Organic Chemistry

Ronald F. Brown
University of Southern California

Wadsworth Publishing Company, Inc.
Belmont, California

Chemistry Editor: Jack Carey

Copy Editor: Edna Ilyin Miller

Production Editor: Susan Yessne

Technical Illustrator: Judith McCarthy

Design: Steve Renick and Robert Gross

© 1975 by Wadsworth Publishing Company, Inc., Belmont, California 94002. All rights reserved. No part of this book may be reproduced, stored in a retrieval system or transcribed, in any form or by any means, electronic, mechanical, photocopying, recording or otherwise, without the prior written permission of the publisher.

ISBN 0-534-00352-4

L. C. Cat. Card No. 74-79856
Printed in the United States of America

2 3 4 5 6 7 8 9 10 — 79 78 77 76 75

Acknowledgments

Chapter 10.   Figures 10.3, 10.5, 10.9, 10.10, 10.11, 10.12, 10.13: From Daniel J. Pasto and Carl R. Johnson, *Organic Structure Determination* © 1969. Reprinted by permission of Prentice-Hall, Inc., Englewood Cliffs, New Jersey.
   Figures 10.4, 10.14, 10.15, 10.16 (adapted), 10.17: From M. W. Hanna, *Quantum Mechanics in Chemistry*, copyright © 1965 by W. A. Benjamin, Inc., Menlo Park, California.
   Figures 10.6, 10.7a, 10.8a,c: Permission for the publication herein of Sadtler Standard Spectra ® has been granted, and all rights are reserved, by Sadtler Research Laboratories, Inc.
   Figure 10.7b,c: From J. D. Roberts and M. C. Caserio, *Basic Principles of Organic Chemistry*, Copyright © 1964, W. A. Benjamin, Inc., Menlo Park, California.
   Figures 10.8b,d: From John R. Dyer, *Applications of Absorption Spectroscopy of Organic Compounds* © 1965. Reprinted by permission of Prentice-Hall, Inc., Englewood Cliffs, New Jersey.

Chapter 11. Figure 11.1: From R. M. Silverstein and G. C. Bassler, *Spectrometric Identification of Organic Compounds,* 1967, John Wiley and Sons, Inc., by permission.

Figures 11.2, 11.26, 11.27, 11.28a,b, 11.29a,b: From Daniel J. Pasto and Carl R. Johnson, *Organic Structure Determination* © 1969. Reprinted by permission of Prentice-Hall, Inc., Englewood Cliffs, New Jersey.

Figures 11.3, 11.5: From R. T. Morrison and R. N. Boyd, *Organic Chemistry,* 2nd ed., 1966, Allyn and Bacon, Inc., by permission.

Figures 11.4, 11.9a,b, 11.11a, 11.12, 11.13a, 11.17, 11.18, 11.19, 11.20, 11.21, 11.22: From N. S. Bhacca, D. P. Hollis, L. F. Johnson, and E. A. Pier, NMR Spectra Catalog, Vols. 1 and 2, Varian Associates, by permission.

Figures 11.9c, 11.10a, 11.11b, 11.14: From John R. Dyer, *Applications of Absorption Spectroscopy of Organic Compounds* © 1965. Reprinted by permission of Prentice-Hall, Inc., Englewood Cliffs, New Jersey.

Figures 11.24, 11.29c, 11.30a, 11.32a, 11.37a: From *Spectral Exercises in Structural Determination of Organic Compounds* by Robert H. Shapiro. Copyright © 1969 by Holt, Rinehart and Winston, Inc. Reprinted by permission.

Figure 11.25: From M. W. Hanna, *Quantum Mechanics in Chemistry,* copyright © 1965 by W. A. Benjamin, Inc., Menlo Park, California.

Figures 11.13b, 11.30b,c, 11.31a,b, 11.32b,c, 11.33, 11.34b,c, 11.35a,b, 11.36a,b, 11.37b,c: Permission for the publication herein of Sadtler Standard Spectra ® has been granted, and all rights are reserved, by Sadtler Research Laboratories, Inc.

Chapter 12. Figure 12.3: From R. T. Morrison and R. N. Boyd, *Organic Chemistry,* 2nd ed., 1966, Allyn and Bacon, Inc., by permission.

Chapter 14. Figures 14.3, 14.5: From J. Pasto and Carl R. Johnson, *Organic Structure Determination* © 1969. Reprinted by permission of Prentice-Hall, Inc., Englewood Cliffs, New Jersey.

Figure 14.4: From J. D. Roberts and M. C. Caserio, *Basic Principles of Organic Chemistry,* copyright © 1964, W. A. Benjamin, Inc., Menlo Park, California.

Figures 14.6, 14.7: From John R. Dyer, *Applications of Absorption Spectroscopy of Organic Compounds* © 1965. Reprinted by permission of Prentice-Hall, Inc., Englewood Cliffs, New Jersey.

Exercise 14.7, Figures 14.8, 14.9, Exercise 14.10: Permission for the publication herein of Sadtler Standard Spectra ® has been granted, and all rights are reserved, by Sadtler Research Laboratories, Inc.

Chapter 16. Exercise 16.3: Permission for the publication herein of Sadtler Standard Spectra ® has been granted, and all rights are reserved, by Sadtler Research Laboratories, Inc.

Chapter 18. Figure 18.1: From John R. Dyer, *Applications of Absorption Spectroscopy of Organic Compounds* © 1965. Reprinted by permission of Prentice-Hall, Inc., Englewood Cliffs, New Jersey.

Chapter 20. Figures 20.1, 20.2: Permission for the publication herein of Sadtler Standard Spectra ® has been granted, and all rights are reserved, by Sadtler Research Laboratories, Inc.

Figure 20.3: From *Spectral Exercises in Structural Determination of Organic Compounds* by Robert H. Shapiro. Copyright © 1969 by Holt, Rinehart and Winston, Inc. Reprinted by permission.

Chapter 21. Figures 21.2, 21.5, 21.6, 21.7, 21.9: Reprinted with permission of Macmillan Publishing Co., Inc. from *Modern Topics in Biochemistry* by T. P. Bennett and E. Frieden. © copyright, The Macmillan Company, 1966.

Figures 21.3, 21.4: From I. T. Millar and H. D. Springall, *A Shorter Sidgwick's Organic Chemistry,* The Clarendon Press, Oxford, 1969, by permission.

Chapter 28. Figure 28.2: From "Enzymatic Modification of Transfer RNA," Dieter Soll, *Science,* Vol. 173, pp. 293–299, July 23, 1971. Copyright 1971 by the American Association for the Advancement of Science.

Figure 28.3: Reprinted with permission from *Modern Topics in Biochemistry* by T. P. Bennett and E. Frieden. © copyright, The Macmillan Company, 1966.

To Allie

and to my students: past, present, and future

Wadsworth Series in Chemistry

Organic Chemistry by Bernard Baker
Chemistry: Basic Concepts and Contemporary Applications by Leonard S. Wasserman
Laboratory Manual for Chemistry: Basic Concepts and Contemporary Applications
  by Leonard S. Wasserman
Energetics, Kinetics, and Life: An Ecological Approach by G. Tyler Miller, Jr.
General Chemistry by Gordon Barrow
Fundamentals of Physical Chemistry by Wayne Wentworth and Jules Ladner

---

Other Books of Interest

Living in the Environment: Concepts, Problems, and Alternatives by G. Tyler Miller, Jr.
Replenish the Earth: A Primer in Human Ecology by G. Tyler Miller, Jr.

Contents

For a detailed table of contents for each chapter, please turn to the beginning of the chapter.

1	Organic Chemistry: What It Is and What It Is About	1
2	Organic Reactions of Simple Organic Compounds	11
3	More Reactions of More Compounds	34
4	Some Compounds with Three and Four Carbon Atoms: Naming	65
5	Atomic and Molecular Structure	86
6	Stereochemistry	118
7	Some More Organic Reactions of Compounds	142
8	About Equilibria and Rates of Reactions	180
9	Alcohols	199
10	Spectroscopy I	239
11	Spectroscopy II	264
12	Nonaromatic Hydrocarbons I	319
13	Alkanes, Alkenes, Alkynes II	344
14	Conjugation and Aromatic Character	392
15	Arenes	439
16	Aldehydes and Ketones	479
17	Acids, Diacids, and Derivatives	536
18	Substituted Acids and Derivatives	588
19	Carbohydrates	613
20	Nonaromatic Nitrogen Compounds	647
21	Amino Acids, Peptides, and Proteins	689
22	Aromatic Nitrogen Compounds	736
23	The Organic Halogen Compounds	776
24	Phenols and Quinones	788
25	Polycyclic Aromatic Compounds	823
26	Terpenes	846
27	Steroids	865
28	Heterocyclic Compounds	885
29	Second-Row-Element Organic Compounds	940
30	Synthetic Polymers	969
	Selected Answers to Exercises and Problems	994
	Index	1009
	Tables	

Guest Essays

John G. Backus: *Enemies are useful!* 10a
Saul G. Cohen: *From freshman to dean* 64a
Herbert C. Brown: *The discovery of hydroboration* 238a
William S. Johnson and John D. Roberts: *Early days with nuclear magnetic resonance* 318a
Jack Hine: *Halogen substituents* 343a
Arie J. Haagen-Smit: *From terpenes and essential oils to smog* 391a
William M. Jones: *The professor knows best* 438a
Clair J. Collins: *How to use isotopic tracers* 535a
D. Stanley Tarbell: *A tough antibiotic to unravel* 612a
Melvin Calvin: *One thing leads to another* 646a
H. Marjorie Crawford: *A lifelong interest* 822a
Charles C. Price: *From antimalarials to involvement* 939a
H. Harry Szmant: *Surprising sulfur radicals* 968a
John E. Leffler: *Why peroxides? It worked!* 993a

Preface

A core enrichment format. Organic chemistry, like most of the other disciplines encompassed by the word "science," has developed in new directions and with new techniques in recent years. Many topics, once reserved for graduate study, are now taken up routinely in the first course in organic chemistry. Other topics, formerly postponed to subsequent courses in biological chemistry, now appear in organic chemistry texts. This presents an author with the task of not just selecting and giving the needed information, but of doing his best to make it understandable and interesting. The student, just introduced into organic chemistry, finds it difficult enough to keep the simple generalities straight and is bewildered and depressed by the mass of detail that must be mastered. I have attempted to write a text in which—without trying to be comprehensive in all things, nor trying to oversimplify—the subject matter is covered in a reasonably thorough fashion and in a manner that will lead the student from the simple to the complex.

In addition, I believe that the textbook should offer more assistance in clearly indicating which material is basic to an understanding of organic chemistry at an introductory level and which material extends both the level and the interest of the basic text. My response to this need has been to organize the book into three levels: the basic text, lined material within the book, and an optional supplement. The basic text covers material that most instructors would feel all students should master. The lined material (identified by a vertical line in the left margin) provides material that some instructors and students may want to cover. This material is not referred to later in the book and, thus, is truly optional.

The *Supplement to Brown's Organic Chemistry: Extensions, the Literature of Organic Chemistry, and Solutions to Extension Exercises and Problems*, is a separate supplement cross-referenced with the text and offering additional material that may be used at the discretion of the instructor. The extensions section of the supplement contains material of two kinds: that which is an expansion of discussion of topics in the text and that which contains topics not covered in the text. With this material, you can extend the boundaries of the course without having to refer to the more complicated advanced texts or research articles. The text and the supplement allow an instructor a wealth of choice for a given course. (The symbol ☐☐ in the text indicates that there is related material in the supplement.)

In addition to the extensions sections, the supplement contains a section on the literature of organic chemistry and detailed solutions to the exercises and problems appearing in the supplement sections.

A thorough treatment of bio-organic chemistry. Another difficulty facing today's author is the necessity for creating a text that is suited to the diverse needs of premedical, predental, and prepharmacy students as well as of the chemistry, biochemistry, biology, and chemical engineering majors. A thorough treatment of bio-organic chemistry is provided for those professors who wish to cover this subject matter more completely. Cardohydrates, amino acids, peptides, and proteins are presented in Chapters 19 and 21 instead of at the end of the text. To make the organic chemistry more meaningful to life science students, numerous examples of the simpler biological structures, reactions, and concepts are integrated throughout the first eighteen chapters. I have been fortunate to have the help of William Beranek, Jr., a biochemist at Duke University Medical Center, who provided many of the examples.

The organization: a balanced view from the beginning. I am adding yet another textbook to the many excellent texts now available for use in the first year course in organic chemistry because this text offers a new and useful organization of the subject material. Other authors have preceded me in offering their versions of how to organize the subject, but none of them comes close to what is offered here.

I was never completely happy with the organization of most organic chemistry texts, because the early chapters give the student the impression that organic chemistry is primarily concerned with atomic and molecular structure and free radical halogenation of alkanes. Later in the semester, a drastic reorganization of priorities becomes necessary on the part of the students in order to cope with the flood of reactions and compounds with which they were faced. Another problem with such an organization is the difficulty of coordinating the laboratory with the lecture when the emphasis in the first part of the text is upon structure and free radical reactions.

In the initial chapters of my book, I introduce both simple structures and simple reactions, including *simple* mechanisms, so that the student will find it easier to cope with the many reactions that follow in the chapters covering the classes of compounds. *Yet only the essentials are given in the first few chapters so that the student will be able to understand these basic concepts.* Structure and reactions are developed gradually, and initially the classes of reactions are introduced by very simple uncomplicated examples. The early introduction of reactions also permits a coordinated laboratory, because after Chapter 3 the laboratory work may begin with almost any of the reactions of aliphatic chemistry. The organization of the text is described in more detail in the prelude.

I have tried to write informally and for the student. To that end, those concepts that in my experience have given difficulty were given more attention than certain others that have enjoyed more time and space in the past. Understanding of and the use of principles and guidelines were the goal. Although qualitative in tone, the quantitative and practical aspects of the subject have not been ignored.

Detailed solutions. Some answers to problems are given in the text. A second supplement entitled *Detailed Solutions to Exercises and Problems in Organic Chemistry* provides solutions to a vast majority of the exercises and problems in the text *in detail.* This was done because the questions and answers are considered by me to be a crucial part of the learning process.

Guest essays. I have been most fortunate to have the help of fourteen prominent organic chemists who contributed essays to the book. These essays illustrate the diverse ways scientists operate, in contrast to the rigid mold of the "scientific method." For example, William Johnson and John Roberts discuss their early days with nuclear magnetic resonance. William Jones writes about the time a student of his made an important discovery, even though Jones was allowing the student to try something the professor "knew" wouldn't work. In another essay Melvin Calvin describes the importance of coordination chemistry with his research.

Other essayists who have contributed of their time and experience with generosity and kindness are: Drs. John Backus, Herbert C. Brown, Saul G. Cohen, Clair J. Collins, W. Marjorie Crawford, Arie J. Haagen-Smit, Jack Hine, William S. Johnson, John F. Leffler,

Charles C. Price, Harry Szmant, and Stanley Tarbell. To these scientists I owe a very real debt of gratitude. Dr. John D. Roberts very kindly gave his permission for the use of Dr. Johnson's essay.

Extensive manuscript review. I am particularly indebted to three of the nineteen reviewers, Professor James A. Campbell (El Camino College), Dr. Frank L. Lambert (Occidental College), and Dr. John J. Uebel (University of New Hampshire), for their incredible patience, diligence, and tact. They read the entire original manuscript and the several revisions thereof. Furthermore, they gave their time to meet with me for six days to discuss the manuscript on a page by page basis. In the process, they found innumerable errors and contributed many, many suggestions for improvement, not only in detail but in general as well.

I also greatly appreciate the following reviewers' assistance in developing the manuscript: Jay Martin Anderson (Bryn Mawr College), A. R. Ballentine (Citadel Military College of South Carolina), Sidney W. Benson (Stanford Research Institute), Edward A. Caress (George Washington University), Leallyn B. Clapp (Brown University), Charles R. Dawson (Columbia University), Louis E. Friedrich (University of Rochester), John Idoux (Florida Tech University), Paul Klinedinst (California State University, Northridge), James A. Marshall (Northwestern University), James D. Morrison (University of New Hampshire), H. LeRoy Nyquist (California State University, Northridge), M. G. Reinecke (Texas Christian University), Kenneth L. Servis (University of Southern California), Charles Stammer (University of Georgia), and Katherine E. Weissman (Charles Stewart Mott Community College).

We would also like to extend our appreciation to the 241 professors who gave us their advice by responding to a twenty-one-page questionnaire.

Acknowledgments

It is impossible to acknowledge all those who have contributed directly and indirectly to the writing and publication of this text. Any author reflects his background and training, his instructors and his colleagues, his reading and previous writing, and the experience of being an instructor. Much as I would like to name everyone, I must desist. However, it would be churlish not to acknowledge the cooperation of all of those at Wadsworth Publishing Company who did their best to bring the project to fruition. Jack Carey, chemistry editor, with his unfailing enthusiasm and supervision kept things on the track over the years. Susan Yessne, production editor, with humor and efficiency coordinated the many facets of production. Edna Ilyin Miller, copy editor, with unbelievable speed turned a beautiful manuscript into an ocean of red pencil marks and corrections decipherable only by compositors. And Judith McCarty turned my scrawls into the excellent art work. I would also like to thank Wadsworth Publishing Company for allowing me to use some excellent exercises from Baker, *Organic Chemistry*.

I was blessed with the services of two exceptionally skilled typists, Maud Bobkowski, who did an original version of the first nine chapters, and Sharon Cutri, who did all the rest of the work including the final complete manuscript. Last, but not least, the encouragement and enthusiasm of the students, who endured the class use of several crude preliminary versions of the first nine chapters, served to give me the final push to go on and complete what seemed to be a never ending task.

Any errors of fact and interpretation and all other shortcomings that may show up, I claim as my responsibility. Any features that may be thought worthy of merit should be shared by the many who helped in the development and preparation of this text and its publication.

Prelude

Chapter 1 is a basic introduction to the organic chemistry text. Chapter 2 begins the reactions of organic chemistry. The reactions are classified as acid-base, substitution, addition-elimination, and oxidation-reduction. Each class is further subdivided into heterolytic, namely nucleophilic and electrophilic, and homolytic or radical types as may be appropriate. This classification is maintained as much as possible throughout the rest of the book as the reactions and preparations of the various classes of compounds are taken up. However, in the second chapter, only the reactions of some of the compounds containing a single carbon are described. This simplification results in, for example, the addition of water to formaldehyde as the introduction to the subject of addition-elimination reactions.

Chapter 3 goes on to some of the two carbon atom compounds. For example, addition-elimination is expanded to include addition to ethylene and to acetylene, and to the production of ethylene from ethanol, from ethyl bromide, and so on. Thus, complications are avoided and the basic ideas are implanted and repeated at the very beginning of the course.

Other aspects of the subject, such as naming, atomic and molecular structure, and stereochemistry, which appear in Chapters 4, 5, and 6 are discussed before addition-elimination and substitution are returned to in detail in Chapter 7. In Chapter 8, reaction rates and equilibria are discussed in elementary terms. Then, in Chapter 9, the alcohols are described as the first class of compounds to be looked at in a comprehensive fashion. Because the students are acquainted by this time with examples of most of the other classes of non-aromatic compounds, the chapter is much more comprehensive than the usual early chapters concerning classes of compounds.

Beginning on page 239, Chapters 10 and 11 cover spectroscopy in enough detail to allow the average student to begin the interpretation of spectra, ir, uv, nmr, and mass. Its use is continued throughout the text. Chapters 12 and 13 take up alkanes, alkenes, alkynes, and the cyclic counterparts together; and petroleum technology is described briefly.

Chapter 14 develops the subject of conjugation from both the MO and VB viewpoints and leads directly into the Huckel rule and aromatic stability. Cycloaddition reactions are described. In Chapter 15, the chemistry of the arenes is discussed including electrophilic, nucleophilic, and homolytic substitution reactions; and coal and coal tar are described.

Chapters 16 to 30 were written with the idea that no single course would require complete coverage of each and every chapter. However, some, or most, of Chapters 16 to 22 would be discussed in any one year course in organic chemistry. Chapters 23 to 30 present a choice of subject matter and may be taken up in toto, in part, omitted, or rearranged as may be needed.

As presently arranged, the organization provides natural break points for both a course of three quarters (before or after spectroscopy and before or after carbohydrates) or of two semesters (before aldehydes and ketones). Also, by the end of Chapter 3, it is possible to introduce reactions in the laboratory with only a small amount of discussion of extension from the reactions already given in lecture.

1 Organic Chemistry: What It Is and What It Is About

In this opening chapter, we look at the subject of organic chemistry from a variety of viewpoints. We hope to convey something of the fascination of the chemistry of the compounds of carbon. The chapter ends with a brief description of the organization of the book and suggestions for using it.

1.1 What is organic chemistry? Some answers   2
1. The origin of the name.   2. A new language.   3. A laboratory science.
4. People in organic chemistry.

1.2 What is organic chemistry? More answers   3
1. Structure.   2. Reactions.   3. Analysis.   4. Identification.   5. Synthesis.   6. Interactions with radiant energy.   7. Physical organic chemistry.
8. Bio-organic chemistry.

1.3 What is organic chemistry about? Some answers   7
1. Evolution and organic chemistry.   2. Development of organic chemistry.
3. Organic products and industries.

1.4 What we shall do with organic chemistry   9
1. The instruction team.   2. Chapter chronology.   3. Study hints.

1.5 Summary   10

1.6 Bibliography   10

1.7 John G. Backus: Enemies are useful!   10a

1.1 What is organic chemistry? Some answers

The origin of the name. The term <u>organic chemistry</u> had its origin in the word <u>organism,</u> a living entity. In the early days of chemistry, a surprising variety of substances were isolated from living organisms—acetic acid, ethyl alcohol, the fats and oils, sugars, fragrant oils, and some alkaloids and drugs. The constitution of such compounds and substances remained a mystery until analytical methods developed enough to allow quantitative analytical data to be gathered. It soon became apparent, though, that all such compounds contained carbon and hydrogen. Oxygen and nitrogen also were frequently present. Little was known (or would be known for a number of years) about the reactions of organic chemistry. No one knew how to make such compounds without starting from some other organic compound; even then the results were poor. Thus, it was thought that the production of organic compounds required a <u>vital force</u> and that they could be put together only by nature, by living organisms. In 1828 Wohler found that by heating ammonium cyanate, an inorganic compound, he could make urea, an organic compound. Although some chemists had suspected for some time that the idea was false, many others remained

$$NH_4^{\oplus} + OCN^{\ominus} \xrightarrow{heat} O=C{\begin{smallmatrix}NH_2\\NH_2\end{smallmatrix}}$$

adherents of the vital force theory. It took many more examples of conversions and many years before the vital force idea was abandoned. By then, chemists had become so used to calling such substances organic compounds (as opposed to inorganic compounds) that the name stuck. Today, organic chemistry is defined as the chemistry of the hydrocarbons and their derivatives—or, which is almost the same, the chemistry of carbon compounds—whether found in organisms or not.

A new language. A major problem throughout our study of organic chemistry will be that of communication. The definition of organic chemistry is not very helpful unless we understand the words used in it. Already we have used some words that are new to some of us, such as <u>constitution, alkaloid,</u> and <u>derivative.</u> As we progress, new concepts, compounds, and reactions will be encountered, named, used, and explained.

If we look up the word <u>constitution,</u> we find it defined as the structure of a compound, as determined by the kind, number, and arrangement of atoms in its molecule. What does <u>structure</u> mean? The dictionary says it is the arrangement and mode of union of the atoms of a chemical compound. Thus, we see that constitution is more than the kind and number of atoms in a molecule. The arrangement of the atoms in space and of the bonds between the atoms also must be considered. Therefore, constitution introduces the idea of the three dimensional arrangement, the architecture, of the atoms and bonds of molecules. Because organic molecules occur in many sizes and shapes, we shall be involved continually with the constitution of organic molecules. Thus, the chemistry of carbon compounds includes the constitution of carbon compounds, but constitution is only one facet of the meaning of the word <u>chemistry.</u>

Organic chemistry is like learning a new language. We learn new words (vocabulary), reactions (sentence structure), and the synthesis of new compounds (speaking and writing).

A laboratory science. Organic chemistry is a laboratory science as opposed to an observational science. The term laboratory science implies that the scientist rests his conclusions on experimental factual data. An observational science, such as astronomy, is helpless when it comes to experimentation. The astronomer can only observe and record what is going on among the sun and planets, the stars, the galaxies, and the universe.

In all sciences, rigorous honesty in experimentation, observation, and recording of results is required. Reliability of data is the first requirement of any science. Honest mistakes are made in all the sciences, but they are found and corrected as more data and more exact data have accumulated. Although personalities and opinions may clash, no scientific scholar will tolerate any deviation from the truth in the data with which he works. One of Webster's definitions of science deserves quotation: "Accumulated and accepted knowledge which has been systematized and formulated with reference to the discovery of general truths or the operation of general laws; knowledge classified and made available in work, life, or the search for truth; . . ."

People in organic chemistry. Organic chemistry is a human activity like any other, and the use of the results of scientific investigation is not always thoughtful. Hasty application of a science can lead to environmental problems. For example, during the development and early use of synthetic detergents, nonbiodegradable substances were introduced because they were cheaper to make than biodegradable detergents. However, when we look askance at such mistakes, we should keep one thing in mind. The supply and distribution of water and food and the collection and disposal of sewage and solid waste have been urban problems since the days of the first city of ancient times. On the other hand, popular concern with the effects of such activities upon the environment has arisen only since about 1950. The men and women of organic chemistry have permitted errors in the application of their science in industry and in agriculture. The fault lies in the common human tendency to oversimplify and ignore the complex consequences of the introduction and use of the products of industry.

Organic chemistry has contributed much that has been beneficial to life itself as well as valuable in making life more pleasant. Industrial organic and pharmaceutical chemistry began with synthetic dyes and today turns out a myriad of products—vitamins, hormones, drugs, antiseptics, antibiotics, rubber, fibers, molded plastics, finishes, solvents, detergents, insecticides, inks, and so on.

Organic chemistry is only one of the tools we shall continue to use in our effort to alleviate our massive ignorance of nature. This tool will remain one of the most useful, but our success will be equally dependent on our vigilance against shortsighted error.

1.2 What is organic chemistry? More answers

Organic chemistry is concerned with the study of these aspects of the compounds of carbon: structure, reactions, analysis, identification, synthesis, and interactions with radiant energy. Let us describe these concerns in a little bit more detail.

Structure. The most impressive thing about organic chemistry is the tremendous number of carbon compounds that are known. More than two million organic compounds have been described, and the number is growing at the rate of two to three hundred new compounds per day. The ability of atoms of carbon to form bonds with other atoms of carbon, as well as with atoms of hydrogen, oxygen, nitrogen, the halogens, and practically all the other elements, is a primary reason for such diversity in the structure of organic compounds. The variations possible in the arrangement of atoms in a molecule is another major cause of diversity.

It is most usual for a carbon atom to be linked or bonded to four other atoms (as in methane, CH_4), although carbon is frequently found to be bonded to only three other atoms (as in formaldehyde, $H_2C=O$). In some instances, carbon is bonded to only two other atoms (as in carbon dioxide, CO_2 or $O=C=O$). In relatively rare cases, a carbon atom may be bonded to only a single atom or to five or six other atoms.

This is not the time or place to discuss bond formation, but we must introduce here the convention that a dash or short line between the atomic symbols represents a shared pair of electrons and is called a covalent bond. There are various types of covalent bonds; we shall discuss them later. For the present, let us remind ourselves that an atom uses only those electrons in the outermost unfilled shell for bond formation. The filled shells closer to the nucleus of the atom remain largely unaffected by what the situation may be in the outermost shell, the valence shell. Thus, carbon has four electrons available for bonding, hydrogen has one, nitrogen has five, oxygen has six, and the halogens have seven. By sharing with four other atoms, carbon is able to form four covalent bonds and have a stable octet of electrons in the valence shell. Hydrogen is able to form only one covalent bond, the first shell being able to accommodate only two electrons. By sharing only three of the total of five electrons to form three covalent bonds, nitrogen has a pair of electrons left over and has an octet. Similarly, oxygen achieves an octet by sharing only two electrons, and a halogen does so by sharing only a single electron. Some examples are shown in Table 1.1. (A dash is also used to denote a pair of electrons not used in bonding.)

Even from the seven simple examples shown (Table 1.1), we can see what a variety of organic compounds are possible. Complexity and variety increase rapidly as the number of carbon atoms in a molecule increases, and we shall find it most convenient to use condensed structural formulas like those at the right of the table.

We should take notice of the formation of double bonds in which four electrons are used to bind two atoms together. Triple bonds involve six electrons in the sharing process.

Various spatial arrangements of the atoms within a molecule also are possible. For the architecture of molecules we use the term stereochemistry, the branch of organic chemistry concerned with spatial effects in structure, stability, and reactivity.

Reactions. The reactions of organic compounds vary from extremely slow to explosive. Much effort has been expended in the study of organic reactions. Today it is possible to classify most of the reactions as follows:

A. Acid-base reactions
B. Substitution reactions
  1. Nucleophilic substitutions
  2. Electrophilic substitutions
  3. Homolytic or free-radical substitutions
C. Addition-elimination reactions
  1. Nucleophilic
  2. Electrophilic
  3. Homolytic
D. Oxidation-reduction reactions

1.2 What is organic chemistry?

Table 1.1 *Bonding in some simple compounds*

hydrogen	H:H	H—H	H$_2$
methane	H:C:H with H above and H below	H—C—H with H above and H below	CH$_4$
ammonia	H:N:H with H below (lone pair above)	H—N—H with H below	NH$_3$
water	H:O: with H below (two lone pairs)	H—O| with H below	H$_2$O
hydrogen fluoride	H:F: (three lone pairs)	H—F|	HF
methanol	H:C:O: with H's (lone pairs on O)	H—C—O| with H's	CH$_3$OH
methyl-amine	H:C:N:H with H's	H—C—N—H with H's	CH$_3$NH$_2$
ethane	H:C:C:H with H's	H—C—C—H with H's	CH$_3$CH$_3$
formaldehyde	H:C::O: with H	H$_2$C=O|	H$_2$C=O or HCHO
ethylene	H:C::C:H with H's	H$_2$C=CH$_2$	H$_2$C=CH$_2$
acetylene	H:C:::C:H	H—C≡C—H	HC≡CH

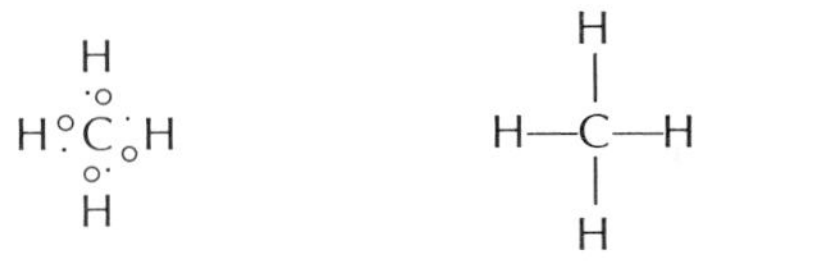

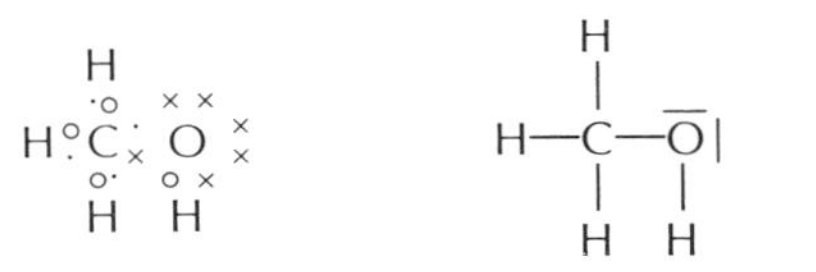

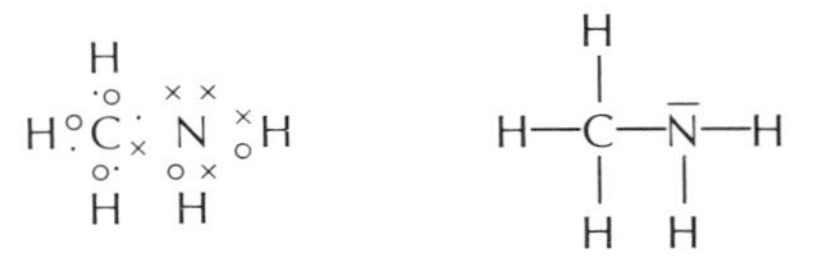

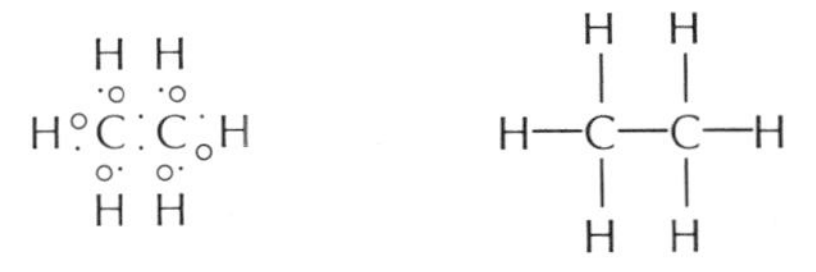

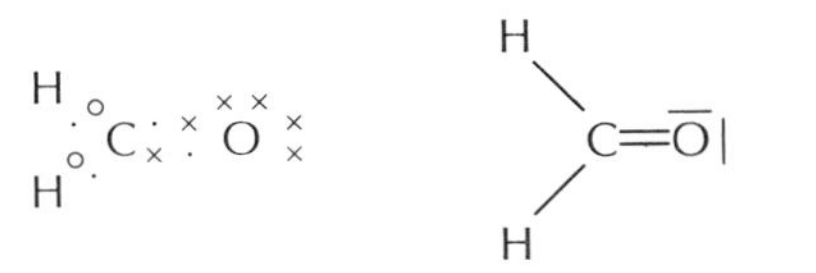

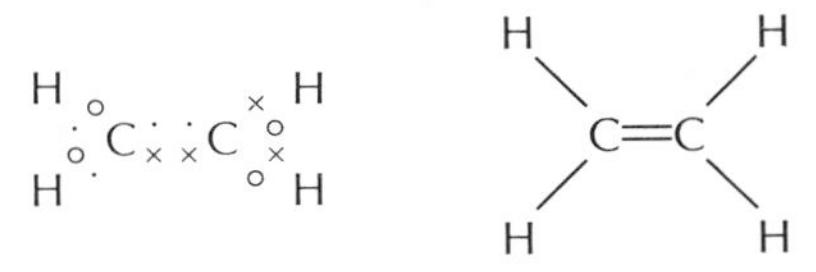

Examples of these reactions will be given in the next chapter. We list the classifications here to emphasize the similarities that we shall encounter again and again in the reactions of organic compounds as we proceed. Certain other classes of reactions exist, but we need not discuss them until we are further along. Later on, we also shall discuss catalysis, the speeding up of otherwise slow reactions.

Analysis. The analysis of organic compounds is divided into qualitative and quantitative aspects. The determination of molecular weight is a necessary quantitative technique. Qualitative analysis usually is concerned with identification of the presence of elements other than carbon, hydrogen, and oxygen—namely, nitrogen, sulfur, the halogens, phosphorus, and others. Analytical techniques have been both automated and refined so that samples of 1 to 5 mg may be used to give the percent composition of any element present.

Identification. Is this compound I have just isolated and purified (from a natural source or from a reaction mixture in the laboratory) already known or is it a new compound? In either case, what is it? Sometimes the answers are easily obtained in a few minutes. In other cases, it takes years of work. The methods used to identify carbon compounds require a good knowledge of organic chemistry.

Synthesis. Synthesis of organic compounds has two aspects. One is the development of industrially suitable methods for the synthesis of certain compounds on a large scale. Examples are: aspirin and many other drugs; dyes; monomers for the synthesis of polymers for use in making rubbers, fabrics, plastics, paints, and so on; and many intermediate compounds to be used in the manufacture of all these products.

3-hydroxy-2-butanone
(odor and flavor in butter)

nylon

The other aspect is the laboratory synthesis of a given compound on a relatively small scale. We may wish to synthesize a new compound for its own sake. In addition, the structure of a compound is not considered proved until it has been synthesized from known compounds by known reactions. Such a synthesis may be incredibly difficult and require a large number of discrete reactions. Ability in synthesis is part training and part talent. Most of the great organic chemists have been virtuosos in synthetic organic chemistry. They have excelled in the art of organic synthesis. To call synthesis an art may seem bizarre now, but during our study you may come to appreciate the talent required to devise and execute a beautiful synthesis.

Enormous families of compounds modeled after penicillin, sulfa, and morphine have been synthesized in hopes both of making an improved drug and of understanding which part of the molecule is essential for drug action. Currently there is much activity in the synthesis of insect hormones to act as nonpersistent, specific insecticides and in the preparation of analogs of human hormones to be better oral contraceptives.

Interactions with radiant energy. Organic compounds interact with radiant energy by absorption and by reradiation (luminescence, phosphorescence, and fluorescence). Depending upon the characteristics of the compound and the energy (frequency) of the radiation, certain consequences ensue. Low energy radiation (microwave) affects the internal rotations within a molecule. Going up the scale of energy to infrared (ir) and Raman spectroscopy, we find it possible to examine a compound for the internal vibratory motions of the atoms. Visible-light absorption (dyes, pigments) causes electron excitation in such molecules, and ultraviolet (uv) light extends the range of such excitation into all organic compounds. Bombardment with electrons energetic enough to knock other electrons out of molecules allows us, in mass spectrometry, to observe what happens to the ions so produced.

A major development has been the use of strong magnetic fields to influence the spin distribution of nuclei as observed in nuclear magnetic resonance (nmr). If we select the proper magnetic field and radio frequency, only the hydrogen nuclei in a compound will respond. Very similar in technique is electron spin resonance (esr), in which only unpaired electrons respond. Both nmr and esr enable us to look (figuratively) into the interior of a compound and see the protons or electrons and the influences to which they are subjected.

The various spectroscopic techniques (there are still more that have not been mentioned here) are invaluable not only as analytical tools and as aids in identification, but also as adjuncts to understanding the influences within a molecule on bonding, structure, and stability.

Physical organic chemistry. Physical organic chemistry is concerned primarily with attempts to understand the structure and reactivity of organic compounds. Correlations of physical properties with either reactivity or stability have been helpful. Rates of reaction, positions of equilibria, effects of irradiation, effects of variations in substituents, and effects of isotopes are only a few of the types of experimentation that have been used in this line of research.

Physical organic chemistry also is concerned with the mechanism of reactions; that is, with the pathway by which the reactants are transformed into products. Mechanisms will provide us with a thread of continuity as we progress from the simple to the complex in the chapters to come.

Bio-organic chemistry. Bio-organic chemistry is devoted to the study of the structure and reactivity of biological compounds. A major concern of scientists in this field is to identify the different factors that biological organisms use to increase the rates of their chemical reactions. Some topics of interest are transport across membranes, mechanism of antibiotic action, and synthetic pathways used by living organisms.

1.3 What is organic chemistry about? Some answers

Evolution and organic chemistry. Organic chemistry is about us: you and me and everybody else. It is about every living thing, which brings us back to organisms.

Some scientists think that in the very early history of the earth, compounds of carbon evolved. Over millions of years, the varieties of these compounds increased until one day an asymmetric molecule was formed. (An asymmetric molecule is one that is not superposable upon its mirror image. We shall take up this and other new ideas in greater detail in subsequent chapters.) After eons of time, these asymmetric molecules became catalysts for some of the existing reactions. Still later, asymmetric catalysts arose that would operate only with certain specific asymmetric molecules. These catalysts were of such high molecular weight that the reactions occurred on their surfaces. Their shape made them behave in a specific (selective) manner, catalyzing only certain reactions. All this development was very slow.

Finally, in some way a collection of these catalysts was enclosed within a membrane of polymeric substances. Thus, the prototype of a cell was created. The reactions continued to occur over more eons, but they did not yield a continuous process. Eventually, the right combination of compounds and catalysts was found, and a cell was created. For all we know, it may have happened countless times until cell division evolved and cell reproduction became a reality. Thereafter, development was more rapid—although still on the millions-of-years time scale—and a myriad of tiny living things evolved. The genetic code ensured the continuation of species, and variations in the coding (mutations) ensured the continual experimentation whereby new species and variants arose. Ultimately, the larger species of life evolved. Thus, any living organism is an incredibly complex system of reactants, catalysts, and products undergoing countless reactions. Organic chemistry is the chemistry of life.

Development of organic chemistry.   Organic chemistry developed with reasonable speed along with the other branches of chemistry until 1856. The preparation of mauve dye by Perkin (England) in that year and of fuchsin in France in 1859 touched off explosive activity in the chemistry of coal tar and its components. Thus organic chemistry became commercially valuable and raced ahead in those parts of the subject not related to living organisms. It was not until the period from about 1920 to 1940, when the science had advanced enough to allow better understanding of the molecules of life, that the emphasis began to swing back to the compounds of carbon derived from living systems.

Organic products and industries.   It is probably easier to make a list of what organic chemistry is not about than of what it is about. However, here is a partial list of products of organic chemical origin:

Antibiotics and drugs
Anesthetics
Vitamins, enzymes, and nucleic acids
Hormones and steroids
Proteins, fats, oils, and carbohydrates
Insecticides and herbicides

Paints and varnishes
Dyes
Synthetic polymers as fibers, sheets, articles, elastics, and plastics
Perfumes and flavors

Industries based on technology derived from organic chemistry are many and varied. We list a few:

Petroleum refining
Coal tar products
Sugar refining
Pulp and paper
Fermentation

Agriculture and agricultural by-products
Pharmaceuticals
Polymers
Leather
Paints and varnishes

1.4 What we shall do with organic chemistry

The instruction team. This textbook is part of the team that will help you learn some of the facts and theories of organic chemistry. The team consists of the lecturer in the course, the discussion leader, the laboratory assistants, this book, the laboratory manual, and a set of models. Among them, they do their best to help you along.

The lectures will supplement and clarify what is given in the text. Your lecture notes, taken in very abbreviated form, should be rewritten as soon as possible and amplified by reference to the textbook. Mark those points on which you need help. It is very helpful to read the proper chapter in the text before the lecture. Certain of the exercises and problems may be assigned as homework. In any case, it will not be amiss to work out as many as time permits.

The laboratory portion of a course in organic chemistry is more than an aid to remembering the lecture material. There we learn something of the techniques and manipulations needed to carry out reactions, isolation, purification, and identification of organic compounds.

In any case, always remember that your instructors are dedicated to helping you learn some organic chemistry. Above all, do not hesitate to ask questions.

Chapter chronology. In the next chapter, we plunge directly into a variety of organic reactions, but we use the very simple organic compounds. We shall maintain our emphasis upon reaction until we come to the slightly more complicated compounds, when we shall need to look into naming, atomic and molecular structure, and stereochemistry. After a relatively more detailed discussion of reactions, we shift our attention to a class of organic compounds, the alcohols. Thereafter, except for the chapters on spectroscopy, we shall find it more convenient to complete our survey of organic chemistry by classes of compounds.

Study hints. The study of organic chemistry requires a certain amount of memorization—new words for compounds, reactions, effects, and principles. Some of us fall into the trap of considering memorization to be sufficient, whereas it is only the beginning. Memorization is like learning the rules of a game; it is only the prelude to learning how to play the game itself. Our game, organic chemistry, consists of reactions (or chemical properties), synthesis, analysis (including the deduction of structure), and identification.

Some of us have poor study habits. A fatal sin in organic chemistry is to fall behind. Attempts at cramming a night or two before a test bring the ghastly realization that the material cannot be quickly learned and used. Furthermore, the pace of the course does not allow much breathing space to catch up once one is behind. So, keep up!

It is necessary to become proficient in the writing of structural formulas and reactions. Our fingers as well as our minds must be educated. In this subject, learning comes through writing, writing, writing, and more writing. Rewrite the lecture notes and incorporate material from the text. Write out the answers to exercises and problems. Write, write, and write.

1.5 Summary

Each chapter will conclude with a summary, which will, in most instances, provide a brief list of the items in the chapter to be kept in mind.

In this chapter, the best summary is probably the table of contents at the head of the chapter. (In some later chapters, the same situation will hold.)

1.6 Bibliography

Most of the chapters will contain a bibliography in which a few references provide starting points for additional information and reading for those interested in pursuing a given topic.

1. MacDonald, Malcolm M., and Robert E. Davis, *Chemistry and Society*. Boston: Willard Grant, 1972. A fascinating little paperback consisting of reprints of articles on various topics related to the title.

2. Maxwell, Kenneth E., *Chemicals and Life*. Belmont, Calif.: Dickenson, 1970. A larger paperback with an extensive collection of reprints.

1.7 John G. Backus: Enemies are useful!

Professor John G. Backus earned his Ph.D. in physics in 1940 at the University of California, Berkeley. After a wartime stint at the Radiation Laboratory, he came to the University of Southern California in 1945 as an associate professor. His interest in music turned his studies toward acoustics. After years of teaching a special course in acoustics for music majors, he wrote a book on it. The preface of his book was so unusual and interesting that most of it is reprinted here for your enjoyment.*

The scientist is generally motivated by the same drive that made him a scientist in the first place: namely, an intense curiosity concerning the behavior of things, and a great delight in discovering the laws governing their behavior. He must make the assumption, rather new in human history but amply justified by this time, that the behavior of things is not decided by certain gods who (like people) are capricious, arbitrary, and unpredictable, but is governed by laws that are based on a few very fundamental principles. He assumes that these principles and laws can be worked out from observations on the way things behave under conditions that he can, to some extent, control and vary in known ways. Observations under such conditions constitute what is called experiment. The working out of laws and principles governing the behavior of things utilizes a complex process of experimentation, interpreting the results, constructing theories to explain the results, using the theories to suggest new experiments, and so on. This process is called the scientific method.

The application of the scientific method has in a very short span of human history given us an immense amount of knowledge about the things around us. This body of knowledge has been worked over and tested by a great many people, so that much of it can, with considerable confidence, be accepted as true. Much more of it can be taken as provisionally true, depending on the outcome of further scientific investigation. However, there is a great deal we do not know about the behavior of things, and the more questions we answer, the more we find remaining to be answered. Nature is infinitely complex and as our body of knowledge increases, our body of ignorance—if we may use the term—increases even faster.

The proper utilization of the scientific method to increase our body of knowledge is not easy. Significant experiments are difficult to devise; a great deal of experimentation turns out ultimately to have been useless, although undertaken for what seemed at the time to be good reasons. Even when soundly conceived, experiments are plagued by technical difficulties of various kinds; like all human efforts, they are subject to what in some circles has become known as "Murphy's Law": If anything can go wrong, it will.

Interpretation of experimental results is equally difficult. There are a great many wrong ways of explaining a given observation, and very few right ways. For example, it is the practice of some aboriginal tribes to beat on drums and other noisemakers when an eclipse of the sun occurs, in the hope of frightening off the god or demon that is swallowing the sun. Since the practice appears to be invariably successful, it might be concluded that hundreds of experiments have demonstrated that an eclipse of the sun may be brought to an end by beating on drums.

As an additional complication, it is necessary to keep in mind that what may appear at one time to be a reasonably correct explanation of certain experimental results may as a

---

*Reprinted from *The Acoustical Foundations of Music* by John Backus. By permission of W. W. Norton & Company, Inc. © 1969 by W. W. Norton & Company, Inc.

result of later research be found inadequate or erroneous; the history of science is full of examples. At one time it was certainly reasonable to suppose that the sun revolved around the earth; today this is not a very popular idea, at least among people with some education.

Because of these various problems, the scientific investigator sometimes gets the feeling that Nature enjoys placing pitfalls for the unwary, traps for the unskilled, and obstacles for everyone, to make understanding her as difficult as possible. This stubbornness develops in the investigator a tremendous skepticism. Things are seldom what they seem; what appeared to be true yesterday may not be true today and must be tested. This skepticism must be extended to include people, especially other scientists. Anyone may be wrong about anything; no statement, made by however high an authority, is to be accepted as true simply because such a high authority has pronounced it so. Such statements may be provisionally accepted as starting points for further work, but must be abandoned without compunction if it becomes necessary.

It is in the areas where new knowledge is sought that one should most distrust Nature and prior authority. Here appearances can be most deceiving; experimental results must be checked, rechecked, and checked again, as time permits. The scientist's distrust must include even himself; he is in the unfortunate position of never being sure he is right. To guard against being led astray by preconceived notions and misconceptions, he must continually submit his work to the scrutiny of his colleagues and must at least try to appear to be grateful if it is torn to shreds. One of the best expressions of this point of view is given by a recent Nobel Laureate in physiology:

> [One] way of dealing with errors is to have friends who are willing to spend the time necessary to carry out a critical examination of the experimental design beforehand and the results after the experiments have been completed. An even better way is to have an enemy. An enemy is willing to devote a vast amount of time and brain power to ferreting out errors both large and small, and this without any compensation. The trouble is that really capable enemies are scarce; most of them are only ordinary. Another trouble with enemies is that they sometimes develop into friends and lose a good deal of their zeal. It was in this way that the writer lost his three best enemies.*

Everyone, not just scientists, needs a few good enemies!

---

* From Georg von Bekesy, *Experiments in Hearing.* New York: McGraw-Hill, 1960, p. 8.

2 Organic Reactions of Simple Organic Compounds

In this chapter we introduce a few of the compounds of carbon, briefly consider physical properties, initiate the method of naming, and begin the study of the reactions of organic chemistry. By restricting ourselves to the very simple compounds, we are able to typify the reactions without the complications that arise with the more complex compounds.

2.1 Introduction   12

2.2 Some compounds with a single carbon atom; elements of structural theory   12
    1. A few compounds and their physical properties.   2. The covalent bond; tetrahedral carbon.   3. Sigma and pi bonds. Exercises.

2.3 Introduction to naming   16
    1. IUPAC naming: hydrocarbons, alcohols, aldehydes, acids, and amines.
    2. Summary of rules. Exercises.

2.4 Reactions   19
    1. Introduction.   2. Four classes of reactions.

2.5 Acid-base reactions   20
    1. Brønsted-Lowry and Lewis acids and bases.   2. Acid-base equilibria.
    3. Control of pH.   4. Use of pK_a values. Exercises.

2.6 Nucleophilic substitution reactions   23
    1. The ABC reaction; S_N2 defined.   2. Comparison of nucleophilic strengths.
    3. Acid-base equilibria and S_N2 reactions.   4. Summary. Exercises.

2.7 Electrophilic substitution reactions   25
    Definition and examples.

2.8 Homolytic substitution reactions   26
    Definition and examples.

12 Chapter two

2.9   Addition and elimination reactions   26
       1. Polarizability; resonance.   2. Addition of water to formaldehyde.
       3. Formaldehyde. Exercises.

2.10   Oxidation and reduction reactions   28
       1. Oxidation; combustion.   2. Partial combustion.   3. Reduction; catalytic
       hydrogenation.   4. Oxidation of formaldehyde by very mild reagents.
       5. The reduction series.   6. Alpha and beta eliminations.

2.11   Summary   31

2.12   Problems   32

2.1 Introduction

Early organic chemistry was hampered by the inability to purify organic compounds. Many compounds now known and described remained as intractable mixtures of gases, liquids, or solids. A number of substances were isolated from natural sources as pure gaseous, liquid, or crystalline compounds; but they defied early attempts to describe their structures. This difficulty forced attention to the simple compounds of carbon, and working out the structures of these relatively simple compounds was difficult enough. Even now there is an amazing amount of research still to be done on the compounds of low and intermediate molecular weight. With refinements of old methods and the introduction of new techniques, it has become possible to begin to unravel the intricacies of substances of higher molecular weight.

We begin our study of organic chemistry with some of the simple substances and their reactions. In this chapter we look at compounds containing only one atom of carbon. In later chapters we study those with two, then three and four atoms of carbon. Later still, we shift our emphasis and consider classes of compounds—compounds of similar reactivity. We shall find many simplifying relationships that will help us understand organic chemistry.

2.2 Some compounds with
a single carbon atom; elements of structural theory

A few compounds and their physical properties.   A selected list of representative compounds that contain a single atom of carbon is given in Table 2.1.

The gases carbon monoxide and carbon dioxide are familiar to us from previous courses in chemistry. We see that all the compounds in the table are gases at room temperature except for methanol, formic acid, and iodomethane, which are liquids, and urea, which is a solid. Only one of the liquids, formic acid, has a freezing point above 0°C. Most organic compounds of low and moderate molecular weight are gases or liquids at room temperature or solids of melting point less than 250°C. Unlike ionic substances, which have strong electrostatic forces between the ions that give rise to high melting points, organic molecules must depend upon relatively weak intermolecular forces for crystal formation.

Of the compounds listed, only methane, carbon dioxide, carbonic acid (carbonated drinks), and urea are nontoxic. Methanol is very toxic, seriously affecting the eyes and frequently causing blindness. Ingestion of methanol may be fatal. Formaldehyde and formic acid, both with repulsive odors, are toxic. A solution of formaldehyde in water (known as formalin) is used as a preservative solution for biological specimens. Formic acid blisters skin on contact. The methyl halides are, perhaps, less toxic even though methyl bromide is a widely used fumigant. Methylamine has a strong ammoniacal odor and is toxic. All organic compounds should be handled with caution and respect until their toxicities are known.

The covalent bond; tetrahedral carbon. For the present, the structure of the molecules will be described simply as consisting of covalent bonds. A covalent bond is directed in space and contains a pair of electrons. A covalent single bond is also called a σ (sigma) bond. Two hydrogen atoms, each with a single electron in its 1s orbital, are able to share the electrons by an overlap of atomic orbitals. Thus, two hydrogen atoms can form a molecule of hydrogen, written as H_2 or H—H or H:H, by overlap of their 1s orbitals. The hydrogen chloride molecule in the gaseous state has a covalent bond formed by overlap of a hydrogen 1s and a chlorine 3p orbital. It is written as HCl or H—$\overline{\text{Cl}}$l or H—Cl or H:$\ddot{\text{Cl}}$:. The shared pair of electrons experiences a greater attraction to the chlorine atom than to the hydrogen atom. Such a covalent bond is said to be polarized. The polarization may be indicated as $\overset{\delta+}{\text{H}}$—$\overset{\delta-}{\text{Cl}}$ or H→Cl. The arrow always points in the direction in which the electron pair is shifted. In a single sodium chloride molecule, the polarization becomes so great that the bond is called an ionic bond, $Na^{\oplus}Cl^{\ominus}$.

Table 2.1 Single carbon atom compounds

Name	Formula	mw	mp, °C	bp, °C
Methane	CH_4	16.04	−182.5	−161.5
Methanol (methyl alcohol)	CH_3OH	32.04	−97.8	64.6
Methanal (formaldehyde)	HCHO	30.03	−92.0	−21.0
Methanoic acid (formic acid)	HCOOH	46.03	8.4	100.7
Carbonic acid	HOCOOH	62.03	(exists only in H_2O)	
Carbon dioxide	CO_2	44.01	−57.0	−78.5 (subl.)
Carbon monoxide	CO	28.01	−207.0	−190.0
Fluoromethane (methyl fluoride)	CH_3F	34.03	−141.8	−78.6
Chloromethane (methyl chloride)	CH_3Cl	50.49	−97.7	−24.2
Bromomethane (methyl bromide)	CH_3Br	94.95	−93.7	3.6
Iodomethane (methyl iodide)	CH_3I	141.95	−66.1	42.5
Methylamine	CH_3NH_2	31.06	−92.5	−6.5
Carbamide (urea)	H_2NCONH_2	60.06	132.7	dec.

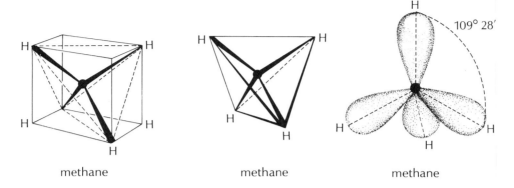

Figure 2.1 Structure of methane; C—H σ bonds are shown

A carbon atom has four electrons in the valence shell (Sec. 1.2) and is able to share them all in various ways. With four hydrogen atoms, four sigma bonds may form to give CH_4. Mutual repulsion between the electrons in these bonds is only one of the factors that influence the arrangement in space of the atoms of methane, CH_4. The arrangement, however, is easiest to describe in terms of the electron repulsion.

The four bonds in methane are all alike, so they must be of equal length. Only a tetrahedral arrangement meets the requirements of electron repulsion under these circumstances. (See Figure 2.1.) Each hydrogen is equally distant from each of the other three hydrogens (the diagonal across a face of the cube). Any movement of one of the hydrogens from the location at a corner of the cube would cause an inequality in C—H bond length and distance among all four hydrogens. This spatial arrangement is called tetrahedral because if we remove those parts of the cube that are not enclosed by the diagonal lines between the hydrogens, we have a tetrahedron, a four-faced structure, each face of which is an equilateral triangle.

A compound such as chloromethane, CH_3Cl, would have three carbon-hydrogen bonds slightly different from those of methane and a carbon-chlorine bond of a still different length and polarity. The tetrahedral arrangement is distorted to some extent. A dihalomethane such as methylene chloride (CH_2Cl_2) would have a slightly different distortion; but a trihalomethane such as trichloromethane ($CHCl_3$) would have a distortion similar in some respects to that in chloromethane. Tetrachloromethane (CCl_4) would be an undistorted tetrahedron.

A tetrahedral arrangement does not correspond with the spatial arrangement of the orbitals, one 2s and three 2p, for the atomic neon configuration. The carbon atom, to form four covalent bonds, has undergone a reorganization of the one 2s and three 2p orbitals into four equivalent orbitals that allow four equivalent covalent bonds in the tetrahedral arrangement. The reorganized orbitals are known as sp^3 orbitals. The reorganization of the orbitals is known as hybridization.

Sigma and pi bonds.   In formaldehyde, only the 2s, $2p_x$, and $2p_y$ carbon orbitals have been reorganized to give three equivalent orbitals, known as sp^2 orbitals. Again, the orbitals repel each other as well as the remaining fourth atomic orbital, the $2p_z$. The repulsion causes the three sp^2 orbitals to lie in the xy plane at 120° angles to each other, and the $2p_z$ orbital remains along the z axis at right angles to the plane of the sp^2 orbitals.

2.2 Some compounds with a single carbon atom 15

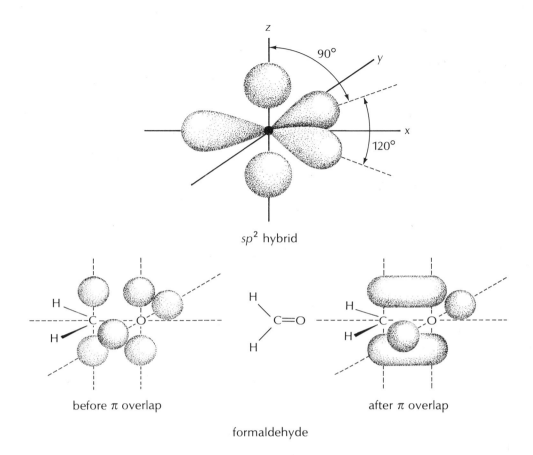

Figure 2.2 Formaldehyde; sp² hybridization. The p orbitals are shown schematically as pairs of spheres.

Two of the sp^2 orbitals are bonded to hydrogens; the third is bonded to oxygen by overlap with the p_x orbital of oxygen. Now there are left over a p_z orbital on carbon with one electron and a p_z orbital on oxygen with one electron. The orbitals, if lined up side by side, overlap sideways with each other to some extent to give rise to what is called a π (pi) bond. A π bond, with less orbital overlap, is a weaker bond than is a σ bond. The combination of a σ bond and a π bond between two atoms is called a double bond, and the formaldehyde structure may be abbreviated $H_2C=O$, which is frequently written as HCHO.

Exercises

2.1 Work out the structure of carbon dioxide. (Hint: it is possible to reorganize the carbon 2s and $2p_x$ orbitals to give two sp orbitals with the $2p_y$ and $2p_z$ orbitals remaining unhybridized.)

2.3 Introduction to naming

Since 1892, the International Union of Pure and Applied Chemistry has charged various committees and commissions with the task of devising a system of nomenclature for the compounds of chemistry. Their recommendations for organic chemistry, as modified and added to occasionally, have been accepted throughout the world and are known as IUPAC names. There are many idiosyncracies left, but the written IUPAC names are recognized by any organic chemist, whatever his language. However, pronunciation is something else again. For example, for methylamine American usage favors methill (rhymes with Ethel) ameen, while the British use mee-thile ameen. The names used in Table 2.1 are IUPAC names, and the older common names are given in parentheses.

IUPAC naming: hydrocarbons, alcohols, aldehydes, acids, and amines. The IUPAC system of naming becomes fairly complicated; but if we understand and keep in mind the first few rules from which the later rules are derived, the system remains comprehensible. The purpose of the system of rules is to build up names so that every pure compound will have a distinctive identification not shared by any other substance. Each name should be so descriptive that one may easily write the structure of the compound from it. Although the objective has not been reached in every instance, the rules have proved very successful. We have to learn to translate from structure to name and back again with ease. We shall define structure for the time being as the order and way in which the individual atoms are linked to form a molecule. For the simple molecules shown here, changing atoms constitute the main variable in structure. We shall assume that we recognize that carbon requires four covalent bonds; oxygen, two; hydrogen and halogens, one. Of course, the four bonds to carbon may all be σ, or three σ and one π, or two σ and two π. Carbon monoxide (CO) is a special case: $|C\equiv O|$.

The IUPAC rules evolved from a naming system already in use. Many of the older names and methods were simply taken over by the new system. For example, those hydrocarbons—compounds containing only carbon and hydrogen—that have σ bonds only had been designated by the ending -ane, as in methane, CH_4. The number of carbon atoms in the molecule was designated by the principal part of the name. Methane (CH_4, one carbon), ethane (CH_3CH_3, two carbons), propane ($CH_3CH_2CH_3$, three carbons), and butane ($CH_3CH_2CH_2CH_3$, four carbons) were well known, so these size names were adopted. From five carbons on up, Greek number names were adopted, so pentane, hexane, heptane, and so on are obvious. Our first rule is to use a size designation and the suffix -ane to name these hydrocarbons. The general name for a hydrocarbon with only σ bonds is alkane. As the number of carbon atoms increases, other complications in naming arise that will be taken up in Chapter 4. For the time being, our rules will be paraphrased from the IUPAC rules.

With the alkane names established, all other substances are named as their relatives or substituted products, according to a second rule. Thus, if a hydrogen in methane is replaced by a halogen, the resulting molecule is named by use of a prefix to the alkane name. The prefix used for a halogen is derived by dropping the ine ending of the element name and adding an o for ease in pronunciation, as in fluoromethane for CH_3F. By the third rule, if more than one substituent is present, all are named as prefixes in alphabetical order, as in dibromochloromethane, $CHBr_2Cl$. Note that the order of the prefixes is alphabetical for the substituents (b before c), ignoring any number designation (di). Even if all the hydrogen atoms are replaced, the alkane name is retained, as in tetrachloromethane (CCl_4) or bromotrifluoromethane ($CBrF_3$).

If a hydrogen atom is replaced by an oxygen atom, a stable molecule is not formed unless the oxygen is in turn linked to something else. A very common "something else" is a hydrogen atom. In other words, we can think of a hydrogen atom in methane as being replaced by an -OH group (hydroxy) to give CH_3OH (hydroxymethane), if we follow the prefix naming. Another way of looking at what we have just done is to think of replacing one of the hydrogen atoms in a molecule of water with a methyl (CH_3-) group. Here we use groupings of atoms related to an alkane as a replacing group. In the IUPAC rules, such groups take their names from parent alkanes. We change the alkane suffix designation of -ane to -yl. Thus ethane becomes ethyl, CH_3CH_2-; propane becomes propyl, $CH_3CH_2CH_2$-; and so on.

Substances such as CH_3OH, related to water in the way we have been discussing, had been recognized long before IUPAC names came into existence. They were called alcohols, and CH_3OH was known as methyl alcohol. The use of two words for a single substance as simple as CH_3OH seemed excessive, so an alternative rule recommends retaining the attention to the class of compound being considered by using the suffix -ol (from alcohol) to replace the final -e in the hydrocarbon name: thus, methanol. It took a long time, but the alkanol names finally are beginning to displace the older alkyl alcohol names.

If both hydrogen atoms in water are replaced with methyl groups, CH_3OCH_3 is formed. This substance, in which an oxygen is bonded to two alkyl groups, belongs to a class of compounds known as ethers. The rules consider such a compound to have a hydrogen of methane substituted by a methyloxy group (usually abbreviated to methoxy), so the name is methyloxymethane, or methoxymethane. The common name is dimethyl ether.

If two hydrogen atoms in methane are replaced by -OH groups, $CH_2(OH)_2$ should be formed. Efforts to isolate such a compound have failed; it is as elusive as carbonic acid, $O=C(OH)_2$, and exists only in aqueous solution. The name would be dihydroxymethane, or methanediol. The molecule is unstable and breaks down to give water and formaldehyde, HCHO. The IUPAC rules recommend the name methanal for HCHO, the ending -al designating the presence of -CHO in the molecule. However, the old common name of formaldehyde has remained in popular usage.

In a similar way, trihydroxymethane or methanetriol may exist in aqueous solution, but it also breaks apart into water and a substance with one C—H bond, one C—OH bond, and one C=O bond, written as HCOOH. The IUPAC rules recommend the name methanoic acid for this compound, -oic acid showing the presence of the -COOH group. Popular usage has ignored the recommendations for the simple acids, and the old common names remain in use. The old name of formic acid is almost universally used. Compounds that contain a —COOH group are acidic (in contrast to methane, methanol, and formaldehyde, which are neutral), readily losing protons to water to give solutions with pH < 7.

Replacement of all four hydrogen atoms with hydroxys would give methanetetrol; but loss of a molecule of water would give HOCOOH, or carbonic acid, which in turn easily loses a molecule of water to give carbon dioxide, CO_2.

Just as we considered methanol as being derived by the use of a methyl group to replace a hydrogen in water, so too can a methyl replace a hydrogen in ammonia to give aminomethane, more commonly called methylamine, CH_3NH_2. (The rules call for methanamine.) Methylamine also may be thought of as being obtained from methane by the replacement of a hydrogen with an amino group, $-NH_2$. Such substances resemble ammonia in the same way that alkanols (alcohols) resemble water. Urea or carbamide, like formaldehyde, has a carbon-oxygen double bond, but it contains two $-NH_2$ groups.

18   Chapter two

The IUPAC system of naming is not followed by everyone all of the time, but everyone with a knowledge of organic chemistry must know the rules. We shall use both IUPAC and common names; the less popular name usually will be given in parentheses.

Summary of rules.   Let us summarize our paraphrased IUPAC rules:

1. Hydrocarbons with only single bonds are named with a size designation and end in -ane. Exceptions are methane, ethane, propane, and butane.
2. Substituents, such as halogens, are given as prefixes: fluoro-, chloro-, bromo-, and iodo-.
3. Substituents are given in alphabetical order. More than one substituent is shown by the use of di-, tri-, tetra-, and so on, as in bromotrifluoromethane, $BrCF_3$.
4. The presence of a hydroxy group in a compound is indicated by the suffix -ol, as in methanol, ethanol, and so on. (Common names utilize the name of the alkyl group with alcohol.)
5. If both hydrogens of water are replaced by alkyl groups, the resulting compound belongs to the class of compounds known as ethers. Methoxyethane is $CH_3OCH_2CH_3$. The common name is methyl ethyl ether.
6. A compound with a $-C\!\!\stackrel{H}{=}\!\!O$, or -CHO, group is classified as an aldehyde. The IUPAC rules designate such a group with the suffix -al, as in methanal for HCHO and ethanal for CH_3CHO.
7. A compound with a $-C\!\!\stackrel{OH}{=}\!\!O$, or -COOH, group is classified as a carboxylic acid. The IUPAC rules designate such a group with the suffix -oic acid, as in methanoic acid for HCOOH and ethanoic acid for CH_3COOH. Common names are formic acid and acetic acid.
8. A compound with an $-NH_2$ group is classified as an amine and named by use of the alkane group name with -amine as a suffix. Thus, CH_3NH_2 is methanamine and $CH_3CH_2NH_2$ is ethanamine. These names are seldom seen, methylamine and ethylamine being the popular names.

Methane and ethane are the first and second members of a series of compounds with increasing molecular weight known as alkanes. Other series exist. A given series in which there is a difference of CH_2 between members is known as a homologous series. For example, propane ($CH_3CH_2CH_3$) is the next higher homolog of ethane (CH_3CH_3). Thus, there are homologous series of alkyl halides, alcohols (or alkanols), ethers, aldehydes, acids, and amines. Other series, or classes, of compounds will be discussed later.

Exercises

2.2   Write formulas for:
    a. Aminomethanol
    b. Dimethyl peroxide
    c. Hexachloroethane
    d. Bromoiodomethane
    e. Propanal
    f. Methoxyethane
    g. Trimethylmethanol
    h. Ethyl alcohol
    i. Propanoic acid

2.3   Give acceptable names, either IUPAC or common, for:
    a. $BrCH_2OH$
    b. $CH_3CH_2NH_2$
    c. FBrCHI
    d. ClCOOH
    e. $BrCH_2CHO$
    f. CI_4
    g. $ClCH_2COOH$
    h. CF_3CH_2OH
    i. $CHBr_3$

2.4 Reactions

Introduction. Let us set aside for the moment the problem of preparation, purification, and storage of organic compounds. In a way, we shall be following in the footsteps of the early investigators in organic chemistry. One of the first things they looked at was the reactivity of an organic compound with the common inorganic reagents such as hydrochloric, sulfuric, and nitric acids; the bases, such as sodium, potassium, and ammonium hydroxides; the oxidizing agents, such as potassium permanganate; and the reducing agents, such as the metals iron, copper, tin, zinc, and sodium. Later on, interactions with an expanded list of reagents became possible, including other organic compounds. If reaction occurred, the scientist had to find out what had happened. As the science developed, it became possible to predict with fair accuracy what would happen, although a large number of reactions still were puzzling. Even today new reactions are found and more and more problems are uncovered—more research is needed and is being done. For the time being, we shall adopt an authoritative stance and simply describe what happens without going into all the details of a given reaction.

Four classes of reactions. Roughly, reactions of organic compounds can be classified as: acid-base, substitution, addition-elimination, and oxidation-reduction reactions. Some reactions may fall into more than one category, but when we get to one of these, it will be apparent why we classify it as we do. The rates of reactions may be explosive, or at the other extreme, they may be very slow. In the course of these reactions, electrons may be transferred from one reagent to another. In a <u>heterolytic</u> reaction, bonds or linkages may be broken or made in such a manner that a pair of electrons goes with one group or the other. On the other hand, if an electron pair is split, each group retaining one electron,

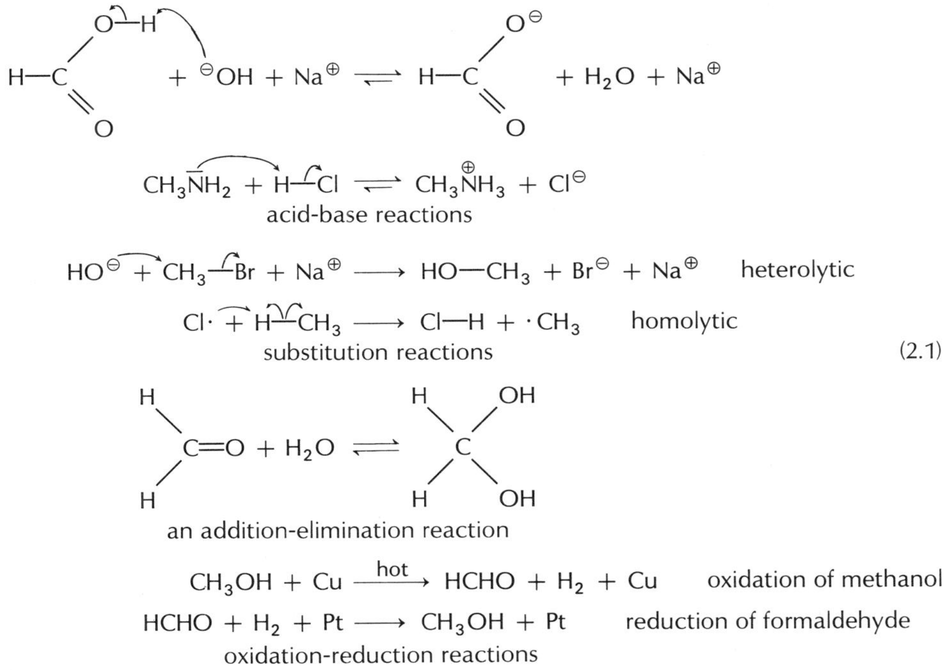

$$\text{H-C(=O)-O-H} + {}^{\ominus}\text{OH} + \text{Na}^{\oplus} \rightleftharpoons \text{H-C(=O)-O}^{\ominus} + \text{H}_2\text{O} + \text{Na}^{\oplus}$$

$$\text{CH}_3\text{NH}_2 + \text{H-Cl} \rightleftharpoons \text{CH}_3\overset{\oplus}{\text{N}}\text{H}_3 + \text{Cl}^{\ominus}$$

acid-base reactions

$$\text{HO}^{\ominus} + \text{CH}_3\text{-Br} + \text{Na}^{\oplus} \longrightarrow \text{HO-CH}_3 + \text{Br}^{\ominus} + \text{Na}^{\oplus} \quad \text{heterolytic}$$

$$\text{Cl} \cdot + \text{H-CH}_3 \longrightarrow \text{Cl-H} + \cdot \text{CH}_3 \quad \text{homolytic} \tag{2.1}$$

substitution reactions

$$\underset{\text{H}}{\overset{\text{H}}{>}}\text{C=O} + \text{H}_2\text{O} \rightleftharpoons \underset{\text{H}}{\overset{\text{H}}{>}}\text{C}\underset{\text{OH}}{\overset{\text{OH}}{<}}$$

an addition-elimination reaction

$$\text{CH}_3\text{OH} + \text{Cu} \xrightarrow{\text{hot}} \text{HCHO} + \text{H}_2 + \text{Cu} \quad \text{oxidation of methanol}$$

$$\text{HCHO} + \text{H}_2 + \text{Pt} \longrightarrow \text{CH}_3\text{OH} + \text{Pt} \quad \text{reduction of formaldehyde}$$

oxidation-reduction reactions

20 *Chapter two*

we are dealing with a homolytic cleavage or free-radical reaction (Eq. 2.1). (We illustrate the electron pair destination with a curved arrow, ⌢, and the destination of a single electron with an arrow with only half a head, ⌢, a fish hook.)

2.5 Acid-base reactions

Brønsted-Lowry and Lewis acids and bases. In the Brønsted-Lowry terminology, an acid (a proton donor) reacts with a base (a proton acceptor) to give a new acid (the conjugate acid of the base used) and a new base (the conjugate base of the acid used). Such a transfer of a proton from a relatively polar bond to oxygen or nitrogen is fast. A transfer from a relatively nonpolar bond to carbon usually is much slower. Depending upon the relative strengths of the acid and of the base, the transfer of the proton may be complete or incomplete.

A more general definition of acids and bases is possible in terms of electrons. A Lewis acid is an electron pair acceptor; a Lewis base is an electron pair donor. The Brønsted-Lowry acid-base concept is a special case of the Lewis acid-base definition. Thus, a base may act in either the Brønsted-Lowry sense and accept a proton or the Lewis sense and donate an electron pair to any acceptor (a Lewis acid). A Brønsted-Lowry acid can only be a proton donor, but a Lewis acid may be any seeker of an electron pair. All Brønsted-Lowry acids are Lewis acids, but not all Lewis acids are Brønsted-Lowry acids. Equations 2.2 illustrate the concepts of the Brønsted-Lowry and the Lewis acid-base reactions.

$$A-H + |\overset{\ominus}{O}-H \rightleftharpoons A^{\ominus} + |\overset{H}{\underset{|}{O}}-H$$

generalized base conjugate conjugate
acid   base   acid

(formic acid) + $H_2\bar{N}CH_3$ ⇌ (formate) + $H_3\overset{\oplus}{N}CH_3$

acid   base   conjugate   conjugate
                                 base   acid

$(CH_3O)_3Al + |\bar{O}-CH_3 \rightleftharpoons (CH_3O)_3\overset{\ominus}{Al}-\overset{\oplus}{O}-CH_3$
            | |
            H   H

acid   base

(2.2)

Acid-base equilibria. Because we shall be using acid-base relationships constantly, let us review briefly acid-base equilibria with a few examples. Follow the discussion with paper and pencil to be sure you understand it.

If we measure the pH of a 0.1 M aqueous solution (we use many solvents other than water in organic chemistry, so we find it necessary to specify the solvent used) of formic acid (4.6 g/liter), we find a value of 2.39. Because pH = $-\log [H^{\oplus}]$, the concentration of hydrogen ion is equal to $10^{-2.39}$ M. A little mathematics shows the hydrogen ion concentration to be 0.00408 M.

$$10^{-2.39} = (10^{0.61})(10^{-3}) = [H^\oplus]$$

$$\text{but } 10^{0.61} = \text{antilog } (0.61) = 4.08$$

$$\text{and } 10^{-3} = \text{antilog } (-3) = 0.001$$

$$\text{and } [H^\oplus] = (4.08)(0.001) = 0.00408 \text{ or } 4.08 \times 10^{-3} \text{ M}$$

which is much less than the total concentration (determined by titration) of formic acid, 0.1 M. The concentration of formate ion, $HCOO^\ominus$, is the same as the hydrogen ion concentration. The equilibrium constant for the reaction (Eq. 2.3) may be calculated as shown. More accurate work gives $K_a = 1.77 \times 10^{-4}$ or $pK_a = 3.75$ ($pK_a = -\log K_a$), and we see that formic acid is a weak acid.

$$HCOOH \rightleftharpoons HCOO^\ominus + H^\oplus$$

$$\text{or, } HCOOH + H_2O \rightleftharpoons HCOO^\ominus + H_3O^\oplus$$

$$\text{from which, } K_a = \frac{[HCOO^\ominus][H^\oplus]}{[HCOOH]} = \frac{(0.00408)(0.00408)}{(0.1 - 0.00408)} \tag{2.3}$$

$$= \frac{(0.00408)^2}{(0.0959)} = \frac{(4.08)^2(10^{-6})}{(9.59)(10^{-2})} = \frac{(16.7)(10^{-4})}{9.59}$$

$$= 1.74 \times 10^{-4}$$

If we measure the pH of a 0.1 M aqueous solution of methylamine (3.10 g/liter), we find a value of 11.81, from which we may calculate a hydrogen ion concentration of 1.55×10^{-12} M or a hydroxide ion concentration of 6.46×10^{-3} M. The equilibrium constant for the reaction may be calculated (Eq. 2.4) to be 4.4×10^{-4} or $pK_b = 3.36$, and methylamine is a weak base.

$$CH_3NH_2 + H_2O \rightleftharpoons CH_3NH_3^\oplus + HO^\ominus$$

$$K_b = \frac{[CH_3NH_3^\oplus][HO^\ominus]}{[CH_3NH_2]} = \frac{(6.46 \times 10^{-3})^2}{(0.1 - 0.00646)} = 4.4 \times 10^{-4} \tag{2.4}$$

If we examine a dilute aqueous solution of methylammonium chloride (Eq. 2.5), we find $K_a = 2.27 \times 10^{-11}$ or $pK_a = 10.64$. We note that $pK_a + pK_b = 10.64 + 3.36 = 14$, which is pK_w for water. The methylammonium ion is a very weak acid, much weaker than formic acid.

$$CH_3NH_3^\oplus + Cl^\ominus \rightleftharpoons CH_3NH_2 + H^\oplus + Cl^\ominus \tag{2.5}$$

Control of pH. The pH of human blood is maintained at 7.4 largely by the bicarbonate-carbonic acid buffer system. Two physiological processes work to hold the $[HCO_3^-]/[H_2CO_3]$ ratio at the necessary 20:1. The carbonic acid concentration is regulated by the lungs. When the concentration of bicarbonate falls below normal (resulting in a lower pH), respiration increases in order to lower the concentration of H_2CO_3.

On the other side, the kidneys aid in control of pH by regulating the bicarbonate concentration. This control is slower acting than the lungs but more sensitive to restoring the pH to exactly 7.4. An individual who hyperventilates will lower the concentration of carbon dioxide in his blood and thus lower the H_2CO_3 level. The pH will rise. His kidneys will then be working to lower the bicarbonate concentration until it is again just twenty times that of the carbonic acid and thus again the pH is normal.

Use of pK_a values. It has become conventional to express both acid and base strength (acidity and basicity) as pK_a values. As with pH values, low pK_a values indicate stronger acids (or weaker bases) than do high pK_a values (weaker acids or stronger bases). For a series of acids, as pK_a approaches zero, the concentration of undissociated acid becomes so small and so difficult to measure that proton transfer to water is considered complete. For a series of bases, as pK_a approaches 14, the concentration of unprotonated base becomes so small (Eq. 2.5) and so difficult to measure that the proton transfer to water is considered negligible. In other words, water behaves as such a strong base toward strong acids that the proton transfer from the acid to water is complete. Towards strong bases, water behaves as such a strong acid that proton transfer from water to the base is complete. This effect is known as the leveling effect and limits the range within which K_a can be determined. If we limit the concentration to 0.1 M to avoid salt effects and other complications in more concentrated solutions, the most acidic solution possible in water would be 0.1 M in hydrogen ion; the most basic solution would be 0.1 M in hydroxide ion or 10^{-13} M in hydrogen ion.

In organic chemistry many solvents other than water are used. If the solvent is capable of proton transfer as in water or methanol, it is called a protic solvent; if not, as in carbon tetrachloride (CCl_4), it is an aprotic solvent. In protic solvents, the positive ion is termed a lyonium ion and the negative ion, a lyate ion. The product constant obtained by multiplying the concentration of the lyonium ion by the concentration of the lyate ion is known as the autoprotolysis constant.

$K_w = (H^\oplus)(HO^\ominus) = 10^{-14}$ is the autoprotolysis constant of water. Table 2.2 gives the negative logarithm of the autoprotolysis constant values for a few protic solvents. The available range for measurement in water at 25° is many powers of 10, actually 10^{12} in 0.1 M solution; but ammonia is even better with a range of 10^{20}; and sulfuric acid, by comparison, has almost no range at all.

If a solvent such as formic acid is used, the pK_a values of some acids that appear to be strong acids in aqueous solution may be measured. By use of ammonia as a

Table 2.2 Some autoprotolysis constants (given as the negative logarithm)

Solvent	Temperature	$pK = -\log K_{auto}$
Sulfuric acid	10°	3.24
Formic acid	25°	6.2
Water	100°	12.3
Acetic acid	25°	12.6
Water	25°	14.0
Methanol	25°	16.7
Ethanol	25°	19.1
Ammonia (liquid)	−33°	22.0

solvent, the pK_a values of compounds that appear to be strong bases in aqueous solution may be measured. By comparing pK_a values for a set of acids or bases for one solvent with those for a second and third solvent, we find that the acids and bases usually fall into the same order of strength. The relative acidity or basicity of a given compound is a specific property of that compound. There are exceptions, but they need not bother us at this point.

Exercises

2.4 Formic acid ionizes in water with a small K_a to give an excess of hydrogen ions. Methylamine ionizes in water to give an excess of hydroxide ions. Calculate K_{eq} for the reaction shown:

$$HCOOH + CH_3NH_2 \rightleftharpoons HCOO^\ominus + CH_3NH_3^\oplus$$

(Hint: use values of K_a given in the text.)

2.5 In a mixture of methanol and water, what ions are present?

2.6 Nucleophilic substitution reactions

The ABC reaction; S_N2 defined. Equation 2.6 illustrates a substitution reaction in

$$A + BC \longrightarrow AB + C \qquad (2.6)$$

which A represents the substituting agent and BC represents the molecule in which C is to be replaced by A. There are three forms that this substitution reaction may take: nucleophilic, electrophilic, and homolytic. If reagent A has a pair of electrons that may form a bond, it is known as a nucleophile (nucleus loving), and the reaction is known as a nucleophilic substitution (a heterolytic reaction). Equation 2.7 shows the use of hydroxide ion as a nucleophile to attack a molecule of bromomethane. We frequently omit the counter ion (in this instance Na$^\oplus$ or K$^\oplus$) to avoid cluttering up an equation.

$$HO^\ominus + CH_3{-}Br \longrightarrow HO{-}CH_3 + Br^\ominus \qquad (2.7)$$

Studies of the rates of such reactions have shown that the speed at any instant is proportional to the concentration at that instant of both A and BC. These reactions are symbolized as S_N2 reactions: S for substitution, N for nucleophilic, and 2 for bimolecular involvement or simultaneous bond changes.

Comparison of nucleophilic strengths. These ABC reactions have many aspects that deserve consideration: the effects of change of structure and character of A, B, and C; the effects of solvent; the effects of temperature and pressure; and the effects of added substances such as neutral salts or acids and bases. For the time being, let us keep B constant as the methyl group and look at the nucleophilic strength of A.

Because a nucleophile attacks with a pair of electrons, it is similar to a base. However, relative basicity is measured by comparing equilibria involving proton shifts. Relative nucleophilicity is measured by comparing rates of reaction involving the making and breaking of bonds to carbon. Relative basicity is related to proton transfer, and relative nucleophilicity is related to the formation of a bond to a carbon. In other words, the displacing ability of nucleophiles is in the order of nucleophilic strength, which usually lies in the order of the strength of the nucleophiles as bases. A strong nucleophile will displace a weaker nucleophile.

The leaving group, C, also affects the reaction. For example, consider the reverse of Equation 2.7, in which a bromide ion is shown attacking methanol (Eq. 2.8), displacing a

$$Br^\ominus + CH_3{-}OH \longrightarrow BrCH_3 + OH^\ominus \qquad (2.8)$$

hydroxide ion, to give bromomethane. It looks reasonable, but the reaction does not happen. The weaker nucleophile, bromide ion, is unable to displace the strong nucleophile, hydroxide ion. In terms of base strength, $OH^\ominus > Br^\ominus$. Let us repeat: in the reaction shown in Equation 2.7, the stronger nucleophile, $HO^\ominus$, displaces the weaker nucleophile, $Br^\ominus$, with ease; but the weak nucleophile, $Br^\ominus$, is unable to displace the stronger nucleophile, $HO^\ominus$.

Acid-base equilibria and S_N2 reactions. Other reactions may compete with or modify an ABC S_N2 reaction by removing or changing A and C in various ways. Consider methanol and its resemblance to water as shown in Equations 2.9. These acid-base equilibria

$$\begin{array}{l} CH_3O{-}H + {}^\ominus OH \rightleftharpoons CH_3O^\ominus + H_2O \\ CH_3\overline{O}H + H_3O^\oplus \rightleftharpoons CH_3OH_2{}^\oplus + H_2O \end{array} \qquad (2.9)$$

are established rapidly and give rise to methoxide ion in the one and to the conjugate acid of methanol, an oxonium ion, in the other. With methoxide ion, an attempt to displace the $O^\ominus$ would give $O^{2\ominus}$, and would require a nucleophile so very much stronger than $CH_3O^\ominus$ that the reaction does not go. With methyloxonium ion, an attempt to displace $OH_2{}^\oplus$ would give water as the product. The reaction with concentrated hydrobromic acid goes well (Eq. 2.10). Water is a nucleophile comparable to the bromide ion. If

$$Br^\ominus + CH_3{-}OH_2{}^\oplus \longrightarrow BrCH_3 + H_2O \qquad (2.10)$$

bromomethane is allowed to stand with water, bromide ion and methyloxonium ion are formed. The oxonium ion reacts with water by proton transfer to produce dilute hydrobromic acid and methanol (Eq. 2.11).

$$\begin{array}{l} H_2\overline{O}| + CH_3{-}Br \longrightarrow H_2\overset{\oplus}{O}{-}CH_3 + Br^\ominus \\[4pt] H_2\overline{O}| + H{-}\underset{|}{\overset{\oplus}{O}}{-}CH_3 \rightleftharpoons H_2\overset{\oplus}{O}{-}H + |\underset{|}{\overline{O}}{-}CH_3 \\ \phantom{H_2\overline{O}| + H{-}O{-}CH_3}H \phantom{\rightleftharpoons H_2O{-}H + |O{-}CH}H \end{array} \qquad (2.11)$$

Summary. The better A is as a nucleophile, the better it is as a displacing agent; and the weaker the free C is as a nucleophile, the easier it is to displace from BC. A stronger nucleophile simply displaces a weaker nucleophile. Bromine may be displaced by water or, more easily, by hydroxide ion. The hydroxide ion is very difficult to displace, but water is displaceable.

Methane is inert to the action of nucleophiles because the extremely strong base hydride ion, $H^{\ominus}$, would have to be displaced. In other words, the C—H bond is inert to nucleophilic substitution, and alkyl groups remain intact during such reactions.

Exercises

2.6 Why do the following reactions not go?
   a. $H_2N^{\ominus} + CH_3OH \longrightarrow H_2NCH_3 + OH^{\ominus}$
   b. $H_3N + CH_3\overset{\oplus}{O}H_2 \longrightarrow H_3\overset{\oplus}{N}—CH_3 + H_2O$

2.7 In general $Y^{\ominus} + CH_3X \rightarrow YCH_3 + X^{\ominus}$ if $X^{\ominus}$ is one of the three halogen ions $Cl^{\ominus}$, $Br^{\ominus}$, or $I^{\ominus}$, because $X^{\ominus}$ is a relatively weak nucleophile. Name five possible ions that could be used as the attacking nucleophile $Y^{\ominus}$ other than $HO^{\ominus}$, NH_3, H_2O, and $NH_2^{\ominus}$. (Hint: use anions of weak acids. The acids do not have to be organic!)

2.8 Complete and balance these reactions:
   a. $CH_3OH + HBr \longrightarrow$
   b. $CH_3NH_2 + CH_3Br \longrightarrow$
   c. $CH_3O^{\ominus} + CH_3Br \longrightarrow$
   d. $Mg + CH_3I \longrightarrow$

2.7 Electrophilic substitution reactions

Definition and examples. Another heterolytic type of substitution that is much less common than the nucleophilic displacements is <u>electrophilic substitution</u>. Substitution (S) by an <u>electrophile</u> (E) with bimolecular dependence (2) is symbolized as S_E2. Making use of the ABC reaction terminology, A is a reagent in quest of electrons, and C is a grouping willing to leave behind a pair of electrons as it departs from B (Eq. 2.12). An electrophile is

$$A^{\oplus} + B—C \longrightarrow A—B + C^{\oplus} \qquad (2.12)$$

analogous to a Lewis acid in the same way that a nucleophile is analogous to a base. To seek electrons, an electrophile must have an incomplete octet of electrons (or reach this condition in the course of the reaction). A proton in transit from an acid to a base is an electrophile seeking a more available electron pair. We shall introduce a new single carbon compound, methylmagnesium iodide (without saying anything yet about how it is prepared), to show two S_E2 reactions (Eqs. 2.13). The stronger electrophile, H of HCl, or X of X_2,

$$\begin{aligned} Cl—H + CH_3—MgI &\longrightarrow Cl^{\ominus} + CH_4 + MgI^{\oplus} \\ X—X + CH_3—MgI &\longrightarrow X^{\ominus} + X—CH_3 + MgI^{\oplus} \end{aligned} \qquad (2.13)$$

is able to displace the weaker electrophile, here $MgI^{\oplus}$. The cations of metals are Lewis acids, the strengths of which vary depending upon the electropositive nature of the metal in question.

2.8 Homolytic substitution reactions

Definition and examples. Homolytic (H) bimolecular (2) substitution (S) is symbolized as S_H2. Such reactions are commonly called <u>free-radical</u> reactions. Homolytic displacements on carbon in a single step are so rare as to be almost unknown. Nevertheless, the concept is useful if we allow B of an ABC reaction to be an atom other than carbon, such as a hydrogen or a halogen.

If we let A be a chlorine atom, B a hydrogen, and C a methyl group, then the reaction to produce hydrogen chloride and a methyl radical meets the classification (Eq. 2.14). This

$$Cl\cdot + H{-}CH_3 \longrightarrow Cl{-}H + \cdot CH_3 \qquad (2.14)$$

step is one of the many involved in the reaction of chlorine with methane to give chloromethane and hydrogen chloride as the principal products (Eq. 2.15). Methane also reacts

$$CH_4 + Cl_2 \longrightarrow CH_3Cl + HCl \qquad (2.15)$$

with bromine, but less energetically, to give bromomethane and hydrogen bromide. Iodine does not react, and fluorine reacts so vigorously with methane with the evolution of so much heat that it is extremely difficult to control. These halogenation reactions usually are carried out at relatively high temperatures (250°C or higher) in the gas phase or in solution near room temperature under the influence of light. It is also possible to cause multiple substitution products to be formed. Thus chloromethane may be converted into bromochloromethane or into dichloromethane by reaction with bromine or with chlorine. Direct halogenation reactions with other single-carbon compounds—methanol, formaldehyde, and so on—are not successful because other reactions intervene.

Examining the formula for the methyl radical, we find the distinctive feature to be the absence of a fourth hydrogen and of the eighth electron. The presence of the odd electron defines a radical. Hence, a chlorine atom ($|\underline{Cl}\cdot$) is a radical, as is a methoxy radical ($CH_3{-}\underline{O}\cdot$) or a hydroxymethyl radical ($HO\dot{C}H_2$). Although their reactivity is usually high, many types of radicals are possible, even in living systems.

2.9 Addition and elimination reactions

Polarizability; resonance. Addition and elimination reactions usually are discussed together. They involve addition to a multiple bond or elimination to form a multiple bond. The reactions may be heterolytic or homolytic.

In Section 2.2 we first encountered a multiple bond, the carbon-oxygen double bond in formaldehyde. Being less confined or constrained, the pair of electrons in the π bond is easily polarizable, or pushed or pulled from one end of the bond to the other by external influences. Strictly speaking, we should describe the situation by saying that the center of density of the electron pair in the orbital may be shifted by external influence. The expression is so clumsy, however, that we easily fall into the usage of the not quite correct but shorter statement.

Other things being equal, in a carbon-oxygen double bond the preference is for oxygen to gain greater control of the π electrons. Oxygen is more electronegative than is carbon. As a result, the oxygen gains a partial negative charge, and the carbon gains a partial positive charge. If the oxygen atom were to gain complete control and acquire a full negative charge, the carbon atom would acquire a full positive charge and share in only six electrons (in the three σ bonds). Such a complete polarization of the π bond is resisted by the increase in attraction for the electron pair as the positive charge on carbon increases and the repulsion of the electron pair as the negative charge on oxygen increases. Thus, polarization or shift in center of density of the electron cloud of the π bond is incomplete. We may picture the situation as shown by a and b, which depict the two extremes.

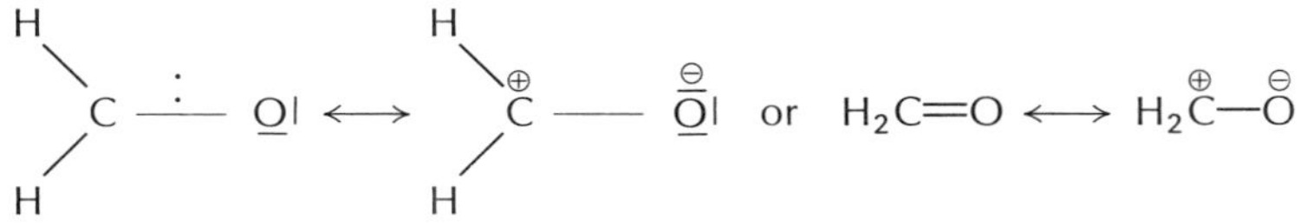

$$\underset{a}{\overset{H}{\underset{H}{\diagup}}}C = \overset{..}{\underset{..}{O}} \longleftrightarrow \underset{b}{\overset{H}{\underset{H}{\diagup}}}\overset{\oplus}{C} — \overset{\ominus}{\underset{..}{O}} \quad \text{or} \quad \underset{c}{H_2C=O} \longleftrightarrow \underset{d}{H_2\overset{\oplus}{C}-\overset{\ominus}{O}}$$

formaldehyde

The in-between situation is awkward to draw, and we make use of c and d with the double-headed arrow between the structures shown. The double-headed arrow symbol may be thought of as a confession of ignorance or of inability to draw or both. We are reduced to this means of indicating that the electron pair of the π bond is imperfectly shared between carbon and oxygen. The arrow is a symbol for <u>resonance,</u> which means that the electron pair location is not known exactly. It does not mean that the electron pair is flapping back and forth between the two structures.

Addition of water to formaldehyde. Formaldehyde dissolves in water and reacts to give methanediol, an addition reaction to the carbon-oxygen double bond (Eq. 2.16).

$$H_2C=O + H_2O \rightleftharpoons H_2C(OH)_2 \qquad (2.16)$$

The equation as written indicates an equilibrium between addition (going to the right) and elimination (going to the left). This statement gives no hint as to how the reaction takes place or how many discrete steps may be involved.

A <u>mechanism</u> is a more elaborate explanation of how a reaction occurs; and it should be able to account for all the step reactions, side-reactions, and products, as well as for the rate of the reaction. A mechanism is a pathway by which reactants go over to products. Obviously, a mechanism can help us see what is going on in this reaction. A possible mechanism might consist of three steps. The first step is the acceptance of a proton from an acid, in this case probably from a hydronium ion (Eq. 2.17). The second step would then be the attachment of a molecule of water to the positive ion a. The loss of a proton to the solvent water by b gives the product, methanediol, c. The proton transfers are rapid (acid-base reactions). The formation of the new C—O bond to give b is thought to be the slowest step. We see that it consists of the reaction of a Lewis acid, $H_2\overset{\oplus}{C}OH$, with water acting as a base. Many organic reactions consist of a series of acid-base (or nucleophile-electrophile) reaction steps. Any attempt in this instance to isolate the product fails because of the ease of reversal of the reaction. The reverse reaction is an example of an <u>elimination reaction</u> and may be recognized by the formation of a multiple bond and by the presence of a relatively simple molecule as the eliminated substance, a molecule of water in this case. Addition-elimination reactions are often, but not always, easily reversible. In many cases, an elimination must be carried out with reagents that cause it to go and that disfavor the addition reaction, as we shall see in the next section.

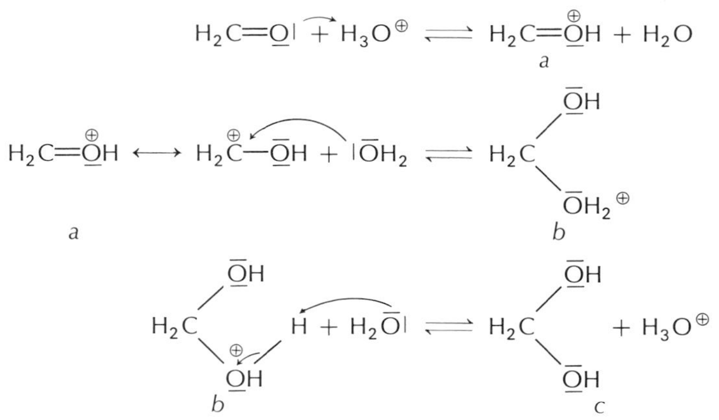

28 Chapter two

$$H_2C=\underline{O}| \stackrel{\frown}{+} H_3O^\oplus \rightleftharpoons H_2C=\overset{\oplus}{\underline{O}}H + H_2O$$
a

$$H_2C=\overset{\oplus}{\underline{O}}H \longleftrightarrow H_2\overset{\oplus}{C}\overset{\frown}{-\underline{O}H} + |\overline{O}H_2 \rightleftharpoons H_2C\overset{\overline{O}H}{\underset{\overline{O}H_2{}^\oplus}{<}}$$

(2.17)

a   b

$$H_2C\overset{\overline{O}H}{\underset{\underset{b}{\overset{\oplus}{\underline{O}}H}}{<}}\stackrel{\frown}{H} + H_2\overline{O}| \rightleftharpoons H_2C\overset{\overline{O}H}{\underset{\underset{c}{\overline{O}H}}{<}} + H_3O^\oplus$$

Formaldehyde. Formaldehyde (methanal) is a colorless gas with an irritating odor familiar to generations of biology students. Its water solution is used as an embalming fluid due largely to its high reactivity with proteins.

Although associated for years with its preservative properties and its crosslinking properties, formaldehyde has recently emerged as the glamour molecule of the new field of astrochemistry. In the winter of 1968–1969, it was discovered by radio telescope, along with ammonia and water, to be present in interstellar clouds. Previously it was thought only simple diatomic molecules could exist in the outer space environment of low density and of harsh exposure to ultraviolet and cosmic radiation. Formaldehyde seems to exist throughout our galaxy, especially in dusty regions. This finding, together with the subsequent identification of other organic compounds, suggests the possibility of a link between interstellar clouds and the origin of life.

Exercises

2.9 Write plausible steps for these reactions:

a. $H_2C=O + H_2O \xrightleftharpoons{\text{dil. NaOH}} H_2C(OH)_2$
b. $CH_2=NH + 3H_2O \longrightarrow CH_2(OH)_2 + NH_4^\oplus + OH^\ominus$

2.10 Oxidation and reduction reactions

Oxidation; combustion. Oxidation and reduction reactions of organic compounds are conveniently considered together. One of the simplest to write out, but with a very intricate mechanism that we shall not go into here, is direct oxidation with oxygen, combustion (Eq. 2.18). The chemistry of a flame or of an explosion is still a long way from being

$$CH_4 + 2O_2 \longrightarrow CO_2 + 2H_2O \tag{2.18}$$

completely understood. These combustions represent the most widespread usage of hydrocarbons because of the large amount of heat given off.

Most of the known organic compounds will burn. Because carbon dioxide is the ultimate product of combustion, the closer the structure of the molecule approaches that of carbon dioxide, the less flammable the substance becomes. Thus, methanoic acid (formic acid) differs from carbonic acid (hydrated CO_2) only slightly; yet it is flammable. Formaldehyde is more flammable than formic acid. The presence of halogens decreases flammability: although trichloromethane, like methanoic acid, may be caused to burn, tetrachloromethane is nonflammable.

Partial combustion. As we mentioned in Section 2.3, we may consider the changes from methane through methanol, formaldehyde, and formic acid to carbonic acid as successive replacements of the methane hydrogens by hydroxy groups. On the other hand, methane can be burned, and the product carbon dioxide can be dissolved in water to give carbonic acid. If the latter is an oxidation, then the former sequence must represent oxidation in four stages. Each replacement of a hydrogen by a hydroxy group is an oxidation. Each stage can be caused to occur separately, but with varying success. The first, the conversion of methane to methanol, is the most difficult. If methane is partially burned in the presence of a catalyst in a restricted amount of air, and the temperature is kept between 350° and 450°C, the principal products are methanol and formaldehyde (Eq. 2.19). The difficulty in trying to prepare methanol appears to be the instability of

$$2CH_4 + O_2 \xrightarrow[\text{cat.}]{350°} 2CH_3OH$$

$$2CH_3OH + O_2 \xrightarrow[\text{cat.}]{350°} 2HCHO + 2H_2O$$

(2.19)

methanol to oxygen so that the second stage of oxidation, the formation of formaldehyde, cannot be completely repressed. The conversion of methanol to formaldehyde is somewhat easier to manage. Depending upon the ratio of methanol to air and the catalyst and temperature used, some methanol may escape reaction; or excess oxygen or even hydrogen may appear.

Reduction; catalytic hydrogenation. Since the conversion of methanol to formaldehyde is an oxidation step, the reverse is a reduction. Thus, under certain conditions it is possible to cause hydrogen to undergo an addition reaction with the carbon-oxygen double bond of formaldehyde to give methanol (Eq. 2.20). A hydrogenation proceeds

$$HCHO + H_2 \xrightarrow[25°]{\text{cat.}} CH_3OH$$

(2.20)

readily in solution at room temperature if a finely divided catalyst—such as platinum, palladium, or nickel—is present and the pressure of hydrogen is greater than one atmosphere. Such an addition of hydrogen is a reduction reaction. It is also an addition reaction that is not as easily reversible as the example given in Section 2.9. However, if methanol vapor is passed over chips of copper at 300° to 325°C, the reverse reaction occurs, although it does not go to completion (Eq. 2.21). In other words, methanol will undergo an elimination

$$CH_3OH \underset{300°}{\overset{Cu}{\rightleftarrows}} HCHO + H_2$$

(2.21)

reaction that is also an oxidation, even though no oxygen or oxidizing agent is present. Elimination reactions are commonly denoted by the prefix de-, so the loss of hydrogen in this way is called a dehydrogenation reaction. Loss of water is a dehydration reaction, and loss of hydrogen chloride is a dehydrochlorination reaction. The word <u>al</u>dehy<u>de</u> was coined from <u>al</u>cohol and <u>de</u>hydrogenation to describe the class of compounds obtainable from alcohols by loss of hydrogen.

In Equations 2.19, 2.20, and 2.21 we introduced the accepted practice of specifying reagents, solvents (if needed), and temperature necessary for the success of the reaction by writing them above and below the arrow or arrows in the chemical equation. When we do so, no attempt is made to balance the equation. It is assumed that we are able to balance such equations by simple inspection.

Oxidation of formaldehyde by very mild reagents. Formaldehyde is the easiest of all in the oxidation series to oxidize: that is, formaldehyde is a good reducing agent. In the process of reducing something else, formaldehyde is oxidized. Even though strong oxidizing agents such as chromic anhydride, sodium dichromate, or potassium permanganate may be used in aqueous solution, it is unnecessary to be so drastic: surprisingly mild conditions may be used. A complexed silver ion in aqueous solution is able to do the job (Eq. 2.22). An aqueous solution of formaldehyde gives no visible reaction with a solution

$$H_2C=O + 2Ag(NH_3)_2^{\oplus} + 3HO^{\ominus} \longrightarrow {}^{\ominus}O\text{-}CH=O + 2Ag\downarrow + 4NH_3 + 2H_2O \tag{2.22}$$

of silver nitrate; but if the silver ion is complexed with ammonia in alkaline solution, the reaction proceeds to give a black precipitate of silver. At suitable dilutions, and if the glass surface of the containing vessel is clean, the silver deposits on the glass to form a mirror.

Formic acid is an interesting substance: it is both an acid and an aldehyde and exhibits reactions characteristic of both. It is easily oxidized, but less so than formaldehyde (Eq. 2.23). The mild oxidation of formaldehyde to formate ion may be managed without

$$^{\ominus}O\text{-}CH=O + 2Ag(NH_3)_2^{\oplus} + 3HO^{\ominus} \longrightarrow {}^{\ominus}O\text{-}C(=O)\text{-}O^{\ominus} + 2Ag\downarrow + 4NH_3 + 2H_2O \tag{2.23}$$

oxidizing formate ion (as was done in Equation 2.23) by suitable control of the amount of the oxidizing agent present. Formic acid is occasionally used as a reducing agent of some versatility.

The reduction series. The reverse sequence, the stepwise reduction from carbon dioxide to methane, is more difficult to carry out in a clean-cut fashion. We shall omit details for the present and only make the statement that CO_2 may be reduced to HCOOH. Formic acid, in turn, is reducible to formaldehyde, and formaldehyde may be reduced to methanol. The reduction of methanol to methane is possible but difficult.

Alpha and beta eliminations. Carbon monoxide is related to formic acid in a way that has some similarity to the relation of carbon dioxide to carbonic acid. Just as carbon dioxide is formed by the loss of water from carbonic acid, carbon monoxide is formed by the loss of water from formic acid. Both reactions are speeded up in the presence of a strong acid such as sulfuric acid (Eqs. 2.24). There the resemblance ends. Both are elimination

$$H_2CO_3 \xrightarrow{H_2SO_4} CO_2 + H_2O$$
$$HCOOH \xrightarrow{H_2SO_4} CO + H_2O$$
(2.24)

reactions; but if we examine carbonic acid more closely, we see that the bonds being broken are to C and O atoms that are bonded to each other. The elimination forms a bond between C and O (Fig. 2.3). In formic acid, the bonds being broken are attached to

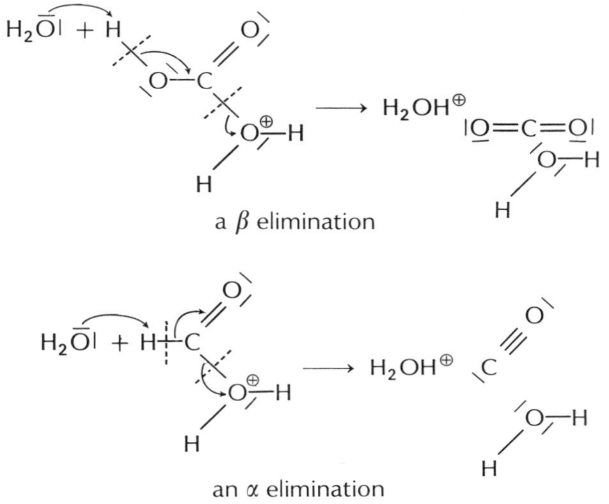

Figure 2.3 Elimination of water from carbonic acid, a β elimination, and from formic acid, an α elimination

the same atom, the carbon; and a molecule of carbon monoxide results. The two eliminations are differentiated by calling the one in which a double bond is established between adjacent atoms a β (beta) elimination; and the one in which both groups leave the same atom, an α (alpha) elimination. The β elimination is by far the more common reaction.

If carbon monoxide is heated under pressure with sodium hydroxide, an addition occurs, and the salt of formic acid (sodium formate) results (Eq. 2.25).

$$CO + NaOH \xrightarrow[\text{6-10 atm.}]{200°} HCOO^{\ominus} + Na^{\oplus}$$
(2.25)

2.11 Summary

Just because almost every reaction mentioned so far has involved an inorganic reagent, it must not be supposed that all reactions of organic compounds require inorganic substances for reaction to occur. We have restricted our examples in order to focus attention on the organic component. Organic compounds may interact with other organic compounds, and most of such reactions will fall into one or another of the four classes that have been illustrated.

It has already become apparent that the classification of reaction types is not exclusive: that is, a given reaction may be both an addition reaction and a reduction, or it may be both a substitution reaction and an oxidation or a reduction. Thus, is the classification of reaction types a real help to us or not? The answer, that the classifications will be of help, can only be appreciated as we proceed. Each homologous series undergoes reactions characteristic of that series of compounds.

We have raised many more questions than answers. This process will continue throughout our study. The business of proper questions and answers may serve as a definition of research or of the learning process, or as an explanation of the fascination of science. As we explore organic chemistry, we shall begin to comprehend interrelationships not previously suspected and become able to ask more and more sophisticated questions. As we try to clarify a given point, we shall continually need to pause and take up another subject to provide a better background for discussion.

For convenience, some sample reactions from the chapter are reproduced here.

Substitution, $S_N 2$
$$HO^\ominus + CH_3Br \longrightarrow HO-CH_3 + Br^\ominus \qquad (2.7)$$

Substitution, $S_E 2$
$$Cl-H + CH_3MgI \xrightarrow{\text{ether}} Cl^\ominus + CH_4 + MgI^\oplus \qquad (2.13)$$

Substitution, $S_H 2$
$$Cl\cdot + H-CH_3 \longrightarrow Cl-H + \cdot CH_3 \qquad (2.14)$$

Addition
$$HCHO + H_2O \rightleftharpoons H_2C(OH)_2 \qquad (2.16)$$

β Elimination
$$H_2C(OH)_2 \xrightarrow{\text{warm}} H_2C=O + H_2O \qquad \text{(similar to 2.24)}$$

α Elimination
$$HCOOH \xrightarrow{H_2SO_4} CO + H_2O \qquad (2.24)$$

Oxidation
$$HCOO^\ominus + 2Ag(NH_3)_2^\oplus + 3HO^\ominus \longrightarrow CO_3^{2-} + 2Ag\downarrow + 4NH_3 + 2H_2O \qquad (2.23)$$

Reduction
$$H_2C=O + H_2 \xrightarrow[25°]{Pt} H_3COH \qquad (2.20)$$

2.12 Problems

The problems given at the end of each chapter will be largely concerned with the material presented throughout that chapter. Some of them will be drills, testing memory only. Others will require some thinking, and as such are a learning device that helps us to see relationships not otherwise brought out. Some will require looking up subjects from previous courses. Some will be concerned with a combination of material from the current chapter and from previous chapters. A few will anticipate material to be presented in future chapters. Some will be difficult; others, easy. No attempt has been made to order the problems in difficulty.

Before doing the problems, go over the chapter at least twice. A good scheme for most of us (study habits vary, and you might find a better scheme for yourself) is to read the chapter first, not idly, but without any attempt to commit anything to memory. Then go through the chapter again, more seriously this time, trying to fix the main points in mind. Now do it a third time, trying to bear details in mind. Prepare a written summary of the reactions given in the text and incorporate any additional reactions and explanations from your lecture notes. Then work on the exercises and problems. Do not look at the answers until and unless nothing more can be done.

A good idea is to purchase an inexpensive, full-size notebook and work out the exercises and problems in it. In this way, you will have a record, easily available, for future study and review—a super answer book. Write, write, and write. Do not try to work out a problem in your mind and then go on to the next one. Write it down, and write it down completely. Organic chemistry requires active and educated fingers. While you are at it, write down questions to be put to the instructor. Cultivate your curiosity as well as your memory.

2.1 Trichloromethane reacts with hot concentrated sodium hydroxide to produce :CCl$_2$, a reactive entity, the fate of which need not concern us here. Devise a reasonable explanation for the reaction, and classify it. Why would you not expect chloromethane to give a similar reaction?

2.2 Define or explain these terms and phrases:
   a. homolytic
   b. addition
   c. substitution
   d. equilibrium
   e. mechanism
   f. S$_N$2
   g. S$_E$2
   h. S$_H$2
   i. reduction
   j. combustion
   k. chlorination
   l. aldehyde
   m. decomposition
   n. acid
   o. polarizable

2.3 Complete the reactions:
   a. HCOOH + NaOH $\longrightarrow$
   b. HCOOH + CH$_3$O$^\ominus$ $\longrightarrow$
   c. HCOOH + NH$_3$ $\longrightarrow$
   d. NH$_4^\oplus$ + CH$_3$OH $\longrightarrow$
   e. NH$_3$ + CH$_3$Br $\longrightarrow$

2.4 If CH$_3$Cl may be called methyl chloride, give names to:
   a. CH$_3$OSO$_3$H
   b. CH$_3$ONO$_2$
   c. CH$_3$ONO
   d. CH$_3$OSO$_2$OCH$_3$
   e. CH$_3$SCH$_3$
   f. CH$_3$CN
   g. CH$_3$SCN
   h. CH$_3$OSO$_2$H
   i. CH$_3$OCOOCH$_3$

2.5 Complete the reactions, giving reasonable steps when possible:
   a. CH$_3$OH + K$_2$Cr$_2$O$_7$ + H$_2$SO$_4$ $\longrightarrow$
   b. HCHO + CH$_3$NH$_2$ $\longrightarrow$
   c. CH$_3$OH$_2^\oplus$ + HSO$_4^\ominus$ $\longrightarrow$
   d. HCHO + CH$_3$OH $\xrightarrow{H^\oplus}$

3 More Reactions of More Compounds

In this chapter we consider organic compounds with two atoms of carbon, and we build on what we have already learned. The reactions from the preceding chapter are applicable to these new compounds; in addition, we shall see some new reactions. We shall pay the most attention to expanding the subject of addition-elimination reactions. We shall make some application of thermodynamics to help our understanding of the energy factors involved in reactions. We shall learn enough naming to accustom us to the new compounds. From here on, we shall not consider acid-base reactions as a separate subject. We have seen how intertwined all other reactions are with acid-base reactions, and it would be tedious to dissect out every acid-base reaction for separate examination.

3.1   Some compounds with two atoms of carbon   35
Two-carbon compounds; physical properties.

3.2   More about naming   37
1. Common and IUPAC names; multiple substituents; rings.   2. Isomers; structural isomerism. Exercises.

3.3   Nucleophilic substitution reactions   39
1. S_N2 reactions of alkyl halides; ether formation.   2. Steric effects.
3. Formation of amines.   4. Inertness of vinyl halides.   5. Organic chemistry in water. Exercises.

3.4   Electrophilic substitution reactions   43
Displacement of metals from metal organic compounds. Exercises.

3.5   Homolytic substitution reactions   43
1. Homolytic displacement on halogen; Grignard reagents.   2. Homolytic displacement on hydrogen: chain reactions. Exercises.

3.6   Addition reactions   45
1. Carbon-carbon double bonds; addition by electrophiles; by free radicals; by catalysts.   2. Polar addition to ethylene by halogen.   3. Addition of acids to alkenes and alkynes.   4. Addition of a peracid.   5. Addition to the carbon-

oxygen double bond.   6. Free-radical addition to alkenes and alkynes.
7. Photochemical smog. Exercises.

3.7 Elimination reactions   51
1. Dehalogenation.   2. Dehydrohalogenation.   3. Dehydration. Exercises.

3.8 Oxidation and reduction reactions   53
1. Combustion; the oxidation series.   2. Strong oxidizing reagents.
3. Reduction of aldehydes, acids, and acid chlorides.   4. Oxidation of alkenes and alkynes by permanganate.   5. Catalytic hydrogenation. Exercises.

3.9 Energy changes   56
1. Enthalpy and the first law of thermodynamics.   2. Entropy and the second law; heterogeneous equilibrium.   3. Gibbs free energy and reaction tendency.
4. Interrelationship of enthalpy, entropy, and free energy.   5. Standard-state reactions.   6. Use of heats (enthalpies) of formation and combustion data. Exercises.

3.10 Summary   61

3.11 Problems   63

3.12 Bibliography   64

3.13 Saul G. Cohen: From freshman to dean   64a

3.1 Some compounds with two atoms of carbon

Two-carbon compounds; physical properties.   As soon as we go on to consider molecules containing two atoms of carbon, we are faced with the beginning of that multiplicity of compounds that is so characteristic of organic chemistry. Instead of having only four hydrogens to replace in methane, we may replace six hydrogens in ethane (CH_3CH_3). Furthermore, ethane has two sets of three hydrogens, so that if two hydrogens are replaced they may be on the same carbon or on separate carbons, as in CH_3CHF_2 and FCH_2CH_2F. Replacement with two hydroxy groups on the same carbon gives the instability we observed with the addition of water to formaldehyde (Eqs. 2.16, 2.17). Equation 3.1 is the reverse of the addition of water to a carbon-oxygen double bond.

$$CH_3CH(OH)_2 \rightleftharpoons CH_3CHO + H_2O \qquad (3.1)$$

If they are on separate carbons, the hydroxys are stable. A selection of two-carbon compounds is listed in Table 3.1.

Table 3.1 Some compounds with two carbon atoms

Name	Formula	mw	mp, °C	bp, °C
ethane	CH_3CH_3	30.07	−172.0	−88.3
ethanol	CH_3CH_2OH	46.07	−114.5	78.3
ethyl fluoride (fluoroethane)	CH_3CH_2F	48.06	−143.2	−37.7
ethyl chloride (chloroethane)	CH_3CH_2Cl	64.52	−138.7	12.2
ethyl bromide (bromoethane)	CH_3CH_2Br	108.98	−119.0	38.0
ethyl iodide (iodoethane)	CH_3CH_2I	155.98	−108.5	72.2
ethylamine (ethanamine)	$CH_3CH_2NH_2$	45.08	−80.6	16.6
dimethylamine	$(CH_3)_2NH$	45.08	−96.0	7.4
acetaldehyde (ethanal)	CH_3CHO	44.05	−123.5	21.0
acetic acid (ethanoic acid)	CH_3COOH	60.05	16.6	118.1
acetyl chloride (ethanoyl chloride)	CH_3COCl	78.50	−112.0	51.0
ethylene (ethene)	$CH_2{=}CH_2$	28.05	−169.4	−103.9
vinyl chloride (chloroethene)	$CH_2{=}CHCl$	62.50	−159.7	−13.9
acetylene (ethyne)	$HC{\equiv}CH$	26.04	−81.8	−83.6
chloroacetylene (chloroethyne)	$HC{\equiv}CCl$	60.49	—	−32.0
glycol (ethanediol)	$HOCH_2CH_2OH$	62.07	−17.4	197.2
glyoxal (ethanedial)	OHCCHO	58.04	15.0	50.4
oxalic acid (ethanedioic acid)	HOOCCOOH	126.07	189.0	150.0
methyl ether (methoxymethane)	CH_3OCH_3	46.07	−138.5	−23.6
methyl formate (methyl methanoate)	$HCOOCH_3$	60.05	−99.0	31.5
ethylene oxide (epoxyethane)	$H_2C\underset{\diagdown O \diagup}{\!-\!\!-\!\!-\!}CH_2$	44.05	−111.3	10.7

Of the twenty-one substances listed, only one is a solid at room temperature. Of the remainder, eight are liquids and twelve are gases. With some significant exceptions, there is a tendency for the mp and bp to increase as the mw increases. Compare the mp and bp of the series of ethane through the ethyl halides to ethyl iodide. On the other hand, comparison of the compounds with similar mw, say about 45, shows a variation from a low of −143° in mp for ethyl fluoride to a high of −81° in mp for ethylamine; and a low of −38° in bp for ethyl fluoride to a high of 78° in bp for ethanol. Such wide variations are usually attributed to hydrogen bonding or other associations that cause the substance in question to act as though the molecular weight were higher than that for the monomolecular species.

Of the compounds listed, only acetic acid (vinegar) may be considered safe for ingestion. Even so, glacial acetic acid (pure enough to freeze on a cold day) will raise welts on the skin. Ethanol (alcohol) has been used since antiquity in various diluted forms as a beverage (beer, ale, wines, and so on). Alcoholism is the tragic result, however, of overindulgence by those who are susceptible to alcohol addiction. Ethane and acetylene may be regarded as physiologically inert. Ethylene has been used as an anesthetic and has been shown to be an important growth hormone. All other compounds in the list are toxic in varying degrees. Methyl formate is particularly dangerous. It is reported to be lethal if a

1% vapor concentration is breathed for only two and a half hours. Oxalic acid is present in plants as diverse as rhubarb and spinach. The two carboxylic acid groups of oxalic acid complex with calcium. At high enough concentrations of the two compounds, crystals of the salt form that can block the function of living tissue. Ethylene oxide is an eye irritant and has been used as a fumigant.

Halogenated hydrocarbons have been put to many uses requiring chemical stability. Teflon, a polymer of repeating —CF_2— units, is a good example. Small fluoro, chloro compounds known as Freons have been used extensively as refrigerants. The gas bromotrifluoromethane is being studied as a clean, nontoxic fire extinguishant. Halothane (2-bromo-2-chloro-1,1,1-trifluoroethane) is one of the common gaseous anesthetics.

Chlorofluoro-carbons have been released over the years by users of aerosol cans in large enough quantities to be measured in the atmosphere. These nontoxic gases have an estimated residence time in the atmosphere of greater than ten years, but their ultimate fate in the environment has yet to be determined.

```
        F              Br  F
        |              |   |
   Cl—C—Cl         H—C—C—F
        |              |   |
        F              Cl  F

     Freon 12         Halothane
```

3.2 More about naming

Common and IUPAC names; multiple substituents; rings. The names given in Table 3.1 are common except for ethane and ethanol. The IUPAC names for these simple substances, given in parentheses, are not used much. Ethanol is beginning to replace the common name, ethyl alcohol, for CH_3CH_2OH. Fluoroethane, chloroethane, bromoethane, and iodoethane are obvious IUPAC names for the ethyl halides.

Ethanal is the IUPAC name for acetaldehyde. The common names for aldehydes are derived from the common name of the related acid, not from the related alcohol. Thus, acetaldehyde is related to acetic acid.

Acetic acid is universally used, not the IUPAC name ethanoic acid. Acetyl chloride is used instead of ethanoyl chloride.

IUPAC rules designate the presence of a carbon-carbon double bond by changing the ending -ane to -ene, and a triple bond by the ending -yne (rhymes with pine). Ethene instead of ethylene ($H_2C\!=\!CH_2$) and ethyne instead of acetylene (HC≡CH) follow the IUPAC rules but are seldom used. However, we frequently use alkane to refer to a member of the C_nH_{2n+2} homologous series, alkene to refer to a member of the C_nH_{2n} homologous series that contains one carbon-carbon double bond, and alkyne to refer to a member of the C_nH_{2n-2} homologous series that contains one carbon-carbon triple bond. The grouping $CH_2\!=\!CH$- is known as the vinyl group, so vinyl chloride for $H_2C\!=\!CHCl$ is a common name.

If there is more than one substituent present in a compound, the locations are indicated by position numbers. The position numbers are obtained by numbering the carbon chain in the direction that will give the substituents the lowest possible numbers. Thus, CH_3CHF_2 is 1,1-difluoroethane, not 2,2-difluoroethane; FCH_2CH_2F can only be 1,2-difluoroethane. Each substituent is given a position number that precedes the name of the substituent. The number and the substituent name are separated by a dash. If more than one position number is needed for a given substituent, the numbers are separated by commas.

The alphabetical order for substituents is retained by disregarding the prefixes di-, tri-, tetra-, and so on, used for denoting the number present of a certain substituent. Thus, 2,2-dibromo-1,1,1-trifluoroethane for Br_2CHCF_3 is preferred to 1,1 and 2,2,2 because the latter adds up to eight as compared to seven. The alphabetical order for substituents has to do with the writing of the name and nothing to do with a preference in numbering the carbon chain.

Multiple substitution names that follow the IUPAC rules are in wide use today, except for the two-carbon compounds. Thus glycol, also known as ethylene glycol, becomes 1,2-ethanediol if the rules are followed. Similarly, the rules indicate the name ethanedial instead of glyoxal for OHCCHO. No numbers are used in this instance because the aldehyde groupings can come only at the end of a chain, and the numbers can be only 1 and 2. Likewise, the name ethanedioic acid follows the rules; but oxalic acid remains in use for HOOCCOOH.

Dimethyl ether (sometimes methyl ether) is considered to arise from a substitution of methane: the grouping CH_3O-, known as methyloxy but contracted to methoxy, is the substituent. Methoxymethane is clear and unambiguous. Ethers, as a class of molecules, have an oxygen linked to two carbon atoms.

Methyl formate, like dimethyl ether, is a new type of molecule to us. According to the rules, it should be methyl methanoate. Although both names suggest a similarity to a salt of the acid, the only similarity is in the name. Molecules like this one are combinations of an alcohol and an acid, and they are known as esters. We refer to this specific ester as the methyl ester of formic acid or as the formic ester of methanol. Esters as a class have an oxygen linking two carbon atoms, one of which has a double bond to another oxygen atom.

$$HC\begin{matrix}\nearrow O \\ \searrow O-CH_3\end{matrix}$$
methyl formate

Ethylene oxide is our first example of a cyclic molecule, a ring instead of a chain of atoms. Rings containing only carbon atoms are known as carbocycles. If an atom other than carbon is present in the ring, as in ethylene oxide, the ring is a heterocycle. Sugars, most alkaloids, and many proteins are heterocyclic compounds. Ethylene oxide may be considered an ether. Such a three-membered oxygen heterocycle is also called an epoxide as a class designation. The IUPAC rules for the naming of heterocyclic compounds are so complicated that they will be postponed until Chapter 28.

Multiple substitution is not limited to the examples given so far. Chloroacetic acid, $ClCH_2COOH$; aminoethanol, $H_2NCH_2CH_2OH$; trichloroacetaldehyde, CCl_3CHO; and glyoxalic acid, OHCCOOH, will give an idea of the variety of possibilities.

The IUPAC rules and names may be likened to the rules of grammar and to a dictionary. Although the grammar and the rules make sense, not everyone follows correct grammar or rules or uses the dictionary. Scientists are human; and organic chemists, although they try to follow the rules and names in their writing, have a tendency to relax and use the old common names at home and in talking shop. So the common names survive.

Isomers; structural isomerism. In Table 3.1 there are two compounds with the molecular formula C_2H_4O, two with $C_2H_4O_2$, two with C_2H_6O, and two with C_2H_7N.

Molecules that have the same general formula but are different substances are known as isomers, and the phenomenon is called isomerism. If the isomerism is caused by a difference in the way the atoms are linked together, as in these instances, it is known as structural isomerism. There are other types of isomerism that will be described later. As chains or rings increase in size, the possibilities for structural isomerism increase many times over. Structural isomers usually are distinguishable by differences in physical and chemical properties.

For a molecular formula of $C_{10}H_{22}$ there are 75 isomers. For $C_{20}H_{42}$, 366,319 isomers are possible!

Exercises

3.1 Give IUPAC names for:
   a. $BrCH_2CHO$
   b. $ClBrCHCH_2F$
   c. ICH_2CH_2OH
   d. H_2NCH_2COOH
   e. $Cl_2CHCOOCH_3$
   f. $Cl_2C=CH_2$
   g. $CH_3COOCH_2CH_3$
   h. CH_3OCH_2CHO
   i. $CH_3CH(OCH_3)_2$

3.2 There are four isomeric chlorobutanes, C_4H_9Cl. Write the structural formulas and names.

3.3 Identify the four pairs of isomers mentioned in Section 3.2.

3.3 Nucleophilic substitution reactions

S_N2 reactions of alkyl halides; ether formation. The S_N2 reactions of the ethyl halides are like those of the methyl compounds, except that the rates are somewhat slower and other reactions that are not possible with the methyl compounds may compete with the S_N2 reaction. Sodium iodide, dissolved in methanol, reacts cleanly with ethyl chloride in an S_N2 reaction to give ethyl iodide and sodium chloride, which, being less soluble than sodium iodide, crystallizes from solution as the reaction proceeds. (Compare Eq. 3.2 with

$$Na^{\oplus} + I^{\ominus} + \overset{CH_3}{\underset{}{CH_2}}-Cl \xrightarrow{25°, CH_3OH} CH_3CH_2I + \underline{NaCl\downarrow} \qquad (3.2)$$

the similar S_N2 reaction of CH_3I, Eq. 2.7.) As the base strength of the nucleophile is increased and as the temperature is increased, an elimination reaction begins to compete and then to dominate. In aqueous acetone at room temperature, hydroxide ion reacts to displace a halogen by an S_N2 reaction and form ethanol (Eq. 3.3). (Acetone is $(CH_3)_2C=O$, a liquid,

$$HO^{\ominus} + \overset{CH_3}{\underset{}{CH_2}}-Br \xrightarrow[\text{acetone}]{25°, H_2O} CH_3CH_2OH + Br^{\ominus} \qquad (3.3)$$

bp 56°. It is an aprotic solvent that is completely miscible with water.) At a higher temperature in a higher boiling solvent, the basic ion is increasingly able to attack a hydrogen on the carbon atom next to the carbon atom bonded to the bromine so that the halogen departs as the ion, leaving behind a carbon-carbon double bond (Eq. 3.4). Such a β elimination

$$\text{HO}^{\ominus} + \text{H—CH}_2\text{—CH}_2\text{—Br} \xrightarrow{100°,\ \text{H}_2\text{O}} \text{HOH} + \text{CH}_2\text{=CH}_2 + \text{Br}^{\ominus} \quad (3.4)$$

reaction, impossible with a methyl halide, is called a <u>dehydrohalogenation</u>. We shall return to eliminations later in the chapter (Sec. 3.7).

There is another competition with the S_N2 reaction of hydroxide ion with both methyl and ethyl halides that we purposely omitted from the previous chapter to keep the picture as simple as possible. Let us suppose that the reaction of hydroxide ion with ethyl bromide is halfway to completion; the concentration of ethanol in the mixture is now equal to that of hydroxide ion. Ethanol and hydroxide ion equilibrate by a rapid acid-base reaction (Eq. 3.5). Ethoxide ion, the lyate ion of ethanol, is a strong base and a good nucleophile.

$$\text{CH}_3\text{CH}_2\text{O—H} + \text{HO}^{\ominus} \rightleftharpoons \text{CH}_3\text{CH}_2\text{O}^{\ominus} + \text{H}_2\text{O} \quad (3.5)$$

Ethoxide ion is able to attack ethyl bromide to form diethyl ether (Eq. 3.6). The same thing

$$\text{CH}_3\text{CH}_2\text{O}^{\ominus} + \underset{\underset{\text{CH}_3}{|}}{\text{CH}_2}\text{—Br} \longrightarrow \text{CH}_3\text{CH}_2\text{OCH}_2\text{CH}_3 + \text{Br}^{\ominus} \quad (3.6)$$

happens in the reaction of hydroxide ion with methyl halides, and dimethyl ether is produced. Of course, if an ether is to be prepared, one would begin with the alkoxide ion and alkyl halide to avoid the formation of the alcohol. If the alcohol is to be prepared, and the formation of the ether is to be avoided, the equilibrium of Equation 3.5 must be pushed and kept to the left to keep the concentration of hydroxide ion as high as possible. It is obvious that the presence of water is required for this purpose, and that is why water is mixed with the acetone as a solvent (as shown in Eq. 3.3). Pure water is not used as the solvent because ethyl bromide is not very soluble in water (only 0.9 g in 100 ml at 20°), and a concentration of 0.008 M is not likely to give a very fast reaction.

For the preparation of mixed ethers, such as methyl ethyl ether, two routes are possible (Eqs. 3.7). Of the two reactions, the use of ethoxide ion on methyl iodide is preferred

$$\begin{array}{l} \text{CH}_3\text{CH}_2\text{O}^{\ominus} + \text{CH}_3\text{—I} \longrightarrow \text{CH}_3\text{CH}_2\text{OCH}_3 + \text{I}^{\ominus} \\ \phantom{\text{CH}_3\text{O}^{\ominus} + } \underset{\underset{\text{CH}_3}{|}}{} \\ \text{CH}_3\text{O}^{\ominus} + \text{CH}_2\text{—I} \longrightarrow \text{CH}_3\text{OCH}_2\text{CH}_3 + \text{I}^{\ominus} \end{array} \quad (3.7)$$

to the use of methoxide ion on ethyl iodide. Not only is the former somewhat faster, but also there is no opportunity provided for the other competing reaction, the elimination reaction (to give CH_3OH, $\text{H}_2\text{C=CH}_2$, and $\text{I}^{\ominus}$).

Because reactions in organic chemistry seldom are free of competition, it is not enough to devise a reaction for the preparation of a substance. The chosen reaction must be carried out with the best selection of reactants (if alternatives are possible) and with the best combination of concentrations, solvent, temperature control, and time in order to obtain the highest possible yield. All this attention to detail may seem discouraging at first, but we shall see that generalizations work quite well.

Steric effects. Nucleophilic substitution reactions (S_N2) are slower with ethyl halides than with methyl halides because the nucleophile is less able to collide properly with an ethyl halide. Previously, we implied that A, in attacking BC, came in on the side of B opposite to C. At the midpoint of the reaction, A is not yet fully bonded to B, and C is not yet completely released. If B is methyl, the midpoint (known as a transition state) would appear to be as shown for the reaction of iodide ion with methyl chloride. The transition

$$I^\ominus + CH_3-Cl \longrightarrow {}^{\delta-}I\cdots\underset{\underset{H}{|}}{\overset{\overset{H}{|}}{C}}\cdots\overset{\delta-}{Cl} \longrightarrow I-CH_3 + Cl^\ominus$$

state, if B is ethyl, would be as shown for the reaction given as Equation 3.2. The iodide ion

$$I^\ominus + \underset{}{\overset{CH_3}{\underset{|}{CH_2}}}-Cl \longrightarrow {}^{\delta-}I\cdots\underset{\underset{H}{|}}{\overset{\overset{CH_3}{|}}{C}}\cdots\overset{\delta-}{Cl} \longrightarrow I-\underset{}{\overset{CH_3}{\underset{|}{CH_2}}} + Cl^\ominus$$

has an easier task in approaching the methyl group than the ethyl group, because in the latter it must avoid a large methyl to gain access to the back side of the carbon atom attached to the chlorine. As a result, at a given temperature and other things being equal, more of the collisions with CH_3Cl are successful than of those with CH_3CH_2Cl; and the rate of reaction is greater with CH_3Cl than with CH_3CH_2Cl. Such an effect, ascribed to spatial causes, is known as a steric effect. The steric effect here caused a decrease in the rate of the reaction and is called steric hindrance.

Formation of amines. An ethyl halide will react with ammonia to give the ethylammonium halide. Treatment of the salt with a strong base will free the amine. Likewise, methylamine will react with a methyl halide to give dimethylamine (Eqs. 3.8). However,

$$H_3N + \overset{CH_3}{\underset{|}{CH_2}}-X \longrightarrow H_3\overset{\oplus}{N}-\overset{CH_3}{\underset{|}{CH_2}} + X^\ominus \qquad (3.8)$$

$$CH_3NH_2 + CH_3-X \longrightarrow CH_3\overset{\oplus}{N}H_2CH_3 + X^\ominus$$

this is beginning to sound like the alcohol-ether situation, and it is. Again, let us assume that a reaction is halfway along in a solvent that holds the salt in solution. Examination of acid-base equilibrium between ammonia and the alkylammonium ion (Eq. 3.9) shows that

$$R-\overset{\oplus}{N}H_2-H + \overline{N}H_3 \rightleftharpoons R-\overline{N}H_2 + NH_4^\oplus \qquad (3.9)$$

the free amine and ammonium ion are present. (R is frequently used as a symbol to represent any alkyl group such as CH_3-, CH_3CH_2-, and so on.) The acids and bases in the solution are not too different in strength; K_{eq} will not be far from one. The amine is able to compete with ammonia in two ways. By lowering the ammonia concentration, it diminishes the rate of attack by ammonia on the remaining alkyl halide. The amine also attacks the alkyl halide to give a dialkylammonium iodide (Eq. 3.8). To prevent the formation of the dialkylammonium halide in solution, the equilibrium (Eq. 3.9) must be pushed to the left, and ammonium ion will be needed in the reaction mixture to do so. The best way to carry out the reaction is to use an aprotic solvent, such as an ether or an alkane, in which the nonionic reactants are soluble and the product salts are insoluble. Then, in the reactions shown in Equations 3.8, ethylammonium halide (or dimethylammonium halide) crystallizes from the solution as the reaction proceeds.

Inertness of vinyl halides. A halogen attached to a carbon atom that has a double bond to another carbon atom is difficult to displace. If the use of the nucleophilic reagent is replaced by the use of a stronger base, the elimination reaction takes place instead of the S_N2 reaction (Eqs. 3.10). Thus, acetylene is formed from vinyl chloride and $NaNH_2$, not vinylamine.

$$I^\ominus + \underset{\underset{CH_2}{\|}}{HC}-Cl \xrightarrow{X} CH_2=CHI + Cl^\ominus \quad (3.10)$$

$$H_2N^\ominus + H-CH=CH-Cl \longrightarrow H_2NH + HC\equiv CH + Cl^\ominus$$

Organic chemistry in water. Proteins, fats, and carbohydrates, which by weight account for practically all of the organic portion of any biochemical substance, are each very large molecules made of smaller molecules.

Proteins are long chains of amino acids connected to one another by amide linkages. The fats all use ester bonds to bind long chain fatty acids to the polyalcohol core. Carbohydrates are chains of simple sugars bonded together with a type of carbon-oxygen bond related to the ester called an acetal link. This bond, with one carbon involved in two ether bonds, has chemical properties different from a single ether bond.

Each of these bonds can be broken by hydrolysis, that is by addition of a water molecule to replace the existing C—N or C—O bond by nucleophilic attack. As you would expect, the rate of reaction is accelerated by acid (the bond to be attacked is made more susceptible by protonation) or in base (the hydroxide ion is a better nucleophile than water). Because water is ubiquitous in biochemical systems, organisms digest these big molecules by catalyzing the hydrolysis reaction. Commercial preparations of meat tenderizer use these catalysts to hydrolyze the "tough" protein fibers in the meat.

Exercises

3.4 Given ammonia, methyl and ethyl iodides, and any solvents needed, use equations to give a reasonable preparation of ethylmethylamine, $CH_3CH_2NHCH_3$.

3.5 Show the major product to be expected from the reactions given:

a. $CH_3CH_2OH + HBr \xrightarrow{25°}$

b. $CH_3CH_2Br + CH_3CH_2NH_2 \xrightarrow[25°]{ether}$

c. $CH_3CH_2Br + HO^\ominus \xrightarrow[H_2O, 25°]{acetone}$

d. $CH_3CH_2Cl + NaI \xrightarrow[25°]{acetone}$

e. $CH_3I + CH_3O^\ominus \xrightarrow[25°]{CH_3OH}$

3.4 Electrophilic substitution reactions

Displacement of metals from metal organic compounds. The reaction of bromine with ethylmagnesium bromide in ether solution to give ethyl bromide and magnesium bromide (Eq. 3.11) parallels the similar reaction of methylmagnesium iodide (Eq. 2.13). The

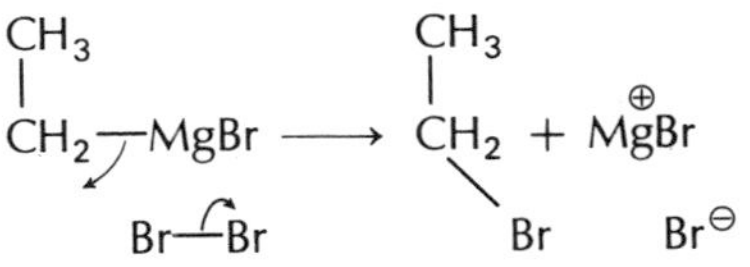

$$\underset{\underset{Br-Br}{}}{\overset{CH_3}{\underset{|}{CH_2-MgBr}}} \longrightarrow \underset{Br}{\overset{CH_3}{\underset{|}{CH_2}}} + M\overset{\oplus}{g}Br + Br^{\ominus} \quad (3.11)$$

electrophile, bromine, attacks the carbon of the ethyl group to displace the magnesium. The electrophile has broken up in the reaction, one atom of bromine gaining the electrons, the other changing partners. We may view the reaction in a different way by saying that the incipient nucleophile, the ethyl group, attacks the bromine molecule and displaces a bromide ion. So the classification of the reaction as an electrophilic substitution, S_E2, is ambiguous unless we also specify substitution on carbon.

Because the displaced group in an electrophilic substitution leaves a pair of electrons behind, it departs with a positive charge. Metals, as a class, are stable with positive charges and constitute most of the known examples of suitable leaving groups. Another example of an electrophilic displacement is given in Equation 3.12. The electrophilic hydrogen of

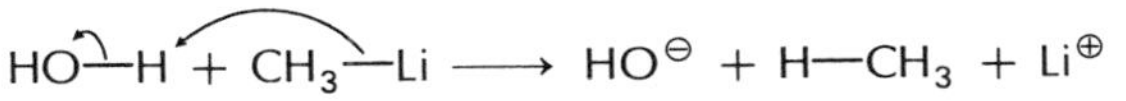

$$HO-H + CH_3-Li \longrightarrow HO^{\ominus} + H-CH_3 + Li^{\oplus} \quad (3.12)$$

water displaces lithium as the positive ion, and the water in turn breaks up to give a hydroxide ion.

Exercises

3.6 Complete the equations:
   a. $I_2 + CH_3MgBr \longrightarrow$
   b. $CH_3OH + CH_3CH_2MgBr \longrightarrow$
   c. $CH_3COOH + CH_3CH_2Li \longrightarrow$
   d. $Br_2 + CH_3CH_2HgCl \longrightarrow$
   e. $ICH_3 + CH_3Na \longrightarrow$

3.5 Homolytic substitution reactions

Homolytic displacement on halogen; Grignard reagents. A useful S_H2 reaction is that of metallic lithium added to a solution of ethyl bromide in anhydrous ether. The reaction occurs with the evolution of heat (Eq. 3.13), and the products remain in solution.

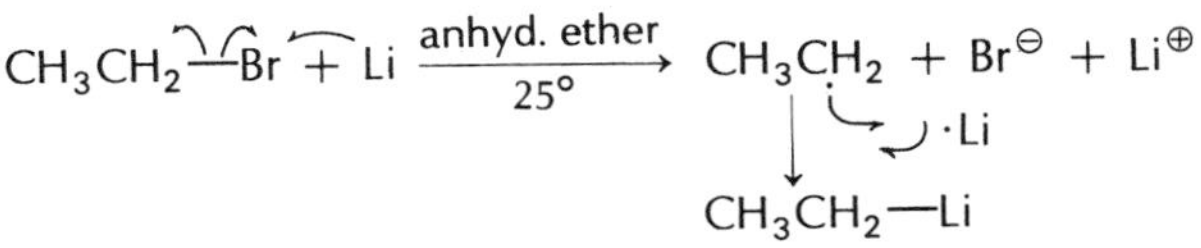

$$CH_3CH_2-Br + Li \xrightarrow[25°]{\text{anhyd. ether}} CH_3CH_2 \cdot + Br^{\ominus} + Li^{\oplus} \quad (3.13)$$
$$\downarrow \cdot Li$$
$$CH_3CH_2-Li$$

or

$$CH_3CH_2Br + 2Li \xrightarrow[25°]{\text{anhyd. ether}} CH_3CH_2Li + LiBr$$

Because of the reactivity of ethyl lithium, it is not isolated but is used immediately in a chosen subsequent reaction, such as an electrophilic substitution reaction. Because each atom of lithium has but one electron to lose, the first step on the surface of the lithium metal seems to be a homolytic electron transfer to give an ethyl radical, a bromide ion, and a lithium ion. Before the radical can diffuse away from the surface, another atom of lithium releases an electron to the ethyl radical, forming a bond between the carbon and lithium atoms.

$$\text{surface layer} \longrightarrow \underset{\text{Li Li Li Li Li Li Li}}{\overset{\overset{\displaystyle CH_3}{\underset{\displaystyle CH_2-Br}{|}} \quad \overset{\displaystyle Li}{}}{}} \quad \overset{\overset{\displaystyle CH_3}{\underset{\displaystyle CH_2}{|}} + \overset{\displaystyle Br}{\underset{\displaystyle Li}{|}}}{}$$

An alkyl halide also will react with magnesium in anhydrous ether, but it does so presumably by a two-electron transfer (Eq. 3.14) to give an alkylmagnesium halide (Sec. 2.7),

$$CH_3CH_2-Br + Mg \xrightarrow[25°]{\text{ether}} CH_3CH_2^{\ominus} + BrMg^{\oplus} \longrightarrow CH_3CH_2-MgBr \quad (3.14)$$

also known as a <u>Grignard reagent</u> (pronounced green-yard). The alkyl lithiums are more reactive than the Grignard reagents but are more expensive as well. Both types are extremely useful in the laboratory because of the wide variety of reactions they undergo. Only a few of these reactions were given in the preceding section; others will come later.

The structure of the reagents has been written in the simplest way, although the actual structure in solution is somewhat different. Some lithium reagents are known to be dimeric (composed of two identical parts). Some are known to have a molecule of solvent bound between the ions. Similar uncertainties exist with respect to the Grignard reagents. Fortunately, we can write almost all the known reactions using the structural formulas as given above.

Homolytic displacement on hydrogen: chain reactions. Ethane, like methane (Eqs. 2.14 and 2.15) undergoes homolytic substitution on hydrogen (Eqs. 3.15). The mechan-

$$Cl-Cl \xrightarrow{\text{heat or light}} 2Cl\cdot \qquad\qquad a \quad \text{initiation}$$

$$\left.\begin{array}{l} CH_3CH_2-H + Cl\cdot \longrightarrow CH_3\dot{C}H_2 + H-Cl \quad b \\ CH_3\dot{C}H_2 + Cl-Cl \longrightarrow CH_3CH_2-Cl + Cl\cdot \quad c \end{array}\right\} \text{propagation}$$

$$\left.\begin{array}{l} Cl\cdot + Cl\cdot \longrightarrow Cl-Cl \qquad\qquad\qquad\qquad\qquad d \\ CH_3\dot{C}H_2 + Cl\cdot \longrightarrow CH_3CH_2-Cl \qquad\qquad\quad e \\ CH_3\dot{C}H_2 + \dot{C}H_2CH_3 \longrightarrow CH_3CH_2-CH_2CH_3 \quad f \end{array}\right\} \text{termination}$$

(3.15)

or

$$CH_3CH_3 + Cl_2 \xrightarrow{\text{heat or light}} CH_3CH_2Cl + HCl \qquad g$$

ism for the reaction with chlorine is shown as an initial rupture of a molecule of chlorine, a, to give two atoms of chlorine. Each atom of chlorine attacks a molecule of ethane, S_H2 reaction, to give HCl and an ethyl radical, b. An ethyl radical attacks a molecule of chlorine, S_H2 on chlorine, to give ethyl chloride and a new atom of chlorine, c. The new chlorine atom, if it is not removed by collision with another chlorine atom, d, or with an ethyl radical, e, sets up another sequence of b, c reactions. Ethyl radicals may be removed from a b, c sequence by reaction with a chlorine atom, e, or by another ethyl radical, f. Bromine also reacts, but at a slower rate; iodine does not react; and fluorine reacts explosively in this sequence.

The reaction is called a free-radical chain reaction. The chain is the self-perpetuating sequence of b, c steps, known as propagation. The chain is initiated by the formation of atom radicals in a. The chain continues until the supply of ethyl radicals and chlorine atoms is so depleted by the termination steps, d, e, and f, and a decreasing rate of formation of chlorine atoms by a that the reaction comes to an end. That the mechanism is satisfactory may be shown by the observation that small amounts of butane, f, have been isolated from the products of the chlorination of ethane. Methane reacts by a similar mechanism, and ethane is present in the product mixture.

Complications arise: ethyl chloride from ethane and methyl chloride from methane may enter into the b, c sequence. Methyl chloride gives dichloromethane, and ethyl chloride gives a mixture of 1,1-dichloroethane and 1,2-dichloroethane (Eqs. 3.16).

$$CH_3Cl + Cl_2 \longrightarrow CH_2Cl_2 + HCl$$
$$CH_3CH_2Cl + Cl_2 \longrightarrow CH_3CHCl_2 + HCl \quad (3.16)$$
$$CH_3CH_2Cl + Cl_2 \longrightarrow ClCH_2CH_2Cl + HCl$$

Exercises

3.7 Devise a mechanism for the bromination of propane, $CH_3CH_2CH_3$. Predict the structures of the possible isomeric C_6H_{14} compounds that will be formed in small amounts.

3.6 Addition reactions

Carbon-carbon double bonds; addition by electrophiles; by free radicals; by catalysts. With the two-carbon compounds we encounter the first examples of carbon-carbon multiple bonds. Although they are nonpolar, they are polarizable. Thus, in ethylene or acetylene there is little or no influence *within* the molecule to pull the less constrained pair of electrons in the π bond to one end or the other of the double bond. However, if a charge approaches from *outside* the molecule (or as the molecule approaches a charge, such as an ion), polarization will occur. Nevertheless, nucleophiles tend not to react with carbon-carbon double bonds, probably because the π electron cloud itself is nucleophilic and repulsive to other nucleophiles. However, an electrophile adds readily to a double bond to give a cation (Eq. 3.17).

$$HO_3SO-H + H_2C=CH_2 \longrightarrow CH_3CH_2^{\oplus} + HSO_4^{\ominus} \longrightarrow CH_3CH_2OSO_3H \quad (3.17)$$

46   Chapter three

A cation by definition has a positive charge. A cation like the one in Equation 3.17 is also called a <u>carbonium ion</u> and contains a carbon atom with three atoms or groups attached by single bonds and sharing only six electrons. A <u>carbanion</u> by definition has a negative charge and contains a carbon atom with three atoms or groups attached by single bonds (sharing six electrons) and has an extra pair of electrons. A free radical is neutral and has an odd electron.

$$\underset{\text{ethyl carbanion}}{\overset{CH_3}{\underset{HH}{\overset{|}{C:^{\ominus}}}}} \quad \underset{\text{ethyl radical}}{\overset{CH_3}{\underset{HH}{\overset{|}{C\cdot}}}} \quad \underset{\text{ethyl cation}}{\overset{CH_3}{\underset{HH}{\overset{|}{C^{\oplus}}}}}$$

Alkenes exhibit three types of addition reaction: polar addition by way of a cation, free-radical addition, and catalytic addition on the surface of a catalyst. The first two will be described here; the third will be described in Section 3.8 on oxidation and reduction.

Polar addition to ethylene by halogen.   In acetic acid, a polar solvent inert to halogen, bromine will add to ethylene in the light or dark by a polar mechanism (Eq. 3.18).

$$H_2C=CH_2 + Br-Br \xrightarrow[25°]{CH_3COOH} H_2\overset{\oplus}{C}-CH_2-Br + Br^{\ominus}$$
$$\longrightarrow Br-CH_2CH_2-Br \quad (3.18)$$

For the time being, we shall write the intermediate ion as the 2-bromoethyl cation, even though we are oversimplifying the structure to avoid certain complexities better left until later chapters. In any event, the cation is a reactive entity, an electrophile, and it reacts rapidly with any nucleophile in the solution. If bromide ion is present, the addition will be completed as shown. A useful trick is to dissolve some sodium bromide in the mixture. The high concentration of bromide ion speeds up the second step of the reaction.

The technique of using added ions may be exploited to guide the addition reaction into giving other products. If 1-bromo-2-chloroethane is wanted, the solution is saturated with chloride ions. Because sodium chloride is relatively insoluble in acetic acid, the much more soluble lithium chloride is used. If 2-bromoethyl acetate is wanted, the solution is saturated with sodium acetate (Eqs. 3.19). Similar results are obtained with acetylene to

$$H_2C=CH \xrightarrow[25°]{\overset{Br-Br}{}\;CH_3COOH} Br-CH_2\overset{\oplus}{C}H_2 \xrightarrow{Br^{\ominus}} \begin{cases} \xrightarrow{\text{NaBr present}} Br-CH_2CH_2-Br \\ \xrightarrow{\text{LiCl present}} Br-CH_2CH_2-Cl \\ \xrightarrow[\text{present}]{CH_3COONa} Br-CH_2CH_2-O-\overset{\overset{O}{\|}}{C}CH_3 \end{cases} \quad (3.19)$$

give the products BrCH=CHBr, ClCH=CHBr, and CH$_3$COOCH=CHBr. Excess bromine will add to BrCH=CHBr to give Br$_2$CHCHBr$_2$. Additions to acetylene are slower than to ethylene.

Addition of acids to alkenes and alkynes. Now let us return to the addition of acids (Eq. 3.17). If ethylene gas is bubbled through a strong acid, some of the gas will dissolve and react (Eq. 3.20). Because alkenes are nucleophilic, the electrophilic proton leads the attack on the π bond (Eq. 3.21). The addition is completed by the coming together of the

$$HX + H_2C=CH_2 \longrightarrow CH_3CH_2X \qquad (3.20)$$

$$X\frown H + H_2C=CH_2 \longrightarrow X^\ominus + H_3C-CH_2^\oplus \qquad (3.21)$$

positive and negative ions (Eq. 3.22). If an aqueous strong acid is used, water is able to com-

$$CH_3\overset{\frown}{CH_2}^\oplus + X^\ominus \longrightarrow CH_3CH_2X \qquad (3.22)$$

pete somewhat with the weakly basic negative ions present (strong acids have weak conjugate bases), and the conjugate acid of ethanol is formed (Eq. 3.23). However, if an

$$CH_3\overset{\frown}{CH_2}^\oplus + H_2\overline{O}| \longrightarrow CH_3CH_2OH_2^\oplus \qquad (3.23)$$

excess of strong acid is present, the conditions for the S_N2 displacement reaction are favorable, and the major product still is an ethyl halide (Eq. 3.24).

$$H\overset{\frown}{-}X + \overset{CH_3}{\underset{|}{CH_2}}\overset{\frown}{-}\overset{\oplus}{OH_2} \longrightarrow H^\oplus + X-\overset{CH_3}{\underset{|}{CH_2}} + H_2O$$

or
$$\qquad (3.24)$$

$$X^\ominus \overset{\frown}{+} \overset{CH_3}{\underset{|}{CH_2}}\overset{\frown}{-}\overset{\oplus}{OH_2} \longrightarrow X-\overset{CH_3}{\underset{|}{CH_2}} + H_2O$$

With acetylene, a vinyl halide is the product ($H_2C=CHX$). A second addition of HX to a vinyl halide is a slower process than the addition to acetylene, and it goes in such a way as to form a 1,1-dihaloethane, not a 1,2-dihaloethane (Eq. 3.25). Directive influences on

$$HC\equiv CH + 2HX \longrightarrow CH_3CHX_2 \qquad (3.25)$$

addition to an unsymmetrical alkene, such as a vinyl halide, $H_2C=CHX$, will be discussed in a later chapter.

If ethylene is bubbled into a really strong acid, such as cold concentrated sulfuric acid, the product of the addition is ethylsulfuric acid (Eq. 3.26). The bisulfate ion, $HSO_4^\ominus$

$$H_2SO_4 + H_2C=CH_2 \xrightarrow{0°} CH_3CH_2OSO_3H \qquad (3.26)$$

is a very good leaving group, better than iodide in an $A + BC$ reaction. Simply pouring the sulfuric acid solution of the ethylsulfuric acid into water (better over ice to take up the large amount of heat evolved in the dilution of sulfuric acid) leads to the formation of ethanol (Eqs. 3.27). The total effect is that of adding a mole of water to a mole of ethylene,

$$H_2O \overset{\frown}{+} \overset{CH_3}{\underset{|}{CH_2}}\overset{\frown}{-}OSO_3H \xrightarrow{0°} CH_3CH_2OH_2^\oplus + HSO_4^\ominus \qquad (3.27)$$

$$CH_3CH_2\overset{\frown}{OH_2}^\oplus + H_2\overset{\frown}{O} \rightleftharpoons CH_3CH_2OH + H_3O^\oplus$$

a reaction that cannot be done directly because of the low acidity of water. The ethanol is isolated by distillation of the dilute aqueous mixture.

With acetylene, again, the vinyl alcohol produced by the addition of one mole of water, rearranges quickly into acetaldehyde, presumably by the addition of a proton from some donor in the solution followed by the loss of a proton from the conjugate acid of acetaldehyde to some acceptor in the solution (Eq. 3.28). The overall equilibrium lies far to

$$H^{\oplus} + H_2C=CH-\overset{..}{O}H \rightleftarrows H_3CC\overset{OH^{\oplus}}{\underset{H}{\diagdown}} + H_2O \rightleftarrows H_3CC\overset{O}{\underset{H}{\diagdown}} + H_3O^{\oplus} \quad (3.28)$$

the right. In fact, we do not have to use concentrated sulfuric acid for this reaction because, in the presence of a reasonable concentration of mercuric ion, acetaldehyde may be formed from acetylene by the action of dilute sulfuric acid (Eq. 3.29).

$$HC\equiv CH + H_2O \xrightarrow[\text{dil. } H_2SO_4,\ 25°]{HgSO_4,} CH_3CHO \quad (3.29)$$

Addition of a peracid. Another interesting addition reaction of alkenes involves a peracid. A peracid is related to hydrogen peroxide in the same way that an ordinary carboxylic acid is related to water. Thus, if we think of a hydrogen of water as being replaced by a $CH_3\overset{|}{C}=O$ group, acetic acid (CH_3COOH) is formed; if a hydrogen of hydrogen peroxide is replaced by a $CH_3\overset{|}{C}=O$ group, peracetic acid ($CH_3\overset{O}{\overset{\|}{C}}OOH$ or CH_3CO_3H) is formed. In a peracid, as in hydrogen peroxide, the weak oxygen-oxygen bond is easily ruptured. If a nucleophile (here, an alkene) were to attack one of the oxygens, a nucleophilic substitution on oxygen would occur. With peracetic acid, the leaving group is the acetate ion (Eq. 3.30). The conjugate acid of ethylene oxide, which is the immediate product,

$$CH_3COO-OH + H_2C=CH_2 \xrightarrow{0°} CH_3COO^{\ominus} + H_2C\overset{\overset{\oplus}{OH}}{\underset{}{\triangle}}CH_2 \xrightarrow{H_2O}$$
$$H_2C\overset{O}{\underset{}{\triangle}}CH_2 + H_3O^{\oplus} \quad (3.30)$$

loses the proton to an acceptor in the solution, and the epoxide is formed. Alternatively, the proton may be transferred to the incipient acetate ion during the addition of the —OH group to the alkene. This path leads directly to the products. Acetylene is strangely resistant to the action of peracids and does not react at 0°.

Addition to the carbon-oxygen double bond. Before leaving the subject of addition reactions, we should remind ourselves that addition at the carbon-oxygen double bond also occurs. In Section 2.9, the addition of water to formaldehyde was described. Acetaldehyde also will add water in aqueous solution; but the equilibrium (a) does not lie as far to the right as with formaldehyde (Eqs. 3.31). If the carbon atom of a carbon-oxygen

$$CH_3C\overset{O}{\underset{H}{\diagdown}} + H_2O \rightleftharpoons CH_3C\overset{OH}{\underset{H}{\diagdown}}OH \qquad a$$

$$Cl_3CC\overset{O}{\underset{H}{\diagdown}} + H_2O \rightleftharpoons Cl_3C-C\overset{OH}{\underset{H}{\diagdown}}OH \qquad b$$

(3.31)

double bond is bonded to a strong electron-pulling group such as the trichloromethyl, the equilibrium lies so far to the right as to give rise to an isolable substance, b. Thus, choral (a common name for trichloroacetaldehyde or trichloroethanal), bp 98°, forms a crystalline substance, mp 52°, called chloral hydrate. Chloral hydrate is a powerful soporific better known as "knock-out drops."

Free-radical addition to alkenes and alkynes. A free-radical or homolytic addition may be illustrated by the addition of a halogen to ethylene or acetylene. Excluding the fiercely reactive fluorine, a mixture of ethylene and halogen vapors is inert in the dark. If the mixture is heated or exposed to light, a reaction ensues (for all but iodine), the color of the halogen disappears, and a liquid is formed. In an inert solvent such as carbon tetrachloride, the same sequence may be noted, except that the liquid formed is soluble in the solvent.

The name <u>olefin</u>, meaning oil forming, is frequently used to refer to alkenes because the halogen addition reaction was one of the first to be observed for unsaturated compounds. The requirement of heat or light means that a molecule of halogen must be broken up into atoms (as in Eq. 3.15a) to start the addition reaction. If the atoms recombine (as in Eq. 3.15d), nothing more will happen. However, the concentration of halogen atoms is so small and the concentration of ethylene is so high that there is an excellent chance for each halogen atom to collide with a molecule of the olefin before another atom is found (Eq. 3.32). The radicals are in small concentration also, but should have no difficulty in

$$H_2C=CH_2 + X\cdot \longrightarrow H_2\dot{C}-CH_2-X \qquad (3.32)$$

finding a halogen molecule to give a homolytic displacement on halogen (Eq. 3.33). The

$$XHC_2CH_2\cdot + X-X \longrightarrow XCH_2CH_2-X + X\cdot \qquad (3.33)$$

new halogen atom repeats the cycle. Depending upon the concentrations, temperature, and pressure, many such cycles occur before an atom or radical meets another atom or radical and the process terminates. In short, free-radical addition is like free-radical substitution (Eq. 3.15) in being a chain reaction (Eq. 3.34). All chain reactions have an initiation step (a), propagation steps (b and c), and one or more termination steps (d, e, and f). Depending upon the number of chains started (how high the temperature, or how intense the light, if it is not strong enough to go completely through the solution, or the wavelength of the light, etc.), a chain reaction may be fast or slow. With all alkenes other than ethylene, free-radical substitution competes with the free-radical addition.

$$X\!-\!X \xrightarrow[\text{or light}]{\text{heat}} 2X\cdot \qquad\qquad a \quad \text{initiation}$$

$$H_2C\!=\!CH_2 + X\cdot \longrightarrow H_2\dot{C}\!-\!CH_2\!-\!X \qquad b \;\Big\}$$
$$X\!-\!X + H_2\dot{C}\!-\!CH_2X \longrightarrow X\cdot + X\!-\!CH_2\!-\!CH_2X \qquad c \;\Big\} \text{propagation}$$

$$\qquad\qquad\qquad\qquad\qquad\qquad\qquad\qquad\qquad\qquad (3.34)$$

$$X\cdot + X\cdot \longrightarrow X\!-\!X \qquad\qquad\qquad\qquad\qquad d \;\Big\}$$
$$X\cdot + H_2\dot{C}\!-\!CH_2X \longrightarrow X\!-\!CH_2\!-\!CH_2X \qquad e \;\Big\} \text{termination}$$
$$XCH_2\dot{C}H_2 + \dot{C}H_2CH_2X \longrightarrow XCH_2CH_2\!-\!CH_2CH_2X \qquad f \;\Big\}$$

$$H_2C\!=\!CH_2 + X_2 \xrightarrow[\text{gas or CCl}_4]{\text{heat or light}} XCH_2CH_2X \qquad g$$

Chain reactions are susceptible to the presence of other substances that can react with the atoms or radicals and break the chains. Oxygen, for example, is an efficient chain breaker for both bromine substitution and addition (Eq. 3.35). The peroxy radical formed is

$$R\cdot + |\ddot{O}\!-\!\dot{O}| \longrightarrow R\!-\!\ddot{O}\!-\!\dot{O}| \qquad\qquad (3.35)$$

relatively stable and can exist long enough to have a good chance of finding another radical, so that each molecule of oxygen breaks two chains. The action of oxygen in obstructing chain reactions has been called the <u>oxygen effect.</u>

Photochemical smog. Reaction of molecular oxygen with radicals is one of the important chemical reactions occurring in the air trapped above cities such as Mexico City, Denver, and Los Angeles. The radical processes are initiated by sunlight acting on the brown gas in urban skies, nitrogen dioxide. The many reactions involving hydrocarbons, nitrogen oxides, and oxygen then continue until the reactants disperse or the sunlight disappears. Formation of smog is described in Section 13.5.

Exercises

3.8 Explain why a mixture of ethylene and bromine, with a small amount of oxygen present, does not seem to react under strong illumination for a long time, but finally starts and then goes rapidly.

3.9 Predict the major product in the reactions given:

a. $H_2C\!=\!CH_2 + Cl_2 \xrightarrow[\text{CH}_3\text{COONa}]{\text{CH}_3\text{COOH}}$

b. $H_2C\!=\!CH_2 + Br_2 \xrightarrow[\text{NaF}]{\text{CH}_3\text{COOH}}$

c. $H_2C\!=\!CH_2 + Br_2 \xrightarrow[\text{NaOH}]{\text{H}_2\text{O}}$

d. $H_2C\!=\!CH_2 + Cl_2 \xrightarrow[\text{NaBr}]{\text{H}_2\text{O}}$

3.7 Elimination reactions

Elimination reactions that give rise to carbon-carbon double and triple bonds may be thought of as the reverse of addition reactions. Unlike the easily reversible addition of water to a carbon-oxygen double bond, the reverse of addition reactions to carbon-carbon double and triple bonds usually requires special conditions to favor the elimination. We shall look at the elimination reactions in the following order: dehalogenation, dehydrohalogenation, and dehydration.

Dehalogenation. Dehalogenation requires a reagent that is willing to lose electrons to convert the covalently bound halogen atoms into the more stable halide anions. A popular method for dehalogenation is simply to heat the dihalide with zinc dust: the ethylene produced, being a gas, flows out of the apparatus (Eq. 3.36). This reaction can be

$$XCH_2CH_2X + Zn \xrightarrow{\Delta} H_2C=CH_2 + ZnX_2 \qquad (3.36)$$

looked upon as a nucleophilic displacement on halogen to give Zn^{2+}, $X^\ominus$, and $^\ominus CH_2CH_2X$. However, the carbanion probably does not exist as such because the second halogen may fall off more or less simultaneously, taking the electron pair of the C—X bond with it. The extra electron pair on the neighboring carbon immediately forms the double bond. In other words, this β elimination is believed to be a concerted process (Eq. 3.37), all three

$$Zn: \; X\!-\!CH_2\!-\!CH_2\!-\!X \longrightarrow \; ^\oplus Zn\!-\!X + H_2C=CH_2 + X^\ominus \qquad (3.37)$$

bond changes occurring at the same time. With a tetrahaloethane, one or two moles of halogen may be removed. The removal of the second mole of halogen from 1,2-dihaloethene is more difficult than the removal of the first mole from the tetrahaloethane (Eq. 3.38).

$$Zn + X_2CHCHX_2 \xrightarrow{\Delta} ZnX_2 + XCH=CHX \xrightarrow[\Delta]{Zn} ZnX_2 + HC\equiv CH \qquad (3.38)$$

Dehydrohalogenation. Dehydrohalogenation has already been mentioned (Eq. 3.4). However, the conditions used there were not the best if elimination was the objective. Such a β elimination, in which a strong base that is usually a strong nucleophile as well is used, is always accompanied by some S_N2 reaction.

One way to diminish the amount of substitution when elimination is the goal is to increase the size of the base to be used. The larger the attacking group, the more difficulty it has in penetrating to the carbon in a displacement reaction; but the increase in size has little effect on the availability of the projecting neighboring hydrogens. So, instead of hydroxide ion, the much larger ethoxide ion is used to carry out an elimination reaction. There are better alcohols that provide larger lyate ions for the attack. On the other hand, ethoxide ions are easy to prepare. We simply dissolve potassium hydroxide in ethanol (Eq. 3.39), add the alkyl halide, and heat. The equilibrium is pushed to the right by the excess

$$K^{\oplus} + HO^{\ominus} + H-OCH_2CH_3 \rightleftharpoons K^{\oplus} + HOH + {}^{\ominus}OCH_2CH_3 \quad (3.39)$$

of alcohol as compared to the small amount of water formed. The complete reaction may now be written (Eq. 3.40).

$$CH_3CH_2O^{\ominus} + H-CH_2CH_2-X \xrightarrow[\text{ethanol}]{\Delta} CH_3CH_2OH + H_2C=CH_2 + X^{\ominus} \quad (3.40)$$

With a 1,1-dihaloethane, the first product of dehydrohalogenation is a vinyl halide. Removal of the second mole of HX from the vinyl halide is more difficult, but it can be done with sodium amide ($NaNH_2$). Since the intermediate is a vinyl halide, we could start with a 1,2-dihaloethane as well. In fact, the sequence of Equation 3.41 is a good method for the conversion of an alkene to an alkyne on a laboratory scale.

$$H_2C=CH_2 \xrightarrow{X_2} XCH_2CH_2X \xrightarrow[\Delta]{\text{alcoholic KOH}} H_2C=CHX \xrightarrow{NaNH_2} HC\equiv CH \quad (3.41)$$

Dehydration. The reversal of the hydration reactions is possible. When ethanol is passed into sulfuric acid at 180°, ethylene is formed in good yield; but it is accompanied by an objectionable amount of sulfur dioxide, which arises from the reduction of some of the sulfuric acid as it oxidizes some of the ethylene at the high temperature of the reaction. Phosphoric acid, although more expensive, may be used for such dehydrations to avoid the complication. The acid used must be nonvolatile. At still higher temperatures, an alcohol in the gas phase may be passed over alumina to cause the dehydration to take place (Eqs. 3.42).

$$CH_3CH_2OH \xrightleftharpoons{H^{\oplus}} CH_3CH_2\overset{\oplus}{O}H_2 \xrightarrow{-H_2O} CH_3CH_2^{\oplus} \xrightarrow{H_2O} H_2C=CH_2 + H_3O^{\oplus}$$

(3.42)

$$CH_3CH_2OH \xrightarrow[\substack{1.\ H_2SO_4,\ 180°,\ \text{or} \\ 2.\ H_3PO_4,\ 180°,\ \text{or} \\ 3.\ Al_2O_3,\ 250°}]{} H_2C=CH_2 + H_2O$$

Exercises

3.10 Predict the products to be expected from:

a. $HOCH_2CH_2Cl + Zn \xrightarrow{\Delta}$

b. $HOCH_2CH_2Cl + NaOH \longrightarrow$

c. $HOCH_2CH_2OH \xrightarrow[250°]{Al_2O_3}$

d. $CH_3CHBrCH_2Br \xrightarrow[\Delta]{Zn}$

3.8 Oxidation and reduction reactions

Combustion; the oxidation series. Every compound listed in Table 3.1 is flammable. During combustion, any halogens present tend to end up as the free molecular halogen; if nitrogen is present, it ends up as molecular nitrogen.

In the oxidation series ethane, ethanol, acetaldehyde, acetic acid, the same reactivity and methods may be used as in the methane series (Sec. 2.10). However, the first step, ethane to ethanol, is so difficult as not to be used. Ethanol to acetaldehyde by dehydrogenation over hot copper chips, like methanol to formaldehyde (Eq. 2.21), goes well. Acetaldehyde is easily oxidized to acetic acid, as is formaldehyde to formic acid (Eq. 2.22), with ammoniacal silver nitrate solution.

Strong oxidizing reagents. With strong oxidizing reagents, conditions become critical for success. Thus, at room temperature, methanol and ethanol do not react with dilute neutral potassium permanganate solution. Under either acidic or basic conditions, reaction occurs. The use of chromates (CrO_3, $H_2Cr_2O_7$, $Na_2Cr_2O_7$, or $K_2Cr_2O_7$ under acidic conditions at 40° or lower is better for oxidation of alcohols in efforts to prepare aldehydes or acids. The acidic conditions are provided by aqueous sulfuric acid, because any of the halogen acids would be oxidized by these strong reagents (Eqs. 3.43). Sodium

$$3CH_3CH_2OH + Na_2Cr_2O_7 + 4H_2SO_4 \longrightarrow 3CH_3CHO + Cr_2(SO_4)_3 + Na_2SO_4 + 7H_2O$$

$$3CH_3CHO + Na_2Cr_2O_7 + 4H_2SO_4 \longrightarrow 3CH_3COOH + Cr_2(SO_4)_3 + Na_2SO_4 + 4H_2O \quad (3.43)$$

$$3CH_3CH_2OH + 2Na_2Cr_2O_7 + 8H_2SO_4 \longrightarrow 3CH_3COOH + 2Cr_2(SO_4)_3 + 2Na_2SO_4 + 11H_2O$$

dichromate, being much more soluble than potassium dichromate, is a popular reagent. One mole of sodium dichromate will oxidize three moles of ethanol to acetaldehyde, three moles of acetaldehyde to acetic acid, or one and a half moles of ethanol to acetic acid.

Acetic acid is stable to the action of these oxidizing agents and is used frequently as a solvent for oxidations that involve chromates or permanganate. Because of the ease with which aldehydes are oxidized, an attempt to prepare acetaldehyde from ethanol by this method will fail unless the apparatus is set up in a manner that will allow the acetaldehyde to distil as rapidly as it is formed. Aldehydes have lower boiling points than do either of the related alcohols or acids and volatilize preferentially from the hot mixture. All oxidations tend to be <u>exothermic</u> (evolve heat) if water is a product. Thus, temperature must be controlled to prevent the reactions from getting out of hand.

Reduction of aldehydes, acids, and acid chlorides. An aldehyde may be reduced to an alcohol by catalytic hydrogenation (see Eq. 2.20), by lithium aluminum hydride (LAH) in ether, or by sodium borohydride in aqueous or alcoholic solution. Both of the latter two reagents are donors of the very strong base hydride ion ($H^{\ominus}$). LAH is the stronger of the two.

$$\text{Li}^\oplus \begin{bmatrix} \text{H} \\ | \\ \text{H}-\text{Al}-\text{H} \\ | \\ \text{H} \end{bmatrix}^\ominus \qquad \text{Na}^\oplus \begin{bmatrix} \text{H} \\ | \\ \text{H}-\text{B}-\text{H} \\ | \\ \text{H} \end{bmatrix}^\ominus$$

or $\qquad\qquad\qquad$ or

LiAlH_4 $\qquad\qquad\qquad$ NaBH_4

lithium aluminum hydride $\qquad$ sodium borohydride

An acid may be reduced only with LAH in ether, and the product is the alcohol. The intermediate aldehyde is easier to reduce than is the acid (Eqs. 3.44). Acetyl chloride is also

$$\begin{aligned} 4\text{CH}_3\text{CHO} + \text{LiAlH}_4 &\xrightarrow{\text{ether}} \text{LiAl}(\text{OCH}_2\text{CH}_3)_4 \\ 4\text{CH}_3\text{COOH} + 3\text{LiAlH}_4 &\xrightarrow{\text{ether}} \text{LiAl}(\text{OCH}_2\text{CH}_3)_4 + 2(\text{LiAl})\text{O}_2 + 4\text{H}_2 \end{aligned} \quad (3.44)$$

$$4\text{CH}_3\text{COCl} + 2\text{LiAlH}_4 \xrightarrow{\text{ether}} \text{LiAl}(\text{OCH}_2\text{CH}_3)_4 + \text{LiAlCl}_4 \quad (3.45)$$

reducible (Eq. 3.45). After the reaction is over, any excess lithium aluminum hydride is decomposed by the careful addition of water, which gives hydrogen and mixed hydroxides (Eq. 3.46). The hydroxides and alkoxides are decomposed by the addition of a dilute acid,

$$4\text{H}_2\text{O} + \text{LiAlH}_4 \longrightarrow \text{LiOH} + \text{Al}(\text{OH})_3 + 4\text{H}_2 \quad (3.46)$$

usually hydrochloric acid, and the alcohol is recovered (Eq. 3.47). Carbon-carbon double and triple bonds are not affected by LAH or sodium borohydride.

$$\text{LiAl}(\text{OCH}_2\text{CH}_3)_4 + 4\text{HCl} \longrightarrow 4\text{CH}_3\text{CH}_2\text{OH} + \text{LiCl} + \text{AlCl}_3 \quad (3.47)$$

Oxidation of alkenes and alkynes by permanganate. Ethylene and acetylene are susceptible to the action of solutions of potassium permanganate under either acidic, neutral, or basic conditions. Various products are obtained depending upon the conditions used. Neutral or basic solutions of potassium permanganate at 0° to 25°C will oxidize ethylene to glycol (Eq. 3.48). Under acidic conditions at the same temperature, the reaction

$$3\text{H}_2\text{C}=\text{CH}_2 + 2\text{KMnO}_4 + 4\text{H}_2\text{O} \xrightarrow{25°} 3\text{HOCH}_2\text{CH}_2\text{OH} + 2\text{KOH} + 2\text{MnO}_2 \quad (3.48)$$

tends to proceed beyond the glycol stage, but not completely. As a result, acidic permanganate solutions are seldom used at room temperature. Once the high-boiling glycol has been formed, the temperature may be raised; under either acidic, neutral, or basic conditions at 100° the oxidation is complete (Eqs. 3.49). The acidic reaction is not only faster, it is more efficient than the basic reaction. The acidic reaction requires only 2 moles

$$\text{HOCH}_2\text{CH}_2\text{OH} + 2\text{KMnO}_4 + 3\text{H}_2\text{SO}_4 \xrightarrow{100°} 2\text{CO}_2 + \text{K}_2\text{SO}_4 + 2\text{MnSO}_4 + 6\text{H}_2\text{O}$$

$$(3.49)$$

$$3\text{HOCH}_2\text{CH}_2\text{OH} + 10\text{KMnO}_4 + 2\text{KOH} \xrightarrow{100°} 6\text{K}_2\text{CO}_3 + 10\text{MnO}_2 + 10\text{H}_2\text{O}$$

of permanganate per mole of glycol, but the basic reaction requires 3.3 moles of permanganate per mole of glycol. The oxidations as shown are not used on ethylene (there are easier ways of producing CO_2!), but we shall keep them in mind for application to other alkenes.

The oxidation of acetylene is complicated by a vinyl-alcohol type of rearrangement like the one we discussed earlier (Eq. 3.28). With acetylene, the initial product (presumably 1,2-ethenediol or $HOCH=CHOH$) is unstable and rearranges to the more stable hydroxyacetaldehyde, $HOCH_2CHO$. The easy oxidizability of the aldehyde group comes into play, and the first isolable product is hydroxyacetic acid, $HOCH_2COOH$. But the reaction is now complicated by possible oxidation of the alcohol group. At any rate, acetylene (as well as ethylene) causes the purple color of a dilute neutral solution of potassium permanganate to fade out, and dark brown manganese dioxide precipitates. The balanced equation for the oxidation to hydroxyacetic acid (as the potassium salt) shows that 1.3 moles of permanganate are required per mole of acetylene (Eq. 3.50) as compared with 0.67 mole per

$$3HC\equiv CH + 4KMnO_4 + 2H_2O \xrightarrow{25°} 3HOCH_2COOK + 4MnO_2 + KOH \quad (3.50)$$

mole of ethylene (Eq. 3.48). The extra oxidizing reagent, of course, is used up in the oxidation of the aldehyde group to the acid. Under more drastic conditions, namely at a higher temperature, any compound that is partially oxidized at neighboring carbons will be completely oxidized. For example, oxalic acid is oxidized by hot permanganate solutions to give two moles of carbon dioxide.

Catalytic hydrogenation. Catalytic hydrogenation (first mentioned in Eq. 2.20) also occurs with carbon-carbon double and triple bonds. The addition occurs on the surface of the catalyst and is difficult to classify as a polar or a free-radical reaction. We express our ignorance by calling it a third class of addition reaction.

Catalytic hydrogenation of alkenes and alkynes occurs under milder conditions than those required for carbon-oxygen bonds. A mixture of hydrogen and ethylene is inert. Addition of a catalyst causes a reaction to ensue, and ethane is produced (Eq. 3.51).

$$H_2C=CH_2 + H_2 \xrightarrow[25°,\ alcohol]{Pt} CH_3CH_3 \quad (3.51)$$

Acetylene reacts similarly. If the amount of hydrogen is restricted, ethylene is the product (Eq. 3.52). If excess hydrogen is used, ethylene, in turn, is hydrogenated (Eq. 3.53), and

$$HC\equiv CH + H_2 \xrightarrow[25°,\ alcohol]{Pd} H_2C=CH_2 \quad (3.52)$$

$$HC\equiv CH + 2H_2 \xrightarrow[25°,\ alcohol]{Pt} CH_3CH_3 \quad (3.53)$$

ethane may be produced directly from acetylene. This ability to take up hydrogen has led to the use of the term <u>unsaturated</u> to refer to the presence of multiple bonds.

The mechanism of addition reactions such as these hydrogenations that are strongly catalyzed by the presence of heterogeneous catalysts is still not understood completely. Platinum or palladium deposited on finely divided carbon are widely used catalysts. The process of adsorption on the surface of the catalyst by one or the other of the reactants,

56 Chapter three

followed by reaction and desorption of the product, is not easy to describe in complete detail despite the large amount of work done on the subject.

Catalytic hydrogenation is used extensively by the chemical industry and in the refining of petroleum (see Sec. 13.7).

Exercises

3.11 Give conditions and reagents for carrying out these reactions:
   a. $CH_3CH_2OH \longrightarrow CH_3COOH$
   b. $HOCH_2COOH \longrightarrow HOCH_2CH_2OH$
   c. $CH_3CH{=}CH_2 \longrightarrow CH_3CH_2CH_3$
   d. $O{=}CHCH{=}O \longrightarrow HOCH_2CHO$

3.9 Energy changes

Enthalpy and the first law of thermodynamics. Ordinary physical and chemical changes (meaning that we are excluding nuclear interchanges of mass and energy) are accompanied by changes of energy. However, the first law of thermodynamics may be stated thus: energy is neither created nor destroyed in such changes, it is converted from one form to another. Thus, if methane is burned, oxygen and methane are consumed, water and carbon dioxide are formed, and heat is released to the surroundings. We interpret this emission of heat to mean that the energy content of oxygen and methane is greater than that of water and carbon dioxide. The reaction releases the extra energy as heat.

To put the first law in another way, the energy of a system and of its surroundings remain constant. The system may release energy to or withdraw energy from the surroundings, but the total energy is unchanged. The energy or heat change of a chemical reaction at constant pressure is called the enthalpy change. The symbol for enthalpy is H, and ΔH is the change in enthalpy of the system, or $\Delta H = H(\text{products}) - H(\text{reactants})$. A change in enthalpy is measured in calories or kilocalories. If enthalpy is released by the system, we say that the reaction is exothermic, and the sign of ΔH is minus. If enthalpy is absorbed by the system, we say that the reaction is endothermic, and the sign of ΔH is plus.

Entropy and the second law; heterogeneous equilibrium. Entropy (symbol S) may be thought of as a measure of disorder. Another way of looking at entropy is to say that it is an indirect measure of randomness or of the probability of a given state or system. The symbol ΔS is defined as $S(\text{products}) - S(\text{reactants})$ and may be either positive or negative. Entropy is measured in entropy units, abbreviated as eu, with the dimensions of calories per degree. The second law of thermodynamics is given thus: for any spontaneous process, the total entropy increases. Total entropy means the entropy of the system plus that of the surroundings.

If a liquid freezes (crystallizes), the solid is more ordered than the liquid, and both ΔH and ΔS for the system are negative. The surroundings take up the enthalpy that is released by the system; and ΔS for the surroundings is positive and is greater numerically than the negative ΔS for the system if the freezing occurs in such a way that the process is spontaneous. A liquid at its bp is converted into vapor by absorption of heat (enthalpy) from the surroundings. (ΔH is positive, or the change in the system is endothermic.) The entropy change, ΔS, for the system also increases (is positive) because disorder in the gas phase is greater than in the liquid. At the bp of a liquid, the two phases may be in equilibrium at a constant pressure of 1 atm, depending upon the conditions: specifically, upon control of heat flow. If no enthalpy change is allowed (the system is completely insulated from the surroundings, a condition described as adiabatic), no change in the relative amounts of the two phases is possible. A heterogeneous equilibrium exists. However, addition of enthalpy to the system (now no longer adiabatic) will not cause a change in temperature (until the liquid phase disappears) but will cause a shift in the direction that will use up the enthalpy: namely, the relative amount of vapor will increase, and the relative amount of liquid phase will decrease. Withdrawal of enthalpy causes a shift in the opposite direction: liquid phase is increased, gas phase is decreased. For changes in phase, the change in entropy of the system is related to the enthalpy change by the very simple relationship in Equation 3.54, in which T is the absolute or Kelvin temperature expression of the bp or mp.

$$\Delta H = T\Delta S \qquad (3.54)$$

Gibbs free energy and reaction tendency. Gibbs free energy, symbol G (some books use F), may be thought of as the available energy from a system, or as the ability of a system to do work. The symbol ΔG is defined as $G(\text{products}) - G(\text{reactants})$ and may be either positive or negative. Free energy is measured in calories or kilocalories. If ΔG is negative for a reaction, the reaction has a tendency to give products. If ΔG is positive, the reaction has no tendency to proceed, but the products of the reaction as written have a tendency to react and give reactants. Nothing is said about the speed of reactions by ΔG: it says only whether the reaction will be spontaneous or not. A spontaneous reaction is called an exergonic reaction, and ΔG is negative. A nonspontaneous reaction is called an endergonic reaction, and ΔG is positive.

What may we expect if ΔG is zero? The system (or reaction) is at equilibrium. At equilibrium, the speed of the reaction going to the right is exactly equal to the speed of the reverse reaction going to the left. Thus, the concentrations of reactants and products do not change.

Interrelationship of enthalpy, entropy, and free energy. Free energy, enthalpy, entropy, and the absolute temperature are interrelated as shown by Equation 3.55. Free

$$\Delta G = \Delta H - T\Delta S \qquad (3.55)$$

energy, enthalpy, and entropy are thermodynamic functions of state. A function of state is independent of the past history of a system, but it does depend upon the conditions under which the system exists. For example, let us specify a system of a mole of ethane at 25°C and 1 atm pressure. The conditions fix the volume at 24.4 liters. The conditions also fix the free energy, the enthalpy, and the entropy of the system. If we heat the sample of ethane to 100°C, keeping the pressure constant, the volume will increase, the free-energy content will increase, the enthalpy content will increase, and the entropy will increase. In short, ΔG, ΔH, and ΔS for the change to 100° all will be positive. We may continue with other changes to the system, as many as we may wish, but a return to 25° and 1 atm pressure will also return the free energy, the enthalpy, and the entropy to the same values they had originally.

The actual content of free energy or enthalpy of a system under a specified set of conditions is unknown. However, because it is the magnitude of the *changes* in free energy

(ΔG) and in enthalpy (ΔH) that we are interested in, any arbitrary zero point may be set. An internationally agreed upon convention assigns a value of zero for the enthalpy of each element at 25°C (298.15°K) and 1 atm pressure, the <u>standard state.</u> Each element is taken in its familiar state: gaseous for hydrogen, fluorine, and chlorine; liquid for bromine and mercury; and solid for sodium, boron, iron, and so on. For carbon, the graphite form is taken; for sulfur, the rhombic crystalline form.

The actual entropy, however, *may* be calculated for many substances. The <u>third law of thermodynamics</u> states that the entropy of a perfect crystal at absolute zero is zero. Because entropy increases with temperature, entropy content of any substance above absolute zero must be a positive quantity. Thus, unlike enthalpy, entropy at 25° and 1 atm pressure is known, not only for the elements, but for many compounds. The values for entropy under such conditions are known as <u>absolute entropies.</u>

The free energy of the elements now may be evaluated at 25°C by the use of Equation 3.55 as shown in Equations 3.56.

$$G = H - TS$$
$$G(H_2, \text{gas}) = 0 - 298 (31.2) = -9300 \text{ cal}$$
$$G(C, \text{graphite}) = 0 - 298 (1.4) = -420 \text{ cal}$$
$$G(Br_2, \text{liq.}) = 0 - 298 (36.4) = -10,800 \text{ cal}$$

(3.56)

Standard-state reactions. Because it is a function of state, we can measure the change in enthalpy of a given reaction. Thus, by starting with the reactants at 25° and 1 atm pressure, carrying out the reaction, and measuring the heat evolved or absorbed by the system, including that needed to return the products to 25° and 1 atm, we obtain the <u>standard enthalpy of reaction.</u> The use of <u>standard</u> means that our conditions are 25°C and 1 atm pressure. Similarly, the <u>standard entropy of reaction</u> and the <u>standard free energy of reaction</u> are defined for the same conditions. The standard condition is shown by the use of a superscript ° to the symbol: $\Delta H°$, $\Delta S°$, $\Delta G°$.

Let us look at some examples: first, the formation of water from hydrogen and oxygen (Eqs. 3.57). The reaction of one mole of hydrogen with one half mole of oxygen to give one

$$H_2 + \tfrac{1}{2}O_2 \longrightarrow H_2O(\text{liq}) \qquad a$$
$$\Delta H° = H°(H_2O, \text{liq}) - H°(H_2, \text{gas}) - H°(\tfrac{1}{2}O_2, \text{gas}) \qquad b$$
$$\Delta H° = -68.3 - 0 - 0 = -68.3 \text{ kcal}$$
$$\Delta S° = S°(H_2O, \text{liq}) - S°(H_2, \text{gas}) - S°(\tfrac{1}{2}O_2, \text{gas}) \qquad c \qquad (3.57)$$
$$\Delta S° = 16.7 - 31.2 - \tfrac{1}{2}(49.0) = -39.0 \text{ eu}$$
$$\Delta G° = \Delta H° - T\Delta S° \qquad d$$
$$\Delta G° = -68,300 - 298 (-39.0)$$
$$\Delta G° = -68,300 + 11,600 = -56,700 \text{ cal}$$
$$\text{or} = -56.7 \text{ kcal}$$

mole of liquid water (*a*) is exothermic and evolves 68.3 kcal of enthalpy (heat) to the surroundings (*b*). The conversion of one and a half moles of gaseous reactants into one mole of liquid product is a change from a disordered state to a much more organized state, and the entropy of the system decreases (c). The free-energy change (d) is large and negative (exergonic reaction) and is a measure of the tendency of the reaction to be spontaneous.

The standard changes in the state functions for the formation of a compound from the elements are called the standard enthalpy of formation, the standard absolute entropy of formation, and the standard Gibbs free energy of formation. The symbols for the thermodynamic function of formation are identified by a subscript f: thus ΔH_f°, ΔS_f°, ΔG_f°. For example, ΔH_f° (H_2O, liq) $= -68.3$ kcal.

Use of heats (enthalpies) of formation and combustion data. Let us see how the thermodynamic functions of formation are obtained for a compound like methane, which cannot be prepared directly from graphite and hydrogen.

The ΔH_f° (CO_2, gas) $= -94.1$ kcal may be obtained in a manner similar to that used for finding ΔH_f° (H_2O, liq). Also, the heat of combustion of methane is found to be -212.8 kcal in Equations 3.58, which also show that the exothermic reaction may be expressed in

$$CH_4(\text{gas}) + 2O_2(\text{gas}) \xrightarrow[1 \text{ atm}]{25^\circ} CO_2(\text{gas}) + 2H_2O(\text{liq})$$

$$\Delta H^\circ = \Delta H_f^\circ(CO_2, \text{gas}) + 2\Delta H_f^\circ(H_2O, \text{liq}) - \Delta H_f^\circ(CH_4, \text{gas}) - 2\Delta H_f^\circ(O_2, \text{gas})$$

but $\Delta H^\circ = -212.8$ kcal; $\Delta H_f^\circ(CO_2, \text{gas}) = -94.1$ kcal

and, $\quad \Delta H_f^\circ(H_2O, \text{liq}) = -68.3$ kcal; $\Delta H_f^\circ(O_2, \text{gas}) = 0$ \hfill (3.58)

Substituting the numerical values gives:
$-212.8 = (-94.1) + 2(-68.3) - \Delta H_f^\circ(CH_4, \text{gas}) - 2(0)$

Rearranging the equation gives:
$\Delta H_f^\circ(CH_4, \text{gas}) = -94.1 - 136.6 + 212.8 = -17.9$ kcal

terms of the standard enthalpies of formation. Numerical values are available for all terms in the equation except for $\Delta H_f^\circ(CH_4, \text{gas})$. Solution of the equation gives $\Delta H_f^\circ(CH_4, \text{gas}) = -17.9$ kcal, which also is the standard enthalpy change for the reaction shown as Equation 3.59. Similar procedures were used to obtain the values given in Table 3.2 (Pg. 60).

$$C(\text{graphite}) + 2H_2(\text{gas}) \xrightarrow[1 \text{ atm}]{25^\circ} CH_4(\text{gas}) \hfill (3.59)$$

Values for the three thermodynamic functions of each of the four reactions are given in Table 3.3. A study of Table 3.3 shows that ΔH° and ΔG° are roughly comparable for a given reaction and that the $T\Delta S^\circ$ term is not comparable in size to either ΔH° or ΔG°. Such is not always the case, especially for reactions in which ΔH° is small. In such instances, the $T\Delta S^\circ$ term may be large enough to cause a change in sign between ΔH° and ΔG°. In general, because ΔH° values are more available than ΔS° or ΔG° for the tremendous variety of reactions in organic chemistry, it has become customary to use ΔH° instead of ΔG° to make rough estimates of whether or not a given reaction will be spontaneous.

Table 3.2 Some selected enthalpies and entropies of formation[a]

Elements	ΔH_f° kcal	ΔS_f° eu	Simple oxides and hydrides	ΔH_f° kcal	ΔS_f° eu
Br_g	26.8	41.8	CH_{4g}	−17.9	44.5
Br_{2g}	7.5	58.6	NH_{3g}	−11.0	45.9
Br_{2l}	0.0	36.4	HO_g	10.1	43.9
C_g	170.9	37.8	H_2O_g	−57.8	45.1
$C_{graphite}$	0.0	1.4	H_2O_l	−68.3	16.7
Cl_g	28.9	39.5	H_2O_{2g}	−32.5	55.6
Cl_{2g}	0.0	53.3	HF_g	−64.2	41.5
F_g	18.9	37.9	HCl_g	−22.1	44.6
F_{2g}	0.0	48.4	HBr_g	−8.7	47.5
H_g	52.1	27.4	HI_g	6.2	49.3
H_{2g}	0.0	31.2	HS_g	32.0	46.8
I_g	25.5	43.2	H_2S_g	−4.8	49.2
I_{2g}	14.9	62.3	PH_{3g}	2.2	50.2
I_{2s}	0.0	27.8	CO_g	−26.4	47.2
N_g	113.0	36.6	CO_{2g}	−94.1	51.1
N_{2g}	0.0	45.8	CS_{2g}	27.6	56.8
O_g	59.6	38.5	N_2O_g	19.5	52.6
O_{2g}	0.0	49.0	NO_g	21.6	50.3
S_g	66.4	40.1	NO_{2g}	8.1	57.5
S_{2g}	30.8	54.5	N_2O_{4g}	2.3	72.7
$S_{s,\,rhombic}$	0.0	7.6	SO_{2g}	−71.0	59.3
			SO_{3g}	−94.4	61.2
Carbon compounds (gases)					
CH_3Cl	−19.6	55.8			
CH_2Cl_2	−21	64.6	Carbon compounds (gases)		
$CHCl_3$	−24	70.7	$HC\equiv CH$	54.2	48.0
CCl_4	−25.5	73.9	$H_2C=CH_2$	12.5	52.5
CF_4	−218	62.5	CH_3CH_3	−20.2	54.9
$COCl_2$	−52.5	67.8	CH_3CH_2OH	−56.6	67.6
CH_3OH	−48.1	57.3	CH_3CH_2SH	−11.0	70.8
CH_3SH	−3.7	60.9	CH_3SCH_3	−9.0	68.3
$HCHO$	−27.7	52.3	CH_3CHO	−39.7	63.2
$HCOOH$	−90.4	59.4	CH_3COOH	−103.8	67.6
CH_3NC	35.9	58.8	CH_3CN	21.0	58.2

[a] H_f° is the standard enthalpy of formation at 25°C, 1 atm; and ΔS_f° is the absolute entropy at 25°C, 1 atm. Values are for one mole of the indicated substance.

Table 3.3 Enthalpy, entropy, and free-energy changes of four selected reactions

Reaction	ΔH° kcal	ΔS° eu	$-T\Delta S^\circ$ kcal	ΔG° kcal
$C + O_2 \longrightarrow CO_2$	−94.1	0.7	−0.2	−94.3
$H_2 + \tfrac{1}{2}O_2 \longrightarrow H_2O$	−68.3	−39.0	11.6	−56.7
$CH_4 + 2O_2 \longrightarrow CO_2 + 2H_2O$	−212.8	−58.0	17.3	−195.5
$C + 2H_2 \longrightarrow CH_4$	−17.9	−19.3	5.8	−12.1

At first thought, life seems to move in a direction contrary to thermodynamic principles. Complex organisms have evolved from simple organisms. The decrease in entropy involved in ordering the molecules in the many tissues and many precise locations is enormous. Although the entropy change between the raw materials and the completed organism may be large and negative, the entropy changes from all the reactions within the organism *and in the environment* is even larger and positive. Living systems couple spontaneous chemical reactions to the endergonic reactions needed for the essential life processes.

One of the spontaneous reactions used by most organisms is the oxidation of glucose.

$$C_6H_{12}O_6 + 6O_2 \rightleftharpoons 6CO_2 + 6H_2O \quad \Delta G = -686 \text{ kcal}$$

Under aerobic conditions glucose can be metabolized to carbon dioxide and water. This overall exergonic reaction proceeds in a controlled, stepwise fashion. Each stepwise reaction is coupled to a reaction that has a positive change in free energy. The sun provided the original energy, driving the reverse reaction to synthesize glucose from carbon dioxide and water in a plant.

Exercises

3.12 Use data from Table 3.2 to calculate $\Delta H°$, $\Delta S°$, $-T\Delta S°$, and $\Delta G°$ for the gas phase reactions given:
  a. $CH_4 + 4Cl_2 \longrightarrow CCl_4 + 4HCl$
  b. $CH_3Cl + H_2S \longrightarrow CH_3SH + HCl$
  c. $CH_3OH + NO_2 \longrightarrow HCHO + NO + H_2O$
  d. $CH_2Cl_2 + 3H_2O_2 \longrightarrow CO_2 + Cl_2 + 4H_2O$
  e. $CO + Cl_2 \longrightarrow COCl_2$

3.10 Summary

In this chapter, the classes of reactions introduced in the previous chapter have been expanded to include the double and triple bonds of carbon. In Section 3.3, the competitions during an ordinary S_N2 reaction were shown to be elimination and other possible S_N2 reactions. At the same time, proper conditions for the preparation of alcohols, ethers, and amines were pointed out.

Any reaction may be viewed from several angles. One way is to look at it as a manifestation of the reactivity of the substances involved. Another way of looking at it is as an example of one of the types of reactions which have been discussed. A third viewpoint would be a concern for the types of products being formed. It is the entanglement of viewpoints as well as the number of compounds and reactions that serve to confuse the beginning student of organic chemistry. By taking things a step at a time we can keep things sorted out, but failure to do so at this point only leads to hopeless disorder later on.

In Section 3.4, we looked at only a few electrophilic displacements, but we can use them to convert alkyl halides through the metal organic compounds to hydrocarbons or to other halides. In Section 3.5, we described the formation of the lithium and magnesium reagents as well as free-radical chain halogenation of alkanes. Then, in Section 3.6, we saw a number of addition reactions of the multiple bonds, mainly the carbon-carbon double

62   Chapter three

bond. Homolytic and polar additions were described. In Section 3.7, several elimination reactions were illustrated. Section 3.8 expanded our knowledge of oxidation-reduction reactions.

In no sense have we finished with any reaction given in this chapter. Just as this chapter was an expansion of the previous chapter, so we shall find future chapters to contain both repetition and new material as we develop and build upon what has already been covered.

The use of thermodynamic functions was introduced, and the calculation of the standard enthalpy, entropy, and free-energy changes of reactions was illustrated.

In the next chapter, we shall introduce some compounds with three and four atoms of carbon and take up naming in a more systematic manner.

S_N2 $\quad HO^\ominus + CH_3CH_2Br \xrightarrow[\text{acetone}]{25°, H_2O} CH_3CH_2OH + Br^\ominus \quad$ (3.3)

$\quad CH_3CH_2O^\ominus + CH_3CH_2Br \longrightarrow CH_3CH_2OCH_2CH_3 + Br^\ominus \quad$ (3.6)

S_E2 $\quad H_2O + CH_3Li \longrightarrow HO^\ominus + CH_4 + Li^\oplus \quad$ (3.12)

S_H $\quad CH_3CH_2Br + 2Li \xrightarrow[25°]{\text{ether}} CH_3CH_2Li + LiBr \quad$ (3.13)

Addition-elimination to carbon-carbon multiple bonds:

H_2 $\quad H_2C{=}CH_2 + H_2 \xrightarrow[25°]{Pd} CH_3CH_3 \quad$ (3.51)

X_2 $\quad H_2C{=}CH_2 + Br_2 \xrightarrow{CH_3COOH} BrCH_2CH_2Br \quad$ (3.18)

$\quad H_2C{=}CH_2 + X_2 \xrightarrow[CH_3COONa]{CH_3COOH} CH_3COOCH_2CH_2X \quad$ (3.19)

Elim. $\quad XCH_2CH_2X \xrightarrow{Zn} ZnX_2 + H_2C{=}CH_2 \quad$ (3.37)

HX $\quad H_2C{=}CH_2 + HX \longrightarrow CH_3CH_2X \quad$ (3.20)

Elim. $\quad CH_3CH_2X + CH_3CH_2O^\ominus \longrightarrow H_2C{=}CH_2 + X^\ominus + CH_3CH_2OH \quad$ (3.40)

H_2O $\quad H_2C{=}CH_2 + H_2SO_4 \xrightarrow[\text{ice}]{0°} CH_3CH_2OH \quad$ (3.27)

$\quad HC{\equiv}CH + \text{dil. } H_2SO_4 \xrightarrow{Hg^{2+}} CH_3CHO \quad$ (3.29)

$\quad CH_3CH_2OH + H_2SO_4 \xrightarrow{180°} H_2C{=}CH_2 + H_2O \quad$ (3.42)

Addition-elimination to carbon-oxygen double bonds:

H_2 $\quad CH_3CHO + H_2 \xrightleftharpoons[Cu, \Delta]{Pt} CH_3CH_2OH \quad$ (similar to 2.20)

$\quad CH_3CHO + LiAlH_4 \xrightarrow{\text{ether}} (CH_3CH_2O)_4LiAl \quad$ (3.44)

H_2O $\quad\quad\quad\quad$ $CH_3CHO + H_2O \rightleftharpoons CH_3CH(OH)_2$ $\quad\quad\quad\quad$ (3.31)

Oxidation-reduction (see also addition of hydrogen):

$$CH_3CH_2OH + Na_2Cr_2O_7 \xrightarrow[100°]{\text{dil. } H_2SO_4} CH_3COOH \quad\quad (3.43)$$

$$H_2C=CH_2 + KMnO_4 \xrightarrow{0°} HOCH_2CH_2OH \quad\quad (3.48)$$

$$HOCH_2CH_2OH + KMnO_4 \xrightarrow[100°]{\text{dil. } H_2SO_4} 2CO_2 \quad\quad (3.49)$$

$$H_2C=CH_2 + CH_3COOOH \longrightarrow H_2C\underset{O}{\overset{}{\diagdown\!\diagup}}CH_2 + CH_3COOH \quad\quad (3.30)$$

3.11 Problems

3.1 Starting with methanol and ethanol as the only organic substances, but having available any solvent and inorganic reagents needed, give a laboratory synthesis of:
  a. glycol
  b. ethane
  c. methyl iodide
  d. 1,1,2,2-tetrabromoethane

Problems like these are best worked out backwards and then put together. Example: synthesize 1,1,2-trichloroethane. First, we have to be able to write the formula for the substance, or $ClCH_2CHCl_2$. Second, what reactions have been given that could form this substance? An addition to a double bond looks good: we could try either the addition of HCl to $ClCH=CHCl$ or the addition of Cl_2 to $H_2C=CHCl$. Third, $ClCH=CHCl$ could be obtained if we added Cl_2 to $HC\equiv CH$. Acetylene, however, is still some way from ethanol. On the other hand, there is a fourth way. Vinyl chloride may be formed from CH_3CHCl_2 or $ClCH_2CH_2Cl$ by treatment with hot alcoholic KOH. Of these dichloroethanes, the latter is available from ethylene by addition of Cl_2, and ethylene is easily produced from ethanol by dehydration. So a correct answer would be written as shown.

$$CH_3CH_2OH \xrightarrow[H_2SO_4]{180°} H_2C=CH_2 \xrightarrow[CCl_4]{Cl_2} ClCH_2CH_2Cl \xrightarrow[\text{hot}]{\text{alc. KOH}} H_2C=CHCl \xrightarrow[CCl_4]{Cl_2} ClCH_2CHCl_2$$

In general, a synthetic sequence with the fewest number of steps (reactions) is the best. Such problems may have more than one answer that could be considered to be correct.

3.2 Given the additional $\Delta H_f°$ values in kcal: $CH_3CH_2OH(\text{liq}) = -66.4$, $CH_3CH_2Cl(\text{gas}) = -25.7$, $CH_3CH_2Br(\text{gas}) = -15.3$, $CH_3CH_2NH_2(\text{gas}) = -11.6$, $(CH_3)_2NH = -6.6$, $CH_3CHO(\text{gas}) = -39.7$, $CH_3COOH(\text{liq}) = -116.4$, calculate $\Delta H°$ for the reactions shown:
  a. $CH_3CH_2OH(\text{liq}) + HCl(\text{gas}) \longrightarrow CH_3CH_2Cl(\text{gas}) + H_2O(\text{liq})$
  b. $CH_3CH_2OH(\text{liq}) + HBr(\text{gas}) \longrightarrow CH_3CH_2Br(\text{gas}) + H_2O(\text{liq})$
  c. $H_2C=CH_2(\text{gas}) + H_2O(\text{liq}) \longrightarrow CH_3CH_2OH(\text{liq})$

64 Chapter three

3.3 Give conditions (other reagents, solvent, catalyst, if needed, and temperature) for these reactions:
a. HC≡CH ⟶ CH_3CHCl_2
b. CH_3CHO ⟶ CH_3CH_2OH
c. OHCCHO ⟶ HOOCCOOH
d. CH_3Cl ⟶ CH_3I
e. CH_3CH_2Br ⟶ $(CH_3CH_2)_2NH$
f. $H_2C{=}CH_2$ ⟶ $CH_3COOCH_2CH_2Cl$
g. $H_2C{=}CH_2$ ⟶ CH_3CH_2Br
h. $BrCH_2CH_2Br$ ⟶ $H_2C{=}CHBr$

3.4 Write equations for the reactions of the substances listed: 1. with $LiAlH_4$ in ether, followed by hydrolysis; and 2. with H_2 in the presence of a catalyst. If no reaction, write NR.
a. CH_3COOH
b. CH_3CHO
c. OHCCHO
d. $H_2C{=}CH_2$
e. OHCCOOH
f. $HOCH_2CHO$
g. $H_2C{=}CHCHO$
h. $HOCH_2COOH$

3.5 Calculate, assuming H_2O to be a gas, the heat of combustion of:
a. ethane   b. ethylene   c. acetylene
On a heat-per-gram basis, which gives the maximum amount of heat?

3.6 If the combustion of one mole of urea, H_2NCONH_2, a crystalline solid, evolves 151.6 kcal of heat, calculate ΔH_f° of urea. Assume the products to be $N_2(g)$, $CO_2(g)$, and $H_2O(l)$.

3.12 Bibliography

It is not easy to find suitable reading material at this early stage in our developing knowledge of organic chemistry. However, a few paperbacks are suggested.

1. Brown, Theodore L., *Energy and the Environment.* Columbus, Ohio: Charles E. Merrill, 1971. Concentrates upon pollution and the energy balance of our planet; non-mathematical.

2. Miller, G. Tyler, Jr., *Energetics, Kinetics, and Life: An Ecological Approach.* Belmont, Calif.: Wadsworth, 1971. A fascinating discussion of thermodynamics at an elementary level with applications to only eleven reactions. The reactions chosen form a starting point for examination of pollution, food supply, mineral resources, energy supply, overpopulation, etc.

3. Nash, Leonard K., *Elements of Chemical Thermodynamics.* Reading, Mass.: Addison-Wesley, 1962. An excellent introduction to chemical thermodynamics at a more sophisticated and mathematical level than that of the preceding reference.

3.13 Saul G. Cohen: From freshman to dean

University Professor Saul G. Cohen of Brandeis is one of the rare individuals who has shown that it is possible to be both a chemist and an administrator. Here, based upon his Norris Award address of 1972, he shares with us some of his reminiscences.

I came to Harvard in 1933, not knowing anything about chemistry. As a student in Professor Lamb's freshman course, I thought him elderly and benign, fascinated by the phenomena and regularities of chemistry, and happy to share his pleasure in the subject with us. But there was a remarkable dichotomy in the course. The section was taught by bright, tense, young postdoctorals. They showed intense interest in the developing science and pleasure in bringing us into this new world. Their deep interest in the subject and in teaching us seemed to be in contrast with the attitudes I had observed in my brief exposure to other areas that I had considered; and I opted to concentrate in chemistry.

The problems and difficulty in teaching and learning chemistry became apparent to me early: the basically abstract nature of the subject, abstract notions of atoms and molecules, and the observable phenomena to be described and dealt with in terms of abstract formulations. In the advanced organic chemistry course, Professor Kohler gave a masterful series of lectures describing, it seemed, a mature body of knowledge, admirably developed, establishing for us that chemistry, with its ideas of structure and change, was a fundamental, powerful, intellectual achievement. A few years later the same course was taught by Paul Bartlett. Not neglecting the past, it emphasized current problems, bringing to the fore the developing character of chemistry. In this, Paul Bartlett made a most important contribution to the teaching and practice of organic chemistry. In teaching, past accomplishments must be selected carefully to illustrate how problems are recognized, perceived differently at different times, and developed. The difficult teaching of the techniques of contemporary problem solving must also be dealt with, but this may not become the entire task.

I remember, in another course, the young professor confided to us in the opening lecture that we would not be able to understand the subject. At Christmas time I decided I had better understand the course. There is a difference between understanding a course and understanding a subject. The course is what is covered and can be understood. The subject has an infinite aspect to it.

In teaching over a long period, we sense more deeply the essence, the complexity, the nuances. What we thought and said with some confidence at an earlier date we are no longer sure of, and what we seem to be sure of will probably not be valid later. Teaching becomes learning for the teacher as for the student, but the student should not be overly confused by the teacher's uncertainty. Our understanding changes, perhaps by a cessation of stupidity; we derive pleasure from a new insight and from the opportunity to share it, and we recognize that it is merely the best view that we may have at the time, and that it too will change.

As a graduate student at Harvard, 1937–1940, I learned to draw my own conclusions and not to accept prior authority, an attitude which seems to come to me easily, but which I nevertheless have to relearn continuously. It is important to teach students to make their own observations, describe them accurately, draw independent inferences, act on them, and accept proof or disproof of the inferences.

This kind of training need not be restricted to the natural sciences. The need to try to see and formulate problems and consequences of recommended action, as they are—

and not as you wish them to be, not as someone else says they are, or wishes them to be, or wishes you to think they are—that need is common to our condition, and the ability to carry out this formulation is essential for solution of any problem. It's not an easy process. Perception and choice of a problem, setting of a goal, are strongly influenced by values and emotion. The solution is largely a rational process, with contributions from experience, which we often term intuition.

My first opportunity to teach was given me in 1940 by Paul Bartlett. I prepared a whole series of lectures, appeared for the first class, and began to talk. When I looked up at the dazed students an hour later I realized I had completed the entire course, the students were lagging behind. I explained, cooly I hope, that I had summarized the material to be covered and would elaborate in succeeding meetings. From this I learned that repetition might be useful in teaching, and that I should conserve what little capital I might have. I am still nervous at the first meeting of a course.

This position also gave me the opportunity for my first independent research. Cleavage of the alkyl-oxygen bond in hydrolysis of esters had been shown not to occur generally. That was a proven, accepted mechanistic conclusion, rare for those days. Showing as we did that it did occur where you might expect it to, and that this too would be general, was to me, what Abraham Maslow later termed a Peak Experience. It established for me, within me, an identity—in which previously I could not have been confident—that I would be able to do original scientific work. The need to reestablish this confidence recurs however.

After a period working on war gasses, in which I developed respect for the physiological action of simple chemical compounds, I went to U.C.L.A. in 1943 as a National Research Council Fellow. I did some experiments and published a paper showing that the acetolysis of certain simple esters could lead to carbonium ions. I may have been ahead of the times or perhaps beyond my depth. Many years later Saul Winstein kindly reminded me of the impact of this work on him.

My education was furthered by a period in industry, 1944–1950. Many of my friends at Polaroid look on me as a teacher in my relation to them, but I have learned far more from them than I taught. It is fascinating to observe invention, ingenious, original daring study and use of nature, to be involved in an attitude that you can solve any problem. Describe it accurately and operationally, think carefully, believe it is important, work without surcease, accept that there are many ways to arrive at the solution. Nature is subtle, resistant but not hostile, neutral, and will reveal its secrets to the worthy suitor.

Think what we learn when we teach about some problems in stereochemistry. The tetrahedral carbon atom was accepted almost one hundred years ago. Ball and stick models were available to anybody. The model of cyclohexane stood before one's eyes, puckered. Yet it took years for it to be accepted that cyclohexane was not planar. The axial and equatorial substituents stood there for all to see. Yet it took many years and sophisticated electron diffraction studies by Hassel to teach this lesson; and more years for Barton to learn and teach the relation of chemical reactivity to this structural feature.

Why does this lag occur? Whence the resistance to drawing the inference from the observation? There seems to be a block, a mysterious difficulty in proclaiming consequences that are logical but elude us for long periods. It appears that chemistry does not develop in a linear fashion. Experience may help one avoid the dead-end paths and encourage a bold stride down the fruitful way, but experience may also be very inhibiting for a long period. What appears to be fresh, daring vision is usually the result of much laborious thought, which may involve the slow discarding of earlier strongly held views. In the latter aspect of the process the inexperienced may have a big advantage.

It was probably the optimistic notion that one could solve any problem that allowed me to move on to three-year-old Brandeis University in 1950 as chairman of the School of Science. Perhaps I was influenced by the admonition to the young of a sheltered philosophy professor that for self-fulfillment one should associate with a self-transcending movement or institution. I subsequently realized that institutional authorities don't always believe that that is a major raison d'être of the institution—they attribute an independent life to it.

I set out to attract faculty who would be teachers and scholars in four disciplines: biology, chemistry, physics, and mathematics. Scholarship might be defined relatively clearly, and so excellence in it might be achieved, while less readily defined goals might well remain elusive. Later on, 1955–1959, while I was dean of faculty, departmental structure and a faculty senate were established, I found the experience a maturing one, which left me fewer illusions about administrators, faculty, students, and myself. I was confirmed in my view that the major task, contribution, and satisfaction lay in learning and teaching, in office, library, classroom, and laboratory. Returning full time to the chemistry department in 1959 I found satisfaction in teaching and in the continuity and development of my research.

4 Some Compounds with Three and Four Carbon Atoms: Naming

We now proceed to a listing of compounds containing three and four carbon atoms. Homologous series are presented, and a discussion of structural isomerism is given. Cyclic compounds are introduced. Systematic naming is presented in much greater detail than was possible before. Thus we can go on to include examples of compounds with more than four carbon atoms.

4.1 Some compounds with three and four carbon atoms   66
 1. How the homologous series of alkanes is built up.   2. Primary, secondary, tertiary, and quaternary.   3. Generalized molecular formulas.   4. Heterocycles and carbocycles.   5. Ketones; isomerism with aldehydes.   6. Some three- and four-carbon-atom compounds. Exercises.

4.2 Naming: chains, substituents, and prefixes   72
 1. Introduction to IUPAC rules.   2. Alkanes.   3. Substituents and prefixes.   4. How to derive a structural formula from a name.   5. Systematic names of alkyl substituents.   6. Hydrocarbons.   7. Heavy metals. Exercises.

4.3 Naming: groups designated by suffixes   77
 1. Double and triple bonds.   2. Reactive substituents; prefixes and suffixes; priority scale. Exercises.

4.4 Naming: cyclic compounds   79
 1. Rings with no double bonds; substituents.   2. Rings with double bonds; priorities in numbering.   3. Rings as substituents on chains.   4. Carboxylic acid method of naming for rings.   5. Carboxylic acid names for chains. Exercises.

4.5 Summary   82

4.6 Problems   83

4.7 Bibliography   85

4.1 Some compounds with three and four carbon atoms

How the homologous series of alkanes is built up. If any of the six equivalent hydrogens in a molecule of ethane is replaced by a methyl group, the next member of the saturated hydrocarbon series, propane, is formed. A replacement of this type, easily done on paper, impossible in the laboratory, is known as a paper reaction (Eq. 4.1).

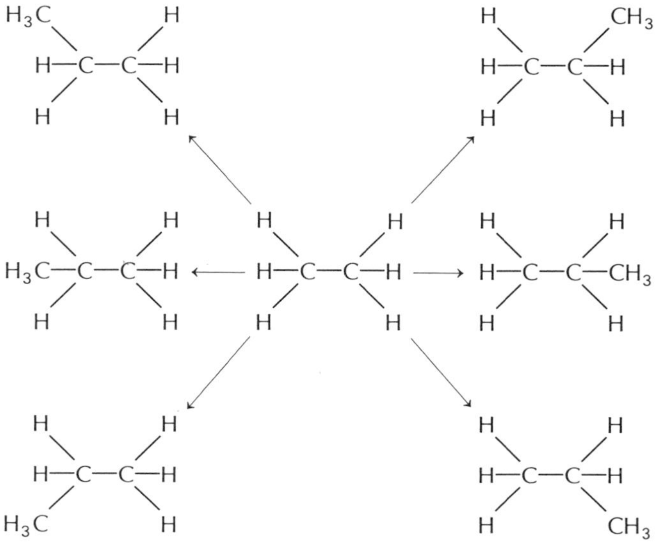

$$\text{(4.1)}$$

In the formula for propane, $CH_3CH_2CH_3$, we recognize two kinds of hydrogen in the molecule: six on the two methyl groups, and two on the central carbon atom, called a methylene group. Replacement of any of the equivalent six hydrogens with a methyl group gives rise to butane, $CH_3CH_2CH_2CH_3$. Replacement of either of the two hydrogens in the methylene group by a methyl gives a different hydrocarbon, $(CH_3)_2CHCH_3$, called isobutane to distinguish it from butane. It is the first structural isomer in the hydrocarbon series.

Primary, secondary, tertiary, and quaternary. We shall use, from time to time, primary, secondary, tertiary, and quaternary to describe various structures and structural features. A primary carbon atom is one that is singly bonded to only one other carbon atom. The other three single bonds may be to any atoms other than carbon. A primary hydrogen, halogen, or hydroxy group is one that is bonded to a primary carbon atom.

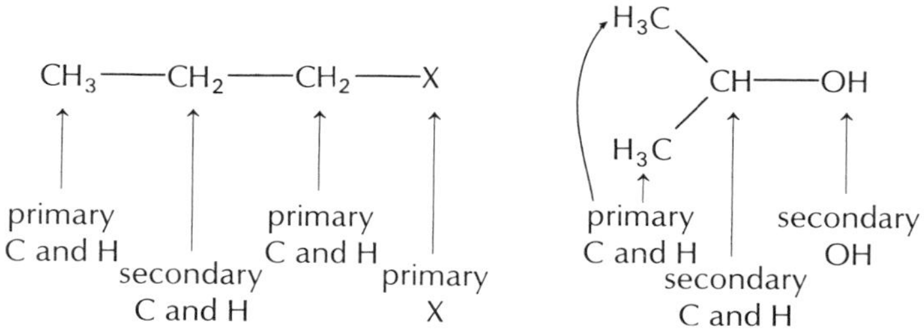

4.1 Some compounds with three and four carbon atoms

A secondary carbon atom is singly bonded to two other carbon atoms. Other atoms or groups bonded to a secondary carbon atom are also classified as secondary atoms or groups.

A tertiary carbon atom is singly bonded to three other carbon atoms, and a quaternary carbon atom is bonded to four other carbon atoms.

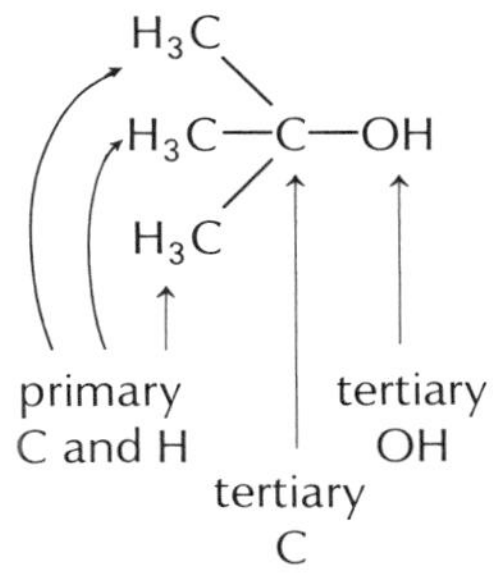

$$H_3C-\underset{\underset{H_3C}{|}}{\overset{\overset{H_3C}{|}}{C}}-OH$$

primary C and H — tertiary C — tertiary OH

Generalized molecular formulas. If any of the four equivalent hydrogens in a molecule of ethylene is replaced by a methyl group, the next member of the alkene or unsaturated hydrocarbon series, propene ($CH_3CH=CH_2$) is formed. Examination of the propene formula shows that three kinds of hydrogens are present: the three in the methyl group, the one adjacent to the methyl group, and the two at the other end of the double bond. Replacement of any of the three hydrogens in the methyl group by a methyl group gives 1-butene ($CH_3CH_2CH=CH_2$). Replacement of the single hydrogen gives 2-methylpropene (($CH_3)_2C=CH_2$), and replacement of either of the two end hydrogens in propene gives 2-butene ($CH_3CH=CHCH_3$). Thus, there are three isomeric structures of molecular formula C_4H_8.

Again, we recognize two kinds of hydrogen in any of the ethyl compounds, such as an ethyl halide, ethanol, or ethylamine (neglecting the hydrogens attached to the oxygen or the nitrogen). Each can give rise to two isomeric propyl compounds. For example, ethyl bromide could give 1-bromopropane ($CH_3CH_2CH_2Br$) and 2-bromopropane ($CH_3CHBrCH_3$). Then 1-bromopropane will give three different bromobutanes: 1-bromobutane ($CH_3CH_2CH_2CH_2Br$), 2-bromobutane ($CH_3CH_2CHBrCH_3$), and 1-bromo-2-methylpropane (($CH_3)_2CHCH_2Br$). On the other hand, upon replacement of hydrogen atoms by methyl groups, 2-bromopropane gives rise to only two bromobutanes, and one of them is identical with one obtained from 1-bromopropane: namely, 2-bromobutane. The new isomer is 2-bromo-2-methylpropane, ($CH_3)_3CBr$ (Eqs. 4.2).

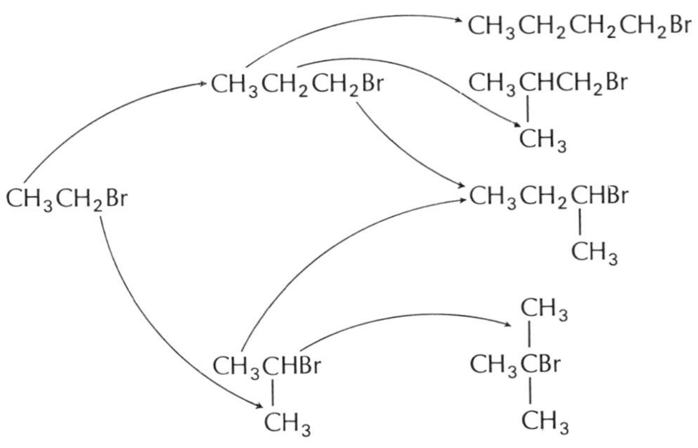

$$CH_3CH_2Br \longrightarrow \begin{matrix} CH_3CH_2CH_2Br \\ CH_3CHBr \\ | \\ CH_3 \end{matrix} \longrightarrow \begin{matrix} CH_3CH_2CH_2CH_2Br \\ CH_3CHCH_2Br \\ | \\ CH_3 \\ CH_3CH_2CHBr \\ | \\ CH_3 \\ CH_3 \\ | \\ CH_3CBr \\ | \\ CH_3 \end{matrix} \quad (4.2)$$

68   Chapter four

These examples show the large number of structural isomers that occur as the number of carbon atoms is increased. Also, we see that any organic compound can be placed in a series of compounds that differ, neighbor to neighbor, by a CH_2 group. To go up the series, we remove a hydrogen from a compound and add a CH_3. Such a series is known as a homologous series. All members of the series can be expressed by a generalized molecular formula. Thus, the alkane series is given by C_nH_{2n+2} and the alkene series by C_nH_{2n}, in which n represents the number of carbon atoms in the compound.

Heterocycles and carbocycles.   Beginning with ethylene oxide, a three-membered ring, it is possible to set up another homologous series of heterocycles. Thus, the next member is trimethylene oxide, a four-membered ring; then tetramethylene oxide, a five-membered ring, and so on.

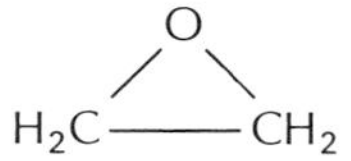

ethylene oxide 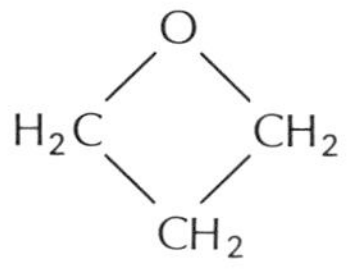   trimethylene oxide 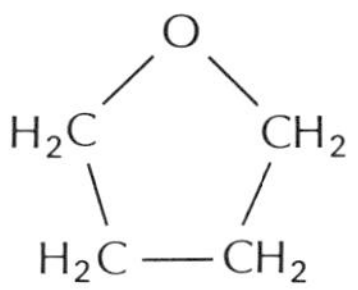   tetramethylene oxide

The first possible carbocyclic ring is cyclopropane, followed by cyclobutane, cyclopentane, and so on. Carbocyclic ring compounds usually are written as simple regular

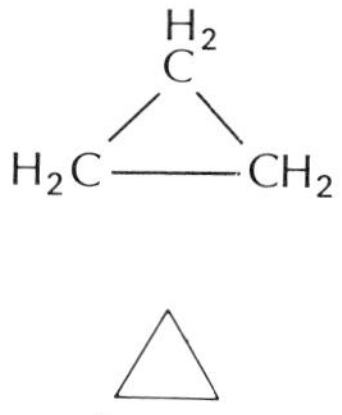

cyclopropane 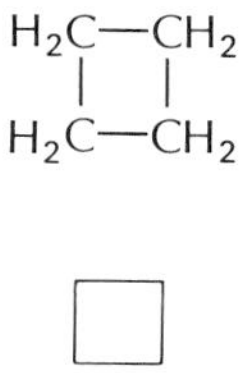   cyclobutane 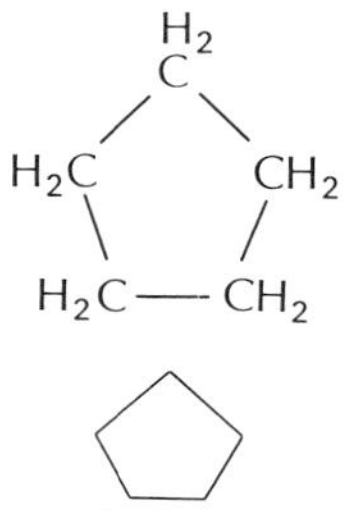   cyclopentane 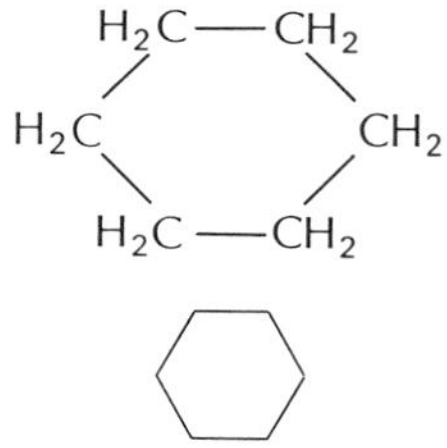   cyclohexane

polygons, and it is understood that each corner represents a -CH_2- group. The physical properties of the carbocyclic hydrocarbons do not differ greatly from those of the corresponding open-chain compounds. There are marked differences in the chemical properties of the smaller rings: they vary depending upon the size of the ring. The polymethylene oxides (which may be regarded as cyclic ethers) also have different reactivities, dependent in part on the size of the ring. The cyclic compounds also may have substitutions on the ring or may be linked to open chains. Double bonds may be present in the ring, but no triple bonds unless the ring has at least eight atoms.

Ketones; isomerism with aldehydes.   The only new type of compound we need to introduce is the ketone. A ketone has a carbon-oxygen double bond: the carbon atom is singly bonded to two other carbon atoms. The simplest ketone is acetone (CH_3COCH_3), which may be thought of as being derived from acetaldehyde (CH_3CHO) by methyl replacement of the single hydrogen. Acetone (C_3H_6O) is isomeric with propionaldehyde (CH_3CH_2CHO). The physical properties are similar, but ketones are far more stable to oxidation and resist addition to the carbon-oxygen double bond more than do aldehydes.

Some three- and four-carbon-atom compounds. Table 4.1 gives a sampling of three-carbon-atom compounds, and Table 4.2 gives a longer list of four-carbon-atom compounds. Of course many compounds have been omitted in both tables. A few esters that have more than the listed number of carbon atoms have been given to familiarize us with that type of compound. Note the proliferation of possibilities for the existence of compounds with more than one reactive center: unsaturated alcohols, aldehydes, ketones, and acids; hydroxy and keto acids; cyclic compounds; and others.

The prefixes ($\pm$) and *meso* designate a type of isomerism known as stereoisomerism. The prefixes *cis-* and *trans-* designate a kind of stereoisomerism known as geometrical isomerism. These forms of isomerism will be described in Chapter 6. The amino acids are described in Chapter 21.

Table 4.1 Some three-carbon-atom compounds

Name[a]	Formula	mw	mp[b]	bp[b]
propane*	$CH_3CH_2CH_3$	44.1	−190	−44
1-propanol* (propyl alcohol)	$CH_3CH_2CH_2OH$	60.1	−127	97
2-propanol* (isopropyl alcohol)	$(CH_3)_2CHOH$	60.1	−89	82
1-chloropropane* (propyl chloride)	$CH_3CH_2CH_2Cl$	78.5	−123	47
2-chloropropane* (isopropyl chloride)	$(CH_3)_2CHCl$	78.5	−117	35
1-propylamine* (propylamine)	$CH_3CH_2CH_2NH_2$	59.1	−83	49
2-propylamine* (isopropylamine)	$(CH_3)_2CHNH_2$	59.1	−101	33
trimethylamine*	$(CH_3)_3N$	59.1	−117	3
propionaldehyde (propanal*)	CH_3CH_2CHO	58.1	−81	49
acetone (2-propanone*)	CH_3COCH_3	58.1	−95	56
propionic acid (propanoic acid*)	CH_3CH_2COOH	74.1	−21	141
ethyl acetate (ethyl ethanoate*)	$CH_3COOCH_2CH_3$	88.1	−84	77
ethyl propionate (ethyl propanoate*)	$CH_3CH_2COOCH_2CH_3$	102.1	−74	99
propylene (propene*)	$CH_3CH=CH_2$	42.1	−185	−48
allene (propadiene*)	$CH_2=C=CH_2$	40.1	−146	−34
propyne* (methylacetylene)	$CH_3C\equiv CH$	40.1	−103	−23
allyl alcohol (2-propen-1-ol*)	$CH_2=CHCH_2OH$	58.1	−129	97
propylene glycol (1,2-propanediol*)	$CH_3CHOHCH_2OH$	76.1	...	189
1,3-propanediol* (trimethylene glycol)	$HOCH_2CH_2CH_2OH$	76.1	...	213
($\pm$)-lactic acid (($\pm$)-2-hydroxypropanoic acid*)	$CH_3CHOHCOOH$	90.1	18	dec.
acrylic acid (propenoic acid*)	$CH_2=CHCOOH$	72.1	13	142
malonic acid (propanedioic acid*)	$CH_2(COOH)_2$	104.1	136	dec.
ethyl malonate (diethyl propanedioate*)	$CH_2(COOCH_2CH_3)_2$	160.2	−50	199
methoxyethane* (ethyl methyl ether)	$CH_3OCH_2CH_3$	60.1	...	7
methoxyethanol* (ethylene glycol monomethyl ether)	$CH_3OCH_2CH_2OH$	76.1	−85	125
propylene epoxide (1,2-epoxypropane*)	$CH_3HC\overset{\displaystyle O}{\underset{}{\triangle}}CH_2$	58.1	...	35
trimethylene oxide (oxetane*)	$CH_2CH_2CH_2$ (with O bridge)	58.1	...	48
cyclopropane*	$CH_2\text{—}CH_2$ (with CH_2 bridge)	42.1	−127	−33
cyclopropanol*	$CH_2\text{—}CHOH$ (with CH_2 bridge)	58.1	...	103

Table 4.1 (continued)

Name[a]	Formula	mw	mp[b]	bp[b]
cyclopropene*	CH₂ bridging CH=CH (cyclopropene structure)	40.1	...	−36
alanine (2-aminopropanoic acid*)	CH₃HC(⁺NH₃)(COO⁻)	89.1	d	...
serine (2-amino-3-hydroxypropanoic acid*)	HOCH₂HC(⁺NH₃)(COO⁻)	105.1	d	...
glycerol (1,2,3-propanetriol*)	HOCH₂CHOHCH₂OH	92.1	20	290 d

[a] The most popular name (a matter of opinion) is given first and an alternate name in parentheses. The IUPAC names are marked with an asterisk.
[b] In °C, rounded off to the nearest degree.

Table 4.2 Some four-carbon-atom compounds

Name[a]	Formula	mw	mp[b]	bp[b]
butane*	CH₃CH₂CH₂CH₃	58.1	−138	−1
isobutane (methylpropane*)	(CH₃)₂CHCH₃	58.1	−145	−10
1-butanol* (butyl alcohol)	CH₃CH₂CH₂CH₂OH	74.1	−90	117
(±)-2-butanol* (sec-butyl alcohol)	CH₃CH₂CHOHCH₃	74.1	−100	99
2-methyl-1-propanol* (isobutyl alcohol)	(CH₃)₂CHCH₂OH	74.1	−108	108
2-methyl-2-propanol* (tert-butyl alcohol)	(CH₃)₃COH	74.1	25	82
1-methoxypropane* (methyl propyl ether)	CH₃OCH₂CH₂CH₃	74.1	...	39
2-methoxypropane* (isopropyl methyl ether)	(CH₃)₂CHOCH₃	74.1	−116	35
ether (ethoxyethane*)	CH₃CH₂OCH₂CH₃	74.1	−116	35
butyl bromide (1-bromobutane*)	CH₃CH₂CH₂CH₂Br	137.0	−112	101
(±)-sec-butyl bromide ((±)-2-bromobutane*)	CH₃CH₂CHBrCH₃	137.0	−112	91
isobutyl bromide (1-bromo-2-methylpropane*)	(CH₃)₂CHCH₂Br	137.0	−118	91
tert-butyl bromide (2-bromo-2-methylpropane*)	(CH₃)₃CBr	137.0	−20	73
butylamine (1-butylamine*)	CH₃CH₂CH₂CH₂NH₂	73.1	−50	78
(±)-sec-butylamine ((±)-2-butylamine*)	CH₃CH₂CH(NH₂)CH₃	73.1	−104	63
isobutylamine (2-methyl-1-propylamine*)	(CH₃)₂CHCH₂NH₂	73.1	−85	68
tert-butylamine (2-methyl-2-propylamine*)	(CH₃)₃CNH₂	73.1	−68	45
tetramethylammonium chloride*	(CH₃)₄NCl	109.6	230 d	...
butyraldehyde (butanal*)	CH₃CH₂CH₂CHO	72.1	−99	76
isobutyraldehyde (2-methylpropanal*)	(CH₃)₂CHCHO	72.1	−66	61
2-butanone* (ethyl methyl ketone)	CH₃CH₂COCH₃	72.1	−87	80
butyric acid (butanoic acid*)	CH₃CH₂CH₂COOH	88.1	−6	163
isobutyric acid (2-methylpropanoic acid*)	(CH₃)₂CHCOOH	88.1	−47	154
butylene (1-butene*)	CH₃CH₂CH=CH₂	56.1	−185	−6
cis-2-butene*	CH₃CH=CHCH₃	56.1	−139	4
trans-2-butene*	CH₃CH=CHCH₃	56.1	−106	1
isobutylene (2-methylpropene*)	(CH₃)₂C=CH₂	56.1	−141	−7
1-butyne* (ethylacetylene)	CH₃CH₂C≡CH	54.1	−122	8

Table 4.2 (continued)

Name[a]	Formula	mw	mp[b]	bp[b]
2-butyne* (dimethylacetylene)	$CH_3C{\equiv}CCH_3$	54.1	−32	27
crotyl alcohol (trans-2-buten-1-ol*)	$CH_3CH{=}CHCH_2OH$	72.1	−30	121
(±)-methylvinylcarbinol ((±)-3-buten-2-ol*)	$CH_2{=}CHCHOHCH_3$	72.1	−100	97
crotonaldehyde (trans-2-butenal*)	$CH_3CH{=}CHCHO$	70.1	−74	104
crotonic acid (trans-2-butenoic acid*)	$CH_3CH{=}CHCOOH$	86.1	72	185
methacrylic acid (2-methylpropenoic acid*)	$CH_2{=}C(CH_3)COOH$	86.1	16	163
succinic acid (butanedioic acid*)	$HOOCCH_2CH_2COOH$	118.1	182	235 d
succinic anhydride	$O{=}\overset{\diagup O\diagdown}{C}CH_2CH_2C{=}O$	100.1	120	261
(±)-malic acid ((±)-hydroxybutanedioic acid*)	$HOOCCH_2CHOHCOOH$	134.1	130	d
(±)-tartaric acid ((±)-2,3-dihydroxybutanedioic acid*)	$HOOCCHOHCHOHCOOH$	150.1	160	d
(±)-tartaric acid hydrate	$HOOCCHOHCHOHCOOH \cdot H_2O$	168.1	203	...
meso-tartaric acid	$HOOCCHOHCHOHCOOH$	150.1	140	...
acetoacetic acid (3-ketobutanoic acid*)	CH_3COCH_2COOH	102.1	...	100 d
acetoacetic ester (ethyl acetoacetate)	$CH_3COCH_2COOCH_2CH_3$	130.1	−80	180
(±)-2,3-butanediol*	$CH_3CHOHCHOHCH_3$	90.1	...	180
meso-2,3-butanediol*	$CH_3CHOHCHOHCH_3$	90.1	25	...
1,4-butanediol*	$HOCH_2CH_2CH_2CH_2OH$	90.1	19	235
isobutylene oxide (1,2-epoxy-2-methylpropane*)	$(CH_3)_2C\overset{\diagup O\diagdown}{}CH_2$	72.1	...	56
tetrahydrofuran* (tetramethylene oxide)	$\overset{\diagup O\diagdown}{CH_2CH_2CH_2CH_2}$	72.1	−65	64
methylcyclopropane*	$CH_2{-}CHCH_3 \diagdown CH_2 \diagup$	56.1	...	4
cyclopropylmethanol*	$CH_2{-}CHCH_2OH \diagdown CH_2 \diagup$	72.1	...	123
cyclobutane*	$CH_2{-}CH_2 \mid \mid CH_2{-}CH_2$	56.1	−50	13
cyclobutene*	$CH{=}CH \mid \mid CH_2{-}CH_2$	54.1	...	2
cyclobutanol*	$CH_2{-}CHOH \mid \mid CH_2{-}CH_2$	72.1	...	123
cyclobutanone*	$CH_2{-}C{=}O \mid \mid CH_2{-}CH_2$	70.1	...	100
aspartic acid (aminobutanedioic acid*)	$HOOCCH_2HC\overset{\diagup \overset{\oplus}{N}H_3}{\diagdown COO^{\ominus}}$	133.1	d	...
threonine (2-amino-3-hydroxybutanoic acid*)	$CH_3CHOHHC\overset{\diagup \overset{\oplus}{N}H_3}{\diagdown COO^{\ominus}}$	119.1	d	...

[a] The most popular name is given first and an alternate name in parentheses. The IUPAC names are marked with an asterisk.
[b] In °C, rounded off to the nearest degree.

72 Chapter four

Exercises

4.1 What are the formulas for the homologous series beginning with:
a. CH_3Cl b. CH_3OH c. $HCHO$ d. $HC{\equiv}CH$ e. $HCOOH$

4.2 Work out the structural formulas for the structurally isomeric:
a. pentanols   b. pentenes

4.2 Naming: chains, substituents, and prefixes

Introduction to IUPAC rules. The time has come to expand on the rules of naming. We do not yet know enough organic chemistry simply to quote the rules laid down by IUPAC, so we shall paraphrase them. We shall not restrict ourselves to the compounds of three and four carbon atoms, because some of the illustrations of the rules require larger compounds. Some of what we say will be a repetition of what was said about naming in Section 2.3.

Alkanes. The first four saturated hydrocarbons are called methane, ethane, propane, and butane. To name the higher members of this series, we use a numerical prefix designating the number of carbons in the chain and the suffix -ane. Thus, we have pentane, hexane, heptane, octane, nonane, decane, undecane, dodecane, tridecane, and so on. The generic name of saturated branched or unbranched hydrocarbons that are not rings is alkane. All other compounds are named as relatives of these hydrocarbons.

Substituents and prefixes. If substituents are present, they are designated by a prefix to the hydrocarbon name, and the position is designated by a location number. The longest chain is numbered from one end to the other, and the direction is chosen so as to give the lowest numbers possible to the substituents. If two or more substituents are present, they are cited in alphabetical order. The alphabetical order is decided by the first letter of the substituent before the multiplying prefix, if any, is inserted. The multiplying prefixes are di-, tri-, tetra-, penta-, hexa-, and so on. If we find two or more chains of equal length as we look for the longest chain, the choice goes in order to: 1. the chain that has the greatest number of substituents; 2. the one with substituents and side chains having the smallest location numbers; 3. the one with the greatest number of carbon atoms in the substituents; and 4. the one with the least branched side chains.

If a substituent is an alkyl group, we name it by changing the saturated hydrocarbon ending from -ane to -yl. For example, methane becomes methyl (CH_3-) and ethane becomes ethyl (CH_3CH_2-). If an alkyl substituent itself is substituted, the longest chain in the grouping is numbered from the point of attachment to the main chain. Substituted substituents are named in the same way as the main chain, but the complete complex substituent name is enclosed in parentheses to indicate that any numbers shown within the parentheses apply only to the substituent. The name of a complex substituent is considered to begin with the first letter of its complete name, even if the first letter is that of a multiplier.

Here are some examples of the application of the rules:

$$\overset{1}{C}H_3\overset{2}{C}H-\overset{3}{C}H\overset{4}{C}H_2-\overset{5}{C}H\overset{6}{C}H_3 \quad \text{or} \quad (CH_3)_2CHCH(CH_3)CH_2CH(CH_3)_2$$
$$\qquad\;\; |\qquad\; |\qquad\qquad |$$
$$\qquad CH_3\;\; CH_3\qquad CH_3$$

4.2 Naming: chains, substituents, and prefixes

We number as shown to give 2,3,5 for the location numbers. Numbering from right to left would give 2,4,5 for the location numbers. The name is 2,3,5-trimethylhexane.

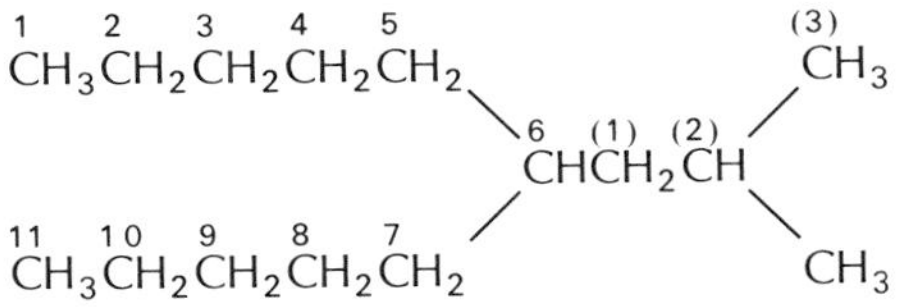

```
 1    2    3    4    5            (3)
CH₃CH₂CH₂CH₂CH₂           CH₃
                   \6  (1) (2) /
                    CHCH₂CH              or   (CH₃CH₂CH₂CH₂CH₂)₂CHCH₂CH(CH₃)₂
                   /           \
CH₃CH₂CH₂CH₂CH₂              CH₃
 11  10  9   8   7
```

The only substituent on the longest chain of eleven atoms is the three-carbon chain at carbon 6, with a methyl group at position 2. The complex substituent, 2-methylpropyl, is located at carbon 6 in the main chain. Thus, the name is 6-(2-methylpropyl)undecane.

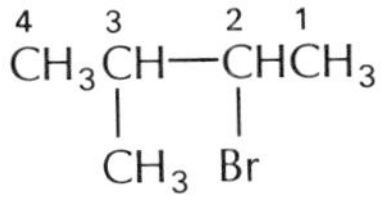

$$\underset{4}{CH_3}\underset{3}{CH}-\underset{2}{CH}\underset{1}{CH_3} \quad \text{or} \quad (CH_3)_2CHCHBrCH_3$$
$$\quad\quad\; |\quad\; |$$
$$\quad\quad CH_3\; Br$$

We number from right to left to give 2 to bromo and 3 to methyl because (only when other things are equal) the substituent that precedes the other in the alphabet is given the smaller location number. Thus, the name is 2-bromo-3-methylbutane, not 3-bromo-2-methylbutane.

How to derive a structural formula from a name. An example that illustrates most of the rules given to this point is 4-(1,2-dimethylpropyl)-6-ethyl-2,2-dimethyl-5-propyl-octane. We shall show how to write a structural formula from the name.

1. Write the main chain with a dash between each pair of carbon atoms.

    C- C- C- C- C- C- C- C

2. Number from left to right and attach the substituents.

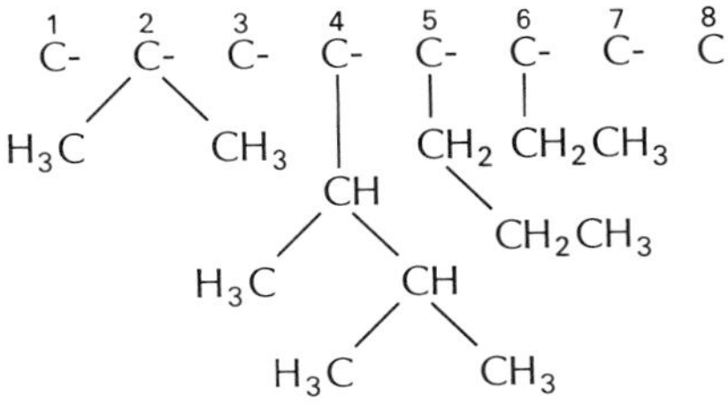

```
  1    2    3    4    5    6    7    8
  C-   C-   C-   C-   C-   C-   C-   C
 / \        |         |    |
H₃C  CH₃    CH        CH₂  CH₂CH₃
           / \        \
         H₃C  CH       CH₂CH₃
             / \
           H₃C  CH₃
```

3. Convert the dashes in the main chain to the required number of hydrogens by adding vertical lines to the dashes to form H's, and add subscript numbers as needed.

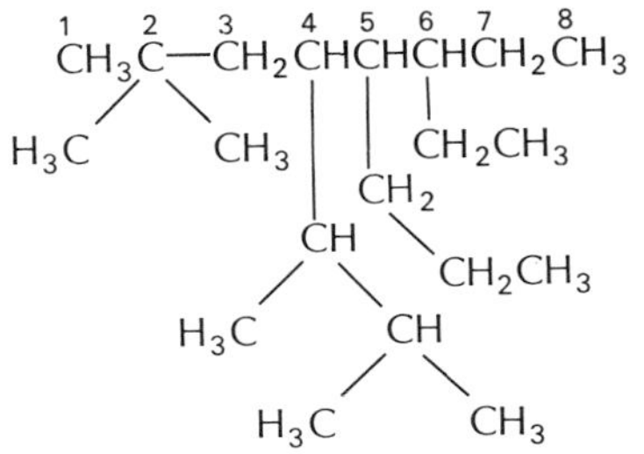

```
  1    2       3    4  5 6 7   8
  CH₃  C   —   CH₂  CHCHCHCH₂  CH₃
      / \          |     |
    H₃C  CH₃       |     CH₂CH₃
                   CH₂
                   |
                   CH     
                  / \   \
                H₃C  CH   CH₂CH₃
                    / \
                  H₃C  CH₃
```

4. Rewrite the structural formula in a more condensed form.

$$(CH_3)_3CCH_2\underset{\underset{CH_3CHCH(CH_3)_2}{|}}{\overset{\overset{CH_2CH_2CH_3}{|}}{C}}HCHCH(CH_2CH_3)_2$$

If the main chain had been selected and numbered as shown, the name would have

$$\overset{1}{C}H_3\overset{2}{\underset{\underset{}{|}}{C}}H-\overset{3}{\underset{\underset{}{|}}{C}}H-\overset{4}{\underset{\underset{}{|}}{C}}H-\overset{5}{\underset{\underset{}{|}}{C}}H-\overset{6}{\underset{\underset{}{|}}{C}}H-\overset{7}{C}H_2\overset{8}{C}H_3$$

with substituents CH₃, CH₃, CH₂, CH₂, CH₂CH₃, and C(CH₃)₂ / CH₂CH₃ / CH₃

been 4-(2,2-dimethylpropyl)-6-ethyl-2,3-dimethyl-5-propyloctane. The name is incorrect because substituents are located at positions 2, 3, 4, 5, and 6, whereas the correct name locates substituents at positions 2, 2, 4, 5, and 6 of the octane chain selected.

The dimethylpropyl substituent comes first in the name because, as a complex substituent, it begins with d. It is followed by ethyl (with e) and then dimethyl, in which the d is ignored and the m is used. Each substituent is designated by a number and a hyphen before the prefix, and each prefix is followed by a hyphen unless it leads into the main chain name, as in propyloctane.

Both 2-dimethyl and 2,2-methyl are incorrect, because dimethyl must be used to specify the presence of two methyl substituents, and the position of each methyl must be shown by a location number. The correct form is 2,2-dimethyl. If complex substituents enclosed in parentheses are present more than once, we use the special multiplying prefixes of bis, tris, tetrakis, and so on, instead of di, tri, tetra, and so on.

Systematic names of alkyl substituents. The substituent 1,2-dimethylpropyl is known as a systematic substituent because it was derived according to the rules. Because of widespread usage, however, the rules allow some specific exceptions for the use of common names for alkanes and alkyl groups. Thus, for the unsubstituted hydrocarbons only, isobutane is allowed for 2-methylpropane, $(CH_3)_3CH$; isopentane for 2-methylbutane, $(CH_3)_2CHCH_2CH_3$; neopentane for 2,2-dimethylpropane, $(CH_3)_4C$; and isohexane for 2-methylpentane, $(CH_3)_2CHCH_2CH_2CH_3$. In Table 4.3 the common substituents are listed in alphabetical order, including the eight nonsystematic names allowed for use by IUPAC. The prefixes sec- and tert- are not treated alphabetically, although iso- is. The groups listed in the table are always used as prefixes. Alkoxy substituents also are always named as prefixes and follow immediately behind the corresponding alkyl group where the order of substituents is concerned (example: 2-methyl-3-methoxybutane for $(CH_3)_2CHCH(OCH_3)CH_3$).

The abbreviation symbols are very useful for note taking and blackboard talk and are entirely unofficial despite widespread use. For example, 2-bromobutane written as sBuBr instead of $CH_3CH_2CHBrCH_3$ or sec-butyl bromide, and tBuCl instead of $(CH_3)_3CCl$ or tert-butyl chloride for 2-chloro-2-methylpropane, represent quite a saving in time and effort. We seldom see such abbreviations in print, especially in textbooks. However, we shall use them frequently.

4.2 Naming: chains, substituents, and prefixes

Table 4.3 Commonly used substituents, IUPAC

Name	Formula	Abbreviation symbol
bromo	Br—	Br—
butyl	$CH_3CH_2CH_2CH_2$—	Bu—
sec-butyl[a]	$CH_3CH_2CHCH_3$	sBu-
tert-butyl[a]	$(CH_3)_3C$—	tBu-
chloro	Cl—	Cl—
ethyl	CH_3CH_2—	Et—
fluoro	F—	F—
heptyl	$CH_3(CH_2)_6$—	Hept—
hexyl	$CH_3(CH_2)_5$—	Hex—
iodo	I—	I—
isobutyl[a]	$(CH_3)_2CHCH_2$—	iBu
isohexyl[a]	$(CH_3)_2CH(CH_2)_3$—	iHex—
isopentyl[a]	$(CH_3)_2CHCH_2CH_2$—	iPent—
isopropyl[a]	$(CH_3)_2CH$—	iPr—
methyl	CH_3—	Me—
neopentyl[a]	$(CH_3)_3CCH_2$—	$tBuCH_2$—
nitro	—NO_2	—NO_2
nitroso	—NO	—NO
pentyl	$CH_3(CH_2)_4$—	Pent—
tert-pentyl[a]	$CH_3CH_2C(CH_3)_2$—	$EtC(Me)_2$—
propyl	$CH_3CH_2CH_2$—	Pr—

[a] These nonsystematic names are specifically allowed for use by IUPAC provided that they are not further substituted.

In the common naming system, the prefix n-, (for normal) was used to specify a straight chain; and it is still seen once in a while, as in n-propyl, to make sure that we all understand that $CH_3CH_2CH_2$- is meant, not isopropyl, $(CH_3)_2CH$-. The prefix iso- originated from isobutane and has now come to mean that the far end of a chain (see isopentyl, isohexyl, and so on) terminates as a $(CH_3)_2CH$- group.

An old scheme of naming, which is still occasionally seen, is the substituted methane system. The most highly substituted carbon atom in the molecule is thought of as the carbon atom of a molecule of methane that has been substituted with alkyl groups. Some examples will show how the scheme works. The method is limited to relatively simple alkanes.

$CH_3CH_2CH_3$ or $(CH_3)_2CH_2$ dimethylmethane
$CH_3CH_2CH_2CH_2CH_3$ or $(CH_3CH_2)_2CH_2$ diethylmethane
CH_3CH_2CH—CH_2CH_3 or $(CH_3CH_2)_3CH$ triethylmethane
      |
   CH_2CH_3
$(CH_3)_2CHCH_2CH_2CH_3$ or $(CH_3)_2CH(CH_2CH_2CH_3)$ dimethylpropylmethane

Similar schemes based on ethylene and acetylene also are still seen. Some examples are shown. The prefixes *sym-* and *as-* are abbreviations of *symmetric* and *asymmetric*.

$$CH_3CH_2CH=CH_2 \quad \text{ethylethylene}$$
$$CH_3CH_2C\equiv CH \quad \text{ethylacetylene}$$
$$CH_3CH=CHCH_3 \quad sym\text{-dimethylethylene}$$
$$(CH_3)_2C=CH_2 \quad as\text{-dimethylethylene}$$
$$(CH_3)_2C=CHCH_3 \quad \text{trimethylethylene}$$
$$(CH_3)_2CHC\equiv CCH_3 \quad \text{isopropylmethylacetylene}$$

The methane, ethylene, and acetylene naming schemes are not approved by IUPAC.

Hydrocarbons. Hydrocarbon isomers often have quite similar chemical properties. Whether the methyl group is on the third or on the fourth carbon makes little difference in most common reactions. One property, a property of hydrocarbons that is of great importance to societies dependent on the internal combustion engine, is, however, quite sensitive to the isomer configuration. This is the burning time of the compound. Branched isomers burn more slowly than the straight chain configurations, and thus engines burning mixtures rich with these isomers have less tendency to knock.

The octane-rating scale is used to compare the burning properties of different gasolines (mixtures of hydrocarbons that distill from crude oil before kerosene). Normal heptane has been assigned zero on this scale and iso-octane a value of 100. In the 1930s the average gasoline had an octane number of 65. Cars today require numbers approaching 100, pressuring oil chemists to devise economical processes to reform the low-octane oil fraction. See Supplement for more details.

Heavy metals. An organic compound that includes a metal atom is usually named as a derivative of the metal. An example is tetraethyl lead, an efficient gasoline antiknock additive that has recently been blamed for the high concentrations of lead in urban air. Organo mercury compounds are used as slimicides in pulp mills and as fungicides for seed dressings. Methyl mercury chloride is the compound that when accidentally released into Minamata Bay in Japan during the 1950s caused extensive poisoning of the local fishing community.

Increased sensitivity to organo metal pollution of the environment has spurred research in the neglected field of detection of these compounds and in the treatment of poisoning. Unfortunately, the only treatment for all types of metal poisoning is the same as it has been for centuries, namely to help the body excrete the metal by natural processes before much tissue damage (especially damage to nervous tissue) has occurred. The two chelating compounds still most commonly used for this are ethylene diamine tetraacetic acid (EDTA) and 1,2-dithioglycerol (BAL).

$$\begin{array}{c} \text{Et} \\ | \\ \text{Et}-\text{Pb}-\text{Et} \\ | \\ \text{Et} \end{array} \quad \text{Me}-\text{Hg}^{\oplus}\text{Cl}^{\ominus} \quad \begin{array}{ccc} \text{SH} & \text{SH} & \text{OH} \\ | & | & | \\ \text{CH}_2-\text{CH}-\text{CH}_2 \\ & \text{BAL} \end{array}$$

Exercises

4.3 Give IUPAC names for:
a. $CH_3CH_2CHClCH_2OCH_3$
b. $(CH_3)_3CCHBrCH_3$
c. $CF_3CH_2CHClCH_2Cl$
d. $(CH_3)_2CHCH(CH_3)_2$
e. $(CH_3)_2CClC(CH_3)CHCHClCH_3$
   with CH_3CH_2 and CH_2CH_3 substituents
f. $(CH_3)_2CHCH-CHC(CH_3)_3$
   with $CH_3CH_2CH_2$ and $CH_2CH_2CH_3$ substituents
g. $CH_3CH_2CHCH_2CHBrCF_3$
   with $CHClCH_3$ substituent

4.3 Naming: groups designated by suffixes

Double and triple bonds. The presence of one carbon-carbon double bond is indicated by changing the ending from -ane to -ene, two double bonds by changing -ane to -adiene, three to -atriene, and so on. The position of a double bond is shown by a single number, that of the carbon at which it begins. Two exceptions to the rules are allowed: ethylene ($H_2C=CH_2$) and allene ($H_2C=C=CH_2$). Triple bonds are shown in a similar manner with -yne, -adiyne, and so on. The chain is numbered so as to give the lowest possible numbers to the double or triple bonds. If both types of bonds are present, the endings change to -enyne, -adienyne, -enediyne, and so on. The selection of the parent chain shifts to that chain which contains the maximum number of double and triple bonds. In other words, the longest possible chain is no longer chosen. Numbers as low as possible are given to double and triple bonds, even though this practice may at times give -yne a lower number than -ene. When there is a choice in numbering, the double bonds are given the lowest numbers. Some examples follow.

$\overset{5}{C}H_3\overset{4}{C}H_2\overset{3}{C}H_2\overset{2}{C}H=\overset{1}{C}H_2$ 1-pentene

$\overset{5}{C}H_3\overset{4}{C}H_2\overset{3}{C}H=\overset{2}{C}H\overset{1}{C}H_3$ 2-pentene

$\overset{6}{C}H_3\overset{5}{C}H=\overset{4}{C}H\overset{3}{C}H_2\overset{2}{C}H=\overset{1}{C}H_2$ 1,4-hexadiene

$H\overset{6}{C}\equiv\overset{5}{C}\overset{4}{C}H_2\overset{3}{C}H_2\overset{2}{C}H=\overset{1}{C}H_2$ 1-hexen-5-yne

The location number preceding hex refers to the first suffix, here en, and later suffixes are immediately preceded by the location number, here -5-yne.

$\overset{6}{C}H_3\overset{5}{C}H=\overset{4}{C}H\overset{3}{C}H_2\overset{2}{C}\equiv\overset{1}{C}H$ 4-hexen-1-yne (not 2-hexen-5-yne)

$H\overset{6}{C}\equiv\overset{5}{C}-\overset{4}{C}H-\overset{3}{C}H\overset{2}{C}H=\overset{1}{C}H_2$ 3,4-dipropyl-1-hexen-5-yne (not 4-ethynyl-5-vinyloctane)
with $CH_3CH_2CH_2$ and $CH_2CH_2CH_3$ substituents

Vinyl, $H_2C=CH-$, allyl, $H_2C=CHCH_2-$, and isopropenyl, $H_2C=C(CH_3)-$, are allowed exceptions as names for substituents which contain a double bond.

78   Chapter four

$$\overset{6}{C}H_3\overset{5}{C}HBr\overset{4}{C}Br\overset{3}{C}H=\overset{2}{C}H\overset{1}{C}H_3$$
$$\underset{CH_3CHCH_3}{|}$$

4,5-dibromo-4-isopropyl-2-hexene (not 2,3-dibromo-3-isopropyl-4-hexene)

Double and triple bonds take precedence for the small location number over all alkyl substituents and others listed in Table 4.3.

Reactive substituents; prefixes and suffixes; priority scale. A reactive substituent (other than those in Table 4.3, which are always named as prefixes) is named as a suffix, according to the rules. However, too much piling up of suffixes is frowned upon. Thus, 4-hydroxybutanal (HOCH$_2$CH$_2$CH$_2$CHO) is preferred to 4-butanolal. The aldehyde grouping can only be at the end of a chain, and it takes precedence over hydroxy for the low number. On the other hand, 4-hydroxy-2-butenal for HOCH$_2$CH=CHCHO is all right. In sum, reactive substituents take one of two different parts in a name, depending upon which substituent takes precedence: they are denoted by prefixes or by suffixes. However, double and triple bonds are always named as suffixes.

Table 4.4   Order of precedence of reactive substituents

No.	Function and formula	Prefix	Suffix	Example
1.	-onium, R$_4$N$^\oplus$	—	-onium	(2-carboxyethyl)trimethyl-ammonium bromide
2.	peroxide, ROOR'	alkylperoxy	peroxide	2,2'-dihydroxydiethyl peroxide
3.	hydroperoxide, ROOH	hydroperoxy	hydroperoxide	4-keto-2-methyl-2-pentyl-hydroperoxide
4.	acid, RCOOH	carboxy	-oic acid	3-ketobutanoic acid
5.	acid halide, RCOX	chloroformyl	-oyl halide	3-methoxybutanoyl chloride
6.	amide, RCONH$_2$	amido	-amide	4-cyanobutanamide
7.	imide, (RCO)$_2$NH	—	-imide	4-hydroxybutanimide
8.	aldehyde, RCHO	aldo or oxo	-al	3-ketobutanal
9.	nitrile, RC≡N	cyano	-nitrile	3-ketobutanenitrile
10.	ketone, RCOR'	keto or oxo	-one	4-amino-2-butanone
11.	alcohol, ROH	hydroxy	-ol	3-buten-2-ol
12.	thiol, RSH	mercapto	-thiol	3-buten-1-thiol
13.	amine, RNH$_2$	amino	-amine	3-buten-2-amine
14.	imine, =NH	imino	-imine	2-butanimine
15.	double bond, \C=C/	—	-ene	1-buten-3-yne
16.	triple bond, =C≡C=	—	-yne	3-butyn-1-ol
17.	other substituents	halo	—	1-iodobutane
		alkyl	—	2-methylbutane
		alkyloxy	—	2-ethoxyethane
		alkylthio	—	2-ethylthiobutane
		epoxy	—	1,2-epoxybutane
		nitro	—	2-nitrobutane
		nitroso	—	2-nitrosopropane

Reactive substituents get preference for low numbers and take precedence in determining the main chain. Although there are no IUPAC rules on the subject, popular usage has set up an order of preference, shown in Table 4.4. The table also gives the prefixes and suffixes that are suggested by the rules and gives examples of names in which two or more reactive substituents are present. Strictly speaking, reactive substituents should be called functional groups, because most of the reactivity of the molecules in which they are present is determined by their own reactivity. Many of the small polyfunctional compounds have durable common (also known as trivial) names. For examples, see Table 4.2 for methacrylic acid, succinic acid, tartaric acid, acetoacetic acid, and others.

Exercises

4.4 Write structural formulas for the compounds named in Table 4.4.

4.5 Write structural formulas for the compounds given:
   a. 2,3-dimethyl-1,3-butadiene
   b. 3-bromo-3-isobutyl-1,4-pentadiene
   c. 2,3,4-trifluoro-2-pentene
   d. 5,5,5-trichloro-2-isopropyl-1-penten-3-yne

4.6 Give IUPAC names for:
   a. $CH_3CH{=}CHC{=}CHCH(CH_3)_2$
      $\quad\ \ |$
      $\quad\ \ CH_2C(CH_3)_3$

   b. $CH_3C{\equiv}CC{-}CH(CH_3)_2$ with CH_2CH_3 above and $CH{=}CHCH_2CH_3$ below

   c. $CH_3CH{=}CHCOCH_2CH_3$
      $\qquad\qquad\ \ |$
      $\qquad\qquad\ \ C{\equiv}N$

   d. $(CH_3)_2CClCH_2C{=}CHCHO$
   e. $HOCH_2COCH_2OH$

4.4 Naming: cyclic compounds

Rings with no double bonds; substituents. Ring hydrocarbons are named by use of the prefix cyclo before the number designation of the carbons forming the ring, as in cyclopropane, cyclobutane, and so on. Because each position in a ring is equivalent to any other position, numbering of a ring for location of substituents may begin at any position. A single substituent is regarded as being located on carbon 1 and is not specified (examples: methylcyclopropane, bromocyclobutane). If more than one substituent is present, numbering begins at one of the substituents and proceeds around the ring (clockwise or counterclockwise) in the direction that gives the substituents the smallest set of location numbers. Some examples are shown.

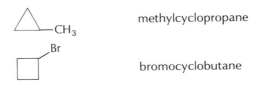

methylcyclopropane

bromocyclobutane

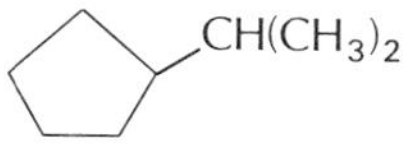

isopropylcyclopentane

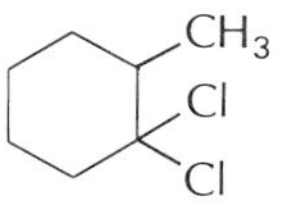

1,1-dichloro-2-methylcyclohexane
(not 2,2-dichloro-1-methylcyclohexane)

Rings with double bonds; priorities in numbering. A double bond, if present, takes precedence and is given location number 1, the double bond ending on 2.

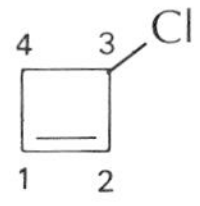

3-chlorocyclobutene (the double bond is not numbered in the name when it takes precedence)

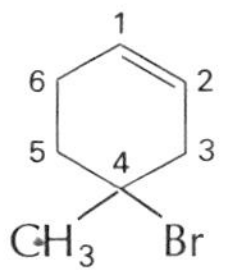

4-bromo-4-methylcyclohexene

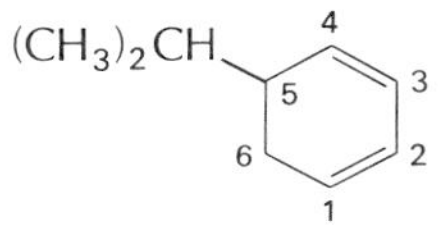

5-isopropyl-1,3-cyclohexadiene (not 6-isopropyl-1,3-cyclohexadiene)

Rings as substituents on chains. If a ring is attached to a chain, the ring may be thought of and named as a substituent on the chain (1-cyclopropylpropane for propylcyclopropane). The choice to use the ring or the chain as the main part of the name usually is made because of convenience or preference. There is something to be said in favor of using the greater number of substituents, because the smaller substituents are easier to name.

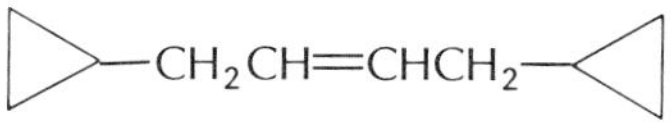

1,4-dicyclopropyl-2-butene (preferable to (4-cyclopropyl-2-buten-1-yl)cyclopropane)

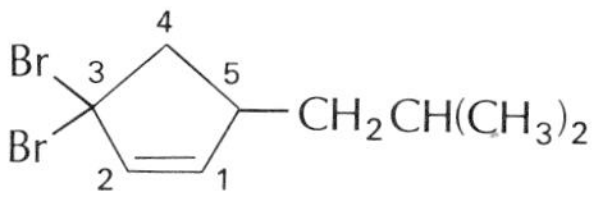

3,3-dibromo-5-isobutylcyclopentene (preferable to 1-(4,4-dibromo-2-cyclopenten-1-yl)-2-methylpropane)

If a ring is named as a substituent, the ring is numbered from the point of attachment to the chain in the direction that gives double bonds the lowest numbers, or, if no double bonds, that gives substituents on the ring the lowest numbers. When necessary, or to avoid any ambiguity, the point of attachment is indicated as -1-yl.

Also, we keep in mind that a ring is shown simply as the proper geometrical figure. Each corner is understood to represent a —CH_2— group unless a double bond is indicated, in which case the corner is a —CH= group.

4.4 Naming: cyclic compounds

Carboxylic acid method of naming for rings. A special difficulty arises in cases such as those in which an acid, ester, acid halide, amide, aldehyde, or nitrile group is linked directly to a ring. In such instances, the carboxylic acid names are used. The carbon atom of the ring to which the carboxyl group is attached is numbered 1. Some examples will show how the system works out.

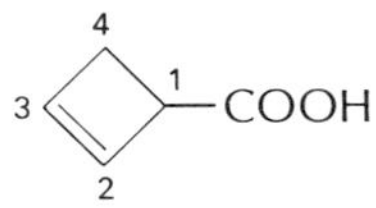

[cyclobutene ring with COOH]   2-cyclobutene-1-carboxylic acid

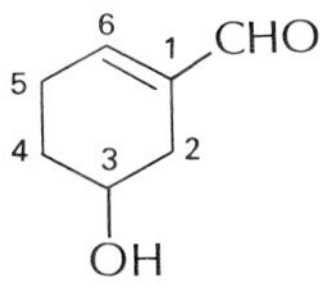

[cyclohexene ring with CHO and OH]   3-hydroxy-6-cyclohexene-1-carboxaldehyde
(not 5-hydroxy-1-cyclohexene-1-carboxaldehyde, because hydroxy has priority over the double bond for the smaller location number)

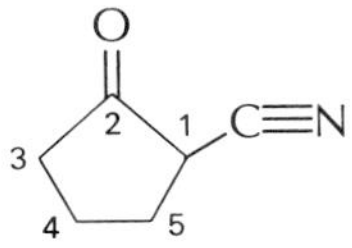

[cyclopentanone ring with C≡N]   2-ketocyclopentane-1-carbonitrile (however, 2-cyano-1-cyclopentanone is permissible in this case)

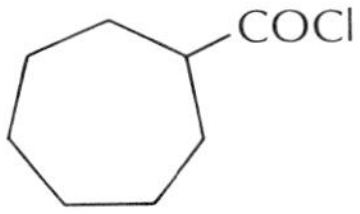

[cycloheptane ring with COCl]   cycloheptanecarbonyl chloride

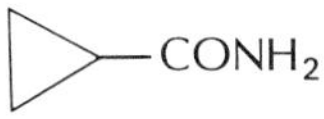

[cyclopropane ring with CONH$_2$]   cyclopropanecarboxamide

Carboxylic acid names for chains. We should make it clear that when we use -oic acid, -oyl halide, -amide, -al, and -nitrile endings, the carbon atom is numbered as part of a chain, but the use of carboxylic acid, carboxaldehyde, carbonitrile, and so on indicates nonparticipation in numbering of a chain or ring. A situation that does not involve a ring but is conveniently handled as a carboxylic acid name would be propane-1,1,2,3-tetracarboxylic acid. Otherwise it would be 2,3-dicarboxypentanedioic acid, a name that does not immediately give as vivid a picture of the structure as does the tetracarboxylic acid name.

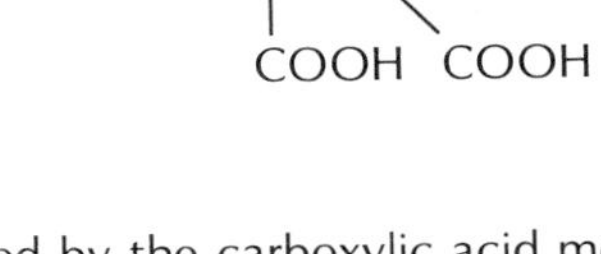

$$\text{HOOCCH}_2^3\text{CHCH}^1\begin{matrix}\text{COOH}\\|\\\text{COOH}\end{matrix}\text{COOH}$$

If an ester group is named by the carboxylic acid method, the term to use is <u>alkyl carboxylate</u>. Thus the product of esterification with ethanol of the acid just mentioned is known as tetraethyl propane-1,1,2,3-tetracarboxylate. An ester group used as a prefix is called an alkoxycarbonyl as in 3-ethoxycarbonyl-1-propanehydroperoxide for $CH_3CH_2OCOCH_2CH_2CH_2OOH$.

Exercises

4.7 Write the structural formulas of:
  a. 2-sec-butyl-3-*tert*-butyl-1-isopropyl-1-cyclobutene
  b. 1-bromo-3-butyl-2-vinyl-1-cyclopentene
  c. 2,4-dichloro-3,3-diethyl-1,4-cycloheptadiene
  d. 4-chloro-2-cyclohexen-1-one
  e. 3,4,5-trimethoxy-1-cyclopentene

4.8 Give IUPAC names to the compounds listed:

a. cyclohexene—CH=CH—cyclohexene

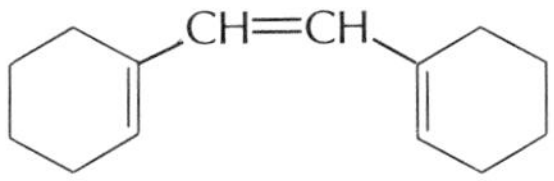

b. cyclopentadiene—CH=CHCH$_2$CH$_3$

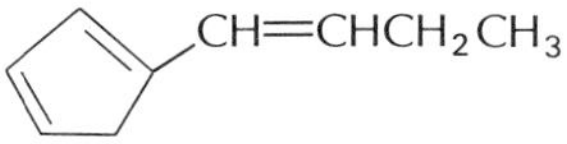

c. cyclooctatetraene with CH$_2$Cl and CH$_3$ substituents

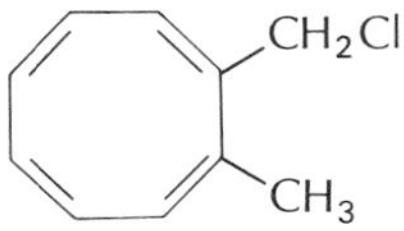

d. cyclohexane with CH$_2$CH=CH$_2$, CH$_3$, and HO substituents

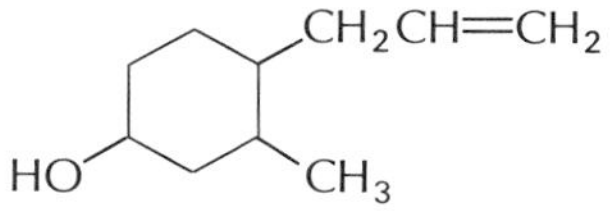

e. cyclopentane—CH$_2$CH$_2$CHOHCH$_2$Cl

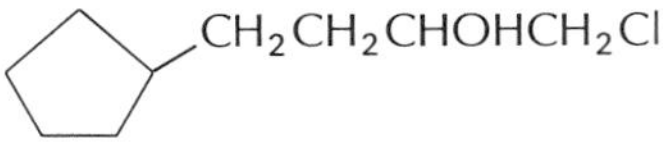

f. cyclohexane—CHCH=CHC≡CCH(CH$_3$)$_2$ with CH$_3$ branch

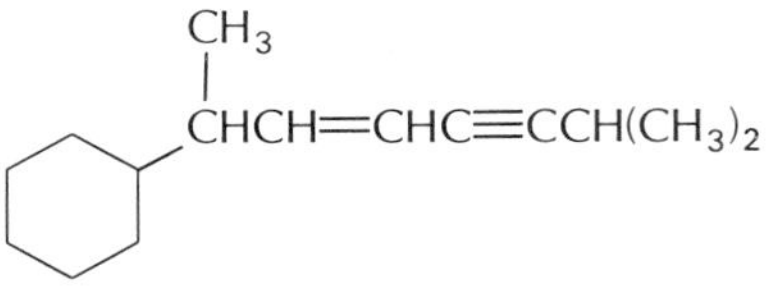

4.5 Summary

In this short chapter, a sampling of compounds with three and four atoms of carbon has been given, not with the expectation that they will be committed immediately to memory, but rather as a matter of reference and illustration of the variety of compounds that exists. The method used to build up a homologous series has been pointed out, and a method for finding the number of isomers of a given molecular formula was given.

Having become familiar with a variety of compounds, we learned many of the rules of systematic naming and saw some examples. One of the common failings in deriving names from structural formulas is lack of attention to the proper use of commas, hyphens, parentheses, brackets, and periods. Another cause of difficulty is failure to follow the order of precedence as given in Table 4.4. Proficiency in systematic naming comes with practice and as familiarity is gained with a variety of compounds.

In the meantime, it is fairly easy to keep in mind the order of precedence of acid derivatives, aldehyde, nitrile, ketone, alcohol, amine, double bond, triple bond, and then the prefixed substituents. Such a list will cover most of the naming problems likely to be encountered.

4.6 Problems

4.1 Write structural formulas and give the correct name for all the isomers of:
   a. $C_4H_8Cl_2$
   b. C_3H_3Br
   c. C_3H_6O
   d. $C_4H_{10}O$
   e. C_4H_6
   f. C_4H_4

4.2 Give a correct name for each compound:
   a. $(CH_3)_2CHCOCH_2CH_2CH_3$
   b. $CH_3CH_2CH_2COCl$
   c. $CH_3COCH_2CH_2COOCH_2CH_3$
   d. $BrCH_2CH=CHCHOHCH_3$
   e. $CH_3OCH=CHCH_2COCH_3$
   f. $HOCH_2C{\equiv}CCH_2CH_2CHO$
   g. 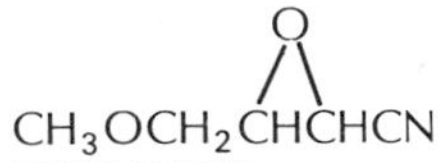 $CH_3OCH_2CHCHCN$ (with C=O substituent)
   h. $HC(COOH)_3$
   i. $HC(CH_2COOH)_3$
   j. $H_2NCH_2COCH_2CONH_2$

4.3 Write a structural formula for each compound:
   a. 2-penten-1-amine
   b. 3-isopropyl-3-methyl-1-hexanol
   c. 5-bromo-4-pentenal
   d. 3-hexyne-1,6-diol
   e. cyclopentanecarboxamide
   f. 2,3-epoxycyclohexanecarboxaldehyde
   g. 3,3-diethylcyclobutanecarbonyl chloride
   h. 2-methoxy-1,3-cyclopentadiene
   i. 2-bromo-2-cyclobutylbutane
   j. 6-bromo-4-butyl-5-keto-2-hexenal

4.4 Give a correct name for each compound:
   a. $(CH_3)_2C=CHCHO$
   b. 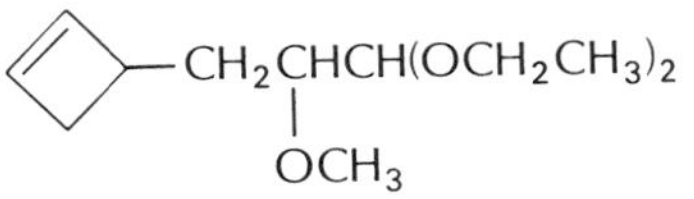 cyclobutyl$-CH_2CHCH(OCH_2CH_3)_2$ with OCH_3
   c. $HC{\equiv}CCH_2CCl_2CH_2CONH_2$
   d. 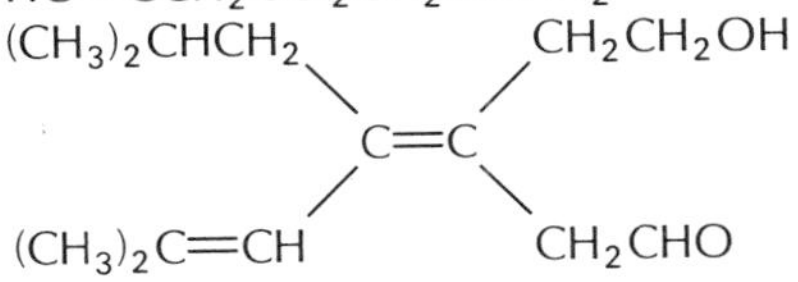 $(CH_3)_2CHCH_2$ and CH_2CH_2OH on C=C, with $(CH_3)_2C=CH$ and CH_2CHO
   e. 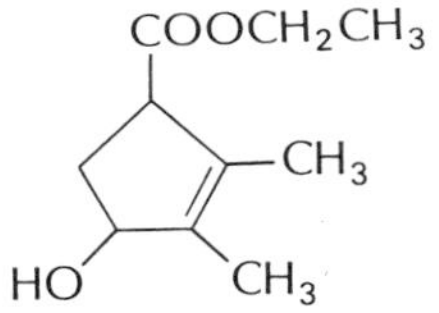 cyclopentene with COOCH$_2$CH$_3$, CH$_3$, CH$_3$, HO substituents
   f. $CH_3CH_2CH_2CH_2CH(CH_2COOH)_2$
   g. 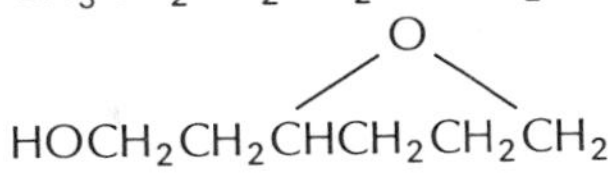 $HOCH_2CH_2CHCH_2CH_2CH_2$ with epoxide O

4.5 Write a structural formula for each compound:
   a. tetrafluoroethylene
   b. tetrabromoallene
   c. isobutylcyclopentane
   d. 3-*sec*-butylcyclopentene
   e. 1,4-epoxycyclohexane
   f. 2-butenedioic acid
   g. 3-keto-2-(2-pentenyl)butanal
   h. 4-(3-methyl-1-cyclopenten-1-yl)-3-propyl-3-buten-2-one

84 Chapter four

4.6 Name the following compounds.

a. CH$_3$CHC≡CCH$_3$
   |
   CH$_3$

b. CH$_3$CH=CHC≡CH

c. CH$_3$CCH$_3$ with CH$_3$ above and CH$_3$ below the central C

d. CH$_3$CHCH$_2$CH$_2$CH$_3$
   |
   CH$_3$

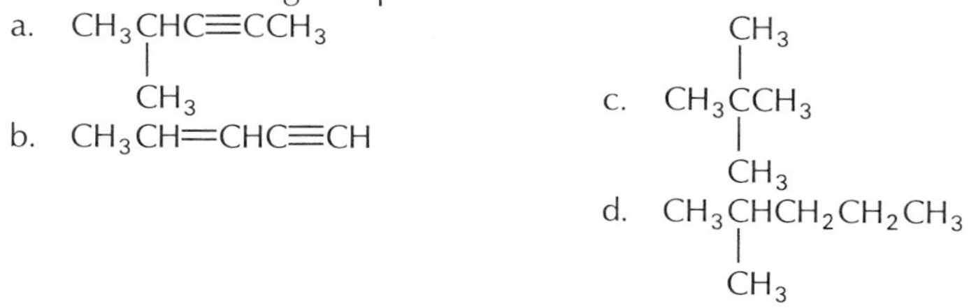

4.7 Give the IUPAC name for the following compounds.

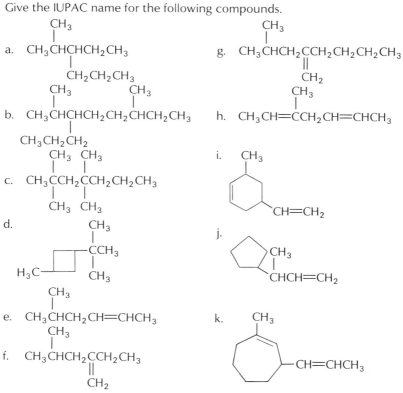

a. CH$_3$CHCHCH$_2$CH$_3$ with CH$_3$ on first CH and CH$_2$CH$_2$CH$_3$ on second CH

b. CH$_3$CHCH$_2$CH$_2$CHCH$_2$CH$_3$ with CH$_3$ on first CH and CH$_3$CH$_2$CH$_2$ on second CH

c. CH$_3$CCH$_2$CCH$_2$CH$_3$ with CH$_3$, CH$_3$ above and CH$_3$, CH$_3$ below

d. (cyclobutane with H$_3$C and substituents CH$_3$, CCH$_3$ with CH$_3$)

e. CH$_3$CHCH$_2$CH=CHCH$_3$ with CH$_3$ substituent

f. CH$_3$CHCH$_2$CCH$_2$CH$_3$ with CH$_3$ on CH and =CH$_2$ on C

g. CH$_3$CHCH$_2$CCH$_2$CH$_2$CH$_2$CH$_3$ with CH$_3$ on CH and =CH$_2$ on C

h. CH$_3$CH=CCH$_2$CH=CHCH$_3$ with CH$_3$ substituent

i. (cyclohexene with CH$_3$ and CH=CH$_2$ substituents)

j. (cyclopentane with CH$_3$ and CHCH=CH$_2$ substituents)

k. (cycloheptene with CH$_3$ and CH=CHCH$_3$ substituents)

4.8 Among the structures in Problem 4.7, pick out and show specifically two compounds that have: a. a <u>primary</u> carbon; b. a <u>secondary</u> carbon; c. a <u>tertiary</u> carbon; and d. a <u>quaternary</u> carbon.

4.9 Draw the structure of each of the following:

a. 2-methyl-1,3-pentadiene
b. 2-methyl-1,3-cyclopentadiene
c. 4-ethyl-6-methyl-4-hepten-1-yne
d. 3,4-diethyl-2,4-hexadiene

4.7 Bibliography

Among the many aids published to help the study and understanding of the naming of organic chemical compounds, we shall mention only three paperbacks, selected at random.

1. Banks, James E., *Naming Organic Compounds*. Philadelphia: Saunders, 1967.
2. Benfey, Otto T., *The Names and Structures of Organic Compounds*. New York: Wiley, 1966.
3. Traynham, James G., *Organic Nomenclature: A Programmed Introduction*. Englewood Cliffs, N.J.: Prentice-Hall, 1966.

For those who wish to see for themselves what the actual IUPAC rules look like, we give either of the following convenient sources.

4. *J. Amer. Chem. Soc. 82*, 5545 (1960).
5. Weast, Robert C., ed., *Handbook of Chemistry and Physics*. Cleveland: The Chemical Rubber Co. The Handbook is revised yearly; the 54th edition is a 1973 copyright. No page reference is given because it varies from one edition to another. However, the IUPAC rules appear just before the extensive table of physical constants (about 14,000 entries) of organic compounds.

5 Atomic and Molecular Structure

Before we describe any more reactions, it will be helpful for us to review and describe some aspects of atomic and molecular structure. Our intention is to show how the physical and chemical properties of a given compound are influenced by the consequences of the various kinds of bonds and atoms present in the molecule. Our presentation will be descriptive rather than rigorous.

5.1   Atomic structure   87
   1. Introduction; hydrogen atom orbitals.   2. Shapes of orbitals.

5.2   Bonds: molecular orbitals   90
   1. Introduction.   2. Hydrogen molecule cation; hydrogen molecule.
   3. Bonding and antibonding orbitals.   4. Nitrogen; oxygen; sigma and pi bonds.   5. Carbon; hybridization; valence and atomic states.   6. MO's of several molecules. Exercises.

5.3   Bonds: valence bonds, resonance   96
   1. Valence bonds and atomic orbitals.   2. Hydrogen fluoride: polarity, resonance.   3. Comparison of VB and MO approaches.   4. Rules for resonance structures.   5. Application of resonance theory. Exercises.

5.4   Bond distances, atomic radii, bond angles   101
   1. Relative constancy of bond distances and angles.   2. Atomic covalent radii.
   3. Bond angles. Exercises.

5.5   Electronegativity: polarity and polarizability   103
   1. Electronegativity.   2. Polarity and dipole moment.   3. Analgesics.
   4. Polarizability.   5. Bond moments; polarizable bonds.

5.6   Intermolecular forces: hydrogen bonds, complexes   105
   1. Intermolecular attraction.   2. Hydrogen bonds.   3. Complex formation: donor-acceptor complex.   4. Intermolecular bonding in biochemistry. Exercises.

5.7   Inductive and resonance effects   108
   1. Bond polarity and inductive effects, I.   2. The R effect. Exercises.

5.8   Bond energies and bond dissociation energies   110
      1. Heats of formation of atomic states; average bond energies, BE's.   2. Bond dissociation energies, BDE's.   3. Uses of BE and BDE.   4. Potential-energy diagram of bond formation. Exercises.

5.9   Summary   115

5.10 Problems   115

5.11 Bibliography   116

5.1 Atomic structure

Introduction; hydrogen atom orbitals.   The early organic chemists developed the science with only a vague idea of what a covalent bond was. They used a single line for a single bond (or two or three lines for double and triple bonds) to represent the mysterious force that held the atoms together. Beginning about 1890, rather vague ideas of an electrical theory of reactivity in organic chemistry began to appear. The electron pair covalent bond was postulated before 1920; but it was not until quantum mechanics clarified atomic structure that our understanding of bonding became better than hazy.

The behavior of an electron may be described by the behavior of the wave associated with it. An electron under the influence of a proton (a hydrogen atom) obeys a wave equation. With wave phenomena, there is no single solution to the equations that describe the waves, and the different solutions differ by integral values such as 0, 1, 2, 3, and so on. The integral values are called quantum numbers. The solution of the wave equation for the hydrogen atom requires four quantum numbers: n, the principal quantum number, which determines the size and energy of the orbital; l, the azimuthal or angular momentum quantum number, which controls the shape of the orbital; m, the magnetic quantum number, which decides the orientation in space; and m_s, the spin quantum number. The quantum numbers are interrelated as follows: $n = 1, 2, 3$, etc., $l = 0, 1, 2$, up to $n - 1$, $m = 0, \pm 1, \pm 2$, up to $\pm l$ for a total of $2l + 1$ values of m, and $m_s = \pm \frac{1}{2}$. The energy of a solution (or orbital, defined by n, l, and m) increases with n and with l. The hydrogen atom orbitals are listed in Table 5.1 in order of increasing energy.

As soon as we look at a system with two or more electrons, so many complications ensue that only approximate mathematical solutions may be found. The difficulty lies in the repulsions between the electrons: each electron requires three coordinates to describe its position. A great simplification is Pauli's exclusion principle, which states that no two electrons in the same system may have the same set of four quantum numbers. Thus, only two electrons are permitted in an orbital, with $m_s = \frac{1}{2}$ for one and $m_s = -\frac{1}{2}$ for the other. We say such electrons are paired.

The periodic table is built up as we add electrons to nuclei of increasing nuclear charge (atomic number). Each electron is placed in an orbital at the lowest possible energy level. When it comes to filling the three p orbitals, all of which are at equal energy levels, one electron is added to each p orbital so that the first three electrons have parallel spins before pairing begins with the fourth electron. We use a special symbolism in describing the electronic structure of atoms. Thus, $2s$ means a single electron in the s orbital ($l = 0$) of the second shell ($n = 2$); and $2p_x^2$ indicates a pair of electrons in the p_x orbital ($l = 1$) of the second shell ($n = 2$).

Table 5.1 Hydrogen atom orbitals

Increasing energy	n	l	m	m_s	Orbitals
↓	1	0	0	$\pm \frac{1}{2}$	$1s$
	2	0	0	$\pm \frac{1}{2}$	$2s$
	2	1	$0, \pm 1$	$\pm \frac{1}{2}$	$2p_x, 2p_y, 2p_z$
	3	0	0	$\pm \frac{1}{2}$	$3s$
	3	1	$0, \pm 1$	$\pm \frac{1}{2}$	$3p_x, 3p_y, 3p_z$
	3	2	$0, \pm 1, \pm 2$	$\pm \frac{1}{2}$	$3d_{z^2}, 3d_{xz}, 3d_{xy}, 3d_{yz}, 3d_{x^2-y^2}$

Table 5.2 shows the electronic distribution in the atoms of the elements up to argon.

Table 5.2 Electronic distribution of some atoms

Atom	Description
H	$1s$
He	$1s^2$
Li	$1s^2 2s$
Be	$1s^2 2s^2$
B	$1s^2 2s^2 2p_x$
C	$1s^2 2s^2 2p_x 2p_y$
N	$1s^2 2s^2 2p_x 2p_y 2p_z$
O	$1s^2 2s^2 2p_x^2 2p_y 2p_z$
F	$1s^2 2s^2 2p_x^2 2p_y^2 2p_z$
Ne	$1s^2 2s^2 2p_x^2 2p_y^2 2p_z^2$
Na	$1s^2 2s^2 2p_x^2 2p_y^2 2p_z^2 3s$
Mg	$1s^2 2s^2 2p_x^2 2p_y^2 2p_z^2 3s^2$
Al	$1s^2 2s^2 2p_x^2 2p_y^2 2p_z^2 3s^2 3p_x$
Si	$1s^2 2s^2 2p_x^2 2p_y^2 2p_z^2 3s^2 3p_x 3p_y$
P	$1s^2 2s^2 2p_x^2 2p_y^2 2p_z^2 3s^2 3p_x 3p_y 3p_z$
S	$1s^2 2s^2 2p_x^2 2p_y^2 2p_z^2 3s^2 3p_x^2 3p_y 3p_z$
Cl	$1s^2 2s^2 2p_x^2 2p_y^2 2p_z^2 3s^2 3p_x^2 3p_y^2 3p_z$
A	$1s^2 2s^2 2p_x^2 2p_y^2 2p_z^2 3s^2 3p_x^2 3p_y^2 3p_z^2$

Shapes of orbitals. To picture an orbital properly is a difficult task. Intensity is related to the probability of finding the electron at a given point or to the density of the electron cloud at a given point. The terms <u>electron density</u> and <u>electron probability</u> are interchangeable. In the 1s one-electron hydrogen orbital there is a large probability of finding the electron in a tiny volume element near the nucleus, but the probability decreases rapidly as the distance from the nucleus is increased. At the same time, the number of volume elements at a constant distance from the nucleus increases rapidly as the distance is increased. (The volume of a layer in an onion is larger the greater the distance from the center.) Between the two effects—the decreasing probability of finding an electron in a given volume and the increasing volume of each successive layer—the total probability of finding the electron close to the nucleus is small, increases rapidly as the distance from the nucleus increases, reaches a maximum, and then falls more slowly as the very low probability per unit of volume approaches a very small value. Figure 5.1 gives a schematic picture of the situation.

The 1s orbital is spherical without a definite boundary. We can represent the orbital by drawing a sphere of radius large enough so that the electron spends 90 or 95% of the time within the arbitrarily selected boundary. The 2s orbital also is spherical but with two regions of relatively high probability or density, as shown in Figure 5.1.

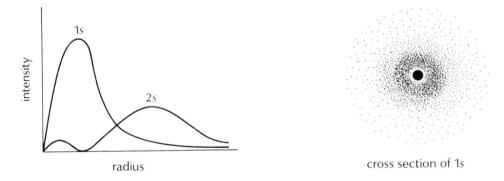

Figure 5.1 The 1s and 2s orbitals

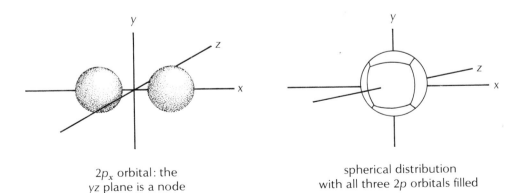

Figure 5.2 A schematic picture of p-orbitals

Each of the 2p orbitals may be described as dumbbell shaped and lying along an axis. There is a node through the nucleus in the plane of the remaining two axes drawn through the nucleus (see Fig. 5.2). The density maxima lie outside the maximum for the 2s orbital, so that the furthest extent of a 2p orbital is greater than that of the 2s orbital. If all three 2p orbitals are filled, the total distribution of the six electrons is spherical, with additional nodes between the orbitals. These are not the only solutions to the wave equation, but they do give the lowest energy state for an atom.

5.2 Bonds: molecular orbitals

Introduction. There are two theories concerning molecular structure and bonding: the molecular orbital theory and the valence bond theory. We begin with the molecular orbital theory, which says that a molecule is built up very much as an atom is built up. The difference is that whereas an atom is built up by addition of electrons to a single nucleus, in a molecule electrons are added to an arrangement of nuclei. Because it is the electrons in the outermost shell of an atom that participate in bonding, the first electrons introduced fill the lower-lying atomic orbitals. (Strictly speaking, the statement is not exact, but it will serve our purpose.) The outermost occupied shell of an atom is known as the valence shell. Depending upon the atoms present, the last electrons used to complete a neutral molecule or an ion are used to fill σ and π molecular orbitals (somewhat analogous to s and p atomic orbitals).

Hydrogen molecule cation; hydrogen molecule. Let us look at the simplest molecule, the hydrogen molecule cation, $H_2^{\oplus}$ or $(H \cdot H)^{\oplus}$. The nuclei in this case are two protons located the proper distance apart. The valence shells are both 1s, each with a capacity for two electrons. To the bare nuclei add the single electron, which goes into the $(\sigma 1s)$ molecular orbital (frequently abbreviated to MO).

If the nuclei are pushed closer to each other, the repulsion of the nuclei exceeds the attraction of the electron and the potential energy is increased. If the nuclei are pulled further apart, the attraction of the electron exceeds the repulsion of the nuclei and the potential energy again is increased. Thus the molecule ion has a minimum of energy when the nuclei are the proper distance apart. We say that the molecule ion has a stable bond distance.

An interesting analogy for visualizing the distribution of electron density of the one-electron bond is as follows. Imagine a camera mounted over a basketball court. We take a picture every second during a basketball game that ends in a tie score. From the pictures, we plot all the positions of the basketball upon a diagram of the court. Where will the center of density of *all* the points lie? At or very close to the center of the court. Where will the points cluster the most? Around the two baskets. Will the ball appear outside the court? Yes, but not very often. Thus, in the one-electron bond the electron density centers at the middle of the bond, but only because the density is equally great around each nucleus.

Now let us add a second electron with spin opposite to that of the electron already present. We now have an electron pair in the orbital, or $(\sigma 1s)^2$. The attraction of the electrons now exceeds the repulsion of the protons. The protons move closer together, and a new balance is struck between attraction and repulsion at a new shorter stable bond distance in the hydrogen molecule, H_2.

Bonding and antibonding orbitals. Because two atomic 1s orbitals are contributing to the molecular orbitals, two molecular orbitals are formed. The one just described, the (σ1s) is a bonding orbital. The other, labeled as (σ*1s), is an antibonding orbital and, like other orbitals, may be occupied by one electron or by an electron pair. Each electron in an antibonding orbital almost cancels the bonding effect of an electron in the corresponding bonding orbital. Thus, the hydrogen molecule anion, with three electrons ($H_2^\ominus$) is about as stable as is the hydrogen molecule cation.

Figure 5.3 illustrates the relative energy levels in the formation of $H_2^\oplus$, H_2, and $H_2^\ominus$ and gives a schematic idea of the shapes of the (σ1s) and (σ*1s) orbitals.

In the $He_2^\oplus$ cation molecule, there are two paired electrons in (σ1s)^2 and one electron in the antibonding orbital, (σ*1s). The net effect is roughly that of one bonding electron, and $He_2^\oplus$ is about as stable as is $H_2^\oplus$. The helium molecule would have a pair of electrons in (σ*1s)^2, which would balance the bonding effect of (σ1s)^2. Thus there is no bonding effect, helium is monatomic, and He_2 has not been observed.

Nitrogen; oxygen; sigma and pi bonds. Let us skip to an examination of the nitrogen molecule, N_2. The valence shell molecular orbitals (MO's) are designated as (σ2s)^2(σ*2s)^2 (σ2p)^2(π2p)^4. With (σ2s)^2 effectively balanced by (σ*2s)^2, there remain six electrons in bonding orbitals, equivalent to what we have been calling a triple bond. We may think of the (σ2p)^2 orbital (or bond) as being made up of good overlap of the atomic p orbitals that lie along the bond. If we label the bond axis as the z coordinate, each nitrogen atom also

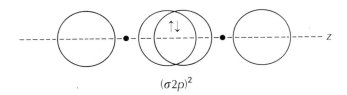

$(\sigma 2p)^2$

has one electron in each of the p orbitals along the x and y axes. These orbitals are able to overlap only sideways in the formation of two equivalent π2p MO's, or (π2p)^4. The sideways overlap is not as complete as is the head-on overlap in a (σ2p)^2 MO, and bonding is weaker.

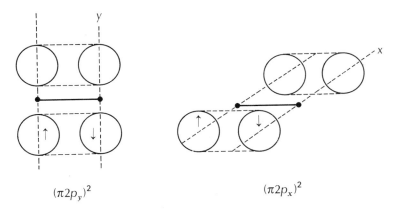

$(\pi 2p_y)^2$ $(\pi 2p_x)^2$

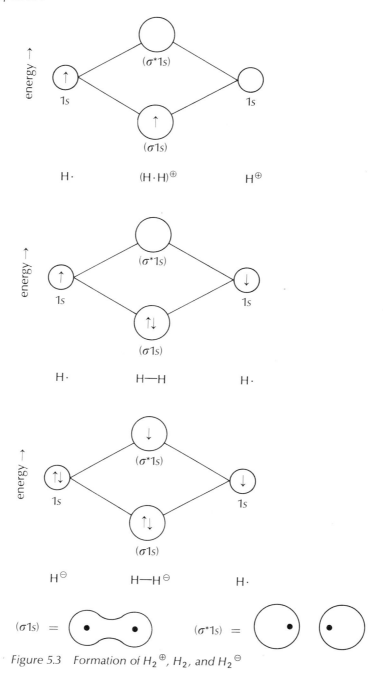

Figure 5.3 Formation of $H_2^{\oplus}$, H_2, and $H_2^{\ominus}$

Of course, there are σ^*2p and π^*2p antibonding MO's that are not occupied in nitrogen. The oxygen molecule, O_2, with two more electrons, is a diradical because the electrons are unpaired, with one electron in one of the π^*2p orbitals and one in the other. The σ^*2p antibonding MO is at the highest energy level of all. It would not fill until neon, another monatomic gas, is reached.

5.2 Bonds: molecular orbitals

Carbon; hybridization; valence and atomic states. When we come to carbon and carbon compounds, we find that the valence-shell atomic-orbital electron distribution of $2s^2 2p_x 2p_y$ would allow only two of the strong ($\sigma 2p$) orbital overlaps with two adjacent atoms. The remaining two electrons could only participate in the formation of ($\sigma 2s$) and ($\sigma^* 2s$) MO's.

So that the atom can participate in the formation of more bonds, the atomic orbitals are reorganized into a less stable (higher energy content) distribution, a process known as hybridization. Another way of looking at the result is to regard it as an alternate solution of the wave equations, a solution that is not at the lowest energy possible. The new result is known as the valence state of the atom to distinguish it from the atomic state (Table 5.2), the most stable state (lowest energy state) of the atom. The less stable valence state is able to form more and better MO's (good overlap) with other atoms, to release more energy in the process, and to arrive at a more stable combination than the atomic state is able to contrive.

The hybrid orbitals differ from the atomic orbitals in shape and distribution in space about the nucleus. The hybrids may take one of three different forms depending upon the number of other atoms to be combined with the carbon atom. Complete hybridization leads to four equivalent sp^3 orbitals and the ability to bond to four other atoms. Partial hybridization leads to three equivalent sp^2 orbitals with one p atomic orbital unaffected or to two equivalent sp orbitals with two p atomic orbitals unaffected (Fig. 5.4). The greater extension in space of the hybrid orbitals gives greater overlap than that obtained with the pure p orbitals.

The hybrids have characteristic angles between the lines along which the hybrid orbitals extend (and the remaining p orbitals, if any). Thus, in the atomic state the three p orbitals are at right angles to each other.

In the sp hybrid, the two sp hybrid orbitals are opposed to each other and lie along the axis of the hybridized p orbital and at right angles to the two remaining p orbitals.

In the sp^2 hybrid, the three sp^2 orbitals lie in the plane determined by the two hybridized p orbitals and at right angles to the remaining, unhybridized p orbital. The three sp^2 orbitals are as far apart from each other as possible at 120° angles.

In the sp^3 hybrid, the four equivalent orbitals form equal-sized angles with each other. The only possibility in three dimensions is the tetrahedral angle of 109°28'.

Actually, because any of the hybrid orbitals has p orbital participation, each hybrid orbital has a node at the nucleus and a small tail (which was omitted in Figure 5.4). Of more significance is the participation of the s orbital, which tends to bring the electron density

closer to the nucleus in the hybrid orbital than in a p orbital. The greater the s participation (50% for sp, 33% for sp^2, and 25% for sp^3) the greater is the tendency. We shall point out the pronounced effects as we go along.

MO's of several molecules. Let us see how the MO theory applies to such molecules as ethane, ethylene, acetylene, and formaldehyde.

Ethane ($CH_3 CH_3$) requires four atoms—three hydrogen atoms and a carbon atom—to be attached to each carbon atom. Therefore, sp^3 hybridization is required for each carbon atom. There are six $(\sigma 1s 2sp^3)^2$ filled MO's and one $(\sigma 2sp^3 2sp^3)^2$ filled MO in each

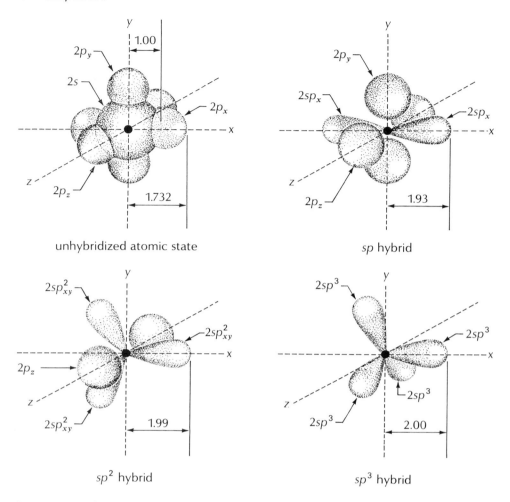

Figure 5.4 Hybridized orbitals; the various orbitals are schematic in shape, but the relative extensions are drawn to an arbitrary scale in units of the assumed radius of the 2s atomic orbital.

molecule. The valence shell antibonding MO's are empty. The overlapping orbitals in ethane are shown in Figure 5.5.

Ethylene ($H_2C=CH_2$) requires three atoms—two hydrogen atoms and a carbon atom—to be attached to each carbon atom. Therefore, sp^2 hybridization is required for each carbon atom. There are four $(\sigma 1s2sp^2)^2$ filled MO's and one $(\sigma 2sp^2 2sp^2)^2$ filled MO in the molecule. There are two atomic 2p orbitals left over, each with a single electron. If they are aligned parallel to each other, they will form a $(\pi 2p2p)^2$ MO. The requirement that the 2p orbitals be parallel to each other to form a π bond is met when all five of the σ MOs (all six nuclei) lie in the same plane. (See Fig. 5.6.)

Acetylene (HC≡CH) requires only two atoms—a hydrogen atom and the other carbon atom—to be attached to each carbon atom. Therefore, the sp hybrid state is adopted by each carbon atom. There are two $(\sigma 1s2sp)^2$ filled MO's and one $(\sigma 2sp2sp)^2$ filled MO in the molecule. There are four atomic 2p orbitals left over, each with a single

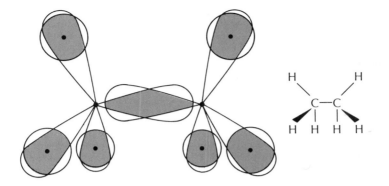

Figure 5.5   Ethane

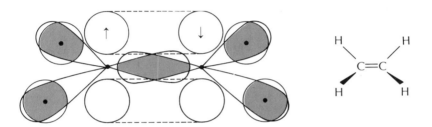

Figure 5.6   Ethylene

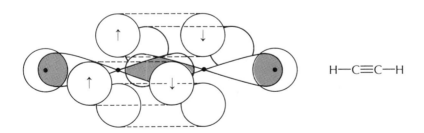

Figure 5.7   Acetylene

electron. The 2p orbitals are aligned parallel to each other to give the maximum possible overlap, and they form two $(\pi 2p2p)^2$ MO's. Because the two $2sp$ orbitals lie in a line, the whole molecule is linear (Fig. 5.7).

Formaldehyde (HCHO) has three atoms—two hydrogen atoms and an oxygen atom—attached to the carbon atom. The oxygen atom is attached to only one atom, the carbon atom. The carbon atom adopts the sp^2 hybrid state and forms two $(\sigma 1s2sp^2)^2$ filled MO's.

96 *Chapter five*

The remaining $2sp^2$ orbital overlaps an oxygen orbital. If the oxygen atom remains in the atomic state, the MO takes the form of $(\sigma 2sp^2 2p)^2$. If the oxygen atom assumes the sp hybrid state, the MO takes the form of $(\sigma 2sp^2 2sp)^2$, which allows greater overlap. The other $2sp$ hybrid orbital on oxygen is filled with a pair of electrons. The $2p$ orbital with one electron left on the carbon atom forms a π MO with a $2p$ orbital with one electron from the oxygen atom. The remaining $2p$ orbital on oxygen is filled with a pair of electrons. The molecule is planar because the three $2sp^2$ orbitals lie in a plane (Fig. 5.8).

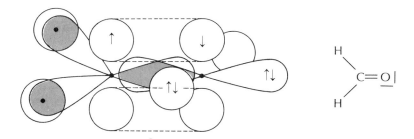

Figure 5.8   Formaldehyde

Exercises

5.1   Refer to Figure 5.3 and set up a similar set of diagrams for the π MO's for $(C_2H_4)^\oplus$, (C_2H_4), and $(C_2H_4)^\ominus$.

5.3   Bonds: valence bonds, resonance

Valence bonds and atomic orbitals.   The valence bond (VB) approach represents a covalent bond as being made up of varying contributions of atomic orbitals. When needed, hybridization is invoked and hybrid atomic orbitals are used.

Let us examine the VB method as applied to the hydrogen molecule, an example that will help us to understand the approach. We bring each hydrogen atom into place. Then we think of each atom as distinct: that is, we label one nucleus A and the electron in the $1s$ orbital a, the other nucleus B and electron in that $1s$ orbital b. The electrons must have opposing (paired) spins. Next, we make the electrons trade places so that b is in the A $1s$ orbital and a is in the B $1s$ orbital. In addition, we think of both electrons as being in the A $1s$ orbital and the B $1s$ orbital as being empty. Similarly, we think of the A $1s$ orbital as being empty and the B $1s$ orbital as containing both electrons. The four situations may be symbolized as shown.

$$A^a\ B^b \qquad A^b\ B^a \qquad A^a_{\ b}{}^{\ominus}\ B^{\oplus} \qquad A^{\oplus}\ B^a_{\ b}{}^{\ominus}$$
$$1 \qquad\qquad 2 \qquad\qquad 3 \qquad\qquad 4$$

Taken together, states 1 and 2 are thought of as a nonpolar covalent bond and symbolized with the dash: A—B or H—H. States 3 and 4 represent ionic bonds, symbolized as $A^{\ominus}B^{\oplus}$ and $A^{\oplus}B^{\ominus}$ or $H^{\ominus}H^{\oplus}$ and $H^{\oplus}H^{\ominus}$. The total bond is contributed to by all four states, but varying weights are assigned to each of the four. States 1 and 2 are assigned equal weights. If the atoms are alike, as in H—H, F—F, Cl—Cl, and so on, states 3 and 4 are assigned equal weights as well. However, the weights of 1 and 2 are greater than those of 3 and 4. We say that the covalent bond is made up of all four atomic states, the major contribution coming from 1 and 2, and only a minor contribution from 3 and 4. Only atomic orbitals are used.

Hydrogen fluoride: polarity, resonance. If we go on to look at a hydrogen halide, say hydrogen fluoride, we find it unreasonable to think of 6, with a positive fluorine, as

$$\underset{5}{\text{H—}\overline{\underline{\text{F}}}|} \qquad \underset{6}{\overset{\ominus}{\text{H}}|\ \overset{\oplus}{\overline{\text{F}}}|} \qquad \underset{7}{\overset{\oplus}{\text{H}}\ \overset{\ominus}{\overline{|\text{F}|}}}$$

making a contribution equal to that of 7. (The states analogous to 1 and 2 for HF have been telescoped into 5.) Consequently, we eliminate 6 as a contributor to the bond between hydrogen and fluorine. Now we symbolize the situation as shown:

$$\underset{5}{\text{H—F}} \longleftrightarrow \underset{7}{\text{H}^{\oplus}\ \text{F}^{\ominus}}$$

The contribution of 7 accounts for the polarity of the H—F bond. The double-headed arrow is the symbol for <u>resonance.</u> The double-headed arrow must not be confused with the two arrows used to represent an equilibrium.

Comparison of VB and MO approaches. Resonance is built into the VB approach to bonding. In MO terminology, resonance does not exist. The valence bond method describes molecules in terms of electron pair valence bonds localized between the atoms involved and uses atomic orbitals to do so. In MO's, the bonding is considered to be delocalized from the beginning, and there is no need to specify atomic orbitals except as they are used to describe the origin of the MO's. In VB theory, no single structure with the electrons assigned to definite positions is a satisfactory representation of a bond. To improve the representation, we are forced in the VB method to write other unsatisfactory structures and then say that the electrons are <u>delocalized</u> or that <u>resonance occurs among the contributing structures.</u> In the language of resonance, one speaks of contributing structures that are imaginary and of resonance among them. A picture that is rather time consuming to write out and still crude may be obtained by averaging the contributing structures.

However, the energy is not an average. The energy of a resonance system is less than that of any of the imaginary contributing structures. The decrease in energy is called the <u>resonance energy,</u> sometimes called the <u>delocalization energy.</u> More exactly, the resonance energy is the difference between the most stable imaginary contributing structure and the actual energy of the molecule, ion, or radical being described.

Let us repeat that resonance contributing structures are imaginary. When we write such structures, only shifts of the positions of electrons are allowed. In other words, no nuclear motion is allowed. The nuclei must be positioned as they exist in the actual molecule. Then, as the electrons are shifted about from one structure to another, the internuclear distances remain the same.

We have been talking of shifting electrons about as though they were discrete particles instead of correctly saying that the filling (or emptying) of the atomic orbitals of the different contributing structures leads to different locations of electron density. Because of the wordiness involved in the correct statement, we all have a tendency to use the "electron shift" expressions. These statements, however, are misleading unless the underlying meaning is understood.

The concept of resonance is *necessary* to the VB approach and *does not exist* in the molecular orbital method. The MO's of the valence shell electrons are spread over the molecule to begin with and delocalization is built in, as it were. The VB theory overemphasizes localization, and MO theory overemphasizes delocalization. Both theories are approximations. The MO method is easier to use mathematically for energy calculations, especially for spectroscopic purposes, because the excited (antibonding and nonbonding) states are part of the theory. The VB method is superior in envisaging structures in the ground state (stereochemistry), but the excited states are very difficult to handle mathematically. The theories should be regarded as complementary, not competitive. Until a better approximation is thought up, we shall be forced to muddle along with both theories.

Rules for resonance structures. A set of rules has been developed for use in writing resonance contributing structures. We list them here along with some illustrations.

1. All electrons are paired, except in radicals. (We use $>$ here to show order of stability.)

$$H_2C=\underline{O}| \gg H_2\dot{C}-\dot{\underline{O}}|$$

$$H_2C=CH_2 \gg H_2\dot{C}-\dot{C}H_2$$

2. Hydrogen prefers to share one electron pair.
3. All other elements (up to chlorine) prefer to share or own four electron pairs.
4. That structure with the maximum number of covalent bonds is the more stable.

$$H_2C=\underline{O}| > H_2\overset{\oplus}{C}-\overset{\ominus}{\underline{O}}|$$

$$H_2C=CH_2 > H_2\overset{\oplus}{C}-\overset{\ominus}{C}H_2$$

5. Like charges on adjacent atoms are so unstable that such structures may be regarded as noncontributors.

$$\underset{CH_3}{\underset{|}{CH_3-\overset{Cl}{\underset{|}{\overset{|}{C}}}-\overset{\oplus}{N}(CH_3)_3}} \gg \underset{CH_3}{\underset{|}{CH_3-\overset{Cl^\ominus}{\underset{|}{\overset{\oplus}{C}}}-\overset{\oplus}{N}(CH_3)_3}}$$

6. A small separation of unlike charges is preferable to a large separation of unlike charges (or to an increase in size of charge).

$$H_2C=CH-\underset{H}{\underset{|}{C}}=\underline{O}| > H_2C=CH-\underset{H}{\underset{|}{\overset{\oplus}{C}}}-\underline{\underline{O}}|^\ominus > \overset{\oplus}{H_2C}-CH=\underset{H}{\underset{|}{C}}-\underline{O}|^\ominus$$

7. The polarity of other bonds affects the relative stability of location and kind of charge.

$$Cl_3C-H > Cl_3C^\ominus H^\oplus \gg Cl_3C^\oplus H^\ominus$$

8. The more electronegative elements are more stable with a negative charge than with a positive charge.

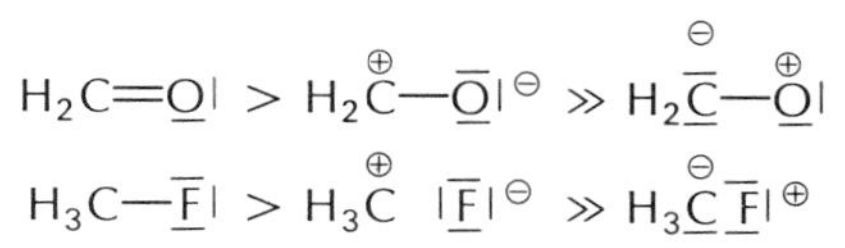

$$H_2C=\underline{O}| > H_2\overset{\oplus}{C}-\overline{\underline{O}}|^{\ominus} \gg H_2\overset{\ominus}{\overline{C}}-\overset{\oplus}{\underline{O}}|$$

$$H_3C-\overline{\underline{F}}| > H_3\overset{\oplus}{C}\ |\overline{\underline{F}}|^{\ominus} \gg H_3\overset{\ominus}{C}\ \overline{\underline{F}}|^{\oplus}$$

The greatest stabilization by means of resonance occurs if two or more contributing structures of equal energy may be written, as is shown in the next subsection.

Application of resonance theory. Let us consider a few examples to help our understanding of resonance.

Why is nitric acid a strong acid? If we consider the structure of nitric acid, we find that of the three contributing structures shown, c may be disregarded (rule 5). Structures

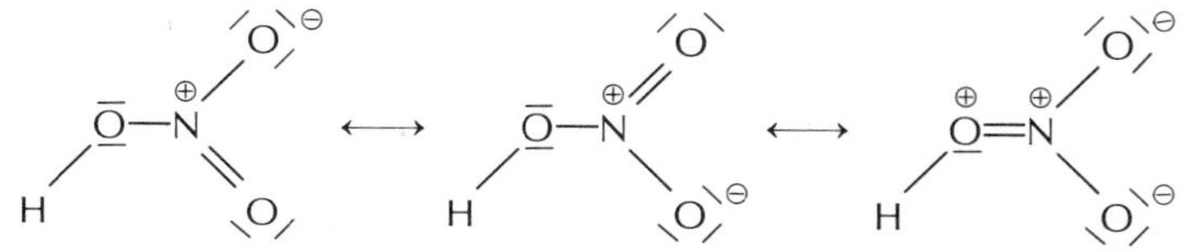

a, b, c (nitric acid resonance structures)

a and b contribute equally, and resonance stabilization is considerable.

Now, if we look at the nitrate ion, we find that the situation is quite different. Structures

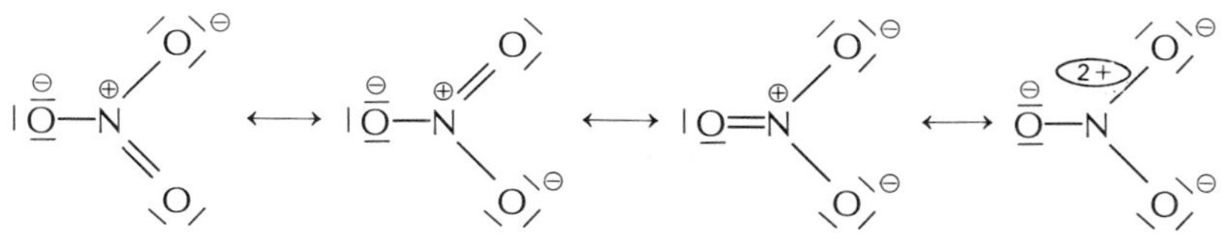

d, e, f, g (nitrate ion resonance structures)

d, e, and f contribute equally, and resonance stabilization is increased dramatically. Thus, the nitrate ion is so stable (such a weak base) that nitric acid will force its proton to move to almost any base (proton acceptor). Structure g is shown to illustrate violations of rules 4 and 6: g is a noncontributor.

Why is acetic acid an acid? If we write out the possible contributing structures of acetic acid, we see that a > b ≫ c. Structures b and c both require charge separation

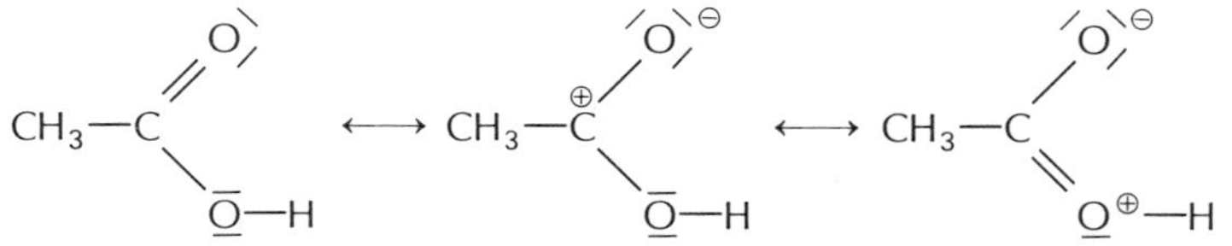

a, b, c (acetic acid resonance structures)

(rule 6), with b > c also, by rule 8. Structure a > b by rule 4. Acetate ion may be represented by structures d = f ≫ e. With d = f, resonance stabilization of the ion is so great, as compared with the undissociated acid, that the puzzle is not why acetic acid is an acid but why it is only a weak acid. Other effects come into the picture, and we shall discuss them later.

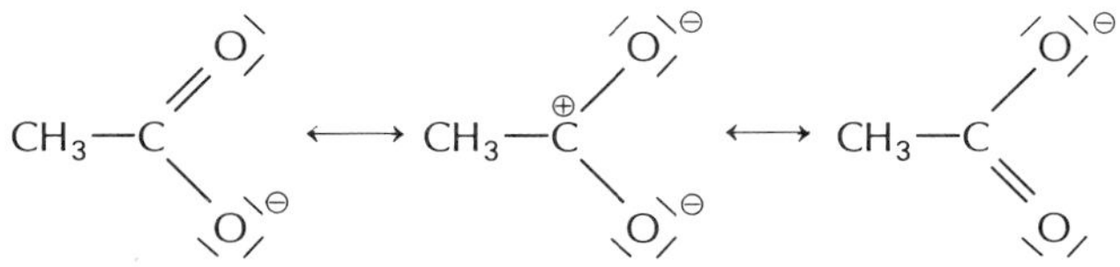

Molecules and ions for which no single structure is satisfactory, such as nitric acid, nitrate ion, and acetate ion, are known as <u>resonance hybrids</u>. Instead of writing out all the contributing structures of a resonance hybrid, we sometimes use a compromise. At other times, we use the very oversimplified abbreviated structures such as HNO_3, $NO_3^{\ominus}$, and $CH_3COO^{\ominus}$, with the understanding that we keep in mind all the implications that they cannot show explicitly.

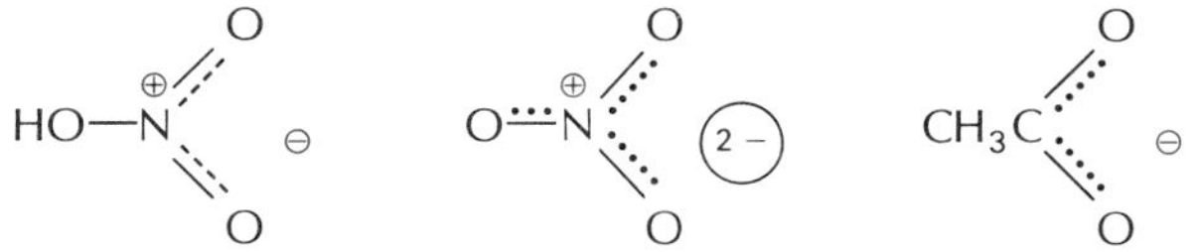

Even though a carbon-hydrogen bond and a carbon-carbon bond may be represented as shown, such single bonds are usually represented as pure covalent bonds. In other words, we usually regard the σ bond framework of a molecule as remaining intact. Resonance theory and structures are used most often in helping us to understand the stability (or reactivity) of molecules that contain π bonds.

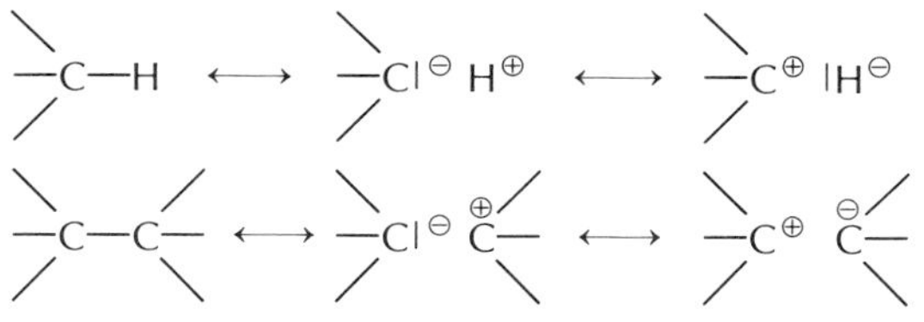

Exercises

5.2 Write out electron pair structural formulas for the compounds given (example: $F_2 = |\overline{F}-\overline{F}|$):
    a. CH_3Cl c. CH_3COCH_3 e. CH_3CH_2CHO
    b. $H_2C=CHCH_2OH$ d. CH_3COCl

5.3 Write out MO descriptions of the compounds given:
    a. CH_4 c. CH_3CHO e. CH_3OH
    b. $CH_3CH=CH_2$ d. $H_2C=CCl_2$

5.4 Write out contributing resonance structures of the compounds and ions given:
a. $H_2C{=}CHCHO$ c. $H_2C{=}CH\overset{\oplus}{C}H_2$ e. $H_2C{=}CHCH{=}CH_2$
b. CH_3NO_2 d. CH_3COCl

5.4 Bond distances, atomic radii, bond angles

Relative constancy of bond distances and angles. From the examination of many compounds (gaseous ones by electron diffraction, crystalline ones by the diffraction of X rays), it has become apparent that covalent bonds are relatively constant in internuclear distance and that the angles between adjacent bonds tend to be constant. From infrared and Raman spectra, it has become equally apparent that the internuclear distances and bond angles are capable of vibratory motions that obey the laws of quantum mechanics. It has been calculated that the energy required to stretch a C—H bond 0.1 A (10^{-9} cm) is about 3.6 kcal/mole. To bend a bond angle, say a H—C—H, by 1° requires only 0.1 kcal/mole. Consequently, a polyatomic molecule is a quivering, jiggling entity at room temperature, as the various nuclei approach and retreat and the bonds between groups of atoms scissor and twist.

The vibrations within a molecule are "followed" by the electronic orbitals: a change in internuclear distance causes changes in the amount of overlap as well as in the internuclear repulsion. A change in bond angle is followed by changes in hybridization, which also lead to changes in overlap and bond strength during the vibration. Despite all the internuclear motion, the magnitudes of shift from positions of equilibrium are relatively small. Thus, there is rather widespread agreement concerning bond distances and angles, whether determined by diffraction or by spectroscopic techniques. On the other hand, the various techniques may, and frequently do, disagree on the exact values.

Atomic covalent radii. Many atoms have relatively constant covalent bond radii. For example, in chlorine (Cl_2), the internuclear distance (bond length) is about 1.98 A. Dividing by 2 gives 0.99 A as the bond radius of chlorine in a covalent bond. In diamond, the C—C bond length is about 1.54 A, as it is in ethane. Thus, the sp^3 bond radius of carbon is 0.77 A. If we add 0.99 A for Cl to 0.77 A for C, we find 1.76 A as the predicted bond distance for a C—Cl bond. In methyl chloride, the measured distance is 1.78 A, in ethyl chloride it is 1.77 A, and in carbon tetrachloride it is 1.76 A. In this manner, covalent bond radii have been developed and used with remarkable success in predicting bond length. However, the exceptions have been of great interest.

For example, in H_2 the bond length is 0.74 A, which gives a radius of 0.37 A. In all other compounds, covalent bonds to hydrogen have a hydrogen atom radius somewhat less than 0.37 A. The effect probably is caused by greater overlap of the 1s orbital of H with all other orbitals than is possible with another 1s orbital. For example, the H—Cl bond distance is 1.27 A. If the Cl radius of 0.99 A is constant, the H radius is only 0.28 A. Similar calculations give 0.20 A for the H radius in H—F, which has a bond distance of 0.92 A; 0.35 A in H—OR, with 1.09 A; and 0.32 A in H—CR_3, with 1.09 A. For general use, 0.32 A is adopted as the H radius.

Changes in hybridization in atoms affect the bond radii. Thus, an sp^3 orbital on carbon seems to have a single bond radius of 0.77 A, but the sp^2 orbital single bond radius is about 0.74 A, and the sp orbital single bond radius is about 0.69 A. Table 5.3 gives some commonly used bond radii.

Table 5.3 Some covalent bond radii (in angstroms)

Atom	Single bond	Double bond	Triple bond
H	0.32	—	—
B	0.81	0.71	—
C	0.77[a]	0.67	0.60
N	0.74	0.62	0.55
O	0.74	0.62	—
S	1.04	0.94	—
F	0.72	0.60	—
Cl	0.99	0.89	—
Br	1.14	1.04	—
I	1.33	1.23	—

[a] For sp^3. Use 0.74 for sp^2, 0.69 for sp.

All the multiple bond radii are shorter than those for single bonds. We can explain this tendency by saying that p orbital sideways overlap helps to strengthen the bonding. In general, the shorter a bond is, the stronger it is, although it is difficult to set up a quantitative relationship because of other effects on bond length and bond strength.

Bond angles. Bond angles usually correspond well with those predicted by hybridization: 109°28' for sp^3, 120° for sp^2, and 180° for sp. Certain exceptions may be noted, such as water, in which the H—O—H angle is about 104°, whereas the H—S—H angle in H_2S is about 92°. In the latter, the sulfur atom evidently prefers to remain in the atomic configuration and bond to the hydrogen atoms by 3p-1s overlap. In water, the oxygen valence shell must undergo some hybridization to follow the angle as it widens and approaches the sp^3 tetrahedral angle of 109°28'. The widening relieves some of the repulsion between the hydrogen atoms. However, the repulsion between the remaining filled nonbonding electron pair orbitals on oxygen should not be forgotten. The 104° angle ends up as the best compromise among the contending repulsions and attractions. In H_2S, the S—H bonds are enough longer (1.33 A) than the O—H bonds in water (0.96 A) so that the nonbonded repulsion between the hydrogen atoms becomes almost negligible, as shown by the tiny distortion of only 2° from the normal 90° p orbital angles. As the groups attached to oxygen increase in size, the angle widens. The C—O—H angle in methanol is 109°, and the C—O—C angle in dimethyl ether is 111°.

Exercises

5.5 Predict the structure of ammonia, including bond lengths and angles.

5.6 Predict the structures (bond lengths and bond angles) of the substances given:
  a. CH_3CHO b. $HCOO^{\ominus}$ c. $CH_3CH=CH_2$

5.5 Electronegativity: polarity and polarizability

Electronegativity. Electronegativity is the tendency for an atom to attract the electron pair in a covalent bond. Because scales of electronegativity have been calculated in several different ways, it is difficult to be more precise without becoming involved in a prolonged description of the various methods used. The originally proposed scale ran from 4.0 for fluorine, the most electronegative element, down to 0.7 for cesium, the least electronegative element. This unique scale remains in use. The values for those elements of most interest to us are shown in Table 5.4.

Table 5.4 Electronegativity values of some elements[a]

H						
2.1						
(2.10)						
Li	Be	B	C	N	O	F
1.0	1.5	2.0	2.5	3.0	3.5	4.0
(0.97)	(1.47)	(2.01)	(2.50)	(3.07)	(3.50)	(4.10)
Na	Mg	Al	Si	P	S	Cl
0.9	1.2	1.5	1.8	2.1	2.5	3.0
(1.01)	(1.23)	(1.47)	(1.74)	(2.06)	(2.44)	(2.83)
						Br
						2.8
						(2.74)
						I
						2.5
						(2.21)

[a] The original scale by Pauling. The values in parentheses are from E. Little and M. Jones, *J. Chem. Educ.* **37**, 231 (1960), by permission.

The difference in electronegativities between two atoms joined by a covalent bond is a very rough representation of the contribution of the ionic structure to the covalent bond, in resonance terminology. Thus, for HCl, the difference is $3.0 - 2.1 = 0.8$, and the H—Cl bond is polar. For CsF, the difference is $4.0 - 0.7 = 3.3$, and the bond is ionic. In general, we may assume that the breaking point is about 2.0. If the difference is less than 2.0, the bond is covalent (with more or less ionic character). If the difference is greater than 2.0, the bond is ionic (with more or less covalent character).

It has been estimated that if a positive charge is present, the electronegativity of a given atom is increased by about 0.5. Thus, the nitrogen atom in $NH_4^{\oplus}$ would have an electronegativity toward hydrogen of $3.0 + 0.5 = 3.5$. On the other hand, if a negative charge is present, the electronegativity is decreased by about 0.5. Thus, the oxygen atom in $CH_3O^{\ominus}$ would have an electronegativity toward carbon of $3.5 - 0.5 = 3.0$.

We shall make use of electronegativity in a qualitative sense; we shall use only relative magnitudes, the order of which is listed here:

$$F > O > N, Cl > Br > C, S, I > H, P > B > Si > \text{metals}$$

Polarity and dipole moment. Polarity was mentioned earlier when we discussed the contribution of ionic structures to the VB formulation of bonding. Polarity exists in a molecule if the center of density of all the negative charges (the electrons) and the center of density of all the positive charges (the nuclei) do not coincide. If the separation of the centers is large enough, we can speak of an ionic bond. Polarity is measurable as a dipole moment, which may be thought of as the turning force that causes a polar molecule to line up with an electrical field. Of course, thermal agitation opposes the process, so that an increase in temperature decreases the lining-up effect. Mathematically, the dipole moment (μ) is equal to the size of the charge in electron units (x) times the distance in Å between the centers (d) times the charge of the electron in electrostatic units, or $\mu = (x)(d)(4.8 \times 10^{-18})$. One unit of 10^{-18} is called a debye (abbreviated as D) in honor of Peter Debye, who did much of the work on the theory involved.

For example, the dipole moment of HCl has been reported: $\mu_{HCl} = 1.03$ D. With a bond distance in HCl of 1.28 Å, and if $x = 1$, we calculate $\mu = (1)(1.28)(4.8) = 6.14$ D. Using $x = 1$ means that the molecule is completely ionic. At 1.28 Å, the situation is unreal because the ionic radius of $Cl^{\ominus}$ is 1.81 Å, which is more than enough to engulf the proton. Nevertheless, $1.03/6.14 \times 100 = 17\%$ ionic character, which means the charge on hydrogen is $+0.17$ and that on chlorine is -0.17 in units of electronic charge.

Analgesics. Acetylsalicylic acid (aspirin), Section 24.2, happens to be a good analgesic. Why is still only a guess. A major part of medical research is devoted to learning what part (or parts) of the structure of a compound account for that compound's action as an analgesic, a stimulant, an antibiotic, a sedative, or whatever else. Traditionally, the investigation involves lengthy chemical and biological investigations with occasional successes.

The story of opium is an example of the tedious procedures now necessary to find a new drug. Opium has been used for several millenia for all types of ailments. In the early nineteenth century, an alkaloid was isolated from opium that had the same sleep-inducing property. It was named morphine after the god of sleep, Morpheus. Chemical derivatives of morphine were synthesized and tested for activity, in hopes of finding a pain-killing drug that was not addictive.

When diacetylmorphine was tested, doctors thought they had found the answer. Amid much publicity as a cure for morphine addiction, clinical use was begun in 1898. The tradename of the new drug was Heroin.

After the chemical structure of morphine was finally established in 1925, many more compounds were synthesized and tested. In the late 1930s a compound was synthesized that looked nothing like morphine but acted as a morphine-like drug. The few structural similarities between this compound (pethidine or, by a tradename, Demerol) and morphine provided a breakthrough leading to the understanding we have today about the opiate drugs. Methadone is a narcotic that can be administered orally (and less frequently) as a substitute for heroin and morphine, with less debilitating effects and with less severe withdrawal difficulties.

The search for a strong analgesic continues. In theory, quantum chemical calculations could be a powerful tool in this search as well as the understanding of many other chemical interactions in biology. One of the first efforts in this direction was a correlation of the effectiveness of a series of anti-tumor compounds with their calculated dipole moments. Much work remains to be done.

Polarizability. Placing a molecule in an electric field, whether or not the molecule is polar, causes the center of density of the electrons to be pulled toward the positive end of the field and the nuclei to be pulled toward the negative end of the field, no matter what the orientation of the molecule in the field may be. The shifts within the molecule affect the polarity. Such a change in polarity is called induced polarization or polarizability. Polarizability is a temporary effect, dependent upon the environment of the molecule at a given time. The environment is made up of other molecules and ions at varying distances, velocities, and so on. During a reactive collision, the polarizability of the participants may become the dominant influence upon the result.

Bond moments; polarizable bonds. In general, the polarity of a molecule (the dipole moment) is a summation of the polarities of the individual bonds. The summation is complicated by the various angles involved. In turn, the polarity of an individual bond (bond moment) is roughly proportional to the difference in electronegativity of the atoms forming the bond. Thus, a given bond may be relatively polar or nonpolar.

The small atoms (up to fluorine) and bonds between them tend to be low in polarizability. The larger atoms and their bonds are more polarizable. The larger polarizability results from the greater distance between the nucleus and the centers of electron density in the orbitals of the valence shell. The valence shell also is screened from the highly charged nucleus by the intervening filled orbitals of the lower-lying shells. Thus, the more weakly held electrons in the valence shell of a larger atom are easily pushed or pulled about by external influences. A big, mushy iodide ion is easy to displace in an S_N2 reaction, but it also is a good nucleophile because the cloud of electron density in the valence shell is so easy to influence.

5.6 Intermolecular forces: hydrogen bonds, complexes

Intermolecular attraction. Besides the strong covalent bonds that atoms form, there are relatively weak attractive forces between molecules. Such weak forces are lumped together and called van der Waals' forces. They all have their origins in dipole effects, either permanent or temporary. In energy terms, they seldom exceed 6 kcal/mole. However, such forces account for the ultimate liquefaction and crystallization of gases such as methane, ethane, and so on.

Hydrogen bonds. Hydrogen bonds are stronger than van der Waals' forces, ranging from 2 to 10 kcal/mole in energy. Some representative values are given in Table 5.5. Aside from the symmetrical and exceptionally strong hydrogen bond in the bifluoride ion ($HF_2^{\ominus}$), hydrogen bonds are unsymmetrical. The hydrogen atom remains closer to the atom to which it is covalently bonded. Hydrogen bonds may be thought of as being formed by the reduction in repulsion between two filled orbitals by the presence of a proton between them. The attraction of a partially ionic hydrogen for the electrons in a filled orbital on another atom is another explanation. Neither of the explanations is completely satisfactory, and our understanding of a hydrogen bond remains limited. However, when possible, a hydrogen bond will form with a reduction in energy of the system and a reduction in the distance between the atoms being hydrogen bonded.

Hydrogen bonding occurs most often between atoms that are highly attractive for electrons: fluorine, oxygen, and nitrogen. There must be a delicate balance of forces to form a hydrogen bond, because if the receiving atom is basic enough relative to the donor,

Table 5.5 Energies of some hydrogen bonds

Bond	Substance	Bond energy, kcal per mole	Bond length, A
F—H—F	KHF_2	27	2.25
F—H··F	HF, gas	6.7	—
OH··O	H_2O, ice	4.5	2.76
O—H··O	CH_3OH	6.2	2.7
O—H··O	$(HCOOH)_2$	7.1	2.67
O—H··O	$(CH_3COOH)_2$	7.6	—
O—D··O	$(CH_3COOD)_2$	7.9	—
O—H··N	average[a]	7	—
N—H··O	average[a]	2	—
N—H··N	average[a]	2–4	—

[a] The average of many such mixtures. The hydrogen bonds occur in almost all proteins as well as in DNA and RNA and play a significant role in the stereochemistry of such compounds.

proton transfer will occur. A consequence of hydrogen bonding is the association of the molecules in all solvents that have H—F, H—O—, and H—N bonds. The association causes the bp and mp of such liquids to be higher than normal. These protic solvents also are able to solvate anions by weak hydrogen bonding.

Complex formation: donor-acceptor complex. Related to hydrogen bonding is donor-acceptor complex formation, in which an electron pair donor operates upon a Lewis acid instead of a proton. An electron donor such as a Lewis base donates or shares a pair of electrons with an acceptor, a molecule or ion with a vacant orbital such as a Lewis acid. Depending upon the extent of transfer of the electron pair, a weak, moderate, or strong complex may be formed with a large or small transfer of charge. For example, the $F_3\overset{\ominus}{B}{:}\overset{\oplus}{N}H_3$ complex has a bond energy of 42 kcal/mole, strong enough to be considered a weak covalent bond. The bond in such a complex is accompanied by a relatively large increase in charge: negative on BF_3 as the electron pair is accepted, and positive on NH_3 as the electron pair is donated. Such a complex is also known as a charge transfer complex.

A shift from BF_3 to the weaker Lewis acid $AlCl_3$ gives the complex $Cl_3\overset{\ominus}{Al}{:}\overset{\oplus}{N}H_3$, with a bond energy of 40 kcal. Aluminium bromide gives a complex with ammonia (38 kcal) and aluminium iodide (only 30 kcal). If we keep the strong Lewis acid BF_3, but shift to the weaker base ethyl ether, the complex $F_3B{:}\overset{\oplus}{O}(CH_2CH_3)_2$ is formed, with a strength of only 14 kcal/mole. Boron trifluoride and aluminum chloride also form complexes with carbon-oxygen and carbon-carbon double bonds. The complex with oxygen of the carbon-oxygen double bond probably is formed by donation of a pair of unshared electrons from the oxygen instead of from the π bond. Complexes with the carbon-carbon double bond can originate only in donation of electrons from the π bond. A π complex, then, is but a special case of the general subject of complex formation. Formed from a relatively weak base, the π complex is weak and dissociates easily.

Intermolecular bonding in biochemistry. Biochemical compounds are often classified as to whether or not they form hydrogen bonds to water. The "water-loving" compounds are called hydrophilic, and those preferring oil solvents to water are called hydrophobic. Sugars are highly hydrophilic. When present as a simple sugar or as a short saccharide chain, they are soluble in water. As a component of a very long, water insoluble polysaccharide such as starch, they absorb much water.

Fats are the opposite. Since little energy is released by the very weak fat-water interaction, the system is at its lowest energy when the water molecules are arranged so they can form hydrogen bonds with each other. Thus the large, nonpolar molecules are forced out of the water solution. Oil and water do not mix.

Every protein has both hydrophobic and hydrophilic side chains. The solubility properties of a certain protein depends to some degree on its total amino acid composition (hence the number of hydrophobic and of hydrophilic side chains) but to an even greater degree on the location of the side chains. For example, in water soluble proteins, the hydrophilic side chains tend to be concentrated on the outer surface, in contact with the water solvent, while the hydrophobic groups are mostly folded inside.

Many lipids belong to an interesting class of molecules with a long, very hydrophobic tail connected to a polar head. The tails can interact with hydrophobic material; the heads, with water. In a water solution, when a certain lipid concentration is exceeded, they aggregate in spheres called micelles. All the tails point inward, away from the water. Other hydrophobic compounds can be absorbed inside the micelle and therefore, in essence, be dissolved in this water solution. Part of the cleaning action of soap is based on this principle. Some lipids, such as lecithin, are added to food products as emulsifying agents to hold other components of the food in suspension.

Lipids are present in the lung where they lessen the surface tension of the water (decrease the number of water–water hydrogen bonds) allowing the alveoli sacs to expand. The major function of the lipids in biochemistry is in the membrane around each cell—thought by many to be a lipid bilayer. These membranes separate the operations in the aqueous solution inside the cells from the aqueous solution, e.g. blood, outside the cells. The structure and mechanism of action of these membranes are currently the object of intense investigation.

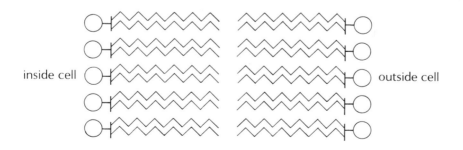

108   Chapter five

Exercises

5.7 Show the structures of
   a. (HCOOH)$_2$ d. (AlCl$_3$·CH$_3$OH)
   b. ((CH$_3$)$_2$NH)$_2$ e. (AlCl$_3$·H$_2$C=CH$_2$)
   c. (HF, H$_2$O)

5.7 Inductive and resonance effects

Bond polarity and inductive effects.   Let us return to the subject of bond polarity and examine some of the effects that a polar bond has on the properties of the molecule in which it is located. We shall use the VB approach in our description.

In methane, each C—H σ bond has an electronegativity difference of 0.4 in favor of carbon. Thus, we expect each C—H σ bond to be polarized, with the hydrogen somewhat positive (say with a charge of $+x$) and carbon negative, $-4x$ to balance the four $+x$ charges. Now, let us replace a hydrogen with a chlorine atom and look at methyl chloride, CH$_3$Cl. In the C—Cl σ bond, we would predict that chlorine would be partially negative, with an equal positive charge on the carbon atom. But this new positive charge (or decrease of negative charge) on carbon will increase the electronegativity of the carbon atom and result in a shift in electron density in each C—H bond toward carbon. Thus, the extra positive charge is transmitted partially to the three hydrogens. In a multiatomic molecule, such polar effects are transmitted and distributed in such a way as to reduce the magnitude of the induced charge on a given atom—in this case, the carbon of the C—Cl bond. Methyl chloride has μ = 1.86 D, which with a C—Cl distance of 1.78 A would give a calculated 22% ionic character to the C—Cl bond if all of the positive charge remained on the carbon atom. This amount is more charge than the 17% in H—Cl! In this diagram of H$_3$CCl$_1$ we use →— to symbolize the direction of shift of the center of electron density in a covalent bond.

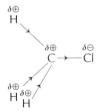

In going from CH$_4$ to CH$_3$Cl, the carbon atom becomes more positive, and we say that chlorine has a —I effect. That is, when we compare C—Cl with C—H, chlorine is more negative than hydrogen. In going from HCl to CH$_3$Cl, chlorine becomes more negative. That is, comparing H—Cl with CH$_3$—Cl, we find the methyl group more positive than hydrogen. We say that methyl has a weak +I effect. I stands for <u>inductive.</u> The effect is given the sign that compares the direction of the shift of the center of electron density in the covalent bond with that of the bond to hydrogen that is replaced.

Now let us compare some pK_a values for some acids. Formic acid (HCOOH) has pK_a = 3.7. Replacing the hydrogen with a methyl group gives acetic acid (CH$_3$COOH), which has pK_a = 4.8, a weaker acid. The methyl releases electrons more strongly than hydrogen, we say, by the +I effect. The release stabilizes the acid molecule more than it does the negative carboxylate ion. Now, if we replace a hydrogen in the methyl group of acetic acid with a chlorine, the −I effect of the chlorine should reverse the electron release of the methyl group. The negative ion is stabilized more than is the acid molecule, and the result is a stronger acid. Chloroacetic acid (ClCH$_2$COOH) has pK_a = 2.8, in accord with prediction. Continuing the substitutions, dichloroacetic acid (Cl$_2$CHCOOH) has pK_a = 1.3, and trichloroacetic acid (Cl$_3$CCOOH) has pK_a = 0.1.

Inductive effects are transmitted throughout the molecule. However, the transmission decreases rapidly with distance or number of bonds to be traversed. It is thought that only about a third of the induced charge is transmitted through the next σ bond. Thus 3-chloropropanoic acid (ClCH$_2$CH$_2$COOH) has pK_a = 4.1 and is stronger than propanoic acid (pK_a = 4.9) but weaker than 2-chloropropanoic acid (pK_a = 2.8). Furthermore, we can now say that chlorine has a strong −I effect and that methyl has a weak +I effect.

The R effect.   The resonance effect, symbolized as +R or −R, is the other major influence a substituent may exert (again, compared to hydrogen). For example, ethyl chloride has μ = 2.05 D, vinyl chloride has μ = 1.44 D, and chloroacetylene has μ = 0.44 D. The decrease in μ in this series cannot be accounted for entirely by the bond shortening of the C—Cl bond that is caused by changes in hybridization of carbon. Neither is it explained by the increasing electronegativity of carbon, caused by hybridization changes of carbon. If we attribute a +R effect to chlorine as shown, we arrive at a better explanation.

The resonance structure *b*, with a double bond to and a positive charge on chlorine, looks unreasonable at first sight. However, the chlorine atom does possess unshared electron pairs in *p* orbitals. One of them may be oriented so as to provide good sideways overlap with the *p* orbital on the neighboring carbon atom if that orbital is vacated by a shift of the electron pair of the π bond to the far carbon atom. Furthermore, the strong −I effect of chlorine provides a partial negative charge on chlorine and a partial positive charge in the vinyl group. The higher-energy resonance structure *b* provides a means of partially counteracting the inductive effect, and μ falls from 2.05 to 1.44 D. We say that chlorine has a weak +R effect.

The R effect requires the presence of a multiple bond in the molecule. The resonance structure with a double bond to chlorine also helps to explain the shortening of the carbon-chlorine bond distance from 1.77 Å in CH$_3$CH$_2$Cl to 1.69 Å in H$_2$C=CHCl.

The I and R effects of substituents are of great help in the correlation and classification of physical and chemical properties. We shall be using I and R over and over in later chapters.

Chapter five

Exercises

5.8 The series EtF, EtCl, EtBr, and EtI have μ = 1.9, 2.0, 2.0, and 1.9 D, even though F, Cl, Br, and I have decreasing electronegativities of 4.0, 3.0, 2.8, and 2.5. Explain.

5.9 The series HCHO, CH_3CHO, CH_3COCH_3 have μ = 2.3, 2.7, and 2.9 D. Explain.

5.10 What may be said about the sign and magnitude of the R effect of H_2C=CH- and HC≡C-?

5.8 Bond energies and bond dissociation energies

Heats of formation of atomic states; average bond energies, BE's. Hydrogen may be heated from the standard state to such a high temperature that it dissociates into hydrogen atoms, not completely but enough to allow a reasonable estimate of the equilibrium constant for the dissociation at several temperatures. By use of van't Hoff's law (Eq. 5.1), $\Delta H°$ may be calculated. Thus, $\Delta H_f°$ may be found for one mole of hydrogen atoms

$$\log K_{eq_2} - \log K_{eq_1} = \frac{\Delta H°}{2.3\,R}\left(\frac{1}{T_1} - \frac{1}{T_2}\right) \tag{5.1}$$

at 25°C and 1 atm to be 52.1 kcal. In the same way, $\Delta H_f°$ may be found for other atoms in the gaseous atomic state even though the usual stable state under standard conditions is liquid or solid (see Table 3.2).

With such enthalpies of formation available, it becomes possible to evaluate the change in enthalpy for reactions like those shown in Equations 5.2 and 5.3. All $\Delta H_f°$ values are taken from Table 3.2. The enthalpy needed to break the two O—H bonds in water is 221.6 kcal, and that needed to break the four C—H bonds in methane is 397.2 kcal.

$$H_2O(gas) \longrightarrow 2H(gas) + O(gas) \tag{5.2}$$
$$\Delta H° = 2\Delta H_f°(H, gas) + \Delta H_f°(O, gas) - \Delta H_f°(H_2O, gas)$$
$$\Delta H° = 2(52.1) + (59.6) - (-57.8) = 221.6 \text{ kcal}$$

$$CH_4(gas) \longrightarrow C(gas) + 4H(gas) \tag{5.3}$$
$$\Delta H° = \Delta H_f°(C, gas) + 4\Delta H_f°(H, gas) - \Delta H_f°(CH_4, gas)$$
$$\Delta H° = (170.9) + 4(52.1) - (-17.9) = 397.2 \text{ kcal}$$

It is convenient to adopt a new standard state in which gaseous atoms under standard conditions are defined as having zero enthalpy. The advantage of using the new standard state is that the $\Delta H°$ for the formation of H_2O(gas) from the atoms is simply the sum of the average bond energies. An average bond energy (BE) is the average $\Delta H°$ necessary to break a bond homolytically to give neutral fragments or atoms, everything being in the gaseous state at 25°C and 1 atm. Similarly, for the reaction in Equation 5.3, methane is 397.2 kcal more stable than the constituent atoms and the BE of a carbon-hydrogen bond is 99.3 kcal.

In methanol there are three C—H bonds, one O—H bond, and one C—O bond.

5.8 Bond energies and bond dissociation energies

If we assume that there is no variation in BE between one molecule and another, we can proceed from the heats of formation to find the BE for the C—O bond (Eq. 5.4), to be 78.3 kcal.

$$
\begin{array}{c}
\text{C(gas)} \quad + 4\text{H(gas)} + \quad \text{O(gas)} \\
\uparrow 170.9 \qquad \uparrow 4(52.1) \qquad \uparrow 59.6 \qquad \searrow {-3(99.3)\ -\ 110.8\ -\ \text{BE(C—O)}} \\
\text{C(graphite)} + 2\text{H}_2\text{(gas)} + \tfrac{1}{2}\text{O}_2\text{(gas)} \xrightarrow{-48.1} \text{CH}_3\text{OH(gas)}
\end{array} \quad (5.4)
$$

$\Delta H^\circ_f(\text{CH}_3\text{OH}) = \Delta H^\circ_f(\text{C}) + 4\Delta H^\circ_f(\text{H}) + \Delta H^\circ_f(\text{O}) - 3\text{BE(C—H)} - \text{BE(O—H)} - \text{BE(C—O)}$

$-48.1 = 170.9 \ + 4(52.1) \ + 59.6 \ - 3(99.3) \ - 110.8 \ - \text{BE(C—O)}$

$\text{BE(C—O)} = 170.9 \ + 208.4 \ + 59.6 \ - 297.9 \ - 110.8 \ + 48.1$

$= 78.3 \text{ kcal}$

Now let us assume nonvariance in the bond energies in formaldehyde, formic acid, and carbon dioxide and calculate BE(C=O). Using ΔH°_f for formaldehyde as -27.7 kcal, for formic acid as -90.4 kcal, and for carbon dioxide as -94.1 kcal, and using BE(C—H) = 99.3, BE(C—O) = 78.3, and BE(O—H) = 110.8 kcal, we find BE(C=O) = 163.8 kcal in formaldehyde, 196.3 kcal in formic acid, and 192.1 kcal in carbon dioxide. We can see that the assumption that bond energies are invariant from one substance to another cannot be correct. However, it is close enough so that we are annoyed that it is not better. Various authors have tinkered with the bond energy values so that for most substances, the sum of the bond energies for a given compound comes within a kcal or two of the correct value. We shall adopt the table of bond energies worked out by J. D. Roberts and M. C. Caserio (Table 5.6), but with all values rounded off. The individual bond energies are derived from thermodynamic data, but additional assumptions are made, and the results obtained by their use are approximations.

Bond dissociation energies, BDE's. If one atom of hydrogen is allowed to combine with one atom of oxygen to form a hydroxyl radical, 102.4 kcal of heat is liberated. Combining this with the total bond energy of water vapor, 221.6 kcal, we see that the formation of the second oxygen-hydrogen bond liberates 119.2 kcal. These accurate values, which depend upon the order of making or breaking the bonds in a molecule or radical, are known as <u>bond dissociation energies,</u> BDE. Thus, to break either one of the two bonds in water vapor (a reaction in the gaseous state at 25° and 1 atm) requires a BDE of 119.2 kcal. To break the remaining bond in the hydroxyl radical requires a BDE of 102.4 kcal. Similarly, breaking one of the four C—H bonds in methane requires 104 kcal. To break the next two bonds, the energy needed for each step is about 106 kcal. But to break the last C—H bond, it seems to be accepted that about 81 kcal is required.

In any case, a BDE, if available (unfortunately, not enough are), is to be preferred over a BE. Some bond dissociation energies are listed in Table 5.7.

Uses of BE and BDE. Of what use are BE and BDE values?

First, BE's are approximate, and BDE's are quantitative measures of the strengths of bonds. Single covalent bonds range in strength from a high of 135 kcal for H—F to a low of 35 kcal for the O—O bond of peroxides. Single bonds to carbon range from a high of 108 kcal for CH_3—F to a low of 50 kcal for tBu—I. Bonds to carbon vary in strength

112   Chapter five

Table 5.6   Bond energies (BE)[a]

Diatomic molecules					
H—H	104	F—F	37	H—F	135
O=O	119	Cl—Cl	58	H—Cl	103
N≡N	226	Br—Br	46	H—Br	87
C≡O[b]	257	I—I	36	H—I	71

Polyatomic molecules					
C—H	99	C—C	83	C—F	116
N—H	93	C=C	146	C—Cl	81
O—H	111	C≡C	200	C—Br	68
S—H	83	C—N	73	C—I	51
P—H	76	C=N	147	C—S	65
N—N	39	C≡N	213	C=S[c]	128
N=N	100	C—O	86	N—F	65
O—O	35	C=O[d]	192	N—Cl	46
S—S	54	C=O[e]	166	O—F	45
N—O	53	C=O[f]	176	O—Cl	52
N=O	145	C=O[g]	179	O—Br	48

[a] These bond energies in kcal/mole at 25° are from J. D. Roberts and M. C. Caserio, *Basic Principles of Organic Chemistry*, New York: Benjamin, 1964, by permission.
[b] Carbon monoxide.
[c] For carbon disulfide.
[d] For carbon dioxide, acids, esters.
[e] For formaldehyde.
[f] For other aldehydes.
[g] For ketones.

Table 5.7   Some bond dissociation energies (BDE)[a]

	—H	—F	—Cl	—Br	—I	—OH	—NH$_2$	—CH$_3$
CH$_3$—	104	108	83.5	70	56	91.5	79	88
CH$_3$CH$_2$—	98	106	81.5	69	53.5	91.5	78	85
CH$_3$CH$_2$CH$_2$—	98	106	81.5	69	53.5	91.5	78	85
(CH$_3$)$_2$CH—	94.5	105	81	68	53	92	77	84
(CH$_3$)$_3$C—	91	—	78.5	63	49.5	90.5	77	80
$\overset{\overset{\displaystyle O}{\|}}{CH_3C}$—	87.5	119	83.5	—	52.5	109	96	82
CH$_2$=CH—	103	—	84	—	—	—	—	92

[a] S. W. Benson, *J. Chem. Educ.* 42, 502 (1965). BDE = $\Delta H°$, 25°C, 1 atm, gas phase, the energy to break the bond between the fragment listed at the left and the atom or fragment listed at the top. Example:

$$\text{BDE} = 69 \text{ kcal for } CH_3CH_2\frown Br \longrightarrow CH_3CH_2\cdot + Br\cdot.$$

depending upon the type of structure. Thus H₂C=CH- > Me- > Et- or primary > iPr- or secondary > tBu- or tertiary. Bonds to CH₃CO- may be either stronger or weaker than those to Me-. Multiple bonds are stronger than single bonds between the same atoms: compare C—N (73 kcal) to C≡N (147 kcal) for example.

Second, they are useful for calculation of $\Delta H°$ of reactions for which $\Delta H°_f$ values of reactants or products or both are not available. A simple example is shown in Equation 5.5.

$$3C + 8H + 2N$$

$$CH_3CH_2CH_3 + N_2 \longrightarrow CH_3C\equiv CH + H_2NNH_2$$

$\Delta H° = 2BE(C—C) + 8BE(C—H) + BE(N\equiv N) - 4BE(C—H) - BE(C—C) - BE(C\equiv C)$
$ - 4BE(N—H) - BE(N—N)$

$\Delta H° = 2(83) + 8(99) + (226) - 4(99) - (83) - (200) - 4(93) - (39)$ \hfill (5.5)

$\Delta H° = 166 + 792 + 226 - 396 - 83 - 200 - 372 - 39$

$\Delta H° = 1184 - 1090 = 94$ kcal

Will propane react with nitrogen to give propyne and hydrazine? The reaction is very endothermic and will not go. The reverse reaction is exothermic and should go.

Potential-energy diagram of bond formation. Let us pull together some of the concepts given in this chapter in a schematic potential-energy (enthalpy) diagram of a covalent bond (Fig. 5.9).

In Section 5.4, vibratory motion was mentioned in connection with bond distance. The energy of vibratory motion is given as $E_{vib} = (v + \frac{1}{2})h\nu_0$, in which v is the vibrational quantum number, with values of 0, 1, 2, 3, etc., and ν_0 is the fundamental frequency, calculable from theory. Thus, the energy of the vibrational levels is quantized, and only certain discrete frequencies occur. At $v = 0$, $E_{vib} = \frac{1}{2}h\nu_0$. At this level, the energy is known as the <u>zero-point energy,</u> because even at absolute zero the residual vibratory motion cannot be frozen out. The first five vibratory levels are shown at the bottom of the <u>potential-energy well,</u> which is practically parabolic in that region. The nuclei at the $v = 3$ level vibrate between the points a and b. In plotting such a diagram (known as a Morse curve) it is convenient to think of the nucleus of the atom on the left hand side as being clamped in place, so that all the motion to and fro is carried out by the nucleus of the atom on the right hand side. Actually, both nuclei move. At a, the nuclei have approached each other, have come to a standstill, and possess potential energy because repulsion exceeds attraction. As the nuclei begin to move away from each other, the potential energy is being converted into kinetic energy. As the nuclei pass the equilibrium distance from each other (the bond distance), the potential energy is at a minimum and the kinetic energy is at a maximum. The repulsive and attractive forces are in balance. Immediately, the attractive force begins to exceed the repulsive force, the nuclei begin to slow down, the kinetic energy begins to decrease, and the potential energy begins to increase. Finally, the nuclei come to rest at b, and all the kinetic energy of vibration has been changed into potential energy. Then the process is repeated in the reverse direction.

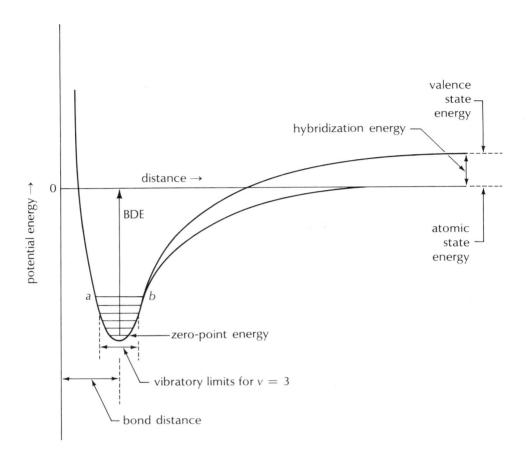

Figure 5.9   Schematic diagram of potential energy of a covalent bond

If the potential well is parabolic, the vibration is harmonic (equal swings to either side of the bond distance). The first few vibrational levels are close to being harmonic, but the higher levels (the bond stretch is greater than the bond compression) become more and more anharmonic (the bond distance is no longer the midpoint between the extremes).

We can think of the bond as acting like a helical spring, with the nuclei at either end vibrating back and forth. Furthermore, the stiffer the spring, the smaller the displacement for the same amount of energy and the greater the number of vibrations per second. Hooke's law applies to such a case: $F = k(r - r_e)$, which shows that the restoring force, F, is proportional to the displacement from the rest position, r_e. The force constant, k, is the proportionality factor. The stiffer the spring, the larger is k. Bond force constants per single bond fall between 3 and 7×10^5 dynes/cm. (The very strong H—F bond has a constant of 9.7×10^5.) Double bonds have constants in the range 7 to 15×10^5, and triple bonds greater than 15×10^5 dynes/cm.

If enough energy is supplied (BDE), the bond will be broken into the component atoms. Conversely, if we start out with the atoms at an infinite distance apart (potential energy of zero) and allow them to approach each other, sooner or later the potential energy of the system will begin to decrease. The energy continues to decrease more and more rapidly until the equilibrium distance (the bond distance) is reached. Any closer approach is opposed by the rapid increase in repulsive force between the nuclei. The decrease in potential energy in formation of the bond is disposed of as heat in the exothermic reaction.

If the atom must be hybridized before bond formation occurs, the hybridization energy must be supplied, after which bond formation follows the upper curve of Figure 5.9.

Exercises

5.11 Calculate $\Delta H°$ for each of the following reactions, all gas phase: 1. by use of BE's (Table 5.6); and 2. by use of $\Delta H°_f$ values (Table 3.2). Compare results.
 a. $CH_4 + \frac{1}{2}O_2 \longrightarrow CH_3OH$
 b. $CH_2Cl_2 + H_2O \longrightarrow HCHO + 2HCl$
 c. $CH_3Cl + 2HI \longrightarrow CH_4 + HCl + I_2$

5.9 Summary

Beginning with atomic structure, this chapter describes the elements of MO and VB approaches to molecular structure. Resonance is introduced as a necessary consequence of the valence bond approach. Bond lengths and angles, vibrations, and atomic radii are discussed. Electronegativity, polarity, polarizability, dipole moments, van der Waals' forces, hydrogen bonds, and donor-acceptor or charge transfer complexes were introduced next. The chapter continued with a description of inductive and resonance effects (I and R) and concluded with bond energies (BE) and bond dissociation energies (BDE), including a potential-energy bond-distance diagram.

Throughout the chapter, the approach has been more qualitative than quantitative, with emphasis on the concepts, and the reasons behind the concepts, involved in molecular structure. Perhaps the most important things for us to keep in mind are the complementary natures of the MO and VB theories of molecular structure and the fact that molecules have definite sizes, shapes, polarities, and stabilities, which determine and influence reactivity. With an understanding of the ideas introduced here, we shall be better prepared to grasp the meaning of descriptions of organic structure and reactions yet to come, especially in Chapters 10, 11, and 14.

5.10 Problems

5.1 a. Arrange water, methanol, and dimethyl ether in order of decreasing dipole moment.
 b. Arrange ammonia, methylamine, dimethylamine, and trimethylamine in order of decreasing dipole moment.

c. Arrange formaldehyde, acetaldehyde, and acetone in order of decreasing dipole moment.

5.2 Arrange the halogens in order of magnitude of the $-I$ effect.

5.3 Why does CH_3COOH partially ionize in aqueous solution, while CH_3CH_2OH and CH_3CHO do not?

5.4 Use hydrogen bonding to show the structure of:
   a. $HOOCCH_2CH_2COO^{\ominus}$
   b. CH_3COCH_2COOH
   (Hint: the use of structural models is very helpful.)

5.5 Verify that the sum of these three equations (all gas phase) gives $\Delta H°$ for the reaction $HCHO + 2H_2 \longrightarrow CH_4 + H_2O$:
   a. $HCHO + H_2 \xrightarrow{cat.} CH_3OH$
   b. $CH_3OH + HCl \longrightarrow CH_3Cl + H_2O$
   c. $CH_3Cl + H_2 \xrightarrow{cat.} CH_4 + HCl$

5.6 Use Table 5.6 to calculate and compare the enthalpy changes in the gas phase for:
   a. $HCHO + H_2O \longrightarrow H_2C(OH)_2$
   b. $CO_2 + H_2O \longrightarrow H_2CO_3$
   What information is needed to compare the reactions in water?

5.7 Use Tables 5.6 and 5.7 to calculate $\Delta H°$ for these reactions (all gases):
   a. $CH_3OH + HCl \longrightarrow CH_3Cl + H_2O$
   b. $(CH_3)_2CHOH + HCl \longrightarrow (CH_3)_2CHCl + H_2O$
   c. $(CH_3)_3COH + HCl \longrightarrow (CH_3)_3CCl + H_2O$

5.8 Show the electronic distribution, including any formal charge, for each of the following molecules.
   a. CH_3ONO_2 c. CH_3OH e. CH_3CNO g. $CH_2=C=O$
   b. CH_3CN d. CH_3NCO f. CH_3NO_2

5.9 If any of the following compounds have a dipole, show the positive and negative ends.
   a. CH_3Br c. Br_2 e. CH_3OH g. ICl
   b. HCl d. NH_3 f. NBr_3 h. H_2O

5.11 Bibliography

Most of the recent textbooks for freshman chemistry have more extensive treatments of atomic structure than we have given in this chapter, and some are very good about molecular structure. Treatments of bond energies and bond dissociation energies are available along with atomic and molecular structure in all the recent organic chemistry texts. We give some references that are somewhat specialized but are useful for those with questions that cannot be answered with the information given in this chapter.

1. Benson, Sidney W., *Atoms, Molecules, and Chemical Reactions: Chemistry from a Molecular Point of View.* Reading, Mass. Addison-Wesley, 1970. A paperback aimed at those who have had a year of general chemistry.

2. Ferguson, Lloyd N., *The Modern Structural Theory of Organic Chemistry*. Englewood Cliffs, N.J.: Prentice-Hall, 1963. A senior level text, packed with data and references.

3. Royer, Donald J., *Bonding Theory*. New York: McGraw-Hill, 1968. A junior-senior level text, mathematical but readable; problems and bibliography. Excellent on MO and VB theories.

4. Wahl, Arnold C., and Maria T. Wahl, *Atomic and Molecular Structure: 4 Wall Charts*. New York: McGraw-Hill, 1970. A set of computer-calculated electron-density contours drawn to scale for the various orbitals; the total picture for atoms, H to Ne; some selected covalent and ionic diatomic molecules.

6 Stereochemistry

A logical extension of our short excursion into molecular structure in the previous chapter is to expand our knowledge of the geometry or architecture of organic molecules. In this chapter, we shall look at the consequences of symmetry or lack of it in molecules. As usual, energy considerations determine the spatial arrangement of the constituent atoms. The result is a compromise of contending forces within the molecule. We shall explore free rotation, touch on group theory, and look into optical activity. In later chapters we shall apply the knowledge gained here to the chemistry of various natural products.

6.1 Stereochemistry, free rotation, nonbonded interactions   119
   1. Stereochemistry defined.   2. Representation of three dimensional structures.
   3. Conformations, conformers, rotamers.   4. Nonbonded repulsion.
   5. Internuclear distances. Exercises.

6.2 Rigidity, isomerism about the double bond   123
   1. Nonrotation about double bonds.   2. Use of cis and trans.   3. Relative stability of alkenes.   4. Geometry of alkynes.   5. Visual system. Exercises.

6.3 Symmetry and chirality   126
   1. Elements of symmetry.   2. Dissymmetric and asymmetric molecules.
   3. Naming: the (RS) rules.   4. Configuration and conformation.
   5. Substance and molecular conformations.   6. Enantiomeric and meso structures.   7. Conformational stability.   8. Helices. Exercises.

6.4 Optical activity   133
   1. Optical activity.   2. Measurement of rotation.   3. Configuration and optical activity.   4. Optical isomerism; racemic mixtures. Exercises.

6.5 Resolution of racemates   137
   1. Resolution: the four steps.   2. Resolution of an acid.   3. Resolution of a base.   4. The age of modern man.

6.6 Summary   139

6.7   Problems   140

6.8   Bibliography   141

6.1   Stereochemistry, free rotation, nonbonded interactions

Stereochemistry defined.   Stereochemistry is the study of molecular shape, size, and symmetry. In dynamic stereochemistry we would study the effect of such considerations upon reaction; static or structural stereochemistry, which we shall discuss here, is concerned with stereoisomerism. In a sense, stereochemistry is the solid geometry of molecules. To master it, we shall need to add more new words to our vocabulary and study some of the conventions used for reducing three-dimensional structures to two-dimensional pictures.

Representation of three-dimensional structures.   Representation of three-dimensional structures on paper is difficult. We show four methods (Fig. 6.1). We shall use the saw-horse and pedestal-flagpole structures more than the others. A set of models will be very helpful to you in visualizing the structures represented. Model sets are available in many scales of sophistication. In an emergency, gumdrops and toothpicks provide satisfactory models.

Conformations, conformers, rotamers.   Methane is a regular tetrahedron. (See Section 2.2.) In ethane, because the C—C σ bond is cylindrically symmetrical, the carbon atoms are free to rotate in relation to each other. Returning to Figure 6.1, we note that of the infinite number of possibilities in the rotation of one methyl group in relation to the other, two are identified and named. In the staggered conformation, a hydrogen atom on one carbon is as far as possible from the nearest two hydrogen atoms on the neighboring carbon. All six hydrogens are similarly located because of the geometry of the system. In the eclipsed conformation, a hydrogen on one carbon atom is brought as close as possible to one of the neighboring carbon's hydrogen atoms, giving rise to three such pairs. Starting with a molecule of ethane in a staggered conformation and rotating one carbon atom in relation to the other, we can see that a rotation of 60° in either direction will bring the molecule into an eclipsed conformation. Continuing the rotation another 60° will bring the molecule back to a staggered conformation but with different atoms of hydrogen as nearest partners. Another 60° rotation, and the conformation again is eclipsed. Any conformation other than eclipsed or staggered is called skewed. The fact that these different conformations are produced by simple rotation leads to the use of the terms rotamers or conformers for the more stable but easily interconvertible structures.

Nonbonded repulsion.   Nonbonded repulsion is repulsion between filled orbitals of neighboring atoms. It decreases rapidly with distance. Such repulsions are countered to some extent by the attractive van der Waals' forces. A complete explanation of the effects has yet to be made. In Figure 6.2 the changes in energy content (potential energy) in ethane

120   Chapter six

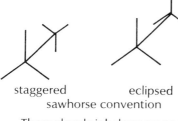

staggered   eclipsed
     sawhorse convention

The molecule is below you and usually to your left, although it is necessary at times to show the molecule as below and to your right. The C—C bond stretches away from you.

staggered   eclipsed
     dotted-line wedge

The molecule is level with and in front of the eye with the C—C bond perpendicular and horizontal to the line of sight. The dotted bonds are behind the paper and the wedges are in front of the paper.

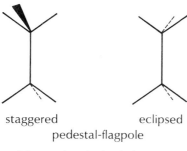

staggered   eclipsed
     pedestal-flagpole

The molecule is sitting on a three footed base. The C—C bond is vertical and in the plane of the paper, the wedge in front, the dotted bond behind. The almost horizontal bonds project slightly above or below the paper depending on whether the third bond is dotted or wedged.

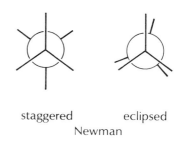

staggered   eclipsed
         Newman

The C—C bond is along the line of sight, the front carbon atom has full length bonds, and the rear carbon atom is shown as lying behind the σ bond (the circle), with the bonds to it partially hidden.

Figure 6.1 Conventions for drawing three-dimensional structures: ethane (CH_3CH_3)

and in butane for a 360° rotation are shown starting from a stable staggered conformation. (In butane the rotation is about the C_2—C_3 bond.) The energy barriers for a shift from one conformation to another are not great, and all conformers exist with an equilibrium distribution at a given temperature. The rotamers with the lower energies tend to dominate the population distribution.

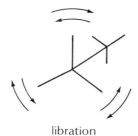

libration

6.1 Stereochemistry, free rotation, nonbonded interactions   121

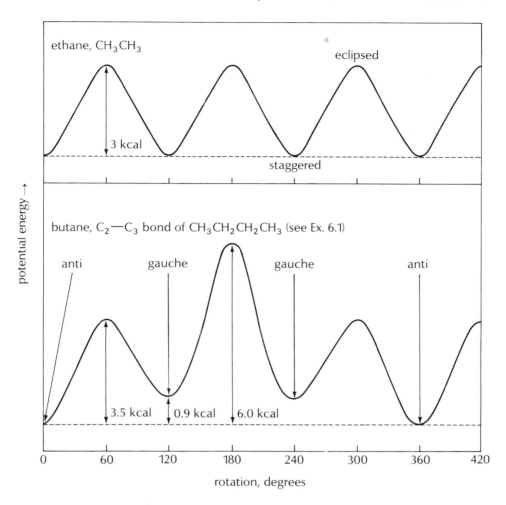

Figure 6.2 Energies of rotation

It is clear that <u>free rotation</u> is not really free in the sense that the groups held by a σ bond are spinning in relation to each other. The situation is more like that in a three-position rotary electric switch in which the handle tends to stay in a given position but may be turned easily to the next position. Actually, a conformer experiences <u>libration,</u> the torsional oscillation back and forth about the relatively stable staggered resting position. As collisions occur and energy is transferred between molecules (at a subreaction level), the librations increase or decrease and frequently become violent enough to cause a shift to an adjacent staggered conformation (or even to an actual spin).

Internuclear distances.   The <u>van der Waals' radii</u> of atoms are approximately 0.8 to 0.9 Å larger than the single-bond covalent radius. Keeping the van der Waals' radii in mind, let us examine Figure 6.3, which shows some internuclear distances calculated by the use of geometry, the tetrahedral angle of 109°28' and bond distances of 1.54 Å for C—C and 1.10 Å for C—H.

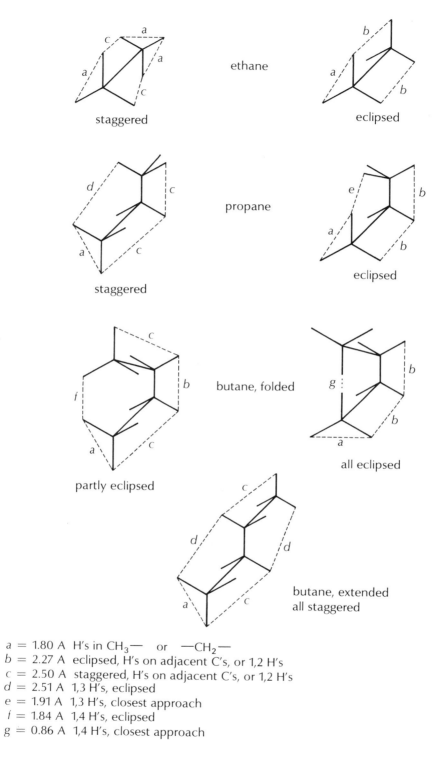

a = 1.80 Å H's in CH_3- or $-CH_2-$
b = 2.27 Å eclipsed, H's on adjacent C's, or 1,2 H's
c = 2.50 Å staggered, H's on adjacent C's, or 1,2 H's
d = 2.51 Å 1,3 H's, eclipsed
e = 1.91 Å 1,3 H's, closest approach
f = 1.84 Å 1,4 H's, eclipsed
g = 0.86 Å 1,4 H's, closest approach

Figure 6.3 Some hydrogen-hydrogen interatomic distances

The first striking thing we may notice is that hydrogens attached to the same carbon are only 1.80 A apart (a) whereas the hydrogen van der Waals' radius is 1.2 A, which gives 2.4 A as the distance of approach at collision. In other words, the carbon-hydrogen σ bond overcomes some of the interatomic repulsion between hydrogen atoms if they are attached to the same carbon atom. The stabilization achieved by C—H bond formation pulls the hydrogen nucleus in to the 1.10 A bond distance and overpowers the weaker repulsive forces between hydrogens.

In the staggered conformation of ethane, the hydrogen atoms on the adjacent carbon atoms are about 2.5 A apart (c), just outside the estimated van der Waals' distance of 2.4 A. There are six such interactions. In the eclipsed conformation of ethane, the eclipsed hydrogen atoms are 2.27 A apart (b), but there are only three such interactions. This difference explains, in part, why the energy barrier between the staggered conformers is only 3 kcal. In propane, only two of the conformations are shown, one in which the hydrogen atoms on adjacent carbon atoms are all staggered, and one in which they are all eclipsed. In the former, two of the hydrogen atoms on one end carbon atom may be considered to be in a 1,3 (for carbons 1 and 3) eclipsed position with their opposites on the other end carbon, with a distance of 2.51 A (d), surprisingly close to c. In the other conformation, along with the eclipsing (b) of adjacent hydrogen atoms, two of the end hydrogen atoms are forced close to one another (1.91 A, e). We should not speak of such a conformation as a conformer, because it corresponds to an unstable transition from one conformer to another. The two unstable folded forms of butane illustrate the closeness of approach of the hydrogens on the C_1 and C_4 carbons, 1.84 A (f) for a 1,4 eclipse and 0.86 A (g) at closest approach. These interactions show why there is a high energy barrier of 5.1 kcal for C_2—C_3 bond rotation when a *gauche* conformer goes to the other gauche conformer by forcing the methyl groups past each other (Figure 6.2). The extended *anti* conformer, the most stable, of butane is also shown. There are four 1,3 eclipsed interactions (d) in the molecule. Remembering the small energy needed to deform a bond angle by 1° (Section 5.4), it is no surprise to find that the C—C—C bond angle in butane is 112.4°, 3° greater than the tetrahedral angle.

Exercises

6.1 Draw structural formulas of the *anti* and *gauche* (pronounced gōsh) rotamers of butane and the conformations of the high energy forms between *anti* and *gauche*, and between *gauche* and *gauche*. Define *anti* and *gauche*.

6.2 Rigidity, isomerism about the double bond

Nonrotation about double bonds. Rotation about a π bond is not possible unless the π bond is broken. For C=C → Ċ—Ċ, the enthalpy change is 146 − 83 = 63 kcal (from Table 5.6). Once the π bond has been broken, rotation about the remaining sigma bond is possible. If a double bond is present in a molecule, two sp^2 hybrid atoms are present, and an sp^2 hybrid has three orbitals lying in a plane with 120° bond angles. The π bond lies above and below the plane of the six atoms involved in the sp^2 bonds. In Figure 6.4 is shown the spatial geometry of ethylene, propylene, 1-butene, and the isomeric 2-butenes. Ethylene, propylene, trimethylethylene, and tetramethylethylene exist in one form only. In 2-butene, on the other hand, we see that the methyl groups (or the hydrogens attached to

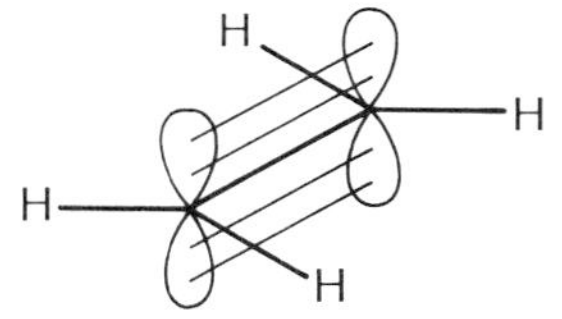

ethylene, $H_2C=CH_2$

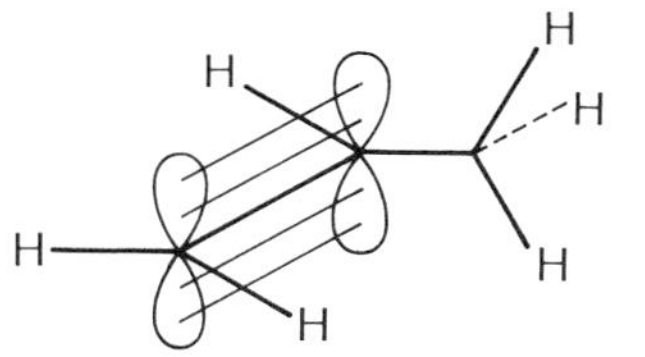

propylene, $H_2C=CHCH_3$

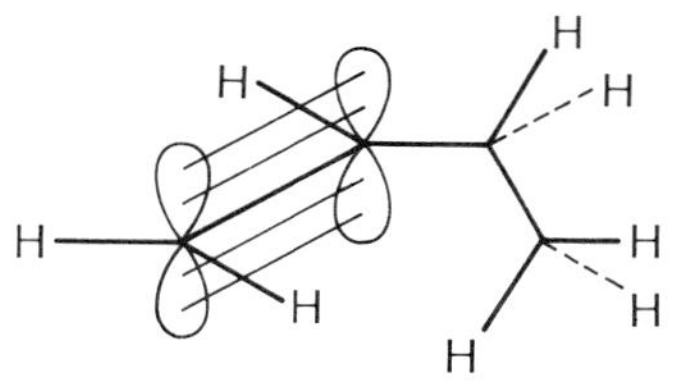

1-butene, $H_2C=CHCH_2CH_3$

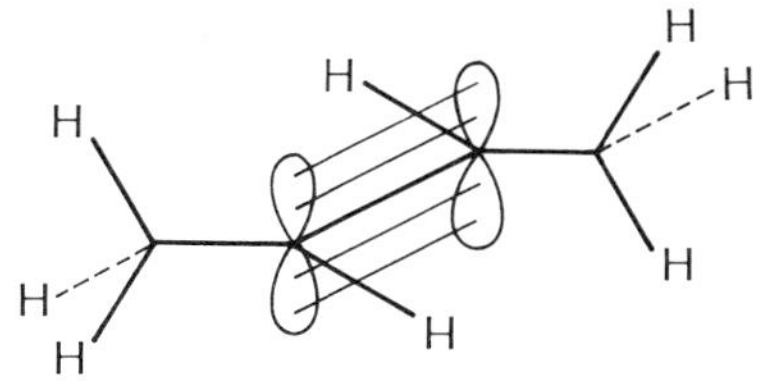

trans-2-butene,

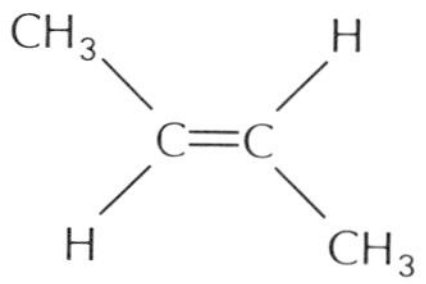

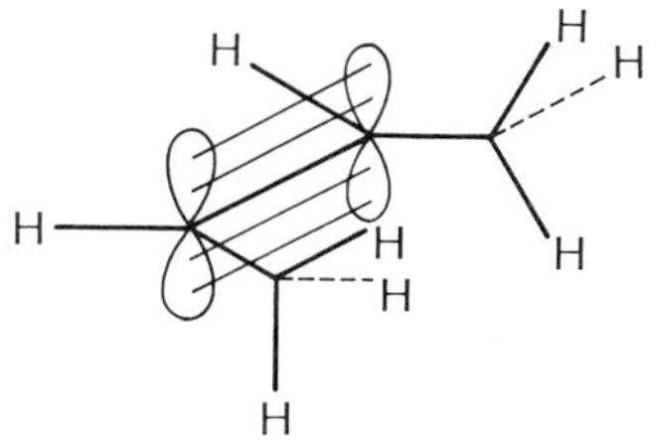

$$\begin{array}{c} CH_3 \\ \end{array} C=C \begin{array}{c} H \\ CH_3 \end{array}$$

cis-2-butene,

$$\begin{array}{c} H \\ CH_3 \end{array} C=C \begin{array}{c} H \\ CH_3 \end{array}$$

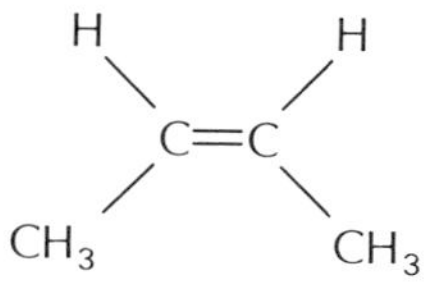

the unsaturated carbon atoms), which lie in the plane of all the sp^2 orbitals, are either on the same side or on opposite sides of the double bond. Groups on the same side of the double bond are known as *cis*; on the opposite sides, as *trans*. Rotation about the double bond does not occur, and *cis*- and *trans*-2-butene exist as distinct substances with different physical and chemical properties. (See Table 4.2.) The substances are rigid and isomeric. This type of isomerism about a double bond is called geometrical isomerism.

Geometrical isomerism is but one type of isomerism of the more general class called stereoisomerism, which includes all forms of isomerism caused by differing arrangements in space of otherwise identical structures.

Different arrangements of the same structure are called configurations and conformations. In general, configuration refers to isolable, or stable, arrangements, and conformation to nonisolable, or unstable, arrangements. Conformations are usually in mobile equilibrium with each other, but configurations are not. A given configuration may exist in many conformations. For example, *cis*-2-butene has many possible conformations caused by rotation of the methyl groups about the σ bonds C_1-C_2 and C_3-C_4. However, any of the conformations is still in a *cis* configuration about the $C_2=C_3$ double bond. Some instances are not clear cut, but the terminology remains useful. Here, it is clear that 2-butene has only the two configurations *cis* and *trans*.

Use of cis and trans. Geometrical isomerism is not possible if two identical atoms or groups are attached to either carbon of the double bond, as in $H_2C=$, $(CH_3)_2C=$, $Br_2C=$, and so on. Therefore, different configurations become possible only if each carbon atom is attached to two different groups. The two different groups may be the same as, or different from, the unlike groups attached to the other end of the double bond: for example, in the 2-butenes and in $CH_3CBr=CHCH_2CH_3$. The prefixes *cis* and *trans* used in the name of a molecule refer to the configuration of the main chain as it passes through the double bond. (Refer to the isomeric 2-butenes in Figure 6.4.)

Also, *cis* and *trans* are used to refer to configuration about a ring. The ring is assumed to be planar, and the substituent groups are called *cis* if they are on the same side and *trans* if they are on opposite sides of the plane. Here we show *cis* and *trans*-2-methyl-1-cyclopentanol. More complicated molecules require different naming rules, the *EZ* system, described in Section 14.1.

Relative stability of alkenes. The *cis* configuration is usually less stable than the *trans*. If we refer back to Figure 6.3, *cis*-2-butene has a configuration forced upon it by the double bond so that it resembles the folded forms of butane, with *f* and *g* interactions that are absent in *trans*-2-butene (Fig. 6.2). Also, the double bond is shorter than the C_2—C_3 bond in butane, so the interaction distances are even shorter. Therefore the *cis* isomer has a higher energy content than the *trans* and should release more energy than the *trans* isomer in a reaction in which the crowding is released. This guess is verified by hydrogenation. It is found that $\Delta H°$ for the addition of hydrogen to the double bond is -28.6 kcal/mole for *cis*-2-butene and -27.6 kcal for *trans*-2-butene. Because the product is butane in both cases, the *cis* isomer must have been 1 kcal higher in energy content than the *trans* isomer.

Geometry of alkynes. Whereas the presence of a double bond requires six atoms to lie in a plane, the triple bond, with two *sp* hybrid carbon atoms, requires four atoms to lie in a line. As a result, it becomes impossible to detect rotation about a triple bond. Examination of CH_3—C≡C—CH_3 shows that the methyl groups are free to rotate about the σ bonds and that they may do so with less hindrance than in ethane because they are farther apart. Also, there is no possibility of different configurations, so there is no isomerism other than structural isomerism, as shown by 1-butyne and 2-butyne.

Visual system. The process of *cis-trans* isomerization is used in our visual system. Vitamin A_1 or retinol is a long chain alcohol with four double bonds attached to a substituted cyclohexene. The four double bonds all are *trans*. Retinol is converted in our bodies to a protein complex, rhodopsin, that can absorb light (absorption maximum at about 500 nm). When light energy is absorbed, a *cis* double bond overcomes its pi bond energy barrier to form the more stable *trans* double bond. (See Eq. 26.16.)

Exercises

6.2 Draw structural formulas of:
    a. *cis*-4-methyl-1-cyclohexanol
    b. *trans*-2-butene-1-ol
    c. 2,4-dimethyl-1,*cis*-3,*trans*-5-heptatriene
    d. *cis*-1,2,3,4-tetrachlorocyclobutene
    e. *trans*-crotonic acid. (Hint: see Table 4.2.)

6.3 Symmetry and chirality

Elements of symmetry. In the previous section, we examined free rotation and the lack of it, and the difference between conformation and configuration. Here we wish to take up some of the stereochemical effects of symmetry or the lack of it in a molecule.

Symmetry may be defined in several ways. Instead of trying to define symmetry, we shall show how it may be described. We shall use four elements of symmetry: 1. point of symmetry; 2. axis of symmetry; 3. interior plane of symmetry; and 4. reflection or exterior mirror symmetry.

By a point of symmetry, we mean a point (center of an object) through which any straight line may be drawn and which will intersect the surface of the object at equal distances on either side of the point. These points of intersection are the inverses of each other. A cube has a point of symmetry, but a tetrahedron does not.

An axis of symmetry may be illustrated by a line through the center of a cube and perpendicular to a pair of opposing faces. Now rotate the cube about the line as an axis. Rotation of 90° brings the cube back to coincidence (or congruence) with the original cube (angle on angle, side on side and face on face). Another 90° rotation brings about another coincidence, and a third and fourth bring the cube back to the starting point. Such a line is said to be a four-fold axis of rotation. A cube has a total of 13 axes of symmetry. (Not all are four fold.)

The third symmetry element, an internal plane of symmetry, is a mirror plane that bisects an object so that the reflected image of one side coincides with the other side of the object. Finding mirror planes in solid figures is difficult and usually requires the use of models. A cube has nine such internal mirror planes.

The fourth and final symmetry element is reflection or external-mirror symmetry, defined as congruence (superposability) of the mirror image of the complete object with the complete object itself. The external mirror image of a cube is superposable on the cube. The mirror image of a cylinder is superposable on the cylinder, but a right hand glove held up to a mirror is reflected as a left hand glove. The image is nonsuperposable: hence a glove has no external-mirror symmetry. In applying this concept to the structure of a molecule, we may place the object in any position with any conformation in relation to the mirror plane, as shown in Figure 6.5, for ethane and ethanol. To test the image for congruence with the object, we may turn and transport the image as needed. The result must be yes or no: either there is external-mirror symmetry and congruence, or there is none and no superposability.

If there is no external-mirror symmetry for a given molecule, chirality or dissymmetry exists, and the molecular structure examined will exhibit stereoisomerism. If, also, the object and the image are capable of independent existence, they are known as enantiomers. Enantiomers are nonsuperposable mirror images of each other.

Examination of a number of cases shows that if an object does not have the first three types of elements of symmetry, it has chirality, or no external-mirror symmetry. It also turns out that rotameric changes usually are without effect. Such rotamers are not isolable, are readily interconverted, and do not exhibit stereoisomerism. If possible, a relatively stable staggered conformer is used in the test for nonsuperposability. (However, see the discussion of 2,3-butanediol later in this section.)

Dissymetric and asymmetric molecules. Absence of a point of symmetry and of an internal plane of symmetry in a structure is enough to cause a lack of superposability on the mirror image even if one or more axes of symmetry are present. See

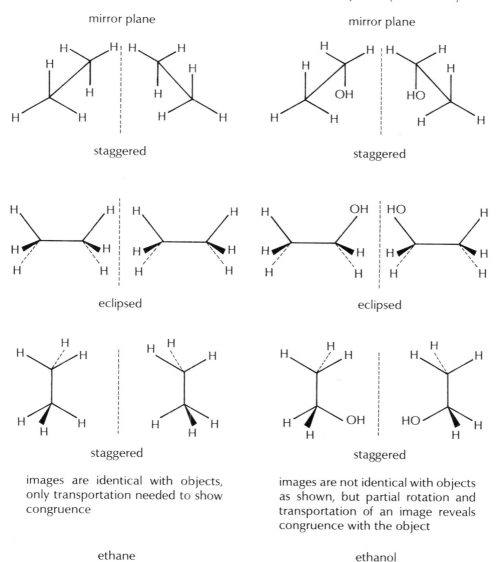

Figure 6.5 Mirror symmetry: ethane and ethanol

CH$_3$CH=C=CHCH$_3$ in Table 6.1, which gives the various symmetry elements of some molecular structures. Let us repeat that molecules that have no external-mirror symmetry exhibit chirality and are described as being dissymmetric or chiral. If, in addition, there is no axis of symmetry in such a chiral molecule, the term <u>asymmetric</u> is used to describe the lack of any element of symmetry.

If we examine the four asymmetric compounds listed in Table 6.1, we see that the common characteristic is the existence of a carbon atom with four different groups attached to it. It is common practice to designate such an atom with an asterisk to call attention to the lack of symmetry, as in CH$_3$CH$_2\overset{*}{\text{C}}$HBrCH$_3$ and CH$_3\overset{*}{\text{C}}$HClCOOH. Such a

Table 6.1 Symmetry elements in various structures

Structure	Point of symmetry present	Number of axes of symmetry present	Number of interior planes of symmetry present	External mirror symmetry present
C_4H_8, cyclobutane	yes	5	5	yes
CCl_4	no	7	6	yes
H_2, Cl_2, CO_2, $HC\equiv CH$	yes	∞	∞	yes
CO_3^{2-}, NO_3^{-}	yes	4	4	yes
$CH_3C\equiv CH$, $CH_3C\equiv N$	no	1	3	yes
$CH_3CH_2CH=CH_2$	no	0	1	yes
$CH_3CH=CHCH_3$, cis	no	1	2	yes
$CH_3CH=CHCH_3$, trans	yes	1	1	yes
CH_4	no	7	6	yes
CH_3Br	no	1	3	yes
CH_2Cl_2	no	1	2	yes
CH_2BrCl	no	0	1	yes
CH_3CH_2Br	no	0	1	yes
$(CH_3)_2CHBr$	no	0	1	yes
$\overset{*}{C}HBrClF^a$	no	0	0	no
$CH_3\overset{*}{C}HBrCl^a$	no	0	0	no
$CH_3CH_2\overset{*}{C}HBrCH_3{}^a$	no	0	0	no
$CH_3\overset{*}{C}HClCOOH^a$	no	0	0	no
$H_2C=C=CH_2$	no	2	0	yes
$CH_3CH=C=CHCH_3$	no	1	0	no

a These molecules are asymmetric.

carbon atom is called an <u>asymmetric carbon atom</u>. Certain other conditions within a molecule may give rise to chirality without the presence of an asymmetric carbon atom. For example, 2,3-pentadiene has a single axis of symmetry even though the mirror images are nonsuperposable.

Naming: the (RS) rules. In Figure 6.6, the enantiomers of 2-butanol ($CH_3CH_2\overset{*}{C}HOHCH_3$) are shown and are labeled as (R) and (S). These designations of configuration are the result of the use of a set of rules devised by Cahn, Ingold, and Prelog and adopted by IUPAC. There are two steps.

The first step is to assign a set of priorities to the four different groups and atoms attached to the asymmetric carbon.

The second step is to think of the molecule as a flower with three petals, the petals represented by the bonds from the asymmetric carbon to the groups with priorities 1, 2, and 3. The stem is the bond to the lowest priority group, or 4. Hold the flower by the stem

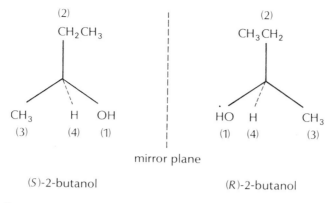

Figure 6.6 Asymmetry of 2-butanol

and look at the three petals. If the priority sequence 1, 2, 3 is clockwise, the designation is (R) (from *rectus*, right). If the sequence 1, 2, 3 is counterclockwise, the designation is (S) (from *sinister*, left).

The priorities are assigned by a set of sequence rules.

1. Only those atoms directly bound to the asymmetric carbon are considered under rule 1. First priority goes to the atom with the highest atomic number. If isotopes are involved, the heavier isotope takes precedence. Second place goes to the atom with the next highest atomic number, and so on. (For the commonly encountered atoms, this gives the sequence, I, Br, Cl, S, P, F, O, N, C, H).

2. If rule 1 gives rise to a tie, consider the atoms bonded to those that were tied. The atom substituted with other atoms of higher atomic number takes precedence. If they are still tied, the one with more atoms of higher number wins. If still tied, go on to the third atom and consider the atoms joined to it.

    For example, in 2-butanol, as shown in Figure 6.6, O of HO is first, H is fourth, C of CH_3CH_2— is second because it is attached to a C and two H's, and C of CH_3— is third because it is attached to three H's. A rather extreme example of priority assignment would be $(CH_3)_2CHCH_2CH_2CH_2$—, which takes precedence over $CH_3CH_2CH_2CH_2CH_2$— because the fourth C in the chain of the former is attached to two C's and an H, whereas in the latter the fourth C is attached to a C and two H's.

3. Other tie-breaking decisions are as follows:

    a. Consider —CH=CH_2 to be —CH$\begin{smallmatrix}CH_2\\ \diagdown\\ C\end{smallmatrix}$ —C≡CCH_3 to be —C$\begin{smallmatrix}C\\ \diagdown\\ \diagup\\ C\end{smallmatrix}$$\begin{smallmatrix}\diagup\\ CCH_3\\ \diagdown\end{smallmatrix}$

    —$COCH_3$ to be —C$\begin{smallmatrix}CH_3\\ \diagup\\ —O\\ \diagdown\\ O \quad C\end{smallmatrix}$. In other words, a multiple bond is imagined as having a multiple of the atom at the other end of the bond. However, a true multiple, as in rule 2, takes precedence over the imaginary multiple from a multiple bond.

    b. Other things being equal, a *cis* configuration takes precedence over a *trans*.

130   *Chapter six*

c. Other things being equal, an (R) configuration takes precedence over an (S). (For the frequently occurring carbon-oxygen double bond the sequence is —COCl, —COOCH$_3$, —COOH, —CON(CH$_3$)$_2$, —CONH$_2$, —COCH$_3$, —CHO.)

Configuration and conformation.   If chirality is caused by the presence of an asymmetric carbon atom, the exchange in location of *any two* groups attached to an asymmetric carbon atom changes the configuration from (R) to (S) or from (S) to (R). For example, try exchanging CH$_3$— and HO— in 2-butanol in Figure 6.6 (or any other pair of groups).

There may be more than one asymmetric carbon atom in a molecule such as in 2,3-butanediol, CH$_3$C̊HOHC̊HOHCH$_3$, which has two. In Figure 6.7 are shown four configurations (ABCD) of 2,3-butanediol in the unstable eclipsed conformations, and the

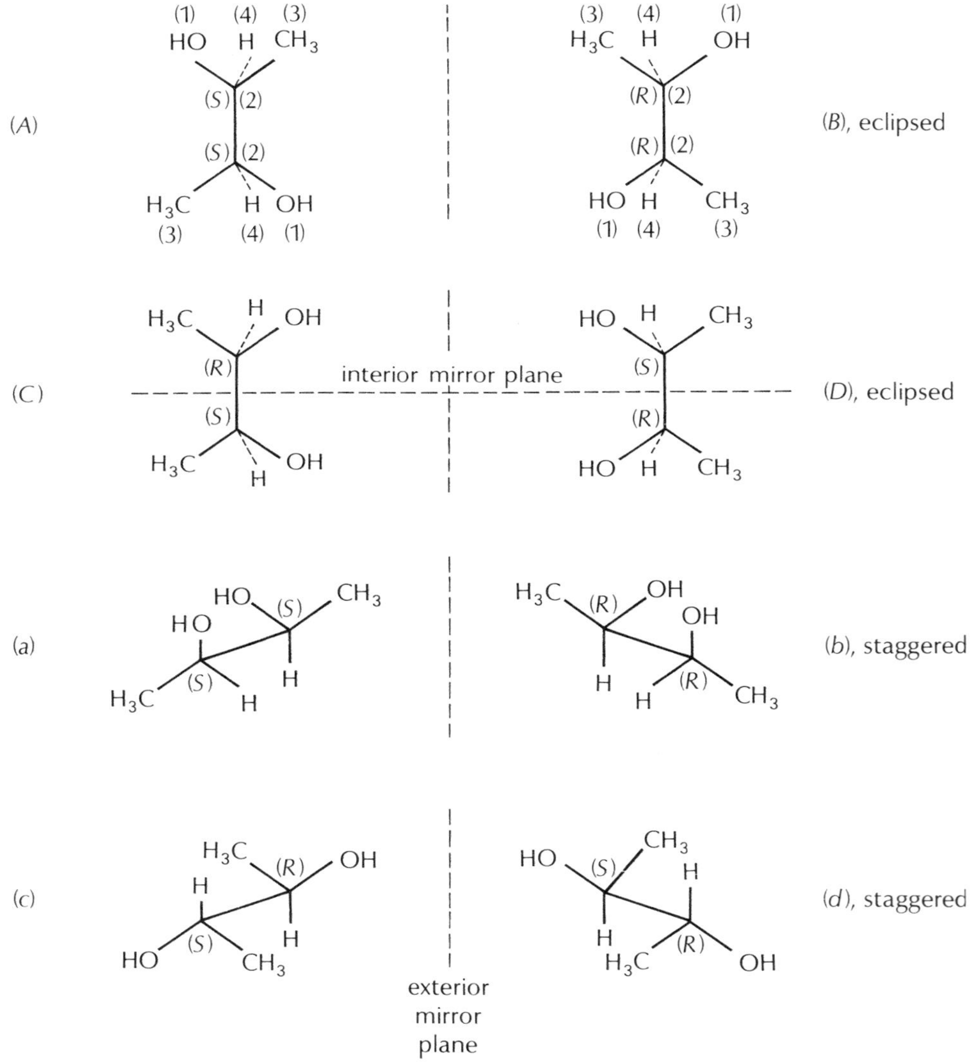

Figure 6.7   Asymmetry in 2,3-butanediol

same four configurations (*abcd*) as stable staggered conformers about the C_2—C_3 bond. (*ABCD*) are superior for deducing configurations; (*abcd*) are better for showing conformational relationships.

In (*A*) and (*B*), the sequence numbers are shown to illustrate the ease of assignment of configuration. The configurational assignment, (*R*) or (*S*), for the C_2 and C_3 atoms is shown in all eight pictures. (*A*) and (*B*) form a pair of enantiomers, as do (*a*) and (*b*). (*C*) and (*D*), as well as (*c*) and (*d*), are not enantiomeric. They are identical (superposable). In crossing a mirror plane, (*R*) becomes (*S*) and vice versa.

Substance and molecular conformations. A point of some delicacy arises when we consider the relationship of (*A*) to (*b*) and of (*a*) to (*B*). Because (*A*) and (*a*) represent the same substance, one might argue, and (*b*) is the enantiomer of (*a*), then (*A*) and (*b*) also are enantiomers. On the other hand, (*A*) and (*b*) definitely are not mirror images of each other. What is the discrepancy?

The discrepancy lies in confusing a substance with a molecule. A sample of 2(*S*),3(*S*)-2,3-butanediol would contain a tremendous number of molecules (the tiny 0.00009 mg = 10^{-9} mole = 6×10^{14} molecules, for example). Thus, an equilibrium distribution of all possible conformations would exist in the sample, of which (*A*) and (*a*) are only two. A like-sized sample of 2(*R*),3(*R*)-2,3-butanediol would have a similar distribution, of which (*B*) and (*b*) are only two possible conformations. For samples of equal size, the number of molecules of a particular conformation in the (*S*),(*S*)-diol is equal to the number of molecules of the mirror image conformation in the (*R*),(*R*)-diol. Thus, the samples are enantiomeric, even though one hypothetically could pick out one molecule from one sample and show that it was not the mirror image of a selected molecule from the other sample.

Consequently, we often speak of (*A*) and (*b*) as being enantiomeric, the statement carrying the unspoken assumption of a sample rather than of single molecules that (*A*) and (*b*) represent. Also, since (*A*) and (*a*) are rotameric forms of molecules that are easily converted to an enantiomeric relationship, (*a*) and (*b*), the relation between (*A*) and (*b*) may be thought of as a mere temporary change from the enantiomeric relationship.

Enantiomeric and meso structures. Inspection of (*C*) and (*D*), or (*c*) and (*d*), shows mirror image relationships (any structure has a mirror image), but there is no enantiomeric relationship because (*C*) and (*D*) are superposable (rotate one or the other 180° in the plane of the paper), as are (*c*) and (*d*) (tumble one or the other end over end). Therefore (*C*) and (*D*) are identical, as are (*c*) and (*d*). We notice again that crossing the mirror plane inverts (*R*) to (*S*) or (*S*) to (*R*). Also, (*C*) and (*D*) both have interior mirror planes

Such structures, with opposite configurations of the equivalent asymmetric carbon atoms in one molecule, are called <u>meso</u> compounds. (If one of the methyl groups were changed to an ethyl, as in 2,3-pentanediol, no interior mirror plane would exist, the equivalency would be lost, and C and D would become nonsuperposable enantiomers of each other.)

Thus, it is possible to have samples of 2,3-butanediol existing in one of three forms, the (*R*),(*R*), the (*S*),(*S*), and the (*R*),(*S*) configurations. An (*R*),(*S*) structure with an interior mirror plane is known as a <u>meso</u> configuration. The relationship of the *meso* form to either of the other two configurations is that of a <u>diastereoisomer</u> (also called by the shorter names <u>diastereomer</u> or <u>diamer</u>). Diastereomers have the same structure but have a nonenantiomeric relationship to each other.

Conformational stability. The question often arises as to which of the conformations is the most stable in a molecule like 2,3-butanediol. A convenient method for answering the question is to assign relative sizes to the three groups attached to each of the neighboring carbon atoms. This task is easily done with *s* (for small, obviously H), *m* (for medium, here OH), and *l* (for large, here CH$_3$). Now nestle *l* on one carbon between *s* and *m* on the other carbon, in a staggered conformation. This arrangement automatically places the other *l* in the same relationship. Furthermore, *m* on one C is staggered between *l* and *m* on the other C, and *s* is between *s* and *l* in the enantiomers (a) and (b). In the *meso* configuration, each *l* rests between *m* and *s* on the neighboring carbon atom, *m* between *l* and *s* on the neighboring carbon atom, and *s* between *l* and *m*. It follows that the most stable conformation is that in which the *l* groups are *anti* to one another. However, other nonbonded interactions may influence the result. Let us consider 2,3-dimethylbutane in the conformation in which the *s* groups are *anti*. The conformation is the most stable because there is no methyl group (*m* and *l*) between methyl groups on the adjacent carbon atom.

Looking at the conformations of (c) and (d) again, we see that if the *l* groups are *anti*, so are the *m* and *s* groups. This occurrence is another characteristic difference between an (R),(R) or (S),(S) configuration and a *meso* configuration in which the asymmetric carbon atoms are adjacent to each other. In other words, a *meso* structure may have all groups *anti*; whereas an (R),(R) or (S),(S) structure can have only one pair of groups *anti* at one time in a staggered conformation. Also, in the all-*anti* conformation of a *meso* configuration, the interactions are *s-ml*, *m-sl*, and *l-sm*, the best possible conformation for minimizing the repulsions between neighboring groups. As a result, *meso* structures should be and are a bit more stable than an (R),(R) or (S),(S) configuration.

A *meso* configuration is also possible with the asymmetric carbon atoms separated from each other by more than one bond, as in CH$_3\overset{*}{\text{C}}$HOHCH$_2\overset{*}{\text{C}}$HOHCH$_3$ or CH$_3\overset{*}{\text{C}}$HOHCH$_2$CH$_2\overset{*}{\text{C}}$HOHCH$_3$.

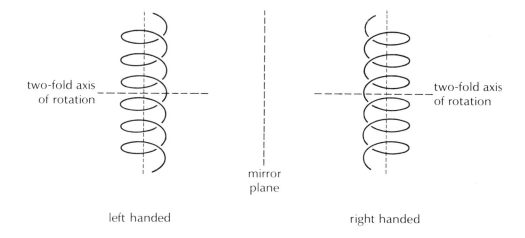

Figure 6.8 *Enantiomeric helices*

Helices. The helix (spiral or coil) may at first glance seem to be a special case of stereoisomerism, but it is really rather common. Some helices may be rotated end over end (Fig. 6.8) and be superposed upon the original. By convention, a right handed helix is defined as the one that when viewed from either end recedes from the eye by following a clockwise rotating path.

Helical conformations are widespread in nature. Many proteins (Chap. 21) take on a structure known as an α-helix. DNA forms a double spiral, (Chap. 28), the winding and unwinding of which is essential to protein synthesis.

Exercises

6.3 Use appropriate structures to draw a stable conformer of each substance given:
 a. (R)-lactic acid
 b. (R)-malic acid
 c. *meso*-tartaric acid
 d. (1R,3R)-cyclopentanediol
 e. (2R,3S,4S)-pentanetriol

6.4 Optical activity

Optical activity. Enantiomers (and diastereomers) are sometimes called optical isomers because the only trustworthy observable physical difference between a pair of enantiomers is the rotation of a plane of polarized light.

We think of light as that portion of the electromagnetic spectrum of radiant energy that our eyes can perceive. More broadly, light is a form of energy travelling at about $c = 3 \times 10^{10}$ cm/sec, characterized by waves and photons. There is no limitation on the frequency, v, and wavelength, λ, other than $c = \lambda v$. A beam or ray of light is a transporter of energy, the amount being proportional to the frequency ($E = hv$) and the intensity (amplitude). Light from most sources vibrates in the infinite number of planes about the direction of propagation. If the light vibrating in all the planes but one is removed, the remaining uniplanar light is called plane polarized. □ □

The index of refraction, n, of a gas, liquid, or solid is the ratio of the velocity of light in a vacuum to the velocity in the medium. The index decreases with increasing wavelength and temperature and varies somewhat depending upon the density and composition of the medium, but is always greater than unity. The D line (589.3 nm) of sodium is the usual convenient monochromatic light used, and the temperature is specified as a superscript, thus: H_2O, n_D^{25} 1.3325; CS_2, n_D^{25} 1.6214. Gases have values close to unity; most liquids and solutions, 1.3 to 1.6; crystals and glasses, 1.4 to 1.9.

Plane polarized light is the resultant of two circularly polarized electric field vectors of the same frequency and phase. Each of the vectors traces a helical path: one clockwise, the other counterclockwise. They are designated as right or (+), and left or (−).

If either the (+) or the (−) component be separated from a ray of plane polarized monochromatic light (which can be done, to give a ray of right or left circularly polarized light), we find that the index of refraction measured for a sample of one liquid enantiomer or a solution of one enantiomer of a pair differs depending upon which light component is used for the measurement. In other words, a collection of molecules of a given chirality will slow down one of the spiral components more than the other. If plane polarized light is passed through such a medium, the emerging light will appear to have turned away from its original plane toward that circular component with the greater velocity (smaller n_D).

Such a rotation of the plane of plane polarized light is called optical activity and is a physical property of the medium causing the rotation. The angle of such a rotation is symbolized as α.

The circular components of plane polarized light also are absorbed to different extents by dissymetric molecules, and the effect is called circular dichroism.

Measurement of rotation.   Optical activity is relatively easy to measure. An instrument called a polarimeter is used. In the polarimeter, polarized light from a sodium

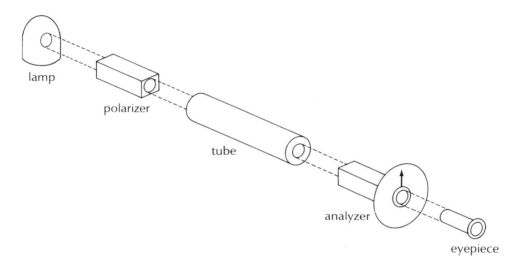

Figure 6.9   Schematic diagram of a polarimeter

lamp passes through the sample whose rotation is to be measured. An analyzer is used to find the new plane in which the polarized light emerges, and a dial shows through how many degrees the plane of the light was rotated. This result might be expressed, for example, as α_D^{25} 10.0° where D stands for the wavelength of light used, 25 denotes the temperature, and the plane of the light was rotated through an angle of 10.0°. □ □

The direction of rotation has been symbolized in various ways. Thus, an observed clockwise rotation has been designated as (+), d, or *d*. We shall use only (+) for clockwise and (−) for counterclockwise rotation. The direction of rotation, (+) or (−), as observed with a polarimeter is in an opposite sense to our previous assignment of right and left handed helices. The polarimeter result is obtained with the helices of the circularly polarized components spiraling toward the observer.

The observed rotation is proportional to the number of chiral molecules in the light path. Thus, the magnitude of the observed rotation is proportional to the concentration, the path length, and the inherent ability of the optically active molecules to cause rotation. The proportionality constant [α] is known as the specific rotation, and $\alpha = [\alpha]cl$. The path length (*l*) is customarily expressed in decimeter (10 cm) units, and the concentration (c) as g/100 ml of solution or as the density (g/ml) if the substance is liquid (Eq. 6.1). Occasionally,

$$[\alpha]_D^{25} = \frac{100\alpha}{cl} \text{ for solution}; \frac{\alpha}{cd} \text{ for liquid} \qquad (6.1)$$

we see a reference to molecular rotation. Molecular rotation is defined as the specific rotation multiplied by 0.01 mw of the optically active substance.

Configuration and optical activity. From studies with crystals, it has been found that if the ray of light enters the crystal in a direction perpendicular to an interior mirror plane, the circular components of plane polarized light are slowed down or absorbed equally. If the light enters from another direction, the circular components are affected unequally, and rotation of the plane of polarized light results. If the crystal has a center of symmetry, the plane polarized light ray from any direction will traverse the crystal without any rotation being set up.

If we carry the result over to molecules, it is apparent that there are many, many ways in which a molecule in a liquid may be oriented such that none of the interior mirror planes of the molecule (if it has them) is perpendicular to the direction of travel of the ray. Thus, most of the molecular orientations imparts a rotation to the plane of the ray. But for every molecular orientation there exists a mirror image of that molecular orientation that is a real molecule in a sample of the substance if the molecules are superposable upon each other. Hence all the (+) rotations are exactly balanced by all the (−) rotations in the sample, and the polarimeter shows no rotation.

If the mirror image oriented molecules are absent from a sample (the molecules are nonsuperposable), then the plane of the ray of polarized light is turned away from the plane in which it entered the polarimeter tube.

The sign of the observed rotation has no relation to the (R),(S) assignments of configuration. Despite many efforts to correlate the two, no satisfactory general relationship has been established. Thus an (R) configuration may exhibit (+) or (−) rotation. Conversely, a (+) rotation may be obtained from either an (R) or (S) configuration, depending upon the molecular structure.

Once the configuration and the sign of rotation have been determined, both are incorporated into the name, as in (S)-(+)-lactic acid. The enantiomer is (R)-(−)-lactic acid, and a 50-50 mixture of the two is racemic lactic acid or (RS)-(±)-lactic acid. If no configuration or sign of rotation is given, we assume that the racemic mixture of enantiomers is what is meant. In other words, lactic acid refers to the racemic mixture or (RS)-(±)-lactic acid.

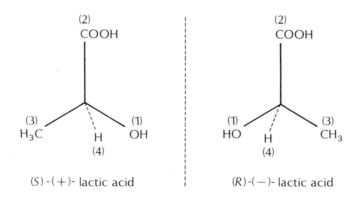

Optical isomerism; racemic mixtures. Optical activity is displayed only by those molecules that are nonsuperposable on their mirror images. As we saw in Section 6.3, this means that the structure, to be optically active, must have no external mirror plane of symmetry. Since optical activity is the only physical property that distinguishes a pair of enantiomers (one rotates (+), the other rotates (−); the specific rotations are equal and opposite), the use of the term optical isomerism to describe this branch of stereoisomerism is obvious. However, optical isomerism is somewhat of a misnomer, because some members of the class are optically inactive ($[\alpha] = 0$). A *meso* structure is an example.

Because pairs of enantiomers have identical physical properties, it is not surprising to us to learn that mixtures of equal amounts of (+) and (−) enantiomers not only are optically inactive, but also require special techniques (to be described later) to bring about a separation of the mixture. Sometimes a racemic mixture crystallizes to give a substance of different physical properties from either of the component enantiomers (differing mp, bp, solubility, and so on). Such a crystalline form is given the name of a racemate or racemic compound. All racemic forms in solution are racemic mixtures.

A pure dissymmetric compound has a rotation larger (either (+) or (−)) than any admixture with its enantiomer. Such a compound is optically pure. A racemic mixture has zero optical purity. Optical purity, expressed in percent, is given by the formula

$$\text{optical purity} = 100 \frac{(+) - (-)}{(+) + (-)} \quad \text{or} \quad 100 \frac{(-) - (+)}{(+) + (-)}$$

depending upon which enantiomer is in excess. For example, a mixture of three parts of (−) with two parts of (+) has an optical purity of 20%.

If there are n asymmetric carbon atoms in a molecule, each of which has a different set of four groups attached to it, the number of optically active isomers is 2^n. One half of 2^n are (+) forms, the other half being (−) forms, each of which is an enantiomer of a (+) form. Thus, the number of possible racemic mixtures is $\frac{1}{2}(2^n)$ or $2^{(n-1)}$. Any two optical isomers that are not enantiomeric are called diastereomers.

If a molecule contains asymmetric carbon atoms attached to identical sets of four groups (as in 2,3-butanediol, Fig. 6.7), the total number of stereoisomers is decreased because of the presence of *meso* structures. Formulas are available for calculating the number of active (enantiomers) and inactive (*meso*) structures, but the formulas are complicated and we shall not be delving that far into the subject.

In any case, the ultimate test for the presence and number of stereoisomers is the absence or presence of superposability for each structural arrangement (configuration) and its mirror image.

Exercises

6.4 Draw structures to satisfy yourself that 2,3,4-pentanetriol has two *meso* structures.

6.5 Work out the number of optically active isomers of $CH_3CH_2\overset{*}{C}HCH=CHCH_3$. (Hint: remember that the chain may be *cis* or *trans*.)
$$\underset{CH_3}{|}$$

6.6 Work out the stereoisomerism of $CH_3CH{=}CH\overset{*}{C}HOHCH{=}CHCH_3$.

6.7 Work out the stereoisomerism of 2,4-dimethyl-1-cyclobutanol.

6.5 Resolution of racemates

Resolution: the four steps. Resolution is any process whereby a racemic mixture may be separated into the component enantiomers.

Because a racemic mixture consists of a pair of enantiomers with the same physical and chemical properties (except for reactions with some other dissymetric or asymmetric compound), how is the separation to be accomplished? The method used most (first used by Pasteur in about 1848) involves the steps shown:

1. Conversion of the racemic mixture by means of a resolving agent into a mixture of diastereomers which differ in physical properties.
2. Separation of the diastereomers by their physical property differences.
3. Each diastereomer, after purification, is decomposed into a mixture of one of the enantiomers of the original racemate and the resolving agent.
4. Separation of each mixture of enantiomer and resolving agent then gives the separated enantiomer and recovered resolving agent.

The resolving agent, we see, is required to react with the enantiomeric mixture and form diastereomers that are stable enough to withstand the separation procedure (step 2) without decomposition. Nevertheless, each purified diastereomer should be easily and conveniently decomposed (step 3) and then easily and conveniently separated into enantiomer and resolving agent (step 4). The resolving agent must be rather versatile to meet so many requirements. The use of acid-base reactions is the answer.

Resolution of an acid. To illustrate the process of resolution, let us start with a racemic carboxylic acid. The conversion of the racemate into a mixture of diastereomers (step 1) is accomplished by salt formation with an optically active base. Nature provides us with a variety of optically active amines of moderately high molecular weight, such as (−)-brucine, (+)-cinchonine, (−)-morphine, (+)-quinidine, (−)-quinine, and (−)-strychnine, which are classified as alkaloids. The alkaloids are crystalline substances that, as bases, react with acids to take up a proton and form crystalline salts. Depending upon the racemic acid to be resolved, some alkaloids are more suitable than others for producing diastereomeric salts with enough difference in solubility to allow good separation by crystallization (step 2). The diastereomeric salts are not mirror images of each other and differ in solubilities, as well as in other physical properties.

After isolation of each impure diastereomic salt, recrystallization from suitable solvents is carried out the number of times necessary to produce each salt in a pure form (step 2). Each salt is then treated separately with a strong mineral acid, usually hydrochloric acid. The carboxylate ion takes up the proton and precipitates from solution as the carboxylic acid while the positive alkaloidal ion remains in solution as a chloride salt (step 3). The carboxylic acid enantiomer is collected by filtration and purified by recrystallization (step 4). The process is outlined in Equations 6.2. The alkaloid hydrochloride salt is decomposed with dilute sodium hydroxide. The free alkaloid precipitates. Filtration allows the alkaloid to be recovered for later reuse.

$$\frac{(+)-\text{RCOOH}}{(-)-\text{RCOOH}} + \frac{(-)-\text{R'NH}_2}{(-)-\text{R'NH}_2} \xrightarrow{\text{step 1}} \frac{(+)-\text{RCOO}^\ominus\ (-)-\text{R'NH}_3^\oplus}{(-)-\text{RCOO}^\ominus\ (-)-\text{R'NH}_3^\oplus} \quad (6.2)$$

racemic mixture   resolving agent   solution of mixed diastereomeric salts

step 2 ↙ recrystallization

(+)—RCOO$^\ominus$(−)—R'NH$_3^\oplus$ +   (−)—RCOO$^\ominus$(−)—R'NH$_3^\oplus$

step 3 │ HCl   step 3 │ HCl

(+)—RCOOH + (−)—R'NH$_3^\oplus$ + Cl$^\ominus$ (−)—RCOOH + (−)—R'NH$_3^\oplus$ + Cl$^\ominus$

step 4 │ recrystallize   step 4 │ recrystallize

(+)—RCOOH   (−)—RCOOH

Natural products capable of chirality almost inevitably occur as a single stereoisomer (enantiomer) and participate in reactions as such. Resolution of laboratory-made compounds is necessary to duplicate natural products.

Resolution of a base. The process may be used in reverse to resolve a racemic amine with an optically active carboxylic acid serving as the resolving agent. After separation of the diastereomeric salts, each salt is decomposed with sodium hydroxide. The resolving agent remains in solution as the sodium salt and the enantiomeric amine separates as an oil or as a precipitate.

Other means of resolution exist, such as the use of enzymes. They will be described in Section 21.2. However, we should mention here Pasteur's original resolution, a happy accident. He was studying the various mixed salts of tartaric acid, and he allowed the sodium ammonium salt of (±)-tartaric acid to crystallize, luckily at a temperature below 27.7°. Under such conditions, the enantiomers crystallize separately into crystals that are themselves nonsuperposable mirror images. With a pair of tweezers and immense patience, Pasteur separated the crystals into two piles, and the first resolution was effected.

The age of modern man. The age of organic materials thousands of years old can be measured approximately using carbon 14. The age of rocks millions of years old can be approximated with methods based on uranium decay. But there is a gap of time that cannot be measured well by either technique. Since this includes the time *Homo sapiens* developed into modern man, it is an interesting period. Recently, a new method, based on asymmetric carbons and biochemistry, has been proposed to date organic material from these ancient millenia.

The principle is that racemization of the asymmetric carbon in the protein amino acids takes a long time. Since changing the configuration about these asymmetric carbons involves breaking and reforming carbon-carbon bonds and carbon-hydrogen bonds, each isomer would be expected to be very stable under ordinary conditions. In fact, the half-life of conversion of isoleucine to the enantiomer at 20°C is something greater than one hundred thousand years.

Plants and animals make and use only specific optically active amino acids in their proteins. At death, the amino acids are all (S) configuration. If the material is preserved in some fashion so the amino acids do not react to form other molecules, the enantiomers will gradually appear in the mixture. The ratio of (R) to (S) will increase as time goes on. Thus the value of this ratio is always an indication of the time from the death of the plant or animal.

This dating method assumes that the environment of the amino acid has remained constant, allowing the isomerization to occur at a constant rate. We must interpret these numbers with great caution; a lot can happen to the environment in a hundred thousand years.

6.6 Summary

The chapter opened with a description of free rotation, conformers, nonbonded interactions, and van der Waals' radii and went on to the rotameric forms of butane. We hope that $CH_3CH_2CH_2CH_3$ now will represent much more than just a symbol of butane to us. Although the discussion was centered on hydrogen interactions, the increased steric influences of larger atoms and groups should be kept in mind.

The consequences of rigidity were taken up and the use of *cis* and *trans* in the naming of unsaturated and cyclic compounds was illustrated. Although a triple bond also is rigid, the linear *sp* hybridization does not allow stereoisomerism to arise.

The four symmetry elements were introduced and the presence or absence of chirality and asymmetry shown. The (RS) rules for designating configuration were described and illustrated. The application of the terms conformer, configuration, enantiomer, sample, substance, and mirror image was explained.

The polarimeter was described briefly, and the meaning of $[\alpha]_D^{25}$, (+), and (−) were given. *Meso* structures and racemates were introduced, and the calculation of the possible number of stereoisomers was shown. Throughout the discussions, the absence or presence of superposability of mirror images was stressed in relation to optical activity.

Because we shall continually be using the concepts and terminology introduced in this chapter, it will be necessary for us to understand and use correctly the words and phrases as listed below:

stereochemistry	stereoisomer, stereoisomerism
free rotation	libration
conformation	conformer, rotamer
staggered	eclipsed, *anti*, *gauche*, skew
geometrical isomerism	*cis*, *trans*
symmetry	dissymmetric, asymmetric
chirality	diastereomers
enantiomers	*meso*, (+), (−)
racemic	(R) and (S)
configuration	polarimeter, α, $[\alpha]_D^{25}$
optical activity	resolution, resolving agent

Finally, it will not be amiss for us to reread Section 1.3. All living matter depends upon the reactions of dissymmetric molecules. Stereochemistry is not an idle diversion. It is fundamental to the existence of life.

6.7 Problems

6.1 Is $CH_3CHOHCD_3$ capable of chirality?

6.2 A compound has $\alpha_D^{25} = 1.52°$ for $c = 2.63$ g/100 ml, $CHCl_3$, in a 2 dm tube. Calculate $[\alpha]_D^{25}$.

6.3 Calculate the number of stereoisomers of 4,5-dichloro-2,3-hexanediol.

6.4 Specify the configuration, (R) or (S), for each asymmetric carbon atom in the structures shown:

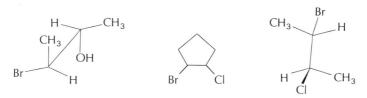

6.5 Name each of the stereoisomers of $CH_3CH=CH\overset{*}{C}HOHCH_3$.

6.6 Draw the appropriate stereoisomers for each structure, and indicate which are enantiomers and which are *meso*:
   a. CH_3—CHOH—CHOH—CH_3
   b. $CH_3CHNH_3^{\oplus}{}^{\ominus}OOCCHOHCH_3$
       |
       CH_2CH_3
   c. HOOCCHCl—CHClCOOH
   d. CH_3—CHCl—CHOH—CH_3

   e. 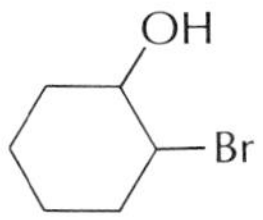 cyclohexane with OH and Br
   f. 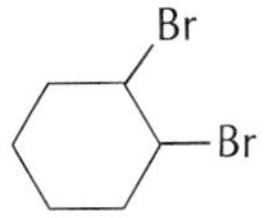 cyclohexane with Br and Br (1,2)
   g. 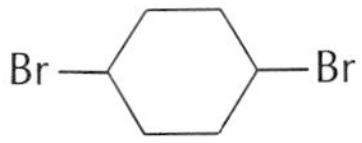 Br—cyclohexane—Br (1,4)
   h. 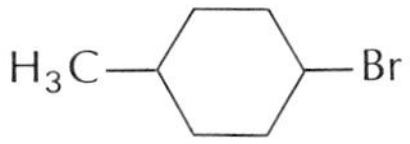 H_3C—cyclohexane—Br

6.7 Draw the isomers of the following compounds, and give each geometric isomer an appropriate prefix.
   a. $CH_3CH=CH—CH_2CH=CHCOOH$
   b. cyclodecene
   c. 1,3,5-cyclooctatriene
   d. $CH_3CH=CHCHCH_3$
                |
                OH

6.8 Draw suitable projection formulas for:
   a. two enantiomers
   b. two diastereomers
   c. a *meso* form

6.9 Draw projection formulas of $(CH_3)_3CCH_2CH_2CH_3$ in the following conformations:
   a. eclipsed
   b. staggered
   c. gauche
   Which is the least stable conformation and why?

6.10 Draw the R and S configurations for the following molecules:
   a. 1-penten-3-ol
   b. $HOOCCH_2CHNH_2COOH$ (aspartic acid)

c. $CH_3CHOHCOOH$ (lactic acid)
d. $C_2H_5CHCl-CH(CH_3)_2$

e. 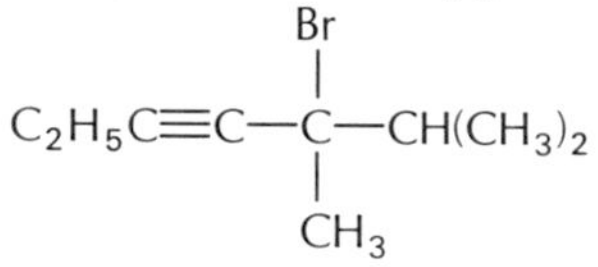 $C_2H_5C\equiv C-\underset{\underset{CH_3}{|}}{\overset{\overset{Br}{|}}{C}}-CH(CH_3)_2$

f. 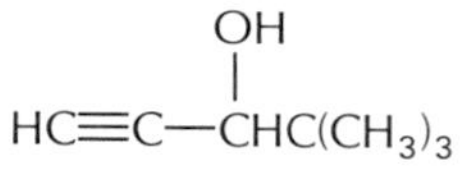 $HC\equiv C-\underset{}{\overset{\overset{OH}{|}}{C}}HC(CH_3)_3$

6.8 Bibliography

At this point in our progress, there are no really suitable references to give other than other textbooks. However, for future reference, we give the following:

1. IUPAC Tentative Rules for the Nomenclature of Organic Chemistry. Section E. Fundamental Stereochemistry, *J. Org. Chem.* **35**, 2849 (1970). The article contains much more than the bare rules. Numerous examples with notes and discussion are included.

2. Bernal, Ivan, Walter C. Hamilton, and John S. Ricci, *Symmetry*. San Francisco: Freeman, 1972.

3. Eliel, Ernest L., *Stereochemistry of Carbon Compounds*. New York: McGraw-Hill, 1962. A comprehensive treatment. Excellent for reference.

4. Lowe, J. P., "The Barrier to Internal Rotation in Ethane," *Science* **179**, 527 (1973).

5. Mislow, Kurt, *Introduction to Stereochemistry*. New York: Benjamin, 1965. An excellent paperback, but much more advanced in treatment than required in a first year course in organic chemistry. Has some problems, no references.

7 Some More Organic Reactions of Compounds

Returning to reactions in this chapter, we shall restrict our study to nucleophilic substitution and to the related heterolytic elimination reactions. In general, we shall try to limit the discussion to compounds with up to four carbon atoms. However, when it is necessary to illustrate a point, we shall introduce larger molecules. Because we have built up a background about molecular structure and stereoisomerism, the details of the reactions we discuss will be more sophisticated than has been possible up to now. When necessary, we shall refer to pertinent sections of previous chapters to maintain and foster our understanding of each topic.

With the presentation of the reactions in this chapter, we shall be able to work reasonable synthetic problems. Consequently, the chapter ends with a short description of these problems and how to solve them.

7.1   Nucleophilic substitution   143
1.   Effects of varying B in the A + BC reaction.   2.   The $S_N 2$ transition state; energy diagrams of $S_N 1$ and $S_N 2$ reactions.   3.   Hydrolysis.   4.   Steric effects.
5.   Characteristics of the two reaction pathways.   6.   Effects of chain lengthening.
7.   The enzyme. Exercises.

7.2   Carbonium ions or alkyl cations   149
1.   Naming.   2.   Need for the $S_N 1$ mechanism; steric effects; $+I$ and $+R$ effects.
3.   Planarity of a cation.   4.   Possible reactions of a carbonium ion. Exercises.

7.3   Solvents and nucleophilic substitution   153
1.   Solvent polarity and $S_N 1$ rates.   2.   Effect of solvent polarity upon $S_N 2$ rates.
3. $S_N 2cA$ and $S_N 1cA$ reactions; solvent effects.   4.   Solvent effects upon the S_N reactions summarized.

7.4   Elimination versus substitution; E2 and E1 reactions   156
1.   The E1 reaction: competition with $S_N 1$.   2.   Relative rates and products. Exercises.

7.5   Stereochemistry of nucleophilic substitution   158
1.   Introduction.   2. $S_N 2$ inversion.   3. $S_N 1$ racemization; ion pairs. Exercises.

7.6   Stereochemistry of elimination reactions   161
   1.   Direction and stereochemistry of elimination.   2.   Effects of structure, configuration, and conformation upon E2.   3.   E1 eliminations. Exercises.

7.7   Elimination reactions; structural isomerism; alkene stability   164
   1.   Saytzeff rule.   2.   Hydrogenation as a measure of stability of alkenes.
   3.   Polar effects and stability.   4.   Rearrangements in E1cA reactions.
   5.   Exhaustive methylation; Hofmann elimination. Exercises.

7.8   Applications to synthesis   169
   1.   Introduction.   2.   Alcohols to alkyl chlorides.   3.   Alcohols to alkyl bromides and iodides.   4.   Alkyl halides to other halides.   5.   Alkyl halides to alcohols.   6.   Alcohols to alkenes; alkyl halides to alkenes.   7.   Alcohols from aldehydes and ketones.   8.   Synthesis.

7.9   Summary   175

7.10   Problems   176

7.11   Bibliography   178

7.1   Nucleophilic substitution

Effects of varying B in the A + BC reaction.   As soon as we consider compounds other than the methyl and ethyl alcohols and halides, we are faced with a variety of structures, rates, conditions, and results for substitution reactions. Let us narrow the field to a discussion of the nucleophilic substitutions that we have met before as S_N2 reactions. We have looked briefly at the effects of changes in A and C in the A + BC reaction. Here, to begin, we shall describe the effect of changes in B.

Preparation of butyl bromide (1-bromobutane) requires a reflux (heating at the boiling point) period of about 1 hr for a mixture of butyl alcohol with hydrobromic and sulfuric acids. But *tert*-butyl chloride preparation requires only a 10 min mixing (or shaking) of *tert*-butyl alcohol (2-methyl-2-propanol) with concentrated hydrochloric acid at room temperature to give a good yield of product. Even though reaction rates will not be discussed until the next chapter, the striking difference in rates just mentioned would convince anyone of the difference in reactivity between a primary and a tertiary alcohol.

Alkyl halides also show marked differences in rates. In a solvent that is made up by mixing 80 ml of ethanol and 20 ml of water (known as 80% aqueous ethanol) and that is 0.01 M in sodium hydroxide, at 55°C, it is found that methyl bromide reacts at a rate such that 0.0214% of the substance remaining reacts each second. For ethyl bromide, isopropyl bromide, and *tert*-butyl bromide, the rates are 0.0017, 0.00029, and 1.010%/sec. Furthermore, the rate for *tert*-butyl bromide is not affected by the absence or presence of this low concentration of sodium hydroxide. In the absence of sodium hydroxide, the rate for isopropyl bromide is diminished to 0.00024%/sec. To express the results in a more comprehensible form, we say that the relative rates are 100, 7.9, 1.4, and 4720 for MeBr, EtBr, iPrBr, and tBuBr.

If we shift to formic acid as solvent with some water present, at 100°, relative rates for the disappearance of the same compounds are 100, 171, 447, and about 10^{10}. In acetone at 25°, relative rates for reaction with radioactive bromide ion are 100, 1.24, 0.013, and 0.0039. We can rationalize these divergent results by recognizing a new nucleophilic substitution mechanism, S_N1, different from S_N2. The symbolism is S for substitution, N for nucleophilic, and 1 for unimolecular. Before discussing the S_N1 reaction, let us review and expand on the S_N2 reaction.

The S_N2 transition state; energy diagrams of S_N1 and S_N2 reactions. An S_N2 reaction (Secs. 2.6 and 3.3) was defined as one in which the replacement of C by A in an $A + BC$ reaction was characterized by the interaction of two molecules (or a molecule and an ion). In an S_N2 reaction (for example, that in Equation 7.1, the replacement of bromine by radio-

$$\overset{*}{Br}{}^{\ominus} + CH_3CH_2—Br \xrightarrow[\text{acetone}]{25°} \overset{*}{Br}CH_2CH_3 + Br^{\ominus} \tag{7.1}$$

active bromine isotope), we can guess that the easiest collision pathway for direct replacement is from the rear of the molecule. At some point, if the energy of the collision is great enough, the incoming ion will have penetrated close enough to the carbon atom to begin bonding, and the leaving ion will have receded far enough to begin a bond breaking. Figure 7.1 depicts the situation, called the <u>transition state.</u> The carbon atom involved may be thought of as being forced from sp^3 to sp^2 hybridization, with both bromine atoms partially overlapping the carbon p orbital of the sp^2 hybrid. The two hydrogens and the carbon of the methyl group have moved into coplanarity with the carbon atom.

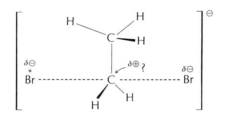

Figure 7.1 The S_N2 transition state, CH_3CH_2Br

A transition state is not a stable situation. This one may be likened to the state of affairs in a head-on collision at the moment when the participants are at a standstill. Kinetic energy is at a minimum and potential energy at a maximum. But the transition state here, described as a <u>back-side collision,</u> is more stable than any that may be described as a front-side collision—one in which $\overset{*}{Br}{}^{\ominus}$ approaches the Br end of the C—Br bond. Consequently, the reasonable back-side attack transition state is the most stable of the many that may be envisaged.

As we arrive at the transition state, the incoming $\overset{*}{Br}{}^{\ominus}$ will lose part of the negative charge as covalent bonding begins. At the same time, the receding Br will gain more control of the pair of electrons from the original C—Br σ bond and become more ionic. In the transition state, each of the bromine atoms is equally charged and equally distant from the carbon atom. The carbon atom, meanwhile, may or may not be more positive than it was before reaching the transition state. In resonance language, the situation may be represented as:

$$\overset{*}{\text{Br}}{}^{\ominus} \quad \underset{\diagdown}{\overset{|}{\underset{}{\text{C}}}}{-}\text{Br} \quad \longleftrightarrow \quad \overset{*}{\text{Br}}{-}\underset{\diagdown}{\overset{|}{\underset{}{\text{C}}}} \quad \text{Br}^{\ominus} \quad \longleftrightarrow \quad \overset{*}{\text{Br}}{}^{\ominus} \quad \underset{\diagdown}{\overset{|}{\underset{}{\text{C}}}}{}^{\oplus} \quad \text{Br}^{\ominus}$$

The carbon-bromine bonds shown in the resonance structures are not normal. They are abnormal, stretched partial bonds. Remember that no atomic motion is allowed in the resonance language picture.

If the attacking ion or group ($HO^{\ominus}$ or NH_3, for example) differs from the leaving group, the contributions of the various resonance structures in the transition state vary, and the negative charge no longer is equally divided between the entering and leaving ions or groups. Also, only if the entering and leaving groups are identical will the distances from the reacting carbon atom be identical.

To reach the transition state, the incoming ion must penetrate close enough to begin bonding and the departing atom must recede enough to begin bond breaking. The energies involved are those for cleavage of covalent bonds to ionic constituents and are greater than bond dissociation energies. Thus, even for the polar covalent molecule hydrogen chloride, 333.2 kcal is required to cleave the bond to give ions, as compared to 103.1 kcal to form the atoms. Of course, solvation must be taken into account in the ionic cleavage, and we shall do so shortly.

For the present, the bond-cleavage energy relationship may be represented schematically by a diagram such as Figure 5.9. Bond making may be represented by the reversed diagram (Figure 7.2). In the graph, Br and $\overset{*}{\text{Br}}$ are fixed in space and far apart. Starting from the left side, the ethyl group is bound to Br and is so far away from $\overset{*}{\text{Br}}{}^{\ominus}$ that there is little inter-

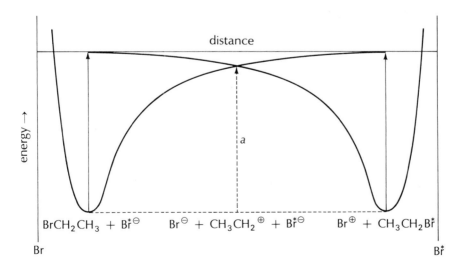

Figure 7.2 S_N1 reaction, energy (not to scale)

146 *Chapter seven*

action. As we move the ethyl group to the right, away from Br and toward $\overset{*}{Br}{}^{\ominus}$, energy must be supplied to break the Br—C bond. Halfway across, the Br—C bond has broken to form the ions $Br^{\ominus}$ and $CH_3CH_2{}^{\oplus}$. The C—$\overset{*}{Br}$ bond has yet to form at this point. The energy requirement a must be supplied. It is practically the same as that needed to break the covalent C—Br bond into the ions. Continuing to the right, the C—$\overset{*}{Br}$ bond is formed, and energy is given up. The process may be represented by Equation 7.2; and $\Delta H = 0$ for the complete reaction.

$$BrCH_2CH_3 + \overset{*}{Br}{}^{\ominus} \longrightarrow Br^{\ominus} + CH_3CH_2{}^{\oplus} + \overset{*}{Br}{}^{\ominus} \longrightarrow Br^{\ominus} + CH_3CH_2\overset{*}{Br} \quad (7.2)$$

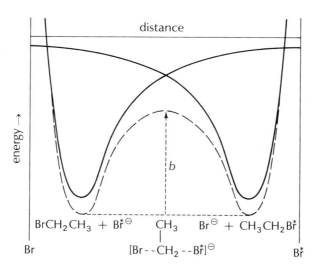

Figure 7.3 $S_N 2$ *reaction, energy (not to scale)*

The graph in Figure 7.3 represents the situation at the transition state, when the $\overset{*}{Br}{}^{\ominus}$ penetrates to a partial state of bonding and the C—Br bond is only partially broken. With $Br^{\ominus}$ and $\overset{*}{Br}{}^{\ominus}$ held much closer together, the energy diagrams for the two C—Br bonds also are pushed together and now cross much lower in the middle of the graph. Again, starting from the left, we represent the Br—C bond as it is affected by the nearness of the $\overset{*}{Br}{}^{\ominus}$. As the Br—C bond is stretched, the energy rises; but the bonding to $\overset{*}{Br}{}^{\ominus}$ has already begun. The energy needed is less than in Figure 7.2 for a given amount of lengthening (dashed line). At the midpoint, the energy requirement (b) of the situation is markedly less than that for a complete ionic dissociation (a in Fig. 7.2). The old bond is not entirely broken, and the new bond being formed is contributing to a lowering of the energy requirement. The lowered energy peak is enough to favor the $S_N 2$ displacement reaction over other pathways for this reaction.

Hydrolysis. Most of the nucleophilic substitution reactions in life are hydrolysis reactions. The digestion of proteins, fats, and carbohydrates down to component amino acids, fatty acids, and simple sugars is exclusively a process of adding a water molecule across a bond. As another example, the ubiquitous energy carrier in each cell, adenosine triphosphate (ATP), releases its energy by the hydrolysis of one of its phosphate ester bonds. This reaction is coupled enzymatically to many of the biosynthetic processes in biochemistry.

$$\text{Adenosine}-O-\underset{\underset{O^{\ominus}}{|}}{\overset{\overset{O}{\|}}{P}}-O-\underset{\underset{O^{\ominus}}{|}}{\overset{\overset{O}{\|}}{P}}-O-\underset{\underset{O^{\ominus}}{|}}{\overset{\overset{O}{\|}}{P}}-OH + H_2O \rightleftharpoons \text{Adenosine}-O-\underset{\underset{O^{\ominus}}{|}}{\overset{\overset{O}{\|}}{P}}-O-\underset{\underset{O^{\ominus}}{|}}{\overset{\overset{O}{\|}}{P}}-OH + H_2PO_4^-$$

Steric effects. Returning to Figure 7.1, we see that the presence of the methyl group interferes with the approach of the incoming $\overset{*}{Br}{}^{\ominus}$ by nonbonded repulsions. It also favors the departure of the $Br^{\ominus}$ by nonbonded repulsion, but this effect is smaller than the interference with the aim of the attacking group. The absence of methyl groups (as in CH_3Br) should favor reaction, and the presence of two or three groups (as in $(CH_3)_2CHBr$ and $(CH_3)_3CBr$) should hinder the reaction. The steric effects predict the observed order of rates (Me > Et > iPr > tBu) for $\overset{*}{Br}{}^{\ominus}$ + RBr, the S_N2 reaction in acetone; but they do not necessarily account for the surprising magnitude of the differences in rate.

If the bromides react in formic acid, an inverse order of reactivity is observed: Me < Et < iPr ≪ tBu, the greatest difference being between iPr and tBu. It is postulated that the easier pathway here for tBuBr is an ionization of the Br—C bond (Fig. 7.2) followed by reaction of the cation (carbonium ion) with solvent and other nucleophiles present. The rate depends upon the single molecule and solvent and is independent of an added reactant; in other words, it is an S_N1 reaction. Only one molecule is involved in the transition state. We may now say that an alkyl halide can react by an S_N2 or an S_N1 pathway. Sometimes, if the rates are not too different, the two pathways may be followed simultaneously: for example, by isopropyl bromide in aqueous alcohol with dilute sodium hydroxide.

Characteristics of the two reaction pathways. Let us summarize some of the characteristics of the two mechanisms.

1. Steric crowding hinders the S_N2 reaction and favors the S_N1. Thus, rates follow the sequence primary > secondary > tertiary in S_N2 reactions and primary < secondary < tertiary for S_N1 reactions. Also, the smaller the entering group is, the more the S_N2 reaction is favored.
2. Because the S_N1 reaction involves ionic dissociation, a polar solvent is required. Because the S_N2 reaction (as discussed so far) does not involve ionic dissociation, a nonpolar solvent will do. The solvent, however, must be polar enough to dissolve a reasonable amount of the ionic nucleophile.
3. The leaving halogen becomes a halide ion, a weak base. Thus, the basicity has only a minor effect on the preference for the S_N1 or S_N2 pathway.
4. The stronger the attacking group is as a nucleophile, the more the S_N2 reaction will be preferred.
5. In nucleophilic substitution, methyl compounds always react by an S_N2 pathway. Tertiary compounds always react by an S_N1 mechanism. Primary compounds nearly always react by S_N2. Secondary compounds generally favor S_N2 but are easily pushed

into an S_N1 scheme. Thus methyl and ethyl bromides in 80% aqueous ethanol reacted by the S_N2 mechanism with 0.01 M hydroxide ion; *tert*-butyl bromide, by S_N1; and isopropyl bromide, by S_N1 83% and S_N2 17% of the time.

Effects of chain lengthening. Aside from the initial drop from methyl to ethyl, the other straight-chain halides show fairly small changes in the S_N2 reaction. One experimentally observed sequence of relative rates is Me 100, Et 5.7, Pr 1.6, Bu 1.3, Pent 1.2. Branching at the adjacent carbon, however, retards the substitution strongly. Compare the relative rates of CH_3-, 100; CH_3CH_2-, 5.7; $CH_3CH_2CH_2-$, 1.6; $(CH_3)_2CHCH_2-$, 0.17; and $(CH_3)_3CCH_2-$, 0.000024. Isobutyl, even though it is a primary group, reacts at a rate only 0.17% that of methyl or 3% that of ethyl. Furthermore, neopentyl, with the bulky *tert*-butyl group blocking the back side, is unreactive by S_N2 substitution unless very drastic conditions are used (high temperature, high concentration of the substituting reagent).

Secondary alkyl halides show little variation if the chain is extended, as shown by CH_3CHCH_3, 100; $CH_3CH_2CHCH_3$, 129; $CH_3CH_2CH_2CHCH_3$, 116; and $CH_3CH_2CHCH_2CH_3$, 93 for S_N2 substitution. For S_N1, the series is very similar, relative rates being 100, 104, 79, and 85.

Tertiary alkyl halides exhibit a slight acceleration as a chain is lengthened. Compare $(CH_3)_3C-$, 100; $CH_3CH_2C(CH_3)_2$, 161; $CH_3CH_2CH_2C(CH_3)_2$, 141; $(CH_3CH_2)_2CCH_3$, 223; and $(CH_3CH_2)_3C-$, 260.

The enzyme. The reaction of carbamyl phosphate with aspartic acid to form N-carbamylaspartic acid is the first step in a biosynthetic pathway to make the pyrimidines of DNA. It proceeds by a nucleophilic substitution attack on a carbonyl carbon by the amine group of aspartic acid. The phosphate ion is the leaving group.

$$NH_2-\overset{\overset{O}{\|}}{C}-OPO_3H_2 + HOOC-CH_2-\overset{\overset{NH_3^{\oplus}}{|}}{CH}-COO^{\ominus} \rightleftharpoons$$

[structure of N-carbamylaspartic acid]

$+ H_3PO_4$

carbamyl phosphate   aspartic acid

Many nucleophilic substitution reactions are involved in the chemistry of life. All of the reactions need a protein catalyst—an enzyme—to proceed at a reasonable rate under physiological conditions. In fact, every chemical reaction in the body is catalyzed by its own specific enzyme. The organism controls the chemistry of the body by controlling its enzymes—either by varying the enzyme concentration or by affecting its catalytic ability. For example, the enzyme aspartate transcarbamylase, which catalyzes the reaction discussed above, has its catalytic activity inhibited in the presence of the pyrimidine that this biosynthetic process is manufacturing. This feedback inhibition insures that the product will only be made if it is needed—if its concentration is too low. Proteolytic digestive enzymes are sent from the pancreas into the small intestine in an inactive state (otherwise they would digest themselves before they reached the food!) and are activated once there.

Medical treatment often is directed knowingly or unknowingly to the action of an enzyme. Penicillin, for example, closely resembles a starting material in the biosynthesis of bacterial cell wall material. An enzyme binds to it instead of the proper sugar compound. With enough penicillin blocking enough enzymes, the bacterial colony dies. Penicillin-resistant strains of bacteria usually have an enzyme that catalyzes a reaction transforming penicillin into a different substance, thus protecting their cell wall manufacturing process.

Genetic diseases are more difficult to treat. The patient cannot simply be rid of another infecting organism and return to normal. For the individual with an inherited affliction, the "normal" state often means an abnormal enzyme. Genetic engineering is a rapidly developing discipline aimed at alleviating the suffering of these people. The techniques of resupplying the deficient enzyme to an individual and of introducing into the patient a gene that codes for the correct enzyme have both been used successfully to lengthen the lives of mortally ill children. Organic chemists have already synthesized a small gene and are working on the difficult problem of sequencing DNA. It cannot be overemphasized that the moral and ethical consequences of genetic engineering to society require much careful consideration.

Exercises

7.1 If in the transition state A---B---C the A—B bond is only one-third formed and the B—C bond is only one-third broken, what would you deduce about the bond stretching and the relative bond strengths of A—B and B—C?

7.2 Carbonium ions or alkyl cations

Naming. The usage of carbonium ions and alkyl cations as names has become blurred. We define our use of the terms as follows. An alkyl cation is an alkyl group with a positive charge, such as $(CH_3)_3C^\oplus$, *tert*-butyl cation. A carbonium ion is a methyl cation, $CH_3^\oplus$. Thus, $(CH_3)_3C^\oplus$ is a trimethylcarbonium ion. In some books, $(CH_3)_3C^\oplus$ is called a *tert*-butylcarbonium ion. By our definitions, *tert*-butylcarbonium ion refers to $(CH_3)_3CCH_2^\oplus$, the neopentyl cation. The IUPAC rules state, "When a cation can be considered as formed by loss of an electron from a radical at the free valence position, it may be named (a) by adding the word 'cation' to the radical name or (b) by replacing the suffix '-yl' of the univalent radical by '-ylium' . . ." We use option (a).

Need for the S_N1 mechanism; steric effects; $+I$ and $+R$ effects. The S_N1 reaction may be pictured as shown in Equation 7.3. Why should a tertiary alkyl halide prefer this pathway? Several causes suggest themselves.

$$BC \underset{\text{slow}}{\rightleftharpoons} B^\oplus + C^\ominus \xrightarrow[\text{fast}]{A^\ominus} AB + C^\ominus \qquad (7.3)$$

First, as we have already pointed out, the back side of a tertiary halide is so blocked off that it is very difficult for the entering group to get within bonding distance of the carbon atom to be substituted. The energy hill for S_N2 reaction is high.

Second, methyl groups are better than hydrogen in releasing electrons by the inductive effect (Sec. 5.7). The $+I$ effect of the three methyl groups stabilizes the *tert*-butyl cation more than it does *tert*-butyl chloride. If methyl has a $+I$ effect, then ethyl should have a

$$CH_3 \rightarrow \underset{\underset{CH_3}{\uparrow}}{\overset{\overset{CH_3}{\downarrow}}{C}} \rightarrow Cl \qquad CH_3 \rightarrow \underset{\underset{CH_3}{\uparrow}}{\overset{\overset{CH_3}{\downarrow}}{C}} ^{\oplus}$$

slightly greater $+I$ effect because another hydrogen has been replaced by a methyl. It turns out that the relative rates are 100 for $(CH_3)_3CX$ and 161 for $CH_3CH_2C(CH_3)_2X$.

Third, methyl groups have a $+R$ effect, <u>hyperconjugation</u>, called into being by the ion.

Hyperconjugation is an extension of resonance theory. In fact, other names for hyperconjugation are <u>second-order resonance</u> and <u>no-bond resonance</u>.

Hyperconjugation is a response to a demand for electrons by a carbon atom with a positive charge. Such a carbon atom does not have an octet of electrons: it has only six. This carbon atom exerts a strong $-I$ effect upon the three covalent bonds left to it; but even so, the electron-hungry carbon atom is still unsatisfied.

If we examine the situation at a second carbon atom bonded to the positively charged carbon atom, we see another way in which electrons may be furnished to the electron-deficient atom. We recognize *b* as an ionic contributor to the C—H bond. However, with

$$\underset{a}{\overset{H}{\underset{H}{\overset{|}{\underset{|}{H-C-C-R}}}}\overset{\oplus}{\underset{R}{\overset{|}{C}}}} \longleftrightarrow \underset{b}{\overset{\oplus H}{\underset{H}{\overset{|}{\underset{|}{H-C^{\ominus}-C-R}}}}\overset{\oplus}{\underset{R}{\overset{|}{C}}}} \longleftrightarrow \underset{c}{\overset{\oplus H}{\underset{H}{\overset{|}{\underset{|}{H-C=C-R}}}}\overset{}{\underset{R}{\overset{|}{}}}}$$

a positive charge right next door, the electron pair may shift so as to occupy the empty p orbital of the sp^2 hybrid positive carbon atom as in *c*. We call the resonance structure *c* a hyperconjugation structure.

The effect of a single hydrogen is not very strong for several reasons. The bare proton, as written in *c*, exerts a very powerful pull on the electrons in the double bond. The double bond is not a full-fledged one because it is not a *p-p* orbital overlap, but an sp^3-p sideways overlap. Furthermore, the sp^3 orbital should be at least roughly aligned with the vacant p orbital. For these reasons (and several others with which we need not be concerned here), the concept of hyperconjugation has not received universal acceptance. Nevertheless, hyperconjugation does provide a reasonable rationale for the distribution of positive charge among the hydrogen atoms attached to the carbon atoms that are bonded to the electron-deficient carbon atom.

If we accept hyperconjugation structure *c*, we also accept that there are three hyperconjugatable hydrogens in the methyl group of *a*. An ethyl group, when bonded to a positive carbon, has two hyperconjugatable hydrogens, and so on. Also, hyperconjugation is an explanation of a $+R$ effect of a methyl group.

A *tert*-butyl cation has nine hyperconjugatable hydrogens, as compared to six for an isopropyl cation, three for an ethyl cation, two for a propyl cation, and only one for an isobutyl cation. Tertiary cations are stabilized by hyperconjugation more than are secondary ions, which, in turn, are more stable than are primary ions.

$$\overset{\oplus}{\underset{H_2C=C}{H}}\diagdown_{CH_3}^{CH_3} \longleftrightarrow H_3C-\overset{\oplus}{C}\diagdown_{CH_3}^{CH_3} \qquad t\text{Bu}^\oplus$$

$$H^\oplus$$
$$(CH_3)_2C=CH_2 \longleftrightarrow (CH_3)_2\overset{H}{\underset{|}{C}}CH_2^\oplus \qquad i\text{Bu}^\oplus$$

A fourth cause is that ionization is favored by polar solvents, and the S_N1 reaction is favored by solvent polarity. Under favorable conditions, an S_N1 reaction is much faster than an S_N2 reaction (even under conditions favorable to S_N2 reaction). A polar solvent aids an S_N1 reaction by the stabilization achieved by the solvation of the ions produced in the initial ionization step.

Planarity of a cation. A tertiary carbonium ion is most likely to be a planar structure, the carbon atom being an sp^2 hybrid with the three sp^2 orbitals engaged in bonding to neighboring carbon atoms. The empty p orbital is partially occupied by hyperconjugation electron pairs. If the cation remained as an sp^3 hybrid with an empty sp^3 orbital, certain stereochemical consequences would ensue that we shall describe later on in this chapter.

The positive charge of the ion is centered on the sp^2 carbon atom but is somewhat dispersed by both the $+I$ and $+R$ effects of the attached alkyl groups. Solvation of a tertiary ion is limited to the front and back, all other approaches being blocked by the bulky adjacent groups. A secondary ion is somewhat more open but does not have as many stabilizing effects available to it as a tertiary ion has. Because the ion is positive, solvation involves the negative end of the dipole in a molecule of solvent. Because the ion is large, the necessary solvation should be less than that needed for a small ion. Because of the relatively intense charge on a small atom, a higher degree of solvation is required.

The steric restriction of alkyl groups to solvation means that most tertiary carbonium ions cannot acquire much additional stability. They have only a transient existence except in certain special solvents, such as a mixture of fluorosulfonic acid, antimony pentafluoride, and sulfur dioxide at $-60°$. (This mixture has been called "magic acid" because of its remarkable stabilization of carbonium ions.)

The transition state for a unimolecular ionization of an alkyl halide may be represented by Equation 7.4. The transition state lies at the peak of the lowered energy hill between R—X and the solvated ions.

$$R-X \rightleftharpoons \left[\overset{\delta\oplus\delta\ominus}{R-X}\right] \rightleftharpoons R^\oplus + X^\ominus \qquad (7.4)$$

Possible reactions of a carbonium ion. Carbonium ions may react: with solvent; with nucleophiles present in the solution to give the same or a new molecule; to lose a proton to a base to produce an olefin; to add to a multiple bond to form a new ion; or to rearrange to give a more stable carbonium ion, which may proceed to react by any of the reactions mentioned. Some possibilities are shown in Figure 7.4 for 3-chloro-2,2-dimethylbutane. In subsequent sections of this chapter, we shall describe the various reactions of carbonium ions.

152   Chapter seven

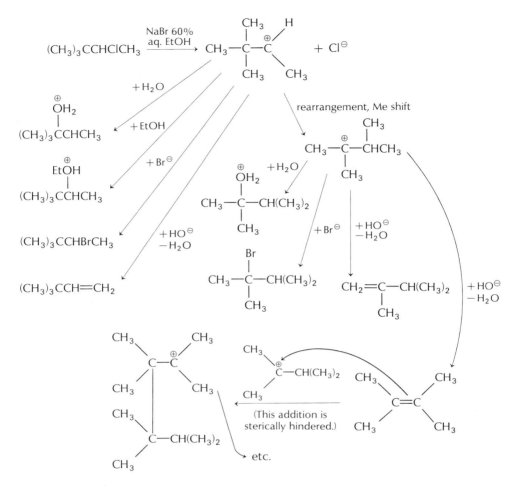

Figure 7.4   Carbonium ion reactions

The rearrangement, originally called the <u>Wagner-Meerwein rearrangement,</u> may involve the migration from a neighboring carbon of any group that may carry an electron pair to the positive carbon. Hydrogen (as hydride, $H^\ominus$) transfers readily, as do most alkyl groups. If there is a choice, that group migrates which will produce the most stable ion. If possible, primary ions always rearrange to secondary or preferably to tertiary ions. Tertiary ions do not rearrange except to more stable tertiary ions. For example, the isobutyl cation $((CH_3)_2CHCH_2^\oplus)$ rearranges by a hydride shift to give $(CH_3)_3C^\oplus$, not by a methyl shift, which would give the secondary ion $CH_3\overset{\oplus}{C}HCH_2CH_3$. Rearrangements are quite common in organic chemistry; the Wagner-Meerwein rearrangement is only the first of many rearrangements we shall encounter.

Exercises

7.2 List the cations in each group in order of stability:
   a. propyl, isobutyl, neopentyl

b. sec-butyl, isopropyl, 3-methyl-2-butyl
c. 2,3,3-trimethyl-2-butyl, tert-pentyl, tert-butyl

7.3 Solvents and nucleophilic substitution

Solvent polarity and S_N1 rates. One extensive study of the relative rates of an S_N1 reaction in a series of solvents showed a range of rates of about 5 million. The data are shown in Table 7.1. We see that the best solvents are the protic solvents, the first six entries in the table. The rates in aprotic solvents roughly parallel the solvent dielectric constant, which may be regarded as a measure of solvent polarity. Other studies have given similar results, with a few changes in order: the rate in water above that in formic acid, and the rate in methanol better than that in acetic acid, for example.

The superiority of the protic solvents probably stems from their ability to solvate the anion in the reaction better than do the aprotic solvents. For those solvents miscible with water, the addition of water has an accelerating effect upon the rates of reaction. Thus, for

Table 7.1 Relative rates of an S_N1 reaction in various solvents[a]

Solvent	$\log k_1$	Relative rate	D^b
Formic acid	−0.93	15300	58
Water	−1.18	3900	78
80% aqueous ethanol	−2.50	185	67
Acetic acid	−2.77	100	6
Methanol	−2.80	95	33
Ethanol	−3.20	37	24
Dimethyl sulfoxide	−3.74[c]	10.8	45
Nitromethane	−3.92	7.3	37
Acetonitrile	−4.22	3.6	37
Dimethyl formamide	−4.30	2.9	37
Acetic Anhydride	−4.47	2.0	21
Pyridine	−4.67	1.27	12
Acetone	−5.07	0.51	21
Ethyl acetate	−5.95	0.067	6.0
Tetrahydrofuran	−6.07	0.050	7.4
Ether	−7.3	0.003	4.2
Chloroform		very small	4.6
Benzene		very small	2.3
Hexane		very small	1.9

[a] Solvolysis of p-methoxyneophyl p-toluenesulfonate at 75°. Some values extrapolated. Adapted from S. G. Smith, A. H. Fainberg, and S. Winstein, J. Amer. Chem. Soc. 83, 618 (1961). Reprinted by permission.
[b] Dielectric constant.
[c] To give an idea of the speed of reaction, for $\log k_1 = -3.74$, the reaction is half completed in 63 min or 97% complete in 5.25 hr.

tert-butyl chloride at 25°, relative rates of appearance of chloride ion of 1, 5, 24, 74, 210, and 760 were found for 90, 80, 70, 60, 50, and 40% aqueous ethanol. We see a large increase in S_N1 rate as the solvent becomes more polar with added water.

Effect of solvent polarity upon S_N2 rates. In the S_N2 reaction, the effect of solvent polarity is more complex and depends upon the changes in charge and dispersion of charge as the system progresses from the reactants to the transition state. For example, in the reaction of hydroxide ion with an alkyl bromide, one reactant is the unchanged alkyl bromide, which has a small solvation requirement, and the other is an ion with a great need of solvation. At the transition state, the hydroxide ion is partially bonded, the incipient bromide ion is not yet fully formed, and both entities bear only partial charges. The entire transition state has a unit negative charge, but it is dispersed. The solvation requirement has been decreased because a large ion with a less intense (dispersed) charge needs less solvation than does a small ion, in which the charge must be more intense (less dispersed). The decrease in need for solvation of the transition state means that a polar solvent would favor the reactant ion, say a hydroxide ion, more than it favors the transition state. Thus, the reaction would go better in a less polar solvent although the effect would probably be small. Experimentally, it has been shown that the relative S_N2 rates are 100, 82, and 50 for the hydroxide ion reacting with isopropyl bromide at 55° in 100, 80, and 60% aqueous ethanol. The rates, we see, decrease as the solvent becomes more polar with added water.

If the substituting group is not an ion, but a neutral species such as water, then the transition state also is neutral but has developed a strong dipole with a partial positive charge on the oxygen and a partial negative charge on the halogen of an alkyl halide.

$$\begin{bmatrix} \overset{H}{\underset{H}{\diagdown}} \overset{\delta\oplus}{} \overline{O} \mathrel{|} ----- \overset{\overset{CH_3}{|}}{\underset{\underset{CH_3}{\diagdown}}{C}} ----- \overset{\delta\ominus}{} \overline{Br} \mathrel{|} \end{bmatrix}$$

$\longmapsto$ strong dipole

S_N2 transition state for H_2O + iPrBr

Now the transition state requires more solvation than do the reactants, and polarity of the solvent should favor the reaction. Isopropyl bromide shows, in the absence of hydroxide ion, relative rates of 3, 9, and 11 (on the same scale as for the hydroxide ion reaction just mentioned) for 100, 80, and 60% aqueous ethanol. The increase in solvation capability caused by the addition of water to the ethanol is more effective in accelerating the rate of substitution in this case (almost four times) than in the deceleration of the ionic attack on isopropyl bromide (one half).

S_N2cA and S_N1cA reactions; solvent effects. As we pointed out previously (Sec. 2.6), the —OH group in an alcohol is very difficult to displace directly, but an —$OH_2^\oplus$ group is easily replaced. An alcohol requires the donation of a proton before an S_N2 or S_N1 replacement can occur. Since the replacement step occurs on the conjugate acid of the alcohol, a refinement of the symbolism designates such substitutions as S_N2cA and S_N1cA (Eqs. 7.5) for attack by a nucleophilic anion, $Nu^\ominus$. The symbol S_N2cA stands for bimolecular nucleophilic substitution upon the conjugate acid of the reactant, and S_N1cA stands for unimolecular nucleophilic substitution upon the conjugate acid of the reactant.

7.3 Solvents and nucleophilic substitution

$$\text{ROH} + \text{HA} \xrightleftharpoons{\text{fast}} \text{ROH}_2^\oplus + \text{A}^\ominus \quad \text{pre-equilibrium}$$

$$\text{Nu}^\ominus + \text{ROH}_2^\oplus \xrightarrow{\text{slow}} \text{NuR} + \text{H}_2\text{O} \quad S_N2\text{cA}$$

$$\text{ROH}_2^\oplus \xrightleftharpoons{\text{slow}} \text{R}^\oplus + \text{H}_2\text{O} \quad S_N1\text{cA} \quad (7.5)$$

$$\text{R}^\oplus + \text{Nu}^\ominus \xrightarrow{\text{fast}} \text{RNu}$$

In the S_N2cA reaction, the leaving group has a positive charge, and the attacking nucleophile may have a negative charge or be neutral. The transition states, therefore, are neutral or have a positive charge. Both systems involve a decrease in requirement for

$$\text{Nu}^\ominus + \text{ROH}_2^\oplus \longrightarrow \left[\overset{\delta\ominus}{\text{Nu}}-\text{R}-\overset{\delta\oplus}{\text{OH}_2}\right] \longrightarrow \text{Nu}-\text{R} + \text{H}_2\text{O}$$

$$\text{Nu} + \text{ROH}_2^\oplus \longrightarrow \left[\overset{\delta\oplus}{\text{Nu}}-\text{R}-\overset{\delta\oplus}{\text{OH}_2}\right]^\oplus \longrightarrow \overset{\oplus}{\text{Nu}}-\text{R} + \text{H}_2\text{O}$$

solvation in going to the transition state and thus would be favored by nonpolarity of solvent. The neutral transition state would be favored more by solvent nonpolarity than the one with a dispersed positive charge.

In the S_N1cA reaction, as in S_N1, the charge or lack of it on the nucleophile is without effect, because the slow step is the ionization step (Eqs. 7.5). The transition state has the charge dispersed, needs less solvation, and would be favored by nonpolarity of solvent. However, the solvent must be polar enough to allow the formation in solution of the positive conjugate acid of the alcohol in the pre-equilibrium step.

Solvent effects upon the S_N reactions summarized. Table 7.2 summarizes the effects of solution polarity upon the various substitution reactions. Aside from using solvents with different polarities, it is also possible to increase the polarity by adding neutral salts to a given solution. The presence of the ions from the salt, with their strong solvation requirements, enhances the polarity of the solution. The effect is stronger the lower the polarity (dielectric constant) of the solvent used.

Table 7.2 Solvent effects upon S_N reactions

Symbol	Reactants	Transition state	Change in charge, amount, and location	Effect of increased polarity of solvent
S_N2	$\text{Nu}^\ominus + \text{RX}$	$[\text{Nu}\text{---}\text{R}\text{---}\text{X}]^\ominus$ with $\delta\ominus, \delta\ominus$	none, dispersed	mild decrease
S_N2	$\text{Nu} + \text{RX}$	$[\text{Nu}\text{---}\text{R}\text{---}\text{X}]$ with $\delta\oplus, \delta\ominus$	none, separation	large increase
S_N2cA	$\text{Nu}^\ominus + \text{RX}^\oplus$	$[\text{Nu}\text{---}\text{R}\text{---}\text{X}]$ with $\delta\ominus, \delta\oplus$	decrease, separation	large decrease
S_N2cA	$\text{Nu} + \text{RX}^\oplus$	$[\text{Nu}\text{---}\text{R}\text{---}\text{X}]^\oplus$ with $\delta\oplus, \delta\oplus$	none, dispersed	mild decrease
S_N1	RX	$[\text{R}\text{---}\text{X}]$ with $\delta\oplus, \delta\ominus$	none, separation	large increase
S_N1cA	$\text{RX}^\oplus$	$[\text{R}\text{---}\text{X}]^\oplus$ with $\delta\oplus, \delta\oplus$	none, dispersed	mild decrease

A striking influence of solvent upon S_N2 reactions is shown by aprotic solvents such as dimethylsulfoxide, dimethylformamide ($\text{HCON}(\text{CH}_3)_2$), acetone, and so on, in contrast to protic solvents. For the S_N2 reaction of azide ion (from NaN_3) with

156 Chapter seven

dimethylsulfoxide, (DMSO or Me$_2$SO)

dimethylformamide (DMF or HCONMe$_2$)

butyl bromide at 25°, the relative rates are 0.076 in methanol, 0.51 in water, 100 in dimethylsulfoxide, 207 in dimethylformamide, 378 in acetonitrile (methyl cyanide, CH$_3$C≡N), and 15,400 in hexamethylphosphoramide, ((CH$_3$)$_2$N)$_3$PO.

In protic solvents, I$^\ominus$ > Br$^\ominus$ > Cl$^\ominus$ is the usual order of attacking reagents, but it reverses to Cl$^\ominus$ > Br$^\ominus$ > I$^\ominus$ in aprotic solvents.

There is no general agreement as to how the aprotic solvents exert their effect; and as is usual when experts disagree, many reasons for this or that explanation have been presented. One explanation involves the fact that the effective aprotic solvents all have strong dipoles with the positive end buried within the molecule and the negative end projecting out. As a result, solvation of cations is good but solvation of anions is poor. Protic solvents solvate anions very well by hydrogen bonding. An anion (even a neutral nucleophile) must be disengaged from solvent molecules before an attack on the molecule to be substituted can occur. The aprotic solvent, by not solvating the nucleophile to begin with, does not have to be stripped from the nucleophile; and substitution may proceed apace. But in opposition to this viewpoint, the group being replaced is not likely to be solvated very well either, and it should be more difficult to push off. Also, the addition of even small amounts of DMSO (dimethylsulfoxide) to an alcohol solvent increases the rate of an S$_N$2 reaction even though hydrogen bonded solvation of the nucleophile cannot be very much diminished. There are other explanations and objections thereto, but we shall omit them.

7.4 Elimination versus substitution; E2 and E1 reactions

The E1 reaction: competition with S$_N$1.   So far in this chapter we have touched very lightly upon the products obtained in substitution reactions. In Sections 3.3 and 3.7 the competition of β elimination with substitution was introduced and conditions favoring S$_N$2 or E2 were discussed. Here it is appropriate to bring up the E1 reaction as well.

The first step of the E1 reaction of an alkyl halide is the first step of the S$_N$1 reaction. The divergence lies in the fate of the carbonium ion (Eq. 7.6). The base may be an added

$$RX \underset{\text{slow}}{\rightleftharpoons} R^{\oplus} + X^{\ominus} \begin{cases} \xrightarrow[\text{fast}]{N^{\ominus}} RN & S_N1 \\ \\ \xrightarrow[\text{fast}]{B^{\ominus}} BH + \text{Alkene} & E1 \end{cases} \quad (7.6)$$

7.4 Elimination versus substitution; E2 and E1 reactions

entity other than the nucleophile. The two may be and often are the same: for example, the solvent. For the same solvent and temperature, the ratio of alkene to substitution product should be the same no matter from what the ion is formed. At 65° in 80% aqueous ethanol, *tert*-butyl chloride gave 36% olefin (isobutylene, $(CH_3)_2C=CH_2$) and 64% *tert*-butyl alcohol and ether mixture. Under the same conditions, but at a slower rate (about one-eighth that of the chloride), the *tert*-butyldimethyl sulfonium ion also gave 36% olefin and 64% alcohol plus ether (Eqs. 7.7). In both instances, S_N1 predominated almost two to one over E1.

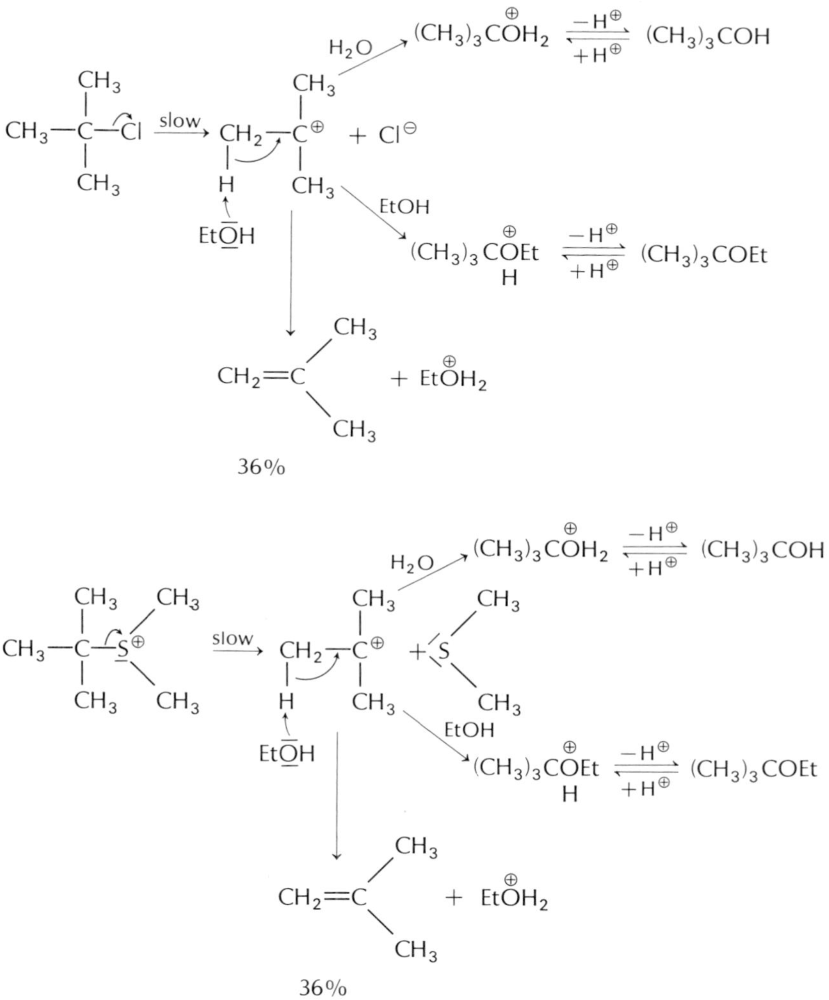

(7.7)

The E1 reaction is always a concomitant of an S_N1 reaction, becoming greater at higher temperatures or smaller at lower temperatures. Thus at 50°, 24% olefin is formed in 80% aqueous ethanol from *tert*-butyl chloride, and at 25°, only 17% olefin is formed.

Because increasing polarity of solvent favors the first step (the slow ionization) of the S_N1-E1 reaction, the competition between the subsequent more rapid steps is a matter of nucleophilicity versus basicity of the solvent, as well as steric effects and so on. Generalities should be stated with caution. However, experimentally it is safe to say that increasing the water content of aqueous ethanol as a solvent while keeping other factors constant tends to decrease the proportion of olefin formed. ☐ ☐

Relative rates and products. Although *tert*-butyl chloride solvolyzes (reacts with solvent) readily by the S_N1, E1 mechanism and the rate is independent of the presence of low concentrations of base, if high concentrations of base are used (or a stronger base) the E2 reaction may dominate. Alcohols, under conditions favorable to the S_N1cA or S_N2cA reactions (presence of acid; low concentration of or weak base), have a tendency to favor E1 mechanisms. Secondary halides and alcohols have not been explicitly discussed because, as might be guessed, they tend to react by all the mechanisms presented.

A primary halide tends to react by S_N2 and E2 mechanisms; a tertiary halide by S_N1 and E1 mechanisms. A secondary halide lies between the extremes. Depending upon the structure and upon conditions, it may be forced toward either of the mechanisms. In 60% aqueous ethanol at 80°, the secondary bromides $CH_3CHBrCH_3$, $CH_3CH_2CHBrCH_3$, $CH_3CH_2CHBrCH_2CH_3$, and $CH_3CH_2CH_2CHBrCH_3$ react with relative unimolecular rates of 100, 105, 85, and 79, to give 5, 9, 15, and 7% alkenes and 95, 91, 85, and 93% alcohols plus ethers. On the other hand, with $NaOCH_2CH_3$ in ethanol at 25°, the same compounds give predominantly alkenes at E2 relative rates of 100, 147, 169, and 117. Under the same conditions, the E2 rates for *tert*-butyl bromide and 2-bromo-2-methylbutane (*tert*-pentyl bromide) are 1270 and 2500.

Exercises

7.3 Predict the products and relative amounts to be expected if $(CH_3)_3CCHClCH_3$ is:
 a. heated with 90% formic acid
 b. heated with excess $NaOCH_2CH_3$ in ethanol

7.4 Predict the products and relative amounts to be expected if (1-methylcyclopentyl) methanol is heated with 90% sulfuric acid.

7.5 Stereochemistry of nucleophilic substitution

Introduction. In Section 7.1, the back-side attack characteristic of an S_N2 reaction and the polar factors favoring such a mechanism were described. In Section 7.2, the S_N1 reaction and the stabilization of carbonium ions was presented. Now we shall examine the stereochemistry of these mechanisms as well as that of the closely related S_N1cA and S_N2cA mechanisms.

S_N2 inversion. If we examine the transition state for the S_N2 reaction (Figure 7.1), we see that if the system reverts to reactants ($Br^{*\ominus}$ and CH_3CH_2Br), there will be no change in the symmetry of the ethyl bromide. If the transition state decomposes in a forward direction, the ethyl bromide ($CH_3CH_2Br^*$) immediately formed is the mirror image (except for the isotopic difference) of the reactant molecule. Because ethyl bromide has other elements of symmetry, the reactant molecule and the product molecule may be superposed and shown to be congruent (again excepting conformational differences). The same

stereochemical results obtain with an isopropyl halide; but consider 2-bromobutane (sec-butyl bromide, $CH_3CH_2\overset{*}{C}HBrCH_3$, Eq. 7.8). If the reactant is the (S)-(+)-2-bromobutane,

$$\overset{*}{Br}^{\ominus} + \underset{(S)-(+)}{\overset{CH_2CH_3}{\underset{H\quad CH_3}{\diagdown\!\!\diagup}}}\!\!-Br \xrightarrow[\text{acetone}]{S_N2} \overset{*}{Br}-\underset{\text{inversion, }(R)-(-)}{\overset{CH_2CH_3}{\underset{CH_3\quad H}{\diagdown\!\!\diagup}}} + Br^{\ominus} \quad (7.8)$$

we see that the product is the enantiomer, (R)-(−)-2-bromobutane. We say that <u>optical inversion</u> has occurred.

When we introduce a radioactive bromide ion, we are at the same time causing a loss of one molecule of the (S)-(+)- enantiomer and a gain of one molecule of the (R)-(−) enantiomer. When the number of the (S)-(+) enantiomers equals the number of (R)-(−) enantiomers, a racemic mixture has been formed. The optical activity of the solution then is zero and remains so thereafter.

The actual course of the reaction is more complicated than what we show in Equation 7.8, because as soon as the solution contains some $Br^{\ominus}$ and (R)-(−) enantiomers, all the possible displacements begin to operate. The initial rate of reaction, however, does show that each incorporation of a radioactive bromine is accompanied by an inversion at the asymmetric carbon. We can follow this process by the decrease in the rotation of plane polarized light by the solution. An S_N2 reaction, in all those cases in which inversion of configuration may be observed by optical activity, proceeds by back-side attack and inversion. By extension, back-side attack and inversion are thought to occur in all S_N2 reactions, even on those primary, secondary, and tertiary carbons without optical activity.

S_N1 racemization; ion pairs. S_N1 reactions, proceeding by a slow ionization step to produce a more or less free carbonium ion, should give rise to a planar ion in which symmetry is present. The planar cation will be superposable on its mirror image even if it is generated from an optically active center. Subsequent reaction of the cation with solvent or nucleophiles would be equally probable from either side (front or back) and lead to racemic products. Such a pure S_N1 reaction is seldom observed. Optically active 2-bromooctane, for example, upon S_N1 hydrolysis or alcoholysis undergoes only about 40 ± 10% loss of activity, the other 60 ± 10% being inverted alcohol or ether. An optically active tertiary alkyl compound (3-chloro-3,7-dimethyloctane, for example) gives a 75 ± 5% loss of activity under similar conditions.

An explanation for these results has been offered (Eq. 7.9). The idea here is that for ionization to occur, help in the form of solvation is needed. The solvent acts as a nucleophile (S_N) to solvate the incipient carbonium ion, and it also behaves as an electrophile (S_E) to solvate the incipient halide ion.

> Electrophilic substances may be added to such mixtures to enhance the ionization step. Sometimes silver ($Ag^{\oplus}$) or mercuric (Hg^{2+}) ions are deliberately added. Their great affinity for halide ions helps to insure that ionization becomes the predominant reaction in solvents of low polarity or ionizing power.
>
> It may be argued that acceleration of the rate of reaction by added nucleophiles, to the extent that it is proportional to the concentration of added nucleophile, is no longer a unimolecular ionization, but a bimolecular one, even if an ion pair is

the immediate result. Our understanding shall be that if the crucial step is unimolecular (aside from participation by solvent), the reaction is S_N1, E1, or S_N1cA. If the critical step is bimolecular, the reaction is S_N2, E2, or S_N2cA.

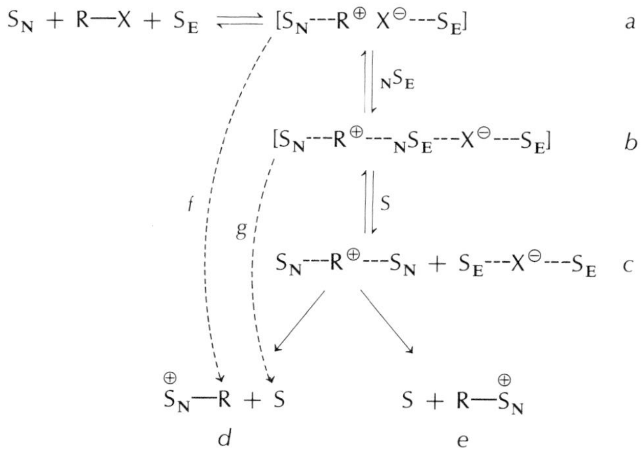

$$S_N + R\text{—}X + S_E \rightleftharpoons [S_N\text{---}R^\oplus\ X^\ominus\text{---}S_E] \qquad a$$

$$\Big\updownarrow {}^NS_E$$

$$[S_N\text{---}R^\oplus\text{---}{}_NS_E\text{---}X^\ominus\text{---}S_E] \qquad b \qquad (7.9)$$

$$\Big\updownarrow S$$

$$S_N\text{---}R^\oplus\text{---}S_N + S_E\text{---}X^\ominus\text{---}S_E \quad c$$

$$\overset{\oplus}{S_N}\text{—}R + S \qquad\qquad S + R\text{—}\overset{\oplus}{S_N}$$
$$d \qquad\qquad\qquad\qquad e$$

The first step shown in Equation 7.9 is ionization, but there is only a small separation: in other words, ion-pair formation. It has been estimated than an increase in bond distance of only 0.4 Å would produce the transition state for ionization. The tight ion pair (a) obviously has room for solvation only at the front on $X^\ominus$ and at the back on $R^\oplus$. The second step is a separation into a loose ion pair (b) in which a molecule of solvent has managed to squeeze in between the ions. The third step is the separation of the completely solvated ions into the bulk of the solution (c). Last of all, the free solvated carbonium ion may go from solvation to σ bond formation from the rear ($\overset{\oplus}{S_N}$—R, d) or from the front (R—$\overset{\oplus}{S_N}$, e) with equal probability, giving a racemic mixture. But all these steps take time, and if the ion is not stable enough to survive long enough, it may collapse into d from any of the preceding stages (f and g). If such early collapse occurs, inversion of configuration will happen more often than will front-side attack with retention of configuration. A secondary halide, which would give rise to a less stable ion, gives more inversion than does a tertiary halide, which gives rise to a more stable ion. □ □

Exercises

7.5 Show the structure of the product to be expected:

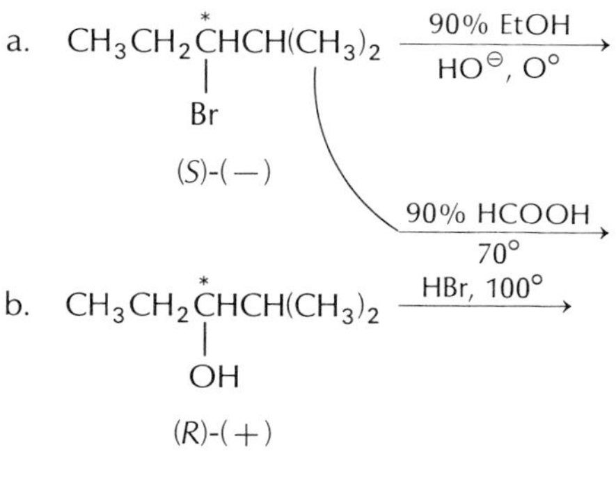

a. $CH_3CH_2\overset{*}{C}HCH(CH_3)_2$ with Br
   (S)-(−)

   $\xrightarrow{\text{90\% EtOH}}{HO^\ominus,\ 0°}$

   $\xrightarrow[70°]{\text{90\% HCOOH}}$

b. $CH_3CH_2\overset{*}{C}HCH(CH_3)_2$ with OH
   (R)-(+)

   $\xrightarrow{\text{HBr, 100°}}$

7.6 Stereochemistry of elimination reactions

Direction and stereochemistry of elimination. A primary halide may eliminate in only one direction to give a 1-alkene, unless rearrangement intervenes (Eq. 7.10). A secondary

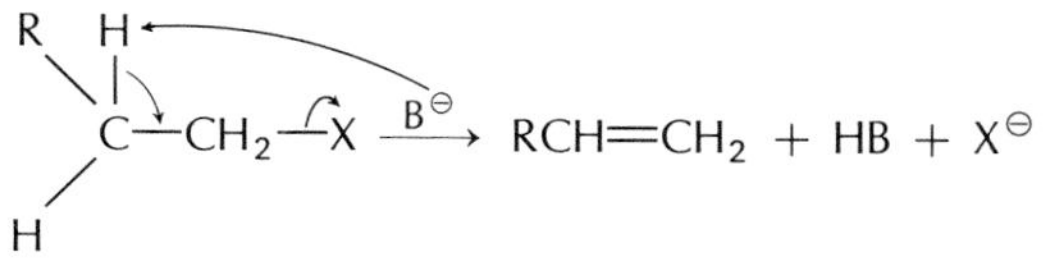

$$\underset{H}{\overset{R}{\diagdown}}\!\!\!\underset{}{\overset{H}{\diagup}}\!C-CH_2-X \xrightarrow{B^{\ominus}} RCH=CH_2 + HB + X^{\ominus} \tag{7.10}$$

or tertiary halide may eliminate in two or three directions, depending upon its structure (Eqs. 7.11). Furthermore, a given halide may give rise to *cis* and *trans* products for a given

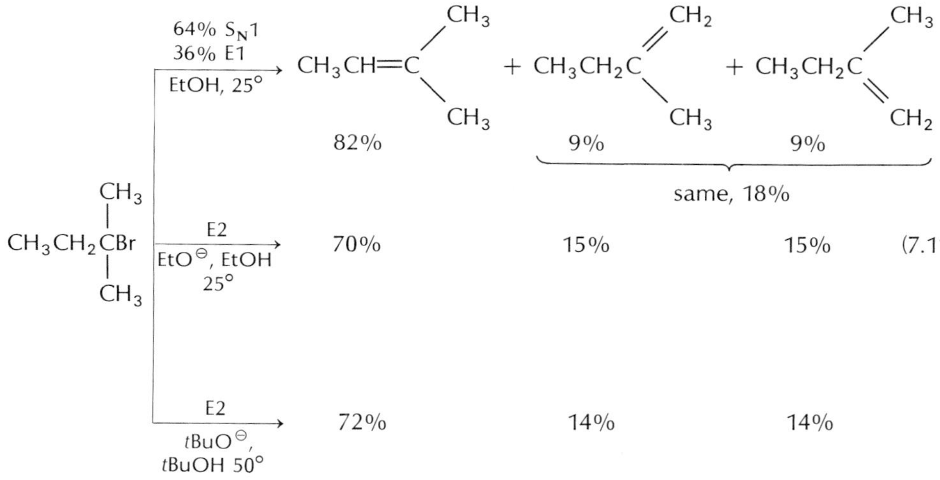

$$CH_3CH_2\underset{\underset{CH_3}{|}}{\overset{\overset{CH_3}{|}}{C}}Br$$

conditions	$CH_3CH=C(CH_3)_2$	$CH_3CH_2C(=CH_2)CH_3$ (cis)	$CH_3CH_2C(CH_3)=CH_2$ (trans)
64% S$_N$1, 36% E1, EtOH, 25°	82%	9%	9%
		same, 18%	
E2, EtO$^{\ominus}$, EtOH, 25°	70%	15%	15%
E2, tBuO$^{\ominus}$, tBuOH 50°	72%	14%	14%

(7.11)

direction (Eq. 7.12). Under E1 conditions, in acetic acid at 118°, a *sec*-butyl compound

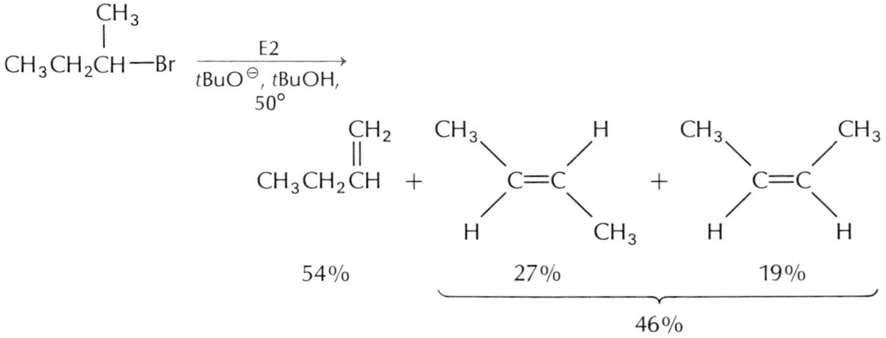

$$CH_3CH_2\underset{\underset{CH_3}{|}}{\overset{}{C}H}-Br \xrightarrow[50°]{E2, \; tBuO^{\ominus}, \; tBuOH,}$$

$CH_3CH_2CH=CH_2$ + *cis*-2-butene + *trans*-2-butene (7.12)

54%   27%   19%
              46%

yielded about 25% olefin, which consisted of 10% 1-butene, 47% *trans*-2-butene, and 43% *cis*-2-butene. To these confusing results may be added a few acid-catalyzed dehydration reactions of alcohols (Eqs. 7.13).

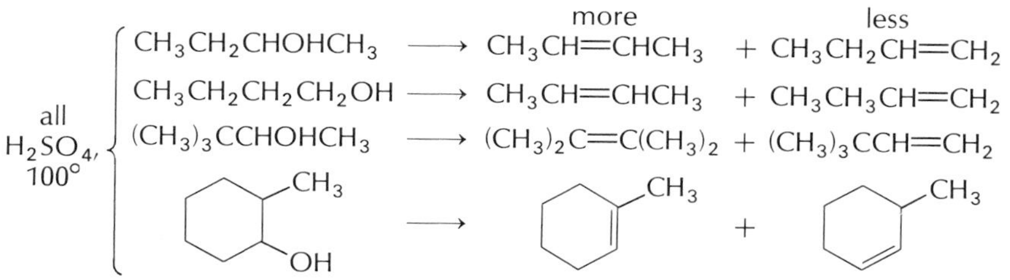

$$\text{all } H_2SO_4, 100° \begin{cases} CH_3CH_2CHOHCH_3 \longrightarrow \overset{\text{more}}{CH_3CH=CHCH_3} + \overset{\text{less}}{CH_3CH_2CH=CH_2} \\ CH_3CH_2CH_2CH_2OH \longrightarrow CH_3CH=CHCH_3 + CH_3CH_3CH=CH_2 \\ (CH_3)_3CCHOHCH_3 \longrightarrow (CH_3)_2C=C(CH_3)_2 + (CH_3)_3CCH=CH_2 \\ \text{cyclohexyl-CH}_3\text{-OH} \longrightarrow \text{cyclohexenyl-CH}_3 + \text{cyclohexenyl-CH}_3 \end{cases} \quad (7.13)$$

The varying direction of elimination leads to structural isomerism (1-butene and 2-butene), whereas the formation of *cis* and *trans* isomers is stereochemical isomerism. In this section, only the stereochemistry of elimination will be discussed.

Effects of structure, configuration, and conformation upon E2. If *cis-trans* isomerism is possible in the products, the *trans* isomer predominates in almost all cases. In Section 6.2, we saw that *trans*-2-butene was more stable than *cis*-2-butene by 1 kcal. This extra stability is reflected in the lowered maximum of the energy hill for an elimination reaction that leads to the *trans* as compared to that for the *cis* isomer. However, the structure, configuration, and conformation of the reacting molecule often determine the structure and configuration of the product. Consider *meso* and ($\pm$)-2,3-dibromobutane and the reaction with a base in an E2 type of reaction: let us examine the possible pathways that lead to 2-bromo-2-butene (Fig. 7.5). (Other products are formed by elimination, such as 3-bromo-1-butene, 1,3-butadiene, and so on; but for the present it will be the stereochemistry of the 2-butenes that will be of interest to us.)

The three relatively stable staggered rotamers of the (RS)-meso configuration of 2,3-dibromobutane are shown on the left hand side of the figure, as well as one of the eclipsed conformers (in brackets). In the center are shown the transition states in which the departing H and Br are coplanar with C_2 and C_3.

Coplanarity is needed in the transition state to accommodate the partially formed π bond as well as the partially broken C—H and C—Br σ bonds. The requirement of coplanarity allows only two possible transition states: one in which H and Br leave from opposite sides of the double bond being formed, and one in which H and Br leave from the same side. Opposite-side elimination is called *trans* elimination, and same side elimination is called *cis* elimination. The use of *cis* and *trans* to describe the orientation of the elimination has nothing to do with the name of the alkene that is produced. Completion of the E2 reaction gives rise to *cis*-2-bromo-2-butene from (RS)-meso-2,3-dibromobutane if elimination is *trans*, and it gives *trans*-2-bromo-2-butene if elimination is *cis*.

For the (RR) enantiomer of the racemate of ($\pm$)-2,3-dibromobutane, only one staggered rotamer is shown (the one that can give rise to *trans* elimination—either Br may leave but only one is shown). The eclipsed form that can give *cis* elimination is also shown. Racemic 2,3-dibromobutane gives *trans*-2-bromo-2-butene by way of *trans* elimination and the *cis* alkene by *cis* elimination.

If we examine the transition states a little more closely, we see that the pair of electrons being released by the proton moves into the forming double bond on the side opposite to the departing bromide ion in *trans* elimination. In a sense, the electrons are executing an S_N2 reaction on the C—Br bond. In *cis* elimination, the electron pair is required to move into place on the same side of the incipient double bond that the bromine is still partially occupying. Therefore, *trans* elimination should be favored. In addition, even if the energy hill to be climbed from the ground state (the starting conformation) were the

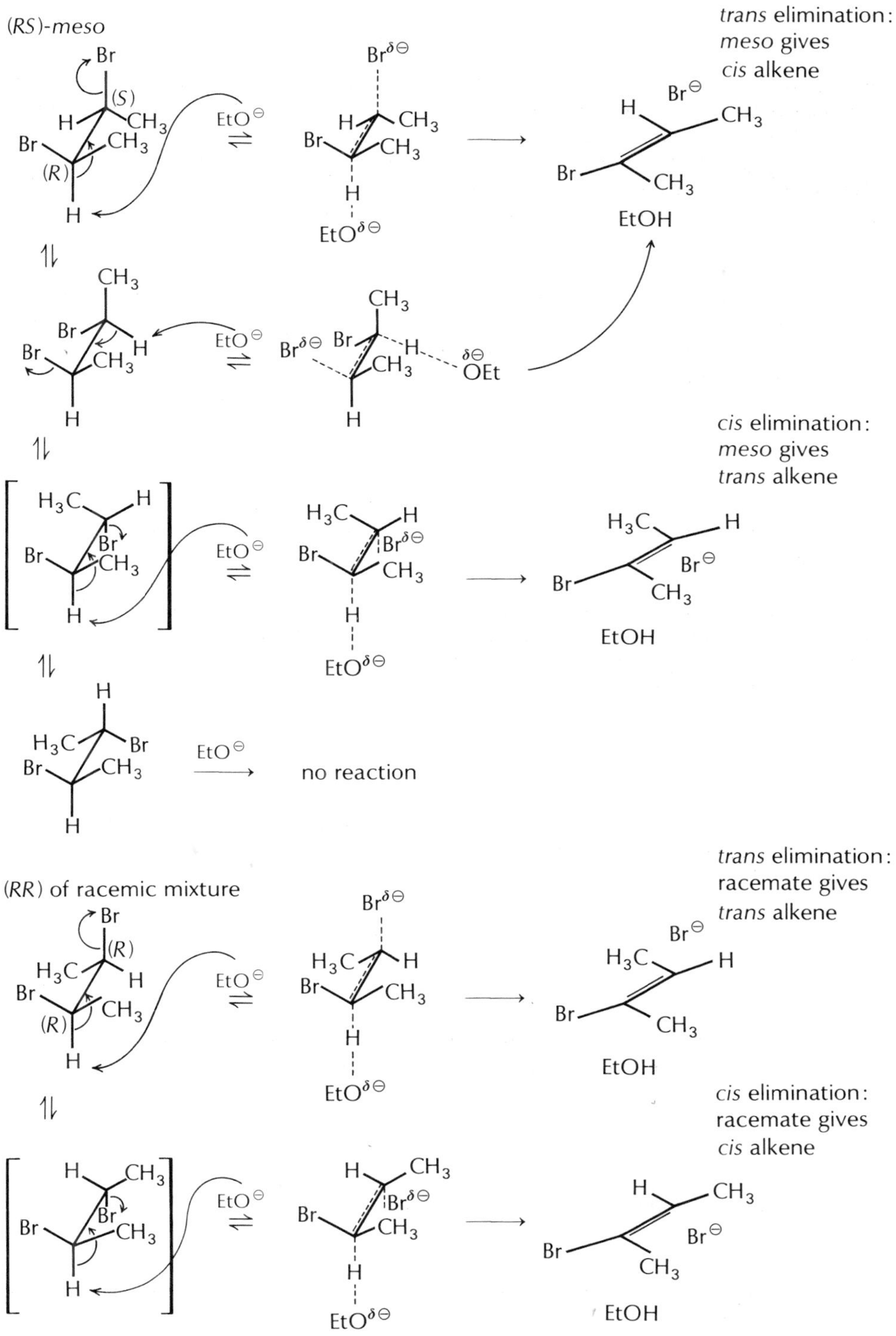

Figure 7.5 Pathways for E2 elimination

same for both *cis* and *trans* elimination, the concentration of staggered conformers is so great relative to the eclipsed form necessary for *cis* elimination that *trans* elimination should be overwhelmingly favored—and it is.

The *meso* compound gives the *cis* olefin, and the racemic mixture (or either of the enantiomers) gives the *trans* olefin exclusively. Many other cases have been examined, and *trans* elimination has been shown to be the rule for E2 reactions.

> A few instances of *cis* elimination have been shown to exist, but they can all be thought of as special cases in which the structures either cannot give a coplanar *trans* elimination transition state or tend to favor a shift to an E1cB (☐ ☐) or other mechanism. By inversion of the rule, we may say that all E2 reactions are considered to proceed by *trans* elimination. We say that the reaction is <u>stereoselective.</u>

E1 eliminations. The E1 mechanism is much less stereoselective than the E2. In the E1 mechanism, an ion has left before a stereochemical decision must be made by the carbonium ion. To some extent, ion-pair intervention favors *trans* over *cis* elimination in the same way that inversion is somewhat favored over retention in S_N1 reactions. The more stable the cation formed, the more chance for rotation about the σ bonds in the ion, and the more chance for scrambling the conformations of the proton to be lost. Thus the ratio of *trans* to *cis* alkene approaches that predictable from the relative stabilities of the olefins: more *trans* than *cis* isomer.

Exercises

7.6 For an E2 elimination reaction (hot alcoholic potassium hydroxide) on 1-bromo-1-ethylcyclopentane, which product is favored, *a* or *b*? Why?

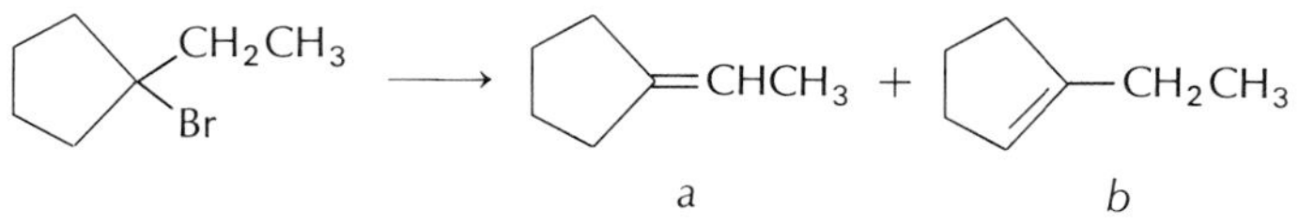

7.7 One of the products obtained from an E2 elimination on 4-bromo-2,3,4-trimethylhexane is 2,3,4-trimethyl-3-hexene. If the halide has the (3S,4R) configuration, will the alkene be *cis* or *trans*?

7.7 Elimination reactions; structural isomerism; alkene stability

Saytzeff rule. As we pointed out in Section 7.6, the direction of elimination may lead to the formation of two or more alkenes of differing structure (Eqs. 7.11–7.13). The olefins whose double bond is not at the end of a chain tend to predominate. This correlation was first noted by Saytzeff almost a hundred years ago. He suggested a rule: the hydrogen atom to be removed would come preferably from that neighboring carbon atom with the fewest number of hydrogens. The statement may be reworded to say that the alkene with

7.7 Elimination reactions; structural isomerism; alkene stability

the greatest number of substituents about the double bond will be preferred; or, very simply, the most stable alkene will be favored.

Hydrogenation as a measure of stability of alkenes. One method for comparing the stability of alkenes is to compare the heats of hydrogenation (defined as $-\Delta H$ for the reactions, all of which are exothermic). Some selected values are given in Table 7.3.

Table 7.3 Heats of hydrogenation of some alkenes[a]

	Alkene	Product	$-\Delta H$
1.	$H_2C=CH_2$	CH_3CH_3	32.8
2.	$CH_3CH=CH_2$	$CH_3CH_2CH_3$	30.1
3.	$CH_3CH_2CH=CH_2$	$CH_3CH_2CH_2CH_3$	30.3
4.	cis-$CH_3CH=CHCH_3$	$CH_3CH_2CH_2CH_3$	28.6
5.	trans-$CH_3CH=CHCH_3$	$CH_3CH_2CH_2CH_3$	27.6
6.	$(CH_3)_2C=CH_2$	$(CH_3)_2CHCH_3$	28.4
7.	$(CH_3)_2CHCH=CH_2$	$(CH_3)_2CHCH_2CH_3$	30.3
8.	$CH_2=C(CH_3)CH_2CH_3$	$(CH_3)_2CHCH_2CH_3$	28.5
9.	$(CH_3)_2C=CHCH_3$	$(CH_3)_2CHCH_2CH_3$	26.9
10.	$(CH_3)_3CCH=CH_2$	$(CH_3)_3CCH_2CH_3$	30.3
11.	$(CH_3)_2C=C(CH_3)_2$	$(CH_3)_2CHCH(CH_3)_2$	26.6
12.	$(CH_3)_3CCH_2C(CH_3)=CH_2$	$(CH_3)_3CCH_2CH(CH_3)_2$	25.5
13.	$(CH_3)_3CCH=C(CH_3)_2$	$(CH_3)_3CCH_2CH(CH_3)_2$	26.8
14.	cis-$(CH_3)_3CCH=CHC(CH_3)_3$	$(CH_3)_3CCH_2CH_2C(CH_3)_3$	36.2
15.	trans-$(CH_3)_3CCH=CHC(CH_3)_3$	$(CH_3)_3CCH_2CH_2C(CH_3)_3$	26.9

[a] At 82° over Pt, 1 atm, gas phase, for compounds 1–11; at 25° in acetic acid for compounds 12–15. The gas-phase values are about 1.5–2.0 kcal greater than solution-phase values.

We see that as the number of alkyl substituents about the double bond increases, the heats of hydrogenation decrease. Compare ethylene at 32.8 kcal with compounds 2, 3, 7, and 10 in Table 7.3 at 30.1 to 30.3 kcal, all of which are $RCH=CH_2$ with R = Me, Et, iPr, and tBu. For compounds with two R groups present on the same carbon or trans at either end of the double bond, compare compounds 5, 6, 8, and 15, at 26.9 to 28.5 kcal.

We see again that trans-2-butene is 1 kcal more stable than cis-2-butene (compare 5 and 4). However, as we increase the size of the R groups from methyl to tert-butyl, trans-2,2,5,5-tetramethyl-3-hexene (15) becomes $36.2 - 26.9 = 9.3$ kcal more stable than the cis isomer (14). The steric crowding of the cis tert-butyl groups leads to the marked instability of compound 14. The crowding is relieved in the saturated hydrocarbon by free rotation about the C_3-C_4 bond.

Three and four alkyl groups (compounds 9, 11, and 13) increase the stability only slightly (26.6 to 26.9 kcal).

Polar effects and stability. Alkyl groups stabilize an alkene relative to the saturated hydrocarbon by the +I effect (the sp^2 carbon atom is more electronegative than an sp^3

166   Chapter seven

carbon atom) and by the +R effect (hyperconjugation). The +R effect is much smaller in an alkene than in a carbonium ion, because in the former, charge is being separated, whereas in the latter it is dispersed.

$$\overset{H}{\underset{|}{CH_2}}\text{—}CH\text{=}CH_2 \longleftrightarrow CH_2\text{=}CH\text{—}\overset{\ominus}{CH_2} \quad (\text{with } \overset{\oplus}{H})$$

Compare compounds 12 and 13 in Table 7.3. Normally, we would expect the trisubstituted ethylene (13) to be more stable than the isomeric disubstituted ethylene (12). The heat of hydrogenation of compound 13, 26.8 kcal, is not out of line, but the heat of hydrogenation of compound 12 is unexpectedly low (stable) at 25.5 kcal. Compare compounds 6 and 8, at 28.4 and 28.5 kcal. Compound 13 possesses six hydrogens capable of hyperconjugation (the two CH_3— groups) and compound 12 has five. But compound 12 also has a *tert*-butyl group in the correct position to con-

$$(CH_3)_3C\text{—}H_2C\diagdown\diagup CH_2 \quad \longleftrightarrow \quad (CH_3)_3\overset{\oplus}{C}H_2C\diagdown\diagup \overset{\ominus}{CH_2}$$
$$H_3C\diagup \phantom{C\text{=}CH_2} \qquad\qquad H_3C\diagup \phantom{C\text{—}CH_2}$$

tribute. Because a *tert*-butyl group is relatively good at accepting a positive charge, it is not too bad an assumption to say that a tBu > H in hyperconjugation effect. This order could be offered as an explanation for the unexpected stability of compound 12. Some authors have ascribed the greater stability of 12 compared to 13 to the greater nonbonded repulsions in 13 (tBu vs Me). Although steric interaction undoubtedly is a factor, it cannot explain the stability of compound 12 as compared to compounds 6 and 8.

In general, then, the order of stability is $R_2C\text{=}CR_2 > R_2C\text{=}CHR > R_2C\text{=}CH_2$, *trans*-$RCH\text{=}CHR$ > *cis*-$RCH\text{=}CHR$ > $RCH\text{=}CH_2$ > $H_2C\text{=}CH_2$. It also is the Saytzeff order of products of elimination. Put very simply, the Saytzeff rule predicts that the double bond prefers not to be at the end of a chain.

Rearrangements in E1cA reactions. The Saytzeff rule applies to all E1 eliminations. Even when rearrangement intervenes, completion of elimination by loss of a proton from the carbonium ion gives the more stable alkene (the most substituted). Rearrangement is more common in E1cA reactions of alcohols than in alkyl halides (Eqs. 7.13). If 1-butene is to be prepared, an E2 reaction on butyl bromide is preferred to a dehydration of butyl alcohol (Eqs. 7.14). If a 2-alkene is wanted, one begins with a 2-alkyl halide.

Exhaustive methylation; Hofmann elimination. Other eliminations follow the Hofmann rule, which was formulated even before the Saytzeff rule. The Hofmann rule states that the least substituted alkene will be formed in the decomposition of a quaternary ammonium hydroxide.

If a tertiary amine is allowed to react with methyl iodide, a quaternary ammonium iodide is formed (Eqs. 7.15) by an S_N2 displacement of iodide ion. The process of converting

7.7 Elimination reactions; structural isomerism; alkene stability

$$CH_3CH_2CH_2CH_2Br \xrightarrow[\Delta]{NaOH \atop EtOH} CH_3CH_2CH=CH_2$$

$$CH_3CH_2CH_2CH_2OH \xrightarrow[\Delta]{H_2SO_4} [CH_3CH_2CH_2\overset{\oplus}{C}H_2] \quad (7.14)$$

$$\downarrow$$

$$[CH_3CH_2\overset{\oplus}{C}HCH_3] \quad \text{small} \atop -H^{\oplus}$$

$$\overset{-H^{\oplus}}{\swarrow} \qquad \overset{-H^{\oplus}}{\searrow}$$

$$CH_3CH=CHCH_3 \qquad CH_3CH_2CH=CH_2$$

$$trans > cis \qquad \qquad small$$

$$R_3\overset{\frown}{N} + CH_3\overset{\frown}{-I} \longrightarrow R_3\overset{\oplus}{N}CH_3 \; I^{\ominus}$$

$$R_2NH \xrightarrow{CH_3I} R_2\overset{\oplus}{N}HCH_3 I^{\ominus} \xrightarrow{HO^{\ominus}} R_2NCH_3 \xrightarrow{CH_3I} R_2\overset{\oplus}{N}(CH_3)_2 I^{\ominus} \quad (7.15)$$

$$RNH_2 \xrightarrow[\text{2. NaOH}]{\text{1. } CH_3I} RNHCH_3 \xrightarrow[\text{2. NaOH}]{\text{1. } CH_3I} RN(CH_3)_2 \xrightarrow{CH_3I} \overset{\oplus}{R}N(CH_3)_3 I^{\ominus}$$

an amine into a quaternary salt by successive treatments with CH_3I and NaOH is known as <u>exhaustive methylation</u>. Quaternary ions are quite stable and do not react with water in the same way that ordinary ammonium and amine salts do. Because there is no hydrogen bonded to nitrogen, the quaternary ion is unable to donate a proton to the solvent (Eqs. 7.16).

$$R\overset{\oplus}{N}H_3 + H_2O \rightleftharpoons R\overline{N}H_2 + H_3O^{\oplus} \quad (7.16)$$

$$R_3\overset{\oplus}{N}CH_3 + H_2O \rightleftharpoons \text{no reaction}$$

Also, because the quaternary ion is stable, the replacement of the counter ion (iodide ion, for example) by hydroxide ion gives rise to a base so strong that ordinary strong hydroxides such as sodium and potassium hydroxides are unable to push the reaction to completion. Moist silver oxide causes the reaction to go to completion by precipitation of the very insoluble silver iodide (Eq. 7.17).

$$2I^{\ominus} + Ag_2O + H_2O \longrightarrow 2AgI + 2OH^{\ominus} \quad (7.17)$$

At moderately high temperatures (100°) quaternary ammonium hydroxides decompose by an S_N2 reaction if a methyl group is present (Eq. 7.18). If an ethyl or larger alkyl

$$R_3\overset{\oplus}{N}\overset{\frown}{-}CH_3 + HO^{\ominus} \xrightarrow{100°} R_3N + CH_3OH \quad (7.18)$$

group is present, an E2 reaction takes place (Eq. 7.19). If the elimination is possible, it

$$R_3\overset{\oplus}{N}\overset{\frown}{-}CH_2-CH_2\overset{\frown}{-}H + HO^{\ominus} \xrightarrow{100°} R_3N + H_2C=CH_2 + H_2O \quad (7.19)$$

always exceeds the substitution. Only a methyl is reactive enough in substitution to be able to compete at all with the E2 reaction.

Other examples of the Hofmann rule are shown in Equations 7.20.

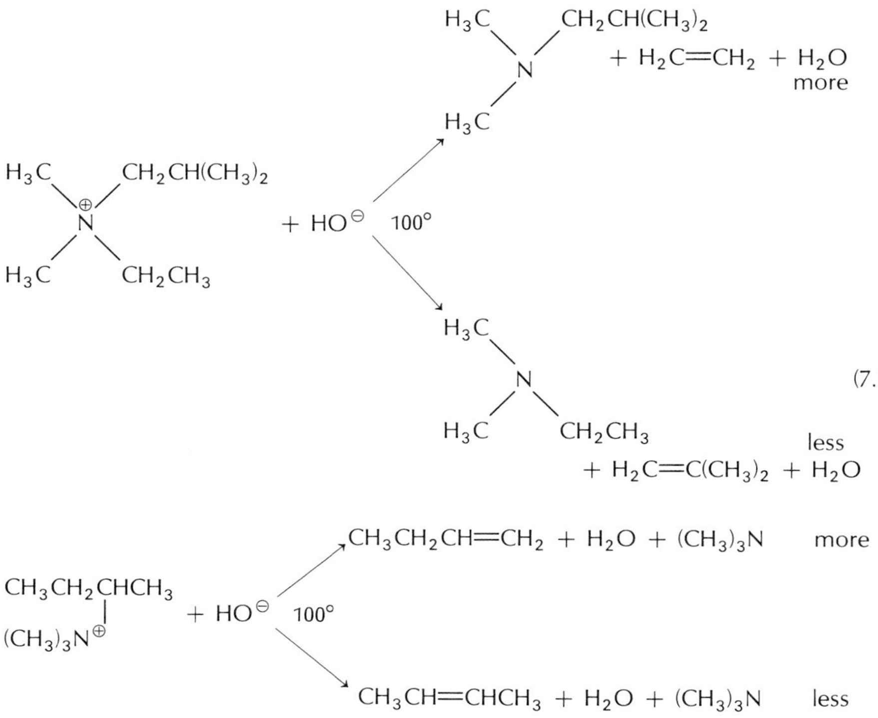

$$\begin{array}{c}
\underset{H_3C}{H_3C}\!\!\diagdown\!\!\underset{|}{\overset{CH_2CH(CH_3)_2}{N}}\!\!\diagup + H_2C=CH_2 + H_2O \\
\text{more}
\end{array}$$

$$\underset{H_3C}{\overset{H_3C}{\diagdown}}\!\overset{\oplus}{N}\!\underset{CH_2CH_3}{\overset{CH_2CH(CH_3)_2}{\diagup}} + HO^{\ominus} \quad 100°$$

$$\begin{array}{c}
\underset{H_3C}{H_3C}\!\!\diagdown\!\!\underset{CH_2CH_3}{\overset{N}{\diagup}} \\
+ H_2C=C(CH_3)_2 + H_2O \quad \text{less}
\end{array} \quad (7.20)$$

$$\begin{array}{c}
CH_3CH_2CHCH_3 \\
| \\
(CH_3)_3\overset{\oplus}{N}
\end{array} + HO^{\ominus} \quad 100°$$

$$\nearrow CH_3CH_2CH=CH_2 + H_2O + (CH_3)_3N \quad \text{more}$$

$$\searrow CH_3CH=CHCH_3 + H_2O + (CH_3)_3N \quad \text{less}$$

The leaving groups in these reactions are a tertiary amine and a proton. The tertiary amine is a poor leaving group, so the carbon-hydrogen bond is presumed to be rather more broken at the transition state than is the carbon-nitrogen bond.

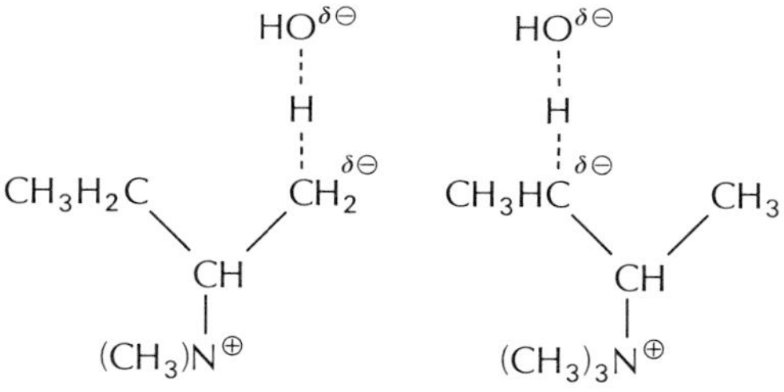

$$\begin{array}{cc}
HO^{\delta\ominus} & HO^{\delta\ominus} \\
\vdots & \vdots \\
H & H \\
CH_3H_2C\diagdown\underset{\delta\ominus}{CH_2} & CH_3HC\underset{\delta\ominus}{\diagdown}CH_3 \\
CH & CH \\
| & | \\
(CH_3)\overset{\oplus}{N} & (CH_3)_3\overset{\oplus}{N} \\
a & b
\end{array}$$

Attack of the base on the methyl group of sBu leads to a partial negative charge on a primary carbon (a). Attack on a hydrogen in the ethyl group leads to a partial negative charge on a secondary carbon (b). Because the order is tertiary > secondary > primary in the stabilization of a positive charge, it follows that primary > secondary > tertiary in toleration of a negative charge. Thus, the less substituted alkene is favored the more anionic character there is to the elimination reaction. Otherwise, the elimination tends to follow the Saytzeff rule.

Steric effects can alter the relative amounts of Saytzeff or Hofmann elimination. (Here we use the rules loosely as synonymous with direction of elimination, into the chain or end of the chain.) In general, E2 reactions of alkyl halides with ethoxide ion give Saytzeff elimination with some concomitant Hofmann elimination (Eq. 7.21). Compare the results

$$\underset{\text{CH}_3\text{CH}_2\text{CHBr}}{\overset{\text{CH}_3}{|}} \xrightarrow[\text{EtO}^\ominus,\ \text{EtOH}]{\text{E2}} \underset{19\%}{\overset{\text{CH}_2}{\overset{||}{\text{CH}_3\text{CH}_2\text{CH}}}} + \underset{81\%\ (cis\ \text{and}\ trans)}{\overset{\text{CH}_3}{\overset{|}{\text{CH}_3\text{CH}=\text{CH}}}} \quad (7.21)$$

for ethoxide on 2-bromobutane (Eq. 7.21) with those for *tert*-butoxide (Eq. 7.12), in which 1-butene exceeded the 2-butenes in yield. □ □

To summarize this section, we may say that Saytzeff elimination (alkene stability) governs the direction of elimination in all E1 reactions and tends to predominate in all E2 reactions unless steric factors (increasing size of base, increased need of solvation, and lack of bond stretch of the departing negative ion) and polar factors (anionic character imparted by a poor leaving group) intervene to favor Hofmann elimination.

Exercises

7.8 Show the expected Saytzeff and Hofmann products and relative amounts from the E2 dehydrobromination by EtO$^\ominus$ in EtOH of these compounds:

   a. (S)-CH$_3$ĊHBrCH$_2$CH$_3$
   b. (R)-(CH$_3$)$_2$CHĊHBrCH$_3$
   c. BrCH$_2$CH$_2$CH$_2$CH$_2$Br
   d. (RS)-CH$_3$ĊHBrĊHBrCH$_3$
   e. 1-bromo-1-methylcyclopentane
   f. (R)-1-bromo-1-cyclopentylethane
   g. (RR)-1-bromo-2-methylcyclopentane

7.8 Applications to synthesis

Introduction. After the preceding discussions, we must clarify the methods used in the laboratory to carry out in reasonable yield any of these five reactions:

1. alcohol to alkyl halide (substitution)
2. alkyl halide to new halide (substitution)
3. alkyl halide to alcohol (substitution)
4. alcohol to alkene (elimination by dehydration)
5. alkyl halide to alkene (elimination by dehydrohalogenation)

Alcohols to alkyl chlorides. Alcohols are converted to the corresponding alkyl chlorides with HCl, PCl$_3$, POCl$_3$, PCl$_5$ or SOCl$_2$. Concentrated hydrochloric acid by itself is useful only with tertiary alcohols (Sec. 7.1). Primary and secondary alcohols react so slowly that zinc chloride must be used with concentrated hydrochloric acid to speed up the reaction. Even so, there is so much S$_N$1cA reaction competing—with carbonium ion formation, rearrangement, and elimination going on—that other methods are preferable.

Thionyl chloride ($SOCl_2$) is widely used for the conversion of primary and secondary alcohols to the chlorides. Tertiary alcohols dehydrate with $SOCl_2$. The reaction usually goes readily and with a minimum of rearrangement or elimination. Purification of the product is easy because the HCl and SO_2 produced are gases, and they escape from the reaction mixture as the reaction proceeds.

The phosphorus compounds phosphorus trichloride, oxychloride, and pentachloride also are used on primary and secondary alcohols. Tertiary alcohols dehydrate with these reagents. The reagents promote the S_N2 reaction and are accompanied by some elimination. The first step of the mechanism involves bonding of phosphorus of PCl_3 to oxygen (Eq. 7.22). Subsequent steps are shown. The $HOPCl_2$ reacts with more alcohol

$$ROH + SOCl_2 \longrightarrow RCl + SO_2 + HCl$$

$$R\overset{\frown}{OH} + PCl_3 \longrightarrow \underset{H}{RO\overset{\oplus}{-}\overset{\ominus}{PCl_3}} \xrightarrow{-HCl} RPOCl_2 + HCl \tag{7.22}$$

(with subsequent branching to $Cl-R + HOPCl_3$ and $Cl-R + \overset{\ominus}{O}PCl_2$, both leading to $HOPCl_2$ via $-Cl^{\ominus}$ and $+H^{\oplus}$ respectively)

and ends up as phosphorus acid, H_3PO_3. Phosphorus oxychloride and pentachloride end up as phosphoric acid, H_3PO_4.

Alcohols to alkyl bromides and iodides. Alkyl bromides are preparable by the action of HBr (or H_2SO_4 + NaBr) on primary and secondary alcohols. The more expensive PBr_3 or $SOBr_2$ may be used on both primary and secondary alcohols. Iodides are not usually made from alcohols with HI. Hydriodic acid is the only easily available reagent; and although the alkyl iodide may be obtained, the yield is low if excess concentrated hydriodic acid is used. The acid is a good reducing agent (Eq. 7.23). Unless the alkyl iodide is removed from

$$I\overset{\frown}{-}H \overset{\frown}{+} R-\overset{\frown}{I+}I^{\ominus} \longrightarrow I^{\ominus} + HR + I_2 \tag{7.23}$$

the reaction mixture as rapidly as it is formed, it suffers reduction to the hydrocarbon. A better method is use of PI_3, which is prepared in the reaction mixture by addition of phosphorus and iodine as needed. Happily, iodides are available from other halides.

Alkyl halides to other halides. Sodium iodide is reasonably soluble in acetone or methanol, whereas sodium chloride and bromide are almost insoluble. An alkyl chloride or bromide dissolved in a solution of sodium iodide in acetone or methanol undergoes S_N2 reaction readily. The reverse reaction does not occur because of the very low concentration of chloride or bromide ions (Eq. 7.24). Of course, this conversion works best on primary halides, less so on secondary, and very poorly on tertiary halides.

$$I^{\ominus} \overset{\frown}{+} R\overset{\frown}{-}Cl \xrightarrow{acetone} IR + Cl^{\ominus} \xrightarrow{Na^{\oplus}} \underline{NaCl\downarrow} \tag{7.24}$$

A roundabout but excellent method of changing one halide into another is via the Grignard reagent (Eq. 7.25). This method works well on primary, secondary, and tertiary

$$RX + Mg \xrightarrow{\text{ether}} RMgX \xrightarrow{X_2'} RX' + MgXX' \qquad (7.25)$$

halides, whatever X and X' may be. Alkyl halides (other than fluorides) react readily with magnesium in ether to give the Grignard reagent without rearrangement. If the halogen is attached to an asymmetric carbon, an optically active halide is racemized in the process. Other asymmetric centers are unaffected.

Alkyl halides to alcohols. We do not often need to convert a halide into an alcohol in a synthetic sequence in the laboratory, because alcohols are usually the precursors of alkyl halides. If it is necessary, a highly aqueous medium is used with only enough of an inert, water-miscible solvent present (acetone) to bring the alkyl halide into at least partial solubility in the mixture. Although the presence of hydroxide ion speeds up the reaction, it also increases the competing elimination reactions, which cannot be completely avoided under any conditions. In any case, S_N1 conditions should be minimized for the hydrolysis of primary and secondary halides to prevent carbonium ion formation, rearrangement, and elimination. Thus, a base must be present. Sodium or potassium hydroxide usually is used, but only in slight excess. With tertiary halides, S_N1 and E1 reactions are almost impossible to avoid. At high concentrations of base, the E2 reaction intervenes, and elimination becomes the principal reaction.

Alcohols to alkenes; alkyl halides to alkenes. The dehydration of alcohols to produce alkenes is accomplished by heating with sulfuric acid, phosphoric acid, or potassium bisulfate. The alkene, with a boiling point less than that of the alcohol from which it is produced, distils from the reaction mixture as it is formed, along with the accompanying water. An alternative is to pass the alcohol, as a vapor, over alumina (Al_2O_3) at 350° to 450°. The ease of dehydration varies with the structure of the alcohol: tertiary > secondary > primary. See Equations 7.26, in which only the principal product is shown. If re-

$$CH_3CH_2OH \xrightarrow[170°]{95\% \ H_2SO_4} CH_2{=}CH_2 + H_2O$$

$$CH_3CH_2CH_2CHOHCH_3 \xrightarrow[95°]{60\% \ H_2SO_4} CH_3CH_2CH{=}CHCH_3 + H_2O$$

$$\underset{\underset{CH_3}{|}}{\overset{\overset{CH_3}{|}}{CH_3CH_2\underset{}{C}OH}} \xrightarrow[95°]{45\% \ H_2SO_4} CH_3CH{=}C\overset{CH_3}{\underset{CH_3}{\diagdown}} + H_2O \qquad (7.26)$$

$$\text{cyclohexanol} \xrightarrow[140°]{95\% \ H_2SO_4} \text{cyclohexene} + H_2O$$

arrangement to a more stable carbonium ion is possible, the major product is the most stable alkene, formed by loss of a proton from the cation according to the Saytzeff rule (Eqs. 7.27).

$$CH_3CH_2CH_2CH_2OH \xrightarrow[140°]{75\% \ H_2SO_4} CH_3CH=CHCH_3$$

$$(CH_3)_3CCHOHCH_3 \xrightarrow[95°]{60\% \ H_2SO_4} (CH_3)_2C=C(CH_3)_2$$

$$(CH_3)_2CHCH(CH_3)CH_2OH \xrightarrow[140°]{75\% \ H_2SO_4} (CH_3)_2C=C(CH_3)_2 \quad (7.27)$$

$$(CH_3)_2CHCH_2CH_2OH \xrightarrow[140°]{75\% \ H_2SO_4} (CH_3)_2C=CHCH_3$$

[cyclohexanol with two CH₃ groups] $\xrightarrow[95°]{60\% \ H_2SO_4}$ [1,2-dimethylcyclohexene] + [isopropylidenecyclopentane]

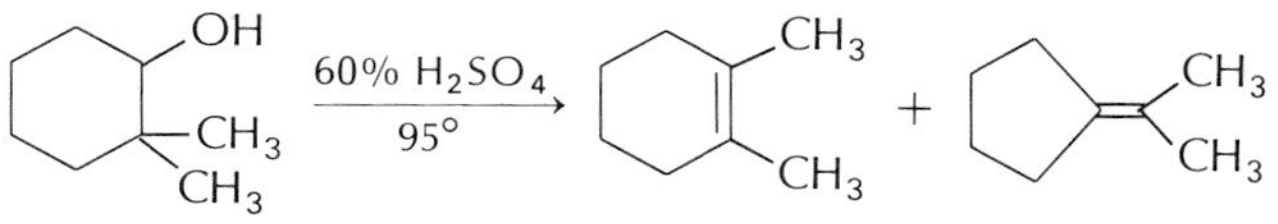

The preparation of 1-alkenes by dehydration of primary alcohols in good yield is not possible. (Ethylene and propylene are exceptions.) Thus, dehydrohalogenation (E2) should be used on the proper 1-haloalkane. Primary halides usually are easily prepared from the corresponding alcohol by $SOCl_2$ or a phosphorus halide (Eqs. 7.28).

$$CH_3CH_2CH_2CH_2OH \xrightarrow[\Delta]{H_2SO_4} CH_3CH=CHCH_3$$

$$\downarrow SOCl_2 \qquad\qquad\qquad\qquad\qquad\qquad (7.28)$$

$$CH_3CH_2CH_2CH_2Cl \xrightarrow[80°]{alc.\ NaOH} CH_3CH_2CH=CH_2$$

Under E2 conditions, a secondary alkyl halide usually follows the Saytzeff rule if sodium hydroxide in ethanol is used. Rearrangement is not a problem in E2 reactions, so we need worry only about the direction of elimination. To increase the yield of the Hofmann product, we shift to the use of potassium *tert*-butoxide in *tert*-butyl alcohol (Sec. 7.7) and preferably use the proper alkyl chloride. There is no advantage in preparing a *tert*-alkyl halide for E2 elimination with ethoxide, because Saytzeff elimination will occur to give the same result as that of the easy E1cA dehydration of the *tert*-alkanol (Eqs. 7.29).

$$(CH_3CH_2)_3COH \xrightarrow[90°]{20\% \ H_2SO_4} \begin{array}{c} CH_3CH_2 \\ \diagdown \\ C=CHCH_3 \\ \diagup \\ CH_3CH_2 \end{array} \quad (7.29)$$

$$\downarrow conc.\ HCl \qquad \nearrow NaOH/EtOH$$

$$(CH_3CH_2)_3CCl$$

Alcohols from aldehydes and ketones. Because alcohols are the key to the preparation of alkyl halides and alkenes, we shall introduce a method for the preparation of alcohols: the addition of Grignard reagents (or of alkyl lithium) to the carbon-oxygen double bond of aldehydes and ketones (Eqs. 7.30). The reactions are carried out by addition of a dry ether solution of the aldehyde or ketone to the already prepared Grignard reagent solution. The first product formed is written as a tight ion pair, a mixed magnesium salt. The

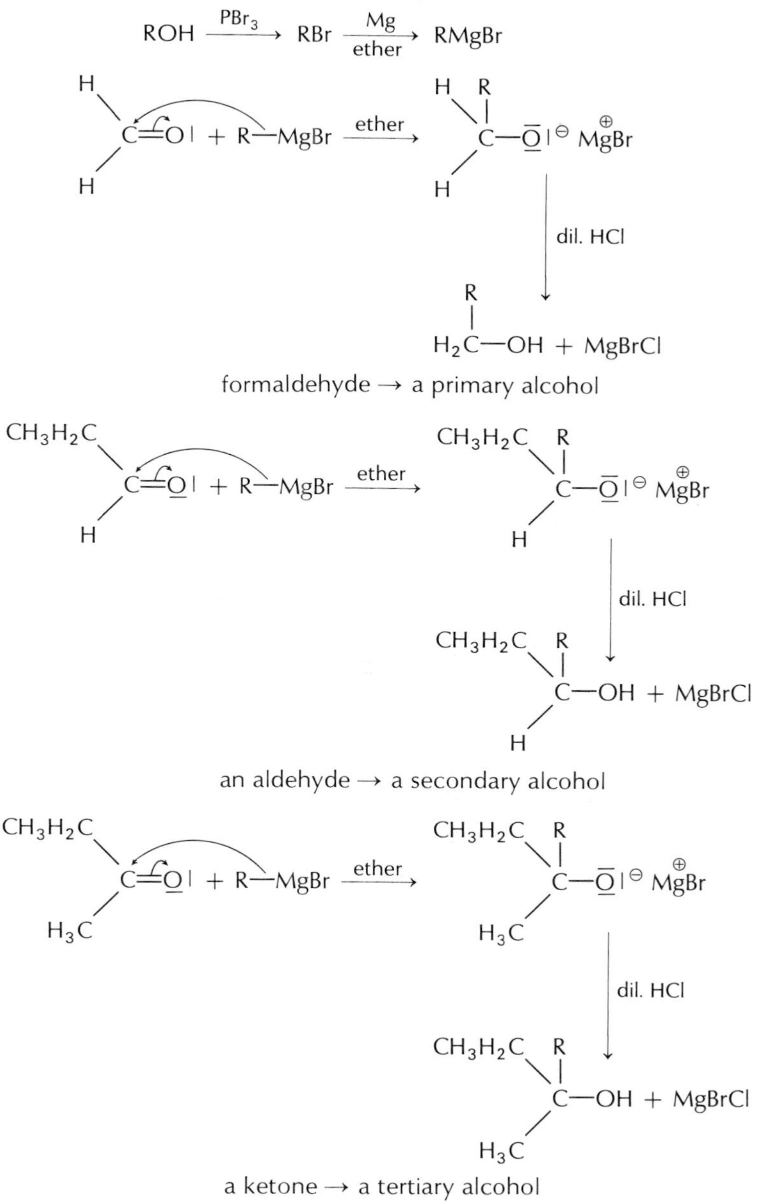

7.8 Applications to synthesis

$$\text{ROH} \xrightarrow{\text{PBr}_3} \text{RBr} \xrightarrow[\text{ether}]{\text{Mg}} \text{RMgBr} \qquad a$$

$$\underset{H}{\overset{H}{\diagdown}}\text{C}{=}\ddot{\text{O}}| + \text{R}{-}\text{MgBr} \xrightarrow{\text{ether}} \underset{H}{\overset{H}{\diagdown}}\underset{|}{\overset{R}{\text{C}}}{-}\bar{\underline{\text{O}}}|^{\ominus}\ \overset{\oplus}{\text{MgBr}}$$

$$\downarrow \text{dil. HCl}$$

$$\text{H}_2\underset{|}{\overset{R}{\text{C}}}{-}\text{OH} + \text{MgBrCl} \qquad b$$

formaldehyde → a primary alcohol

$$\underset{H}{\overset{CH_3H_2C}{\diagdown}}\text{C}{=}\ddot{\text{O}}| + \text{R}{-}\text{MgBr} \xrightarrow{\text{ether}} \underset{H}{\overset{CH_3H_2C}{\diagdown}}\underset{|}{\overset{R}{\text{C}}}{-}\bar{\underline{\text{O}}}|^{\ominus}\ \overset{\oplus}{\text{MgBr}} \qquad (7.30)$$

$$\downarrow \text{dil. HCl}$$

$$\underset{H}{\overset{CH_3H_2C}{\diagdown}}\underset{|}{\overset{R}{\text{C}}}{-}\text{OH} + \text{MgBrCl} \qquad c$$

an aldehyde → a secondary alcohol

$$\underset{H_3C}{\overset{CH_3H_2C}{\diagdown}}\text{C}{=}\ddot{\text{O}}| + \text{R}{-}\text{MgBr} \xrightarrow{\text{ether}} \underset{H_3C}{\overset{CH_3H_2C}{\diagdown}}\underset{|}{\overset{R}{\text{C}}}{-}\bar{\underline{\text{O}}}|^{\ominus}\ \overset{\oplus}{\text{MgBr}}$$

$$\downarrow \text{dil. HCl}$$

$$\underset{H_3C}{\overset{CH_3H_2C}{\diagdown}}\underset{|}{\overset{R}{\text{C}}}{-}\text{OH} + \text{MgBrCl} \qquad d$$

a ketone → a tertiary alcohol

addition of dilute sulfuric or hydrochloric acids decomposes the alkoxide into water-soluble magnesium salts. The product alcohol remains in the ether layer. The ether layer is separated and dried, and the ether is removed from the alcohol by distillation.

Except for formaldehyde, which gives a primary alcohol (b) with one more carbon atom than in the R group of the starting alcohol, aldehydes give secondary alcohols (c), and ketones give tertiary alcohols (d). By selection of the proper alcohol (a) to begin with, and of the proper aldehyde or ketone, a variety of alcohols may be prepared. If necessary, the new alcohol may be used in a second sequence of reactions to give a second new alcohol. When we have obtained the proper alcohol, we may prepare the corresponding alkyl halide or alkene.

Of course, we should keep in mind that aldehydes may be reduced to primary alcohols, and ketones to secondary alcohols. Either LiAlH$_4$ in ether, NaBH$_4$ in water, or catalytic hydrogenation may be used.

Synthesis. The game of <u>Synthesis</u> may now be set up. The rules usually will supply a variety of simple organic compounds (mainly alcohols) from which we are to devise a set of equations, utilizing any inorganic reagents and any solvents needed, to prepare some given compound. To begin, let us make available any alcohol, aldehyde, and ketone with up to four carbon atoms. Now suppose the problem is to prepare 3-methylpentane.

Until we have gained experience, we can waste much time and effort in trying out various routes from the starting materials. *Begin at the end,* or near it.

What reactions are available that produce saturated hydrocarbons? Reference to Chapter 3 reminds us of two such reactions, Equations 3.12 and 3.51. Which is better? The first reaction requires an alkyl halide from which to prepare a lithium or a Grignard reagent. The halogen could be on any of the six carbon atoms of 3-methylpentane. Tentatively, let us select a tertiary position, or 3-methyl-3-halopentane. The other reaction is the hydrogenation of an alkene. Which of the three possible alkenes will be the most highly substituted? The trisubstituted ethylene, 3-methyl-2-pentene, is the answer. But both the alkene and the alkyl halide would require the same precursor, 3-methyl-3-pentanol. Thus, we arrive at two possible pathways (Eqs. 7.31). Either route from the alcohol would be satisfactory.

$$\begin{array}{c}
\text{CH}_3 \\
| \\
\text{CH}_3\text{CH}_2\text{CHCH}_2\text{CH}_3
\end{array} \xleftarrow{\text{dil. HCl}} \begin{array}{c}
\text{CH}_3 \\
| \\
\text{CH}_3\text{CH}_2\text{CCH}_2\text{CH}_3 \\
| \\
\text{MgBr}
\end{array} \xleftarrow[\text{ether}]{\text{Mg}} \begin{array}{c}
\text{CH}_3 \\
| \\
\text{CH}_3\text{CH}_2\text{CCH}_2\text{CH}_3 \\
| \\
\text{Br}
\end{array}$$

$$\uparrow \text{H}_2, \text{Pt} \qquad \begin{array}{c} \text{CH}_3 \\ | \\ \text{CH}_3\text{CH}=\text{CCH}_2\text{CH}_3 \end{array} \xleftarrow{\text{conc. HBr}} \uparrow \qquad (7.31)$$

$$\text{H}_2\text{SO}_4 \Big| \Delta$$

$$\begin{array}{c}
\text{CH}_3 \\
| \\
\text{CH}_3\text{CH}_2\text{CCH}_2\text{CH}_3 \\
| \\
\text{OH}
\end{array}$$

Now, to prepare the alcohol, we note that it is tertiary: a ketone is needed. Only two are available by the rules of the game, acetone and 2-butanone. The latter, with EtMgBr, would give the alcohol. (Acetone would give an alcohol with two methyl groups, (CH$_3$)$_2$C(R)OH, and cannot be used.) Such step by step analysis, *from end to beginning,* will lead to an answer: Equations 7.32 followed by Equations 7.31.

There may be a variety of answers. The best answer is the one that involves the smallest number of operations or reactions, because every manipulation incurs a loss of material. However, the reactions chosen must give reasonable yields at each step; otherwise, a longer sequence with better yields may be preferable.

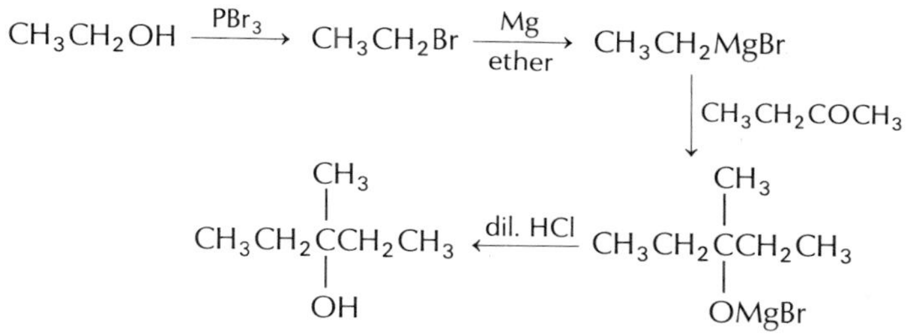

$$CH_3CH_2OH \xrightarrow{PBr_3} CH_3CH_2Br \xrightarrow[\text{ether}]{Mg} CH_3CH_2MgBr$$

$$\downarrow CH_3CH_2COCH_3$$

$$\underset{\underset{OH}{|}}{\overset{\overset{CH_3}{|}}{CH_3CH_2CCH_2CH_3}} \xleftarrow{\text{dil. HCl}} \underset{\underset{OMgBr}{|}}{\overset{\overset{CH_3}{|}}{CH_3CH_2CCH_2CH_3}} \quad (7.32)$$

7.9 Summary

In this chapter, nucleophilic substitution reactions and those elimination reactions that accompany substitution have been discussed in some detail. The polar and steric effects of the attacking reagent and of the substance being attacked have been pointed out. The influence of these effects upon the transition state and the stability of the products were also examined. The object of going into such detail is to learn to apply postulated influences in explaining and understanding organic reactions. In later chapters, we shall consider other reactions in a similar manner, but with less verbosity. We shall assume that as a result of this chapter, we shall recognize the effects of certain influences without having to have it all spelled out again and again.

The chapter ended with the introduction of the addition of Grignard reagents to carbon-oxygen double bonds. These reactions, when combined with reactions learned earlier, are useful in synthesis. Knowing them, we can use simple synthetic problems as a device to help us fix the reactions in mind. Some new reactions given in this chapter are as follows.

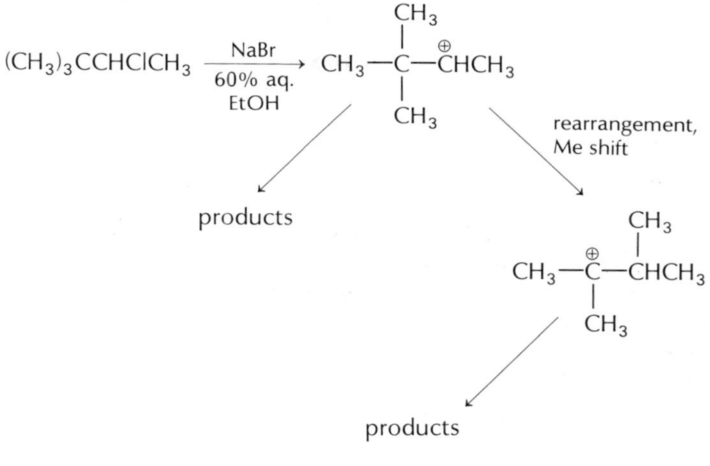

$$(CH_3)_3CCHClCH_3 \xrightarrow[\text{EtOH}]{NaBr \atop 60\% \text{ aq.}} CH_3-\underset{\underset{CH_3}{|}}{\overset{\overset{CH_3}{|}}{C}}-\overset{\oplus}{C}HCH_3$$

Fig. 7.4

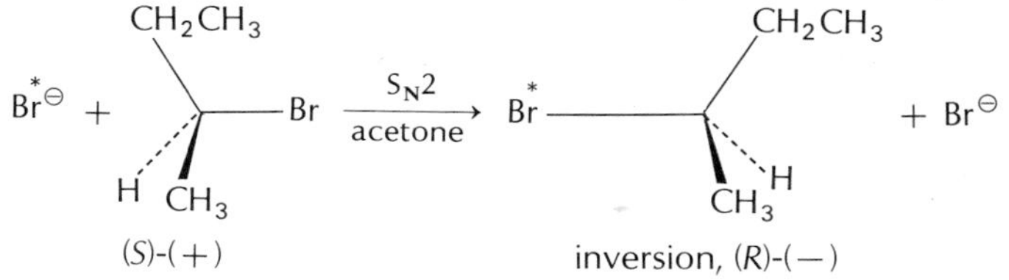

products

rearrangement, Me shift

$$CH_3-\underset{\underset{CH_3}{|}}{\overset{\overset{CH_3}{|}}{\overset{\oplus}{C}}}-CHCH_3$$

products

$$Br^{\ominus} + \underset{\underset{H\ CH_3}{}}{\overset{CH_2CH_3}{\diagup}}-Br \xrightarrow[\text{acetone}]{S_N2} \overset{*}{Br}-\underset{\underset{CH_3}{}}{\overset{CH_2CH_3}{\diagup}}H + Br^{\ominus} \quad (7.8)$$

(S)-(+)   inversion, (R)-(−)

$$R_3\overset{\oplus}{N}CH_2CH_3 + OH^{\ominus} \xrightarrow{100°} R_3N + H_2C=CH_2 + H_2O \quad (7.19)$$

$$ROH + SOCl_2 \longrightarrow RCl + SO_2 + HCl$$

$$3ROH + PX_3 \longrightarrow 3RX + H_3PO_3 \quad (7.22)$$

$$RX + Mg \xrightarrow{ether} RMgX \xrightarrow{X_2'} RX' + MgXX' \quad (7.25)$$

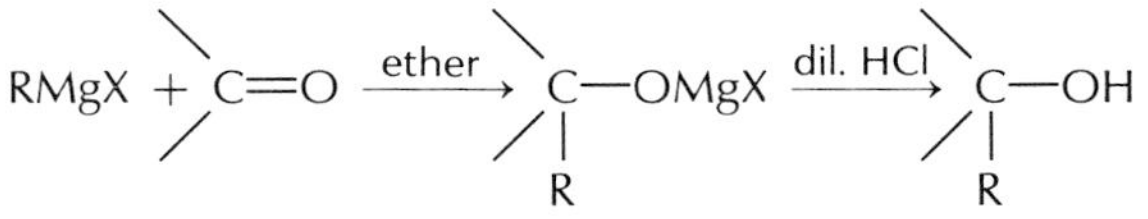

$$RMgX + \overset{\diagdown}{\underset{\diagup}{C}}=O \xrightarrow{ether} \underset{R}{\overset{\diagdown}{\underset{\diagup}{C}}-OMgX} \xrightarrow{dil.\ HCl} \underset{R}{\overset{\diagdown}{\underset{\diagup}{C}}-OH}$$

7.10 Problems

7.1 Write structural formulas for all the isomeric alkenes with molecular formula C_6H_{12}. Which is the most stable?

7.2 Using exhaustive methylation and Hofmann elimination, show the structure of the alkene product to be obtained from:

a. 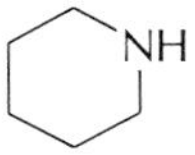 [piperidine structure]

b. 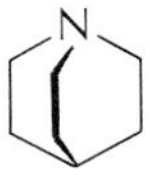 [quinuclidine structure]

c.  [bicyclic NH structure]

d. $(CH_3)_2CHCH_2CH_2Br$

e. $CH_3CH_2CHBrCH_3$

f. 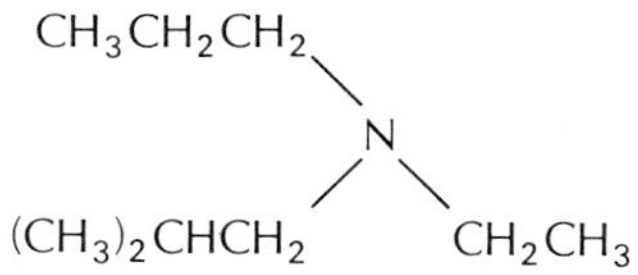 $CH_3CH_2CH_2\diagdown$
N
$(CH_3)_2CHCH_2\diagup\diagdown CH_2CH_3$

7.3 Starting with any alcohol, aldehyde, or ketone with up to four carbons, any inorganic reagents, and any solvents needed, give a good laboratory synthesis of:
 a. 2,2-dimethyl-3-hexanol   e. 4,4-dimethyl-2-pentene
 b. 3-methyl-1-butene   f. 3-pentanol
 c. 3-heptene   g. octane
 d. 2,5-dimethyl-2-hexene

7.4 Which compound in each of the following pairs reacts more rapidly in an S_N2 reaction?
 a. $CH_3CHBrCH_3$ or $CH_3CH_2CH_2Br$
 b. $(CH_3)_3CCl$ or $CH_3CHClCH_2CH_3$

7.5 Which compound in each of the following pairs reacts more rapidly in an S_N1 reaction?

a. 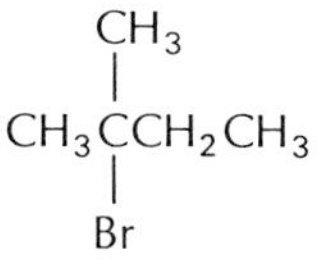 $CH_3\underset{Br}{\overset{CH_3}{\underset{|}{\overset{|}{C}}}}CH_2CH_3$ or $CH_3\overset{CH_3}{\underset{|}{C}}HCHBrCH_3$

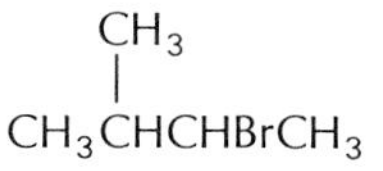

b. 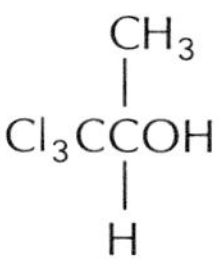 $Cl_3\underset{H}{\overset{CH_3}{\underset{|}{\overset{|}{C}}}}OH$ or $CH_3\underset{H}{\overset{CH_3}{\underset{|}{\overset{|}{C}}}}OH$

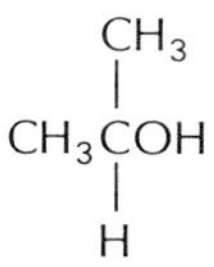

7.6 Which compound in each of the following pairs reacts more rapidly in an E2 reaction? 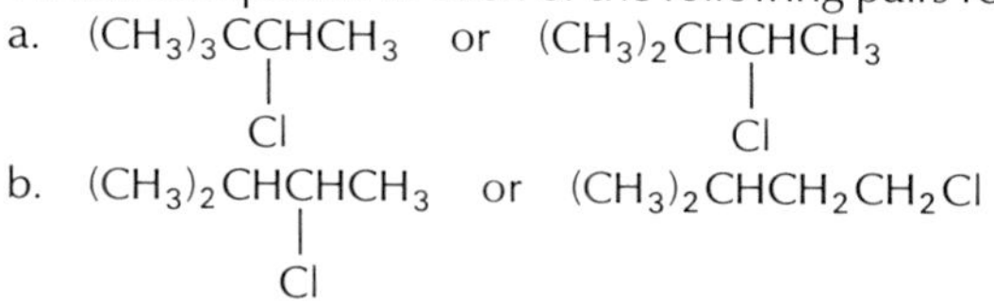
a. $(CH_3)_3CCHCH_3$ or $(CH_3)_2CHCHCH_3$
      | |
      Cl   Cl
b. $(CH_3)_2CHCHCH_3$ or $(CH_3)_2CHCH_2CH_2Cl$
      |
      Cl
c. CH_3CH_2I or $CH_2{=}CHI$

7.7 Which compound in each of the following pairs dehydrates with H_2SO_4 more rapidly by the E1cA reaction?

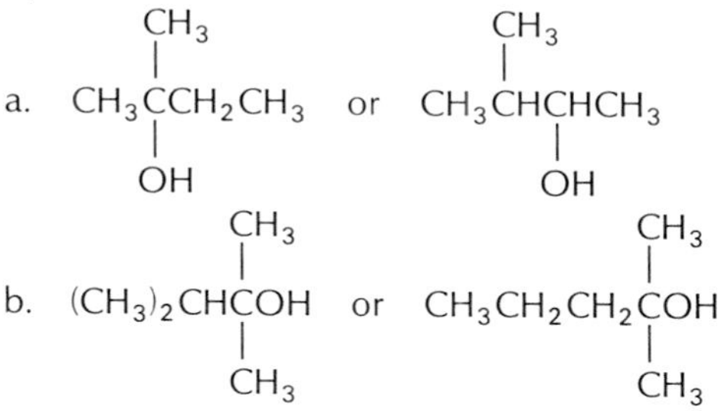

a.
$\quad\quad\;\;\;CH_3 \quad\quad\quad\quad\quad CH_3$
$\quad\quad\;\;\;\;|\quad\quad\quad\quad\quad\quad\;\;\;|$
$CH_3CCH_2CH_3$ or $CH_3CHCHCH_3$
$\quad\quad\;\;\;\;|\quad\quad\quad\quad\quad\quad\;\;\;|$
$\quad\quad\;\;\;OH \quad\quad\quad\quad\quad OH$

b.
$\quad\quad\quad\quad\;\; CH_3 \quad\quad\quad\quad\quad CH_3$
$\quad\quad\quad\quad\;\;\;|\quad\quad\quad\quad\quad\quad\;\;\;|$
$(CH_3)_2CHCOH$ or $CH_3CH_2CH_2COH$
$\quad\quad\quad\quad\;\;\;|\quad\quad\quad\quad\quad\quad\;\;\;|$
$\quad\quad\quad\quad\;\; CH_3 \quad\quad\quad\quad\quad CH_3$

7.8 What is the major product in each of the following reactions? By which type of mechanism is it formed?

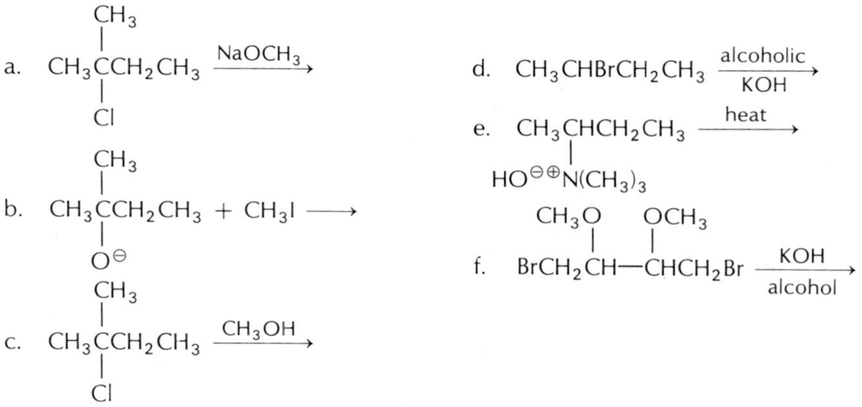

a.
$\quad\quad\;\;\;CH_3$
$\quad\quad\;\;\;\;|$
$CH_3CCH_2CH_3 \xrightarrow{NaOCH_3}$
$\quad\quad\;\;\;\;|$
$\quad\quad\;\;\;Cl$

b.
$\quad\quad\;\;\;CH_3$
$\quad\quad\;\;\;\;|$
$CH_3CCH_2CH_3 + CH_3I \longrightarrow$
$\quad\quad\;\;\;\;|$
$\quad\quad\;\;\;O^\ominus$

c.
$\quad\quad\;\;\;CH_3$
$\quad\quad\;\;\;\;|$
$CH_3CCH_2CH_3 \xrightarrow{CH_3OH}$
$\quad\quad\;\;\;\;|$
$\quad\quad\;\;\;Cl$

d. $CH_3CHBrCH_2CH_3 \xrightarrow[KOH]{alcoholic}$

e.
$CH_3CHCH_2CH_3 \xrightarrow{heat}$
$\quad\;\;\;|$
$HO^{\ominus\oplus}N(CH_3)_3$

f.
$\quad\quad CH_3O \quad\;\; OCH_3$
$\quad\quad\;\;\;|\quad\quad\quad\;\;\;|$
$BrCH_2CH{-}CHCH_2Br \xrightarrow[alcohol]{KOH}$

7.9 The following reactions proceed by molecular rearrangement of intermediate carbonium ions. What is the major rearrangement product, and what are the intermediate carbonium ions in each case?

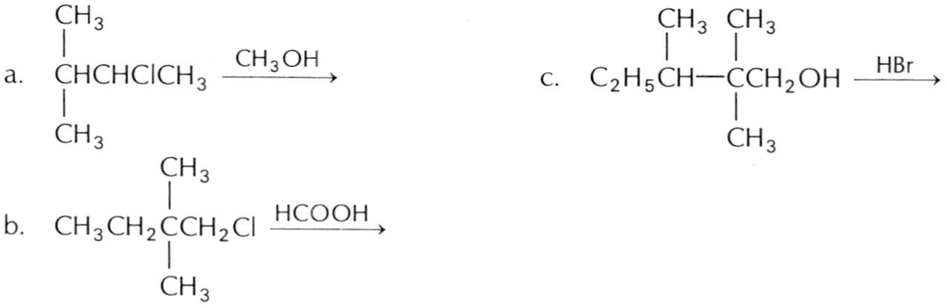

a.
$\quad CH_3$
$\quad\;\;|$
$CHCHClCH_3 \xrightarrow{CH_3OH}$
$\quad\;\;|$
$\quad CH_3$

b.
$\quad\quad\quad\quad\;\; CH_3$
$\quad\quad\quad\quad\;\;\;|$
$CH_3CH_2CCH_2Cl \xrightarrow{HCOOH}$
$\quad\quad\quad\quad\;\;\;|$
$\quad\quad\quad\quad\;\; CH_3$

c.
$\quad\quad\;\;\;CH_3\;\; CH_3$
$\quad\quad\;\;\;\;|\quad\quad\;|$
$C_2H_5CH{-}CCH_2OH \xrightarrow{HBr}$
$\quad\quad\quad\quad\;\;\;|$
$\quad\quad\quad\quad\;\; CH_3$

7.11 Bibliography

We list here some general organic chemistry textbooks that may now be consulted concerning the topics covered in this chapter. In addition, a few paperbacks are listed. The titles chosen are not recommended more than those that are not included. Rather, the list should be considered a random sampling.

1. Allinger, N. L., M. P. Cava, D. C. De Jongh, C. R. Johnson, N. A. Lebel, and C. L. Stevens, *Organic Chemistry*. New York: Worth, 1971. A recent big textbook, excellently done.
2. Bordwell, Frederick G., *Organic Chemistry*. New York: Macmillan, 1963. A good general textbook.
3. Butler, G. B., and K. D. Berlin, *Fundamentals of Organic Chemistry, Theory and Application*. New York: Ronald, 1972. An up-to-date general textbook.
4. Cason, James, *Principles of Modern Organic Chemistry*. Englewood Cliffs, N.J.: Prentice-Hall, 1966. A good general textbook.
5. Corwin, Alsoph H., and Maurice M. Bursey, *Elements of Organic Chemistry*. Reading, Mass.: Addison-Wesley, 1966. A good general textbook with a very unusual arrangement of topics. Easy to read.
6. Ferguson, Lloyd N., *Textbook of Organic Chemistry*, 2nd ed. Princeton: Van Nostrand, 1965. A good general textbook; better than most on petroleum.
7. Fieser, Louis F., and Mary Fieser, *Organic Chemistry*. Boston: Heath, 1956. A deservedly popular text in its time, it is still very good for extensive coverage. Very readable. No problems.
8. Geissman, T. A., *Principles of Organic Chemistry*, 3rd ed. San Francisco: Freeman, 1968. A good textbook. Chapters 7, 8, and 9 are related to substitution, alcohols, and unsaturation.
9. Hendrickson, James B., Donald J. Cram, and George S. Hammond, *Organic Chemistry*, 3rd ed. New York: McGraw-Hill, 1970. A big, modern textbook written at a rigorous level.
10. Morrison, Robert T., and Robert N. Boyd, *Organic Chemistry*, 3rd ed. Boston: Allyn and Bacon, 1973. A very good textbook, the most popular one for a number of years.
11. Noller, Carl R., *Chemistry of Organic Compounds*, 3rd ed. Philadelphia: Saunders, 1965. A giant of a book, about as comprehensive as a single volume on organic chemistry can be.
12. Rakoff, Henry, and Norman C. Rose, *Organic Chemistry*. New York: Macmillan, 1966. A good general textbook.
13. Roberts, John D., and Marjorie C. Caserio, *Basic Principles of Organic Chemistry*. New York: Benjamin, 1964. A good big textbook, the first to introduce nmr in the course for the first year of organic chemistry.
14. Smith, L. Oliver, Jr., and Stanley J. Cristol, *Organic Chemistry*. New York: Reinhold, 1966. A good general textbook.
15. Weininger, Stephen J., *Contemporary Organic Chemistry*. New York: Holt, Rinehart and Winston, 1972. A good modern general textbook.

There are a number of paperbacks, each of which covers a limited area. Some are better than others for use in the first course in organic chemistry—but it depends a lot on the individual's reaction. All have useful bibliographies and reading lists.

16. Allinger, Norman L., and Janet Allinger, *Structures of Organic Molecules*. Englewood Cliffs, N.J.: Prentice-Hall, 1965. One of the Foundations of Modern Organic Chemistry Series by Prentice-Hall.

17. Breslow, Ronald, *Organic Reaction Mechanisms*, 2nd ed. New York: Benjamin, 1969. One of the Organic Chemistry Monograph Series by Benjamin.

18. Payne, Charles A., and Lamar B. Payne, *How To Do an Organic Synthesis*. Boston: Allyn and Bacon, 1969. Practically a "must."

19. Pryor, William A., *Introduction to Free Radical Chemistry*. Englewood Cliffs, N.J.: Prentice-Hall, 1966. Excellent.

20. Saunders, William H., *Ionic Aliphatic Reactions*. Englewood Cliffs, N.J.: Prentice-Hall, 1965. Excellent.

21. Stille, John K., *Industrial Organic Chemistry*. Englewood Cliffs, N.J.: Prentice-Hall, 1968. Very good for information about industrial processes and reactions.

22. Weiss, Howard D., *Guide to Organic Reactions*. Minneapolis: Burgess, 1969. More than an outline: a good compilation of reactions and conditions.

A few articles of more than ordinary interest are as follows.

23. Lloyd, W. G., Free Radical Addition Reactions, *Chem. Technology 1*, 687 (1971).

24. Parker, A. J., Generation of Olefins via Elimination Promoted by Weak Bases, *Chem. Technology 1*, 297 (1971).

8 About Equilibria and Rates of Reactions

Having shown the usefulness of equilibria and rates of reactions in understanding the mechanisms of reactions, we pause and examine some of the underlying concepts. We shall not describe methods of measurement of rates of reaction, being more interested in the uses of rate constants. More and more material about equilibria and reaction rates is appearing in the newer texts for the first year course in chemistry. Those of us who have been through such a text will find some repetition here, because this chapter assumes no previous knowledge of the subject.

8.1 Equilibria and free energy   181
   1. Standard enthalpy, entropy, and free-energy changes.   2. The standard free-energy change and the equilibrium constant.   3. Effect of temperature.
   4. In cold blood. Exercises.

8.2 Transition-state theory   183
   1. The transition state.   2. Rate of reaction and the rate constant.
   3. Examples.   4. Transition-state rate constants.

8.3 Arrhenius equation   186
   1. Arrhenius equation.   2. Magnitudes of terms; Arrhenius plots.
   3. Calculation of E_a, A, $\Delta H^\ddagger$, $\Delta S^\ddagger$, $\Delta G^\ddagger$. Exercises.

8.4 Energy-reaction-coordinate diagrams   189
   1. Description.   2. Examples.   3. Competitive S_N2 and E2 reactions.
   4. Energetics of enzyme catalysis.

8.5 Mechanisms; steady states   192
   1. Stoichiometry and reaction rates.   2. First- and second-order rates of reaction.   3. First order and second order vs. unimolecular and bimolecular.
   4. Mechanism of hydrolysis of tBuCl.   5. Mechanism of conversion of tBuOH into tBuCl.   6. The steady state.

8.6 Kinetic and thermodynamic control of reactions   194
   1. Kinetic control of competing reactions.   2. Thermodynamic control of competing reactions.   3. Thermodynamic versus kinetic control.
   4. Oxygen transport in the blood.

8.7   Summary   196

8.8   Problems   197

8.9   Bibliography   197

8.1   Equilibria and free energy

Standard enthalpy, entropy, and free-energy changes. In Section 3.9 we mentioned that in a system at equilibrium, $\Delta G = 0$. Let us examine this situation in more detail (Eq. 8.1). We have seen that if the reactants and products are under standard-state con-

$$aA + bB \rightleftharpoons mM + nN \qquad (8.1)$$

ditions (1 atm and 25°), the values of $\Delta H°$, $\Delta S°$, and $\Delta G°$ may be calculated for the reaction from the standard enthalpies, entropies, and free energies of formation (Eqs. 8.2). In the

$$\begin{aligned} \Delta H° &= m\Delta H°_f(M) + n\Delta H°_f(N) - a\Delta H°_f(A) - b\Delta H°_f(B) \\ \Delta S° &= m\Delta S°_f(M) + n\Delta S°_f(N) - a\Delta S°_f(A) - b\Delta S°_f(B) \\ \Delta G° &= m\Delta G°_f(M) + n\Delta G°_f(N) - a\Delta G°_f(A) - b\Delta G°_f(B) \end{aligned} \qquad (8.2)$$

process, m moles of M and n moles of N are formed, and a moles of A and b moles of B disappear.

However, if the reaction approaches an equilibrium, A and B do not disappear entirely, and the formation of M and N is incomplete. At equilibrium, the equilibrium

$$K_{eq} = \frac{[M]^m[N]^n}{[A]^a[B]^b} \qquad (8.3)$$

constant K_{eq} is formulated as shown in Equation 8.3. The brackets are used to indicate that the *concentrations at equilibrium* of the products and reactants are to be inserted.

The thermodynamic relationship between ΔG (the actual free-energy change experienced by a system in reaching equilibrium under any specified conditions), $\Delta G°$ (the standard free-energy change if the reaction were to go to completion under standard conditions), and K_{eq} (the equilibrium constant) is given by Equation 8.4. Thus any reaction may be thought of as an equilibrium-seeking system.

$$\Delta G = \Delta G° + 2.3\, RT \log K_{eq} \qquad (8.4)$$

The standard free-energy change and the equilibrium constant. If ΔG is negative, a reaction tends to be spontaneous. If ΔG is positive, a reaction will not be spontaneous. In fact, if ΔG is positive, a reaction should be spontaneous in the reverse direction. In short, the free-energy content of a system (of a reaction) tends toward a minimum. Once the minimum content has been reached, no further change is possible within the system unless there is a change in the surroundings (in pressure, volume, temperature) or of concentration. Thus, when the free-energy content has reached a minimum, $\Delta G = 0$ thereafter.

When $\Delta G = 0$, a rearrangement of Equation 8.4 gives Equations 8.5. Thus, a

$$\Delta G° = -2.3\, RT \log K_{eq} = 2.3\, RT(pK_{eq}) \tag{8.5}$$

measurement of K_{eq} allows us to calculate $\Delta G°$ for the reaction. We see that the equilibrium constant is related in magnitude to the standard free-energy change $\Delta G°$, not to ΔG, which is zero at equilibrium.

Because of the logarithmic relationship, K_{eq} is very sensitive to changes in $\Delta G°$. This dependence is shown in Table 8.1, which covers a range of 60 kcal in $\Delta G°$, but in which

Table 8.1 Relationship of $\Delta G°$, log K_{eq}, and K_{eq} at 25°C

$\Delta G°$, kcal	log K_{eq} or $-pK_{eq}$	K_{eq}
30	-21.989	1.03×10^{-22}
20	-14.659	2.19×10^{-15}
10	-7.330	4.68×10^{-8}
1	-0.733	1.85×10^{-1}
0	0	1.00
-1	0.733	5.41
-10	7.330	2.14×10^{7}
-20	14.659	4.56×10^{14}
-30	21.989	9.74×10^{21}

K_{eq} varies over a range of 10^{44}. If $\Delta G°$ is negative, K_{eq} is greater than one and the position of equilibrium lies to the right. If $\Delta G°$ is zero, K_{eq} is one. If $\Delta G°$ is positive, K_{eq} is less than one and the position of equilibrium lies to the left.

If K_{eq} is very large, the reaction goes essentially to completion, and the measurement of K_{eq} becomes difficult because it is almost impossible to find an accurate value for the concentrations of the tiny amounts of reactants left. If K_{eq} is very small, the problem is reversed, essentially no reaction takes place, and it becomes difficult to measure the traces of products present.

Effect of temperature. To consider the effect of temperature on a reaction at equilibrium, we use Equation 8.6, in which $T_2 > T_1$ represent temperatures that are close enough together so that $\Delta H°$ and $\Delta S°$ may be regarded as constants.

$$\Delta(\log K_{eq}) = \log K_{2_{eq}} - \log K_{1_{eq}} = \frac{\Delta H°}{2.3\, R}\left(\frac{1}{T_1} - \frac{1}{T_2}\right) \tag{8.6}$$

The equation, known as van't Hoff's law (Section 5.8), is obeyed surprisingly well. If the equilibrium constants at the two temperatures can be determined with some precision, we have an independent method for the calculation of $\Delta H°$ and a check upon the calorimetric method for measurement of standard enthalpy changes. Reversing the process, and knowing only standard enthalpies, we can calculate the change in equilibrium constant for a given reaction over a reasonable range of temperatures, even though the reaction may not have been carried out. The greater $\Delta H°$ is, the greater the difference in log K_{eq} will be for a given temperature change.

In cold blood. Many of the major thermodynamic questions in biology have been answered by chemical investigations of photosynthesis, sugar metabolism, replication, and oxidative phosphorylation. Neither the "vital force" nor new scientific laws (e.g., the idea of quantum mechanics in physics) are needed to explain the chemistry of life. But our understanding is incomplete for a number of phenomena. One is the phenomenon of cold-bloodedness.

We can now describe to a good approximation many of the complicated systems of biochemical reactions at a certain temperature. We know that as we lower the temperature, we decrease the rate at which a reaction proceeds to equilibrium. Morever, each reaction has an equilibrium constant with its own unique temperature dependence. A decrease in the temperature then not only slows all of the reactions in an organism but it changes all of the equilibrium constants. Binding constants, ionization constants, solubility constants—all change with temperature. This difficulty is avoided in birds and mammals by maintaining a constant body temperature of 37°C. Some bacteria live only in a place with constant temperature, such as the human gut (Escherichia coli) or hot springs (Thermobacillus). But what of the life forms such as those at the sea shore that experience daily body temperature fluctuations? The chemistry of cold-blooded animals is not yet understood.

Exercises

8.1 Use Equation 8.6 to calculate log K_{eq} at 40°, 50°, and 60°C if log $K_{25°} = -2.00$ and $\Delta H° = 4$ kcal. Does K_{eq} increase or decrease with temperature? If $\Delta H°$ were negative, what would happen to K_{eq} as the temperature increased?

□ □

8.2 Transition-state theory

The transition state. The transition-state theory begins with the idea that for a reaction to occur, the species involved must accumulate enough energy to get into a transition state, or activated complex, that is so unstable that it decomposes spontaneously into reactants or products.

At first glance, it does not seem that thermodynamics, with its concern for systems at equilibrium and functions of state, would help us to understand the factors that influence the rate of a reaction, either to completion or to equilibrium. Nevertheless, we shall find that the language and techniques of thermodynamics will help us.

Rate of reaction and the rate constant. A rate, like a speed or a velocity, is expressed as the amount of change of something per unit of time. In the study of reaction rates, kinetics, the change is in the concentration of a reactant or product per unit of time, or moles per liter per second. Although it is possible in principle, there is no simple device (like a speedometer) that may be inserted into a reaction mixture to give a direct reading of the rate of change in moles per liter per second. Certain methods have been developed that, by direct linkage to a computer, allow all the necessary computation to be done as the reaction proceeds. These elaborate techniques are beyond the scope of our study.

For the reaction $A + B \rightarrow$ products, the rate of change in concentration per second at any given instant is expressed as a constant multiplied by the concentrations of the reactants at that instant (Eq. 8.7). The rate of a reaction (except for a special case known

$$\text{rate (moles/liter/sec)} = k[A][B] \tag{8.7}$$

as a zero-order reaction) is constantly changing as the concentrations of reactants change.

The rate constant is the rate of reaction at the instant when each reactant is at a concentration of 1 M. The rate constant is used in comparing the speed of one reaction to another. This constant may be regarded as a conglomeration of all the factors that affect the rate of reaction except for the concentrations of reactants.

In the transition-state theory, the rate constant is related to these other factors by Equation 8.8. The symbols are as follows: R is the gas constant, N is Avogadro's number,

$$\log k = \log \frac{R}{Nh} + \log T - \frac{\Delta G^{\ddagger}}{2.3\, RT} \tag{8.8}$$

h is Planck's constant, T is the absolute temperature, and $\Delta G^{\ddagger}$ is the free energy of activation. The superscript $\ddagger$ is used to show that $\Delta G^{\ddagger}$ is a transition-state free energy only. Now we can use thermodynamic language and set $\Delta G^{\ddagger} = \Delta H^{\ddagger} - T\Delta S^{\ddagger}$ and speak not only of the free energy of activation, but also of the enthalpy of activation and the entropy of activation.

Examples. Let us concentrate for the moment on $\Delta G^{\ddagger}$ and look at a few examples, the S_N2 (and E2) reactions of ethoxide ion with methyl bromide and with ethyl bromide in ethanol as solvent (Eqs. 8.9).

$$CH_3CH_2O^{\ominus} + CH_3-Br \longrightarrow CH_3CH_2OCH_3 + Br^{\ominus}$$
$k(25°) = 1.55 \times 10^{-3};\quad \log k(25°) = -2.810 \qquad a$
$k(55°) = 3.40 \times 10^{-2};\quad \log k(55°) = -1.469 \qquad b$

$$CH_3CH_2O^{\ominus} + \underset{\underset{CH_3}{|}}{CH_2}-Br \longrightarrow \underset{\underset{CH_3}{|}}{CH_3CH_2OCH_2} + Br^{\ominus} \tag{8.9}$$

$k(55°) = 1.18 \times 10^{-3};\quad \log k(55°) = -2.928 \qquad c$

$$CH_3CH_2O^{\ominus} + H-CH_2-CH_2-Br \longrightarrow CH_3CH_2OH + CH_2=CH_2 + Br^{\ominus}$$
$k(55°) = 1.2 \times 10^{-5};\quad \log k(55°) = -4.921 \qquad d$

Rearrangement of Equation 8.8 allows us to evaluate $\Delta G^{\ddagger}$ from $\log k$ and T (Eqs. 8.10). Making the substitutions from the data given in Equations 8.9 gives the results shown in Equations 8.11.

$$\Delta G^{\ddagger} = 2.3\,RT\,(\log \frac{R}{Nh} + \log T - \log k)$$

and

$$\Delta G^{\ddagger} = 4.576\,T\,(\log (2.083 \times 10^{10}) + \log T - \log k) \quad (8.10)$$

or

$$\Delta G^{\ddagger} = 4.576\,T\,(10.319 + \log T - \log k)$$

For

a $\Delta G^{\ddagger} = (4.576)(298)(10.319 + 2.474 + 2.810)$
= 21 280 cal for S_N2, 25°, MeBr, EtO$^{\ominus}$

b $\Delta G^{\ddagger} = (4.576)(328)(10.319 + 2.516 + 1.469)$
= 21 470 cal for S_N2, 55°, MeBr, EtO$^{\ominus}$

(8.11)

c $\Delta G^{\ddagger} = (4.576)(328)(10.319 + 2.516 + 2.928)$
= 23 660 cal for S_N2, 55°, EtBr, EtO$^{\ominus}$

d $\Delta G^{\ddagger} = (4.576)(328)(10.319 + 2.516 + 4.921)$
= 26 650 cal for E2, 55°, EtBr, EtO$^{\ominus}$

Comparison of a and b shows that although k increases about 22 times from 25° to 55°, $\Delta G^{\ddagger}$ increases by about 1%. Comparison of b and c shows that MeBr is about 29 times more reactive than EtBr at 55°, and the free energy of activation is about 2 kcal (about 10%) greater for EtBr than for MeBr. In turn, comparison of c and d shows that under the same conditions at 55°, EtBr undergoes the S_N2 reaction about 98 times more often than it does the E2 reaction, and that $\Delta G^{\ddagger}$ for the E2 reaction is about 3 kcal greater than for the S_N2 reaction.

Transition-state rate constants. As we might expect from the presence of $\Delta G^{\ddagger}$, the transition-state theory envisions an equilibrium between the reactants and the transition state (Eq. 8.12). The rate of formation of the transition state from the reactants must

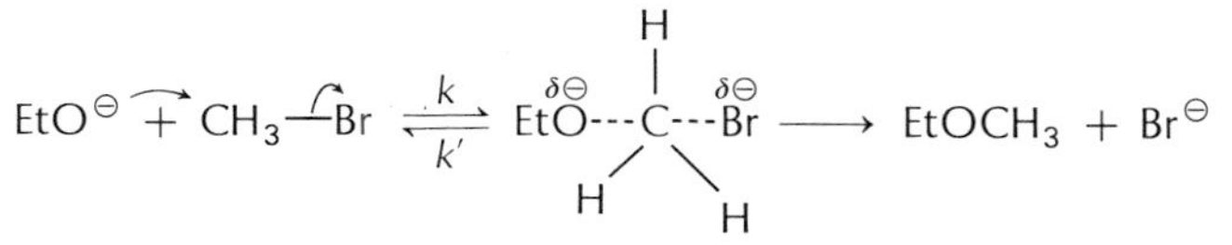

$$\text{EtO}^{\ominus} + \text{CH}_3\text{—Br} \underset{k'}{\overset{k}{\rightleftarrows}} \overset{H}{\underset{H}{\text{EtO}\cdots\overset{|}{\underset{H}{C}}\cdots\text{Br}}}{}^{\delta\ominus\;\;\;\delta\ominus} \longrightarrow \text{EtOCH}_3 + \text{Br}^{\ominus} \quad (8.12)$$

transition state

be equal to the rate of decomposition of the transition state into reactants for an equilibrium to exist (Eqs. 8.13). A little rearranging shows us that the equilibrium constant for reactants

forward rate = $k[\text{EtO}^{\ominus}][\text{CH}_3\text{Br}]$

reverse rate = $k'[\text{EtO}\text{---}\text{CH}_3\text{---}\text{Br}]^{\ominus} = k'[\text{TS}]$

forward rate = reverse rate = $k[\text{EtO}^{\ominus}][\text{CH}_3\text{Br}] = k'[\text{TS}]^{\ominus}$ (8.13)

and

$$\frac{[\text{TS}]^{\ominus}}{[\text{EtO}^{\ominus}][\text{CH}_3\text{Br}]} = \frac{k}{k'} = K^{\ddagger}_{eq}$$

going to the transition state is $K_{eq}^{\ddagger}$. Also, $K_{eq}^{\ddagger}$ is equal to the ratio of the forward rate constant to the reverse rate constant.

We previously found that $\Delta G^{\ddagger}$ is around 20 kcal. Table 8.1 shows that $K_{eq}^{\ddagger}$ is about 2×10^{-15}. Such a minute value indicates that $[TS]^{\ominus}$ is at an exceedingly small concentration in relation to reactants. Also, $k' \gg k$. Furthermore, without delving into the mathematics, let us state that the rate constant for the decomposition of the transition state into products also is k'. Thus, we envisage the reaction as proceeding by a conversion (with a small rate constant) of reactants into a transition state that decomposes (at very large rate constants) equally into products and reactants. The free energy of activation is the free-energy change of the equilibrium between the reactants and the transition state.

Before taking up the enthalpy of activation ($\Delta H^{\ddagger}$) and the entropy of activation, let us turn back and look at a predecessor of the transition-state theory.

8.3 Arrhenius equation

Arrhenius equation. Soon after rates of reaction began to be measured, it was found that an increase in temperature had a remarkably large accelerating effect upon a rate. Attempts to put the effect into an exact relationship resulted in what we now call the Arrhenius equation, which is given in the exponential and logarithmic forms in Equation 8.14. The logarithmic form is a linear equation that says that a plot of log k against $1/T$ has an intercept of log A and a slope of $-E_a/2.3\, R$.

$$k = A e^{-(E_a/RT)}; \quad \log k = \log A - \left(\frac{E_a}{2.3\, R}\right)\left(\frac{1}{T}\right) \tag{8.14}$$

The temperature affects the rate constant (k) for the reaction, which means that equal concentrations at a higher temperature react faster because the rate constant is larger. The A term is known as the Arrhenius A factor, the pre-exponential factor, or more simply, just as the A factor. The E_a is called the activation energy, measured in cal. R is the gas constant, 1.987 cal/deg/mole, and T is the absolute temperature. In the Arrhenius Equation, k must always be used with seconds as the unit of time. □ □

Magnitudes of terms; Arrhenius plots. To give us a feeling for the sizes of the numbers involved, Table 8.2 presents the data for five hypothetical reactions arbitrarily

Table 8.2 Activation energies and A factors

Reaction	A	B	C	D	E
log k(300°)	−3.00	−4.00	−5.00	−6.00	−7.00
log k(400°)	−0.25	−0.75	−1.25	−1.75	−2.25
log k(500°)	1.40	1.20	1.00	0.80	0.60
log k(600°)	2.50	2.50	2.50	2.50	2.50
log k(700°)	3.29	3.43	3.57	3.71	3.86
log k(∞°) = log A	8.00	9.00	10.00	11.00	12.00
E_a, cal	15,180	17,940	20,700	23,460	26,220

8.3 Arrhenius equation 187

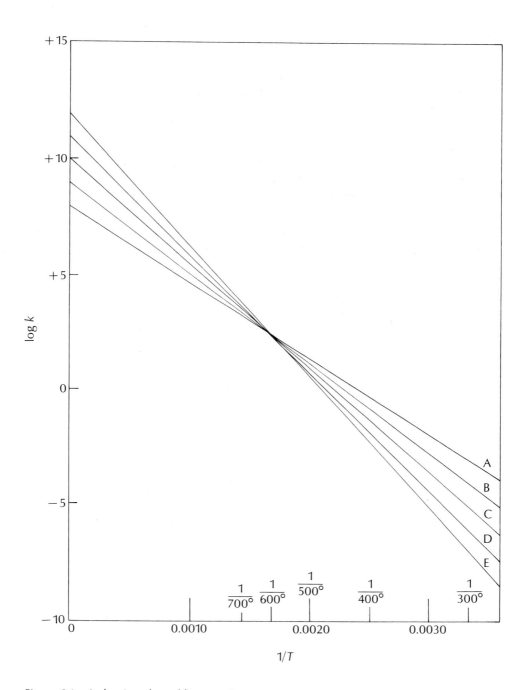

Figure 8.1 Arrhenius plots of five reactions

chosen so that log k (300°) = −3, −4, −5, −6, and −7, with log A = 8, 9, 10, 11, and 12. All other values were calculated. Figure 8.1 shows the corresponding Arrhenius plots, log k versus 1/T.

Several items need attention. At room temperature (300°K = 27°C), reaction A is the fastest and reaction E is the slowest. The fastest reaction has the smallest E_a, and the slowest has the largest E_a. For some reason that has never had a satisfactory explanation, reactions with large E_a usually have large log A. As the temperature increases, the reaction with the larger E_a increases more rapidly, and if log A also is larger, the slower reaction may overtake and surpass the faster reaction. (The fact that the reactions shown here all cross at 600°K is a coincidence that resulted from the choices made for log k(300°) and log A in setting them up.) In short, a large E_a means a large temperature effect. A fast reaction means a reaction with a large rate constant, and a slow reaction means a reaction with a small rate constant. The actual rate of a reaction depends upon both the size of the rate constant and the concentration of the reactants.

In any case, the Arrhenius equation expresses log k as a relatively small number that is the difference between two large numbers. Because log k is the experimentally determined number, log A and $E_a/2.3$ R magnify any error in log k many times. Notice the distance we must extrapolate to obtain log A if the highest experimental temperature is 333°K.

Calculation of E_a, A, $\Delta H^\ddagger$, $\Delta S^\ddagger$, $\Delta G^\ddagger$. Because E_a and log A do not vary with temperature (which is not always true), it is a simple matter to calculate their values if the rate constants at two or more temperatures are known (Eqs. 8.15).

$$\log k_2 - \log k_1 = \frac{E_a}{2.3\ R}\left(\frac{1}{T_1} - \frac{1}{T_2}\right)$$

or

$$E_a = 2.3\ R\left(\frac{T_1 T_2}{T_2 - T_1}\right)(\log k_2 - \log k_1)$$

and (8.15)

$$\log A = \frac{(T_2 \log k_2) - (T_1 \log k_1)}{T_2 - T_1}$$

Once E_a and log A have been evaluated, we can calculate $\Delta H^\ddagger$ and $\Delta S^\ddagger$ (Eqs. 8.16).

$$\Delta H^\ddagger = E_a - RT$$
$$\Delta S^\ddagger = 4.58\ (\log A - \log T) - 49.21$$ (8.16)

If we return to Equations 8.9, we may use the data of a and b and Equations 8.15 and 8.16 to calculate that

$$E_a = 20\ 000\ \text{cal} \qquad \log A = 11.857$$

and that, at 25°,

$$\Delta H^\ddagger = 19\ 400\ \text{cal} \qquad \Delta S^\ddagger = -6.3\ \text{eu}$$
$$\Delta G^\ddagger = 21\ 300\ \text{cal} \qquad T\Delta S^\ddagger = -1900\ \text{cal}$$

Exercises

8.2 Calculate E_a and log A for a reaction that has log $k(300°) = -4.00$ and log $k(310°) = -3.70$.

□ □

8.4 Energy-reaction-coordinate diagrams

Description.   An energy-reaction-coordinate diagram is a representation of the potential-energy changes as reactants are converted into products. Depending upon the availability of data, these diagrams range from fairly accurate representations to schematic pictures.

The ordinate is energy, either E_a, G, or H, in cal or kcal. The abscissa is the reaction coordinate, a vague expression meaning the progress of the reaction being depicted. The reactants are to the left, the transition state or activated complex in the center, and products to the right of the usual diagram. We first illustrated this kind of diagram in Figures 7.2 and 7.3.

Examples.   Let us use the reactions *a* and *b* of Equations 8.9 as illustrations of these diagrams. The data from *a* and *b* allow us to calculate E_a, log A and $\Delta H^{\ddagger}$, $T\Delta S^{\ddagger}$, and $\Delta G^{\ddagger}$ at 25°. We construct the diagram shown in Figure 8.2. Because we do not know the values of the various energy levels for the solution in ethanol of the reactants, it is customary to assign a value of zero as starting points for E_a, $\Delta G^{\ddagger}$, $\Delta H^{\ddagger}$, and $T\Delta S^{\ddagger}$. Also, because in this case we do not know the values, we assign $\Delta G = -10$ kcal and $\Delta H = -11.5$ kcal, from which $\Delta S = -5$ eu.

Examination of the diagram shows that the extra height of the $\Delta G^{\ddagger}$ hill above the $\Delta H^{\ddagger}$ hill is the same as the dip in the curve for $T\Delta S^{\ddagger}$. Also, the E_a hill is only RT above the $\Delta H^{\ddagger}$ hill. However, all three energy humps are close together in height and shape. Consequently, in drawing schematic energy-reaction-coordinate diagrams, we usually label the ordinate as "energy" and let it go at that.

Competitive S_N2 and E2 reactions.   Equations *c* and *d* of 8.9 represent competitive reactions, S_N2 and E2, taking place simultaneously in the same flask. The reaction-coordinate-energy diagram is shown in Figure 8.3. With only $\Delta G^{\ddagger}$ (S_N2) and $\Delta G^{\ddagger}$ (E2) available, the other values shown [ΔG (S_N2) and ΔG (E2)] are convenient guesses. However, the higher E2 energy hill is shown clearly in comparison to the S_N2 hill. A return to Table 8.2 and Figure 8.1 reminds us that a reaction with a higher energy hill increases in rate (rate constant) with increase in temperature more rapidly than does a reaction with a lower energy hill. Thus, an increase in temperature favors a competitive E2 reaction more than it does an S_N2 reaction.

Energy-reaction-coordinate diagrams help us to visualize the energy effects that accompany the reactions we have been concerned with. We shall be using such diagrams frequently in the chapters to come.

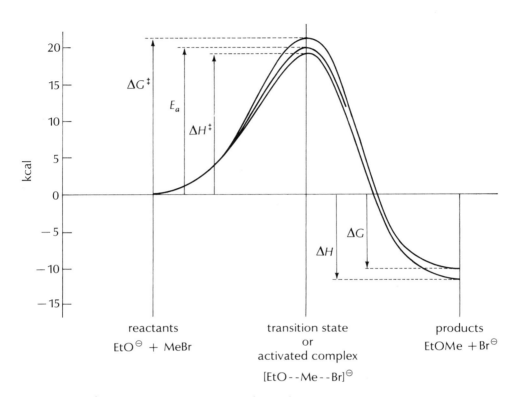

Figure 8.2 Schematic energy-reaction coordinate diagram

Energetics of enzyme catalysis. A catalyst lowers the activation energy of a reaction, allowing the reaction to proceed to its (unaltered) equilibrium state more quickly. Mechanisms for chemical catalysts have been studied for years in order to improve industrial processes. Of the catalysts developed thus far, however, few exceed the efficiency of enzymes. For instance, liver catalase, one of the first enzymes discovered, can accelerate the breakdown of hydrogen peroxide into oxygen and water to a rate of several million molecules per minute per molecule of enzyme. As enzyme extraction techniques improve, industrial use, especially of enzymes made insoluble by binding to polymers, will expand rapidly in areas such as waste recycling, food production, and drug synthesis.

The energetics of enzyme catalysis are still the subject of much speculation. That some enzymes appear to bind the transition state more strongly than they do the starting material or product is predicted to be a rate-enhancing property to apply to enzymes in general. In any event, enzymes do tend to bind the starting materials at the active site in an orientation for reaction. All such effects that essentially increase the number of productive collisions during a period of time can be lumped together as entropy effects. By binding

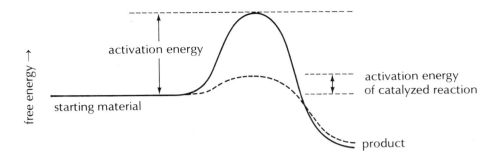

the reactants properly, an enzyme can eliminate the entropy of activation. The enthalpy of activation may be altered by allowing a covalent intermediate between the enzyme and one of the starting materials or by providing at the active site an environment with a dielectric constant more favorable to reaction than that of the solution. There have been suggestions but as yet it has proven difficult to attach accurate measures to the possible contributing factors in enzyme catalysis.

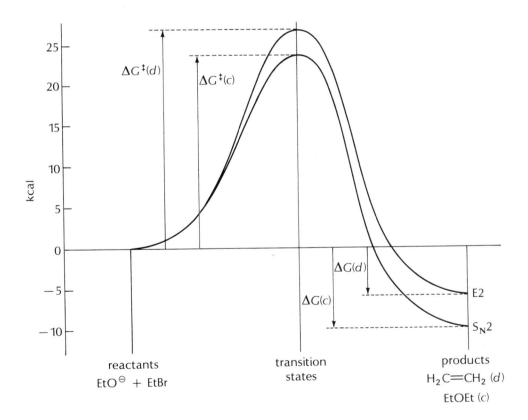

Figure 8.3 Energy-reaction-coordinate diagram, competitive reactions

8.5 Mechanisms; steady states

Stoichiometry and reaction rates. Stoichiometry, the weight relations in chemical reactions, must be kept in mind in the study of rates of reaction. For example, if we were studying the rate of the S_N2 reaction of methyl iodide with sodium ethoxide to give methoxyethane and sodium iodide, we could safely assume that the numbers of moles of sodium iodide and of the ether (methoxyethane) formed at any instant are the same. Because iodide ion concentration is easily found and the concentration of ether is difficult to determine, the iodide concentration is measured as the reaction proceeds. Furthermore, the formation of each mole of iodide ion implies that a mole of methyl iodide and a mole of ethoxide ion have reacted. Alternatively, we could elect to follow the concentration of ethoxide ion as the reaction proceeds, because it has the properties of a strong base and is used up. Each mole consumed uses up one mole of methyl iodide, and one mole each of the ether and iodide ion are formed. In this instance we easily follow the reaction by keeping track of a reactant (the base) or of a product (iodide ion).

Other reactions may not be so easily studied because of the difficulty of finding a way to analyze for any of the reactants or products. Thus, in the acid-catalyzed dehydration of an alcohol, although the rate is dependent upon the concentration of acid used, the acid is not used up and remains at constant concentration throughout the reaction. We must follow the concentration of the alcohol as it is used up or the concentration of alkenes or water as they are formed.

First- and second-order rates of reaction. Most reactions fall into either of two classes of rate of reaction, first order and second order (Eqs. 8.17). The rate constants are

rate = k_1[reactant]	first order, a	
rate = k_2[reactant A][reactant B]	second order, b	(8.17)
rate = k_2[reactant A]^2	second order, c	

identified by a subscript numeral. A <u>first-order</u> reaction (a) is one whose rate depends upon the concentration (denoted by brackets [ ], in moles per liter, M) of a single reactant. An example is the hydrolysis of tBuCl (Eq. 8.18). A <u>second-order</u> reaction is one whose rate

$$tBuCl + H_2O \xrightarrow{k_1} tBuOH + HCl$$
$$\text{rate} = k_1[tBuCl] \tag{8.18}$$

depends upon the product of the concentrations of two reactants (b) or upon the product of the concentration of a single reactant with itself (c). Examples are the reaction of ethoxide ion with ethyl bromide (Eq. 8.19) and the reaction of a radical with itself (Eq. 8.20).

$$EtO^\ominus + EtBr \longrightarrow EtOEt + Br^\ominus$$
$$\text{rate} = k_2[EtO^\ominus][EtBr] \tag{8.19}$$

$$CH_3\dot{C}H_2 + \dot{C}H_2CH_3 \longrightarrow CH_3CH_2CH_2CH_3$$
$$\text{rate} = k_2[CH_3\dot{C}H_2]^2 \tag{8.20}$$

8.5 Mechanisms; steady states

Many reactions exhibit other simple rate patterns (zero order, third order) as well as much more complicated expressions. However, we shall restrict our discussion to the first- and second-order reactions.

First order and second order vs. unimolecular and bimolecular. We should not confuse the expressions <u>unimolecular</u> with <u>first order</u> and <u>bimolecular</u> with <u>second order</u>. The <u>order</u> expressions refer to the form of the rate equation and the <u>molecular</u> expressions refer to how a given reaction is taking place. For example, Equation 8.19 clearly is an S_N2 reaction. (We ignore for the moment the small amount of concurrent E2 reaction.) Now let us set up the reaction with $[EtO^\ominus] = 1\ M$ and $[EtBr] = 0.01\ M$. When the reaction is over, $[EtO^\ominus]$ will have decreased only to 0.99 M. The rate of the reaction appears to depend only upon [EtBr] and to be first order with $k_1(\text{observed}) = k_2[EtO^\ominus]$. Nevertheless, the reaction is still bimolecular, an S_N2 reaction.

Mechanism of hydrolysis of tBuCl. Consider the hydrolysis of tBuCl (Eq. 8.18), which, written as a balanced equation, looks as though it should be a second-order reaction. The explanation lies in the mechanism, which, to be satisfactory, should account for the first-order rate (Eqs. 8.21). If k_1 is smaller than any of the other rate constants (except k_{-W},

$$tBuCl \underset{k_{-1}}{\overset{k_1}{\rightleftarrows}} tBu^\oplus + Cl^\ominus \qquad a$$

$$tBu^\oplus + H_2O \underset{k_{-W}}{\overset{k_W}{\rightleftarrows}} tBu\overset{\oplus}{O}H_2 \qquad b \qquad (8.21)$$

$$tBu\overset{\oplus}{O}H_2 + H_2O \underset{k_{-H}}{\overset{k_H}{\rightleftarrows}} tBuOH + H_3O^\oplus \qquad c$$

see the next paragraph), then steps *b* and *c* may proceed to the right no faster than step *a*. In other words, the slow step *a* limits the concentration of $tBu^\oplus$, which reacts swiftly with the excess water present to give the conjugate acid (*b*) of tBuOH. The proton transfer *c* is very fast, of the order of 10^{10}. The reverse step in *a*, even though k_{-1} is comparable to k_W, is limited by the small concentration of $Cl^\ominus$. Thus, the S_N1 reaction of *a*, we say, is the <u>rate-determining step.</u> The following fast steps (*b* and *c*) do not affect the rate of the first-order reaction. The fast steps, we repeat, can go no faster than the slow step (*a*) will allow them to go.

Mechanism of conversion of tBuOH into tBuCl. Will the reverse of Equations 8.21 offer an explanation of the formation of tBuCl from tBuOH by reaction with concentrated hydrochloric acid? Let us see.

Step *c* is the very rapid proton-transfer equilibrium to give the conjugate acid of tBuOH. The next step (*b*) is what we have classified as an S_N1cA reaction, the relatively slow dissociation of $tBu\overset{\oplus}{O}H_2$ to give $tBu^\oplus$. The *tert*-butyl cation has a choice of reacting with $Cl^\ominus$ or with H_2O. With concentrated hydrochloric acid, the high concentration of $Cl^\ominus$ favors *a* to give tBuCl at a relatively rapid rate. The rate-determining step in this reverse sequence is that set by $k_{-W}[tBu\overset{\oplus}{O}H_2]$. However, $[tBu\overset{\oplus}{O}H_2]$ must also conform to the requirements of the fast equilibrium of *c*. Nevertheless, the rate remains first order and depends only upon [tBuOH] when an excess of the acid is used. The first-order rate constant k_{obs} in this instance is a conglomeration of terms.

194 Chapter eight

The mechanism (Eqs. 8.21), we see, is satisfactory for both the hydrolysis and the formation of *t*BuCl. (Actually, the mechanism is not completely satisfactory, because we have ignored the competing E1 and E1cA reactions as well as the tight ion pairs, and so on. A more complete discussion, however, would take us too far afield and involve us in a lot of mathematics.)

The steady state. We should mention that there is a mathematical approximation known as the steady state that is of use in the manipulation of rate equations. Although we shall make no use of the steady state in that sense, we shall occasionally use the concept in some descriptions in chapters to come. For our purposes, we shall define the steady state as the tiny concentration of a radical or ion (sometimes a molecule) in a mechanism.

For example, the *tert*-butyl cation in the mechanism shown by Equations 8.21 may be assumed to be in a steady state. The cation is formed slowly and is so reactive that it is gobbled up as soon as it is formed. The concentration of the cation never rises above miniscule. Consequently, the rate of change in concentration of the cation is practically zero. Because it is practically zero, we assume that it is zero. Hence the name steady-state approximation is used to denote something that remains steady in the midst of change, a very reactive intermediate in a mechanism of reaction.

The concept of the steady state leads to considerable simplification of the mathematics involved in the derivation of satisfactory expressions for the rates of many reactions, including enzymatic reactions.

8.6   Kinetic and thermodynamic control of reactions

Kinetic control of competing reactions. We already have seen an illustration of kinetic control of competitive reactions in the S_N2, E2 competition (Eqs. 8.9 *c* and *d*). The ratio of the products formed was determined by the relative sizes of the rate constants of the competing reactions.

Thermodynamic control of competing reactions. Another form of competition occurs when we heat one mole of acetic acid with a large excess of a one-to-one molar mixture of methanol and ethanol in the presence of some sulfuric acid. The two alcohols compete with each other for the limited amount of acetic acid to form the esters methyl acetate and ethyl acetate (Eqs. 8.22). At the end of the mixed reaction, the concentration of

$$\text{CH}_3\text{C}(=\text{O})\text{OH} + \text{CH}_3\text{OH} \underset{\text{H}^\oplus}{\overset{K_M}{\rightleftarrows}} \text{CH}_3\text{C}(=\text{O})\text{OCH}_3 + \text{H}_2\text{O}$$

$$\text{CH}_3\text{C}(=\text{O})\text{OH} + \text{CH}_3\text{CH}_2\text{OH} \underset{\text{H}^\oplus}{\overset{K_E}{\rightleftarrows}} \text{CH}_3\text{C}(=\text{O})\text{OCH}_2\text{CH}_3 + \text{H}_2\text{O}$$

(8.22)

methyl ester present must meet the requirements of the equilibrium constant K_M for that esterification, as must the concentrations of acetic acid, methanol, and water. The concentration of ethyl ester present must satisfy the equilibrium constant K_E for that reaction, as must the concentrations of acetic acid, ethanol, and water. The sulfuric acid may be regarded as a catalyst for speeding up the attainment of equilibrium. The strong mineral acid has no effect on the position of equilibrium. The mechanism is given in the next chapter.

In the mixed reaction, the concentrations of acetic acid and of water at equilibrium must be identical in both equilibria. Because the reactions were started with equal concentrations of methanol and ethanol, it follows that the concentrations of the two esters will be determined by the values of the equilibrium constants for the two reactions (Eqs. 8.23). The ratio of the equilibrium concentrations of the two esters (c) is equal to the ratio

and

$$[CH_3COOCH_3] = K_M \frac{[CH_3COOH][CH_3OH]}{[H_2O]} \qquad a$$

$$[CH_3COOCH_2CH_3] = K_E \frac{[CH_3COOH][CH_3CH_2OH]}{[H_2O]} \qquad b$$

then

(8.23)

$$\frac{[CH_3COOCH_3]}{[CH_3COOCH_2CH_3]} = \frac{K_M[CH_3OH]}{K_E[CH_3CH_2OH]} \qquad c$$

of the equilibrium constants multiplied by the ratio of the equilibrium concentrations of the two alcohols. Because a large excess (at equal concentrations) of the two alcohols was used to begin with, the ratio of the concentrations of the alcohols remains close to unity even after equilibrium has been reached. Thus, the ratio of the equilibrium concentrations of the esters is equal to the ratio of the equilibrium constants. This relationship is thermodynamic control of competitive reactions.

Thermodynamic versus kinetic control. If competitive reactions are essentially irreversible, the rate constants determine the ratio of products formed, and the reactions are said to be under kinetic control. If competitive reactions are reversible and the system is carried to equilibrium, the equilibrium constants (other things being equal) determine the ratio of products formed, and the reactions are said to be under thermodynamic control.

Let us consider a competition between an irreversible and a reversible reaction.

If the irreversible reaction is fast and the reversible reaction is slow in reaching equilibrium, kinetic control dominates throughout the period of time needed to carry out the reaction.

If the irreversible reaction is slow and the reversible reaction is fast in coming to equilibrium, thermodynamic control dominates during the first phase of the reaction. As time goes on, the irreversible reaction begins to compete and finally to dominate the competition. If the product of kinetic control is wanted, the reaction should not be terminated until the reaction is well into the second phase. Knowing the length of time required for the first phase, for the transition to the second phase, and for the second phase to become established requires a knowledge of the relative rate constants, the concentrations involved, and the temperature.

Often when the rate constants are not known, an organic chemist must make a few trial runs with small amounts of material to find conditions appropriate for the reaction. With experience, the chemist comes to sense what to do and frequently is able to carry out reactions that are the despair of the novice in the organic chemistry laboratory.

196 Chapter eight

If the irreversible and reversible reactions are comparable in rate, the product will be a mixture until enough time has elapsed to allow the irreversible reaction to win out.

Of course, the greater the disparity of rates involved in the competition, the more that reaction with the greater rate dominates.

Oxygen transport in the blood. Oxygen transport in the blood is a fairly complicated system of equilibria and rates. Blood leaves the lungs with oxygen associated with the iron of the porphin ring of hemoglobin with an oxygen tension of 80 mmHg (a method for expressing the amount of oxygen present by specifying the pressure of oxygen required to maintain equilibrium with blood under a given set of conditions). Upon arriving in a capillary requiring oxygen, hemoglobin enters into an equilibrium (with a cytochrome at the site?) and takes up a bicarbonate ion, $HCO_3^{\ominus}$, or a chloride ion. Returning to the lungs, blood in the veins has about 35 mmHg oxygen tension. In the lungs, $HCO_3^{\ominus}$ is released and discharged as CO_2; and a fresh supply of O_2 is taken up by hemoglobin—the equilibria now favoring CO_2 release and O_2 uptake.

8.7 Summary

In this chapter, we have tried to cover the rudiments of equilibria and of the collision and transition-state theories of reaction rates. In Section 8.1, equilibria and $\Delta G°$ were related and Table 8.1 gave an idea of the range of values to be expected. The transition-state equation (Sec. 8.2) and the Arrhenius (or collision-theory) equation (Sec. 8.3) were described and illustrated with examples. Energy-reaction-coordinate diagrams were introduced in Section 8.4, and the usefulness of the diagrams in helping to understand kinetic control of competitive reactions was illustrated. The next section went into the interrelationships of mechanism, rates, and equilibria. Section 8.6 returned to kinetic control of reactions, went on to thermodynamic control, and discussed briefly the consequences of mixed or competing control of reactions.

We now can appreciate that many, many reactions in organic chemistry are competitive, with only slight differences in activation energy. If we set up the proper conditions, one or another of the competing processes may be caused to predominate.

Many reactions are so slow that they require an appreciable amount of time to approach completion. Often, catalysts are added to speed up a recalcitrant reaction. Thus, every reaction ends up with a purification problem in which the remaining reactants and the unwanted products of competing reactions, as well as any added catalysts, must be removed from the wanted product. This problem leads to somewhat different emphases in laboratory work and in industrial processes.

An industrial process is concerned with the day-after-day, month-after-month repetition of a few reactions with tons of material. The process is kept as continuous as possible. It pays to take the time to work out the very best conditions for a given reaction and to devise means to make every process continuous. Continuous processes usually involve the use of easily handled gases, liquids, and solutions rather than solids, which require much more equipment and labor to handle.

Laboratory bench work is concerned with a great variety of batch processes on a molar scale. This work requires versatility, the ability to carry out a whole spectrum of reactions. Gases, liquids, solutions, and solids are handled indiscriminately but carefully. It seldom is worth the bother and time to find the very best conditions for a given reaction

8.8 Problems

(unless the problem is to find such conditions for industrial production). Nevertheless, the bench chemist is always looking for better reactions and better ways of doing things. A knowledge of equilibria and rates is part of the repertoire that the chemist uses to attack the various chemical problems that may arise. This knowledge has practical usefulness, whether in the laboratory or in the plant.

8.8 Problems

8.1 In methanol (as both solvent and reactant), acetic acid (0.5 M) and dry hydrogen chloride (0.005 M) react to give methyl acetate with log k_2 = 6.10 − 10 000/2.3 RT. The rate constant k_2 was calculated on the basis of complete ionization of the hydrogen chloride so that rate = $k_2[CH_3COOH][CH_3OH_2^{\oplus}]$.
   a. If pure methanol is 24.7 M, and assuming that k_2 = 24.7 k_2', what is k_2' at 25°C?
   b. If K_{eq} = 4.0, what is the concentration of acetic acid remaining at equilibrium?

8.2 Construct an energy-reaction-coordinate diagram for these competing reactions:
   a. $\Delta H^{\ddagger}$ = 20 kcal and $\Delta H°$ = −20 kcal
   b. $\Delta H^{\ddagger}$ = 15 kcal and $\Delta H°$ = −15 kcal
   Which will give the major product, a or b?

8.3 Construct an energy-reaction-coordinate diagram for $A + B \rightleftharpoons C \rightarrow D$ at 25°C. For $A + B \rightleftharpoons C$, $\Delta H^{\ddagger}$ = 13 kcal and $\Delta H°$ = 10 kcal. For $C \rightarrow D$, $\Delta H^{\ddagger}$ = 6 kcal and $\Delta H°$ = −15 kcal.
   a. For the complete reaction, $A + B \rightarrow D$, what would $\Delta H^{\ddagger}$ and $\Delta H°$ be?
   b. What would $\Delta H^{\ddagger}$ be for $D \rightarrow C$?
   c. Could C be assumed to be in a steady state for either the forward or the reverse reaction?
   d. If it were possible to start out with pure C, which would form faster, $A + B$ or D?

8.9 Bibliography

Recent textbooks of organic chemistry (see Bibliography, Ch. 7) all have sections or chapters that describe equilibria and rates of reaction (kinetics). Particularly good in this respect are Allinger et al; Butler and Berlin; Hendrickson, Cram, and Hammond; and Weininger. The paperback by Breslow also is helpful.

We list a few advanced texts and a few articles for those who wish a more detailed treatment.

1. Frost, Arthur A., and Ralph G. Pearson, *Kinetics and Mechanisms*, 2nd ed. New York: Wiley, 1961.

2. Gilliom, Richard D., *Introduction to Physical Organic Chemistry*. Reading, Mass.: Addison-Wesley, 1970.

3. Gould, Edwin S., *Mechanism and Structure in Organic Chemistry*. New York: Holt, Rinehart and Winston, 1959.

The articles listed below all appeared in a single issue, June, 1968, of the *Journal of Chemical Education*, volume 45. Therefore, we give only the page number.

4. Hammond, George S., and Harry B. Gray, *Chemical Dynamics for College Freshmen,* page 354.

5. Wolfgang, Richard, *The Revolution in Elementary Kinetics and Freshman Chemistry,* page 359.

6. Halpern, Jack, *Some Aspects of Chemical Dynamics in Solution,* page 372.

7. Edwards, John O., Edward F. Greene, and John Ross, *From Stoichiometry and Rate Law to Mechanism,* page 381.

8. Edwards, John O., *Bimolecular Nucleophilic Displacement Reactions,* page 386.

9. Taube, Henry, *Mechanisms of Oxidation-Reduction Reactions,* page 452.

The following articles will be of interest to those who are interested in automation for the determination of rate constants.

10. Malmstadt, H. V., E. A. Cordos, and C. J. Delaney, Automated Reaction-Rate Methods of Analysis. *Analytical Chemistry* 44 (12), 26A (1972).

11. Malmstadt, H. V., C. J. Delaney, and E. A. Cordos, Instruments for Rate Determinations. *Analytical Chemistry* 44 (12), 79A (1972).

9 Alcohols

In this chapter we shall look at alcohols (alkanols) and some of their physical properties, reactions, and preparations. We choose them rather than another class as the first class of compounds to examine because of the variety of reactions by which alcohols may be converted into alkanes, alkenes, alkyl halides, aldehydes, ketones, acids, esters, and ethers. We shall discuss the reactions before the methods of synthesis so that we can better understand the preparations. The reactions will be taken up in some detail, whereas syntheses will be given in brief form. Any reaction may be viewed either as a chemical property of a reactant or as a method of preparation of a product. We shall study the preparations of alcohols in detail later, considering these reactions as properties of the reactants. A large number of reactions and methods of preparation is given, but it is only a reasonable selection from the large number known.

9.1   Physical properties   201
  1. Polar and nonpolar characteristics.   2. Boiling points.   3. Melting points.
  4. Uses.

9.2   Types of reaction   204
  Exercises.

9.3   Reactions at the oxygen-hydrogen bond, a   204
  1. Hydrogen bonding; deuterium exchange.   2. Reaction with active metals.
  Exercises.

9.4   Acid-base properties; azeotropes   206
  1. Oxonium ions.   2. Azeotropes.

9.5   Esterification   207
  1. Acid-catalyzed esterification.   2. Benzyl alcohol in perfume.   3. Tertiary alcohols.   4. Other strong acids.   5. Energy-reaction-coordinate diagram of esterification.   6. Esterification of inorganic acids.   7. Inorganic esters from halides of sulfur and phosphorus.   8. Ester formation in biology.   9. Esters from acid chlorides and acid anhydrides. Exercises.

200   Chapter nine

9.6   Reactions at the carbon-oxygen bond, b: alkyl halides   214
      Necessity of acidic conditions. Exercises.

9.7   Ether formation   214
      1. Williamson ether synthesis.   2. S_N1cA method with tertiary alcohols.
      3. Sulfuric acid method. Exercises.

9.8   Reactions involving a carbon-hydrogen bond, c: oxidation   216
      1. Oxidation of primary and secondary alcohols.   2. Catalytic dehydrogenation.
      3. Chromium reagents.   4. Potassium permanganate.   5. Oppenauer oxidation.   6. Biological oxidation. Exercises.

9.9   Balancing oxidation-reduction equations   220
      1. Introduction.   2. Hydrogen method.   3. Examples.

9.10 Alkene formation, d   223
      1. Acidic conditions.   2. Chugaev reaction. Exercises.

9.11 Preparation of alcohols   225
      Introduction.

9.12 Preparation by substitution reactions   225
      1. Formation of the b bond; of the a bond.   2. From a Grignard reagent and an epoxide.   3. Relative uselessness of a bond formation. Exercises.

9.13 Preparation by addition reactions   227
      1. Addition of Grignard reagents to aldehydes, ketones, and esters.
      2. Hydration of alkenes.   3. Hydroboration of alkenes. Exercises.

9.14 Preparation by oxidation-reduction reactions   229
      1. Utility of reduction.   2. Catalytic hydrogenation.   3. Acidic sodium borohydride with a catalyst: direct reduction.   4. Use of lithium aluminum hydride.   5. Use of diborane.   6. Meerwein-Ponndorf-Verley reduction.
      7. Cannizzaro and Tishchenko reactions. Exercises.

9.15 Industrial preparation of alcohols   233
      1. Methanol.   2. Ethanol.   3. Other alcohols.

9.16 Summary   234

9.17 Problems   236

9.18 Bibliography   238

9.19 Herbert C. Brown: The discovery of hydroboration   238a

9.1 Physical properties

Polar and nonpolar characteristics.   Alcohols, possessing both a nonpolar hydrocarbon group and polar carbon-oxygen and oxygen-hydrogen bonds, have properties intermediate between those of water and the corresponding hydrocarbon. The polar group dominates in the low molecular weight alkanols, which are all completely miscible with water (Table 9.1). As the hydrocarbon group is increased in size, the solubility in water (and of water in the alcohol) decreases rapidly up to the butyl and pentyl alcohols. Thereafter, the small water solubility decreases more slowly.

The solubility for a given set of isomers is greatest for the tertiary alcohols, less for a secondary alcohol or a branched-chain primary alcohol, and least for the straight-chain primary isomer. Conversely, solubility in nonpolar liquids such as hexane is least for methanol but increases rapidly to complete miscibility for the alcohols of higher molecular weights.

Boiling points.   Boiling points increase with molecular weight, as expected, but the increase is surprisingly constant. Aside from the small jump from methanol to ethanol, there is close to a 20° increase for each additional CH_2, whether one compares the straight-chain primary alcohols, the 2-alkanols, or the tertiary alcohols. Within a set of isomers, the 1-alkanol boils about 19° higher than the 2-alkanol and about 36° higher than the tertiary alcohol. However, a cycloalkanol invariably boils at a higher temperature than does a corresponding 1-alkanol, by 3° to 10°. (Remember that a cycloalkanol is not an isomer of an alkanol. The former are members of the $C_nH_{2n}O$ homologous series, and the latter belong to the $C_nH_{2n+2}O$ series.)

Melting points.   The melting points of the alcohols tend to be low so that, except for methanol and ethanol, the liquid phase has a range of existence of about 200° for the 1-alkanols. As branching increases within a set of isomers, there is tendency (with some exceptions) for the melting point to rise. This increase, along with a general tendency for the boiling point to decrease, leads to a shortening of the temperature range for the existence of the liquid phase. Thus, *tert*-butyl alcohol (2-methyl-2-propanol) has a liquid phase range of only $82 - 25 = 57°$ and neopentyl alcohol (2,2-dimethyl-1-propanol) has a range of $114 - 53 = 61°$. Crudely, the more spherical the molecule becomes, the lower the boiling point and the higher the melting point of the substance. The more like a ball (or less like a

202 Chapter nine

Table 9.1 Alcohols

Number of carbons	Structural formula	Name[a]	Class[b]		
			Primary	Secondary	Tertiary
1	CH_3OH	methanol	−98°, 65°, ∞		
2	CH_3CH_2OH	ethanol	−115°, 78°, ∞		
3	$CH_3CH_2CH_2OH$	1-propanol	−126°, 97°, ∞		
	$CH_3CHOHCH_3$	2-propanol		−88°, 82°, ∞	
	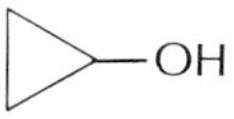 ▷—OH	cyclopropanol		100°	
4	$CH_3CH_2CH_2CH_2OH$	1-butanol	−90°, 118°, 7.9		
	$(CH_3)_2CHCH_2OH$	2-methyl-1-propanol	−108°, 108°, 10.0		
	$CH_3CH_2CHOHCH_3$	2-butanol[c]		−114°, 99°, 12.5	
	$(CH_3)_3COH$	2-methyl-2-propanol			25°, 82°, ∞
	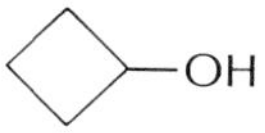 ◇—OH	cyclobutanol		123°	
5	$CH_3(CH_2)_3CH_2OH$	1-pentanol	−78°, 138°, 2.3		
	$CH_3CH_2\!\!\setminus\!\!CHCH_2OH / CH_3$	2-methyl-1-butanol[d]	128°, 3.6		
	$(CH_3)_2CHCH_2CH_2OH$	3-methyl-1-butanol	−117°, 131°, 2.0		
	$(CH_3)_3CCH_2OH$	2,2-dimethyl-1-propanol	53°, 114°		
	$CH_3(CH_2)_2CHOHCH_3$	2-pentanol[e]		119°	
	$(CH_3CH_2)_2CHOH$	3-pentanol		116°	
	$CH_3CH_2C(CH_3)_2OH$	2-methyl-2-butanol			−12°, 102°, 12.5
	$(CH_3)_2CHCHOHCH_3$	3-methyl-2-butanol		112°	
	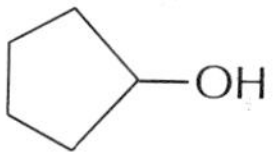 ⬠—OH	cyclopentanol		−19°, 140°	
6[f]	$CH_3(CH_2)_4CH_2OH$	1-hexanol	−52°, 157°, 0.6		
	$CH_3(CH_2)_2\!\!\setminus\!\!CHCH_2OH / CH_3$	2-methyl-1-pentanol	148°		
	$(CH_3)_2CH(CH_2)_2CH_2OH$	4-methyl-1-pentanol	148°		
	$(CH_3CH_2)_2C(CH_3)OH$	3-methyl-3-pentanol			122°
	$CH_3(CH_2)_2C(CH_3)_2OH$	2-methyl-2-pentanol			122°
	$(CH_3)_2CHC(CH_3)_2OH$	2,3-dimethyl-2-butanol			121°

9.1 Physical properties

Table 9.1 (continued)

Number of carbons	Structural formula	Name[a]	Class[b]		
			Primary	Secondary	Tertiary
	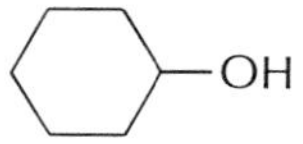 (cyclohexane)—OH	cyclohexanol		24°, 161°	
7[g]	CH$_3$(CH$_2$)$_5$CH$_2$OH	1-heptanol	−35°, 176°, 0.2		
	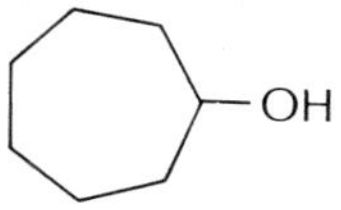 (cycloheptane)—OH	cycloheptanol		2°, 185°	
8[h]	CH$_3$(CH$_2$)$_6$CH$_2$OH	1-octanol	−17°, 195°, 0.05		
	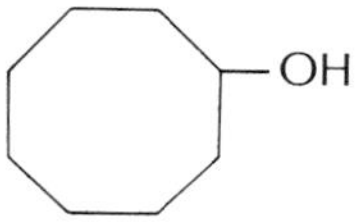 (cyclooctane)—OH	cyclooctanol		15°, 100° (15 mm)	

[a] IUPAC names.
[b] Under each class designation (primary, secondary, and tertiary) are given, in order, melting point, boiling point, and solubility in g/100 ml of water.
[c] Data given for the racemic mixture; the enantiomeric forms do not differ appreciably, $[\alpha]_D^{20}$ 13.9°.
[d] As in note c, $[\alpha]_D^{20}$ 5.9°.
[e] As in note c, $[\alpha]_D^{20}$ 13.7°.
[f] Only a few of the 17 structurally isomeric hexanols are given.
[g] There are 39 structurally isomeric heptanols.
[h] There are 89 structurally isomeric octanols.

string) the alcohol is, the easier it is to fit into a crystalline lattice: thus, the higher its melting point. The more like a ball (or less like a string), the less the possibility for intermolecular van der Waals' attractive forces to operate, and the lower the boiling point. It is not just alcohols that behave this way; it is a consequence of the general shape of the molecule, whether it is an alcohol, a hydrocarbon, an acid, or a ketone.

Uses. Alcohols are used extensively as solvents. Being both polar and nonpolar, they dissolve a wide variety of substances. Convenient boiling points and low melting points allow easy removal of solvent alcohols by boiling or evaporation of a solution or mixture to recover the solutes.

Even though we shall not go into the subjects in this chapter, we should mention that diols, triols, and polyols exist, in open chain, cyclic, and heterocyclic forms. Ethylene glycol, HOCH$_2$CH$_2$OH (and formerly, glycerol, CH$_2$OHCHOHCH$_2$OH) is used extensively as an antifreeze for internal combustion engine coolant. Glycerol and propylene glycol (CH$_2$OHCHOHCH$_3$) are included in foods such as bakery products to maintain moisture content. They are put to the same use in hair conditioners. Sugars and other carbohydrates are heterocyclic polyols. (See Chap. 19.)

9.2 Types of reaction

Alcohols characteristically undergo heterolytic rather than homolytic reactions. Like water, alcohols act as bases toward strong acids and as acids toward strong bases. Many of the reactions of alcohols are preceded by the acceptance or donation of a proton (or by the donation of an electron pair to a Lewis acid or an electrophile). The alkyl portions of an alkanol tend to remain inert during reaction, except for dehydration.

If we look at the partial structure of an alcohol, we can identify four bonds that may participate in its reaction. Some reactions are characterized by a rupture of the oxygen-hydrogen bond at *a*. In others, the carbon-oxygen bond is broken at *b*. These reactions

$$\begin{array}{c} | | \mathrm{O{-}H} \\ -\mathrm{C{-}C} {}^{b} {}_{a} \\ | | \\ \mathrm{H} \mathrm{H} \\ d c \end{array}$$

ordinarily are preceded by electron pair donation by oxygen to a proton or to a Lewis acid. In other instances, the carbon-hydrogen bond at *c* is involved, but only in conjunction with a change at *a*. Reaction at *d* occurs only if *b* also participates, as in dehydration. We shall organize our perusal of the reactions of alcohols by these letters.

Exercises

9.1 All alcohols are capable of reaction at *a* and *b*. Using Table 9.1, list those alcohols incapable of reaction at *c*. Also list those incapable of reaction at *d*.

9.3 Reactions at the oxygen-hydrogen bond, *a*

Hydrogen bonding; deuterium exchange. Like water, alcohols are strongly hydrogen bonded in the liquid state. The association into clumps of molecules, which are constantly changing in size and shape, requires energy for dissociation and for separation into a gas. As a result, the heats of vaporization and the boiling points are abnormally high. The effect diminishes in magnitude with increase in molecular weight because of increasing preponderance of the bulky alkyl group, which not only decreases the polarity of the liquid, but also decreases the number of hydroxy groups per unit of volume. Compare water (mw 18, bp 100°) and methane (mw 16, bp −184°) which have a difference of 284° in bp. Methanol and ethane give a difference of 154°; 1-decanol and undecane have a difference of only 32°.

Again like water, the lower alcohols are able to dissolve most acids and bases because of the ability to form lyonium and lyate ions. The lower dielectric constant of an alcohol results in a lowered ability to dissolve salts. Thus, methanol and ethanol can reasonably solubilize sodium iodide, but the other sodium halide salts are practically insoluble.

9.3 Reactions at the oxygen-hydrogen bond, a

The lability of the oxygen-hydrogen bond at *a* may be illustrated by the rapid equilibration with deuterium oxide (heavy water) (Eq. 9.1). Even high molecular weight,

$$ROH + DOD \rightleftarrows \begin{array}{c} R\overset{\oplus}{O}H + OD^{\ominus} \\ | \\ D \\ \\ RO^{\ominus} + D\overset{\oplus}{O}D \\ | \\ H \end{array} \rightleftarrows RO + OD \quad \begin{array}{cc} | & | \\ D & H \end{array} \qquad (9.1)$$

water-insoluble alcohols in the presence of a strong acid or base equilibrate rapidly at the interface between the two liquid phases.

Reaction with active metals. Alcohols, like water, are acidic enough to react with active metals to produce hydrogen and the alkoxide of the metal. Again, the reactivity is influenced by the bulk of the alkyl group. Methanol reacts with sodium: not as violently as water does, but extreme caution is necessary to prevent fire. Ethanol is less reactive, 1-propanol still less, and 2-propanol reacts rather gently. With *tert*-butyl alcohol, refluxing is usually necessary to complete the reaction. The alkoxides are reasonably soluble in the alcohols from which they are made (Eq. 9.2) and are widely used as strong bases.

$$2ROH + 2Na \longrightarrow 2RO^{\ominus} + 2Na^{\oplus} + H_2 \qquad (9.2)$$

A solution of potassium hydroxide in 95% ethyl alcohol frequently is used to carry out dehydrohalogenation reactions. As we saw in Chapter 7, the bulkier strong base, potassium *tert*-butoxide in *tert*-butyl alcohol, tends to favor Hofmann elimination as compared to alcoholic potassium hydroxide.

The less reactive metals magnesium and aluminum also react readily with alcohols, because the magnesium and aluminum alkoxides are reasonably soluble in the alcohols from which they are prepared. These metals fail to react with water under ordinary conditions because of the adherent coating of oxide-hydroxide formed on the surface of the metals, which prevents a continuation of the reaction.

> The coating presents a practical difficulty as we try to get the reaction with an alcohol to start. Once started, the reaction continues because the coating is gradually stripped from the remaining metal by the action of the increasing concentration of alkoxide ion. The presence of moisture in the alcohol counteracts the effect by reformation of the coating, so the alcohol must be reasonably dry at the start. Secondary and tertiary alcohols usually require a period of heating under reflux if the reaction is to go to completion.

Once prepared, solutions of magnesium or aluminum alkoxides are useful for the removal of the last traces of moisture from the corresponding alcohol (Eq. 9.3). The reaction

$$(RO)_3Al + 3H_2O \longrightarrow \underline{Al(OH)_3} + 3ROH \qquad (9.3)$$

goes to completion because the hydroxide is insoluble. The remaining anhydrous alcohol is distilled from the mixture as needed.

The aluminum alkoxides have a metal-oxygen bond that is less polar than those in the magnesium alkoxides. This lower polarity leads to greater solubility of the aluminum alkoxides in the corresponding alcohol. The aluminum alkoxides also may be isolated as solids with definite melting points, and they may be distilled under reduced pressure without decomposition.

Aluminum isopropoxide is the reagent most used in the Meerwein-Ponndorf-Verley reduction of ketones and aldehydes, and aluminum *tert*-butoxide is most used in the reverse of the Meerwein-Ponndorf-Verley reduction, the Oppenauer oxidation. These reactions will be described later in this chapter.

Exercises

9.2 Give a method for preparing CD_3OD.

9.4 Acid-base properties, azeotropes

Oxonium ions.   Before continuing with reactions at a, let us pause to look at the part played by the nonbonding electron pairs of oxygen. We have already seen several instances of the participation by these electrons: the formation of a lyonium ion, $ROH_2^\oplus$, and (in deuterium exchange) of $ROHD^\oplus$. An oxygen covalently bound to three atoms and bearing a positive charge is known as an oxonium ion. In principle, protons are not necessary; they have merely furnished us with examples up to now. Any electrophile or Lewis acid also should be able to form a bond to the oxygen of water, an alcohol, or an ether. Thus, alcohols (like water) are able to act as Brønsted acids toward bases and as Lewis or Brønsted bases toward Lewis or Brønsted acids.

Azeotropes.   In an ideal liquid mixture of components A and B, the attraction between the molecules of component A does not differ greatly from that between the molecules of component B. The attraction between a molecule of A and a molecule of B is about the same as that of A for A or of B for B. A mixture of heptane and octane or of 1-heptanol and 1-octanol is an example. If such a mixture is heated, the vapor pressure increases until, at a pressure of 1 atm, boiling begins. The more volatile component exerts a partial vapor pressure in excess of its concentration in the liquid phase, so it is removed at a greater rate than is the less volatile component. (In a 50-50 mixture, the more volatile component exerts something more than 50% of the total pressure.) The boiling liquid left behind becomes richer in the higher-boiling component as the lower-boiling component is removed. □ □

Except for methanol, most of the lower alcohols form azeotropes with water. Ethanol boils at 78.5°, but a mixture of ethanol and water (95.6% to 4.4% by weight) has a bp of 78.2°. A propanol (bp 97.2°) and water mixture (81.8% and 18.2%) has bp 88.1°, and a

2-propanol (bp 82.3°) and water mixture (87.8% and 12.2%) has bp 80.4°. These azeotropes all have boiling points less than those of either component. This phenomenon leads to an awkward situation: any attempt to obtain the alcohol free from water by distillation fails because the most volatile component is the low-boiling azeotrope. Other means must be used to free the alcohols from water. □ □

9.5 Esterification

Acid-catalyzed esterification. An ester may be regarded as an alcohol that has had the hydrogen of the oxygen-hydrogen bond replaced by a moiety (a group) derived from an acid (either inorganic or organic) by apparent loss of an OH group from the acid. The definition excludes an alkyl halide from being classed as an ester. If a strong acid, such as sulfuric acid, is added to a mixture of an alcohol and an acid like acetic acid, an equilibrium is set up (shown in Eq. 8.22, repeated here as Eq. 9.4).

$$CH_3COOH + ROH \underset{}{\overset{H_2SO_4}{\rightleftharpoons}} CH_3COOR + H_2O \qquad (9.4)$$

The mechanism is complicated by the large number of steps necessary. The first step, proton transfer, may occur in any of three ways (Eqs. 9.5) as a rapid reversible equilibrium.

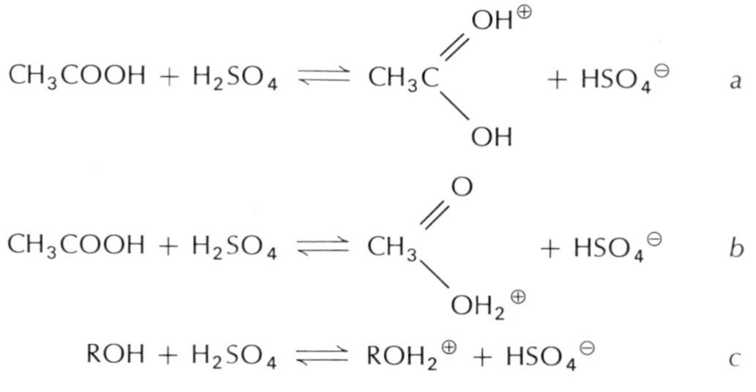

$$
\begin{aligned}
CH_3COOH + H_2SO_4 &\rightleftharpoons CH_3C(OH^+)(OH) + HSO_4^- \qquad &a\\
CH_3COOH + H_2SO_4 &\rightleftharpoons CH_3C(=O)(OH_2^+) + HSO_4^- \qquad &b\\
ROH + H_2SO_4 &\rightleftharpoons ROH_2^+ + HSO_4^- \qquad &c
\end{aligned}
\qquad (9.5)
$$

Of the two conjugate acids formed from acetic acid, the symmetrical ion $CH_3C(OH)_2^{\oplus}$ (a) has the greater stabilization by resonance between two equivalent structures. Although the alcohol is the stronger base, enough $CH_3C(OH)_2^{\oplus}$ is formed to allow the next steps to occur (Eq. 9.6). B is any proton acceptor in the mixture: water, acetic acid, bisulfate ion, alcohol, ester, or even the tetrahedral intermediate, the product in Equation 9.6.

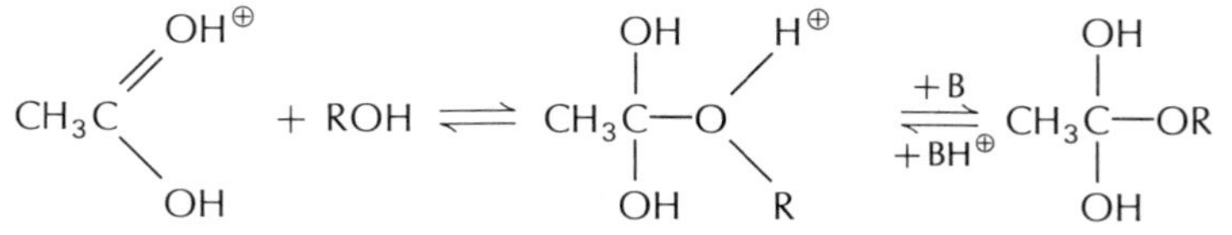

$$CH_3C(OH^+)(OH) + ROH \rightleftharpoons CH_3C(OH)(OH)(O^+HR) \underset{+BH^\oplus}{\overset{+B}{\rightleftharpoons}} CH_3C(OH)(OH)(OR) \qquad (9.6)$$

In any event, the tetrahedral intermediate will accept a proton from some donor (symbolized as $BH^{\oplus}$) in the mixture. If the proton is taken on by the oxygen of the OR group, the reaction is the reversal of Equation 9.6. However, the OH groups also are available for accepting a proton, in which case (Eq. 9.7) a molecule of water is lost and the conjugate acid

$$\underset{\underset{OH}{|}}{\overset{\overset{OH}{|}}{CH_3C-OR}} \underset{+B}{\overset{+BH^\oplus}{\rightleftharpoons}} \underset{\underset{OH}{|}}{\overset{\overset{OH_2^\oplus}{|}}{CH_3C-OR}} \underset{+H_2O}{\overset{-H_2O}{\rightleftharpoons}} CH_3C\overset{OR}{\underset{OH^\oplus}{\diagdown\!\!\diagup}} \qquad (9.7)$$

of the ester results. The conjugate ester then is able to lose a proton to any proton acceptor (Eq. 9.8), and the ester is formed.

$$CH_3C\overset{OR}{\underset{OH^\oplus}{\diagdown\!\!\diagup}} + B \rightleftharpoons CH_3C\overset{OR}{\underset{O}{\diagdown\!\!\diagup}} + BH^\oplus \qquad (9.8)$$

The presence of the strong acid is necessary to produce enough of the conjugate acid form of the organic acid to hasten the attainment of equilibrium. If the reaction is carried out in an excess of the alcohol, the rate is second order, dependent upon the concentration of the strong acid and of the organic acid. The equilibrium constant for the reaction of acetic acid with ethanol has a value near 4. Branching in the alcohol slows the rate of reaction: secondary alcohols esterify much more slowly than do primary alcohols. Tertiary alcohols, however, esterify as rapidly as does methyl alcohol.

Benzyl alcohol in perfume. The perfume industry is part science, part art, and perhaps another part marketing. The chemistry questions it tries to answer are demanding. Foremost is the analysis and synthesis of fragrances. Each fragrance is a complex mixture of organic compounds, predominately alcohols and esters. Benzyl acetate, for instance, was found to be a major component of jasmin. It is now synthesized (it is the ester of acetic acid and benzyl alcohol) for use in synthetic fragrances.

$$\underset{}{\text{C}_6\text{H}_5}-CH_2OH$$

Often the odor strength, physical properties, or expense of the essential compounds of a fragrance are such that it is sold as an alcoholic tincture or is diluted with other materials such as benzyl alcohol. A 50 percent benzyl alcohol solution is, in fact, required as solvent for phenylacetaldehyde. Pure phenylacetaldehyde tends to polymerize.

By their nature fragrant compounds are volatile but each with a different vapor pressure. This is the root of another problem. Since the bouquet of a perfume depends on the ratio of its many components, the fragrance must change as the bottle is opened to the atmosphere. The perfume makers, using the principle of Raoult's Law, try to minimize this by adding another liquid as a fixative. The fixative retards the volatility of all the other liquids. A good choice is an oil soluble compound with low vapor pressure and low molecular weight, for example benzyl benzoate.

Tertiary alcohols. In the mechanism described above, called the Fischer method of esterification, the fission of the oxygen-hydrogen bond (a) did not occur until the second step of Equation 9.6. With a tertiary alcohol, a different mechanism comes into operation. The conjugate acid of the alcohol (for example, *tert*-butyl alcohol) loses water easily to form a *tert*-butyl cation (rupture at bond b). The ion, a Lewis acid and an electrophile, equilibrates with the organic acid to give the conjugate acid of the ester (Eq. 9.9). The conjugate acid of

$$CH_3COOH + CH_3-\overset{CH_3}{\underset{CH_3}{\overset{|}{C}}}{}^{\oplus} \rightleftharpoons CH_3C\overset{\overset{\oplus}{O}-C(CH_3)_3}{\underset{OH}{\diagup\!\!\!\diagdown}} \underset{+BH^{\oplus}}{\overset{+B}{\rightleftharpoons}} CH_3COOC(CH_3)_3 \qquad (9.9)$$

the ester then equilibrates with any base (proton acceptor) in the mixture to give the product.

In Fischer esterification, the oxygen initially present in the alcohol is present in the ester. This fact has been demonstrated by use of the heavy oxygen isotope ^{18}O as a tracer. The alcohol is <u>labeled,</u> we say: it is synthesized so as to have ^{18}O present. The presence of the heavy isotope may be followed or traced from reactant to product (Eq. 9.10). Because

$$RCOOH + CH_3-{}^{18}OH \overset{H_2SO_4}{\rightleftharpoons} RC\overset{{}^{18}OCH_3}{\diagup\!\!\!\diagdown_O} + H_2O \qquad (9.10)$$

the ^{18}O appears in the ester and not in the water, we know that the $CH_3-{}^{18}O$ bond is maintained during the reaction. Likewise, the retention of optical activity (configuration), if the carbon holding the OH group is asymmetric, demonstrates the persistence of the carbon-oxygen bond during the reaction. If the alcohol is tertiary and the tertiary carbon atom is asymmetric, extensive racemization is observed in the ester, as we expect if a cation is an intermediate.

Other strong acids. Strong acids other than sulfuric acid, such as gaseous hydrogen chloride or a Lewis acid, may be used to catalyze the Fischer esterification, but an excess should be avoided. Excess strong acid complicates the situation by tying up the alcohol as the conjugate acid, which decreases the amount of alcohol available for the reaction shown in Equation 9.6. The reaction is slowed down as a result. If the alcohol is tertiary, excess acid causes the carbonium ion to be formed too rapidly to be accommodated by the organic acid. Then dehydration competes with esterification. Certain organic acids are strong enough to catalyze their own esterification. Formic and oxalic acids are examples.

Energy-reaction-coordinate diagram of esterification. A schematic energy-reaction-coordinate diagram for the six steps in a Fischer esterification is shown in Figure 9.1. The slow steps (large activation energy) are those for the association of ROH with $CH_3C(OH)_2{}^{\oplus}$ and of H_2O with $CH_3C(OR)OH^{\oplus}$. (See Ex. 9.3.) The proton transfer reactions are fast, with small activation energies. With an equilibrium constant not far from unity, the energy levels at start and finish are nearly the same. The position of equilibrium is easily shifted one way or the other by use of an excess of a reactant or product or by removal of one of the reactants or products as the reaction proceeds.

Esterification of inorganic acids. Esters of inorganic acids may be prepared. Two groupings of acids may be set up: 1. acids (and the oxides from which they are derived) that esterify directly, and 2. acids that do not esterify directly.

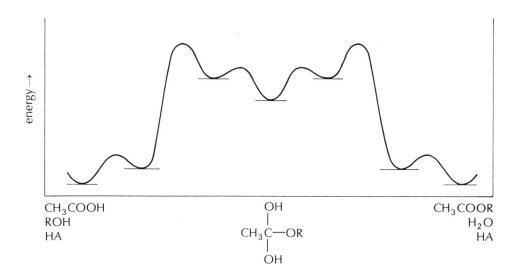

Figure 9.1 Schematic energy-reaction-coordinate diagram for Fischer esterification

Group 1 includes sulfuric, nitric, chromic, boric, and nitrous acids, and so on (Eqs. 9.11).

$$H_2SO_4 + ROH \rightleftharpoons ROSO_3H + H_2O$$
$$HNO_3 + ROH \rightleftharpoons RONO_2 + H_2O$$
$$H_2Cr_2O_7 + ROH \rightleftharpoons ROCr_2O_6H + H_2O \text{ (or } ROCrO_3H) \quad (9.11)$$
$$H_3BO_3 + ROH \rightleftharpoons ROB(OH)_2 + H_2O$$
$$HNO_2 + ROH \rightleftharpoons RONO + H_2O$$

If the acid is weak, the presence of a strong catalyzing acid is required. The mechanisms are roughly similar to that of the Fischer esterification. With sulfuric acid a pentacoordinate intermediate that is analogous to the tetrahedral one in Fischer esterification (Eq. 9.12) is formed. Then a molecule of water is lost, and the alkyl hydrogen sulfate is formed.

$$ROH + OS(OH)_3^{\oplus} \rightleftharpoons \begin{array}{c} R \quad O^{\ominus} \, OH \\ \diagdown \oplus \quad | \diagup \\ O-S^{\oplus} \\ \diagup \quad | \diagdown \\ H \quad OH\,OH \end{array} \rightleftharpoons H_3O^{\oplus} + ROS(O)_2OH \quad (9.12)$$

The second alkyl group of a sulfate ester is introduced via an S_N2cA reaction (Eq. 9.13).

$$ROS(O)_2O^{\ominus} + R-\overset{\oplus}{O}H_2 \rightleftharpoons ROS(O)_2OR + H_2O \quad (9.13)$$

This path has been shown by inversion of configuration and by loss of ^{18}O from the alcohol. Methyl and ethyl sulfates are prepared by distillation of the alkyl hydrogen sulfate at low pressure. Sulfates of the higher alcohols are prepared by less direct methods to avoid the decomposition that sets in at the higher bp of the alkyl hydrogen sulfates.

Nitric acid esterifies rapidly because the lyonium ion (ON(OH)$_2^{\oplus}$) loses a molecule of water to form the nitronium ion (NO$_2^{\oplus}$), which has little steric hindrance to bond formation with a molecule of alcohol. Nitrous acid also esterifies with astonishing speed. This weak acid is so unstable that it is prepared as needed from sodium or potassium nitrite and sulfuric acid. At 0°, the esterification is complete as soon as all the reagents have been combined. Whether the nitrosonium ion (NO$^{\oplus}$) or the lyonium ion (N(OH)$_2^{\oplus}$) or one of the nitrogen oxides related to nitrous acid is the reactive entity that attacks the oxygen of the alcohol is not clear.

Boric acid, even though weak as a proton acid, esterifies quickly because it is a Lewis acid and associates readily with an electron pair of the oxygen in ROH.

Inorganic esters from halides of sulfur and phosphorus. Group 2 acids are, for example, sulfurous, phosphoric, phosphorous, arsenic, arsenous, and silicic acids. Although these acids may not be esterified directly, the esters may be prepared by interaction of an alcohol with the chloride or oxychloride of the element. (Bromides and iodides cannot be used: the esters are easily substituted by these more nucleophilic halide ions. Even with the chlorides, some conversion of the esters into alkyl chlorides takes place.)

The reaction of an alcohol with thionyl chloride (SOCl$_2$) has been discussed. By using an excess of the alcohol, adding the thionyl chloride slowly, and keeping the temperature below 45° during the first half of the reaction, we can cause the intermediate alkyl chlorosulfite ester to react with the excess alcohol to form the dialkyl sulfite, rather than with chloride ion to give alkyl chloride (Eqs. 9.14).

$$2ROH + SOCl_2 \xrightarrow{<45°} (RO)_2SO + 2HCl$$

or

$$ROH + SOCl_2 \longrightarrow \underset{H}{\overset{R}{\underset{|}{\overset{\oplus}{O}}}}\text{—}\underset{Cl}{\overset{O^{\ominus}}{\underset{|}{S}}}\text{—}Cl \xrightarrow{-HCl} RO\text{—}\underset{|}{\overset{O^{\ominus}}{\underset{|}{S^{\oplus}}}}\text{—}Cl \quad (9.14)$$

$$\downarrow ROH$$

$$(RO)_2SO \xleftarrow{-HCl} RO\text{—}\underset{ROH^{\oplus}}{\overset{O^{\ominus}}{\underset{|}{S}}}\text{—}Cl$$

When similar techniques are used, phosphorus trichloride reacts with alcohols to give trialkyl phosphites; phosphorus pentachloride and phosphorus oxychloride (POCl$_3$) react to give trialkyl phosphates; and silicon tetrachloride gives tetra-alkoxysilanes. Sometimes a tertiary amine such as triethylamine (Et$_3$N) is added to the reaction mixture to bind the liberated hydrogen chloride and keep it from protonating the esters as they are formed. The protonated esters are more subject to S$_N$2 displacement by chloride ion than are the unprotonated esters. Sometimes, too, the aluminium or sodium alkoxides are prepared from the alcohols first, and the alkoxides are allowed to react with inorganic chloride to produce the esters and the chloride salt of the metal.

Esters of phosphoric acid are widespread in nature and are necessary to life (Chaps. 21, 28, and 29).

Ester formation in biology. Under neutral aqueous conditions, at body temperature, the alcohol is simply not a strong nucleophile. Even if it were, under these conditions, the carbon of the resonance stabilized carboxylate anion is not very susceptible to nucleophilic attack. Living systems have evolved to include a special molecule called coenzyme A (CoA) to activate the carboxyl group for ester formation. CoA is a long molecule with a sulfhydryl group on one end. Thus alcohol attacks the carbonyl carbon of a thioester, rather than that of a carboxylate anion.

The formation of phosphatidic acid, the initial step in the biosynthesis of lipids and fats is illustrated below.

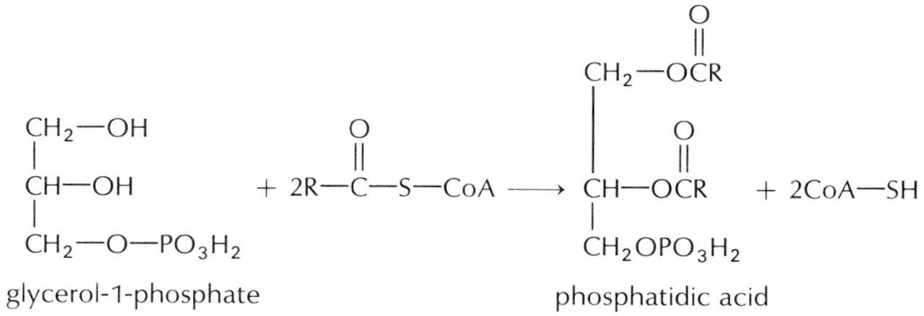

$$\begin{array}{c} CH_2-OH \\ | \\ CH-OH \\ | \\ CH_2-O-PO_3H_2 \end{array} + 2R-\overset{O}{\underset{\|}{C}}-S-CoA \longrightarrow \begin{array}{c} CH_2-O\overset{O}{\underset{\|}{C}}R \\ | \\ CH-O\overset{O}{\underset{\|}{C}}R \\ | \\ CH_2OPO_3H_2 \end{array} + 2CoA-SH$$

glycerol-1-phosphate phosphatidic acid

where R is hydrocarbon chain

Esters from acid chlorides and acid anhydrides. The preparation of esters of carboxylic acids is not restricted to the Fischer esterification. Alcohols react with acid chlorides and with acid anhydrides to form esters. Acid chlorides result from replacement of the hydroxy group of an acid by chloride. Just as an alkyl halide is obtained by proper treatment of an alcohol with thionyl chloride or one of the phosphorus chlorides, an acid reacts with the same reagents to give an acid chloride. Acid anhydrides in turn result from the interaction of an acid chloride with the salt of the acid, or with the acid in the presence of a tertiary amine in a nonpolar solvent (Eqs. 9.15).

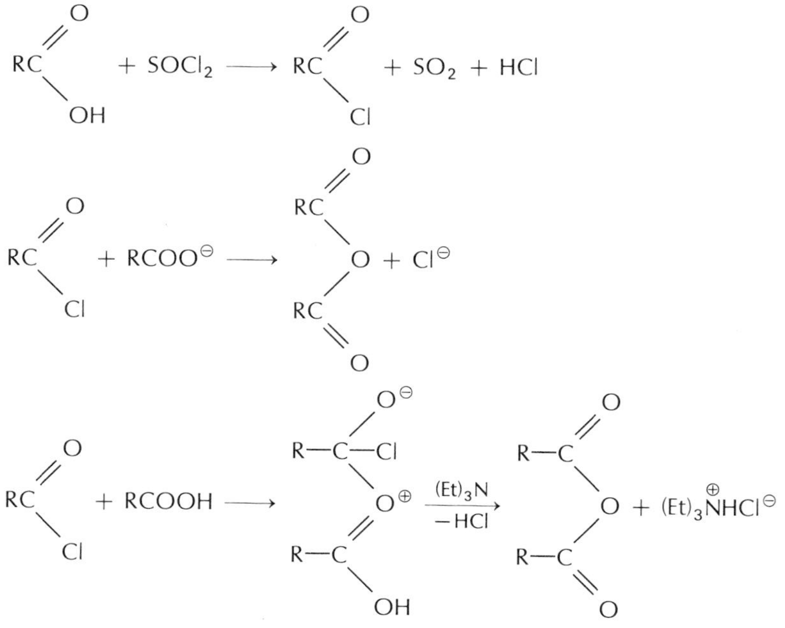

$$RC\overset{O}{\underset{OH}{\diagup}} + SOCl_2 \longrightarrow RC\overset{O}{\underset{Cl}{\diagup}} + SO_2 + HCl$$

$$RC\overset{O}{\underset{Cl}{\diagup}} + RCOO^{\ominus} \longrightarrow \begin{array}{c} RC\overset{O}{\diagup} \\ \diagdown O \\ \diagup \\ RC \\ \diagdown O \end{array} + Cl^{\ominus} \qquad (9.15)$$

$$RC\overset{O}{\underset{Cl}{\diagup}} + RCOOH \longrightarrow \begin{array}{c} R-C-Cl \\ | \\ O^{\ominus} \\ O^{\oplus} \\ \| \\ R-C \\ | \\ OH \end{array} \xrightarrow[-HCl]{(Et)_3N} \begin{array}{c} R-C\overset{O}{\diagup} \\ \diagdown O \\ \diagup \\ R-C \\ \diagdown O \end{array} + (Et)_3\overset{\oplus}{N}HCl^{\ominus}$$

9.5 Esterification

An acid chloride or anhydride has a more firmly defined carbon-oxygen double bond than does an acid; and, especially with a chlorine present, the carbon end of the double bond is more positive. Acid chlorides react vigorously with primary alcohols and more moderately with secondary alcohols, and they dehydrate tertiary alcohols unless special techniques are used. The acid anhydrides are more gentle in these reactions. The first step in the mechanism for ester formation is coordination of the oxygen of the alcohol with the carbon of the carbon-oxygen double bond. Loss of a proton and of Y (Y = $Cl^{\ominus}$ or $RCOO^{\ominus}$) follows (Eq. 9.16). The reaction goes to completion instead of to an equilibrium.

$$RCOY + EtOH \longrightarrow R-\underset{\underset{H}{\overset{\oplus}{\text{OEt}}}}{\overset{O^{\ominus}}{C}}-Y \xrightarrow{-HY} R-C\overset{O}{\underset{OEt}{\diagup}} + HY \qquad (9.16)$$

A tertiary amine acts as a catalyst for the reaction whenever one is needed. Because amines are much better nucleophiles than are alcohols, it is thought that the amine is able to coordinate with carbon faster than does the alcohol in the rate-determining first step.

The intermediate loses $Y^{\ominus}$ to give the ion $RC\overset{O}{\overset{\|}{-}}\overset{\oplus}{N}Et_3$. This new intermediate, despite increased crowding about the carbon, is even more electrophilic and reacts quickly with alcohol, the amine probably leaving as the alcohol oxygen bonds to the carbon to give the ester.

Although the tetrahedral intermediate of Equation 9.16 is formed if a tertiary alcohol is used, and the ester results, this ester of a tertiary alcohol is unstable in the presence of HY. Thus, if the ester accepts a proton from HY, it may dissociate in a rate-determining step to give the acid RCOOH and a tertiary carbonium ion, the reverse of Equation 9.9. The answer to the problem of persuading the reaction to give an ester of a tertiary alcohol is to use an excess of a base to take care of all the HY that is formed. A tertiary amine should be used.

We should note that the esters produced by the action of acid chlorides and acid anhydrides still possess the oxygen of the alcohol used. Also, formyl chloride (HCOCl) is so unstable that it cannot be used for esterifications. It falls apart into HCl and CO by α elimination. However, the diacid chloride ($COCl_2$) of carbonic acid (H_2CO_3) is stable and may be used to prepare dialkyl carbonates (($RO)_2C=O$). Carbonyl chloride ($COCl_2$) is also known as phosgene.

Exercises

9.3 If an ester of *tert*-butyl alcohol labeled with ^{18}O is hydrolyzed under acidic conditions what becomes of the labeled oxygen?

9.6 Reactions at the carbon-oxygen bond, b: alkyl halides

Necessity of acidic conditions. As we have said many times already, the hydroxy group is difficult to remove, but a protonated hydroxy group is easy to remove. Therefore, all reactions of alcohols in which the carbon-oxygen bond (b) is broken must be preceded by protonation or by association with a Lewis acid or electrophile. Thus, reactions at b all take place under acidic conditions. Depending upon the class of alcohol, the reactions are S_N1cA, S_N2cA, E1cA, or E2cA. The substitution reactions by which alcohols may be converted into alkyl halides were discussed in Section 7.8 and need not be repeated here.

Exercises

9.4 Give a reasonably good method for carrying out these conversions:
   a. tBuOH ⟶ tBuBr
   b. nBuOH ⟶ nBuCl
   c. iBuOH ⟶ iBuBr
   d. 2,2-dimethyl-1-propanol ⟶ 1-chloro-2,2-dimethylpropane
   e. cyclopentyl-OH ⟶ cyclopentyl-Br

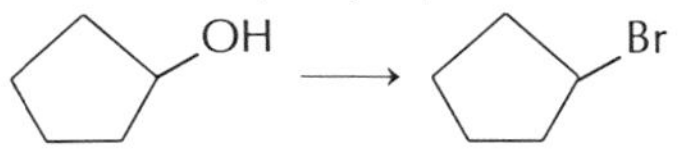

9.7 Ether formation

An ether looks as if it is derived from water by the replacement of both hydrogens with alkyl groups. Its close relationship to an alcohol is obvious. Alcohols may serve as the starting points for the preparation of ethers in two ways: the Williamson ether synthesis or any of the sulfuric acid methods.

Williamson ether synthesis. The Williamson ether synthesis is nothing more than the use of an alkoxide ion as the nucleophile in an S_N2 reaction (Eq. 9.17). The departing

$$RO^\ominus + CH_3CH_2-I \longrightarrow ROCH_2CH_3 + I^\ominus \qquad (9.17)$$

group usually is a halide ion. The reaction is encumbered by the usual competitions (E2 for S_N2, E1 for S_N1) and the usual unsuitability of tertiary halides for the reaction. However, a tertiary alkoxide ion is perfectly suitable as the nucleophile. Thus, *tert*-butyl ethyl ether (2-ethoxy-2-methyl propane) may be prepared in good yield by the reaction of sodium *tert*-butoxide with ethyl bromide or iodide in *tert*-butyl alcohol as the solvent. An attempt to obtain the ether by the action of sodium ethoxide on *tert*-butyl chloride or bromide in ethyl alcohol as solvent gives more isobutylene than ether. Strictly speaking, the Williamson ether synthesis is a reaction not of an alcohol, but of an alkoxide ion.

S_N1cA method with tertiary alcohols. If *tert*-butyl alcohol is dissolved in an excess of methyl alcohol and a small amount of sulfuric acid is added, a good yield of methyl

tert-butyl ether may be isolated after a short period of heating under reflux. The tertiary alcohol readily accepts a proton and loses water to form *tert*-butyl cations by an S_N1cA reaction. Of course, some of the ions continue along the E1 path to give some isobutylene, but most of them associate with the solvent methanol to produce the conjugate acid of the product ether (Eqs. 9.18). Water is the other product.

$$tBuOH + MeOH \underset{}{\overset{H_2SO_4}{\rightleftharpoons}} tBuOMe + H_2O$$

$$\text{or} \quad tBuOH + H_2SO_4 \rightleftharpoons tBu^{\oplus} + H_2O + HSO_4^{\ominus} \quad (9.18)$$

$$tBu^{\oplus} + MeOH \rightleftharpoons tBu\overset{\oplus}{O}\text{—Me} \underset{}{\overset{MeOH}{\rightleftharpoons}} tBuOMe + MeOH_2^{\oplus}$$
$$\quad\quad\quad\quad\quad\quad\quad\quad\quad\quad\quad |$$
$$\quad\quad\quad\quad\quad\quad\quad\quad\quad\quad\quad H$$

The complete reaction is a dehydration in which the components of the water molecule come from two molecules of alcohol: scissions at both *a* and at *b* occur. Using a tertiary alcohol to give a bond breakage at *b* and a primary alcohol to give a bond breakage at *a* works well. A tertiary alcohol would not associate well with the tertiary carbonium ions (steric hindrance) and dehydration to an alkene predominates. Secondary alcohols with tertiary alcohols do not react very well either.

Sulfuric acid methods. If ethanol is heated to about 120° to 140° with concentrated sulfuric acid, diethyl ether is the major constituent of the distillate, along with some ethylene and unreacted ethanol. We find that there are several pathways to the product. The first step, again, is protonation. As a second step, an S_N2 displacement of water by an attacking molecule of ethanol leads to product (Eqs. 9.19). As another second step, an S_N2

$$EtOH \underset{HSO_4^{\ominus}}{\overset{H_2SO_4}{\rightleftharpoons}} Et\overset{\oplus}{O}H_2 \overset{EtOH, S_N2}{\rightleftharpoons} Et\overset{\oplus}{O}Et + H_2O \rightleftharpoons EtOEt + H_3O^{\oplus}$$
$$\quad\quad\quad\quad\quad\quad\quad\quad\quad\quad\quad\quad\quad\quad\quad\quad |$$
$$HSO_4^{\ominus}, S_N2 \quad\quad\quad\quad EtOH, S_N2 \quad H$$
$$\quad H_2O + EtOSO_3H \quad\quad\quad\quad\quad\quad\quad\quad\quad (9.19)$$

displacement of water by bisulfate ion gives ethyl sulfuric acid. Then S_N2 displacement of bisulfate ion by attacking ethanol leads to product. Complicating the series of equilibria involved are the possible E2cA reactions of $EtOH_2^{\oplus}$ and $Et\overset{\oplus}{O}Et$ with any weak bases
$$\quad\quad\quad\quad\quad\quad\quad\quad\quad\quad\quad\quad\quad\quad\quad\quad\quad\quad\quad\quad\quad |$$
$$\quad\quad\quad\quad\quad\quad\quad\quad\quad\quad\quad\quad\quad\quad\quad\quad\quad\quad\quad\quad\quad H$$
to give *ethylene*.

> Diethyl ether has bp 35°, much lower than that of ethanol, bp 78°. If we run ethanol through a narrow tube below the surface into sulfuric acid heated to 130°, the substitution reactions predominate and the volatile ether is expelled from the mixture as rapidly as it is formed. The success of this method of operation depends upon the difference in bp of the alcohol and the ether. It works well only for methanol and ethanol. With 1-propanol, the ether (bp 90°) is too close to the bp of the alcohol (97°) to be separated easily. With 1-butoxybutane (bp 142°) the corresponding alcohol, 1-butanol, has bp 118°, so the ether would be left behind.

216   *Chapter nine*

In such cases, only a little sulfuric acid is mixed with the alcohol, and the mixture is heated under reflux. The condensed vapor contains both water and 1-butanol. The apparatus is designed so that the two liquids of the condensate separate and only the alcohol is returned to the heated flask. In this manner, the equilibria are shifted in favor of the ether. After the calculated amount of water is collected, the operation is stopped and the ether is isolated.

The sulfuric acid method is good only for the preparation of symmetrical ethers, and only primary alcohols may be used. Even so, sizable amounts of the dehydration products (alkenes) are formed. Any alkene that does not escape from the reaction mixture undergoes some cationic polymerization (Chap. 30) to give substances of high bp, which are retained in the mixture. At the temperatures involved, sulfuric acid can act as a mild oxidizing agent toward such substances, giving rise to sizable amounts of sulfur dioxide. Phosphoric acid is a suitable (if more expensive) substitute for sulfuric acid in dehydration reactions for the preparation of alkenes or ethers.

Exercises

9.5 Starting with any alkanols with up to four carbon atoms, give a reasonable synthesis of:
   a. $CH_3CH_2OC(CH_3)_3$ c. $CH_3OCH_2CH_2OCH_3$
   b. $(CH_3)_2CHOCH(CH_3)_2$ d. $CH_3OC(CH_3)_2CH_2OCH_3$

9.8   Reactions involving a carbon-hydrogen bond, c: oxidation

Oxidation of primary and secondary alcohols.   In a primary or secondary alcohol a hydrogen is bonded to the carbon that holds the hydroxy group. A heterolytic fission of the carbon-hydrogen bond (c), in which the pair of electrons of the bond is removed with the proton, gives rise to a positive ion that is a resonance structure of the conjugate acid of an aldehyde or ketone. Loss of the proton (a) gives the aldehyde or ketone. In effect, the operation is a dehydrogenation; it is thus an oxidation (Eq. 9.20). A primary alcohol is

$$\begin{array}{c} \diagdown \\ \overset{\oplus}{C}\!\!-\!\!OH \\ \diagup \end{array} \longleftrightarrow \begin{array}{c} \diagdown \\ C\!=\!\overset{\oplus}{O}H \\ \diagup \end{array}$$

$$-H^{\ominus} \nearrow \qquad\qquad \searrow -H^{\oplus}$$

$$\begin{array}{c}\diagdown \\ C \\ \diagup\diagdown \\ H\end{array} \overset{O-H}{\underset{a}{\curvearrowleft}} \quad \underset{+H_2, \text{ cat.}}{\overset{-H_2, \text{ cat.}}{\rightleftarrows}} \quad \begin{array}{c}\diagdown \\ C\!=\!O \\ \diagup\end{array} \qquad (9.20)$$

oxidized to an aldehyde, a secondary alcohol to a ketone. A tertiary alcohol does not have a hydrogen bonded to the carbon that holds the hydroxy group. Therefore, this kind of oxidation of a tertiary alcohol is impossible.

The list of usable oxidizing agents is impressively long, and for each agent there exist a variety of conditions and methods. Here, we shall restrict the discussion to the use of

9.8 Reactions involving a carbon-hydrogen bond, c: oxidation

catalytic methods, chromium trioxide and compounds derivable from it, potassium permanganate, and the Oppenauer method.

Catalytic dehydrogenation. The catalytic method involves passing an alcohol as a vapor at a temperature of 330° to 350° over copper chips or a copper chromium oxide catalyst. A reversal of the catalytic addition of hydrogen to a carbon-oxygen double bond occurs. The amount of aldehyde or ketone formed in a single pass over the catalyst usually is low. Aldehydes and ketones all have boiling points lower than those of the related alcohols (because of lack of hydrogen bonding in the oxidized products), so the product may be isolated by distillation and the recovered alcohol sent through the apparatus again. This procedure is a tiresome one in the laboratory, but it presents no difficulty in an industrial continuous process. Through recycling of the alcohol, the yield of product exceeds 90%.

A distinction is made between yield and conversion in a process that involves some recycling. The yield is calculated as the total number of moles of product obtained (times 100 to change to percent) divided by the total number of moles of reactant consumed. A conversion is calculated as 100 times the number of moles of product obtained in a single pass or charge divided by the number of moles of reactant used in the single pass or charge. If only a single operation is carried out (one pass, or one charge to a reactor) and none of the reactant is recovered for future use, then the conversion and the yield are the same. Otherwise, yield is greater than conversion.

Because it is not always convenient to use the catalytic method, and because of the low conversion, other methods are more popular in the laboratory.

Chromium reagents. In chromium trioxide, the oxidation number of chromium is $+6$. This does not mean that chromium bears a charge of $6+$; it means that a chromium atom is using all six of its electrons in the $3d4s$ atomic orbitals for bonding. The compound is not monomeric, but polymeric, with chromium hexacoordinated. It is soluble not only in water, but also in pyridine (an aromatic tertiary amine, C_5H_5N), acetic anhydride, and *tert*-butyl alcohol. In solution, the polymer is destroyed and monomeric soluble species such as $(HO)_2CrO_2$, $C_5H_5N^{\oplus}\text{—}CrO_2(\overset{\ominus}{O})$, $(CH_3COO)_2CrO_2$, and $(tBuO)_2CrO_2$ are formed. Related compounds are potassium and sodium chromates and dichromates, as well as chromyl chloride (CrO_2Cl_2). Chromium with oxidation number $+6$ is a strong oxidizing agent and is reduced finally to oxidation number $+3$. Because the hydroxide or oxide of the $+3$ chromium is insoluble, an acid such as acetic or sulfuric usually is used to prevent hydroxide formation. Sodium dichromate dihydrate is much more soluble in water at $0°$ (238 g/100 ml) than is potassium dichromate (4.9 g/100 ml). The former is the preferred reagent for laboratory use, although the latter is often used as well. Other, more specialized reagents are described in laboratory manuals.

The mechanism of oxidation of an alcohol by a chromium reagent is complex because of the formation of intermediate oxidation number forms of chromium. A schematic mechanism designed to show the intermediate oxidation number forms of $+4$ and $+5$ is given in Equations 9.21 for the oxidation of a primary alcohol to an aldehyde. Actually, various esters of chromium are involved in the mechanism. □ □

$$\begin{aligned} RCH_2OH + CrO_3 &\longrightarrow RCHO + CrO_2 + H_2O \\ CrO_2 + CrO_3 &\longrightarrow Cr_2O_5 \\ 2RCH_2OH + Cr_2O_5 &\longrightarrow 2RCHO + Cr_2O_3 + 2H_2O \\ Cr_2O_3 + 3H_2O &\longrightarrow 2Cr(OH)_3 \\ 2Cr(OH)_3 + 6H_3O^{\oplus} &\longrightarrow 2Cr^{3+} + 12H_2O \end{aligned} \quad (9.21)$$

Addition of the five equations gives the complete reaction:

$$3RCH_2OH + 2CrO_3 + 6H_3O^{\oplus} \longrightarrow 3RCHO + 2Cr^{3+} + 12H_2O$$

Chromium +6 is the usual choice of an oxidizing agent for alcohols. Depending upon the conditions, the reaction may be rapid or slow. There is usually no complication in the oxidation of a secondary alcohol to a ketone. The oxidation of a primary alcohol to an aldehyde is complicated by the oxidation of the aldehyde to an acid, both directly and indirectly (the ester of the alcohol used and of the acid produced is formed). Primary alcohols also may be oxidized so as to produce the acid as the major product, but if acidic chromium +6 is used, some of the ester is always produced as well.

Potassium permanganate. Potassium permanganate ($KMnO_4$) is a powerful oxidizing agent but reacts much more slowly with alcohols at room temperature than do the chromium reagents. The manganese in $MnO_4^{\ominus}$ has oxidation number +7. Under alkaline or neutral conditions, three electrons are gained, and manganese dioxide (MnO_2) precipitates from the mixture. Under acidic conditions (sulfuric acid), five electrons are taken up, and the manganous ion (Mn^{2+}) remains in solution (Eqs. 9.22). The mechanism is similar

$$3CH_3CH_2CHOHCH_3 + 2KMnO_4 \longrightarrow 3CH_3CH_2COCH_3 + 2MnO_2 + 2KOH + 2H_2O$$

$$5CH_3CH_2CHOHCH_3 + 2KMnO_4 + 3H_2SO_4 \quad (9.22)$$
$$\longrightarrow 5CH_3CH_2COCH_3 + K_2SO_4 + 2MnSO_4 + 8H_2O$$

to that of the chromium reagents for oxidations of alcohols, going through a manganese ester with oxidative cleavage. However, direct transfer of hydrogen at c to the oxidant is a possibility that has not been ruled out. Rates are greater at high pH than for mildly acidic pH values. Rates again become fast in 10 to 30% sulfuric acid. Alkaline $KMnO_4$ solutions at room temperature oxidize primary alcohols to acids in good yields, and secondary alcohols to ketones. If the alcohol to be used is so insoluble as to give very slow reactions, 60 to 80% aqueous acetone or acetic acid may be used as the solvent. Reaction times vary from 2 to 12 hr. Permanganate oxidations of alcohols may be heated to boiling after an hour or two to complete the reaction.

Oppenauer oxidation. The Oppenauer method of oxidation and the reverse reaction, Meerwein-Ponndorf-Verley reduction, both make use of the mobile equilibrium that may be set up under anhydrous conditions with a suitable catalyst between an alcohol and its oxidation product (aldehyde or ketone) and some other aldehyde or ketone and its reduction product (primary or secondary alcohol). (See Eq. 9.23.)

$$RCH_2OH + CH_3COCH_3 \underset{}{\overset{cat.}{\rightleftharpoons}} RCHO + CH_3CHOHCH_3 \quad (9.23)$$

The nearly ideal catalyst for the system is an aluminum alkoxide. An aluminum alkoxide keeps the system strictly anhydrous; and, by ready exchange of alkoxy groups with alcohols, it is able to bring the reactants into close proximity. Acetone is widely used as the oxidant for the oxidation of primary and secondary alcohols, although any other ketone (or aldehyde) could be used. The equilibrium reaction is carried to the right by the excess of acetone. Isopropyl alcohol generally is used in the reverse reaction for the reduction of an aldehyde or ketone, although any other alcohol (except a tertiary alcohol) could be used.

9.8 Reactions involving a carbon-hydrogen bond, c: oxidation

In a reduction, isopropyl alcohol is used as the solvent and aluminum isopropoxide as the catalyst. The large excess of isopropyl alcohol pushes the equilibrium to the left. In addition, the more volatile acetone may be distilled from the boiling mixture to aid in pulling the reaction to the left.

In an oxidation, an excess of acetone is used, and an inert but miscible solvent such as toluene is added to raise the boiling point of the mixture. The alcohol to be oxidized and the aluminum *tert*-butoxide catalyst complete the system. The excess acetone pushes the equilibrium to the right. Slow distillation of the acetone carries away some of the isopropyl alcohol and pulls the reaction to the right. Because we are dealing with an essentially neutral anhydrous mixture of a mild oxidation-reduction system, no other part of the alcohol and aldehyde or ketone is interfered with, and yields approach 100% in most instances. □ □

Before leaving the subject of oxidation, we should emphasize that tertiary alcohols are inert to oxidation. A tertiary alcohol under acidic conditions is readily dehydrated via the S_N1cA route. The alkene that results is subject to oxidation. Thus, *tert*-butyl alcohol in the presence of both an acid and an oxidizing agent reacts to give carbon dioxide and acetone, but it is isobutylene that is oxidized, not *tert*-butyl alcohol.

Biological oxidation.   Electron transfer molecules coordinate oxidation-reduction reactions in living systems. An enzyme catalyzing an oxidation may use the oxidizing power of a molecule abbreviated $NAD_{(oxidized)}$. For example, the conversion of malic acid into oxaloacetic acid involves the reduction of $NAD_{oxidized}$ to $NAD_{reduced}$. The $NAD_{reduced}$ can

$$\underset{\text{malic acid}}{\begin{array}{c}\text{O}\\\|\\\text{C}\\/\backslash\\ \text{OH}\\\text{CHOH}\\|\\\text{CH}_2\\|\\\text{C}=\text{O}\\\backslash\\\text{OH}\end{array}} \xrightarrow[NAD_{ox}]{\text{L-malate dehydrogenase}}[NAD_{red}] \underset{\text{oxaloacetic acid}}{\begin{array}{c}\text{O}\\\|\\\text{C}\\/\backslash\\ \text{OH}\\\text{HC}=\text{O}\\|\\\text{CH}_2\\|\\\text{C}=\text{O}\\\backslash\\\text{OH}\end{array}}$$

now be used to donate an electron in a reduction reaction. In our body, there are many more oxidation reactions than reductions. Excess $NAD_{reduced}$ is ultimately oxidized indirectly by oxygen.

The final reaction in the biosynthesis of ethyl alcohol by brewer's yeast is catalyzed by an enzyme called alcohol dehydrogenase. This same enzyme in the liver catalyzes the reverse reaction whenever (if ever?) an excess of alcohol is ingested.

$$H_3C-\overset{H}{C}=O \xrightarrow[NAD_{red}]{\text{alcohol dehydrogenase}}[NAD_{ox}] H_3C-CH_2-OH$$

Exercises

9.6 Predict the products (organic and inorganic) of these reactions:

a. $(CH_3)_2CHCH_2OH \xrightarrow[\text{dil. } H_2SO_4]{Na_2Cr_2O_7,\ 25°}$

b. $CH_3CH_2CHOHCH_3 \xrightarrow[\text{dil. KOH}]{KMnO_4,\ 25°}$

c. cyclopentyl-OH $\xrightarrow[\text{dil. } H_2SO_4]{CrO_3,\ 25°}$

d. cyclobutyl-$CH_2OH \xrightarrow{Cu,\ \Delta}$

e. $CH_3CH=CHCH_2OH \xrightarrow[\text{acetone}]{Al(OtBu)_3,\ \text{reflux}}$

9.9 Balancing oxidation-reduction equations

Introduction. The balancing of an oxidation-reduction equation involving organic compounds may be done in any of several ways. Methods have been devised for adapting the balancing methods learned in previous courses in chemistry to the oxidation-reduction equations in organic chemistry. The use of oxidation numbers in inorganic chemistry in electron-transfer reactions was a help in understanding oxidation and reduction. The assignment of oxidation numbers to atoms of carbon, however, is at best an artificial concept. Also, as we have seen, electron transfers tend to involve one or two electrons at a time in mechanisms that are much more complicated than a balanced stoichiometric equation suggests.

However, we shall assume that we are still able to find the oxidation state number of an oxidizing agent when we need it, as for chromium in $K_2Cr_2O_7$, for example (K is $+1$, O is -2, and $K_2O_7 = -12$. Therefore, $2\ Cr = -(-12)$, or $Cr = +6$).

More difficult is the prediction of the product or products that arise by oxidation of the organic substance. For the time being, these predictions will be easy. Primary alcohols are oxidized to aldehydes or acids, depending upon conditions; secondary alcohols, to ketones; and tertiary alcohols are inert under neutral and basic conditions. If an organic acid is the product of an oxidation under neutral or basic conditions, it is produced as a salt instead of the free acid.

Oxidations carried out in partly or completely aqueous solutions should always have water written as a product. Neutral or basic oxidations with potassium or sodium permanganate should have potassium or sodium hydroxide written as a product along with manganese dioxide. Acidic oxidations should have sulfuric acid written as a reactant and potassium or sodium sulfate and chromic or manganous sulfate written as products.

We shall describe a successful method here for balancing oxidation-reduction reactions (sometimes called redox reactions). ☐☐

Hydrogen method. The hydrogen method for balancing a redox equation depends upon writing balanced half-reactions in terms of the oxidant, $H^\oplus$, $HO^\ominus$, H_2O,

9.9 *Balancing oxidation-reduction equations* 221

and H (hydrogen atoms) and of the reductant, $H^\oplus$, $HO^\ominus$, H_2O, and H. The hydrogen atoms of the two half-reactions are then balanced and the two half-reactions added. (This method is similar to the one in which half-reactions in terms of electron transfer are used.)

The steps necessary for the use of the hydrogen method are as follows.

1. Select the oxidizing agent and the ion or product to which it is reduced.
2. Balance the charges by adding the proper number of $H^\oplus$ if the reaction is acidic or of $HO^\ominus$ if the reaction is neutral or basic.
3. Balance oxygen by adding the proper number of molecules of H_2O.
4. Balance hydrogen by adding the proper number of H atoms. The half-reaction of the oxidizing agent should now be in balance.
5. Repeat steps 1 through 4 for the reducing agent.
6. Equalize the H atoms between the half-reactions by multiplication by the proper factors.
7. Add the equalized half-reactions to give a balanced redox equation. (The H atoms should cancel out in the addition.)
8. To obtain a balanced equation for the substance as weighed out in the laboratory, supply $Na^\oplus$, $K^\oplus$, $SO_4^{2\ominus}$, and other ions as may be necessary.

As usual, the description of the method is longer than its performance. Some examples will help to clarify the steps, most of which are condensed in the process of balancing the equation.

Examples. Consider the preparation of butyric acid by the alkaline potassium permanganate oxidation of 1-butanol (Eq. 9.24).

$$CH_3CH_2CH_2CH_2OH + KMnO_4 \longrightarrow \quad (9.24)$$
$$CH_3CH_2CH_2COOK + KOH + MnO_2 + H_2O$$

We do the eight steps as follows:

1. $MnO_4^\ominus \longrightarrow MnO_2$
2. $MnO_4^\ominus \longrightarrow MnO_2 + HO^\ominus$ charge balance
3. $MnO_4^\ominus \longrightarrow MnO_2 + HO^\ominus + H_2O$ O balance
4a. $3H + MnO_4^\ominus \longrightarrow MnO_2 + HO^\ominus + H_2O$ H balance

Now we repeat steps 1 through 4 for the reducing agent (step 5), in this case, butyl alcohol:

1. $CH_3CH_2CH_2CH_2OH \longrightarrow CH_3CH_2CH_2COO^\ominus$
2. $HO^\ominus + CH_3CH_2CH_2CH_2OH \longrightarrow CH_3CH_2CH_2COO^\ominus$ charge balance
3. No change in 2 because O is in balance
4b. $HO^\ominus + CH_3CH_2CH_2CH_2OH \longrightarrow CH_3CH_2CH_2COO^\ominus + 4H$ H balance

222 Chapter nine

We see that 4a requires 3H on the left and that 4b requires 4H on the right. Therefore we multiply 4a by 4 and 4b by 3:

6. $12H + 4MnO_4^\ominus \longrightarrow 4MnO_2 + 4HO^\ominus + 4H_2O$

  $3HO^\ominus + 3CH_3CH_2CH_2CH_2OH \longrightarrow 3CH_3CH_2CH_2COO^\ominus + 12H$

Adding the two equalized half-reactions gives 7:

7. $3CH_3CH_2CH_2CH_2OH + 4MnO_4^\ominus \longrightarrow 3CH_3CH_2CH_2COO^\ominus + 4MnO_2 + HO^\ominus + 4H_2O$

Inspection of 7 shows that $4K^\oplus$ are needed on the left and on the right to give the final balance in step 8 (Eq. 9.25).

$$3CH_3CH_2CH_2CH_2OH + 4KMnO_4 \longrightarrow 3CH_3CH_2CH_2COOK + KOH + 4MnO_2 + 4H_2O \quad (9.25)$$

For our second example, let us look at the chromium trioxide oxidation of 1-butanol to give butyraldehyde, an acidic oxidation (Eq. 9.26).

$$CH_3CH_2CH_2CH_2OH + CrO_3 + H_2SO_4 \longrightarrow CH_3CH_2CH_2CHO + Cr_2(SO_4)_3 + H_2O \quad (9.26)$$

We write:

1. $CrO_3 \longrightarrow Cr^{3+}$
2. $3H^\oplus + CrO_3 \longrightarrow Cr^{3+}$ charge balance
3. $3H^\oplus + CrO_3 \longrightarrow Cr^{3+} + 3H_2O$ O balance
4a. $3H + 3H^\oplus + CrO_3 \longrightarrow Cr^{3+} + 3H_2O$ H balance

and 5 becomes:

1. $CH_3CH_2CH_2CH_2OH \longrightarrow CH_3CH_2CH_2CHO$
2. Not needed   charge balance
3. Not needed   O balance
4b. $CH_3CH_2CH_2CH_2OH \longrightarrow CH_3CH_2CH_2CHO + 2H$ H balance

then:

6. $6H + 6H^\oplus + 2CrO_3 \longrightarrow 2Cr^{3+} + 6H_2O$

  $3CH_3CH_2CH_2CH_2OH \longrightarrow 3CH_3CH_2CH_2CHO + 6H$

7. $3CH_3CH_2CH_2CH_2OH + 2CrO_3 + 6H^\oplus \longrightarrow 3CH_3CH_2CH_2CHO + 2Cr^{3+} + 6H_2O$

Addition of $3SO_4^{2\ominus}$ to the left and the right of 7 (step 8) gives the final balance (Eq. 9.27).

$$3CH_3CH_2CH_2CH_2OH + 2CrO_3 + 3H_2SO_4 \longrightarrow 3CH_3CH_2CH_2CHO + Cr_2(SO_4)_3 + 6H_2O \quad (9.27)$$

9.10 Alkene formation, d

Acidic conditions. As we said in Chapter 7, alcohols may be dehydrated readily under acidic conditions to give alkenes. Ruptures at *b* and at *d* occur in an E2cA reaction typical of primary alcohols. However, even with primary alcohols, not to mention secondary and tertiary, the E1cA mechanism tends to intrude. The intermediate carbonium ion may rearrange. If rearrangement occurs, the proton that is ultimately lost may or may not be the one we have designated as *d*. In general, primary alcohols cannot be dehydrated to give 1-alkenes in good yield because rearrangement predominates. Secondary alcohols, of course, rearrange if a more stable carbonium ion will result. In any case, the Saytzeff rule is followed in acidic dehydration. Tertiary alcohols seldom rearrange, because seldom is a more stable tertiary carbonium ion available by rearrangement. Tertiary ions have a choice of direction for loss of a proton, and although the most stable alkene is the major product, the other possible alkenes are also formed. As a result of the difficulties imposed by the E1cA mechanism, some alkenes are not often prepared by direct dehydration of an alcohol.

If rearrangement is not likely, or if rearrangement does not lead to changes in structure (as in unsubstituted cycloalkanols), dehydration by phosphoric or sulfuric acids or in the vapor phase over alumina is an excellent way of preparing alkenes (Eqs. 9.28).

$$CH_3CH_2OH \xrightarrow[H_3PO_4]{180°} H_2C=CH_2$$

$$CH_3CHOHCH_3 \longrightarrow CH_3CH=CH_2$$

$$CH_3CHOHCH_2CH_3 \longrightarrow CH_3CH=CHCH_3 \qquad (9.28)$$

$$(CH_3)_3COH \longrightarrow (CH_3)_2C=CH_2$$

$$CH_3CH_2CHOHCH_2CH_3 \longrightarrow CH_3CH_2CH=CHCH_3$$

cyclohexanol → cyclohexene   cyclopentanol → cyclopentene

Chugaev reaction. Conversion of an alcohol to the corresponding alkyl halide followed by a dehydrohalogenation is an indirect but preferable route from an alcohol to an alkene in many instances (1-butanol to 1-butene, for example). Another indirect route, in which no rearrangement at all is possible, is the Chugaev reaction (sometimes spelled Tschugaeff). Salts and esters of dithiocarbonic acid, $O=C(SH)_2$, are known as <u>xanthates.</u> If a primary or secondary alcohol is allowed to react with potassium hydroxide and carbon disulfide, a potassium alkyl xanthate is formed (Eqs. 9.29). The alkoxide ion first formed adds to the carbon of the carbon-sulfur double bond to give a potassium xanthate (a), an isolable crystalline compound. Being a negative ion, it is an excellent nucleophile and readily displaces iodide ion from methyl iodide in an S_N2 reaction to give an O-alkyl S-methyl xanthate (b). When it is heated, this compound decomposes by a cyclic mechanism into the alkene, methylthiol, and carbon oxysulfide. The geometry of the decomposition is *cis*. At no time is a carbonium ion formed. By the Chugaev reaction, 3,3-dimethyl-1-butene may be prepared in good yield from 3,3-dimethyl-2-butanol (Eq. 9.30).

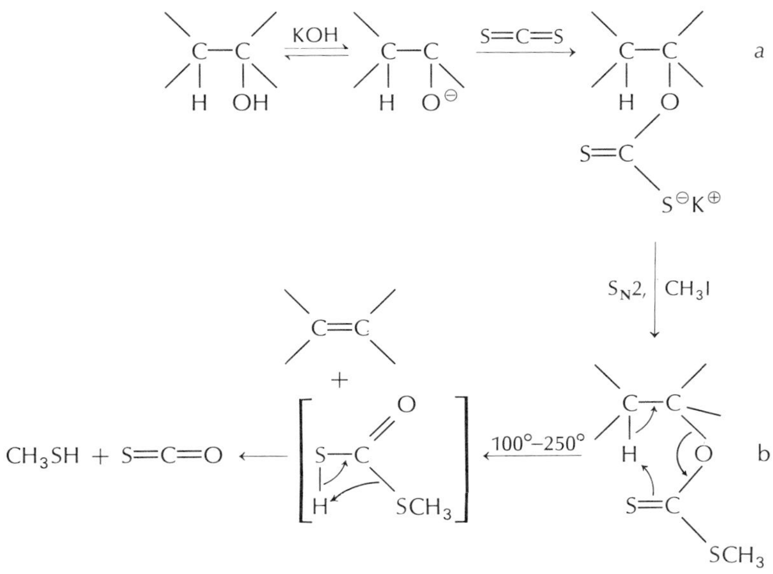

$$(CH_3)_3CCHOHCH_3 \xrightarrow[\substack{1.\ KOH,\ CS_2 \\ 2.\ CH_3I \\ 3.\ heat}]{} (CH_3)_3CCH=CH_2 \quad (9.30)$$

Exercises

9.7 Predict the products and relative amounts to be expected from:

a. $(CH_3)_2CHCHOHCH_3 \xrightarrow[180°]{H_3PO_4}$

b. 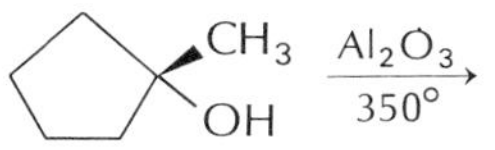 [cyclopentane with CH$_3$ and OH substituents] $\xrightarrow[350°]{Al_2O_3}$

c. 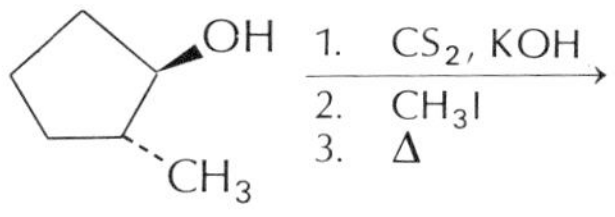 [cyclopentane with OH and CH$_3$ substituents] $\xrightarrow[\substack{1.\ CS_2,\ KOH \\ 2.\ CH_3I \\ 3.\ \Delta}]{}$

d. $(CH_3)_2\underset{\underset{OH}{|}}{\overset{\overset{OH}{|}}{C}}-C(CH_3)_2 \xrightarrow[350°]{Al_2O_3}$

9.11 Preparation of alcohols

Introduction. We take up the preparation of alcohols under three headings: substitutions, additions, and reductions. In most cases, the reactions have been discussed in previous chapters. The new reactions will be discussed later in the text at greater length as properties of the substances used, but enough detail will be given here so that we can understand what is going on. The most widely applicable of all the methods for producing alcohols is the Grignard reaction with compounds containing a carbon-oxygen double bond. The other methods are not without value and are frequently used.

Now that we know something of the properties of alcohols, we shall recognize the preparative methods, in many cases, as reversals of reactions of alcohols.

9.12 Preparation by substitution reactions

Formation of the b bond; of the a bond. Under this heading, we recognize two types of substitution. One is the replacement of a group attached to carbon by a hydroxy group, or *b* formation. The other is the replacement of a group attached to an oxygen (in turn, attached to carbon) with a hydrogen, or *a* formation. Another way of looking at it is to say that in the first case, a carbon-oxygen bond must be introduced; and in the second, an oxygen-hydrogen bond needs to be made.

Under the first classification, we list those methods that begin with an alkyl halide. An alkyl halide may be treated with hydroxide ion in an effort to carry out an S_N2 displacement. This approach works fairly well with simple primary halides; but, as we have seen, it is seldom clean cut.

If the alkyl halide is first converted into a Grignard reagent, many possibilities open up; but only two may be thought of as substitutions. If oxygen is bubbled through a Grignard reagent at $-40°$ or lower, a good yield of a hydroperoxide is obtained after hydrolysis (RX → RMgX → ROOMgX → ROOH). At a higher temperature, 0° or higher, the product is an alcohol. Presumably, something like the speculative mechanism shown in Equation 9.31 happens at the higher temperature. In any case, the reactions are not very useful, because the principal sources for the preparation of alkyl halides are alcohols.

$$R\text{—}O\text{—}O^{\ominus} \overset{\oplus}{M}gX + R\text{—}Mg\text{—}X \longrightarrow R\text{—}O\overset{\frown}{\text{—}}O\text{—}\overset{|}{\underset{R}{Mg}}\text{—}X$$

$$\longrightarrow RO^{\ominus} + \underset{R}{\underset{|}{O}}\text{—}MgX + \overset{\oplus}{M}gX \tag{9.31}$$

From a Grignard reagent and an epoxide. The other substitution is that with an epoxide. A Grignard reagent acts as a nucleophile towards an epoxide and displaces the oxygen from one of the two carbon atoms involved. It may be thought of as an S_N2 reaction, because the displacement is directed more at a primary carbon atom than at a secondary or tertiary carbon atom (Eqs. 9.32). The reaction with ethylene oxide is often used to prepare

$$H_2C\overset{O}{-\!\!\!-\!\!\!-}CH_2 + R\!-\!MgX \longrightarrow H_2\overset{OMgX}{\underset{|}{C}}-CH_2-R$$
$$\xrightarrow{\text{dil. HCl}} H_2\overset{OH}{\underset{|}{C}}-CH_2R \quad \text{primary}$$

$$CH_3CH\overset{O}{-\!\!\!-\!\!\!-}CH_2 + R\!-\!MgX \longrightarrow CH_3\overset{OMgX}{\underset{|}{CH}}-CH_2R \quad (9.32)$$
$$\xrightarrow{\text{dil. HCl}} CH_3\overset{OH}{\underset{|}{CH}}-CH_2R \quad \text{secondary}$$

$$(CH_3)_2C\overset{O}{-\!\!\!-\!\!\!-}CH_2 + R\!-\!MgX \longrightarrow (CH_3)_2\overset{OMgX}{\underset{|}{C}}-CH_2R$$
$$\xrightarrow{\text{dil. HCl}} (CH_3)_2\overset{OH}{\underset{|}{C}}-CH_2R \quad \text{tertiary}$$

primary alcohols that have a chain with two more carbon atoms than were present in the RX used to prepare the Grignard reagent. One can envision the reaction proceeding by the attack of an $R^\ominus$ (or an incipient $R^\ominus$), with displacement of the oxygen as $O^\ominus$. Although oxygen is reluctant to depart with a negative charge (ethers are stable to bases), it is faced here with a much stronger nucleophile. Also, there probably is some association of the ring oxygen with magnesium in the reaction mixture, which would help the departure of the oxygen from the carbon. The strongest influence undoubtedly is the lowering of the activation energy of the reaction by the weakened C—O bonds in the strained three-membered ring (bond angles are 60°, 49° less than the normal 109° 28' tetrahedral angle).

Relative uselessness of a bond formation. Under the second classification, the introduction of an O—H bond, we find it necessary to use a compound that already has a carbon-oxygen bond with something other than hydrogen as the other group attached to oxygen. A little thought reduces the possibilities rapidly to alkoxides, ethers, esters, and acetals. However, all of these compounds except the alkoxides require alcohols in their own preparation. Consequently, the preparation of alcohols by introduction of an O—H bond is limited to the hydrolysis of the products of addition of Grignard reagents to $\underset{/}{\overset{\backslash}{C}}=O$ compounds and to the hydrolysis of naturally occurring esters.

Exercises

9.8 Show which Grignard reagents and which epoxides are needed to prepare:
    a. $(CH_3)_2CHCH_2CH_2OH$ c. $CH_3CH_2CH_2CHOHCH_3$
    b. $(CH_3)_2CHCH_2CHOHCH_3$ d. $(CH_3)_3CCH_2CH_2OH$

9.13 Preparation by addition reactions

Because there are so many addition reactions, we shall delegate the addition of hydrogen to carbon-oxygen double bonds to the next section, on reductions.

Addition of Grignard reagents to aldehydes, ketones, and esters. The addition of Grignard reagents to aldehydes and ketones has previously been described. Esters also may be used (Eqs. 9.33). The unstable intermediate breaks apart in the ether solution

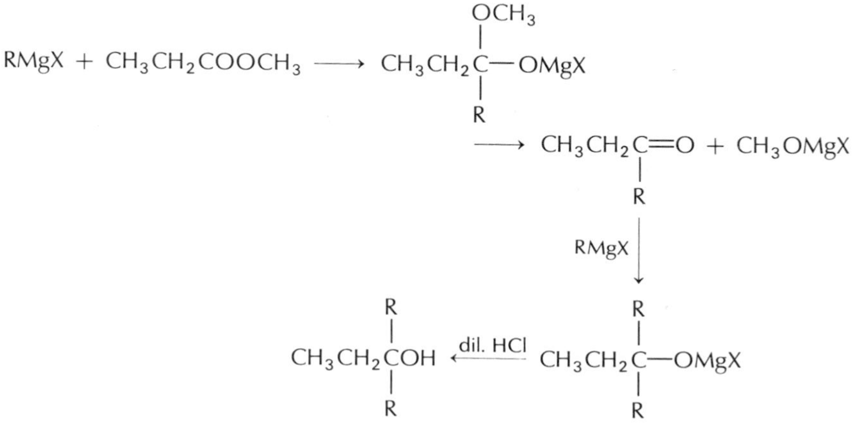

$$RMgX + CH_3CH_2COOCH_3 \longrightarrow CH_3CH_2\underset{R}{\overset{OCH_3}{\underset{|}{\overset{|}{C}}}}\text{—}OMgX$$

$$\longrightarrow CH_3CH_2\underset{R}{\overset{|}{C}}\text{=}O + CH_3OMgX$$

$$\downarrow RMgX$$

$$CH_3CH_2\underset{R}{\overset{R}{\underset{|}{\overset{|}{C}}}}OH \xleftarrow{\text{dil. HCl}} CH_3CH_2\underset{R}{\overset{R}{\underset{|}{\overset{|}{C}}}}\text{—}OMgX \tag{9.33}$$

or

$$CH_3CH_2COOCH_3 \longrightarrow CH_3CH_2CR_2OH$$

in which the reaction is done, forming an alkoxymagnesium halide and a ketone. The ketone reacts with another mole of Grignard reagent to give a tertiary alcohol. A formic ester gives a secondary alcohol. The alcohols formed from esters contain two groups from the Grignard reagent. The alcoholic moiety of the ester is discarded in the process, and only methyl and ethyl esters are ever used.

Hydration of alkenes. We have already seen that alkenes add sulfuric acid and that the alkyl hydrogen sulfates that form may be hydrolyzed to alcohols. Only ethylene gives a primary alcohol by this process. All other alkenes give secondary or tertiary alcohols, because a carbonium ion is an intermediate. This reaction is used on a tremendous scale in industry, but only seldom in the laboratory, where it serves mostly as a test for the presence of an alkene rather than for preparative purposes.

Hydroboration of alkenes. An addition reaction that has proved very useful is hydroboration. Borane is boron hydride (BH_3), a gas. It is so reactive that it dimerizes to diborane (B_2H_6), also a gas. Diborane may be prepared as needed by the reaction of sodium borohydride with boron trifluoride, usually in ether as a solvent (Eq. 9.34). In the presence of

$$3NaBH_4 + BF_3 \xrightarrow{\text{ether}} 2B_2H_6 + 3NaF \tag{9.34}$$

an alkene, diborane dissociates into borane, which adds rapidly to the carbon-carbon double bond. The first product formed, an alkylborane, adds to another molecule of alkene to give a dialkylborane. The dialkylborane repeats the process to form a trialkylborane (Eq. 9.35). The trialkylboranes also are reactive. Upon hydrolysis, they give alkanes (reduction

$$6H_2C=CH_2 + B_2H_6 \longrightarrow 2(CH_3CH_2)_3B \tag{9.35}$$

products of alkenes) and boric acid. Of interest to us here is the reaction with alkaline hydrogen peroxide to form alcohols and borate ion (Eqs. 9.36).

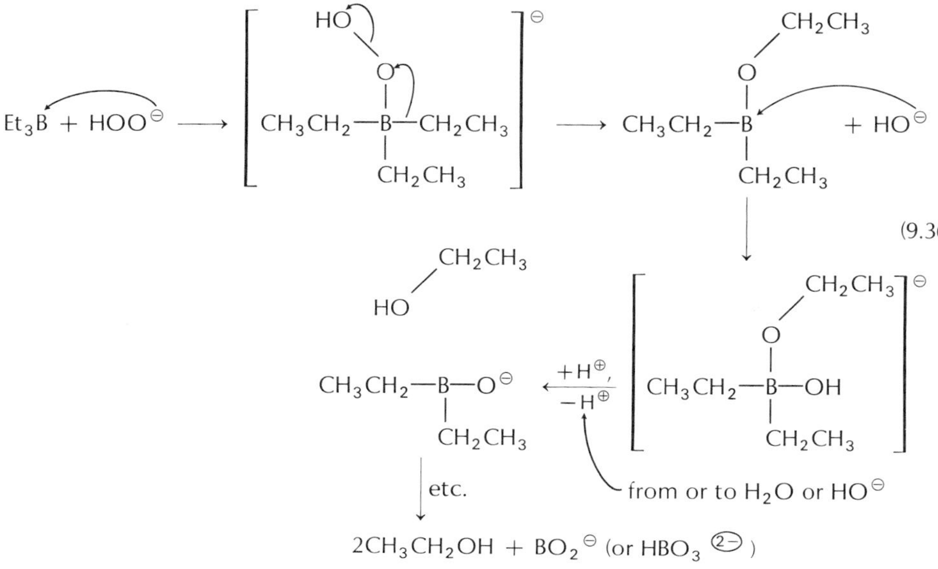

$$Et_3B + HOO^\ominus \longrightarrow \begin{bmatrix} HO \\ \backslash \\ O \\ | \\ CH_3CH_2-B-CH_2CH_3 \\ | \\ CH_2CH_3 \end{bmatrix}^\ominus \longrightarrow CH_3CH_2-B\begin{matrix} CH_2CH_3 \\ \diagdown \\ O \\ \end{matrix} + HO^\ominus$$

$$\begin{matrix} CH_2CH_3 \\ HO \\ \end{matrix}$$

$$CH_3CH_2-B-O^\ominus \xleftarrow[-H^\oplus]{+H^\oplus} \begin{bmatrix} CH_2CH_3 \\ | \\ O \\ | \\ CH_3CH_2-B-OH \\ | \\ CH_2CH_3 \end{bmatrix}^\ominus$$

(9.36)

$$\downarrow \text{etc.} \qquad \text{from or to } H_2O \text{ or } HO^\ominus$$

$$2CH_3CH_2OH + BO_2^\ominus \text{ (or } HBO_3^{\;2\ominus}\text{)}$$

The mechanism is speculative but reasonable. The value of the reaction is increased by the fact that if there is a choice of direction for the addition, reaction goes in the direction that will give primary in preference to secondary or secondary in preference to tertiary trialkyl boranes. Also, the geometry of the addition is *cis*. Finally, the hydroperoxide ion bonds exclusively to the boron, so that the migration of the alkyl group from boron to oxygen is a front-side displacement on the carbon atom of the migrating group. Thus, the alcohol produced is the result of the *cis* addition of a hydrogen and an OH group to the double bond, in what we call anti-Markovnikoff addition. Since no carbonium ion is formed, no rearrangement occurs; and the yields are high.

A few examples are shown (Eqs. 9.37). Although BH_3 is a reducing agent, hydrogen

$$(CH_3)_2C=CH_2 \xrightarrow[2.\ H_2O_2,\ HO^\ominus]{1.\ B_2H_6} (CH_3)_2CHCH_2OH$$

$$(CH_3)_3CCH=CH_2 \longrightarrow (CH_3)_3CCH_2CH_2OH$$

$$(CH_3)_2C=CHCH_3 \longrightarrow (CH_3)_2CHCHOHCH_3$$

(9.37)

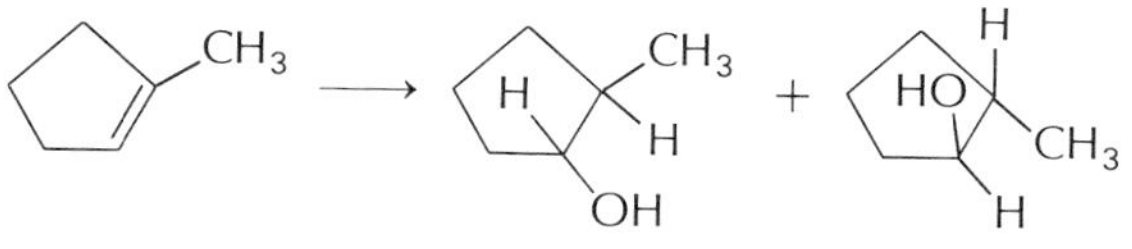

racemic mixture

9.14 Preparation by oxidation-reduction reactions

peroxide is an oxidizing agent, and the net result is no oxidation or reduction of the alkene (addition of water to an alkene does not change the oxidation state). Borane is oxidized; hydrogen peroxide is reduced.

Exercises

9.9 Give the product, after hydrolysis, of reaction of $(CH_3)_2CHMgBr$ with:
- a. HCHO
- b. $CH_3CH_2COCH_3$
- c. $(CH_3)_2CHCHO$
- d. $HCOOCH_2CH_3$
- e. $(CH_3)_2CHCH_2COOCH_2CH_3$
- f. $O_2, 0°$
- g. dil. HCl
- h. I_2
- i. CO_2
- j. 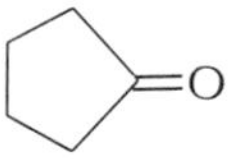 cyclopentanone

---

9.14 Preparation by oxidation-reduction reactions

Utility of reduction. Because an alcohol is only one stage above an alkane in oxidation state, and because there are no satisfactory methods of oxidizing an alkane to an alcohol, the oxidation-reduction reactions referred to here for the preparation of alcohols all will be concerned with the reduction of aldehydes, ketones, acids, esters, and other acid derivatives. In contrast to the oxidation reactions of alcohols, the reduction reactions for the synthesis of alcohols have equations that are easy to balance by inspection.

Catalytic hydrogenation. The direct addition of hydrogen to a carbon-oxygen double bond may be carried out with a variety of catalysts, solvents, temperatures, and pressures. It is not our intent here to go into all the possible combinations of conditions. Let us simply say that finely divided platinum (from $PtO_2 + 2H_2 \rightarrow Pt + 2H_2O$) deposited on carbon is a popular catalyst and that ethyl acetate, ethanol, water, and acetic acid are frequently used solvents for catalytic hydrogenations of carbon-oxygen double bonds at room temperature and pressures up to 4 atm. Most reactive are acid chlorides, which rapidly undergo hydrogenolysis to give the aldehyde. Unless special precautions are taken, the reaction does not stop, but proceeds to take up another mole of hydrogen to give the

$$CH_3CH_2CH_2COCl \xrightarrow{H_2, Pt} CH_3CH_2CH_2CHO + HCl$$

$$CH_3CH_2CH_2CHO \longrightarrow CH_3CH_2CH_2CH_2OH$$

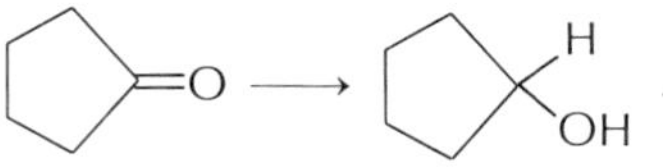

cyclohexanone $\longrightarrow$ cyclohexanol (9.38)

$$CH_3CH_2CH_2COOCH_3 \longrightarrow CH_3CH_2CH_2CH_2OH + CH_3OH$$

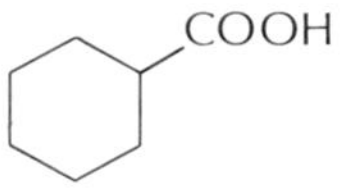

cyclohexane-COOH $\longrightarrow$ no reaction

primary alcohol. Halogens also are easily removed by hydrogenolysis of most carbon-halogen bonds (alkyl halides). Next in order of reactivity are aldehydes, followed by ketones; and then, a long way behind, are esters. Acids or salts of acids are extraordinarily inert to hydrogenation. Equations 9.38 give some examples.

Acidic sodium borohydride with a catalyst: direct reduction. An alternative to using hydrogen from a tank is to use a solution of sodium borohydride ($NaBH_4$) in water or ethanol. The solution is allowed to drip into an acidified mixture of the compound, solvent, and hydrogenation catalyst in a closed system. Under acidic conditions sodium borohydride reacts to produce hydrogen (Eq. 9.39).

$$NaBH_4 + CH_3COOH + 3H_2O \longrightarrow 4H_2 + CH_3COONa + H_3BO_3 \quad (9.39)$$

Instead of using sodium borohydride to generate hydrogen, we may use it to react directly with the carbonyl group (a carbon-oxygen double bond, Eq. 9.40). The immediate

$$CH_3CH_2CHO + BH_4^\ominus \longrightarrow CH_3CH_2CH_2OBH_3^\ominus \quad (9.40)$$

product is an alkoxyborohydride, formed by donation of a hydride ion to the carbon of the carbonyl group to give an alkoxide ion, which immediately bonds to the borane. The reaction continues until a tetraalkoxyborate ion, $(RO)_4B^\ominus$, is formed. Addition of water gives the alcohol and a salt of boric acid (Eq. 9.41). This use of sodium borohydride is limited to the

$$4CH_3CH_2CHO + NaBH_4 + 3H_2O \longrightarrow 4CH_3CH_2CH_2OH + NaOB(OH)_2 \quad (9.41)$$

reduction of acid chlorides, aldehydes, and ketones. Esters react very slowly, and acids and salts are not reduced.

Use of lithium aluminum hydride. A more powerful reducing agent is lithium aluminum hydride ($LiAlH_4$), which is so reactive that it must be used in anhydrous ether, as though it were a Grignard reagent. Just as a Grignard reagent adds an R group to carbon and MgX to oxygen, $LiAlH_4$ adds H to carbon and either Li or Al to oxygen. The reaction is completed by hydrolysis of the mixture. (Extreme caution must be used because of the vigorous reaction of any excess $LiAlH_4$ with water.) The organic product remains in the ether layer, and the inorganic products are taken into the aqueous layer as salts of the hydrochloric acid that is now added. The reaction in ether usually is complete in 15 to 30 min at room temperature. Acid chlorides react so explosively that they are not used in this reaction. Aldehydes, ketones, and esters follow in order of reactivity; but now acids and soluble salts of acids also react. Epoxides also are reduced: hydride ion attacks the least substituted carbon of the three-membered heterocyclic ring in the same manner that R of a Grignard reagent attacks. Equations 9.42 give some examples.

Use of diborane. As we mentioned in Section 9.13, diborane may be used as a reducing agent for alkenes. It is of interest that acids are more reactive than aldehydes, ketones, or alkenes. The initial step with an acid seems to be the formation of $(RCOO)_3B + 3H_2$, and it is the triacylborane that is reduced to a primary alcohol. Also, acid chlorides are inert to the action of diborane. Whereas sodium borohydride and lithium aluminum hydride function by donation of a hydride ion, borane acts as a Lewis acid and interacts

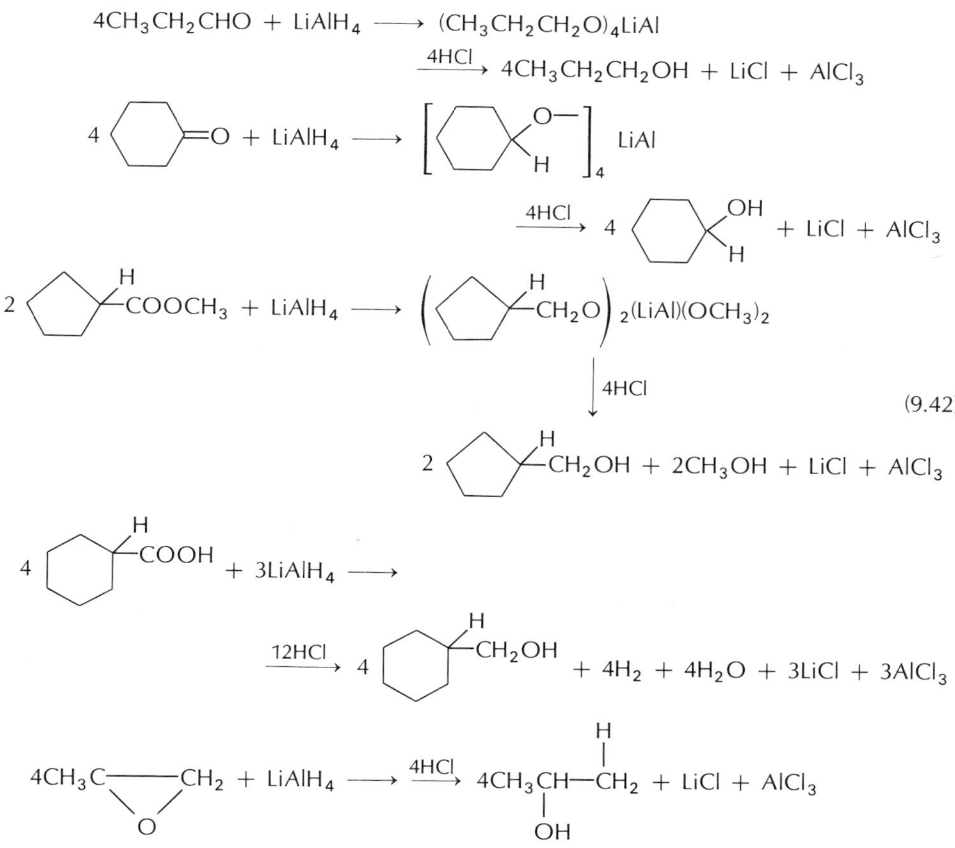

$$4CH_3CH_2CHO + LiAlH_4 \longrightarrow (CH_3CH_2CH_2O)_4LiAl$$
$$\xrightarrow{4HCl} 4CH_3CH_2CH_2OH + LiCl + AlCl_3$$

$$4 \langle\text{C}_6H_{10}\rangle=O + LiAlH_4 \longrightarrow \left[\langle\text{C}_6H_{10}\rangle\text{CH(O-)(H)}\right]_4 LiAl$$
$$\xrightarrow{4HCl} 4\ \langle\text{C}_6H_{10}\rangle\text{CH(OH)(H)} + LiCl + AlCl_3$$

$$2\ \langle\text{cyclopentyl}\rangle\text{CH(H)-COOCH}_3 + LiAlH_4 \longrightarrow \left(\langle\text{cyclopentyl}\rangle\text{CH(H)-CH}_2\text{O}\right)_2 (LiAl)(OCH_3)_2$$
$$\downarrow 4HCl \qquad\qquad (9.42)$$
$$2\ \langle\text{cyclopentyl}\rangle\text{CH(H)-CH}_2OH + 2CH_3OH + LiCl + AlCl_3$$

$$4\ \langle\text{C}_6H_{10}\rangle\text{CH(H)-COOH} + 3LiAlH_4 \longrightarrow$$
$$\xrightarrow{12HCl} 4\ \langle\text{C}_6H_{10}\rangle\text{CH(H)-CH}_2OH + 4H_2 + 4H_2O + 3LiCl + 3AlCl_3$$

$$4CH_3\overset{\triangle}{C\text{—}}CH_2 + LiAlH_4 \xrightarrow{4HCl} 4CH_3\underset{OH}{\overset{H}{CH}}\text{—}CH_2 + LiCl + AlCl_3$$

with a lone pair on the carbonyl oxygen, after which a hydride is shifted from boron to carbon (Eq. 9.43).

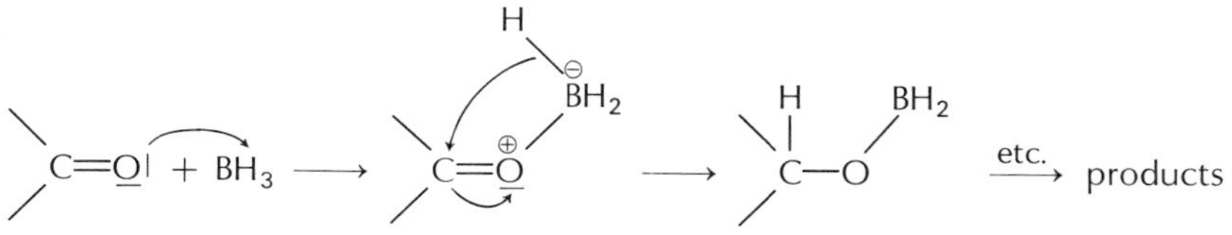

$$\text{>}C=\underset{..}{\overset{..}{O}}| + BH_3 \longrightarrow \text{>}C=\overset{\oplus}{O}\text{—}\overset{H}{\underset{\ominus}{BH_2}} \longrightarrow \text{>}\overset{H}{C}\text{—}O\text{—}BH_2 \xrightarrow{etc.} \text{products} \quad (9.43)$$

Meerwein-Ponndorf-Verley reduction. The Meerwein-Ponndorf-Verley reduction of aldehydes and ketones is the reverse of the Oppenauer oxidation (Eq. 9.23). In other words, the reaction is carried out in an isopropyl alcohol solvent, and aluminum isopropoxide is used as the catalyst. Acetone, with a lower bp than that of isopropyl alcohol, may be distilled from the boiling mixture to help shift the equilibrium toward the reduced product (Eq. 9.44). □ □

$$RCOR' + CH_3CHOHCH_3 \underset{}{\overset{cat.}{\rightleftharpoons}} RCHOHR' + CH_3COCH_3 \quad (9.44)$$

Cannizzaro and Tishchenko reactions. A rather specialized method of preparing alcohols is restricted to the use of aldehydes that have no hydrogen on the carbon adjacent to the carbonyl group. It is known as the Cannizzaro reaction. The aldehyde is treated with sodium hydroxide and disproportionates into equal amounts of alcohol and acid. It is not very useful, because the maximum amount of alcohol obtained is only 50% of the aldehyde used. However, crossed Cannizzaro reactions are possible whereby two aldehydes are employed in equal amounts. It turns out that if the other aldehyde is formaldehyde, the acid produced is not a mixture but is formic acid (Eq. 9.45).

$$(CH_3)_3CCHO + HCHO + NaOH \longrightarrow (CH_3)_3CCH_2OH + HCOONa \quad (9.45)$$

The mechanism of the Cannizzaro reaction, with a hydride transfer to the carbonyl group is reminiscent of the Meerwein-Ponndorf-Verley reduction, as shown by Equations 9.46.

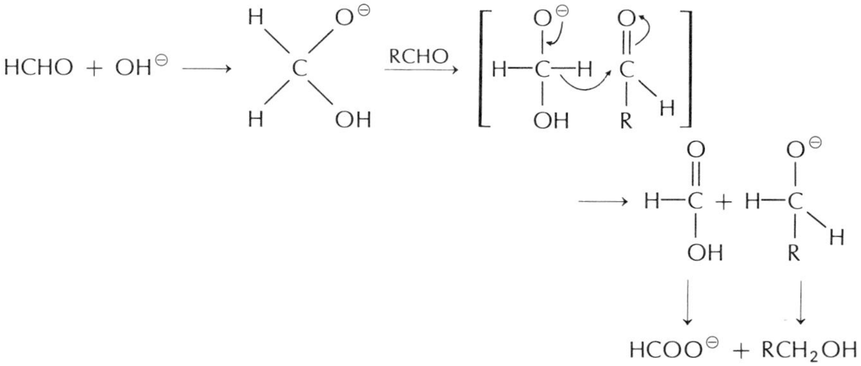

$$HCHO + OH^{\ominus} \longrightarrow \underset{H}{\overset{H}{\underset{OH}{>\!\!\!C\!\!\!<}}}\overset{O^{\ominus}}{} \xrightarrow{RCHO} \left[H-\underset{OH}{\overset{O^{\ominus}}{\underset{|}{C}}}-H \cdots \underset{R}{\overset{O}{\underset{|}{C}}}\overset{}{\underset{H}{}} \right]$$

$$\longrightarrow H-\underset{OH}{\overset{O}{\underset{|}{C}}} + H-\underset{R}{\overset{O^{\ominus}}{\underset{|}{C}}}\!\!\!-H \quad (9.46)$$

$$\downarrow \qquad \downarrow$$

$$HCOO^{\ominus} + RCH_2OH$$

In fact, under anhydrous conditions, an aluminum alkoxide may be used instead of hydroxide ion (the Tishchenko reaction), in which case the product is the ester of the alcohol and acid that are produced in the Cannizzaro reaction (2RCHO $\longrightarrow$ RCOOR).

Exercises

9.10 Predict the product of each reaction:

a. $(CH_3)_2CHCH_2CHO \xrightarrow[\text{iPrOH}]{Al(OiPr)_3,\ 80°}$

b. $CH_3CH_2CHO \xrightarrow[\text{dil. } CH_3COOH]{NaBH_4,\ 25°,\ cat.,}$

c. 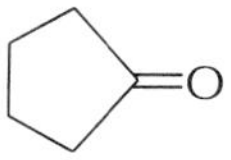 <chemical structure: cyclopentanone> $\xrightarrow[\text{2.   dil. HCl}]{\text{1.   LiAlH}_4,\ 25°,\ \text{ether}}$

d. $(CH_3)_3CCHO \xrightarrow[\text{dil. NaOH}]{HCHO,\ 40°}$

e. $(CH_3)_3CCHO \xrightarrow[Pt]{H_2,\ 25°}$

f. $(CH_3)_2CHCOOH \xrightarrow[\text{2.   dil. HCl}]{\text{1.   LiAlH}_4,\ 25°,\ \text{ether}}$

9.15 Industrial preparation of alcohols

Methanol. Methanol, the first member of the homologous series of alkanols, is prepared in tremendous amounts by hydrogenation of carbon monoxide (Eq. 9.47). An old

$$CO + 2H_2 \xrightarrow[350°, \ 200 \ atm.]{Zn \ and \ Cr \ oxides,} CH_3OH \tag{9.47}$$

common name of methanol is wood alcohol. This name derives from the former major source, the dry distillation of hardwoods (charcoal manufacture). The major uses are dehydrogenation to formaldehyde (about 50% of the total) and other synthetic applications (about 25%). Methanol is poisonous. Ingestion or excessive exposure leads to blindness and death.

Ethanol. Ethanol has been known at least since the time of the cave dwellers. How much further back in prehistory it was produced by deliberate fermentation seems likely to remain an unanswered question until and unless time travel is discovered and perfected. All beverage alcohol (for human consumption) still must be made by fermentation. Almost every possible substance from the vegetable world has been used or tried as a source for fermentation. Any starch may be hydrolyzed to give maltose and glucose. Glucose, in the presence of the enzyme zymase (present in growing yeast), is decomposed into ethanol and carbon dioxide (Eq. 9.48). Minor products are acetaldehyde and fusel oil,

$$C_6H_{12}O_6 \xrightarrow[32°]{zymase} 2CH_3CH_2OH + 2CO_2 \quad \Delta H° = -26 \ kcal \tag{9.48}$$

a mixture of higher alcohols: propyl, isobutyl, isopentyl, and other pentyl alcohols that arise from amino acids from proteins present in the original grain, fruit, or potatoes used. The maximum concentration of alcohol (the word alcohol, used alone, refers to ethanol) attainable by fermentation is about 15%. A higher concentration harms the yeast organisms. Distillation of the fermentation product is used to produce whiskey (from grains), brandy (from wines), and so on.

Industrial alcohol, 95%, is produced by fermentation and distillation of blackstrap molasses, the noncrystallizable residue from the refining of sucrose (cane or beet sugar). The major source is from hydration of ethylene.

Ethanol is used extensively as a solvent and for synthetic purposes. Ethanol is taxed heavily unless it is denatured by addition of any of many legally specified denaturants, which make the liquid unfit for human consumption and which are very difficult to remove from the denatured alcohol.

Other alcohols. Other alcohols are prepared by hydration of specific alkenes and by hydrolysis of alkyl chlorides available from chlorination of mixtures of butanes and of pentanes.

A special fermentation of starch gives butyl alcohol.

Fats and oils (esters of glycerol with long-chain acids) may be hydrogenated to give long-chain alkanols.

9.16 Summary

In this chapter we have been assuming inertness of the alkyl groups in the reactions and preparations given. The activity has been centered on the reactive groups: the —OH, the \C=O, the —COOR, and so on. Appreciation of the integrity of alkyl groups and the reactivity of the bonds of carbon to elements other than carbon and hydrogen is fundamental to understanding organic chemistry. Of course, if a carbonium ion is formed, rearrangements occur and the original alkyl group then participates in determining the course of the reaction.

The chapter begins with a list of some typical alcohols and considers a few of the physical properties. Reactions at the oxygen-hydrogen bond, such as hydrogen and deuterium exchange and reaction with metals, are described. The use of magnesium and of aluminium alkoxides as drying reagents is mentioned. Azeotropes are described. Esterification is discussed in greater detail than before, and inorganic esters are introduced. Conversion of alcohols to alkyl halides is given as the first example of reactions that cleave the carbon-oxygen bond. Ether syntheses are described. Oxidation is considered at some length. The distinction between conversion and yield is made clear. A method for balancing oxidation-reduction equations is given with illustrations. Dehydration is taken up and the Chugaev reaction is introduced.

The preparation of alcohols is considered as an application of substitution, addition, and reduction reactions. Under substitution are given two new reactions of the Grignard reagent, oxygenation and reaction with epoxides. Under addition, the reaction of Grignard reagents with esters is added to the list of Grignard reactions. Hydroboration of the carbon-carbon double bond is introduced. Reduction reactions of aldehydes, ketones, acids, and acid derivatives by hydrogen, lithium aluminum hydride, diborane, and sodium borohydride are given. The Meerwein-Ponndorf-Verley reduction is described. The Cannizzaro and Tishchenko reactions are mentioned. The industrial methods used for the preparation of MeOH and EtOH are outlined.

Despite the numerous reactions and preparations given in this chapter, the list is not complete by any means. Many properties, reactions, and syntheses remain to be mentioned. This chapter might be a prototype of the descriptions needed for each class of compound. However, no class of compound can be treated in a comprehensive manner in a first year textbook of organic chemistry. Only a sampling of reactions and properties is possible.

Some new reactions given in this chapter are as follows.

$$-\underset{H}{\overset{|}{C}}-\underset{}{\overset{|}{C}}\overset{b}{\underset{d}{\diagdown}}\overset{O\text{—}H}{\underset{c}{\diagup}}a$$

Reactions at a:

$$ROH + D_2O \rightleftharpoons ROD + DOH \qquad (9.1)$$

$$ROH + Na \longrightarrow RONa + \tfrac{1}{2}H_2 \qquad (9.2)$$

$$Al(OR)_3 + 3H_2O \longrightarrow \underline{Al(OH)_3} + 3ROH \qquad (9.3)$$

$$CH_3COOH + tBuOH \xrightleftharpoons{H^{\oplus}} CH_3COOtBu + H_2O \qquad (9.9)$$

$$ROH + H_2SO_4 \rightleftharpoons RO(SO_2)OH \xrightleftharpoons{+ROH} RO(SO_2)OR \qquad (9.11, 9.13)$$

$$ROH + SOCl_2 \longrightarrow (RO)_2SO + 2HCl \qquad (9.14)$$

$$RCOOH + SOCl_2 \longrightarrow RCOCl + SO_2 + HCl \qquad (9.15)$$

$$RCOCl + RCOONa \longrightarrow (RCO)_2O + NaCl \qquad (9.15)$$

$$RCOCl + EtOH \longrightarrow RCOOEt + HCl \qquad (9.16)$$

Reactions at b:

$$tBuOH + MeOH \xrightarrow{H^{\oplus}} tBuOMe + H_2O \qquad (9.18)$$

$$2EtOH \xrightarrow[H_2SO_4]{130°} EtOEt + H_2O \qquad (9.19)$$

Reactions at a and c:

$$RCHO + H_2 \xrightleftharpoons[Cu, \Delta]{Pt} RCH_2OH \qquad (9.20)$$

$$RCOR' + H_2 \xrightleftharpoons[Cu, \Delta]{Pt} \underset{R'}{\overset{R}{>}}CHOH \qquad (9.20)$$

$$RCH_2OH + CrO_3 \xrightarrow[hot]{H^{\oplus}} RCHO + Cr^{3+} + H_2O \qquad (9.21)$$

$$R_2CHOH + CrO_3 \xrightarrow[40°]{H^{\oplus}} R_2C=O + Cr^{3+} + H_2O \qquad (9.21)$$

$$RCH_2OH + KMnO_4 \xrightarrow[40°-60°]{dil.\ OH^{\ominus}} RCOOK + MnO_2 + KOH + H_2O \qquad (9.22)$$

$$RCH_2OH + KMnO_4 \xrightarrow[25°-100°]{dil.\ H_2SO_4} RCOOH + MnSO_4 + K_2SO_4 + H_2O \qquad (9.22)$$

$$RCH_2OH + CH_3COCH_3 \xrightarrow[]{\text{excess}\ Al(OtBu)_3} RCHO + CH_3CHOHCH_3 \qquad (9.23)$$

Reactions at b and d:

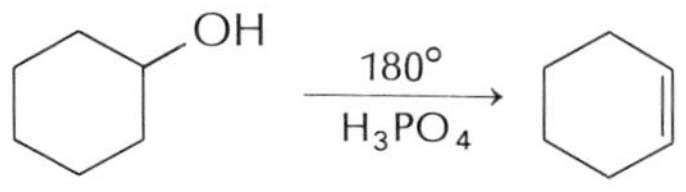

$$\text{cyclohexanol} \xrightarrow[H_3PO_4]{180°} \text{cyclohexene} \qquad (9.28)$$

$$(CH_3)_3CCHOHCH_3 \xrightarrow{\begin{array}{l}1.\ KOH,\ CS_2\\2.\ CH_3I\\3.\ 100°\end{array}} (CH_3)_3CCH=CH_2 \qquad (9.30)$$

Preparations:

$$\text{RMgX} \xrightarrow[\text{2. dil. HCl}]{\text{1. O}_2, 0°} \text{ROH} \qquad (9.31)$$

$$\text{RMgX} + \text{CH}_3\text{CH}\underset{\text{O}}{-\!-\!-\!-}\text{CH}_2 \longrightarrow \underset{\text{OMgX}}{\text{CH}_3\text{CHCH}_2\text{R}} \xrightarrow{\text{dil. HCl}} \text{CH}_3\text{CHOHCH}_2\text{R} \qquad (9.32)$$

$$\text{RMgX} + \text{CH}_3\text{CH}_2\text{COOCH}_3 \longrightarrow \underset{\underset{\text{R}}{|}}{\overset{\overset{\text{R}}{|}}{\text{CH}_3\text{CH}_2\text{COMgX}}} \xrightarrow{\text{dil. HCl}} \underset{\underset{\text{R}}{|}}{\overset{\overset{\text{R}}{|}}{\text{CH}_3\text{CH}_2\text{COH}}} \qquad (9.33)$$

$$3\text{CH}_3\text{CH}=\text{CH}_2 + \text{BH}_3 \longrightarrow (\text{CH}_3\text{CH}_2\text{CH}_2)_3\text{B} \qquad (9.35)$$

$$(\text{CH}_3\text{CH}_2\text{CH}_2)_3\text{B} \xrightarrow{\text{H}_2\text{O}_2, \text{HO}^\ominus} 3\text{CH}_3\text{CH}_2\text{CH}_2\text{OH} + \text{H}_3\text{BO}_3 \qquad (9.36)$$

$$\text{CH}_3\text{CH}_2\text{COCl} \xrightarrow{\text{H}_2, \text{Pt}} \text{CH}_3\text{CH}_2\text{CHO} \xrightarrow{\text{H}_2, \text{Pt}} \text{CH}_3\text{CH}_2\text{CH}_2\text{OH} \qquad (9.38)$$

$$\text{CH}_3\text{CH}_2\text{CHO} \xrightarrow[\text{alc.}]{\text{NaBH}_4} (\text{CH}_3\text{CH}_2\text{CH}_2\text{O})_3\text{B} \xrightarrow{\text{H}_2\text{O}} \text{CH}_3\text{CH}_2\text{CH}_2\text{OH} \qquad (9.41)$$

$$\text{cyclohexanone} \xrightarrow[\text{2. dil. HCl}]{\text{1. LiAlH}_4, \text{ether}} \text{cyclohexanol} \qquad (9.42)$$

$$(\text{CH}_3)_3\text{CCHO} + \text{HCHO} \xrightarrow{\text{NaOH}} (\text{CH}_3)_3\text{CCH}_2\text{OH} + \text{HCOONa} \qquad (9.45)$$

$$(\text{CH}_3)_3\text{CCHO} \xrightarrow[\text{2. dil. HCl}]{\text{1. Al(OiPr)}_3} (\text{CH}_3)_3\text{CCOOCH}_2\text{C}(\text{CH}_3)_3 \qquad \text{(Tishchenko)}$$

9.17 Problems

9.1 Starting with an alkanol with up to three carbon atoms, cyclopentanol, any solvent, and any inorganic reagents needed, give a good, reasonable synthesis of:
    a. 2-methyl-3-pentanone
    b. 4-methyl-1-pentene
    c. 2-methoxy-1-ethanol
    d. 2,3-pentanediol
    e. 2,2-dimethylpropanoic acid
    f. propylcyclopentane
    g. 2,2-dimethylbutanal
    h. 1-chloro-3-methylbutane
    i. *trans*-2-bromocyclopentyl propanoate
    j. *cis*-1,2-dimethyl-1,2-cyclopentanediol

9.2 Show how to convert isobutyl chloride into each of the following products. Other reagents and solvents may be used as required. More than one step may be necessary.
    a. isobutylene
    b. *tert*-butyl chloride
    c. isobutyl methyl ether
    d. isobutyl acetate
    e. isobutyric acid
    f. methyl isobutyrate
    g. 3-methyl-2-butanol
    h. 3-methyl-1-butanol
    i. trimethylacetic acid
    j. 4-methylpentanal

9.3 Substance A ($C_6H_{14}O$), when heated to 180° with phosphoric acid, gives a good yield of B (C_6H_{12}). Drastic oxidation of B gives acetone as the only product. B reacts with diborane and then with alkaline hydrogen peroxide to give C, an isomer of A. Dehydration of C gives B as the major product. Catalytic hydrogenation of B gives D. B reacts with bromine in acetic acid with dissolved sodium bromide to give E. A, upon treatment with CS_2, KOH, and methyl iodide, followed by heat, gives F, an isomer of B. Catalytic hydrogenation of F gives H, an isomer of D. Drastic oxidation of F gives carbon dioxide and G, ($C_5H_{10}O_2$). F with diborane, then alkaline hydrogen peroxide, gives A. A with thionyl chloride at 80° forms I. Under S_N2 conditions, I reacts with hydroxide ion to give A and a small amount of F. Catalytic hydrogenation of I forms H. Drastic oxidation of A by alkaline $KMnO_4$ gives J ($C_6H_{12}O_2$). J, with $LiAlH_4$ followed by dilute HCl, gives A. Give a structural formula for each lettered substance.

9.4 Substance A ($C_7H_{14}O$) is inert to the action of H_2 over Pt. A heated with phosphoric acid gives a mixture of three isomeric products: B and C in large amounts and D, barely detectable, each one of formula C_7H_{12}. Upon oxidation with warm acidic $KMnO_4$, B forms E (C_5H_8O) and acetic acid. Similar treatment of C gives only F ($C_7H_{12}O_3$); but D gives G ($C_6H_{10}O_2$) and carbon dioxide. A, with alkaline CS_2 followed by CH_3I and heating, gives a good yield of D with some B. B, C, and D all react with hydrogen over Pt to give H (C_7H_{14}). A is easily oxidized by CrO_3 in acetic acid to give I ($C_7H_{12}O$). I with CH_3MgI forms J. Reflux of G in CH_3OH with some sulfuric acid present gives a product that also reacts with CH_3MgI to give J. Give a structural formula for each lettered substance.

9.5 Substance A ($C_4H_8Br_2$), heated with zinc dust, forms B. B with bromine in CCl_4 gives A. Oxidation of B yields acetic acid as the only product. Mild oxidation of B with $KMnO_4$ at 0° gives C, a racemate. Hydrogenation of B gives D, with evolution of about 28 kcal of heat per mole. A, with an equimolar amount of hot alcoholic KOH, gives racemic F and a single substance E (both C_4H_7Br) plus other compounds. B with bromine in acetic acid with excess sodium acetate present undergoes *trans* addition to give a single racemate, the enantiomers being G and H. I and J are the enantiomers that would compose the other possible racemate (which is not formed from B). Give a structural formula for each lettered substance.

9.6 Give the IUPAC name for the following compounds.

a. $HOCH_2CH=CHCH_2OH$

b. $FCH_2\overset{Cl}{\underset{|}{C}}HCH_2\underset{|}{C}HCH_2CH_3$
     CH_2OH

c. $HOCH-C\equiv CCH_3$ attached to cyclohexane with Br

d. cyclohexane with OH, Cl, CH_3CHCH_3

e. cyclohexane with Br, Cl, $CH_3\underset{|}{C}CH_3$ / CH_3

f. cyclohexene with OH and cyclopentyl

g. cyclopentane with CH_3, OH, CH_3

9.7 Write the major organic reaction product for each of the following.

a. $CH_3CH_2\underset{\underset{CH_3}{|}}{\overset{\overset{CH_3}{|}}{C}}Cl + H_2O \longrightarrow$

b. 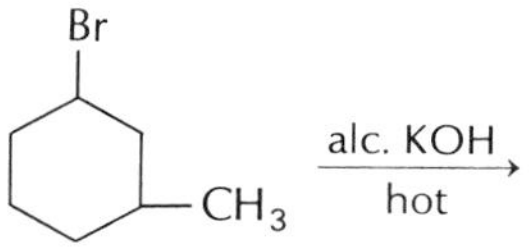 [1-bromo-2-methylcyclohexane] $\xrightarrow[\text{hot}]{\text{alc. KOH}}$

c. 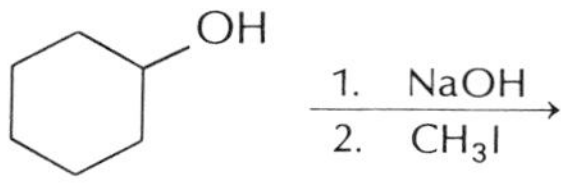 [cyclohexanol] $\xrightarrow[\text{2. } CH_3I]{\text{1. NaOH}}$

d. 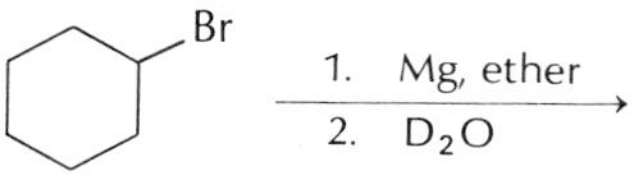 [bromocyclohexane] $\xrightarrow[\text{2. } D_2O]{\text{1. Mg, ether}}$

e. $CH_3CH_2CH_2OH \xrightarrow[H^\oplus, \text{ hot}]{K_2Cr_2O_7}$

f. $CH_3\underset{\underset{CH_3}{|}}{CHOH} \xrightarrow[H^\oplus]{K_2Cr_2O_7}$

g. 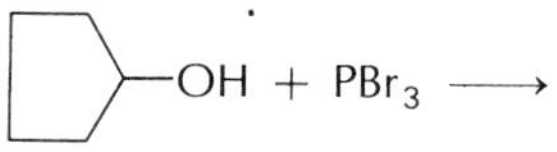 [cyclopentanol] $-OH + PBr_3 \longrightarrow$

h. 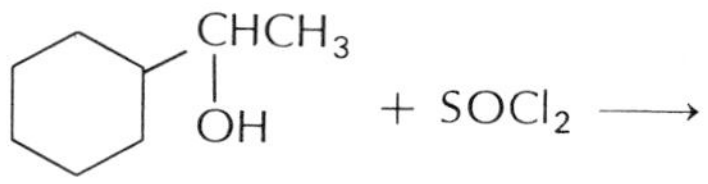 [cyclohexyl-CH(CH_3)-OH] $+ SOCl_2 \longrightarrow$

9.18 Bibliography

Any of the textbooks given in the bibliographies of Chapters 7 and 8 are good for additional information about alcohols. The paperbacks by Breslow (Chap. 7, ref. 17), Payne and Payne, (Chap. 7, ref. 18), Saunders (Chap. 7, ref. 20), and Weiss (Chap. 7, ref. 17) all are helpful. Stille (Chap. 7, ref. 16) is particularly valuable concerning industrial syntheses.

9.19 Herbert C. Brown: The discovery of hydroboration

Professor Herbert C. Brown describes the steps leading to the discovery of hydroboration at Purdue.* Professor Brown started out as an inorganic chemist at Chicago, Ph.D. 1938, moved to Wayne State in 1943, and on to Purdue in 1947. A prolific investigator, a recipient of many awards, he is well known for his research in many facets of organic chemistry.

Early observations on the reaction of diborane with alkenes indicated that the reaction was quite slow, requiring elevated temperatures and long reaction periods. These results certainly were not promising for achieving a fast quantitative addition of diborane to olefins in ether solvents at 0°.

So, how was the discovery made?

Dr. B. C. Subba Rao received his Ph.D. degree from Purdue University in 1955, working with me, and decided to remain with me for two years of postdoctoral work before returning to India. He was in the midst of a study of the reducing powers of sodium borohydride, enhanced by aluminum chloride. He established that representative aldehydes and ketones utilized 1 "hydride" per mole. Obviously, reduction was occurring to the alcohol stage. Nitriles utilized 2 moles of "hydride," corresponding to reduction to the amine. Esters, such as ethyl acetate and ethyl stearate, revealed the uptake of 2 moles of hydride. Evidently, reduction to the alcohol was occurring here also. One of the compounds Subba Rao examined was ethyl oleate. It showed "hydride" uptake of 2.4 moles per mole of compound.

I asked how we might account for this nonstoichiometric value. The ethyl oleate he had obtained from the stockroom had been somewhat discolored, but he had used it anyhow for this first exploratory experiment. He suggested that the material might contain sufficient impurities, possibly peroxides, to account for the high results. He recommended that we drop ethyl oleate. Its addition to the list had really been an afterthought.

Fortunately, the research director is in an enviable position to insist on high standards—he does not have to do the actual experimental work! I persuaded Subba Rao to return to the bench and repeat the experiment with purified ethyl oleate.

The repetition yielded the same result! It required only a little more effort to establish that we were achieving the simultaneous reduction and hydroboration of ethyl

* Based on Herbert C. Brown, *Boranes in Organic Chemistry*. Cornell University Press, 1972, pp. 256–261.

oleate. But we were still not home safe. We argued that the successful hydroboration of alkenes must involve aluminum borohydride. It had been previously established that diborane adds to aluminum hydride to form aluminum borohydride. It then occurred to us that we might develop this into a catalytic process. All we had to do was place the alkene in the ether solvent together with a small catalytic quantity of aluminum hydride. Then, diborane could be introduced. They would form aluminum borohydride, and the latter would react with the alkene present to form the desired organoborane, regenerating the aluminum hydride. This catalytic process was tested. It worked perfectly.

We were on the point of submitting a publication reporting this new catalytic process when we decided to run a blank experiment. We knew that diborane could not possibly react with olefins under these mild conditions. The previous work made that quite clear. However, we thought it dramatic to include an experiment showing the remarkable change in the rate of reaction of diborane with alkenes achieved by small catalytic quantities of aluminum hydride. We tried the experiment. The addition of the diborane to the olefin proceeded just as rapidly in the absence of the aluminum hydride!

$$6 RCH\!\!=\!\!CH_2 + B_2H_6 \longrightarrow 2(RCH_2CH_2)_3B$$

Investigation soon revealed the reaction to be rapid, quantitative, and of wide generality. The reaction appeared to be complete as fast as the two reagents could be brought into contact. In its speed, generality, and quantitative nature, the reaction resembles the addition of bromine to simple alkenes.

These observations were exceedingly puzzling in view of the earlier reports. Accordingly, we undertook a reexamination of the reaction of alkenes with diborane in a high-vacuum apparatus. It was observed that the reaction of diborane with pure alkenes is indeed quite slow, extending over periods of many hours. However, the addition is markedly catalyzed by ethers and similar weak bases. Addition of mere traces of ethers changed the initially slow reaction to a fast one. Consequently, the inadvertent use of the ether solvents for the reactions involving sodium borohydride appears to have been responsible for the fast, quantitative hydroboration procedures.

10 Spectroscopy I

In this chapter, we shall examine the various interactions of radiant energy and organic compounds, known as <u>spectroscopy.</u> The whole subject has been expanding rapidly in recent years, and we shall need two chapters to cover the topics. A bibliography will be found at the end of Chapter 11.

Certain forms of spectroscopy are of more use to an organic chemist than are others. We shall spend most of our time on infrared, visible, and ultraviolet spectra in this chapter and go on to nuclear magnetic resonance and mass spectroscopy in the next.

Then, having added the very powerful probe of spectroscopy to our background, we shall be in a position in the chapters to follow to discuss and understand the properties of the various classes of compounds more comprehensively than heretofore.

10.1   Background; types of spectroscopy   240
   1. Radiant energy; sunlight.   2. Units: frequency, wavelength, energy.
   3. Electromagnetic spectrum.   4. Color.   5. Definitions.   6. Ranges of microwave, ir, visible, and uv spectroscopy. Exercises.

10.2   Microwave spectroscopy   245
   1. Kinds of kinetic energy.   2. Transitions.   3. Uses of microwave spectroscopy.

10.3   Theory of infrared spectroscopy   246
   1. Vibrational energy.   2. Triatomic molecules.   3. Polyatomic molecules.

10.4   Interpretation of infrared spectroscopy   248
   1. Absorption frequencies.   2. Alkenes.   3. Alkynes.   4. Alkyl groups: the fingerprint region.   5. Carbon-oxygen bonds: alcohols, ethers.
   6. Carbon-oxygen double bonds: aldehydes, ketones, acids, acid derivatives.
   7. Chemical communication. Exercises.

10.5   Raman spectroscopy   257

10.6   Theory of visible and ultraviolet spectroscopy   258
   1. Electronic transitions.   2. Selection rules; energy diagrams.   3. Singlet and triplet states.   4. The σ, σ^*, n, π, and π^* states.

10.7 Interpretation of visible and ultraviolet spectroscopy 261
   1. Chromophores.   2. Effect of conjugation.   3. Extended conjugation.
   4. Uses.   5. Extinction coefficients.

10.8 Summary   263

10.1 Background; types of spectroscopy

Radiant energy; sunlight.   The universe is sparsely populated by two types of bodies: those that primarily emit energy (stars) and those that primarily absorb energy (planets, asteroids, dust, and so on). From the point of view of the earth, the sun is the major source of radiant energy. A tiny amount of radiation is received on earth by reflection of sunlight from the moon and from the planets. Starlight contributes even less energy. In turn, the earth reradiates some of this energy into space. This chapter will be concerned with the interaction of radiant energy and matter: in particular, organic compounds.

Sunlight, as received at the surface of the earth, has been filtered through the atmosphere and has its maximum intensity in the yellow-green visible region. The intensity trails off slowly into the infrared but drops off quickly in the opposite direction beyond the violet. Sunlight is the radiant energy absorbed by the leaves of plants: it allows photosynthesis to occur. Thus, carbon dioxide and water interact to give oxygen and organic compounds. This endothermic reduction process maintains the oxygen content of the atmosphere and is the starting point for the maintenance of life on earth.

Units: frequency, wavelength, energy.   Radiant energy traverses a vacuum at 2.998×10^{10} cm/sec. The velocity in a region of density greater than zero is lower and is expressed as a ratio, the refractive index, n_D^{25} (see Sec. 6.4). Although the velocity is constant (in a vacuum), the wavelength and frequency cover a tremendous range of values. The product of the wavelength (λ) and the frequency (v) is equal to the velocity of light (c, Eq. 10.1). The frequency (v) is the number of waves passing a fixed point in a second, or

$$c = v\lambda \tag{10.1}$$

the number of cycles per second. The term hertz (Hz) is now used for a cycle/sec. A million Hz is called a megahertz (MHz); a thousand, a kilohertz (kHz). At long wavelengths, the frequency is small, and at short wavelengths the frequency is large.

The energy transmitted by a beam of light is proportional to both the frequency and the intensity. The proportionality constant of 6.626×10^{-27} erg sec, known as Planck's constant, h (Eq. 10.2), relates the energy and the frequency. Because energy is also pro-

$$E_{\text{erg}} = hv = \frac{hc}{\lambda} = \frac{1.986 \times 10^{-16} \text{ (erg cm)}}{\lambda \text{ (cm)}} \tag{10.2}$$

portional to the reciprocal of the wavelength, a new term, the wavenumber ($\bar{v}$, called "nu bar") is defined as equal to $1/\lambda$. It has units of cm^{-1} (called a reciprocal centimeter) and may be pictured as the number of waves of wavelength λ needed to add up to 1 cm. Substitution into Equation 10.2 gives Equation 10.3. These tiny packages of energy are

$$E_{\text{erg}} = hc\bar{v} = (1.986 \times 10^{-16})\bar{v} \tag{10.3}$$

known as <u>quanta</u> or <u>photons</u>. A photon (quantum) originates by emission or ends in absorption by a single atom or molecule. Because chemists feel much more at home with the concept of the mole (6.023×10^{23} molecules), we may also speak of one mole of photons, which is given the special name of 1 Einstein and is equal to 2.859 $\bar{v}$ in cal. Thus, energy is proportional to the frequency and to the number of photons arriving at or passing by a given point, the <u>intensity</u>.

Electromagnetic spectrum. To avoid excessively small or large numbers, we use various units of measurement for different regions of the spectrum of electromagnetic radiation. For short wavelengths, Angstrom units (Å) are used. For the ultraviolet and visible ranges, nanometers (nm) are used. The micrometer (μm) usually is employed in the infrared region. Numerically, 1 cm = 10^4 μm = 10^7 nm = 10^8 Å. Also, 1 m = 10^6 μm = 10^9 nm = 10^{10} Å.

> The words and symbols used here are those that have been approved by the Nomenclature Committee of the Division of Analytical Chemistry of the American Chemical Society. Some of the older symbols are still in use and can cause confusion. Thus, the old term for micrometer (μm) is micron (μ, simply called "mu"). The old expression for nanometer (nm) is millimicron (mμ, commonly called "millimu").

Energies are expressed as ergs, calories, kilocalories, or electron volts (eV). One calorie equals 4.184×10^7 ergs. An electron volt is the energy imparted to a mole of electrons when accelerated through a potential drop of 1 volt. An electron volt is equal to 23.06 kcal.

Table 10.1 displays the range of values encountered in terms of the various units of wavelength, frequency, and energy. The boxed-in areas show the convenient ranges for use of the various units. The first column shows the names given to certain portions of the spectrum. The table is arranged so that the long wavelength (small frequency) and small energy portion of the spectrum is at the top. Short wavelength (high frequency) and high energy appear at the bottom.

Emission of radiant energy may be thought of as the result of a drop from a higher energy level to a lower level by an electric charge (an electron, a proton) or of changes in vibrational or rotational energies, and so on. Absorption of radiant energy in the microwave region results in an increase in rotational energy of molecules. Frequencies in the infrared region are characteristic of bond-length and bond-angle vibrations. The visible and ultraviolet absorption regions result from excitation (absorption) or de-excitation (emission) of electrons. The far ultraviolet and X-ray regions originate from bond breaking, ionization reactions, or expulsion of electrons. In Table 10.2, the region for $\bar{v}$ from 33 to 1,000,000 cm^{-1} is presented in more detail than in Table 10.1. We see that the energy of an Einstein of radiation increases through the range of activation energies (10 to 30 kcal) in the near infrared, into the weaker bond energies in the visible, and on into the stronger bond energies and ionization energies in the ultraviolet.

Table 10.1 The electromagnetic spectrum

Approximate region	Wavelength, λ				Frequency	
	cm	μm	nm	Å	$\bar{\nu}$ cm^{-1}	ν^a Hz
AM radio	10^6	10^{10}	10^{13}	10^{14}	10^{-6}	3×10^4
nmr	10^4	10^8	10^{11}	10^{12}	10^{-4}	3×10^6
TV	10^2	10^6	10^9	10^{10}	10^{-2}	3×10^8
radar, microwave	10^0	10^4	10^7	10^8	10^0	3×10^{10}
far ir	10^{-2}	10^2	10^5	10^6	10^2	3×10^{12}
ir, vis, uv	10^{-4}	10^0	10^3	10^4	10^4	3×10^{14}
far uv, X rays	10^{-6}	10^{-2}	10^1	10^2	10^6	3×10^{16}
X rays, γ rays	10^{-8}	10^{-4}	10^{-1}	10^0	10^8	3×10^{18}
γ rays, cosmic rays	10^{-10}	10^{-6}	10^{-3}	10^{-2}	10^{10}	3×10^{20}
cosmic rays	10^{-12}	10^{-8}	10^{-5}	10^{-4}	10^{12}	3×10^{22}

Approximate region	Energy			
	photon,^{b} ergs	Einstein		
		cal^c	kcal^c	eV^d
AM radio	2×10^{-22}	2.9×10^{-6}	2.9×10^{-9}	1.2×10^{-10}
nmr	2×10^{-20}	2.9×10^{-4}	2.9×10^{-7}	1.2×10^{-8}
TV	2×10^{-18}	2.9×10^{-2}	2.9×10^{-5}	1.2×10^{-6}
radar, microwave	2×10^{-16}	2.9×10^0	2.9×10^{-3}	1.2×10^{-4}
far ir	2×10^{-14}	2.9×10^2	2.9×10^{-1}	1.2×10^{-2}
ir, vis, uv	2×10^{-12}	2.9×10^4	2.9×10^1	1.2×10^0
far uv, X rays	2×10^{-10}	2.9×10^6	2.9×10^3	1.2×10^2
X rays, γ rays	2×10^{-8}	2.9×10^8	2.9×10^5	1.2×10^4
γ rays, cosmic rays	2×10^{-6}	2.9×10^{10}	2.9×10^7	1.2×10^6
cosmic rays	2×10^{-4}	2.9×10^{12}	2.9×10^9	1.2×10^8

^{a} To save space, 3 is entered in the table instead of the more exact 2.998.
^{b} 2 is entered instead of 1.986.
^{c} 2.9 is entered instead of 2.859.
^{d} 1.2 is entered instead of 1.240.

Table 10.2 The infrared, visible, and ultraviolet spectrum

Region		Wavelength		Frequency	Energy
		μm	nm	$\bar{\nu}$, cm^{-1}	Einstein, kcal
ir	far ir	300 — 300,000 — 33 — 0.10			
	most used ir range	25 25,000 400 1.14			
		2.5 2,500 4,000 11.4			
	near ir	0.78 — 780 — 12,820 — 36.6			
vis	red	0.65 650 15,380 44.0			
	orange	0.59 590 16,950 48.5			
	yellow	0.54 540 18,520 53.0			
	green	0.49 490 20,410 58.4			
	blue	0.42 420 23,810 68.1			
	violet	0.38 — 380 — 26,320 — 75.2			
uv	most used uv range	0.20 200 50,000 143			
	vacuum uv	0.01 — 10 — 1 × 10^6 — 2,860			

Color. The human eye is a marvelous organ: it is a physiological device that detects and discriminates radiation from 12,820 cm^{-1} (780 nm) to 26,320 cm^{-1} (380 nm). The ability to discriminate in this range is what we know as color. Obviously, this spread of frequencies (or wavelengths) is called the visible range. (A color-blind person is still able to detect light in the visible range but it appears to him as light and dark, or grayness, the way a color program on television appears in a noncolor receiving set.) The eye is so sensitive that only about 2 million quanta/cm^2/sec of the yellow sodium line are required to give a detectable sensation. This reception is equivalent to only 1.6 × 10^{-13} cal/sec.

Definitions. We should consider a few definitions at this point. Spectrometry is the branch of science that treats the measurement of spectra. A spectrometer is an instrument with a radiation source, slits, and a dispersing device (a prism or grating) whereby measurements may be made at selected wavelengths. A detector responds to the radiant power (ergs per sec) transmitted by the sample, which is contained in a cell. The instrument also may scan over the frequency range for which it was built. A spectrograph is an instrument like a spectrometer, except that the detector is a photographic plate and the response is in terms of energy (think of taking a time exposure with a camera). A spectrophotometer is a spectrometer that utilizes two beams, and the detector and associated devices give a ratio of the power in one beam to that in the other.

The sample and reference cells vary in size and material depending upon the portion of the spectrum to be scanned. The beam that strikes a cell is partially reflected from the cell wall and partially transmitted through the wall into the interior. As the beam leaves the cell, the same effects occur. We are interested in the amount of radiation that gets through the medium in the cell. The <u>transmittance</u> (T) (old terms used were transmittancy and transmission) is defined as the ratio of the radiant power (energy per second, intensity) transmitted by a sample or reference (I) to the radiant power (I_0) incident on the sample. Thus $T = I/I_0$. It would be zero for complete absorption, or unity for no absorption. The <u>absorbance</u> (A) (old terms used were optical density, absorbancy, and extinction) is the logarithm of the reciprocal of the transmittance. Thus $A = \log(1/T) = \log(I_0/I)$, and A is infinite for complete absorption and zero for no absorption.

If there is any absorption of the light passing through the sample, the intensity diminishes as more and more molecules are encountered. Thus, the amount of light absorbed depends on the concentration and the path length. We may express this dependence as $A = abc$, in which b is the internal cell dimension or sample path length in cm and c is the concentration in g/liter. The third factor, a, is a proportionality constant called the <u>absorptivity.</u> (Old terms were absorbancy index, specific extinction, and extinction coefficient. Sometimes k was used as the symbol instead of a). If the concentration is expressed in moles/liter, $A = \varepsilon bM$, where ε is called the <u>molar absorptivity.</u> (Old terms were molar absorbancy index, molar extinction coefficient, and molar absorption coefficient). Numerically $\varepsilon = a(\text{mw})$. Also, a and ε usually are constant for dilute solutions, but they are inclined to shift as the concentration increases. The expression $A = \varepsilon bm$ is a mathematical statement of the <u>Beer-Lambert Law,</u> sometimes simply called <u>Beer's Law</u>. The molar absorptivity may vary over quite a range, from 2 or 3 to over 200,000. The molar absorptivity at the frequency (or wavelength) at which the maximum of an absorption occurs is symbolized as ε_{max}. Sometimes the expression $\log \varepsilon_{max}$ is used if ε_{max} is large. For example, if $\varepsilon_{max} = 100,000$, $\log \varepsilon_{max} = 5.00$.

Ranges of microwave, ir, visible, and uv spectroscopy. We shall describe the various types of spectroscopy in order of increasing energy. Microwave spectroscopy, making use of rotational energies, is arbitrarily defined as being in the cm wavelength region ($\bar{v}$ from 0.1 to 10 cm^{-1}, λ from 10 to 0.1 cm). Infrared spectroscopy, used mostly in the 400 to 4000 $\bar{v}$ region (25–2.5 µm), gives information concerning vibrational energies. As we mentioned previously, visible and ultraviolet spectroscopy (12,000–50,000 $\bar{v}$ or 780–200 nm) lies in the energy region of electron shifts. The far ultraviolet, beyond 50,000 $\bar{v}$, is well into energies of ionization and bond rupture.

Exercises

10.1 a. Calculate the wavelength of the sodium line at 5896 A in nm and µm.
    b. Calculate $\bar{v}$ and v.
    c. Calculate the energy in kcal and eV per Einstein.

10.2 Calculate the frequency ($\bar{v}$) and the wavelength (λ) that correspond to the energy needed to cause a mole of bromine gas to dissociate into bromine atoms (see Table 5.6).

10.2 Microwave spectroscopy

Kinds of kinetic energy. The total energy of a molecule nearly isolated in the gas state at low pressure is the sum of all the component energies. The electronic energy consists of binding energies, atomic energies (for the electrons lying below the valence shell), and so on, which in turn consist of the several components of electronic energy in the various orbitals. Vibrational energy is present as bond stretching, as angular stretching, and in librations. Rotational energy is present as the spin of the entire molecule and of any portions of the molecule that rotate with respect to the rest of the molecule. Finally, translational energy is present as kinetic energy, $(3RT)/2$ for a gas. Each of the component energies is quantized. When molecules become more closely associated in the liquid or solid state, each molecule is affected by the nearest neighbors through the electromagnetic fields involved. The involvement complicates and increases the number of energy levels to such an extent that bands rather than lines characterize the spectrum.

Microwave spectroscopy is concerned with the changes in the closely spaced, quantized rotational energy levels of molecules. Radar-sized waves are employed; and, because radar waves are reflected or absorbed by liquids and solids, only gases may be used in the spectrometer.

Transitions. In all types of spectroscopy, a transition by a molecule from one energy level to a lower level results in emission of the energy difference as radiation (a photon) of equivalent energy (frequency) (Fig. 10.1). The interaction is reversible, and a photon of equivalent energy may be absorbed at the lower level, exciting the molecule to the upper level. Photons of the incorrect energy equivalent cannot be absorbed. The interaction (for rotation, vibration, or an electronic state) of a changing dipole moment with the electric field component of the wave is the cause or result of a transition (Fig. 10.2). The strength of absorption depends upon the transition moment and the difference in population between the two states. We may think of the transition moment as related to the magnitude of the difference in electric dipole moment between the two energy states of the molecule and to a probability factor for the absorption. There also must be a smaller population in the upper state than in the lower state for observable absorption to occur.

We have said that only a photon of the correct energy may be absorbed by a molecule undergoing a transition from a lower to a higher energy level. The implication is that only certain energy levels exist for the molecule, and we find that it is so for the rotational, vibrational, and electronic energies. There are quantum numbers that describe these energies.

For example, the rotational levels are designated by values of the quantum number J. <u>Selection rules</u> state the preferred changes in value of quantum numbers. (A transition in which J changes by ± 1 is favored.) Transitions that do not obey these rules are called <u>forbidden</u>. □ □

Uses of microwave spectroscopy. Although microwave spectroscopy has been known since 1933, only recently has it been used to any extent in organic chemistry. The large number of lines, the weakness of the intensity of the lines, and the need for gaseous samples has restricted the use to small molecules. However, the results are of value for conformational studies, dipole-moment measurement, isotope effects, and structure determination. Microwave spectroscopy is the technique used to identify carbon monoxide, ammonia, and formaldehyde, among other substances, in interstellar space.

246   Chapter ten

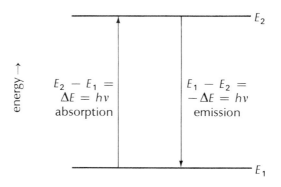

Figure 10.1   Energy transitions

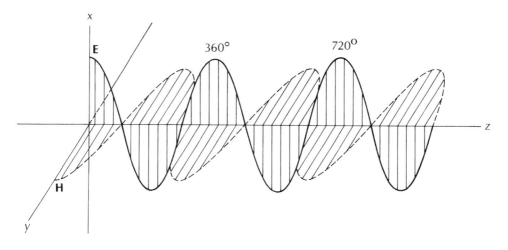

*Figure 10.2   Electric (**E**) and magnetic (**H**) components of plane polarized light. If **E** is in the xz plane, **H** is in the yz plane for a beam traveling from left to right along the z axis. (By convention, the three axes used for a three-dimensional coordinate system follow the <u>right hand rule</u>. Hold the right forearm horizontal, palm of the hand up, and the thumb pointing right. Let the other fingers curl up. The origin is taken at the base of the thumb, the x axis is positive up along the line to which the fingers point, the y axis is positive along the forearm from wrist to elbow, and the z axis is positive to the right along the thumb.)*

10.3   Theory of infrared spectroscopy

Vibrational energy.   The vibrational quantum number v has values of 0, 1, 2, ... The vibrational energy levels are equally spaced. The selection rule calls for $\Delta v = \pm 1$ for

10.3 Theory of infrared spectroscopy

allowed transitions. For a diatomic molecule with only one stretching mode, only one transition line should be observed in the spectrum, and then <u>only if the stretching vibration produces a change in the dipole moment.</u>

Triatomic molecules. For a linear triatomic molecule (for example, O=C=O), there are four modes of vibration (Fig. 10.3). Of the four modes, only v_1 does not give rise

$$O=C=O \qquad O=C=O \qquad O=C=O \qquad O=C=O$$
$$v_1 \qquad\qquad v_2 \qquad\qquad \sigma_1 \qquad\qquad \sigma_2$$
$$\text{inactive} \qquad \text{active} \qquad \text{active} \qquad \text{active}$$

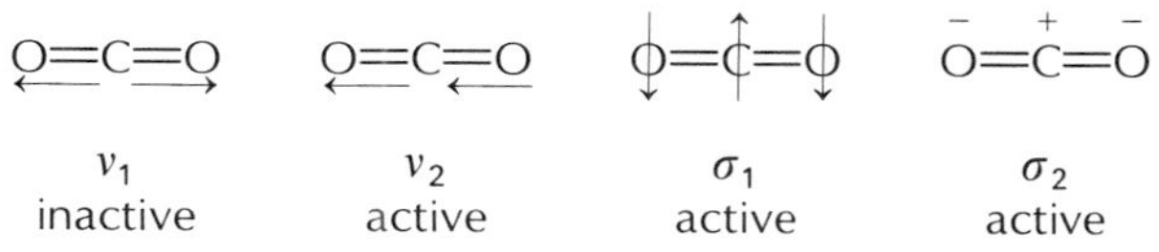

Figure 10.3 Vibrational modes of carbon dioxide. Arrows indicate the direction of movement of the atoms. in the plane of the page; + and − indicate movement up and down, respectively, out of the plane of the paper.

to a change in dipole moment. The other three modes do cause changes in dipole moment and are observable in the infrared region. If a triatomic molecule is nonlinear, there are only three modes of vibration (Fig. 10.4).

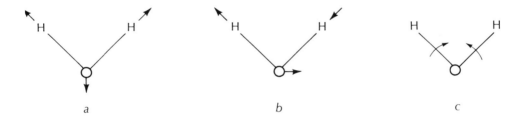

Figure 10.4 Vibrational modes of the water molecule. The frequencies are 3652, 3756, and 1545 cm^{-1} for the symmetric stretching (a), asymmetric stretching (b), and symmetric bending (c).

Polyatomic molecules. Most substances in organic chemistry are polyatomic and have many vibrational modes. Thus, methane has nine vibrational modes, and ethane has eighteen. Not all of the vibrational modes cause changes in dipole moment and appear in the infrared. However, enough of the modes are infrared active to give a spectrum rich in lines and bands. Most groups give characteristic spectra caused by certain persistent vibrational modes (see Fig. 10.7). By persistent modes, we mean the typical vibrational modes of a given group. The modes shown for methylene (—CH$_2$—) in Figure 10.5 appear in the spectrum of any molecule that contains a —CH$_2$— group. Infrared spectra of organic compounds display a seemingly bewildering variety of troughs. Our task now is to sort things out.

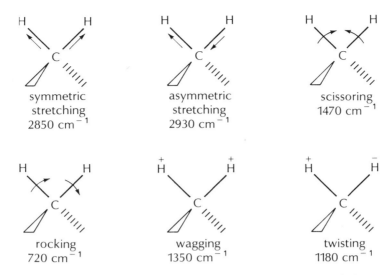

Figure 10.5 Vibrational and bending deformations of the methylene group. Arrows indicate movement in the plane of the page and the + and − signs indicate movement to positions above and below the plane of the paper.

10.4 Interpretation of infrared spectroscopy

Absorption frequencies. Infrared spectrophotometry gives more information about organic compounds than does any other single technique. Every bond and every bond angle vibration contributes an absorption unless the vibration causes no change in dipole moment. The restriction is minor, and even a highly symmetrical triatomic molecule like carbon dioxide has some absorption bands in the infrared (Fig. 10.3).

Elaborate tables have been published that give the position and relative intensities of the bands characteristic of certain groups. We give a very simplified table here (Table 10.3).

Both frequency and wavelength have been used to specify location of a peak. We shall use cm^{-1} only. Because adjacent bonds and groups affect the frequencies involved, many entries are needed to cover most of the contingencies we are likely to meet.

The greatest value of ir spectra in organic chemistry probably is the relatively easy identification of the various reactive groups in a molecule. The presence or absence of a C=C, C≡C, C≡N, C—O, O—O, O—H, C=O is indicated quickly and easily by a perusal of the ir spectrum of a compound. We shall begin with the alkenes. Groups other than those listed here will be taken up at appropriate points later in the book.

Alkenes. The C=C bond stretch may appear at 1430 to 2000 cm^{-1}, depending upon substitution about the double bond. In the simpler alkenes, a weak to medium absorption is seen from about 1640 to 1680 cm^{-1}.

Propylene (Figure 10.6) displays both the methyl hydrogens at 2950 cm^{-1} and the alkene hydrogens at 3100 cm^{-1}. The 1800 cm^{-1} weak absorption is an overtone. An overtone is a harmonic of a strong absorption at a smaller frequency. The C=C stretch appears at 1650 cm^{-1} in the unsymmetrical molecule. The 1470, 1440, and 1420 cm^{-1}

10.4 Interpretation of infrared spectroscopy

Table 10.3 Some characteristic infrared frequencies

Frequency, cm^{-1}	Group
3600	O—H, monomeric alcohol or acid
3300	≡C—H or hydrogen bonded alcohol
3050	=C—H
2950	C—H, CH_2, and CH_3
2840 and 2740	C—H in aldehyde, —CHO
2600	O—H, hydrogen bonded acid
2230	—C≡N or R—C≡C—R
2120	RC≡CH
1820 and 1760	C=O in acid anhydride
1800	C=O in acid chloride
1740	C=O in ester^a or dilute acid
1720	C=O in aldehyde, ketone, or dimeric acid
1660	C=C
1600 and 1400	RCO_2^- in salts
1475	CH_2
1475 and 1380	CH_3
1410, 990 and 910	$RCH=CH_2$
1410 and 890	$R_2C=CH_2$
1395 and 1370	$(CH_3)_3C$, unequal
1385, 1370, and 1160	$(CH_3)_2CH$, equal
1360	tertiary O—H
1340	tertiary C—H
1300	primary and secondary O—H
1300 and 960	trans—RCH=CHR
1250	multiple CH_2 or CH_3
1250	epoxide
1200	tertiary C—O, alcohol, ether, or ester
1200	C—F, variable
1160	$(CH_3)_2CH$
1100	secondary C—O
1050	primary C—O
810	$R_2C=CHR$
700	C—Cl
690	cis-RCH=CHR
550	C—Br
500	C—I

^{a}An ester also has two strong bands, C—O, in the 1250–1050 cm^{-1} region.

peaks are characteristic of alkene C—H. The 1380 cm^{-1} band, even though weak and appearing only as a shoulder, is characteristic of a methyl group. The 990 and 910 cm^{-1} bands are characteristic of a —CH=CH_2 group.

Without going on with more examples, let us summarize here the assignments of the alkenes. RCH=CH_2 has bands at 1410, 990, and 910 cm$^{\ominus 1}$. RCH=CHR' has a band at about 690 cm^{-1} if cis, or at about 1300 and 960 cm^{-1} if trans. $R_2C=CH_2$ has bands at 1410 and 890 cm^{-1}. $R_2C=$CHR' also has a somewhat variable band at about 810 cm^{-1}, and the C=C band at about 1660 cm^{-1} usually is weak.

250   *Chapter ten*

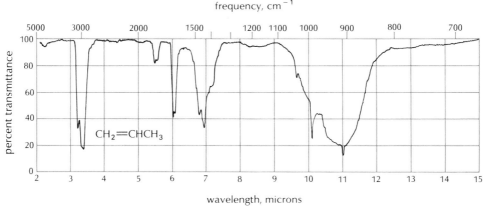

Figure 10.6   Infrared spectrum of propylene

Alkynes.   An alkyne hydrogen (—C≡C—H) absorbs at a high frequency, about 3300 cm^{-1}. We can see that there is a progression from about 2950 cm^{-1} for an sp^3 C—H to about 3050 cm^{-1} for an sp^2 C—H and about 3300 cm^{-1} for an sp C—H. The C≡C bond stretch occurs at about 2120 cm^{-1} for RC≡CH and at about 2230 cm^{-1}, with less intensity, for RC≡CR'. The —C≡N stretch falls at about 2230 cm^{-1} and is much more intense than the C≡C stretch.

Alkyl groups: the fingerprint region.   Before going ahead, let us summarize what to expect from the presence of some typical alkyl groups. Both CH$_3$— and —CH$_2$— absorb in the 1430 to 1470 cm^{-1} region, but the 1375 to 1380 cm^{-1} peak is characteristic of a methyl group only. If an isopropyl group is present, it is usual to see two absorptions of about equal intensity at 1385 and 1370 cm^{-1}. The *tert*-butyl group shows a doublet at 1395 and 1370 cm^{-1}, the latter being very strong. A 1340 cm^{-1} absorption is characteristic of a C—H bond with atoms other than hydrogen forming the other three bonds to carbon (a tertiary hydrogen).

That part of the spectrum with frequencies less than 1200 cm^{-1} is known as the fingerprint region because of the great variety of absorptions that may be observed and are characteristic of the substance being scanned. Some of the absorptions in this region have been identified and assigned to a specified molecular motion, but most remain unassigned.

We should mention here that carbon-carbon single bonds show weak and variable absorptions and are seldom useful for identification. Carbon-halogen bonds tend to appear at the low frequency end of the spectrum: C—F at 1000 to 1400 cm^{-1}, C—Cl at 600 to 800 cm^{-1}, C—Br at 500 to 600 cm^{-1}, and C—I at about 500 cm^{-1}. Depending upon the range of the spectrometer, they may or may not be seen.

Carbon-oxygen bonds: alcohols, ethers.   Turning to C—O bonds, let us look at dimethyl ether (CH$_3$OCH$_3$) first (Fig. 10.7a). The 2900, 1450, and 1410 cm^{-1} bands, although displaced somewhat, are the methyl absorptions. The absorptions at 1170 cm^{-1}, 1160 cm^{-1}, and so on, are in the region of C—O stretching; but not all of these bands are assignable. In general, a strong absorption in the 1250 to 1000 cm^{-1} region (usually about 1110 cm^{-1}) is typical of the presence of a C—O bond. For an epoxide, a band at 1250 cm^{-1} is characteristic.

The hydrogens in the O—H bonds are of the most help in identifying alcohols. Before looking at examples, we should make clear that the C—O bond absorption in primary alcohols appears at 1050 cm^{-1}, in secondary alcohols at 1100 cm^{-1}, and in tertiary alcohols at 1200 cm^{-1}.

The position of the O—H bond stretch varies depending upon the amount of hydrogen bonding present. Hydrogen bonding is at a maximum for the pure liquid. As the concentration of an alcohol is diminished by dilution with an inert solvent, such as CCl_4 or CS_2, the absorption shifts from a broad band at 3300 ± 100 cm^{-1} to higher frequencies and becomes sharper, approaching about 3600 cm^{-1} for a non-hydrogen-bonded O—H at high dilution. (See Figs. 10.7a, b, and c.) The O—H bending frequencies occur at 1300 ± 50 cm^{-1} for primary and secondary alcohols and at 1360 ± 50 cm^{-1} for tertiary alcohols.

For the dilution effect, Figure 10.7b for ethanol as a vapor (not associated) shows a sharp line at 3600 cm^{-1}. In Figure 10.7c, the appearance of the broad band at 3300 cm^{-1} is typical of association (hydrogen bonding). The 3600 cm^{-1} line is still present in the solution in carbon tetrachloride, although it now is weak.

The spectra in Figures 10.7b and c show the differences in appearance of the other bands in the vapor state and in solution. The differences in intensity are marked in some of the bands (for instance, at 1300 cm^{-1}) and negligible in others. We may ascribe these effects vaguely to solvation and leave it at that.

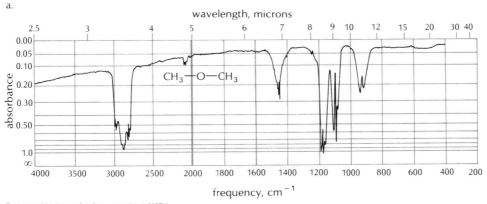

© 1969 Sadtler Research Laboratories, Inc. (15307K)

Figure 10.7 Infrared spectra of a. dimethyl ether, b. ethanol as a vapor, and c. ethanol in dilute solution

252 Chapter ten

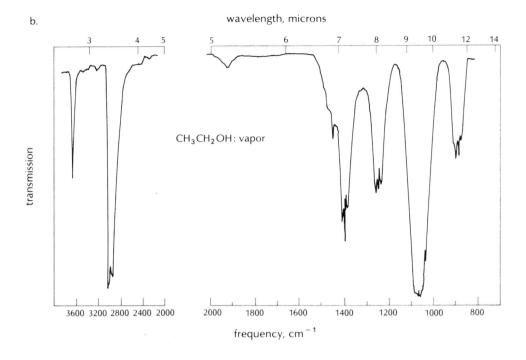

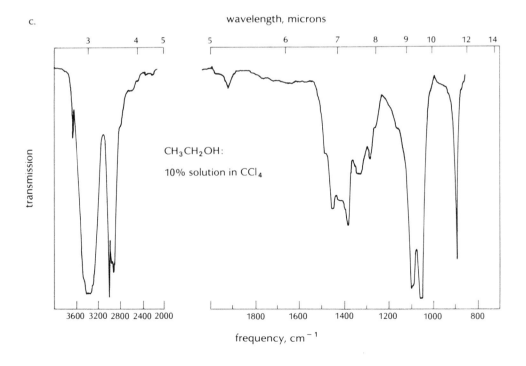

Figure 10.7 (continued)

Carbon-oxygen double bonds: aldehydes, ketones, acids, acid derivatives. The presence of a C=O bond, with its strong dipole, gives rise to a strong bond in the 1750 to 1670 cm^{-1} region (Fig. 10.9). The exact position is influenced by the groups bonded to the carbon atom and by solvation. An aldehyde or ketone displays the C=O stretch at 1720 ± 20 cm^{-1} (Figs. 10.8a and b). The aldehydic group (—CHO) also gives rise to two weak C—H stretching frequencies at about 2840 cm^{-1} and at 2740 ± 50 cm^{-1} (Fig. 10.8a). The higher-frequency band may be lost (or covered over) by the usual C—H absorptions in this range.

Conjugation with a double bond $\left(\begin{array}{c}\diagdown\\C=C-C=O\\\diagup\end{array}\right)$ causes a decrease of 20 to 30 cm^{-1} in the C=O bond stretch frequency.

If the C=O is part of an ester group, the frequency is increased by 15 to 20 cm^{-1} to 1740 ± 20 cm^{-1} (Fig. 10.8c). If the C=O is part of an acid chloride, the increase is still more marked, to 1800 ± 20 cm^{-1}. In an acid anhydride, the absorption not only is increased but is split into two peaks at about 1820 and 1760 cm^{-1}. An acid, unless it is in very dilute solution in a nonpolar solvent, absorbs as a dimeric hydrogen-bonded form at the usual C=O position of 1720 ± 20 cm^{-1}. The monomeric form, if present, may be detectable at about +20 cm^{-1} with reference to the strong dimeric absorption. In addition, the dimeric acid shows a broad hydrogen-bond stretch at 2600 ± 200 cm^{-1} (Fig. 10.10d). The monomeric acid gives a sharp absorption at about 3570 cm^{-1}. A salt of an acid, which has the form $\left[-C\begin{array}{c}\diagup\!\!\!\!\diagup O\\\diagdown\!\!\!\!\diagdown O\end{array}\right]^{-}$, does not exhibit a C=O absorption. Instead there are two bands (symmetric and antisymmetric stretches) at about 1400 and 1600 cm^{-1}.

Chemical communication. The use of chemicals to transport messages between organisms is under intense investigation. Some messages are direct and uncomplicated, such as that carried by butanethiol, the defensive weapon of the skunk. Other messages are more subtle. Many organisms secrete substances called pheromones that affect the behaviour of other organisms. Sex pheromones, for instance, attract members of the opposite sex of the same species and stimulate sexual activity. An understanding of the chemistry of sex pheromones of pest insects could lead to the development of highly specific, nonpersistent pesticides. Incidentally, a search is also underway for human sex pheromones.

The dilemma of relating molecular characteristics to biological activity in pheromones is the same problem of relating molecular characteristics to odor. Some workers emphasize the molecular shape as being the dominant factor; others, the molecular vibrations. Many molecules of a similar shape do have similar odors; and it is easy to imagine a detection system whereby the receptor protein for, for example, benzaldehyde (bitter almond flavor) will bind to molecules with a similar shape and not to others. However, there are a number of exceptions to this hypothesis.

Proponents of the molecular vibration hypothesis claim a closer correlation can be made to the infrared spectra than to the shape, although a mechanism is obscure. One of the systems under investigation is that of the alarm signal of the ant species *Iridomyrmex pruinosus*. The alarm pheromone has been identified as 2-heptanone. By comparison of the far infrared spectra of a variety of compounds to the ant alarm activity, ir frequencies could be identified as favorable or adverse. The high activity of compounds as dissimilar in shape to the natural pheromone as triethylamine and heptyl butyrate could be accounted for by favorable ir spectra!

254 Chapter ten

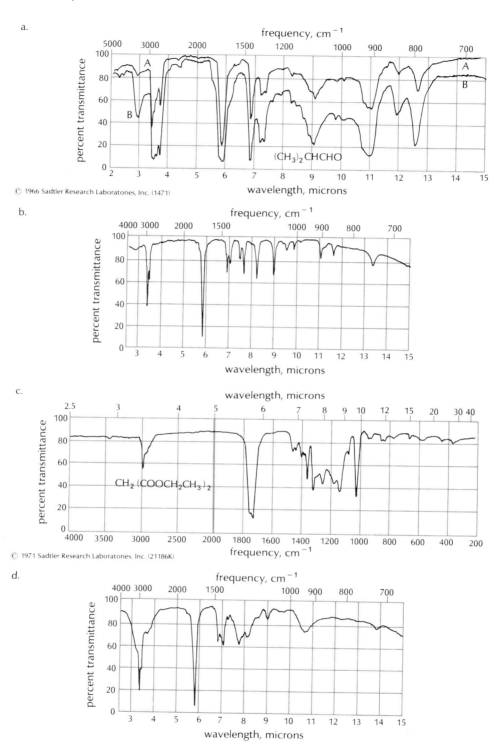

Figure 10.8 Infrared spectra of a. isobutyraldehyde, b. cyclohexanone, c. diethyl malonate, and d. nonanoic acid

10.4 Interpretation of infrared spectroscopy

Exercises

In these exercises, the infrared spectrum and the molecular formula of a substance are given. You are asked to identify the substance in as much detail as is possible by giving the correct structural formula. (Even a partial answer should be attempted: you should at least say that the compound has a methyl, has an isopropyl, has a C=C, has a C=O, etc.)

10.3 Spectrum with formula, Figure 10.9.
10.4 Spectrum with formula, Figure 10.10.
10.5 Spectrum with formula, Figure 10.11.
10.6 Spectrum with formula, Figure 10.12.
10.7 Spectrum with formula, Figure 10.13.

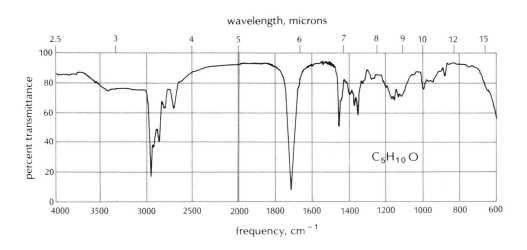

Figure 10.9 For Exercise 10.3

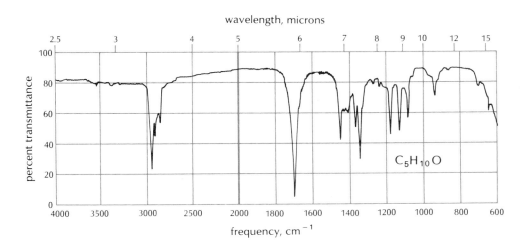

Figure 10.10 For Exercise 10.4

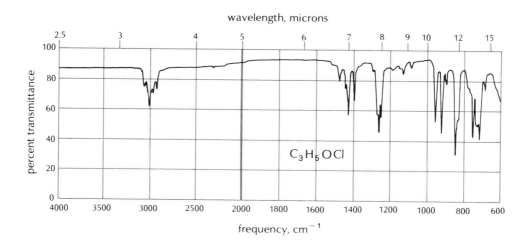

Figure 10.11   For Exercise 10.5

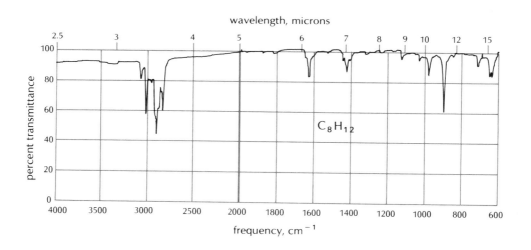

Figure 10.12   For Exercise 10.6

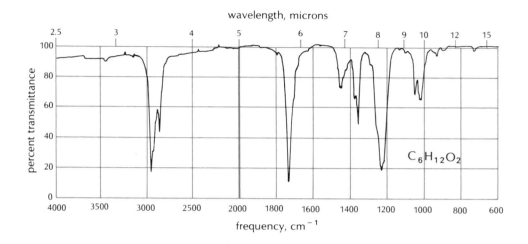

Figure 10.13   For Exercise 10.7

10.5 Raman spectroscopy

Raman spectroscopy, which has been known since the 1920's, has never become as useful in the organic laboratory as infrared for several reasons. The principal reason is cost: a Raman spectrophotometer is much more expensive than any of the relatively simple infrared spectrophotometers now available. The Raman spectrum is of the light scattered (actually absorbed and re-emitted) by a substance. An intense monochromatic source is needed to produce a Raman spectrum bright enough for easy scanning. Only with the use of a laser as a source has this requirement been met satisfactorily. Experimentally, the sample (a liquid or a solution) must be carefully filtered and handled to remove and keep out dust particles, which also scatter the incident light.

The Raman spectrum has its origin in a change of polarizability during absorption, in contrast to infrared, which requires a change in dipole moment. A Raman spectrum complements the infrared spectrum. Referring to Figure 10.3, the v_1 vibrational mode, although inactive in the infrared, is Raman active. The other three modes are Raman inactive. In general, the Raman spectrum exhibits those absorptions that are infrared inactive. Thus it is of value in the study of substances that have high symmetry. Even so, many bands appear in both the Raman and the infrared. Despite continuing theoretical interest in Raman spectroscopy, infrared spectroscopy will remain the technique of greater use in organic chemistry.

258   Chapter ten

10.6   Theory of visible and ultraviolet spectroscopy

Electronic transitions. Visible and ultraviolet spectra are the result of electronic transitions between a ground state and an excited state. Such a transition usually needs so large an energy that it requires light in the ultraviolet region. For an electronic transition to occur in the visible region, the energy difference between the ground and excited states must be reduced by structural modifications in the substance: for example, by charge delocalization (resonance).

Just as vibrational spectra are complicated by rotational transitions, so electronic spectra are complicated by vibrational and rotational transitions (Fig. 10.14). Under very high resolution in the gas phase, the spectrum may be seen as a mass of discrete sharp lines. In solution or liquid, however, line broadening occurs and the spectrum coalesces into rather wide bands (Fig. 10.15).

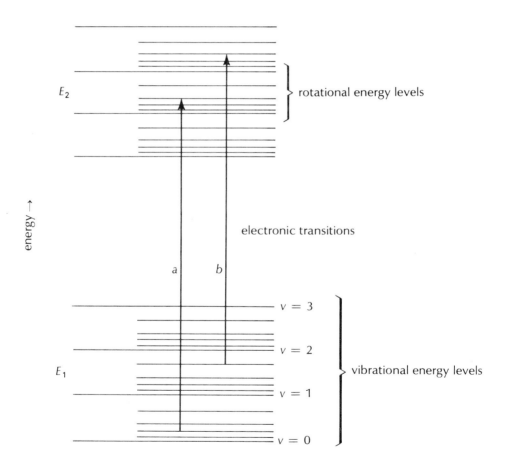

Figure 10.14   Schematic representation of electronic, vibrational, and rotational energy levels. The arrows show transitions a, from $E_1 = 1$, $v = 0$, $J = 2$ to $E_2 = 2$, $v = 1$, $J = 3$ and b, from $E_1 = 1$, $v = 1$, $J = 4$ to $E_2 = 2$, $v = 2$, $J = 3$

10.6 Theory of visible and ultraviolet spectroscopy 259

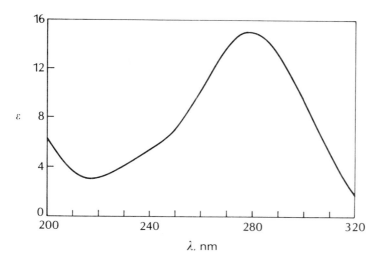

Figure 10.15 The ultraviolet spectrum of acetone in cyclohexane

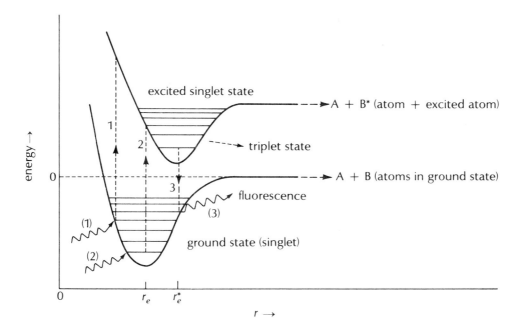

Figure 10.16 Schematic potential-energy diagram for ground and excited electronic single states of a diatomic molecule, A—B and A—B*, respectively. The horizontal lines represent vibrational energy levels. The wavy lines represent the arrival or departure of light quanta. In the lowest vibrational state, the most probable distance between the nuclei corresponds to the center of the vibrational energy level. In the higher vibrational states, the most probable distances approach the extremes of the vibrations. The curves shown here are schematic, and for actual systems the positions and shapes of the curves may be quite different from those shown.

260   Chapter ten

Selection rules; energy diagrams.   The selection rules are not simple. The situation is clarified by the application of the Franck-Condon principle, which states that an electron jump takes place with such speed that the nuclei may be thought of as stationary. In other words, vibratory and rotatory motion are so slow in comparison to an electronic transition that there is no change in bond length, twist, and so on, during the transition. Figure 10.16 illustrates the situation for a diatomic molecule, simplified by the omission of the rotational energy levels. An excited state always lies above (higher energy) and to the right of a ground state (longer equilibrium bond distance). □ □

Singlet and triplet states.   Among other items in Figure 10.16 that require explanation are the terms singlet and triplet. The multiplicity of a molecule is given by $2S + 1$, where S = sum of m_s of all the electrons in the molecule (see Sec. 5.1). Because $m_s = \pm 1/2$, if all the electrons are paired $S = 0$, the multiplicity is one, and the state is designated as a singlet. If all the electrons are paired except two that have parallel spins, $S = \pm 1$, and the multiplicity is 3 or -1. Taking the positive value, the multiplicity is termed a triplet state. Excitation from a singlet state to a triplet is not allowed. Therefore the excited state in Figure 10.16 is that of a singlet. □ □

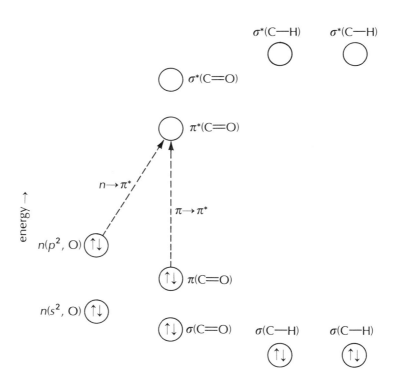

Figure 10.17   Schematic representation of the molecular orbitals and bonding electrons of formaldehyde and the $n \rightarrow \pi^*$ and $\pi \rightarrow \pi^*$ transitions. Only the outer-shell electrons are shown for carbon and oxygen. (The unshared electrons on oxygen may be formulated as being in sp^2 orbitals instead of s^2 and p^2 as shown.)

The σ, σ*, n, π, and π* states. Simplification is achieved in other energy diagrams by omission of the vibration levels and the triplet states. The ground state of formaldehyde is shown in Figure 10.17. The *n* orbitals are nonbonding and represent the 2p orbitals of oxygen occupied by electron pairs. Two of the possible absorption transitions are shown, $n \to \pi^*$ and $\pi \to \pi^*$, the latter requiring a higher frequency (more energy change). Other possible transitions at still higher frequencies are the $n \to \sigma^*$, $\pi \to \sigma^*$, $\sigma \to \pi^*$, and $\sigma \to \sigma^*$, all in the carbonyl (C=O) bond. The $\sigma \to \sigma^*$ absorption for a C—H bond is at such a high energy that it is outside the range of C=O absorptions. The $n(p^2, O)$ and $n(s^2, O)$ orbital energy levels do not differ significantly from the level in the isolated atomic orbitals. Whereas π is higher than σ in the ground state, σ* is greater than π* in energy in the excited state. Thus, any absorption leading to a π* state requires less energy (longer wavelength) than any absorption leading to a σ* state.

The $n \to \pi^*$ absorption is forbidden, and the molar absorptivity (ε_{max}) is weak. In other words, the absorption peak at 280 nm is small. In this usage, we should remember that forbidden means low probability. The $\pi \to \pi^*$ absorption is allowed, so ε_{max} is large, and the peak at 185 nm is huge. □ □

10.7 Interpretation of visible and ultraviolet spectroscopy

Chromophores. A grouping that gives rise to an absorption in the visible and near uv region is known as a chromophore. Thus we can speak of the C=O bond as a chromophore. Because we are limited to wavelengths above 200 nm and there are no $\sigma \to \pi^*$ nor $\sigma \to \sigma^*$ transitions in this region, we may say that a spectrum is observed only if a multiple bond is present. Therefore, saturated hydrocarbons, alkyl halides, alcohols, and ethers are transparent down to 200 nm because only σ bonds are present. The C=O bond in either an aldehyde or ketone shows a broad, weak band at 280 to 290 nm (ε_{max} 15 to 20), which, as we have seen, is the $n \to \pi^*$ transition. (See Figure 10.15 for acetone.) If the C=O bond is flanked by an electron-withdrawing group, the band shifts to a shorter wavelength (hypsochromic effect) with a larger ε_{max}. Thus, the ester ethyl acetate absorbs at 211 nm (ε_{max} 57); the amide acetamide absorbs at 220 nm (ε_{max} 63); the acid chloride acetyl chloride absorbs at 220 nm (ε_{max} 100); and acetic acid absorbs at 208 nm (ε_{max} 32).

An isolated C=C bond has some strong absorption peaks below 200 nm but is transparent at wavelengths above 200 nm because no $n \to \pi^*$ transition is available. The smallest possible energy shift is the $\pi \to \pi^*$.

Effect of conjugation. If two or more chromophores are present in a molecule and conjugation is absent, the spectrum is simply characteristic of each chromophore independent of the other. However, if the chromophores are conjugated (alternating multiple and single bonds as in 1,3-butadiene, CH_2=CHCH=CH_2, or 3-butene-2-one, CH_2=CHCOCH$_3$), an interaction between the π bonds in the molecule (resonance) leads to a pronounced decrease in the energy levels for both the $n \to \pi^*$ and $\pi \to \pi^*$ shifts (about 30 nm longer wavelength than normal) and to an increase in ε_{max}. Thus butadiene absorbs at 217 nm (ε_{max} 20 900) and 324 nm (ε_{max} 24). (See Table 10.4.)

Extended conjugation. If the conjugation is extended, the shift becomes more pronounced. Thus 1,3,5-hexatriene absorbs at 248 nm (ε_{max} 56,000, 257 nm (ε_{max} 79,000), and 267 nm (ε_{max} 68,000).

Table 10.4 Some electronic transitions of simple organic molecules [a]

Compound	Type	λ_{max}, nm	ε_{max}	Solvent [b]
$(CH_3)_2C=O$	$n \to \pi^*$	280	15	cyclohexane
	$\pi \to \pi^{*c}$	190	1,100	
	$n \to \sigma^{*c}$	156	strong	
$CH_2=CH_2$	$\pi \to \pi^*$	162	10,000	vapor
$CH_2=CH-CH=CH_2$	$\pi \to \pi^*$	217	20,900	hexane
$CH_3-CH=CH-CH=CH-CH_3$	$\pi \to \pi^*$	227	22,500	hexane
$CH_2=CH-CH_2-CH_2-CH=CH_2$	$\pi \to \pi^*$	185	20,000	alcohol
$CH_3-C\equiv CH$	$\pi \to \pi^*$	186	450	cyclohexane

$CH_2=CH-C=O$
             $\vert$
             CH_3
	$n \to \pi^*$	324	24	alcohol

CH_4	$\sigma \to \sigma^*$	122	strong	vapor
CH_3-CH_3	$\sigma \to \sigma^*$	135	strong	vapor
CH_3-Cl	$n \to \sigma^*$	172	weak	vapor
CH_3-Br	$n \to \sigma^*$	204	200	vapor
CH_3-I	$n \to \sigma^*$	258	365	pentane
CH_3-O-H	$n \to \sigma^*$	184	150	vapor
CH_3-O-CH_3	$n \to \sigma^*$	184	2,520	vapor
$(CH_3)_3N$	$n \to \sigma^*$	227	900	vapor

[a] From M. W. Hanna, *Quantum Mechanics in Chemistry*, © 1965 by W. A. Benjamin, Inc., Menlo Park, California.
[b] It is necessary to specify the solvent because λ_{max} and ε_{max} vary somewhat with solvent.
[c] These assignments are not certain.

The carotenoid β-carotene, the major constituent of the coloring matter of carrots, occurs widely in other representatives of the vegetable kingdom (green leaves) as well. It has a system of eleven conjugated double bonds, which absorb in the visible: 452 nm (ε_{max} 139,000) and 478 nm (ε_{max} 122,000).

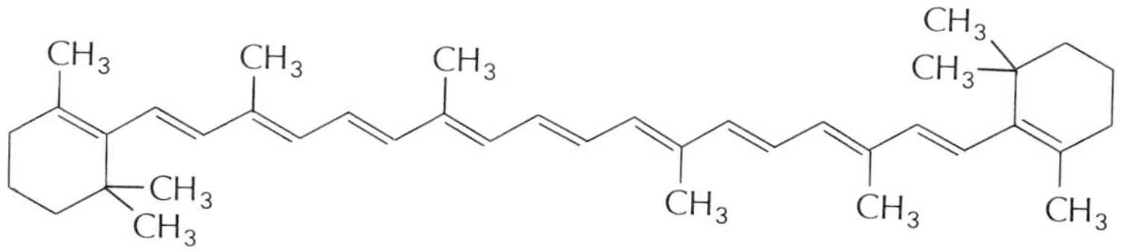

The ubiquitous green plant pigment necessary for photosynthesis, chlorophyll, is known in four structures. Chlorophyll a (Sec. 28.9) has absorption peaks in ether solution at 660, 613, 577, 531, 498, 429, and 409 nm. A solution in alcohol is blue-green with a deep red fluorescence.

Uses. Although visible-ultraviolet spectroscopy is not as widely useful as infrared because of the relatively few chromophores available to us so far, we shall find more applications in later chapters, when we discuss amino acid analysis, proteins, heterocycles, dyes, and pigments.

Finally, the quantitative use of visible-ultraviolet spectroscopy is readily accomplished because the path lengths and very dilute solutions (10^{-5} to 10^{-2} M) used allow us to apply the Beer-Lambert law with ease.

Extinction coefficients. Extinction coefficients are often used to measure the concentration of a particular biological compound. Nucleic acids of DNA and RNA are measured by the absorbance at 260 nm. Protein concentrations, on the other hand, are usually proportional to the absorbance at 280 nm. This peak at 280 nm is the sum of contributions from each of the aromatic side chains on the protein. The peptide bond itself absorbs at 215 nm, and so the concentration of even those proteins without aromatic side chains can be measured.

Enzyme-catalyzed oxidation-reduction reactions using NAD are readily monitored by a spectrophotometer. Thus the rate of appearance or disappearance of reduced NAD is rate of change in the absorbance at 340 nm. The flavin oxidation-reduction enzymes are even easier to observe because the flavin molecule absorbs in the visible part of the spectrum. In fact, the first flavin enzyme discovered, despite attempts to standardize enzyme nomenclature, is still often called by its earlier name, Old Yellow Enzyme.

10.8 Summary

We have by no means exhausted the subject of radiation spectroscopy. For example, we have not even mentioned the far-ultraviolet, X-ray, or gamma-ray effects. Our objective has been to build up our background in units, in terminology, and in understanding of the processes involved on the molecular scale in each region of the spectrum. The time spent on microwave spectroscopy served mainly as an introduction whereby selection rules, absorption processes, and other topics used later in ir and uv were brought up. The sections on ir and uv were the heart of the chapter. The material will be used extensively in later chapters.

There is no summary for this chapter. A valuable exercise is to make a list of new words encountered in this chapter with your own definitions and examples of usage.

No problems are given at the end of this chapter, because we shall want to use ir, uv, nmr, and mass spectra all together in the solution of structural and identification problems at the end of the next chapter.

11 Spectroscopy II

We continue our examination of the various types of spectroscopy in this chapter by going on to those forms that are operationally more complicated than those of the previous chapter. The interpretation of the results is, however, relatively simple in many instances.

11.1 Nuclear magnetic resonance spectroscopy   265

11.2 Theory of nuclear magnetic resonance   265
   1. Magnetic field alignment of certain nuclei.   2. Scanning.   3. Saturation; relaxation.   4. Shielding; deshielding; the chemical shift.   5. Scales.
   6. Integration: counting protons.   7. Summary.

11.3 Practical aspects of nuclear magnetic resonance   269
   1. The nmr spectrometer.   2. Operational techniques.   3. Solvents.
   4. Chemical shift reagents.

11.4 Spin-spin splitting   271
   1. Spin-spin splitting with nearby protons.   2. The coupling constant.
   3. The $n + 1$ rule.   4. Distortion of patterns.   5. Summary. Exercises.

11.5 Interpretation of nuclear magnetic resonance   276
   1. The A_2X_3 system.   2. Spin decoupling.   3. Pure ethanol.
   4. The AX_6 and $A_2B_2C_3$ systems.   5. The A_4X_2, $A_2B_2M_2$, and ABC systems.
   6. Acetylenic and ethylenic protons.   7. Chemical shifts.   8. Coupling constants.   9. Unsaturation centers. Exercises.

11.6 Electron spin resonance   292

11.7 Theory of mass spectrometry   294
   1. Gaseous ions and ion radicals.   2. Types of fragmentation.
   3. Rearrangement.   4. Base and parent peaks; isotope effects.   5. The nitrogen rule.

11.8 Practical aspects of mass spectrometry   299

11.9   Interpretation of mass spectrometry   300
       1. The mass spectrum of octane.   2. Frequently occurring ions and ion radicals.   3. Alkenes.   4. McLafferty rearrangement.   5. Alkyl halides.   6. Alcohols and ethers.   8. Ketones and aldehydes; acids and esters.

11.10 Summary   307

11.11 Problems   307

11.12 Bibliography   317

11.13 William S. Johnson and John D. Roberts: Early days with nuclear magnetic resonance   318a

11.1   Nuclear magnetic resonance spectroscopy

Nuclear magnetic resonance is a phenomenon observable only in the presence of a strong magnetic field. Because of the electronics and fine control needed for observation, the first signals were not seen until 1945. In 1949 it was found that resonance frequencies depended upon the compound being used. Thereafter, nuclear magnetic resonance spectroscopy (nmr) developed rapidly. The principal advantage of nmr to organic chemistry is that hydrogen nuclei give transition peaks, but the rest of the molecule (carbon, oxygen, nitrogen, sulfur and other atoms) is transparent. This effect has led to some use of the term proton magnetic resonance spectroscopy (pmr). Nuclear magnetic resonance is the more general term and, as we shall see, some other atoms do affect the proton signals and also may give rise to signals under the proper conditions. We shall use nmr instead of pmr to identify this type of spectroscopy.

11.2   Theory of nuclear magnetic resonance

Magnetic field alignment of certain nuclei.   Nuclear magnetic resonance is concerned with the effect of a magnetic field upon the spins of nuclei. Certain nuclei (for example, a proton) have a magnetic moment, accompanied by a magnetic field set up by the spinning charge. The nuclear magnetic field is directed along the axis of the spin. In effect, the spinning nucleus acts like a small bar magnet (Fig. 11.1). In the absence of a strong magnetic field, the spins are randomly oriented. If a molecule with hydrogen nuclei is placed in a strong magnetic field, the protons tend to be oriented with or against the field: they are symbolized as ↑ and ↓.

Quantum mechanics introduces the nuclear spin quantum number I and the statement that a spinning nucleus assumes $2I + 1$ spin orientations. It turns out that although I may have values such as $0, \frac{1}{2}, 1, \frac{3}{2}, 2\ldots$, the I for a particular nucleus must be determined by experiment. Certain trends are noticeable.

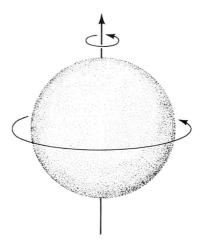

Figure 11.1 Spinning charge in proton generates magnetic dipole

1. If the numbers of protons and neutrons in a given nucleus both are even, $I = 0$, as in ^{4}He, ^{12}C, ^{16}O, ^{20}Ne. Presumably, the protons and neutrons both are paired and there is only $2(0) + 1 = 1$ spin orientation.

2. If the number of either protons or neutrons is even (the other number being odd), $I = \frac{1}{2}, \frac{3}{2}, \frac{5}{2}\ldots$. In ^{1}H, ^{13}C, ^{15}N, ^{19}F, $I = \frac{1}{2}$; these atoms have $2(\frac{1}{2}) + 1 = 2$ spin orientations; and when $I = \frac{3}{2}$, $2(\frac{3}{2}) + 1 = 4$ spin orientations are present.

3. If the number of protons and the number of neutrons both are odd, $I = 1, 2, 3, \ldots$, as in ^{2}H, ^{14}N, with $I = 1$; or ^{10}B, with $I = 3$.

Because we are concerned with ^{1}H, which has only two spin orientations in the presence of a strong magnetic field, our discussion is simplified considerably.

As the magnetic field is increased in strength, those protons with a spin orientation aligned with the field experience a slight decrease in energy, and those with a spin orientation opposed to the field undergo a slight increase in energy (Fig. 11.2). The energy difference

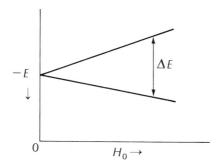

Figure 11.2 Spin energy-level separation for a nucleus with I of $\frac{1}{2}$ as a function of the applied field H_0

leads to a slight excess of the more stable spin state. For hydrogen in a magnetic field of 10,000 gauss, the distribution ratio is 0.99999:1.00001. Out of every 200,000 nuclei, there are only 2 more nuclei in the lower energy state than in the upper. □ □

Scanning. If the sample, immersed in a magnetic field, is bathed at a right angle to the field with radiation of the proper frequency, there is absorption of energy from the radiation. The absorption causes a spin flip from the lower energy level to the higher energy level. The loss in energy from the radiation is detected by what is essentially a radio antenna and receiver and displayed on a recorder. The frequency (at a preset magnetic field) is slowly increased through the range in which absorption will occur. Alternatively, the magnetic field strength may be slowly increased (at a preset frequency) through the range in which absorption occurs. The latter method is easier to control electronically, and most nmr spectrometers are operated in this fashion. The receiver needs to be tuned only to the preset frequency. However, it has become conventional to record on a frequency scale, no matter which scanning method is employed.

Unfortunately, absorption, which occurs only under the proper conditions of field strength and frequency, has also been called resonance, not to be confused with resonance of resonance theory. Resonance, as used here, refers to being in tune (as a radio set may be tuned to a station to be in resonance with the broadcast wave).

Saturation; relaxation. Returning to our sample, and remembering the very small population difference between the two energy levels, we can see that the absorption probability also is very small. The receiver must be carefully tuned to the preset frequency and must be capable of great amplification to detect the very small change in signal strength when absorption occurs. Also, since absorption depletes the excess population of the lower level and adds to the population at the upper level, saturation (populations equal in the two levels) would soon be achieved if there did not exist pathways whereby the nuclei in the upper level could return to the lower level. These relaxation processes are of two types, spin lattice and spin-spin relaxation. Spin lattice relaxation is a process whereby the excess energy of the upper level is dissipated among all the neighboring molecules as kinetic energy (heat). Spin-spin relaxation refers to the effect of a neighboring nucleus in taking up energy from a nucleus in the upper level.

Shielding; deshielding; the chemical shift. So far, we have introduced nothing to show that protons in differing parts of a molecule appear to absorb at differing frequencies on a recorder scale. A given nucleus sees the applied magnetic field through a haze of tiny fields set up by electrons and other nuclei close by. Some of these tiny fields enhance the field produced by the machine magnet; others detract. The major factor seems to be the effect of nearby electrons, which for protons in organic compounds means those electrons in the H—C, H—N, H—O bonds. If any influence within a molecule tends to cause the H—C bond to become more like $\overset{\delta\oplus}{H}$—$C^{\delta\ominus}$, the proton becomes deshielded. The opposite effect, the bond becoming more like $\overset{\delta\ominus}{H}$—$C^{\delta\oplus}$, is known as shielding. If we add up all the effects about a given proton and call this sum the shielding constant (σ), we may write Equation 11.1, in which H is the magnetic field felt by the proton, and H_m is the field set up

$$H = H_m(1 - \sigma) \tag{11.1}$$

by the machine. As we have seen, once a given fixed frequency (often 60 MHz) has been established by the machine, then H, the field that must be seen by the proton for absorption to occur, is invariant at 14,092 gauss. Thus, a proton that is shielded so that σ has a positive value requires H_m to be larger than otherwise in order for $H_m(1 - \sigma)$ to equal 14,092 gauss. Deshielding results in a requirement for H_m to be smaller than otherwise. So we come to speak of deshielding as giving a <u>downfield shift</u> and shielding as giving an <u>upfield shift</u>.

The appearance of nmr signals at various values of H_m gives rise to an nmr spectrum. However, because of difficulties in reproducing spectra, a reference standard is needed. Tetramethylsilane (TMS, $(CH_3)_4Si$) is universally used for this purpose: a drop or two is added to the sample or solution of the sample. All twelve protons in TMS are alike and a single sharp peak is observable. Silicon, which is less electronegative than carbon, allows an electron shift towards the carbons in the C—Si bonds, and the carbons in turn allow the protons to become shielded with electrons more than in most other organic compounds. Thus, σ is large. All other compounds with hydrogen present are deshielded with respect to TMS and give nmr signals downfield from TMS. The distance downfield from the TMS signal is known as the <u>chemical shift</u>.

In Figure 11.3, the nmr spectrum (for 60 MHz and 14,092 gauss, as are all nmr spectra in this text unless otherwise noted) of 1,2-dibromo-2-methylpropane $((CH_3)_2CBrCH_2Br)$ is shown. The largest peak is that given by the six identical protons in the two methyl groups. The medium peak is given by the two protons on C_1. The smallest peak, at the right, is given by the added TMS (the magnitude of this peak depends upon the amount of TMS added). The two protons on C_1 are deshielded more than the six on the methyl carbons because they are closer to an electron-withdrawing C—Br bond. Their signal appears farther downfield than does that for the six protons, which in turn are not as well shielded as are those in TMS.

Scales. Other features of Figure 11.3 that need explanation are the scales used. The magnitude of the chemical shift depends upon the strength of the magnetic field, as we have already seen. For a 60 MHz instrument, the change in magnetic field strength is represented at the top of the spectrum chart as the equivalent cps or Hz downfield from TMS. For another instrument, operating at 100 MHz, the separation of the peaks would be 100/60 times that of the 60 MHz machine if measured in Hz. To convert to a scale in-

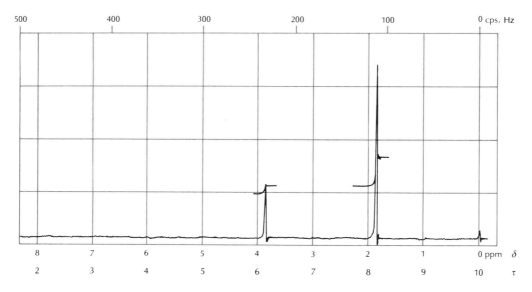

Figure 11.3 The nmr spectrum of $(CH_3)_2CBrCH_2Br$

dependent of the machine frequency, we divide the chemical shift in Hz from the TMS peak by the machine frequency and then multiply by 10^6 to avoid the continual writing of 10^{-6}. Thus, the largest peak (113 Hz) from Figure 11.3 becomes $113/(60 \times 10^6) = 1.88 \times 10^{-6}$. Multiplied by 10^6, this figure becomes 1.88 ppm (parts per million) and is known as δ. An alternate scheme for expressing chemical shifts is to use $\tau = 10 - \delta$. We shall use only δ hereafter for chemical shift values. Both δ and τ are scaled at the bottom of the chart.

Integration: counting protons. The fine lines that appear in Figure 11.3 above or crossing the nmr signal lines are from an integrator. They are used to measure the relative areas under the peak curves. Thus, for the large peak at δ 1.88 ppm, the integrator curve runs horizontally at about 11.2 and then rises suddenly and levels out at about 16.6. The rise, $16.6 - 11.2 = 5.4$ units, is proportional to the number of protons giving the signal. Examination of the integration of the peak at δ 3.9 shows a rise of $11.1 - 9.4 = 1.7$ units. From the molecular formula of the compound used, $C_4H_8Br_2$, we know that there are eight protons per molecule. Therefore, there are $(5.4 + 1.7)/8 = 0.89$ units per proton, and $5.4/0.89 = 6.07$ or 6 protons gave the signal at δ 1.88, and $1.7/0.89 = 1.91$ or 2 protons gave the signal at δ 3.9. In short, the area under a signal is proportional to the number of protons giving that signal. This correlation is a unique advantage of nmr.

Summary. To summarize for the moment, nmr allows us 1. to look at the hydrogen nuclei (protons) in a compound, 2. to see separately those protons with different environments within the molecule (chemical shift), and 3. to count the protons in each environment (integration). These tools alone would be valuable enough, but we shall see later on that there is still more information to be obtained from nmr.

11.3 Practical aspects of nuclear magnetic resonance

The nmr spectrometer. An nmr spectrometer is a complex and relatively expensive piece of equipment. Figure 11.4 is a schematic diagram of an nmr spectrometer. The sample liquid or solution is contained in a thin-walled, narrow tube, which is spun about its vertical axis. The spinning averages out any inhomogeneities in the magnetic field and in the sample.

The magnet of either the permanent or the electromagnetic type produces the intense magnetic field through the sample along a horizontal axis. On the pole faces of the magnet are a few turns of a coil used to vary the strength of the field: hence the name sweep coils. The sweep generator supplies the current to give a slow, linear field increase.

The R-F (radiofrequency) transmitter supplies the steady 60 MHz signal to the broadcast antenna (transmitter coil) mounted at right angles to both the other axes. The receiving antenna (receiver coil) is placed around the sample tube so as to be at right angles to both the transmitter coil and the magnetic field. The nmr signals are detected and amplified by the R-F receiver. The recorder displays the amplitude of the signal in a vertical direction and the field strength (in Hz) increasing to the right along the horizontal direction.

Operational techniques. The actual instrument is far more complicated, because many interlocking devices are needed to provide stability. In addition, coarse and fine controls are included to adjust the sweep time, length of sweep in Hz, the

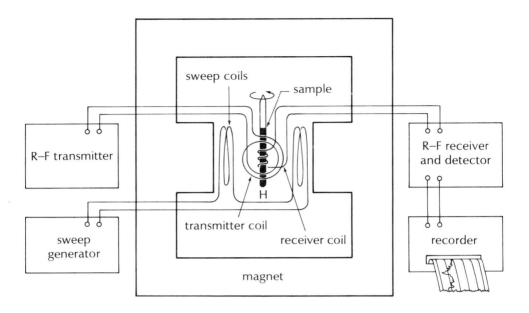

Figure 11.4 Schematic diagram of an nmr spectrometer (courtesy of Varian Associates)

zero of sweep to correspond with the TMS signal, the height of the baseline on the chart, the signal amplification, the strength of the R-F field, the R-F frequency, the integrator, and so on. Accessories are available for controlling the temperature of the sample and that of the magnet and for other purposes too complicated to be included here. We make one exception and mention the CAT (computer averaging of transients). The CAT is a small computer that controls the instrument so as to repeat a scan of a signal or spectrum up to a 100 or more times. The computer adds the results as they accumulate and, on command, displays the result on the recorder. This technique is especially valuable when the signal is very weak (very dilute solution or other cause) and the electronic noise arising from extreme amplification and other causes obscures the signal. Noise, being random, tends to cancel out as the repetitive scans are added, whereas the weak signal adds up to become stronger and stronger.

Solvents. Neat liquids (undiluted pure substances) require little amplification to give sharp, strong signals. Solids must be observed as solutions. Carbon tetrachloride is a widely used solvent. Other solvents in which deuterium has replaced hydrogen are also used: $CDCl_3$, CD_3COCD_3, CD_3COOD, D_2O, $(CD_3)_2SO$, $CD_3C{\equiv}N$, and so on. At times, a solvent containing hydrogen may be used if it is known that the strong signals from the solvent will not cover any signals from the solute. Solutions used usually are of about 10% concentration. Sometimes a solvent (usually a polar one) causes a small shift (1 to 5 Hz) of some of the peaks from the position in a nonpolar solvent.

Chemical shift reagents. Recently, complexes of certain metals of the lanthanide series (rare earths), which act as Lewis acids toward such classes of compounds as alcohols, ethers, epoxides, esters, and others, have been found to modify strongly the chemical shift values of such compounds. Such chemical shift reagents are useful in the separation of peaks that normally fall so close together that identification and integration are difficult.

11.4 Spin-spin splitting

Spin-spin splitting with nearby protons. Before going ahead, we should introduce the subject of spin-spin splitting (sometimes called spin-spin coupling). If we examine the nmr spectrum of 1,1,2-tribromoethane (Br_2CHCH_2Br, Fig. 11.5), we find a set of three peaks centered at δ 5.76 ppm and a set of two peaks centered at 4.15 δ ppm. But we also note that there are only three hydrogen nuclei in the molecule: one on C_1 (b) and two on C_2 (a). As we might deduce, the single proton on C_1 is more deshielded by the electron-withdrawing effect of the two bromine atoms than are the two protons on C_2 by the effect of only one bromine atom. Therefore, the triplet (set of three peaks) at b is the signal from C_1—H, and the doublet (set of two peaks) at a is the signal from the two like protons on C_2. The triplet and doublet are the observable result of spin-spin coupling between the single proton and the two like protons.

The term AX system is a way of designating the coupling between two protons that are far apart in chemical shift. An AB system is one in which the two chemical shifts are close together. An intermediate case is designated as AM. In any case, the proton most downfield (largest δ) is designated A. The number of protons is designated by subscripts as usual. Thus, $BrCH_2CH_3$ gives an A_2X_3 pattern, $(CH_3)_2CHBr$ an AX_6 pattern, CH_3CHO an AX_3 pattern, and $CHBr_2CH_2Br$ an AX_2 pattern.

The coupling constant. We said earlier that a given nucleus sees the applied magnetic field H_m through a haze of tiny fields set up by electrons and other nuclei close

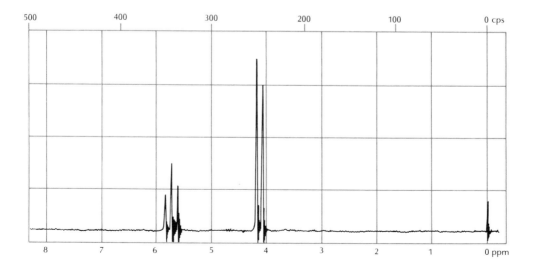

Figure 11.5 Nmr spectra: splitting of signals for 1,1,2-tribromoethane

by. We have already discussed the effect of electrons (or rather, electronegativity effects transmitted by electrons) on the chemical shift (δ). Now we come to the effects of nearby protons.

A proton in a given molecule may be oriented with ($\uparrow$) or against ($\downarrow$) the machine field. As we have seen, almost exactly half the molecules have the proton oriented with the field and half have the proton oriented against the field. If the proton is oriented with the field, the tiny magnetic field set up by the proton reinforces the field seen by a nearby proton. If the orientation is against the machine field, the field seen by the neighbor is decreased. The first proton is similarly affected by the possible orientations of the neighbor. Because the neighboring proton effect is the same in both directions, all the protons involved feel an identical effect. If we set up a diagram (Fig. 11.6), we see that the nmr

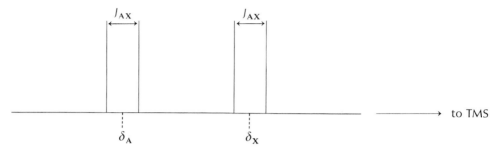

Figure 11.6 The AX system coupling. In order for absorption (resonance) of proton A to occur: a. Proton X is $\uparrow$ in 50% of the molecules and reinforces H_m as seen by A. Resonance (absorption) of A in these molecules occurs at a point ($\frac{1}{2}$)J less than the chemical shift of A. b. Proton X is $\downarrow$ in 50% of the molecules and reduces H_m as seen by A. Resonance (absorption) of A in the molecules occurs at a point ($\frac{1}{2}$)J more than the chemical shift of A. In order for absorption (resonance) of proton X to occur: a. Proton A is $\uparrow$ in 50% of the molecules and reinforces H_m as seen by X. Resonance (absorption) of X in these molecules occurs at a point ($\frac{1}{2}$)J less than the chemical shift of X. c. Proton A is $\downarrow$ in 50% of molecules and reduces H_m as seen by X. Resonance (absorption) of X in these molecules occurs at a point ($\frac{1}{2}$)J more than the chemical shift of X.

spectrum of an AX two-proton system shows proton A as split by X into two peaks, each of which is ($\frac{1}{2}$)J from the chemical shift position of A. Likewise, proton X appears as a doublet, each peak ($\frac{1}{2}$)J from the chemical shift position of X.

J is defined as the coupling constant. Because J is a characteristic of the situation within the molecule, it is independent of the instrument frequency or field. J is relatively constant for certain atomic systems within a molecule and serves as an aid in identification, as we shall soon see. J is always evaluated in Hz and is seldom greater than 14 Hz in organic compounds.

The n + 1 rule.   We used an AX system for illustration with a purpose in mind. If the chemical shifts of the two protons are identical, they are in the same magnetic environment, the effect of J disappears, and there is no splitting pattern. Returning to the discussion of Figure 11.3 for a moment, note that we specified six identical protons in the two methyl groups. The six protons all have the same chemical shift value and do not split each other. However, multiple protons have a multiple effect upon neighboring protons that have different chemical shift values.

11.4 Spin-spin splitting 273

Now we return to Figure 11.5 and 1,1,2-tribromoethane, an example of an AX$_2$ system. The identical two X protons (a) see A as ↑ in half the molecules, as ↓ in the other half. Thus the X$_2$ part of the spectrum appears as a doublet with J_{AX} about 7 Hz. The A proton (b) in a single molecule sees the effect of the two X protons in one of four ways: ↑↑, ↑↓, ↓↑, or ↓↓. Thus, in 25% of the molecules, the two X protons are ↑↑; in another 25%, ↑↓; and so on. The effect of ↑↑ on A is $(\frac{1}{2})J + (\frac{1}{2})J$, or J. The effect of ↑↓ on A is $(\frac{1}{2})J - (\frac{1}{2})J$, or 0. The effect of ↓↑ on A is $-(\frac{1}{2})J + (\frac{1}{2})J = 0$. The effect of ↓↓ on A is $-(\frac{1}{2})J - (\frac{1}{2})J = -J$. Therefore, 50% of the molecules show no effect on A and give resonance at the chemical shift, 25% give resonance at J less than the chemical shift, and 25% give resonance at J greater than the chemical shift. The result is a triplet whose central peak is twice as large as the outer peaks, each of which is J distant from the center, about 7 Hz. One neighbor splits a signal

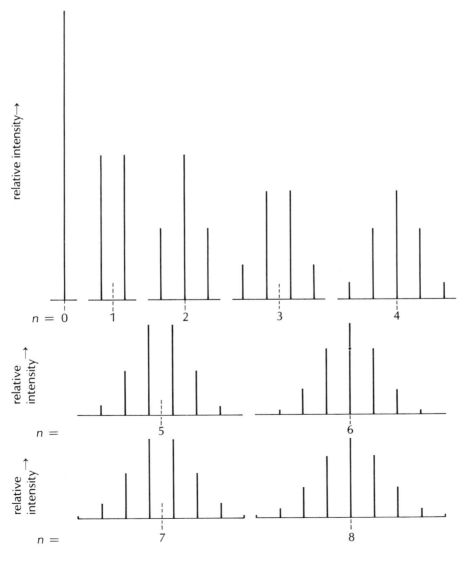

Figure 11.7 Splitting and intensity patterns for $n = 0$ to 8, $J = 10$ Hz

into two peaks. Two neighbors split a signal into three peaks. Thus *n* identical neighbors split a peak into n + 1 peaks.

The intensities (areas, approximate heights) of the n + 1 peaks are in the ratio of the coefficients of $(a + b)^n$ when expanded. The total area is unchanged when splitting occurs. In Figure 11.7, the relative intensities are shown for the given values of *n*. As *n* increases, the intensities of all the peaks diminish, and the outermost peaks decrease to such an extent that only close inspection of a given spectrum shows the difference between n = 5 or 7, or n = 6 or 8. However, the problem seldom arises.

Distortion of patterns. A useful fact is that the nmr spectral pattern of splitting is always distorted from the ideal patterns shown in Figure 11.7. The peaks in a splitting pattern always are more intense on the side closer to the pattern that is doing the splitting (Fig. 11.8). The figure shows the result of change in *J* for an AX system where the chemical shift difference has been held constant at 100 Hz, as calculated from the theoretical equations. With *J* = 0 Hz, there is no splitting and each proton gives a single peak (relative height = 2) at the assumed chemical shift values of 300 and 400 Hz. At *J* = 10 Hz, there is doublet splitting, but with the distortion already present (inner peaks larger than outer peaks). At *J* = 20 Hz, the distortion is greater. At $J = (1/\sqrt{3})(100) = 57.7$ Hz, the distortion has reached such a point that the spacing is equal between all four peaks and the intensitites reproduce the 1:3:3:1 pattern of a three-neighbor splitting pattern (compare Figure 11.7). As *J* continues to increase, the inner peaks come closer to each other and increase in intensity. The outer peaks decrease and recede. Ultimately, there is a single peak (relative height = 4) and no outer peaks. The critical factor to note is the magnitude of the ratio

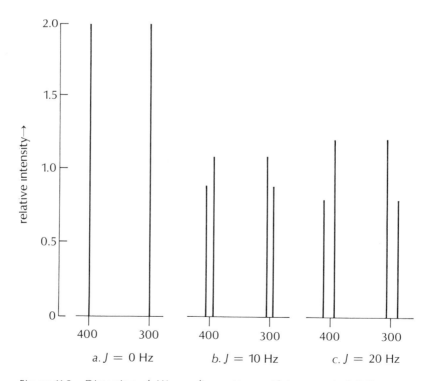

Figure 11.8 Distortion of AX coupling pattern with increase in *J*. Difference in chemical shift = 100 Hz

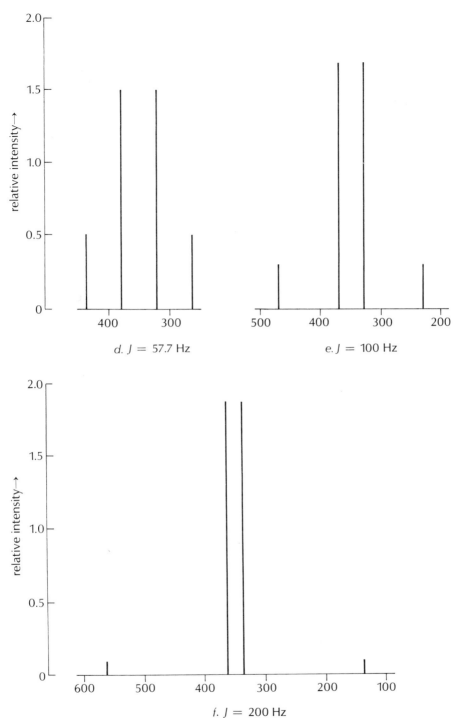

Figure 11.8 (continued)

of J to the difference in chemical shift values, which must be small (less than 0.2) to give the recognizable pattern of doublets. The patterns shown in Figure 11.8 would also occur if we had used a constant value for J and varied the chemical shift difference from a large number toward zero. ☐ ☐

276   Chapter eleven

Summary.   To summarize this section, we shall group the points to remember as follows:
1. The protons in a given molecule tend to fall into one or another of several different arrangements, such as AB, AX, AB_2, and so on.
2. If the chemical shift difference between neighboring protons is much greater than the coupling constant (J) the splitting follows the $n + 1$ rule.
3. If the ratio of J to the chemical shift difference is not small, the $n + 1$ rule no longer holds.
4. The inner peaks of a pattern are nearly always more intense than the outer peaks.

Exercises

11.1 Predict the nmr splitting patterns and relative areas to be expected from the compounds listed.
   a. CH_3CH_2COCl
   b. $(CH_3)_3CCl$
   c. $CH_3COOCH_2CH_3$
   d. $BrCH_2CH_2CH_2Br$
   e. $(CH_3)_2NCH_2CH_3$
   f. $CH_3CHBrCHBr_2$
   g. $(CH_3CH_2)_2CHOCH(CH_2CH_3)_2$
   h. 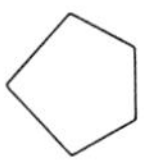 cyclopentane
   i. 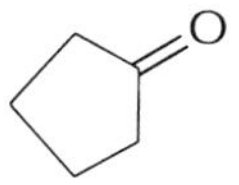 cyclopentanone
   j. 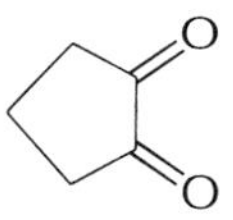 1,2-cyclopentanedione

11.5   Interpretation of nuclear magnetic resonance

The A_2X_3 system.   The nmr spectrum of ethyl chloride is shown in Figure 11.9a. The A_2X_3 system is clearly obvious. The methylene protons (A_2) are deshielded so as to appear at δ 3.57 ppm and exhibit the 1:3:3:1 quartet splitting pattern caused by the three identical methyl protons (X_3). The methyl protons, being shielded more than the methylene protons, appear at δ 1.48 ppm as a 1:2:1 triplet caused by the neighboring methylene protons, $J_{AX} = 7.0$ Hz. Ethyl bromide exhibits a similar picture at δ 3.43 and δ 1.67 ppm; ethyl iodide has δ 3.20 and δ 1.83 ppm. As expected, the decreasing electronegativity in the sequence Cl > Br > I gives a decreasing δ for the methylene protons. However, δ increases for the methyl protons in this same sequence, which remains something of a puzzle.

The small peak that appears at δ 7.27 ppm in many spectra is caused by the presence of a small amount of chloroform ($CHCl_3$) present in the solvent $CDCl_3$ when it is used.

Spin decoupling.   Ethanol in dilute solution (Fig. 11.9b) gives the same ethyl picture, this time with δ 3.70 and δ 1.22 ppm. In addition, the proton in the —OH group appears at δ 2.58 ppm as a rather broad peak. Because of hydrogen bonding and rapid transfer (about 0.0008 sec) of a proton from one molecule to another, coupling between

11.5 Interpretation of nuclear magnetic resonance 277

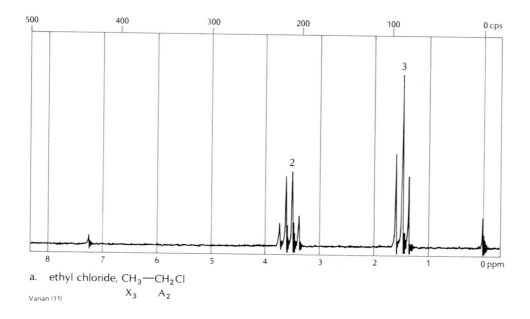

a. ethyl chloride, CH_3—CH_2Cl
   X_3 A_2

Varian (11)

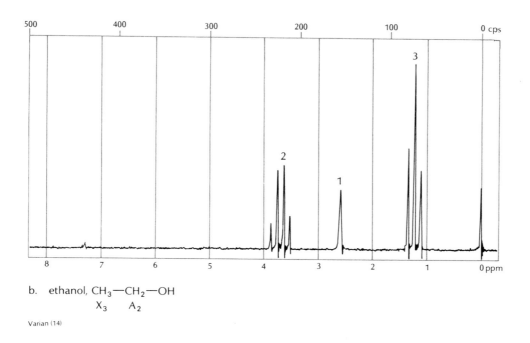

b. ethanol, CH_3—CH_2—OH
   X_3 A_2

Varian (14)

Figure 11.9 The nmr spectra of a. ethyl chloride, b. ethanol, and c. methanol. The numerals above the patterns represent the relative areas.

278   Chapter eleven

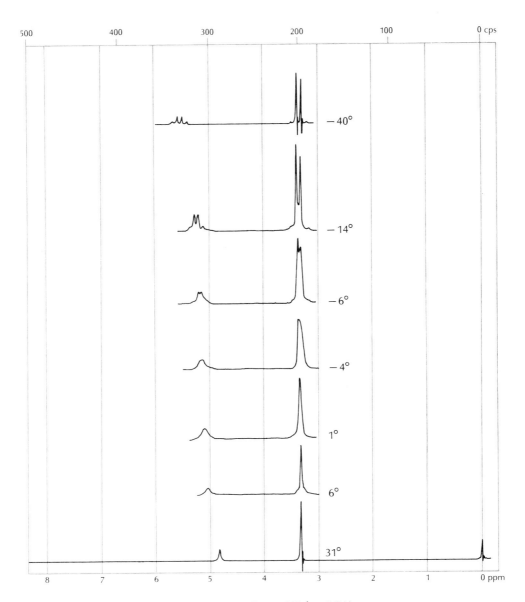

c. methanol, neat, at varying temperatures, CH$_3$—OH, J = 5.2 Hz
   A$_3$ X

Figure 11.9 (continued)

—OH and —CH$_2$— does not occur. We say that the exchange causes spin decoupling. If the rapid transfer can be slowed down enough so that a given proton remains on a given molecule long enough (about 0.003 sec), coupling is observed. One way of achieving this situation is to lower the temperature.

Figure 11.9c shows that pure methanol at 31°C gives only two sharp peaks, at δ 4.85 ppm for the hydroxylic proton and δ 3.33 ppm for the methyl protons. As the temperature is lowered, the position of the hydroxylic proton shifts to higher δ and rapidly broadens. At -4°C an indication of splitting begins to appear. A decrease of only 2°C more reveals splitting in both peaks. Finally, at -40°C, the complete AX$_3$ pattern of a sharp quartet ($J = 5.2$ Hz, now at δ 5.50 ppm) of the hydroxylic proton coupled to the doublet of the three methyl protons (still at δ 3.33 ppm) is seen.

As we have shown, the location of a hydroxylic proton (chemical shift) varies widely with solvent and with concentration. Unless special care is taken, it appears as a single peak.

Pure ethanol. Now let us look at pure neat ethanol (Figure 11.10a), in which the hydroxylic proton coupling is observable as a triplet (split by the neighboring methylene group) at δ 5.28 ppm. The methyl protons at δ 1.17 ppm show the expected triplet of an AM$_2$X$_3$ pattern. The methylene protons at δ 3.62 ppm, however, now consist of eight peaks, clearly seen in the expansion of the trace in the upper left corner. What is happening is a splitting of a split pattern. The values are far enough apart so that we can approximate the system at δ 3.62 ppm for the M$_2$ protons with $J_{AM} = 5.0$ Hz and $J_{MX} = 7.2$ Hz, as shown in Figure 11.10b. The first splitting, using J_{MX}, gives the quartet with reduced intensities but the

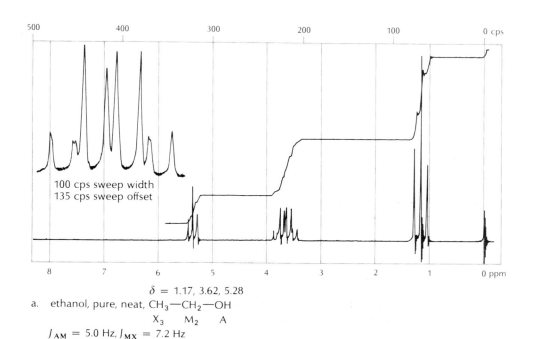

Figure 11.10 a. The nmr spectrum of pure 100% ethanol, and b. a splitting chart of the M$_2$ protons

280 Chapter eleven

b.

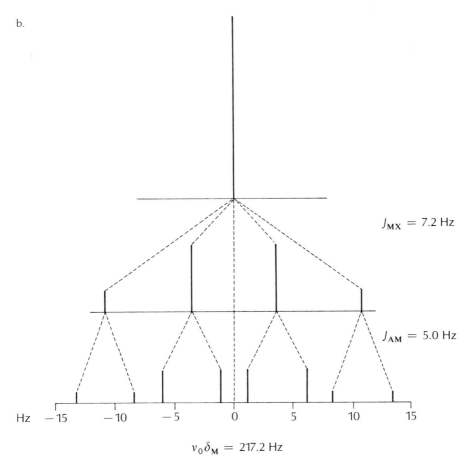

$J_{MX} = 7.2$ Hz

$J_{AM} = 5.0$ Hz

$v_0 \delta_M = 217.2$ Hz

Figure 11.10 (continued)

same total area. The second splitting, using J_{AM}, causes each line of the quartet to divide into two lines of reduced but equal height. The final octet clearly reproduces the result seen in Figure 11.10a. Note that if J_{AM} and J_{MX} had been equal, lines 2 and 3, 4 and 5, and 6 and 7 would have coalesced to give a five-line spectrum of a 1:4:6:4:1 pattern, characteristic of four identical neighboring protons. Such coincidence occurs at times in nmr spectra of other compounds.

The AX_6 and $A_2B_2C_3$ systems. Isopropyl alcohol (Fig. 11.11a), an example of an AX_6 system, shows the A proton at δ 4.00 ppm split by the six identical neighboring protons into a septet. As we have mentioned before, the outermost peaks in such an extensive splitting pattern are so weak that they are almost invisible unless amplified. The six X protons at δ 1.20 ppm appear as a strong doublet because there is only the single A neighbor ($J_{AX} = 7.0$ Hz). Just as the quartet-triplet is characteristic of an ethyl group, so the septet-doublet is characteristic of an isopropyl group. This time the hydroxylic proton appears at δ 1.60 ppm, another example of the variability of the chemical shift of a hydroxylic proton.

The propyl group, an example of an $A_2B_2C_3$ system, is seen in Figure 11.11b, the nmr spectrum of propyl iodide. The A_2 triplet (δ 3.17 ppm), split by B_2 ($J_{AB} = 6.8$ Hz), is the most deshielded set. The C_3 triplet (δ 1.02 ppm), split by B_2 ($J_{BC} = 7.3$ Hz), is the most shielded

11.5 Interpretation of nuclear magnetic resonance 281

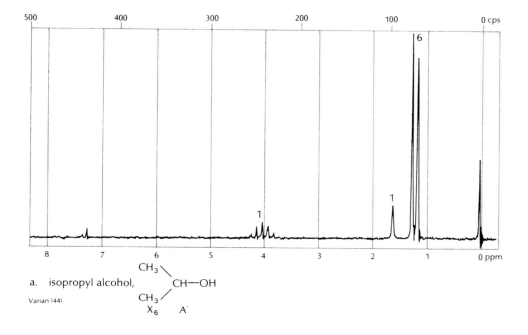

a. isopropyl alcohol, $\underset{\underset{X_6}{CH_3}}{\overset{CH_3}{\diagdown}} CH-OH$

Varian (44)

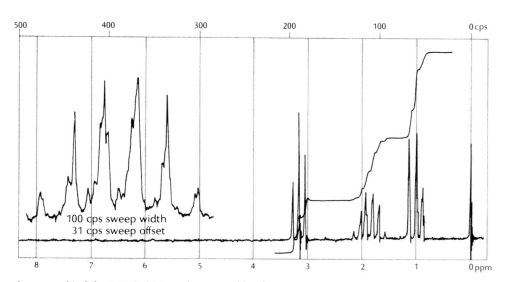

b. propyl iodide, 15% (w/v) in carbon tetrachloride,
 δ 1.02, 1.86, 3.17 ppm
 $\underset{C_3}{CH_3}-\underset{B_2}{CH_2}-\underset{A_2}{CH_2}-I$ $J_{AB} = 6.8$ Hz, $J_{BC} = 7.3$ Hz

Figure 11.11 The nmr spectra of a. isopropyl alcohol and b. propyl iodide; and c. the splitting chart of B_2 of the propyl $A_2B_2C_3$ system

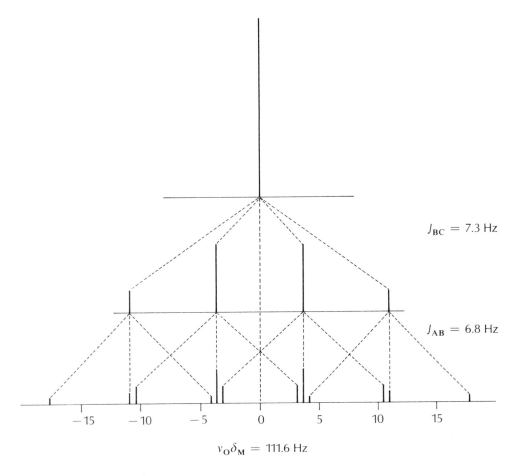

$J_{BC} = 7.3$ Hz

$J_{AB} = 6.8$ Hz

$\nu_0 \delta_M = 111.6$ Hz

Figure 11.11 (continued)

set. The pattern of B_2 (δ 1.86 ppm), is barely interpretable in terms of the $n + 1$ rule. The quartet formed by coupling with C_3 is coupled to A_2 as well to give a triplet of each line of the quartet. The result is a twelve-line pattern. However, some of the lines are so close together that there is partial consolidation into only six lines. Under increased amplification of the signal and scale expansion, a partial resolution may be seen (on the left of Fig. 11.11b). Comparison with the chart in Figure 11.11c shows how the complex pattern arises because of the quartet and triplet splittings. Extra complexity occurs in the actual spectrum because of increased splitting as a result of a sizable $J/(\Delta \nu_0 \delta)$ ratio.

The A_4X_2, $A_2B_2M_2$, and ABC systems. Figure 11.12a (1,3-dichloropropane, $ClCH_2CH_2CH_2Cl$) shows the simplification in the spectrum when the system is made more symmetrical, an A_4X_2 system. Note that the two protons on C_1 are identical to those on C_3. Thus each of the two pairs of protons is split into identical triplets (δ 3.70 ppm, $J_{AX} = 6.25$ Hz) that are superposed to give a single large triplet. The $n + 1$ rule predicts five lines for the X_2 protons (δ 2.20 ppm), which are distorted somewhat from the 1:4:6:4:1 pattern, but are still recognizable.

Unsymmetrical substitution, as in 1-bromo-3-chloropropane ($ClCH_2CH_2CH_2Br$), an $A_2B_2M_2$ system (Fig. 11.12b), has little effect on M_2 (δ 2.28 ppm) because $J_{AM} = J_{BM} =$

11.5 Interpretation of nuclear magnetic resonance   283

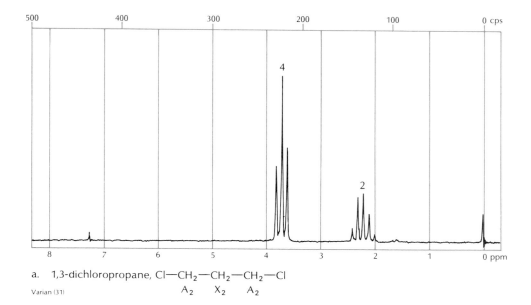

a. 1,3-dichloropropane, Cl—CH$_2$—CH$_2$—CH$_2$—Cl
                         A$_2$ X$_2$ A$_2$

Varian (31)

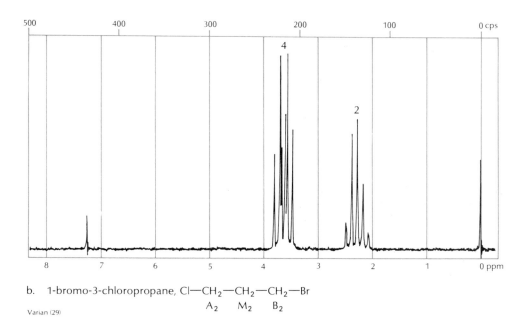

b. 1-bromo-3-chloropropane, Cl—CH$_2$—CH$_2$—CH$_2$—Br
                              A$_2$ M$_2$ B$_2$

Varian (29)

Figure 11.12 The nmr spectra of a. 1,3-dichloropropane, b. 1-bromo-3-chloropropane, and c. propylene oxide

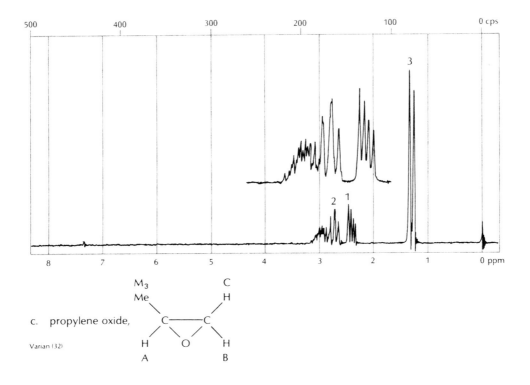

c. propylene oxide,

Varian (32)

Figure 11.12 (continued)

6.25 Hz. A_2 (δ 3.70 ppm) and B_2 (δ 3.55 ppm) give separate but equal triplets by coupling with M_2. The two triplets overlap each other because $\Delta v_0 \delta = (3.70 - 3.55)(60) = 9$ Hz. On a 100 MHz machine, $\Delta v_0 \delta = 15$ Hz, and the triplets would be separated.

Now let us look at 1,2-epoxypropane (or propylene oxide, $CH_3CH\overset{O}{\overset{\diagup\diagdown}{\text{———}}}CH_2$, Fig. 11.12c), an ABC system complicated by an X_3 that couples only with A. The protons B and C are not identical. B is on the same side of the three-membered ring as is A (B is cis to A) and trans to the methyl group. C is trans to A and cis to the methyl. Thus, B and C experience different environments and exhibit different chemical shifts. B (δ 2.72 ppm) is split to a doublet by C (J_{BC} about 5 Hz), and the doublet is again split by A (J_{AB} about 4.5 Hz) to give the awkward-looking triplet (the inner lines of the doublets almost coincide). C (δ 2.42 ppm) is split to a doublet by B, which is split by A (J_{AC} about 2.5 Hz) to give what would be four equally spaced and equally intense peaks except for the usual distortion. M_3 couples with A (J_{AM} about 5 Hz) to give the strong doublet at δ 1.32 ppm. Thus A (δ 2.98 ppm) is coupled with B (a doublet), with C (a doublet), and with X_3 (a quartet), or (2)(2)(4) = 16 lines if there are no coincidences. The actual result (Fig. 11.12c) is a mass of small peaks. □ □

Acetylenic and ethylenic protons. The protons about a double bond would be expected to show some deshielding because of the increased electronegativity of the atom to which they are attached (sp^2 is more electronegative than sp^3 because of increased s character). Similarly, we would expect an acetylenic proton to be even more deshielded; but if we examine Figure 11.13a, we find that the ethylenic protons are surprisingly far

11.5 Interpretation of nuclear magnetic resonance 285

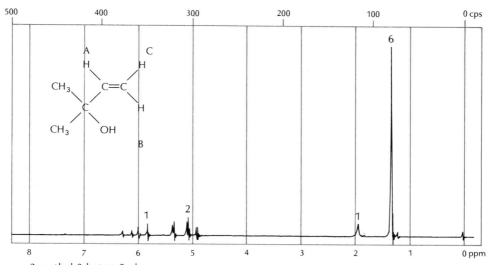
a. 2-methyl-3-buten-2-ol

Varian (444)

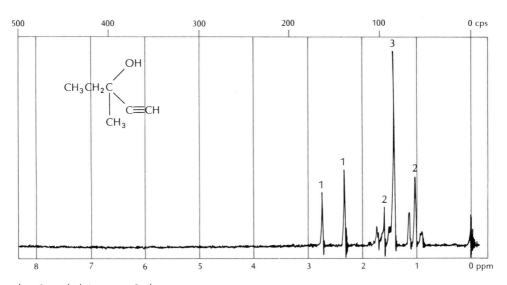
b. 3-methyl-1-pentyn-3-ol

© 1969 Sadtler Research Laboratories, Inc. (6032M)

Figure 11.13 The nmr spectra of a. 2-methyl-3-buten-2-ol and b. 3-methyl-1-pentyn-3-ol

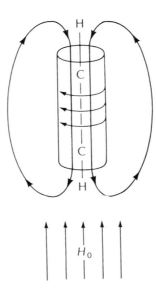

a. shielding of an acetylenic proton, H—C≡C—H

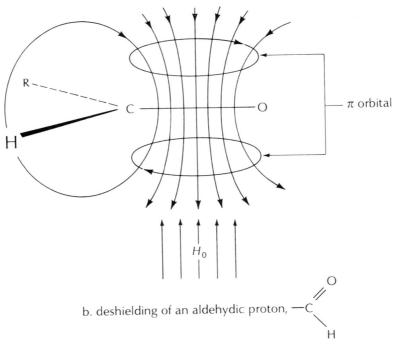

b. deshielding of an aldehydic proton, —C(=O)H

Figure 11.14 a. Shielding of acetylenic protons, b. deshielding of an aldehydic proton, and c. the nmr spectrum of acetaldehyde

downfield as an ABC system (δ 6.04, 5.20, 4.99 ppm) in 3-methyl-1-buten-3-ol. The singlets at δ 1.32 and δ 1.92 ppm obviously are the 6 identical methyl protons and the hydroxylic proton. We shall not go into the coupling pattern at this point.

Looking at Figure 11.13b, 3-methyl-1-pentyn-3-ol, we find nothing beyond δ 2.8 ppm.

11.5 Interpretation of nuclear magnetic resonance   287

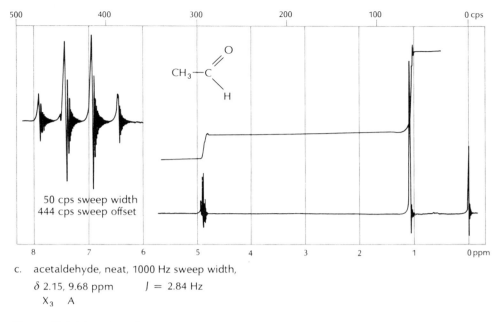

c. acetaldehyde, neat, 1000 Hz sweep width,
    δ 2.15, 9.68 ppm   J = 2.84 Hz
    X₃   A

Figure 11.14 (continued)

Evidently the acetylenic proton is at δ 2.35 or δ 2.72 ppm. The sharpness of the peak at δ 2.35 ppm and the broadening of the peak at δ 2.72 ppm leads to the assignment of —C≡C—H to the former and of —O—H to the latter. Something other than the deshielding caused by the inductive effect must be operating.

The explanation is as follows. The electrons in a π bond are loosely held as compared to those of a σ bond. The magnetic field applied to the sample by the instrument encounters the individual molecules in all possible orientations. The charged electrons respond to the magnetic field by circulating to produce a local magnetic field that reinforces (paramagnetic effect) or opposes (diamagnetic effect) the applied magnetic field in the region of the electronic field (the bond orbital). In Figure 11.14a, we see the induced diamagnetic effect upon neighboring protons as the lines of the magnetic field are closed. In acetylene, the protons are immersed in the opposing field and require a stronger machine field (upfield toward TMS, about δ 2 to δ 3) to reach resonance. In aldehydes and alkenes, the protons are immersed in the local induced field at a point at which the machine field is reinforced. They require a smaller machine field (downfield, about δ 2 to δ 3) to reach resonance (Fig. 11.14). Thus, ethylenic protons appear more downfield than expected, and acetylenic protons appear more upfield than expected. The aldehydic proton (Fig. 11.14c), under the influence of a strong inductive effect as well as the diamagnetic effect from the double bond, is pushed all the way to δ 9.68 in acetaldehyde (CH₃CHO). (The scale is 1000 Hz, not the usual 500 Hz. The splitting patterns are almost invisible unless the scale is expanded.)

Acidic protons (RCOOH) appear far downfield (δ 10 to δ 13 ppm), not only because of inductive and field effects but also because hydrogen bonding exerts a downfield (deshielding) effect. □ □

Chemical shifts.   In Table 11.1 are given some approximate values of δ for some of the more common groupings encountered in nmr spectroscopy. (Aromatic compounds will be described in Chapter 14.) We see that a given group is reasonably consistent in chemical shift, especially if the protons are not bound to the carbon atom adjacent to the perturbing group. For example, the methyl protons in CH₃CH₂CH₂— show a chemical shift of δ 0.9 to δ 1.0. In CH$_3$CH$_2$—, the values range from δ 1.0 to δ 1.7, and in CH$_3$—, δ 1.6 to δ 4.3. Thus the closer a given proton is to the perturbing group, the more downfield (larger δ) is the shift. Also, for a given location, a methine proton $\left(\begin{array}{c}\diagdown\\-CH\\\diagup\end{array}\right)$ appears at a larger δ than do the methylene protons $\left(\begin{array}{c}\diagdown\\\diagup CH_2\end{array}\right)$; and the methylene protons in turn appear at a larger δ than do methyl protons (—CH$_3$). For example, compare (CH$_3$)$_2$CHOH at δ 3.9 with CH$_3$CH$_2$OH at δ 3.6 or CH$_3$OH at δ 3.3. An interesting regularity appears when we compare the proton chemical shifts just mentioned in the correspond isopropyl, ethyl, and methyl esters; δ 5.0, 4.1, and 3.6. As we go from alcohol to ester, the methine proton shift is increased by 5.0 − 3.9 = 1.1 ppm, methylene by 4.1 − 3.6 = 0.5 ppm, and methyl by 3.6 − 3.3 = 0.3 ppm. Thus, those protons with the larger δ increase still more as we progress from alcohol to ester.

If more than one perturbing group is present in a molecule, we must not assume that the total effect is the sum of the group effects. There is a downfield shift of the affected protons, but the extent of the shift is variable. Predictive methods have been devised for calculating such cases, but we shall not go into them here.

The presence of a double bond in a molecule is easily recognized because it causes some complexities in nmr spectra (Fig. 11.15). Alkenyl protons are deshielded so that most appear at δ 5.0 ± 0.6 ppm, a region not inhabited by other types of protons (unless strongly deshielded by multiple substitution, as in Cl$_2$CHCOOCH$_2$CH$_3$, with δ 5.9 ppm). In the RCH=CH$_2$ molecule, the proton *trans* to R usually has the smallest δ, the *cis* proton usually has δ a bit larger, and the *vic* proton (from *vicinal*, bonded to the same carbon as the

Table 11.1 Some chemical shift values (δ, ppm)

	CH$_3$—	CH$_3$—CH$_2$—		CH$_3$—CH$_2$—CH$_2$—			(CH$_3$)$_2$CH—	
RF	4.3	1.6	4.5	0.9	1.8	4.3		4.8
RCl	3.0	1.5	3.5	0.9	1.8	3.3		4.0
RBr	2.7	1.7	3.4	1.0	1.9	3.3	1.7	4.2
RI	2.2	1.7	3.2	1.0	1.9	3.2		4.2
ROH[a]	3.3	1.2	3.6	0.9	1.6	3.6	1.2	3.9
ROR'	3.2	1.1	3.4	1.0	1.6	3.6	1.0	3.6
R'COOR	3.6	1.3	4.1	0.9	1.6	4.2	1.3	5.0
RCH=CH$_2$[b]	1.6	1.0	1.9	0.9	1.3	1.9	0.9	2.4
RC≡CH[c]	1.8	1.1	2.1	0.9	1.5	2.2		2.8
RCHO[d]	2.2	1.1	2.3	1.0	1.7	2.4		2.4
RCOR'	2.2	1.0	2.5	0.9	1.5	2.3	1.1	2.5
RCOOH[e]	2.1		3.2	1.0	1.5	2.4	1.2	2.7
RCOOR'	2.0	1.2	2.2	1.0	1.5	2.3		2.5
RNH$_2$[f]	2.3	1.0	2.5	0.9	1.4	2.6		2.9

[a] Hydroxylic protons exhibit strong temperature and concentration effects, δ 0.5–δ 6.0.
[b] See Figure 11.15 for alkenyl proton shifts.
[c] Acetylenic protons appear at δ 2.4–2.8.
[d] Aldehydic protons appear at δ 9.7–10.0.
[e] Acidic protons appear at δ 10–δ 13.
[f] Amine protons appear at δ 0.5–δ 3.0.

11.5 Interpretation of nuclear magnetic resonance

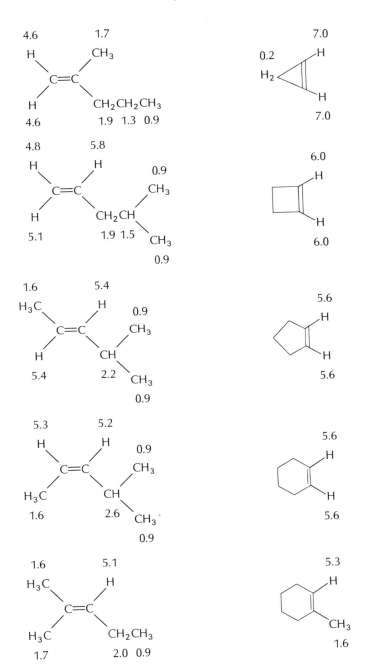

Figure 11.15 Some alkenyl chemical shift values (δ, ppm)

R group) has the largest δ. For example, the values of δ are 4.8, 5.1, and 5.8 ppm in 4-methyl-1-pentene. In the $H_2C=CR_2$ molecule, the gem (from geminal, meaning twinned) protons appear at the same value of δ, but if the R groups differ markedly (methyl and chloro; ethyl and alkoxy) the proton cis to the more electronegative group has the larger δ. In the RCH=CHR molecule, cis and trans isomers exist. If the protons are cis, they appear to be less deshielded (smaller δ) than if they are *trans*. In the R_2C=CHR molecule, the lone alkenyl proton is easily spotted at δ 5 to δ 6 ppm, depending upon the nature of the *vic* R group. If the double bond is part of a ring system, the alkenyl protons appear at the normal δ values unless the ring is strained, in which case more deshielding comes into play and δ increases to 6.0 in cyclobutene and to 7.0 in cyclopropene. In contrast, the methylenic protons in cyclobutane are strongly deshielded to δ 1.97. In cyclopropane, the methylenic protons are even more strongly shielded to δ 0.22. (Compare cyclopentane at δ 1.51 and cyclohexane at δ 1.43.)

Coupling constants. Spin-spin coupling constants (J, in Hz) are summarized and illustrated in Figure 11.16 for some frequently met systems. We need to keep in mind that a substantial difference in chemical shift is necessary before we can observe the $n + 1$ splitting rule. In general, too, it is helpful to remember that coupling is limited to transmission through only three σ bonds. If a π bond is present, transmission of the effect is increased: coupling can occur through four σ with a centrally located π bond. The negative sign of J for *gem* coupling of methylene protons need not bother us: the effect of the sign change is to invert the intensities of certain lines when the $J/(\Delta v_0 \delta)$ ratio is large. The large variation in magnitude of J with bond angle is a consequence of the location of the nodal planes of the bonds involved, but we shall not go into the details of this effect. We should note that *trans* alkenyl protons interact more strongly than do *cis* protons and that *gem* alkenyl protons couple only weakly. In isomeric *cis* and *trans* alkenes, J_{trans} is about 1.5 times as large as J_{cis}.

Unsaturation centers. Now we are ready to show how it is possible in certain cases to work out the complete structure of a molecule from a knowledge of only the molecular formula and the nmr spectrum.

To find out how much unsaturation or how many rings are present in a molecule, we need only four simple operations on the molecular formula, as long as the substance is not a salt.
1. Replace all halogens with hydrogen atoms.
2. Subtract all oxygen (and sulfur) atoms.
3. If nitrogen is present, subtract the nitrogen atoms and an equal number of hydrogen atoms.
4. Add enough atoms of hydrogen to bring the formula up to C_nH_{2n+2}. One half the number of hydrogen atoms necessary is the number of π bonds or rings in the molecule.

For example, 2-chlorocyclopentanone has the molecular formula C_5H_7ClO. Replacing Cl with H and subtracting O gives C_5H_8. The formula C_nH_{2n+2} for five atoms of carbon is C_5H_{12}. Thus, four atoms of hydrogen are needed to bring C_5H_8 up to C_5H_{12}. Half of four is two, which indicates that the molecule has two double bonds, two rings, a triple bond, or one double bond and one ring. Of course, we knew the answer to begin with: one C=O and one ring. □ □

Exercises

In the following exercises, the nmr spectrum and the molecular formula are given. You are asked to analyze the spectrum in as much detail as is necessary to give the correct structural formula. The relative areas of each set of peaks are given on each spectrum.

11.2   Figure 11.17.   11.4   Figure 11.19.   11.6   Figure 11.21.

11.3   Figure 11.18.   11.5   Figure 11.20.   11.7   Figure 11.22.

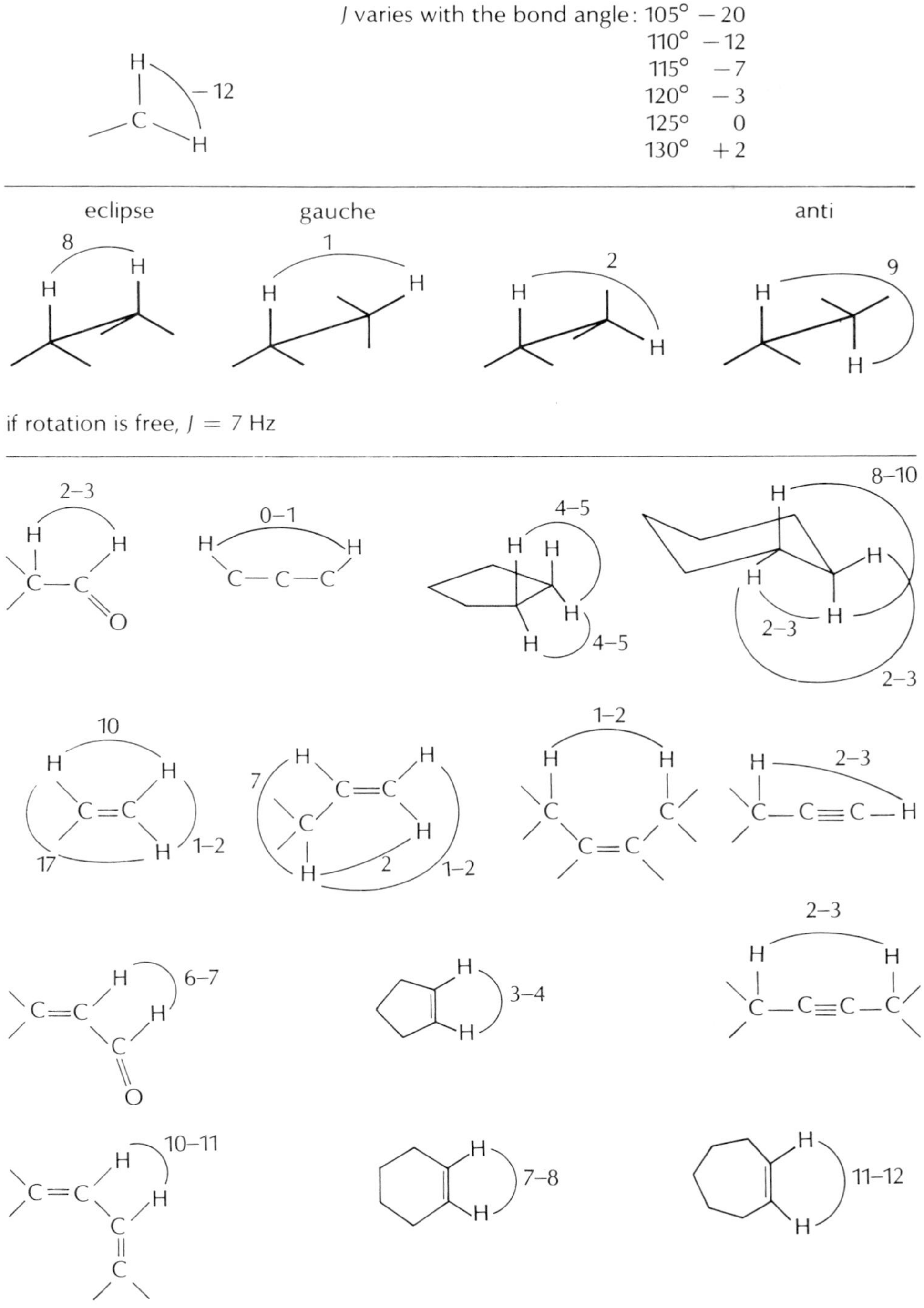

11.5 Interpretation of nuclear magnetic resonance

J varies with the bond angle:
105° −20
110° −12
115° −7
120° −3
125° 0
130° +2

if rotation is free, $J = 7$ Hz

Figure 11.16 Some proton spin coupling constants (J, Hz)

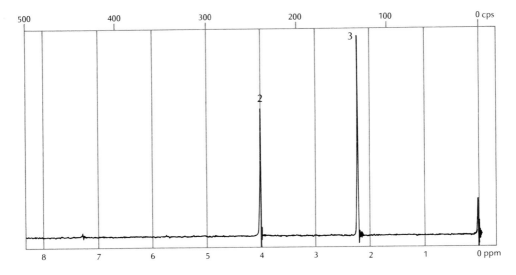

Varian (383)

Figure 11.17 For Exercise 11.2, $C_3H_5Cl_3$

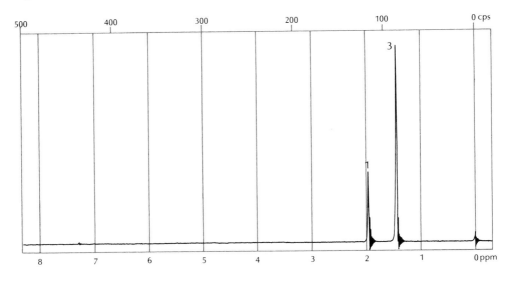

Varian (141)

Figure 11.18 For Exercise 11.3, $C_6H_{12}O_2$

11.6 Electron spin resonance

Electron spin resonance (esr), also known as electron paramagnetic resonance (epr), is analogous to nmr. We shall not go into the subject in any detail, only enough for purposes of recognition.

A molecule or ion with all electrons paired does not exhibit esr. With an odd electron

11.6 Electron spin resonance   293

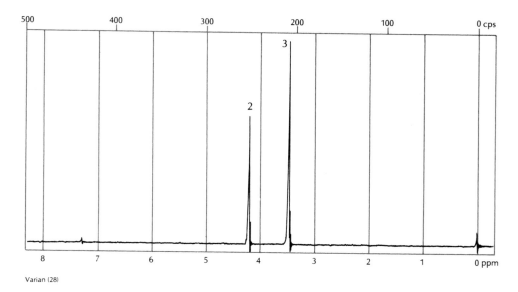

Figure 11.19 For Exercise 11.4, C_3H_5NO

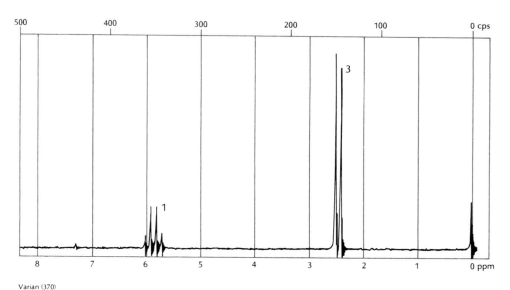

Figure 11.20 For Exercise 11.5, $C_2H_4Br_2$

present as in a free radical (say $(CH_3)_3C\cdot$) or a radical ion (say $H_2\dot{C}-O^{\ominus}$), the odd electron has $I = \pm\frac{1}{2}$. In the presence of a magnetic field, the two spin states separate in energy, and spin transitions become observable. The scale is quite different from that for protons. For a field of 10,000 gauss, $\nu = 28.0246$ GHz (GHz = gigahertz = 10^9 Hz). In esr instruments the usual field is 3400 gauss, which requires 9.5 GHz. Sensitivity is such that we may routinely work with 10^{-10} to 10^{-12} mole per liter of free radical. Unlike the case with nmr, there are some advantages to the use of solid samples, enhanced by lower temperatures.

The use of esr is restricted, of course, to radicals and ion radicals and is of limited

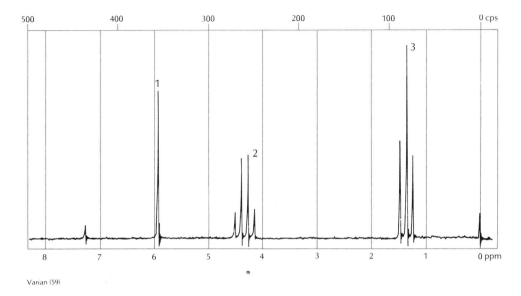

Figure 11.21 For Exercise 11.6, $C_4H_6O_2Cl_2$

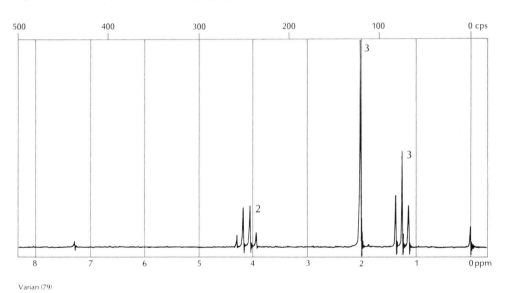

Figure 11.22 For Exercise 11.7, $C_4H_8O_2$

interest to us at this point. Nevertheless, the technique is extremely valuable in the detection and study of radicals and homolytic reactions.

11.7 Theory of mass spectrometry

Gaseous ions and ion radicals. Mass spectrometry, unlike the forms of spectroscopy previously discussed, is a destructive technique: the sample is destroyed in the process.

Fortunately, the amount of material required is miniscule, so no practical difficulty is introduced. In principle, mass spectrometry depends upon some means of producing ions in the gaseous state and on subjecting the ions to acceleration by an electric field and to curvature in a magnetic field so as to cause the ions of different mass to separate and be detectable. Actually, the result of the focusing is to cause ions of differing mass-to-charge ratios (m/e) to be detected. Because e is unity for almost every ion, the detector signals the arrival of ions, the mass of which is the sum of the atomic mass units, not atomic weights. We shall return to this point later.

The most popular type of operation is to bombard the substance being examined with a stream of energetic electrons. Under the conditions in the instrument, a collision knocks out an electron from the molecule (Eq. 11.2). The positive ion radical produced is a

$$\text{molecule} + e^{\ominus} \longrightarrow \text{molecule}^{\dot{\oplus}} + 2e^{\ominus} \qquad (11.2)$$

very unstable entity and begins to fragment or to rearrange in an attempt to find a more stable situation. Under the low pressures used (about 10^{-5} torr or less), bimolecular collisions are not frequent, so the major pathway is unimolecular fragmentation in the ionization chamber. After the ions have been accelerated by the strong electric field, there is little time for anything to happen before they are neutralized at the ion collecter or by the walls of the system.

During fragmentation, a bond in the ion radical is ruptured to produce two fragments, one of which has an odd electron but no charge (a radical) and the other of which is a carbonium or other type of positive, electron-deficient ion. The neutral radical fragments do not respond to the accelerating potential and remain in the ionization chamber to interact and to be swept out by the evacuating pumps. The neutral radicals are not detected and are of interest to us here only as a means of keeping track of the mass of the ionic fragments that are focused and detected. The carbonium and carbonium-like ions in turn may rearrange or fragment again to produce a new ion and a new neutral entity, usually a molecule. Thus, a given substance in the mass spectrometer gives rise to a characteristic pattern of detected ions.

Types of fragmentation. Because none of the neutral radicals or molecules are recorded, the fragmentations are inferred from the ions that are detected. In other words, we are dealing with a highly hypothetical type of chemistry in the mass spectrometer. Much empirical correlation took place before any hypothetical ideas evolved. We bring up the point here to emphasize the lack of proof of the types of fragmentation shown in Figure 11.23.

Depending upon the substance used, an electron may be expelled in the ion-radical-forming collision from 1. a nonbonded n pair, 2. a π bond, or 3. a σ bond. The listing is in order of probability of occurrence and is the same order of increasing energy as is seen in uv. Thus, in Figure 11.23a we see the ion radicals that result from the removal of an n or π electron from a ketone. Actually, the same ion radical would be formed in either case, and the structures shown are only two of the possible resonance structures that can be written. We use the single barbed curved line, as before, to keep track of single electrons and help us see how the bond disruption gives a radical and an ion. No matter which structure we write, the n or the π, the hypothetical homolytic bond rupture of $[R—COR']^{\oplus}$ leads to the same products, an ion and a radical.

Because only the ion is detectable, we need to look at the type of m/e pattern likely to result. Because R· may be $\dot{H}$, $\dot{C}H_3$, $CH_3\dot{C}H_2$, or other radical, the m/e of the fragment ion is 1, 15, 29,..., or $1 + 14c$ units less than that of the parent molecular ion radical. We use c to designate the number of carbon atoms in the fragment. Also, either R or R' may undergo scission. For example, consider methyl isopropyl ketone (3-methyl-2-butanone, $(CH_3)_2CHCOCH_3$), which gives somewhat more isopropyl radical ($c = 3$) than methyl

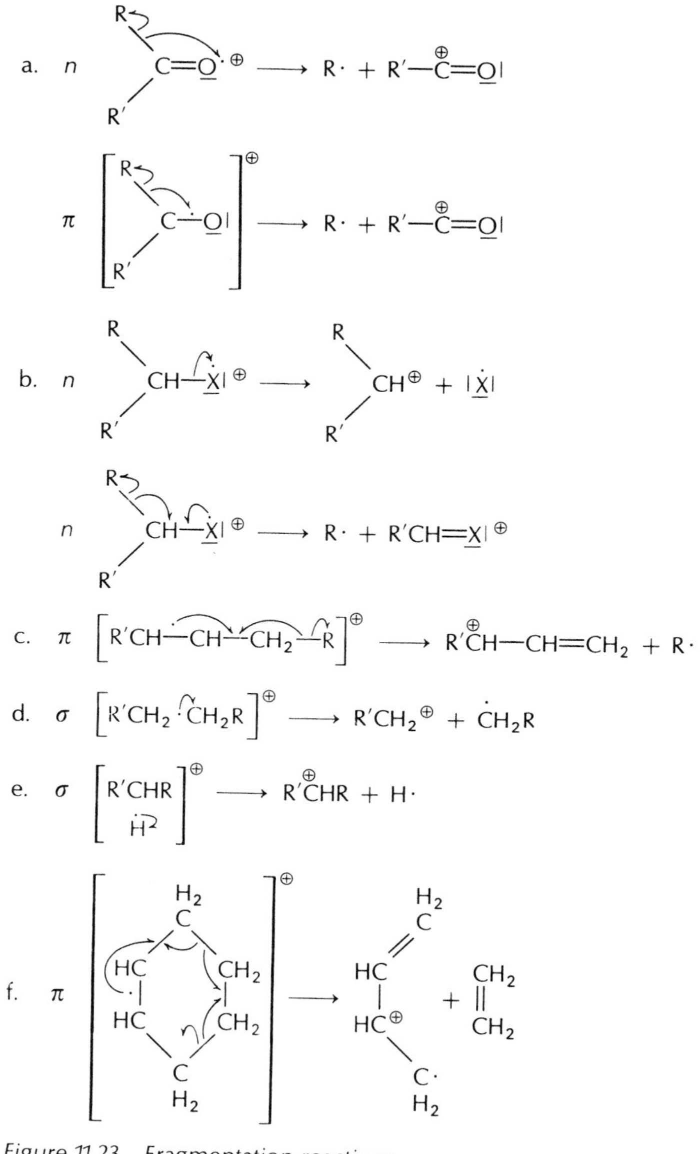

a. n $\quad R'\!-\!\overset{R}{\underset{}{C}}\!=\!\overset{\cdot\oplus}{\underline{O}}\ \longrightarrow\ R\cdot\ +\ R'\!-\!\overset{\oplus}{C}\!=\!\underline{O}|$

$\pi\quad \left[R'\!-\!\overset{R}{\underset{}{C}}\!\doteq\!\underline{O}|\right]^{\oplus}\ \longrightarrow\ R\cdot\ +\ R'\!-\!\overset{\oplus}{C}\!=\!\underline{O}|$

b. n $\quad R'\!-\!\overset{R}{\underset{}{CH}}\!-\!\underline{\dot{X}}|^{\oplus}\ \longrightarrow\ R'\!-\!\overset{R}{\underset{}{CH}}{}^{\oplus}\ +\ |\underline{\dot{X}}|$

$n\quad R'\!-\!\overset{R}{\underset{}{CH}}\!-\!\underline{\dot{X}}|^{\oplus}\ \longrightarrow\ R\cdot\ +\ R'CH\!=\!\underline{X}|^{\oplus}$

c. $\pi\quad \left[R'CH\!-\!CH\!-\!CH_2\!-\!R\right]^{\oplus}\ \longrightarrow\ R'\overset{\oplus}{C}H\!-\!CH\!=\!CH_2\ +\ R\cdot$

d. $\sigma\quad \left[R'CH_2\!-\!CH_2R\right]^{\oplus}\ \longrightarrow\ R'CH_2{}^{\oplus}\ +\ \dot{C}H_2R$

e. $\sigma\quad \left[\begin{array}{c}R'CHR\\ \dot{H}\end{array}\right]^{\oplus}\ \longrightarrow\ R'\overset{\oplus}{C}HR\ +\ H\cdot$

f. π [cyclohexene radical cation] $\longrightarrow$ [butadiene cation + ethylene]

Figure 11.23 Fragmentation reactions

radical in such a fragmentation. Thus we expect to find signals at $m/e = 86 - 43 = 43$ and at $86 - 15 = 71$ (Eqs. 11.3). Carbonylic ions, also called acylium ions ($R\overset{\oplus}{C}\!=\!O$),

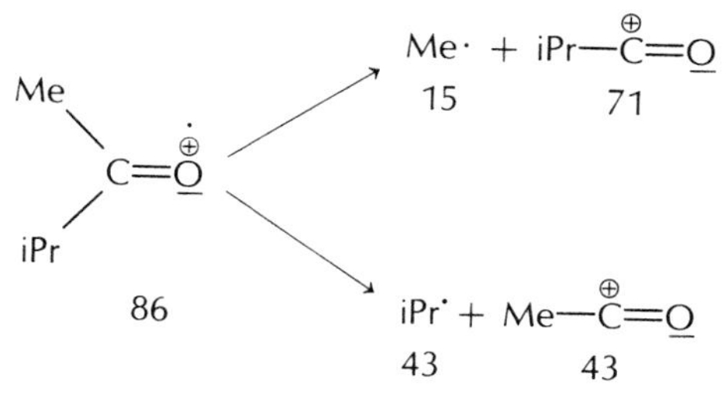

$$\underset{86}{\overset{Me}{\underset{iPr}{\diagdown}}\!C\!\overset{\cdot\oplus}{=}\!\underline{O}}\ \begin{array}{c}\nearrow\ Me\cdot\ +\ iPr\!-\!\overset{\oplus}{C}\!=\!\underline{O}\\ \quad\ \ 15\qquad\ \ 71\\ \\ \searrow\ iPr\cdot\ +\ Me\!-\!\overset{\oplus}{C}\!=\!\underline{O}\\ \quad\ \ 43\qquad\ \ 43\end{array}\qquad (11.3)$$

11.7 Theory of mass spectrometry 297

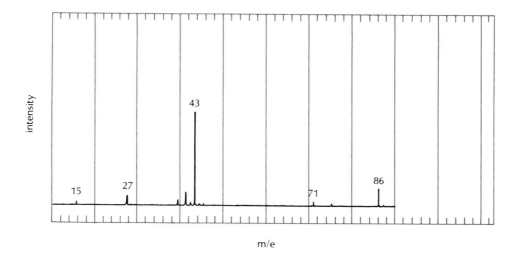

Figure 11.24 Mass spectrum of $(CH_3)_2CHCOCH_3$

readily decompose to give $R^{\oplus}$ and CO, so we also expect to find signals at $m/e = 43 - 28 = 15$ and at $71 - 28 = 43$ (Fig. 11.24). The peak at $m/e = 27$ and numerous other small peaks will not be discussed at this point. Note, however, that enough of the molecular ion radicals survived to give a *parent peak* at 86 m/e.

Returning to Figure 11.23b, we see that two common types of fragmentation are shown: loss of X as an atom or group (X here represents a halogen, OH, OR, NH_2, NHR, SR, or other: any atom capable of the loss of an *n* electron), or the retention of X in the ion.

In Figure 11.23c is shown a frequent decomposition of the ion radical formed from an alkene. The remnant of the double bond produces a rupture to give a radical and a stable allylic carbonium ion.

A saturated hydrocarbon hypothetically should exhibit fragmentation by either of the routes shown in Figure 11.23d and e. Practically, the loss of a hydrogen is not observed because the formation of a polyatomic radical (Figure 11.23d) is energetically favored. Also, because the methyl is less stable than other alkyl radicals, the breaking of a straight-chain alkyl ion radical always gives an ethyl or larger radical.

Figure 11.23f shows a decomposition of cyclohexene. This fragmentation differs from that of Figure 11.23c in that a neutral molecule instead of a radical is one of the products, and a new ion radical is left to undergo another fragmentation.

Rearrangement. Besides the usual 1,2 migration of H or R that we have learned to expect in carbonium ions, several other shifts occur in the rupture of certain structural types in the mass spectrometer. One of them is shown in Equation 11.4, in which a carbon-

(11.4)

ium ion breaks up into an alkene and a new smaller carbonium ion, which appears at 28, 42, 56 or more m/e units less than the position of the original ion (if ethylene is the alkene split off).

Base and parent peaks; isotope effects. Mass spectra consist of lines of varying intensity separated by units of m/e. The most intense line is called the base peak. The intensities of all the other lines are expressed as percentages of the base peak. This method of displaying intensities does not require absolute measurements and offers certain advantages in the determination of molecular weights.

As we have seen (Figure 11.24) the molecular ion radical is observable in some cases. The molecular peak is known as the parent peak. Identification of the parent peak gives the molecular weight of the substance with better accuracy than any other method is able to do. However, the parent peak often is very weak or appears difficult to place because of the presence of isotopes. Naturally occurring carbon and hydrogen contain measurable amounts of the heavier isotopes ^{13}C and ^2H. Any substance that contains the natural distribution of these elements shows, in the mass spectrometer, a small peak at $P + 1$ (one mass unit greater than the location of the parent peak, P) caused by the presence of one ^{13}C or ^2H atom in some of the molecules. Likewise, there is a smaller peak at $P + 2$ which is partly caused by the presence of two isotopic atoms (2 ^{13}C, 2 ^2H, or $^{13}C + ^2H$) in the same molecule.

Another cause for $P + 1$ and $P + 2$ peaks (limiting ourselves to carbon and hydrogen only for the moment) is the abstraction of a hydrogen atom by an ion radical as a result of collision in the ionization chamber with a neutral molecule (Eq. 11.5). The $P + 1$ and

$$M^{\oplus} + H\frown M \longrightarrow MH^{\oplus} + M\cdot$$
$$\left\{\begin{array}{l} \text{thus } P \longrightarrow P + 1 \\ \text{and } P + 1 \longrightarrow P + 2 \end{array}\right\} \tag{11.5}$$

$P + 2$ peaks may be enhanced by a lower electron accelerating potential, which decreases the number of ion radicals formed and increases the probability of collisions between molecules and ion radicals. Actually, if the parent peak is tall enough so that the relative $P + 1$ and $P + 2$ peak heights can be measured, the result is of help in determination of the molecular formula of an unknown compound.

In Table 11.2 are given some data on the isotopic composition of the elements most likely to be encountered in organic chemistry. For use in mass spectrometry, the isotopic compositions are recalculated to a basis of 100% for the lightest isotope as a means of comparing $P + 1$ and $P + 2$ heights relative to P. For example, an alkyl bromide (RBr) exhibits a $P + 2$ peak about 98% as tall as P. No other element comes close to such an effect. An alkyl chloride shows $P + 2$ about 33% as high as P. Fluorine, iodine, and phosphorus have no naturally occurring isotopes. Thus, if present, they serve to decrease the effect of other constituents on the $P + 1$ and $P + 2$ peaks relative to P. □ □

The nitrogen rule. The nitrogen rule sometimes is of help in determination of a parent peak. The rule states that a molecule with a molecular weight that is an even number has either no nitrogen atoms or an even number of nitrogen atoms. A molecular weight with an odd number has an odd number of nitrogen atoms. The rule leads to another statement: a fragmentation of a single bond in a molecular ion radical gives an even-numbered ion from an odd-numbered ion radical, or an odd-numbered ion from an even-numbered ion radical. Returning to Figure 11.24, we find that all the sizable peaks have odd mass numbers and come from a molecule with an even mass number.

Table 11.2 Isotope abundances[a]

Element	Atomic weight	Isotope	% of total	% of lightest isotope	Mass
Hydrogen	1.008	^{1}H	99.9844	100.0	1.00783
		^{2}H	0.0156	0.016	2.01410
Carbon	12.011	^{12}C	98.892	100.0	12.00000
		^{13}C	1.108	1.120[b]	13.00335
Nitrogen	14.0067	^{14}N	99.635	100.0	14.00307
		^{15}N	0.365	0.366	15.00011
Oxygen	15.9994	^{16}O	99.761	100.0	15.99491
		^{17}O	0.037	0.037	16.99914
		^{18}O	0.202	0.202	17.99916
Silicon	28.086	^{28}Si	92.21	100.0	27.97693
		^{29}Si	4.70	5.10	28.97649
		^{30}Si	3.09	3.35	29.97376
Sulfur	32.06	^{32}S	95.08	100.0	31.97207
		^{33}S	0.74	0.78	32.97146
		^{34}S	4.18	4.40	33.96786
Chlorine	35.453	^{35}Cl	75.53	100.0	34.96886
		^{37}Cl	24.47	32.6	36.96590
Bromine	79.904	^{79}Br	50.57	100.0	78.91835
		^{81}Br	49.43	97.8	80.91635

[a] Assembled from various sources.
[b] Beynon uses 1.08 in the construction of his tables: J. H. Beynon, Mass Spectrometry and Its Application to Organic Chemistry. Amsterdam: Elsevier, 1960. His tables give calculated $P + 1$ and $P + 2$ values for possible isomeric ions for each mass number up to 250.

11.8 Practical aspects of mass spectrometry

Many variations of design, geometry, construction, and control exist in the commercially available mass spectrometers. A simplified schematic diagram of a mass spectrometer is shown in Figure 11.25.

The most popular types of instrument utilize a beam of electrons accelerated to about 70 eV, which shoots through a thin stream of gaseous molecules. The gaseous molecules are introduced into the ionization chamber, kept at about 10^{-5} torr by pumps, through a molecular leak. If the vapor pressure of the substance is too low, provision is made for heating liquid samples or for the insertion of solid samples mounted on a probe directly into the path of the electron beam. Any positive ions formed are accelerated toward and through slits maintained at a high negative potential into a curved tube that is immersed in a magnetic field. A collector slit allows only properly focused ions through to the ion collecter. The rate at which the ions are collected is determined by the amount of current required by the collector plate to neutralize the ions.

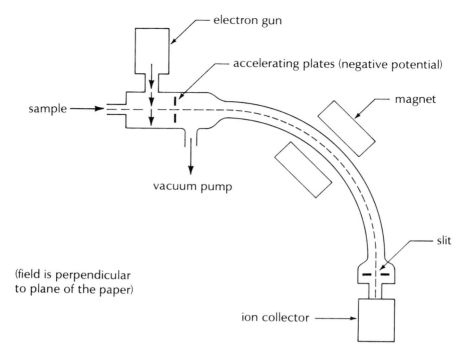

Figure 11.25   Schematic diagram of a mass spectrometer

Because peaks usually are very sharp, results often are charted in the form of vertical lines located at the various m/e values. It is convenient in such graphs to draw the base peak as 100% so that all other peaks may be read directly in percent of the base peak. Scans seldom are carried below 20 m/e, and charts usually are cut off at 20 m/e or higher.

> Other means for producing ions have been and are used. Photons may be utilized (photoionization) or alpha rays (radiation-induced ionization) or high temperature, either thermally induced ionization, as in flames, or produced by shock waves. An indirect technique is to add a gas to the sample, the gas being a substance (such as isobutane) that is very good at producing *tert*-butyl cations. The ions, in turn, abstract a proton from the sample, which then undergoes fragmentation (chemically induced ionization). By operation at low electron accelerating potentials (about 2 eV or less), electron capture occurs, and negative-ion mass spectrometry gives rise to a different fragmentation scheme for study and experimentation.

11.9   Interpretation of mass spectrometry

The mass spectrum of octane.   Beginning with a straight-chain alkane, octane (Fig. 11.26), we note first that the base peak is at 43 m/e, equivalent to a $C_3H_7^{\oplus}$ ion. Next, we easily locate the parent peak at 114 m/e and make note of prominent peaks at 85, 71, 57, and 29 m/e, which correspond with the ions $C_6H_{13}^{\oplus}$, $C_5H_{11}^{\oplus}$, $C_4H_9^{\oplus}$, and $C_2H_5^{\oplus}$. These ions result from σ bond rupture as shown in Figure 11.23d. Such a sequence of ions

11.9 Interpretation of mass spectrometry   301

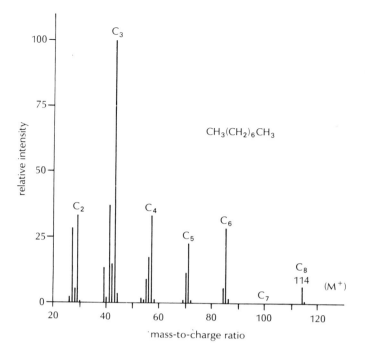

Figure 11.26   Mass spectrum of octane

is a common occurrence for a hydrocarbon chain. The absence of the $P - CH_3$ peak is also typical. Each of the prominent peaks is accompanied by a tiny isotopic $I + 1$ (I for a fragment ion) peak and up to three or four peaks of varying intensity on the low side of I. Thus the peaks at 56, 55, and 54 m/e may be ascribed to the presence of $C_4H_8^{\oplus}$, $C_4H_7^{\oplus}$, and $C_4H_6^{\oplus}$. The smaller carbonium ions, $C_3H_7^{\oplus}$ and $C_2H_5^{\oplus}$ at 43 and 29 m/e, commonly are accompanied by prominent peaks at 41 and 27 m/e for the $C_3H_5^{\oplus}$ and $C_2H_3^{\oplus}$ ions. In fact, the 41 and 27 peaks at times may overshadow the 43 and 29 peaks. We have now accounted for the presence of the previously unexplained 27 m/e peak in Figure 11.24.

Frequently occurring ions and ion radicals.   Table 11.3 gives a list of fragment ion weights that may be referred to as needed for ready identification of the type of ion that peaks at a given m/e.

Alkenes.   If a double bond is present, as in 3-ethyl-2-methyl-1-pentene, we should immediately suspect π bond participation with fragmentation (as in Fig. 11.23c) to give rise to a $P - C_2H_5$ allylic ion at $112 - 29 = 83$ m/e (Fig. 11.27). But other prominent peaks appear at 41, 55, and 84 m/e. How do they arise? The answer requires the introduction of a new fragmentation reaction.

McLafferty rearrangement.   The McLafferty rearrangement was first proposed for ketone radical ions (Eq. 11.6). Such cyclic mechanisms, in which a molecule (or ion or radical) may, by free rotation, assume an effective position for internal reaction, tend to

302   Chapter eleven

Table 11.3   Commonly occurring fragment weights

Number of carbons	Alkane	Alkyl	Alkene[a]	Alkenium	Alkyne[b]
	Ion radical C_nH_{2n+2}	Carbonium ion C_nH_{2n+1}	Ion radical C_nH_{2n}	Ion C_nH_{2n-1}	Ion radical C_nH_{2n-2}
1	16	15	14	13	12
2	30	29	28	27	26
3	44	43	42	41	40
4	58	57	56	55	54
5	72	71	70	69	68
6	86	85	84	83	82
7	100	99	98	97	96
8	114	113	112	111	110
9	128	127	126	125	124
10	142	141	140	139	138

[a] Or a single ring.
[b] Or two double bonds, a double bond and a ring, or two rings.

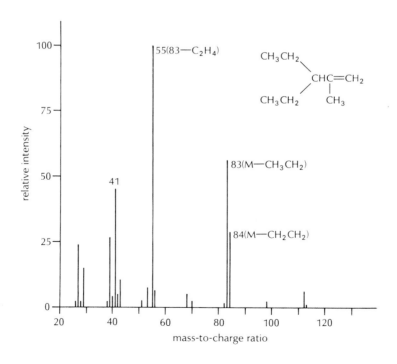

Figure 11.27   Mass spectrum of 3-ethyl-2-methyl-1-pentene

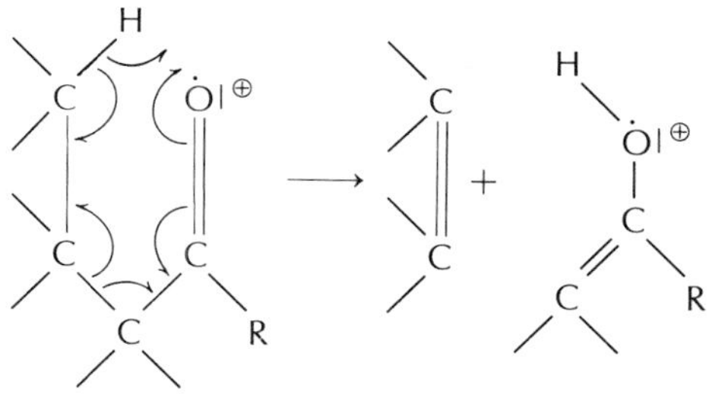

$$\text{(11.6)}$$

have lower activation energies than usual and become favored pathways for reaction. Cyclic mechanisms, especially those that may be written involving six atoms, have been postulated in many instances since the early days of organic chemistry.

The McLafferty rearrangement requires a chain of at least three carbon atoms on one side of the carbonyl group (C=O) and at least one hydrogen atom on the third carbon. Otherwise the proper cycle to allow the hydrogen transfer and alkene elimination cannot be written. The presence of a carbonyl group is not always necessary (Eq. 11.7). The base

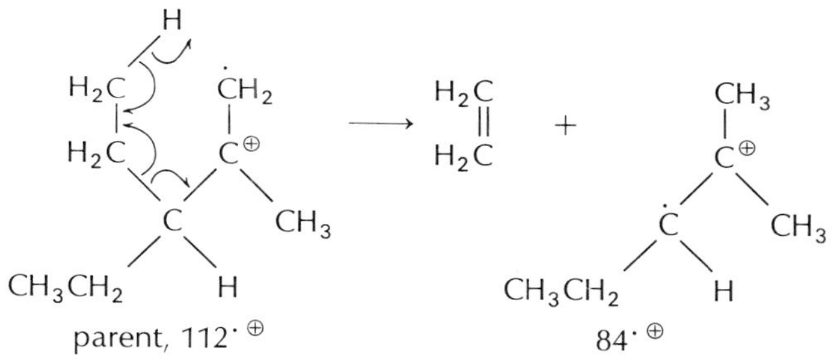

$$\text{(11.7)}$$

peak at 55 m/e is the result of a McLafferty-like rearrangement of the 83 m/e allylic ion (Eq. 11.8).

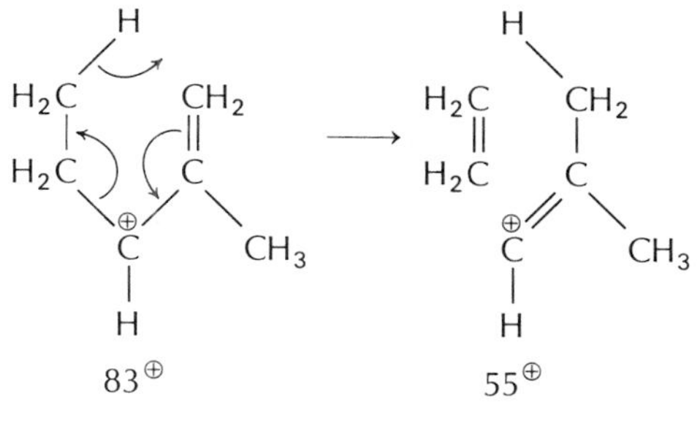

$$\text{(11.8)}$$

Alkyl halides. If a halogen atom is present in a molecule, we assume that the ion radical is formed by loss of one of the nonbonding p electrons on the halogen. The parent ion radical then fragments by one or several of the schemes shown in Equations 11.9.

The presence of chlorine in the parent or any daughter ions is recognizable by a peak at $+2$ m/e that is about one-third as intense as that of the parent or daughter ion (Table 11.2). Bromine, if present, exhibits a $+2$ m/e peak of almost equal intensity (Fig. 11.29c, ethyl 3-bromopropanoate).

$$R-\underset{|}{\overset{H}{C}}-\underset{|}{\overset{|}{C}}-X^{\oplus} \longrightarrow R-\underset{|}{\overset{H}{C}}-\underset{|}{\overset{|}{C}}^{\oplus} + X\cdot \quad (F, Cl < Br, I)$$

$$\longrightarrow R-\overset{\oplus}{\underset{|}{C}}-\underset{|}{\overset{|}{C}}\cdot + HX \quad (F, Cl > Br, I) \quad (11.9)$$

$$\longrightarrow R-\underset{|}{\overset{H}{C}}\cdot + \underset{|}{\overset{|}{C}}=X^{\oplus}$$

Alcohols and ethers. Surprisingly, alcohols seldom give a $P - 17$ peak from $RO\dot{H}^{\oplus} \rightarrow R^{\oplus} + \dot{O}H$. The most frequent fragmentation is the loss of a radical from the carbon atom to which the oxygen is bonded (Eq. 11.10). The parent peak usually is very small or invisible. Figure 11.28a shows the mass spectrum of 2-butanol.

$$\underset{H}{\overset{R}{\underset{|}{\diagdown}}}\underset{}{\overset{}{C}}OH^{\oplus} \longrightarrow \underset{H}{\overset{R\cdot}{\underset{|}{\diagup}}}CH_3-C=OH^{\oplus}, \quad \text{more than } CH_3\cdot \quad \underset{H}{\overset{R}{\diagdown}}C=OH^{\oplus}$$

$$\text{more than } CH_3-\underset{H\cdot}{\overset{R}{\diagdown}}C=OH^{\oplus} \quad (11.10)$$

An ether is stable enough to give an identifiable parent peak, as seen in Figure 11.28b (1-ethoxybutane). The fragmentation is like that of an alcohol, but complicated by the ability to give bond scission on either side of the oxygen.

Ketones and aldehydes; acids and esters. If a carbonyl group is present in a molecule, the major fragmentation is always to one side or the other of the carbon atom of the C=O, with the larger radical (smaller ion) predominating. The parent peak of an aldehyde is almost invisible, but the $P - 1$ loss-of-H peak is present. A sizable $P - 1$ peak is always a sign of the presence of an aldehyde (RCHO, Fig. 11.29a). A ketone (4-methyl-2-pentanone, Fig. 11.29b) exhibits a prominent parent peak. The base peak at 43 m/e is the result of the loss of an isobutyl radical to give the $CH_3\overset{\oplus}{C}=O$ ion. Acids (RCOOH) show peaks for $R^{\oplus}$, $COOH^{\oplus}$, and $RCO^{\oplus}$, as well as a McLafferty ion if R is long enough. The parent peak is weak.

Esters ($RCOOCH_3$, for example) show $RCO^{\oplus}$, $^{\oplus}OCOCH_3$, $^{\oplus}OCH_3$, and $R^{\oplus}$ ion peaks of varying intensity. Figure 11.29c shows the mass spectrum of ethyl 3-bromopropanoate. The parent peak at 180 m/e is apparent and the $P + 2$ ion peak is of equal intensity because of the presence of the bromine isotope. The 107 and 109 m/e peaks are the result of $BrCH_2CH_2^{\oplus}$, and 135 and 137 m/e are the result of $BrCH_2CH_2CO^{\oplus}$.

Pursuing the subject beyond this point would take us too far afield. We shall return to mass spectrometry from time to time in later chapters.

11.9 Interpretation of mass spectrometry 305

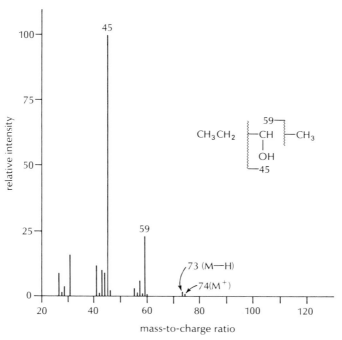

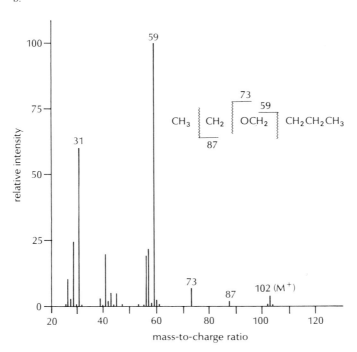

Figure 11.28 Mass spectra of a. 2-butanol and b. 1-ethoxybutane

306 Chapter eleven

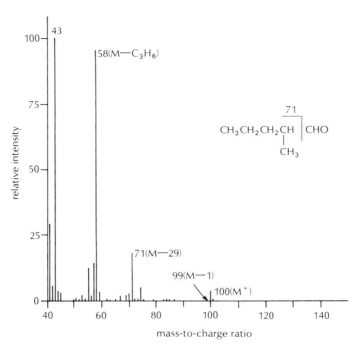

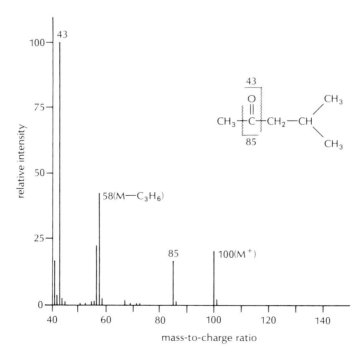

Figure 11.29 Mass spectra of a. 2-methylpentanal, b. 4-methyl-2-pentanone, and c. ethyl 3-bromopropanoate

c.

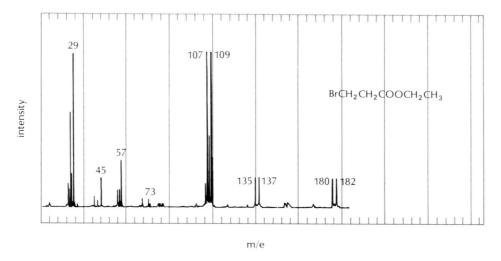

Figure 11.29 (continued)

11.10 Summary

Having been through an introduction to ir, uv, nmr, and mass spectrometry, we are in a position to begin to use all four as a means of identification. Throughout our discussion we have assumed the purity of the substances being examined. Once the spectrum of a pure compound has been determined, the process can be turned around and the presence, as well as the amount, of the compound may be determined in mixtures. Such analysis is most easily performed by uv and nmr, with more difficulty by ir and mass spectrometry.

Again, we shall not attempt a summary. Instead, the points to be kept in mind will be brought out as the problems are worked out. We should emphasize again that all of organic chemistry, including spectroscopy, is a matter of details. A general idea of a topic is only the barest of a beginning. We should strive for facility by practice, not by memorization alone. By now we should begin to feel the peculiar satisfaction that comes when nature yields a secret as the result of our efforts to see through the clues provided.

In succeeding chapters, as we encounter new classes of compounds, we shall enlarge on certain aspects of spectroscopy.

11.11 Problems

11.1 Predict the approximate λ_{max} in the ultraviolet spectrum region (170–400 mμ) for each of the following molecules.

a. $CH_2=C(CH_3)-C(CH_3)=CHCH_3$

b. cyclohexene with $-CH=CH_2$ substituent

c. $CH_2=C(CH_3)-CH=C(CH_3)-CH=CHCH_3$

d. $CH_2=CH-CH_2CH=CH_2$

e. $CH_2=CHCH_2CH=CH-CH=CH_2$

308   Chapter eleven

11.2 Two compounds, A and B, have the formula C_6H_8. Each absorbs 2 moles of H_2 on catalytic reduction, yielding cyclohexane. The λ_{max} for A and B are 190 and 227 mμ, respectively. What are the structures of A and B?

11.3 Each of the following molecules was observed to have infrared bands at the frequencies indicated. What bond is responsible for each band?
  a. Ethyl acetate: 1220, 1400, 1720, 2900 cm^{-1}
  b. Isobutyric acid: 1380, 1720, 2500–3000 (broad) cm^{-1}
  c. $HOCH_2C\equiv N$: 1400, 2250, 2900, 3250 cm^{-1}

11.4 Suggest an infrared band that would distinguish between the following pairs.

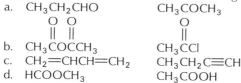

11.5 Suggest a likely structure for the following compound based on the given infrared absorption data; indicate by an arrow which peak goes with which bond.
  C_5H_{10}: 3050, 2950, 1640, 1430, 990, 900 cm^{-1}

11.6 Indicate the relative field position and strength of the signal of each class of protons in the following compounds; also indicate the multiplets that can be expected by spin-spin coupling.

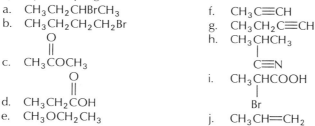

11.7 Indicate a difference in the NMR spectra of A and B that would allow you to distinguish between them.

	A	B
a.	$ClCH_2CH_2COOH$	$CH_3CHClCOOH$
b.	$HCOCH_2CH_3$ (with C=O)	CH_3COCH_3 (with C=O)
c.	$CH_3CH_2C\equiv CH$	$CH_2=CH-CH=CH_2$
d.	⌂ (cyclopentane)	$CH_3CH=CHCH_2CH_3$
e.	$(CH_3)_3CH$	$CH_3CHCH_2CH_3$ with CH_3 branch
f.	$CH_3CH_2CH_2C\equiv N$	CH_3CHCH_3 with $C\equiv N$

11.8 The mass spectrum of 2,3-dimethyl-1-butene shows a molecular ion of m/e 84 with relative intensity of 24 and a base peak at m/e 41. Other peaks with relative intensities in parentheses occur at m/e 69 (82), 39 (41), 29 (6), 27 (21), and 15 (7). Assuming that each species has only one positive charge, draw logical structures for the seven compounds.

11.11 Problems 309

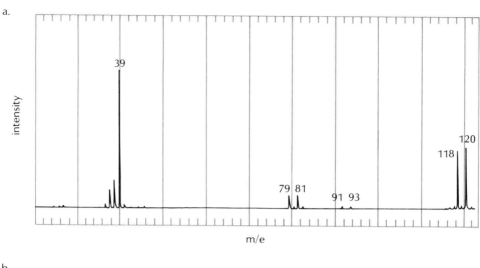

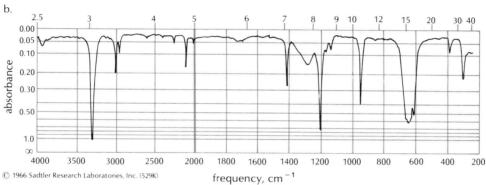

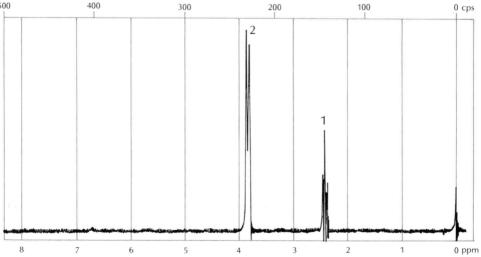

Figure 11.30 For Problem 11.9, C_3H_3Br

310   Chapter eleven

In each of the eight problems that follow, a set of spectra of a substance is given along with a molecular formula. The correct answer is a structural formula. For our satisfaction, we should verify that each spectrum fits the answer. As we have seen from time to time, any one spectrum may not allow a unique structure to be assigned. With the information available from a set of spectra, the correct structure usually can be deduced without too much difficulty.

11.9   Figure 11.30.   11.13   Figure 11.34.

11.10 Figure 11.31.   11.14   Figure 11.35.

11.11 Figure 11.32.   11.15   Figure 11.36.

11.12 Figure 11.33.   11.16   Figure 11.37.

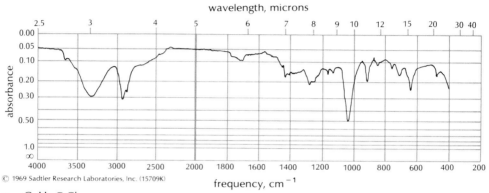

a. C_3H_7OCl

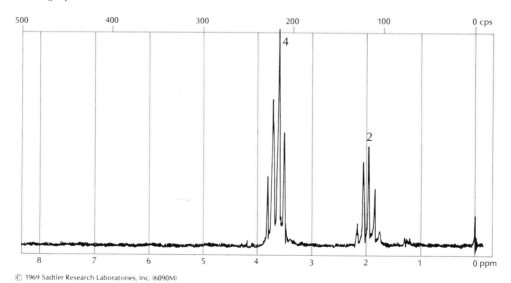

b. C_3H_6DOCl

Figure 11.31   For Problem 11.10

11.11 Problems 311

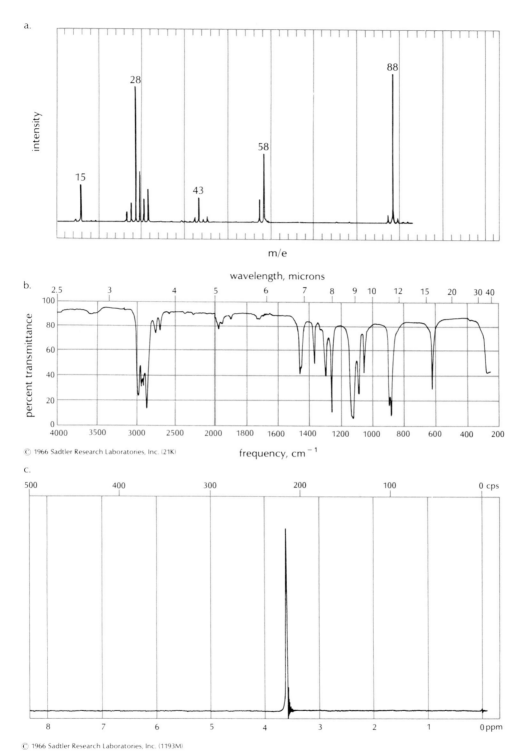

Figure 11.32 For Problem 11.11, $C_4H_8O_2$

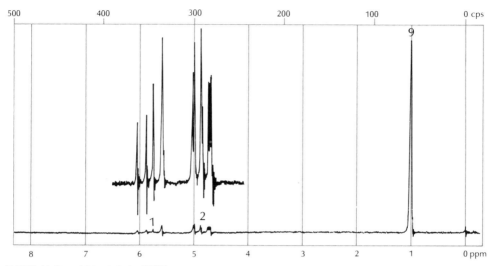

Figure 11.33   For Problem 11.12, C_6H_{12}

11.11 Problems 313

a.

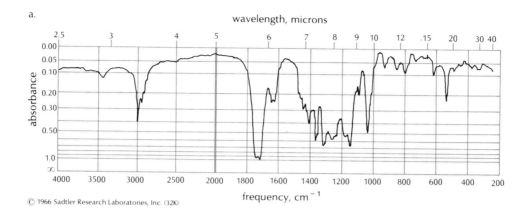

b.

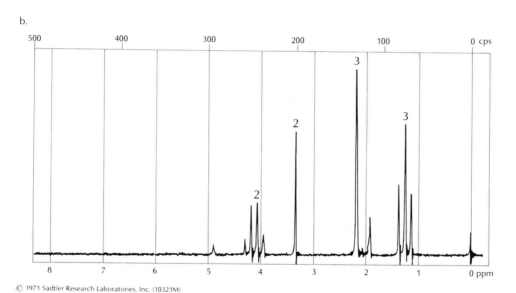

Figure 11.34 For Problem 11.13, $C_6H_{10}O_3$

314 Chapter eleven

a.
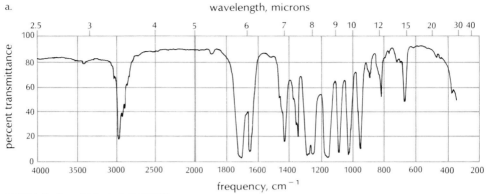

b.
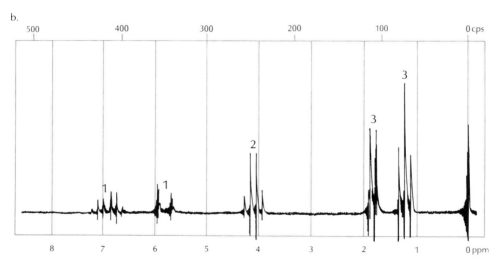

Figure 11.35   For Problem 11.14, $C_6H_{10}O_2$

11.11 Problems 315

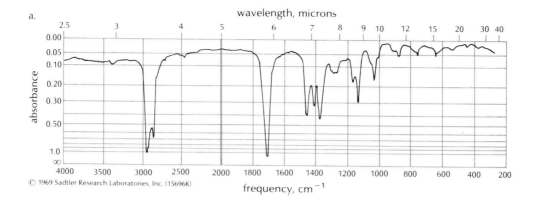

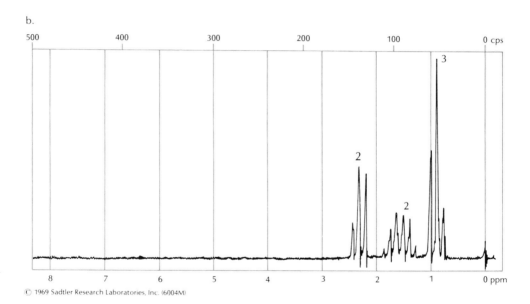

Figure 11.36 For Problem 11.15, $C_7H_{14}O$

316   Chapter eleven

a.

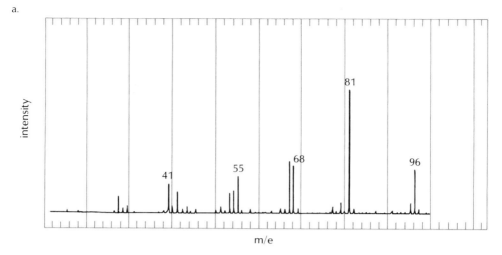

b.

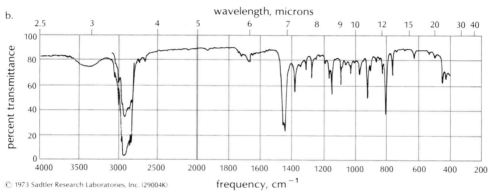

c.

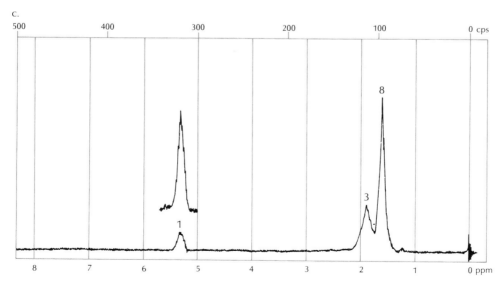

Figure 11.37   For Problem 11.16, C_7H_{12}

11.12 Bibliography

Most of the recent textbooks have treatments of spectroscopy: see references 1, 3, 9, 10, 13, and 15 in Section 7.12, for example. Some are more extensive than others.

There are a number of excellent books about spectroscopy at all levels of sophistication. The list given here is a sampling and is intended only as a starting point for anyone interested in going into the subject more extensively than we were able to do here.

1. Ault, Addison, *Problems in Organic Structure Determination*. New York: McGraw-Hill, 1967. A paperback of problems selected and designed for students who have completed a year of organic chemistry.

2. Bhacca, N. S., D. P. Hollis, L. F. Johnson, and E. A. Pier, *NMR Spectra Catalog*, Vol. 1. Palo Alto: Varian Associates, 1962. This and the following volume are paperbacks that contain nmr spectra of a large variety of organic compounds.

3. Bhacca, N. S., D. P. Hollis, L. F. Johnson, and E. A. Pier, *NMR Spectra Catalog*, Vol. 2. Palo Alto: Varian Associates, 1963.

4. Creswell, Clifford J., Olaf Runquist, and Malcolm M. Campbell, *Spectral Analysis of Organic Compounds, An Introductory Programmed Text*, 2nd ed. Minneapolis: Burgess, 1972. Excellent.

5. De Puy, Charles H., and Orville L. Chapman, *Molecular Reactions and Photochemistry*. Englewood Cliffs, N.J.: Prentice-Hall, 1971. A paperback. No spectra. For those interested in photochemistry.

6. Dyer, John R., *Applications of Absorption Spectroscopy of Organic Compounds*. Englewood Cliffs: Prentice-Hall, 1965. An excellent little paperback that covers uv, ir, and nmr with numerous illustrations. There are a few problems and references.

7. Hanna, Melvin W., *Quantum Mechanics in Chemistry*. New York: Benjamin, 1965. For those with enough mathematical background, this paperback provides an introduction to the theory of bonding and spectra.

8. Ionin, B. I., and B. A. Ershov, *NMR Spectroscopy in Organic Chemistry*, 2nd ed. New York: Plenum, 1970. Many examples and references.

9. Nakanishi, Koji, *Infrared Absorption Spectroscopy—Practical*. Tokyo and San Francisco: Nankodo and Holden-Day, 1962. Old but excellent, with many examples and problems.

10. Pasto, Daniel J., and Carl R. Johnson, *Organic Structure Determination*. Englewood Cliffs, N. J.: Prentice-Hall, 1969. A textbook for a second course in organic chemistry, excellent for reference and explanation concerning spectra. Covers ir, uv, nmr, esr, and mass spectra.

11. Paudler, William W., *Nuclear Magnetic Resonance*. Boston: Allyn and Bacon, 1971. A paperback. More depth than Dyer, reference 6. Many references and problems.

12. Roberts, Royston M., John C. Gilbert, L. B. Rodewald, and Alan S. Wingrove, *An Introduction to Modern Experimental Organic Chemistry*. New York: Holt, Rinehart and Winston, 1969. An excellent example of the several new laboratory manuals that use ir and nmr. Numerous examples.

13. Sadtler Research Laboratories, *Sadtler Standard Spectra*. Issued annually. Philadelphia: Sadtler Research Laboratories. The most extensive catalog of ir, nmr, and uv spectra published. Also, work books are published annually.

14. Shapiro, Robert H., *Spectral Exercises in Structural Determination of Organic Compounds*. New York: Holt, Rinehart and Winston, 1969. A collection of one hundred problems in which ir, uv, nmr, and mass spectra are provided along with other information needed to identify each compound. The level of difficulty in this paperback is appropriate for beginners.

15. Silverstein, Robert M., G. Clayton Bassler, and Terence C. Morrill, *Spectrometric Identification of Organic Compounds*, 3rd ed. New York: Wiley, 1974. Universally referred to for introduction to and interpretation of ir, uv, nmr, and mass spectra.

16. Swinehart, James S., *Organic Chemistry, An Experimental Approach*. New York: Appleton-Century Crofts, 1969. Another of the new laboratory manuals. This one is exceptionally detailed and good about ir. The other types of spectra, however, are neglected.

A few articles of more than ordinary interest are given.

17. Gevantman, Lewis H., Survey of Analytical Spectral Data Sources and Related Data Compilation Activities, *Analytical Chemistry 44*, 30A (1972).

18. Johnson, Leroy F., Nuclear Magnetic Resonance with Superconducting Magnets, *Analytical Chemistry 43*, 28A (1971).

19. Wilson, E. Bright, Jr., Microwave Spectroscopy in Chemistry, *Science 162*, 59 (1968).

11.13 William S. Johnson and John D. Roberts: Early days with nuclear magnetic resonance

William S. Johnson

Professor William S. Johnson of Stanford was head of the department of chemistry there from 1960 until 1969. Previously, after earning his Ph.D. at Harvard in 1940, he had gone to Wisconsin as an instructor where he rose rapidly to the rank of professor and remained until 1960. He is well known for his research work in the fields of steroids and alkaloids.

Professor John D. Roberts, Ph.D. 1944 at U.C.L.A., remained at U.C.L.A. as an instructor for a year, spent a year at Harvard as a National Research Council Fellow, and moved on to M.I.T. in 1946 as an instructor. He advanced through the ranks there until 1953 when he moved to Cal Tech where he is now Institute Professor of Chemistry. A prolific research worker and author in many fields, he probably is best known for his expertise in the applications of nmr to organic chemistry. Both men have received many honors and awards, too many to list here. Professor Johnson relates the beginning of a close friendship.*

Of course I knew of Jack Roberts by reputation and through his publications, but I had never met him until Monday, July 22, 1957, when he came to the Chemistry Department of the University of Wisconsin to deliver the Folkers Lectures. He stayed for twelve days. When he left I was a changed man, invigorated but exhausted.

Soon after his arrival I asked Jack if he would show me how to operate our new nuclear magnetic resonance spectrograph. He was one of the few experts in the field and was, in fact, delivering the Folkers Lectures on this subject. Jack agreed to take me on as a student, and thus was established a relationship that has remained essentially unchanged to this day: I always feel a bit like a student who must measure up.

For almost two weeks I was busy day and night. Jack was in my office early each morning to find out what I had accomplished and to quiz me. He is a dynamic man of

---

* Taken from William S. Johnson, "My First Encounter with John D. Roberts," from *John D. Roberts: On Thirty Years of Teaching and Research.* Menlo Park, Calif.: W. A. Benjamin, Inc., 1970.

action who has bulldog tenacity and is not satisfied with anything much short of perfection; he cannot tolerate laziness or sloppiness. This has a very tonic effect on any discussion and when he brings his penetrating insight to bear on a complex problem, it is often like a master magician opening locks. In dealing with Jack, whether it's a matter of chemistry or a game of Tiddlywinks, you immediately realize that you are involved with a fierce competitor and one of the great gamesmen. I can hear him saying in the plaintive tone of a true gamesman, "What do you mean?—You know I'm usually one down."

Those who remember the early 40 Mc Varian Associates V-4300-B spectrometer realize that the matter of obtaining satisfactory spectra in those days was really tricky, more of an art than a science. You had to learn how to recycle the magnet and then to search, sometimes for long periods of time, with the probe for a position where the field had some reasonable stability. Finally, after making numerous routine adjustments while scanning on the oscilloscope, you would start serious runs. With luck two or three satisfactory spectra might be obtained without interruption, but more often than not, in the middle of an early scan, someone would let the door to the nmr room slam, or there would be a power fluctuation, or a truck would go by, and the spectrometer would give out mostly noise; you were back at the beginning again. In such instances, long after a normal person was ready to give up for the day, Jack would show extraordinary patience, repeatedly coaxing the instrument into performance.

On the morning of July 26, 1957, Jack was in my office as usual, and we were discussing the matter of obtaining nmr spectra of some of the cevine alkaloids, which have numerous hydroxyl groups. I was deploring the fact that there was no really elegant way of exhaustively methylating all of the hydroxyl groups of such compounds, the classical method, i.e., prolonged treatment with dimethyl sulfate and sodium hydroxide, being inapplicable because of the sensitivity of these substances to base. I said something to the effect that what chemists really need is a way of getting diazomethane to react readily with alcoholic groups, similarly to the way this reagent reacts with acidic hydroxyl groups (as with carboxylic acids) to effect methylation. Jack said he thought this might be possible. The posing of the problem brought to his mind the fact that diazoacetic ester, under the influence of acid catalysts like p-toluenesulfonic acid, reacts with alcohols to form the alkoxyacetic esters and nitrogen. After recalling this information, Jack proceeded to advance the following argument to me:

"In principle, diazomethane should methylate alcohols under acid catalysis. However, the usual protonic acid, such as hydrochloric acid, is unsatisfactory, being itself consumed by reaction with diazomethane; the intermediate methyldiazonium cation combines more rapidly with the acid anion than with the alcohol. Fluoroboric acid, on the other hand, should serve as a useful methylation catalyst, because it would produce a relatively long-lived diazonium cation that would react with an alcohol as the nucleophile, and it would be consumed in reaction with diazomethane only by some process involving rupture of the relatively strong B-F bond."

I was tremendously impressed and said he should try out his hypothesis. Jack's reply was typical. In a very loud and firm tone that one might use to a misbehaving child, he said, "What are we waiting for?—Let's do it now!" Thus, in a typical display of generosity, he invited me to share in the fruits of his idea.

So on the afternoon of July 26, having located some fluoroboric acid, we ran two experiments which are recorded in my laboratory notebook, No. 5, in my handwriting. One of these involved the gradual addition of an ethereal solution of a large excess of diazomethane to a solution of 2 ml of n-octanol in 15 ml of ether containing 2 drops of 48–50% aqueous fluoroboric acid. The addition was started at 3:25 P.M., and immediate evolution of nitrogen was noted. The addition was completed in 2 minutes, and the yellow color from the excess diazomethane persisted. After an additional 2 minutes, 2 more drops of catalyst were added, a vigorous evolution of nitrogen ensued, and the mixture became colorless. After treatment with anhydrous potassium carbonate, the supernatant liquid was

11.13 Early days with nuclear magnetic resonance

analyzed on Harlan Goering's vapor phase chromatograph. The results are recorded in my notebook in Jack's handwriting, namely, that there was no vpc peak corresponding to the original alcohol. Instead there was a shorter retention time peak which we assumed was due to the methyl ether, and the hoped-for conversion appeared to have occurred completely. The other experiment was with t-butyl alcohol, but the reaction was not complete.

As a consequence of these preliminary exploratory experiments, Jack and I laid plans for putting the method on a sound basis: he and Marjorie Caserio in his laboratory examined the reaction with low molecular weight alcohols, and I with Moshe Neeman at Wisconsin looked at the behavior of several steroidal alcohols. A number of joint publications arose from this study.

In the course of applying the new methylation method to testosterone to give the then hitherto unknown methyl ether, Moshe Neeman made an interesting observation. He noted the formation of a by-product which had lost the α, β-unsaturated ketonic function of testosterone. This observation was exciting because it had been well known that this type of steroidal α, β-unsaturated ketone system was inert to diazomethane. Thus the acid-catalyzed diazomethane ring enlargement of α, β-unsaturated ketones was discovered and exploited. Surely an impressive number of significant results developed from the idea Jack had in my office on the morning of July 26, 1957.

But this is not all that evolved from that visit of Jack's. Jack and I were poring over some spectra at my home after dinner on July 31. When our attention came to a certain spectrum, our conversation went something like this:

Jack: Those peaks certainly look like two ⟩CCH$_3$ groups.

Bill: But that's impossible; only one of them can be such a methyl.

Jack: Well, if one of those peaks isn't due to a methyl, it is the result of an exceptionally sharp, unsplit high-field methylene signal and I've never seen one like it. I still think it's a methyl.

Bill: But, there just is no way to formulate this ketol with two methyl groups attached to a quaternary carbon.

Thus the conversation proceeded, until suddenly, by virtue of Jack's insistence, another possible structure for the compound did occur to me, namely, the structure shown (left) which does contain two of such methyl groups.

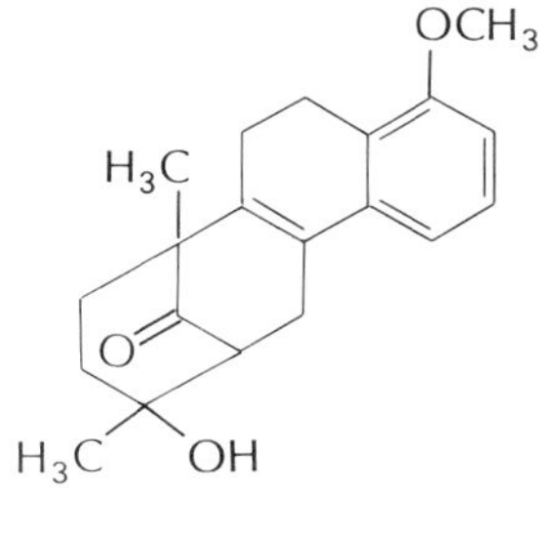

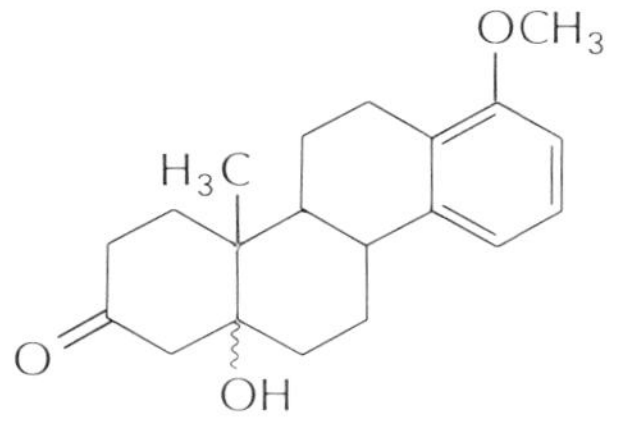

This episode prompted us to undertake a thorough re-examination of the chemistry of synthetic ketols related to the structures shown above.

In conclusion I would like to mention that I am the proud owner of a complimentary copy of Jack's book, Nuclear Magnetic Resonance. Application to Organic Chemistry, published early in 1959, which carries the inscription "To Bill Johnson—my best pupil of NMR—Jack Roberts 1/7/59." It is obvious that since then Jack has had many pupils who are my superiors, but I doubt if he has ever had a more enthusiastic one.

12 Nonaromatic Hydrocarbons I

In this chapter, we return to the classes of compounds and begin a study of the structure of cycloalkanes, cycloalkenes, and cycloalkynes and the substitution reactions of open-chain and cyclic alkanes, alkenes, and alkynes. Although we shall find a number of new reagents, most of the reactions are very similar to many we have already seen. The next chapter will continue with the hydrocarbons and take up the addition reactions of the π bond and the competition that ensues with the substitution reactions.

For a bibliography, see the end of chapter 13.

12.1 Physical properties   320
   1. Boiling points.   2. Melting points, densities.   3. Oil and water.
   4. Dipole moments.   5. Baeyer's strain theory.   6. Strain energy.
   7. Cyclohexane: conformations.   8. Substituted rings.   9. Larger cycloalkanes.   10. Cycloalkenes.   11. Highly strained ring systems.
   12. Occurrence. Exercises.

12.2 Spectroscopy   331

12.3 Nucleophilic substitution reactions   331
   1. Inertness of alkyl groups.   2. Acidity of 1-alkynes; Grignard reagents.
   3. Isomerization of alkenes by bases. Exercises.

12.4 Electrophilic substitution reactions   334
   1. Hydride removal by electrophiles.   2. Carbene insertion. Exercises.

12.5 Homolytic substitution reactions   335
   1. Attack upon a C—H or a π bond.   2. Halogenation of methane: mechanism.
   3. Comparison of ΔH and E_a for halogenation chains.   4. Relative activity of hydrogen atoms at various positions.   5. Uses of halogenation.   6. Organic liquid breathing.   7. Allylic substitution.   8. Summary. Exercises.

12.6 Summary   342

12.7 Problems   343

12.8 Jack Hine: Halogen substituents   343a

12.1 Physical properties

Boiling points. Alkanes, cycloalkanes, alkenes, cycloalkenes, alkynes, and cycloalkynes all have C—H bonds and, with only the three exceptions of methane, ethylene, and acetylene, C—C single bonds. Because of the low polarity of the individual bonds, molecules of these types of compounds have little attraction for each other unless the molecular weight is large. The larger molecules exert a sizable attraction for each other by van der Waals' forces. Consequently, the smaller members of each family (C_1 to C_4) are gases at room temperature. Beginning with the straight-chain or cyclic C_5 compounds, which are low-boiling liquids, the bp rises, but less and less rapidly as the CH_2 increment between members of a family becomes a smaller proportion of the whole. Branching, as we have seen (Sec. 9.1), causes a decrease in bp. Table 12.1 gives a sampling of alkanes, alkenes, and alkynes with some data on mp, bp, density, and dipole moments. Table 12.2 gives similar data of some cycloalkanes, cycloalkenes, and cycloalkynes.

Melting points, densities. Melting points never get very high for the alkanes or for those compounds with only one ring or one double or triple bond. Because crystal formation requires a decrease in entropy from the liquid state, the $-T\Delta S$ term is positive for crystallization. This effect works against a moderate negative ΔH in the equation $\Delta G = \Delta H - T\Delta S$, and the two effects must balance for $\Delta G = 0$, which must be true for the phase change from liquid to solid at the freezing point. At a higher temperature, the positive $-T\Delta S$ term exceeds the negative ΔH, ΔG is positive, and any solid phase melts spontaneously. Conversely, no solid may form from the liquid above the melting point.

The densities of the liquids and solids all are less than that of water, in which they are only very slightly soluble. This combination creates a messy situation when an oil well at sea springs a leak or a tanker is wrecked.

The refractive indices tend to increase with unsaturation and with molecular weight. Because of ease of measurement, the refractive index is one of the three physical properties classically used to characterize a liquid, the other two being bp and density.

Oil and water. As long as oil remains an important fuel transported across the ocean, hydrocarbons will be mixing with water on a geologically significant scale. The denser material sinks to the bottom; and, in spills close to shore, marine bottom life may be smothered. Most hydrocarbons are to some slight extent soluble in water. In this form they have been shown to have no effect on the marine mussel. On the other hand, even minute quantities of the water-soluble fraction of kerosene inhibits the food-gathering ability of the marine snail, *Nassarius obsoletus*. The partial dissolution of fuel oil in sea water is important to its eventual degradation.

The large fraction of the spilled oil goes neither under the ocean nor into it, but on top of it. Being less dense than water and only sparingly soluble, it exists as a long-lived film covering the surface. This film can interfere with the life processes of the plankton, which is a major source of molecular oxygen for our atmosphere.

Dipole moments. Alkanes and cycloalkanes have zero dipole moment and are nonpolar. Yet an alkyl group and a hydrogen atom both differ a bit in polarity upon being bonded to an sp^2 or sp hybridized carbon atom. A methyl group has a +I effect as compared with hydrogen. The difference is detectable as a weak dipole moment. Thus, propylene has

Table 12.1 Some alkanes, alkenes, and alkynes

Formula	mp, °C	bp, °C	Density,[a] g/ml	Refractive index[b]	μ, debyes
$CH_3C{\equiv}CH$	−103	−23	$0.706^{-50°}$	$1.3863^{-40°}$	0.75
$CH_3CH_2CH_2CH_3$	−138	0	$0.601^{0°}$	$1.3543^{-13°}$	0
$(CH_3)_2CHCH_3$	−159	−12	$0.603^{-20°}$	—	0
$CH_3CH_2CH{=}CH_2$	−185	−6	0.595	$1.3962^{-20°}$	0.30
cis-$CH_3CH{=}CHCH_3$	−139	4	$0.621^{0°}$	$1.3931^{-25°}$	0.33
trans-$CH_3CH{=}CHCH_3$	−106	1	$0.604^{0°}$	$1.3848^{-25°}$	0
$(CH_3)_2C{=}CH_2$	−141	−7	$0.640^{-20°}$	$1.3727^{-12.7°}$	0.49
$CH_3CH_2C{\equiv}CH$	−122	8	$0.678^{0°}$	$1.3962^{0°}$	0.80
$CH_3C{\equiv}CCH_3$	−32	27	0.691	1.3921	0
$CH_3(CH_2)_3CH_3$	−130	36	0.626	1.3579	0
$CH_3(CH_2)_4CH_3$	−95	68	0.659	1.3749	0
$CH_3(CH_2)_5CH_3$	−91	98	0.684	1.3876	0
$CH_3(CH_2)_6CH_3$	−57	126	0.703	1.3975	0
$CH_3(CH_2)_8CH_3$	−30	174	0.730	1.4119	0
$CH_3(CH_2)_{10}CH_3$	−10	216	0.749	1.4216	0
$CH_3(CH_2)_{12}CH_3$	6	254	0.763	1.4290	0

[a] At 20°C unless superscript indicates otherwise.
[b] $n_D^{20°}$ unless another temperature is indicated.

Table 12.2 Some cycloalkanes, cycloalkenes, and cycloalkynes

Name	mp, °C	bp, °C	Density, g/ml[b]	$n_D^{20°}$
cyclopropane	−127	−33	$0.720^{-79°}$	
cyclopropene		?		
cyclobutane	−50	13	0.703	1.426
cyclobutene		2	0.733	
cyclopentane	−94	49	0.751	1.406
cyclopentene[a]	−135	44	0.774	1.422
cyclohexane	6	81	0.779	1.427
cyclohexene[a]	−104	83	0.811	1.446
cycloheptane	−12	118	0.811	1.445
cycloheptene		115	0.824	1.454
cyclooctane	13	149	0.834	1.457
cis-cyclooctene	−12	138	0.845	1.468
trans-cyclooctene	−6	143	0.846	1.474
cyclononane		171		1.466
cis-cyclononene		167	0.867	1.480
trans-cyclononene		$73^{20\,mm}$	0.862	1.480
cyclononyne		177	0.897	1.489
cyclodecane		199	0.857	1.472
cis-cyclodecene		194	0.877	1.485
trans-cyclodecene		194	0.867	1.482
cyclodecyne		203	0.897	1.490

[a] Cyclopentene has μ = 0.55 D; cyclohexene, 0.61 D.
[b] At 20°C unless superscript indicates otherwise.

a small dipole moment (μ = 0.35 D). The C—H bonds *trans* to each other balance their polar effects, but CH$_3$—C is not balanced by the C—H *trans* to it, and the dipole results.

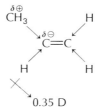

Propyne has μ = 0.75 D. Two methyl groups, as in isobutylene, give μ = 0.49 D. *Trans*-2-butene has zero dipole moment, but *cis*-2-butene has μ = 0.33 D. Cyclic *cis*-alkenes have larger dipole moments: cyclohexene has μ = 0.61 D, and cyclopentene has μ = 0.98 D.

Baeyer's strain theory. We should consider the properties and conformations of cyclic compounds in more detail than we have in previous discussions. Cyclopropane, the simplest ring compound, is striking because of its 60° bond angles.

> This sharp angle surely must introduce some differences in properties. The idea of angle strain was applied to ring compounds as long ago as 1885 by Baeyer. He erred by assuming that all rings were planar. However, it was only in 1874 that the tetrahedral carbon atom had been proposed independently by van't Hoff and Le Bel, and the tetrahedral angle had been thought of as rigid. Until 1881, all efforts to synthesize a saturated ring other than cyclohexane had failed. In that year, a substituted cyclobutane was reported. Cyclopropane was made in 1882. Thus, Baeyer assumed that the tetrahedral angle (109° 28′) could be altered, but that, as it was done, strain was introduced into the molecule proportional to the deviation from 109° 28′. This idea explained the instability of the small rings and predicted the stability of cyclopentane, with 108° bond angles. Later on, in 1885, a substituted cyclopentane was reported and found to be very stable to ring opening. Still, no ring larger than cyclohexane had been made or found in nature and <u>Baeyer's strain theory</u> predicted that such rings would be unstable because of strain by angle expansion.
> It was not until 1890 that Sachse proposed strainless ring puckering for cyclohexane and larger rings. Even so, it was not until 1918, when Mohr revived Sachse's ideas, that Baeyer's unfortunate assumption concerning large rings was finally laid aside and the stereochemistry of cyclohexane and larger rings was firmly established. □ □

Strain energy. The strain of the C—C bonds in cyclopropane appears both in the tendency for the ring to open up (to be described later in the chapter) and in the heat of combustion (Eq. 12.1). If we divide the heat of combustion (which is ΔH for the reaction

$$C_3H_6 + \tfrac{9}{2}O_2 \longrightarrow 3CO_2 + 3H_2O, \quad \Delta H = -499.8 \text{ kcal} \tag{12.1}$$

with the sign reversed) by 3, we get a value of 166.6 kcal per CH$_2$ group. Because all the cycloalkanes may be represented as (CH$_2$)$_n$, where n is the number of carbon atoms in the ring, and because the products are the same in every instance from every CH$_2$ group, the heat of combustion per methylene (CH$_2$) group serves as a convenient measure of the instability to oxidation. Table 12.3 shows the heats of combustion of some cycloalkanes,

Table 12.3 Heats of combustion, cycloalkanes

	$-\Delta H$, kcal per mole	$-\Delta H$, kcal per CH_2	Destabilization energy, kcal[a]	Total strain energy, kcal[b]
Cyclopropane	499.8	166.6	9.2	27.6
Cyclobutane	656.0	164.0	6.6	26.4
Cyclopentane	793.5	158.7	1.3	6.5
Cyclohexane	944.4	157.4	—	—
Cycloheptane	1108.1	158.3	0.9	6.3
Cyclooctane	1268.8	158.6	1.2	9.6
Cyclodecane	1586.0	158.6	1.2	12.0
Cyclododecane	1891.2	157.6	0.2	2.4
Cyclotetradecane	2203.6	157.4	0.0	0.0

[a] Destabilization or strain energy per CH_2 as compared to cyclohexane.
[b] The destabilization energy per CH_2 multiplied by the number of CH_2 groups in the molecule.

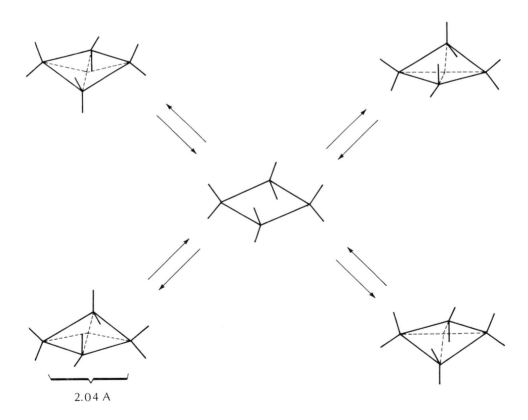

2.04 Å

Figure 12.1 Cyclobutane vibrations

as well as the heats per methylene group and comparisons with cyclohexane. We see that cyclopropane is less stable than cyclohexane by 9.2 kcal per CH_2, or a total of 27.6 kcal. Cyclobutane also has a sizable strain energy (destabilization energy) of 6.6 kcal per CH_2, or a total of 26.4 kcal. (All the other rings have smaller strain energies.) However, cyclobutane has more eclipsed C—H nonbonded repulsions than does cyclopropane. To reach a more stable level, the planar ring folds slightly and vibrates through the plane. As shown in Figure 12.1, the fold of the ring (to about 150°–160° from the planar 180°) relieves the eclipsing by allowing an approach to a staggered conformation of the C—H bonds around the ring.

Cyclopentane, with an interior angle of 108°, has practically no angle compression, but the planar form suffers from even more C—H bond eclipsing than does cyclobutane. In ethane (Fig. 6.2) the eclipsed conformer is 3 kcal per mole above the staggered form in energy. Thus, one eclipse of C—H bonds costs 1 kcal of stabilization. Planar cyclopentane has 10 eclipses and should have a total of 10 kcal of strain energy. From the heat of combustion, we find only 6.5 kcal. Therefore, by doing some folding, the ring relieves 3.5 kcal of eclipsing strain (Fig. 12.2). As with cyclobutane, all the carbon atoms may shift out of the ring plane at one time or another. Only two of the twenty possible vibrational conformations are shown in Figure 12.2.

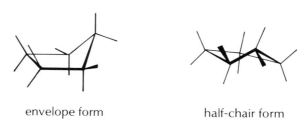

envelope form   half-chair form

Figure 12.2 Two cyclopentane vibrational conformations

Cyclohexane: conformations. Cyclohexane is able to assume a completely staggered conformation without any bond-angle compression, and for this reason we may assume that no strain energy is present. Furthermore, no cycloalkane has a smaller heat of combustion per CH_2 unit. The most stable conformation of cyclohexane is known as the chair form (Fig. 12.3). By the application of 11 kcal of activation energy, the fairly rigid chair form may pass through the unstable half-chair to the twist boat and boat forms. The boat (sometimes called a bed form) is rather floppy; and by suitably twisting a model, we can bring any pair of carbon atoms, in a 1,4 relationship to each other to the prow and stern positions of the boat, right side up or upside down. (Use models.)

The boat itself is 1.6 kcal less stable than a twist boat, because the twist boat partially relieves some of the eclipsing of C—H bonds that is present in a boat form. In any case, a twist boat has 5.5 kcal of strain energy relative to the chair form, and it has an activation energy of only 5.5 kcal for return to the more stable chair form. At room temperature there is enough thermal energy so that an equilibrium is rapidly attained. Thus there is no possibility of separating the chair and twist boat conformers at room temperature.

If we examine the structure of a cyclohexane chair form more closely, we recognize three C—H bonds pointing up and three down from the ring. These six bonds are called axial (a) because they are parallel to the vertical three-fold axis of the ring. The other six C—H bonds, which project outward from the ring and alternate slightly up and down, are

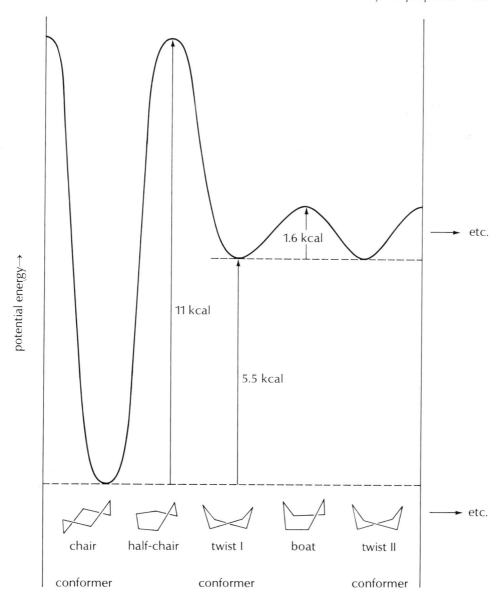

Figure 12.3 Potential-energy relationships among conformations of cyclohexane

called <u>equatorial</u> (e) because of their positions near the rim of the ring. If we allow a given boat form to go over to a chair form, we see that each set of six C—H bonds has an equal chance of becoming axial or equatorial (Fig. 12.4). In the figure we have used circles to point out only one hydrogen from each set of six to simplify the picture. For the same reason, a boat instead of a twist boat form was used as the intermediate. The equilibrium between the two chair forms gives a mixture in which the two sets of six hydrogen atoms that are axial and equatorial in one form become equatorial and axial in the other form.

326   Chapter twelve

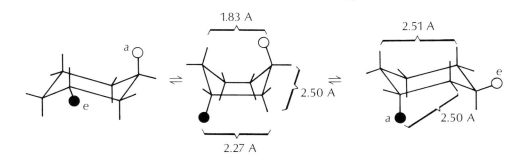

Figure 12.4   Interconversion of axial and equatorial conformations in cyclohexane

At room temperature it has been calculated that there are 160,000 chair-to-chair conversions per sec. This rate is so rapid that, as with the alcohols (Figure 11.9b and c), spin averaging occurs and the nmr of cyclohexane protons is a singlet at δ 1.43. If the temperature of a CS_2 solution of cyclohexane is lowered, the peak broadens and then separates into two peaks at $-70°$ that do not sharpen much more at $-106°$ (δ_a 1.20 and δ_e 1.66). Thus the interconversions are slowed enough to allow the nmr of the a and e protons to be seen, even though the spin-spin splitting cannot be resolved.

Substituted rings.   Now let us envision one of the circles in Figure 12.4 as a substituent, say a methyl group. If the methyl group is in an axial position, there are nonbonded repulsions with the two axial hydrogen atoms on the same side of the ring as well as with e protons on adjacent carbon atoms. But if the methyl group is in an equatorial position, the only nonbonded interactions are with the hydrogen atoms on the adjacent carbon atoms. The result is to stabilize the conformer in which the substituent is an equatorial position. The equilibrium between the conformers shifts to favor the substituent in the equatorial position. (A single substituent in cyclopropane has no choice concerning a favored location. A single substituent in cyclobutane or cyclopentane favors the bent form that allows the substituent to become most nearly equatorial.)

Two substituents, whether in a 1,2, a 1,3, or a 1,4 relationship, are either *cis* or *trans* to each other (Sec. 6.2). This arrangement is easy to see in the more planar rings. In cyclohexane, however, we need to be careful in deciding which isomer is *cis* and which is *trans*. If we look at Figure 12.4 once more and let the circles represent any substituents we choose, we have an example of a 1,3-disubstituted cyclohexane. We also see that the substituents are *trans* to each other and that the conformation is e and a in equilibrium with the a and e conformation. Going through all six of the possibilities (Table 12.4), we find, if the two substituents are alike, that all the ae $\rightleftharpoons$ ea equilibria have K_{eq} = 1 because both chair forms have equal energy in the cis-1,2-, trans-1,3-, and cis-1,4-isomers. The equilibria between aa and ee all favor the ee conformer. We should note that the cis-1,3-isomer is extra crowded in the aa conformer because the two substituents are axial on the same side of the ring, whereas the other aa conformers have the substituents on opposite sides of the ring.

Table 12.4 Disubstituted cycloalkane conformations[a]

Substituent locations	Geometry	Stereoisomerism	Conformations		K_{eq}
1,2	cis	meso	ae	ea	1
	trans	(RS)	aa	ee	large, ee favored
1,3	cis	meso	aa	ee	very large, ee favored
	trans	(RS)	ae	ea	1
1,4	cis		ae	ea	1
	trans		aa	ee	large, ee favored

[a] Dimethyl, diethyl, dibromo, and the like: both substituents the same.

Larger cycloalkanes.   The rings with 7, 8, 9, and more carbon atoms all are buckled in a manner similar to cyclohexane. The buckling, however, cannot rid the molecules of all nonbonded repulsions, and some strain energy is present. Models and the heats of combustion show that cyclononane has a maximum of 1.4 kcal per CH_2 of strain energy, caused in part by crowding of hydrogen atoms in the center of the ring. If we think of the rings as being like doughnuts, then the hole in the center does not become big enough to allow much flexibility in the ring until cyclotetradecane is reached. Thereafter, the larger cycloalkanes may assume many conformations.

All the cycloalkanes from cyclohexane up have normal C—C and C—H bond distances and angles, close to 1.54 and 1.09 Å and 109° 28'. (Cyclohexane has been reported to have the C—C—C angle of 111.5°.) Cyclopentane also has normal bond distances. However, cyclobutane has a longer than normal 1.57 Å C—C bond distance.

Cycloalkenes.   A double bond has a normal bond distance of 1.33 Å, and a triple bond has a distance of 1.21 Å. Because a double bond imparts coplanarity to six atoms (Sec. 6.2), cyclohexene is half planar and may be semi-chair or semi-boat, with the former predominating in the equilibrium between the two. Bond distances and angles are normal, as is also true for the larger cycloalkenes. Cyclopentene has a bit of angle compression strain of the C=C—C 120° angle and some expansion of the C—C—C tetrahedral angle. Cyclobutene is about as might be expected, the principle strain arising from angle compression (hydrogen atoms omitted from the diagram).

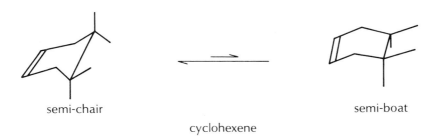

semi-chair   semi-boat

cyclohexene

328 Chapter twelve

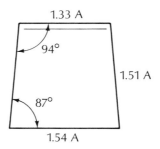

Cyclopropene is so highly destabilized by strain that it is reported to explode spontaneously at room temperature and to polymerize slowly at −78°. It must be stored as a solid at liquid nitrogen temperature, −196°. The heat of formation has been calculated as 48 kcal (positive, not negative). The structure exhibits the bond shortening also found in cyclopropane (only one hydrogen atom shown in the diagram) as well as a shortened C=C bond.

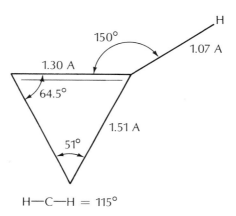

H—C—H = 115°

Highly strained ring systems. Many other highly strained ring systems have been prepared by organic chemists, especially in the last ten to twenty years, when it was realized that such structures would hang together if they could be put together. A few of these compounds are shown in Figure 12.5.

Some of the substances have been given obvious trivial names, some of which are rather frivolous. Thus: *f*, cubane; *g*, prismane; *h*, birdcage hydrocarbon; *i*, basketane; and *j*, apartmenthousane! (It may be of interest that the systematic name of the birdcage hydrocarbon is hexacyclo[6.2.1.1^{3,6}.0^{2,7}.0^{4,10}.0^{5,9}]dodecane, which is not a name used conveniently in conversation.)

Cycloalkenes with *trans* C=C bonds are possible beginning with *trans*-cyclooctene. Beginning at about *trans*-cyclododecene, the *trans* isomer is more stable than is the *cis* isomer. Cyclononyne is strainless, and the strained cyclooctyne has been made.

Figure 12.6 shows some of the ways of drawing the cyclic structures named in Table 12.2. The use of models is of great help as we visualize what the drawings represent.

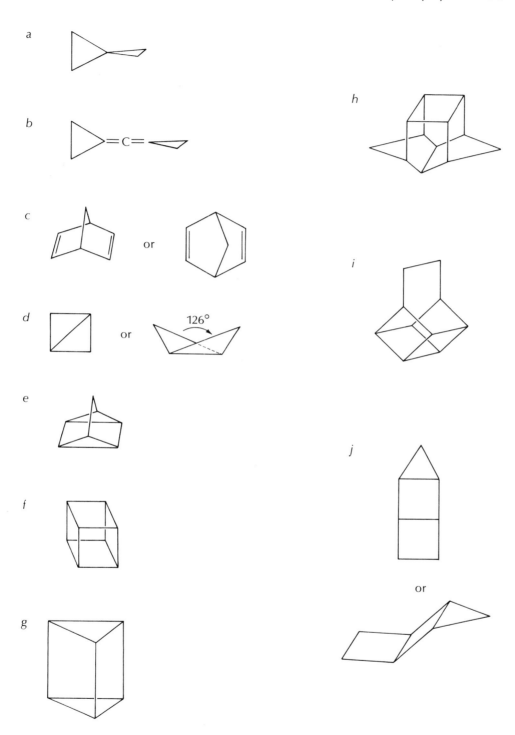

Figure 12.5 Some highly strained ring systems. Hydrogen atoms have been omitted.

330 Chapter twelve

Occurrence.   A tremendous number of hydrocarbons occur in nature, in addition to those in coal and petroleum. For example, ethylene is known to be produced by most fruit trees and hastens ripening of the fruit. Probably the most abundant naturally occurring hydrocarbons other than from coal and petroleum are the terpenes, which occur in units of five carbon atoms. Terpene chemistry is too complex to go into here, but we shall take it up in Chapter 26.

Exercises

12.1   Draw *cis*- and *trans*-4-*t*-butyl-1-cyclohexanol. Why are both structures incapable of being optically active?

12.2   Draw *cis*- and *trans*-3-methyl-1-cyclohexanol. Why are both structures capable of optical activity?

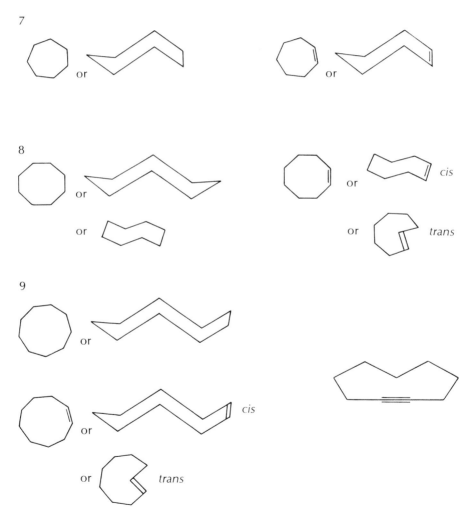

Figure 12.6   Various representations of cycloheptane, cyclooctane, and cyclononane, as well as some unsaturated cyclic structures

12.2 Spectroscopy

Because we are limiting ourselves in this chapter to hydrocarbons with no more than one double or triple bond present in the molecule, and because the spectrometry of these compounds was covered in Chapters 10 and 11, we shall note only that nmr is the most useful technique for the study of alkanes, alkenes, and alkynes.

We have observed that the coupling constant between ^{13}C and H in a ^{13}C—H bond depends almost linearly upon the percentage of s character in the hybridized orbital. Thus, in methane and ethane $J(^{13}CH)$ has been found to be 125 Hz; in ethylene, 156 Hz; and in acetylene and methyl acetylene, 248 Hz, for 25, 33, and 50% s character in sp^3, sp^2, and sp orbital bonds. The interesting result is that the $J(^{13}CH)$ in cyclopropane is 162 Hz, which may be assumed to mean that the carbon-hydrogen bonds in cyclopropane are sp^2 hybrids! Furthermore, in 3,3-dimethylcyclopropene, the vinylic coupling constant is 221 Hz, or close to the value for an sp hybrid.

Microwave spectrometry favors the rotamer shown here for propene. The

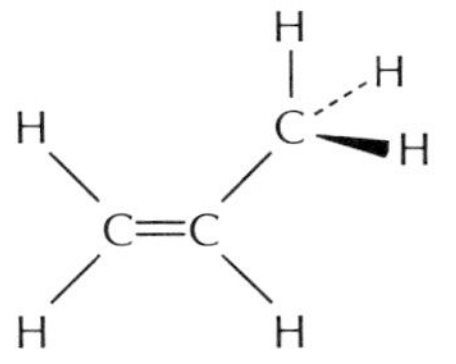

vinylic hydrogen and a hydrogen that is part of a *cis*-CH$_3$ group prefer the eclipsed conformation from among the available rotamers.

12.3 Nucleophilic substitution reactions

Inertness of alkyl groups. As we have seen previously, nucleophilic substitution on an alkane is a difficult reaction (Eqs. 12.2) because it is not easy to find a nucleophile stronger

$$N^\ominus + R\text{—}H \rightleftharpoons N\text{—}R + H^\ominus$$
$$N^\ominus + R'\text{—}R \rightleftharpoons N\text{—}R' + R^\ominus$$
(12.2)

than hydride ion or a carbanion. The reverse reaction is much easier (Eq. 12.3). (The very

$$BH_4^- + CH_3CH_2Br \longrightarrow BH_3 + CH_3CH_3 + Br^\ominus$$
(12.3)

reactive borane continues the reaction by electrophilic attack on bromine of the C—Br bond.) The inertness of alkyl groups is a characteristic which we have come to take for granted.

Acidity of 1-alkynes; Grignard reagents. If a strong nucleophile (a strong base) is used on acetylene or a 1-alkyne, a proton is easily removed (Eq. 12.4). In other words,

$$R-C\equiv C-H + H_2N^\ominus \rightleftharpoons R-C\equiv C^\ominus + NH_3 \tag{12.4}$$

the increased s orbital content of the *sp* orbital (as compared to sp^3) increases the electronegativity of the carbon atom so that the C—H bond is polarized as $\overset{\delta\ominus}{C}-H^{\delta\oplus}$. Also, the anion of acetylene is more stable than any simple alkyl carbanion. In short, 1-alkynes are weak acids. It has been estimated that pK_a for acetylene is 25, that of CH_3OH is 16, and that of NH_3 is 36 on the scale that puts H_2O at 15.7 (or 10^{-14} divided by the concentration of water in liquid water: (1000 g/liter)/(18 g/mole H_2O) = 55.6 M). Ethylene is thought to have a pK_a of about 36; ethane, about 42. Thus 1-alkynes, including acetylene, undergo acid-base reactions that are denied to alkenes and alkanes.

A useful test-tube reaction used as a test for the presence of a 1-alkyne is addition of a drop or two of the alkyne to ammoniacal aqueous silver nitrate or ammoniacal aqueous cuprous chloride, or bubbling a gaseous alkyne through it (Eqs. 12.5). These two metallic 1-alkyne salts are insoluble. Because alkanes, alkenes,

$$\begin{array}{l}RC\equiv CH + Ag(NH_3)_2NO_3 \longrightarrow RC\equiv C^\ominus Ag^\oplus + NH_4NO_3 + NH_3 \\ RC\equiv CH + Cu(NH_3)_2Cl \longrightarrow RC\equiv C^\ominus Cu^\oplus + NH_4Cl + NH_3\end{array} \tag{12.5}$$

and alkynes (with no acetylenic hydrogen atoms) do not give a precipitate, the reactions serve as positive tests for 1-alkynes. The solid acetylenic salts react with dilute nitric acid to regenerate the alkyne. On a larger scale, precipitation, washing, and regeneration may be used to purify a 1-alkyne, because any alkanes or alkenes present as impurities do not react and are not present when the 1-alkyne is reformed. The heavy metal salts must not be allowed to become dry because they are easily exploded by shock or by heat.

The sodium salt of a 1-alkyne is easily prepared by reaction with sodium amide in liquid ammonia (Eq. 12.4). The carbanion is a good nucleophile (Eq. 12.6), although the strong

$$RC\equiv C^\ominus + R'X \longrightarrow RC\equiv CR' + X^\ominus \tag{12.6}$$

basicity limits the S_N2 reaction to primary halides because otherwise the E2 reaction of the alkyl halide predominates. Even so, the reaction is good for the preparation of alkynes in which the triple bond no longer is at the end of a chain.

Even more useful is the reaction of a 1-alkyne with a Grignard (or a lithium) reagent (Eq. 12.7) to produce an acetylenic Grignard reagent. Acetylenic Grignard reagents react

$$RC\equiv C-H + R'-Mg-X \longrightarrow R'H + RC\equiv CMgX \tag{12.7}$$

as ordinary Grignard reagents do. For example, reaction with an aldehyde gives an alkynol (Eq. 12.8) after hydrolysis of the intermediate magnesium alkoxide. (Often, to reduce

$$RC\equiv CMgX + CH_3CHO \longrightarrow RC\equiv CCHOHCH_3 \tag{12.8}$$

repetitious statements after we have become used to what is required, we shall not mention the hydrolysis step or the isolation and purification operations, merely referring to them as the work-up.)

Because the original alkyl Grignard reagent is destroyed in the preparation of the acetylenic Grignard reagent, it is convenient to use a cheap alkyl halide to prepare the reagent. Ethyl bromide of a technical grade is relatively cheap, and the by-product alkane (R'H) is ethane, which simply evaporates and does not interfere in the work-up.

Isomerization of alkenes by bases. A strong base causes isomerization of an alkene or alkyne if an α hydrogen is present (Eq. 12.9). Propylene has $pK_a = 35$, about the

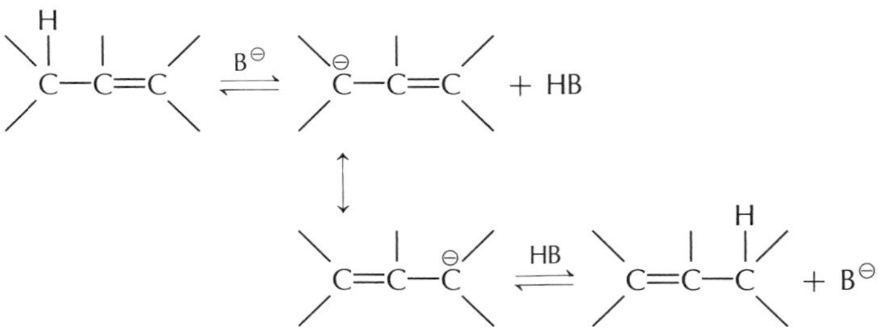

$$\underset{H}{\overset{}{\diagdown}}\!\!\overset{|}{\underset{|}{C}}\!-\!\overset{|}{\underset{|}{C}}\!=\!\overset{|}{\underset{|}{C}}\!\diagup \quad \overset{B^{\ominus}}{\rightleftharpoons} \quad \overset{\ominus}{\diagdown}\!\overset{|}{\underset{|}{C}}\!-\!\overset{|}{\underset{|}{C}}\!=\!\overset{|}{\underset{|}{C}}\!\diagup \;+\; HB$$

$$\updownarrow \tag{12.9}$$

$$\diagdown\!\overset{|}{\underset{|}{C}}\!=\!\overset{|}{\underset{|}{C}}\!-\!\overset{\ominus}{\underset{|}{C}}\!\diagup \quad \overset{HB}{\rightleftharpoons} \quad \diagdown\!\overset{|}{\underset{|}{C}}\!=\!\overset{|}{\underset{|}{C}}\!-\!\overset{H}{\underset{|}{C}}\!\diagup \;+\; B^{\ominus}$$

same as that of ammonia. Thus sodium amide ($NaNH_2$) causes isomerization of an alkene to the more stable structure.

Allylic resonance, as shown in Equation 12.9 for an allylic carbanion, also is exhibited by allylic free radicals and allylic carbonium ions. Long before the resonance explanation for the stabilization of the ion was given, it was recognized by organic chemists as an empirical rule: any atom or group attached to a carbon (or other atom) that is α to a double bond exhibits enhanced reactivity. Equation 12.10 shows removal of A as a cation, although

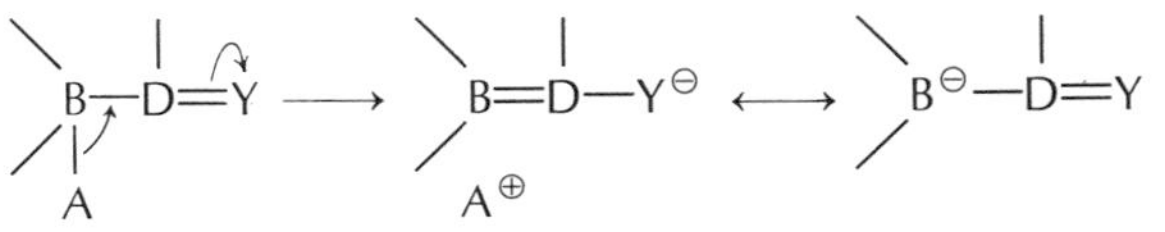

$$\underset{A}{\overset{}{\diagdown}}\!B\!-\!D\!=\!Y \;\longrightarrow\; \underset{A^{\oplus}}{\overset{}{\diagdown}}\!B\!=\!D\!-\!Y^{\ominus} \;\longleftrightarrow\; \overset{}{\diagdown}\!B^{\ominus}\!-\!D\!=\!Y \tag{12.10}$$

the reaction could have been written with equal ease with A as a radical or as an anion. The nature of Y influences the ease of a given reaction. Thus, an acid is much more apt to lose a proton than a hydride ion (Eqs. 12.11).

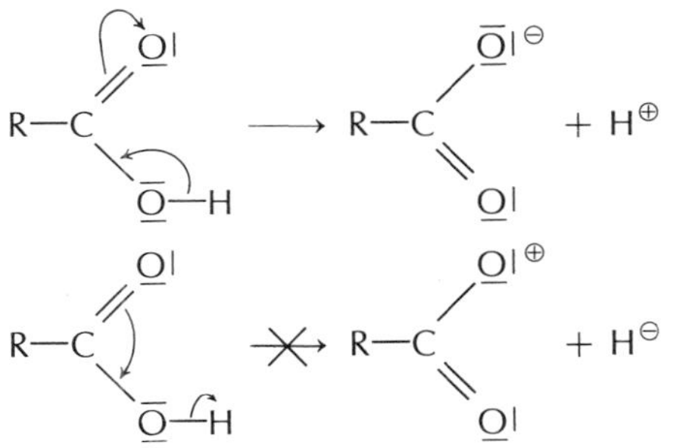

$$R-C\underset{\overline{\underline{O}}-H}{\overset{\overline{\underline{O}}|}{\diagup\!\!\!\!\diagdown}} \;\longrightarrow\; R-C\underset{\underline{\underline{O}}|}{\overset{\overline{\underline{O}}|^{\ominus}}{\diagup\!\!\!\!\diagdown}} \;+\; H^{\oplus}$$

$$\tag{12.11}$$

$$R-C\underset{\overline{\underline{O}}-H}{\overset{\underline{\underline{O}}|}{\diagup\!\!\!\!\diagdown}} \;\xrightarrow{\;\;\;\;}\!\!\!\!\!\!\!\!\!\!\!/\!\!\!\!\!\!\;\; R-C\underset{\underline{\underline{O}}|}{\overset{\underline{\underline{O}}|^{\oplus}}{\diagup\!\!\!\!\diagdown}} \;+\; H^{\ominus}$$

Exercises

12.3 Predict the nmr spectrum of $HC\!\equiv\!CCH_2CH_2C\!\equiv\!CH$.

334 Chapter twelve

12.4 How may $CH_3CH_2CH_2C\equiv CH$ be purified (1-pentene and pentane are the principal impurities) and transformed into 2-pentyne?

12.5 A Grignard reagent is prepared from $CH_3CH=CHCH_2Br$ and allowed to react with acetone. Give the structure of the product.

12.4 Electrophilic substitution reactions

Hydride removal by electrophiles. An electrophile, if strong enough, removes a hydride ion from an alkane and produces a carbonium ion (Eq. 12.12). This step is essential

$$R^{\oplus} + H-R' \rightleftharpoons RH + R'^{\oplus} \tag{12.12}$$

in the alkylation reaction to be described in the next chapter. However, electrophilic attack on carbon to displace a hydrogen ion or carbonium ion is not often observed. Interesting rearrangements are involved in such reactions, but they will be postponed to the next chapter.

If a double or triple bond is present, an electrophile attacks the easily accessible sources of electrons in the π bonds in preference to anything else to give an addition reaction (Chap. 13).

Carbene insertion. The very strong electrophile $:CH_2$, a <u>carbene</u> with only six electrons, readily attacks alkanes to give an <u>insertion</u> reaction. The reaction may be envisioned as occurring in either of two ways (Eqs. 12.13). The first way is a <u>three-center</u>

$$\begin{aligned} R\text{—}H + :CH_2 &\longrightarrow R\text{—}CH_2\text{—}H \\ R\text{—}H + :CH_2 &\longrightarrow R\cdot + \cdot CH_3 \longrightarrow R\text{—}CH_3 \end{aligned} \tag{12.13}$$

reaction in which the very reactive carbene pulls the electrons and the proton of the R—H bond to itself, at the same time allowing the electrons originally owned by the carbene to form a bond to R. The second way is a form of free-radical reaction.

Carbene is so reactive that there is no selectivity in the attack. If it is allowed to attack pentane, the products are as shown in Equation 12.14. Statistically, there are six primary

$$\begin{aligned} C_5H_{12} \xrightarrow{:CH_2} &\; CH_3CH_2CH_2CH_2CH_2 \quad 48\% \\ &\qquad\qquad\qquad\;\; | \\ &\qquad\qquad\qquad CH_3 \\ &+ CH_3CH_2CH_2CHCH_3 \quad 35\% \\ &\qquad\qquad\qquad\;\; | \\ &\qquad\qquad\qquad CH_3 \\ &+ CH_3CH_2CHCH_2CH_3 \quad 17\% \\ &\qquad\qquad\; | \\ &\qquad\qquad CH_3 \end{aligned} \tag{12.14}$$

hydrogen atoms in C_5H_{12}, so we expect $(6/12) \times 100 = 50\%$ of hexane. There are four secondary hydrogen atoms on C_2 and C_4, so we expect $(4/12) \times 100 = 33\%$ of 2-methylpentane. The two hydrogen atoms on C_3 lead us to expect $(2/12) \times 100 = 17\%$

of 3-methylpentane. The observed yields correspond almost exactly. In other words, towards :CH$_2$, all C—H bonds are equally reactive. The reaction is not very useful because of the mixtures formed with most reactants. If π bonds are present, the principal reaction is addition (next chapter).

Exercises

12.6 Predict the products and yields from the reaction of carbene with 2,2,4,4-tetramethylpentane.

12.7 Predict the products and yields from the reaction of carbene with methylcyclobutane.

12.5 Homolytic substitution reactions

Attack upon a C—H or a π bond. In contrast to nucleophilic and electrophilic attack, alkanes, alkenes, and alkynes all yield readily to homolytic attack (Eq. 12.15). The

$$Q\cdot + H-R \rightleftharpoons Q-H + R\cdot \qquad (12.15)$$

attacking reagent must have an odd electron. Again, as with the nucleophiles and electrophiles, the attack by the homolytic reagent is almost always upon a C—H rather than a C—C bond. If a double or triple bond is present, the attacking radical may seek an odd electron from the π bond, but if an α hydrogen is present, the formation of the allylic radical may take precedence (Eq. 12.16).

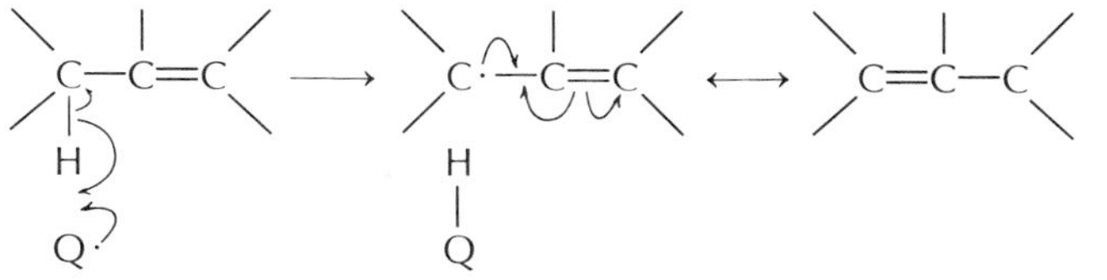

$$\underset{\underset{Q\cdot}{\overset{H}{|}}}{\overset{}{\underset{}{\text{C}-\text{C}=\text{C}}}} \longrightarrow \underset{\underset{Q}{\overset{H}{|}}}{\text{C}\cdots\text{C}=\text{C}} \longleftrightarrow \text{C}=\text{C}-\text{C} \qquad (12.16)$$

Halogenation of methane: mechanism. Let us reconsider the halogenation of methane (Eq. 12.17; review Sections 2.8 and 3.5). Careful study of the reactions shows the

$$CH_4 + X_2 \longrightarrow CH_3X + HX \qquad (12.17)$$

presence of some di-, tri-, and tetrahalomethanes as well as some ethane and haloethanes in the products. Also, if any oxygen is present, the reactions (except for fluorination) are slowed down to a very low rate. Fluorination may be studied by dilution of the mixtures with inert gases such as argon or nitrogen.

The mechanism postulated for the halogen substitution is shown (Eqs. 12.18; compare Eq. 3.15). The reaction is begun by a homolytic rupture of the X—X bond (a) caused by heat or by radiation at room temperature. The halogen atoms may recombine (d), but only if a third body is present (walls of the container) to take up the energy released by bond formation. The atoms have a greater opportunity to collide with those molecules present in highest concentration, X$_2$ and CH$_4$. Collision with an X$_2$ molecule produces only a new

336 *Chapter twelve*

$$X\text{—}X \xrightarrow[\text{or heat}]{h\nu} 2X\cdot \qquad a \quad \text{initiation}$$

$$\begin{array}{rl} X\cdot + H\text{—}CH_3 &\longrightarrow X\text{—}H + \cdot CH_3 \qquad b \\ X\text{—}X + \cdot CH_3 &\longrightarrow X\cdot + X\text{—}CH_3 \qquad c \end{array} \Bigg\} \text{propagation}$$

$$\begin{array}{rl} X\cdot + \cdot X &\xrightarrow{M} X\text{—}X \qquad d \\ X\cdot + \cdot CH_3 &\longrightarrow X\text{—}CH_3 \qquad e \\ H_3C\cdot + \cdot CH_3 &\longrightarrow H_3C\text{—}CH_3 \qquad f \end{array} \Bigg\} \text{terminations} \qquad (12.18)$$

$$\begin{array}{rl} X\cdot + H\text{—}CH_2CH_3 &\longrightarrow X\text{—}H + \cdot CH_2CH_3 \qquad g \\ X\cdot + H\text{—}CH_2X &\longrightarrow X\text{—}H + \cdot CH_2X \qquad h \\ X\text{—}X + \cdot CH_2X &\longrightarrow X\cdot + X\text{—}CH_2X \qquad i \end{array} \Bigg\} \text{secondary reactions}$$

X_2 and $X\cdot$; it is a <u>nonproductive reaction</u>. Collision of $X\cdot$ with CH_4 (b) gives the hydrogen halide and a new <u>methyl radical</u> ($CH_3\cdot$). This radical, in turn, collides more often with X_2 and CH_4 molecules than with anything else. Collision with X_2 (c) leads to CH_3X and $X\cdot$. The new atom ($X\cdot$), upon collision with another CH_4 (b), starts another cycle. Collision of $CH_3\cdot$ with CH_4 can lead only to CH_4 and $CH_3\cdot$, another nonproductive reaction.

The cyclic system (b, c; b, c; b, c; etc.) is a <u>free-radical chain reaction</u>. As shown, CH_3X is the major product. Strictly speaking, we should include a, the <u>initiation step</u>; b, c, the <u>propagation steps</u>; and d, e, f, the <u>termination steps</u> in our meaning of a chain reaction. However, without the propagation steps, in which a new radical is produced each time one disappears, there would be no chain.

Occasionally, a radical finds another radical, as in ,d e, or f. When this happens, two of the chains are terminated, and continuation of the reaction depends upon the formation of new initiators (a) and the presence of other b, c chains. Reaction e gives more product, but f gives ethane. The mechanism does not predict that much ethane is formed, and thus it fits the experimental evidence. What little is formed is subject to attack by $X\cdot$ (g), and the ethyl radical thus produced is a momentary member of a chain when it encounters an X_2 to give a c reaction to form CH_3CH_2X. Even less likely is a collision with a $CH_3\cdot$ to give propane, $CH_3CH_2CH_3$. There is a remote possibility of finding another $CH_3CH_2\cdot$ to give butane, $CH_3CH_2CH_2CH_3$. Both propane and butane may react with $X\cdot$ just as ethane does—but we are close to zero probability now and had better turn to reactions h and i.

As the concentration of methyl halide builds up, more and more often a halogen atom enters into reaction h. Especially near the end of the reaction, when the concentration of methyl halide is greater than that of methane, the major chains become h, i; h, i; etc., rather than b, c; b, c; etc. As the concentration of CH_2X_2 builds up, it enters into reaction to give some CHX_3. Then CHX_3 gives some CX_4.

If any oxygen is present, it reacts very rapidly with any of the carbon-containing radicals to give $ROO\cdot$ (Eq. 12.19). The peroxyradical is relatively stable and reacts so slowly

$$O_2 + \cdot CH_3 \longrightarrow \cdot OOCH_3 \qquad (12.19)$$

with more hydrocarbon (and other radicals) that it breaks the b, c chain. The chain-breaking activity slows down the halogen substitution reaction until all the oxygen has been consumed.

Comparison of ΔH and E_a for halogenation chains. The mechanism seems to fit the experimental data quite well. Let us go on and consider the enthalpy and activation data (Table 12.5). For a, the rate order is $I_2 > F_2 \gg Br_2 \gg Cl_2$ if we go by the E_a values.

Table 12.5 Enthalpy and activation energies for $X_2 + CH_4 \rightarrow CH_3X + HX$[a]

Reaction from Eq. 12.18	X = F		Cl		Br		I	
	ΔH	E_a^b	ΔH	E_a^b	ΔH	E_a^b	ΔH	E_a^b
a	37	37	58	58	46	46	36	36
b	−31	1	1	4	16	18	33	35
c	−71	1	−25.5	2	−24	2	−20	2
d	−37	0	−58	0	−46	0	−36	0
e	−108	0	−83.5	0	−70	0	−56	0
f	−88	0	−88	0	−88	0	−88	0
$CH_4 + X_2 \rightarrow CH_3X = HX$	−102		−24.5		−7		13	

[a] As calculated by use of Tables 3.2, 5.6, and 5.7 for ΔH. All values in kcal/mole.
[b] Very approximate, primarily for illustration.

Thus a cannot be the rate-determining step. Furthermore, the very vigorous fluorination requires an $E_a = 37$ kcal, which does not seem available in the dark at −80°C, under which conditions fluorination still proceeds. Thus, the mechanism must be modified for the fluorine-methane reaction. We do so by proposing molecule-induced homolysis (Eq. 12.20), in which the interaction of two molecules produces two radicals. The ΔH

$$F-F + H-CH_3 \longrightarrow F\cdot + F-H + \cdot CH_3 \tag{12.20}$$

calculated for this reaction is only 6 kcal/mole. The activation energy cannot be much greater, so it seems to be a reasonable modification for fluorination. Chlorination does not require such modification. The molecule-induced homolysis calculates ΔH to be 59 kcal/mole, which is greater than the BDE for Cl_2 (a). From Table 10.2, we see that the light of $\lambda = 490$ nm or less is sufficient to cause a for Cl_2 to occur.

If we pass over reaction b for the moment and look at c, we again find poor correlation with reactivity, although the order is correct. Chlorine, bromine, and iodine are quite close together in enthalpy change for the reaction. Now if we go back to b, we find a parallel in E_a. With the smallest E_a, $F\cdot$ has no difficulty in causing reaction at almost every collision; whereas $I\cdot$ ($E_a = 35$ kcal/mole) has almost no chance and wanders about until it undergoes reactions d or e. In short, it is b that determines the reactivity. This fact may be seen more clearly, perhaps, in Figure 12.7, in which the activation hill that must be climbed in b definitely is the rate-determining factor that fixes the reactivity.

Before going on, we notice that iodination is endothermic (ΔH = 13 kcal/mole), so that $CH_3I + HI \rightarrow CH_4 + I_2$ is exothermic with an E_a of 20 kcal/mole. But I_2 is almost as good as Cl_2 and Br_2 in c. Iodine is a frequently used free-radical trap because of its chain-breaking reactivity and because the iodine atoms are so unreactive that they cannot initiate new chains.

Relative activity of hydrogen atoms at various positions. As we go on to the halogenation of ethane, propane, and others, we must take into account the differing reactivities of hydrogen atoms in primary, secondary, and tertiary positions. If a small amount of halogen is allowed to react with an excess of an alkane, analysis of the products gives the

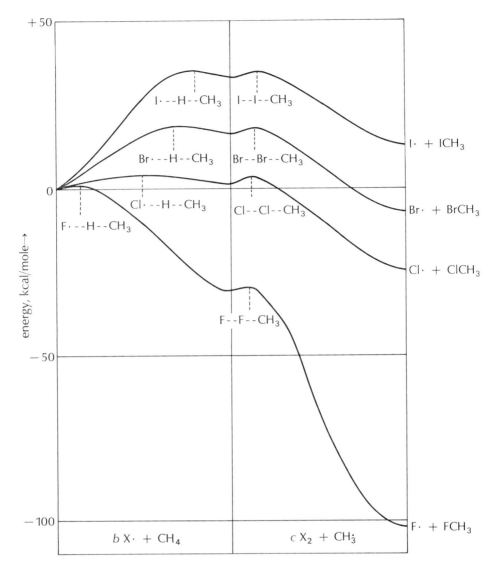

Figure 12.7 Energy-reaction-coordinate diagram for $X\cdot + CH_4 \rightarrow XH + \cdot CH_3$ and for $X_2 + \cdot CH_3 \rightarrow X\cdot + XCH_3$ for F_2, Cl_2, Br_2, and I_2

information needed to find the relative reactivity at each position in the alkane. Propane, upon chlorination, gives about 46% of 1-chloropropane ($ClCH_2CH_2CH_3$), which could arise by replacement of any of the six primary hydrogen atoms. The 54% of 2-chloropropane (isopropyl chloride, $CH_3CHClCH_3$) arises by replacement of either of the two secondary hydrogen atoms. Therefore, each primary atom accounts for 46/6 = 7.7% of product, and each secondary atom accounts for 54/2 = 27% of product. Thus, each secondary atom is 27/7.7 = 3.5 times as active as a primary hydrogen.

In Table 12.6 are given some reactivity ratios, all adjusted to a primary hydrogen atom as unity. Despite the variations, it is clear that tertiary > secondary > primary ≫ methane is the order of reactivity. It is plain that reactivity for chlorination is not very selective,

Table 12.6 Relative rates of removal of hydrogen atoms $X\cdot + H{-}M \rightarrow X{-}H + M\cdot$

M—H	Attacking radical, 40°	
	Cl·	Br·
CH_3—H	0.004	0.0007
RCH_2—H	1	1
R_2CH—H	4.3	220
R_3C—H	6.0	19,400
$(CH_3)_2CClCH_2$—H	0.12	
$ClCH_2$—H	0.07	
Cl_2CH—H	0.01	
Cl_3C—H	0.005	
cyclohexyl-H	2.6	

whereas bromination is. At 40°, we see that isobutane gives 40% of *tert*-butyl chloride (from 1 H × 6) and 60% of isobutyl chloride (from 9 H × 1) upon chlorination but gives 99 + % of *tert*-butyl bromide upon bromination. Similarly, butane gives 26% of butyl chloride (from 6 H × 1) and 74% of *sec*-butyl chloride (from 4 H × 4.3) upon chlorination but 99 + % of *sec*-butyl bromide upon bromination.

The selectivity of bromine (and of chlorine) decreases with increasing temperature. This effect is to be expected, because the slowest reactions near room temperature have the highest activation energies. Selectivity here means selectivity by kinetic control, the influence of E_a upon the reaction. Very reactive attacking groups (F·, Cl·) with small E_a are not selective. They pick off the first hydrogen they run into.

Up to now we have concentrated on the reactivity of the hydrogen atoms and ignored the radical being formed. For each C—H bond being broken, a gaseous radical is formed. The methyl radical is the most reluctant to be formed: it is the least stable of the alkyl radicals. We have seen also that the methyl cation (carbonium ion) is so unstable that it seldom is formed. Consequently, there is a parallel between the stability of radicals and that of carbonium ions: allyl > tertiary > secondary > primary > methyl.

Uses of halogenation. Practically, halogenation of alkanes and cycloalkanes is a poor reaction for laboratory synthetic reactions. Only in a few instances is selectivity great enough to give good yields in chlorination reactions. Bromination is selective enough, but the reactions are slow. In any case, primary products are difficult to come by. Neopentane chlorination is an exception, because only primary hydrogen atoms are present (Eq. 12.21).

$$(CH_3)_4C + Cl_2 \xrightarrow[25°]{h\nu} (CH_3)_3CCH_2Cl + HCl \tag{12.21}$$

On the other hand, fluorination and chlorination are practiced extensively on a commercial scale. Fluorination is modified by the use of a diluent gas (nitrogen) and the presence of metallic chips or gauzes to help carry off the heat evolved. Another modification is to pass fluorine over a bed of a metallic fluoride such as cobaltous fluoride at about 250°, then to flush the reactor with nitrogen and pass in the alkane until the cobaltic fluoride is exhausted. The cycle is repeated (Eqs. 12.22). In this

$$F_2 + 2CoF_2 \longrightarrow 2CoF_3$$
$$R\text{—}H + 2CoF_3 \longrightarrow R\text{—}F + 2CoF_2 + HF \tag{12.22}$$

manner, about half of the total heat evolved comes during the preparation of the cobaltic fluoride, and the temperature of the second step is much easier to control.

Because of the unusual properties of fluorinated alkanes (nontoxicity, inertness, low solubility in both polar and nonpolar solvents) most fluorination reactions are directed toward the formation of the completely fluorinated product (fluorocarbons). Teflon is a polymer of tetrafluoroethylene, $F_2C\text{=}CF_2$ (Chap. 30).

Chlorination is easier to control and usually is done thermally, at 250° to 400°. The chlorination of methane may be carried out, by varying the ratio of reactants, to give CH_3Cl, CH_2Cl_2, $CHCl_3$, or CCl_4 as the major product. In any case, a mixture always is formed. However, it is easily separated by distillation. All four products are useful (solvents, reagents). Other hydrocarbons usually are chlorinated to give only a single substitution as the major product.

Organic liquid breathing. Perfluorinated liquids are used commercially because they are electrically nonconductive, chemically inert, and stable to heat. An additional property of these compounds is high solubility of gases. Although it is undesirable in industry, this property is being used to advantage in medicine in the preparation of artificial blood.

The liquids are insoluble in water and thus must be added as tiny spheres of oil suspended in the solution. These spheres can dissolve large amounts of oxygen and carbon dioxide. The major obstacle to this approach has been that within several days the emulsions aggregate irreversibly in the liver and spleen. This problem may now be solved by the discovery that several perfluorinated compounds, including perfluoromethylcyclohexane, do not accumulate in the liver irreversibly. Artificial blood may be with us soon.

Allylic substituion. If a π bond is present in the molecule, the hydrogen atom α to the π bond is preferentially attacked in free-radical halogen substitution. This kind of reaction is known as <u>allylic substitution</u>. However, the π bond also is attacked, and addition to the double or triple bond occurs. We shall postpone discussion of the situation to the next chapter.

A rather specific reagent for allylic bromination is N-bromosuccinimide (NBS), which has been shown to react as in Equations 12.23. The role of N-bromosuccinimide is to

$$\text{succinimide-N-Br} + \text{Br—H} \longrightarrow \text{succinimide-N}^{\ominus} + Br_2 + H^{\oplus}$$

$$\text{succinimide-N}^{\ominus} + H^{\oplus} \longrightarrow \text{succinimide-N—H} \tag{12.23}$$

$$\diagup_{C=C-\underset{|}{\overset{|}{C}}-H}^{|} + Br\frown Br \longrightarrow \diagup_{C=C-\underset{|}{\overset{|}{C}}\cdot}^{|} + HBr + Br\cdot$$

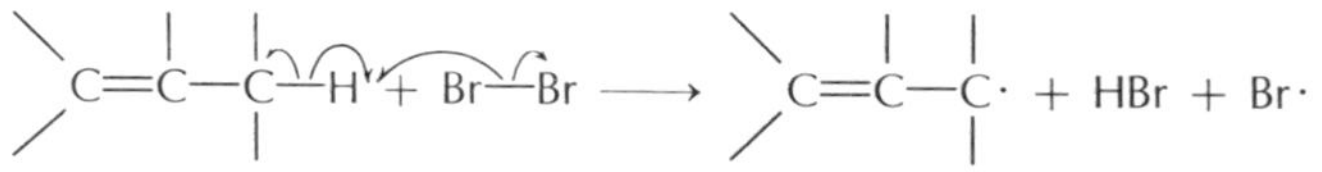

(molecule-induced-homolysis, perhaps)

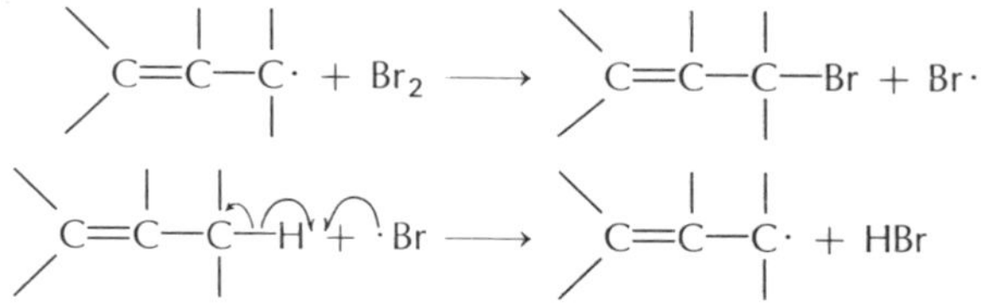

$$\diagup_{C=C-\underset{|}{\overset{|}{C}}\cdot}^{|} + Br_2 \longrightarrow \diagup_{C=C-\underset{|}{\overset{|}{C}}-Br}^{|} + Br\cdot$$

$$\diagup_{C=C-\underset{|}{\overset{|}{C}}-H}^{|} + \cdot Br \longrightarrow \diagup_{C=C-\underset{|}{\overset{|}{C}}\cdot}^{|} + HBr$$

regenerate bromine from the hydrogen bromide being formed. The chain-propagation steps are the last two of the reactions shown. Competition is provided by the π bond for Br·, but the radical so formed, Br—$\overset{|}{\underset{|}{C}}$—$\overset{|}{\underset{|}{\overset{\cdot}{C}}}$—CH, is not allylic and so has a higher activation energy of formation. N-chlorosuccinimide is not as selective as is the bromo analog. If there is a choice, a *tert*-allylic hydrogen atom is replaced in preference to a *sec*-allylic hydrogen, which in turn takes precedence over a primary allylic hydrogen. Because of resonance in the allylic radical, rearranged products are possible (Eq. 12.24). The geometry

$$CH_3CH_2CH=CHCH_3 \xrightarrow{Br\cdot} \begin{array}{c} CH_3\overset{\cdot}{C}HCH=CHCH_3 \\ \updownarrow \\ CH_3CH=CH\overset{\cdot}{C}HCH_3 \end{array} \xrightarrow{Br_2} CH_3CH=CHCHBrCH_3 + Br\cdot \text{ (most)}$$

$$\xrightarrow{Br\cdot} \begin{array}{c} CH_3CH_2CH=CH\overset{\cdot}{C}H_2 \\ \updownarrow \\ CH_3CH_2\overset{\cdot}{C}HCH=CH_2 \end{array} \xrightarrow{Br_2} \begin{array}{c} CH_3CH_2CH=CHCH_2Br + Br\cdot \text{ (some)} \\ \\ CH_3CH_2CHBrCH=CH_2 + Br\cdot \text{ (least)} \end{array}$$

(12.24)

of the double bond is retained in the allylic radical: *cis*-2-pentene gives *cis*-1-bromo-2-pentene. Both *cis*- and *trans*-3-bromo-2-pentene are formed, because the end methyl group (the C_5) may assume either a *cis* or a *trans* position in the allylic radical. □ □

Summary. Although we have spent some time on the homolytic substitution reaction of alkanes, alkenes, and so on, we may summarize briefly by saying that by and large, alkanes are stable to nucleophilic or electrophilic attack and that alkyl groups maintain their identity unless attacked by energetic radicals. The allylic positions have been shown to be the susceptible points of attack in alkenes and in alkynes.

Exercises

12.8 Calculate the percent of each isomer formed in the reaction at 40° of 2-methylbutane with a. Cl_2 and b. Br_2.

342 Chapter twelve

12.9 N-bromosuccinimide is a good electrophile (Eq. 12.23). Why? Succinimide is fairly acidic. Why?

12.10 Thermal chlorinations require fairly high temperatures, yet cooling is required to prevent the temperature from going higher. Why not let the temperature increase without limit?

12.11 How should we carry out the chlorination of methane to obtain the maximum possible yield of methyl chloride?

12.6 Summary

In this chapter we have begun a tabulation of the reactions (specifically, the substitution reactions) of alkanes, alkenes, and alkynes. Because we also wished to include the cycloalkanes and cycloalkenes, we spent some time on the stereochemistry of the cyclic compounds. We should never allow ourselves to think of molecules as the flat symbols we are accustomed to seeing on the page before us. We also introduced ring strain as a factor to be considered along with nonbonded repulsions in any discussion of reactivity of cyclic compounds.

We have seen that the substitution reactions of alkanes, alkenes, and alkynes are relatively few. Nucleophiles do not affect alkanes unless the basicity is high. In other words, we are dealing with very, very weak acids and very, very strong bases. Alkenes, we found, are acidic enough to be isomerized by bases, the equilibrium favoring the formation of the most highly substituted alkene. Allylic resonance stabilized the anion enough to allow the reaction to occur. Alkynes with the triple bond at the end of the chain, we learned, are good enough acids to cause decomposition of Grignard reagents and to form salts.

Electrophilic substitution is limited to the carbene reaction or insertion. Perhaps it should have been listed under homolytic reactions, but the characteristics and results favor our thinking of carbene as an electrophilic reagent.

Homolytic substitution is the major pathway for conversion of an alkane into a more reactive substance. Halogenation is a generally useful commercial process.

The presence of a π bond introduces the possibility for addition to occur as well as substitution. This subject will be discussed in the next chapter.

Some of the new reactions given in this chapter are listed here.

$$RC\equiv CH + Ag(NH_3)_2NO_3 \longrightarrow RC\equiv CAg \qquad (12.5)$$

$$RC\equiv CH + EtMgBr \longrightarrow RC\equiv CMgBr + EtH \qquad (12.7)$$

$$(CH_3)_2CHCH=CH_2 \xrightarrow[\Delta]{NaNH_2} (CH_3)_2C=CHCH_3 \qquad (12.9)$$

$$RH + :CH_2 \longrightarrow RCH_3 \qquad (12.13)$$

$$RH + X_2 \xrightarrow{h\nu} RX + HX \qquad (12.17)$$

$$RH + 2CoF_3 \longrightarrow RF + HF + 2CoF_2 \qquad (12.22)$$

$$(CH_3)_2CHCH=CH_2 \xrightarrow{NBS} (CH_3)_2CBrCH=CH_2 \qquad (12.24)$$

12.7 Problems

Show how to carry out the conversions shown. More than one step may be necessary. Assume that any reagents and solvents needed are available.

12.1 $(CH_3)_2CHCH=CH_2 \longrightarrow CH_2=\underset{\underset{CH_3}{|}}{C}-CH=CH_2$

12.2 cyclohexane $\longrightarrow$ 1-(prop-1-yn-1-yl)cyclohexan-1-ol (cyclohexane ring with OH and C≡CCH$_2$ substituents on the same carbon)

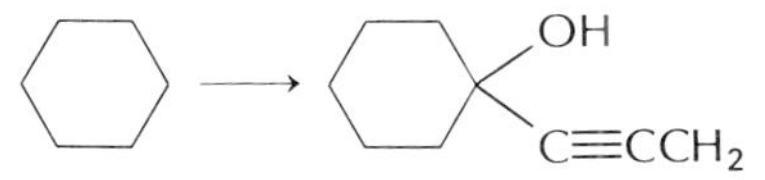

12.3 $(CH_3)_4C \longrightarrow (CH_3)_3CCH_2COOH$

12.4 $(CH_3)_3CH \longrightarrow (CH_3)_3CCH=CH_3$

12.5 Assume that $tBu \gg iPr > Et > Me > Cl$ in size. Draw the most stable conformation of:
  a. trans-1,2-dimethylcyclohexane
  b. cis-1-tert-butyl-4-chlorocyclohexane
  c. trans-1-methyl-3-ethylcyclohexane
  d. cis-1-chloro-2-methylcyclohexane.

12.6 Give the major organic product to be expected in the reactions given:
  a. $\underset{H_3C}{\overset{H_3C}{\diagdown\diagup}}CHCH=CH_2 + NBS \longrightarrow$
  b. $tBuCH_2CH_3 + :CH_2 \longrightarrow$
  c. $(CH_3)_2CHCH_2CH_3 + Br_2 \xrightarrow{heat}$
  d. $(CH_3)_4C + 2Br_2 \xrightarrow{heat}$
  e. $CH_3CH_2C{\equiv}CH + CH_3CH_2MgBr \xrightarrow{ether}$

12.8 Jack Hine: Halogen substituents

Professor Jack Hine probably is best known for his research work on divalent carbon intermediates. Now at Ohio State, he first went to Georgia Tech from postdoctoral experience at M.I.T. and then at Harvard after earning his Ph.D. at Illinois. He is the author of several books and many research papers.

   Although the assignment of laying out a worthwhile original research project is nowadays a common part of the examination system for the doctoral degree in many schools, it was very uncommon in my graduate student days. Consequently, I had given little thought to a research project of my own until I was a first-year postdoctorate with John D. Roberts at M.I.T. and was applying to various agencies for a second-year postdoctorate to work independently. When I remember the struggles I went through trying to decide which idea was more significant, more tractable, etc., I feel much more sympathetic to the graduate students around me who are going through similar struggles. The problem I finally settled on for my application for a Du Pont Postdoctoral Fellowship at Harvard seemed solid but not exciting. I proposed to study the effect of halogen substituents on reactivity by the S_N1 and S_N2 mechanisms. I thought that the relative simplicity of halogen substituents might make them particularly valuable in shedding more light on the effect of structure on reactivity in nucleophilic substitution reactions. The problem seemed worth exploring, but its proposal certainly required no great originality. Happily, I was awarded the fellowship anyway.
   When I got to Harvard and settled in Paul Bartlett's group, I began work by fleshing out the rather incomplete literature search I had made on the project at M.I.T. Very few relevant reliable kinetic studies had been made, but I did run across some interesting, if crude, studies by Petrenko-Kritschenko and coworkers. A number of the halides, including polyhalides, of methane had been treated under various conditions and the extent of liberation of halide ions measured. The reaction mixtures were sometimes heterogeneous, no rate constants were determined, and the nature of the organic reaction products was not investigated, but the relative reactivities showed some interesting patterns. Under fairly nonbasic conditions the chlorides showed reactivity patterns of the type $CH_3Cl > CH_2Cl_2 > CHCl_3 < CCl_4$. This suggested that, like α-methyl substituents, α-halogen substituents decrease S_N2 reactivity and increase S_N1 reactivity. The marked S_N2 reactivity of the methyl halide decreases steadily on going to CH_2X_2 and then to CHX_3, but by the

time CX_4 is reached the S_N1 reactivity has become sufficient to make the S_N1 mechanism dominant. Under strongly basic conditions, however, this simple pattern was broken, and the reactivity sequence $CH_3Cl > CH_2Cl_2 < CHCl_3 > CCl_4$ was observed; that is, the haloform had become unexpectedly reactive. These observations suggested that the haloforms were being transformed to carbanions, which were losing halide ions to give derivatives of divalent carbon as reaction intermediates. There had been good evidence

$$CHCl_3 \underset{}{\overset{-H^\oplus}{\rightleftharpoons}} CCl_3^\ominus \xrightarrow{-Cl^\ominus} CCl_2 \xrightarrow[H_2O]{OH^\ominus} CO + HCO_2^\ominus$$

for the formation of divalent carbon species in the homolysis of diazo compounds but not as intermediates in polar reactions.

The work in the following months to establish the proposed divalent carbon mechanism had its ups and downs. The idea of capturing the intermediate with sodium thiophenoxide to obtain triphenyl orthothioformate seemed a good one, but I could not isolate the product in decent yield. In my wife's hands the yield rose to a respectable 85 percent. The product of the first capturing experiment had been diphenyl disulfide; I hadn't removed the peroxide impurities from the dioxane in the solvent, and they had oxidized the sodium thiophenoxide. I try to remember such stupidities on such occasions as when I found a graduate student drying chloroform (bp 61°C) by putting it on a watch glass in a desiccator.

13 Alkanes, Alkenes, Alkynes II

This long chapter is the equivalent of two ordinary chapters. The first half is concerned with addition reactions; the second half, with oxidation, reduction, sources, and preparation of alkanes, alkenes, and alkynes.

There are many new reactions; however, most of them involve the use of new reagents in ways we are already familiar with: namely, additions to π bonds. We have looked over the variety of possible addition reagents as a background to understanding the action of the relatively few reagents to which we pay most of our attention.

To use a rough analogy, we have reached a point in learning to play the piano (organic chemistry) when we begin to use both hands and some of the keys outside the middle octave.

13.1 Nucleophilic addition reactions   345
   1. Difficulty of carbanion formation; addition of tBuLi.   2. HCN addition.
   3. Alcohols to alkynes.   4. Negatively substituted alkenes.

13.2 Electrophilic addition reactions   347
   1. AA and AB reagents; Markovnikov's rule.   2. Stereochemistry: trans addition.   3. AA reagents: bromonium ions.   4. Chlorine; fluorine; iodine.   5. Competition between addition and substitution.   6. Alkynes and X_2.   7. AB reagents: acids.   8. Isomerization; solvent participation.   9. Hydration of alkenes and alkynes.   10. Dimerization.   11. Alkylation; hydride transfer.   12. Alkyne addition to alkyne.   13. Carbene addition. Exercises.

13.3 Hydroboration and similar addition reactions   356
   1. Hydroboration; isomerization.   2. Hydrolysis: reduction of an alkene.
   3. Oxidation: hydration of an alkene.   4. Displacement. Exercises.

13.4 Homolytic addition reactions   360
   1. Introduction.   2. Halogen addition.   3. Alkynes.   4. AB reagents and the peroxide effect.   5. Cyclopropane reactions.   6. Competition between addition and allylic substitution.   7. Summary. Exercises.

13.5 Oxidation reactions   365
   1. Oxidation by substitution.   2. Carbon dioxide in the atmosphere; air

pollution; smog.   3. Oxidation by addition: ozone.   4. Epoxidation.
5. Halohydrins; dihalides.   6. Osmium tetroxide; alkaline potassium permanganate.   7. Mild and drastic oxidations. Exercises.

13.6   Reduction reactions   374
1. Catalytic hydrogenation.   2. Use of sodium borohydride.
3. Reduction by sodium or lithium.   4. Diimide. Exercises.

13.7   Sources: petroleum   377
1. Natural gas and petroleum.   2. Refining.   3. Summary.

13.8   Preparation of alkanes   379
1. Substitution reactions.   2. Addition reactions: Wolff-Kishner and Clemensen reductions.   3. Kolbe electrolysis: fusion of salts of acids.
4. Summary. Exercises.

13.9   Preparation of alkenes   383
1. Summary of reactions previously given.   2. Ester pyrolysis.   3. Cope reaction.   4. Summary. Exercises.

13.10   Preparation of alkynes   384
1. Acetylene: preparation, uses.   2. Dehydrohalogenation. Exercises.

13.11   Summary   386

13.12   Problems   389

13.13   Bibliography   390

13.14   Arie J. Haagen-Smit: From terpenes and essential oils to smog   391a

13.1   Nucleophilic addition reactions

Difficulty of carbanion formation; addition of tBuLi.   Many reagents add to alkenes and alkynes. The mechanisms allow us to classify the reactants and products.

Nucleophilic addition to double and triple bonds is not a common reaction (Eqs. 13.1). The first step is attack by the nucleophile, which must force the electron pair from a

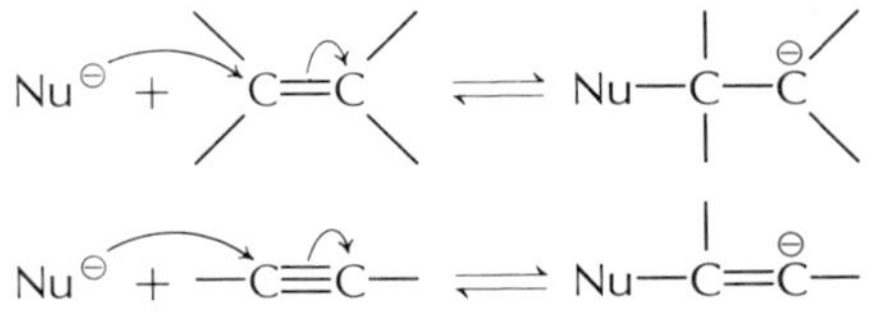

$$Nu^{\ominus} + \ \ \mathrm{C}{=}\mathrm{C} \ \ \rightleftharpoons \ \ Nu-\underset{|}{\overset{|}{\mathrm{C}}}-\underset{|}{\overset{\ominus}{\mathrm{C}}}$$

$$Nu^{\ominus} + -\mathrm{C}{\equiv}\mathrm{C}- \ \ \rightleftharpoons \ \ Nu-\underset{|}{\mathrm{C}}{=}\overset{\ominus}{\mathrm{C}}-$$

(13.1)

π bond on to a carbon atom to form a carbanion. Given the known reactivity of carbanions, the reversability of the reaction is no surprise. If the reaction goes, the nucleophile bonds to the carbon atom that allows the more stable anion to be formed (Eq. 13.2).

$$Nu^\ominus + CH_3CH=CH_2 \begin{array}{c} \nearrow CH_3CH-CH_2^\ominus \quad \text{primary anion, more} \\ | \\ Nu \\ \\ \searrow CH_3\overset{\ominus}{C}H-CH_2 \quad \text{secondary anion, less} \\ | \\ Nu \end{array} \quad (13.2)$$

One of the few nucleophiles able to add to alkenes and nonterminal alkynes is tBuLi (Eq. 13.3). (If a 1-alkyne is used, the nucleophile simply takes away the acidic

$$Li^\oplus tBu^\ominus + H_2C=CH_2 \rightleftharpoons tBuCH_2CH_2^\ominus Li^\oplus \quad (13.3)$$

alkynyl proton to give a salt of the 1-alkyne.) The tBuLi is written as an ion pair to emphasize the nucleophilic strength of the anion $tBu^\ominus$. The new carbanion is primary and more stable (less basic and less nucleophilic) than the tertiary nucleophile, and the equilibrium lies to the right. Once formed, the new lithium reagent may be used like any other lithium reagent to add to a $\diagdown C=O \diagup$ or to attack a weak acid (Eqs. 13.4).

$$\begin{array}{l} tBuCH_2CH_2Li + CO_2 \longrightarrow tBuCH_2CH_2COO^\ominus Li^\oplus \\ tBuCH_2CH_2Li + H_2O \longrightarrow tBuCH_2CH_3 + LiOH \end{array} \quad (13.4)$$

In complete reaction sequence it appears that tBuCOOH or tBu—H has been added to ethylene, which, of course, is what has happened in a roundabout way. The Grignard reagent (tBuMgX) is less reactive than tBuLi and will not add to ethylene. □ □

HCN addition. Hydrogen cyanide adds to alkenes in the presence of dicobalt octacarbonyl ($Co_2(CO)_8$), presumably by nucleophilic attack (Eq. 13.5). Conversions

$$CH_3CH_2CH=CH_2 + HCN \xrightarrow[\substack{130° \\ \text{pressure}}]{\text{cat.}} CH_3CH_2CHCH_3 \atop | \atop CN \quad (13.5)$$

decrease if the chain is increased in length or if the double bond is not at the end of a chain. Hydrogen cyanide adds even more readily to alkynes, and the addition to acetylene is one of the reactions used industrially to produce acrylonitrile (Eq. 13.6). Acrylonitrile is the

$$HC\equiv CH + HCN \xrightarrow[\substack{NH_4Cl \\ HCl \ 90°}]{CuCl} H_2C=CHCN \quad (13.6)$$

monomer used for polymerization and formation of acrylic fibers (rugs, etc.).

Alcohols to alkynes. Bases catalyze nucleophilic addition of alcohols to alkynes to give vinyl ethers (Eq. 13.7). Vinyl ethers are polymerizable also, but the polymers are not as useful as those obtained from acrylonitrile.

$$CH_3C\equiv CH + ROH \xrightarrow[150°]{OH^{\ominus}} \underset{RO}{\overset{CH_3}{>}}C=CH_2 \qquad (13.7)$$

Negatively substituted alkenes. We should mention that alkenes with strong −I groups substituted about the double bond force the molecule to undergo nucleophilic additions (Eqs. 13.8).

$$F_2C=CF_2 + Cl_2 \longrightarrow F_2C-\overset{\ominus}{C}F_2 \longrightarrow F_2C-CF_2$$
$$\underset{}{\overset{Cl}{|}} \qquad \underset{}{\overset{Cl}{|}} \quad \underset{Cl}{}$$

$$\underset{NC}{\overset{NC}{>}}C=C\underset{CN}{\overset{CN}{<}} + HF \longrightarrow \underset{NC}{\overset{NC}{>}}\overset{F}{\underset{|}{C}}-\overset{\ominus}{C}\underset{CN}{\overset{CN}{<}} \longrightarrow \underset{NC}{\overset{NC}{>}}\overset{F}{\underset{|}{C}}-\overset{|}{\underset{H}{C}}\underset{CN}{\overset{CN}{<}} \qquad (13.8)$$

$$F_2C=CF_2 + HCN \longrightarrow F_2C-\overset{\ominus}{C}F_2 \longrightarrow F_2C-CHF_2$$
$$\overset{CN}{\underset{}{|}} \qquad \overset{CN}{\underset{}{|}}$$

13.2 Electrophilic addition reactions

AA and AB reagents; Markovnikov's rule. Electrophilic reagents in embarrassing profusion add readily to alkenes and alkynes. Happily, all may be thought of as A—A or A—B reagents, where A or A$^{\oplus}$ is the electrophilic portion of the molecule. The electrophile may be written as leading the attack upon the electrons in the π bond to form a carbonium ion (Eq. 13.9). If the alkene or alkyne is substituted in such a way that two carbonium ions

$$RCH\overset{\frown}{=}CH_2 + A^{\oplus} \underset{}{\overset{slow}{\rightleftharpoons}} R\overset{\oplus}{C}H-CH_2-A \qquad (13.9)$$

are possible ($R\overset{\oplus}{C}HCH_2A$ or $RCHA\overset{\oplus}{C}H_2$), the more stable ion is formed almost exclusively. Knowing that the most stable ion is an intermediate, we can predict the orientation of the product (Eq. 13.10). When the addend is AB, the product is RCHBCH$_2$A, not RCHACH$_2$B.

$$R\overset{\oplus}{C}H-CH_2-A + B^{\ominus} \xrightarrow{fast} R\overset{|}{\underset{}{C}}H-CH_2-A \qquad (13.10)$$
$$\phantom{R\overset{\oplus}{C}H-CH_2-A + B^{\ominus} \xrightarrow{fast} R}\overset{B}{\underset{}{|}}$$

This orientation was stated as a much more limited rule in 1870 by Markovnikov. The original rule applied to the addition of HX to an alkene, stating that "the halogen adds itself to the less hydrogenated carbon atom." Thus we have come to speak of Markovnikov and anti-Markovnikov addition with reference to the direction of addition. This distinction arises only in the case of the addition of an unsymmetrical reagent AB to an unsymmetrical alkene or alkyne.

348   Chapter thirteen

Stereochemistry: trans addition. The four possible situations are illustrated in Figure 13.1 for the additions of AA and of AB to cyclopentene and to 1-methylcyclopentene. The cyclic compounds clearly show steric consequences. Only *trans* addition is shown—it is the predominate result for electrophilic addition reactions. We see that *a* and *b* are enantiomers and that *c* is identical with *a* and *d* with *b*. Therefore, addition of AB to cyclopentene gives a racemic mixture of *a* and *b*. The addition of AA to cyclopentene gives a racemic mixture of the enantiomers *e* and *f*.

The addition of AA to 1-methylcyclopentene gives the racemic mixture *k* and *l*. The addition of AB to 1-methylcyclopentene gives a mixture of the racemates of *g* and *h* and of *i* and *j* if no directive influences are operating. Application of our principle of the more stable carbonium ion predicts that *i* and *j* will be the major racemic product. Only in the addition of AB to the unsymmetrical cycloalkene does the possibility arise of forming two racemic products. The major product is easily predicted by use of the concept of the more stable carbonium ion.

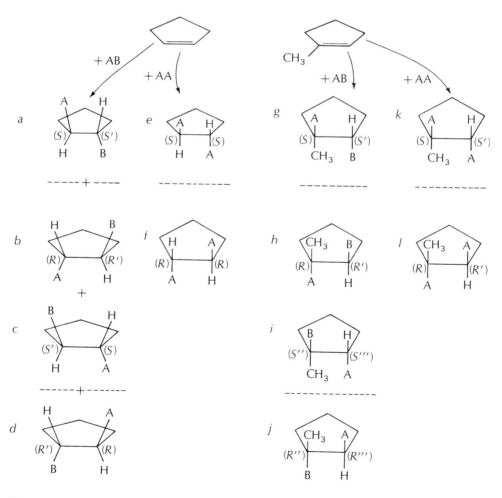

Figure 13.1 Stereochemistry of trans addition of symmetrical and unsymmetrical reagents to symmetrical and unsymmetrical alkenes. For determination of (R) and (S), assume that A and B take priority over other groups present.

13.2 Electrophilic addition reactions

AA reagents: bromonium ions. Aside from hydrogenation and oxidation (discussed in Secs. 13.5 and 13.6), there are only a few AA reagents. The most studied AA reagents have been the halogens. As we pointed out earlier, X_2 adds *trans* to an alkene or an alkyne in almost all instances to give *e* and *f* or *k* and *l* types of racemates. Thus *trans*-2-butene takes up bromine in polar solvents (acetic acid, ethers, etc.) to produce *meso*-(RS)-2,3-dibromobutane. The isomer, *cis*-2-butene, gives equal amounts of 2,3-(RR)-dibromobutane and 2,3-(SS)-dibromobutane.

Because only *trans* addition is observed, the intermediate carbonium ion cannot be a complete answer. If the ion exists for any time at all, free rotation should come into play. The ion from *cis*-2-butene should rapidly equilibrate with the more stable rotamer (Eq. 13.11).

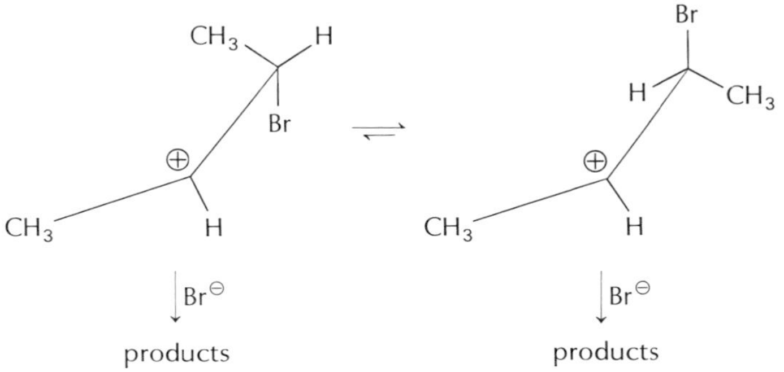

$$\text{(13.11)}$$

Either rotamer should collapse with $Br^{\ominus}$ coming in from above or below the ion to give about equal amounts of *cis* and *trans* addition.

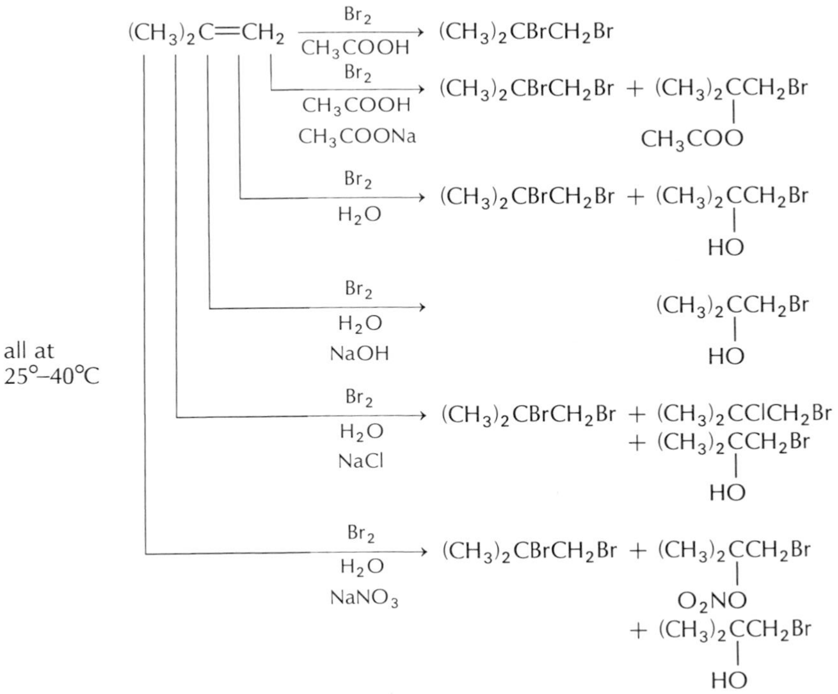

$$(CH_3)_2C{=}CH_2 \xrightarrow[CH_3COOH]{Br_2} (CH_3)_2CBrCH_2Br$$

$$\xrightarrow[\substack{CH_3COOH \\ CH_3COONa}]{Br_2} (CH_3)_2CBrCH_2Br + (CH_3)_2CCH_2Br$$
$$\qquad\qquad\qquad\qquad\qquad\qquad\qquad\qquad\qquad |$$
$$\qquad\qquad\qquad\qquad\qquad\qquad\qquad\qquad\qquad CH_3COO$$

all at 25°–40°C

$$\xrightarrow[H_2O]{Br_2} (CH_3)_2CBrCH_2Br + (CH_3)_2CCH_2Br$$
$$\qquad\qquad\qquad\qquad\qquad\qquad\qquad\qquad\qquad |$$
$$\qquad\qquad\qquad\qquad\qquad\qquad\qquad\qquad\qquad HO$$

$$\xrightarrow[\substack{H_2O \\ NaOH}]{Br_2} (CH_3)_2CCH_2Br$$
$$\qquad\qquad\qquad\qquad\qquad\qquad\qquad\qquad\qquad |$$
$$\qquad\qquad\qquad\qquad\qquad\qquad\qquad\qquad\qquad HO \qquad\qquad\text{(13.12)}$$

$$\xrightarrow[\substack{H_2O \\ NaCl}]{Br_2} (CH_3)_2CBrCH_2Br + (CH_3)_2CClCH_2Br$$
$$\qquad\qquad\qquad\qquad\qquad\qquad + (CH_3)_2CCH_2Br$$
$$\qquad\qquad\qquad\qquad\qquad\qquad\qquad\qquad\qquad |$$
$$\qquad\qquad\qquad\qquad\qquad\qquad\qquad\qquad\qquad HO$$

$$\xrightarrow[\substack{H_2O \\ NaNO_3}]{Br_2} (CH_3)_2CBrCH_2Br + (CH_3)_2CCH_2Br$$
$$\qquad\qquad\qquad\qquad\qquad\qquad\qquad\qquad\qquad |$$
$$\qquad\qquad\qquad\qquad\qquad\qquad\qquad\qquad\qquad O_2NO$$
$$\qquad\qquad\qquad\qquad\qquad\qquad + (CH_3)_2CCH_2Br$$
$$\qquad\qquad\qquad\qquad\qquad\qquad\qquad\qquad\qquad |$$
$$\qquad\qquad\qquad\qquad\qquad\qquad\qquad\qquad\qquad HO$$

The bromonium ion was suggested as a way out of the quandary. We imagine that

$$\underset{\substack{|\overset{\oplus}{Br}| \\ \diagup \diagdown \\ C\text{———}C \\ \diagup \quad \diagdown}}{} \longleftrightarrow \underset{\substack{|\overline{Br}| \\ \diagup \diagdown \\ C\text{—}\overset{\oplus}{C} \\ \diagup \quad \diagdown}}{} \longleftrightarrow \underset{\substack{\diagdown Br \diagdown \\ \overset{\oplus}{C}\text{—}C \\ \diagup \quad \diagdown}}{} \longleftrightarrow \underset{\substack{|\overline{Br}|^{\oplus} \\ \diagdown \quad \diagup \\ C\!=\!C \\ \diagup \quad \diagdown}}{}$$

the positive charge is distributed among all three atoms. The resonance energy helps the stability and the ease of formation of the ion. The bromine atom is bonded somewhat to each carbon and the stereoisomerism of the alkene is preserved. Subsequent reaction with $Br^{\ominus}$ now must occur from the side opposite the Br already present in the cation to give *trans* addition only. The $Br^{\ominus}$ attaches to the more positive carbon of the ion if direction of addition is in question. The last point may be shown by the participation of added nucleophiles (Eqs. 13.12). The added nucleophiles and the solvent compete for the intermediate: the higher the concentration of a given anion, the more of that product is formed. □ □

Chlorine; fluorine; iodine. The addition of chlorine to alkenes in polar solvents is similar to that of bromine, though faster. Here again we need to postulate a chloronium ion to account for *trans* addition. The chloronium ion must be less stable than the bromonium ion, because more carbonium ion reactivity has been observed. Thus, addition of hypochlorous acid (Eq. 13.13) to isobutylene gives about 85% of $(CH_3)_2C(OH)CH_2Cl$ and

$$Cl_2 + H_2O \rightleftharpoons HOCl + HCl \rightleftharpoons H_2\overset{\oplus}{O}Cl + Cl^- \qquad (13.13)$$

about 10% of $CH_2\!=\!C(CH_3)CH_2Cl$ by loss of a proton from the presumed $(CH_3)_2\overset{\oplus}{C}CH_2Cl$. The same products in the same amounts resulted from the hydrolysis at the same temperature (45°) of the dichloro compound, $(CH_3)_2C(Cl)CH_2Cl$: the reaction must have been of the E1 type. □ □

In general, the order is $Cl_2 > BrCl > Br_2 > ICl > IBr > I_2$ in rate for polar addition. In other words, it seems that BrCl, for example, approaches the π bond so that a probable transition state is as shown in Equation 13.14. The more electronegative halogen atom

$$\underset{\substack{\diagdown \diagup \\ C \\ \| \quad + Br\text{—}Cl \\ C \\ \diagup \diagdown}}{} \rightleftharpoons \delta\oplus \underset{\substack{\diagdown \diagup \\ C \\ \vdots\diagdown \\ \quad Br\text{—}\overset{\delta\ominus}{Cl} \\ \vdots\diagup \\ C \\ \diagup \diagdown}}{} \longrightarrow \underset{\substack{\diagdown \diagup \\ C \\ \diagdown \\ \quad Br^{\oplus} + Cl^{\ominus} \\ \diagup \\ C \\ \diagup \diagdown}}{} \qquad (13.14)$$

becomes the negative ion, leaving the less electronegative halogen atom to form the positive ion. There probably is an association complex between the alkene and halogen before the displacement reaction. The reaction may be classified as an S_N2 attack by the weakly nucleophilic π bond upon a halogen molecule, displacing a halide ion.

As we have seen, the bromonium ion does not seem to allow much carbonium ion formation, but the chloronium ion equilibrates to some extent with a carbonium ion. The second step may involve any nucleophile (usually an anion) that can exist in solution with the halogen. In general, the major product is that predicted by the bromonium (or chloronium ion), unless steric factors intervene.

Fluorine is so reactive that the usual reagent is CoF_3 in ether. Iodine is so unreactive that iodination is an unusual reaction. Also, di-iodides are so unstable that they must be treated gently (Eq. 13.15). Solutions of sodium iodide in methanol or acetone frequently are

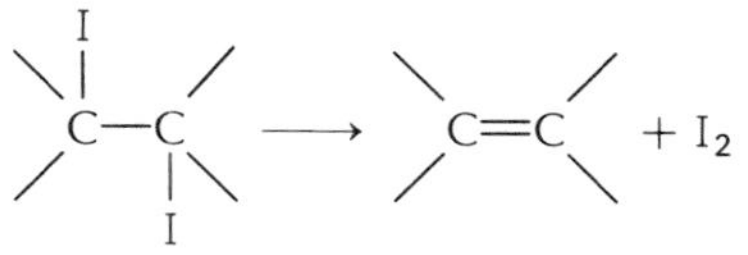

$$\diagdown_{\text{C}-\text{C}}\diagup^{\text{I}}_{\text{I}} \longrightarrow \diagdown_{\text{C}=\text{C}}\diagup + I_2 \qquad (13.15)$$

used to cause dehalogenation of dihalides (Eq. 13.16) instead of the zinc dust method

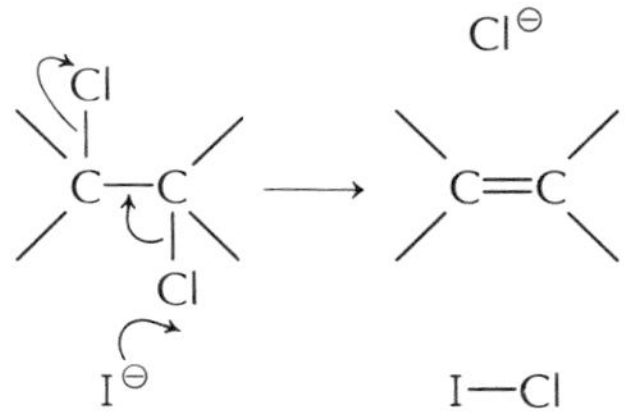

$$\diagdown_{\text{C}-\text{C}}\diagup^{\text{Cl}}_{\text{Cl}} \longrightarrow \diagdown_{\text{C}=\text{C}}\diagup + \text{Cl}^{\ominus} \qquad (13.16)$$

$$\text{I}^{\ominus} \qquad \qquad \text{I}-\text{Cl}$$

and

$$\text{I}^{\ominus} + \text{I}-\text{Cl} \longrightarrow \text{I}-\text{I} + \text{Cl}^{\ominus}$$

(Eqs. 3.36, 3.37, and 3.38). The alkene formed does not react with the iodine produced as the reaction proceeds.

> An old laboratory method used to purify alkenes contaminated with alkanes of similar bp was to add bromine to the mixture (polar solvent), distil the alkanes from the now much higher boiling dibromide, add zinc dust to the distilling flask, and distil the pure alkene as the dehalogenation reaction took place. Iodide ion dehalogenation also is useful in the last step.

Competition between addition and substitution. Bromination and chlorination substitution reactions, as we have seen, are catalyzed by light. In polar solvents, the ionic electrophilic addition usually goes so rapidly that competition with a light-catalyzed substitution or addition reaction is hardly a problem. Nevertheless, careful workers carry out the reaction in the dark to lessen any such competition. Likewise, to reduce the activity of the solvent in the second step, it is good practice to add LiCl to chlorine addition mixtures. A test-tube reaction that will show the presence of a double or triple bond is to see if a few drops of a bromine solution are decolorized rapidly without HBr evolution.

Alkynes and X_2. Alkynes react slowly as compared with alkenes for most electrophilic addition reactions, perhaps because something like a $R\overset{\oplus}{C}=CHX$ ion is less stable than an ordinary carbonium ion. Halogens add *trans*, and the direction is Markovnikov (Eq. 13.17) for AB addition.

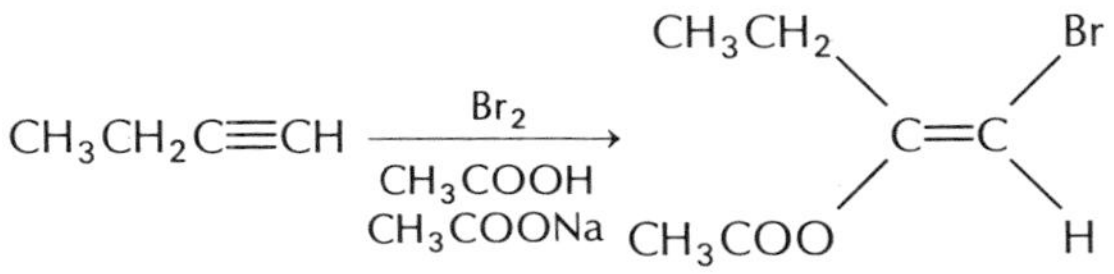

$$CH_3CH_2C\equiv CH \xrightarrow[\substack{CH_3COOH \\ CH_3COONa}]{Br_2} \underset{CH_3COO}{\overset{CH_3CH_2}{\diagdown}} C=C \overset{Br}{\underset{H}{\diagup}} \qquad (13.17)$$

AB reagents: acids. Going on to the AB unsymmetrical reagents, we have already mentioned the interhalogen compounds and the hypohalous acids, as well as the variety of products obtainable by variation of solvent and nucleophile during chlorination and bromination.

Strong acids constitute another class of AB reagents. An alkene or alkyne is attacked by a proton from the acid ($H_3O^\oplus$, HX, $HONO_3$, H_2SO_4, etc.) to give a carbonium ion. Some evidence exists for a rapid, reversible preliminary complex formation (Eq. 13.18). The

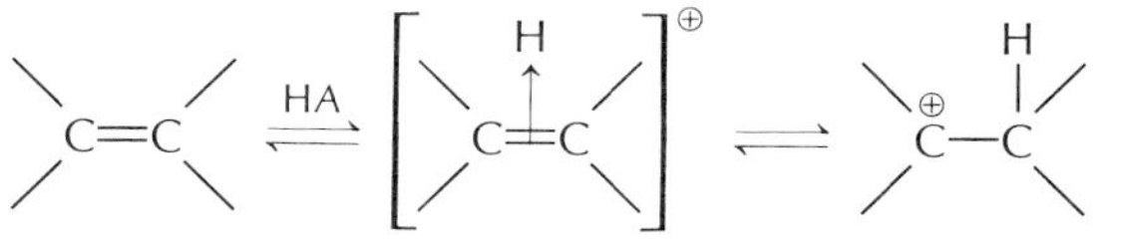

$$\diagdown C = C \diagup \; \underset{}{\overset{HA}{\rightleftarrows}} \; \left[\diagdown \underset{}{\overset{H}{C \overset{\uparrow}{=} C}} \diagup \right]^{\oplus} \; \rightleftarrows \; \diagdown \overset{\oplus}{C} - \overset{H}{\underset{|}{C}} \diagup \qquad (13.18)$$

complex should not be confused with what looks like the similarly formulated chloronium or bromonium ion. The subsequent reactions are all those we have come to expect of carbonium ions. Rearrangements and loss of a proton to give a new alkene frequently occur. The addition reactions are completed by the second step of an S_N1 reaction: namely, bond formation of the carbonium ion with the best available nucleophile.

Practically any substance with even a moderately active hydrogen may be added to an alkene or alkyne. A short list includes H_2SO_4, $HONO_2$, HX, HOX, RCOOH, H_2O, ROH, H_2S, RSH, and even RH. Careful control of conditions and amounts is required in most instances.

Isomerization; solvent participation. Let us begin with an example of excess alkene in momentary contact with solid acidic materials under fairly drastic conditions: for example, with H_3PO_4 on alumina as the acidic material at 250° to 400°C (Eqs. 13.19).

Passage of 1-butene over the catalyst at 400° gives almost complete conversion to a mixture of 52% *trans-* and 48% *cis-*2-butene. Recycling of the mixture gives increasing amounts of isobutylene, approaching 50% of the total. Other alkenes isomerize in a similar way. The unstable primary carbonium ion serves only as a pathway between the stable

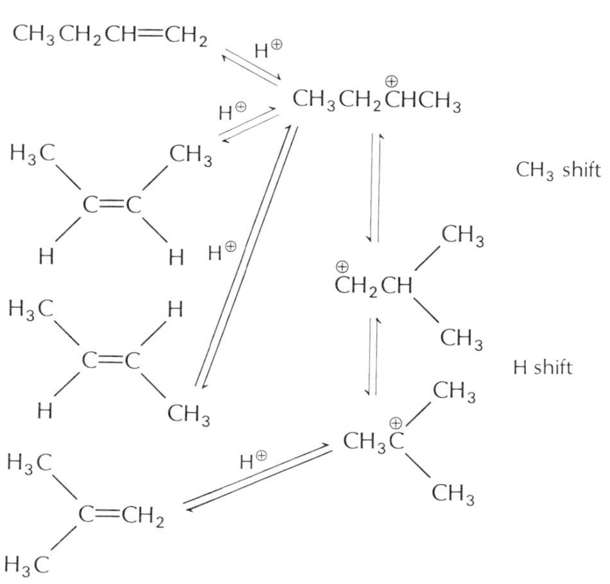

(13.19)

secondary and the more stable tertiary carbonium ions. Acidic isomerization of alkenes and alkynes is more rapid and requires less drastic conditions than does basic isomerization.

A liquid-phase example is the conversion of $t\text{BuCH}=CH_2$ (bp 41°) into a mixture of $(CH_3)_2C=C(CH_3)_2$ (bp 73°) and $(CH_3)_2CHC(CH_3)=CH_2$ (bp 56°) by heating under reflux with 30% sulfuric acid. The $t\text{BuCH}=CH_2$ concentration at equilibrium is negligible, because it is only a monosubstituted alkene in competition with the more stable di- and tetra-substituted alkenes. By fractional distillation of the mixture, the more volatile, but less stable 2,3-dimethyl-1-butene is removed first. The equilibrium in the mixture is restored by isomerization of more 2,3-dimethyl-2-butene into 2,3-dimethyl-1-butene. In this manner, a good yield of the less stable isomer may be won.

Isomerization, then, is a possible complication in all HA addition reactions. If a strong acid (HA) is used with more than 1 mole of water present, hydrogen ion really is the proton donor, with subsequent competition between water and $A^\ominus$ as nucleophiles (Eq. 13.20).

$$R^\oplus \underset{A^\ominus}{\overset{H_2O}{\rightleftarrows}} \begin{matrix} ROH_2^\oplus \\ \\ RA \end{matrix} \qquad (13.20)$$

If a protic solvent other than water is present, the competition in the second step is between the solvent and $A^\ominus$. For example, if the solvent is methanol, both $RCHACH_3$ and $RCH(OCH_3)CH_3$ are possible addition products from $RCH=CH_2$ (assuming that no isomerization occurs). To minimize solvent attack in HA additions, the reaction is carried out with no solvent or an aprotic solvent that dissolves both the alkene (or alkyne) and anhydrous HA. The solvent, however, also should be polar to allow carbonium-ion formation to occur. The reaction is not very useful because it is usually more convenient to prepare alkyl halides by other routes.

However, the addition to alkynes is useful. One route to vinyl chloride ($H_2C=CHCl$), the monomer from which most vinyl plastics (polymer) are made, is by addition of HCl to acetylene (Eq. 13.21). Addition is *trans* and follows the Markovnikov rule when required.

$$HC\equiv CH \xrightarrow[\substack{HgCl_2 \text{ on } C \\ 200°}]{HCl} H_2C=CHCl \qquad (13.21)$$

Hydration of alkenes and alkynes. Sulfuric acid (and other strong acids) add to alkenes, depending upon the proportions used, to give either the mono- or the diester (Eqs. 13.22). The concentration of the acid and the temperature are critical. Industrially, the

$$CH_3CH=CH_2 + H_2SO_4 \longrightarrow (CH_3)_2CHOSO_3H$$

$$CH_3CH=CH_2 + (CH_3)_2CHOSO_3H \longrightarrow (CH_3)_2CHO\diagdown_{\underset{O_2}{S}}\diagup OCH(CH_3)_2 \qquad (13.22)$$

reaction is used on a large scale to prepare alcohols. Dilute acid is used so that water may intervene in the second step (Eqs. 13.23). Phosphoric acid may be used in place of sulfuric

$$H_2C=CH_2 \xrightarrow[\substack{240° \\ \text{pressure}}]{10\% H_2SO_4} CH_3CH_2OH$$

$$(CH_3)_2C=CH_2 \xrightarrow[25°]{10\% H_2SO_4} (CH_3)_3COH \qquad (13.23)$$

acid. Milder conditions are used if the carbonium ion formed is secondary, and very mild if it is tertiary. Because of carbonium-ion formation, addition is both *trans* and *cis*. □ □

Alkynes are hydrated with ease if mercuric ion is present (Eqs. 13.24). The enol first formed tautomerizes (isomerization by relocation of H) to the ketone. Acetylene gives acetaldehyde (CH_3CHO). □ □

$$CH_3(CH_2)_3C\equiv CH \xrightarrow[\substack{\text{aq. }CH_3COOH \\ 1\text{ g }HgSO_4 \\ 1\text{ g }H_2SO_4 \\ 60°-80°,\ 3\text{ hr}}]{\text{aq. MeOH or}} \left[CH_3(CH_2)_3\underset{\underset{OH}{|}}{C}=CH_2 \right] \longrightarrow CH_3(CH_2)_3\overset{O}{\overset{\|}{C}}CH_3 \quad (13.24)$$

$>80\%$

Dimerization. A complication that may be turned to advantage at times is the tendency for the carbonium ion, in the absence of a good nucleophile, to attack a π bond. The ion is strongly electrophilic, and the π bond is a relatively weak nucleophile. The result is a new carbonium ion made up of two of the original alkene molecules: a dimeric ion (Eq. 13.25). The dimeric ion may lose a proton to any base in the mixture and become an

$$(CH_3)_3C^\oplus + H_2C=C(CH_3)_2 \longrightarrow (CH_3)_3C-CH_2\overset{\oplus}{C}(CH_3)_2 \quad (13.25)$$

alkene (($CH_3)_3CCH=C(CH_3)_2$), or it may go on to add to another molecule of isobutylene to give a trimeric ion, and so on. The reaction is favored by the use of Lewis acids (BF_3, $AlCl_3$, etc.) or small amounts of a strong protic acid (to hold down the concentration of $A^\ominus$). Under special conditions, cationic polymerization is successful (Sec. 30.3). Tertiary ions are more prone to give dimers and trimers than are secondary and primary ions because of the stability needed to keep the loss of a proton to a minimum before an alkene molecule is encountered. Even so, there seems to be an unavoidable amount of dimerization in any HA addition reaction.

Alkylation; hydride transfer. Before leaving the subject of HA additions, we should look into the alkylation reaction. Overall, it has the deceptively simple appearance (Eq. 13.26) of addition of RH to an alkene. Although almost any alkene may be used, RH

$$RH + \overset{\diagdown}{\underset{\diagup}{C}}=\overset{\diagup}{\underset{\diagdown}{C}} \xrightarrow[-20° \text{ to } 10°]{HF \text{ or } H_2SO_4} R\overset{|}{\underset{|}{C}}-\overset{|}{\underset{|}{C}}-H \quad (13.26)$$

must be tertiary. The accepted mechanism is shown in Equations 13.27.

The first step is the formation of a carbonium ion by addition of a proton. The ion adds to a π bond: in this case, another molecule of isobutylene. This step is the unwanted side-reaction we have just described. It is perhaps most annoying during the addition of water to a double bond (hydration). Here, it is the reaction we wish to favor. At the low temperatures used, the tertiary dimeric carbonium ion undergoes an equilibrium reaction with isobutane to give the monomeric ion (*tert*-butyl) and the dimeric alkane. We have seen intermolecular transfers of hydride ion before in the Oppenauer oxidation (Eqs. 9.23 and 9.24) and in the Cannizzaro reaction (Eqs. 9.53). The equilibrium here favors the smaller monomeric cation to some extent. Because the smaller ion is constantly disappearing by addition to more isobutylene to give more dimeric ion, the dimeric ion must continually be attacking more isobutane to maintain the equilibrium. □ □

$$\begin{array}{c}
\text{H}_3\text{C} \\
\diagdown \\
\text{C}=\text{CH}_2 \xrightarrow{\text{HA}} \\
\diagup \\
\text{H}_3\text{C}
\end{array}
\begin{array}{c}
\text{H}_3\text{C} \\
\diagdown \\
\overset{\oplus}{\text{C}}-\text{CH}_3 \\
\diagup \\
\text{H}_3\text{C}
\end{array}$$

(13.27)

With sizable quantities of butanes, propanes, butenes, and propenes on hand as a result of cracking of larger hydrocarbons (to be discussed in Sec. 13.7), petroleum refiners are able to use alkylation to convert the relatively unusable gaseous substances into liquid heptanes and octanes for gasoline. The highly branched compounds give excellent anti-knock performance in gasoline. The product shown in Equation 13.27 is the misnamed iso-octane, the 100-octane gasoline.

Alkyne addition to alkyne. We have seen that addition of an alkene to an alkene may be induced by acids. An alkyne adds to an alkyne in the presence of cuprous chloride in ammonium chloride solution (Eq. 13.28). Addition of HCl to the product, vinylacetylene,

$$2\text{HC}{\equiv}\text{CH} \xrightarrow[\substack{\text{NH}_4\text{Cl} \\ \text{H}_2\text{O} \\ 25°}]{\text{CuCl}} \text{HC}{\equiv}\text{C}-\text{CH}=\text{CH}_2 \qquad (13.28)$$

gives 2-chloro-1,3-butadiene ($H_2C{=}CClCH{=}CH_2$), the monomer from which the oil-resistant rubber neoprene is made. □ □

Carbene addition. Carbene may be generated as a singlet ($CH_2\uparrow\downarrow$) or as a triplet ($CH_2\uparrow\uparrow$). The singlet state carbene adds to a π bond with only a small E_a and with ΔH of about -80 kcal. Consequently, there is only a small difference in relative rates: $CH_2{=}CH_2$, 1; $(CH_3)_2C{=}CH_2$, 2; $(CH_3)_2C{=}C(CH_3)_2$, 2.2. Triplet carbene shows more discrimination, not only in addition but also in insertion reactions (relative rates of 1, 1.5, and 4.3 for primary, secondary, and tertiary C—H). Singlet carbene addition is stereospecific: it is exclusively cis. Thus cis-2-butene gives cis-1,2-dimethylcyclopropane as the only product of addition. Insertion products also are produced, from which we conclude that singlet carbene shows little choice of location as long as it finds something to react with, be it a C—H bond or a C=C bond. Triplet carbene, in reaction with a π bond, is nonstereospecific. Reaction with cis-2-butene gives a mixture of cis- and trans-1,2-cyclopropane and presumably a smaller amount of insertion products. The singlet is thought to add in a concerted mechanism; but the triplet must give rise to a triplet diradical, which has to wait so long for spin inversion to occur that free rotation about the σ bond of the original double bond occurs (Eq. 13.29). This explanation seems to be satisfactory for the stereospecificity of singlet carbene addition and the lack of it in triplet carbene addition.

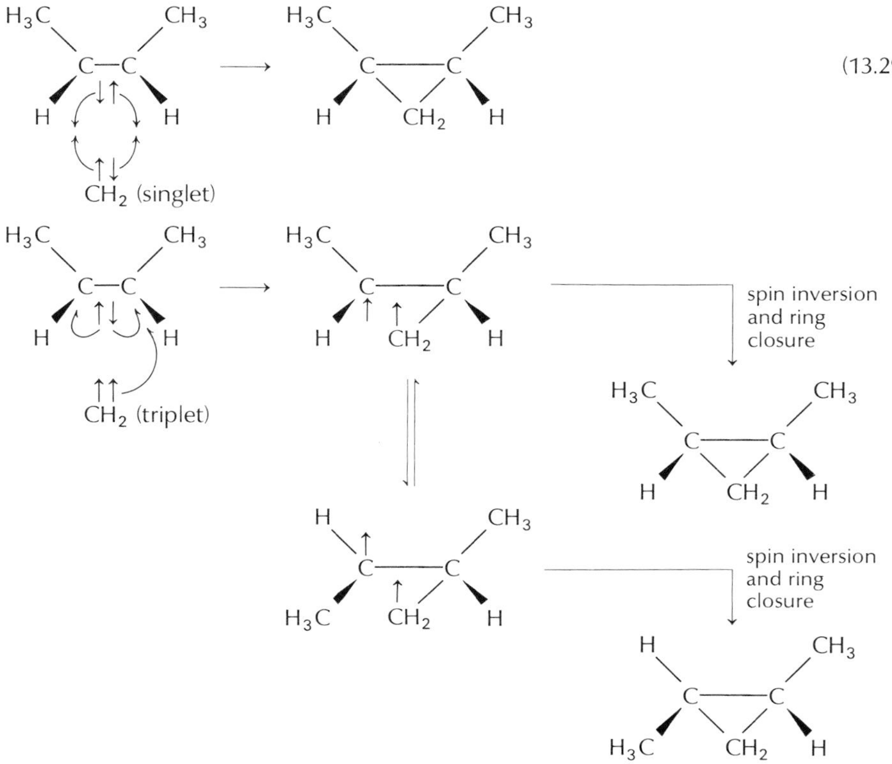

(13.29)

Exercises

13.1 Starting with 2-methyl-2-pentene, show the product to be expected under the conditions shown.
   a. Br₂ in CH₃COOH, CH₃COONa present
   b. 10% H₂SO₄
   c. HF at −20° and 2-methylpentane
   d. HBr, polar aprotic solvent
   e. Cl₂ and dilute NaOH

13.2 Give the products to be expected if each alkyne is treated with Hg²⁺ and dilute H₂SO₄.
   a. CH₃CH₂C≡CH
   b. tBuC≡CH
   c. CH₃C≡CCH₂CH₃

13.3 Hydroboration and similar addition reactions

Hydroboration; isomerization. In Section 9.13, the hydroboration reaction was introduced. Although it is classified as an electrophilic addition reaction, the details are

13.3 Hydroboration and similar addition reactions

unusual enough so that it merits a separate section of this chapter. Borane is an intense electrophile because the fourth orbital of BH_3 is vacant. It attacks a π bond readily at room temperature. The solvent of choice is diglyme (diethylene glycol methyl ether, $CH_3OCH_2CH_2OCH_2CH_2OCH_3$, bp 162°). The three ether linkages provide good solvation for the reactants and products. As we have pointed out previously, addition is exclusively cis.

Although the mechanism definitely does not go via a carbonium ion, the concept of going through a carbonium ion can help us predict the direction of addition (Eq. 13.30).

$$RCH=CH_2 \xrightarrow{BH_3} \left[\begin{array}{c} R\overset{\oplus}{C}H-CH_2 \\ | \\ ^{\ominus}BH_3 \end{array} \text{ or } \begin{array}{c} RCH-\overset{\oplus}{C}H_2 \\ | \\ ^{\ominus}BH_3 \end{array} \right] \quad (13.30)$$

$$\downarrow$$

$$\begin{array}{c} RCH-CH_2 \\ | \quad | \\ H \quad BH_2 \end{array} \xrightarrow{RCH=CH_2} \text{etc. } (RCH_2CH_2)_3B$$

We think of the actual mechanism as a <u>four center</u> transition state (<u>concerted addition</u>) to explain the cis addition, the direction, and the reversibility (Eq. 13.31). At room temperature,

$$\mathrm{C=C} + \mathrm{H-B} \rightleftharpoons \left[\begin{array}{c} \overset{\delta\oplus}{C\cdots C} \\ \vdots \quad \vdots \\ H\cdots B \\ ^{\delta\ominus} \end{array} \right] \rightleftharpoons \begin{array}{c} C-C \\ | \quad | \\ H \quad B \end{array} \quad (13.31)$$

transition state

the equilibrium lies far to the right and the relative rates of addition are rapid.

Equation 13.31 is reversible. If we use a slight excess of BH_3 (generated according to Eq. 13.32 and heat the reaction mixture to 162° for an hour, we find isomerization to the

$$3NaBH_4 + 4BF_3 \longrightarrow 4BH_3 + 3NaBF_4 \quad (13.32)$$

primary positions. Equation 13.33 shows this isomerization for the addition to 3-hexene.

$$\left(\begin{array}{c} CH_3CH_2CH- \\ | \\ CH_3CH_2CH_2 \end{array} \right)_3 B \underset{\Delta}{\rightleftharpoons} \left(\begin{array}{c} CH_3CH- \\ | \\ CH_3CH_2CH_2CH_2 \end{array} \right)_3 B \underset{\Delta}{\rightleftharpoons} \left(\begin{array}{c} CH_2- \\ | \\ CH_3CH_2CH_2CH_2CH_2 \end{array} \right)_3 B$$

(13.33)

The isomerization cannot pass a quaternary carbon atom. Compare the results shown in Equations 13.34. Also, isomeric alkenes give the same equilibrium mixture. Thus

$$\left(\begin{array}{c} (CH_3)_2CHCH- \\ | \\ CH_3CH_2 \end{array} \right)_3 B \xrightarrow{\Delta} B\left(\begin{array}{c} -CH_2CHCH_2 \\ | \quad | \\ H_3C \quad CH_2CH_3 \end{array} \right)_3 \quad 39\%$$

$$+ \left(\begin{array}{c} (CH_3)_2CHCH_2 \\ | \\ CH_2CH_2- \end{array} \right)_3 B \quad 59\% \quad (13.34)$$

$$\left(\begin{array}{c} (CH_3)_3CCH_2 \\ | \\ CH_3CH- \end{array} \right)_3 B \xrightarrow{\Delta} \left(\begin{array}{c} (CH_3)_3CCH_2 \\ | \\ CH_2CH_2- \end{array} \right)_3 B \quad >98\%$$

$CH_3CH_2C(CH_3)=CH_2$, $(CH_3)_2CHCH=CH_2$, and $(CH_3)_2C=CHCH_3$ all give 40% of $(CH_3CH_2CH(CH_3)CH_2-)_3B$ and 58% of $[(CH_3)_2CHCH_2CH_2-]_3B$ after addition and heating. If we start with 1-, 3-, or 4-methylcyclohexene, we may obtain at least 50% of $[(C_6H_{11})CH_2-]_3B$ by the shifting about the ring and into the side-chain primary position.

Hydrolysis: reduction of an alkene. A trialkylboron, upon treatment with water, is remarkably stable to hydrolysis. However, carboxylic acids are effective: heating under reflux with propionic acid added to the diglyme mixture is the recommended procedure (Eq. 13.35). The effect is that of *cis* addition of hydrogen to the π bond, a useful alternative to catalytic hydrogenation.

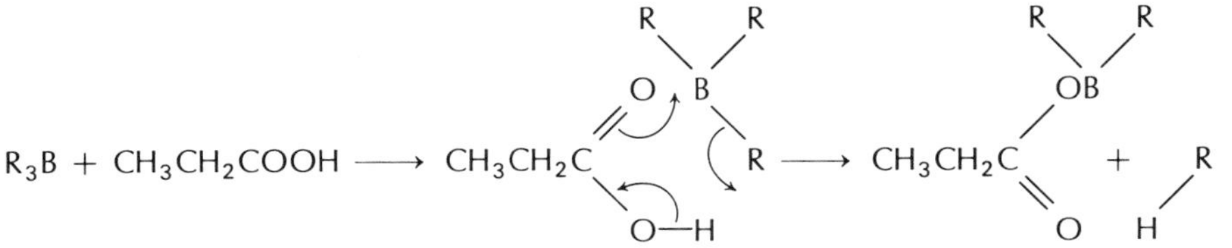

$$R_3B + CH_3CH_2COOH \longrightarrow CH_3CH_2C\begin{smallmatrix}O\\O-H\end{smallmatrix}\overset{R\diagup B\diagdown R}{\underset{}{}} \longrightarrow CH_3CH_2C\begin{smallmatrix}OB(R)_2\\O\end{smallmatrix} + R-H \tag{13.35}$$

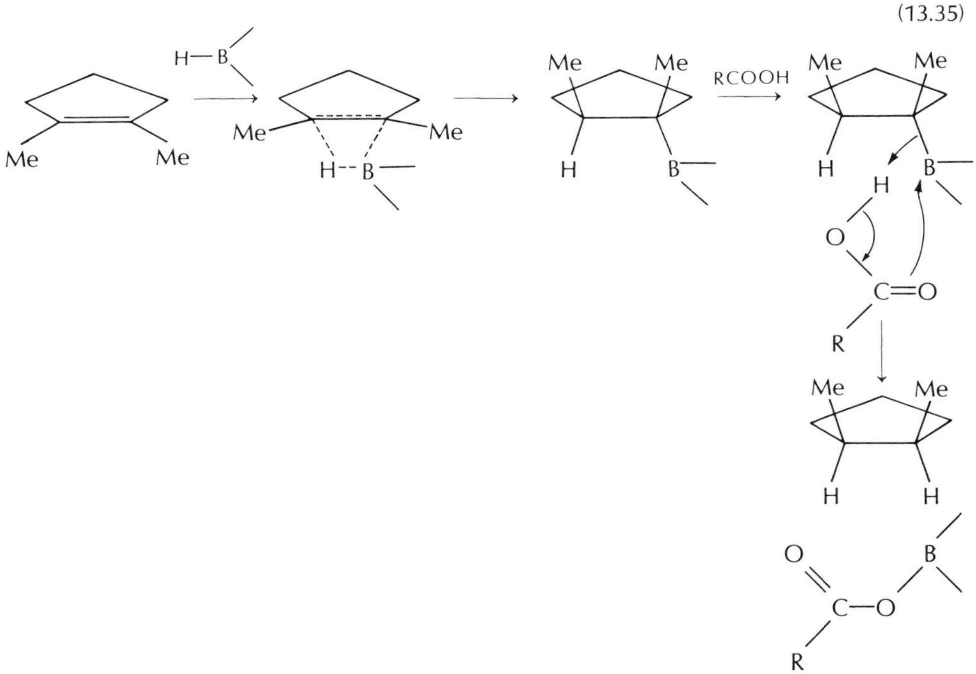

Oxidation: hydration of an alkene. Treatment of a trialkylboron with alkaline hydrogen peroxide at room temperature gives a fast reaction even at 0°. The mechanism probably is as shown (Eqs. 13.36). The original BH_3 addition is *cis* and the shift of the R group from B to O occurs with retention of configuration of the R group. Therefore, the final product alcohol is the result of *cis* addition to the π bond, but it is anti-Markovnikov as compared to an acid-catalyzed addition of water.

Combined with the isomerization reaction, the decomposition of trialkylborons allows the synthesis of primary alcohols difficult to make by other means (Eq. 13.37).

Displacement. We also may make use of displacement. If another alkene is added to a solution of R_3B in diglyme and the mixture is heated, an equilibrium with the new

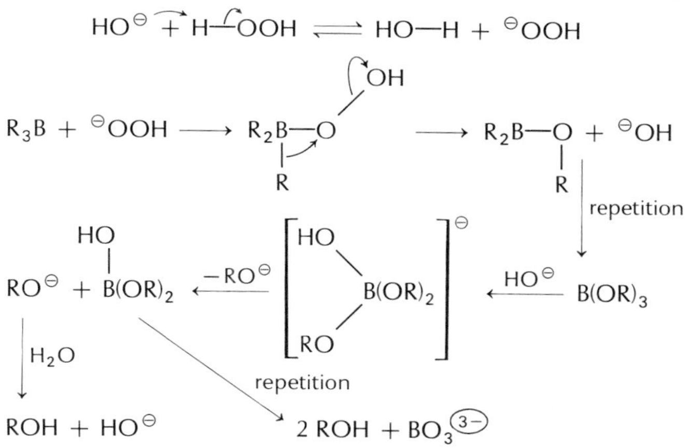

$$HO^\ominus + H\text{—}OOH \rightleftharpoons HO\text{—}H + {}^\ominus OOH$$

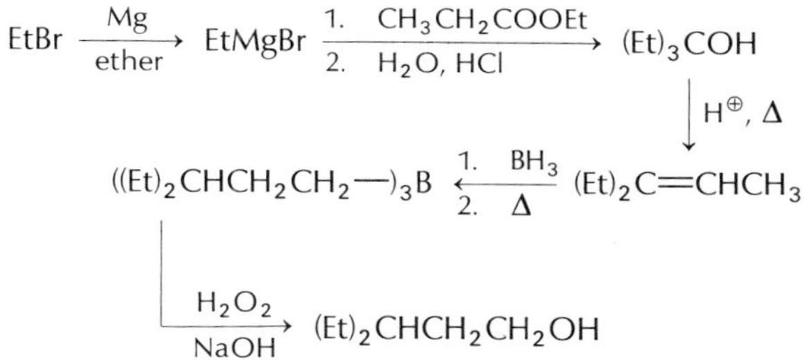

$$R_3B + {}^\ominus OOH \longrightarrow R_2B\text{—}O\underset{R}{|}\overset{\overset{\displaystyle OH}{\diagup}}{} \longrightarrow R_2B\text{—}O\underset{R}{|} + {}^\ominus OH$$

$$\underset{\underset{\displaystyle ROH + HO^\ominus}{\Big\downarrow H_2O}}{RO^\ominus + B(OR)_2\overset{\displaystyle HO}{|}} \xleftarrow{-RO^\ominus} \left[\underset{RO}{\overset{HO}{\diagdown}}B(OR)_2\right]^\ominus \xleftarrow{HO^\ominus} B(OR)_3 \quad \Big\downarrow \text{repetition}$$

$$\text{repetition} \searrow 2\,ROH + BO_3{}^{3\ominus} \quad (13.36)$$

$$\text{EtBr} \xrightarrow[\text{ether}]{\text{Mg}} \text{EtMgBr} \xrightarrow[\text{2. } H_2O, HCl]{\text{1. } CH_3CH_2COOEt} (Et)_3COH$$

$$\Big\downarrow H^\oplus, \Delta$$

$$((Et)_2CHCH_2CH_2\text{—})_3B \xleftarrow[\text{2. } \Delta]{\text{1. } BH_3} (Et)_2C\text{=}CHCH_3 \quad (13.37)$$

$$\Big\downarrow \xrightarrow[\text{NaOH}]{H_2O_2} (Et)_2CHCH_2CH_2OH$$

alkene comes into play. Rearrangement does not occur, because the excess of the new alkene destroys all the remaining BH_3, the presence of which is essential for rapid rearrangement. If the alkene from R is more volatile, it may be distilled from the mixture. If an excess of the new, higher-boiling alkene is supplied, the equilibria shift so as to form more alkene from R. An example is shown in Equation 13.38. The more stable pentene

$$3CH_3(CH_2)_7CH\text{=}CH_2 + ((CH_3)_2CHCH\underset{|}{\overset{CH_3}{|}}\text{—})_3B \rightleftharpoons$$

$$(CH_3(CH_2)_7CH_2CH_2\text{—})_3B + 3(CH_3)_2C\text{=}CH\underset{|}{\overset{CH_3}{|}} \quad (13.38)$$

$((CH_3)_2C\text{=}CHCH_3$, the most substituted) is released, instead of the less stable $(CH_3)_2CHCH\text{=}CH_2$, from the R_3B given in the equation. In other words, whether combined with a preceding isomerization step or not, the addition of an alkene with a high bp allows the separation of the more stable alkene. This operation is of help at times in the synthesis of alkenes otherwise difficult to make. □ □

Exercises

13.3 Starting with 1-methylcyclohexene in each case, give a good method of preparation of:
   a. 2-methylcyclohexanol
   b. cyclohexylmethanol
   c. 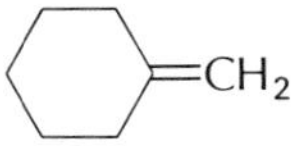 ⬡=CH$_2$

13.4 Homolytic addition reactions

Introduction. The homolytic addition reactions of alkenes and alkynes, like the electrophilic reactions, involve A—A or A—B reagents. A major difference lies in whether the radical A· or B· initiates the attack. Also, there is competition by allylic substitution. As in the homolytic substitution reactions, the reaction must be initiated by a radical that has been produced thermally, by peroxides, or by photoexcitation.

Halogen addition. Let us begin with the halogens again and go back to Equations 3.34, 12.18, and 12.19. Once the halogen atoms are formed, a chain propagation reaction for addition to an alkene may be written (Eqs. 13.39b and c). The intermediate radical can get into all sorts of trouble, especially if oxygen is present (d). The usual termination reactions

$$X_2 \xrightarrow[\text{or heat}]{hv} 2X\cdot \qquad a$$

$$R\overset{\frown}{CH\!=\!CH_2} + X\cdot \longrightarrow R\dot{C}H\!-\!CH_2X \qquad b$$

$$R\dot{C}HCH_2X + X_2 \longrightarrow R\overset{X}{\underset{|}{C}H}CH_2X + X\cdot \qquad c \qquad (13.39)$$

$$R\dot{C}HCH_2X + O_2 \longrightarrow R\overset{O\!-\!O\cdot}{\underset{|}{C}H}CH_2X \qquad d$$

are going on as well. Competing with the free-radical chain addition reaction are the electrophilic addition and free-radical chain substitution reactions. The electrophilic addition reaction is so favored by polar solvents that little free-radical halogen addition occurs. In nonpolar solvents, such as CCl_4 or Cl_2FCCF_2Cl, the polar electrophilic additions with the necessary intermediate cations are less favored, and the free-radical processes compete more successfully. If the radical chains are discouraged by the presence of oxygen, the polar electrophilic reaction still is able to operate in such nonpolar solvents, although with less facility. The intermediate cation, if branched, becomes so unstable under these conditions that proton elimination, rather than completion of the addition, becomes the favored process. Thus, isobutylene with chlorine (no solvent) under oxygen at $-9°C$ gives 87% of $ClCH_2C(CH_3)\!=\!CH_2$ and only 13% of the addition product, $(CH_3)_2CClCH_2Cl$. Straight-chain alkenes, however, give excellent yields of *trans* addition product (*trans*-$CH_3CH\!=\!CHCH_3 \rightarrow$ *meso*-$CH_3CHClCHClCH_3$ in 98% yield). The radical pathway for chlorination, on the other hand, gives somewhat more addition (>50%) than substitution with isobutylene and gives 18% substitution plus 82% addition with *trans*-2-butene. The addition is nonstereoselective, with both *meso* and racemic products formed.

Alkynes. With an alkyne (example: 1-butyne), free-radical chlorination gives a mixture of products, but the *trans*-addition product, $CH_3CH_2CCl\!=\!CHCl$, is formed in 65 to 90% yield. The reaction in the absence of light is much slower than the reaction with an alkene. The usual rule is that the more substitution there is about a π bond, the faster is the rate of addition. However, chlorine atoms are so reactive (E_a about 1 to 2 kcal) that little selectivity is shown, and the rates do not vary much within a given class of compound. Bromination is more selective in rates of addition and substitution and is very sensitive to

13.4 Homolytic addition reactions

the presence of oxygen. Thus, addition of bromine to an alkene in CCl_4 as solvent should be done under nitrogen. Room light catalyzes the free radical reaction. Free-radical fluorination and iodination addition reactions are little known—there is too much reactivity in the one and too little in the other.

AB reagents and the peroxide effect. As in the electrophilic additions, there are a large number of AB reagents that add to π bonds by the homolytic mechanism. For each of them, an initiator must produce the radicals to start the chain reaction. The initiators usually are peroxides such as tBuOO-tBu, tBuOOH, and $(CH_3COO)_2$. With some AB compounds, bond rupture may be induced by light, usually in the uv.

As an illustration, let us look at the addition of HBr by the homolytic mechanism. The other hydrogen halides are reluctant to add by such a pathway, for reasons we shall explore soon. The accepted mechanism for the homolytic addition of HBr is given in Equations 13.40.

$$
\begin{aligned}
&\text{Br-H} + \text{O}_2 \xrightarrow{\Delta} \text{Br}\cdot + \text{HOO}\cdot &&a \\
&\text{ROOR} \xrightarrow{\Delta} 2\text{RO}\cdot &&b \quad \text{initiation} \\
&\text{Br-H} + \cdot\text{OR} \longrightarrow \text{Br}\cdot + \text{HOR} &&c \\
&\text{RCH=CH}_2 + \text{Br}\cdot \longrightarrow \text{R}\dot{\text{C}}\text{H-CH}_2\text{Br} &&d \\
&\text{R}\dot{\text{C}}\text{HCH}_2\text{Br} + \text{H-Br} \longrightarrow \text{RCH}_2\text{CH}_2\text{Br} + \text{Br}\cdot &&e \quad \text{propagation} \\
&\text{R}\dot{\text{C}}\text{HCH}_2\text{Br} + \text{O}_2 \longrightarrow \text{R}\overset{\overset{\text{O-O}\cdot}{|}}{\text{C}}\text{HCH}_2\text{Br} &&f \quad \text{a termination}
\end{aligned}
\qquad (13.40)
$$

There is the usual hodgepodge of termination reactions. The reaction was a real puzzle to early research workers because the direction of the addition is anti-Markovnikov, opposite to that of the electrophilic addition. It now seems to us that at times, alkenes containing traces of autooxidation products were used. The peroxides could set off the homolytic addition in competition with the electrophilic addition. Depending upon the amount of peroxides present, the addition product formed would contain more or less of the anti-Markovnikov-oriented result. The presence of oxygen did not help matters, because a small amount served as an initiator (a), but an excess could act as a chain-terminating step (f). When the homolytic mechanism finally was suspected, the answer to the puzzling results was quickly found. The best conditions for free-radical addition require the use of a peroxide initiator or light of wavelength less than 290 nm, nonpolar conditions, and oxygen exclusion (to prevent chain breaking by f) or the presence of an antioxidant.

Why is it that the other hydrogen halides do not add to alkenes by this free-radical mechanism? If we tabulate the ΔH values for the various steps (Table 13.1), we see that only

Table 13.1 ΔH (kcal) for the hydrogen halides (Eq. 13.40)

Step X =	F	Cl	Br	I
c	24	−8	−24	−40
d	−53	−18	−5	12
e	36	4	−12	−28
d + e	−17	−14	−17	−16

X = Br has a negative value for ΔH for steps c, d, and e. HF is blocked from the homolytic addition by the large positive ΔH values for initiation (c) as well as for propagation (e). Hydrogen chloride is blocked by ΔH of 4 kcal in e. This amount is small enough so that it may be overcome at high temperatures. A few instances of peroxide-catalyzed addition of HCl at 400° to 600°C have been reported. For hydrogen bromide, each ΔH is negative and the steps are downhill all the way. Although HI is easily broken up to produce iodine atoms, the first addition step (d) is uphill by 12 kcal (E_a about 14 kcal), and it is effectively blocked. So it is that of the four HX compounds, only HBr exhibits the peroxide effect, or the ability to add by the homolytic mechanism, even though ΔH for a chain cycle (d + e) is about the same for all four compounds. The chain length for HBr homolytic addition is so large that it is susceptible to variation in rate by added substances capable of chain-termination reactions.

With cycloalkenes, it has been shown that addition is *trans* (Eq. 13.41). (Let us not be misled by the name of the product, which is *cis*-1-bromo-2-methylcyclohexane for R =

$$\text{(13.41)}$$

R = CH$_3$, Cl, or Br

methyl.) With open-chain alkenes, addition is usually *trans*, although the reactions may be complicated by free rotation of the radical formed in d and reversibility of d so that an equilibrium between the *cis* alkene and the *trans* alkene is achieved. With alkynes the situation is not clear: both *cis* and *trans* additions have been reported. □ □

Cyclopropane reactions. In many respects, cyclopropane resembles an alkene. Because ring opening is favored by 27.6 kcal of strain energy (Table 12.3), cyclopropane undergoes addition reactions with a number of reagents, both heterolytic and homolytic (Eqs. 13.42).

Why free-radical chlorination favors substitution while bromination and iodination favor addition by ring opening is a puzzle. The answer may lie in the higher bond energy of HCl, which causes ΔH for substitution to become negative and thus lowers E_a sufficiently to allow substitution to be the favored reaction. The enthalpy changes show that bromine or iodine substitution are both quite endothermic, so addition with ring opening is the major result. Aqueous acids must produce a $C_3H_7^{\oplus}$ ion of uncertain structure, which completes ring opening with $X^{\ominus}$.

Cyclobutane opens only upon drastic (200°) hydrogenation conditions, despite a total strain energy that is almost as large as that of cyclopropane. However, the strain energy per angle is less than that of cyclopropane by enough to allow the ring to maintain its structure. □ □

13.4 Homolytic addition reactions

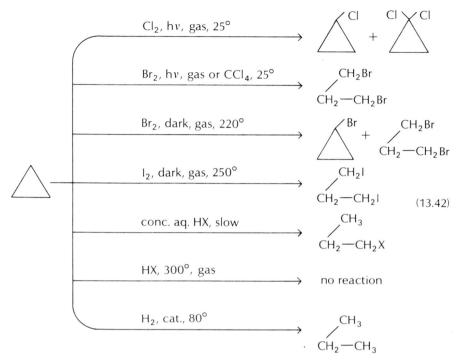

(13.42)

Competition between addition and allylic substitution. All the homolytic additions suffer competition by substitution at the allylic position if a hydrogen is present at that location (Eqs. 13.43). In many instances, this substitution completely overcomes the

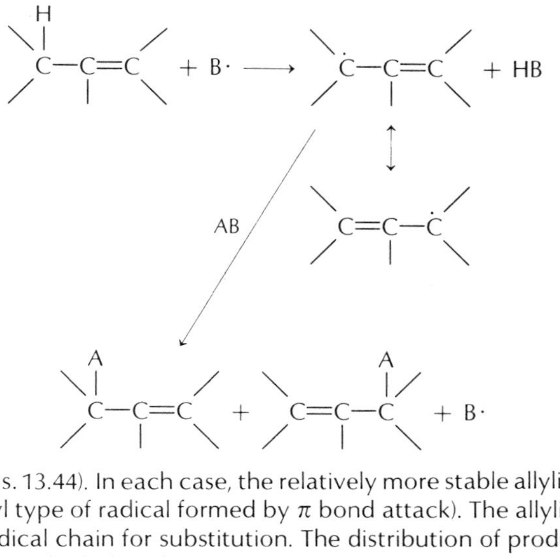

(13.43)

addition reaction (Eqs. 13.44). In each case, the relatively more stable allylic radical is formed (rather than the alkyl type of radical formed by π bond attack). The allylic radical then sets up the usual free-radical chain for substitution. The distribution of products is determined by the relative rates at which the allylic positions react with the AB reagent (kinetic control of reaction). NBS is thought to participate only indirectly in the reactions, by a fast polar reaction with HBr to give a low, steady concentration of Br_2. The same effect may be obtained by careful control of the direct addition of bromine to the reaction mixture. At high concentrations of bromine, free-radical addition becomes competitive with substitution.

$$CH_3CH=CH_2 \xrightarrow[400°]{Cl_2} ClCH_2CH=CH_2 + HCl$$
$$80\%$$

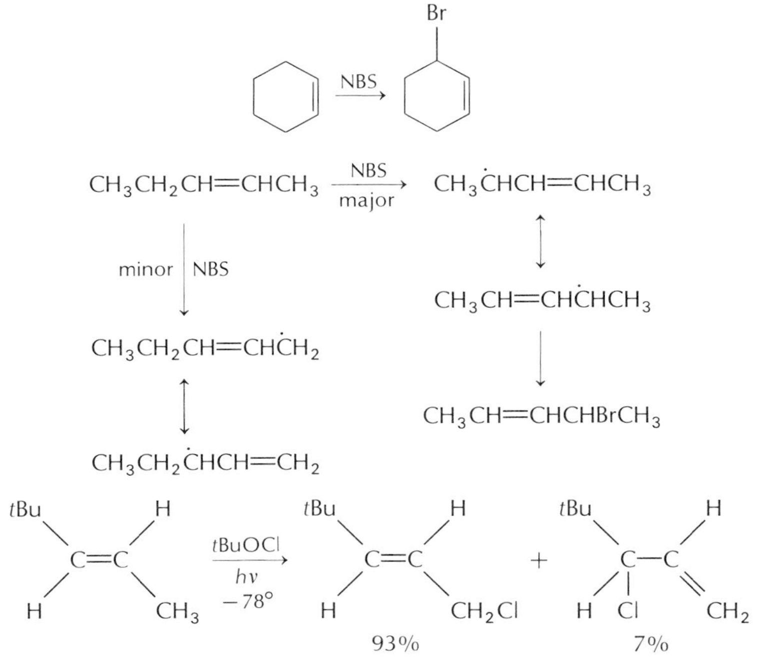

[cyclohexene] $\xrightarrow{NBS}$ [3-bromocyclohexene]

$$CH_3CH_2CH=CHCH_3 \xrightarrow[\text{major}]{NBS} CH_3\dot{C}HCH=CHCH_3$$

minor $\downarrow$ NBS $\qquad\qquad\qquad \updownarrow$

$$CH_3CH_2CH=CH\dot{C}H_2 \qquad CH_3CH=CH\dot{C}HCH_3 \qquad (13.44)$$

$\downarrow \qquad\qquad\qquad \downarrow$

$$CH_3CH_2\dot{C}HCH=CH_2 \qquad CH_3CH=CHCHBrCH_3$$

$$\begin{array}{c} tBu \\ \diagdown \\ H \end{array} C=C \begin{array}{c} H \\ \diagup \\ CH_3 \end{array} \xrightarrow[\substack{hv \\ -78°}]{tBuOCl} \begin{array}{c} tBu \\ \diagdown \\ H \end{array} C=C \begin{array}{c} H \\ \diagup \\ CH_2Cl \end{array} + \begin{array}{c} tBu \\ \diagdown \\ H \end{array} \begin{array}{c} H \\ | \\ C-C \\ | \\ Cl \end{array} \begin{array}{c} \\ \diagup \\ CH_2 \end{array}$$
$$93\% \qquad\qquad 7\%$$

Summary. In general, it appears that electrophilic addition of halogens as a procedure is to be preferred to homolytic addition. On the other hand, homolytic addition is capable of a wider diversity of AB reagents. If we operate with oxygen purposely present, the homolytic substitution and addition reactions are suppressed sufficiently to allow electrophilic addition to proceed even in relatively nonpolar aprotic solvents.

Exercises

13.4 Predict the principal product to be expected from:

    a. 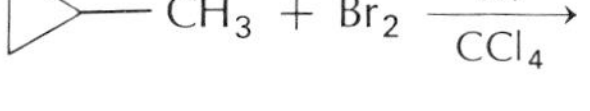 [cyclopropyl]—CH_3 + Br_2 $\xrightarrow[CCl_4]{hv}$

    b. 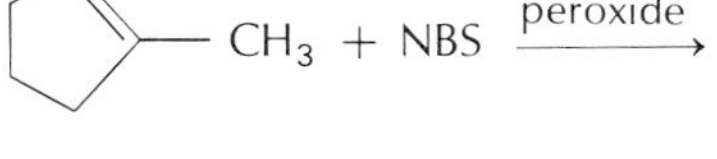 [cyclopentenyl]—CH_3 + NBS $\xrightarrow{peroxide}$

13.5 Show how to carry out the conversions shown. More than one step may be necessary.
    a. $CH_3CH=CH_2 \longrightarrow ClCH_2CHClCH_2Cl$
    b. $tBuCH=CH_2 \longrightarrow tBuCH_2CH_2CCl_3$
    c. $tBuC\equiv CH \longrightarrow tBuCH=CHBr$
    d. $Et_2C=CHCH_3 \longrightarrow Et_2CHCHBrCH_3$

13.5 Oxidation reactions

Oxidation by substitution. We may break down the oxidation reactions into two classes: oxidation by substitution and oxidation by addition.

Taking up substitution oxidation first, let us look at a probable course of oxidation of the very simple methane molecule. Methane, the principal constituent of natural gas, is consumed in tremendous quantity as a source of energy by combustion. It is generally agreed that the reaction is initiated by attack of oxygen with a fairly high E_a, so that high temperatures (a spark, a flame) are needed. In the presence of a catalyst (Pt), the E_a may be lowered to a point where exposure of the alkane to oxygen adsorbed on the surface of the catalyst at room temperature starts the reaction. Thereafter, reactions proceed in many directions (Eqs. 13.45), and we have only begun. Another set of equations involving CH_3OH

$$
\begin{aligned}
CH_4 + O_2 &\longrightarrow CH_3\cdot + HOO\cdot & a \\
CH_4 + HOO\cdot &\longrightarrow CH_3\cdot + HOOH & b \\
HOOH &\longrightarrow 2HO\cdot & c \\
CH_4 + HO\cdot &\longrightarrow CH_3\cdot + H_2O & d \\
CH_3\cdot + O_2 &\longrightarrow CH_3OO\cdot & e \\
CH_4 + CH_3OO\cdot &\longrightarrow CH_3\cdot + CH_3OOH & f \\
CH_3OOH &\longrightarrow CH_3O\cdot + HO\cdot & g \\
CH_4 + CH_3O\cdot &\longrightarrow CH_3\cdot + CH_3OH & h
\end{aligned}
\qquad (13.45)
$$

and $HOCH_2\cdot$ and ending with $HCHO$ and water may be written, and so on until CO_2 is reached. The mechanism, we see, is a turmoil of chain reactions.

As soon as we go on to ethane, things become much worse. We did not go into a complication of the methane combustion, bond rupture at the high temperatures involved (CH_3—$H \rightarrow CH_3\cdot + H\cdot$). Ethane possesses a C—C bond, easier to break than a C—H bond; and $CH_3CH_3 \rightarrow 2CH_3\cdot$ and $CH_3CH_3 \rightarrow CH_3CH_2\cdot + H\cdot$ become factors in the combustion of ethane. There is no need to go on. Combustion is not a simple process. □ □

Carbon dioxide in the atmosphere; air pollution; smog. In 1957, it was estimated that about 10^{16} moles of carbon (coal, peat, petroleum) had been removed from the earth and used as a source of energy (about 130 billion tons), enough to have caused an 18% increase in the amount of carbon dioxide in the atmosphere. One series of measurements showed an increase of 1.4% during the five-year period between 1958 and 1963. Another estimate, from 1966 data, comes up with a figure of about 100 million tons of carbon monoxide added to the atmosphere of the United States. It was also estimated that 63% came from transportation sources (automobiles, trucks, railroads, aircraft, ships). Still another estimate gives a figure of 10^{12} tons of gases and dusts from all sources added to the atmosphere, world-wide, per year. Surprisingly, only 0.05% (500 million tons) come from human activities. Vegetation is a major contributor of pollen and organic vapors to the atmosphere (odor of flowers, aroma of a pine forest, terpenes and ethylene from trees, etc.), not to mention dust and salt particles and other debris arising from storms at sea, volcanic activity, and so on.

Relatively minor components of organic compounds in the atmosphere (in terms of amounts) are the substances used by insects to communicate with each other. An example is the simple straight-chain alkene cis-9-tricosene ($C_{23}H_{46}$), also called muscalure. It is the sex attractant pheromone of the female of the common

$$CH_3(CH_2)_{11}CH_2 \overset{H}{\underset{}{\diagdown}} C=C \overset{H}{\underset{}{\diagup}} CH_2(CH_2)_6CH_3$$

muscalure

house fly. The search for pheromones, despite the large number already known, is an active field of present-day research. (Each species of insect utilizes a specific compound to carry signals—hence the name used for such compounds, pheromone. Phero- is from the Greek, meaning carrier).

Air pollution, if we define it as the presence of substances in the atmosphere other than N_2, O_2, H_2O, CO_2, H_2, and the rare gases, is global. Problems arise when pollutants in a given locality exceed certain concentrations. No longer can we discharge waste materials into the air and water and depend upon simple dilution for dispersion. Every living thing exerts a change upon the environment, be it a seed, an elephant, a virus, a coral, a blue jay, or a bee; but man is the one who really has the ability to affect the surface of the earth and the atmosphere as well. The management disposal of waste materials and objects, both industrial and domestic, urban and rural, has been a neglected endeavor. In any event, air pollution is here to stay for a while.

Smog (the photochemical variety) was first noticed in Los Angeles in the 1940's. Since then, it has been experienced in almost every large city in the world. The chemistry of smog is not simple, and many details remain to be worked out.

The nitrogen in the fuel mixture of gasoline and air is partially oxidized during the explosion in the cylinder of an engine. Thus there emerges from the exhaust pipe a witch's brew of CO_2, CO, N_2, NO, NO_2, O_2, unburned fuel, partially oxidized fuel, H_2O, lead compounds, and various halogen compounds. Thereafter, the NO is rapidly oxidized to NO_2. This oxidation, it seems, is catalyzed by light of about 3000 to 4200 A: sunlight. It is possible that singlet oxygen is involved. Equations 13.46 show the possible beginning of what happens next. Essentially, a catalyzed autoxida-

$$NO_2 \xrightarrow{h\nu} NO + O$$
$$O_2 + O \longrightarrow O_3$$
$$RH + O \longrightarrow R\cdot + HO\cdot$$
$$RH + O_3 \longrightarrow R\cdot + HO\cdot + O_2 \quad (13.46)$$
$$NO + O_3 \longrightarrow NO_2 + O_2$$
$$R\cdot + O_2 \longrightarrow ROO\cdot$$

tion of alkanes and alkenes is the trouble. Among the products are peroxides, aldehydes, ketones, and $RCOONO_2$ (peroxyacyl nitrates, PAN's for short). It is the PAN's that are most potent in damaging vegetation, bringing tears, clogging sinuses, and so on. The NO_2 is responsible for the yellow to orange color of a smog blanket viewed from above. The products are sensitive to radiation, and once into an excited state, they are able to cause triplet oxygen to rise to the singlet state. Ozone concentration serves as a convenient marker for measuring the intensity of a smog attack.

An atmospheric condition favorable for smog is a quiet, sunny day with a temperature inversion. An inversion is a quiet layer of warmer air over a quiet ground layer of cooler air. The cool layer cannot rise, and with sunlight from above and plenty of raw materials from active sources below (cars, trucks, smokestacks, and so on) the smog rapidly accumulates and stays. One of the surprising things about smog is the rapidity of formation from reactants that are at ppm levels of concentration. Other pollutants are SO_2 from sulfur-containing fuels, which eventually is oxidized to SO_3 and hydrated to H_2SO_4, carbon (soot) particles, compounded rubber particles (tire wear), asbestos (brake linings), plus any number of items from the entire spectrum of manufacturing, distribution, and service industries.

Oxidation by addition: ozone. Let us return to the laboratory and oxidation by addition reactions. Ozone has been used for a long time as a reagent for the location of double bonds in a molecule (Eq. 13.47). With the rise in availability of spectroscopic

$$\mathrm{\diagdown_{\diagup}C{=}C\diagdown^{\diagup}} \xrightarrow[\text{2. Zn, CH}_3\text{COOH}]{1.\ O_3} \mathrm{\diagdown_{\diagup}C{=}O + O{=}C\diagdown^{\diagup}} \qquad (13.47)$$

methods, this use of ozone has declined in popularity. Ozonization is occasionally used as a synthetic cleavage reaction. Depending upon the structure of the alkene and upon the method used to decompose the intermediate ozonide, aldehydes or acids and ketones are produced (Eqs. 13.48). If zinc dust with acetic acid or catalytic hydrogenation with a limited

$$\begin{aligned}
RCH{=}CH_2 &\xrightarrow{O_3,\ etc.} RCHO\ or\ RCOOH\ +\ HCHO\ or\ HCOOH \\
R_2C{=}CH_2 &\longrightarrow R_2CO \qquad\qquad\ \ +\ HCHO\ or\ HCOOH \\
RCH{=}CHR' &\longrightarrow RCHO\ or\ RCOOH\ +\ R'CHO\ or\ R'COOH \qquad (13.48) \\
R_2C{=}CHR' &\longrightarrow R_2CO \qquad\qquad\ \ +\ R'CHO\ or\ R'COOH \\
R_2C{=}CR'_2 &\longrightarrow R_2CO \qquad\qquad\ \ +\ R'_2CO
\end{aligned}$$

supply of hydrogen is used to decompose the ozonide, aldehydes and ketones are formed. If hydrogen peroxide is used, acids and ketones are formed. If sodium borohydride or lithium aluminum hydride is used, the alcohols related to the aldehydes or acids and ketones result.

A double bond may be located in a chain, as shown in Equations 13.49. In short,

$$\begin{aligned}
CH_3CH_2CH_2CH{=}CHCH_3 &\xrightarrow{O_3,\ etc.} CH_3CH_2CH_2CHO \\
&\qquad\qquad\quad +\ CH_3CHO \qquad (13.49) \\
CH_3CH_2CH{=}CHCH_2CH_3 &\xrightarrow{O_3,\ etc.} 2CH_3CH_2CHO
\end{aligned}$$

aldehydes or acids and ketones are easy to separate and identify. Thus, of the possible isomeric hexenes (of which we have shown only two, 2- and 3-hexene) ozonization clearly shows that if acetaldehyde and butyraldehyde are formed, the alkene had to be 2-hexene. If only propionaldehyde is formed, the alkene must have been 3-hexene. □ □

Ozonolysis of alkynes has not been used very much. Sometimes diketones may be isolated, but acids are the usual products.

Epoxidation. Epoxidation of alkenes was introduced in Equation 3.30, and a use of epoxides was given in Section 9.12. Here we need to develop the subject in more detail

(Eqs. 13.50). A peracid is the reagent of choice (occasionally CrO_3 has been known to give an epoxide). Both schemes, a and b, fit the experimental requirement for a second-order

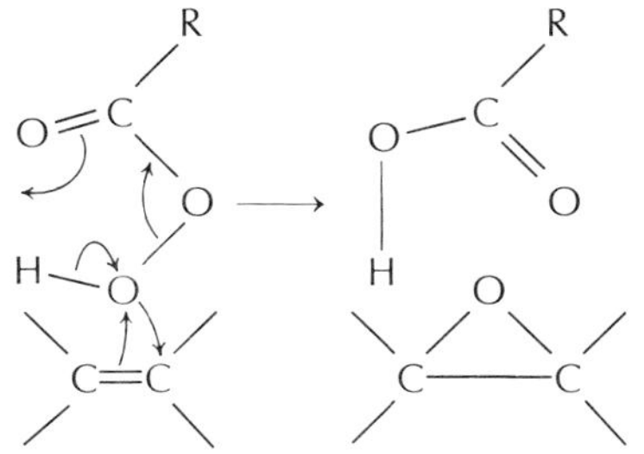

$$\text{(scheme a and b shown, Eq. 13.50)}$$

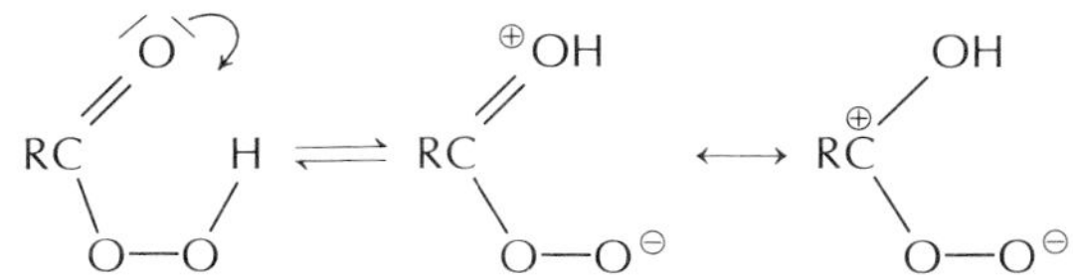

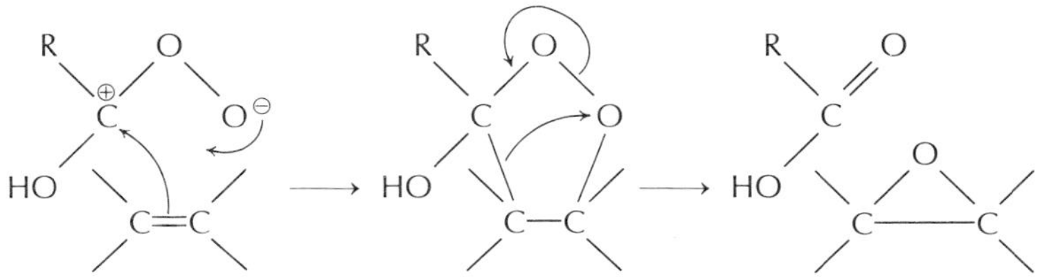

reaction, ease of reaction in nonpolar solvents, and stereospecific cis addition (a trans-alkene gives only a trans-epoxide and a cis-alkene gives only a cis-epoxide). Which mechanism is correct is still under investigation. The more substituents there are about the double bond, the faster is the reaction, as is true for other electrophilic additions.

Peracids may be formed in several ways. One of the simplest is to allow an acid to interact with hydrogen peroxide in the presence of a bit of strong acid (say H_2SO_4) to speed up the reaction (Eq. 13.51). Formic acid and trifluoroacetic acid are strong enough to catalyze their own conversion to the peracid.

$$RCOOH + H_2O_2 \underset{}{\overset{H^\oplus}{\rightleftharpoons}} RC\overset{O}{\underset{OOH}{\diagdown}} + H_2O \quad (13.51)$$

Peracids are weaker than the corresponding acid: pK_a for HCOOH is 3.6; for HCO_3H, 7.1; for CH_3COOH, 4.8; and for CH_3CO_3H, 8.2. Several causes may contribute to the effect, one of which undoubtedly is the loss of symmetrical resonance structures for the peracid anion ($RCO_3^\ominus$) as compared to $RCO_2^\ominus$.

With the weaker peracids, the epoxides may be obtained in good yield even though by slow reaction. Performic and trifluoroperacetic acids (the latter even in a nonpolar solvent such as CH_2Cl_2) cause a decomposition of the highly strained three-membered ring (Eq. 13.52). The alkanediol (or half-ester of the diol) is a trans addition product in relation

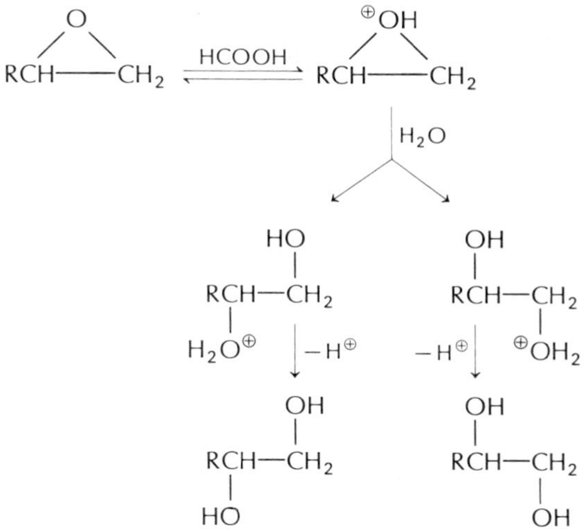

$$\text{RCH}\underset{O}{\overset{}{\triangle}}\text{CH}_2 \xrightleftharpoons{\text{HCOOH}} \text{RCH}\underset{\overset{\oplus}{O}H}{\overset{}{\triangle}}\text{CH}_2$$

$$\Big\downarrow \text{H}_2\text{O}$$

(13.52)

to the original alkene, which is clearly seen in the reaction of cyclohexene with a mixture of formic acid and 30% hydrogen peroxide at 40° to 45° (Eq. 13.53). The initially formed

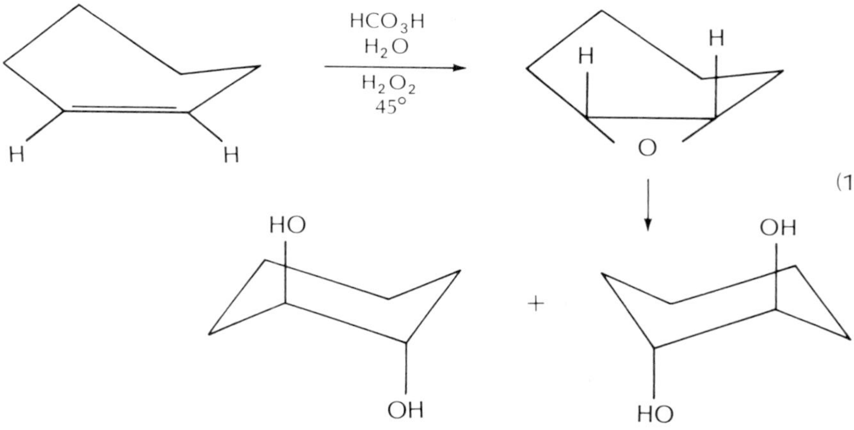

(13.53)

diaxial diols quickly go over to the more stable diequatorial conformers. Any epoxide may be hydrolyzed under either dilute acidic or basic conditions to the diol. Thus epoxidation and hydrolysis is equivalent to the *trans* addition of hydrogen peroxide to an alkene.

> Commercially, ethylene oxide (but no other epoxide) is prepared by a catalytic method, the passage of a mixture of ethylene and air over silver at 250°. Epoxides are formed in sizable but relatively low yields by what may be a free-radical reaction of an alkene with hydroperoxides or under autoxidation conditions.
> An alkyne, it is thought, may react slowly to give an unstable unsaturated epoxide (an oxirene), which may react in different ways to give a variety of products (Eq. 13.54). The yields of the several products vary markedly as either the peracid or solvent is varied.

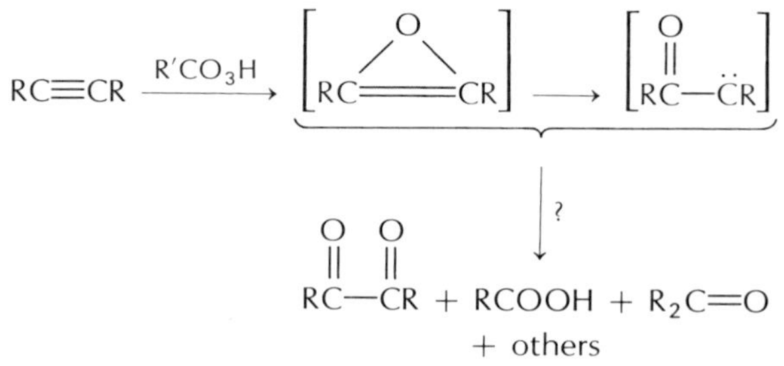

$$RC\equiv CR \xrightarrow{R'CO_3H} \left[RC\underset{O}{=\!=\!=}CR\right] \longrightarrow \left[\underset{O}{\overset{\parallel}{RC-\ddot{C}R}}\right] \quad (13.54)$$

$$\downarrow ?$$

$$\underset{O\ \ O}{\overset{\parallel\ \ \parallel}{RC-CR}} + RCOOH + R_2C=O$$
$$+ \text{others}$$

Halohydrins; dihalides. An alternate route to an epoxide is the electrophilic addition of HOX to an alkene, followed by treatment of the halohydrin with dilute base (Eq. 13.55). Similarly, hydrolysis of a dihalide gives rise to an oxide (Eq. 13.56). If the concen-

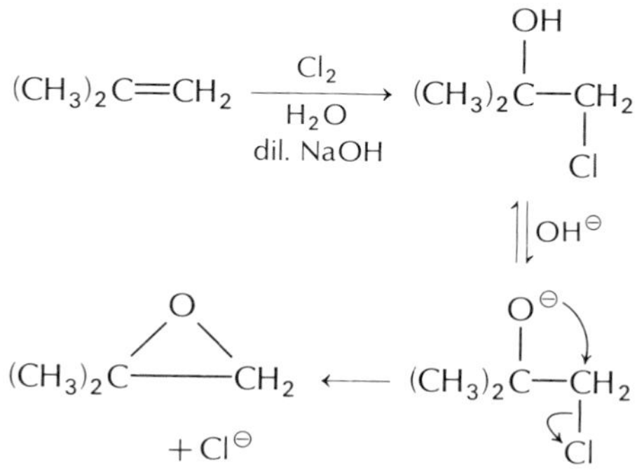

$$(CH_3)_2C=CH_2 \xrightarrow[\substack{H_2O \\ \text{dil. NaOH}}]{Cl_2} (CH_3)_2\underset{Cl}{\overset{OH}{\underset{|}{C}-CH_2}}$$

$$\updownarrow OH^\ominus$$

$$(CH_3)_2C\underset{O}{\overset{}{\triangle}}CH_2 \longleftarrow (CH_3)_2\underset{Cl}{\overset{O^\ominus}{\underset{|}{C}-CH_2}} \quad (13.55)$$
$$+ Cl^\ominus$$

tration of base is too great, the epoxide reacts to give a diol. Both routes are equivalent to an oxidation of an alkene. Of course, if we wish, we may hydrolyze the epoxide by dilute acid to give the diol, a process equivalent to a *trans* addition.

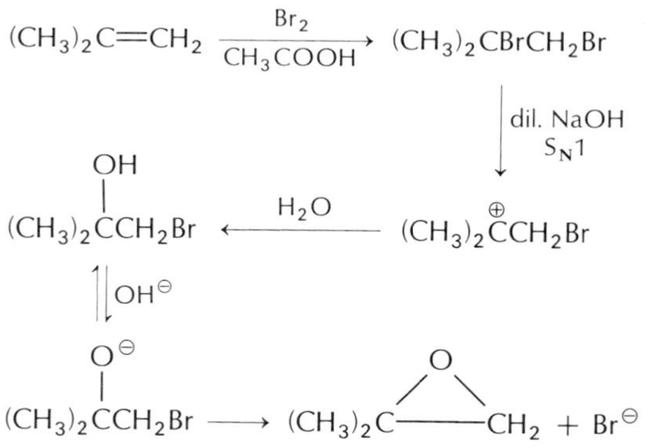

$$(CH_3)_2C=CH_2 \xrightarrow{Br_2}{CH_3COOH} (CH_3)_2CBrCH_2Br$$

$$\downarrow \substack{\text{dil. NaOH} \\ S_N1}$$

$$(CH_3)_2\overset{OH}{\underset{|}{C}CH_2Br} \xleftarrow{H_2O} (CH_3)_2\overset{\oplus}{C}CH_2Br \quad (13.56)$$

$$\updownarrow OH^\ominus$$

$$(CH_3)_2\overset{O^\ominus}{\underset{|}{C}CH_2Br} \longrightarrow (CH_3)_2C\overset{O}{\underset{}{\triangle}}CH_2 + Br^\ominus$$

Osmium tetroxide; alkaline potassium permanganate. Several other routes are available for the oxidation of alkenes to diols. One is the use of osmium tetroxide (OsO_4, mp 42°, bp 130°, soluble in CCl_4, ether, alcohols, water; very poisonous—vapors attack skin, eyes, and respiratory tract). Despite the difficulties involved, osmium tetroxide is the best reagent known for the conversion of alkenes to diols by *cis* attack (Eqs. 13.57). The

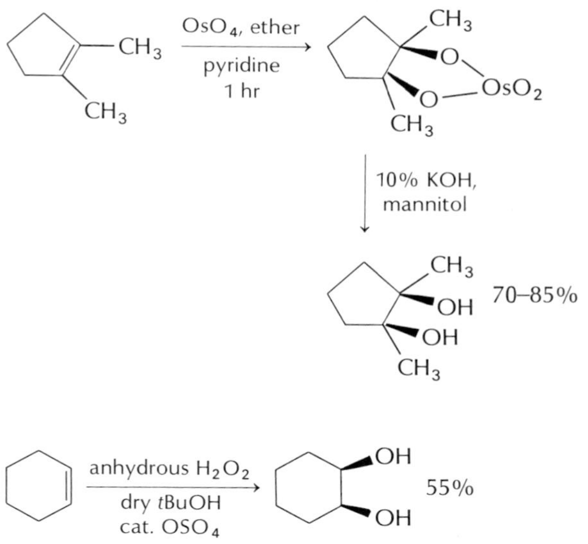

[Reaction 13.57: 1,2-dimethylcyclopentene + OsO₄, ether, pyridine, 1 hr → cyclic osmate ester intermediate; then 10% KOH, mannitol → cis-1,2-dimethylcyclopentane-1,2-diol, 70–85%]

or

[Cyclohexene + anhydrous H₂O₂, dry tBuOH, cat. OsO₄ → trans/cis-cyclohexane-1,2-diol, 55%]

reagent may be used on a mole-per-mole basis or in tiny amounts as a catalyst, in which case the ester presumably is decomposed by hydrogen peroxide with regeneration of the osmium oxide, which is used over and over.

Potassium permanganate at 0° under basic conditions in aqueous tBuOH also oxidizes an alkene to a *cis*-diol. Yields generally are low (cyclohexene, 38%; cyclopentene, 55%), although several exceptions are known in which yields of 70 to 80% are obtained. It has been shown by the use of ^{18}O labeled potassium permanganate that the oxygen atoms in the product diol come from the reagent, not the solvent. Thus, an intermediate ester similar to that with OsO$_4$ must be involved (Eq. 13.58). Unless the mixture is basic

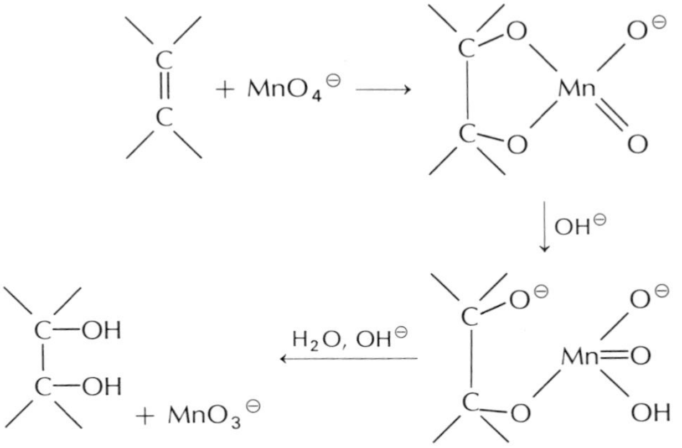

$$\text{C=C} + \text{MnO}_4^\ominus \longrightarrow \text{cyclic manganate ester} \xrightarrow{\text{OH}^\ominus} \text{hydroxylated Mn intermediate} \xrightarrow{\text{H}_2\text{O, OH}^\ominus} \text{diol} + \text{MnO}_3^\ominus$$

(13.58)

$$3\text{MnO}_3^\ominus + \text{H}_2\text{O} \longrightarrow 2\text{MnO}_2 + \text{MnO}_4^\ominus + 2\text{OH}^\ominus$$

to begin with, the oxidation does not stop at the diol stage. Diols are very easily oxidized even if both —OH are tertiary. A variety of products is possible.

Mild and drastic oxidations. Let us classify those oxidation reactions that give epoxides or diols from alkenes as <u>mild oxidations</u> and those that go beyond the epoxide or diol stage as <u>drastic oxidations.</u> Both potassium permanganate and potassium dichromate may be used as drastic reagents. Although alcohols are more easily attacked by the chromium compound, the π bond is more subject to attack by permanganate. However, dichromate with sulfuric acid present always is a drastic reagent, especially at temperatures of 50° or more. Under anhydrous conditions (for example, CrO_3 in acetic acid), mixtures of products often result (allylic attack, epoxides, hydroxyketones). Permanganate, except at temperatures less than 25° and in initially basic solutions, is a drastic reagent, particularly under initially acidic conditions. We summarize possible intermediates and products in Equations 13.59 and look at the final products obtainable at either end of a π bond for preparative drastic oxidations in Table 13.2.

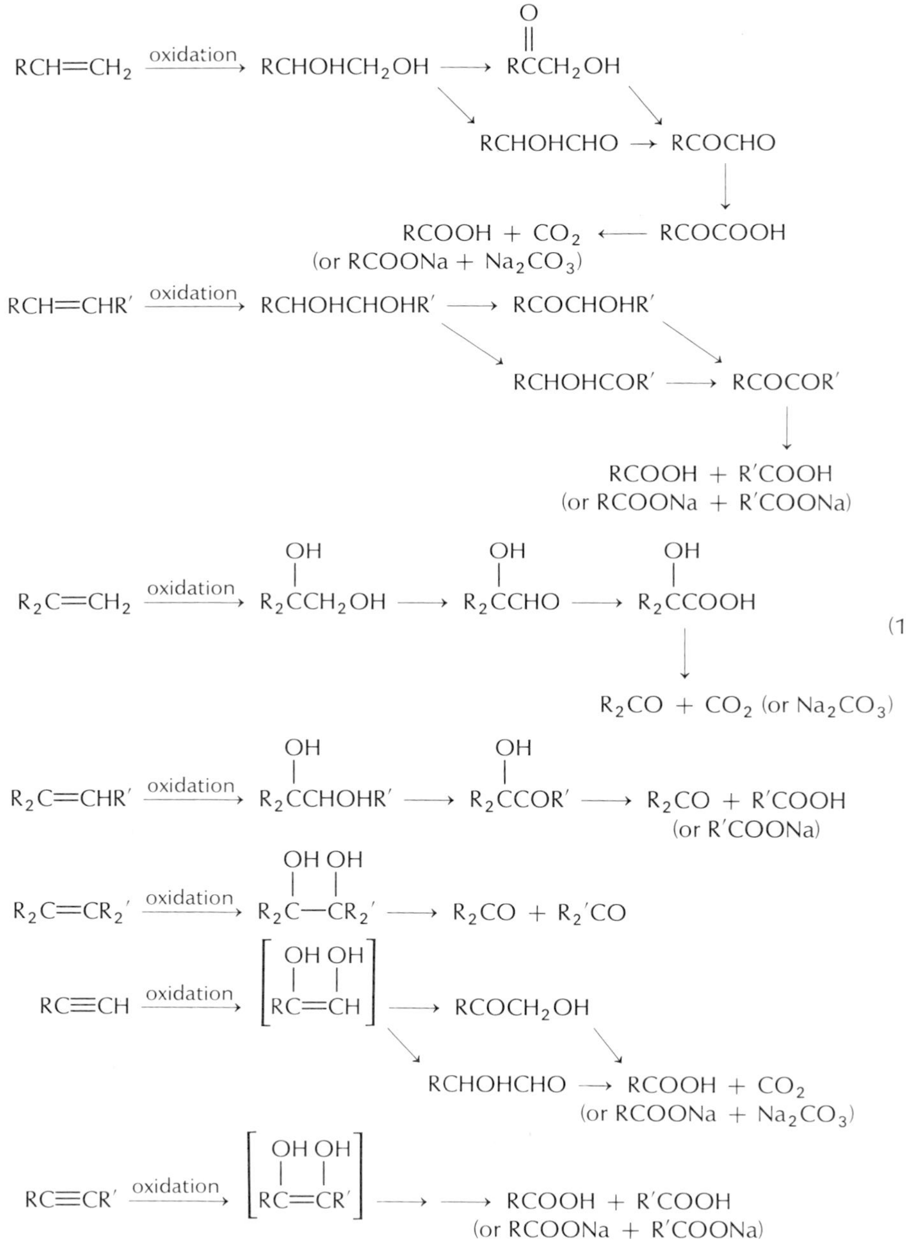

$$RCH=CH_2 \xrightarrow{\text{oxidation}} RCHOHCH_2OH \longrightarrow RCCH_2OH \text{ (=O)}$$
$$\searrow RCHOHCHO \rightarrow RCOCHO$$
$$\downarrow$$
$$RCOOH + CO_2 \longleftarrow RCOCOOH$$
$$(\text{or } RCOONa + Na_2CO_3)$$

$$RCH=CHR' \xrightarrow{\text{oxidation}} RCHOHCHOHR' \longrightarrow RCOCHOHR'$$
$$\searrow RCHOHCOR' \longrightarrow RCOCOR'$$
$$\downarrow$$
$$RCOOH + R'COOH$$
$$(\text{or } RCOONa + R'COONa)$$

$$R_2C=CH_2 \xrightarrow{\text{oxidation}} R_2\underset{OH}{C}CH_2OH \longrightarrow R_2\underset{OH}{C}CHO \longrightarrow R_2\underset{OH}{C}COOH$$
$$\downarrow$$
$$R_2CO + CO_2 \text{ (or } Na_2CO_3)$$

(13.59)

$$R_2C=CHR' \xrightarrow{\text{oxidation}} R_2\underset{OH}{C}CHOHR' \longrightarrow R_2\underset{OH}{C}COR' \longrightarrow R_2CO + R'COOH$$
$$(\text{or } R'COONa)$$

$$R_2C=CR_2' \xrightarrow{\text{oxidation}} R_2\underset{OH\ OH}{C-CR_2'} \longrightarrow R_2CO + R_2'CO$$

$$RC\equiv CH \xrightarrow{\text{oxidation}} \left[\underset{OH\ OH}{RC=CH} \right] \longrightarrow RCOCH_2OH$$
$$\searrow RCHOHCHO \longrightarrow RCOOH + CO_2$$
$$(\text{or } RCOONa + Na_2CO_3)$$

$$RC\equiv CR' \xrightarrow{\text{oxidation}} \left[\underset{OH\ OH}{RC=CR'} \right] \longrightarrow \longrightarrow RCOOH + R'COOH$$
$$(\text{or } RCOONa + R'COONa)$$

13.5 Oxidation reactions

Table 13.2 Products of drastic oxidations

Reactant	Product	
	Acidic conditions	Basic or initially neutral conditions
=CH$_2$	CO$_2$	Na$_2$CO$_3$
=CHR	RCOOH	RCOONa
=CR$_2$	RCOR	RCOR
≡CH	CO$_2$	Na$_2$CO$_3$
≡CR	RCOOH	RCOONa
KMnO$_4$	K$_2$SO$_4$ + MnSO$_4$	KOH + MnO$_2$
K$_2$Cr$_2$O$_7$	K$_2$SO$_4$ + Cr$_2$(SO$_4$)$_3$	-----

The products of drastic oxidations are acids, ketones, and carbon dioxide under acidic conditions or salts of acids, ketones, and sodium carbonate under basic conditions (NaOH added initially). Intermediate alcohols and aldehydes seldom can be isolated. Ketones are more stable, but under strong acidic or alkaline conditions they may be attacked, especially if excess oxidizing agent is present. Acids and salts of acids are stable to oxidation. (We shall discuss exceptions to these general statements later.) In acidic drastic oxidations, intermediate alcohols are prone to dehydration and rearrangement, especially if they are tertiary. Any new double bonds are subject to more oxidation (Eq. 13.60).

$$
\begin{array}{c}
\text{CH}_3\text{CH}_2 \\
\diagdown \\
\text{C}=\text{CH}_2 \\
\diagup \\
\text{CH}_3
\end{array}
\xrightarrow[\text{oxidation}]{\text{acidic}}
\begin{array}{c}
\text{CH}_3\text{CH}_2 \text{OH} \\
\diagdown | \\
\text{C}-\text{CH}_2\text{OH} \\
\diagup \\
\text{CH}_3
\end{array}
\downarrow
$$

$$
\begin{array}{c}
\text{CH}_3\text{CH} \\
\diagdown\!\!\!\diagdown \\
\text{C}-\text{CH}_2\text{OH} \\
\diagup \\
\text{CH}_3
\end{array}
\xleftarrow{-\text{H}^\oplus}
\begin{array}{c}
\text{CH}_3\text{CH}_2 \\
\diagdown \\
\overset{\oplus}{\text{C}}-\text{CH}_2\text{OH} \\
\diagup \\
\text{CH}_3
\end{array}
\quad (13.60)
$$

$$
\downarrow \text{oxidation}
$$

$$
\begin{array}{c}
\text{CH}_3\text{CHOH} \text{OH} \\
\diagdown | \\
\text{C}-\text{CH}_2\text{OH} \\
\diagup \\
\text{CH}_3
\end{array}
\xrightarrow{\text{etc.}}
\begin{array}{c}
\text{CH}_3\text{CHOH} \\
\diagdown \\
\text{C}-\text{CH}_2\text{OH} \\
\diagup\!\!\!\diagup \\
\text{CH}_2
\end{array}
\xrightarrow{\text{etc.}} 3\text{CO}_2 + \text{CH}_3\text{COOH}
$$

Ozonization is superior to a drastic oxidation of small amounts of alkenes for the purpose of locating the position of the π bond, because ozonization is much more clean cut. For preparative purposes, drastic oxidation is preferred because larger quantities may be handled and the process is more rapid. On the other hand, uses of ozone are not unknown on an industrial scale. □□

374   Chapter thirteen

Many other reactions of alkanes, alkenes, and alkynes may be considered oxidations. We have omitted many oxidations involving metal catalysts.

Exercises

13.6 What alkene or cycloalkene, upon ozonization and Zn dust work-up, will give the products or product shown?
  a. $CH_3CH_2COCH_2CH_3 + CH_3CH_2CHO$
  b. $(CH_3)_2CHCHO + CH_3CHO$
  c. $CH_3CH_2COCH_3$
  d. $CH_3COCH_2CH_2CH_2COCH_3$
  e. $CH_3CH_2COCH_2CH_2CH_2CHO$

13.7 Given cyclopentene, cyclohexene, and 3-ethyl-2-methyl-2-pentene, and any other needed reagents and solvents, show how to prepare:

  a. 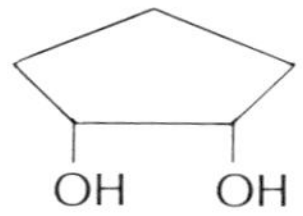 [cyclopentane with two adjacent OH groups]

  c. $HOOCCH_2CH_2CH_2COOH$

  b. 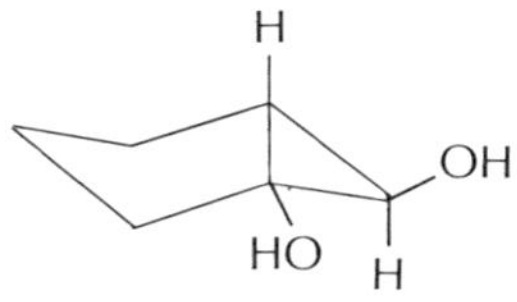 [cyclohexane with OH groups, HO and H stereochemistry shown]

  d. $OHC(CH_2)_4CHO$

13.8 Predict the products of drastic oxidation by $KMnO_4$ (acidic and basic) and by $K_2Cr_2O_7$ (acidic) of the alkenes shown, and balance the equations.

  a. 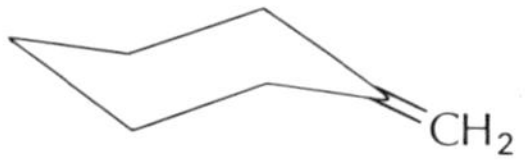 [alkene structure with =CH$_2$]

  b. $(CH_3CH_2)_2C=CHCH_2CH_3$

13.6   Reduction reactions

We shall look at four of the general methods for carrying out the reduction (hydrogenation) of π bonds: catalytic hydrogenation, reduction by metal hydrides, reduction by active metals, and reduction by diimide. Each general method has many variations in procedure, solvent, temperature, and the like.

Catalytic hydrogenation.   Catalytic hydrogenation (heterogeneous catalysis) of alkenes and alkynes depends upon the reduction in bond strength of the hydrogen molecule and of the alkene or alkyne. It is not known whether hydrogen is adsorbed on the surface of the catalyst followed by a π bond, the alkene or alkyne precedes hydrogen on the surface, or both creep about on the catalyst surface after being adsorbed. The reactants combine with release of heat at rates that vary from fast to very slow. Addition is *cis*, but isomerization may occur. Although an alkane is less strongly adsorbed than is an alkene, experiments with alkanes and deuterium have clearly shown the reversibility of

$$H_2 + \text{cat} \rightleftharpoons \underset{|\quad\;|}{H\quad H} \qquad\qquad a$$

$$D_2 + \text{cat} \rightleftharpoons \underset{|\quad\;|}{D\quad D} \qquad\qquad b$$

$$\diagdown_{\displaystyle C}\!\!=\!\!\diagup^{\displaystyle C}\diagdown + \text{cat} \rightleftharpoons \diagdown_{\displaystyle C}\!\!-\!\!\diagup^{\displaystyle C}\diagdown \qquad\qquad c$$

$$\diagdown\!C\!-\!C\!\diagup + H\;H \rightleftharpoons \diagdown\!C\diagup\;\;C\!\diagdown^{H}\;\;H \qquad d \qquad (13.61)$$

$$\diagdown\!C\diagup\;C\!\diagdown^{H} + H\;H \rightleftharpoons \;\;\;\;\;H + \diagdown\!C\!-\!C\!\diagup \qquad e$$
$$\qquad\qquad\qquad\qquad\qquad\qquad\qquad H\;H$$

the reaction (Eqs. 13.61). In a or b, the reverse reaction does not occur (is inhibited) in the presence of alkenes or alkynes, which leads us to believe that c, d, and e (and analogous reactions with D) predominate. Adsorption of a π bond is stronger than that of H_2 or D_2. The reversal of e is slow at less than 150°, but even if the equilibrium lies far to the right at all temperatures, an alkane in the presence of excess deuterium exhibits an exchange reaction. Combining the reverse of e with the reverse of d, we see that either one hydrogen or two neighboring hydrogen atoms may be exchanged, but not three or more at a time unless they are neighbors. Thus neopentane gives $(CH_3)_3CCH_2D$ as the initial product, but butane gives all products from C_4H_9D to C_4D_{10}. Also, cyclopentane gives more $C_5H_5D_5$ and C_5D_{10} than any other initial products. This result has been interpreted to mean that c, d, and e involve adsorption of eclipsed conformers. Cyclopentane, once adsorbed, easily exchanges all five hydrogen atoms on that side of the ring. A desorption by e and a subsequent resorption on the other side of the ring is required before exchange of the last five atoms can take place.

In industrial gaseous hydrogenation, a mixture of the reactants is allowed to flow through a bed of catalyst with temperature, pressure, and rate of flow carefully controlled to obtain the maximum conversion per pass.

In the laboratory and in industry, when the continuous-flow process is impractical, hydrogenations are carried out as batch processes with or without a solvent. Catalysts used are platinum, palladium, rhodium, ruthenium, or nickel, finely divided or deposited on inert carriers such as carbon, alumina, silica, or calcium carbonate. Many of these powdered metallic substances are pyrophoric (ignite spontaneously) if exposed to air oxygen. The activity of the catalyst increases as the acidity or polarity of the solvent is increased. Thus hydrochloric or perchloric acids sometimes are added to hydrogenation mixtures. Sulfuric acid (and all sulfur compounds) are avoided because of preferential adsorption on the surface of the catalyst (poisoning). Sometimes traces of ferrous chloride help to promote a reduction. Commonly used solvents are cyclohexane, ethyl acetate, ethanol, water, and acetic acid.

Temperatures may range up to 300° and pressures up to 300 atm. Many hydrogenations in the laboratory are carried out at room temperature and 1 to 2 atm pressure. At the higher temperatures and pressures, a mixture of copper and chromium oxides called copper chromite is a frequently used catalyst.

Rates of hydrogenation are difficult to compare because of differences in activity of catalysts from run to run. However, a general rule (with many exceptions) is that the ease of catalytic hydrogenation goes as follows:

$$RC\equiv CH > RC\equiv CR' > RCH=CH_2 > RCH=CHR',$$
$$RR'C=CH_2 > R_2C=CHR' > R_2C=CR_2'$$

Heats of hydrogenation have been given (Table 7.3).

Use of sodium borohydride. Several methods have been devised for carrying out hydrogenation in simple glassware instead of the usual pressure and agitation equipment. One such variation has been to drip a solution of $NaBH_4$ into an aqueous (or alcoholic) mixture of solvent and dilute hydrochloric acid (which reacts with $BH_4^{\ominus}$ to produce hydrogen), catalyst (usually Pt or Pd on carbon), and the substance to be reduced. A very active nickel catalyst, perhaps Ni_2B, is produced if $NiCl_2$ is present in the mixture. This combination reduces 1-octene in 15 min.

A fuel cell (ordinarily used with H_2 or CH_4 with oxygen) has been found to operate for hydrogenation of alkenes also.

Complexes of transition metals (Rh) have been found that are soluble and facilitate hydrogenation (homogeneous hydrogenation).

Reduction by metal hydrides already has been discussed in Section 13.3 (B_2H_6, AlH_3, etc.).

Reduction by sodium or lithium. Reduction by active metals refers to the use of lithium or sodium dissolved in liquid ammonia or ethylamine. These processes have been shown to be free radical reactions in some instances. A metal atom donates an electron to the π bond to form a radical anion, which takes up a proton from the solvent (more rapidly from purposely added alcohols such as CH_3OH or $tBuOH$) to give a radical. The radical takes up an electron from another metal atom to give an anion, which abstracts another proton (Eqs. 13.62).

$$RCH=CH_2 + Na\cdot \longrightarrow R\dot{C}HCH_2^{\ominus} + Na^{\oplus}$$
$$R\dot{C}HCH_2^{\ominus} + CH_3OH \longrightarrow R\dot{C}HCH_3 + CH_3O^{\ominus}$$
$$R\dot{C}HCH_3 + Na\cdot \longrightarrow R\overset{\ominus}{C}HCH_3 + Na^{\oplus} \qquad (13.62)$$
$$R\overset{\ominus}{C}HCH_3 + CH_3OH \longrightarrow RCH_2CH_3 + CH_3O^{\ominus}$$

A side reaction (other than reaction of metal directly with alcohol) is a coupling of radicals to give, for example, $R(CH_3)CHCH(CH_3)R$. The reaction cannot be used with terminal alkynes because of the ease of formation of the salt of the alkyne. Mono- and disubstituted alkenes react at a reasonable rate; tetrasubstituted alkenes react so slowly that they are classified as inert.

In a variation of the reaction, electrons are furnished at a plate in an electrolysis cell. However, the coupling reaction rather than reduction may give the major product.

13.7 Sources: petroleum

Diimide. Diimide, HN=NH, though known for a long time, has only recently been looked at thoroughly enough to be recognized as particularly useful in reductions. Of the various methods for preparation and use of diimide (which must be used as it is formed), we show only a possible mechanism for the copper-catalyzed oxidation of hydrazine, H_2NNH_2 (Eqs. 13.63). The cyclic addition gives exclusively *cis* addition. The

$$Cu^{\oplus} + H_2O_2 \longrightarrow Cu^{2\oplus} + HO^{\ominus} + HO\cdot$$

$$HO\cdot + H_2NNH_2 \longrightarrow H_2O + H\dot{N}NH_2$$

$$H\dot{N}NH_2 + H_2O_2 \longrightarrow HN=NH + H_2O + HO\cdot$$

$$H\dot{N}NH_2 + Cu^{2+} \longrightarrow HN=NH + Cu^{\oplus} + H^{\oplus} \qquad (13.63)$$

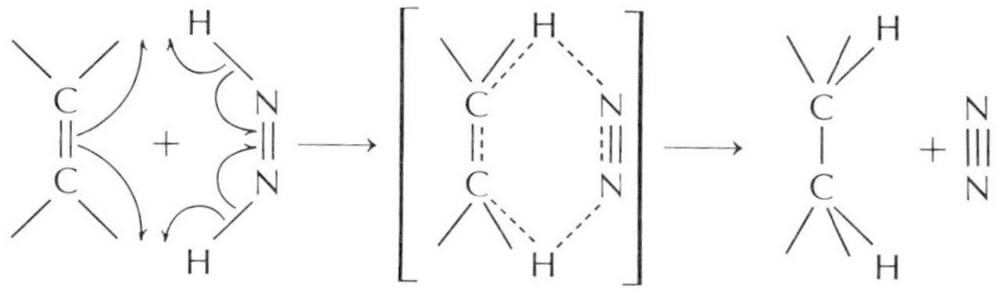

reaction is rapid, and it does better the less polar is the π bond to be reduced. The reagent seems to be generally applicable to the reduction of alkenes and alkynes. In use, the alkene with an excess of hydrazine and a small amount of a dilute solution of cupric sulfate is dissolved in alcohol, and a slight excess of 30% hydrogen peroxide is added dropwise at 0°.

Exercises

13.9 Show how to convert 2-butyne into a. *cis*-2-butene; b. *trans*-2-butene.

13.7 Sources: petroleum

Natural gas and petroleum. The major sources of the hydrocarbons are natural gas and petroleum. Their greatest use is for the production of energy by combustion. However, a substantial amount is used for the production of solvents and organic compounds.

The total energy for all heat, light, and power used in the United States for 1969 was about 16.5×10^{15} kcal, equivalent to the burning of about 2 billion tons of carbon or 1.5 billion tons of ethane. Not all of this energy requirement is furnished by natural gas and petroleum. For example, figures projected up to 1975 are shown in Table 13.3, which demonstrates the dependence upon coal as the principal source for the production of electrical energy. We should remember in this connection that carbon is only the fourteenth most abundant element, about 0.08% of the earth's crust and oceans. Note as well the doubling in demand in 1975 over that of 1965. The demand for energy is real.

Despite heavy use of gas and petroleum products as the principal energy source for transporation and as a major one for electrical energy, we shall find the amounts of simple organic compounds derived from petroleum to be almost unbelievable (Table 13.4).

Table 13.3 Electrical energy requirements and forecast[a]

Source	1950			1965			1970			1975		
	kWh	C	%	kWh	C	%	kWh	C	%	kWh	C	%
Coal	155	17.0	47	573	62.9	54	825	90.6	55	1152	126.5	56
Fuel oil	34	3.7	10	65	7.1	6	108	11.9	7	130	14.3	6
Gas	44	4.8	14	219	24.0	21	287	31.5	19	310	34.0	15
Hydroelectric	96	10.5	29	194	21.3	18	226	24.8	15	261	28.6	13
Nuclear	0	0	0	4	0.4	<1	50	5.5	3	196	21.5	10
Total	329	36.0		1055	115.8		1496	164.2		2049	224.9	

[a] Energy given as billions of kilowatt hours and in the equivalent of combustion of millions of metric tons of carbon at 100% efficiency. A metric ton = 1000 kg = 1.1023 short tons. Adapted from *Chemical and Engineering News,* June 15, 1970, p. 16. Copyright 1970 by the American Chemical Society. Reprinted by permission.

Table 13.4 Production of some organic compounds, 1969[a]

Compound	Produced, millions of tons	Shipped, %
ethylene	7.7	28
propylene	4.0	46
1,2-dichloroethane	2.9	10
formaldehyde	2.1	36
methanol	2.0	39
vinyl chloride	1.8	49
ethylene oxide	1.6	15
carbon black	1.5	—
ethylene glycol	1.2	80
cyclohexane	1.1	96
ethanol	1.0	59
2-propanol	1.0	39
acetic acid	0.9	22
acetaldehyde	0.8	—
acetic anhydride	0.8	8
acetone	0.7	75

[a] Selected from *Chemical and Engineering News,* May 25, 1970, p. 18. Copyright 1970 by the American Chemical Society. Reprinted by permission.

For comparison, ammonia production was 12.7 million tons and nitric acid was 6.2 million tons in 1969. The percentage shipped gives an idea of how much of the substances made was sold to customers. Big sellers were ethylene glycol (used as an antifreeze) and cyclohexane (a starting point for the manufacture of nylon). The amount retained gives an indication of use as solvents and for conversion to other products at the point of manufacture. □ □

Refining. The original refining of petroleum consisted of nothing more than a crude distillation into fractions with different ranges of bp. Until the development of the internal combustion engine, kerosene was the most used fraction. At present, efficient fractionation allows separation into cuts of variable bp ranges depending upon the crude oil being refined: petroleum ether, bp 20 to 60°; ligroin, 60 to 100°; gasoline, 40 to 205°; kerosene, 175 to 325°; gas oil, 275 to 350°. Vacuum distillation yields fractions that after further treatment become lubricating oil, paraffin oil, petrolatum, and paraffin wax. The residue is tar (asphalt) or petroleum coke, depending again upon the source of the crude oil and upon the final temperature.

Because of the demand for gasoline much effort has gone into finding ways of building up the propanes and butanes, breaking down the higher boiling substances, and converting straight-chain to branched-chain or cyclic or aromatic compounds (all of which have high octane numbers) of the proper range of bp. As a side line, a few companies supply petroleum ether, bp 30 to 60°; ligroin, bp 60 to 90°; 63 to 75°, 100 to 115°; and petroleum octane, bp 125 to 127°, for use as solvents.

Other companies as well as refiners have found it profitable to make <u>petrochemicals</u>: compounds largely derived from acetylene, ethylene, propylene, and butylenes. □ □

Summary. The refining of petroleum has become an operation no longer concerned only with gasoline, kerosene, lubricating oils, and greases (an oil with an aluminum or lithium soap dispersed in it). Profitable use is made of ethylene, propylene, and butadiene. As a result, not only all the alkanes, alkenes, and alkynes up to C_6, but also many of the larger compounds are easily available.

13.8 Preparation of alkanes

Substitution reactions. The laboratory methods for the synthesis of specific alkanes all may be classed as reductions. We shall subclassify the methods as substitution and addition-elimination reactions. Most of the methods are applicable to the preparation of cycloalkanes as well, although methods of ring closure will be postponed until a later chapter.

Most of the substitution methods have been described (Eqs. 13.64). Only very active

$$RX \begin{cases} \xrightarrow[\text{ether}]{Mg} RMgX \xrightarrow{\text{dil. HCl}} RH + MgClX & a \\ \xrightarrow[\text{ether}]{Li} RLi \xrightarrow{\text{dil. HCl}} RH + LiCl & b \\ \xrightarrow[\text{ether}]{Mg} RMgX \xrightarrow{CH_2=CHCH_2Cl} RCH_2CH=CH_2 + MgXCl \\ \qquad\qquad\qquad\qquad \downarrow \text{reduction} \\ \qquad\qquad\qquad\quad RCH_2CH_2CH_3 & c \quad (13.64) \\ \xrightarrow{HI} RH + X^\ominus + I_2 & d \\ \xrightarrow[\text{cat.}]{H_2} RH + HX & e \\ \xrightarrow[Zn]{HCl} RH + ZnX_2 & f \\ \xrightarrow{LiAlH_4} RH + LiX + AlX_3 & g \end{cases}$$

halides such as allyl chloride may be used in c for synthetic work. A little of c occurs in the preparation of any Grignard or lithium reagent ($2RX + Mg \rightarrow R_2 + MgX_2$) and lowers the yield of reagent. Alkyl halides are easily reduced by HI, HCl + Zn, and LiAlH$_4$. Reduction by catalytic hydrogenation is called hydrogenolysis, and we must keep it in mind when we are considering a synthetic sequence that involves a reduction.

Certain alcohols (for example, allylic alcohols) are subject to hydrogenolysis (Eq. 13.65). If the double bond to be hydrogenated is removed from the allylic position,

$$CH_3CH=CHCH_2OH \xrightarrow[\text{cat.}]{H_2} CH_3CH_2CH_2CH_3 + H_2O \qquad (13.65)$$

the alcohol resists hydrogenolysis (Eq. 13.66).

$$H_2C=CHCH_2CH_2OH \xrightarrow[\text{cat.}]{H_2} CH_3CH_2CH_2CH_2OH \qquad (13.66)$$

A new (to us) substitution reaction is the use of sulfonate esters and LiAlH$_4$ (Eq. 13.67).

$$(RO)_2SO_2 \xrightarrow{LiAlH_4} 2RH + SO_4^{2-}$$

or $\qquad (13.67)$

$$ROSO_2C_6H_5 \xrightarrow{LiAlH_4} RH + C_6H_5SO_3^{-}$$

This reaction, like the reduction of alkyl halides by LiAlH$_4$, is thought to be an S_N2 reaction in which hydride ion acts as the nucleophile.

Another new reaction is the Wurtz reaction (Eq. 13.68), which is also an S_N2 reaction

$$2RX \xrightarrow{Na} R-R + NaX \qquad (13.68)$$

(Eq. 13.69). As such, it works only with primary alkyl halides. Secondary and tertiary halides

$$RX + Na\cdot \longrightarrow R\cdot + X^{\ominus} + Na^{\oplus}$$
$$\downarrow Na\cdot$$
$$R^{\ominus}Na^{\oplus} \qquad (13.69)$$
$$\downarrow RX$$
$$R-R + NaX$$

are subject to easy dehydrohalogenation by the strongly basic carbanion $R^{\ominus}$, so that an alkene and the alkane RH become the principal products. Even with primary halides, the elimination reaction is a sizable part of the total reaction, and yields are low: 50% for the conversion of BuBr to octane.

In every class in organic chemistry, there are a few students who use the Wurtz reaction as the answer to all their difficulties with any synthetic reaction or problem in the course. The reason for this fascination with the Wurtz reaction is a mystery. In any case, no problem or exercise in this book will use it.

Addition reactions: Wolff-Kishner and Clemmensen reductions. Addition reactions for the synthesis of alkanes have already been described: catalytic hydrogenation of alkenes and alkynes, hydroboration, use of active metals, and diimide. Alkylation also has been covered.

Another addition-elimination reaction is the reduction of aldehydes and ketones to alkanes without formation of the intermediate alcohol. There are two ways of carrying out such a reaction.

The <u>Wolff-Kishner reduction</u> as modified by Huang-Minlon consists of mixing the aldehyde or ketone with aqueous hydrazine and sodium (or potassium) hydroxide in a high-boiling solvent (diethylene glycol, $HOCH_2CH_2OCH_2CH_2OH$, bp 245°, is frequently used), heating, and distilling the water until the temperature of the mixture is 180° to 200°. Heating under reflux is continued for two to five hours. The alkane separates as an upper layer (Eq. 13.70). The mechanism will be discussed in Section 20.6. Most ketones may be

$$\diagup\!\!\!\!\diagdown\!\!C=O + H_2NNH_2 \xrightarrow{NaOH} \diagup\!\!\!\!\diagdown\!\!CH_2 + N_2 + H_2O \qquad (13.70)$$

reduced to alkanes by this procedure; but aldehydes are troublesome because of their sensitivity to base, which competes successfully in many instances. In such cases, a hydrazone may be prepared and isolated. Then, at room temperature, the hydrazone is dissolved

$$\diagup\!\!\!\!\diagdown\!\!C=O + H_2NNH_2 \longrightarrow \diagup\!\!\!\!\diagdown\!\!C=NNH_2 + H_2O$$

in dimethyl sulfoxide (($CH_3)_2SO$) containing sodium *tert*-butoxide and is allowed to stand for an hour.

The other reaction is known as the <u>Clemmensen reduction.</u> The aldehyde or ketone, dissolved in a solvent immiscible with water (toluene or other), is heated under reflux with concentrated hydrochloric acid and zinc that has been amalgamated with mercury. (The amalgam provides a liquid on the surface of the zinc so that the atoms of zinc in the surface are easily attacked.) The reaction is slow and may take as long as one to two days for completion. The concentrated acid enhances the solubility of the aldehyde or ketone in

$$C=O + H_3O^\oplus \rightleftharpoons C=OH^\oplus \xrightarrow{Zn} \underset{Zn^\oplus}{\overset{OH}{\diagup\!\!\!\!C\!\!\!\diagdown}}$$

$$\underset{Zn^\oplus}{\overset{Zn^\oplus}{\diagup\!\!\!\!C\!\!\!\diagdown}} \xleftarrow{Zn} \underset{Zn^\oplus}{\overset{}{\diagup\!\!\!\!C^\oplus\!\!\!\diagdown}} \xleftarrow{-H_2O} \underset{Zn^\oplus}{\overset{OH_2^\oplus}{\diagup\!\!\!\!C\!\!\!\diagdown}} \qquad (13.71)$$

$$\downarrow -Zn^{2+}$$

$$\underset{Zn^\oplus}{\overset{}{\diagup\!\!\!\!C^\ominus\!\!\!\diagdown}} \xrightarrow{H^\oplus} \underset{Zn^\oplus}{\overset{H}{\diagup\!\!\!\!C\!\!\!\diagdown}} \xrightarrow{-Zn^{2+}} \overset{H}{\diagup\!\!\!\!C^\ominus\!\!\!\diagdown} \xrightarrow{H^\oplus} \overset{H}{\diagup\!\!\!\!C\!\!\!\diagdown_H}$$

the aqueous layer. (Only the simplest aldehydes and ketones are completely miscible with water.) The bulk of the carbonyl compound remains in the upper toluene layer. The mechanism of the reaction is in doubt, but a possible mechanism is shown in Equation 13.71. Some reduction to the alcohol followed by dehydration to an alkene competes with the main reaction. The alkane produced returns to and remains in the toluene layer, from which it must be separated during work-up. Ketones and aldehydes that are sensitive to concentrated hydrochloric acid cannot be reduced by the Clemmensen procedure.

Kolbe electrolysis: fusion of salts of acids. In 1849, Kolbe passed an electric current through a solution of acetic acid and obtained some ethane. The reaction, now known as the Kolbe electrolysis, gives fair to good yields of the coupled alkane only in simple cases (Eq. 13.72). The reaction is thought to pass through a radical

$$\text{RCOO}^{\ominus} \xrightarrow{\oplus \text{ pole}} \text{RCOO} \cdot + e$$
$$\downarrow$$
$$\text{R} \cdot + \text{CO}_2 \qquad (13.72)$$
$$\downarrow \text{R} \cdot$$
$$\text{R}-\text{R}$$

(adsorbed on the positive pole) because by-products that could have come from radicals accompany the alkane.

Another early reaction was discovered in the drastic treatment of acetic acid with sodium hydroxide (Eq. 13.73). However, the reaction is poor for salts of other

$$\text{CH}_3\text{COONa} + \text{NaOH} \xrightarrow{350°} \text{CH}_4 + \text{Na}_2\text{CO}_3 \qquad (13.73)$$

acids. Although sodium acetate gives a 99% yield of methane, CH_3CH_2COONa gives 33% of hydrogen, 20% of methane, and only 44% of ethane. Other salts are even worse. An exception is sodium formate (HCOONa), which, upon being heated to 400°, decomposes in a different manner to give hydrogen and sodium oxalate in good yield (Eq. 13.74).

$$2\text{HCOONa} \xrightarrow{400°} \text{H}_2 + \begin{array}{c} \text{COONa} \\ | \\ \text{COONa} \end{array} \qquad (13.74)$$

Summary. The most useful methods for alkane synthesis are substitution reactions of alkyl halides or sulfonates and hydrogen addition reactions of alkenes and alkynes or of aldehydes and ketones.

Exercises

13.10   Show how to convert cyclohexanol into cyclohexane by four different routes: a. via cyclohexene, b. via cyclohexanone, c. via chlorocyclohexane, and d. via a cyclohexyl sulfonate.

13.11   Show a method for the preparation of pentane from 3-ethylpentane.

13.9 Preparation of alkenes

Summary of reactions previously given. The synthesis of specific alkenes may be done by a variety of methods, most of which have been described already.

Applicable elimination reactions are as follows:

1. dehydration of alcohols, either acidic or by the Chugaev reaction (Secs. 7.6 and 9.10)
2. dehydrohalogenation (Secs. 7.6 and 7.7)
3. dehalogenation (Eqs. 13.15, 13.16, and 3.36, 3.37, and 3.38)
4. the Hofmann elimination (Eq. 7.20)

Addition reactions are the following:

1. hydrogenation of alkynes (Sec. 13.6)
2. hydroboration of alkynes (Sec. 13.3)
3. diimide reaction with alkynes (Sec. 13.6)
4. active metals with alkynes (Sec. 13.6)

We should keep in mind as well the utility of the isomerization reactions, either acidic, basic, or via the heating of trialkylboranes (Eq. 13.33).

Ester pyrolysis. An ester, heated to 300° to 550° in the gas phase, loses the elements of the acid. The elimination is *cis* and has a good yield (Eq. 13.75). A cyclic elimination

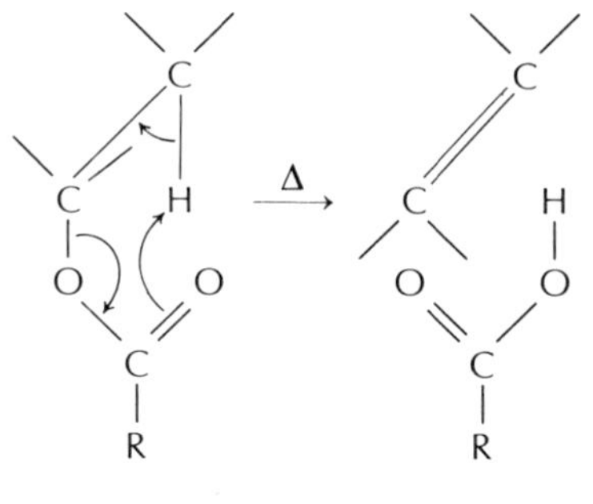

$$\text{(13.75)}$$

mechanism like that shown is known as Ei (i for internal). Like the other elimination mechanisms, the Ei may tend to proceed to an ion pair (bond breaking preceding bond making) on the one hand, or bond making may precede bond breaking. Elimination at a high temperature in the absence of specific reagents is known as a pyrolytic elimination. The Chugaev reaction is an example of this kind of elimination (even though the temperature used is rather low because the xanthates are sensitive to heat), as is the Hofmann elimination.

Cope reaction. Another pyrolytic elimination is the Cope reaction. If a tertiary amine is allowed to react with hydrogen peroxide or a peracid, the product is an amine oxide (Eq. 13.76). Although the hydrogen bonded complex $R_3N\text{---}HOOH$ may be isolated,

$$R_3N + H_2O_2 \longrightarrow R_3\overset{\oplus}{N}\text{---}O^{\ominus} + H_2O \qquad (13.76)$$

it is easier to think of the reaction as an S_N2 displacement on oxygen (Eq. 13.77) followed by

$$R_3N: + HO-OH \longrightarrow R_3\overset{\oplus}{N}-OH + OH^{\ominus} \qquad (13.77)$$

loss of a proton from the conjugate acid of the amine oxide. The nitrogen-oxygen bond is covalent, but it is formed with a pair of electrons from nitrogen. Hence, the nitrogen atom is positive, the oxygen is negative, and the bond is very polar. Such a bond is called a <u>coordinate covalent</u> bond. Amine oxides hydrogen bond strongly to proton donors: water, alcohols, and others.

When heated to 120° to 150° in water or without any solvent, a tertiary amine oxide undergoes an Ei reaction (Eq. 13.78). In the absence of special effects, the direction of

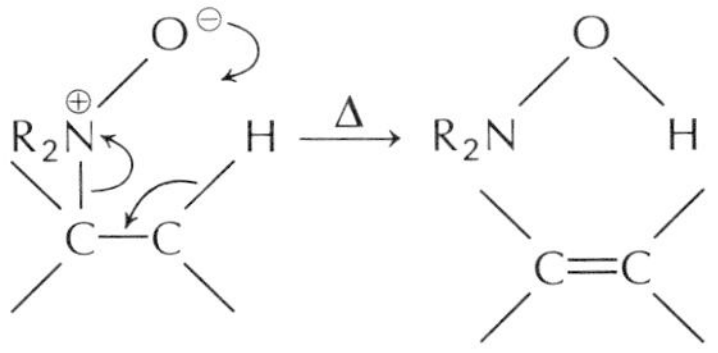

$$\begin{array}{c}\text{(cyclic transition state)} \xrightarrow{\Delta} \text{(products)}\end{array} \qquad (13.78)$$

elimination follows the Hofmann rule. If the amine oxide is isolated, it decomposes at room temperature in dry dimethyl sulfoxide or tetrahydrofuran. The Cope reaction is an alternative to the Hofmann elimination reaction of quaternary ammonium hydroxides.

Summary. Possibly the most used reactions for the preparation of alkenes are dehydration, dehydrohalogenation, and any of the amine eliminations. However, we must not be misled by this statement, because any of the other reactions may be more convenient and efficient for the synthesis of a given alkene.

Exercises

13.12 Give two methods for converting 3,3-dimethyl-1-butanol into 3,3-dimethyl-1-butene.

13.13 Show how to prepare 1-octene from:
a. acetylene and allyl chloride
b. butyl bromide, propionaldehyde, and formaldehyde.
(Assume the availability of all needed solvents, inorganic and organic substances, etc.)

13.10 Preparation of alkynes

Acetylene: preparation, uses. The methods for synthesis of an alkyne are limited in number. Of a large number of possible elimination reactions, only dehydrohalogenation (Sec. 7.7) is of much utility. In many cases, the starting compound is most easily obtained by an addition reaction to the alkyne that is to be prepared!

Acetylene itself is made in vast quantities (as is ethylene) by cracking of various petroleum fractions including methane. An older method of preparation, which is still competitive, is the action of water on calcium carbide, which in turn is made from lime and coke in an electric furnace (Eqs. 13.79). Calcium carbide thus could be considered the calcium salt of acetylene.

$$CaO + 3C \xrightarrow{3000°} CaC_2 + CO$$
$$CaC_2 + 2H_2O \longrightarrow Ca(OH)_2 + HC{\equiv}CH$$
(13.79)

Other carbides are MgC_2, BaC_2, and CrG_2. Some carbides, such as Al_4C_3 and Be_2C, give methane when allowed to react with water and are thought of as methanides. Still other carbides must be covalent because of inertness to water, acids, and bases. Many of the covalent carbides (like a diamond) are very hard. Silicon carbide (carborundum) and boron carbide are examples. An unusual carbide is Mg_2C_3, which reacts with water to give propyne.

As we have seen, acetylene serves as the starting point for many substitution and addition reactions. Commercially, a major use is in welding and metal cutting by the oxyacetylene torch, whose flame is one of the hottest known, 2500° to 2700°. Compressed acetylene is explosive. However, it also is very soluble in acetone. Thus, under moderate pressure, acetylene is dissolved in acetone in gas cylinders and shipped to the point of use.

The substitution reactions of acetylene and 1-alkynes by way of sodium amide or Grignard reagents for the synthesis of alkynes have been discussed (Sec. 12.3), as have isomerizations.

Dehydrohalogenation. Dehydrohalogenation of vinyl halides is more difficult than that of alkyl halides. A halogen bonded to a carbon atom at one end of a double bond is relatively inert, and the strong base sodium amide is required if elimination is to occur. Isomerization of the product under the strongly basic conditions also takes place.

Suppose that in answer to a problem that asks us to prepare 2-butyne, we try to start from 2-butene (Eqs. 13.80). It would be wishful thinking to suppose that a worthwhile

$$CH_3CH{=}CHCH_3 \xrightarrow{Br_2} CH_3CHBrCHBrCH_3$$

$$\Delta \downarrow alc.\ KOH$$

$$\begin{array}{c} CH_3C{\equiv}CCH_3 \\ + \\ CH_3CH{=}C{=}CH_2 \end{array} \xleftarrow{NaNH_2} CH_3CH{=}CBrCH_3\ (+\ some\ CH_2{=}CHCHBrCH_3 \rightarrow etc.)$$
(13.80)

$$NaNH_2 \downarrow warm$$

$$CH_3CH_2C{\equiv}C^{\ominus}\ Na^{\oplus}$$

yield of 2-butyne could be obtained by dehydrohalogenation from 2,3-dibromobutane. If there is no reasonable alternative and the starting material is abundant and easily available, the sequence shown in Equation 13.81 may be used for the conversion of methyl

$$CH_3(CH_2)_7CH{=}CH(CH_2)_7COOCH_3$$

$$1.\ Br_2 \downarrow 2.\ alc.\ KOH\ 150°$$
(13.81)

$$CH_3(CH_2)_7C{\equiv}C(CH_2)_7COOK$$
35–40%

oleate to stearolic acid. Alkynes usually are easier to prepare from acetylene by substitution reactions (Eq. 13.82).

$$HC\equiv CH \xrightarrow{EtMgBr} BrMgC\equiv CMgBr \xrightarrow{CH_3I} CH_3C\equiv CCH_3 \qquad (13.82)$$

Exercises

13.14 Starting with acetylene, show how to prepare a. 1-pentyne, b. 1,7-octadien-4-one, c. cyclohexylethyne. As usual, assume that other needed substances are available.

13.11 Summary

In this chapter some of the addition and oxidation-reduction reactions of alkanes, alkenes, and alkynes were described. We looked briefly at the petroleum industry and outlined the synthesis or preparation of specific hydrocarbons. Although we wrote many new reactions most of them are variations of reactions first described in Chapters 2 and 3. Looking back for a moment, we have now accumulated a large number of interconversion reactions among the alkanes, alkenes, alkynes, alkyl halides, and alcohols as well as a number of reactions that involve oxidation and reduction of these and related compounds. Along the way, we have increased our understanding of steric and polar influences upon stability and reactivity.

Some of the reactions given in this chapter are listed here.

$$HC\equiv CH + HCN \xrightarrow[\substack{NH_4Cl\\HCl\\90°}]{CuCl} H_2C=CHCN \qquad (13.6)$$

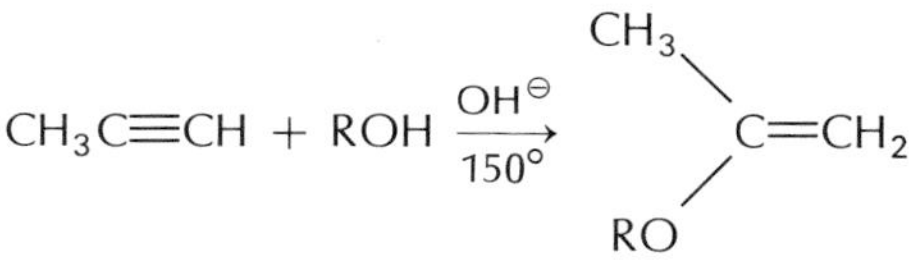

$$CH_3C\equiv CH + ROH \xrightarrow[150°]{OH^{\ominus}} \underset{RO}{\overset{CH_3}{>}}C=CH_2 \qquad (13.7)$$

$$(CH_3)_2C=CH_2 + Br_2 \xrightarrow{CH_3COOH} (CH_3)_2CBrCH_2Br \qquad (13.12)$$

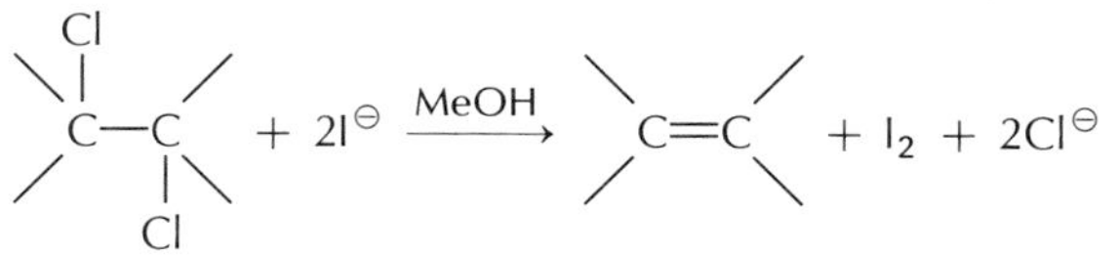

$$\underset{Cl}{\overset{Cl}{>}}C-C\underset{}{<} + 2I^{\ominus} \xrightarrow{MeOH} >C=C< + I_2 + 2Cl^{\ominus} \qquad (13.16)$$

$$CH_3CH_2CH=CH_2 \xrightarrow[400°]{cat.} CH_3CH=CHCH_3 + (CH_3)_2C=CH_2 \qquad (13.19)$$

$$(CH_3)_2C=CH_2 \xrightarrow[25°]{10\% H_2SO_4} (CH_3)_3COH \qquad (13.23)$$

$$RC\equiv CH \xrightarrow[H^{\oplus}, H_2O]{HgSO_4} RCOCH_3 \qquad (13.24)$$

$$tBu^{\oplus} + CH_2=C(CH_3)_2 \longrightarrow tBuCH_2\overset{\oplus}{C}(CH_3)_2 \qquad (13.25)$$

$$(CH_3)_3CH + H_2C=C(CH_3)_2 \xrightarrow[-20°]{HF} (CH_3)_3CCH_2CH(CH_3)_2 \qquad (13.27)$$

$$2HC\equiv CH \xrightarrow[\substack{NH_4Cl \\ H_2O \\ 25°}]{CuCl} HC\equiv CCH=CH_2 \qquad (13.28)$$

$$cis\text{-}CH_3CH=CHCH_3 + :CH_2 \underset{\text{singlet}}{\longrightarrow} \begin{array}{c} CH_3 \quad\quad CH_3 \\ \triangle \\ H \quad\quad\quad H \end{array} \qquad (13.29)$$

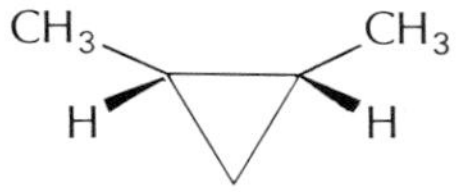

$$3RCH=CH_2 + BH_3 \longrightarrow (RCH_2CH_2)_3B \qquad (13.30)$$

$$\left(\begin{array}{c} CH_3CH_2 \\ \diagdown \\ CH\text{—}B \\ \diagup \\ CH_3CH_2CH_2 \end{array}\right)_3 \xrightarrow[BH_3]{\Delta} (CH_3CH_2CH_2CH_2CH_2CH_2)_3B \qquad (13.33)$$

$$R_3B \xrightarrow[\text{diglyme}]{CH_3CH_2COOH} 3RH + H_3BO_3 \qquad (13.35)$$

$$R_3B \xrightarrow[NaOH]{H_2O_2} 3ROH + BO_3{}^{3-} \qquad (13.36)$$

$$RCH_2CH=CH_2 \xrightarrow[h\nu]{X_2} RCHCH=CH_2 + RCH_2CHXCH_2X \qquad (13.39)$$
$$\phantom{RCH_2CH=CH_2 \xrightarrow[h\nu]{X_2} R}|$$
$$\phantom{RCH_2CH=CH_2 \xrightarrow[h\nu]{X_2} R}X$$

$$RCH=CH_2 \xrightarrow[\text{peroxide}]{HBr} RCH_2CH_2Br \qquad (13.40)$$

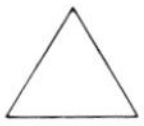

$$\triangle \xrightarrow[\text{cat.}]{H_2, 80°} CH_3CH_2CH_3 \qquad (13.42)$$

$$CH_4 + O_2 \longrightarrow CH_3OH \longrightarrow \text{etc.} \longrightarrow CO_2 + H_2O \qquad (13.45)$$

$$NO \xrightarrow{O_3} NO_2 \xrightarrow{h\nu} NO + O \xrightarrow{RH} R\cdot + HO\cdot \qquad (13.46)$$

$$\diagdown\diagup \diagdown\diagup$$
$$C=C \xrightarrow{O_3} \text{ozonide} \xrightarrow[\substack{H_2O \text{ or} \\ CH_3COOH}]{Zn} C=O + O=C \qquad (13.49)$$
$$\diagup\diagdown \diagup \diagdown$$

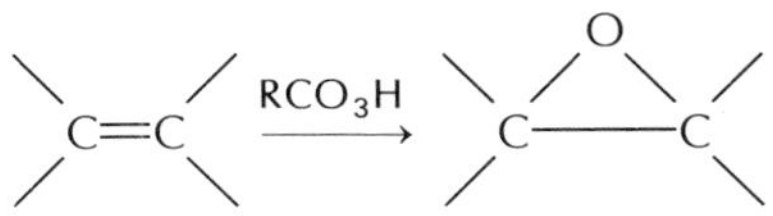

$$\diagdown\diagup O$$
$$C=C \xrightarrow{RCO_3H} \overset{\diagdown\diagup}{C\text{—}C} \qquad (13.50)$$
$$\diagup\diagdown \diagup\diagdown$$

$$RCH=CHR' \xrightarrow{HCO_3H} RCH\text{—}CHR' \quad trans\text{ addition} \qquad (13.52)$$
$$||$$
$$HOOH$$

(with OH above right carbon and HO below left carbon)

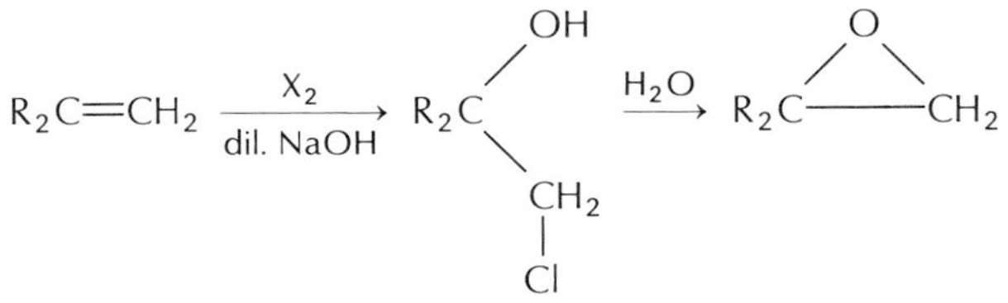

$$R_2C{=}CH_2 \xrightarrow[\text{dil. NaOH}]{X_2} R_2C\begin{matrix}OH\\|\\\\CH_2\\|\\Cl\end{matrix} \xrightarrow{H_2O} R_2C\overset{O}{\underset{}{\diagup\!\!\!\diagdown}}CH_2 \quad (13.55)$$

$$RCH{=}CHR' \xrightarrow[\text{2. KOH}]{\text{1. OsO}_4} \underset{OH\ OH}{RCH{-}CHR'} \quad \text{cis addition} \quad (13.57)$$

$$RCH{=}CHR' \xrightarrow[\text{KOH, 0°}]{\text{KMnO}_4} \underset{OH\ OH}{RCH{-}CHR'} \quad \text{cis addition} \quad (13.58)$$

$$R_2C{=}CHR' \xrightarrow[\text{oxidation}]{\text{drastic}} R_2CO + R'COOH \quad (13.59)$$

$$RCH{=}CH_2 \xrightarrow[\text{cat.}]{H_2} RCH_2CH_3 \quad (13.61)$$

$$RCH{=}CH_2 \xrightarrow[\substack{NiCl_2\\H^{\oplus}}]{NaBH_4} RCH_2CH_3$$

$$RCH{=}CH_2 \xrightarrow[NH_3]{Na} RCH_2CH_3 \quad (13.62)$$

$$RCH{=}CH_2 \xrightarrow[H_2O_2,\ Cu^{\oplus}]{H_2NNH_2} RCH_2CH_3 \quad (13.63)$$

$$RX \longrightarrow RH \quad \text{(many methods)} \quad (13.64)$$

$$ROSO_2C_6H_5 \xrightarrow{LiAlH_4} RH \quad (13.67)$$

$$2RX \xrightarrow{Na} R{-}R \quad (13.68)$$

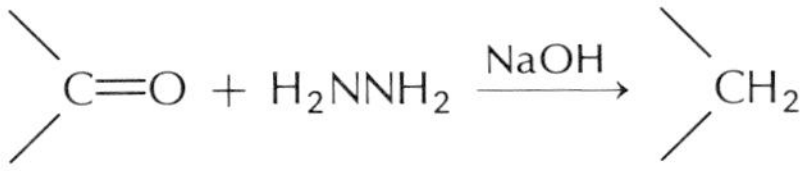

$$\diagdown\!\!C{=}O + H_2NNH_2 \xrightarrow{NaOH} \diagdown\!\!CH_2 \quad (13.70)$$

$$\diagdown\!\!C{=}O \xrightarrow[\text{conc. HCl}]{ZnHg_x} \diagdown\!\!CH_2 \quad (13.71)$$

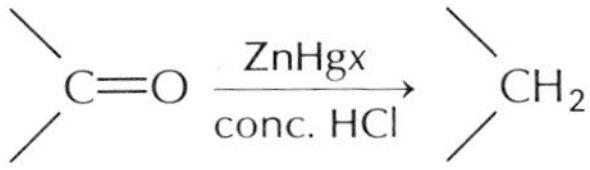

$$RCOO^{\ominus} \xrightarrow{\text{electrolysis}} R{-}R \quad (13.72)$$

$$CH_3COOCH_2CH_2R \xrightarrow{\Delta} H_2C{=}CHR \quad (13.75)$$

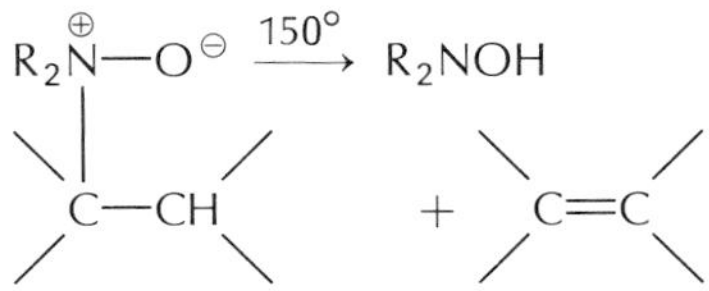

$$R_2\overset{\oplus}{N}{-}O^{\ominus} \xrightarrow{150°} R_2NOH \quad (13.78)$$

$$\underset{}{\diagdown\!\!\underset{|}{C}{-}\underset{}{CH}\diagup} \quad + \quad \diagdown\!\!C{=}C\diagup$$

$$CaC_2 + H_2O \longrightarrow HC{\equiv}CH \qquad (13.79)$$

$$RCH{=}CHR' \xrightarrow[\substack{2.\ \text{alc. KOH} \\ 150°}]{1.\ Br_2} RC{\equiv}CR' \quad 40\% \text{ or less} \qquad (13.81)$$

13.12 Problems

13.1 Substance A (C_5H_8) reacts with EtMgBr in ether to give off a gas. Addition of acetone to the solution and work-up gives B. Catalytic hydrogenation of B gives C. Heating C with H_3PO_4 gives D. D with BH_3, followed by heating with CH_3CH_2COOH, gives E (C_8H_{18}). The nmr of E displays a multiplet at 1.8 ppm (area 2), a doublet at 1.3 ppm (area 4), and a doublet at 0.9 ppm (area 12). A with dilute H_2SO_4 and $HgSO_4$ gives F. F with H_2NNH_2, base, and Me_2SO gives G. G with Br_2 and light gives H. F with $LiAlH_4$ gives I, which with concentrated HBr also gives H. I heated with H_3PO_4 gives J. Addition of HBr (no peroxides present) to J gives H.
Give structural formulas for the lettered substances.

13.2 Substance A (C_7H_{14}) reacts with excess BH_3. Heating the solution before oxidation with H_2O_2 and NaOH gives B. B undergoes a Chugaev dehydration to give C, an isomer of A. Ozonization of C gives formaldehyde and D ($C_6H_{12}O$), the nmr of which shows a triplet at 9.8 ppm (area 1), a doublet at 2.4 ppm (area 2), and a singlet at 1.0 ppm (area 9). Treatment of C with N-bromosuccinimide gives E, which with sodium amide and heat gives F. E reacts readily with water to give G ($C_7H_{14}O$) as the major product. A reacts with moderately concentrated H_2SO_4 at 40° to 60° to give H. H dehydrates readily with hot H_3PO_4 to give I. Ozonization of I gives a mixture of two ketones. Treatment of I with either performic acid or OsO_4 gives the same product, J, a racemic mixture.
Give structural formulas for the lettered substances.

13.3 Starting with any alcohol, $C_nH_{2n+2}O$, with $n = 1,2,3,$ or 4; any solvent; and any inorganic or special reagents, give a good laboratory synthesis of the following:

a. $(CH_3)_2CHC{\equiv}CCH_3$
b. $(CH_3)_2CHCH_2C{\equiv}CH$
c. $(CH_3)_2C{=}CHCH_2CH_3$
d. $(CH_3)_2CHCH{=}CHCH_3$
e. $(CH_3)_2CHCH_2CH{=}CH_2$
f. $H_2C{=}C\begin{smallmatrix}\diagup CH_2CH_2CH_3 \\ \diagdown CH_3\end{smallmatrix}$
g. $(CH_3)_2CHCH_2CH_2CH_3$

13.4 What are the organic reaction products in each of the following reactions?

a.  ⬠ + $Cl_2 \longrightarrow$

b. $CH_3CH_2CH_2CH_3 \xrightarrow[\text{"cracking" catalyst}]{700°}$

c. $CH_3CH_2CH_2OH \xrightarrow[H_2SO_4]{96\%}$

d. $CH_3CH{=}CH_2 \xrightarrow[H_2SO_4]{H_2O}$

e.  ⬡ + $2H_2 \xrightarrow{Pt}$

f. $CH_3CH_2CH{=}CH_2 \xrightarrow{HBr}$

390 Chapter thirteen

g. $CH_3CH_2CH_2CH=CH_2 \xrightarrow[\text{alkaline KMnO}_4]{\text{hot}}$

h. $CH_3CH_2CH_2CH=CH_2 \xrightarrow[\text{2. Zn—HOAc}]{\text{1. O}_3}$

i. $CH_3C{\equiv}CCH_3 + 2HBr \longrightarrow$

j. 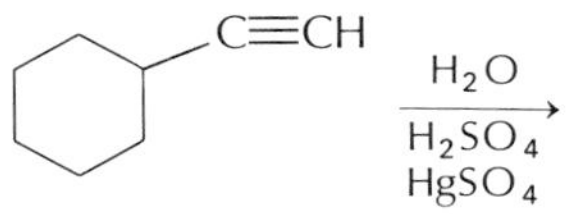 (cyclohexyl)—C≡CH $\xrightarrow[\text{HgSO}_4]{\underset{\text{H}_2\text{SO}_4}{\text{H}_2\text{O}}}$

13.5 Give a chemical reaction that will distinguish between the following pairs. Draw the reactions products.

a. (cyclobutane) and $CH_3CH_2CH=CH_2$

b. $CH_3\underset{\underset{CH_3}{|}}{C}=CH_2$ and $CH_3CH=CHCH_3$

c. $CH_3C{\equiv}CH$ and $CH_3CH=CH_2$

d. 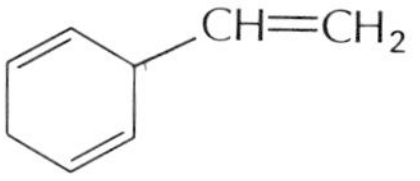 (cyclohexadiene with CH=CH_2 substituent) and (benzene with CH_2CH_3 substituent)

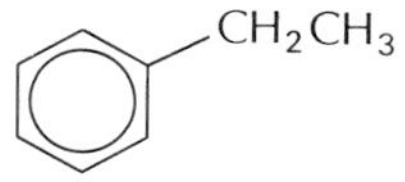

13.6 Draw the possible isomers obtained by monochlorination of the following compound catalyzed by light.

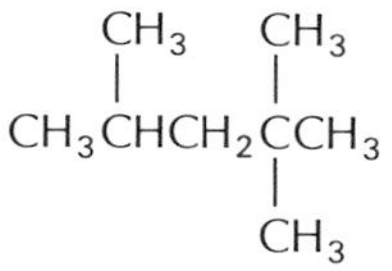

$$CH_3\underset{\underset{}{|}}{\overset{\overset{CH_3}{|}}{C}H}CH_2\underset{\underset{CH_3}{|}}{\overset{\overset{CH_3}{|}}{C}}CH_3$$

13.7 A compound C_5H_8 (A) absorbed one mole of H_2 in the presence of Pt to give cyclopentane. A reacted with Br_2 at 25° to give $C_5H_8Br_2$ (B). When A was treated with O_3, then Zn and acetic acid, $C_5H_8O_2$ (C) was obtained. When A was treated with hot alkaline permanganate, $C_5H_8O_4$ (D) was obtained. Write the structures for A, B, C, D.

13.8 A hydrocarbon, C_8H_{14}, gave CH_3CH_2CHO and $O=\underset{\underset{H}{|}}{C}-\underset{\underset{H}{|}}{C}=O$ on treatment with ozone, then Zn in acetic acid. What is the structure of the hydrocarbon?

13.13 Bibliography

Some general organic chemistry textbooks that are helpful about certain aspects of the alkanes, alkenes, and alkynes are listed in Chapter 7.

In addition, all the paperbacks listed in Chapter 7 are useful, as is reference 5 in Chapter 6. We list one additional textbook here.

1. House, Herbert O., *Modern Synthetic Reactions*, 2nd ed. Menlo Park, Calif.: Benjamin, 1972. The first eight chapters are particularly pertinent to the subjects of this chapter. The rather specialized articles here pursue a few subjects in more depth.

2. Cadle, Richard D., and Eric R. Allen, Atmospheric Photochemistry, *Science* 167, 243 (1970).

3. Carlson, D. A., M. S. Mayer, D. L. Silhacek, J. D. James, Morton Beroza, and B. A. Bierl,

Sex Attractant Pheromone of the House Fly: Isolation, Identification and Synthesis, *Science* 174, 76 (1971).

4. Fahey, Robert C., and C. Allen McPherson, Kinetics and Stereochemistry of the Hydrochlorination of 1,2-Dimethylcyclohexene, *Journal of the American Chemical Society* 93, 2445 (1971).

5. Greek, Bruce F., Gasoline, *Chemical and Engineering News*, Nov. 9, 1970, p. 52.

6. Lloyd, William G., The Use of Free Radical Reactions. Part 2, Substitution Reactions, *Chemical Technology* 1, 371 (1971).

7. O'Sullivan, Dermot A., Air Pollution, *Chemical and Engineering News*, June 8, 1970, p. 38.

8. Pryor, William A., Daniel L. Fuller, and J. P. Stanley, Reactivity Patterns of the Methyl Radical, *Journal of the American Chemical Society* 94, 1632 (1972).

9. Squires, Arthur M., Clean Power from Coal, *Science* 169, 821 (1970).

10. Story, Paul R., John A. Alford, John R. Burgess, and Wesley C. Ray, Mechanisms of Ozonolysis, *Journal of the American Chemical Society* 93, 3042, 3044 (1971).

13.14 Arie J. Haagen-Smit: *From terpenes and essential oils to smog*

Professor Emeritus Arie J. Haagen-Smit of the California Institute of Technology also is chairman of the National Air Quality Criteria Advisory Committee for the Environmental Protection Agency. A native of the Netherlands, awardee of many prizes and honors, he describes for us the research that led him to smog.

I was born in Utrecht, a small town in the center of the Netherlands where my father was chief chemist at the Royal Mint. Watching the pouring of molten gold and silver was a common occurrence, and my two sisters and I used to play hide-and-seek between stacks of gold and silver bars. My early experience in the laboratory was watching my father carry out the Volhard gold and silver analysis, dissolving money in a liquid which turned a beautiful blue and gave off brown, sweet-smelling vapors. It took several years before I understood what was going on in the laboratory. In high school I had learned some chemistry, but I thought that the chemistry of gold, silver, and copper was rather dull.

In 1918, I left the high school with top grades in science; the only failing grade was in the Dutch language. I entered the University of Utrecht that year and enrolled as a chemist. Organic chemistry as taught by Professor P. Van Romburgh showed an orderly structure that I liked. "Once you know the structures of the compounds, the properties can be predicted."

In 1928, Professor L. Ruzicka succeeded Van Romburgh as professor of organic chemistry at Utrecht, and I started to work on my thesis, which consisted of a structure determination of the sesquiterpenes guaiol and copaene.

After finishing my thesis and receiving my Ph.D., I became chief assistant in the laboratory supervising laboratory courses and continuing research in the field of natural products. In 1933, Professor Kögl became director of the laboratory; and, upon his suggestion, I started working on the chemical nature of the cell elongation hormone: the "Wuchs stoff" that F. Went had postulated to be present in plants. Went had developed a test method that was a prequisite for the isolation. Highly potent crystallisates called auxins and an active heteroauxin identified as 3-indoleacetic acid were isolated.

In 1936, Harvard University requested a visit to Cambridge, Massachusetts, to give a series of lectures on natural products. During this time, a research project was undertaken with Dr. K. V. Thimann to investigate the active components responsible for the excystment of protozoa. After one year at Harvard I went as an associate professor in bio-organic chemistry to the California Institute of Technology.

13.14 From terpenes and essential oils to smog 391b

I continued my work on the nature and chemistry of natural-occurring substances of physiological significance. Some of the substances studied were traumatic acid, drosophyla V^+ hormone, and xanthurenic acid. In continuation of my earlier work on essential oils, I investigated oils from desert plants, various species of pine, and the volatile components of fruits, such as pineapple. The investigation of the active ingredients present in minute quantities in the natural products required tons of starting material that yielded only milligrams of final pure product. It was, therefore, necessary to conduct a microchemical analytical laboratory that I operated for the Institute. During the war years when the German analytical laboratories were inaccessible, it served chemical researchers nationally.

In the course of time (especially after the war), the character of the State of California changed rapidly, and one of the most obvious signs of the change was a deterioration of the quality of the air. Every morning the smog came rolling into Pasadena from Los Angeles. It had a peculiar odor, damaged plants, and irritated the eyes. The intriguing odor aroused my curiosity and made me look for the "flavor or bouquet of Los Angeles smog." Since my previous work on the volatile trace materials of fruits had given me the techniques, it was not difficult to conclude that smog was the result of an oxidation of organics, mostly gasoline emitted by the petroleum industry and the millions of automobiles.

A preliminary publication of these findings resulted in a one year's leave-of-absence from the Institute to take over the direction of research from the Los Angeles County Air Pollution District. The continuation of these studies established the photochemical nature of the transformation and showed clearly that the irritation was caused by an atmospheric conversion from primary pollutants to highly irritating compounds. The primary mover in this process was the oxides of nitrogen that originates in all high-temperature combustions from a combination of oxygen and nitrogen in the air. The results of these laboratory conclusions were not readily accepted, and it took several years—until 1954—before several laboratories confirmed the photochemical nature of smog.

In the meantime, I had become interested in some of the technical aspects of control and participated in the studies of the emissions of the petroleum industry and automobiles. Since the production of oxides of sulfur, oxides of nitrogen, and smoke from power plants are ingredients in the formation of smog symptoms, I took a leave-of-absence in 1957 to direct some of the control work at the Southern California Edison El Segundo plant.

At the time I published my first findings on the nature and origin of the smog, I thought I could return to my academic interests. However, this turned out to be an illusion. The problems of air pollution are linked with social and economic problems that determine success or failure of the control program. This realization brought me more and more into the field of governmental control of pollution: first as a member of various local and governmental committees dealing with the environment, later in 1960 as a member of the State of California Motor Vehicle Pollution Control Board where I chaired the criteria committee until 1962. After a brief intermission of taking over the directorship of the controlled environment plant laboratories at the Institute, I was appointed as chairman of the newly formed State Air Resources Board. This board has the responsibility for the control of emissions from stationary, as well as mobile, sources. The duties as chairman of the board and participation in various federal committees have become nearly a full-time occupation, and my office is just as busy as before my retirement as emeritus professor of the California Institute of Technology.

14 Conjugation and Aromatic Character

We continue the examination of the properties of the π bond by looking at the properties of a molecule with two or more π bonds. In some instances we find enhanced reactivity, and in others enhanced stability. The stability reaches a maximum in cyclic molecules that contain $4n + 2$ electrons in π orbitals. We use both resonance and MO theories to help our understanding of the effects upon spectra and reactivity. We shall look at new addition reactions called <u>concerted reactions</u>. They form a new class of reactions, neither heterolytic nor homolytic. We shall mention the Woodward-Hoffman rules and illustrate their use with cycloaddition reactions.

14.1   Dienes and conjugation; naming   393
  1. Conjugated dienes.   2. Naming: EZ system.   3. Conformations, s.
  4. Multiple substituents on a ring.   5. Benzene. Exercises.

14.2   Resonance and MO theory of conjugation   398
  1. Resonance in butadiene.   2. Energies: hydrogenation of butadiene.
  3. Summary. Exercises.

14.3   Rings, conjugation, and the $4n + 2$ rule   399
  1. Cyclopentadienyl ions and radical.   2. The Hückel $4n + 2$ rule.
  3. Benzene: stability, heats of hydrogenation.   4. Aromaticity; resonance.
  5. Symbols.   6. MO of benzene. Exercises.

14.4   Spectrometry of conjugated systems   404
  1. The ir spectra of conjugated systems.   2. The uv spectra.   3. The coal tar dyes.   4. The nmr spectra.   5. Mass spectrometry. Exercises.

14.5   Aromatic stability   416
  1. Ozonization of benzene and o-xylene.   2. Inertness to oxidation of benzene; side-chain oxidation.   3. Drastic hydrogenation and metal reductions.   4. SE, RE, and DE. Exercises.

14.6   Diene reactivity: 1,2 and 1,4 additions   419
  1. Addition of HCl to 1,3-pentadiene.   2. Addition to butadiene.

3. Kinetic and thermodynamic control. 4. Homolytic additions.
5. Benzyl cation, anion, and radical. 6. Styrene: addition.
7. 1-Phenylbutadiene and 1,2 addition. 8. Summary. Exercises.

14.7 Oxidation and reduction   427
1. Summary of reactions; ozonization.   2. Mild and drastic oxidation.
3. Reduction by active metals. Exercises.

14.8 Cycloaddition reactions   428
1. Concerted reactions.   2. Conservation of orbital symmetry.
3. Diels-Alder reaction.   4. Definitions of exo and endo. Exercises.

14.9 Preparation of dienes   432
1. Summary of methods.   2. Catalytic methods.   3. Commercial routes to butadiene.   4. Laboratory preparation of isoprene. Exercises.

14.10 Summary   434

14.11 Problems   437

14.12 Bibliography   438

14.13 William M. Jones: The professor knows best   438a

14.1 Dienes and conjugation; naming

Conjugated dienes. If two or more double and triple bonds are present in a molecule, three possibilities exist for structure. In *a*, an <u>allenic</u> system, the arrangement of

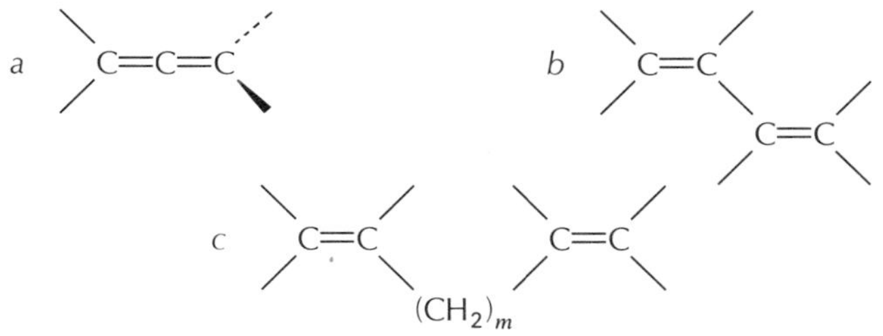

a   C=C=C

b   C=C
         C=C

c   C=C
         $(CH_2)_m$
         C=C

the double bonds is described as <u>twinned</u> or <u>cumulative</u>. In *b*, an alternation of double with single bonds is described as <u>conjugated</u>. In *c*, in which two or more single bonds intervene, the double bonds are <u>isolated</u>. We shall not pursue the allenic system. As for the isolated systems, it is safe to say that each π bond reacts as though the other were not present. Thus, 1,5-hexadiene undergoes reactions characteristic of terminal bonds, and we need not discuss such cases (except for naming).

A conjugated system, however, has physical and chemical properties somewhat different from those we would expect of a molecule with two or more double bonds. (See the spectra in Sec. 10.8, for example.) Conjugated cyclic systems approach extremes in stability: they are either very stable, as is benzene (C_6H_6), or very unstable, as is cyclobutadiene (C_4H_4).

Naming: EZ system.   We have used the prefixes *cis-* and *trans-* (abbreviations *c-* and *t-*), and they are preferred for those molecules with only one π bond or for disubstituted ring compounds. If there are two or more double bonds in a chain, difficulties arise. Consider 2,4-hexadienoic acid (CH_3CH=$CHCH$=$CHCOOH$), for example. There are four possible isomeric forms: *c,c; c,t; t,c;* and *t,t*, reading the bonds from left to right. In principle, we could use this designation to distinguish by name any one of the four. This scheme runs into confusion with more complex examples. The sequence rule scheme is never ambivalent, however, and is easily adapted and learned.

We look at the two groups attached to one end of a double bond, and using the sequence rule of Sec. 6.3, we assign priority numbers. We do the same for the other end. If the low-numbered groups are on the same side (*cis*) of the double bond, they are zusammen (German, *together*), or (Z). If the preferred groups are on opposite sides (*trans*) of the double bond, they are entgegen (German, *opposite*), or (E). Thus,

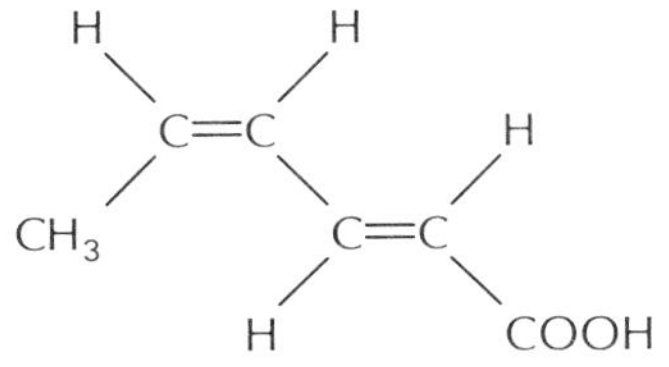

is (2E,4Z)-2,4-hexadienoic acid, but

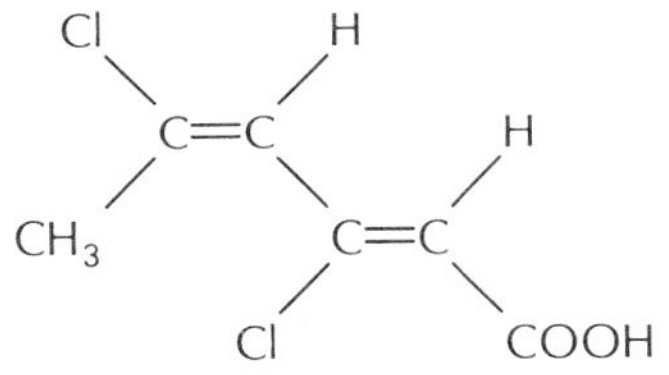

is (2Z,4E)-3,5-dichloro-2,4-hexadienoic acid, because the chlorine atoms take precedence in the sequences. The (E,Z) scheme is unambiguous and allows clear-cut decisions. When necessary, (Z) takes precedence over (E), (R) over (S), and *cis* over *trans*, as in

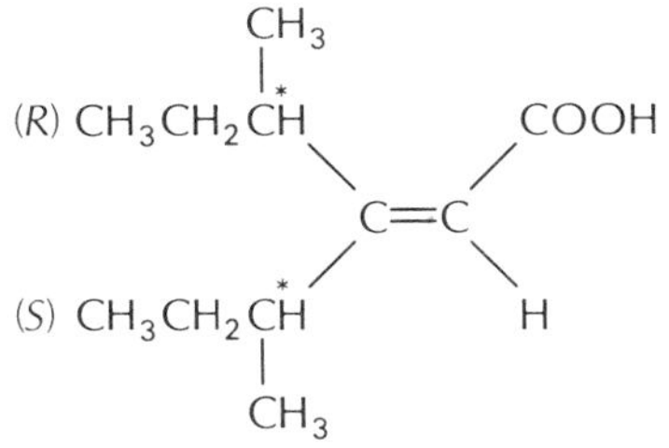

which is named as (Z)-(4R)-3-[(S)-*sec*-butyl]-4-methyl-2-hexenoic acid, and

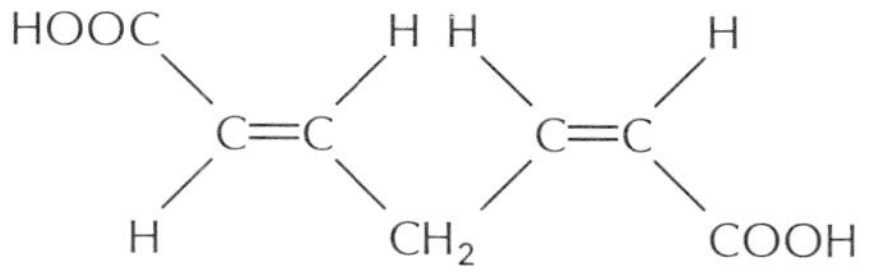

14.1 Dienes and conjugation; naming

[Structure: HOOC-CH=CH-CH₂-CH=CH-COOH with specified geometry]

which is named as (2Z,5E)-2,5-heptadienoic because the numbering is from right to left, not (2E,5Z), as it would be if numbered left to right.

Conformations, s. Occasionally it becomes necessary to distinguish between rotamers of conjugated systems. As we shall see, conjugated systems prefer coplanarity for the double bonds. Thus, the two conformations shown are favored for butadiene. The

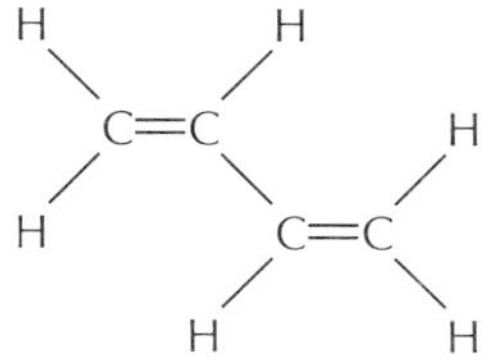

[Structures a and b of butadiene conformations]

a 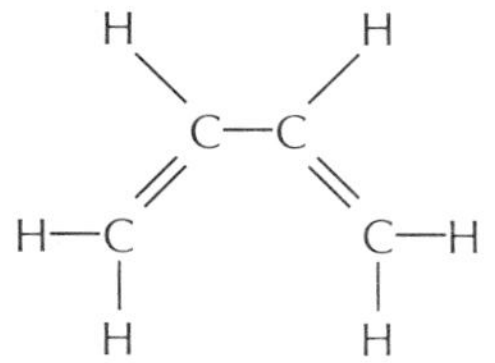   b

distinction between a and b lies in the conformation about the single bond between the two double bonds. In a, the groupings are *trans* and in b, *cis* to each other, and a is termed the *s-trans* form and b is called the *s-cis* form. The new prefix, *s*, (from single) tells us that the *cis* or *trans* following refers to conformation about the single bond. Some people use *syn* and *anti* or *cisoid* and *transoid* along with s to emphasize that only the conformation about the single bond is being referred to.

Multiple substituents on a ring. If three or more substituents are present on a ring, the principal substituent is labeled *r* (from relative) and each other substituent is labeled *c* or *t* in relation to the *r* substituent. Thus,

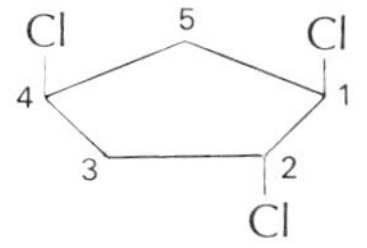

[Structure: cyclopentane with Cl at positions 1, 2, 4]

is a 1,2,4-trichlorocyclopentane whether numbering is clockwise or counterclockwise. The preferred name is *r*-1,*t*-2,*c*-4-trichlorocyclopentane (not *r*-1,*t*-2,*t*-4 from counterclockwise numbering, because other things being equal, *c* takes precedence over *t*).

Sometimes a ring is drawn so that the plane of the ring is that of the paper. A heavy line to a substituent means that the substituent is above the ring; a broken line means that the substituent is below the ring. Thus

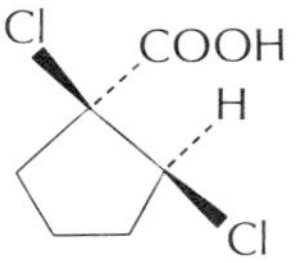

[Structure: cyclopentane with Cl, COOH, H, Cl substituents]

is 1,*t*-2-dichloro-*r*-1-cyclopentanecarboxylic acid.

Note that hydrogen atoms normally are omitted in the drawing of cyclic structures unless they are required to illustrate a point.

Benzene. Of the completely conjugated ring structures cyclobutadiene, cyclohexatriene, and cyclooctatetraene, the name cyclohexatriene is used only for a

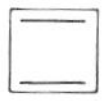

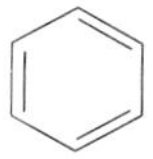

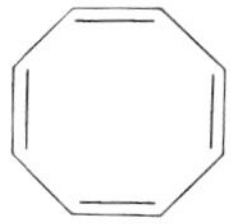

nonexistent molecule that is one of two major resonance structures (Sec. 14.3). The substance itself is called benzene (C_6H_6).

Many mono- and disubstituted benzenes have common names that we shall memorize later. Here we wish to pursue the naming and numbering of the fused benzenoid ring systems and the hydrogenated products. As usual, common names persist. The IUPAC rules list thirty-five exceptions, of which fifteen include fused rings other than benzenoid ones. Some examples are shown here.

naphthalene

anthracene

phenanthrene

indene

fluorene

azulene

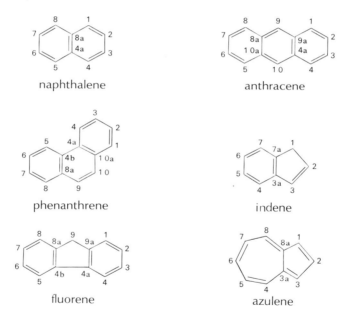

The numbering should begin at the first angle in the upper right hand ring (with the most rings possible in a line) and proceed clockwise about the ring. The interior positions (or angles) are given letters following the preceding numbered exterior angle (for example, 8a follows 8). Note that anthracene and phenanthrene are exceptions to the clockwise rule. Each structure as given here is only one of the possible resonance structures that can be written. Writing in the double bonds in fixed locations is a bookkeeping device used to help keep track of the number of pairs of π orbital electrons present in the molecule.

The reduced (hydrogenated) structures are named with dihydro-, tetrahydro-, and so on, followed by the parent name. The extra hydrogen atoms are given the lowest possible numbers. Some examples follow.

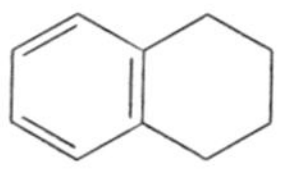

14.1 Dienes and conjugation; naming

1,2,3,4-tetrahydronaphthalene
(common name: tetralin)

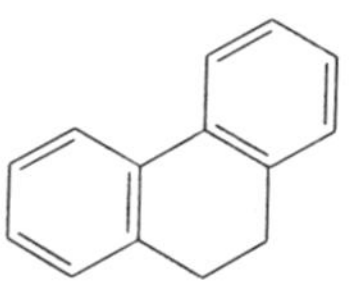

9,10-dihydrophenanthrene

If hydrogenation is complete, the prefix perhydro- is used, as in perhydronaphthalene (the common name is decalin).

Exceptions are the hydrogenated benzenes, which are named as 1,4-cyclohexadiene or 1,3-cyclohexadiene instead of dihydrobenzenes; cyclohexene instead of 1,2,3,4-tetrahydrobenzene; and, of course, cyclohexane. Another exception is the use of indane for 2,3-dihydroindene (hydrindene).

Exercises

14.1 Write the structural formulas for:
   a. 2,6-dibromophenanthrene
   b. 1,10-dibromoanthracene
   c. 2-methyl-1,3-cyclohexadiene
   d. s-trans-(3E)-2-chloro-3-methyl-1,3-pentadiene
   e. cis-3,8-dimethylcyclooctyne

14.2 Give names for:

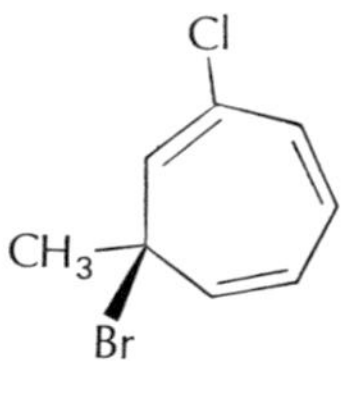

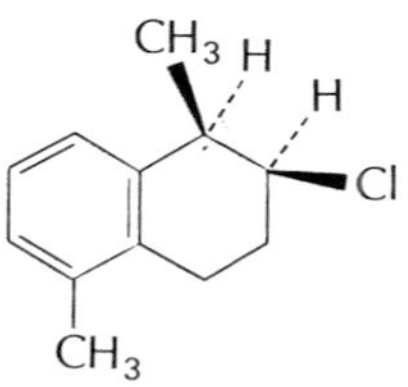

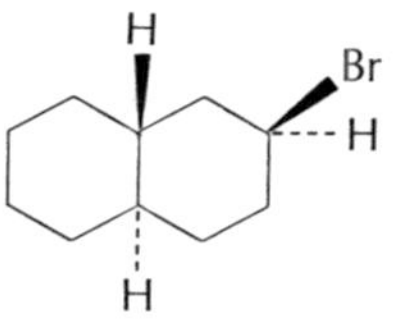

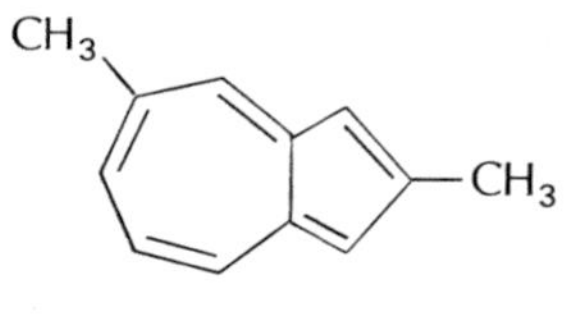

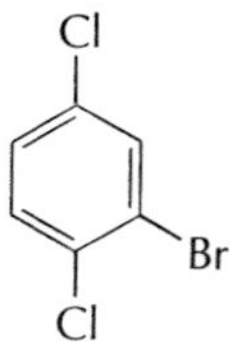

14.2 Resonance and MO theory of conjugation

Resonance in butadiene. Four of the contributing resonance structures of the simplest conjugated diene, 1,3-butadiene, are shown here.

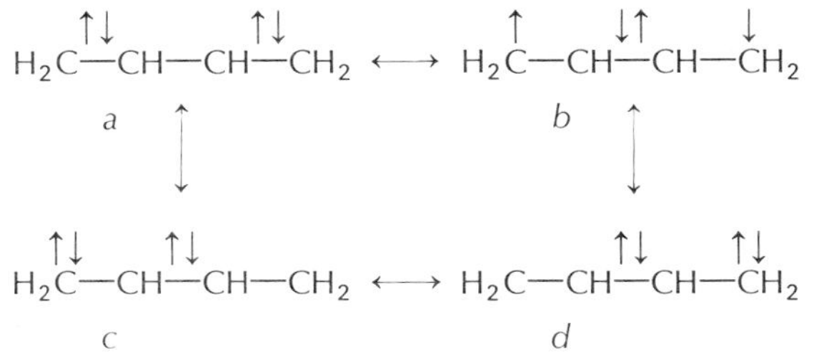

$$H_2\overset{\uparrow\downarrow}{C}-CH-\overset{\uparrow\downarrow}{CH}-CH_2 \longleftrightarrow H_2\overset{\uparrow}{C}-CH-\overset{\downarrow\uparrow}{CH}-\overset{\downarrow}{CH_2}$$
$$a \updownarrow \qquad\qquad b \updownarrow$$
$$H_2\overset{\uparrow\downarrow}{C}-\overset{\uparrow\downarrow}{CH}-CH-CH_2 \longleftrightarrow H_2C-CH-\overset{\uparrow\downarrow}{CH}-\overset{\uparrow\downarrow}{CH_2}$$
$$c \qquad\qquad d$$

or

$$H_2C=CH-CH=CH_2 \longleftrightarrow H_2\dot{C}-CH=CH-\dot{C}H_2$$
$$a \updownarrow \qquad\qquad b \updownarrow$$
$$H_2\overset{\ominus}{C}-CH=CH-\overset{\oplus}{C}H_2 \longleftrightarrow H_2\overset{\oplus}{C}-CH=CH-\overset{\ominus}{C}H_2$$
$$c \qquad\qquad d$$

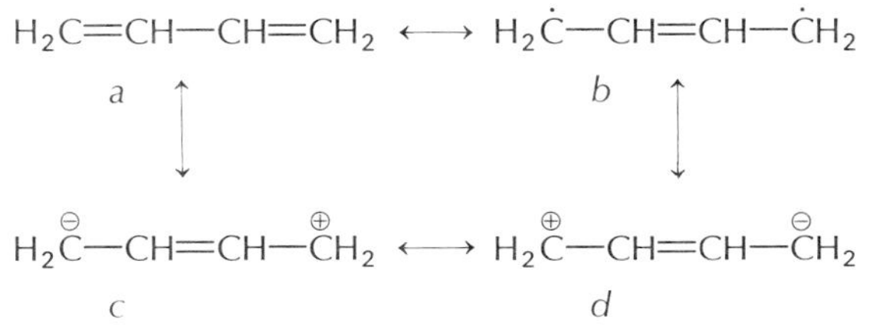

The four p electrons may be shared as shown and in other ways: for example, $H_2\overset{\ominus}{C}-\overset{\oplus}{CH}-CH=CH_2$. At first sight, b does not appear to be a valid structure, but the odd electrons on C_1 and C_4 are paired. In any event, the structures b, c, and d should cause the single C_2-C_3 bond to become partially π-bonded and the double bonds $C_1=C_2$ and $C_3=C_4$ to become less than completely double-bonded. Thus, the shortening of the single bond from the normal 1.54 A to 1.46 A and the slight lengthening of the double bonds from 1.34 A to 1.37 A is explained if we take a as the major structure with small contributions from the other three. On the other hand, the assumption that 1.54 A is the normal single-bond distance is in error, because the σ bond is formed by overlap of sp^2 orbitals, not sp^3 orbitals. From Table 5.3 we take 0.74 A as the covalent single-bond radius of the sp^2 orbital and calculate the bond distance as 1.48 A, which is within experimental error of 1.46 A, the measured distance. So, as is often the case, the experimental evidence (the measured bond lengths) can be explained by either of two theories, and we are left in the uncomfortable position of saying that both are involved and that the resonance shortening effect seems not as strong as the orbital radius effect. Resonance theory does, however, predict an interaction between the p orbitals in a conjugated system. □ □

Energies: hydrogenation of butadiene. We may see a practical effect of conjugation by comparing heats of hydrogenation. From Table 7.3 we find that 30.3 kcal of heat is liberated in the hydrogenation of 1-butene. Therefore, we expect two isolated terminal double bonds to evolve 2(30.3) = 60.6 kcal of heat during hydrogenation. It has been found that 1,4-pentadiene gives off 60.8 kcal and that 1,5-hexadiene gives off 60.5 kcal of heat, and these results are close enough to expectation to provide corroboration.

However, 1,3-butadiene gives off only 57.1 kcal, or 60.6 − 57.1 = 3.5 kcal less than expected. Thus we find that 1,3-butadiene is 3.5 kcal more stable than a molecule with isolated terminal double bonds.

Summary. We use stability in the thermodynamic meaning of the term—a more stable compound has a lower energy content than does a less stable isomer or other compound. Strictly speaking, we should talk in terms of free energy, although we often shall use enthalpy and energy interchangeably. Reactivity refers to the relative activation energies (or free energies of activation): a more reactive substance has a smaller activation energy than does a less reactive substance. Thus, reactivity and stability are independent terms—a stable compound may be reactive or unreactive: likewise, an unstable compound may be unreactive or reactive.

Exercises

14.3 Write the three principal contributing resonance structures of the pentadienyl cation.

14.3 Rings, conjugation, and the 4n + 2 rule

Cyclopentadienyl ions and radical. The cyclopentadienyl cation, according to simple resonance theory, should be very stable because we can write five equivalent structures.

However, the cation is very unstable. On the other hand, the corresponding anion is so stable that cyclopentadiene is more acidic than is acetylene. One source lists pK_a of cyclopentadiene as 16 and that of acetylene as 26. In any case, if cyclopentadiene is added to a Grignard reagent, it reacts as an acid (Eq. 14.1). MO theory predicts that the order of

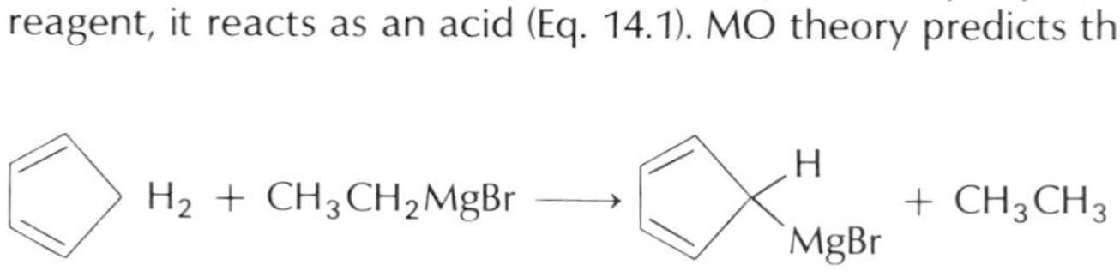

$$\text{C}_5\text{H}_6 + \text{CH}_3\text{CH}_2\text{MgBr} \longrightarrow \text{C}_5\text{H}_5\text{MgBr} + \text{CH}_3\text{CH}_3 \quad (14.1)$$

stability should be anion > radical > cation, which is the order found by experience. Resonance theory predicts equal stability for the three species. □ □

The Hückel 4n + 2 rule. Hückel proposed that any completely conjugated monocyclic ring that contained 4n + 2 electrons would be more stable than other ring systems. The variable n has the values 0, 1, 2, ... Thus, the Hückel rule predicts that monocyclic conjugated rings with 2, 6, 10, 14, . . . π electrons are stable.

In accord with the rule, each member atom of the ring must provide a p orbital for incorporation in the MO's. Also, we should realize that the stabilization provided by the application of the rule is only one factor in overall stability. Angle strain and other steric considerations must be taken into account as well as the energy of ion or radical formation, when needed. For example, we might expect the cyclobutadiene dication with only two electrons to show enhanced stability. However, the dication, with a $2^{\oplus}$ charge, would be very strongly electrophilic (a very strong acid). A still stronger acid would have been necessary to provide the energy needed to form the dication in the first place. The cyclobutadienyl dication is of lower energy than other dications, but all dications are difficult to produce.

Benzene: stability, heats of hydrogenation. Benzene (C_6H_6), is a molecule in which the most favorable stereochemistry for conjugative stability is found. Each carbon atom in this conjugated system is an sp^2 hybrid with 120° bond angles, exactly as in a hexagon: there is no strain. Because it is a molecule, not an ion or a radical, energy of formation is the lowest possible.

Let us turn to heats of hydrogenation again, as we did for 1,3-butadiene, to estimate the stabilization of benzene (Fig. 14.1). The data for cyclooctatetraene are included for comparison. The conversion of benzene to cyclohexane by reaction with 3 moles of hydrogen releases only 49.8 kcal of heat, less than the 55.4 kcal released by 2 moles of hydrogen reacting with 1,3-cyclohexadiene. Cyclohexene releases 28.6 kcal, the same amount as *cis*-2-butene. If we use twice that figure ($28.6 \times 2 = 57.2$) as the value expected for cyclohexadiene, we find that DE (delocalization energy, a measure of stabilization) is $57.2 - 55.4 = 1.8$ kcal for cyclohexadiene. This amount is half that of 1,3-butadiene, which, however, did not have an *s-cis* conformation built in. For three double bonds, we calculate that $28.6 \times 3 = 85.8$ kcal should be released: then $85.8 - 49.8 = 36.0$ kcal is the DE for benzene, a different order of magnitude altogether.

Cyclooctatetraene, on the other hand, shows little conjugative stabilization. The hydrogenation of cyclooctene releases only 24.5 kcal of heat, probably because unlike cyclohexane, the product cyclooctane has a sizable strain energy arising from nonbonded interactions. Using 4×24.5 kcal, we calculate 98.0 kcal as the expected heat of hydrogenation of cyclooctatetraene. The actual measured value is 99.5 kcal, or DE = -1.5 kcal. Other estimates of DE give 2.4 to 4.8 kcal. The bond distances have been reported as C—C, 1.46 A and C=C, 1.35 A; and the C—C=C angle as 125°. Benzene, however, has only one carbon-carbon bond distance, 1.39 A. Thus the structure of cyclooctatetraene must explain why the molecule has no conjugation energy even though it contains alternating single and double bonds. A nonplanar ring, the "tub" conformation, shows how the overlap of the p orbitals between carbon atoms 2 and 3, 4 and 5, 6 and 7, and 8 and 1 is largely

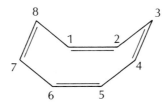

destroyed in the preferred nonplanar system. However, cyclooctatetraene reacts spontaneously with lithium in a 90% ether-10% tetrahydrofuran solvent to give the dianion with 10 electrons (Eq. 14.2).

$$C_8H_8 + 2Li \longrightarrow C_8H_8^{\;2\ominus} + 2Li^{\oplus} \qquad (14.2)$$

14.3 Rings, conjugation, and the 4n + 2 rule   401

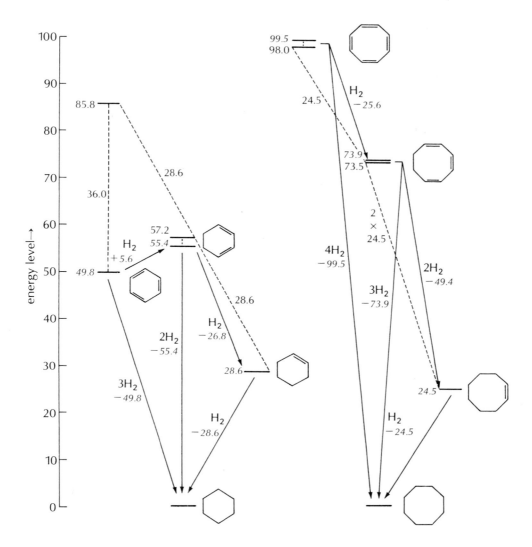

Figure 14.1 Energy levels for the various hydrogenation products of benzene and of cyclooctatetraene; experimental results in italic and calculated values in roman type, in kcal

Benzene is difficult to hydrogenate. The first mole of hydrogen is used in an endothermic step to break into the conjugative system ($\Delta H = 5.6$ kcal) to form cyclohexadiene, after which the rest of the way is downhill.

An interesting reaction is the treatment of cyclohexene with a hydrogenation catalyst (Pd on charcoal is good) in a closed system but in the absence of any hydrogen. On the catalyst surface, the loss of a mole of hydrogen costs 26.8 kcal to give cyclohexadiene, but the addition of the mole of hydrogen to another cyclohexene releases 28.6 kcal, so there is a net release of 1.8 kcal. Furthermore, cyclohexadiene loses a mole of hydrogen to form benzene and give off 5.6 kcal. The new mole of hydrogen with another cyclohexene releases another 28.6 kcal. The total reaction is the disproportionation of cyclohexene to cyclohexane and benzene (Eq. 14.3). To put it another way, 2 moles of cyclohexane and 1 of benzene are more stable than 3 moles of cyclohexene by 36 kcal.

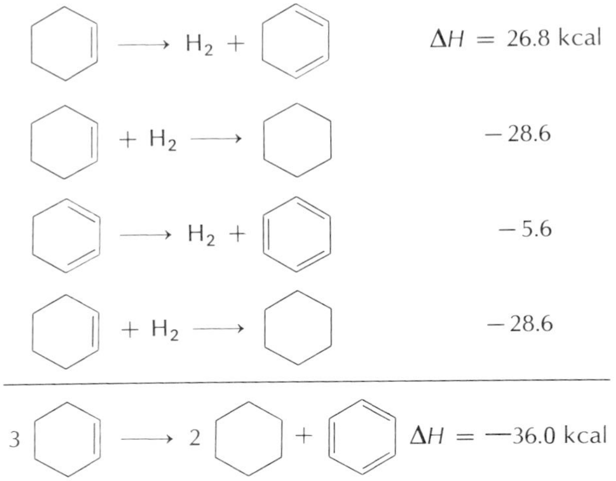

$$\text{cyclohexane} \longrightarrow H_2 + \text{cyclohexene} \qquad \Delta H = 26.8 \text{ kcal}$$

$$\text{cyclohexene} + H_2 \longrightarrow \text{cyclohexane} \qquad -28.6$$

$$\text{cyclohexadiene} \longrightarrow H_2 + \text{benzene} \qquad -5.6$$

$$\text{cyclohexene} + H_2 \longrightarrow \text{cyclohexane} \qquad -28.6$$

$$3 \text{ cyclohexene} \longrightarrow 2 \text{ cyclohexane} + \text{benzene} \qquad \Delta H = -36.0 \text{ kcal} \qquad (14.3)$$

Aromaticity; resonance. Benzene, the prototype of the magic number of six, is remarkably resistant to addition reactions. Instead, electrophilic reagents give substitution products, as we shall see in Chapter 15.

Benzene was discovered by Faraday in 1825. Later on, when it was found that the benzene ring system usually was present in substances with pleasing aromas, the word aromatic came into use as a descriptive term for all compounds with benzene or fused benzene rings present. Then it evolved into a word applied to all substances that show marked conjugative stability. Thus we speak of the cyclooctatetraenyl dianion as a non-benzenoid aromatic ion.

Resonance theory works well in giving a reasonable qualitative feel for aromatic compounds. Benzene and pyridine, each with two equivalent principal contributing

benzene

pyridine

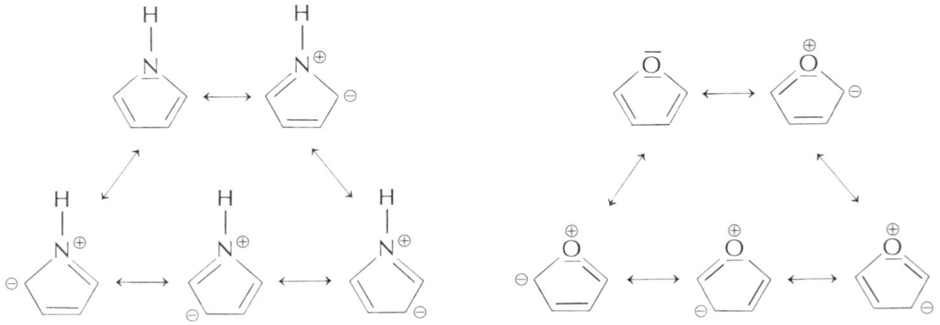

pyrrole

furan

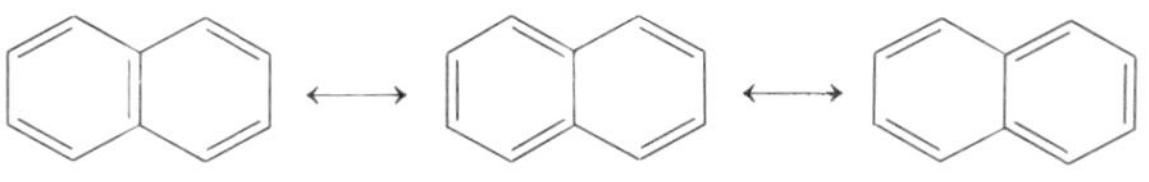

naphthalene

structures, we say are typically aromatic. Pyrrole and furan are examples of heterocycles other than pyridine. By using an electron pair from a p orbital on nitrogen and oxygen, each has six electrons for conjugation. Their resonance structures, which require charge separation (higher energy structures), do not contribute equally, and the substances are only partially aromatic. Naphthalene, with three resonance structures, is an example of the fused benzene ring molecule. Again we emphasize that the resonance structures written are imaginary because no atomic motion is allowed. Thus, the representation of benzene is not a picture of resonance between two cyclohexatriene molecules with single and double bond lengths. Instead, the resonance structures represent localization of the electrons between adjacent p orbitals in a fixed σ-bonded ring. The combination of resonance structures represents delocalization.

Symbols. Various symbolic representations are in use for aromatic compounds, as shown for toluene ($C_6H_5CH_3$). In symbol a, one of the resonance structures, a Kekulé

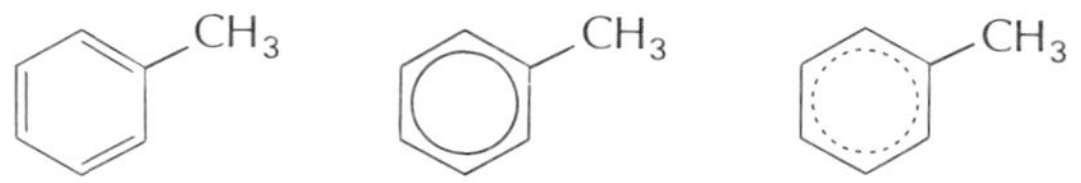

a   b   c ϕCH_3 $PhCH_3$
                              d   e

structure, represents the benzene ring. Symbols b and c may be thought of as attempts to pull either all the resonance structures or the occupied MO's together into a single symbol. Symbols d and e are merely shorthand, where ϕ and Ph represent C_6H_5—, which is called phenyl (not benzyl, as one might logically assume: benzyl is used instead for ϕCH_2—). We shall use all the symbols from time to time. Let us keep in mind that a symbol is a symbol: none of them is a realistic picture of the molecule in question.

MO of benzene. The occupied MO's of benzene as pictured in Figure 14.2 give an idea, in contrast to the resonance picture, of the equalized bonding when the lowest three MO's are occupied by six electrons. It is obvious that no single symbol can present either the MO or the resonance picture of the benzene molecule with any accuracy.

Exercises

14.4 Styrene, $\phi CH=CH_2$, prefers a planar conformation. Why?

14.5 Benzoquinone, $O=\!\!\bigcirc\!\!=O$ is easily reduced to 1,4-dihydroxybenzene (common name is hydroquinone). Why?

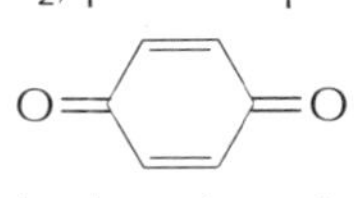

14.6 Draw orbital pictures of pyridine, pyrrole, and furan to show why six, and only six, p electrons are available for π bonding.

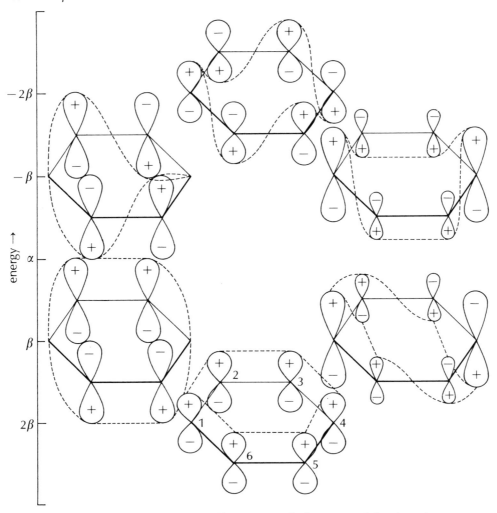

Figure 14.2 Diagram of the six MO's of benzene. Only the positive lobes have been joined with dashed lines to give an idea of the shapes.

14.4 Spectrometry of conjugated systems

The ir spectra of conjugated systems. In the ir, the presence of conjugation is shown by replacement of the usual vinylic absorption at 1650 ± 10 cm^{-1} with two weak peaks (1600 and 1650 cm^{-1} for 1,3-pentadiene) for unsymmetrical conjugation, or with a single band (1600 cm^{-1} for butadiene) if the conjugation is symmetrical.

Substituted benzenes display strong bands in the 700 to 900 cm^{-1} region, as shown in Figure 14.3. These bands are so common that lack of absorption in this region may be taken to mean that aromatic rings are absent. Substitution on a benzene (and, to a certain extent, on other aromatic rings) gives rise to weak peaks in the 1650 to 2000 cm^{-1} region. Unfortunately, they are not always as easily seen as those shown in Figures 14.3 and 14.4. The set of four peaks at 1450 to 1600 cm^{-1} sometimes is covered by other peaks. Figure 14.5 shows the type of variation to be expected. The aromatic C—H bond stretch at about 3030 cm^{-1} is variable in strength and may be a multiple band.

14.4 Spectrometry of conjugated systems 405

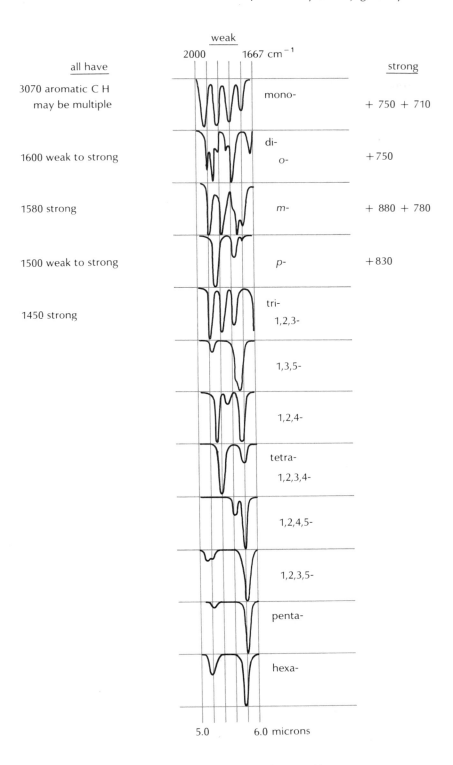

Figure 14.3 Typical ir absorption patterns of substituted benzenes

406 Chapter fourteen

a.

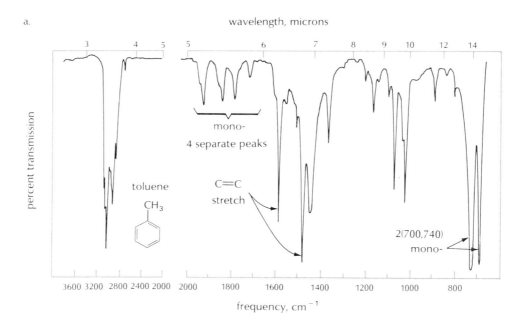

b.

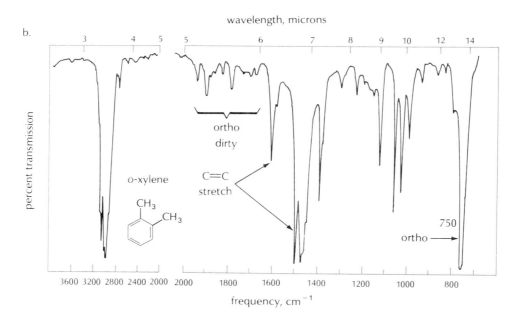

Figure 14.4 Infrared spectra of toluene, o-, m-, and p-xylenes. The number and positions of ring substituents and the pattern of the weak bands in the region 2000 to 1650 cm^{-1} and the pattern of the strong bands in the region 800 to 690 cm^{-1} are shown.

14.4 Spectrometry of conjugated systems 407

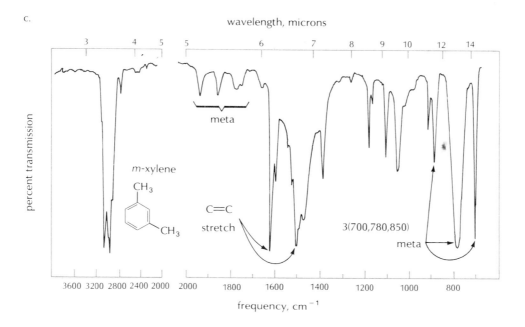

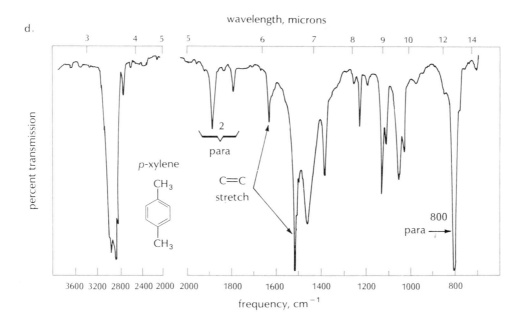

Figure 14.4 (continued)

408   Chapter fourteen

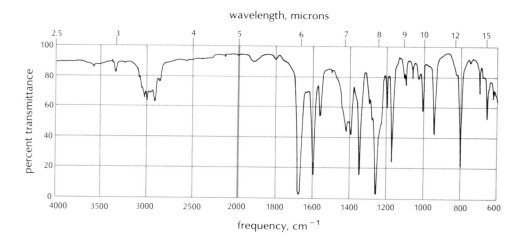

Figure 14.5   Infrared spectrum of methyl p-tolyl ketone as a neat liquid film

The uv spectra.   The visible and uv spectra of conjugated systems are characterized by a <u>bathochromic shift</u> (towards longer wavelengths) because a smaller energy is needed for the excitation of an electron in a $\pi \rightarrow \pi^*$ shift from the highest occupied to the lowest unoccupied molecular orbitals (HOMO to LUMO). Butadiene absorbs at 217 nm in hexane solvent. The ε_{max} for such a shift usually increases as well, to 21,000 for butadiene. Increasing conjugation lowers the energy of the $\pi \rightarrow \pi^*$ transition still more, as in 1,3,5-hexatriene with $\lambda_{max} = 258$ nm, $\varepsilon_{max} = 35,000$. Cyclopentadiene has $\lambda_{max} = 239$ nm, $\varepsilon_{max} = 3400$.

Benzene and substituted benzenes in solution show a broad band at 230 to 270 nm of low intensity: $\lambda_{max} = 255$ nm, $\varepsilon_{max} = 215$ for benzene; 262 nm and 174 nm for toluene. (In the vapor state the absorption shows a large number of sharp peaks, which are assigned as vibrational sublevels of the $\pi \rightarrow \pi^*$ transition.) In addition, other bands of stronger intensity appear at shorter wavelengths, as shown by λ_{max} at 200 nm, $\varepsilon_{max} = 8000$, and λ_{max} at 180 nm, $\varepsilon_{max} = 60,000$ for benzene; and λ_{max} at 208 nm, $\varepsilon_{max} = 2460$ for toluene. Stilbene (ϕCH=CHϕ), in which the aromatic rings are conjugated with a double bond, has 268 nm and 18,500 for *cis* and 280 nm and 29,000 for *trans*. Both isomers have very strong bands as well in the 200 to 230 nm region. Styrene (ϕCH=CH$_2$) has 282 nm and 450, and 244 nm and 12,000.

The coal tar dyes.   Benzene, naphthalene, and anthracene are colorless products of coal tar distillation. By simple chemical addition of chromophoric groups, beautiful dyes can be created. This finding in the nineteenth century excited the fashion world to colorful fantasies, marked the beginnings of chemical industry, and led to the development of organic chemistry. Mauve, the first coal tar dye, was synthesized by oxidizing aniline with sulfuric acid and nitric acid. (See Sec. 15.7.) It was the sensation of the 1862 World Exhibition in London. Martius yellow, Caro's induline blue, and Hoffman's violet were synthesized and widely accepted. Soon thereafter the bright red dye fuchsin was prepared in France, making an even greater impact than mauve.

The artificial dyes, although more brilliant, were less permanent than the natural dyes. Chemists set to work analyzing the structure of the natural compounds. The structure of alizarin, the popular red dye from madder root, was quickly determined to be a derivative of anthracene. Shortly, the dyestuff industry was producing alizarin from anthracene, as well as a family of alizarin-like dyes with different colors. By the turn of the century, the oldest known coloring matter, indigo, was being synthesized commercially. The economic value of organic chemistry had been recognized and the chemical industry born.

The nmr spectra. The nmr of open-chain conjugated systems presents no special problems other than those already considered for vinylic protons. Benzene and other aromatic systems have protons that are strongly deshielded. The cause is the ring current, which is set up by exposure of the electrons in the π network to the strong magnetic field in the spectrometer (Fig. 14.6). The circulation of the electrons, as in a coil, sets up a magnetic field in opposition to that of the machine. Closing the lines of force exposes the protons attached to the ring to a field greater than that imposed by the machine. These protons come into resonance at machine fields less than those expected. Thus, benzene shows a single peak, far downfield at δ 7.27 ppm.

Substituted benzenes are more complex. In toluene ($C_6H_5CH_3$), for example, we find the methyl protons shifted downfield to about δ 2.2 ppm and the five aromatic protons at about δ 7.2 ppm. If the substituent on the ring has a strong $-I$ effect ($-NO_2$ or $-COOEt$), the chemical shift, besides increasing because of deshielding, no longer is the same for all five aromatic protons. We see that the protons at positions 2 and 6 (the *ortho* protons) are

identical and have the same chemical shift; those at positions 3 and 5 (the *meta* protons) have identical but different shifts from the *ortho* protons; and the lone proton at 4 (the *para* proton) shows a still different shift. It turns out that for ethyl benzoate, the two H_o have

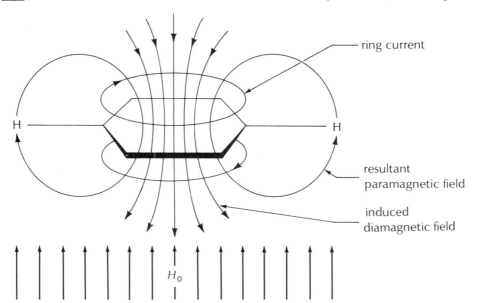

Figure 14.6 The deshielding of aromatic protons caused by a ring-current effect

410   Chapter fourteen

a.

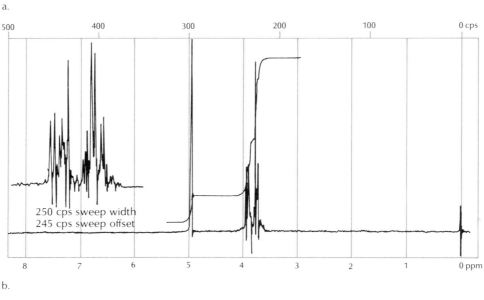

b.

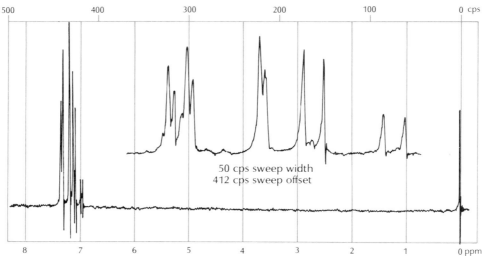

Figure 14.7 a. Benzaldehyde, 15% (w/v) in carbon tetrachloride, 1000 cps sweep width;

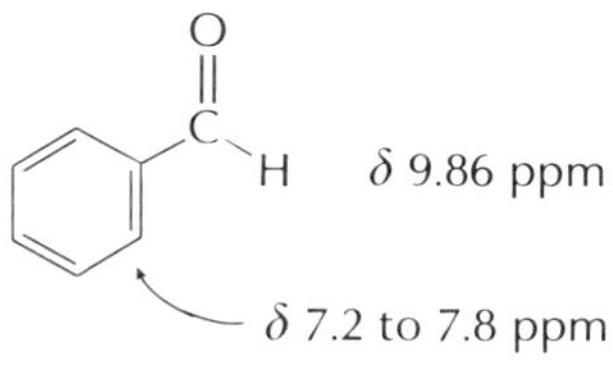

δ 9.86 ppm

δ 7.2 to 7.8 ppm

b. 1,2,4-trichlorobenzene, 15% (w/v) in carbon tetrachloride, 500 cps sweep width.

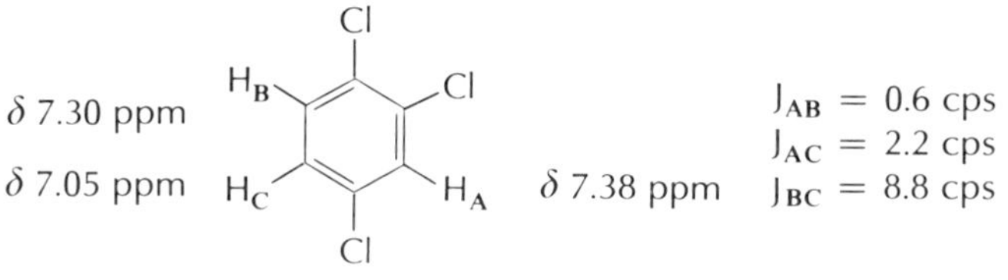

δ 7.30 ppm
δ 7.05 ppm
δ 7.38 ppm

J_{AB} = 0.6 cps
J_{AC} = 2.2 cps
J_{BC} = 8.8 cps

δ 8.2 ppm, the two H_m have δ 7.4 ppm, and H_p has δ 7.6 ppm. The electron-withdrawing tendency of the substituent appears to affect the protons so that the order is $o > p > m$ in δ. If the substituent is a strong electron donor ($-NH_2$), the effect on the protons falls into the same order of magnitude. However, because electrons are being furnished to the ring, the *ortho* protons become more shielded than the others, decreasing δ so that the H_o now have δ 6.5 ppm, the H_m have δ 7.2 ppm, and H_p has δ 6.8 ppm. The order in δ now is $m > p > o$, the reverse of the order for $-I$ substituents.

With the differing chemical shifts caused by substituents, splitting patterns of the A_2BC_2 type emerge. Because the chemical shift differences are small, the patterns are complex indeed. For protons *p* to each other, J_p is 0 to 1 cps; for protons *m* to each other, J_m is 1 to 3 cps; and for protons adjacent (*o*) to each other, J_o is 6 to 10 cps, about the same as for *cis* ethylenic protons. An example of the complexity that may arise is shown in Figure 14.7a for benzaldehyde. The nmr spectrum of 1,2,4-trichlorobenzene (Fig. 14.7b), with only three protons (hence an ABC system), shows the splitting pattern involving the three coupling constants.

We should work through the nmr spectra of disubstituted and trisubstituted benzenes using the principles already given. If the substituents are identical, great simplification is possible in some instances. Consider the three dibromobenzenes, for example. The *ortho* compound gives an A_2B_2 pattern involving only J_o and J_m. The *meta* compound gives a complex AB_2C pattern; but the *para* compound, with four identical protons, gives a single line.

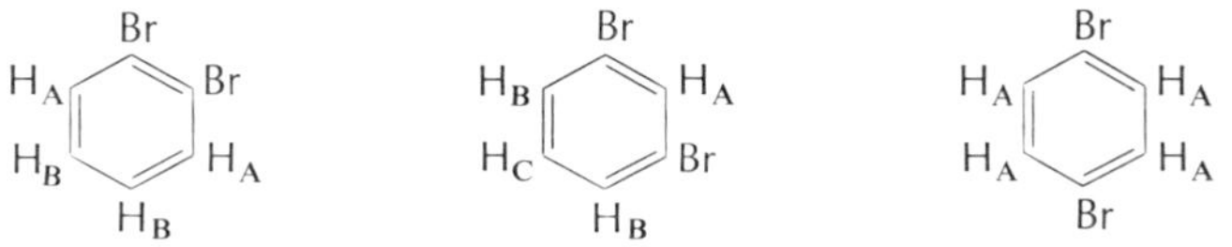

Mass spectrometry. Mass spectrometry of conjugated systems usually gives easily identifiable parent peaks. Aromatic compounds characteristically tend to fragment outside the ring as a first step rather than in the ring. There is a high probability that the second bond from the ring, called a β bond, will be broken (Eqs. 14.4). The benzylic cation

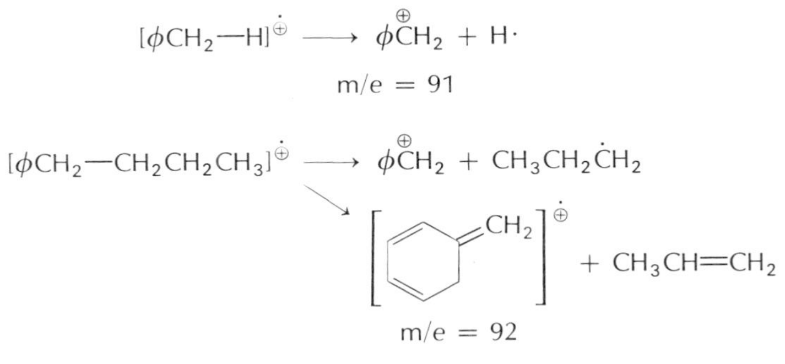

$$[\phi CH_2-H]^{\oplus \cdot} \longrightarrow \phi \overset{\oplus}{CH_2} + H\cdot$$
$$m/e = 91$$

$$[\phi CH_2-CH_2CH_2CH_3]^{\oplus \cdot} \longrightarrow \phi \overset{\oplus}{CH_2} + CH_3CH_2\dot{C}H_2 \quad (14.4)$$

$$\left[\begin{array}{c}\text{=CH}_2\\ \end{array}\right]^{\oplus \cdot} + CH_3CH=CH_2$$
$$m/e = 92$$

($\phi \overset{\oplus}{CH_2}$) is comparable to the allylic cation in stability when in solution. In the mass spectrometer, there is evidence (from isotope scrambling) that a rearrangement occurs to give a tropylium cation, which by loss of acetylene gives a cyclopentadienyl cation with $m/e = 65$ (Eq. 14.5). Thus a peak at 91 is always accompanied by one at 65 m/e.

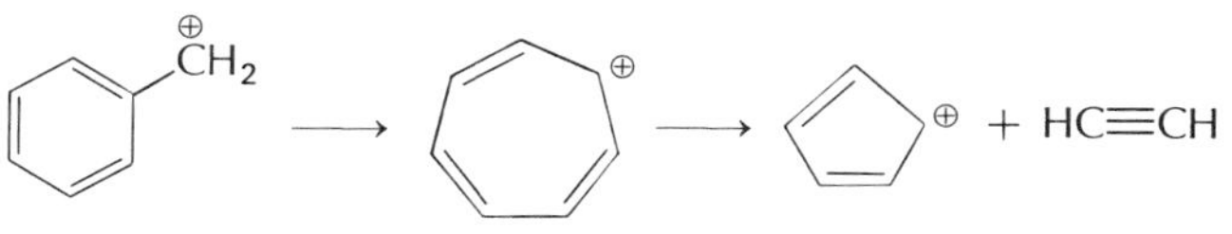

412   Chapter fourteen

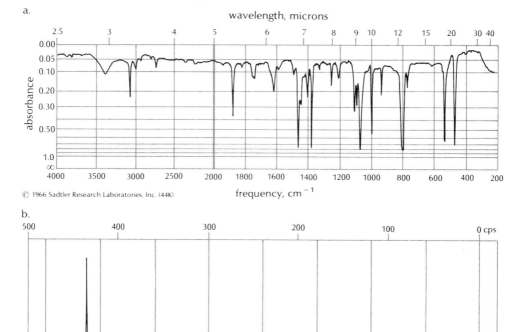

If an oxygen atom is attached to a carbon atom that is bonded to the ring, as in ϕCH_2OH, ϕCHO, ϕCOR, $\phi COOH$, $\phi COOR$, and so on, the species $[\phi C{=}O]^{\oplus}$, with m/e = 105, and $[\phi]^{\oplus}$, with m/e = 77, give strong lines.

Exercises

In each exercise, the molecular formula and the spectra of a compound are given. Give the structural formula of each compound.

Ex. 14.7 $C_6H_4Cl_2$

a.

b.

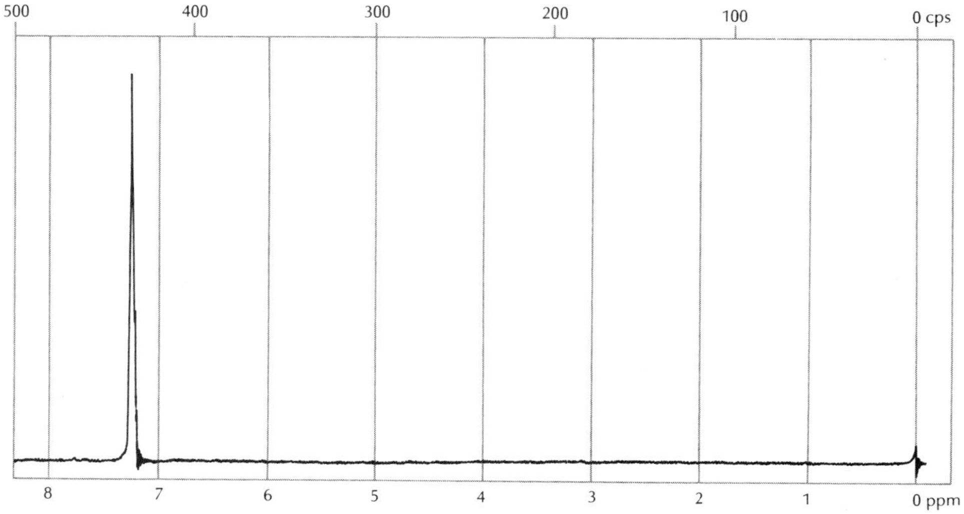

Ex. 14.8 C_8H_7OCl

a.
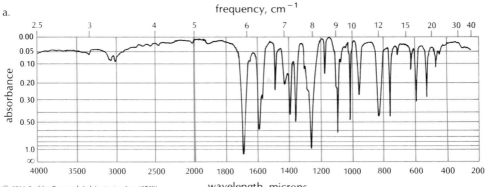

b.
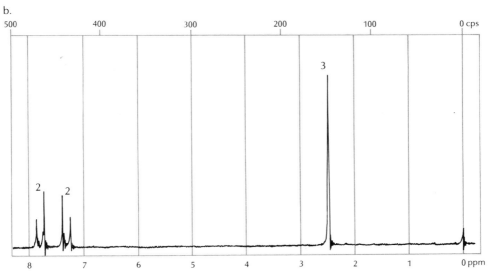

c.
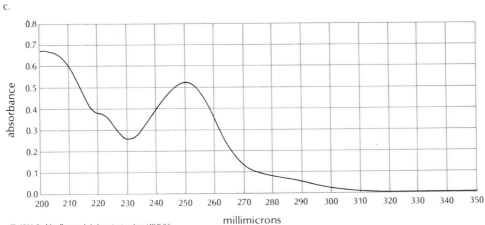

Ex. 14.9 C_7H_7Br

a.
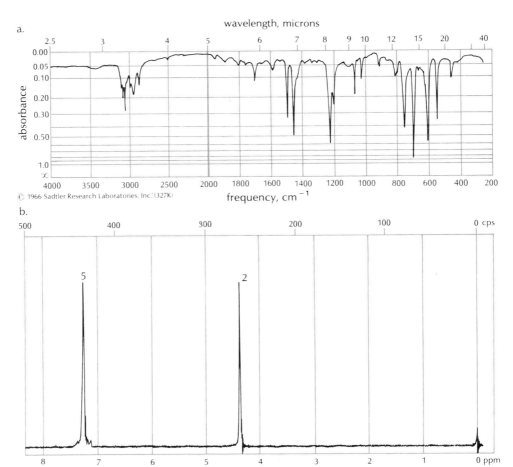

b.

c.
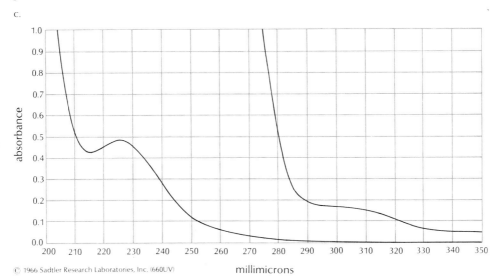

Ex. 14.10 $C_9H_{10}O$

a.

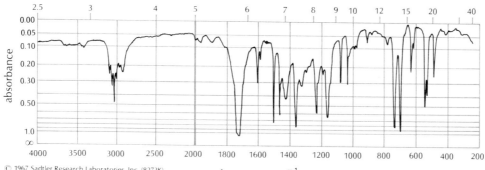

b.

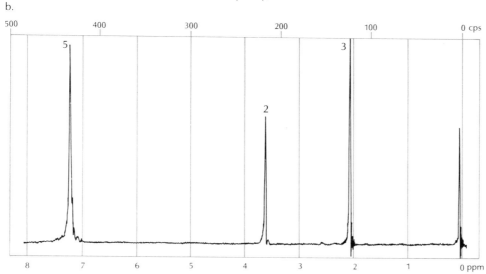

c.

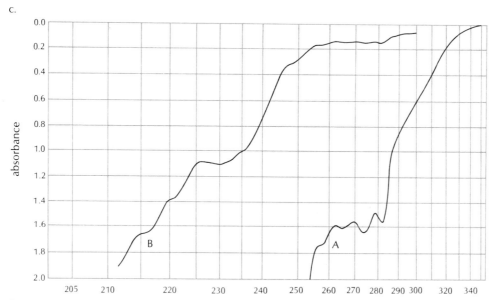

14.5 Aromatic stability

Ozonization of benzene and o-xylene. Aromatic compounds undergo addition reactions to the ring with difficulty. Substituents influence the ease of hydrogenation, and any general statement is subject to many reservations.

Benzene reacts slowly with ozone to form a triozonide, which gives glyoxal (OHC—CHO) as the final product. The ozonization of o-xylene is interesting because the two possible triozonides should be formed in equal amounts (Eq. 14.6). After hydrolysis, we see

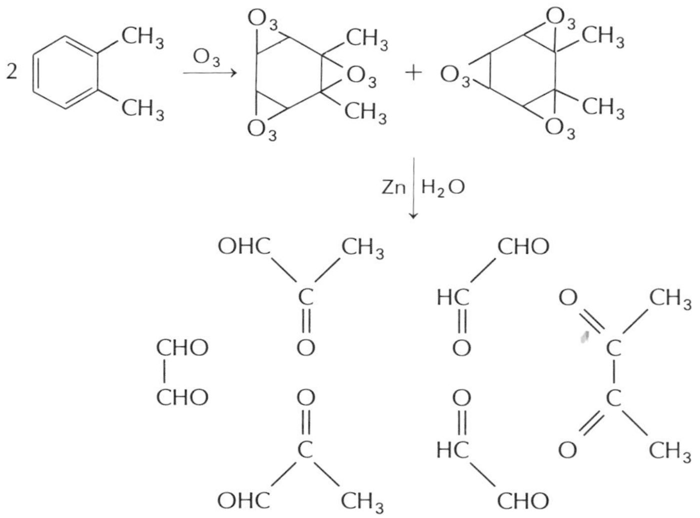

(14.6)

that for every two moles of o-xylene reacting, the products should be three moles of glyoxal, two moles of 2-ketopropanal (CH_3COCHO), and one mole of butanedione ($CH_3COCOCH_3$). Confirmation of this prediction before the advent of resonance or MO theory was used by organic chemists as evidence for a very rapid equilibrium between the two Kekulé structures.

Inertness to oxidation of benzene; side-chain oxidation. Benzene is stable to the action of the ordinary oxidizing agents ($KMnO_4$, CrO_3, HNO_3). Substituted benzenes may be oxidized with survival of the ring (Eqs. 14.7). The initial attack is thought to occur at the carbon atom adjacent to the ring, and we can rationalize a sequence as shown in Equation 14.8. If there is no hydrogen or double bond at the reactive position, oxidation does not occur ($\phi C(CH_3)_3$ is not easily oxidizable) or the reaction stops, as in $\phi CO\phi$ or $\phi_3 COH$. Fused ring systems are subject to oxidative ring attack, but a benzene ring survives. Depending upon conditions and structure, either a side chain or one of the fused rings is oxidized. Under very drastic conditions, benzene may be oxidized (Eq. 14.9). These catalyzed air oxidations are used extensively by industry. The point of our discussion is that the benzene ring is so stable that very drastic conditions are required to break into the π electron system.

14.5 Aromatic stability

$$\phi CH_3 \xrightarrow[\text{or } CrO_3]{KMnO_4} \phi COOH$$

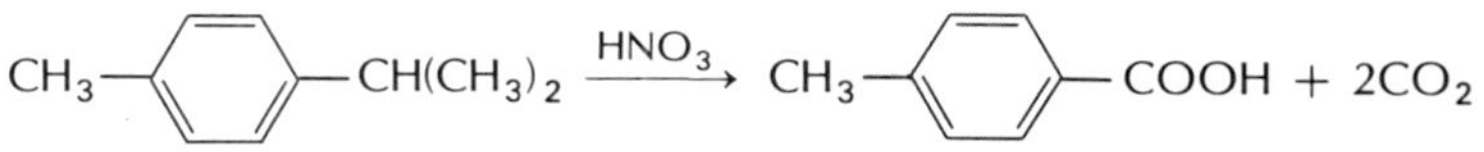

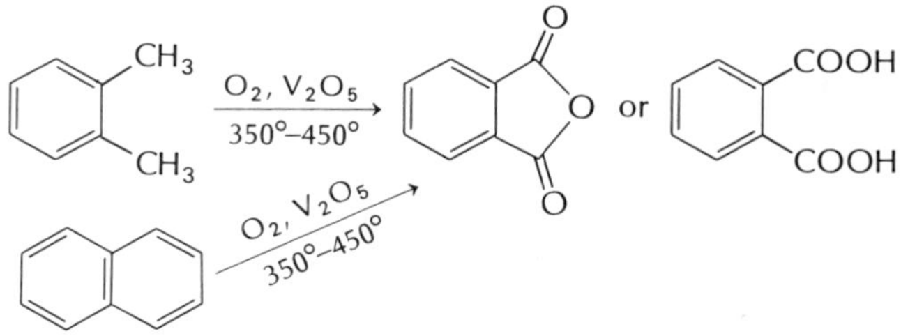

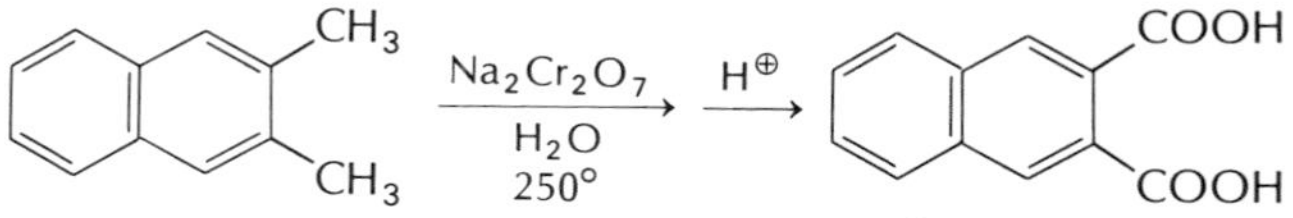

$$CH_3\text{-}C_6H_4\text{-}CH(CH_3)_2 \xrightarrow{HNO_3} CH_3\text{-}C_6H_4\text{-}COOH + 2CO_2$$

o-xylene $\xrightarrow[350°-450°]{O_2, V_2O_5}$ phthalic anhydride or phthalic acid

naphthalene $\xrightarrow[350°-450°]{O_2, V_2O_5}$ phthalic anhydride

2,3-dimethylnaphthalene $\xrightarrow[\substack{H_2O \\ 250°}]{Na_2Cr_2O_7} \xrightarrow{H^\oplus}$ naphthalene-2,3-dicarboxylic acid

2-methylnaphthalene $\xrightarrow[CH_3COOH]{CrO_3}$ 2-methyl-1,4-naphthoquinone

1-methylnaphthalene $\xrightarrow[CH_3COOH]{CrO_3}$ 8-methyl-1,4-naphthoquinone

(14.7)

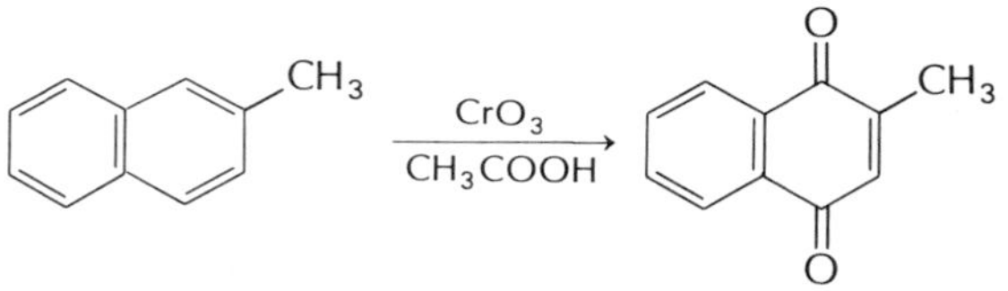

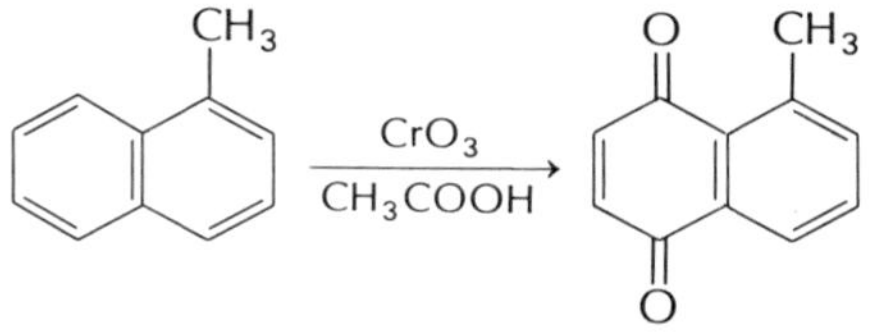

$$\phi CH_2 \phi \xrightarrow[\text{or } CrO_3]{KMnO_4} \phi CO \phi$$

$$\phi_3 CH \xrightarrow[\text{or } CrO_3]{KMnO_4} \phi_3 COH$$

$$\phi CH_2CH_2CH_2CH_3 \xrightarrow{\text{oxidation}} [\phi CHOHCH_2CH_2CH_3] \longrightarrow [\phi COCH_2CH_2CH_3]$$

$$\phi COOH + HOOCCH_2CH_3 \longleftarrow [\phi C{=}CHCH_2CH_3] \atop \underset{OH}{|}$$

(14.8)

benzene $\xrightarrow[450°-500°]{O_2, V_2O_5}$ maleic anhydride or maleic acid (HC-COOH, HC-COOH)

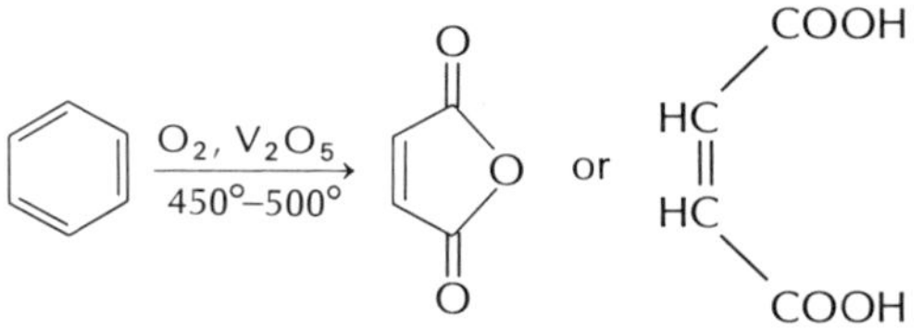

(14.9)

Drastic hydrogenation and metal reductions. Benzene is inert to the usual reducing agents ($NaBH_4$, $LiAlH_4$, B_2H_6, H_2), so compounds that have phenyl groups along with reducible functions may be reduced without effect on the ring (Eqs. 14.10). However, if conditions are drastic enough, catalytic hydrogenation of benzene rings may be carried out (Eqs. 14.11).

$$\phi CH_2COOH \xrightarrow[\text{ether}]{LiAlH_4} \phi CH_2CH_2OH$$

$$\phi CH=CHCO\phi \xrightarrow[\substack{CH_3COOEt \\ 25°}]{H_2, Pt} \phi CH_2CH_2CO\phi \quad (14.10)$$

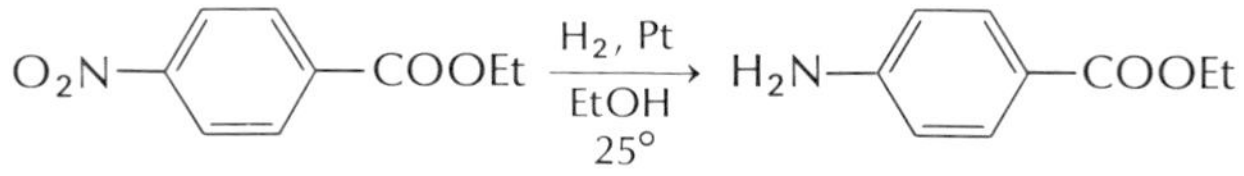

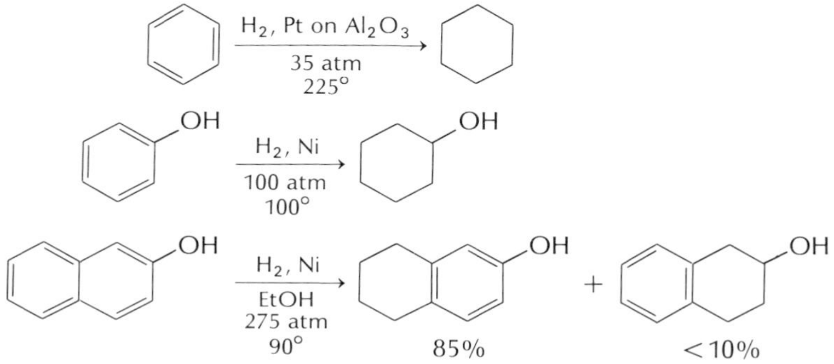

$$O_2N\text{-}\!\!\left\langle\!\!\!\bigcirc\!\!\!\right\rangle\!\!\text{-}COOEt \xrightarrow[\substack{EtOH \\ 25°}]{H_2, Pt} H_2N\text{-}\!\!\left\langle\!\!\!\bigcirc\!\!\!\right\rangle\!\!\text{-}COOEt$$

$$\text{benzene} \xrightarrow[\substack{35 \text{ atm} \\ 225°}]{H_2, \text{Pt on }Al_2O_3} \text{cyclohexane}$$

$$\text{phenol} \xrightarrow[\substack{100 \text{ atm} \\ 100°}]{H_2, Ni} \text{cyclohexanol} \quad (14.11)$$

$$\text{2-naphthol} \xrightarrow[\substack{EtOH \\ 275 \text{ atm} \\ 90°}]{H_2, Ni} \text{(6-hydroxytetralin, 85\%)} + \text{(2-hydroxytetralin, <10\%)}$$

Effective reducing reagents for aromatic systems are the active metals. Naphthalene, for example, with sodium in dimethoxyethane, gives rise to the anion radical (Eq. 14.12).

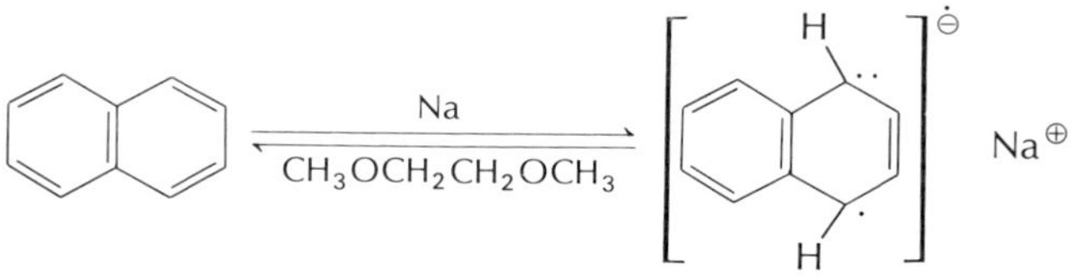

$$\text{naphthalene} \xrightleftharpoons[CH_3OCH_2CH_2OCH_3]{Na} \left[\text{1,4-dihydronaphthalenyl}\right]^{\ominus} Na^{\oplus} \quad (14.12)$$

$$\text{naphthalene} \xrightarrow[EtOH]{Na^{\cdot}} \left[\text{anion}\right] Na^{\oplus} \xrightarrow{EtOH} \text{1,4-dihydronaphthalene} \xrightarrow{Na^{\cdot}} \text{anion}^{\ominus} Na^{\oplus} \xrightarrow{EtOH} \text{1,2,3,4-tetrahydronaphthalene} \quad (14.13)$$

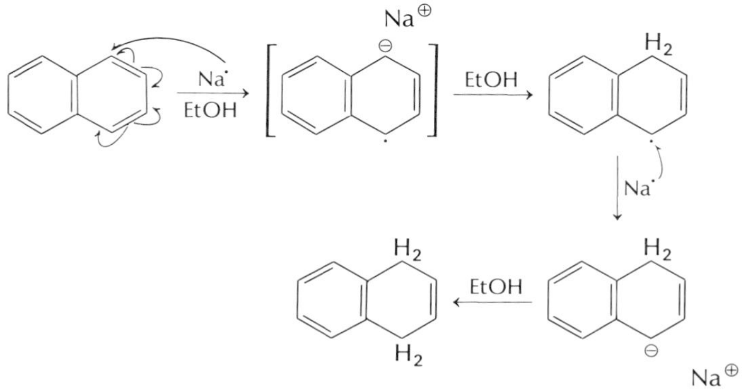

If a solvent (such as ethanol) that is capable of losing a proton to an anion radical is used, the reaction continues (Eq. 14.13) to give 1,4-dihydronaphthalene. Using the higher boiling pentyl alcohol ($C_5H_{11}OH$, mixture of isomers) at the boiling point with sodium and naphthalene gives 1,2,3,4-tetrahydronaphthalene (tetralin). Substituted benzenes may be reduced with lithium in liquid ammonia or ethylamine and an alcohol (Sec. 14.7).

SE, RE, and DE. As we have seen, aromatic rings have a great tendency to maintain their identity. To break into an aromatic ring, we must use very reactive agents or drastic conditions to overcome the large activation energies that must be surmounted in the endothermic destruction of the delocalization energy of the six-electron conjugation. Other terms in use besides DE are resonance energy (RE) and stabilization energy (SE). In magnitude, they are roughly equivalent. When necessary, we shall use DE as the MO-evaluated energy, RE as the similar energy thought of as resonance stabilization, and SE as the energy term evaluated from heats of hydrogenation or of combustion. According to these definitions, SE encompasses more than DE or RE because it includes steric effects, conformational differences, and so on. □ □

Exercises

14.11 A modern test for aromaticity is use of nmr to probe for the existence of a ring current. Explain.

14.12 The SE of biphenyl is greater than that of two benzene molecules. Explain.

14.6 Diene reactivity: 1,2 and 1,4 additions

Addition of HCl to 1,3-pentadiene. If we examine the various cations that may be formed in the first step of the addition of hydrogen chloride to 1,3-pentadiene (Eq. 14.14),

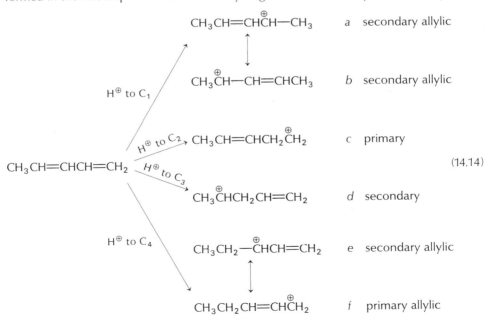

(14.14)

we find four possibilities. Of the four, the two allylic cations ($a \leftrightarrow b$ and $e \leftrightarrow f$) are the most stable because of the conjugative stability or delocalization of charge, as we have seen. Of the two, the $a \leftrightarrow b$ ion, both forms of which are secondary (and equivalent), is more stable than $e \leftrightarrow f$, which consists of a secondary and a primary form. Of the remaining two, d (as a secondary ion) is more stable than c (a primary ion).

Because the first step of the addition reaction is endothermic (uphill)—rupture of H—Cl and C=C, formation of C—H—we may safely assume that the activation energies parallel the positive ΔH of the step (Fig. 14.8). Thus the activation energy for going to the allylic ion $a \leftrightarrow b$ is less than that for going to $e \leftrightarrow f$, which is less than that for going to d, which is less than that for going over the highest activation energy hill to c. Putting it another way, the rates of formation of the ions are $a \leftrightarrow b > e \leftrightarrow f \gg d \gg c$; or we may generalize by saying that the attacking electrophile (here the proton) finds it easier to attack either end of the conjugated system at C_1 or C_4, rather than C_2 or C_3. The attacking electrophile goes to the carbon atom of a conjugated system that will result in the formation of the more stable allylic cation.

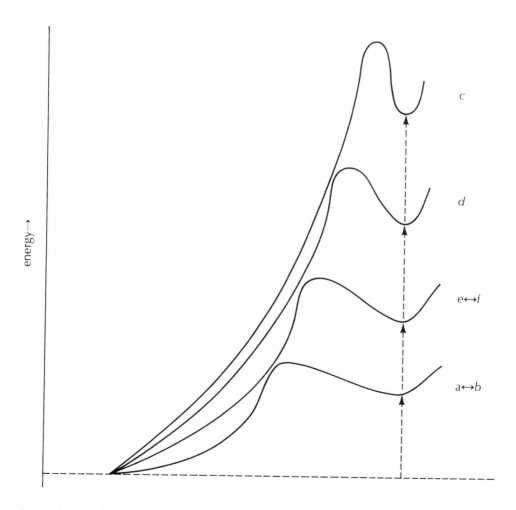

Figure 14.8 Schematic energy diagram for the first step in the addition of HCl to 1,3-pentadiene (Eq. 14.14)

14.6 Diene reactivity: 1,2 and 1,4 additions

Addition to butadiene. Having settled on a ↔ b as the preferred cation, we come to the problem of finding out whether the reaction is completed by chloride ion attachment at C_2 or at C_4. In the present case it makes no difference, because attack at either position would give the same product, 4-chloro-2-pentene ($CH_3CH=CHCHClCH_3$).

So let us shift the reaction to the addition of hydrogen chloride to 1,3-butadiene. Now we see that if $Cl^\ominus$ draws near to C_2, the cation reacts as structure a' and gives rise to

$$H_2C=CH-\overset{\oplus}{C}HCH_3 \longleftrightarrow H_2\overset{\oplus}{C}-CH=CHCH_3$$

a' $Cl^\ominus \downarrow\uparrow$ $Cl^\ominus \downarrow\uparrow$ b'

$H_2C=CH-CHClCH_3$ *trans*-$ClCH_2-CH=CHCH_3$

3-chloro-1-butene, a product that appears to be the result of addition of HCl to the 1 and 2 positions in butadiene. (We could call them positions 3 and 4, but we tend to use the lower numbers if it makes no difference.) If, however, $Cl^\ominus$ reacts at C_4, the cation reacts as structure b' and forms 1-chloro-2-butene, a product that is the result of addition to the 1 and 4 positions. Actually, both products are formed. The problem now becomes how to account for why 1,2 or 1,4 addition predominates under a given set of conditions.

Returning to the ion a' ↔ b', we see that the allylic ion is a resonance hybrid of a secondary and a primary structure. Because a secondary cation is more stable than is a primary, we can say in a simple-minded way that on the average, the positive charge is more intense at the secondary position than at the primary. Opposing that idea is the fact that when the charge is at the secondary position (a'), the double bond is terminal, which is a less stable situation than that of a nonterminal double bond, as in b'; but in b' the charge is at a primary position. So, as is usual in human relations as well, we are up against an evaluation of counterbalancing effects. Let us proceed on the assumption that the electrical effect outweighs the steric effect of the location of the double bond and write an "average" ion. We predict that $Cl^\ominus$ will more readily find the more intense positive center, so the 1,2 addition

product will be the major product. An equivalent statement is that the activation energy for interaction of $Cl^\ominus$ and the cation is less for the production of the 1,2 product than for the production of the 1,4 product. However, there can be little doubt that the 1,4 product is more stable (nonterminal double bond) than is the 1,2 product (terminal double bond); so we are not through yet.

Kinetic and thermodynamic control. Activation energy favors 1,2 addition; and stability favors 1,4 addition. What we have is one of the classic examples of kinetic *versus* thermodynamic control of a reaction (Fig. 14.9).

At room temperature or below, most of the reactants pass over the lower activation energy hill $\Delta G^\ddagger_{1,2}$, and the major product is the 1,2 isomer (75 to 80%) accompanied by 20 to 25% of the 1,4 isomer, for a ratio of 3 to 4 in favor of 1,2 addition. Under equilibrium control ($ZnCl_2$ present) at 15°, the 1,4 product is favored by 3.5 to 1. At higher temperatures, the 1,2 product undergoes an S_N1 dissociation into allylic ion and chloride ion faster and

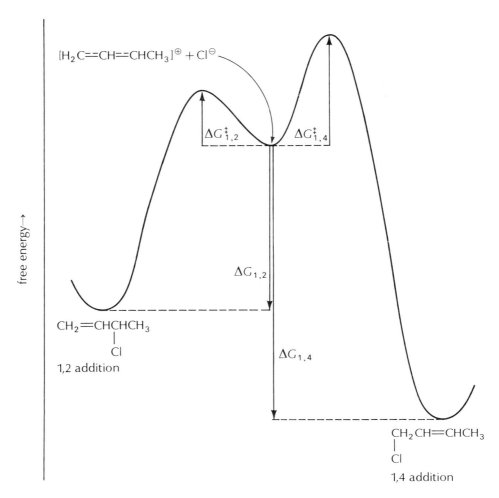

Figure 14.9 Schematic energy diagram for the second step in the addition of HCl to butadiene

faster. Also, at a higher temperature a larger fraction of the ions pass over the higher activation hill ($\Delta G^{\ddagger}_{1,4}$) and fall to the more stable 1,4 product. If the reaction mixture is allowed to stand long enough at any temperature, an equilibrium between the 1,2 and 1,4 products is established. In the absence of catalyst, $K_{eq} = 2.34$ at 80° in toluene; the rate of establishment of equilibrium is so slow that it requires weeks. At low temperatures, the time to reach equilibrium is so long that interruption of the reaction after a reasonable time gives a good yield of the less stable, kinetically controlled 1,2 product. At higher temperatures, in the presence of a catalyst such as $ZnCl_2$, which assists the S_N1 dissociation, equilibrium is established in so short a time that it becomes very difficult to isolate products in any ratio other than that determined by the equilibrium constant at that temperature. □ □

Homolytic additions.   A homolytic pathway favors 1,4 addition, probably because the intermediate allyl radical is at a relatively high energy level and the activation energy for 1,4 product formation becomes less than that for 1,2. It is also possible that there is less

14.6 Diene reactivity: 1,2 and 1,4 additions

difference in energy level between the primary and secondary allylic radical structures than there is between the allylic cations. If so, the primary radical structure would contribute more to the resonance hybrid of the radical than would the primary ionic structure to the hybrid of the ion.

As we have seen, catalytic hydrogenation tends to give the most stable of a set of possible isomers by isomerization on the catalyst surface. At room temperature, butadiene takes up 1 mole of hydrogen to give 60 to 70% of *trans*-2-butene, the more stable 1,4 product. The reaction may be stopped at that point or continued with more hydrogen to give butane.

Dienes also react with active metals (Eq. 14.15) to give a radical anion or a dianion. If a proton donor such as an alcohol is present, the result is equivalent to a 1,4 hydrogen addition.

$$RCH=CH-CH=CHR \xrightarrow{Na \cdot} R\dot{C}H-CH=CH-\overset{\ominus}{C}HR \quad Na^{\oplus}$$
$$\downarrow Na \cdot$$
$$RCH_2CH=CHCH_2R \xleftarrow{EtOH} \overset{\ominus}{R C}H-CH=CH-\overset{\ominus}{C}HR \quad 2Na^{\oplus}$$

(14.15)

Other electrophilic and radical reagents add to dienes to give varying proportions of 1,2 and 1,4 products.

Benzyl cation, anion, and radical. What happens if an aromatic ring is present in conjugation with an extra orbital, as in the benzyl cation ($\phi CH_2^{\oplus}$), or a double bond, as in styrene ($\phi CH=CH_2$)? Formally, at least, we may write a set of resonance structures for the benzyl cation in which three forms appear with the positive or negative charge or odd electron in the ion or radical at positions 2, 4, and 6 in the ring (or *o*, *p*, and *o*). It is possible

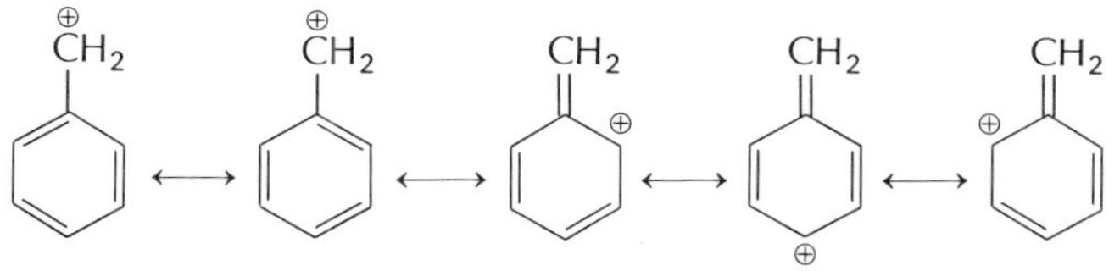

for conjugation to occur between the aromatic ring and adjacent orbitals. The lowest occupied MO would appear as shown, with only the overlap above the ring drawn in. The contribution of the ring to stabilization of the charge or odd electron lends stability to the

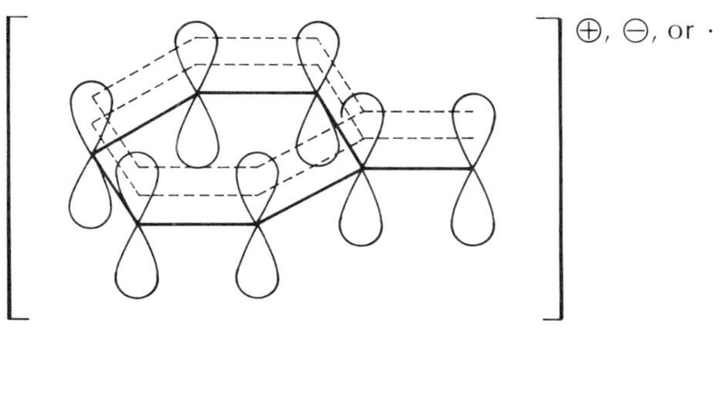

$\oplus, \ominus,$ or $\cdot$

ion, and a benzyl ion or radical is comparable to an allyl ion or radical. We also could say that a benzene ring shows either a +R or a −R effect, depending upon the demand. The presence of more than one ring bonded to the reactive carbon atom imparts additional stability, as we can see in the effect upon activation energies shown in relative rates for S_N1 reactions (Table 14.1). The allyl and benzyl halides also become more susceptible to

Table 14.1 Conjugation effects upon relative rates of reaction

R—Cl	S_N1, 80% EtOH, 50°	S_N2, KI, acetone, 50°
CH_3CH_2Cl	—	2.5
$(CH_3)_2CHCl$	0.7	0.02
$CH_2=CHCH_2Cl$	52	79
ϕCH_2Cl	100	200
$\phi(CH_3)CHCl$	7400	—
$(CH_3)_3CCl$	8200	—
ϕ_2CHCl	$\sim 10^7$	—

S_N2 attack, as shown. Benzhydryl ($\phi_2CH^\oplus$) and triphenylmethyl ($\phi_3C^\oplus$) are relatively stable carbonium ions. □ □

Styrene: addition. If we look at styrene, we see a stabilization as shown. The charge-separated forms, as in butadiene, contribute only a little to the ground state, so the

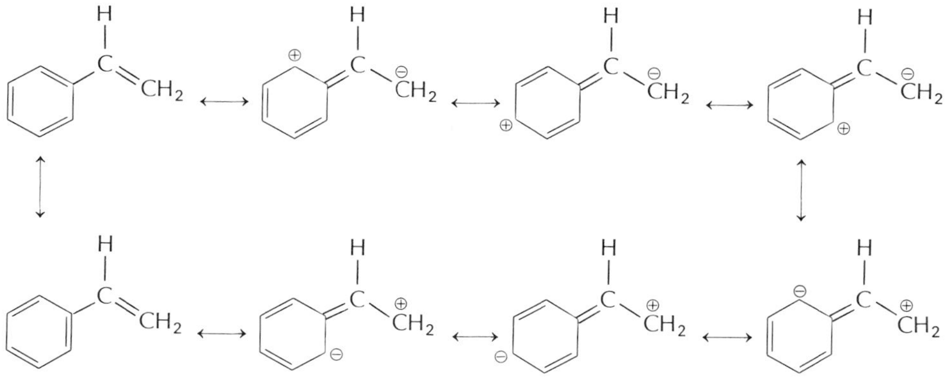

heat of hydrogenation of styrene, −77.5 kcal, gives DE = 38.6 kcal, only 2.6 kcal greater than for benzene alone. However, the extra stabilization is comparable to the 3.5 kcal of butadiene.

Styrene reacts with electrophilic reagents. Let us examine Figure 14.10, the first step possible for acceptance of a proton, as we did in Equation 14.14. We show five of the possible locations that the entering proton may take up: 1. at the β position as in a to e, to give a substituted benzyl cation; 2. at the α position as in f, to give a substituted ethyl cation; 3. at either of the *ortho* positions as in g; 4. at either of the *meta* positions as in h; or 5. at the *para* position as in i. In both a and f, the Kekulé resonance is preserved. In g,

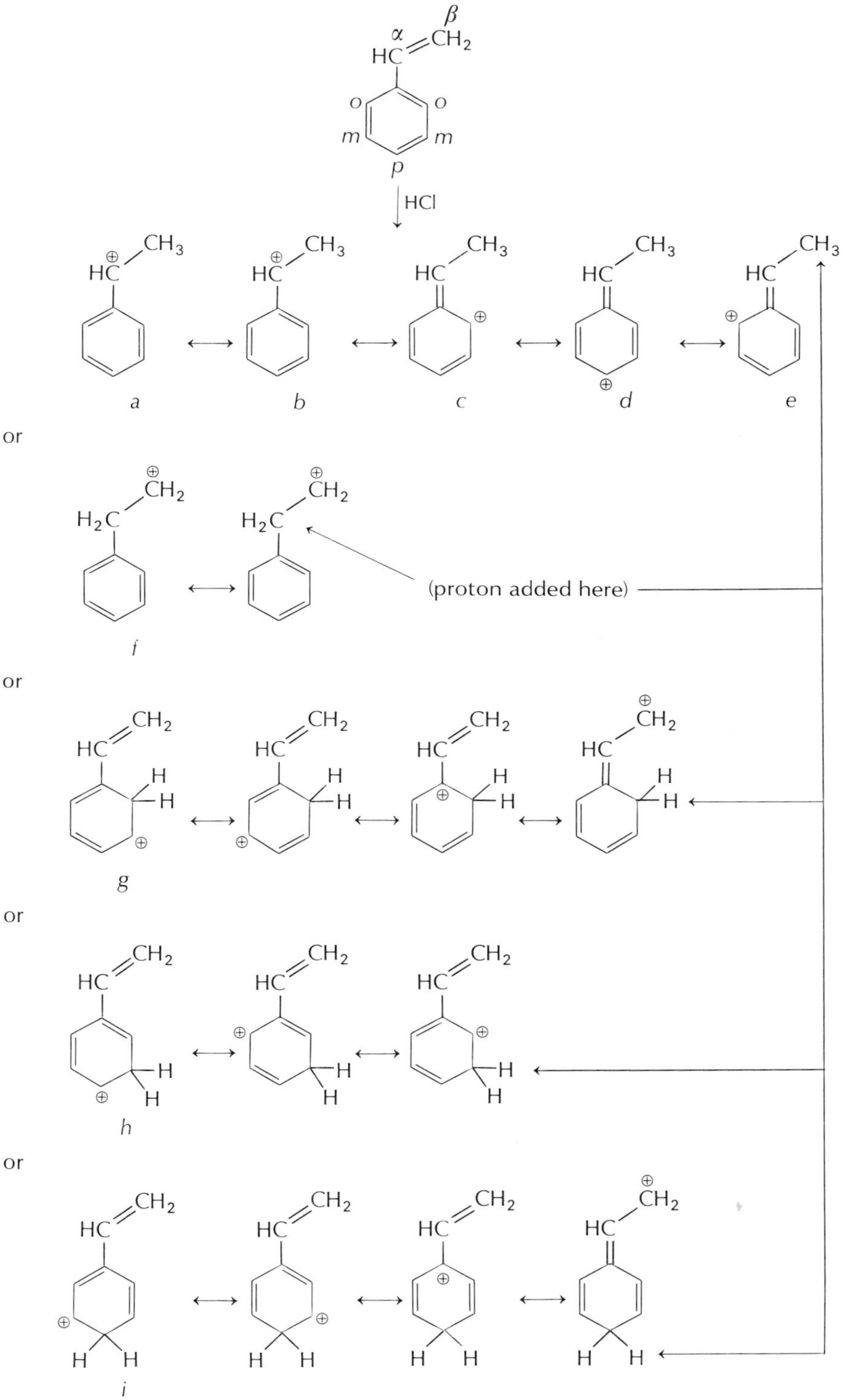

14.6 Diene reactivity: 1,2 and 1,4 additions 425

Figure 14.10 First step in the reaction of styrene with HCl

h, and *i*, the Kekulé resonance has been destroyed, and the ions must be at a higher energy level than in *a* or *f*, despite the extensive conjugation remaining. Comparing *a* with *f*, we see that *a* is more stable (has less energy) than *f*, in which the positive charge is isolated at a primary position.

Thus, initial attack upon a double bond conjugated with a benzene ring always proceeds to place the charge upon the α carbon atom, which allows the ring to participate in stabilizing (delocalizing) the charge. The Kekulé structures *a* and *b* remain the major contributors to the hybrid structure of the cation. However, *c*, *d*, and *e* contribute enough to lower the energy level of the hybrid far enough below those of *f*, *g*, *h*, or *i* to encourage exclusive attack by HCl at the β position. A discussion in MO terms comes to the same conclusion, but the resonance method is easier to see and follow.

The second step is disposed of quickly: chloride ion completes the addition at the α position to give ϕCHClCH$_3$. If chloride ion completed the addition by locating at an *o* or *p* position (from *c*, *d*, or *e*), the product would be a substituted cyclohexadiene. This product would require the loss of almost all the 36 kcal of DE of the benzene ring: ΔH for the reaction would become positive. Thus, an aromatic ring is able to affect strongly a reaction in a side chain while preserving its own identity.

1-Phenylbutadiene and 1,2 addition. We could work through in detail the course of addition of HCl to 1-phenylbutadiene, but let us be content to look at only two of the resonance structures of the cation that is formed by proton bonding to C$_4$. We see that we have an allylic resonance as well as a benzylic resonance (from *b*). Without thinking about it,

$$\phi\text{CH}=\text{CH}\overset{\oplus}{\text{CH}}\text{CH}_3 \longleftrightarrow \overset{\oplus}{\phi\text{CH}}\text{CH}=\text{CHCH}_3$$
$$\qquad\qquad a \qquad\qquad\qquad\qquad\qquad b$$

we might expect chloride ion to attack the ion so as to give ϕCHClCH=CHCH$_3$, the 1,4 product, rather than ϕCH=CHCHClCH$_3$. However, we should think about it, because ϕCH=CHCHClCH$_3$ is almost the exclusive product. Why? Because the 1,2 (the 3,4) product leaves a double bond conjugated with the ring, but the double bond is isolated from the ring in the less stable 1,4 product.

Summary. Conjugated systems undergo both 1,2 and 1,4 addition; the proportions vary with the substrate, the reagent, and conditions so that prediction is not invariably successful; and the presence of an aromatic ring exerts a powerful effect upon the course of an addition reaction.

Exercises

14.13 Predict the major organic product or products of the reactions given:
   a. $\phi\text{CH}=\text{CHCH}=\text{CH}_2 + \text{HBr} \xrightarrow{\text{peroxide}}$
   b. $\phi\text{CH}=\text{CHCH}=\text{CH}\phi + \text{Br}_2 \xrightarrow[\text{CH}_3\text{COONa}]{\text{CH}_3\text{COOH}}$
   c. $\phi_2\text{C}=\text{CH}_2 + \text{Br}_2 \xrightarrow{\text{MeOH}}$
   d. $\text{CH}_3\text{CH}=\text{CHCH}=\text{CH}_2 + \text{Cl}_2 \xrightarrow{hv}$
   e. $\text{H}_2\text{C}=\text{CHCH}=\text{CHCH}=\text{CH}_2 + \text{Br}_2 \longrightarrow$

14.14 Show how to carry out the conversion between these compounds:

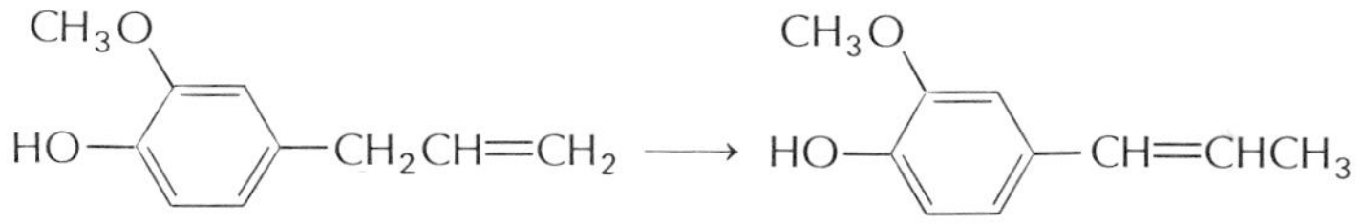

HO—[C₆H₃(OCH₃)]—CH₂CH=CH₂ ⟶ HO—[C₆H₃(OCH₃)]—CH=CHCH₃

14.7 Oxidation and reduction

Summary of reactions; ozonization. We have already considered several oxidation and reduction reactions of conjugated systems: see hydrogenation of dienes (Sec. 14.2), of rings, (Sec. 14.3); ozonization of benzene and o-xylene, drastic oxidation of substituted benzenes and naphthalenes, hydrogenation of benzene and naphthalene, active metal reduction (Sec. 14.5); and hydrogenation and active metal reduction of dienes (Sec. 14.6). Here we shall enlarge on a few of the reactions.

Ozone has no respect for conjugation, as we have seen from the destruction of benzene. Hence, an ordinary conjugated diene reacts with ozone as though the double bonds were isolated (Eq. 14.16).

$$CH_3CH=CHCH=CH_2 \xrightarrow[\text{2. Zn, H}_2\text{O}]{\text{1. O}_3} CH_3CHO + OHCCHO + HCHO \quad (14.16)$$

Mild and drastic oxidation. Mild oxidation of dienes is difficult to carry out and is seldom attempted because of the host of possible products. Drastic oxidation generally destroys the conjugated system (Eq. 14.17). In the example, oxalic acid does not survive. A single bond between carbon atoms that are partially oxidized is susceptible to cleavage by oxidation, and the C—C bond in oxalic acid is an example.

$$\phi CH=CHCH=CHCH_3 \xrightarrow[\text{drastic}]{\text{oxidation}} \phi COOH + [HOOCCOOH] + HOOCCH_3 \quad (14.17)$$
$$\downarrow$$
$$2CO_2 + 2H_2O$$

Reduction by active metals. Reduction by active metals deserves a bit more attention. Isolated double bonds resist reduction by these means, even though conjugated systems are easily reducible. Liquid ammonia (bp −33°) frequently is used alone or with additional aprotic solvents such as ether. The most used metals are Li, Na, and K. Two examples are shown in Equations 14.18. If a protic solvent more acidic than ammonia

$$\phi_2C=CH_2 \xrightarrow[\text{ether, }-33°]{\text{Na}/\text{NH}_3} \xrightarrow{NH_4Cl} \phi_2CHCH_3 + \phi_2CHCH_2CH_2CH\phi_2$$
$$\qquad\qquad\qquad\qquad\qquad\qquad 67\% \qquad 17\%$$

$$H_2C=CHCH=CH_2 \xrightarrow[\text{2. NH}_4\text{Cl}]{\text{1. Na, NH}_3} \underset{H}{\overset{CH_3}{>}}C=C\underset{H}{\overset{CH_3}{<}} + \underset{H}{\overset{CH_3}{>}}C=C\underset{CH_3}{\overset{H}{<}} \quad (14.18)$$

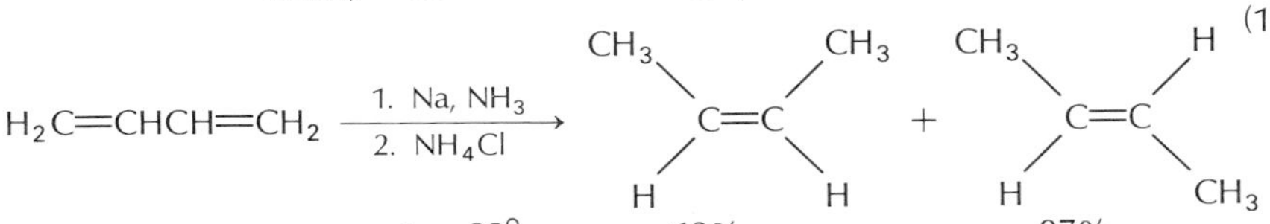

at −33°   13%   87%
at −78°   50%   50%

is present in the mixture, the radical dimerization product (ϕ_2CHCH$_2$CH$_2$CHϕ_2) is completely suppressed. ☐ ☐

Exercises

14.15 Predict the product to be expected after work-up from:

a. ϕ_2CHCH=CH$_2$ $\xrightarrow{\frac{K_2Cr_2O_7}{H_2SO_4}}$

b. 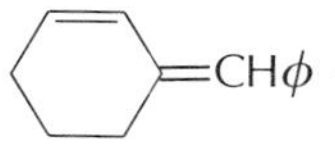 (cyclohexenyl)=CHϕ $\xrightarrow[\text{2. Zn, CH}_3\text{COOH}]{\text{1. O}_3}$

c. trans-ϕCH=CHϕ $\xrightarrow{\text{OsO}_4}$

d. 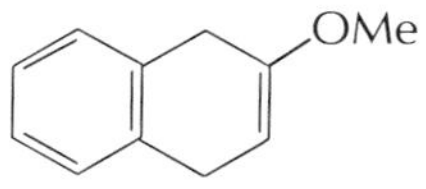 (dihydronaphthalene with OMe) $\xrightarrow[\text{2. CH}_3\text{CH}_2\text{COOH}]{\text{1. BH}_3}$

14.8 Cycloaddition reactions

Concerted reactions. There exists a class of reactions that we have avoided until now. All members of the class involve a ring or cycle as reactant, transition state, or product. The general name for these reactions is <u>pericyclic</u>. (The prefix <u>peri</u> means around

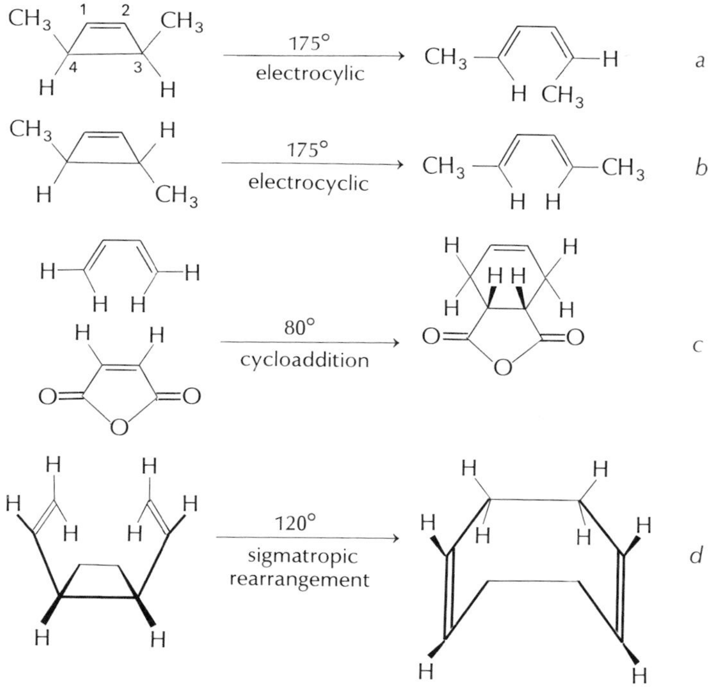

(14.19)

or about.) Pericyclic reactions may be divided into three classes: electrocyclic reactions, cycloaddition reactions, and sigmatropic rearrangements. They all have certain common features that are not immediately apparent from a glance at a simple example of each class of reaction (Eqs. 14.19).

The reactions shown all are brought about by heat. The presence or absence of cations, anions, radicals, radical traps, catalysts, or solvents has little or no effect upon the rates or positions of equilibrium. In each instance, although the number of bonds does not change, several bonds are made or broken. It seems that each reaction is a single concerted change, not a stepwise series of changes. These reactions may be brought about by uv (photolysis, a photochemical reaction) as well as by the heat. The stereochemistries of the thermal and photolytic reactions differ, however, and the stereochemical course of a given reaction is quite specific. Yields are high.

The electrocyclic reactions shown (a and b of Eqs. 14.19) illustrate the stereochemical specificity involved in the ring openings of cis- and trans-3,4-dimethyl-1-cyclobutene to give (Z-2,E-4)-hexadiene (a) and (E-2,E-4)-2,4-hexadiene (b).

The cycloaddition reaction between butadiene and maleic anhydride (c) is an example of a [4 + 2] cycloaddition, better known as the Diels-Alder reaction, discovered in 1928. It has been explored in many ways and has been very useful in the synthesis of cyclic structures.

The sigmatropic rearrangement is characterized by the loss of a single bond at one location in the molecule and the creation of another at some other location, with a change in position of the double bonds.

Conservation of orbital symmetry. It was not until 1965 that the common features of these reactions were given a satisfactory explanation by Woodward and Hoffman, who emphasized the part played by conservation of orbital symmetry. We shall not describe conservation of orbital symmetry in this text because, for the time being, it is more properly a subject for advanced study in organic chemistry. As a result, we cannot easily describe electrocyclic reactions and sigmatropic rearrangements. Information on these topics is easily available elsewhere: see the bibliography. However, we shall describe cycloaddition reactions. □ □

Diels-Alder reaction. The Diels-Alder reaction, a [4 + 2] cycloaddition, goes so readily that many of these reactions are carried out at room temperature, or only slightly above it. Because the reaction is between an alkene and a diene, the alkene component of a given reaction is called the dienophile. It turns out that if the dienophile is substituted with electron-withdrawing substituents, the reaction proceeds more easily than otherwise. The presence of electron-withdrawing substituents in the diene slows the reaction down, and the presence of electron-donating substituents in the diene speeds it up. What may be thought of as the parent reaction, the production of cyclohexene from ethylene and butadiene, requires drastic conditions (Eq. 14.20). The diene must be in the s-cis conformation to react, and the reaction may also be thought of as a stereospecific cis-1,4 addition to the diene and as a simultaneous cis addition to the alkene.

$$\text{butadiene} + \text{ethylene} \xrightarrow[300 \text{ atm}]{200°} \text{cyclohexene} \quad 18\% \tag{14.20}$$

430   Chapter fourteen

Because an *s-cis* conformation is built into cyclic compounds, a diene like cyclopentadiene is particularly reactive (Eqs. 14.21). We show one *p* orbital in each reactant and

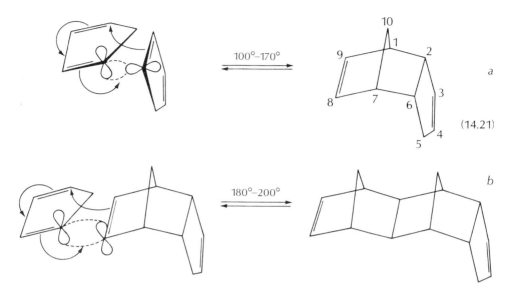

(14.21)

use dashed lines to illustrate the geometry of the orbital overlap of σ bond formation. One molecule acts as a dienophile to the other, slowly at room temperature, more quickly as the temperature is raised, to give tricyclo[5.2.1.0^{2,6}]3,8-decadiene, better known as dicyclopentadiene (a). At 170°, the bp of dicyclopentadiene, the equilibrium with the cyclopentadiene monomer is set up rather quickly; it is reached more slowly at lower temperatures. By fractional distillation of the dimer-monomer mixture the small amount of monomer (bp 41°) present at equilibrium may be removed. Then more of the dimer dissociates to maintain the equilibrium. In this manner, the reverse Diels-Alder reaction may be carried out in good yield.

Because the monomer dimerizes slowly at room temperature, it cannot be stored for any length of time. However, if the monomer is collected at 0°, it may be used the same day for synthetic purposes. At still higher temperatures, the monomeric diene undergoes a second Diels-Alder reaction with the strained and more reactive 8,9 double bond instead of the unstrained 3,4 double bond to give a trimer (b). At 200°, subsequent Diels-Alder reactions continue on to give pentamers [(C$_5$H$_6$)$_5$] and hexamers [(C$_5$H$_6$)$_6$].

Because dienes show such great variation in reactivity, many different dienophiles have been tested. Probably the most reactive dienophile known is tetracyanoethylene, (N≡C)$_2$C=C(C≡N)$_2$. In general, any system that has a C=O or C≡N conjugated with C=C is an active dienophile: some examples are C=C—CHO, C=C—COR, C=C—CN, C=C—COOH, C=C—COOR, and C=C—CONH$_2$. Also, in general, the presence of two electron-withdrawing groups enhances the activity.

Definitions of exo- and endo-.   Two geometries are possible in a Diels-Alder reaction if the diene is part of a ring. Equation 14.22 shows the two possible products from cyclopentadiene and diethyl maleate (the diethyl ester of ethylenedicarboxylic acid). The bicyclic ring system, bicyclo[2.2.1]heptane, usually is referred to as norbornane. The unsaturated compound then is norbornene, and it is numbered as shown. The substituents at the 2, 3, 5, and 6 positions in norbornane may be either equatorial about the boat-shaped

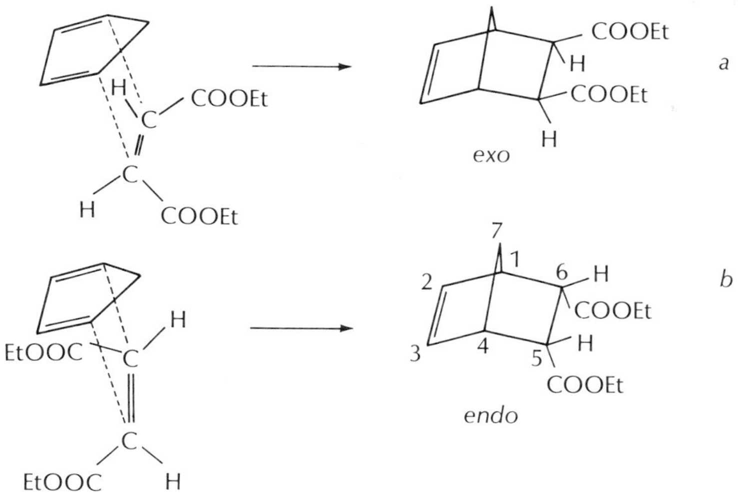

14.8 Cycloaddition reactions

(14.22)

six-membered ring, known as *exo*, or they may be axial, known as *endo*. Substituents at the 2 or 3 positions in norbornene have no descriptive prefix, because they must be coplanar with atoms 1, 2, 3, and 4 in any case. Thus the product in a is diethyl *exo-cis*-5,6-norbornenedicarboxylate; and in b, diethyl *endo-cis*-5,6-norbornenedicarboxylate. The dicyclopentadiene shown in Equation 14.21a is the *endo* isomer. In Equation 14.21b, the norbornene ring system of dicyclopentadiene undergoes the Diels-Alder reaction to give the *exo* isomer. The reaction to give the *exo* isomer is an exception, because the *endo* isomer is the usual major product (if not the only one) when the dienophile is highly polar. ☐ ☐

Exercises

14.16 Complete the Diels-Alder reactions:

a. 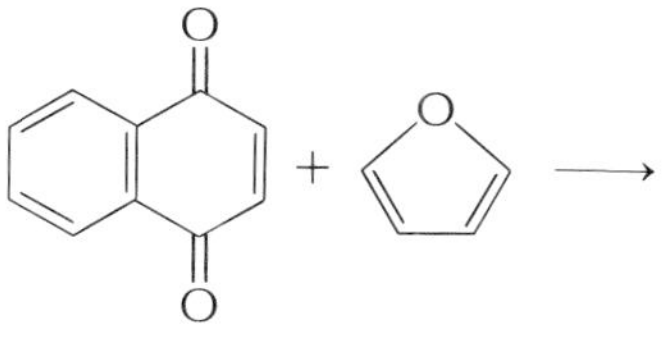 [naphthoquinone] + [furan] ⟶

b. 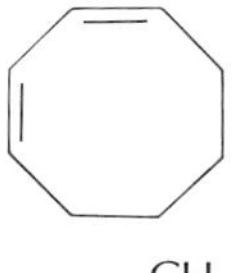 [cyclooctatetraene] + $H_2C=CHCHO$ ⟶

c. $H_2C=\underset{CH_3}{C}-\underset{CH_3}{C}=CH_2$ + *cis*-$CH_3CH=CHCN$ ⟶

d. 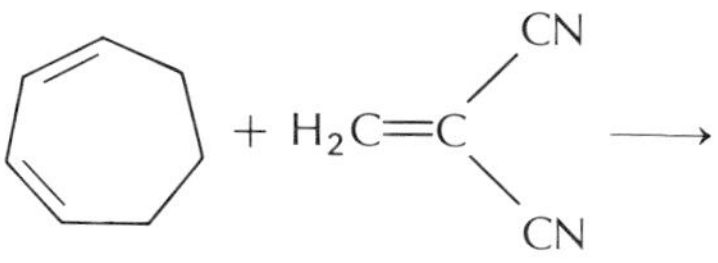 [cycloheptatriene] + $H_2C=C(CN)_2$ ⟶

14.17 Complete the reaction:

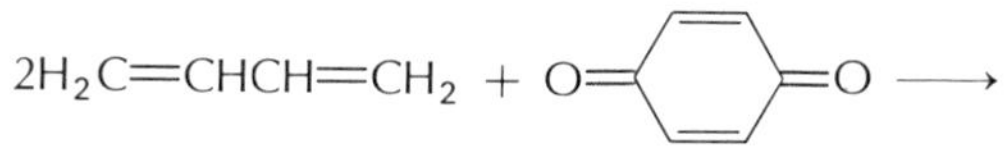

$$2H_2C{=}CHCH{=}CH_2 + O{=}\underset{}{\bigcirc}{=}O \longrightarrow$$

14.9 Preparation of dienes

Summary of methods. To prepare dienes in the laboratory, we usually adapt a method for the synthesis of an alkene. Thus, dehydration of an alcohol, either acidic or Chugaev; pyrolysis of esters; dehalogenation of dihalides by zinc or iodide ion; dehydrohalogenation by bases; thermal decomposition of quarternary ammonium hydroxides (Hofmann) or tertiary amine oxides; and reduction of alkynes with catalytic hydrogenation or borane, with or without isomerization, all have been used.

An example of a commonly used sequence is given in Equations 14.23.

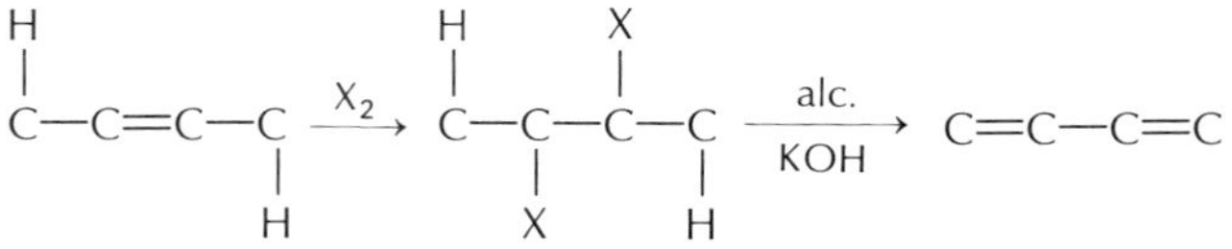

$$\underset{\underset{H}{|}}{\overset{\overset{H}{|}}{C}}{-}C{=}C{-}\underset{\underset{H}{|}}{C} \xrightarrow{X_2} \underset{\underset{X}{|}}{\overset{\overset{H}{|}}{C}}{-}\underset{\underset{X}{|}}{C}{-}\underset{\underset{H}{|}}{C}{-}C \xrightarrow[KOH]{alc.} C{=}C{-}C{=}C$$

$$\phi CH_2CH{=}CHCH_2CH_3 \xrightarrow[CHCl_3]{Br_2} \phi CH_2 CHBrCHBrCH_2 CH_3 \qquad (14.23)$$

$$\downarrow \begin{array}{c} alc. \\ KOH \end{array} 80°$$

$$\phi CH{=}CHCH{=}CHCH_3$$

Because of the stability to be gained by conjugation with an aromatic ring, a double or triple bond has a great tendency (especially under thermodynamic control) to be formed or to migrate in a side chain to a position adjacent to the ring (Eq. 14.24).

$$\phi CH_2CH_2 CHBrCHBrCH_3 \xrightarrow[80°]{alc.\ KOH} [\phi CH_2 CH{=}CHCHBrCH_3]?$$

$$\downarrow -Br^{\ominus}$$

$$[\phi CH_2 \overset{\oplus}{C}H{-}CH{=}CHCH_3 \longleftrightarrow \phi CH_2 CH{=}CH\overset{\oplus}{C}HCH_3]? \qquad (14.24)$$

$$\downarrow -H^{\oplus}$$

$$\phi CH{=}CHCH{=}CHCH_3$$

Even if $\phi CH_2CH{=}CHCH{=}CH_2$ were formed, in the presence of base some $\phi \overset{\ominus}{C}HCH{=}CHCH{=}CH_2 \leftrightarrow \phi CH{=}CHCH{=}CH\overset{\ominus}{C}H_2$ would be present in the solution, so thermodynamic control still is possible.

Catalytic methods. A method new to us but involving no new principles is oxidative dehydrogenation. Previously (Eq. 14.3), we observed the disproportionation of cyclohexene in a closed system. If the system is open so that hydrogen may leave the mixture, the hydrogenation catalyst serves as a dehydrogenation catalyst, and cyclohexene is converted completely into benzene (Eq. 14.25). Also, any cyclohexane or cyclohexadiene undergoes the reaction, which is usually heated to speed it up. Actually, we are disturbing an equilibrium by removing one of the reactants.

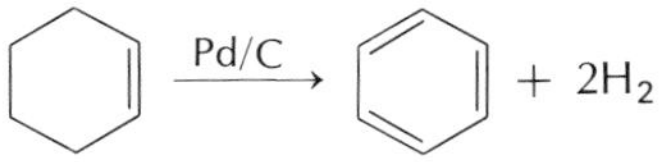

$$\text{cyclohexane} \xrightarrow{Pd/C} \text{benzene} + 2H_2 \qquad (14.25)$$

If an oxidizing agent is present to combine with hydrogen, the same principle (removal of a reactant) is used. In the laboratory, sulfur, selenium, and various quinones are used at 230° to 250° for S, 300° to 330° for Se, and 100° to 250° for quinones. (A <u>quinone</u> is an unsaturated cyclic diketone that is easily reducible to a diphenol, a dihydroxybenzene as in Equation 14.26. (See Chapter 24.) Some examples of oxidative dehydrogenation are shown in Equations 14.27.

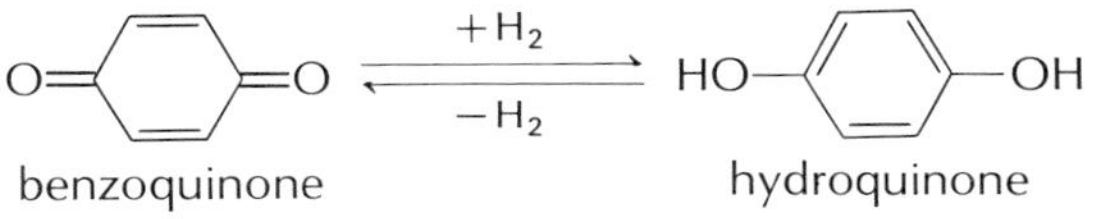

$$\text{benzoquinone} \underset{-H_2}{\overset{+H_2}{\rightleftharpoons}} \text{hydroquinone} \qquad (14.26)$$

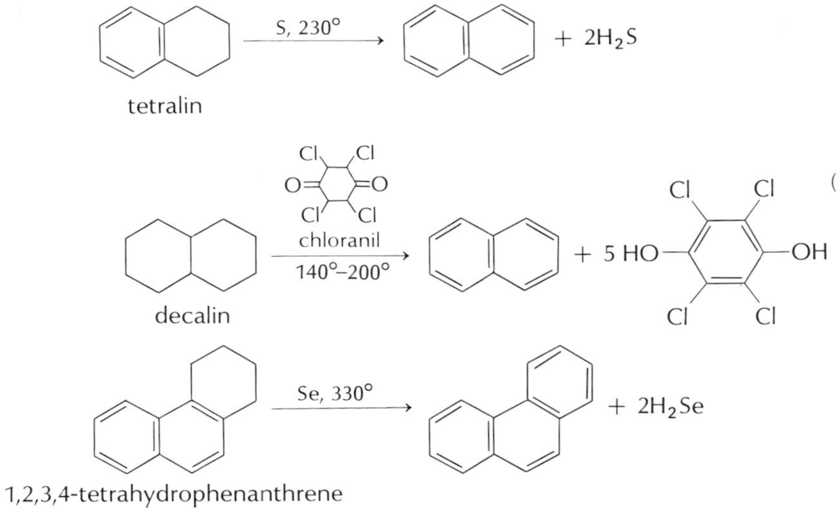

tetralin $\xrightarrow{S, 230°}$ naphthalene $+ 2H_2S$

decalin $\xrightarrow[140°-200°]{\text{chloranil}}$ naphthalene $+ 5\,HO\text{-}C_6Cl_4\text{-}OH$ (14.27)

1,2,3,4-tetrahydrophenanthrene $\xrightarrow{Se, 330°}$ phenanthrene $+ 2H_2Se$

Commercial routes to butadiene. Commercially, the demand for butadiene and isoprene (2-methyl-1,3-butadiene) is high for use in the preparation of polymers. From 1000 tons of naptha feedstock, the industry expects to obtain (by cracking, steam cracking, reforming, etc.) about 300 tons of ethylene, 160 tons of propylene, 45 tons of butylenes, 45 tons of butadiene, and 200 tons of pentanes, pentenes, pentadienes, and so on. Several methods have been developed for converting butenes to butadiene by oxidative dehydrogenation utilizing oxygen or air as the oxidant. The reactions become exothermic because water is one product, so the energy requirement is small. However, control of the reactions becomes difficult.

434 Chapter fourteen

They may be moderated by the presence of halogens (iodine is very good, but so expensive that complete recovery and recycling is necessary) or of sulfur from H_2S (Eq. 14.28). The best ratio of reactants is given, and a 52% conversion per pass over the catalyst is reached.

$$C_4H_8 + 1.2H_2S + 0.75O_2 \xrightarrow[\Delta]{\text{5% cerium oxide on alumina}} 0.52C_4H_6 \quad (14.28)$$

Laboratory preparation of isoprene. A laboratory synthesis of isoprene is given in Equation 14.29. The first step is the addition of $HC{\equiv}\overset{\ominus}{C}K^{\oplus}$ to a $C{=}O$. The hydrogenation of $C{\equiv}C$ to a $C{=}C$ and the dehydration of the *tert*-allylic alcohol are familiar reactions.

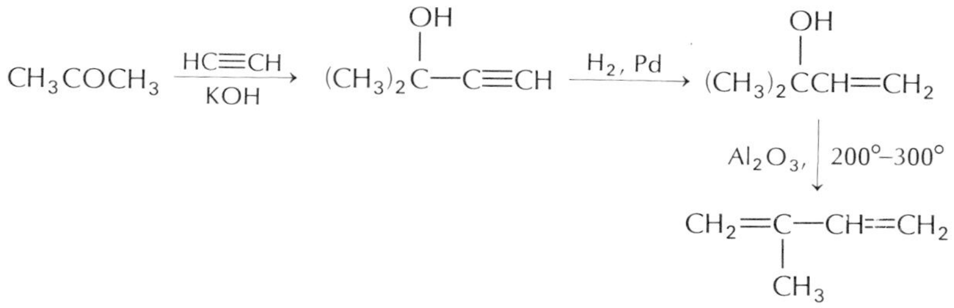

$$CH_3COCH_3 \xrightarrow[KOH]{HC{\equiv}CH} (CH_3)_2\underset{\underset{OH}{|}}{C}{-}C{\equiv}CH \xrightarrow{H_2,\ Pd} (CH_3)_2\underset{\underset{OH}{|}}{C}CH{=}CH_2$$

$$\xrightarrow[]{Al_2O_3,\ 200°-300°} CH_2{=}\underset{\underset{CH_3}{|}}{C}{-}CH{=}CH_2 \quad (14.29)$$

Exercises

14.18 Give a suitable laboratory preparation of:

a. 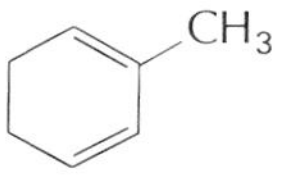 [structure: 1-methyl-1,3-cyclohexadiene] from 1-methylcyclohexanol

b. $(CH_3)_2C{=}CHCH{=}C(CH_3)_2$ from 2,5-dimethyl-3-hexanol

c. 2,3-dimethyl-1,3-butadiene from 2,3-dimethyl-2-butanol

14.10 Summary

After some preliminaries about the naming of systems containing two or more double bonds, we looked at the effects of conjugation from both a resonance and an MO viewpoint. The concepts of delocalization and of delocalization energy were developed and applied to cyclic conjugated systems. The $4n + 2$ rule was introduced, and its application to various sizes of rings and ions was shown.

We went on to see how the concepts are used to increase understanding of the spectral behavior of conjugated systems and arrived at the aromatic compounds, exemplified by benzene. Other benzenoid and some nonbenzenoid aromatic substances were included in the discussion.

Moving on into the reactivity of conjugated compounds, we examined 1,2 and 1,4 additions and brought in several factors for consideration: kinetic and thermodynamic control, reagents, ions, radicals, and acidity. Oxidation and reduction were examined: specifically, drastic oxidation products and the utility of active metal reductions.

The pericyclic reactions—electrocyclic reactions and sigmatropic rearrangements—were only mentioned, but cycloaddition reactions, with emphasis upon the Diels-Alder reaction, were described in some detail.

We ended the chapter with a few remarks concerning the preparation of dienes.

Some new reactions given in this chapter are repeated here, sometimes with different reactants to show that a given reaction need not be limited to the reactant first given in an example or section.

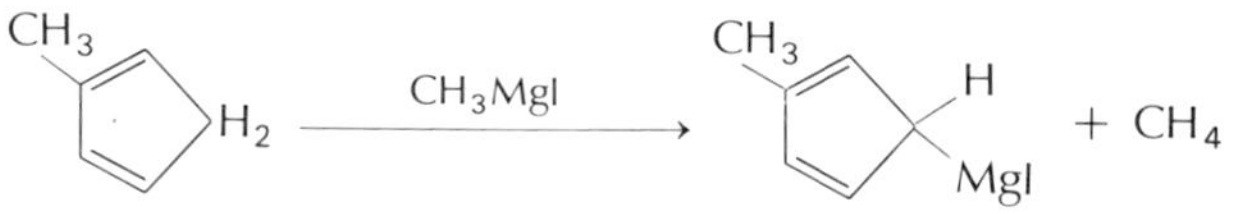

$$\text{methylcyclopentadiene} + \text{CH}_3\text{MgI} \longrightarrow \text{methylcyclopentadienyl-MgI} + \text{CH}_4 \tag{14.1}$$

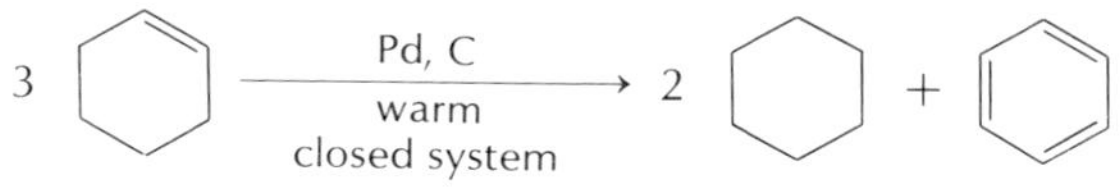

$$3\,\text{cyclohexene} \xrightarrow[\text{closed system}]{\text{Pd, C, warm}} 2\,\text{cyclohexane} + \text{benzene} \tag{14.3}$$

$$2\,\text{(1,2-diethylbenzene)} \xrightarrow[\text{2. Zn, H}_2\text{O}]{\text{1. O}_3} 3\,\text{OCH—CHO} + 2\,\text{EtCOCHO} + \text{EtCOCOEt} \tag{14.6}$$

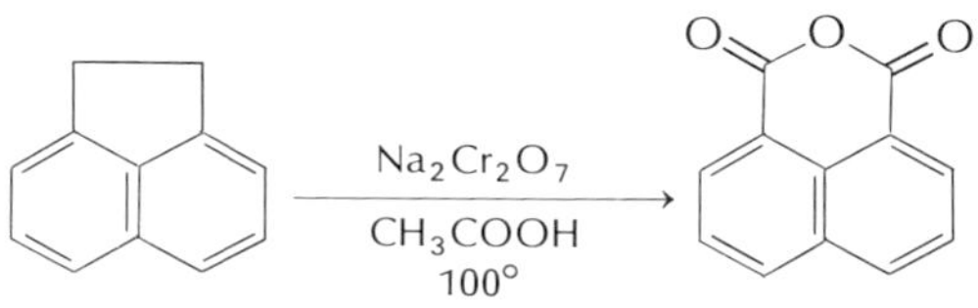

$$\text{acenaphthylene} \xrightarrow[\substack{\text{CH}_3\text{COOH} \\ 100°}]{\text{Na}_2\text{Cr}_2\text{O}_7} \text{naphthalic anhydride} \tag{14.7}$$

$$\text{benzene} \xrightarrow[450°-500°]{\text{O}_2, \text{V}_2\text{O}_5} \text{cis-HOOCCH=CHCOOH} \tag{14.9}$$

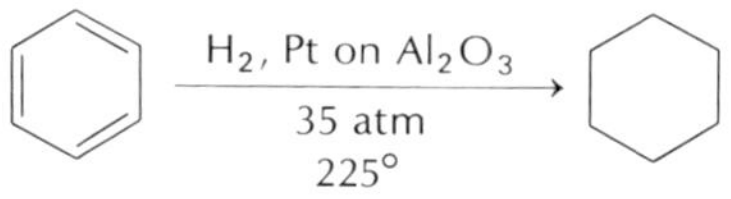

$$\text{benzene} \xrightarrow[\substack{35\text{ atm} \\ 225°}]{\text{H}_2, \text{Pt on Al}_2\text{O}_3} \text{cyclohexane} \tag{14.11}$$

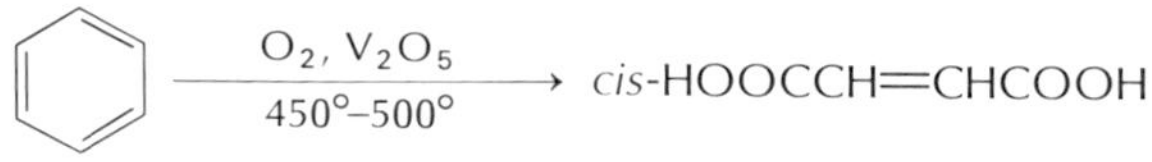

$$\text{RCH=CHCH=CH}_2 \xrightarrow[\text{R'OH}]{\text{Na}} \text{RCH}_2\text{CH=CHCH}_3 \tag{14.15}$$

$$\phi\text{CH=CHCH}_3 \xrightarrow{\text{HCl}} \phi\text{CHClCH}_2\text{CH}_3 \tag{Fig. 14.10}$$

$$\phi\text{CH=CHCH=CH}\phi \xrightarrow{\text{HCl}} \phi\text{CH=CHCHClCH}_2\phi \tag{Fig. 14.10}$$

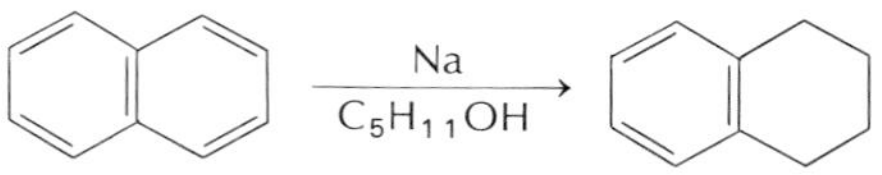

$$\text{naphthalene} \xrightarrow[\text{C}_5\text{H}_{11}\text{OH}]{\text{Na}} \text{tetralin} \tag{14.13}$$

$$\text{(2-methyl-1-(cinnamyl)benzene)} \xrightarrow{\text{oxidation}} \text{phthalic acid} + \text{CO}_2 + \phi\text{COOH} \tag{14.17}$$

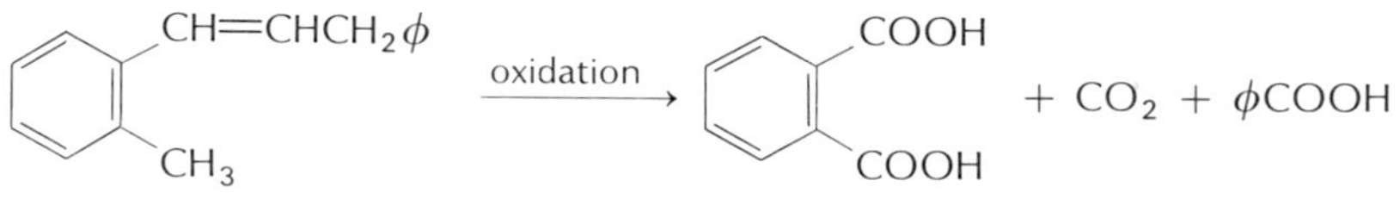

$$\text{H}_2\text{C=CHCH=CH}_2 \xrightarrow[\substack{\text{2. NH}_4\text{Cl} \\ -33°}]{\text{1. Na, NH}_3} \text{trans-CH}_3\text{CH=CHCH}_3 \quad 87\% \tag{14.18}$$

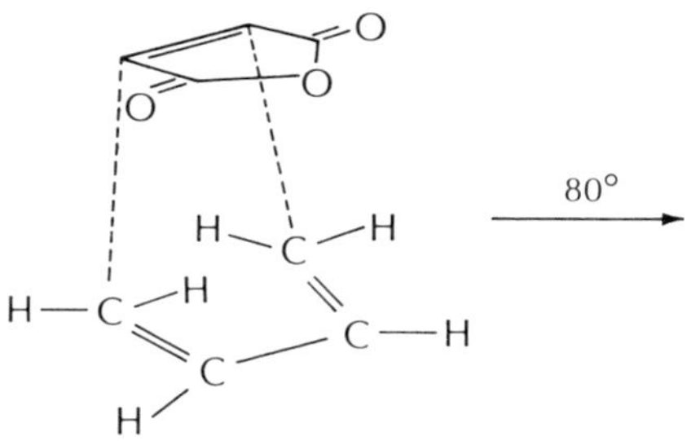

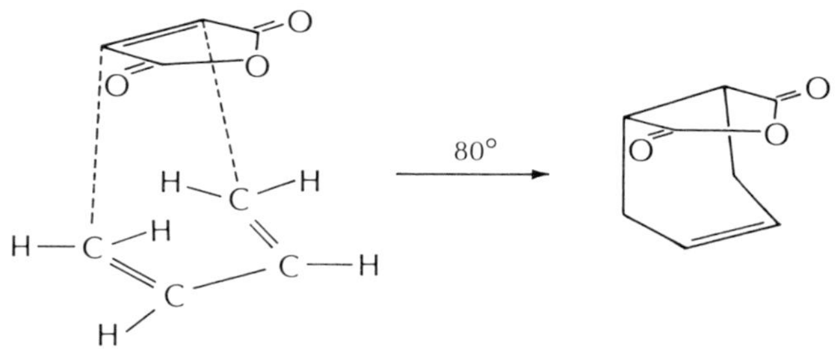

(14.19a)

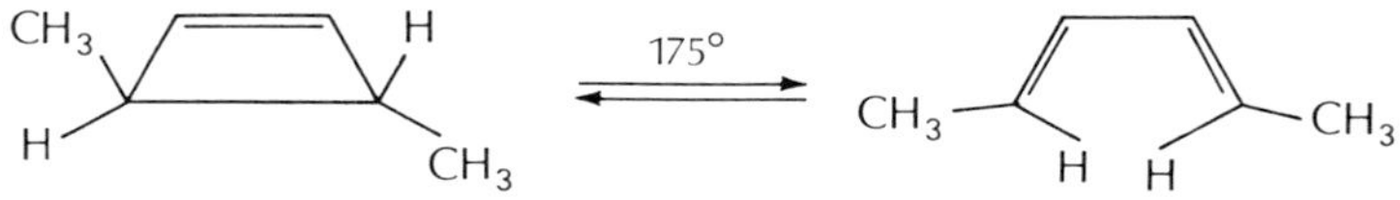

(14.19b)

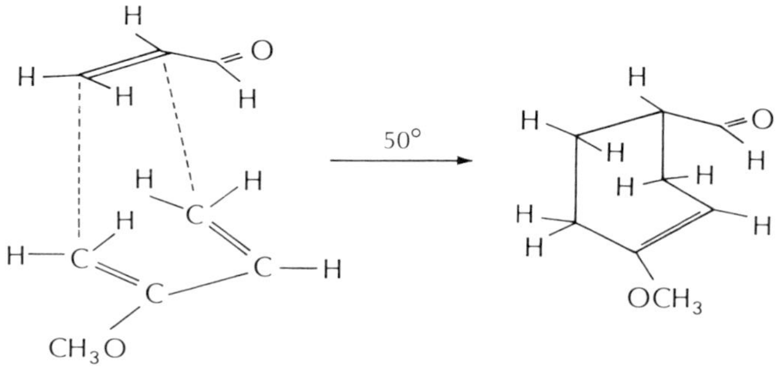

(14.19c)

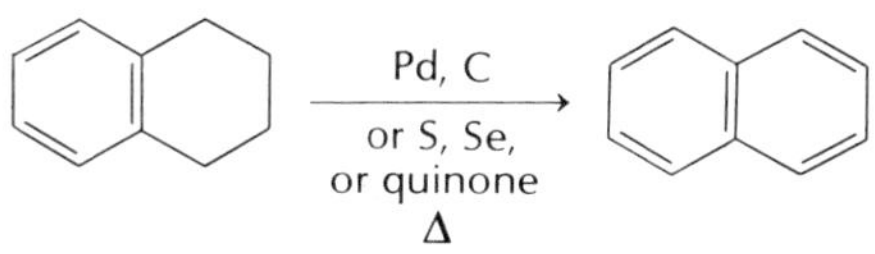

(14.27)

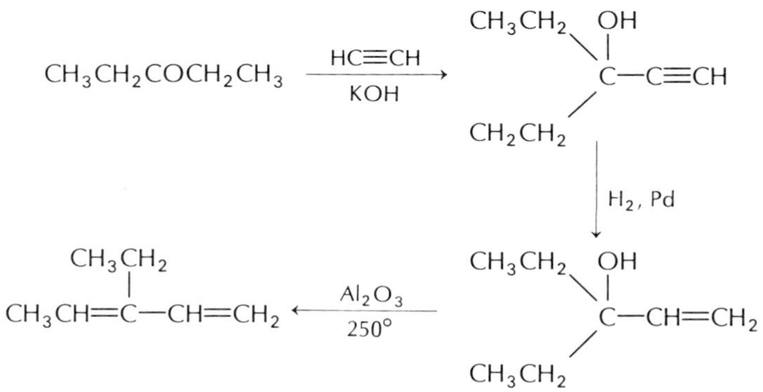

(14.29)

14.11 Problems

14.1 Substance A ($C_{10}H_{14}O$) is formed by reaction of C_4H_6 with C_6H_8O. Treatment of A with LAH gives B. B, with CS_2 and NaOH followed by CH_3I and heat, gives a mixture of C and D. Heating the mixture with Pd on C or with sulfur gives E ($C_{10}H_8$). Oxidation of A gives F ($C_{10}H_{14}O_5$). F with LAH gives G. Treatment of G with CS_2, NaOH, etc. as was done with B gives rise to H ($C_{10}H_{14}$). H with H_2 and Pt gives I. Heating with Pd on C or with sulfur gives J ($C_{10}H_{14}$). The nmr of J has an A_2B_2 pattern at about δ 7 ppm (area 4), a quadruplet at about δ 2.5 ppm (area 4), and a triplet at about δ 1.2 ppm (area 6).

Give a structural formula for each lettered substance.

14.2 Specify distinctly what is wrong with the structures shown:

a. 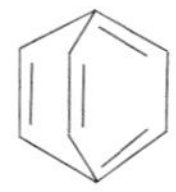 [bicyclic diene structure]   c.  [cyclopentene with triple bond]   e. 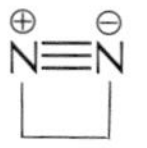 $\overset{\oplus}{N}\equiv\overset{\ominus}{N}$ [with ring]

b. 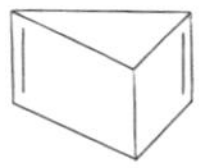 [bicyclic structure]   d. 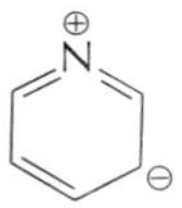 [pyridine N-oxide anion structure]   f. 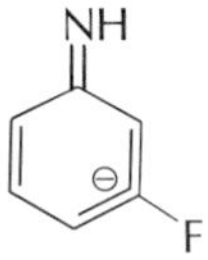 [benzene ring with NH and F substituents, anion]

14.3 What is the major product in each of the following reactions? Write the mechanism.

a. $CH_2=CH-CH=CH_2 + HBr \xrightarrow{40°}$

b. $CH_2=CH-CH=CH_2 + HBr \xrightarrow{-80°}$

c. $\underset{\underset{CH_3}{|}}{CH_2=C}-CH=CH_2 + HBr \xrightarrow{-80°}$

d. 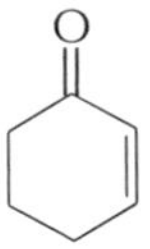 [cyclohexenone] $+ CH_2=CH-CH=CH_2 \longrightarrow$

e. $\underset{\underset{CH_3}{|}}{CH_2=C}-CH=CH_2 + \begin{array}{c} H \quad\quad COOCH_3 \\ \diagdown \quad \diagup \\ C \\ \| \\ C \\ \diagup \quad \diagdown \\ H \quad\quad COOCH_3 \end{array} \longrightarrow$

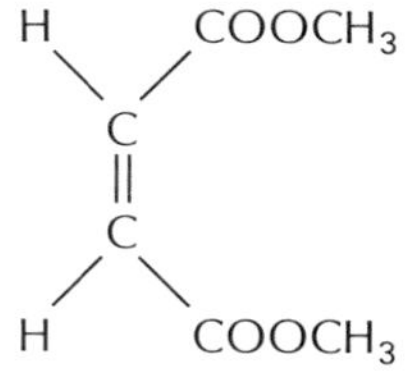

f.  [benzene] $+ CH_3OOC-C\equiv C-COOCH_3 \longrightarrow$

g. $CH_2=CH-CH=CH_2 + C_6H_5CH=CHCCH_3 \underset{(trans)}{\overset{O}{\|}} \longrightarrow$

438 Chapter fourteen

h. 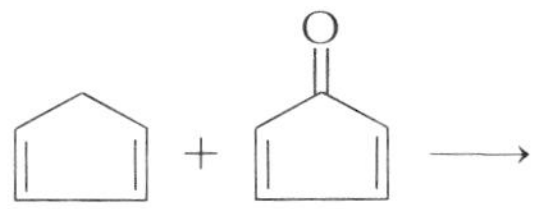 (cyclopentadiene) + (cyclohexenone) ⟶

(excess)

14.4 Starting with any unsaturated hydrocarbon(s) containing no more than five carbon atoms, how would you synthesize the following? Where it is critical, indicate solvent or temperature; any other appropriate organic molecule not containing a C=C or C≡C linkage can be employed.

a. 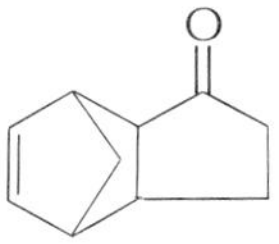 (tricyclic ketone structure)

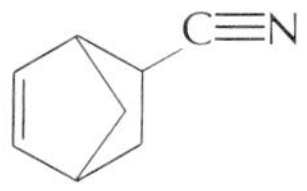

b. 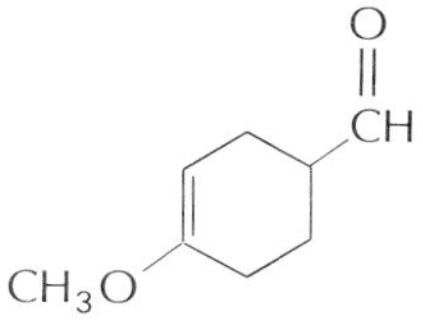 (6-methoxycyclohex-3-enyl carbaldehyde)

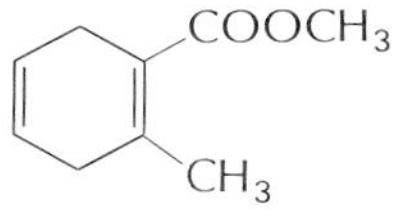

c. 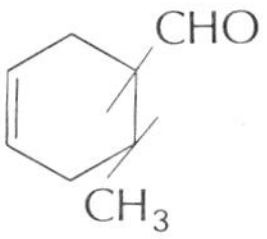 (methylcyclohexene with CHO group)

d. (norbornene-2-carbonitrile)

e. (methyl 6-methylcyclohexa-1,3-diene-1-carboxylate)

f. $BrCH_2CH(OCH_3)-CH=CH_2$

14.12 Bibliography

Beside the general organic textbooks and paperbacks already given in Section 7.12, several other references pertinent to the subjects discussed in this chapter are as follows:

1. March, Jerry, *Advanced Organic Chemistry*. New York: McGraw-Hill, 1968. A good advanced text book with concentration on reactions, mechanisms, and structure. We have reached the point in our study of organic chemistry at which such texts may be referred to with profit. Clearly written and with extensive documentation.

2. Vollmer, J. J., and K. L. Servis, *Journal of Chemical Education* 45, 214 (1968) and 47, 491 (1970). Cover the Woodman-Hoffman rules as applied to electrocyclic and cycloaddition reactions with excellent illustrations.

3. Hoffman, R., and R. B. Woodward, Orbital Symmetry Control of Chemical Reactions, *Science* 167, 825 (1970). A short review of the subject; should not be attempted unless reference 2 has been read first.

4. Pearson, R. G., Molecular Orbital Symmetry Rules, *Chemical and Engineering News*, Sept. 28, 1970, p. 66. Discusses the application of MO to reactions other than pericyclic.

14.13 William M. Jones: The professor knows best

Professor William M. Jones remained at the University of Southern California for one year after earning the Ph.D. there in 1955. Going to the University of Florida in 1956, he became professor in 1965 and chairman of the department in 1968. The youngest of our essayists, he has already begun to collect several honors and awards. Here, he tells of an instance in which he learned something.

In my years of research perhaps our most exciting discovery was the rearrangement of phenylcarbene I to cycloheptatrienylidene II. This came about in a rather unusual way, and its recounting may have a moral worth sharing.

By mid-1969, our research group had spent some time exploring the so-called aromatic carbene—cycloheptatrienylidene (II). Since, among its other unusual properties, this was expected to be a relatively stable carbene, we thought it would be interesting to see if it would be possible to deposit it on a cold surface for direct observation. After a number of failures, we decided to try to generate the carbene from the sodium salt of its tosylhydrazone at about 250° in a stream of dry nitrogen with the hope that the nitrogen would carry the carbene into a trap cooled with liquid nitrogen where it might be stable for a reasonable period of time. It was known that the carbene normally dimerizes to the fulvalene III, which has a very characteristic brownish-black color, so we felt it would be pretty easy simply to determine visually if the reaction had gone too far. Indeed, when the reaction was carried out as planned, the cold trap was found to contain a nearly white substance, which, upon warming to a bit below $-100°$, turned brownish-black. Needless to say, we were very excited because the dark colored material proved to be III, and we therefore thought we had successfully isolated the carbene (incidentally, we hadn't; and we still don't know what the white material is, but that's another story).

Now, before we determined that we had not actually isolated the carbene, my young collaborator, who was actually doing the laboratory work, thought this might be a general way to isolate carbenes and therefore decided to study phenylcarbene I in the same way.

When I learned of this, I strongly discouraged him because, with my years of experience in the carbene field, I was quite positive that such a reaction had no chance of success, that the products would be completely trivial, and that it would be a general waste of time. Fortunately, he was ready to do the experiment before telling me of his plans; and, since it was set up and ready to go, I decided it might be worthwhile as an object lesson to let him run the experiment and discover for himself his folly. As expected, upon removal of the liquid nitrogen from the trap, the receiver contained a light colored product. However, to my complete amazement, upon warming, it turned to a brownish-black. In fact, the color change was so characteristic that we knew at once that the carbene must have rearranged to cycloheptatrienylidene. Thus we were introduced to the area of carbene-carbene rearrangements that has been one of our most exciting adventures.

If this story has a moral, perhaps it is the value of surrounding oneself with bright, young coworkers who have enough gumption to carry out the experiment that their experienced (biased?) research director is sure isn't worthwhile.

15 Arenes

Here we take up the naming and reactions of compounds that contain the aromatic benzenoid ring. The electrophilic and nucleophilic substitution reactions of aromatic compounds make up the bulk of the chapter.

15.1 Naming of arenes and some common substituted benzenes 440
1. Arenes and substituted arenes. 2. Common names.
3. Disubstituted compounds. 4. Naphthalenes: common names.
5. Cigarette smoke. Exercises.

15.2 Substitution reactions: introduction 445
1. Use of resonance theory. 2. Electrophilic attack. 3. DNA bases.
4. Bromination: the rate step. 5. Mechanism; energy diagram.
6. Substituent effects: three classes. 7. Class 1. 8. Class 2.
9. Class 3. 10. I and R effects. Exercises.

15.3 Electrophilic substitution reactions 452
1. Activation and deactivation. 2. Halogenation. 3. Nitration.
4. Sulfonation. 5. Friedel-Crafts alkylation and acylation. 6. Mercuration.
7. Phenyl mercury fungicide. 8. Partial rate factors. 9. Other substituents; steric effects. 10. Hyperconjugation. 11. Effect of two substituents.
12. Polycyclics. Exercises.

15.4 Homolytic aromatic substitution reactions 458
1. Addition of radicals. 2. Hexabromobenzene. 3. Hydrogen abstraction in side chains. 4. Thyroxine.

15.5 Nucleophilic aromatic substitution reactions 460
1. Modified $S_N 2$ mechanism. 2. Effect of substituents. 3. Trinitro compounds. 4. Use of 2,4-dinitrofluorobenzene. 5. Role of solvent.
6. The benzyne pathway. 7. Commercial preparations. Exercises.

15.6 Oxidation and reduction of arenes 465
1. Indirect oxidation of side chains. 2. Autoxidation. Exercises.

440   *Chapter fifteen*

15.7   Sources of arenes   467
       1. Coal tar; coal.   2. Perkin's mistake.

15.8   Preparation and synthesis of arenes   468
       1. Control of electrophilic substitution.   2. Sulfonation.   3. Syntheses: the proper sequence of reactions.   4. Aromatic compounds. Exercises.

15.9   Summary   472

15.10 Problems   475

15.11 Bibliography   478

15.1 Naming of arenes and some common substituted benzenes

Arenes and substituted arenes. The naming of benzene compounds is not very systematic because of the persistence of many common names, not only for compounds but also for radical names. Hydrocarbons containing a benzenoid ring, as a class, are known as <u>arenes</u>. Although many names have been mentioned previously, let us put them down here in one spot, along with some examples of substituted arenes. We note that phenol

ϕH	ϕCH_3	ϕCH_2CH_3	$\phi CH(CH_3)_2$	$\phi C(CH_3)_3$
benzene	toluene	ethylbenzene	cumene	*tert*-butylbenzene

ϕCl	ϕCH_2Cl	$\phi CHCl_2$	ϕCCl_3
chlorobenzene	benzyl chloride	benzal chloride	benzo trichloride

ϕOH	ϕCH_2OH	ϕCHO	$\phi COOH$
phenol	benzyl alcohol	benzaldehyde	benzoic acid

$\phi_2 CHCl$	$\phi_2 CHOH$	$\phi_3 CCl$	$\phi_3 COH$
benzhydryl chloride	benzhydrol	triphenylmethyl chloride	triphenylmethanol

contains a hydroxy group, as the ending -ol indicates. However, phenols are different enough from alcohols in their reactions so that they are considered a new class.

The radical names are illustrated by the chloride names just listed: phenyl, ϕ—; benzyl, ϕCH_2—; benzhydryl, $\phi_2 CH$—; triphenylmethyl, $\phi_3 C$—; benzal, $\phi CH\diagdown$; and benzo, $\phi C\diagdown$.

15.1 Naming of arenes and some common substituted benzenes

Common names. If a phenyl is substituted into an open chain or another cyclic compound, the most convenient name is used. Thus $\phi CH_2CH_2CH_2OH$ is 3-phenylpropanol, $\phi CHOHCH_2CH_3$ is 1-phenylpropanol, and $\phi CH_2CHOHCH_3$ is 1-phenyl-2-propanol. In ketones a peculiarity in naming arises. Ethyl phenyl ketone ($CH_3CH_2CO\phi$), for example, is also known as propiophenone. The prefix propio- shows that the chain is related to propionic acid (CH_3CH_2CO-OH). The rest of the name, -phenone, shows the presence of the phenyl ketone grouping. Thus, $\phi COCH_3$ is acetophenone, $\phi COCH(CH_3)_2$ is isobutyrophenone, and $\phi CO\phi$ is benzophenone.

Disubstituted compounds. If two substituents are present on the ring, the prefixes ortho-, o-, or 1,2- are used interchangeably, as are meta-, m-, or 1,3-; and para-, p-, or 1,4-. Again, common names abound and are approved by IUPAC in many instances. We just have to memorize them as we encounter them later on. The listing given is not intended to be complete, even though a few new polycyclic compounds have been included. Those

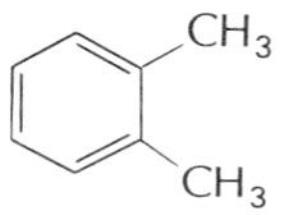

o-xylene* 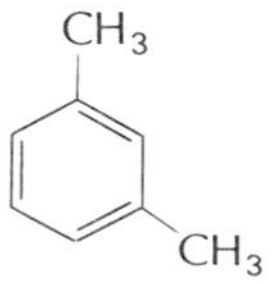 m-xylene* 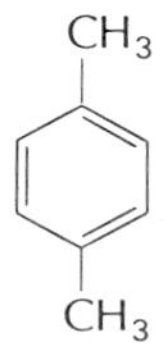 p-xylene* 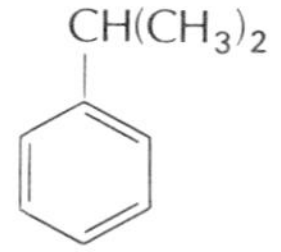 cumene*

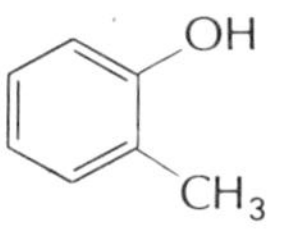

o-cresol 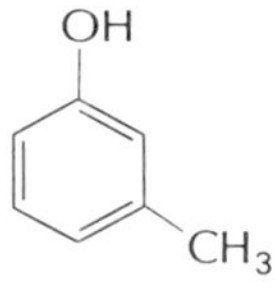 m-cresol 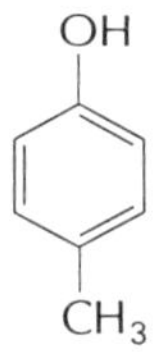 p-cresol 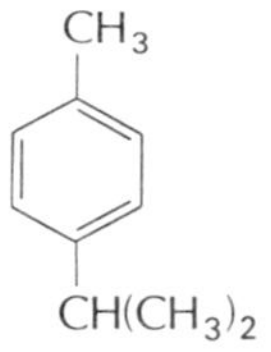 p-cymene*

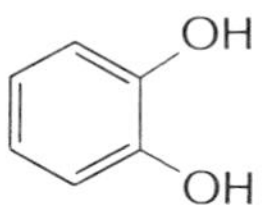

catechol or pyrocatechol resorcinol  hydroquinone 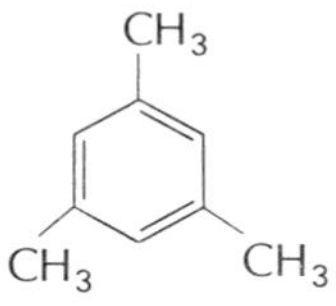 mesitylene*

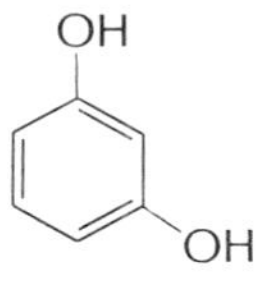

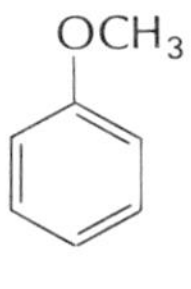

anisole 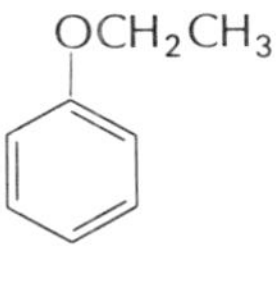 phenetole 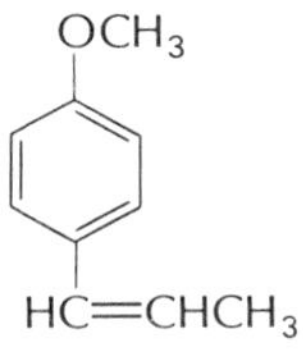 anethole 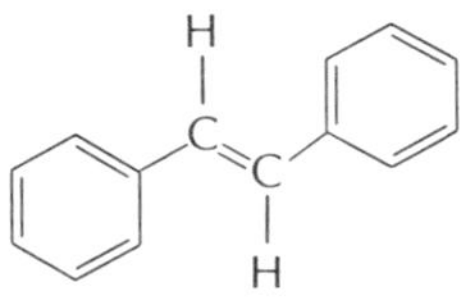 trans-stilbene*

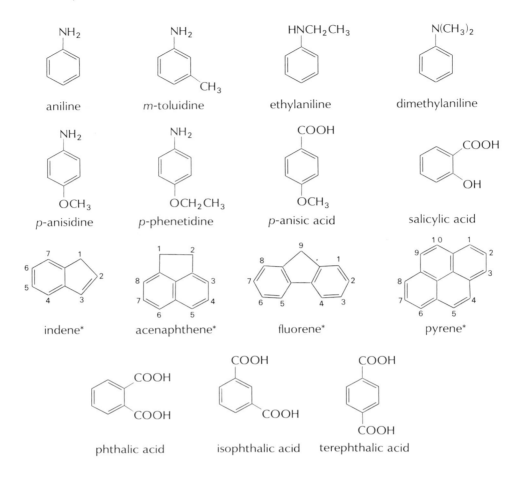

names with an asterisk are approved by IUPAC as exceptions to systematic naming. We shall be running into other compounds with common names as we go along. Names like o-chlorotoluene, m-methylbenzoic acid, or p-ethylphenol should be obvious.

When we encounter tri- and tetrasubstituted benzenes, we shall use numbers to designate positions as shown, except that number 1 is omitted.

A substituted phenyl as a substituent in an open-chain molecule is symbolized as Ar- (from aryl) instead of by ϕ-. Thus, if instead of phenylacetone (ϕCH$_2$COCH$_3$, also known as benzyl methyl ketone) we wish to refer to p-chlorophenylacetone in a general way, we shall call it an arylacetone, or ArCH$_2$COCH$_3$.

15.1 Naming of arenes and some common substituted benzenes

Naphthalenes: common names. Monosubstituted naphthalenes are named as α- or β- because only the two possibilities exist.

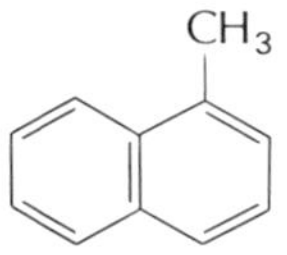

α-methylnaphthalene 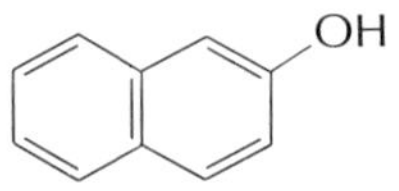 β-naphthol 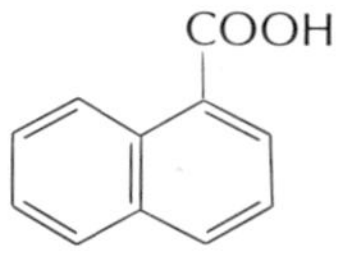 α-naphthoic acid

For di- and trisubstituted naphthalenes and so on, we shall use numbers.

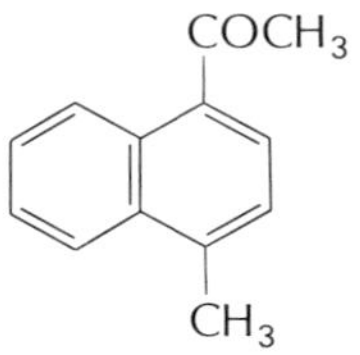

4-methyl-1-aceto-naphthone 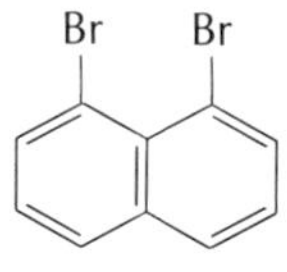   1,8-dibromonaphthalene 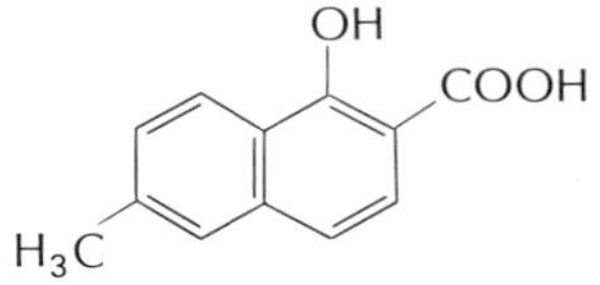   1-hydroxy-6-methyl-2-naphthoic acid

For all substituted anthracenes and phenanthrenes we shall use numbers only, even though other systems of naming exist.

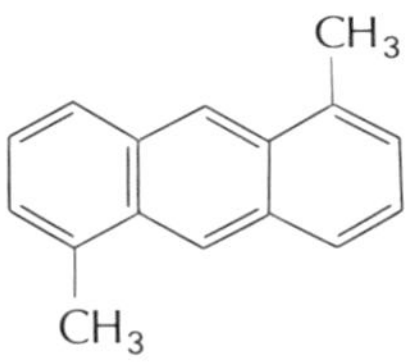

1,5-dimethylanthracene 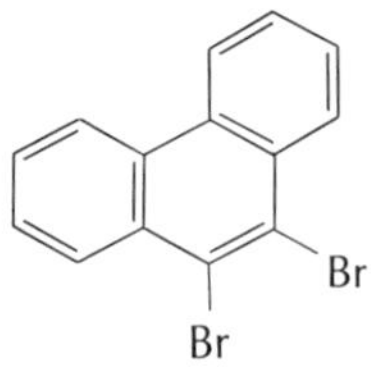   9,10-dibromophenanthrene

Table 15.1 lists some arenes to give an idea of the range of mp and bp to be expected.

Table 15.1 Arenes

Name	Formula	mp, °C	bp, °C
benzene	C_6H_6	5.4	80.1
toluene	$C_6H_5CH_3$	−93	110.6
o-xylene	$1,2\text{-}(CH_3)_2C_6H_4$	−28	144
m-xylene	$1,3\text{-}(CH_3)_2C_6H_4$	−54	139
p-xylene	$1,4\text{-}(CH_3)_2C_6H_4$	13	138
mesitylene	$1,3,5\text{-}(CH_3)_3C_6H_3$	−57	165
durene	$1,2,4,5\text{-}(CH_3)_4C_6H_2$	80	195
pentamethylbenzene	$C_6H(CH_3)_5$	53	231
hexamethylbenzene	$C_6(CH_3)_6$	166	265
ethylbenzene	$C_6H_5CH_2CH_3$	−93	136
n-propylbenzene	$C_6H_5CH_2CH_2CH_3$	−100	159.5

Table 15.1 (continued)

Name	Formula	mp, °C	bp, °C
cumene	$C_6H_5CH(CH_3)_2$	−96	152
tert-butylbenzene	$C_6H_5C(CH_3)_3$	−58	169
p-cymene	$p\text{-}CH_3C_6H_4CH(CH_3)_2$	−68	177
styrene	$C_6H_5CH{=}CH_2$	−31	145
allylbenzene	$C_6H_5CH_2CH{=}CH_2$	−40	156
stilbene (trans)	$C_6H_5CH{=}CHC_6H_5$	124	307
diphenylmethane	$(C_6H_5)_2CH_2$	27	262
triphenylmethane	$(C_6H_5)_3CH$	92.5	359
biphenyl	$C_6H_5\text{—}C_6H_5$	70.5	254
indene	C_9H_8	−2	181
naphthalene	$C_{10}H_8$	80	218
acenaphthene	$C_{12}H_{10}$	95	278
fluorene	$C_{13}H_{10}$	114	295
anthracene	$C_{14}H_{10}$	216	354
phenanthrene	$C_{14}H_{10}$	101	340
pyrene	$C_{16}H_{10}$	151	393

Cigarette smoke. Some of the aromatic compounds in cigarette smoke are known to cause cancer in mice and are thought by some to be responsible for the greater than average incidence of lung cancer among smokers. The compounds are adsorbed on the particulate matter, some of which becomes trapped in the lung tissue. The most studied is benzo[a]pyrene. Another is 5-methylchrysene.

Activated charcoal tends to adsorb aromatic compounds. In theory, at least, it would be possible to design a charcoal filter to remove all these aromatics. Unfortunately, nicotine, the component that the smokers crave, is also an aromatic compound.

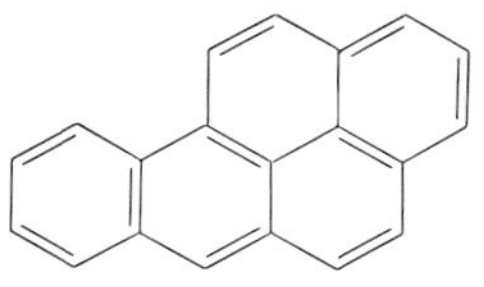

benzo[a]pyrene 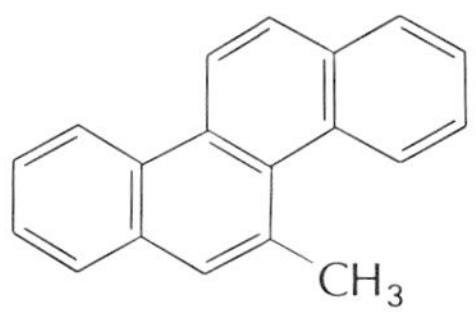   5-methylchrysene 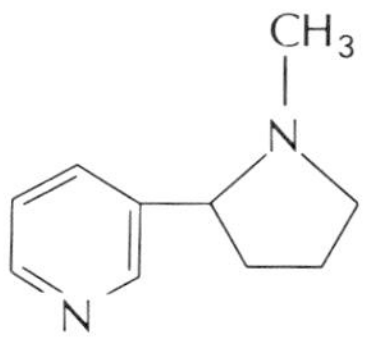   nicotine

Exercises

15.1 Write structural formulas for:
 a. 2,3′-dinitro-*trans*-stilbene
 b. 3-bromo-1-(3,4-dimethoxyphenyl)-1-propanone
 c. β-naphthaldehyde
 d. 1,1-dimethyl-4-acenaphthenol
 e. 3-bromo-5-chloro-4-nitroaniline

15.2 Give names for:

a. 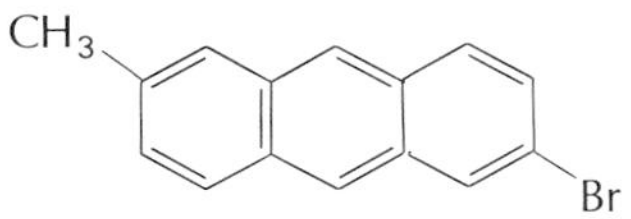 [structure: anthracene with CH₃ and Br substituents]

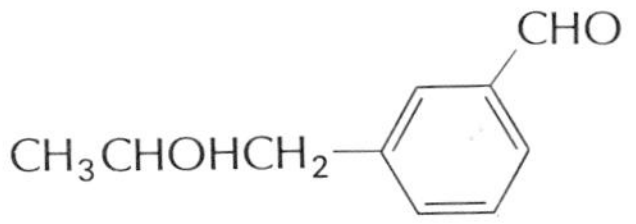

b. 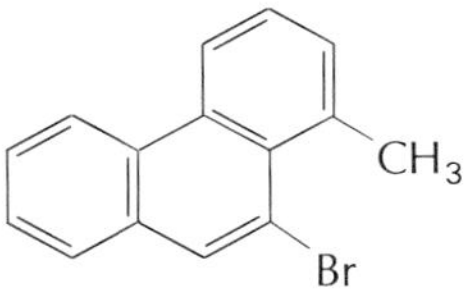 [structure: phenanthrene with CH₃ and Br substituents]

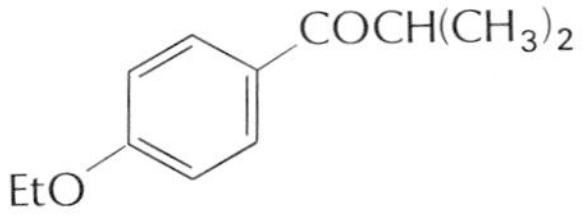

c. 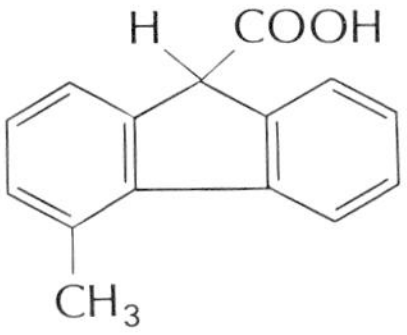 [structure: fluorene with H, COOH, and CH₃ substituents]

d. CH₃CHOHCH₂—[benzene ring with CHO]

e. EtO—[benzene ring]—COCH(CH₃)₂

15.2 Substitution reactions: introduction

Use of resonance theory. We have seen that benzene undergoes a few addition reactions, but the usual result of attack by nucleophilic or electrophilic reagents or by radicals is a substitution. If a side chain or other substituents are already present, complications arise. We shall postpone a description of these situations.

Benzene has a cloud of electrons above and below the ring. Like ethylene, it is a weak nucleophile. Unlike the case of ethylene, attack by a reagent on benzene must break into the aromatic delocalization energy. At some point in the reaction, we come to a consideration of the resonance structures shown in Equations 15.1, either as intermediates or as

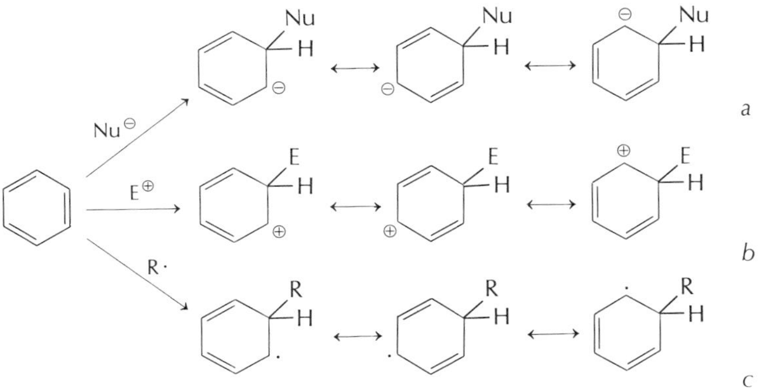

[reaction scheme showing benzene attacked by Nu⁻, E⁺, or R· giving three sets of resonance structures labeled a, b, c] (15.1)

transition states. As the reaction goes from benzene to a pentadienyl resonance, we see that resonance theory predicts that each one should require about the same energy. Actually, it is found that, as with ethylene, electrophilic attack is easier than radical attack and that nucleophilic attack is rather difficult. We shall discuss only electrophilic attack (*b*) before going on to the others. The benzenonium ion (b) is also called a σ complex and is sometimes symbolized as shown.

Electrophilic attack. Iodine vapor is violet, and a solution of iodine in carbon tetrachloride also is violet. The addition of benzene to such a solution causes the color to turn to brown, an indication of reaction. The product has a different absorption spectrum from that of molecular iodine. The product, $\phi H \cdot I_2$, is called a π complex or a charge-transfer complex. This complex is held together somewhat more strongly than if only the London attractive forces were in operation. We can explain spectra in terms of a mixing (or a partial transfer) of the electrons from the π cloud of benzene into the influence of the I_2 molecule. This particular complex is easily separated and benzene and iodine recovered. Many electrophiles and many arenes form charge-transfer complexes. The structure of the π complexes varies: that of benzene with silver ion from $AgClO_4$ has the ion above one of the bonds between adjacent carbon atoms of the ring, and that with bromine has the molecule lying on the axis of the ring (above the center of the ring).

DNA bases. The code of life is contained in giant DNA molecules in the chromosomes. The backbone of the DNA is a polymer of alternating deoxyribose and phosphate units. In the chromosome, two DNA strands spiral around each other in the famous double helix. (See Fig. 28.1.) Each deoxyribose carries one of four possible aromatic bases, which serve as the information bits in the code. Three successive bases have been shown to code for one amino acid. (See Table 28.2.) Thus 600 bases code for a protein with 200 amino acids.

Each base on one DNA strand is hydrogen bonded to a base on the sister strand of the double helix. However, because only adenine-thymine and cytosine-guanine base pairing occur, the sequences of the two strands complement each other. During DNA replication, the strands are thought to unwind and each to determine the pattern of a new complement strand for itself (Section 28.7).

A large number of aromatic compounds are carcinogens or mutagens. They must interact with the genetic machinery in some manner. One proposal is that the aromatic compounds are binding between two aromatic bases, disrupting the native structure of the molecule significantly.

Bromination: the rate step. It seems reasonable to postulate that the electrophilic substitution reaction of bromine with benzene proceeds as shown in Equations 15.2. In *a* the steps are emphasized, while in *b* the structures are emphasized. The atomic *p*

$$\phi H + Br_2 \rightleftharpoons \phi H \cdots Br_2$$
π complex
$$+ Br_2$$

$$C_6H_6Br^{\oplus} + Br_3^{\ominus} \quad (15.2)$$
σ complex

$$\phi Br \xleftarrow{-HBr, -Br_2} \phi Br \cdots \overset{\oplus}{H} \cdots Br_3^{\ominus} \qquad a$$
π complex

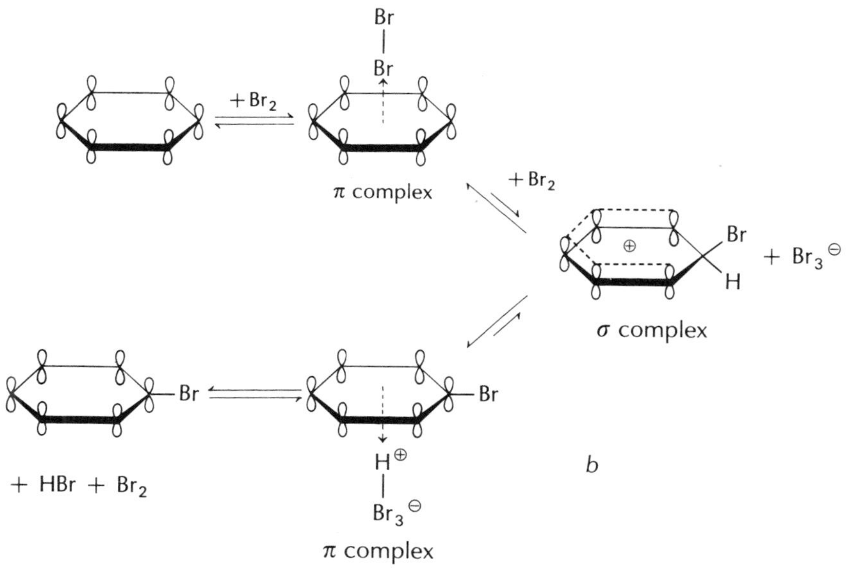

orbitals in *b* are a schematic representation. Overlap is partially shown only in the σ complex. The slow step is the conversion of the initial π complex into the σ complex. The second molecule of bromine seems necessary to aid the departure of the $Br^{\ominus}$. It is known that Lewis acids such as $FeBr_3$ accelerate the reaction, in which case the initial π complex may be formulated as $\phi H \cdots Br—Br \cdots FeBr_3$, which goes on to give $C_6H_6Br^{\oplus} + FeBr_4^{\ominus}$

Mechanism; energy diagram. It is known that a C—H bond-breaking reaction proceeds about five to seven times faster than does a C—D bond-breaking reaction under identical conditions (the <u>kinetic isotope effect</u>). Therefore, the decomposition of the σ complex in the third step cannot be rate determining, because both C_6H_6 and CD_6 react at the same rate. Thus we see that either the first step leading to the π complex or the second step leading to the σ complex must be the rate-determining step, the slowest in the sequence of steps. The first step cannot be the slow step because the π complex between benzene and bromine is formed rapidly at low temperatures, under which condition little or no substitution is observed. We come to the conclusion that the reaction proceeds as follows: an equilibrium concentration of the π complex is quickly reached, and the complex undergoes a slow conversion into the σ complex. The σ complex may revert to the initial π complex or decompose into the second π complex, in which the displaced proton is associated with the π cloud. The proton π complex then falls apart to give the products: bromobenzene, HBr, and $FeBr_3$.

The energy diagram of this reaction sequence is shown in Figure 15.1. Using bond-energy values from Table 5.6, we calculate that ΔH for $\phi H + Br_2 \rightarrow \phi Br + HBr$ is −10 kcal. The rest of the diagram is schematic and designed to show the small activation energies for the equilibria involving benzene and bromobenzene with the π complexes and the large activation energies needed for either π complex to climb to the σ-complex level. The reaction is irreversible because bromobenzene and hydrogen bromide cannot be induced to give benzene and bromine.

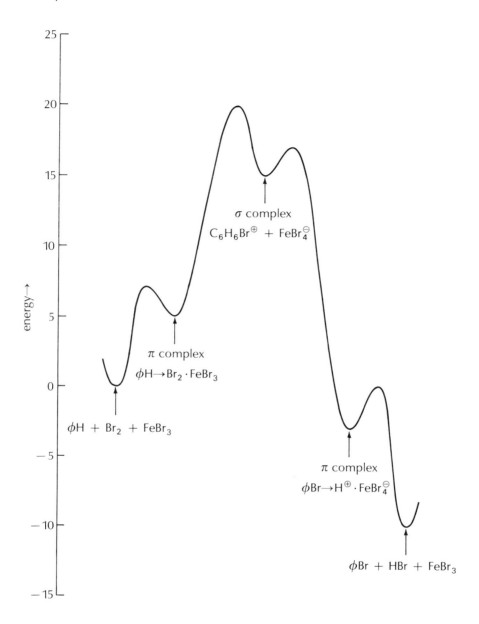

Figure 15.1 Schematic energy-reaction-coordinate diagram for $\phi H + Br_2 \rightarrow \phi Br + HBr$ catalyzed by $FeBr_3$

Substituent effects: three classes. Now, we ask ourselves, what if there is a substituent already on the ring? Will the reaction go faster or slower? Will the entering substituent end up *o, m,* or *p* to the original substituent? Will the original substituent affect the reaction in some unexpected fashion? There is no simple answer: we did not pose the questions properly. There is no such thing as a substituent in a general sense, and we must ask these questions for each different substituent. Qualitatively, we find that substituents may be classified into three groups based on their effects upon electrophilic substitution reactions.

1. *Substituents that cause the reaction to go more rapidly than with benzene.* The entering substituent ends up o or p more often than m to the original substituent. The ratio of o to p is variable.
2. *Those that cause the reaction to go more slowly than with benzene.* The entering substituent ends up m more often than o or p to the original substituent.
3. *The halogens and a few other substituents, which cause the reaction to go more slowly than with benzene.* The entering substituent ends up more o or p than m to the original halogen substituent.

There are no substituents that cause the reaction to go rapidly and cause the entering substituent to go to the m rather than to the o or p positions.

Some examples of substituents in each of the three classes are as follows:

1. o, p, activating: R—, ϕ—, H$_2$N—, R$_2$N—, RCONH—, HO—, RO—, RCOO—;
2. m, deactivating: —CHO, —COR, —COOH, —COOR, —CX$_3$, —C≡N, —NO$_2$, —SO$_3$H, —NH$_3^{\oplus}$, —NR$_3^{\oplus}$;
3. o, p, deactivating: F—, Cl—, Br—, I—, —CH$_2$Cl, —CH=CHCOOH.

Class 1. Let us examine an example from each of the classes to see whether we can make sense of these effects. The methoxyl group is in Class 1, so let us use resonance theory to look at the various possibilities in the bromination of anisole. We remember that the influence must come to bear on the conversion of the π complex to the σ complex. As usual, we shall try to estimate the relative stabilities of the product of this step, the σ complex. Because the transition state lies closer to the σ complex than to the π complex, we then apply the estimate to make an educated guess about the relative magnitude of the activation energies.

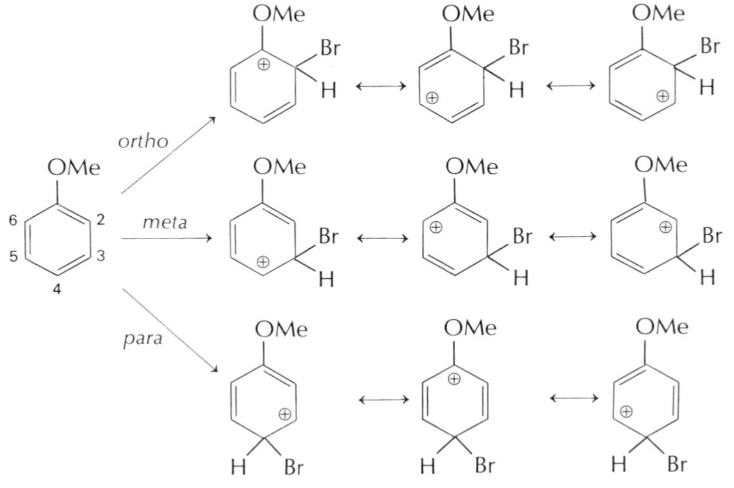

In Equations 15.3 are shown the resonance structures of the σ complexes for attack at one of the two *ortho* positions, at one of the two *meta* positions, and at the *para* position to the methoxyl group. We see that attack at either the o or p position places the positive charge at the 1, 3, and 5 positions. Attack at the m position places the positive charge at the 2, 4, and 6 positions. So far, there is little to choose among the three possibilities. Looking at the methoxyl group, however, we see that in a C—O bond the oxygen atom exerts a −I effect, pulling the electrons of the σ bond away from the carbon atom. In addition, if there is a vacant p orbital adjacent to and parallel to one of the filled p orbitals on the oxygen atom, a double bond forms (C=O$^\oplus$). This effect, we remember, is called a +R effect. Any atom bonded to the ring (O in φOMe) and having a nonbonded pair of electrons (again, like O) exerts a +R effect.

In *meta* attack, the p orbital on the carbon atom adjacent to the oxygen atom (C_1) is occupied, so only the −I effect of the oxygen atom can operate. Because a positive charge already is present in the pentadienyl system, the −I effect is in opposition to the formation of the σ meta complex. The +R effect can operate in this case only with charge separation, which, as we have seen in other cases, is not a serious contributor to the ground state. One possible structure is shown.

meta attack

In *o* or *p* attack, a positive charge appears at C_1 in one of the three resonance structures. Although this form is disfavored by the −I effect, it is strongly favored by the +R effect as shown. This contributing structure lowers the energy of the σ complex and thus

ortho attack   *para* attack

lowers the activation energy leading to it. It follows that the activation energy for *o, p* attack is less than that for *m* attack. Thus, more *o, p* product than *m* product is formed. Furthermore, the overall rate should be greater than for attack on benzene. Also, because the *o* position is adjacent to the original substituent, it is more sterically crowded than is the *p* position. Therefore *p* is favored over one *o* position. Because there are two *ortho* positions, the final proportion is in doubt. We can say, though, that the larger the substituent, the more hindered is *ortho* attack. Qualitatively, we may conclude that Class 1 substituents all have a sizable +R effect, larger in magnitude than the I effect, whether positive or negative.

Class 2. Class 2 substituents are of two types: one is exemplified by a keto group (say CH_3CO—), the other by $Me_3\overset{\oplus}{N}$—. Again we make use of Equation 15.3, imagining the —OMe replaced by CH_3CO— (which also exerts a −I effect) or by —$\overset{\oplus}{N}(Me)_3$ (with a very strong −I effect). The keto group, unlike methoxyl, exerts a −R effect. The —$\overset{\oplus}{N}(Me)_3$ is incapable of an R effect, either positive or negative. In *meta* attack, as before, the positive charge does not appear at C_1, whereas in *o, p* attack, the positive charge may appear there.

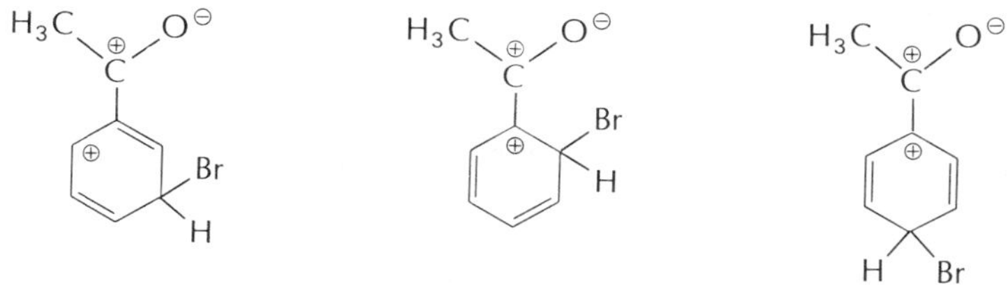

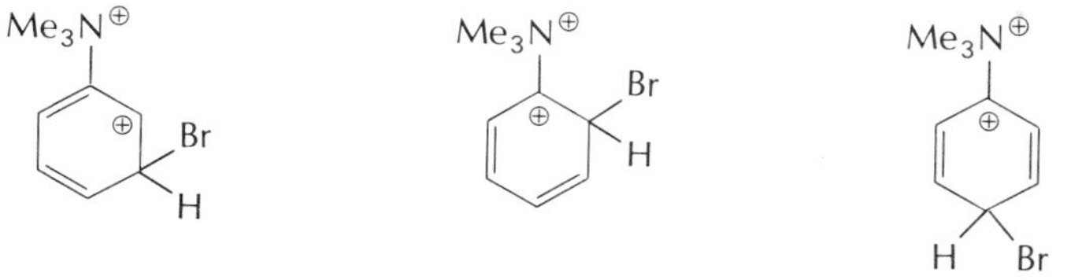

[Structures showing σ-complex intermediates for o, m, p attack with acetyl (top row) and trimethylammonium (bottom row) substituents]

Therefore, o, p attack is disfavored more than is m attack, and the reaction as a whole is slower than that with benzene. Class 2 substituents all have a −I effect, and some have a −R effect as well.

Class 3. Class 3 substituents are like those of Class 1 in being +R and −I, but they differ in that the magnitude of −I is greater than that of +R. The overall effect, like that of the Class 2 substituents, is a withdrawal of electrons from the ring in the ground state, a shift of the equilibrium between reactants and the π complex to decrease the concentration of π complex, and a slowing down of the rate of conversion to the σ complex. On the other hand, as with Class 1 substituents, the +R effect favors o, p orientation over m orientation in the transition state for the conversion of the π to the σ complex.

I and R effects. The resonance interpretation leads us to these general conclusions:

1. If the overall effect of I plus R is negative, electron withdrawal from the ring is sufficient to cause the rate of reaction to be slower than that with benzene.
2. If the overall effect of I plus R is positive, electron donation to the ring is sufficient to cause the rate of reaction to be faster than that with benzene.
3. Orientation of the product is determined by R. If R is positive, electron donation is sufficient to favor o, p orientation. If R is negative, electron withdrawal favors m orientation. If R is absent, the I effect determines the orientation (usually m because of a strong −I effect caused by the presence of a positive charge, as in $-\overset{\oplus}{\text{N}}(\text{Me})_3$).

Exercises

15.3 Show the major product to be expected from bromination in the presence of FeBr$_3$ of:

a. 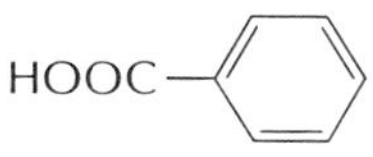 HOOC—⟨benzene ring⟩

b. 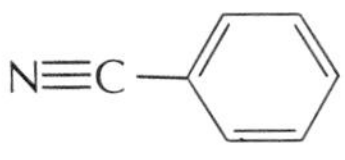 N≡C—⟨benzene ring⟩

c. OHCCH$_2$—⟨benzene ring⟩

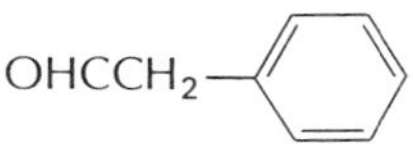

d. 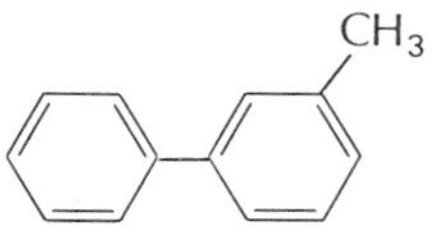 ⟨benzene ring⟩—⟨benzene ring with CH$_3$⟩

452   Chapter fifteen

e. 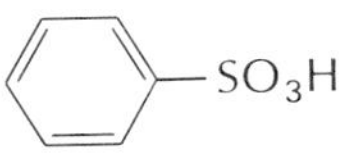 ⌬—SO₃H

f. φCH=CHCOOH

g. φCHCl₂

h. 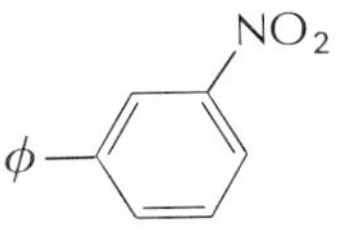 φ—⌬-NO₂ (meta)

i. 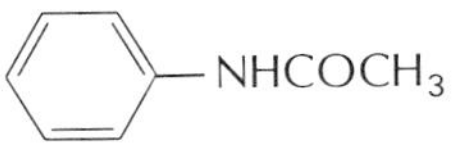 ⌬—NHCOCH₃

j. 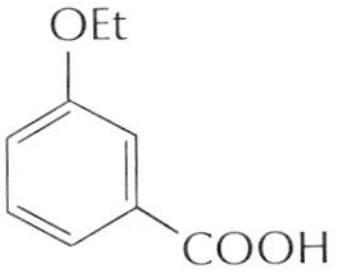 3-OEt, 1-COOH benzene

---

15.3   Electrophilic substitution reactions

Activation and deactivation.   Benzene reacts with various electrophilic reagents to give φCl, φBr, φNO₂, φSO₃H, φR, φCOR, and φHgOOCCH₃. Activated aromatic rings react with vigor, and deactivated rings react slowly or not at all in a reasonable time. By activation we mean the presence of strong *o*, *p*-directing substituents (large +R); deactivation is the presence of strong *m*-orienting substituents (large −I or large −I and −R). Activated rings also react with milder electrophilic reagents, a subject we shall discuss later in the book under the specific substituents (phenols, amines, and so on).

The many products mentioned above are obtained by the choice of reagent and conditions for five electrophilic reactions. With one exception, they seem to follow a mechanism similar to that outlined in the previous section. The five reactions are halogenation, nitration, sulfonation, Friedel-Crafts alkylation and acylation, and mercuration.

Halogenation.   Halogenation of benzene is slow unless a polar solvent and a Lewis acid catalyst (FeX₃, AlX₃) are present. Usually, chlorine or bromine is added directly to the reaction mixture. Fluorination is carried out by indirect routes, because direct fluorination has a tendency to tear the ring apart. Iodine is so weakly electrophilic that it is usually regarded as nonreactive. If an oxidizing agent is present (usually nitric acid or hydrogen peroxide) to convert iodine to the more electrophilic HOI (Eq. 15.4), iodination

$$I_2 + H_2O_2 \longrightarrow 2HOI \qquad (15.4)$$

may be accomplished (Eq. 15.5).

$$\phi H + IOH_2^{\oplus} \longrightarrow \phi I + H_3O^{\oplus} \qquad (15.5)$$

Nitration.   It seems certain that nitration is the result of attack by the nitronium ion (NO₂⁺). The usual procedure is to use a mixture of nitric and sulfuric acids. In the absence of water (or with water in such small amounts that it is all converted to H₃O⁺), nitric acid acts as a base to the stronger sulfuric acid (Eqs. 15.6). Although *a* appears to be

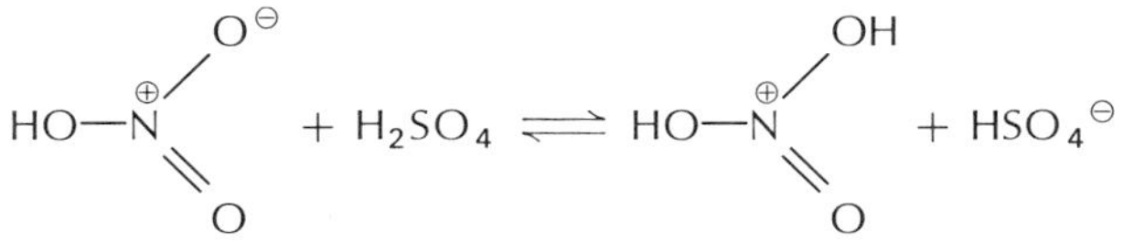

$$HO-\overset{\oplus}{N}(O^{\ominus})=O + H_2SO_4 \rightleftharpoons HO-\overset{\oplus}{N}(OH)=O + HSO_4^{\ominus} \qquad a \qquad (15.6)$$

and

$$HO-NO_2 + H_2SO_4 \rightleftharpoons H_2\overset{\oplus}{O}-NO_2 + HSO_4^{\ominus} \qquad b$$

$$H_2\overset{\oplus}{O}-NO_2 \rightleftharpoons H_2O + NO_2^{\oplus} \qquad c$$

$$H_2O + H_2SO_4 \rightleftharpoons H_3O^{\oplus} + HSO_4^{\ominus} \qquad d$$

or, adding b, c, and d,

$$HNO_3 + 2H_2SO_4 \rightleftharpoons NO_2^{\oplus} + H_3O^{\oplus} + 2HSO_4^{\ominus} \qquad e$$

perfectly reasonable, it is swamped by e. To see why, we remember that water is a strong base to sulfuric acid, so much so that d lies far to the right. This reaction removes water from the right hand side of c, causing all the $H_2\overset{\oplus}{O}-NO_2$ to decompose into H_2O and $NO_2^{\oplus}$. Loss of $H_2\overset{\oplus}{O}-NO_2$ causes b in turn to shift to the right. Then a shifts to the left to maintain an equilibrium concentration of HNO_3. Thus the magnitude of K_a for sulfuric acid (d) causes e to lie far to the right. The melting-point depressions of solutions of nitric acid in sulfuric acid are twice those expected from a, which shows that one mole of nitric acid produces four ions in the solution instead of two.

There are other ways of producing $NO_2^{\oplus}$ (Eqs. 15.7), and nitration can be

$$N_2O_5 + H^{\oplus} \longrightarrow HNO_3 + NO_2^{\oplus} \qquad a$$

$$CH_3COONO_2 + H^{\oplus} \longrightarrow CH_3COOH + NO_2^{\oplus} \qquad b$$

$$NO_2Cl + AgBF_4 \longrightarrow AgCl + BF_4^{\ominus} + NO_2^{\oplus} \qquad c \qquad (15.7)$$

$$\phi H + NO_2^{\oplus} \longrightarrow \phi NO_2 + H^{\oplus} \qquad d$$

observed with these reagents. Some nitrations seem to be governed by the rate of π-complex formation instead of the rate of σ-complex formation, as we shall see later.

Sulfonation. Sulfonation may be carried out with sulfuric acid or, more vigorously, with fuming sulfuric acid (SO_3 dissolved in H_2SO_4). Sulfuric acid undergoes a series of equilibria (Eqs. 15.8). They indicate that SO_3 is the active electrophilic reagent, although

$$H_2SO_4 + H_2SO_4 \rightleftharpoons H_3SO_4^{\oplus} + HSO_4^{\ominus} \qquad a$$

$$H_3SO_4^{\oplus} \rightleftharpoons H_2O + HSO_3^{\oplus} \qquad b$$

$$H_2SO_4 + H_2O \rightleftharpoons H_3O^{\oplus} + HSO_4^{\ominus} \qquad c$$

$$HSO_3^{\oplus} + HSO_4^{\ominus} \rightleftharpoons SO_3 + H_2SO_4 \qquad d \qquad (15.8)$$

adding gives

$$H_2SO_4 + H_2SO_4 \rightleftharpoons H_3O^{\oplus} + HSO_4^{\ominus} + SO_3 \qquad e$$

$HSO_3^{\oplus}$ may take part as well. The product is a sulfonic acid (Eqs. 15.9). If one of the acid

$$\phi H + H_2SO_4 \rightleftharpoons \phi SO_3H + H_2O$$
$$\phi H + SO_3 \rightleftharpoons \phi SO_3H \quad (15.9)$$

halides of sulfuric acid is used instead of H_2SO_4, the sulfonic acid interchanges with the acid halide to give a sulfonyl halide (Eqs. 15.10). Unlike halogenation or nitration, sulfonation

$$\phi SO_3H + FSO_3H \longrightarrow \phi SO_2F + H_2SO_4$$
$$\phi SO_3H + ClSO_3H \longrightarrow \phi SO_2Cl + H_2SO_4 \quad (15.10)$$

is reversible. Boiling a sulfonic acid with 80% sulfuric acid or treating it with superheated steam (200° or higher) causes the replacement of —SO_3H by a proton (reversal of Eq. 15.9).

Friedel-Crafts alkylation and acylation. The Friedel-Crafts reaction is a substitution reaction with an alkyl halide or an acid chloride or anhydride. Aluminum chloride is the most frequently used catalyst, although many others have been used: BF_3, $SnCl_4$, $FeCl_3$ with alkyl halides and acid chlorides; H_2SO_4, HF, $BF_3 + H_2O$ with alcohols; and $HF + BF_3$, $HCl + AlCl_3$ with alkenes (Eqs. 15.11).

$$\phi H \begin{cases} \xrightarrow{CH_3CH_2Cl,\ AlCl_3} \phi CH_2CH_3 + HCl & a \\ \xrightarrow{CH_3CH_2COCl,\ AlCl_3} \phi COCH_2CH_3 + HCl & b \\ \xrightarrow{CH_3CHOHCH_3,\ AlCl_3} \phi CH(CH_3)_2 + H_2O & c \\ \xrightarrow{(CH_3)_2C=CH_2,\ AlCl_3,\ HCl} \phi C(CH_3)_3 & d \end{cases} \quad (15.11)$$

It seems certain that the electrophilic reagent is a carbonium ion (or an incipient carbonium ion), either $R^{\oplus}$ or $R\overset{\oplus}{C}=O$. In *a*, the Lewis acid $AlCl_3$ is thought to complex with EtCl to form Et---Cl---$AlCl_3$. Attack by benzene (a weak nucleophile) then gives the π complex and $AlCl_4^{\ominus}$. Formation of a similar complex, CH_3CH_2CO---Cl---$AlCl_3$, followed by attack by benzene gives the π complex, which leads to *b*. In *c* the free isopropyl cation is thought to be present, and in *d* HCl supplies a proton to form $tBu^{\oplus}$. The fact that the use of $(CH_3)_2CHCH_2Cl$ gives little $\phi CH_2CH(CH_3)_2$ and a very good yield of $\phi C(CH_3)_3$ reinforces the evidence for the presence of a carbonium ion: we have frequently encountered the rearrangement of the isobutyl cation into the *tert*-butyl cation.

Alcohols, alkenes, and alkyl halides react to give arenes; acid chlorides and acid anhydrides react to give ketones. The arene reaction is reversible; the ketone reaction presumably is not. Because a carbonium ion is required (or an easily displaceable halogen, as in MeBr, EtCl, and so on), unreactive halides like ϕCl or $H_2C=CHCl$ are not usable as the nonaromatic reactant in the Friedel-Crafts reaction.

Mercuration. Mercuric acetate is surprisingly reactive with arenes (Eq. 15.12)

$$\phi H + (CH_3COO)_2Hg \xrightarrow[HClO_4]{CH_3COOH} CH_3COOHg\phi + CH_3COOH \quad (15.12)$$

when catalyzed by the presence of a strong acid. However, C_6D_6 under the same conditions reacts at a slower rate than does C_6H_6—actually only 1/2.4 times as fast. Thus, the decomposition of the σ complex becomes slow enough to begin to complicate the mechanism. We include mercuration here to offset the impression that only relatively vigorous reagents (Cl_2, HNO_3, H_2SO_4, $RCl + AlCl_3$) can substitute arenes by an electrophilic mechanism.

Phenyl mercury fungicide. One of the few clear-cut effects of a man-made chemical on the environment involved the phenyl mercury compound. In the 1950's the death rate of birds in Sweden increased markedly. Investigation of the feathers of birds such as the eagle owl from the early 1800s to the present revealed that the mercury content was uniformly low until a big jump in the 1940's. The mercury-content curve seemed to parallel the bird fatalities. It happened that during these years the seed dressings used to prevent fungal growth were changed from phenyl mercury and inorganics such as arsenic to the more potent alkyl mercury compounds. A causal relation was suspected, and in 1965 the Swedish legislature banned the use of alkyl mercury compounds. The return to the phenyl mercury fungicides resulted in a return to the low mercury concentrations in feathers and, most importantly, to the normal death rate of birds.

Partial rate factors. Let us now consider the effects of substituents on a more quantitative level. The relative rate constants for the reactions of ϕCH_3, ϕH, ϕF, ϕCl, and ϕBr with Cl_2 in acetic acid at 25° are 340, 1.0, 0.74, 0.10, and 0.072. Other things being equal, each of the five replaceable aromatic hydrogens in toluene, for example, is reacting at a rate of $340/5 = 68$, as compared with one of the six hydrogens in benzene, at $1/6 = 0.17$. Other things are not equal, however, because the reaction with toluene gives 60% of *o*-chlorotoluene, 0.5% of *m*-chlorotoluene, and 39.5% of *p*-chlorotoluene.

A partial rate factor (prf) is defined as the ratio of a rate constant at a particular position to the rate constant at one of the six positions in benzene. The relative rate constant at one of the *ortho* positions in toluene may be set equal to half of the percent of the *ortho* product divided by 100, with the result multiplied by the relative rate constant for the total reaction (Eq. 15.13). Then prf for the *ortho* position in toluene for this particular reaction

$$k(CH_3, o) = \frac{(\% \text{ ortho}) \times k(CH_3, \text{rel})}{100 \times 2} \tag{15.13}$$

and conditions may be set up as Equation 15.14, in which $k(\phi H) = 1/6$ as before.

$$\text{prf}(CH_3, o) = \frac{k(CH_3, o)}{k(\phi H)} \tag{15.14}$$

Combining the equations and carrying out the calculations gives

$$\text{prf}(CH_3, o) = 3(0.60)(340) = 610$$
$$\text{prf}(CH_3, m) = 3(0.005)(340) = 5$$
$$\text{prf}(CH_3, p) = 6(0.40)(340) = 810$$

The results show that an *ortho* position in toluene reacts at a relative rate 610 times greater than at a single position in benzene. The *para* position is even more reactive, by a factor of 810. Even though only 0.5% of the product was *m*-chlorotoluene, the *meta* position was five times as reactive as a position in benzene. The methyl substituent, we see, activates the entire ring to electrophilic substitution and orients a single substituent according to the order of preference $p > o \gg m$.

Other substituents; steric effects. Similar calculations from the data for other substituents and other reactions give the results shown in Table 15.2. The

Table 15.2 *Some prf factors for various compounds and substitution reactions*

Compound	Reaction	prf			Ratio, p/o
		ortho	meta	para	
PhMe	Cl_2, CH_3COOH, 25°	610	5	810	1.3
PhF	Cl_2, CH_3COOH, 25°	0.2	0.0056	3.9	19.5
PhCl	Cl_2, CH_3COOH, 25°	0.01	0.0023	0.4	40
PhBr	Cl_2, CH_3COOH, 25°	0.08	0.0032	0.3	3.8
PhMe	HNO_3, $(CH_3CO)_2O$, 25°	38	3	49	1.3
PhF	HNO_3, $(CH_2CO)_2O$, 25°	0.4	—	0.8	2
PhCl	HNO_3, $(CH_3CO)_2O$, 25°	0.03	0.00084	0.13	4.3
PhBr	HNO_3, $(CH_3CO)_2O$, 25°	0.03	0.00098	0.10	3.3
PhMe	HOBr, $HClO_4$, aq. dioxane	76	2.5	59	0.78
PhMe	Br_2, CH_3COOH, 25°	800	5.5	2400	3.0
PhtBu	Br_2, CH_3COOH, 25°	50	7.0	800	16

deactivation by the halogen substituents is clearly shown for all positions, but m deactivated is much greater than o and p. (Chlorination p to F is an exception.) Thus, a halogen deactivates but orients to the o, p positions. Note the variability of the ratio of prf for *para* to prf for *ortho* for chlorination.

How do the prf values look for another reaction? For nitric acid in acetic anhydride at 25°, the results show that the strong electrophilic reagent $NO_2^\oplus$ is less subject than is the weaker Cl_2 to variation caused by the activating and orienting influences of Me. The deactivating and orienting influences of the halogens remain about the same for chlorination and nitration. However, in a catalyzed halogenation (HOBr in 50:50 dioxane-water with $HClO_4$ present), the halogen becomes a strong electrophile. This change is shown by the prf values for PhMe, which are not much different from the values for nitration. In contrast, if we look at uncatalyzed bromination (Br_2 is a weaker electrophile than is Cl_2), we find prf values for PhMe that bear out our prediction that the activating substituent has a greater effect.

To examine the effect of the size of the substituent, we look at the prf values for *t*Bu in uncatalyzed bromination. The much larger *t*Bu crowds the adjacent *ortho* position more than does Me, and the *ortho* prf is reduced more than is the *para* prf (the ratio increases). Also, we see that *meta* prf for *t*Bu (7.0) is larger than that for Me (5.5). This difference may be accounted for if the $+I$ effect for *t*Bu is greater than that for Me. After all, *t*Bu$^\oplus$ is more stable than Me$^\oplus$. The effect is attenuated in the transfer from C_1 to C_3 or C_5 through the framework of the ring, so the difference is quite small.

15.3 Electrophilic substitution reactions

Hyperconjugation. It is surprising, at first, that prf for the position *para* to *t*Bu is so small (800 as compared with 2400 for *para* to Me). Hyperconjugation supplies an answer. We have not yet accounted for the +R effect of methyl, and now it is convenient to include *t*Bu as well. Let us look at the σ complex for *para* attack by bromine on toluene (a) and on *tert*-butylbenzene (b). If we make the assumption that H$^\oplus$ is more stable than $^\oplus$CH$_3$, it

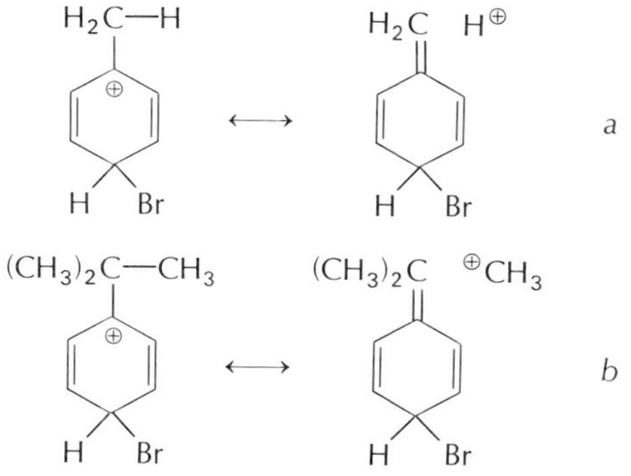

follows that the no-bond structure in *a* (hyperconjugation) contributes more to the stabilization of the σ complex than does the no-bond structure in *b* and lowers the activation energy for the conversion of the π complex to the σ complex. Thus, Me activates *p* more than *t*Bu activates *p*. A more realistic illustration shows that the proton participating in hyperconjugation must lie above or below the ring. (Overlap is shown to correspond with *a*.) □ □

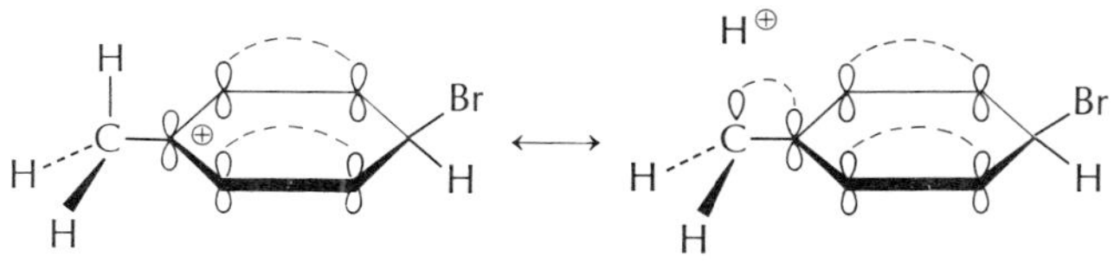

Effect of two substituents. Let us pose a problem. What is the major product in the nitration of *m*-chlorotoluene? Qualitatively, we can set up a diagram as shown. We see that C$_2$, C$_4$, and C$_6$ are activated, and C$_5$ is deactivated strongly by being *m* to Cl.

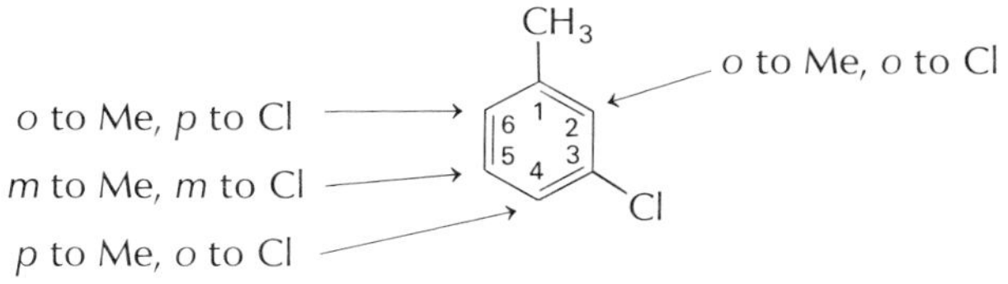

o to Me, *p* to Cl
m to Me, *m* to Cl
p to Me, *o* to Cl
o to Me, *o* to Cl

Because C$_2$ is *o* to both Me and Cl, it is crowded more than C$_4$ or C$_6$. Therefore, the entering NO$_2$$^\oplus$ goes to C$_4$ or C$_6$ rather than to C$_2$ or C$_5$. Now if we happen to remember that prf *o* to Me is almost as large as prf *p* and that prf *o* to Cl is small in relation to prf *p*, we predict more C$_6$ product than C$_4$. The experimental result is 59% of 5-chloro-2-nitrotoluene, 32% of 3-chloro-4-nitrotoluene, and 9% of 3-chloro-2-nitrotoluene, in accord with our prediction. It is possible to make more precise predictions by use of prf values (see Chapter 4 in Stock, this chapter, ref. 1). In general, we may conclude that the effect of an activating substituent drowns out the orienting effect of a deactivating (*m*-directing) substituent.

458   Chapter fifteen

Thus we would predict that *m*-nitrotoluene would react predominantly at the 6 position, just as we predicted for *m*-chlorotoluene.

Polycyclics.   The polycyclic compounds all are more reactive than benzene. For bromination, the prf value for α or C_1 substitution in naphthalene is 200,000 (with reference to benzene); it is 2000 for β or C_2. However, the β position is less crowded than the α, so the reversible substitution reactions (sulfonation and Friedel-Crafts) give the β product under thermodynamic control. Kinetic control (the usual conditions) gives good yields of the α product (nitration, halogenation).

Anthracene reacts almost exclusively by addition to the 9 and 10 positions. Heating the addition product or carrying out an elimination reaction then leads to the 9-substituted anthracene (Eq. 15.15).

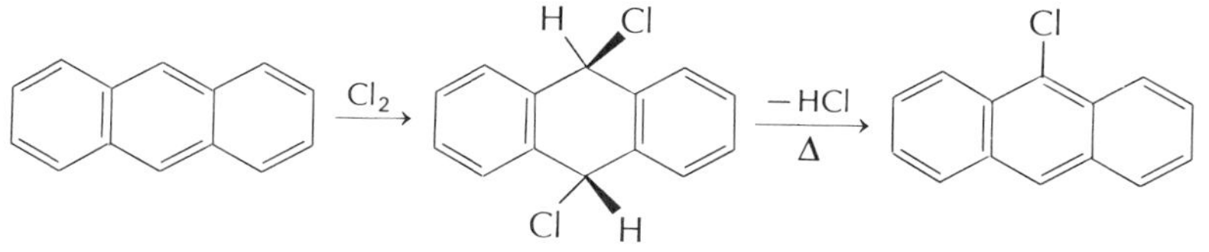

$$\text{anthracene} \xrightarrow{Cl_2} \text{9,10-dihydro-9,10-dichloroanthracene} \xrightarrow[\Delta]{-HCl} \text{9-chloroanthracene} \quad (15.15)$$

Phenanthrene reacts with halogens by addition at the 9 and 10 positions, but nitration substitutes at every available position in the order 9 > 1 > 3 > 2, 4.

Exercises

15.4   Predict the major product to be expected upon nitration of:

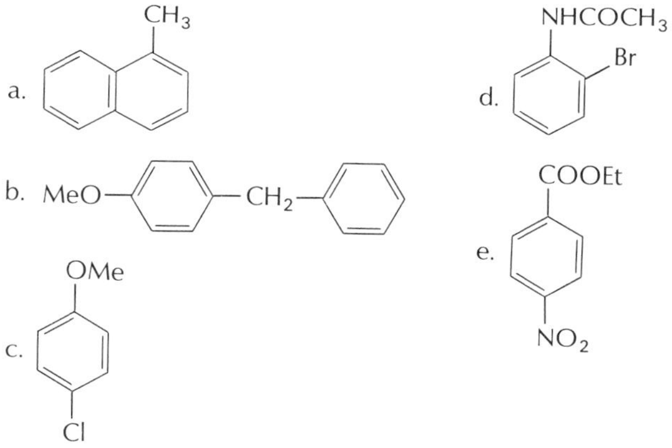

a. 1-methylnaphthalene

b. MeO—C₆H₄—CH₂—C₆H₅ (4-methoxydiphenylmethane)

c. 4-chloroanisole (OMe para to Cl)

d. 2-bromoacetanilide (NHCOCH₃ with ortho Br)

e. ethyl 4-nitrobenzoate (COOEt para to NO₂)

15.4   Homolytic aromatic substitution reactions

Addition of radicals.   Free radicals attack aromatic rings by addition rather than by abstraction of a hydrogen atom (Eqs. 15.16). If R· is Cl·, a is thought to have $\Delta H = -8 \pm 1$ kcal. In a, a σ bond to carbon is formed and some conjugation is preserved in the

$$\phi H + R\cdot \longrightarrow \text{[cyclohexadienyl radical resonance structures with R and H]} \quad a$$

$$\phi H + R\cdot \not\longrightarrow R-H + \text{[phenyl]} \quad b \tag{15.16}$$

pentadienyl radical system. The radical is delocalized. In b, the abstraction of a hydrogen atom leads to a localized radical. In it, the odd electron is isolated in the sp^2 orbital that formerly overlapped with the hydrogen s orbital.

If benzene is exposed to chlorine and light, the chlorine atoms attack as shown in Equation 15.16a. The subsequent step ($Cl\phi\cdot + Cl_2 \rightarrow C_6H_6Cl_2 + Cl\cdot$) is the second step in the propagation of the chain. The dichlorocyclohexadiene is very subject to free-radical addition as compared to benzene, and the subsequent steps lead rapidly to completion to give hexachlorocyclohexane. It is an interesting exercise to work out the possible stereoisomerism.

Only one of the isomers has insecticidal activity: axial-1,2,3-equatorial-4,5,6-hexachlorocyclohexane. The mixture of isomers has been sold under the names gammexane and lindane. This molecule is only one of the many halogenated compounds that have striking physiological activity.

Hexabromobenzene. In contrast, if benzene is dissolved in bromine and some $FeBr_3$ is added (or iron filings: $2Fe + 3Br_2 \rightarrow 2FeBr_3$) and the mixture is allowed to stand for a week, C_6Br_6 is produced in good yield. Polar conditions favor electrophilic substitution and radical conditions favor addition to benzene.

Hydrogen abstraction in side chains. If hydrogen atoms are available on a carbon atom adjacent to the ring in a substituted benzene, hydrogen-atom abstraction by the attacking radical occurs at that point instead of attack upon the ring (Eq. 15.17). The

$$\phi CH_2CH_3 + R\cdot \longrightarrow \phi\dot{C}HCH_3 + R-H \tag{15.17}$$

reason, of course, is that the new radical is a benzyl type, stabilized by delocalization. (Compare with benzyl cation and benzyl anion stability.) However, the rate of reaction of ethylbenzene is not markedly different from that of an alkane like ethane or cyclohexane with reactive radicals such as $Cl\cdot$, $tBuO\cdot$, $\phi\cdot$, and $Me\cdot$. As we saw previously (halogenation of alkanes and alkenes, Sec. 12.5), $Br\cdot$ and NBS are much more subject to the influence of the substance being attacked. It has been estimated that ΔH for $\phi CH_3 + X\cdot \rightarrow \phi CH_2\cdot + HX$ is about -10 kcal for $Br\cdot$ and about -25 kcal for $Cl\cdot$. Thus the transition state for the $Cl\cdot$ reaction is of less energy and has much less C—H breaking and H—X making than does the $Br\cdot$ reaction. Again, bromination is much more selective than is chlorination. □ □

Thyroxine. Thyroxine and triiodothyronine, two iodinated aromatic compounds, are secreted by the thyroid gland to influence the rate of metabolic activity of practically every tissue in the body. The compounds (Table 21.2) are uncommon amino acids synthesized from tyrosine while it is a part of the large protein thyroglobulin. Certain enzymes catalyze the iodination of the phenolic side chain on the intact protein; and others, the subsequent conversion to L-thyronine. It is stored on thyroglobulin until secreted.

The absence of iodine from the diet causes goiter, a condition in which the thyroid gland expands in size in an attempt to produce the necessary levels of thyroxine. In the early nineteenth century, when iodine deficiency was first associated with goiter, seaweed became a popular, effective dietary supplement. In this century, when health officials recognized that addition of iodine to the diets could eliminate goiter in the entire population, they recommended it be put in salt. Together with publicity encouraging people to include iodine in their diets, iodized salt (usually .01 percent potassium iodide or cuprous iodide) has proved valuable in the prevention of goiter.

15.5 Nucleophilic aromatic substitution reactions

Modified S_N2 mechanism. Although phenyl cations may be formed in several ways, their reactivity is so great that they are quite unselective in their reactions. As in the phenyl radical, the positive charge of a phenyl cation is isolated in an sp^2 orbital, with no chance for delocalization. Thus S_N1 reactions of aryl halides are negligible.

Aryl halides react under S_N2 conditions with strong nucleophiles or under drastic conditions. There is no pentacoordinate transition state: instead, the sp^2 hybridization of an aryl halide causes a modification whereby an unstable sp^3 hybrid intermediate is formed (Eq. 15.18). The negative charge is accommodated by the conjugated pentadienyl anion

$$\phi Br + \phi CH_2NH_2 \xrightarrow[HCONMe_2]{160°} \left[\text{intermediate}\right] \xrightarrow{} \phi \overset{\oplus}{N}H_2CH_2\phi \quad Br^{\ominus} \quad (15.18)$$

system. We do not say that the intermediate is always isolable: usually it is not, and it goes on to product by what seems like an S_N1 reaction, the ionization of a tertiary halide. However, it is even more than that, because the ionizing second step is favored by the regeneration of the aromatic stabilization of the ring. Aromatic nucleophilic substitution may be thought of as an addition-elimination reaction.

It seems quite certain that the formation of the intermediate is a slow step, the second step being a fast reaction. The reaction is bimolecular. A schematic energy diagram is shown in Figure 15.2 for $Y^{\ominus} + \phi X \rightarrow Y\phi + X^{\ominus}$. As shown, the intermediate $C_6H_5XY^{\ominus}$ reposes at the bottom of a relatively shallow well and may easily revert to reactants or go on to products.

Effect of substituents. Whether the nucleophile is a negative ion or not, the intermediate must take care of a negative charge in the pentadienyl system (Eq. 15.18). Therefore, we expect electron-withdrawing substituents ($-I$, $-R$) to stabilize the system, especially if they are located at C_2, C_4, or C_6, where the negative charge is greatest. Electron-donating substituents ($+I$, $+R$) should destabilize the system—again, especially if they are located at C_2, C_4, or C_6, o and p to the group being replaced. Substituents in the m positions (C_3 or C_5) exert their influence largely through the I effect.

A particularly effective substituent for favoring nucleophilic displacement is the nitro group, which has a powerful $-I$ effect as well as a strong $-R$ effect. The $-R$ effect resonance structures in the intermediate for $Y^{\ominus}$ attack upon p-nitrochlorobenzene are shown.

15.5 Nucleophilic aromatic substitution reactions

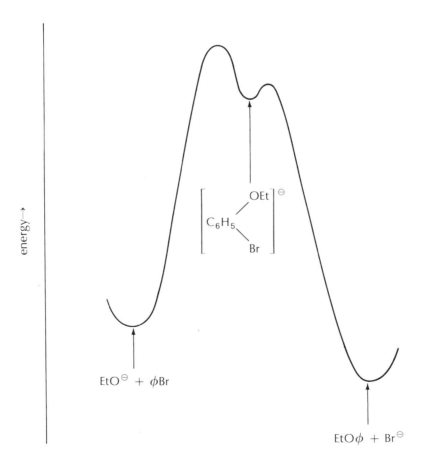

Trinitro compounds. The effect is so powerful that 2,4,6-trinitrobenzene compounds easily accommodate a negative charge in the ring. Thus, 1-chloro-2,4,6-trinitrobenzene is so reactive that the weak nucleophile water reacts to replace the chlorine (Eq. 15.19). The product, picric acid (strictly speaking, a phenol), is a strong acid ($K_a = 0.42$) because of the stabilization of the anion. The trinitro compounds (picric acid; 1,3,5-trinitrobenzene) so deplete the aromatic ring of electrons that they form stable, crystalline charge-transfer complexes with arenes such as naphthalene, or [$C_{10}H_8 \rightarrow C_6H_3(NO_2)_3$].

Trinitroanisole (we have left off the numbers) reacts with EtOK to produce a stable

Figure 15.2 Schematic energy-reaction-coordinate diagram of a nucleophilic aromatic displacement, $EtO^\ominus + \phi Br \rightarrow EtO\phi + Br^\ominus$

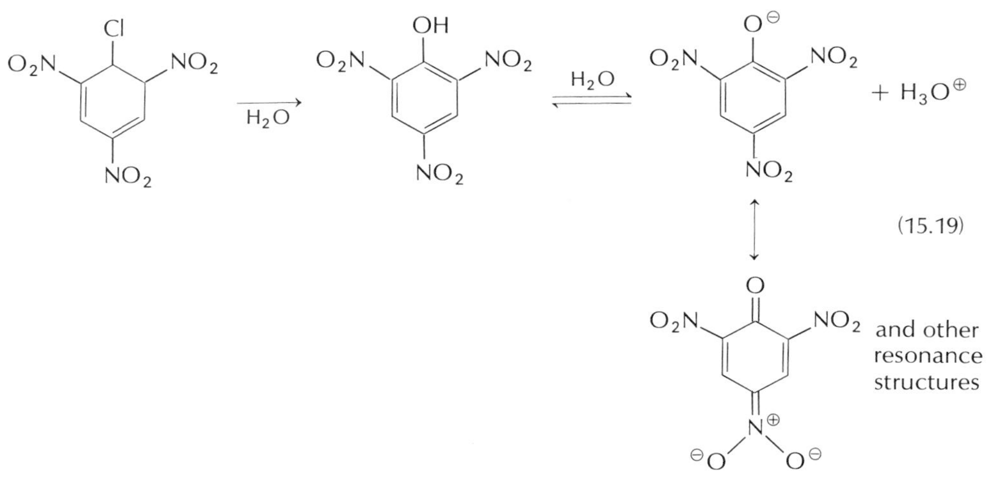

[Eq. 15.19 shows the hydrolysis of picryl chloride to 2,4,6-trinitrophenol (picric acid) and its deprotonation to the picrate anion, with a resonance structure shown.]

intermediate. The same intermediate is formed if we allow trinitrophenetole to react with MeOK. In the presence of water, the intermediate decomposes to give the salt of picric acid (Eq. 15.20). Nevertheless, the red crystalline intermediate is isolable as the relatively stable potassium salt.

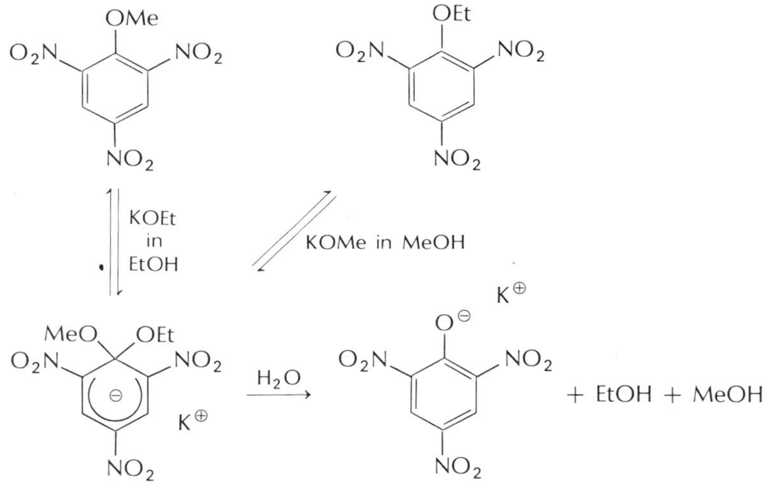

[Eq. 15.20 shows trinitroanisole and trinitrophenetole reacting with KOEt in EtOH or KOMe in MeOH to form a Meisenheimer complex (with both OMe and OEt on the sp³ carbon, K⁺ counterion), which on hydrolysis gives the potassium picrate salt + EtOH + MeOH.]

One series of studies has shown that a single p-nitro group gives a rate 41,000 times that of the unsubstituted compound for the replacement of chlorine by an amine. Other substituents that are m-orienting and deactivating for electrophilic substitution also accelerate the nucleophilic substitution from the p position (for example, p-CN, 6000; and p-COOEt, 920 times). These groups also speed up the reaction from the m position, but because only the −I effect operates, the rate acceleration is much smaller (m-CN, 60; and m-COOEt, 5 times). The −I > +R effect of the halogens is shown by their rate accelerations (m-Cl, 32; and p-Cl, only 6 times). Oxygen, with a much stronger +R effect, is deactivating at the p position (0.025) but is activating from the m position (4 times). The consistency of the qualitative predictions from the resonance theory for both electrophilic and nucleophilic substitutions is very satisfying. □ □

15.5 Nucleophilic aromatic substitution reactions

Use of 2,4-dinitrofluorobenzene. The presence of nitro groups in a molecule causes a marked increase in bp and mp and a decrease in solubility. This property has been used to tag and identify the location of free amino groups in proteins and other naturally occurring substances. A protein is a polyamide chain of many different amino acids (Chap. 21). At one end of a chain there is a free amino group ($H_2NCHRCONHCHR'$—). Treatment of a protein with 1-fluoro-2,4-dinitrobenzene results in a displacement of fluorine by the amino group (the amide groups are unreactive). Hydrolysis of the protein allows the tagged amino acid at the end of the chain to be isolated and identified with ease (Eq. 15.21).

$$O_2N\text{-}C_6H_3(NO_2)\text{-}F + H_2NCHRCONH\text{-} \xrightarrow{-HF} O_2N\text{-}C_6H_3(NO_2)\text{-}NHCHRCONH\text{-}$$

$$\xrightarrow{H^{\oplus}, H_2O} O_2N\text{-}C_6H_3(NO_2)\text{-}NHCHRCOOH + \text{amino acids} \quad (15.21)$$

Role of solvent. Because the reaction is difficult to carry out under ordinary conditions in the usual protic solvents without an activating substituent, it was thought for a long time that weak nucleophiles (bases) could not be used on unactivated halobenzenes. Then it was discovered that aprotic solvents, particularly DMSO (dimethylsulfoxide, $(CH_3)_2S\text{=}O$), DMF (dimethylformamide, $HCONMe_2$), and DMA (dimethylacetamide, CH_3CONMe_2) caused the aromatic nucleophilic substitution reactions to proceed so much faster that ordinary elevated temperatures were possible (Eq. 15.18). The explanation given is that the highly polar but aprotic solvents are excellent at solvating cations but are very poor at solvating anions. In

$$\underset{\text{strong attraction}}{Me_2S^{\oplus}\text{-}O^{\ominus}\text{---}K^{\oplus}} \qquad \underset{\text{strong repulsion}}{Me_2S^{\oplus}\text{-}O^{\ominus} \quad OH^{\ominus}}$$

effect, the concentration of anion is increased in these solvents because, in entering into a reaction, the nucleophilic anion does not have to fight its way out of a clinging crowd of solvent molecules as it would in protic solvents.

The benzyne pathway. If a strong base ($H_2N^{\ominus}$) is used, the reaction with a halobenzene follows another path that involves a <u>benzyne</u> as an intermediate. That a benzyne is a satisfactory explanation is shown in a number of ways.

First, it is known that benzene is attacked by such strong bases (Eq. 15.22), because deuterium may be replaced by hydrogen.

$$\phi D \xrightarrow[NH_3]{NaNH_2} \underbrace{\phi^{\ominus} + NH_2D}_{?} \xrightarrow{NH_3} \phi H + NH_2^{\ominus} \quad (15.22)$$

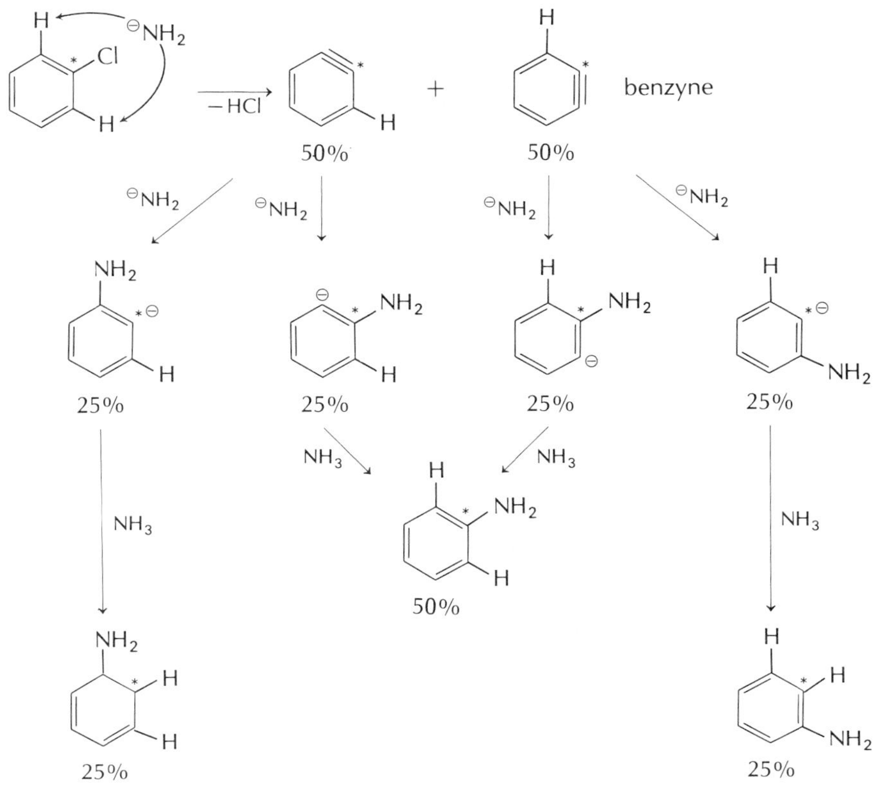

(15.23)

Second, if we use chlorobenzene in which a ^{14}C atom is present at C_1, we can analyze the aniline produced and find that the carbon to which the amino group is bonded is only half as radioactive as we expected. The rest of the radioactivity is found in the carbons at the positions *ortho* to the amino group (Eq. 15.23: the asterisk is used to identify ^{14}C). The elimination of HCl in the first step must be close to concerted (E2), because if the *ortho* positions are blocked by substituents, as in this example, the reaction does not go.

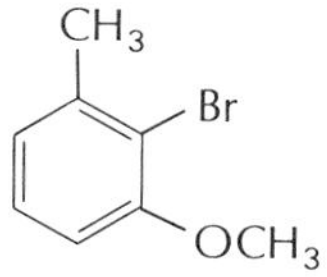

Benzyne is so unstable that it has not been isolated. (The third bond is a very weak overlap of two sp^2 orbitals that lie in the plane of the ring and are directed somewhat away from each other.) On the other hand, if furan (a strong enophile) is added to a reaction mixture in which a benzyne is present, furan is able to compete with other reagents for the benzyne by a Diels-Alder reaction (Eq. 15.24). Other reactive dienes may be used as well.

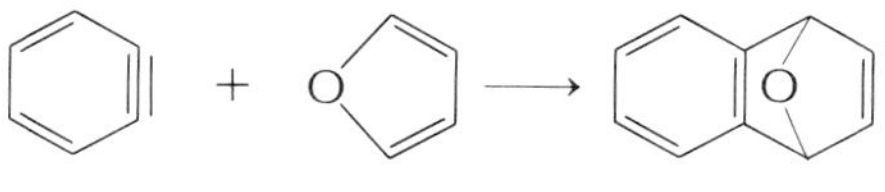

(15.24)

The Diels-Alder reactions of benzynes are so useful that other and more convenient methods for the production of benzynes have been developed (Eqs. 15.25).

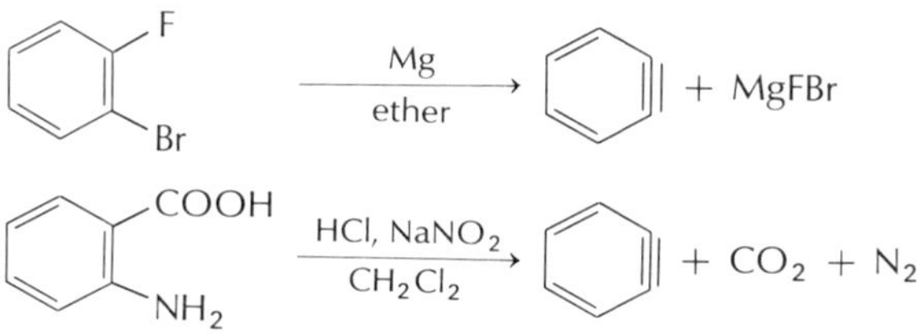

$$\text{o-F,Br-C}_6\text{H}_4 \xrightarrow{\text{Mg, ether}} \text{benzyne} + \text{MgFBr}$$

$$\text{o-COOH,NH}_2\text{-C}_6\text{H}_4 \xrightarrow[\text{CH}_2\text{Cl}_2]{\text{HCl, NaNO}_2} \text{benzyne} + CO_2 + N_2 \quad (15.25)$$

Commercial preparations. Commercially, nucleophilic reactions have been used for a long time for the preparation of ϕOH, ϕCN, and ϕNH_2 (Eqs. 15.26).

$$\phi SO_3Na \xrightarrow[300°]{\text{NaOH}} \phi ONa + Na_2SO_3 + H_2O$$

$$\phi Cl \xrightarrow[350°]{\text{NaOH, H}_2\text{O}} \phi ONa + NaCl + H_2O$$

$$\phi Cl \xrightarrow[\Delta,\text{ pressure}]{H_2O} \phi OH + HCl \quad (15.26)$$

$$\phi Cl \xrightarrow[Cu^\oplus,\ 250°]{NH_3,\ H_2O} \phi NH_2 + NH_4Cl$$

Another reaction, which also is sometimes used in the laboratory, gives low yields (5 to 50%) of a nitrile (Eq. 15.27).

$$\phi SO_3Na \xrightarrow[\Delta]{\text{NaCN}} \phi CN + Na_2SO_3 + H_2O \text{ (less than 50\%)} \quad (15.27)$$

Exercises

15.5 Show how to prepare, from benzene or naphthalene:
 a. p-nitroaniline
 b. 2,4-dinitrophenetole
 c. α-naphthol
 d. picric acid

15.6 Oxidation and reduction of arenes

Indirect oxidation of side chains. Oxidation and reduction of aromatic compounds were covered in Chapter 14. However, a few items remain to be described.

Side-chain halogenation is a useful indirect method of oxidation. Because a second halogen is harder to introduce than the first, the reaction is easily controlled (Eqs. 15.28).

$$\phi CH_3 \begin{cases} \xrightarrow[hv]{Cl_2} \phi CH_2Cl \xrightarrow{H_2O} \phi CH_2OH \\ \xrightarrow[hv]{2Cl_2} \phi CHCl_2 \xrightarrow{H_2O} \phi CHO \\ \xrightarrow[3Cl_2]{hv} \phi CCl_3 \xrightarrow{H_2O} \phi COOH \end{cases} \quad (15.28)$$

By free-radical chlorination followed by easy hydrolysis of the various chlorides, toluene is converted in excellent yields into benzyl alcohol, benzaldehyde, or benzoic acid. For longer chains, the more selective bromine is used (Eqs. 15.29), although the products are usually

$$\phi CH_2CH(CH_3)_2 \begin{array}{c} \xrightarrow[hv]{Br_2} \phi CHBrCH(CH_3)_2 \xrightarrow{H_2O} \phi CHOHCH(CH_3)_2 \\ \xrightarrow[hv]{2Br_2} \phi CBr_2CH(CH_3)_2 \xrightarrow{H_2O} \phi COCH(CH_3)_2 \end{array} \qquad (15.29)$$

more easily obtained by the Friedel-Crafts acylation reaction (Sec. 15.8).

With a xylene, the stepwise chlorination alternates between the methyl groups (Eqs. 15.30).

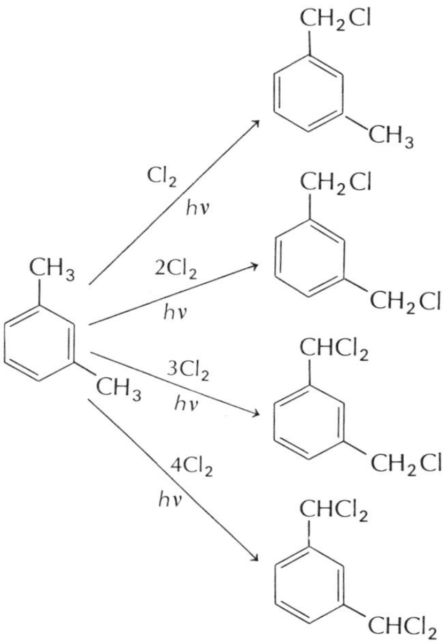

(15.30)

Autoxidation. The tertiary benzylic hydrogens are so reactive (or the tertiary benzylic radicals are so stable) that autoxidation is a practical process (Eq. 15.31). Once

$$\phi CH(CH_3)_2 \xrightarrow[100°]{O_2,\ air} \phi - \underset{\underset{CH_3}{|}}{\overset{\overset{CH_3}{|}}{C}} - OOH \qquad (15.31)$$

90%

the reaction is begun, the radical speedily takes up oxygen (R· + O_2 → ROO·), and the peroxy radical continues the chain (ROO· + HR → ROOH + R·).

The reaction is used commercially because it provides an economical pathway to phenol, with acetone as a bonus (Eq. 15.32). The acid-catalyzed rearrangement resembles a Wagner-Meerwein rearrangement in that a tertiary cation is formed by the migration of phenyl to oxygen (which is like a primary carbon). The tertiary cation picks up a molecule of water to become a hemiacetal of acetone, which hydrolyzes rapidly.

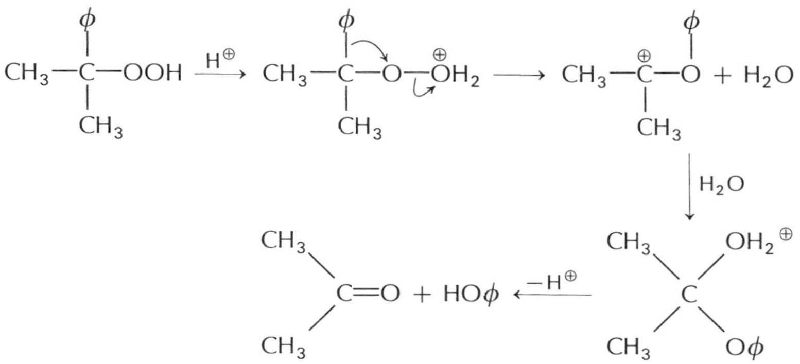

$$CH_3-\underset{CH_3}{\underset{|}{C}}-OOH \xrightarrow{H^\oplus} CH_3-\underset{CH_3}{\underset{|}{\overset{\phi}{C}}}-O-\overset{\oplus}{O}H_2 \longrightarrow CH_3-\underset{CH_3}{\underset{|}{\overset{\phi}{\overset{\oplus}{C}}}}-O + H_2O$$

$$\Big\downarrow H_2O \qquad (15.32)$$

$$\underset{CH_3}{\overset{CH_3}{\diagdown}}C=O + HO\phi \xleftarrow{-H^\oplus} \underset{CH_3}{\overset{CH_3}{\diagdown}}\underset{O\phi}{\overset{\overset{\oplus}{O}H_2}{\diagup}}C$$

Many other autoxidation reactions are known. Two more examples emphasize that benzylic hydrogens are involved even if they are not tertiary (Eqs. 15.33).

$$\phi CHO \xrightarrow{O_2} \phi COOH \xrightarrow{\phi CHO} 2\phi COOH \qquad (15.33)$$

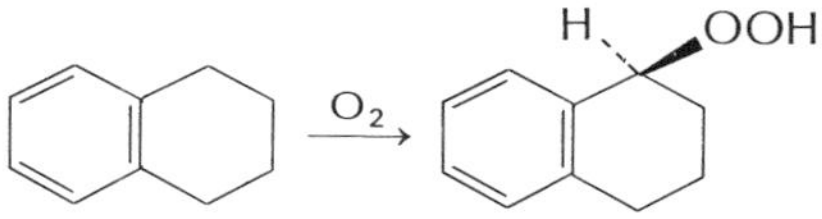

Exercises

15.6 Show how to carry out these conversions:
   a. p-cymene → p-cresol
   b. o-xylene → o-methylbenzaldehyde
   c. $\phi CH_2CH_2CH_3 \rightarrow \phi CHBrCHBrCH_3$

15.7 Sources of arenes

Coal tar; coal. Following the early discovery of benzene by Faraday, production of iron and steel rose throughout the 1800's, requiring ever increasing amounts of coke. Coal tar, a by-product of the manufacture of coke (pyrolysis of bituminous coal), became a constant problem in disposal. Distillation of coal tar, it was soon found, yielded a whole catalog of aromatic compounds, the most abundant of which was naphthalene. Other easily separable arenes are benzene, toluene, xylene, other alkyl benzenes, alkylnaphthalenes, acenaphthene, fluorene, indene, anthracene, phenanthrene, and pyrene, as well as basic nitrogen-containing substances (pyridine and quinoline) and oxygen- and sulfur-containing substances in great variety.

Coal is thought to be of vegetable origin. The various types (peat, lignite, and bituminous and anthracite coals) are formed by bacterial decomposition and compression after burial, the harder coals resulting from greater pressure.

Coal tar remains the principal source of aromatic compounds other than benzene, toluene, and ethylbenzene, which are available today from the petroleum industry. After

Perkin's mistake. Perkin's mistake came about as follows. The antimalarial drug quinine had been shown to be $C_{20}H_{24}N_2O_2$. Perkin, who had begun training under Hofmann, was looking into the possibilities of preparing quinine by an oxidation reaction like $2C_{10}H_{13}N + 3[O] \rightarrow C_{20}H_{24}N_2O_2 + H_2O$. Casting about for a $C_{10}H_{13}N$ compound, he came upon N-allyl-p-toluidine. Treatment with dichromate did not give any quinine but did produce a tarlike substance that was soluble enough in water to produce a colored solution. Following up this observation, Perkin oxidized aniline and found that the tarlike product was soluble in alcohol, giving a purple color. The aniline of the time was obtained from coal tar and was so impure that it contained sizable amounts of the toluidines. Perkin soon found that the toluidines were essential to the process. At the age of eighteen, he applied for and obtained a patent. A year later the dye mauveine, was on the market. The first coal-tar dye caught on and the race for new dyes and colors began. From a waste product, coal tar was transformed into a useful commodity. Quinine was not synthesized in the laboratory until 1944 by Woodward and Doering. The mistaken idea for the preparation of quinine, by brilliant observation and diligence, led to a successful conclusion in an entirely unexpected direction.

15.8 Preparation and synthesis of arenes

Control of electrophilic substitution. Let us look at a few compounds other than arenes to illustrate certain points. We begin by making certain that we understand some of the consequences of electrophilic substitution.

Of the five electrophilic substitution reactions (omitting mercuration for the moment), we must adopt more drastic measures in four to persuade a second substituent to enter. Thus, the halogenation of benzene produces a halobenzene, which is more reluctant than benzene to undergo a second halogenation to give o- and p-dihalobenzene: halogen deactivates. Nitration and sulfonation produce substances that contain a deactivating, m-orienting group. To introduce a second nitro or sulfonic acid group, we need an increase in temperature or more reactive reagents (fuming acids) or both. Friedel-Crafts reactions are particularly subject to activating and deactivating groups on the ring, and the acylation reaction to give a ketone (deactivating, m orienting) stops dead. These four reactions give monosubstitution without any particular effort at control.

The Friedel-Crafts alkylation reaction produces a substance that is more reactive than the starting material. For example, if we start with one mole each of EtCl and ϕH and some $AlCl_3$, at the halfway mark (assuming that nothing else happens in the meantime) there is half a mole of benzene left and half a mole of ethylbenzene present. The ethyl group, like methyl, is $+I$ and $+R$, activating, and an o, p director. Thus, from this point on, ethylbenzene is consumed faster than is the remaining benzene. Actually, under these conditions we get a mixture of benzene, ethylbenzene, diethylbenzenes, some triethylbenzenes, and small amounts of tetra- and pentasubstituted benzenes. To some extent, the difficulty may be circumvented by use of a large excess of benzene (twenty moles to one mole of EtCl) to favor monosubstitution. This approach may be satisfactory with cheap

15.8 Preparation and synthesis of arenes 469

starting materials, but it is wasteful and laborious to operate with an expensive or rare arene in this manner. We also have seen that alkylation proceeds by way of a carbonium ion, so that benzene and PrCl produce iPrϕ, iBuCl and benzene give tBuϕ, and so on.

To avoid these difficulties, we can use the acylation reaction. For example, if iBuϕ is required, we start with ϕH and iPrCOCl (Eq. 15.34) and reduce the ketone by the Clemmensen procedure to the required arene. In this instance, the Wolff-Kishner reduction would be

$$\phi H \xrightarrow[\text{iPrCOCl}]{\text{AlCl}_3} \phi COiPr \xrightarrow[\text{HCl}]{\text{ZnHg}_x} \phi CH_2CH(CH_3)_2$$

$$\begin{array}{c} CH_3 \\ \\ CH_3 \end{array} \Bigg\uparrow SOCl_2$$

$$\begin{array}{c} CH_3 \\ \diagdown \\ CHCOOH \\ \diagup \\ CH_3 \end{array}$$

(15.34)

just as useful: perhaps more so with the use of DMSO. The acylation reaction stops with one group on the ring, and no rearrangements interfere. □ □

Sulfonation. Sulfonation is reversible and the ratios of o, m, and p are subject to thermodynamic control. At 100°, sulfonation of toluene gives a 73% yield of p-toluenesulfonic acid. At 0°, o-toluenesulfonic acid is obtained in a yield of 45%, along with 52% of the p isomer.

Syntheses: the proper sequence of reactions. It is critical not only to select the proper reactions for a synthesis, but also to put them in the proper sequence. We shall use examples to illustrate some of the difficulties. In every case we shall assume that benzene and toluene are available as starting materials and that the usual assortment of ordinary reagents and solvents is at hand.

1. Prepare p-nitrobenzoic acid. As we work backwards, the last step must be either the nitration of benzoic acid or the introduction of the acid group p to the nitro group. Nitration of benzoic acid gives m-nitrobenzoic acid as the major product, so we must use the other reaction. Oxidation of methyl to an acid group is possible, so a last step may be the oxidation of p-nitrotoluene. Unfortunately, nitration of toluene gives more o and p, but we just take our loss: there is no other convenient route. Equation 15.35 is the answer.

$$O_2N-\langle\rangle-COOH \xleftarrow[\substack{H_2SO_4 \\ 100°}]{KMnO_4} O_2N-\langle\rangle-CH_3 \xleftarrow[H_2SO_4]{HNO_3} \phi CH_3 \quad (15.35)$$

2. Prepare m-bromopropiophenone. Only two reactions are required. Starting from benzene, we can either use bromination and then the Friedel-Crafts reaction with CH$_3$CH$_2$COCl or use the Friedel-Crafts reaction first and then bromination. If we brominate first, the Friedel-Crafts reaction will give the p isomer as the major product. If the Friedel-Crafts is done first, the keto group orients m, and bromination, though slow, will give the wanted product (Eq. 15.36).

$$\underset{Br}{\underset{}{\langle\rangle}}\!\!-\!\!COCH_2CH_3 \xleftarrow[FeBr_3]{Br_2} \langle\rangle\!\!-\!\!COCH_2CH_3 \xleftarrow[AlCl_3]{\phi H} CH_3CH_2COCl \quad (15.36)$$

3. Prepare *m*-bromobenzal chloride. Again the product is two reactions away from the starting material, toluene: chlorinate, then brominate; or brominate, then chlorinate. Bromination of toluene goes *p*: therefore we chlorinate first under radical conditions to attack the methyl group instead of the ring. Then bromination goes *m* to the deactivating —CHCl$_2$ group (Eq. 15.37).

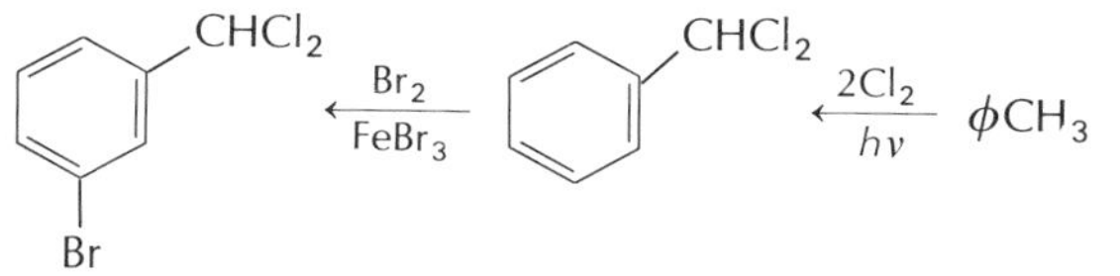

$$\text{m-Br-C}_6\text{H}_4\text{CHCl}_2 \xleftarrow{\text{Br}_2 / \text{FeBr}_3} \text{C}_6\text{H}_5\text{CHCl}_2 \xleftarrow{2\text{Cl}_2 / h\nu} \phi\text{CH}_3 \qquad (15.37)$$

4. Prepare 4-bromo-3-nitro-1-ethylbenzene. There are three possible last steps. Bromination of *m*-nitroethylbenzene probably would give a mixture of the 4- and 6-bromo compounds. Nitration of *p*-bromoethylbenzene would give the 2-nitro compound as the major product; ethyl activates *o*, *p*; bromine deactivates; ethyl will win. Friedel-Crafts ethylation will not work on a ring containing a deactivating group, such as nitro in *o*-nitrobromobenzene. The alternative is to reduce the keto group in —COCH$_3$ by some means. The Clemmensen reduction of the ketone will also cause a reduction of the nitro group to —NH$_2$. The Wolff-Kishner reduction in an alkaline medium will cause a nucleophilic replacement of bromine *ortho* to the nitro group. An alternative is to reduce a vinyl group with BH$_3$ (the nitro group is inert). In turn, we can prepare the vinyl group by simple distillation from a trace of acid of the —CHOHCH$_3$ group. Treatment with NaBH$_4$ will reduce the keto group of —COCH$_3$ without affecting the nitro group.

Now we are back to 4-bromo-3-nitroacetophenone. Nitration of *p*-bromoacetophenone will introduce the nitro group *ortho* to the bromine (also *meta* to the keto group, a favorable factor but not the deciding one). Bromination of *m*-nitroacetophenone will put any entering bromine in the 5 position, but the two deactivating groups, nitro and keto, will make the reaction very slow. As before, the Friedel-Crafts ketone synthesis will not proceed with a nitro group present.

So we need to make *p*-bromoacetophenone. Now (see example 2) we are home free by a ketone synthesis on bromobenzene, which we make from benzene (Eq. 15.38).

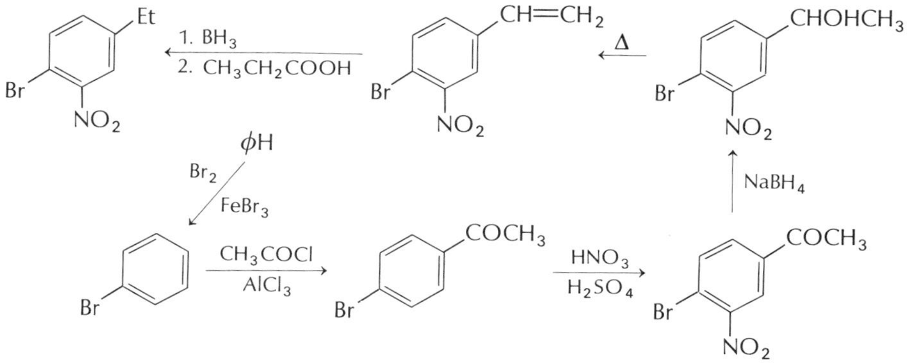

$$(15.38)$$

This last sequence was not easy. We had to know the properties of the nitro group with respect to the various reducing agents, a subject yet to be discussed. Nevertheless, we include the problem here to demonstrate the analytical process of working out a synthesis—how it is possible to work around a seemingly decisive roadblock.

5. Prepare *trans*-stilbene. The last step seems to be the easy dehydration of $\phi CH_2CHOH\phi$, a secondary alcohol that would result from $\phi CH_2MgCl + \phi CHO$. The Grignard reagent can be made from benzyl chloride, and the aldehyde from benzal chloride. Toluene is the starting point (Eq. 15.39).

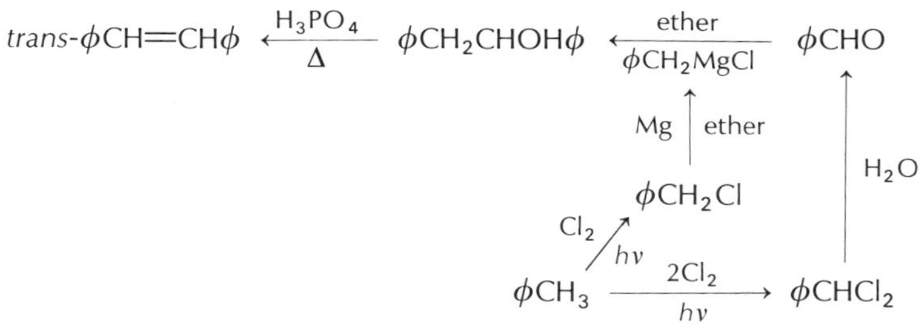

$$trans\text{-}\phi CH=CH\phi \xleftarrow[\Delta]{H_3PO_4} \phi CH_2CHOH\phi \xleftarrow[\phi CH_2MgCl]{ether} \phi CHO$$

$$\phi CH_2MgCl \xleftarrow[]{Mg \mid ether} \phi CH_2Cl \xleftarrow[hv]{Cl_2} \phi CH_3 \xrightarrow[hv]{2Cl_2} \phi CHCl_2 \xrightarrow{H_2O} \phi CHO \quad (15.39)$$

This synthesis is included to remind us that we must remember most of the reactions of nonaromatic compounds for use in manipulating side chains.

6. Prepare 1-phenyl-1-*p*-tolylethylene. The last step again is a dehydration, preferably of the tertiary rather than the primary alcohol. There are three ways of getting to the tertiary alcohol: a. TolCOϕ + MeMgBr, b. TolCOMe + ϕMgBr, and c. ϕCOMe + TolMgBr. In making tertiary alcohols, it is usually best to add the smallest group last, so we choose a. The required ketone is available by one of two Friedel-Crafts ketone syntheses: either TolCOCl + ϕH or ϕCOCl + ϕCH$_3$. Of these, the first requires several operations on ϕCH$_3$ to get to TolCOCl, so we choose the second and use the sequence in Equations 15.40.

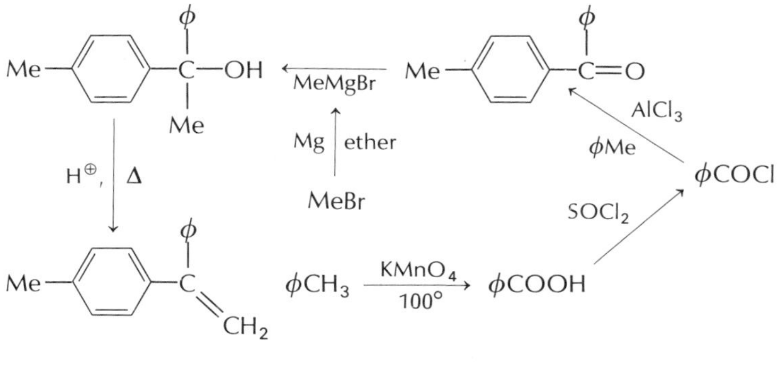

[Scheme showing synthesis via Me-C$_6$H$_4$-C(OH)(Me)(ϕ) ← MeMgBr — Me-C$_6$H$_4$-C(=O)ϕ ← AlCl$_3$, ϕCOCl from ϕMe via SOCl$_2$; with H$^\oplus$, Δ giving Me-C$_6$H$_4$-C(ϕ)=CH$_2$; ϕCH$_3$ →(KMnO$_4$/100°) ϕCOOH; MeBr via Mg/ether] (15.40)

7. Prepare ethyl *p*-ethoxybenzoate. The only way we have met for the introduction of the ethoxy group is nucleophilic displacement of a halogen. It should

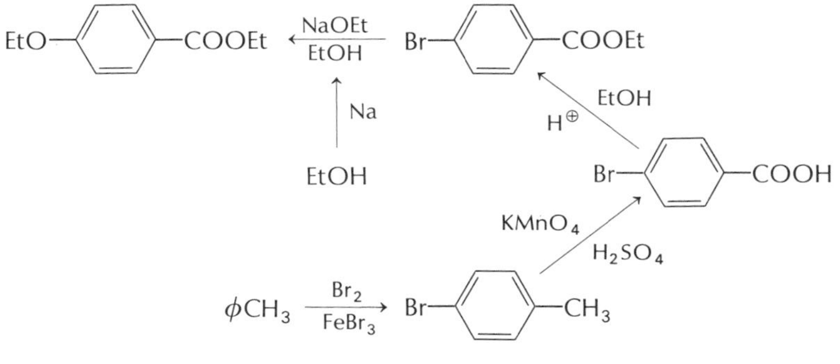

[Scheme: EtO-C$_6$H$_4$-COOEt ← NaOEt/EtOH — Br-C$_6$H$_4$-COOEt ← EtOH, H$^\oplus$ — Br-C$_6$H$_4$-COOH ← KMnO$_4$/H$_2$SO$_4$ — Br-C$_6$H$_4$-CH$_3$ ← Br$_2$/FeBr$_3$ — ϕCH$_3$; Na/EtOH] (15.41)

work here, because there is an electron-withdrawing group *p* to the point of reaction. Getting the *p*-bromobenzoic acid ester from the acid is easy, as is getting the acid from *p*-bromotoluene (Eqs. 15.41).

Aromatic compounds. A constant tension exists in pharmaceutical and agricultural research to develop a compound that interferes with some life forms but not others—something that kills only the "bad" or helps only the "good." However, the biochemical reactions of different organisms are so similar and the lives themselves often so intertwined ecologically that magic bullets are difficult to produce. Here are two examples of aromatic compounds that were thought for a long time to do only good but that are now virtually banned from use in the United States.

After its discovery in 1941, hexachlorophene came to be widely used as an antibacterial agent in soaps, and later on in deodorants and other cosmetics. As a 3% solution used to wash babies in hospitals, it prevented staphylococcus outbreaks. Workers investigating its possible use as a fungicide first noted in 1971 that hexachlorophene could cause brain damage. Warnings were sent to hospitals. The following year, when the deaths of French babies were traced to talcum powder consisting of 6 percent hexachlorophene, a virtual ban was placed on it. For instance, it now may be only used as a preservative in drugs and cosmetics at concentrations not greater than 0.1 percent.

DES was administered to farm animals to fatten them up for market. Consequently, the meat from such animals contained DES, now known to be a carcinogen. In 1971 just as a study with 1000 college women was being completed that demonstrated DES to be a "safe and effective" morning-after pill to prevent pregnancy, another study was uncovering evidence that DES had caused vaginal cancer in the daughters of women who had used it during their pregnancy.

hexachlorophene

DES
diethylstilbestrol

Exercises

15.7 Synthesize:
   a. *o*-nitrobenzaldehyde
   b. *m*-ethylphenol
   c. *p*-methylallylbenzene
   d. 4-methoxy-3-nitroacetophenone
   e. naphthalene

15.9 Summary

After we described the naming of aromatic compounds, we took up the energetics of substitution reactions. The σ and π complexes were introduced, and we examined the mechanism of electrophilic substitution. We discussed the effect of substituents upon the reaction (activation or deactivation, and orientation of the entering group) and emphasized the recognition of three classes of substituents. The reasons for the effects of substituents were clarified in terms of resonance theory and the I and R effects.

15.9 Summary

The electrophilic reagents were discussed in more detail in Section 15.3, and substituent effects were taken up in a more quantitative way by the use of partial rate factors. We briefly mentioned the problems of orientation in substitution in disubstituted benzenes and in polycyclic compounds.

Homolytic substitution was confined to a consideration of the halogens and of the phenyl radical. Nucleophilic substitution was described in some detail, and the marked effect of electron-withdrawing substituents was accounted for in terms of I and R effects. The benzyne intermediate and its synthetic usefulness were mentioned.

Under oxidation and reduction, only the indirect route by halogenation and hydrolysis and autoxidation were added to the methods already described in Chapter 14.

After a mention of coal and coal tar as the principal source of aromatic compounds, we discussed the preparation of arenes and certain substituted arenes.

We have seen that the aromatic substitution reactions differ from all the reactions we had previously encountered in terms of the influence of the aromatic system upon the mechanism and of substituents upon the rates and orientations. Aromatic chemistry tends to fall into a class of its own. However, the reactions of side chains provide a link to the other material we have considered.

Some new reactions presented in this chapter are listed here.

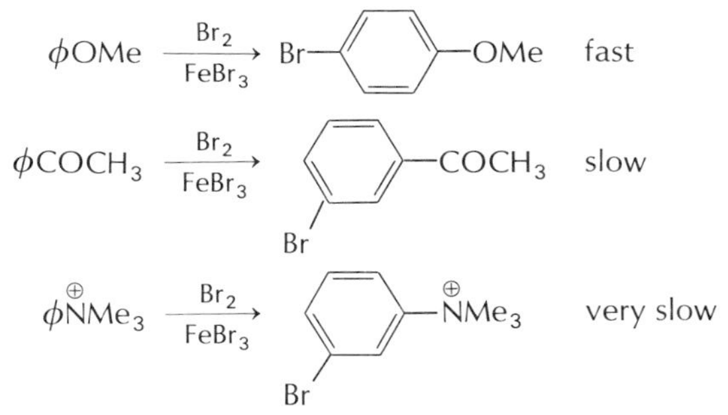

$$\phi OMe \xrightarrow{Br_2/FeBr_3} Br\text{-}C_6H_4\text{-}OMe \quad \text{fast} \tag{15.3}$$

$$\phi COCH_3 \xrightarrow{Br_2/FeBr_3} \text{m-Br-}C_6H_4\text{-}COCH_3 \quad \text{slow}$$

$$\phi \overset{\oplus}{N}Me_3 \xrightarrow{Br_2/FeBr_3} \text{m-Br-}C_6H_4\text{-}\overset{\oplus}{N}Me_3 \quad \text{very slow}$$

$$CH_3\phi \xrightarrow{HNO_3/H_2SO_4} \text{o-}CH_3\text{-}C_6H_4\text{-}NO_2 \text{ (more)} + \text{p-}CH_3\text{-}C_6H_4\text{-}NO_2 \text{ (less)} \tag{15.6}$$

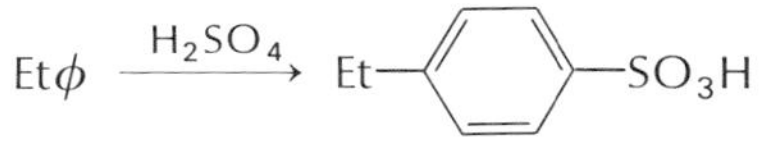

$$Et\phi \xrightarrow{H_2SO_4} \text{p-}Et\text{-}C_6H_4\text{-}SO_3H \tag{15.9}$$

$$\phi NO_2 \xrightarrow{ClSO_3H} \text{m-}O_2N\text{-}C_6H_4\text{-}SO_2Cl \tag{15.10}$$

$$\text{p-}CH_3\text{-}C_6H_4\text{-}SO_3H \xrightarrow[\text{boil}]{80\% \text{ H}_2SO_4} CH_3\text{-}C_6H_5$$

$$CH_3\phi \xrightarrow{CH_3COCl/AlCl_3} \text{p-}CH_3\text{-}C_6H_4\text{-}COCH_3 \tag{15.11}$$

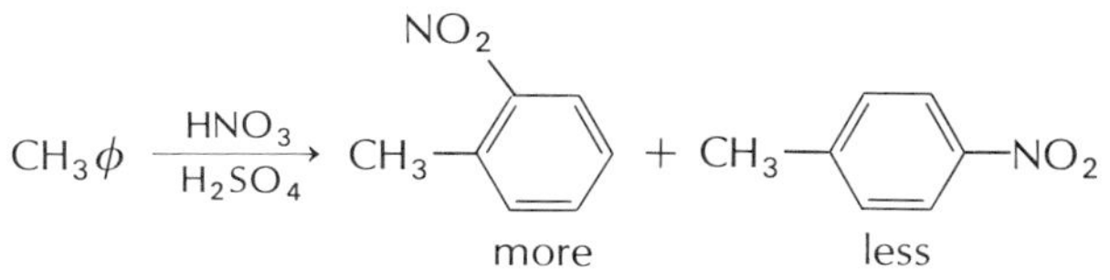

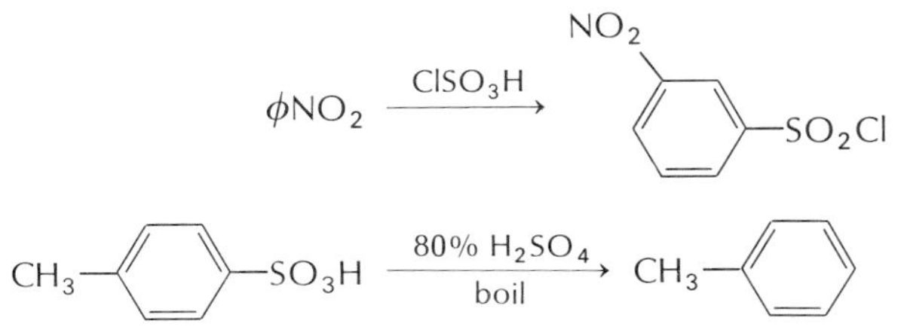

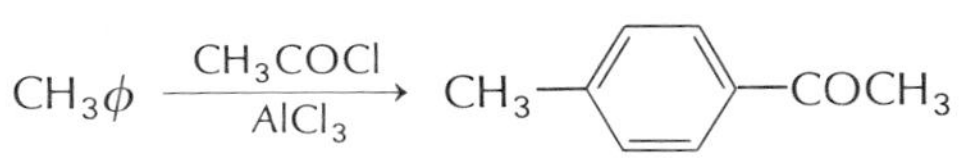

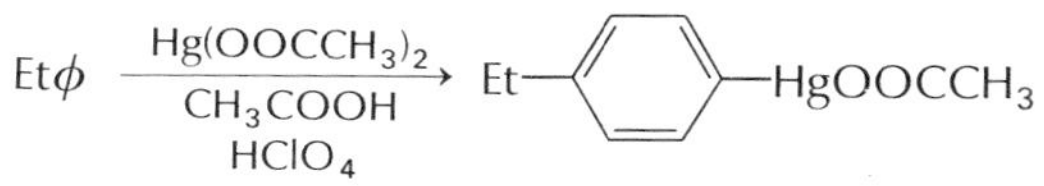

$$Et\phi \xrightarrow[\substack{CH_3COOH \\ HClO_4}]{Hg(OOCCH_3)_2} Et\text{—}C_6H_4\text{—}HgOOCCH_3 \quad (15.12)$$

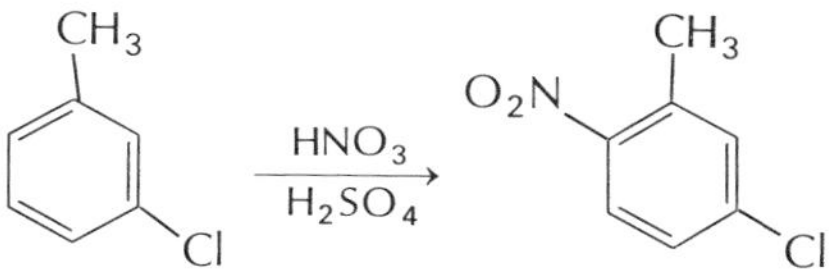

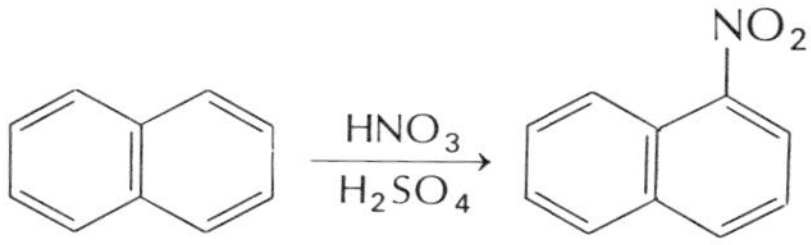

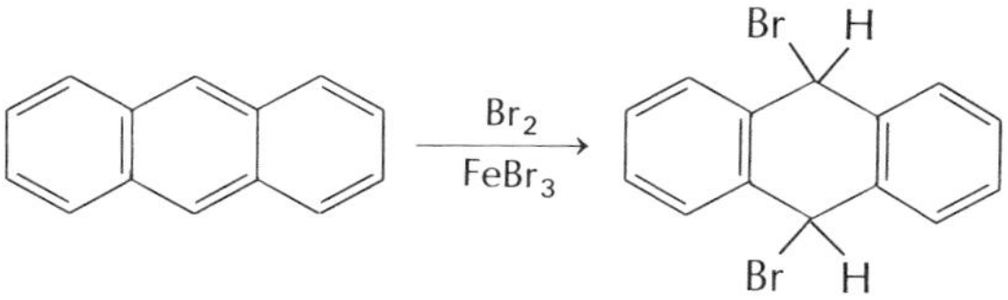

$$\underset{Cl}{\overset{CH_3}{C_6H_4}} \xrightarrow[H_2SO_4]{HNO_3} \underset{Cl}{\overset{CH_3,\ O_2N}{C_6H_3}}$$

$$\text{naphthalene} \xrightarrow[H_2SO_4]{HNO_3} \text{1-nitronaphthalene}$$

$$\text{anthracene} \xrightarrow[FeBr_3]{Br_2} \text{9,10-dibromo-9,10-dihydroanthracene} \quad (15.15)$$

$$\phi CH(CH_3)_2 \xrightarrow[h\nu]{Cl_2} \phi C(CH_3)_2Cl$$

$$CH_3\text{—}C_6H_4\text{—}Br + \phi CH_2CH_2NH_2 \xrightarrow{HCONMe_2} \phi CH_2CH_2\overset{\oplus}{N}H_2\text{—}C_6H_4\text{—}CH_3 \quad (15.18)$$

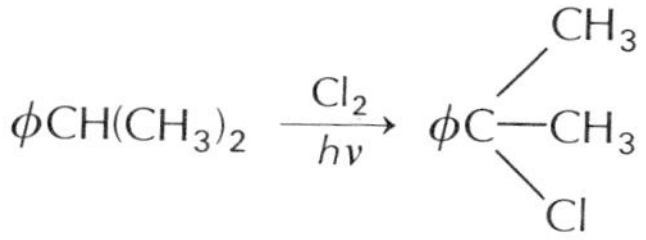

$$C_6H_2(NO_2)_3Cl \xrightarrow{H_2O} C_6H_2(NO_2)_3OH \quad (15.19)$$

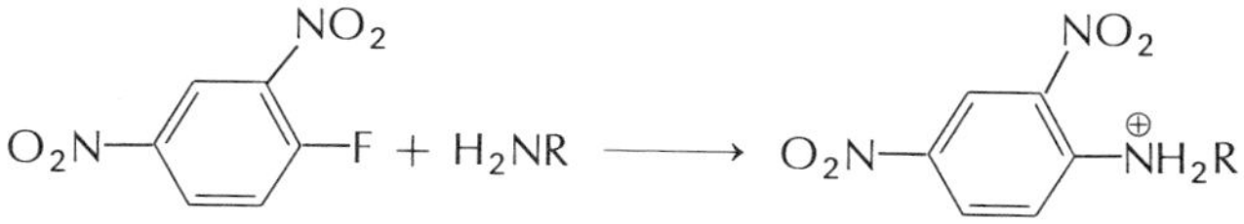

$$\underset{NO_2}{\underset{|}{O_2N\text{—}C_6H_3\text{—}F}} + H_2NR \longrightarrow \underset{NO_2}{\underset{|}{O_2N\text{—}C_6H_3\text{—}\overset{\oplus}{N}H_2R}} \quad (15.21)$$

$$\phi Cl + \overset{\ominus}{N}H_2 \longrightarrow \phi NH_2 \text{ (benzyne reaction)} \quad (15.23)$$

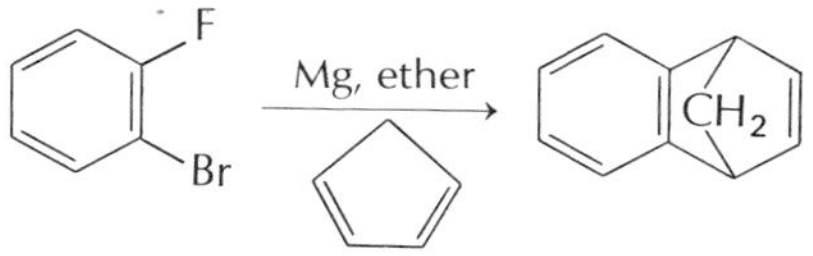

$$\underset{Br}{\overset{F}{C_6H_4}} \xrightarrow[\text{Mg, ether}]{\text{+ cyclopentadiene}} \text{benzonorbornene} \quad (15.24,\ 15.25)$$

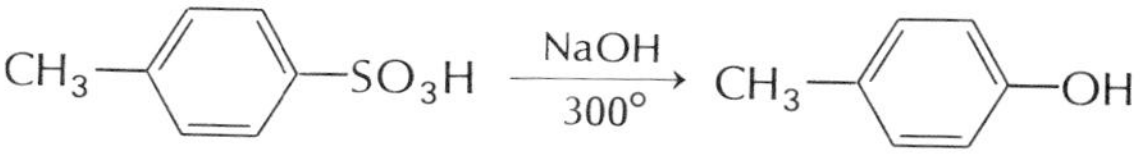

$$CH_3\text{—}C_6H_4\text{—}SO_3H \xrightarrow[300°]{NaOH} CH_3\text{—}C_6H_4\text{—}OH \quad (15.26)$$

$$\text{2-naphthalenesulfonic acid} \xrightarrow[\Delta]{NaCN} \text{2-naphthonitrile} \quad (15.27)$$

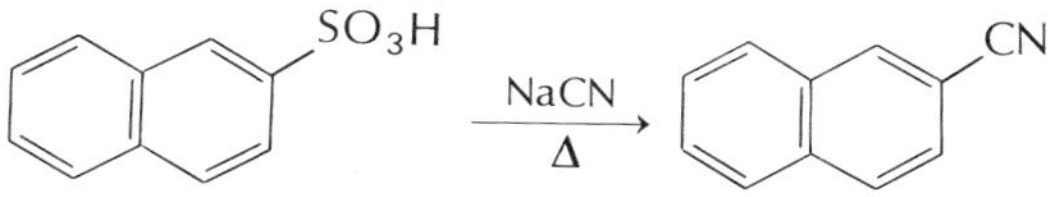

$$\phi CH_2CH_3 \xrightarrow[h\nu]{Br_2} \phi CHBrCH_3 \xrightarrow{H_2O} \phi CHOHCH_3 \quad (15.29)$$

$$\phi CH(CH_3)_2 \xrightarrow{O_2} \phi-\underset{\underset{CH_3}{|}}{\overset{\overset{CH_3}{|}}{C}}-OOH \quad (15.31)$$

$$\phi-\underset{\underset{CH_3}{|}}{\overset{\overset{CH_3}{|}}{C}}-OOH \xrightarrow{H^{\oplus}} \phi OH + CH_3COCH_3 \quad (15.32)$$

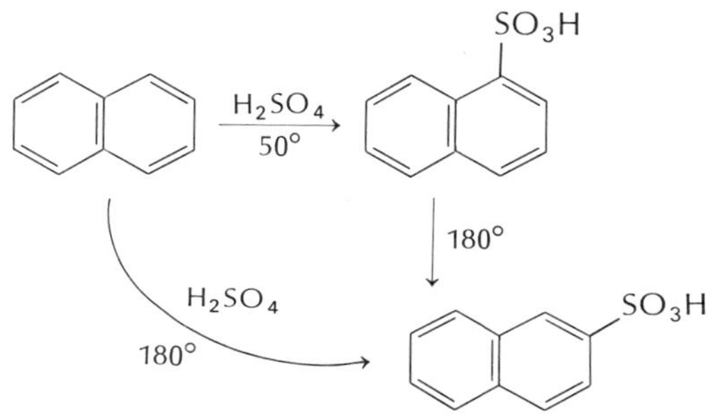

15.10 Problems

15.1 Substance A ($C_{10}H_{10}$) gives a precipitate with ammoniacal silver nitrate solution. A reacts with MeMgBr in ether to give off a gas. Addition of acetone to the ethereal mixture followed by work-up gives B. B with one mole of H_2 over Pt gives C, which, with another mole of H_2, gives D. D heated with H_2SO_4 or H_3PO_4 gives E. E with one mole of H_2 over Pt gives F ($C_{13}H_{20}$). C with hot H_3PO_4 gives G. G will take up two moles of H_2 to give F. G with one mole of dry HCl gives H as the major product and a small amount of an isomer I. A reacts with two moles of H_2 over Pt to give J. Nitration of J gives a single mononitro product.
Give a structural formula for each lettered substance.

15.2 Substance A (C_9H_{10}), upon drastic oxidation, gives one mole of benzoic acid and two moles of CO_2. A with BH_3 followed by alkaline H_2O_2 gives B. B reacts with $SOCl_2$ to give C. C with $AlCl_3$ gives D and HCl. Oxidation of D gives one mole of phthalic acid and one mole of CO_2. A with HBr under polar conditions and in the absence of peroxides gives E as the major product and a small amount of F. Treatment of E or F with hot alcoholic KOH gives G, an isomer of A. E with dilute NaOH at 25° quickly gives H. H with Al(OtBu)$_3$ and excess acetone gives I. I with ZnHg$_x$ and concentrated HCl gives J, which is also obtained if A or G is hydrogenated.
Give structural formulas for the lettered substances.

15.3 Which member of each of the following pairs will react more rapidly with the given reagent?

	Reagent	A	B
a.	Br_2—$FeBr_3$	C_6H_5Cl	$C_6H_5COOCH_3$
b.	Br_2—$FeBr_3$	C_6H_5Cl	$C_6H_5OCH_3$
c.	Cl_2—$FeCl_3$	$C_6H_5CH_3$	$C_6H_5CH_2Cl$
d.	H_2SO_4—SO_3	C_6H_6	C_6H_5OH
e.	HNO_3—H_2SO_4	$C_6H_5COOCH_3$	$C_6H_5OCOCH_3$
f.	Cl_2—$AlCl_3$	$C_6H_5CCl_3$	$C_6H_5CH_2Cl$
g.	CH_3Cl—$AlCl_3$	$C_6H_5NHCOCH_3$	$C_6H_5CONHCH_3$
h.	CH_3CH_2Br—$AlCl_3$	C_6H_6	$C_6H_5CH_3$
i.	NaOH	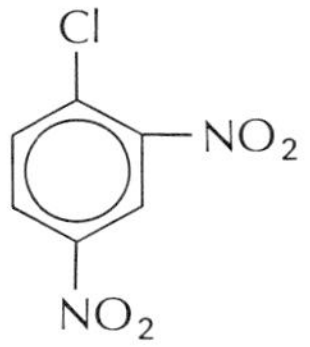 (2,4-dinitrochlorobenzene: Cl, NO₂ at ortho, NO₂ at para)	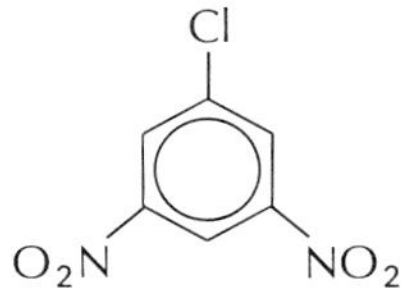 (3,5-dinitrochlorobenzene: Cl with two NO₂ groups meta)

15.4 Write the major product(s) for each of the following reactions.

a. $C_6H_5CH_3 \xrightarrow{Cl_2,\ FeCl_3}$

b. $C_6H_5CH_2Cl \xrightarrow{Cl_2,\ FeCl_3}$

c. $C_6H_5CCl_3 \xrightarrow{Cl_2,\ AlCl_3}$

d. $C_6H_5CH_2Cl + 2Cl_2 \xrightarrow{light}$

e. $C_6H_5COOH \xrightarrow[H_2SO_4]{HNO_3}$

f. $C_6H_5CH_2COOH \xrightarrow[H_2SO_4]{HNO_3}$

g. $C_6H_5Cl \xrightarrow[H_2SO_4]{HNO_3}$

h. $C_6H_5NO_2 \xrightarrow[AlCl_3]{Cl_2}$

i. 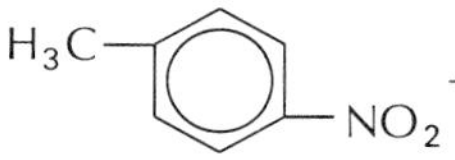 H_3C—(C₆H₄)—NO_2 + $Cl_2 \xrightarrow{FeCl_3}$

j. 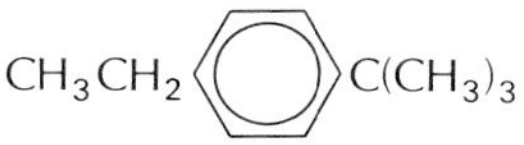 CH_3CH_2—(C₆H₄)—$C(CH_3)_3 \xrightarrow{ClSO_3H}$

k. 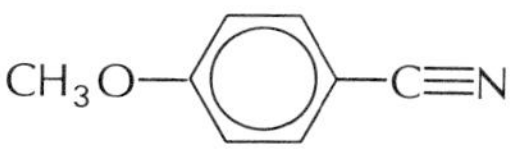 CH_3O—(C₆H₄)—$C{\equiv}N \xrightarrow[H_2SO_4]{HNO_3}$

l. 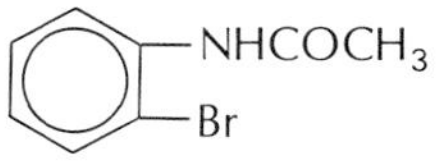 (C₆H₄ with NHCOCH₃ and Br) $\xrightarrow[SO_3]{H_2SO_4}$

m. $C_6H_5SO_2Cl \xrightarrow{ClSO_3H}$

n. $C_6H_5CH_2SO_2Cl \xrightarrow{ClSO_3H}$

o. Excess $C_6H_6 + CHCl_3 \xrightarrow{AlCl_3}$

p. $C_6H_5CH_3 + CH_2=\underset{\underset{CH_2CH_3}{|}}{C}CH_3 \xrightarrow{H_2SO_4}$

q. $C_6H_5NHCOC_6H_5 + (CH_3CO)_2O \xrightarrow{AlCl_3}$

r. $C_6H_5COOCH_3 + C_6H_5COCl \xrightarrow{AlCl_3}$

s. 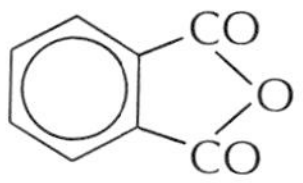 [phthalic anhydride] + $C_6H_5Br \xrightarrow{AlCl_3}$

t. $CH_3OOC(CH_2)_5COCl + C_6H_6 \xrightarrow{AlCl_3} A \xrightarrow[\text{2. } H^\oplus]{\text{1. KOH}} B \xrightarrow{Zn(Hg)/HCl} C$

u. $C_6H_5CH_3 + (CH_3)_2CHCH_2Br \xrightarrow{AlCl_3}$

v. $C_6H_5CH_3 + (CH_3)_2CHCOCl \xrightarrow{AlCl_3} A \xrightarrow{N_2H_4/KOH} B$

w. $C_6H_5COCH_2CH_2COCl \xrightarrow{AlCl_3}$

x. $C_6H_5CH_2CH_2CH_2COCl \xrightarrow{AlCl_3}$

y. 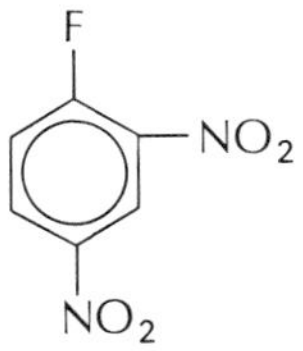 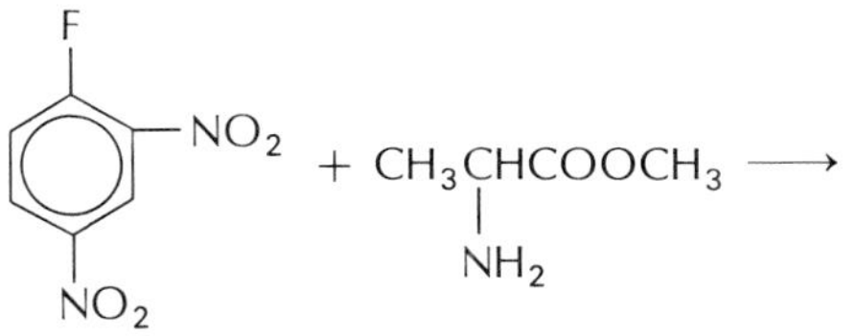 2-fluoro-1,4-dinitro... $+ CH_3\underset{\underset{NH_2}{|}}{CH}COOCH_3 \longrightarrow$

z. 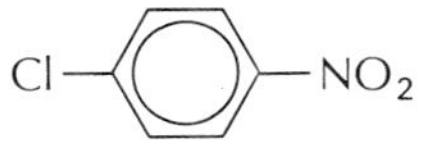 $Cl-\text{C}_6H_4\text{-}NO_2 \xrightarrow[160°]{NH_4OH}$

15.5 Write a mechanism for each of the following reactions.

a. 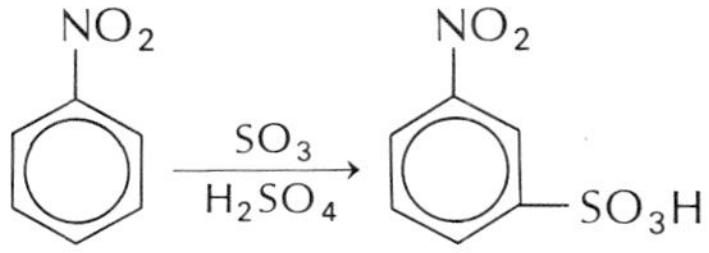 [nitrobenzene] $\xrightarrow{SO_3 / H_2SO_4}$ [m-nitrobenzenesulfonic acid, SO_3H]

b. $C_6H_5Cl + (CH_3)_3CCH_2Cl \xrightarrow{AlCl_3} Cl-\text{C}_6H_4\text{-}\underset{\underset{CH_3}{|}}{\overset{\overset{CH_3}{|}}{C}}CH_2CH_3$

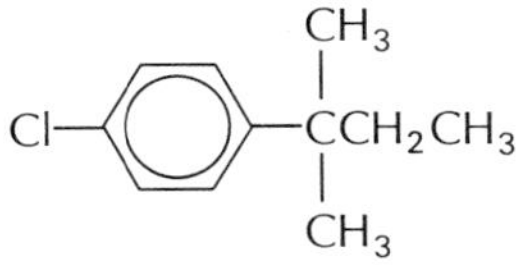

c. 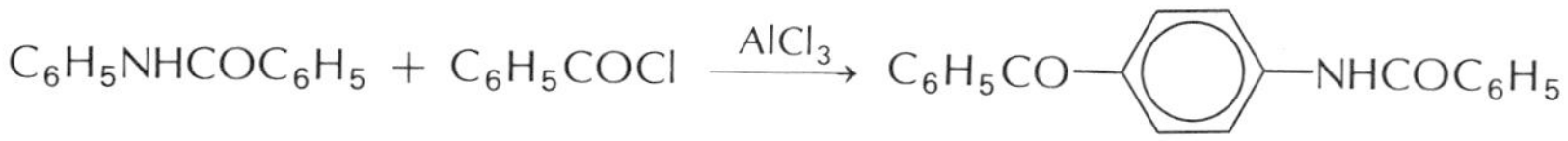 $C_6H_5NHCOC_6H_5 + C_6H_5COCl \xrightarrow{AlCl_3} C_6H_5CO\text{-}\langle\text{C}_6\text{H}_4\rangle\text{-}NHCOC_6H_5$

d. 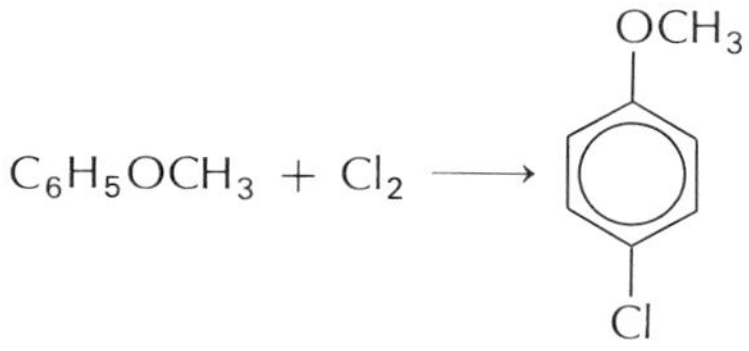 $C_6H_5OCH_3 + Cl_2 \longrightarrow$ *p*-chloroanisole (OCH₃ and Cl para on benzene ring)

e. 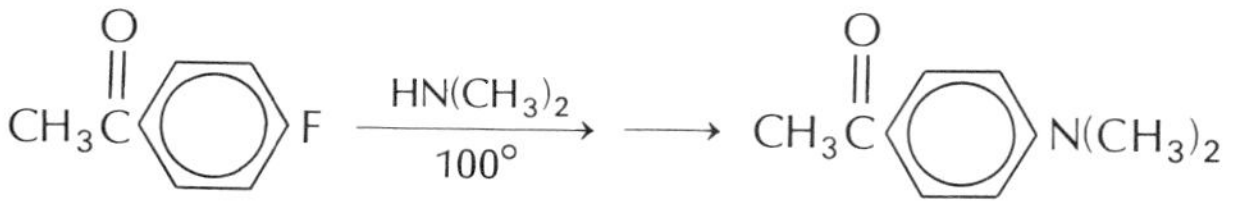 $CH_3CO\text{-}C_6H_4\text{-}F \xrightarrow[100°]{HN(CH_3)_2} CH_3CO\text{-}C_6H_4\text{-}N(CH_3)_2$

15.11 Bibliography

Beside the general organic textbooks and paperbacks previously mentioned in Chapter 7, all of which discuss arenes and the aromatic substitution reactions, a very good paperback is the one by Stock.

1. Stock, Leon M., *Aromatic Substitution Reactions*. Englewood Cliffs, N.J.: Prentice-Hall, 1968.

   Some excellent articles are the following.

2. Bunnett, Joseph F., The Base-Catalyzed Halogen Dance, and Other Reactions of Aryl Halides, *Accounts of Chemical Research 5*, 139 (1972).

3. Hill, George R., Some Aspects of Coal Research, *Chemical Technology 2*, 292 (1972).

4. Olah, George A., Mechanism of Electrophilic Aromatic Substitutions, *Accounts of Chemical Research 4*, 240 (1971).

5. Ridd, John H., Mechanism of Aromatic Nitration, *Accounts of Chemical Research 4*, 248 (1971).

6. Roberts, Royston M., Friedel-Crafts Chemistry, *Chemical and Engineering News*, Jan. 25, 1965, p. 96.

16 Aldehydes and Ketones

Besides having its own reactions, the double bond between carbon and oxygen influences its neighboring atoms in aldehydes and ketones. We may also consider this subject as the chemistry of organic anions, which we have rather neglected in our pursuit of cations. Even though we shall encounter many reactions in this chapter, there are relatively few new principles: addition and substitution are the key words that describe most of the chapter.

16.1   Introduction; physical properties; equilibria   480
    1. Occurrence.   2. Physical properties.   3. The carbon-oxygen double bond.   4. Acid-base equilibria: addition, substitution.   5. Enols and enolate ions.   6. Ketosis. Exercises.

16.2   Spectrometry   487
    1. The carbonyl stretching frequency.   2. Absorption in the uv. Exercises.

16.3   Addition reactions   490
    1. Hydrates; acetals.   2. Acid-catalyzed trimerization.   3. Sodium bisulfite.   4. Hydrogen cyanide.   5. Grignard reagents; the Reformatsky reaction.   6. Wittig reaction: ylides.   7. Ammonia, amines, and substituted amines.   8. Schiff base in biochemistry.   9. Chloromethylation. Exercises.

16.4   Substitution and condensation reactions   501
    1. Basic and acidic halogenation.   2. Aldol condensation: base catalysis, acid catalysis.   3. Mixed aldol condensations.   4. β-diketones.   5. C and O acylation.   6. Michael reaction.   7. Knoevenagel reaction.   8. Benzoin condensation. Exercises.

16.5   Rearrangements   515
    1. Benzilic acid rearrangement.   2. Baeyer-Villiger reaction. Exercises.

16.6   Oxidation and reduction   517
    1. Oxidations: $KMnO_4$, the chromium reagents.   2. Special reagents.   3. Previously described reductions.   4. Bimolecular reduction. Exercises.

480   Chapter sixteen

16.7   Preparation of aldehydes and ketones   520
       1. Substitutions.   2. Additions.   3. Rearrangements.
       4. Oxidations.   5. Reductions. Exercises.

16.8   Summary   531

16.9   Problems   531

16.10 Bibliography   535

16.11 Clair J. Collins: How to use isotopic tracers   535a

16.1 Introduction; physical properties; equilibria

Occurrence. Aldehydes and ketones are widespread in nature, not only as simple compounds, but also as derivatives and as intermediates in many of the reactions in plants and animals. Aromatic aldehydes in particular occur frequently in odoriferous and flavoring substances such as bitter almond, cinnamon, vanilla, and so on.

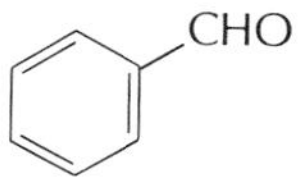

benzaldehyde   cinnamaldehyde   vanillin
oil of bitter almonds oil of cinnamon

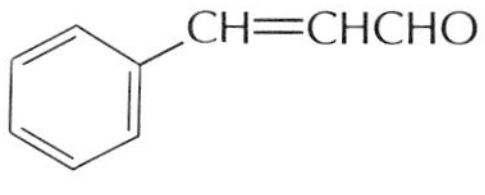

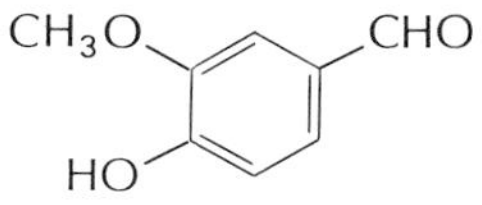

Aldehydes and ketones undergo a variety of reactions, probably more than any other class of compounds. Standing between the saturated hydrocarbons (C_nH_{2n+2}) and the acids ($C_nH_{2n}O_2$), aldehydes and ketones ($C_nH_{2n}O$) are at an intermediate stage of oxidation and are subject to both oxidation and reduction. We need only recall the disproportionation of formaldehyde in the Cannizzaro reaction (Eq. 9.53) to set the stage for one of the several classes of reactions we shall meet.

Physical properties. Some representative aldehydes and ketones are listed in Table 16.1. The melting points tend to be low and variable. The boiling points are greater than those of the corresponding hydrocarbons and ethers and lower than those of the corresponding alcohols and acids. Compare C_4H_8O (butyraldehyde, bp 76°; 2-butanone, bp 80°) with C_5H_{12} (pentane, bp 36°) and with $C_4H_{10}O$ (1-butanol, bp 118°) and $C_3H_6O_2$ (propionic acid, bp 141°). The aldehydes and ketones are unable to hydrogen bond with each other, but they do possess strong dipoles, and some dipolar association occurs. The liquid range is large, on the order of 200°.

Solubility in water is limited roughly to $C_5H_{10}O$ or less, and only acetaldehyde and acetone are completely miscible with water.

If we consider formaldehyde to be the parent, replacement of one of the hydrogen

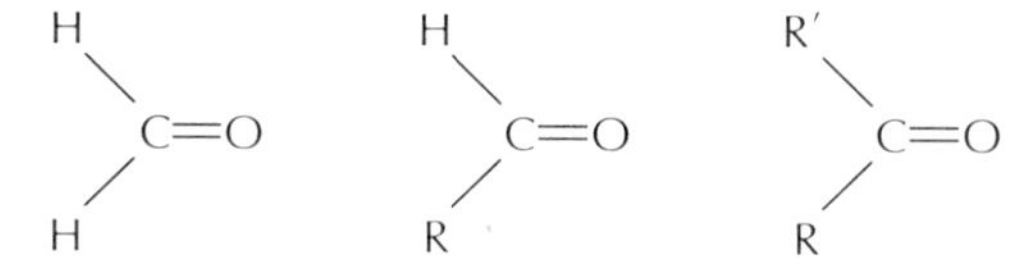

$$\underset{\text{formaldehyde}}{\overset{H}{\underset{H}{\diagdown}}C=O} \qquad \underset{\text{aldehyde}}{\overset{H}{\underset{R}{\diagdown}}C=O} \qquad \underset{\text{ketone}}{\overset{R'}{\underset{R}{\diagdown}}C=O}$$

atoms with an alkyl or aryl group gives rise to an aldehyde, and replacement of both hydrogen atoms with alkyl or aryl groups gives rise to a ketone.

Table 16.1 Aldehydes and ketones

	m.p., °C	b.p., °C	Solubility, g/100 g H_2O, 25°
Formaldehyde	−92	−21	v. sol.
Acetaldehyde	−121	20	∞
Propionaldehyde	−81	49	16
n-Butyraldehyde	−99	76	7
n-Valeraldehyde	−91	103	sl. s
Caproaldehyde	−56	131	sl. s
Heptaldehyde	−42	155	0.1
Phenylacetaldehyde	34	194	sl. s
Benzaldehyde	−26	178	0.3
o-Tolualdehyde		200	
m-Tolualdehyde		199	
p-Tolualdehyde		205	
Salicylaldehyde (o-Hydroxybenzaldehyde)	−7	197	1.7
p-Hydroxybenzaldehyde	116		1.4
Anisaldehyde	3	248	0.2
Vanillin	82	285	1
Piperonal	37	263	0.2
Acetone	−94	56	∞
Methyl ethyl ketone	−86	80	26
2-Pentanone	−78	102	6.3
3-Pentanone	−41	101	5
2-Hexanone	−35	150	2.0
3-Hexanone		124	sl. s
Methyl isobutyl ketone	−85	119	1.9
Acetophenone	21	202	
Propiophenone	21	218	
n-Butyrophenone	11	232	
Benzophenone	48	306	

482 Chapter sixteen

The carbon-oxygen double bond.   The carbon-oxygen double bond in the ground state consists of a σ and a π bond. The carbon atom is an sp^2 hybrid, and the carbon atom and the three atoms bound to it all lie in the same plane.

The oxygen atom may be in the atomic configuration $2s^2 p_x^2 p_y^1 p_z^1$, the p_z orbital

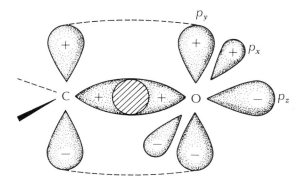

overlapping with a carbon sp^2 orbital to form a σ bond, and the p_y orbital overlapping the carbon p_y orbital to give a π bond. This formulation leaves two electrons buried in the $2s$ oxygen orbital and two in the p_x orbital, which lies in the same plane as the two groups (H or R) bonded to the carbon atom.

On the other hand, the oxygen atom may be hybridized into sp^2 (or into any intermediate hybrid). In this case the four electrons on oxygen are divided between the two nonbonded sp^2 orbitals, which lie in the same plane (the xz plane) as the groups bonded

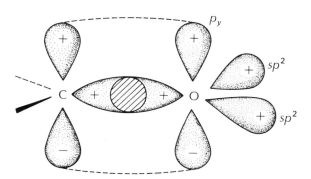

to carbon. In any case, besides the easily available electrons of the π bond, there are one or two pairs of available electrons on oxygen. Therefore, we expect the ╲C═O bond to be more nucleophilic than a ╲C═C╱ bond.

If we look at the resonance structures, we expect little or no contribution from c, but

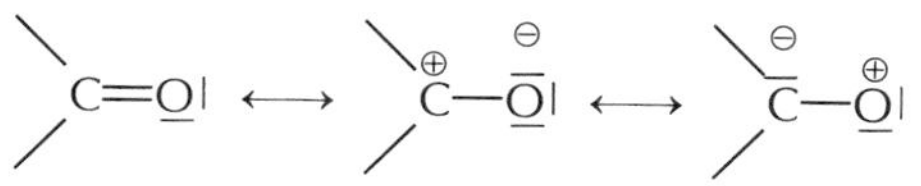

$$\diagdown\!\!\!\diagup \text{C}=\overline{\text{O}}| \;\longleftrightarrow\; \diagdown\!\!\!\diagup \overset{\oplus}{\text{C}}-\overset{\ominus}{\overline{\underline{\text{O}}}}| \;\longleftrightarrow\; \diagdown\!\!\!\diagup \overset{\ominus}{\text{C}}-\overset{\oplus}{\overline{\text{O}}}|$$

a   b   c

16.1 Introduction; physical properties; equilibria

a strong contribution of *b* because the electronegativity of oxygen is greater than that of carbon. A strong contribution of *b* suggests that *d* would be a rough picture of the carbon-

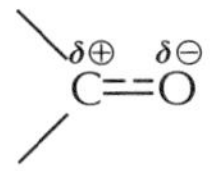

$$\text{C}\overset{\delta+}{=}\overset{\delta-}{\text{O}}$$

d

oxygen double bond and that any molecule containing it will have a sizable dipole moment. Aldehydes and ketones with R and R' alkyl groups do have μ about 2.5 D.

This large dipole moment suggests that *a* and *b* contribute about equally. Thus, *b* would be as good a single structure to use as would *a*. It is conventional, however, to use $\text{C}=\text{O}$. We should keep in mind that the double bond is much less double and much more ionic than the $\text{C}=\text{C}$ with which we have become familiar.

The increased ionic character of $\text{C}=\text{O}$ causes an increase in the bond energy (176 to 179 kcal) as compared to that of $\text{C}=\text{C}$ (146 kcal). We may also compare the process $\text{C}=\text{C} \rightarrow \text{C}-\text{C}$, which requires $146 - 83 = 63$ kcal, to $\text{C}=\text{O} \rightarrow \text{C}-\text{O}$, which requires at least $175 - 86 = 89$ kcal. (We may consider that the extra 26 kcal is needed to transfer an electron from oxygen to carbon in order to destroy the contribution of *b*.) Also, the alkyl aldehydes and ketones have a carbon-oxygen bond length of 1.22 Å, shorter than the 1.33 Å length of a $\text{C}=\text{C}$.

The contribution of *b* causes the $\text{C}=\text{O}$, called the <u>carbonyl group</u>, to exert a $-I$ effect and, when possible, a $-R$ effect as well. Thus, a $\text{C}=\text{C}$ that is conjugated with a

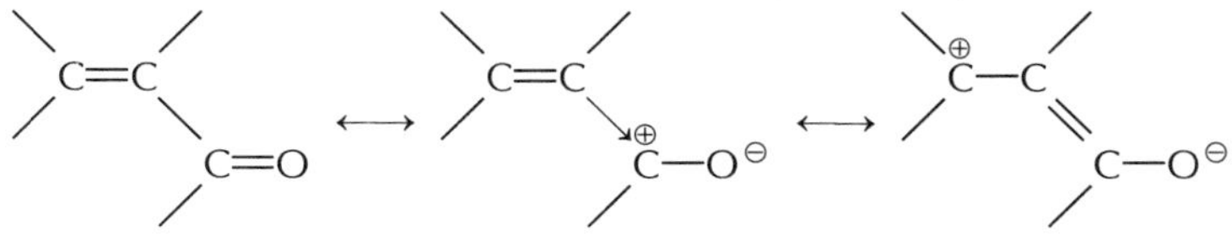

$$\text{C}=\text{C}-\text{C}=\text{O} \longleftrightarrow \text{C}=\text{C}-\overset{+}{\text{C}}-\text{O}^{-} \longleftrightarrow \overset{+}{\text{C}}-\text{C}=\text{C}-\text{O}^{-}$$

$\text{C}=\text{O}$ is polarized in the ground state. In effect, the positive charge is delocalized and the $\text{C}=\text{C}$ becomes less nucleophilic (more electrophilic). If an aryl group is bonded to $\text{C}=\text{O}$, the conjugation leads to partial removal of electrons (both $-I$ and $-R$ in operation) from the ring. The result is deactivation to electrophilic substitution and *meta* orientation and activation to nucleophilic substitution at the positions *ortho* and *para* to the carbonyl group.

484 Chapter sixteen

Perhaps it is surprising that relatively few of the reactions of aldehydes involve the carbon-hydrogen bond of the aldehyde group, $R-C\overset{\displaystyle O}{\underset{H}{\diagdown}}$. (Oxidation and radical attack are exceptions.) Therefore, we consider the reactions of aldehydes and ketones side by side and concentrate our attention on the three points of initial attack: oxygen, carbon, and an α hydrogen atom $\overset{H}{\underset{\diagup}{\diagdown}}C-C=O$. Let us look at the $\diagdown\!\!C=O\!\!\diagup$ first.

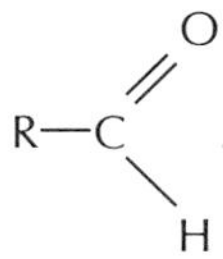

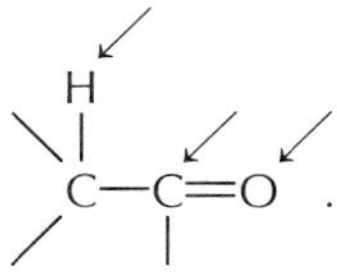

Acid-base equilibria: addition, substitution. Because of the polarization of $\diagdown\!\!C=O\!\!\diagup$, the oxygen is subject to attack by acids (both protic and aprotic), and the carbon can be attacked by bases or nucleophiles. Let us use HB to represent a protic acid and $B^{\ominus}$ to represent a base. Equation 16.1 is a generalized picture of the possible interactions of a carbonyl group with an acid and with a base.

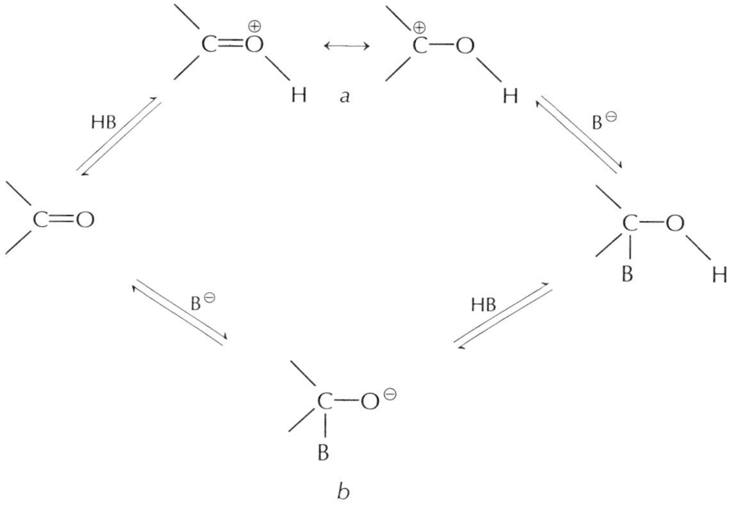

$$(16.1)$$

In essence, we have an acid-base equilibrium in which the carbonyl group may act as a base towards an acid (HCl or $H_3O^{\oplus}$, as examples) to give a, a cation. The cation, with the positive charge distributed between C and O, reacts with the conjugate base $B^{\ominus}$ to lose a proton and return to $\diagdown\!\!C=O\!\!\diagup$. The base also may react to form a σ bond to C, in which case the result is an addition to the carbonyl group. If $B^{\ominus}$ dissociates from the product, a is reformed. The cation a is an intermediate in the equilibrium set up between $\diagdown\!\!C=O\!\!\diagup$ + HB and the addition product, $\overset{OH}{\underset{B}{\diagdown\!\!C\!\!\diagup}}$

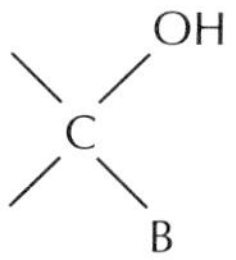

If $B^{\ominus}$ is a good nucleophile (for example, cyanide ion, $C{\equiv}N^{\ominus}$) the first step may be the formation of a σ bond to C to give the anion b. Reaction of b with the acid HB leads to the addition product $\underset{B}{\overset{OH}{\underset{|}{C}}}$. If the product loses a proton to $B^{\ominus}$, b is again formed. The anion may return to the reactant $\overset{}{\underset{}{C}}{=}O$ by loss of $B^{\ominus}$. The anion b is an intermediate in the equilibrium between reactants and the addition product.

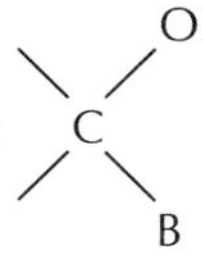

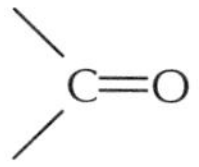

Whether a given reaction of a carbonyl group with HB proceeds by way of a or b may be difficult to decide. Either route is possible in a given mixture. In general, under acidic conditions the major intermediate is a, and under basic conditions, b. Both routes require participation by both HB and $B^{\ominus}$. The difference lies in the order of their participation. The position of the equilibrium varies and depends upon the structure of the aldehyde or ketone as well as upon the nature of B.

The ability of the carbonyl group to act as an acid or base is strongly affected by the groups to which it is attached. Thus, if the relative K_{eq} for the acceptance of a proton by acetophenone is 1.00, K_{eq} for p-nitroacetophenone is 0.02, and that for p-methoxyacetophenone is 25. The strong $-I$, $-R$ effect of the p-nitro group weakens the ability of the carbonyl group to accept a proton because of the positive charge adjacent to the carbonyl. The moderate $-I$, strong $+R$ effect of the p-methoxy strengthens the ability to take up a proton.

If an α hydrogen is present, it is strongly activated, as an allylic hydrogen would be. Another set of equilibria enter the picture (Eq. 16.2). The product b, known as an <u>enol</u>

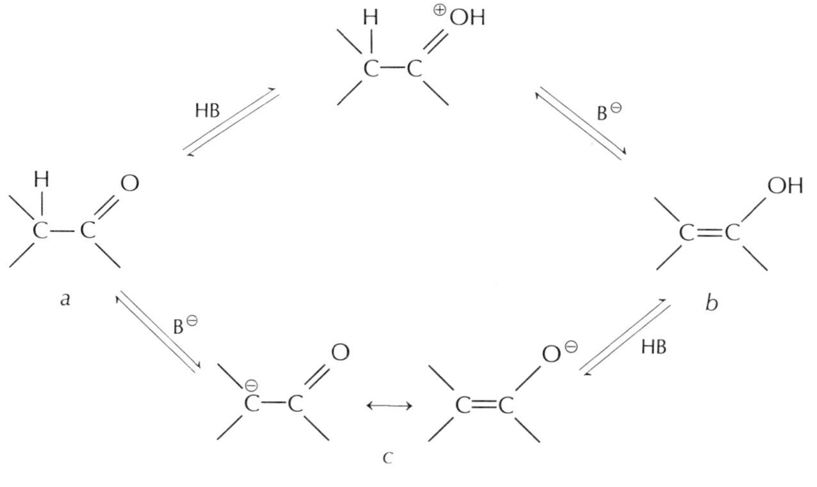

(16.2)

(from alk<u>ene</u> plus alcoh<u>ol</u>), is a <u>tautomer</u> of the original aldehyde or ketone a. Enols (b) and enolate ions (c, the anionic intermediate) are bases (nucleophiles) and are very reactive with acids (electrophiles). Thus, if α hydrogens are present, we have competition between the acid-base equilibria of Equation 16.1 and substitution at the α position (Eq. 16.3).

$$\phi COCH_3 \underset{}{\overset{HO^{\ominus}}{\rightleftharpoons}} \phi COCH_2^{\ominus} \xrightarrow{Br_2} \phi COCH_2Br + Br^{\ominus} \qquad (16.3)$$

Actually, the base-catalyzed substitution reaction shown is only the first in a sequence of reactions that we shall examine later in more detail.

Enols and enolate ions. If we replace the general base $B^{\ominus}$ with water, the equilibrium constant becomes K_a (Eq. 16.4). The magnitude of K_a may be taken as a measure

$$CH_3COCH_3 + H_2O \rightleftharpoons CH_3COCH_2^{\ominus} + H_3O^{\oplus} \qquad (16.4)$$

of the stability of the enolate ion. Table 16.2 gives pK_a for a few selected compounds and

Table 16.2 Stability of enolate ions and rates of formation

Compound	pK_a	k
CH_3COCH_2—H	20	4.7×10^{-10}
$CH_3COCHCl$—H	16.5	5.5×10^{-8}
CH_3COCCl_2—H	15	7.3×10^{-7}
$(CH_3CO)_2CH$—H	9	1.7×10^{-2}
$(OHC)_2CH$—H	5	—
C_6H_5O—H	10	fast
CH_3CH_2O—H	18	fast

the rate constant for the forward reactions. We see that acetone is only a hundred times less acidic than ethanol. If chlorine is substituted, the electron-withdrawing effect increases the acidity of the remaining protons. The rate of removal of the proton from carbon is slow, but it increases as the acidity increases. This parallel holds only for similarly constructed compounds. If we compare C—H bond breaking with O—H bond breaking, for example, we find no correlation with acidity. A C—H bond acid ionizes slowly, whereas the O—H acids ionize rapidly.

If a hydrogen is α to two carbonyl groups, as in $CH_3COCH_2COCH_3$ or $OHCCH_2CHO$, it is more than doubly activated. In fact, the dialdehyde is as acidic as acetic acid: both have pK_a of 5. The explanation lies in the extensive delocalization that is possible in the anion.

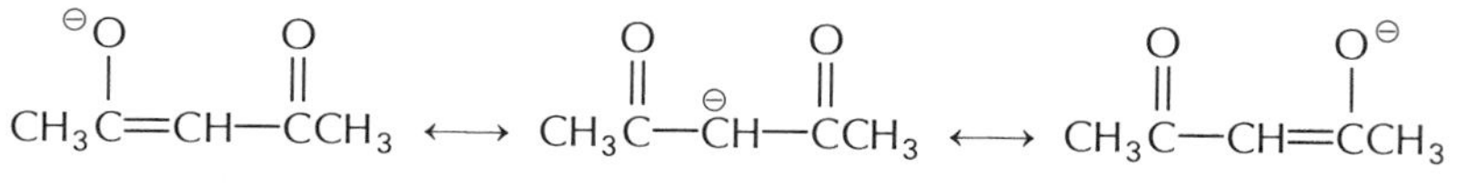

$$\underset{CH_3C=CH-CCH_3}{\overset{\ominus O \quad\quad O}{|\quad\quad\quad ||}} \longleftrightarrow \underset{CH_3C-CH-CCH_3}{\overset{O \quad\ominus\ O}{||\quad\quad ||}} \longleftrightarrow \underset{CH_3C-CH=CCH_3}{\overset{O \quad\quad O^{\ominus}}{||\quad\quad |}}$$

We can measure the equilibrium on the acidic side (between the tautomers) as well. At equilibrium, pure acetone is only 0.00025% in the enol form, but acetylacetone $(CH_3COCH_2COCH_3)$ is 80% enol. The latter enol is stabilized by an internal hydrogen bond in what we call a <u>chelate</u>.

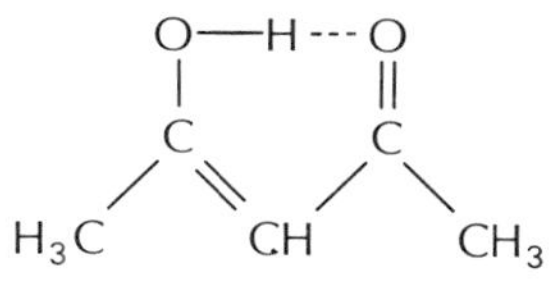

$$\underset{H_3C}{\overset{O-H\cdots O}{\underset{\diagup\diagdown}{\underset{C}{}\underset{C}{}}}}\underset{CH_3}{}$$
$$H_3CCHCH_3$$

Another extreme is found in phenol, which is the enol of 2,4-cyclohexadienone.

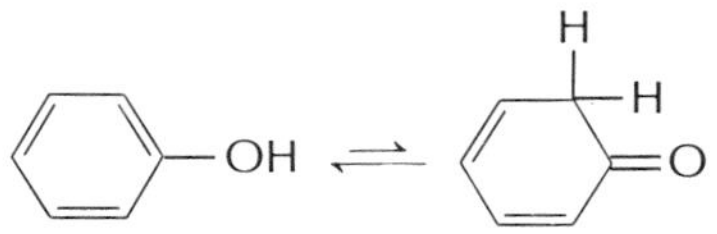

[structure: phenol ⇌ 2,4-cyclohexadienone]

The carbonyl structure usually is the stable form, but in this instance the keto form does not have the aromatic delocalization energy that is present in the enol. Thus phenol exists as a stable enol, not as a ketone.

Most of the reactions of aldehydes and ketones involve as the first step the equilibria of Equations 16.1 and 16.2 along with the equilibria of all other acids and bases present in the mixture.

Ketosis. Under certain kinds of stress, fatty-acid degradation in the liver occurs so fast that products accumulate. These are acetone, acetoacetic acid, and β-hydroxybutyric acid. The primary symptom of this condition is ketosis—the presence of acetone in the breath. Ketosis is often noted in starvation when carbohydrate is absent and the body is forced to break down fatty acids for energy. This occurrence is most common among infants with digestive-tract malfunction and among pregnant women. In the general population, the most important cause of ketosis is diabetes.

Exercises

16.1 The aqueous solubilities of salicylaldehyde and of *p*-hydroxybenzaldehyde are not very different, but the mp are very different. Why?

16.2 Both $(CH_3)_2C$═$CHCOCH_3$ and H_2C═$CHCOCH_3$ add HBr (polar conditions) in the same direction. Explain.

16.2 Spectrometry

Much could be added to the little we have presented on nmr and mass spectrometry of aldehydes and ketones, but for details we must refer to the more extensive treatments that are available. However, we shall add a little bit to the ir and uv of the carbonyl compounds.

The carbonyl stretching frequency. The ir of the carbonyl group exhibits a $\diagdown\!\!C$═$O\!\diagup$ stretching frequency of close to 1730 cm^{-1} for aldehydes, 1715 cm^{-1} for ketones.

Electronic effects cause a shift in the position of the band: electron withdrawal causes an increase in the frequency; donation of electrons, a decrease. Compare acetone at 1715 cm^{-1}, chloroacetone at 1724 cm^{-1}, and hexafluoroacetone at 1801 cm^{-1}.

Steric effects that cause an expansion or contraction of the normal 120° sp^2 bond angle also decrease or increase the frequency of absorption. Compare hexamethylacetone (di-*tert*-butyl ketone) at 1686 cm^{-1}; cyclohexanone, 1715 cm^{-1}; cyclopentanone, 1745 cm^{-1}; and cyclobutanone, 1784 cm^{-1}. A diminution in frequency corresponds to a decrease in the force constant for stretching the bond, which means that the bond is weaker and has more single-bond than double-bond character. Thus, conjugation with a $\diagdown\!\!\!\!\diagup\!\!\!\!\! C\!\!=\!\!C \diagdown\!\!\!\!\diagup$ causes a decrease of 25 to 40 cm^{-1} in frequency because the increase in delocalization increases the single-bond character of $\diagdown\!\!\!\! C\!\!=\!\!O$.

Absorption in the uv. In the uv, the $\diagdown\!\!\!\! C\!\!=\!\!O$ is capable of both the lower energy forbidden $n \rightarrow \pi^*$ and the higher energy allowed $\pi \rightarrow \pi^*$ transitions. An ordinary aldehyde or ketone is essentially transparent above 200 nm. Conjugation with a $\diagdown\!\!\!\!\diagup\!\!\!\!\! C\!\!=\!\!C \diagdown\!\!\!\!\diagup$ or aryl group lowers the energy of the $\pi \rightarrow \pi^*$ absorption so that the band moves above 200 nm. Compare $CH_3CH\!\!=\!\!CHCOCH_3$ (224 nm) and $\phi COCH_2CH_3$ (244 nm). Thus, the position of the uv band is an indication of the extent of conjugation of the carbonyl group.

Exercises

16.3 The spectra of $C_{10}H_{12}O$ are given. What is the compound?

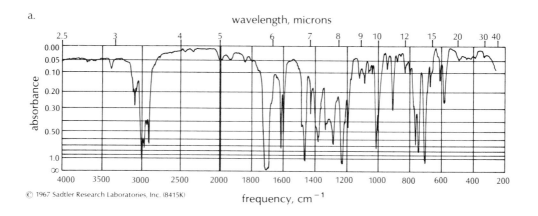

b.

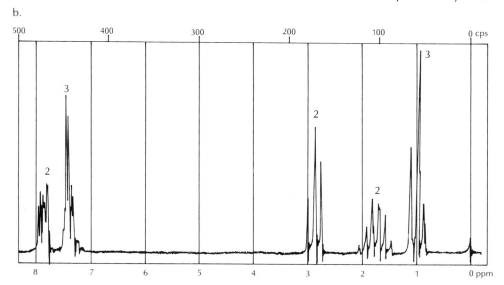

© 1967 Sadtler Research Laboratories, Inc. (3199M)

c.

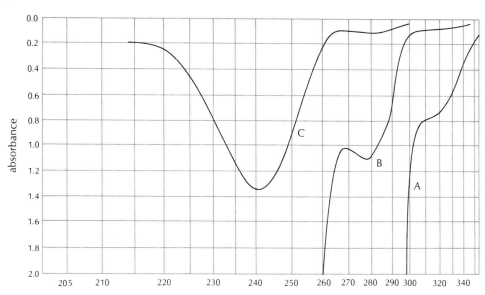

© Sadtler Research Laboratories, Inc. (1613UV)

16.3 Addition reactions

In this section we shall describe those addition reactions that are largely uncomplicated by the intervention of enols and enolate ions.

Hydrates; acetals. Let us begin with the additions to the carbonyl group of HCl, water (Eqs. 2.16, 2.17, 3.31), and alcohols. Since we first saw these reactions in Chapters 2 and 3, our knowledge of organic chemistry has grown and we are now able to ask much more sophisticated questions about them.

In view of the acid-base relationships described in Section 16.1, are the additions of HCl, H_2O, and ROH subject to both acid and base catalysis? The answer is no for HCl and yes for H_2O and ROH.

For the addition of dry HCl, the cation is a more stable intermediate than is the anion (Eq. 16.5) because of charge delocalization in the cation. Also, let us look at the reverse

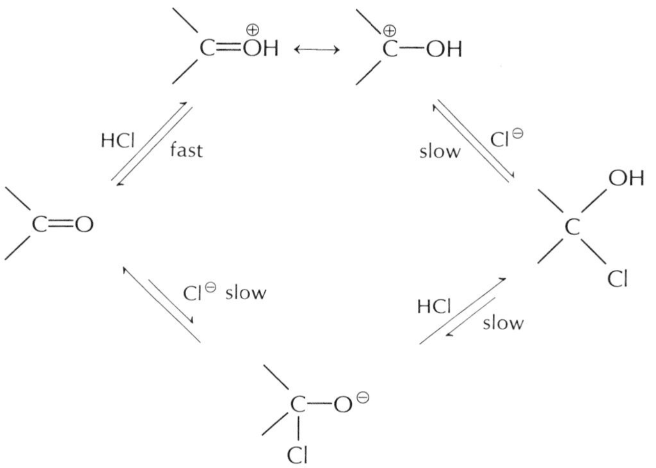

$$\begin{array}{c} \diagup \\ C = \overset{\oplus}{O}H \longleftrightarrow \overset{\oplus}{C} - OH \\ \diagup \end{array} \qquad (16.5)$$

reactions. The removal of $Cl^\ominus$ from a quasi-tertiary position to reproduce the cation is a much more likely prospect than the loss of a proton to a $Cl^\ominus$ or other weak base present in the anhydrous system.

The addition of water to acetaldehyde (hydration) proceeds very slowly at pH 7 and much more rapidly at pH 4 and 11. Under acidic conditions, $H_3O^\oplus$ is faster than H_2O in forming the cation, and the equilibrium lies more to the right. The slow step is the reversible addition of water to the cation. The last step is the fast proton transfer (Eq. 16.6). Under

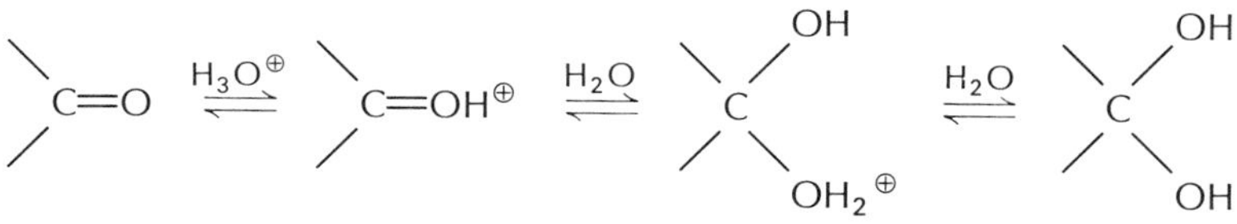

$$\diagup C=O \underset{}{\overset{H_3O^\oplus}{\rightleftharpoons}} \diagup C=OH^\oplus \underset{}{\overset{H_2O}{\rightleftharpoons}} \diagup\!\!\!\!C\!\!\!\!\diagdown \begin{smallmatrix} OH \\ \\ OH_2^\oplus \end{smallmatrix} \underset{}{\overset{H_2O}{\rightleftharpoons}} \diagup\!\!\!\!C\!\!\!\!\diagdown \begin{smallmatrix} OH \\ \\ OH \end{smallmatrix} \qquad (16.6)$$

basic conditions, hydroxide ion and water are the reactants present in high concentration, and the slow step again is the formation of the new C—O σ bond (Eq. 16.7). At pH 7, both

$H_3O^{\oplus}$ and $HO^{\ominus}$ are at low concentration, and water is slower than either $H_3O^{\oplus}$ or $OH^{\ominus}$ in the initial step (protonation or addition to carbon). □ □

$$\diagup_{\diagdown}C=O \underset{}{\overset{HO^{\ominus}}{\rightleftarrows}} \diagup_{\diagdown}\overset{O^{\ominus}}{\underset{OH}{C}} \underset{}{\overset{H_2O}{\rightleftarrows}} \diagup_{\diagdown}\overset{OH}{\underset{OH}{C}} \qquad (16.7)$$

Even though little hydrate exists at equilibrium for many aldehydes and most ketones, the reaction does occur. Acetone reacts with water containing the heavy oxygen isotope ^{18}O to form a trace of hydrate. The reverse reaction regenerates $(CH_3)_2C=O$ and $(CH_3)_2C\overset{18}{=}O$ in equal amounts. At equilibrium, the percent of ^{18}O in acetone and solvent water is the same. The attainment of equilibrium is catalyzed by acids and by bases. The tracer isotope shows that the reaction occurs even though no hydrate is detectable at any time.

An alcohol, in the presence of acids and bases, adds to the carbonyl group to give a hemiacetal (from an aldehyde) or a hemiketal (from a ketone). (See Eq. 16.8.) The mechan-

$$\diagup_{\diagdown}C=O + HOR \rightleftarrows \diagup_{\diagdown}\overset{OR}{\underset{OH}{C}} \qquad (16.8)$$

ism is thought to be the same as that for hydration. Under acidic conditions, either the —OH or —OR group of a hemiacetal or hemiketal may accept a proton. (Hereafter we shall use only one name, because the reactions apply to both.) If the —OR accepts the proton, the reverse of a reaction analogous to Equation 16.6 is set on its way to regenerate the carbonyl compound and alcohol. If the —OH accepts the proton, subsequent reaction in the presence of excess alcohol leads to an acetal (Eq. 16.9). Under basic conditions, a

$$\diagup_{\diagdown}\overset{OH}{\underset{OR}{C}} \overset{HB}{\rightleftarrows} \diagup_{\diagdown}\overset{OH_2^{\oplus}}{\underset{OR}{C}} \overset{-H_2O}{\rightleftarrows} \diagup_{\diagdown}\overset{C^{\oplus}}{\underset{OR}{}} \leftrightarrow \diagup_{\diagdown}\overset{C}{\underset{\overset{\oplus}{O}R}{\diagdown\!\!\!\!=}}$$

$$\Updownarrow ROH$$

$$\diagup_{\diagdown}\overset{\overset{H}{\underset{|}{OR^{\oplus}}}}{\underset{OR}{C}} \overset{-H^{\oplus}}{\rightleftarrows} \diagup_{\diagdown}\overset{OR}{\underset{OR}{C}} \qquad (16.9)$$

hemiacetal, with only a single —OH, can only undergo the reverse of the reaction analogous to Equation 16.7 and regenerate the carbonyl compound and alcohol. An acetal is stable under basic conditions: there is no proton to be lost. Acetal formation may be reversed only under acidic conditions.

Numerous studies of the rates and equilibria of acetal formation have shown the order primary > secondary > tertiary for the alcohol used and that the rates for aldehydes are greater than those for ketones. Equation 16.10 for the complete reaction shows

$$RCHO + 2EtOH \overset{H^{\oplus}}{\rightleftarrows} RCH(OEt)_2 + H_2O \qquad (16.10)$$

that water is a product. Use of an excess of alcohol and removal of water as the reaction proceeds help to shift reactants into products. Although the presence of an acid hastens the attainment of equilibrium, we avoid an excess to prevent other reactions that we shall describe later. Aluminum chloride or gaseous HCl are commonly used acids for this purpose. ☐ ☐

Acid-catalyzed trimerization. In the absence of water, a strong acid can cause aldehydes, especially the simple ones, to polymerize. Addition of a few drops of sulfuric acid to acetaldehyde gives rise to <u>paraldehyde</u>. The cation, having nothing else with which to react, attacks another carbonyl group. Although the reaction proceeds through a number of intermediates, we show the cyclization as though it were concerted (Eq. 16.11).

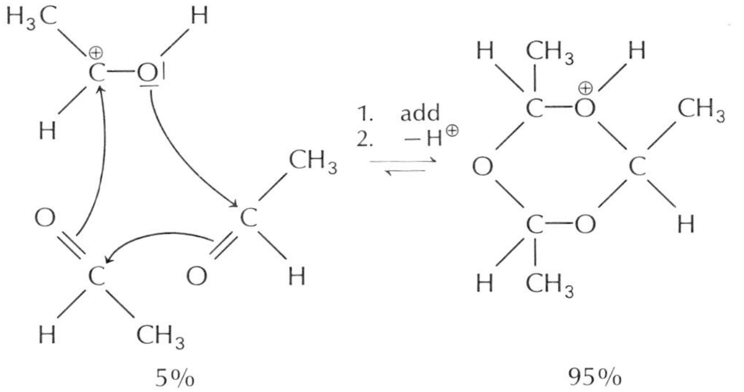

$$\text{(structures shown)} \quad \xrightleftharpoons[\text{2. } -H^{\oplus}]{\text{1. add}} \quad \text{(paraldehyde)} \qquad (16.11)$$

5%   95%

The product, paraldehyde (2,4,6-trimethyl-1,3,5-trioxane), is favored by the equilibrium and is relatively insoluble in water. Thus, washing the reaction mixture with dilute base neutralizes the acid, freezes the equilibrium, and removes the small remaining amount of acetaldehyde to the aqueous layer. The much less volatile paraldehyde (bp 124°) is easier to purify, store, ship, and use. When dry, pure acetaldehyde is needed, a few drops of sulfuric acid are added to paraldehyde and the mixture is distilled. The equilibrium is re-established and continually shifts to the left as the volatile acetaldehyde distils from the mixture.

At 0° or less, acetaldehyde with traces of strong acids reacts in a sealed tube to give a tetramer, <u>metaldehyde,</u> which is an eight-membered ring of mp 246°. Metaldehyde sublimes readily. The vapor attracts snails and slugs; and, because it is also toxic to them, metaldehyde is widely used as a specific poison. Paraldehyde also is physiologically active and has been used as a powerful inducer of sleep (a hypnotic, sedative, or soporific).

Formaldehyde is more active than is acetaldehyde in these polymerizations (a type of acetal formation) because water need not be excluded. If formalin (a solution of formaldehyde in water, about 37% HCHO w/w) is evaporated, some of the formaldehyde escapes but most precipitates as <u>paraformaldehyde,</u> an amorphous solid of high and variable molecular weight (Eq. 16.12: the mechanism shown is speculative). Paraformaldehyde

$$H_2C=O + HOCH_2OH \rightleftharpoons H_2\overset{\overset{O^{\ominus}}{|}}{\underset{\underset{H}{|}}{C}}-\overset{\oplus}{O}-CH_2OH \rightleftharpoons HOCH_2OCH_2OH$$

$$\downarrow \text{repeat} \qquad (16.12)$$

$$HOCH_2(OCH_2)_n OH$$

decomposes into formaldehyde upon being heated. When the polymerization is carried out under carefully controlled conditions to cause n to become as large as possible and the hydroxyl groups at the ends of the chains are esterified, the product is stable to depolymerization. This polymer is known as delrin.

If a 60% solution of formaldehyde in water with 2% sulfuric acid is distilled, the principal product in the distillate is the cyclic trimer of formaldehyde, called trioxane or metaformaldehyde (mp 64°). Trioxane, like paraldehyde, may be caused to decompose in the presence of acid catalysts.

Sodium bisulfite. Aldehydes, most methyl ketones, and some cyclic ketones react with a saturated aqueous solution of sodium bisulfite to give an addition product (Eq. 16.13). The products are crystalline solids. They are soluble in water, from which they

$$\phi CHO + HSO_3^{\ominus} Na^{\oplus} \rightleftharpoons \phi CH\begin{matrix}OH\\SO_3^{\ominus}Na^{\oplus}\end{matrix} \quad (16.13)$$

$$\xrightarrow{HCl} \phi CHO + SO_2 + H_2O + NaCl$$

$$\xrightarrow{NaOH} \phi CHO + Na_2SO_3 + H_2O$$

may be recrystallized and purified. Treatment of an aqueous solution with either acid or base regenerates the pure carbonyl compound by a nonreversible destruction of bisulfite ion.

The position of the equilibrium is markedly affected by the size of the groups attached to the carbonyl carbon atom. Because an aldehyde always has a small hydrogen attached, the alkyl or aryl group may become quite bulky before the reaction fails. A methyl ketone is more sensitive: the other group must not become too large. Even a phenyl is too large: acetophenone fails to give the reaction. If both groups are larger than methyl, the reaction fails: diethyl ketone (3-pentanone) does not react.

Hydrogen cyanide. The addition of hydrogen cyanide to a carbonyl group has been studied in detail. The reaction is catalyzed by bases (amines, cyanide ion, and others). It seems likely that $B^{\ominus} + HCN \rightarrow BH + CN^{\ominus}$ is a rapid preliminary step; and cyanide ion, a strong nucleophile, then attacks the carbonyl carbon in a rate-determining slow step (Eq. 16.14). The anion rapidly accepts a proton from any handy proton donor to give an

$$\diagdown C=O + CN^{\ominus} \rightleftharpoons \diagup_{\diagdown}^{\diagup} C \diagdown_{C\equiv N}^{O^{\ominus}} \underset{HB}{\rightleftharpoons} \diagup_{\diagdown}^{\diagup} C \diagdown_{C\equiv N}^{OH} \quad (16.14)$$

α-hydroxynitrile, known as a cyanohydrin. The complete reaction may be written as shown in Equation 16.15.

$$\diagdown C=O + HCN \underset{B^{\ominus}}{\rightleftharpoons} \diagup_{\diagdown}^{\diagup} C \diagdown_{C\equiv N}^{OH} \quad (16.15)$$

Table 16.3 Cyanohydrin equilibria

Compound	K_{eq}
CH_3CHO	very large
$CH_3COCH_2CH_3$	38
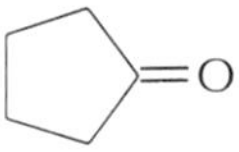 cyclopentanone	500
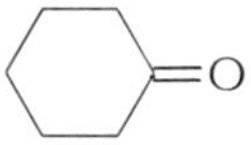 cyclohexanone	10,000
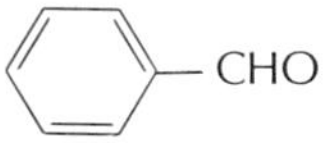 ϕ—CHO	210
 MeO—ϕ—CHO	32
 O_2N—ϕ—CHO	1,430
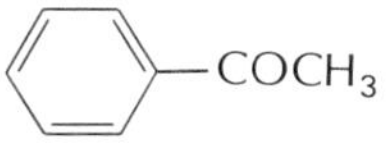 ϕ—COCH$_3$	0.77

Let us look at the equilibrium constants of Equation 16.15 for a few compounds to gain an idea of the magnitude of the steric and electronic effects involved (Table 16.3). As we go from acetaldehyde to ethyl methyl ketone, there is a large decrease in K_{eq}. This decrease may be ascribed partly to steric causes (extra crowding in the ketone) and partly to electronic causes (less double-bond character in the ketone). If we now consider cyclopentanone (equivalent to pinning back the floppy alkyl groups of diethyl ketone), K_{eq} goes back up. In a cyclopentane ring, however, there is still considerable nonbonding interference between the new sp^3 carbon and its neighbors. Going on to cyclohexanone, we see another jump up in K_{eq}. Steric factors are a major influence affecting these K_{eq}.

If we go back and contrast benzaldehyde with acetaldehyde, we find that K_{eq} for ϕCHO has decreased (even less double-bond character, and phenyl is larger than methyl). With methoxy *para* to the aldehyde group, K_{eq} decreases still more (less double-bond character because of increased conjugation). With *para*-nitro, K_{eq} goes back up (more double-bond character). However, the changes caused by the electronic factors alone, though not without influence, do not compare with steric effects in magnitude. Finally, with acetophenone K_{eq} drops to less than one.

Cyanide ion in alcoholic solution causes aldehydes with no α hydrogens to undergo the benzoin condensation. (We shall describe this reaction in the next section.) To prevent the benzoin condensation, the addition of HCN (from NaCN) to benzaldehyde is carried out in the presence of sodium bisulfite. The sodium bisulfite serves several purposes: it increases the acidity of the medium, forms the addition product with benzaldehyde and releases benzaldehyde as needed, and prevents the Cannizzaro reaction.

Grignard reagents; the Reformatsky reaction. The addition of hydrogen cyanide gives a new carbon-carbon σ bond. There is a far more useful nucleophile for producing

16.3 Addition reactions

these bonds: the Grignard reagent. Although we already have met the Grignard and lithium reagents, we have not yet given much attention to the steric and electronic aspects of the reaction with the carbonyl group.

Although the R in RMgX is a very strong base (and nucleophile) no matter what it is, there is reason to believe that the initial step is coordination of magnesium with the carbonyl oxygen. In the Grignard reagent, the magnesium is coordinated with two molecules of ether. Therefore, it seems reasonable to postulate the mechanism shown in Equations 16.16

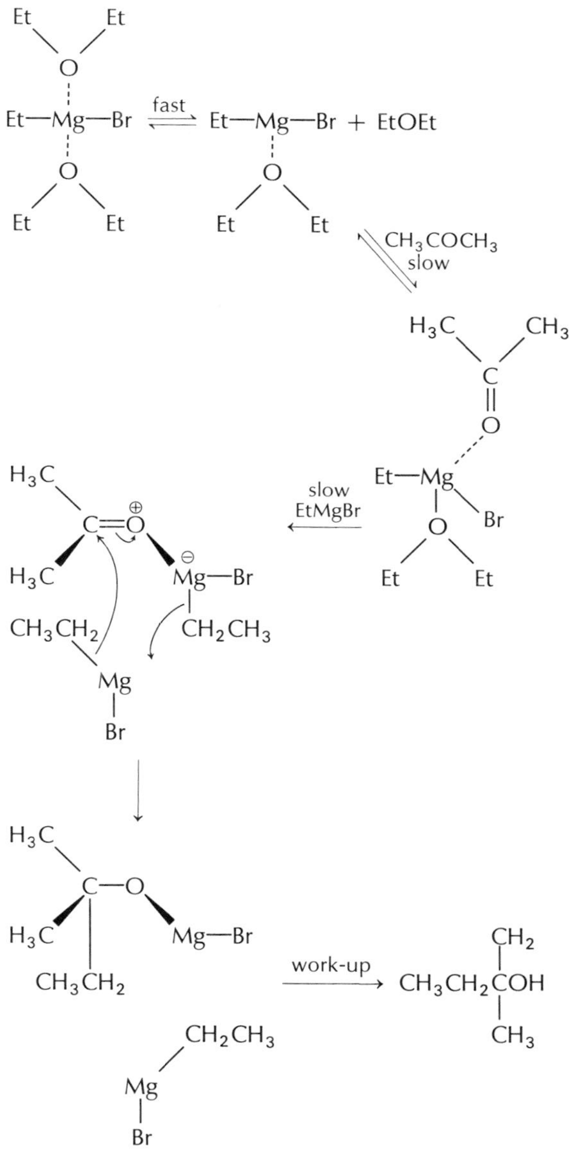

$$(16.16)$$

for the addition of EtMgBr to acetone. (The ether molecules are omitted after the second step in Eq. 16.16 for clarity.) □ □

Rather similar to a Grignard reaction is the Reformatsky reaction. Before the discovery of the Grignard reagent, it was known that alkyl halides could react with zinc powder to give substances that combusted spontaneously in air. It was found that if the work was done under CO_2 gas, the substances were reactive towards aldehydes and ketones, but not towards esters or CO_2. The less reactive zinc reagents, which are more selective than Grignard reagents, may be safely prepared if they are used immediately (Eq. 16.17) by

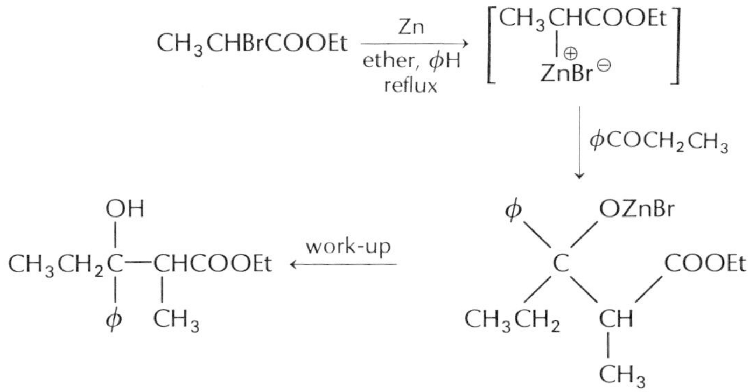

$$CH_3CHBrCOOEt \xrightarrow[\text{ether, }\phi H]{\text{Zn}} \begin{bmatrix} CH_3CHCOOEt \\ | \oplus \\ ZnBr^\ominus \end{bmatrix}$$

$$\downarrow \phi COCH_2CH_3 \qquad (16.17)$$

$$\underset{\underset{\phi \quad CH_3}{|\quad\quad|}}{CH_3CH_2\overset{OH}{\underset{|}{C}}-CHCOOEt} \xleftarrow{\text{work-up}} \underset{CH_3CH_2 \quad\quad CH}{\overset{\phi\quad\quad OZnBr}{\underset{|}{C}}}\underset{CH_3}{\overset{}{\underset{|}{}}}COOEt$$

reaction with a ketone (or aldehyde) present in the mixture from the beginning. The product (a tertiary β-hydroxy ester from a ketone, secondary from an aldehyde) is very easily dehydrated to give an α,β unsaturated ester (CH_3CH_2C=C—COOEt from the ester shown, with φ and CH_3 on the second C). Because a >C=C< is easy to hydrogenate as compared to >C=O, especially if the >C=O is part of an ester group, hydrogenation of the unsaturated ester gives the saturated ester ($CH_3CH_2CH(\phi)CH(CH_3)COOEt$). Thus the Reformatsky reaction provides a route to rather complex esters from a ketone or aldehyde and an α-bromo ester. If the halogen is located at a position other than α to the >C=O of the ester, the reaction fails. For this reason, we may consider that the zinc reagent has some characteristics of an

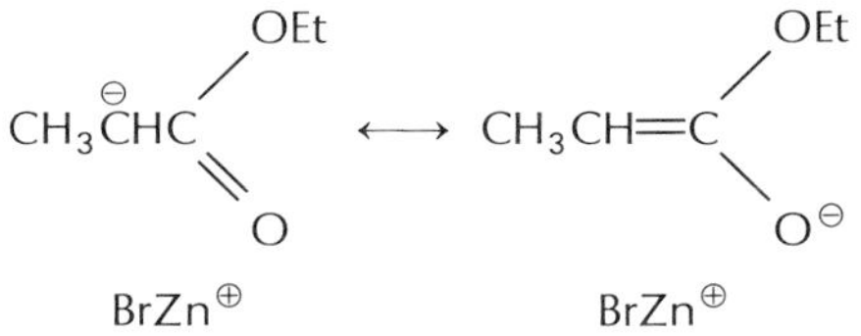

$$CH_3\overset{\ominus}{C}HC\overset{OEt}{\underset{\underset{BrZn^\oplus}{O}}{\diagup}} \quad\longleftrightarrow\quad CH_3CH=C\overset{OEt}{\underset{\underset{BrZn^\oplus}{O^\ominus}}{\diagdown}}$$

enolate ion. If the ketone is highly hindered with bulky groups, successful addition is hard to achieve.

In addition reactions to a carbonyl group, the secondary alcohol that results from an aldehyde is capable of asymmetry (unless the two R groups are identical): RR′C̊HOH. The tertiary alcohol from a ketone is capable of asymmetry if all three R groups differ: RR′R″C̊OH. As the reaction is ordinarily carried out, there is equal opportunity for the approaching RMgX (or RLi or Zn reagent) to come in on either side of the carbonyl group of an aldehyde, ketone, or ester. The newly created asymmetric center thus becomes 50% (R) and 50% (S). ☐ ☐

Wittig reaction: ylides. The Wittig reaction is another way of joining a carbon atom to the carbon of a carbonyl. The reaction makes use of a phosphorus ylide. We may think of an ylide as a substance in which a negative carbon is bonded to a hetero atom (usually P). The ylide is stabilized by partial overlap of the filled carbon *p* orbital with an empty *d* orbital of positive P. To prepare a phosphorus ylide, we start with a trialkyl or preferably a triarylphosphine and carry out an S_N2 reaction with an alkyl halide. (Primary alkyl halides react faster than secondary; tertiary ones are not used.) The quaternary salt is treated with a strong base (butyl lithium, sodium amide, sodium hydride, and $CH_3\overset{\overset{O}{\|}}{S}-CH_2^{\ominus}Na^{\oplus}$ all have been used) to give the ylide (Eq. 16.18).

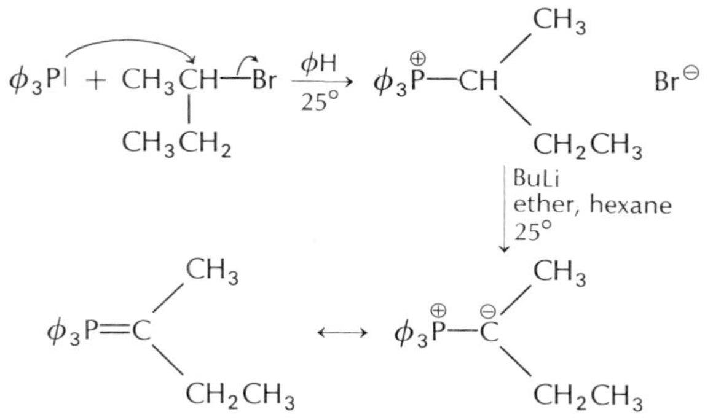

$$\phi_3P: + CH_3CH-Br \xrightarrow[25°]{\phi H} \phi_3\overset{\oplus}{P}-CH\begin{matrix}CH_3\\ \\ CH_2CH_3\end{matrix} \quad Br^{\ominus} \tag{16.18}$$

$$\downarrow \begin{matrix}BuLi\\ether, hexane\\25°\end{matrix}$$

$$\phi_3P=C\begin{matrix}CH_3\\ \\ CH_2CH_3\end{matrix} \longleftrightarrow \phi_3\overset{\oplus}{P}-\overset{\ominus}{C}\begin{matrix}CH_3\\ \\ CH_2CH_3\end{matrix}$$

an ylide

The carbon atom bonded to phosphorus bears enough negative charge to become a strong nucleophile. If an aldehyde or ketone is added to the solution, the nucleophilic carbon attacks the carbonyl carbon and forms a betaine. The betaine is unstable and decomposes into an alkene and a phosphine oxide. Depending upon the steric hindrance involved, either the addition or the decomposition may be the rate-determining step in the Wittig reaction (Eq. 16.19). It usually is carried out under an inert gas such as nitrogen to

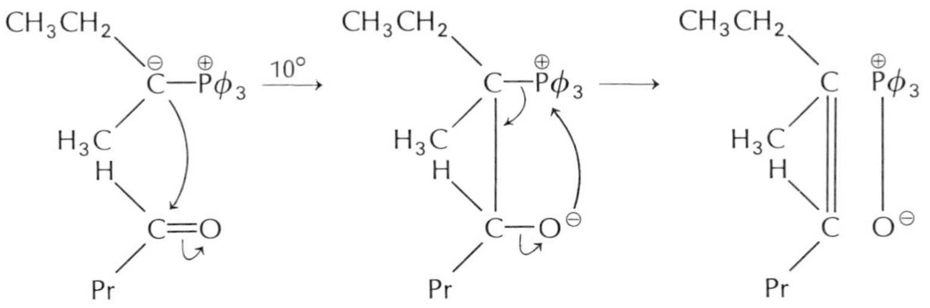

$$(16.19)$$

a betaine

prevent a side reaction of the ylide with oxygen of air. The complete, generalized reaction may be written as shown in Equation 16.20. The Wittig reaction can be used in conversion

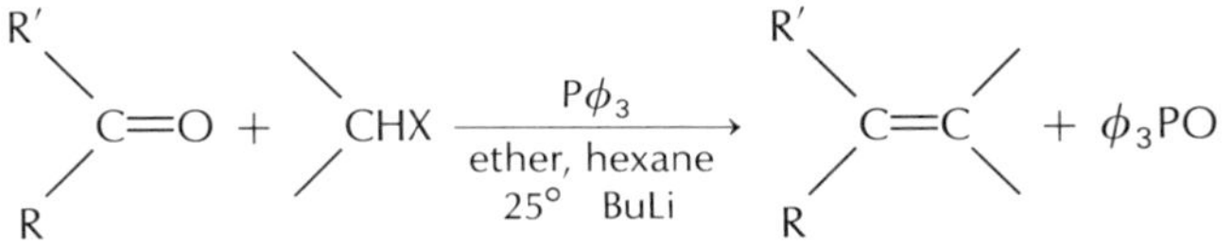

$$\begin{matrix}R'\\ \\R\end{matrix}C=O + \begin{matrix}\\ \\ \end{matrix}CHX \xrightarrow[\text{ether, hexane}]{P\phi_3}_{25° \text{ BuLi}} \begin{matrix}R'\\ \\R\end{matrix}C=C\begin{matrix}\\ \\ \end{matrix} + \phi_3PO \tag{16.20}$$

of an aldehyde or ketone into an alkene, the structure of which depends upon the carbonyl compound and alkyl halide chosen. Thus, the Wittig reaction is an excellent method for the synthesis of alkenes.

Ammonia, amines, and substituted amines. Turning now to ammonia and various derivatives of ammonia (primary amines, hydroxylamine, hydrazine, substituted hydrazines), we find another group of addition reactions in which, in general, a carbonyl group reacts with —NH_2 to give an <u>imine</u> $\left(\diagdown_{\diagup}C{=}N{-}\right)$.

A primary amine, which is a nucleophile, attacks the carbonyl carbon as the first step. Subsequent steps are catalyzed by acids. Consequently, several equilibria are involved (Eqs. 16.21).

$$\diagdown_{\diagup}C{=}O + RNH_2 \rightleftharpoons \underset{\overset{\oplus}{NH_2R}}{\overset{O^{\ominus}}{\diagdown C \diagup}} \underset{HB}{\rightleftharpoons} \underset{\overset{\oplus}{NH_2R}}{\overset{OH}{\diagdown C \diagup}}$$

$$B^{\ominus} \updownarrow \quad (16.21)$$

$$\underset{NHR}{\overset{\oplus}{\diagdown C \diagup}} \underset{-H_2O}{\rightleftharpoons} \underset{NHR}{\overset{\overset{\oplus}{OH_2}}{\diagdown C \diagup}} \underset{HB}{\rightleftharpoons} \underset{NHR}{\overset{OH}{\diagdown C \diagup}} \quad \text{(proton exchanges)}$$

$$\updownarrow$$

$$\underset{\overset{\oplus}{NHR}}{\diagdown C \diagdown} \underset{B^{\ominus}}{\rightleftharpoons} \diagdown_{\diagup}C{=}N\diagdown_R$$

Reaction depends upon having the maximum possible concentrations of $\diagdown_{\diagup}C{=}O$ and of RNH_2 in solution. Accordingly, the solution should be basic or neutral to prevent formation of $R\overset{\oplus}{N}H_3$, which cannot add. Nevertheless, the solution should be acidic enough to facilitate the numerous proton transfers (as in acetal formation), which are accelerated by the presence of an acid.

At a high pH, the reaction is slow because the proton transfers are slow. At a small pH, the reaction is slow because the concentration of RNH_2 is small. There is an intermediate pH at which the rate is a maximum: it depends upon the base strength of the amine on the one hand and the optimization of the rate on the other. In general, the intermediate pH falls at about 5, close to the pH of a buffer mixture of equal parts of acetic acid and sodium acetate. In practice, the same effect can be obtained by addition of hydrochloric acid to a sodium acetate solution.

The product of reaction of a primary amine and an aldehyde or ketone is known as an imine, and also as a Schiff base. The complete reaction (Eq. 16.22) shows that the

$$\diagup\!\!\!\!\!\!\diagdown C=O + H_2NR \rightleftharpoons \diagup\!\!\!\!\!\!\diagdown C=N\diagdown_R + H_2O \qquad (16.22)$$

equilibrium lies to the right because of the energy of formation of water. As might be expected, the order is HCHO > RCHO > R_2CO in ease of formation of Schiff bases.

Schiff base in biochemistry. The occurrence of a Schiff base in an enzyme mechanism is detected by addition of borohydride to the incubation mixture to reduce the labile bond to a stable ketimine. The trapped intermediate can be isolated and identified.

An example of a Schiff base is the important biochemical reaction of transfer of amino groups to an acceptor molecule. In animal tissues, the amino group is almost always donated by glutamic acid, leaving behind α-ketoglutaric acid. The amino group replaces a carbonyl oxygen on the acceptor molecule. Two of the compounds of vitamin B_6 are used by the enzymes that catalyze this complicated reaction. Although the mechanism is incompletely understood, at least four different Schiff base intermediates have been identified.

The products from reaction of carbonyl compounds with ammonia ($\diagup\!\!\!\!\!\!\diagdown C=NH$) are unstable and undergo reaction with more ammonia or carbonyl to give other products. For example, formaldehyde and ammonia in the proper proportions react to give hexamethylene tetramine *a*, and benzaldehyde and ammonia react to give *b* (Eqs. 16.23). Hexamethylene tetramine is of interest because of its

$$6HCHO + 4NH_3 \longrightarrow a + 6H_2O$$

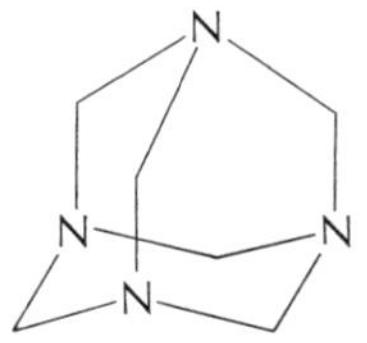

$$3\phi CHO + 2NH_3 \longrightarrow \phi CH\diagup^{N=CH\phi}_{\diagdown N=CH\phi} + 3H_2O$$

$$b \qquad (16.23)$$

very stable adamantane-like structure. It still is used as an orally administered urinary antiseptic.

Table 16.4 lists various amines and the structures and class names of products that they form with aldehydes and ketones.

Because most aldehydes and ketones are liquids, whereas most of the derivatives are solids with sharp melting points, extensive mp tables of oximes, semicarbazones, and phenylhydrazones have been set up as an aid in the identification of carbonyl compounds. Although phenylhydrazine was used extensively in the past, a recent trend has been to

Table 16.4 Aldehyde and ketone amine derivatives

Amine	Name	Product	Name of derivative
RNH$_2$	amine	\C=NR/	imine, Schiff's base
H$_2$N—NH$_2$	hydrazine	\C=NNH$_2$/	hydrazone
		\C=NN=C/	azine
ϕNHNH$_2$	phenylhydrazine	\C=NNHϕ/	phenylhydrazone
H$_2$NCONHNH$_2$	semicarbazide	\C=NNHCONH$_2$/	semicarbazone
NH$_2$OH	hydroxylamine	\C=NOH/	oxime

use 2,4-dinitrophenylhydrazine, which is easier to prepare (by nucleophilic aromatic substitution of H$_2$NNH$_2$ on the activated chlorine of 2,4-dinitrochlorobenzene).

The amine derivatives (except for those of symmetrical ketones) may exist in two forms, as shown for benzaldoxime. The isomer in which the group on nitrogen is *cis* to the

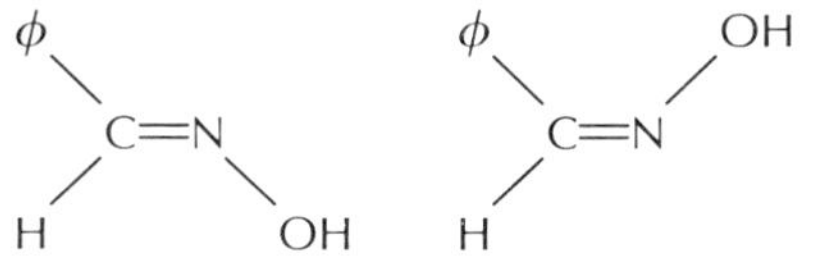

$$\begin{array}{cc} \phi\diagdown \quad \diagup \\ C=N \\ \diagup \quad \diagdown OH \\ H \\ a \end{array} \qquad \begin{array}{cc} \phi\diagdown \quad \diagup OH \\ C=N \\ \diagup \\ H \\ b \end{array}$$

smaller group on carbon usually is the more stable. However, the *syn* (a) and *anti* (b) isomers usually are easy to interconvert by a variety of means. In naming these derivatives of ketones, we must specify the group to which the substituent on nitrogen is *syn* or *anti*. Thus, *syn*-methyl ethyl ketone phenylhydrazone refers to the same isomer as does *anti*-ethyl methyl ketone phenylhydrazone. ☐ ☐

Chloromethylation. A reaction that is restricted to formaldehyde as the carbonyl component is chloromethylation of aromatic compounds. It is a variety of the Friedel-Crafts reaction (Eq. 16.24). A mixture of the aromatic compound to be substituted and trioxane is treated with ZnCl$_2$ and HCl. Trioxane decomposes to formaldehyde in the presence of HCl, and the cation enters into a Friedel-Crafts type of reaction to give a benzyl alcohol. This alcohol is rapidly and easily converted in the reaction mixture into the benzyl chloride. The reaction is subject to all the restraints and influences that we discussed concerning the Friedel-Crafts reaction. A complete reaction is given in Equation 16.25. ☐ ☐

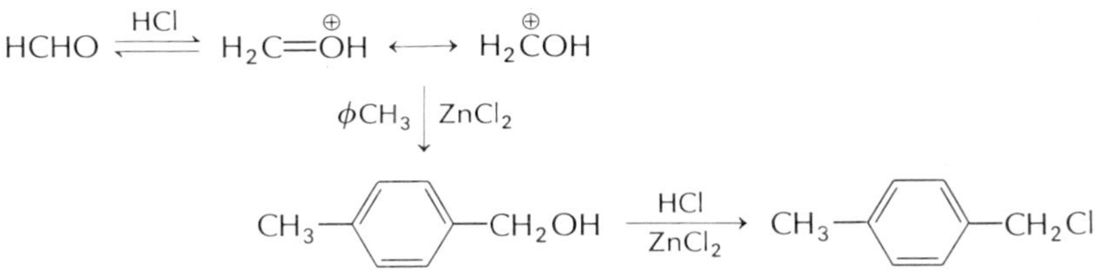

$$\text{HCHO} \xrightleftharpoons{\text{HCl}} \text{H}_2\text{C}=\overset{\oplus}{\text{OH}} \longleftrightarrow \text{H}_2\overset{\oplus}{\text{C}}\text{OH}$$

$$\downarrow \phi\text{CH}_3 \mid \text{ZnCl}_2 \qquad (16.24)$$

$$\text{CH}_3-\phi-\text{CH}_2\text{OH} \xrightarrow[\text{ZnCl}_2]{\text{HCl}} \text{CH}_3-\phi-\text{CH}_2\text{Cl}$$

$$3\phi\text{CH}_3 + \text{(trioxane)} + 3\text{HCl} \xrightarrow{\text{ZnCl}_2} 3\text{CH}_3-\phi-\text{CH}_2\text{Cl} + 3\text{H}_2\text{O} \qquad (16.25)$$

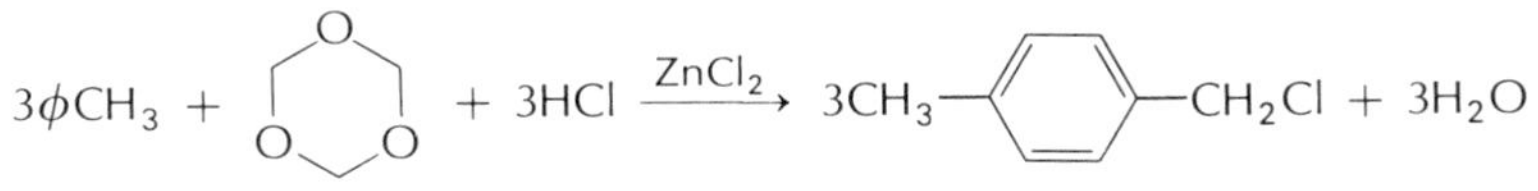

Exercises

16.4 Give the major products to be expected:

$$\phi\text{COCHMe}_2 + \phi\text{NHNH}_2 \xrightarrow[\substack{\text{CH}_3\text{COOH} \\ \text{CH}_3\text{COONa}}]{\text{H}_2\text{O, EtOH}}$$

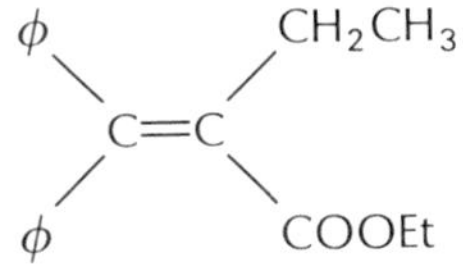

16.5 Show how to carry out the conversions shown:

a. $\text{CH}_3\phi \longrightarrow \text{CH}_3-\phi-\underset{\underset{\text{H}_3\text{C}}{|}}{\overset{\overset{\text{CH}_3}{|}}{\text{C}}}=\text{C}-\text{CH}_3$

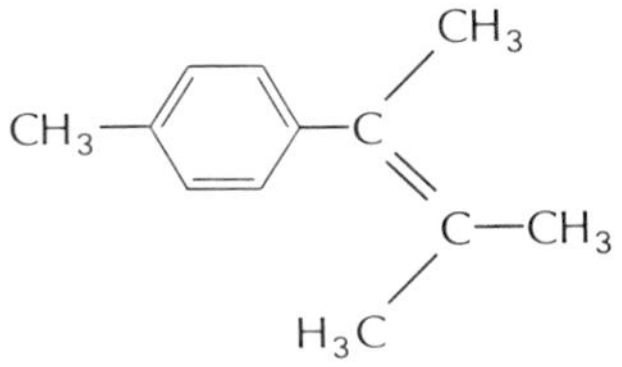

b. $\phi\text{CO}\phi \longrightarrow \underset{\phi}{\overset{\phi}{\diagdown}}\text{C}=\text{C}\underset{\text{COOEt}}{\overset{\text{CH}_2\text{CH}_3}{\diagup}}$

c. $\text{CH}_3\phi \longrightarrow \text{HOOC}-\phi-\overset{\overset{\text{O}}{\|}}{\text{C}}-\phi-\text{COOH}$

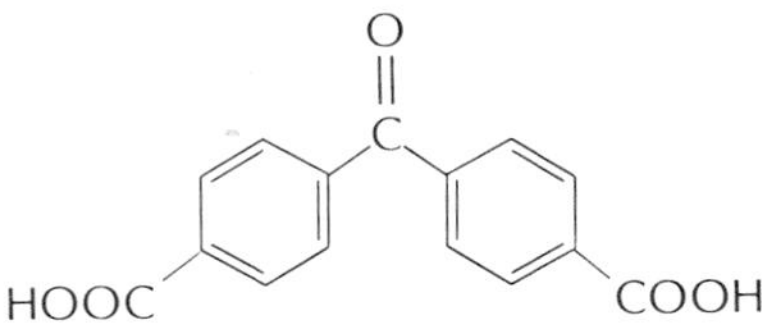

16.6 Show how to separate and obtain pure each component of a mixture of equal parts of ϕCH_3, ϕCHO, and $\phi\text{CH}_2\text{OH}$.

16.4 Substitution and condensation reactions

Basic and acidic halogenation. Halogen substitution of an α hydrogen in an aldehyde or ketone proceeds by way of an enol under acidic conditions and by way of an enolate ion under basic conditions. In basic solution, the rate of reaction is proportional to the concentration of carbonyl compound and of base. The concentration of halogen does not appear at all (Eq. 16.26). In fact, the rates of chlorination, bromination, and iodination

$$CH_3COCH_3 \underset{slow}{\overset{HO^\ominus}{\rightleftarrows}} CH_3COCH_2^\ominus \xrightarrow[fast]{X_2} CH_3COCH_2X + X^\ominus \qquad (16.26)$$

$$\text{rate} = k_B[CH_3COCH_3][HO^\ominus]$$

are the same. The rate constant k_B reflects the slow production of enolate ion, which is immediately snapped up by halogen before the return step can take place.

If no halogen is present, the equilibrium is established. Because K_{eq} is less than one, the reverse step (ion + water) is faster than the forward step. If the equilibrium is set up in water containing deuterium, it is found that deuterium is incorporated (the reverse step) at a rate determined by the slower forward step. In other words, the same k_B is found.

If an optically active ketone with the asymmetric carbon atom bonded to the carbonyl group is used, it is found that racemization also proceeds at the same rate (Eq. 16.27): namely, the rate at which enolate ion is formed. The enolate ion is not dissymmetric and only racemates may be produced from it.

$$(R)-\phi CO\overset{*}{C}H\begin{smallmatrix}CH_3\\CH_2CH_3\end{smallmatrix} \xrightarrow{HO^\ominus, k_B} \phi CO\overset{\ominus}{C}\begin{smallmatrix}CH_3\\CH_2CH_3\end{smallmatrix} \begin{array}{c}\xrightarrow[fast]{X_2} (RS)-\phi CO\overset{*}{C}X\begin{smallmatrix}CH_3\\CH_2CH_3\end{smallmatrix}\\ \xrightarrow[fast]{H_2O} (RS)-\phi CO\overset{*}{C}H\begin{smallmatrix}CH_3\\CH_2CH_3\end{smallmatrix}\\ \xrightarrow[fast]{D_2O} (RS)-\phi CO\overset{*}{C}D\begin{smallmatrix}CH_3\\CH_2CH_3\end{smallmatrix}\end{array} \qquad (16.27)$$

The kinetic studies of halogenation all were performed with a large excess of ketone (or aldehyde) because the product is able to react as well (Eq. 16.28). Of the two possible

$$CH_3COCH_2X \begin{array}{c}\xrightarrow{HO^\ominus} CH_3\overset{O}{\underset{\|}{C}}-\overset{\ominus}{C}HX \longleftrightarrow CH_3\overset{O^\ominus}{\underset{|}{C}}=CHX \quad a\\ \xrightarrow{HO^\ominus} \overset{\ominus}{C}H_2\overset{O}{\underset{\|}{C}}CH_2X \longleftrightarrow CH_2=\overset{O^\ominus}{\underset{|}{C}}-CH_2X \quad b\end{array} \qquad (16.28)$$

enolate ions, a not only is the more stable, but also is formed faster. The halogen atom, with a strong $-I$ effect, polarizes the remaining C—H bonds next to the C—X bond so that a proton is more easily removed. The halogen atom also helps to stabilize a by increasing the contribution of the carbon anion resonance structure. Consequently, when competing for remaining halogen in the mixture, CH_3COCH_2X reacts with a larger k_B than does CH_3COCH_3. In turn, the new product, CH_3COCHX_2, has an even larger k_B in going on to CH_3COCX_3. Thus we see that under basic conditions, it is difficult to monohalogenate an aldehyde or ketone. Once started, the reaction is more easily allowed to proceed until all the α hydrogens on a given carbon have been replaced.

16.4 Substitution and condensation reactions

What happens if R_1 is different from R_2 in R_1COR_2? We already have seen that an α halogen eases the formation of enolate ion. Thus, an alkyl group with a +I effect should act in the opposite sense (Eq. 16.29). In other words, the less alkyl-substituted group is the more reactive.

$$(CH_3)_2CHCOCH_3 \xrightarrow[\text{dil. NaOH}]{Br_2} \begin{array}{l} (CH_3)_2CBrCOCH_3 \quad \text{small} \\ (CH_3)_2CHCOCBr_3 \quad \text{large} \end{array} \qquad (16.29)$$

Under basic conditions, a trihalomethyl group is easily cleaved. It is thought that an addition is followed by the breaking of the most favored bond (Eqs. 16.30). Of the three

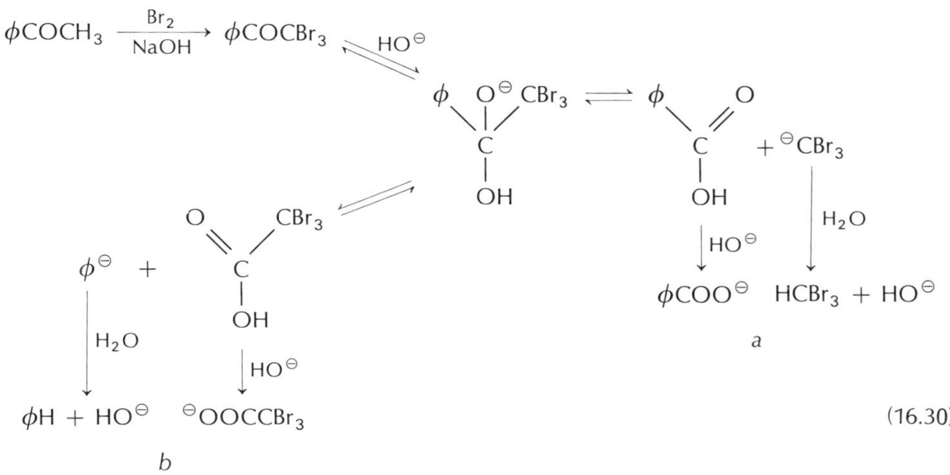

$$\phi COCH_3 \xrightarrow[NaOH]{Br_2} \phi COCBr_3 \xrightarrow{HO^\ominus} \cdots$$

(scheme showing tetrahedral intermediate with ϕ, $O^\ominus$, CBr_3, OH substituents, which proceeds via pathway a to give $\phi COO^\ominus + HCBr_3 + HO^\ominus$, or via pathway b to give $\phi H + HO^\ominus + {}^\ominus OOCCBr_3$)

(16.30)

possibilities, reversal of the addition is nonproductive. Loss of $Br_3C^\ominus$ instead of $\phi^\ominus$ is favored because the three halogens help to stabilize the anion in a as compared to that in b. After cleavage, the benzoic acid and the anion undergo rapid proton transfers to give benzoate ion and bromoform. Because of the generality of the complete reaction to give HCX_3 with chlorine, bromine, or iodine, it is known as the <u>haloform</u> reaction (Eq. 16.31).

$$R-COCH_3 \xrightarrow[NaOH]{X_2} RCOOH + HCX_3 \qquad (16.31)$$

The haloform reaction is restricted to compounds that have a —$COCH_3$ group present. Consequently, aside from methyl ketones (R may be almost any group), acetaldehyde is the only aldehyde to undergo the reaction. Because alkaline solutions of halogens are oxidizing agents, those alcohols that may be oxidized to acetaldehyde (ethanol) or methyl ketones (isopropyl alcohol, 2-butanol, α-phenethyl alcohol, etc.) also undergo the reaction (Eqs. 16.32). For testing small amounts of a

$$CH_3CH_2OH \xrightarrow[NaOH]{Cl_2} HCCl_3 + HCOOH$$

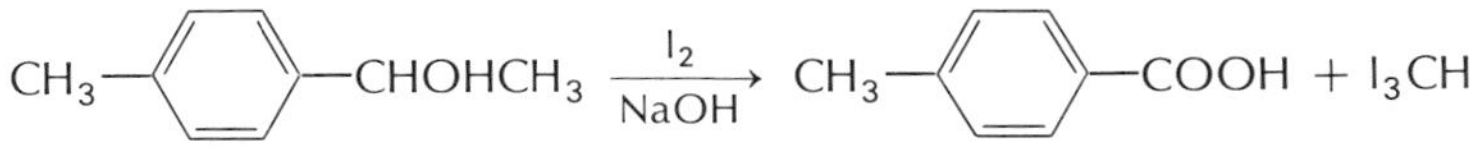

$$CH_3\text{-}C_6H_4\text{-}CHOHCH_3 \xrightarrow[NaOH]{I_2} CH_3\text{-}C_6H_4\text{-}COOH + I_3CH$$

(16.32)

substance, iodine is the halogen used because the product, iodoform, is an easily isolated, pale yellow solid (mp 119°) with a characteristic odor. Because of the oxidizing character of the test mixture and the presence of base, which can cause numerous rearrangements, there is a facetious saying that "if it is organic, it will give a positive iodoform test."

The haloform reaction sometimes is used in synthesis because it provides a mild method for the removal of a methyl group and for conversion of the carbonyl to an acid (Eqs. 16.33).

$$tBuCOCH_3 \xrightarrow[NaOH]{Cl_2} tBuCOOH + HCCl_3$$

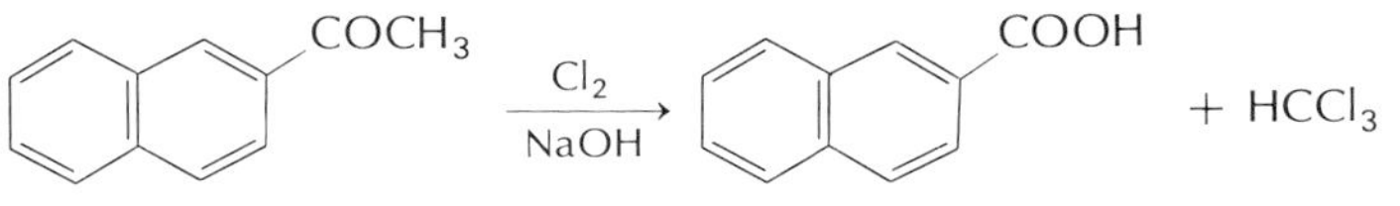

$$\text{Naphthyl-COCH}_3 \xrightarrow[NaOH]{Cl_2} \text{Naphthyl-COOH} + HCCl_3 \quad (16.33)$$

Under acidic conditions, the rate of halogenation is proportional to the concentration of ketone and of $H_3O^\oplus$ if a strong acid is used (Eqs. 16.34). Here, the enol is the intermediate;

$$CH_3COCH_3 \underset{slow}{\overset{H_3O^\oplus}{\rightleftharpoons}} CH_3\overset{OH}{C}=CH_2 \xrightarrow[fast]{X-X} CH_3\overset{OH^\oplus}{\underset{\|}{C}}-CH_2Br \quad (16.34)$$

$$\text{rate} = k_A[CH_3COCH_3][H_3O^\oplus]$$

and, if R and R' of RR'C=O are different, the more highly substituted R group gives the more highly substituted and more stable enol. The acidic halogenation of methyl isopropyl

$$(CH_3)_2C=\overset{OH}{\overset{|}{C}}CH_3 > (CH_3)_2CH\overset{OH}{\overset{|}{C}}=CH_2$$

$$\downarrow Br_2 \qquad\qquad \downarrow Br_2 \quad (16.35)$$

$$(CH_3)_2CBrCOCH_3 \qquad (CH_3)_2CHCOCH_2Br$$

$$\text{more} \qquad\qquad \text{less}$$

ketone places the halogen preferentially in the isopropyl group. The presence of a halogen slows the subsequent rates of enol formation. As a result, it is possible, under acidic control, to introduce a single halogen.

Aldol condensation: base catalysis, acid catalysis. Now we come to the prototype of many condensation reactions, the aldol condensation. Acetaldehyde, in the presence of dilute sodium hydroxide, reacts to give 3-hydroxybutanal (common name aldol, Eq. 16.36). One possible initial reaction of a base with acetaldehyde is the removal of an α

$$2CH_3CHO \xrightarrow{\text{dil. NaOH}} CH_3CHOHCH_2CHO \quad (16.36)$$

$$\text{aldol}$$

16.4 Substitution and condensation reactions 505

hydrogen to produce an enolate ion. Another, a nucleophilic reaction, is addition of the base to the carbon of the carbonyl group (Eq. 16.37). Of the two, the nucleophilic step (the

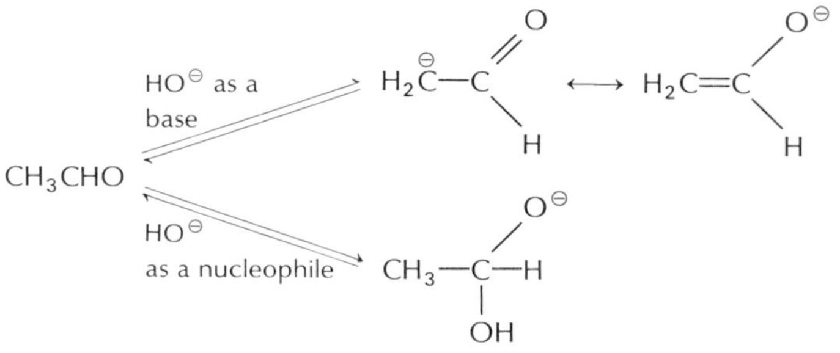

$$CH_3CHO \begin{array}{c} \xrightarrow{HO^\ominus \text{ as a base}} \\ \xleftarrow{} \\ \xrightarrow{HO^\ominus \text{ as a nucleophile}} \end{array} \begin{array}{c} H_2\overset{\ominus}{C}-\overset{O}{\underset{H}{C}} \longleftrightarrow H_2C=\overset{O^\ominus}{\underset{H}{C}} \\ \\ CH_3-\overset{O^\ominus}{\underset{OH}{C}}-H \end{array}$$ (16.37)

initial step in hydration of a carbonyl) is the more easily reversible. The proton removal, also reversible, is less facile. In either case, a new nucleophile (and base) is formed that also can react with another carbonyl group (Eqs. 16.38). We see that if the enolate ion acts as a

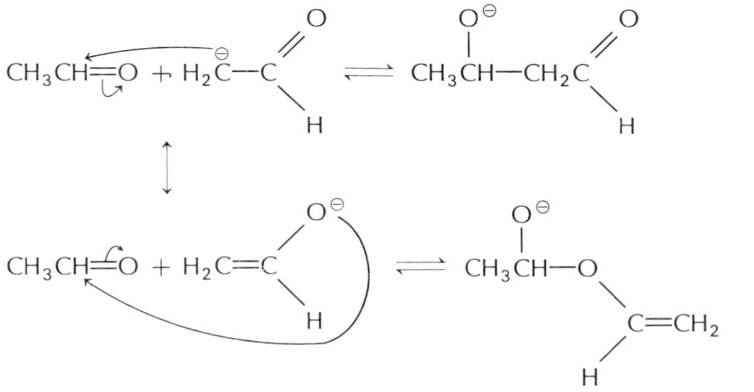

$$CH_3CH=O + H_2\overset{\ominus}{C}-\overset{O}{\underset{H}{C}} \rightleftharpoons CH_3\overset{O^\ominus}{\underset{}{CH}}-CH_2\overset{O}{\underset{H}{C}} \quad a$$

$$CH_3CH=O + H_2C=\overset{O^\ominus}{\underset{H}{C}} \rightleftharpoons CH_3\overset{O^\ominus}{\underset{}{CH}}-O\underset{H}{\overset{}{\diagdown}}C=CH_2 \quad b$$ (16.38)

nucleophile (an anion of carbon), a is the product; if it acts as an anion of oxygen, b is the product. The b product is on the way to becoming a hemiacetal of vinyl alcohol, a readily reversible reaction under basic conditions. The product a has a new C—C σ bond, much less reversible. Thus, after looking at some of the possible equilibria, we find that a wins out by thermodynamic control and goes on to add a proton and give aldol as the product. (See eq. 16.39, in which we have simply omitted the myriad of other equilibria in the system.)

$$CH_3CHO \xrightleftharpoons{HO^\ominus} \overset{\ominus}{C}H_2CHO \xrightleftharpoons{CH_3CHO} CH_3\overset{O^\ominus}{\underset{}{CH}}CH_2CHO \xrightleftharpoons{H_2O} CH_3CHOHCH_2CHO$$ (16.39)

Any aldehyde with an α hydrogen can undergo the aldol condensation (Eqs. 16.40).

$$2CH_3CH_2CHO \xrightleftharpoons{HO^\ominus} CH_3CH_2CHOHCHCHO$$
$$\phantom{2CH_3CH_2CHO \xrightleftharpoons{HO^\ominus} CH_3CH_2CHOHCH}|$$
$$\phantom{2CH_3CH_2CHO \xrightleftharpoons{HO^\ominus} CH_3CH_2CHOHCH}CH_3$$

$$2(CH_3)_2CHCHO \xrightleftharpoons{HO^\ominus} (CH_3)_2CHCHOH\overset{CH_3}{\underset{CH_3}{\overset{|}{C}}}CHO$$ (16.40)

The product in every case is a β-hydroxyaldehyde. The aldol condensation, we see, is a substitution in one molecule of reactant and an addition to the other molecule. If no α hydrogens are present, as in φCHO or tBuCHO, the Cannizzaro reaction takes place instead of the aldol condensation.

The addition of an enolate ion to a carbonyl group is so strongly influenced by steric factors that the simplest ketone, acetone, forms only 2% of product (diacetone alcohol) at equilibrium (Eq. 16.41). We may persuade an equilibrium of this type to give a good yield by exposing acetone to an insoluble base, then removing the mixture, allowing fresh acetone to be exposed, and repeating the operation over and over.

$$2CH_3COCH_3 \underset{}{\overset{HO^\ominus}{\rightleftharpoons}} (CH_3)_2\underset{|}{\overset{OH}{C}}-CH_2COCH_3 \quad (16.41)$$
$$98\% \qquad\qquad\qquad 2\%$$

A base-catalyzed aldol condensation should be carried out at or only slightly above room temperature, because at higher temperatures a base-catalyzed dehydration occurs (Eq. 16.42). Of course, most of the aldol reacts with hydroxide ion to equilibrate with

$$CH_3\underset{|}{\overset{OH}{C}}HCH_2CHO \overset{HO^\ominus}{\rightleftharpoons} CH_3\overset{OH}{\overset{|}{C}}HCHCHO \overset{-HO^\ominus}{\longrightarrow} CH_3CH=CHCHO \quad (16.42)$$

$CH_3\overset{O^\ominus}{\underset{|}{C}}HCH_2CHO$; but any enolate ion formed readily loses hydroxide ion to give crotonaldehyde ($CH_3CH=CHCHO$), which is stabilized by conjugation. If the α,β-unsaturated aldehyde or ketone is wanted, it may be preferable to carry out the condensation reaction at a higher temperature. However, if a γ hydrogen is available in an α,β-unsaturated aldehyde or ketone, a dienolate ion may be formed (Eqs. 16.43). It, in turn, may add to another

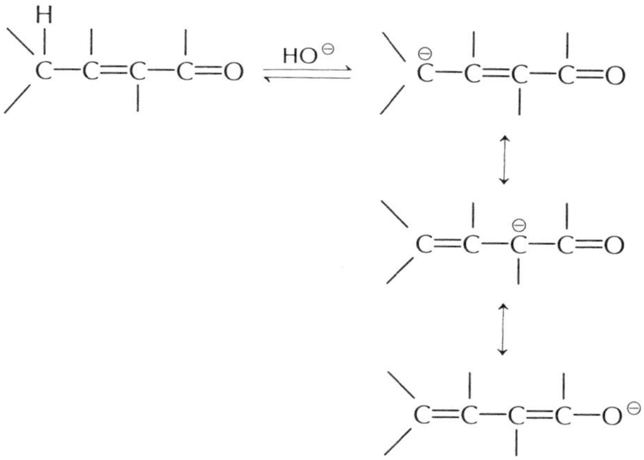

$$\overset{H}{\underset{}{>}}C-C=C-C=O \overset{HO^\ominus}{\rightleftharpoons} \overset{\ominus}{>}C-C=C-C=O$$
$$\updownarrow$$
$$>C=C-\overset{\ominus}{C}-C=O$$
$$\updownarrow \qquad\qquad (16.43)$$
$$>C=C-C=C-O^\ominus$$

$$2CH_3CHO \overset{HO^\ominus}{\underset{hot}{\longrightarrow}} CH_3CH=CHCHO \overset{HO^\ominus}{\underset{\underset{hot}{CH_3CHO}}{\longrightarrow}} [CH_3CHOHCH_2CH=CHCHO]$$
$$\downarrow -H_2O$$
$$CH_3CH=CHCH=CHCHO$$

16.4 Substitution and condensation reactions

molecule of acetaldehyde. On the other hand, another enolate ion from acetaldehyde may add to crotonaldehyde. Thus, acetaldehyde with hot concentrated sodium hydroxide reacts to give a mixture of products: crotonaldehyde; 2,4-hexadienal; 2,4,6-octatrienal; and so on.

Acids also catalyze the aldol condensation (Eq. 16.44). However, under acidic con-

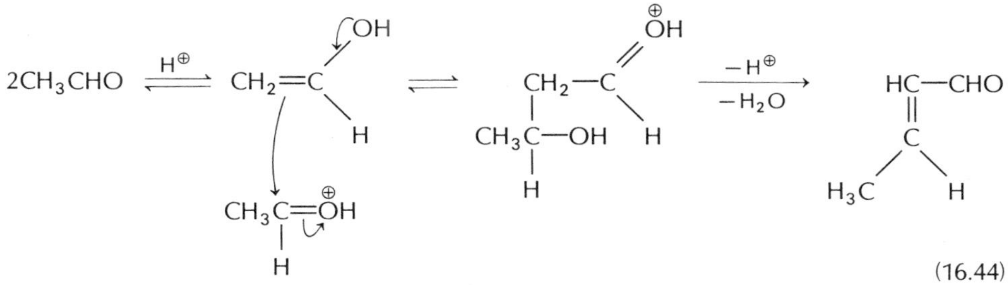

$$2CH_3CHO \xrightleftharpoons{H^\oplus} CH_2=C\begin{smallmatrix}OH\\H\end{smallmatrix} \rightleftharpoons CH_2-C\begin{smallmatrix}\overset{\oplus}{O}H\\H\end{smallmatrix} \xrightarrow[-H_2O]{-H^\oplus} \underset{H_3C\quad H}{HC-CHO} \quad (16.44)$$

ditions, it is almost impossible to stop the reaction at the β-hydroxy stage, because acid-catalyzed dehydration proceeds so readily to give the α,β-unsaturated product, which, of course, may condense again. Because the dehydration step pulls the reaction to the right, ketones become susceptible to condensation under acidic conditions. Thus, acetone, in the presence of dry HCl and $ZnCl_2$, reacts to give phorone (Eq. 16.45). Phorone also is

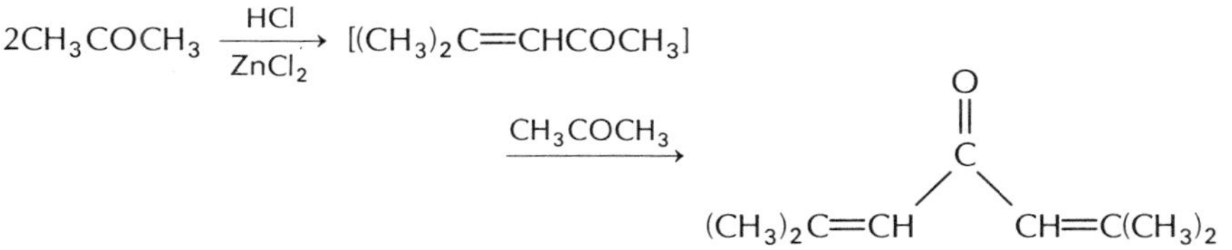

$$2CH_3COCH_3 \xrightarrow[ZnCl_2]{HCl} [(CH_3)_2C=CHCOCH_3] \xrightarrow{CH_3COCH_3} (CH_3)_2C=CH-\underset{\text{phorone}}{C(=O)}-CH=C(CH_3)_2 \quad (16.45)$$

available from the reaction of <u>mesityl oxide</u> $((CH_3)_2C=CHCOCH_3$, obtained from diacetone alcohol by very mild acidic dehydration) with acetone.

Mixed aldol condensations. Aldol condensations with mixed reactants are most successful if one of the reactants has no α hydrogens: for example, an aromatic aldehyde (Eq. 16.46). Again, dehydration pulls the reaction to completion. Condensation of an

$$\phi CHO + CH_3COtBu \xrightarrow[\substack{H_2O\\EtOH, 25°}]{NaOH} \phi CH=CHCOtBu \quad (16.46)$$

aromatic aldehyde with an aliphatic aldehyde or ketone under strongly basic conditions is known as a <u>Claisen-Schmidt condensation.</u> We should remember that under basic conditions, the order of contribution to the stability of an enolate ion is $\overset{\ominus}{C}H_2$—COR > $R\overset{\ominus}{C}H$—COR > $R_2\overset{\ominus}{C}$—COR. Thus, if methyl ethyl ketone is used, condensation occurs preferentially at the methyl group (Eq. 16.47).

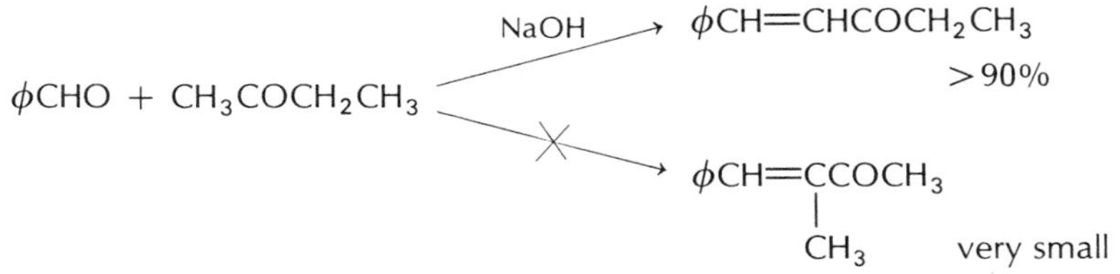

$$\phi CHO + CH_3COCH_2CH_3 \underset{\xcancel{\quad}}{\overset{NaOH}{\longrightarrow}} \begin{array}{l} \phi CH=CHCOCH_2CH_3 \quad >90\%\\ \phi CH=CCOCH_3\\ \quad\quad |\\ \quad\quad CH_3 \quad\text{very small}\end{array} \quad (16.47)$$

An enolate ion is reluctant to add to the carbonyl group of a ketone. Let us look at the reaction of acetophenone with acetaldehyde, in which there are two enolate ions (Eqs. 16.48). The larger anion finds it more difficult to add, so aldol is the major product. Thus, in

$$CH_3CHO + \overset{\ominus}{C}H_2CHO \longrightarrow CH_3CHOHCH_2CHO \quad (16.48)$$

$$CH_3CHO + \overset{\ominus}{C}H_2CO\phi \longrightarrow CH_3CHOHCH_2CO\phi$$

the Claisen-Schmidt condensation, the aldehyde used must have no α hydrogens. To prevent aldol formation in the mixed reaction of two aldehydes ($\phi CHO + CH_3CHO \rightarrow \phi CHOHCH_2CHO$), acetaldehyde is dripped into a mixture of benzaldehyde and base. The enolate ion from acetaldehyde thus has little chance of finding an acetaldehyde molecule to which to add.

An interesting variation is the use of another aldehyde without any α hydrogens, formaldehyde. Equation 16.49 shows the preparation of <u>pentaerythritol</u> from $CH_3CHO + 4HCHO$. The last step is a mixed Cannizzaro reaction.

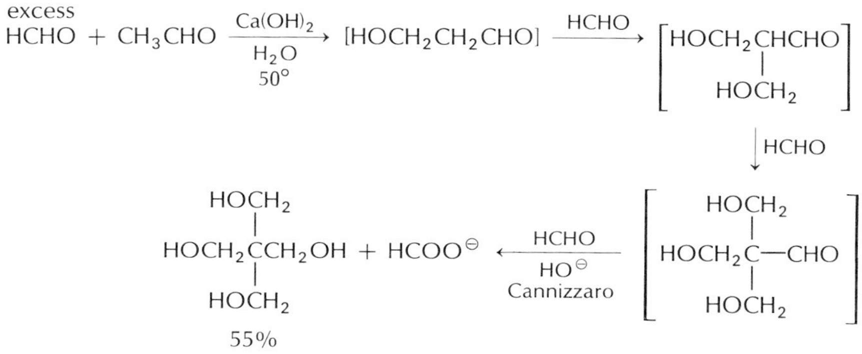

$$\text{excess HCHO} + CH_3CHO \xrightarrow[\substack{H_2O \\ 50°}]{Ca(OH)_2} [HOCH_2CH_2CHO] \xrightarrow{HCHO} \begin{bmatrix} HOCH_2CHCHO \\ | \\ HOCH_2 \end{bmatrix}$$

$$\downarrow HCHO \quad (16.49)$$

$$\underset{\substack{| \\ HOCH_2 \\ 55\%}}{\overset{HOCH_2}{HOCH_2CCH_2OH}} + HCOO^\ominus \xleftarrow[\substack{HO^\ominus \\ \text{Cannizzaro}}]{HCHO} \begin{bmatrix} HOCH_2 \\ | \\ HOCH_2C-CHO \\ | \\ HOCH_2 \end{bmatrix}$$

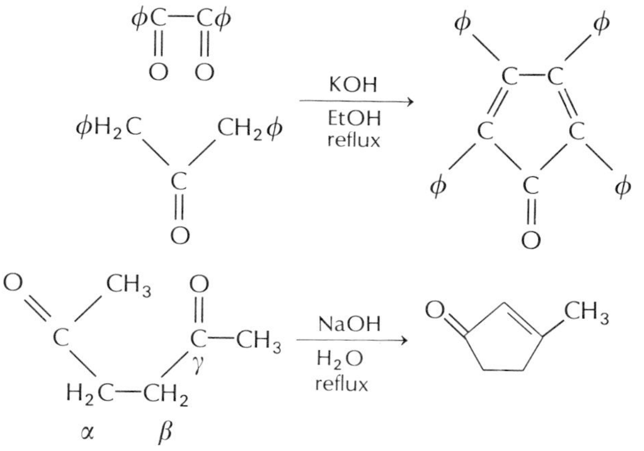

Another possibility is the use of the aldol or Claisen-Schmidt condensations to prepare cyclic compounds (Eqs. 16.50). In a, an α-diketone is condensed on both sides of another ketone, here dibenzyl ketone, to give a substituted cyclopentadienone. In b, a γ-diketone (2,5-hexadione) is shown going to a cyclopentenone, and c shows a δ-diketone (2,6-heptadione) forming a cyclohexenone. The cyclization of a δ-diketone takes place so easily that many such ketones are unknown.

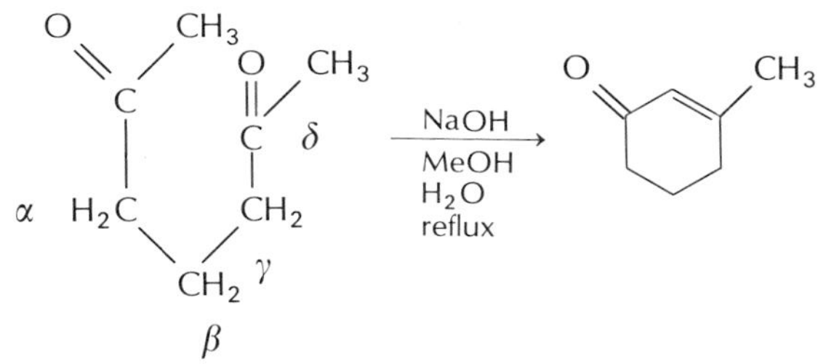

β-diketones. We have omitted β-diketones. Why? Let us look at the pK_a of $CH_3COCH_2COCH_3$ (acetylacetone), which is 9, and of propanedial ($OHCCH_2CHO$), which is 5. They are much more acidic than are the monocarbonyl compounds. If we examine the enolate ion, we find that the answer is increased stabilization of the ion by charge delocalization. The readiness with which enolization occurs is shown by the failure of Grignard

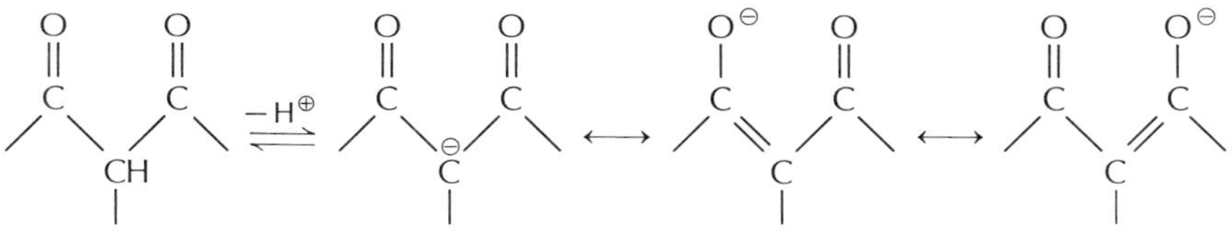

reagents to add to either of the carbonyl groups. Instead, 100% enolization results. The enol too is stable, not necessarily by conjugation so much as by strong internal hydrogen bond formation, called chelation. Liquid acetylacetone is 80% enol and only 20% dicarbonyl at 25°. The liquid also has no ir absorption at 1715 cm^{-1}, only a strong, broad band at 1540 to 1640 cm^{-1}.

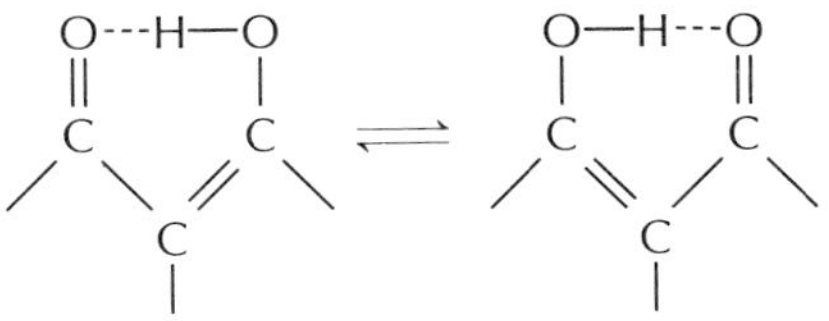

The enolate ions of β-dicarbonyl compounds are able to participate as nucleophiles in S_N2 reactions with alkyl halides (sometimes as bases for dehydrohalogenation). Although most of the charge resides on the oxygen atoms, most reaction takes place at the partially negative carbon atom (Eq. 16.51). Because of the basic media used, secondary alkyl halides are less useful than primary, and tertiary halides cannot be used at all.

In a diketone, C alkylation occurs in preference to O alkylation. The C-alkylated product may be subjected to a second alkylation (Eq. 16.52).

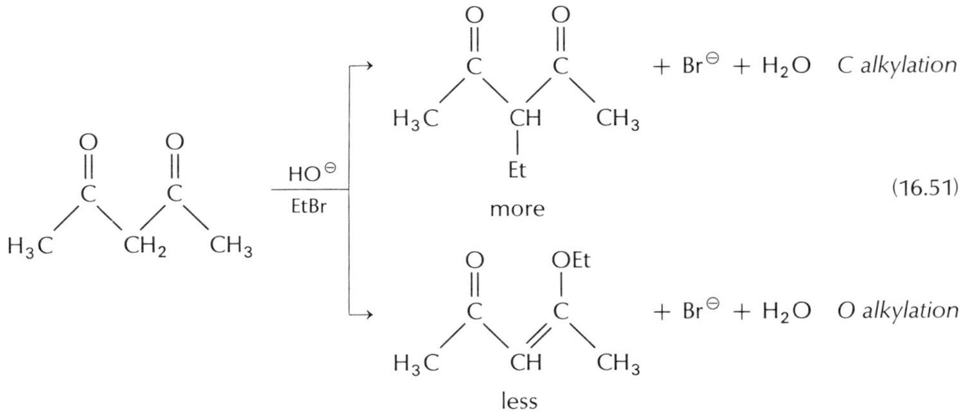

$$\underset{H_3C}{\overset{O}{\overset{\|}{C}}}\underset{CH_2}{\diagdown}\underset{CH_3}{\overset{O}{\overset{\|}{C}}}\xrightarrow[EtBr]{HO^{\ominus}}\begin{cases}\underset{H_3C}{\overset{O}{\overset{\|}{C}}}\underset{\underset{Et}{|}}{\diagdown}\underset{CH}{\overset{O}{\overset{\|}{C}}}\underset{CH_3}{\diagup} + Br^{\ominus} + H_2O \quad \text{C alkylation}\\[2em]\text{more}\\[2em]\underset{H_3C}{\overset{O}{\overset{\|}{C}}}\underset{CH}{\diagdown}\underset{CH_3}{\overset{OEt}{\overset{|}{C}}}\underset{CH_3}{\diagup\!\!\!\diagup} + Br^{\ominus} + H_2O \quad \text{O alkylation}\\[1em]\text{less}\end{cases}$$

(16.51)

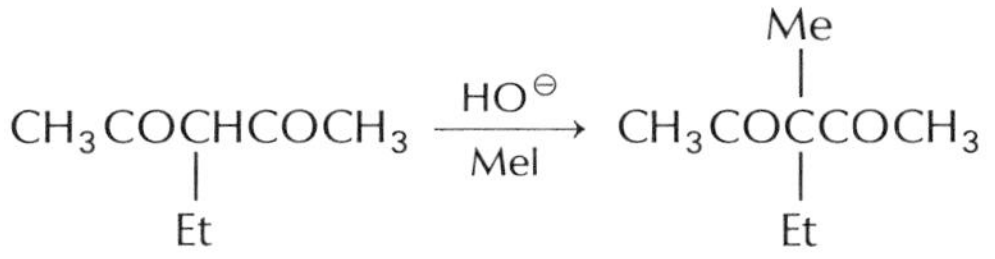

$$CH_3COCHCOCH_3 \xrightarrow[MeI]{HO^{\ominus}} CH_3CO\underset{\underset{Et}{|}}{\overset{\overset{Me}{|}}{C}}COCH_3 \qquad (16.52)$$
$$\underset{Et}{|}$$

The β-dicarbonyl compounds are rather more susceptible to base cleavage than are other dicarbonyl or monocarbonyl compounds (except of the haloform type). The enolate ion formed lowers the transition state for reaction in that direction (Eq. 16.53). When

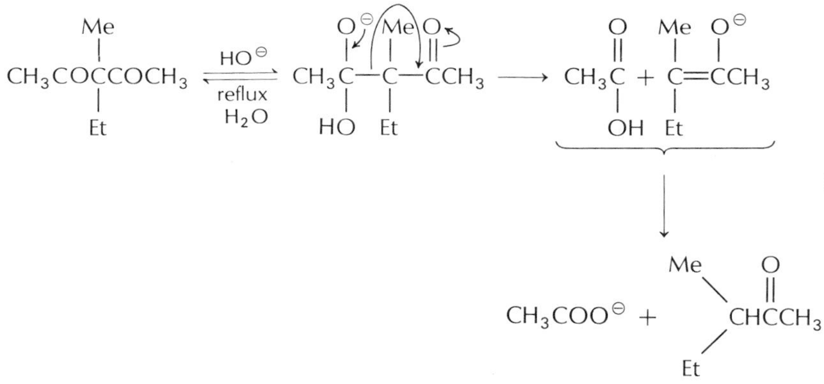

$$CH_3CO\underset{\underset{Et}{|}}{\overset{\overset{Me}{|}}{C}}COCH_3 \underset{\underset{H_2O}{reflux}}{\overset{HO^{\ominus}}{\rightleftharpoons}} CH_3\underset{\underset{HO}{|}}{\overset{\overset{O^{\ominus}}{|}}{C}}\!\!-\!\!\underset{\underset{Et}{|}}{\overset{\overset{Me}{|}}{C}}\!\!-\!\!\overset{\overset{O}{\|}}{C}CH_3 \longrightarrow CH_3\overset{\overset{O}{\|}}{C}\underset{OH}{} + \underset{\underset{Et}{|}}{\overset{\overset{Me}{|}}{C}}\!\!=\!\!\overset{\overset{O^{\ominus}}{|}}{C}CH_3$$

$$\downarrow$$

$$CH_3COO^{\ominus} + \underset{Et}{\overset{Me}{\diagdown}}CH\overset{\overset{O}{\|}}{C}CH_3$$

(16.53)

combined with previous alkylation steps, the cleavage provides a pathway to the synthesis of substituted ketones. In later sections and chapters we shall find even more useful schemes for the preparation of substituted ketones.

C and O acylation. We have pointed out that in alkylation, the C-alkyl product generally is preferred. In contrast, in acylation the O-acyl product dominates (Eq. 16.54).

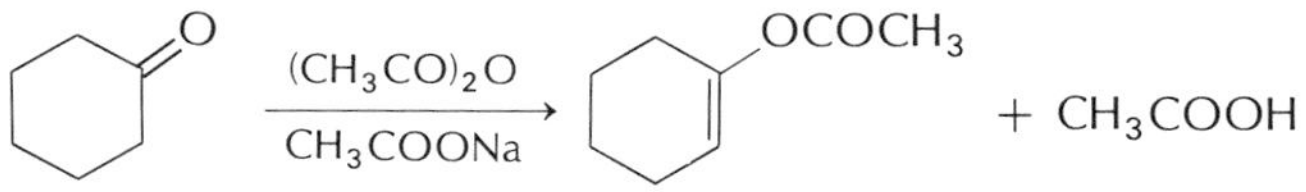

cyclohexanone $\xrightarrow[CH_3COONa]{(CH_3CO)_2O}$ cyclohexenyl-OCOCH$_3$ + CH$_3$COOH (16.54)

An enol acetate (a vinyl ester) frequently is useful in a transesterification (Eq. 17.48), especially for tertiary alcohols, because very mild conditions allow the equilibrium in Equation 16.55 to be set up.

$$CH_2=\underset{CH_3}{\underset{|}{C}}-OCOCH_3 + tBuOH \xrightleftharpoons{\text{trace } H^\oplus} CH_3-\underset{CH_3}{\underset{|}{C}}=O + CH_3COOtBu \quad (16.55)$$

There are many exceptions to these generalities. Sometimes only a change in solvent or a change from a homogeneous solution to a heterogeneous mixture causes a change from C to O alkylation or acylation or the reverse. However, another general rule is that acylation (with an acid chloride or acid anhydride) tends to give the O-acyl product if the acylating agent is used in excess, or to give the C-acyl product if the enolate ion is in excess. The C-acylation is valuable as a method for getting to β-carbonyl compounds (Eqs. 16.56).

$$CH_3COCH_3 \xrightarrow[\text{ether}]{NaH} \underset{\text{excess}}{CH_3COCH_2^\ominus} \xrightarrow{(CH_3CO)_2O} CH_3COCH_2COCH_3 + CH_3COONa$$

$$tBuCOCH_3 \xrightarrow[\text{ether}]{NaNH_2} \underset{\text{excess}}{tBuCOCH_2^\ominus} \xrightarrow{tBuCOCl} tBuCOCH_2COtBu + NaCl$$
$$(16.56)$$

Michael reaction. The Michael reaction is the addition of nucleophiles (activated β-carbonyl enolate ions were used originally) to the β position of α,β-unsaturated aldehydes or ketones. It is a 1,4-addition reaction in competition with a 1,2-addition to the carbonyl carbon (Eq. 16.57). The 1,2-addition may be recognized as the addition step of an aldol

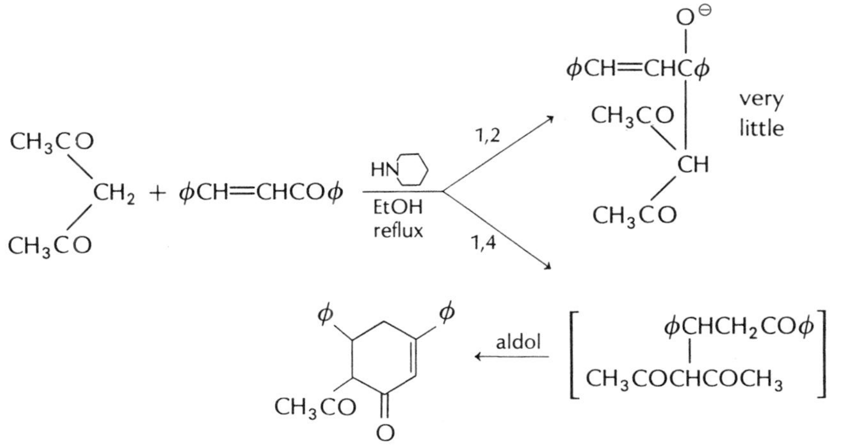

$$\underset{CH_3CO}{\overset{CH_3CO}{\diagdown}}CH_2 + \phi CH=CHCO\phi \xrightarrow[\text{EtOH reflux}]{HN\bigcirc} \begin{array}{c} \overset{1,2}{\nearrow} \phi CH=CHC\phi \text{ with } \underset{CH_3CO}{\overset{CH_3CO}{\diagdown}}CH \text{ very little} \\ \overset{1,4}{\searrow} \end{array} \quad (16.57)$$

$$\xleftarrow{\text{aldol}} \begin{bmatrix} \phi CHCH_2CO\phi \\ | \\ CH_3COCHCOCH_3 \end{bmatrix}$$

(cyclohexenone product with φ, φ, CH₃CO substituents)

$$R^\ominus + \underset{\diagdown}{\overset{\diagup}{C}}=\underset{\diagdown}{\overset{\diagup}{C}}\underset{C=O}{\overset{|}{\diagdown}} \begin{array}{c} \overset{1,4}{\nearrow} \underset{R}{\overset{|}{C}}-\underset{\diagdown}{\overset{\diagup}{C}}-\underset{\diagdown}{\overset{C=O}{\diagup}} \longleftrightarrow \underset{R}{\overset{|}{C}}-\underset{\diagdown}{\overset{\diagup}{C}}=\underset{\diagdown}{\overset{C-O^\ominus}{\diagup}} \\ \overset{1,2}{\searrow} \underset{\diagdown}{\overset{\diagup}{C}}=\underset{\diagdown}{\overset{\diagup}{C}}\underset{R}{\overset{C-O^\ominus}{\diagdown}} \end{array} \quad (16.58)$$

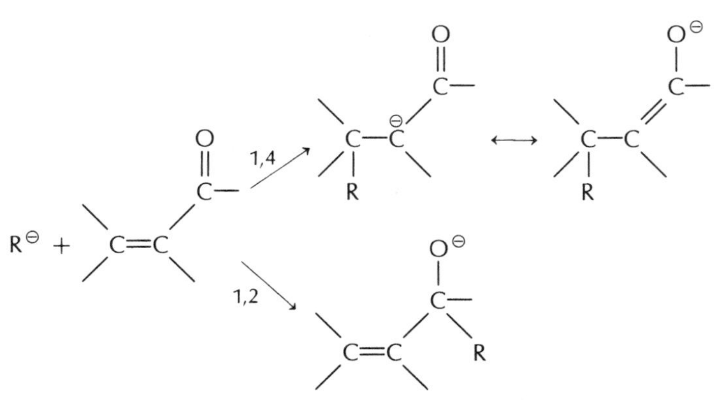

condensation. The 1,4-addition is preferred here (the Michael reaction) because the result of the first step is the formation of an enolate ion that is more stable than the alcoholate ion that results from 1,2-addition (Eq. 16.58). The Michael reaction product in most cases is capable of other reactions (aldol, Claisen-Schmidt, alkylation), so conditions are kept as mild as possible (weak base, low temperature, and short reaction times) to maximize the kinetically controlled product. Piperidine, pyridine, triethylamine, and benzyltrimethylammonium hydroxide ($\phi CH_2 \overset{\oplus}{N}(CH_3)_3 HO^\ominus$, Triton B) are frequently used catalysts. Because the base is regenerated as the reaction proceeds (enolate ion + piperidine $H^\oplus$ → ketone + piperidine), only catalytic amounts are used (0.1 to 0.3 equivalent).

A very useful application of the Michael reaction employs acrylonitrile as the unsaturated component (Eq. 16.59). Also, the anion of a β-diketone is not necessary.

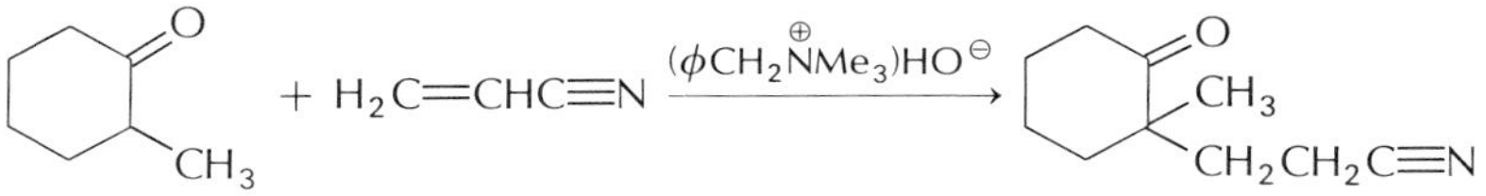

$$\text{(cyclohexanone-CH}_3\text{)} + H_2C=CHC\equiv N \xrightarrow{(\phi CH_2 \overset{\oplus}{N}Me_3)HO^\ominus} \text{(product with CH}_3\text{ and CH}_2CH_2C\equiv N\text{)} \tag{16.59}$$

In other words, if a stronger base and an unhindered but activated double bond (as in acrylonitrile) are used, the enolate ion of a simple ketone will react. Because the product has gained a —CH_2CH_2CN group, the use of $H_2C=CHC\equiv N$ has given rise to the name <u>cyanoethylation</u> for the reaction. A cyano group can be hydrolyzed to a carboxylic acid group (—$C\equiv N$ + $2H_2O$ → —$COOH$ + NH_3). Thus, cyanoethylation followed by hydrolysis is an excellent method for introducing a —CH_2CH_2COOH group into a ketone.

Another useful substance is 3-butene-2-one (vinyl methyl ketone), which allows the synthesis of cyclohexenones (Eq. 16.60) in a subsequent aldol condensation.

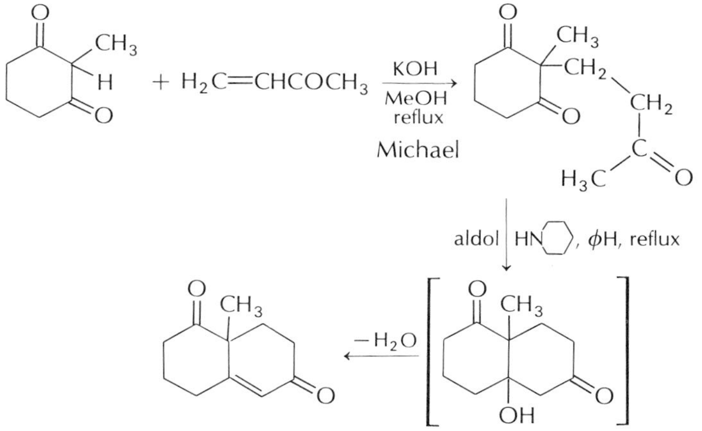

$$\tag{16.60}$$

This reaction is known as the <u>Robinson annelation reaction</u>.

Knoevenagel reaction. The <u>Knoevenagel reaction</u> refers to the facile aldol type of reaction of active methylene compounds (β-diketones, etc.) with aldehydes (and some ketones) in the presence of mild bases. A trace of acid has been found beneficial (Eq. 16.61).

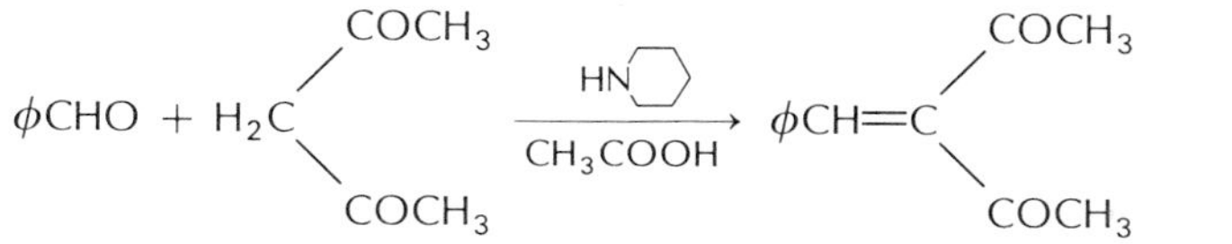

16.4 Substitution and condensation reactions 513

$$\phi CHO + H_2C\begin{matrix}COCH_3\\ \\COCH_3\end{matrix} \xrightarrow[CH_3COOH]{HN\bigcirc} \phi CH=C\begin{matrix}COCH_3\\ \\COCH_3\end{matrix} \quad (16.61)$$

If an excess of the β-diketone is used, a Michael reaction takes place with the Knoevenagel product in the reaction mixture (Eq. 16.62). The Michael product cyclizes as in Equations 16.57 and 16.60. ☐ ☐

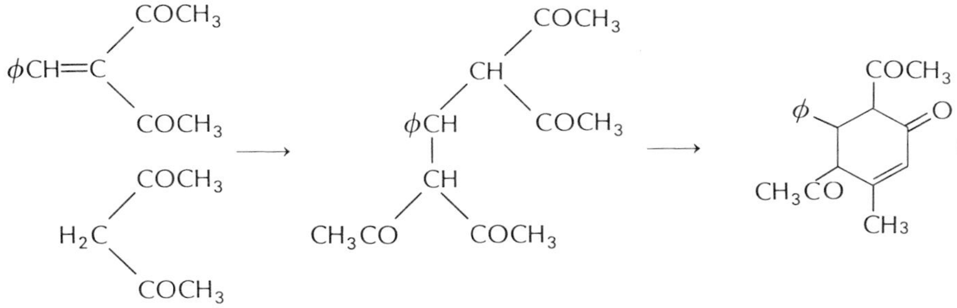

(16.62)

Benzoin condensation. The benzoin condensation is a reaction of aldehydes with no α hydrogens. Cyanide ion plays a decisive role. The complete reaction is shown for the production of benzoin. As with the aldol condensation, this product name has been used for the reaction (Eq. 16.63).

$$2\phi CHO \xrightarrow[H_2O]{CN^\ominus, EtOH} \phi CHOHCO\phi \quad (16.63)$$

The mechanism, which fits the rate expression, rate = $k[\phi CHO]^2[CN^\ominus]$, is a series of equilibria (Eqs. 16.64) in which the slow step probably is the aldol addition

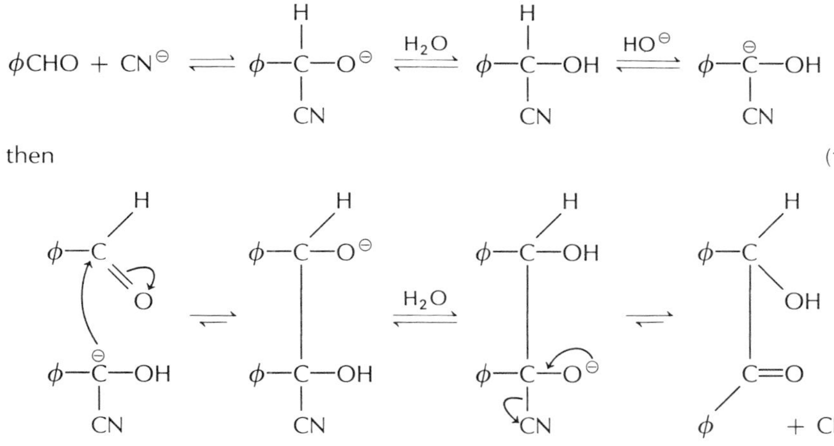

$$\phi CHO + CN^\ominus \rightleftharpoons \phi-\underset{CN}{\overset{H}{\underset{|}{C}}}-O^\ominus \xrightarrow{H_2O} \phi-\underset{CN}{\overset{H}{\underset{|}{C}}}-OH \xrightleftharpoons{HO^\ominus} \phi-\underset{CN}{\overset{\ominus}{\underset{|}{C}}}-OH$$

then (16.64)

step. Sodium cyanide hydrolyzes in solution to produce enough hydroxide ion to cause "enolate ion" formation in the cyanohydrin. The triple bond in —C≡N activates neighboring α hydrogen atoms in the same way that a carbonyl group does

by delocalization of the charge of the anion $\left(\begin{array}{c} \\ \\ \end{array} \overset{\ominus}{C}-C\equiv N \longleftrightarrow \begin{array}{c} \\ \\ \end{array} C=C=N^{\ominus} \right)$.
Strictly speaking, we should not call the anion an enolate ion, but the analogy is apt. After addition of the anion to the carbonyl carbon of another molecule of benzaldehyde, there are several proton transfers. Then the cyanide ion falls away and takes part again by attack on another benzaldehyde molecule. The Cannizzaro reaction is the principal competing reaction, and that is minimized by use of a rather high concentration of sodium cyanide.

Benzoin, as an α-hydroxyketone, is easily oxidized by a number of reagents to the α-diketone benzil (ϕCOCOϕ) which was used in Equation 16.50 to demonstrate ring formation by double aldol condensation with dibenzyl ketone.

Exercises

16.7 Complete the reactions:

a. ϕCOCH$_3$ $\xrightarrow{\text{HCl}}$

b. 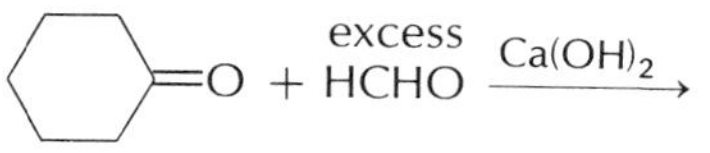 (cyclohexanone)=O + HCHO $\xrightarrow[]{\text{excess Ca(OH)}_2}$

c. CH$_3$COCH$_2$CH$_2$CHO $\xrightarrow[\text{warm}]{\text{NaOH}}$

d.  ϕCHO + CH$_3$CH$_2$COCH$_2$COCH$_2$CH$_3$ $\xrightarrow[\text{CH}_3\text{COOH}]{\text{excess HN}\bigcirc}$

e. 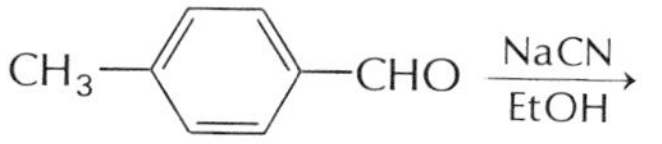 CH$_3$—⟨⟩—CHO $\xrightarrow[\text{EtOH}]{\text{NaCN}}$

f. (CH$_3$)$_2$CHCOCH$_3$ + ϕCHO $\xrightarrow{\text{NaOH}}$

16.8 Show how to carry out the conversions:

a. CH$_3$CH$_2$CHO ⟶ CH$_3$CH$_2$CH$_2$CH(CH$_3$)CH$_2$OH

b. CH$_3$CH$_2$COCH$_2$CH$_3$ ⟶ Et$_3$CCOϕ

c. 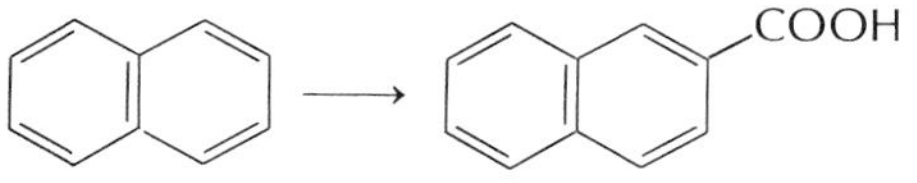 (naphthalene) ⟶ (2-naphthoic acid, COOH)

d. 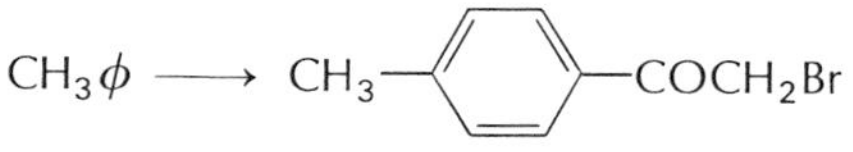 CH$_3\phi$ ⟶ CH$_3$—⟨⟩—COCH$_2$Br

16.9 Under drastic acidic conditions (warm to hot concentrated H$_2$SO$_4$), acetone trimerizes and dehydrates to give mesitylene (1,3,5-trimethylbenzene). Write out a reasonable mechanism for the reaction.

16.5 Rearrangements

We shall take up only two of the many rearrangement reactions: the benzilic acid rearrangement of α-diketones; and the Baeyer-Villiger reaction, in which ketones rearrange to esters.

Benzilic acid rearrangement. Benzil undergoes an interesting reaction with hydroxide ion known as the <u>benzilic acid rearrangement</u>. It contains elements of resemblance to base cleavage of a ketone, to the Cannizzaro reaction, and to nucleophilic addition to a carbonyl carbon (Eq. 16.65). The absence of α hydrogens helps this base-catalyzed

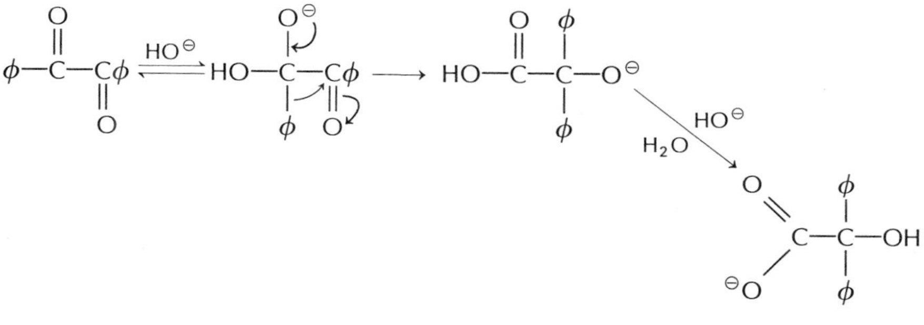

$$\phi-\overset{O}{\underset{\|}{C}}-\overset{}{\underset{\|}{C}}\phi \xrightleftharpoons{HO^\ominus} HO-\overset{O^\ominus}{\underset{\phi}{\underset{|}{C}}}-\overset{}{\underset{O}{\underset{\|}{C}}}\phi \longrightarrow HO-\overset{O}{\underset{\|}{C}}-\overset{\phi}{\underset{\phi}{\underset{|}{C}}}-O^\ominus$$

$$\xrightarrow[H_2O]{HO^\ominus} \overset{O}{\underset{\ominus O}{\underset{\|}{C}}}-\overset{\phi}{\underset{\phi}{\underset{|}{C}}}-OH \qquad (16.65)$$

rearrangement of α-diketones to succeed. Another example shows a way to go easily from phenanthrene quinone to the fluorene ring system (Eq. 16.66).

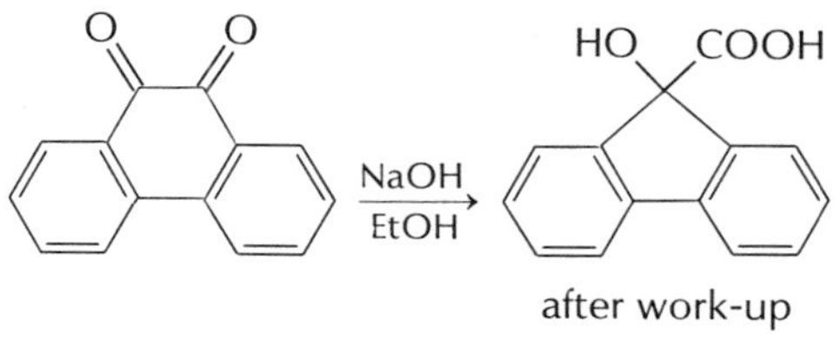

[phenanthrenequinone] $\xrightarrow[\text{EtOH}]{\text{NaOH}}$ [9-hydroxyfluorene-9-carboxylic acid] (16.66)

after work-up

Baeyer-Villiger reaction. The <u>Baeyer-Villiger reaction</u> usually is carried out with a reactive peracid such as CF_3CO_3H in acetic acid. A strong acid (H_2SO_4, $ArSO_3H$) is present to cause a ketone or aldehyde to rearrange to an ester. At times, $CHCl_3$ and CH_2Cl_2 are used as solvents. The reaction is believed to go as shown in Equation 16.67.

If the two R groups are different, which R migrates to oxygen and which remains bonded to the carbonyl carbon? If the R groups are substituted phenyls, results show clearly that electron-withdrawing substituents (—NO_2, —C≡N, —COOEt) retard migration and that electron-donating substituents (—OH, —OMe) provide aptitude for migration. In general, it may be said that the order of migratory aptitudes is *tert*-R > cyclohexyl, *sec*-R, ϕCH_2—, allyl, ϕ— > primary —R > Me, not only here but in all rearrangements of the Wagner-Meerwein type, the benzilic acid rearrangement, and so on. Aldehydes react to give esters of formic acid (R > H in migration) or to give acids (H > R in migration) depending on the R group. Some other examples are shown in Equations 16.68. ☐☐

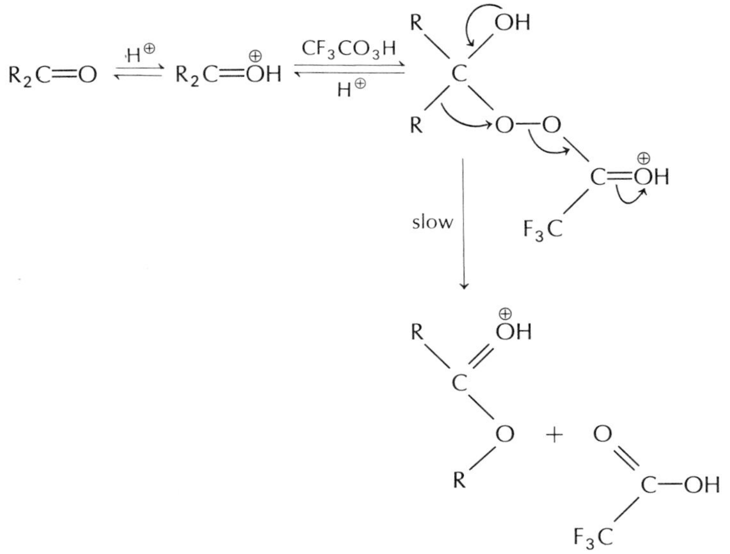

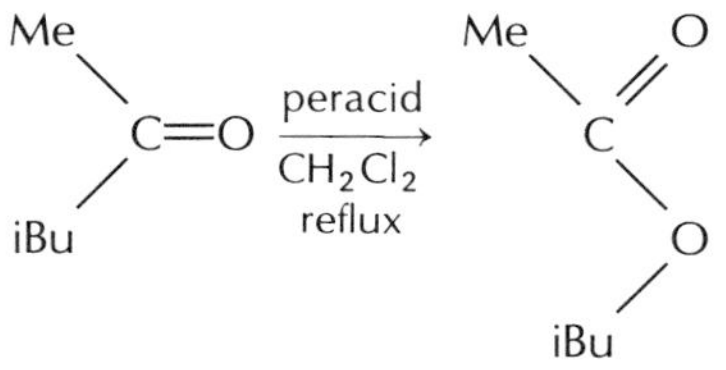

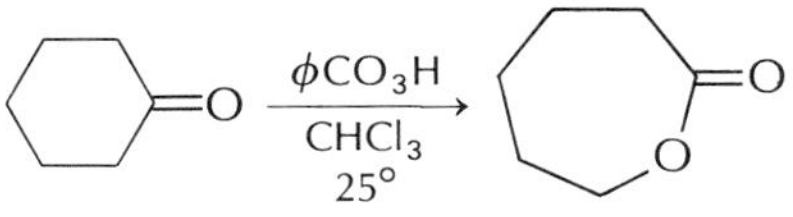

Exercises

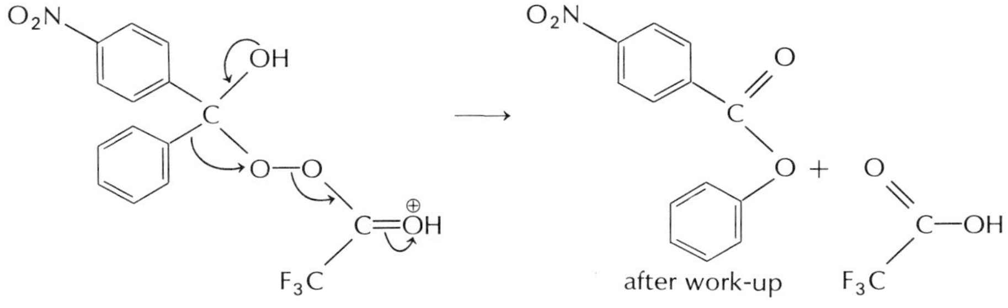

16.10 Complete the reactions:

a. $\phi\text{COCHO} \xrightarrow{\text{NaOH}}$

b. 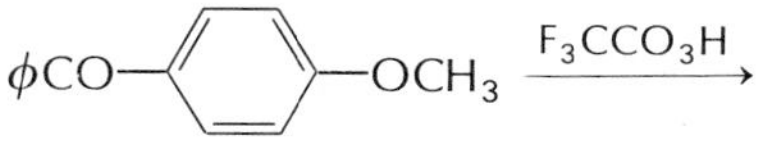 $\phi\text{CO}\text{–}\langle\text{C}_6\text{H}_4\rangle\text{–OCH}_3 \xrightarrow{\text{F}_3\text{CCO}_3\text{H}}$

16.6 Oxidation and reduction

Many reagents are used in many ways to carry out oxidation and reduction reactions. Some were described in Chapter 9 (alcohols) and elsewhere. In this section, we shall give quite a list of these reactions, even though the description may lie in some previous section or chapter. Let us take up oxidations first.

Oxidations: KMnO₄, the chromium reagents. Aldehydes are easily oxidized to the corresponding acids by many reagents. Potassium permanganate under acidic, neutral, or basic conditions probably is the most widely used reagent. Acidic conditions are convenient because the lengthy and difficult filtration of manganese dioxide (formed under neutral and basic conditions) is avoided. In any case, dilute solutions of acid or base are used to avoid the unwanted aldol condensation, Cannizzaro, and other possible side reactions. The chromium reagents also are used (acid conditions only). Some examples are shown in Equations 16.69. Note that conditions are mild.

$$\phi CHO \xrightarrow[\substack{H_2O \\ 80°}]{KMnO_4} \phi COOH \quad \text{after work-up}$$

$$CH_3CH_2CH_2CHO \xrightarrow[\substack{dil.\ H_2SO_4 \\ 40°}]{KMnO_4} CH_3CH_2CH_2COOH \tag{16.69}$$

$$\phi CH_2CH_2CHO \xrightarrow[\substack{dil.\ H_2SO_4 \\ 40°}]{KMnO_4} \phi CH_2CH_2COOH$$

Enolizable ketones are subject to oxidation either as enols or as enolate ions. Because the ketone is broken up in the process, the reaction is not often used in preparative work unless a single enol is possible. Strongly acid or basic conditions are used to catalyze the slow step, enolization or enolate ion formation (Eqs. 16.70). Ketones are oxidized much

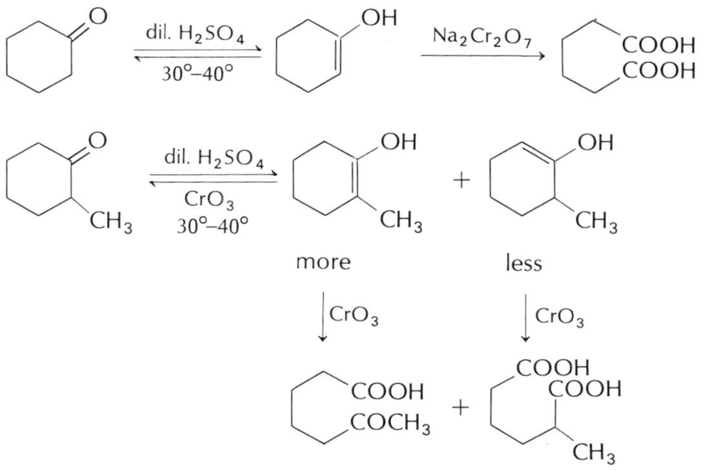

(16.70)

more slowly than are alcohols, alkenes, and so on; and they usually survive these oxidations in some degree. However, aryl alkyl ketones are readily cleaved by oxidation (Eq. 16.71).

$$\phi COCH(CH_3)_2 \xrightleftharpoons[CrO_3]{dil. H_2SO_4, 40°} \phi\overset{OH}{\underset{|}{C}}=C(CH_3)_2 \xrightarrow{CrO_3} \phi COOH + O=C(CH_3)_2 \quad (16.71)$$

Special reagents. The ease of oxidation of aldehydes is demonstrated by the very mild reagents cupric and silver ion. The reactions are used as tests for the presence of —CHO in a molecule. Because of several factors (one being the necessity of keeping the oxidized product, an acid, in solution), the tests require an alkaline medium. To maintain high concentrations of cupric or silver ion in solutions of high pH ($Cu(OH)_2$ and Ag_2O are not very soluble), complexing agents are used. They reduce the concentration of hydrated Cu^{2+} or $Ag^{\oplus}$ to levels less than those required for precipitation of the hydroxides. □ □

Previously described reductions. Reduction reactions of aldehydes and ketones have been taken up in many previous instances. The uses of catalytic hydrogenation, lithium aluminum hydride, sodium borohydride, and borane have been discussed. Likewise, we should remember the Clemmensen and the Wolff-Kishner reductions and the Meerwein-Ponndorf-Verley reduction. A mixed Cannizzaro reaction with formaldehyde is a reduction of the other aldehyde used.

Bimolecular reduction. A method we have not seen is the use of reactive metals to bring about <u>bimolecular reduction</u>. Under the proper conditions (control of potential,

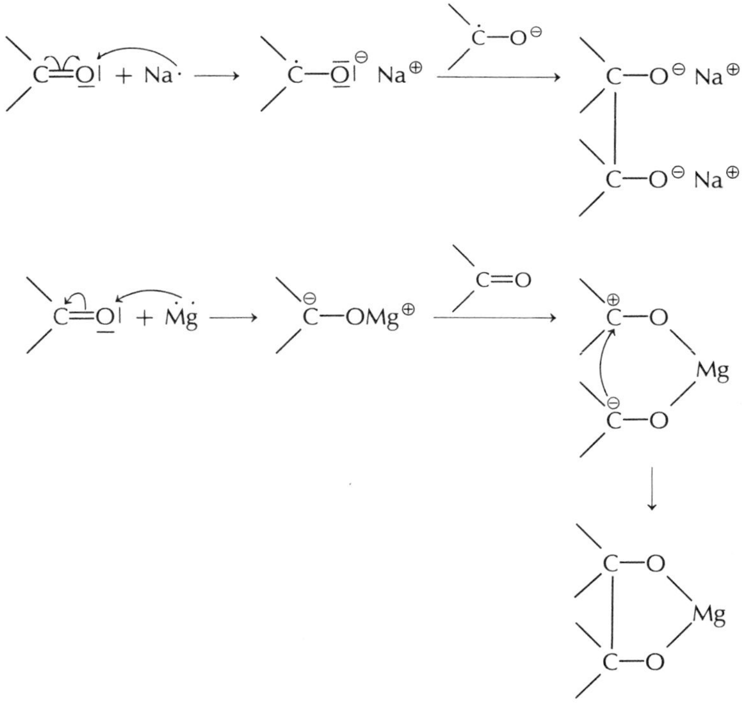

(16.72)

composition of electrodes, etc.), bimolecular reductions also may be carried out by electrolytic means. Sodium, magnesium, and aluminum have been the most used metals. The usual conditions are sodium with moist ether, or sodium amalgam with more active protic solvents at 25°; amalgamated magnesium or aluminum in dry benzene under reflux; or an equimolar mixture of Mg + $MgBr_2$ or MgI_2 in dry ether or dry ether and benzene under reflux (for diaryl ketones only).

The reduction may be shown as taking place in either of two different ways (Eqs. 16.72). The evidence seems to favor the anion-radical pathway (a) rather than the anion-cation pathway (b). In any case, the reaction is thought to occur on the surface of the metal. The presence of mercury helps the reaction by providing a liquid surface on the magnesium, by providing an inert surface for electron transfer to the aldehyde or ketone, or by both. The α-diol produced is called a <u>pinacol</u> because the product formed from acetone crystallizes as a hexahydrate in large platelike crystals. Some other examples are shown in Equations 16.73.

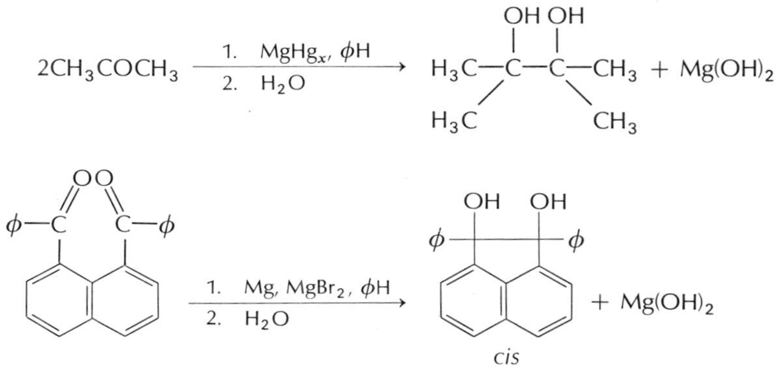

$$2CH_3COCH_3 \xrightarrow[\text{2. }H_2O]{\text{1. }MgHg_x, \phi H} H_3C\underset{H_3C}{\overset{OH}{\underset{|}{C}}}-\underset{CH_3}{\overset{OH}{\underset{|}{C}}}-CH_3 + Mg(OH)_2 \quad a$$

$$\text{[1,8-diacylnaphthalene]} \xrightarrow[\text{2. }H_2O]{\text{1. Mg, MgBr}_2, \phi H} \text{[cis diol]} + Mg(OH)_2 \quad b \qquad (16.73)$$

$$\phi CHO \xrightarrow{\text{Mg, MgHg}_x} \phi CHOHCHOH\phi \quad c$$

Pinacols, especially ditertiary alcohols prepared from ketones, readily undergo an acid-catalyzed dehydration and rearrangement known as the <u>pinacol rearrangement</u> (Eq. 16.74). The product from pinacol is called pinacolone (methyl *tert*-butyl ketone).

$$Me_2\underset{|}{\overset{OH}{C}}-\underset{|}{\overset{OH}{C}}Me_2 \xrightarrow[70-100°]{\text{dil. }H_2SO_4} Me_3CCOMe + H_2O \qquad (16.74)$$

Because the product of the pinacol rearrangement is a ketone, we shall discuss the ramifications of the reaction in Section 16.7 when we describe the preparation of aldehydes and ketones. □ □

Exercises

16.11 Show how to carry out these conversions:

a. $\phi CH=CHCHO \longrightarrow \phi CH=CHCOOH$

b. $\phi CH_2CO\phi \longrightarrow \phi CH_2CH_2\phi$

520   Chapter sixteen

c. 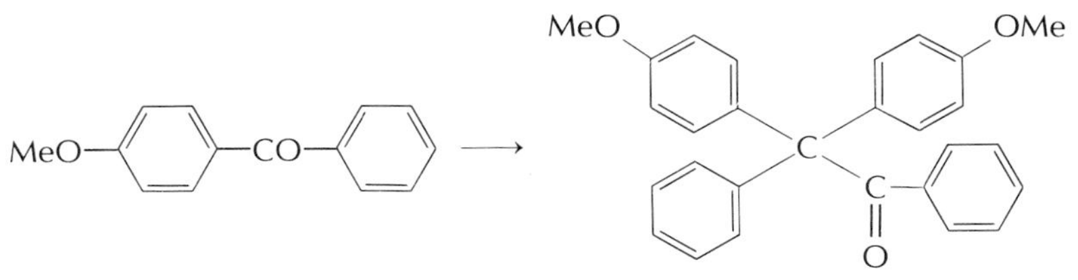 MeO—⟨C₆H₄⟩—CO—⟨C₆H₅⟩ ⟶ (MeO—C₆H₄)₂C(C₆H₅)—C(=O)—C₆H₅

☐ ☐

16.7 Preparation of aldehydes and ketones

There are a great many preparative reactions for synthesizing the carbonyl compounds. One advanced text lists forty-four methods for the preparation of aldehydes and fifty-seven for ketones. There is some duplication between the lists, but even so, the variety is impressive. We shall tabulate over twenty methods, most of which we already have discussed in connection with other topics. We shall arbitrarily classify them as substitutions, additions, rearrangements, oxidations, and reductions. In later chapters we shall encounter still more preparative reactions.

Substitutions.   Under substitution reactions, we may recall the aldol condensation (Sec. 16.4) and the Friedel-Crafts acylation reaction (Sec. 15.3).

In addition to the straightforward preparation of aromatic ketones, two variations have been used to introduce an aldehyde group. The Gattermann-Koch reaction utilizes carbon monoxide, HCl, and AlCl₃ under pressure (100–250 atm) or, if CuCl is added to coordinate (complex) the CO, under atmospheric pressure. The mixture of gases may be conveniently prepared from formic acid (Eq. 16.75). The

$$HCOOH + ClSO_3H \longrightarrow CO + HCl + H_2SO_4 \qquad (16.75)$$

gaseous mixture is conducted into an excess of the arene mixed with AlCl₃ and CuCl (Eq. 16.76). The resemblance to the Friedel-Crafts reaction extends to the fact that

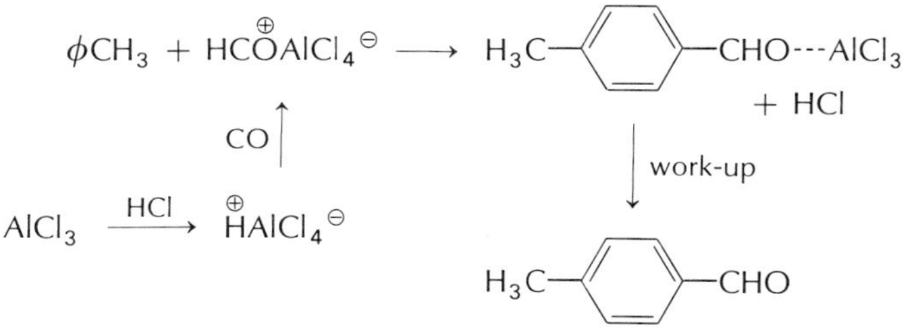

$$\phi CH_3 + HC\overset{\oplus}{O}AlCl_4{}^{\ominus} \longrightarrow H_3C\text{—}C_6H_4\text{—}CHO\cdots AlCl_3 + HCl$$

$$CO \uparrow$$

$$AlCl_3 \xrightarrow{HCl} H\overset{\oplus}{A}lCl_4{}^{\ominus}$$

$$\downarrow \text{work-up}$$

$$H_3C\text{—}C_6H_4\text{—}CHO \qquad (16.76)$$

the reaction does not work if meta-directing substituents are present. Likewise, phenols and phenyl ethers do not react successfully, possibly because AlCl₃ complexes with oxygen strongly enough to cause inactivation of the ring.

16.7 Preparation of aldehydes and ketones 521

The other variation is called the Gattermann reaction and employs $Zn(CN)_2$ and HCl. It is, in turn, a special case of the Houben-Hoesch reaction. The latter reaction may be formulated as shown (Eq. 16.77) for the preparation of a ketone. The Gattermann reaction is analogous (Eq. 16.78), although the reactive ion may be $HCN + H\overset{\oplus}{C}=NH \rightarrow H\overset{\oplus}{C}=N-CH=NH$. The milder catalyst ($ZnCl_2$) allows phenols and phenyl ethers to be used as well as arenes in both the Gattermann and Houben-Hoesch reactions.

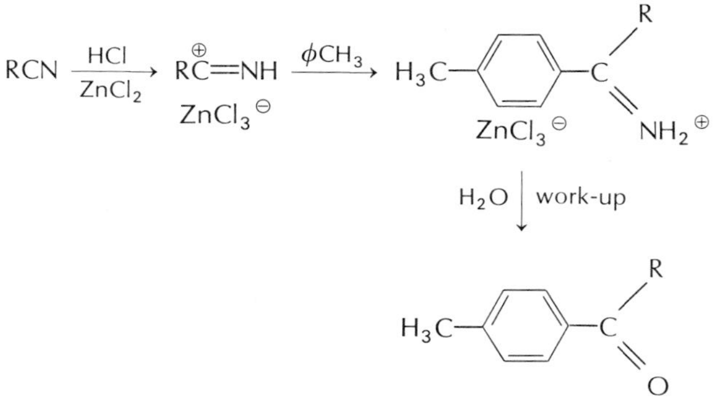

$$RCN \xrightarrow[ZnCl_2]{HCl} R\overset{\oplus}{C}=NH \; ZnCl_3^{\ominus} \xrightarrow{\phi CH_3} H_3C-C_6H_4-\underset{NH_2^{\oplus}}{\overset{R}{C}}=... \;\; ZnCl_3^{\ominus}$$

$$\downarrow H_2O \; \text{work-up}$$

$$H_3C-C_6H_4-\underset{O}{\overset{R}{C}}= \qquad (16.77)$$

$$HCN \xrightarrow[ZnCl_2]{HCl} H\overset{\oplus}{C}=NH \; ZnCl_3^{\ominus} \xrightarrow{\phi CH_3} H_3C-C_6H_4-\underset{NH_2^{\oplus}}{\overset{H}{C}}=... \;\; ZnCl_3^{\ominus}$$

$$\uparrow$$
$$Zn(CN)_2 + HCl$$

$$\downarrow H_2O \; \text{work-up}$$

$$H_3C-C_6H_4-\underset{O}{\overset{H}{C}}= \qquad (16.78)$$

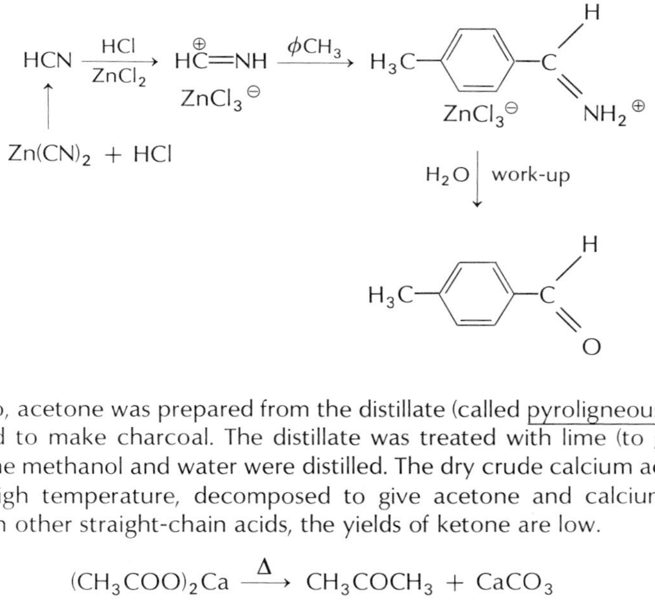

Years ago, acetone was prepared from the distillate (called pyroligneous liquid) from hardwood used to make charcoal. The distillate was treated with lime (to give calcium acetate), and the methanol and water were distilled. The dry crude calcium acetate, when heated to a high temperature, decomposed to give acetone and calcium carbonate (Eq. 16.79). With other straight-chain acids, the yields of ketone are low.

$$(CH_3COO)_2Ca \xrightarrow{\Delta} CH_3COCH_3 + CaCO_3 \qquad (16.79)$$

The reaction has been used to produce cyclic ketones from diacids, and (as we might guess) it is very successful for the preparation of 5- and 6-membered rings (Eq. 16.80).

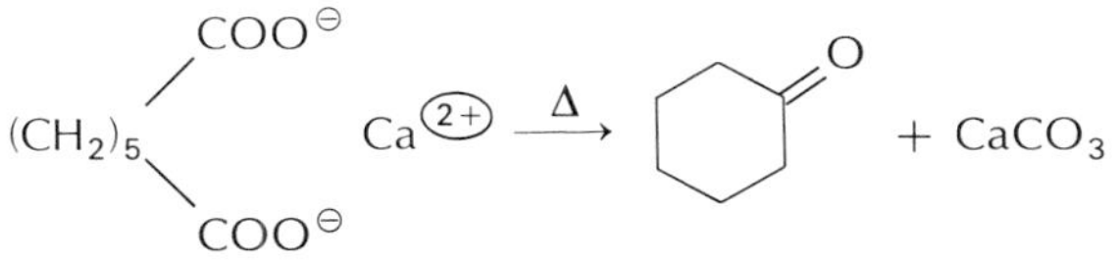

$$(CH_2)_5\underset{COO^{\ominus}}{\overset{COO^{\ominus}}{<}} Ca^{2+} \xrightarrow{\Delta} \text{cyclohexanone} + CaCO_3 \qquad (16.80)$$

We have classified this reaction (a decarboxylation) as a substitution (an α hydrogen is replaced by another group), but it could just as well be called a fragmentation, a disproportionation, or an oxidation-reduction.

Additions. Addition reactions are given here, although you may wish to reclassify them into other categories. Addition reactions already discussed elsewhere are hydration of alkynes (Sec. 13.2) and addition of a nucleophile to an α,β-unsaturated aldehyde or ketone (Michael reaction, Sec. 16.4). To these we shall add some Grignard reactions, the cadmium reagents, and the use of ylides.

A Grignard reagent adds to a cyano group rather less readily than to a carbonyl group, but the reaction provides another route to ketones (Eq. 16.81). The carbon-nitrogen

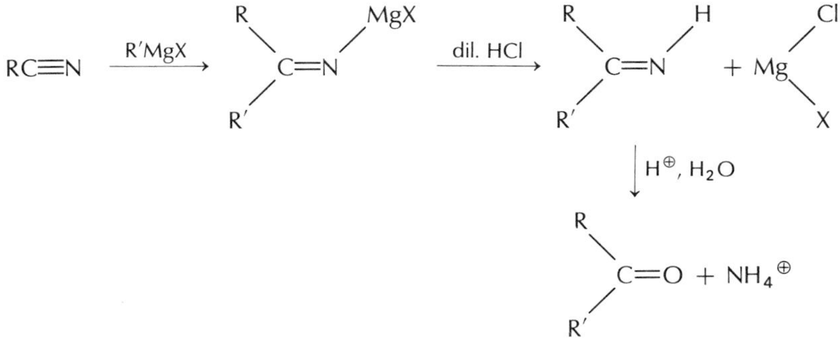

$$RC\equiv N \xrightarrow{R'MgX} \underset{R'}{\overset{R}{>}}C=N\underset{}{\overset{MgX}{<}} \xrightarrow{\text{dil. HCl}} \underset{R'}{\overset{R}{>}}C=N\underset{}{\overset{H}{<}} + Mg\underset{X}{\overset{Cl}{<}}$$

$$\downarrow H^{\oplus}, H_2O \qquad (16.81)$$

$$\underset{R'}{\overset{R}{>}}C=O + NH_4^{\oplus}$$

triple bond is less polar than a carbon-oxygen double bond, and the carbon-nitrogen double bond is even more nonpolar. The addition of a second mole of R'MgX does not occur. Hydrolysis leads first to the ketimine, which in turn is readily hydrolyzed to the ketone.

Grignard reagents add to α,β-unsaturated ketones to give both 1,2 and 1,4 (Michael) addition. The proportions seem to be governed by the size of the groups involved. Unsaturated aldehydes give only 1,2 addition; and alkyl and aryl lithium reagents give more 1,2 and less 1,4 addition than do the corresponding Grignard reagents. An example is shown in Equation 16.82. The 1,4 product is an enolate salt which, upon work-up, reverts to a ketone. The 1,4 product then looks as though R— and H— had been added to the conjugated C=C bond.

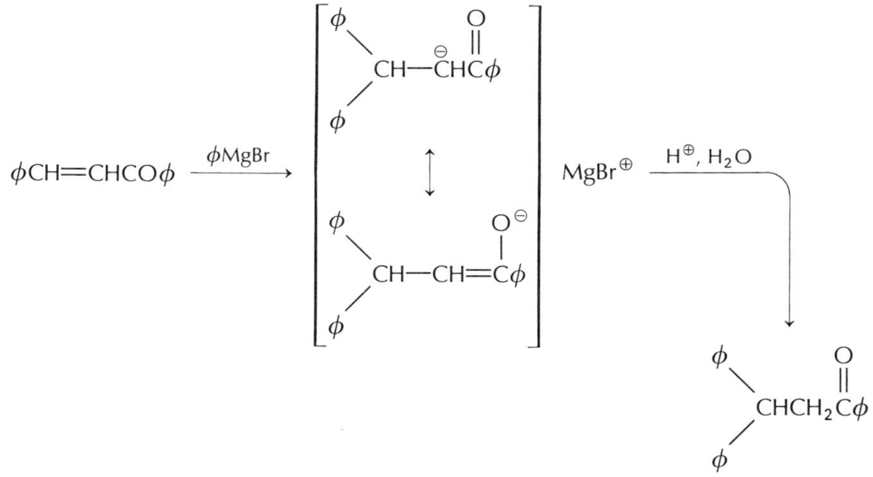

$$\phi CH=CHCO\phi \xrightarrow{\phi MgBr} \left[\begin{array}{c}\underset{\phi}{\overset{\phi}{>}}CH-\overset{\ominus}{C}HC\phi\overset{O}{\underset{\parallel}{}} \\ \updownarrow \\ \underset{\phi}{\overset{\phi}{>}}CH-CH=C\phi\overset{O^{\ominus}}{\underset{|}{}}\end{array}\right] MgBr^{\oplus} \xrightarrow{H^{\oplus}, H_2O} \underset{\phi}{\overset{\phi}{>}}CHCH_2C\phi\overset{O}{\underset{\parallel}{}} \qquad (16.82)$$

16.7 Preparation of aldehydes and ketones

Crowding about the carbonyl group increases the proportion of 1,4 product, and crowding about the C=C bond increases the proportion of 1,2 product (a tertiary alcohol). Thus, in ϕCH=CHCOR, addition of ϕMgBr gives these yields of 1,4 product as R is varied: R = H, 0%; Me, 12%; and ϕ, 94%. Addition of EtMgBr gives 0%, 60%, and 99% with the same compounds. Crowding about the C=C, as in ϕ_2C=CHCOϕ, gives 100% of 1,2 addition with ϕMgBr. Thus, Michael addition of a Grignard is useful in the synthesis of ketones.

A good method for preparation of ketones is the reaction of acid chlorides with cadmium reagents (RCdCl). We saw in the Reformatsky reaction that zinc reagents were less reactive than the magnesium and lithium reagents and did not add to esters. The cadmium reagents are still less reactive and do not add to ketones, although they react with acid chlorides, especially in the presence of anhydrous magnesium halide. Thus the easy preparation of the cadmium reagent from a Grignard reagent makes possible a convenient laboratory procedure (Eq. 16.83). The cadmium reagents are not stable to heat,

$$\text{RMgCl} + \text{CdCl}_2 \xrightarrow[0°]{\text{ether}} \text{RCdCl} + \text{MgCl}_2 \xrightarrow[\phi\text{H, warm}]{\text{R'COCl}} \text{R'COR} + \text{CdCl}_2 \qquad (16.83)$$

which limits R to a primary alkyl or aryl group (Eq. 16.84). A secondary or tertiary alkyl cadmium reagent decomposes even at 0°. An advantage of the reaction for the preparation of ketones is that R' may contain almost any other substituent, including those that would react with Grignard reagents, such as aldehyde, ketone, ester, and even nitro groups (Eq. 16.85).

$$\text{CH}_3(\text{CH}_2)_7\text{CdCl} \xrightarrow{80°–100°} \text{CH}_3(\text{CH}_2)_6\dot{\text{C}}\text{H}_2 + \dot{\text{C}}\text{dCl}$$
$$\qquad\qquad\qquad\qquad\qquad\qquad \downarrow \qquad\qquad\qquad\qquad\qquad (16.84)$$
$$\qquad\qquad\qquad\qquad\qquad \text{CH}_3(\text{CH}_2)_6\text{CH}_3 + \text{CH}_3(\text{CH}_2)_5\text{CH}=\text{CH}_2$$

$$\underset{\text{NO}_2}{\phi}\text{—COCl} \xrightarrow[\phi\text{H}]{\underset{\text{MgX}_2}{\text{EtCdCl}}} \underset{\text{NO}_2}{\phi}\text{—COCH}_2\text{CH}_3 \qquad (16.85)$$

An ylide (Sec. 16.3) reacts readily with acid chlorides (less so with esters) to give a product that may be hydrolyzed under basic conditions or may be reduced with zinc and acetic acid to give ketones. The ylide used must have a hydrogen present on the α carbon if the reaction is to go (Eq. 16.86). Thus, the ylide method for conversion of an acid chloride

$$\phi_3\text{P} + \text{RCH}_2\text{Br} \xrightarrow[25°]{\phi\text{H}} \phi_3\overset{\oplus}{\text{P}}\text{—CH}_2\text{R Br}^{\ominus} \xrightarrow[\substack{\text{ether, hexane} \\ 25°}]{\text{BuLi}} \phi_3\overset{\oplus}{\text{P}}\text{—}\overset{\ominus}{\text{C}}\text{HR}$$

$$\qquad\qquad\qquad\qquad\qquad\qquad\qquad\qquad\qquad\qquad\qquad\qquad\qquad \Bigg\downarrow \substack{\text{R'COCl} \\ \text{Et}_3\text{N}}$$

(16.86)

has the same limitation as does the cadmium reagent: the group to be introduced must be from a primary halide. In fact, the limitation is worse, because an aryl group may be used in the cadmium reagent.

The keto ylide, upon being heated, decomposes to give an alkyne (Eq. 16.87).

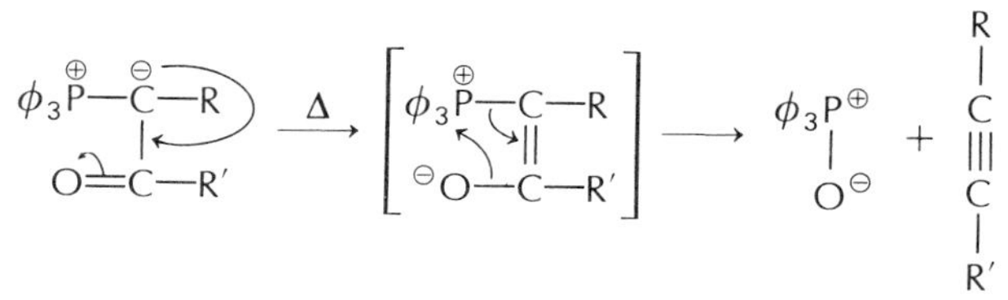

$$\phi_3\overset{\oplus}{P}-\overset{\ominus}{C}-R \atop O=C-R' \xrightarrow{\Delta} \left[\phi_3\overset{\oplus}{P}-C-R \atop \ominus O-C-R' \right] \longrightarrow \phi_3\overset{\oplus}{P} \atop O^\ominus + \begin{matrix} R \\ | \\ C \\ ||| \\ C \\ | \\ R' \end{matrix} \quad (16.87)$$

This reaction should be added to our list of methods for the preparation of non-terminal alkynes (Eq. 16.88).

$$RCH_2X + R'COCl \xrightarrow[\text{heat}]{\phi_3P, \text{BuLi}} RC\equiv CR' \quad (16.88)$$

Rearrangements. Here we shall give only one method for the preparation of aldehydes and ketones.

The pinacol rearrangement is a very useful reaction in some cases for the preparation of highly substituted and branched aldehydes and ketones. The name is derived from the rearrangement of pinacol to pinacolone (methyl *tert*-butyl ketone, Eqs. 16.89). The initial

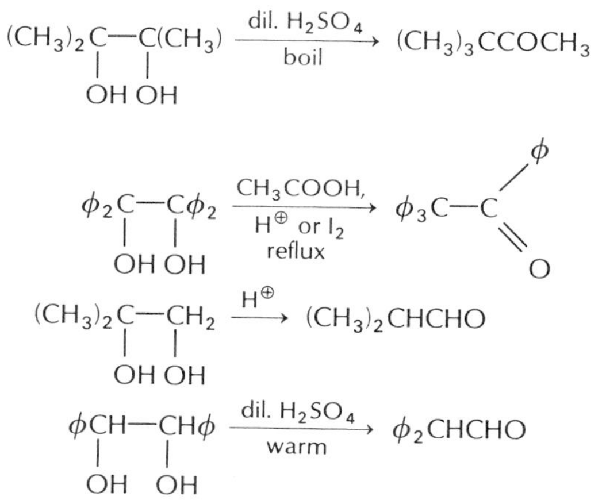

$$(CH_3)_2\underset{OH}{C}-\underset{OH}{C}(CH_3) \xrightarrow[\text{boil}]{\text{dil. } H_2SO_4} (CH_3)_3CCOCH_3$$

$$\phi_2\underset{OH}{C}-\underset{OH}{C}\phi_2 \xrightarrow[\text{reflux}]{\text{CH}_3\text{COOH,} \atop H^\oplus \text{ or } I_2} \phi_3C-C\overset{\phi}{\underset{O}{\diagdown}}$$

$$(CH_3)_2\underset{OH}{C}-\underset{OH}{CH_2} \xrightarrow{H^\oplus} (CH_3)_2CHCHO \quad (16.89)$$

$$\phi CH-CH\phi \atop \underset{OH}{|} \underset{OH}{|} \xrightarrow[\text{warm}]{\text{dil. } H_2SO_4} \phi_2CHCHO$$

formation of a carbonium ion seems reasonable. Like other carbonium ions, it has several options for subsequent reaction. Thus, at high temperatures pinacol dehydrates readily to give 2,3-dimethyl-2,3-butadiene (Eq. 16.90), just as one would expect a tertiary alcohol

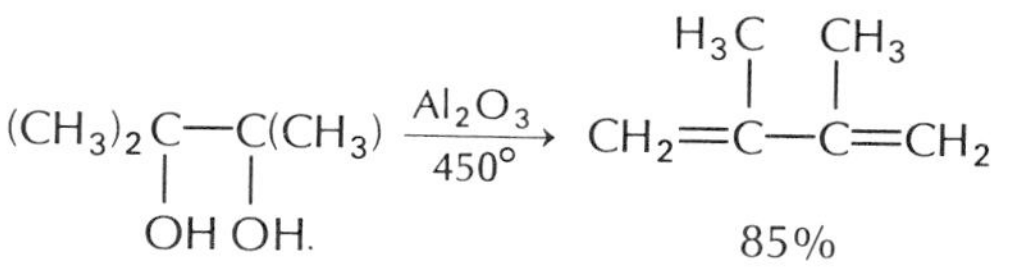

$$(CH_3)_2\underset{OH}{C}-\underset{OH.}{C}(CH_3) \xrightarrow[450°]{Al_2O_3} CH_2=\underset{|}{\overset{H_3C}{C}}-\underset{|}{\overset{CH_3}{C}}=CH_2 \quad (16.90)$$
$$85\%$$

16.7 Preparation of aldehydes and ketones

to do. The rearrangement is favored at lower temperatures, and ordinary acidic dehydration at higher temperatures.

If we look at the initial carbonium ion, we find it possible for each of the groups on the neighboring carbon atom to migrate (Eq. 16.91). We do not show the ion itself,

and the equation seems to say that the migration is concerted, which it is not. It is shown this way to stress the consequences of conformer population. Depending upon the bulk of the groups at the migration terminus, either a, b, or c is favored. However, a is reversible, and only b and c lead to the more stable ketones. During the migration of R_1 or R_2, a transition state (d) resembling a cyclopropane must be attained in

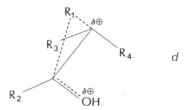

which R_2 (the nonmigrating group in b) becomes eclipsed with R_3 (one of the groups at the terminus). If R_2 migrates, as in c, R_1 becomes eclipsed with R_4. The migration that proceeds through the transition state of lowest energy is favored. The migrating group tends to approach from the rear: the released molecule of water probably interferes with approach from the front side. In other words, the change at the migration terminus, even though we think of it as an S_N1cA reaction, has the stereochemical characteristics of a back-side internal S_N2 substitution.

Studies with substituted benzopinacols have given results as shown in Equation 16.92. In these symmetrical pinacols there is equal opportunity for loss of either —OH group, and the migratory aptitude of the two groups may be compared. Because 94% of p-tolyl group migrated as compared with only 6% phenyl group, we may calculate the migratory aptitude as $(94/6) = 15.7$. From the result (because p-Me is $+I$, $+R$), we can say that electron donation enhances migratory aptitude. The conclusion is not surprising; because, from the viewpoint of the migrating benzene ring, an electrophilic displacement is occurring. Other substituents gave migratory aptitudes as follows: p-OCH$_3$, 500; p-iPr, 9; p-Et, 5; m-CH$_3$, 2; m-OCH$_3$, 1.6; p-Cl, 0.7; and m-Cl, 0.

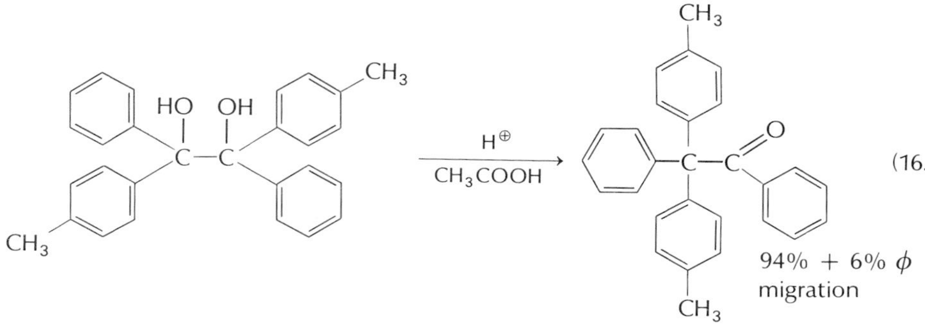

$$\text{(16.92)}$$

94% + 6% ϕ migration

In unsymmetrical pinacols of the type shown in Equation 16.93, another

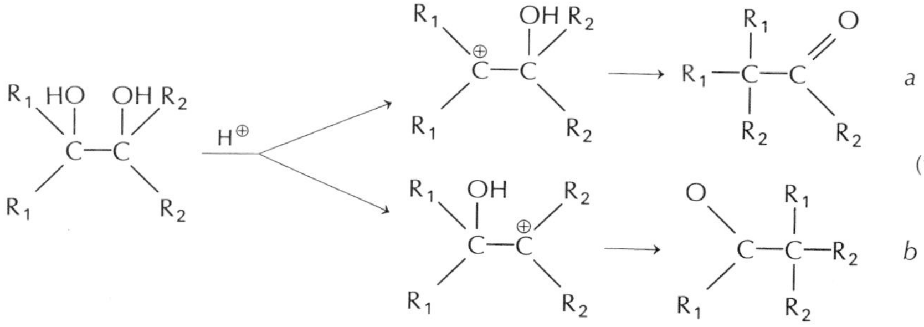

$$\text{(16.93)}$$

determining factor is the relative stability of the cations that rearrange to a and b. If the cation in a is more stable (assuming the energy of activation to be decreased by the stability of the cation), then only R$_2$ migrates. If the cation in b is more stable, then only R$_1$ migrates. If stabilities are comparable, both products are formed. In other words, that group migrates which is the less effective in stabilizing the cation (neglecting the effect that —COH(R)$_2$ may have). Some examples are shown in Equations 16.94.

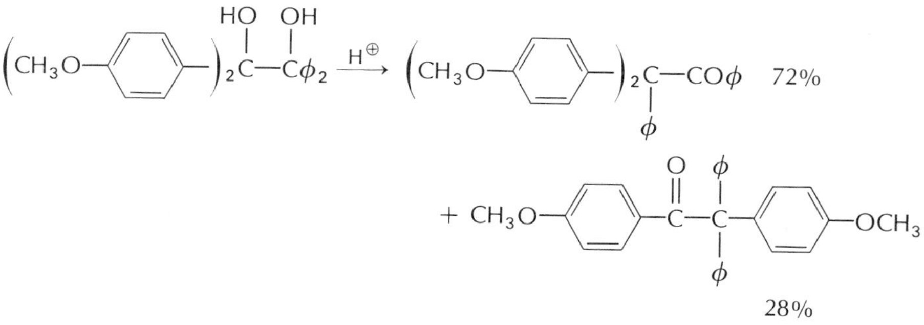

$(CH_3O-\phi-)_2C(OH)-C(OH)\phi_2 \xrightarrow{H^\oplus} (CH_3O-\phi-)_2C(\phi)-CO\phi$ 72%

$+ CH_3O-\phi-CO-C(\phi)_2-\phi-OCH_3$

28%

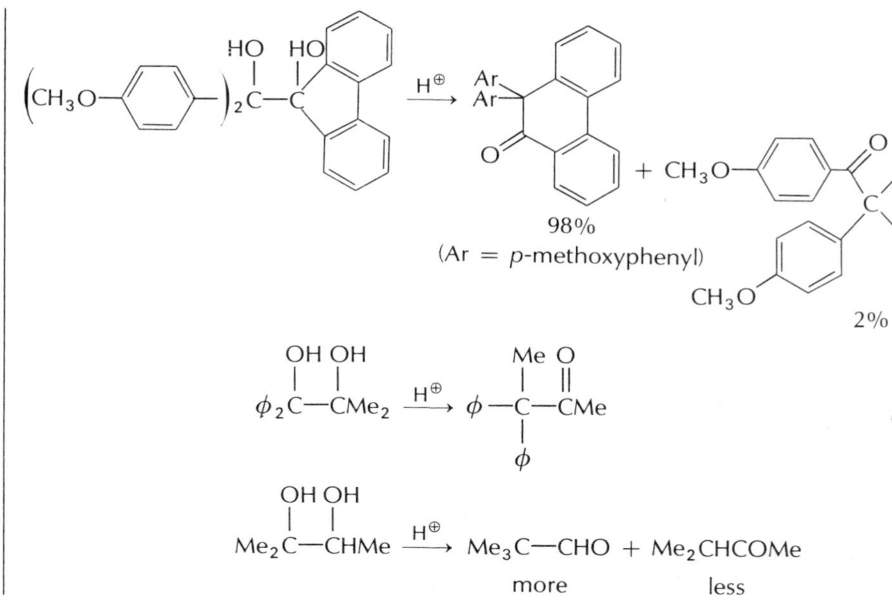

16.7 Preparation of aldehydes and ketones

[Diol rearrangement: (CH$_3$O—C$_6$H$_4$)$_2$C(OH)—C(OH)(fluorenyl) $\xrightarrow{H^\oplus}$ Ar,Ar-substituted phenanthrone (98%) + CH$_3$O—C$_6$H$_4$—C(O)—C(fluorenyl)(C$_6$H$_4$—OCH$_3$) (2%); Ar = p-methoxyphenyl]

$$\phi_2\overset{\underset{|}{OH}}{C}-\overset{\underset{|}{OH}}{C}Me_2 \xrightarrow{H^\oplus} \phi-\overset{\underset{|}{Me}}{\underset{|}{C}}-\overset{O}{\overset{\|}{C}}Me \quad\quad (16.94)$$
$$\phi$$

$$Me_2\overset{\underset{|}{OH}}{C}-\overset{\underset{|}{OH}}{C}HMe \xrightarrow{H^\oplus} Me_3C-CHO + Me_2CHCOMe$$
$$\text{more} \qquad\qquad \text{less}$$

The rearrangement is not limited to diols. What is required is the proper cation, and the cation may be formed in numerous ways (Eqs. 16.95).

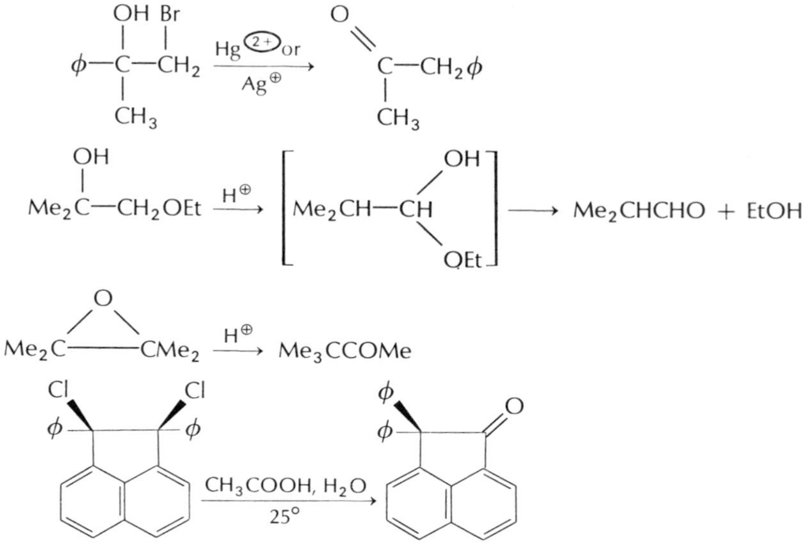

$$\phi-\overset{\underset{|}{OH}}{\underset{\underset{|}{CH_3}}{C}}-\overset{\underset{|}{Br}}{C}H_2 \xrightarrow{Hg^{2+} \text{or} \atop Ag^\oplus} \overset{O}{\overset{\|}{C}}-CH_2\phi$$
$$\underset{CH_3}{|}$$

$$Me_2\overset{\underset{|}{OH}}{C}-CH_2OEt \xrightarrow{H^\oplus} \left[Me_2CH-\overset{OH}{\underset{OEt}{CH}}\right] \longrightarrow Me_2CHCHO + EtOH$$

$$Me_2C\overset{O}{\underset{}{\diagdown\!\!\diagup}}CMe_2 \xrightarrow{H^\oplus} Me_3CCOMe \quad\quad (16.95)$$

[Acenaphthylene-derived: 1,2-dichloro-1,2-diphenyl acenaphthene $\xrightarrow{CH_3COOH, H_2O \atop 25°}$ 2-phenyl-2-phenylacenaphthenone]

Oxidations. The preparation of ketones by oxidative methods generally gives good yields because ketones resist further oxidation (Sec. 16.6). Aldehydes generally require special conditions or special reagents because it is difficult to stop the oxidation at the aldehyde stage. Of the methods of oxidative preparation listed, several have already been discussed elsewhere.

The chromium and manganese reagents were discussed in Sections 9.8 and 16.6. For the preparation of aldehydes, a chromium reagent usually is chosen for oxidation of the primary alcohol. Provision is made for removal of the more volatile aldehyde by distillation (fractionation) as rapidly as it is formed. Another technique used with aryl-methyl groups is to carry out the oxidation in the presence of acetic anhydride (Eq. 16.96).

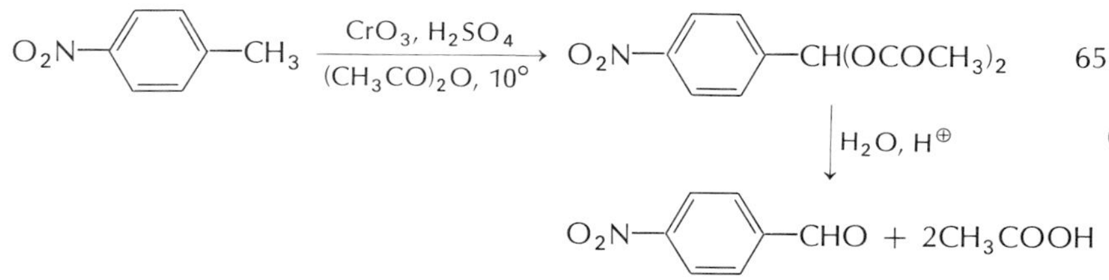

$$O_2N-C_6H_4-CH_3 \xrightarrow[(CH_3CO)_2O,\ 10°]{CrO_3,\ H_2SO_4} O_2N-C_6H_4-CH(OCOCH_3)_2 \quad 65\%$$

$$\downarrow H_2O,\ H^\oplus \quad (16.96)$$

$$O_2N-C_6H_4-CHO + 2CH_3COOH$$

A reagent of special utility is active manganese dioxide (MnO_2), prepared by precipitation from an alkaline solution after interaction of $KMnO_4$ and $MnSO_4$. It is used to oxidize allylic and benzylic alcohols (primary and secondary) to the corresponding aldehyde and ketone without affecting the C=C bonds (Eqs. 16.97) or other hydroxyl groups that might be present.

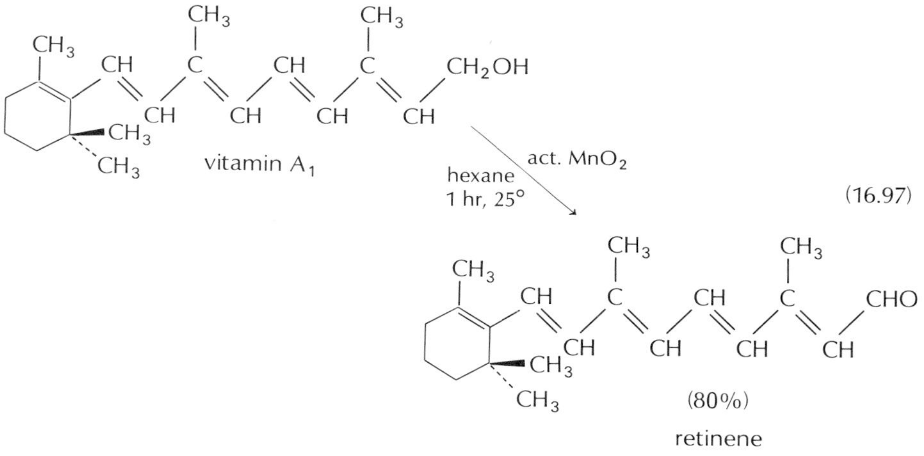

vitamin A₁ $\xrightarrow[\text{hexane, 1 hr, 25°}]{\text{act. MnO}_2}$ retinene (80%) (16.97)

$$CH_3O-C_6H_3(OCH_3)-CH(OH)CH_2CH_2OH \xrightarrow[\text{hexane, 5 hr}]{\text{act. MnO}_2} CH_3O-C_6H_3(OCH_3)-CCH_2CH_2OH\ (=O)$$

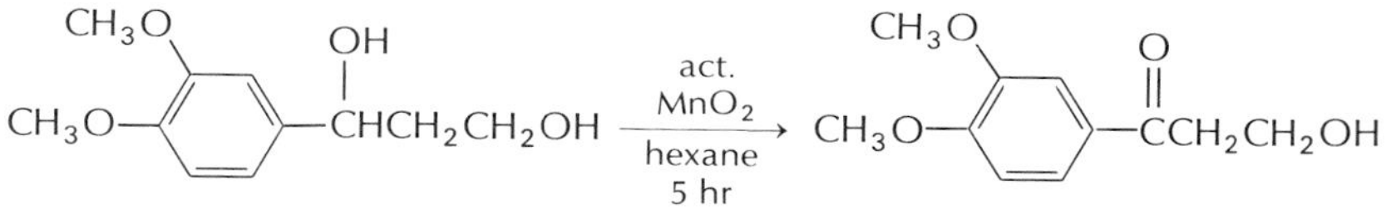

Other oxidations include the Oppenauer oxidation of alcohols (discussed in Sec. 9.8), dehydrogenative oxidation of alcohols by means of hot copper (Sec. 9.8), and indirect oxidation by means of halogenation of side chains (Sec. 15.6).

Diols in which the hydroxy groups are on adjacent carbon atoms may be oxidized quantitatively with either periodic acid (useful in water) or lead tetraacetate (useful in acetic acid, benzene, methanol, etc.), depending upon the solubility characteristics of the diol (Eqs. 16.98). The reagents attack not only diols, but also α-hydroxycarbonyl and

$$\underset{\underset{\diagdown}{\overset{HO\;\;OH}{\underset{|}{C}-\underset{|}{C}}}\diagup}{\overset{\diagup}{}} \begin{cases} \xrightarrow[H_2O,\ 25°]{H_5IO_6} \diagdown C{=}O + O{=}C \diagup + 3H_2O + HIO_3 \\ \\ \xrightarrow[\phi H,\ 25°]{(CH_3COO)_4Pb} \diagdown C{=}O + O{=}C \diagup + 2CH_3COOH + (CH_3COO)_2Pb \end{cases}$$

(16.98)

($H_5IO_6 = HIO_4 \cdot 2H_2O = KIO_4 + H_2SO_4 + H_2O$, sometimes called Malaprade's reagent.)

α-dicarbonyl compounds. They have been very useful in studies on sugars (Ch. 19). In such instances, we think of the carbonyl group as being hydrated. ☐ ☐

A primary hydroxy (—CH_2OH) is oxidized to formaldehyde, a secondary hydroxy (—CHROH) to an aldehyde (RCHO), and a tertiary alcohol (—CR_2OH) to a ketone (RCOR). The usefulness for preparative purposes, as with the pinacol rearrangement, is limited by the availability of the proper diol.

Reductions. Ketones cannot be made by reduction methods because the only higher oxidation state requires a fragmentation of the ketone (or an adjacent oxidation state, as in α-dicarbonyl compounds). Aldehydes, however, can be made by reduction of acid derivatives. Several methods are known, of which we list three.

The Rosenmund reduction makes use of the fact that acid chlorides react readily with hydrogen, even with a poisoned catalyst, whereas aldehydes are more stable. The usual catalyst is palladium deposited on barium sulfate. Sulfur, in almost any form, is held so tenaciously on a catalyst surface that it effectively prevents the usual hydrogenation and hydrogenolysis reactions. The Rosenmund reduction is generally applicable (Eq. 16.99): R— may be alkyl, allyl, aryl, and so on. Yields are good.

$$RCOCl \xrightarrow[\text{xylene, } 150°,\ 2\ hr]{H_2,\ Pd\ /\ BaSO_4} RCHO + HCl \qquad (16.99)$$

The Stephen reduction utilizes stannous chloride to reduce nitriles (usually aryl or straight-chain alkyl nitriles) to imines, which complex with stannic chloride and precipitate from the reaction mixtures. Subsequent hydrolysis of the complex generates the aldehyde (Eq. 16.100). Yields are excellent.

$$RC{\equiv}N \xrightarrow[\substack{\text{ether, HCl}\\25°,\ 15\ min}]{\substack{\text{anhydrous}\\SnCl_2}} (RCH{=}NH_2)_2^{\oplus} SnCl_6^{2-} \xrightarrow[100°]{H_2O} RCHO + NH_4Cl + SnCl_4 \qquad (16.100)$$

Various metal hydrides under special conditions may be used on acids, acid chlorides, esters, N,N-dialkyl amides, or nitriles to prepare aldehydes. It seems, however, that sodium bis (2-methoxyethoxy)aluminum hydride ($Na^{\oplus}(CH_3OCH_2CH_2O)_2AlH_2^{\ominus}$) will become the reagent of choice for the reduction of esters. At temperatures less than $-50°$, aldehydes are stable in the presence of the reagent, so esters, at $-70°$ (the temperature of solid CO_2 cooled baths) in solvents such as toluene, ether, or tetrahydrofuran, are reduced to the aldehydes in the course of two to eight hours. The ester should be methyl or ethyl (branching,

530 Chapter sixteen

as in secondary and tertiary alkyls, lowers the yield dramatically). Yields are about 80% unless R— in RCOOMe is Ar— or Ar$_2$CH— (Eq. 16.101).

$$tBuCOOMe \xrightarrow[\substack{\text{ether} \\ -70°}]{Na^{\oplus}(RO)_2AlH_2^{\ominus}} \xrightarrow{\text{work-up}} tBuCHO \quad (81\%) \qquad (16.101)$$

Exercises

16.12 Complete the reactions:

a. $\phi CH_2CH_2CHO \xrightarrow[25°]{NaOH}$

b. $CH_3\phi \xrightarrow{\text{Gatterman}}$

c. 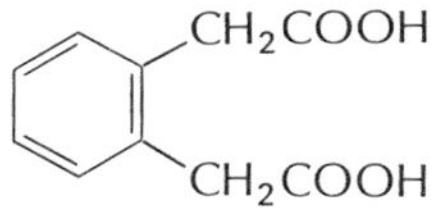 [ortho-disubstituted benzene with CH$_2$COOH groups] $\xrightarrow[\text{2. }\Delta]{\text{1. CaO}}$

d. $\phi C\equiv C\phi \xrightarrow[Hg^{2+}]{H^{\oplus}, H_2O}$

e. $CH_3CH=CHCO\phi + H_2C(COCH_3)_2 \xrightarrow{\text{cat.}}$

f. $\phi MgBr + \phi CH_2CN \longrightarrow$

g. $H_2C=CHCO\phi + \phi MgBr \longrightarrow$

h. $\phi CH=CHCOCl + \phi CH_2CH_2CdCl \longrightarrow$

i. $(CH_3)_2CHCH_2COCl + \phi_3\overset{\oplus}{P}\!\!-\!\!\overset{\ominus}{C}HCH_2CH(CH_3)_2 \xrightarrow[\text{2. HO}^{\ominus}]{\text{1. Et}_3N}$

16.13 Carry out the conversions:

a. 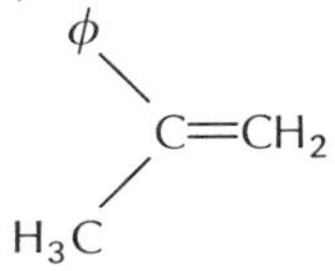 $\underset{H_3C}{\overset{\phi}{\diagdown}}C=CH_2 \longrightarrow \phi CH_2COCH_3$

b. $\phi COCH_2COOEt \longrightarrow \phi COCH_2CH_2OH$

c. $\phi CH_2CH_3 \longrightarrow O_2N\!\!-\!\!\!\bigcirc\!\!\!-\!\!COCH_3$

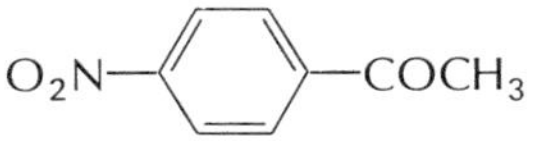

d. 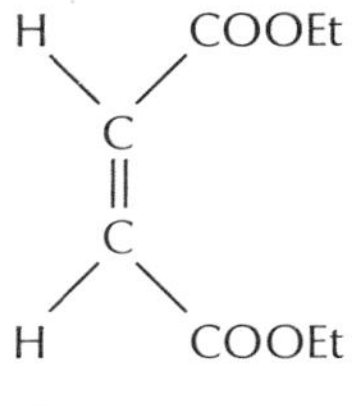 $\underset{H}{\overset{H}{\diagdown}}\!\!\underset{COOEt}{\overset{COOEt}{\diagup}} C=C \longrightarrow 2\,OHCCOOEt$

e. $\phi CH_2Cl \longrightarrow \phi CH_2CHO$

f. $CH_3CH=CHCH=CHCOOEt \longrightarrow CH_3CH=CHCH=CHCHO$

16.8 Summary

"So what do we do now, coach?"

No one expects anyone to master all the material in this chapter the first, third, or tenth time through it. You and your instructor will decide what parts should be stressed, depending upon the objectives of the course for which you are using this book. Enough material has been included to allow a choice.

In a sense, aldehydes and ketones epitomize organic chemistry, and we have just crossed the high point in our journey through that subject. No other class of compounds encompasses so many reactions. From here on, although there are many topics yet to be taken up, we shall more and more frequently find similarities to what already has been covered.

We shall not attempt even a brief summary of this chapter. It is time for you to begin, if you have not already done so, to draw up your own summaries. In so doing, you will find them peculiarly suited to your own needs.

We have not drawn up a list of equations, as we have done in previous chapters; but do not despair. The exercises are designed so that the answers provide the list.

16.9 Problems

16.1 Substance A is neutral, does not give an aldehyde test, but does form an oxime. A with Br_2 under acidic conditions and exclusion of Fe gives B. B with hot alcoholic KOH gives C. Drastic oxidation of A or C gives the same products, 1 mole each of CO_2, benzoic acid, and terephthalic acid. B reacts with water to give a substance that upon very gentle oxidation gives D. A with SeO_2 and heat gives D directly. D reacts with lead tetraacetate to give benzoic acid and E ($C_9H_{10}O_2$). Refluxing of E with acidified methanol yields a product that with sodium bis(2-methoxyethoxy)-aluminum hydride at $-70°$ gives F. F reacts with ϕCH_2MgCl to give G. E with $SOCl_2$ gives H, and H with ϕCH_2CdCl gives I. I reacts with $LiAlH_4$ to give G; and G, under mild oxidation, gives I. I, with benzil at 100° and a mild base, reacts to give J ($C_{30}H_{22}O$).

Give a structural formula for each lettered substance.

16.2 Benzaldehyde reacts in the presence of NaCN to give A. Oxidation of A by a variety of schemes gives B. A or B with $NaBH_4$ gives more *meso*-C than racemic D. C or D with lead tetraacetate gives benzaldehyde. C or D with warm dilute H_2SO_4 gives E. A, under Clemmensen reduction conditions, gives F. F with OsO_4 gives D. A with $SOCl_2$ gives G. G with $NaBH_4$ followed by Zn powder also gives F. G with HI gives H, identical with the product from $\phi CH_2COCl + \phi H + AlCl_3$. B with KOH gives I. I with $LiAlH_4$ gives J. J with warm dilute H_2SO_4 gives E.

Write the reactions given and show the structural formula of each lettered substance.

532   Chapter sixteen

16.3   Give the IUPAC names for the following compounds.

a. $C_6H_5OCH_2CH=CHCHO$

b. 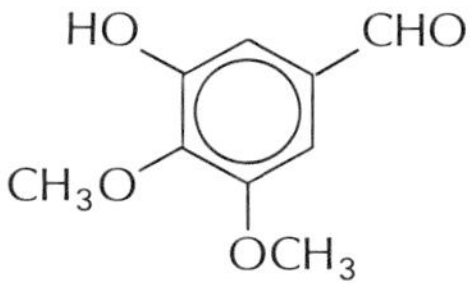 HO—[benzene ring with CHO, OCH$_3$, OCH$_3$ substituents]

 (structure: HO and CHO on benzene with CH$_3$O and OCH$_3$)

c. 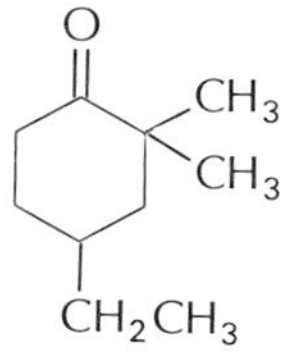 [cyclohexanone with two CH$_3$ groups and CH$_2$CH$_3$ on α-carbon]

d. $C_6H_5\underset{\underset{O}{\|}}{C}CH=CHC_6H_5$

16.4   Draw the structures for the following compounds:
a. isopropyl phenyl ketone
b. 3,4-dimethoxyacetophenone
c. m-nitrocinnamaldehyde
d. 2,2-dimethyl-1,3-cyclohexanedione

16.5   Write the organic reaction product(s):

a. $CH_3CH=CHCHO \xrightarrow{Ag(NH_3)_2^{\oplus}}$

b. 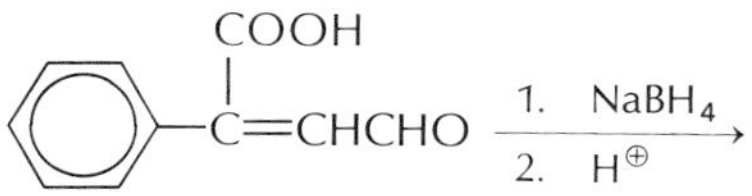 [phenyl]—$\underset{\underset{COOH}{|}}{C}=CHCHO \xrightarrow[2.\ H^{\oplus}]{1.\ NaBH_4}$

c. [cyclohexane with $CH_3C(=O)$— and —$COOCH_2CH_2C_6H_5$ substituents] $\xrightarrow[2.\ H_3O^{\oplus}]{1.\ LiAlH_4}$

d. $CH_2=CHCH_2MgCl + CH_2=O \xrightarrow[2.\ H_3O^{\oplus}]{1.\ dry\ ether}$

e. $CH_3CH_2MgBr + C_6H_5CHO \xrightarrow[2.\ H_3O^{\oplus}]{1.\ dry\ ether}$

f. 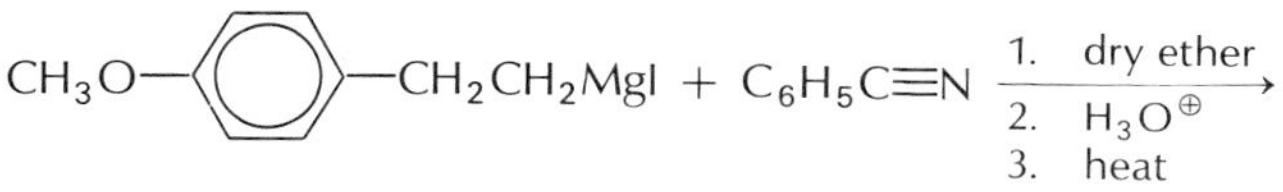 CH_3O—[benzene]—$CH_2CH_2MgI + C_6H_5C\equiv N \xrightarrow[\substack{2.\ H_3O^{\oplus} \\ 3.\ heat}]{1.\ dry\ ether}$

g. 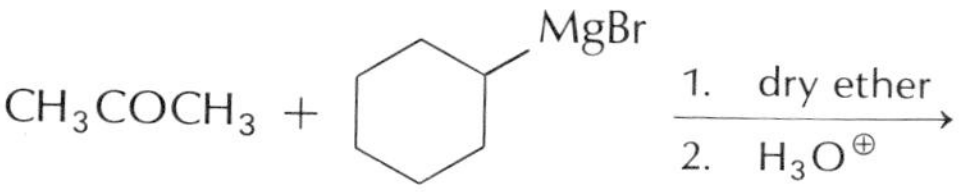 $CH_3COCH_3 +$ [cyclohexyl-MgBr] $\xrightarrow[2.\ H_3O^{\oplus}]{1.\ dry\ ether}$

16.6   Starting with ethyl magnesium iodide and a suitable benzene derivative containing no more than nine carbon atoms, synthesize the following:

a. 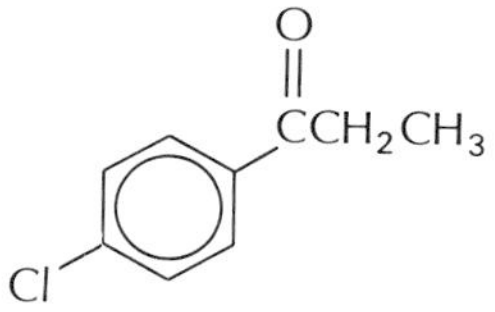 [4-chlorophenyl]—$\underset{\underset{O}{\|}}{C}CH_2CH_3$

b. 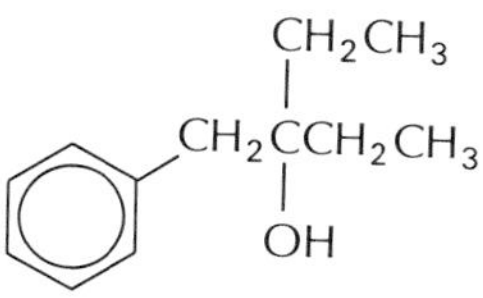 [phenyl]—$CH_2\underset{\underset{OH}{|}}{\overset{\overset{CH_2CH_3}{|}}{C}}CH_2CH_3$

c. CH_3CH_3

d. 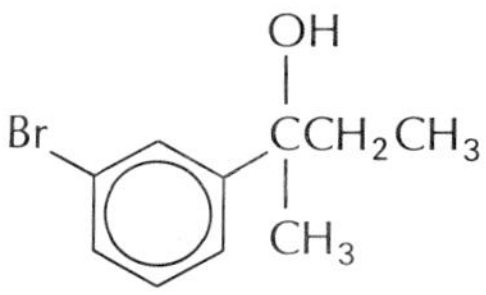 [3-bromophenyl]—$\underset{\underset{CH_3}{|}}{\overset{\overset{OH}{|}}{C}}CH_2CH_3$

e. 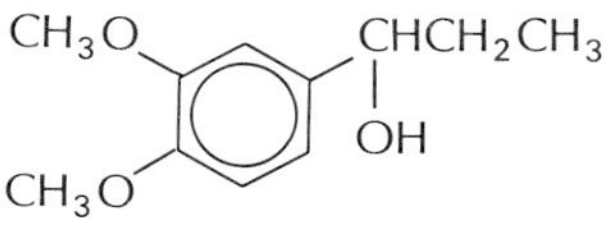 [3,4-dimethoxyphenyl]—$\underset{\underset{OH}{|}}{C}HCH_2CH_3$

16.7 A compound C_3H_6O (A) gave $C_3H_6O_2$ (B) and a silver mirror with Tollens' reagent. Treatment of A with benzyl magnesium chloride, then acid, gave $C_{10}H_{14}O$ (C). When C was allowed to react with $SOCl_2$, $C_{10}H_{13}Cl$ (D) was formed. Reaction of D with 1. Mg in dry ether, 2. formaldehyde, and 3. $H_3O^\oplus$ gave $C_{11}H_{16}O$ (E). What are the structures of A–E?

16.8 Will A or B react faster with the given reagent? Why? Write the reaction product.

	Reagent	A	B
a.	$^\ominus CN$	CH_3COCH_3	$(CH_3)_2CHCOCH_3$
b.	$NaHSO_3$	CH_3CHO	CH_3COCH_3
c.	$^\ominus CH_2CHO$	CH_3CHO	C_6H_5CHO
d.	Br_2 (basic solution)	$BrCH_2COCH_3$	CH_3COCH_3
e.	Br_2 (acid solution)	$BrCH_2COCH_3$	CH_3COCH_3
f.	I_2 (basic solution)	CH_3COCH_3	$CH_3COCH_2COC_6H_5$
g.	$^\ominus CH_2CHO$	CH_3CHO	CH_3COCH_3

16.9 Write the major product for the following reactions:

a. $CH_3CH_2CHO + CH_2O \xrightarrow[\text{cold}]{OH^\ominus}$

b. $(CH_3)_2CHCHO + CH_2O \xrightarrow[\text{cold}]{OH^\ominus}$

c. $(CH_3)_3CCHO + CH_2O \xrightarrow[\text{cold}]{OH^\ominus}$

d. $(CH_3)_3CCHO + CH_2O \xrightarrow[\text{hot}]{OH^\ominus}$

e. $C_6H_5CH_2COCH_3 + Br_2 \xrightarrow{OH^\ominus}$

f. 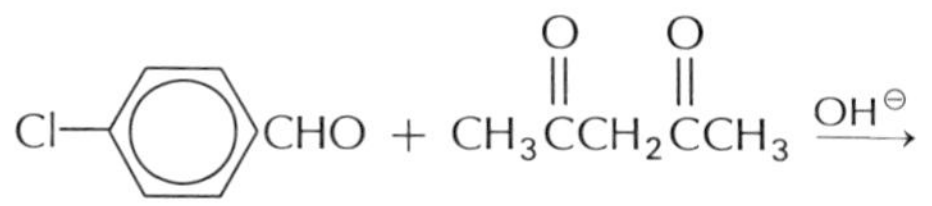 $Cl\text{-}\underset{}{\bigcirc}\text{-}CHO + CH_3\overset{O}{\overset{\|}{C}}CH_2\overset{O}{\overset{\|}{C}}CH_3 \xrightarrow{OH^\ominus}$

g. 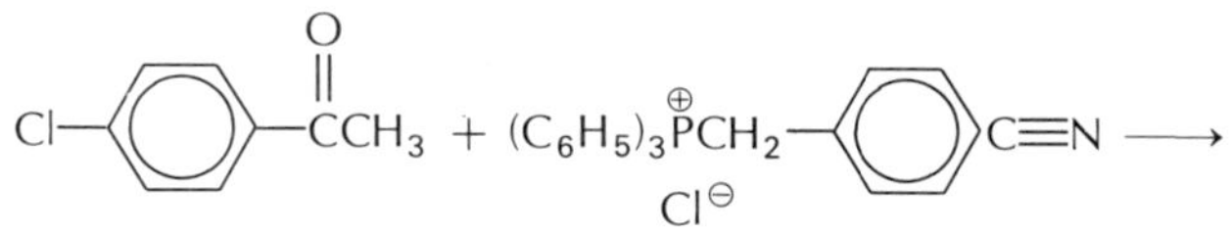 $Cl\text{-}\underset{}{\bigcirc}\text{-}\overset{O}{\overset{\|}{C}}CH_3 + (C_6H_5)_3\overset{\oplus}{P}CH_2\text{-}\underset{}{\bigcirc}\text{-}C\equiv N \longrightarrow$
  $Cl^\ominus$

h. 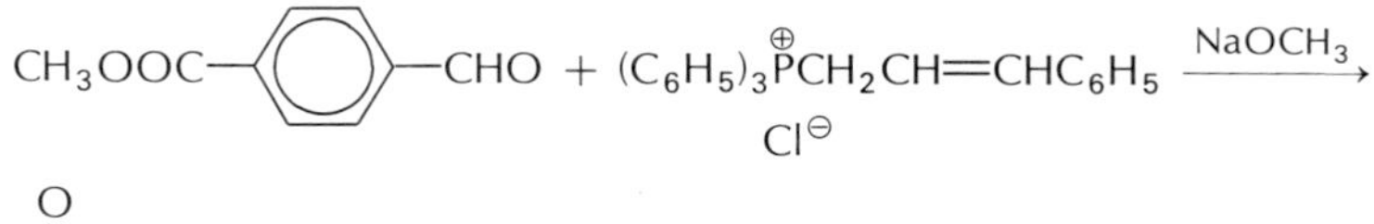 $CH_3OOC\text{-}\underset{}{\bigcirc}\text{-}CHO + (C_6H_5)_3\overset{\oplus}{P}CH_2CH=CHC_6H_5 \xrightarrow{NaOCH_3}$
  $Cl^\ominus$

i. 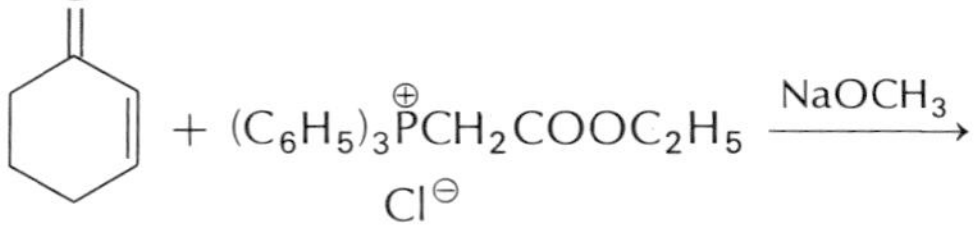 $\underset{}{\bigcirc}=O + (C_6H_5)_3\overset{\oplus}{P}CH_2COOC_2H_5 \xrightarrow{NaOCH_3}$
  $Cl^\ominus$

j. 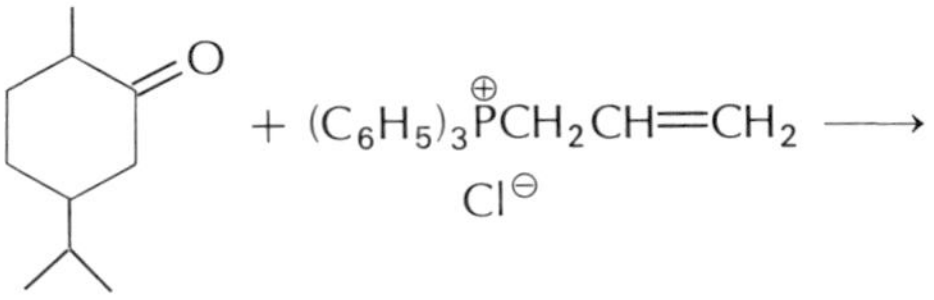 [cyclohexanone with methyl and isopropyl substituents] $+ (C_6H_5)_3\overset{\oplus}{P}CH_2CH=CH_2 \longrightarrow$
  $Cl^\ominus$

k. $C_6H_5CH=CHCHO + CH_3OH \xrightarrow{H^\oplus}$

l. $Br_3CCHO + C_2H_5OH \longrightarrow$

m. $C_6H_5CH_2COCH_2C_6H_5 + HOCH_2CH_2OH \xrightarrow{H^\oplus}$

n.

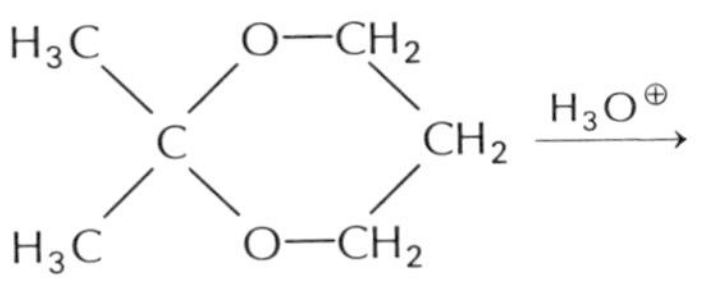

$$\begin{array}{c} H_3C \\ \diagdown \\ C \\ \diagup \\ H_3C \end{array} \begin{array}{c} O-CH_2 \\ \diagdown \\ CH_2 \\ \diagup \\ O-CH_2 \end{array} \xrightarrow{H_3O^{\oplus}}$$

o. (2-naphthyl)–CCH$_2$C$_6$H$_5$ $\xrightarrow{N_2H_4}{KOH}$

(with C=O between naphthyl and CH$_2$C$_6$H$_5$)

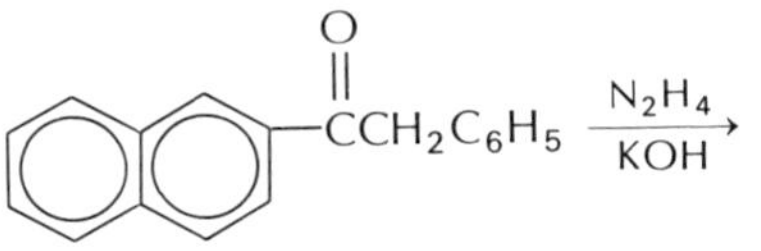

p. $(-)-C_6H_5\overset{O}{\underset{\|}{C}}\overset{}{\underset{CH_3}{C}}HC_6H_5 \xrightarrow{D_2O}{OD^{\ominus}}$

q. $C_6H_5COCH_3 + I_2 \text{ (excess)} \xrightarrow{OH^{\ominus}}$

r. $Cl_3CCH_2COCH_3 + I_2 \text{ (excess)} \xrightarrow{OH^{\ominus}}$

s. $C_6H_5CHO + CH_3CH_2CHO \xrightarrow{OH^{\ominus}}$

t. $(CH_3)_3CCHO + CH_3COC_6H_5 \xrightarrow{OH^{\ominus}}$

u. 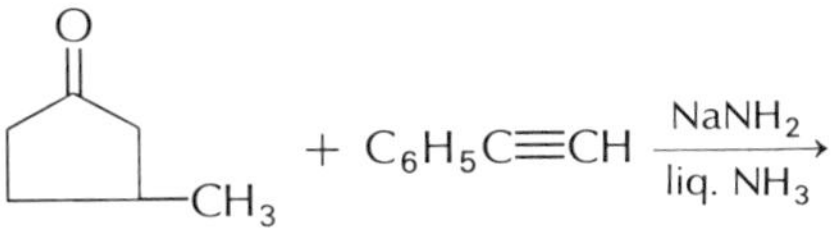 cyclopentanone-2-CH$_3$ + $C_6H_5C\equiv CH$ $\xrightarrow{NaNH_2}{\text{liq. NH}_3}$

v. $C_6H_5\overset{O}{\underset{\|}{C}}CHO \xrightarrow{OH^{\ominus}}$

16.10 The following aldol or reverse aldol reactions are catalyzed by enzymes within the living cell. Write the product(s).

a. $^{\ominus}O\overset{O}{\underset{\|}{P}}OCH_2CHOHCHO + CH_3COCOOH \longrightarrow$
$\phantom{^{\ominus}OP}|$
$\phantom{^{\ominus}OPOCH_2}OH$

(with OH on P)

b. $HOOCCH_2CH_2COCOOH + CH_3\overset{O}{\underset{\|}{C}}-S-CoA \longrightarrow$
(a step in mevalonic acid biosynthesis)

c. $CH_3\overset{O}{\underset{\|}{C}}-S-CoA + CH_3COCH_2\overset{O}{\underset{\|}{C}}-S-CoA \longrightarrow$
(a step in cholesterol biosynthesis)

d. $CH_3\overset{O}{\underset{\|}{C}}-S-CoA + (CH_3)_2CH\overset{O}{\underset{\|}{C}}COOH \longrightarrow$

e. $CH_3\overset{O}{\underset{\|}{C}}-S-CoA + CO_2 \longrightarrow$

16.11 The following can be synthesized from *n*-butyraldehyde. Use any other organic or inorganic reagents of your choosing.

a. 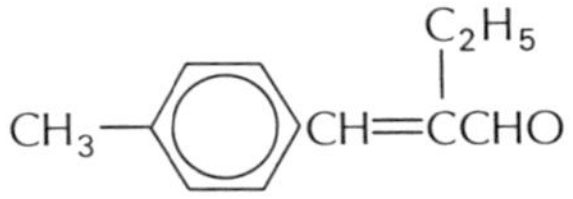 CH$_3$—C$_6$H$_4$—CH=C(C$_2$H$_5$)CHO

b. C$_6$H$_5$CH=CHCH=CH(CH$_2$)$_2$CH$_3$

c. HOOCCH=CH(CH$_2$)$_2$CH$_3$

d. 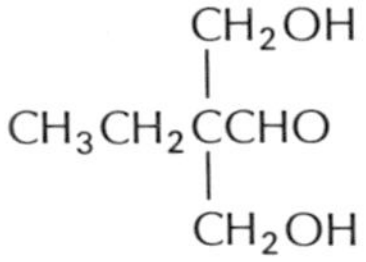 CH$_3$CH$_2$C(CH$_2$OH)$_2$CHO

e. 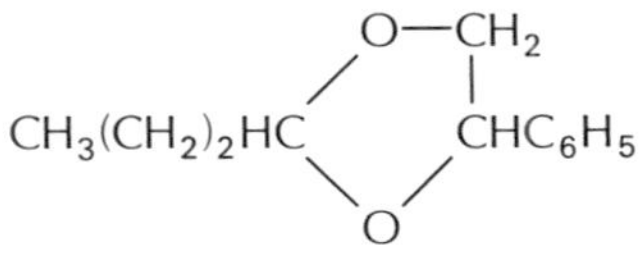 CH$_3$(CH$_2$)$_2$HC$\underset{O}{\overset{O-CH_2}{\diagup \diagdown}}CHC_6H_5$

f. 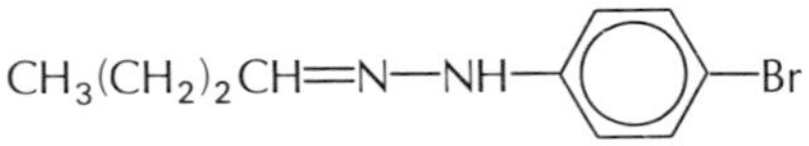 CH$_3$(CH$_2$)$_2$CH=N—NH—C$_6$H$_4$—Br

g. CH$_3$CH$_2$CBr$_2$CHO

h. CH$_3$(CH$_2$)$_2$CHOHC≡CCHOH(CH$_2$)$_2$CH$_3$

16.10 Bibliography

Aside from the textbooks previously mentioned, the paperback by Weiss (Ch. 7, ref. 22) and the text by House (Ch. 13, ref. 1) will be found helpful. We add another good paperback here.

1. Gutsche, C. David, *The Chemistry of Carbonyl Compounds.* Englewood Cliffs, N.J.: Prentice-Hall, 1967. The book considers acids and acid derivatives to be carbonyl compounds as well as aldehydes and ketones. It contains some good problems. Many references.

16.11 Clair J. Collins: How to use isotopic tracers

Professor Clair J. Collins of Tennessee is also a group leader of the chemistry division at Oak Ridge National Laboratory. After obtaining his Ph.D. at Northwestern in 1944, Professor Collins was drafted into the army and later, after being commissioned, was assigned to the Isotopes Branch of the Manhattan District at Oak Ridge. Separated from the army in 1946, he continued on as a civilian at Oak Ridge. In 1964 he was appointed as a professor of chemistry at the University of Tennessee. He and his coworkers have had vast experience with organic syntheses using isotopes and with the use of isotopes in the study of organic reaction mechanisms. Here he relates a memorable experience.

When Professor Ronald Brown asked me if I would like to contribute some special, personal research experience to his book, the answer was easy. My most satisfying scientific memory is that of our work on the nitrous acid deamination of 1,1-diphenyl-2-amino-1-propanol. I value this particular problem above all others, because in a unique moment of recognition my coworkers and I deduced the correct answer to a chemical riddle before we did the crucial experiments.

16.11 How to use isotopic tracers

Prior to our explanation it was commonly held that all 1,2-shifts—for example, of hydrogen, alkyl, or aryl during Wagner-Meerwein, pinacol, Demjanov rearrangements and the like—took place with inversion of configuration at the migration terminus. The nitrous acid deamination of diaryl-substituted α-aminoalcohols was thus visualized with both anisyl (An) and phenyl (Ph) groups finally occupying positions on the opposite side of the carbon from the original bond to nitrogen. In a series of such reactions, it was always that group which migrated through the trans-transition state which yielded about 90 percent of the product, the steric properties of the molecule being more important here than the electron-donating properties of the individual groups (the "cis-effect" of D. Y. Curtin).

Two observations led us to believe, however, that the mechanism shown is only partially correct. For example, the classic deamination of optically active 1,1-diphenyl-2-amino-1-propanol yielded predominately inverted product. It was clearly stated in the

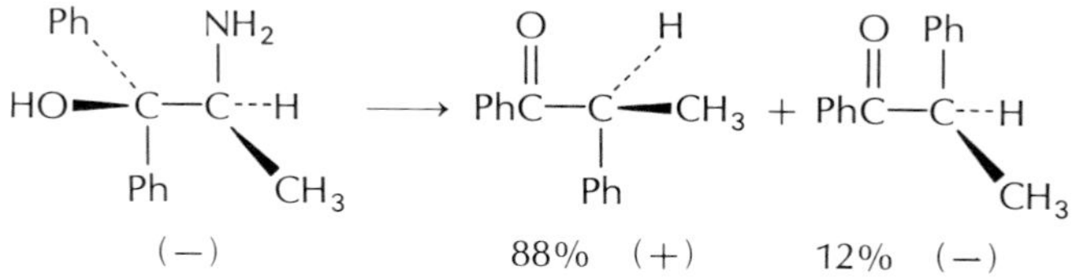

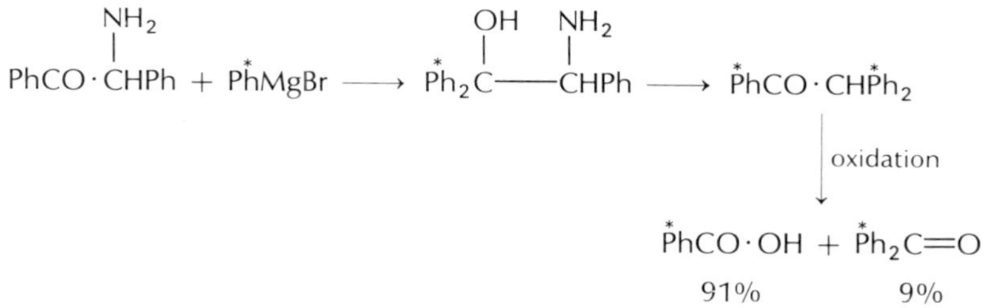

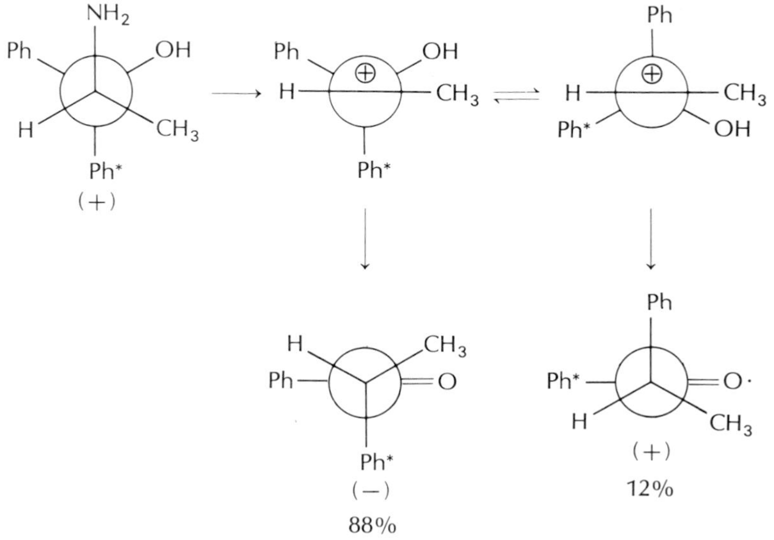

experimental section of the paper, however, that 24 percent racemization had occurred, meaning that 12 percent of the product was of retained configuration. A second interesting observation was made in 1956 by my coworker, Dr. Howard J. Schaeffer, when he added carbon-14 labeled phenylmagnesium bromide to aminodesoxybenzoin and subjected the aminoalcohol so obtained to nitrous acid deamination. After oxidation and radioactivity analysis of the products, Dr. Schaeffer showed that the unlabeled phenyl had undergone 91 percent migration, whereas the labeled phenyl migrated to the extent of only 9 percent.

It was then that we were struck with two obvious facts. When we placed these three equations side by side, it was clear that: the addition of Ph*MgBr to aminodesoxybenzoin must be about 99 percent (or more) stereospecific, and the mechanism for the deamination of 1,1-diphenyl-2-amino-1-propanol had to be approximately as shown in the last equation. From this equation, we see that both phenyls migrate through trans-transition states, one (Ph*) with inversion and the other (Ph) with retention of configuration. It remained only to synthesize the starting material optically active and labeled with carbon-14 in Ph*. From the third equation we knew this synthesis was feasible; we had only to treat optically active α-aminopropiophenone-phenyl-^{14}C with phenylmagnesium bromide! The synthesis was accomplished, optically active, stereospecifically phenyl-labeled (+)-1,1-diphenyl-2-amino-1-propanol was prepared and treated with nitrous acid in aqueous solution. The product ketone was re-resolved. The (−)-ketone (88 percent) had its labeled phenyl adjacent to the —CH— group, whereas in the (+)-ketone the labeled phenyl was adjacent to carbonyl!

I have participated in many other research problems since that time in 1957, but the plain fun and exhilaration of knowing the answer to the mechanism of these reactions and knowing with near certainty how to solve the problem, before the fact, has never since been matched in my experience.

17 Acids, Diacids, and Derivatives

Here we bring together the reactions and preparations of saturated acids, unsaturated acids, and diacids, along with the compounds closely related to acids: the acid halides, acid anhydrides, and esters. We shall find that the reactions of the derivatives are similar in many respects to those of aldehydes and ketones. We already have met many of the reactions in one way or another.

Also, we shall learn something of the widespread occurrence of esters in nature. We shall find their reactivity centered upon the carbonyl group and the adjacent atoms. The alkyl and aromatic systems present survive most of the reactions without change.

17.1   Occurrence   537
 1. Occurrence; trivial names.   2. Waxes, fats, and oils.   3. Some unusual acids.   4. B vitamins.   5. Food values.   6. Autoxidation: rancidity, drying of paints.   7. Soaps and detergents.   8. Diacids.

17.2   Physical properties   542
 1. The mp of acids and derivatives; odors.   2. The bp: hydrogen bonding.   3. Solubility.

17.3   Spectrometry   544
 Carbonyl stretching frequency.

17.4   Reactions of acids   546
 1. Acidity: effect of dimerization.   2. Effect of substituents upon acidity.   3. Replacement of —OH: conversion of acids into derivatives.   4. Olysis reactions.   5. Peracids and derivatives.   6. Decarboxylation.   7. Reduction.   8. α-Hydrogen activity.   9. The four-carbon plants. Exercises.

17.5   Preparation of acids   557
 1. Oxidative methods.   2. Hydrolytic methods.   3. Disproportionation and addition methods.   4. Rearrangements.   5. Synthesis of diacids. Exercises.

17.6   Reactions of acid chlorides and acid anhydrides   560
 1. Olysis reactions.   2. Exchange and addition reactions.   3. Reduction reactions.   4. Condensation. Exercises.

17.7   Preparation of acid chlorides and acid anhydrides   565
        1. Acid chlorides.   2. Acid anhydrides. Exercises.

17.8   Reactions of esters   566
1. Reaction sites.   2. Additions to the carbonyl group: olysis reactions
and others.   3. α-Hydrogen reactions: Claisen condensation.   4. Malonic
ester synthesis.   5. Reactions in the alkyl group of an ester. Exercises.

17.9   Preparation of esters   580
        1. Olysis reactions.   2. Substitution and addition reactions.   3. Oxidation
        and reduction reactions.   4. Ortho esters. Exercises.

17.10   Summary   583

17.11   Problems   584

17.12   Bibliography   587

17.1   Occurrence

Occurrence; trivial names.   Carboxylic acids as such are not widespread in nature. Derivatives, however, especially esters and amides of a wide variety of acids, occur naturally throughout all forms of organic matter.

A few well known acids are acetic acid, in vinegar; lactic acid ($CH_3CHOHCOOH$), in sour milk; malic acid ($HOOCCH_2CHOHCOOH$), in sour apples; and citric acid ($HOOCC(OH)(CH_2COOH)_2$, in citrus fruits. Other acids occur as participants in metabolic pathways. Amino acids will be discussed with the proteins in Chapter 21. In this chapter we shall restrict ourselves to the simple mono- and diacids and derivatives and a few unsaturated acids. Table 17.1 lists some representative monoacids. Many unsubstituted straight chain acids are known by their trivial names, which usually come from the original natural source. Substituted acids and derivatives are becoming known more and more by the systematic IUPAC names.

Waxes, fats, and oils.   Waxes, fats, and oils occur as the saponifiable portion of a larger class of substances known as lipids. (We shall examine saponification, an alkaline hydrolysis, in Section 17.8.) A lipid is a naturally occurring substance that is more soluble in ether than in water. Waxes are arbitrarily defined as esters of long-chain acids (C_{16} or more) with long-chain monoalcohols (C_{16} or more). Actually, other structures also are present. Without exception, waxes, fats, and oils are mixtures of components. Most waxes are solids that soften and melt at temperatures above 37° and less than 100°. Fats and oils all are esters of aliphatic saturated and unsaturated acids with the triol glycerol, ($HOCH_2CHOHCH_2OH$). Fats are solids that soften and melt at temperatures above 20° to 25°; oils are liquid at room temperature. Saponification of waxes, fats, and oils gives rise to mixtures of the component acids and alcohols (only glycerol from fats and oils). The proportions of the different acids vary not only from source to source, but also with the season, time of harvest, age, diet, and so on of a single source. □ □

Table 17.1 Some monocarboxylic acids

Number of C atoms	Name	mp, °C	bp, °C	Water solubility[a]	pK_a[b]
1	formic	8.4	100.5	∞	3.75
2	acetic	16.6	118	∞	4.75
3	propionic	−21	141	∞	4.87
4	butyric	−5	164	∞	4.82
	isobutyric	−47	154	∞	4.85
5	valeric	−34	187	3.7	4.81
	trimethylacetic	−35	164		5.02
6	caproic	−2	205	1.0	4.85
7	enanthic	−8	223		4.89
	cyclohexanecarboxylic	31	233	0.2	
	benzoic	122	250	0.34	4.17
8	caprylic	16	239	0.7	4.89
	phenylacetic	77	266	1.66	4.28
9	pelargonic	12	254		4.96
10	capric	31	269	0.2	
11	undecanoic	28			
12	lauric	44		insol.	
13	tridecanoic	42		,,	
14	myristic	58		,,	
15	pentadecanoic	52		,,	
16	palmitic	63		,,	
17	margaric	61		,,	
18	stearic	70		,,	
19	nonadecanoic	69		,,	
20	arachidic	75		,,	
22	behenic	80		,,	
24	lignoceric	84		,,	
26	cerotic	88		,,	

[a] Solubility in g/100 g H_2O, 20°–25°.
[b] In water, 25°.

Some unusual acids. A host of acids have been identified as occurring in derivatives isolated from various sources. A few of the unusual samples are shown here along with some more common ones.

$(CH_3)_2CHCH=CH(CH_2)_4COOH$ from capsaicin (red pepper)

$CH_3(CH_2)_{10}C\equiv C(CH_2)_4COOH$ tariric acid (tropical American shrub)

$HC\equiv CC\equiv CCH=C=CHCH=CHCH=CHCH_2COOH$ mycomycin
(microorganisms)

$CH_3(CH_2)_7\overset{\underset{\displaystyle CH_3}{|}}{C}HCH_2(CH_2)_7COOH$ tuberculostearic acid (capsule of tuberculosis bacillus)

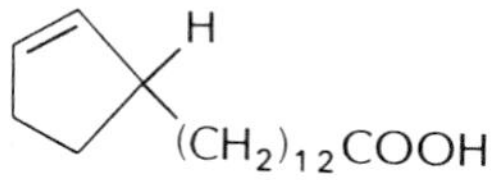

chaulmoogric acid (seeds of chaulmoogra tree, East India)

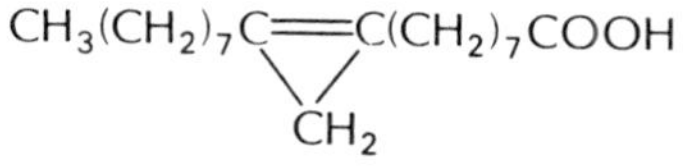

sterulic acid (seed oil of a shrub of the cola family)

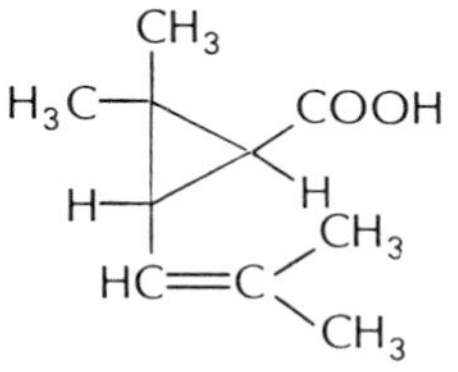

chrysanthemic acid (chrysanthemum)

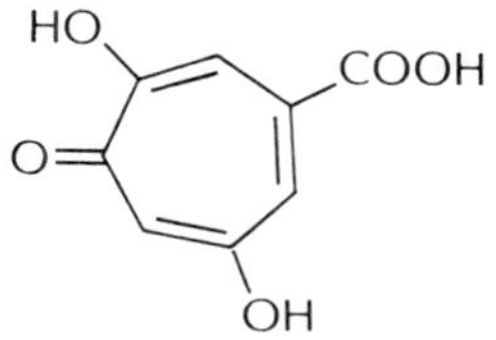

stipitatic acid (penicillium mold)

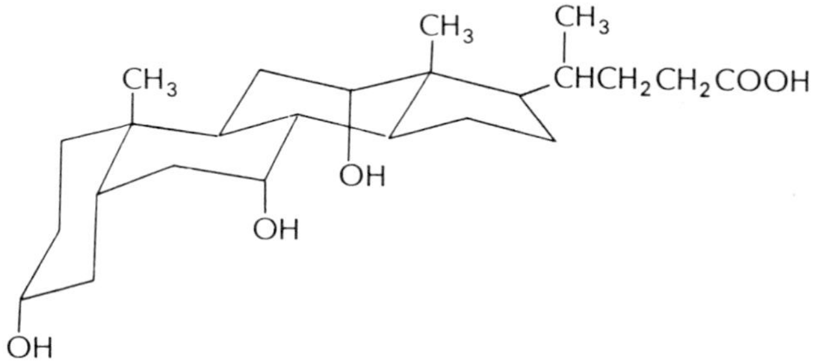

cholic acid (bile)

ϕCOOH benzoic acid (cranberries, etc.)

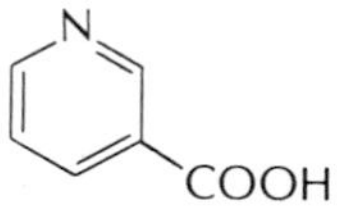

nicotinic acid (nicotine, tobacco)

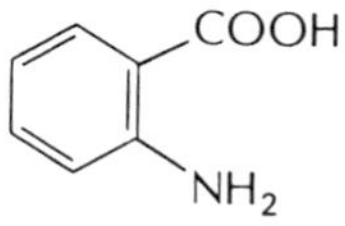

anthranilic acid (indigo)

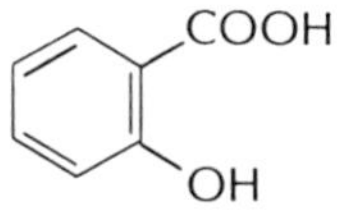

salicylic acid (willow trees, oil of wintergreen)

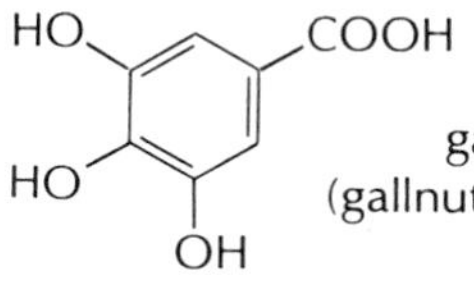

gallic acid (gallnuts on oak trees)

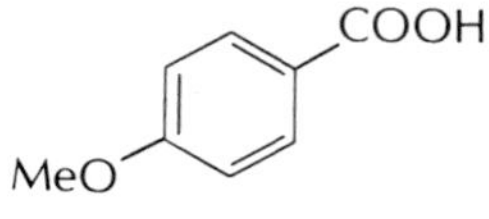

anisic acid (oil of cinnamon)

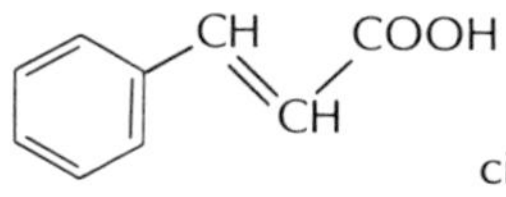

cinnamic acid

B vitamins. Two B vitamins, which themselves contain carboxylic acid groups, are essential for the metabolism of acids in the body. The level of the human requirement is unknown, but apparently practically any diet supplies an adequate amount of pantothenic acid. Biotin is constantly provided by the intestinal bacteria. (See Table 21.4.)

Food values. Fats and oils are essential components of the food intake of all animal life, as are proteins and carbohydrates. They serve (among other functions) as the principal source of energy: 1 g of fat provides about 9.5 kcal (9500 cal), 1 g of protein about 4.4 kcal, and 1 g of carbohydrate about 4.2 kcal.

It has been shown that diets containing fats and oils with a high percentage of saturated acids tend to increase cholesterol in the blood, whereas a high percentage of "polyunsaturated" acids (linoleic > linolenic ≫ oleic in effect) tends to lower blood cholesterol. Deposition of cholesterol within the arteries leads to increased blood pressure and to "hardening" of the arteries. The factors that influence this deposition are not entirely clear, although the concentration of cholesterol in the blood is thought by some to be correlated with deposition.

Fats and oils, aside from their presence in foods, are used extensively as shortening, cooking and salad oils, and the like. Oleomargarine, a butter substitute, is a carefully contrived mixture of partially hydrogenated oils, natural oils, emulsifiers, milk, coloring, flavor agents, and vitamins.

Autoxidation: rancidity; drying of paints. The unsaturated fats and oils are subject to oxidative attack, either at a double bond or at an allylic position. In edible fats and oils, the aldehydes, ketones, and acids that are formed produce odors and off-tastes: we call the effect rancidity. Hydrolysis (perhaps by microorganisms) also leads to rancidity, especially if any of the C_4–C_{10} acids are formed, all of which have unpleasant odors. Antioxidants are added sometimes to prevent autoxidative rancidity from developing. A pile of oily rags and other flammable oily waste may develop enough heat (the result of the autoxidative chain reactions) to burst into flame (spontaneous combustion).

A practical use of autoxidation is the employment of the more unsaturated oils (linseed, tung, and dehydrated castor oil) as the vehicle in paints, varnishes, and enamels. In a thin film, the autoxidation (already rapid because of the presence of the doubly allylic hydrogen atoms: $C=C-CH_2-C=C + O_2 \rightarrow C=C-\dot{C}H-CH=C$) is catalyzed by a drier (a Co, Mn, Pb, or Zn salt of naphthenic acids, which are obtained by alkaline extraction of petroleum fractions). The "drying" of paint actually is autoxidation that creates linkages between the oil molecules within the film. The increase in molecular weight at first increases the viscosity of the film and ultimately causes solidification. In addition, a paint, varnish, or enamel contains a thinner (turpentine or mineral spirits, a petroleum fraction), which lowers the viscosity to permit easy application and helps to give initial setting of the film by evaporation. A paint contains pigment, a varnish contains rosin (and resins), and an enamel is a varnish with pigment.

Soaps and detergents. Another major use of fats and oils is the manufacture of soap and detergents. Alkaline hydrolysis of a fat or oil leads to the formation of glycerol and the sodium salts of the component fatty acids. The alkali metal and ammonium salts of the fatty acids are much more soluble in water than are the free acids. If the chain is ten or more carbons long, the salts no longer are soluble in the usual sense. Instead (depending upon the concentration, chain length, and temperature), colloidal dispersions are formed. These dispersions can act as wetting agents, which concentrate at the interface between water and a hydrophobic surface and lower the surface tension between the two. They also can be emulsifying agents, which have the ability to disperse a hydrophobic substance in water (or vice versa), creating a colloidal mixture. Also, they can behave as detergents, substances with wetting and emulsifying action that are able to remove dirt from fabrics and other surfaces by emulsification of oil films surrounding it. Foaming is a result of emulsification of air and water and may or may not be related to cleansing ability.

Soaps have the disadvantage of forming insoluble scums with any Ca^{2+}, Mg^{2+}, or Fe^{2+} ions that are present in surface and ground waters. Cleansing action may be enhanced by builders, usually alkaline substances such as sodium carbonate (washing soda), sodium silicates, and sodium phosphates. Cold water soaps are made from fats with a high proportion of lauric acid, hot water soaps from fats high in palmitic and stearic acids. In any case, the fats or oils used are hydrogenated to prevent deterioration of the product by autoxidation. Hand or face soaps (toilet soaps) have just about every imaginable substance added—color, perfumes, air (to float), creams, strange oils, and so on—none of which contribute to the purpose of a soap, to cleanse. Pumice is added to some hand soaps to help scrub away imbedded soil from the skin.

Soaps are one of a class of agents known as anionic surfactants. Their hydrophilic (water-liking) part is the negative carboxylate ion, and the hydrophobic (water-avoiding) part is the hydrocarbonlike chain. Cationic surfactants have a positive quarternary nitrogen group for the hydrophilic ion, as in $CH_3(CH_2)_{15}\overset{\oplus}{N}(CH_3)_3$ $Cl^{\ominus}$. Nonionic surfactants (the hydrophilic part is a polyether or polyol) and amphoteric surfactants (containing both anionic and cationic centers) are known and are useful under alkaline, neutral, and acidic conditions.

Synthetic detergents for laundry use have become popular because they do not precipitate in hard waters (containing Ca^{2+}, Mg^{2+}, Fe^{2+}). Most often, they are sodium salts of long-chain alcohol esters of sulfuric acid or of long-chain alkylsulfonic acids ($ROSO_3^{\ominus}Na^{\oplus}$ or $RSO_3^{\ominus}Na^{\oplus}$). It is now possible to tailor a detergent or surfactant for specific purposes: to foam or to suppress foam under given conditions, to cause flotation of a given ore, and so on.

Diacids. The diacids also are known by common names, some of which are listed in Table 17.2.

> The origins of some names are interesting: oxalic from oxalis, in which the monopotassium salt occurs; malonic from malic acid (malum = apple); succinic from distillation of succinum (amber); glutaric from glutamic plus tartaric; adipic and pimelic from oxidation of fats (adipis = fat, pimele = fat); suberic from oxidation of suber (cork). Azelaic is from oxidation of olive oil by nitric acid (azote = nitro-

Table 17.2 Dicarboxylic acids

Name	Formula	mp, °C	Water solubility[a]	pK_{a_1}	pK_{a_2}
Oxalic	HOOC—COOH	189	9	1.23	4.19
Malonic	HOOCCH$_2$COOH	136	74	2.83	5.69
Succinic	HOOC(CH$_2$)$_2$COOH	185	6	4.19	5.48
Glutaric	HOOC(CH$_2$)$_3$COOH	98	64	4.34	5.41
Adipic	HOOC(CH$_2$)$_4$COOH	151	2	4.43	5.41
Pimelic	HOOC(CH$_2$)$_5$COOH	105	5	4.51	5.43
Suberic	HOOC(CH$_2$)$_6$COOH	144	0.2	4.52	5.41
Azelaic	HOOC(CH$_2$)$_7$COOH	106	0.3	4.54	5.41
Sebacic	HOOC(CH$_2$)$_8$COOH	134	0.1	4.58	5.40
Maleic	cis-HOOCCH=CHCOOH	130.5	79	1.92	6.23
Fumaric	trans-HOOCCH=CHCOOH	302	0.7	3.02	4.38
Phthalic	1,2-C$_6$H$_4$(COOH)$_2$	231	0.7	2.89	5.51
Isophthalic	1,3-C$_6$H$_4$(COOH)$_2$	348.5	0.01	3.54	4.60
Terephthalic	1,4-C$_6$H$_4$(COOH)$_2$	300 (subl.)	0.002	3.51	4.82
Trimesic	1,3,5-C$_6$H$_3$(COOH)$_3$	380	2	3.12	3.89

[a] g/100 g H$_2$O at 20°.

gen, <u>elaion</u> = olive oil); sebacic is from <u>sebum</u> (tallow). The names of the first six acids are remembered easily by use of a mnemonic: <u>O</u>h <u>M</u>y, <u>S</u>uch <u>G</u>ood <u>A</u>pple <u>P</u>ie.

As we have seen, esters are widespread in nature. Amides also occur frequently. Acid halides and anhydrides, as well as the free acids, seldom occur as such because of their reactivity.

17.2 Physical properties

The mp of acids and derivatives; odors. The mp of acids and diacids show an interesting behavior: each acid or diacid with an even number of carbon atoms in the chain has a higher mp than does the next higher homolog with an odd number of carbon atoms in the chain. This effect is caused by a better packing of the crystal if the extended chain has an even number of carbon atoms. The greater stability of a crystalline form frequently is reflected in a higher heat of fusion and a lowering of solubility.

The simple esters all have lower mp than do the acids. The ethyl ester of a given acid has a lower mp than that of the methyl ester. The simple esters thus are liquids at room temperature.

The esters of the repulsive-smelling C$_4$–C$_8$ acids are fragrant and flavorful. Artificial flavors usually are mixtures of esters (raspberry, strawberry, cherry, and so on). Some specific esters have odors that are specifically identified with a given source.

CH₃COOCH₂CH₂CH(CH₃)₂
banana oil

CH₃CH₂CH₂CH₂COOCH₂CH₂CH(CH₃)₂
apple

CH₃CH₂CH₂COOCH₂CH₂CH₂CH₃
pineapple

CH₃CH₂COOCH₂CH(CH₃)₂
rum

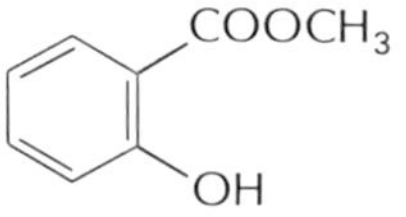

[structure: benzene ring with COOCH₃ and OH substituents]
wintergreen

CH₃COO(CH₂)₇CH₃
oranges

Acid chlorides and anhydrides all tend to have a lower mp than that of the free acid. Amides almost always have a higher mp than that of the acid. Diacids all are solids at room temperature, and only one has mp under 100°.

The bp: hydrogen bonding. The bp of acids are somewhat higher than those of the corresponding alcohols. The high bp of the lower members of the series is caused by dimerization. The two hydrogen bonds provide enough stability so that the bp of an acid is roughly that of a polar substance of twice the mw. In diacids, the dimerization can take place at each end of the chain and produce stable crystals of high mp (Eq. 17.1). Even the

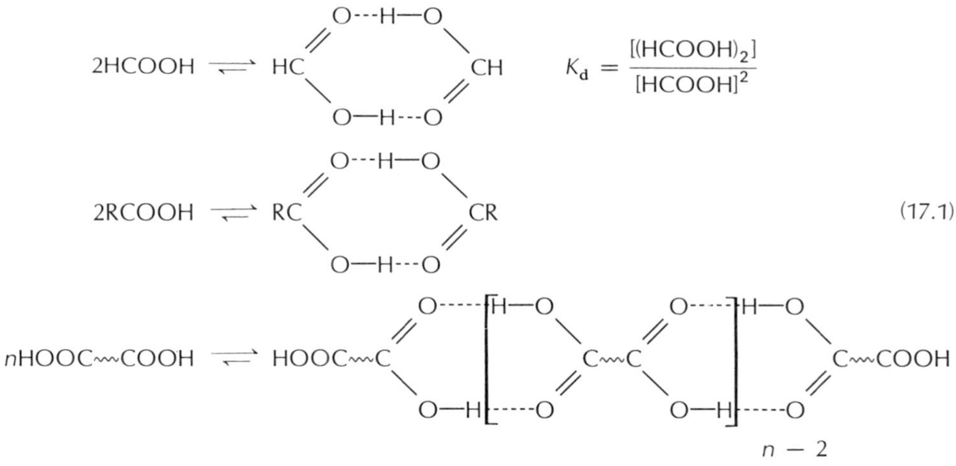

$$2\text{HCOOH} \rightleftharpoons \text{dimer} \qquad K_d = \frac{[(\text{HCOOH})_2]}{[\text{HCOOH}]^2}$$

$$2\text{RCOOH} \rightleftharpoons \text{dimer} \qquad (17.1)$$

$$n\text{HOOC}\mathtt{\sim}\text{COOH} \rightleftharpoons \text{HOOC}\mathtt{\sim}\text{C}[\cdots\text{C}\mathtt{\sim}\text{C}\cdots]_{n-2}\text{C}\mathtt{\sim}\text{COOH}$$

vapour of formic and acetic acids at the bp consists largely of dimers. At higher temperatures (and at the higher bp of the higher members of the homologous series), the dimers dissociate into the monomeric acids.

Methyl esters usually have bp about 62° less than those of the acids; ethyl esters, about 42° less. Because they are not hydrogen bonded, there is only a weak dipolar attraction for association, so esters have bp only slightly higher than those of the corresponding saturated hydrocarbons. For example, $CH_3CH_2COOCH_2CH_3$ has bp 99°; $CH_3(CH_2)_5CH_3$ has bp 98°.

Amides all have high bp unless both hydrogen atoms on the nitrogen are substituted. Compare $HCONH_2$ (bp 193°) and $HCONMe_2$ (bp 153°); also, CH_3CONH_2 (bp 222°) and CH_3CONMe_2 (bp 165°). Obviously, unsubstituted amides are strongly hydrogen bonded. Acid halides and anhydrides tend to have bp characteristic of polar, non-hydrogen-bonded substances of similar mw.

544   *Chapter seventeen*

Solubility.   Acids tend to be more soluble in water than are the corresponding alcohols: two points in the molecule are capable of hydrogen bonding with the solvent. However, solubility decreases rapidly once the chain reaches a length of four carbons. Amides have water solubilities similar to those of the acids. Esters are insoluble in water. Acid chlorides and anhydrides are insoluble in water, but they react with it to produce the corresponding acids.

$$ROH\cdots O\begin{matrix}H\\ \diagup\\ \diagdown\\ H\end{matrix} \qquad RC\begin{matrix}O\cdots H-O\diagdown H\\ \diagup\\ \diagdown\\ O-H\cdots O\diagdown H\end{matrix}$$

The ammonium, sodium, and potassium salts of slightly soluble acids are uniformly much more soluble in water. Thus, most acids may be transferred back and forth at will between an aqueous layer and an immiscible organic solvent like ether, petroleum ether, benzene, or carbon tetrachloride. Lowering the pH of the aqueous layer with a strong acid (hydrochloric, sulfuric, etc.) repels the organic acid from the aqueous layer. If a strong base is added (ammonium, sodium, or potassium hydroxide; carbonate or bicarbonate), the organic acid is removed from the nonaqueous layer. For example, benzoic acid transfers from an aqueous layer at low pH into an ether layer and stays there. At high pH, the aqueous layer removes benzoic acid from an ether layer and retains the acid as the soluble salt.

17.3 Spectrometry

Carbonyl stretching frequency.   There is little to be added here to what has already been said about uv, visible, nmr, or mass spectrometry of acids and acid derivatives. The ir deserves a few comments. Like aldehydes and ketones, acids and derivatives show characteristic carbonyl absorptions. As we pointed out in Section 16.2, the presence of an electron-withdrawing group near or on the carbon of a $\diagdown C=O \diagup$ decreases the contribution of $\diagdown \overset{\oplus}{C}-\overset{\ominus}{O} \diagup$ to the resonance hybrid. In effect, if double-bond character increases, the ir frequency increases. Likewise, electron-donating groups cause a decrease in the frequency. In acids and derivatives, the presence of $-I$, $+R$ groups produces partially compensating effects upon the double-bond character of the carbonyl group.

$$R-C\begin{matrix}O\\ \diagup\\ \diagdown\\ OH\end{matrix} \longleftrightarrow R-C\begin{matrix}O^{\ominus}\\ \diagup\\ \diagdown\\ \overset{\oplus}{OH}\end{matrix} \qquad R-C\begin{matrix}O\\ \diagup\\ \diagdown\\ O^{\ominus}\end{matrix} \longleftrightarrow R-C\begin{matrix}O^{\ominus}\\ \diagup\\ \diagdown\\ O\end{matrix}$$

$\quad\quad\quad -I \quad\quad\quad\quad\quad +R \quad\quad\quad\quad\quad -I \quad\quad\quad\quad\quad +R$

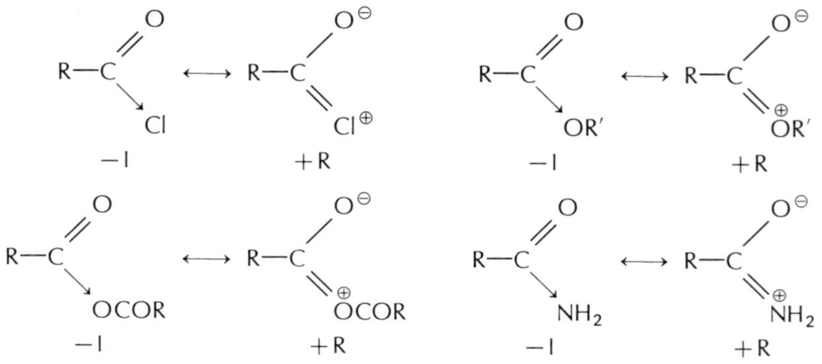

$$\underset{-I}{\overset{O}{\underset{Cl}{R-C}}} \longleftrightarrow \underset{+R}{\overset{O^{\ominus}}{\underset{Cl^{\oplus}}{R-C}}} \qquad \underset{-I}{\overset{O}{\underset{OR'}{R-C}}} \longleftrightarrow \underset{+R}{\overset{O^{\ominus}}{\underset{OR'^{\oplus}}{R-C}}}$$

$$\underset{-I}{\overset{O}{\underset{OCOR}{R-C}}} \longleftrightarrow \underset{+R}{\overset{O^{\ominus}}{\underset{OCOR^{\oplus}}{R-C}}} \qquad \underset{-I}{\overset{O}{\underset{NH_2}{R-C}}} \longleftrightarrow \underset{+R}{\overset{O^{\ominus}}{\underset{NH_2^{\oplus}}{R-C}}}$$

If we take 1715 cm^{-1} as representative of the C=O stretching frequency (acetone), replacement of a methyl by a hydrogen gives an increase of 15 cm^{-1}. Replacement of a methyl by hydroxy (to form an acid) gives a slight decrease of 5 cm^{-1}, an indication that −I of HO— is slightly overbalanced by +R. Ionization of the acid effectively cancels the −I effect, leaving a powerful +R (balanced by the other resonance structure), which results in a decrease in frequency of 135 cm^{-1}. Replacement of a methyl in acetone by a chlorine (to form an acid chloride) gives an increase of 80 cm^{-1}, consonant with a large −I overbalancing a small +R. Acid anhydrides, esters, and amides fall into line about as expected. Table 17.3 summarizes the data.

If the remaining R group becomes capable of conjugation (by unsaturation, either C=C or Ar), an additional decrease in frequency occurs. The magnitude of the extra decrease is variable and depends inversely upon the size of the +R effect already

Table 17.3 Some carbonyl stretching frequencies[a]

Compound	R		
	alkyl	RCH=CH—	Ar
RCOR'[b]	1715	1675	1690
RCHO	1730	1690	1705
RCOOH	1710	1700	1690
RCOO$^-$	1580[c]	1580	1580
RCOCl	1795	1765[d]	1765[d]
(RCO)$_2$O	1760[e]	1750[f]	1750[f]
RCOOR'[b]	1740	1720	1720
RCONH$_2$	1670	—	—

[a] In cm^{-1} ± 10.
[b] R' is alkyl.
[c] And at 1350 cm^{-1}.
[d] And at 1735 cm^{-1}.
[e] And at 1820 cm^{-1}.
[f] And at 1800 cm^{-1}.

present. Thus, when +R is very large, as in RCOO⁻, conjugation in the R group is without effect. In RCOCl, with a small +R effect, conjugation in the R group has a sizable effect upon the frequency.

The frequency of absorption by the O—H group of an acid is strongly influenced by hydrogen bonding. If the acid is very dilute in an aprotic solvent, the O—H absorbs at 3600 cm^{-1}. If the acid concentration is large enough to favor some dimerization by hydrogen bonding, an absorption band appears at 2600 cm^{-1} (Table 10.3).

17.4 Reactions of acids

We shall abandon our orderly arrangement of reactions in describing the acids and their derivatives because, to a great extent, we would be adding complications to a subject already confused by overlapping and by the duplications that exist.

Acidity: effect of dimerization. Organic carboxylic acids are protic acids. They all ionize to some extent in aqueous solution. They vary widely in strength (extent of ionization). Several factors are involved when we look into the relationship between structure and acid strength. A problem common to all the acids is the change in measured K_a with concentration, which is caused by the competing equilibrium for dimerization (Eqs. 17.2). As

$$2CH_3COOH \underset{}{\overset{K_d}{\rightleftharpoons}} CH_3C\begin{matrix}O\cdots H\text{—}O\\ \diagup \quad \diagdown\\ \diagdown \quad \diagup\\ O\text{—}H\cdots O\end{matrix}CCH_3 \quad a$$

$$CH_3COOH + H_2O \overset{K_a}{\rightleftharpoons} CH_3COO^{\ominus} + H_3O^{\oplus} \quad b \qquad (17.2)$$

measurements are made at ever increasing dilution, the equilibrium *a* shifts to the left. As infinite dilution is approached, the true value of K_a is also approached. Thus, the use of K_a or pK_a to calculate pH at usable concentrations gives only an approximate result unless dimerization is taken into account.

This effect may be seen more clearly if we look at the usual definition of K_a, in which *c* is total amount of carboxylic acid in the solution and α is the degree of ionization (Eq. 17.3). Now, if we let β be the degree of dimerization, the amount of

$$K_a = \frac{[H_3O^{\oplus}][CH_3COO^{\ominus}]}{[HAc]} = \frac{(c\alpha)^2}{c(1-\alpha)} = \frac{c\alpha^2}{1-\alpha} \qquad (17.3)$$

unionized monomeric acid becomes $c(1 - \alpha - 2\beta)$, the denominator is decreased, and the numerator (hydrogen-ion concentration squared) must decrease as well if K_a is to remain constant. Some values of K_d (Eq. 17.2a) are as follow: formic acid, 0.04; acetic, 0.16; propionic, 0.23; butyric, 0.36; chloroacetic, 0.40; and benzoic, 0.75. Strong solvent effects must be acting on the dimer and the monomer to cause such a range of values.

Effect of substituents upon acidity. Let us begin our examination of structure and acidity with formic acid (pK_a = 3.75), the strongest of the unsubstituted monoacids. The transfer of a proton from acid to solvent increases the number of ions in solution (favored by an increase in dielectric constant and polarity of solvent). The ions require solvation for stabilization, and the coordination of solvent molecules about the ions causes a decrease in entropy of the solvent. With only a tiny hydrogen atom attached to —COOH or —COO$^\ominus$, formic acid is completely miscible with water, and there is little to hinder solvation of either the undissociated acid or the anion. A balance of all effects leads to a moderate pK_a value.

If H of H—COOH is replaced by CH$_3$, the change to acetic acid is not great enough to affect solubility much. The major effects are upon the extent of dimerization (an increase) and of ionization (a decrease: pK_a increases to 4.75). The acidity (extent of ionization) decreases because the +I effect of methyl is more in demand in the acid than in the anion. Continued increase in chain length has only a minor effect upon pK_a (Table 17.1). A *tert*-butyl group, with a still larger +I effect than methyl, gives the still weaker trimethylacetic acid, at pK_a = 5.02.

Replacement of H in formic acid by an electron-withdrawing group should cause an increase in acidity. Unfortunately, we cannot test the possibility in ClCOOH, because chloroformic acid is too unstable, decomposing to CO$_2$ and HCl. We can, however, look at oxalic acid (Table 17.2); and sure enough, we find a marked decrease in pK_{a_1} to 1.23. The second ionization constant (pK_{a_2} = 4.19) may be regarded as the effect of replacing H in formic acid by the anionic group $^\ominus$OOC—, which should be acid weakening. The data show that the acid-weakening effect of $^\ominus$OOC— is less than that of the methyl. Note, however, that we must consider solubility and solvation effects in making this comparison.

If the new —COOH, with a strong −I effect, is insulated by a —CH$_2$—, the effect is lessened, and malonic acid has pK_{a_1} of 2.83, which is still stronger than formic acid. Two CH$_2$ groups, as in succinic acid, insulate still more, and pK_{a_1} of 4.19 shows that the acid is now intermediate in strength between formic and acetic acids. Increasing the chain length between the —COOH groups has little further effect upon pK_{a_1}. If we look at butyric acid (pK_a = 4.82) and compare the acid-strengthening effect of chlorine in the α position (CH$_3$CH$_2$CHClCOOH, pK_a = 2.86), the β position (CH$_3$CHClCH$_2$COOH, pK_a = 4.05), and the γ position (ClCH$_2$CH$_2$CH$_2$COOH, pK_a = 4.52), the steadily diminishing effect shows the insulating resistance of σ bonds for the transmission of inductive (I) effects. □ □

Although it would be interesting to pursue the subject of the effect of structure upon acidity, we must be brave and turn to a contemplation of other reactions of acids.

Replacement of —OH: conversion of acids into derivatives. Acids may be converted directly into salts, acid chlorides, and esters and only indirectly (with some exceptions) into acid anhydrides, amides, and substituted amides. Our subject here is the replacement of the —OH group.

The choice of reagent for the conversion of an acid into an acid chloride is largely governed by the relative bp of reagent and products (Eqs. 17.4). Of the five methods given,

$$\text{RCOOH} \begin{cases} \xrightarrow{\text{SOCl}_2} \text{RCOCl} + \text{HCl} + \text{SO}_2 & a \\ \xrightarrow{\text{PCl}_3} \text{RCOCl} + \text{H}_3\text{PO}_3 & b \\ \xrightarrow{\text{PCl}_5} \text{RCOCl} + \text{POCl}_3 + \text{HCl} & c \\ \xrightarrow{\phi\text{COCl}} \text{RCOCl} + \phi\text{COOH} & d \\ \xrightarrow{\underset{|}{\text{COCl}}}_{\text{COCl}} \text{RCOCl} + \text{CO}_2 + \text{CO} + \text{HCl} & e \end{cases} \quad (17.4)$$

548 Chapter seventeen

a and *e* have the advantage of volatile products that are easily removed from the reagents used. Thionyl chloride ($SOCl_2$, bp 77°) reacts quantitatively, is catalyzed by a drop or two of pyridine, and may be used as the solvent. Benzene and carbon tetrachloride are useful solvents as well. Oxalyl chloride (*e*), the double acid chloride $(COCl)_2$ of oxalic acid (bp 63°), has the disadvantage of high cost or of preparation by *a*, *b*, or *c*. However, the lower bp allows the use of the reagent in situations unsuitable for thionyl chloride. Benzoyl chloride (*d*, bp 197°) like oxalyl chloride must be prepared by some other method, but it is useful in the preparation of acid chlorides of low bp (acetyl chloride, bp 52°) because separation is easy. The phosphorus reagents (PCl_5, mp 167°; PCl_3, bp 76°) and their products (H_3PO_3, decomposes at about 200°; $POCl_3$, bp 107°) provide two other methods (*b* and *c*), so that a satisfactory choice of methods is easily made. Other methods exist but will not be considered here.

Let us look more closely at *d*, an <u>exchange reaction</u> (Eq. 17.5). The equilibria are dis-

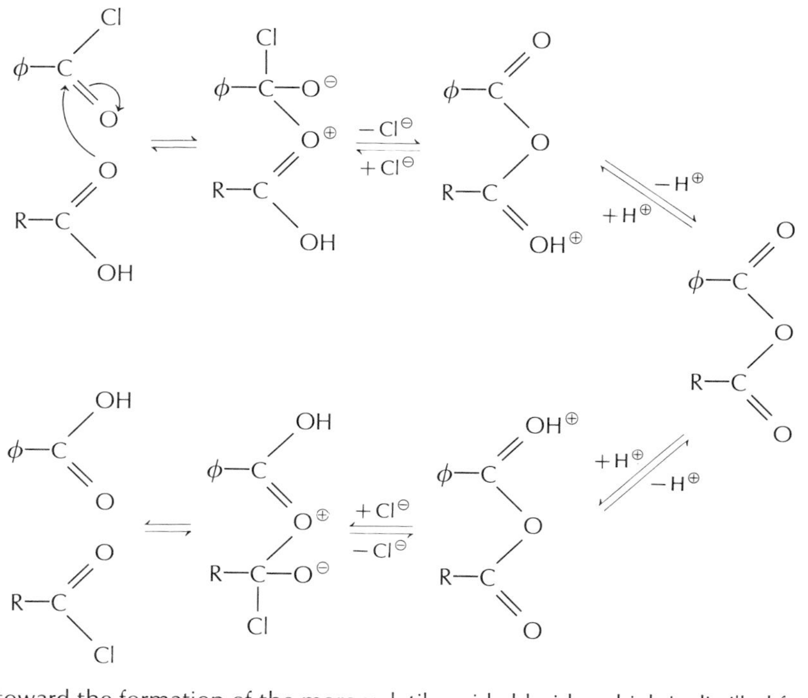

$$\text{(17.5)}$$

placed toward the formation of the more volatile acid chloride, which is distilled from the reaction mixture. The intermediate is the mixed acid anhydride. The reaction passes through structures, on either side of the anhydride, in which a carbonyl carbon is rehybridized to sp^3. Alternatively, we may regard the reaction as the attack of a nucleophile upon the relatively positive carbonyl carbon of the acid chloride. We have met such additions many times before, beginning with additions to $\diagdown C=O \diagup$ in Chapter 2. We recall that the similarity to Fischer esterification (Sec. 9.5) is notable. These equilibria are ubiquitous in the interreactions of acids and derivatives.

We can make the reaction of Equation 17.5 irreversible by the simple expedient of using an anion of the acid if we wish to prepare an acid anhydride (Eq. 17.6). In this fashion, mixed (R ≠ R') or symmetrical anhydrides (R = R') may be prepared. Acetic anhydride (bp 140°) may be utilized in a reaction similar to that of Equation 17.5 to prepare symmetrical anhydrides of other acids (Eqs. 17.7). Acetic acid, the most volatile of the substances

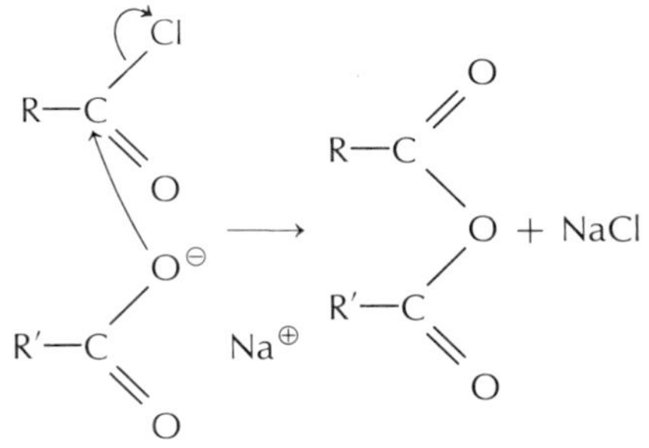

17.4 Reactions of acids 549

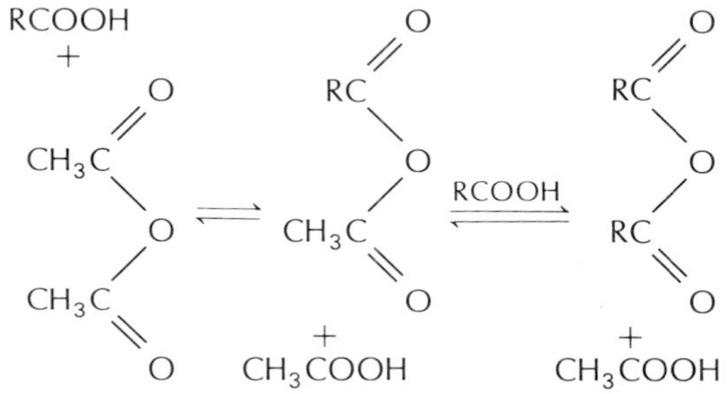

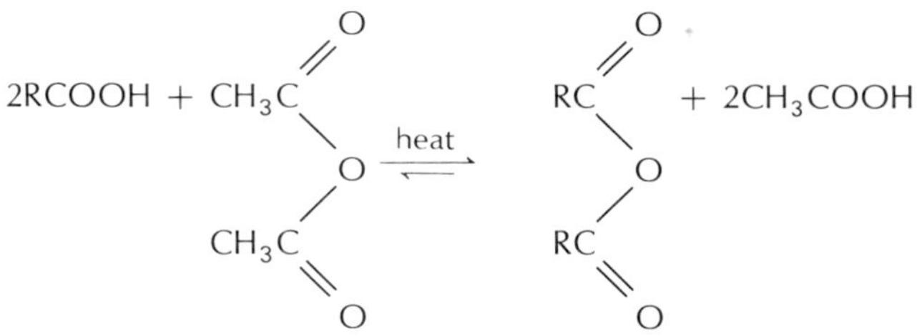

$$R-C(Cl)=O \cdots R'-C(O^-)=O \cdot Na^+ \longrightarrow (RCO)O(COR') + NaCl \tag{17.6}$$

$$RCOOH + (CH_3CO)_2O \rightleftharpoons RCO-O-COCH_3 + CH_3COOH \xrightarrow{RCOOH} (RCO)_2O + CH_3COOH \quad a$$

or

$$2RCOOH + (CH_3CO)_2O \underset{heat}{\rightleftharpoons} (RCO)_2O + 2CH_3COOH \quad b \tag{17.7}$$

and

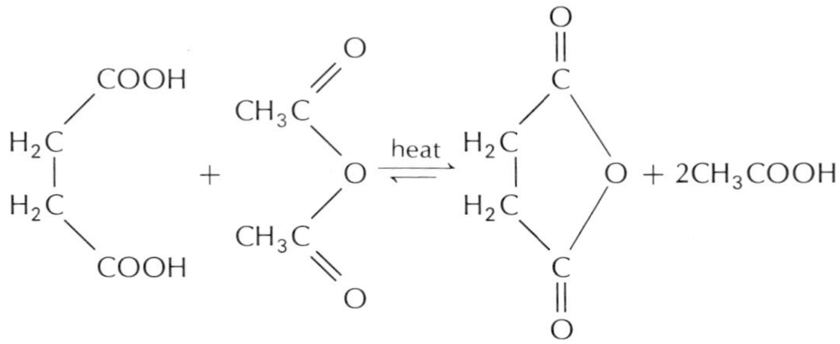

$$\underset{\substack{H_2C-COOH\\|\\H_2C-COOH}}{} + (CH_3CO)_2O \underset{heat}{\rightleftharpoons} \underset{\substack{H_2C-C(=O)\\|\quad\quad\quad\;\;\;\,\diagdown\\H_2C-C(=O)}}{\;\;O} + 2CH_3COOH \quad c$$

present, is distilled, causing the reactions to shift to the right. Notice c, the easy preparation of cyclic anhydrides if the ring has five or six members. Acid anhydrides cannot be made by simple dehydration of acids (heating with sulfuric acid, passage over hot Al_2O_3).

Esterification by strong-acid-catalyzed reaction between a carboxylic acid and an alcohol has been described. <u>Transesterification</u> (Secs. 17.8 and 17.9) may be used as an exchange reaction. Thus, dimethyl phthalate (bp 282°), upon being heated with a carboxylic acid (in the presence of a little sulfuric acid), undergoes ester interchange; and the much more volatile new methyl ester is distilled from the mixture. Valeric acid (bp 187°) is converted to methyl valerate (bp 127°), leaving behind the nonvolatile phthalic acid (mp 231°, Eq. 17.8).

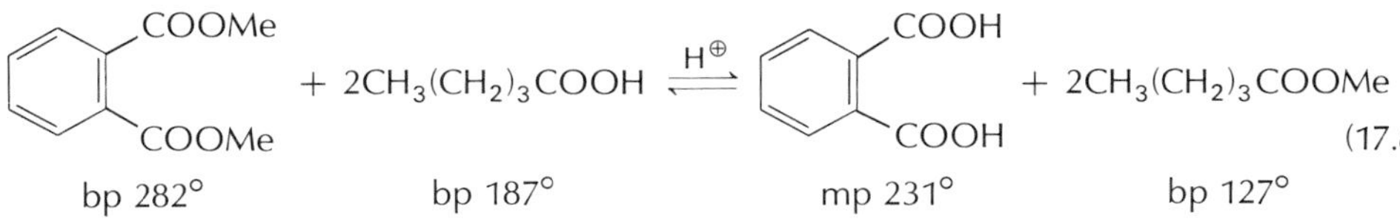

$$\text{o-C}_6\text{H}_4(\text{COOMe})_2 + 2\text{CH}_3(\text{CH}_2)_3\text{COOH} \underset{}{\overset{\text{H}^\oplus}{\rightleftharpoons}} \text{o-C}_6\text{H}_4(\text{COOH})_2 + 2\text{CH}_3(\text{CH}_2)_3\text{COOMe} \quad (17.8)$$

bp 282° bp 187° mp 231° bp 127°

A rarely used alternative (because of expense and trouble) is to allow reaction under anhydrous conditions of a silver salt with an alkyl halide (Eq. 17.9). The reaction may be cyclic, as shown, because even tertiary halides may be esterified. The reaction also is irreversible because of the precipitation of the insoluble silver halide.

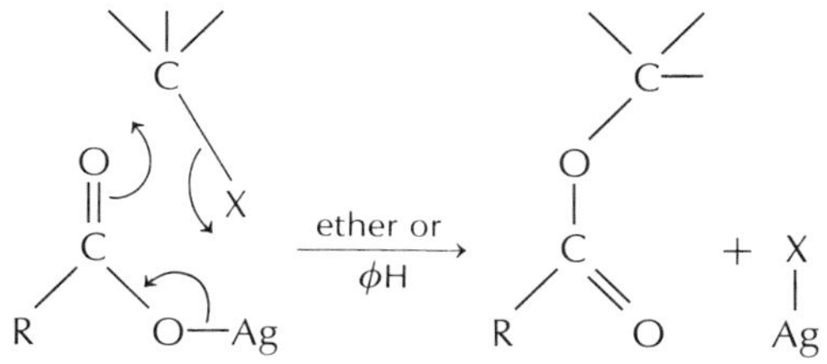

$$\text{(acyl-O-Ag + R'X)} \xrightarrow{\text{ether or } \phi\text{H}} \text{ester} + \text{AgX} \quad (17.9)$$

By the use of diazomethane, methyl esters may be prepared in such purity that subsequent purification is unnecessary (Eq. 17.10). The only other substance involved is nitrogen, which bubbles from the solution as the reaction proceeds.

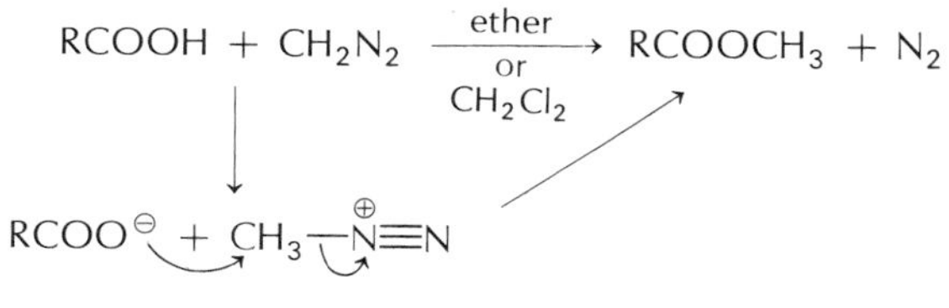

$$\text{RCOOH} + \text{CH}_2\text{N}_2 \xrightarrow[\text{CH}_2\text{Cl}_2]{\text{ether or}} \text{RCOOCH}_3 + \text{N}_2 \quad (17.10)$$

$$\text{RCOO}^\ominus + \text{CH}_3-\overset{\oplus}{\text{N}}\equiv\text{N}$$

$$\text{RCOONH}_4 \underset{}{\overset{\text{heat}}{\rightleftharpoons}} \text{RCOOH} + \text{NH}_3 \rightleftharpoons \text{R}-\underset{\underset{\text{H}_3\text{N}^\oplus}{|}}{\overset{\overset{\text{O}^\ominus}{|}}{\text{C}}}-\text{OH}$$

$$\Updownarrow \text{H}^\oplus \text{ in and out}$$

$$\text{R}-\underset{\underset{\text{H}_2\text{N}}{|}}{\overset{\overset{\text{O}^\ominus}{|}}{\text{C}}}-\text{OH}_2^\oplus \quad (17.11)$$

$$\underset{-\text{H}_2\text{O}}{\overset{}{\Updownarrow}} +\text{H}_2\text{O}$$

$$\text{R}-\overset{\overset{\text{O}}{\|}}{\text{C}}-\text{NH}_2$$

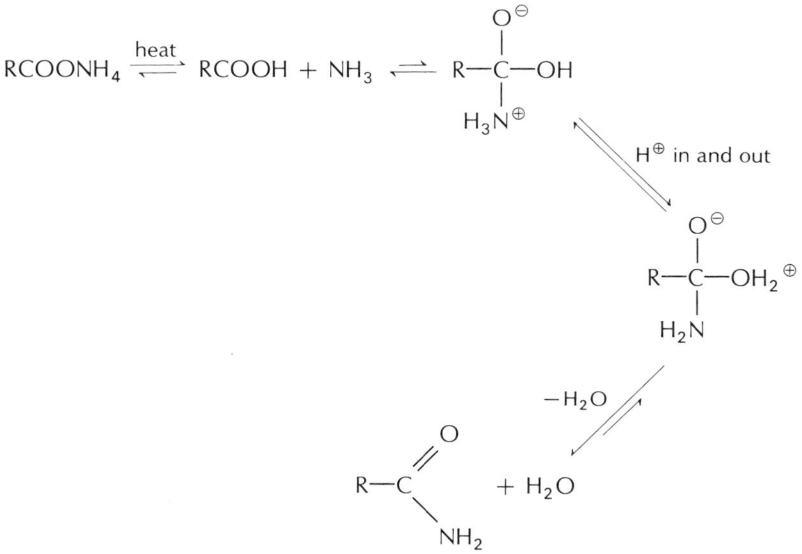

Amides may be formed from acids by heating of the ammonium salts of acids. The reaction probably proceeds by the mechanism shown in Equation 17.11. An alternative method is to take advantage of another exchange reaction, this time acid exchange of an amide (Eq. 17.12). The amide used here is urea, the diamide of carbonic acid. It is stable alone but reacts with an acid as shown.

$$\text{RCOOH} + \text{H}_2\text{NCONH}_2 \overset{\text{heat}}{\rightleftharpoons} \underset{\underset{\underset{O}{\overset{\|}{C}}}{\overset{|}{\underset{H_2N-C}{}}}}{\overset{O}{\overset{\|}{RC}}}\overset{}{\underset{NH}{}} + \text{H}_2\text{O} \rightleftharpoons \text{RC}\overset{O}{\overset{\|}{}}\underset{NH_2}{} + \begin{bmatrix} \text{H}_2\text{NCOH} \\ \| \\ \text{O} \end{bmatrix}$$

$$\text{NH}_3 + \text{CO}_2 \qquad (17.12)$$

Olysis reactions. Acids, then, may be used in exchange reactions for the preparation of acid chlorides, anhydrides, esters, and amides. We shall dub such reactions olysis reactions. Thus, the use of benzoyl chloride to prepare an acid chloride of another acid (Eq. 17.4d) becomes an acid chloridolysis; Equations 17.7 become acid anhydridolysis reactions; transesterification, as described for dimethyl phthalate and valeric acid, becomes esterolysis; and Equation 17.12 is an amidolysis. We are already familiar with other olysis terms, such as hydrolysis and alcoholysis. Soon we shall become acquainted with ammonolysis and aminolysis. Our purpose in adopting olysis as a general class of reaction is to emphasize that acids and acid derivatives all undergo the olysis reactions and that all the olysis reactions pass through a tetrahedral sp^3 intermediate.

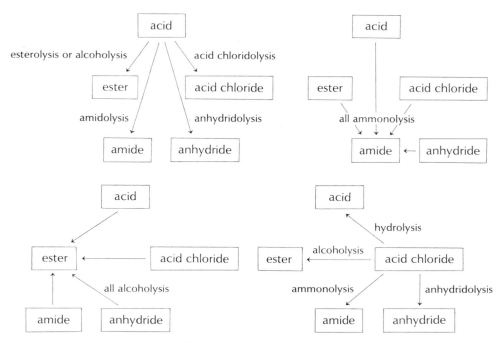

Figure 17.1 Diagrams of some relationships between acids and acid derivatives

The prefix used with olysis designates the reagent used: hydrolysis refers to water, alcoholysis to an alcohol, ammonolysis to ammonia, and so on. If we set up a diagram with each type of acid and acid derivative at the corner of a pentagon, we have a device for showing the olysis reactions (Fig. 17.1). The olysis reactions divide into two sets: the acid or acid-portion reagents (esterolysis, amidolysis, etc.) and the nonacid reagents (hydrolysis, alcoholysis, etc.). A given reaction (for example, ammonolysis) converts any of the other four into a given compound (in this instance, an amide). Any four may be converted into any other one by use of the proper olysis reaction. Some of the reactions are vigorous and others are slow, as we shall learn shortly.

Peracids and derivatives. So far we have been concerned in the reactions of acids with the breaking of the O—H bond, and of the C—OH bond after addition to the carbonyl carbon. The peracids and derivatives are available by similar reactions (Eqs. 17.13). We have

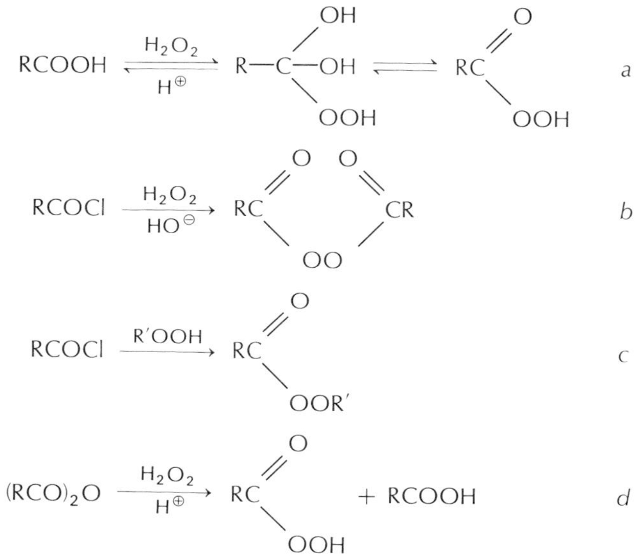

$$
\begin{array}{ll}
\text{RCOOH} \xrightleftharpoons[\text{H}^\oplus]{\text{H}_2\text{O}_2} \text{R—C(OH)(OH)(OOH)} \rightleftharpoons \text{RC(=O)(OOH)} & a \\[6pt]
\text{RCOCl} \xrightarrow[\text{HO}^\ominus]{\text{H}_2\text{O}_2} \text{RC(=O)—OO—C(=O)R} & b \\[6pt]
\text{RCOCl} \xrightarrow{\text{R'OOH}} \text{RC(=O)(OOR')} & c \\[6pt]
(\text{RCO})_2\text{O} \xrightarrow[\text{H}^\oplus]{\text{H}_2\text{O}_2} \text{RC(=O)(OOH)} + \text{RCOOH} & d
\end{array}
\quad (17.13)
$$

referred often to the use of peracids in the epoxidation of alkenes, and we have seen benzoyl peroxide (R = ϕ in b) as a free-radical initiator. Performic acid is used as in a, which is an equilibrium because formic acid is strong enough to catalyze the attainment of equilibrium. Other peracids are prepared and isolated as in d, an irreversible reaction. All peracids, as well as any compound with an O—O bond, should be handled with fear and respect. They are sensitive to shock (shock here meaning something as simple as removing a cap from a bottle) and they explode very readily.

Decarboxylation. Reactions in which the carboxylic acid group is replaced are not numerous. Because we may consider a salt of an acid to be formed by $R^\ominus + CO_2 \rightarrow RCOO^\ominus$, the reverse reaction, decarboxylation, should be possible if $R^\ominus$ is reasonably stable. The stability of CO_2 should aid the process. On the other hand, the anion $RCOO^\ominus$ is stabilized by a symmetrical resonance. Therefore, in general, the free acid is easier to decarboxylate than is the salt, especially if some strong acid also is present (Eq. 17.14).

$$H^\oplus + R\text{—COOH} \xrightarrow{\Delta} H\text{—R} + O\text{=}C\text{=}OH^\oplus \quad (17.14)$$

17.4 Reactions of acids

Except for acetic acid ($CH_3COONa + NaOH \rightarrow CH_4 + Na_2CO_3$), the aliphatic acids do not decarboxylate well in either the acid or the salt forms. If, however, there is a substituent on the α carbon atom, or if some other structural feature present in the molecule or ion acts to stabilize $R^\ominus$, then decarboxylation becomes a useful reaction (Eqs. 17.15).

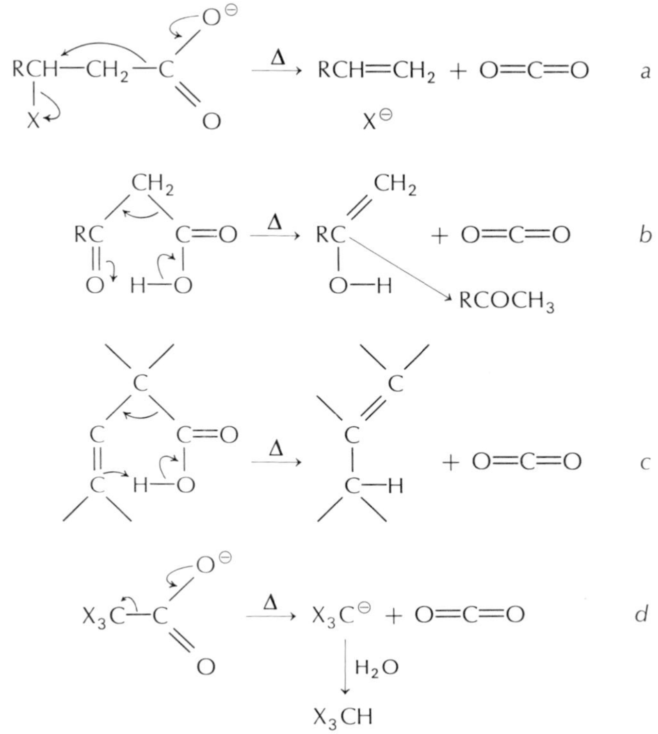

$$
\begin{array}{ll}
RCH{-}CH_2{-}COO^\ominus \xrightarrow{\Delta} RCH{=}CH_2 + O{=}C{=}O & a \\
\quad | \\
\quad X & X^\ominus \\
\\
RC(=O){-}CH_2{-}C(=O){-}O{-}H \xrightarrow{\Delta} RC(=CH_2){-}O{-}H + O{=}C{=}O \rightarrow RCOCH_3 & b \\
\\
C{=}C{-}C{-}C(=O){-}O{-}H \xrightarrow{\Delta} C{-}C{=}C{-}C{-}H + O{=}C{=}O & c \\
\\
X_3C{-}COO^\ominus \xrightarrow{\Delta} X_3C^\ominus + O{=}C{=}O \xrightarrow{H_2O} X_3CH & d
\end{array}
$$
(17.15)

Heating the salt of a β-halo acid (a) leads to elimination, as does the heating of a β,γ-unsaturated acid (c). Heating a β-carbonyl-containing acid (b) or a trihaloacetate salt (d) leads to replacement of —COOH with —H. Decarboxylation of salts follows noncyclic mechanisms, whereas decarboxylation of acids goes by cyclic mechanisms. Again, we notice the prevalence of six-membered cyclic transition states. Some specific examples are shown in Equations 17.16. If possible, an α,β-unsaturated acid rearranges to the more

$$
\begin{array}{l}
(CH_3)_2CHCH{=}CHCOOH \xrightarrow{\Delta} [(CH_3)_2C{=}CHCH_2COOH] \\
\qquad\qquad\qquad\qquad\qquad\qquad\qquad \searrow \\
\qquad\qquad\qquad\qquad\qquad\qquad\qquad (CH_3)_2CHCH{=}CH_2 + CO_2 \\
\\
\phi CH{=}CHCOOH \xrightarrow{\Delta} \phi CH{=}CH_2 + CO_2 \\
\\
\begin{array}{l} COOH \\ | \\ COOH \end{array} \xrightarrow{\Delta} HCOOH + CO_2 \\
\\
\begin{array}{l} COOH \\ | \\ COOH \end{array} \xrightarrow[H_2SO_4]{\Delta} H_2O + CO + CO_2 \\
\\
H_2C\begin{array}{l} {\diagup}COOH \\ {\diagdown}COOH \end{array} \xrightarrow{140°} CH_3COOH + CO_2
\end{array}
$$
(17.16)

easily decarboxylated β,γ-unsaturated acid. Cinnamic acid decarboxylates easily because of the stabilization of the anion, $\phi CH=CH^{\ominus}$.

Aromatic acids decarboxylate easily. One scheme is to heat the free acid in quinoline (bp 238°) with copper powder. Another is simply to heat a salt of the acid, or in some cases to heat the acid itself (Eqs. 17.17). Electron-donating substituents aid a, and electron-withdrawing substituents favor b.

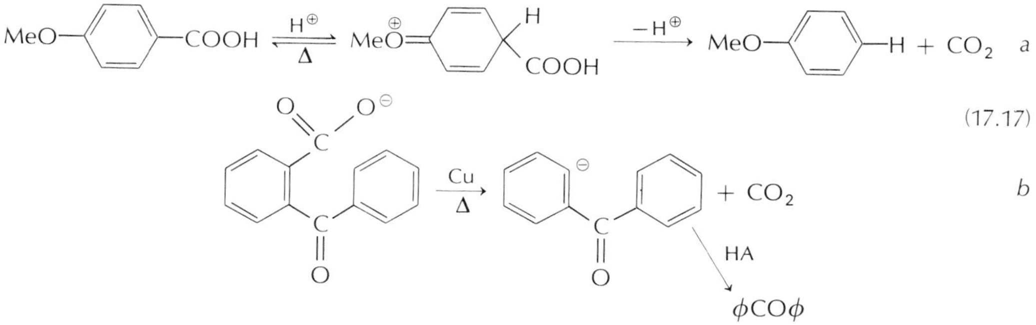

$$\text{MeO}-\text{C}_6\text{H}_4-\text{COOH} \underset{\Delta}{\overset{H^{\oplus}}{\rightleftharpoons}} \text{MeO}=\underset{\text{COOH}}{\overset{H}{\bigcirc}} \xrightarrow{-H^{\oplus}} \text{MeO}-\text{C}_6\text{H}_4-\text{H} + CO_2 \quad a$$

(17.17)

$$\text{(diphenic acid anion)} \xrightarrow{Cu, \Delta} \text{(benzophenone carbanion)} + CO_2 \xrightarrow{HA} \phi CO\phi \quad b$$

The diacids split into groups of two if we consider only the effects of heat (Eqs. 17.18). Oxalic and malonic acids decarboxylate (a). Succinic and glutaric acids dehydrate to give cyclic anhydrides—they are exceptions to our previous statement that anhydrides cannot be prepared this way (b). Adipic and pimelic acids tend to decarboxylate and dehydrate to give cyclic ketones (c). The longer chain acids simply dehydrate to give polymeric anhydrides.

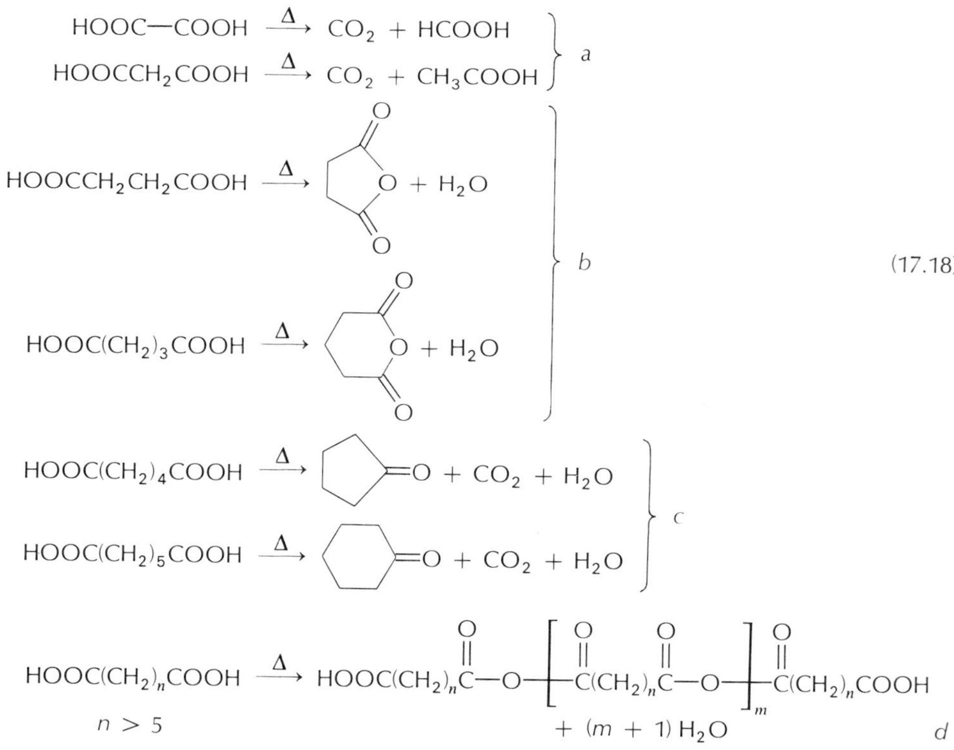

$$\left. \begin{array}{l} \text{HOOC—COOH} \xrightarrow{\Delta} CO_2 + HCOOH \\ \text{HOOCCH}_2\text{COOH} \xrightarrow{\Delta} CO_2 + CH_3COOH \end{array} \right\} a$$

$$\left. \begin{array}{l} \text{HOOCCH}_2\text{CH}_2\text{COOH} \xrightarrow{\Delta} \text{succinic anhydride} + H_2O \\ \\ \text{HOOC(CH}_2)_3\text{COOH} \xrightarrow{\Delta} \text{glutaric anhydride} + H_2O \end{array} \right\} b \qquad (17.18)$$

$$\left. \begin{array}{l} \text{HOOC(CH}_2)_4\text{COOH} \xrightarrow{\Delta} \text{cyclopentanone} + CO_2 + H_2O \\ \text{HOOC(CH}_2)_5\text{COOH} \xrightarrow{\Delta} \text{cyclohexanone} + CO_2 + H_2O \end{array} \right\} c$$

$$\begin{array}{l} \text{HOOC(CH}_2)_n\text{COOH} \xrightarrow{\Delta} \text{HOOC(CH}_2)_n\overset{O}{\underset{\|}{C}}-O\left[\overset{O}{\underset{\|}{C}}(CH_2)_n\overset{O}{\underset{\|}{C}}-O\right]_m \overset{O}{\underset{\|}{C}}(CH_2)_n\text{COOH} \\ \quad n > 5 \qquad\qquad\qquad + (m+1) H_2O \end{array} \quad d$$

Decarboxylation to form hydrocarbons by electrolysis (the Kolbe electrolysis reaction) has been described (Eq. 13.72). ☐ ☐

Many microorganisms contain enzymes catalyzing the decarboxylation of amino acids. The decarboxylation of lysine forms cadaverine, a foul-smelling compound released especially during the putrification of animal flesh.

$$H_2NCH_2CH_2CH_2CH_2CH\underset{COO^{\ominus}}{\overset{\overset{\oplus}{N}H_3}{\diagup}} \qquad H_2NCH_2CH_2CH_2CH_2CH_2NH_2$$

lysine   cadaverine

Reduction. The direct reduction of acids or salts is limited to the use of lithium aluminium hydride (Eq. 17.19) or BH_3 as quick, convenient methods in the laboratory.

$$4RCOOH + 3LiAlH_4 \xrightarrow{\text{ether}} (RCH_2O)_4LiAl + 2LiAlO_2 + 4H_2 \qquad (17.19)$$

Diacids tend to give trouble because the initial reaction is release of hydrogen to form the dinegative anion, which precipitates as the lithium aluminium salt. In these cases, the diesters are reduced instead.

α-Hydrogen activity. We have not brought up the subject of α-hydrogen reactivity in acids because it is practically negligible. Treatment of an acid with a halogen under acidic, neutral, or basic conditions gives little reaction. Free-radical halogenation gives mixtures of products even with acetic acid, and longer chains are worse. However, acids do display some α-hydrogen activity in deuterium exchange and racemization (Eqs. 17.20).

$$CH_3COOD + D_2O \xrightarrow[\text{slow}]{\text{KOD}} CD_3COOD$$

$$(R)\ \underset{R_2}{\overset{R_1}{\diagdown}}\overset{*}{C}HCOOH \underset{\text{slow}}{\overset{H^{\oplus}}{\rightleftarrows}} \left[\underset{R_2}{\overset{R_1}{\diagdown}}C=C\underset{OH}{\overset{OH}{\diagup}}\right]^{\oplus} \rightleftarrows (RS)\ \underset{R_2}{\overset{R_1}{\diagdown}}\overset{*}{C}HCOOH \qquad (17.20)$$

The acid derivatives all display more α-hydrogen activity than do the acids, which, strictly speaking, do not possess a full-fledged carbonyl group. It is possible to bring about an acid-catalyzed chlorination or bromination at the α position by the Hell-Volhard-Zelinsky reaction. The actual enolization and substitution is with the acid halide. In the reaction, some red phosphorus is added to the acid, and the halogen is added as rapidly as it is taken up. Some PX_3 may be added instead of phosphorus, and sometimes some $RCOX$, all of which speed up the reaction (Eqs. 17.21).

Reactions 17.21a and b show the formation of some acid bromide. (Only catalytic amounts of P, PBr_3, or RCOBr are necessary.) Reaction c shows the enolization and bromination of the acid bromide. Unlike the case with halogenation of a ketone, the reaction rate here does depend upon the concentration of halogen. The bromo-enol presumably is so much less reactive than is a simple enol (because of reluctance to act as a nucleophile) that the displacement of $Br^\ominus$ from Br_2 becomes the rate-limiting step. Then, in d, we have an acid bromidolysis occurring, which permits c to operate over again. The operation of d sets up a chain reaction. As with acidic halogenation of ketones, single substitution takes place. A second substitution (to form $RCBr_2COOH$) is difficult.

$$2P + 3Br_2 \longrightarrow 2PBr_3 \qquad a$$

$$3RCH_2COOH + PBr_3 \longrightarrow 3RCH_2COBr + P(OH)_3 \qquad b$$

$$RCH_2COBr + HBr \rightleftharpoons RCHC\begin{smallmatrix}H\\ \\Br\end{smallmatrix}\!\!\begin{smallmatrix}OH^\oplus\\ //\\ \end{smallmatrix} \rightleftharpoons RCH{=}C\begin{smallmatrix}OH\\ \\Br\end{smallmatrix} + H^\oplus \qquad (17.21)$$

$$\downarrow Br_2 \qquad c$$

$$RCHC\begin{smallmatrix}Br\\ |\\ \\Br\end{smallmatrix}\!\!\begin{smallmatrix}O\\ //\\ \end{smallmatrix} \xleftarrow{-H^\oplus} RCH{-}C\begin{smallmatrix}Br\\ |\\ \\Br\end{smallmatrix}\!\!\begin{smallmatrix}OH^\oplus\\ //\\ \end{smallmatrix} + Br^\ominus$$

$$RCHC\begin{smallmatrix}Br\\ |\\ \\Br\end{smallmatrix}\!\!\begin{smallmatrix}O\\ //\\ \end{smallmatrix} + RCH_2COOH \rightleftharpoons RCHCOOH\begin{smallmatrix}Br\\ |\\ \\ \end{smallmatrix} + RCH_2C\begin{smallmatrix}O\\ //\\ \\Br\end{smallmatrix} \qquad d$$

and

$$RCH_2COOH \xrightarrow[Br_2]{P} RCHBrCOOH + HBr$$

We should mention, in concluding this section, that many enzymes have been devised in nature to carry out specific reactions with acids and acid derivatives. Oxidations, reductions, decarboxylations, dehydrations, isomerizations, and the like occur, as in the Krebs cycle, for example.

The four-carbon plants. Some three- and four-carbon compounds are playing an important role in the research leading to what could be the next stage of the Green Revolution. This is the study of why some plant species are two to three times as efficient as others in the photosynthetic fixation of carbon dioxide into organic compounds. Most important crops, such as cereal grains, peanuts, and soybeans are among the less efficient plants. Corn and sugar cane, however, are high-efficiency plants. A number of biochemical differences have been identified between the two categories, but the reason for the difference in efficiency is still unknown.

As high-efficiency plants assimilate the carbon dioxide, four-carbon acids (including oxaloacetic, malic, and aspartic) are the first products. Hence, these species are called C_4 plants. The low efficiency plants are referred to as C_3 plants because as they take up CO_2, 3-phosphoglyceric acid with only three carbon atoms is predominantly the first product. If the secret of the C_4 plants could be understood and incorporated into the low efficiency plants, the yields of these crops could probably be increased by up to 50 percent.

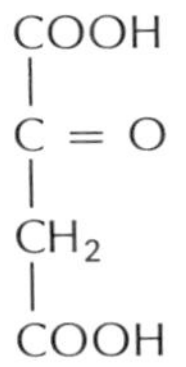

COOH	COOH	COOH	COOH
C=O	HOCH	H$_2$NCH	HCOH
CH$_2$	CH$_2$	CH$_2$	H$_2$COPO$_3$ ②⁻
COOH	COOH	COOH	
oxaloacetic acid	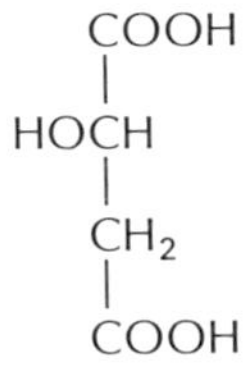 malic acid	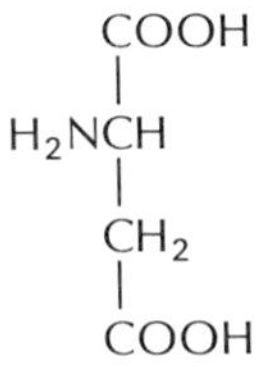 aspartic acid	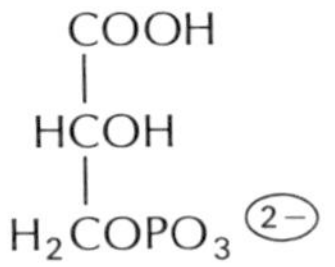 3-phosphoglyceric acid

Exercises

17.1 Show how to complete these conversions, using any other substances needed:

a. ϕCH$_2$COOH $\longrightarrow$
$$\begin{array}{c} Et \\ \diagdown \\ \end{array} C=C \begin{array}{c} \phi \\ \diagup \\ \end{array}$$
$$\begin{array}{c} \diagup \\ Et \end{array} \quad \begin{array}{c} \diagdown \\ COOEt \end{array}$$

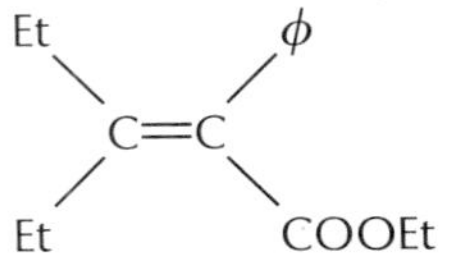

b. CH$_3$CH=CHCOOH $\longrightarrow$ CH$_3$CH(OCH$_2$CH=CHCH$_3$)$_2$
c. Et$_3$CCOOH $\longrightarrow$ Et$_3$CCOOEt
d. CH$_3$CH$_2$COOH $\longrightarrow$ CH$_3$CH$_2$CONH$_2$
e. HOOCCH$_2$CH$_2$COOH $\longrightarrow$ EtOOCCH$_2$CH$_2$COCl
f. ϕCOOH $\longrightarrow$ ϕCOOCϕ
        (with O=C—O—C=O linkage)

17.2 Explain why *o*-hydroxybenzoic acid is stronger and *m*- and *p*-hydroxybenzoic acids are weaker than benzoic acid.

17.5 Preparation of acids

Many methods are available for the preparation of acids. Happily, most of the routes have been described elsewhere: a list will serve to gather them together in one place.

Oxidative methods.
1. Alkenes (RCH=CH$_2$ and RCH=CHR), Section 13.5.
2. Alkynes (RC≡CH and RC≡CR), Section 13.5.
3. Primary alcohols (RCH$_2$OH), Section 9.8.
4. Aldehydes (RCHO), Section 16.6.
5. Ketones (RCOR), Section 16.6.
6. Methyl ketones (RCOCH$_3$), the haloform reaction, Section 16.4.
7. α-Diketones (RCOCOR), periodic acid or lead tetraacetate, Section 16.7.
8. Ketones (RCOR), Baeyer-Villiger reaction, Section 16.5.
9. Side chains of aromatic rings (Ar—R, etc.), Section 14.5. ☐☐

Hydrolytic methods.
1. Esters, Section 17.1: primarily used for the recovery of acids from naturally occurring oils, fats, and waxes.

2. Amides, Section 17.4, and Equation 17.22.

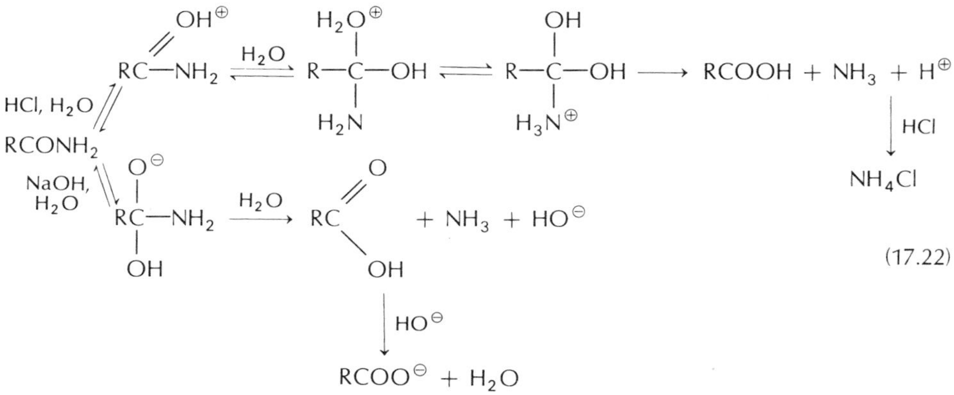

$$\begin{array}{c} \text{RCONH}_2 \xrightarrow[\text{H}_2\text{O}]{\text{HCl, H}_2\text{O}} \text{RC(=OH}^\oplus\text{)NH}_2 \xrightleftharpoons[]{\text{H}_2\text{O}} \text{R–C(OH}_2^\oplus\text{)(H}_2\text{N)–OH} \rightleftharpoons \text{R–C(OH)(H}_3\text{N}^\oplus\text{)–OH} \longrightarrow \text{RCOOH} + \text{NH}_3 + \text{H}^\oplus \xrightarrow{\text{HCl}} \text{NH}_4\text{Cl} \\ \text{RCONH}_2 \xrightarrow[\text{H}_2\text{O}]{\text{NaOH}} \text{RC(O}^\ominus\text{)(OH)NH}_2 \xrightarrow{\text{H}_2\text{O}} \text{RC(=O)OH} + \text{NH}_3 + \text{HO}^\ominus \xrightarrow{\text{HO}^\ominus} \text{RCOO}^\ominus + \text{H}_2\text{O} \end{array}$$

(17.22)

3. Nitriles. The carbon-nitrogen triple bond may be thought of as a dehydrated amide on the one hand or as analogous to a terminal alkyne on the other. Hydrolysis proceeds under both acidic and basic conditions. However, concentrated hydrochloric acid is the usual choice for reagent (Eq. 17.23). Once the amide is formed, continued hydrolysis gives the acid (Eq. 17.22).

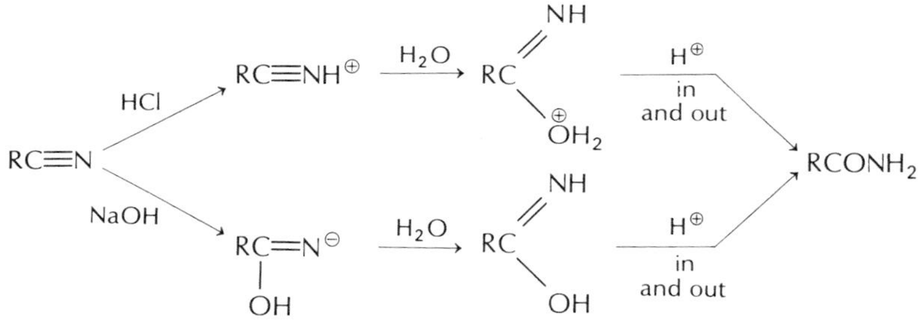

$$\text{RC}{\equiv}\text{N} \xrightarrow[\text{NaOH}]{\text{HCl}} \begin{array}{c} \text{RC}{\equiv}\text{NH}^\oplus \xrightarrow{\text{H}_2\text{O}} \text{RC(=NH)OH}_2^\oplus \xrightarrow[\text{in and out}]{\text{H}^\oplus} \\ \text{RC}{\equiv}\text{N}^\ominus\text{–OH} \xrightarrow{\text{H}_2\text{O}} \text{RC(=NH)OH} \xrightarrow[\text{in and out}]{\text{H}^\oplus} \end{array} \text{RCONH}_2 \quad (17.23)$$

4. Acid chlorides. The substances react vigorously if they are soluble in water, as is CH$_3$COCl. If the substance is insoluble in water (ϕCOCl), the reaction is somewhat slower (Eq. 17.24).

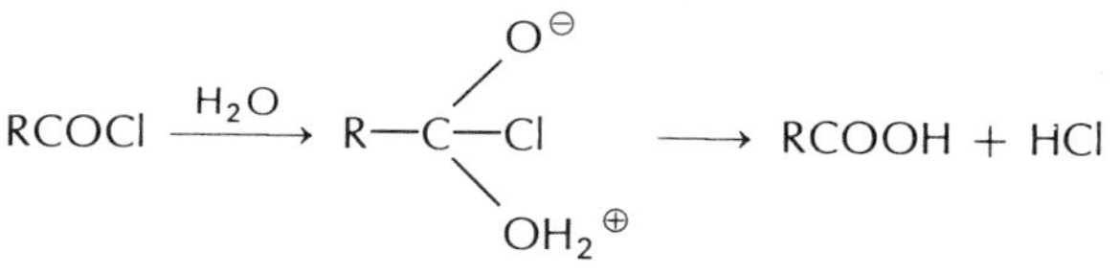

$$\text{RCOCl} \xrightarrow{\text{H}_2\text{O}} \text{R–C(O}^\ominus\text{)(Cl)(OH}_2^\oplus\text{)} \longrightarrow \text{RCOOH} + \text{HCl} \quad (17.24)$$

5. Acid anhydrides. Anhydrides react with water much more gently than do acid chlorides, and bases may be required to hasten the hydrolysis, as with phthalic anhydride and succinic anhydride (Eq. 17.25).

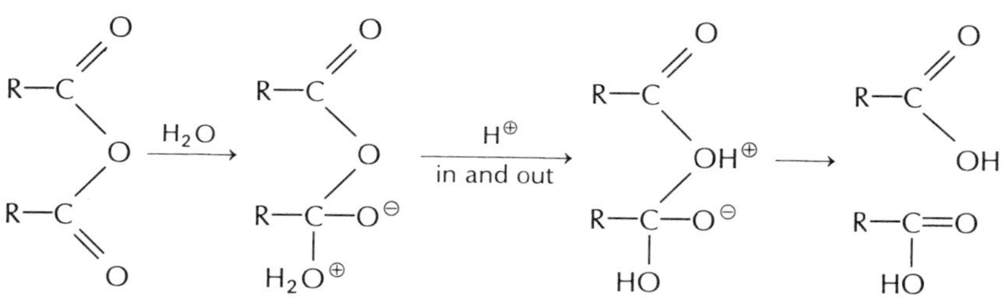

$$(\text{RCO})_2\text{O} \xrightarrow{\text{H}_2\text{O}} \text{RC(=O)–O–C(R)(O}^\ominus\text{)(OH}_2^\oplus\text{)} \xrightarrow[\text{in and out}]{\text{H}^\oplus} \text{RC(=O)–O–C(R)(O}^\ominus\text{)(OH}^\oplus\text{)(HO)} \longrightarrow \text{RCOOH} + \text{RC(=O)OH} \quad (17.25)$$

17.5 Preparation of acids

6. Trihalomethyl groups, Section 15.6: refers to the hydrolysis of benzotrichloride; aromatic side chains.

Disproportionation and addition methods.
1. Grignard reaction: addition of RMgX or RLi to CO_2, Section 16.7.
2. Cannizzaro reaction, Equation 9.46.
3. Diels-Alder reaction of unsaturated acids, Section 14.8.

Rearrangements. Benzilic acid rearrangement, Section 16.5.

Synthesis of diacids. Several methods are used to synthesize diacids.

Oxalic acid can be made in numerous ways, one of which is the careful oxidation of ethylene glycol ($HOCH_2CH_2OH$). Another method is the fusion of sodium formate (Eq. 17.26).

$$2HCOONa \xrightarrow{400°} H_2 + \begin{array}{c} COONa \\ | \\ COONa \end{array} \xrightarrow{H_2SO_4} \begin{array}{c} COOH \\ | \\ COOH \end{array} \quad (17.26)$$

Malonic acid and esters are made from sodium bromo- or chloroacetate and sodium cyanide (Eqs. 17.27). In Section 17.8, we shall go into the usefulness of malonic ester for the synthesis of acids.

$$ClCH_2COONa + NaCN \longrightarrow NCCH_2COONa \begin{array}{c} \xrightarrow{H^\oplus, H_2O} HOOCCH_2COOH \\ \xrightarrow{H^\oplus, EtOH} EtOOCCH_2COOEt \end{array} \quad (17.27)$$

Succinic acid may be put together in several convenient ways, depending upon the availability of starting materials (Eqs. 17.28).

Glutaric acid, like succinic acid, can be made from malonic ester, as we shall soon see. Other methods are shown in Equations 17.29.

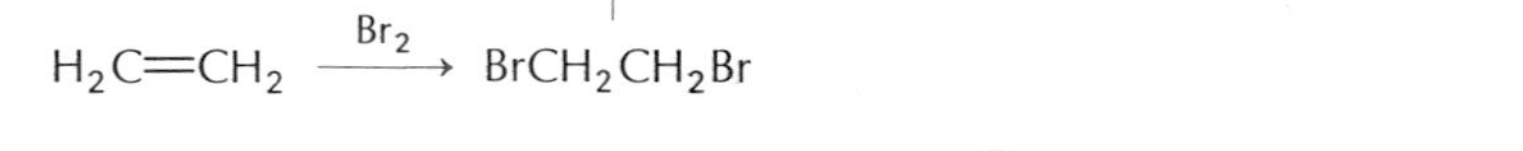

$$\begin{array}{c} HOOCCH=CHCOOH \\ \text{maleic or fumaric} \end{array} \xrightarrow{H_2, Pt}$$

$$H_2C=CHCN \xrightarrow{HCN} \begin{array}{c} H_2CCH_2CN \\ | \\ CN \end{array} \xrightarrow{H^\oplus, H_2O} HOOCCH_2CH_2COOH \quad (17.28)$$

$$\uparrow NaCN$$

$$H_2C=CH_2 \xrightarrow{Br_2} BrCH_2CH_2Br$$

cyclopentanone $\xrightarrow{H^\oplus}$ cyclopentenol $\xrightarrow{HNO_3}$ $HOOCCH_2CH_2CH_2COOH$

$$HOCH_2CH_2CH_2OH \xrightarrow{HBr} BrCH_2CH_2CH_2Br \xrightarrow{H^\oplus, H_2O} \quad (17.29)$$

$$\downarrow NaCN$$

$$NCCH_2CH_2CH_2CN$$

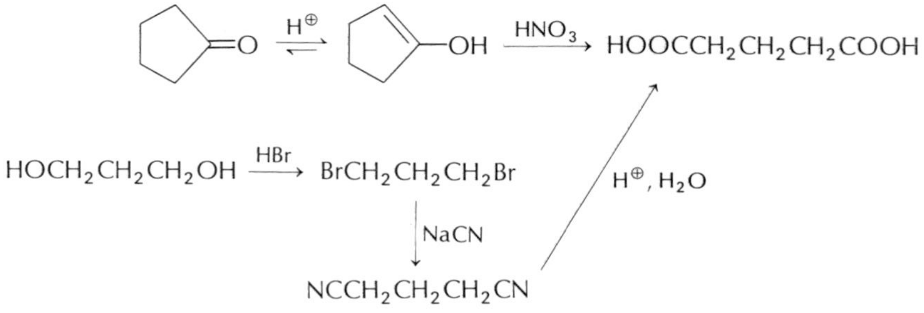

Adipic acid is commonly made in the laboratory by cyclohexanol or cyclohexanone oxidation (Eq. 17.30). Commercially, the acid is made by a catalytic (($CH_3COO)_2Co$ in CH_3COOH) air oxidation of cyclohexane. ☐ ☐

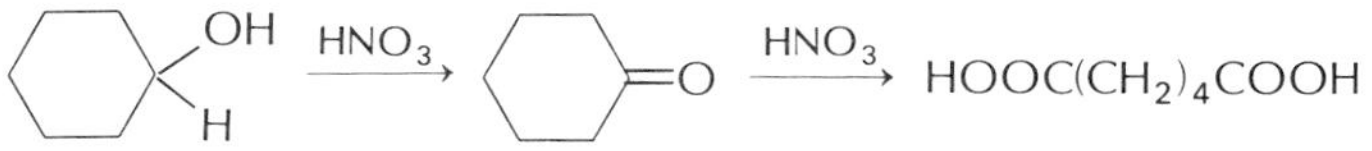

$$\text{cyclohexanol} \xrightarrow{HNO_3} \text{cyclohexanone} \xrightarrow{HNO_3} HOOC(CH_2)_4COOH \quad (17.30)$$

Exercises

17.3 Show how to carry out the conversions:

a. $CH_3CH_2CHOHCH_3 \longrightarrow \begin{array}{c} CH_3CH-COOH \\ | \\ CH_3CH-COOH \end{array}$

b. 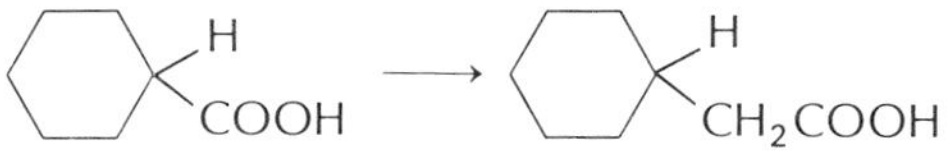 cyclohexane-H,COOH ⟶ cyclohexane-H,CH$_2$COOH

c. 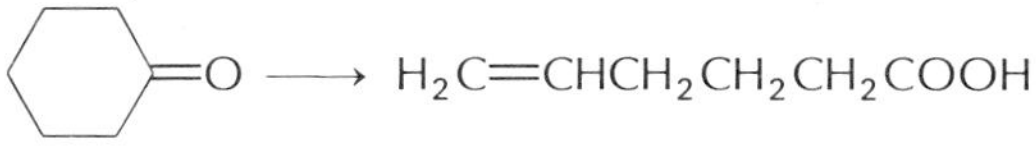 cyclohexanone ⟶ $H_2C=CHCH_2CH_2CH_2COOH$

d. 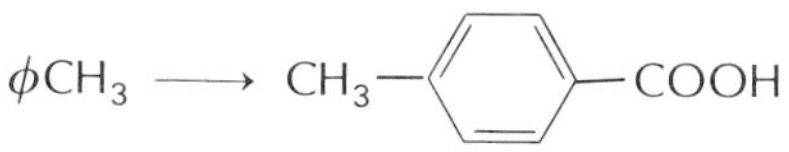 $\phi CH_3 \longrightarrow CH_3-\text{C}_6H_4-COOH$

e.  $H_2C=CHCOOH \longrightarrow$ cyclohexane-H,COOH

17.6 Reactions of acid chlorides and acid anhydrides

Acid chlorides and anhydrides undergo very similar reactions, although at different rates. Comparison of RCO—Cl with RCO—OCOR shows that an acid chloride has somewhat more double-bond character than does an acid anhydride because of the somewhat larger −I effect and smaller +R effect of Cl.

Olysis reactions. The olysis reactions have been mentioned. Acid chlorides and anhydrides provide useful, convenient reagents for the preparation of esters, amides, and substituted amides (Eqs. 17.31). Acetylation is carried out with either acetyl chloride or acetic anhydride, but benzoylation usually is done with benzoyl chloride only.

$$\left.\begin{array}{c} RCOCl \\ \text{or} \\ RCO\diagdown\\ O \\ RCO\diagup \end{array}\right\} + \left\{\begin{array}{c} R'OH \longrightarrow RCOOR' \\ \text{or} \\ NH_3 \longrightarrow RCONH_2 \\ \text{or} \\ R'_2NH \longrightarrow RCONR'_2 \end{array}\right\} + \left\{\begin{array}{c} HCl \\ \text{or} \\ \text{or} \\ RCOOH \end{array}\right. \quad (17.31)$$

17.6 Reactions of acid chlorides and acid anhydrides

Benzoyl chloride is insoluble in water and thus reacts slowly with it. The alcoholysis, ammonolysis, and aminolysis reactions may be carried out in aqueous mixtures with dilute sodium hydroxide. The alcohol or amine tends to be less soluble in the aqueous layer than in the immiscible benzoyl chloride in which reaction occurs. The hydrogen chloride is neutralized as rapidly as it is formed by the sodium hydroxide at the interface of the immiscible liquids. The newly formed ester or amide remains in the excess benzoyl chloride layer until the reaction reaches the point at which the amide crystallizes or the liquid ester takes over as the principal constituent of the immiscible layer. This use of dilute sodium hydroxide with benzoyl chloride is known as the <u>Schotten-Baumann reaction.</u>

A useful variation is to employ the tertiary amine pyridine as a basic solvent for both polar and nonpolar substances (Eqs. 17.32) in olysis reactions. As a good nucleophile,

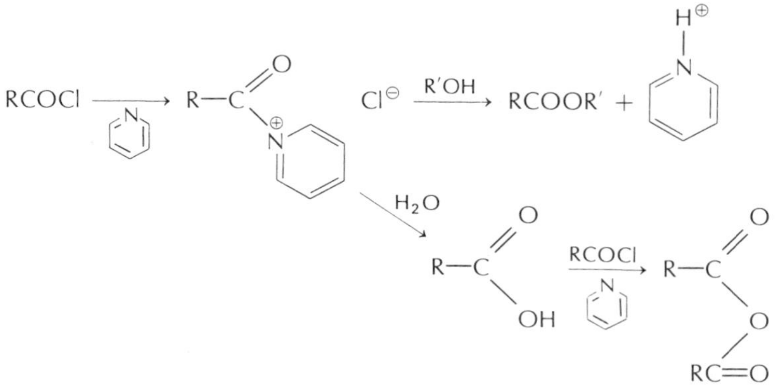

$$RCOCl \xrightarrow{\text{pyridine}} R-C(=O)-N^+\text{pyridine} \; Cl^- \xrightarrow{R'OH} RCOOR' + \text{pyridine} \cdot H^+ \tag{17.32}$$

$$\downarrow H_2O$$

$$R-C(=O)OH \xrightarrow{RCOCl, \text{pyridine}} R-C(=O)-O-C(=O)R$$

pyridine displaces chlorine readily and, in turn, is easily displaced by other olysis reagents. An excellent preparation of anhydrides consists of treating two moles of acid chloride with one mole of water in excess pyridine. The hydrogen chloride evolved is neutralized by the excess of amine used.

We have seen that the reaction of an acid chloride with an ylide is a route to ketones, or alkynes (Eq. 16.86).

The cyclic anhydrides succinic anhydride, maleic anhydride, and phthalic anhydride react slowly in the alcoholysis reaction. Thus, it sometimes is more convenient to use an alkoxide in excess alcohol to prepare a monoester from such unreactive anhydrides (Eq. 17.33). If the diester is wanted, the cyclic anhydride simply is heated under reflux with

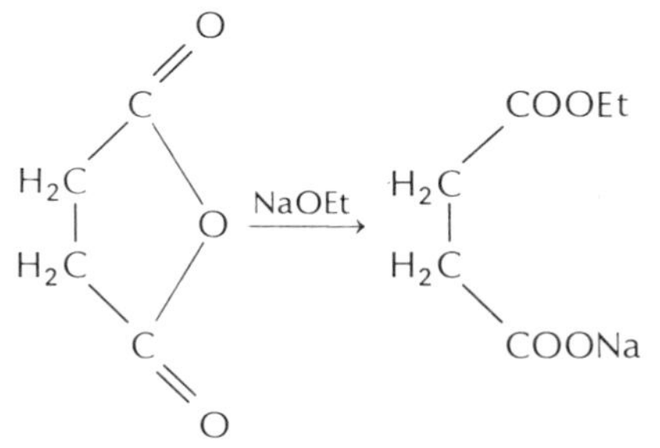

$$\text{succinic anhydride} \xrightarrow{\text{NaOEt}} \begin{array}{c} H_2C-COOEt \\ | \\ H_2C-COONa \end{array} \tag{17.33}$$

excess alcohol in the presence of a strong acid, just as though a Fischer esterification were being carried out. In fact, the second step is a Fischer esterification, the slow step in the overall reaction (Eq. 17.34).

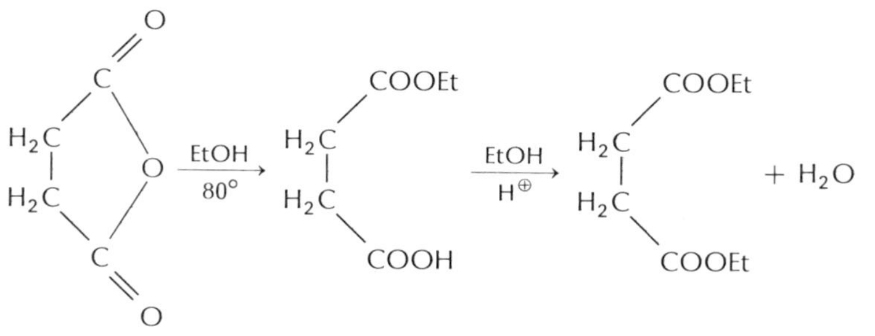

$$\text{[succinic anhydride]} \xrightarrow[80°]{\text{EtOH}} \underset{\underset{\text{COOH}}{\overset{\text{COOEt}}{|}}}{\overset{H_2C}{\underset{H_2C}{|}}} \xrightarrow[H^\oplus]{\text{EtOH}} \underset{\underset{\text{COOEt}}{\overset{\text{COOEt}}{|}}}{\overset{H_2C}{\underset{H_2C}{|}}} + H_2O \qquad (17.34)$$

From the standpoint of an acid chloride or anhydride, the Friedel-Crafts reaction with an aryl compound in the presence of $AlCl_3$ (or BF_3 or HF, etc.) may be thought of as an arylolysis (Eqs. 17.35).

$$ArH + RCOX \xrightarrow{AlCl_3} ArCOR + HX$$

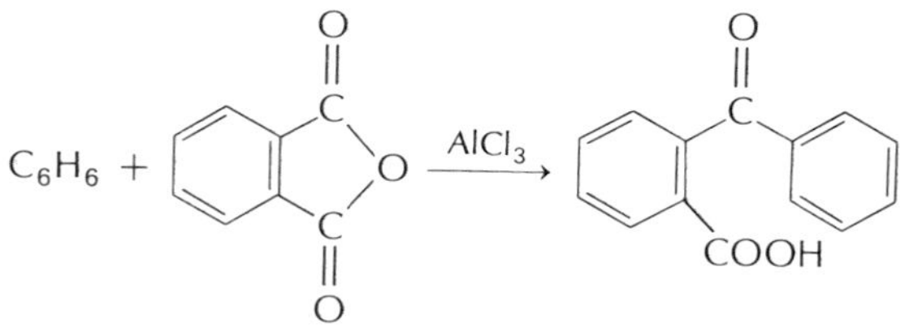

$$C_6H_6 + \text{[phthalic anhydride]} \xrightarrow{AlCl_3} \text{[2-benzoylbenzoic acid]} \qquad (17.35)$$

Exchange and addition reactions. Acid chlorides undergo exchange reactions with anhydrous HF, HBr, or HI to give the corresponding acid fluoride, bromide, or iodide. Both acid chlorides and anhydrides react with Grignard and lithium reagents to give tertiary alcohols. The reactions are vigorous compared to the relatively mild effect of the reagents with esters. The use of the cadmium reagents with acid chlorides to prepare ketones has been described (Eq. 16.83).

Reduction reactions. The Rosenmund reduction of acid chlorides to give aldehydes (Eq. 16.99) also has been described. Acid chlorides react with lithium aluminium hydride and with sodium borohydride to give primary alcohols. Acid chlorides seem to be inert to reduction by borane. Acid anhydrides are inert to sodium borohydride, but lithium aluminium hydride reduces them.

Cyclic anhydrides may be partially reduced with zinc dust and acetic acid or with catalytic hydrogenation to give lactones (Eq. 17.36). Lithium aluminium hydride may be

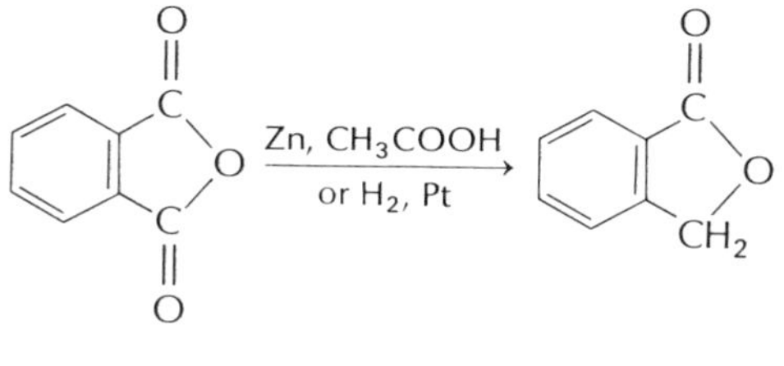

$$\text{[phthalic anhydride]} \xrightarrow[\text{or } H_2, \text{Pt}]{Zn, CH_3COOH} \text{[phthalide]} \qquad (17.36)$$

used to accomplish the same result, although it usually is used in excess to give complete reduction to a diol (Eq. 17.37).

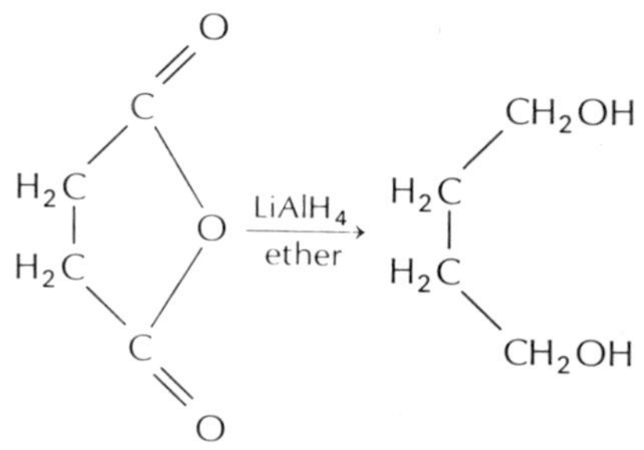

17.6 Reactions of acid chlorides and acid anhydrides

$$\text{(glutaric anhydride)} \xrightarrow[\text{ether}]{\text{LiAlH}_4} \text{HOCH}_2\text{CH}_2\text{CH}_2\text{CH}_2\text{CH}_2\text{OH} \qquad (17.37)$$

Condensation. Acid anhydrides have enough α-hydrogen activity to undergo an aldol-like condensation called the Perkin reaction with aldehydes that have no α hydrogens (aromatic aldehydes). The reaction usually is carried out in the presence of the sodium or potassium salt of the acid corresponding to the anhydride used (Eq. 17.38). At times it is

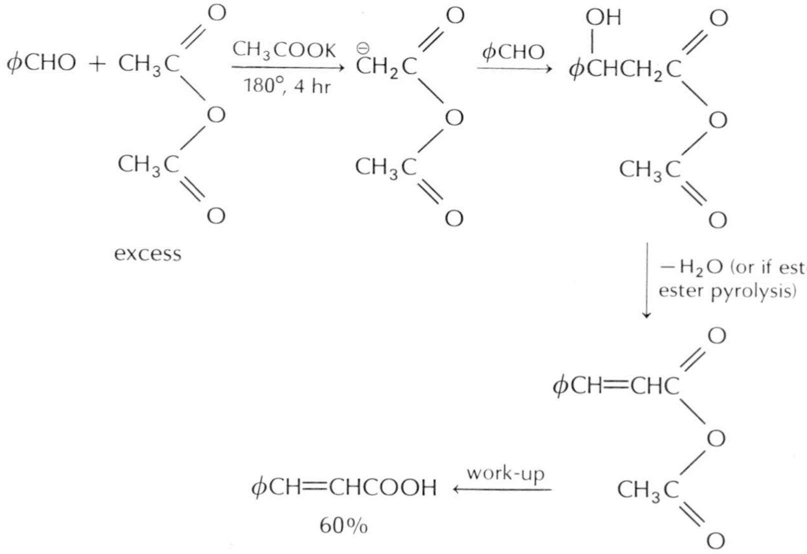

$$\phi\text{CHO} + (\text{CH}_3\text{CO})_2\text{O} \xrightarrow[180°, 4\text{ hr}]{\text{CH}_3\text{COOK}} {}^{\ominus}\text{CH}_2\text{C(O)OC(O)CH}_3 \xrightarrow{\phi\text{CHO}} \phi\text{CH(OH)CH}_2\text{C(O)OC(O)CH}_3 \qquad (17.38)$$

excess

$$\downarrow -\text{H}_2\text{O} \text{ (or if esterified, ester pyrolysis)}$$

$$\phi\text{CH}=\text{CHCOOH} \xleftarrow{\text{work-up}} \phi\text{CH}=\text{CHC(O)OC(O)CH}_3$$

60%

possible to use an acid that gives rise to an anhydride that is more reactive than acetic anhydride and to use a tertiary amine as the base (Eq. 17.39).

$$\phi\text{CHO} + \phi\text{CH}_2\text{COOH} \xrightarrow[\text{Et}_3\text{N reflux}]{(\text{CH}_3\text{CO})_2\text{O}} \phi\text{CH}=\overset{\phi}{\text{C}}-\text{COOH} \qquad (17.39)$$

25%

The acetic anhydride reacts to give rise to some phenylacetic anhydride, which condenses with benzaldehyde.

The addition of the anion to the carbonyl carbon of the aldehyde is subject to steric and electronic effects. Thus, the yield with phenylacetic acid (bulky anion) is decreased. Electron-donating substituents on the benzene ring of benzaldehyde decrease the yield; electron-withdrawing substituents increase the yield. Thus, under conditions that give a 45 to 50% yield of cinnamic acid from benzaldehyde, p-methylbenzaldehyde gives 33% and p-chlorobenzaldehyde gives 52% of the substituted cinnamic acid. More striking are o-methylbenzaldehyde, which gives 15%, and o-chlorobenzaldehyde, which gives 71%.

Surprising, too, is the reactivity of phthalic anhydride, which can take the place of the aldehyde in the Perkin reaction (Eq. 17.40).

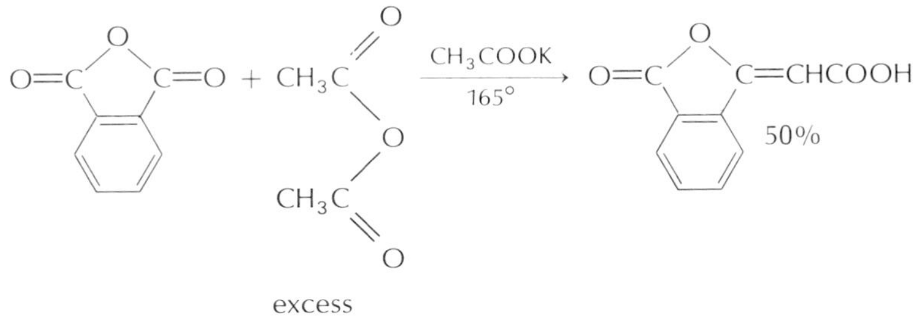

$$\text{phthalic anhydride} + CH_3C(O)OC(O)CH_3 \text{ (excess)} \xrightarrow[165°]{CH_3COOK} \text{phthalide-C=CHCOOH} \quad 50\% \quad (17.40)$$

Exercises

17.4 Complete the equations:

a. $CH_3COCl + \phi CH_2OH \longrightarrow$

b. $(CH_3CO)_2O + \phi NH_2 \longrightarrow$

c. 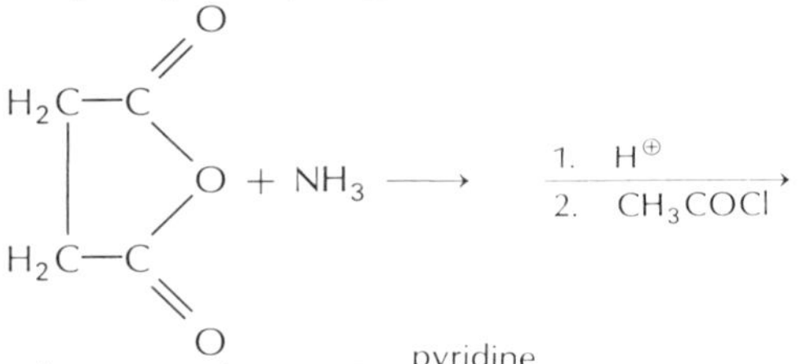 succinic anhydride $+ NH_3 \longrightarrow \xrightarrow[\text{2. } CH_3COCl]{\text{1. } H^{\oplus}}$

d. $2\phi CH_2COCl + H_2O \xrightarrow{\text{pyridine}}$

e. 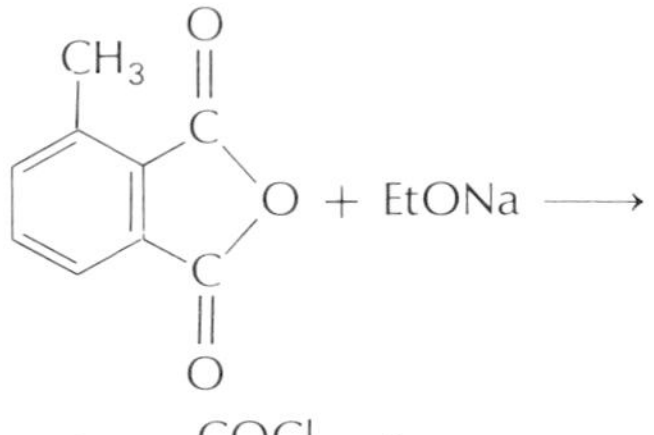 3-methylphthalic anhydride $+ EtONa \longrightarrow$

f. 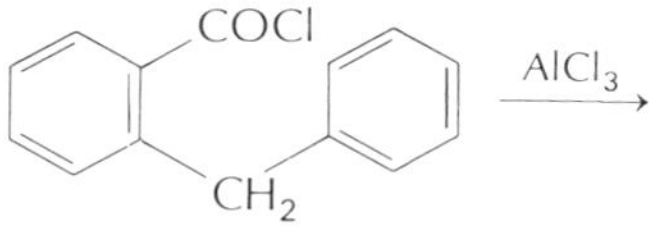 2-(phenylmethyl)benzoyl chloride $\xrightarrow{AlCl_3}$

g. $(CH_3)_2CHCOCl + \phi CH_2CH_2CH_2CdCl \longrightarrow$

h.  $CH_3\text{-}\phi\text{-}CHO + (CH_3CH_2CO)_2O \xrightarrow[\Delta]{CH_3CH_2COONa}$

i. $\phi CH_2COCl + \phi_3\overset{\oplus}{P}\text{—}\overset{\ominus}{C}HCH_3 \longrightarrow \xrightarrow{\text{alc. KOH}}$

17.7 Preparation of acid chlorides and acid anhydrides

Most of the preparative reactions already have been described. Nevertheless, we shall list them here for convenience.

Acid chlorides.
1. The reaction of acids with $SOCl_2$, PCl_3, PCl_5, or oxalyl chloride. Anhydrides are less reactive; esters, still less so. Thus, it is possible to prepare an acid chloride at one location in a molecule without affecting an ester group at some other location (Eq. 17.41).

$$EtOOC(CH_2)_4COOH \xrightarrow{SOCl_2} EtOOC(CH_2)_4COCl \qquad (17.41)$$

2. The addition of anhydrous HX to a ketene (Eq. 17.42).

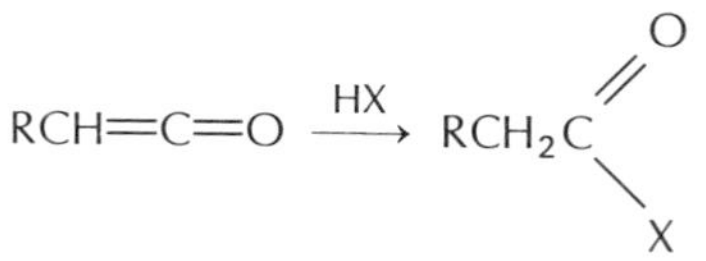

$$RCH=C=O \xrightarrow{HX} RCH_2C{\overset{\displaystyle O}{\underset{\displaystyle X}{\Big\langle}}} \qquad (17.42)$$

Acid anhydrides.
1. The reaction of an acid chloride with a salt of a carboxylic acid.
2. The reaction of an acid with an anhydride. A variation is the reaction of an acid with an enol ester (Eq. 17.43).

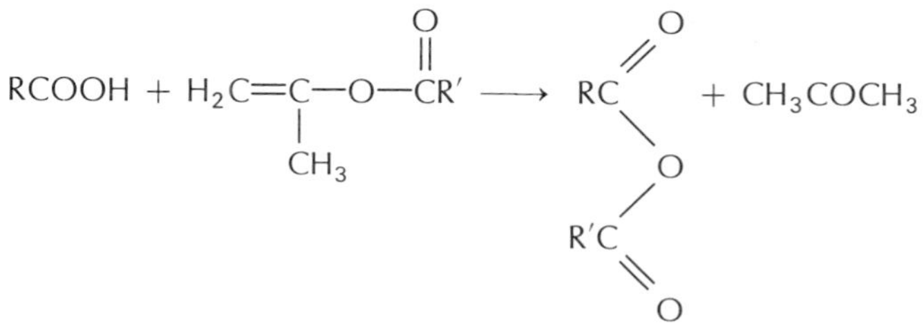

$$RCOOH + H_2C=\underset{CH_3}{\overset{\displaystyle }{C}}-O-\overset{\displaystyle O}{\overset{\displaystyle \|}{C}}R' \longrightarrow RC{\overset{\displaystyle O}{\underset{\displaystyle O-CR'\!=\!O}{\Big\langle}}} + CH_3COCH_3 \qquad (17.43)$$

3. The Baeyer-Villiger reaction with an α-diketone (Eq. 17.44).

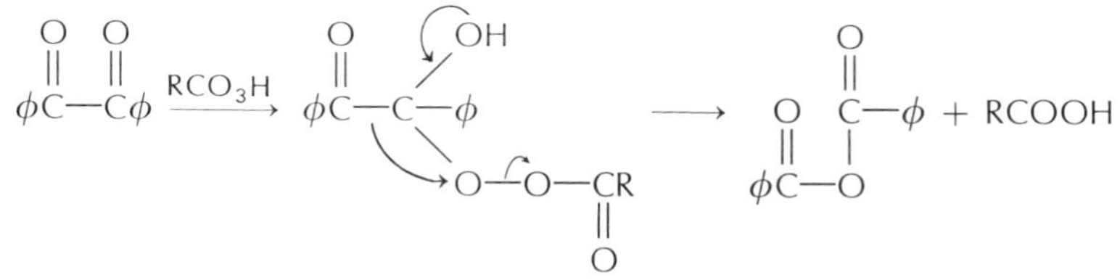

$$\underset{}{\overset{O\ \ O}{\underset{}{\overset{\|\ \ \|}{\phi C-C\phi}}}} \xrightarrow{RCO_3H} \phi C\overset{O}{\overset{\|}{-}}\overset{OH}{\underset{O-O-CR\ \ \|\ \ O}{\overset{\displaystyle }{C}}}\!\!\!-\!\phi \longrightarrow \overset{O}{\underset{\phi C-O}{\overset{\|}{O\ \ C}}}\!\!-\phi + RCOOH \qquad (17.44)$$

4. The addition of an acid to a ketene (Eq. 17.45).

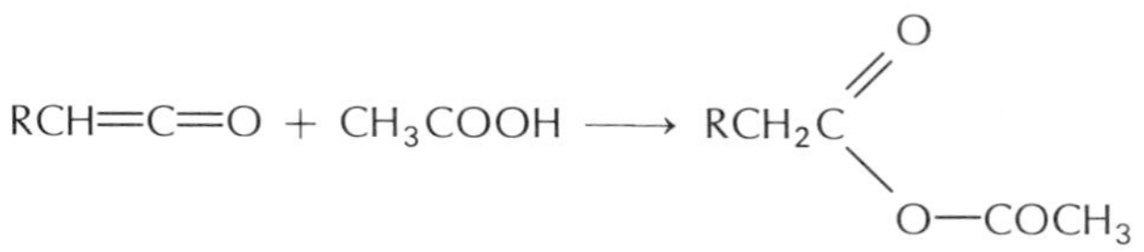

$$RCH=C=O + CH_3COOH \longrightarrow RCH_2C{\overset{\displaystyle O}{\underset{\displaystyle O-COCH_3}{\Big\langle}}} \qquad (17.45)$$

5. The free-radical reaction of an aldehyde with a *tert*-butyl perester. Cuprous ion is a good catalyst (Eq. 17.46).

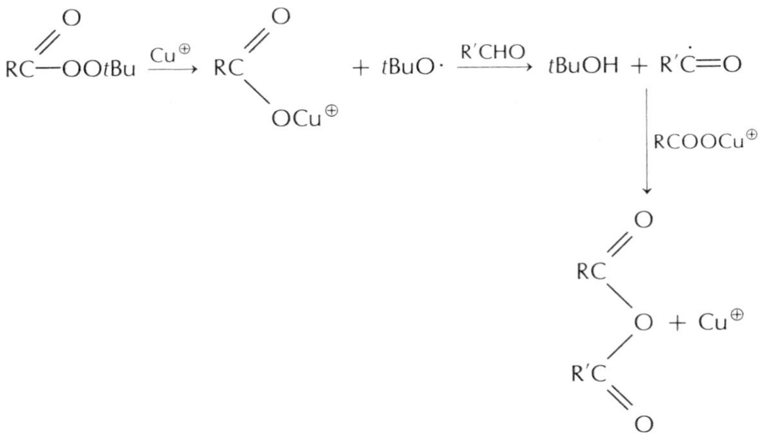

$$RC(=O)-OOtBu \xrightarrow{Cu^{\oplus}} RC(=O)OCu^{\oplus} + tBuO\cdot \xrightarrow{R'CHO} tBuOH + R'\dot{C}=O$$

$$\downarrow RCOOCu^{\oplus}$$

$$RC(=O)-O-C(=O)R' + Cu^{\oplus} \quad (17.46)$$

6. Catalytic oxidation of benzene (to maleic anhydride) and of naphthalene (to phthalic anhydride).

Exercises

17.5 Complete these reactions or conversions (more than one step may be needed):
  a. $2(CH_3)_2C=C=O + H_2O \longrightarrow$

  b. 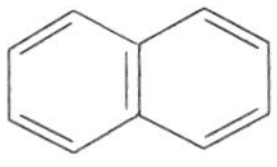 [naphthalene] $+ O_2 \xrightarrow[\Delta]{cat.}$

  c. $EtOOCCH_2CH_2COOEt \longrightarrow EtOOCCH_2CH_2COCl$

17.8 Reactions of esters

Reaction sites. Esters contain a variety of positions that are subject to attack by various reagents. An ester may be viewed as a special type of ether or as a special type of

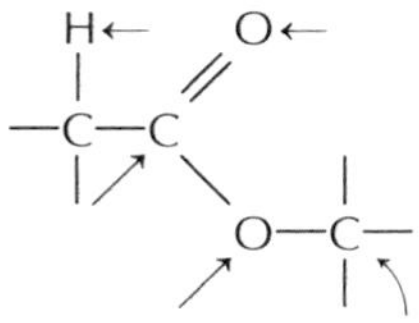

ketone that has a carbonyl group and α-hydrogen activity. On the other hand, we may regard an ester as having an acyl group (RCO—) and an alkyl group (R') in RCOOR'. Most of the reactions of esters involve changes brought about by initial attack upon the carbonyl group. We shall organize the reactions of esters into sections built around the subsequent fate of the intermediate that results from the initial attack by acid or base. We shall conclude with a few of the reactions of ortho esters.

Additions to the carbonyl group: olysis reactions and others. Esters hydrolyze more rapidly under basic conditions (saponification) than under acidic conditions. Because esters are insoluble in water, addition of a solvent miscible with water (ethanol, for example) to the mixture generally speeds up the reaction by increasing the solubility of the ester (Eqs. 17.47). The abbreviated mechanism shows that acidic hydrolysis leads to an equilibrium

$$\phi\text{COOEt} \underset{\text{slow}}{\overset{H_3O^\oplus/\text{fast}}{\rightleftharpoons}} \phi-\overset{\overset{HO^\oplus}{\|}}{C}-OEt \underset{\text{slow}}{\overset{+H_2O,\text{ etc.}}{\rightleftharpoons}} \phi-\overset{\overset{HO}{|}}{\underset{HO}{C}}-OEt \overset{+H^\oplus, -EtOH,\text{ etc.}}{\rightleftharpoons} \phi\text{COOH} + \text{EtOH}$$

$$\phi\text{COOEt} \overset{HO^\ominus}{\underset{\text{slow}}{\rightarrow}} \phi-\overset{\overset{O^\ominus}{|}}{\underset{OH}{C}}-OEt \overset{\text{fast}}{\rightleftharpoons} \phi-\overset{O}{\underset{OH}{C}} + \text{EtO}^\ominus \longrightarrow \phi\text{COO}^\ominus + \text{EtOH}$$

(17.47)

(pushed to the right by excess water), whereas saponification is irreversible because of salt formation by the acid. Substitution in the aromatic ring has a greater effect upon the alkaline rate than upon the acidic rate. Electron-withdrawing substituents such as nitro speed up the reaction. The greater susceptibility of the basic hydrolysis to electronic effects suggests that the attack by hydroxide ion upon the ester is the slow step in the sequence. Substituents in the *ortho* position hinder reaction (formation of a tetrahedral intermediate) whether the electronic effect is accelerating or decelerating. The combined effects usually lead to a rate less than that of ethyl benzoate, although there are some exceptions.

The evidence that the tetrahedral intermediate exists in both acidic and basic mechanisms and that the alkyl-oxygen bond remains intact may be presented thus:

1. Hydrolysis of $RCOOCH_2C(CH_3)_3$ gives only neopentyl alcohol $((CH_3)_3CCH_2OH)$, showing that no $(CH_3)_3CCH_2^\oplus$ is involved. (If it were involved, rearrangement would intervene.)
2. Esters of optically active alcohols $(R'COO\overset{*}{C}HR_1R_2)$ hydrolyze to give optically active alcohols (retention of configuration).
3. Use of $H_2^{18}O$ in the solvent gives rise to the appearance of the heavy oxygen isotope in the acid, with none in the alcohol.
4. An allylic ester gives an unrearranged allylic alcohol. □ □

Esters undergo alcoholysis and acidolysis reactions to give transesterification (Eqs. 17.48, previously met in Eq. 16.55). A competing reaction in acidolysis of esters is formation of an anhydride and alcohol. The competition is slight, however, unless the ester is that of an enol, in which case the reaction is useful for the preparation of mixed anhydrides (Eq. 17.49).

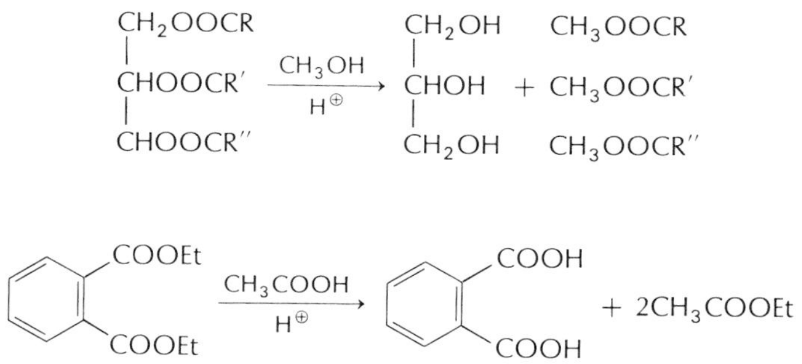

568   Chapter seventeen

$$\begin{array}{c}CH_2OOCR\\|\\CHOOCR'\\|\\CHOOCR''\end{array} \xrightarrow[H^{\oplus}]{CH_3OH} \begin{array}{c}CH_2OH\\|\\CHOH\\|\\CH_2OH\end{array} + \begin{array}{c}CH_3OOCR\\CH_3OOCR'\\CH_3OOCR''\end{array}$$

(17.48)

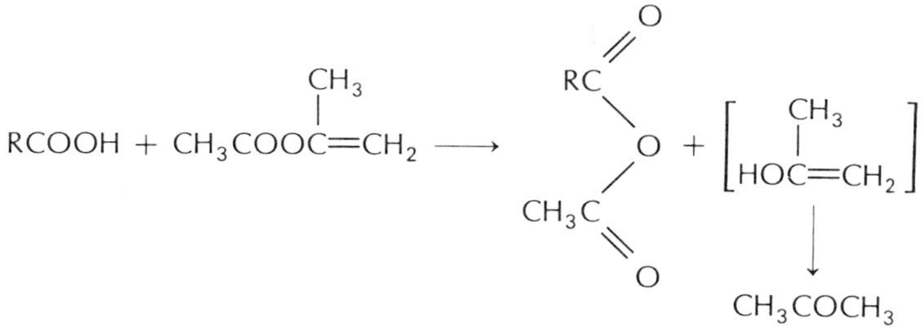

[phthalate diester] $\xrightarrow[H^{\oplus}]{CH_3COOH}$ [phthalic acid] $+ 2CH_3COOEt$

$$RCOOH + CH_3COOC(CH_3)=CH_2 \longrightarrow \begin{array}{c}RC(=O)\\ \diagdown \\ O\\ \diagup \\ CH_3C(=O)\end{array} + \left[\begin{array}{c}CH_3\\|\\HOC=CH_2\end{array}\right] \quad (17.49)$$

$$\downarrow$$
$$CH_3COCH_3$$

Alcoholysis has been useful in exploration of the mechanisms of olysis reactions of esters. Thus, heating methyl benzoate in methanol with sodium methoxide for 52 hr at 100° gives 74% of dimethyl ether, an S_N2 reaction. Of course, the major (but unproductive) reaction is transesterification, but the slow irreversible reaction is shown to exist. Other methyl esters give the same slow reaction (Eq. 17.50). A form of S_N1 mechanism is displayed by *tert*-butyl esters (Eqs. 17.51).

$$ArCOO\frown Me + {}^{\ominus}OMe \longrightarrow ArCOO^{\ominus} + MeOMe \quad (17.50)$$

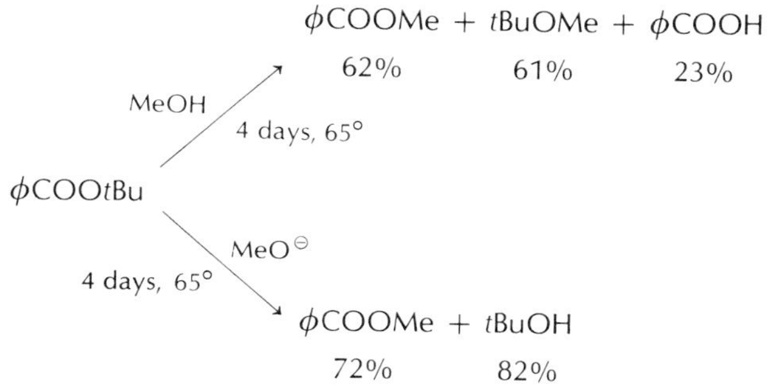

$$\phi COOtBu \begin{array}{c} \xrightarrow[\text{4 days, 65°}]{MeOH} \phi COOMe + tBuOMe + \phi COOH \\ 62\% 61\% 23\% \\ \\ \xrightarrow[\text{4 days, 65°}]{MeO^{\ominus}} \phi COOMe + tBuOH \\ 72\% 82\% \end{array} \quad (17.51)$$

Esters react slowly with ammonia and amines (ammonolysis and aminolysis) to give amides and N-substituted amides (Eq. 17.52). Even the much less nucleophilic substances, amides, give amidolysis reactions (Eq. 17.53) in the presence of an anhydrous base. The reaction may be carried out in reverse also (Eq. 17.54).

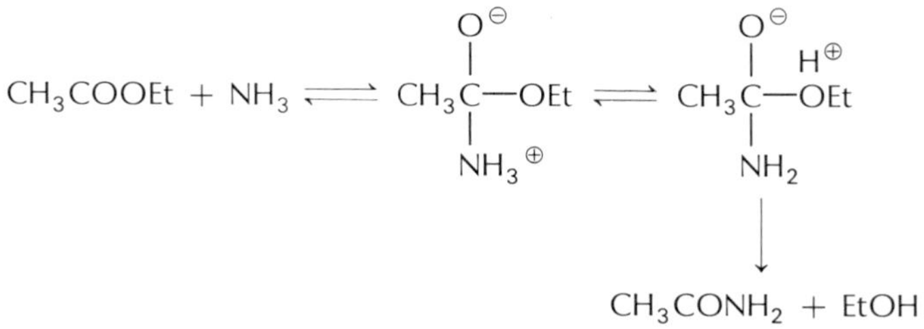

17.8 Reactions of esters

$$CH_3COOEt + NH_3 \rightleftharpoons CH_3\underset{NH_3^{\oplus}}{\overset{O^{\ominus}}{\underset{|}{C}}}-OEt \rightleftharpoons CH_3\underset{NH_2}{\overset{O^{\ominus}}{\underset{|}{C}}}-OEt \overset{H^{\oplus}}{}$$

$$\downarrow$$

$$CH_3CONH_2 + EtOH \tag{17.52}$$

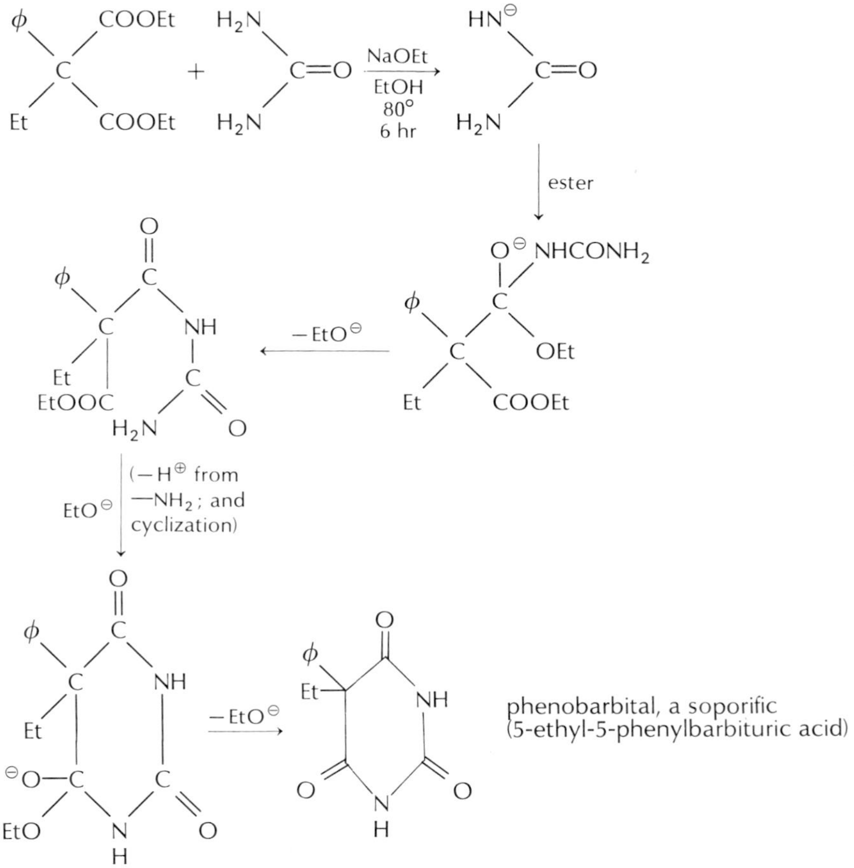

phenobarbital, a soporific
(5-ethyl-5-phenylbarbituric acid)

(17.53)

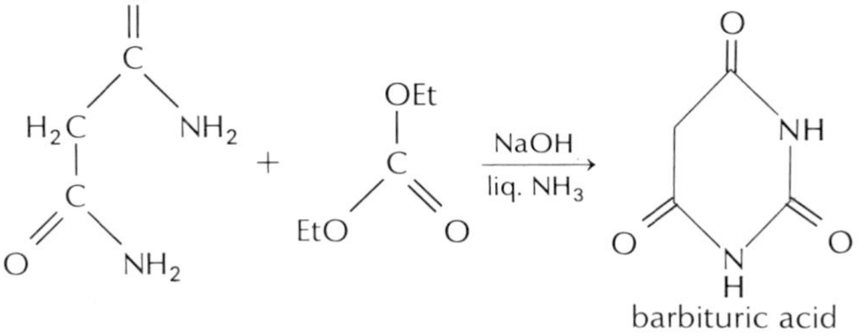

barbituric acid

(17.54)

Esters are much more inert to the action of $SOCl_2$, PCl_3, PCl_5, and acid chlorides than are acids (Eq. 17.55). Under drastic conditions, PCl_5 with an ester gives the acid chloride.

$$CH_3OOCCH_2CH_2COOH \xrightarrow[80°]{SOCl_2} CH_3OOCCH_2CH_2COCl \qquad (17.55)$$

Esters are not usually employed in a Friedel-Crafts reaction. The catalyst ($AlCl_3$) causes a cleavage of the ester, and alkylation instead of acylation of the aromatic component occurs.

Other addition reactions to the carbonyl group to be considered are the reactions of Grignard and lithium reagents with esters (Eqs. 17.56). Little difficulty is encountered in the initial attack of the reagent unless the ester is very hindered.

$$RCOOMe \xrightarrow[\text{ether}]{\phi MgBr} \underset{\phi}{\overset{\phi}{RC}}-OH \quad \text{tertiary alcohol} \qquad (17.56)$$

$$HCOOEt \xrightarrow[\text{ether}]{\phi MgBr} \phi_2CHOH \quad \text{secondary alcohol}$$

Esters may be used in the ylide reaction instead of acid chlorides (Eq. 17.57), although the rate of reaction is slower.

$$R'COOMe + R\overset{\ominus}{C}H-\overset{\oplus}{P}\phi_3 \xrightarrow{-MeO^\ominus} R'CO-\underset{R}{\overset{H}{C}}-\overset{\oplus}{P}\phi_3 \xrightarrow{H_2O} R'COCH_2R \qquad (17.57)$$

The reduction reactions of the esters have been described. We should remember that catalytic hydrogenation is on the difficult side, so other possible reductions occur first. Also, reduction is accompanied by hydrogenolysis of the ester (Eq. 17.58). Sodium boro-

$$\phi CH=CHCOOEt \xrightarrow[Pt]{H_2} \phi CH_2CH_2CH_2OH + EtOH \qquad (17.58)$$

hydride under the usual conditions (ethanol solution, acetic acid) does not reduce esters. Lithium aluminium hydride reduces esters to primary alcohols. Sodium bis(2-methoxyethoxy) aluminium hydride, at $-70°$, reduces an ester to an aldehyde. An old method, the Bouveault-Blanc reduction, used sodium and an alcohol with a bp near 130° to reduce esters to alcohols. □ □

α-Hydrogen reactions: Claisen condensation. Now we come to a very useful group of reactions, all of which involve the α-hydrogen activity of the ester. Just as the aldol condensation of aldehydes is the prototype for several other condensations, so is the Claisen condensation for several others involving esters (Eq. 17.59).

We see that the Claisen condensation parallels the aldol condensation. A base (usually EtONa) establishes an equilibrium by the removal of an α hydrogen from the ester, after which the ester anion adds to the carbonyl carbon of another molecule of ester. The new anion may reverse or go forward by ejection of ethoxide ion to give acetoacetic ester, AAE (also called ethyl acetoacetate and ethyl 3-ketobutanoate). In the presence of base, the relatively acidic AAE ($pK_a = 10.2$—compare pK_a of 2,4-pentanedione, 9.0; and of ethanol, 17) loses a proton to form the anion of AAE (b from a). During work-up, the reaction mixture is acidified and AAE is isolated.

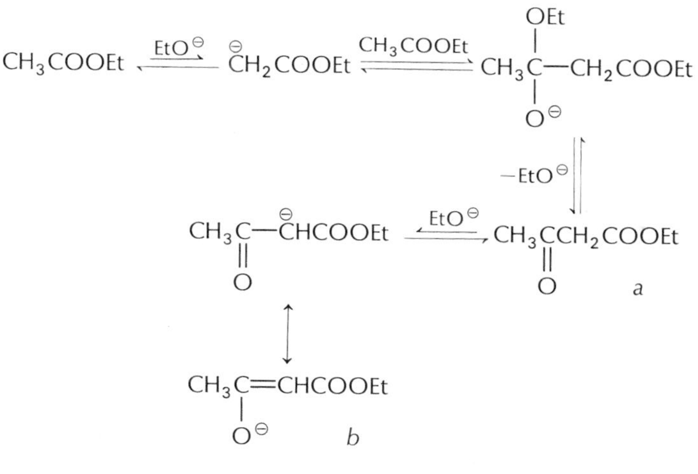

$$CH_3COOEt \xrightleftharpoons{EtO^\ominus} \overset{\ominus}{C}H_2COOEt \xrightleftharpoons{CH_3COOEt} CH_3\underset{\underset{O^\ominus}{|}}{\overset{\overset{OEt}{|}}{C}}-CH_2COOEt$$

$$\Bigg\updownarrow -EtO^\ominus \quad (17.59)$$

$$CH_3\underset{\underset{O}{\|}}{C}-\overset{\ominus}{C}HCOOEt \xleftarrow{EtO^\ominus} CH_3\underset{\underset{O}{\|}}{C}CH_2COOEt \quad a$$

$$\updownarrow$$

$$CH_3\underset{\underset{O^\ominus}{|}}{C}=CHCOOEt \quad b$$

If we write the complete reaction as though one mole of EtONa were prepared and added to an excess of ethyl acetate (Eq. 17.60), we find that two moles of ethanol are needed

$$2CH_3COOEt + EtONa \longrightarrow [CH_3COCHCOOEt]^\ominus Na^\oplus + 2EtOH \quad (17.60)$$

on the right to balance the equation. In other words, ethanol is a product of the reaction. Therefore, if only a trace of ethanol is present, it is possible to add metallic sodium to the mixture to start the reaction. The EtONa produced then would give rise to the AAE anion (as the sodium salt) and produce even more ethanol, which would react with more sodium to give more EtONa in a cyclic process. Thus, we now write the reaction with sodium as shown in Equation 17.61.

$$2CH_3COOEt + 2Na \xrightarrow{reflux} [CH_3COCHCOOEt]^\ominus Na^\oplus + EtONa + H_2 \quad (17.61)$$

It is quite difficult to remove the last traces of ethanol (bp 78.5°) from ethyl acetate (bp 77.1°), and the ester usually contains up to 2 to 3% of ethanol. The reaction usually is carried out by addition of sodium to an excess of ethyl acetate; and the yield, calculated from Equation 17.61 and based upon the sodium added, is 75 to 80%.

Acetoacetic ester is very useful in the synthesis of ketones, a subject we shall defer until the next chapter.

The Claisen condensation, we see, is an equilibrium pulled to the right by salt formation of the acidic product. Other esters may be used if two α hydrogens are present (Eqs. 17.62). If only a single α hydrogen is present, the reaction with NaOEt fails because the

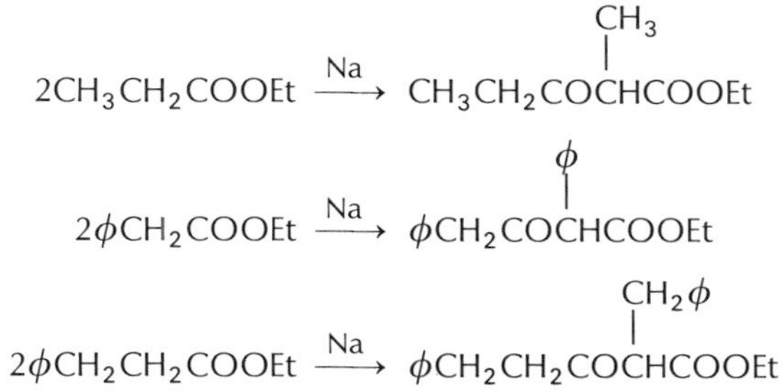

$$2CH_3CH_2COOEt \xrightarrow{Na} CH_3CH_2CO\underset{\underset{CH_3}{|}}{C}HCOOEt$$

$$2\phi CH_2COOEt \xrightarrow{Na} \phi CH_2CO\underset{\underset{\phi}{|}}{C}HCOOEt \quad (17.62)$$

$$2\phi CH_2CH_2COOEt \xrightarrow{Na} \phi CH_2CH_2CO\underset{\underset{CH_2\phi}{|}}{C}HCOOEt$$

product no longer has an acidic proton present (Eq. 17.63).

$$2(CH_3)_2CHCOOEt \xrightarrow[\text{no reaction}]{Na} (CH_3)_2CHCOCCOOEt(CH_3)_2 \quad (17.63)$$

In such a case, a stronger base will induce the reaction to proceed (Eq. 17.64),

$$2(CH_3)_2CHCOOEt + 2\phi_3CNa \longrightarrow (CH_3)_2\overset{\ominus}{C}COC(CH_3)_2COOEt\;Na^{\oplus} + EtONa + 2\phi_3CH \quad (17.64)$$

because now the strong base (triphenylmethyl anion) is able to cause the product to act as an acid.

If the product before work-up is an anion, why does the product anion not add to another carbonyl carbon and continue the condensation to give ever lengthening chains? A little bit of this reaction goes on, but it never gets very far for two reasons. First, the product anion is more stable and less reactive than is the reacting anion. Second, the addition to the carbonyl carbon is subject to steric influences, as are all other nucleophilic additions to the carbonyl carbon; and, even in the simplest case, the product anion is AAE anion, which is so big and bulky that it adds with difficulty as compared with $\overset{\ominus}{C}H_2COOEt$.

Diesters undergo the Claisen condensation. If the chain is of the proper length, the reaction may lead to cyclic keto esters, in which case it is known as the <u>Dieckmann condensation</u> (Eqs. 17.65). Practically, the Dieckmann condensation is limited to the esters

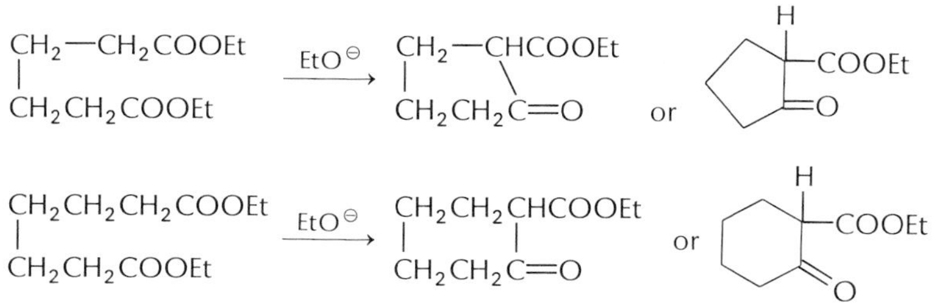

$$\begin{array}{c} CH_2-CH_2COOEt \\ | \\ CH_2CH_2COOEt \end{array} \xrightarrow{EtO^{\ominus}} \begin{array}{c} CH_2-CHCOOEt \\ | \quad \backslash \\ CH_2CH_2C=O \end{array} \text{ or } \begin{array}{c} \text{cyclopentanone with } H, COOEt, =O \end{array} \quad (17.65)$$

$$\begin{array}{c} CH_2CH_2CH_2COOEt \\ | \\ CH_2CH_2COOEt \end{array} \xrightarrow{EtO^{\ominus}} \begin{array}{c} CH_2CH_2CHCOOEt \\ | \quad | \\ CH_2CH_2C=O \end{array} \text{ or } \begin{array}{c} \text{cyclohexanone with } H, COOEt, =O \end{array}$$

of adipic and pimelic acids, which give five and six membered rings. The other diesters tend to give intermolecular rather than intramolecular condensation. Diethyl succinate is the only other ester that provides a useful reaction (Eq. 17.66).

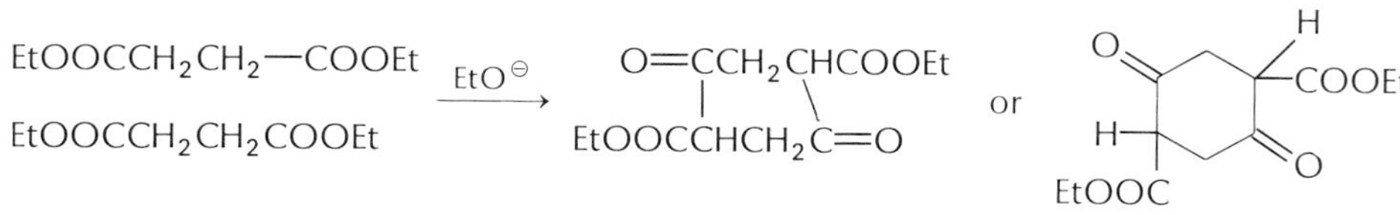

$$\begin{array}{c} EtOOCCH_2CH_2-COOEt \\ EtOOCCH_2CH_2COOEt \end{array} \xrightarrow{EtO^{\ominus}} \begin{array}{c} O=CCH_2CHCOOEt \\ | \quad \backslash \\ EtOOCCHCH_2C=O \end{array} \text{ or } \begin{array}{c} \text{cyclohexanedione with } H, COOEt, EtOOC \end{array}$$

$$(17.66)$$

17.8 Reactions of esters

If we employ two different esters in a <u>crossed Claisen condensation</u>, we may expect to find four different products (Eqs. 17.67); and separation of the two crossed products will

$$CH_3COOEt + \phi CH_2COOEt \xrightarrow{Na} \begin{cases} CH_3COCH_2COOEt \\ CH_3COCH\phi COOEt \\ \phi CH_2COCH_2COOEt \\ \phi CH_2COCH\phi COOEt \end{cases} \quad (17.67)$$

be difficult. However, if one of the esters has no α hydrogens (ethyl formate, oxalate, benzoate, carbonate), only two products are possible. If we add the ester with α hydrogens last to the basic reaction mixture, the ester anion has little choice but to add to the other ester to form the crossed product (Eqs. 17.68). Thus, in every crossed Claisen condensation

$$HCOOEt + CH_3CH_2COOEt \xrightarrow{EtO^\ominus} HCOCHCOOEt \underset{|}{\overset{CH_3}{}} \longrightarrow \text{trimer}$$

$$EtOOCCOOEt + \phi CH_2COOEt \xrightarrow{EtO^\ominus} EtOOCCOCHCOOEt \underset{|}{\overset{\phi}{}}$$

$$\phi COOEt + CH_3COOEt \xrightarrow{EtO^\ominus} \phi COCH_2COOEt \quad (17.68)$$

$$EtOCOOEt + CH_3CH_2CH_2COOEt \xrightarrow{EtO^\ominus} EtOOCCHCOOEt \underset{|}{\overset{CH_2CH_3}{}}$$

$$t\text{BuCOOEt} + CH_3COOEt \xrightarrow{EtO^\ominus} t\text{BuCOCH}_2\text{COOEt}$$

of this type, RCH_2COOEt gives $R'COCHRCOOEt$ as a product from $R'COOEt$ if R' has no α hydrogens ($R' = H-$, $EtOOC-$, $\phi-$, $EtO-$, and $t\text{Bu}-$).

Crossed Dieckmann condensations are possible as well (Eq. 17.69). The reaction is favored by the reluctance of diethyl glutarate to undergo an internal Dieckmann condensation to form a cyclobutanone derivative. Another interesting cyclization is a means of entry into the indane ring system (Eq. 17.70).

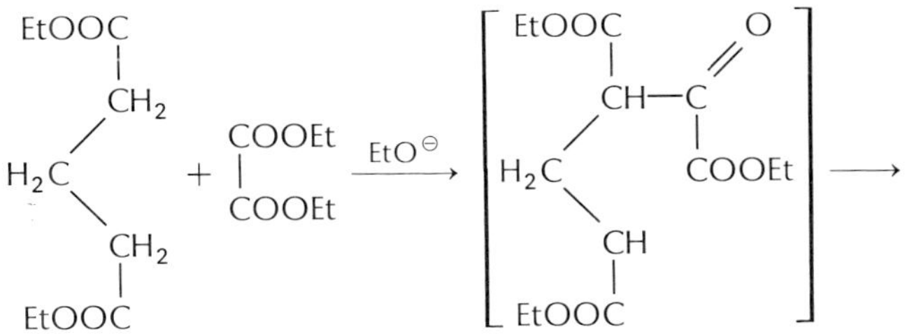

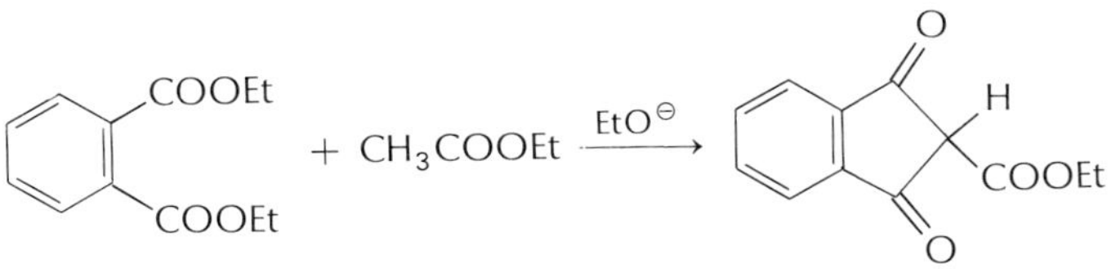

If we go on and ask about the possibility of a mixed condensation of an aldehyde or ketone with an ester, we recognize two possibilities: addition of an aldehyde or ketone anion to the ester carbonyl, which we shall call a mixed aldol condensation; and addition of an ester anion to an aldehyde or ketone, which we shall call a mixed Claisen condensation. Our definition implies that a mixed aldol condensation gives a β-diketone as product and a mixed Claisen condensation gives an α,β-unsaturated ester.

In mixed aldol condensation we can use only methyl ketones with ethyl acetate and with alkoxide as the base, unless we use aromatic compounds. The mixed condensations involve competition, in the general case, between the ester anion and the keto anion for attack on the carbonyl carbons of ester and ketone. Ketones usually are more acidic than are esters, and addition to ester carbonyls is less hindered than that to ketone carbonyls. Thus, mixed aldol condensation is more prevalent than mixed Claisen condensation. Stronger bases ($NaNH_2$ or NaH) allow other ketones and esters to be used. Aldehydes cannot be used successfully, because any anion finds it easier to add to an aldehydic carbon than to an ester carbonyl carbon. Some examples of mixed aldol condensations are shown in Equations 17.71.

$$CH_3COOEt + CH_3COCH_3 \xrightarrow{EtO^\ominus} CH_3COCH_2COCH_3$$

$$\phi COOEt + CH_3CO\phi \xrightarrow{EtO^\ominus} \phi COCH_2CO\phi$$

$$\phi COOEt + CH_3COCH_2CH_2CH_3 \xrightarrow{NaH} \phi COCH_2COCH_2CH_2CH_3$$

$$CH_3CH_2COOEt + \text{(cyclohexanone)} \xrightarrow{NaH} \text{(2-propanoylcyclohexanone)} \quad (17.71)$$

$$CH_3CH_2COOEt + CH_3COCH_2CH_3 \xrightarrow{NaNH_2} CH_3CH_2COCH_2COCH_2CH_3$$

$$EtOOCCOOEt + CH_3COC(CH_3)_3 \xrightarrow{MeO^\ominus} MeOOCCOCH_2COC(CH_3)_3$$
(ester interchange, Me for Et in MeOH)

$$EtOCOOEt + CH_3CH_2CO\phi \xrightarrow{EtO^\ominus} EtOOCCH(CH_3)CO\phi$$

A mixed Claisen condensation of an ester with an aldehyde that has no α hydrogens is easily accomplished (Eq. 17.72), as is shown for the preparation of ethyl cinnamate and

$$ArCHO + CH_3COOEt \xrightarrow[0°]{Na} ArCH=CHCOOEt \quad (17.72)$$

substituted ethyl cinnamates. The Perkin reaction, for example, fails if 2,4,6-trimethylbenzaldehyde is used, but the mixed Claisen condensation gives 70% of ethyl 2,4,6-trimethylcinnamate. Ketones give poor yields, mainly because the order of ease of addition to the carbonyl carbon is aldehyde > ester > ketone. □ □

Malonic ester synthesis. The malonic ester synthesis is primarily a method for putting together acids. However, the variety that is possible requires a bit of description. Malonic ester refers to diethyl malonate. The ester, like acetylacetone, has hydrogens that are α to two carbonyl groups. The effect, while not as great as that in acetylacetone (pK_a 9.0), is still great enough (pK_a 13.5) to cause malonic ester to act as a relatively strong acid to ethoxide ion (Eq. 17.73). The anion is nucleophilic and attacks alkyl halides (allylic, benzylic,

$$\text{EtOOC}\diagdown_{\text{CH}_2}\diagup^{\text{EtOOC}} \xrightarrow{\text{EtO}^\ominus} \begin{array}{c}\text{EtO-C}^{\diagup\!\!\!\!\diagdown O}\\ \text{CH}^\ominus\\ \text{EtOC}_{\diagdown\!\!\!\!\diagup O}\end{array} \longleftrightarrow \begin{array}{c}\text{EtO-C}^{\diagup\!\!\!\!\diagdown O^\ominus}\\ \text{CH}\\ \text{EtOC}_{\diagdown\!\!\!\!\diagup O}\end{array} \longleftrightarrow \begin{array}{c}\text{EtO-C}^{\diagup\!\!\!\!\diagdown O}\\ \text{CH}\\ \text{EtOC}_{\diagdown\!\!\!\!\diagup O^\ominus}\end{array} \quad (17.73)$$

primary > secondary ≫ tertiary), carbonyl carbon atoms, and conjugated double bonds. Each of the reactions leads to a substituted malonic ester. We can then take advantage of the reaction shown as Equation 17.18, the easy decarboxylation of malonic acid.

Let us follow a sequence of reactions that leads to 4-pentenoic acid (Eqs. 17.74).

$$\text{H}_2\text{C(COOEt)}_2 \xrightleftharpoons{\text{EtONa}} \text{H}\overset{\ominus}{\text{C}}\text{(COOEt)}_2 \xrightarrow[a]{\text{CH}_2=\text{CHCH}_2\text{Cl}} \text{H}_2\text{C}=\text{CHCH}_2\text{CH(COOEt)}_2$$

$$\text{H}_2\text{C}=\text{CHCH}_2\text{CH}\diagdown_{\text{COOH}}^{\text{COOH}} \xleftarrow{\text{H}^\oplus} \text{H}_2\text{C}=\text{CHCH}_2\text{CH(COONa)}_2$$

(via dil. H$_2$SO$_4$ reflux, or dil. NaOH reflux *b*)

$$c \downarrow 140° \; -\text{CO}_2$$

$$\text{H}_2\text{C}=\text{CHCH}_2\text{CH}_2\diagup^{\text{COOH}} + \text{CO}_2 \qquad \text{or} \quad \text{RX} + \text{H}_2\text{C(COOEt)}_2$$

$$\downarrow 4 \text{ steps} \quad d.$$

$$\text{RCH}_2\text{COOH}$$

(17.74)

Step *a* is an S$_N$2 substitution reaction. Step *b* may be carried out in two steps by saponification of the ester followed by acidification, which is actually faster than the single-step acidic hydrolysis of the ester. The substituted malonic acid is reasonably stable at room temperature but decarboxylates smoothly at temperatures above 100°. The decarboxylation usually is carried out as a distillation: the product has a lower bp than that of the substituted malonic acid.

Except for step *a*, yields are very good. To see the trouble with the alkylation step, let us look at the situation in the flask when the reaction is halfway through. At the halfway point we have NaCH(COOEt)$_2$, CH$_2$=CHCH$_2$Cl, CH$_2$=CHCH$_2$CH(COOEt)$_2$, and NaCl present. There is nothing to prevent an equilibrium from being set up (Eq. 17.75). The new anion, being a big, bulky tertiary one, is not as good a nucleophile as is the secondary anion. Nevertheless, it begins to compete for the allyl chloride remaining and forms enough (EtOOC)$_2$C(CH$_2$CH=CH$_2$)$_2$ to be a nuisance, both in purification and in lowering the yield of the monoallyl malonic ester.

$$\begin{array}{c} \text{EtOOC} \\ \diagdown \\ \text{CH}^{\ominus} \\ \diagup \\ \text{EtOOC} \end{array} + \begin{array}{c} \text{EtOOC} \quad \text{H} \\ \diagdown \diagup \\ \text{C} \\ \diagup \diagdown \\ \text{EtOOC} \quad \text{CH}_2\text{CH}=\text{CH}_2 \end{array}$$

$$\rightleftharpoons \begin{array}{c} \text{EtOOC} \\ \diagdown \\ \text{CH}_2 \\ \diagup \\ \text{EtOOC} \end{array} + \begin{array}{c} \text{EtOOC} \\ \diagdown \\ \text{C}^{\ominus} \\ \diagup \\ \text{EtOOC} \quad \text{CH}_2\text{CH}=\text{CH}_2 \end{array} \qquad (17.75)$$

On the other hand, the presence of a remaining α hydrogen allows us to introduce a second alkyl group into the sequence. Let us synthesize 2-methylpentanoic acid (Eq. 17.76). We introduce the methyl group last to avoid competition by elimination of HX from the alkyl halide.

$$\text{H}_2\text{C(COOEt)}_2 \xrightarrow{\text{EtONa}} \text{H}\overset{\ominus}{\text{C}}(\text{COOEt})_2 \xrightarrow{\text{PrBr}} \text{PrCH(COOEt)}_2$$

$$\downarrow \text{EtONa}$$

$$\begin{array}{c} \text{Me} \quad \text{COOEt} \\ \diagdown \diagup \\ \text{C} \\ \diagup \diagdown \\ \text{Pr} \quad \text{COOEt} \end{array} \xleftarrow{\text{MeI}} \text{Pr}\overset{\ominus}{\text{C}}(\text{COOEt})_2 \qquad (17.76)$$

$$\downarrow \begin{array}{l} 1. \text{ hydrolysis} \\ 2. \ \Delta, \ -\text{CO}_2 \end{array}$$

$$\text{CH}_3\text{CH}_2\text{CH}_2\text{CHCOOH}$$
$$|$$
$$\text{CH}_3$$

Other variations are possible. One is the use of α-bromoesters (Eq. 17.77) to produce substituted succinic acids.

$$\text{H}_2\text{C(COOEt)}_2 \xrightleftharpoons{\text{EtONa}} \text{H}\overset{\ominus}{\text{C}}(\text{COOEt})_2 \xrightarrow{\text{CH}_3\text{CHBrCOOEt}} \text{CH}_3\text{CHCH(COOEt)}_2$$
$$|$$
$$\text{COOEt}$$

$$\downarrow \begin{array}{l} 1. \text{ hydrolysis} \\ 2. \ \Delta, \ -\text{CO}_2 \end{array} \qquad (17.77)$$

$$\text{CH}_3\text{CHCH}_2\text{COOH}$$
$$|$$
$$\text{COOH}$$

Another is to use an epoxide instead of an alkyl halide to give unsaturated acids (Eq. 17.78).

$$\text{H}_2\text{C(COOEt)}_2 \xrightarrow{\text{EtONa}} \text{H}\overset{\ominus}{\text{C}}(\text{COOEt})_2 \xrightarrow{\text{CH}_3\text{HC}\overset{\text{O}}{\underset{}{\triangle}}\text{CH}_2} \text{CH}_3\text{CHCH}_2\text{CH(COOEt)}_2$$
$$|$$
$$\text{OH}$$

$$\downarrow \begin{array}{l} 1. \text{ hydrolysis} \\ 2. \ \Delta, \ -\text{CO}_2, \ -\text{H}_2\text{O} \end{array} \qquad (17.78)$$

$$\text{CH}_3\text{CH}=\text{CHCH}_2\text{COOH}$$
$$+ \ \text{CH}_3\text{CH}_2\text{CH}=\text{CHCOOH}$$

17.8 Reactions of esters

Still another method is use of a dihalide (Eq. 17.79). If excess EtONa is used, the second step leads to ring closure (even to a cyclopropane ring). If only one mole of EtONa is used per mole of malonic ester, then the substituted alkyl halide may be added to another mole of malonic ester anion; and, as shown, adipic acid is produced after decarboxylation.

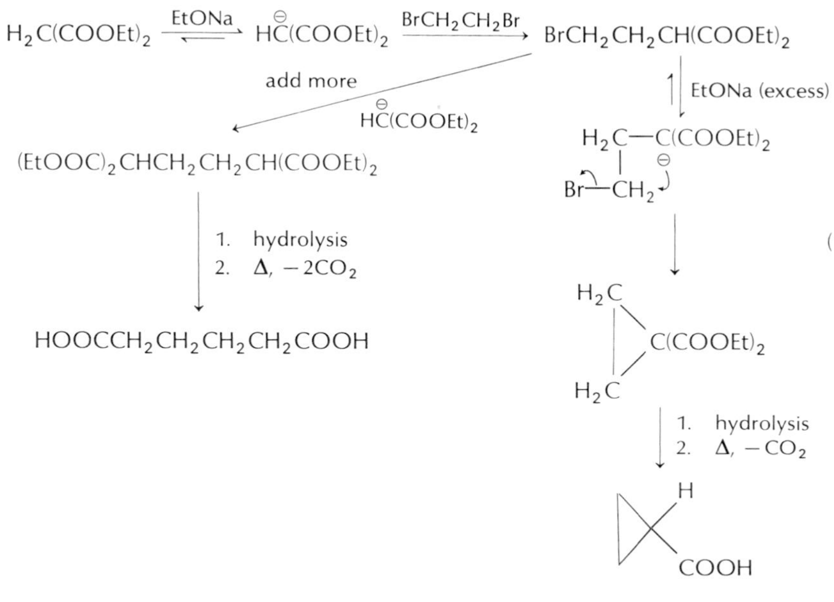

$$H_2C(COOEt)_2 \xrightarrow{EtONa} H\overset{\ominus}{C}(COOEt)_2 \xrightarrow{BrCH_2CH_2Br} BrCH_2CH_2CH(COOEt)_2$$

add more $H\overset{\ominus}{C}(COOEt)_2$ ← | ↑ EtONa (excess)

$(EtOOC)_2CHCH_2CH_2CH(COOEt)_2$

$$\begin{array}{c} H_2C-C(COOEt)_2 \\ | \quad\quad \ominus \\ Br-CH_2 \end{array}$$

(17.79)

1. hydrolysis
2. Δ, −2CO₂

↓

$HOOCCH_2CH_2CH_2CH_2COOH$

$$\begin{array}{c} H_2C \\ \diagdown \\ C(COOEt)_2 \\ \diagup \\ H_2C \end{array}$$

1. hydrolysis
2. Δ, −CO₂

↓

(cyclopropane-COOH with H)

Yet another variation is to use methylene iodide (CH_2I_2; the other methylene halides are less reactive) to give glutaric acid, or acrylic acid.

The last variation in substitutions that we shall give is the use of bromo or iodomalonic ester to produce succinic acid or substituted succinic acids (Eq. 17.80).

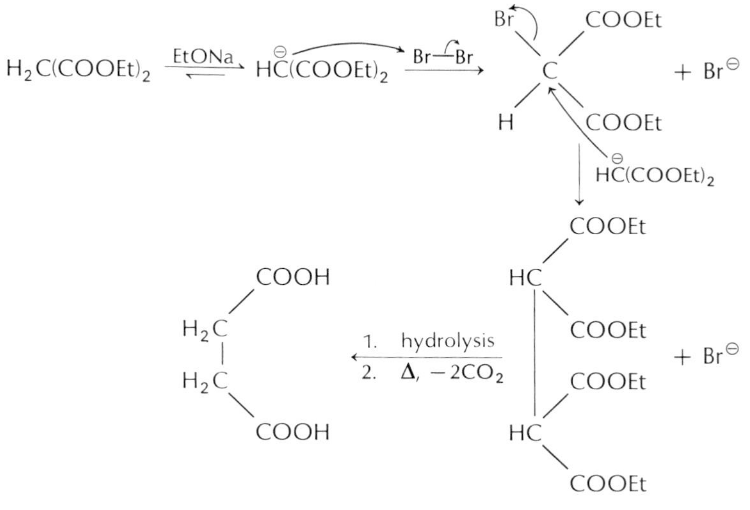

$$H_2C(COOEt)_2 \xrightarrow{EtONa} H\overset{\ominus}{C}(COOEt)_2 \xrightarrow{Br-Br} \begin{array}{c} Br \quad COOEt \\ \diagdown \diagup \\ C \\ \diagup \diagdown \\ H \quad COOEt \end{array} + Br^{\ominus}$$

↓ $H\overset{\ominus}{C}(COOEt)_2$

$$\begin{array}{c} COOEt \\ \diagup \\ HC \\ \diagdown COOEt \\ | \\ \diagup COOEt \\ HC \\ \diagdown \\ COOEt \end{array} + Br^{\ominus}$$

(17.80)

$$\begin{array}{c} COOH \\ \diagup \\ H_2C \\ | \\ H_2C \\ \diagdown \\ COOH \end{array} \xleftarrow[\text{2. } \Delta, -2CO_2]{\text{1. hydrolysis}}$$

Upon bromination, esters ordinarily react in the alkyl rather than in the acyl portion of the molecule. However, the anion of malonic ester (like the anion in basic halogenation of ketones) reacts rapidly to give bromomalonic ester. The first-formed bromomalonic ester then suffers S_N2 attack by the remaining malonic ester anion to give the tetraester (Eq. 17.81). □ □

$$2HC(COOEt)_2 + Br_2 \longrightarrow (EtOOC)_2CHCH(COOEt)_2 \qquad (17.81)$$

Malonic acid derivatives may be used in separately contrived Michael reactions (Eqs. 17.82).

$$\phi CH{=}CHCO\phi \xrightarrow[EtONa]{H_2C(COOEt)_2} \phi CH{-}CH_2CO\phi \atop \underset{HC(COOEt)_2}{|}$$

$$CH_3CH{=}CHCOOEt \xrightarrow[EtONa]{H_2C(COOEt)_2} CH_3CH{-}CH_2COOEt \atop \underset{HC(COOEt)_2}{|} \qquad (17.82)$$

Thus, by proper choice of malonic acid derivative, alkyl halide, aldehyde, ketone, and unsaturated compound with which to carry out a synthesis, a great variety of substituted esters and other derivatives may be constructed. Strictly speaking, <u>malonic ester synthesis</u> refers only to reactions like those shown in Equations 17.74 to 17.80. However, we shall include in the meaning of the term any reaction in which a malonic ester or derivative is a reactant.

Reactions in the alkyl group of an ester. Heated to 300° to 500°, an ester pyrolyzes with loss of an acid to produce an alkene from the alkyl group (Eq. 17.83; see Eq. 13.75).

$$\phi CH_2COOCH_2CH_3 \xrightarrow{500°} \phi CH_2COOH + H_2C{=}CH_2 \qquad (17.83)$$

The mechanism involves a cyclic transition state (Eq. 17.84) and may be symbolized as Ei.

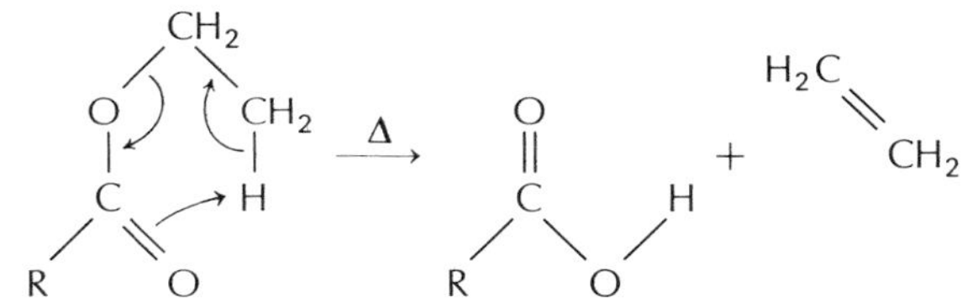

(17.84)

If there is a choice of direction (as there is if the alkyl group is secondary or tertiary), the distribution of products tends to be statistical or to follow the Hofmann rule (Eq. 17.85). The cyclic mechanism requires that a β hydrogen in the alkyl group be able to come into an

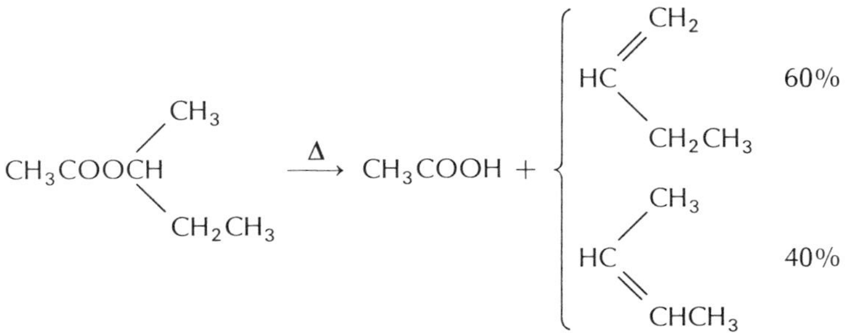

(17.85)

eclipsed conformation (or close to it) to form the lowest energy transition state. Thus, in cyclohexyl esters we find the results to be as shown in Equations 17.86. If the ester group is axial or equatorial, neighboring equatorial hydrogens are subject to attack, whether they are *cis* or *trans* in configuration.

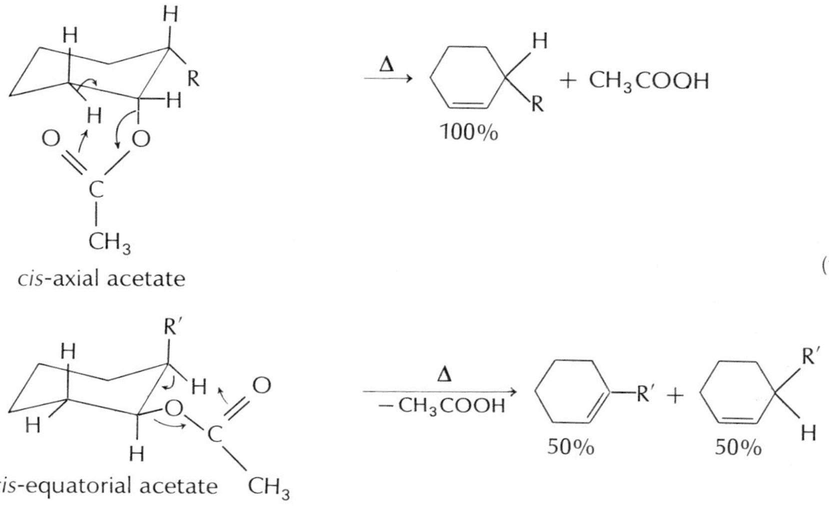

cis-axial acetate → (100%) cyclohexene with R + CH$_3$COOH

(17.86)

cis-equatorial acetate → −CH$_3$COOH → (50%) + (50%) isomers with R'

The pyrolysis of esters, because of the cyclic mechanism, frequently gives a distribution of product alkenes different from the products of acidic dehydration of alcohols or the alkaline dehydrohalogenation of alkyl halides. The Chugaev reaction (Sec. 9.10) is a special case of ester pyrolysis. □ □

Exercises.

17.6 Show how to carry out the conversions:

a. Et$_2$C(COOEt)$_2$ ⟶ [5,5-diethylbarbituric acid: ring with Et, Et on one carbon, two NH groups, three C=O]

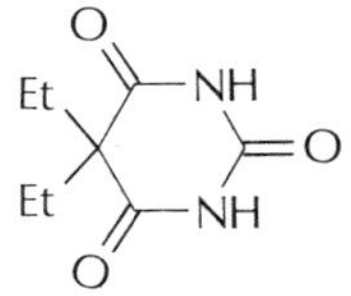

b. φCOOMe ⟶ φCONH$_2$

c. φCOOCH$_2$CH=CHCH$_3$ ⟶ φCOOH + HOCH$_2$CH=CHCH$_3$

17.7 Complete the reactions:
a. φCH$_2$CH$_2$COOEt + CH$_3$MgI ⟶
b. φCH$_2$CH$_2$COOEt + CH$_3$$\overset{\ominus}{C}H\overset{\oplus}{P}$φ$_3$ ⟶
c. CH$_3$CH=CHCOOEt $\xrightarrow{\text{H}_2}{\text{Ni or Pt}}$
d. φCH=CHCOOEt $\xrightarrow{(\text{MeOCH}_2\text{CH}_2\text{O})_2\text{AlH}_2{}^{\ominus}\text{Na}^{\oplus}}{-70°}$

17.8 A Claisen condensation of 275 ml of CH$_3$COOEt with 23 g of sodium gave 51 ml of CH$_3$COCH$_2$COOEt. Calculate the percent yield.

580   Chapter seventeen

17.9   Use a Claisen (or a crossed Claisen) condensation to prepare:

a. tBuCOCHCOOEt
        |
     CH$_2$CH$_3$

b. EtOOCCHCOOEt
      |
   CH$_2\phi$

c. OHCCHCOOEt
    |
 CH$_3$CHCH$_3$

d. EtOOCCOCHCOOEt
          |
     CH$_2$tBu

e. ϕCOCHCOOEt
      |
    ϕ

17.10   Write equations for:
a. a mixed aldol condensation to give ϕCOCHϕCOOEt
b. a mixed Claisen condensation to give tBuCH=CϕCOOEt
c. a malonic ester synthesis of ϕCH=CHCH$_2$COOH
d. a malonic ester synthesis of 1,2-cyclopentanedicarboxylic acid

17.11   Using CH$_3$CH$_2$CHBrCOOEt, write equations illustrating:
a. Reformatsky reaction
b. ylide reaction

17.9   Preparation of esters

Almost all the methods for the preparation of esters have already been described. A widely used method, the acetoacetic ester synthesis, will be taken up in the next chapter. We shall list the reactions under separate headings.

Olysis reactions.

1. The Fischer esterification (alcoholysis of an acid) has been described. Being the reverse of hydrolysis of an ester, it is subject to the same influences in coming to equilibrium (Eqs. 17.47).
2. Transesterification has been given (Eqs. 17.48).
3. Esters are prepared quickly and conveniently by alcoholysis of acid chlorides and anhydrides. Even *tert*-butyl mesitoate may be formed (Eq. 17.87).

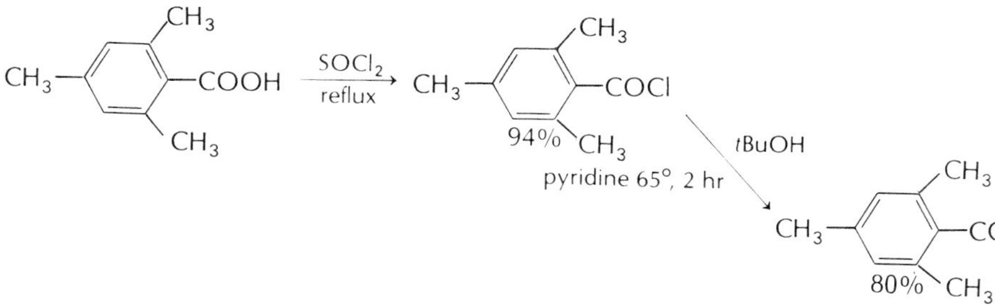

$$\text{mesitylene-COOH} \xrightarrow[\text{reflux}]{\text{SOCl}_2} \text{mesitylene-COCl} \xrightarrow[\text{pyridine 65°, 2 hr}]{\text{tBuOH}} \text{mesitylene-COO}t\text{Bu} \quad (17.87)$$

94%, 80%

4. Nitriles undergo alcoholysis to give esters (Eq. 17.88).

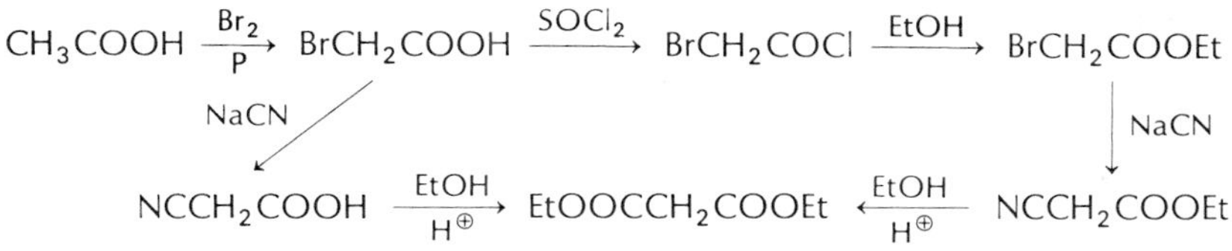

17.9 *Preparation of esters* 581

$$CH_3COOH \xrightarrow[P]{Br_2} BrCH_2COOH \xrightarrow{SOCl_2} BrCH_2COCl \xrightarrow{EtOH} BrCH_2COOEt$$

$$\downarrow NaCN \qquad\qquad\qquad\qquad\qquad\qquad\qquad \downarrow NaCN \qquad (17.88)$$

$$NCCH_2COOH \xrightarrow[H^\oplus]{EtOH} EtOOCCH_2COOEt \xleftarrow[H^\oplus]{EtOH} NCCH_2COOEt$$

5. New esterification reactions are the use of dicyclohexylcarbodiimide (DCC) or of carbonyldiimidazole (CDI) to bring about esterification. Both reagents may be thought of as mild dehydrating substances that cause esterification to become irreversible. A probable mechanism for dicyclohexylcarbodiimide is shown in Equation 17.89. The dicyclohexylurea is a very good leaving group, so the process of taking up the molecule of alcohol may be thought of as concerted.

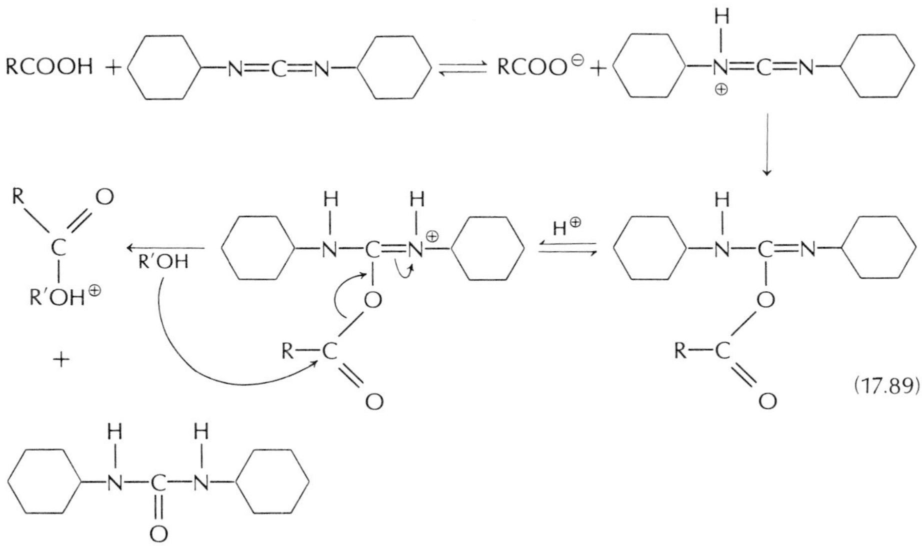

(17.89)

The action of CDI is similar in that imidazole also is a very good leaving group (Eq. 17.90).

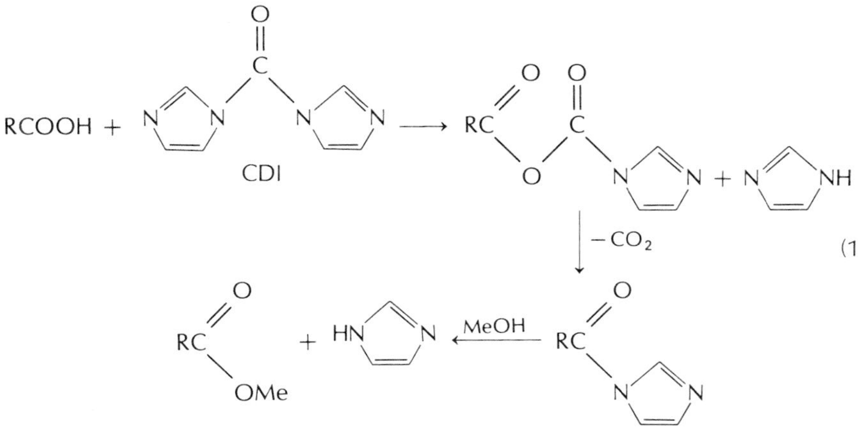

(17.90)

Both of these gentle newer methods have been adopted quickly by chemists working in the field of proteins, amino acids, and other sensitive naturally occurring substances.

Substitution and addition reactions. The substitution reactions may be listed as follows:

1. Diazomethane with acids (Eq. 17.8).
2. The silver salt method (Eq. 17.9) is nothing more than an assisted S_N2 displacement in which alkyl halides primary > secondary ≫ tertiary are used. It was originally devised for the preparation of esters of highly hindered acids, because it has no tetrahedral intermediate (Eq. 17.91) and is irreversible.

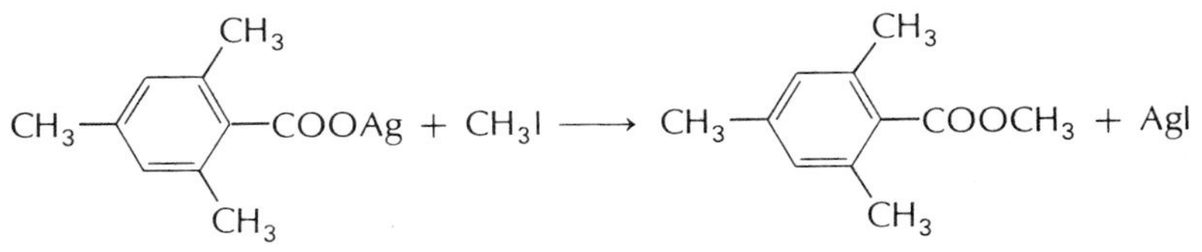

CH3
CH3
COOAg + CH3I
CH3
CH3
CH3
COOCH3 + AgI
CH3

(17.91)

3. The malonic ester synthesis (Sec. 17.8).

Addition reactions include the various condensation reactions, such as the Claisen, crossed Claisen, Dieckmann, Reformatsky, Knoevenagel, Michael, and malonic ester syntheses (Sec. 17.8).

Oxidation and reduction reactions. Under oxidations we mention the Baeyer-Villiger reaction of ketones to give esters or lactones (Eq. 16.67).

Reductions include:

1. Treatment of cyclic anhydrides with zinc dust and acetic acid, the calculated amount of $LiAlH_4$, or hydrogen over Pt (Eq. 17.92).

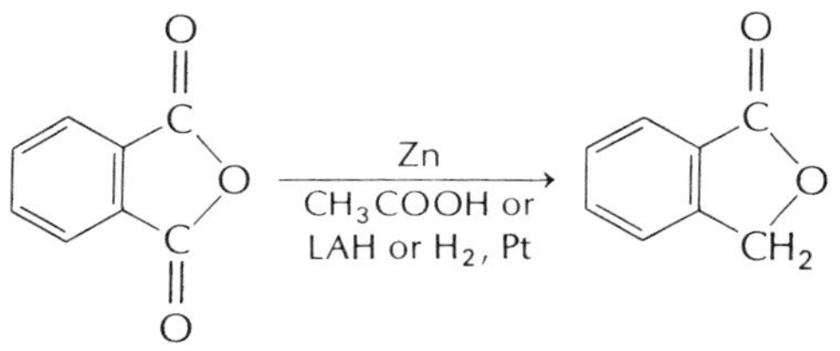

Zn
CH3COOH or
LAH or H2, Pt
CH2

(17.92)

2. The Tishchenko reaction of aldehydes with aluminum ethoxide. The reaction formally resembles a Cannizzaro reaction, except that the mild reagent allows aldehydes with α hydrogens to be used, and the acid and alcohol are obtained combined as the ester (Eq. 17.93).

$$2(CH_3)_2CHCHO \xrightarrow{Al(OEt)_3} (CH_3)_2CHCOOCH_2CH(CH_3)_2 \qquad (17.93)$$

Ortho esters.

1. The Williamson ether synthesis may be used in the syntheses of those ortho esters in which no α hydrogen is present. If hydrogen is present on the carbon

atom adjacent to the trichloromethyl group ($\diagdown$CHCCl$_3\diagup$), the presence of the strong base (MeO$^\ominus$, EtO$^\ominus$) is likely to cause dehydrohalogenation rather than replacement. Equation 17.94 shows the preparation of triethyl orthoformate.

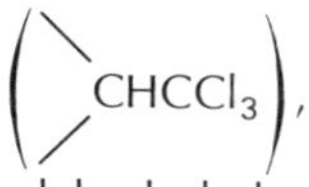

$$HCCl_3 + EtOH \xrightarrow{Na} HC(OEt)_3 + NaCl \qquad (17.94)$$

2. A more generally applicable synthesis starts with a nitrile. An anhydrous mixture of a nitrile, an alcohol, and HCl gives an imino ester salt, which, in the presence of excess alcohol, reacts to form an ortho ester (Eq. 17.95).

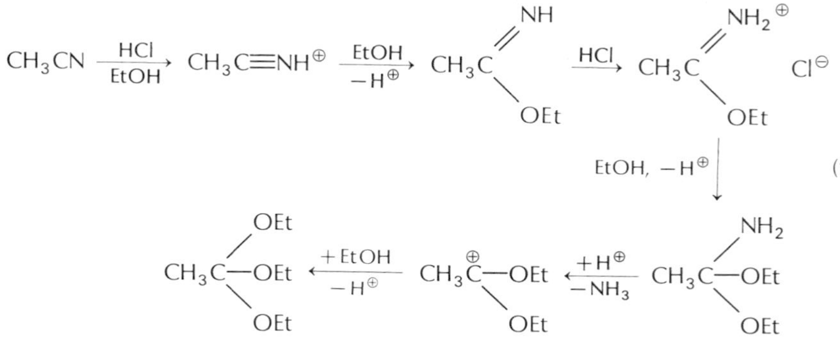

$$CH_3CN \xrightarrow[EtOH]{HCl} CH_3C\!\equiv\!NH^\oplus \xrightarrow[-H^\oplus]{EtOH} CH_3C\!\!\begin{array}{c}\diagup NH\\ \diagdown OEt\end{array} \xrightarrow{HCl} CH_3C\!\!\begin{array}{c}\diagup NH_2^\oplus\\ \diagdown OEt\end{array} Cl^\ominus$$

$$\Big\downarrow EtOH, -H^\oplus \qquad (17.95)$$

$$CH_3C\!\!\begin{array}{c}\diagup OEt\\ -\!OEt\\ \diagdown OEt\end{array} \xleftarrow[-H^\oplus]{+EtOH} CH_3\overset{\oplus}{C}\!\!\begin{array}{c}-OEt\\ \diagdown OEt\end{array} \xleftarrow[-NH_3]{+H^\oplus} CH_3C\!\!\begin{array}{c}\diagup NH_2\\ -OEt\\ \diagdown OEt\end{array}$$

Exercises

17.12 Prepare a methyl ester from each of the substances given.
  a. ϕCOOH
  b. MesCOOH
  c. ϕCOCl
  d. (CH$_3$CO)$_2$O
  e. CH$_3$CH$_2$CH$_2$COOEt
  f. CH$_3$CH$_2$CH=CH$_2$

17.10 Summary

Again we are faced with the task of summarizing a chapter with many diverse subjects and reactions. Although it is not as long as the preceding chapter, we still must choose what to stress in our study and what to go over lightly. Naturally, those majoring in chemistry, biochemistry, and chemical engineering will expect to master more of the subject matter than will others. On the other hand, we are beginning to reach the parts of organic chemistry that are of greater applicability and interest to biology, premedical, and predental majors.

In constructing your own summary and list of reactions, a good starting point is to go back and go through the table of contents at the head of the chapter. This table is the shortest possible outline and may be expanded at will by inclusion of as many equations and notes as needed.

584   Chapter seventeen

17.11   Problems

17.1 Substance A is prepared by a crossed Dieckmann condensation from diethyl oxalate and diethyl 3,3-dimethylglutarate. A with one mole each of EtONa and CH_3I gives B. Reduction of B with sodium and alcohol gives, after work-up, C ($C_{10}H_{16}O_6$). Treatment of C with hydriodic acid causes a dehydration, replacement of an HO— with I—, and reduction of the C—I to C—H to give D ($C_{10}H_{14}O_4$). D adds HBr, and C—Br is reduced to C—H with Zn dust in acetic acid to give E ($C_{10}H_{16}O_4$). Treatment of E with acetyl chloride gives F ($C_{10}H_{14}O_3$). Treatment of F with sodium amalgam and water reduces the least hindered carbonyl group to give G ($C_{10}H_{16}O_2$). G reacts with NaCN to give H ($C_{11}H_{17}O_2N$), which upon acidic hydrolysis gives I ($C_{11}H_{18}O_4$). Pyrolysis of the calcium salt of I gives J ($C_{10}H_{16}O$), a well known natural product.
Write a structural formula for each lettered substance.

17.2 Indicate which member of each pair is more stable (lower energy) and why.

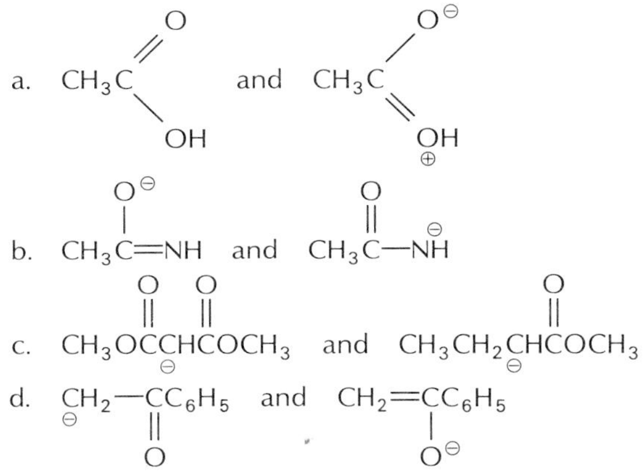

a. $CH_3C(=O)OH$ and $CH_3C(=\overset{\oplus}{O}H)\overset{\ominus}{O}$

b. $CH_3C(\overset{\ominus}{O})=NH$ and $CH_3C(=O)\overset{\ominus}{NH}$

c. $CH_3OCCHCOCH_3$ (with O's double-bonded and $\ominus$ on central CH) and $CH_3CH_2\overset{\ominus}{C}HCOCH_3$

d. $CH_2\overset{\ominus}{-}\underset{\parallel}{C}C_6H_5$ and $CH_2=CC_6H_5$
       $\quad\quad O$ $\quad\quad\overset{\ominus}{O}$

17.3 Which is the stronger acid or stronger base of the following pairs? Use the appropriate dipole or resonance structure to explain why.

a. $ClCH_2CH_2COOH$ and $CH_3CHClCOOH$

b. NO_2CH_2COOH and CH_3COOH

c. $CH_3COO^\ominus$ and $HCOO^\ominus$

d. $CH_3CO\overset{\ominus}{N}H$ and $CH_3CH_2\overset{\ominus}{N}H$

e. $CH_3\underset{\parallel}{C}CH_2C_6H_5$ and $CH_3\underset{\parallel}{C}CH_2CH_2C_6H_5$
     $\quad O$ $\quad O$

17.4 Write the major organic products in the following reactions:

a. $(CH_3)_2CHCOOH + HOCH_2CH(CH_3)_2 \xrightarrow{H^\oplus}$

b. $(CH_3)_2CHCOOH + HOC(CH_3)_3 \xrightarrow{H^\oplus}$

c. $(CH_3)_2CHCOOH + CH_2=C(CH_3)_2 \xrightarrow{H^\oplus}$

d. $C_6H_5COOH + H_2NC_2H_5 \xrightarrow{220°}$

e. $C_6H_5CH_2CH_2OH$ + [3,5-dinitrobenzoyl chloride] $\xrightarrow{\text{pyridine}}$

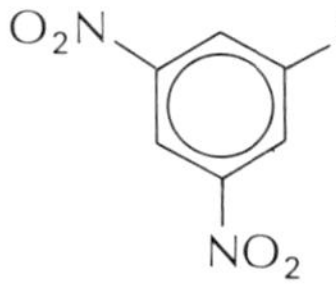

f. 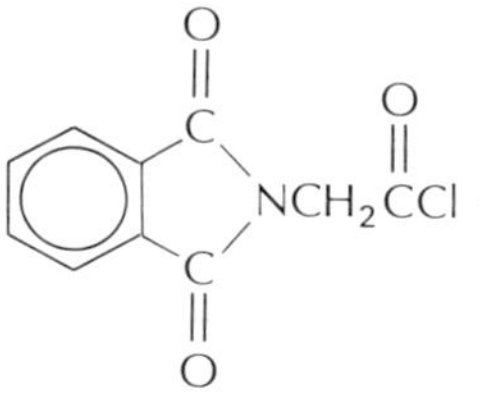 [phthalimide]-$NCH_2\overset{O}{\overset{\|}{C}}Cl$ + $HOC(CH_3)_3$ $\xrightarrow{\text{pyridine}}$

g. $C_6H_5CH_2\underset{\underset{\overset{\|}{O}}{(CH_3)_3CO\overset{}{C}NH}}{CH}\overset{O}{\overset{\|}{C}}NHCH_2COOH$ $\xrightarrow{H^\oplus}$

h. $C_2H_5OOC(CH_2)_4COCl$ + H_2 $\xrightarrow[\text{"quinoline-S"}]{\text{Pd-BaSO}_4}$

i. $C_6H_5CH_2CH_2COOH$ + Br_2 $\xrightarrow{\text{trace P}}$

j. $(CH_3)_2CHCH_2COOC_2H_5$ + $C_6H_5COOC_2H_5$ $\xrightarrow{{}^\ominus OC_2H_5}$

k. $(CH_3)_2CHCOOC_2H_5$ + $C_6H_5COOC_2H_5$ $\xrightarrow{{}^\ominus OC_2H_5}$

l. $C_6H_5CH_2C{\equiv}N$ + $HCOOC_2H_5$ $\xrightarrow{{}^\ominus OC_2H_5}$

m. $CH_3(CH_2)_2COOC_2H_5$ + $CH_3COC_6H_5$ $\xrightarrow{{}^\ominus OC_2H_5}$

n. $2CH_3COC_6H_5$ $\xrightarrow{{}^\ominus OC_2H_5}$

o. $CH_3OOC(CH_2)_5COOCH_3$ $\xrightarrow{{}^\ominus OCH_3}$

p. 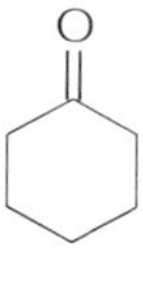 [cyclohexanone] + $CH_3O\overset{O}{\overset{\|}{C}}OCH_3$ $\xrightarrow{{}^\ominus OCH_3}$

17.5 Using any organic or inorganic reagents of your own choosing, convert ethyl acetate to the following products:
a. CH_3COOH
b. $(CH_3)_3COH$
c. $CH_3CONHCH_2C_6H_5$
d. $\begin{array}{l}CH_2OCOCH_3\\|\\CHOCOCH_3\\|\\CH_2OCOCH_3\end{array}$
e. $C_2H_5OOCCOCH_2COOC_2H_5$
f. $CH_2(COOC_2H_5)_2$

g. $Cl-\underset{}{\bigcirc}COCH_2COCH_3$

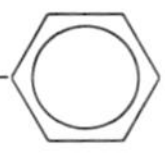

586 Chapter seventeen

17.6 Write a plausible mechanism for each of the following reactions.

a. 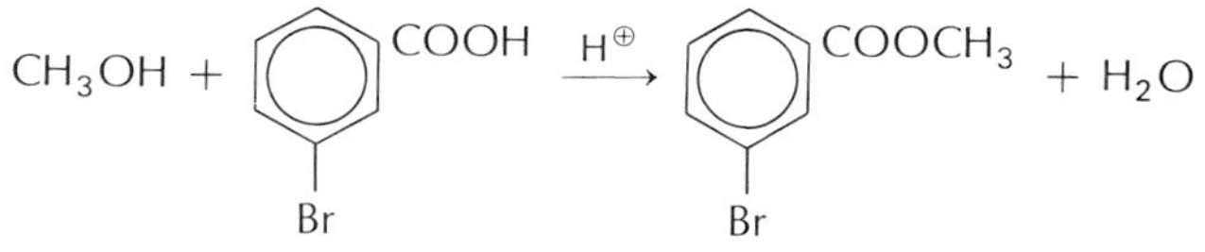 CH_3OH + (3-bromobenzoic acid, COOH) $\xrightarrow{H^{\oplus}}$ (3-bromo methyl benzoate, COOCH$_3$) + H_2O

b. (methyl 3-bromobenzoate, COOCH$_3$) + $^{\ominus}OH$ ⟶ (3-bromobenzoate, COO$^{\ominus}$) + CH_3OH

c. (methyl 3-bromobenzoate, COOCH$_3$) + $(CH_3)_2NH$ ⟶ (3-bromo-N,N-dimethylbenzamide, CON(CH$_3$)$_2$) + CH_3OH

d. (methyl 3-bromobenzoate, COOCH$_3$) + $C_6H_5CH_2OH$ $\xrightarrow[\ominus OCH_3]{trace}$ (benzyl 3-bromobenzoate, COOCH$_2C_6H_5$) + CH_3OH

e. $(CH_3)_2N$—(C$_6$H$_4$)—$COO^{\ominus}$ + $ClCOC_2H_5$ (with C=O)

⟶ $(CH_3)_2N$—(C$_6$H$_4$)—$COCOC_2H_5$ (with two C=O) + $Cl^{\ominus}$

f. $CH_3OPOPO^{\ominus}$ (with two C=O, $^{\ominus}O$, OH substituents) + $CH_3CO^{\ominus}$ (with C=O) $\xrightarrow{enzyme}$ CH_3COP—OH (with two C=O, O$^{\ominus}$) + $CH_3OPO^{\ominus}$ (with C=O, OH)

g. CH_3COCl + HO—(3-nitrophenyl, NO$_2$) ⟶ CH_3CO—O—(3-nitrophenyl, NO$_2$) + HCl

h. $C_6H_5CH_2COOCH_3$ + $CH_3OOCCOOCH_3$
$\xrightarrow[2. \ H_3O^{\oplus}]{1. \ ^{\ominus}OCH_3}$ $C_6H_5CHCOOCH_3$ + CH_3OH
with substituent O=CCOOCH$_3$

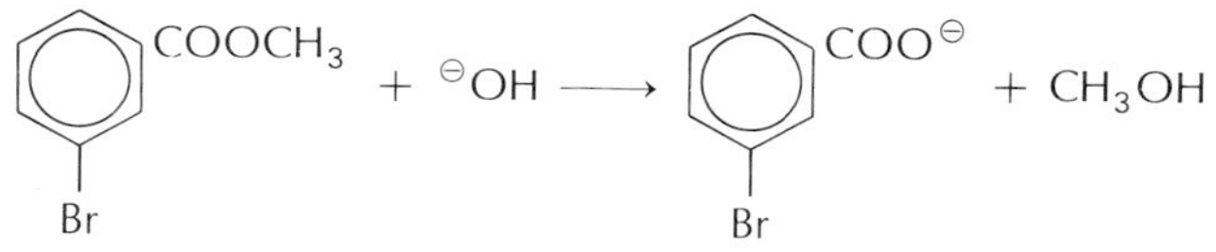

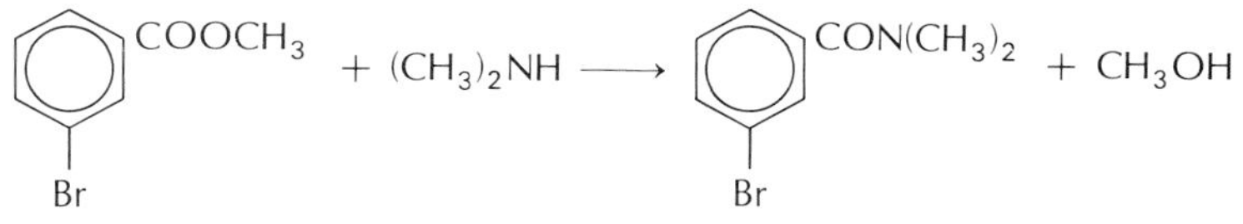

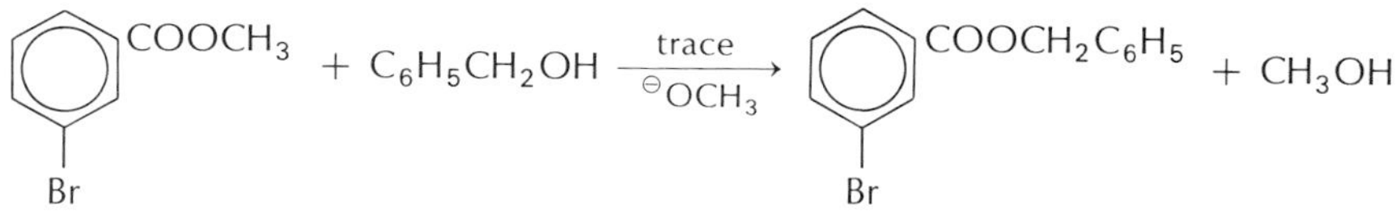

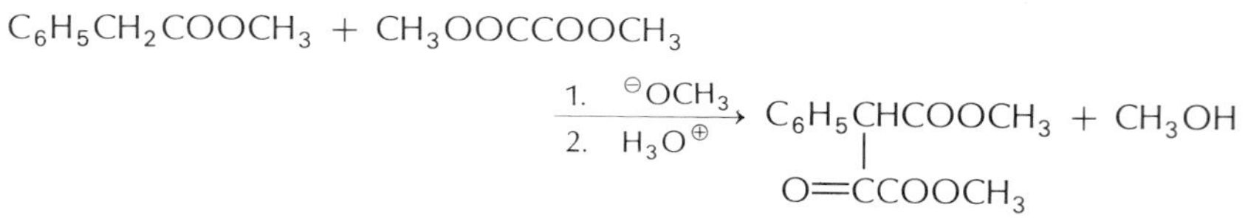

i. 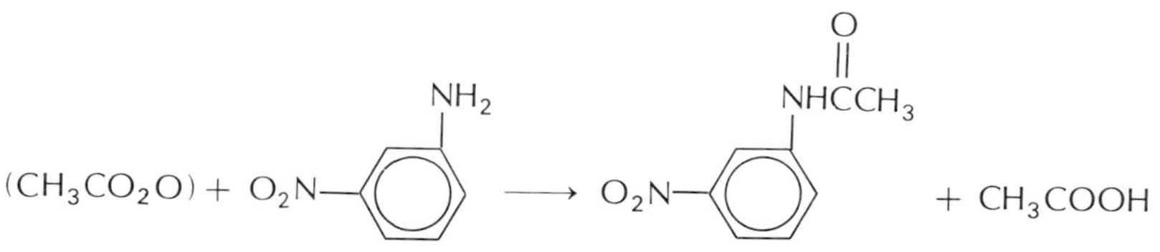 (CH_3CO_2O) + O_2N—(aniline with NH$_2$) ⟶ O_2N—(with NHCCH$_3$, C=O) + CH_3COOH

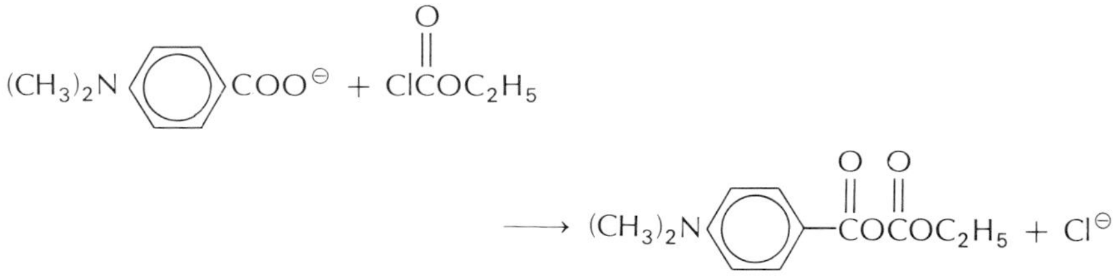

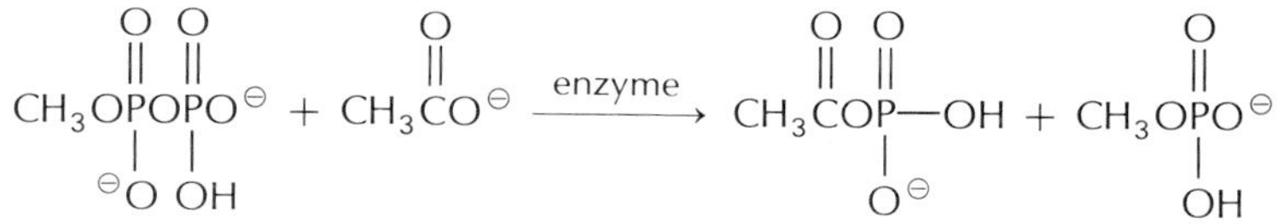

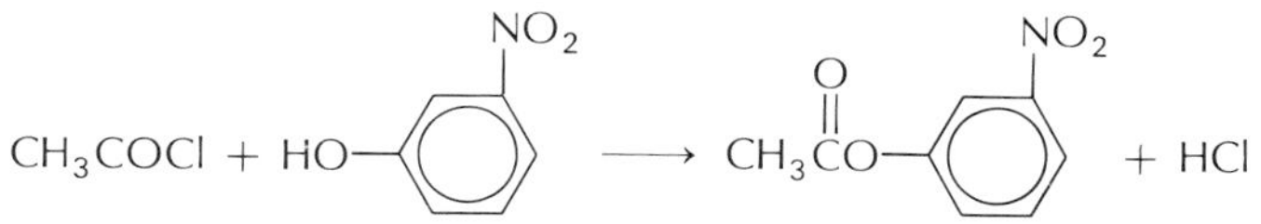

17.7 In Problem 17.6a, 80% of the *m*-bromobenzoic acid is converted to the methyl ester at equilibrium when the starting ratio of the carboxylic acid to methanol is 1:4.
    a. What is the equilibrium constant?
    b. What percentage of the carboxylic acid is converted to ester if the ratio of the acid to methanol is 1:10?

17.8 In each of the following pairs, which one will react more rapidly with the given reagent? Why?

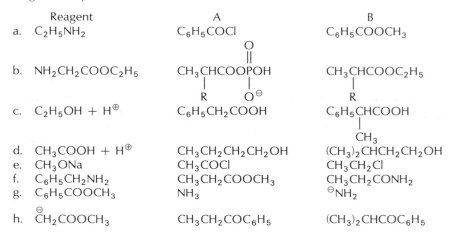

17.12 Bibliography

See Gutsche (Chap. 16, ref. 1), House (Chap. 13, ref. 1), and Gould (Chap. 8, ref. 3), in particular. The general textbooks also are helpful. An excellent review article is given here.

1. Wolff, Ivan A., Seed Lipids, *Science* **154**, 1140 (1966).

18 Substituted Acids and Derivatives

Having been through the acids and derivatives, we go on to look at hydroxy, unsaturated, and keto acids. We shall emphasize either the independence of the reactive centers or the effects of interaction between them. We shall spend more time on the synthetic uses of acetoacetic ester than on any other single topic. However, we shall describe a few compounds of natural origin and physiological effects as well.

18.1   Hydroxy acids and their reactions   589
    1. Introduction; α-hydroxy acids and lactides.   2. β-Hydroxy acids: dehydration.   3. γ- and δ-hydroxy acids: lactones. Exercises.

18.2   Preparation of hydroxy acids   593
    Natural occurrence; list of syntheses. Exercises.

18.3   Reactions of unsaturated acids   594
    1. Occurrence.   2. Sorbic acid.   3. Addition reactions.   4. Stereochemistry. Exercises.

18.4   Preparation of unsaturated acids   596
    Summary of synthetic methods. Exercises.

18.5   Reactions of keto acids   597
    1. α-Keto acids.   2. Acetoacetic ester: ketone and acid cleavages.   3. Alkylation and cleavage.   4. Variations.   5. Use of dihalides.   6. Acylation.   7. Enolization. Exercises.

18.6   Preparation of keto acids   604
    1. Preparation of α-keto acids.   2. Preparation of β-keto acids. Exercises.

18.7   Several interesting esters   606
    1. Pyrethrin.   2. Juvenile hormone.   3. Queen bee substance.   4. Schizophrenic sweat odor; prostaglandins.

18.8   Summary   608

18.9   Problems   609

18.10 Bibliography   612

18.11 D. Stanley Tarbell: A tough antibiotic to unravel   612a

18.1   Hydroxy acids and their reactions

Introduction; α-hydroxy acids and lactides.   Various hydroxy, unsaturated, and keto acids and their derivatives are widespread in nature. Many of the naturally occurring substances combine several substituents in one molecule. It will be convenient, in this chapter, to describe derivatives along with the acids and to show certain syntheses at the same time.

Hydroxy acids may be classified according to the location of the hydroxy group with relation to the carboxyl group. An α-hydroxy acid reacts differently from a β- or δ-hydroxy acid. Table 18.1 shows a random selection of hydroxy acids, many of which have asymmetric centers.

The α-hydroxy acids undergo a bimolecular esterification. The cyclic diester is called a lactide, after lactic acid (Eq. 18.1). The six-membered ring is quite stable. However, continued application of heat causes ring opening to give polymeric esters such as

Table 18.1   Some hydroxy acids

Name	Structure	mp, °C	bp, °C	pK_a
glycolic	$HOCH_2COOH$	80	dec.	3.83
(+)-lactic	$CH_3\overset{*}{C}HOHCOOH$	53	dec.	3.79
(±)-lactic		18	dec.	3.86
(±)-glyceric	$HOCH_2\overset{*}{C}HOHCOOH$	syrup	dec.	3.55
(±)-β-hydroxybutyric	$CH_3\overset{*}{C}HOHCH_2COOH$	50	dec.	4.41
γ-hydroxybutyric	$HOCH_2CH_2CH_2COOH$	< −17	dec.	4.72
(−)-malic	$HOOCCH_2\overset{*}{C}HOHCOOH$	100	140 dec.	3.40, 5.11
(+)-tartaric	$HOOC\overset{*}{C}HOH\overset{*}{C}HOHCOOH$	170	—	2.93, 4.23
(±)-tartaric		206	—	2.96, 4.24
meso-tartaric		140	—	3.11, 4.80
mevalonic	$HOCH_2CH_2\underset{\underset{CH_3}{\mid}}{C(OH)}CH_2COOH$	oil	dec.	—
citric	$HOOCCH_2\underset{\underset{COOH}{\mid}}{C(OH)}CH_2COOH$	153	dec.	3.09, 4.75, 5.41
(+)-mandelic	$\phi\overset{*}{C}HOHCOOH$	134	—	—
(±)-mandelic		119	—	3.37
benzilic	$\phi_2C(OH)COOH$	150	180 dec.	3.04

$HO(CH_2COO)_nCH_2COOH$ (mp 223°), polyglycolide from glycolic acid. Distillation of these polymeric esters under reduced pressure gives back the dimeric cyclic lactides.

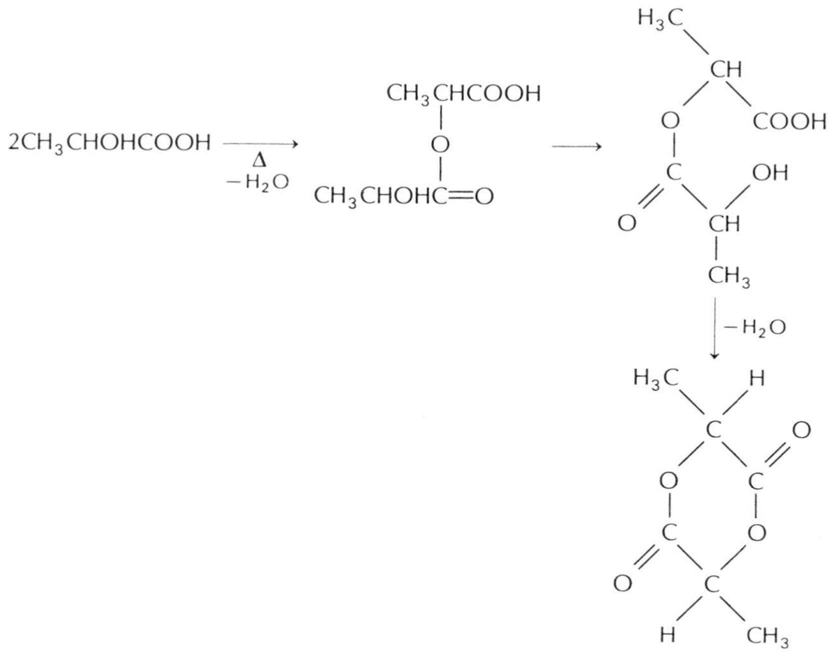

$$2CH_3CHOHCOOH \xrightarrow[-H_2O]{\Delta} \begin{matrix} CH_3CHCOOH \\ | \\ O \\ | \\ CH_3CHOHC\!=\!O \end{matrix} \longrightarrow \begin{matrix} H_3C\diagdown \\ CH \\ \diagup \diagdown \\ O \quad COOH \\ | \quad \diagup \\ C \quad OH \\ \diagup\!\!\!= \diagdown \diagup \\ O \quad CH \\ | \\ CH_3 \end{matrix}$$

$\Big\downarrow -H_2O$ (18.1)

(±) form, mp 125° lactide

Acids with an α-hydroxy group decarbonylate in the presence of strong acids, probably by the mechanism shown in Equations 18.2. The loss of CO gives rise to a stable

$$RCHOHCOOH \xrightarrow[0°]{H_2SO_4} RCHO + CO + H_2O$$

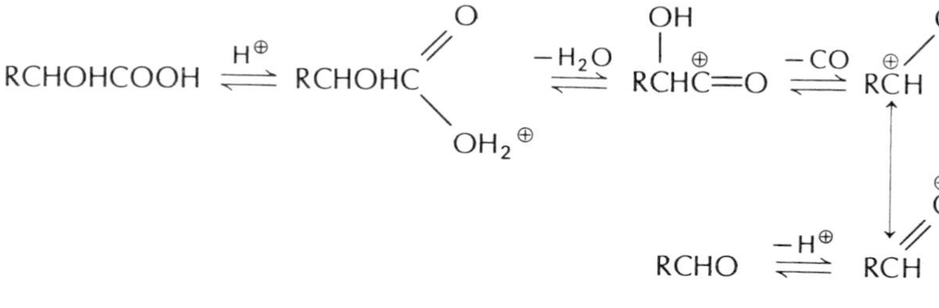

$$RCHOHCOOH \underset{}{\overset{H^\oplus}{\rightleftharpoons}} RCHOHC\!\!\diagup\!\!\overset{O}{\underset{OH_2^\oplus}{\diagdown}} \overset{-H_2O}{\rightleftharpoons} R\overset{OH}{\underset{}{C}}H\!\!\overset{\oplus}{=}\!O \overset{-CO}{\rightleftharpoons} R\overset{OH}{\underset{}{C}}H^\oplus \qquad (18.2)$$

$$RCHO \overset{-H^\oplus}{\rightleftharpoons} R\overset{\oplus}{C}H\!\!=\!\!\overset{OH}{}$$

ion, a protonated aldehyde or ketone, which helps to lower the activation energy for the breaking of the C—C σ bond. The reaction is not very satisfactory if the product is an aldehyde with α hydrogens, because acid-catalyzed cyclization and condensation reactions of the aldehyde come into play (Chap. 16).

β-Hydroxy acids: dehydration. Whereas the α-hydroxy acids resist dehydration to give α,β-unsaturated acids (or prefer cyclic and polymeric ester formation), the β-hydroxy acids are unusually susceptible to dehydration (Eq. 18.3).

$$CH_3CHOHCH_2COOH \xrightarrow[H^\oplus, HO^\ominus \text{ or } \Delta]{-H_2O} CH_3CH\!=\!CHCOOH \qquad (18.3)$$

18.1 Hydroxy acids and their reactions

A relatively rare β-hydroxy acid, 3-hydroxydecanoic acid, is secreted by some strains of ants. The substance is active as a herbicide and antibiotic. One of the puzzles of pharmacology is the relationship between structure and physiological activity. Thus, 3-hydroxyoctanoic and 3-hydroxyhexanoic acids seem to be inactive. On the other hand, 10-undecenoic acid (undecylenic acid) is used to combat fungus skin infections (athlete's foot). Is the β-hydroxy acid the active entity, or is it the corresponding α,β-unsaturated acid? (See the discussion of queen bee substance in Sec. 18.7.)

The internal cyclic ester of a β-hydroxy acid, called a β-lactone, is an easily opened four-membered ring (Eq. 18.4). Depending upon the reagent and conditions, the ring is

$$
\begin{array}{c}
H_2C-C=O \\
| \quad | \\
H_2C-O
\end{array}
\begin{cases}
\xrightarrow[CH_3CN]{NH_3} & H_3\overset{\oplus}{N}-CH_2-CH_2COO^{\ominus} \xrightarrow[S_N2]{warm, -NH_3} HC-COOH \\
& \| \\
& CH_2 \\
\xrightarrow{conc.\ HCl} & Cl-CH_2-CH_2COOH \xrightarrow[S_N2]{warm, -HCl} CH_2 \\
\xrightarrow[CH_3OH]{CH_3O^{\ominus}} & HO-CH_2-CH_2COOCH_3 \xrightarrow{warm, -H_2O} HC-COOCH_3 \\
& \| \\
& CH_2
\end{cases}
\quad (18.4)
$$

opened by S_N2 attack at the β position or by attack on the carbonyl carbon. The β-lactone is made by the cyclization of ketene with formaldehyde.

γ- and δ-hydroxy acids: lactones. The γ- and δ-hydroxy acids react to give the five- and six-membered rings of γ- and δ-lactones so readily that it is difficult to isolate the free acids (Eqs. 18.5). If the hydroxyl group is still more distant from the carboxylic acid

$$
\begin{array}{c}
RCHOH\ COOH \\
| \quad \quad | \\
H_2C-CH_2
\end{array}
\xrightarrow{-H_2O}
\begin{array}{c}
RCH \quad \quad C \\
\diagdown O \diagup \diagdown O \\
H_2C-CH_2
\end{array}
\xrightarrow{NaOH}
\begin{array}{c}
RCOH\ COO^{\ominus} \\
| \quad \quad | \\
H_2C-CH_2
\end{array}
$$

$$
\begin{array}{c}
RCHOH\ COOH \\
| \quad \quad | \\
H_2C \quad CH_2 \\
\diagdown CH_2 \diagup
\end{array}
\xrightarrow{-H_2O}
\begin{array}{c}
RCH \quad \quad C \\
| \quad O \quad \| O\\
H_2C \quad CH_2 \\
\diagdown CH_2 \diagup
\end{array}
\xrightarrow{NaOH}
\begin{array}{c}
RCHOH\ COO^{\ominus} \\
| \quad \quad | \\
H_2C \quad CH_2 \\
\diagdown CH_2 \diagup
\end{array}
\quad (18.5)
$$

group, polymeric esters are formed rather than a seven-, eight-, or nine-membered lactone. Distillation of the polymers under reduced pressure produces the cyclic lactones.

If a choice is available, a γ-lactone (five-membered ring) always is formed in preference to a δ-lactone or other possibility (Eq. 18.6). A γ-lactone also is formed by cyclization of certain unsaturated and keto acids (Eqs. 18.7).

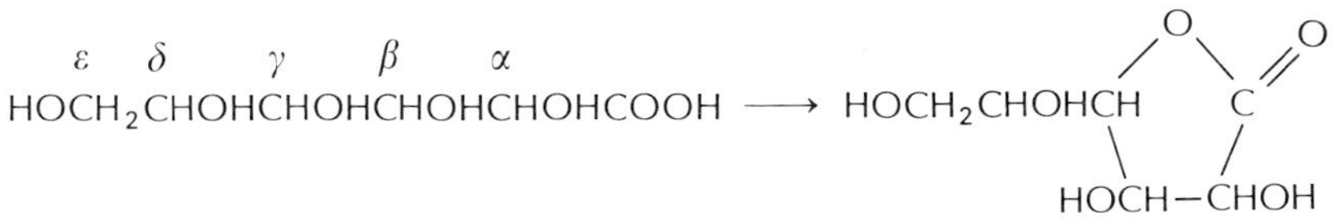

$$\underset{\varepsilon\quad\delta\quad\gamma\quad\beta\quad\alpha}{HOCH_2CHOHCHOHCHOHCHOHCOOH} \longrightarrow HOCH_2CHOHCH\underset{HOCH-CHOH}{\overset{O\diagdown\quad\diagup\overset{O}{\diagup}}{\diagdown\quad C\diagup}} \qquad (18.6)$$

$$\begin{matrix} H_2C=CHCH_2CH_2COOH \\ CH_3CH=CHCH_2COOH \end{matrix} \xrightarrow[H^\oplus]{H^\oplus} CH_3HC\underset{H_2C-CH_2}{\overset{O\diagdown\quad\diagup\overset{O}{\diagup}}{\diagdown\quad C\diagup}}$$

$$CH_3COCH_2CH_2COOH \xrightarrow{H^\oplus} CH_3C\underset{H_2C-CH_2}{\overset{HO\quad O\diagdown\quad\diagup\overset{O}{\diagup}}{\diagup\diagdown\quad C\diagup}} \xrightarrow{-H_2O} CH_3C\underset{HC-CH_2}{\overset{O\diagdown\quad\diagup\overset{O}{\diagup}}{\diagdown\diagup\quad C\diagup}} \qquad (18.7)$$

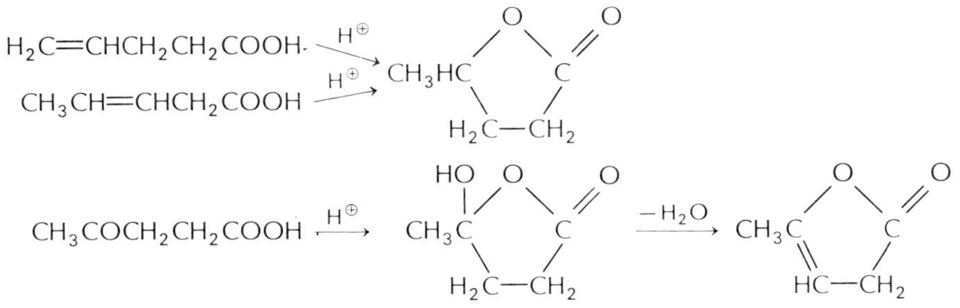

Ascorbic acid (vitamin C) is a γ-lactone with a unique enediol structure in the ring.

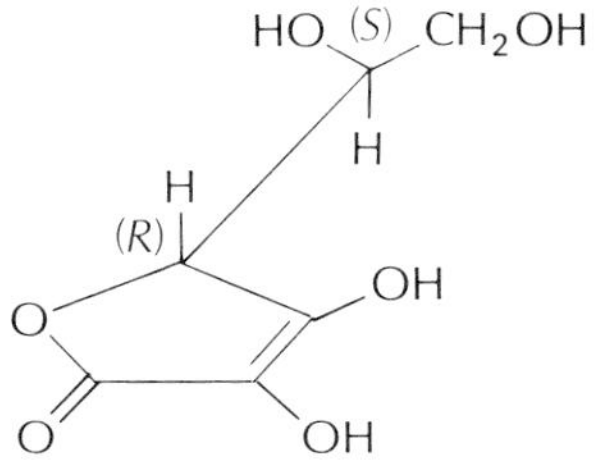

vitamin C
(ascorbic acid)

Hydroxy acids (and salts) undergo the usual reactions of acids (esterification, etc.), but the complication of lactone formation must be kept in mind. Reduction by lithium aluminum hydride provides an example (Eqs. 18.8).

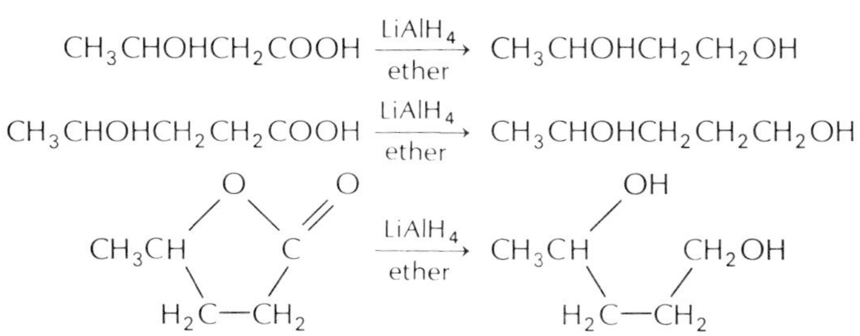

$$CH_3CHOHCH_2COOH \xrightarrow[ether]{LiAlH_4} CH_3CHOHCH_2CH_2OH$$

$$CH_3CHOHCH_2CH_2COOH \xrightarrow[ether]{LiAlH_4} CH_3CHOHCH_2CH_2CH_2OH \qquad (18.8)$$

$$CH_3CH\underset{H_2C-CH_2}{\overset{O\diagdown\quad\diagup\overset{O}{\diagup}}{\diagdown\quad C\diagup}} \xrightarrow[ether]{LiAlH_4} CH_3CH\underset{H_2C-CH_2}{\overset{OH}{\diagup}\diagdown\ CH_2OH}$$

Exercises

18.1 Complete the reactions:

a. $\phi CHOHCOOH \xrightarrow{\Delta}$

b. $\phi\underset{OH}{\overset{CH_3}{\underset{|}{\overset{|}{C}}}}-COOH \xrightarrow[0°]{H_2SO_4}$

c. $CH_3CHOHCH_2COOH \xrightarrow{warm}$

d. $CH_3CHOHCH_2CH_2COOH \xrightarrow{warm}$

e. $CH_3CHOHCH_2CH_2CH_2COOH \xrightarrow{\Delta}$

f. $CH_3CHOH(CH_2)_5COOH \xrightarrow{\Delta}$

18.2 Complete the reactions:

a. 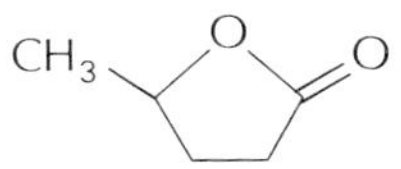 γ-valerolactone (CH₃-substituted γ-butyrolactone) + φMgBr ⟶

b. 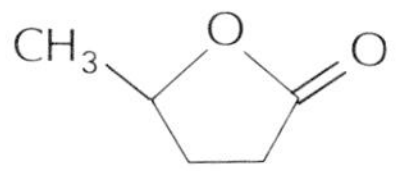 γ-valerolactone + 2CH₃MgI ⟶

c. 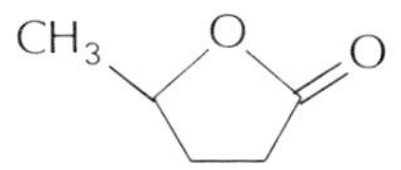 γ-valerolactone + excess EtOH $\xrightarrow[reflux]{H^\oplus}$

18.2 Preparation of hydroxy acids

Natural occurrence; list of syntheses. Many hydroxy acids are available from natural sources. Potassium hydrogen tartrate, for example, is the principal component of argol, the crude product that separates in the lees of wine. The purified material, cream of tartar, when neutralized with sodium hydroxide, is known as Rochelle salt. Others, such as lactic, glyceric, malic, and citric acids, participate in the metabolism of carbohydrates.

The α-hydroxy acids can be made from acids by the Hell-Volhard-Zelinsky bromination (Eq. 17.21) followed by replacement of the halogen by hydroxide. Another route is formation of a cyanohydrin followed by hydrolysis of the nitrile group.

The β-hydroxy acids can be made by reduction of β-ketoacids or esters and by the Reformatsky reaction of α-bromoesters (zinc, ether) with aldehydes and ketones. The aldol condensation of aldehydes followed by gentle oxidation (Tollens' reagent, etc.) of the aldehyde group is also useful.

The γ-hydroxy acids (and those with the hydroxy group farther removed) can be obtained by hydrolysis of the corresponding lactone, which in turn may be prepared by the Baeyer-Villiger reaction on cyclic ketones (Eq. 16.68).

Exercises

18.3 Show how to carry out the conversions:

a. $CH_3CH_2CH_2COOH \longrightarrow CH_3CH_2CHOHCOOH$

b. $CH_3COCH_3 \longrightarrow (CH_3)_2\underset{OH}{\overset{|}{C}}COOH$

c. $CH_3CH_2CO\underset{}{\overset{CH_3}{\underset{|}{C}}}HCOOEt \longrightarrow CH_3CH_2CHOH\underset{}{\overset{CH_3}{\underset{|}{C}}}HCOOH$

d. $CH_3CH_2COCH_2CH_3 \longrightarrow$

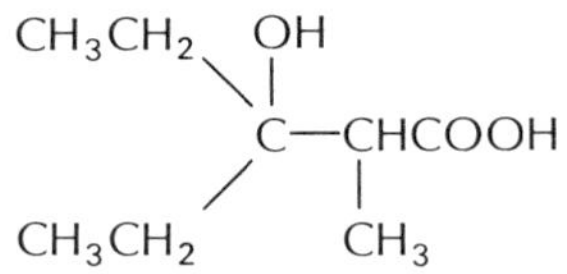

$$\begin{array}{c} CH_3CH_2 \\ \diagdown | \\ C-CHCOOH \\ \diagup | \\ CH_3CH_2 CH_3 \end{array} \begin{array}{c} OH \\ | \\ \\ \end{array}$$

e. $CH_3CHO \longrightarrow CH_3CHOHCH_2COOH$

f. 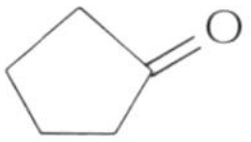 cyclopentanone $\longrightarrow HOCH_2CH_2CH_2CH_2COOH$

g. fumaric acid $\longrightarrow$ *meso*-tartaric acid

h. fumaric acid $\longrightarrow$ ($\pm$)-tartaric acid

18.3 Reactions of unsaturated acids

Occurrence. Unsaturated acids are common in nature: not only the derivatives of the fatty acid group, but many of short chain length as well. Some representative examples of unsaturated acids are given in Table 18.2. The smaller acids, usually in the form of derivatives (nitrile, amide, methyl ester) have been used to make some polymers that are widely used: examples are polyacrylonitrile (Acrilan, Orlon) and poly(methyl methacrylate) (Lucite, Plexiglas). (For others, see Table 30.1.)

Aside from the fatty-acid group of unsaturated acids, more is known about the α,β- and β,γ-unsaturated acids and derivatives than about those whose double bond is elsewhere in the molecule. Depending upon the number of alkyl groups present, the

Table 18.2 Some unsaturated acids

Name	Structure	mp, °C	bp, °C	pK_a
acrylic	$H_2C=CHCOOH$	13	141	4.26
methacrylic	$H_2C=C(CH_3)COOH$	15	163	
crotonic	*trans*-$CH_3CH=CHCOOH$	72	185	4.70
isocrotonic	*cis*-$CH_3CH=CHCOOH$	15	168	4.44
angelic	(Z)-$CH_3CH=C(CH_3)COOH$	45	185	4.30
tiglic	(E)-$CH_3CH=C(CH_3)COOH$	64	198	5.02
sorbic	*trans, trans*-$CH_3CH=CHCH=CHCOOH$	134	228 dec.	4.76
cinnamic	*trans*-$\phi CH=CHCOOH$	133	300 dec.	4.44
allocinnamic	*cis*-$\phi CH=CHCOOH$	68	—	3.96
phenylpropiolic	$\phi C\equiv CCOOH$	137	subl.	2.23
fumaric	*trans*-$HOOCCH=CHCOOH$	287	subl.	3.0; 4.5
maleic	*cis*-$HOOCCH=CHCOOH$	130	135 dec.	1.9; 6.5
mesaconic	(E)-$HOOCCH=C(CH_3)COOH$	204	250 dec.	3.1; 4.8
citraconic	(Z)-$HOOCCH=C(CH_3)COOH$	91 dec.	—	2.4; —
itaconic	$HOOCCH_2C(=CH_2)COOH$	162 dec.	—	3.8; 5.4
glutaconic	*trans*-$HOOCCH_2CH=CHCOOH$	138	—	
acetylenedicarboxylic	$HOOCC\equiv CCOOH$	179	—	

α,β-unsaturated isomer (in conjugation with the acid group) is not necessarily the most stable one. Crotonic acid is stable as compared with 3-butenoic acid (the β,γ isomer, $H_2C=CHCH_2COOH$): heating a mixture of their salts results in 100% crotonic acid at equilibrium. On the other hand, 4-methyl-2-pentenoic acid ((CH_3)$CH_2CH=CHCOOH$) is present only as 5% of the equilibrium mixture with 95% of 4-methyl-3-pentenoic acid ((CH_3)$_2C=CHCH_2COOH$).

Sorbic acid. 2,4-Hexadienoic acid is added to many processed foods to prevent the growth of mold and fungi. In the list of ingredients on packages it is denoted as sorbic acid. Because it shows little bacteriocidal activity, it is ideal for use in the control of mold growth during the preparation of cheese. Moreover, because its chemical structure is similar to that of natural fatty acids, sorbic acid is metabolized by normal processes. Instead of being harmful, it is, in fact, a food.

Addition reactions. The α,β- and β,γ-unsaturated acids decarboxylate readily (Eqs. 17.16). The α,β-unsaturated acids add nucleophilic reagents in only the one direction (Eq. 18.9). Their derivatives do so even more readily. The double bond is easily hydrogenated,

$$RCH=CHCOOH \xrightarrow{HY} RCHCH_2COOH \atop |\atop Y \quad (18.9)$$

and the acid group or derivative (except for acid chlorides) is not affected. The α,β-unsaturated acids and derivatives have been used extensively as dienophiles in Diels-Alder reactions (Eqs. 14.22).

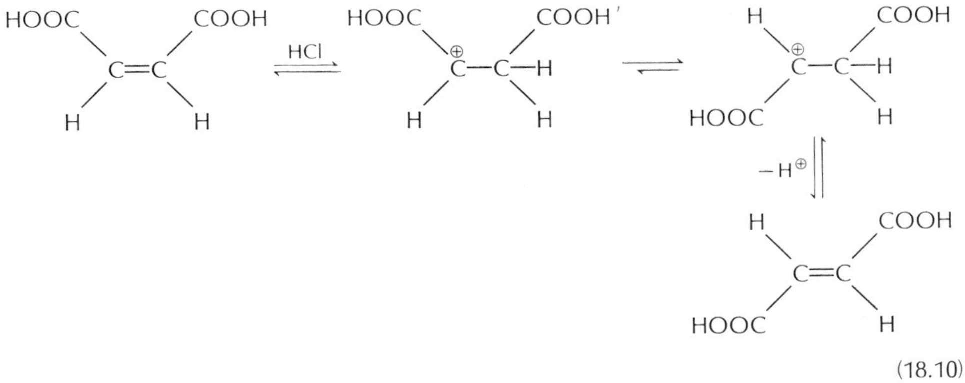

(18.10)

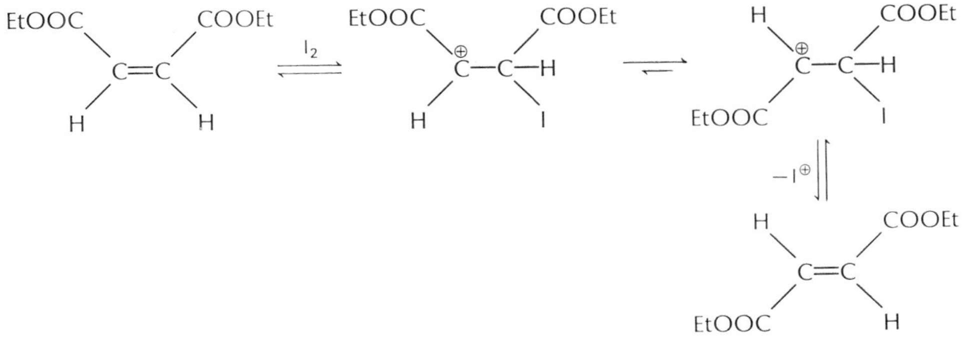

596 Chapter eighteen

Stereochemistry. Fumaric and maleic acids and mesaconic and citraconic acids were studied extensively as the relationships between *trans* and *cis* isomers were worked out. The *cis* diacids have lower mp, are more soluble in water, are stronger acids, and have higher heats of combustion than the *trans* isomers. The *cis* diacids and derivatives may be isomerized to the *trans* isomers by treatment with catalytic amounts of a halogen or of an acid (Eqs. 18.10). On the other hand, fumaric acid may be converted into maleic anhydride (Eq. 18.11).

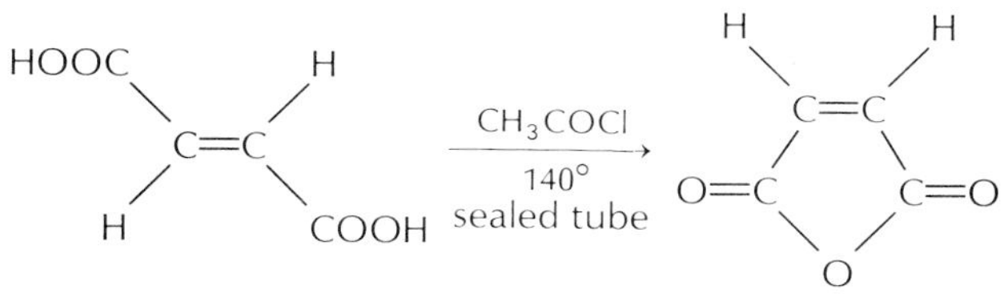

$$\underset{H}{\overset{HOOC}{\diagdown}}C=C\underset{COOH}{\overset{H}{\diagup}} \xrightarrow[\text{sealed tube}]{\text{CH}_3\text{COCl} \atop 140°} \text{maleic anhydride} + \text{HCl} + \text{CH}_3\text{COOH} \qquad (18.11)$$

The unsaturated acids and derivatives undergo the usual olysis and other reactions, except for an occasional intrusion of the C=C or C≡C into the reaction (for instance, a Michael addition to an unsaturated ester).

Exercises

18.4 Complete the reactions:
   a. $CH_3CH=CHCOOEt + HBr \xrightarrow{\text{polar}}$
   b. $CH_3CH_2CH=CHCOOEt + H_2C(COOEt)_2 \xrightarrow{EtO^\ominus}$
   c. $H_2C=CHCOOEt + H_2C=CHCH=CH_2 \longrightarrow$
   d. $\phi CH_2CH_2CH=CHCOOH \xrightarrow{\Delta}$
   e. $CH_3CH=CHCH_2CH_2COOH \xrightarrow{H^\oplus}$
   f. $\phi CH=CHCOOCH_3 + HOCH_2CH_2OH \xrightarrow{H^\oplus}$
   g. $\phi C\equiv CCOOEt + H_2O \xrightarrow[H^\oplus]{Hg^{2+}}$

18.4 Preparation of unsaturated acids

Summary of synthetic methods. The α,β-unsaturated acids and derivatives usually can be made by way of the acid or ester from a Reformatsky or ylide reaction or by a malonic ester synthesis. Other useful schemes go by way of a cyanohydrin or a Perkin reaction. Occasionally the Hell-Volhard-Zelinsky bromination of an acid followed by dehydrohalogenation is of use. If the double bond is not α,β to the carboxylic acid group, the problem must be worked out in each individual case.

Exercises

18.5 Show how to carry out the conversions:
   a. $\phi H \longrightarrow \phi CH=CHCH_2COOH$

b. 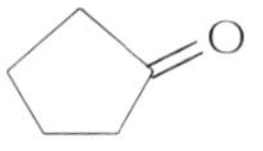 [cyclopentanone] ⟶ $H_2C=CHCH_2CH_2COOEt$

c. 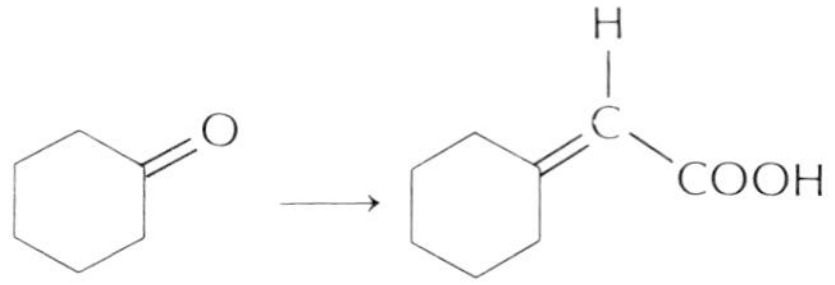 [cyclohexanone] ⟶ [cyclohexylidene-CH-COOH]

d. $(CH_3)_2CHCOOH \longrightarrow (CH_3)_2C=CHCOOH$

e. $\phi COOH \longrightarrow \phi CH=CHCOOH$

18.5 Reactions of keto acids

α-Keto acids. Glyoxylic acid (OHCCOOH), the simplest acid with a carbonyl group, acts like chloral in forming an isolable hydrate. Pyruvic acid (α-ketopropionic acid, $CH_3COCOOH$) and α-ketosuccinic acid (oxaloacetic acid, $HOOCCH_2COCOOH$) are intermediates in the metabolism of carbohydrates.

The special properties of α-keto acids (other than the usual reactions as a ketone or as an acid) are <u>easy oxidation</u> and <u>decarbonylation.</u> Pyruvic acid, for example, reacts with the very mild Tollens' reagent (Eq. 18.12). An α-keto acid or ester loses carbon monoxide

$$CH_3COCOOH \xrightarrow{Ag(NH_3)_2^{\oplus}} CH_3COOH + CO_2 + Ag + 2NH_3 \qquad (18.12)$$

upon being heated (Eqs. 18.13). However, decarboxylation of the malonic acid type is more

$$CH_3COCOOH \xrightarrow{170°} CH_3COOH + CO$$
$$CH_3COCOOEt \xrightarrow[\text{distil}]{\Delta} CH_3COOEt + CO \qquad (18.13)$$

facile than is decarbonylation, as oxaloacetic acid and ester demonstrate (Eq. 18.14). The acid decarboxylates, but the ester decarbonylates.

$$HOOCCH_2COCOOH \xrightarrow{heat} CO_2 + CH_3COCOOH$$
$$\downarrow EtOH, H^{\oplus} \qquad (18.14)$$
$$EtOOCCH_2COCOOEt \xrightarrow{distil} EtOOCCH_2COOEt + CO$$

Acetoacetic ester: ketone and acid cleavages. Acetoacetic ester, like malonic ester, is extraordinarily useful in organic synthesis. The key reactions are easy substitution (alkylation and acylation) and the ability to be hydrolyzed to a β-keto acid, which is easily decarboxylated. Actually, acetoacetic ester may be cleaved in two different ways, the <u>ketone</u> and <u>acid cleavages.</u> The ketone cleavage is analogous to the hydrolysis and decarboxylation of malonic ester (Eq. 18.15). The alkaline hydrolysis and subsequent acidification are faster and cleaner than the one-step acid hydrolysis of the ester. The product from acetoacetic ester itself and from alkylated esters is a methyl ketone.

The acid cleavage results from a reversal of a Claisen condensation under the action of concentrated NaOH (Eq. 18.16). Although this method could be a synthesis of acids, it seldom is used, because the malonic ester synthesis is just as convenient to carry out, if not more so.

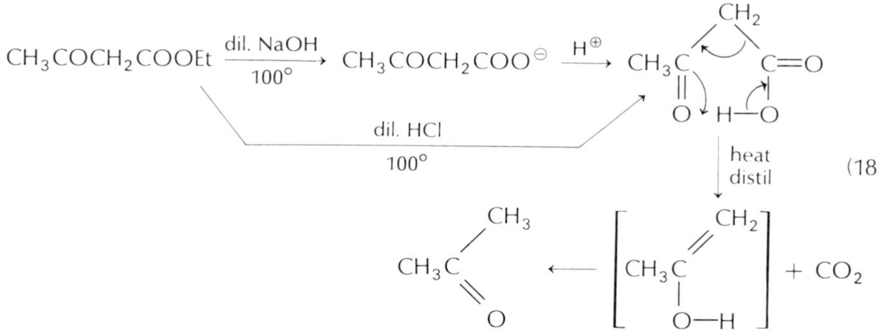

$$CH_3COCH_2COOEt \xrightarrow[100°]{\text{dil. NaOH}} CH_3COCH_2COO^{\ominus} \xrightarrow{H^{\oplus}} \cdots \xrightarrow{\text{heat, distil}} [\text{enol}] \longrightarrow CH_3COCH_3 + CO_2 \quad (18.15)$$

$$CH_3COCH_2COOEt \xrightarrow{\text{conc. NaOH}} \cdots \xrightarrow{100°} CH_3COO^{\ominus} + {}^{\ominus}CH_2COOEt \xrightarrow{+H_2O, \text{etc.}} CH_3COO^{\ominus} + CH_3COO^{\ominus} + EtOH \quad (18.16)$$

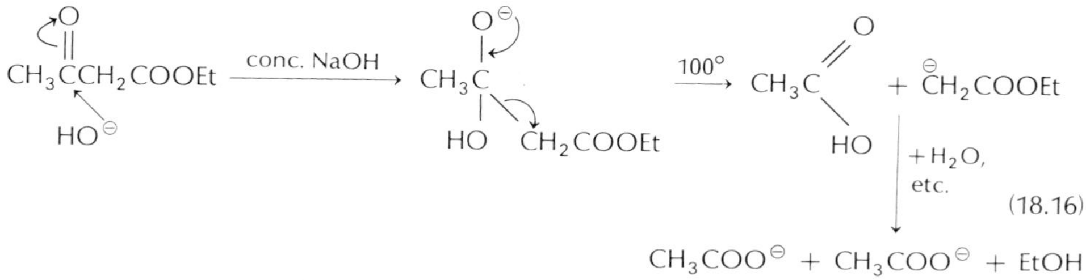

Alkylation and cleavage. Acetoacetic ester (AAE), with pK_a of 10.7, is more acidic than ethanol and even more acidic than malonic ester. Thus, the anion or ion pair is easily formed by reaction with sodium ethoxide. Reaction of the anion with an alkyl halide then gives a substituted acetoacetic ester, which in turn may be alkylated a second time. Ketone cleavage completes the operation of an acetoacetic ester synthesis (Eqs. 18.17). As with the malonic ester synthesis, a primary alkyl halide should react more readily than a secondary one, and tertiary ones cannot be used.

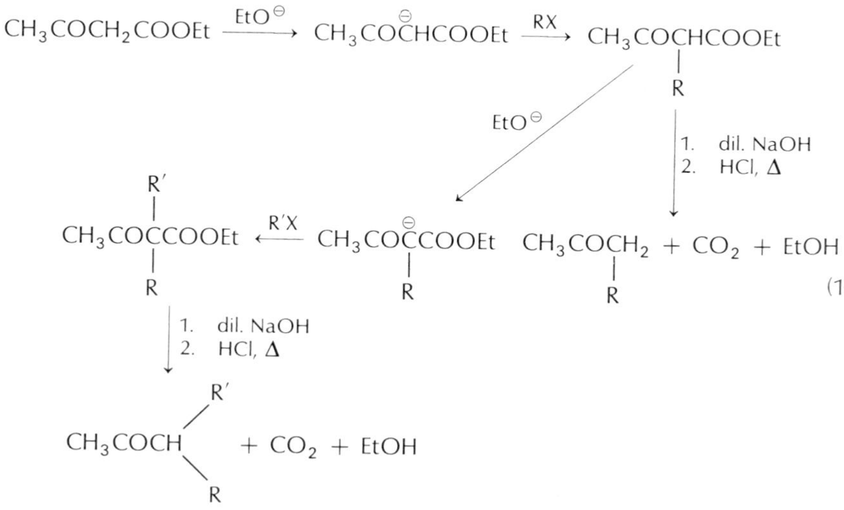

$$CH_3COCH_2COOEt \xrightarrow{EtO^{\ominus}} CH_3CO\overset{\ominus}{C}HCOOEt \xrightarrow{RX} CH_3COCHCOOEt \mid R$$

$$\xrightarrow{EtO^{\ominus}} CH_3CO\overset{\ominus}{C}COOEt \xrightarrow{R'X} CH_3COCCOOEt \quad\quad \xrightarrow[\text{2. HCl, }\Delta]{\text{1. dil. NaOH}} CH_3COCH_2 + CO_2 + EtOH$$

(with R, R' substituents; final product $CH_3COCHRR'$ + CO_2 + EtOH)

(18.17)

18.5 Reactions of keto acids

Variations. If a ketone other than a methyl ketone is wanted, some variation is possible. Use of the Claisen condensation can produce a β-keto ester that gives the desired group at the end of the chain. For example, suppose the ketone EtCOCH(CH$_3$)CH$_2$CH(CH$_3$)$_2$ is needed. Equation 18.18 shows how to do it. Of course, we see the limitation of the scheme: the β-keto ester already had a methyl group at the α position. We can get around that problem by using a crossed Claisen condensation. For example, if tBuCOBu is needed, we do it as shown in Equation 18.19. Again we are limited, this time by the use of esters that have no α hydrogens for the crossed Claisen condensation (for example, HCOOEt, φCOOEt, EtOOCCOOEt, besides the one shown: see Eq. 17.68).

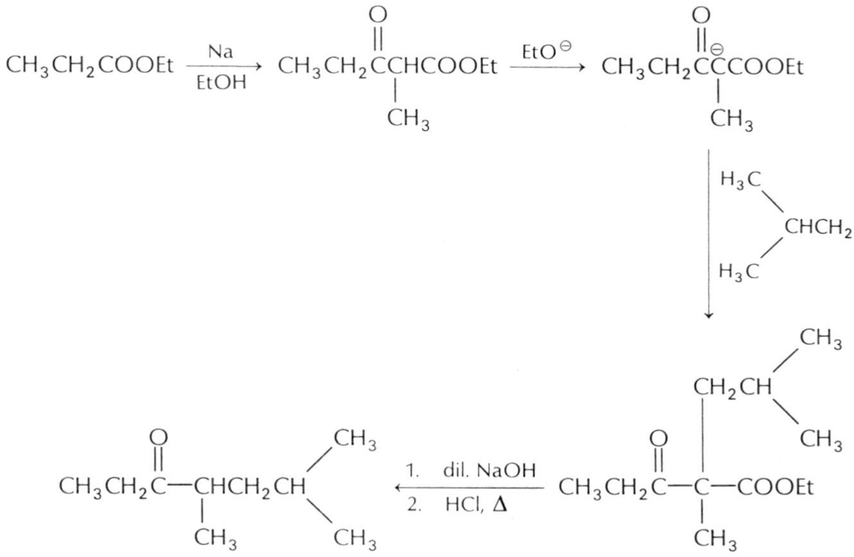

$$CH_3CH_2COOEt \xrightarrow[EtOH]{Na} CH_3CH_2\overset{O}{\underset{CH_3}{\overset{\|}{C}}}CHCOOEt \xrightarrow{EtO^\ominus} CH_3CH_2\overset{O}{\underset{CH_3}{\overset{\|}{C}}}\overset{\ominus}{C}COOEt$$

$$\downarrow (CH_3)_2CHCH_2Br$$

$$CH_3CH_2\overset{O}{\overset{\|}{C}}-\underset{CH_3}{\overset{CH_3}{\underset{|}{C}H}}CH_2CH\overset{CH_3}{\underset{CH_3}{\diagdown}} \xleftarrow[\text{2. HCl, }\Delta]{\text{1. dil. NaOH}} CH_3CH_2\overset{O}{\overset{\|}{C}}-\underset{CH_3}{\overset{CH_2CH(CH_3)_2}{\underset{|}{C}}}-COOEt \quad (18.18)$$

$$(CH_3)_3CCOOEt + CH_3COOEt \xrightarrow[EtOH]{Na} (CH_3)_3C\overset{O}{\overset{\|}{C}}CH_2COOEt$$

$$\downarrow EtO^\ominus$$

$$(CH_3)_3C\overset{O}{\overset{\|}{C}}\underset{CH_2CH_2CH_3}{\overset{|}{C}}HCOOEt \xleftarrow{CH_3CH_2CH_2Br} (CH_3)_3C\overset{O}{\overset{\|}{C}}\overset{\ominus}{C}HCOOEt$$

$$\downarrow \begin{array}{l}\text{1. dil. NaOH}\\\text{2. HCl, }\Delta\end{array}$$

$$(CH_3)_3C\overset{O}{\overset{\|}{C}}\underset{CH_2CH_2CH_3}{\overset{|}{C}}H_2 \quad + CO_2 + EtOH \quad (18.19)$$

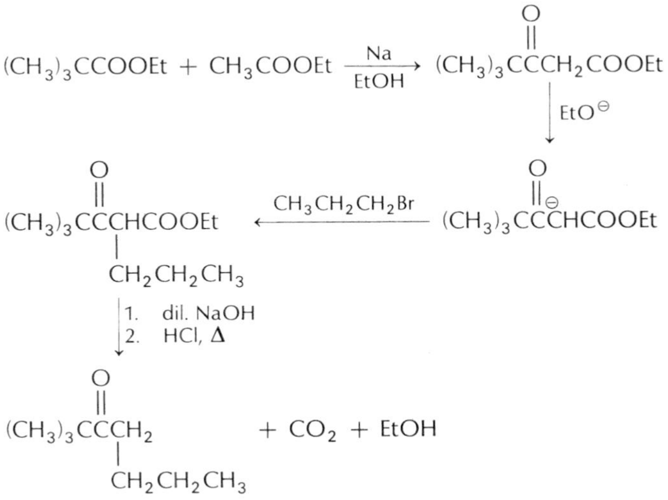

However, we still have room to maneuver a bit. We can make use of mixed condensations in two ways. One way is to use diethyl carbonate (available from COCl$_2$ + 2EtOH → EtOCOOEt) with a methyl ketone, as shown in Equation 18.20. The other way is to use

$$\text{RCOCH}_3 \xrightarrow{\text{NaH}} \text{RCOCH}_2^{\ominus} \xrightarrow{\text{EtOCOOEt}} \left[\text{RCOCH}_2-\underset{\underset{\text{OEt}}{|}}{\overset{\overset{\text{O}^{\ominus}}{|}}{\text{C}}}-\text{OEt} \right]$$

$$\downarrow -\text{EtO}^{\ominus} \quad (18.20)$$

$$\text{RCOCH}_2\text{COOEt}$$

diethyl oxalate with a methyl ketone (Eq. 18.21). The α-keto ester, we remember, is subject to decarbonylation (Eq. 18.14) upon distillation, and the β-keto ester is stable to heat (even if the acid is not). So because almost any methyl ketone we want may be synthesized from AAE ester itself, we now are in a position to vary the end group in RCOCH$_2$COOEt at will. Just as we used the term <u>malonic ester synthesis</u> to include reactions of the ester other than just alkylation and decarboxylation, we shall use the term <u>acetoacetic ester synthesis</u> to include the many reactions of the ester other than alkylation and cleavage. Furthermore, we also shall include in the meaning the use of any β-keto ester, not just AAE itself.

$$\text{RCOCH}_3 \xrightarrow{\text{NaH}} \text{RCOCH}_2^{\ominus} \xrightarrow{\overset{\text{COOEt}}{\underset{}{|}}\text{COOEt}} \text{RCOCH}_2\underset{\underset{\text{COOEt}}{|}}{\overset{\overset{\text{O}}{\parallel}}{\text{C}}}$$

$$\downarrow \Delta, -\text{CO} \quad (18.21)$$

$$\text{RCOCH}_2\text{COOEt}$$

Use of dihalides. If a dihalide is used, either rings or diketones may be made (Eqs. 18.22). The rings are produced by treatment of the monosubstituted ester with a second mole of ethoxide ion to cause the internal ring-closing substitution. The use of CH$_2$I$_2$ provides a route to a vinyl ketone, as shown. Side reactions always are dehydrohalogenations of the halides. The diketones undergo internal aldol condensations to form cyclic ketones (Eq. 16.50).

Treatment of two moles of AAE anion with one mole of halogen leads to a γ-diketone (Eq. 18.23).

Acylation. Acylation of AAE is carried out in aprotic solvents to prevent reaction of the acid chloride with the solvent. The AAE anion, it has been found, may be formed and stabilized as an ion pair by the use of magnesium. The use of sodium or sodium hydride for this purpose is possible with MeOCH$_2$CH$_2$OMe as the solvent, because it keeps the sodium enolate in solution. An example is shown (Eq. 18.24) in which several products may be obtained, depending upon conditions and/or where the sequence is stopped.

Now that we can add the malonic ester and acetoacetic ester syntheses to our ability along with aldol, Claisen, Michael, Perkin, ylide, and other reactions and condensations, we can construct ketones, aldehydes, and acids of almost any complexity desired.

Enolization. Let us return for a moment to the structure of acetoacetic ester and the subject of tautomerism. Historically, AAE was the subject of a long controversy. The

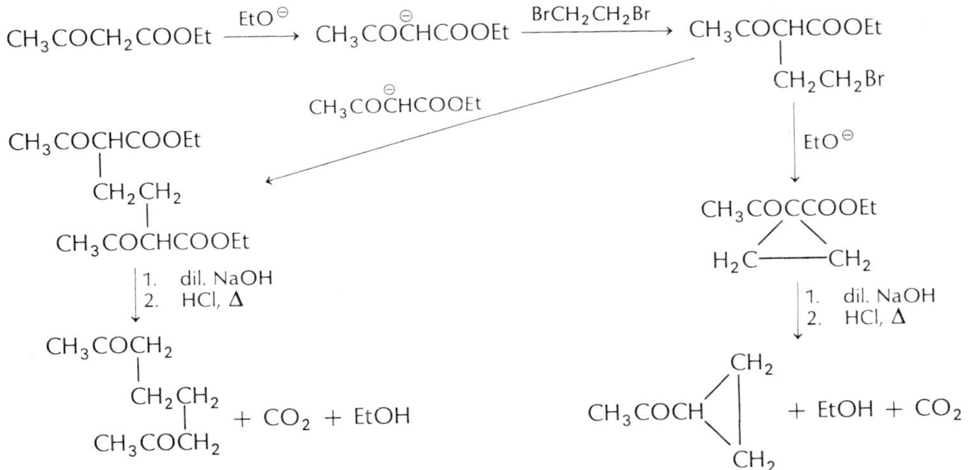

18.5 Reactions of keto acids

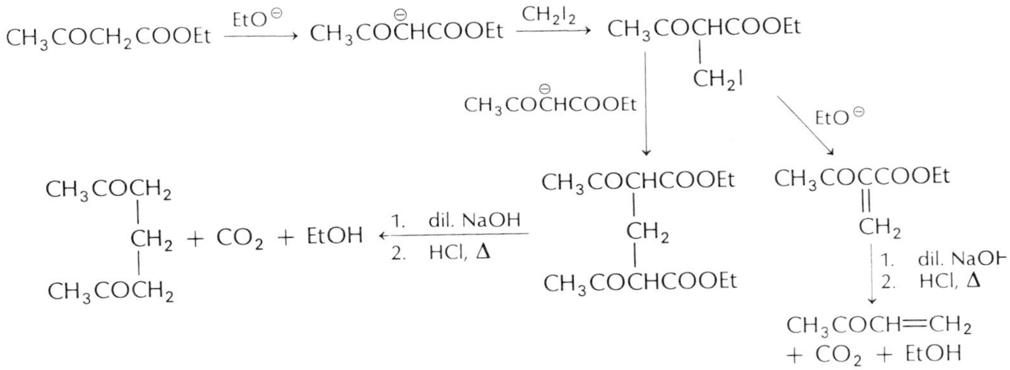

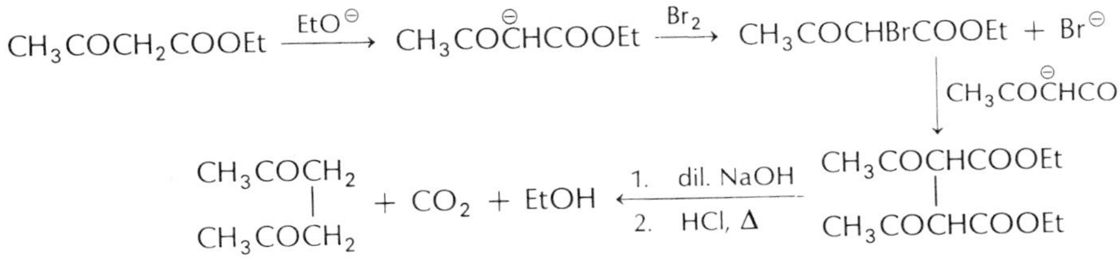

$$CH_3COCH_2COOEt \xrightarrow{EtO^{\ominus}} CH_3CO\overset{\ominus}{C}HCOOEt \xrightarrow{BrCH_2CH_2Br} \underset{\underset{CH_2CH_2Br}{|}}{CH_3COCHCOOEt}$$

$$\xrightarrow{CH_3CO\overset{\ominus}{C}HCOOEt}$$

$$\underset{\underset{CH_3COCHCOOEt}{|}}{\underset{CH_2CH_2}{|}}{CH_3COCHCOOEt} \quad \xleftarrow{} \quad$$

$$\downarrow \begin{array}{l}1.\ \text{dil. NaOH}\\ 2.\ \text{HCl, } \Delta\end{array}$$

$$\underset{\underset{CH_3COCH_2}{|}}{\underset{CH_2CH_2}{|}}{CH_3COCH_2} + CO_2 + EtOH$$

Right branch:

$$\xrightarrow{EtO^{\ominus}} CH_3COCCOOEt \text{ (cyclopropane with } H_2C\text{—}CH_2)$$

$$\downarrow \begin{array}{l}1.\ \text{dil. NaOH}\\ 2.\ \text{HCl, } \Delta\end{array}$$

$$CH_3COCH\underset{CH_2}{\overset{CH_2}{\diagup\!\!\diagdown}} + EtOH + CO_2 \qquad (18.22)$$

$$CH_3COCH_2COOEt \xrightarrow{EtO^{\ominus}} CH_3CO\overset{\ominus}{C}HCOOEt \xrightarrow{CH_2I_2} \underset{\underset{CH_2I}{|}}{CH_3COCHCOOEt}$$

$$\xrightarrow{CH_3CO\overset{\ominus}{C}HCOOEt}$$

$$\underset{\underset{CH_3COCHCOOEt}{|}}{\underset{CH_2}{|}}{CH_3COCHCOOEt}$$

$$\xleftarrow{\begin{array}{l}1.\ \text{dil. NaOH}\\ 2.\ \text{HCl, } \Delta\end{array}} \underset{\underset{CH_3COCH_2}{|}}{\underset{CH_2}{|}}{CH_3COCH_2} + CO_2 + EtOH$$

And:

$$\xrightarrow{EtO^{\ominus}} \underset{\underset{CH_2}{||}}{CH_3COCCOOEt}$$

$$\downarrow \begin{array}{l}1.\ \text{dil. NaOH}\\ 2.\ \text{HCl, } \Delta\end{array}$$

$$CH_3COCH=CH_2 + CO_2 + EtOH$$

$$CH_3COCH_2COOEt \xrightarrow{EtO^{\ominus}} CH_3CO\overset{\ominus}{C}HCOOEt \xrightarrow{Br_2} CH_3COCHBrCOOEt + Br^{\ominus}$$

$$\downarrow CH_3CO\overset{\ominus}{C}HCOOEt \qquad (18.23)$$

$$\underset{\underset{CH_3COCH_2}{|}}{CH_3COCH_2} + CO_2 + EtOH \xleftarrow{\begin{array}{l}1.\ \text{dil. NaOH}\\ 2.\ \text{HCl, } \Delta\end{array}} \underset{\underset{CH_3COCHCOOEt}{|}}{CH_3COCHCOOEt}$$

idea of fairly rapid shift of a proton from one position to another was very slow to be accepted. We shall not pursue the twists and turns of the arguments against the idea, nor the ingenious experimentation that finally led to the isolation of the pure keto and enol forms of AAE. (Other keto and enol tautomers had been isolated long before: $CH_3COCH(CO\phi)_2$ and $CH_3\overset{OH}{\underset{|}{C}}=C(CO\phi)_2$, for example).

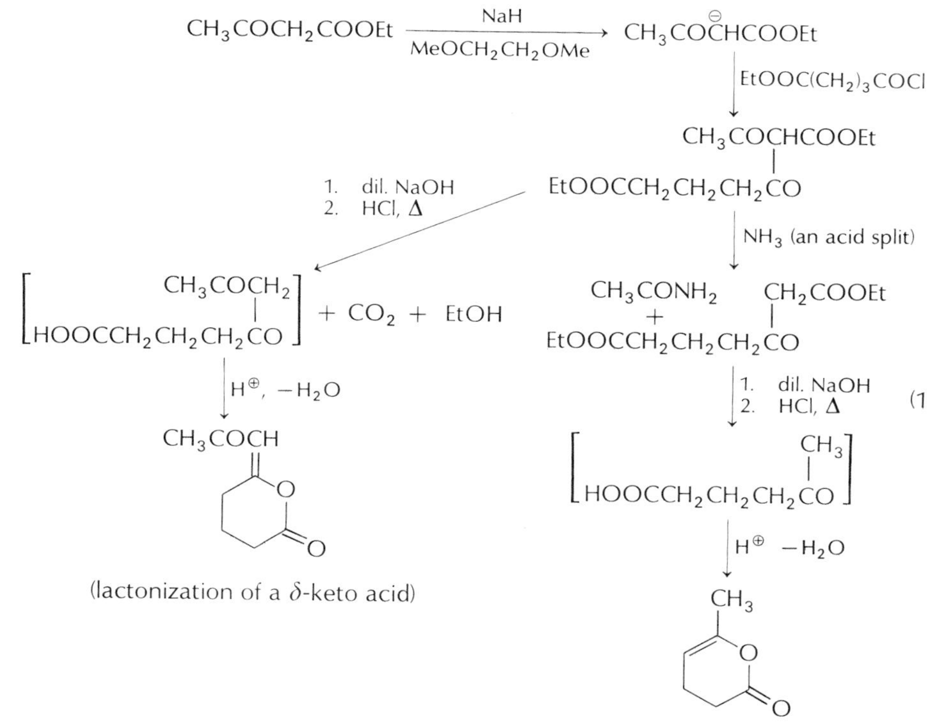

$$CH_3COCH_2COOEt \xrightarrow[MeOCH_2CH_2OMe]{NaH} CH_3CO\overset{\ominus}{C}HCOOEt$$

$$\downarrow EtOOC(CH_2)_3COCl$$

$$\begin{array}{c} CH_3COCHCOOEt \\ | \\ EtOOCCH_2CH_2CH_2CO \end{array}$$

$$\downarrow NH_3 \text{ (an acid split)}$$

1. dil. NaOH
2. HCl, Δ

$$\begin{array}{cc} CH_3CONH_2 & CH_2COOEt \\ + & | \\ EtOOCCH_2CH_2CH_2CO \end{array}$$

$$\left[\begin{array}{c} CH_3COCH_2 \\ | \\ HOOCCH_2CH_2CH_2CO \end{array} \right] + CO_2 + EtOH$$

$$\downarrow \text{1. dil. NaOH} \atop \text{2. HCl, Δ} \quad (18.24)$$

$$\downarrow H^{\oplus}, -H_2O$$

$$\left[\begin{array}{c} CH_3 \\ | \\ HOOCCH_2CH_2CH_2CO \end{array} \right]$$

CH₃COCH (lactone structure)

(lactonization of a δ-keto acid)

$$\downarrow H^{\oplus} -H_2O$$

CH₃ (lactone structure)

Today, it is an easy task to estimate the position of a keto-enol equilibrium from nmr. Figure 18.1 shows the nmr of acetylacetone and of acetoacetic ester, from which the positions of the equilibria are easily calculated (80% enol for acetylacetone and 7.5% enol for acetoacetic ester as pure liquids). Also, the ir of liquid AAE has a band at 1650 cm^{-1} (α,β-unsaturated carbonyl) as well as at 1718 cm^{-1} (ketone) and 1742 cm^{-1} (ester). Furthermore, it also is known that the amount of enol in acetylacetone (15% enol in water) and in AAE increases as the polarity of the solvent decreases (0.4% in H_2O, 16.2% in ϕH, 46.4% in hexane). Many studies have been made of the effects of substitution upon the keto-enol equilibria in β-diketones and β-ketoesters. For example, liquid $CH_3COCH\phi COOEt$ is 30% enol.

The speed of shift of a proton is catalyzed by both acids and bases. Thus, by operating at low temperatures and keeping acids and bases to a minimum, we may freeze the shift or position of equilibrium. Enols react rapidly with halogens, whereas the keto forms react only as rapidly as the shift to enol occurs. Enols form colored complexes with Fe^{3+} and Cu^{2+}. An enol, being a compound with an active hydrogen, does not allow a Grignard reaction to proceed by direct addition (Eq. 18.25).

$$CH_3COCH_2COOEt \xrightarrow{MeMgI} CH_3\overset{OMgI}{\underset{|}{C}}=CHCOOEt + CH_4 \quad (18.25)$$

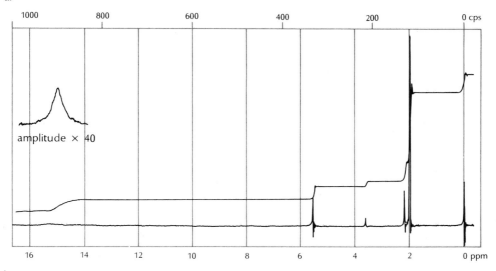
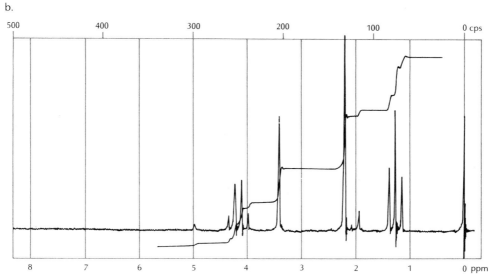

Figure 18.1 a. Acetylacetone, pure liquid, 1000 cps sweep width

b. Acetoacetic ester, 20% (w/v) in carbon tetrachloride, 500 cps sweep width

Excess Grignard reagent may now add 1,4 to the conjugated ester and 1,2 to the ester group (Eq. 18.26).

$$CH_3C(OMgI)=CHCOOEt \xrightarrow[\text{2. work-up}]{\text{1. excess MeMgI}} CH_3-\underset{CH_3}{\underset{|}{C}}(OH)-CH_2COOEt + CH_3COCH_2\underset{CH_3}{\underset{|}{C}}(OH)-CH_3$$

$$\text{1,4-product} \qquad\qquad \text{1,2-product}$$

(18.26)

The haloform reaction fails with acetoacetic ester because the active hydrogens are replaced first and the alkali present causes a reverse Claisen condensation (alkaline cleavage by base, Eq. 18.27).

$$CH_3COCH_2COOEt \xrightarrow[\text{NaOH}]{I_2} CH_3COCl_2COOEt \xrightarrow{HO^\ominus} CH_3COOH + I_2\overset{\ominus}{C}COOEt$$

$$\downarrow HO^\ominus, H_2O$$

$$CH_3COO^\ominus + CHI_2COO^\ominus$$

(18.27)

As we have seen, γ, δ, and higher keto acids can be made by a number of synthetic routes. Both γ and δ keto acids form unsaturated lactones (Eq. 18.24: only δ keto acids are shown).

Exercises

18.6   a. Use Figure 18.1b and identify the source of each peak or set of peaks.
       b. Calculate the percent of enol present.

18.7   Show how to carry out the conversions:
       a. EtOOCCOCH$_2$COOEt $\longrightarrow$ CH$_3$CH$_2$CH$_2$CH$_2$COOH
       b. EtOOCCOCH$_2$COOEt $\longrightarrow$ CH$_3$CH$_2$CH$_2$CH$_2$COCOOH

18.8   Use the acetoacetic synthesis to prepare:
       a. CH$_3$COCH$_2$CH$_2$CH$_3$ d. ϕCOCH$_2$CH$_2$COϕ
       b. CH$_3$CH$_2$COCHCH$_2$CH$_3$
           $\quad\quad\quad\quad\;\;|$
           $\quad\quad\quad\;\;$CH$_3$
       c. CH$_3$COCH$_2$CH$_2\phi$ e. 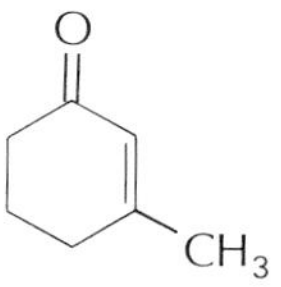 [3-methylcyclohex-2-enone structure]

18.6   Preparation of keto acids

Preparation of α-keto acids.   The α-keto acids can be made in several ways. The simplest, glyoxylic acid, may be made by:

1. treatment of maleic or fumaric acids with ozone, with Zn dust work-up;

2. hydrolysis of dichloroacetic acid; or
3. hydrolysis of the ester obtained by the action of lead tetraacetate upon diethyl tartrate (Eq. 18.28).

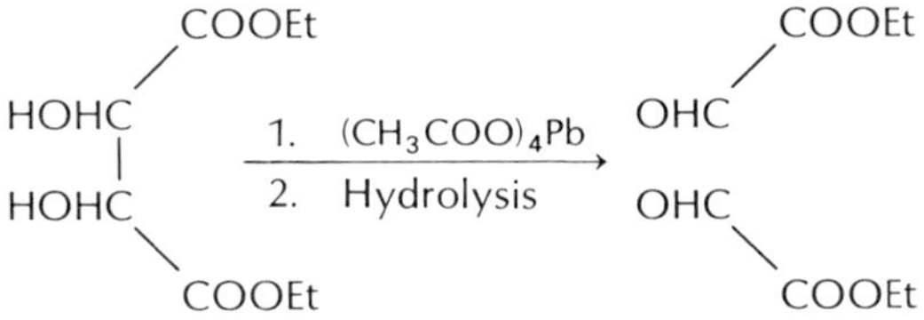

$$\begin{array}{c}\text{HOHC}\diagup^{\text{COOEt}}\\|\\\text{HOHC}\diagdown_{\text{COOEt}}\end{array}\xrightarrow[\text{2. Hydrolysis}]{\text{1. }(CH_3COO)_4Pb}\begin{array}{c}\text{OHC}\diagup^{\text{COOEt}}\\\\\text{OHC}\diagdown_{\text{COOEt}}\end{array} \quad (18.28)$$

Pyruvic acid may be made by pyrolysis of tartaric acid. The first step is a dehydration, and it is followed by decarboxylation of the β-keto acid (Eq. 18.29). A more general method is

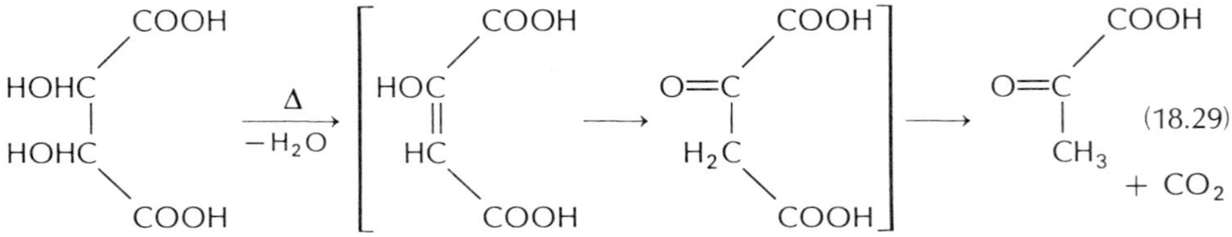

$$\begin{array}{c}\text{HOHC}\diagup^{\text{COOH}}\\|\\\text{HOHC}\diagdown_{\text{COOH}}\end{array}\xrightarrow[-H_2O]{\Delta}\left[\begin{array}{c}\text{HOC}\diagup^{\text{COOH}}\\\|\\\text{HC}\diagdown_{\text{COOH}}\end{array}\longrightarrow\begin{array}{c}O=C\diagup^{\text{COOH}}\\|\\H_2C\diagdown_{\text{COOH}}\end{array}\right]\longrightarrow\begin{array}{c}O=C\diagup^{\text{COOH}}\\|\\CH_3\end{array}+CO_2 \quad (18.29)$$

to add a carbon atom to an acid chloride (Eq. 18.30). Previously, we have mentioned the

$$\text{RCOCl}\xrightarrow{\text{NaCN}}\text{RC}\overset{O}{\overset{\|}{C}}\text{CN}\xrightarrow[\text{hydrolysis}]{\text{mild}}\text{RC}\overset{O}{\overset{\|}{C}}\text{COOH} \quad (18.30)$$

formation of α-keto esters by the reaction of diethyl carbonate or oxalate in crossed and mixed Claisen condensations (Eq. 18.21, for example).

Preparation of β-keto acids. The β-keto acids usually are prepared by alkaline hydrolysis of β-keto esters, followed by acidification of the mixture at 0°. We already have covered the formation of β-keto esters, and we shall add nothing new here. (See Secs. 17.8, 17.9, and 18.5.)

The other keto acids can be prepared by way of the malonic ester (Sec. 17.8) or acetoacetic ester syntheses (Eq. 18.24).

Exercises

18.9 Show how to prepare:
    a. ϕCOCOOH
    b. ϕCH$_2$COCOOH
    c. ϕCOCH$_2$CH$_2$CH$_2$COOEt
    d. CH$_3$COCCOOH
                ||
                ϕCH

18.7 Several interesting esters

Here are a few esters and acids of physiological interest.

Pyrethrin. Pyrethrin I has the structure shown. It is a substituted cyclopentenyl

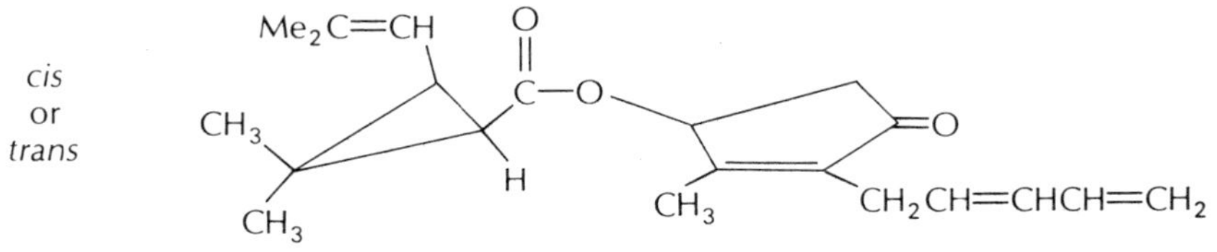

cis
or
trans

Pyrethrin I

ester of chrysanthemic acid. It is a component of natural pyrethrins from pyrethrum flowers (relatives of chrysanthemums) and is an effective, nonpersistent insecticide.

A promising synthetic substitute is put together as shown in Equation 18.31. It

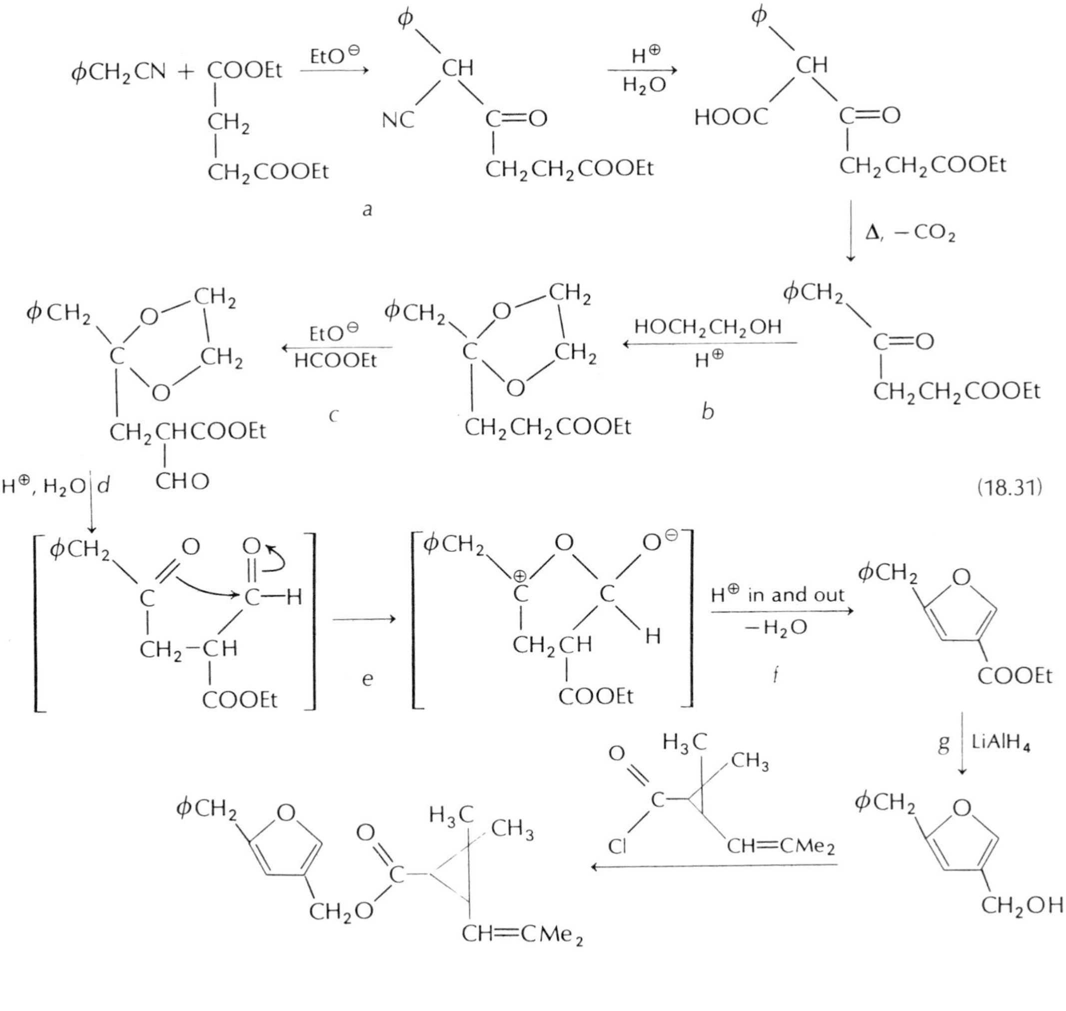

(18.31)

illustrates many of the reactions we have been discussing. Thus, a shows the phenylacetonitrile as having more reactive α hydrogens than does diethyl succinate. After hydrolysis of the cyano group and decarboxylation of the β-keto acid, the keto group is protected by the formation of the cyclic ketal with ethylene glycol (b). With the keto group protected (meaning a lack of α-hydrogen activity as well as inertness to addition), the crossed Claisen condensation with ethyl formate gives a β-aldehydo ester (c). Acidification breaks open the very susceptible acetal, freeing the keto group (d). The γ-keto aldehyde cyclizes (e) and dehydrates (f) to give the substituted furan. Reduction of the ester with LiAlH$_4$ (g) gives the primary alcohol, which is esterified with the acid chloride of chrysanthemic acid. The synthetic substance is reported to be five times more effective than the natural product as an insecticide and to be nonpersistent. The toxicity of the ester (and of the many possible metabolic breakdown products) to animal life remain to be measured.

Juvenile hormone. Juvenile hormone (JH) is secreted by insects and is essential for their normal growth in the larval stage. Continued presence of the hormone hinders proper metamorphosis into the next stage of the life cycle. The structure is shown. Such

minute amounts are required by an insect that exposure to the substance (and to analogous synthetic substances: for example, an analog with a methyl instead of an ethyl at C$_7$) brings about sterility and other harmful effects. Many laboratories are engaged in research on JH and compounds that will mimic JH activity for possible use as insecticides.

Queen bee substance. Queen bee substance is *trans*-9-keto-2-decenoic acid. Secreted by a queen bee, it is a sex attractant and an inhibitor of the ovaries in worker bees. It may be synthesized by a series of familiar steps (Eq. 18.32).

We start with a mixed aldol condensation of cyclohexanone with ethyl acetate (a) to produce the β-diketone 2-acetocyclohexanone. A reverse aldol condensation (b, also an alkaline cleavage of a diketone) opens the ring to give 7-ketooctanoic acid. Thus a chain of eight carbon atoms with a keto group located adjacent to the terminal carbon is constructed in just two steps. The acid group is converted into an aldehyde by way of the acid chloride (c, other methods could have been used) in preparation for extending the chain by two more carbon atoms. The more reactive aldehyde group reacts with malonic acid, while the less reactive keto group remains inert (d). Malonic acid is used instead of the ester to prevent a Michael reaction with the new double bond. Heat causes the usual malonic acid decarboxylation to give the product (e).

Schizophrenic sweat odor; prostaglandins. Schizophrenic patients have sweat of a peculiar odor. The odor is caused by the presence of *trans*-3-methyl-2-hexenoic acid.

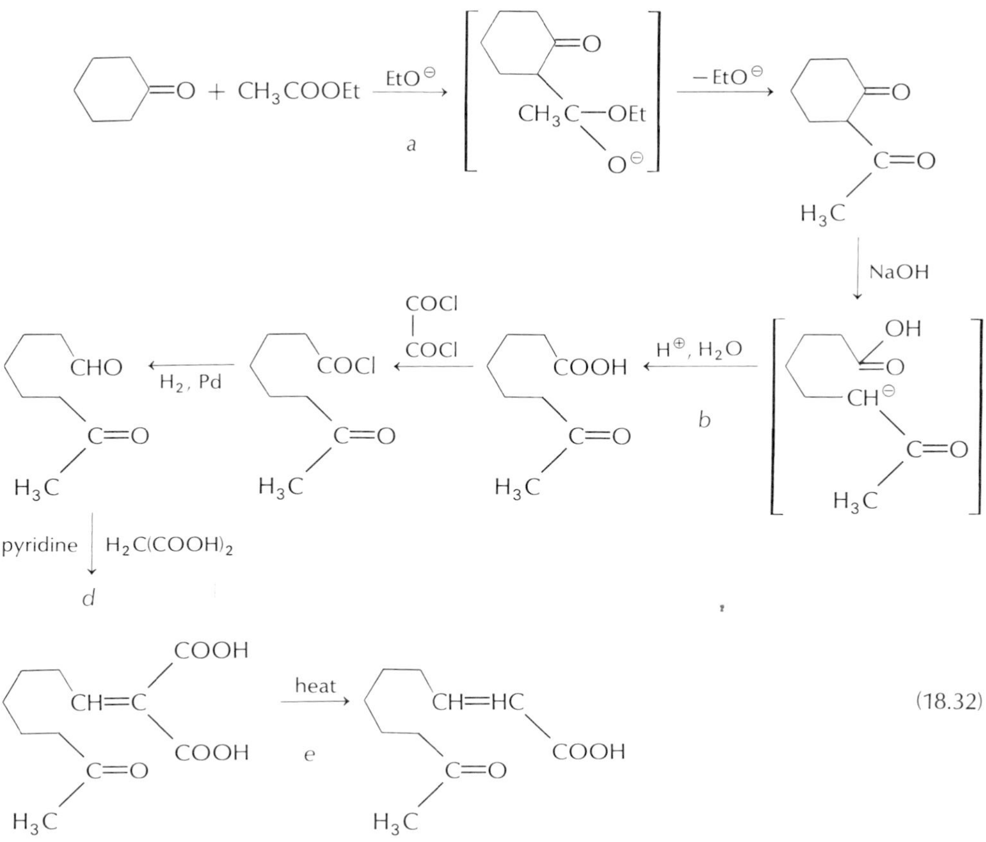

Over sixteen prostaglandins have been isolated from a variety of tissues and biological liquids in very small amounts. Prostaglandins are thought to affect many functions in the body, such as smooth muscle function, lipid metabolism, and reproductive physiology.

The structure of prostaglandin E_3 is shown. Other prostaglandins have various double bonds hydrogenated.

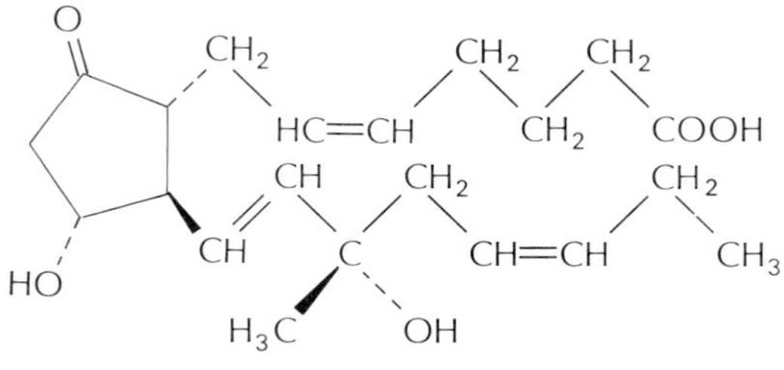

(18.32)

18.8 Summary

In this brief chapter, we have sampled polyfunctional compounds by way of substituted acids and derivatives, concluding with a few examples of the synthesis of naturally occurring substances of physiological interest. To pursue these topics rigorously or comprehensively is not our objective. Much more information is available in advanced textbooks and monographs on the subjects mentioned here.

Acetoacetic ester and malonic ester syntheses provide versatile approaches to the synthesis of ketones and acids. They supplement the Grignard reactions. We are now in a position to construct many, many organic compounds. Essentially, we have completed our survey of the C,H,O, and X compounds, except for some topics concerning phenols, quinones, and heterocyclics.

18.9 Problems

18.1 Substance A ($C_{15}H_{22}O_8$) may be prepared from chloroform and excess malonic ester and sodium ethoxide. Treatment of A with dilute NaOH, followed by acidification and heat, gives B ($C_5H_6O_4$). Treatment of B with acetyl chloride gives C ($C_5H_4O_3$), a compound that gives a color reaction with Fe^{3+}. B also may be prepared in a sequence that begins with the product (D) from the crossed Claisen condensation of ethyl acetate with diethyl oxalate. D reacts with Zn and ethyl bromoacetate to give E. Careful hydrolysis of E with dilute NaOH and work-up gives F ($C_6H_8O_7$). Treatment of F with concentrated H_2SO_4 at 0° gives G. Upon being heated, G decomposes into one mole of acetone and two of CO_2. G reacts with sodium borohydride to give H. H, when gently warmed, gives B. E, when heated, gives I, which when treated with dilute NaOH and work-up gives J ($C_6H_6O_6$).

Write the structural formula of each lettered substance.

18.2 Write the major product for each of the following reactions:

a. $HOCH_2CH_2CHCH_2COO^{\ominus} \xrightarrow{H^{\oplus}}$
$\qquad\qquad\qquad |$
$\qquad\qquad\;\; C_6H_5$

b.  [benzene ring]—COONa $\xrightarrow[> 300°]{\text{soda lime}}$

c. 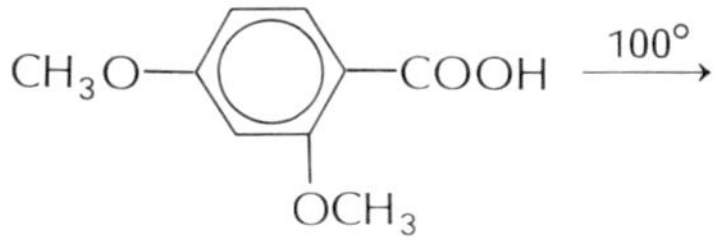 CH_3O—[benzene ring with OCH_3]—COOH $\xrightarrow{100°}$

d. $CH_3CH=CCOOH \xrightarrow{150°}$
$\qquad\quad\;\; |$
$\qquad\; C\equiv N$

e. $CH_3(CH_2)_7CH_2Br + CH_2(COOC_2H_5)_2 \xrightarrow{^{\ominus}OC_2H_5} A \xrightarrow[H_2SO_4]{\text{strong}} B$

f. $CH_3(CH_2)_7COCl + CH_2(COOC_2H_5)_2 \xrightarrow{^{\ominus}OC_2H_5} A \xrightarrow{\Delta, H_2O} B$
$\qquad\qquad\qquad\qquad\qquad\qquad\qquad\qquad\qquad\;\; \searrow \Delta, H_3O^{\oplus}$
$\qquad\qquad\qquad\qquad\qquad\qquad\qquad\qquad\qquad\qquad\quad C$

g. 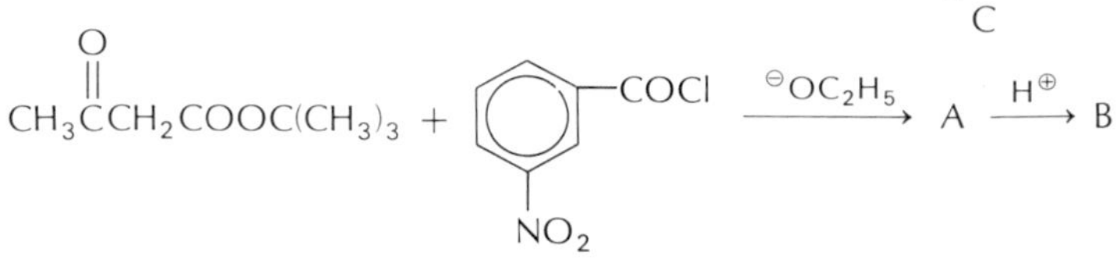 $CH_3\overset{O}{\underset{\|}{C}}CH_2COOC(CH_3)_3 +$ [benzene ring with NO_2]—COCl $\xrightarrow{^{\ominus}OC_2H_5} A \xrightarrow{H^{\oplus}} B$

610   Chapter eighteen

18.3   In each of the following equilibria indicate whether the equilibrium will be mainly to the left or to the right. Why?

a. 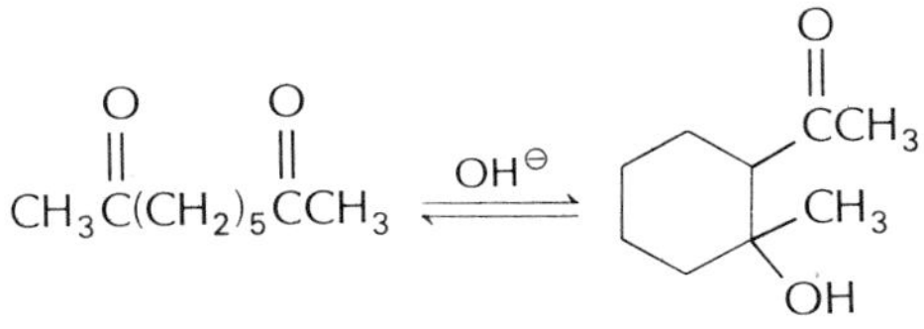 $CH_3\overset{O}{\overset{\|}{C}}(CH_2)_5\overset{O}{\overset{\|}{C}}CH_3 \; \underset{}{\overset{OH^\ominus}{\rightleftharpoons}}$ [cyclohexane with $\overset{O}{\overset{\|}{C}}CH_3$, CH_3, and OH substituents]

b. 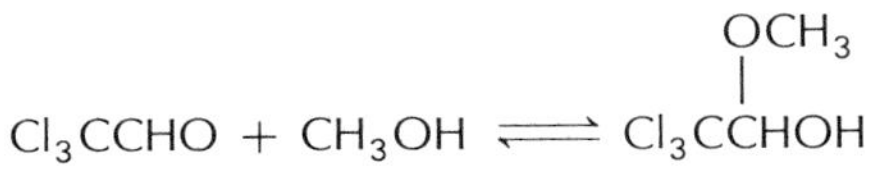 $Cl_3CCHO + CH_3OH \rightleftharpoons Cl_3C\underset{\underset{OCH_3}{|}}{C}HOH$

c. $CH_3CH_2CHO + 2CH_3OH \overset{H^\oplus}{\rightleftharpoons} CH_3CH_2CH(OCH_3)_2 + H_2O$

d. $CH_3COCH_3 + 2CH_3OH \overset{H^\oplus}{\rightleftharpoons} (CH_3)_2CH(OCH_3)_2 + H_2O$

e. 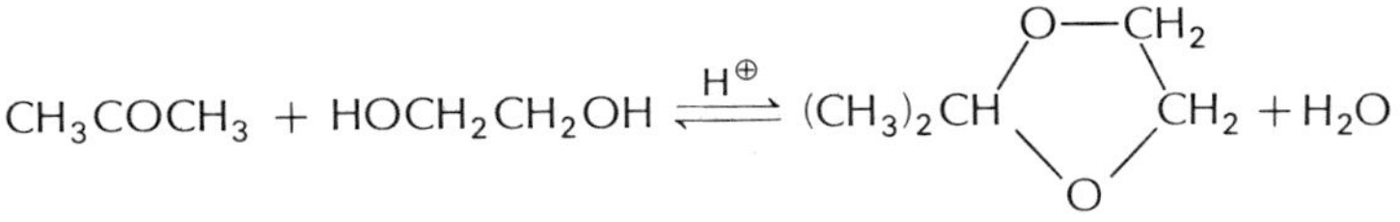 $CH_3COCH_3 + HOCH_2CH_2OH \overset{H^\oplus}{\rightleftharpoons} (CH_3)_2C\underset{O}{\overset{O-CH_2}{\diagup\diagdown}}CH_2 + H_2O$

f. [cyclopentanone]$=O + HO(CH_2)_3OH \overset{H^\oplus}{\rightleftharpoons}$ [spiro ketal] $+ H_2O$

g. $CH_3\underset{\underset{OH}{|}}{\overset{\overset{CH_3}{|}}{C}}CH_2CH_2COOH \overset{H^\oplus}{\rightleftharpoons} CH_3\overset{\overset{CH_3}{|}}{C}\underset{O\diagdown\diagup}{\overset{-CH_2}{}}\underset{\overset{\|}{O}}{\underset{C}{}}CH_2 + H_2O$

h. $C_6H_5\underset{\underset{OH}{|}}{C}H(CH_2)_3COOH \overset{H^\oplus}{\rightleftharpoons} C_6H_5\underset{\underset{O}{|}}{C}H\text{---}(CH_2)_3 + H_2O$
$\phantom{h.\ C_6H_5\underset{\underset{OH}{|}}{C}H(CH_2)_3COOH \overset{H^\oplus}{\rightleftharpoons} C_6H_5\underset{\underset{O}{|}}{C}} \overset{|}{\underset{C=O}{}}$

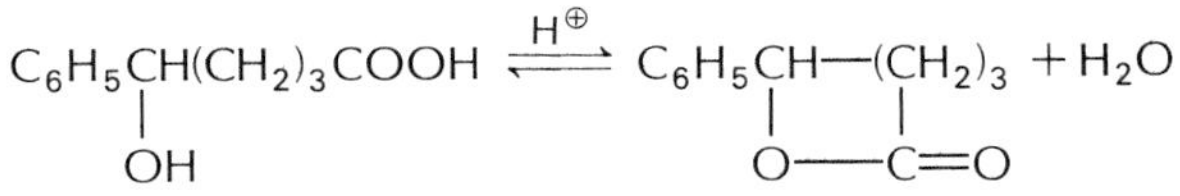

i. 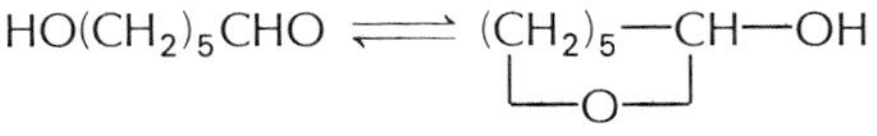 $HO(CH_2)_5CHO \rightleftharpoons (CH_2)_5\underset{\underset{O}{\llcorner\text{---}\lrcorner}}{\text{---}CH\text{---}OH}$

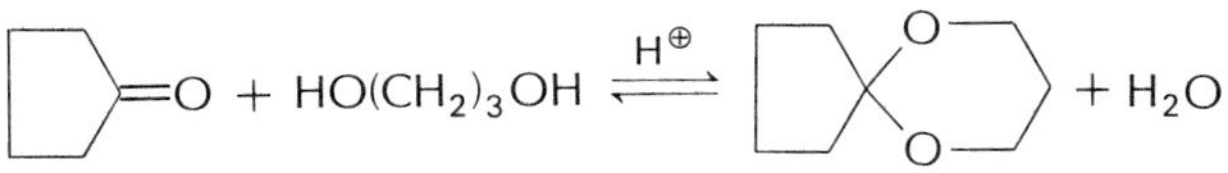

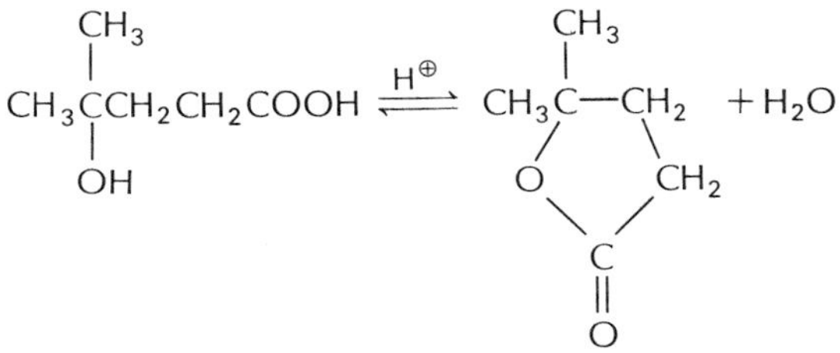

j. 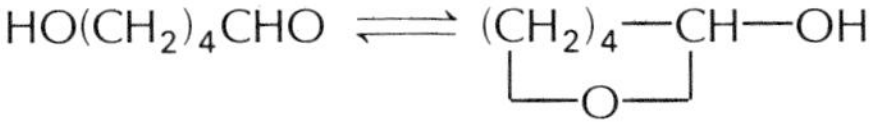 $HO(CH_2)_4CHO \rightleftharpoons (CH_2)_4\underset{\underset{O}{\llcorner\text{---}\lrcorner}}{\text{---}CH\text{---}OH}$

k. 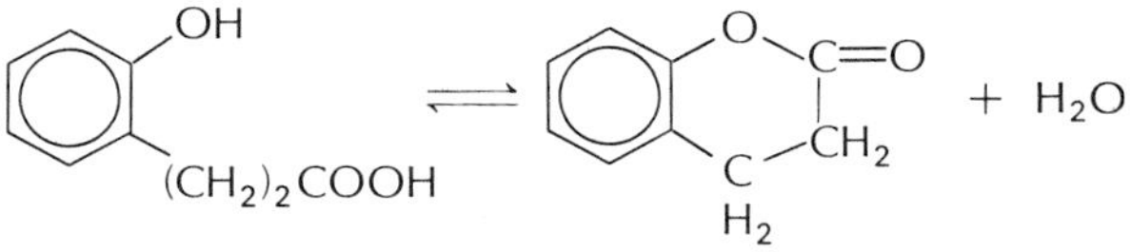 [benzene-OH, (CH₂)₂COOH] $\rightleftharpoons$ [benzo-lactone ring] $+ H_2O$

18.4   What is the major product of each of the following reactions?

a. 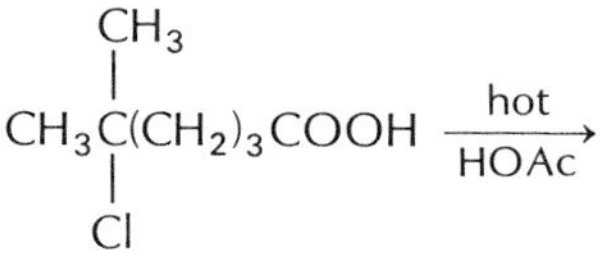 $CH_3\underset{\underset{Cl}{|}}{\overset{\overset{CH_3}{|}}{C}}(CH_2)_3COOH \xrightarrow[HOAc]{hot}$

b. $CH_3\underset{Cl}{\overset{CH_3}{\underset{|}{\overset{|}{C}}}}(CH_2)_4COOH \xrightarrow[HOAc]{hot}$

c. $HOOC(CH_2)_3COOH \xrightarrow{Ac_2O}$

d. 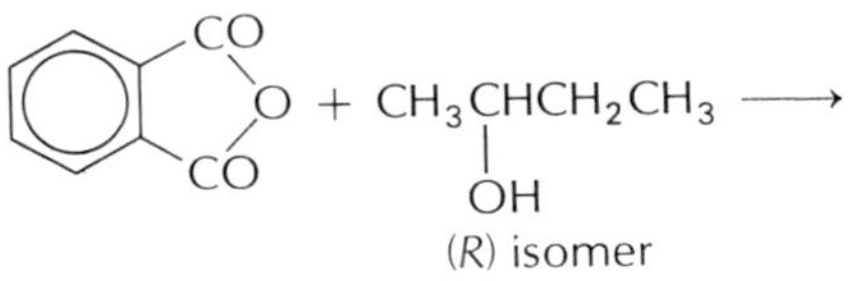
$\begin{array}{c}\text{phthalic anhydride} \\ \end{array} + CH_3\underset{OH}{\underset{|}{CH}}CH_2CH_3 \longrightarrow$
(R) isomer

e. 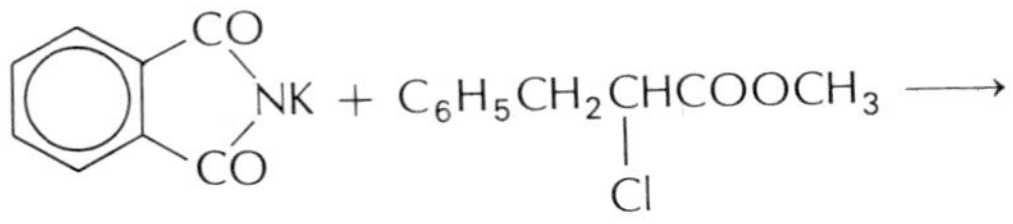
$\begin{array}{c}\text{phthalimide-NK}\end{array} + C_6H_5CH_2\underset{Cl}{\underset{|}{CH}}COOCH_3 \longrightarrow$

f. $CH_3O\underset{\underset{O}{\|}}{C}(CH_2)_5COOCH_3 \xrightarrow{^{\ominus}OCH_3}$

g. $CH_3OOC\underset{CH_3}{\underset{|}{CH}}(CH_2)_4COOCH_3 \xrightarrow{^{\ominus}OCH_3}$

h. $(C_2H_5OOC)_2C\begin{matrix}C_2H_5 \\ \diagdown \\ \diagup \\ CH(CH_2)_3CH_3 \\ | \\ CH_3\end{matrix} + NH_2\overset{\overset{O}{\|}}{C}NH_2 \xrightarrow{^{\ominus}OC_2H_5}$ (Nembutal)

i. 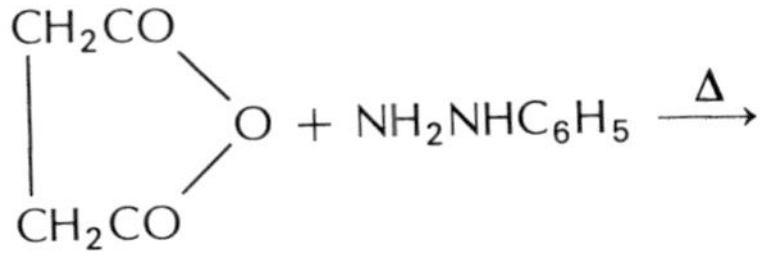 $\begin{matrix}CH_2CO \\ | \quad \diagdown \\ \quad \quad O \\ | \quad \diagup \\ CH_2CO\end{matrix} + NH_2NHC_6H_5 \xrightarrow{\Delta}$

j. $CH_3\overset{\overset{O}{\|}}{C}(CH_2)_3\overset{\overset{O}{\|}}{C}OC_2H_5 \xrightarrow{^{\ominus}OC_2H_5}$

k. $CH_3\overset{\overset{O}{\|}}{C}(CH_2)_2\overset{\overset{O}{\|}}{C}OC_2H_5 \xrightarrow{N_2H_4}$

l. $C_6H_5CH=CHCOCH_3 + N\equiv CCH_2COOC_2H_5 \xrightarrow{^{\ominus}OC_2H_5}$

m. $CH_3CO(CH_2)_5COCH_3 \xrightarrow{^{\ominus}OH}$

n. $C_6H_5\underset{\underset{CH_2}{\|}}{C}(CH_2)_2COOH \xrightarrow{^{\oplus}H}$

18.10 Bibliography

The references mentioned in Chapter 17 are applicable to this chapter as well. In addition, it may be of interest to read some reports that describe the isolation and identification of a few substances from natural sources.

1. Boylan, David B., and Paul J. Scheuer, Pahutoxin: A Fish Poison, *Science* 155, 52 (1967).

2. Jacobson, Martin, Robert E. Redfern, William A. Jones, and Mary H. Aldridge, Sex Pheromones of the Southern Armyworm Moth: Isolation, Identification, and Synthesis, *Science* 170, 542 (1970).

3. Meinwald, J., Y. C. Meinwald, A. M. Chalmers, and T. Eisner, Dihydromatricaria Acid: Acetylenic Acid Secreted by Soldier Beetle, *Science* 160, 890 (1968).

18.11 D. Stanley Tarbell: A tough antibiotic to unravel

D. Stanley Tarbell, Distinguished Professor of Chemistry at Vanderbilt since 1967, earned his Ph.D. at Harvard in 1937. After a postdoctoral fellowship year at Illinois, he went on to Rochester where he stayed until 1966, his last years there as chairman of the department. Professor Tarbell has been the recipient of several honors and is well known for his research work on phenolic ethers, organic sulfur compounds, natural products, and in theoretical organic chemistry. His short essay conveys some of the characteristics of research.

I suppose that the satisfaction in my job of teaching and doing research in organic chemistry has come from two sources: from people and from solving scientific problems. The people whom I remember with particular pleasure are Louis Fieser, who first interested me in organic chemistry; E. P. Kohler, who gave a masterly advanced course; and Paul D. Bartlett, whose first Ph.D. student I was. In addition to these, there have been many graduate students and postdoctoral fellows who have worked with me, some of whom were rather amateurish and lacked confidence when they started, but who developed into expert original chemists who could work on their own. I think that I got as much satisfaction out of these people as out of anything connected with research.

In research on many problems, one thing that stands out in my mind particularly as one of my greatest satisfactions was the final solution of the problem of the structure of an antibiotic, fumagillin.

$$HOOC(CH=CH)_4COO-\text{[cyclohexane with spiro epoxide CH}_2\text{]}-\overset{OCH_3}{\underset{CH_3}{C}}-\text{[epoxide]}-CHCH_2CH=C\overset{CH_3}{\underset{CH_3}{}}$$

fumagillin

The work on this compound, which had interesting medicinal properties, had been frustrating; the compounds formed were difficult to separate, they frequently were not formed reproducibly, and very few of them were crystalline. The work continued over a period of several years, eight or nine I believe; and finally I began to think that we were unusually stupid and that someone else would solve the problem by some simple fashion that we had overlooked. However, so much time and money had been invested in the problem that I did not feel we could give it up.

Finally after many ups and downs, we arrived at a structure for the compound; and it remained only to synthesize a product obtained from the antibiotic in order to establish its structure conclusively. I put a postdoctoral fellow and a graduate student on this problem together, working from different angles; and the graduate student reached the solution first in 1961. I had been staying at home for a couple of days with a touch of flu when the graduate student came over to see me to tell me that he had completed a successful synthesis of the degradation product and that, therefore, the problem was solved.

A good deal of the thrill of research is in pitting one's energy, experimental skill, and knowledge against a difficult problem, and then, by a combination of good luck and hard work, obtaining a satisfactory solution. I believe that all research scientists feel this to a very high degree.

19 Carbohydrates

Carbohydrate is the class name for a naturally occurring compound of formula $C_m(H_2O)_n$. However, there are so many exceptions that it may be better to say that the word carbohydrate applies to glucose and all its friends and relations, its cousins, its uncles and aunts, and its ancestors and progeny, even unto nitrogen-containing compounds. Perhaps we can improve the definition by saying that a carbohydrate is a polyhydroxy compound that usually has an aldehyde or a ketone function present, either free (unusual) or as a hemiacetal or acetal. Carbohydrates are sugars, starches, and cellulose; they are a lot more besides.

19.1   Glucose   614
     1. Occurrence; mutarotation; epimers and anomers.   2. Oxidation of glucose: gluconic acid, glucaric acid.   3. Reduction: glucitol.   4. Cyanohydrins: up the chain.   5. Down the chain.   6. Osazones.   7. Glucosides: proof of pyranose ring.   8. Other representations of glucose. Exercises.

19.2   Aldoses   622
     1. Fischer's proof of structure.   2. Epimer interconversion.   3. The D and L family assignments.   4. The D family of aldoses. Exercises.

19.3   Ketoses and vitamin C   627
     1. The D family of 2-ketoses: fructose.   2. Vitamin C or ascorbic acid: synthesis.   3. Scurvy. Exercises.

19.4   Synthesis of sugars   630
     Lobry de Bruyn-van Eckenstein transformation.

19.5   Disaccharides   632
     1. Cellobiose.   2. Use of periodic acid.   3. Gentiobiose.   4. Lactose.   5. Lactase.   6. Maltose.   7. Melibiose.   8. Sucrose.   9. Trehalose. Exercises.

19.6   Trisaccharides   637
     1. Gentianose.   2. Raffinose. Exercises.

614 Chapter nineteen

19.7   Polysaccharides   637
       1. Cellulose.   2. Chitin.   3. Hemicelluloses.   4. Starch: amylose, amylopectin.   5. Glycogen.   6. Other polysaccharides.
       7. Physiological polysaccharides.   8. Blood group substances.

19.8   Sugar alcohols   642
       Mannitol.

19.9   Photosynthesis and metabolism of carbohydrates   643
       1. Photosynthesis.   2. Pathways for utilization of glycogen.

19.10 Summary   645

19.11 Problems   645

19.12 Bibliography   646

19.13 Melvin Calvin: One thing leads to another   646a

19.1   Glucose

Occurrence; mutarotation; epimers and anomers.   Glucose (also called dextrose), alone and in polymeric form as starch and cellulose, is the most abundant organic compound on earth. It is present throughout the plant kingdom and the animal world. It is the product of photosynthesis and the precursor of every other naturally occurring organic compound, simple and complex. Let us begin our examination of glucose by looking at the physical properties.

Glucose ($C_6H_{12}O_6$) exists in two crystalline forms known as α-D-glucopyranose and β-D-glucopyranose (X-ray data). The α-glucose is obtained by crystallization from aqueous solution below 30° as a hydrate of mp 83°; the β form, by crystallization from aqueous solution above 98°. When the anhydrous crystalline forms are dissolved, the initial specific rotations are 112° for α and 19° for β. The rotations begin to change immediately: that of α decreases and that of β increases until the final equilibrium value of 52.7° is reached. The phenomenon is called mutarotation (Eq. 19.1).

Examine the configurations at carbon atom 1 in the α and β forms. An inversion is taking place between the axial hydroxy (equatorial hydrogen) in the α form and the equatorial hydroxy (axial hydrogen) in the β form. We see also that the carbon atom at 1 is linked not only to a hydroxy group, but also to another oxygen (through the ring oxygen to 5). It is a hemiacetal. Hemiacetals (Eq. 16.9) are subject to both acid and base catalysis in reaching equilibrium. The equilibria here favor the cyclic hemiacetal formation, even in aqueous solution. However, the open-chain aldehyde structure is intermediate between the two cyclic forms. The structures *a* and *b* are open-chain ionic intermediates in the acid- and base-catalyzed reactions. At equilibrium, when the specific rotation is 52.7°, the solution contains 36% of α, 64% of β, and about 0.02% of the aldehyde (as the aldehyde + *a* + *b*). From the equilibrated solution (or even before equilibrium is reached), the α or β

19.1 Glucose

α-D-glucose
$[\alpha]_D 112°$
mp 146°

(most of the C—H bonds omitted for clarity)

β-D-glucose
$[\alpha]_D 19°$
mp 148°–155°

(19.1)

crystalline stereoisomer may be obtained as we have described. These epimers (epimers differ in configuration at only one asymmetric center) are given the special name of anomers (epimeric at the 1 position only) in the sugar series.

Because the amount of free aldehyde is as small as it is during mutarotation or at equilibrium, it is not surprising that glucose does not undergo all the reactions of an aldehyde. However, if the reaction in question is irreversible or has an equilibrium constant greater than that for the hemiacetal formation, then it will take place.

Let us stop and state some definitions. The α and β refer to the location of the hydroxyl group at 1: α is below the plane of the ring and β is above. The small capital D designates the configuration at 5. Gluco- refers to the presence of six carbon atoms and to the configuration of the hydroxy groups at 2, 3, and 4: namely, all equatorial. Pyran- refers to a six-membered ring with one oxygen atom as part of the ring; and -ose means that we are referring to a saccharide (a sugar). Glucose is a monosaccharide: it cannot be hydrolyzed into smaller saccharides. We shall elaborate each of these points as we go on.

Oxidation of glucose: gluconic acid, glucaric acid. Glucose reacts with Fehling's solution or Tollens' reagent (Sec. 16.6) and is oxidized to gluconic acid. Because it does so, glucose is termed a reducing sugar. (Some sugars are not reducing.) For preparative purposes, gluconic acid is obtained by oxidation with bromine water. The reagents differ in pH. Both Fehling's solution and Tollens' reagent are basic, and under basic conditions glucose undergoes extensive rearrangement by way of the enolate anion (to be discussed later). Bromine water is an acidic reagent, strong enough to oxidize an aldehyde or hemiacetal group but not an alcohol (Eq. 19.2). For clarity, we have shown the reaction as though the aldehyde were being oxidized. The gluconic acid may close (in the presence of hydrogen ion) to give either the γ- or the δ-lactone. Of the two, the γ-lactone is the more stable, so the δ-lactone isomerizes to give the γ-lactone. The oxidation actually seems to proceed by way of the β-anomer (the α-anomer is oxidized at the rate at which it mutarotates) through a hypobromite decomposition $\left(\begin{array}{c} \diagdown \diagup \\ \diagup C \diagdown \\ H \end{array} \begin{array}{c} OBr \\ \\ \end{array} \rightarrow \begin{array}{c} \diagdown \\ C=O + HBr \\ \diagup \end{array} \right)$ to give the

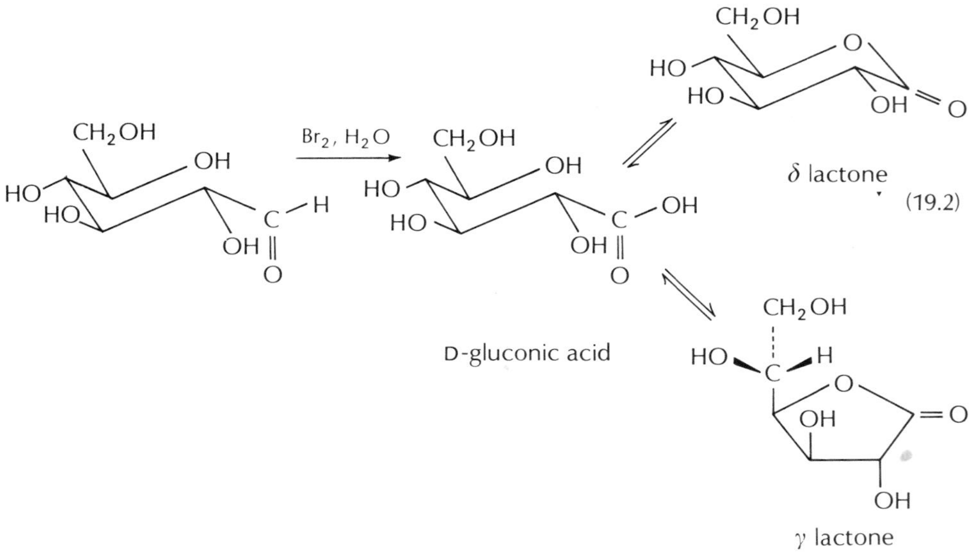

$$\text{[D-glucose]} \xrightarrow{Br_2, H_2O} \text{D-gluconic acid} \rightleftharpoons \delta \text{ lactone} \rightleftharpoons \gamma \text{ lactone} \qquad (19.2)$$

δ-lactone as the first product. We obtain gluconic acid by opening the lactone ring with dilute NaOH, followed by acidification at low temperature.

A stronger oxidizing agent, moderately concentrated nitric acid oxidizes glucose to D-glucaric acid (a tetrahydroxy adipic acid) in which the primary alcohol group has been oxidized as well as the free aldehyde group (Eq. 19.3). (Let us not get into the subject of the

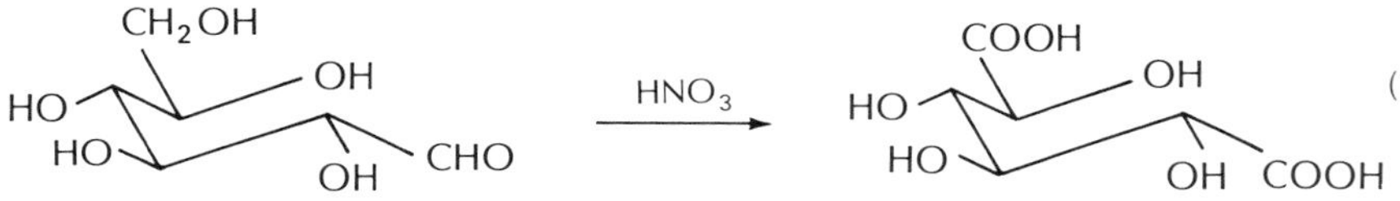

$$\text{[D-glucose]} \xrightarrow{HNO_3} \text{[D-glucaric acid]} \qquad (19.3)$$

possible lactones and dilactones.) If the reaction becomes too drastic, oxalic acid becomes the major product (other than CO_2).

Reduction: glucitol. Glucose may be reduced by catalytic hydrogenation, by sodium amalgam (used extensively before catalytic hydrogenation became a reliable technique), or by $NaBH_4$ with a hydrogenation catalyst (the present popular method). The product is called D-glucitol (common name, sorbitol). It is an intermediate in the synthesis of vitamin C (Sec. 19.3).

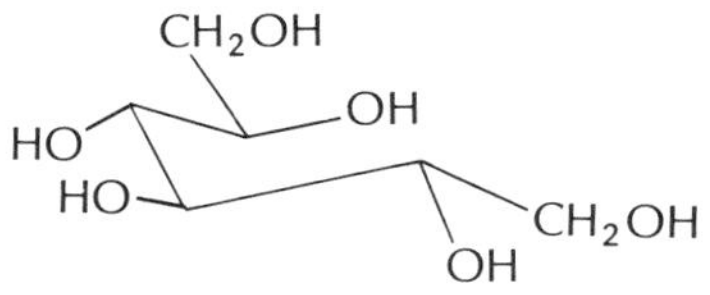

D-glucitol (sorbitol)

Cyanohydrins: up the chain. Glucose as an aldehyde, will form two cyanohydrins in uneven amounts; the asymmetry of the chain prevents complete randomness in the addition to the carbonyl group. Hydrolysis under acidic conditions of the separated cyanohydrins (which are not enantiomers, but epimers) gives heptonic acids, which are converted into the γ-lactones. Reduction of the lactones at pH 3 with NaBH$_4$ or sodium amalgam gives epimeric D-heptoses (Eqs. 19.4). The series of reactions is known as the Kiliani-Fischer synthesis. The products have been written as aldehydes instead of hemiacetals for simplicity and clarity. The sequence of reactions has given rise to the epimeric pair (at C$_2$ of the longer chain) of sugars that have one additional carbon in the chain. We shall call this process "going up the chain."

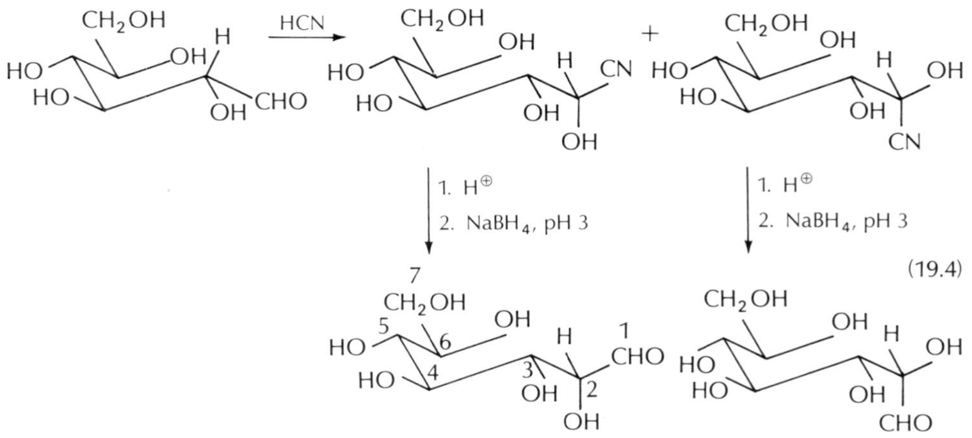

Down the chain. "Going down the chain," or degradation, also is possible. Several methods have been used. We shall give two.

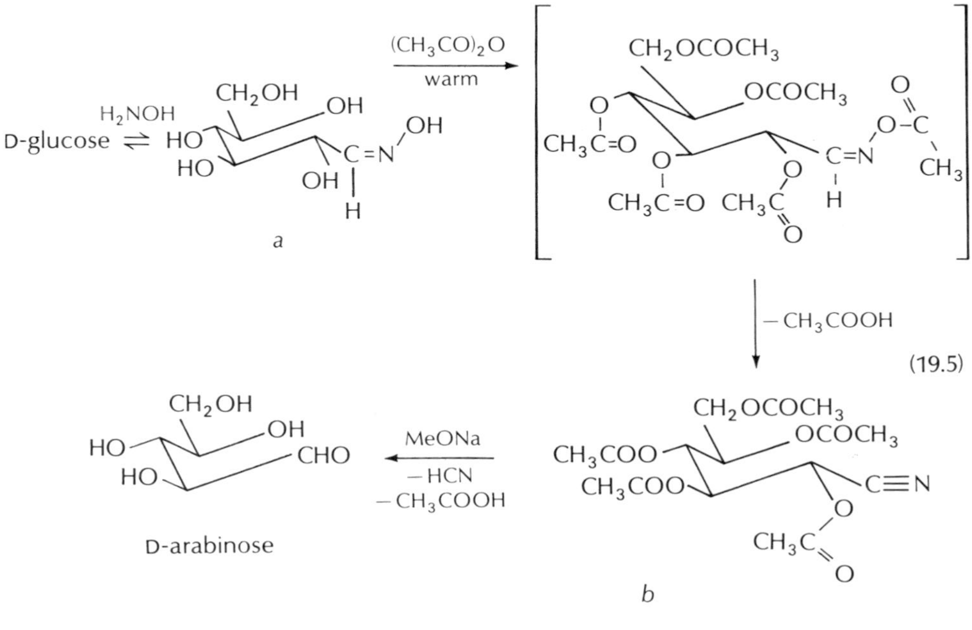

The Wohl degradation is the reverse of the cyanohydrin synthesis. The first step is the formation of the oxime (a, favored by the equilibrium) of the aldehyde group. Next, the oxime acetate is prepared by the action of acetic anhydride. At the same time, the hydroxy groups also are partly esterified. Thus, we must use an excess of the anhydride to get a pure product. An aldoxime acetate is unstable and decomposes by loss of acetic acid to give a nitrile (b). All the ester groups are decomposed by warming with anhydrous base (which shifts the cyanohydrin-carbonyl-group equilibrium towards the free carbonyl by converting HCN into CN$^\ominus$) or by the presence of Ag$_2$O (removes HCN as insoluble AgCN). The product is the aldose that has one less carbon atom in the chain (Eq. 19.5). As we go down the chain, only a single product is formed: from D-glucose, the product is D-arabinose.

The Ruff degradation consists of oxidation of an aldose to the glyconic acid (D-glucose to D-gluconic acid) and treatment of the calcium salt of the acid with Fenton's reagent (H_2O_2 + Fe^{2+}), which oxidizes the acid to an α-keto acid. This product decarboxylates under acidic conditions (Eqs. 19.6).

$$\text{D-glucose} \xrightarrow[H_2O]{Br_2} \text{D-gluconic acid} \xrightarrow{CaCO_3} \text{Ca(gluconate)}_2$$

$$\downarrow H_2O_2, Fe^{2+}, H^{\oplus} \quad (19.6)$$

D-arabinose ← (−CO_2) ← [α-keto acid intermediate]

While we are looking at arabinose, we had better point out that the aldopentoses tend to favor the furanose ring structure rather than the pyranose ring. Furan- refers to a five-membered ring with one oxygen atom.

α-D-arabinofuranose ⇌ (open chain) ⇌ β-D-arabinofuranose

Osazones. Phenylhydrazine reacts with a hydroxy aldehyde or a hydroxy ketone to give a double phenylhydrazone. These products, which were first seen with sugars, are known as osazones. Because we are interested for the moment only in what happens at positions 1 and 2, let us simplify matters and use R— for the rest of the chain. The balanced equation (a) and a possible mechanism (b) are shown in Equations 19.7, as well as the formation of the osazone of benzil (c) by reaction of benzoin with phenylhydrazine. The reaction is a reduction of phenylhydrazine by the oxidation of the hydroxy group α to the carbonyl.

For our purposes, osazone formation destroys the configuration at position 2 in the chain: the tail (positions 3, 4, etc.) retains configuration. Osazones crystallize nicely and serve as useful derivatives of the sugars for identification.

Glucosides: proof of pyranose ring. The hemiacetals, the α and β forms of a sugar, are easily converted into mixed acetals by reaction with an alcohol in the presence of acid (Eq. 19.8). Either α or β glucose gives the same mixture of α- and β-methyl glucosides (separable by crystallization). No acyclic dimethyl acetal is formed. The methyl glucosides hydrolyze readily with dilute acids to give back a solution of glucose and methanol. Under basic conditions, the acetal is stable: it does not react with Fehling's solution or Tollens' reagent, it does not give an oxime or an osazone under alkaline conditions, and it does not mutarotate.

The methyl glucosides (as well as those from other alcohols) are hydrolyzed with specific enzyme catalysts. Thus the enzyme α-glucosidase (old name, maltase) catalyzes

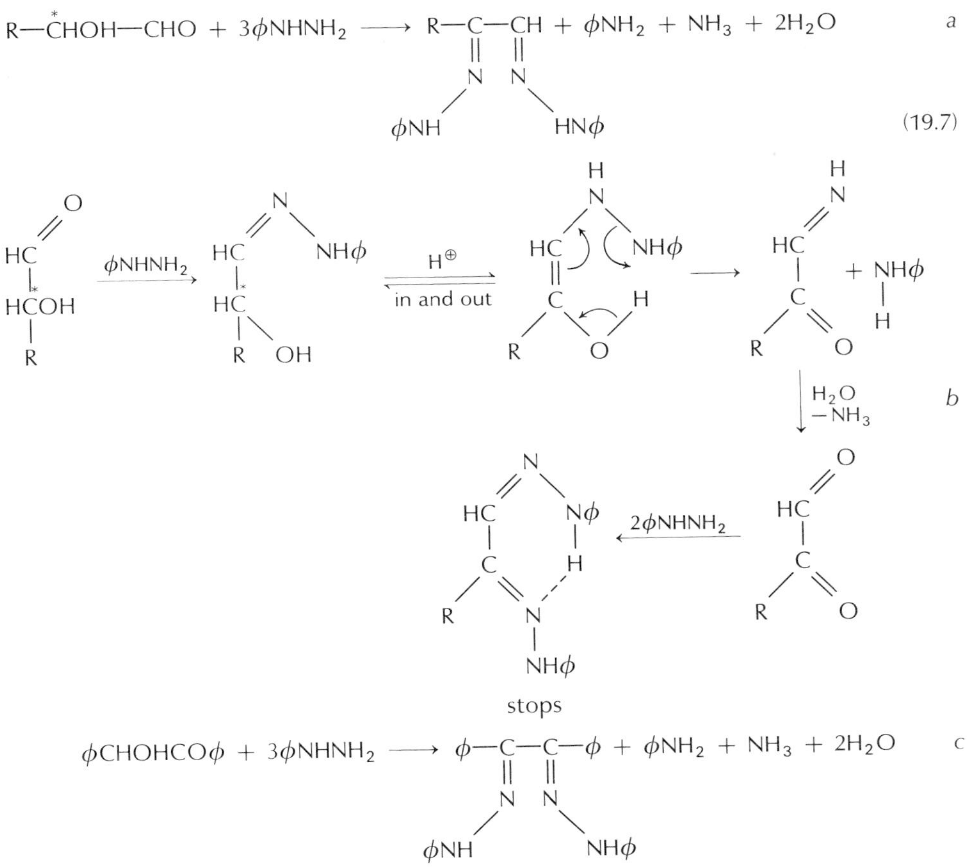

$$\text{R}-\overset{*}{\text{CHOH}}-\text{CHO} + 3\phi\text{NHNH}_2 \longrightarrow \text{R}-\underset{\underset{\phi\text{NH}}{\overset{\|}{\text{N}}}}{\text{C}}-\underset{\underset{\text{HN}\phi}{\overset{\|}{\text{N}}}}{\text{CH}} + \phi\text{NH}_2 + \text{NH}_3 + 2\text{H}_2\text{O} \qquad a$$

(19.7)

b

stops

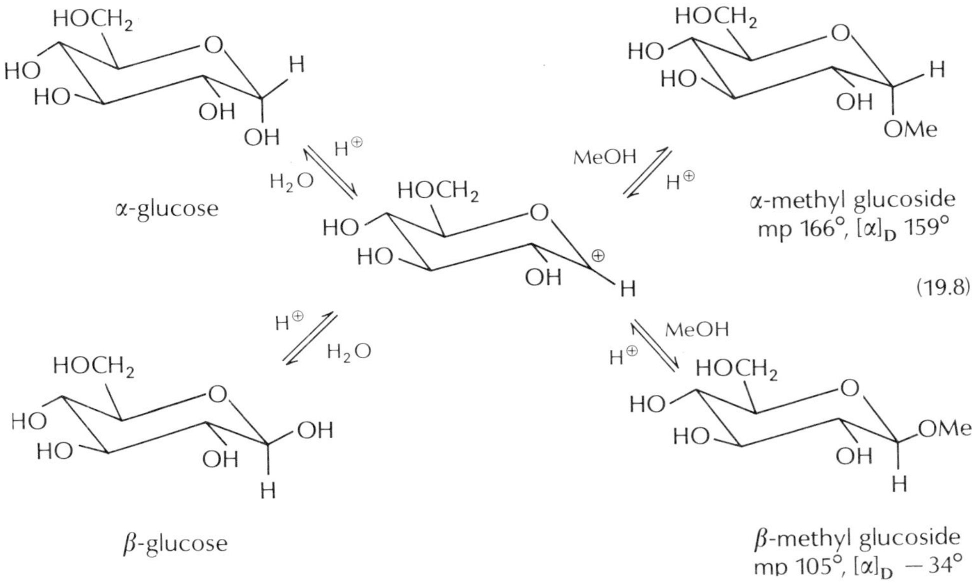

$$\phi\text{CHOHCO}\phi + 3\phi\text{NHNH}_2 \longrightarrow \phi-\underset{\underset{\phi\text{NH}}{\overset{\|}{\text{N}}}}{\text{C}}-\underset{\underset{\text{NH}\phi}{\overset{\|}{\text{N}}}}{\text{C}}-\phi + \phi\text{NH}_2 + \text{NH}_3 + 2\text{H}_2\text{O} \qquad c$$

α-glucose

α-methyl glucoside
mp 166°, [α]$_D$ 159°

(19.8)

β-glucose

β-methyl glucoside
mp 105°, [α]$_D$ −34°

the hydrolysis of only α-glucosides; and β-glucosidase (old name, emulsin) will hydrolyze only β-glucosides. Most of the naturally occurring glucosides are β (thus, equatorial).

Once a glucoside has been formed by an acid-catalyzed reaction, the remaining hydroxy groups may be methylated with $(MeO)_2SO_2$ and alkali or by MeI and Ag_2O. Once the remaining four hydroxy groups have been converted into methoxyl groups (methyl ethers), the methoxyl group at position 1 may be easily hydrolyzed by dilute acid to give a tetramethylglucose (Eq. 19.9). The name tetramethylglucose implies that α and β forms exist and mutarotation occurs.

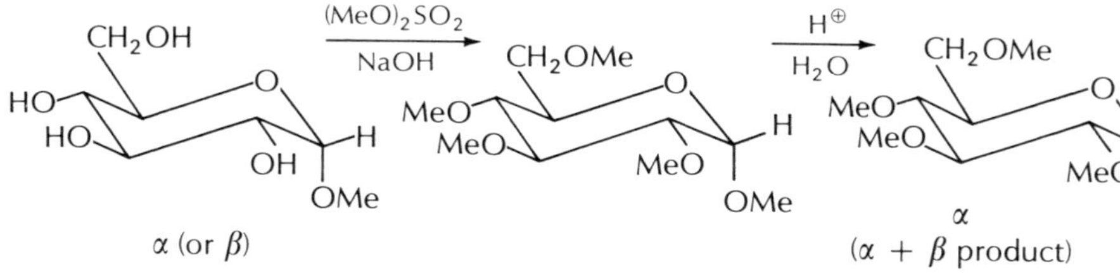

Nitric acid oxidation of either α- or β-tetramethylglucose acts on the free aldehyde form by dehydration. The free hydroxy group at 5 reacts to give a mixture of dimethoxy-succinic acid (a, also called di-O-methyltartaric acid) and trimethylxylaric acid (b, or tri-methoxyglutaric acid), as seen in Equation 19.10. These two acids can be formed only if

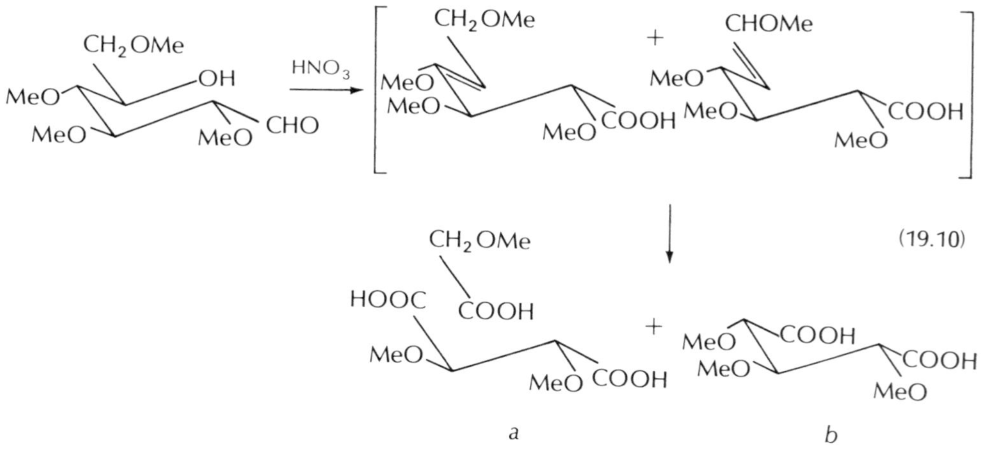

the free hydroxy group is at position 5. Therefore, the ring must be a pyranose, and not a furanose, which cannot give b.

Other representations of glucose. We have used the perspective structural formulas for glucose and the products of reactions in this section for a purpose: we must learn to think in three dimensions when we discuss sugars.

Other formulations have been used. The immediate predecessor of the perspective structural formula was the Haworth formulation, in which the pyranose ring was shown as planar. It has the advantage of showing clearly which groups are on the α and which are on the β side of the molecule. However, we shall continue to use the perspective formulas when we need them.

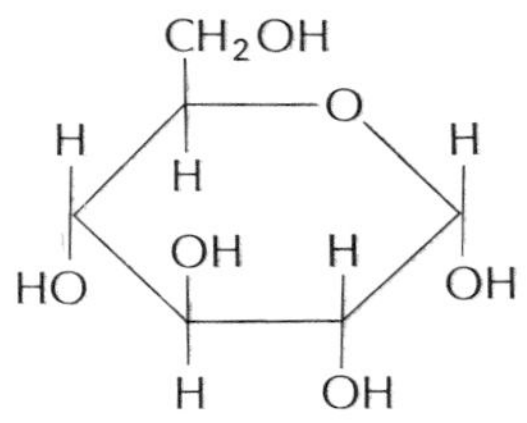

19.1 Glucose 621

Haworth formula of α-D-glucose

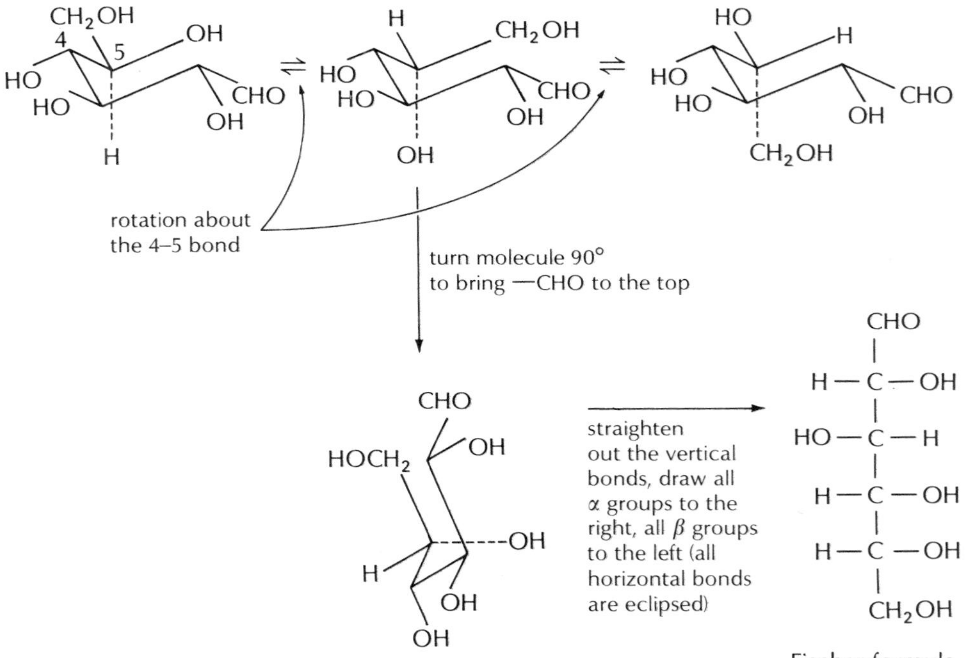

rotation about
the 4–5 bond

turn molecule 90°
to bring —CHO to the top

straighten
out the vertical
bonds, draw all
α groups to the
right, all β groups
to the left (all
horizontal bonds
are eclipsed)

Fischer formula,
D-glucose

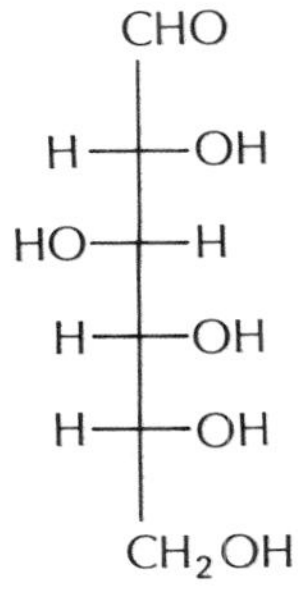

simplified
Fischer
formula,
D-glucose

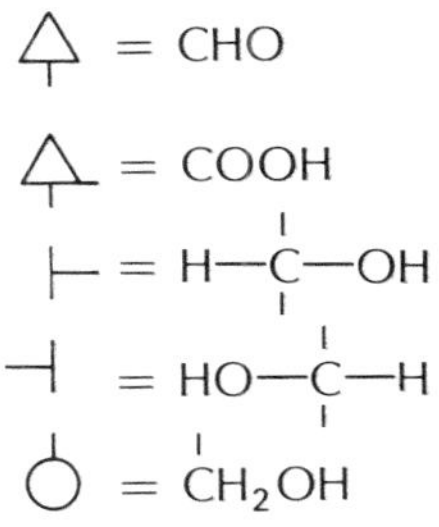

Rosanoff
symbols

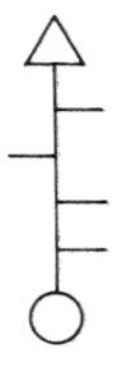

symbolic
formula,
D-glucose

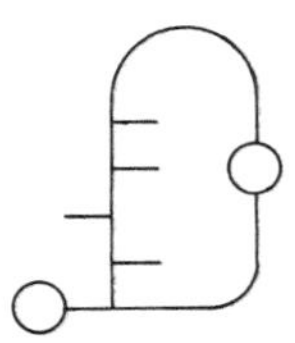

symbolic
formula,
α-D-glucose

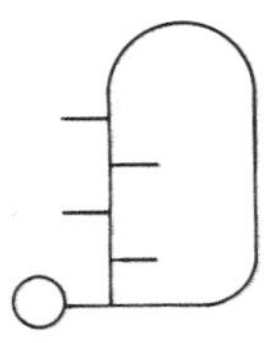

symbolic
formula,
β-D-glucose

The original Fischer planar projection formula was a convention for representation of tetrahedral structure. The chain is written vertically, with the aldehyde group at the top. Each vertical bond (now a bond in the chain) is thought to project below the plane of the paper it is written on. The horizontal bonds are thought of as projecting above the paper. This convention is most useful in the writing of open-chain structures: the aldehyde form is easily represented, but the hemiacetal forms are less realistic.

We can simplify the Fischer formula by omitting the carbon atoms of the chain. Rosanoff introduced additional symbolism for further simplification. The modified symbolism as we shall use it is shown. The symbolic structures for α- and β-D-glucose show the —CH_2OH group on the β side of the ring (or chain), as it is in the perspective formulas for α and β glucose. We must keep both the Fischer and symbolic structures in the plane of the paper to preserve the convention adopted in writing them.

Exercises

19.1 Rewrite the equations in this section using symbolic formulas instead of perspective formulas. (Omit Eq. 19.7.)

19.2 Aldoses

We have been using a given specific structure for glucose. How was the structure worked out originally, before the days of X-ray crystal-structure determination? In 1888, Emil Fischer began work on the problem. In those days, only four monosaccharide sugars were known: glucose, mannose, fructose, and arabinose. How did he go about it? (We shall set fructose aside for the time being.)

Fischer's proof of structure. It was known that glucose, mannose, and arabinose gave <u>some</u> reactions characteristic of aldehydes. Oxidation to the glyconic acids and reduction by prolonged heating with hydriodic acid gave hexanoic acid from either glucose or mannose and pentanoic acid from arabinose. Therefore, the compounds were straight-chain aldehydes. (The troublesome α and β forms are not in question for the moment.)

Glucose and mannose gave the same osazone. Therefore, glucose and mannose were epimeric at position 2 and had the same configuration for the rest of the chain.

Going down the chain, glucose and mannose gave a sugar that was the enantiomer of the known arabinose, as was shown by formation of a racemate if the two forms were mixed. Going up the chain from arabinose gave enantiomers of glucose and mannose. Given the formula $C_6H_{12}O_6$, the hexoses had to be pentahydroxyaldohexoses; and arabinose ($C_5H_{10}O_5$) had to be a tetrahydroxyaldopentose. The formulas showed that a hexose had four asymmetric centers and a pentose had three asymmetric centers. Therefore, there were $2^4 = 16$ stereoisomers (8 enantiomeric pairs) of aldohexoses and $2^3 = 8$ stereoisomers (4 enantiomeric pairs) of aldopentoses. To simplify matters, Fischer hit upon the scheme of classifying all the sugars that could be related directly to glucose as belonging to one set of compounds, which he called the D family. Their enantiomers, then, belonged to the L family. Thus D-glucose, D-mannose, and L-arabinose were the known sugars.

He made the arbitrary decision that the configuration of the bottom asymmetric hydroxy group of D-glucose, that at C_5, would be to the right. The symbol D was chosen simply because natural glucose had a positive rotation.

19.2 Aldoses

Without going down the chain completely, and with a minimum of experimental work, he solved the immediate problem as follows:

1. Arabinose could be oxidized to an optically active arabinaric acid. Therefore, two of the four possible structures (Eq. 19.11) were ruled out for D-arabinose.

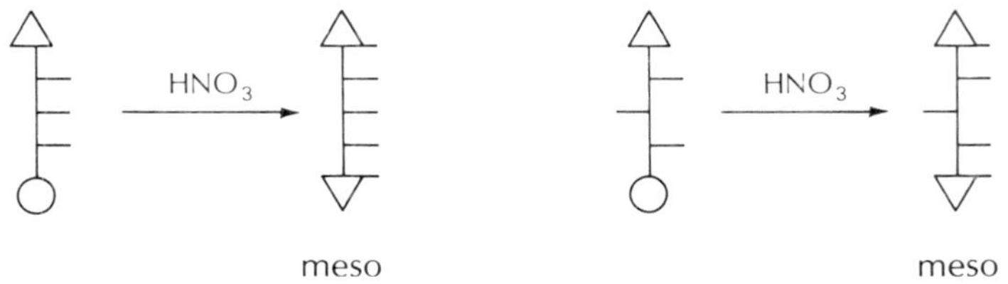

$$\text{(structures)} \xrightarrow{HNO_3} \text{meso} \qquad \text{(structures)} \xrightarrow{HNO_3} \text{meso} \qquad (19.11)$$

2. Oxidation of glucose and of mannose gave optically active glucaric acid and optically active mannaric acid. Therefore, the third structure for arabinose may be eliminated, because one of the hexoses to which it would give rise would produce a *meso*-glycaric acid (Eqs. 19.12).

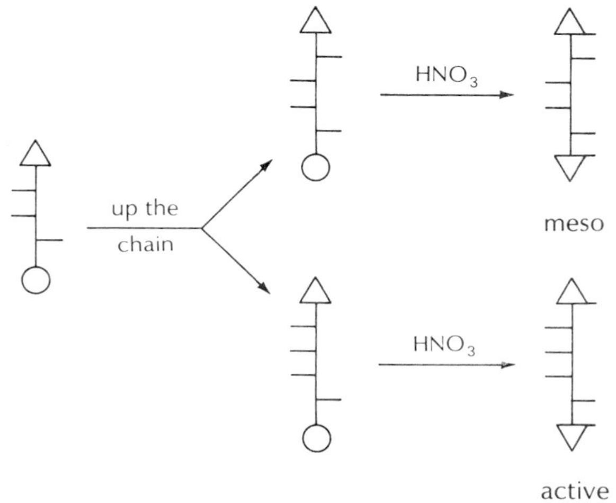

$$\xrightarrow{\text{up the chain}} \begin{cases} \xrightarrow{HNO_3} \text{meso} \\ \xrightarrow{HNO_3} \text{active} \end{cases} \qquad (19.12)$$

3. By logical examination of the situation, and knowing that only optically active glycaric acids were formed, Fischer deduced that D-arabinose must have the structure shown, because three of the possible four had been eliminated (Eq. 19.13). So he came to the

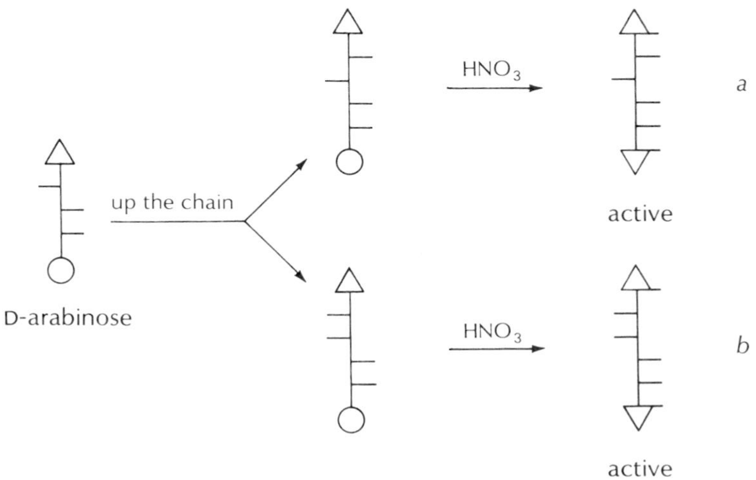

$$\text{D-arabinose} \xrightarrow{\text{up the chain}} \begin{cases} \xrightarrow{HNO_3} \text{active} \quad a \\ \xrightarrow{HNO_3} \text{active} \quad b \end{cases} \qquad (19.13)$$

final problem: which pair is glucose and glucaric acid, and which is mannose and mannaric acid, *a* or *b*?

4. What Fischer did next solved the problem very neatly. He interchanged the groups at the ends of the chains (by a series of reactions we omit). The sugar of formula *a* gives an L sugar, whereas the sugar of formula *b* gives back *b*. Of course, *a* is D-glucose and *b* is D-mannose (Eq. 19.14). Fischer had proved the structures: the year was 1891.

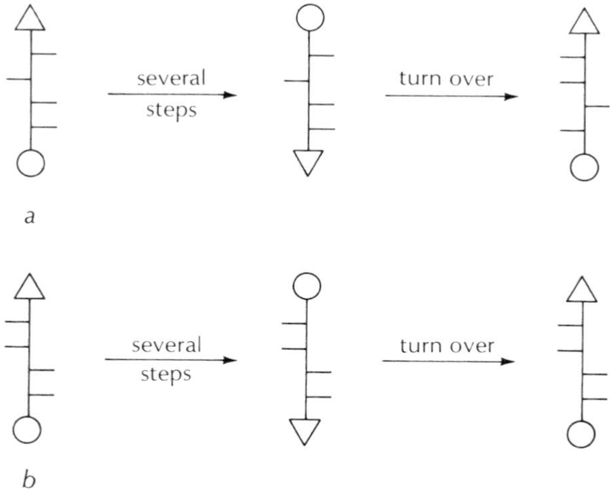

$$(19.14)$$

a

b

Epimer interconversion. With the help of an additional reaction, the structures of the entire D family of sugars were worked out. The reaction, although it was not necessary, saved much experimental work. Heating a glyconic acid with a mild base (pyridine is commonly used) causes epimerization at the carbon adjacent to the carboxylic acid group (enolization). The reaction is an equilibration, so the conversion may or may not be appreciable, depending upon the position of equilibrium and which epimeric glyconic acid is used (Eqs. 19.15). The mixture of lactones is separable by crystallization, and the

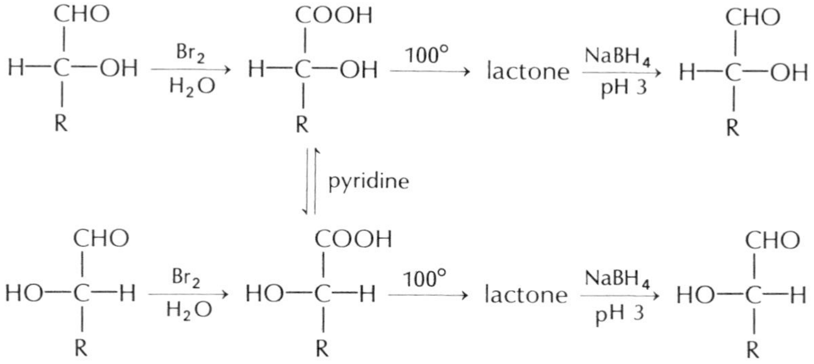

$$(19.15)$$

separate aldoses are obtained by reduction. The reaction is a short cut conversion between epimers, which otherwise requires the relatively long sequence of one step down the chain and one back up.

The D and L family assignments. Although Fischer related the other sugars to D-glucose, the family progenitor was changed to D-(+)-glyceraldehyde, and all sugars up the chain from D-(+)-glycerose (glyceraldehyde) were called D sugars. The single asymmetric hydroxy group in D-glycerose was written to the right. That configuration is retained adjacent to the —CH₂OH group in all the other sugars of the D family. The enantiomer of D-(+)-glycerose, L-(−)-glycerose, is the progenitor of all the enantiomers of the D family: namely, the L family. The designation of family, D or L, has nothing to do with the signs of rotation of the individual sugars, because the original assignment of the hydroxy group to the right was based only on the fact that natural glucose has a positive rotation.

Later on, in 1951, Bijvoet determined the absolute configuration of (+)-tartaric acid by a special X-ray diffraction method on the crystalline sodium rubidium salt. Previously, it had been known that D-(−)-threose could be oxidized to give (−)- tartaric acid and that the enantiomer could be oxidized to give (+)-tartaric acid (Eq. 19.16). Thus, the original

$$\text{L-(+)-threose} \xrightarrow{HNO_3} \text{L-(+)-tartaric acid} \qquad \text{D-(−)-tartaric acid} \xleftarrow{HNO_3} \text{D-(−)-threose} \tag{19.16}$$

arbitrary assignment of configuration happened to be correct. The (RS) convention was designed to fit the absolute configurations as far as possible. Thus, D-(+)-glyceraldehyde is (R), but the (RS) convention should not be confused with the D and L family designation. For example, D-(−)- threose is (2S,3R) and D-(−)-tartaric acid is (2S,3S).

The D family of aldoses. The complete D family of aldoses is shown in Figure 19.1. We illustrate the process of going up the chain from D-(+)-glycerose by placing the new hydroxy group on the right and writing the new sugar above and to the right, or by placing the new hydroxy group on the left and writing the new sugar above and to the left. Knowing which name goes with which formula is a matter of memory.

The most abundant aldoses in nature, either free or as glycosides, are glucose, mannose, galactose, xylose, arabinose, and ribose. Most of the naturally occurring sugars belong to the D family, although L-arabinose and L-galactose are well known. The other sugars either are of rare occurrence or have not yet been identified in nature. All members in both the D and L families have been synthesized. A few aldoheptoses have been isolated. In addition, branched-chain, amino, and deoxysugars are known, as well as the ketoses (Sec. 19.3). A few examples are shown.

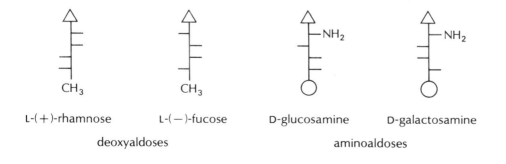

L-(+)-rhamnose   L-(−)-fucose   D-glucosamine   D-galactosamine

deoxyaldoses   aminoaldoses

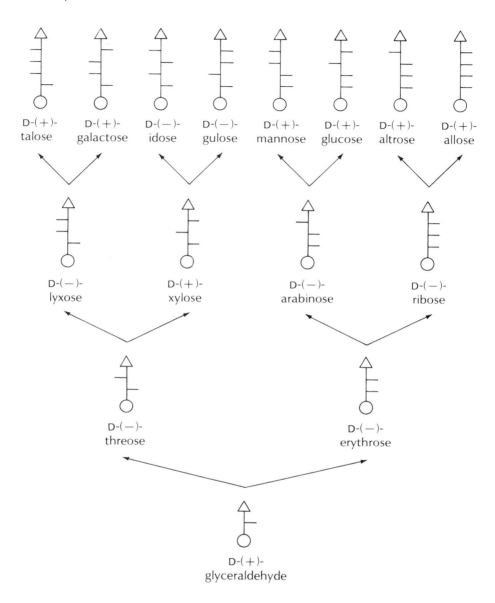

Figure 19.1 The D-aldoses: configurations, rotations

Exercises

19.2 A D-aldoheptose gives a *meso*-heptaric acid upon oxidation with nitric acid. On going down the chain, the aldohexose gives an optically active glycaric acid upon oxidation. The epimer of the aldohexose, when oxidized, gives a *meso*-glycaric acid. Give the structures of the compounds and name the aldohexoses involved.

19.3 Write a perspective structural formula for the β-anomer of each of the D-aldohexoses.

19.4 D-Mannose, after mutarotation in solution, consists of 69% of the α-anomer and 31% of the β. Why is the proportion different from that of D-glucose?

19.3 Ketoses and vitamin C

The D family of 2-ketoses: fructose. Another sugar sequence is that of the ketoses. Although it is no simple matter in the laboratory to go up or down the chain when dealing with ketoses, we can at least do it easily on paper (Fig. 19.2).

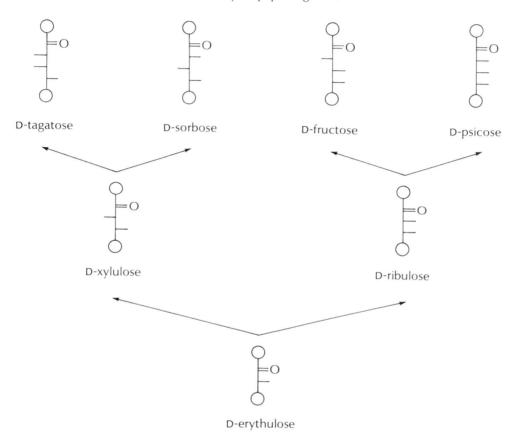

Figure 19.2   The D-ketoses: configurations

The most abundant of the ketoses is fructose, present in honey, fruit juices, and various glycosides. It is the sweetest of the sugars. (Not all sugars are sweet: many are tasteless, some are bitter.) Fructose is also known as levulose because of its negative rotation.

> The ending -<u>ulose</u> is the ending used to denote a ketose in the systematic naming. All the common ketoses have the keto group at position 2, and position 1 is a —CH_2OH group. The prefix, from the aldose system, denotes the number of carbons in the chain and the configuration of the chain with the keto group written to the right: hence, D-ribulose, not D-arabinulose. Another method is to name the tail the same as the corresponding tail in an aldose, insert the number of carbon atoms in the molecule, and end up with -ulose. Thus D-ribulose also is named as

D-erythropentulose, and D-fructose as D-arabohexulose. The names of the ketohexoses in Figure 19.2 are common names. Dihydroxyacetone (HOCH$_2$COCH$_2$OH) may be thought of as the progenitor of the ketoses; but since it has no asymmetric center, the family progenitors are D- and L-erythulose. Ketoses with more than six carbon atoms are known: D-sedoheptulose is part of the photosynthetic cycle, and D-mannoheptulose is present in avocados.

D-sedoheptulose D-mannoheptulose

Fructose crystallizes as a furanose and as a pyranose. It exhibits a mutarotation that is complicated by the participation of five structures in the equilibrium (Eq. 19.17). The

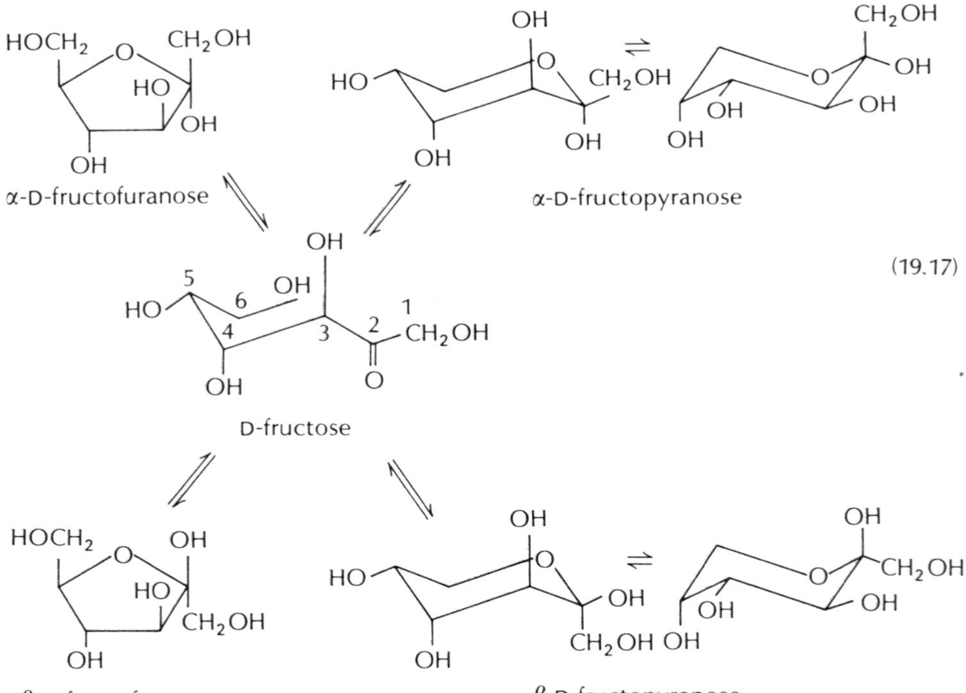

(19.17)

usual crystalline form of fructose decomposes at 103° to 105° and shows mutarotation from $-132°$ to $-92°$. Fructose gives an osazone identical to that from glucose and mannose. Therefore, the configuration of the chain is known to be identical to that of positions 3, 4, and 5 in those aldoses.

19.3 Ketoses and vitamin C

Vitamin C or ascorbic acid: synthesis. Vitamin C or ascorbic acid is the γ-lactone of the enediol form of 2-keto-L-gulonic acid. It is synthesized on a large scale. One route is shown in Equation 19.18. Glucose is reduced to the hexahydroxyhexane sorbitol (systematic name, glucitol). The next step (a) is critical. It saves many steps otherwise needed to introduce a keto group at position 5. As it is, the bacterium *acetobacter suboxydans* is able to accomplish the specific oxidation in 60 to 70% yield. The product is L-sorbose, the enantiomer of

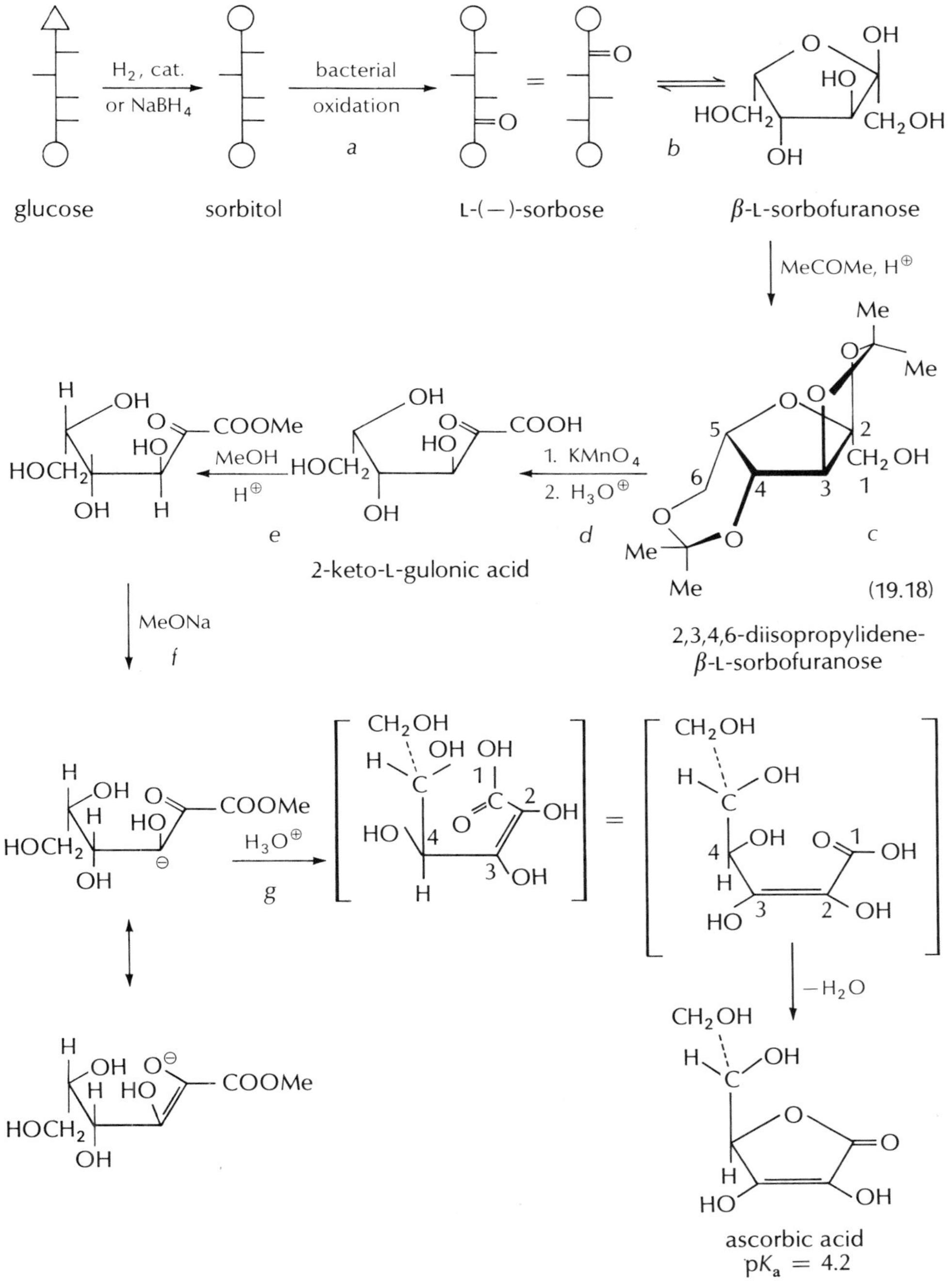

(19.18)

ascorbic acid
pK_a = 4.2

D-sorbose, which cyclizes to the furanose ring. The β-furanose reacts with two moles of acetone to give two acetone ketals (c). Hydroxy groups that are *cis*-1,2 or *cis*-1,3, as they are in β-sorbofuranose, are able to form acetals and ketals with ease. The reaction releases two moles of water, so better yields are obtained if *trans*-ketalization is used. Thus, if the preformed diethylketal of acetone is used with a trace of acid, ethanol is released in the reaction instead of water, and the conversion is almost quantitative.

The purpose of the ketalization is to protect the hydroxy groups at 2, 3, 4, and 6 so that the oxidation (at 1) to the carboxylic acid by neutral $KMnO_4$ will be clean (d). Hydrolysis of the ketals then gives 2-keto-L-gulonic acid, which is esterified (e) so that the enolization (f) will not be complicated by the carboxylate ion. Acidification of the enolate ion (g) gives the enediol, which cyclizes spontaneously to give ascorbic acid.

Ascorbic acid is easily oxidized in biochemical systems to the 2,3-keto structure. In turn the diketo form is easily reduced back to ascorbic acid. In other words, ascorbic acid is a good reducing agent (hydrogen donor) and the diketo form is a good oxidizing agent (hydrogen acceptor). By good, we mean that the activation energies are small and reaction is facile.

Scurvy. Scurvy, or scorbutus, has been known for centuries. In about the sixteenth century, the symptoms (tendency to hemorrhage; degeneration of bones, teeth, etc.) began to be treated in various localities with infusions of twigs, herbs, and leaves. About 1750 the prevention of scurvy among sailors on extended voyages by the use of fresh vegetables or fruit juices began on a small scale. Captain Cook, on his 1772 voyage, avoided scurvy entirely. In 1795, the British navy began the use of lime juice to prevent scurvy (hence, "limey" for a British sailor). However, not until 1917 was it recognized that the disease is caused by lack of a specific substance in the diet. Isolation of the antiscorbutic factor was not accomplished until 1932. The structure was worked out in 1933, and in 1937 synthetic ascorbic acid became available. Many claims and counterclaims have been made concerning the efficacy of excess vitamin C in the diet for the prevention of colds, maintenance of general health, healing of wounds, and so on.

Exercises

19.5 The fact that fructose is a 2-ketohexose was shown by the addition of HCN, hydrolysis, and HI reduction. Write the equations and show the product that proves the structure to be a 2-ketohexose.

19.6 D-sorbose gives an osazone identical with those from which pair of epimeric aldoses?

19.4 Synthesis of sugars

Lobry de Bruyn-van Eckenstein transformation. Fischer's first attempts to use sodium amalgam to reduce glucose gave mannitol instead of glucitol (sorbitol). This product resulted from a Lobry de Bruyn-van Eckenstein rearrangement or transformation caused by the sodium hydroxide that was formed as the sodium reacted. The Lobry de Bruyn-van Eckenstein rearrangement strictly refers only to isomerizations (epimerizations) induced by removal of an α hydrogen to give an enolate ion. We shall use it to include all other alkali-induced reactions of aldoses and ketoses. Although all the steps are reversible,

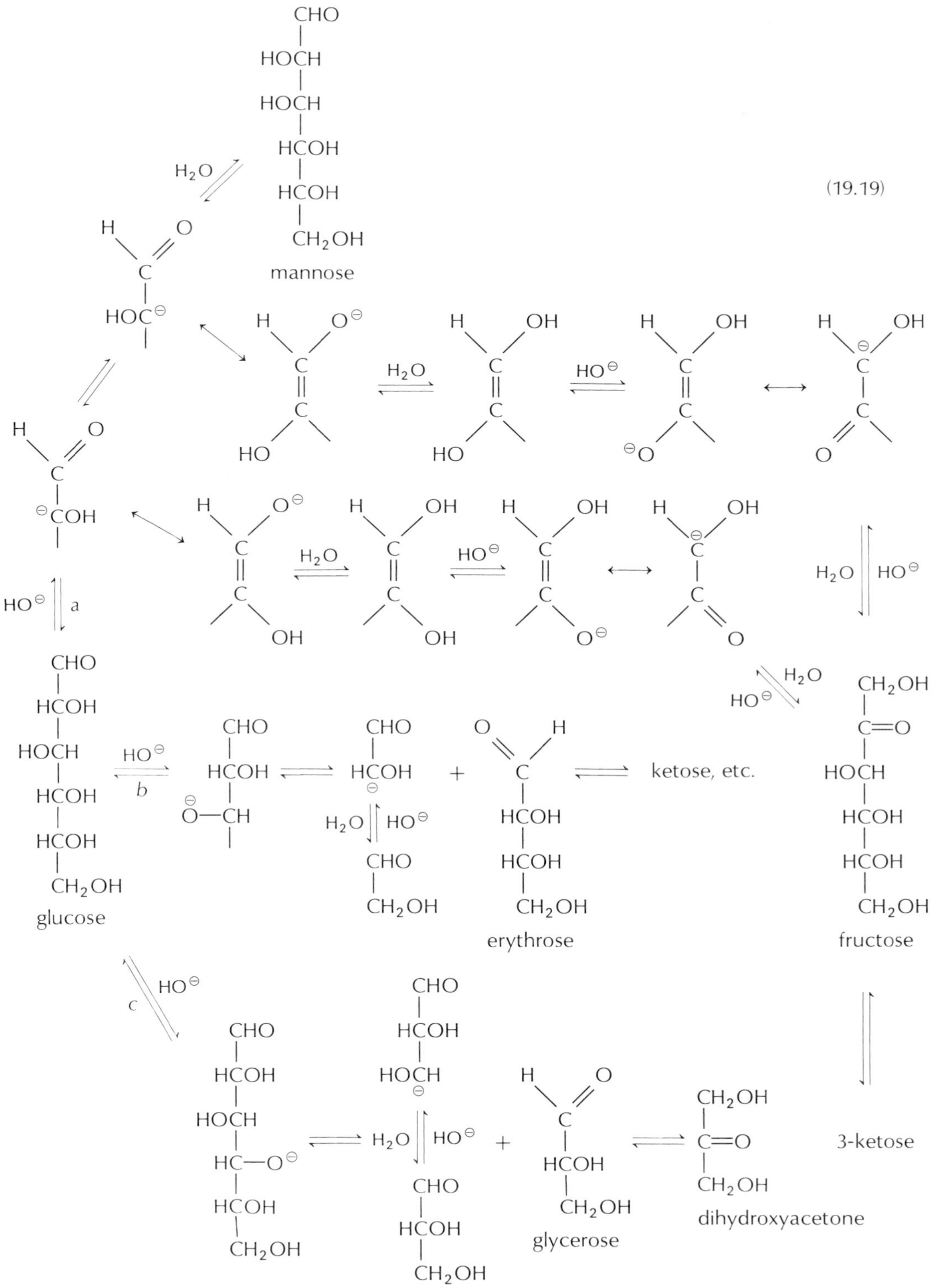

19.4 Synthesis of sugars 631

(19.19)

the composition of the mixture of products is very sensitive to conditions (concentrations, temperature) and to starting material. The process undergone by glucose is shown in Equations 19.19. Sequence *a* leads through enolate ions to mannose and fructose as well as other ketoses. Sequence *b* leads to a reverse aldol condensation to hydroxyacetaldehyde and erythrose, which may either remain or go on to isomerize to threose or a ketose or may be split to give more hydroxyacetaldehyde. Sequence *c* leads to another reverse aldol condensation to give glycerose and dihydroxyacetone. In short, aldoses and ketoses are unstable to the presence of base and slowly isomerize and break down. For example, glucose in Ca(OH)$_2$ solution after several days becomes a mixture of 63% glucose, 21% fructose, 3% mannose, and 13% other compounds.

Now we see the reason for not carrying out an epimerization directly by the action of alkali on an aldose, and why the route through the glyconic acids gives better results (Eq. 19.15). Also, we see why reduction of aldoses must be done under neutral or mildly acidic conditions for the glycitol to retain configuration. □ □

19.5 Disaccharides

The general term for a saccharide that may be hydrolyzed to a few (usually two to six) moles of monosaccharides is oligosaccharide. A disaccharide is the simplest of the oligosaccharides.

A characteristic of disaccharides is acidic or enzymatic hydrolysis to yield monosaccharides. How are the monosaccharides linked together? At what positions? Are there ring structures? If so, are they furanose, pyranose, or others?

Cellobiose.   Cellobiose (dec. 225°) is obtained by hydrolysis of cellulose. It does not occur naturally as a disaccharide. It is a reducing sugar; therefore, at least one of the constituent monosaccharides is a hemiacetal in a furanose or pyranose structure in equilibrium with an aldehyde or ketone form. It exhibits mutarotation, going from 14° to 35°. Upon hydrolysis, cellobiose yields two moles of D-glucose. Hydrolysis also occurs if the enzyme emulsin (β-glucosidase) is used; therefore, one of the glucose units is linked to the other as a β-glucoside. Methylation of β-methyl cellobioside gives an octamethyl cellobiose, which on acidic hydrolysis gives equal quantities of 2,3,4,6-tetramethylglucose and 2,3,6-trimethylglucose. Therefore, the bond between the glucose units is from the 4 position of one to the 1 position of the other (Eq. 19.20). The systematic name for cellobiose is 4-O-(β-D-glucopyranosyl)-D-glucopyranose.

Use of periodic acid.   Periodic acid is a specific reagent for 1,2-diols and is of use in the determination of the structure of sugars. (Periodic acid is useful for water-soluble diols, lead tetraacetate for water-insoluble diols.) We shall use $IO_4^\ominus$ ion to designate the oxidizing agent that is reduced to $(HO)_2IO_2^\ominus$ or $H_2O + IO_3^\ominus$. The α-hydroxy carbonyl compounds also are cleaved (Eqs. 19.21). A mole of diol (a) consumes one mole of $IO_4^\ominus$ and gives two moles of carbonyl compounds (formaldehyde from —CH$_2$OH, aldehyde from —CRHOH, ketone from —CR$_2$OH). An α-hydroxy carbonyl compound (b) consumes one mole of reagent to give a mole of acid and a mole of carbonyl compound. A triol (c— the hydroxy groups must be adjacent) consumes two moles of reagent to give two moles of carbonyl compounds and one mole of acid.

Let us see how this reaction works out for β-D-glucose and for methyl β-D-glucoside (Eqs. 19.22). The former consumes three moles of reagent and gives two moles of formic

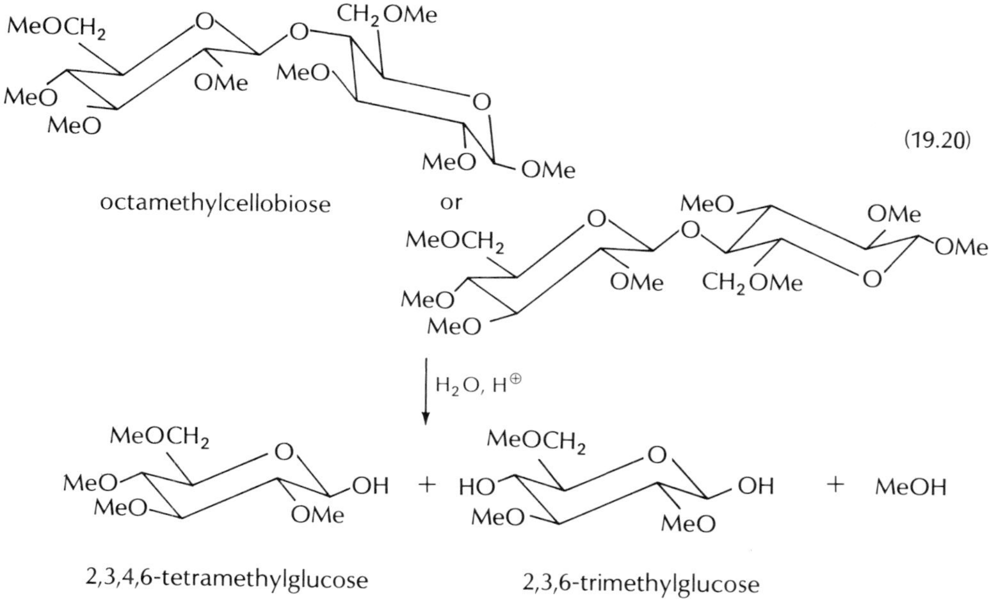

19.5 Disaccharides

$$\text{octamethylcellobiose} \quad \text{or} \quad (19.20)$$

$$\downarrow \text{H}_2\text{O, H}^\oplus$$

2,3,4,6-tetramethylglucose + 2,3,6-trimethylglucose + MeOH

$$\begin{array}{l}\text{—C—OH}\\\text{—C—OH}\end{array} + \text{IO}_4^\ominus \longrightarrow \begin{array}{l}\text{—C=O}\\\text{—C=O}\end{array} + \text{IO}_3^\ominus + \text{H}_2\text{O} \qquad a$$

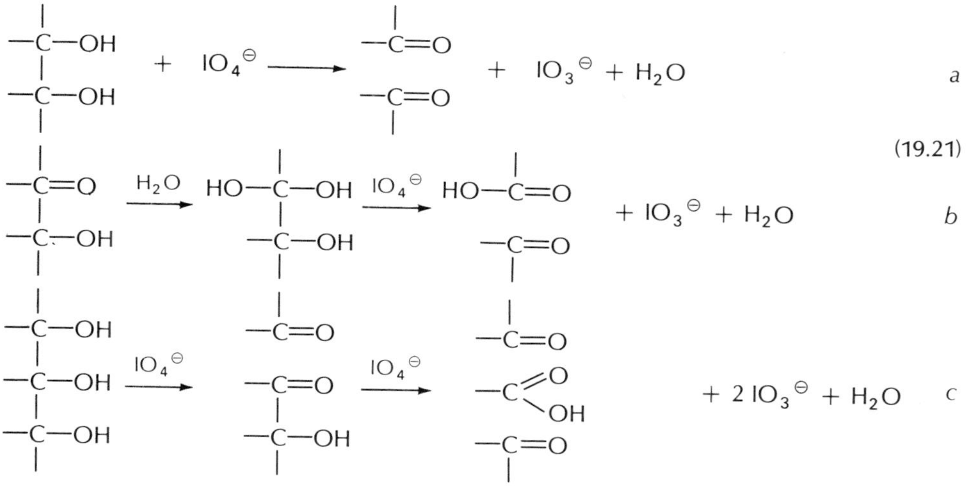

(19.21)

$$\begin{array}{l}\text{—C=O}\\\text{—C—OH}\end{array} \xrightarrow{\text{H}_2\text{O}} \begin{array}{l}\text{HO—C—OH}\\\text{—C—OH}\end{array} \xrightarrow{\text{IO}_4^\ominus} \begin{array}{l}\text{HO—C=O}\\\text{—C=O}\end{array} + \text{IO}_3^\ominus + \text{H}_2\text{O} \qquad b$$

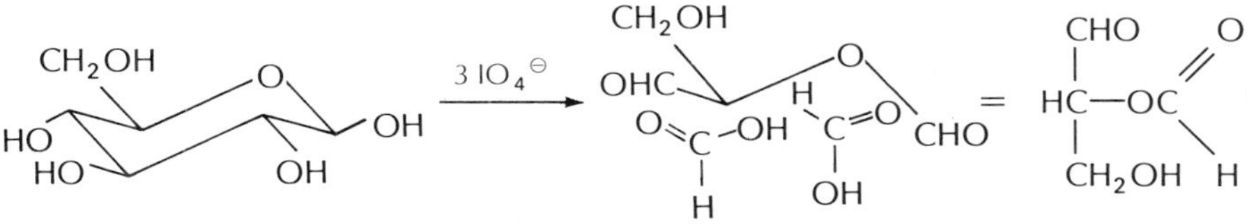

$$\begin{array}{l}\text{—C—OH}\\\text{—C—OH}\\\text{—C—OH}\end{array} \xrightarrow{\text{IO}_4^\ominus} \begin{array}{l}\text{—C=O}\\\text{—C=O}\\\text{—C—OH}\end{array} \xrightarrow{\text{IO}_4^\ominus} \begin{array}{l}\text{—C=O}\\\text{—C}\begin{array}{l}\diagup\text{O}\\\diagdown\text{OH}\end{array}\\\text{—C=O}\end{array} + 2\,\text{IO}_3^\ominus + \text{H}_2\text{O} \qquad c$$

$$\xrightarrow{3\,\text{IO}_4^\ominus} \qquad = \qquad (19.22)$$

$$\xrightarrow{2\,\text{IO}_4^\ominus} \qquad =$$

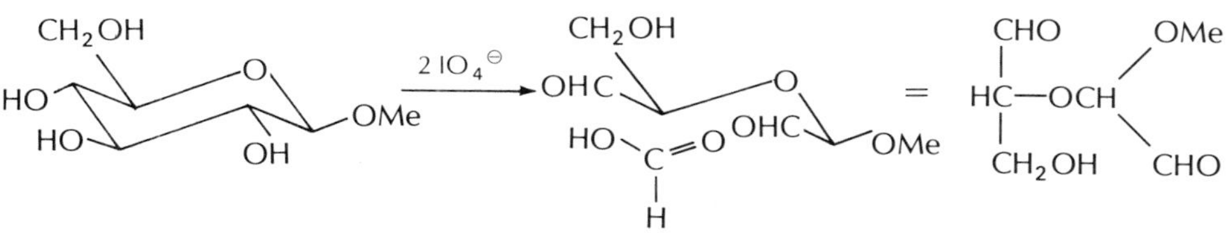

acid and one of the 2-formate ester of D-glyceraldehyde. The methyl β-D-glucoside consumes only two moles of reagent and gives one mole of formic acid and the D-glyceraldehyde derivative shown.

Returning to cellobiose, we see that the methyl cellobioside consumes only three moles of reagent but gives only one mole of formic acid. If the glycoside link were at 2 or 3, only two moles of reagent would be required. If the link were to the 6 position, four moles of reagent would be required and two moles of formic acid would be formed.

Gentiobiose. Gentiobiose is a hydrolytic product of the trisaccharide gentianose, which is present in gentian. Gentiobiose may be obtained as the α-anomer (mp 86°, mutarotation 16° → 9°) and as the β-anomer (mp 190°–195°, mutarotation −6° → 9°). It is a reducing sugar. Gentiobiose can be hydrolyzed by emulsin (or acid) to give two moles of glucose; therefore, it is a glucosyl β-glucoside. Methyl gentioside consumes four moles of periodic acid and gives two moles of formic acid. Therefore, there must be three adjacent hydroxy groups in each ring, and the glucosidic link must be 1β to 6. Methylation and hydrolysis to give 2,3,4,6-tetramethylglucose and 2,3,4-trimethylglucose confirm the point. The structure of the β-anomer is shown.

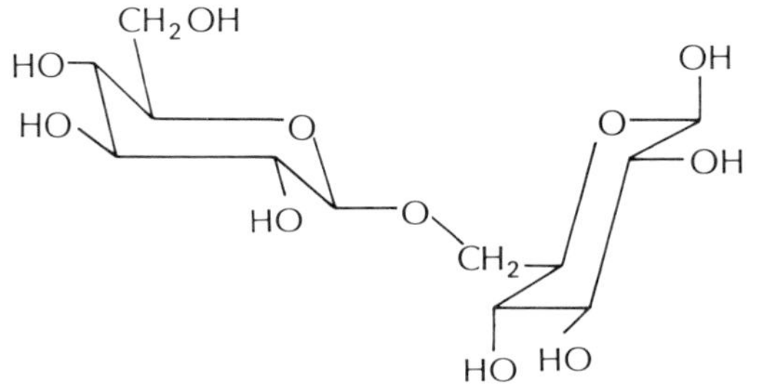

β-gentiobiose, or 6-O-(β-D-glucopyranosyl)-β-D-glucopyranose

Lactose. Lactose, or milk sugar, is present in milk in about 5% concentration. Lactose crystallizes as a hydrated α-anomer or an anhydrous β-anomer. Mutarotation of the anhydrous forms is α, 93° → 52°, and β, 34° → 52°. Hydrolysis requires the enzyme β-galactosidase and gives a mole each of galactose and glucose. Oxidation of lactose with bromine water gives the -onic acid of lactose. Upon hydrolysis, this acid gives galactose and gluconic acid. Thus, the molecule is a β-galactoside. Methylation or periodic acid treatment both indicate a 1 to 4 linkage. The structure of lactose then differs from cellobiose only in the single configurational change at 4 in the galactose ring.

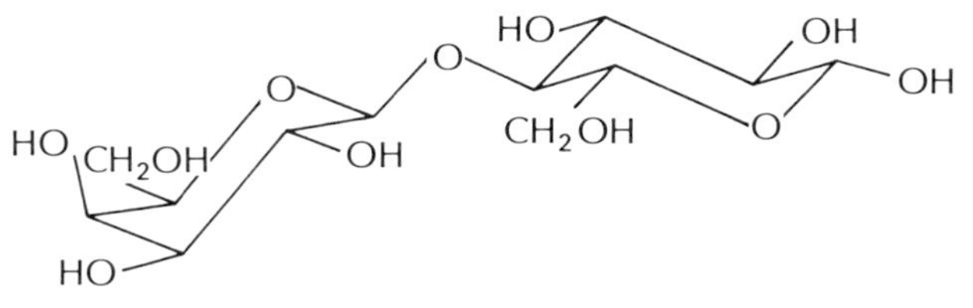

β-lactose, or 4-O-(β-D-galactopyranosyl)-β-D-glucopyranose

Lactase. Milk is no longer considered a universal complete food. The use of powdered milk to supplement the diets of the undernourished has recently been questioned by the finding that only a small group of mankind, notably those of northern European stock, can digest lactose. Although some Africans can and do digest fresh milk, the blacks taken as slaves were almost exclusively from intolerant tribes. The addition of European genes over the years has increased the tolerance, but still today it has been estimated that 70% of adult black Americans cannot digest lactose.

The reason is the amount of a single enzyme present in the gut. Sugars must be hydrolyzed to free simple sugars to pass through the intestinal wall into the blood. Because lactose is a double sugar, an enzyme (called <u>lactase</u>) is required for its digestion. In the case of intolerant individuals, there is simply not enough lactase. Much of the lactose then passes intact through the small intestine into the large intestine. The high concentration of the sugar here causes water to enter osmotically. Moreover, bacteria in the large intestine ferment some of the lactose to carbon dioxide gas. This combination is responsible for diarrhea, the major symptom of intolerance. Human milk contains especially high concentrations of lactose (7.5% compared to 4.5% for the cow). It can overload normal lactase concentrations, and occasional digestive disorders of the baby are the result.

Maltose. Maltose itself is found occasionally in nature, but it is easily available from acidic or enzymatic (diastase) hydrolysis of starch. Maltose, as the β-anomer, has mp 102° to 103° and mutarotation from 112° to 130°. Maltose is hydrolyzed under acidic conditions or in the presence of maltase (α-glucosidase) to give two moles of glucose. The usual reactions indicate a 1α-to-4 linkage.

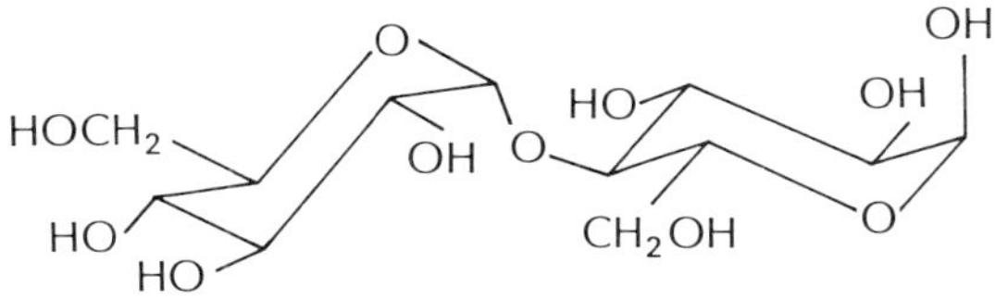

α-maltose or 4-O-(α-D-glucopyranosyl)-α-D-glucopyranose

Melibiose. Melibiose is a mutarotating disaccharide obtained by partial hydrolysis of the trisaccharide raffinose. Like lactose, it is a galactylglucose, but it is linked 1α to 6 (similarly to α-gentiobiose, which has a 1α-to-6 linkage).

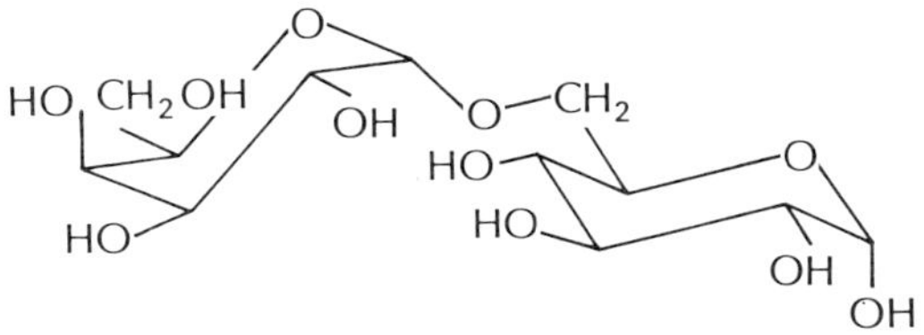

α-melibiose, or 6-O-(α-D-galactopyranosyl)-α-D-glucopyranose

Sucrose. Sucrose (cane sugar, beet sugar, table sugar, etc.) occurs in up to about 15% concentration in the juice of sugar cane, a little more in the juice of some varieties of sugar beets. Purification is carried out by recrystallization: decolorizing carbon is used to remove color.

Refined sucrose is the champion. No other pure organic compound comes close to sucrose in amount produced per year throughout the world. No other sugar crystallizes as easily as sucrose.

Sucrose is unusual in not being a reducing sugar; neither does it exhibit mutarotation. Upon hydrolysis, either acidic or by the enzyme invertase (sucrase), sucrose gives a mole each of glucose and fructose. Sucrose, which decomposes at 160° to 186° to caramel, has $[\alpha]_D$ of 66°. Fructose has an equilibrium rotation of $-92°$; glucose, 53°. Thus, the hydrolysis of sucrose may be followed by the change in rotation from 66° to $-20°$. Because of the change in sign, the hydrolysis is called inversion and the mixture of fructose and glucose is called invert sugar. Invert sugar is about as sweet as sucrose (fructose is sweeter, glucose is not as sweet) but has little tendency to crystallize. Thus, invert sugar is used to sweeten those products (ice cream, certain candies, soft drinks) in which crystallization is not wanted.

Sucrose also may be hydrolyzed by maltase, and therefore we know it is an α-glucoside. Sucrose also may be hydrolyzed by the enzyme β-fructosidase, and therefore it is a β-fructofuranoside. Thus, fructose is linked β at the 2 position to the 1α position of glucose. Methylation (sucrose does not form methyl sucrosides) gives an octamethylsucrose that hydrolyzes to give 1,3,4,6-tetramethylfructose and 2,3,4,6-tetramethylglucose. Sucrose consumes only three moles of periodate and forms only one mole of formic acid, which shows that if glucose is in the pyranose structure, fructose is in the furanose structure.

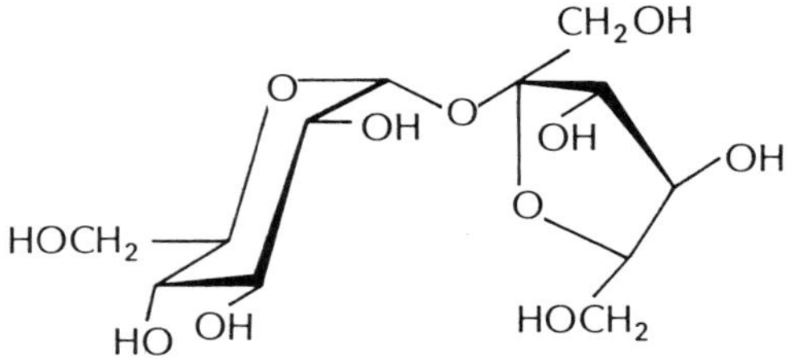

sucrose, or α-D-glucopyranosyl-β-D-fructofuranoside

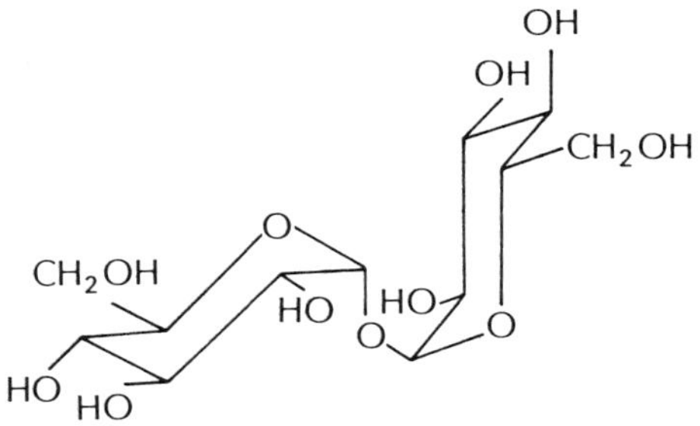

trehalose, or α-D-glucopyranosyl-α-D-glucopyranoside

Trehalose. Trehalose occurs in bacteria, yeasts, insects, and so on. Like sucrose, it is nonreducing and nonmutarotating, does not form an osazone, and does not form methyl glycosides. (The prefix glyco- indicates a general sugar.) Methylation and hydrolysis gives two moles of 2,3,4,6-tetramethylglucose. The α-methyl glycosides of the D sugars all have higher positive rotations than do the sugars from which they are derived, and the β-methyl derivatives are more negative. (The reverse is true for the L sugars.) Thus, the very high positive rotation of 178° suggests that trehalose is linked 1α to 1α.

Many other disaccharides are known. The linkage between the rings has been found to be 1α or 1β to 2 and 3 as well as to 4, 6, and 1, as in the ones we have seen.

Exercises

19.7 Nigerose (found in beer) is 3-O-(α-D-glucopyranosyl)-D-glucopyranose. Draw the perspective structural formula.

19.6 Trisaccharides

Gentianose. Gentianose is present in the tuberous roots (rhizomes) of gentian. Complete acidic hydrolysis yields one mole of fructose and two moles of glucose. Hydrolysis by sucrase gives fructose and gentiobiose; that by β-glucosidase (emulsin) gives glucose and fructose. Therefore, we may represent gentianose as glucose-1β-6-glucose-1α-2β-fructose.

Raffinose. Raffinose occurs in many plants in small quantities. It is obtained as a by-product in the refining of beet sugar. Complete acidic hydrolysis of raffinose gives one mole each of fructose, galactose, and glucose. Partial enzymatic hydrolysis gives, with α-galactosidase (emulsin), one mole of galactose and one mole of sucrose; with sucrase, one mole of fructose and one mole of melibiose. Thus, we may represent raffinose as galactose-1α-6-glucose-1α-2β-fructose.

Exercises

19.8 Write perspective structures of gentianose and raffinose.

19.7 Polysaccharides

The polysaccharides, upon complete hydrolysis, yield a large number of monosaccharides. Partial hydrolysis gives a mixture of mono- and oligosaccharides. The most plentiful polysaccharides in nature are cellulose, chitin, glycogen, and starch. There is a host of others, only a few of which we shall mention.

Cellulose. Cellulose occurs in practically pure form in cotton, flax, and ramie; and it is present in all plants. Complete hydrolysis gives only glucose; partial hydrolysis allows cellobiose to be isolated. Thus, cellulose consists of long chains of glucose units linked 1β to 4 in an extended conformation. Unfortunately, humans do not tolerate enzymes capable of hydrolyzing β-glucosides, so cellulose cannot serve as a food. Some animals (grazing animals, termites) harbor bacteria in the intestinal tract or in a forestomach (rumen) that does possess these enzymes. These animals are able to eat cellulose and utilize the glucose provided from it.

The principal use of cotton and other forms of native cellulose is for fabrics; impure cellulose (wood pulp) is used for paper. Large quantities of waste cellulose (cotton linters, the fuzz left on cottonseed after ginning) are used in making rayon and cellulose acetate.

The viscose rayon process consists of soaking cellulose in CS_2 in the presence of NaOH solution. The cellulose xanthate salt that results is soluble in the aqueous system (see Eq. 9.37). The solution, called viscose, is pumped through very fine holes into $NaHSO_4$ solution, whereupon the cellulose is regenerated as very fine filaments. These filaments are twisted together to make thread.

Cellulose acetate also is used for making thread and fabrics, in addition to sheets, as is viscose (cellophane). Ethers and mixed esters also are manufactured for special purposes (laminate for safety glass, for example). Nitration leads to nitrocellulose or guncotton, depending upon the extent of nitration. Each glucose unit in the chain of cellulose has three free hydroxy groups available for reaction at positions 2, 3, and 6. The properties of the product depend upon the average degree of reaction per glucose unit: in other words, whether one, two, or three of the hydroxy groups react.

Pure cellulose is a high polymer with a molecular weight of about 10^6, having 3000 to 6000 glucose units per chain. Its solubility is negligible in most solvents. Exposure to concentrated NaOH while under tension causes a change in cotton thread that is known as mercerization. The tensile strength is increased and the thread is given a lustrous finish. On the other hand, dipping paper of good quality in 80% sulfuric acid and then washing it thoroughly produces a stiff, tough substance known as parchment paper.

Chitin. Chitin is the shell substance, a polymeric β-D-glucosamine in which the amino groups are acetylated. The hard, horny substance is used by crustaceans (crab, lobster, shrimp, etc.), insects, and some fungi as rigid structural material instead of bones. Chitin is very insoluble and is difficult to hydrolyze, even with concentrated hydrochloric acid. X-ray data indicate that chitin has the same repeating unit of 10.3 Å as does cellulose.

chitin

Hemicelluloses. Hemicelluloses may be defined as noncellulosic polysaccharides present in plants. Cornstalks, straw, grain hulls, and so on, may consist of up to 40% hemicelluloses. Some examples are:

1. <u>Araban</u>, from peanut shells, is a poly-L-arabinoside, linked 1 to 4.
2. <u>Xylan</u>, from corn cobs, is a poly-D-xyloside, linked 1β to 4.
3. <u>Mannan</u>, from yeast and ivory nuts, is a poly-D-mannopyranoside, linked 1β to 4.
4. <u>Galactans</u> and arabinogalactans, from larch wood, are poly-D-galactosides with both 1-to-6 and 1-to-3 links. More than an occasional unit of L-arabinose is present in the branched chains.

Starch: amylose, amylopectin. Starch is a polyglucoside that has mostly 1α to 4 linkages. Hydrolysis by acid leads to maltose and to glucose. These statements are greatly oversimplified, because starch is widely distributed in the plant world and each variety produces a starch of slightly different properties, ranging from size of the granules to proportions of the various fractions present. Starches from corn, potato, wheat, rice, tapioca, and sago are commercially available. Boiling a suspension of starch granules causes the granules to swell and burst. The boiled suspension sets on cooling to a gel whose thickness depends upon the concentration. Such a boiled suspension of starch may be separated into two fractions; impure <u>amylose</u>, which has diffused from the granules, and a residue of impure <u>amylopectin</u> retained by the granules. Pure amylose, obtained by complexation with thymol and regeneration at room temperature and absence of air, is relatively insoluble in water. Pure amylopectin is more soluble.

Amylose and amylopectin vary in molecular weight depending upon the source. Amylose, mw about 300,000 $\pm$ 100,000 (about 1800 α-D-glucose units linked 1α to 4 as in maltose), is the fraction that forms the intense blue color with iodine. The linear polymer coils about the iodine in such a manner that the photochemistry of the complex differs drastically from the brown color of iodine in polar solvents. Unlike amylose, amylopectin forms a red complex with iodine. Amylopectin, mw about 100 $\times$ 10^6 (about 600,000 α-D-glucose units), has a branch, about every 25th unit, formed by a 1α to 6 link. The branching gives a three-dimensional tangle of chains. A very simple picture of a portion of amylopectin is shown with a single branch.

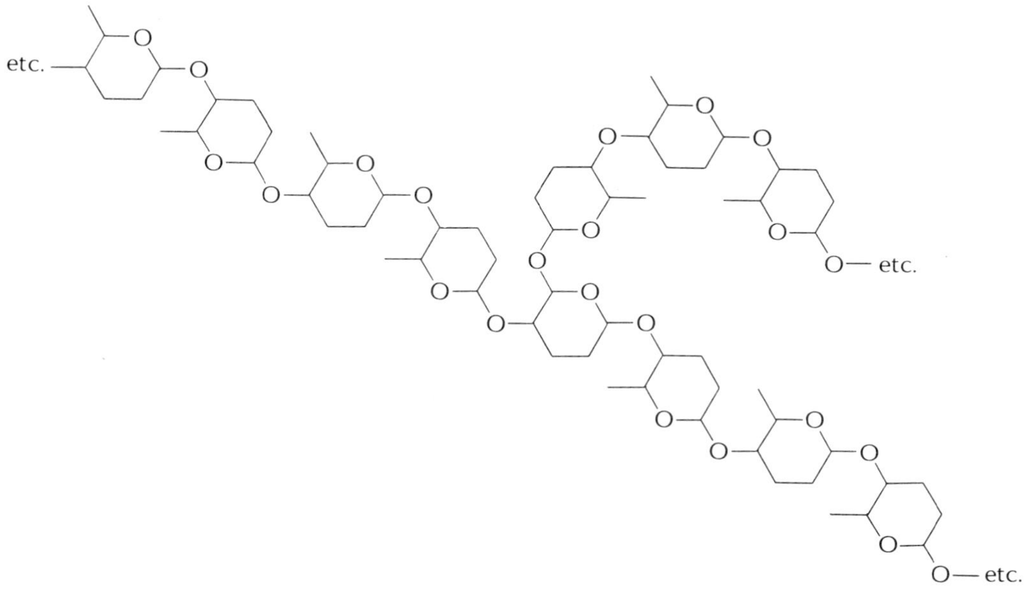

Starch is one of the energy storehouses of the plant world. Metabolic enzymatic breakdown furnishes glucose within the plant as needed. Starch, as the source of glucose, is also the source of almost all beverage alcohol (wines excepted). Diastase (from malt, which is sprouted barley, dried and ground) converts starch into maltose, after which yeast is added as a source of the enzymes maltase (α-glucosidase) and zymase, which catalyzes the fermentation of glucose to two moles of EtOH and two moles of CO_2. The reaction is exothermic ($\Delta H = -26$ kcal), so that large fermentation vats require cooling to keep the temperature below 32°. Other enzymes (present in certain strains of bacteria) cause the fermentation to take a different course. One such process gives butyl alcohol, acetone, and ethanol in about a six to three to one ratio. In any case, small amounts of other alcohols are formed, including PrOH, iBuOH, iBuCH$_2$OH, and sBuCH$_2$OH. The mixture is known as fusel oil.

Glycogen. Glycogen is the animal form of starch. It is synthesized within the body after glucose has passed through the intestinal wall. Glycogen is present throughout the body, sometimes free, sometimes bound to protein, the highest concentrations being found in liver and muscle.

Glycogen is similar to amylopectin in structure but is much more highly branched and of a much greater molecular weight. Some estimates of mw have been as high as 100,000,000 (or 60,000 glucose units).

Other polysaccharides. There are many other polysaccharides with properties more or less like those of starch. Some examples are:

1. Pectins and pectic acids, which are produced by partial hydrolysis of protopectins present in fruits and berries. They are gelling agents. Exact structures are difficult to specify, but they seem to be polygalacturonic acids. The glycuronic acids are aldoses in which the primary alcohol group (position 6 in aldohexoses) is oxidized to an acid group. The linkage is thought to be 1α to 4.

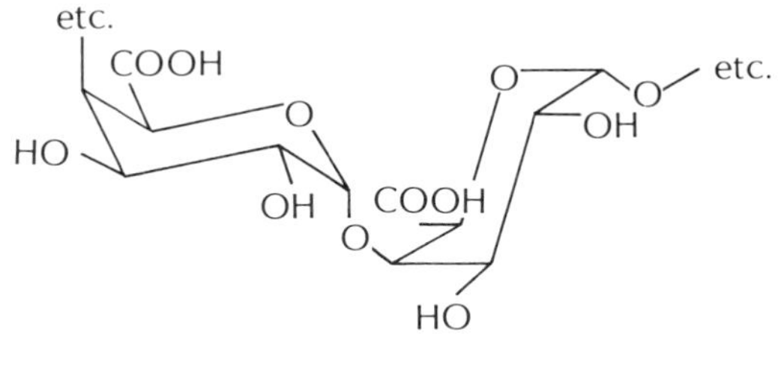

pectic acid, or poly-D-galacturonic acid

2. Inulin is present in such plants as dahlia, chicory, and Jerusalem artichoke. It is a polymeric fructose with some glucose units present. The linkage is 2β to 1 with fructofuranoside units.
3. Algin, from algae and certain seaweeds (kelp), is a salt of polyglycuronic acids. It is used as a thickening and emulsifying agent in creams, lotions, and the like.

Hydrolysis gives two moles of D-mannuronic acid to one mole of L-guluronic acid. Thus, an alginic acid resembles pectic acid in being a polyglycuronic acid, but differs in being composed of D-mannuronic and L-guluronic acids, linked probably 1β to 4.

4. Agar, a gel-forming substance obtained from certain East Indian seaweeds by extraction with hot water, seems to be made up of two components. One, called agarose, is a polysaccharide containing D-galactose and 3,6-anhydro-L-galactose units. (An anhydro sugar is one in which an internal ether has been formed by elimination of a molecule of water between the positions designated.) The other component, called agaropectin, is less well characterized. It contains D-glucuronic acid as well as D-galactose and 3,6-anhydro-L-galactose units.

5. Gums are known in great variety, and they have a variety of compositions. Gum arabic is typical. It is the calcium salt of the polysaccharide arabic acid, which gives L-arabinose, L-rhamnose, D-galactose, and D-glucuronic acid upon hydrolysis.

6. Dextran is a polymer formed from sucrose by fermentation by certain bacteria. It has been used as a blood plasma extender. Dextrins are breakdown products formed by the heating of starch and are used for mucilage and other adhesives.

Physiological polysaccharides. Other polysaccharides play roles within the body. Some examples are given here.

1. Hyaluronic acid is a component of connective tissue and of the fluid that lubricates the joints. It has a mw of between 400,000 and several million. It seems to be made up of equal amounts of N-acetylated D-glucosamine and D-glucuronic acid with 1β to 3 links. Closely related is chondroitin sulfate, a component of cartilage, in which it is present as an anion where collagen, a protein, is the cation. The structure is similar to that of hyaluronic acid, with D-galactosamine replacing D-glucosamine and with the 6-hydroxy group in the acetylated amine esterified by sulfuric acid. The mw seems relatively small, about 100,000. Heparin is present in blood and acts to prevent coagulation. It is similar to chondroitin sulfate, except that the acetylated amino group ($-NHCOCH_3$) is replaced by a $-NHSO_3H$ group and some of the hydroxy groups (beside the hydroxy group at 6) are esterified with sulfuric acid.

2. A protein foreign to a given person or animal stimulates the production of antibodies that are specific in combining with the foreign protein or antigen. The foreign protein, for example, from the capsule coating of a variety of pneumococcus usually is made up of a proteinoid part and a polysaccharide portion. It seems clear that it is the polysaccharide portion, called a hapten, that acts as the antigen.

3. Antibiotics frequently are glycosides of hitherto unknown sugars. Many have been uncovered as the search for new antibiotics goes on. An example is streptomycin.

Blood group substances. Sugars covalently bound to protein or to lipid molecules coat the surface of cells. In mammals, the oligosaccharides seem to serve a function analogous to the identifying flags of a ship. All cells that the body's immune system recognizes as carrying foreign colors are removed. This protective device must be considered when blood transfusions are administered. The tolerance when an entire organ is transplanted is extremely slight, and a near-identical match of donor and recipient cell types is necessary.

The first blood group system discovered was described by Karl Landsteiner in 1900. People were shown to have blood of type A, B, AB, or O. Blood cells of A individuals contain

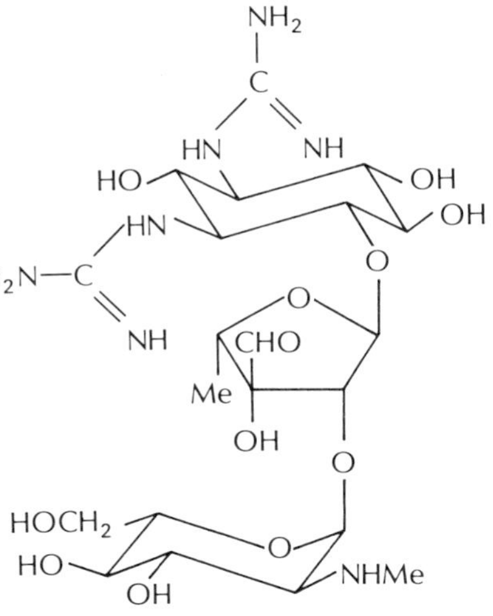

streptidine

streptose

N-methyl-L-glucosamine

streptomycin

the A oligosaccharides on their surface. Type B individuals have cells with the B oligosaccharide, type O cells contain neither the A nor the B glycosyl chains, and AB contain both. The immune system of type A people recognizes B red blood cells as foreign and causes them to precipitate. It does not recognize the type O cells as foreign, and so they are not attacked. The corresponding situation occurs for the type B people. Type A blood can be accepted by type A people, but not by type B. Thus O group blood is the universal donor. AB individuals, who consider both the A and B as friendly, can accept blood from anybody. In the United States 45% of the population is O group, 40% A, 9% B, and 3% AB. The sole difference between the A and B oligosaccharides is the terminal residue—in A it is N-acetyl-D-galactosamine, in B it is D-galactose. This terminal group is missing in the O group oligosaccharide.

19.8 Sugar alcohols

The noncarbohydrate reduction products of the common aldoses and ketoses occur in nature. As we have mentioned previously, they also are available by catalytic reduction of aldoses under conditions (absence of alkali) that do not cause epimerization. The glycitols are frequently used as thickeners, humectants, emulsifiers, and so on.

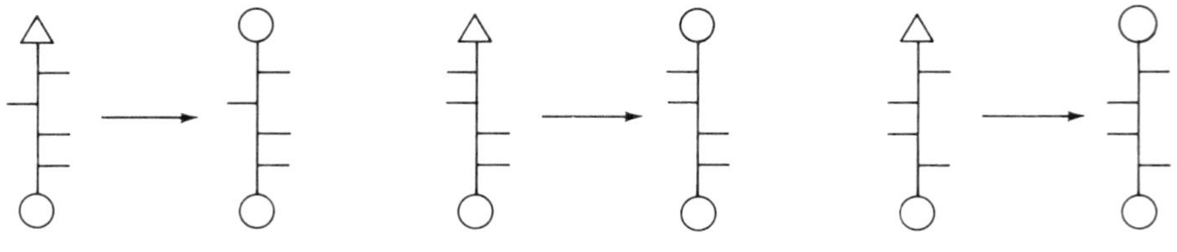

D-glucose → D-glucitol, or sorbitol

D-mannose → D-mannitol

D-galactose → galactitol, or dulcitol

19.9 Photosynthesis and metabolism of carbohydrates

Although they are not carbohydrates, the hexahydroxycyclohexanes, known as inositols, have physical properties similar to those of the sugar -itols, and they also occur in nature. Many stereoisomers are possible. One in particular, *myo*-inositol (also called *meso*-inositol, even though seven of its other stereoisomers are *meso*), is part of the vitamin B complex and probably is present in every living cell. The hexaester with phosphoric acid is known as phytic acid.

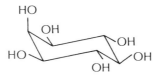

myo-inositol

Mannitol. The chemistry of many simple sugars that are hemiacetals and acetals is similar to the chemistry of aldehydes and ketones. For instance, mannose, like most of these sugars, can be detected in solution by a color test based on its ability to reduce the ferric ion. The sugar carbonyl can itself be reduced to give a sugar alcohol (glycitol).

The sugar alcohol of mannose has several striking properties different from those of the parent sugar. For some reason it is poorly digested by the body. Only half its energy value is recovered. This property is put to commercial advantage by the use of mannitol to sweeten low-calorie diet foods. Cariogenic bacteria on teeth also have much more difficulty digesting mannose than, for example, sucrose, and for this reason mannitol is used to sweeten sugarless gum. Mannitol is important in the manufacture of regular chewing gum as well, not because it is sweet, but rather because it does not absorb water. It is dusted on each stick of gum to prevent the gum from absorbing water and becoming tacky.

19.9 Photosynthesis and metabolism of carbohydrates

Photosynthesis. Photosynthesis is a complex process that we shall describe in a very simplified way.

The primary event is absorption of light by chlorophyll a, whereby an electron is promoted to an excited level. Two pathways are followed thereafter. (See Chapters 21 and 28 for structures and other information about the many compounds named in the discussion that follows.)

One pathway, a cyclic process called cyclic photophosphorylation, involves a number of electron transfers through ferrodoxin to cytochromes and back to chlorophyll. In the process, a number of adenosine diphosphonucleotides (ADP) are converted by reaction with phosphoric acid (or an ion) into adenosine triphosphonucleotides (ATP). The net result is the conversion of the light energy absorbed into ATP synthesis.

The other process, noncyclic photophosphorylation, is a disproportionation in which water, nicotinamide adenine dinucleotide phosphate (NADP), and ADP are converted into NADPH$_2$ (reduced NADP), O$_2$, and ATP (Eq. 19.23). The reaction is

$$2NADP + 2ADP + 2H_2O + 2P \longrightarrow 2NADPH_2 + 2ATP + O_2 \qquad (19.23)$$

(P is symbolic of H$_3$PO$_4$ and the ions)

endergonic (like endothermic, but referring to Gibbs free energy instead of enthalpy), the required energy being furnished by the excited chlorophyll. The light energy in this case is stored as both ATP and NADPH$_2$.

Now the dark reactions of photosynthesis follow, in which the energy stored in ATP and NADPH$_2$ operate through a cycle whereby CO$_2$ is converted into 3-phosphoglyceraldehyde, or <u>triose phosphate</u> (Eq. 19.24).

$$3CO_2 + 9ATP + 6NADPH_2 + 5H_2O \longrightarrow \begin{array}{c} \triangle \\ \vdash \\ OP \end{array} + 9ADP + 6NADP + 8P \quad (19.24)$$

In Figure 19.3 a representation of the dark-reaction cycle is shown. In brief, CO$_2$ is added to the reactive ribulose diphosphate, which is cleaved and oxidized to give two moles of phosphoglyceric acid (PGA). PGA also serves as the point of departure for the synthesis of fatty acids and amino acids. Phosphorylation of PGA by ATP followed by reduction by NADPH$_2$ gives triose phosphate. Triose phosphate is the origin of glycerol phosphate, from which fats and oils are formed. The following steps, whereby ribulose 1,5-diphosphate is formed, involve many operations. We need to keep in mind that each individual step requires an enzyme (and a coenzyme in most instances).

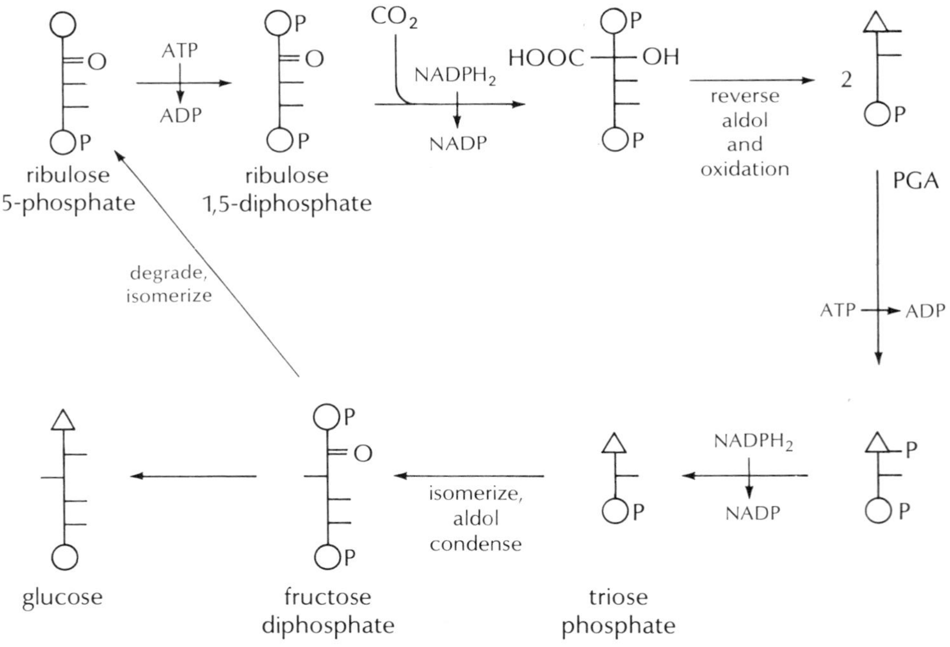

Figure 19.3 Photosynthetic dark-reaction cycle

Pathways for utilization of glycogen. Glucose is stored as glycogen. As needed for metabolism, glucose is split off by hydrolysis or by phosphorylation to give glucose-1-phosphate, which is isomerized to glucose-6-phosphate. Glucose-6-phosphate may follow three different but interrelated schemes of oxidative breakdown in which each step releases a portion of the energy originally stored in glycogen. Not all the steps are oxidative: energy is transferred in conversion of ADP to ATP in some of the steps. NADP is used freely as an oxidizing agent (hydrogen acceptor), and $NADPH_2$ is used as a reducing agent (hydrogen donor). We briefly describe the three pathways.

The Emden-Meyerhof glycolysis leads through fructose to pyruvic acid, lactic acid, ethanol, and CO_2.

The Krebs citric acid cycle takes pyruvic acid from the Emden-Meyerhof pathway, decarboxylates it, and combines acetyl with oxaloacetic acid (2-ketosuccinic acid) to give citric acid. Isomerization, oxidation, and decarboxylation lead to 2-ketoglutaric acid. Another decarboxylation and reduction give succinic acid. Dehydrogenation to fumaric acid, hydration to malic acid, and oxidation give back oxaloacetic acid, ready for another cycle. In the process, three moles of CO_2 are evolved, equivalent to oxidation of pyruvic acid ($CH_3COCOOH$).

The phosphogluconate oxidative pathway begins with an oxidation of glucose-6-phosphate to 6-phosphogluconic acid. Oxidation and decarboxylation lead to the pentose system, whereby ribose, ribulose, and xylulose are formed. Erythrose and sedoheptulose also become involved. Essentially, this pathway is a means for pentose synthesis.

Details may be found in any textbook of biochemistry.

19.10 Summary

For many years, sugar or carbohydrate chemistry was a field of research work that was so specialized that there was little interaction between it and other research in organic chemistry. However, as usually happens, a coalescence is taking place between carbohydrate chemistry and molecular biology on the one hand and photo-organic chemistry on the other.

We have been getting more and more into physiologically active compounds. Future chapters also will be filled with these substances. In the meantime, if we pause and review this chapter, we find that most of the reactions have been those of aldehydes, ketones, hemiacetals, acetals, diols, hydroxy acids, and lactones, all of which were shown in the preceding chapters.

19.11 Problems

19.1 A D-aldoheptose, A, when oxidized with nitric acid, gives an optically inactive glycaric acid, B. Going up the chain from A gives the epimeric pair of D-aldooctoses C and D. Oxidation with nitric acid converts C and D to optically active glycaric acids, E and F. Oxidation of A with bromine water and heating the product with pyridine give a separable mixture, the components of which after reduction with $NaBH_4$ give A and G. G, upon going up the chain, gives H and I. H gives an optically active glycaric acid J upon nitric acid oxidation, but I gives an optically inactive glycaric acid, K.

Write structural formulas (symbolic will be satisfactory) of the lettered substances.

646   *Chapter nineteen*

19.2 How can the following pairs be distinguished by a physical or chemical test?

	A	B
a.	L-xylose	L-arabinose
b.	D-ribose	2-deoxy-D-ribose
c.	lactose	sucrose
d.	methyl α-D-ribofuranoside	methyl α-D-ribopyranoside
e.	methyl β-D-ribofuranoside	methyl α-D-ribofuranoside
f.	D-glucose	methyl α-D-glucopyranoside
g.	D-arabinose	D-arabitol
h.	D-glucose	sorbitol
i.	D-glucose	D-galactose
j.	threitol	erythritol

19.3 Show the chemical transformations needed to convert D-ribose to the following carbohydrate derivatives.
   a. D-allose
   b. D-altrose
   c. D-erythrose
   d. D-ribonolactone
   e. methyl α-D-ribopyranoside
   f. 2,3,4-tri-O-methyl-D-ribose
   g. *meso*-ribaric acid
   h. D-arabinose
   i. α-D-ribopyranose tetraacetate
   j. *meso*-ribitol

19.4 What are the structures of the following sugars?
   a. A sugar gives D-glucosazone with phenylhydrazine but is not oxidized by bromine water.
   b. An aldopentose not identical with D-arabinose gives D-arabinitol on reduction with $NaBH_4$.
   c. An aldohexose not identical with D-glucose gives D-glucaric on oxidation with HNO_3.
   d. An aldopentose gives D-xylosazone with phenylhydrazine and an optically active dicarboxylic acid with HNO_3.
   e. An amino sugar gives D-ribosazone on reaction with phenylhydrazine.
   f. A ketose gives D-allitol and D-altritol on reduction with H_2–Ni.
   g. A methyl pentoside gives the same oxidation product with $NaIO_4$ as methyl α-D-ribofuranoside. Acid hydrolysis of the methyl pentoside followed by oxidation with HNO_3 gives an optically inactive dicarboxylic acid.

19.12   Bibliography

Fieser and Fieser (Chap. 7, ref. 7) and especially Noller (Chap. 7, ref. 11) are much more complete concerning carbohydrates than are other texts. Numerous advanced books are available. Textbooks of biochemistry usually are concerned more with carbohydrate metabolism than with carbohydrates in general.

1. Bassham, J. A., The Control of Photosynthetic Carbon Metabolism, *Science* 172, 526 (1971). A good brief article on photosynthesis is given here.

2. Sugars Made from Water, Carbon Dioxide, *Chemical and Engineering News* Sept. 29, 1969, p. 40. A revival of interest in carbohydrate synthesis from simple substances as part of a self-sustaining food cycle for extended space flights is the subject of this short article.

19.13 Melvin Calvin: One thing leads to another

Professor Melvin Calvin, director of the Laboratory of Chemical Biodynamics at the University of California, Berkeley and Nobel Prize winner on top of numerous awards, started out with a Ph.D. in 1935 at Minnesota. After a two-year stay at Manchester on a Rockefeller grant, he returned in 1937 to an instructorship at Berkeley where he has remained. Here he describes how the thread of coordination chemistry runs through his research program.

My activity in coordination chemistry is really the seed of practically everything I've done. Perhaps not everything, but a good bit of what I have done in the succeeding thirty years branched out from metal complexes. The work on photosynthesis arose because of the fact that chlorophyll is a complex. I was interested in how it worked and the reason for its stability. In a sense my activity in coordination chemistry and catalytic function is still central to almost everything we have done in the thirty years I have been in Berkeley. I started out in Berkeley on homogeneous catalytic hydrogenation which came from metal complex systems. Thus, hydrogen is activated to reduce other reducible compounds in presence of a coordination compound of copper. This was the first demonstration of our ability to activate molecular hydrogen by molecules in solution. Hydrogen activation had been performed only on the surfaces of metal or metal oxide catalysts prior to my work in 1938. It was, then, the first demonstration of a homogeneous catalytic facility for metal coordination compounds. Since that time it has developed into a very important commercial application with books written on the subject. I haven't been in the field since about 1940. Actually I initiated some early studies on coordination chemistry in England as a postdoctoral fellow.

I began in my early postdoctoral work in England with Professor Polanyi at the University of Manchester. It was the mid-summer of 1936, I believe, that the discovery of phthalocyanine, a compound resembling in its general structure chlorophyll and heme, was made by Linstead. His work was the result of some observations noted at Imperial Chemical Industries. They were making phthalonitrile, ortho-dicyanobenzene, in glass-lined iron kettles. Well, one batch turned out to be blue instead of colorless. R. P. Linstead, professor of organic chemistry at Imperial College, successfully undertook the project of finding out what the source of blue color was. Polanyi recognized that this class of compounds was structurally very similar to the structure of chlorophyll and of heme but was very much more stable. In fact it is one of the most stable dye stuffs known. He thought that maybe we could use this synthetic compound (coming from a "cracked *pot")* with its high stability as a model of chlorophyll and heme of red blood cells. We hopefully would learn something about the fundamental nature of that peculiar structure, one so important in biology and obviously one with very important catalytic functions—chlorophyll in the plant and heme in the animal. I went to London and got a small sample of phthalocyanine from Linstead and took it back to Manchester for work on hydrogenation. After using up Linstead's supply I had to make it myself. It was my first real effort in organic synthesis. You see the chain. That's where the work on coordination chemistry began, that's where the work on biological catalysts began, that's what led to photosynthesis, that's what led to my interest in chemical evolution. It began as a result of an accidental cracking of the glass pot in the Imperial Chemical Industries factory.

The tools that we have available for exploring chemistry, biochemistry, molecular biology, and cellular biology—that whole sequence—are increasing in sophistication in their capabilities for examining things at molecular and molecular aggregate levels all the time—every day we get new tools. There were new tools available to us at the end of 1945 for measuring radioactivity. When we applied them to biological problems such as photosynthesis we were successful. I think now the general turn toward the use of instruments and concepts—which come, of course, from physical science, physics, and chemistry—in biological problems will allow us to make a dent in the major problem of growth and differentiation. We don't really know yet how a cell knows when to stop growing or when to continue growing or when to change its form.

20 Nonaromatic Nitrogen Compounds

We met our first nitrogen-containing organic compound almost at the beginning of our study and have seen several others since then. We need an organized discussion of the chemistry of the nitrogen containing compounds. Nitrogen has many oxidation states and can form single, double, and triple bonds with itself and with carbon. Its chemistry is very extensive, and we must omit many aspects of it and be very brief about much of what we do say. (We shall postpone the aromatic nitrogen compounds, the amino acids (except for lactam formation), and the nitrogen heterocycles.)

We must know some nitrogen chemistry to be able to understand biological chemistry, as we saw in Section 19.9 concerning photosynthesis and metabolism. This chapter also will prepare us for a later look at amino acids and proteins.

20.1 Introduction to the nitrogen-containing compounds   648
Classes of nitrogen compounds.

20.2 Reactions of amines   649
1. Occurrence; structure; physical properties.   2. Spectroscopy.
3. Basicity of amines: effects of substitution.   4. Nucleophilicity: alkylation of amines.   5. Addition of amines to carbonyl groups.   6. Hinsberg reaction.
7. Reaction with nitrous acid.   8. Preparation and properties of quaternary salts.   9. Exhaustive methylation: Hofmann elimination.
10. Stereochemistry of quaternary salts.   11. Oxidation of amines.
12. Cope elimination. Exercises.

20.3 Preparation of amines   660
1. Substitution reactions.   2. Olysis reactions.   3. Reductions.
4. Rearrangements. Exercises.

20.4 Amides and imides; lactams   664
1. Occurrence and physical properties.   2. Structure and acidity.
3. Spectroscopy.   4. Reactions.   5. Preparation.   6. Lactams. Exercises.

20.5 Nitriles and cyanohydrins   671
1. Nitriles: properties, reactions.   2. Preparation of nitriles: reactions.
3. Cyanohydrins: reactions and preparation. Exercises.

648   Chapter twenty

20.6   Hydroxylamine and hydrazine; oximes and hydrazones   676
       1. Hydroxylamine and hydrazine.   2. Oximes.   3. Hydrazones. Exercises.

20.7   Nitroso and nitro groups; nitrites and nitrates   679
       1. Nitrosoalkanes.   2. Nitroalkanes.   3. Alkyl nitrites and nitrates.
       Exercises.

20.8   Azo, diazo, azide, and cyanate groups   682
       1. Azoalkanes.   2. Diazoalkanes: diazomethane.   3. Azides.
       4. Cyanates. Exercises.

20.9   Summary   684

20.10 Problems   685

20.11 Bibliography   688

20.1 Introduction to the nitrogen-containing compounds

Classes of nitrogen compounds.   Nitrogen-containing organic compounds have some parallels to the alkanes on one hand and to water on the other. Compare CH_4, NH_3, and H_2O. Each is hybridized to some extent: CH_4 completely, H_2O partially, and NH_3 between the two. Methane has four σ bonds, ammonia has three σ bonds and a pair of electrons in the fourth orbital, and water has two σ bonds and two pairs of electrons in two orbitals. Each of the three compounds may lose a proton and form a negative ion: the order of base strength of these ions is $CH_3^{\ominus} > H_2N^{\ominus} > HO^{\ominus}$. However, if a proton is gained, we find that $CH_5^{\oplus}$ (which has only transitory existence) and $H_3O^{\oplus}$ are greater than $NH_4^{\oplus}$ in acid strength. Furthermore, although $NH_4^{\oplus}$ is an ion, it has a tetrahedral structure like that of CH_4.

If a hydrogen atom in each compound is replaced with an HO—, we find CH_3OH, H_2NOH (hydroxylamine), and HOOH. If H_2N— is the replacing group, we have CH_3NH_2, H_2NNH_2, and H_2NOH. If CH_3— is the replacing group, we have CH_3CH_3, CH_3NH_2, and CH_3OH, which have properties similar to those in the CH_4, NH_3, H_2O sequence. If we replace a hydrogen with an acetyl, we have this sequence: CH_3COCH_3, a ketone; CH_3CONH_2, an amide; and CH_3COOH, an acid.

Nitrogen forms a variety of oxides (N_2O, NO, N_2O_3, NO_2, N_2O_4, and N_2O_5). By hydration, they give a variety of acids. By bonding in one way or another to carbon or oxygen of an organic compound, nitrogen forms an impressive array of compounds. Table 20.1 shows a few of them. We shall not discuss naming. Do not memorize the list given in Table 20.1. Some of the names and structures will become familiar as we use them. We have used only familiar alkyl groups; but it should not be too difficult to see how other alkyl, acyl, and aryl groups could fit into the names or give rise to more highly substituted relatives of the compounds shown. Notice the relationships among compounds: amines, imines, amides, imides, nitriles, and isonitriles all may be regarded as substituted ammonias.

Most of this chapter will be devoted to the amines and amides. We shall take a brief look at some of the other classes to get some idea of properties and reactivity without becoming burdened by too much detail.

Table 20.1 Some representative nitrogen compounds

Name	Structure	Name	Structure
ethylamine	EtNH$_2$	ethylhydrazine	EtNHNH$_2$
diethylamine	Et$_2$NH	1,1-diethylhydrazine	
triethylamine	Et$_3$N	(N,N-diethyl-	
		hydrazine)	Et$_2$NNH$_2$
N-ethyl-N-	Me	acetohydrazide	MeCONHNH$_2$
methylpropylamine	PrN⟨	acetone hydrazone	Me$_2$C=NNH$_2$
	Et	acetone azine	Me$_2$C=NN=CMe$_2$
tetraethylammonium		semicarbazide	H$_2$NCONHNH$_2$
chloride	Et$_4$N$^{\oplus}$Cl$^{\ominus}$	acetone	
		semicarbazone	Me$_2$C=NNHCONH$_2$
triethylamine oxide	Et$_3$N$^{\oplus}$—O$^{\ominus}$	ethyldiimide	EtN=NH
trimethylammonium		azoethane	EtN=NEt
methylide	Me$_3$N$^{\oplus}$—CH$_2^{\ominus}$	diazoethane	MeCH=N$^{\oplus}$=N$^{\ominus}$
acetamide	CH$_3$CONH$_2$	N-ethylhydroxylamine	EtNHOH
N-ethylacetamide	CH$_3$CONHEt	ethoxyamine	H$_2$NOEt
N,N-diethylacetamide	CH$_3$CONEt$_2$	acetohydroxamic acid	MeCONHOH
proprionitrile		O-acetylhydroxylamine	MeCOONH$_2$
(propanenitrile)	EtC≡N	acetone oxime	Me$_2$C=NOH
ethylcarbylamine		proprionitrile oxide	EtC≡N$^{\oplus}$—O$^{\ominus}$
(ethyl isonitrile)	EtN$^{\oplus}$≡C$^{\ominus}$	ethyl cyanate	EtOC≡N
urea (carbamide)	H$_2$NCONH$_2$	ethyl isocyanate	EtN=C=O
ethyl carbamate		ethyl azide	EtN=N$^{\oplus}$=N$^{\ominus}$
(urethane)	H$_2$NCOOEt	acetyl azide	MeCON=N$^{\oplus}$=N$^{\ominus}$
biuret	H$_2$NCONHCONH$_2$	nitrosoethane	EtNO
guanidine		ethyl nitrite	EtONO
(carbamidine)	(H$_2$N)$_2$C=NH	nitroethane	EtNO$_2$
cyanamide	H$_2$NC≡N	ethyl nitrate	EtONO$_2$
diethylcarbodiimide	EtN=C=NEt		
ethylidenimine			
(aldimine)	MeCH=NH		
2-propylidenimine			
(ketimine)	Me$_2$C=NH		
succinimide	(succinimide ring structure)		

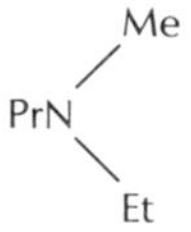

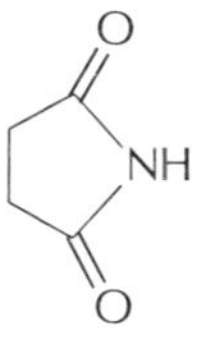

20.2 Reactions of amines

Occurrence; structure; physical properties. Amines are widespread in living organisms, and many of them provoke major physiological effects. The smaller compounds are so common in marine life that we usually say they have fishlike odors. Others, such as putrescine (H$_2$N(CH$_2$)$_4$NH$_2$) and cadaverine (H$_2$N(CH$_2$)$_5$NH$_2$) possess appropriate names. They are products of bacterial decomposition of proteins. At the opposite end of the life cycle, spermine ($H_2N(CH_2)_3NH(CH_2)_4NH(CH_2)_3NH_2$) occurs in human semen (and also in yeast).

Amines are classified as primary (RNH_2), secondary (R_2NH), or tertiary (R_3N), depending upon how many of the hydrogen atoms in ammonia have been replaced by alkyl or aryl groups. The amines resemble ammonia in structure about the nitrogen atom. The structure is very close to tetrahedral, the fourth sp^3 orbital being occupied by a pair of electrons. It is unstable and vibrates (about 10^{10} Hz, E_a about 6 kcal) through the planar

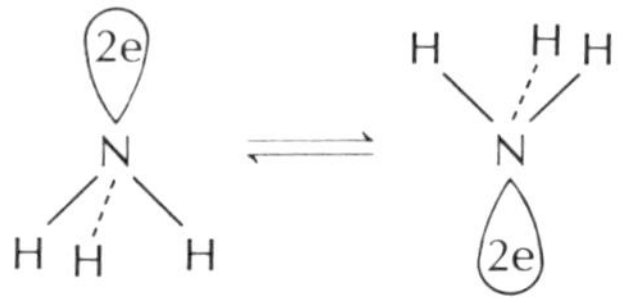

sp^2 hybridized nitrogen atom. The covalent radii of nitrogen (Table 5.3) are 0.3 to 0.5 Å less than those of carbon; the N—H bond length in ammonia is 1.01 Å, and the H—N—H bond angle is 107°; the C—N bond length in trimethylamine is 1.47 Å, and the C—N—C bond angle is 109°. The dipole moments of ammonia and of methyl, dimethyl, and trimethylamine are 1.5, 1.3, 1.0, and 0.6. They illustrate the electron-donating effect of methyl in opposition to the large partial dipole of the occupied orbital.

Table 20.2 gives a sample of some typical amines. The primary and secondary amines are able to hydrogen bond each other weakly, but enough so that the bp are greater than those of the corresponding alkanes (compare pentane, bp 36°, with diethylamine, bp 56°). The tertiary amines, having no N—H bonds, do not hydrogen bond among themselves, and their bp are not far from those of the corresponding alkanes (compare 3-ethylpentane, bp 93°, with triethylamine, bp 89°). Thus primary and secondary amines are analogous to alcohols, and tertiary amines, to ethers (Secs. 9.1 and 9.7).

Table 20.2 Some typical amines

Name	Structure	mp, °C	bp, °C	pK_a
methylamine	$MeNH_2$	−94	−6	10.6
ethylamine	$EtNH_2$	−81	17	10.6
propylamine	$PrNH_2$	−83	49	10.7
butylamine	$BuNH_2$	−49	77	10.8
tert-butylamine	$tBuNH_2$	−67	44	10.8
allylamine	$H_2C{=}CHCH_2NH_2$		58	9.5
dimethylamine	Me_2NH	−93	7	10.8
diethylamine	Et_2NH	−48	56	11.0
diallylamine	$(H_2C{=}CHCH_2)_2NH$		111	9.3
trimethylamine	Me_3N	−117	3	9.8
triethylamine	Et_3N	−115	89	10.6
triallylamine	$(H_2C{=}CHCH_2)_3N$		155	8.3
guanidine	$(H_2N)_2C{=}NH$	~50		13.7
benzylamine	ϕCH_2NH_2		184	9.4
cyclohexylamine	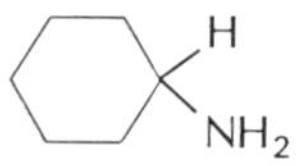 C₆H₁₁NH₂	−18	134	10.6

The lower molecular weight amines are all more soluble in water than are the corresponding alcohols, probably because of the greater tendency of the amine to accept a hydrogen bond from water.

> This trend gives rise to some peculiar solubility behavior. Triethylamine, for example, is completely miscible with water at low temperatures. As the temperature is raised, a point is reached at which miscibility ceases and the mixture separates into two phases, one of the amine dissolved in water, the other of water dissolved in the amine. Presumably, this situation is caused by a marked decrease in the proportion of hydrogen-bonded water to amine species (as compared with normal solvation). Increasing the temperature still more increases solubility of the less soluble non-hydrogen-bonded amine, and complete miscibility returns.

Spectroscopy. The ir of amines show N—H stretching at 3330 to 3570 cm^{-1}. Primary amines have two sharp peaks in this region, secondary amines only one, and tertiary amines none (no N—H bond). Amine salts have the bands reduced to about 3080 cm^{-1}. The N—H bending frequency occurs at 1550 to 1650 cm^{-1}, weak in secondary, medium in primary amines. Amine salts show two strong bands, 1480 to 1520 and 1610 to 1660 cm^{-1}. The C—N vibration bands are weak at about 1000 to 1200 and 1400 cm^{-1}. Imines have bands at 3300 to 3400 and at 1640 to 1690 unless conjugation is present (C=C—C=N), which causes a lowering to 1630 to 1660 cm^{-1}.

In the uv, an amine, like an alcohol, is transparent down to the usual 200 to 220 nm cut-off.

In nmr, amine protons, usually undergoing rapid exchange, give a single peak at δ 0.3 to 2.2 ppm. The nitrogen atom, with spin number (*I*) of one, should interact by spin-spin splitting and cause any protons bound to nitrogen to give a triplet. Amine salts (or amines in strongly acid solutions) show this behavior, with a broad triplet (J_{N-H} = 52 cps) as well as spin interaction with protons on neighboring carbon (absent in the rapidly exchanging amines).

The mass spectra of amines usually show a weak (or absent) parent peak because of a facile cleavage of a group bonded to an α carbon R—C—N$^{\oplus}$ → R· + C=N$^{\oplus}$

If more than one R group is bonded to the α carbon, the larger group departs. A primary amine (RCH$_2$NH$_2$) thus gives H$_2$C=NH$_2^{\oplus}$ (m/e = 30), which is the base peak. Depending upon substitution, m/e peaks at 44 and 58 also may be prominent.

Basicity of amines: effects of substituion. Amines, like ammonia, are basic (Sec. 2.5). In general, amines are more basic than ammonia (pK_a = 9.2), another indication of the +I effect of alkyl groups. However, the presence of unsaturation (greater electronegativity of sp^2 than sp^3 hybridized carbon) almost counteracts the +I effect, so that allyl and benzyl-amines are only slightly more basic than ammonia (Table 20.2).

If we remember that as pK_a increases, basicity increases, and look at the pK_a values for the sequence NH$_3$, RNH$_2$, R$_2$NH, R$_3$N, we find for R = Me, 9.2, 10.6, 10.8, and 9.8; for R = Et, 9.2, 10.6, 11.0, and 10.6; for R = allyl, 9.2, 9.5, 9.3, and 8.3. How do we explain the relatively large base-strengthening effect of the first R, small effect of the second R, and reversal of effect for the third R group? Something other than a polar effect must be in operation, and immediately we suspect something steric in nature. However, for pK_a measured in water, there cannot be much steric difference between the hydrogen bonded amine and the hydrated ion, $R_3N\text{---}HOH$ and $R_3NH^{\oplus} \cdot H_2O$. Steric effects do operate, as we can show by the use of a bulky Lewis acid. Using $(tBu)_3B$ as the acid, we find the order $Et_3N < Et_2NH < EtNH < NH_3$ in base strength. The answer to the problem of protonation seems to lie in the different stabilizations of amine and protonated amine by differences in solvation of the two, even in aprotic solvents. To the extent that differences in solvation result from steric effects, our first impression is correct; but polarity, dielectric constant, and other factors of the solvent also participate. The systems find the best compromise among the contending influences, whether or not we can understand how that compromise is reached.

Guanidine is a strong base, about three pK_a units stronger than a common alkylamine. The reason is simple. The protonated structure is capable of resonance stabilization between three equivalent structures (Eq. 20.1), whereas free guanidine cannot form equivalent

$$\begin{array}{c}
H_2N-C\begin{array}{c}{\nearrow NH}\\{\searrow NH_2}\end{array}\\
\updownarrow\\
H_2\overset{\oplus}{N}=C\begin{array}{c}{\nearrow \overset{\ominus}{N}H}\\{\searrow NH_2}\end{array} + H_3O^{\oplus}\\
\updownarrow\\
H_2N-C\begin{array}{c}{\nearrow \overset{\ominus}{N}H}\\{\searrow NH_2^{\oplus}}\end{array}
\end{array}
\rightleftharpoons
\begin{array}{c}
H_2N-C\begin{array}{c}{\nearrow NH_2^{\oplus}}\\{\searrow NH_2}\end{array}\\
\updownarrow\\
H_2\overset{\oplus}{N}=C\begin{array}{c}{\nearrow NH_2}\\{\searrow NH_2}\end{array} + H_2O\\
\updownarrow\\
H_2N-C\begin{array}{c}{\nearrow NH_2}\\{\searrow NH_2^{\oplus}}\end{array}
\end{array}
\quad (20.1)$$

resonance structures. The effect is so great that it overwhelms all other influences within the system. Aromatic amines (Chap. 22) are very susceptible to resonance effects of ring substituents.

> The hydrochloride and sulfate salts of amines that are not miscible with water are more soluble in water than are the amines themselves. An amine can be extracted from dilute sodium hydroxide solutions into ether (or other water-immiscible solvent). An amine can be extracted from ether into dilute hydrochloric acid. Unlike ammonium chloride, the amine hydrochlorides are soluble in ethanol. Some salts are quite insoluble and once were used as quantitative precipitants for amines. Platinic chlorides $((RNH_3)_2^{\oplus}PtCl_6^{2\ominus})$ are examples. Picrates have been useful in the identification of tertiary amines because the mp of the salts $(R_3NH^{\oplus}\text{ picrate}^{\ominus})$ and purification by recrystallization from alcohol is easy.

Nucleophilicity: alkylation of amines. Amines, being basic, also are nucleophiles (see Secs. 2.6 and 3.3); and the reaction of ammonia or a primary, secondary, or tertiary amine with an alkyl halide, sulfate, or sulfonate gives rise to primary, secondary, tertiary, and quaternary ammonium salts. Primary amines react more quickly by S_N2 than do secondary; tertiary amines eliminate. As we know, the reactions are not clean-cut because of competing equilibria. Industrially, alkylation of ammonia is useful because the products are easily separable by distillation and are salable. When a specific amine is to be prepared in the laboratory, we use a reaction that gives a good yield of the specific amine rather than the alkylation reaction, with its lower yield and separation problem. An exception is the preparation of quaternary ammonium salts, in which there is no competing equilibrium (Eq. 20.2). If it is carried out in ether (which is polar enough to favor the charge separation

$$R_3N + CH_3I \longrightarrow R_3\overset{\oplus}{N}CH_3 \quad I^{\ominus} \tag{20.2}$$

in the transition state), the salt simply crystallizes as the reaction proceeds. □ □

Ammonia and amines react quickly with halogens. For preparative purposes, NaOCl and NaOBr usually are used ($RNH_2 + NaOCl + H_2O \rightarrow RNH_2Cl^{\oplus} + Na^{\oplus} + 2HO^{\ominus}$). The quaternary ion readily loses a proton to give an N-chloroamine ($RNH_2Cl^{\oplus} + HO^{\ominus} \rightarrow RNHCl + H_2O$). If excess halogen (or hypohalite) is used, all the hydrogen atoms on nitrogen are replaced. The chloroamines hydrolyze under acidic conditions to give the conjugate acid of the amine and hypochlorous acid.

Addition of amines to carbonyl groups. Being nucleophiles, amines add to the carbonyl carbon atom of $\diagup \overset{\diagdown}{C}=O$ and also undergo a Michael type of addition to α,β-unsaturated carbonyl systems. These reactions may be thought of as aminolysis reactions. The ultimate course of such a reaction depends on several factors, not the least being the class of amine employed (Eqs. 20.3). A primary or secondary amine allows the reaction to

$$CH_3COOEt + NH_3 \longrightarrow CH_3CONH_2 + EtOH$$

$$(CH_3CO)_2O + 2RNH_2 \longrightarrow CH_3CONHR + CH_3COO^{\ominus} + R\overset{\oplus}{N}H_3$$

$$\phi CHO + RNH_2 \rightleftharpoons \phi CH=NR + H_2O \tag{20.3}$$

$$CH_3COCl + 2R_2NH \longrightarrow CH_3CONR_2 + R_2\overset{\oplus}{N}H_2 + Cl^{\ominus}$$

$$\phi COCl + R_3N \rightleftharpoons \phi\overset{O^{\ominus}}{\underset{R_3N^{\oplus}}{C-Cl}} \rightleftharpoons \phi\overset{O}{\underset{R_3N^{\oplus}}{C}} + Cl^{\ominus}$$

go to completion by loss of a proton; but a tertiary amine, having no proton to lose, does no better than to go into an equilibrium situation. However, the equilibrium is mobile enough to allow easy replacement of the tertiary amine. Thus, tertiary amines are frequently used catalysts for reactions susceptible to mild bases. (The preparation of derivatives of aldehydes and ketones will be discussed in Sec. 20.6.)

Hinsberg reaction. A laboratory procedure that is useful for identification of the class of an amine as well as for preparation of solid derivatives of amines and for the separation of amines is the Hinsberg reaction. The original reagent used was benzenesulfonyl chloride (ϕSO_2Cl), an oily liquid, the acid chloride of benzenesulfonic acid. Also frequently used is *p*-toluenesulfonyl chloride, a solid. □ □

Use of the Hinsberg reaction for testing or preparation of derivatives is done as follows. A few drops of the unknown amine (or a small amount of solid) is mixed with a few drops of the Hinsberg reagent in a few ml of dilute sodium hydroxide. After the sulfonyl chloride has reacted, there are three possibilities: 1. The reaction mixture contains no oil or precipitate (the amine is I, and the I sulfonamide is in solution in the form of a sodium salt, $Na^\oplus\ \phi SO_2\overset{\ominus}{N}R$). Acidification with dilute HCl gives a precipitate of the primary sulfonamide. (The sulfonic acid, also formed, is soluble in water.) 2. The reaction mixture contains a precipitate of a secondary sulfonamide. Or 3. the reaction contains an oil or solid that appears to be unchanged amine. If acidification causes solution to occur, the amine is tertiary. If acidification causes no change in the precipitate, the amine is secondary.

Reaction with nitrous acid. Before describing the reactions of amines with nitrous acid, we need to look at the reactions of nitrous acid in aqueous solution (Eqs. 20.4). Nitrous

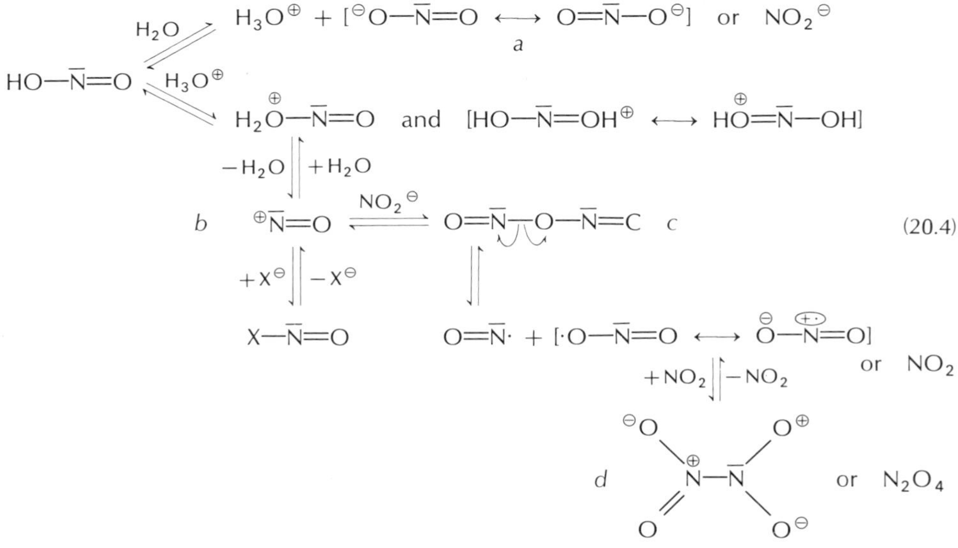

$$\text{(20.4)}$$

acid, like carbonic acid, is unstable and cannot be isolated from aqueous solution (Eq. 20.5).

$$3HNO_2 \rightleftharpoons HNO_3 + 2NO + H_2O \quad (20.5)$$

Nitrous acid is prepared by the addition of dilute sulfuric acid to a cold aqueous solution of sodium nitrite and used immediately. (Sodium or potassium nitrite are made by heating of the nitrate salts with carbon: $2NaNO_3 + C \rightarrow 2NaNO_2 + CO_2$). Nitrous acid ($pK_a = 3.35$) is not a strong acid. Thus, the presence of a strong acid decreases the extent of its acid ionization (nitrite ion (a) concentration decreases) and formation of nitrosonium ion (b, $NO^\oplus$) and blue dinitrogen trioxide (c) is favored. The nitrosonium ion reacts with all the various nucleophiles present: water, nitrite ion, nitrous acid (not shown), and chloride ion (if hydrochloric acid is used as the strong acid in place of sulfuric acid). The trioxide (c) equilibrates as shown with nitric oxide (NO), nitrogen dioxide (brown), and dinitrogen tetroxide.

It seems to be generally agreed that dinitrogen trioxide is the reactive entity toward amines. (The same results may be obtained, in paper reactions, by use of undissociated nitrous acid, as well as the other oxides, and ions.) The initial step is nucleophilic attack of the amine upon nitrogen of a —N=O bond (Eq. 20.6) to give a nitrosoamine.

$$R\overset{\frown}{\bar{N}H_2} + O\overset{\frown}{=}N-O-N=O \rightleftharpoons \left[\begin{array}{c} \overset{\oplus}{R{N}H_2} \\ | \\ {}^{\ominus}O-N-O-N=O \end{array} \right]$$

$$\begin{array}{c} RNH \\ | \\ O=N \end{array} \underset{+H^\oplus}{\overset{-H^\oplus}{\rightleftharpoons}} \begin{array}{c} \overset{\oplus}{R{N}H_2} \\ | \\ O=N \end{array} + NO_2{}^\ominus \tag{20.6}$$

What happens next depends upon the class of amine used, temperature, and stability and reactivity of the intermediates and products. To prevent decomposition of nitrous acid (Eq. 20.4), the reaction of an amine with nitrous acid usually is carried out in ice-cold (0° to 5°) solutions. In the presence of a strong acid (pH < 2), the alkylnitrosoamine from a primary amine is unstable and decomposes (Eq. 20.7), by proton transfer, into the hypo-

$$R\bar{N}H-\bar{N}=O \overset{H^\oplus}{\rightleftharpoons} \left[\begin{array}{c} R\overset{\frown}{N}H-\bar{N}\overset{\frown}{=}OH^\oplus \\ \updownarrow \\ \overset{\oplus}{R{N}H}=\bar{N}-OH \end{array} \right] \overset{-H^\oplus}{\rightleftharpoons} [R\bar{N}=\bar{N}-OH]$$

$$+H^\oplus \updownarrow -H^\oplus \tag{20.7}$$

$$R^\oplus \overset{-N_2}{\longleftarrow} \left[\begin{array}{c} R-\overset{\frown}{N}\equiv\bar{N}^\oplus \\ \updownarrow \\ R-\overset{\oplus}{N}\equiv\bar{N} \end{array} \right] \underset{+H_2O}{\overset{-H_2O}{\rightleftharpoons}} [R\bar{N}=\bar{N}-\overset{\oplus}{O}H_2]$$

thetical diazohydroxide (RN=NOH), which accepts a proton and dehydrates to give a diazonium ion, $R\overset{\oplus}{N}\equiv N$.

A diazonium ion is not very stable and is subject to irreversible loss of a molecule of nitrogen to give a carbonium ion. The carbonium ion is subject to all the possibilities of reaction we have previously discussed. This kind of ion, formed by the "explosion" of a diazonium ion, has at times been referred to as a "hot carbonium ion," because the proportion of the various products formed sometimes differs from those formed when the ion is produced by one of the S_N1 (or E1) pathways. We may now write the equation for the reaction of butylamine with sodium nitrite and hydrochloric acid (Eq. 20.8). ☐ ☐

$$CH_3CH_2CH_2CH_2NH_2 \xrightarrow[HCl, 0°]{NaNO_2}$$
$$CH_3CH_2CH_2CH_2OH + CH_3CH_2CH_2CH_2Cl + CH_3CH_2CH=CH_2 +$$
$$CH_3CH=CHCH_3 + CH_3CH_2CHOHCH_3 + CH_3CH_2CHClCH_3 + N_2 \tag{20.8}$$

A primary aromatic amine gives a relatively stable diazonium ion (Chap. 22).

An amino alcohol rearranges to give a ketone (Eq. 20.9) in good yield. The reaction is sometimes called the <u>semipinacol rearrangement.</u>

A secondary amine reacts with nitrous acid to give a stable nitrosoamine (Eq. 20.10).

$$R_2\overset{\overset{OH}{|}}{C}-CH_2NH_2 \xrightarrow[H^\oplus]{NaNO_2} R_2\overset{\overset{OH}{|}}{C}-\overset{\oplus}{C}H_2 \xrightarrow{shift} R\overset{\overset{O}{\|}}{C}-CH_2R \qquad (20.9)$$

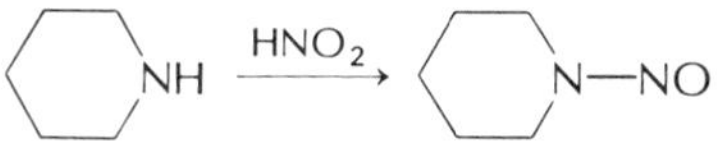

$$\underset{}{\bigcirc}\!\!\text{NH} \xrightarrow{HNO_2} \underset{}{\bigcirc}\!\!\text{N}-\text{NO} \qquad (20.10)$$

Most secondary nitrosoamines are nonbasic, water-insoluble, oily liquids. (The very simple nitrosoamines, such as Me_2N-NO, are soluble in water.) Reduction (catalytic hydrogenation, $LiAlH_4$, etc.) regenerates the secondary amine.

A tertiary amine, under acidic conditions (pH < 1), does not react with nitrous acid and remains in solution as the ammonium salt. □ □

A supplement to the Hinsberg test for the class of an amine is the nitrous acid test. A little bit of the amine is dissolved in dilute hydrochloric acid, the solution is chilled, and a few drops of $NaNO_2$ solution are added. If nitrogen is evolved, the amine is primary (Eq. 20.8); if no gas is evolved, the amine is secondary or tertiary. If a yellow oil separates, the amine is secondary (Eq. 20.10). If there is no apparent change, the amine is tertiary.

Preparation and properties of quaternary salts. A tertiary amine reacts with an alkyl halide, sulfate, or sulfonate to form a quaternary ammonium salt (Eqs. 20.2). The rates of formation depend upon the bulk of the groups in the tertiary amine and the usual S_N2 activity for the halides (Me > primary > secondary ≫ tertiary). Thus, the last step in the preparation of a quaternary salt is chosen so that the alkyl group is the smallest primary of the four alkyl groups (Eq. 20.11). As shown, the order is $a > b > c \gg d$ in the

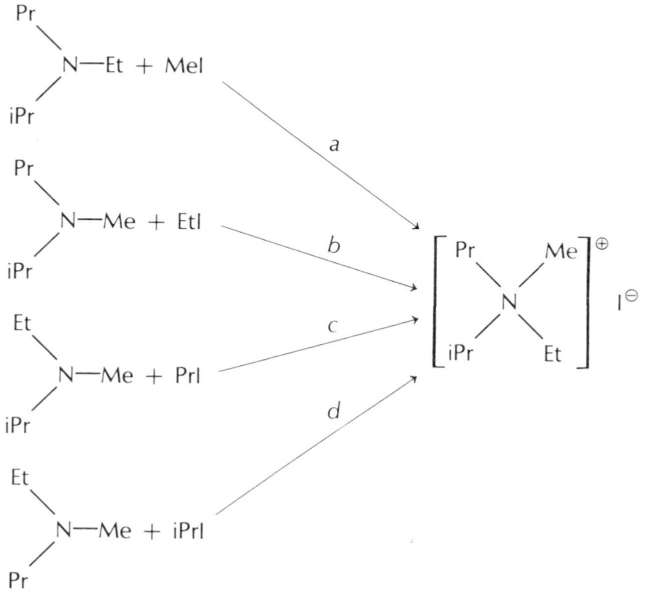

$$\begin{array}{c}
\underset{iPr}{\overset{Pr}{\diagdown}}N-Et + MeI \\
\underset{iPr}{\overset{Pr}{\diagdown}}N-Me + EtI \\
\underset{iPr}{\overset{Et}{\diagdown}}N-Me + PrI \\
\underset{Pr}{\overset{Et}{\diagdown}}N-Me + iPrI
\end{array}
\begin{array}{c} a \\ b \\ c \\ d \end{array}
\left[\begin{array}{c} Pr \diagdown \diagup Me \\ N \\ iPr \diagup \diagdown Et \end{array}\right]^{\oplus} I^{\ominus} \qquad (20.11)$$

selection of the best possible last step for the preparation of the quaternary ammonium salt (starting with ammonia).

The quaternary ammonium salts are completely ionic (shown by X-ray crystal studies), and their aqueous solutions conduct an electric current. The large cations are remarkably tolerant of large anions, as is shown by salts like $Et_4N^\oplus I_7^\ominus$ and $Et_4N^\oplus I_9^\ominus$. Anions such as $Cl_3^\ominus$, $Br_5^\ominus$, and so on have been observed as well. The long-chain quarternary salts are used as cationic surfactants (Sec. 17.1).

The quaternary ammonium hydroxides require special techniques for preparation because the hydroxides are strong bases, resembling NaOH and KOH. One method is to add an excess of a suspension of Ag_2O (from $AgNO_3$ + NaOH) to an aqueous solution of a quaternary ammonium iodide or bromide (Eq. 7.17). Removal of the precipitated silver halide by filtration gives a solution from which the hydroxide may be obtained by evaporation of water (though not by boiling, as we shall see; Eq. 20.12). A second method, useful

$$2R_4N^\oplus I^\ominus + Ag_2O \xrightarrow{H_2O} SR_2N^\oplus OH^\ominus + 2\underline{AgI} \qquad (20.12)$$

for chlorides, depends upon the insolubility of KCl in methanol (Eq. 20.13). The most general

$$R_4N^\oplus Cl^\ominus + KOH \xrightarrow{MeOH} R_4N^\oplus OH^\ominus + \underline{KCl} \qquad (20.13)$$

method is to pass an aqueous solution of the salt through a column of an anion exchange resin, whereby the halide ion is replaced by hydroxide ion.

Exhaustive methylation: Hofmann elimination. The quaternary ammonium hydroxides are thermally unstable. The reaction is known as the Hofmann elimination and was discussed in Section 7.7. Combined with <u>exhaustive methylation,</u> the Hofmann elimination is of use in the elucidation of the structure of naturally occurring nitrogenous compounds. A simple example is shown in Equation 20.14. The process of methylation is

repeated until the quaternary salt is obtained. Conversion to the quaternary hydroxide, elimination, and repetition until $N(CH_3)_3$ is obtained gives a non-nitrogen-containing fragment that, after identification, provides evidence for the structure of the original amine.

The quaternary ammonium salts are less subject to thermal decomposition but will fragment if temperatures are high enough ($Cl^\ominus$ is less basic than $HO^\ominus$). If no β hydrogen atom is present, a quaternary hydroxide undergoes an S_N2 reaction at the α carbon atom (Eq. 20.15). The alkyl group that is most susceptible to an S_N2 displacement is removed as an alcohol. Substitution on methyl is the usual result of the reaction.

$$Me_3\overset{\oplus}{N}CH_2\phi \xrightarrow[\Delta]{HO^{\ominus}} \left[\underset{HH}{Me_3\overset{\overset{\phi}{|}}{\underset{|}{N}}\text{---}\overset{(\delta+)}{\underset{}{C}}\text{---}\overset{(\delta-)}{OH}} \right] \longrightarrow Me_3N + \phi CH_2OH \quad (20.15)$$

Stereochemistry of quaternary salts. The quaternary salts offer an opportunity to test the stereochemistry of nitrogen. If the four groups bonded to nitrogen are all different, the compound may be separated into the two enantiomeric components. This resolution was first accomplished in 1899, when allylbenzylmethylphenylammonium iodide was obtained in (+) and (−) forms. Amines are incapable of resolution because the rate of

$$H_2C=CHCH_2-\overset{\overset{CH_3}{|}}{\underset{\underset{\phi}{|}}{\overset{\oplus}{N}}}-CH_2\phi \quad I^{\ominus}$$

vibration through the planar sp^2 hybrid is so great that the lifetime of a given enantiomer is not long enough to allow a separation. The salts of amines also cannot be resolved because of rapid proton transfers. Even though the amount of free amine is negligible (high acid concentration), racemization is so rapid that resolution fails. Even an optically active quaternary ammonium salt undergoes racemization at a reasonable rate.

By tying down the bonds to nitrogen so that inversion cannot take place, it has been possible to produce a few amines in optically active forms. An example is Troger's base, in which the bridgehead tertiary nitrogens are incapable of inversion.

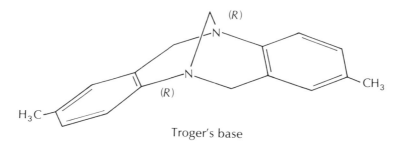

Troger's base

Imines are capable of geometrical isomerism, but the facile proton transfer allows only the more stable of the two possible structures to be isolated (Eq. 20.16).

$$\underset{H_3CCH_3}{\overset{\phi}{\underset{}{\diagdown}}C=N\overset{}{\diagup}} \underset{a}{\overset{+H^{\oplus}}{\rightleftarrows}} \underset{H_3CCH_3}{\overset{\phi}{\diagdown}C=\overset{\oplus}{\underset{}{N}}\overset{H}{\diagup}} \longleftrightarrow \underset{H_3CCH_3}{\overset{\phi}{\diagdown}\overset{\oplus}{C}-N\overset{H}{\diagup}} \rightleftarrows$$

$$\underset{H_3CH}{\overset{\phi}{\diagdown}\overset{\oplus}{C}-N\overset{CH_3}{\diagup}} \longleftrightarrow \underset{H_3CH}{\overset{\phi}{\diagdown}C=\overset{\oplus}{N}\overset{CH_3}{\diagup}} \overset{-H^{\oplus}}{\rightleftarrows} \underset{H_3C}{\overset{\phi}{\diagdown}C=N\overset{CH_3}{\diagup}} \quad (20.16)$$

The more stable form, of course, is that in which the smaller group on carbon is syn to the group on nitrogen (a).

Oxidation of amines. Amines are subject to oxidation in neutral and basic solution, whereas amine salts (acidic solutions) are stable. Peracids and hydrogen peroxide react by attack on the free electron pair of nitrogen to give a tertiary amine oxide (Eq. 20.17). An amine oxide has a σ bond between nitrogen and oxygen, but the bond is

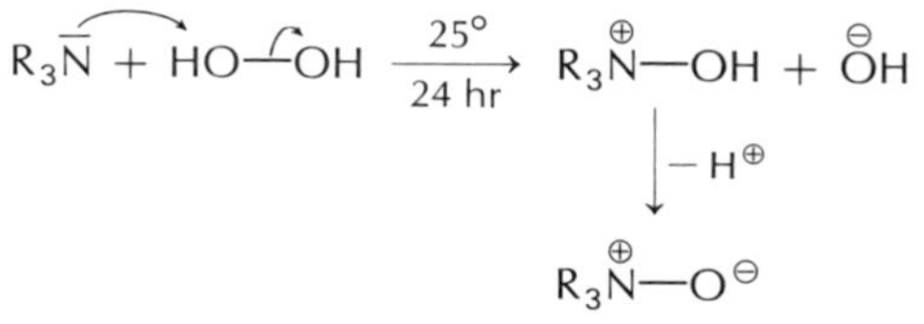

$$R_3N + HO-OH \xrightarrow[24\ hr]{25°} R_3\overset{\oplus}{N}-OH + \overset{\ominus}{O}H$$

$$\downarrow -H^{\oplus}$$

$$R_3\overset{\oplus}{N}-O^{\ominus} \qquad (20.17)$$

polarized by the formal charges on both atoms. This coordinate covalent bond is formed when one of the atoms furnishes both electrons to the bond. The bond usually is shorter than normal (1.27 Å) and has a large dipole (5.0 D for Me_3NO). The tertiary amine oxides are relatively stable, with a strong tendency to form hydrates ($[Me_3NOH]^{\oplus}$ $HO^{\ominus} \cdot H_2O$, mp 96°). They are tetrahedral; and, when properly substituted, they may be resolved into the constituent enantiomers. The tertiary amine oxides are basic, but they are weaker than

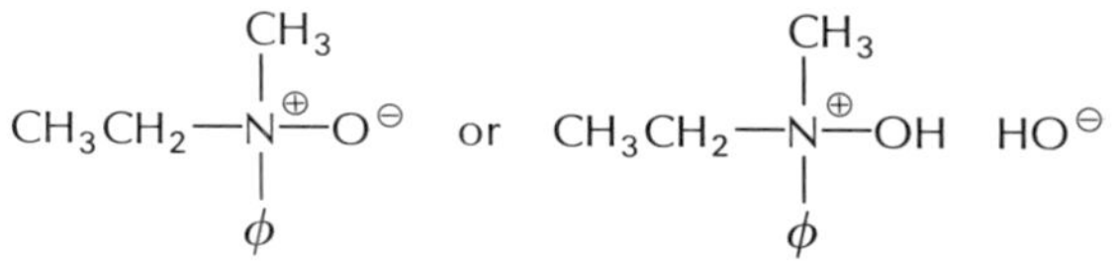

$$CH_3CH_2-\underset{\underset{\phi}{|}}{\overset{\overset{CH_3}{|}}{\overset{\oplus}{N}}}-O^{\ominus} \quad \text{or} \quad CH_3CH_2-\underset{\underset{\phi}{|}}{\overset{\overset{CH_3}{|}}{\overset{\oplus}{N}}}-OH \quad HO^{\ominus}$$

the corresponding tertiary amines (Me_3N, $pK_a = 9.8$; Me_3NO, $pK_a = 4.65$).

Cope elimination. Like the quaternary ammonium hydroxides, amine oxides are thermally unstable. If β hydrogen atoms are present, a cyclic elimination pathway gives exclusively *cis* elimination (Eq. 20.18). This reaction is known as the *Cope reaction* or

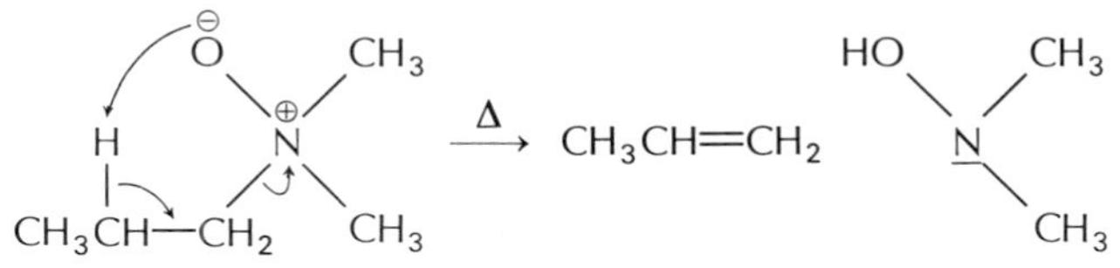

(20.18)

elimination (see Sec. 13.9). The oxidation of the tertiary amine and the Cope reaction may be carried out in the same flask without isolation of the amine oxide.

Exercises

20.1 Using cyclohexylamine (A), N-methylcyclohexylamine (B), and N,N-dimethylcyclohexylamine (C), complete the reactions.
   a. $C + CH_2=CHCH_2Cl \longrightarrow$
   b. $B + NaOCl + H_2O \longrightarrow$

c. B + φCOCl + NaOH ⟶

d. A + φCHO ⟶

e. B + φSO₂Cl + NaOH ⟶

f. B + NaNO₂ + HCl $\xrightarrow{0°}$

g. B + MeI $\xrightarrow[2.\ \Delta]{1.\ Ag_2O}$

h. B + φMgBr ⟶

i. A ⟶ a racemic salt

j. C + H₂O₂ $\xrightarrow{\Delta}$

20.3 Preparation of amines

An amine may be prepared in many different ways. We shall classify the various methods as substitutions, olysis reactions, reductions, and rearrangements. Still other methods are used for the synthesis of amino acids. As usual, our classification of reactions is arbitrary because more than one step may be required.

Substitution reactions. Alkylation of ammonia and amines has been described (Sec. 3.3 and Eq. 20.2).

The Gabriel synthesis makes use of the acidity of phthalimide. The nucleophilicity of nitrogen is increased by the negative charge that arises from salt formation of the imide with a base. Also, the method is specific for the preparation of primary amines. The first step is the reaction of phthalimide with potassium carbonate or potassium hydroxide in dimethylformamide (DMF), in which potassium phthalimide is soluble (Eq. 20.19). The anion of phthalimide displaces halogen (or sulfate or sulfonate) from an alkyl halide (primary > secondary ≫ tertiary, as usual) to give a substituted imide. The substituted imide may be hydrolyzed to give phthalic acid and the hydrochloride salt of RNH₂ (a). However, unless R is one of the simple alkyl groups, the hydrolysis is slow. Hydrazinolysis

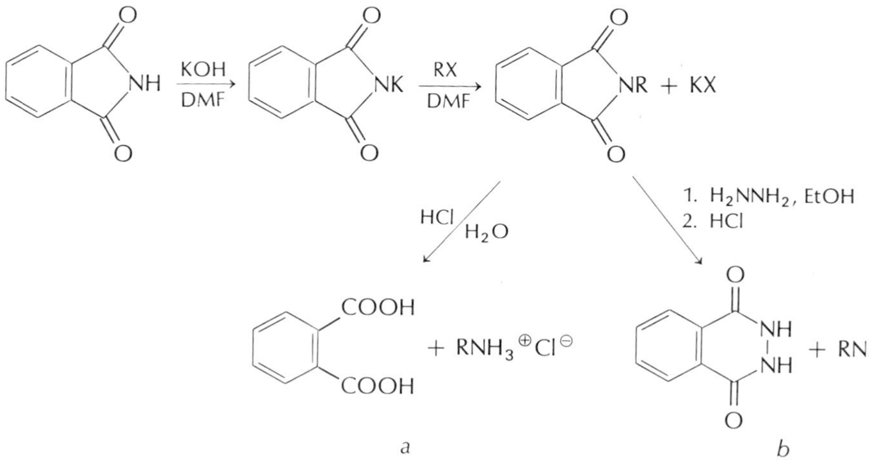

(20.19)

(warming with H_2NNH_2 in alcohol) is gentle and rapid. When the reaction is complete, addition of hydrochloric acid holds the amine in solution so that the insoluble phthalhydrazide may be removed by filtration. The Gabriel synthesis is used widely for the preparation of RNH_2 in which R is primary or secondary. □ □

Olysis reactions. Olysis reactions for the production of amines are limited to the hydrolysis of N-substituted amides and imides. We could have listed the Gabriel synthesis as an olysis method because the second step may be the hydrolysis of an N-substituted imide. Because amides and imides usually are prepared by acid chloridolysis or anhydridolysis of amines, hydrolysis of a synthetic amide or imide is of little preparative value unless the reaction is preceded by an alkylation of the amide or imide, as in the Gabriel synthesis.

Another route to a substituted amide is the Ritter reaction, which is limited to the introduction of alkyl groups that form relatively stable carbonium ions, such as $tBu^{\oplus}$ (Eq. 20.20). The carbonium ion, which may be produced in any convenient way (usually

$$tBuOH \xrightarrow{H_2SO_4}$$
$$Me_2C{=}CH_2 \xrightarrow{H_2SO_4} tBu^{\oplus} \begin{array}{c} \xrightarrow{HCN} tBuN{\equiv}CH \xrightarrow{H_2O} tBuNHCHO \\ \\ \xrightarrow{RCN} tBuN{\equiv}CR \xrightarrow{H_2O} tBuNHCR(O) \end{array}$$
$$tBuX \xrightarrow{H_2SO_4}$$

$$HCONHtBu \xrightarrow[H_2O]{NaOH} HCOONa + tBuNH_2$$

(20.20)

from the alcohol in the laboratory), adds to the nitrogen of a nitrile group (or of $HC{\equiv}N$ formed by $H_2SO_4 + NaCN$) to form a nitrilium salt. The unstable nitrilium salt hydrates in the reaction mixture to give the N-substituted amide. The amide, upon hydrolysis, gives the *tert*-alkylamine. Alkaline hydrolysis is preferred because acidic hydrolysis nearly always gives a type of reaction that is essentially a reversal of the Ritter reaction. In any case, a formamide hydrolyzes faster than any of the other possible amides.

Reductions. In general, all nitrogen compounds that are not already amines are reducible to amines—but there are many exceptions. We shall give only the more popular methods.

Nitro, nitroso, hydroxylamine, azo, diazo, and hydrazino compounds all may be reduced to an amine. Popular reagents are $LiAlH_4$, H_2 with a catalyst, BH_3, and a metal (Sn, Zn, or Fe) with an acid (Eqs. 20.21).

$$(CH_3)_2CHNO_2 \xrightarrow{Sn,\ HCl} (CH_3)_2CHNH_2$$
nitro

$$CH_3CH_2NHOH \xrightarrow{LiAlH_4} CH_3CH_2NH_2$$
hydroxylamine

$$CH_3CH_2N{=}NCH_2CH_3 \xrightarrow{H_2}{Pt} CH_3CH_2NH_2$$
azo

$$CH_3NHNHCH_3 \xrightarrow{BH_3} CH_3NH_2$$
hydrazine

(20.21)

Compounds in which nitrogen is double or triple bonded to carbon also are reducible (Eqs. 20.22), as are amides. Catalytic hydrogenation, although generally useful, suffers from

$$(CH_3)_2C=NOH \xrightarrow{Na + EtOH} (CH_3)_2CHNH_2$$

$$CH_3CH_2CH=NNH\phi \xrightarrow{LiAlH_4} CH_3CH_2CH_2NH_2 + H_2N\phi$$

$$\phi_2C=NNHCONH_2 \xrightarrow[Pt]{H_2} \phi_2CHNH_2 + H_2NCONH_2 \qquad (20.22)$$

$$(CH_3)_2CHCONH_2 \xrightarrow{LiAlH_4} (CH_3)_2CHCH_2NH_2$$

$$\phi CH_2CH_2C\equiv N \xrightarrow[Pt, (CH_3CO)_3O]{H_2} \phi CH_2CH_2CH_2NHCOCH_3 \xrightarrow[H_2O]{HO^{\ominus}} \phi CH_2CH_2CH_2NH_2$$

a side reaction that reduces the yield of primary amine (amides require high pressures and temperatures). A nitrile is reduced first to an imine ($RC\equiv N \rightarrow RCH=NH$), which is capable of aminolysis with some of the amine product to give a new imine ($RCH=NH + H_2NCH_2R \rightarrow RCH=NCH_2R + NH_3$). The new imine then is reduced to a secondary amine (RCH_2NHCH_2R). A similar difficulty is encountered if $LiAlH_4$ is used, unless the substance to be reduced is added to a solution of $LiAlH_4$ so that an excess of reducing agent is always present. The difficulty is avoided during the catalytic hydrogenation of a nitrile by having acetic anhydride present, which acylates the amine as rapidly as it is formed. The free amine is obtained in a later hydrolysis of the N-substituted acetamide.

Imides are reduced to secondary amines by $LiAlH_4$. This method provides a pathway to some of the nitrogen-containing ring compounds (Eq. 20.23).

$$(CH_3)_2C\begin{matrix}CH_2COOH\\ \\CH_2COOH\end{matrix} \xrightarrow[\substack{1.\ SOCl_2\\2.\ NH_3\\3.\ \Delta}]{} (CH_3)_2C\begin{matrix}H_2C-C(=O)\\ \\H_2C-C(=O)\end{matrix}NH \xrightarrow{LiAlH_4} \begin{matrix}H_3C\\ \\H_3C\end{matrix}\!\!\!\bigcirc\!\!\!NH \qquad (20.23)$$

Reductive alkylation is a means of converting an aldehyde or ketone into an amine without isolating the intervening imine or carbonyl derivative (oximes, and so on, as in Eq. 20.22). As the method was originally devised, the carbonyl compound is hydrogenated in the presence of ammonia or an amine. The intermediate hydroxylamine is subjected to hydrogenolysis, or it eliminates water to form an imine that is hydrogenated (Eq. 20.24).

$$\!\!>\!\!C=O + NH_3 \rightleftharpoons \!\!>\!\!C\!\!\begin{matrix}NH_2\\ \\OH\end{matrix} \xrightarrow{-H_2O} \!\!>\!\!C=NH \xrightarrow[cat.]{H_2} \!\!>\!\!CHNH_2 \qquad (20.24)$$

(with H_2/cat. pathway directly from hydroxylamine)

Ammonia gives rise to primary amines, primary amines give secondary amines, and secondary amines give tertiary amines as products.

Rearrangements. An amide, treated with halogen and alkali (bromine and dilute sodium hydroxide), undergoes the Hofmann rearrangement to give an amine by what appears to be decarbonylation ($RCONH_2 \rightarrow RNH_2$). However, the stoichiometry of the

20.3 Preparation of amines

reaction shows that bromine is acting as an oxidant (Eq. 20.25). Although it was discovered

$$RCONH_2 + Br_2 + 4NaOH \longrightarrow RNH_2 + Na_2CO_3 + 2NaBr + 2H_2O \quad (20.25)$$

in 1882 by Hofmann, and despite many investigations, a few points in the mechanism still await clarification. The first step is a halogenation of the amide to give a bromoamide (Eq. 20.26). With RCO on one side of N and Br on the other, each exerting a $-I$ effect,

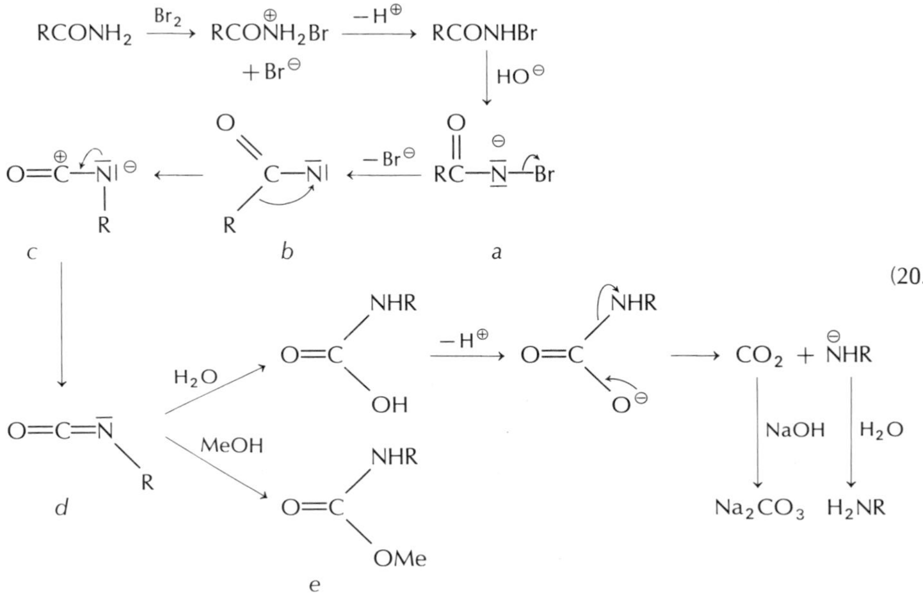

$$RCONH_2 \xrightarrow{Br_2} RCO\overset{\oplus}{N}H_2Br \xrightarrow{-H^\oplus} RCONHBr$$
$$+ Br^\ominus \qquad \downarrow HO^\ominus$$

(20.26)

the bromoamide is acidic enough to lose a proton to give the anion a. Subsequent steps to b (loss of $Br^\ominus$), migration of R from C to N (c), and formation of isocyanate (d) are shown as separate steps for clarification. However, this point is not clear: the steps are thought by many to be concerted. That is, R migrates as $Br^\ominus$ departs and the carbon-nitrogen double bond is formed. An alternative point of view is that b, a <u>nitrene</u> (analogous to a carbene), is formed before R migrates. In any case, R is never free, because a chiral R has been shown to retain configuration during the rearrangement (R = ϕ—$\overset{*}{C}H$—, for example).
$$\qquad\qquad\qquad\qquad\qquad\qquad\qquad\qquad\qquad\qquad\qquad\qquad\quad |$$
$$\qquad\qquad\qquad\qquad\qquad\qquad\qquad\qquad\qquad\qquad\qquad\qquad\; CH_3$$

The isocyanate may be isolated in certain cases, but the reaction usually is carried out in water, and hydrolysis leads to the products shown. In most instances, yields are improved by use of methanol as the solvent to isolate the urethane (a substituted methyl carbamate) as product (e). Subsequent easy hydrolysis of the urethane gives the amine.

Imides react to give amino acids (Eq. 20.27) if hydrolysis is carried out first. □ □

The <u>Beckmann rearrangement</u> is useful at times because amines can be prepared by hydrolysis of the substituted amide obtained as a result of the rearrangement. Oximes of

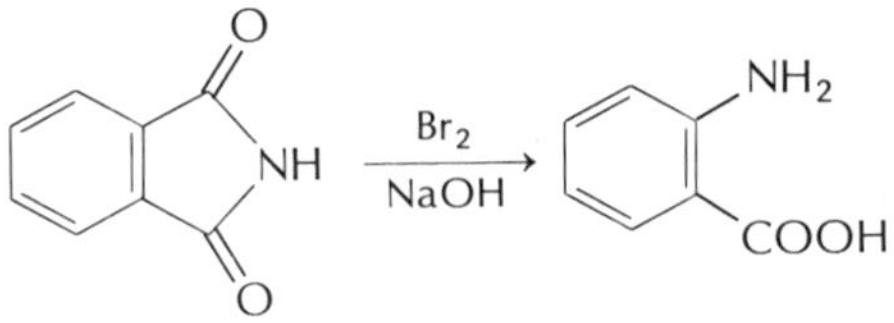

(20.27)

aldehydes and ketones are readily prepared (Sec. 16.3). The stable form of an oxime is that in which the —OH group on nitrogen is *syn* to the smaller group on carbon. However, interconversion of the *syn* and *anti* isomers is easy. (In some instances, filtration of a solution of an isomer through decolorizing carbon causes isomerization.)

A ketoxime reacts in ether solution with PCl$_5$ to undergo rearrangement. Subsequent treatment with water gives the amide (Eq. 20.28). The bulky group *anti* to the —OH on

$$\begin{array}{c}H_3C\\ \diagdown\\ C=N\\ \diagup\hspace{1em}\diagdown\\ \phi\hspace{2em}OH\end{array} \xrightarrow[\text{ether}]{PCl_5} \left[\begin{array}{c}HO\\ \diagdown\\ C=N\\ \diagup\hspace{1em}\diagdown\\ H_3C\hspace{2em}\phi\end{array}\right] \xrightarrow{H_2O} \begin{array}{c}O\\ \|\\ C-NH\phi\\ \diagup\\ H_3C\end{array} \quad a \quad (20.28)$$

nitrogen migrates. It took long and ingenious research work to prove that it is the *anti* group that migrates. The point has been proved, and the reaction is considered stereospecific. Hydrolysis of the amide shown gives aniline (ϕNH$_2$) and acetic acid. A disadvantage of the Beckmann rearrangement is that only one of the R groups in a ketone migrates, whereas the other R group remains attached to the carbonyl carbon and ends up as an acid (acetic acid from *a*). Thus, only a part of the molecule of ketone is used in amine formation.

Other than PCl$_5$ in ether, reagents, that cause the rearrangement are sulfuric acid, polyphosphoric acid, and anhydrous HCl in acetic acid and acetic anhydride mixture. Rearrangement also is caused by warming of the aqueous solution of the sulfonate ester of an oxime formed by ϕSO$_2$Cl or other arylsulfonyl chloride. ☐ ☐

Exercises

20.2 Show how to prepare the primary amines by the methods given:
  a. iBuNH$_2$ by the Gabriel synthesis
  b. tBuNH$_2$ by the Ritter reaction
  c. iBuNH$_2$ by reduction of a nitrile
  d. BuNH$_2$ by reduction of an oxime
  e. ϕCHMeNH$_2$ by the Hofmann rearrangement
  f. tBuCH$_2$NH$_2$ by the Beckmann rearrangement

20.4 Amides and imides; lactams

Occurrence and physical properties. An <u>amide</u> may be thought of as an amine (or ammonia) in which one hydrogen has been replaced by an acyl group (RCONH$_2$). An <u>imide</u> may be considered an amine in which two hydrogens have been replaced by two acyl groups: it may be cyclic (phthalimide, succinimide) or noncyclic. A <u>lactam</u> is a cyclic amide, and it is related to an amino acid in the same way that a lactone is related to a hydroxy acid.

Amides are widespread in nature, and formamide has been identified even in outer space. All proteins are polyamides that can be hydrolyzed to amino acids. Many amides have pronounced physiological effects. Some examples are shown in Table 20.3.

Some of the simple amides and lactams and an imide are shown in Table 20.4. All the amides have bp greater than those of their constituent acids and amines (and ammonia). Amides that have hydrogen atoms attached to nitrogen (especially the unsubstituted

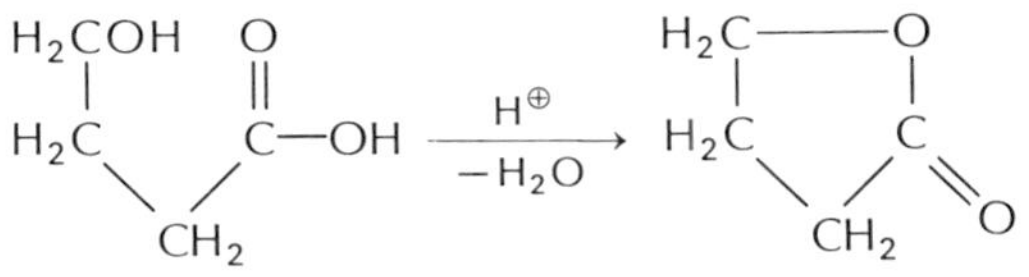

20.4 Amides and imides; lactams   665

$$\underset{\text{CH}_2}{\overset{\text{H}_2\text{COH}}{\underset{|}{\text{H}_2\text{C}}}} \overset{\text{O}}{\underset{}{\overset{\|}{\text{C}}}} - \text{OH} \xrightarrow[-\text{H}_2\text{O}]{\text{H}^\oplus} \underset{\underset{\gamma\text{-lactone}}{\text{CH}_2}}{\text{H}_2\text{C} - \text{O}}\text{C}=\text{O}$$

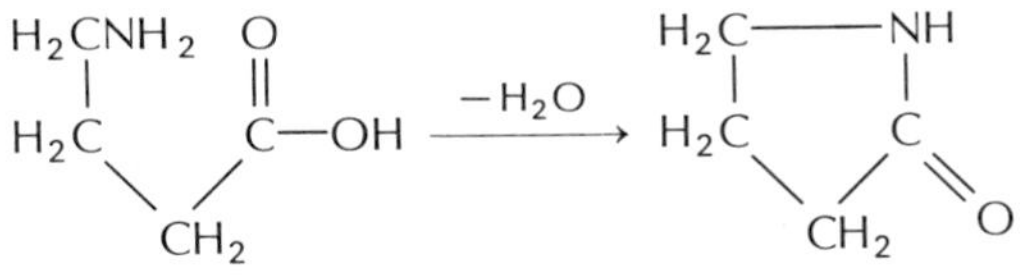

$$\underset{\text{CH}_2}{\overset{\text{H}_2\text{CNH}_2}{\underset{|}{\text{H}_2\text{C}}}} \overset{\text{O}}{\underset{}{\overset{\|}{\text{C}}}} - \text{OH} \xrightarrow{-\text{H}_2\text{O}} \underset{\underset{\gamma\text{-lactam}}{\text{CH}_2}}{\text{H}_2\text{C} - \text{NH}}\text{C}=\text{O}$$

Table 20.3 Some physiologically active lactams, amides, and imides

Name	Structural formula	Physiological effect
penicillin G	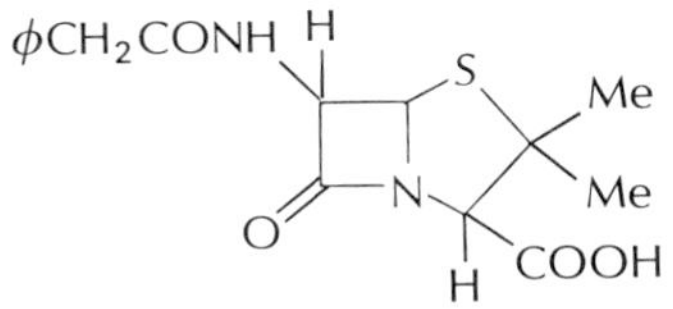 $\phi\text{CH}_2\text{CONH}$ — (β-lactam fused thiazolidine with Me, Me, COOH)	antibiotic
piperine	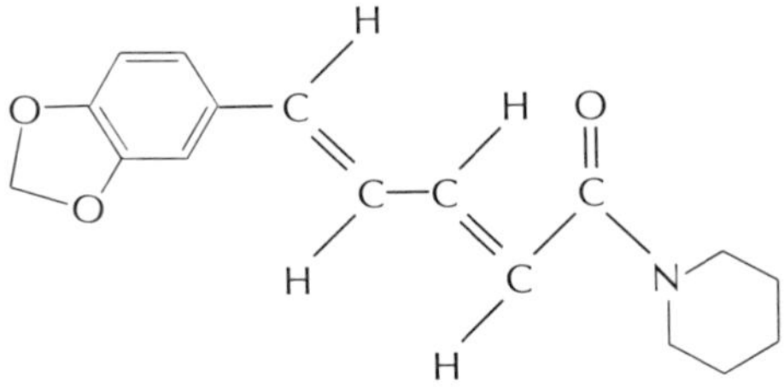 (methylenedioxyphenyl)–CH=CH–CH=CH–C(=O)–N(piperidine)	active principle of black pepper
riboflavin	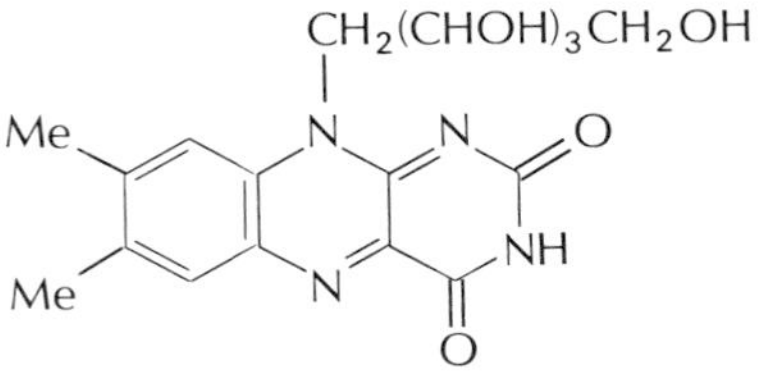 isoalloxazine with $\text{CH}_2(\text{CHOH})_3\text{CH}_2\text{OH}$, two Me groups	vitamin B_2
diethylamide of lysergic acid	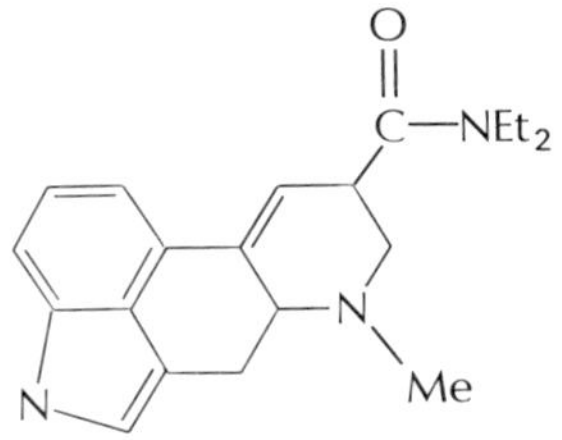 lysergic acid structure with $\text{C}(=\text{O})-\text{NEt}_2$ and N–Me	hallucinogen

Table 20.4 Some simple amides, lactams, and imides

Name	Structure	mp, °C	bp, °C
formamide	HCONH$_2$	2.5	211
methylformamide	HCONHMe		185
dimethylformamide	HCONMe$_2$	−60	153
acetamide	CH$_3$CONH$_2$	82	222
N-methylacetamide	CH$_3$CONHMe	28	206
N,N-dimethylacetamide	CH$_3$CONMe$_2$	−20	165
propionamide	CH$_3$CH$_2$CONH$_2$	79	213
isobutyramide	(CH$_3$)$_2$CHCONH$_2$	129	220
2-pyrrolidone	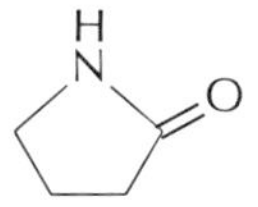 (5-membered ring lactam)	25	251
2-piperidone	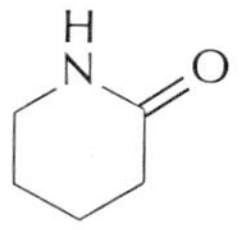 (6-membered ring lactam)	39	256
succinimide	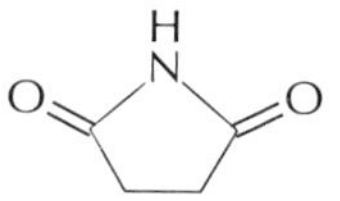 (5-membered ring imide)	126	287 dec.
oxamide	CONH$_2$–CONH$_2$	419 dec.	

amides) tend to be solids with respectable mp. Oxamide behaves as though it were a polymer, with a very high mp (above 400°). The reason for the high mp and bp of amides, imides, and lactams is the presence of hydrogen bonding. The ability to hydrogen bond is reflected in the relatively large solubilities of amides in water and in alcohol. The amides from secondary amines, N,N-dimethylformamide (DMF), N,N-dimethylacetamide (DMA), and N-methyl-2-pyrrolidone (bp 202°) are good solvents for many reactions, especially those in which an anion is the reactant. DMF, for example, solvates cations and does not solvate anions well. Thus, the anion has enhanced nucleophilic activity. Also, DMF is miscible with water as well as with most organic solvents. The solvent power of DMF is such that many polymeric substances are dissolved by it.

Structure and acidity. Amides are neutral as compared to the amines from which they are formed (pK_a of CH$_3$CONH$_2$ is 0.1, that of ϕCONH$_2$ is −2). Equation 20.29 shows that oxygen is more likely to accept a proton than is nitrogen of the amide group (resonance delocalization is enhanced by the cation a). Furthermore, the conjugate acid is a strong acid, and the amide is reluctant to accept a proton. Imides are even less willing to accept a proton and instead are weak proton donors (weakly acidic). Succinimide has pK_a of 9.6, about the same acid strength as RNH$_3^{\oplus}$, of pK_a about 10.

$$RC\begin{matrix}O\\\parallel\\\diagdown\\NH_3^{\oplus}\end{matrix} + H_2O \rightleftharpoons \left[RC\begin{matrix}O\\\parallel\\\diagdown\\NH_2\end{matrix} \longleftrightarrow RC\begin{matrix}O^{\ominus}\\\diagup\\\diagdown\\\diagdown\\NH_2^{\oplus}\end{matrix}\right] + H_3O^{\oplus}$$

$$\left[RC\begin{matrix}OH^{\oplus}\\\parallel\\\diagdown\\NH_2\end{matrix} \longleftrightarrow RC\begin{matrix}OH\\\diagup\\\diagdown\\\diagdown\\NH_2^{\oplus}\end{matrix}\right] \underset{a}{\rightleftharpoons} \quad (20.29)$$

Spectroscopy. The ir of amides are characterized by a strong carbonyl stretch at 1690 cm^{-1} and variable N—H absorptions. For example, —NH$_2$ gives two sharp absorptions at 3400 and 3500 cm^{-1} as well as a broad hydrogen bond absorption at the lower frequency 3100 to 3300 cm^{-1}. In contrast, an —NHR shows a single absorption at 3440 cm^{-1}.

In the nmr, protons attached to nitrogen exchange so slowly that coupling with protons on a neighboring carbon may sometimes be observed. The major effects, however, are a broadening of the proton peak (and a lowering of the peak height) caused by the quadrupole moment of ^{14}N (a nonspherical nucleus) and the spin number of ^{14}N, which is 1. In other words, a nonexchanging proton signal is split into a triplet (J = 52 cps) by the nitrogen to which it is bonded, and each peak is broadened by the quadrupole moment effect. The result usually is a single broad (100 Hz or more), weak signal for an amide hydrogen peak, which may appear in the δ5 to δ8.5 ppm region. Imides and lactams behave in a similar fashion.

Reactions. The reactions of the amides may be classified as olysis, reduction, dehydration, and rearrangement reactions.

Amides undergo the usual <u>olysis reactions,</u> catalyzed by acids or bases. Hydrolysis is the most used of these reactions (Eqs. 20.30). Hydrolysis also may be catalyzed by an

$$RCONHEt \begin{matrix} \overset{H^{\oplus}}{\nearrow} & R-C\begin{matrix}OH^{\oplus}\\\parallel\\\diagdown NHEt\end{matrix} \xrightarrow{etc.} RCOOH + EtNH_3^{\oplus} \\ \underset{HO^{\ominus}}{\searrow} & R-C\begin{matrix}O^{\ominus}\\\diagup\\-NHEt\\\diagdown OH\end{matrix} \xrightarrow{etc.} RCOO^{\ominus} + EtNH_2 \end{matrix} \quad (20.30)$$

enzyme such as pepsin (from gastric juice). The rates of hydrolysis are strongly affected by steric hindrance in the same way that ester hydrolysis is affected. The details of amide hydrolysis form a subject of active research because of the direct relationship of this reaction to the breakdown and formation of proteins and enzymes.

Unsubstituted amides react with nitrous acid (or alkyl nitrites) in the same manner as do primary amines to give what is in effect a hydrolysis product, the acid (Eqs. 20.31). A substituted amide such as RCONHMe, analogous to a secondary

$$RCONH_2 \xrightarrow[\substack{NaNO_2 \\ HCl}]{} RCOOH + N_2 + H_2O$$
$$RCONH_2 \xrightarrow{R'ON=O} RCOOH + N_2 + R'OH \qquad (20.31)$$

amine, gives an N-methyl-N-nitrosoamide (RCON(NO)Me). Nitrosoamides are used in the preparation of diazo compounds like diazomethane (CH_2N_2, Sec. 20.8).

Amides may be <u>reduced</u> in several ways. Catalytic hydrogenation, LAH, and BH_3 reduce amides to amines. Sodium and alcohol cause reduction to an alcohol and an amine. Sodium borohydride does not react with an amide group. Equations 20.32 summarize the possibilities.

$$\phi CH_2CONHEt \xrightarrow[cat.]{H_2} \phi CH_2CH_2NHEt$$

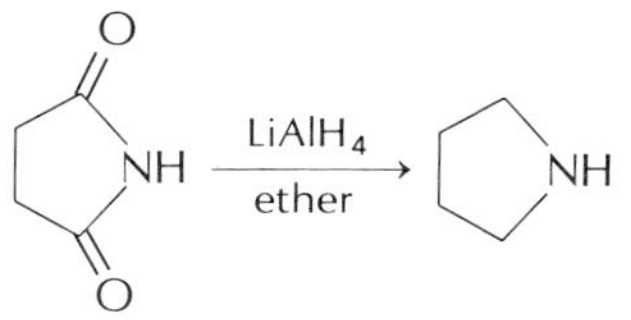

$$\text{succinimide} \xrightarrow[ether]{LiAlH_4} \text{pyrrolidine} \qquad (20.32)$$

$$CH_3CONEt_2 \xrightarrow{BH_3} CH_3CH_2NEt_2$$

$$CH_3CH_2CONH_2 \xrightarrow[EtOH]{Na} CH_3CH_2CH_2OH + NH_3$$

Unsubstituted amides are easily <u>dehydrated</u> with $SOCl_2$, PCl_5, or P_2O_5 to give nitriles (Eqs. 20.33). Strictly speaking, this reaction is not a reduction, but an elimination.

$$tBuCONH_2 \xrightarrow[80°]{SOCl_2} tBuC \equiv N + SO_2 + 2HCl$$
$$3N \equiv CCH_2CONH_2 + PCl_5 \xrightarrow{heat} 3N \equiv CCH_2C \equiv N + HPO_3 + 5HCl \qquad (20.33)$$
$$Me_2CHCONH_2 \xrightarrow[\Delta]{P_2O_5} Me_2CHC \equiv N$$

The Hofmann rearrangement of amides may be regarded as a reduction and has been described in Section 20.3. Imides, such as succinimide, do not rearrange. N-bromosuccinimide is relatively stable, and its use for allylic bromination has been described (Sec. 12.5).

Preparation. Almost all the reactions that lead to an amide have been described. We list them here.

Olysis reactions
Aminolysis (ammonolysis) of acid chlorides, acid anhydrides, and esters (Eqs. 20.3) and the effect of heat upon ammonium salts (to be discussed later in this section).
Hydrolysis of a nitrile is useful when the reaction may be stopped at the amide (Eq. 17.23).

20.4 Amides and imides; lactams

Substitution reactions
The Gabriel (phthalimide) synthesis gives a substituted imide as the initial product (Eq. 20.19).

Addition reactions
The Ritter reaction is useful for the preparation of N-substituted amides. The limitation is that the substituent must be capable of relatively stable carbonium ion formation (Eq. 20.20). ☐ ☐

Rearrangements
The Beckmann rearrangement of oximes was described in Equations 20.28, 20.29, and 20.31. ☐ ☐

Lactams. The lactams are closely related to the corresponding amino acids in the same way that the lactones are related to the hydroxy acids. Otherwise, lactams behave as ordinary amides. The chemistry of amino acids, especially that of the α-amino acids, will be discussed in Chapter 21. Here we shall deal with the effect of heat upon the various classes of amino acids.

An amino acid, containing both an acidic group and a basic group, exists as a dipolar ion or zwitterion. The terminology is misleading because, strictly speaking, this substance is not an ion. It is an internal salt, and it is nearly neutral. Like an amine oxide, it is highly polar and strongly associated. Thus, amino acids have no sharp mp and decompose upon being heated. They are somewhat soluble in water and other highly polar solvents. An amino acid always should be written as the dipolar structure. Thus, glycine, (aminoacetic acid) should be written as a, not b.

$$H_3\overset{\oplus}{N}CH_2COO^{\ominus} \qquad H_2NCH_2COOH$$
$$a \qquad\qquad\qquad b$$

If glycine or another α-amino acid is heated to 160° in a sealed tube or to 170° in glycerol, dehydration occurs to produce a six-membered cyclic diamide, a diketopiperazine (Eq. 20.34) as the major product and a polymeric amide as a minor product. (Compare Eq. 18.1 for the similar reaction of an α-hydroxy acid to form a lactide.)

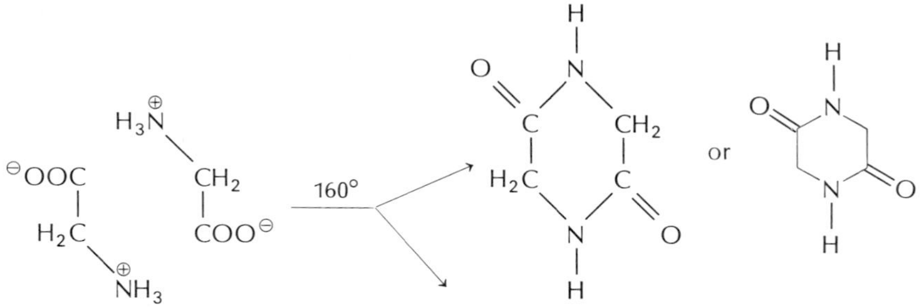

$$\text{(20.34)}$$

$$H_3\overset{\oplus}{N}CH_2CO(NHCH_2CO)_nNHCH_2COO^{\ominus} + (n+1)H_2O$$

A β-amino acid, when heated, simply loses a molecule of ammonia to give an α,β-unsaturated acid (Eq. 20.35).

The γ- and δ-amino acids, when subjected to heat, give γ- and δ-lactams (Eqs. 20.36).

If the amino group is more distant than δ from the acid group, a polymeric amide results from heating the amino acid. Again, notice the similarity to the reactions of the hydroxy acids (Sec. 18.1).

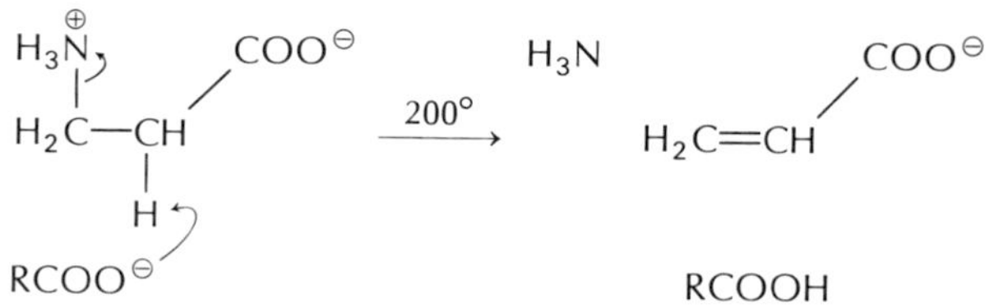

670   Chapter twenty

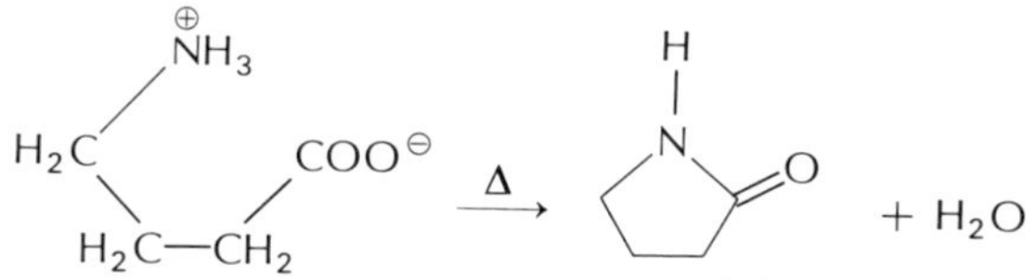

$$\underset{RCOO^{\ominus}}{\overset{H_3\overset{\oplus}{N}}{\underset{H}{\underset{|}{H_2C-CH}}}}\hspace{-2em}\diagdown^{COO^{\ominus}} \xrightarrow{200°} \underset{RCOOH}{\overset{H_3N}{H_2C=CH}}\diagdown^{COO^{\ominus}} \quad (20.35)$$

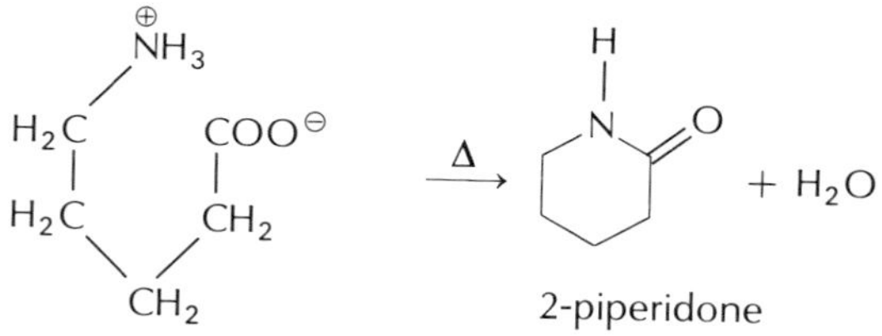

$$\underset{H_2C-CH_2}{\overset{\overset{\oplus}{NH_3}}{H_2C}}\diagdown^{COO^{\ominus}} \xrightarrow{\Delta} \underset{\text{2-pyrrolidone}}{\begin{array}{c}H\\|\\N\end{array}\diagup\hspace{-0.5em}=\hspace{-0.5em}O} + H_2O$$

(20.36)

$$\underset{\underset{CH_2}{H_2C-CH_2}}{\overset{\overset{\oplus}{NH_3}}{H_2C}}\diagdown^{COO^{\ominus}} \xrightarrow{\Delta} \underset{\text{2-piperidone}}{\begin{array}{c}H\\|\\N\end{array}\diagup\hspace{-0.5em}=\hspace{-0.5em}O} + H_2O$$

All lactams (as well as diketopiperazines and polymeric amides) are hydrolyzed to the constituent amino acids when heated with dilute hydrochloric or other strong acid.

Exercises

20.3   Show the products to be expected:

a. $(CH_3)_2CHCONH_2$
   - $\xrightarrow{HCl, H_2O}$
   - $\xrightarrow{NaNO_2, HCl}$
   - $\xrightarrow{LiAlH_4}$
   - $\xrightarrow{SOCl_2}$
   - $\xrightarrow{Br_2, NaOH}$

b. $CH_3CONHEt \xrightarrow[HCl]{NaNO_2}$

c. $\phi CH_2CN + H_2O_2 \longrightarrow$

d. $CH_3COCl + Et_2NH \longrightarrow$

e. $H_2C=CHCH_2OH + CH_3CN \xrightarrow{H_2SO_4}$

f. $CH_3CH_2N=C=O + sBuMgBr \longrightarrow$

g. $H_3\overset{\oplus}{N}CH_2CH_2CH_2COO^{\ominus} \xrightarrow{\Delta}$

h. $HOOC(CH_2)_4COOH + H_2N(CH_2)_6NH_2 \xrightarrow{\Delta}$

i. $HOOC(CH_2)_2COOH + NH_3 \xrightarrow{\Delta}$

20.5 Nitriles and cyanohydrins

Nitriles (RCN) and cyanohydrins (RCHOHCN), provide us with an opportunity to look at nitrogen with multiple bonds.

Nitriles: properties, reactions. Nitriles are named in relation to the acid that would be formed upon hydrolysis. Thus, CH_3CN (→ CH_3COOH) is acetonitrile; ϕCN (→ $\phi COOH$) is benzonitrile. On occasion they may be referred to as cyanides. Thus, CH_3CN also would be methyl cyanide, and $(CH_3)_2CHCN$ would be isopropyl cyanide (but isobutyronitrile). In naming as a nitrile, we take the carbon of the —C≡N group as number 1.

Nitriles tend to be liquids with bp less than those of the corresponding acids. Nitriles lack hydrogen bonding but have a sizable dipole moment (about 3.5 D), which causes the liquids to be associated. Acetonitrile is used as a polar aprotic solvent, and it is miscible with water. The carbon-nitrogen triple bond may be considered a resonance hybrid. As in

$$R-C\equiv \underline{N} \longleftrightarrow R-\overset{\oplus}{C}=\overset{\ominus}{\underline{N}}|$$

a carbonyl group, the carbon atom is electrophilic. The nitrogen atom, of course, is equally nucleophilic.

A nitrile, like a carbon-carbon triple bond, absorbs in the ir at about 2250 cm^{-1}. Unlike the —C≡C— bond, however, —C≡N gives a strong band (a strong dipole is affected). The nitriles are transparent in the uv down to 167 nm unless conjugated with some other multiple bond.

We shall take up the reactions of the nitriles in three groups: additions, conjugated additions, and condensation reactions.

Nitriles undergo hydrolysis to amides (which react further to form acids) and alcoholysis to form esters (Eqs. 17.22, 17.23, and 17.88). These olysis reactions involve addition to the carbon-nitrogen triple bond. The reaction of a nitrile with H_2O_2, which gives an amide, is described in the supplement.

A Grignard reagent adds to a nitrile to give a ketone (Eq. 16.81).

Nitriles may be reduced catalytically with hydrogen or with LAH to give amines (Eq. 20.22). The use of stannous chloride allows an aldehyde to be formed from a nitrile by the Stephen reduction (Eq. 16.100).

If a nitrile is conjugated, 1,4 addition becomes possible (Michael addition). Acrylonitrile undergoes Michael addition with so many nucleophilic reagents that the reaction has come to be known as cyanoethylation (Eq. 16.59). Conjugated nitriles are excellent dienophiles in the Diels-Alder reaction (Sec. 14.8). Tetracyanoethylene not only is one of the best dienophiles known but is an excellent electrophile as well and readily forms charge-transfer complexes with electron donors. (Sec. 15.2 described the $\phi H \cdot I_2$ complex.)

The carbon-nitrogen triple bond, like the carbon-oxygen double bond, activates any α hydrogens. The activity of these hydrogens is then intermediate between those of the active aldehydes and ketones and those of the less active esters. Thus, nitriles may be alkylated (Eq. 20.37).

$$\phi CH_2CN \xrightarrow{EtO^{\ominus}} \phi \overset{\ominus}{C}H-C\equiv N \xrightarrow{RX} \underset{\phi}{\overset{R}{\diagdown}}CH-C\equiv N \qquad (20.37)$$
$$\updownarrow$$
$$\phi CH=C=\overset{\ominus}{N}$$

672   Chapter twenty

A competing reaction that is brought on by the small steric hindrance to addition to R—C≡N as compared with R—CHO, RCOR, or RCOOR' is self-condensation of the nitrile. It is similar to the aldol condensation (Eq. 20.38). The iminonitrile, upon complete hydrolysis,

$$RCH_2CN \xrightarrow{EtO^\ominus} R\overset{\ominus}{C}HCN \xrightarrow{RCH_2C\equiv N} \underset{\underset{RCHCN}{|}}{RCH_2C=N^\ominus} \xrightarrow{H_2O} \underset{\underset{RCHCN}{|}}{RCH_2C=NH}$$

$$\Big\downarrow H^\oplus, H_2O \quad (20.38)$$

$$\underset{\underset{RCH_2}{|}}{RCH_2C=O} \xleftarrow{\Delta} \underset{\underset{RCHCOOH}{|}}{RCH_2C=O}$$

gives a β-keto acid, which decarboxylates when heated to give a symmetrical ketone. Yields are low because the intermediate anion of the imine is able to add to another cyano group (Eq. 20.39) to give a substituted pyrimidine. Other reactions compete as well.

(20.39)

Compounds such as β-ketobutyronitrile (CH₃COCH₂C≡N), cyanoacetic ester (N≡CCH₂COOEt), and malononitrile (N≡CCH₂C≡N) all display the enhanced reactivity characteristic of the β-carbonyl compounds. In other words, a cyano group in place of a carbonyl group gives about the same α-hydrogen activity and allows mixed condensations to occur.

> The Thorpe reaction, analogous to the Dieckmann condensation of diesters (Eqs. 17.65), gives excellent yields of cyclized products from dinitriles if the conditions are carefully controlled (Eqs. 20.40). The product has been shown by nmr and ir to be a, the conjugated enamine-nitrile, rather than b, the imine-nitrile. This result contrasts with the keto-nitrile (d), which is more stable than c, the enol-nitrile. Hydrolysis of the ketonitrile gives a β-keto acid, which decarboxylates upon being heated.
> (One of the unexplained footnotes in the history of organic chemistry is a report by Thorpe in 1909 that he was able to prepare a, then thought to be b, in 84% yield from adiponitrile by reaction with a small amount of sodium ethoxide in dry ethanol. No one has ever been able to repeat the preparation as described by Thorpe.)
> The method of preparation was modified by Ziegler (Thorpe-Ziegler reaction or condensation) in a successful effort to prepare large rings (fourteen or more carbon atoms). To favor intramolecular condensation (cyclization) and to prevent intermolecular condensation (dimeric products), he used very high dilution. The very small concentration of anion, under these dilute conditions, slows the second-order dimerization to a negligible rate, whereas the first-order cyclization rate is un-

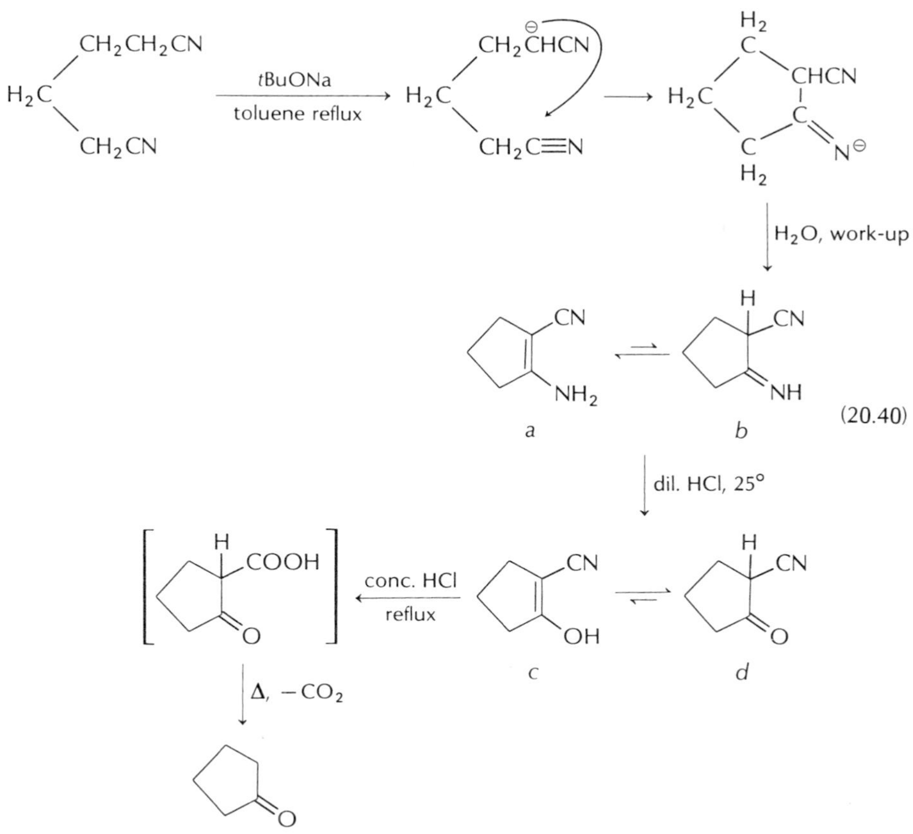

$$\text{H}_2\text{C} \underset{\text{CH}_2\text{CN}}{\overset{\text{CH}_2\text{CH}_2\text{CN}}{\diagup}} \xrightarrow[\text{toluene reflux}]{t\text{BuONa}} \text{H}_2\text{C} \underset{\text{CH}_2\text{C}\equiv\text{N}}{\overset{\text{CH}_2\overset{\ominus}{\text{CHCN}}}{\diagup}} \longrightarrow$$

[cyclopentane intermediate with CHCN and C=N⁻]

↓ H₂O, work-up

a (cyclopentene with CN and NH₂) ⇌ b (cyclopentane with H, CN, =NH)   (20.40)

↓ dil. HCl, 25°

[cyclopentanone with H, COOH] ← conc. HCl, reflux ← c (cyclopentene with CN, OH) ⇌ d (cyclopentanone with H, CN)

↓ Δ, −CO₂

cyclopentanone

affected. It was also found that the use of lithium diethylamide (LiNEt$_2$) or of lithium methylphenylamide (LiN(Me)ϕ) in anhydrous ether gave better yields than those obtained with other bases tested (Eq. 20.41).

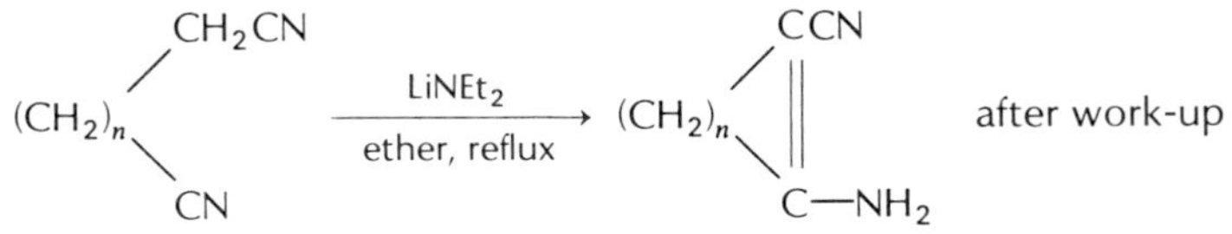

$$(\text{CH}_2)_n \underset{\text{CN}}{\overset{\text{CH}_2\text{CN}}{\diagup}} \xrightarrow[\text{ether, reflux}]{\text{LiNEt}_2} (\text{CH}_2)_n \underset{\text{C—NH}_2}{\overset{\text{CCN}}{\diagup}} \quad \text{after work-up} \quad (20.41)$$

Preparation of nitriles: reactions. The preparation of nitriles will be described under substitution, addition, and elimination reactions.

The S_N2 reaction is widely used for the preparation of nitriles (Eq. 20.42). Dimethyl-

$$\text{N}\equiv\text{C}^{\ominus} + \text{R—Br} \xrightarrow{\text{Me}_2\text{SO}} \text{N}\equiv\text{C—R} + \text{Br}^{\ominus} \quad (20.42)$$

sulfoxide is the solvent of choice, yields are much better, and the reaction time (at 80°) may be reduced from the three to four days required for aqueous ethanol to one to one and a half hours in DMSO. Other leaving groups, such as those in alkyl benzenesulfonates and alkyl *p*-toluenesulfonates, react easily and well. The sequence that takes alcohol (ROH) to alkyl halide or sulfonate (RX or ROSO$_2\phi$) and to nitrile (RCN) and acid (RCOOH) is a frequently used alternative to the route *via* Grignard reagent and carbon dioxide to acid, especially if other reactive groups are present. Equation 20.43 shows a good laboratory procedure for the preparation of diethyl malonate (malonic ester).

$$NaC\equiv N + BrCH_2COOEt \xrightarrow{Me_2SO} N\equiv C-CH_2COOEt + NaBr$$
$$\downarrow HCl, EtOH \quad (20.43)$$
$$EtOOC-CH_2-COOEt$$

Sometimes it is advantageous to use a Grignard reagent and cyanogen chloride—for example, if R is tertiary (Eq. 20.44).

$$tBuCl \xrightarrow[ether]{Mg} tBuMgCl \xrightarrow{Cl-C\equiv N} tBuC\equiv N + MgCl_2 \quad (20.44)$$

Addition reactions for the preparation of nitriles include several variations of the Michael reaction (Eqs. 20.45). We should remind ourselves of a method of preparation of

$$\phi CH=CHCO\phi \xrightarrow{HCN} \phi CHCH_2CO\phi$$
$$\qquad\qquad\qquad\qquad |$$
$$\qquad\qquad\qquad\qquad C\equiv N$$

$$H_2C(COOEt)_2 + H_2C=CHCN \xrightarrow{piperidine} H_2C-CH_2CN \quad (20.45)$$
$$\qquad\qquad\qquad\qquad\qquad\qquad\qquad | \searrow COOEt$$
$$\qquad\qquad\qquad\qquad\qquad\qquad\qquad HC$$
$$\qquad\qquad\qquad\qquad\qquad\qquad\qquad\qquad \searrow COOEt$$

acrylonitrile and of succinonitrile (Eq. 20.46).

$$HC\equiv CH + HCN \xrightarrow[\substack{NH_4Cl \\ HCl \\ 80°}]{CuCl} H_2C=CHCN$$
$$\qquad\qquad\qquad\qquad\qquad\qquad \downarrow HCN, \text{ excess} \quad (20.46)$$
$$\qquad\qquad\qquad\qquad\qquad\qquad NCCH_2CH_2CN$$

Several elimination reactions are useful for making nitriles. The dehydration of amides with thionyl chloride has been described (Eqs. 20.33). The tendency of aldoximes to undergo dehydration with acetic anhydride to give a nitrile (S-181) provides a route for converting an aldehyde into a nitrile.

Of commercial interest is the dehydrogenation of primary amines over copper or nickel catalysts at high temperatures (Eq. 20.47).

$$RCH_2NH_2 \xrightarrow[500°]{Cu \text{ or } Ni} RC\equiv N + 2H_2 \quad (20.47)$$

Cyanohydrins: reactions and preparation. Cyanohydrins (α-hydroxy nitriles) are worthy of a little attention. The cyano group undergoes the usual reactions of hydrolysis and reduction (Eqs. 20.48). The hydrolysis must be acidic, because basic conditions catalyze

$$\underset{\underset{Me_2C-C\equiv N}{|}}{OH} \xrightarrow[reflux]{HCl} [\underset{\underset{Cl}{|}}{Me_2C-C=NH_2^{\oplus}}] \longrightarrow \underset{\underset{Me_2C-COOH}{|}}{OH} + NH_4Cl$$
$$\qquad\qquad\qquad\qquad\qquad\qquad\qquad\qquad\qquad (20.48)$$

$$\phi CHOHC\equiv N \xrightarrow[excess]{LiAlH_4} \phi CHOHCH_2NH_2$$

the decomposition of a cyanohydrin into cyanide ion and the carbonyl-containing component. Cyanohydrins thus serve as intermediates for the preparation of α-hydroxy acids and α-hydroxy primary amines.

Reactions at the hydroxy group of cyanohydrins are those characteristic of secondary and tertiary alcohols (Eqs. 20.49).

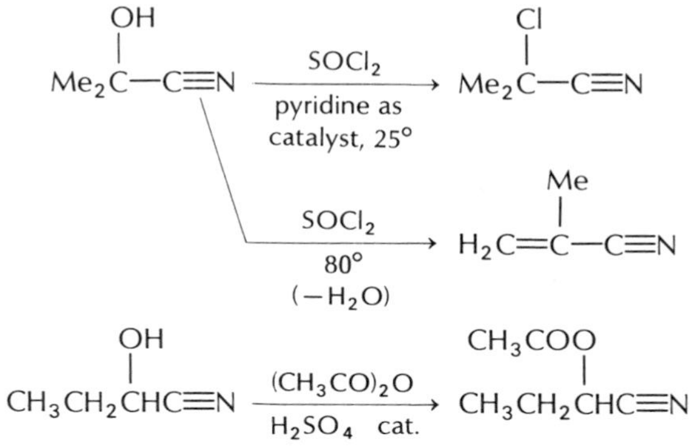

$$\text{Me}_2\overset{\overset{\text{OH}}{|}}{\text{C}}-\text{C}\equiv\text{N} \xrightarrow[\text{pyridine as catalyst, 25°}]{\text{SOCl}_2} \text{Me}_2\overset{\overset{\text{Cl}}{|}}{\text{C}}-\text{C}\equiv\text{N}$$

$$\xrightarrow[80° \ (-\text{H}_2\text{O})]{\text{SOCl}_2} \text{H}_2\text{C}=\overset{\overset{\text{Me}}{|}}{\text{C}}-\text{C}\equiv\text{N} \quad (20.49)$$

$$\text{CH}_3\text{CH}_2\overset{\overset{\text{OH}}{|}}{\text{CH}}\text{C}\equiv\text{N} \xrightarrow[\text{H}_2\text{SO}_4 \text{ cat.}]{(\text{CH}_3\text{CO})_2\text{O}} \text{CH}_3\text{CH}_2\overset{\overset{\text{CH}_3\text{COO}}{|}}{\text{CH}}\text{C}\equiv\text{N}$$

The preparation of cyanohydrins was described in Section 16.3. Here we should introduce a variation of the addition of HCN to a carbonyl group whereby an α-amino nitrile is formed. Acidic hydrolysis then leads to the α-amino acid. The sequence of reactions is known as the <u>Strecker synthesis</u> (Eq. 20.50). The Strecker synthesis is only one method for the synthesis of amino acids (Sec. 21.2), but a useful one.

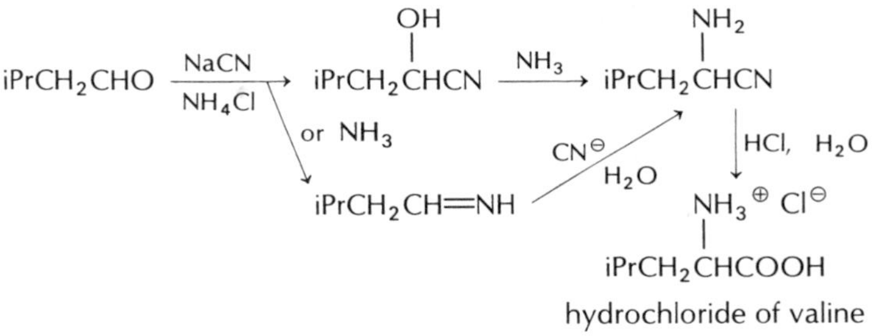

$$\text{iPrCH}_2\text{CHO} \xrightarrow[\text{NH}_4\text{Cl}]{\text{NaCN}} \text{iPrCH}_2\overset{\overset{\text{OH}}{|}}{\text{CH}}\text{CN} \xrightarrow{\text{NH}_3} \text{iPrCH}_2\overset{\overset{\text{NH}_2}{|}}{\text{CH}}\text{CN}$$

or NH$_3$ ↘ ↓ HCl, H$_2$O

$$\text{iPrCH}_2\text{CH}=\text{NH} \xrightarrow[\text{H}_2\text{O}]{\text{CN}^{\ominus}} \quad \overset{\overset{\text{NH}_3^{\oplus} \text{ Cl}^{\ominus}}{|}}{\text{iPrCH}_2\text{CHCOOH}} \quad (20.50)$$

hydrochloride of valine

Exercises

20.4 Complete the reactions:

a. N≡CCH$_2$CH$_2$C≡N + EtOH $\xrightarrow{\text{H}^{\oplus}}$

b. NCCH$_2$COOEt + H$_2$C=CHCN $\xrightarrow[\text{base}]{\text{mild}}$

c. H$_2$C=CHCH$_2$Cl + NaCN $\xrightarrow{\text{DMSO}}$

d. φCH$_2$CN + SnCl$_2$ $\xrightarrow[\text{HCl}]{\text{ether}}$

e. (CH$_3$)$_2$CHCN + CH$_3$MgI $\longrightarrow$

f. EtOOCCH$_2$CONH$_2$ + SOCl$_2$ $\longrightarrow$

g. (CH$_3$)$_2$$\overset{}{\underset{\underset{\text{OH}}{|}}{\text{C}}}$CN $\xrightarrow[\text{warm}]{\text{SOCl}_2}$

h. CH$_3$CH$_2$CHO + NaCN $\xrightarrow{\text{NH}_4\text{Cl}}$

i. $t\text{BuCHO} + \text{CH}_3\text{CH}_2\text{CN} \xrightarrow{\text{EtONa}}$

j. $\text{NCCH}_2\text{CN} + \text{BrCH}_2\text{COOEt} \xrightarrow{\text{EtONa}}$

k. $\text{NCCH}_2\text{CH}_2\text{C(Me)}_2\text{CH}_2\text{CH}_2\text{CN} \xrightarrow[\phi\text{CH}_3]{t\text{BuONa}} \xrightarrow[\text{reflux}]{\text{conc. HCl}}$

l. 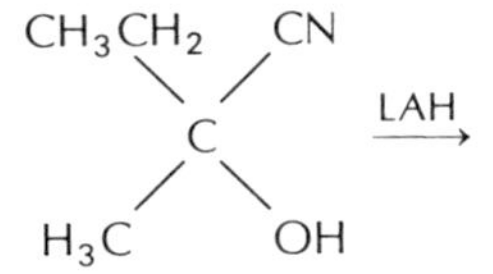
$$\begin{array}{c} \text{CH}_3\text{CH}_2 \quad \text{CN} \\ \diagdown \quad \diagup \\ \text{C} \\ \diagup \quad \diagdown \\ \text{H}_3\text{C} \quad \text{OH} \end{array} \xrightarrow{\text{LAH}}$$

20.6 Hydroxylamine and hydrazine; oximes and hydrazones

Hydroxylamine and hydrazine. Hydroxylamine and hydrazine are related to ammonia in the same way that methanol and methylamine are related to methane, or hydroxylamine and hydrogen peroxide to water. The effects of the substitutions may be

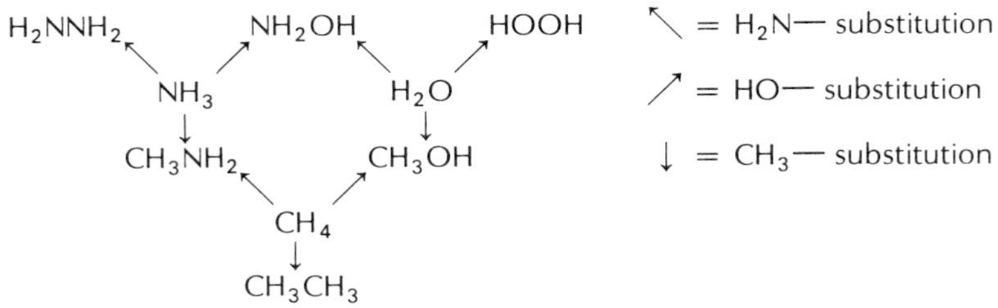

$$\begin{array}{ccc}
\text{H}_2\text{NNH}_2 & \text{NH}_2\text{OH} & \text{HOOH} \\
\nwarrow \nearrow & \nwarrow \nearrow & \\
\text{NH}_3 & \text{H}_2\text{O} & \\
\downarrow & \downarrow & \\
\text{CH}_3\text{NH}_2 & \text{CH}_3\text{OH} & \\
\nwarrow \nearrow & & \\
\text{CH}_4 & & \\
\downarrow & & \\
\text{CH}_3\text{CH}_3 & &
\end{array}$$

$\diagdown$ = H_2N— substitution

$\diagup$ = HO— substitution

$\downarrow$ = CH_3— substitution

seen by comparison of the pK_a values of ammonia (9.2), hydrazine (8.1), and hydroxylamine (6). Thus the order is $\text{NH}_3 > \text{H}_2\text{NNH}_2 > \text{H}_2\text{NOH}$ in base strength. In many respects, the properties of hydrazine are those of any diamine, and those of hydroxylamine are like those of an amino alcohol. The reactions with acids and bases are shown in Equations 20.51.

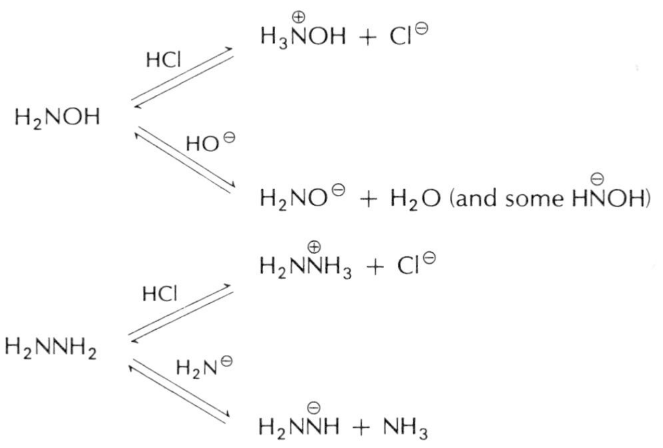

$$\text{H}_2\text{NOH} \xrightleftharpoons[\text{HO}^\ominus]{\text{HCl}} \begin{array}{c} \text{H}_3\overset{\oplus}{\text{N}}\text{OH} + \text{Cl}^\ominus \\ \\ \text{H}_2\text{NO}^\ominus + \text{H}_2\text{O} \text{ (and some } \overset{\ominus}{\text{H}}\text{NOH)} \end{array} \qquad (20.51)$$

$$\text{H}_2\text{NNH}_2 \xrightleftharpoons[\text{H}_2\text{N}^\ominus]{\text{HCl}} \begin{array}{c} \text{H}_2\text{N}\overset{\oplus}{\text{N}}\text{H}_3 + \text{Cl}^\ominus \\ \\ \text{H}_2\text{N}\overset{\ominus}{\text{N}}\text{H} + \text{NH}_3 \end{array}$$

Hydroxylamine is acidic enough to form the anion with sodium hydroxide, but hydrazine requires a much stronger base, amide ion.

Many alkylated hydroxylamines and hydrazines are known, but we shall not consider

20.6 Hydroxylamine and hydrazine; oximes and hydrazones

them. We shall look at arylhydroxylamines and arylhydrazines in Chapter 22. Also, we shall limit ourselves to one reaction of hydrazine and hydroxylamine.

The addition of these two substances to a carbonyl group was described in part in Section 16.3. The addition of hydrazine to a carbonyl group goes readily to give a hydrazone (Eq. 20.52). If the reaction is pushed, with the calculated amount of aldehyde or ketone, a

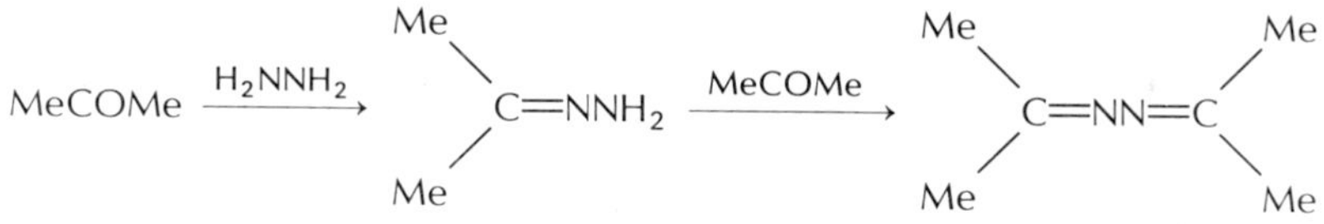

$$\text{MeCOMe} \xrightarrow{H_2NNH_2} \underset{Me}{\overset{Me}{\diagdown}}C=NNH_2 \xrightarrow{\text{MeCOMe}} \underset{Me}{\overset{Me}{\diagdown}}C=N-N=C\underset{Me}{\overset{Me}{\diagup}} \quad (20.52)$$

dihydrazone (known as an <u>azine</u>) is formed. Both hydrazones and azines may be hydrolyzed under acidic conditions to give back the aldehyde or ketone and the salt of hydrazine.

The rate of formation of acetoxime from acetone and hydroxylamine is greatest at pH 4.7; that of hydrolysis is greatest at pH 2.3. At pH > 5, hydrolysis is negligible. At high pH, the rate of formation also is rapid, probably because of the greater nucleophilicity of $\overset{\ominus}{\text{HNOH}}$. Thus, oximes are best prepared at about pH 5 (an acetic acid and sodium acetate buffer) or in the presence of excess NaOH.

The α-diketones (or ketoaldehydes) react easily to form the less hindered monoximes. The dioxime preparation requires the presence of excess NH_2OH and heat. The β-diketones (and β-keto esters) give monoximes with difficulty because of a pronounced tendency to close to a five-membered ring known as an <u>isoxazole</u> (Eqs. 20.53). Hydrazine reacts similarly to give a pyrazole ring. The corresponding six-membered heterocyclic rings may be obtained from γ-diketones and γ-keto esters. □ □

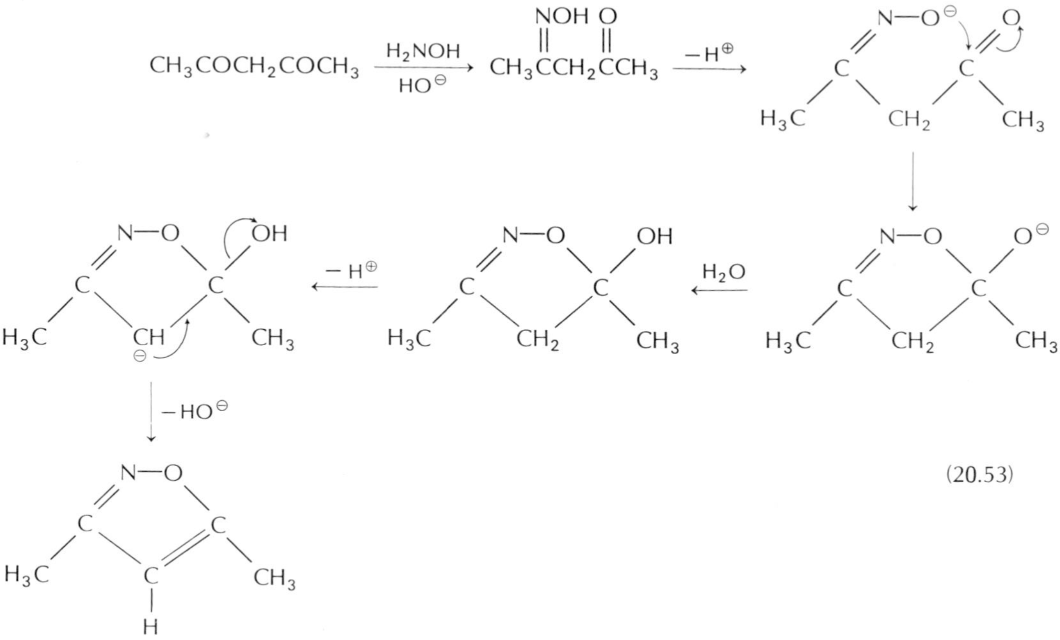

(20.53)

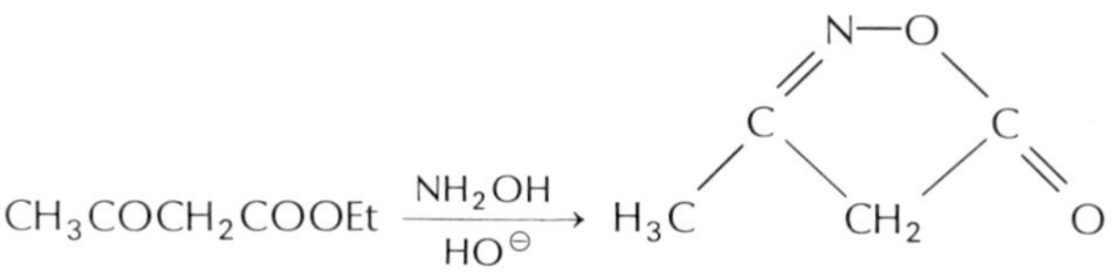

Oximes. The structure of oximes and the Beckmann rearrangement of oximes have been discussed (Sec. 20.3).

Oximes react as both weak acids and weak bases and are more soluble in low and high pH solutions than in water (Eq. 20.54).

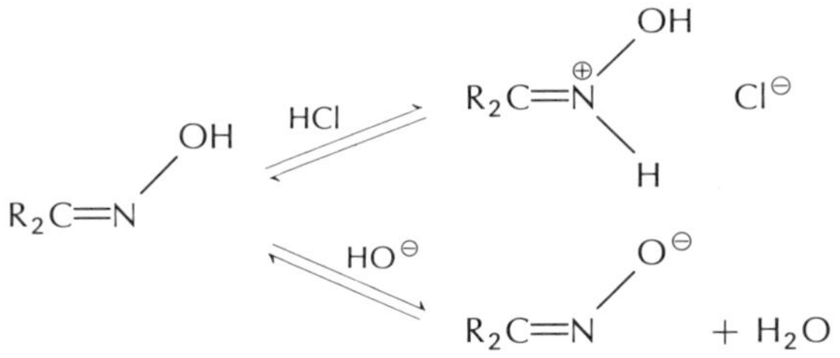

$$R_2C=N-OH \underset{HO^\ominus}{\overset{HCl}{\rightleftarrows}} R_2C=\overset{\oplus}{N} \begin{smallmatrix}OH \\ \\ H\end{smallmatrix} \quad Cl^\ominus$$
$$R_2C=N-O^\ominus + H_2O \tag{20.54}$$

Oximes are oxidizable: the peracid of trifluoroacetic acid often is used (Eq. 20.55),

$$R_2C=N-OH \xrightarrow{F_3CCO_3H} R_2CHNO_2 + F_3CCOOH \tag{20.55}$$

and the reaction is a good one for the synthesis of specific nitroalkanes.

Reduction of oximes may be accomplished with LiAlH$_4$, Zn and CH$_3$COOH, or catalytic hydrogenation to give primary amines (Eq. 20.56). If diborane is used, the reduction may be stopped at the hydroxylamine stage (R$_2$CHNHOH). □ □

$$R_2C=N-OH \xrightarrow[\text{ether}]{LiAlH_4} R_2CHNH_2 \tag{20.56}$$

Hydrazones. Hydrazones and the related compounds semicarbazones, phenylhydrazones, and 2,4-dinitrophenylhydrazones react in many ways, only a few of which we shall describe. Hydrazones are capable of *syn* and *anti* isomerism in the same way that the oximes are. However, because the hydrazones do not undergo a rearrangement analogous to the Beckmann rearrangement of oximes, the subject is not essential to the chemistry of the hydrazones.

Azines and hydrazones are unstable to heat and decompose with evolution of nitrogen. In the presence of a strong base, the decomposition may be carried out at a lower temperature than otherwise. A possible mechanism is shown (Eq. 20.57). We recognize the reaction

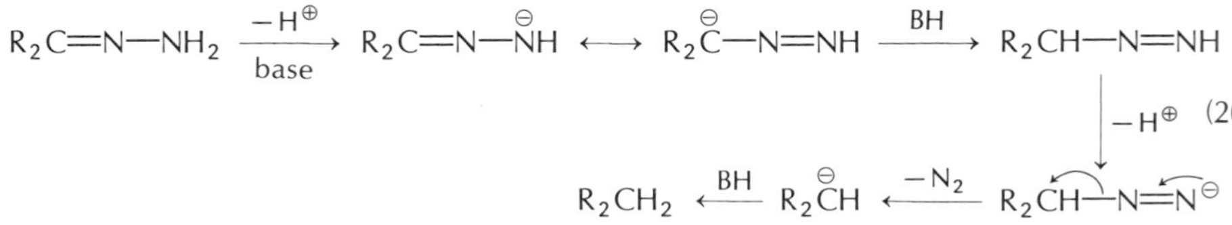

$$R_2C=N-NH_2 \xrightarrow[\text{base}]{-H^\oplus} R_2C=N-\overset{\ominus}{N}H \longleftrightarrow R_2\overset{\ominus}{C}-N=NH \xrightarrow{BH} R_2CH-N=NH$$
$$\downarrow -H^\oplus \quad (20.57)$$
$$R_2CH_2 \xleftarrow{BH} R_2\overset{\ominus}{C}H \xleftarrow{-N_2} R_2CH-N=N^\ominus$$

as the Wolff-Kishner reduction (Eq. 13.70). There is some evidence for homolytic cleavage instead of heterolytic, but the ionic fragmentation as shown is easier to keep in mind.

Hydrazones possess a free —NH$_2$ group, are basic, and undergo reactions characteristic of a primary amine. Thus, alkylation to give R$_2$C=NNHMe and acylation to give R$_2$C=NNHCOϕ proceed normally. Hydrazones are useful precursors for the preparation of diazo compounds (Eq. 20.58). Hydrazones may be reduced to hydrazines (R$_2$CHNHNH$_2$)

$$R_2C=NHNH_2 \xrightarrow{HgO} R_2C\overset{\oplus}{=}N\overset{\ominus}{=}N + Hg + H_2O \qquad (20.58)$$

with mild reagents. Stronger reducing agents, such as LiAlH$_4$, cleave the N—N bond as well to give a primary amine and ammonia (R$_2$CHNH$_2$ + NH$_3$).

Exercises

20.5 Use equations to describe the reaction of Me$_2$C=NOH with:
    a. PCl$_5$ in ether   c. LiAlH$_4$
    b. dilute NaOH, 25°   d. ϕCOCl

20.6 Use equations to describe the reaction of Me$_2$C=NNH$_2$ with:
    a. NaOH in DMSO, heat   d. LiAlH$_4$
    b. ϕCOCl   e. HCl, reflux
    c. HgO

20.7 Using H$_2$NOH or H$_2$NNH$_2$ and any other necessary substances, show how these compounds may be obtained:

    a. 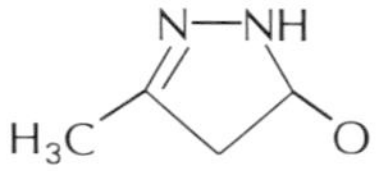 [structure: ring with N—NH, H$_3$C, O]

    b. 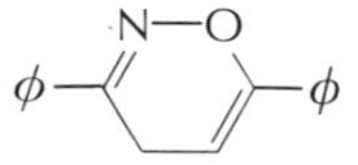 ϕ—[ring with N—O]—ϕ

    c. CH$_3$CH$_2$CH$_2$CONHOH

    d. ϕCONHNH$_2$

20.7 Nitroso and nitro groups; nitrites and nitrates

Nitrosoalkanes. The nitroso group (—N̈=Ö|) is at an intermediate stage of oxidation of nitrogen and is subject to both oxidation and reduction reactions. Moreover, the nitroso group is unstable with respect to tautomerization if a hydrogen atom is available on neighboring carbon (Eq. 20.59) to give the more stable oxime.

$$\diagdown\text{CH}-\text{N}=\text{O} \rightleftharpoons \diagdown\text{C}=\text{N}-\text{OH} \qquad (20.59)$$

The nitroso compounds are unstable with respect to dimerization, which complicates the determination of mp, bp, and solubility behavior. A solid nitroso compound is colorless (or light yellow) in the crystalline dimeric state and blue in the liquid state or in solution.

Nitroalkanes. The nitro group (—NO$_2$) contains nitrogen in the highest oxidation state that it can have and still have a bond to carbon. The structure of the nitro group is dipolar and leads to characteristic large dipole moments (3.2 D for EtNO$_2$). This polar

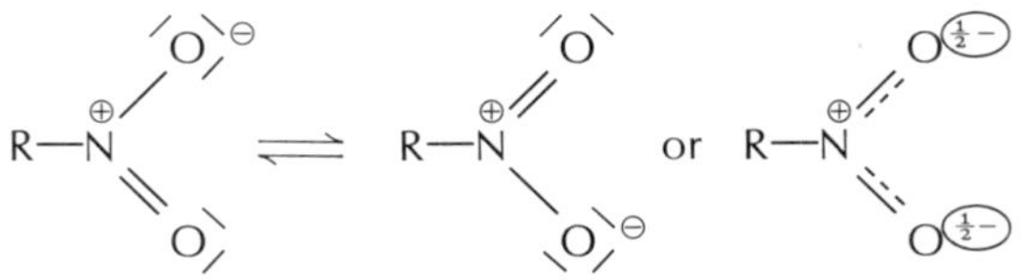

$$R-\overset{\oplus}{N}\begin{matrix}\diagup O^{\ominus}\\ \diagdown O\end{matrix} \rightleftharpoons R-\overset{\oplus}{N}\begin{matrix}\diagup O\\ \diagdown O^{\ominus}\end{matrix} \quad \text{or} \quad R-\overset{\oplus}{N}\begin{matrix}\diagup O^{\ominus\frac{1}{2}}\\ \diagdown O^{\ominus\frac{1}{2}}\end{matrix}$$

character leads to association and higher than normal bp. (MeNO$_2$ has bp 101°; EtNO$_2$, 115°; PrNO$_2$, 131°; iPrNO$_2$, 120°; tBuNO$_2$, 128°.) Strangely enough, despite the high polarity, solubility in water is small.

Nitroalkanes absorb at 1580 and 1375 cm^{-1} in the ir and have a weak $n \rightarrow \pi^*$ transition at 270 nm.

All nitro compounds with α hydrogen dissolve in dilute NaOH, and they give reactions reminiscent of aldehydes and ketones (Eqs. 20.60). The loss of a proton, although not

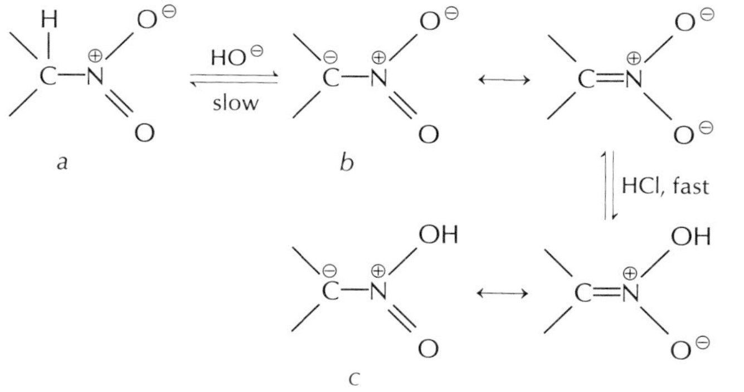

$$(20.60)$$

immediate, is sufficient to cause nitro compounds (a) to be classed as acids unless R in RNO$_2$ is tertiary. (CH$_3$NO$_2$ has a pK_a of 10.2, EtNO$_2$ of 8.6, and iPrNO$_2$ of 5.1.) The anion b accepts a proton on oxygen rapidly to give c, known as the acinitro form (analogous to an enol). Unlike the rapid keto-enol tautomerism, the nitro-acinitro tautomerism is so slow that the acinitro form may be isolated in some instances (Eq. 20.61). An acinitro compound,

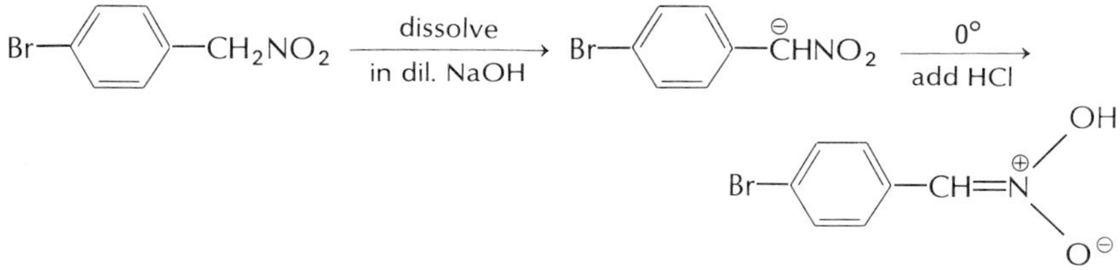

$$Br-\phi-CH_2NO_2 \xrightarrow[\text{in dil. NaOH}]{\text{dissolve}} Br-\phi-\overset{\ominus}{C}HNO_2 \xrightarrow[\text{add HCl}]{0°}$$

$$Br-\phi-CH=\overset{\oplus}{N}\overset{OH}{\underset{O^{\ominus}}{}} \qquad (20.61)$$

also known as a nitronic acid, slowly reverts to the more stable nitro form. Here is another example of kinetic control of the product first formed when the anion accepts a proton, and thermodynamic control ultimately favoring the nitro structure.

Another similarity of nitroalkanes to aldehydes and ketones is shown by the participation of nitroalkanes as the anion in crossed aldol condensations (Eq. 20.62) and in Michael reactions (Eqs. 20.63). Followed by reduction of the nitro group to an amino group (also reductive alkylation), the reactions are of use in the preparation of complex amines and other compounds.

$$CH_3CHO + CH_3CH_2NO_2 \xrightarrow{HO^{\ominus}} \underset{\underset{CH_3}{|}}{CH_3CHOHCHNO_2} \xrightarrow[-H_2O]{\text{if warm}} CH_3CH=C\underset{CH_3}{\overset{NO_2}{\diagup}} \qquad (20.62)$$

$$CH_3CH=CHCO\phi + CH_3NO_2 \xrightarrow{HO^{\ominus}} \underset{\underset{CH_2NO_2}{|}}{CH_3CHCH_2CO\phi}$$

$$H_2C=CHCN + EtNO_2 \xrightarrow{\text{base}} \underset{\underset{CH_3CHNO_2}{|}}{H_2C-CH_2CN} \xrightarrow{\text{repeat}} \underset{\underset{\underset{H_2C-CH_2CN}{|}}{CH_3CNO_2}}{H_2C-CH_2CN} \qquad (20.63)$$

The nitroalkanes, like the nitrosoalkanes, may be reduced to substituted hydroxylamines or to primary amines, depending upon the reagent and conditions.

The simple nitroalkanes are made by the commercial gas-phase nitration of propane (Eq. 20.64). A common laboratory preparation is the use of an alkyl iodide

$$CH_3CH_2CH_3 \xrightarrow[400°]{HNO_3}$$

$$CH_3NO_2 + CH_3CH_2NO_2 + CH_3CH_2CH_2NO_2 + (CH_3)_2CHNO_2 \quad (20.64)$$

$$25\% \qquad\quad 10\% \qquad\qquad 25\% \qquad\qquad 40\%$$

with silver nitrite, even though the reaction is complicated by nitrite and nitrate ester formation, the nitrate ester arising from the presence of $AgNO_3$ in the $AgNO_2$ used (Eq. 20.65). The nitroalkane and the nitrite ester usually are easy to separate

$$CH_3CH_2CH_2I + AgNO_2 \longrightarrow CH_3CH_2CH_2NO_2 + CH_3CH_2CH_2ONO \quad (20.65)$$

because of the greater volatility of the ester. Occasionally the oxidation of a primary amine or the use of a crossed aldol or of a Michael reaction is invoked to prepare a complex nitro compound.

Alkyl nitrites and nitrates. The esters of nitrous and nitric acids are known as alkyl nitrites and nitrates. The nitrites of the simple alcohols are gases or low-boiling liquids. The bp are lower than those of the corresponding alcohols (MeONO, bp $-12°$; EtONO, bp 17°; PrONO, bp 48°; iPrONO, bp 45°). The nitrate esters, however, tend to have bp slightly above those of the corresponding alcohols because of polarity effect upon association (MeONO$_2$, bp 65°; EtONO$_2$, bp 88°; PrONO$_2$, bp 111°; iPrONO$_2$, bp 102°). The nitrites tend to be coplanar, with a small rotational barrier of about 8 kcal. Thus, because a barrier of 18 kcal is necessary for isolation of specific rotamers at room temperature, the equilibrium between *a* and *b* is mobile. If R is primary, both *a* and *b* are present. If R is secondary or tertiary, the tendency is for

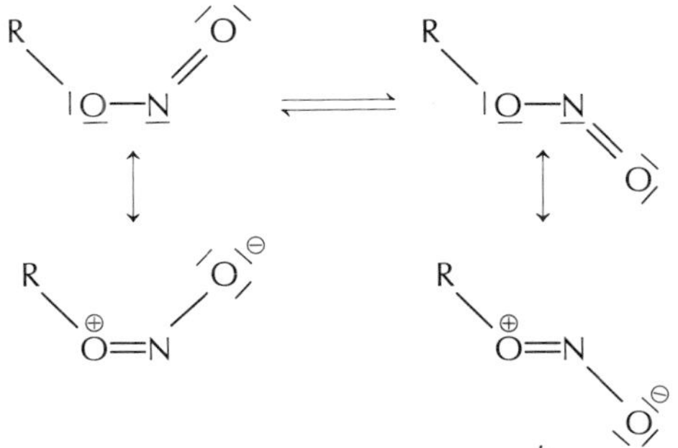

a to decrease and *b* to increase. The nitrates do not have *cis* and *trans* conformations.

The nitrites are toxic substances. Isopentyl nitrite (Me$_2$CHCH$_2$CH$_2$ONO) is used in small amounts as a coronary vasodilator. The nitrates also are toxic, and the high explosive misnamed nitroglycerine (actually glyceryl trinitrate) also is a commonly used coronary vasodilator. ☐ ☐

682 Chapter twenty

Exercises

20.8 Show the reactants needed to prepare:
  a. CH$_3$CH$_2$CHCH$_2$CH$_2$COCH$_3$
        |
        NO$_2$
  b. (CH$_3$)$_3$CCH=CCH$_2$CH$_3$
           |
           NO$_2$

20.9 Write the structures of the products to be expected if the products in Equations 20.63 are subjected to catalytic hydrogenation.

20.8 Azo, diazo, azide, and cyanate groups

Azoalkanes. Azomethane has the structure CH$_3$N=NCH$_3$. The rationale for the name is that each carbon-linked grouping (here methyl) is bound to a single nitrogen atom, hence azo. One may easily work out more logical names for these compounds, but the azo compounds have been known by this name for so long now that we must perpetuate it to prevent confusion.

The azoalkanes are neutral substances and are used principally as sources of alkyl free radicals for chain initiation (free-radical polymerization) and similar purposes. Azomethane requires a temperature of 200° for a reasonable rate of decomposition (Eqs. 20.66),

$$CH_3\!:\!\overset{\frown}{N\!=\!N}\!:\!CH_3 \xrightarrow{200°} CH_3\cdot + N_2 + CH_3\cdot$$

$$\underset{\underset{Me}{|}}{N\!\equiv\!C\!-\!\overset{Me}{\underset{|}{C}}\!:\!\overset{\frown}{N\!=\!N}\!:\!\overset{Me}{\underset{|}{C}}\!-\!C\!\equiv\!N} \xrightarrow{100°} N_2 + 2\,\underset{\underset{Me}{|}}{\overset{Me}{\underset{|}{C}}\!-\!C\!\equiv\!N}\cdot \qquad (20.66)$$

but only 100° is required for the fragmentation of 2,2'-dicyano-2,2'-azopropane (better known as azobisisobutyronitrile) into the more stable 2-cyano-2-propyl radicals.

An azoalkane may be reduced to a substituted hydrazine, but the reaction is seldom carried out because the principal method used for the preparation of an azoalkane is the oxidation of a hydrazine (Eq. 20.67). The required substituted hydrazines usually are made

$$MeNHNHMe \xrightarrow[H_2SO_4]{Na_2Cr_2O_7} MeN\!=\!NMe \qquad (20.67)$$

by reduction of an azine. Thus, we show the preparation of azobisisobutyronitrile from acetone, hydrazine dihydrochloride, and potassium cyanide in a reaction analogous to the Strecker synthesis (Eq. 20.50), the difference being the use of hydrazine instead of ammonia (Eq. 20.68). The reduction of the azine here is replaced by an addition reaction of HCN. The substituted hydrazine, in this instance, usually is oxidized to the azo compound with NaOCl.

$$2CH_3COCH_3 + 2KCN + H_2NNH_2\cdot 2HCl \longrightarrow NC\!-\!\underset{\underset{Me}{|}}{\overset{Me}{\underset{|}{C}}}\!-\!NHNH\!-\!\underset{\underset{Me}{|}}{\overset{Me}{\underset{|}{C}}}\!CN \qquad (20.68)$$

20.8 Azo, diazo, azide, and cyanate groups

Diazoalkanes: diazomethane. The diazoalkanes have two nitrogen atoms per carbon grouping. The structure and use of diazomethane (CH_2N_2) are described in the supplement. □ □

The most popular method for the preparation of diazomethane (which must be prepared immediately before use) is by the reaction of EtONa with N-methyl-N-nitroso-*p*-toluenesulfonamide (Eq. 20.69) in a complicated elimination reaction. Other leaving groups

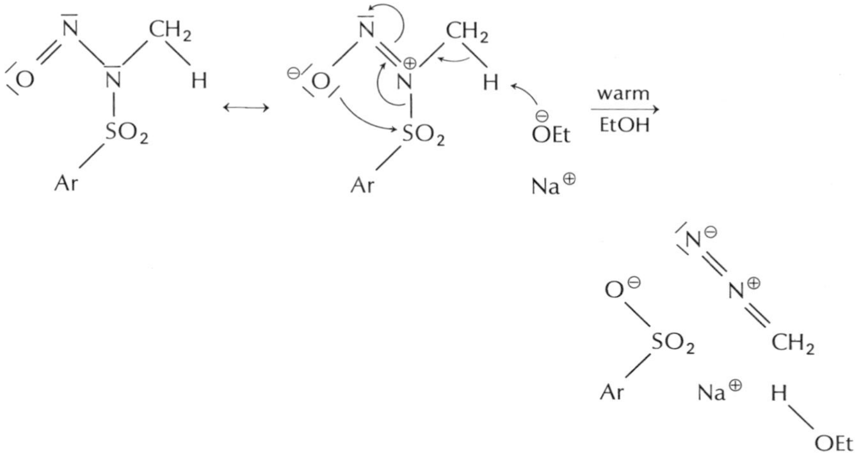

(20.69)

than the *p*-toluenesulfonate ion may be used, but the *p*-toluenesulfonamide has the advantages of being more stable and commercially available. The yellow, toxic, gaseous, and unpredictably explosive diazomethane distils from the mixture as it is formed and is absorbed in CH_2Cl_2; and the solution is used immediately.

Other diazoalkanes are available from primary amines, as shown in Equation 20.70. If the hydrogen on the carbon adjacent to an amino group is activated, nitrosa-

$$R_2CHNH_2 \xrightarrow[NaOH]{ArSO_2Cl} ArSO_2NHCHR_2$$

$$\xrightarrow[HCl]{NaNO_2} ArSO_2NCHR_2 \text{ (NO)} \xrightarrow[\text{warm}]{EtONa, EtOH} ArSO_3Na + \overset{\ominus}{N}=\overset{\oplus}{N}=CR_2 \quad (20.70)$$

tion gives some of the diazo compound (Eq. 20.71) instead of forming a carbonium ion.

$$\underset{NH_2}{CH_2COOEt} \xrightarrow[HCl]{NaNO_2} \underset{\overset{\oplus}{N}\equiv N}{CH_2COOEt} \xrightarrow{-H^{\oplus}} \underset{\overset{\oplus}{N}=\overset{\ominus}{N}}{CHCOOEt} \quad (20.71)$$

Azides. Azides are named as esters of hydrazoic acid (HN_3). The anion, azide ion, is nucleophilic and can displace good leaving groups, such as sulfate (Eq. 20.72). Acyl azides are prepared (Eqs. 20.73) by the reaction of sodium azide with an acid chloride or by

$$\overset{\ominus}{|N}=\overset{\oplus}{N}=\overset{\ominus}{N|} + ROSO_3R \longrightarrow \overset{\ominus}{|N}=\overset{\oplus}{N}=N-R + {}^{\ominus}OSO_3R \quad (20.72)$$

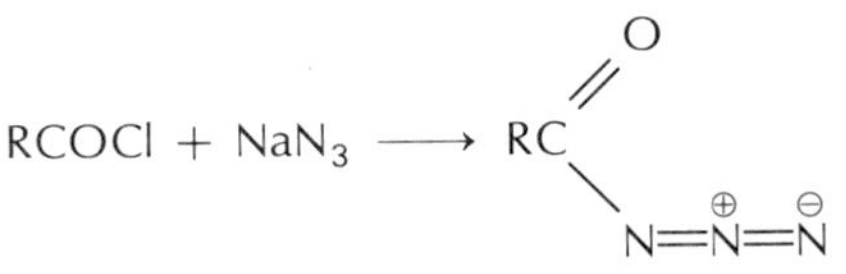

$$RCOCl + NaN_3 \longrightarrow RC\underset{\overset{|}{N=\overset{\oplus}{N}=\overset{\ominus}{N}}}{\overset{\overset{O}{\|}}{}}$$

$$RCONHNH_2 + HNO_2 \longrightarrow RCON_3 + 2H_2O \tag{20.73}$$

nitrous acid with a hydrazide (usually made from an ester and hydrazine). Acyl azides, because $N_3^{\ominus}$ is a fairly good leaving group, are used as gentle acylating agents (instead of acid chlorides) in peptide synthesis (Chap. 21).

Cyanates. Cyanates (ROCN) are relatively uncommon as compared with isocyanates (RNCO). Cyanates tend to trimerize to substituted 1,3,5-triazines unless the R group is bulky. Isocyanates are more stable but react readily with protic substances (Eq. 20.74). We have met the isocyanates before as intermediates in the Hofmann rearrange-

$$R-N=C=O + EtOH \longrightarrow R-NH-\underset{\underset{OEt}{|}}{C}=O \tag{20.74}$$

ment. A description of the preparation and other reactions of isocyanates will be given in Chapter 22 when the aromatic isocyanates are taken up.

Exercises

20.10 Complete the reactions:
  a. $EtNHNHEt \xrightarrow[H_2SO_4]{Na_2Cr_2O_7}$
  b. $EtOOCCHN_2 + CH_3CH_2COOH \longrightarrow$
  c. $CH_3COCl \xrightarrow{H_2NNH_2} \xrightarrow[HCl]{NaNO_2} \xrightarrow{\Delta} \xrightarrow{EtNH_2}$
  d. $\underset{\underset{Me_2CCH_2COMe}{|}}{ON-N-Me} \xrightarrow[warm]{EtONa}$

20.9 Summary

The variety of organic nitrogen compounds ranges from bases to acids and from reduced to oxidized states, with many intermediate forms. We have chosen a path through the jumble of compounds and reactions that shows that our ideas of substitution, addition and elimination, and oxidation and reduction apply to the nitrogen compounds as they do to the carbon, hydrogen, and oxygen compounds.

By now, we hope that the value of your own summary of previous chapters has become evident. However, the answers to the exercises again provide a list of most of the reactions in the chapter.

20.10 Problems

20.1 Substance A ($C_7H_{15}O_2N$) gives a purple color in solution with $FeCl_3$. Upon hydrolysis, A gives B and NH_2OH. B with $SOCl_2$ gives C. C with H_2NNH_2 gives D. D with $NaNO_2$ and HCl gives E. Subjecting E to heat gives rearrangement to F. F reacts with water to give G. C with ammonia gives H. H with bromine in dilute NaOH gives G. G upon exhaustive methylation gives I. I with Ag_2O, then heat, gives J. Ozonolysis of J gives CH_3COCH_3 as the only product. Give structural formulas for the lettered substances.

20.2 In each of the figures a set of spectra and the molecular formula are given. Deduce the structural formula of each substance.

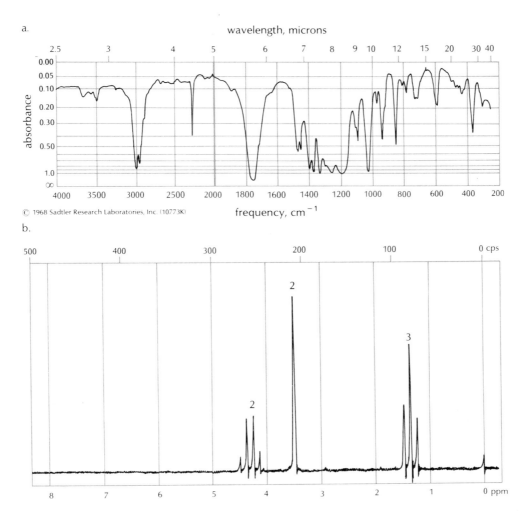

Figure 20.1 For Problem 20.2a. $C_5H_7O_2N$

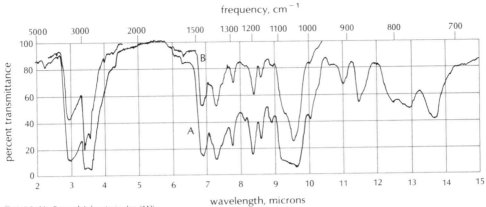

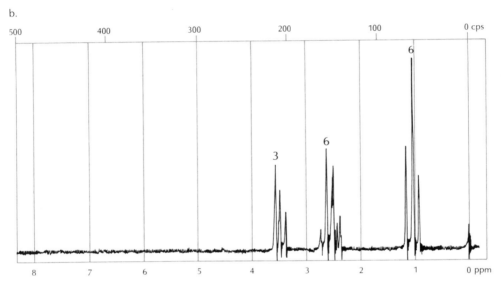

Figure 20.2 For Problem 20.2b. $C_6H_{15}ON$

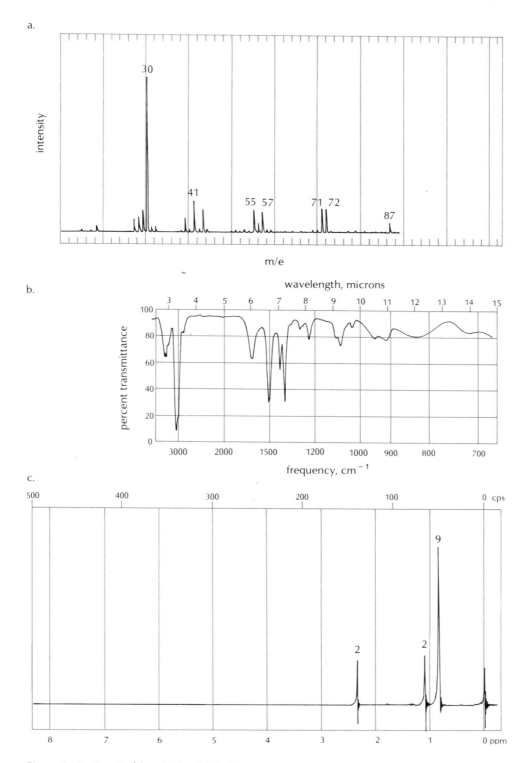

Figure 20.3 For Problem 20.2c. $C_5H_{13}N$

20.11 Bibliography

Coverage of the nonaromatic nitrogen compounds is relatively sparse in most first year textbooks of organic chemistry. Two of the better textbooks in this area are Fieser and Fieser (Chap. 7, ref. 7) and Roberts and Caserio (Chap. 7, ref. 13). The best reference is a paperback:

1. Millar, Ian T., and H. D. Springall, *A Shorter Sidgwick's Organic Chemistry of Nitrogen.* London: Oxford University Press, 1969. As the name describes it, this book is not a condensation of the classic reference work in the field so much as it is a selection of portions of the much larger set. The coverage is spotty in parts, but it is more than adequate as a starting point for additional material on the subjects covered in this chapter (and the following two). Many references. No problems.

2. Moss, Robert A., Deamination Chemistry. *Chemical and Engineering News* Nov. 22, 1971, p. 28. This is a good review article of one aspect of the chemistry of amines.

21 Amino Acids, Peptides, and Proteins

Of the three major classes of organic compounds found in nature, the lipids, carbohydrates, and proteins, the latter have been the most difficult to study. Proteins not only are polymeric, but are made up from about twenty monomeric units, the amino acids. They are very complex. Proteins also are very sensitive substances, and it is not easy to purify them or prevent them from decomposing. Nevertheless, progress in their study has been rapid in recent years.

Proteins (from the Greek proteios, primary or prime) were recognized in the early 1800's as vital components of living organisms. In this chapter we introduce the subject by looking at amino acids first, then progressing to peptides, and then to proteins in general. Finally, we look at some hormones and enzymes to illustrate a few of the many active fields of work within protein chemistry. Protein chemistry is fundamental to our understanding of health and disease.

21.1   Proteins, peptides, and amino acids   690
Definitions; descriptions.

21.2   Amino acids   691
1. The twenty common L-family amino acids.   2. Physical properties; isoelectric points.   3. Less common amino acids.   4. Distribution in various proteins.   5. Reactions.   6. Syntheses.   7. Resolution of synthetic racemic amino acids.

21.3   Peptides   705
1. Properties and representation.   2. Analysis: difficulties.   3. End-group analysis.   4. Chain cleavage.   5. Structural formulas from analytical results.   6. Syntheses. Exercises.

21.4   Proteins   715
1. Conformations; structures.   2. Peptide bonds: geometry.   3. Resonance in proteins.   4. Hydrogen bonding: secondary structures.   5. The α-helix.   6. Tertiary structures.   7. Quaternary structures.   8. Nmr in biochemistry.

21.5   Hormones, antibodies, and genetic defects   722
1. Hormones.   2. Antibody-antigen defense mechanism.   3. Genetic diseases. Exercises.

690 Chapter twenty-one

21.6 Enzymes   725
 1. Definitions.   2. Effects of pH, temperature, and inhibitors.
 3. Coenzymes and vitamins. Exercises.

21.7 Summary   732

21.8 Problems   733

21.9 Bibliography   734

21.1 Proteins, peptides, and amino acids

Definitions; descriptions. What are proteins? Proteins may be defined as substances of molecular weight greater than 7000 to 10,000 that upon complete hydrolysis yield amino acids, either exclusively or in large part. If a given protein contains a non-amino-acid portion, it is classed as a conjugated protein, and the non-amino-acid group is called a prosthetic group. Hemoglobin is an example of a conjugated protein, the iron-containing heme being the prosthetic group (Chap. 28).

Proteins are classified in several different ways. One arbitrary division is made in terms of physical and chemical properties: fibrous proteins and globular proteins.

The fibrous proteins (those in hair, horn, fingernails, and skin) are insoluble in water and resist hydrolysis, especially enzymatic hydrolysis. Leather, furs, wool, and silk are protein containing substances that are used extensively in commerce. Specific proteins present in these substances are sericine (or silk fibroin), hard keratin in hair, nails, claws, and so on, soft keratin in skin, myosin in muscle, and collagen in connective tissue and bones. Fibrous proteins have structural integrity: that is, they tend to persist in their original structure or form.

The globular proteins are soluble (or, if their molecular weight is large, form hydrophilic colloidal dispersions) in water. The name is descriptive: the molecular shape is globular or ellipsoidal. Examples are hemoglobin, enzymes, antibodies, and some hormones.

Partial hydrolysis of a protein gives rise to a mixture of substances of molecular weight intermediate between that of the protein and those of the constituent amino acids. (Some individual amino acids also appear in partially hydrolyzed mixtures.) The intermediate-weight substances are known as peptides. If the exact number of combined amino acids in a peptide is known, the compound is designated with a suitable numerical prefix: tripeptide, decapeptide, and so on. Polypeptide is used to designate compounds with molecular weights that are large but less than 7000. An oligopeptide is a peptide of molecular weight intermediate between those of a polypeptide and a simple peptide. The words peptide and pepsin (an enzyme) are derived from Greek words meaning to cook or to digest.

The expression peptide bond refers to the substituted amide bond whereby the individual amino acids are linked together to form peptides. Thus, a polypeptide is a polyamide of amino acids. Because peptides are more tractable to work with than are proteins, and because they are intermediates in any attempt to hydrolyze or to synthesize a protein, they have been the subject of an immense research effort as a key to the understanding of proteins.

Twenty different α-amino acids are commonly found as complete hydrolysis products of proteins. Many other α-amino acids are known to occur in nature as constituents of proteins but are not as widespread nor as abundant as the common amino acids. Of the twenty amino acids, eight are known to be essential to man. These eight must be present in the diet because the human body cannot synthesize them. The other twelve, if not present in the diet, can be synthesized in the body.

Now let us go over these topics in more detail, beginning with the amino acids and working our way up to the high molecular weight proteins.

21.2 Amino acids

The twenty common L-family amino acids. The twenty common α-amino acids are known by their trivial names. All are alike in possessing an amino group and a hydrogen attached to the α carbon atom to the carboxylic acid group. They differ in the third group attached to the α carbon atom, which then is asymmetric (except in glycine, in which the third group is a hydrogen atom). Alanine (2-aminopropionic acid) was related to lactic acid

and to glyceraldehyde, and it was shown that alanine belongs to the family of compounds derived from L-(−)-glyceraldehyde. The common amino acids all belong to the L family and have the (S) configuration about the α carbon atom. An amino acid, with both basic and acidic groups present, should be written as a dipolar ion (zwitterion) with the proton transferred from the carboxylic acid group to the basic amino group (Sec. 20.4). However, we often see amino acids written as the non-proton-transferred structures, or RCH(NH$_2$)COOH.

The twenty common α-amino acids are shown in Table 21.1. They may be classified in several different ways: neutral, acidic, and basic; containing reactive or unreactive side chains; and so on. We list the acids in alphabetical order, which seems to be as easy a way to remember them as any. The structures are shown in such a way as to emphasize that all the acids except proline may be written as $^\oplus$H$_3$NCHRCOO$^\ominus$ and differ only in the structure of the side chain (R group). The R group may be hydrogen (as in glycine) or a simple alkyl (Me in alanine, iPr in valine, iBu in leucine, sBu in isoleucine). The Et, Bu, and tBu groups are not represented in the common amino acids. Proline differs from the other amino acids by incorporating the α-amino group into a pyrrolidine ring (Chap. 28) as an amino group.

Two of the amino acids have R groups that contain hydroxy groups: serine and threonine. Two amino acids have sulfur present in the R groups: HS— in cysteine and MeS— in methionine. Two more amino acids have aromatic benzene rings: phenyl in phenylalanine and p-hydroxyphenyl in tyrosine. Another two amino acids have extra basic groups: guanadinium in arginine, amino in lysine. Still another two have extra acid groups: aspartic acid (2-aminosuccinic acid) and glutamic acid (2-aminoglutaric acid). Asparagine and glutamine have the extra acid groups of aspartic and glutamic acids

Table 21.1 The common α-amino acids[a]

Name	Abbreviation	mw	Structural formula	Isoelectric point
alanine	Ala	89.09	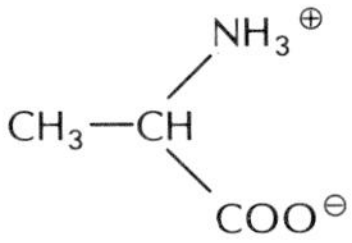 $CH_3-CH(NH_3^{\oplus})COO^{\ominus}$	6.11
arginine	Arg	174.20	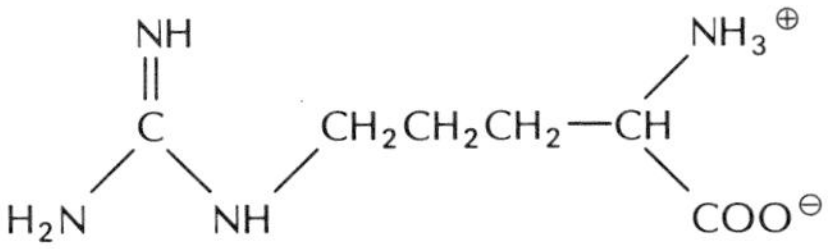 $H_2N-C(=NH)-NH-CH_2CH_2CH_2-CH(NH_3^{\oplus})COO^{\ominus}$	10.77
asparagine	Asn	132.12	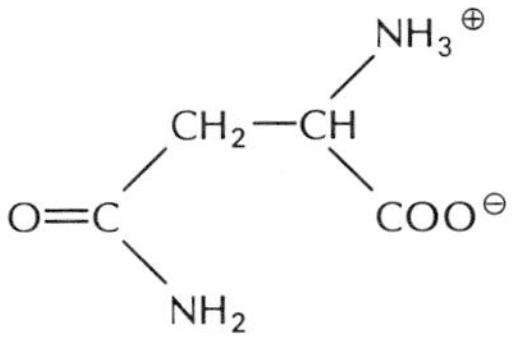 $O=C(NH_2)-CH_2-CH(NH_3^{\oplus})COO^{\ominus}$	5.41
aspartic acid	Asp	133.10	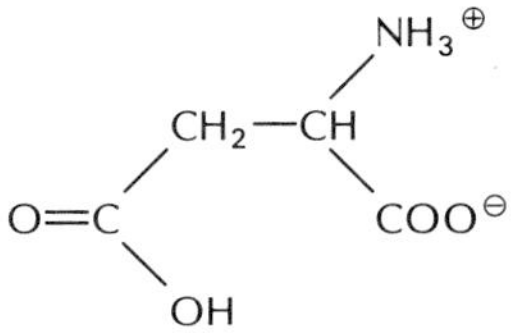 $O=C(OH)-CH_2-CH(NH_3^{\oplus})COO^{\ominus}$	2.98
cysteine	Cys	121.16	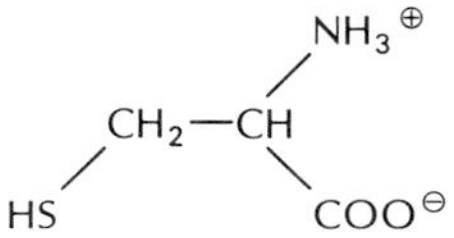 $HS-CH_2-CH(NH_3^{\oplus})COO^{\ominus}$	5.02
glutamic acid	Glu	147.13	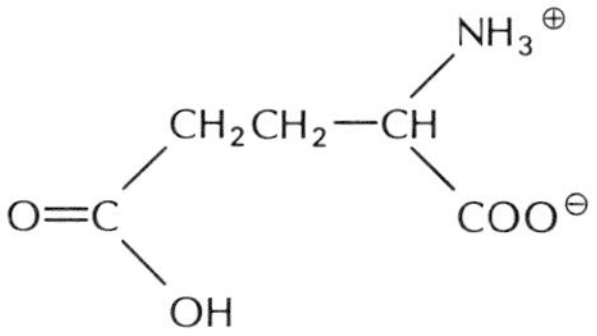 $O=C(OH)-CH_2CH_2-CH(NH_3^{\oplus})COO^{\ominus}$	3.08
glutamine	Gln	146.15	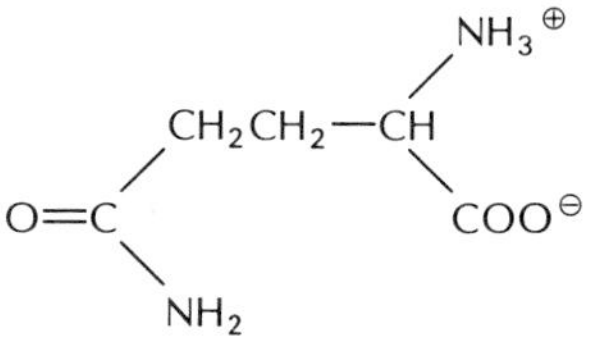 $O=C(NH_2)-CH_2CH_2-CH(NH_3^{\oplus})COO^{\ominus}$	5.65
glycine	Gly	75.07	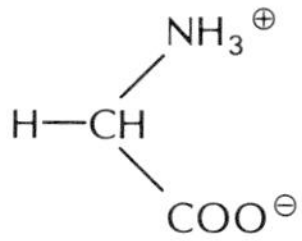 $H-CH(NH_3^{\oplus})COO^{\ominus}$	6.06
histidine	His	155.16	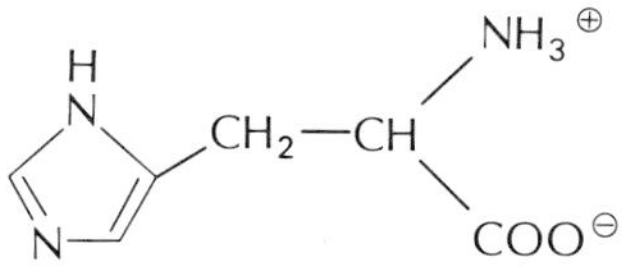 imidazole-$CH_2-CH(NH_3^{\oplus})COO^{\ominus}$	7.64

Table 21.1 (continued)

Name	Abbreviation	mw	Structural formula	Isoelectric point
isoleucine[b]	Ile	131.17	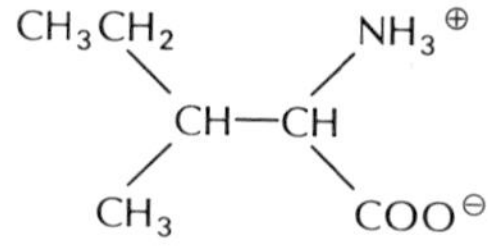 CH_3CH_2–CH(CH$_3$)–CH(NH$_3^\oplus$)–COO$^\ominus$	6.04
leucine[b]	Leu	131.17	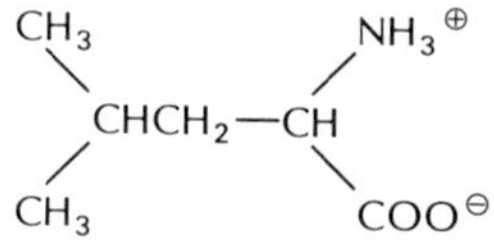 (CH$_3$)$_2$CH–CH$_2$–CH(NH$_3^\oplus$)–COO$^\ominus$	6.04
lysine[b]	Lys	146.19	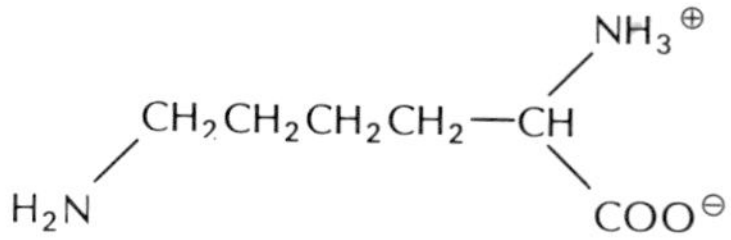 H$_2$N–CH$_2$CH$_2$CH$_2$CH$_2$–CH(NH$_3^\oplus$)–COO$^\ominus$	9.74
methionine[b]	Met	149.21	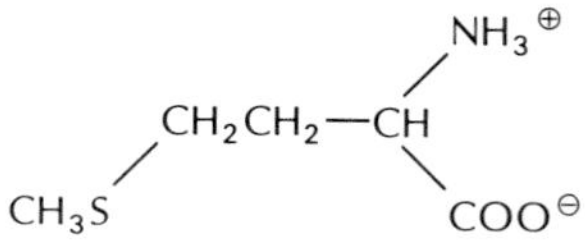 CH$_3$S–CH$_2$CH$_2$–CH(NH$_3^\oplus$)–COO$^\ominus$	5.74
phenylalanine[b]	Phe	165.19	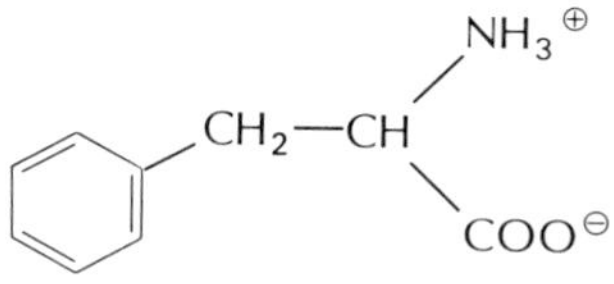 C$_6$H$_5$–CH$_2$–CH(NH$_3^\oplus$)–COO$^\ominus$	5.91
proline	Pro	115.13	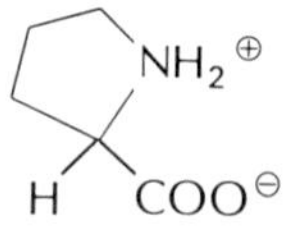 (pyrrolidine)–NH$_2^\oplus$, H, COO$^\ominus$	6.30
serine	Ser	105.09	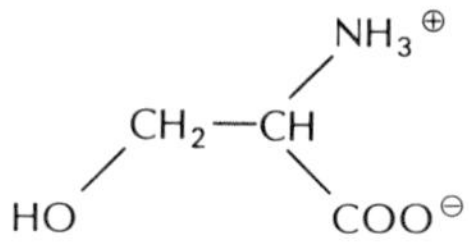 HO–CH$_2$–CH(NH$_3^\oplus$)–COO$^\ominus$	5.68
threonine[b]	Thr	119.12	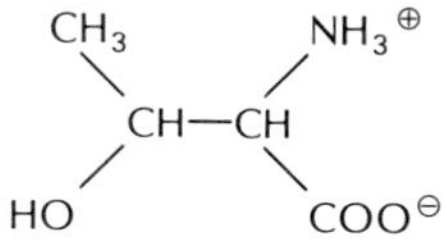 CH$_3$–CH(OH)–CH(NH$_3^\oplus$)–COO$^\ominus$	5.64
tryptophan[b]	Try	204.22	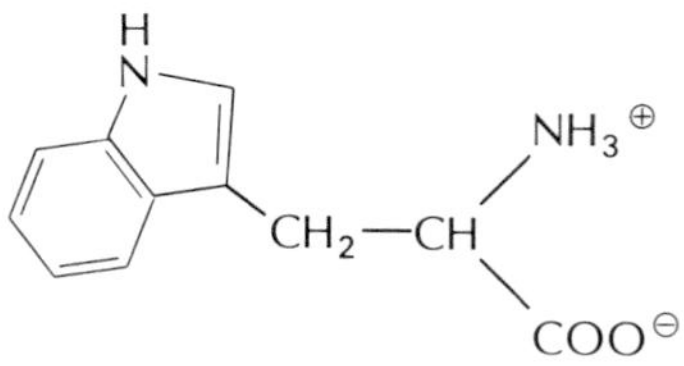 (indol-3-yl)–CH$_2$–CH(NH$_3^\oplus$)–COO$^\ominus$	5.88
tyrosine	Tyr	181.19	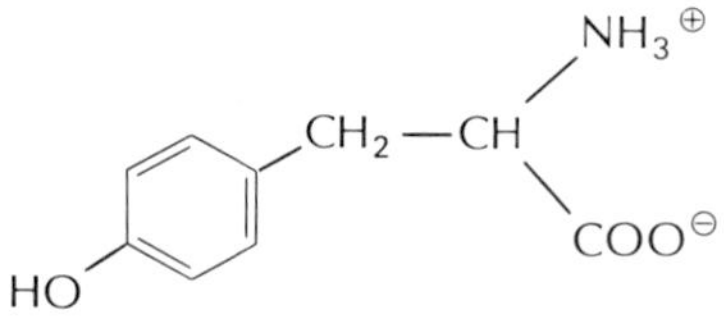 HO–C$_6$H$_4$–CH$_2$–CH(NH$_3^\oplus$)–COO$^\ominus$	5.66

694 Chapter twenty-one

Table 21.1 (continued)

Name	Abbreviation	mw	Structural formula	Isoelectric point
valine[b]	Val	117.15	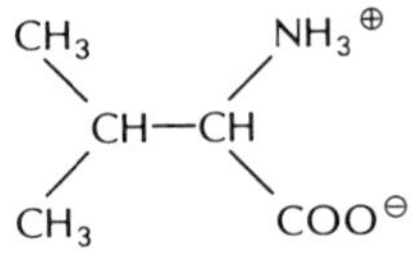 $\text{CH}_3\text{-CH(CH}_3\text{)-CH(NH}_3^{\oplus}\text{)-COO}^{\ominus}$	6.00

[a] Some authors list hydroxylysine and hydroxyproline as common amino acids. Cystine, the disulfide formed by oxidation of cysteine, is usually considered along with cysteine as a single substance.
[b] Essential amino acids.

neutralized by amide formation. The remaining two amino acids have heterocyclic rings: imidazole in histidine and indole in tryptophan. This two-by-two sequence reminds us of Noah's Ark, even though the analogy is far-fetched.

The abbreviations listed in Table 21.1 are useful when we represent peptides. Strictly speaking, an abbreviation symbolizes $-\text{HNCHRC}(=\text{O})-$. Thus, the amino acid alanine by itself would be $^{\oplus}\text{H}_2-\text{Ala}-\text{O}^{\ominus}$ as a zwitterion, or $\text{H}-\text{Ala}-\text{OH}$ as the non-proton-transferred structure. These end groups often are omitted when peptide structures are written out.

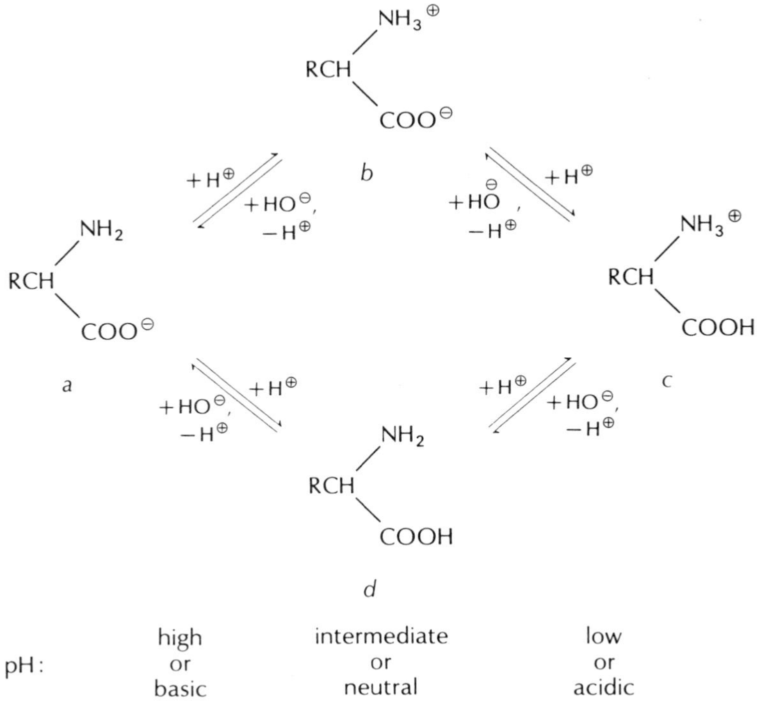

$$\begin{array}{c}
\text{RCH}(\text{NH}_3^{\oplus})\text{COO}^{\ominus} \\
\text{b} \\
\text{RCH}(\text{NH}_2)\text{COO}^{\ominus} \quad \rightleftharpoons \quad \rightleftharpoons \quad \text{RCH}(\text{NH}_3^{\oplus})\text{COOH} \\
\text{a} \qquad\qquad\qquad\qquad\qquad\qquad\qquad\qquad c \\
\text{RCH}(\text{NH}_2)\text{COOH} \\
\text{d}
\end{array} \qquad (21.1)$$

pH: high or basic | intermediate or neutral | low or acidic

Physical properties; isoelectric points. The crystalline amino acids are quite dense (about 1.5 g/cc) and have no sharp melting point. They decompose at 250° or more. X-ray crystal analysis has shown that the bond distances are best explained in terms of hydrogen bonded dipolar ions.

An aqueous solution of an amino acid has properties that vary with the pH (amino acids are only slightly soluble in MeOH and EtOH, and they are practically insoluble in most other organic solvents). The possible proton-transfer equilibria involving an amino acid are shown in Equation 21.1.

Let us dissolve an amino acid in an excess of dilute aqueous sodium hydroxide. All of the acid is in solution as the anion *a* at a high pH. Now, let us add dilute hydrochloric acid dropwise to the solution. There is competition between the bases present (hydroxide ion and *a*) for the available protons. The stronger base, hydroxide ion, wins out, and the hydrogen ion concentration begins to increase (the pH begins to decrease). As the hydroxide ion concentration decreases, we reach a point at which *a* begins to compete for the hydrogen ions being added and the equilibria $a + H^{\oplus} \rightleftharpoons b$ and $a + H^{\oplus} \rightleftharpoons d$ come into play. Which is favored, *b* or *d*?

Actually, we are asking which is the stronger base, the —NH_2 group or the carboxylate ion, —$COO^{\ominus}$. Let us find the answer by starting with a solution of the amino acid as *c* in an excess of dilute hydrochloric acid. Now let us add dilute sodium hydroxide dropwise to the solution. This time, as the pH increases, we reach a point at which the equilibria $c + HO^{\ominus} \rightleftharpoons b$ and $c + HO^{\ominus} \rightleftharpoons d$ come into play. Now the question may be modified: which is the stronger acid, the —$NH_3^{\oplus}$ or the —COOH group?

The equilibrium between *c* and *b* is that for the ionization of a carboxylic acid, which, we learned previously (Table 17.1), has pK_a of about 5. The equilibrium between *c* and *d* is for the ionization of the conjugate acid of an amine, which we know (Sec. 20.2) has pK_a of about 10 to 11. Therefore, we expect the amino acid cation *c* to lose a proton from the carboxylic acid group at a pH of about 5 to give equal concentrations of *b* and *c*. In turn, the dipolar ion *b* loses a proton from the —$NH_3^{\oplus}$ group to give equal concentrations of *a* and *b* at a pH of about 10 to 11. Thus, as we add base to the solution containing *c*, the pH increases, and it reaches 5 before it reaches 10. Therefore, *c* goes into equilibrium with *b* instead of with *d*.

A similar argument applies to *a*. The stronger base, the amino group, is better able to take up a proton to give *b* than is the weaker base, the carboxylate anion, to give *d*. Therefore, at intermediate values of pH, *b* is in equilibrium with both *a* and *c*. Some *d* is present, but in only miniscule concentration as compared to *b*.

What is actually observed? Titration from either direction of a solution of an amino acid gives inflection points that show pK_a 2.35 and 9.78 for glycine, 2.35 and 9.87 for alanine, and 2.21 and 9.15 for serine (Fig. 21.1). The high values are only slightly less than our predicted pK_a value of 10 to 11, but the low values are decidedly less than our predicted pK_a of 5. Why?

Structure *c* has a strong —I group (—$NH_3^{\oplus}$) in the α position. The effect is to strengthen the acid. Compare the pK_a values of acetic acid, 4.75; chloroacetic acid, 2.85; glycine, 2.35; and dichloroacetic acid, 1.48. We see that the —$NH_3^{\oplus}$ has a larger —I effect than that of a single chlorine but less than that of two chlorine atoms.

Another observation of interest is that if the pH of a solution of an amino acid is adjusted to a point midway between the pK_a values for that amino acid, the solubility of the amino acid is at a minimum. The solubility is greater at pH values above or below the midpoint pH. At the midpoint pH, the concentration of *b* is at a maximum, and the concentration of *a* is the same as that of *c*. This fact may be shown if an electric current is passed through such a solution. Although *a* migrates to the positive pole and *c* to the negative pole, there is no net migration, because equal amounts of *a* and *c* migrate in opposite directions. At the same time, *b*, as a neutral molecule, does not migrate. At higher pH, *a* is greater than *c* in concentration, and net migration to the positive pole results.

At lower pH, a is less than c, and net migration to the negative pole results. The pH at which no net migration occurs is known as the isoelectric point. It is identical in value with the average of the pK_a values for a given amino acid. Isoelectric points are listed in Table 21.1 for most of the amino acids.

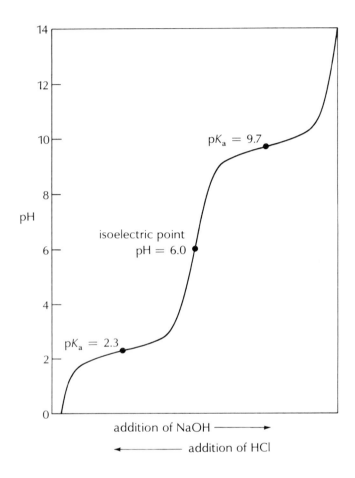

Figure 21.1   Schematic titration curve of an amino acid

Two of the amino acids listed, aspartic and glutamic acids, have two carboxylic acid groups and one amino group per molecule. The isoelectric points listed, 2.98 and 3.08, are the average of the pK_a values for the ionizations of the two carboxylic acid groups: 2.10 and 3.86 for aspartic acid, and 2.10 and 4.07 for glutamic acid. The zwitterions for these substances have unionized carboxylic acid groups present. The loss of a proton from the conjugate acid of the amino group, $-NH_3^{\oplus}$, occurs from the negative ion and does not compete with zwitterion formation. Aspartic acid ionization is shown.

21.2 Amino acids

$$\underset{\text{p}K_a \;=\;}{\underset{\text{HOOC}}{\overset{\text{CH}_2\text{CH}}{\diagup}}\overset{\overset{\oplus}{\text{NH}_3}}{\diagdown}\text{COOH}} \quad \underset{2.10}{\overset{-\text{H}^{\oplus}}{\rightleftharpoons}} \quad \underset{\substack{\text{zwitterion, no net charge,} \\ \text{maximum at pH 2.98}}}{\underset{\text{HOOC}}{\overset{\text{CH}_2\text{CH}}{\diagup}}\overset{\overset{\oplus}{\text{NH}_3}}{\diagdown}\text{COO}^{\ominus}}$$

$$\underset{3.86}{\overset{-\text{H}^{\oplus}}{\rightleftharpoons}} \quad \underset{\text{}^{\ominus}\text{OOC}}{\overset{\text{CH}_2\text{CH}}{\diagup}}\overset{\overset{\oplus}{\text{NH}_3}}{\diagdown}\text{COO}^{\ominus} \quad \underset{9.82}{\overset{-\text{H}^{\oplus}}{\rightleftharpoons}} \quad \underset{\text{}^{\ominus}\text{OOC}}{\overset{\text{CH}_2\text{CH}}{\diagup}}\overset{\text{NH}_2}{\diagdown}\text{COO}^{\ominus}$$

Arginine and lysine have two basic groups present and one carboxylic acid group. The isoelectric points are large, 10.77 and 9.74. The situation with lysine is shown.

$$\underset{\text{p}K_a \;=\;}{\underset{^{\oplus}\text{H}_3\text{N}}{\overset{(\text{CH}_2)_4\text{CH}}{\diagup}}\overset{\overset{\oplus}{\text{NH}_3}}{\diagdown}\text{COOH}} \quad \underset{2.18}{\overset{-\text{H}^{\oplus}}{\rightleftharpoons}} \quad \underset{^{\oplus}\text{H}_3\text{N}}{\overset{(\text{CH}_2)_4\text{CH}}{\diagup}}\overset{\overset{\oplus}{\text{NH}_3}}{\diagdown}\text{COO}^{\ominus}$$

$$\underset{\substack{8.95 \;\; \text{zwitterion, no net charge,} \\ \text{maximum at pH 9.74}}}{\overset{-\text{H}^{\oplus}}{\rightleftharpoons} \;\; \underset{^{\oplus}\text{H}_3\text{N}}{\overset{(\text{CH}_2)_4\text{CH}}{\diagup}}\overset{\text{NH}_2}{\diagdown}\text{COO}^{\ominus}} \quad \underset{10.53}{\overset{-\text{H}^{\oplus}}{\rightleftharpoons}} \quad \underset{\text{H}_2\text{N}}{\overset{(\text{CH}_2)_4\text{CH}}{\diagup}}\overset{\text{NH}_2}{\diagdown}\text{COO}^{\ominus}$$

Confirmation of the dipolar structure of amino acids is found in the ir. Unionized carboxylic acids show absorption at about 1760 cm^{-1}, dimers at about 1710 cm^{-1}, and the carboxylate ions at about 1600 cm^{-1} and 1400 cm^{-1}. Amino acids show only the latter bands unless the solutions are made strongly acidic.

Less common amino acids. Some of the less common amino acids are shown in Table 21.2. We note that D-amino acids are not unknown in nature.

Table 21.2 Some less common amino acids[a]

Name	Structure	Occurrence
alliin	$\text{H}_2\text{C}=\text{CH}\diagdown\text{CH}_2\diagdown\overset{\oplus}{\underset{\underset{\text{O}^{\ominus}}{\|}}{\text{S}}}\diagup\text{CH}_2-\text{CH}\diagup^{\overset{\oplus}{\text{NH}_3}}_{\diagdown\text{COO}^{\ominus}}$	garlic oil

Table 21.2 (continued)

Name	Structure	Occurrence
α-aminoadipic acid	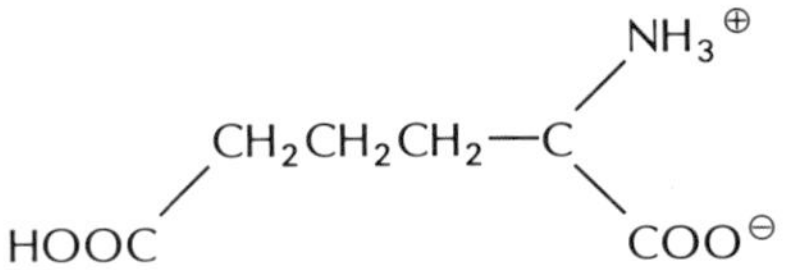 HOOC–CH$_2$CH$_2$CH$_2$–CH(NH$_3^\oplus$)–COO$^\ominus$	corn; other plants
citrulline	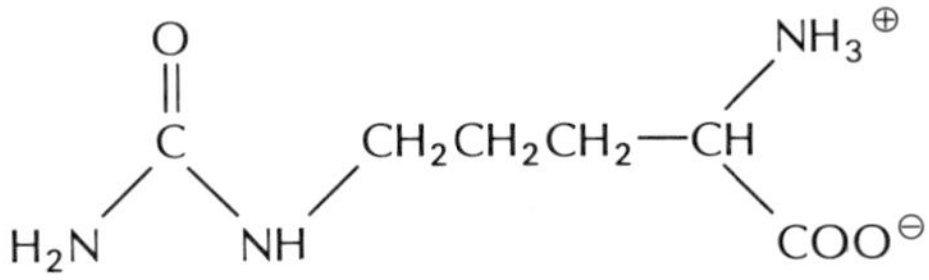 H$_2$N–C(=O)–NH–CH$_2$CH$_2$CH$_2$–CH(NH$_3^\oplus$)–COO$^\ominus$	casein; watermelon
D-glutamic acid	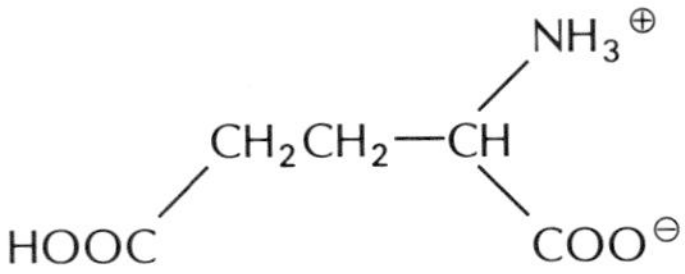 HOOC–CH$_2$CH$_2$–CH(NH$_3^\oplus$)–COO$^\ominus$	Bacillus anthracis
hydroxylysine[b]	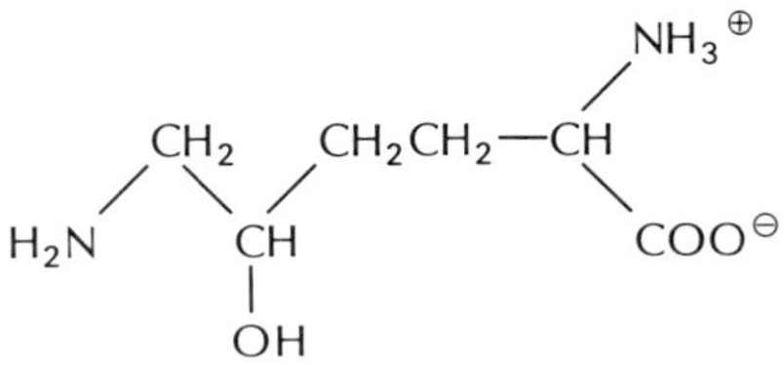 H$_2$N–CH$_2$–CH(OH)–CH$_2$CH$_2$–CH(NH$_3^\oplus$)–COO$^\ominus$	collagen
hydroxyproline[c]	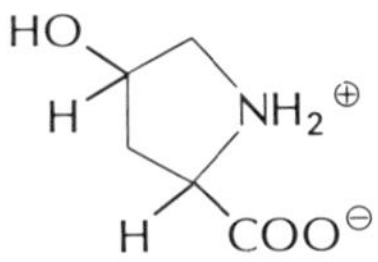 HO–(pyrrolidine ring with NH$_2^\oplus$)–COO$^\ominus$	collagen
lanthionine	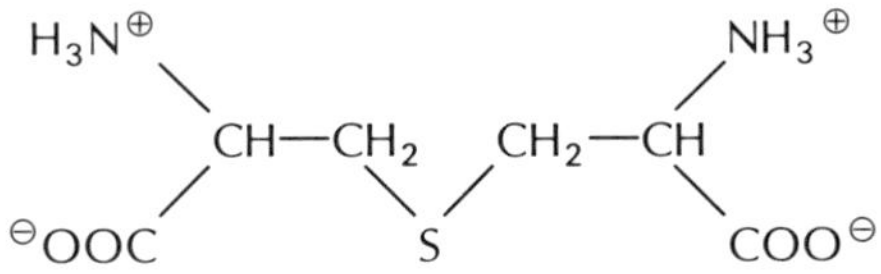 $^\ominus$OOC–CH(N H$_3^\oplus$)–CH$_2$–S–CH$_2$–CH(NH$_3^\oplus$)–COO$^\ominus$	subtilin
ornithine	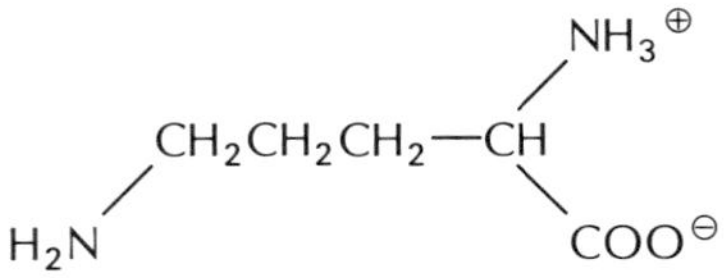 H$_2$N–CH$_2$CH$_2$CH$_2$–CH(NH$_3^\oplus$)–COO$^\ominus$	intermediate in formation of urea
D-phenylalanine	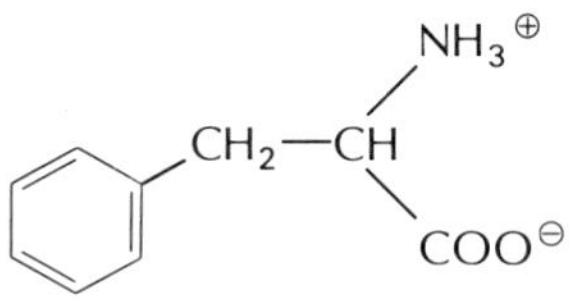 C$_6$H$_5$–CH$_2$–CH(NH$_3^\oplus$)–COO$^\ominus$	gramicidin
sarcosine	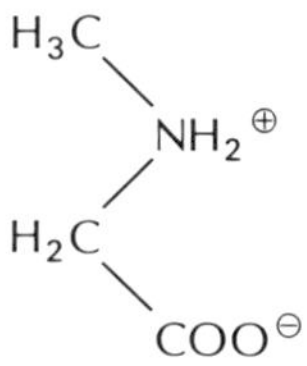 H$_3$C–NH$_2^\oplus$–CH$_2$–COO$^\ominus$	widespread but not in proteins

Table 21.2 (continued)

Name	Structure	Occurrence
thyroxine	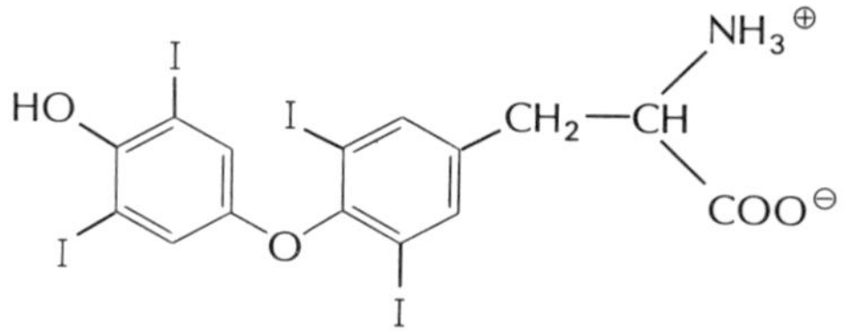 HO–(3,5-diiodo-phenyl)–O–(3,5-diiodo-phenyl)–CH$_2$–CH(NH$_3^{\oplus}$)–COO$^{\ominus}$	thyroglobin, thyroid hormone
triiodothyronine	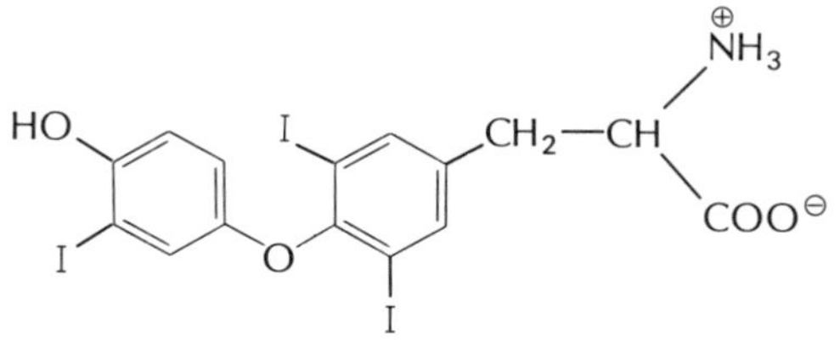 HO–(3-iodo-phenyl)–O–(3,5-diiodo-phenyl)–CH$_2$–CH(NH$_3^{\oplus}$)–COO$^{\ominus}$	thyroglobin, thyroid hormone

[a] All are L family unless labelled otherwise.
[b] Abbreviation is Hylys.
[c] Abbreviation is Hypro.

Antibiotics are a prolific source of rare amino acids. Some non-α-amino acids are known also: for example, γ-aminobutyric acid is present as such in the brain.

Distribution in various proteins. Proteins differ in composition to a large extent. The partial composition of some representative proteins is shown in Table 21.3. Only the

Table 21.3 Partial composition of a few proteins[a]

Amino Acid[b]	Serum Albumin[c]	Collagen[d]	Keratin[e]	Myosin[f]
Ala	8	11	5	9
Asp	9	5	6	10
Cys	6	0	11	1
Glu	13	7	12	18
Gly	3	33	8	5
Leu	11	2	7	9
Lys	10	3	2	11
Pro	5	13	10	4
Ser	5	4	10	4

[a] In mole per cent of the amino acids listed as present in the completely hydrolyzed protein.
[b] Only the nine most abundant amino acids are listed. See text.
[c] From cattle.
[d] From rat tail tendon.
[e] From merino wool.
[f] From skeletal muscle of rabbit.

nine common amino acids that are present in largest amounts are shown. They form about 70 to 80% of the total. The other eleven amino acids are present in all the proteins shown in small and variable amounts (2 to 7%) except for histidine, methionine, and tryptophan, which are present in smaller amounts yet. Collagen is unusual in having no cysteine nor tryptophan present at all, in being very rich in glycine, and in having both hydroxylysine (0.5%) and hydroxyproline (10%) present. Fibroin, the principal component of silk, is even more unusual in that only four amino acids make up 90% of the total products of hydrolysis (glycine 46%, alanine 26%, serine 12%, and tyrosine 6%). During hydrolysis, asparagine and glutamine are converted into aspartic acid and glutamic acid.

Reactions. The amino acids undergo most of the reactions we expect of amines and of acids. We may carry out methyl or ethyl esterification of the carboxylic acid group by suspending an amino acid in the appropriate anhydrous alcohol and passing in hydrogen chloride until the mixture is saturated. Evaporation of the mixture at room temperature or below leaves a residue of the hydrochloride of the amino acid ester (Eq. 21.2). Like other

$$H_3\overset{\oplus}{N}-CH(R)-COO^{\ominus} \xrightarrow[HCl]{\text{excess EtOH}} H_3\overset{\oplus}{N}-CH(R)-COOEt + H_2O + Cl^{\ominus} \tag{21.2}$$

amine hydrochlorides, the hydrochloride of an amino acid ester is soluble in water. Neutralization of this solution with one equivalent of base (at 0°) gives the free amino acid ester, which must be extracted immediately by ether. These esters, even when pure, are unstable and decompose slowly (rapidly upon heating) by aminolysis to give a diketopiperazine (Eq. 21.3; see also Eq. 20.34).

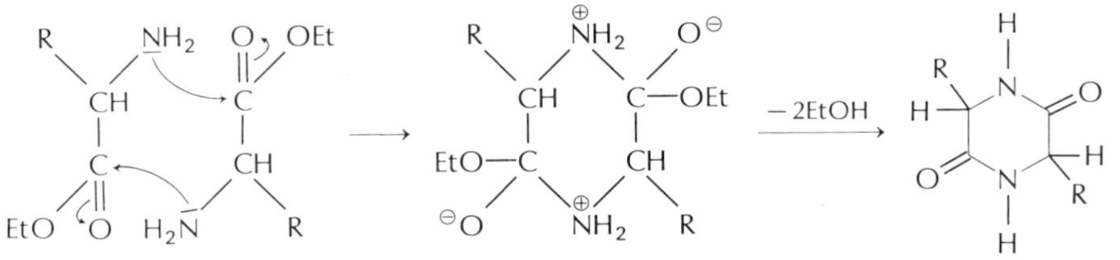

$$(21.3)$$

We see that attempts to carry out a reaction at one site or the other must take into account the possibilities of interaction with the other group. Attempts to prepare an acid chloride of an amino acid are doomed to fail because of interference by the amino group (Eq. 21.4). However, if the amino group is protected by amide formation (Eq. 21.5), which reduces the basicity and reactivity of the group, reactions of the carboxylic acid group may be carried out that otherwise would not be possible. An azlactone (to be discussed later), rather than the acid chloride a, is the real intermediate. In principle, it also is possible

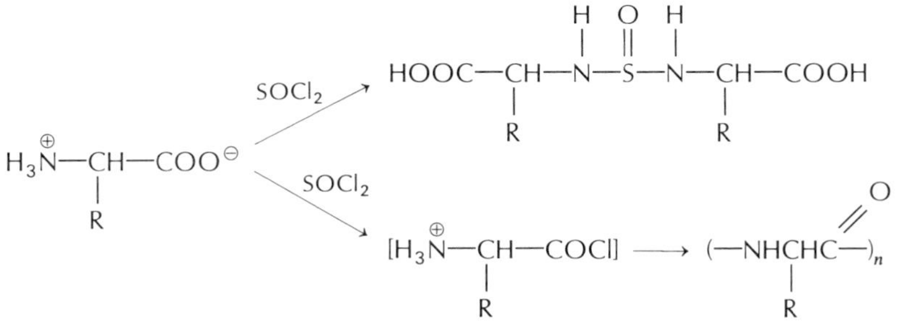

$$(21.4)$$

$$\underset{R}{\overset{\oplus}{H_3N}-CH-COO^{\ominus}} \xrightarrow[\text{dil. NaOH}]{\phi COCl} \underset{R}{\phi CONH-CH-COO^{\ominus}} \xrightarrow[\text{2. SOCl}_2]{\text{1. H}^{\oplus}}$$

$$\begin{bmatrix} \phi CONH-CH-COCl \\ | \\ R \end{bmatrix} \xrightarrow[\substack{\overset{\oplus}{H_3N}-CHCOO^{\ominus} \\ | \\ R' \\ + NaHCO_3}]{} \underset{R}{\phi CONH-CH-CONH-CHCOO^{\ominus}} \underset{R'}{|} \quad (21.5)$$

to protect the acid group in order to carry out specific reactions on the amino group. In practice, it usually is the amino group that is protected. We shall encounter several such reactions when we describe peptide and protein analysis and synthesis. □ □

Syntheses. The synthesis of amino acids has been studied extensively, and many methods have been devised. Some of the reactions used for the synthesis of amines have been modified and used for the preparation of amino acids.

The formerly difficult task of separating an amino acid from a reaction mixture has been simplified by the use of ion exchange resins. All the common amino acids (and many less common ones) and derivatives are available in racemic, D-, and L-forms from biochemical supply houses. Thus, laboratory syntheses these days are employed only for the preparation of unusual amino acids.

Many syntheses require the use of an α-bromo acid. Equation 21.6 shows a synthesis

$$\overset{\oplus}{H_3N} + \underset{COO^{\ominus}}{\overset{CH_3}{\underset{}{\big\backslash}}}CH\overset{\frown}{-}Br \longrightarrow \underset{COO^{\ominus}}{\overset{CH_3}{\underset{}{\big/}}}\overset{\oplus}{H_3N}-CH + Br^{\ominus} \quad (21.6)$$

of alanine. Both glycine and alanine may be prepared as shown, but the simple method fails for most of the other amino acids. The Gabriel (phthalimide) amine synthesis (Eq. 20.19) may be used with esters of α-halo acids to prepare amino acids. A clever variation of the Gabriel synthesis is the <u>phthalimidomalonic ester synthesis</u> (Eq. 21.7). The easily brominated malonic ester is allowed to react with potassium phthalimide. The substituted malonic ester *a* may be alkylated with an alkyl halide of choice (benzyl chloride is shown). The phthalimide group is broken up by formation of the cyclic hydrazide, and the amino alkyl malonic ester *b* is hydrolyzed and decarboxylated to give phenylalanine *c*.

A reaction sequence that employs more gentle conditions is the <u>acetamidomalonic ester synthesis</u> (Eq. 21.8). Nitrous acid (or an alkyl nitrite) reacts readily with malonic ester to give diethyl oximinomalonate *a*. The oximino group is easily reduced by catalytic hydrogenation to give the unstable aminomalonic ester *b*. (The ester groups are more resistant to hydrogenation.) Acetylation of the aminomalonic ester gives *c*, the stable acetamidomalonic ester. Saponification of the ester, followed by heating with acid, hydrolyzes the amide and causes decarboxylation to give glycine (not shown). The remaining α hydrogen can be removed by sodium ethoxide to give the acetamidomalonic ester anion *d*. The anion may be alkylated by a variety of alkyl halides, saponified, and decarboxylated to give amino acids *e* of the $\overset{\oplus}{H_3N}-CHR-COO^{\ominus}$ type. The anion also may be added to the carbonyl group of formaldehyde to give serine f. The anion also adds to methyl acrylate (a Michael addition) to give glutamic acid g. Cyanoacetic ester may be used instead of malonic ester.

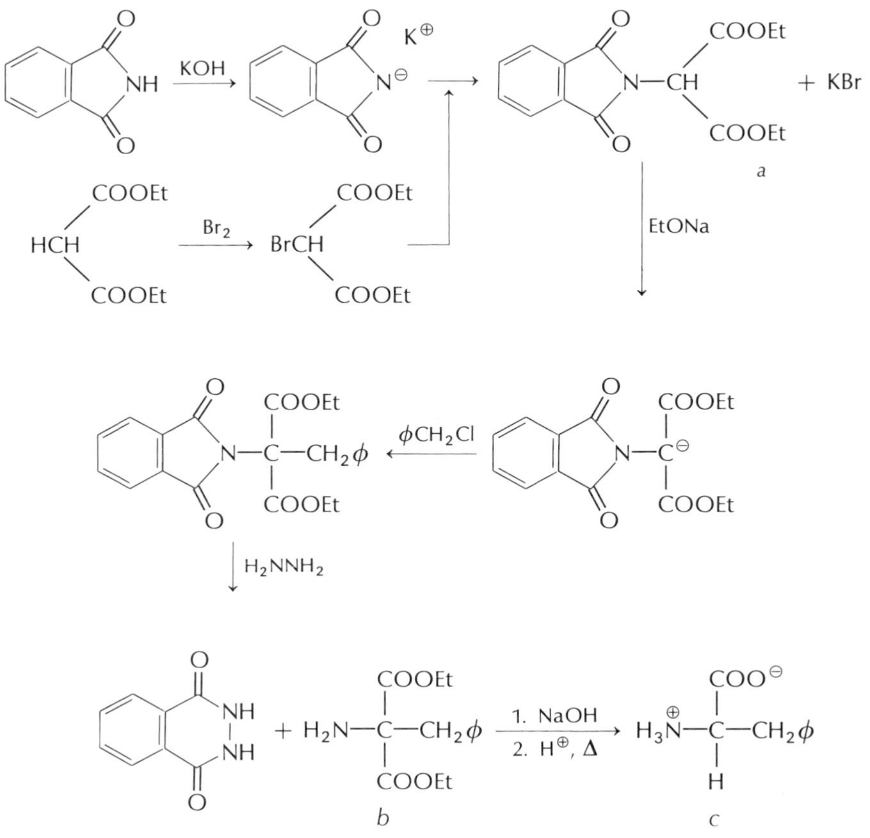

$$(21.7)$$

The azlactone synthesis begins with the condensation of an aldehyde with hippuric acid, N-benzoylglycine (or with N-acetylglycine; see Eq. 21.9). Phenylalanine, tyrosine, and thyroxine have been prepared by this method.

The Strecker synthesis (Eq. 20.50) exists in many modifications and also begins with an aldehyde (Eq. 21.10). It differs from the azlactone synthesis (in which two carbon atoms are added to the aldehyde group) by the addition of only a single carbon atom. The Strecker synthesis also may be regarded as a variant of cyanohydrin formation (see Sec. 16.9). The variations of the Strecker synthesis consist largely of the use of reagents other than NH_3 and HCN. Thus, the Bucherer modification employs NH_4CN and $(NH_4)_2CO_3$.

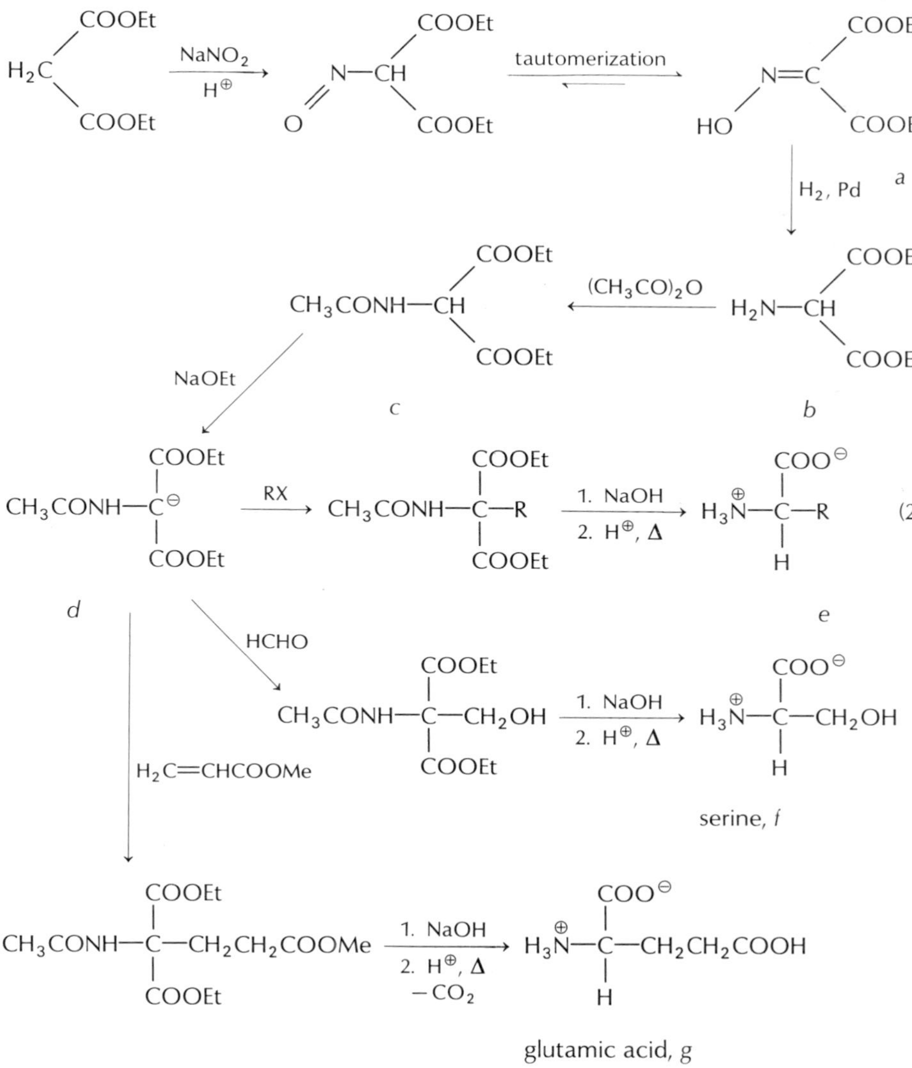

21.2 Amino acids

$$\underset{H_2C}{\overset{COOEt}{\diagdown}}\overset{COOEt}{\diagup} \xrightarrow{\underset{H^\oplus}{NaNO_2}} \underset{O}{\overset{COOEt}{N=CH}}\overset{COOEt}{\diagup} \xrightleftharpoons{\text{tautomerization}} \underset{HO}{\overset{COOEt}{N=C}}\overset{COOEt}{\diagup}$$

$$\downarrow H_2, Pd \quad a$$

$$CH_3CONH-\underset{COOEt}{\overset{COOEt}{CH}} \xleftarrow{(CH_3CO)_2O} H_2N-\underset{COOEt}{\overset{COOEt}{CH}}$$

$$\qquad\qquad\qquad c \qquad\qquad\qquad\qquad b$$

$$\downarrow NaOEt$$

$$CH_3CONH-\underset{COOEt}{\overset{COOEt}{C^\ominus}} \xrightarrow{RX} CH_3CONH-\underset{COOEt}{\overset{COOEt}{C-R}} \xrightarrow[2.\ H^\oplus, \Delta]{1.\ NaOH} \underset{H}{\overset{COO^\ominus}{\underset{H_3\overset{\oplus}{N}}{C-R}}} \quad (21.8)$$

$$d \qquad\qquad\qquad\qquad\qquad\qquad\qquad\qquad\qquad\qquad\qquad e$$

$$\downarrow HCHO$$

$$CH_3CONH-\underset{COOEt}{\overset{COOEt}{C-CH_2OH}} \xrightarrow[2.\ H^\oplus, \Delta]{1.\ NaOH} \underset{H}{\overset{COO^\ominus}{\underset{H_3\overset{\oplus}{N}}{C-CH_2OH}}}$$

$$\qquad\qquad\qquad\qquad\qquad\qquad\qquad\qquad\qquad\qquad \text{serine, } f$$

$$\downarrow H_2C=CHCOOMe$$

$$CH_3CONH-\underset{COOEt}{\overset{COOEt}{C-CH_2CH_2COOMe}} \xrightarrow[\substack{2.\ H^\oplus, \Delta \\ -CO_2}]{1.\ NaOH} \underset{H}{\overset{COO^\ominus}{\underset{H_3\overset{\oplus}{N}}{C-CH_2CH_2COOH}}}$$

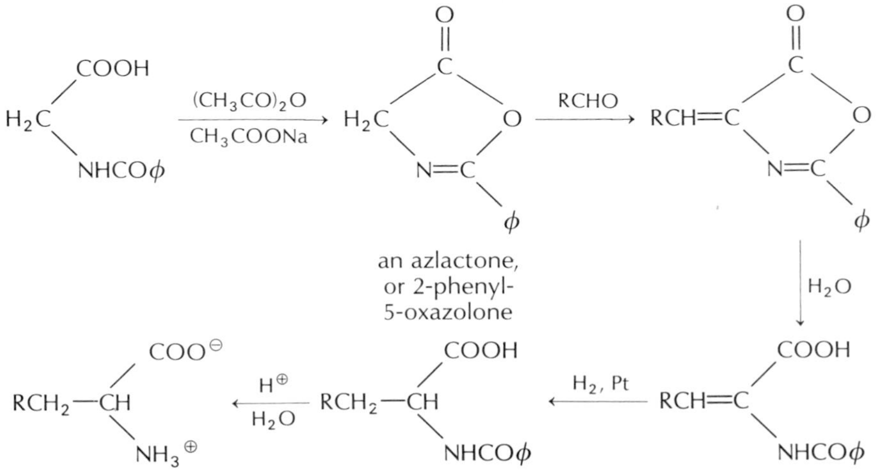

$$\qquad\qquad\qquad\qquad\qquad\qquad\qquad\qquad\qquad \text{glutamic acid, } g$$

$$\underset{H_2C}{\overset{COOH}{\diagdown}}\overset{}{\diagup}_{NHCO\phi} \xrightarrow[CH_3COONa]{(CH_3CO)_2O} \underset{N=C}{\overset{O}{\underset{\phi}{\overset{\|}{C}}\diagdown O}} H_2C\diagup \xrightarrow{RCHO} \underset{N=C}{\overset{O}{\underset{\phi}{\overset{\|}{C}}\diagdown O}} RCH=C\diagup$$

an azlactone,
or 2-phenyl-
5-oxazolone

$$\downarrow H_2O \quad (21.9)$$

$$\underset{NH_3^\oplus}{\overset{COO^\ominus}{RCH_2-CH}} \xleftarrow[H_2O]{H^\oplus} \underset{NHCO\phi}{\overset{COOH}{RCH_2-CH}} \xleftarrow{H_2, Pt} \underset{NHCO\phi}{\overset{COOH}{RCH=C}}$$

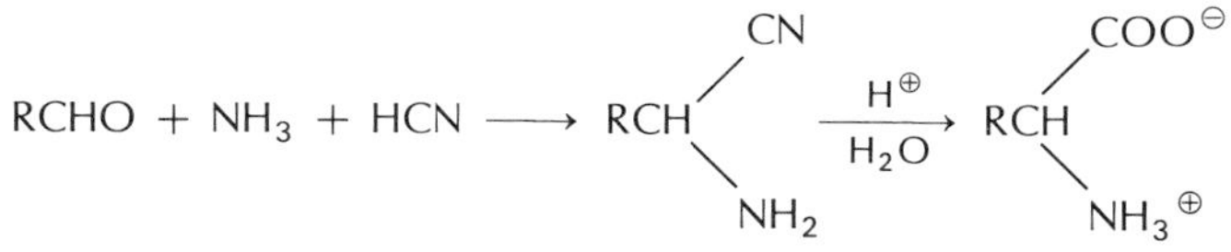

$$RCHO + NH_3 + HCN \longrightarrow RCH\begin{array}{c}CN\\ \diagdown\\ NH_2\end{array} \xrightarrow[H_2O]{H^{\oplus}} RCH\begin{array}{c}COO^{\ominus}\\ \diagdown\\ NH_3^{\oplus}\end{array} \qquad (21.10)$$

Resolution of synthetic racemic amino acids. Synthesis of amino acids in the laboratory always results in a racemic mixture of the enantiomeric forms. Hydrolysis of the naturally occurring proteins, however, gives a mixture of amino acids that are all L or (S) in configuration, unless the hydrolytic conditions are severe enough (basic hydrolysis) to cause partial racemization. Thus, it has become the usual practice to hydrolyze proteins with 6 N hydrochloric acid at 105° for 24 hr.

The resolution of synthetic racemic amino acids is best done by the action of the enzyme pig kidney acylase, which hydrolyzes the L form of an N-acetylamino acid without affecting the D form at all (Eq. 21.11). After hydrolysis, the L-amino acid and the D-N-acetyl-amino acid are easily separated by extraction with ether. Evaporation of the ether leaves

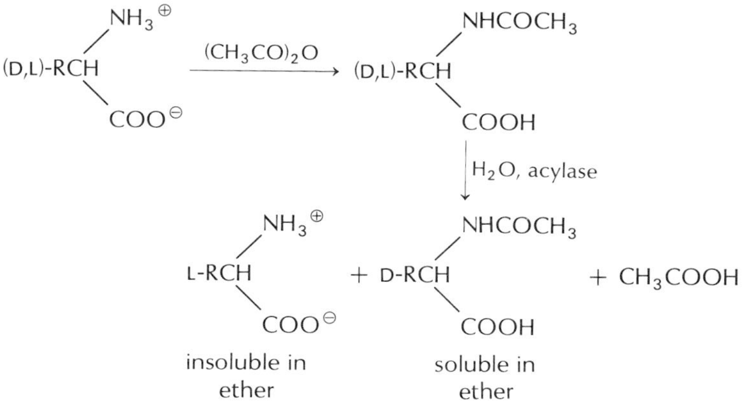

$$(D,L)\text{-}RCH\begin{array}{c}NH_3^{\oplus}\\ \diagdown\\ COO^{\ominus}\end{array} \xrightarrow{(CH_3CO)_2O} (D,L)\text{-}RCH\begin{array}{c}NHCOCH_3\\ \diagdown\\ COOH\end{array}$$

$$\downarrow H_2O, \text{ acylase} \qquad (21.11)$$

$$L\text{-}RCH\begin{array}{c}NH_3^{\oplus}\\ \diagdown\\ COO^{\ominus}\end{array} + D\text{-}RCH\begin{array}{c}NHCOCH_3\\ \diagdown\\ COOH\end{array} + CH_3COOH$$

insoluble in ether   soluble in ether

a residue that, after hydrolysis by 6 N hydrochloric acid, yields the D-amino acid. This resolution, which depends upon differences in rates of reaction, is known as a <u>*kinetic resolution.*</u> Chemical resolution, which is less clean cut, usually is performed by fractional crystallization of the diastereomeric salts formed by the reaction of the racemic (D,L)-N-acylamino acids with asymmetric (optically active) bases such as strychnine, brucine, or cinchonine (Sec. 28.7).

Exercises

21.1 Show why the isoelectric point of an amino acid is equal to the average of the two pK_a values.

21.2 Threonine is so named because of the presence of a *threo* configuration. Draw a perspective structural formula of threonine.

21.3 Give a satisfactory synthesis of:
 a. Sarcosine   c. L-threonine
 b. L-leucine   d. D-glutamic acid

21.3 Peptides

Properties and representation. At one extreme, we may regard peptides as fragments of a protein formed by hydrolysis. At the other, we may see them as low molecular weight polyamino acids, such as oxytocin, a nonapeptide pituitary hormone. Because they have peptide (substituted amide) linkages, simple peptides have been extensively studied from both degradative and synthetic approaches. Thus peptides may provide a means of understanding the polypeptides and the more intractable, high molecular weight proteins.

The isolation and purification of peptides present the same difficulties as do the same processes for amino acids, but they are complicated by the effects of increased molecular weight, including decreased solubilities. Electrophoresis, countercurrent distribution, and column chromatography on ion exchange resins are some of the methods used for purification and for demonstration of homogeneity. Purification is complicated by the solubility behavior of peptides. Because a peptide has a free amino group at one end of the chain and a free carboxylic acid group at the other end, dipolar ion formation occurs. The isoelectric point of a peptide thus is close to pH = 6 unless basic amino acids (Arg, Lys) or acidic amino acids (Asp, Glu) are present in the peptide. If both basic and acidic amino acids are present, the solubility characteristics of a peptide become complex indeed.

A labor-saving device for writing the structures of peptides (and of proteins—much of our discussion of peptides is applicable to proteins as well) is the internationally accepted usage of abbreviations for the amino acids (Table 21.1). The free amino group is always considered to be at the left of the abbreviation, the carboxylic acid group at the right. Horizontal lines represent replacement of hydrogen in the amino group or of —OH in the carboxylic acid group. The end groups usually are omitted. Vertical lines represent substitution of the obvious group in the side chain. A peptide chain is numbered from left to right, each amino acid group being counted as a unit.

An example will help us to see how the scheme works. Thus, a is the side-chain ethyl

$$H_3\overset{\oplus}{N}-CH-C\underset{CH_3}{\overset{O}{\parallel}}\diagdown NH-CH-C\underset{\underset{\underset{COOEt}{|}}{\underset{CH_2}{|}}}{\overset{O}{\parallel}}\diagdown NH-CH-C\underset{\underset{CH(CH_3)_2}{|}}{\overset{O}{\parallel}}\diagdown O^{\ominus}$$

a

ester of the tripeptide alanylglutamylleucine, and it is written in abbreviated form as:

$$\underset{\underset{OEt}{|}}{\overset{1\ \ \ \ \ \ 2\ \ \ \ \ \ 3}{Ala-Glu-Leu}}$$

The dipolar structure for the end groups is implicit if not shown. Certain substituents also have abbreviations, which will be shown as we go along.

Analysis: difficulties. The smaller peptides give meaningful results upon quantitative analysis for C, H, O, N, and S (when present). For example, the tripeptide Ala-Glu-Leu is $C_{14}H_{25}O_6N_3$, which is 50.75% C, 7.60% H, 28.97% O, and 12.68% N; but Ala-Asp-Leu is $C_{13}H_{23}O_6N_3$ (or CH_2 smaller) and is 49.20% C, 7.31% H, 30.25% O, and 13.24% N. Analyses are considered to be good if their reproducibility is within ±0.1 to 0.2%. Thus the two peptides differ enough in composition to be easily distinguished on the basis of C analysis. However, as the number of amino acids in a peptide increases, the molecular weight becomes so large that all analytical differences fall into the range of error of the analyses. With improvement in methods of amino acid analysis it became possible (and easier) to express peptide analysis in terms of the component amino acids formed by complete hydrolysis. Thus, ACTH (adrenocorticotropic hormone) is a polypeptide consisting of thirty-nine amino acid units. Of the fifteen different component amino acids, Gln, His, Leu, and Met occur only once each, and the molecular weight and amino acid composition are relatively easily determined.

After the composition of a peptide is found, the next question is this one: what is the order of appearance of the component amino acids in the peptide chain? The possible arrangements increase enormously as the number of component amino acids increases. Let us look at the possible tripeptides made up of the three amino acids A, B, and C, each being present once. The possibilities are six: ABC, ACB, BAC, BCA, CAB, and CBA. If there are n separate amino acids, each occurring once only in a peptide, then $n!$ (factorial n) isomeric peptide chains are possible. We have seen that for $n = 3$, there are $1 \times 2 \times 3 = 6$ isomeric tripeptides. For $n = 4$, there are 24 tetrapeptide isomers; for $n = 5$, 120 isomers; and for $n = 6$, 720 isomers. If each of the 20 amino acids occurred once only in each possible chain of all 20 amino acids, the number of isomers would be 2,432,902,008,176,640,000, or about 2.43 million million million!

To determine the sequence in which the amino acids appear in a peptide, we need a selective method for breaking off the <u>N-terminal residue</u> (the free amino end) or the <u>C-terminal residue</u> (the free acid group end).

End-group analysis. Analysis for identification of end groups is more satisfactory with chemical reagents for N-terminal groups than with those for C-terminal groups.

If a peptide is allowed to react with 1-fluoro-2,4-dinitrobenzene (FDNB) in aqueous ethanol in the presence of sodium bicarbonate, the terminal amino group of the peptide reacts to displace fluorine in an aromatic nucleophilic substitution (Sec. 15.5). Any other free amino groups present react also. Acidic hydrolysis of the peptide into the component amino acids gives a mixture from which the yellow 2,4-dinitrophenylamino acid may be isolated by extraction into ether, identified, and quantitatively estimated (Eq. 21.12). The method is very sensitive, and less than 10^{-6} mole of peptide is required. Unfortunately, the method is useful only for determination of the N-terminal residue.

$$\overset{\oplus}{H_3N}-CH(R)-CONH-CH(R')\mathbf{\sim}COO^{\ominus} \xrightarrow[NaHCO_3]{EtOH, H_2O} \text{ and } \underset{O_2N}{\text{2,4-(O_2N)_2C_6H_3F}}$$

$$\longrightarrow (O_2N)_2C_6H_3-NH-CH(R)-CONHCH(R')\mathbf{\sim}COO^{\ominus} \quad (21.12)$$

$$\xrightarrow{\text{dil. HCl}} (O_2N)_2C_6H_3-NHCH(R)COOH \text{ (soluble in ether)} + \overset{\oplus}{H_3N}-CH(R')-COOH + \text{etc. (insoluble in ether)}$$

$$\xrightarrow{\text{ether extraction}}$$

21.3 Peptides

A more useful method is the <u>Edman degradation,</u> which does not require a complete hydrolysis of the peptide. It makes possible the removal of amino acids one at a time from the N-terminal end of a peptide. The reaction utilizes phenylisothiocyanate, which reacts with a free amino group by addition at pH 9 to give a substituted thiourea (Eq. 21.13). In the

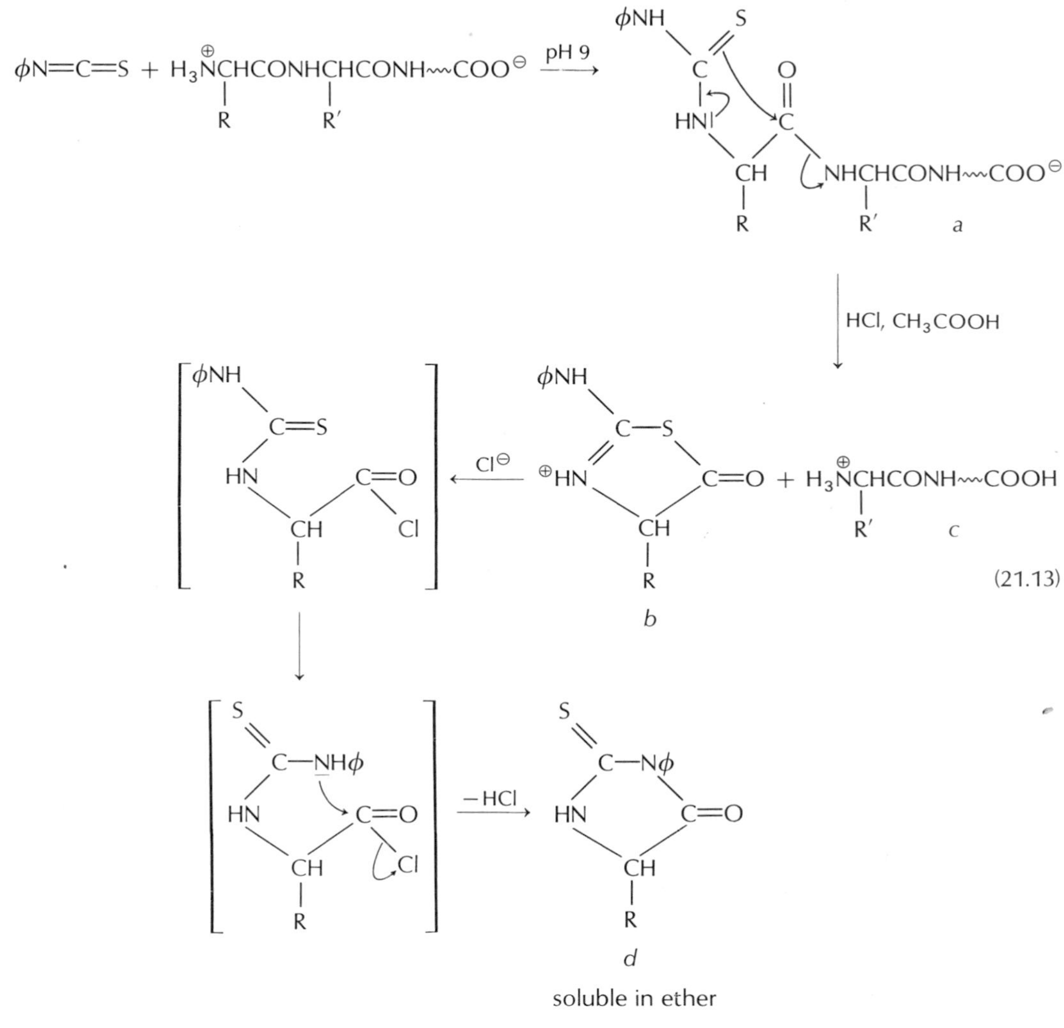

$$\phi N=C=S + H_3\overset{\oplus}{N}CHCONHCHCONH\sim\sim COO^{\ominus} \xrightarrow{pH\ 9}$$

(structures a, b, c, d of reaction scheme 21.13)

(21.13)

soluble in ether

presence of anhydrous HCl in nitromethane or acetic acid, the thiourea (a) undergoes a cyclization with rupture of the adjacent peptide bond to give a substituted thiazolone (b) and the peptide chain (c) of the remaining amino acids. Repetition of the reaction with c allows the new N-terminal amino acid (which was the second one in the original chain) to be identified, and so on. A practical limit is about the first eight to twelve amino acids, counting from the N-terminal end of the original polypeptide. (This limit is caused by accumulation of impurities.) Under the anhydrous acidic conditions used, the thiazolone (b) rearranges to a substituted thiohydantoin (a substituted imidazole, d), which is the product actually isolated by extraction into ether and identified. Each amino acid gives rise to a different substituted thiohydantoin, each of which is easily identified.

The C-terminal amino acid may be found by hydrazinolysis of a sample of the peptide

(12 hr or more, 100°, anhydrous hydrazine). Only the C-terminal amino acid is obtained as such: all the peptide bonds are converted into hydrazides (Eq. 21.14).

$$\overset{\oplus}{H_3N}CHCO\sim\sim NHCHCOO^{\ominus} \xrightarrow[100°]{H_2NHNH_2} H_2NCHCONHNH_2$$
$$\quad\; |\qquad\qquad\quad\; |\qquad\qquad\qquad\qquad\qquad\quad\; |$$
$$\quad\; R\qquad\qquad\quad R'\qquad\qquad\qquad\qquad\qquad\; R$$

$$+ \text{ hydrazides of other amino acids} \qquad (21.14)$$

$$+\; \overset{\oplus}{H_3N}CHCOO^{\ominus}$$
$$\qquad\quad |$$
$$\qquad\quad R'$$

Chain cleavage. Fortunately, it has been possible to find and purify enzymes that can hydrolyze certain peptide bonds without attacking others. Some other enzymes that are not quite so specific are used also. Those that catalyze hydrolysis of peptide bonds are known as proteolytic enzymes and are classified as endopeptidases (attack nonterminal peptide bonds) and exopeptidases (attack terminal peptide bonds).

Among the exopeptidases, we shall mention carboxypeptidase (ox pancreas) and leucine aminopeptidase (kidney). Carboxypeptidase is specific for hydrolytic cleavage of the C-terminal peptide bond. A practical difficulty is that the enzyme does not discriminate: the new peptide chain with one less residue is attacked as well as the original chain; so is the $n-2$ chain, and so on. Also, not all terminal bonds are hydrolyzed at the same rate: in fact, some may appear to be inert. Leucine aminopeptidase is specific for cleavage of the N-terminal peptide bond (unless the N-terminal amino acid is proline).

Among the endopeptidases, we mention three. Papain, from papaya latex, has little specificity and attacks peptide chains more or less at random. It is useful for breaking down peptide chains to a mixture of di-, tri-, and tetrapeptides. Chymotrypsin (pancreas) attacks peptide bonds in which the carboxylic acid component is Phe, Try, or Tyr most readily. More slowly, it also attacks the carboxylic peptide bonds of Asn, Glu, Leu, and Met. Trypsin (pancreas) is the most specific of the three enzymes mentioned. It catalyzes hydrolysis of only the carboxylic peptide bonds of Arg and Lys.

Chemical reagents have been found that are specific for certain peptide bonds: N-bromosuccinimide for Try and Tyr, and BrCN for Met are examples. Details of the reactions may be found in the references given in the bibliography.

Structural formulas from analytical results. From the analytical results, it is possible to deduce the sequence of amino acids in the original peptide. The process is similar to that used in putting together a jigsaw puzzle.

Let us see how a certain heptapeptide sequence would be worked out. Amino acid analysis of our hypothetical peptide gives one mole each of Glu, Gly, Leu, Lys, Phe, Ser, and Try. FDNB shows Phe to be the N-terminal amino acid, and the use of carboxypeptidase shows Leu to be the C-terminal amino acid, with Lys (formed more slowly) likely to be the adjacent amino acid. Trypsin hydrolysis gives Leu and a hexapeptide, which with carboxypeptidase confirms that Lys is the new end group. Summarizing to this point, we can write the sequence as shown.

Phe-(Glu, Gly, Ser, Try)-Lys-Leu

Chymotrypsin hydrolysis yields Phe (as expected) and two tripeptides as the major products. One of the tripeptides contains Glu, Leu, Lys, and we can write the sequence as:

Phe-(Gly, Ser, Try)-Glu-Lys-Leu

Amino acid analysis shows that the second tripeptide contains Gly, Ser, Try, as expected. FDNB shows the N-terminal unit to be Gly, and carboxypeptidase shows Try to be the C-terminal (as expected from the chymotrypsin hydrolysis). Therefore, the tripeptide must be Gly-Ser-Try. Because we know that the other tripeptide has to be on the right and that Phe has to be on the left, the heptapeptide can only be:

Phe-Gly-Ser-Try-Glu-Lys-Leu

Our example has been relatively simple, but it illustrates the time and labor involved in the deduction of the sequence.

Syntheses. A chemist does not consider a structure to be proved until the compound has been synthesized by rational steps: that is, by understood reactions from known reactants. The synthesis of peptides is particularly thorny because of the need to protect all other amino and carboxylic acid groups from reaction at the time when a specific peptide bond formation is planned to occur. Then the protecting groups have to be removed, leaving the newly formed peptide bond unaffected. Also, at times it may be necessary to activate a group to cause reaction to occur under milder conditions than usual.

Let us be more specific and look at the synthesis of a simple dipeptide, Phe-Gly. Of the methods for formation of a peptide bond, $RCOCl + H_2NR''$, $(RCO)_2O + H_2NR''$, or $RCOOEt + H_2NR''$ all seem possible. The trouble is that the acid chloride, anhydride, or ester will react with the amino group of the same molecule. As we showed in Equations 21.3, 21.4, and 21.5, cyclization or polymerization results unless the amino group is protected. Thus, we might try the method shown in Equation 21.15. In step a, the amino group

$$CH_3COCl + H_3\overset{\oplus}{N}CHCOO^{\ominus} \xrightarrow[\substack{-H^{\oplus} \\ -HCl}]{\text{base}} CH_3CONHCHCOO^{\ominus}$$
$$\quad\quad\quad\quad\quad |\quad\quad\quad\quad\quad\quad\quad\quad\quad\quad\quad\quad |$$
$$\quad\quad\quad\quad\quad \phi CH_2 \quad\quad\quad\quad\quad\quad\quad\quad\quad\quad\phi CH_2$$

(21.15)

step a → step b (1. $H^{\oplus}$, 2. $SOCl_2$) → [$CH_3CONHCHCOCl$, ϕCH_2] → step c ($H_3\overset{\oplus}{N}CH_2COO^{\ominus}$ / $NaHCO_3$) → $CH_3CONHCHC(=O)$—$NHCH_2COO^{\ominus}$ with ϕCH_2

step d ($H^{\oplus}$, H_2O) → $CH_3COOH + H_3\overset{\oplus}{N}CHC(=O)$—$NHCH_2COOH$ with ϕCH_2 $\quad$ + $CH_3CONHCHCOOH + H_3\overset{\oplus}{N}CH_2COOH$ with ϕCH_2

of Phe is protected by acetylation because it is to be the N-terminal or unreacted group. In step b, the carboxylic acid of acetylated Phe is converted into the more reactive acid chloride (azlactone) so that step c, the formation of the peptide bond with Gly, will proceed smoothly at room temperature or lower. The acetylated dipeptide may be subjected to step b and allowed to react with a third amino acid, say Val, to give CH$_3$CO-Phe-Gly-Val. By continued repetition of steps b and c, the peptide chain may be extended indefinitely. A practical limit is encountered in the purification of the increasingly more insoluble acetylated peptide chain. The procedure we have given builds the peptide from the N-terminal end. Later we shall see a procedure for building the chain from the C-terminal end.

Step d is the downfall of the use of the acetyl group to protect the N-terminal group. The hydrolysis of the substituted acetamide is no more probable than is the hydrolysis of the newly formed peptide bond! Thus, no more than a 50% yield of Phe-Gly may be expected. The situation becomes even worse in a polypeptide chain, which may be broken initially at any peptide bond. Thus, a major problem was to find a protecting group that could be removed later without effect on the peptide bonds. Several alternative protecting groups have been found.

The first satisfactory N-terminal protecting group to be devised was the carbobenzoxy (or benzyloxycarbonyl) group. The carbobenzoxy group is introduced by reaction of the amino acid with benzyloxycarbonyl chloride, which is prepared from phosgene (carbonyl chloride) and benzyl alcohol (Eq. 21.16). The acylated amino group allows

$$\phi CH_2OH + COCl_2 \longrightarrow \phi CH_2OCCl \overset{O}{\underset{}{\|}}$$

$$\downarrow H_3\overset{\oplus}{N}CHCOO^{\ominus}, NaHCO_3$$
$$ | R$$

$$\phi CH_2OC\overset{\nearrow O}{\underset{\searrow}{}} NHCHCOO^{\ominus}$$
$$ | R$$

(21.16)

steps b and c of Equation 21.15 to be carried out and repeated as we described earlier. The removal of the protecting group was done originally by hydrogenolysis, a reductive reaction peculiar to benzyl groups, particularly benzyl esters (Eq. 21.17). The intermediate product is unstable and loses CO$_2$ to give the peptide. Later on, it was found that the benzyl ester could be decomposed by acidic hydrolysis under conditions that did not affect the peptide bonds: 2 N HBr in an organic solvent such as nitromethane or acetic acid. If the benzyl group is substituted, variations in reactivity and physical properties may be achieved. A p-nitro group improves crystallinity of the protected peptides but causes resistance to hydrolysis. A p-methoxy group increases susceptibility to hydrolytic removal.

$$\phi CH_2OC\overset{\nearrow O}{\underset{\searrow NH\sim COOH}{}} \xrightarrow[Na, liq. NH_3]{H_2, Pd \text{ or}} \phi CH_3 + \left[HOC\overset{\nearrow O}{\underset{\searrow NH\sim COOH}{}} \right]$$

$$\downarrow -CO_2$$

$$ H_3\overset{\oplus}{N}\sim COO^{\ominus}$$

(21.17)

A *p*-phenylazo group introduces color so that chromatography is easier to carry out than it would be with the colorless benzyloxy derivatives.

A helpful development has been the use of the *tert*-butoxycarbonyl group, which is inert to hydrogenolysis but is even more susceptible to acid hydrolysis than is the carbobenzoxy group. Equation 21.18 shows the preparation of the reagent, *tert*-butoxycarbonyl azide (a), and the conditions for use and removal of the group.

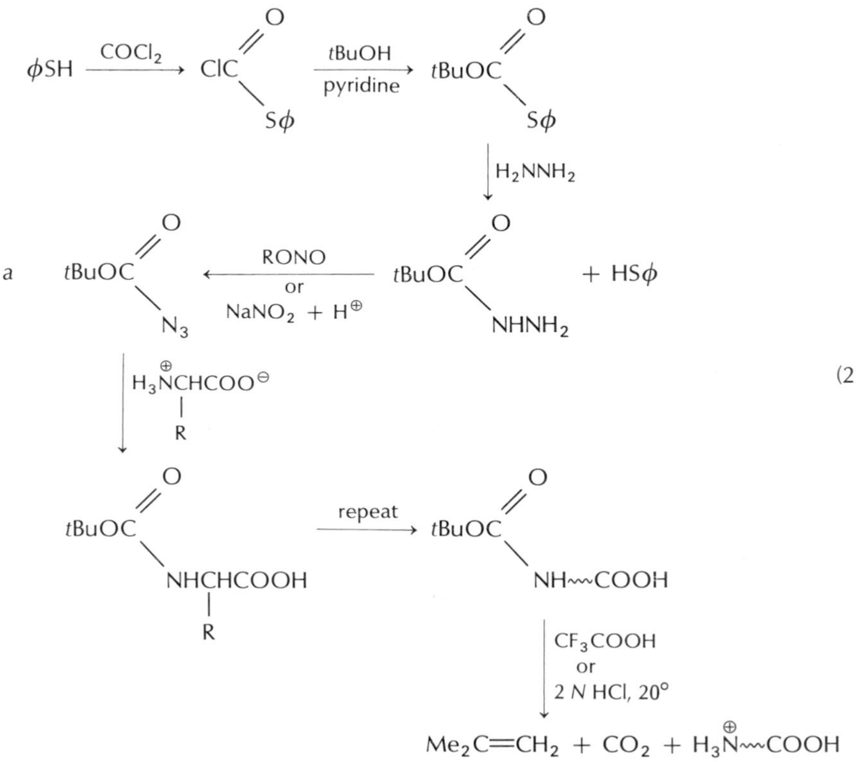

$$\phi SH \xrightarrow{COCl_2} ClC(=O)S\phi \xrightarrow[\text{pyridine}]{tBuOH} tBuOC(=O)S\phi$$

$$\downarrow H_2NNH_2$$

a $\quad tBuOC(=O)N_3 \xleftarrow[\text{NaNO}_2 + H^\oplus]{RONO \text{ or}} tBuOC(=O)NHNH_2 + HS\phi$

$$\downarrow \overset{\oplus}{H_3N}CHCOO^\ominus \text{ with } R$$

$$tBuOC(=O)NHCHCOOH \text{ with } R \xrightarrow{\text{repeat}} tBuOC(=O)NH\sim COOH$$

$$\downarrow \begin{array}{c} CF_3COOH \\ \text{or} \\ 2\ N\ HCl,\ 20° \end{array}$$

$$Me_2C=CH_2 + CO_2 + \overset{\oplus}{H_3N}\sim COOH$$

(21.18)

Other protecting groups of less general use are the toluenesulfonyl, trityl (triphenylmethyl), formamidyl, trifluoracetyl, and phthaloyl groups.

In Equation 21.15, we were shown the preparation (b) of an acid chloride for use in the formation of c, a peptide bond with an amino acid. Acid chlorides of α-amido acids are subject to racemization and to decomposition at room temperature (Eqs. 21.19). Thus, other means of activating the carboxylic acid group have been developed. Acid azides replaced acid chlorides very early (Eqs. 21.18, 21.20) because the acid azides do not racemize as readily as do the acid chlorides.

The most popular method for the formation of peptide bonds is the use of dicyclohexylcarbodiimide (DCC), Sec. 17.9. The method avoids the separate activation of the carboxylic acid (step b of Eq. 21.15). The components are mixed and DCC is added at 0° (Eq. 21.21).

To this point, we have been concerned with the use of the simple amino acids. If one of the more complex amino acids is to be incorporated into a peptide, it is necessary to protect the extra functional group. Thus Asp and Glu require protection of the second carboxylic acid group, Lys of the second amino group, Arg of the guanadyl group, His of the imidazolyl group, Cys of the mercapto group, and so on. Details of this protection would take us too far afield, but they are easily available: see the bibliography.

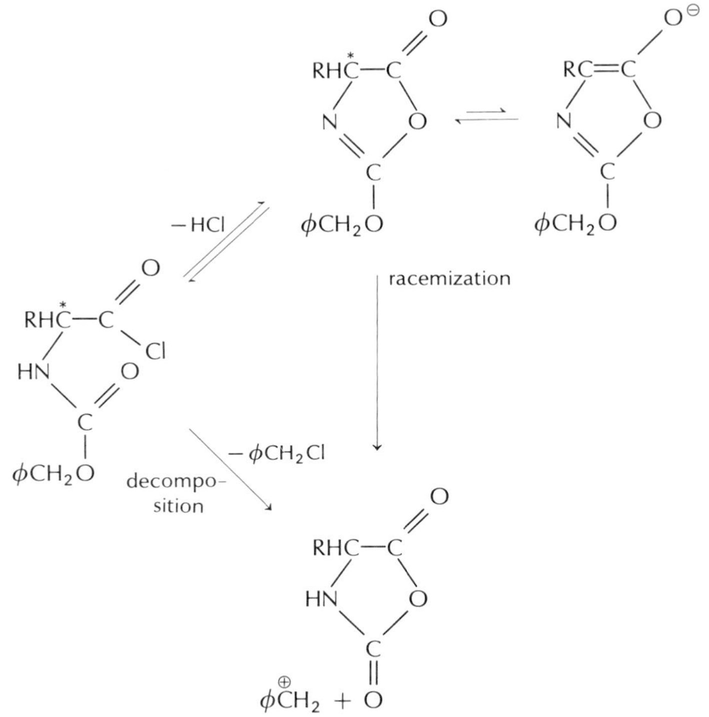

$$\begin{array}{c} \text{RHC*-C} \\ | \quad \backslash \\ \text{HN} \quad \text{O} \\ \backslash \quad / \\ \text{C} \\ | \\ \phi\text{CH}_2\text{O} \end{array} \xrightarrow[-\text{HCl}]{} \begin{array}{c} \text{RHC*-C} \\ / \quad \backslash\backslash\text{O} \\ \text{N} \quad \text{O} \\ \backslash\backslash \quad / \\ \text{C} \\ | \\ \phi\text{CH}_2\text{O} \end{array} \rightleftharpoons \begin{array}{c} \text{RC=C} \\ / \quad \backslash\text{O}^{\ominus} \\ \text{N} \quad \text{O} \\ \backslash\backslash \quad / \\ \text{C} \\ | \\ \phi\text{CH}_2\text{O} \end{array}$$

(the starting structure has Cl attached; racemization via the center form; decomposition with $-\phi\text{CH}_2\text{Cl}$ gives final oxazolone with $\phi\overset{\oplus}{\text{CH}}_2 + \text{O}^\ominus$)

(21.19)

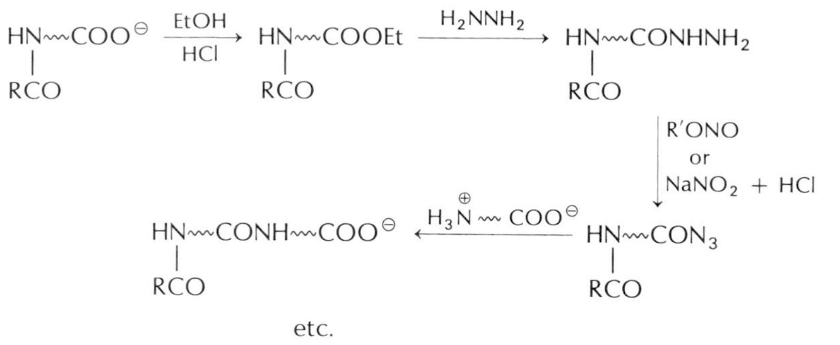

$$\begin{array}{c} \text{HN}\sim\text{COO}^{\ominus} \\ | \\ \text{RCO} \end{array} \xrightarrow[\text{HCl}]{\text{EtOH}} \begin{array}{c} \text{HN}\sim\text{COOEt} \\ | \\ \text{RCO} \end{array} \xrightarrow{\text{H}_2\text{NNH}_2} \begin{array}{c} \text{HN}\sim\text{CONHNH}_2 \\ | \\ \text{RCO} \end{array}$$

$$\downarrow \begin{array}{c} \text{R'ONO} \\ \text{or} \\ \text{NaNO}_2 + \text{HCl} \end{array}$$

$$\begin{array}{c} \text{HN}\sim\text{CONH}\sim\text{COO}^{\ominus} \\ | \\ \text{RCO} \end{array} \xleftarrow{\text{H}_3\overset{\oplus}{\text{N}}\sim\text{COO}^{\ominus}} \begin{array}{c} \text{HN}\sim\text{CON}_3 \\ | \\ \text{RCO} \end{array}$$

(21.20)

etc.

The preparation of polypeptides has been shown so far as a stepwise addition of a single amino acid at a time. A difficulty with the process is the purification of the $n + 1$ peptide formed from the n peptide. Even with moderate values of n, the properties of the two peptides are so similar that complete separation is almost impossible. A better approach is to prepare the needed di-, tri-, and tetrapeptides and then to link the smaller peptides. For example, to synthesize the hypothetical octapeptide ABCDEFGH, it is better to prepare the dipeptides AB, CD, EF, and GH, purify each, and link AB to CD and EF to GH. After purification of ABCD and EFGH, the final peptide bond may be formed; and the purification of the octapeptide from the reacting tetrapeptides is relatively easy.

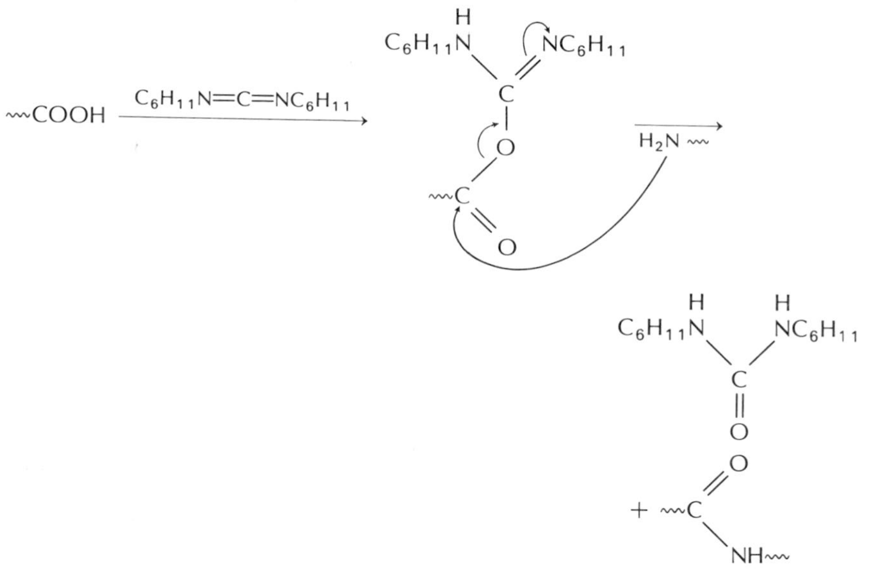

$$\sim\text{COOH} \xrightarrow{C_6H_{11}N=C=NC_6H_{11}}$$

[reaction scheme showing carbodiimide coupling: C₆H₁₁NH–C(O–C(=O)∼)=NC₆H₁₁ intermediate attacked by H₂N∼ to give C₆H₁₁NH–C(=O)–NHC₆H₁₁ (urea byproduct) + ∼C(=O)–NH∼ (amide product)]

(21.21)

Although it usually is easier to begin a peptide synthesis from the N-terminal end, a convenient method for stepwise growth of a chain from the C-terminal end (devised by Merrifield) employs a synthetic polymer to anchor the chains to a solid, easily separated, and easily washed support. Polystyrene is easily prepared in the form of globules. Those phenyl groups at the surface of the globules undergo the usual aromatic electrophilic substitution reactions. If chloromethylation (Sec. 16.3) is used, a benzyl-like ester may be formed with the C-terminal amino acid. The amino group is unblocked (a) and the next peptide bond formed (b, Eq. 21.22). Steps a and b are repeated as often as necessary. The peptide is removed from the polymeric globules by a hydrolysis (c) that ruptures both the *tert*-butyl and the benzyl-like esters. The process has been automated. A chromatographic column is packed with the polymeric beads, which remain in place, and the various reactions and purifications are done as the needed reagents are passed in solution down the column, one after the other.

Exercises

21.4   A certain nonapeptide, upon complete hydrolysis, gave a mixture of Ala, Asp, Glu, Gly, Leu, Lys, Phe, Tyr, and Val. FDBN showed the N-terminal to be Val, and carboxypeptidase showed the C-terminal to be Gly. With chymotrypsin, the nonapeptide gave a mixture of a pentapeptide (which contained Glu, Leu, Lys, Phe, and Val) and two dipeptides (one contained Asp and Tyr; the other, Ala and Gly). With trypsin, the nonapeptide gave a mixture of a tetrapeptide (Glu, Leu, Lys, and Val) and a pentapeptide (Ala, Asp, Gly, Phe, and Tyr). With this information, the primary structure may be deduced except for the position of two units. Give the structure as deduced and describe how to obtain the information needed to complete the sequence.

21.5   Show how to synthesize the tetrapeptide Ala-Gly-Phe-Val from the amino acids by a route in which the last step requires the bonding of two dipeptides. (For the purpose of this question, do not use the solid-support method).

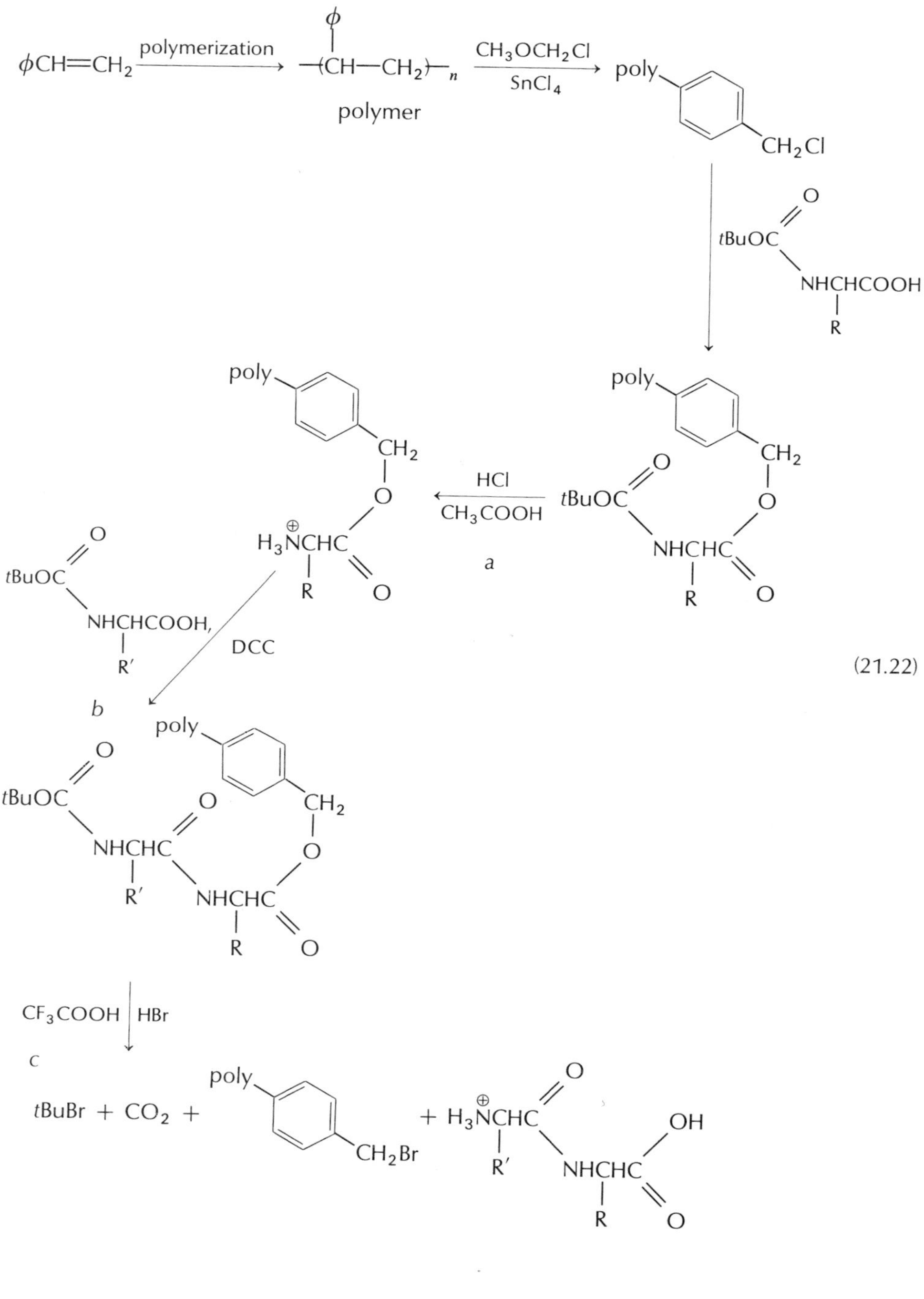

$$\phi CH=CH_2 \xrightarrow{\text{polymerization}} -(CH-CH_2)_n- \xrightarrow[SnCl_4]{CH_3OCH_2Cl} \text{poly-}\phi\text{-CH}_2Cl$$

(21.22)

21.4 Proteins

Conformations; structures.   Proteins are polymeric substances whose properties are mainly determined by the conformations that the polypeptide chains assume. If the conformation of the native state is changed, the biological activity of the protein is altered (usually lost), and the change is known as denaturation. Frying or boiling an egg is a familiar denaturation caused by heat.

The conformations of a polypeptide are classified into three groups. The so called secondary structure (which actually refers to conformation) is concerned with the level of organization caused by near neighbor interactions, including hydrogen bonding. The tertiary structure is the level of organization concerned with folding or changes in direction of the secondary structure. The quaternary structure is the organization of tertiary units into larger structures. The primary structure is the amino acid sequence of a polypeptide chain. Figure 21.2, hemoglobin, shows the distinctions. The quarternary structure consists of four separate units, two of which are identical α chains of 141 amino acid units with a prosthetic heme, and the other two of which are identical β chains of 146 units, also with a prosthetic heme. The α chains are shown in white, the β chains as shaded. The association of the four tertiary units (a) is tetrahedral. The hemes are shown as discs. A single tertiary unit (b), a β chain, shows the location of His in the chain and the association with Fe at the center of a heme. In c is shown part of the secondary helix structure, a straight section. In d, a part of the peptide sequence (primary structure) is shown. (Heme will be described in Chapter 28.)

Peptide bonds: geometry.   Peptide bonds and the constituent amino acid units possess a structure in which the atoms from the α carbon of one unit to the α carbon of the next unit all lie in single plane (Fig. 21.3). The chain in fibrous proteins (keratin, myosin, and so on) assumes a completely planar structure when the fiber is stretched—we know it is so from X-ray diffraction studies. In the unstretched state, the chains fold up somewhat into what is called a pleated structure (Fig. 21.4). The pleated structure shows clearly the folding at only the α carbon atoms and the coplanarity between them. We see also the regularity with which R groups (side chains) of the amino acid units appear above and below the chains.

Resonance in proteins.   The folding of a polypeptide chain into its active (native) three-dimensional configuration is affected by resonance. In order to gain 21 kcal of resonance stabilization energy, the six atoms at the site of the O—C—N bond must be in a plane. Another way to express the same idea is that the energy barrier to alter the planar structure is 21 kcal—large enough to suppose that these bonds are planar in solution. One of the major resonance structures that can be drawn has a C=N double bond, freezing the position of six atoms relative to one another.

Thus the backbone of the polypeptide can twist and turn only by rotation of the bonds to and from the carbon in the middle of each amino acid (C_α, Fig. 21.4).

Hydrogen bonding: secondary structures.   Hydrogen bonding is a strong influence in determining the secondary and tertiary structures of proteins. Even simple amides have strong hydrogen bonding (of the order of $\Delta H = -3$ kcal) in the crystalline state and in

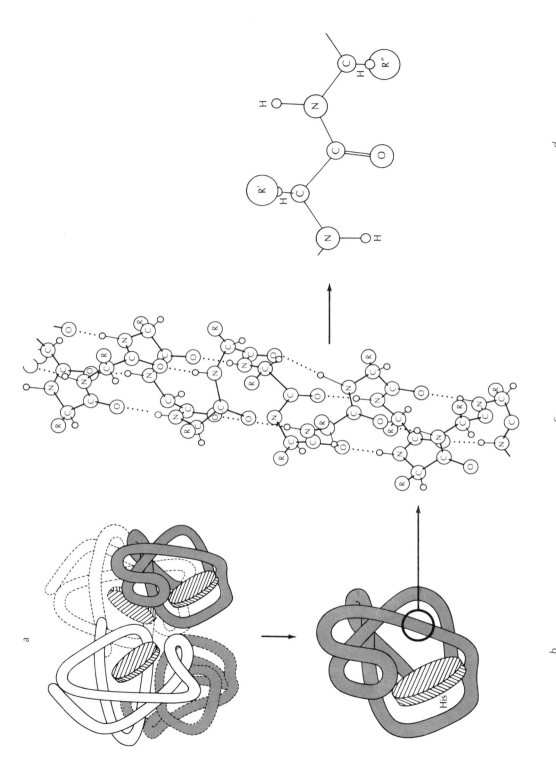

Figure 21.2 Levels of structural organization in hemoglobin: a. quaternary, showing aggregation of two α- and two β-chain subunits; b. tertiary, showing folding of β-chain helix; c. secondary, composition of helical structure; d. primary, peptide-bond structure and sequence

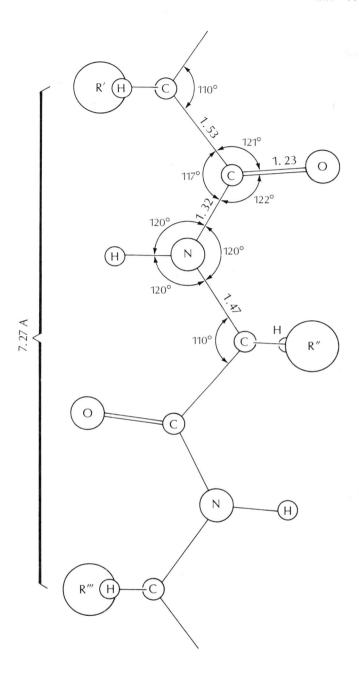

Figure 21.3 Dimensions of a fully extended polypeptide chain

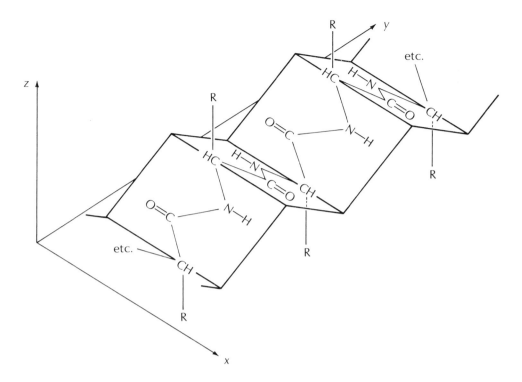

Figure 21.4   Pleated (β) polypeptide chain

aprotic solvents (Eq. 21.23). In a peptide, opportunities for hydrogen bonding exist, not only within a given chain between units, but also between units of separate chains. The system seeks the most stable situation (lowest free energy) as a compromise between the lowest enthalpy and the highest entropy. To do so, it uses nonbonded interactions and hydrogen bonding between the peptide bonds of one or more chains and a solvent (if present).

$$\underset{\underset{H_3C}{|}}{\overset{\overset{CH_3}{|}}{O=C}}\diagdown_{N-H} + \underset{\underset{CH_3}{|}}{\overset{\overset{H_3C}{|}}{N-H}}\diagdown_{O=C} \xrightleftharpoons{CCl_4} \underset{\underset{H_3C}{|}}{\overset{\overset{CH_3}{|}}{O=C}}\diagdown_{N-H\cdots O=C}\diagup^{\overset{H_3C}{|}}_{\underset{CH_3}{|}} \qquad (21.23)$$

The α-helix.   We have seen secondary structure for stretched and unstretched fibrous proteins. Other classes of proteins have much more complex X-ray diffraction patterns. A commonly occurring secondary structure is that of the α-helix. The name refers to a coil with a right handed pitch: the coil recedes from the eye by clockwise rotation when viewed along the axis. There are eighteen amino acid units $\left(-\text{NHCHRC}\overset{\overset{O}{\|}}{-}\right)$ in five turns. They take up 27.0 Å along the axis (Fig. 21.5). Thus there are 3.6 units per turn,

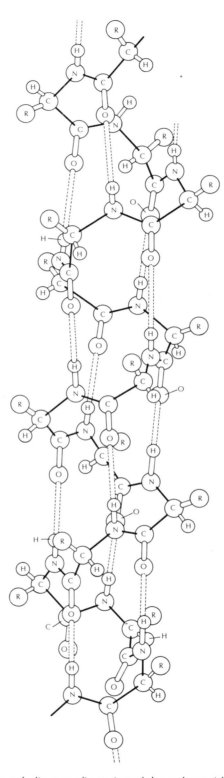

Figure 21.5 Model of an α-helix, a configuration of the polypeptide chain in proteins. The backbone of the chain consists of repeating sequences of C, C, and N. R represents side groups of the different amino acids. The broken lines represent hydrogen bonds that stabilize the helix.

with a pitch of 5.4 A: each unit occupies 100° (equal to 5/18) of a turn, equivalent to 1.5 A along the axis. The figure shows a fifteen-unit portion of an α-helix with the N-terminal amino acid above the top and the C-terminal unit below the bottom. We note that each \C=O/ is hydrogen bonded to the \NH/ of the fourth unit below it. Thus, the \C=O/ of the first unit is hydrogen bonded to the \NH/ of the fifth unit, the second to the sixth, the third to the seventh, and so on. The hydrogen bonds are skewed slightly in relation to the axis because the bonds are between every fourth unit and there are only 3.6 units per turn. Furthermore, such a helix, besides providing for the maximum amount of internal hydrogen bonding, causes every side chain (R group) to be located on the outside surface, where there is the maximum amount of room for any bulky groups.

In a given protein there may be long stretches of α-helical structure, but often a fold intervenes. This folding leads to tertiary structure.

Tertiary structures. Hydrogen bonding stabilizes the α-helix. Solution in water, with displacement of the internal hydrogen bonds by hydrogen bonding to water molecules, causes the helix to uncoil into an unoriented chain. Similarly, if there are other bonding possibilities within a chain, a folding of the helical structure occurs. Thus, other hydrogen bonds may form, such as those of acid dimerization (for example, between the free carboxylic groups of glutamic and aspartic acids), or those between the —OH of tyrosine and the imidazole ring of histidine. Salt formation, with the resulting ionic attraction, also may occur. Bond formation, as in the disulfide link of cystine, is possible. Nonbonded interactions also may contribute, as well as steric effects. Thus, the rigid unit proline is associated with folds in peptide chains, which determine tertiary structure.

The structure of myoglobin from sperm whale has been worked out. Myoglobin contains a sequence of 153 units connected to a prosthetic heme. Myoglobin is the constituent of muscle in vertebrates that accepts, stores, and releases oxygen as required. The tertiary structure has been described as containing eight sections of α-helical secondary structure with intermediate nonhelical sections of various lengths, which constitute the curved and folded portions of the chain. The tertiary structure is not much different from those of the α and β units of hemoglobin, as shown in Figure 21.2a and b.

Quaternary structures. Quaternary structure is the assembly of tertiary units into the protein molecule. Not much is yet known of the various means by which the assemblage is formed and stabilized. Some examples are shown in Figure 21.6. As we see, they range from the relatively simple hemoglobin to tubular tobacco mosaic virus. Some examples of less well defined quaternary structure are shown in Figure 21.7.

Nmr in biochemistry. Nuclear magnetic resonance studies of large biochemical molecules are hampered because of the huge number of nuclei in similar environments. The spectrum of a protein is a composite of the resonances of its hundred or more amino acids. However, the information that can be obtained from protein nmr studies is unique and so valuable that major efforts have been made to use this technique.

Some workers have concentrated on the lines distinguishable either upfield or downfield from the many others. If a line can be identified as being associated with a particular amino acid at an interesting location on the protein, then its chemical shift and line width under a variety of conditions may give information about that location. Other workers

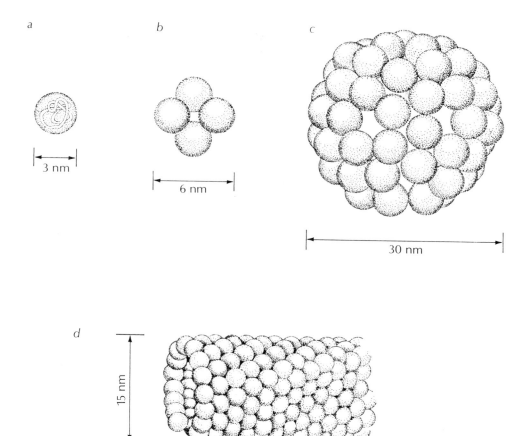

Figure 21.6 Quarternary structure of proteins: a. myoglobin has no quarternary structure; b. hemoglobin, four similar subunits in tetrahedral arrangement; c. polio virus, sixty identical subunits arranged in rings of five to form a hollow sphere, with each subunit built up of two or three smaller subunits comparable in size with a.; d. tobacco mosaic virus, a hollow tube formed by a spiral of subunits, with about sixteen units per turn

have watched the influence that a protein has on the nmr spectrum of a small molecule that is known to interact with the protein. Sometimes in these studies, advantage is taken of the fact that the ^{19}F and ^{13}C nuclei resonate at a much different frequency than does the proton. These lines can therefore be recorded in the absence of the many unwanted proton lines.

The advantage of nmr is that one can focus on one specific part of the large biomolecule and observe its behaviour in solution under conditions closely approximating the natural state. The changes observed in the spectra are interpreted to reflect such behaviour as a conformational change in the polypeptide chain, ionization of a nearby

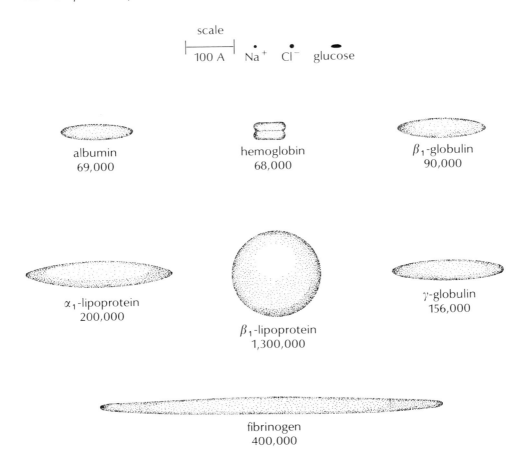

Figure 21.7 Relative dimensions of various proteins

functional group, a shift in the plane of the large side chain of tryptophan, and so forth. In cases where the three-dimensional crystal structure of the protein has been determined, interpretation of the nmr data can lead to more accurate insight to the movement of the side chains responsible for protein activity. Alternatively, it may be possible to observe changes in a small molecule as it is acted upon by an enzyme.

21.5 Hormones, antibodies, and genetic defects

Of the many functions of proteins that we could discuss, we shall choose hormones, antibodies, and genetic diseases in this section and take up enzymes in the next section.

Hormones. A <u>hormone</u> is a substance produced in certain cells of endocrine tissues. Transported in the blood in small concentration, it exerts an enormous effect upon the activity of a specific tissue or organ. For example, only a tiny amount of insulin from the pancreas is required to maintain blood sugar at the proper level by conversion to glycogen. Most hormones have concomitant effects: epinephrine increases blood sugar and heart beat and diverts blood from other areas to arms and legs for use in muscular activity.

Hormones may be proteins or nonproteins—usually the latter are steroidal. The pituitary gland is the source of most of the identified proteinoid hormones of mammals, although the pancreas, adrenals, kidneys, thyroid, and other glands also serve as sources. All hormones are subject to some form of feedback control. An example is the hormone ACTH (β-corticotropin, one of the adrenocorticotropic hormones) from the anterior lobe of the pituitary gland, which stimulates the adrenal cortex to secrete steroid hormones. ACTH is a thirty-nine-unit peptide that has been completely sequenced and synthesized. Species variations occur in units 25 through 32, as shown in Figure 21.8. Tests have shown

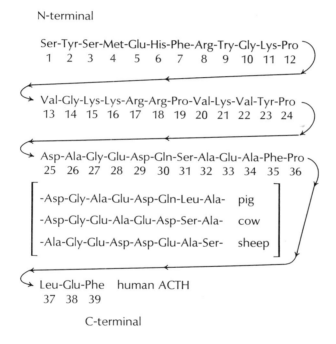

Figure 21.8 The amino acid sequence of human ACTH, with the variation among species at units 25 to 32

that these variations have little effect upon the hormonal activity. A very interesting result, however, appeared when smaller peptides were tested for activity. All the peptides of sequence up to 1 through 16 were inactive, but the 1 through 17 exhibited 15%; the 1 through 19, 50%; and the 1 through 20 and larger peptides, 100% of the activity shown by ACTH.

Antibody-antigen defense mechanism. Despite the statement we just made about ACTH and the lack of effect of species variation upon the hormonal activity, the general situation is far different. One of the defenses used by the body is <u>antibody</u> formation. If a substance not normally found in a human is introduced into the blood by injection or through an open wound, the various enzymes present all are tried out until those are found that are able to metabolize the substance to products that may be eliminated. (Several other mammals also have this capability.) If the foreign substance is large enough (a polypeptide or protein, for example), the substance is removed by precipitation. A foreign substance that induces precipitation is known as an <u>antigen.</u> The substance employed by the body to engulf the antigen and form a precipitate is the antibody. The antibody is a <u>gamma globulin</u> that is synthesized within the body for the specific purpose of reaction with a given antigen. An excess of antibody always is produced. If a second invasion of the same antigen should occur, the antigen-antibody precipitation is immediate, and we say that <u>immunity</u> to the antigen has been gained. The precipitated antigen-antibody is removed from circulation and may be disposed of by metabolic disintegration into smaller peptides.

Immunity is not permanent, even though antibodies persist for months and years. A practical application is vaccination, whereby antibody formation is caused by a weakened or modified antigen (smallpox, diphtheria, etc.) so that upon later exposure, the vaccinated individual is immune to infection (or the virulence is reduced).

Attempts at organ transplantation (kidney, heart, etc.) must overcome antibody formation if the transplant is to be successful. In preparation for the surgery, the patient is placed on a regime of drugs that suppress gamma globulin production by the body. The body accepts the new organ at the cost of losing immunity to all previous infections. When gamma globulin production is allowed to resume in the patient, antibodies to the foreign organ gradually increase, and rejection eventually occurs.

Genetic diseases. A number of diseases are transmitted by inheritance. These genetic diseases include <u>hemophilia</u> (failure of blood to clot) and <u>sickle-cell anemia.</u> These abnormalities arose originally as mutations, which, not being immediately lethal, are perpetuated in progeny following the laws of genetics. Sickle-cell anemia is a disabling disease: the afflicted person is oxygen starved—physical work and high altitudes must be avoided. The deoxygenated red blood cells of hemoglobin-S take on a sickle (crescent) shape as opposed to the disclike shape of the normal cell. The cause has been traced to a single substitution in the β chain of 146 amino acid units: valine is substituted for glutamic acid that is normally at position 6. The effect of the substitution is alteration of the tertiary structure of the chain to such an extent that access by oxygen to the heme is blocked. The quaternary structure also is modified so as to give rise to sickle cells. The oxygen-carrying capacity of the affected hemoglobin is limited to the α-chain units, with the result that a given number of cells are able to transport only half the oxygen that normal cells are capable of carrying. Cyanate has been shown to be effective in treatment of sickle-cell anemia.

Other abnormal hemoglobins are known also. However, we shall close at this point by reemphasizing the extraordinary specificity that nature has built into the countless proteins that help to make up a living organism. We shall see other instances of specificity as we look at enzymes.

Exercises

21.6 A certain graduate student, after thinking about the results obtained on ACTH activity, had a great idea. He synthesized Arg-Arg-Pro-Val and tested it for ACTH activity. To his amazement, activity was zero. What was wrong with the idea? What should he do to salvage something from his work?

21.6 Enzymes

Definitions. A simple definition of an enzyme is that it is a protein that catalyzes (accelerates) a chemical reaction within a living system. An enzyme is not necessarily inactive *in vitro*. An enzyme was developed by nature for use with *in vivo* reactions. Enzymes are essential for the steps involved in the synthesis and breakdown of proteins, lipids, and carbohydrates; in the oxidation and reduction reactions whereby energy is transferred, stored, and utilized; and in all other ancillary reactions needed for the maintenance of life in all its forms. □ □

Enzymes usually are large proteins, much larger than the substrates involved in the reaction being catalyzed. A substrate is the molecule that enters into the reaction on the surface of the enzyme as a reactant. Most enzymes require the presence of a cofactor to catalyze or participate in the reaction in question. If the cofactor is bound to the polypeptide chain, the complete molecule is called a conjugated protein, the cofactor is called a prosthetic group, and the polypeptide portion of the molecule is called the apoenzyme. The apoenzyme is inactive without the cofactor. Hemoglobin or myoglobin are examples. In other instances, the cofactor is not bound to the apoenzyme and enters into the picture at the time of reaction. These independent cofactors are known as coenzymes and may be as simple as a metallic ion, such as Mg^{2+}, Zn^{2+}, and so on. Other reactions may require large organic nonproteinoid molecules, such as NAD. These coenzymes may participate in many different reactions with many different apoenzymes. When the term is necessary for clarity, the temporary combination of coenzyme with a given apoenzyme is known as a holoenzyme. Unfortunately, often no distinction is made, and a holoenzyme is simply called the enzyme for the given reaction.

In any case, the catalysis observed is always an acceleration of the attainment of equilibrium between the reactants (substrates) and products. The cofactor, whether a prosthetic group or a coenzyme, is needed as a participant in the reaction. Later on, we shall see how the cofactors participate in enzyme catalysis. □ □

Effects of pH, temperature, and inhibitors. Most enzymes operate best at or near neutrality, whether in the body or in a test tube. Only in the stomach, with pH of about 2 to 3, is there much deviation from neutrality within the body. Thus, only the enzymes (pepsin) that are found in the stomach operate well under acidic conditions. A given enzyme shows maximum catalytic activity at a given pH, and activity decreases rapidly at pH's much above or below the optimum.

Enzymes work best at body temperature, 37.0° C or 98.6° F. At lower temperatures, the catalytic effect decreases as in any other reaction, the rate decreasing as temperature is decreased. In general, increasing the temperature slightly increases the catalytic effect, but enzymes begin to be destroyed and to lose their effectiveness at 40° and above. However, enzymes of bacteria and algae that have become acclimated to hot springs or to arctic cold have their maximum effect at those temperatures.

Enzymes are very sensitive to inhibitors, which, in extreme cases, are poisons. Inhibition is inversely related to specificity. Specificity may be of any of the following types.

1. Absolute: limited to one substrate only
2. Group: limited to esters only, or amides only, etc.
3. Reaction: limited to a single reaction type, such as hydrolysis, reduction, etc.

Absolute specificity can be inhibited only with extreme difficulty. Thus, the decomposition of urea by urease is limited to urea alone. No other compound is able to compete for the active sites on the urease surface. The only ways to inhibit this kind of enzyme are to decrease the available substrate (urea) or to cause a reaction with the enzyme surface in such a way as to distort (and thus deactivate) the absolutely specific active sites for urea. Group and reaction specificity are more easily interfered with: the inhibitor may act by forming an inactive complex at an active site on the enzyme. The inhibitor denies access of the substrate to the active site.

Inhibition, then, is classified into three types. Competitive inhibition can be observed with the relatively nonspecific enzymes. The inhibitor simply competes with substrate for active sites, and the more stable the enzyme-inhibitor complex, the more the number of catalytically active sites of the enzyme is decreased. Competitive inhibition is reversible (that is, catalytic activity is restored) if substrate concentration is increased.

Noncompetitive inhibition (distortion of active sites by reaction elsewhere) is not reversible and may be remedied only by addition of fresh enzyme (in vitro) or by synthesis of more enzyme (in vivo). In Figure 21.9, a very schematic idea of competitive, noncompetitive, and lack of inhibition is shown. Examples of competitive inhibition are that of malonate ion in competition with the substrate (succinate ion) for the enzyme succinate dehydrogenase and of a sulfa drug (a sulfanilamide) in compeition with p-aminobenzoic acid (PAB) for the enzyme responsible for the incorporation of PAB into folic acid, a coenzyme essential to the growth of bacteria (Eqs. 21.24).

$$\text{succinic acid} \xrightarrow[\text{succinic dehydrogenase}]{\text{FAD}} \text{fumaric acid} + FADH_2$$

$$\downarrow \text{malic acid}$$

enzyme-inhibitor complex

(21.24)

$$\text{PAB} \xrightarrow[\text{enzyme}]{\text{other components}} \text{folic acid (a coenzyme)}$$

$$\downarrow \text{sulfa drug}$$

a modified folic acid in which there is no PAB component and is not a coenzyme

The third type of inhibition is nonreversible. The inhibitor forms a compound with the enzyme at or near the active site and destroys the enzyme as a catalyst. An example is the nerve poison diisopropyl fluorophosphate $(iPrO)_2POF$, or DFP), which acts on acetylcholine esterase. The arrival of an impulse at a nerve cell causes formation of some acetylcholine along the cell. The ester reacts with a receptor protein on the next cell, which, in turn, forms some acetylcholine. In this way, the impulse or signal is transmitted from cell to cell. Once the signal is passed on, acetylcholine is hydrolyzed by the enzyme acetylcholine esterase. If allowed to remain unhydrolyzed, the ester maintains transmission, and the final receptor, be it a muscle, a gland, or a nerve, is in a state of constant tension or stimulation. Convulsions leading to exhaustion and paralysis quickly cause death. DFP reacts with the enzyme, probably at the hydroxyl group of a serine unit. Catalytic activity

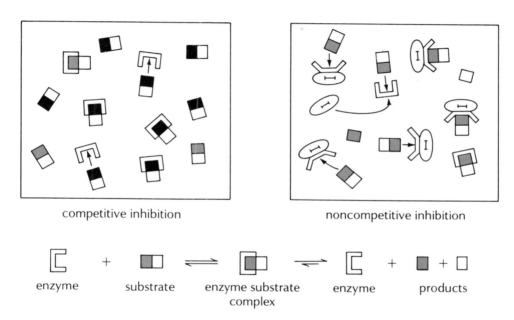

Figure 21.9 Schematic competitive and noncompetitive inhibition. Competitive inhibitors (▮▯) are bound to enzyme at the substrate site. Noncompetitive inhibitors (⊂⊐) influence the catalytic activity of an enzyme, although they are not bound to the active site.

ceases, ester is not hydrolyzed, and poisoning occurs (Eqs. 21.25). A good antidote, if given in time, is 2-formyl-1-methylpyridinium chloride oxime, which has been reported to reactivate the enzyme by removal of the diisopropyl fluorophosphate group.

Because of the catalytic effect of enzymes, reactions that otherwise would require weeks to come to equilibrium go astoundingly fast. The acceleration in rate has been estimated as high as 10^{23} for certain reactions. Thus, inhibition and poisoning produce dramatic effects even if the enzyme is exposed to only tiny amounts of the inhibitor or poison.

Other phosphate esters, such as malathion and parathion, have been used as

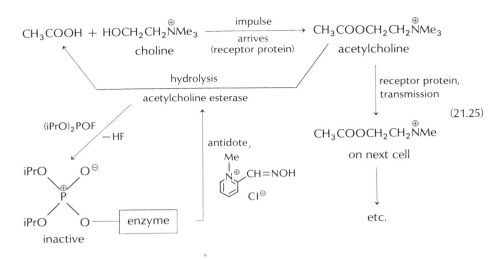

(21.25)

insecticides. They are nonpersistent and decompose within a month. However, because they also operate on acetylcholine esterase, these insecticides are very dangerous for humans to handle. Very special precautions are necessary in the application of these potent substances.

Coenzymes and vitamins. A vitamin may be defined as a substance required in the diet in small amounts for the maintenance of normal growth and well-being. A vitamin exerts an effect by serving as a coenzyme. The absence of a vitamin causes a vitamin deficiency. For example, scurvy results from a lack of vitamin C.

A coenzyme, as we have pointed out, is a nonproteinoid substance that, with the proper apoenzyme (a protein), forms a holoenzyme. It turns out that the B vitamins, which in general are water soluble, act as coenzymes by themselves or in combination with other substances. The oil soluble vitamins A, D, E, and K do not need modification to play roles as coenzymes.

Because only a small amount of a vitamin is effective, it did not take long for the suspicion to arise that vitamins were associated with enzyme activity. Confirmation came quickly, although many details remain to be worked out.

Table 21.4 lists some vitamins and the coenzymes in which they participate.

Table 21.4 Vitamins in coenzymes

Vitamin	Coenzyme	Use
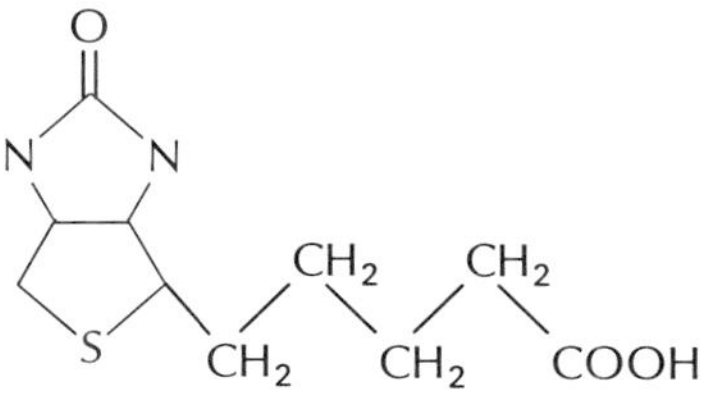 biotin (once called vitamin H, now considered part of the vitamin B complex)	may be a coenzyme in itself, a CO_2 carrier	probably needed to make other enzymes, to synthesize acids, and to metabolize proteins and carbohydrates
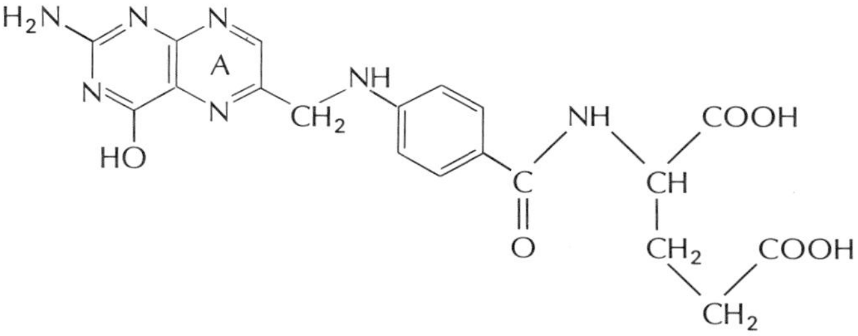 folic acid (member of vitamin B complex, without a number; once called vitamin M)	the coenzyme has a reduced A ring and a peptide chain of 3 to 7 glutamic acid units	used in purine synthesis and in use of amino acids (absence causes anemia)
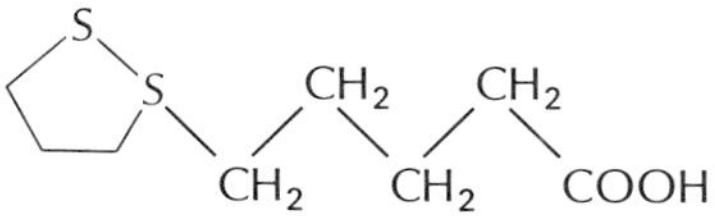 lipoic acid (or thioctic acid; may be a member of vitamin B complex)	may itself be a coenzyme	used in oxidation-reduction reactions: the disulfide acts as a hydrogen acceptor
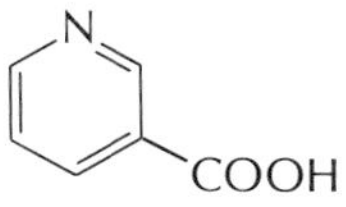 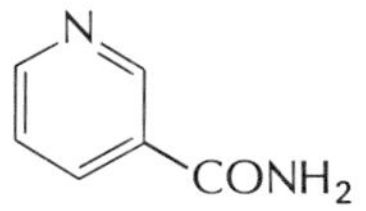 niacin (or nicotinic acid) niacinamide (or nicotinamide) (vitamin B_3)		

Table 21.4 (continued)

Vitamin	Coenzyme	Use
	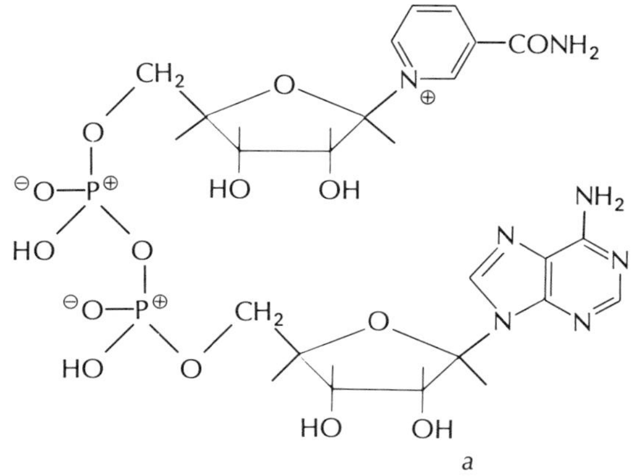 nicotinamide adenine dinucleotide (or NAD): NADP has the hydroxy group at a esterified by a phosphate	used in oxidation-reduction reactions as a hydrogen acceptor to give NADH (absence causes pellagra)
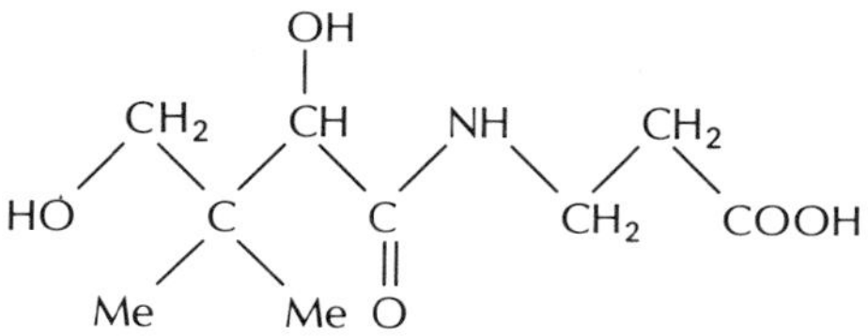 pantothenic acid (vitamin B$_5$)	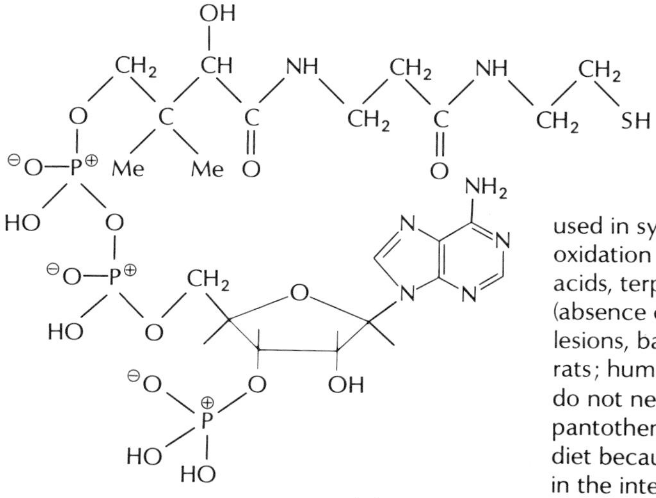 coenzyme A (or CoA—SH)	used in synthesis and oxidation of fatty acids, terpenes, sterols (absence causes skin lesions, baldness in rats; humans probably do not need pantothenic acid in diet because bacteria in the intestine synthesize enough to mitigate requirement)

Table 21.4 (continued)

Vitamin	Coenzyme	Use
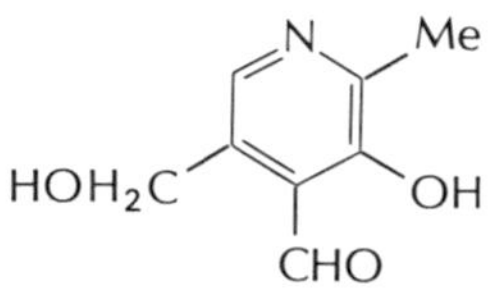 pyridoxal (vitamin B_6)	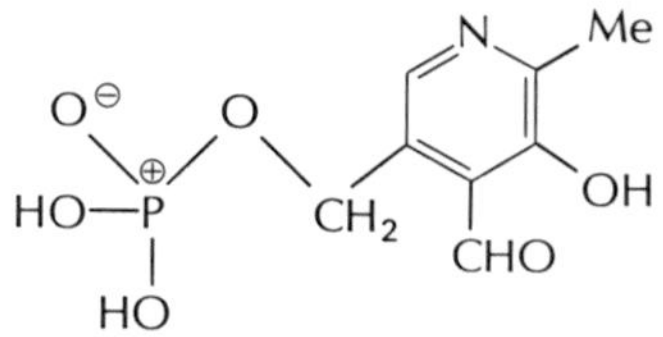 pyridoxal phosphate	use of amino acids (absence causes abnormal nerve function, convulsions)
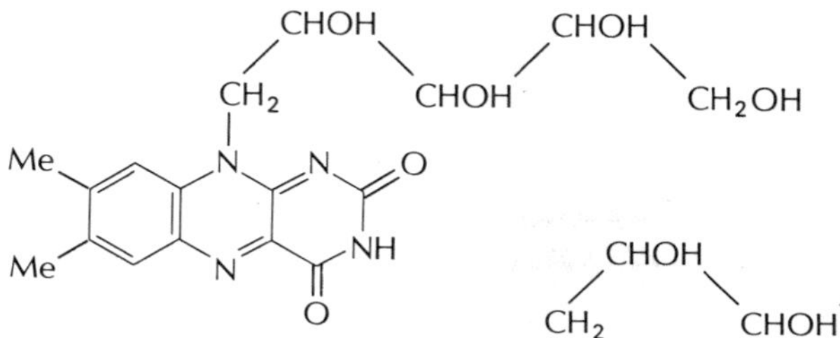 rivoflavin (vitamin B_2)	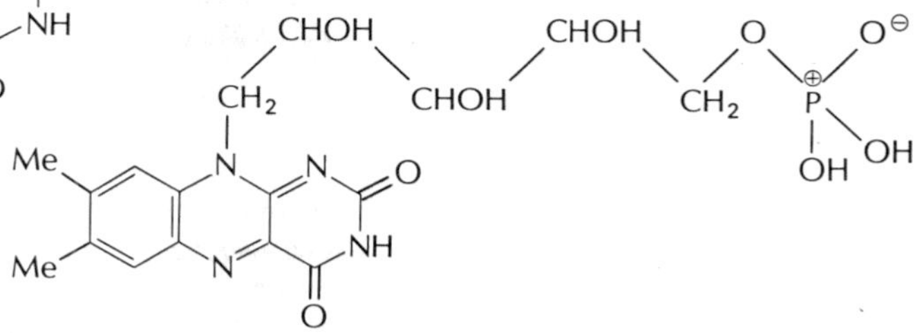 riboflavin phosphate, or flavin mononucleotide, FMN	oxidation-reduction reactions (absence causes skin lesions, general poor health)
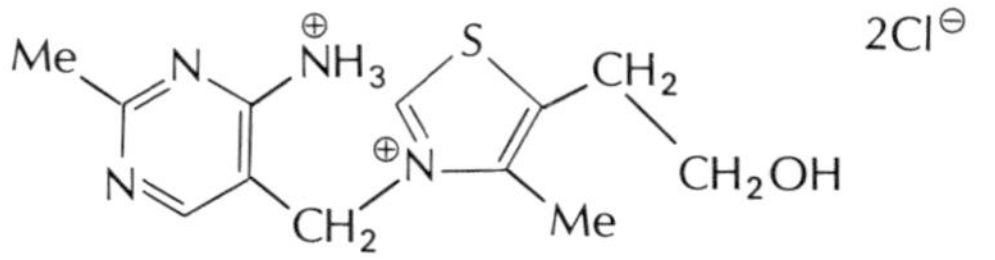 thiamine hydrochloride (vitamin B_1)	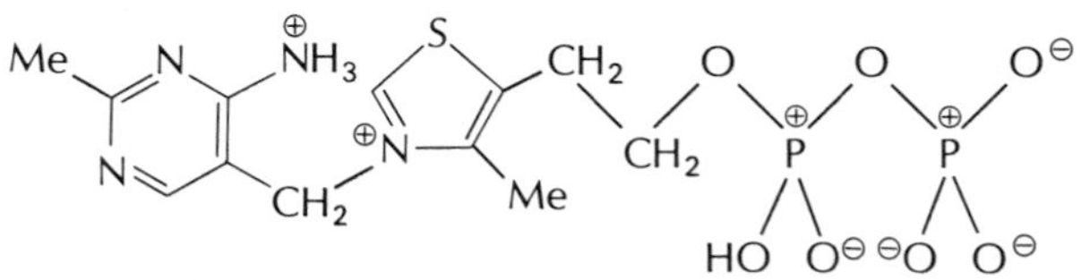 thiamine pyrophosphate or cocarboxylase	decarboxylation (absence from diet causes beri-beri)
cobalamine (vitamin B_{12})	(see Section 28.8)	anti-pernicious-anemia factor

Synthetic vitamins have replaced many that formerly were available only by isolation from natural sources. The best natural sources of the B vitamins are liver and yeast. Syntheses of the B vitamins are given in Chapter 28.

Exercises

21.7 Choline is the quaternized derivative (by methyl groups) of ethanolamine ($HOCH_2CH_2NH_2$), which may arise by decarboxylation of serine. Interactions between fatty acids, glycerol, phosphoric acid, and the nitrogenous compounds just mentioned give rise to phospholipids. Phospholipids occur in all types of cells, most often as constituents of membranes or of the surfaces of membranes. The specific role played by the phospholipids remains a puzzle. Distribution is varied, from as high as 30% of the dry weight of brain and 15% of liver down to much smaller amounts in plant cells.

Draw structural representations of the substances described:
a. Phosphatidyl choline (lecithin) is a diester of phosphoric acid, with the hydroxy group of choline and with the hydroxy at position 1 of glycerol. The remaining hydroxy groups of glycerol are esterified with fatty acids (use palmitic and oleic acids).
b. Phosphatidyl serine (a cephalin) is like phosphatidyl choline, except that serine is incorporated in place of choline.
c. A sphingolipid is a derivative of sphingosine, 2-amino-4-octadecene-1,3-diol. The amino group is acylated with a fatty acid (use linoleic acid) and the hydroxy at 1 is esterified with phosphoric acid, which in turn is also esterified with choline.

Note: Other lipids are glycolipids (cerebrosides), in which glucose has replaced the phosphocholine group of a sphingolipid by glycoside formation, and lipoproteins in which a lipid and a protein are associated or bound together.

21.7 Summary

This chapter is an introduction to amino acids, peptides, and proteins. Research progress in the fields of biochemistry and molecular biology has been great, and no one facet of the subject of proteins could possibly be comprehensive or up to date.

After protein terminology and definitions were given, amino acids were taken up. Following the tabulation of some common and uncommon amino acids, we discussed the interrelationships of pK_a, isoelectric point, dipolar ions, and solubility. Then we saw some of the reactions of amino acids. Various syntheses of amino acids were shown.

We next saw that peptide representation and analysis include the use of FDNB, the Edman degradation, hydrazinolysis, and enzyme selective hydrolysis. Subsequently, the interpretation of results was illustrated. We examined the need for and use of protecting groups in peptide synthesis, including the use of solid support systems.

The problems of the structures of proteins with several examples were shown next. We then concentrated on hormones, antibodies, genetic diseases, and enzymes. We looked at the specificity and effect of inhibitors upon enzyme activity. The B vitamins and relationship to coenzyme activity concluded the section.

Many topics in protein chemistry have not even been mentioned. Nevertheless, enough subject matter has been given to enable us to proceed to more specialized texts and articles, some of which are given in the bibliography.

21.8 Problems

21.1 Match the items in column A with the proper associated items in column B.

	A		B
1.	acidic amino acid	a.	C-terminal
2.	α-helix	b.	prosthetic group
3.	tobacco mosaic virus	c.	NAD
4.	basic amino acid	d.	arginine
5.	cofactor	e.	aspartic acid
6.	isoelectric point	f.	quaternary structure
7.	N-terminal	g.	secondary structure
8.	carboxypeptidase	h.	pH
9.	niacin	i.	DFNB

21.2 Senator E. M. Teahead, upset by hearing that nerve gases are phosphates and that excess phosphates are supposed to ruin lakes and streams, introduces a bill to ban phosphates from all foods, drugs, and detergent formulations. We are assigned the task of explaining the facts of life to the Senator. How do we go about it?

21.3 A group of food faddists claims that plants grown under "natural" conditions are superior in nutritive values to those plants that have had the misfortune to have been subjected to "nonorganic" fertilizers, insect sprays, and so on. An earnest friend seeks our opinion. How should we respond?

21.4 A number of the amino acids, including threonine, in Table 21.1 have more than one asymmetric center.
   a. Draw projection formulas for the L isomers.
   b. Threonine received its name from its *threo* configuration. Which projection isomer is threonine?

21.5 How would you synthesize the following amino acids from the given starting material? Indicate the reagent to be used.
   a. Val from isopropyl bromide
   b. Phe from benzyl chloride
   c. Lys from ε-aminocaproic acid
   d. His from ethyl acetamidomalonate
   e. Asp from fumaric acid
   f. Glu from ethyl acrylate
   g. Tyr from anisaldehyde
   h. Ser from ethyl acetamidomalonate
   i. Met from $CH_3SCH_2CH_2Cl$
   j. Try from gramine methiodide (indole-$CH_2N(CH_3)_3^+ I^-$)
   k. Pro from cyclopentanone

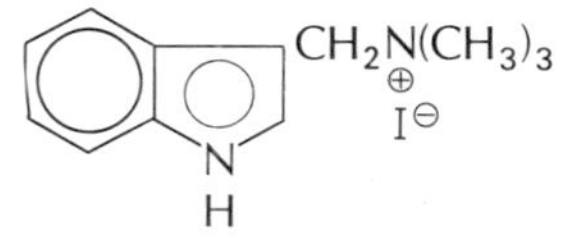

21.6 How would you differentiate between the following pairs of compounds?

	A	B
a.	Pro	Val
b.	Ala.Phe	Phe.Ala
c.	Ala	Lactic acid

	A	B
d.	Val.Gly.Val	Val.Val.Gly

e.
```
     CH₂
    /   \
  NH     C=O
  |       |
O=C       NH
    \   /
     CH₂
```

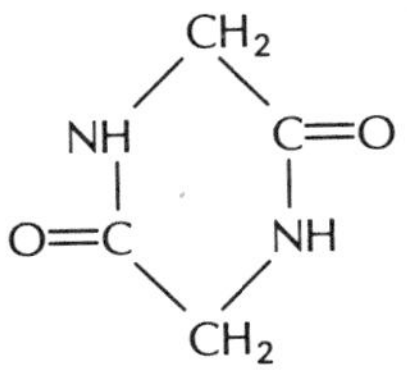

Gly.Gly

21.7 A hexapeptide gave the following products:

a. $\xrightarrow{C_6H_5NCS}$ $\xrightarrow{HF}$ phenylthiohydantoin of Pro

b. $\xrightarrow[100°]{3\,N\,HCl}$ 2 Gly, 1 Leu, 1 Phe, 1 Pro, 1 Tyr

c. $\xrightarrow[80°]{1\,N\,HCl}$ Phe.Gly.Tyr

Gly.Phe.Gly
Pro.Leu.Gly
Leu.Gly.Phe

What is the structure of the hexapeptide?

21.8 A hexapeptide gave the following products:

a. $\xrightarrow{C_6H_5NCS}$ $\xrightarrow{HF}$ phenylthiohydantoin of Phe

b. $\xrightarrow[100°]{3\,N\,HCl}$ 1 Asp, 1 Glu, 2 Phe, 1 Pro, 1 Tyr

c. $\xrightarrow[80°]{1\,N\,HCl}$ Pro.Phe.Tyr

Asp.Pro.Phe

Phe.Asp.Pro

Phe.Tyr.Glu

What is the structure of the hexapeptide?

21.9 Starting with the L-amino acids, how would you synthesize the four tripeptides in Problem 21.8? Indicate reagents, solvents, and temperatures.

21.10 An enzyme can reduce acetoacetate to β-hydroxybutyrate with DPNH as a coenzyme. Write a mechanism.

21.11 An enzyme with pyridoxal phosphate as a cofactor can convert serine to glycine and formaldehyde; the reaction is reversible. Write a mechanism.

21.9 Bibliography

It is difficult to be selective in suggesting additional sources of information about amino acids, peptides, and proteins. The books by Allinger et al. (Chap. 7, ref. 1); Hendrickson, Cram, and Hammond (Chap. 7, ref. 9); Morrison and Boyd (Chap. 7, ref. 10); Roberts and Caserio (Chap. 7, ref. 13); Fieser and Fieser (Chap. 7, ref. 7); Geissman (Chap. 7, ref. 8); Noller (Chap. 7, ref. 11); and Millar and Springall (Chap. 20, ref. 1) are all useful. In addition, seven other books are listed, as well as a number of articles.

1. Baker, B. R., *Organic Chemistry*. Belmont, Calif.: Wadsworth, 1971. One of the recently published shorter texts with extra emphasis upon biological chemistry.
2. Baum, S. J., *Introduction to Organic and Biological Chemistry*. New York: Macmillan, 1970. Another short book.
3. Bennett, T. P., and E. Frieden, *Modern Topics in Biochemistry*. New York: Macmillan, 1966. An excellent paperback.
4. Calvin, M., and W. A. Pryor, comps., *Organic Chemistry of Life*. San Francisco: Freeman, 1973. A collection of 47 articles from the period 1954 to 1973, published in Scientific American divided into four sections: Biological Regulators, Macromolecular Architecture, Cellular Architecture, and Chemical Biodynamics. This book has to be seen to be appreciated.
5. Holum, J. R., *Introduction to Organic and Biological Chemistry*. New York: 1969. Also short. Wonderful illustrations.
6. Kopple, Kenneth D., *Peptides and Amino Acids*. New York: Benjamin, 1966. An excellent paperback. References but no problems.
7. Reithel, F. J., *Concepts in Biochemistry*. New York: McGraw-Hill, 1967.
8. Sutterby, L., Elements in the Chemistry of Bioluminescence, *Chemical Technology 1*, 563 (1971). A student research paper.
9. Witkop, B., Chemical Cleavage of Proteins, *Science 162*, 318 (1968).
10. Fox, Sidney W., Kaoru Harada, Gottfried Krampitz, and George Mueller, Chemical Origin of Cells, *Chemical and Engineering News*, June 22, 1970, p. 80.
11. Fox, Sidney W., Chemical Origins of Cells—2, *Chemical and Engineering News*, Dec. 6, 1971, p. 46.
12. Lipmann, Fritz, Attempts to Map a Process Evolution of Peptide Biosynthesis, *Science 173*, 875 (1971).
13. Wang, Jui H., Oxidative and Photosynthetic Phosphorylation Mechanisms, *Science 167*, 25 (1970).
14. Allergy, A Protective Mechanism Out of Control, *Chemical and Engineering News*, May 11, 1970, p. 84.
15. Harris, Maureen, Interferon: Clinical Application of Molecular Biology, *Science 170*, 1068 (1970).
16. Wang, Jui H., Facilitated Proton Transfer in Enzyme Catalysis, *Science 161*, 328 (1968).
17. Paik, Woon Ki, and Sangduk Kim, Protein Methylation, *Science 174*, 114 (1971).
18. Riordan, James F., and Mordecai Sokolovsky, Chemical Approaches to the Study of Enzymes, *Accounts of Chemical Research 4*, 353 (1971).
19. Bodansky, M., and D. Perlman, Peptide Antibiotics, *Science 163*, 352 (1969).
20. Zaoral, M., and K. Slama, Peptides with Juvenile Hormone Activity, *Science 170*, 92 (1970).
21. Klotz, Irving M., and Dennis W. Darnall, Protein Subunits: A Table (Second Edition), *Science 166*, 126 (1969).
22. Meienhofer, Johannes, Why Peptides are Synthesized and How, *Chemical Technology 3*, 242 (1973).

22 Aromatic Nitrogen Compounds

The aromatic nitrogen compounds are different from the nonaromatic ones. The presence of the aromatic ring and conjugation with it give rise to considerable differences in reactivity of the nitrogen-containing groups. Also, a number of electrophilic and nucleophilic reactions and rearrangements appear that are not present in the reactions of the non-aromatic nitrogen compounds. Another difference is that most of the aromatic nitrogen compounds are derived from nitro compounds.

We shall give more attention to the nitro-, the amino-, and diazoaromatic compounds than to any of the other groups. However, others will be included, and we should not be surprised to find that many of the reactions given in previous chapters are repeated here.

22.1 Aromatic nitro compounds   737
1. Physical properties; spectra; basicity.   2. Reactions.   3. Electrophilic substitution.   4. Nucleophilic aromatic substitution.   5. Vinylogy; condensation reactions.   6. Charge-transfer complexes.   7. Reductions: reduction cycle.   8. Conditions.   9. Ammonium sulfide: the nitro group as an oxidant.   10. Preparation. Exercises.

22.2 Aromatic amines   744
1. Occurrence; naming; spectra.   2. Basicity: conjugation.   3. Effects of substituents.   4. Dipole moments.   5. The ortho effect.   6. Reactions: amine reactions.   7. Electrophilic substitution.   8. Routes to nitroanilines.   9. Toluidines.   10. Sulfonation.   11. Sulfanilamide and sulfa drugs: synthesis.   12. Halogenation.   13. Friedel-Crafts reaction: use of amides.   14. Oxidation.   15. Reaction with nitrous acid: diazotization of primary aromatic amines.   16. Nitrosation of secondary aromatic amines.   17. Nitrosation of tertiary aromatic amines.   18. Preparations: Bucherer reaction. Exercises.

22.3 Aromatic diazonium compounds   757
1. Properties of diazonium compounds.   2. Reactions.   3. Reduction.   4. Coupling reaction.   5. Effects of substituents.   6. Indicators: color change with pH change.   7. S_N1 replacements of nitrogen.   8. Radical reactions: Sandmeyer reaction.   9. Reduction.   10. Anthranilic acid; benzyne.   11. Summary of reactions.   12. Uses in synthesis. Exercises.

22.4 Aromatic nitroso and hydroxylamino compounds   768
   1. Physical properties of nitroso compounds.   2. Addition reactions to the nitroso group.   3. Instability of phenylhydroxylamine. Exercises.

22.5 Other aromatic nitrogen compounds   770
   1. Introduction.   2. Summary of reductions and oxidations.
   3. Arylhydrazines.   4. Aryl isocyanates.   5. Aryl nitriles.
   6. Benzidine rearrangement. Exercises.

22.6 Summary   772

22.7 Problems   773

22.8 Bibliography   774

22.1 Aromatic nitro compounds

We remember from Chapter 15 that the aromatic nitro compounds can be made by nitration of the aromatic hydrocarbons, arenes, and many substituted aromatic compounds. Therefore, they serve as the starting point for the preparation of almost all other aromatic nitrogen compounds, and we shall begin with them. The various aromatic nitrogen compounds are interrelated, and we shall skip about a bit to keep things from becoming too muddled.

Physical properties; spectra; basicity.   The presence of a nitro group invariably causes an increase in the bp of an aromatic compound and also in the mp (unless there is *ortho* substitution). For example, even the simplest nitro compound, nitrobenzene (ϕNO_2), has bp 209°. The o-, m-, and p-nitrotoluenes have bp of 222°, 231°, and 238° and have mp of $-4°$, 15°, and 51°. The presence of more than a single nitro group invariably causes the bp to approach 300° (with danger of explosive decomposition) and the mp to increase. For example, the o-, m-, and p-dinitrobenzenes have mp of 118°, 89°, and 174°. Nitro-substituted reagents are useful for the preparation of crystalline, sharp-melting derivatives of alcohols (the 3,5-dinitrobenzoate esters) and of aldehydes and ketones (the 2,4-dinitrophenylhydrazones).

Nitrobenzene (oil of mirbane) has an almond odor, penetrates the skin, and is very toxic. Its high polarity leads to good solvent properties. Its main industrial use is for reduction to aniline.

Nitrobenzene (in hexane) absorbs in the uv at 252 nm (ε_{max} 10,000), 280 nm (ε_{max} 1000), and 330 nm (ε_{max} 125). In the ir, bands associated with the nitro group appear at about 1350 and 1500 cm^{-1}. In the mass spectrum, the parent peak is usually strong, and the base peak comes at m/e of $P - 46$, which is $P - NO_2$.

The aromatic nitro compounds do not have α hydrogen and do not act as acids (however, see Sec. 15.5). Basicity is weak (pK_a is about -11), so that aside from organic solvents only very strong acids (concentrated H_2SO_4) will dissolve nitrobenzene. Nitrobenzene is a good solvent for Friedel-Crafts reactions because it does not undergo the reaction and because it is able to dissolve not only aluminum chloride but also the various complexes involved.

Reactions. The reactions of the nitro compounds may be classified as 1. those reactions involving the ring (electrophilic and nucleophilic substitution), 2. those that involve other substituents on the ring, and 3. those that involve a change in the nitro group itself, primarily reduction (discussed separately).

Electrophilic substitution. The nitro group is a strongly deactivating group: electrophilic substitution reactions proceed with difficulty unless an activating group also is present. Sulfonation of nitrobenzene requires fuming sulfuric acid (H_2SO_4 containing 20% of SO_3) at 100° for over an hour to give *m*-nitrobenzenesulfonic acid. Nitration of nitrobenzene usually is done with a mixture of fuming nitric acid and sulfuric acid at 90° to give *m*-dinitrobenzene. Bromination of nitrobenzene requires an excess of bromine, a large amount of $FeBr_3$ (0.25 mole to 1 mole of ϕNO_2) at 140° for 4 hr for a 75% yield of *m*-bromonitrobenzene. The Friedel-Crafts reaction does not go at all with nitroaromatic compounds. However, mercuration of nitrobenzene with mercuric acetate proceeds slowly at 95° to 150° to give a statistical distribution of isomers. Mercuric perchlorate in excess perchloric acid at room temperature with nitrobenzene for 10 days reacts to give a 97% yield of crude product, of which 89% is the *meta* isomer. Evidently the ionic mercuric perchlorate gives the usual electrophilic attack, whereas mercuric acetate gives a much more random attack.

Nucleophilic aromatic substitution. Nucleophilic aromatic substitution occurs readily at positions *o* and *p* to the nitro group (Sec. 15.5). In addition to the replacements shown in Section 15.5, it is possible to substitute hydrogen *o* or *p* to a nitro group (Eq. 22.1).

$$\phi NO_2 \xrightarrow{NaNH_2} O_2N{-}\underset{}{\bigcirc}{-}NH_2 + H^\ominus$$

$$\phi NO_2 \xrightarrow[\text{heat}]{\text{KOH fuse by}} \underset{NO_2}{\bigcirc}{-}OH + H^\ominus + K^\oplus \xrightarrow[\text{oxidant}]{KOH} \underset{NO_2}{\bigcirc}{-}OK + H_2O$$

(22.1)

Of course, hydride ion itself is not displaced: it is oxidized by dissolved oxygen from air or by nitrobenzene. Sometimes an added oxidant is used. A nitro group *o* or *p* to another nitro group can be replaced by nucleophiles.

Vinylogy; condensation reactions. Closely related to the activation by nitro groups for nucleophilic substitution is the activation of hydrogen on carbon α to *o* and *p* positions relative to a nitro group (Eq. 22.2b). This activation is similar to that of the nitroalkanes (a). The ring acts as a good transmitter of the influence of the nitro group, and *o*- and *p*-nitrotoluenes show reactivity similar to that of nitromethane. The transmitter effect is not

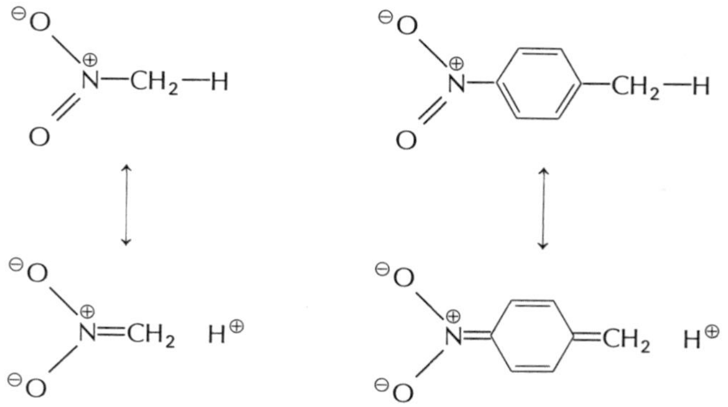

22.1 Aromatic nitro compounds

$$\text{equation 22.2 resonance structures}$$

(22.2)

restricted to nitro groups. For example, compare the reactions shown in Equations 22.3, in which the easy loss of CO_2 from carbonic acid (a) is paralleled by a similar loss from p-hydroxybenzoic acid (b). The transmission effect has been called the principle of vinylogy because a double bond also acts as a transmitter.

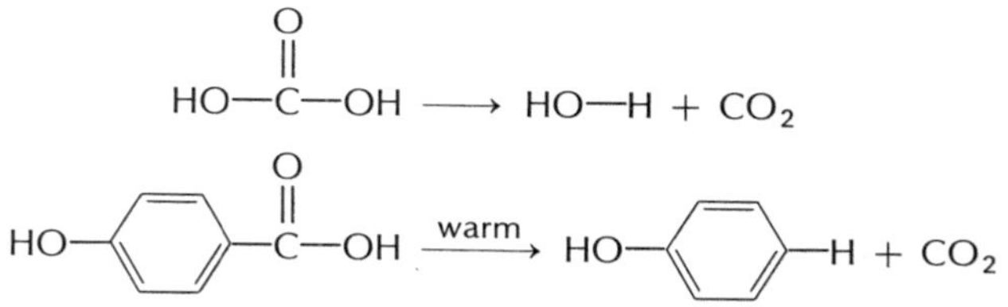

$$HO-\overset{O}{\underset{\parallel}{C}}-OH \longrightarrow HO-H + CO_2 \qquad a$$

$$HO-\text{C}_6H_4-\overset{O}{\underset{\parallel}{C}}-OH \xrightarrow{warm} HO-\text{C}_6H_4-H + CO_2 \qquad b$$

(22.3)

Among the nitroarenes, the condensation reactions shown in Equations 22.4 are possible. Here, a and b again illustrate the parallelism we have mentioned, and c shows a condensation with a nitroso group to give an imine. Hydrolysis of the imine gives 2,4-dinitrobenzaldehyde and constitutes still another aldehyde synthesis.

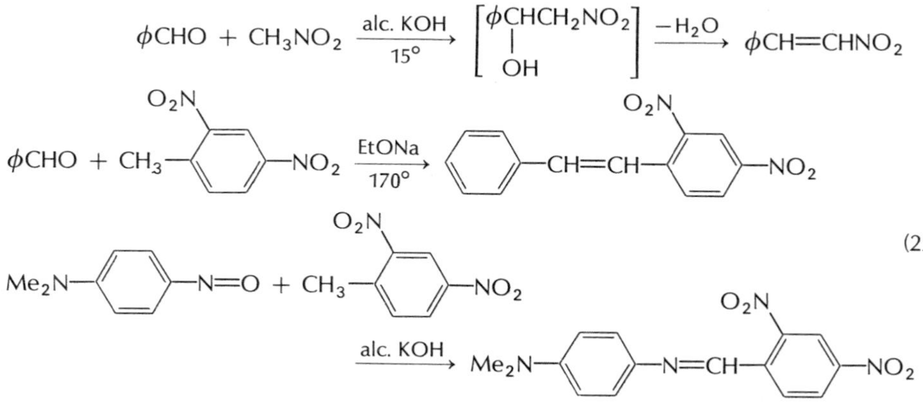

$$\phi CHO + CH_3NO_2 \xrightarrow[15°]{alc.\ KOH} \left[\begin{array}{c}\phi CHCH_2NO_2 \\ | \\ OH\end{array}\right] \xrightarrow{-H_2O} \phi CH=CHNO_2 \qquad a$$

$$\phi CHO + CH_3-\text{C}_6H_3(O_2N)(NO_2) \xrightarrow[170°]{EtONa} \text{Ar}-CH=CH-\text{C}_6H_3(O_2N)(NO_2) \qquad b$$

$$Me_2N-\text{C}_6H_4-N=O + CH_3-\text{C}_6H_3(O_2N)(NO_2) \xrightarrow{alc.\ KOH} Me_2N-\text{C}_6H_4-N=CH-\text{C}_6H_3(O_2N)(NO_2) \qquad c$$

(22.4)

Charge-transfer complexes. The polynitro compounds (picric acid and 1,3,5-trinitrobenzene, TNB, are the most popular) are used to form weak ($-\Delta H \cong 2-4$ kcal), highly colored charge-transfer or π complexes with arenes, polycyclic compounds, phenols, and amines. Most of these crystalline complexes have mp higher than either component, and picrates of amines are salts. Their rings are stacked like pancakes, alternating between the polynitro acceptor molecule and the donor molecule. They are most easily understood, perhaps, in terms of MO theory: there is a partial electron transfer from the highest occupied MO of the donor to the lowest unoccupied MO of the acceptor. Because their components are easily separated by chromatography, these complexes have been of use in the isolation and purification of arenes and polycyclic compounds.

Reductions: reduction cycle. The reduction of aromatic nitro compounds has been studied extensively. The products obtained depend upon the reagent (or the potential in electrolytic reduction) and the conditions used. The products cover the entire list of intermediate states of reduction down to the product of complete reduction, an aromatic primary amine. The products may be separated into monomolecular and bimolecular groups. The monomolecular products are obtained under neutral or acidic conditions. The bimolecular products are obtained under basic conditions, mainly as a result of condensation reactions of certain monomolecular products. We illustrate the situation by looking at the reduction of nitrobenzene (Eq. 22.5). Beginning with nitrobenzene and going clockwise around the diagram, the various stages of reduction are listed, ending with aniline. The dashed line separates the monomolecular from the bimolecular products.

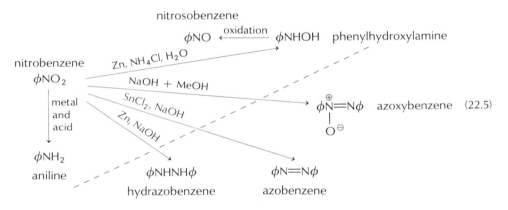

Conditions. Although the evidence is overwhelming that nitrosobenzene, under all conditions, is the initial reduction product of nitrobenzene, it also has been shown that nitrosobenzene is reduced so much more rapidly than nitrobenzene that it cannot be isolated. The reduction of nitrosobenzene leads to phenylhydroxylamine, which, despite its reactivity, is the first isolable reduction product of nitrobenzene. One can obtain phenylhydroxylamine easily by stirring an aqueous mixture of nitrobenzene, ammonium chloride, and zinc dust at 60° for 30 min. Nitrosobenzene may be used instead of nitrobenzene. This method is useful in the case of the easily prepared p-nitrosophenols and p-nitroso-N,N-dialkylanilines.

The next three reduction products are bimolecular and result from condensation reactions in the alkaline environment. Nitrosobenzene and phenylhydroxylamine react to give azoxybenzene at a rate much greater in an alkaline environment than in an acidic one (Eq. 22.6). One of several possible mechanisms is shown. Thus a scheme for a mild reduction of nitrobenzene, which allows a reasonable amount of nitrosobenzene to exist in the

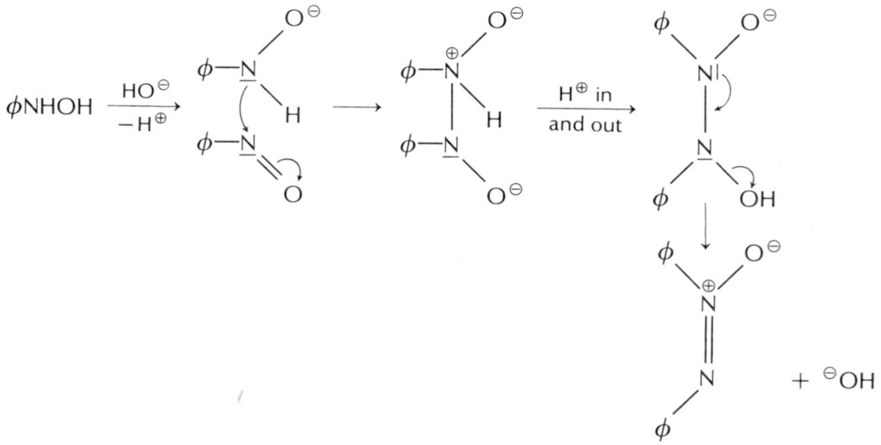

$$\phi\text{NHOH} \xrightarrow[-H^{\oplus}]{HO^{\ominus}} \begin{array}{c} \phi-\underline{N}\diagdown_{O^{\ominus}} \\ \big(\diagdown_{H} \\ \phi-\underline{N}\diagdown_{\diagdown O} \end{array} \longrightarrow \begin{array}{c} \phi-\overset{\oplus}{N}\diagdown_{O^{\ominus}} \\ \diagdown_{H} \\ \phi-N\diagdown_{O^{\ominus}} \end{array} \xrightarrow[\text{and out}]{H^{\oplus} \text{ in}} \begin{array}{c} \phi\diagdown_{N\big|}\diagup^{O^{\ominus}} \\ \big| \\ \phi\diagdown_{N}\diagup^{\curvearrowright}_{OH} \end{array}$$

$$\downarrow$$

$$\begin{array}{c} \phi\diagdown_{\overset{\oplus}{N}}\diagup^{O^{\ominus}} \\ \parallel \\ N \\ \diagup \\ \phi \end{array} \quad + \quad {}^{\ominus}OH \tag{22.6}$$

systems and to react with phenylhydroxylamine, gives azoxybenzene. Three useful schemes have been found (Eq. 22.7). It is useful to think of nitrobenzene as an oxidizing

$$\phi NO_2 \xrightarrow[\substack{\text{1. NaOH, CH}_3\text{OH, reflux} \\ \text{2. Na}_3\text{AsO}_3, \text{H}_2\text{O, boil} \\ \text{3. NaOH, glucose, boil}}]{} \phi-\overset{\oplus}{N}=N-\phi \atop \underset{O^{\ominus}}{|} \tag{22.7}$$

agent for the reagents used. Thus, in scheme 1, the product of oxidation of methanol is sodium formate; in 2, sodium arsenate (Na_3AsO_4); and in 3, several products, such as gluconic, glycaric, tartaric, and oxalic acids (as sodium salts).

If the calculated amount of zinc dust (to remove the oxygen atom from azoxybenzene) is added to a methanolic sodium hydroxide reduction of nitrobenzene, azobenzene is produced. As shown in Equation 22.4, azobenzene is the principal product if an alkaline solution of stannous chloride is used. Only slightly more vigorous conditions (zinc dust with sodium hydroxide solution) reduce nitrobenzene to hydrazobenzene. Azoxybenzene also is reduced to hydrazobenzene with zinc dust and alkali.

Finally, complete reduction of nitrobenzene to aniline may be accomplished in a variety of ways: 1. almost any active metal (Sn or Fe, for example) with an acid, 2. catalytic hydrogenation, 3. $SnCl_2$ and HCl, 4. hydrazine and Raney nickel catalyst, or 5. alkaline sodium hydrosulfite ($Na_2S_2O_4$). Also, any of the intermediate reduction products may be reduced to aniline by any of the five procedures given. Any of the bimolecular intermediate products is reduced to aniline by sodium in ethanol as well. Commercially, aniline is obtained from nitrobenzene either by catalytic hydrogenation or by the use of scrap iron and small amounts of hydrochloric acid (Eq. 22.8). The hydrolysis of the anilinium ion maintains enough acidity for the reaction to continue.

$$\phi NO_2 \xrightarrow[\text{dil. HCl}]{Fe} \phi NH_3^{\oplus}Cl^{\ominus} + Fe_3O_4 \tag{22.8}$$

In the laboratory, the nitro group usually is reduced to an amino group by catalytic hydrogenation or (for small amounts) by stannous chloride solutions. Both methods are relatively clean and rapid compared to acidic reduction with tin and hydrochloric acid (Eq. 22.9), which is the usual introduction to the reaction in the organic chemistry laboratory. Many side reactions are possible, as we shall see when we go into the reactions of the intermediate reduction products.

Unfortunately, $LiAlH_4$ and $NaBH_4$ are of little use for the reduction of aromatic nitro compounds because side reactions and condensations predominate. With nitrobenzene, azobenzene seems to be the major product of $LiAlH_4$ reduction.

$$2\phi NO_2 + 6Sn + 24HCl \xrightarrow[H_2O]{100°} (\phi\overset{\oplus}{N}H_3)_2(SnCl_4)^{2-} + 5H_2SnCl_4 + 4H_2O$$

$$\downarrow \text{excess NaOH}$$

$$\phi NH_2 + Na_2SnO_3 + NaCl + H_2O \tag{22.9}$$

Ammonium sulfide: the nitro group as an oxidant. If more than one nitro group is present, $(NH_4)_2S$ (made from a calculated amount of H_2S and NH_4OH) in aqueous ethanol reduces only one group. Because nitro groups *o* and *p* to each other are obtainable only by roundabout routes, the practical use of the reaction is limited to *m*-dinitro aromatic compounds that can be obtained by direct nitration. The mechanism is in doubt, but it is helpful to regard the reaction as an oxidation of $(NH_4)_2S$ by the dinitro compound, which is a stronger oxidizing agent than the nitroaniline (*a*) that is formed (Eqs. 22.10). Why 2,4-dinitroaniline should be reduced at the 2-NO_2 group (*b*) and 2,4-dinitrotoluene should be reduced at the 4-NO_2 group (*c*) is another puzzle awaiting solution.

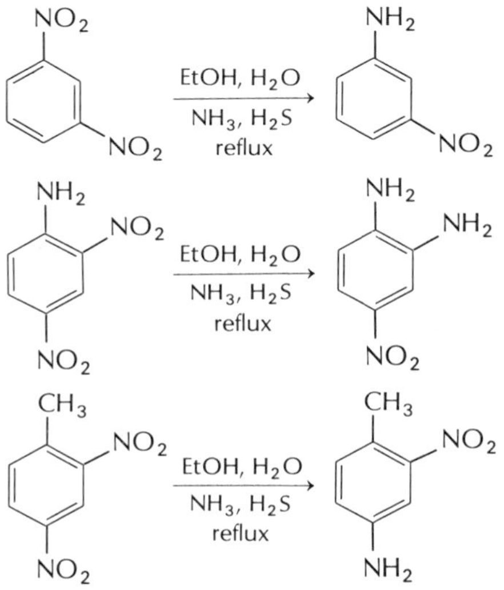

$$\text{(22.10)}$$

From time to time, we have suggested thinking of an aromatic nitro compound as an oxidizing agent. A clear case of internal or <u>intramolecular</u> oxidation-reduction results if *o*-nitrotoluene is heated in alcoholic NaOH solution (Eq. 22.11).

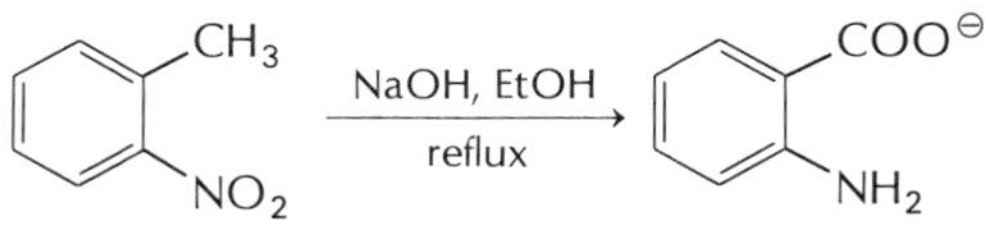

$$\text{(22.11)}$$

Preparation. Direct nitration of arenes and substituted aromatic compounds has been discussed (Sec. 15.3). Here we wish to bring up some of the details of direct nitration. Other methods of preparation of nitro aromatic compounds will come later in the chapter, as we encounter the reactions of other nitrogen-containing compounds.

Nitration usually is carried out with "mixed acid", a mixture of sulfuric and nitric acids. For unreactive compounds, the amount of sulfuric acid may be increased, and fuming nitric acid may be used. For reactive compounds, the amount of sulfuric acid may be decreased (or eliminated entirely) and the nitric acid may be diluted. Some examples will clarify the situation.

The hydroxy group of a phenol is a strong activator and o,p director. The nitration of phenol is carried out at 20° with an aqueous mixture of 1.75 moles of nitric acid and 0.70 moles of sulfuric acid per mole of phenol. It gives about a 35% yield of o-nitrophenol and 25% of p-nitrophenol (Eq. 22.12). The reaction is accompanied by considerable dinitration and oxidation. Attempts to prepare 2,4-dinitrophenol by more drastic conditions fail because of the oxidation that occurs. Therefore, 2,4-dinitrophenol is prepared by nucleophilic displacement of chlorine from 2,4-dinitrochlorobenzene, which may be obtained by dinitration of chlorobenzene without the troublesome oxidation encountered with phenol (Eq. 22.13).

$$\phi\text{-OH} \xrightarrow[20°]{HNO_3, H_2SO_4, H_2O} o\text{-O}_2N\text{-}\phi\text{-OH} + p\text{-O}_2N\text{-}\phi\text{-OH} \quad (22.12)$$

$$\phi H \xrightarrow[FeCl_3]{Cl_2} \phi Cl \xrightarrow[\text{hot}]{HNO_3, \text{fuming } H_2SO_4} \text{2,4-dinitrochlorobenzene} \xrightarrow[100°, 24\ hr]{Na_2CO_3, H_2O} \text{2,4-dinitrophenol} \quad (22.13)$$

Picric acid (2,4,6-trinitrophenol) can be obtained in low yield by the nitration of phenol with concentrated mixed acid. Commercially, nitration of 2,4-dinitrophenol is used. The presence of the two nitro groups reduces the oxidizability of the molecule, prepared as shown in Equation 22.13. A good laboratory preparation begins with reversible disulfonation of phenol, following which the sulfonic acid groups are removed by hydrolysis with added nitric acid and replaced by nitro groups. The final nitration occurs in the same mixture (Eq. 22.14).

$$\phi\text{OH} \xrightarrow[\substack{100° \\ 30\ min}]{H_2SO_4} \text{2,4-disulfonic acid of phenol} \xrightarrow[\substack{100° \\ 2\ hr}]{\text{add } HNO_3} \text{picric acid} \quad (22.14)$$

Trinitrobenzene requires heating (100°) of m-dinitrobenzene with a large excess of fuming nitric and fuming sulfuric acids for five days, and the yield is only 45%. A much better route is trinitration of toluene to give 2,4,6-trinitrotoluene (the explosive, TNT), which is much easier to accomplish because of the presence of the activating and o,p-directing methyl group. Oxidation of the methyl group to an acid and decarboxylation gives the product (Eq. 22.15). The nitration of aniline will be described in the next section.

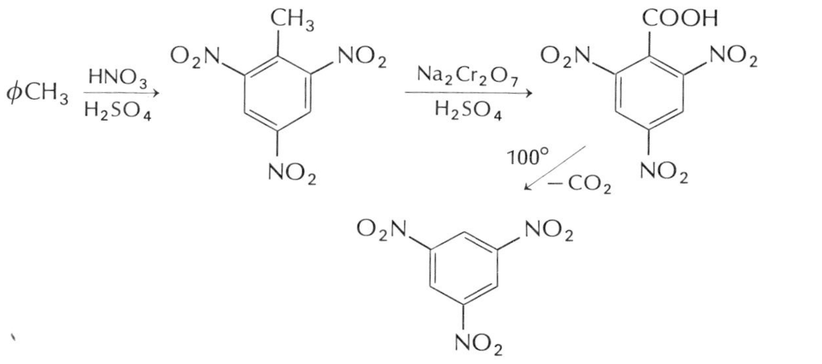

$$\phi\text{CH}_3 \xrightarrow[\text{H}_2\text{SO}_4]{\text{HNO}_3} \underset{\text{NO}_2}{\underset{|}{\text{2,4,6-trinitrotoluene}}} \xrightarrow[\text{H}_2\text{SO}_4]{\text{Na}_2\text{Cr}_2\text{O}_7} \underset{\text{NO}_2}{\text{2,4,6-trinitrobenzoic acid}} \xrightarrow[-\text{CO}_2]{100°} \text{1,3,5-trinitrobenzene} \qquad (22.15)$$

Exercises

22.1 Replacement of a nitro group by nucleophilic substitution (Eq. 22.2) is not very useful. Why not?

22.2 Use the principle of vinylogy to predict some reactions of:

a. 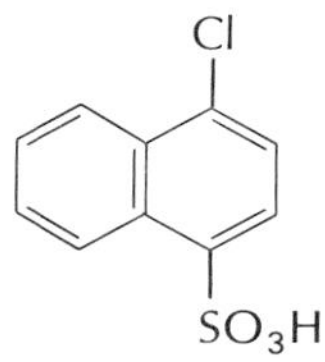 1-chloro-4-sulfonaphthalene   b. 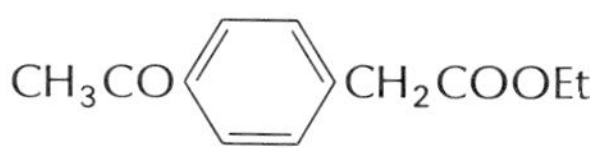 $CH_3CO\text{-}C_6H_4\text{-}CH_2COOEt$ c. $MeO\text{-}C_6H_4\text{-}CO\phi$

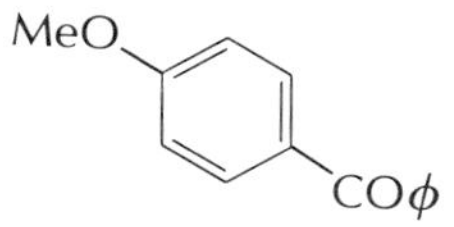

22.3 Starting with ϕH or ϕCH_3, show how to prepare:

a. 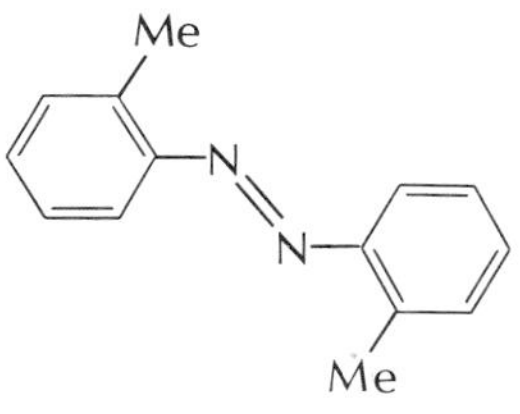 2,2'-dimethylazobenzene   b. 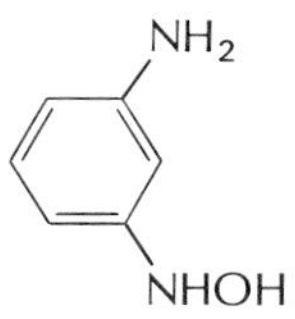 3-(hydroxyamino)aniline   c. 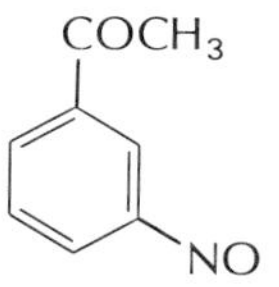 3-nitroso-acetophenone

22.2 Aromatic amines

Occurrence; naming; spectra. Let us skip over the intermediate stages of reduction of nitro compounds for the time being and go directly to a description of aromatic amines, which, considering the ability of living organisms to devise organic compounds, are remarkable because they do not occur in nature except in polycyclic and heterocyclic systems. By our definition, an aromatic amine is exemplified by aniline (ϕNH_2). It is an amine in which the nitrogen is bonded to an aromatic ring. Thus, benzylamine (ϕCH_2NH_2) is a nonaromatic amine. A few examples of naturally occurring or derivable aromatic amines are shown. They include a few of the more common substances in which the aromatic amino group is incorporated into a ring system.

22.2 Aromatic amines

p-aminobenzoic acid

pteroic acid

physostigmine

A few of the aromatic amines and substituted amines have common names: aniline, ϕNH_2; the toluidines, o-, m-, and p-$CH_3C_6H_4NH_2$; the anisidines, $MeOC_6H_4NH_2$; the phenetidines, $EtOC_6H_4NH_2$; anthranilic acid for o-aminobenzoic acid; benzidine for 4,4'-diaminobiphenyl; and picramide for 2,4,6-trinitroaniline. Otherwise, systematic names are used.

The spectra of the aromatic amines offer few surprises. In mass spectra, the parent peak is usually strong and the base peak is usually at $P - 27$ (or $P - HCN$). Aside from the expected N—H stretching vibrations (near 3400 and 3500 cm^{-1} for primary amines, near 3350 cm^{-1} for secondary amines, and broad hydrogen bonded peaks at lower frequencies), the C—N stretch appears at 1300 ± 40 cm^{-1}. The uv is strongly affected by electron-withdrawing substituents. Thus aniline absorbs at 230 nm (ε_{max} 8600) and at 280 nm (ε_{max} 1430), but p-nitroaniline absorbs at 381 nm (ε_{max} 13,500), which spills over into the visible region, causing the substance to be yellow.

The aromatic amines undergo all the reactions of the nonaromatic amines, but the presence of an interaction with the aromatic ring causes some differences. In addition, aromatic amines strongly affect the ring; and, as a result, those reactions in which the ring participates also are affected.

Basicity: conjugation. The influence of the ring (and substituents on the ring) is illustrated if we look at the basicity of the aromatic amines. Aniline has pK_a of 4.60 and is a much weaker base than ammonia (pK_a = 9.25) and the alkylamines (pK_a = 10–11). If we write only one of the Kekulé resonance structures and only the p structure for interaction with the ring, the ionization of the conjugate acid of aniline is shown in Equation 22.16.

$$\text{PhNH}_3^+ \xrightleftharpoons[]{H_2O, -H^+} \text{PhNH}_2 \leftrightarrow \text{(quinoid)} + H_3O^+ \quad (22.16)$$

The unprotonated amino group is able to conjugate with the ring by electron donation to the π system in such a way that the nitrogen accepts a partial positive charge and the ring a partial negative charge (a +R effect of the amino group). The protonated amine does not have a nonbonded pair of electrons to furnish, and the —NH$_3^+$ exerts only a strong −I effect on the ring. As a result, the stability of the amine is increased with respect to the protonated amine: K_a is larger than it otherwise would be (or pK_a is smaller). The acid is stronger and the base is weaker because the +R effect stabilizes only the base. Another way of looking at it is to say that the +R effect decreases the availability of the pair of electrons normally on nitrogen for bonding to an incoming proton. The presence of two rings, as in diphenylamine (ϕ_2NH), enhances the effect (pK_a = 0.79). Although diphenylamine still may be regarded as a base, it requires concentrated hydrochloric acid for solution. Addition of water to this solution causes the amine to precipitate. Triphenylamine (ϕ_3N) may be regarded as a neutral substance.

The basicity of amines contributes to the ability to form charge-transfer complexes. We may think of picrates of aromatic amines as salts rather than complexes.

Effects of substituents. As might be expected, substitution of alkyl groups on nitrogen causes a mild increase in basicity: pK_a values for ϕNHMe and ϕNMe$_2$ are 4.85 and 5.15. On the other hand, substitution on the ring affects the basicity to a larger extent, either increasing or decreasing it, depending upon the electron-donating or electron-withdrawing effect of the substituent and upon location of the substituent. Because the protonated amino group is unable to conjugate with the ring, no substituent is able to affect the stability of the conjugate acid of the amine to a great extent, even if the substituent is able to conjugate with the ring. Any effect within the ring may be transmitted to the protonated amino group only by an inductive effect through the C—N σ bond. The major effect of a substituent, then, is upon the unprotonated amino group by way of the ring and may be relatively large if the substituent is able to conjugate with the amino group. For conjugation, the substituent must be either *o* or *p* to the amino group and must have a −R effect. The pK_a values for some substituted anilines are given in Table 22.1 in order of increasing basicity for the *para*-substituted anilines.

Table 22.1 Substituted anilines: pK_a values at 25°

Substituent	pK_a ortho	pK_a meta	pK_a para	ΔpK_a[a] (p − m)	ΔpK_a[b] (m − H)
—NO$_2$	−0.29	2.46	1.00	−1.46	−2.14
—C≡N			1.74		
—COCH$_3$			2.19		
—COOCH$_3$			2.46		
—Br	2.53	3.58	3.86	0.28	−1.02
—Cl	2.65	3.52	3.98	0.46	−1.08
—H	4.60	4.60	4.60	0	0
—F	3.20	3.50	4.65	1.15	−1.10
—CH$_3$	4.44	4.73	5.08	0.35	0.13
—OCH$_2$CH$_3$	4.43	4.18	5.20	1.02	−0.42
—OCH$_3$	4.52	4.23	5.34	1.11	−0.37
—OH	4.7	4.3	5.6	1.3	−0.3
—NH$_2$	4.5	4.9	6.1	1.2	0.3

[a] The values in this column are obtained by subtraction of pK_a for the *meta* isomer from pK_a for the *para* isomer and serve to give the sign and a rough magnitude of R, the resonance effect of the substituent.

[b] The values in this column are obtained by subtraction of pK_a of aniline from the pK_a of the *meta*-substituted aniline and serve to give the sign and a rough magnitude of I, the inductive effect of the substituent (as felt by the amino group).

22.2 Aromatic amines

Dipole moments. It is interesting to compare some dipole moment data. Aniline (*a*) has μ = 1.55 D, and nitrobenzene (*b*), 3.95 D (at 25° in benzene). If there

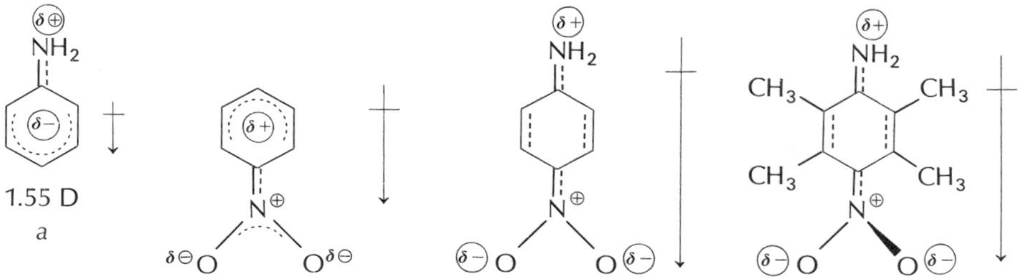

were no interaction (conjugation) between the two groups, *p*-nitroaniline would be expected to have 1.55 + 3.95 = 5.5 D. The compound actually has μ = 6.3 D. The structure shown as *c* must contribute strongly to the overall electron distribution. In the last compound shown, 2,3,5,6-tetramethyl-4-nitroaniline (*d*), the dipole moment (the methyl dipoles cancel each other) again is expected to be 5.5 D. However, the observed moment is only 5.0 D. This result is interpreted as a steric effect. The methyl groups at 3 and 5 force the nitro group at 4 to rotate out of the plane of the benzene ring and decrease the conjugation of the nitro group with the ring (from 3.95 to 3.45 D?) to such an extent that the moment is less than expected. This effect is known as <u>steric inhibition of resonance</u>. In this case, it reduces the magnitude of the −R effect of the nitro group.

The ortho effect. Returning to Table 22.1, we now note that only if oxygen is in an *o* position is the basicity greater than that of the *m* isomer. In all other cases, the *o* isomers are less basic than the *m* isomers and the *p* isomers. We saw the same effect with the *o*-substituted benzoic acids. The *ortho* effect is independent of the I and R effects. It favors the stability of the unprotonated amine. It has been postulated that the *ortho* effect is caused by steric interference with the ground state, the transition state, hydrogen bonding, solvation, and so on. In short, there is no single accepted explanation of the effect.

Reactions: amine reactions. Aside from their lessened basicity, the aromatic amines undergo all the usual amine reactions (Sec. 20.2) except for the reaction with nitrous acid, several rearrangements from nitrogen to the ring, and oxidations. The exceptions all involve the ring in one way or another. Some of the usual amine reactions are shown in Equations 22.17.

Electrophilic substitution. The free amino group is a powerful *o,p* activator and director, exceeded only by —O$^\ominus$, the anion of a phenol. However, electrophilic aromatic substitution usually is carried out in the presence of strong acids so that the concentration of free amine is small. The protonated amino group bears a positive charge (compare —$\overset{\oplus}{\text{N}}Me_3$) and is deactivating for substitution, and thus *m* directing. Let us look at the nitration of aniline as our first example (Eq. 22.18). Depending upon the concentration of mixed

$$\phi NH_2 + (MeO)_2SO_2 \longrightarrow \phi \overset{\oplus}{N}H_2Me + MeOSO_3^{\ominus}$$

$$\phi NH_2 + (CH_3CO)_2O \longrightarrow CH_3CONH\phi + CH_3COOH$$

$$\phi NHMe + \phi SO_2Cl \xrightarrow{NaOH} \phi SO_2 \underset{\underset{Me}{|}}{N}\phi + NaCl$$

$$\phi NH_2 + \phi \overset{\oplus}{N}H_3Cl^{\ominus} \xrightarrow[36 \text{ hr}]{225°} \phi_2NH + NH_4Cl \qquad (22.17)$$

$$\phi NH_2 + \phi CHO \longrightarrow \phi CH{=}N\phi + H_2O$$

$$\phi NH_2 + NaNH_2 \rightleftharpoons \phi \overset{\ominus}{N}H \; Na^{\oplus} + NH_3$$

$$\phi NMe_2 + CH_3CO_3H \longrightarrow \phi -\underset{\underset{Me}{|}}{\overset{\overset{Me}{|}}{\overset{\oplus}{N}}}-O^{\ominus} + CH_3COOH$$

$$\phi NH_2 \xrightarrow[\text{fast}]{\overset{\oplus}{N}O_2} \underset{}{\text{o-nitroaniline}} + \underset{}{\text{p-nitroaniline}} \xrightarrow{H^{\oplus}} \text{protonated amines}$$

$$H^{\oplus} \updownarrow$$

$$\phi \overset{\oplus}{N}H_3 \xrightarrow[\text{slow}]{\overset{\oplus}{N}O_2} \underset{}{\text{m-nitroaniline}} \xrightarrow{H^{\oplus}} \text{protonated amine} \qquad (22.18)$$

acid and the temperature, varying small amounts of unprotonated aniline are present, and it undergoes rapid nitration to give the o and p isomers. The high concentration of protonated aniline is substituted slowly at the m position. Nitration of the quaternary phenyltrimethylammonium ion ($\phi \overset{\oplus}{N}Me_3$) gives 11% p and 89% m substitution. As a result of the competition (small concentration, fast; and large concentration, slow), the nitration gives rise to a mixture of all three isomers. In addition, yields are diminished by oxidation of aniline. Direct nitration of aniline (or of ring-substituted anilines) seldom is carried out because better alternatives are available.

Most of the alternative routes involve the use of an acetylated aniline. The conversion of —NH_2 to —$NHCOCH_3$ decreases the basicity of nitrogen so much that protonation is diminished, and the o,p activation and direction, although also diminished, still are powerful. Acetanilide (b) is less basic than aniline because of the contribution of a and c to the structure. Structure a contributes less to the structure than does c, and the ring positions become less electrophilic than they would be in aniline as a consequence. Nevertheless, the intermediate d in nitration is favored so decisively that nitration of acetanilide gives a 90% yield of p-nitroacetanilide. The acetylation also diminishes the susceptibility to oxidation, and this inertness contributes to the high yield. The size of the acetamido group diminishes the amount of o isomer obtained to a few percent by steric hindrance. Thus p-nitroaniline is best prepared as in Equation 22.19. In this case, three steps in excellent yield are better than a single step that has poor yield and a separation problem.

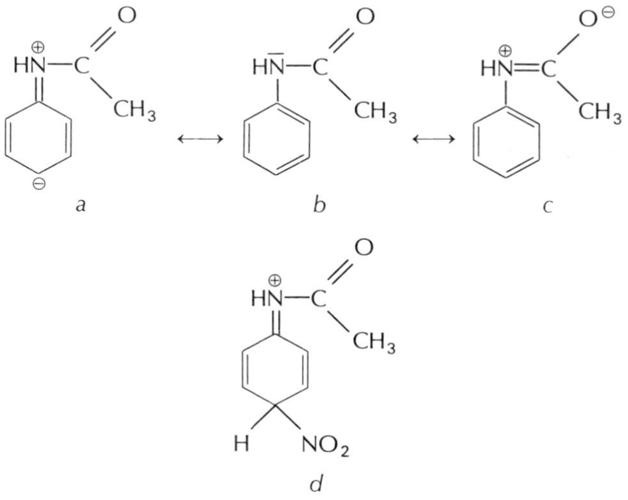

22.2 Aromatic amines

[Structures a, b, c showing resonance structures of acetanilide with HN⁺=C(O⁻)CH₃ and related forms]

[Structure d showing protonated acetanilide with H and NO₂ at para position]

Acid hydrolysis is necessary in the last step. Basic hydrolysis of *p*-nitroacetanilide indeed gives *p*-nitroaniline, but *p*-nitroaniline undergoes aromatic nucleophilic substitution easily with hydroxide ion to give *p*-nitrophenol and ammonia.

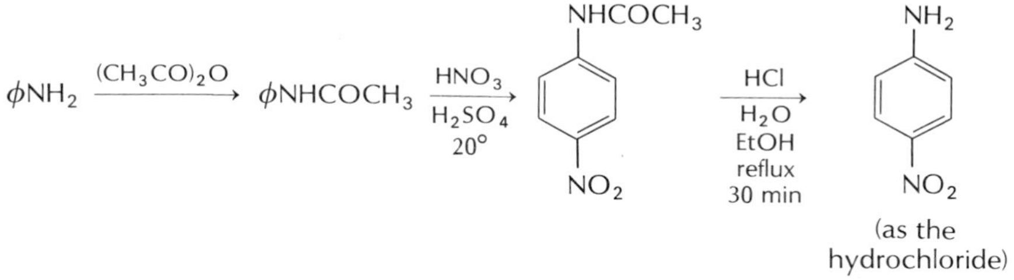

$$\phi NH_2 \xrightarrow{(CH_3CO)_2O} \phi NHCOCH_3 \xrightarrow[H_2SO_4]{HNO_3} \text{(p-NHCOCH}_3\text{-C}_6\text{H}_4\text{-NO}_2\text{)} \xrightarrow[\substack{H_2O \\ EtOH \\ reflux \\ 30 \text{ min}}]{HCl} \text{(p-NH}_2\text{-C}_6\text{H}_4\text{-NO}_2\text{)} \quad (22.19)$$

(as the hydrochloride)

Routes to nitroanilines. How do we prepare *o*-nitroaniline? We block the *p* position with a removable group and force the nitration of acetanilide into the *o* position. A suitable removable group is —SO₃H (the sulfonic acid group), and the sequence of steps is shown in Equation 22.20.

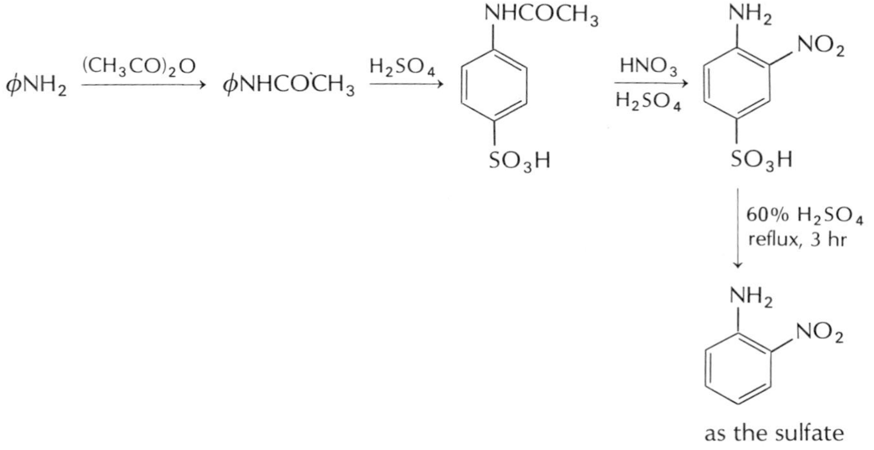

$$\phi NH_2 \xrightarrow{(CH_3CO)_2O} \phi NHCOCH_3 \xrightarrow{H_2SO_4} \text{(p-NHCOCH}_3\text{-C}_6\text{H}_4\text{-SO}_3\text{H)} \xrightarrow[H_2SO_4]{HNO_3} \text{(2-NO}_2\text{-4-SO}_3\text{H-aniline)}$$

$$\xrightarrow[\substack{60\% \text{ H}_2\text{SO}_4 \\ reflux, 3 \text{ hr}}]{} \text{(o-NH}_2\text{-C}_6\text{H}_4\text{-NO}_2\text{)} \quad (22.20)$$

as the sulfate

In the preparation of *m*-nitroaniline, benzene is dinitrated (the second nitro group enters *m* to give *m*-dinitrobenzene) and the reduction with $(NH_4)_2S$ is used (Eq. 22.10). The nitroanilines are reducible by the usual methods to the *o*-, *m*-, and *p*-phenylenediamines (common names).

Toluidines. The various toluidines, after acetylation, nitration, and hydrolysis, give the various nitrotoluidines as shown in Equations 22.21.

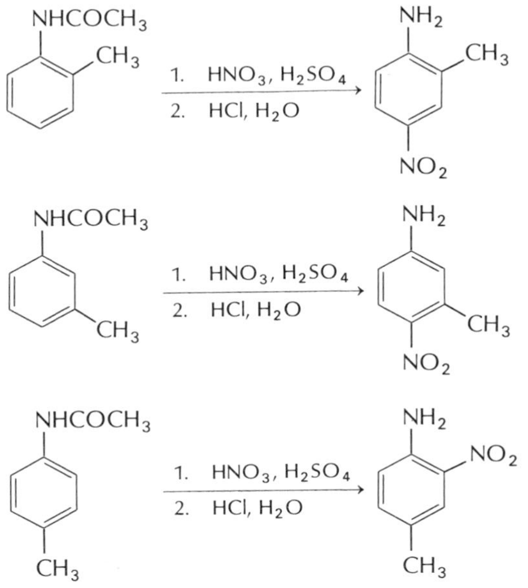

(22.21)

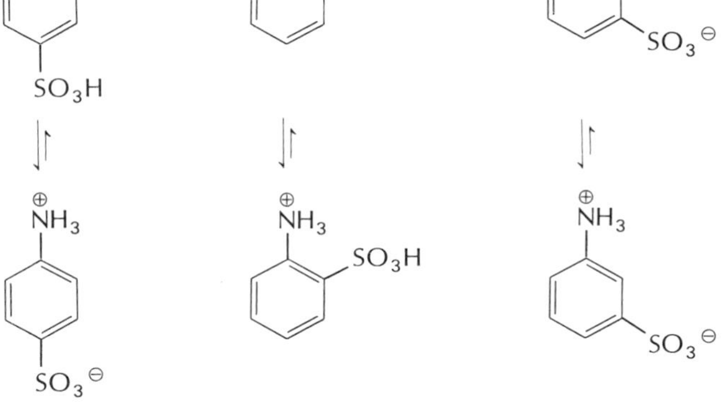

(22.22)

Sulfonation. In contrast to nitration, direct sulfonation of aniline carried out at 180°, or baking of the salt $\phi\overset{\oplus}{N}H_3\ HSO_4^{\ominus}$ at 180° to 190°, gives a reasonable yield of sulfanilic acid (*p*-aminobenzenesulfonic acid). The mechanism is in doubt, but we may rationalize the situation by pointing out that sulfonation is reversible and that if any of the unprotonated amine is sulfonated, the *p* isomer is by far the most stable because of conjugation and lack of steric hindrance. Instead of kinetic control, we have an instance of thermodynamic control (Eq. 22.22). We also should keep in mind that the aminobenzenesulfonic acids have a weak basic group and a very strong acid group, a situation favorable for zwitterion formation. Sulfanilic acid is insoluble in alcohol but will dissolve in hot water. It is inert to hydrochloric acid (the amino group is already protonated and a stronger acid than HCl is required to protonate the very weakly basic sulfonate anion), but it dissolves easily in dilute NaOH (deprotonation of the amino group).

Orthanilic acid is prepared by a sequence of reactions new to us. The third step is a nucleophilic aromatic substitution by disulfide ion ($S_2^{\ 2-}$), and the fourth is a chlorination with rupture of a S—S bond and oxidation by nitric acid (Eq. 22.23). The fifth step is reduction of the nitro group.

$$\phi H \xrightarrow[\text{FeCl}_3]{\text{Cl}_2} \phi Cl \xrightarrow[\text{H}_2\text{SO}_4]{\text{HNO}_3} \underset{+\text{ para}}{o\text{-O}_2N\text{-C}_6H_4\text{-Cl}} \xrightarrow[\text{aq. EtOH}]{\text{Na}_2\text{S}_2} \cdots \tag{22.23}$$

orthanilic acid

The preparation of metanilic acid is straightforward (Eq. 22.24).

$$\phi H \xrightarrow[\text{H}_2\text{SO}_4]{\text{HNO}_3} \phi NO_2 \xrightarrow[\substack{\text{H}_2\text{SO}_4 \\ 60°}]{\text{SO}_3} m\text{-O}_2N\text{-C}_6H_4\text{-SO}_3H \xrightarrow{\text{H}^{\oplus}, \text{Fe}} m\text{-}\overset{\oplus}{N}H_3\text{-C}_6H_4\text{-SO}_3^{\ominus} \tag{22.24}$$

Sulfanilamide and sulfa drugs: synthesis. Sulfanilamide, the first sulfa drug (1935), was the first big breakthrough in chemotherapy, the use of a chemical to inhibit the growth of disease-causing organisms within the body. The major requirement is a maximum of antagonism to the disease-causing organism with a minimum of toxicity and untoward side effects, such as nausea, headache, allergy, and soon, to the host.☐☐

The sulfa drugs are relatively easy to put together (Eq. 22.25). The chlorosulfonation is believed to proceed by way of the sulfonic acid, which then undergoes exchange with chlorosulfonic acid to give the *p*-acetamidobenzenesulfonyl chloride. The formation of the sulfonamide *b* (if R = H, the product is sulfanilamide) is similar to the Hinsberg reaction,

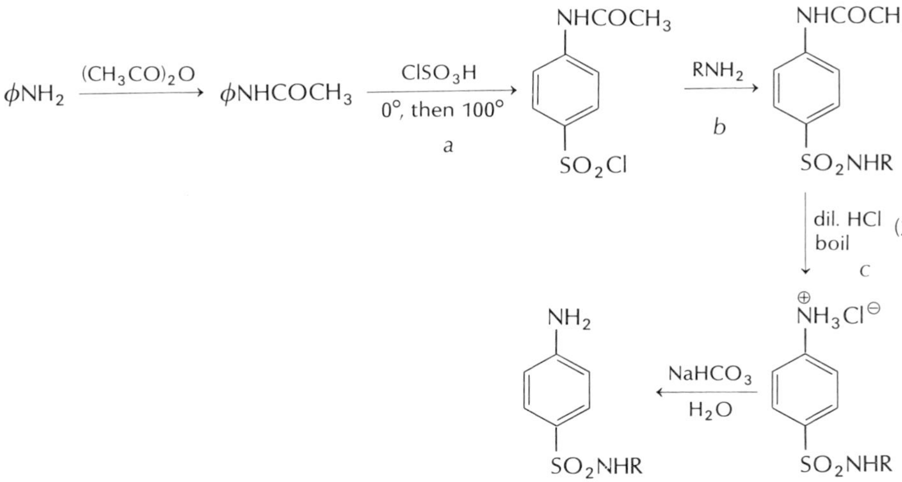

$\phi NH_2 \xrightarrow{(CH_3CO)_2O} \phi NHCOCH_3 \xrightarrow[0°, \text{then } 100°]{ClSO_3H}$ [p-NHCOCH₃-C₆H₄-SO₂Cl] $\xrightarrow{RNH_2}_{b}$ [p-NHCOCH₃-C₆H₄-SO₂NHR]

$\xrightarrow[\text{boil}]{\text{dil. HCl}}$ (22.25)

[p-NH₃⁺Cl⁻-C₆H₄-SO₂NHR] $\xrightarrow{NaHCO_3, H_2O}$ [p-NH₂-C₆H₄-SO₂NHR]

but without the presence of a strong base (NaOH) that would hold the product in solution as the sodium salt. The hydrolysis step (c) must be acidic if we are to avoid aromatic nucleophilic displacement of the amino group. (The —SO₂NHR is moderately activating for nucleophilic displacement, as are other m-directing groups.) The acetamido group is hydrolyzed much more rapidly than is the sulfonamido group. (Sulfonyl chlorides are less reactive than are carboxylic acid chlorides, etc.) A pentacoordinate intermediate (with more crowding) must be passed through with a sulfonyl group reaction, as compared with the tetrahedral intermediate of the usual olysis reactions of acids and acid derivatives. Other synthetic routes have been devised and used.

Halogenation. The amino group is such a good activating group for electrophilic substitution that, under polar aqueous conditions, it is not possible to stop halogenation short of trisubstitution (Eqs. 22.26) unless there are blocking groups in the 2, 4, and 6 positions.

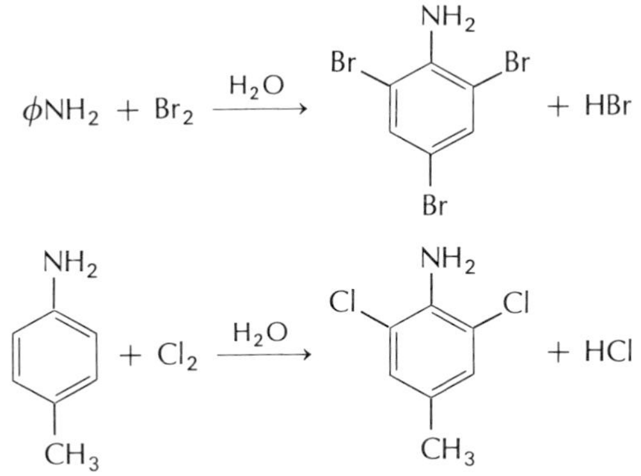

$\phi NH_2 + Br_2 \xrightarrow{H_2O}$ [2,4,6-tribromoaniline] $+ HBr$

(22.26)

[p-toluidine] $+ Cl_2 \xrightarrow{H_2O}$ [2,6-dichloro-4-methylaniline] $+ HCl$

The di- and trihalo aromatic amines are weak bases and precipitate from dilute aqueous acid solutions. By the use of relatively nonpolar (but basic) conditions, it is possible to introduce a single halogen into an aromatic amine in reasonable yield (Eqs. 22.27). Under these conditions, it is possible that the first step is halogenation of the amine group, followed by a rearrangement to the ring by nucleophilic replacement of hydrogen as shown for N-*tert*-butylaniline with calcium hypochlorite in CCl₄. (This reaction however,

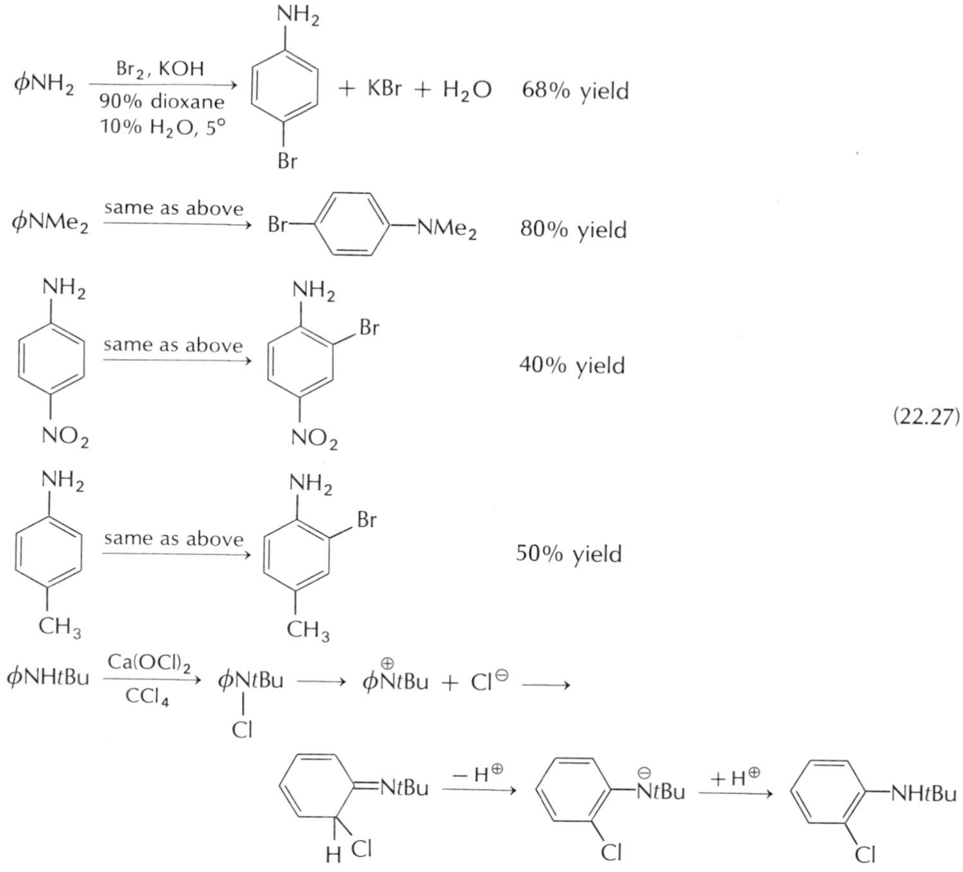

$$\phi NH_2 \xrightarrow[\substack{90\% \text{ dioxane} \\ 10\% H_2O, 5°}]{Br_2, KOH} \underset{Br}{\underset{\big|}{\bigodot}}\text{-}NH_2 + KBr + H_2O \quad 68\% \text{ yield}$$

$$\phi NMe_2 \xrightarrow{\text{same as above}} Br\text{-}\bigodot\text{-}NMe_2 \quad 80\% \text{ yield}$$

[p-nitroaniline] $\xrightarrow{\text{same as above}}$ [2-bromo-4-nitroaniline] 40% yield

[p-toluidine] $\xrightarrow{\text{same as above}}$ [2-bromo-4-methylaniline] 50% yield

(22.27)

$$\phi NHtBu \xrightarrow{\underset{CCl_4}{Ca(OCl)_2}} \phi \underset{\underset{Cl}{\big|}}{NtBu} \longrightarrow \phi \overset{\oplus}{N}tBu + Cl^{\ominus} \longrightarrow$$

[cyclohexadienyl intermediate =NtBu with H, Cl] $\xrightarrow{-H^{\oplus}}$ [o-chloro-N-tBu anilide anion] $\xrightarrow{+H^{\oplus}}$ [o-chloro-NHtBu]

favors *ortho* over *para* by 2 to 1.) An alternative is to acetylate the amine, brominate *para* in acetic acid, and hydrolyze the amide (the same route used for mononitration).

Friedel-Crafts reaction: use of amides. The Friedel-Crafts reaction with an aromatic amine is unsatisfactory for two reasons. The amino group complexes with aluminium chloride. Deactivation of the ring prevents the reaction, which is sensitive to activation influences, from proceeding. When necessary, moderate yields of ketone can be obtained by the use of an amide with an excess of aluminium chloride (Eq. 22.28).

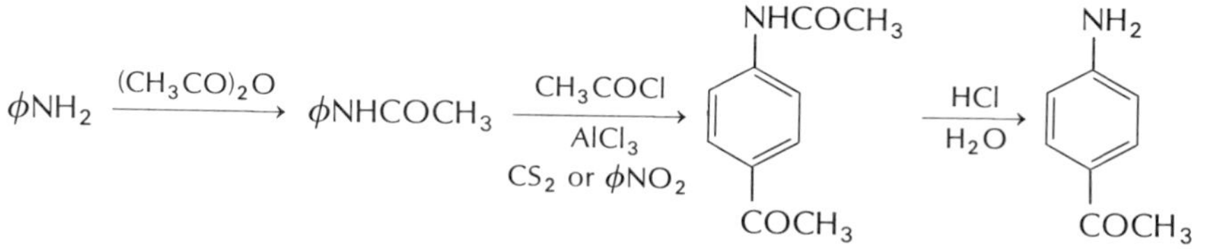

$$\phi NH_2 \xrightarrow{(CH_3CO)_2O} \phi NHCOCH_3 \xrightarrow[\substack{AlCl_3 \\ CS_2 \text{ or } \phi NO_2}]{CH_3COCl} \underset{COCH_3}{\underset{\big|}{\bigodot}}\text{-}NHCOCH_3 \xrightarrow{\underset{H_2O}{HCl}} \underset{COCH_3}{\underset{\big|}{\bigodot}}\text{-}NH_2 \quad (22.28)$$

Oxidation. Aromatic amines are very susceptible to oxidation by air, and they become colored with yellow to red oxidation products. However, a simple distillation with

Reaction with nitrous acid: diazotization of primary aromatic amines. The reactions of nitrous acid with the aromatic amines differ somewhat from the reactions with the nonaromatic amines. If a cold solution of sodium nitrite is added to a solution of aniline in three or more equivalents of hydrochloric acid ($\phi\overset{\oplus}{N}H_3 Cl^{\ominus}$ frequently crystallizes) at 0° to 5°, phenyldiazonium chloride is formed. Unlike the situation with the nitrosation of nonaromatic primary amines, which gives an unstable diazonium ion (Eq. 20.7), the aromatic diazonium ions are relatively stable (Eq. 22.29). If a solution of a diazonium salt (hydro-

$$\phi\bar{N}H_2 + |\overset{\oplus}{N}=\bar{O}| \xrightarrow{0°} \phi\overset{\oplus}{N}(H)(H)-\underline{N}=\underline{O}| \xrightarrow[\text{and out}]{H^{\oplus} \text{ in}} \phi\bar{N}(H)-\underline{N}=\overset{\oplus}{O}H \longleftrightarrow \phi\overset{\oplus}{N}(H)=\underline{N}-\bar{O}H$$

$$+H^{\oplus} \Big\updownarrow -H^{\oplus} \qquad \qquad \Big\downarrow H^{\oplus} \text{ in and out}, -H_2O$$

$$\phi\overset{\oplus}{N}H_3 \qquad\qquad\qquad \phi\overset{\oplus}{N}\equiv\underline{N} \longleftrightarrow \phi\underline{N}=\overset{\oplus}{\underline{N}} \qquad (22.29)$$

chloride or sulfate) is heated, decomposition occurs and the diazonium group is replaced by a hydroxy group (Eq. 22.30). However, the hydrolytic decomposition is only one of many

$$\phi\overset{\oplus}{N}\equiv N + H_2O \xrightarrow{100°} \phi OH + N_2 + H^{\oplus} \qquad (22.30)$$

reactions that the diazonium compounds undergo, as we shall see.

Nitrosation of secondary aromatic amines. Secondary aromatic amines, like nonaromatic amines, react with nitrous acid to give N-nitrosoamines (Eq. 22.31). However,

$$\phi NHCH_3 \xrightarrow[\substack{NaNO_2 \\ 0° \\ 1\,hr}]{HCl,} \phi-\overset{\overset{\displaystyle H}{|}}{\underset{\underset{\displaystyle H_3C}{|}}{N}}{}^{\oplus} \; Cl^{\ominus} \xrightarrow{-H^{\oplus}} \underset{H_3C}{\overset{\phi}{\diagdown}} N-\underline{N}=\bar{O}| \qquad (22.31)$$

aromatic N-nitrosoamines are rather more useful than are the nonaromatic relatives because an acid-catalyzed rearrangement (the Fischer-Hepp rearrangement) is possible. The nitroso group, more labile than alkyl, rearranges to the *p* position (if not blocked) if the N-nitroso-N-alkylaniline is dissolved in a mixture of alcohol and ether to which some concentrated hydrochloric acid is added at room temperature (Eq. 22.32). The value of the reaction lies in the mildness (and good yield) whereby an N-alkylaniline may be converted into the *p*-nitroso derivative. From the latter, a *p*-nitro derivative can be made by mild oxidation or a *p*-amino derivative by reduction.

Nitrosation of tertiary aromatic amines. N,N-dialkylanilines, it is thought, react initially to give an unisolable quaternary nitrosoamino ion, which rearranges rapidly and spontaneously to give the *p*-nitroso product in excellent yield (Eq. 22.33).

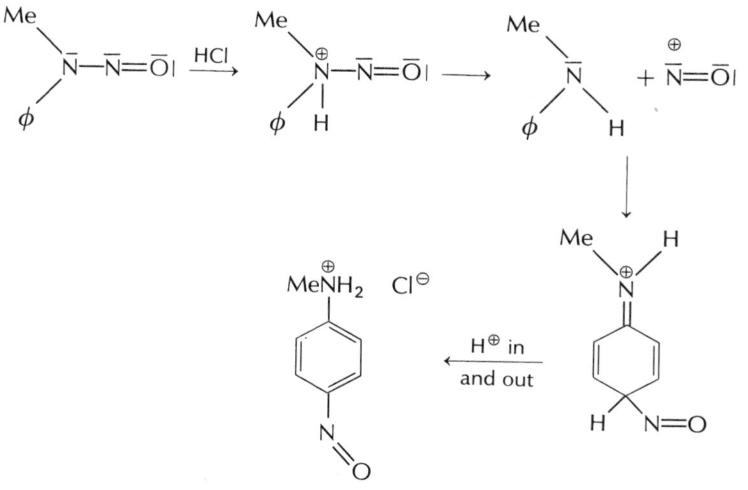

$$\text{Me}-\underset{\phi}{\text{N}}-\overset{..}{\text{N}}=\overset{..}{\text{O}}| \xrightarrow{\text{HCl}} \text{Me}-\overset{\oplus}{\underset{\phi}{\text{N}}}-\overset{..}{\text{N}}=\overset{..}{\text{O}}| \longrightarrow \underset{\phi}{\text{Me}}\text{N}\overset{H}{\underset{}{}} + \overset{\oplus}{\text{N}}=\overset{..}{\text{O}}|$$

(22.32)

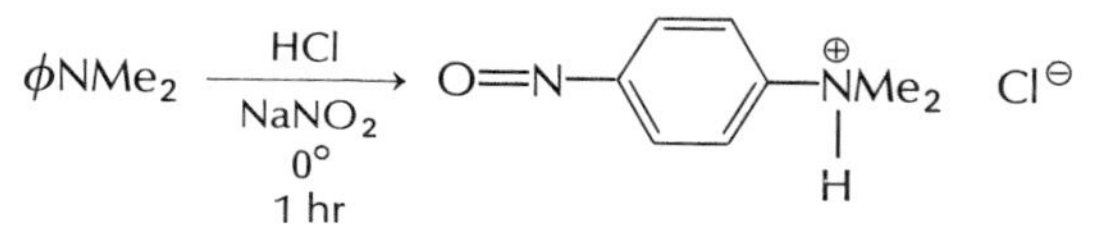

[Mechanism continues: MeNH$_2^\oplus$ Cl$^\ominus$ / p-nitroso-N-methylaniline ⇌ cyclohexadienimine intermediate, H$^\oplus$ in and out]

$$\phi\text{NMe}_2 \xrightarrow[\substack{\text{NaNO}_2 \\ 0° \\ 1\text{ hr}}]{\text{HCl}} \text{O}=\text{N}-\!\!\left\langle\right\rangle\!\!-\overset{\oplus}{\underset{H}{\text{NMe}_2}} \text{Cl}^\ominus \quad (22.33)$$

Preparations: Bucherer reaction. The preparation of certain substituted anilines has been described. Here we list the reactions useful in the preparation of aromatic amines.

1. Reduction of nitro compounds (Section 22.1)
2. Reduction of nitroso (and all other nitrogen compounds at a higher oxidation state than amino (Sec. 22.1)
3. Aromatic nucleophilic substitution reactions (Sec. 15.5)
4. Benzyne reactions (Sec. 15.5)
5. Rearrangements of:
   a. Oximes of alkyl aryl ketones (Beckmann, Sec. 20.3)
   b. Aryl amides, azides, hydroxamic acids (Hofmann, Sec. 20.3)
6. The <u>Bucherer reaction</u> is employed industrially far more than it is in the laboratory. A description follows.

The Bucherer reaction is useful for conversions into amines of those phenols in which the keto tautomeric structure is not too unfavorable. The catalyzing reagent is bisulfite ion ($HSO_3^\ominus$) and the reactant is ammonia or an amine (Eq. 22.34). The reaction mixture is

$$\text{ArOH} \underset{150°, 6\text{ atm} \atop 8\text{ hr}}{\overset{HSO_3^\ominus,\ NH_3}{\rightleftharpoons}} \text{ArNH}_2 \quad (22.34)$$

prepared from concentrated ammonium hydroxide and SO_2 or $NaHSO_3$. The reaction requires a pressure apparatus known as an autoclave. The reaction is useful primarily for the preparation of β-napthylamine (a carcinogen), which is not otherwise conveniently obtainable. (β-Nitronapthalene is not available by the nonreversible nitration of napthalene, whereas β-napthalenesulfonic acid is prepared by a reversible sulfonation at high temperature under thermodynamic control and may be converted into β-napthol by fusion with NaOH.)

The mechanism seems to be as shown in Equation 22.35. The reaction is reversible: β-napthylamine may be converted to β-napthol by prolonged boiling with NaHSO₃ solution. Although the overall reaction as written (Eq. 22.34) appears to be an aromatic nucleophilic substitution, we see that it actually is a series of addition-elimination steps including tautomerization. The reaction does not take place with phenol, although resorcinol reacts well.

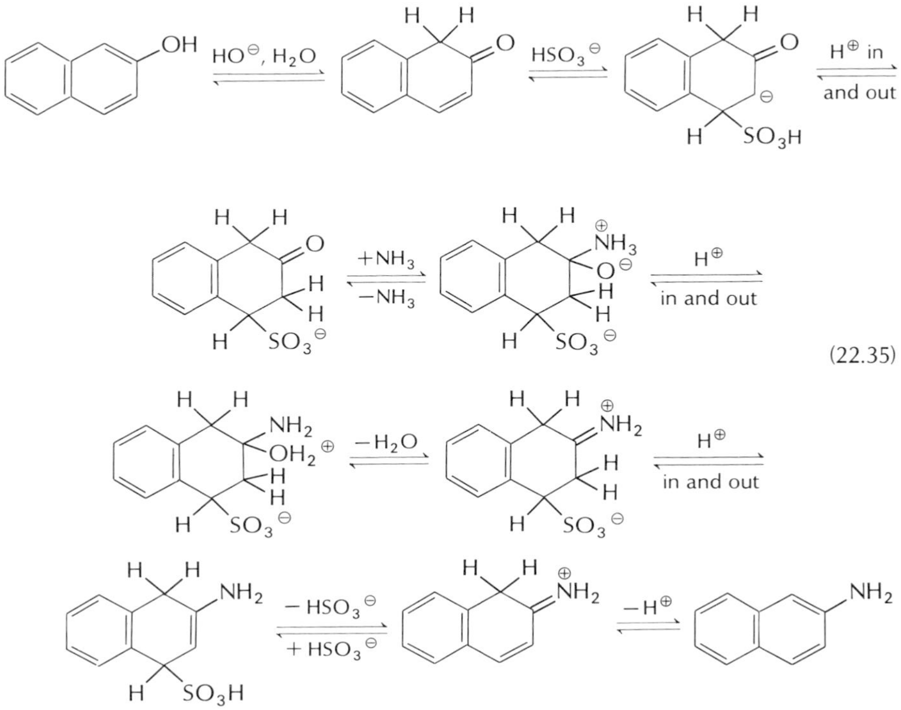

$$(22.35)$$

Exercises

22.4 Why should we expect little or no hyperconjugation in the —$NH_3^{\oplus}$ group (+R), whereas —CH_3 has +R of about 0.35?

22.5 Which is the weaker base? Why?

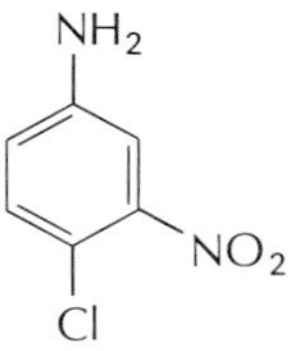

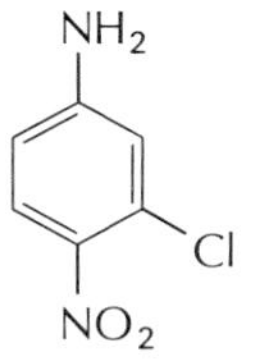

22.6 Show how to prepare the compounds shown, beginning with benzene, toluene, or napthalene and any other reagents needed.

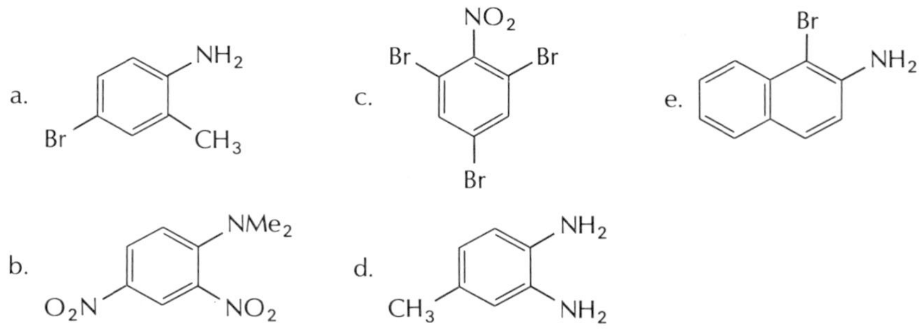

a. [4-bromo-2-methylaniline structure]

b. [4-nitro-2-(N,N-dimethylamino)-3-nitro substituted benzene: NMe₂, NO₂, O₂N substituents]

c. [2-nitro-1,3,5-tribromobenzene structure]

d. [3-methyl-1,2-diaminobenzene structure]

e. [1-bromo-2-aminonaphthalene structure]

22.3 Aromatic diazonium compounds

Properties of diazonium compounds. The aromatic diazonium compounds are relatively unstable substances that usually are prepared in cold, acidic solutions (Eq. 22.29) and used for subsequent reactions without isolation. In a sense, the diazonium compounds are analogous to nitrous acid (Eq. 20.4) and the nitrite salts (Eqs. 22.36). Nitrous acid with

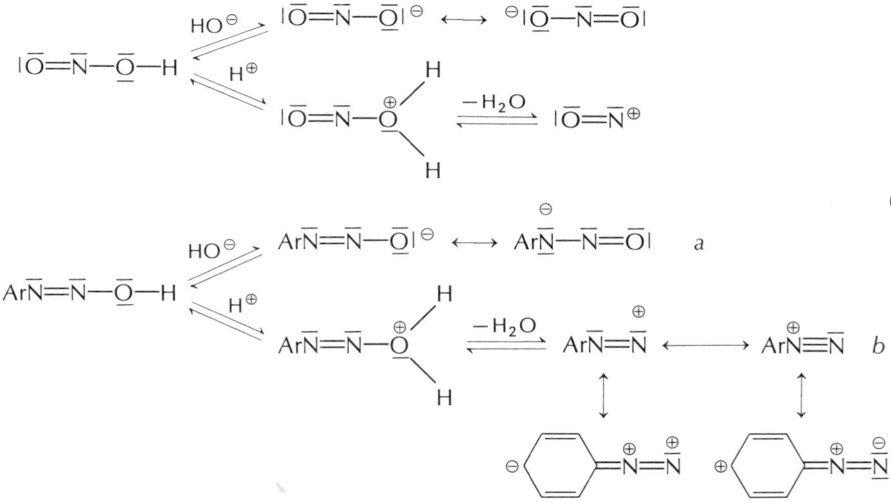

$$|\bar{O}=\bar{N}-\underline{O}-H \; \underset{H^{\oplus}}{\overset{HO^{\ominus}}{\rightleftarrows}} \; |\bar{O}=\bar{N}-\bar{\underline{O}}|^{\ominus} \longleftrightarrow {}^{\ominus}|\underline{\bar{O}}-\bar{N}=\bar{O}|$$

$$|\bar{O}=\bar{N}-\overset{\oplus}{\underline{O}}\overset{H}{\underset{H}{\diagdown}} \; \underset{}{\overset{-H_2O}{\rightleftarrows}} \; |\bar{O}=\bar{N}^{\oplus}$$

$$Ar\bar{N}=\bar{N}-\underline{O}-H \; \underset{H^{\oplus}}{\overset{HO^{\ominus}}{\rightleftarrows}} \; Ar\bar{N}=\bar{N}-\bar{\underline{O}}|^{\ominus} \longleftrightarrow Ar\bar{N}-\bar{N}=\bar{O}| \quad a$$

$$Ar\bar{N}=\bar{N}-\overset{\oplus}{\underline{O}}\overset{H}{\underset{H}{\diagdown}} \; \underset{}{\overset{-H_2O}{\rightleftarrows}} \; Ar\bar{N}=\bar{N}^{\oplus} \longleftrightarrow Ar\bar{N}{\equiv}N \quad b$$

(22.36)

hydroxide ion gives nitrite ion. Addition of hydrogen ion (strong acid) reverses the equilibrium, and nitrosonium ion ($ON^{\oplus}$) is formed. The corresponding diazoacid, like nitrous acid, cannot be isolated. Furthermore, no evidence (uv, nmr) has been found for its existence in solution. Nevertheless, at high pH the anion *a* is known, and metal salts may be isolated as with nitrite ion. At low pH, the cation *b* is formed. Evidently the equilibrium between *a* and *b* at intermediate hydrogen ion concentrations is rapid and consists almost completely of the ions *a* and *b*, with practically none of the substance present as ArN=NOH or ArN=NOH$_2^{\oplus}$. Ion *a* is known as the diazoate ion and ion *b* as the diazonium ion. If KCN is added to a solution of a diazonium compound, the cyanide ion forms a covalent bond, and a nonionic diazocyanide (ArN=N—CN) results that exists in interconvertible *syn* and *anti* geometrical isomeric structures.

The diazonium ion, being a cation, requires an anion as the counterion. As it is usually prepared (NaNO$_2$ + HCl + arylamine), the anion is chloride ion. The diazonium chlorides are very soluble in water, much more so than the arylamine hydrochlorides. The diazonium sulfates are less soluble than the chlorides but still very soluble. The solid salts can be made in several ways, but they must be handled with caution because they are so unstable that they may decompose explosively. However, addition of NaBF$_4$ to a solution of a diazonium chloride (or sulfate) gives a precipitate of the insoluble salt diazonium fluoroborate (ArN≡N$^\oplus$ BF$_4^\ominus$), which is stable and safe. Commercially, the double salt with zinc chloride (ArN≡N$^\oplus$ ZnCl$_3^\ominus$) is used as a safe diazonium salt.

Reactions. The diazonium compounds are of interest because they have many synthetically useful reactions. We group the reactions into five classes. In the first two classes, the two atoms of nitrogen are retained in the products. In the other three, nitrogen gas is evolved in the course of the reactions.

Reduction. A diazonium salt can be reduced with Na$_2$SO$_3$ to give a product that can be hydrolyzed with HCl to give the hydrochloride of an arylhydrazine. A possible mechanism is shown (Eqs. 22.37). An alternate reduction is the use of SnCl$_2$ in HCl, unless

$$\text{ArN}_2^\oplus \text{Cl}^\ominus + 2\text{Na}_2\text{SO}_3 + \text{HCl} + 2\text{H}_2\text{O} \longrightarrow \text{ArNHNH}_3^\oplus \text{Cl}^\ominus + \text{NaCl} + \text{NaHSO}_4 + \text{Na}_2\text{SO}_4$$

$$\phi\text{N}_2^\oplus + \text{SO}_3^{2-} \longrightarrow \phi\text{—N=N—S(=O)(O}^\ominus\text{)(O}^\ominus\text{)} \xrightarrow{\text{SO}_3^{2-}} \underset{\text{O}_3\text{S}}{\phi\text{N—N—SO}_3^\ominus} \xrightarrow{\text{H}^\oplus} \underset{\text{O}_3\text{S}}{\phi\text{N—NH—SO}_3^\ominus} \qquad (22.37)$$

$$\xrightarrow[\text{boil}]{\text{HCl}} \phi\text{NHNH}_3^\oplus \text{Cl}^\ominus + 2\text{HSO}_4^\ominus$$

other reducible groups (nitro, nitroso, etc.) are present, in which case the sulfite reduction is necessary. Although many substituted arylhydrazines may be made by this method from an arylamine, if several —I, —R groups are present it is preferable to use a nucleophilic aromatic substitution. (See Eq. 22.38, in which the preparation of 2,4-dinitrophenylhydrazine is shown.)

$$\text{2,4-(O}_2\text{N)}_2\text{C}_6\text{H}_3\text{Cl} + \text{H}_2\text{NNH}_2 \longrightarrow \text{2,4-(O}_2\text{N)}_2\text{C}_6\text{H}_3\text{NHNH}_2 \qquad (22.38)$$

Coupling reaction. Of great utility is the coupling reaction in which the diazonium ion, like the nitrosonium ion, is able to engage in aromatic electrophilic substitution with aromatic compounds such as phenols and amines, activated by the presence of strong *o,p* director substituents. The products are substituted azo compounds (Eq. 22.39). It is the

22.3 Aromatic diazonium compounds

$$\text{Me}_2\text{N}\phi + \text{N}\overset{\oplus}{\equiv}\text{N}-\phi \xrightarrow{0°-10°} \text{Me}_2\overset{\oplus}{\text{N}}=\underset{\text{N}=\text{N}\phi}{\underset{\text{H}}{\bigcirc}} \quad \left[\text{not Me}_2\text{N}=\underset{\text{H}}{\underset{\phi}{\bigcirc}}\overset{\text{N}\overset{\ominus}{=}\text{N}}{\underset{\oplus}{|}} \right]$$

$$\downarrow -\text{H}^{\oplus}$$

$$\text{Me}_2\text{N}-\bigcirc-\text{N}\underset{\text{N}\phi}{\overset{\nwarrow}{=}}$$

(22.39)

diazonium ion (favored by low pH) that reacts with the free amine (favored by high pH), so a compromise is needed. We remember that an aromatic amine is a relatively weak base, so that even at a pH of 4 to 5 there is enough of the free amine in solution to give a reasonably fast reaction. At a higher pH, the concentration of diazonium ion decreases and at a lower pH, the concentration of free amine decreases so much that the rate of coupling also decreases. Thus it is common practice in carrying out a coupling reaction to add sodium acetate to the acidic diazonium chloride (or sulfate) solution to buffer the mixture to a pH close to what is needed for maximum rate of reaction.

The reaction shown is for coupling with a tertiary amine, and it occurs preferentially at the *p* position to the amino group. If the *p* position is blocked, coupling then goes to the *o* position. With a primary or secondary amine, reaction with a diazonium ion goes first at the amine nitrogen atom, just as it does with nitrosonium ion, to give a diazoamino compound (Eq. 22.40). Warming a diazoamino compound with a strong acid (HCl is commonly

$$\phi\text{NH}_2 + \text{N}\overset{\oplus}{\equiv}\text{N}-\phi \xrightarrow[\text{0°-10°}]{\text{acetate buffer}} \phi\overset{\oplus}{\text{NH}}_2-\text{N}=\text{N}-\phi \xrightarrow{-\text{H}^{\oplus}} \phi\text{NH}-\text{N}=\text{N}-\phi$$

(22.40)

used) causes the usual rearrangement from amino nitrogen to the *p* position. (Compare Eq. 22.41 with Eq. 22.32 for the rearrangement of a nitroso group.) During diazotization

$$\bigcirc-\text{NH}\underset{\text{N}}{\overset{\nwarrow}{\underset{\text{N}}{|}}}\text{N}-\bigcirc \xrightarrow[\text{warm}]{\text{HCl}} \text{H}_2\text{N}-\bigcirc-\text{N}\underset{\text{N}}{\overset{\nwarrow}{=}}-\bigcirc$$

(22.41)

of an aromatic amine, the solution is kept strongly acidic to prevent the formation of the diazoamino compound. In other words, the nitrosonium ion is more reactive than and competes successfully with diazonium ion for attack on aniline at low pH.

Coupling with phenols is relatively slow, but it is rapid with phenoxide ions, which require a relatively high pH. The pH must not be so high, however, as to completely convert the diazonium ion into the diazotate ion, which does not couple. A maximum rate is achieved at pH 9 to 10. In practice, the cold acidic solution of the diazonium ion is added to a weakly alkaline solution of the phenol obtained by the use of Na_2CO_3 (Eq. 22.42). Coupling again goes to the *p* position, unless it is blocked, in which case *o* coupling occurs.

$$\phi\overset{\oplus}{\text{N}}\equiv\text{N} + \phi\text{O}^{\ominus} \xrightarrow[\text{H}_2\text{O}]{\text{cold}} \bigcirc-\text{N}\underset{\text{N}}{\overset{\nwarrow}{=}}-\bigcirc-\text{OH}$$

(22.42)

Effects of substituents. What is the effect of substituents? Any substituent in the phenol or aromatic amine that contributes additional stabilization to the intermediate in the electrophilic substitution reaction helps to lower the activation energy for the reaction and favor coupling. Thus, an *o,p* directing substituent in the 3 position (to the amino or hydroxy at 1) favors coupling. A similar substituent in the 2 position has little influence.

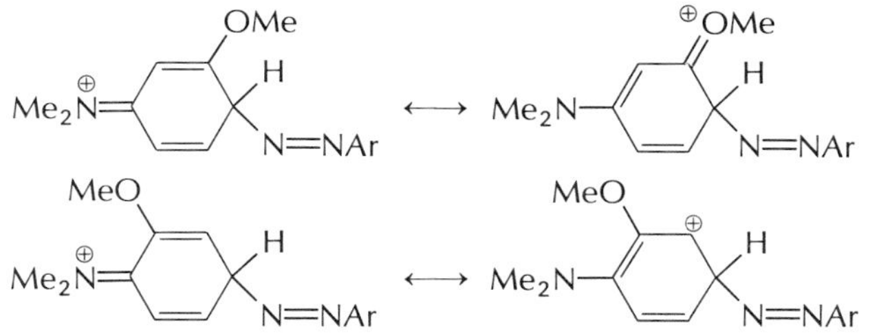

An *m*-directing substituent in the 3 position disfavors coupling more than does one in the 2 position (perhaps a steric effect).

Substituents in the diazonium ion that favor delocalization of the positive charge increase the stability of the ion and decrease the reactivity for coupling. Substituents that tend to remove electrons from the ring ($-I$, $-R$ effects) increase the localization of the positive charge in the diazo group, which increases the reactivity for coupling. Increasing the reactivity of the diazonium ion enables the coupling reaction to take place with aromatic compounds less reactive than the amines or phenoxide ions (Eqs. 22.43).

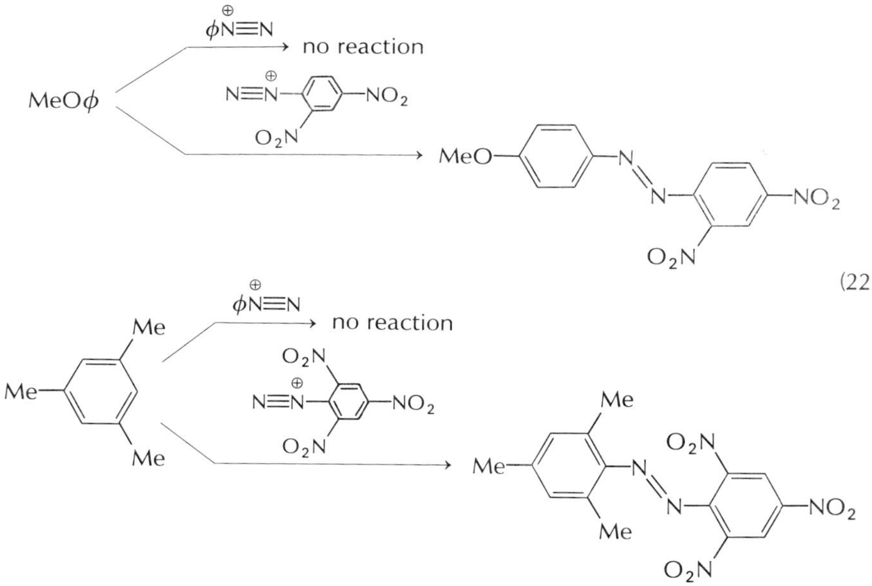

(22.43)

Indicators: color change with pH change. The syntheses of two indicators are shown in Equations 22.44. The indicator action of methyl orange is caused by the presence or absence of a proton. In basic and weakly acidic solution, the yellow anionic azo structure is present. Addition of a strong acid lowers the pH of the solution: a proton is accepted by the azo nitrogen and the conjugation is affected so that the absorption band is shifted to a longer wavelength. At the midpoint of the change, pH 3.8, both forms are present and the

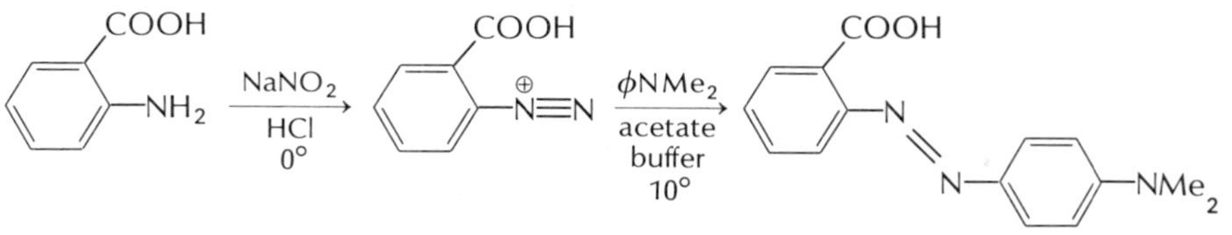

22.3 Aromatic diazonium compounds

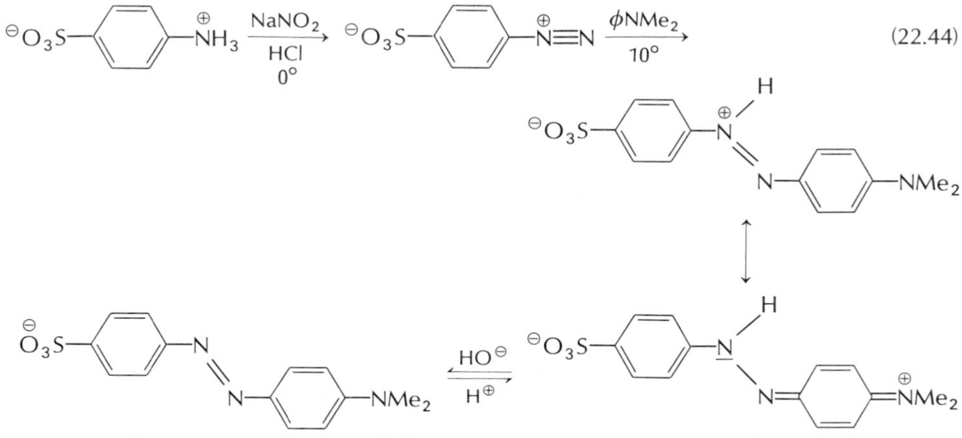

[Scheme: 2-aminobenzoic acid (anthranilic acid) → diazonium salt (NaNO₂/HCl, 0°) → coupling with φNMe₂ (acetate buffer, 10°) → **methyl red** (2-carboxyphenyl-azo-4-dimethylaminobenzene)]

$$\ominus O_3S\text{—}C_6H_4\text{—}\overset{\oplus}{N}H_3 \xrightarrow[\text{HCl, }0°]{\text{NaNO}_2} \ominus O_3S\text{—}C_6H_4\text{—}\overset{\oplus}{N}\!\!\equiv\!\!N \xrightarrow[10°]{\phi NMe_2} \tag{22.44}$$

[Scheme showing coupling product, then acid/base equilibrium:

yellow (pH > 4.4) ⇌ (HO⁻ / H⁺) ⇌ red (pH < 3.2)

methyl orange]

solution is orange. There are many other indicators, and they cover the pH range 1 through 13. For example, methyl red is red at pH < 4.8 and yellow at pH > 6.0. An indicator seldom is used as a dye, because the color change is not usually wanted in a fabric.

Dyes may be classified in several different ways: by structure; acidity or basicity; method of application; suitability for various textiles; fastness (durability) to light, to washing and so on; and other characteristics. The subject is well covered by Fieser and Fieser (Chap. 7, ref. 7) and Noller (Chap. 7, ref. 11). □ □

S_N1 replacements of nitrogen. Here we include those few reactions of the aromatic diazonium compounds that are analogous to the brisk decomposition of the nonaromatic diazonium ions (Eq. 22.30). This decomposition is the reason for keeping a diazotization cold. At 0° to 5° the decomposition is slow enough so that the formation of phenol is not troublesome. Nevertheless, usually enough phenol is formed and enough coupling of the phenol with remaining diazonium ion occurs to give some azo compound colors to the diazonium solution. For preparation of a phenol, the diazotization is done with sulfuric acid. (If HCl is used, the decomposition is accompanied by some replacement by chloride ion.) The cold diazonium solution is added slowly to hot dilute sulfuric acid. If the phenol is volatile with steam, it steam distils as it is formed. Otherwise, the mixture is cooled (Eq. 22.45) and the phenol crystallizes.

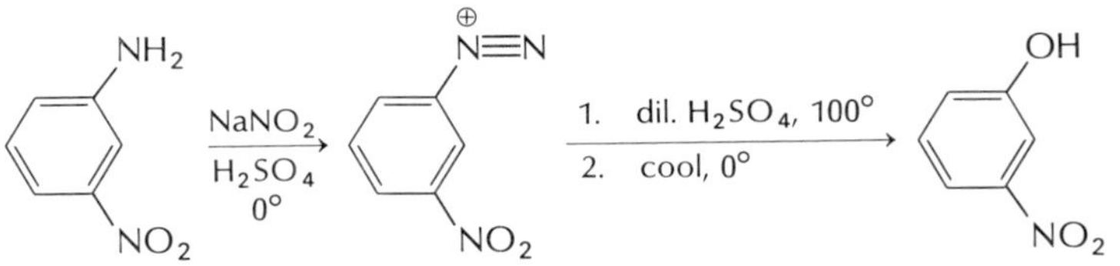

$$\underset{NO_2}{\underset{|}{C_6H_4}}\text{—}NH_2 \xrightarrow[\text{H}_2\text{SO}_4, 0°]{\text{NaNO}_2} \underset{NO_2}{\underset{|}{C_6H_4}}\text{—}\overset{\oplus}{N}\!\!\equiv\!\!N \xrightarrow[\text{2. cool, }0°]{\text{1. dil. H}_2\text{SO}_4,\,100°} \underset{NO_2}{\underset{|}{C_6H_4}}\text{—}OH \tag{22.45}$$

If NaBF$_4$ is added to a diazonium ion solution, the diazonium fluoroborate precipitates. The diazonium fluoroborates are stable and may be separated by filtration and dried with safety. When the dried salt is heated cautiously, a decomposition sets in that is known as the Schiemann reaction (Eq. 22.46). If the salt is not dry, the decomposition is more

$$\phi\overset{\oplus}{N}\equiv N\ BF_4^{\ominus} \xrightarrow{heat} \phi^{\oplus} + N_2 \xrightarrow{BF_4^{\ominus}} \phi F + BF_3 \quad (22.46)$$
$$55\%$$

vigorous. It is thought that the reaction proceeds by an S$_N$1 first step (as in the replacement by an —OH group) to give a phenyl cation, which reacts rapidly in the hot mixture to form a bond to fluorine from a fluoroborate ion.

A nucleophilic aromatic substitution is possible with good nucleophiles such as iodide ion (Eq. 22.47). A solution of KI is added to the cold acidic diazonium ion solution.

$$Ar\overset{\oplus}{N}\equiv N \xrightarrow[\text{2. warm}]{\text{1. KI, H}_2\text{O, 0°}} ArI + N_2 \quad (22.47)$$
$$70\%$$

The mixture is allowed to stand and warm up to room temperature (about 2 to 3 hr) and finally heated slowly to 100° to complete the reaction. If sodium azide is used, the product is phenyl azide ($\phi N\overset{\oplus}{=}N\overset{\ominus}{=}N$).

Radical reactions: Sandmeyer reaction. A variety of radical reactions are observed with the diazonium ions, many of which are catalyzed by or proceed with cuprous complexes (supplied as such or by finely divided copper metal). The Sandmeyer reaction utilizes preformed cuprous salts to replace the diazo group with —Cl, —Br, or —CN. The reaction is shown in Equation 22.48, which is schematic to show electron shifts. The reaction

$$\begin{array}{c}Ar\quad Y^{\ominus}\\ |\\ \overset{\oplus}{N}\leftarrow\overset{(+)}{Cu}\\ |||\\ \underline{N}\end{array} \longrightarrow \begin{array}{c}Ar\curvearrowright Y^{\ominus}\\ \overline{N}\quad (2+)\\ |||\quad Cu\\ \underline{N}\end{array} \longrightarrow \begin{array}{c}Ar-Y\\ \\ N_2\quad \overset{(+)}{Cu}\end{array} \quad (22.48)$$

actually proceeds by way of solid cuprous complexes, which decompose to give the product.

For replacement of the diazo group with chlorine, the cold solution of diazonium chloride is poured into a cold solution of freshly prepared cuprous chloride in concentrated HCl. The insoluble complex ArN$_2$$^{\oplus}$ CuCl$_2$$^{\ominus}$ separates. As the solution stands and slowly warms to room temperature, decomposition begins. The mixture is warmed slowly to maintain a gentle rate of decomposition (evolution of nitrogen) and is completed by heating to 100°. The product usually is isolated by steam distillation (Eqs. 22.49).

$$2CuSO_4 + 2NaCl + NaHSO_3 + H_2O \longrightarrow 2CuCl + 3NaHSO_4$$

$$CuCl + HCl + H_2O \longrightarrow H_3O^{\oplus} + CuCl_2^{\ominus} \quad (22.49)$$

$$Me-\underset{}{\bigcirc}-\overset{\oplus}{N}\equiv N \xrightarrow[\text{HCl, 0°}]{CuCl_2^{\ominus}} complex \xrightarrow{25°} Me-\underset{}{\bigcirc}-Cl + N_2 + CuCl$$
$$75\%$$

22.3 Aromatic diazonium compounds

For replacement with bromine, a cold solution of the diazonium sulfate is dripped into a boiling solution of freshly prepared cuprous bromide in concentrated HBr. The complex decomposes as rapidly as it is formed, and the aryl bromide steam distils from the reaction mixture (Eq. 22.50). The use of HCl is avoided to prevent formation of the chloro

$$\text{o-MeC}_6\text{H}_4\text{N}_2^{\oplus} \xrightarrow[\text{HBr, 100°}]{\text{CuBr}_2^{\ominus}} \text{Complex} \longrightarrow \text{o-MeC}_6\text{H}_4\text{Br (70\%)} + \text{N}_2 + \text{CuBr} \quad (22.50)$$

compound. The method of steam distillation and slow addition may be adapted to the preparation of chloro compounds.

For replacement of the diazo group with a cyano group, a cold solution of the diazonium chloride (or sulfate) is added in portions to a solution of cuprous cyanide in excess sodium or potassium cyanide held at 60° to 70°. The mixture (nearly neutral) is boiled for 15 to 20 min to complete the reaction and then is steam distilled (Eqs. 22.51). □ □

$$2\text{CuSO}_4 + 2\text{NaCN} + \text{NaHSO}_3 + \text{H}_2\text{O} \longrightarrow 2\text{CuCN} + 3\text{NaHSO}_4$$

$$\text{CuCN} + 2\text{NaCN} \longrightarrow \text{Cu(CN)}_3^{2-} + 2\text{Na}^{\oplus}$$

$$\text{Me-C}_6\text{H}_4\text{-N}_2^{\oplus}\,\text{Cl}^{\ominus} \xrightarrow[60°]{\text{Na}_2\text{Cu(CN)}_3} \text{Complex} \xrightarrow{100°} \text{Me-C}_6\text{H}_4\text{-C}\equiv\text{N (65\%)} + \text{N}_2 + \text{CuCN} + \text{NaCl} \quad (22.51)$$

Reduction. A useful reaction is reduction of the diazonium group by replacement with hydrogen. Of the various methods, the most generally useful is reduction by hypophosphorous acid (H_3PO_2) in the presence of a little cuprous salt (Eqs. 22.52). (The rate is

$$\text{ArN}_2^{\oplus} + \text{H}_3\text{PO}_2 + \text{H}_2\text{O} \xrightarrow[0°-10°]{\text{Cu}^{\oplus}} \text{ArH} + \text{H}_3\text{PO}_3 + \text{N}_2 + \text{H}^{\oplus}$$

or

$$\text{ArN}_2^{\oplus} + \text{Cu}^{\oplus} \longrightarrow \text{N}_2 + \text{Cu}^{2+} + \text{Ar}\cdot \longleftarrow \text{Ar}\cdot + \text{H}_3\text{PO}_3 + \text{N}_2 + \text{H}^{\oplus}$$

$$\quad (22.52)$$

accelerated also by traces of $CuSO_4$, $FeSO_4$, $KMnO_4$, or $K_2Cr_2O_7$.) The cold solution of diazonium chloride (or sulfate) is poured into an excess of H_3PO_2 and allowed to stand until the evolution of nitrogen is complete. Other reducible groups do not seem to be affected. Other reagents that have been used are $NaBH_4$, Na_2SnO_2, MeOH, and EtOH.

764 Chapter twenty-two

The alcohols are oxidized to aldehydes in the process and give some ArOMe and ArOEt as side products. The usefulness of reductive replacement will be demonstrated later.

Anthranilic acid; benzyne.   Anthranilic acid, when diazotized, reacts normally with hot water to give salicyclic acid. However, in a dry, warm aprotic solvent (CH$_2$Cl$_2$, MeOCH$_2$CH$_2$OMe, ClCH$_2$CH$_2$Cl, ϕH), decomposition leads to the formation of benzyne (Eq. 22.53). Benzyne may be trapped as it is formed by various enophiles, such as anthracene

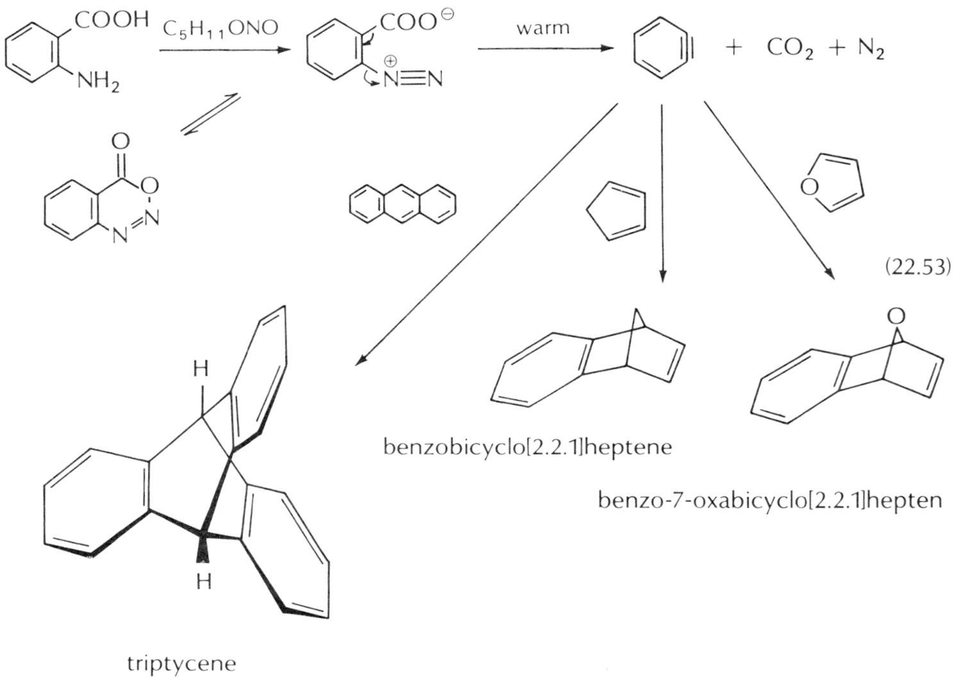

(22.53)

triptycene

benzobicyclo[2.2.1]heptene

benzo-7-oxabicyclo[2.2.1]hepten

(to give triptycene), cyclopentadiene (to give benzobicyclo[2.2.1]heptene), or furan (to give benzo-7-oxabicyclo[2.2.1]heptene).

Summary of reactions.   Let us review and summarize most of the reactions of the diazonium compound reactions. The instability of the diazo group gives rise to high reactivity of its compounds. The sequence nitro → amine → diazo → product enables us to use the reactions for a variety of transformations. In Figure 22.1, we have charted most of the reactions of the diazonium ion. The nitrogen-retaining products are shown on the left; the products from reactions in which nitrogen is evolved are shown on the right.

We should keep in mind that the preparation of the diazonium ion, unless kept cold and strongly acidic, is accompanied by some phenol formation. Also, most of the reactions are accompanied by some phenol and phenolic coupling products. Nevertheless, the reactions provide useful means of synthesis.

Uses in synthesis.   Let us look at a few examples of syntheses that make use of diazonium-ion reactions.

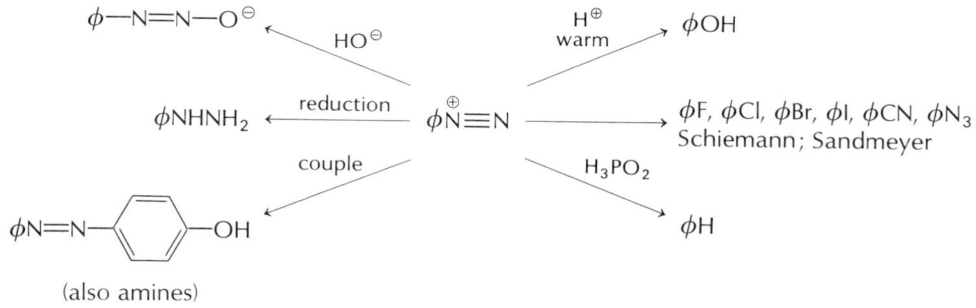

22.3 Aromatic diazonium compounds

$$\phi-N=N-O^{\ominus} \xleftarrow{HO^{\ominus}} \phi\overset{\oplus}{N}\equiv N \xrightarrow[\text{warm}]{H^{\oplus}} \phi OH$$

$$\phi NHNH_2 \xleftarrow{\text{reduction}} \phi\overset{\oplus}{N}\equiv N \xrightarrow{} \phi F, \phi Cl, \phi Br, \phi I, \phi CN, \phi N_3 \quad \text{Schiemann; Sandmeyer}$$

$$\phi N=N-\underset{\text{(also amines)}}{\bigcirc}-OH \xleftarrow{\text{couple}} \phi\overset{\oplus}{N}\equiv N \xrightarrow{H_3PO_2} \phi H$$

Figure 22.1   Chart of diazonium ion reactions

A favorite test question for years has been to show how to prepare *m*-bromotoluene. The answer is shown in Equation 22.54. The transformation of toluene to *p*-toluidine is

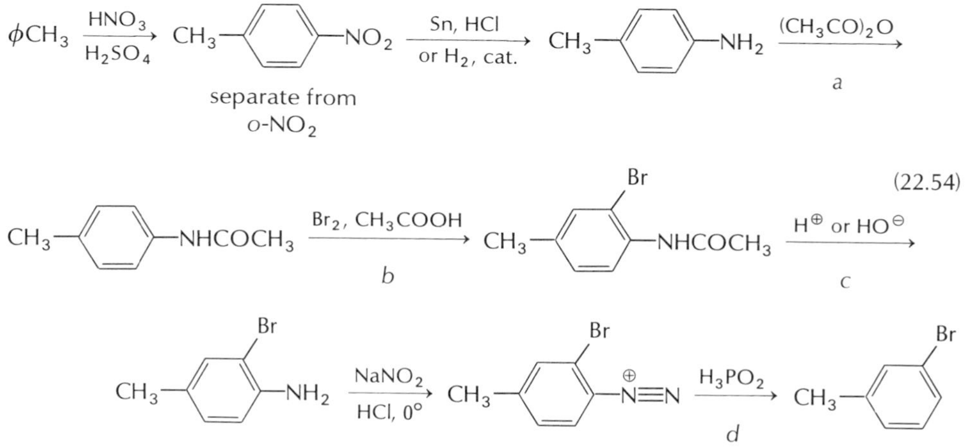

$$\phi CH_3 \xrightarrow[H_2SO_4]{HNO_3} CH_3-\bigcirc-NO_2 \xrightarrow[\text{or } H_2, \text{cat.}]{Sn, HCl} CH_3-\bigcirc-NH_2 \xrightarrow{(CH_3CO)_2O} \quad a$$

separate from
o-NO$_2$

(22.54)

$$CH_3-\bigcirc-NHCOCH_3 \xrightarrow[b]{Br_2, CH_3COOH} CH_3-\underset{Br}{\bigcirc}-NHCOCH_3 \xrightarrow[c]{H^{\oplus} \text{ or } HO^{\ominus}}$$

$$CH_3-\underset{Br}{\bigcirc}-NH_2 \xrightarrow[HCl, 0°]{NaNO_2} CH_3-\underset{Br}{\bigcirc}-\overset{\oplus}{N}\equiv N \xrightarrow[d]{H_3PO_2} CH_3-\bigcirc-Br$$

clear. Acetylation (*a*) of the amino group is necessary because only one bromine is to be substituted in the next step (*b*). The bromine enters *ortho* to the stronger activator (acetamido > methyl for *o,p* activation and direction). In other words, the acetamido group has been used to overpower the methyl in order to introduce bromine *meta* to methyl, as required by the problem. Now, the acetamido group must be removed. Hydrolysis of the amide gives the amine (*c*). (Basic conditions here are satisfactory, because there is no activation within the molecule for nucleophilic aromatic displacement of the amino group.) Diazotization and reduction with H_3PO_2 (*d*) removes the amino group to give the product.

Once the diazonium ion is prepared, many other replacements are possible besides *d*. The products in each instance are difficult to obtain by any other route.

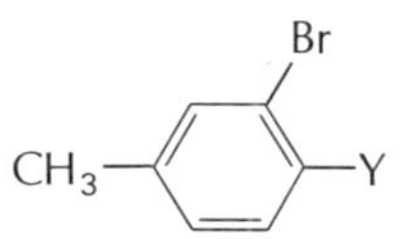

$$CH_3-\underset{Br}{\bigcirc}-Y \quad Y = F, Cl, Br, I, CN, N_3, OH, \text{etc.}$$

If instead of bromination (*b*) some other electrophilic aromatic substitution is used (say nitration), then another series of products becomes available. After the replacement of the amino group, the nitro group may be reduced to amino and diazotized. Then the series shown becomes available.

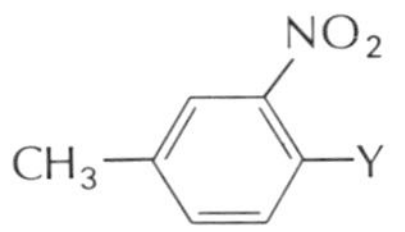

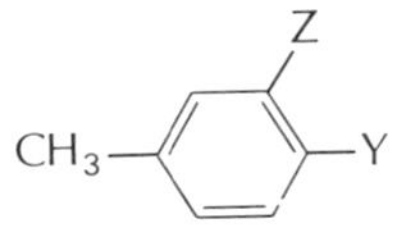

Z = H, F, Cl, Br, I, CN, N$_3$, OH, etc.
Y = H, F, Cl, Br, I, CN, N$_3$, OH, etc.

We are not limited to toluene as the starting material. Thus any alkyl or aryl group, halogen, and so on could have been present in place of methyl to begin with (any o,p director weaker than —NHCOCH$_3$).

Let us prepare o-bromotoluene. Two routes are possible (Eq. 22.55). In route a, the p-NO$_2$ is used to block the *para* position, and the directive influence of Me gives an excellent yield of the o-bromo compound, after which the blocking group is reduced to amino, diazotized, and removed. In route b, the major product of nitration, o-nitrotoluene, is

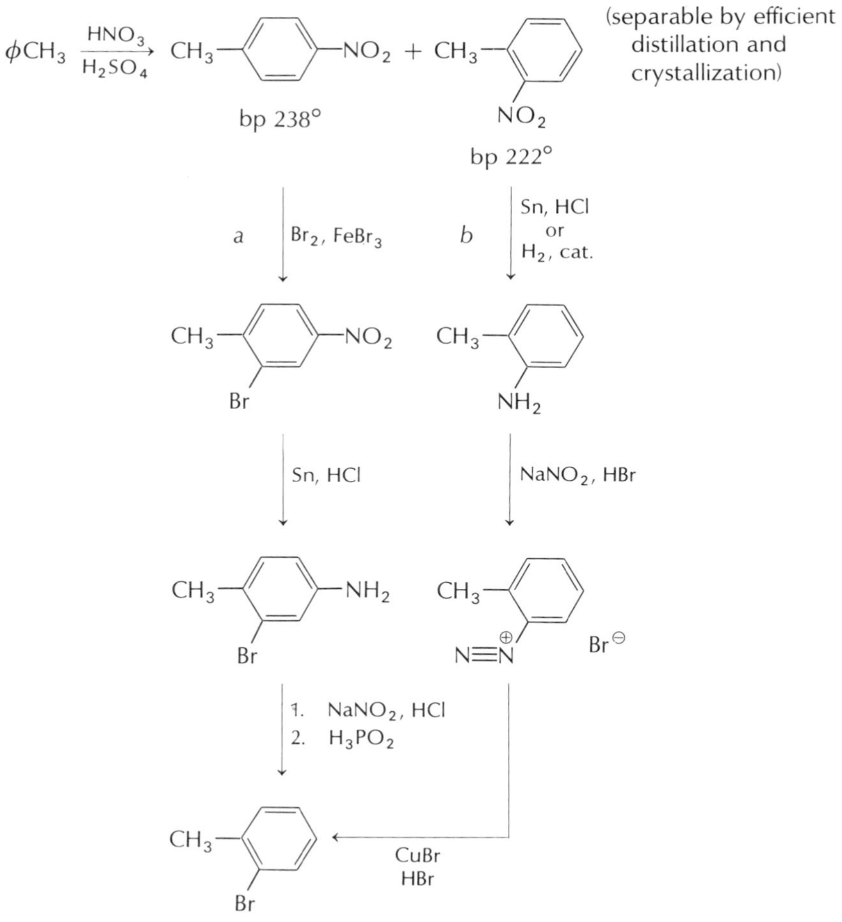

(22.55)

reduced, diazotized, and replaced by bromine in a Sandmeyer reaction. Routes a and b allow other products to be prepared as shown.

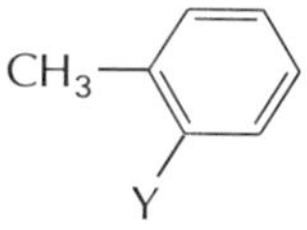

[Structure: CH₃ on benzene ring with Y substituent ortho]

Route a: Y = Cl, Br, SO₃H; not NO₂ nor Friedel-Crafts

Route b: Y = F, Cl, Br, I, CN, N₃, OH, etc.

For the preparation of 2-bromonaphthalene, the Bucherer reaction (Eq. 22.34) must be used. The 2 position of naphthalene is available only by thermodynamic control of electrophilic substitution (sulfonation and Friedel-Crafts reaction). Thus, we utilize the sequence of reactions shown in Equations 22.56. Once 2-naphthylamine is made, the

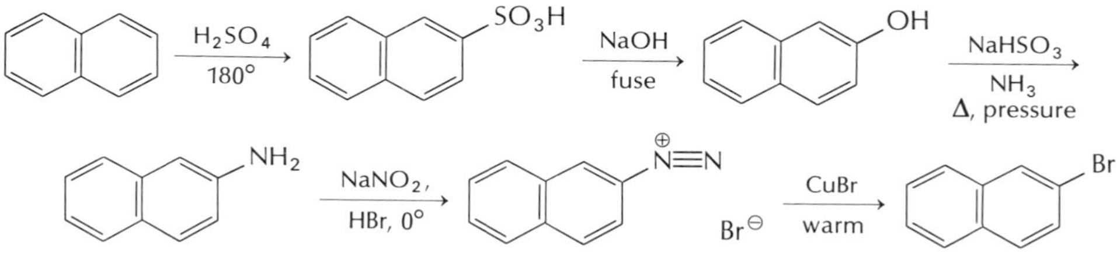

$$\text{naphthalene} \xrightarrow{H_2SO_4, 180°} \text{2-naphthalene-SO}_3\text{H} \xrightarrow[\text{fuse}]{NaOH} \text{2-naphthol} \xrightarrow[\Delta, \text{pressure}]{NaHSO_3, NH_3} \text{2-naphthylamine}$$

$$\text{2-naphthylamine-NH}_2 \xrightarrow{NaNO_2, HBr, 0°} \text{ArN}_2^+ \text{Br}^- \xrightarrow[\text{warm}]{CuBr} \text{2-bromonaphthalene}$$

(22.56)

various groups shown in Figure 22.1 may be introduced into the 2 position by means of the diazonium ion replacement.

Exercises

22.7 Using *p*-toluidine as the only aromatic compound, show how to prepare:

a. 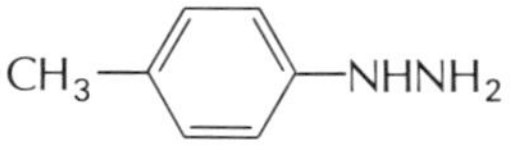 CH₃—C₆H₄—NHNH₂

b. 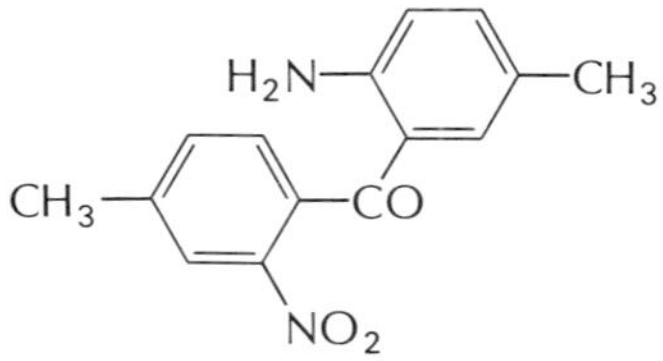 CH₃ substituted benzene with H₂N, CO, NO₂, CH₃ groups

c. 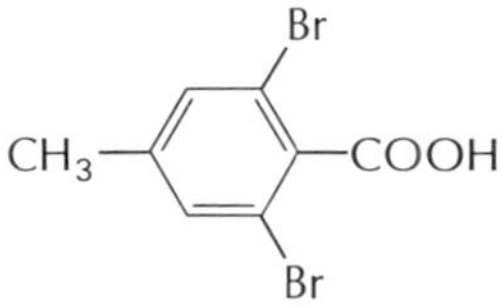 CH₃—C₆H₂(Br)₂—COOH

d. 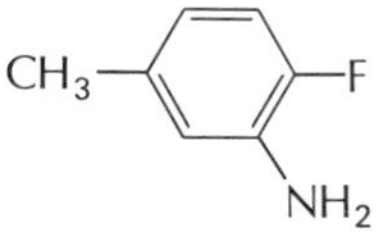 CH₃—C₆H₃(F)(NH₂)

e. NMe₂ and N=N linked to COOH-bearing ring, with CH₃

f. O₂N—C₆H₂(OH)(CH₃)—N=N—C₆H₃(CH₃)(NO₂)

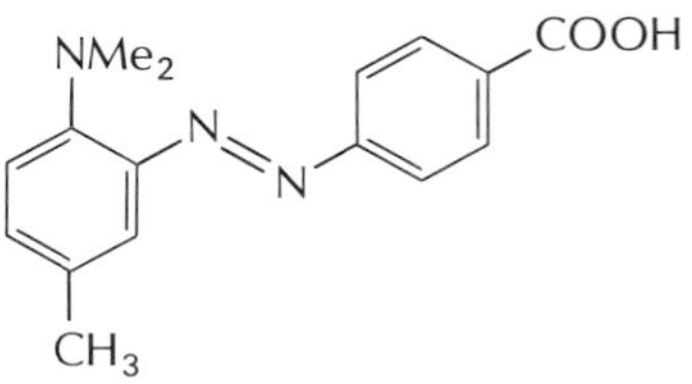

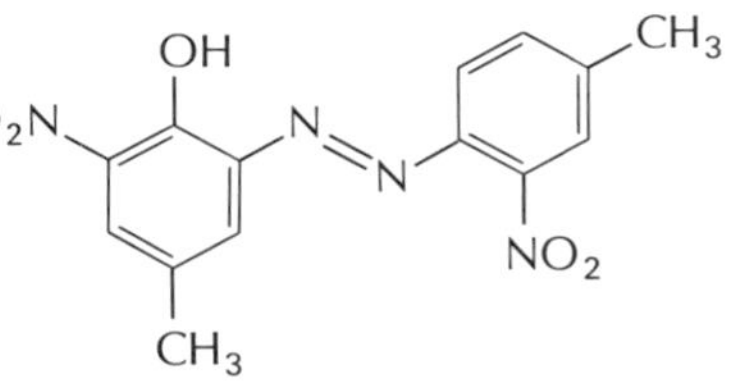

768   Chapter twenty-two

22.8 Using toluene as the only aromatic compound, show how to prepare:

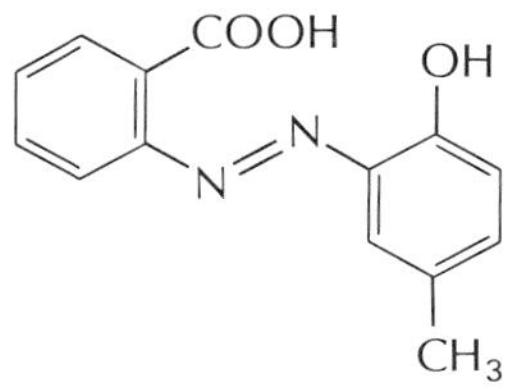

[Structure: 2-carboxyphenyl azo group attached to 2-hydroxy-5-methylphenyl — i.e., benzene ring with COOH and N=N− linked to a phenol ring bearing OH (ortho to N=N) and CH₃ (para to N=N / meta to OH)]

22.9 Give the structure of the compound that would be formed as a result of your method of synthesis if *m*-toluidine had been used instead of *p*-toluidine in Exercise 22.7b, c, d, e, and f.

22.4 Aromatic nitroso and hydroxylamino compounds

Physical properties of nitroso compounds. The aromatic nitroso (ArNO) and hydroxylamino (ArNHOH) compounds are taken up together because of the fairly close relationship between the two classes, which is rather like that between an aldehyde (ArCHO) and the primary alcohol (ArCH$_2$OH).

The aromatic nitroso compounds resemble the aromatic nitro compounds on the one hand and the tertiary alkyl nitroso compounds on the other. Nitrosobenzene is almost colorless as the dimeric solid, has mp 68° (melting to a green liquid), and is volatile with steam. It acts as a base to strong acids. The dipole moment is large (3.2 Debye) and is electron withdrawing from the ring (N,N-dimethylaniline, 1.6 D; *p*-nitrosodimethylaniline, 6.9 D; calculated, 3.2 + 1.6 = 4.8 D).

N-phenylhydroxylamine is an unstable colorless solid of mp 82°, somewhat less basic (pK_a = 3.2) than aniline. (The —OH group is called <u>hydroxyl</u> in the naming of inorganic compounds, as in hydroxylamine, HONH$_2$. If we name the compound ϕNHOH as a substituted aniline, it becomes N-hydroxyaniline.)

Addition reactions to the nitroso group. A nitroso group undergoes addition reactions with ease, the nitrogen atom acting as an electrophile to nucleophiles (Eq. 22.57;

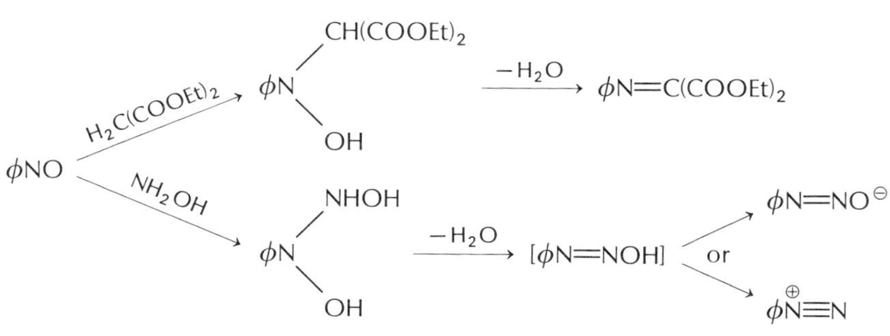

$$\phi NO \xrightarrow{H_2C(COOEt)_2} \phi N\begin{matrix} CH(COOEt)_2 \\ OH \end{matrix} \xrightarrow{-H_2O} \phi N{=}C(COOEt)_2$$

$$\phi NO \xrightarrow{NH_2OH} \phi N\begin{matrix} NHOH \\ OH \end{matrix} \xrightarrow{-H_2O} [\phi N{=}NOH] \begin{matrix} \nearrow \phi N{=}NO^{\ominus} \\ \text{or} \\ \searrow \phi \overset{\oplus}{N}{\equiv}N \end{matrix}$$

(22.57)

see also Eq. 22.4c). With a Grignard reagent, a disubstituted hydroxylamine is formed (Eq. 22.58). The reactions of nitrosobenzene with the other reduction products of nitrobenzene are of interest (Eq. 22.59). With aniline (a) or with phenylhydroxylamine (c, also Eq. 22.6), we see how the azo and azoxy compounds arise in the cycle of reduction products

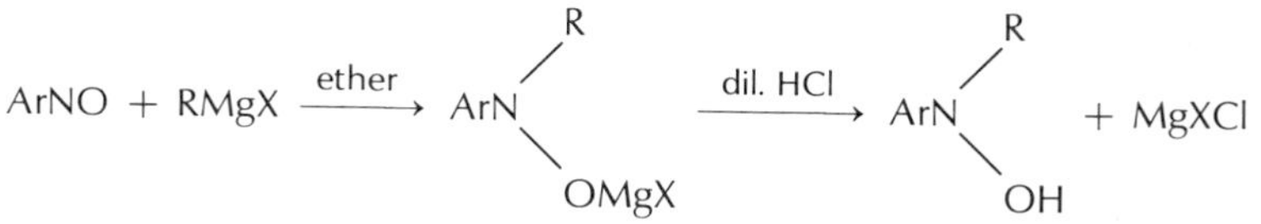

22.4 Aromatic nitroso and hydroxylamino compounds

$$\text{ArNO} + \text{RMgX} \xrightarrow{\text{ether}} \text{ArN}\begin{array}{c}R\\|\\\\OMgX\end{array} \xrightarrow{\text{dil. HCl}} \text{ArN}\begin{array}{c}R\\|\\\\OH\end{array} + \text{MgXCl} \quad (22.58)$$

$$\phi\text{NO} \begin{array}{l} \xrightarrow{\phi\text{NH}_2} \phi\overset{O^{\ominus}}{\underset{|}{N}}-\overset{\oplus}{N}H_2\phi \xrightarrow{-H_2O} \phi N{=}N\phi \quad a \\ \xrightarrow{\phi\text{NHNH}\phi} \phi\text{NHOH} + \phi N{=}N\phi \quad b \\ \xrightarrow{\phi\text{NHOH}} \phi\overset{O^{\ominus}}{\underset{|}{N}}-\overset{\oplus}{N}H\phi \xrightarrow{-H_2O} \phi\overset{O^{\ominus}}{\underset{|}{N}}{=}N\phi \quad c \\ \qquad\qquad\qquad\qquad OH \end{array} \quad (22.59)$$

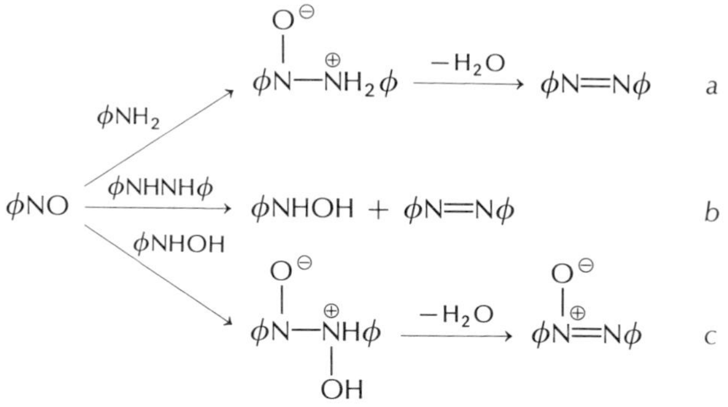

of nitrobenzene. With hydrazobenzene (b), the reaction is a reduction of nitrosobenzene and an oxidation of hydrazobenzene. As b takes place, the phenylhydroxylamine that is formed begins to add to the remaining nitrosobenzene (c) to give azoxybenzene, so b is not a clean-cut reaction.

Instability of phenylhydroxylamine. Phenylhydroxylamine is so unstable that it undergoes slow disproportionation to nitrosobenzene and aniline when stored in the absence of air. We already have seen that phenylhydroxylamine acts like an amine in adding to nitrosobenzene. It also adds to carbonyl groups (Eq. 22.60).

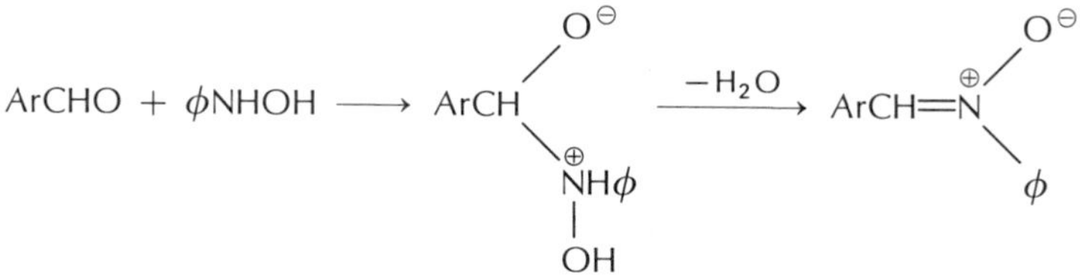

$$\text{ArCHO} + \phi\text{NHOH} \longrightarrow \text{ArCH}\begin{array}{c}O^{\ominus}\\/\\\\\overset{\oplus}{N}H\phi\\|\\OH\end{array} \xrightarrow{-H_2O} \text{ArCH}{=}\overset{\oplus}{N}\begin{array}{c}O^{\ominus}\\/\\\\\phi\end{array} \quad (22.60)$$

Upon standing in dilute sulfuric acid, phenylhydroxylamine rearranges (as usual, to *p* if not blocked, then to *o*) according to Equation 22.61, to give *p*-aminophenol in good yield. Other nucleophiles may compete for the cation *a*. If the rearrangement is carried out in ethanol with HCl, ethanol and chloride ion compete to give *p*-phenetidine and *p*-chloro-aniline, among other products.

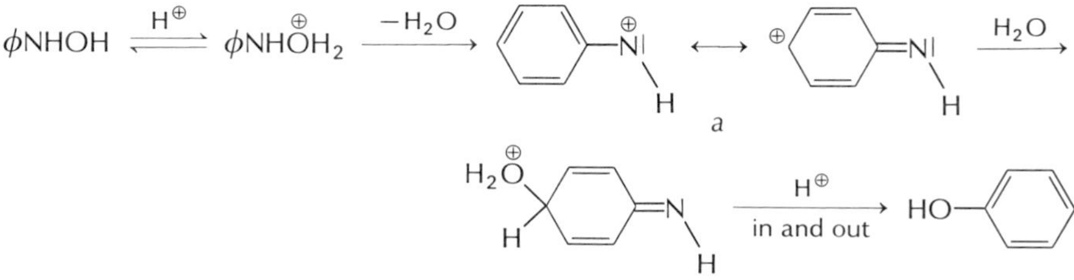

$$\phi\text{NHOH} \xrightleftharpoons{H^{\oplus}} \phi\text{N}\overset{\oplus}{H}\text{OH}_2 \xrightarrow{-H_2O} \left[\text{Ph}{-}\overset{\oplus}{\underset{H}{N}}\text{I} \leftrightarrow \overset{\oplus}{\text{Ph}}{=}\underset{H}{N}\text{I}\right] \xrightarrow{H_2O}$$
$$\qquad\qquad\qquad\qquad\qquad\qquad\qquad a$$
$$\xrightarrow{}\; \overset{\overset{\oplus}{H_2O}}{\underset{H}{\bigcirc}}{=}\underset{H}{N} \xrightarrow[\text{in and out}]{H^{\oplus}} \text{HO}{-}\bigcirc{-}\text{NH}_2 \quad (22.61)$$

Neither nitrosobenzene nor phenylhydroxylamine is useful in electrophilic aromatic substitution reactions, because the groups usually are destroyed by the reagents used. Substituted nitrosobenzenes or phenylhydroxylamines must be made from the corresponding substituted nitrobenzene, except for certain anilines and phenols. □ □

Exercises

22.10 Starting with benzene, toluene, or naphthalene, show how to prepare:

a. 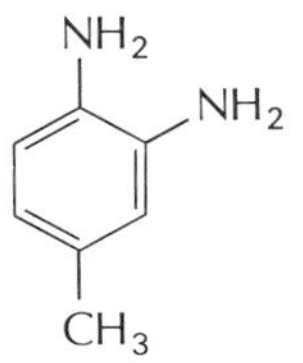 4-methyl-1,2-diaminobenzene (NH$_2$, NH$_2$, CH$_3$)

b. 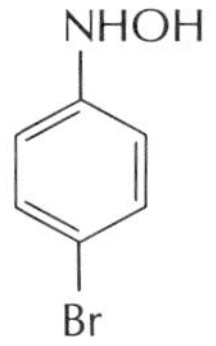 4-bromo-N-phenylhydroxylamine (NHOH, Br)

c. 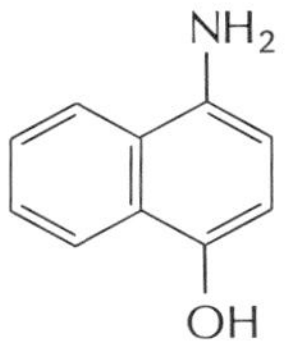 4-amino-1-naphthol (NH$_2$, OH on naphthalene)

d. 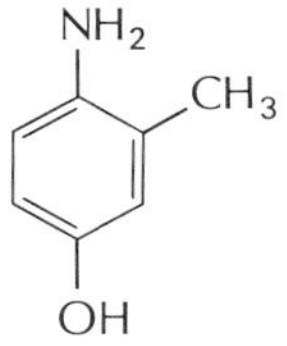 2-methyl-4-hydroxyaniline (NH$_2$, CH$_3$, OH)

e. 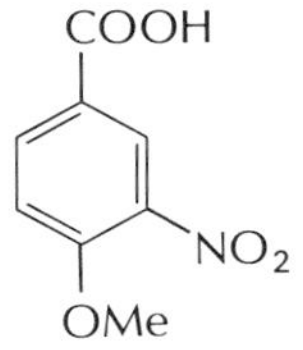 4-methoxy-3-nitrobenzoic acid (COOH, NO$_2$, OMe)

22.11 Using aniline and 1-bromo-5-chloro-3,3-dimethylpentane, show how to prepare and obtain pure a cyclic secondary amine.

22.5 Other aromatic nitrogen compounds

Introduction. In this section, we shall focus on hydrazines, isocyanates, and nitriles. Also, we shall look at the benzidine rearrangement.

Summary of reductions and oxidations. Let us return to the nitrobenzene reduction cycle (Eq. 22.5). We may summarize some of the relationships in more detail (omitting nitrosobenzene and phenylhydroxylamine) in Equation 22.62. We have added several oxidations, some condensations, and a rearrangement (to be described later). Most of the reactions have been described elsewhere.

Arylhydrazines. The arylhydrazines (ArNHNH$_2$) are familiar to us, as is their use to prepare arylhydrazones of aldehydes and ketones and to make substituted acid hydrazides. The nitrogen-nitrogen bond is easily cleaved by reducing agents to give amines (or amides) and ammonia.

Aryl isocyanates. The aryl isocyanates are useful for the preparation of derivatives of alcohols, phenols, and amines (Eq. 22.63). The reaction medium must be dry if c, the formation of diphenylurea, is to be avoided. The products a, a substituted phenylurea, and

22.5 Other aromatic nitrogen compounds

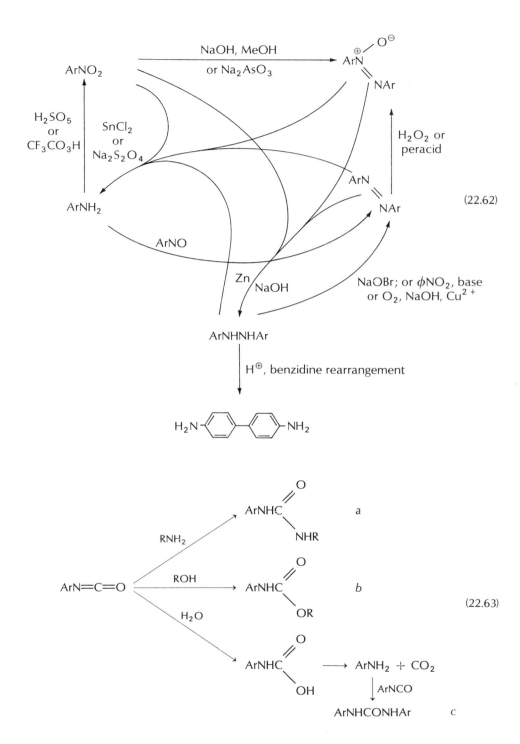

b, an alkyl or aryl N-arylcarbamate, are crystalline solids. Both phenyl and α-naphthyl isocyanate are used.

Isocyanates are prepared either from an amine and phosgene (Eq. 22.64) or

$$\phi NH_2 + O=C\begin{matrix}Cl\\ \\Cl\end{matrix} \longrightarrow \phi NHCOCl \xrightarrow[-HCl]{\Delta} \phi N=C=O \quad (22.64)$$

from an acid derivative by rearrangement under anhydrous conditions.

Aryl nitriles. Aromatic nitriles are obtained by the Sandmeyer reaction or by dehydration of an amide. An aryl halide is unable to undergo replacement by a cyanide ion unless activated by *o* or *p* nitro or nitroso groups.

Benzidine rearrangement. The benzidine rearrangement has been the subject of extensive research because of the usefulness of benzidine (and substituted benzidines) in the manufacture of azo dyes. It is known that, unlike most of the other rearrangements to the *p* position, the benzidine rearrangement is intramolecular: the migrating group is never separated from the rest of the molecule. Hydrazobenzene rearranges in the presence of cold concentrated HCl to give a good yield of benzidine (*p,p*'-diaminobiphenyl, Eq. 22.65).

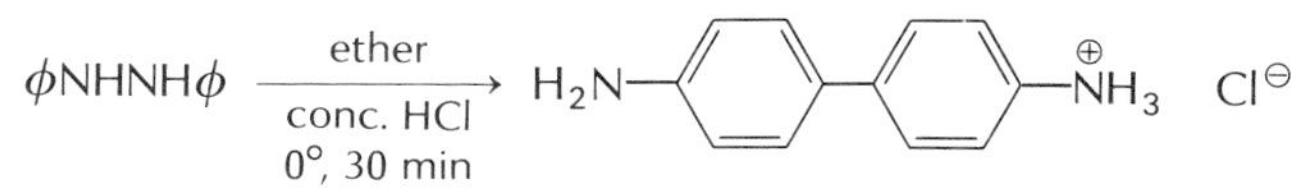

$$\phi NHNH\phi \xrightarrow[\substack{\text{conc. HCl}\\ 0°, 30 \text{ min}}]{\text{ether}} H_2N-\!\!\!\left\langle\!\!\!\bigcirc\!\!\!\right\rangle\!\!\!-\!\!\!\left\langle\!\!\!\bigcirc\!\!\!\right\rangle\!\!\!-\overset{\oplus}{N}H_3 \quad Cl^{\ominus} \quad (22.65)$$

Certain groups, if they are in the *p* position in hydrazobenzene, are displaced in the course of the rearrangement (—COOH, —OCOCH$_3$, —SO$_3$H, —Cl). Others are not displaced (—R, —OR, —NH$_2$, etc.), and a shift to an *o* position occurs. A small amount of *o* shift occurs anyway, as well as a halfway shift to give semidines. □ □

Exercises

22.12 Using benzene as starting material show how to synthesize:

a. 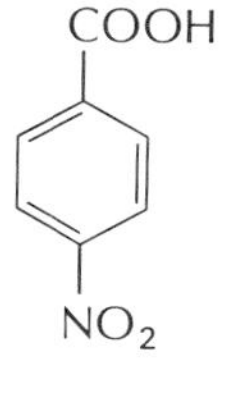 4-nitrobenzoic acid (COOH, NO$_2$)

b. 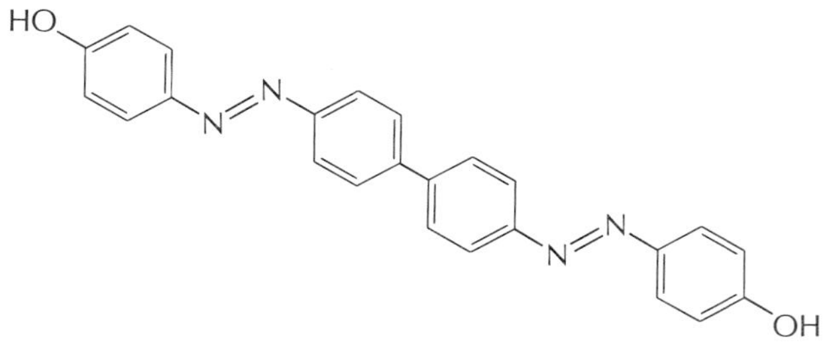 HO—C$_6$H$_4$—N=N—C$_6$H$_4$—C$_6$H$_4$—N=N—C$_6$H$_4$—OH

22.6 Summary

The versatility of nitration, reduction, diazotization, and replacement or coupling in aromatic chemistry has been demonstrated in this chapter. We also have sampled the many other types of aromatic compounds containing nitrogen and a few of the many reactions that they undergo.

22.7 Problems

Along the way, we have seen that there are many rearrangements of groups from nitrogen to the ring, both inter- and intramolecular, both ionic and radical. We have seen how useful and necessary the cuprous-cupric ion one-electron couple is in many of the nitrogen-compound reactions.

A large amount of material has been compressed into this chapter, and not all of it is necessary for understanding the chapters to follow. Consequently, your instructor probably has considered the purpose of your course and deleted parts of this chapter.

We should mention that many nitrogen containing compounds have pronounced physiological activity. Some are potent inducers of cancer (carcinogens). Perhaps the simplest of all is N-nitrosodimethylamine (Me_2N—NO). Aniline is safe, but 1-naphthylamine, 2-naphthylamine, and 2-naphthylhydroxylamine are not. Benzidine, 4-aminobiphenyl, and most of the methyl substituted compounds of these two substances are carcinogenic. A few azo compounds are active cancer inducers. Butter yellow, p-$Me_2NC_6H_4N$=NC_6H_5, has been studied so much that it can be said that the presence of —OH or —SO_3H groups as substituents remove carcinogenicity, but alkyl and —OMe groups increase activity, depending upon the position of substitution in butter yellow. However, if rupture of the azo linkage gives rise to a carcinogen, —OH and —SO_3H elsewhere in the molecule fail to confer safety. Many N-nitrosoamines and amides are carcinogenic.

Carcinogenic activity is not limited to nitrogen compounds. Dioxane and carbon tetrachloride both are dangerous. It behooves us all to be as clean and careful in the laboratory as is possible at all times.

22.7 Problems

22.1 Substance A ($C_{16}H_{18}N_2$) is a neutral red solid. Treatment with Sn and HCl or with $Na_2S_2O_4$ gives B. B with pertrifluoroacetic acid gives C. Drastic oxidation of C gives 4-nitrophthalic acid. C with Sn and HCl gives B. C with Zn powder and NH_4Cl solution gives D. Mild oxidation of D gives E. Treatment of either D or E with Sn and HCl gives B. D, allowed to stand in dilute H_2SO_4, gives F. Treatment of either D or E with CF_3CO_3H gives C. B and E react to give A. Treatment of A, C, D, or E with Zn and NaOH gives G. G, upon standing in cold concentrated HCl, gives H. H, with one equivalent of $NaNO_2$ and HCl, gives a solution that reacts with CuCN to give I. Hydrolysis of I gives a product that cyclizes easily (with loss of water) to give J ($C_{17}H_{17}ON$).

Give a structural formula for each lettered substance.

22.2 Starting with methyl anthranilate, show how to prepare the products shown. Use any organic or inorganic reagents you wish. (More than one synthetic step may be required.)

a. 2-bromo-benzene-COOCH₃ with Br

b. 3-bromo-benzene-COOCH₃ with Br

c. benzene-1,2-dicarboxylic acid (COOH, COOH)

d. 2-(NHCOCH₃)-benzene-COOH

e. 2-COOCH₃-C₆H₄—N=N—C₆H₄—N(C₂H₅)₂

f. 2-OH-benzene-COOH (salicylic acid)

g. 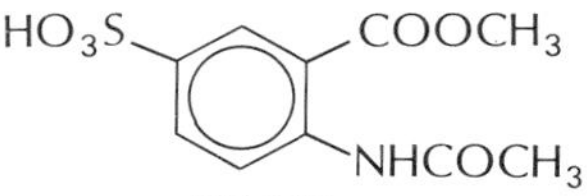 [structure: benzene with HO₃S, COOCH₃, NHCOCH₃]

h. 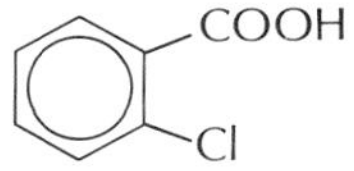 [structure: benzene with COOH, Cl]

i. [structure: benzene with COOCH₃, I]

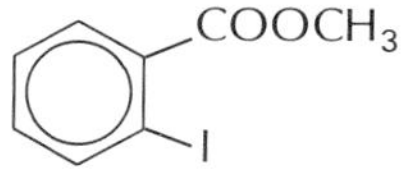

j. 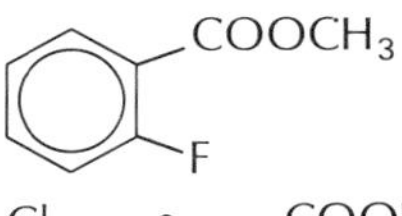 [structure: benzene with COOCH₃, F]

k. 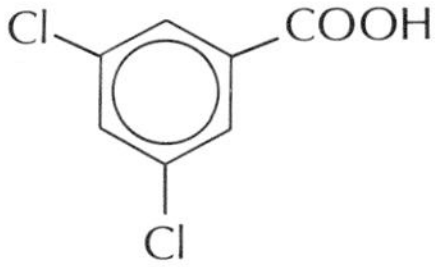 [structure: benzene with Cl, COOH, Cl]

l. 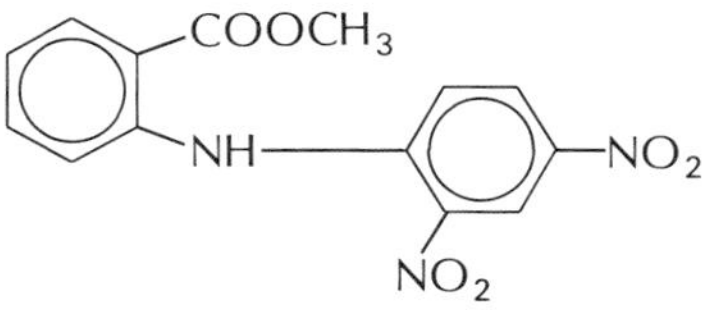 [structure: benzene with COOCH₃, NH—benzene(NO₂)(NO₂)]

22.3 By use of the following reagents, p-toluidine can be converted to a large variety of products; write their structures.

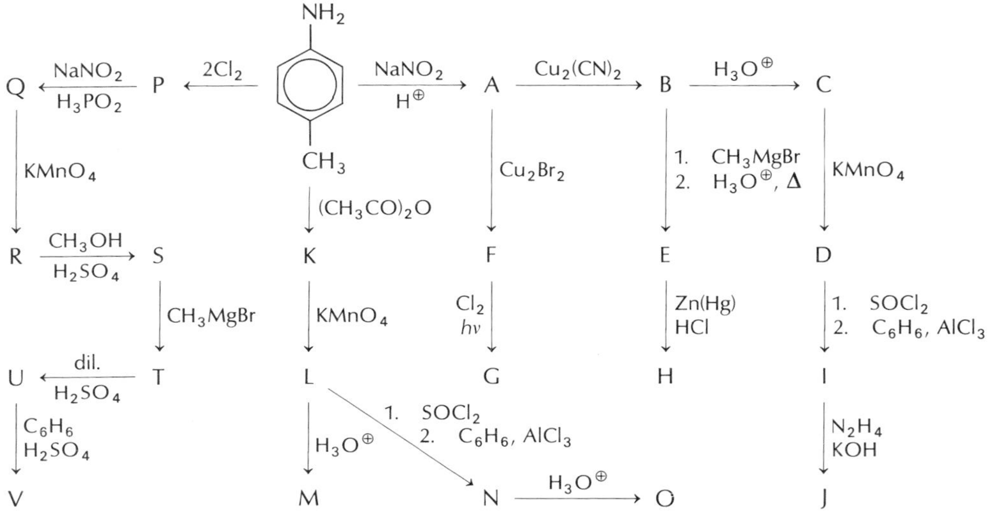

[reaction scheme with p-toluidine center, branches to Q, P, A, B, C with reagents NaNO₂/H₃PO₂, 2Cl₂, NaNO₂/H⊕, Cu₂(CN)₂, H₃O⊕; further to R, S, K, F, E, D via KMnO₄, CH₃OH/H₂SO₄, (CH₃CO)₂O, Cu₂Br₂, 1. CH₃MgBr 2. H₃O⊕,Δ, KMnO₄; further to U, T, L, G, H, I via dil. H₂SO₄, CH₃MgBr, KMnO₄, Cl₂/hv, Zn(Hg)/HCl, 1. SOCl₂ 2. C₆H₆,AlCl₃; further to V, M, N, O, J via C₆H₆/H₂SO₄, H₃O⊕, 1. SOCl₂ 2. C₆H₆,AlCl₃, H₃O⊕, N₂H₄/KOH]

22.8 Bibliography

Millar and Springall (Chap. 20, ref. 1) is excellent for additional material on the subjects taken up in this chapter. Fieser and Fieser (Chap. 7, ref. 7) and Noller (Chap. 7, ref. 11) are excellent, especially on dyes. The list given here is only a sampling of the literature on the subject.

1. Buehler, Calvin A., and Donald E. Pearson, *Survey of Organic Syntheses*. New York: Wiley, 1970. A useful advanced textbook.

2. Hickinbottom, Wilfrid J., *Reactions of Organic Compounds*, 3rd ed. London: Longmans, Green, 1957. An older advanced textbook still in use because of the inclusion of so much detailed information.

3. Ingold, Christopher K., *Structure and Mechanism in Organic Chemistry*, 2nd ed. Ithaca:

Cornell University Press, 1969. A modern classic advanced textbook, very good on mechanisms. Many references.

4. Norman, R. O. C., *Principles of Organic Synthesis*. London: Methuen, 1968. Available as a paperback from Barnes and Noble, Inc. A valuable advanced text with references and problems.

5. Saunders, Kenneth H., *The Aromatic Diazo Compounds and Their Technical Application*, 2nd ed. London: Edward Arnold, 1949. The classic book on the subject. There are several newer books—see reference 6.

6. Zollinger, Heinrich, *Azo and Diazo Chemistry: Aliphatic and Aromatic Compounds*. New York: Interscience, 1961.

7. Weisburger, John H., and Elizabeth K. Weisburger, Chemicals as Causes of Cancer, *Chemical and Engineering News*, Feb. 7, 1966. An excellent starting point for information on carcinogenesis.

23 The Organic Halogen Compounds

We have learned a lot about the halogen compounds as we went along looking at other classes of compounds. We do not want to make a complete summary of this information now, but we do need to add a few more details to our knowledge.

23.1   The organic halogen compounds   776
       Occurrence, uses, reactions, and preparations.

23.2   Haloalkanes and haloalkenes   777
       1. Commercial manufacture.   2. Inertness of polyfluoro compounds.
       3. Relative inertness of CX_2 group and of CX_4; vinyl halides. Exercises.

23.3   Haloarenes and allylic halides   780
       1. Preparation; reactions of haloarenes.   2. Allylic and benzylic Grignard reagents.   3. The hexaphenylethane mistake.   4. Insecticides. Exercises.

23.4   Halohydrins and other halogen-substituted compounds   785
       Exercises.

23.5   Summary   787

23.6   Bibliography   787

23.1   The organic halogen compounds

Occurrence, uses, reactions, and preparations.   The organic halogen compounds are rare in nature. One of the few instances is the presence of fluoroacetate ion in a South African plant. The plant is toxic, and sodium fluoroacetate is used as a rat poison. Another case is the presence of iodine in thyroxine (Chap. 21). Most of the halogen-containing compounds are toxic.

The chloro- and fluoroalkanes are used widely in industry: the chloroalkanes because they are cheaper than the bromo- and iodoalkanes, the fluoroalkanes (especially the polyfluoro compounds) because of special properties. In the laboratory, the bromoalkanes and bromoarenes are encountered frequently because they are easy to make and have a somewhat greater reactivity than do the chloro compounds. The chloro-, dichloro-, and polychloroalkanes are widely used as aprotic solvents. Depending upon the number of atoms of and kinds of halogens present, the halogen-containing compounds range in density from about 0.9 to over 3.0 g/cc, compared to about 0.7 to 0.9 for most other organic liquids.

The reactions of the halogen-containing compounds have been given throughout the preceding chapters. Reactions in which haloalkanes are products also have appeared in many chapters. We shall not repeat all this information. Instead, we shall classify the reactions and preparations of halogen-containing compounds and list the chapters in which most of the information is given.

1. Reactions of haloalkanes appear in Chapters 2, 3, 7, 8, 11, 13, 16, 17, 18, 20, 22.
2. Preparative reactions of haloalkanes were given in Chapters 2, 3, 7, 12, 13.
3. Reactions of vinyl halides appear in Chapters 3, 13, 14.
4. Reactions that give vinyl halides as products appear in Chapters 7, 13, 14.
5. Reactions of aryl halides appear in Chapters 14, 15, 16, 17, 20, 22.
6. Preparations of aryl halides were shown in Chapters 15, 16, 22.
7. Reactions of allylic and benzylic halides were described in Chapters 13 through 18, 20, and 22.
8. Preparative reactions of allylic and benzylic halides were shown in Chapters 13, 15, 16.
9. Reactions of halohydrins, halogen-substituted aldehydes, ketones, acids, esters, amides, and so on, and acid chlorides were given in Chapters 2, 3, 5, 9, 13 through 18, 20, and 22.
10. Preparations of halohydrins, halogen-substituted aldehydes, ketones, acids, esters, amides, and so on, and acid chlorides appeared in Chapters 3, 9, 13, 16, 17, 18, 20, and 22.

Because of their use in the preparation of Grignard and lithium reagents, we have mentioned the organic halogen compounds even more often than the tabulation shows. The reactivity of the halogen compounds makes them very useful in synthetic sequences in nucleophilic, electrophilic, and radical substitution, as well as in addition and elimination reactions.

23.2 Haloalkanes and haloalkenes

Commercial manufacture. Commercially, methyl chloride is made from methanol and HCl. Methylene chloride and chloroform are made chiefly by chlorination of methane or methyl chloride. Carbon tetrachloride is manufactured by three routes (Eqs. 23.1).

$$
\begin{aligned}
&\text{C} + 2\text{S} \xrightarrow{\Delta} \text{CS}_2 \xrightarrow{\text{Cl}_2} \text{CCl}_4 + \text{S} &\quad a\\
&\text{CH}_4 + \text{Cl}_2 \xrightarrow[\text{or } h\nu]{\Delta} \text{CCl}_4 + \text{HCl} &\quad b \\
&\text{CH}_3\text{CH}_2\text{CH}_3 + \text{Cl}_2 \xrightarrow{600°} \text{Cl}_2\text{C}=\text{CCl}_2 + \text{CCl}_4 &\quad c
\end{aligned} \quad (23.1)
$$

Route a is the principal method used. The sulfur is recycled to produce more carbon disulfide (dangerously flammable, flash point less than 100°). Route b is used principally to obtain CH_2Cl_2 and CHCl_3, so that CCl_4 may be considered a by-product. Route c is used as a method of preparation of tetrachloroethylene: again, CCl_4 is a valuable by-product.

The chloroethanes and chloroethylenes are made almost entirely from ethylene and acetylene (Eqs. 23.2). It is interesting to examine the tangle of addition and substitution

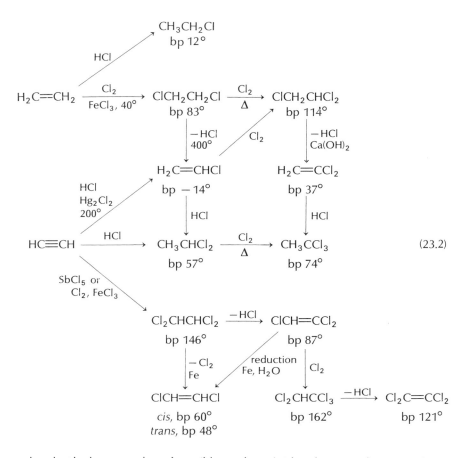

(23.2)

reactions whereby the large number of possible products (with only two carbon atoms) are manufactured. Most of the products are used as solvents (dry cleaning, degreasing metals, etc.). Perchloroethylene ($Cl_2C=CCl_2$) and trichloroethylene ($ClCH=CCl_2$) are most used for these purposes. Others are used for the production of polymers, sometimes alone, but more likely with other monomers. Vinyl chloride ($H_2C=CHCl$) and vinylidene chloride ($H_2C=CCl_2$) are so used (vinyls, sarans).

Perchloroethane (hexachloroethane) is of interest because it forms a liquid phase only under pressure. Thus, it is listed as having bp 185°, but mp 187° in a sealed tube. Practically, the data imply that it sublimes readily.

Inertness of polyfluoro compounds. The fluorine-containing compounds are characterized by extremes as compared with the other halogen-containing compounds. Thus the fluorine compounds are either very reactive or markedly inert, either almost completely nontoxic or very toxic. For example, CF_2Cl_2 is so safe that it is the commonly used refrigerant in domestic refrigerators and air conditioners. It also is useful as the pressurizer for aerosol spray cans. Octafluorocyclobutane is used as the pressurizer for foamed food products (whipped cream). In general, compounds with —CF_2— groups are inert and nontoxic; but not always! Hexafluorocyclopropane and perfluoroisobutylene (($F_3C)_2C=CF_2$) are very toxic.

The largest commercial use of fluorinated compounds is for the production of polymers resembling polytetrafluoroethylene. Teflon is the high molecular weight (1×10^6

and more) polymer of tetrafluoroethylene. This polymer is so inert that it is attacked only by molten sodium, potassium, and the other Group I metals. Its surface is remarkably frictionless toward other substances (it feels slippery). The shorter chain polyfluorinated hydrocarbons (called <u>fluorocarbons</u>) such as C_7F_{16} are insoluble in most other nonpolar and polar solvents (heptane to water) except ethers and other fluorocarbons.

Various modifications have been made in the Teflons. If some hexafluoropropylene is used as a copolymer, the product is known as Teflon 100 and has a lower softening point. As a result, it is easier to extrude or mold. Fluorothene or Kel-F is a polymer of $ClCF=CF_2$ and has similar properties. Viton, an elastomer used to manufacture O-rings, gaskets, and washers resistant to solvents and oils, is a copolymer of vinylidene fluoride ($H_2C=CF_2$) and perfluoropropylene ($CF_3CF=CF_2$).

The preparation of the haloalkanes and haloalkenes has been adequately covered already.

Relative inertness of $\diagdown C X_2 \diagup$ group and of CX_4; vinyl halides. If we compare CH_3X, CH_2X_2, CHX_3, and CX_4 in terms of reactivity with a base, we find the order $CH_3X > CHX_3 \gg CX_4 > CH_2X_2$ for a given halogen. The relative inertness of compounds in which an even number of halogen atoms (2 or 4) are attached to a single carbon has been ascribed to the symmetrical resonance structures that can be written (a and b). The double-bonded structures not only help to disperse the positive charge but also diminish the

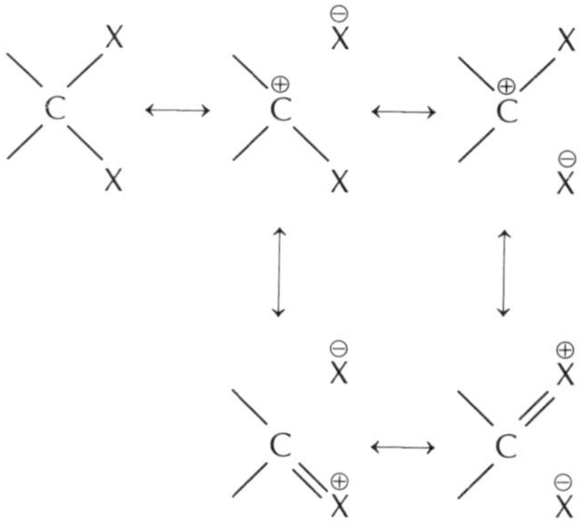

negative charge on each halogen. The result is to lessen the polarity of a given C—X bond as well as to increase the binding energy.

A similar effect may be noted in a vinyl halide, in which the C—X bond also gains some double-bond character, which decreases polarity and increases the binding energy.

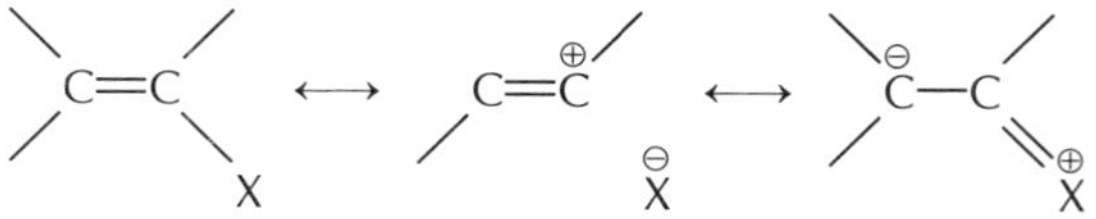

We recognize the effect as the +R effect of a halogen, which we discussed in our description of aromatic electrophilic substitution in Chapter 15. The effect is strong enough to cause vinyl chloride ($H_2C=CHCl$) and vinyl bromide ($H_2C=CHBr$) to be inert to magnesium in ether. To prepare vinyl Grignard reagents, we must use the better ethereal solvent tetrahydrofuran (THF).

The effect is particularly noticeable in the fluorine compounds. The contribution of the double-bond structures a and b is such that bond shortening predominates over the bond-lengthening effect of the noncovalent ionic form of the bond. Thus the C—F bond distance in CH_3F is 1.42 A, but it is only 1.36 A in CH_2F_2. The C—Cl bond length is almost invariant in CH_3Cl, CH_2Cl_2, $CHCl_3$, and CCl_4, at 1.76 A. Yet in CCl_2F_2, not only is the C—F bond length lessened, but the C—Cl as well, down to 1.70 A. In vinyl chloride, in which an sp^2 hybrid single bond also is involved, the C—Cl bond distance is decreased still more, to 1.69 A.

Exercises

23.1 Why should the C—Cl bond length be shortened in CF_2Cl_2 and not in CCl_4?

23.2 Tetrafluoroethylene is reported to be made as shown.

$$2HCF_2Cl \xrightarrow{700°} F_2C=CF_2 + 2HCl$$

Devise a possible radical mechanism.

23.3 Show how to prepare:
   a. $CH_3CH_2CHBrCH_2Br$ from BuOH
   b. $Cl_3CCH_2CBrMe_2$ from CH_4 and tBuOH

23.3 Haloarenes and allylic halides

Preparation; reactions of haloarenes. We have discussed the aromatic chloro and bromo compounds previously, and we also have looked at preparation by direct electrophilic substitution. Side-chain chlorination and bromination have been shown as well. The fluorobenzenes can be made by means of the Schiemann reaction. Iodobenzenes, although electrophilic iodination is possible in the presence of mercuric oxide or nitric acid, usually are prepared by means of iodide displacement of a diazo group. The Sandmeyer reaction may be used to substitute Cl— or Br— for H_2N—.

Two additional reactions of aryl halides should be mentioned here. The Ullmann reaction is useful for the synthesis of substituted biphenyls (Eq. 23.3). It is a complement to

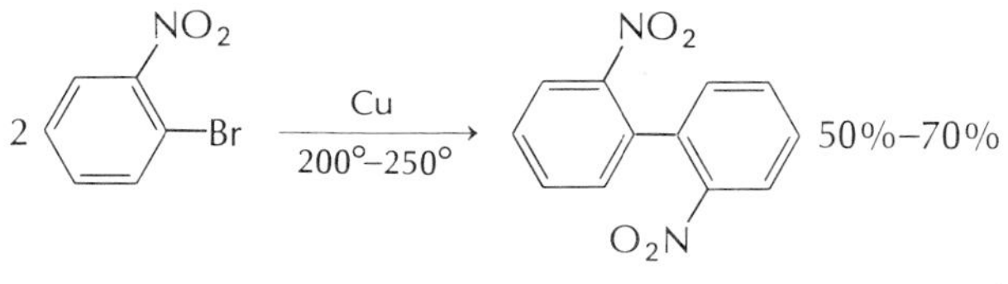

$$2 \underset{}{\text{Ar(NO}_2\text{)—Br}} \xrightarrow[200°-250°]{Cu} \text{2,2'-dinitrobiphenyl} \quad 50\%-70\% \quad (23.3)$$

the benzidine rearrangement for entry into the biphenyl system. As with other systems in which copper and cuprous or cupric salts are involved, the reaction probably is a free-radical one. It proceeds best if the halogen present is iodine, less readily if bromine or chlorine, and not at all if fluorine.

The Wurtz-Fittig reaction sometimes is used for the preparation of an alkylbenzene (Eq. 23.4). Presumably, both φNa and BuNa are formed, but the anions attack

$$CH_3CH_2CH_2CH_2Br \xrightarrow[20°]{Na} \phi CH_2CH_2CH_2CH_3 \quad (23.4)$$
$$60\%$$

BuBr much more readily than ϕBr. Thus, by-product octane exceeds side product biphenyl. Temperature is kept low to minimize the elimination reaction $\phi^\ominus$ or $Bu^\ominus + BuBr \rightarrow \phi H$ or $BuH + CH_3CH_2CH=CH_2 + Br^\ominus$. □ □

Allylic and benzylic Grignard reagents. Allylic and benzylic halides give Grignard reagents that often react as though they had been prepared from the halide that would result from an allylic rearrangement (Eq. 23.5). This kind of reaction sometimes is called

$$CH_3CH=CHCH_2Br \xrightarrow[\substack{1. \text{ Mg, ether} \\ 2. \text{ CO}_2 \\ 3. \text{ dil. HCl}}]{} \underset{\underset{COOH}{|}}{CH_3CHCH=CH_2} \quad (23.5)$$

abnormal addition. The nmr of a solution of $H_2C=CHCH_2MgX$ shows a doublet of area 4 and a triplet of area 1, which means that $H_2\overset{\delta-}{C}\!=\!=\!CH\!=\!=\!\overset{\delta-}{C}H_2$ is symmetrical. The product to be expected from the addition of an allylic Grignard reagent to a carbonyl group (aldehyde, ketone, ester, acid chloride, N,N-dialkylamine, CO_2) should be looked up when needed, because predictions are unreliable. The Grignard reagents prepared from benzyl halides sometimes give the normal product (Eq. 23.6a) and sometimes give abnormal addition at

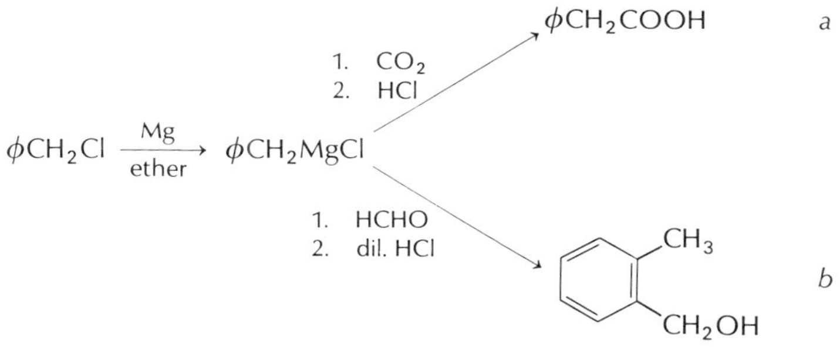

$$\phi CH_2Cl \xrightarrow[\text{ether}]{Mg} \phi CH_2MgCl \begin{array}{c} \xrightarrow[]{1.\ CO_2 \\ 2.\ HCl} \phi CH_2COOH \quad a \\ \\ \xrightarrow[]{1.\ HCHO \\ 2.\ \text{dil. HCl}} \text{o-}CH_3C_6H_4CH_2OH \quad b \end{array} \quad (23.6)$$

the *ortho* position (b) or the *para* position (Eq. 23.7). No satisfactory mechanism has been given for this behavior. Ketones are supposed to react correctly, but one should look up a given reaction to be on the safe side.

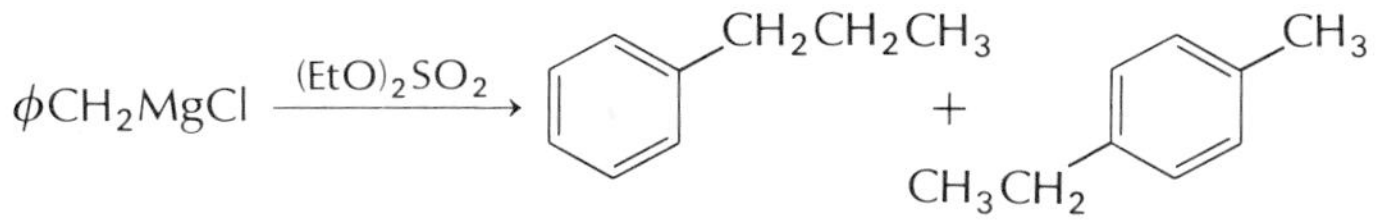

$$\phi CH_2MgCl \xrightarrow{(EtO)_2SO_2} \text{C}_6\text{H}_5CH_2CH_2CH_3 + \text{p-}CH_3CH_2C_6H_4CH_3 \quad (23.7)$$

Diphenylmethyl halides (ϕ_2CHCl, benzhydryl halides) and triphenylmethyl halides (ϕ_3CCl) may be said to show "super" benzylic halide reactivity in their easy S_N1 solvolysis. Grignard reagents may be prepared, and they seem to react normally. (Certainly there must be an abnormal reaction at an *ortho* or *para* position in some instances.)

The hexaphenylethane mistake. In 1900, Gomberg reported the synthesis of hexaphenylethane by a Wurtz reaction in the absence of oxygen (Eq. 23.8) from triphenyl-

$$2\phi_3\text{CCl} \xrightarrow[\phi\text{H no O}_2]{\text{Ag or Zn}} \phi_3\text{CC}\phi_3 \quad (23.8)$$

methylchloride. The product was a colorless crystalline solid; the mw and the C and H analyses were correct. However, a benzene solution of the arene, again free of oxygen, soon became yellow. Treatment of the solution with small portions of O_2, Cl_2, Br_2, or I_2 gave instantaneous decolorization, followed by a return of the yellow color. The first authenticated free radicals had been observed (Eq. 23.9).

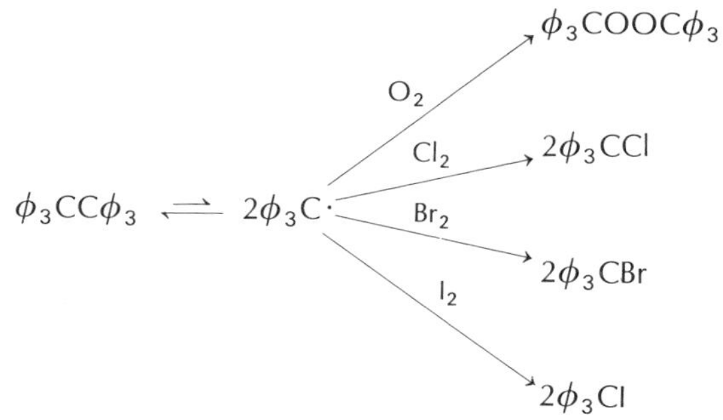

$$\phi_3\text{CC}\phi_3 \rightleftharpoons 2\phi_3\text{C}\cdot \begin{array}{l} \xrightarrow{O_2} \phi_3\text{COOC}\phi_3 \\ \xrightarrow{Cl_2} 2\phi_3\text{CCl} \\ \xrightarrow{Br_2} 2\phi_3\text{CBr} \\ \xrightarrow{I_2} 2\phi_3\text{CI} \end{array} \quad (23.9)$$

Subsequent research showed that $\Delta H = 11$ kcal for the dissociation, and many suggestions were made as to why the C—C bond should be so weak. The effect of substituents was thoroughly looked into. For example, hexa(o-methoxyphenyl)ethane was found to be almost completely dissociated in 0.08 M solution in benzene at 5°.

In all the excitement over the free radicals, it was not until 1968 that Lankamp, Nauta, and Kmet thought to examine the nmr of hexaphenylethane in CCl_4. The protons on the benzene rings appear as expected at $\delta 6.8$ to 7.4 ppm. However, an A_2B_2C pattern (correct by integration) also appears at $\delta 6.4$ to 5.8 and 5 ppm. Hexaphenylethane is not hexaphenylethane: it is a substituted cyclohexadiene.

The structures of the many "hexaarylethanes" remain to be examined,

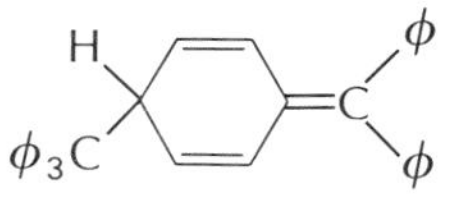

especially those with *para* substituents. For example, "hexa(*p*-biphenyl)ethane" and "hexa(*p*-tert-butylphenyl)ethane" are known to be highly dissociated. It is interesting that it was known in 1939 that radicals of the tri(*p*-alkylphenyl)methyl type disproportionated as shown in Equation 23.10. Even earlier (1904), it was reported that

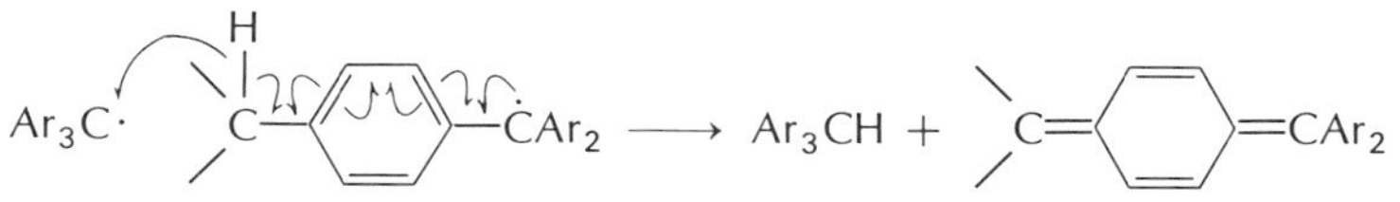

$$Ar_3C\cdot \quad \text{C}\text{—}\text{C}Ar_2 \longrightarrow Ar_3CH + \text{C}=\text{C}Ar_2 \quad (23.10)$$

"hexaphenylethane" in the presence of metallic sodium in ether or of traces of HCl in benzene gave rise to *p*-benzhydryltetraphenylmethane, a stable (nondissociating) substance (Eq. 23.11). The aromatization of the substituted cyclohexadiene ring by transfer of a hydrogen is easily acceptable.

$$\phi_3\text{CCl} \xrightarrow{\text{Ag}} [\phi_3\text{C}-\text{C}\phi_3?] \xrightarrow[\text{Na}]{\text{HCl or}} \phi_3\text{C}-\!\!\left\langle\!\!\!\bigcirc\!\!\!\right\rangle\!\!-\text{CH}\!\!\begin{array}{c}\phi\\ \phi\end{array} \qquad (23.11)$$

The triarylmethyl system offers excellent delocalization and stabilization of radical, anion, or cation (Eqs. 23.12). The structure and reactivity of the triarylmethyl radicals is not questioned.

$$\begin{array}{c}
\phi_3\text{COH} \xrightarrow{\text{HCl}} \phi_3\text{C}^{\oplus} \\
{\scriptstyle\text{AlCl}_3}\nearrow \\
\phi_3\text{CCl} \xrightarrow{\text{Na}} \phi_3\text{C}^{\ominus} \\
{\scriptstyle\text{Na}}\nearrow \\
\phi_3\text{C}\cdot \xrightarrow{\text{FeCl}_3} \phi_3\text{CCl} + \text{FeCl}_2
\end{array} \qquad (23.12)$$

Insecticides. In 1939 it was found that DDT was insecticidal. In 1942 it first was offered for sale by Geigy, a Swiss firm. It was the first effective, wide-spectrum, persistent insecticide. Widespread use soon followed: during World War II, principally against lice (typhus carriers) and mosquitos (malaria carriers).

Enthusiasm led to indiscriminate and overabundant use until the persistent wonder insecticide became distributed throughout the world, accumulating in the fatty tissue of living organisms of the land, sea, and air. In the meantime, the incidence of malaria, typhus, yellow fever, bubonic plague, sleeping sickness, and other pestilences had been dramatically reduced. In Ceylon, where the use of DDT had almost eradicated malaria by 1961

DDT

(110 cases in a population of about 11 million), the spraying was stopped. By 1969, malaria again was a major problem (almost 3 million cases). DDT also has been widely used as an agricultural spray to control insect devastation of food crops. In 1942, no one ever dreamed that DDT would be produced in such quantities and used so indiscriminately that it would become so widely distributed. The persistence of DDT (lack of biodegradation), which was one of the factors favoring its use, has turned into a nuisance. The principal effect of DDT in biological systems seems to be an interference with calcium *in vivo* reactions. It causes thinner than normal egg shells to be produced by certain bird species. It also has been reported that those who work with DDT and who have 100 times more DDT in their fat than does the normal population also have a lower incidence of cancer than normal and no other symptoms after twenty-five years.

Another chlorinated hydrocarbon insecticide introduced in 1945 is γ-hexachlorocyclohexane. Nine isomers are possible (called α-, β-, etc.). Gammexane (or lindane, benzene hexachloride, BHC, 666) is one of the possible isomers produced by the free-radical addition of chlorine to benzene (Eq. 23.13). The γ-isomer has three adjacent chlorine atoms in axial

$$C_6H_6 + 3Cl_2 \xrightarrow[360-400 \text{ nm}]{h\nu} C_6H_6Cl_6 \qquad (23.13)$$

positions, the other three in equatorial positions. The other isomers have no insecticidal activity.

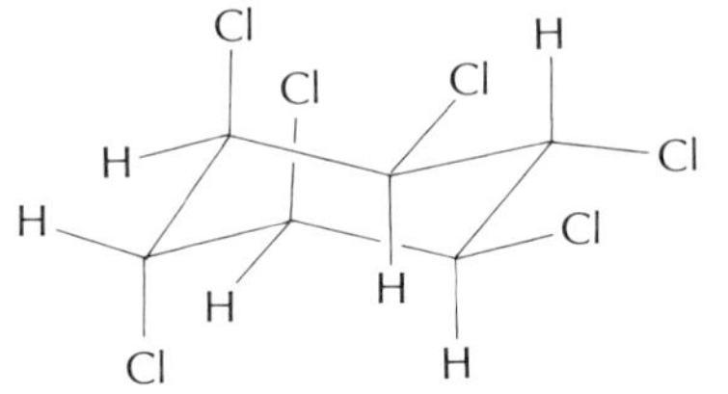

Other chlorinated insecticides that have been used are aldrin, dieldrin, endrin, and chlordane. All four may be considered derivatives of hexachlorocyclopentadiene (Eq. 23.14). It is interesting that the Diels-Alder reactions proceed to give the *endo*

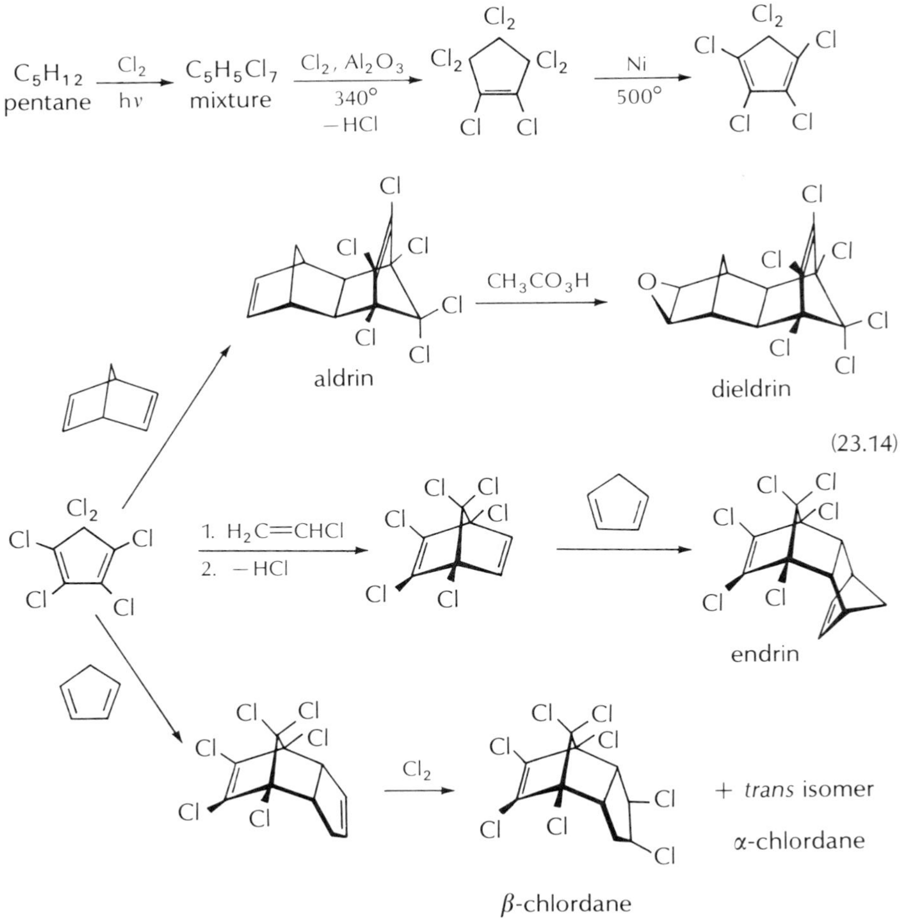

$$C_5H_{12} \xrightarrow[h\nu]{Cl_2} C_5H_5Cl_7 \xrightarrow[340°]{Cl_2, Al_2O_3} \text{(hexachlorocyclopentene)} \xrightarrow[500°]{Ni} \text{(hexachlorocyclopentadiene)}$$

aldrin → dieldrin (via CH₃CO₃H)

(23.14)

β-chlordane → α-chlordane + trans isomer (via Cl₂)

endrin

isomer of the bicyclo ring system in which the bridge is —CCl$_2$—, as in Aldrin and Endrin. On the other hand, the aldrin product is the *exo* isomer of the unchlorinated bicycloheptene system but is the *endo* isomer in the endrin product.

The toxicity and persistence of the chlorinated insecticides has caused rigid controls to be imposed on their use. Other types of insecticides are being explored, as well as other ways to control insect populations, such as biological methods (encouragement of natural predators, diseases, etc.) and sterilization.

Exercises

23.4 If we designate the γ-isomer of hexachlorocyclohexane as *aaaeee* (or the equivalent: *ttcttc*) work out the nine possible isomeric designations.

23.5 Show how to prepare:

a. 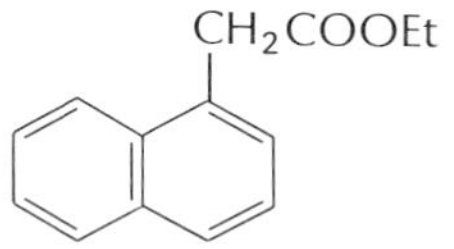 naphthalene with CH$_2$COOEt substituent — from naphthalene

b. 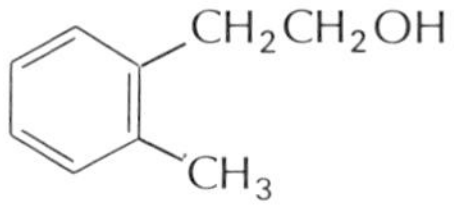 benzene with CH$_2$CH$_2$OH and CH$_3$ substituents (ortho) — from toluene

c. 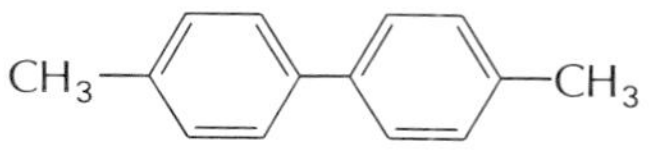 CH$_3$—⟨⟩—⟨⟩—CH$_3$ from toluene

23.4 Halohydrins and other halogen-substituted compounds

The halohydrins, haloaldehydes, ketones, acids, and acid derivatives have been discussed in the chapters on alcohols, carbonyl compounds, and acids.

Halohydrins and β-haloethers result from the addition of halogen (Cl$_2$ or Br$_2$) to an alkene in mildly basic aqueous or alcoholic solution. Under more strongly basic conditions, a halohydrin reacts to give an epoxide (Eq. 23.15). Under acidic conditions, a halohydrin

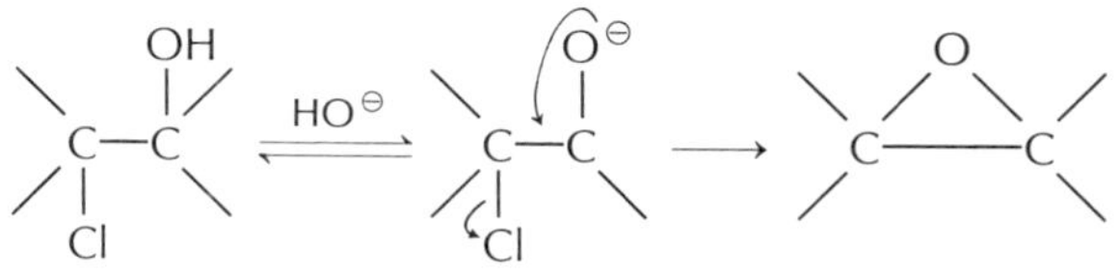

$$\begin{matrix} \text{OH} \\ | \\ \text{C—C} \\ | \\ \text{Cl} \end{matrix} \quad \xrightleftharpoons{\text{HO}^{\ominus}} \quad \begin{matrix} \text{O}^{\ominus} \\ | \\ \text{C—C} \\ | \\ \text{Cl} \end{matrix} \quad \longrightarrow \quad \begin{matrix} \text{O} \\ \diagup \diagdown \\ \text{C——C} \end{matrix} \qquad (23.15)$$

dehydrates or undergoes a pinacol or Wagner-Meerwein rearrangement (Eqs. 23.16). The major products are the aldehyde or ketone. A halohydrin (or an α-haloether) with zinc dust gives the original alkene back.

Multiple halogens on the carbon atom adjacent to the hydroxy, as in 2,2,2-trifluoroethanol (F$_3$CCH$_2$OH), cause the alcohol to become acidic. Hexafluoroisopropyl alcohol ((CF$_3$)$_2$CHOH) has pK_a = 6.7.

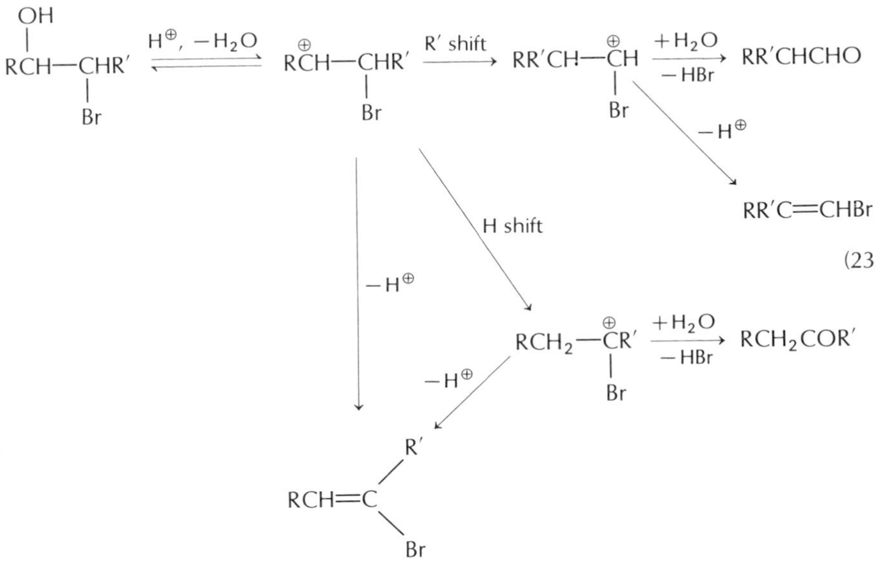

$$\underset{\underset{Br}{|}}{RCH-CHR'}\overset{OH}{\underset{}{|}} \underset{-H_2O}{\overset{H^\oplus,\ -H_2O}{\rightleftharpoons}} \underset{\underset{Br}{|}}{R\overset{\oplus}{C}H-CHR'} \overset{R'\ shift}{\longrightarrow} \underset{\underset{Br}{|}}{RR'CH-\overset{\oplus}{C}H} \overset{+H_2O}{\underset{-HBr}{\longrightarrow}} RR'CHCHO$$

$$\downarrow -H^\oplus$$

$$RR'C=CHBr$$

(23.16)

H shift path → $\underset{\underset{Br}{|}}{RCH_2-\overset{\oplus}{C}R'} \overset{+H_2O}{\underset{-HBr}{\longrightarrow}} RCH_2COR'$

$\downarrow -H^\oplus$

$RCH=\underset{Br}{\overset{R'}{C}}$

The aldehyde chloral (Cl$_3$CCHO) and the trihalomethylketones form stable hydrates. Under alkaline conditions, they are readily cleaved, as in the haloform reaction.

The β-haloaldehydes, ketones, acids, and so on, like the corresponding hydroxy compounds (which lose water), easily lose HX to form the α,β-unsaturated compounds. Similarly, the γ and δ halogen compounds react to give cyclic hemiacetals or lactones (Eqs. 23.17). The α-halogen esters undergo the Reformatsky reaction (Eq. 16.17).

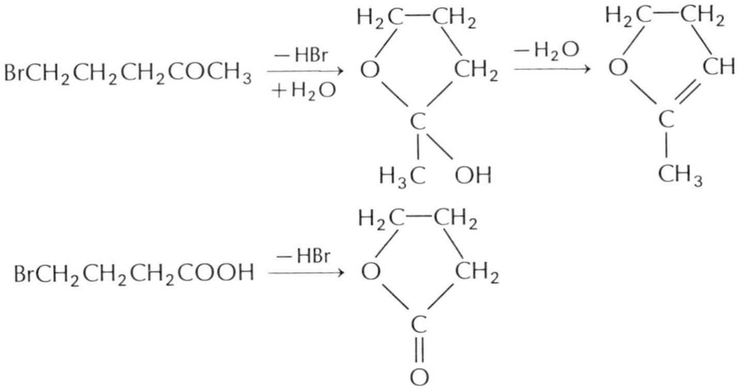

$$BrCH_2CH_2CH_2COCH_3 \overset{-HBr}{\underset{+H_2O}{\longrightarrow}} \text{(cyclic hemiacetal with } H_3C,\ OH) \overset{-H_2O}{\longrightarrow} \text{(dihydrofuran with } CH_3)$$

$$BrCH_2CH_2CH_2COOH \overset{-HBr}{\longrightarrow} \text{(γ-butyrolactone)}$$

(23.17)

Exercises

23.6 Show how to prepare:

a. $\phi_2C=CHCOOEt$

b. $\phi_2CClCOCl$

c. 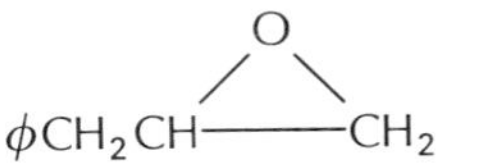 $\phi CH_2CH\overset{O}{\overset{\diagup\diagdown}{-\!-\!-}}CH_2$

d. $\phi CHClCH_2CH_2COOEt$

23.5 Summary

In this chapter we have provided a list of the reactions and preparations of the organic halogen compounds and the chapters in which they have appeared. In so doing, we reviewed many of the topics covered in this course.

We have also taken the opportunity to include some description of the fluorocarbons, triarylmethyl radicals, and some chlorinated insecticides.

The chapter has been short, but only because the subject has been covered in detail throughout the preceding chapters, where the halogen-containing compounds were related to the class of compound or reaction being considered.

23.6 Bibliography

Aside from the general textbooks previously listed, most of which have separate chapters on aliphatic and aromatic halides, do not overlook the paperbacks by Stille (Chap. 7, ref. 21) and Weiss (Chap. 7, ref. 22).

For the hexaphenylethane story see the following:

1. Smith, William B., *Journal of Chemical Education* 47, 535 (1970).

2. Wheland, George W., *Advanced Organic Chemistry*, 3rd ed. New York: Wiley, 1960. Sections 15.9–15.17, pp. 763–793, are very good on the history of the triphenylmethyl radicals.

3. Gilman, Henry, ed., *Advanced Organic Chemistry*, 2nd ed. New York: Wiley, 1943. Chapter 6 (Free Radicals), by W. E. Bachman, is excellent in coverage of the subject up to the time of publication.

For more on insecticides, two articles provide starting points:

4. Bergmann, E. D., The Future of Insect Control—A Problem of Human Environment, *Chemical Technology* 1, 740 (1971).

5. Metcalf, R. L., I. P. Kapoor, and A. S. Hirwe, Development of Biodegradable Analogues of DDT, *Chemical Technology* 2, 105 (1972).

24 Phenols and Quinones

Phenols and quinones are involved in many naturally occurring substances: lignins, vitamins, alkaloids and many other drugs, antioxidants, metabolic products, and flower pigments, to name a few. They also are components of synthetically produced substances such as antiseptics, antipyretics, polymers, flavors and odors, dyes, and so on. The chemistry of phenols is diverse, characterized by ease of aromatic electrophilic substitution, reactions as enols, reactions as ketones, and ease of oxidation. Most quinones also are reactive substances and undergo many addition reactions as well as reduction to diphenols (hydroquinones).

24.1   Phenols   789
    1. Occurrence.   2. Naming.   3. Physical properties.   4. Spectra.

24.2   Phenols: reactions at the oxygen-hydrogen bond   790
    1. Phenols as enols: acidity.   2. Formation of ethers: aryloxyacetic acids.
    3. Hydrolysis of ethers.   4. Esterification.   5. Aspirin, salicylates.
    6. Ferric chloride colors. Exercises.

24.3   Phenols: reactions at the carbon-oxygen bond   795
    1. Replacement of hydroxy.   2. Bucherer reaction.   3. Replacement by hydrogen.

24.4   Phenols: reactions on the ring   795
    1. Sulfonation.   2. Nitration; nitrosation.   3. Coupling: p-aminophenols.
    4. Halogenation: protic and aprotic solvents.   5. Friedel-Crafts reactions: mild catalysts, rearrangements.   6. Reactions with aldehydes and ketones.
    7. Mercuration.   8. Base-catalyzed substitutions.   9. Oxidation.
    10. Reduction. Exercises.

24.5   Di- and trihydroxybenzenes   805
    1. Catechol.   2. Polyethers.   3. Resorcinol.   4. Hydroquinone.
    5. Pyrogallol.   6. Phloroglucinol. Exercises.

24.6   Preparation of phenols   808
    1. The sulfonation route.   2. From chlorobenzene.   3. From cumene.
    4. Cresols.   5. Di- and trihydroxybenzenes.   6. Other routes. Exercises.

24.7   Quinones   811
   1. Origin of the name.   2. Addition reactions.   3. Diels-Alder reactions.
   4. Preparations. Exercises.

24.8   Lignin   814

24.9   Some drugs and vitamins   815
   1. Phenolic phenylethylamine compounds.   2. Chloramphenicol.
   3. Marijuana.   4. Vitamin E: synthesis, antioxidant activity.   5. The aging process.   6. Vitamin K; dicumarol; menadione. Exercises.

24.10   Summary   820

24.11   Problems   821

24.12   Bibliography   821

24.13   H. Marjorie Crawford: A lifelong interest   822a

24.1 Phenols

Occurrence. Phenols and derivatives of phenols are widespread in nature. Lignin, used to bind cellulose into wood, seems to consist mainly of units constructed from coniferyl alcohol (3-(4-hydroxy-3-methoxyphenyl)-2-propenol). Many pheromones (chemi-

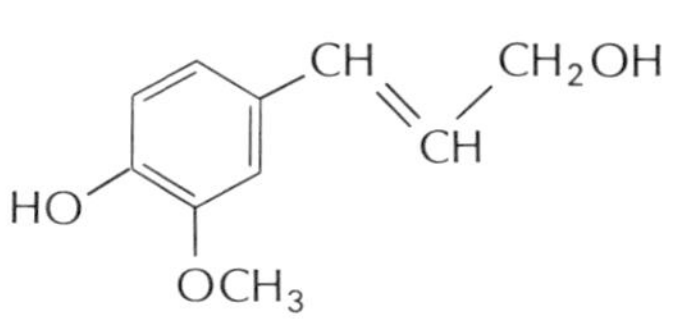

coniferyl alcohol 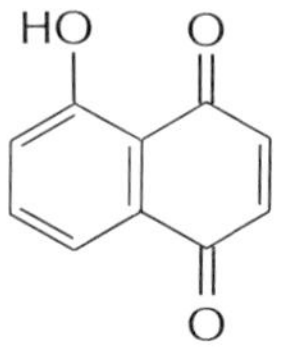   juglone 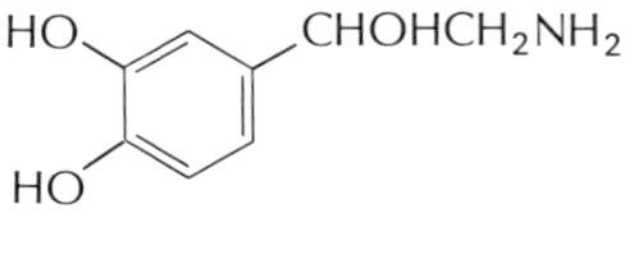   norepinephrine

cals that serve as messengers in and between species) are phenolic. For example, juglone is washed from the leaves of walnut trees into the soil beneath, where it exerts an allelopathic effect. It prevents germination and growth of the seeds of other species. In the class of phenols containing nitrogen, many physiologically active substances are known: for example, norepinephrine, a vasoconstrictor (used in nose drops and nasal sprays). Phenol, the cresols and xylenols, and α- and β-naphthols are obtainable from coal tar in small amounts, a total of about 2.5%.

Naming. The simple phenols have nonsystematic names that must be learned. Many derivatives of phenols have common names as well. Otherwise, systematic names are used.

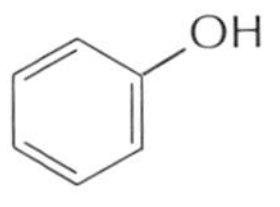

790   Chapter twenty-four

phenol, 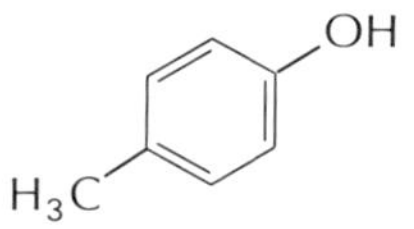 **p-cresol** (also o and m), **catechol**, 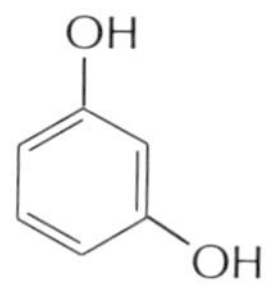 **resorcinol**,  **hydroquinone**

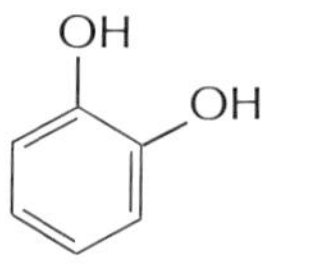

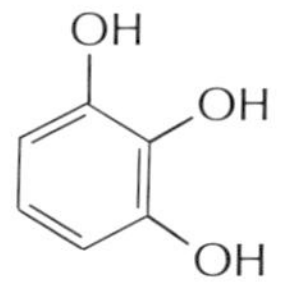

pyrogallol, 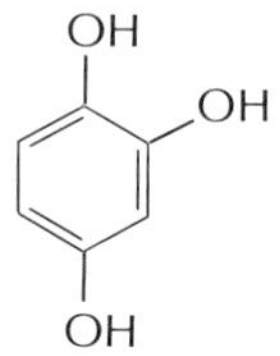 **hydroxyhydroquinone**, 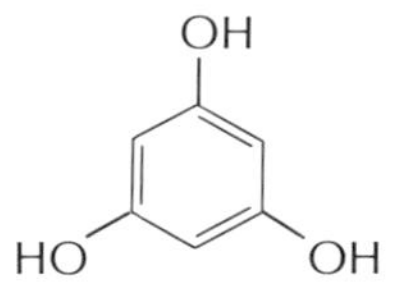 **phloroglucinol**

Physical properties.   Phenols have bp in excess of 200° (hydrogen bonding), except for phenol itself and a few *ortho*-substituted phenols. The mp tend to fall between 10° and 110°, depending upon the symmetry of the molecule. Solubility in water is usually appreciable (for phenol, 9 g/100 g H_2O, 25°). Water is quite soluble in phenol (29 g/100 g phenol, 25°), and it lowers the mp so much that a 5% solution of water in phenol is liquid (carbolic acid). Phenols are toxic to bacteria. Lister introduced the use of carbolic acid as the antiseptic agent in sterile surgery in 1867. Antiseptics are still rated in terms of phenol coefficients—roughly, the ratio of phenol to the substance being tested that produces the same lethal effect on a given bacterium.

Spectra.   The mass spectra of phenols are characterized by a large parent peak. The $P - 28$ and $P - 29$ peaks (loss of CO and CHO) are usually present. The ir is similar to that of alcohols, but with a broad out-of-plane O—H bending at 650 to 770 cm^{-1} and a C—O stretch at 1200 to 1240 cm^{-1}. The phenolic proton appears in nmr at δ 4.0 to 7.5 ppm. The uv is affected strongly by the type and location of substituents and varies with the pH in aqueous solutions.

We shall divide the reactions of phenols into three groups: those at the oxygen-hydrogen bond (ArO—H), those at Ar—OH, and those upon the ring.

24.2   Phenols: reactions at the oxygen-hydrogen bond

Phenols as enols: acidity.   Phenols are enols stabilized by the aromatic ring. A simple calculation (from bond energies) gives about 20 kcal for ΔH for enolization of a ketone. Ketonization of a phenol then would be favored by -20 kcal, disfavored by 36 kcal (loss of stabilization of the aromatic benzene ring), and favored by an unknown amount of stabilization of a cyclohexadienone (perhaps -5 to -10 kcal). The sum, 6 to 11 kcal, indicates that an insignificant amount of tautomeric ketone would be present at equilibrium with phenol (Eq. 24.1). However, as we shall see later on, some phenols have ketonic reactions.

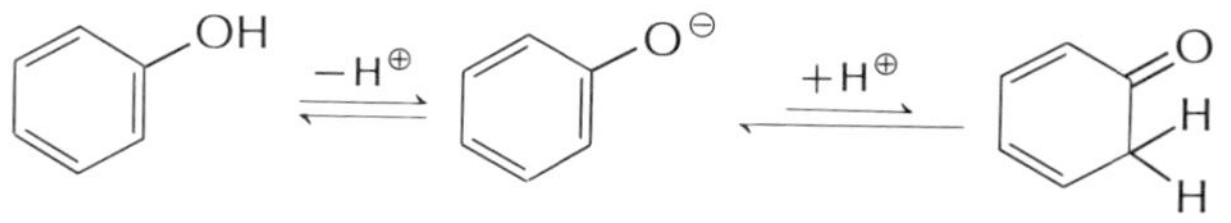

24.2 Phenols: reactions at the oxygen-hydrogen bond

$$\text{PhOH} \underset{+H^\oplus}{\overset{-H^\oplus}{\rightleftharpoons}} \text{PhO}^\ominus \rightleftharpoons \text{cyclohexadienone form} \tag{24.1}$$

The phenoxide anion can be stabilized by delocalization of charge, whereas similar resonance structures of phenol require charge separation. Thus phenols are acidic. Substitution on the ring by electron-withdrawing substituents increases acidity by stabilizing

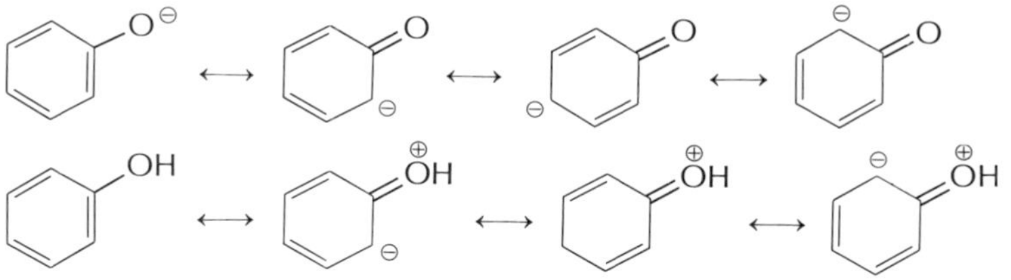

the anion more than the undissociated phenol, especially if conjugation is possible (o and p). Electron-donating substituents are unable to conjugate from any position, and the acid-weakening effect is less marked. Thus phenol has pK_a of 9.9; o-nitrophenol, of 7.2; m-nitrophenol, of 8.3; and p-nitrophenol, of 7.2. The o, m, and p cresols have pK_a of 10.2, 10.0, and 10.2, which illustrate the small acid-weakening effect. The polyhalophenols are surprisingly weak acids: pK_a of 5.5 for C_6F_5OH, as compared with 2,4-dinitrophenol, $pK_a = 4.0$; and 2,4,6-trinitrophenol (picric acid), $pK_a = 0.4$. A phenol is acidic enough to react with diazomethane (CH_2N_2) to give a methyl ether.

Practically, phenols dissolve in dilute NaOH. Solubility in sodium carbonate and sodium bicarbonate solutions is slight but appreciable: greater, of course, for the more acidic phenols.

Formation of ethers: aryloxyacetic acids. Phenols easily form ethers by a Williamson ether synthesis in which the phenoxide ion is always used as the nucleophile (Eq. 24.2). The

$$\text{ArOH} \xrightarrow[\substack{RX \\ Na_2CO_3 \\ (RO)_2SO_2}]{NaOH} \begin{array}{l} \text{ArOR} + \text{NaX} \\ \\ \text{ArOR} + Na_2SO_4 \end{array} \tag{24.2}$$

reaction is used for the preparation of phenoxyacetic acids, which are convenient solid derivatives of phenols (Eq. 24.3).

$$\phi\text{OH} + \text{ClCH}_2\text{COOH} \xrightarrow[\substack{1 \text{ hr} \\ 100°}]{NaOH} \phi\text{OCH}_2\text{COONa} + \text{NaCl} \tag{24.3}$$

Following the discovery that indoleacetic acid was a plant growth hormone, a search for synthetic compounds led to 2,4-dichloro- and 2,4,5-trichlorophenoxyacetic

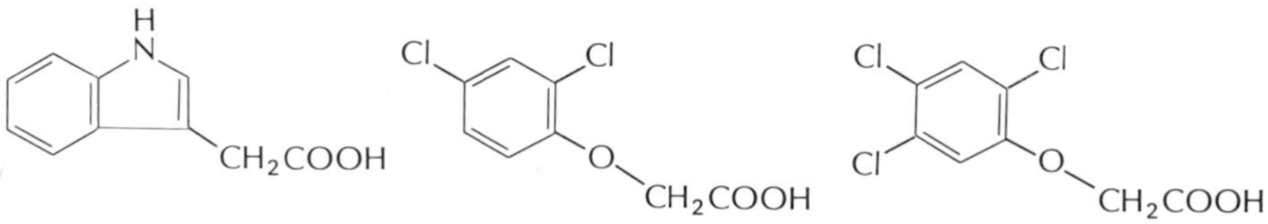

indoleacetic acid   2,4-D   2,4,5-T

acids (known as 2,4-D and 2,4,5-T), which showed pronounced herbicidal effects on broad-leaved plants, but little effect on grasses. Strangely enough, the presence of other halogens, including fluorine, was much less effective. Phenoxyacetic acid is without any effect, and any other substituent on the ring has no (or slight) effect. Both 2,4-D and 2,4,5-T are relatively nontoxic. A controversy concerning the toxicity of 2,4,5-T in 1970 showed that traces of the extremely toxic 2,3,7,8-tetrachlorodibenzo-p-dioxin is the cause (Eq. 24.4). The dioxin is formed in a side reaction during the

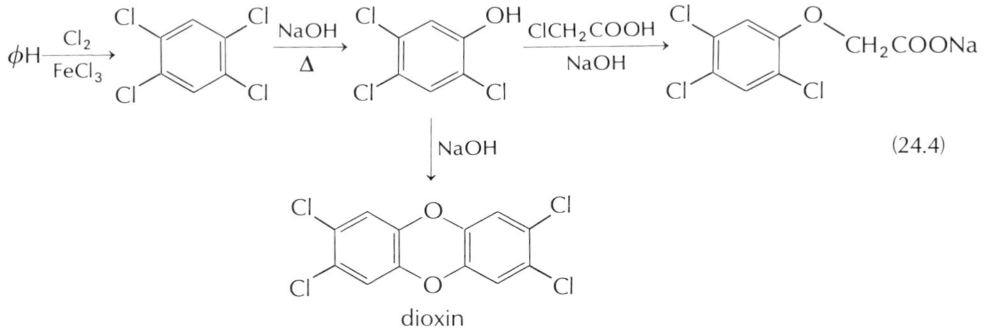

$$\phi H \xrightarrow[\text{FeCl}_3]{\text{Cl}_2} \text{(2,4,5-trichlorobenzene)} \xrightarrow[\Delta]{\text{NaOH}} \text{(2,4,5-trichlorophenol)} \xrightarrow[\text{NaOH}]{\text{ClCH}_2\text{COOH}} \text{(aryloxyacetate)} \quad (24.4)$$

$$\downarrow \text{NaOH}$$

dioxin

preparation of 2,4,5-trichlorophenol and, unless the product is carefully purified, the dioxin survives to contaminate the 2,4,5-T.

Other phenoxy-substituted acids are used in other ways. For example, 2-(p-chlorophenoxy)-2-methylpropionic acid (a) is used as the ethyl ester (Clofibrate or Atromid) or as the basic aluminium salt in the treatment of atherosclerosis and (it is hoped) to decrease cholesterol levels in blood. It has been reported that an aminoether, tilorone hydrochloride (b) is an effective antiviral agent in mice.

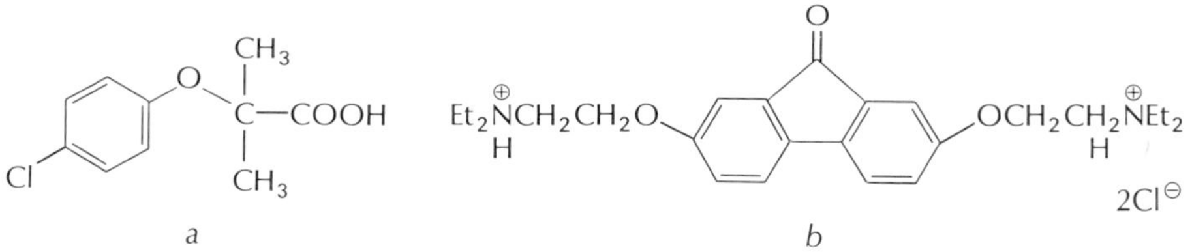

a

b

Hydrolysis of ethers. The aryl ethers are somewhat more difficult to hydrolyze than are the alkyl ethers. The cause lies in the reluctance of the aryl group to undergo nucleophilic substitution. As a result, an aryl alkyl ether reacts as shown in Equation 24.5. Other useful reagents are concentrated hydrobromic acid in acetic acid or AlCl$_3$ in benzene. Diaryl ethers resist this acid-catalyzed cleavage but may be cleaved by nucleophilic aromatic displacement under sufficiently drastic conditions.

Esterification. The esters of phenols are conveniently prepared by reaction with anhydrides or acid chlorides (Eq. 24.6). Fischer esterification is unsuitable because the equilibrium constant is so unfavorable (K_{eq} for CH$_3$COOEt is 4.0; that for CH$_3$COOϕ is 0.009) that it requires the removal of water from the reaction mixture. On the other hand, phenolic esters are easily hydrolyzed and are not prevalent in nature.

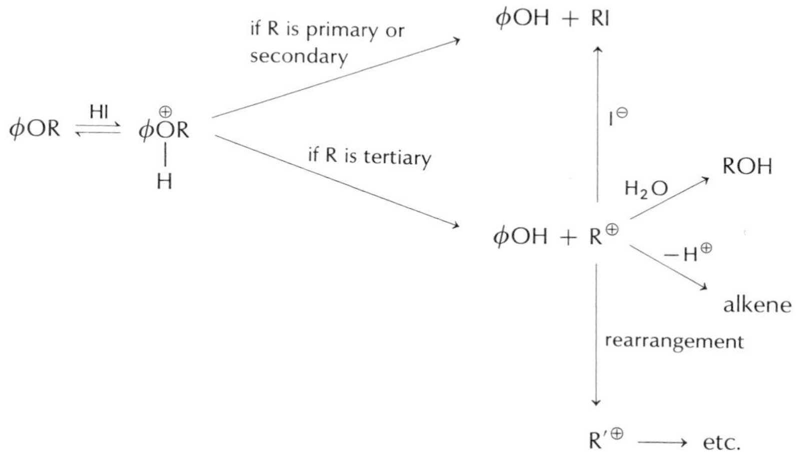

24.2 Phenols: reactions at the oxygen-hydrogen bond

$$\phi OR \underset{}{\overset{HI}{\rightleftarrows}} \phi\overset{\oplus}{\underset{H}{OR}} \begin{array}{c} \xrightarrow{\text{if R is primary or secondary}} \phi OH + RI \\ \xrightarrow{\text{if R is tertiary}} \phi OH + R^{\oplus} \end{array}$$

$$R^{\oplus} \xrightarrow{I^{\ominus}} \text{ } \xrightarrow{H_2O} ROH$$
$$R^{\oplus} \xrightarrow{-H^{\oplus}} \text{alkene}$$
$$R^{\oplus} \xrightarrow{\text{rearrangement}} R'^{\oplus} \longrightarrow \text{etc.}$$

(24.5)

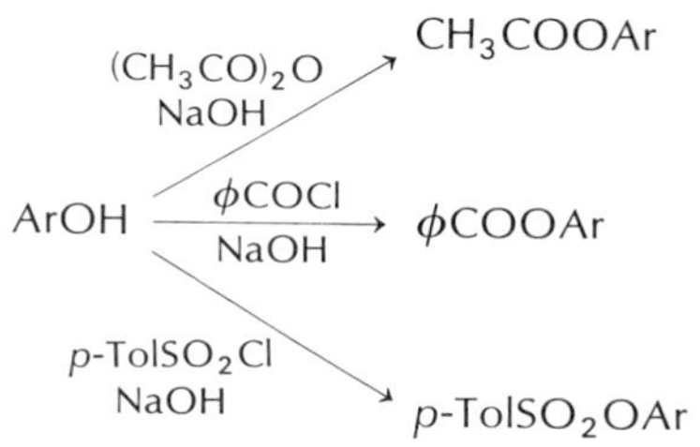

$$\text{ArOH} \begin{array}{c} \xrightarrow{(CH_3CO)_2O, \text{ NaOH}} CH_3COOAr \\ \xrightarrow{\phi COCl, \text{ NaOH}} \phi COOAr \\ \xrightarrow{p\text{-TolSO}_2Cl, \text{ NaOH}} p\text{-TolSO}_2OAr \end{array}$$

(24.6)

Aspirin, salicylates. Aspirin is the acetate ester of salicylic acid, prepared as shown in Equation 24.7. Aspirin is an antipyretic and analgesic, and it is used extensively in the

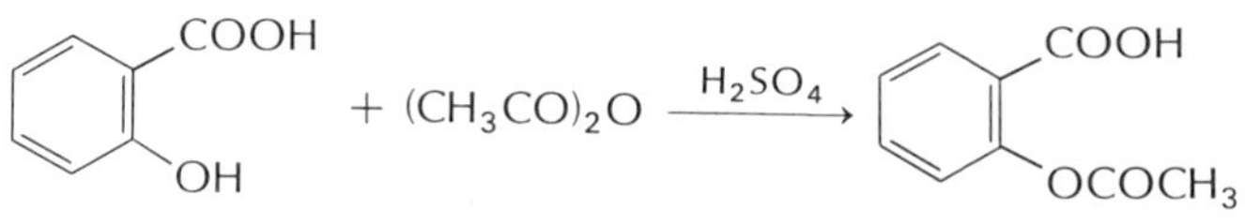

$$\underset{\text{OH}}{\overset{\text{COOH}}{\bigcirc}} + (CH_3CO)_2O \xrightarrow{H_2SO_4} \underset{\text{OCOCH}_3}{\overset{\text{COOH}}{\bigcirc}}$$

(24.7)

treatment of all forms of arthritis for reduction of pain and inflammation. Strangely enough, how aspirin operates is not definitely known. Salicylic acid (o-hydroxybenzoic acid) occurs in nature as several derivatives. The methyl ester is the principal constituent of oil of wintergreen. Salicylates were known to and used by the ancient Greeks.

Salicylic acid is relatively insoluble, and efforts to find a more soluble (and palatable) form led to the use of the sodium salt as an antipyretic in 1875. Upon reaching the stomach, the acid was precipitated and produced great irritation. Other alternatives were looked for. In 1886, the phenyl ester of the acid group was introduced as salol, but the phenol formed in the intestinal tract by hydrolysis was not very conducive of gastric happiness either. It was not until 1898 that acetylsalicylic acid was tried (it was first made in 1853) and found to be efficacious and free of undesirable side effects. (Nothing is perfect—a small percentage of individuals have allergic reactions to aspirin and show other side effects, such as nausea, temporary deafness, etc.). Over the years, no other drug has been so thoroughly proved by so many participants to be as effective and as safe as aspirin. The name was coined from acetyl and spiraeic acid, an old common name for salicylic acid (from the name of a flowering shrub, spiraea). Aspirin passes the stomach unchanged and is hydrolyzed by the mildly alkaline environment of the intestine. The salicylic acid is absorbed through the intestinal membrane. Over 40 tons per day of aspirin is made in the United States.

Ferric chloride colors.   Phenols, like enols, form transient colored complexes with $FeCl_3$ solutions. However, many exceptions are known. For example, although salicylic acid gives a purple color, *m*- and *p*-hydroxybenzoic acids fail to give colors.

Exercises

24.1 Show how to separate a mixture of phenol, benzoic acid, and anisole (ϕOMe) by extraction.

24.2 Many natural products contain aromatic methoxy groups. Zeisel worked out a method for determining the number of these groups by 1. boiling an HI solution with a weighed sample, 2. steam distillation of the mixture, 3. collecting the distillate in alcohol containing $AgNO_3$, and 4. collecting and weighing the precipitate. A 0.0208 g sample of methoxyanthracene will give how much precipitate?

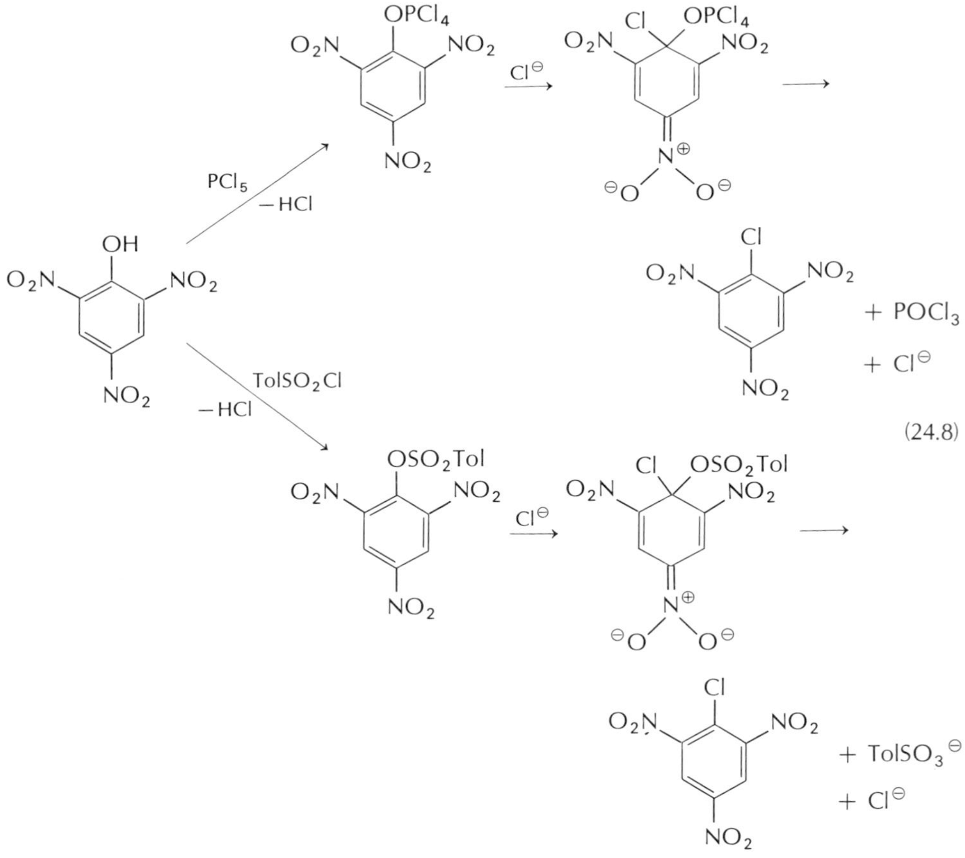

(24.8)

24.3 Phenols: reactions at the carbon-oxygen bond

Replacement of hydroxy. Phenols do not undergo aromatic nucleophilic substitution reactions in the usual sense, because any nucleophile is much more likely to remove a proton from the acidic hydroxy group. It is possible to carry out a replacement of the hydroxy group by certain reagents, however, which might appear to be a direct replacement. Actually, these reactions involve nucleophilic aromatic substitution of an ester (Eqs. 24.8). The usual activation (by −R substituents) is necessary.

Bucherer reaction. The Bucherer reaction, in which hydroxy is replaced by an amino group, takes place by a sequence of addition-elimination steps on those phenols that are able to ketonize to some extent, such as β-naphthol (Eqs. 22.34, 22.35). An amination that might be thought of as a nucleophilic displacement is shown in Equation 24.9, although the conditions are on the drastic side.

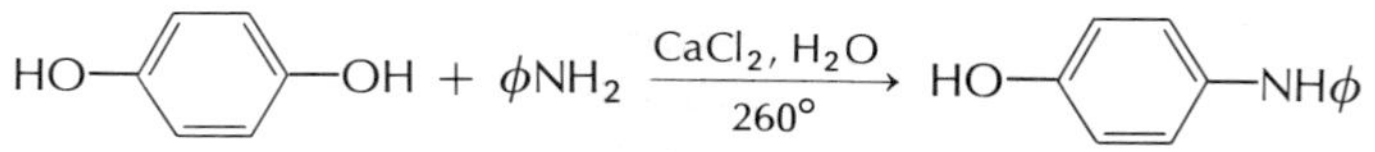

$$HO-C_6H_4-OH + \phi NH_2 \xrightarrow[260°]{CaCl_2, H_2O} HO-C_6H_4-NH\phi \qquad (24.9)$$

Replacement by hydrogen. An old reaction of no synthetic usefulness consists of heating a compound suspected of being a phenol to a high temperature with zinc dust. Oxygen is removed as zinc oxide. A small yield of the compound in which the hydroxy group has been replaced by hydrogen is obtained (Eq. 24.10).

$$ArOH \xrightarrow[400°]{Zn} ArH + ZnO \qquad (24.10)$$

24.4 Phenols: reactions on the ring

The phenols are very reactive in electrophilic substitutions. The hydroxy group is a strong o,p director and activator.

Sulfonation. Phenols are easily sulfonated, and the reaction is subject to temperature effects (Eq. 24.11).

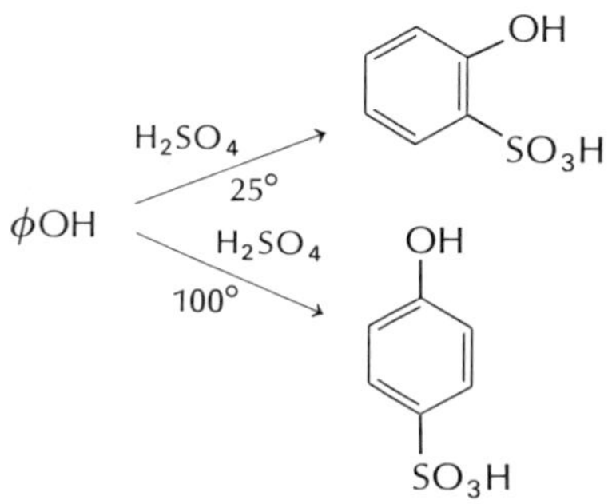

$$\phi OH \begin{array}{c} \xrightarrow[25°]{H_2SO_4} \text{o-HOC}_6H_4SO_3H \\ \xrightarrow[100°]{H_2SO_4} \text{p-HOC}_6H_4SO_3H \end{array} \qquad (24.11)$$

Nitration; nitrosation. Nitration of phenols goes so rapidly that dilute nitric acid at room temperature must be used (Eq. 24.12). There is some oxidation, as well, which causes

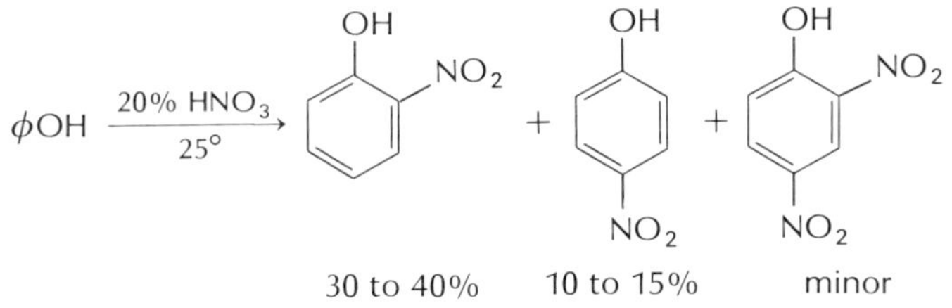

$$\phi OH \xrightarrow[25°]{20\% \ HNO_3} \text{o-nitrophenol} + \text{p-nitrophenol} + \text{2,4-dinitrophenol} \quad (24.12)$$

30 to 40%   10 to 15%   minor

some nitrous acid to be produced. The nitrous acid rapidly nitrosates the phenol, and the nitrosophenol is oxidized by more nitric acid. In other words, nitration by dilute nitric acid is indirect (Eqs. 24.13). If urea is added to the mixture, the nitrous acid is destroyed, and no

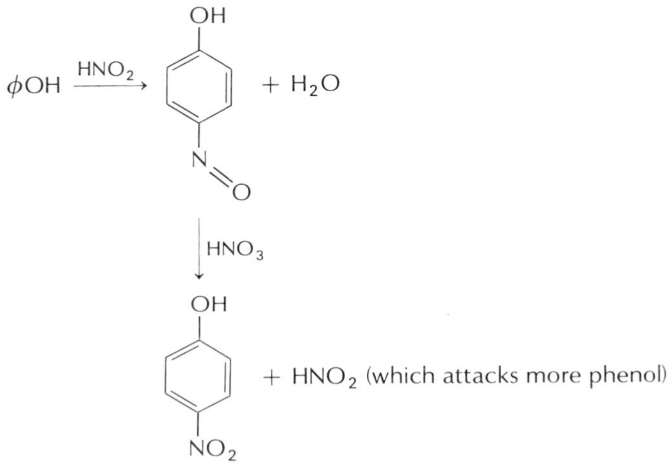

$$\phi OH \xrightarrow{HNO_2} \text{p-nitrosophenol} + H_2O$$

$$\downarrow HNO_3 \quad (24.13)$$

$$\text{p-nitrophenol} + HNO_2 \text{ (which attacks more phenol)}$$

nitration occurs. The mixture of products is easily separated, because o-nitrophenol is volatile with steam (internal hydrogen bonding, chelation) and p-nitrophenol is not.

Nitrosation of phenols, as was just mentioned, is a rapid reaction, and it proceeds well under mild conditions to give the p-nitroso product (unless blocked) in very good yield. The product, a vinylog of nitrous acid, tautomerizes to an oximinoquinone (Eq. 24.14). Reduction of the product offers a good synthetic route to p-aminophenols.

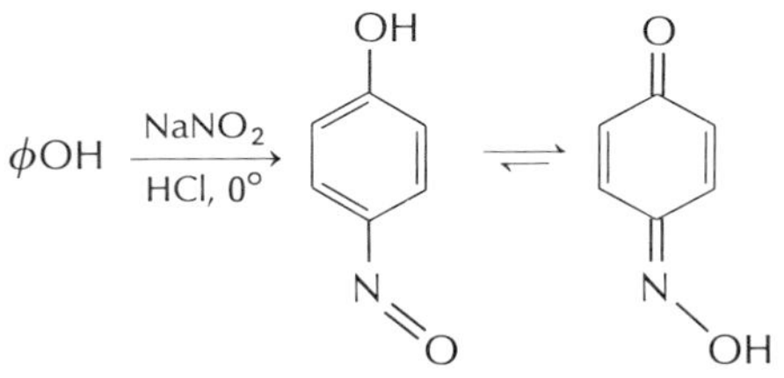

$$\phi OH \xrightarrow[HCl, \ 0°]{NaNO_2} \text{p-nitrosophenol} \rightleftharpoons \text{oximinoquinone} \quad (24.14)$$

Coupling: p-aminophenols. Coupling of phenols with diazonium ions has been discussed (Sec. 22.3). Here we should remind ourselves that coupling is a mild reaction and that success depends upon control of pH: phenoxide ion is even more reactive than phenol to electrophilic substitution.

24.4 Phenols: reactions on the ring

For example, phenyl ethers and esters do not couple, although the $-I$, $+R$ effects are about the same as for the hydroxy group. The pK_a values of substituted benzoic acids are a measure of these effects. Thus, the pK_a values of o-, m-, and p-hydroxybenzoic acids are 3.0, 4.1, and 4.5; those of the corresponding methoxybenzoic acids, 4.1, 4.1, and 4.5. The methoxy and hydroxy groups have identical effects except for the o-hydroxy which shows the influence of hydrogen bonding. The increased reactivity of phenols to electrophilic substitution then must be ascribed to the presence of the very reactive phenoxide ion, even at very small concentrations.

Coupling gives rise to p-hydroxyazo compounds, which, like the nitroso compounds, may be reduced to p-aminophenols (Eq. 24.15). For this purpose sulfanilic acid is used, because it allows of easy separation of the products of reduction.

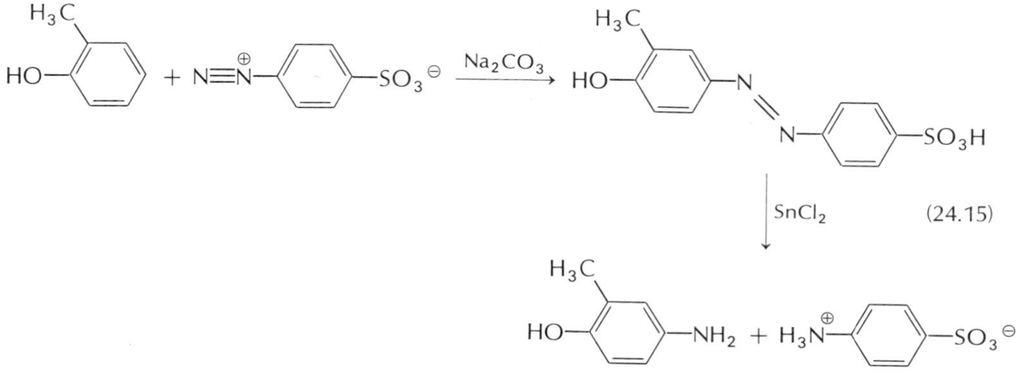

$$\text{(24.15)}$$

Halogenation: protic and aprotic solvents. The products of halogenation of phenols depend upon the conditions used. In aprotic solvents (CCl_4), chlorination gives a mixture of o- and p-chlorophenols. The mixture may be separated by distillation: the bp are 176° and 217°. In CS_2 at 5°, bromine gives up to 85% of p-bromophenol. Pure halophenols can be made from haloarylamines by way of the diazonium ions.

In protic solvents (water especially), phenols are polyhalogenated with such ease that it is impossible to stop at monohalogenation (Eq. 24.16).

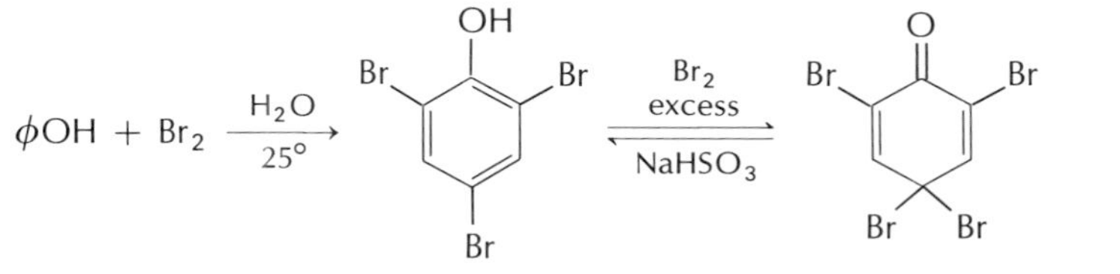

$$\text{(24.16)}$$

Friedel-Crafts reactions: mild catalysts, rearrangements. Friedel-Crafts reactions with phenols usually are modified by the use of milder catalysts. In other instances, rearrangements give better yields than do direct alkylation or acylation.

Alkylation may be achieved by the use of sulfuric acid and a cation source (alcohols and alkenes as well as alkyl halides; Eq. 24.17). Another way of viewing these reactions is to say that phenol rings add to alkenes in the presence of acids.

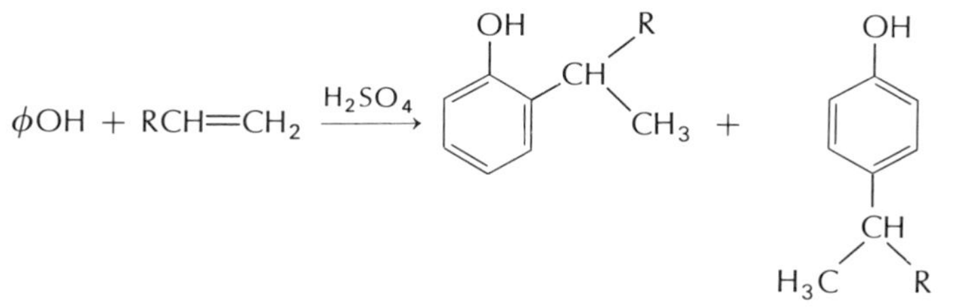

$\phi\text{OH} + \text{RCH}=\text{CH}_2 \xrightarrow{\text{H}_2\text{SO}_4}$ 2-(1-methylalkyl)phenol + 4-(1-methylalkyl)phenol (24.17)

Acylation in the presence of AlCl_3 requires an excess of the catalyst because a phenolic salt is formed first (Eq. 24.18). The aluminum salts of phenols often are so insoluble

$$\text{ArOH} + \text{AlCl}_3 \longrightarrow \text{ArOAlCl}_2 + \text{HCl} \qquad (24.18)$$

in the reaction mixture that they interfere with the smooth progress of the reaction. Several alternatives are possible.

An ether (usually the easily prepared methyl ether) may be used (Eq. 24.19). A low

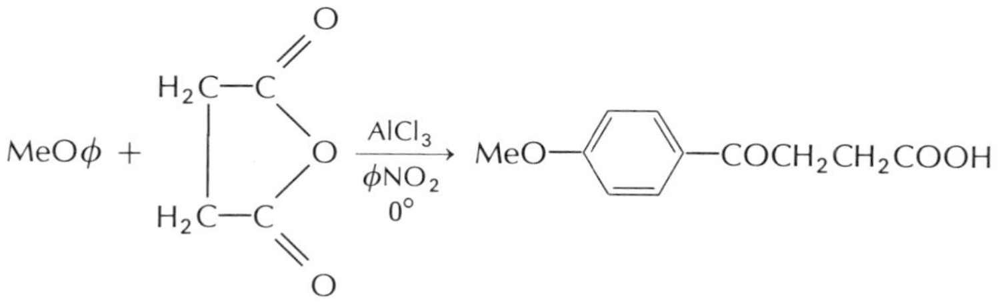

$\text{MeO}\phi + $ succinic anhydride $\xrightarrow[\phi\text{NO}_2]{\text{AlCl}_3}$ $\text{MeO}-\text{C}_6\text{H}_4-\text{COCH}_2\text{CH}_2\text{COOH}$ (24.19)
at $0°$

temperature is used to minimize cleavage of the ether. If we want the phenolic compound, we can add more AlCl_3 after the Friedel-Crafts reaction is complete and warm the mixture to about $80°$ to break the ether link.

Another alternative, which is useful with reactive compounds such as resorcinol, is to use the mild catalyst ZnCl_2. In Equation 24.20 is shown the preparation of hexylresorcinol, a

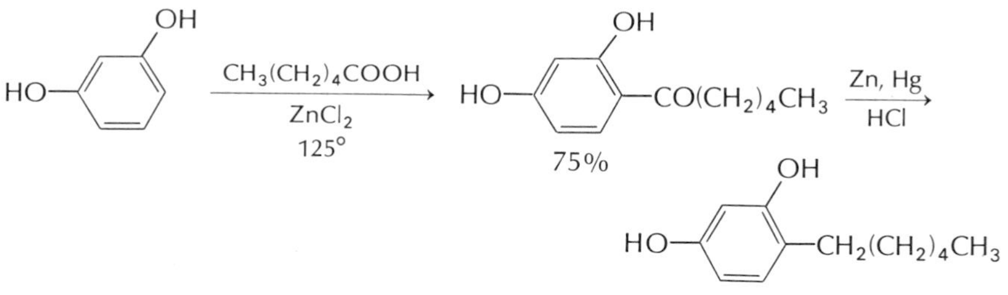

resorcinol $\xrightarrow[\text{ZnCl}_2, 125°]{\text{CH}_3(\text{CH}_2)_4\text{COOH}}$ 2,4-dihydroxyphenyl pentyl ketone (75%) $\xrightarrow[\text{HCl}]{\text{Zn, Hg}}$ 4-hexylresorcinol (24.20)

substance with a phenol coefficient of over 500. This efficiency permits use at higher dilutions and with lowered toxicity. If the alkyl side chain is made shorter or longer, the effectiveness decreases.

If Zn(CN)_2 and HCl gas are used (equivalent to ZnCl_2 and $\text{HC}\equiv\text{N}$), the reaction used most with phenols and phenyl ethers and known as the Gattermann aldehyde synthesis occurs (Eq. 24.21). The Gattermann aldehyde synthesis may be used on less active compounds (such as toluene) also, with AlCl_3 added, if the temperature of the reaction is raised to $100°$. If a nitrile is used instead of HCN, the reaction is known as the Hoesch ketone synthesis (Eq. 24.22).

Another route to the formation of hydroxyarylketones is the Fries rearrangement. We have previously noted the frequency with which groups attached to nitrogen rearrange

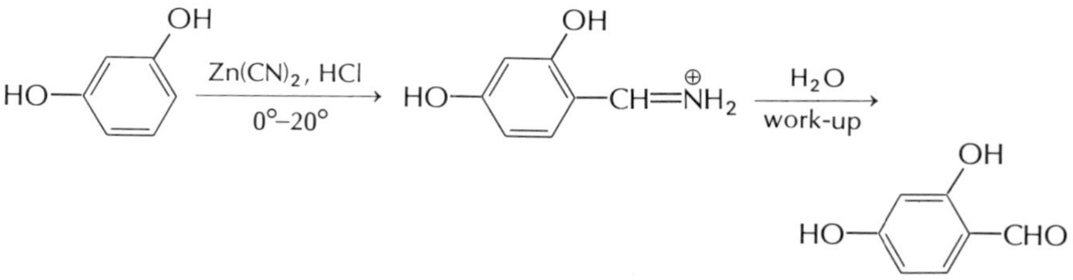

24.4 Phenols: reactions on the ring

[Resorcinol] + Zn(CN)₂, HCl, 0°–20° → [2,4-dihydroxyphenyl-CH=NH₂⁺] →(H₂O work-up)→ 2,4-dihydroxybenzaldehyde (HO-C₆H₃(OH)-CHO) (24.21)

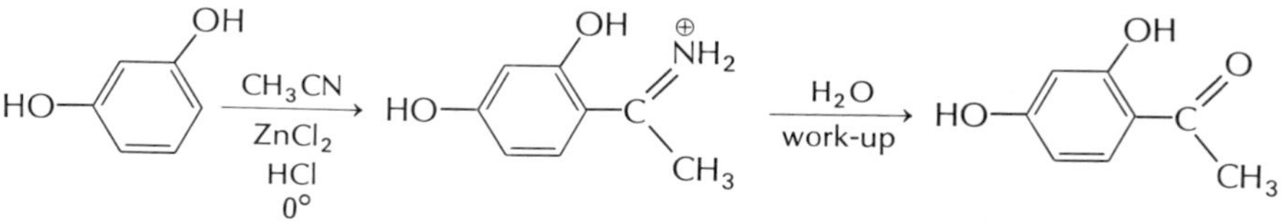

[Resorcinol] + CH₃CN, ZnCl₂, HCl, 0° → [2,4-dihydroxyphenyl-C(=NH₂⁺)CH₃] →(H₂O work-up)→ 2,4-dihydroxyacetophenone (24.22)

to $p > o$ in the ring. The same tendency exists for groups attached to oxygen. The Fries rearrangement refers to the shift of the acyl component of a phenyl ester to the o and p positions if the ester is subjected to the influence of AlCl₃ (Eq. 24.23). The temperature effect is pronounced and is opposite to that usually noted in electrophilic substitutions—an increase in temperature, which usually favors p substitution, here favors o substitution.

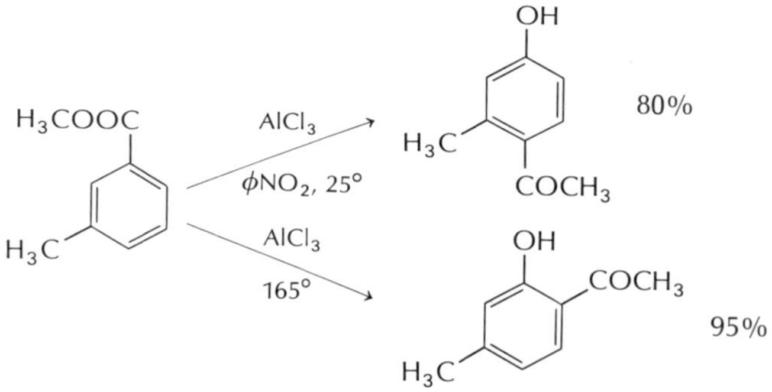

m-cresyl acetate (H₃COOC—C₆H₄—CH₃)
→ AlCl₃, φNO₂, 25° → p-product: 2-hydroxy-4-methylacetophenone (80%)
→ AlCl₃, 165° → o-product: 2-hydroxy-4-methylacetophenone (95%) (24.23)

The reaction, like most of the nitrogen rearrangements, is intermolecular, although one may argue about the identity of the various fragments. It seems clear that the shift to the p-position is faster than that to the o-position, so, at low temperatures, kinetic control is observed. At higher temperatures, the reversible reaction leads to thermodynamic control and to the more stable chelated aluminum salt of the o product.

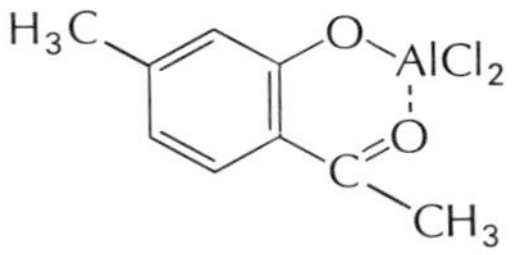

[Chelated structure: H₃C—C₆H₃—O—AlCl₂ with C=O···Al chelation, CH₃]

Phenolic ethers, as we have seen, do not rearrange so much as they are split by AlCl₃. However, allyl phenyl ethers undergo a thermal rearrangement known as the <u>Claisen rearrangement</u> (Eq. 24.24). The reaction is intramolecular and proceeds by way of a cyclic

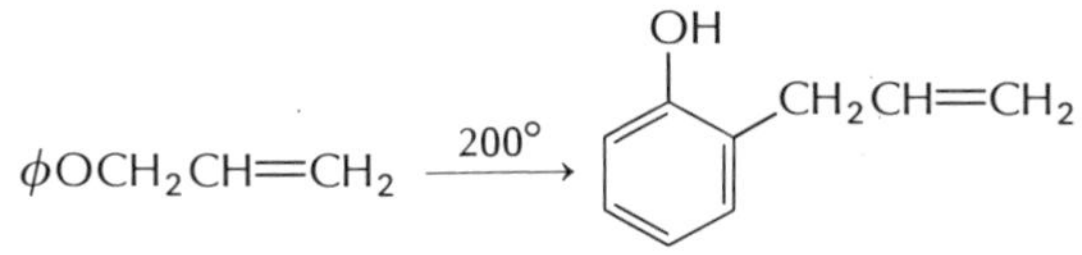

φOCH₂CH=CH₂ —200°→ o-allylphenol (2-CH₂CH=CH₂-C₆H₄-OH) (24.24) transition state and rearrangement of the allyl group (a, Eq. 24.25). If the o positions are

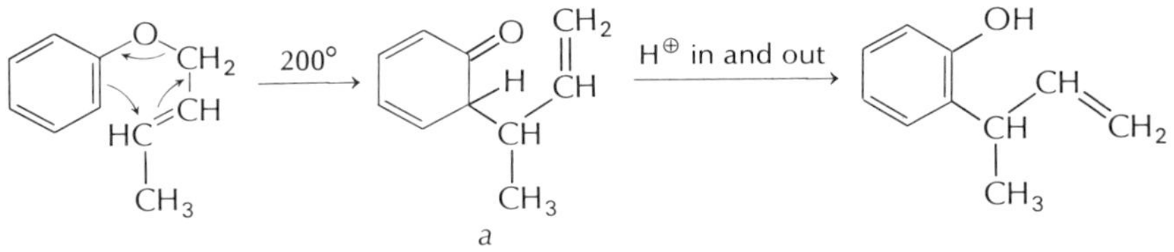

$$\text{(24.25)}$$

blocked, the intermediate (b) corresponding to a cannot tautomerize, and a second shift takes place, this time to the p position (Eq. 24.26).

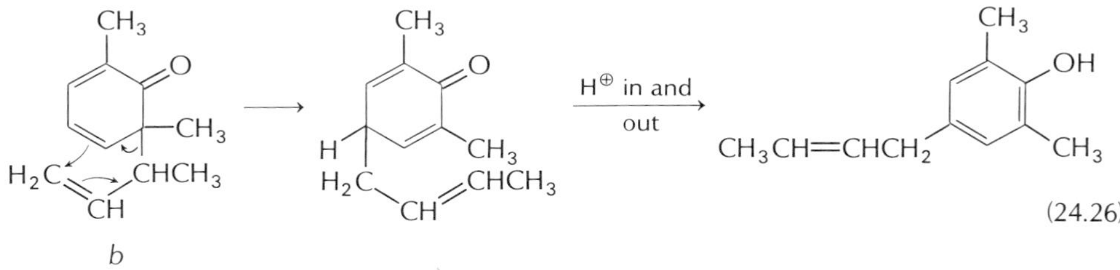

$$\text{(24.26)}$$

Allylation of a phenol proceeds to give the ether in polar solvents and to give a ring-substituted product in nonpolar solvents (Eqs. 24.27). (We recall that the anion of acetoacetic ester undergoes both C and O alkylation and acylation.) We shall find that there are other base-catalyzed substitutions on phenols.

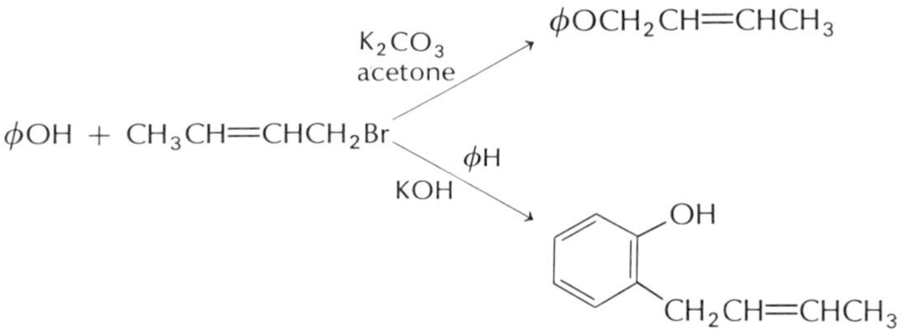

$$\phi OH + CH_3CH=CHCH_2Br \xrightarrow[\text{acetone}]{K_2CO_3} \phi OCH_2CH=CHCH_3$$

$$\xrightarrow[\text{KOH}]{\phi H} \text{o-}HOC_6H_4CH_2CH=CHCH_3 \quad (24.27)$$

Reactions with aldehydes and ketones. Aldehydes and ketones may be used to form cations that react with phenols to give a variety of products (Eq. 24.28). Under alkaline conditions, phenoxide ions are able to attack a carbonyl carbon atom (Eq. 24.29). With formaldehyde, the reaction must be carefully controlled, because the hydroxybenzyl alcohols can react again (o-hydroxybenzyl alcohol also is formed) with more formaldehyde to give 2,4-bis(hydroxymethyl)phenol or with more phenol to give 4,4'-dihydroxydiphenylmethane. The reaction may be purposely continued to give a polymer, originally known as Bakelite, the first thermosetting synthetic polymer.

Heating phenol with phthalic anhydride in the presence of zinc chloride or H_2SO_4 leads to an initial Friedel-Crafts-like reaction. The reaction does not stop at that point but continues as shown in Equation 24.30 to give phenolphthalein, a γ-lactone. Phenolphthalein (a widely used laxative) is an indicator, colorless at pH lower than 8.3, pink at pH 8.3 to 10. At pH above 10, the substance again becomes colorless (Eq. 24.31). Several monoanions may be written, but the dianion is the structure, with the equivalent quinoid resonance structures, that absorbs in the visible at 555 nm. The trianion again is colorless.

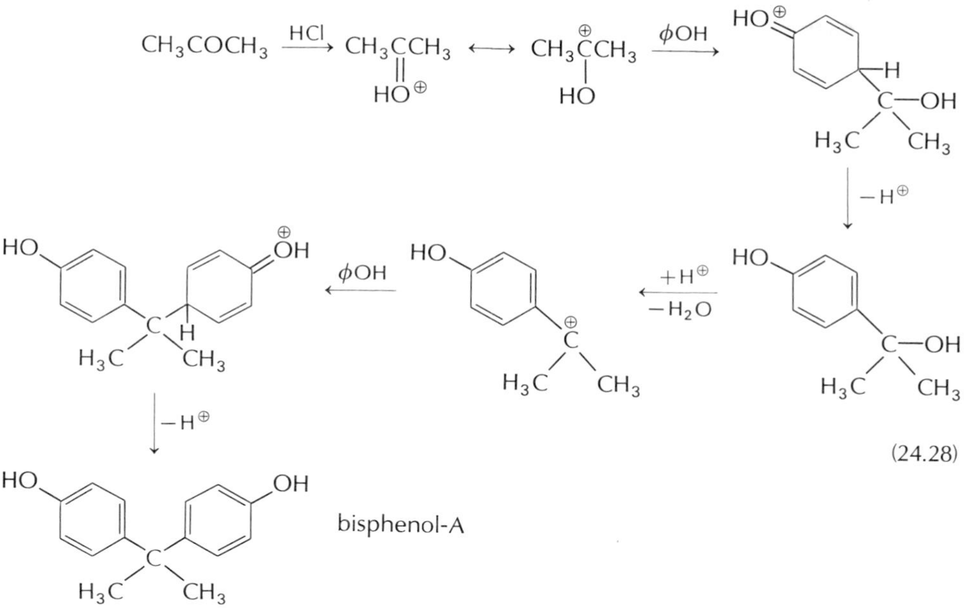

24.4 Phenols: reactions on the ring

(Reaction schemes showing synthesis of bisphenol-A from acetone and phenol with HCl catalyst — equation 24.28; synthesis of 4-hydroxybenzyl alcohol from phenol and formaldehyde with HO⁻ — equation 24.29; and synthesis of phenolphthalein from phthalic anhydride and phenol with ZnCl₂ or H₂SO₄, Δ — equation 24.30.)

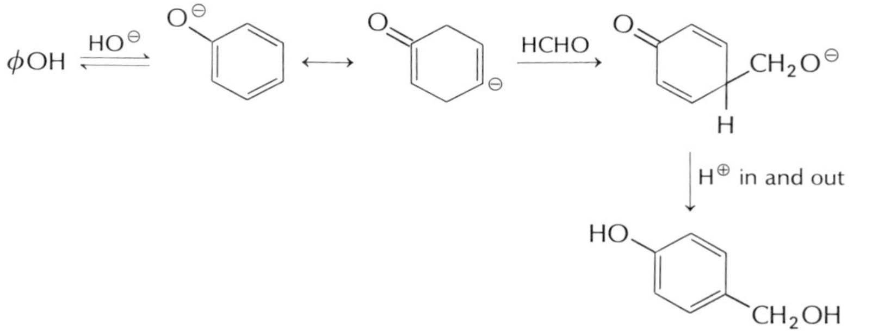

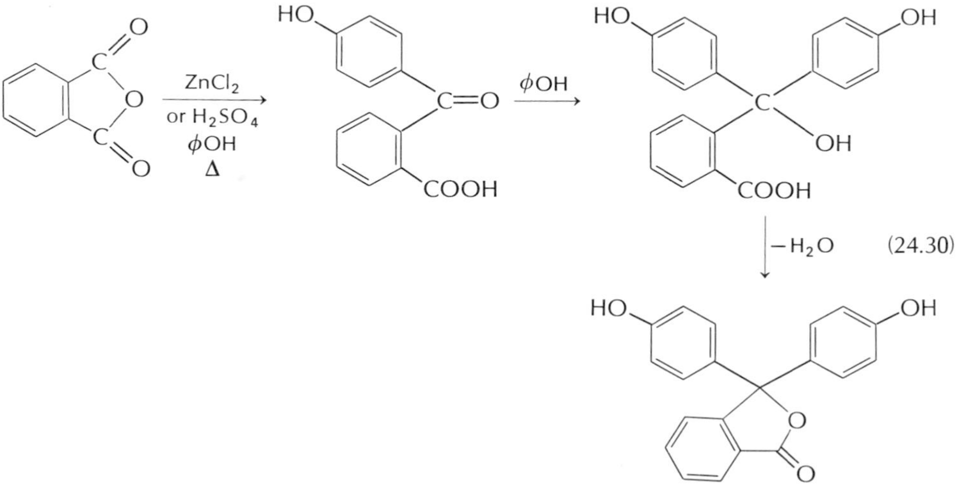

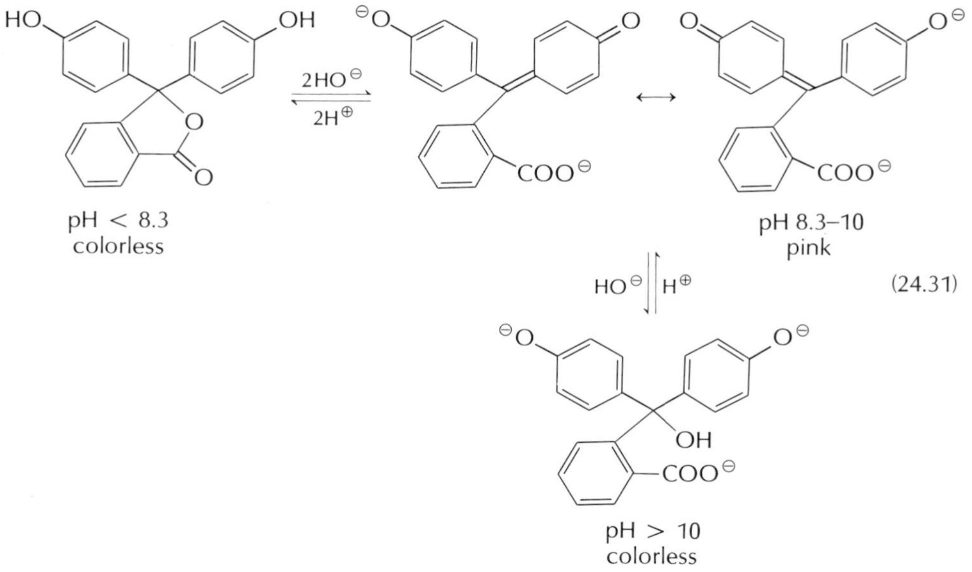

$$\text{(structures shown)} \quad (24.31)$$

pH < 8.3
colorless

pH 8.3–10
pink

pH > 10
colorless

Phenolphthalein belongs to the general class of triphenylmethane dyes, of which there are many subclasses, one being the phthaleins. Another subclass comprises the sulfonphthaleins, to which pyrocatechol violet belongs. Another subclass is the xanthenes, to which fluorescein, eosin, and mercurochrome belong. (Only one of the resonance structures contributing to the colored form is shown.) Pyrocatechol

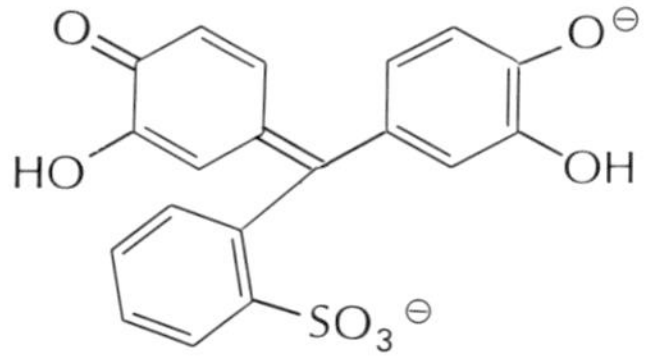

pyrocatechol violet

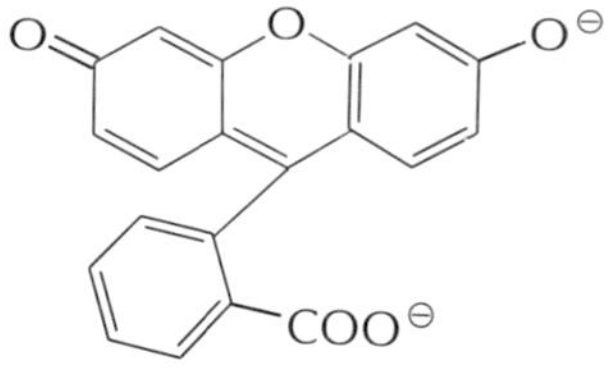

fluorescein

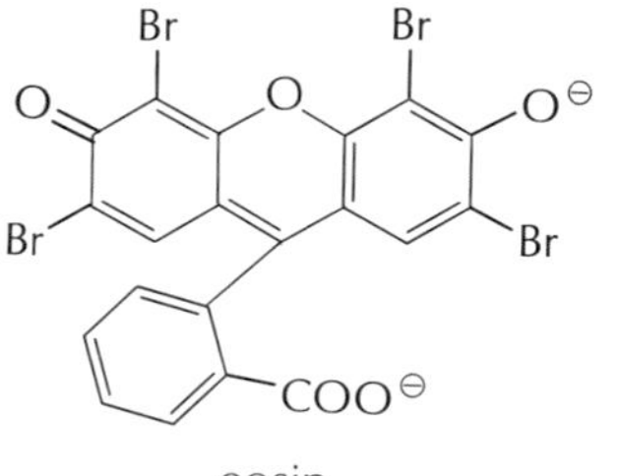

eosin

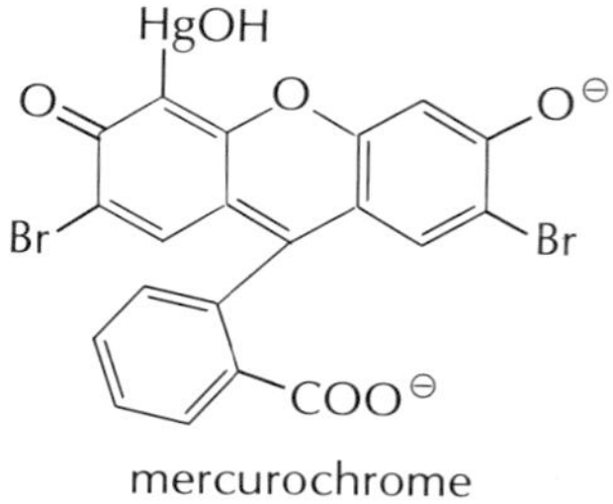

mercurochrome

violet is used as an indicator in chelatometric titrations. Fluorescein dyes wool or silk yellow, but the color fades. In solution, fluorescein is yellow with green fluorescence. Eosin, in solution, is an intense red (red ink) and has a blue fluorescence. Mercurochrome is used as an antiseptic and is bright red.

Mercuration. Phenols react nicely with mercuric acetate. Mercurochrome is made by mercuration of dibromofluorescein followed by treatment with NaOH. The mercuriacetate can be replaced in several ways, one of which is shown in Equation 24.32.

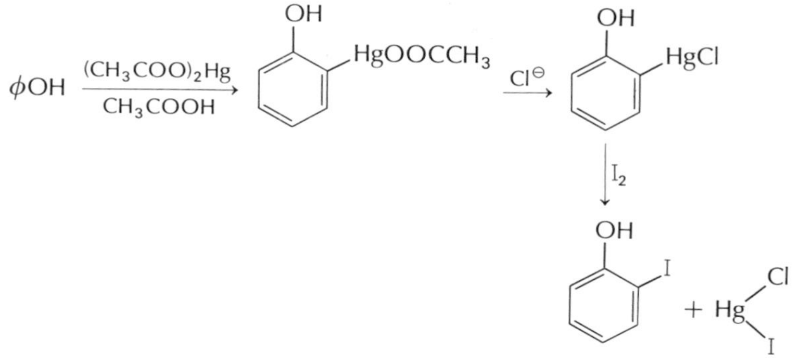

ϕOH $\xrightarrow{\text{(CH}_3\text{COO)}_2\text{Hg}}_{\text{CH}_3\text{COOH}}$ *o*-HO-C$_6$H$_4$-HgOOCCH$_3$ $\xrightarrow{\text{Cl}^\ominus}$ *o*-HO-C$_6$H$_4$-HgCl

$\downarrow$ I$_2$

o-HO-C$_6$H$_4$-I + Hg(Cl)(I)   (24.32)

Base-catalyzed substitutions.

Several base-catalyzed reactions of phenols are useful in synthesis.

The Kolbe reaction is used commercially for the synthesis of salicylic acid. Dry sodium phenoxide is heated under pressure (4 to 7 atm) with carbon dioxide (Eq. 24.33). The *ortho*

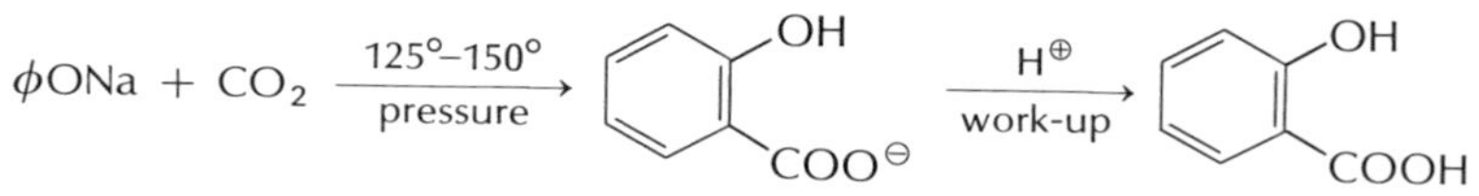

ϕONa + CO$_2$ $\xrightarrow[\text{pressure}]{125°-150°}$ *o*-HO-C$_6$H$_4$-COO$^\ominus$ $\xrightarrow[\text{work-up}]{\text{H}^\oplus}$ *o*-HO-C$_6$H$_4$-COOH   (24.33)

acid is formed almost exclusively because of greater stability of the hydrogen-bonded structure. The reaction may be thought of as that of the very reactive phenoxide ion with a carbonyl carbon, similar to the reaction with formaldehyde (Eq. 24.29). The reaction, as applied to β-naphthol, gives 3-hydroxy-2-naphthoic acid rather than 2-hydroxy-1-naphthoic acid (Eq. 24.34).

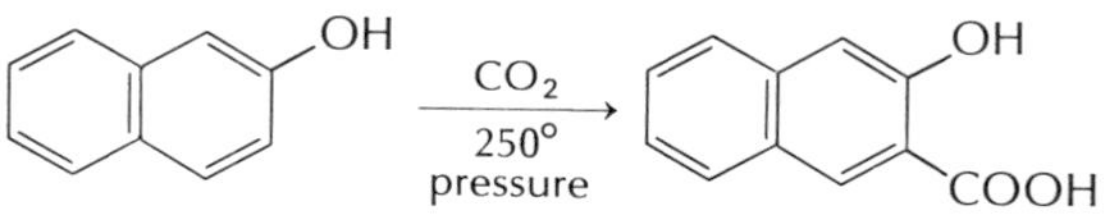

2-naphthol $\xrightarrow[\text{pressure}]{\text{CO}_2, 250°}$ 3-hydroxy-2-naphthoic acid   (24.34)

The Reimer-Tiemann reaction of phenols with HCCl$_3$ and alkali leads to the formation of hydroxyarylaldehydes (Eq. 24.35). The reaction is relatively slow unless an emulsifying agent is present to increase the surface area between the alkaline solution of sodium phenoxide and the insoluble chloroform layer. If potassium hydroxide is used instead of sodium hydroxide, the relative amount of the *para* product is increased. Although yields are low, the products are easily separated. Steam distillation of the acidified reaction mixture volatilizes remaining phenol and the *o*-hydroxyaldehyde; the *p* isomer remains behind.

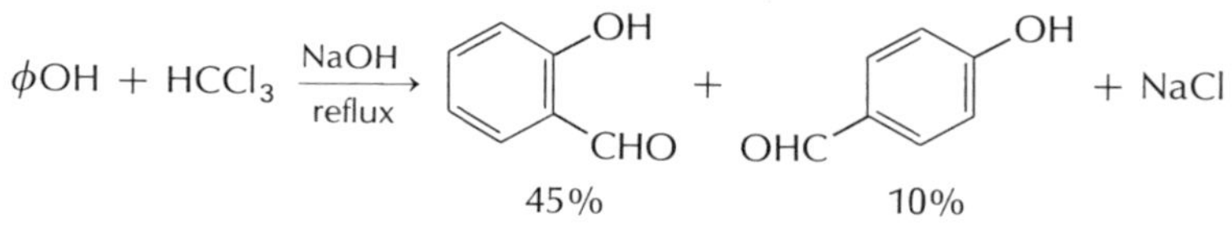

ϕOH + HCCl$_3$ $\xrightarrow[\text{reflux}]{\text{NaOH}}$ *o*-HO-C$_6$H$_4$-CHO (45%) + *p*-OHC-C$_6$H$_4$-OH (10%) + NaCl   (24.35)

The mechanism is thought to involve dichlorocarbene ($\ddot{C}Cl_2$) as shown in Equation 24.36. However, there remain several puzzling features. For instance,

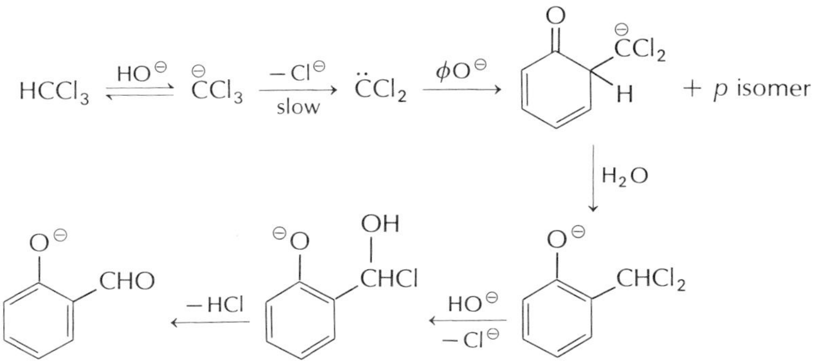

$$HCCl_3 \xrightleftharpoons{HO^\ominus} {}^\ominus CCl_3 \xrightarrow[slow]{-Cl^\ominus} \ddot{C}Cl_2 \xrightarrow{\phi O^\ominus} \text{[cyclohexadienone-CCl}_2\text{H]} + p \text{ isomer} \quad (24.36)$$

[Mechanism showing: cyclohexadienone intermediate → (H₂O) → ortho-phenoxide-CHCl₂ → (HO⁻/−Cl⁻) → ortho-phenoxide-CHCl-OH → (−HCl) → ortho-phenoxide-CHO]

dichlorocarbene is a reactive electrophile and adds readily to alkenes and to primary amines (to give isonitriles). However, the Reimer-Tiemann reaction requires the very reactive phenoxide anion to go as shown. The failure of other reactive substances like ϕOMe and ϕNMe$_2$ to react with $\ddot{C}Cl_2$ is difficult to explain. Also, the effect of changing the counterion from Na$^\oplus$ to K$^\oplus$ is not explained. The presence of alkyl groups *o* or *p* to hydroxy promote reactivity at these positions even though tautomerization cannot take place. In these cases, the products are substituted 2,5-cyclohexadienones (Eqs. 24.37).

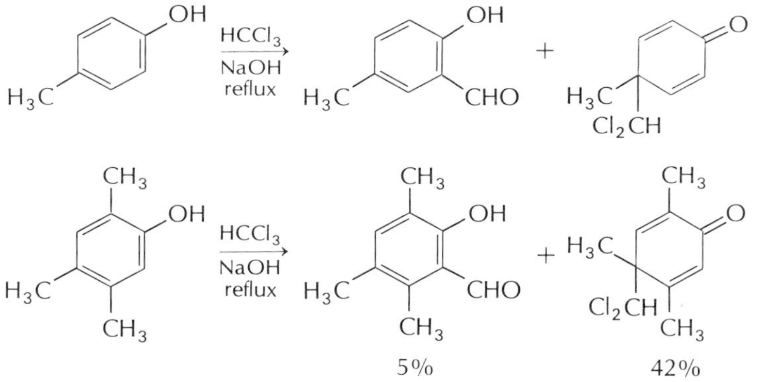

[p-cresol] $\xrightarrow[\text{NaOH reflux}]{HCCl_3}$ [2-hydroxy-5-methylbenzaldehyde] + [4-methyl-4-(dichloromethyl)cyclohexa-2,5-dienone]

[2,3,5-trimethylphenol] $\xrightarrow[\text{NaOH reflux}]{HCCl_3}$ [aldehyde product] 5% + [dienone product] 42%

(24.37)

[2,6-di-tert-butylphenol] $\xrightarrow{O_2 \text{ or } K_3Fe(CN)_6}$ [phenoxyl radical] ⟷ [carbon radical resonance form]

↓

[bisphenol HO-Ar-Ar-OH with tBu groups] ⟵ tautomerization ⟵ [bis-dienone coupled product]

↓
etc.

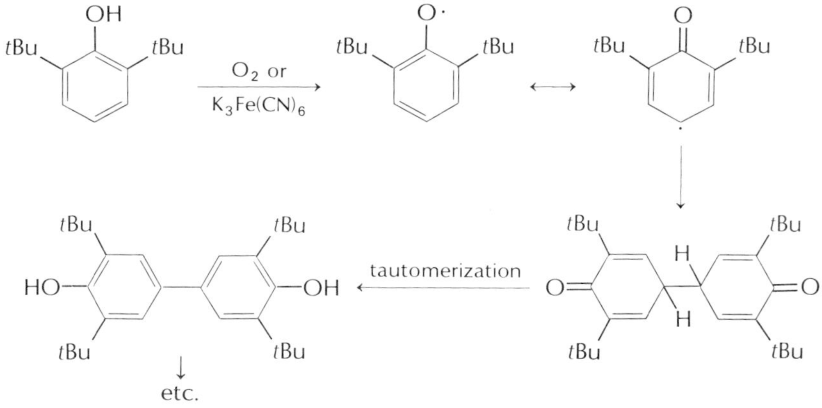

(24.38)

Oxidation. Phenols are very susceptible to oxidation, so much so that some phenols are specifically used as antioxidants (Eq. 24.38).

Reduction. The presence of the electron-donating hydroxy group tends to increase the difficulty of reduction of the aromatic ring. As shown in Equations 24.39, the

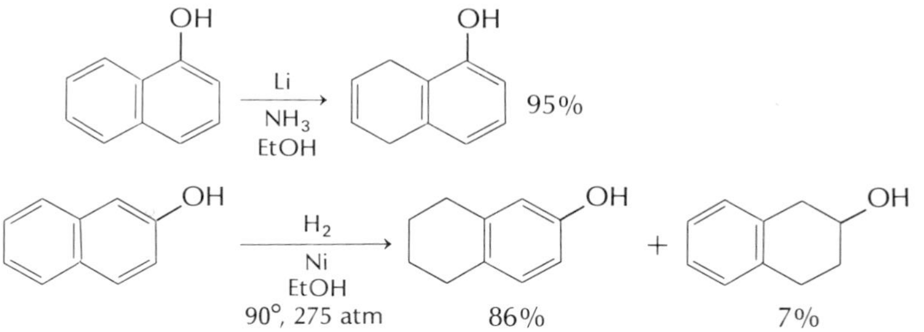

(24.39)

ring that does not have the hydroxy group is the more easily reduced, either by a metal combination or by catalytic hydrogenation. Nevertheless, cyclohexanol usually is prepared by catalytic hydrogenation of phenol. ☐ ☐

Exercises

24.3 Hexachlorophene, a germicide with a phenol coefficient of 125, used in soaps and so on, is bis(2-hydroxy-3,5,6-trichlorophenyl)methane. Suggest a method of manufacture from benzene and other needed reagents.

24.4 Heating salicylic acid with formaldehyde and sulfuric acid with a mild oxidizing agent present gives a dye, Chrome Violet. Suggest a structure for the product.

24.5 Predict the major product formed if *p*-cresol is treated with:

a. H_2SO_4 at 25°
b. 20% HNO_3 at 25°
c. $NaNO_2$, HCl, 0°
d. $\overset{\oplus}{\phi}N\equiv N$, 0°, pH 9
e. Br_2, H_2O, 25°
f. Br_2, CS_2, 5°
g. $Me_2C=CH_2$, H_2SO_4, 25°
h. $Zn(CN)_2$, HCl, 0°
i. $(CH_3)_2CHCN$, $ZnCl_2$, HCl, 0°
j. $(CH_3CO)_2O$; then $AlCl_3$, 165°
k. $CH_2=CHCH_2Cl$, K_2CO_3, acetone; then 200°
l. CH_3COCH_3, HCl
m. $(CH_3COO)_2Hg$, CH_3COOH
n. CO_2, NaOH, heat, pressure
o. $HCCl_3$, NaOH
p. NaOH, Me_2SO_4
q. H_2, Pt
r. ϕCHO, H_2SO_4

24.5 Di- and trihydroxybenzenes

Catechol. Catechol (sometimes called pyrocatechol) is properly pronounced kat-e-chole, reflecting its origin as one of the products of distillation of gum catechu (kat-e-chu). Because of confusion with the word catechism (kat-e-kism), which has an entirely different Greek origin, pronunciation as kat-e-kole has become accepted.

Catechol, its ethers, and its substitution products are common in nature. A few examples are shown.

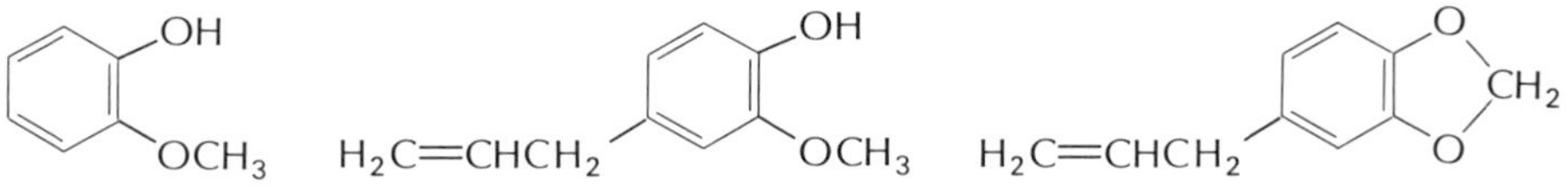

[Structures: guiacol, eugenol, safrole, vanillin, veratric acid, (−)-(R)-epinephrine (adrenaline)]

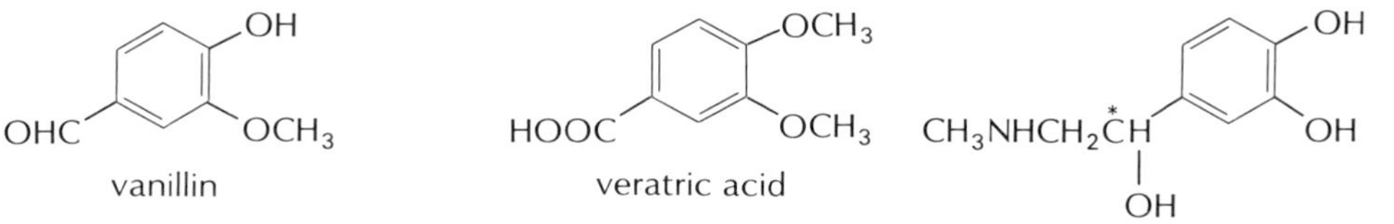

Catechol enters enthusiastically into all the reactions previously given because the remaining positions of the ring are o or p to one or the other of the hydroxy groups.

Polyethers. Catechol is the starting point for the preparation of a series of cyclic polyethers, which have the unique property of being able to preferentially complex the alkali metal ions (Eq. 24.40). Compound *a* bears the awesome systematic

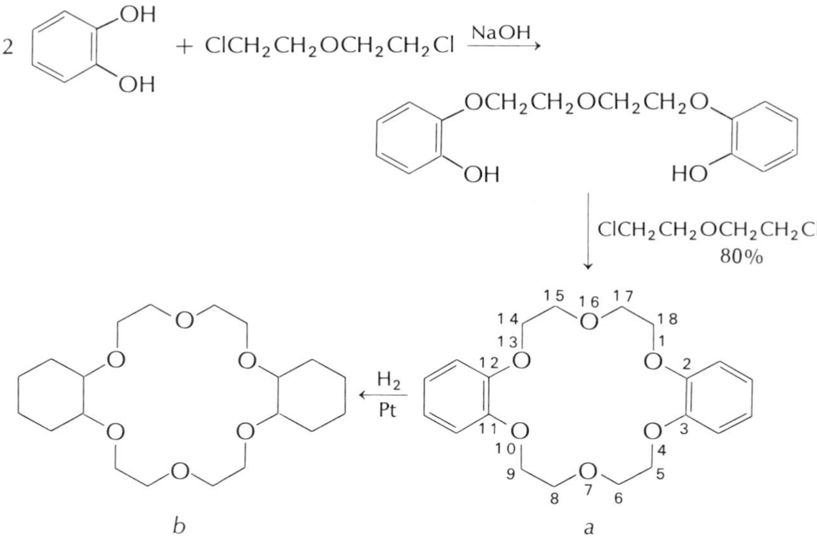

$$2 \text{ catechol} + \text{ClCH}_2\text{CH}_2\text{OCH}_2\text{CH}_2\text{Cl} \xrightarrow{\text{NaOH}}$$

[intermediate structure]

$$\xrightarrow{\text{ClCH}_2\text{CH}_2\text{OCH}_2\text{CH}_2\text{Cl, NaOH}}_{80\%}$$ (24.40)

[Structures of *b* (dicyclohexyl-18-crown-6) and *a* (dibenzo-18-crown-6), with *a* numbered 1–18; *b* obtained from *a* by H₂/Pt]

name of 2,3,11,12-dibenzo-1,4,7,10,13,16-hexaoxacyclooctadeca-2,11-diene. Because of a fancied resemblance to a crown, a trivial naming system has been devised in which *a* is known as dibenzo-18-crown-6. (The ring size is 18, and 6 is the number of oxygens, each oxygen being separated by 2 carbons from nearest oxygens.) Thus *b* is dicyclohexyl-18-crown-6. Crowns of different sizes have been made: 14-crown-4 is best for complexing $Li^\oplus$ (the ion fits into the open center of the crown), 15-crown-5 is best for $Na^\oplus$, and 16-crown-6 is best for $K^\oplus$. In general, the dicyclohexyl crowns are better complexing agents with greater solubility in organic solvents. For example,

24.5 Di- and trihydroxybenzenes

Resorcinol. Resorcinol, because the two hydroxy groups reinforce each other, is most reactive at the 4 and 6 positions (Eq. 24.20). A second substituent enters at the 2 or 6 position depending upon conditions. Resorcinol is relatively easy to reduce, probably because the tautomeric cyclohexendione is relatively stable (Eq. 24.41). The reduction is easily terminated at the diketone stage.

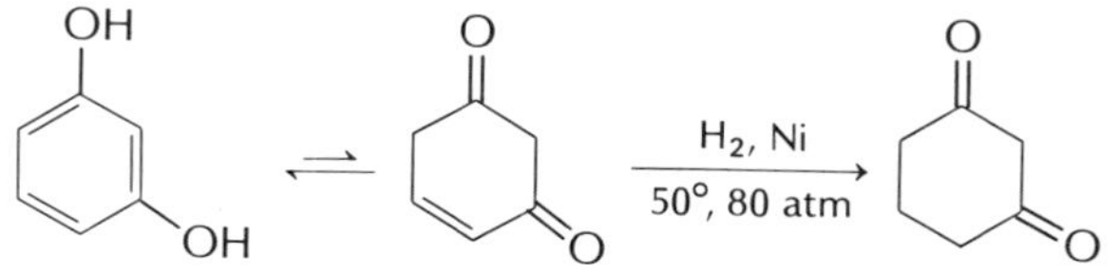

$$\text{resorcinol} \rightleftharpoons \text{cyclohexendione} \xrightarrow[50°, 80 \text{ atm}]{H_2, \text{Ni}} \text{cyclohexanedione} \quad (24.41)$$

Hydroquinone. Hydroquinone, like catechol, is reactive. It is so easily oxidized that it is used as a photographic developer, a reducing agent (Eq. 24.42). The oxidized product, benzoquinone, will be discussed in Section 24.7.

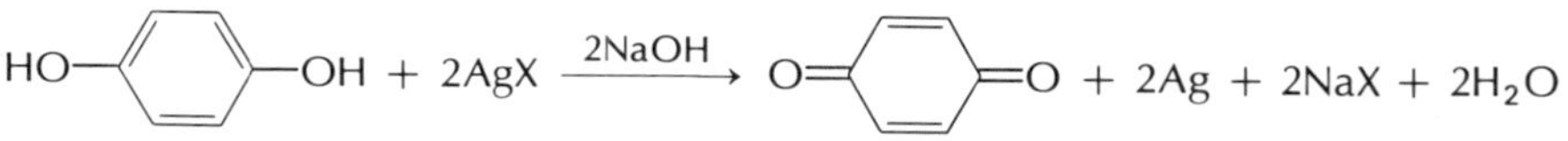

$$HO\text{-}C_6H_4\text{-}OH + 2AgX \xrightarrow{2NaOH} O\text{=}C_6H_4\text{=}O + 2Ag + 2NaX + 2H_2O \quad (24.42)$$

Hydroquinone is so reactive that it acts as a diene in the Diels-Alder reaction (Eq. 24.43).

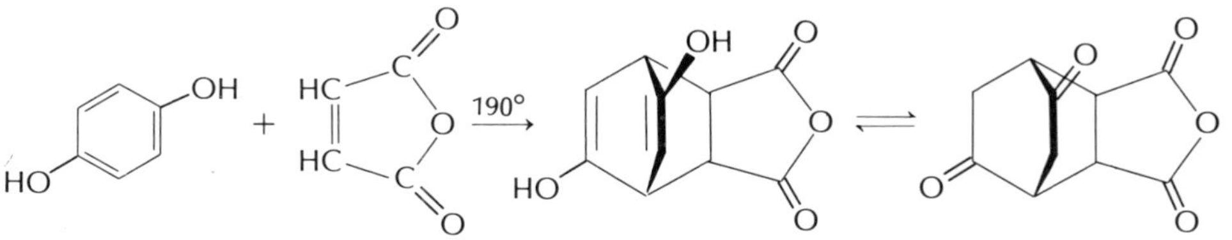

$$(24.43)$$

Pyrogallol. Pyrogallol is such a good reducing agent that an alkaline solution of it absorbs oxygen from the air or other gases. It also is used as a photographic developer. The name is derived from galls, which appear on plants as a result of the activities of certain insects. Those that appear on oak trees are most used and yield tannic acid upon extraction. Tannic acid (also called tannin) is made up of complicated esters of gallic acid with glucose and the hydroxy groups of other molecules of gallic acid. Hydrolysis gives gallic acid, which gives pyrogallol upon decarboxylation (Eq. 24.44). The potent hallucinogen mescaline is a triether of pyrogallol with an aminoethyl group at the 5 position.

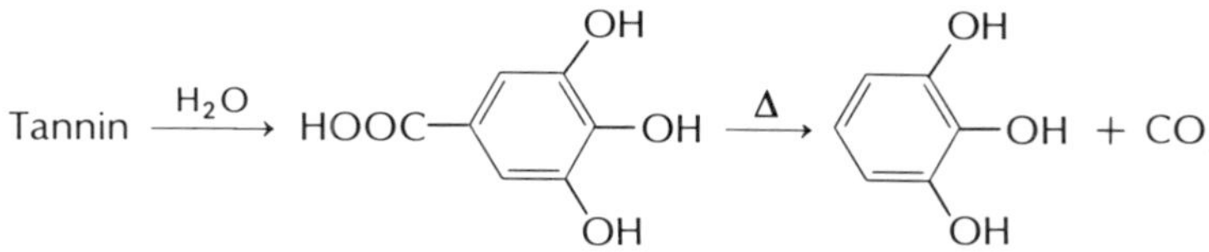

$$\text{Tannin} \xrightarrow{H_2O} \text{HOOC-C}_6H_2(OH)_3 \xrightarrow{\Delta} \text{C}_6H_3(OH)_3 + CO_2 \quad (24.44)$$

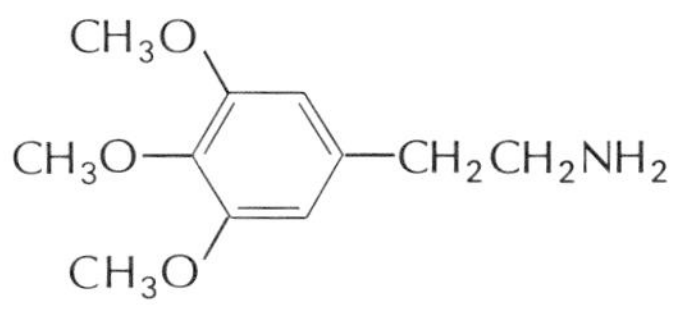

808   Chapter twenty-four

mescaline

Phloroglucinol.   Skipping over hydroxyhydroquinone, we come to the third isomeric trihydroxybenzene, phloroglucinol, which is of interest because of its pronounced tendency to react as a ketone (Eqs. 24.45). It also gives typical phenolic reactions, such as a color with $FeCl_3$, coupling with diazonium ions, ethers with CH_2N_2, and so on.

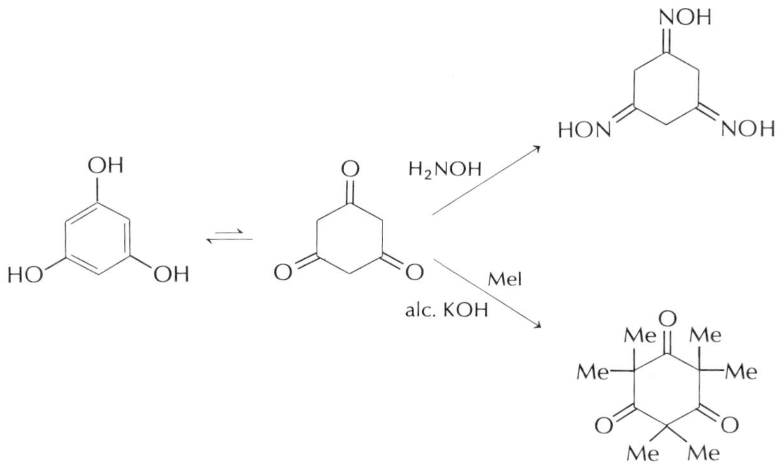

(24.45)

Exercises

24.6   Show how to obtain vanillin: from a. guiacol, and b. eugenol.

24.7   a. Is the product from the Diels-Alder reaction in Equation 24.43 capable of optical activity?   b. Is the product optically active?

24.6   Preparation of phenols

The search for cheaper processes for the manufacture of phenol has led to a diversity of methods and starting materials.

The sulfonation route.   The oldest route (still in use) begins with the sulfonation of benzene (Eq. 24.46). In the laboratory, a mixture of KOH and NaOH is used for the fusion reaction (the mp is lower) and a metallic vessel must be used (the molten alkali attacks glass).

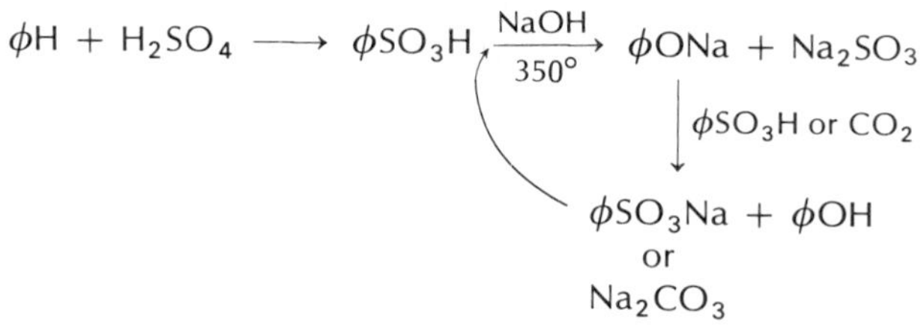

$$\phi H + H_2SO_4 \longrightarrow \phi SO_3H \xrightarrow[350°]{NaOH} \phi ONa + Na_2SO_3$$
$$\downarrow \phi SO_3H \text{ or } CO_2 \quad (24.46)$$
$$\phi SO_3Na + \phi OH$$
$$\text{or}$$
$$Na_2CO_3$$

From chlorobenzene. Next on the scene (1928) was the Dow process, the high-temperature hydrolysis of chlorobenzene. Ideally, the process would look like Equations 24.47, consumption of benzene, water, and electrical energy giving phenol and hydrogen

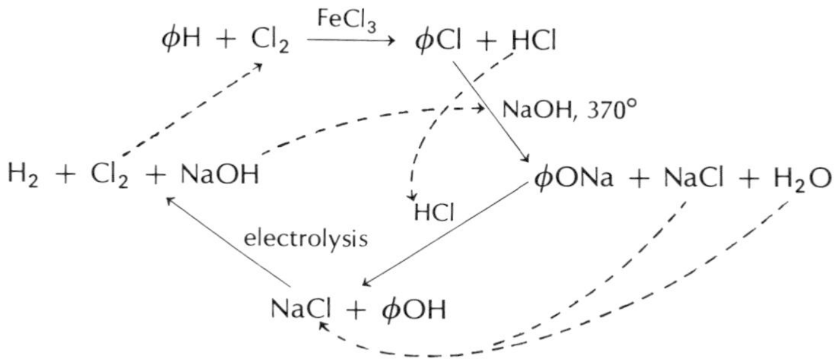

$$\phi H + Cl_2 \xrightarrow{FeCl_3} \phi Cl + HCl$$
$$\xrightarrow{NaOH, 370°}$$
$$H_2 + Cl_2 + NaOH \qquad \phi ONa + NaCl + H_2O \quad (24.47)$$
$$\xleftarrow{electrolysis} \quad HCl$$
$$NaCl + \phi OH$$

as products. Chlorine, HCl, NaOH, and NaCl are not consumed except for unavoidable losses. Among the unavoidable losses are the formation of diphenyl ether ($\phi O \phi$) and small amounts of o- and p-phenylphenol. The losses may be kept to a minimum by recycling of the ether, which at 370° is in equilibrium with 2 moles of sodium phenoxide and water.

The Raschig process (1940) is an all-gas-phase process in which the hydrolysis of chlorobenzene is carried out with steam at 425°.

From cumene. In 1954, the cumene process was introduced. It is of interest because of an ionic rearrangement of cumene hydroperoxide (Eqs. 24.48). The advantages

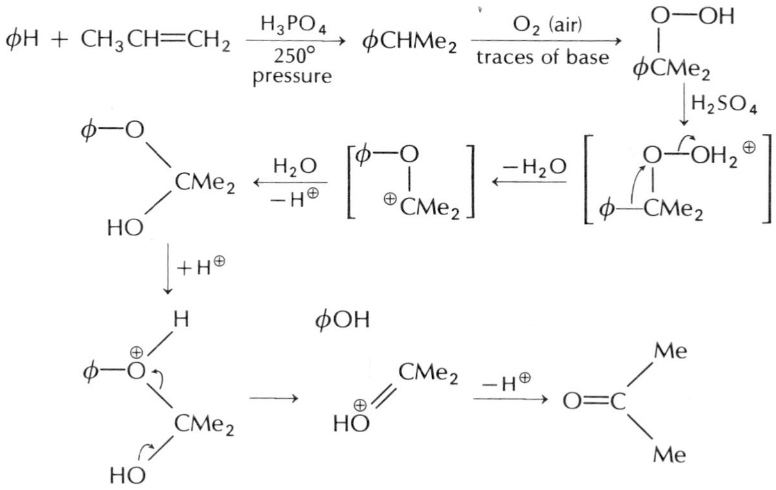

$$\phi H + CH_3CH=CH_2 \xrightarrow[\text{pressure}]{H_3PO_4, 250°} \phi CHMe_2 \xrightarrow[\text{traces of base}]{O_2 \text{ (air)}} \begin{array}{c} O-OH \\ | \\ \phi CMe_2 \end{array}$$

$$\downarrow H_2SO_4$$

$$\begin{array}{c}\phi-O \\ \diagdown \\ CMe_2 \\ HO \end{array} \xleftarrow[-H^\oplus]{H_2O} \left[\begin{array}{c}\phi-O \\ | \\ {}^\oplus CMe_2\end{array}\right] \xleftarrow{-H_2O} \left[\begin{array}{c} O-OH_2^\oplus \\ | \\ \phi-CMe_2 \end{array}\right] \quad (24.48)$$

$$\downarrow +H^\oplus$$

$$\begin{array}{c} H \\ \phi-\overset{\oplus}{O} \\ \diagdown \\ CMe_2 \\ HO \end{array} \longrightarrow \begin{array}{c} \phi OH \\ \overset{\oplus}{}CMe_2 \\ HO \end{array} \xrightarrow{-H^\oplus} O=C\begin{array}{c} Me \\ \diagdown \\ Me \end{array}$$

here are the availability of cheap propylene; the low energy requirement; and two valuable products, phenol and acetone. The cumene process is currently the most popular commercial process.

Cresols. Phenol and the cresols can be made from aniline and the toluidines by way of the diazonium ions in the laboratory. Alkali fusion of *p*-toluenesulfonic acid is a good method for making *p*-cresol. The alkali fusion of either an *o*- or *p*-chlorotoluene probably proceeds, at least in part, by way of a benzyne (Section 15.5), because *m*-cresol is present in the product (Eq. 24.49). Pure *m*-cresol is probably best (if laboriously) prepared from

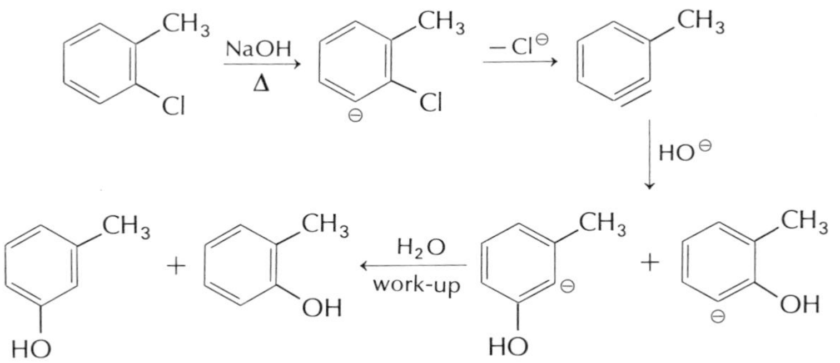

(24.49)

p-toluidine by acetylation, nitration, hydrolysis, diazotization, reduction (H_3PO_2), reduction of nitro, diazotization, and hydrolysis.

Di- and trihydroxybenzenes. Catechol, aside from being obtained by hydrolysis of guiacol or veratrole, can be made by the Dakin reaction, as is shown from salicylaldehyde (Eq. 24.50). Resorcinol is most easily prepared by alkali fusion of *m*-benzenedisulfonic acid. Hydroquinones usually are prepared by reduction of the corresponding quinones.

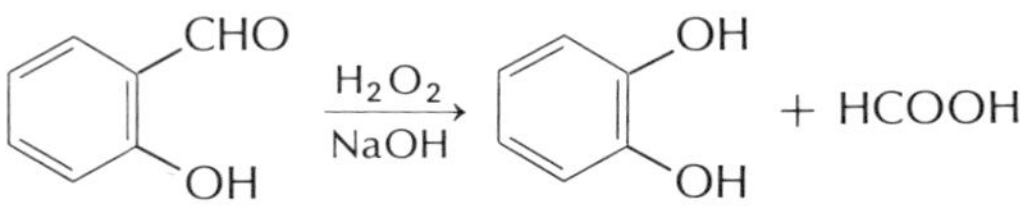

(24.50)

Pyrogallol is most easily obtained from natural sources. Phloroglucinol is troublesome to prepare. Beginning with trinitrotoluene, a synthesis has been devised as shown in Equations 24.51.

Other routes. Substituted phenols can be made by many alternate routes. Nitrosation and coupling reactions are mild methods for introducing groups that are easily reduced to amino groups. Amino groups can be replaced by means of the low-pH diazonium-ion reactions. The Fries and Claisen rearrangements should be kept in mind as well as the special phenolic reactions—the Kolbe, the Reimer-Tiemann, and the Bucherer reactions.

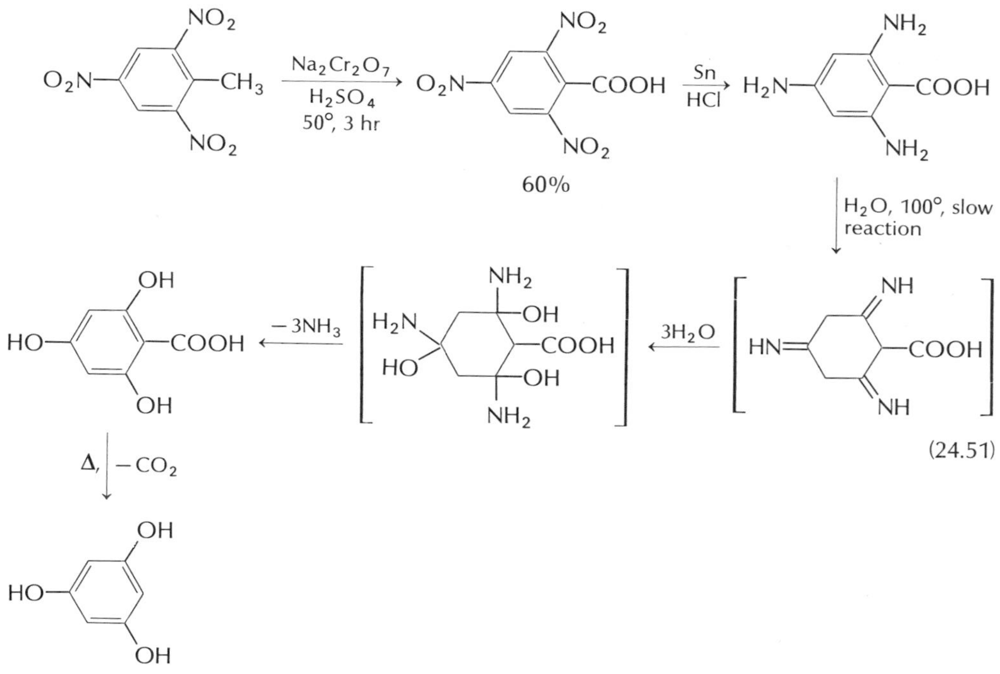

[Reaction scheme: 2,4,6-trinitrotoluene → (Na₂Cr₂O₇/H₂SO₄, 50°, 3 hr) → 2,4,6-trinitrobenzoic acid (60%) → (Sn/HCl) → 2,4,6-triaminobenzoic acid → (H₂O, 100°, slow reaction) → diimine intermediate → (3H₂O) → triamino-trihydroxy intermediate → (−3NH₃) → 2,4,6-trihydroxybenzoic acid → (Δ, −CO₂) → phloroglucinol (1,3,5-trihydroxybenzene)]

(24.51)

Exercises

24.8 Starting with benzene, toluene, aniline, and any necessary reagents and solvents, give a good synthesis of:

    a. *m*-chlorophenol   d. 3-methylcatechol
    b. 4-methylresorcinol   e. 4-methylcatechol
    c. 5-methylresorcinol   f. 2,4-dihydroxyacetophenone

□ □

24.7 Quinones

Origin of the name. Quinones take their name from quinic acid, first reported in 1790 as a constituent of cinchona bark (along with quinine, cinchonine, and many other alkaloids). In 1838, quinic acid was oxidized, and a neutral yellow crystalline product was obtained (Eq. 24.52). The name <u>quinone</u> is used to designate both the compound *p*-benzoquinone (2,5-cyclohexadiene-1,4-dione) and the class of compounds. Quinones may be *ortho* or *para*, never *meta*. Quinones are intensely colored substances. They are not aromatic, and they react according to the nature of conjugated diketones. They can be

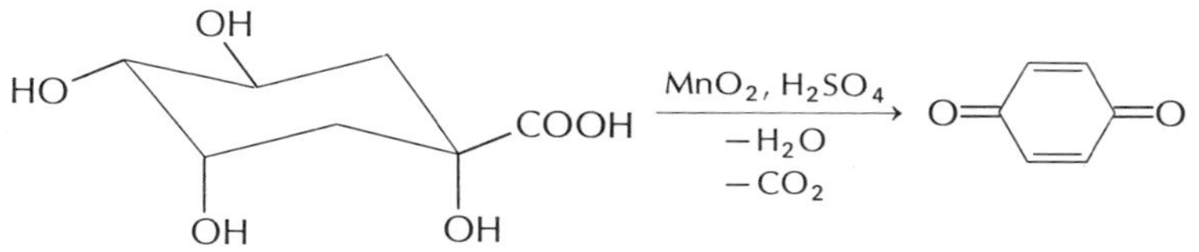

[Reaction: quinic acid (cyclohexane with OH, OH, OH, OH, COOH substituents) → (MnO₂, H₂SO₄, −H₂O, −CO₂) → *p*-benzoquinone]

(24.52)

reduced to dihydroxybenzenes and undergo a number of addition reactions, almost all of which are of the 1,4 type (Michael type). ☐ ☐

Addition reactions. Addition of HCl to benzoquinone gives a substituted hydroquinone (Eq. 24.53). (Benzoquinone is unable to oxidize chlorohydroquinone.)

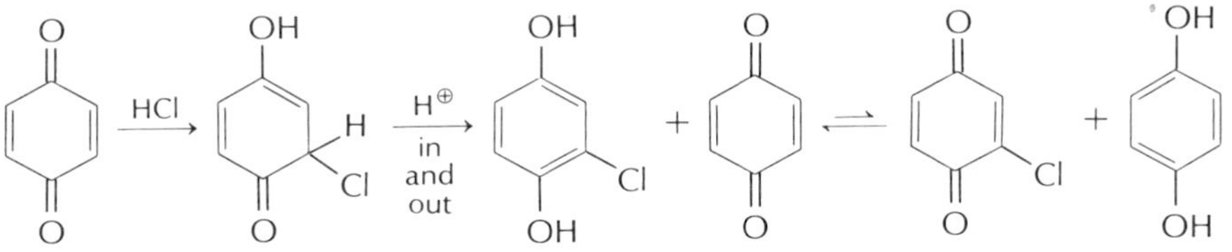

$$ \text{(24.53)} $$

Quinones give similar 1,4-additions with many reagents: for instance, HCN, malonic ester, and Grignard reagents. Quinones also react with substances that prefer to add to the carbon-carbon double bond (Eqs. 24.54). Phthiocol has been obtained also from human tubercle bacilli.

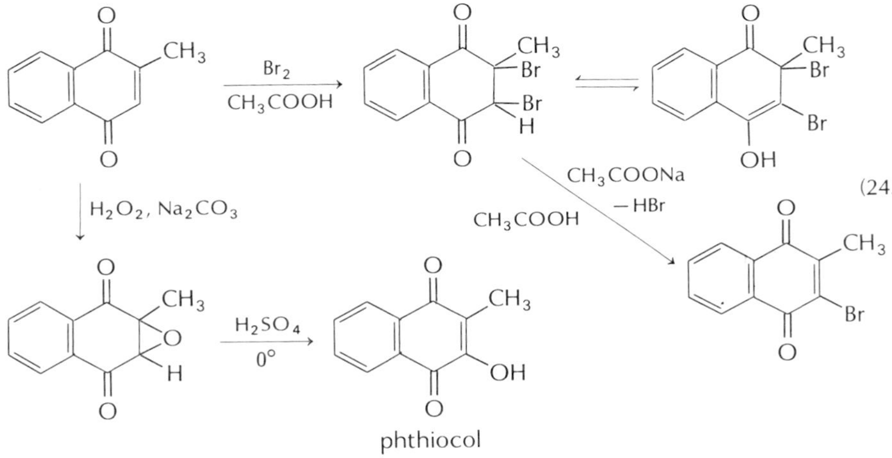

$$ \text{(24.54)} $$

phthiocol

Diels-Alder reactions. Quinones are efficient dienophiles (Eqs. 24.55). Anthra-

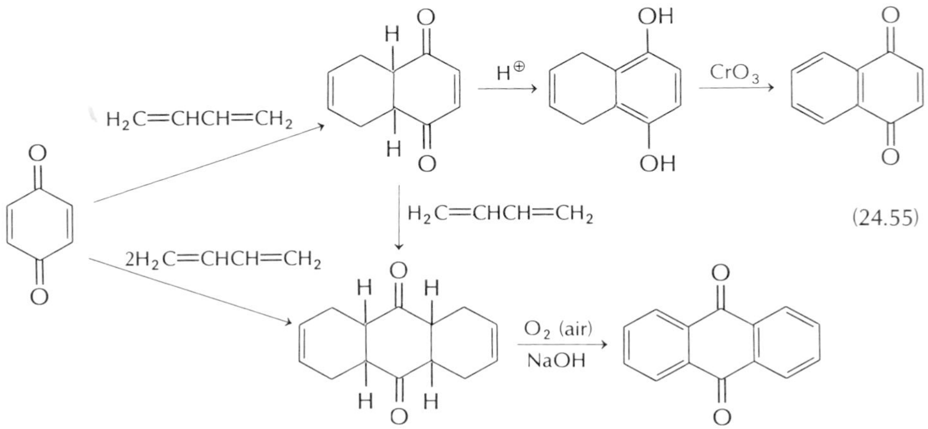

$$ \text{(24.55)} $$

quinone, with two benzene rings fused to benzoquinone is so stable that the hydroquinone can be oxidized in the presence of air, especially when alkali is present. ☐ ☐

Preparations. Quinones may be prepared by oxidation of phenols, anilines, dihydroxybenzenes, aminophenols, or diamines (Eqs. 24.56) or by hydrolysis of a monoxime of a quinone obtained by nitrosation of a phenol.

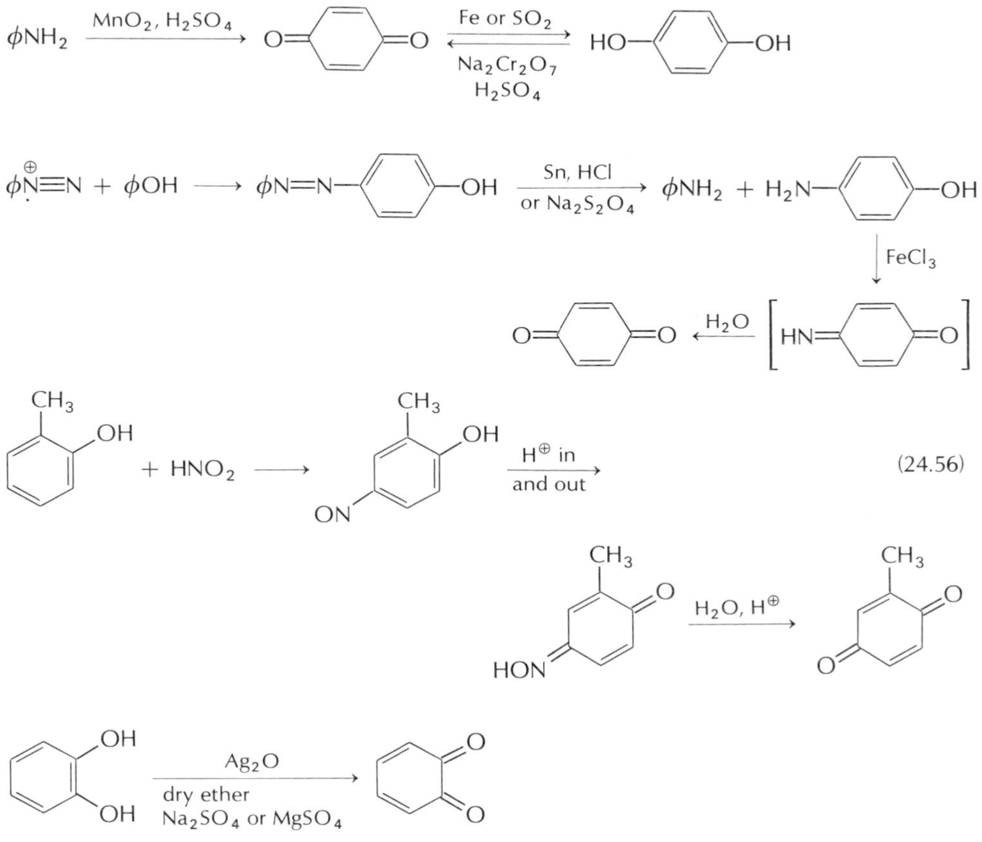

(24.56)

Exercises

24.9 Both *o*- and *m*-cresol give 2-methylhydroquinone by the nitrosation or the coupling route. Starting from toluene, which is the easier route (least steps) to the product?

24.10 Starting with quinone, show how to prepare:
  a. 2,5-dihydroxybenzoic acid
  b. 2,5-dihydroxyphenylacetic acid
  c. 2,5-diphenylhydroquinone
  d. 6-methylnaphthoquinone

☐ ☐

24.8 Lignin

Lignin is the phenolic "glue" that holds the cells of wood in place. The cell wall material is largely cellulose. Our uncertainty about lignin is one of the disappointments of organic chemistry. Despite the effort that has been expended in research on lignin, the structure remains unknown. Part of the problem lies in its polymeric nature. Perhaps we should speak of lignin in the plural, because it is clear that the lignin of the gymnosperms (naked-seed-bearing plants: the cycads, gingkoes, and conifers; conifers include pines, firs, spruces, redwoods) differs from that of the angiosperms (enclosed seeds: flowering plants, shrubs, bushes, trees, etc.; hardwoods).

Wood (lumber) is a remarkable substance because of the unique combination of lignin and cellulosic cell structure. It may be cut, sawed, split, drilled, chipped, planed, and sanded. It holds screws and nails. It may be stained, varnished, waxed, or oiled. It has good strength, may be steamed and bent, and weathers well (some varieties better than others). It varies in density, hardness, and toughness. Failing all else, it serves as a fuel.

Chemically, wood may be decomposed in several ways. The cellulose may be removed by enzymatic hydrolysis (rotting) and the remaining lignin studied. Commercially, lignin (and other components) are removed in the production of wood pulp (partially degraded cellulose), used in vast quantities for the manufacture of paper. Paper mills are faced with disposal problems of a magnitude not faced by other industries—what to do with the solutions of lignin?

The three major processes for lignin removal are the soda pulp method (use of NaOH solutions), the sulfite pulp method (use of a bisulfite, Na, Mg, or Ca), and the sulfide-sulfate pulp (Kraft) method (use of alkaline sodium sulfide and sulfate). In each process, wood chips are heated under pressure (digestion) with the solutions indicated. The spent liquors contain degraded lignin as salts (soda method) or as salts of lignosulfonates. In general, lignins resist acid hydrolysis, are soluble in hot alkaline or bisulfite solutions, and are easily oxidized.

It is generally agreed that softwood lignins are built up from coniferyl alcohol, 3-(4-hydroxy-3-methoxyphenyl)-2-propene-1-ol, a substituted relative of cinnamyl alcohol. Hardwood lignins also contain sinapyl alcohol (3-(3,5-dimethoxy-4-hydroxyphenyl)-2-propene-1-ol) units. The units are linked by means of the hydroxy groups,

MeO—[Ar]—CH=CHCH$_2$OH
HO
coniferyl alcohol

MeO—[Ar]—CH=CHCH$_2$OH
HO OMe
sinapyl alcohol

except that no diphenyl ethers have been found. Many modifications of the propenol chain are found, such as ArCH=CHCHO, ArCH$_2$COCH$_3$, ArCHOHCH$_2$CH$_2$OH, and ArCHOHCHOHCH$_2$OH with formation of acetals and ketals as well as ethers. The hydroxy groups also are found with bonds to molecules of sugars and presumably to cellulose.

The major by-product obtained from softwood sulfite pulp spent liquor is vanillin. The process is simple and cheap (Eq. 24.57). However, only so much vanillin is

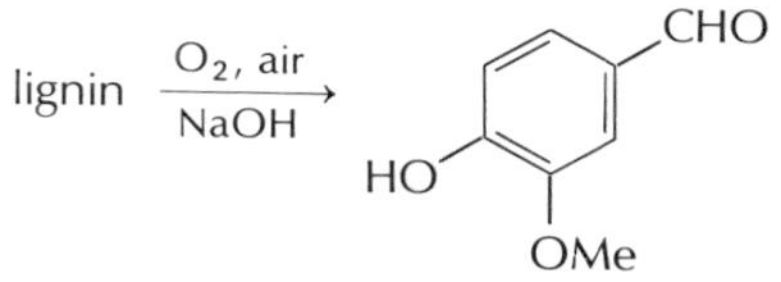

$$\text{lignin} \xrightarrow[\text{NaOH}]{O_2,\text{ air}} \underset{\underset{\text{vanillin}}{\text{OMe}}}{\text{HO}\diagup\!\!\!\diagdown\text{CHO}} \quad (24.57)$$

needed by civilization, so the great mass of extracted lignins continues to be a disposal problem instead of a source of useful aromatic compounds. Despite the attention that has been given to the problem, it remains intractable.

24.9 Some drugs and vitamins

Phenolic phenylethylamine compounds. There is a group of compounds that may be considered as derived from 2-phenylethylamine (phenethylamine). Many of them have pronounced effects upon the central nervous system. One sequence appears to arise from the presence of phenolic hydroxy groups. Tyramine and dopamine are normally occurring

$\phi CH_2 CH_2 NH_2$
2-phenylethylamine
(phenethylamine)

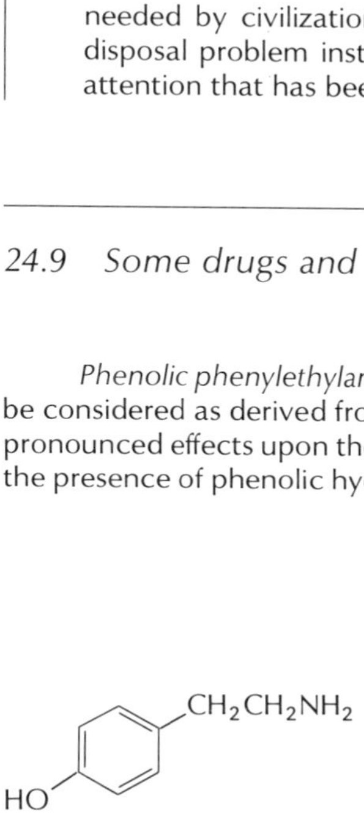

tyramine 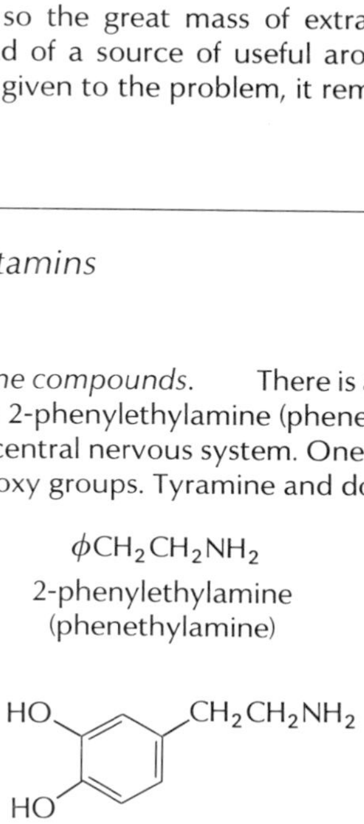   dopamine 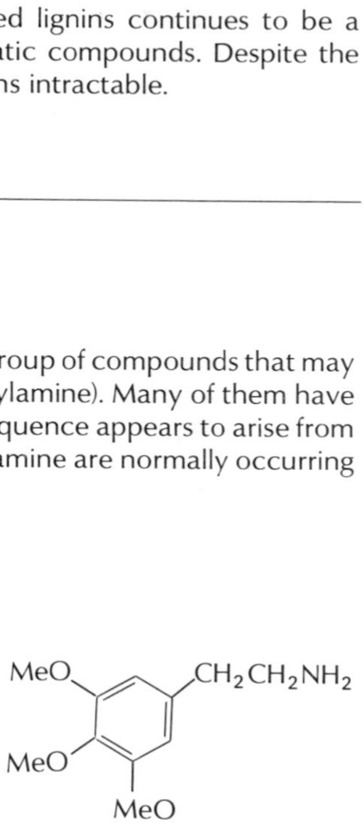   mescaline

amines in the body. Mescaline, a powerful hallucinogen, occurs in the mescal buttons of a variety of cactus indigenous to Mexico and the southwestern area of the United States. Even more powerful is DOM (STP), a synthetic substance, 1-(2,5-dimethoxy-4-methylphenyl)-2-propylamine.

Phenethylamine is a sympathomimetic substance; ingestion results in mydriatic action (dilation of the pupil of the eye) and pressor activity (increase in blood pressure). Tyramine is about as active as phenethylamine. Dopamine is a precursor of norepinephrine, which is epinephrine without the N-methyl group.

A second sequence of compounds is related by having a three-carbon side chain. Tyrosine is one of the common amino acids. It is a precursor of tyramine by decarboxylation, and of DOPA by hydroxylation. DOPA (the letters arose from d̲i̲o̲x̲y̲phenylalanine)

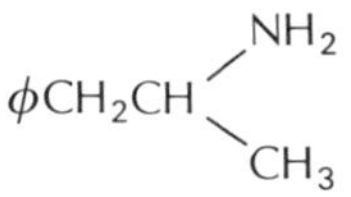

amphetamine
(benzidrine)   tyrosine 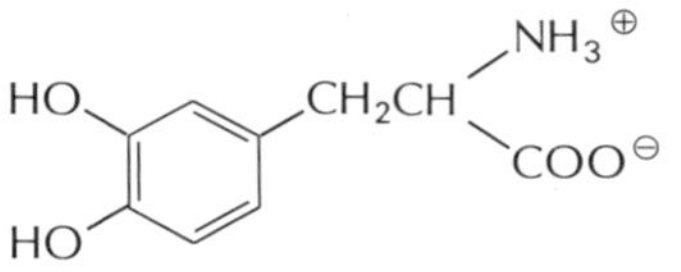   DOPA

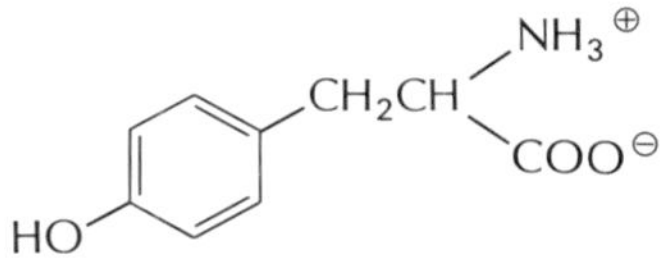

is a precursor of dopamine by decarboxylation. The positive-rotating form of amphetamine is known as <u>dexedrine</u> and is the more active of the enantiomers. Amphetamine is a stimulant: it increases sleeplessness (and irritability). The aftereffect is depression. Hydrogenation of the ring gives the corresponding cyclohexylpropylamine, a compound known as <u>benzedrex</u>, which is used in nasal inhalers to constrict capillaries.

A third sequence of compounds involves the presence of an alcoholic hydroxy group α to the benzene ring and with or without methylation of the amino group. Epinephrine and norepinephrine are powerful stimulants of the central nervous system and are secreted by the adrenal glands for this purpose. Ephedrine and norephedrine are plant products with similar, though less intense, effects. Neosynephrine is popular for use as nose drops, exerting a constrictive effect upon capillaries with a minimum of effect upon the heartbeat and constriction of arteries. Synephrine is similar.

ϕCHOHCH(NHMe)(CH$_3$)
ephedrine

HO-C$_6$H$_4$-CHOHCH$_2$NHMe
synephrine

(HO)$_2$C$_6$H$_3$-CHOHCH$_2$NHMe
epinephrine (adrenaline)

ϕCHOHCH(NH$_2$)(CH$_3$)
norephedrine (propadrine)

HO-C$_6$H$_4$-CHOHCH$_2$NHMe
neosynephrine (phenylephrine)

(HO)$_2$C$_6$H$_3$-CHOHCH$_2$NH$_2$
norepinephrine (noradrenaline)

The 3,4-dihydroxyphenylamines are also known as the catecholamines. The catecholamines, along with serotonin, have been implicated in activity in the brain. A proper balance of dopamine, serotonin, and norepinephrine is necessary for mental health. An

serotonin (5-hydroxyindole with CH$_2$CH$_2$NH$_2$ at position 3, NH indole)

excess or a deficiency of one of the three produces profound effects—hallucination, depression, stimulation (manic activity), crippling effects such as Parkinson's disease, and others. L-DOPA (the active enantiomer) has been found effective in treatment of Parkinson's disease. It crosses from blood to the brain and there acts as a precursor for dopamine. Dopamine cannot cross the barrier. In Equation 24.58 is shown the biochemistry of the DOPA → dopamine → norepinephrine pathway. A specific enzyme is necessary for each step, and each step has been blocked by the presence of a given inhibitor.

The part played by serotonin in brain function remains to be elucidated, even though all research workers in the field of mental health agree that the serotonin level (or balance) is heavily involved. In the meantime, many ignorant persons (as well as many who should know better) have taken to tinkering with hallucinogens and other drugs in an alleged search for "freedom of the mind." Despite the ability of the body to stand up to drug abuse, little is known of the long-term damage that uncontrolled usage of nonaddictive drugs will cause. What is known is that the "drug culture" has given rise to costly civic expense in the form of crime to support the expensive habit and to a needless diversion of time of those in the health fields in taking care of the human misery that results.

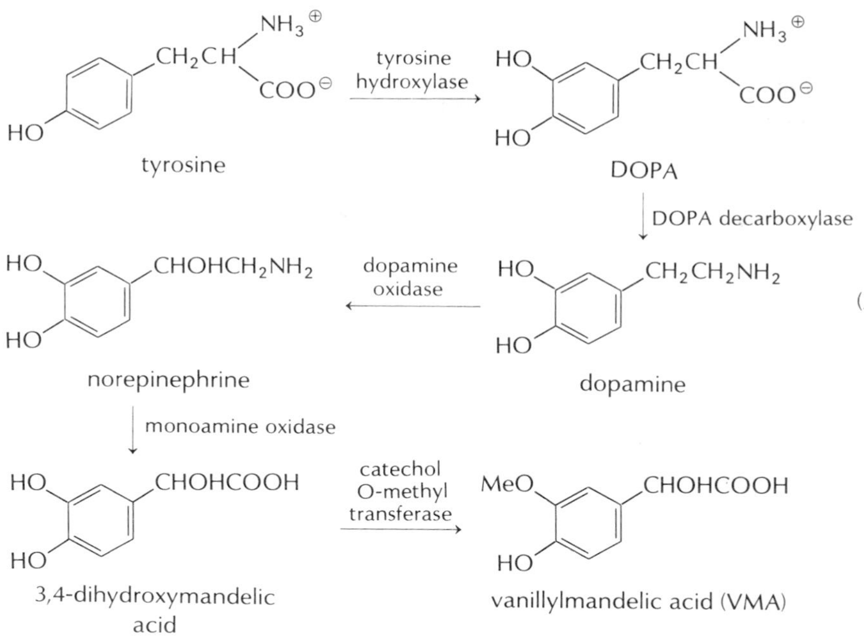

24.9 Some drugs and vitamins

[Structural diagrams showing the biosynthetic pathway:]

tyrosine →(tyrosine hydroxylase)→ DOPA →(DOPA decarboxylase)→ dopamine →(dopamine oxidase)→ norepinephrine →(monoamine oxidase)→ 3,4-dihydroxymandelic acid →(catechol O-methyl transferase)→ vanillylmandelic acid (VMA)

(24.58)

Chloramphenicol. Chloramphenicol (chloromycetin), one of the simplest antibiotics, is unusual in being closely related to the amines we have been looking at. It is one of the very few naturally occurring substances with a nitro group or a dichloroacetyl group.

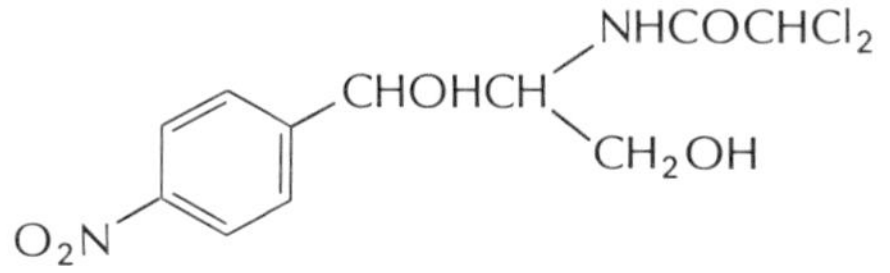

[Structure: O_2N-C$_6$H$_4$-CHOHCH(NHCOCHCl$_2$)CH$_2$OH]

Marijuana. Another series of compounds with psychotomimetic properties occurs in the plant *Cannabis sativa*. Various parts of the plant have been used: the bracts, flowers, and pollen contain more of the hallucinogens than do the stems and leaves. Under various names (marijuana, hashish, pot, grass, dagga, charas, and bhang, to name a few), it has been in use for centuries, either taken orally or by smoking, with debatable results for the individual or for society. The active principle has been shown to be *trans*-10a,6a-(—)-Δ^9-tetrahydrocannabinol, or Δ^9-THC (by pyran numbering). The subject has been thoroughly confused by the use of two alternative numbering systems and of occasional mixing of the two systems in a single name (Δ^9-3,4-*trans*-tetrahydrocannabinol, for example). The other naming system is derived from that used for terpenes (Eq. 24.59). The use of Δ^1, Δ^2, etc., is a method for designating the position of a double bond. It is used in terpene and steroid chemistry. Cannabidiol is one of the inactive compounds present in marijuana. It is partially converted into Δ^9-THC (we use the pyran numbering system) during smoking. A Δ^8-THC also is present but is reported to be less active than Δ^9-THC. Metabolism in the body introduces a hydroxy group at the 11 position.

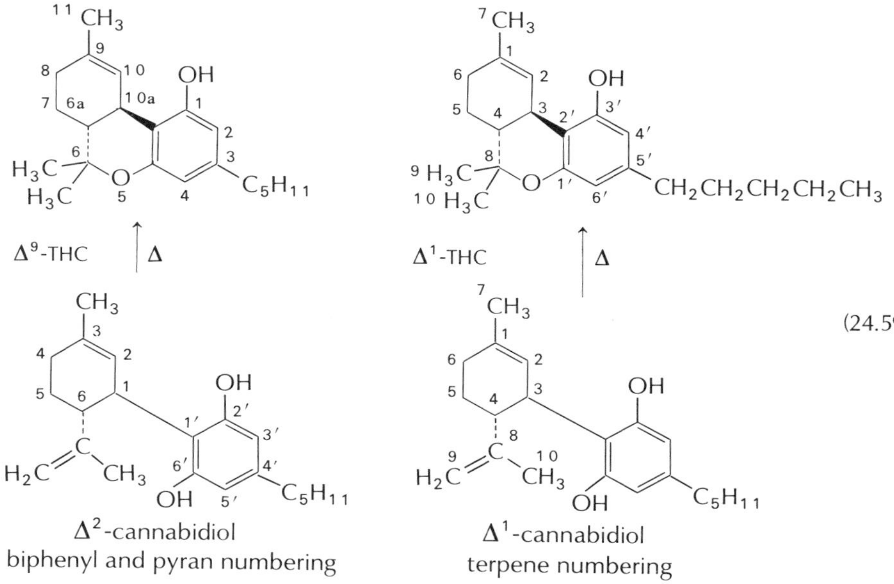

818   Chapter twenty-four

[Structures of Δ⁹-THC, Δ¹-THC, Δ²-cannabidiol (biphenyl and pyran numbering), and Δ¹-cannabidiol (terpene numbering)] (24.59)

Vitamin E: synthesis, antioxidant activity. Vitamin E was isolated from wheat germ oil in 1936 and was found to consist of two substances with phenolic properties. They were called α- and β-tocopherol (from two Greek words meaning childbirth and to bear). Later, the active factors in cottonseed oil and corn oil were isolated and called γ- and δ-tocopherol. The vitamin was discovered as the factor necessary for successful reproduction in rats fed a restricted diet. A synthesis of α-tocopherol is shown in Equation 24.60, as well as an

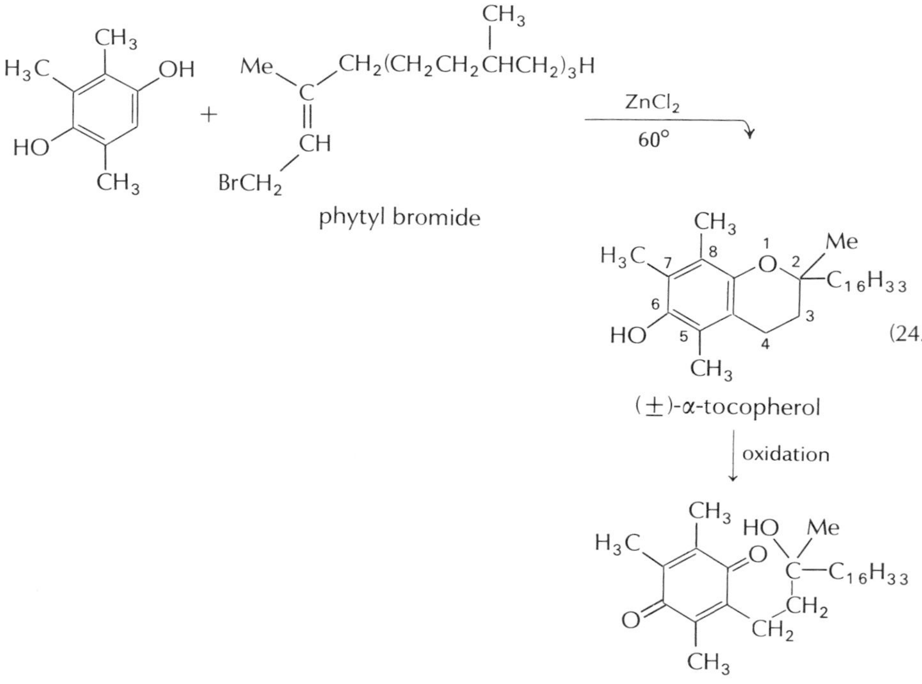

[Reaction scheme: trimethylhydroquinone + phytyl bromide, $ZnCl_2$, 60°, giving (±)-α-tocopherol, which upon oxidation gives the quinone product] (24.60)

oxidation that illustrates the *in vivo* antioxidant properties of the substance. Each of the various tocopherols has methyl groups at positions 2 and 8. As shown, α-tocopherol has two more methyls at positions 5 and 7. The other tocopherols have fewer additional methyls: β has one at position 5, γ has one at 7, and δ has none. The relative activities, based on $(\pm)$-α-tocopheryl acetate as 1.0, are $(+)$-α, 1.5; $(-)$-α, 0.5; β, 0.1; γ, 0.1; and δ, very low.

There is little agreement about human requirements. It seems that much more of the vitamin is needed if the diet is high in polyunsaturated fats and oils. It has been shown, too, that the life span of mice is prolonged if more than the minimum required amount of vitamin E is included in the diet. It has been hypothesized that the vitamin acts as an antioxidant in such instances by inhibiting radical reactions of HOO· and HO·, which destroy cell walls and mitochondrial membranes. By this slowing of the rate of destruction, the repair functions are able to maintain the proper balance and prevent aging. However, other hypotheses about aging do not consider radical reactions (or inhibition of radical reactions) as a factor. It is of interest that coenzyme Q, a quinone, is involved in biochemical oxidation reactions, and it can be employed instead of vitamin E in the alleviation of certain diseases caused by

coenzyme Q

vitamin E deficiency. As we have seen, quinones (and hydroquinones) both are able to intercept radicals.

The aging process. Carbon tetrachloride (CCl_4), a colorless liquid commonly used as a solvent in organic chemistry laboratories, is toxic to man in extremely small doses. When ingested, it is absorbed by the liver where it is induced to dissociate into radicals. The radicals (including Cl·) react and damage molecules important to the liver.

One of the several prominent theories about the process of aging attaches importance to the accumulation of such radical induced damage from many sources over the years. By this reasoning, the aging process would be slowed if the radicals could be removed before doing the damage. In fact, it has been shown that this destruction can be greatly reduced by treatment with radical scavengers, special molecules that react much more quickly with the radicals than do the molecules of the tissue.

Tissue protection of this sort is thought to be a major biological role for vitamin E. Vitamin E does, in fact, decrease the toxicity of carbon tetrachloride. However, much work remains to be done on radical reactions in biological systems, and it is too early to do other than speculate on the implications of this research.

Vitamin K; dicumarol; menadione. Vitamin K is a quinone that is essential for the clotting of blood, a process still not completely worked out. A possible sequence of events is shown in Equation 24.61. Vitamin K, in this scheme, is thought of as the enzyme or

$$\text{biosynthesis} \xrightarrow{\text{vitamin K}} \text{prothrombin} \xrightarrow{\text{injury}}$$
$$\downarrow \text{thromboplastin, } Ca^{2+}\text{, cofactors}$$
$$\text{fibrinogen} \xrightarrow{\text{thrombin}} \text{fibrin} \xrightarrow{\text{polymerization}} \text{clot} \quad (24.61)$$

coenzyme necessary for the biosynthesis of prothrombin, a protein present in blood plasma. A cut or injury somehow causes the release of thromboplastin by the damaged cells. Thromboplastin, along with calcium ions and other cofactors, acts as an enzymatic system for the conversion of prothrombin into the enzyme thrombin, needed for the conversion of fibrinogen (a soluble protein in the plasma) into fibrin (an insoluble protein), which polymerizes (coagulates) into the clot. Lack of vitamin K, or the presence of antagonists or competitors, upsets the clotting mechanism. Heparin, a polysaccharide that consists of alternate glucosamine and glucuronic acid units and has the amino groups mostly converted to —$NHSO_3H$ groups, and dicumarol are two naturally occurring substances that

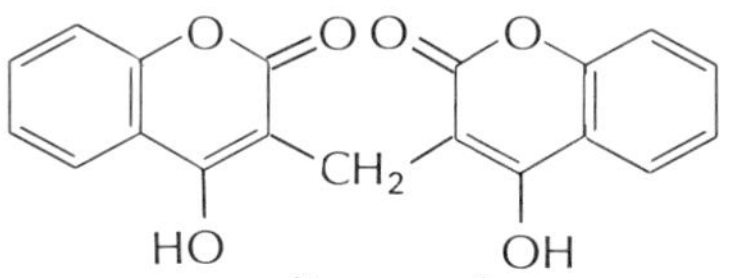

dicumarol

interfere with the role of vitamin K and prevent clotting. They have been used in preparation of a patient for surgery to lessen the danger of a thrombosis during and after the procedure.

There are several vitamin K's. Vitamin K_1, from alfalfa, is 3-phytyl-2-methylnaphthoquinone. Slightly less active is vitamin K_2, from putrefied fish meal. It has two forms, one with a six-unit isoprenoid side chain at position 3 instead of phytyl, the other with a seven-unit isoprenoid chain. It came as something of a surprise that 2-methylnaphthoquinone, known as menadione, gave vitamin K activity. Evidently the body has little difficulty in attaching the proper side chain at position 3. Hemophilia is not caused by lack of vitamin K, but rather by an absent ingredient in thromboplastin formation. (The problem is hardly that simple—one sentence cannot summarize books and symposia on the subject.)

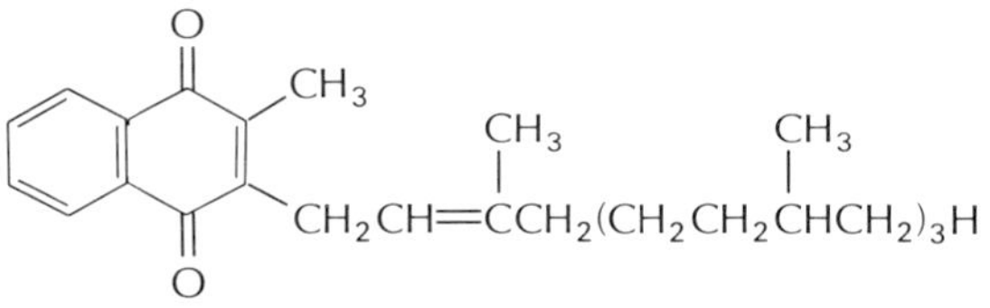

vitamin K_1 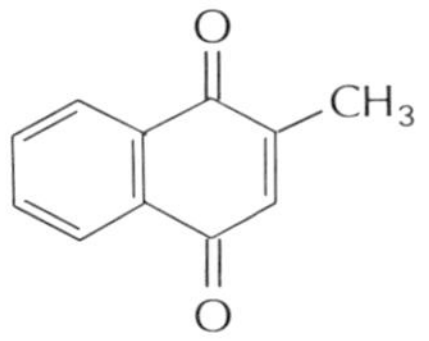   menadione

Exercises

24.11 Show how to prepare:

a. 2-phenylethylamine from toluene
b. dopamine from catechol
c. propadrine from benzaldehyde
d. epinephrine from catechol

24.10 Summary

We have now covered the reactions of phenols and of quinones and have seen how many of them are interlocked with metabolic processes. In coming chapters we shall consider other types of compounds involved in other aspects of metabolism. Along the way, we have learned about the reactivity of the phenols in substitution reactions and oxidation and reduction and about their manufacture and uses. We examined quinones in enough depth to help us understand the results of addition reactions of quinones. Lignin and some drugs and vitamins were described. Phenols and quinones thus are involved in many ways and have diverse effects in nature.

We have given a lot of background material in this chapter—some of it is not organic chemistry at all. Background material may be of interest and of help in understanding and remembering certain items. It is not as pertinent to the subject as the actual chemistry we have discussed, however.

24.11 Problems

24.1 Substance A is soluble in dilute NaOH and gives a $FeCl_3$ color. A with $CHCl_3$ and NaOH gives B. B reacts with MeI and NaOH to give a compound that may be oxidized to C ($C_9H_8O_5$). With allyl chloride and K_2CO_3 in acetone, A gives D. Heating D gives E. E with MeI and NaOH gives a substance that gives C upon oxidation. A with acetic anhydride gives a substance that, with $AlCl_3$ at 25° gives F. F with MeI and NaOH gives a compound that with NaOCl and NaOH gives G. Oxidation of G gives H, an isomer of C. H, unlike C, easily gives an anhydride. A, with nitrous acid gives I. I is easily reduced to J. Mild oxidation of J gives 2-methylbenzoquinone.

Write a structural formula for each lettered compound.

24.12 Bibliography

All the general textbooks previously mentioned are good on the subject of phenols and tropolones. For quinones and for dyes, only Fieser and Fieser (Chap. 7, ref. 7) and Noller (Chap. 7, ref. 11) are good: other texts tend to give little space to the two subjects. Do not miss reference 4 of Chapter 21. Some additional readings on some of the topics covered in this chapter follow.

Crown compounds:
1. Macrocyclic Polyethers Complex Alkali Metal Ions. *Chemical and Engineering News (C & EN)* March 2, 1970, pp. 26, 27.

Dyes:
2. Stinson, Stephen C., Textile Dye Industry, *C & EN* Oct. 26, 1970, pp. 42–50.

Lignin:
3. Pearl, Irwin A., Lignin, Century Old Puzzle, *C & EN* July 6, 1964, pp. 81–93.

Catechol amines:
4. Everett, G. M., and J. W. Borcherding, L-DOPA: Effect on Concentrations of Dopamine, Norepinephrine, and Serotonin in Brains of Mice, *Science 168*, 849 (1970).

5. Schildkraut, Joseph J., and Seymour S. Kety, Biogenic Amines and Emotion, *Science 156*, 21 (1967).

6. Stein, Larry, and C. David Wise, Possible Etiology of Schizophrenia, *Science 171*, 1032 (1971).

Marijuana:
7. Branham, J. M., S. A. Reed, Julie H. Bailey, and J. Caperon, Marihuana Components: Effects of Smoking on Δ^9-Tetrahydrocannabinol and Cannabidiol, *Science* 172, 1158 (1971).
8. Marijuana Program Advances at NIMH, *C & EN* July 6, 1970, pp. 30–33.
9. Mechoulan, Raphael, Marijuana Chemistry, *Science* 168, 1159 (1970). The best place to begin outside reading on the subject.
10. Mechoulan, Raphael, Arnon Shani, Habib Edery, and Yona Grunfeld, Chemical Basis of Hashish Activity, *Science* 169, 611 (1970).

Drugs, vitamins, and allelochemics:
11. Pryor, William A., Free Radical Pathology, *C & EN* June 7, 1971, pp. 34–51. A good summary of free-radical reactions, application to *in vivo* systems, vitamin E and coenzyme Q activity, and so on.
12. Sanders, Howard J., Arthritis Drugs, *C & EN* Aug. 12, 1968, pp. 46–73.
13. Whittaker, R. H., and P. P. Ferry, Allelochemics: Interactions between Species, *Science* 171, 757 (1971).

24.13 H. Marjorie Crawford: A lifelong interest

Professor Emeritus H. Marjorie Crawford of Vassar went there in 1927 as an instructor fresh from Minnesota with her Ph.D. She retired first in 1964 but kept busy by teaching at Marist College for four years. After a year of retirement she returned to Vassar to help with the planning and teaching of the organic chemistry labs when enrollment shot from 35 to 140. She was the first woman to serve on the Council Policy Committee of the American Chemical Society and also has been a member of the Visiting Scientists Committee. Here she tells, all too briefly, of her research work.

Working alone, as I have, I do not have stories about graduate students; but I was lucky in picking a topic to work on that was not "hot news," and so my slow results did not meet with competition from big research groups.

For a long time I have been interested in the reactions of quinones with organometallic reagents, primarily the Girgnard and lithium reagents. The syntheses of the required quinones and related compounds and the identification of the expected as well as some very unexpected products has made this work interesting.

The first quinone studied was duroquinone, 2,3,5,6-tetramethyl-1,4-benzoquinone. Products resulting from the 1,2- and 1,4-additions of phenylmagnesium bromide were isolated and identified. The yields were small; and, because anthraquinone gave good yields of 1,2-addition products only, I turned to the reactions of 2,3-dimethyl-1,4-naphthoquinone. Phenyllithium gave only 1,2-addition, but phenylmagnesium bromide again gave 1,2- and 1,4-monoaddition products, as well as some 1,2-1,2- and 1,2-, 1,4-double addition products. The diaddition products lost water readily and underwent rearrangement. The 2,3-diphenyl-1,4-naphthoquinone was no better and gave the same mixture of products. With acenaphthenequinone, there was the expected trans-1,2-diphenyl-1,2-acenaphthenediol, as well as small amounts of a lactone and a ketone in both of which a phenyl group has gone into the naphthalene ring. This was really unexpected under the mild conditions (bp of ether) under which the reaction took place.

The tert-butyl group should afford great steric hindrance, so I looked into that possibility. Two di-tert-butylnaphthalenes were known, but the locations of the groups were not known. I showed that the groups were in different rings and that the higher melting one was the 2,6- isomer. The di-tert-butylnaphthalenes are very easy to make but difficult to separate. Because I had these compounds, I made the quinones and various derivatives in order to see what they would do with phenyllithium and phenylmagnesium bromide. The expected mono- and di-addition products were obtained as well as more surprises, such as the lactones and traces of the red quinone.

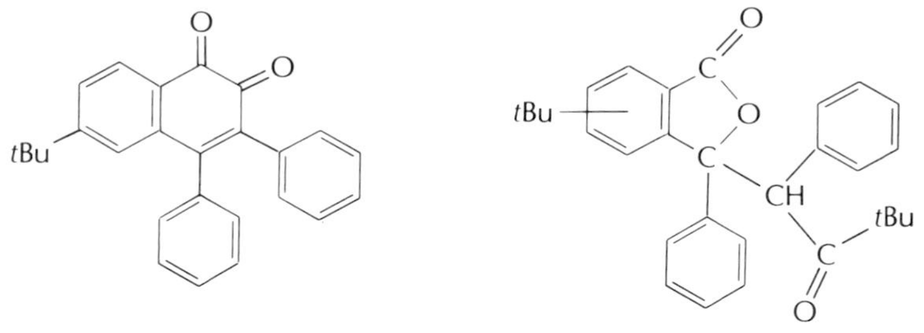

Now I am retired for the third time but still have laboratory space to continue research.

25 Polycyclic Aromatic Compounds

      From time to time in previous chapters we have illustrated certain reactions with naphthalene and anthracene derivatives. Now we shall look into the occurrence, reactions, and synthesis of naphthalene, anthracene, and some other substances that look as though they are made up of fused benzene rings (rings that have at least one side of a hexagon in common). There will be only a few new reactions. However, it will be interesting to see how some familiar reactions proceed with the polycyclic aromatic compounds.

25.1   Naphthalene: occurrence and stability   824
        1.   Occurrence; naming.   2.   Structure; resonance; bond fixation.
        3.   Resonance energy; other polycyclics.

25.2   Naphthalene: substitution   825
        1.   Nitration; halogenation.   2.   o-Xylene comparison: structures of intermediates.   3.   Sulfonation.   4.   Friedel-Crafts reactions.
        5.   Other β substituents.   6.   Disubstitution.   7. β-Naphthol. Exercises.

25.3   Naphthalene: additions   830
        1.   Reductions.   2.   Oxidation. Exercises.

25.4   Naphthalene: synthesis   831
        1.   Synthesis from benzene and succinic anhydride.   2.   Variations, substituted naphthalenes. Exercises.

25.5   Anthracene and anthraquinone   833
        1.   Occurrence; structure; naming; reactivity.   2.   Anthraquinone: properties, stability.   3.   Reduction of anthraquinone.   4.   Anthraquinone: substitution reactions.   5.   Anthraquinone: syntheses. Exercises.

25.6   Phenanthrene   837
        1.   Occurrence; naming; bond fixation.   2.   Reactions.   3.   Synthesis. Exercises.

25.7   Other polycyclic compounds   839
        1.   Fusions with cyclopentane rings.   2.   Polycyclic fused benzene ring systems.

824 Chapter twenty-five

25.8   Biphenyl   842
       1. Resonance energy: conjugation between the rings.   2. Chirality of substituted biphenyls. Exercises.

25.9   Summary   844

25.10 Problems   844

25.11 Bibliography   845

25.1 Naphthalene: occurrence and stability

Occurrence; naming.   Naphthalene ($C_{10}H_8$) is the simplest of the fused benzene ring compounds. It is the most abundant (about 11%) of the substances obtained by the distillation of coal tar, and it was the first pure compound to be isolated from coal tar (1820). It is a crystalline solid (mp 80°) with a high vapor pressure that causes easy sublimation.

The ring is numbered as shown. In addition, because a single substituent at the

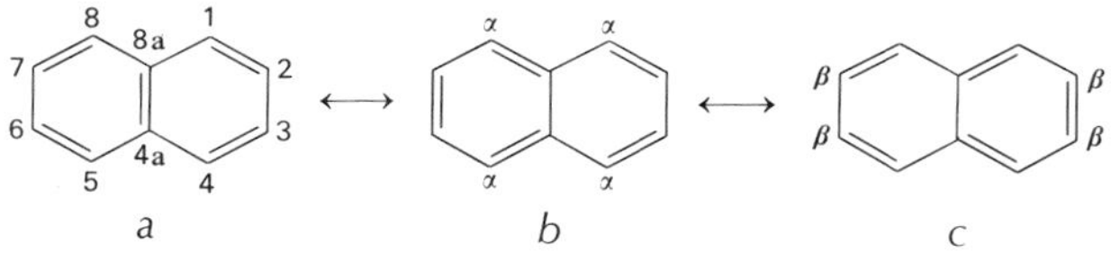

1, 4, 5, or 8 position gives the identical substance, such compounds often are called α, as in α-chloronaphthalene. Similarly, a single substituent at the 2, 3, 6, or 7 position is called β, as in β-methylnaphthalene. If two or more substituents are present, the numbering system must be used. (Occasionally we find the prefixes *peri* for 1,8- and *amphi* for 2,6-disubstituted naphthalenes.)

Structure; resonance; bond fixation.   It was shown early that naphthalene contained two rings (Eq. 25.1). Oxidation of α-nitronaphthalene gives 3-nitrophthalic acid, which shows that ring B has survived. Reduction to α-naphthylamine and oxidation gives phthalic

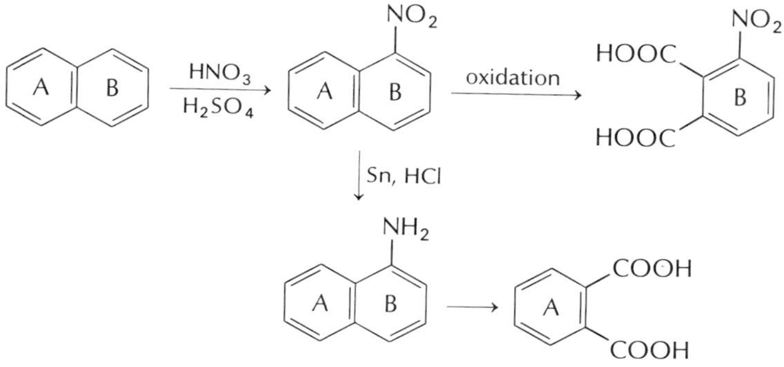

(25.1)

acid, which shows that ring A has survived. Phthalic acids in each instance indicate that remnants of the destroyed ring are *ortho*. Thus $C_{10}H_8$ can be only as shown in resonance theory, with five double bonds. Structure *a* contributes more, with both rings benzenoid; structures *b* and *c* contribute less, with one ring benzenoid and the other o-quinoid (higher energy). Furthermore, bond distances (from X-ray diffraction) show that the α-β bonds at 1.36 A are shorter than normal (all other bond lengths in naphthalene are 1.42 A) or shorter than those of benzene (1.40 A). Thus we come to speak of bond fixation in the polycyclic aromatic hydrocarbons, and we refer to partial double-bond fixation in the four α-β bonds. The effect influences the course of certain reactions, as we shall see shortly.

Resonance energy; other polycyclics. We can calculate what the heat of combustion should be and compare the value with what it actually is to estimate resonance energies. Thus, ΔH for combustion of a mole of benzene is -789 kcal, and that of naphthalene is -1250 kcal. The values for benzene should have been -825 to -827 kcal, and the greater stability of benzene (36 to 38 kcal) is called resonance energy. Naphthalene should have released 1321 or 1311 kcal, which gives 71 or 61 kcal for the resonance energy. Let us adopt the most popular values and make some calculations as shown in Table 25.1,

Table 25.1 Resonance energies of aromatic compounds (kcal)

Compound	Number of rings	Resonance energy[a]	Resonance energy per ring
benzene	1	36	36
naphthalene	2	61	30.5
anthracene	3	84	28
phenanthrene	3	91	30.3

[a] Calculated from heats of combustion.

which includes anthracene and phenanthrene as well. What we wish to bring out is that the polycyclics have less stabilization by resonance energy than we would expect if each ring were completely aromatic (or benzenoid). With higher energy content (smaller resonance energy than if they were perfectly aromatic) the polycyclic compounds are less stable and more reactive than benzene.

25.2 Naphthalene: substitution

Nitration; halogenation. Naphthalene undergoes the substitution reactions (nitration and halogenation) to give almost exclusively the α product (Eqs. 25.2). Care must be taken to prevent disubstitution. Even though the substituted ring is deactivated, the other ring is only slightly affected and reacts easily to give the disubstituted products (Eqs. 25.3). In disubstitution, 1,8-dinitronaphthalene is the major product of dinitration, whereas 1,5-dibromonaphthalene is the major product of dibromination.

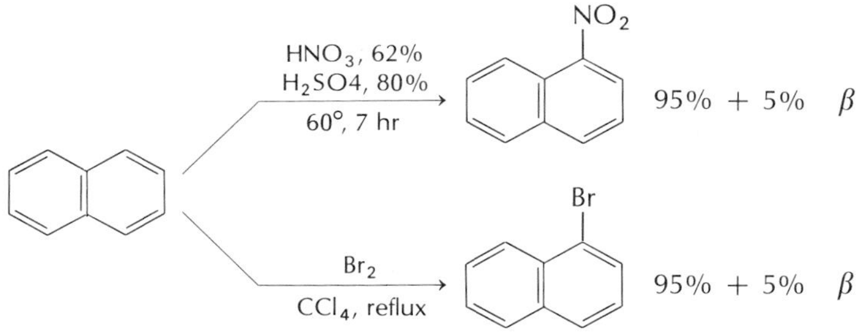

[Reaction scheme 25.2: Naphthalene + HNO₃, 62% / H₂SO₄, 80% / 60°, 7 hr → 1-nitronaphthalene, 95% + 5% β; Naphthalene + Br₂ / CCl₄, reflux → 1-bromonaphthalene, 95% + 5% β]

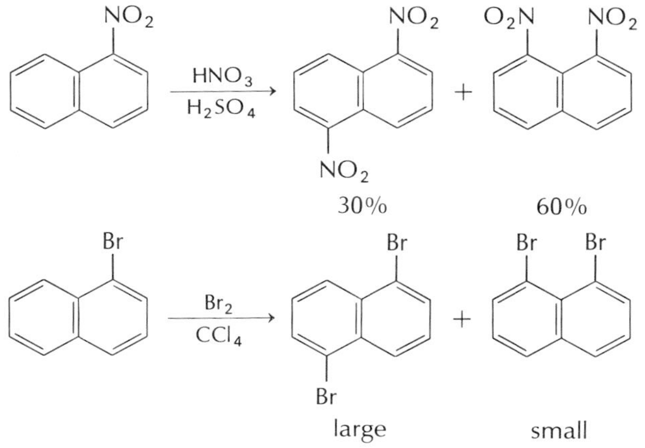

[Reaction scheme 25.3: 1-nitronaphthalene + HNO₃/H₂SO₄ → 1,5-dinitronaphthalene (30%) + 1,8-dinitronaphthalene (60%); 1-bromonaphthalene + Br₂/CCl₄ → 1,5-dibromonaphthalene (large) + 1,8-dibromonaphthalene (small)]

o-Xylene comparison: structures of intermediates. Why α and not β? Offhand, we might expect β to be favored because it is less sterically hindered. For example, compare substitution of o-xylene (Eq. 25.4). The α position in naphthalene may be considered *ortho*

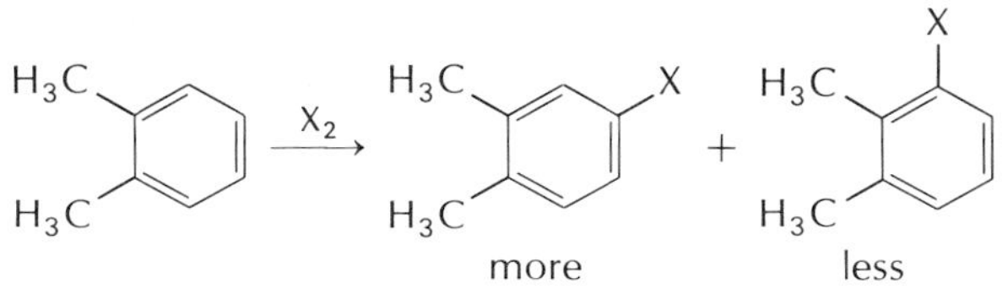

[Equation 25.4: o-xylene + X₂ → 4-substituted o-xylene (more) + 3-substituted o-xylene (less)]

to one side chain and *meta* to the other, and the β position as *meta* to one and *para* to the other. Thus, although about equally activated, the α position (like the 3 position in o-xylene) is more hindered than the β position (like the 4 position in o-xylene). The explanation, we find, lies in the lower activation energy for production of the intermediate for α substitution. The contributing resonance structures for α and β attack are shown. Only the resonance pairs a, b, and c can be written for Kekulé resonance in ring A. All the other structures contribute only slightly, being quinoid in both rings and higher in energy. We see that a and b both stabilize the α intermediate, and only c stabilizes the β intermediate. Thus, α attack is favored by kinetic control.

Sulfonation. The reversible electrophilic substitutions may be carried out so that either kinetic (α) or thermodynamic (β) control governs the result. The β product is more

25.2 Naphthalene: substitution

α intermediate (E = electrophile)

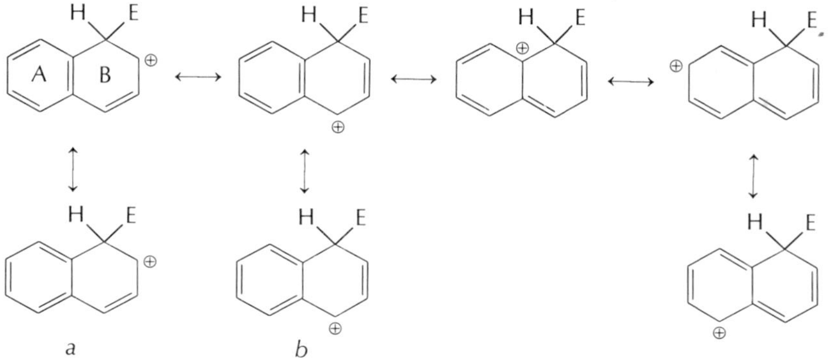

β intermediate

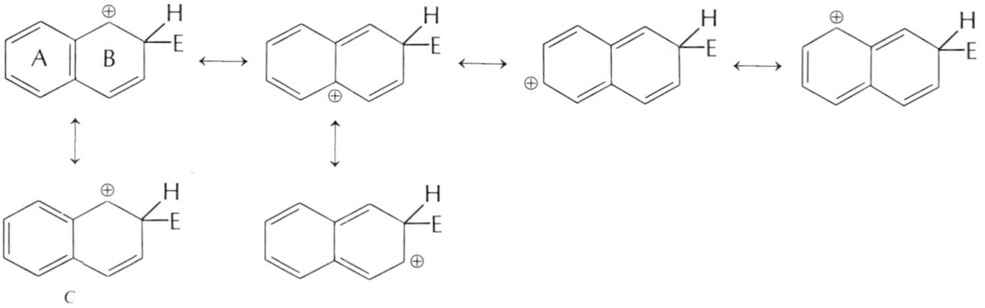

stable than the α product (because the α position is more crowded), but it is formed more slowly. The situation resembles the one that we encountered previously when we looked at 1,2 versus 1,4 addition to dienes. The 1,2 product was formed faster (at low temperature) but the 1,4 product was the more stable. Thus sulfonation at low temperature gives naphthalene-α-sulfonic acid, but it gives naphthalene-β-sulfonic acid at high temperature. The α acid may be rearranged to give the β acid if it is heated in the presence of sulfuric acid (Eq. 25.5). Again, we must be careful to avoid disubstitution.

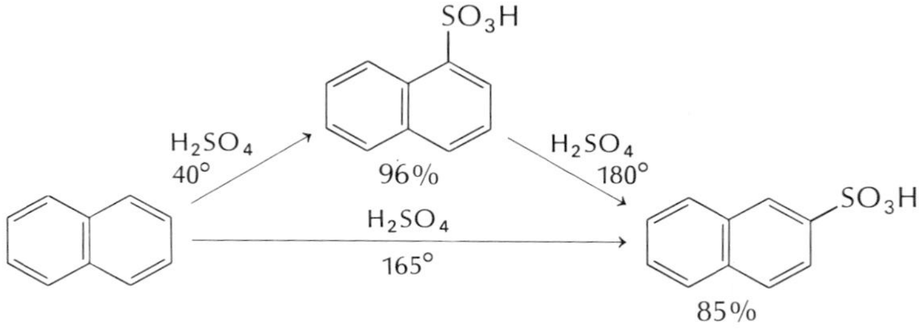

(25.5)

Friedel-Crafts reactions. Friedel-Crafts acylation may be controlled to some extent. Low temperature and relatively nonpolar solvent favor kinetic control to give the α product (Eq. 25.6). However, extrapolation to other cases is unreliable. Thus, naphthalene with succinic anhydride in ϕNO_2 gives 36% α and 47% β, but with phthalic anhydride in tetrachloroethane ($Cl_2CHCHCl_2$; the $-CCl_2-$ group is unreactive) it gives 71% α, or with benzoyl chloride in CS_2, 81% α. If a good yield of the β isomer is needed, tetralin (tetrahydronaphthalene) may be used (Eq. 25.7), after which ring A is easily aromatized (dehydrogenated) if the product is heated with S, Se, Pt, or Pd.

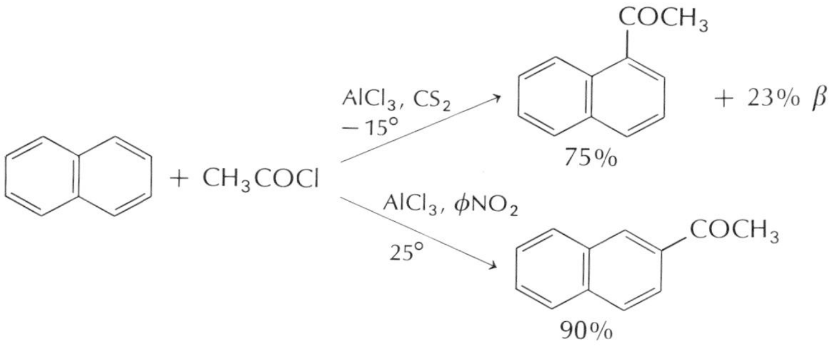

<!-- Equation 25.6: Naphthalene + CH3COCl -->
Naphthalene + CH₃COCl:
- AlCl₃, CS₂, −15°: → 1-acetylnaphthalene (α-COCH₃) 75% + 23% β
- AlCl₃, φNO₂, 25°: → 2-acetylnaphthalene (β-COCH₃) 90%

(25.6)

<!-- Equation 25.7: Tetralin with anhydrides -->
Tetralin (rings A, B):
- succinic anhydride, AlCl₃, φH → 6-(COCH₂CH₂COOH)-tetralin, 76%
- phthalic anhydride, AlCl₃, φH → 6-(2-carboxybenzoyl)-tetralin, 97%

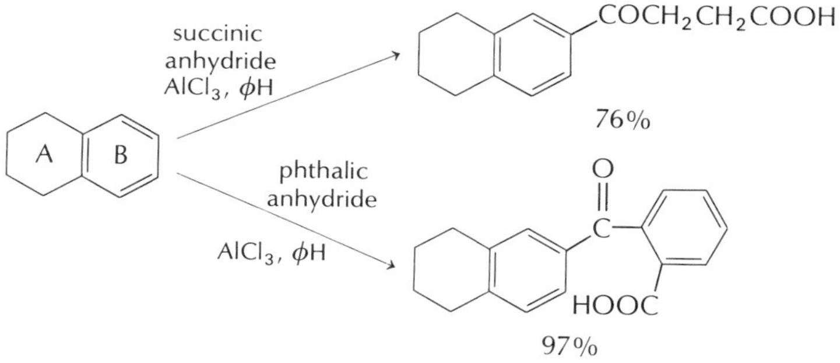

(25.7)

Other β substituents. Both sulfonation and the Friedel-Crafts reaction offer entry to other β substituents. For example, fusion of the β-sulfonic acid with NaOH gives β-naphthol, which may be converted by the Bucherer reaction into β-naphthylamine. Then the β-amine (as well as the α-amine from α-nitronaphthalene) may be replaced by any number of groups (Sandmeyer, Gattermann, and other reactions).

Disubstitution. Directive and activation effects of substituents upon further substitution, in general, are similar to the effects upon corresponding substituted benzenes. The activating and o,p-directing substituents favor substitution in the same ring, but with varying effectiveness, depending upon whether the activating substituent is in the α or β position. Some examples are shown in Equations 25.8. Thus, an o,p director in an α position directs to the 4 (or *para*) in preference to the 2 position, even for those reactions (acylation in ϕNO_2) that prefer a β position. An o,p director in a β position directs to the 1 position (kinetic control), as shown in nitration, but thermodynamic control favors the 6 position. A *meta*-directing, deactivating substituent, whether in an α or a β location, causes substitution to take place in the unsubstituted ring (Eq. 25.3). In any case, all the substitution reactions on naphthalene and substituted naphthalenes proceed much more rapidly than do those with benzene. For example, note the use of benzene as solvent for the Friedel-Crafts reactions in Equation 25.7 (and in place of ϕNO_2 in other acylations).

β-Naphthol. We will not go further into the subject of di- and trisubstitution in naphthalene and substituted naphthalenes except to point out the special reactivity of β-naphthol. Partial bond fixation (Secs. 14.5 and 25.1) leads β-naphthol to react as though it were the enol of a ketone. In the presence of CH₃COONa in acetic acid, bromination (and chlorination) proceed as shown in Equation 25.9. Product a is obtained if HBr is removed by reaction with CH_3COONa. In the absence

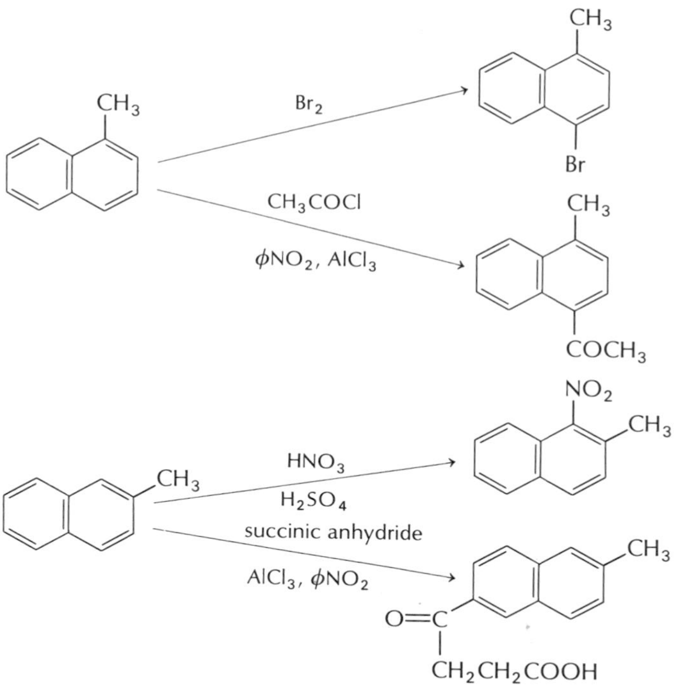

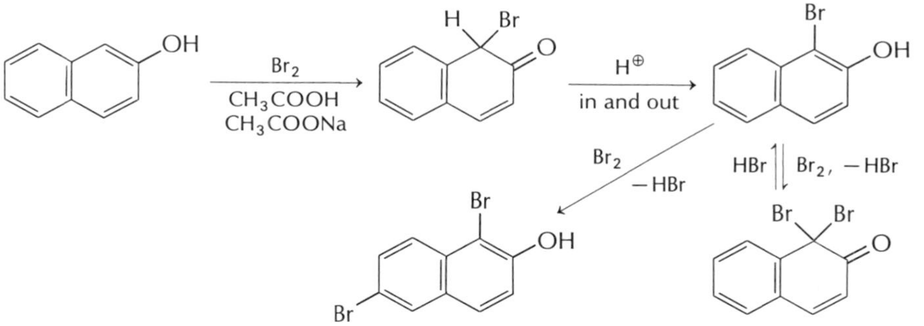

of CH_3COONa, HBr causes a slow reversal of a to 1-bromo-2-naphthol, which undergoes a slow reaction with bromine to give b. Although substitution of a halogen in a β (6) position is not common, a strong electron donor at the 2 position is able to conjugate nicely for substitution at 6 or 8. In brief, β-naphthol shows no tendency to direct to the 3 position (except for the Kolbe reaction of CO_2 with alkaline β-naphthol).

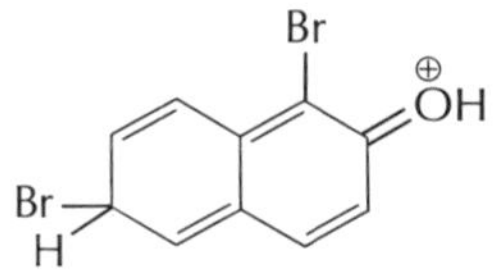

Exercises

25.1 Starting with naphthalene, show how to prepare:
   a. α-naphthoic acid
   b. β-napthoic acid
   c. α-naphthylamine

25.3 Naphthalene: additions

The addition reactions of naphthalene may be regarded as reductions or oxidations.

Reductions. Reduction is much easier than with benzene. Only 25 kcal of resonance energy is lost if one ring is hydrogenated. Sodium in ethanol reduces naphthalene to 1,4-dihydronaphthalene. If the reaction is carried out with sodium in isopentyl alcohol at the bp, the reduction may be continued to the tetrahydro stage, tetralin (Sec. 14.5). Catalytic hydrogenation proceeds easily to give tetrahydronaphthalene and somewhat more slowly thereafter (hydrogenation of a substituted benzene) to give decahydronaphthalene (decalin) (Eq. 25.10). In tetrahydrofuran, sodium or lithium donates an electron to naphthalene to form a radical anion.

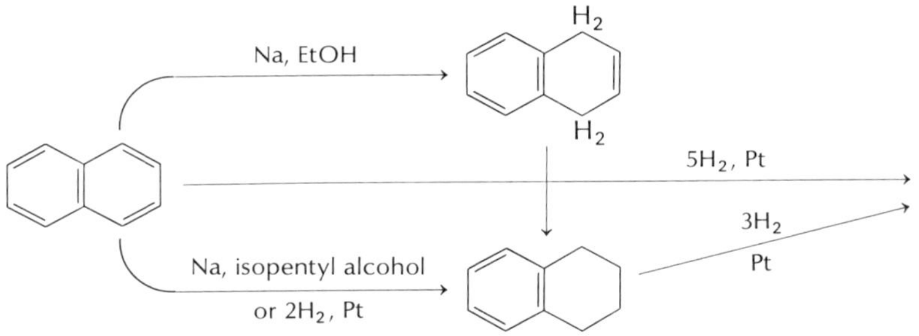

(25.10)

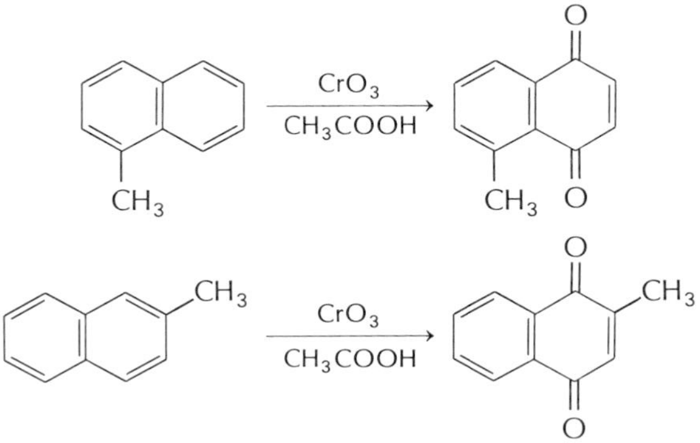

(25.11)

Oxidation. Naphthalene may be oxidized to α-naphthoquinone in low yield by chromic acid in acetic acid. Oxidation of α-methylnaphthalene gives 5-methyl-1,4-

naphthoquinone, and oxidation of β-methylnaphthalene gives 2-methyl-1,4-naphthoquinone (Eqs. 25.11). The reaction obviously proceeds by attack at an α position, presumably by initial addition (Eq. 25.12), although initial abstraction of α hydrogen also is possible.

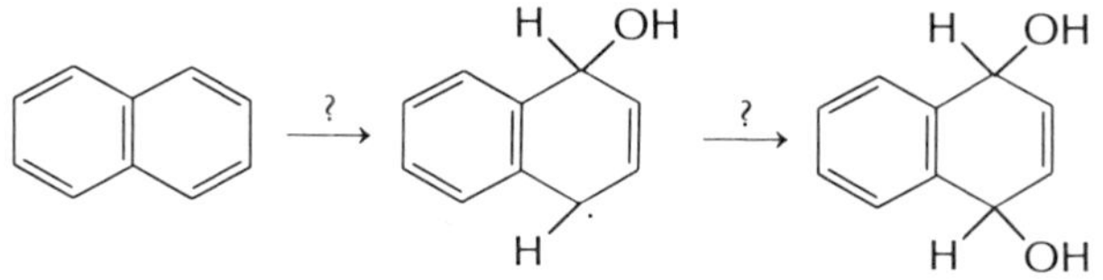

$$\text{naphthalene} \xrightarrow{?} \text{1-hydroxy-1,2-dihydronaphthalenyl radical} \xrightarrow{?} \text{1,4-dihydroxy-1,4-dihydronaphthalene} \quad (25.12)$$

Exercises

25.2 Decalin exists in two forms, *cis* and *trans* fused. Draw the structures.

25.4 Naphthalene: synthesis

Synthesis from benzene and succinic anhydride. Naphthalene may be synthesized from benzene and succinic anhydride as shown in Equation 25.13. All the reactions proceed

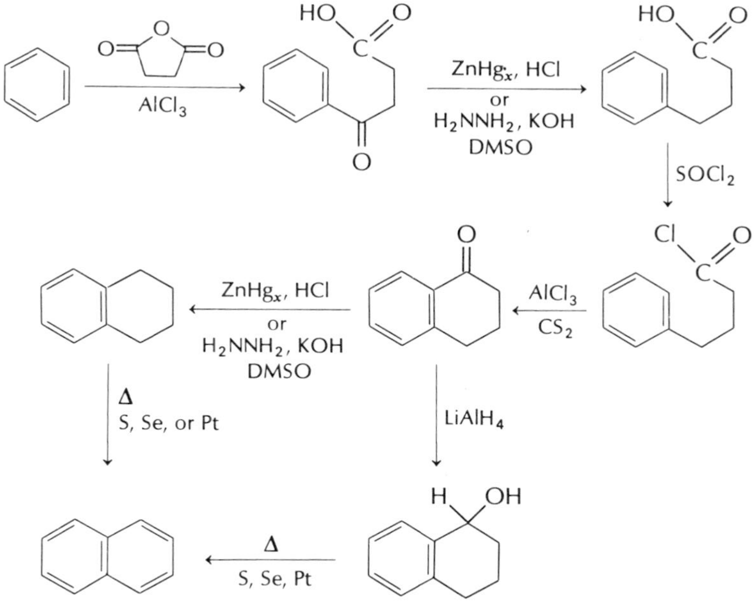

$$\text{(25.13)}$$

in excellent yield. The ketone, which deactivates the ring, is reduced (Clemmensen or Wolff-Kishner) in the second step to the activating —CH₂— to allow the cyclizing Friedel-Crafts reaction (step four) to go.

Variations, substituted naphthalenes. Although naphthalene itself is in good supply and need not be synthesized in this manner, the route is useful for the synthesis of various alkyl and aryl naphthalenes. If toluene is used instead of benzene, the product is β-methylnaphthalene; if biphenyl is used, the product is β-phenylnaphthalene. Another variation gives an α substituent from the intermediate α-tetralone (Eq. 25.14). Less satisfactory is the use of a substituted succinic anhydride,

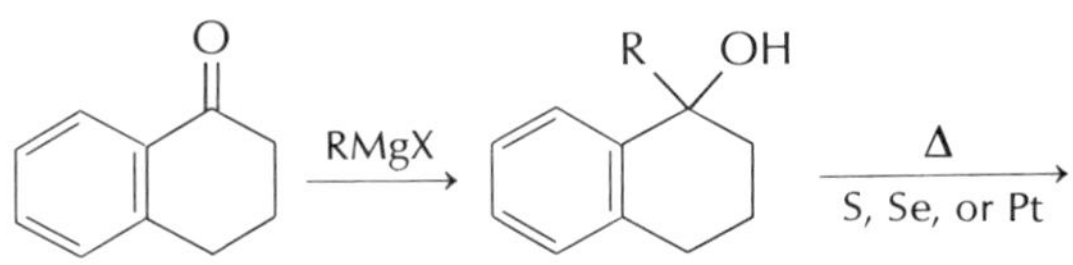

$$\text{α-tetralone} \xrightarrow{\text{RMgX}} \text{R, OH tetralin} \xrightarrow[\text{S, Se, or Pt}]{\Delta} \text{R-naphthalene} \quad (25.14)$$

which gives β substituents. For example, if we wanted 2,6-dimethylnaphthalene, we might try to prepare it as in Equation 25.15, whereupon we would find that both *a* and

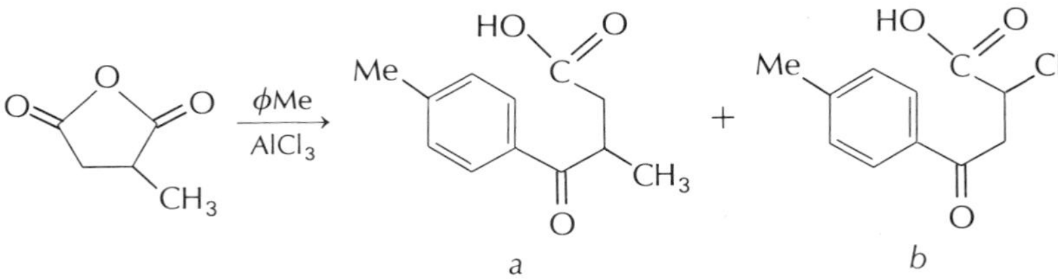

$$\text{2-methylsuccinic anhydride} \xrightarrow[\text{AlCl}_3]{\phi \text{Me}} a + b \quad (25.15)$$

b are formed, with more *b* than *a*. We should note that cyclization may occur in either direction on the ring but that it will not affect the results in this instance. Ketoacid *a* is what is needed to give the 2,6 isomer, whereas *b* gives the 2,7 isomer. In short, the substituted anhydride prefers to react at the less hindered carbonyl group. We can take advantage of the preference and proceed as shown in Equation 25.16. (We are

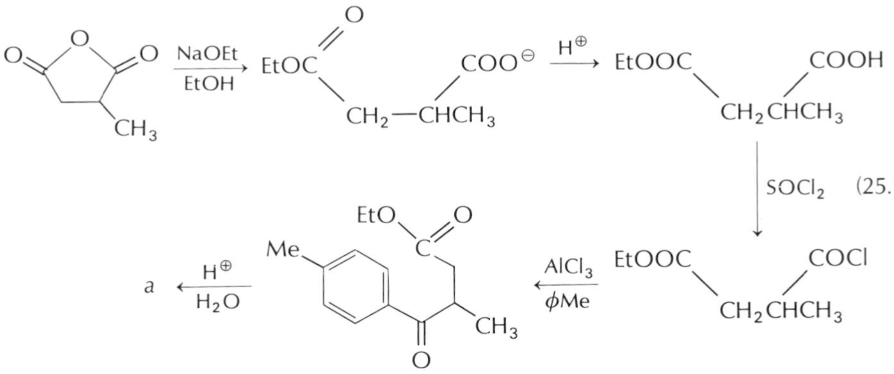

$$\text{anhydride} \xrightarrow[\text{EtOH}]{\text{NaOEt}} \text{EtOC-CH}_2\text{-CHCH}_3\text{-COO}^\ominus \xrightarrow{\text{H}^\oplus} \text{EtOOC-CH}_2\text{CHCH}_3\text{-COOH}$$

$$\downarrow \text{SOCl}_2 \quad (25.16)$$

$$a \xleftarrow[\text{H}_2\text{O}]{\text{H}^\oplus} \text{ketoester} \xleftarrow[\phi\text{Me}]{\text{AlCl}_3} \text{EtOOC-CH}_2\text{CHCH}_3\text{-COCl}$$

assuming that 2-methylsuccinic anhydride is available.) If we block the less hindered carbonyl group as an ester, the hindered acid group may be converted into the necessary acid chloride for the preparation of *a* in good yield. If the 2,7 isomer were wanted, then *b* could be obtained as in Equation 25.15.

Exercises

25.3 Assuming that the necessary substituted benzenes and succinic anhydrides are available, give a usable synthesis of:

a. 1,3-dimethylnaphthalene
b. 1,4-dimethylnaphthalene
c. 1,7-dimethylnaphthalene

25.5 Anthracene and anthraquinone

Occurrence; structure; naming; reactivity. Anthracene ($C_{14}H_{10}$), a linear tricyclic aromatic hydrocarbon, was the second substance to be isolated from coal tar (1832). The pure substance forms colorless crystals with blue fluorescence (mp 216°). The numbering of the molecule is suited to both anthraquinone and an analogy with naphthalene whereby the α and β positions have the same numbers. Bond fixation of the α-β bonds is even more

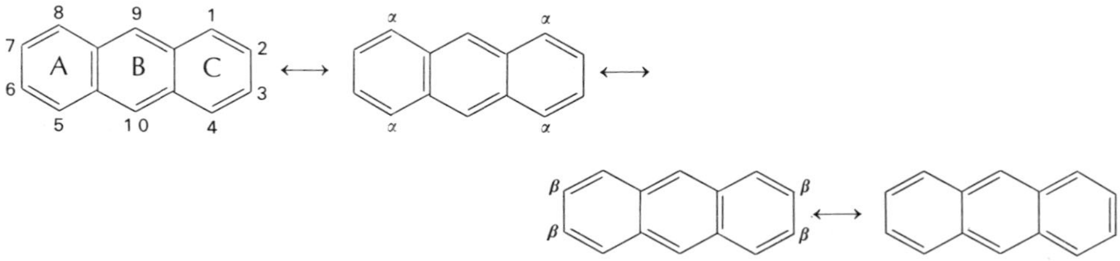

extreme than it is in naphthalene. Referring back to Table 25.1, we find that anthracene has a resonance energy deficit of -24 kcal, more than double that of naphthalene. The destabilization that results is shown by the reactivity of anthracene, particularly at the 9 and 10 positions.

Electrophilic substitution reactions are nearly always accompanied by or superseded by addition to the 9 and 10 positions. The mixtures that are produced cause trouble in purification of a given product, and we simply shall not discuss electrophilic substitution of anthracene. The reason for the reactivity at the 9 and 10 positions lies in the formation of two independent benzene rings with a total resonance energy of 72 kcal, only 12 kcal less than that of anthracene (Eq. 25.17). In essence,

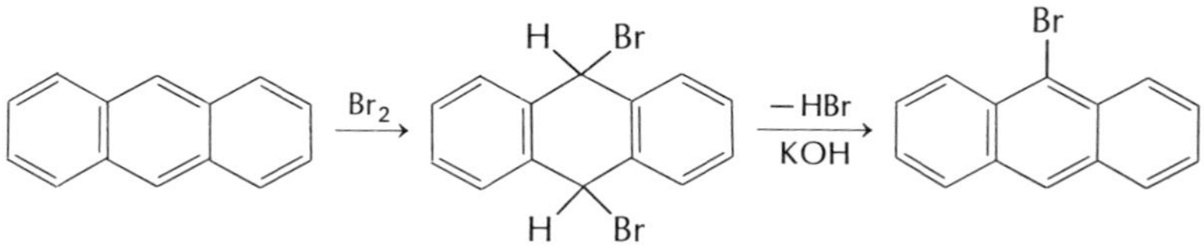

(25.17)

addition of bromine to the 9 and 10 positions is analogous to the 1,4 addition of bromine to butadiene, which has $\Delta H = -27$ kcal (calculated by the use of the BE table), more than enough to take care of the 12 kcal decrease in resonance energy for the 9,10 addition. The addition product, 9,10-dibromo-9,10-dihydroanthracene, may be isolated in good yield. Addition of a base causes a rapid dehydrobromination to give 9-bromoanthracene.

Anthracene may be reduced by sodium amalgam in alcohol to give 9,10-dihydro-anthracene. The reactivity at the 9 and 10 positions is also displayed in Diels-Alder reactions, in which anthracene reacts as a diene to dienophiles. Oxidation of anthracene gives the 9,10-quinone, known simply as anthraquinone (Eq. 25.18).

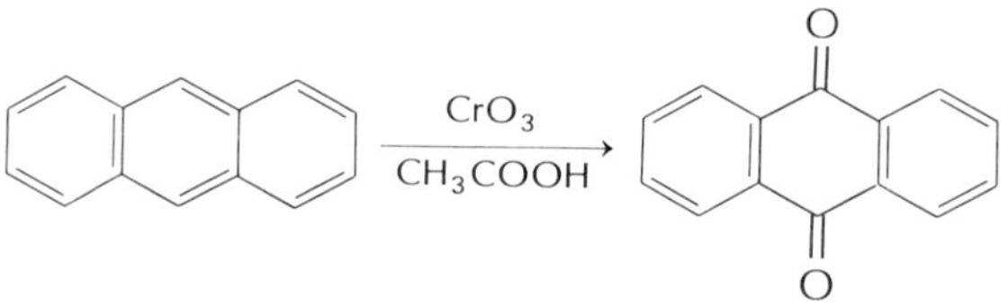

$$\text{anthracene} \xrightarrow[\text{CH}_3\text{COOH}]{\text{CrO}_3} \text{anthraquinone} \tag{25.18}$$

Anthraquinone: properties, stability. We have mentioned anthraquinone in several places previously, but we did not say much about it as a derivative of anthracene. Anthraquinone (mp 286°, bp 377°) is a stable crystalline substance that is very unreactive toward oxidative reagents. Even though quinones, in general, are reactive substances, anthraquinone is not. The end rings may be regarded as benzene rings with *meta*-directing substituents in *ortho* positions. The result is deactivation of all four open positions in each end ring. The central ring is already oxidized and is in a 1,4 quinoid state. Any addition reaction of the Michael type to the quinoid central ring would destroy the Kekulé resonance stability of an end ring, an uphill battle. Thus, anthraquinone resists all oxidations (the end rings are electron poor), electrophilic substitution reactions (the end rings are deactivated), and 1,4 quinone addition reactions.

Reduction of anthraquinone. Reduction of anthraquinone is possible, but only with strong reducing agents (Eqs. 25.19). Under alkaline conditions, the soluble disodium

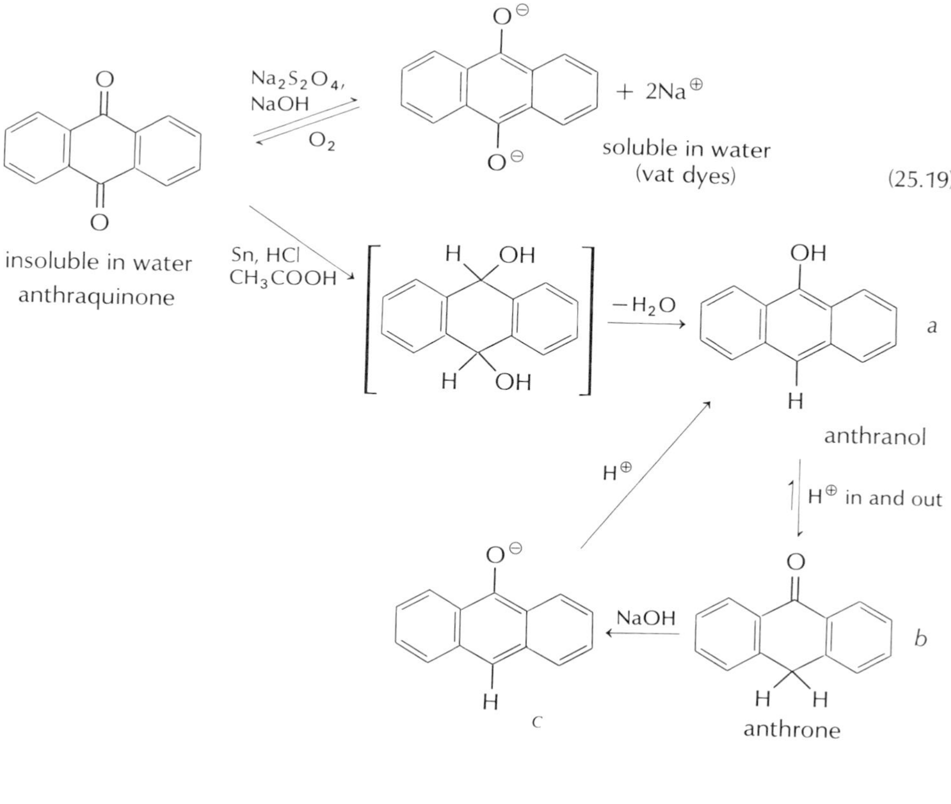

$$\text{anthraquinone (insoluble in water)} \underset{O_2}{\overset{Na_2S_2O_4, NaOH}{\rightleftarrows}} \text{disodium salt} + 2Na^{\oplus} \quad \text{soluble in water (vat dyes)} \tag{25.19}$$

Sn, HCl / CH₃COOH → [9,10-dihydroxy-9,10-dihydroanthracene] → (−H₂O) → anthranol (a)

H⁺ in and out ↕ anthrone (b)

NaOH ← (c)

salt of anthrahydroquinone is formed (vat dyes). It is easily oxidized by oxygen of the air back to the quinone. If zinc and NaOH are used, the disodium salt is further reduced after long reaction, first to the salt of anthranol (c), and finally to anthracene. Under acidic conditions (Sn and HCl in CH_3COOH), the quinone is reduced to anthranol (a), possibly by way of a diol. However, the presence of an acid catalyzes the attainment of the tautomeric equilibrium between anthranol and anthrone (b), in which b is favored. Thus, anthrone is the actual isolated product of the acidic reduction. Anthrone slowly dissolves in hot dilute NaOH to give c. Acidification of the solution at low temperature causes the relatively insoluble anthranol to precipitate. (Protonation at oxygen is faster than at carbon, similar to the protonation of the anion of a nitroalkane.) Anthranol behaves in phenolic fashion (coupling at the 10 position, etc.) whereas anthrone behaves somewhat as does benzophenone (reaction with RMgX, etc.). Both are easily oxidized back to anthraquinone.

Anthraquinone: substitution reactions. Electrophilic substitution of anthraquinone is possible only under drastic conditions. Thus, halogenation and the Friedel-Crafts reactions cannot be used. Under drastic conditions, nitration gives α-nitroanthraquinone in poor yield, because the poor control also allows the formation of 1,5- and 1,8-dinitroanthraquinones. The low solubilities of the high-melting components of the mixture make separation and purification difficult.

Sulfonation is easier to control, and the sulfonic acids are easier to separate and purify than are the nitro compounds. Even so, the reaction is carried out in such a manner that not all the anthraquinone is sulfonated (to avoid disulfonation). As with naphthalene, at an elevated temperature substitution is at the β position (Eq. 25.20). The presence of $HgSO_4$ in

$$\text{anthraquinone} \xrightarrow[\text{then}]{\substack{1.\ 18\%\ SO_3,\ H_2SO_4,\ 135° \\ 2.\ 66\%\ SO_3,\ H_2SO_4,\ 110°}} \text{2-}SO_3H\text{-anthraquinone} + \text{2,6 and 2,7 diacids} \quad (25.20)$$

	conversion	49%	14%
	yield	78%	22%

the sulfonation mixture causes the substitution to take place at an α position, possibly because a somewhat faster mercuration occurs at the α position, followed by a nucleophilic replacement of mercury by SO_3 (Eq. 25.21).

$$\text{anthraquinone} \xrightarrow[\text{then}]{\substack{1.\ 2\%\ HgSO_4,\ 20\%\ SO_3,\ H_2SO_4,\ 135° \\ 2.\ 60\%\ SO_3,\ H_2SO_4,\ 135°}} \text{1-}SO_3H\text{-anthraquinone} \quad (25.21)$$

81% + 19% 1,5 and 1,8 diacids.

$$\text{2-}SO_3Na\text{-anthraquinone} \xrightarrow[\text{NH}_4Cl,\ 200°]{NH_3,\ Na_3AsO_4} \text{2-}NH_2\text{-anthraquinone} \quad (25.22)$$

82%

The sulfonic acid group, either α or β, undergoes a Bucherer reaction (replacement by —NH_2) if an oxidizing agent such as an arsenate is present to oxidize displaced sulfite ion to sulfate (Eq. 25.22).

Anthraquinone: syntheses. Of the possible syntheses of anthraquinone (other than oxidation of anthracene) we shall give two. The first is similar to the synthesis of naphthalene from benzene, but with the use of phthalic anhydride instead of succinic anhydride (Eq. 25.23). As in the naphthalene synthesis, substituted benzenes and phthalic

(25.23)

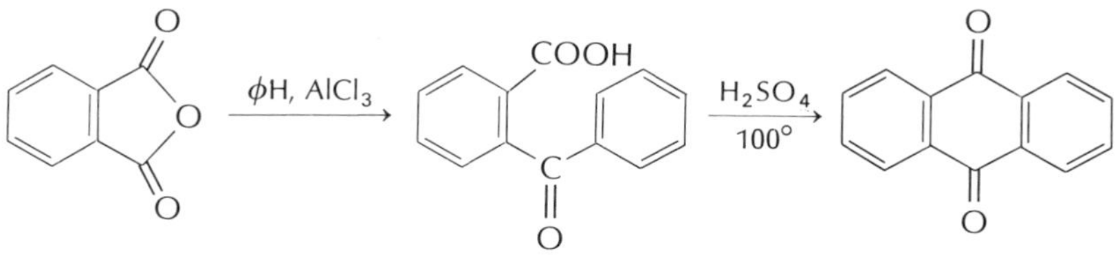

anhydrides lead to substituted anthraquinones. The remarkable cyclization step (an acid-catalyzed internal Friedel-Crafts reaction *ortho* to a *meta*-directing and deactivating keto group) may be catalyzed by other reagents ($AlCl_3$) or by acid chlorides at high temperature (ϕCOCl at 250°). It may well be that, despite a high activation energy (which requires an elevated temperature for the reaction to proceed at a reasonable rate), the stability of the products (anthraquinone and water) is great enough to give good yields. □ □

The second synthesis is by way of the Diels-Alder reaction (Eq. 25.24), starting from either benzoquinone or α-naphthoquinone.

(25.24)

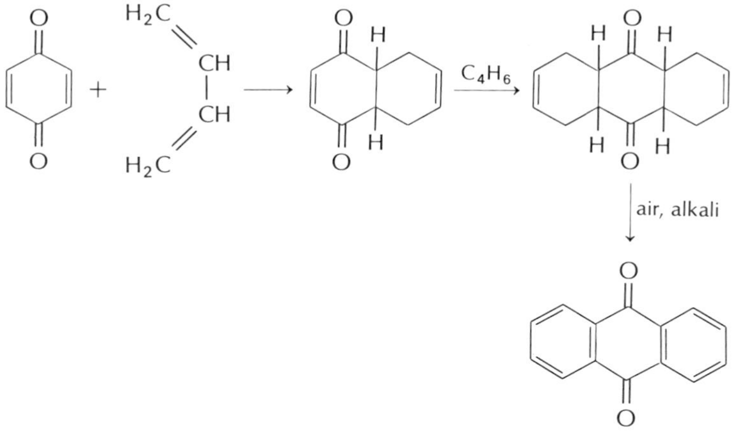

Exercises

25.4 Show how to prepare:
  a. a mixture of 1,6- and 1,7-anthraquinonedisulfonic acids
  b. 1,3-dimethylanthraquinone
  c. 2,3-dimethylanthraquinone

25.6 Phenanthrene

Occurrence; naming; bond fixation. Phenanthrene (from phenyl and anthracene) is the angular tricyclic isomer of anthracene. It is the second most abundant compound in coal tar (about 4%), but it is more difficult to isolate in a pure state because of a relatively low mp (101°) and greater solubility than that of anthracene.

Many uninformed graduate students have spent days and weeks trying to purify crude phenanthrene by recrystallization, only to find the mp getting worse and worse. Unfortunately, anthracene, the principal impurity in phenanthrene (they have bp only 14° apart: phenanthrene, bp 340°; anthracene, bp 354°) precipitates almost quantitatively with the first batch of crystals, which normally is the most pure in recrystallization. Fortunately, anthracene is more easily oxidized than phenanthrene, so a partial dichromate oxidation removes the anthracene as the nearly insoluble anthraquinone (some of the phenanthrene also is lost as phenanthraquinone). The remaining phenanthrene crystallizes satisfactorily.

Phenanthrene is numbered as shown, an exception to the rules sanctioned by

IUPAC. It is easy to remember, even when the compound is written in another orientation, if we note that the 9 and 10 positions are on the outside of the angle.

Resonance theory allows us to write five resonance structures, four of which have a double bond in the 9,10 position. The resulting fixation of the 9,10 bond as being almost a double bond (1.8 bonds) permits phenanthrene to undergo several alkenelike (or *cis*-stilbenelike) reactions. The resonance energy is 91 kcal, with only a −16 kcal deficit. We may thus say that phenanthrene is more aromatic than is anthracene.

Reactions. Electrophilic substitution reactions unfortunately proceed with little selectivity for location of attack and frequently give mixtures of all five possible monosubstituted isomers (the 1, 2, 3, 4, and 9) as well as some of the twenty-five possible disubstituted isomers. The Friedel-Crafts acylation is an exception, and the 3 isomer may be obtained in yields of 50 to 60%.

The reduction, oxidation, and addition reactions of phenanthrene take place at the alkenelike 9,10 bond. Thus, phenanthrene is easily hydrogenated to give 9,10-dihydrophenanthrene or oxidized to give phenanthrenequinone (Eqs. 25.25). Phenanthrenequinone may be oxidized further to diphenic acid in good yield (70%). Although the quinone resembles an *o*-quinone, the reactions resemble those of an α-diketone (benzil). One reaction, the benzilic acid rearrangement (Sec. 16.5) is of use for gaining access to the fluorene system (Eq. 25.26). If phenanthrene is oxidized with $KMnO_4$ in alkaline solution, the final product is fluorenone; the four steps are oxidation to quinone, rearrangement, decarboxylation, and oxidation of fluorenol to fluorenone.

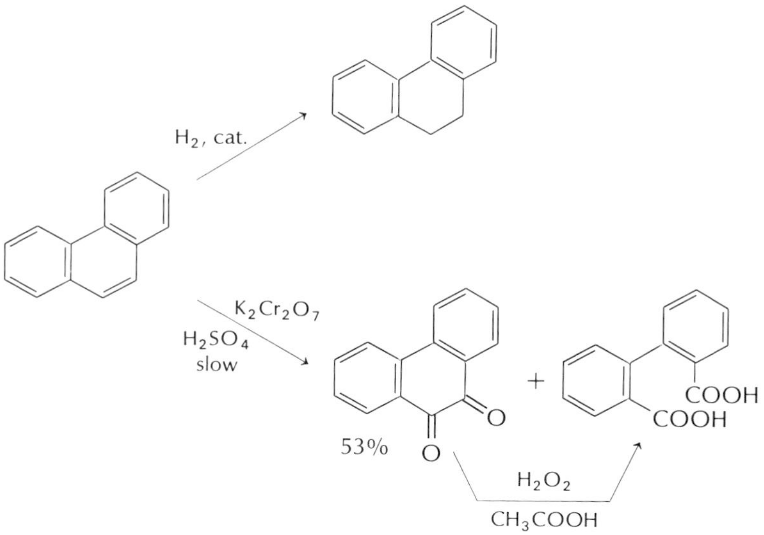

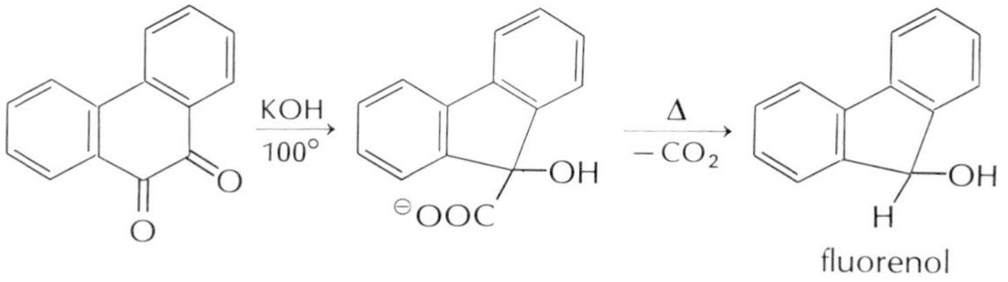

Bromine in acetic acid adds to phenanthrene at the 9 and 10 positions (alkenelike addition). If FeBr$_3$ (or other polar catalyst) is present, substitution takes place instead, and 9-bromophenanthrene is formed. Furthermore, the addition of FeBr$_3$ to a solution of the addition product gives 9-bromophenanthrene (Eqs. 25.27).

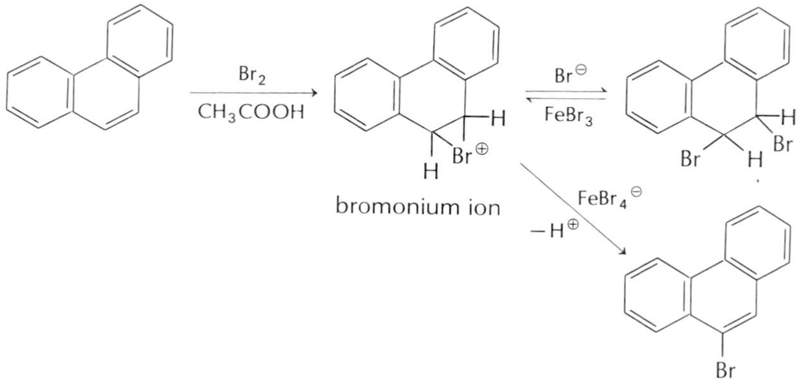

Synthesis. There are several synthetic routes to phenanthrenes. One is a variation of the naphthalene synthesis from succinic anhydride and benzene. If we start with succinic anhydride and naphthalene, phenanthrene is the product (Eq. 25.28). The keto acids can be separated (with some labor) but the following steps are alike. The mixture itself may be put through these steps if purification is delayed to the end of the sequence. The cyclization step always goes in the direction that gives the tetrahydroketophenanthrene, never to the linear isomer (which would lead to anthracene). As with the naphthalene synthesis, many variations can be done to obtain various substituted phenanthrenes.

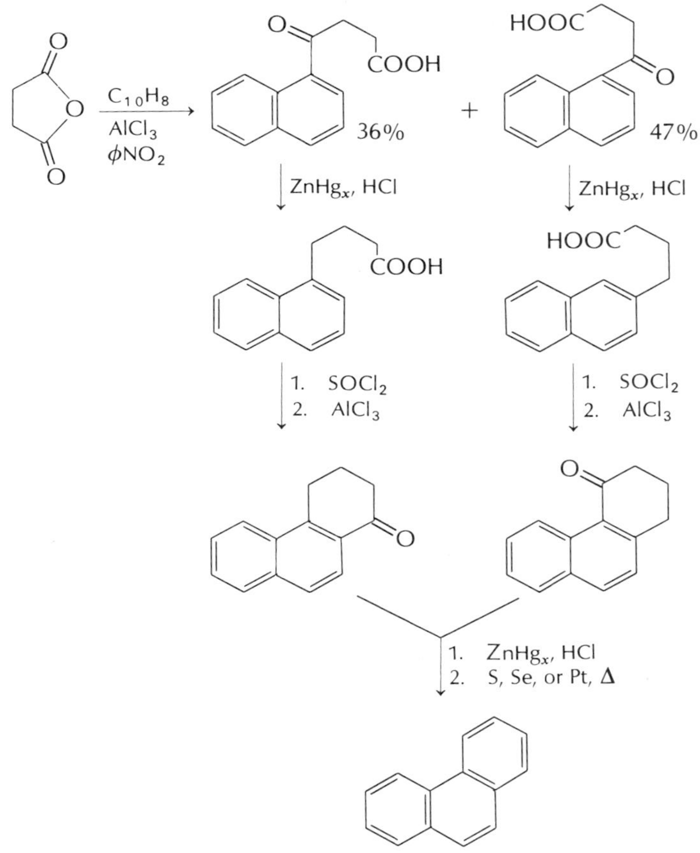

(25.28)

Exercises

25.5 Show how to synthesize 9-methylphenanthrene.

25.6 Starting with phenanthrene and any other needed reagents, show how to prepare:
  a. 9-phenanthrenecarboxylic acid
  b. 9-phenanthrol

25.7 Other polycyclic compounds

In this section, we take a passing glance at some of the other polycyclic compounds isolated from coal tar.

Fusions with cyclopentane rings. Those compounds in which a five-membered ring is present are not completely aromatic and properly should be called arenes. The

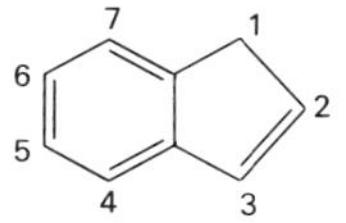

indene 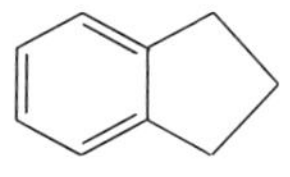   indane

simplest of these is indene (C_9H_8). Indene is easily reduced to indane (also called hydrindene). Oxidation gives phthalic acid. In electrophilic substitution reactions, the compounds resemble *o*-xylene.

A widely used reagent for detection of amino acids is ninhydrin, which gives a blue color. Ninhydrin is 1,2,3-triketoindane hydrate, one preparation of which is shown in Equation 25.29. Dimethylsulfoxide, upon condensation with diethyl

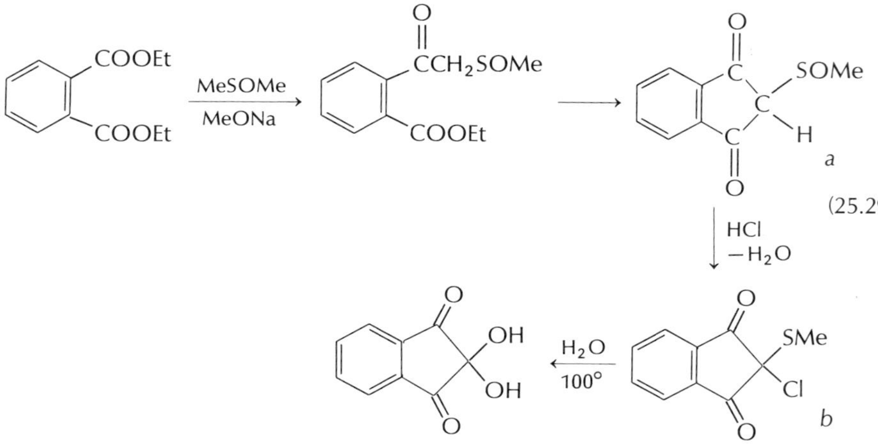

(25.29)

phthalate (mixed Claisen), gives the skeleton of the product (*a*), which enolizes in the presence of HCl and undergoes reduction of the sulfoxide group and takes up $Cl^\ominus$ to give *b*. Hydrolysis of *b* gives ninhydrin.

Fluorene ($C_{13}H_{10}$) is similar to phenanthrene in some respects. It is easily oxidized to fluorenone, which may be compared to benzophenone. The acidity of cyclopentadiene is relatively high (pK_a 16), strong enough to react with a Grignard reagent (cyclopentadienyl

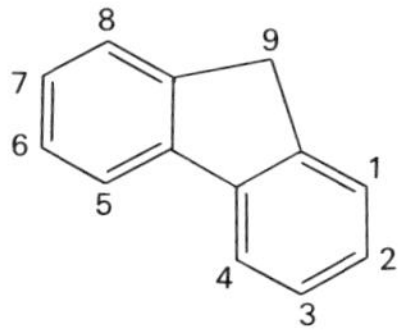

fluorene 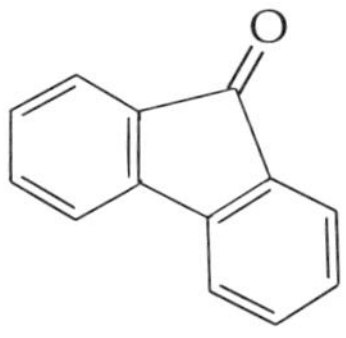   fluorenone

anion is aromatic, Chap. 14). Fusion with one benzene ring (indene) lowers the acidity to pK_a 21; with two rings (fluorene), to pK_a 25 (about the same as HC≡CH or CH$_3$C≡N).

Acenaphthene (C$_{12}$H$_{10}$) resembles naphthalene in many ways, including the

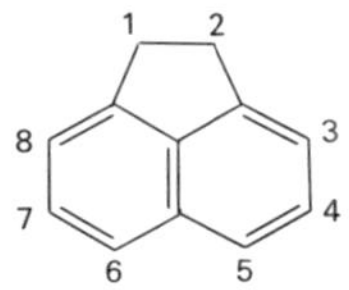

acenaphthene 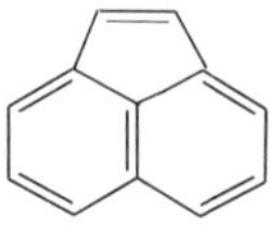   acenaphthylene 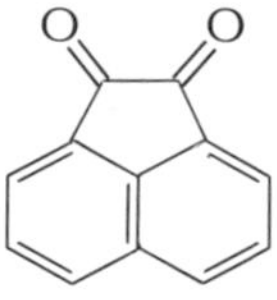   acenaphthoquinone

odor. The fusion of the five-membered ring into the 120° angle of naphthalene (two sides) introduces considerable strain energy into the molecule, and the 1–2 bond is longer than usual. Both the 1 and 2 positions are also benzylic in type. Thus, free-radical bromination goes with ease; and removal of HBr gives acenaphthylene, with the even more strained 1–2 double bond. The orange-red compound is very reactive. Oxidation of acenaphthene with the calculated amount of chromate in acetic acid goes readily to give a fair yield of the brilliant yellow α-diketone called acenaphthoquinone. Unlike benzil or phenanthraquinone, acenaphthoquinone does not give the benzilic acid rearrangement. Oxidation of acenaphthene or of the quinone with excess chromate gives 1,8-naphthalenedicarboxylic acid, called naphthalic acid. Naphthalic acid, though soluble in bases, is otherwise quite insoluble and also dehydrates with ease to give the even more stable anhydride. Naphthalic anhydride

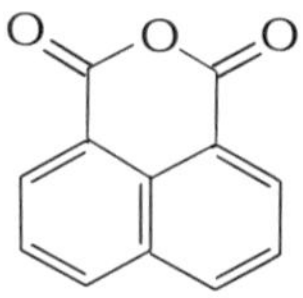

naphthalic anhydride

is one of the few organic compounds that may be dissolved in boiling nitric acid and will then crystallize nicely on cooling.

Polycyclic fused benzene ring systems. There are many fused ring systems consisting of four, five, six, or more benzene rings. A few examples are pictured. Some of the substances are carcinogenic: dibenzanthracene is one. Two others are benzpyrene and methylcholanthrene. These substances seem to act only slowly: prolonged exposure is necessary.

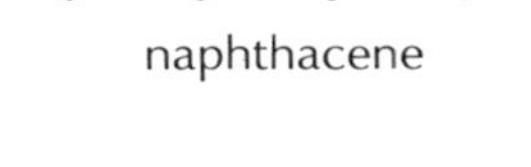

naphthacene 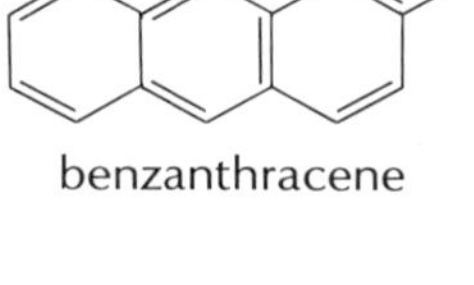   benzanthracene 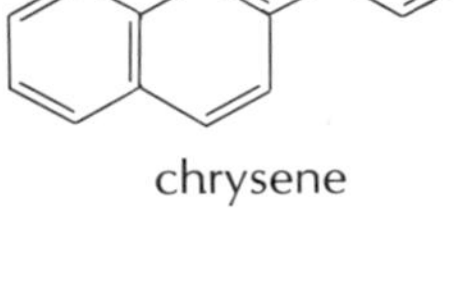   chrysene

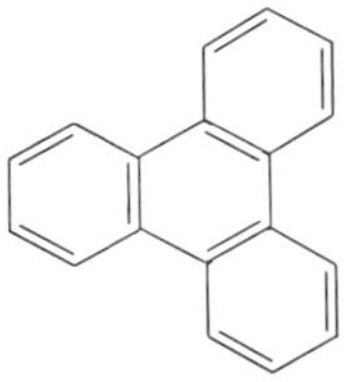

triphenylene

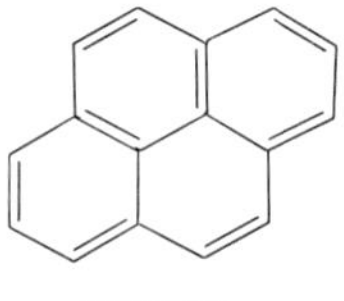

pyrene

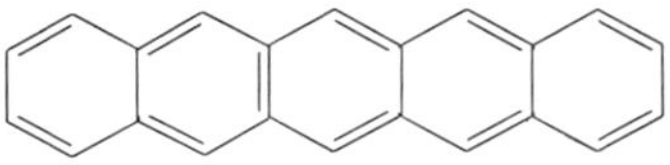

pentacene

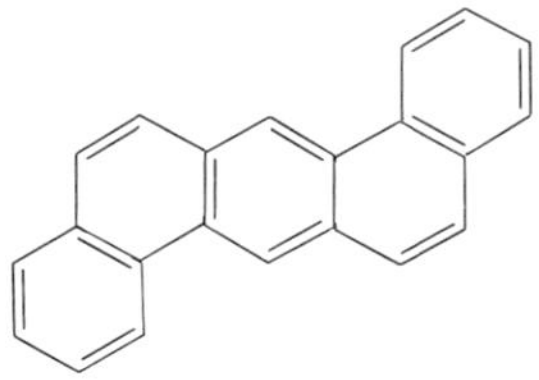

dibenzanthracene

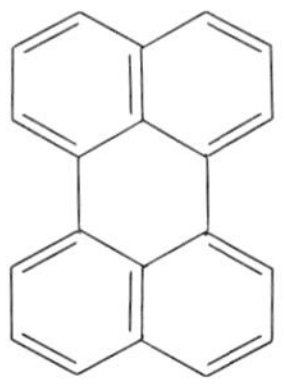

perylene

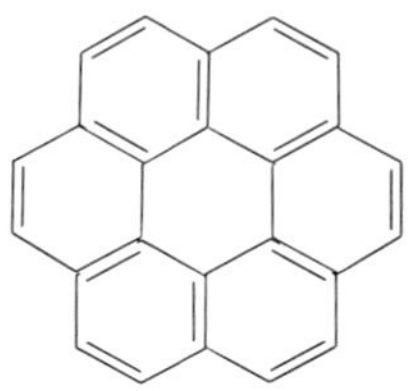

coronene

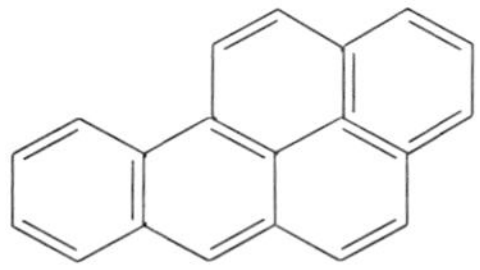

benzpyrene

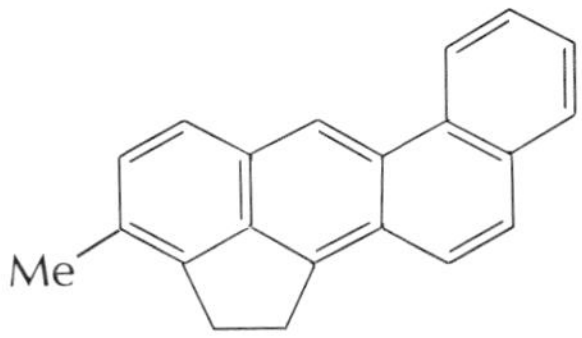

methylcholanthrene

25.8 Biphenyl

Resonance energy: conjugation between the rings. Biphenyl and substituted biphenyls are nonfused aromatic systems with interesting properties. For one thing, biphenyl is more aromatic than two benzene rings if we go by resonance energies: that of biphenyl is about 7 kcal greater than that of two molecules of benzene. This fact has been interpreted to mean that there is some contribution of the conjugated structures shown, which also favor coplanarity of the two rings. In the crystalline state, biphenyl is planar.

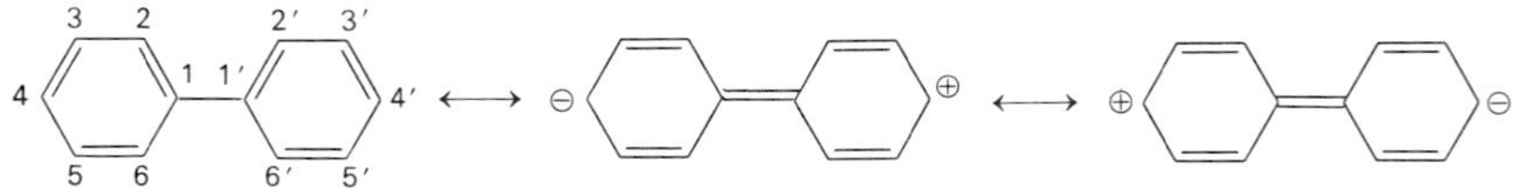

As a vapor, however, the angle between the planes of the two rings (about the long axis of the molecule) is about 40°. Thus there is some repulsion between the 2, 2′, and 6, 6′ hydrogens. The molecule then adopts a conformation that is between maximum relief from repulsion and maximum conjugation of the completely planar structure. In 1922, a substituted biphenyl was resolved into the two component enantiomers.

Chirality of substituted biphenyls. The subject of biphenyl chirality has been extensively studied and shown to be the result of restricted rotation about the axis of the bond between the rings, provided that the rings are unsymmetrically substituted. The compound 6,6'-dinitro-2,2'-biphenyldicarboxylic acid (also known as 6,6'-dinitrodiphenic acid) is shown in the (R) and (S) configurations. (See Sec. 6.3 for naming by the sequence rules.)

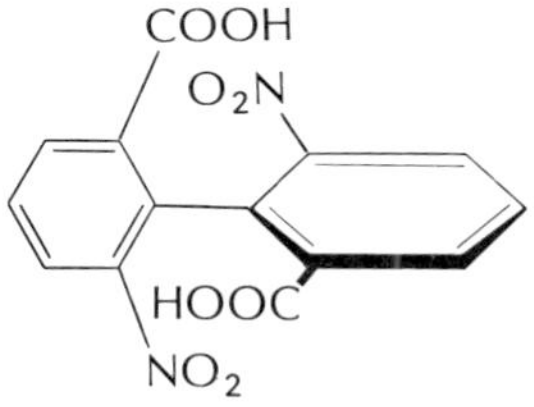

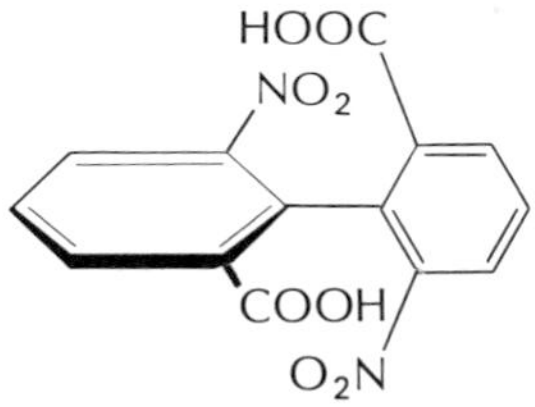

The configuration is determined as follows.

1. View the molecule along the long axis. (It makes no difference from which end.)
2. Assign sequence numbers to the four groups causing the chirality. The nearer pair always takes priority over the pair farther away.
3. If going from 2 to 3 is clockwise, the configuration is (R). If going from 2 to 3 is counter-clockwise, the configuration is (S).

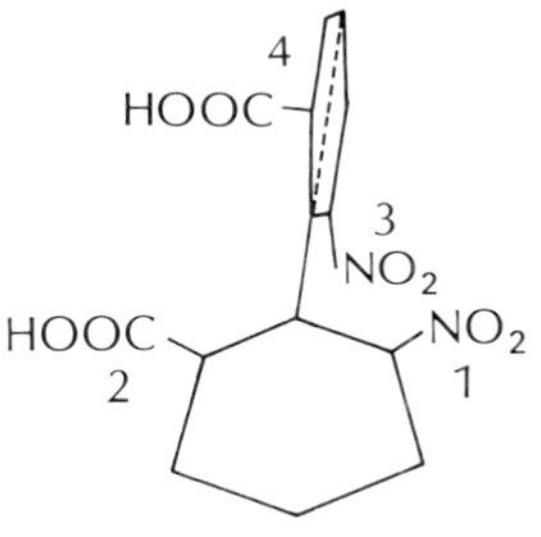

For racemization to occur, a given molecule must be able to assume a coplanar form from which it can rotate with equal probability into the (R) or (S) configuration. (The use of models is a great help here.) Coplanarity, in this case, would require crowding of one —COOH group against a —COOH or —NO$_2$ on the other ring, and simultaneously of —NO$_2$ against —NO$_2$ or —COOH on the other ring. Although difficult, it is not impossible, and the rate of racemization increases with temperature. (Of course, the rate of racemization at room temperature must be slow enough in the first place to allow time for the resolution to take place.)

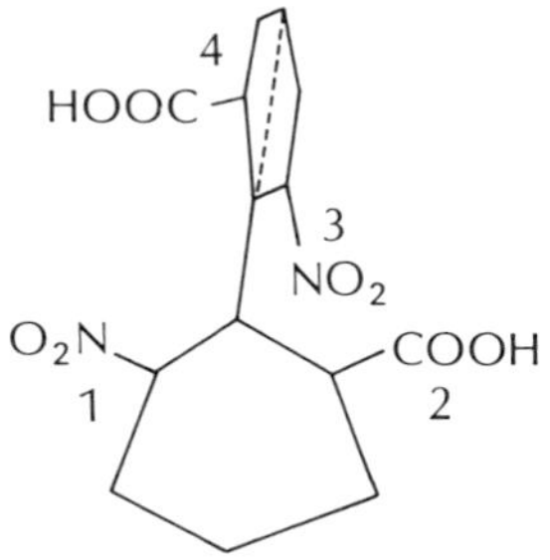

The rates of racemization, assumed to be proportional only to the relative sizes of the groups in the 2, 2', 6, and 6' positions, were used to estimate the size of substituents before the far more accurate X-ray measurements were applied to the problem. It was soon found that certain groups had trouble getting past hydrogen, so that only the 2 and 2' positions needed to be substituted, as in 2,2'-biphenyldisulfonic acid (a). Later on, a single large substituent ($-\overset{\oplus}{\text{A}}\text{sMe}_3$) was found sufficient (b). We must remember the need for unsymmetrical

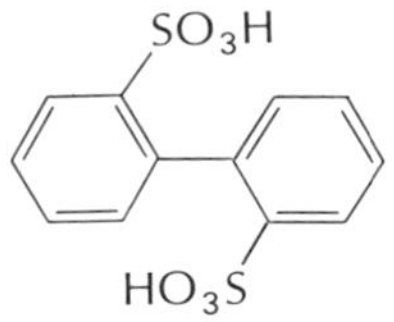

a 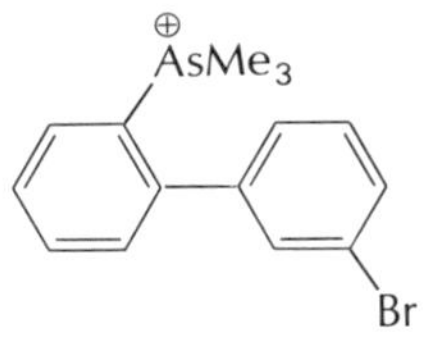   b

substitution in each ring: hence, there must be at least one *m* substituent in the other ring (b). If one of the rings is symmetrically substituted, the plane of the other ring bisects the symmetrically substituted ring, and the molecule is superposable upon the mirror image.

Exercises

25.7 Biphenyl has ε_{max} = 18,000 at 250 nm, but 2,2',6,6'-tetrachlorobiphenyl has no absorption band at 250 nm. What is your interpretation?

25.8 Which member of the pairs shown will racemize faster?
   a. 6,6'-dichlorodiphenic acid or 6,6'-diiododiphenic acid
   b. 6,6'-dimethyldiphenic acid or 6,6'-di-*tert*-butyldiphenic acid
   c. 1-(2,6-dimethylphenyl)-2-naphthoic acid or
      2-(2,6-dimethylphenyl)-1-naphthoic acid

☐ ☐

25.9 Summary

As we suggested in the introduction, very few new reactions were introduced in this chapter, although many familiar ones were used in new applications. We have tried to relate most of the material to what we have already learned in previous chapters.

25.10 Problems

25.1 Substance A ($C_{10}H_8$) reacts with H_2SO_4 at 40° to give B. Heating B with NaOH and work-up gives C. C reacts with diazotized sulfanilic acid to give D. D with Sn and HCl, gives E and sulfanilic acid. E reacts with $NaNO_2$ and concentrated HCl to give a solution that decomposes upon being heated to give F. F reacts with diazotized sulfanilic acid to give G. G with Sn and HCl gives H. H reacts with one mole of acetic anhydride to give I. Mild oxidation of I gives J ($C_{12}H_9O_3N$).
   Give a structural formula for each lettered substance.

25.11 Bibliography

The most complete discussion of the polycyclic aromatic compounds (except for biphenyl, ferrocene, and spectra) is in Fieser and Fieser (Chap. 7, ref. 7). Noller (Chap. 7, ref. 11) also is good. The other texts seldom have more details than we have given here (except, of course, for the advanced texts).

26 Terpenes

Most terpenes occur in plants. Originally, a terpene meant a volatile oil derived from conifers. As knowledge grew, the term was applied to more and more varieties of oils that could be obtained from plant sources by pressure, extraction, or steam distillation. Finally it was recognized that many of the volatile oils, whether hydrocarbon, alcohol, ketone, or aldehyde, had ten carbon atoms. (The aromatic odoriferous substances, such as cinnamaldehyde, methyl salicylate, etc., are exceptions.) The terpenes look as if they are constructed from units of the five carbon diene isoprene (2-methyl-1,3-butadiene). Isoprene itself does not occur in nature, so the terpenes arise from some other source. Nevertheless, once chemists started looking for isoprene units in structures, they found a host of compounds that, on paper at least, could be dissected into them. We use the term terpene to include all isoprenoid compounds.

26.1   Terpenes   847
Classification; terminology; origins.

26.2   Acyclic terpenes   847
1. Open-chain terpenes.   2. The isoprene rule.   3. Alcohols, aldehydes, and ketones.   4. Synthesis of citral.

26.3   Cyclic terpenes   850
1. Cyclization of acyclic terpenes: synthesis of menthol.   2. Limonene; dipentene.   3. Exceptions to the isoprene rule: the ionones and irones. Exercises.

26.4   Bicyclic terpenes   852
1. The bicyclic systems.   2. α-Pinene and rearrangements.   3. Camphor: reactions; synthetic camphor. Exercises.

26.5   Sesquiterpenes   855
1. Acyclic sesquiterpenes: farnesol, cyclization.   2. Azulenes: synthesis of azulene. Exercises.

26.6   Diterpenes   857
1. Acyclic diterpenes: phytol, synthesis.   2. Cyclic diterpenes; Vitamin A, synthesis.   3. Chemistry of vision.   4. Tricyclic diterpenes: abietic acid. Exercises.

26.7   Triterpenes and tetraterpenes   860
       1. Squalene: occurrence, synthesis.   2. Carotenes: occurrence, oxidation.

26.8   Polyterpenes   862
       1. Rubber.   2. Gutta-percha and balata; chicle.   3. Technology of rubber: vulcanization. Exercises.

26.9   Summary   864

26.10 Problems   864

26.11 Bibliography   864

26.1 Terpenes

Classification; terminology; origins.   As we have said, we shall consider any compound that can be dissected into isoprene units as a member of the class of terpenes. We say that these compounds follow the isoprene rule. We shall also include a few exceptions.

Within the classification of terpenes, we call the C_{10} compounds terpenes, the C_{15} compounds sesquiterpenes, the C_{20} compounds diterpenes, the C_{30} compounds triterpenes, and so on. The terpenes, as a class, exist as open-chain compounds and as cyclic, bicyclic, tricyclic, and higher compounds with a variable number of double bonds and with or without an alcohol or carbonyl function present. It all sounds complicated, and it was a nightmare for a long time to the chemists who ventured to do research in the field of terpenes. However, as we pick our way along, many similarities will become apparent. Also, no new reactions will be needed, only applications of some that we already know.

The volatile oils, usually distillable with steam, also are known as essential oils. Their composition is a characteristic of the source and may vary from one part of a plant to another and from season to season. Nevertheless, most essential oils are mixtures of many component terpenes in which one, two, or three may predominate. In many instances the presence of a small amount of a given terpene gives a particular oil its unique odor or other characteristic. Much of the earlier work on the terpenes was hampered by an inability to isolate pure components. The development of preparative vapor-phase chromatography has made isolation of pure components from natural products a relatively simple operation.

26.2 Acyclic terpenes

Open-chain terpenes.   Isoprene has the molecular formula C_5H_8, and a dimer has the formula $C_{10}H_{16}$. The formula for decane is $C_{10}H_{22}$. Thus, the dimer is short of saturation by $3H_2$, which indicates that it has three double bonds, two double bonds and a ring, one double bond and two rings, or three rings. (Triple bonds do not occur often in the isoprenoids.) We begin with the $C_{10}H_{16}$ compounds with three double bonds.

The isoprene rule. Simplified structures for terpenes are shown for ocimene (from basil) and myrcene (from bayberry). Systematic naming and numbering seldom is used in terpene chemistry because so much of the literature on the subject uses common names and numbering that was devised for specific compounds before many structures had been proved. We have written the acyclic trienes in a specific form for a reason. As far as possible, we shall adhere to the form shown for the various cyclic terpenes and for the sesqui- and diterpenes. The dashed line shows dissection according to the isoprene rule. In applying the rule, we look only for the ⟩—⟨ unit, neglecting the number and positions of double bonds and hydroxyl and carbonyl functions.

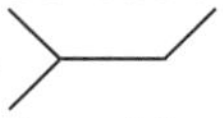

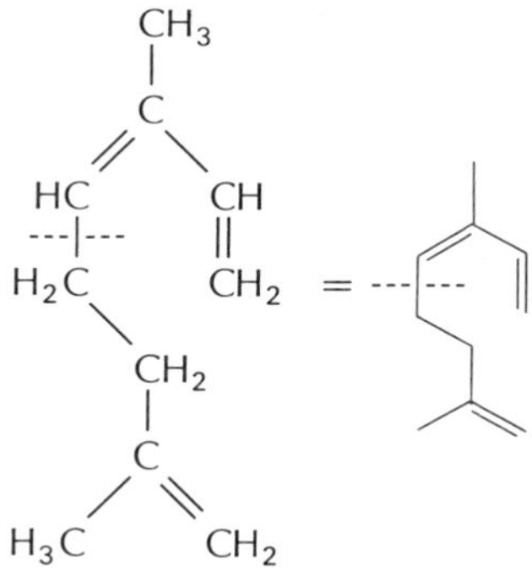

ocimene,
2,6-dimethyl-1,5,7-octatriene, or
3,7-dimethyl-1,3,7-octatriene

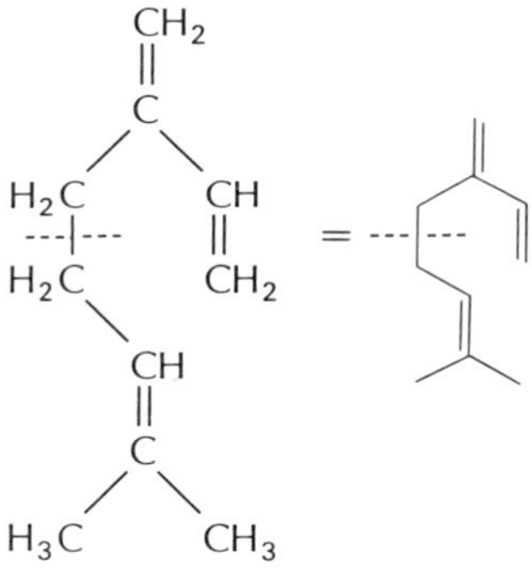

myrcene,
7-methyl-3-methylenyl-1,6-octadiene

Alcohols, aldehydes, and ketones. Geraniol (rose oil, lemon grass oil) and nerol (orange blossom oil) are alcohols with only two double bonds. They differ only in that geraniol is *trans* or (*E*) and nerol is *cis* or (*Z*) about the allylic double bond. The corresponding

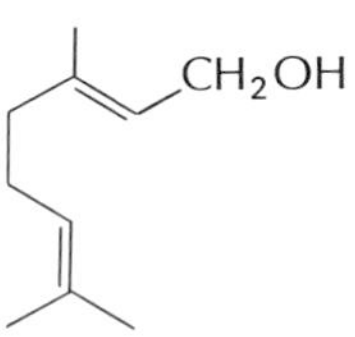

geraniol

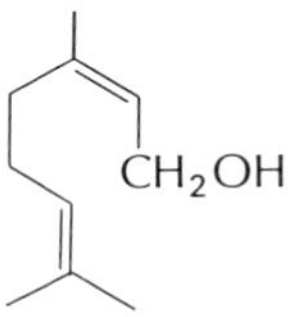

nerol

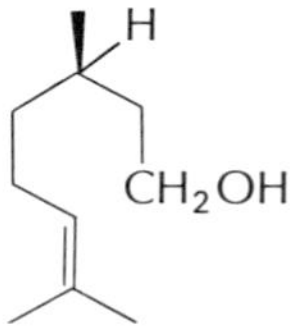

(*R*)-(+)-citronellol

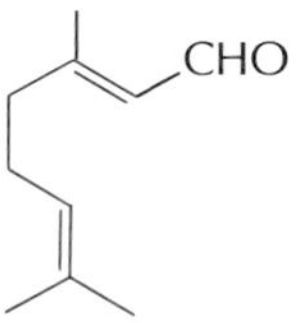

geranial

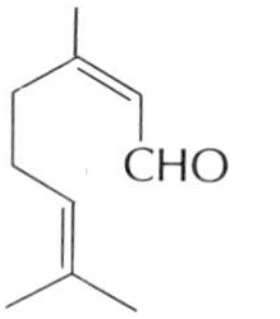

neral

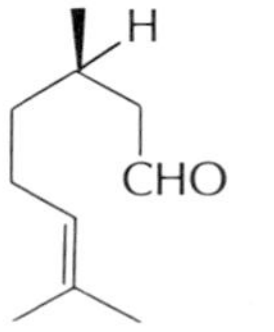

(*R*)-(+)-citronellal

26.2 Acyclic terpenes

aldehydes, geranial and neral, are the principal constituents of lemon grass oil (about 80%). Distillation gives a 90:10 mixture of the two aldehydes, called citral. Geranial is sometimes called citral-a, and neral, citral-b. Very closely related are two of the components of citronella oil. Citronellol and citronellal differ from the preceding alcohols and aldehydes in having the allylic double bond hydrogenated. The C_3 becomes asymmetric. The (R)-(+) aldehyde is the major component of citronella oil; the (S)-(−), of lemon grass oil from Java. The (S)-(−) alcohol is present in rose oil.

Synthesis of citral. A synthesis of citral (geranial) is shown in Equation 26.1. Note that if *a* is dehydrated, isoprene is the product. The rest of the structure is built up with three carbon atoms from the acetoacetic ester (from diketene) and two from acetylene.

26.3 Cyclic terpenes

Cyclization of acyclic terpenes: synthesis of menthol. The cyclic terpenes may be regarded as cyclization products of the open-chain terpenes. Citronellal, upon treatment with an acid, undergoes ring closure to give the cyclic alcohol isopulegol (Eq. 26.2). This compound is present in pennyroyal, a member of the mint family, as is the oxidation product, isopulegone. Catalytic reduction of isopulegol gives menthol (Eq. 26.3), as does reduction of the phenol thymol. With three asymmetric centers in the cyclohexane ring, there are $2^3 = 8$ stereoisomers. Menthol is the racemic pair in which all three groups are equatorial (the Me-, iPr-, and hydroxy).

Limonene; dipentene. Possibly the most widespread of the cyclic terpenes is limonene, both the (+) and (−) forms of which occur naturally. Racemic limonene may be formed by a Diels-Alder reaction of isoprene with itself (Eq. 26.4). It is known as dipentene. As with other cyclohexenes (Eq. 14.3), heating with a hydrogenation catalyst causes limonene to disproportionate into *p*-cymene and *p*-menthane (Eq. 26.5).

Exceptions to the isoprene rule: the ionones and irones. Some exceptions to the isoprene rule are the ionones (with thirteen carbon atoms) and irones (with fourteen carbon atoms). The ionones (in violets) may be synthesized from citral (Eq. 26.6). The aldehyde undergoes an aldol condensation with acetone to give the conjugated ketone (pseudo-ionone). Pseudoionone, in the presence of a strong acid (H_2SO_4 or BF_3 in acetic acid), closes to give a cyclohexene (α-ionone, kinetic control) which isomerizes more slowly to give β-ionone, the thermodynamically favored conjugated product. The most abundant irone

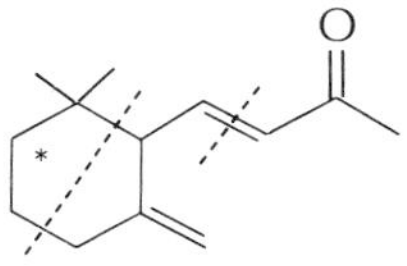

γ-irone

is γ-irone, which has the structure shown. The other irones differ in the location of the double bond and in the asymmetry of the odd methyl group.

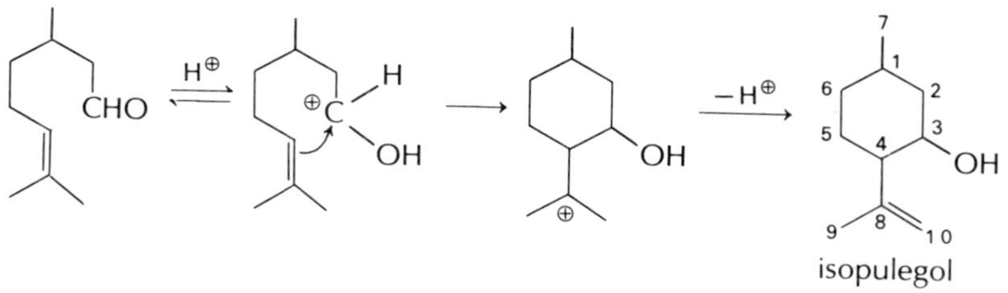

26.3 Cyclic terpenes

(26.2)

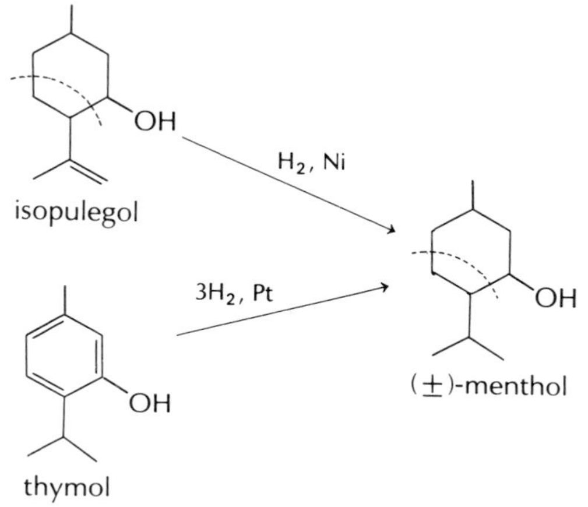

(26.3)

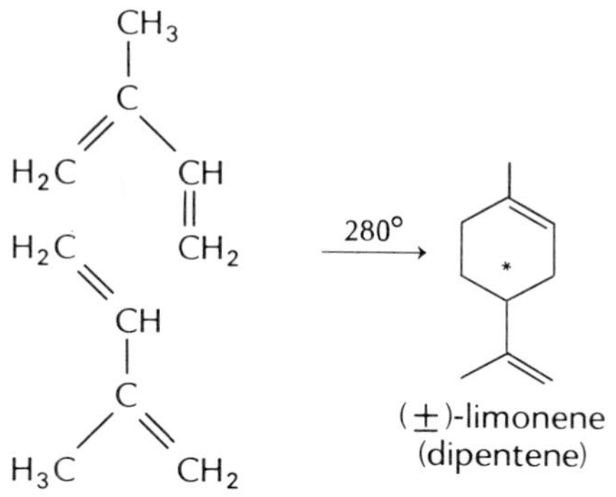

(26.4)

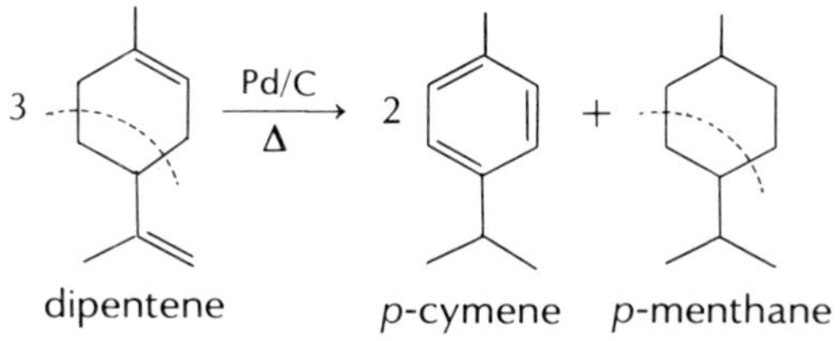

(26.5)

26.4 Bicyclic terpenes

The bicyclic systems. Most of the bicyclic terpenes may be looked at as monocyclic terpenes in which the isopropyl group has been joined to the three possible positions in the cyclohexane ring to give a [4.1.0], [3.1.1.], or [2.2.1] bicycloheptane ring system. (See Supplement for naming of bicyclics.) The other bicyclic terpenes may be regarded as rearrangements of the ring structures shown (camphene, Eqs. 26.7).

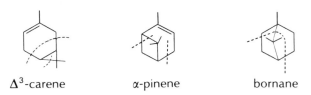

Δ^3-carene   α-pinene   bornane

α-Pinene and rearrangements. Probably the most widespread terpene is α-pinene, which is also the principal constituent of turpentine, the volatile component of conifers. At low temperatures (0°), α-pinene adds HCl without rearrangement to give pinene hydrochloride (Eqs. 26.7). At room temperature, an acid adds a proton to the double bond in pinene to give the cation *a*, which rearranges to the secondary cation *b* (relief of strain in four-membered ring). This species can accept a molecule of water from the *endo* direction to give borneol or from the *exo* direction to give isoborneol. On the other hand, a may open up to give c, which goes on to give α-terpineol, the principal product if α-pinene is treated with dilute sulfuric acid. With hydrogen chloride, α-pinene gives bornyl chloride as the principal product. (The approach of chloride ion from the *endo* direction is less hindered than that from the *exo* because of the presence of methyl groups on the bridge carbon atom.)

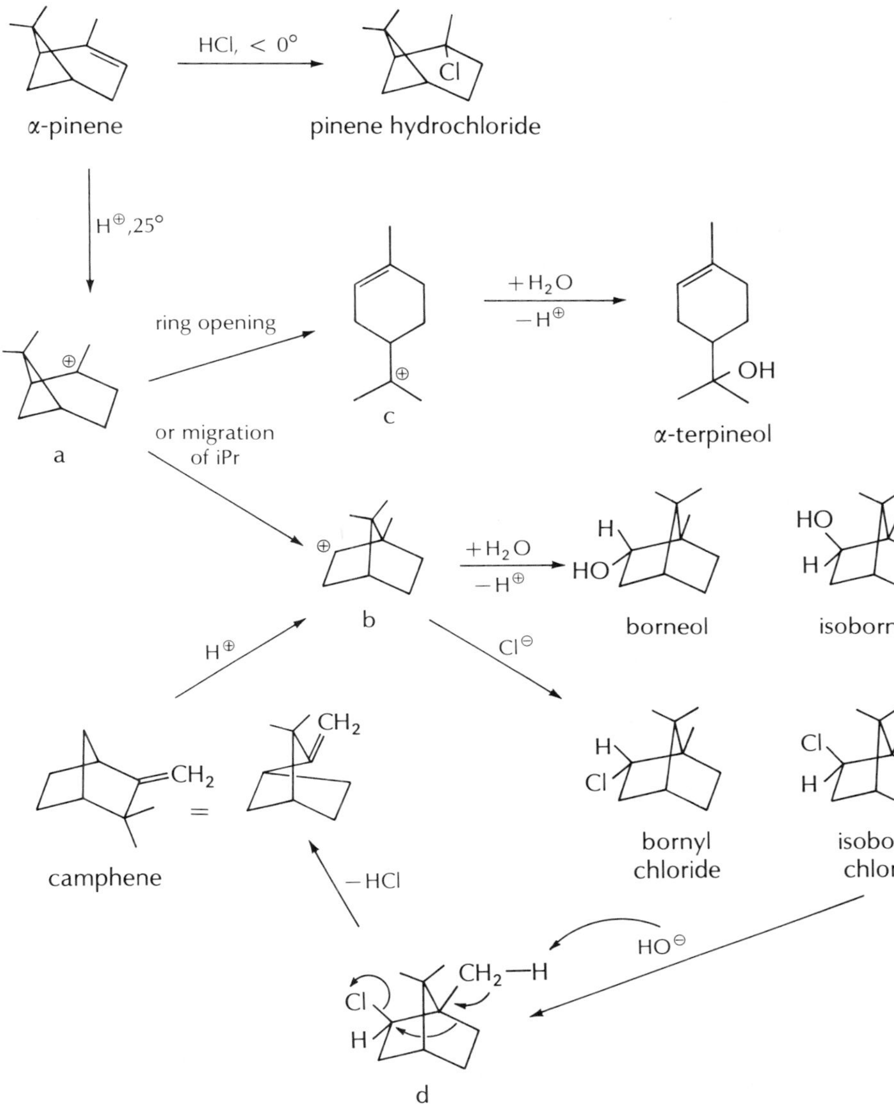

If bornyl or isobornyl chloride is treated with base (or if borneol or isoborneol is dehydrated) another rearrangement occurs, typified by d, and camphene (an isocamphane ring structure) is formed. The *exo* compounds (isobornyl chloride or isoborneol) react much more rapidly than do the *endo* compounds, because the departing group (Cl in d) is being displaced by back-side attack (the migrating bond). At low temperatures, HCl may be added to camphene to give camphene hydrochloride. At higher temperatures, reaction proceeds either through b to give bornyl and isobornyl chlorides or through a transition state (e) that looks like the reverse of d to give only isobornyl chloride. The latter is the predominant product.

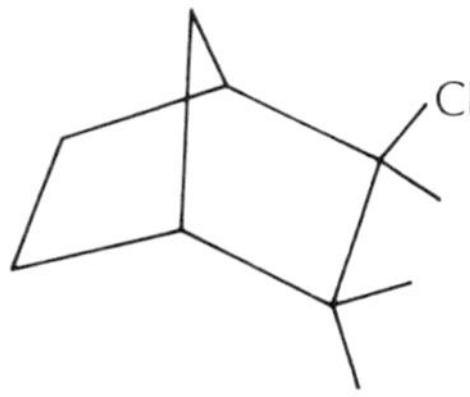

camphene hydrochloride 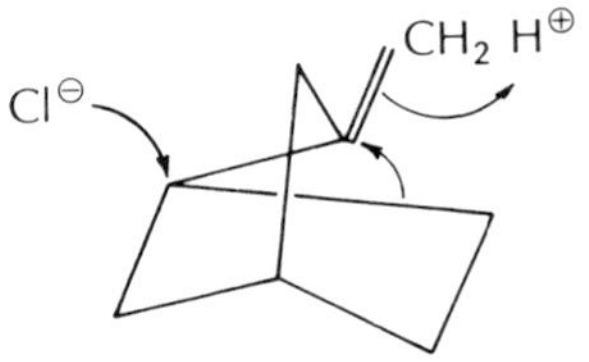   e

Camphor: reactions; synthetic camphor. Camphor occurs in the camphor tree in the dextrorotatory form. Being almost spherical in shape, it has the usual small liquid range (mp 177°, bp 204°) and sublimes readily. It has a very large molal lowering constant of the freezing point, 39.7° per mmole of substance dissolved in 1 g of camphor. For comparison, the value for water is 1.86°.

Oxidation of camphor gives camphoric acid, reduction gives camphane, sulfonation gives camphorsulfonic acid, and SeO_2 (or isopentylnitrite followed by hydrolysis) gives the α-diketone known as camphorquinone (Eqs. 26.8).

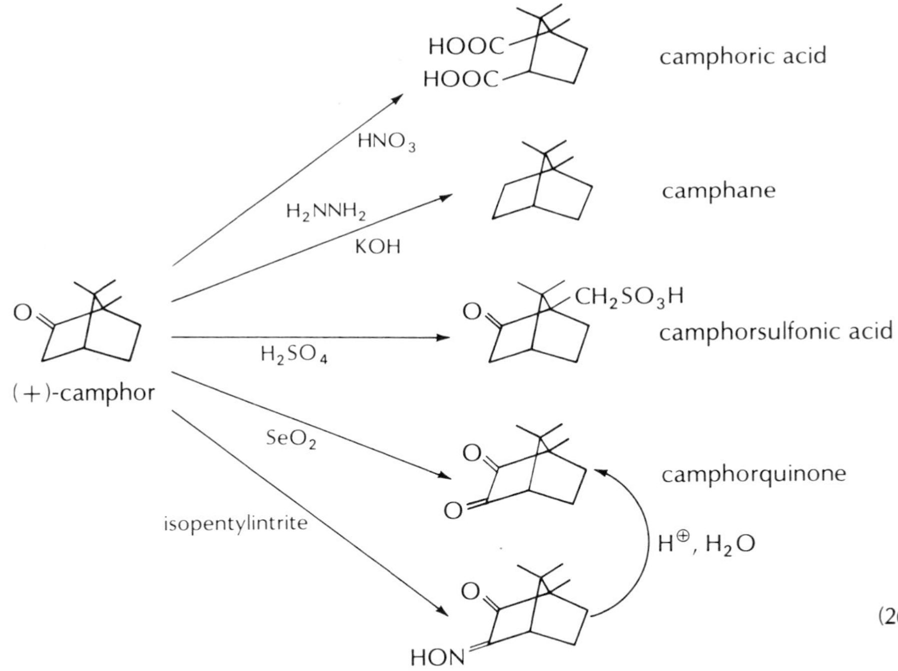

(26.8)

Synthetic (±) camphor is made from α-pinene by conversion to camphene (Eqs. 26.7), addition of acetic acid to give isobornyl acetate, hydrolysis of the ester to give isoborneol, and mild oxidation (ϕNO_2) of isoborneol to racemic camphor. (See also Prob. 17.1.)

An often quoted reaction that illustrates the stability of aromatic compounds is the formation of *p*-cymene from camphor by warming with P_2O_5.

26.5 Sesquiterpenes

Exercises

26.2 Camphorquinone (like other α-diketones and o-quinones) reacts with o-phenylenediamine (1,2-diaminobenzene) to give a stable crystalline derivative. Write the equation for the reaction using structural formulas.

26.5 Sesquiterpenes

Acyclic sesquiterpenes: farnesol, cyclization. Of the acyclic sesquiterpenes (C_{15}), we mention only the widely distributed (in small amounts), fragrant farnesol (citronella, rose, orange blossoms, etc.). Farnesol may be synthesized from acetone, diketene, and acetylene as shown in Equation 26.9. In the presence of an acid, farnesol dehydrates and cyclizes to give bisabolene (Eq. 26.10), a monocyclic sesquiterpene.

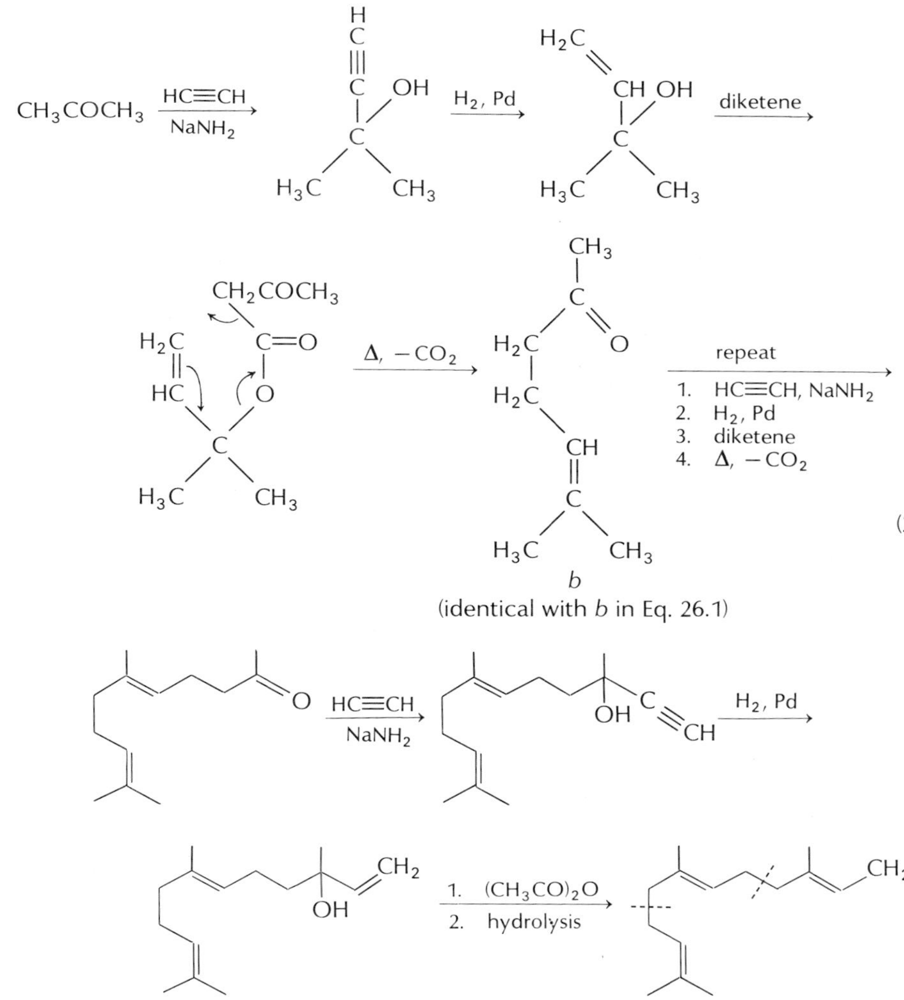

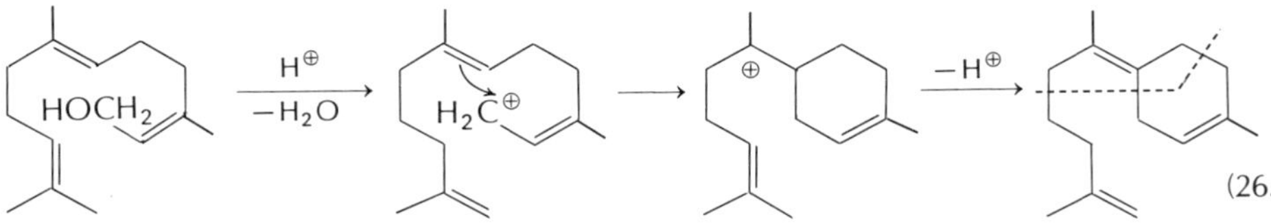

farnesol → [H⁺, −H₂O] → [intermediate] → [intermediate] → [−H⁺] → bisabolene (oil of bergamot)   (26.10)

Interesting sesquiterpenes are the monocyclic α-humulene (cloves, hops) and the closely related bicyclic compound β-caryophyllene (also in cloves). One has an eleven-membered triene ring with all the double bonds *trans* or (*E*); the other has a nine-membered

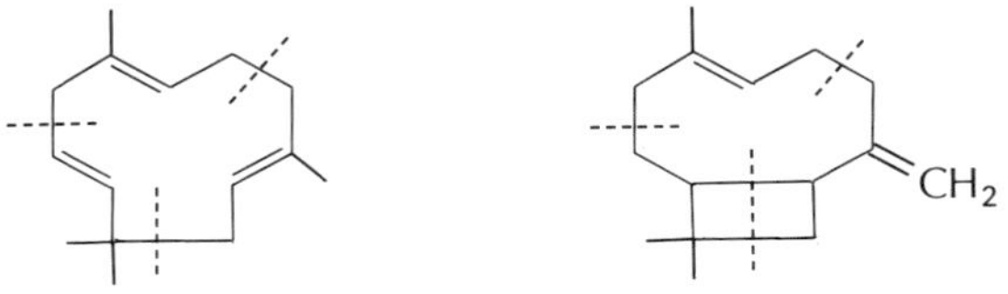

α-humulene

β-caryophyllene

ring (with a *trans* double bond) fused to a cyclobutane ring. The structural relationships to farnesol and to each other are easily seen.

Azulenes: synthesis of azulene. Another way of closing the rings leads to azulene systems (Eqs. 26.11). The development of a blue color (or intensification of the color) remained

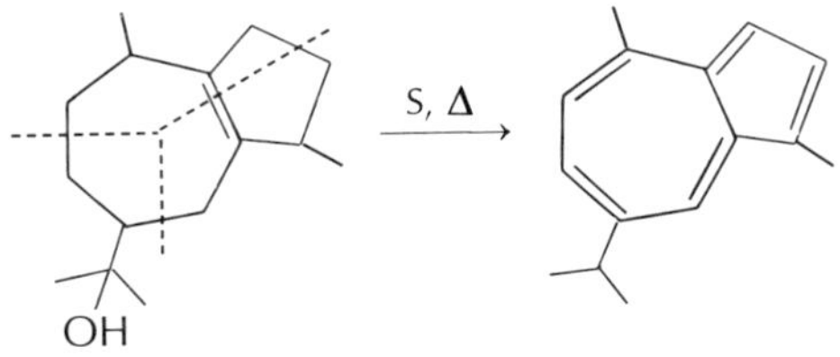

guaiol (guaiacum wood) → [S, Δ] → guaiazulene (geranium)

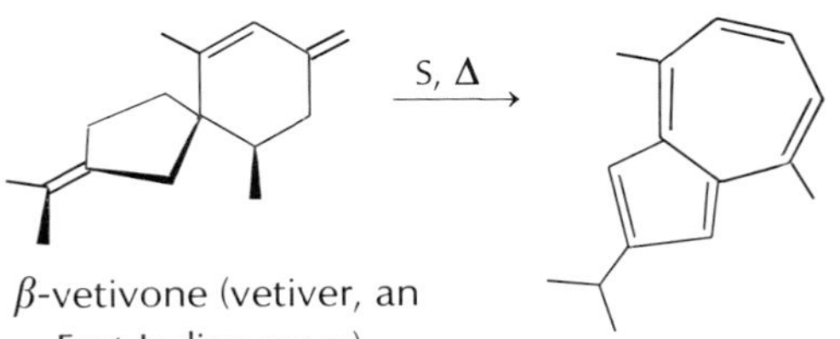

β-vetivone (vetiver, an East Indian grass) → [S, Δ] → vetivazulene (vetiver)

(26.11)

a mystery until the synthesis and chemistry of azulene, the parent blue hydrocarbon, was finally worked out in the 1930's. A synthesis of azulene is shown in Equation 26.12. It is interesting because of the internal aldol condensation of the diketone a.

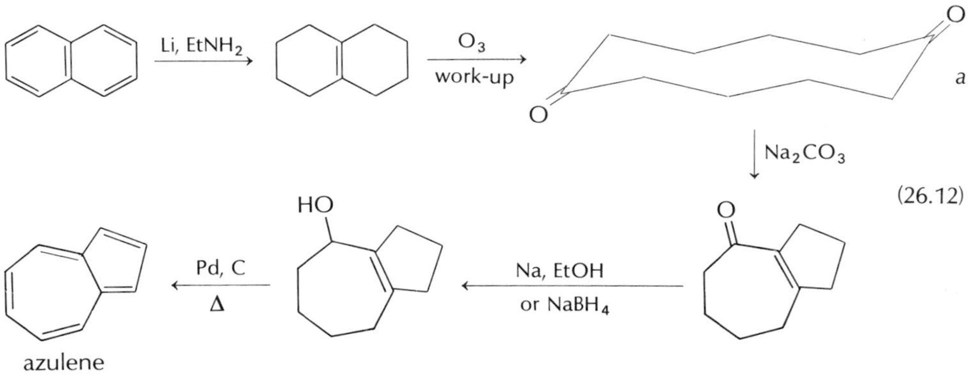

(26.12)

azulene

Exercises

26.3 In the synthesis of farnesol (Eq. 26.9), the hydrogenation of the triple bond in the later steps is complicated by reduction of the double bonds. An alternate scheme uses the Reformatsky reaction to add the two carbon atoms. Write out the equations, using the Reformatsky reaction, for:

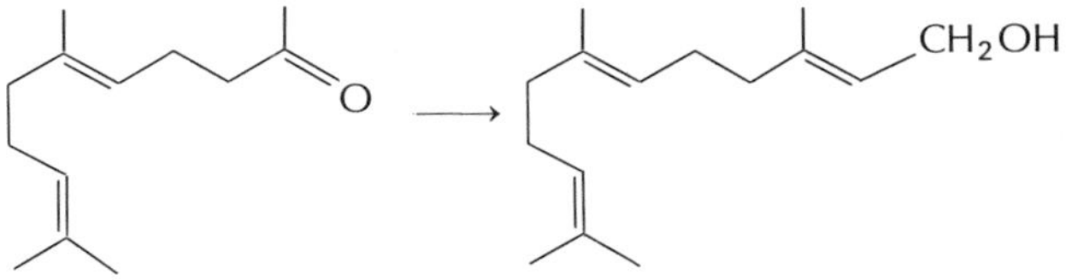

26.6 Diterpenes

Acyclic diterpenes: phytol, synthesis. Chlorophyll, the universal green pigment of all plants that is essential for photosynthesis, may be saponified to give a complex nitrogen-containing acid and an acyclic diterpene alcohol, phytol ($C_{20}H_{39}OH$), with one double bond. Ozonization of phytol gives a methyl ketone ($C_{16}H_{33}COCH_3$) and glycolic aldehyde ($HOCH_2CHO$). Thus, the alcohol is allylic. It was suspected that phytol was related to farnesol, and a synthesis proved the point (Eq. 26.13). The phytyl chain also occurs in vitamin K_1 (Sec. 24.9) and in modified form in vitamin E (α-tocopherol; Eqs. 26.14).

Cyclic diterpenes: vitamin A, synthesis. A monocyclic diterpene is vitamin A, necessary for the growth of mammals. The body is able to cleave β-carotene (Sec. 26.7) to vitamin A aldehyde (called <u>retinal</u>) and to obtain vitamin A by reduction of the aldehyde. Fish liver oils were a widely used source of vitamin A before convenient syntheses became available. One of the possible syntheses starts with β-ionone (Eq. 26.6). Thus vitamin A_1 may be obtained as shown in Equation 26.15. Vitamin A_2 has an additional conjugated double bond, and its biological activity is less than that of vitamin A_1.

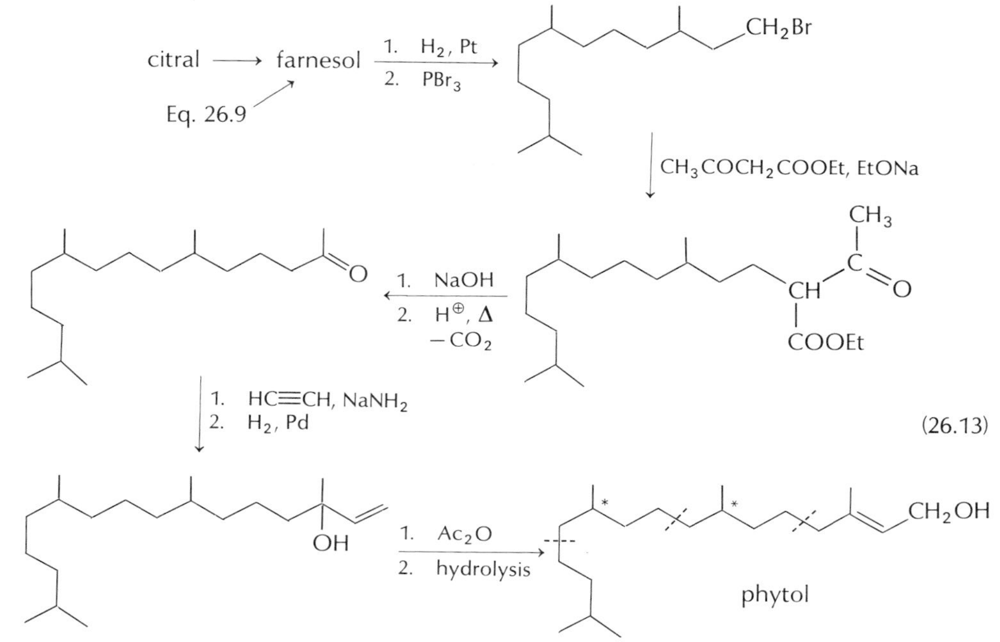

(26.13)

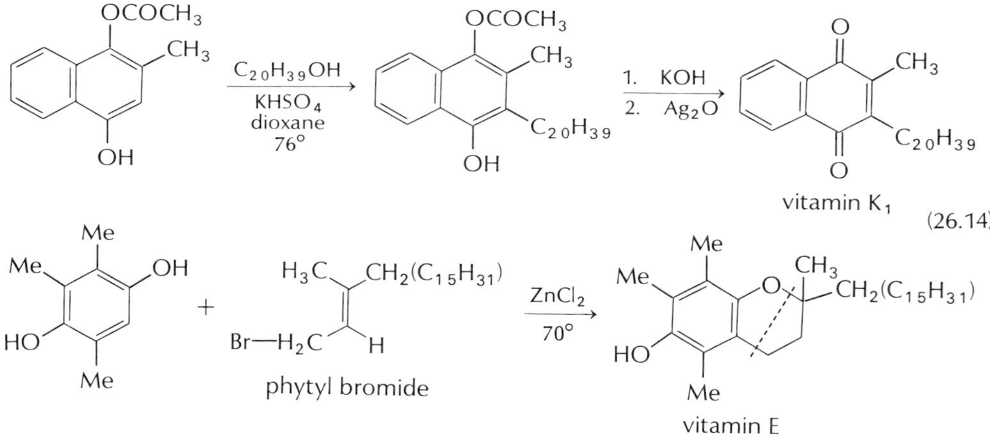

(26.14)

Chemistry of vision. The role of vitamins as coenzymes to catalyze specific biochemical transformations was pointed out in Section 21.6. The part played by vitamin A as a growth factor is not yet clear, but the chemistry of sight is fairly well known. Vitamin A may be oxidized enzymatically to the corresponding aldehyde, now known as retinal (formerly called retinene$_1$). Retinal may be isomerized (the *trans*-4,5 double bond to *cis*) to neoretinal-b (neoretinene-b) by the enzyme called retinene isomerase. Neoretinal-b reacts with a specific protein in the eye, opsin, to give a Schiff base known as rhodopsin (visual purple). In the presence of light, isomerization to *trans*-rhodopsin occurs. Unlike rhodopsin, *trans*-rhodopsin is very unstable with respect to hydrolysis of the Schiff base function. The cycle now is ready for repetition (Eq. 26.16). How the reaction is transformed into a nerve impulse and transmitted to the brain remains to be worked out.

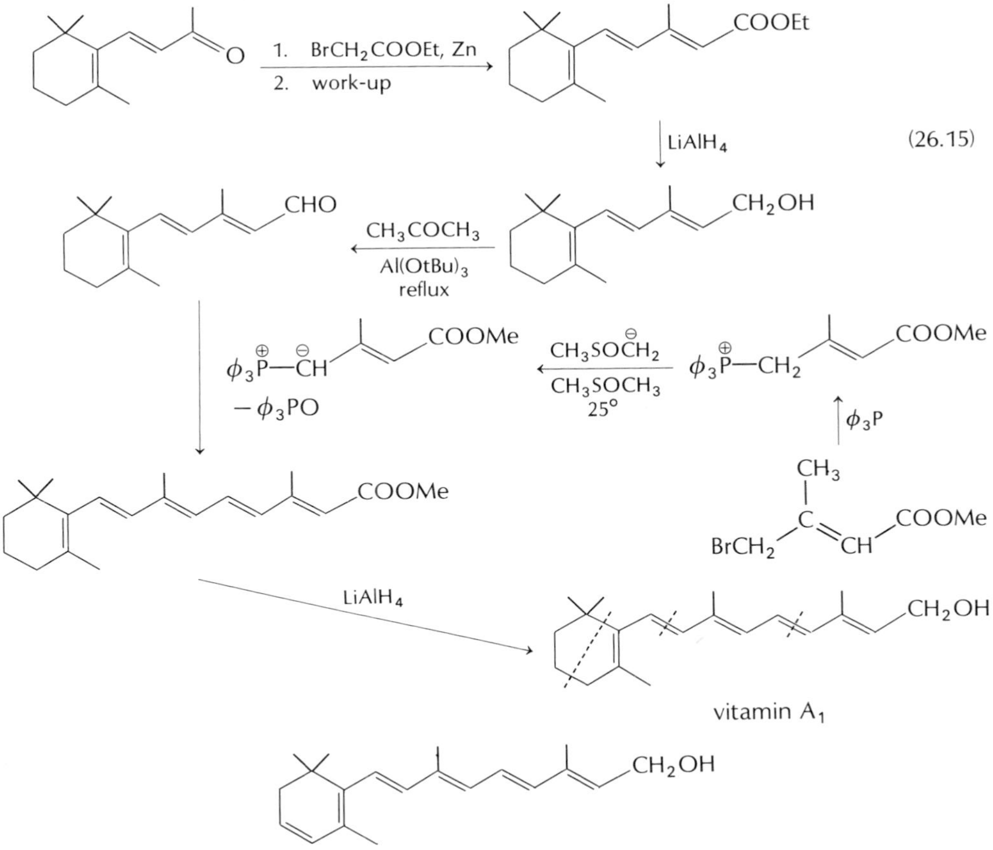

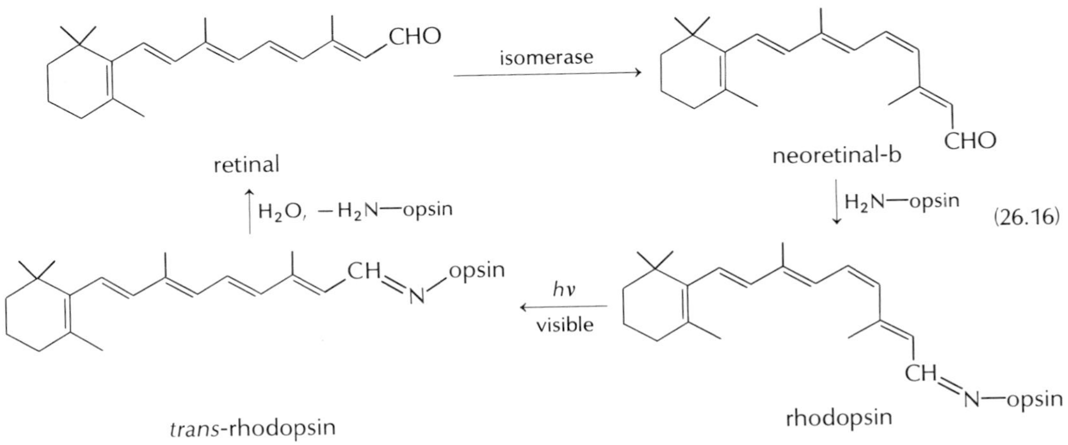

Tricyclic diterpenes: abietic acid. We skip the bicyclic diterpenes and mention only one tricyclic diterpene, abietic acid ($C_{20}H_{30}O_2$), the principal component of rosin, the residue left after distillation of turpentine from extracts of pine wood. Heating with sulfur (Eq. 26.17) causes aromatization to 7-isopropyl-1-methylphenanthrene (common

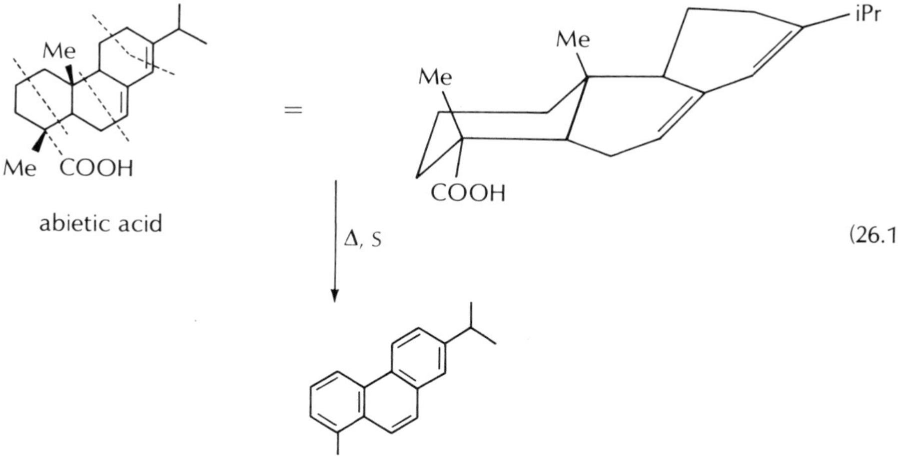

abietic acid

Δ, S

(26.17)

retene

name, retene) by loss of hydrogen, CO_2, and methane (from the angular methyl group). We shall come across many other phenanthrenelike ring structures in the next chapter.

Exercises

26.4 Starting with acetone and ethyl bromoacetate, devise a synthesis of methyl 4-bromo-3-methyl-2-butenoate for use in the ylide in the synthesis of vitamin A_1.

26.7 Triterpenes and tetraterpenes

Squalene: occurrence, synthesis. Squalene ($C_{30}H_{50}$), from shark liver oil, is now recognized as the biochemical precursor of the sterols, specifically of lanosterol (see Chap. 27). Squalene in turn is formed by an enzyme-catalyzed reaction whereby two molecules of the pyrophosphate ester of farnesol are linked together tail to tail (Eq. 26.18: we omit charges in the pyrophosphate ester for simplicity). In the laboratory, squalene may be made from the reactive allylic farnesyl bromide by means of the Grignard reagent.

Carotenes: occurrence, oxidation. The most common tetraterpenes are the carotenes, the yellow and red pigments of many plants. Lycopene ($C_{40}H_{56}$), the almost

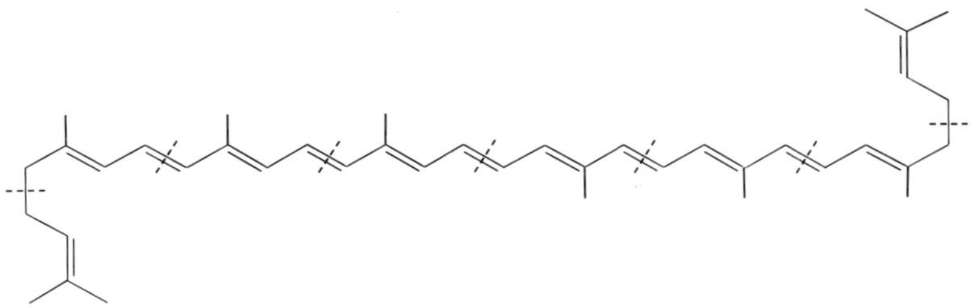

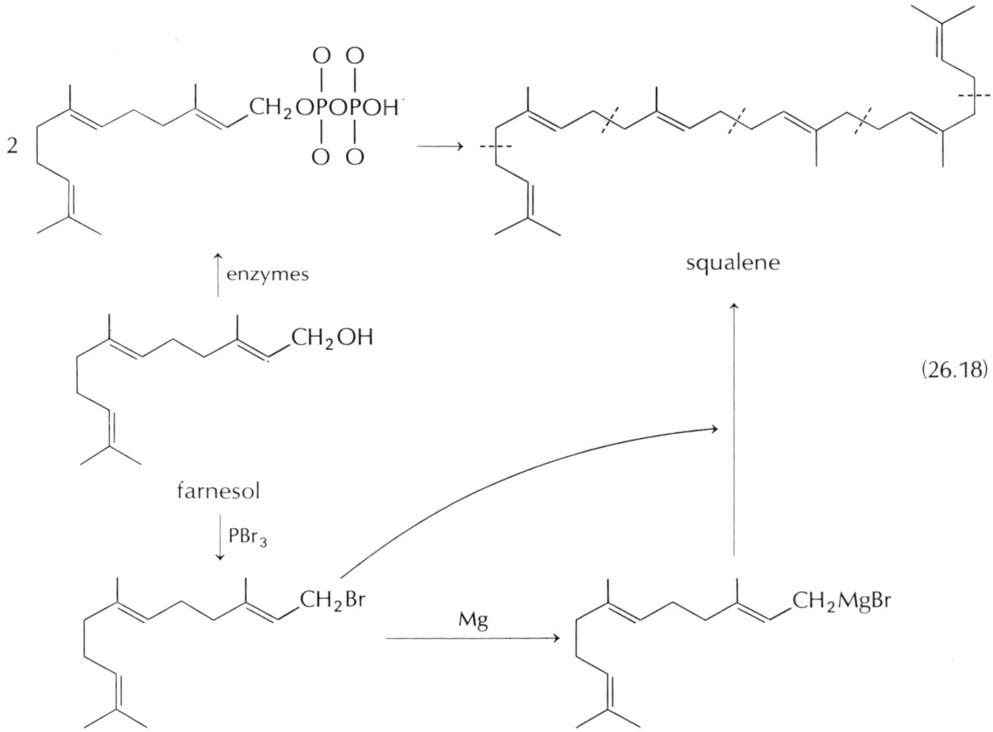

$$\text{(26.18)}$$

completely conjugated acyclic unsaturated tetraterpene, is the red of tomatoes and other fruits. As we see from the structure, lycopene looks like two phytyl groups joined tail to tail and partially dehydrogenated.

Possibly the most interesting carotene is β-carotene (green leaves, carrots), which acts as a precursor of vitamin A_1. Also, the carotenes in general may be thought of as cyclization products of lycopene. Although β-carotene is a "twinned" system (in the same way that squalene and lycopene are twinned), not all the carotenes have this symmetry, and there are many isomers of β-carotene. It is thought that the oxidative cleavage of β-carotene at the center of the molecule occurs on the walls of the intestines (Eq. 26.19).

Many oxygenated carotenes, known generically as xanthophylls or xanthins, also occur in nature (autumn leaves, paprika, egg yolk, oranges, saffron, etc.).

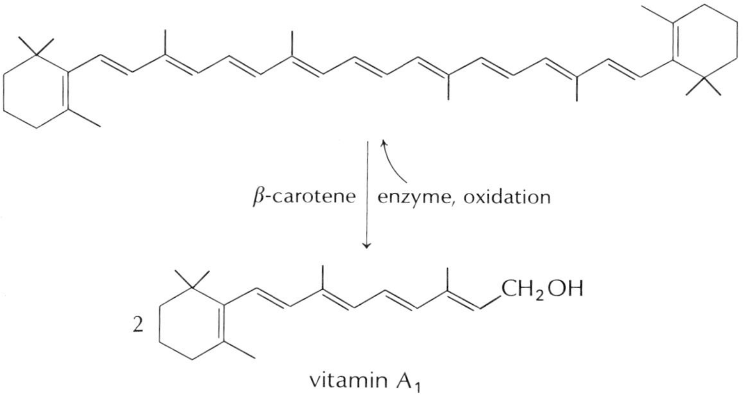

862 Chapter twenty-six

$$\beta\text{-carotene} \xrightarrow{\text{enzyme, oxidation}} 2 \text{ vitamin A}_1 \text{ (CH}_2\text{OH)} \quad (26.19)$$

26.8 Polyterpenes

Rubber.   Rubber $((C_5H_8)_n$, with n distributed over the range 700 to 44,000) is the product obtained upon coagulation (smoking, or acidification) of the colloidal suspension (latex) that exudes from the trunk of the rubber tree if the outer bark is cut. Rubber (also known as <u>caoutchouc</u>) occurs in small amounts in numerous other plants: guayule shrubs, goldenrod, dandelions, and so on. As with most of the terpenes, the part played by rubber in the biochemistry of the plant is not known.

Distillation of rubber causes decomposition, and isoprene is present in the distillate. Ozonization of rubber leads to levulinic aldehyde as the principal product (Eqs. 26.20).

$$+\!\!\left[\!\text{CH}_2\overset{\overset{\displaystyle\text{CH}_3}{|}}{\text{C}}\!\!=\!\!\text{CHCH}_2\!\dotplus\!\text{CH}_2\overset{\overset{\displaystyle\text{CH}_3}{|}}{\text{C}}\!\!=\!\!\text{CHCH}_2\right]_{\frac{n}{2}}\!\!\xrightarrow{\Delta} n\ \text{CH}_2\!=\!\!\overset{\overset{\displaystyle\text{CH}_3}{|}}{\text{C}}\!-\!\text{CH}\!=\!\text{CH}_2$$

$$\downarrow \begin{array}{l}1.\ O_3\\ 2.\ \text{work-up}\end{array}$$

$$(n-1)\ \ O\!\!=\!\!\text{CHCH}_2\!-\!\text{CH}_2\overset{\overset{\displaystyle\text{CH}_3}{|}}{\text{C}}\!\!=\!\!O\ +\ ?\ \text{(end groups)}$$

(26.20)

Raw rubber is characterized by extreme extensibility before rupture (elasticity) and ability to return to approximately the original dimensions when the stress is released. Raw rubber is also very sensitive to temperature. At low temperatures, elasticity is lost (it becomes stiff and even brittle); and at elevated temperatures, raw rubber becomes soft and sticky, and extensibility increases, but ability to return to shape decreases. These properties are characteristic of what are known as <u>thermoplastic polymers.</u> A polymer is a substance composed of many repeated units: in the case of rubber, the repeating unit is the isoprene unit.

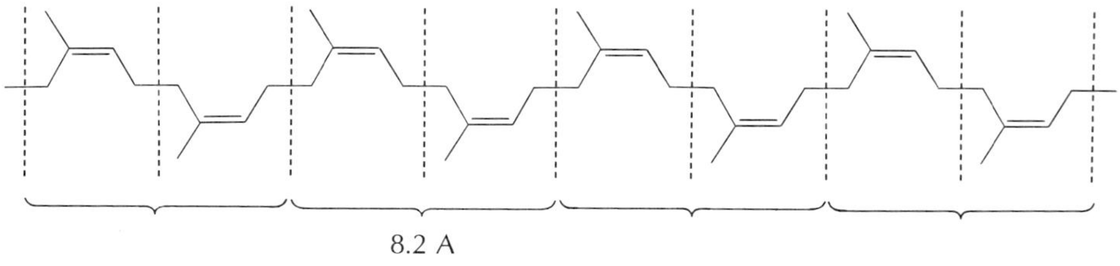

8.2 Å

Raw rubber shows no regular pattern when subjected to X rays (characteristic of amorphous, noncrystalline substances). When rubber is stretched, an X-ray diffraction pattern appears, which indicates that some crystallinity is being induced. A repeating pattern of 8.2 Å is characteristic of a nonplanar, extended, all-*cis* configuration.

We may consider a rubber band as an entropy spring. Application of force causes elongation (work is done on the system) and a decrease in entropy ($-T\Delta S$ is positive and contributes to the increase in free energy of the band). When released, the band contracts and does work (paper clips or spit balls may be propelled quite a distance) by loss of free energy. During contraction, entropy increases (less ordered, more amorphous system), and $-T\Delta S$ is negative and contributes to the loss of ΔG, which is converted into work.

Gutta-percha and balata; chicle. Two other polyisoprenes are gutta-percha (from the leaves of a tree indigenous to southeast Asia and the East Indies) and balata (from a native plant of tropical South America and Panama). Analysis, ozonolysis, and so on indicate that they too are $(C_5H_8)_n$. The two substances are nearly identical in physical properties, but these properties do not resemble those of rubber except for sensitivity to temperature. Balata and gutta-percha are tough, nonelastic substances best described as similar to horn. Examination by X rays shows that the chain has the *trans* configuration.

Chicle (used almost exclusively for chewing gum) is a complex mixture of both *cis*- and *trans*-polyisoprenes and numerous other substances. It is present as a latex in a tree native to Yucatan.

Technology of rubber: vulcanization. Goodyear discovered hot vulcanization of rubber in 1839 when he dropped a mixture of rubber and sulfur on a hot stove. Sulfur was only one of the many substances he had mixed with raw rubber in an effort to lessen the sensitivity to temperature. We now know that heating rubber with sulfur leads to attack at double bonds and at allylic positions and gives sulfide and disulfide linkages between the chains. The tying together of the chains increases the dimensional stability of the product and decreases the sensitivity to temperature. We say that the polymer becomes more like a thermosetting polymer and less like a thermoplastic polymer. The number of interchain sulfur bonds increases with the amount of sulfur incorporated into the mixture. Large proportions of sulfur (30 to 50% by weight) give a hard, stable polymer known as ebonite, with little unsaturation remaining. Intermediate amounts of sulfur (10 to 30%) give a useless product, not hard enough for some uses and not plastic enough for others. Small amounts of sulfur (2 to 3%) give a soft rubber (rubber bands, shoe soles, rubber gloves), and not much more (5 to 8%) is needed for more rigorous uses (tires).

Even a small amount of cross-linking diminishes the extensibility of rubber, but it increases the tensile strength. Certain fillers impart other desirable properties. Carbon black and zinc oxide, within limits, increase tensile strength and resistance to abrasion and tearing. Other substances added to the mixture before vulcanization (usually in molds heated by steam) are accelerators such as 2-mercaptobenzothiazole, which cause vulcanization to proceed at lower temperatures, and antioxidants such as N-phenyl-β-naphthylamine, which increase resistance to oxygen and ozone and thus prolong the useful life of the rubber article. For special purposes, softening agents, abrasives, pigments, and so on, also are added.

Exercises

26.5 Isoprene polymerizes easily in the presence of free-radical initiators (peroxides), but the product is almost not like rubber. Why?

26.9 Summary

We have looked briefly at a class of compounds obtained from living natural sources, largely from the plant kingdom, as opposed to those from petroleum and coal tar. By working from the simple (terpenes) to the complex (polyterpenes) we have used continuity in the application of the isoprene rule and shown how the complex cyclic compounds are related to the acyclic compounds. We also became aware that nature, by working with countless enzymes, each taking care of a single specific reaction in almost quantitative yield, is still miles ahead of what the organic chemist can accomplish in the laboratory.

26.10 Problems

26.1 Substance A ($C_{10}H_{16}$) takes up only one mole of hydrogen upon hydrogenation. Ozonization of A and decomposition of the ozonide without the presence of zinc dust (called oxidative ozonization) gives a single product (B, $C_{10}H_{16}O_3$). B reacts with Cl_2 in NaOH to give $CHCl_3$ and C ($C_9H_{14}O_4$). C reacts with EtOH in the presence of H_2SO_4 to give D. Treatment of D with EtONa and work-up give E ($C_{11}H_{16}O_3$). Hydrolysis of E and heating of the acidified product give CO_2 and F ($C_8H_{12}O$). Heating F with alkaline $KMnO_4$ gives a diacid that is easily dehydrated by heat to give G ($C_8H_{10}O_3$). F with $NaBH_4$ gives H. Warming H with H_2SO_4 gives a mixture of products, one of which, I (C_8H_{12}), takes up two moles of hydrogen during hydrogenation. Oxidative ozonization of I gives one mole of acetone and J ($C_5H_6O_5$). Heating J gives one mole of acetone and two moles of CO_2.

Give a structural formula for each lettered substance.

(This problem is very difficult. Do not attempt to work it out unless it is assigned.)

26.11 Bibliography

The literature concerning the terpenes is immense and still growing.

Fieser and Fieser (Chap. 7, ref. 7) and Noller (Chap. 7, ref. 11) have the largest sections on terpenes of the standard textbooks. Roberts and Caserio (Chap. 7, ref. 13) and Allinger et al. (Chap. 7, ref. 1) have shorter sections, as do the other large texts.

1. Russell, G. F., and J. I. Hills, Odor Differences between Enantiomeric Isomers, *Science* 172, 1043 (1971). Of interest to those wishing to pursue the subject of odors, as is the next reference. Both are concerned with terpenes.

2. Friedman, L., and J. G. Miller, Odor Incongruity and Chirality, *Science* 172, 1044 (1971).

27 Steroids

Another special class of substances are the steroids, naturally occurring substances of which many are essential to life. The chemistry of the steroids has a long history. Despite many years of labor, it was not until 1955 that the structure of cholesterol finally was completely worked out. However, much was known previously, and the sex hormone equilenin was synthesized in 1939. The steroids make a fascinating story which, regretably, we must abridge in many ways.

27.1   Steroids   866
    1. Definition.   2. Cholestane and coprostane: structures, numbering.
    3. The α and β sides: epimeric relationship.

27.2   Biosynthesis   867
    1. The acetate building block; role of coenzyme A.   2. Acetyl-CoA and malonyl-CoA to squalene: labeling.   3. Squalene to lanosterol and to cholesterol.   4. Fatty acid synthesis. Exercises.

27.3   Sterols   872
    1. Cholesterol: occurrence; properties.   2. Cholesterol, coprostanol: reactions.   3. Stigmasterol; ergosterol. Exercises.

27.4   Vitamin D   874
    Occurrence; formation.

27.5   Sex hormones   875
    1. Occurrence; effects.   2. Structures of estrogens and androgens.
    3. Progesterone; the Pill. Exercises.

27.6   Adrenal hormones   877
    1. Occurrence; structures; effects.   2. Cortisone: synthesis. Exercises.

27.7   Bile acids   880
    Occurrence; effects; structures.

866 *Chapter twenty-seven*

27.8   Sapogenins; cardiac glycosides   881
  1. Saponin, sapogenin; classes of sapogenins; tigogenin.   2. Cardiac glycosides: digitalis, digitoxigenin, strophanthidin.

27.9   Toad poisons; insect hormones   882
  1. Bufotalin and batrachotoxin.   2. Ecdysone.

27.10   Summary   883

27.11   Bibliography   883

27.1 Steroids

Definition.   It is difficult to define a steroid: precision is hampered by numerous exceptions and variations. We shall work from a fuzzy statement: a steroid is a substance in which we are able to recognize a more or less hydrogenated tetracyclic ring structure, 1,2-cyclopentanophenanthrene. Various side chains and oxygen functions are usually present.

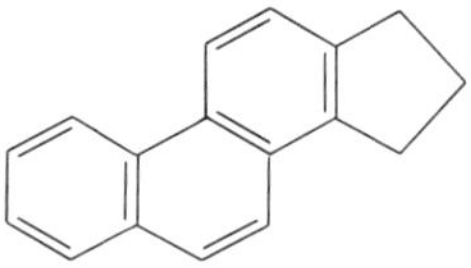

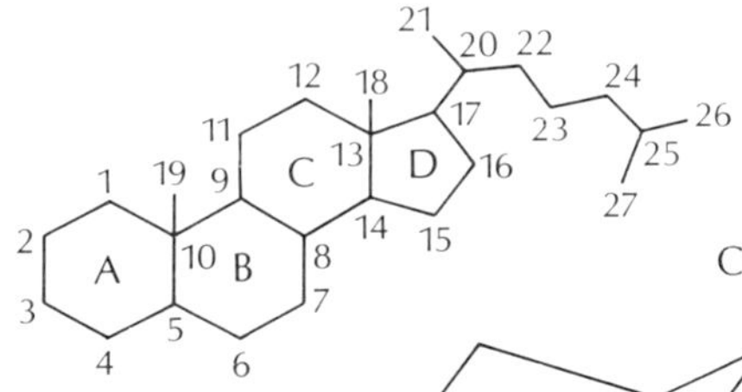

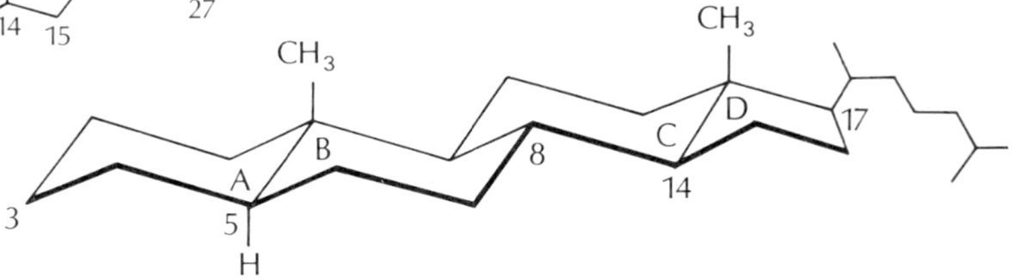

cholestane

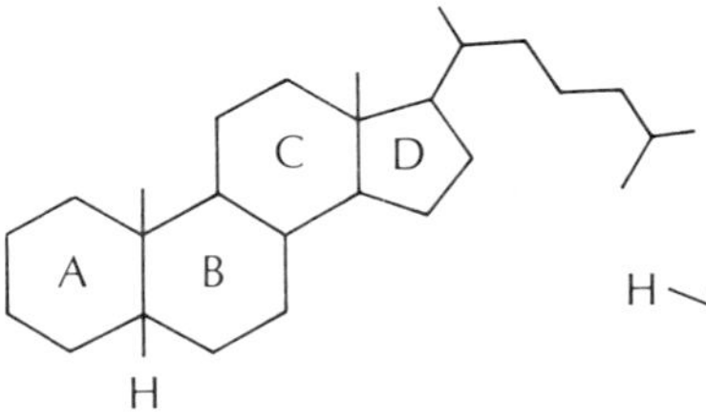

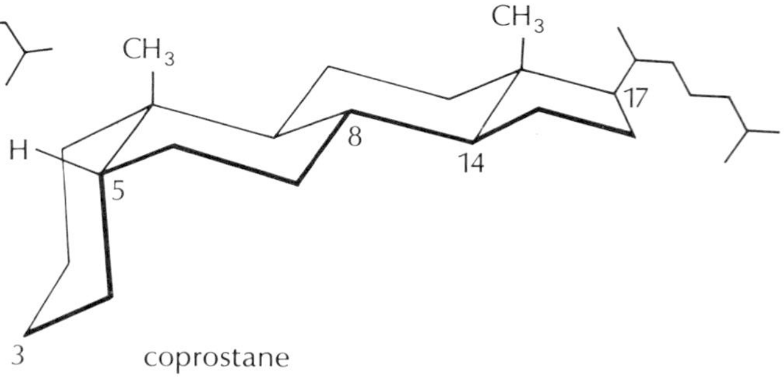

coprostane

Cholestane and coprostane: structures, numbering. Cholestane and coprostane may be considered the parent hydrocarbons for most of the steroids. The numbering system is far from systematic, but it is retained as an official exception to the rules. Rings A and B are numbered counterclockwise, 1 to 10. Then there is a jump to ring C, clockwise; and on into ring D, counterclockwise, ending with 17. Then the angular methyl groups are numbered, 18 and 19. The side chain at 17 begins with 20 and goes up to 27.

Cholestane and coprostane differ only in the A/B ring junction. All the other ring junctions are *trans*. In cholestane, the A/B ring junction is *trans*, as are the hydrogen at 5 and the 19 methyl group at 10. In coprostane, the A/B ring junction is *cis*, as are the hydrogen at 5 and the 19 methyl group at 10. Rings A, B, and C all are chair cyclohexane rings and are rigidly held in that conformation (except for C_2 or C_3, which may flip so as to throw ring A into the unstable boat form: models are helpful).

The α and β sides: epimeric relationship. Substituents on the upper surface of the roughly planar cholestane are designated as $β$, whether axial or equatorial; those on the under side are termed $α$. The same terminology is applied to the coprostane system, even though the rough plane is bent. Thus the methyl groups at 10 and 13 as well as the side chain at 17 are all $β$ substituents. The hydrogens attached at 9, 14, and 17, for example, are all on the $α$ side of both cholestane and coprostane.

With 8 asymmetric carbon atoms present, 128 racemic pairs of enantiomers are possible. Nevertheless, cholestane and coprostane have the absolute configurations shown. The two stereoisomers, differing only in configuration at 5, all the other asymmetric centers being identical, are called epimers—or we say that one is an epimer of the other.

27.2 Biosynthesis

The acetate building block; role of coenzyme A. Let us look at how terpenes, steroids, and fatty acids arise from the simple precursor acetic acid in a living organism. Although the idea that acetic acid could be the building block is an old one, evidence was hard to come by until radioactive carbon-14 became widely available (from atomic reactors) after World War II. Thereafter, the use of ^{14}C as a label enabled the workers in the field to trace the pathway of the atom as it was incorporated into one compound after another. For example, if a buffered solution containing $^{14}CH_3COONa$ is injected into a rat, it is only seconds before ^{14}C is present in cholesterol in the liver. At least thirty-five steps (enzyme-catalyzed reactions) are needed.

Evidently, acetate ion is present at various sites in the body, arising by breakdown of fats and carbohydrates so that a constant level is maintained (acetate metabolic pool) by enzyme-catalyzed reversible reactions. From here on, we must understand that an enzyme enters into every reaction mentioned.

The first step in the biosynthesis of terpenes is the formation of the acetate ester (Eq. 27.1) of the thiol group of coenzyme A. (The structure was shown in Sec. 21.6.) The next

$$CoA-SH + CH_3COO^{\ominus} \xrightarrow{\text{buffer}} CoA-S-\underset{\underset{O}{\|}}{C}CH_3 + H_2O \qquad (27.1)$$

step is condensation of acetyl-CoA with CO_2 in a form of Claisen condensation to give malonyl-CoA (Eq. 27.2).

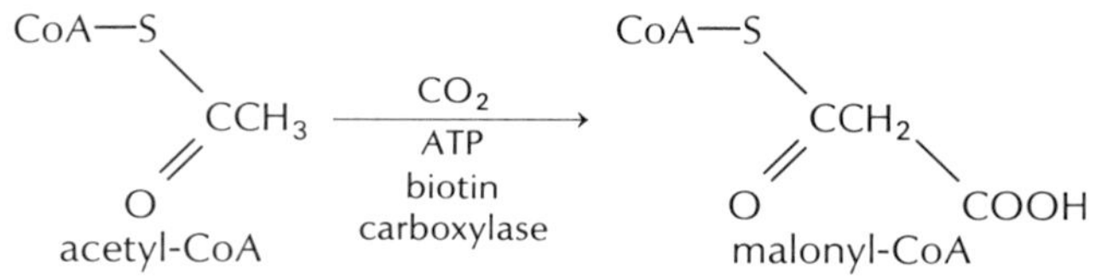

$$\underset{\text{acetyl-CoA}}{\text{CoA—S—C(=O)—CH}_3} \xrightarrow[\substack{\text{ATP}\\\text{biotin}\\\text{carboxylase}}]{CO_2} \underset{\text{malonyl-CoA}}{\text{CoA—S—C(=O)—CH}_2\text{—COOH}} \quad (27.2)$$

Acetyl-CoA and malonyl-CoA to squalene: labeling. With both acetyl-CoA and malonyl-CoA present, reaction to give acetoacetyl-CoA with loss of CO_2 (used to form more malonyl-CoA), further reaction to 3-hydroxy-3-methylglutaryl-CoA, reduction to mevaldic acid, and a second reduction to mevalonic acid take place (Eq. 27.3). NADH is the

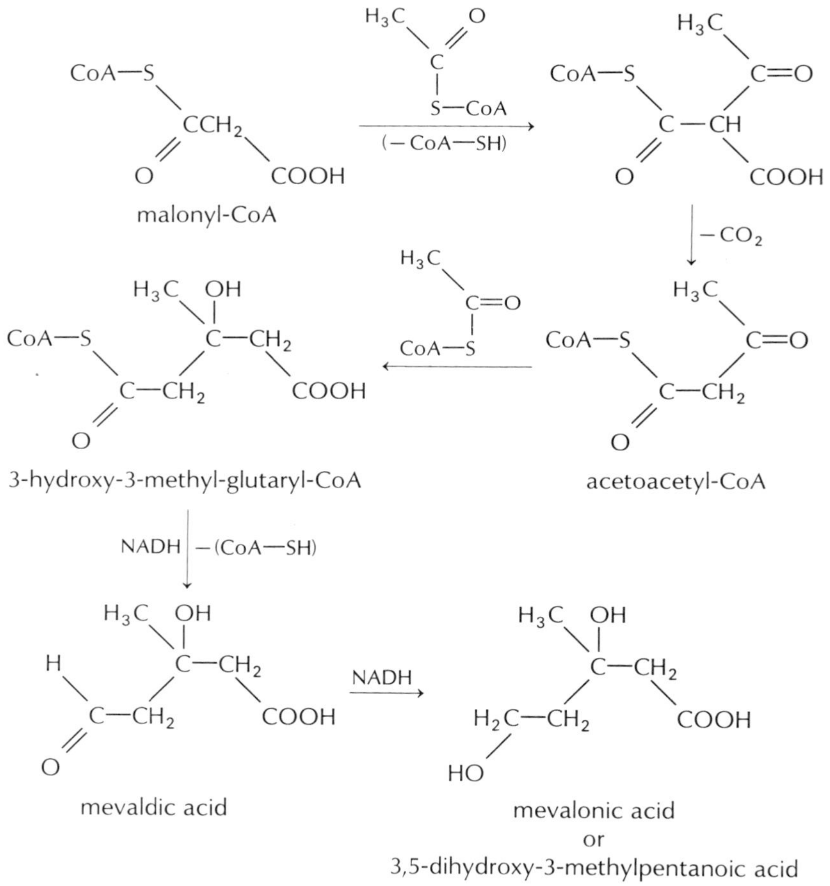

(27.3)

[Reaction scheme: malonyl-CoA + H₃C—C(=O)—S—CoA → (−CoA—SH) → intermediate → (−CO_2) → acetoacetyl-CoA; then combination with H₃C—C(=O)—S—CoA → 3-hydroxy-3-methyl-glutaryl-CoA; then NADH, −(CoA—SH) → mevaldic acid; then NADH → mevalonic acid or 3,5-dihydroxy-3-methylpentanoic acid]

reduced form of nicotinamide adenine dinucleotide, also known as diphosphopyridine nucleotide (DPNH). The structures of these nucleotides are of no concern to us at present; all we need to remember here is that NADH is an efficient hydride donor or reducing agent, widely used for that purpose in biological systems.

Mevalonic acid is the precursor of the five-carbon isoprenoid building block in nature.

How mevalonic acid is transformed into geraniol (or nerol), to farnesol, and to squalene is shown in Equation 27.4. First, mevalonic acid is converted into the dipyrophosphate ester

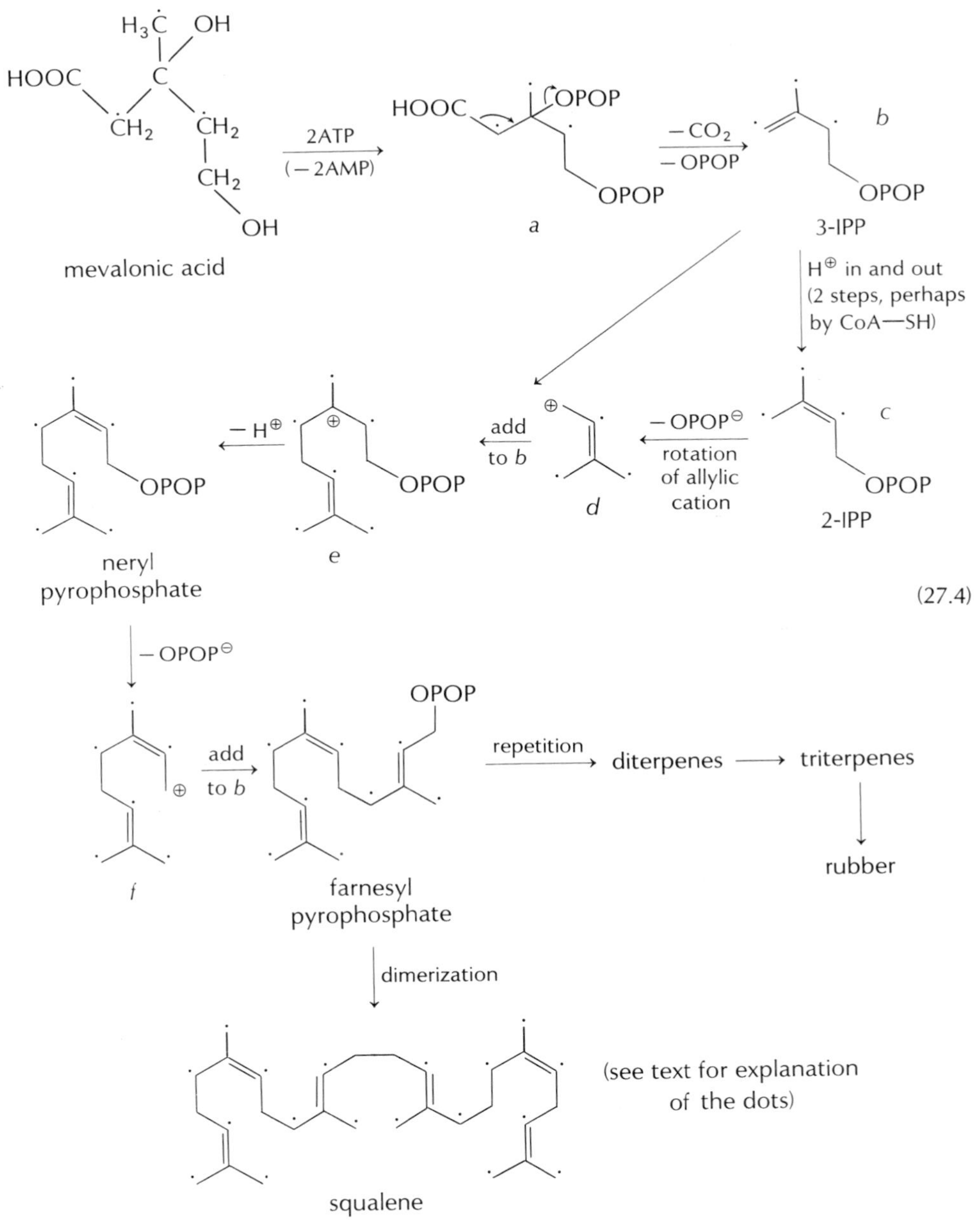

(27.4)

(a) by ATP (adenosine triphosphate) which becomes AMP (adenosine monophosphate). We abbreviate the pyrophosphate group (—OPO_2OPO_2OH) to —OPOP for simplicity and neglect any negative charges caused by ionization. The pyrophosphate is an excellent leaving group (comparable to sulfonates), and in leaving the tertiary carbon atom of a it causes decarboxylation to give b. The CO_2 that is lost comes from the carbonyl group of the last molecule of acetyl-CoA to be incorporated in the production of mevalonic acid

(Eq. 27.3). The pyrophosphate ester of 3-methyl-3-butenol (b) is also called 3-isopentenyl pyrophosphate (3-IPP). Isomerization of b to the allylic ester (c, 2-IPP) takes place by addition of thiol (CoA-SH) to the double bond followed by removal of the thiol to give the more stable (more substituted) location to the regenerated double bond.

The primary pyrophosphate now leaves (ionizes) to give the allylic cation (d), which is added to another molecule of b. Addition occurs to give the more stable tertiary carbonium ion (e), which loses a proton to give neryl (or geranyl) pyrophosphate, a primary allylic ester. Ionization of the ester and addition of the allylic cation (f) to another molecule of b (and loss of a proton) give farnesyl pyrophosphate. Continuation of the process (ionization of the primary allylic ester, addition to another molecule of b, loss of a proton) gives an acyclic diterpenyl ester, and so on. Ultimately, depending upon the enzymes present, all-*trans* balata or gutta-percha or all-*cis* rubber results.

Returning to the cation f for a moment, it is also possible (and easy to see) that reaction with water would give nerol or geraniol. An internal addition (cyclization) would lead to limonene, and on to the various bicyclic terpenes.

The cation from farnesyl pyrophosphate, in the presence of another enzyme and NADH, is able to give the dimer, squalene. If we go back to the formation of mevalonic acid and use $^{14}CH_3COOH$, placing a dot above each radioactive ^{14}C atom, we see that the distribution is as shown in Equation 27.4. If $C\overset{\bullet}{H_3}{}^{14}COOH$ were used, then the undotted carbon atoms would be the radioactive ^{14}C atoms. The experimental distribution of radioactivity in the products is in accord with the predictions.

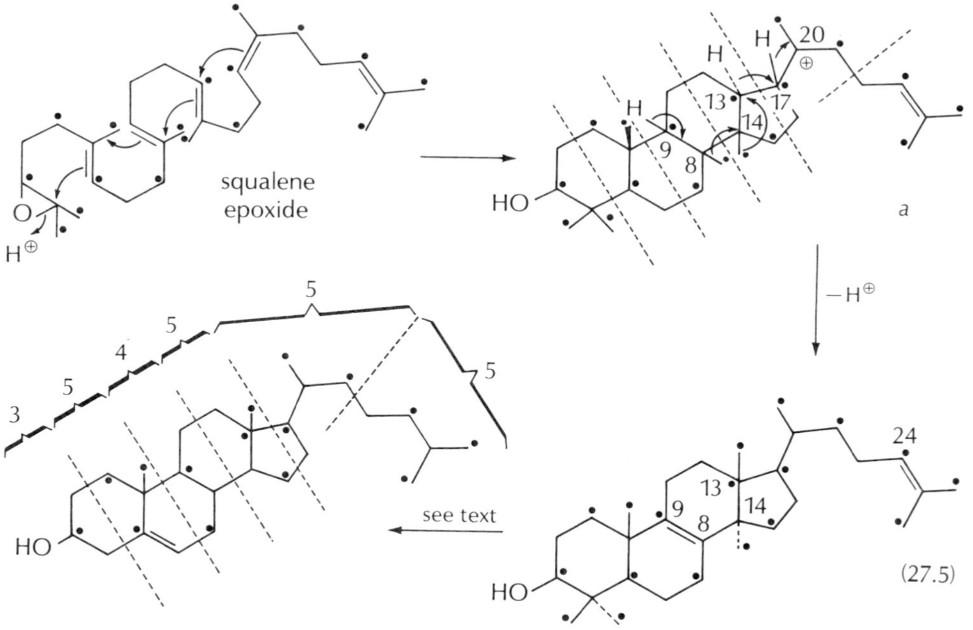

cholesterol
(5-cholestene-3β-ol)
$C_{27}H_{46}O$

lanosterol
(4,4,14α-trimethyl-8,24-cholestadiene-3β-ol)
$C_{30}H_{50}O$

(27.5)

Squalene to lanosterol and to cholesterol. Now we come to a remarkable sequence of reactions in which squalene is "zipped up" into a steroid, lanosterol, which in turn is transformed into cholesterol, which seems to be the precursor of all the other steroids (Eq. 27.5). The first step (not shown) is the formation of an epoxide at either end of the squalene chain. Then comes the opening of the epoxide ring by a proton, whereupon the tertiary cation adds to the next double bond, and so on until the process stops at *a* (perhaps by enzyme control) with the tetracyclic steroid framework in position. The proton at 17 shifts to 20 with an electron pair, the one at 13 shifts to 17, the methyl moves to 13, and the methyl at 8 migrates to 14. The proton at 9 is removed and the electron pair forms a π bond between 8 and 9. The product is lanosterol (present in lanolin).

Lanosterol is converted into cholesterol by a number of discrete steps that involve removal of the three methyl groups (at 4 and 14) as CO_2 by oxidation and decarboxylation, reduction of the double bond at 24, and rearrangement of the double bond at 8 to 5. Several of the intermediates have been isolated.

The isoprene rule is thrown out of kilter once *a* is passed. Comparison of the dashed lines in *a* and in cholesterol shows how parts of the isoprenoid structure of *a* are destroyed during the rearrangements and oxidations on the way to cholesterol. Incidentally, the facts that it is the methyl group at 14 in *a* that moves to 13 in lanosterol and cholesterol and that it is the methyl group at 8 that is present at 14 in lanosterol and removed in cholesterol have been shown by a clever series of double labeling experiments. Again, the dotted carbons in cholesterol designate the surviving carbon atoms from the methyl groups of $^{14}CH_3COOH$ as shown by labeling; the undotted carbon atoms come from the carboxyl group of $^{14}CH_3COOH$.

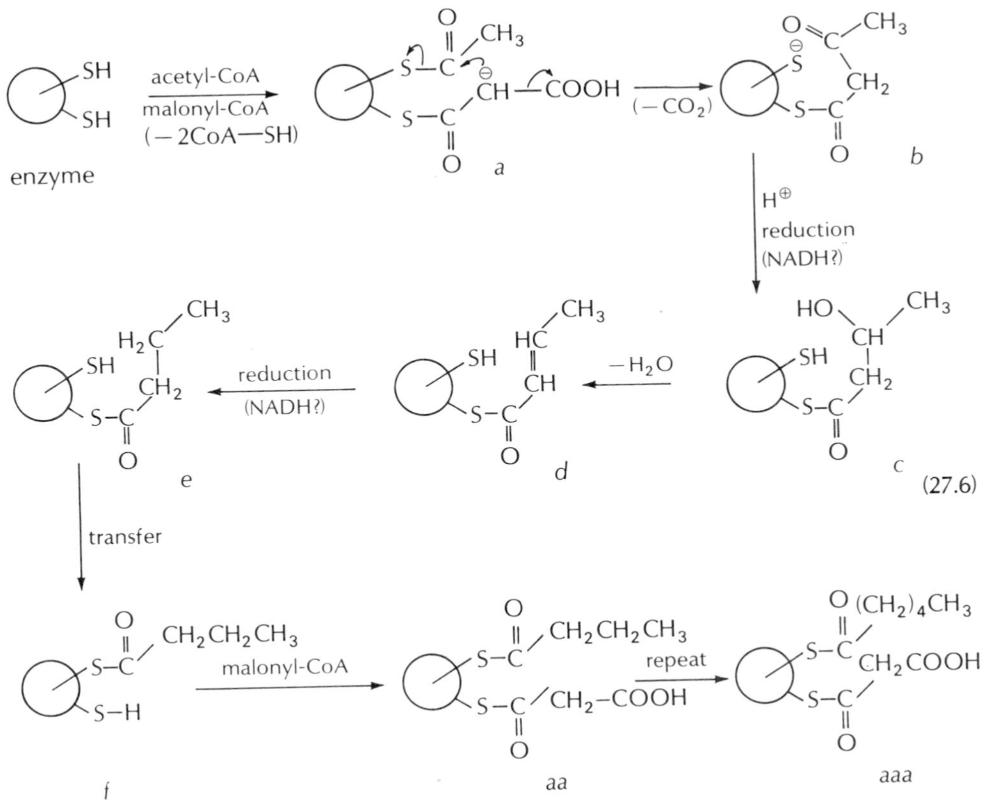

(27.6)

Fatty acid synthesis. Before concluding this section, let us return to acetyl-CoA and malonyl-CoA and look at the biosynthesis of fatty acids. Although many details remain to be elucidated, it seems that a special enzyme, fatty acid synthetase, enables the long-chain fatty acids to be put together on the enzyme surface on a sort of assembly-line basis from acetyl-CoA. There is some evidence that the enzyme possesses two thiol groups so spaced as to allow interaction and transfer of groups between the two (Eq. 27.6). One of the groups is specific in taking up only malonyl-CoA. The stage is set by transfer from acetyl-CoA and malonyl-CoA to the enzyme (a). The malonyl group is acetylated and loses carbon dioxide to give a β-keto ester in b (as shown, an acetoacetic ester). The β-keto ester is reduced to c, dehydrated to d, and reduced to e. The four-carbon chain is shifted over to the other thiol position (f), and another malonyl-CoA is taken up by the malonyl-holding thiol (aa). Repetition of the cycle gives aaa, with a six-carbon chain in position to be increased to an eight-carbon chain in the next succeeding cycle, and so on. When the chain has reached the proper length (10, 12, 14, 16, 18, and so on) as determined by the specific synthetase, it is transferred to CoA-SH, and a new acetyl is accepted to start another chain. The process is very neat and is quickly done at 37°. Our skills in the laboratory still do not match nature: but then we have not been at it quite so long, either.

Exercises

27.1 If mevalonic acid with ^{14}C incorporated only at C_2 is employed in the biosynthesis of cholesterol, at what locations will ^{14}C be found?

27.3 Sterols

Cholesterol: occurrence, properties. The most widely occurring of all the steroids is cholesterol (Eq. 27.5), which occurs in most body tissues, especially in the spinal cord, brain, and marrow, either uncombined or esterified at the 3 position. Cholesterol is not very soluble and tends to deposit in the gall bladder (gall stones) and in the arteries (arteriosclerosis). Consequently, excess cholesterol in the diet should be avoided. However, as we have just seen, the body is able to synthesize cholesterol with great rapidity from the acetate pool, so it may be that high levels of cholesterol in the blood and elsewhere is the result of defective control of the cholesterol level by excessive synthesis or ineffective degradation. Cholesterol does not seem to have any specific metabolic role to play other than as the raw material for the synthesis in the body of vitamin D, sex hormones, adrenal hormones, bile acids, and so on. Cholesterol may have a part to play in transmission of pulses in the nerves. Cholesterol is accompanied by smaller amounts of other sterols.

The term sterol now refers only to the monohydroxy steroids such as lanosterol (lanolin, wool fat) and cholesterol. Other commonly occurring sterols (always mixed with other sterols in the nonsaponifiable fraction of ether extracts of the lipids) are cholestanol, coprostanol, ergosterol, and stigmasterol. Let us return, however, to the structure and reactions of cholesterol.

Cholesterol, coprostanol: reactions. Cholesterol, with the double bond at 5, is somewhat more planar than is cholestane. Cholesterol also has two angular methyl groups and the side-chain on the β side of the molecule, and together these facts imply that almost all attack on the ring structure must be initiated on the less hindered α side. Thus, catalytic hydrogenation gives *cis* addition to the double bond from the α side to give 3β-cholesterol (the hydroxy group at 3 is β and equatorial). Coprostanol (present in feces) results from a bacterial reduction of ingested cholesterol in the intestines. The 3-hydroxy group is still β but is axial in the coprostane configuration.

Addition of bromine to the double bond in cholesterol proceeds by initial attack on the α side (a), followed by bromide ion on the β side (Eq. 27.7: only partial structures are

(27.7)

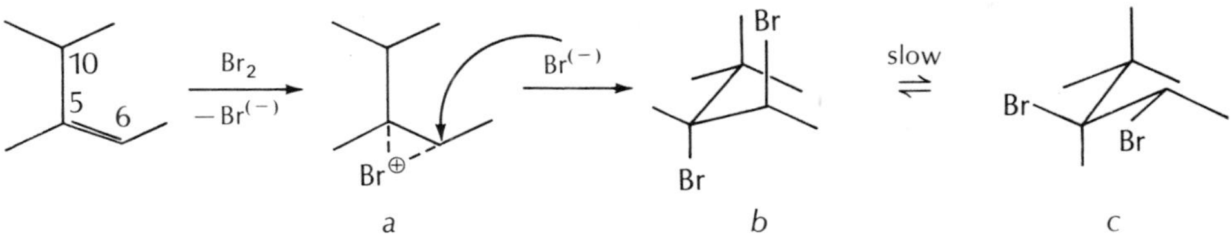

shown) to give the 5α,6β-dibromocholestanol with bromines axial. The structure is unstable (because of axial repulsion of the methyl at 10 with the bromine at 6) and slowly equilibrates to give c, 5β,6α-dibromocoprostanol. (There is dissociation of the bromine at 6, the molecule goes back to a, and ring A flips into the coprostane configuration as $Br^\ominus$ attacks a at 5β.) Treatment of b with NaI in acetone regenerates cholesterol immediately (the bromines are *anti* to each other, so *trans* elimination is possible), but c reacts slowly because the bromines are *gauche* to each other and elimination cannot be concerted. These reactions first showed the importance of stereochemical control of addition-elimination. □ □

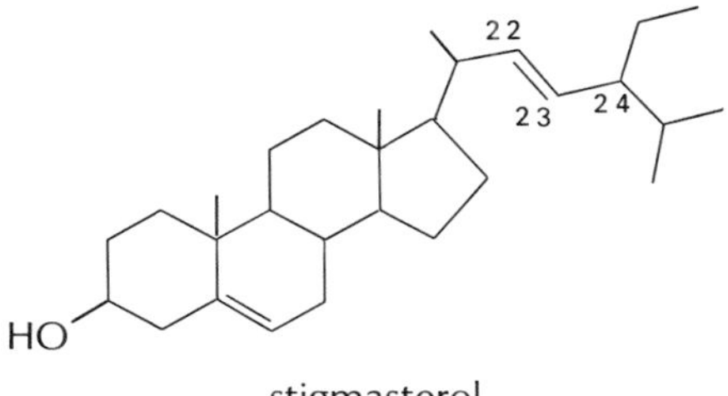

Stigmasterol; ergosterol. Other sterols, especially those of plant origin, have more double bonds and variations in the side chain. For example, stigmasterol (soybean oil) differs from cholesterol only in the side chain. There is a double bond at 22 and an ethyl group at 24. Ergosterol (ergot, yeast) differs from cholesterol in having a methyl at 24 and

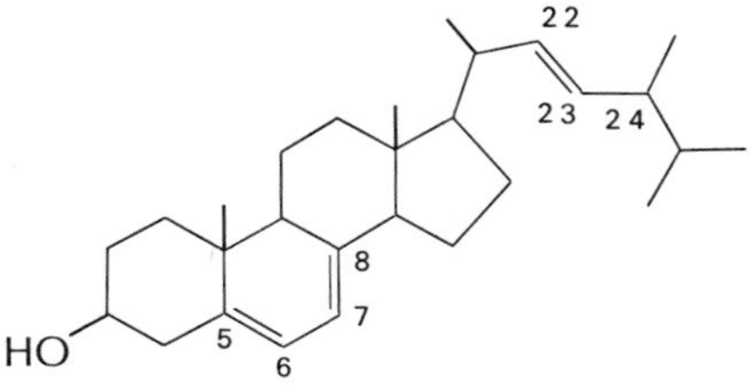

extra double bonds, not only at 22, but also at 7, in conjugation with one at 5. Ergosterol is a precursor of vitamin D_2.

Exercises

27.2 Show how to convert cholesterol into cholestanone.

27.3 Show the structures of the products to be expected:

$$\text{cholestanol} \xrightarrow[\text{pyridine}]{ArSO_2Cl} \xrightarrow[\substack{\text{EtONa}\\S_N2}]{H_2C(COOEt)_2}$$

27.4 Vitamin D

Occurrence; formation. During the search for the antirachitic factor in the 1920's, it turned out that only those sterol fractions, mostly from fish oils, that gave a uv band characteristic of conjugated dienes would acquire antirachitic activity. Soon afterwards, the conversion of ergosterol into vitamin D by uv irradiation was announced. Unfortunately, the announcement was premature, because the vitamin D so formed was a mixture. When the pure vitamin was isolated, it was called vitamin D_2 (or calciferol) and the mixture was called vitamin D_1. Later, an even more potent substance, vitamin D_3, was prepared. The D vitamins are necessary for the regulation of deposition of calcium (bones, teeth).

Ergosterol, upon irradiation (282 nm), absorbs energy and is converted to a singlet state, some of which undergoes intersystem crossing to a triplet, which does not return to the ergosterol ground state. Instead, cis-tachysterol seems to be the first substance formed. However, irradiation of cis-tachysterol probably regenerates the triplet, which also gives rise to trans-tachysterol, lumisterol, and vitamin D_2, which seems to be the end product of this complex photochemical reaction (Eq. 27.8).

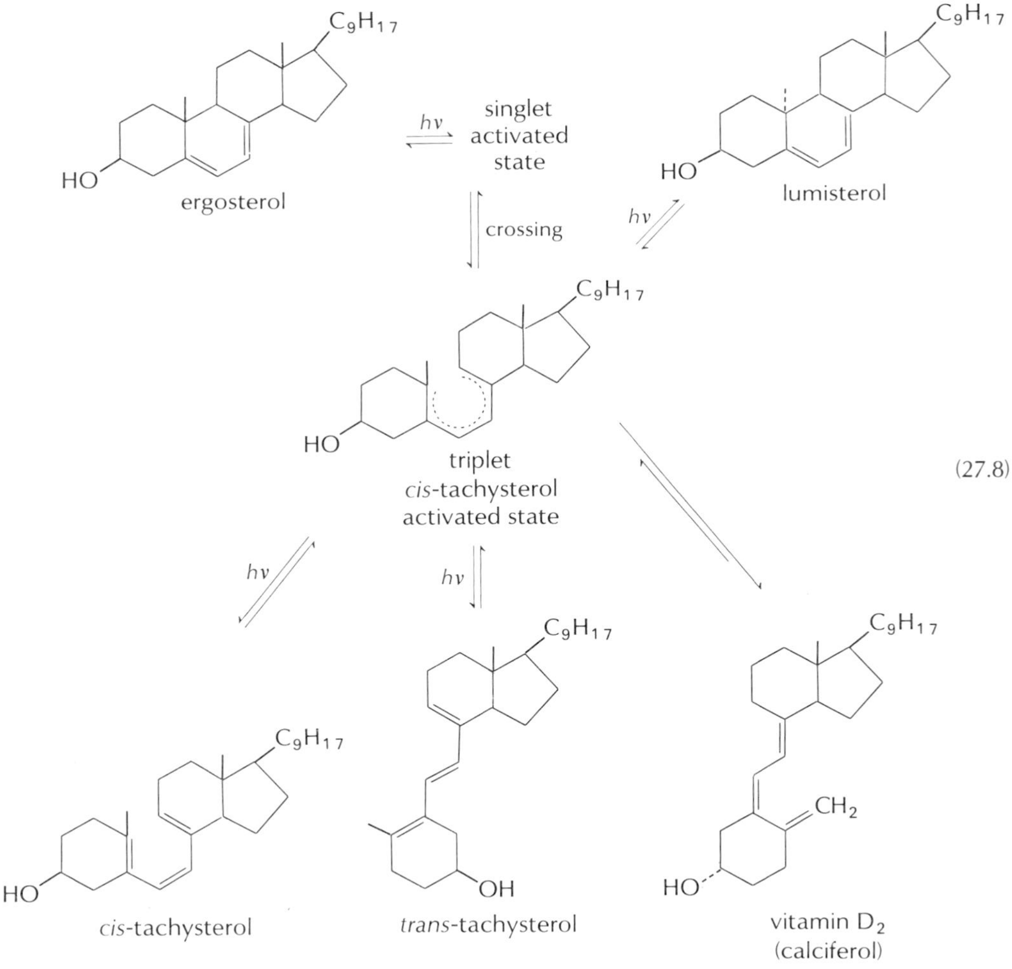

(27.8)

Then it was found that if the double bond at 22 in the side chain was hydrogenated, the activity of the irradiated product was equal to or greater than that of vitamin D_2. Cholesterol already has a saturated side chain, but without the extra methyl. Cholesterol was converted into 7-dehydrocholesterol (5,7-cholestadiene-3β-ol) and irradiated. The result was a similar array of products to those from ergosterol. The final product was called vitamin D_3 and was found to be even more active than vitamin D_2. Afterwards, vitamin D_3 was isolated from cod liver oil and other fish oils.

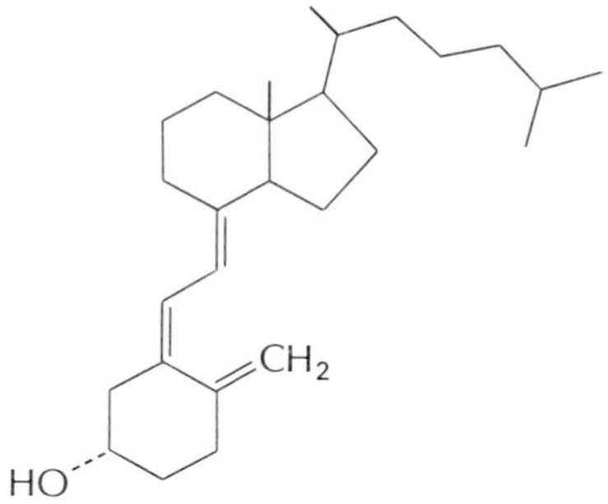

vitamin D_3

27.5 Sex hormones

Occurrence; effects. The sex hormones, formed in the testes or ovaries under the influence of hormones from the pituitary, are steroids that are responsible for development and control of the genitals and other primary sexual characteristics. Secondary sexual characteristics such as high-pitched voice and lack of facial and body hair in the female (presence of estrogens) or low-pitched voice, presence of facial and body hair (and perhaps tendency to baldness) in the male (presence of androgens) also respond to sex hormones.

Structures of estrogens and androgens. The sex hormones are very similar to one another in shape, but they differ somewhat in number of double bonds. The first five compounds shown are female hormones. Their common characteristics are the aromaticity of the A ring (they all are phenols) and the presence of a keto or hydroxy group in place of a side chain at the 17 position. Estrone was the first sex hormone to be isolated in 1929; identification of the others followed quickly in the next few years. Estradiol probably is the hormone itself, being about nine times more active than estrone, which is about twice as active as estriol. Equilenin (first isolated from the urine of pregnant mares, but much more abundant in stallion urine) is only weakly estrogenic. Stilbestrol is a synthetic compound with estrogenic properties: its shape, size, and functions closely resemble those of estradiol.

The male hormone is testosterone: androsterone is weaker in activity and evidently is either an intermediate on the way to testosterone or a metabolic product. The similarities in structure of estradiol and testosterone and of estrone and androsterone are surprising in view of the differences in their effects. The differences in structure are confined to ring A and to the absence of the angular methyl group at 10 in the estrogens.

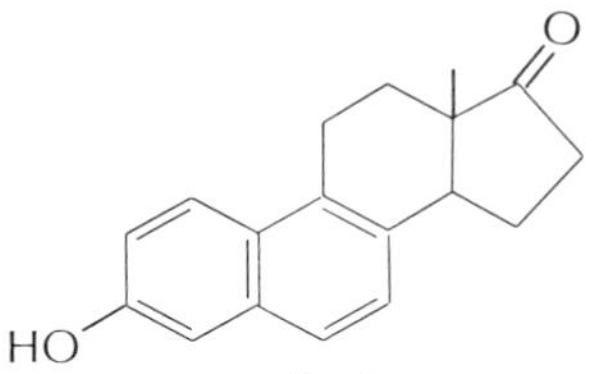

[Structures: equilenin, estrone, estradiol, estriol, stilbestrol, androsterone, testosterone, progesterone]

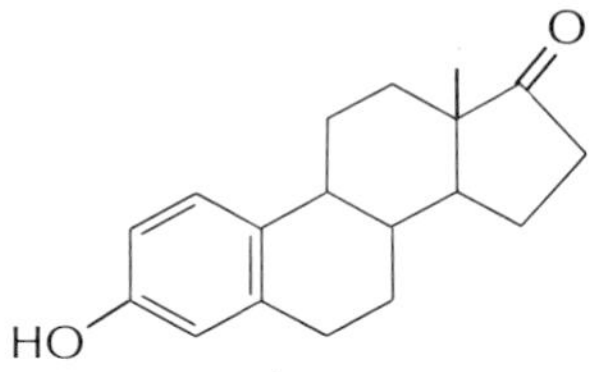

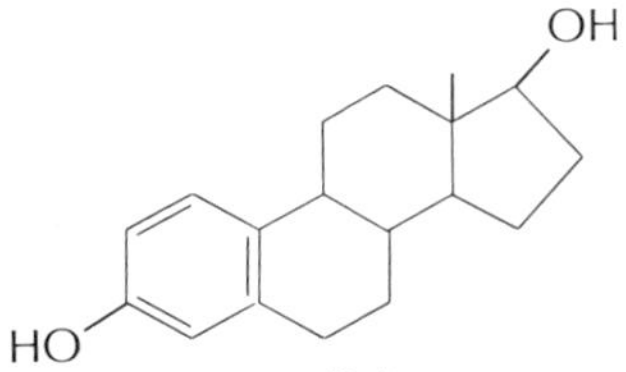

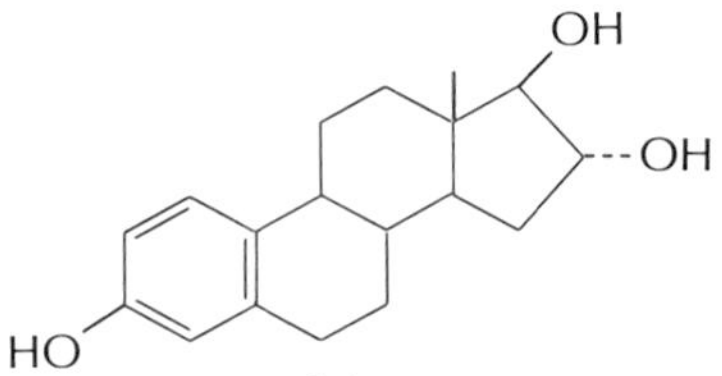

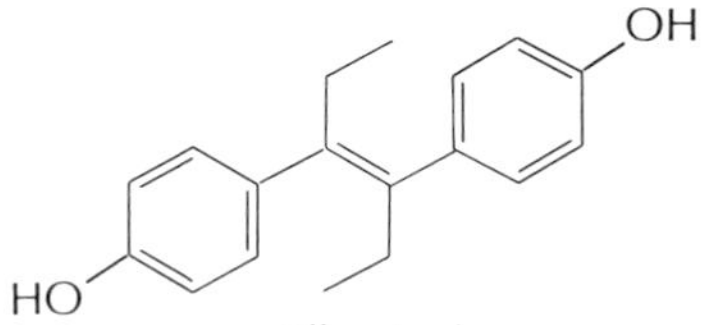

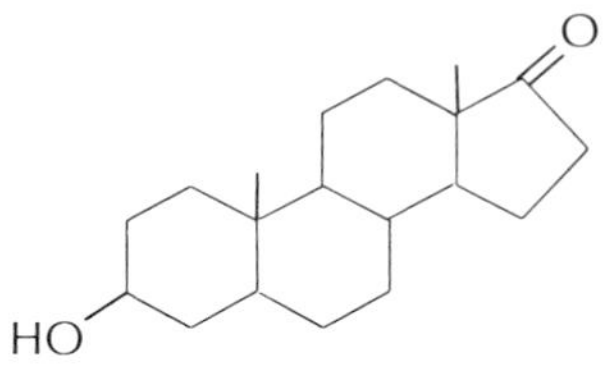

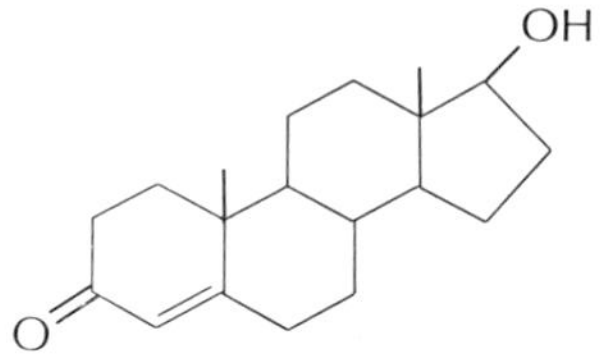

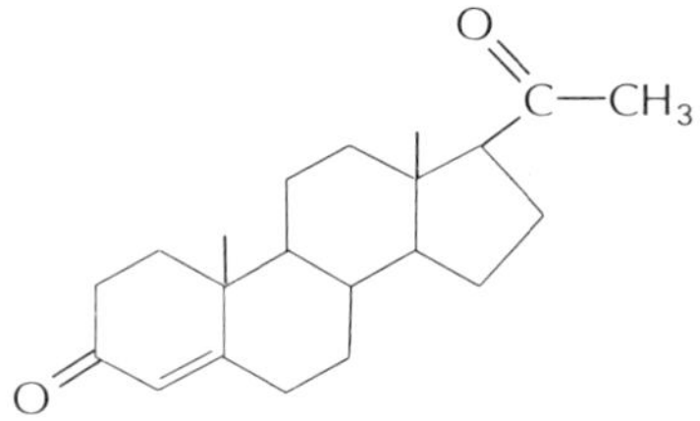

Progesterone; the Pill. Progesterone differs from testosterone only in the replacement of the 17 hydroxy group by an acetyl. Progesterone is secreted by the corpus luteum (a yellow part of the ovary) and operates to prepare the uterus for acceptance of a fertilized ovum. It also acts to maintain pregnancy by suppression of ovulation. Research work on the modification of the sex hormones resulted in the discovery of 17α-ethynyl-estradiol and the related enovid ketone, both effective in suppression of ovulation. When they were put together as "the Pill", oral contraception was made possible and rapidly became popular. The ultimate personal, social, economic, and political impact has not yet been felt.

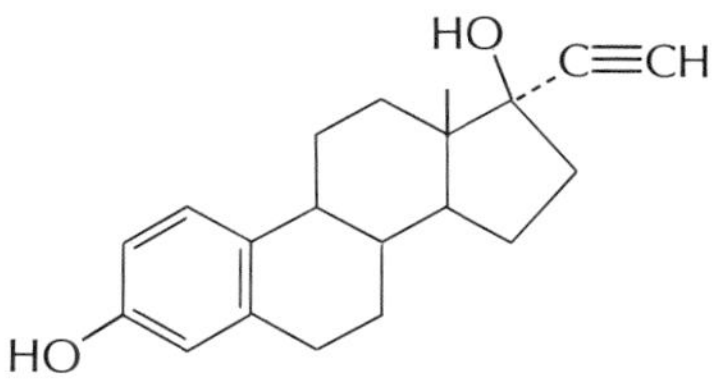

17α-ethynylestradiol

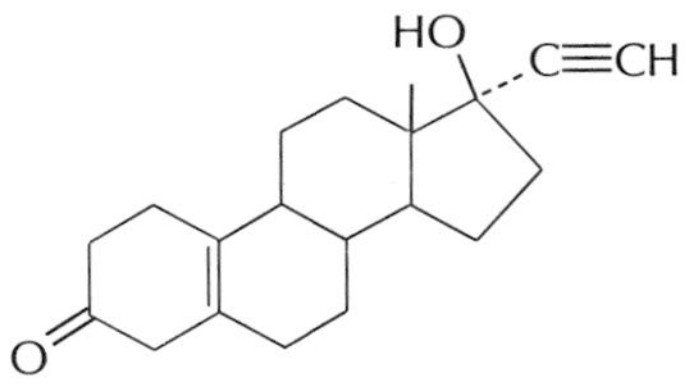

enovid ketone

Exercises

27.4 Show how to convert estrone into 17α-ethynylestradiol.

27.5 Devise a method for obtaining testosterone from progesterone.

27.6 Adrenal hormones

Occurrence; structures; effects. The adrenal glands secrete not only epinephrine and norepinephrine (Sec. 24.9) but also a group of steroids known collectively as cortin (adrenal cortex), which are essential to well-being. A large number of individual steroids from this source have been isolated and identified, but not all are active. (By <u>active</u>, we mean that the steroid can sustain the life of a test animal after removal of the adrenals.) A few of the active cortical hormones are shown. They all have a keto group at 3, a conjugated

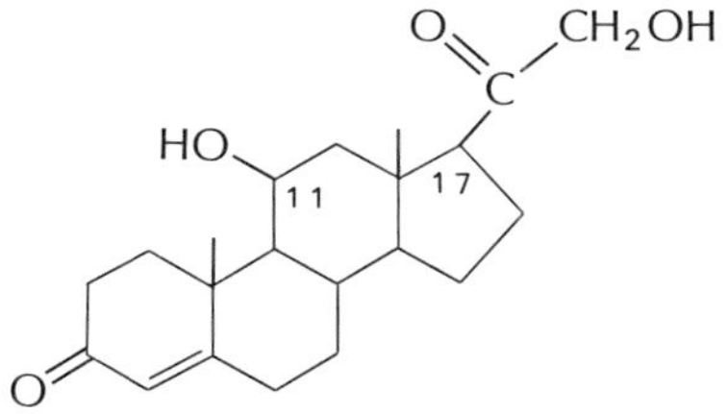

corticosterone

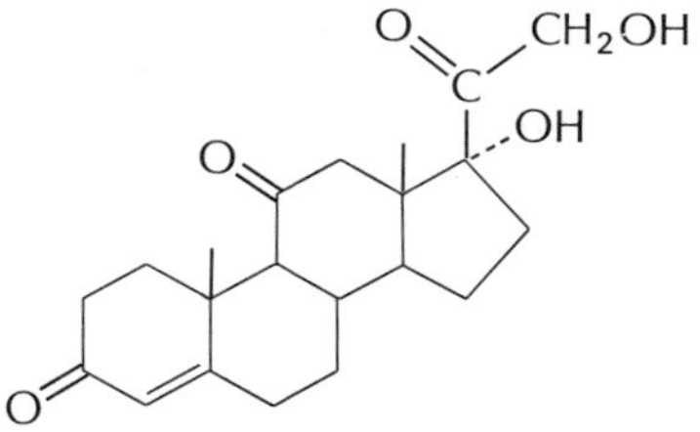

cortisone

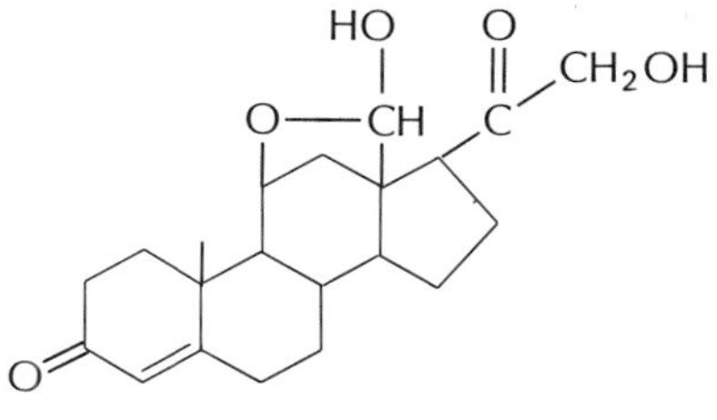

aldosterone

double bond at 4, an angular methyl group at 10, and a —COCH$_2$OH group at 17. Most of the hormones have an angular methyl group at 13 (aldosterone is an exception) and a hydroxy or keto group at 11 (deoxycorticosterone, not shown, is an exception). Some, like cortisone, have an α hydroxy group at 17. Corticosterone differs from progesterone only by having two more hydroxy groups, one at 11, the other in the acetyl group.

The cortical hormones are active in maintaining electrolyte balance, control of carbohydrate and protein metabolism, and skin health. Cortisone, it was discovered, was helpful in rheumatoid arthritis and was thought to be a wonder drug until clinical use uncovered a variety of undesirable side effects in many patients. Several synthetic variations have been found to be superior to the naturally occurring substances for treatment of certain ailments. Thus, prednisone, prednisolone (the keto group at 11 is reduced to a β hydroxy group), and dexamethasone are superior to cortisone in antirheumatic activity.

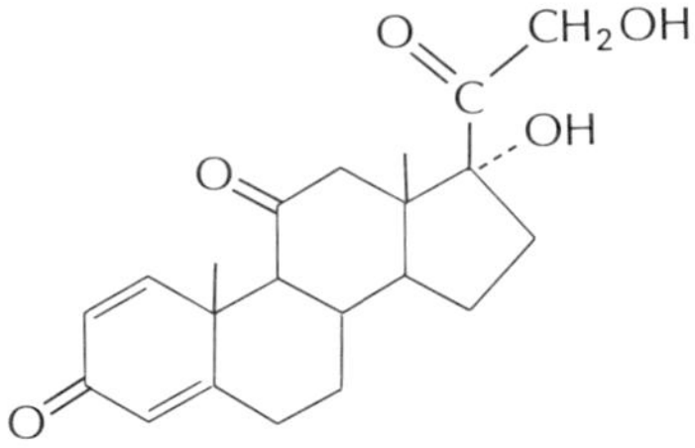

prednisone

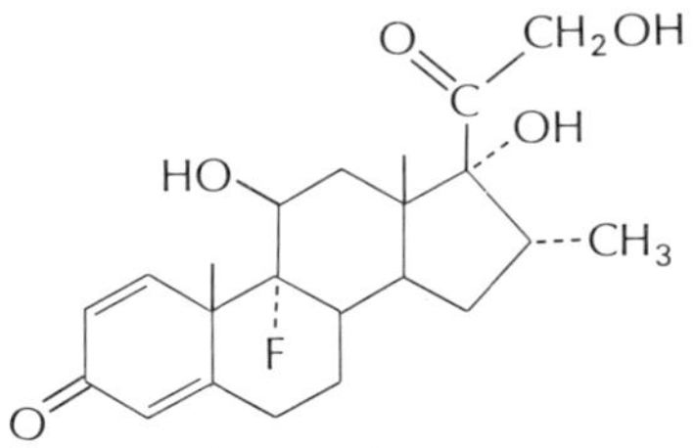

dexamethasone

Cortisone: synthesis. At first, cortisone was practically unavailable because it was very difficult to isolate from cattle adrenal glands, and the quantity present was small to begin with. Attempts at synthesis from other steroids were long and involved because of the

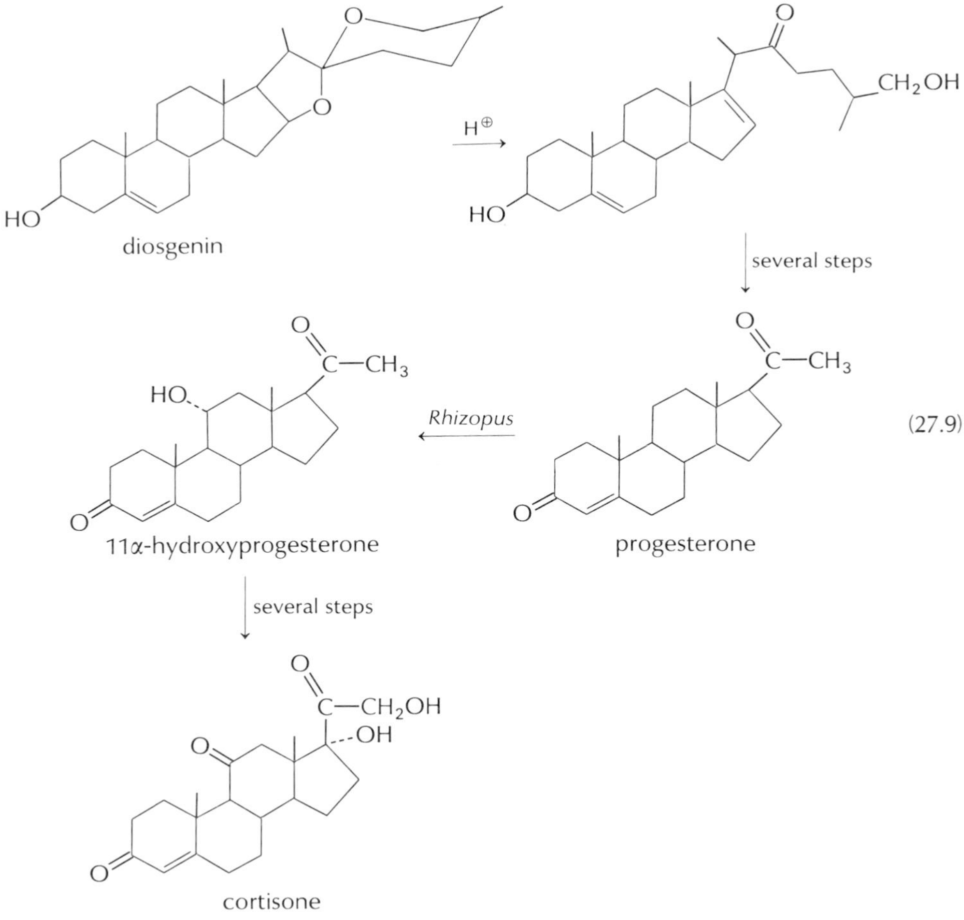

diosgenin

several steps

11α-hydroxyprogesterone ←—Rhizopus—— progesterone (27.9)

several steps

cortisone

difficulty of introducing a hydroxyl or keto group at 11. (One synthesis involved 32 steps from desoxycholic acid.) The solution to the problem came unexpectedly from another area of research: work on molds. It was found that the mold *Rhizopus nigricans* grown upon a medium containing progesterone, introduced a hydroxy group at 11α in yields up to 90%. From 11α-hydroxyprogesterone, only ten steps, in high yield, gave cortisone. In the meantime, efforts to find a suitable source of progesterone had met success. An abundant source of diosgenin is the Mexican yam. Thus, a reliable source and convenient reactions gave access to both the sex hormones and the cortical hormones (Eq. 27.9).

Cholesterol, available in abundant supply from packing houses, is a poor starting material for synthesis of hormones because the long side chain at 17 must be removed. Because it is a saturated alkyl group, there is no specific location for breaking into the chain without disrupting the ring structure. Until the diosgenin route was worked out, cholesterol was used as shown in Equation 27.10. The double bond and the hydroxy group are protected

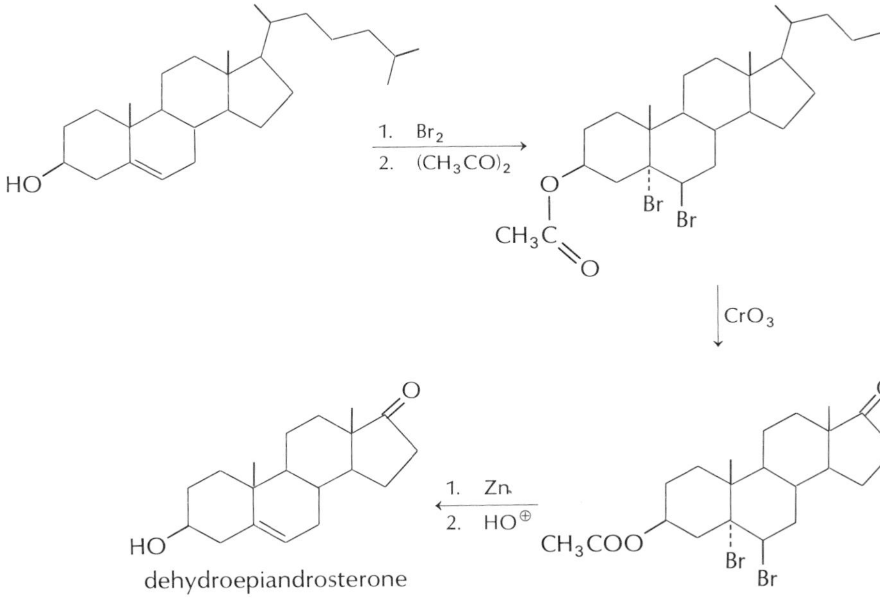

(27.10)

first by bromination and esterification. Drastic oxidation attacks tertiary hydrogens first and breaks off the chain. (The tertiary hydrogens in the ring structure are attacked also.) Removal of bromine with zinc and hydrolysis of the ester gives dehydroepiandrosterone, but in only 5% yield from cholesterol. It is no wonder that the approach was abandoned. Ergosterol and stigmasterol also were considered as starting materials, because the double bond in the side chain offered a point of attack. However, difficulties arose in the attempt to protect the double bonds in ring B without protecting the 22 double bond at the same time.

Exercises

27.6 The first total synthesis of cortisone from nonsteroidal precursors was achieved by Woodward and his students (Harvard University) in 1951. Some of the steps follow; what are the reagents (organic and inorganic) needed for each reaction?

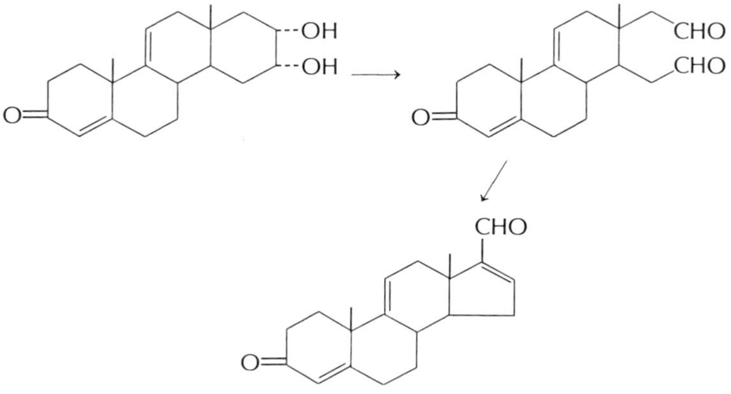

27.7 Bile acids

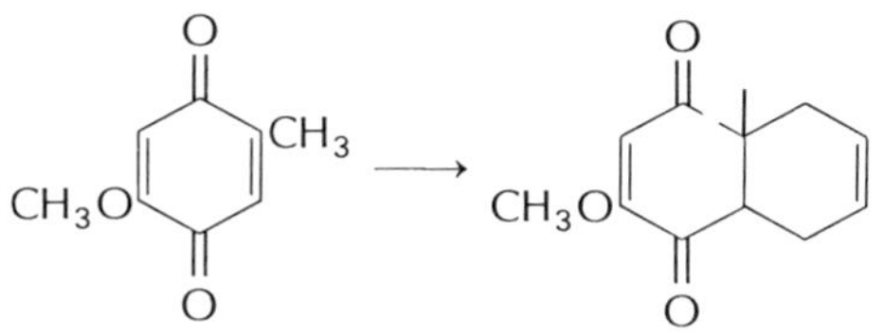

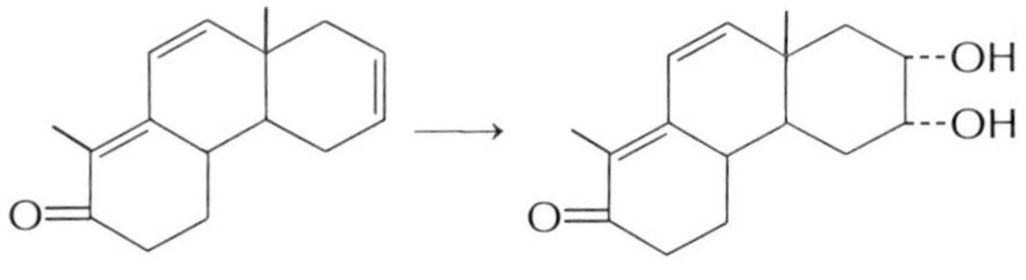

Occurrence; effects; structures. The bile acids occur as bile salts (from bile) in the stomach and intestines. They serve as emulsifying agents for fats and help in the hydrolysis of fats and transfer of fatty acids to the walls of the intestines. A variety of bile acids occur in different ratios in different species. The bile salts all are amides of glycine or taurine. The bile acids are liberated upon alkaline hydrolysis of the salts (Eq. 27.11).

$$\text{Cholic-CONHCH}_2\text{COOH} \xrightarrow{\text{HO}^\ominus} \text{Cholic-COOH} + \text{H}_3\overset{\oplus}{\text{N}}\text{CH}_2\text{COO}^\ominus$$
$$\text{glycocholic acid} \qquad\qquad\qquad\qquad\qquad \text{glycine}$$

$$\text{Cholic-CONHCH}_2\text{CH}_2\text{SO}_3\text{H} \xrightarrow{\text{HO}^\ominus} \text{Cholic-COOH} + \text{H}_3\overset{\oplus}{\text{N}}\text{CH}_2\text{CH}_2\text{SO}_3^\ominus$$
$$\text{taurocholic acid} \qquad\qquad\qquad\qquad\qquad \text{taurine}$$

(27.11)

The bile acids differ from most of the steroids we have seen in that they have a coprostane configuration with α hydroxy groups. Cholesterol has been shown by ^{14}C labeling to be transformed in the liver into cholic acid and the other bile acids. This biosynthesis must require a large number of steps to break the side chain by oxidation, reduce cholesterol to a coprostanol, invert configuration of 3-hydroxy from β to α, and selectively oxidize the 7 and 12 positions to α hydroxy groups.

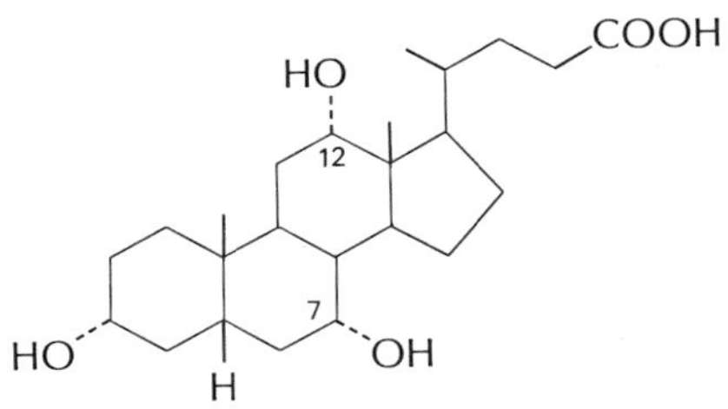

cholic acid

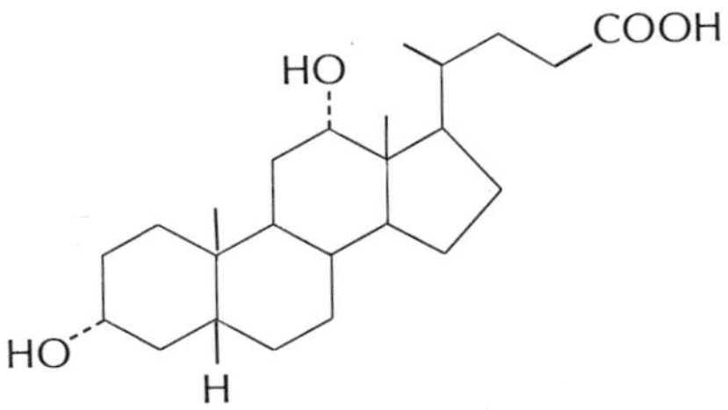

desoxycholic acid

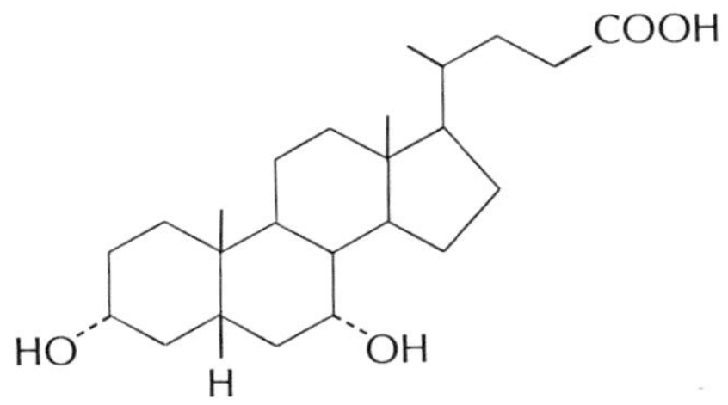

chenodesoxycholic acid

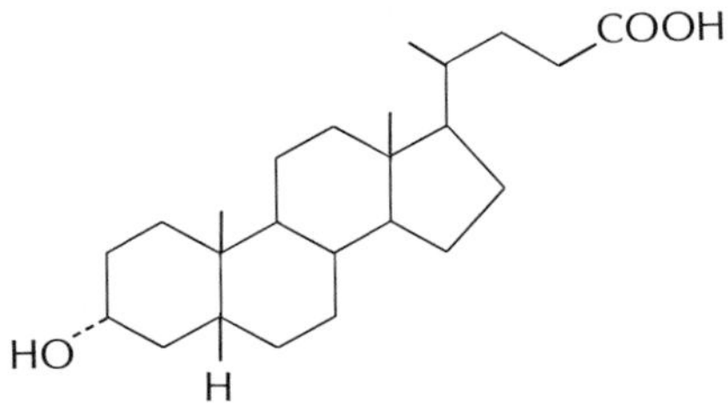

lithocholic acid

27.8 Sapogenins; cardiac glycosides

Saponin, sapogenin; classes of sapogenins; tigogenin. Saponin is the name used for certain glycosides (linked to a molecule of a sugar) that occur in some plants and form soapy, colloidal dispersions in water. Saponins are toxic to fish.

Hydrolysis of a saponin gives a sugar and a nonsugar. The nonsugar, called an aglycone, is also known as a sapogenin. Most sapogenins fall into two categories, the many pentacyclic triterpene sapogenins (not described in this book) and the relatively rare steroid sapogenins. We already have met a steroid sapogenin in diosgenin (Eq. 27.10). Tigogenin (from tigonin in digitalis seeds) has the cholestanol configuration. The carbon atom at 22 is a

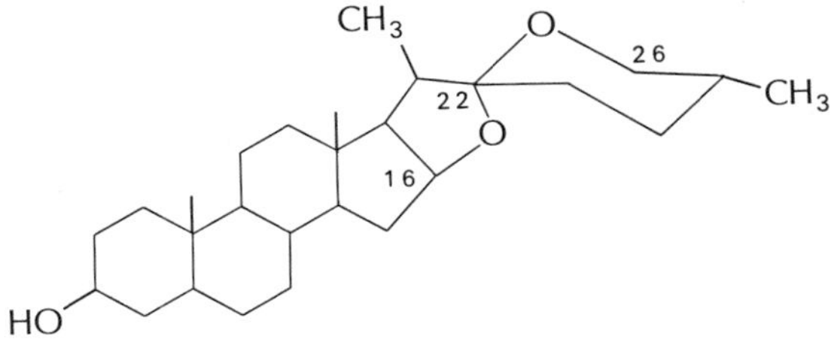

tigogenin

keto group that has formed a ring system by acetal formation with hydroxy groups at 16 and 26. The sugar molecule is linked to the 3-hydroxy group to form the saponin.

Cardiac glycosides: digitalis, digitoxigenin, strophanthidin. Digitalis is a mixture of glycosides derived from the purple foxglove and known as cardiac glycosides. In small doses, digitalis causes heart muscle to contract and has been so used medically for a long time. The aglycones obtained upon hydrolysis are far less active. The aglycones are different from most other steroids in having a *cis* junction not only between rings A and B but also between rings C and D. They also have a hydroxy group at the rarely seen 14 position. The side chain is a lactone of an α,β unsaturated γ-hydroxy acid. Digitoxigenin is shown as an example.

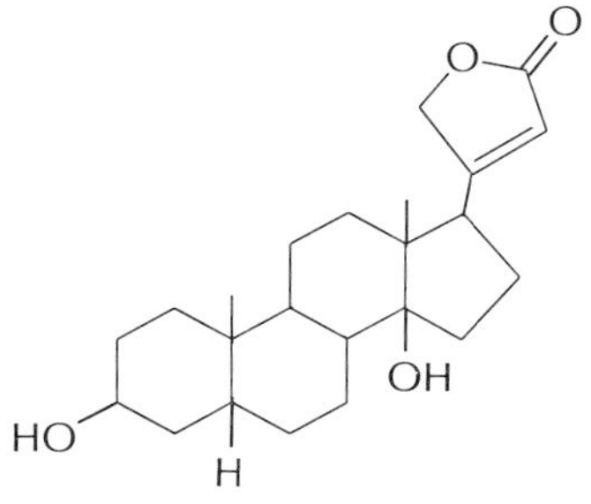

digitoxigenin 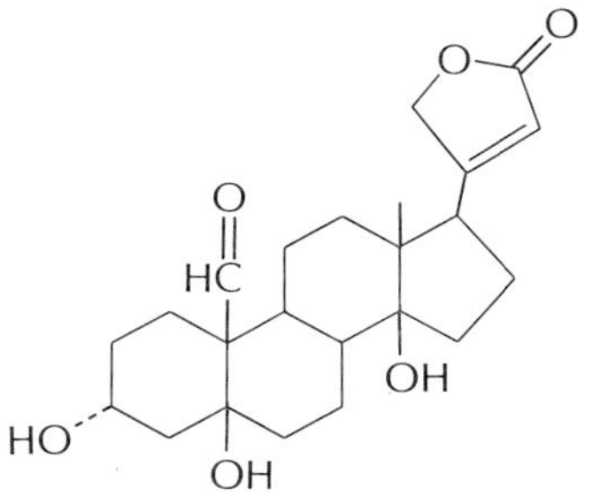   strophanthidin

Strophanthidin differs even more in having the methyl at 10 oxidized to an aldehyde group. Strophanthin (used in some arrow poisons in Africa), the glycoside of strophanthidin, is a virulent heart poison. Many other poisonous steroidal glycosides are known.

27.9 Toad poisons; insect hormones

Bufotalin and batrachotoxin.   Toads and frogs (and the other amphibians) have developed a variety of chemical defenses against their enemies. Unlike the plant toxic saponins, which are glycosides, the toad poisons (only a few of which are steroid) are esters of various acids.

Bufotalin, from the skin glands of the European toad, has been known for some time. Like digitoxigenin, it has *cis* A/B and C/D ring junctions. In fact, it differs from digitoxigenin only in having a slightly larger lactone ring at 17 and a β acetate at 16.

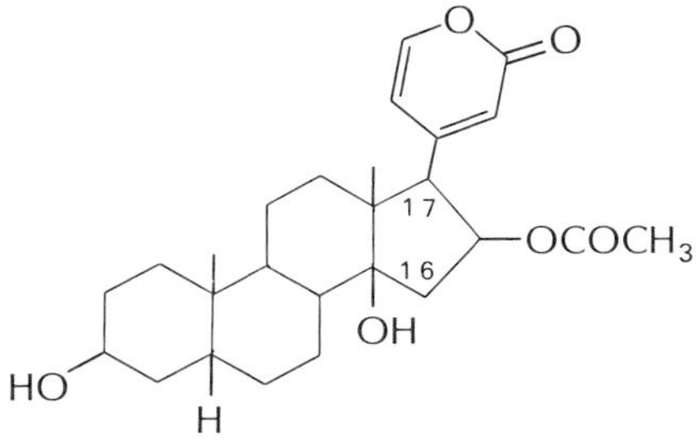

bufotalin

Batrachotoxin, from the skin of a small frog native to the forests of Colombia, is one of the most toxic substances known. On a scale of relative toxicity to mice of strychnine as 1, NaCN is 0.05, bufotalin is 1.25, and batrachotoxin is 250. The venom is used by the Indians as a blowgun dart poison. The structure of batrachotoxin, although of the *cis* A/B and C/D steroid system, also contains nitrogen and is classed as a steroidal alkaloid (Chap. 28). The D ring projects to the α side of the steroid system, and the seven-membered ring containing both N and O projects to the β side. The oxygen link between 3 and 9 on the α side of the system is a hemiacetal formation between a keto group at 3 and an α hydroxy group at 9. (Use models here.)

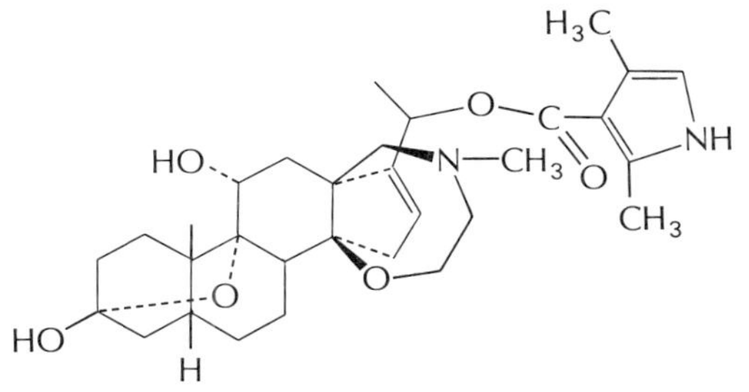

batrachotoxin

Ecdysone.   Ecdysone is an insect hormone that develops in the larval stage of an insect and sets in motion the transformation into the pupal (chrysalis) form. Premature exposure of a larva to ecdysone causes molting to occur before the system is prepared for it, and the larva is killed. Ecdysone does not seem to have any effects on animals and may be useful as a practical larvicide, specific in its effect and biodegradable.

Ecdysone is a pentahydroxy steroidal ketone.

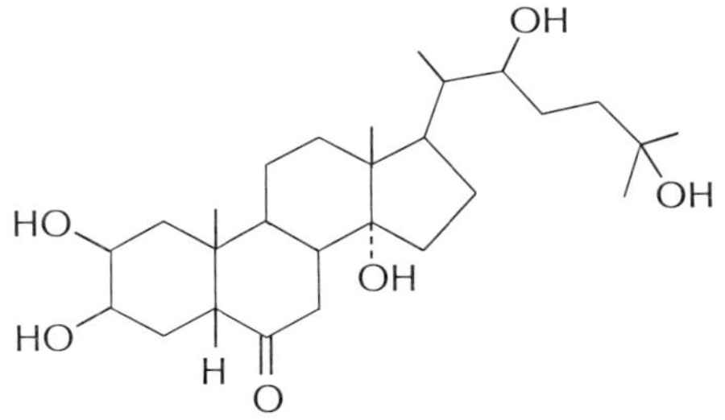

ecdysone

27.10   Summary

We made a start on the steroids with the structure of cholestane and coprostane and the use of α and β to designate on which side of the ring system a given substituent may be. Thereafter, we took up the various steroids class by class, with occasional digressions to nonsteroids.

We have omitted all discussion of the evidence, so painstakingly gathered, for the structures given. For those who are interested in pursuing a topic, the bibliography provides a number of starting points.

27.11   Bibliography

Of the textbooks, Fieser and Fieser (Chap. 7, ref. 7) is the most complete concerning steroids. Noller (Chap. 7, ref. 11) is good. Allinger *et al.* (Chap. 7, ref. 1) is up to date but brief, as is Baker (Chap. 21, ref. 1). Norman (Chap. 22, ref. 4) has excellent summaries concerning some specific syntheses of terpenes and steroids such as vitamin A, β-carotene, cholesterol, estrone, and epiandrosterone. A variety of advanced books deal not just with a single section of this chapter but with parts of parts of sections.

A few specific references follow.

1. Fieser, L. F., in *Bio-organic Chemistry*, M. Calvin and M. J. Jorgenson, Eds. San Francisco: Freeman, 1968, pp. 158–166. A chronological account of steroids, 1770 to 1955. The same account appears in *Organic Chemistry of Life*, M. Calvin and W. A. Pryor, comps. Freeman, 1973, pp. 32–40.

2. Bachman, W. E., W. Cole, and A. L. Wilds, The Total Synthesis of the Sex Hormone Equilenin and Its Stereoisomers, *J. Amer. Chem. Soc.* 62, 824 (1940). If at all possible, this paper should be read as a model, not only as a masterpiece of clear reporting and description of experimental details, but as a readable account of an outstanding achievement in synthetic organic chemistry: the first total synthesis of a sex hormone.

28 Heterocyclic Compounds

Heterocyclic compounds occur in vast array throughout nature. From the beginning of organic chemistry, the heterocycles have posed the greatest challenges to the synthetic skills of organic chemists. The chemistry of the heterocycles brings together all of the reactions and concepts we have considered separately in the preceding chapters. The variety of heterocycles is enormous. We shall adopt a classification depending upon the number and size of rings present, and in each class we shall proceed with increasing number of heteroatoms. We cannot cover all the possible substituents and types of side chains for each ring system, but we shall mention what is necessary for a cursory introduction to the heterocyclic compounds.

28.1 Heterocycles: naming   886
   1. Heterocycles: aromatic character.   2. Oxa-aza naming system; systematic naming.   3. Fused ring systems; common names. Exercises.

28.2 Substituted cyclopentadienes   889
   1. Aromaticity; resonance structures; reactivity.   2. Reactions: acidic polymerization; substitution; Diels-Alder reaction; reduction.   3. Two hetero atoms: pyrazole, imidazole, isoxazole, oxazole, thiazole. Exercises.

28.3 Substituted benzenes   894
   1. Pyran; pyrylium salt; pyrones, synthesis.   2. Pyridine: basicity, solvent properties; substitution reactions.   3. Resistance to oxidation; pyridine oxide; reduction.   4. Substituted pyridines: reactions.   5. Synthesis of substituted pyridines.   6. Diazines; pyrimidines; barbiturates; vitamin B_1. Exercises.

28.4 Bicyclic compounds with a five-membered heterocycle   904
   1. Types; naming.   2. Indole: similarity to pyrrole; substituted indoles.   3. Fischer indole synthesis. Exercises.

28.5 Bicyclic compounds with a six-membered heterocycle   906
   1. Quinoline: properties; reactions.   2. Skraup synthesis.   3. Isoquinoline: properties; reactions; synthesis. Exercises.

28.6 Purines and pteridines   910
   1. Purine and pteridine: structures, numbering.   2. Purines: uric acid.   3. Methylated xanthines.   4. Pteridines: folic acid; riboflavin. Exercises.

886   Chapter twenty-eight

28.7   Alkaloids   916
        1. Definition; examples.   2. Indole alkaloids.   3. Quinoline, isoquinoline, and tropane alkaloids.

28.8   Nucleic acids   920
        1. Discovery: DNA, RNA; structures of nucleosides and nucleotides.
        2. Secondary structure.   3. Tertiary structure: the double helix.
        4. Various RNA's; coding; the genetic code.

28.9   Heterocyclic metal coordination compounds   927
        1. Porphin systems: aromaticity.   2. Hemoglobin.   3. Cytochromes.
        4. Chlorophylls: phorbin.   5. Corrin systems: vitamin B_{12}.

28.10   Summary   933

28.11   Problems   934

28.12   Bibliography   937

28.13   Charles C. Price: From antimalarials to involvement   939a

---

28.1 Heterocycles: naming

Heterocycles: aromatic character.   A heterocycle is a ring compound (it may be polycyclic) in which one or more of the carbon atoms in a carbocycle is replaced by an atom of another element. The most common hetero atoms are nitrogen, oxygen, and sulfur. However, heterocycles that have phosphorus, silicon, tin, boron, and so on, are known. We shall examine the cyclopentadiene and benzene replacements and the di- and tricyclic ring systems formed by fusion with benzene and other heterocycles. Other sizes of heterocycles exist, but we have already looked at epoxides, β, γ, and δ lactones and lactams, cyclic anhydrides, and imides in previous chapters. These compounds, although they are heterocycles, react more in the nature of their open-chain relatives than as heterocycles. A heterocycle, we shall say in this chapter, has aromatic character and ring stability. We really are studying the heterocyclic aromatic compounds.

Oxa-aza naming system; systematic naming.   The naming of heterocycles is a hodgepodge of common and systematic names. Common names have endured for all the well known heterocycles. In general, we shall continue to use the common names. From time to time a systematic name will be used as well. Also, many of the common names are specifically authorized exceptions to the rules.

Two systematic naming systems are in use. The simpler of the two is the oxa-aza scheme, in which the hetero atoms are treated as substitutions in a carbocyclic system. The hetero atoms are named as prefixes: oxa, thia, aza (for O, S, N) in order of preference both for numbering and for position in the name. A few examples show how the system operates.

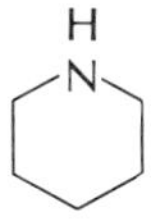

azacyclohexane (piperidine)

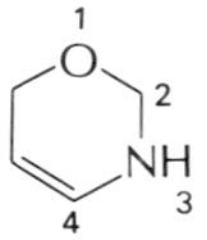

1,3-oxaza-4-cyclohexene

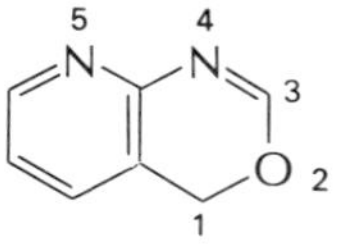

1,2-dihydro-2-oxa-4,5-diazanaphthalene

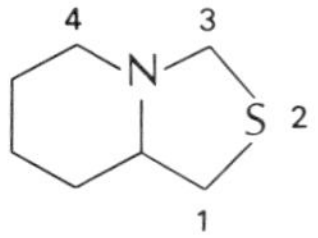

perhydro-2,3a-thiazaindane

The other systematic naming system is somewhat more complicated.
1. Ring size is shown by the use of stems, which may be remembered as the underlined portions of some of the words shown: 3, third; 4, tetra; 5, pyrrole; 6, pyridine; 7, hepta; 8, octa; 9, nona; and 10, deca.
2. Hetero atoms are designated as in the oxa-aza scheme.
3. Unsaturation is specified by a suffix, as shown in Table 28.1.

Table 28.1 Systematic naming of heterocycles

Ring size	Stem	N in ring		No N in ring	
		Unsaturated[a]	Saturated	Unsaturated[a]	Saturated
3	ir	-ine	-idine	-ene	-ane
4	et	-e	-idine	-e	-ane
5	ol	-e	-idine	-e	-ane
6	in	-e	b	c	-ane
7	ep	-ine	b	-in	-ane
8	oc	-ine	b	-in	-ane
9	on	-ine	b	-in	-ane
10	ec	-ine	b	-in	-ane

[a] *Unsaturated* implies the maximum number of double bonds.
[b] In these cases, the prefix perhydro is used.
[c] Sometimes -e; sometimes no suffix is used: see examples.

4. Numbering always begins with the preferred hetero atom (we are speaking of monocyclic rings only) and proceeds in the direction to give a. other hetero atoms, b. substituents, and c. double bonds, the lowest possible numbers.
5. Partial reduction is specified as dihydro, tetrahydro, and so on. In such cases the remaining double bond or bonds are numbered last. (The extra hydrogens are regarded as substituents in rule 4.) In some instances, as in odd-numbered rings, the maximum number of double bonds leaves a single extra hydrogen, which is indicated as shown in the examples.

The examples illustrate the application of the rules. The common names (if any) of the compounds shown are given in parentheses. It has been recommended that heterocyclic rings be oriented so that the number 1 hetero atom is at the top and numbering proceeds clockwise. The old custom was to write the hetero atom at the bottom of the ring. The custom has persisted: do not be surprised to find structures written in that way. We shall follow the recommended orientation.

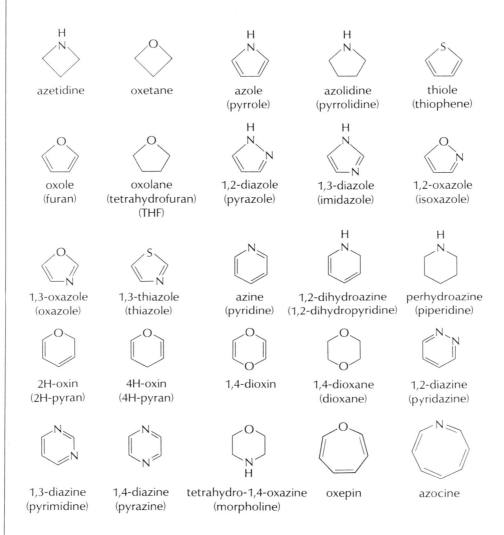

Fused ring systems: common names. Rules exist for the naming of fused ring systems involving heterocycles, but we shall not try to explain them because it would take too long. Besides that, many of the fused ring systems have common names exempted from the naming system, and we shall use them. For those interested in the subject of naming, the bibliography gives a few references.

Here are some of the more common fused ring heterocyclic compounds, some of which are known by more than one name.

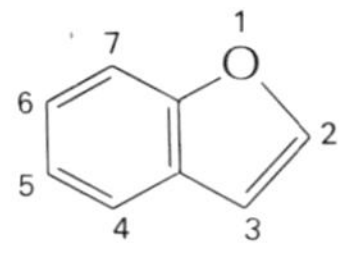

28.2 Substituted cyclopentadienes

benzofuran (coumarone)

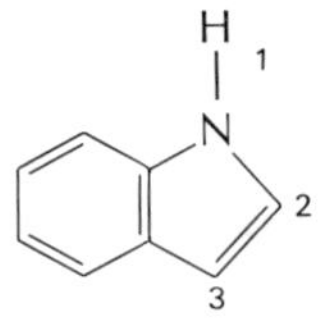

indole

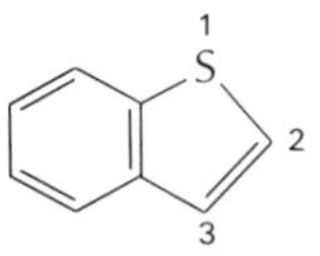

benzothiophene (thianaphthene)

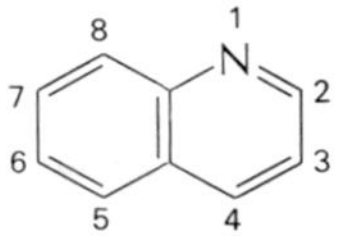

quinoline

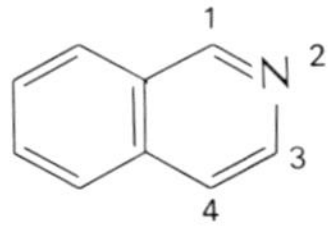

isoquinoline

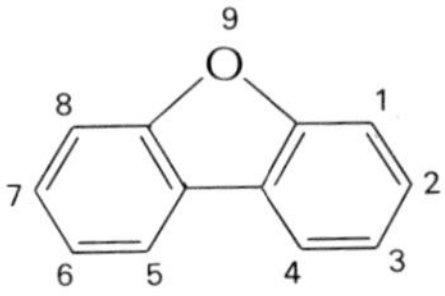

dibenzofuran (diphenylene oxide)

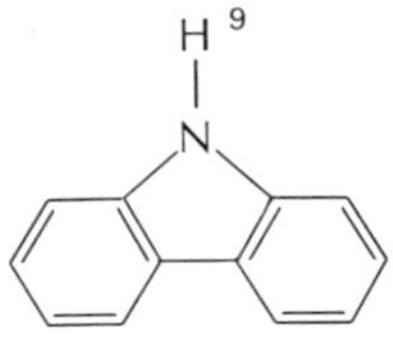

carbazole

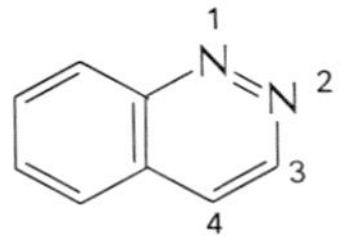

cinnoline

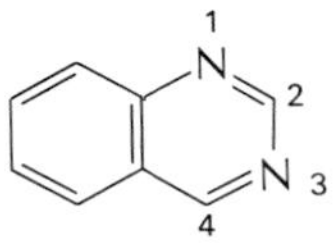

quinazoline

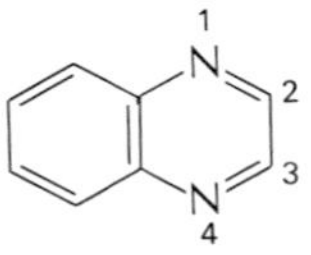

quinoxaline

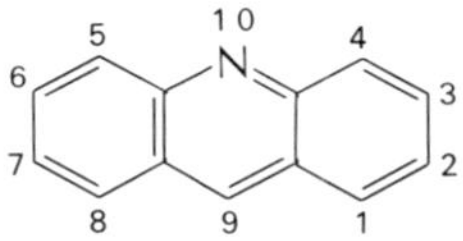

acridine

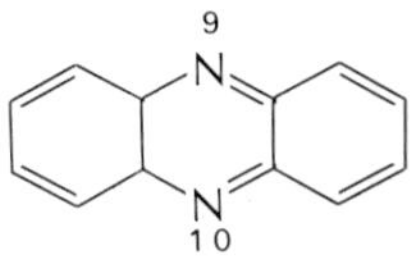

phenazine

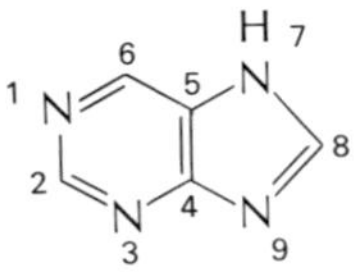

purine

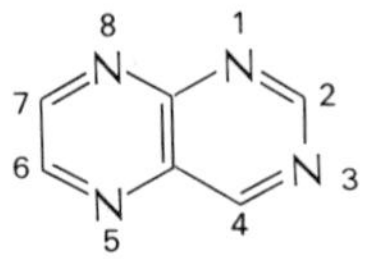

pteridine

Exercises

28.1 Draw structural formulas of:
   a. 2,3-dimethyloxirane
   b. 2-quinolinecarboxylic acid
   c. 8-hydroxyisoquinoline
   d. 2H-1,3-oxazine
   e. 2,3-dihydro-3-ketoindole
   f. 2-acetamino-1,3-thiazole

28.2 Substituted cyclopentadienes

We have titled this section <u>substituted cyclopentadienes</u> to stress the similarities and differences between cyclopentadiene and the five-membered heterocycles, furan, thiophene, and pyrrole.

Aromaticity; resonance structures; reactivity. The aromaticity of the five-membered heterocycles, while not so marked as in benzene, nevertheless is striking. We remember the aromaticity of cyclopentadienyl anion, with five *p* orbitals and six electrons. The heterocycles do not have to be ionized to provide the third pair of electrons needed for aromatic stabilization. The hetero atoms O, S, and N all possess an electron pair that participates in aromaticity. Although we write five participating structures for each molecule (resonance theory) we keep in mind the molecular orbital theory, which explains the need for six electrons.

[resonance structures of cyclopentadienyl anion]

[resonance structures of furan, with positions labeled 1 (O), 2, 3, 4, 5]

furan

[resonance structures of thiophene, with positions labeled α, α′, β, β′]

thiophene

[resonance structures of pyrrole]

pyrrole

Although we have no value for the resonance energy of cyclopentadienyl anion, values for furan (16 kcal), thiophene (11 to 29 kcal), and pyrrole (16 to 21 kcal) have been estimated, roughly about half that of benzene. For one thing, delocalization is accompanied by charge separation, which requires energy. (The cyclopentadienyl anion does not: it delocalizes charge as well as electrons.) The charge separation is in a direction opposite to the inductive effects (O > N > S), and it operates either to diminish the magnitude of the dipole moment or to change the direction of the moment to the opposite direction. Thus furan has $\mu = 0.7$ D, but tetrahydrofuran has 1.6 D; thiophene has 0.6 D, but tetrahydrothiophene has 1.9 D; and pyrrole has 1.8 D, but pyrrolidine has 1.6 D. In the case of thiophene, the π bond to sulfur is a 3p-2p π bond, and we expect such bonding to be weak. However, the gain in stability of the molecule as a whole is more than enough to offset the reluctance of sulfur to undergo this bonding.

If we look at furan for a moment, we can see a resemblance to divinyl ether ($H_2C=CH-O-CH=CH_2$). Therefore, we should see reactivity of the type we have come to expect of activated double bonds. Also, we can extend the analogy to thiophene and pyrrole.

If we look at pyrrole for a moment, we recognize a secondary amine. With a strong positive charge, it is a very weak base ($pK_a = 0.4$) and even is a weak acid ($pK_a = 15$).

In all three heterocyclic ring compounds, with all four positions (2, 3, 4, and 5 or α, α′, β, β′) bearing a partial negative charge, electrophilic substitution is easily accomplished. In fact, electrophilic substitution is not as useful a method for preparing substituted furans,

28.2 Substituted cyclopentadienes

thiophenes, and pyrroles as one might anticipate, because it is difficult to stop at a monosubstituted product. Reactivity is comparable to that of aniline and phenol. Therefore, other methods of synthesis of substituted furans, thiophenes, and pyrroles become useful.

Reactions: acidic polymerization; substitution; Diels-Alder reaction; reduction. Let us take up the reactions of the substituted cyclopentadienes in more detail.

Acids cause decomposition and polymerization of furan and pyrrole, but thiophene is less susceptible to this effect (Eqs. 28.1). The proton is shown as being accepted at an

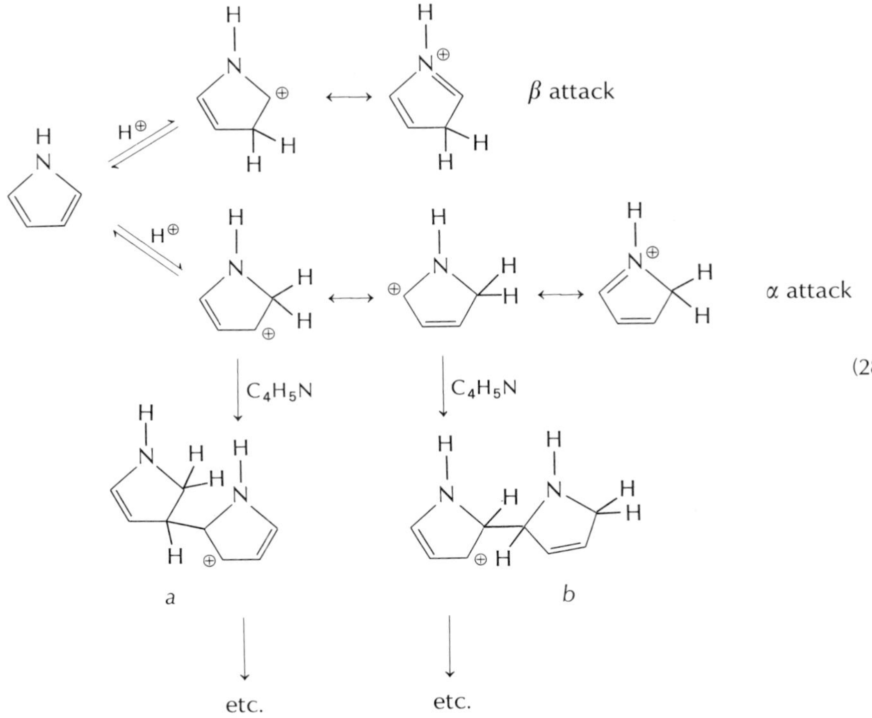

(28.1)

α position of pyrrole rather than a β. For α attack the positive charge in the protonated ion may be delocalized between the 3, 5, and 1 positions as shown, whereas β attack gives an ion in which delocalization is possible only between the 2 and 1 positions. The cation, in turn, attacks another pyrrole molecule at an α position, and so on to give polymeric products. The ion may attack as a β ion to give a or as an α ion to give b. Thus, strongly acidic conditions must be avoided with furans and pyrroles.

Electrophilic substitution reactions must be carried out under nonacidic conditions, if possible. Sulfonation with sulfuric acid and nitration with nitric and sulfuric acids are possible only with the less reactive thiophene. In any case, initial attack is at an α position. It makes little difference if a directive group (o,p or m) is already present at an α position: substitution then occurs at the other α position (α'). In Table 28.2 we summarize the conditions and results of some electrophilic substitution reactions. Other reactions (chloromethylation, the Kolbe, and the Reimer-Tiemann reactions) also may take place.

The relative lack of aromaticity in furan (as compared with thiophene and pyrrole) is evident in the reactivity with dienophiles in the Diels-Alder reaction (Eq. 28.2). The bridging oxygen atom is easily removed by acid and aromatization to give phthalic anhydride (or substituted phthalic acids if substituted furans are used).

Table 28.2 Electrophilic substitution reactions of furan, thiophene, and pyrrole

Reaction	Conditions and results		
	Furan	Thiophene	Pyrrole
1. Nitration	$CH_3COO^\ominus \overset{\oplus}{N}O_2$ α-NO_2	in $(CH_3CO)_2O$ at 5° α-NO_2	α-NO_2
2. Sulfonation	SO_3 in pyridine, warm α-SO_3H	α-SO_3H	α-SO_3H
3. Bromination	Br_2 in dioxane, 0° α-Br + α,α'	Br_2 in ϕH, 0° α,α'-diBr	Br_2 in EtOH, 0° 2,3,4,5-tetraBr
4. Friedel-Crafts	$(CH_3CO)_2O$, $SnCl_4$, 0° α-$COCH_3$	α-$COCH_3$	α-$COCH_3$
5. Coupling	$Ar\overset{\oplus}{N}\equiv N$, pH > 7, 0° α-N=NAr	poor results	α-N=NAr
6. Gattermann	$Zn(CN)_2$, HCl; H_2O α-aldehyde	—	α-aldehyde
7. Mercuration	$(CH_3COO)_2Hg$, CH_3COOH α-HgOOCCH$_3$	α-HgOOCCH$_3$	α-HgOOCCH$_3$

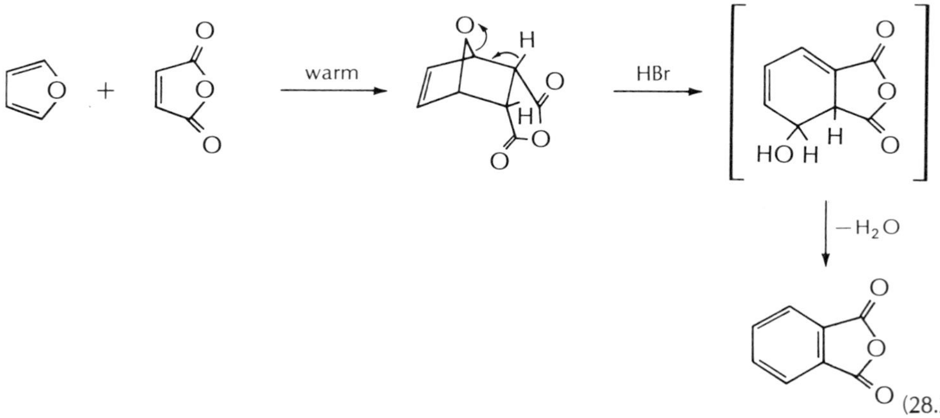

$$\text{(28.2)}$$

Catalytic reduction converts furan into tetrahydrofuran and pyrrole into pyrrolidine. Thiophene (and derivatives), upon treatment with Raney nickel, loses sulfur (Eq. 28.3). However, reduction with sodium and EtOH converts thiophenes into tetrahydrothiophenes.

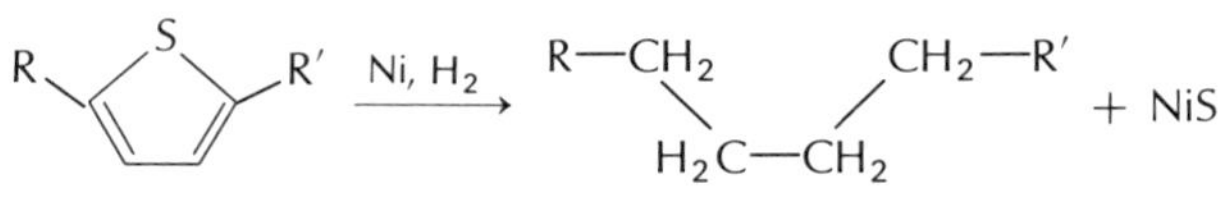

$$\text{(28.3)}$$

Thiophene and pyrrole are present in small amounts in coal tar. Furan is made commercially from pentosans. (Oat hulls are a good source, as are corn cobs, etc.) □ □

Pyrrole and pyrrolidine derivatives are present in nature. Two of the amino acids are pyrrolidines, and pyrrole derivatives make up the porphin and corrin ring systems (hemin, chlorophylls, vitamin B_{12}). □ □

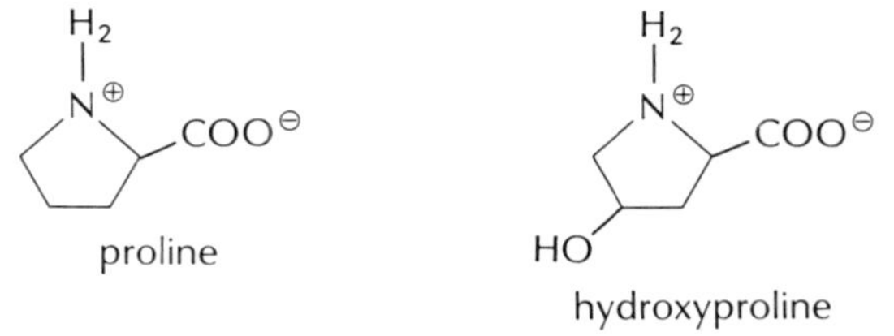

[Structures of proline and hydroxyproline]

Two hetero atoms: pyrazole, imidazole, isoxazole, oxazole, thiazole. There are many five-membered heterocycles with two or more heterocyclic atoms present. We shall confine our attention to the diazoles, oxazoles, and thiazoles. (The dioxoles and dithioles are of either the cyclic peroxide or the acetal types.)

Pyrazoles (1,2-diazoles) do not occur naturally, but imidazoles (1,3-diazoles) occur frequently (the amino acid histidine, and histamine). Both diazoles (especially imidazole)

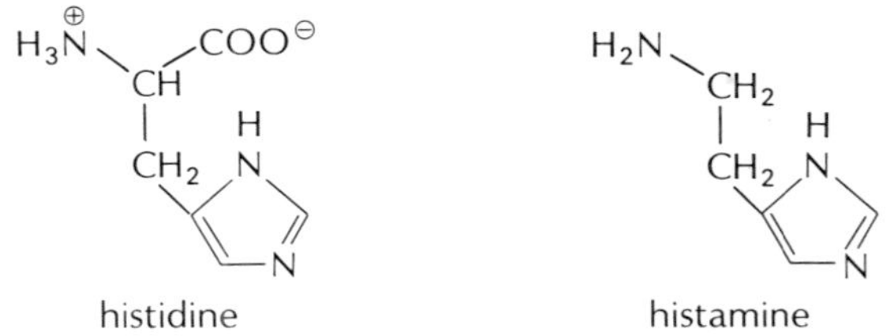

[Structures of histidine and histamine]

are much stronger bases than is pyrrole. The cation, as shown, is stabilized by the contributions of the two equivalent structures a and b.

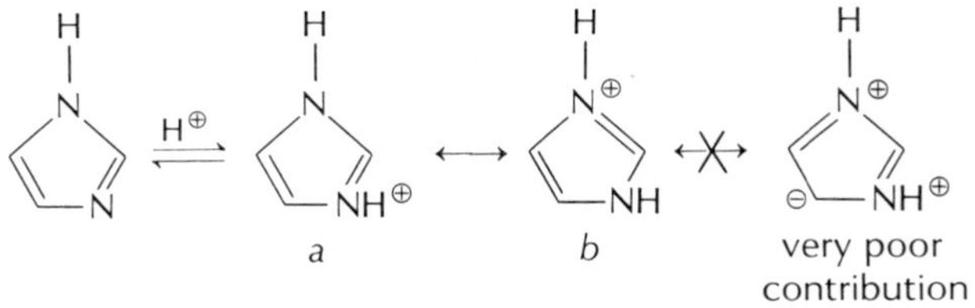

[Resonance structures of protonated imidazole: a, b, and "very poor contribution"]

Pyrazoles result from the reaction of 1,3-dicarbonyl compounds with hydrazine. If aceto acetic ester (AAE) is used as an example (Eq. 28.4), a 3-methyl-5-pyrazolone

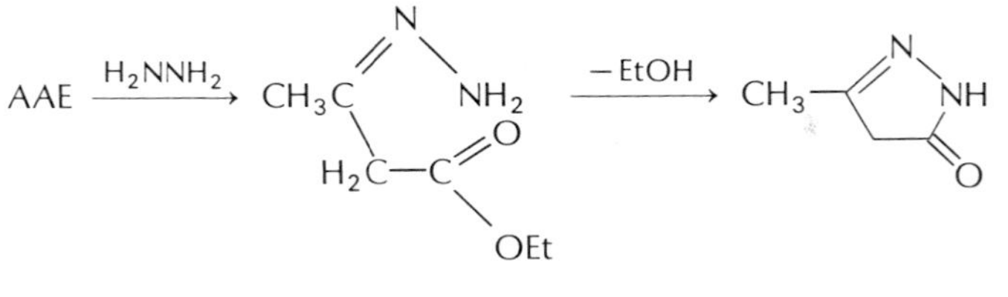

$$\text{AAE} \xrightarrow{H_2NNH_2} \text{intermediate} \xrightarrow{-\text{EtOH}} \text{3-methyl-5-pyrazolone} \qquad (28.4)$$

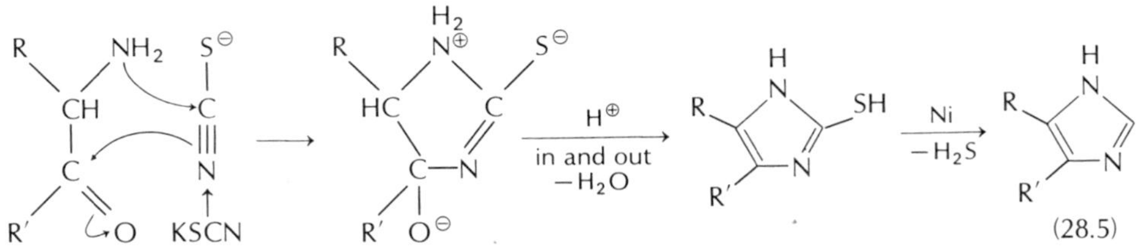

[Mechanism scheme for reaction 28.5 showing formation of imidazole from amino compound, KSCN, with H⁺ (in and out, −H₂O), then Ni/−H₂S desulfurization]

(28.5)

results. Imidazoles may be synthesized in a number of ways. One method is shown in Equation 28.5, beginning with an amino ketone and KSCN.

Isoxazolones are formed if H_2NOH reacts with AAE. Oxazoles are produced if the oxime of an α-diketone is acetylated under reducing conditions to give an acetylated α-amino ketone, which cyclizes with $SOCl_2$ (Eq. 28.6). If the intermediate acetylated α-amino ketone (a) is treated with P_2S_5, a reaction occurs to give a thiazole (b).

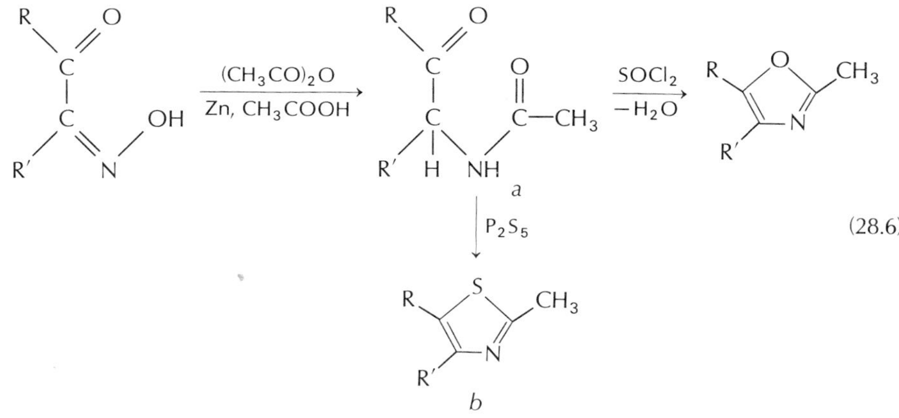

(28.6)

Exercises

28.2 Starting with acetoacetic ester, show how to prepare:
  a. an α-amino ketone
  b. an isoxazolone
  c. 3-methyl-1-phenyl-5-pyrazolone

28.3 Substituted benzenes

The six-membered rings are comparable to benzene. We shall take up only pyridine and pyran as representatives of the compounds with only a single hetero atom, and pyridazine, pyrimidine, and pyrazine to represent those with two hetero atoms.

Pyran; pyrylium salt; pyrones, synthesis. Pyran is nonaromatic and hypothetically could exist in two tautomeric forms, α and γ. Loss of a hydrogen as hydride ion (an oxidation) gives rise to a pyrylium salt, relatively stable because of aromaticity. In the salt, oxygen forms two σ bonds with sp^2 orbitals. The remaining sp^2 orbital is occupied by a pair of electrons. The p orbital furnishes two electrons for the aromatic sextet and causes a positive charge

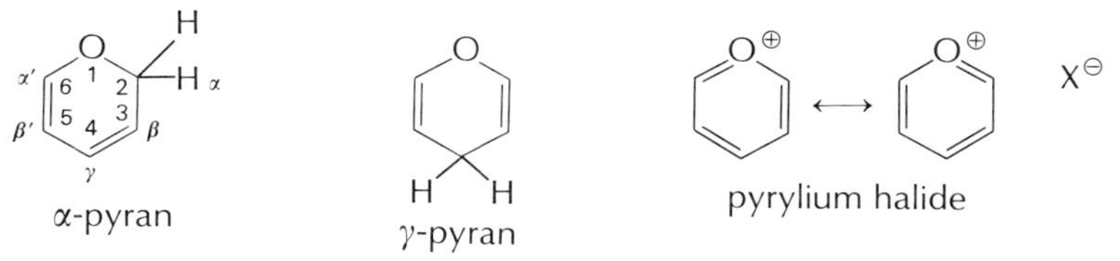

α-pyran   γ-pyran   pyrylium halide

to be localized on oxygen. (A pair of electrons is required to make up for the loss of an electron pair from the α or γ carbon when a hydride ion is lost to form the salt.)

Pyran is not known as such, although the dihydro- and tetrahydropyrans are known. The ketones α- and γ-pyrone are known. The α-ketone is a δ-lactone of the enol form of an unsaturated aldehyde acid (Eq. 28.7). The γ-ketone is of interest because it displays aromatic

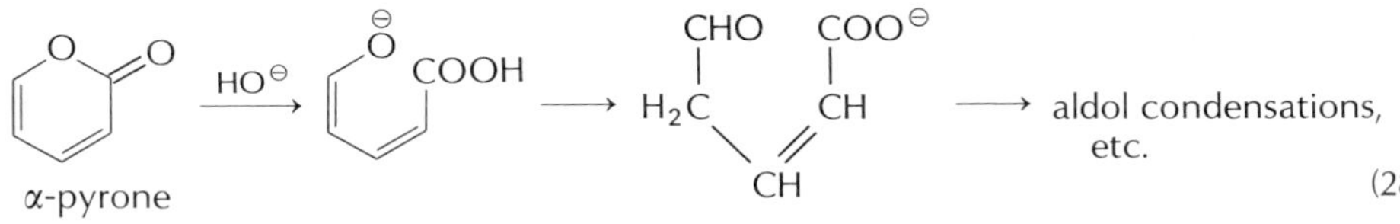

$$\text{α-pyrone} \xrightarrow{HO^\ominus} \text{[enolate-COOH]} \longrightarrow \underset{CH}{\overset{CHO \quad COO^\ominus}{H_2C \quad CH}} \longrightarrow \text{aldol condensations, etc.} \quad (28.7)$$

properties (nitration, bromination) instead of acting like an unsaturated ketone (no derivatives of the keto group, no or little 1,4-addition). The answer is found in a consideration of the contributing resonance structures of γ-pyrone. The dipolar structures resemble the phenoxide ions in appearance.

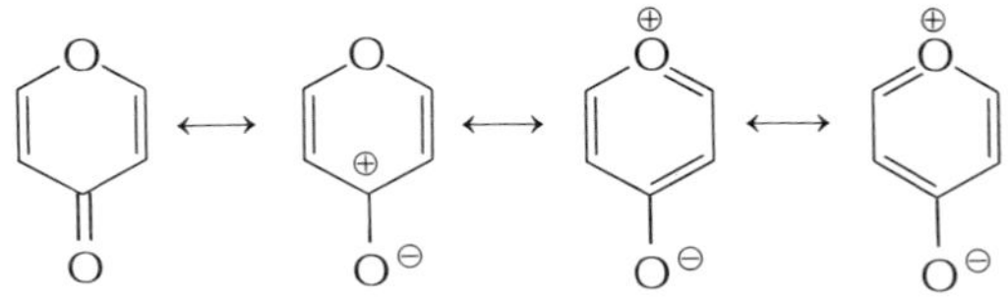

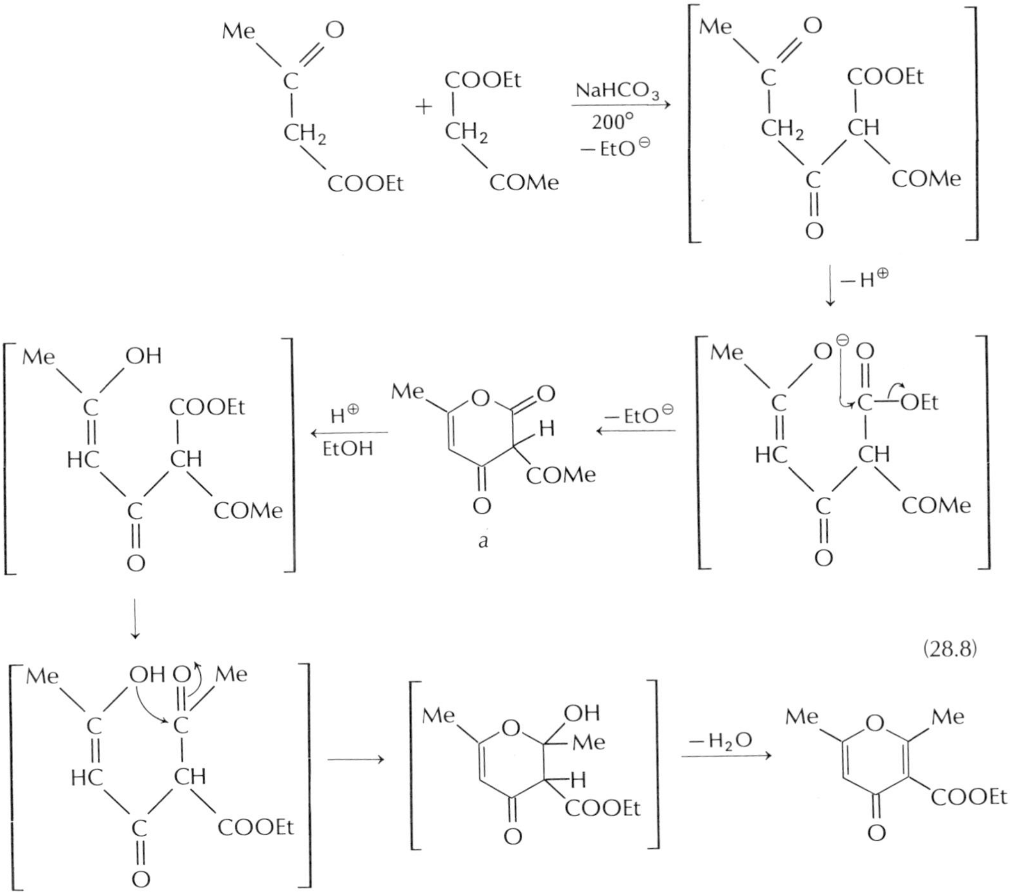

$$(28.8)$$

AAE condenses and cyclizes in the presence of NaHCO$_3$ at 200° to give a diketolactone (a, known as dehydroacetic acid), which rearranges with EtOH and acid to give a substituted γ-pyrone (Eq. 28.8).

Pyridine: basicity, solvent properties; substitution reactions. Pyridine and pyridine derivatives are widespread in nature. Coal tar is a practical source of pyridine, the methylpyridines (α-, β-, and γ-picolines), the various dimethylpyridines (lutidines), and other substituted pyridines.

Pyridine is decidedly aromatic, even though the resonance energy is estimated to be only 21 to 23 kcal. Aside from what we call Kekulé resonance, pyridine has dipolar contributing structures that cause partial positive charges to appear at the 2, 4, and 6 positions.

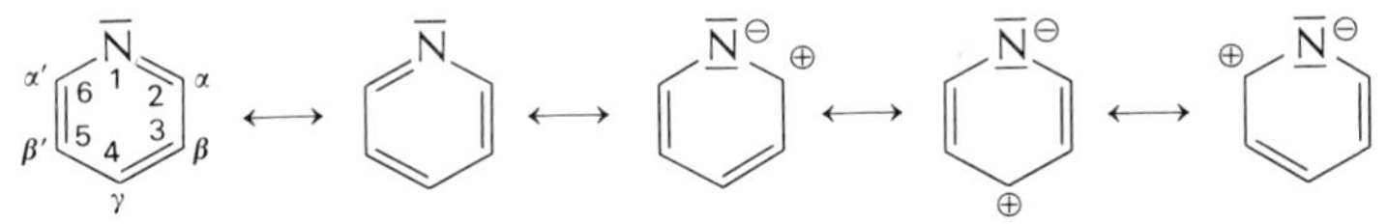

The dipolar structures also contribute to the large dipole moment of 2.3 D. The polarity causes the bp to be relatively high (115°).

Pyridine is only weakly basic, even though it seems that the partial negative charge on nitrogen should enhance the ability to accept a proton. The weak basicity is caused by the sp^2 orbital in which the extra pair of electrons is present (in the plane of the ring). Electrons in such an orbital with increased s character are less available for bonding an incoming proton, or the conjugate acid (C$_5$H$_5$N̈H) is much more willing to lose a proton. Thus, pyridine (pK$_a$ = 5.2) is a weak base as compared to piperidine (hexahydropyridine, pK$_a$ = 11.2). However, the contributions of the dipolar structures appear as a base-strengthening effect, so that pyridine is not as weak a base as is aniline (pK$_a$ = 4.6).

Pyridine is completely miscible with water, as with most organic liquids, polar and nonpolar. It is an excellent solvent for many organic compounds. It is very useful as a basic tertiary amine solvent for many reactions requiring the presence of a base, either as a catalyst or to neutralize acids formed in the reaction (as in the reaction of an acid chloride with an alcohol).

The electron-withdrawing effect of nitrogen causes pyridine to be inert to electrophilic substitution reactions. Furthermore, since most electrophilic reagents are strongly acidic, the conjugate acid of pyridine is even less reactive. In other words, pyridine is much more like nitrobenzene in lack of reactivity than it is like aniline. In fact, pyridine is less reactive than nitrobenzene, and only very drastic conditions cause electrophilic substitution to occur (Eqs. 28.9). The substitutions take place at the 3 or 5 (called β) positions

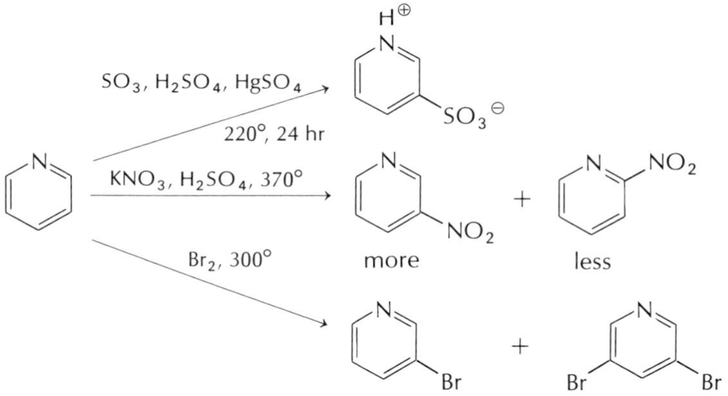

(28.9)

because not only are the 2 and 4 positions deactivated more (partial positive charge), but also the intermediate at the 3 position does not require a positive charge on the already positive nitrogen.

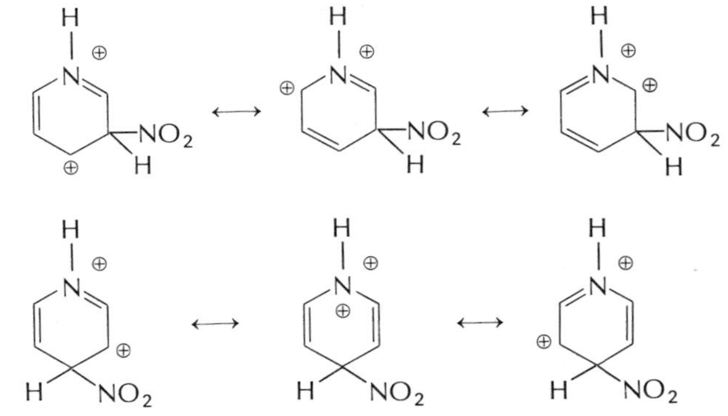

Pyridine undergoes nucleophilic substitution with relative ease (Eqs. 28.10), primarily at an α position (some γ substitution also occurs). The reaction with sodium amide is known as the <u>Chichibabin reaction.</u> The product from the reaction with KOH, 2-hydroxypyridine (2-pyridol or 2-pyridinol) is tautomeric with the more stable lactam structure, which is misnamed as an α-pyridone. A lithium reagent also attacks at the α position. (The intermediate anion allows the negative charge to reside partially on nitrogen, whereas β attack does not.)

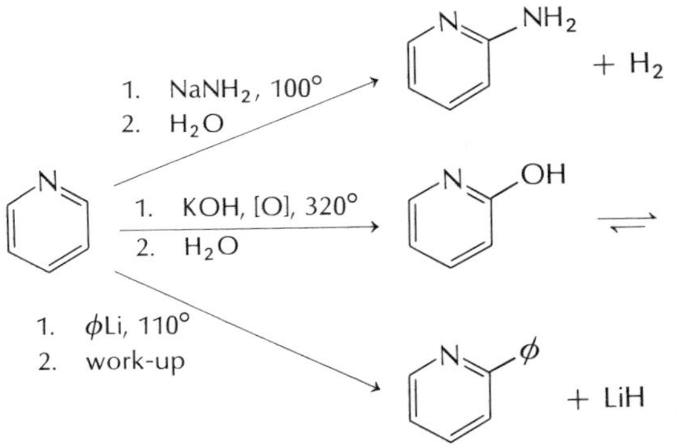

$$(28.10)$$

Resistance to oxidation; pyridine oxide; reduction. Although pyridine is very resistant to oxidation, it does respond to the use of a peracid or H_2O_2 to form a tertiary amine oxide. The presence of the polar $\overset{\oplus}{N}$—$\overset{\ominus}{O}$ bond enhances the reactivity of the ring to electrophilic attack. It also modifies the position of attack so that $\gamma > \alpha \gg \beta$ (Eq. 28.11). Thus 4-nitropyridine (a) is easily made by removal of oxygen with PCl_3. The nitro group may be reduced to an amino group. Also, the nitro group may be displaced by a nucleophilic substitution (b) before removal of oxygen.

Pyridine is relatively easy to reduce to piperidine, either catalytically or by sodium and ethanol.

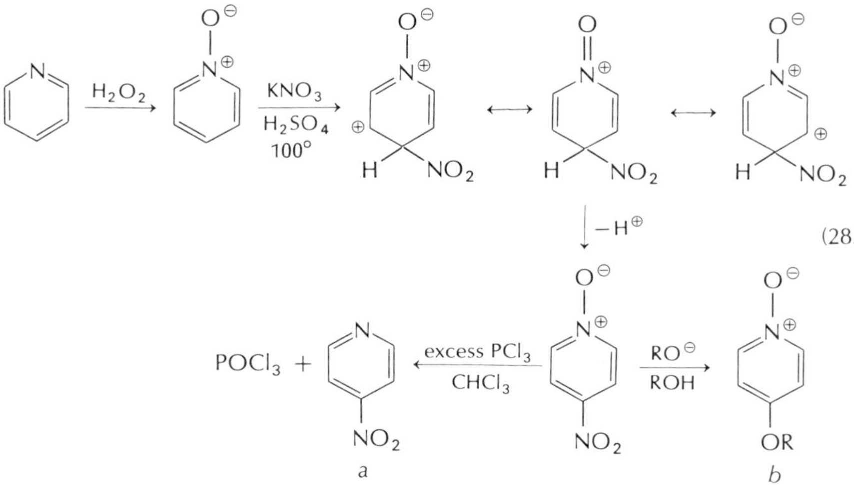

$$\text{pyridine} \xrightarrow{H_2O_2} \text{pyridine N-oxide} \xrightarrow[\substack{H_2SO_4 \\ 100°}]{KNO_3} \cdots \xleftrightarrow{} \cdots \xleftrightarrow{} \cdots$$

$$\downarrow -H^{\oplus} \qquad (28.11)$$

$$POCl_3 + \text{4-nitropyridine} \xleftarrow[CHCl_3]{\text{excess } PCl_3} \text{4-nitropyridine N-oxide} \xrightarrow[ROH]{RO^{\ominus}} \text{4-alkoxypyridine N-oxide}$$

a \qquad\qquad\qquad\qquad b

Substituted pyridines: reactions. Pyridine derivatives fall into two groups. If the substituted group is at the α or γ position, reactivity seems similar to that of the group adjacent to a carbonyl group. This similarity is illustrated by the participation of α-picoline (the common name for 2-methylpyridine) in an aldol condensation (Eq. 28.12). In other words,

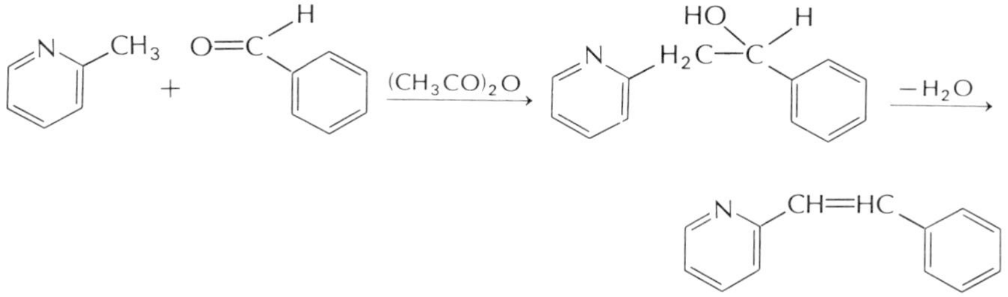

(28.12)

cis < trans

the methyl group at the α position (or the γ position) is activated in such a way that forms analogous to enols and enolate ions are possible (Eqs. 28.13).

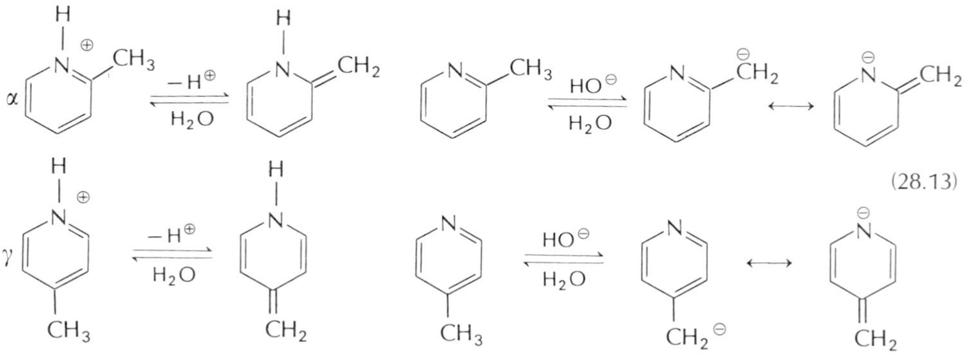

(28.13)

The β-substituted picolines, on the other hand, have reactivity analogous to that of a m-nitrobenzene derivative in which electron withdrawal without conjugation is occurring. Thus, β-picoline does not have any more tendency toward aldol condensation than does toluene or m-nitrotoluene.

28.3 Substituted benzenes

[Resonance structures: β-picoline cation showing N⁻ and CH₃ substituent; nitro-methylbenzene resonance structures with NO₂ and CH₃]

The picolines are all available from coal tar and serve as convenient starting points for the preparation of various monosubstituted pyridines. Oxidation of a picoline with permanganate gives the corresponding pyridine carboxylic acid (Eq. 28.14). Nicotinic acid

$$\text{β-picoline} \xrightarrow{\text{KMnO}_4} \text{nicotinic acid} \quad (28.14)$$

(niacin) is present in all living cells and is a component of NAD and NADH. It was originally isolated as a product of oxidation of the alkaloid nicotine, which is N-methyl-2(3'-pyridyl)pyrrolidine. The amide of nicotinic acid was called vitamin B_3 at one time. Both niacin and the amide are antipellagra agents. The demand for synthetic niacin is so great that an alternate source of supply is needed (Eq. 28.15).

$$\text{pyridine} \xrightarrow[\text{SO}_3\ 220°]{\text{H}_2\text{SO}_4} \text{3-SO}_3\text{H} \xrightarrow[\Delta]{\text{KCN}} \text{3-CN} \xrightarrow[\text{H}_2\text{O}]{\text{H}^\oplus} \text{3-COO}^\ominus \quad (28.15)$$

The γ acid is called isonicotinic acid. The hydrazide of isonicotinic acid, isoniazid, is useful in treatment of tuberculosis. The α acid is named picolinic acid. All the acids probably are zwitterions, because the pK_a of each (α, 5.52; β, 4.85; γ, 4.96) is not far from that of pyridine (4.63). The α and γ acids are easily decarboxylated (similarly to α-keto acids), whereas the β acid is not.

The acids may be converted into acid chlorides with $SOCl_2$, and esters and amides may be made. Treatment of the amides with NaOBr (Hofmann rearrangement) gives the aminopyridines. The α- and γ-aminopyridines react with nitrous acid at 0° to go to the α- and γ-pyridols. The β-aminopyridine, with aromatic character, may be diazotized, and the many possible substitutions and coupling reactions may be carried out. The various aldehydes and ketones can be made by an assortment of methods.

Synthesis of substituted pyridines. We shall illustrate only one of the available methods for the synthesis of substituted pyridines, the <u>Hantzsch synthesis</u>. Acetoacetic ester (AAE) and an aldehyde in the presence of ammonia condense to give either of two products, depending upon which reactant is in excess (Eq. 28.16).

The first step is the addition of AAE anion to the aldehyde carbonyl group to give a. (AAE is more acidic than CH_3CHO, and —CHO is easier to add to than a keto or ester grouping.) The second step is the base-catalyzed elimination of water from a β-hydroxy carbonyl compound to give b. The third step is a Michael addition to the α,β-unsaturated carbonyl compound (b). If the aldehyde is in excess, c is the product; if AAE is in excess, c' is the product. Thereafter, ammonia adds, water is eliminated (to give the imine), and the imino group adds to the other carbonyl group, closing the ring. Loss of water gives the

(28.16)

substituted dihydropyridines *d* and *d'* (or 1,4-dihydro tautomers), the immediate product of allowing a mixture of the reactants to stand at 25° or with moderate warming. The substituted pyridines *e* and *e'*, are obtained in a separate step by mild oxidation of *d* and *d'*. Hydrolysis of the ester groups and decarboxylation of the acids (heating with lime) gives fair yields of 2,4-dimethylpyridine from *e* and of 2,4,6-trimethylpyridine from *e'*. Many variations of the Hantzsch pyridine synthesis are possible by the use of other aldehydes and other β-keto esters (RCOCH$_2$COOEt).

There is an elegant synthesis of pyridoxine or pyridoxol (previously known as vitamin B$_6$), a component of transamination and decarboxylation enzymes. The phosphate ester is a coenzyme for several other reactions. The synthesis begins with a cleverly conceived condensation in which the amide —NH$_2$ group of cyanoacetamide adds to the less hindered methyl keto group of a β-diketone (Eq. 28.17).

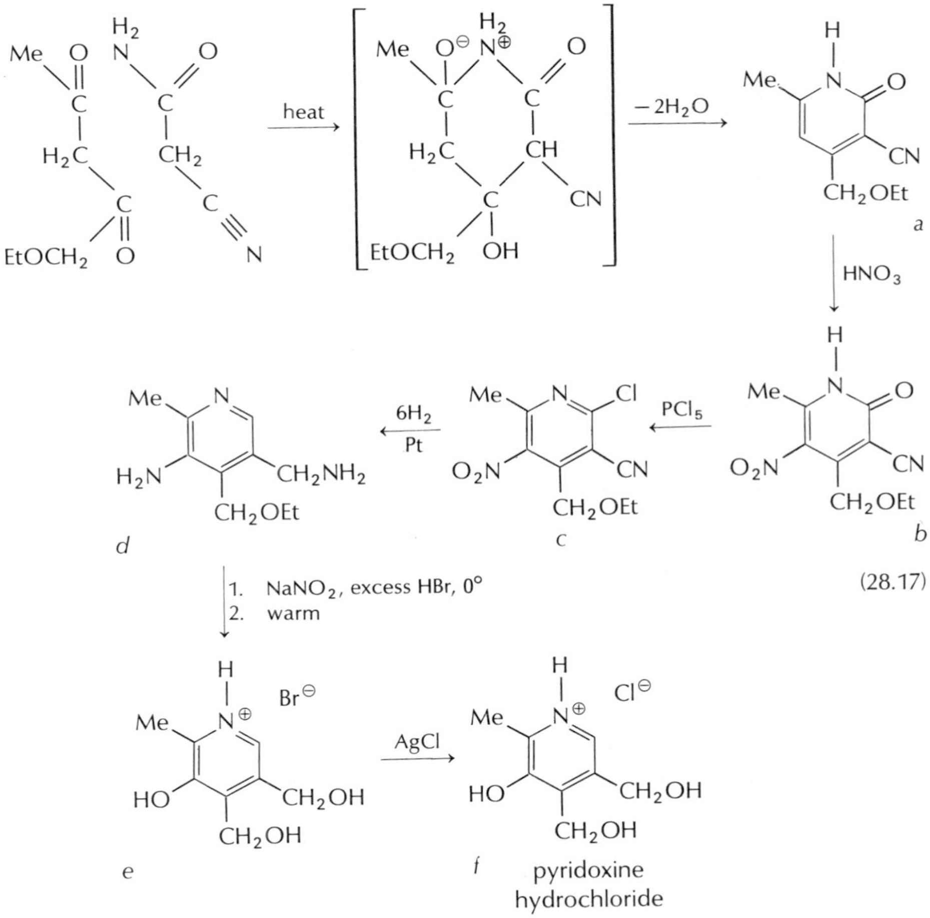

(28.17)

Condensation follows by simple heating to give the α-pyridone (*a*). Nitration goes at the only open β position, 5, activated by a hydroxy (or keto) at 2, to give *b*. Pyridones react well with PCl$_5$ to give, in this case, a chlorine at 2 (*c*). Now comes a hydrogenation, which accomplishes three reactions in a single step: removal of chlorine by reduction, conversion of the cyano group into an aminomethyl (—CH_2NH_2), and reduction of the nitro group. The next step, conversion of d to e, is equally remarkable. Diazotization with $NaNO_2$ and HBr also converts —CH_2NH_2 into —CH_2OH. Warming the mixture hydrolyzes the diazo group to hydroxy, and the HBr is strong enough to hydrolyze the benzylic-like ethyl ether. The product (e) is the HBr salt of pyridoxine, which must be converted with AgCl to the HCl salt to become the synthetic natural product (f).

Diazines; pyrimidines; barbiturates; vitamin B_1. The three isomeric diazines are 1,2-diazabenzene (pyridazine), 1,3-diazabenzene (pyrimidine), and 1,4-diazabenzene (pyrazine). We shall spend most of our time on pyrimidine because of the widespread occurrence of pyrimidine derivatives in nature, especially of cytosine, thymine, and uracil in DNA and RNA.

The diazines are all less basic than pyridine. Think of pyridine with an electron-

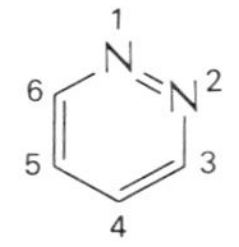

	pyridazine	pyrimidine	pyrazine
pK_a =	2.3	1.3	0.6
bp	207°	124°	116°
mp	−64°	22°	57°

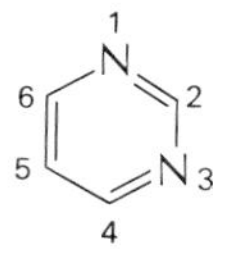

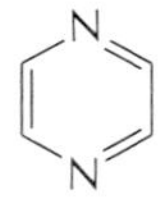

withdrawing substituent (the second atom of nitrogen) present. Pyrazine, with no dipole, has the lowest bp and the highest mp of the three. The positions α, β, and γ relative to each nitrogen have some of the character of the α, β, and γ positions of pyridine. Thus, in pyrazine, each carbon atom is both an α and a β position with respect to a nitrogen atom. The α or γ effect deactivates for electrophilic attack and activates for nucleophilic attack, whereas the β effect is similar to that of an *m*-nitro group in affecting reactivity. Pyrazine, then, is more resistant to electrophilic substitution than is pyridine, and substituted pyrazines (Me—, HO—, H_2N—, etc.) are all subject to easy nucleophilic reactions.

In pyrimidine, position 2 is α to both nitrogen atoms, and positions 4 and 6 both are α and γ. Only position 5 is β to both nitrogens. The contributing resonance structures show that the two nitrogen atoms are competing for negative charge from positions 2, 4, and 6. Only position 5 has no formal positive charge in any of the eight structures. Electrophilic

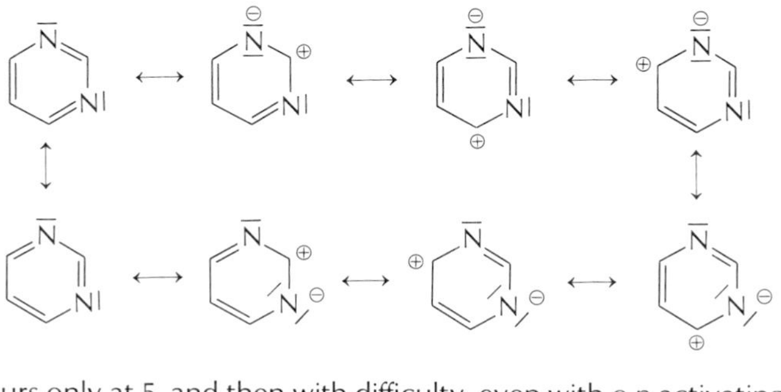

substitution occurs only at 5, and then with difficulty, even with *o,p* activating substituents present. Substituents at the 2, 4, or 6 positions are all subject to nucleophilic substitution or to aldol condensation reactivity (Eq. 28.12).

Substituted pyrimidines usually are made by synthesis instead of by operations on pyrimidine itself. For example, 2,4,6-trihydroxypyrimidine (common name <u>barbituric acid</u>) is easily made from urea and malonic ester (Eq. 28.18). Barbituric acid originally was isolated

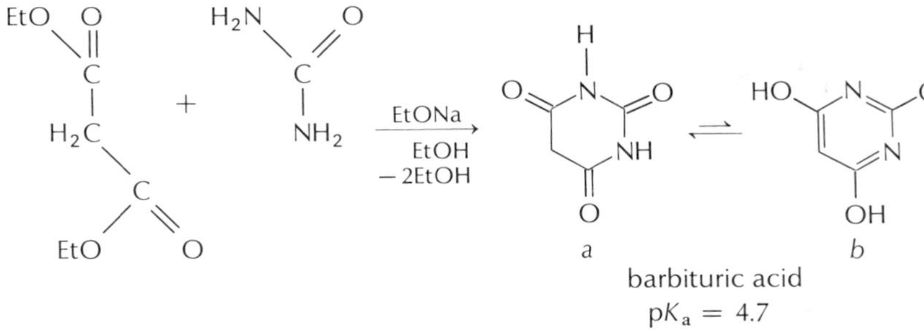

(28.18)

barbituric acid
$pK_a = 4.7$

as a degradation product of uric acid. As with the α- and γ-pyridols, tautomerization to the lactam structure is possible and in fact is favored in barbituric acid. Of course, there are several possible intermediates between the trihydroxypyrimidine structure *b* and the all-keto structure *a*. In writing hydroxy or lactam structures hereafter, we should understand that the structure written stands for all the possible tautomeric structures as well.

Although barbituric acid is without any notable physiological effect, numerous 5,5-disubstituted barbituric acids have pronounced sedative and hypnotic effects. Habitual use tends to create addiction to such substances. The substances are easily made from the properly substituted malonic ester. The most used of the barbiturates are listed.

5,5-diethyl	Barbital or Veronal
5-ethyl-5-phenyl	Phenobarbital or Luminal
5-ethyl-5-(1-methylbutyl)	Pentobarbital or Nembutal
5-ethyl-5-(3-methylbutyl)	Amobarbital or Amytal

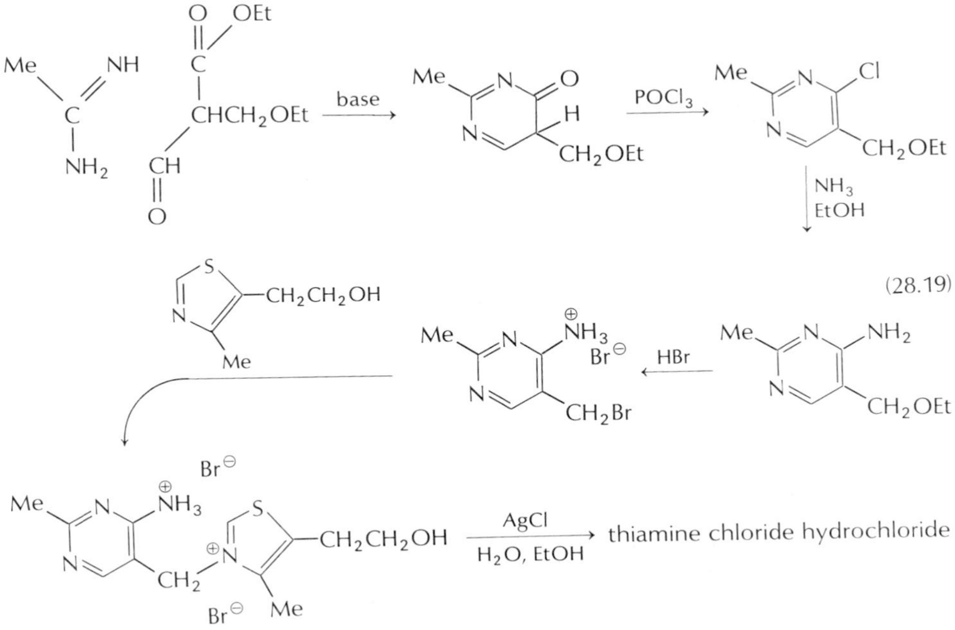

(28.19)

thiamine bromide hydrobromide

Many variations of the malonic ester and urea synthesis of barbituric acid are possible. For example, 2,4-dihydroxypyrimidine (uracil), 4-amino-2-hydroxypyrimidine (cytosine), and 2,4-dihydroxy-5-methylpyrimidine (thymine) may be made. The compounds named are constituents of the nucleic acids RNA (cytosine, uracil) and DNA (cytosine, thymine).

Thiamin (vitamin B_1) is both a substituted pyrimidine and a substituted thiazole. First isolated in 1926, it was not synthesized until 1936 (Eq. 28.19). Diphosphorylation of thiamin (at the primary OH group) gives the coenzyme cocarboxylase.

Exercises

28.3 Devise a reasonable synthesis of:
   a. uracil
   b. cytosine
   c. thymine

28.4 Use of an excess of $CH_3CH_2COCH_2COOEt$ instead of AAE in the Hantzsch pyridine synthesis with ϕCHO instead of CH_3CHO gives what product?

28.4 Bicyclic compounds with a five-membered heterocycle

Types; naming. Other rings, both carbocyclic and heterocyclic, may be fused with a heterocyclic ring in various ways. Thus, the fusion of a benzene ring with pyrrole gives rise to three bicyclic systems. (In systematic naming, the sides of the parent ring are labeled

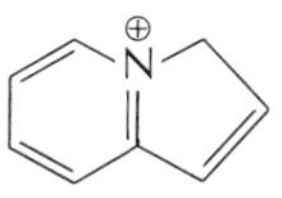

benzo[a]pyrrole 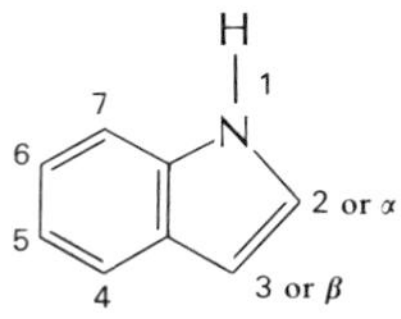   benzo[b]pyrrole   benzo[c]pyrrole
                     or indole

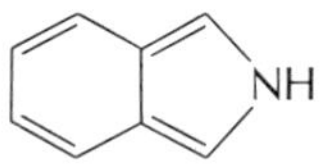

[a] for the 1,2 side; [b] for the 2,3 side; and so on. Selection of the parent ring is taken care of by rules beyond the scope of this book.) We shall discuss only a few of the bicyclic systems, and in this section only one, indole.

Indole: similarity to pyrrole; substituted indoles. Indole has properties similar to those of pyrrole, modified by the presence of the benzene ring. Indole is less basic ($pK_a = -2.3$) than pyrrole ($pK_a = 0.4$) because the benzene ring decreases basicity in the same way that it does in aniline ($pK_a = 4.6$) as compared with ammonia ($pK_a = 9.2$). Reactivity of indole is similar to that of pyrrole in sensitivity to strong acids and in ability to undergo alkaline reactions. Unlike pyrrole, indole reacts primarily at the β position. The cationic intermediate for β attack is stabilized more than that for α attack.

Indole undergoes nucleophilic attack, as shown by the useful Mannich reaction (Eq. 28.20). The product, gramine, is present in grasses. Gramine may be converted into β-indolylacetic acid (also called indoleacetic acid), a plant growth hormone known as an auxin (Eq. 28.21).

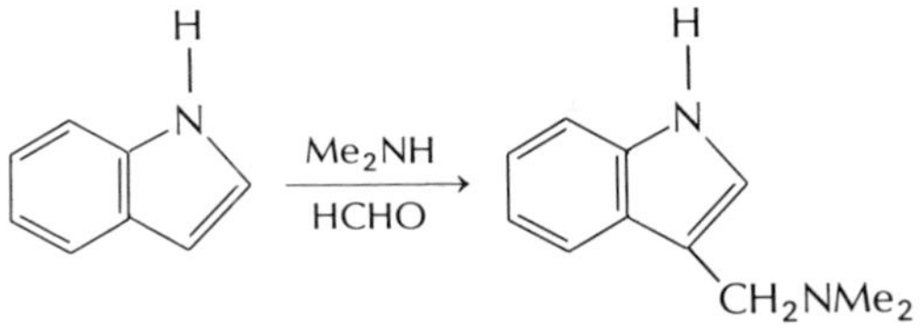

28.4 Bicyclic compounds with a five-membered heterocycle

$$\text{indole} \xrightarrow[\text{HCHO}]{\text{Me}_2\text{NH}} \text{gramine (3-CH}_2\text{NMe}_2\text{-indole)} \tag{28.20}$$

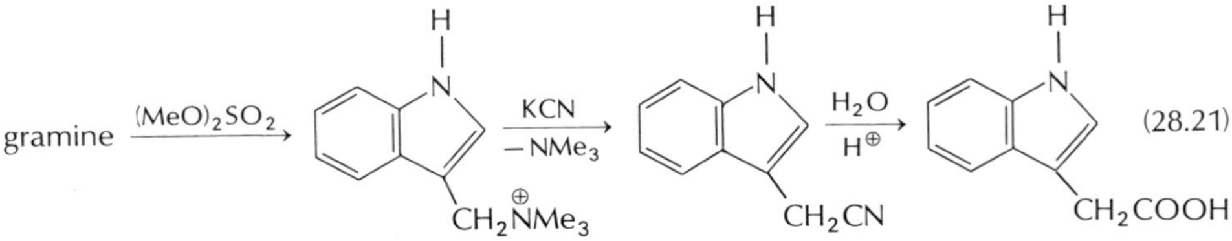

$$\text{gramine} \xrightarrow{(\text{MeO})_2\text{SO}_2} \text{3-CH}_2\overset{\oplus}{\text{N}}\text{Me}_3\text{-indole} \xrightarrow[-\text{NMe}_3]{\text{KCN}} \text{3-CH}_2\text{CN-indole} \xrightarrow[\text{H}^\oplus]{\text{H}_2\text{O}} \text{3-CH}_2\text{COOH-indole} \tag{28.21}$$

Tryptophan, one of the essential amino acids, is an indole derivative that serves as the precursor of such substances as indolylacetic acid, serotonin, and skatole (Eq. 28.22). Skatole occurs along with indole in feces. Serotonin, as we have mentioned, is a factor in the functioning of the brain. It affects blood pressure, peristalsis, and other body functions as well.

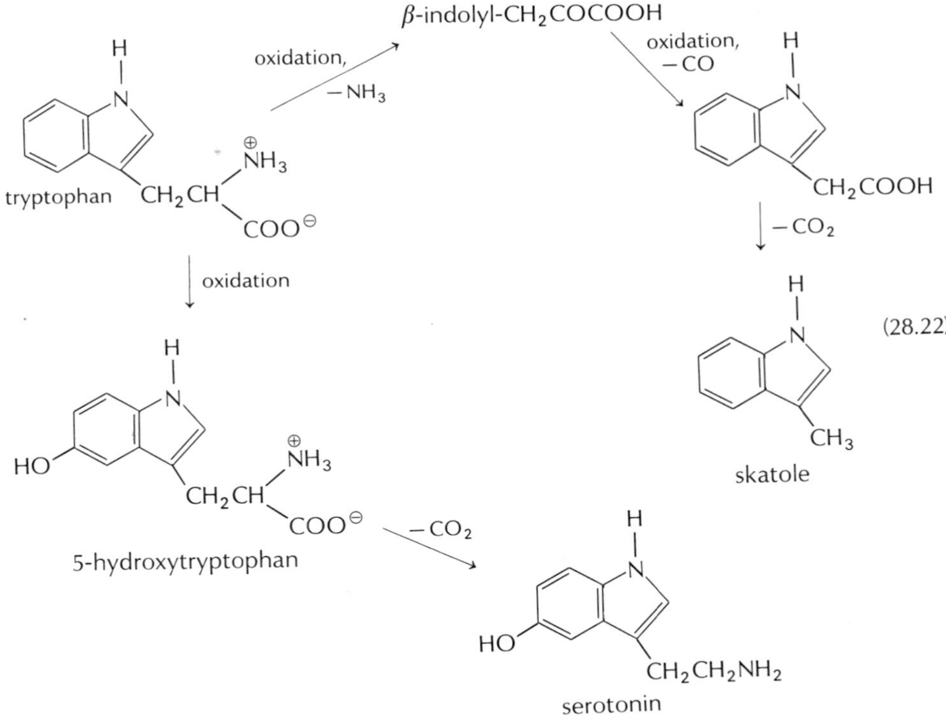

$$\text{tryptophan} \xrightarrow[-\text{NH}_3]{\text{oxidation}} \beta\text{-indolyl-CH}_2\text{COCOOH} \xrightarrow[-\text{CO}]{\text{oxidation}} \text{3-CH}_2\text{COOH-indole} \xrightarrow{-\text{CO}_2} \text{skatole} \tag{28.22}$$

tryptophan $\xrightarrow{\text{oxidation}}$ 5-hydroxytryptophan $\xrightarrow{-\text{CO}_2}$ serotonin

Most of the hallucinogenic drugs are serotonin antagonists and thus are substituted indoles. Psilocin (mushrooms) and lysergic acid (ergot fungus) are examples. Dimethylation of serotonin gives bufotenine (toads). Similar small changes in structure in other substances illustrate the extreme specificity needed in certain vital

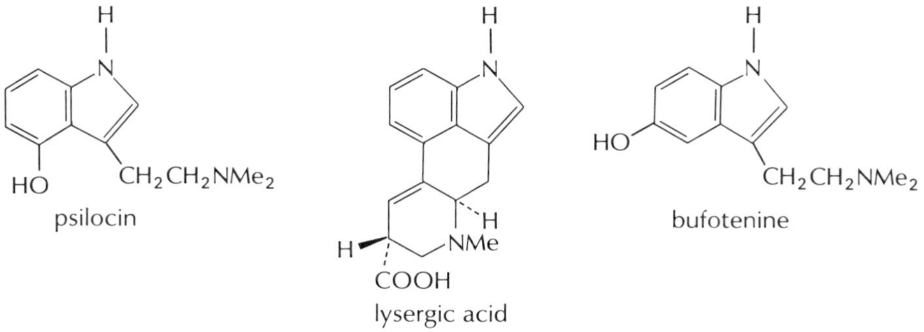

[Structures: psilocin, lysergic acid, bufotenine]

reactions for healthy existence. Many alkaloids (Sec. 28.7) have the indole ring system.

Fischer indole synthesis. Indoles may be synthesized by many routes. One of the versatile methods is the Fischer indole synthesis, in which a phenylhydrazone is heated under anhydrous conditions with a strong acid, such as $ZnCl_2$ or H_3PO_4 and P_2O_5 (polyphosphoric acid; Eq. 28.23). The starting aldehyde or ketone must have two α hydrogen atoms. The R group ends up in the α position (hydrogen if an aldehyde is used) and the group attached to the —CH_2— group ends up in the β position. If pyruvic acid ($CH_3COCOOH$) is used, the product is 2-indole carboxylic acid, which, upon decarboxylation, gives indole.

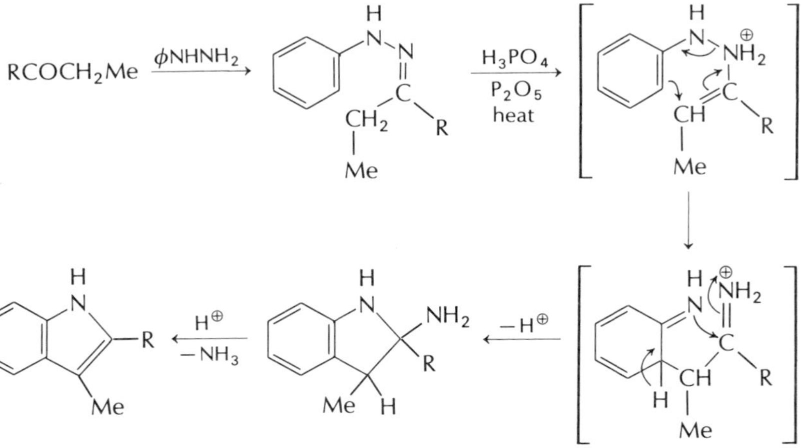

[Reaction scheme 28.23: Fischer indole synthesis mechanism]

(28.23)

Exercises

28.5 Starting with *p*-benzyloxyphenylhydrazine, show how to prepare serotonin.

28.5 Bicyclic compounds with a six-membered heterocycle

We shall limit ourselves in this section to the fusion of benzene with pyridine to produce quinoline and isoquinoline. The various alkaloids and drugs that contain quinoline and isoquinoline ring systems will be discussed elsewhere.

28.5 Bicyclic compounds with a six-membered heterocycle

Quinoline: properties; reactions. Quinoline (or benzo[b]pyridine), in the form of numerous derivatives, is widespread in nature. In many respects, quinoline has properties we have learned to expect from resemblance to naphthalene and to pyridine. Pure quinoline is a colorless liquid (bp 238°) with a pleasant odor, and it is a weak base (pK_a = 4.90).

The resistance to oxidation of pyridine can be seen in the drastic oxidation of quinoline, which gives 2,3-pyridinedicarboxylic acid (quinolinic acid; Eq. 28.24) by destruction of

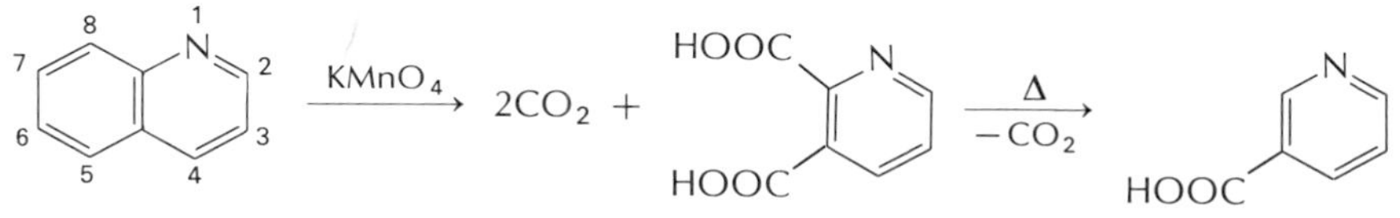

$$\text{quinoline} \xrightarrow{KMnO_4} 2CO_2 + \text{quinolinic acid} \xrightarrow[-CO_2]{\Delta} \text{nicotinic acid} \quad (28.24)$$

the benzene ring. Quinolinic acid is preferentially decarboxylated at the α-COOH group to give nicotinic acid.

Reduction of quinoline is easy: the benzene ring survives, and 1,2,3,4-tetrahydroquinoline is produced (Eq. 28.25).

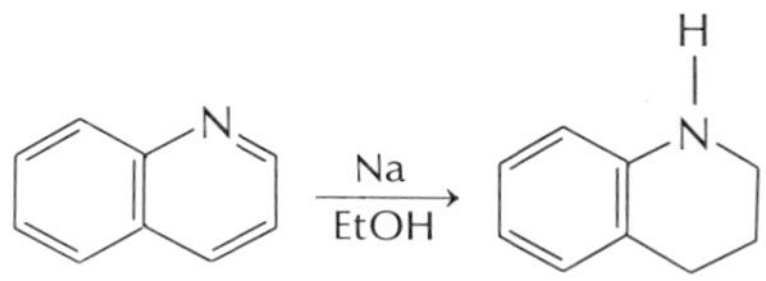

$$\text{quinoline} \xrightarrow[EtOH]{Na} \text{1,2,3,4-tetrahydroquinoline} \quad (28.25)$$

Electrophilic substitution reactions take place on the susceptible benzene ring, not on the resistant pyridine ring. Thus, nitration gives a mixture of 5- and 8-nitroquinoline. The two isomers are separable, and reduction of the nitro group to an amino group allows admission to diazotization and the replacement and coupling reactions of the diazonium ion. Sulfonation at 200° gives a separable mixture of the 5- and 8-quinolinesulfonic acids. At 300°, the 8 acid rearranges to the 6 acid. The Friedel-Crafts reactions do not take place because of complex formation with AlCl$_3$. Bromination in CCl$_4$ under reflux unexpectedly gives 3-bromoquinoline.

Nucleophilic reactions, as with pyridine, proceed nicely at the 2 and 4 positions. The Chichibabin reaction with NaNH$_2$ gives 2-aminoquinoline. Treatment with nitrous acid (remember that α- and γ-aminopyridines do not give diazonium salts) gives 2-hydroxyquinoline, which, like 2-hydroxypyridine, tautomerizes to 2-quinolone (also called carbostyril; Eq. 28.26).

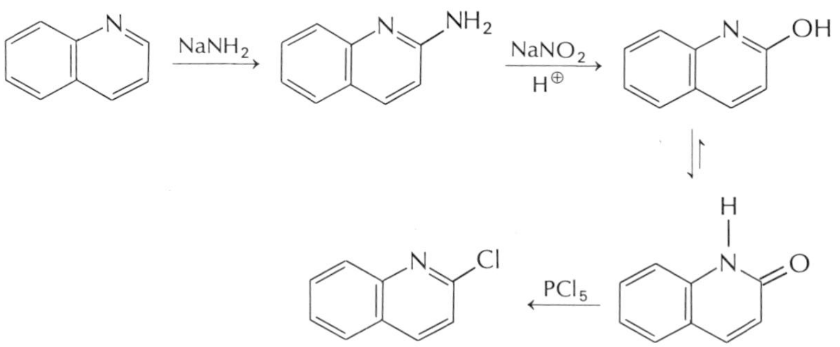

$$\text{quinoline} \xrightarrow{NaNH_2} \text{2-aminoquinoline} \xrightarrow[H^\oplus]{NaNO_2} \text{2-hydroxyquinoline} \quad (28.26)$$

$$\text{2-chloroquinoline} \xleftarrow{PCl_5} \text{2-quinolone (carbostyril)}$$

The 2- and 4-quinolinecarboxylic acids can be made by interesting syntheses. The 2 acid (quinaldinic acid) may be prepared by a nucleophilic reaction activated by a benzoyl group (Eq. 28.27). The 4 acid (cinchoninic acid) is made by a related sequence (Eq. 28.28). Iodine is used as an oxidant to aromatize the dihydro intermediate.

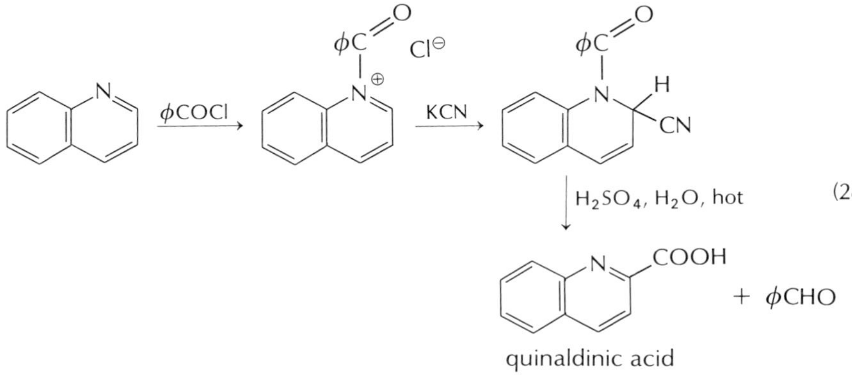

$$(28.27)$$

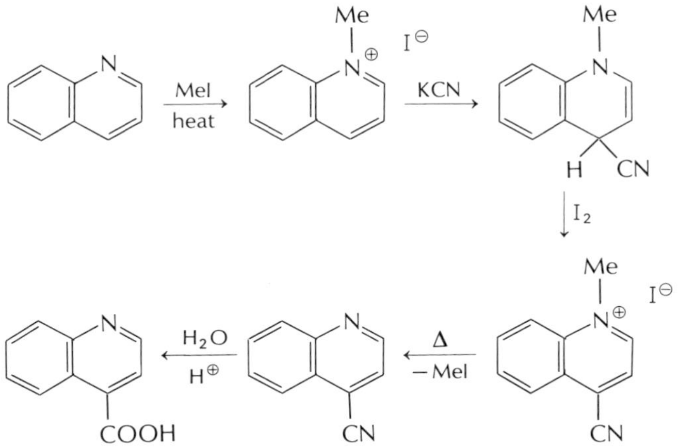

$$(28.28)$$

Skraup synthesis. Quinoline and substituted quinolines may be prepared by many syntheses in which the pyridine ring is closed.

The Skraup synthesis involves a dehydration, a Michael addition, an electrophilic substitution (the ring-closing reaction), and an oxidation, all occurring in one flask during a relatively short period of refluxing (5 hr; Eq. 28.29). By use of substituted anilines, substituted quinolines may be obtained. Thus a *p*-alkylaniline gives a 6-alkylquinoline, an *o*-alkylaniline gives an 8-alkylquinoline, and a *m*-alkylaniline gives a mixture of 5- and 7-alkylquinolines.

Substitution in the pyridine ring of quinoline is achieved by substitution of other α,β-unsaturated aldehydes or ketones for acrolein (glycerol). Thus crotonaldehyde ($CH_3CH=CHCHO$), with aniline, As_2O_5, and sulfuric acid, gives 2-methylquinoline (quinaldine). The reaction shows that the addition to the unsaturated carbonyl system must be 1,4. ☐ ☐

Isoquinoline: properties; reactions; synthesis. Isoquinoline, or benzo[c]pyridine, serves as the framework for many alkaloids. Like quinoline, it is a colorless liquid (bp 240°) and a weak base (pK_a = 5.42). Oxidation of isoquinoline destroys the benzene ring and gives

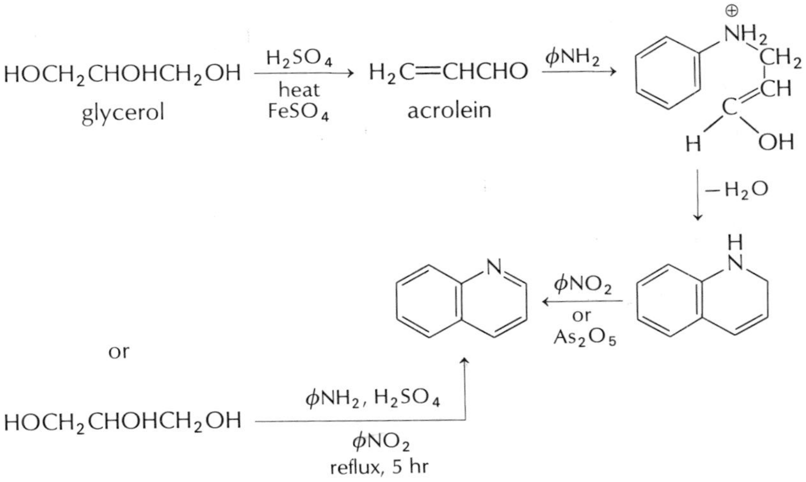

28.5 Bicyclic compounds with a six-membered heterocycle

$$\text{HOCH}_2\text{CHOHCH}_2\text{OH} \xrightarrow[\text{FeSO}_4]{\substack{\text{H}_2\text{SO}_4 \\ \text{heat}}} \text{H}_2\text{C}=\text{CHCHO} \xrightarrow{\phi\text{NH}_2} \text{[anilinium-propenol intermediate]}$$

glycerol → acrolein

$\downarrow -\text{H}_2\text{O}$

[1,2-dihydroquinoline] $\xrightarrow[\text{As}_2\text{O}_5]{\phi\text{NO}_2 \text{ or}}$ quinoline

or

$$\text{HOCH}_2\text{CHOHCH}_2\text{OH} \xrightarrow[\substack{\phi\text{NO}_2 \\ \text{reflux, 5 hr}}]{\phi\text{NH}_2,\ \text{H}_2\text{SO}_4}$$

(28.29)

3,4-pyridinedicarboxylic acid (cinchomeronic acid). Reduction of isoquinoline gives 1,2,3,4-tetrahydroisoquinoline. Electrophilic substitution on isoquinoline takes place on the benzene ring, as with quinoline. The 1 position in isoquinoline is equivalent to an α position in pyridine and is reactive to nucleophilic reagents. Bond fixation results in relative inactivation at the 3 position. Thus, the Chichibabin reaction takes place at the 1 position (Eq. 28.30).

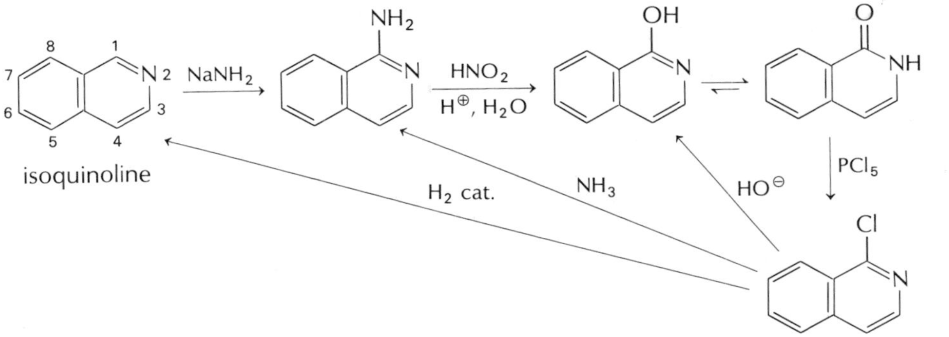

isoquinoline $\xrightarrow{\text{NaNH}_2}$ 1-aminoisoquinoline $\xrightarrow[\text{H}^\oplus,\ \text{H}_2\text{O}]{\text{HNO}_2}$ 1-hydroxyisoquinoline $\rightleftharpoons$ isocarbostyril $\xrightarrow{\text{PCl}_5}$ 1-chloroisoquinoline

1-chloroisoquinoline $\xrightarrow{\text{H}_2 \text{ cat.}}$ isoquinoline; $\xrightarrow{\text{NH}_3}$ 1-aminoisoquinoline; $\xrightarrow{\text{HO}^\ominus}$ 1-hydroxyisoquinoline

(28.30)

$$\phi\text{CH}_2\text{CH}_2\text{NH}_2 \xrightarrow{\text{RCOCl}} \text{[N-acyl-}\beta\text{-phenethylamine]} \xrightarrow[\substack{\text{toluene} \\ \text{reflux} \\ -\text{H}_2\text{O}}]{\text{P}_2\text{O}_5} \text{[3,4-dihydroisoquinoline, 1-R]}$$

$\swarrow \underset{-\text{H}_2}{\overset{\Delta}{\text{Pd}}} \qquad \searrow \text{H}_2, \text{Pd}$

a: 1-R-isoquinoline b: 1-R-1,2,3,4-tetrahydroisoquinoline

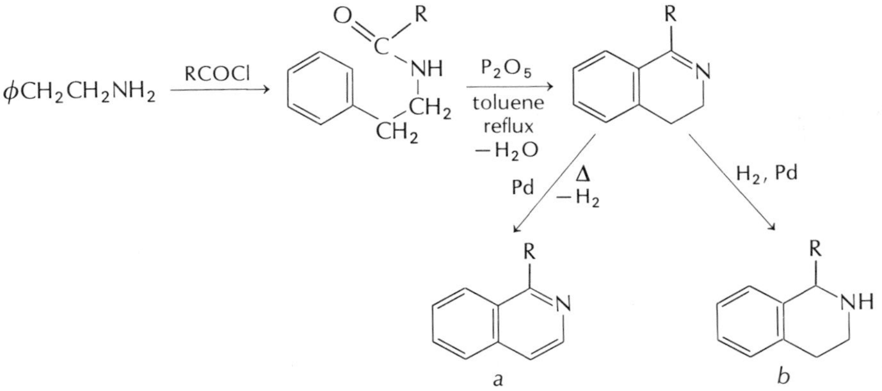

(28.31)

Of the various ways in which the isoquinoline ring system may be formed, we shall look at the Bischler-Napieralski synthesis. The method involves the formation of an amide of a 2-phenylethylamine and ring closure of the amide in the presence of a dehydrating reagent such as P_2O_5 in boiling toluene (Eq. 28.31). ($POCl_3$ in solvents of higher bp also is used.) The product, as shown, is a 1-alkyl-3,4-dihydroisoquinoline, which may be dehydrogenated to give the substituted isoquinoline a or hydrogenated to give a substituted tetrahydroisoquinoline. The 2-phenylethylamines are conveniently made from benzoic acid and substituted benzoic acids as shown in Equation 28.32.

$$\phi COOH \xrightarrow[\text{ether}]{\text{LiAlH}_4} \phi CH_2OH \xrightarrow{SOCl_2} \phi CH_2Cl \xrightarrow[\text{DMSO}]{\text{NaCN}} \phi CH_2CN \xrightarrow{\text{LiAlH}_4} \phi CH_2CH_2NH_2 \quad (28.32)$$

Exercises

28.6 Give a useful synthesis of:
  a. 6-quinolinecarboxylic acid
  b. 1,4-dimethylisoquinoline
  c. 4,6-dichloro-2-methylquinoline

28.6 Purines and pteridines

Purine and pteridine: structures, numbering. Purine and pteridine are our two examples of the very many possible bicyclic ring systems that are the result of fusion of two heterocyclic rings. Although the numbering of pteridine follows the naming rules, the numbering of purine, because of long usage, is an allowed exemption from the rules. Purine is a fusion product of imidazole and pyrimidine; pteridine is a fusion product of pyrazine and pyrimidine.

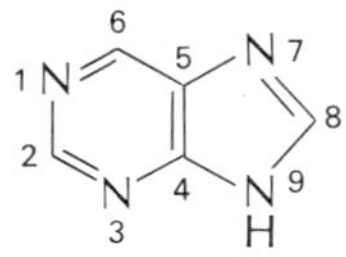

purine 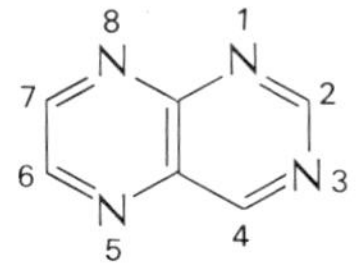   pteridine

The numbering of purine goes back to early work upon the structure, when uric acid was written (as late as the 1920's) as though a central column of three carbon atoms were joined to two molecules of urea.

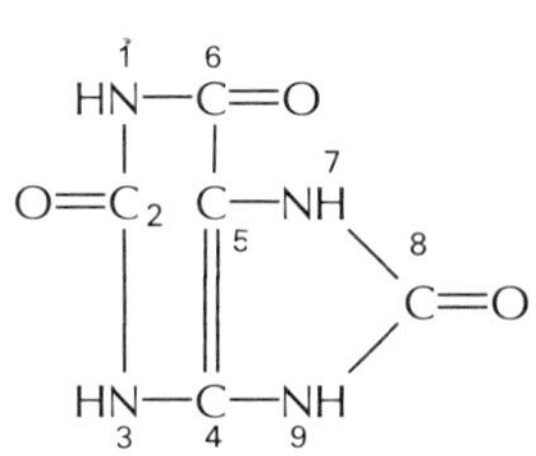

28.6 Purines and pteridines

Purines: uric acid. One of the first organic compounds to be isolated from an animal source was a purine derivative, uric acid, present in small amounts in urine (also gallstones). Later on, it was found that uric acid is the major nitrogen-containing component in the excreta of birds, reptiles, and insects. Many other naturally occurring compounds are purine derivatives also.

We write purine with the N—H bond at position 9 (it could be at position 7 and still preserve the Kekulé resonance of the pyrimidine ring) because the uv spectrum of purine resembles that of 9-methylpurine instead of that of 7-methylpurine, in which tautomerization of hydrogen between positions 7 and 9 cannot occur. Most of the naturally occurring purine derivatives are bonded at 9 to sugars (purine glycosides) or to proteins. Commonly occurring purines, other than uric acid, are caffeine (coffee, tea), theobromine (cocoa bean), theophylline (tea), and xanthine, hypoxanthine, adenine, and guanine in most living cells.

Purine is a weak base ($pK_a = 2.6$). Purine is seldom used as a starting point for preparation of substituted purines. Instead, substituted purines are synthesized directly or by transformations of other substituted purines. Uric acid is a convenient starting point.

Most purine compounds are prepared from a pyrimidine, which has amino groups at the 4 and 5 positions. Let us first see how such a substituted pyrimidine may be formed (Eq. 28.33) and closed to give uric acid (2,6,8-trihydroxypurine). The hydroxypurines exist in

(28.33)

the lactam structures rather than in the hydroxypurine structures. Naming, however, is easier if they are considered to be in the hydroxypurine tautomer. Aminopurines, on the other hand, seem to prefer to remain aminopurines. The pyrimidines (a) may be varied to obtain different substituents at the 2 position in the substituted purine. Thus, use of guanidine (($H_2N)_2C$=NH) instead of urea gives 2-amino-6,8-dihydroxypurine.

The hydroxy (carbonyl) groups in uric acid all are in positions whereby nucleophilic substitution is favored (α and γ to heterocyclic nitrogen). Treatment of uric acid with $POCl_3$ gives 2,6 or 2,6,8-trichloropurine. The chlorine atoms, in turn, may undergo nucleophilic substitution by ammonia, hydroxide, or alkoxide ion or may be removed by HI reduction. The order of reactivity of replacement of Cl is 6 > 2 > 8. A suitable combination of reagents and quantities allows the conversion of uric acid into a number of various substituted purines (Eqs. 28.34). Adenine and guanine are present in DNA and RNA as glycosides, attached to deoxyribose and to ribose. Other synthetic routes are available whereby

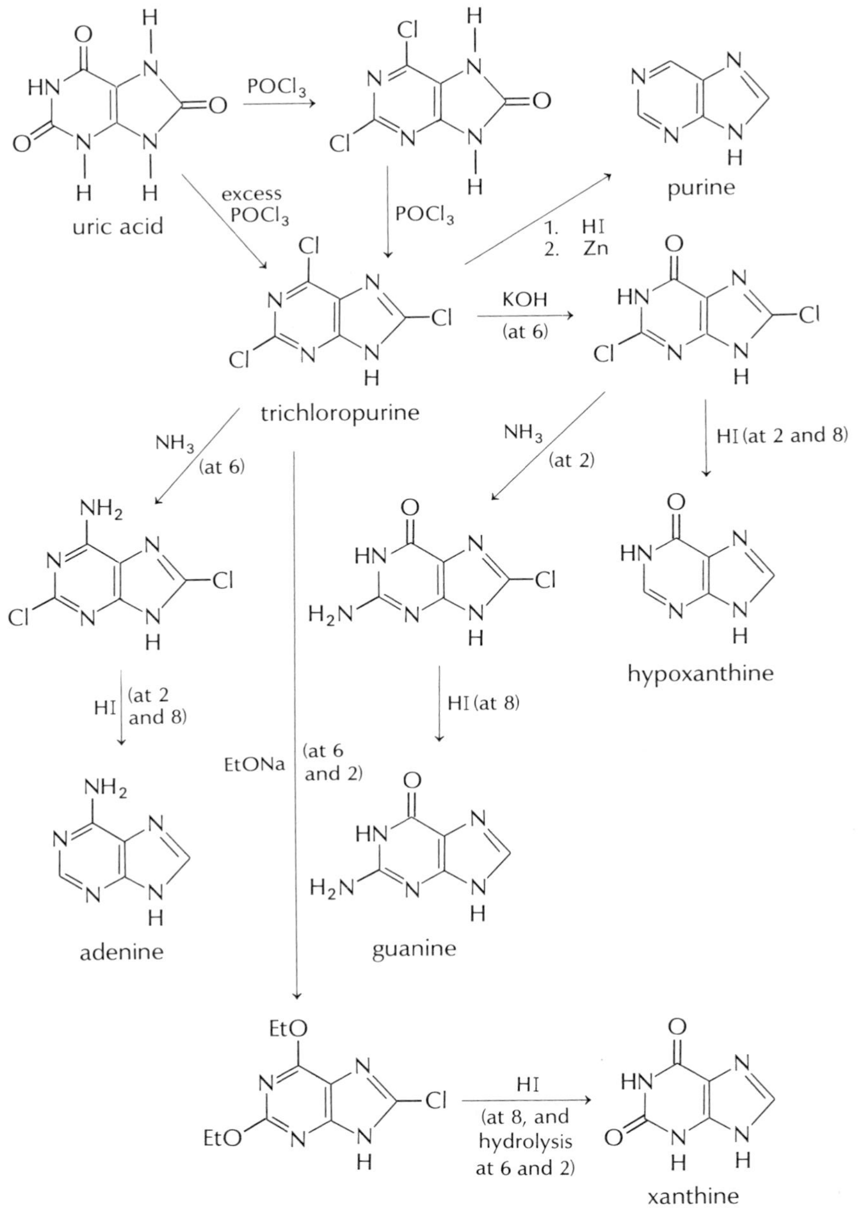

(28.34)

adenine, guanine, hypoxanthine, and xanthine may be formed by ring closure of suitably substituted pyrimidines.

Methylated xanthines. Three well-known mild stimulants are methylated xanthines. Caffeine (1,3,7-trimethylxanthine) may be obtained by trimethylation of 8-bromoxanthine. If xanthine itself is methylated, the methyl groups end up at 1, 3, and 9. Why the presence of bromine modifies the reaction is a puzzle (Eq. 28.35).

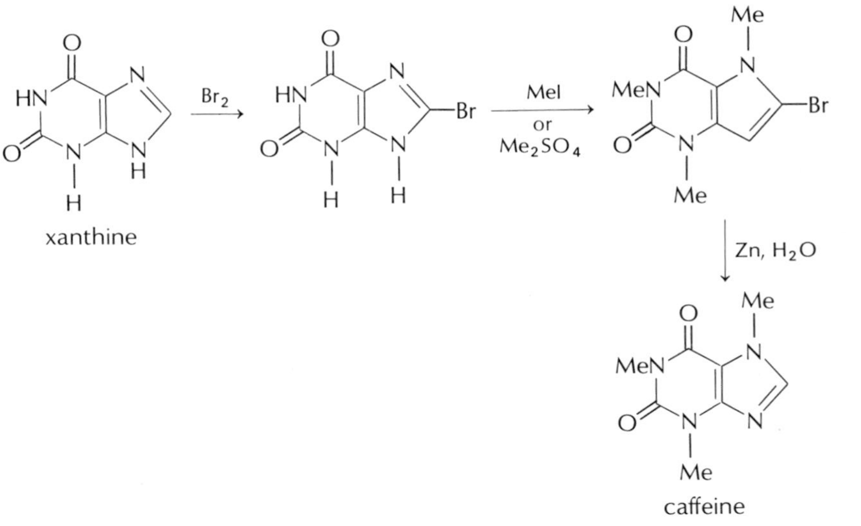

[Scheme showing xanthine → 8-bromoxanthine (Br₂) → methylated bromo intermediate (MeI or Me₂SO₄) → caffeine (Zn, H₂O)] (28.35)

The partially methylated derivatives theophylline (1,3-dimethylxanthine) and theobromine (3,7-dimethylxanthine) are more difficult to obtain by synthesis or conversion.

Pteridines: folic acid, riboflavin. Substituted pteridines were first isolated from nature as butterfly wing pigments. The chemistry of the pteridines, however, was not explored thoroughly until it was found that two of the B vitamins, folic acid and riboflavin, were substituted pteridines.

Pteridine, a fusion of pyrimidine with pyrazine, is not susceptible to electrophilic aromatic substitution, but it does undergo nucleophilic substitution (both rings are electron deficient).

Substituted pteridines are prepared by a combination of ring synthesis and aromatic nucleophilic substitution (Eq. 28.36). Leucopterin (from oxalic acid or ester with 4-hydroxy-2,5,6-triaminopyrimidine) is a colorless pteridine derivative that occurs along with xanthopterin (yellow) in butterfly wings.

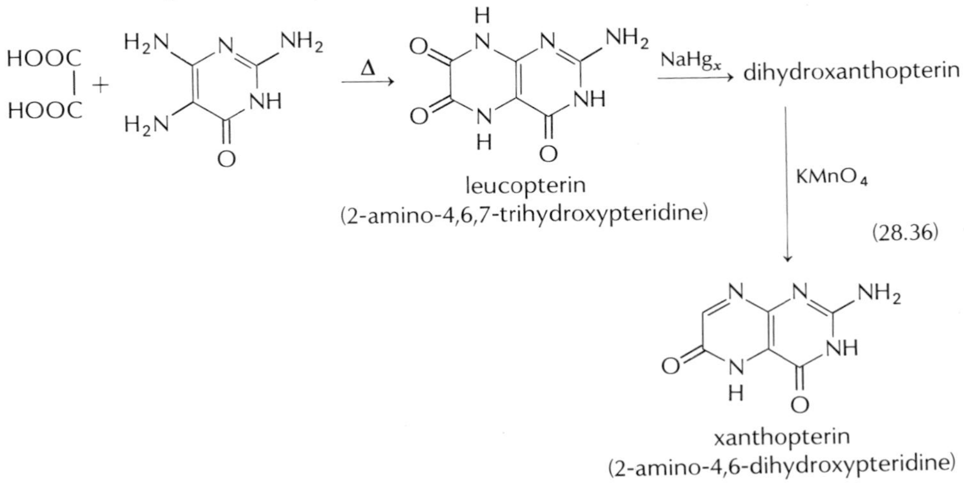

[Scheme: HOOC-COOH + 4-hydroxy-2,5,6-triaminopyrimidine →(Δ) leucopterin (2-amino-4,6,7-trihydroxypteridine) →(NaHg$_x$) dihydroxanthopterin →(KMnO₄) xanthopterin (2-amino-4,6-dihydroxypteridine)]   (28.36)

Folic acid (also called pteroyglutamic acid, PGA, vitamin B_c, vitamin M, etc.) is one of the vitamins of the B complex. It is needed to prevent certain types of anemia. It also is a precursor of the citrovorum factor that is required for cell division. It may be synthesized in about 10% yield by a one-step synthesis (Eq. 28.37).

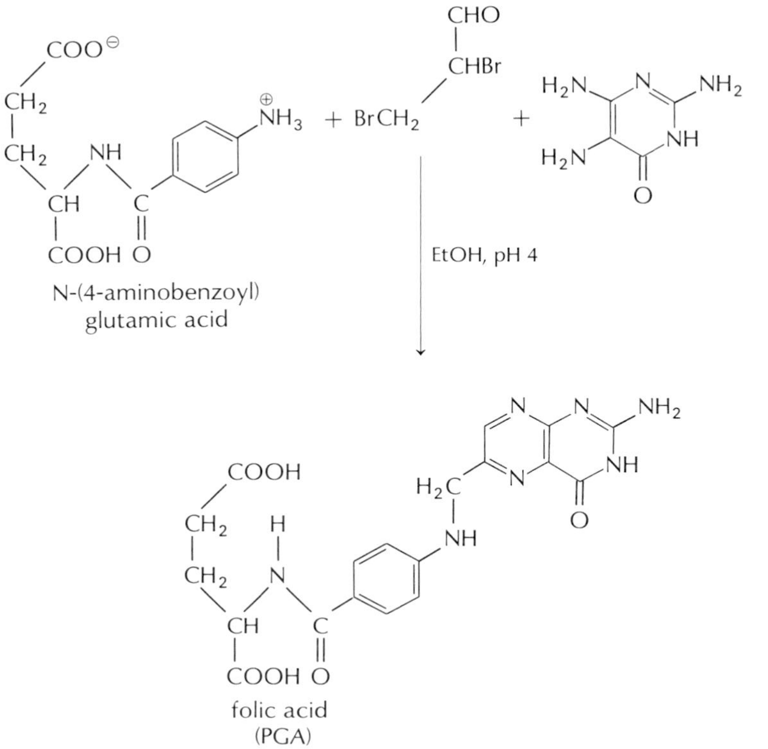

N-(4-aminobenzoyl) glutamic acid + BrCH$_2$CHBr(CHO) + diaminopyrimidinone $\xrightarrow{\text{EtOH, pH 4}}$ folic acid (PGA)

(28.37)

Riboflavin, or vitamin B_2, is present in small amounts in all animal cells and in many plants as well. Absence of the vitamin in the diet causes poor health. One of the symptoms of deficiency of riboflavin is skin sores, usually about the lips. It is a derivative of isoalloxazine (the oxa in the name is not part of the oxa-aza naming system: oxygen is not present in the heterocyclic ring system), a pteridine derivative with a benzene ring fused to the pyrazine

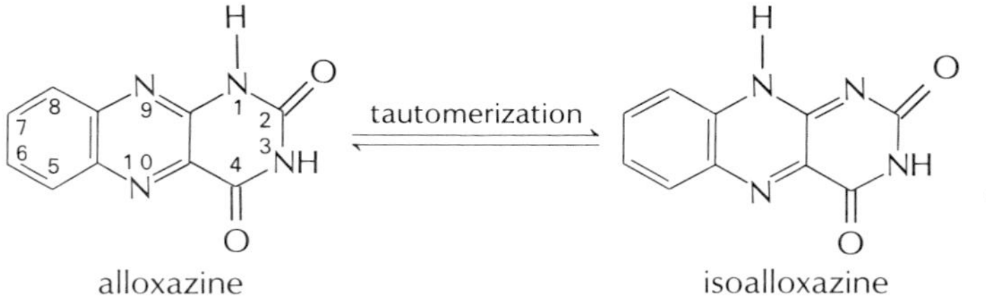

alloxazine $\rightleftharpoons$ isoalloxazine (tautomerization)

ring. Riboflavin may be synthesized as shown in Equation 28.38. The synthetic product is used in the fortification of foods as well as in vitamin tablets. Alloxan (from which the name alloxazine is obtained) is produced by acidic oxidation (dilute HNO_3) of uric acid.

Within the body, riboflavin is easily reduced at the 9 and 10 positions to give a dihydro compound that is easily oxidized. Thus, riboflavin participates as a coenzyme in hydrogen transport. The two components FMN and FAD are shown.

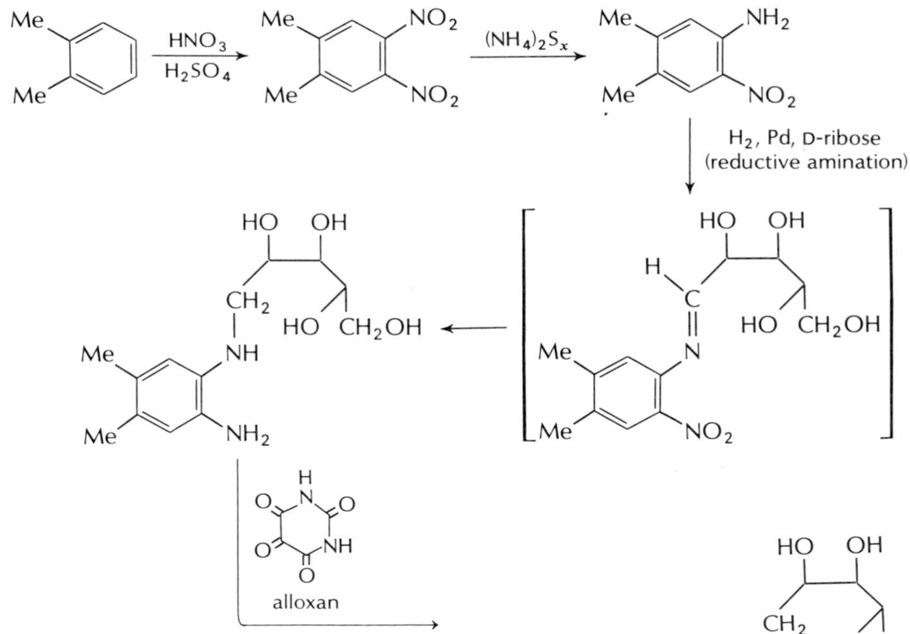

28.6 Purines and pteridines

(28.38)

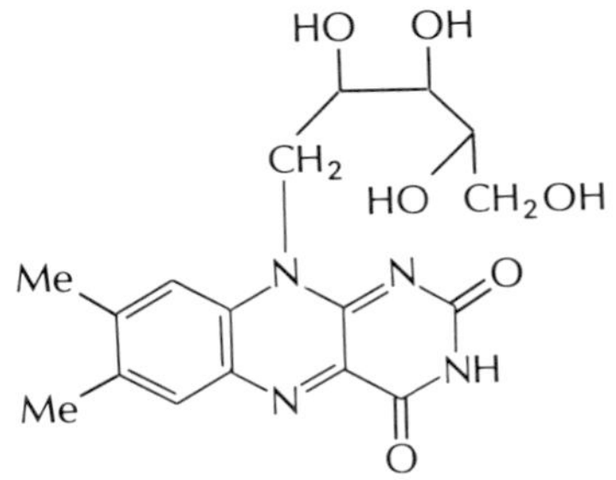

riboflavin
(6,7-dimethyl-9-(D-1-ribityl)isoalloxazine)

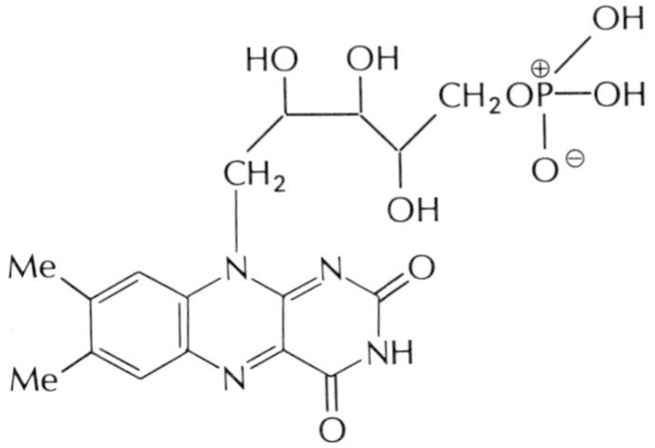

FMN (flavin mononucleotide)

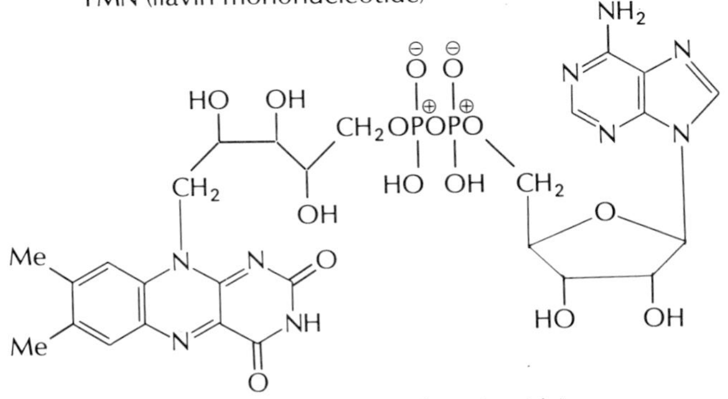

FAD (flavin adenine dinucleotide)

Exercises

28.7 Alloxan may be obtained from barbituric acid. Show how.

28.8 Starting with glutaric acid and toluene, show how to prepare N-(4-aminobenzoyl) glutamic acid.

28.7 Alkaloids

Definition; examples. Alkaloid (similar to an alkali, or a basic substance) is the general name used for a nitrogen-containing product obtainable from a plant and having physiological activity. However, there are numerous exceptions, such as caffeine (usually not considered to be an alkaloid), marijuana (no nitrogen), and meperidine (a synthetic product). Most alkaloids are derivatives of heterocycles and range from relatively simple to very complex structures.

A very simple alkaloid is (+)-coniine, the poisonous substance in hemlock. The racemic substance (2-propylpiperidine) is easily made (Eq. 28.39). Other simple alkaloids are

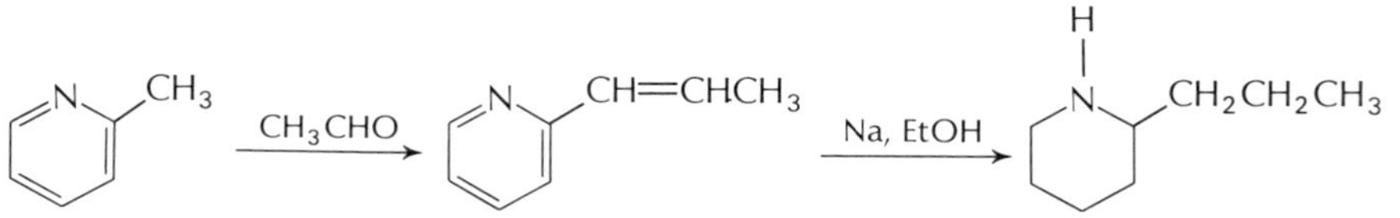

$$\text{2-methylpyridine} \xrightarrow{CH_3CHO} \text{2-(propenyl)pyridine} \xrightarrow{Na, EtOH} (\pm)\text{-coniine} \quad (28.39)$$

nicotine and piperine. Nicotine contains both a pyridine ring and a pyrrolidine ring. Although nicotine is most abundant in tobacco, it is only one of many others occurring in that plant. Piperine occurs in black pepper and is an amide of piperidine.

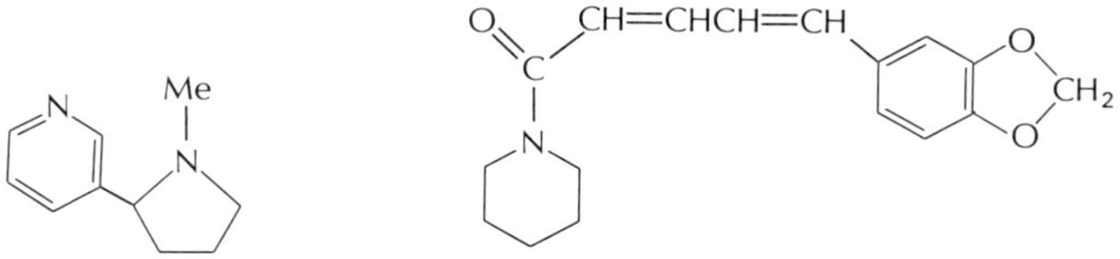

nicotine

piperine

Indole alkaloids. Complexity of structure increases rapidly as soon as we begin to examine the indole alkaloids. The indole alkaloids may be subdivided, for convenience, into: 1. the simple indole alkaloids; 2. the ergot; 3. the harmala; 4. the yohimbe; and 5. the strychnos alkaloids.

The simple indole alkaloids (1.) resemble tryptophan, the essential amino acid, from which they are probably derived. Tryptamine (acacia) is an example.

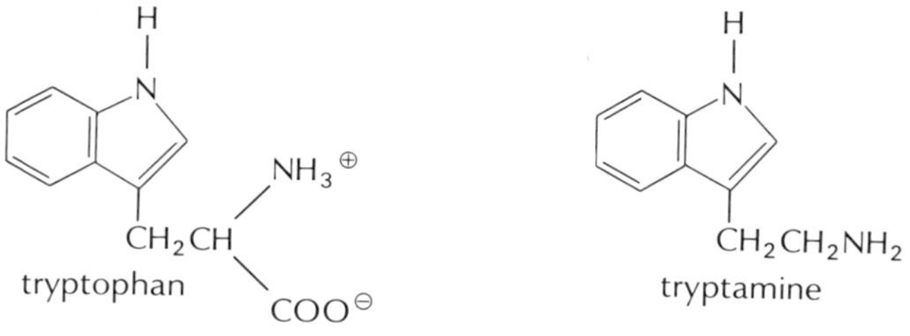

tryptophan

tryptamine

The six alkaloids (2.) from the fungus ergot are amides of lysergic acid. We note that lysergic acid has an isoprene unit fused to a methylated tryptamine. The diethylamide, LSD, is a synthetic product that is widely misused.

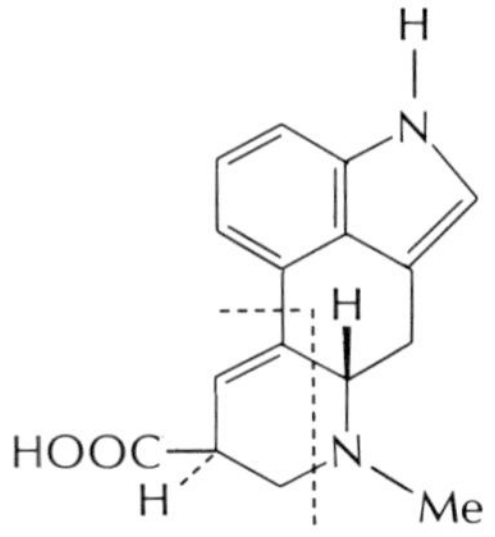

lysergic acid

The harmala alkaloids (3.) all have a pyridine ring fused to the pyrrole ring of indole; the parent ring system is known as β-carboline. An example is harman, from an herb, harmel. The yohimbine alkaloids (4.) have an isoquinoline system fused to the pyridine ring

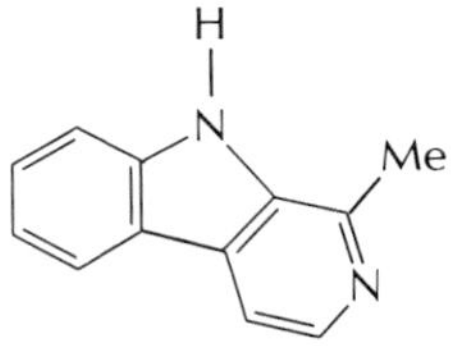

harman

of β-carboline. An example is reserpine from Indian snake root. Reserpine relieves hypertension and is a powerful tranquilizer.

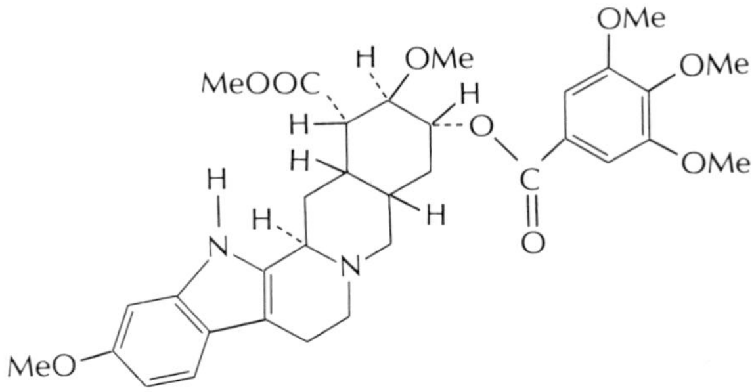

reserpine

The strychnos alkaloids (5.) are most abundant in nux vomica. Strychnine and brucine (10,11-dimethoxystrychnine), both poisonous, are examples. The heptacyclic system proved particularly difficult to pick apart and to synthesize.

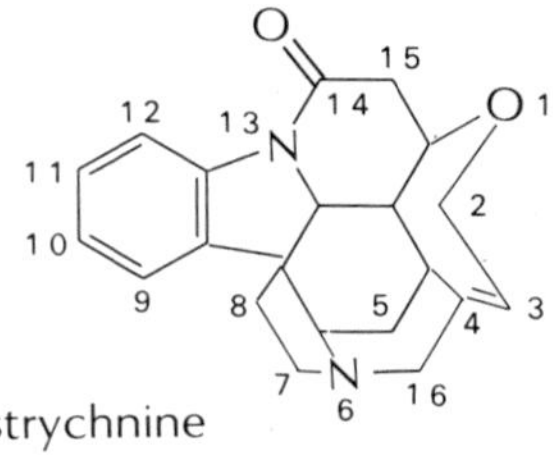

strychnine

Quinoline, isoquinoline, and tropane alkaloids. Of the alkaloids with a quinoline ring system, we give only some of those that occur in the bark of the cinchona tree (pronounced sinkona), native to the Andes. Quinine (70%) and cinchonine are the most abundant of the numerous alkaloids (>25) present (@ 6%) in cinchona bark. There are four asymmetric carbon atoms in each compound, and many of the other alkaloids present

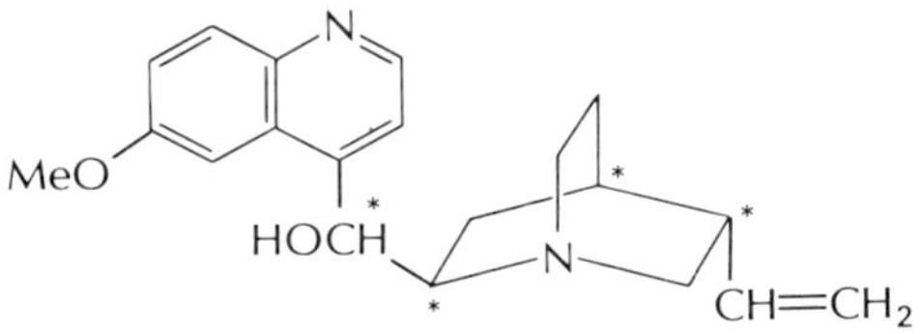

(−)-quinine

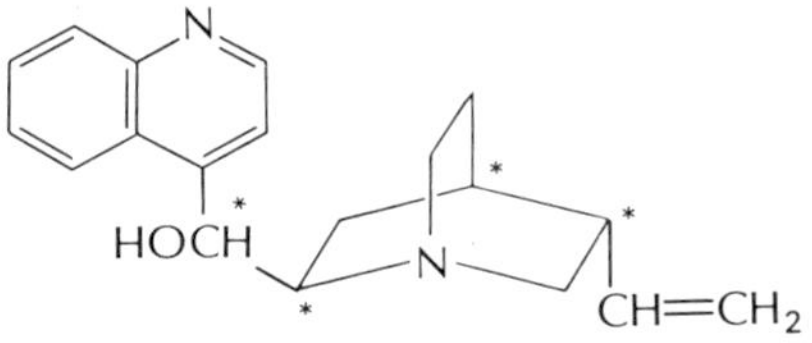

(+)-cinchonine

in the bark are stereoisomers of quinine and cinchonine. Quinine was finally synthesized in 1944, after years of effort, but the route was so long that no commercial use of the method was ever made. ☐ ☐

We shall illustrate the isoquinoline alkaloids by some of the compounds in opium, the exudate of the opium poppy. Of the more than twenty alkaloids present, morphine (7–15%) is the most abundant. It was first isolated in 1803, but the final proof of structure by synthesis was not achieved until 1952.

Laudanosine is N-methyl-1,2,3,4-tetrahydropapaverine (a). If we demethylate laudanosine, hydrogenate a ring, and twist the molecule as shown (b), we can visualize the spatial relationship by which laudanosine might be considered a precursor of morphine (c). Other perspective views of morphine, d and e, are to be compared with the usual representation of the molecule (f), which gives the false impression of a more or less planar hexahydrophenanthrene ring system.

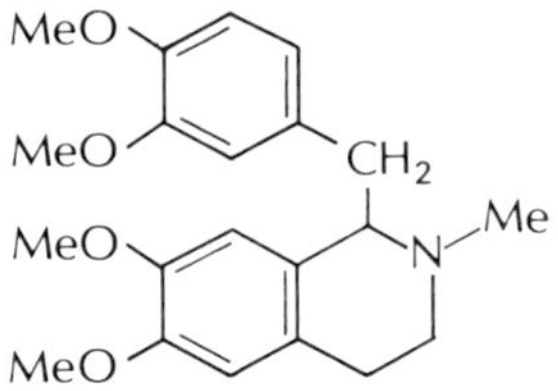

laudanosine
a

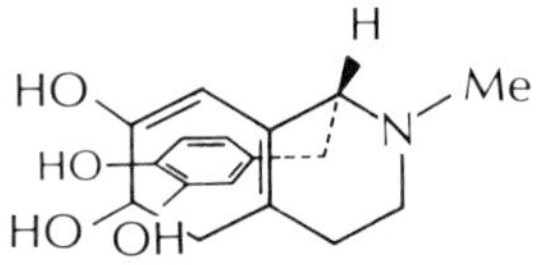

b

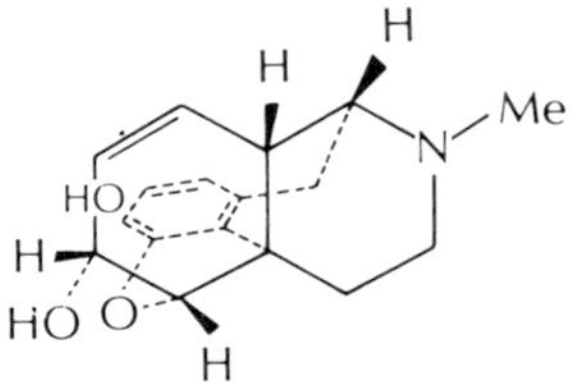

morphine, c

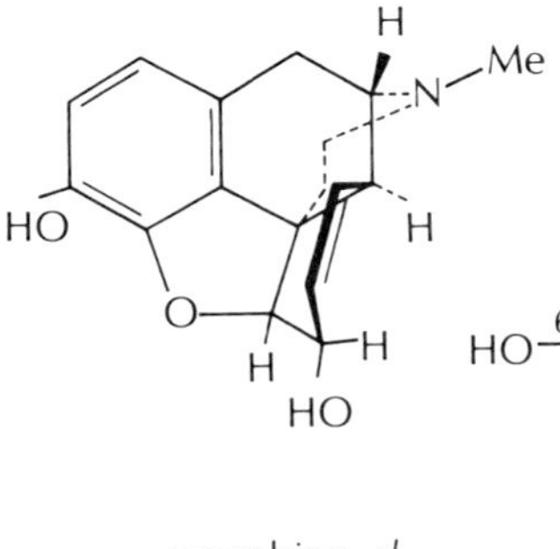

morphine, d

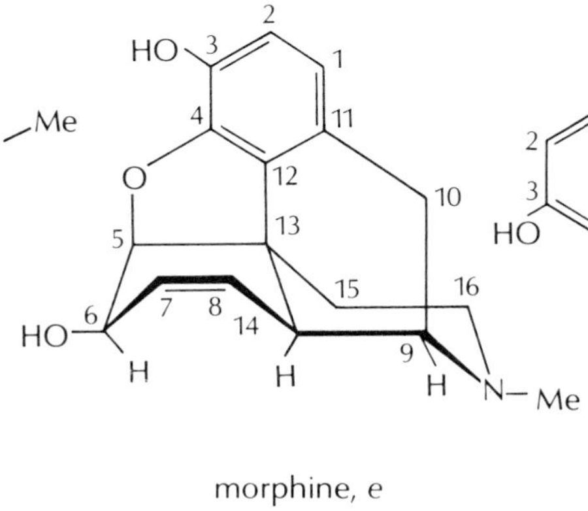

morphine, e

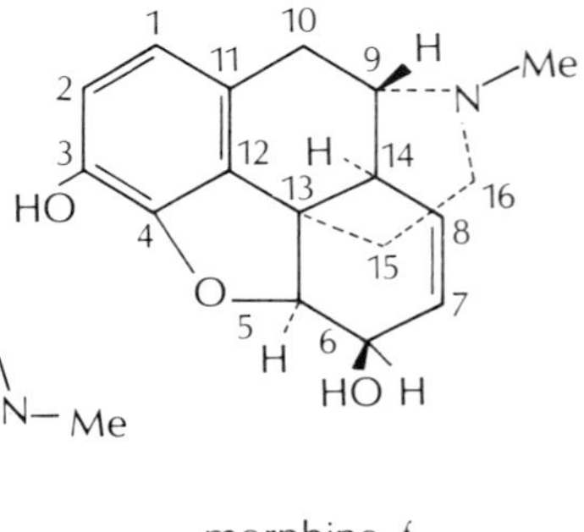

morphine, f

Opium has been known since antiquity as an effective pain reliever. (Aspirin relieves only minor aches and pains except for arthritis.) Morphine is the active ingredient in opium for pain relief, and it remains a popular drug for this purpose. It has two serious drawbacks, however: addiction and depression of respiration. Many, many attempts have been made to modify morphine to find a drug with its analgesic properties but without its side effects. Unfortunately, in most cases analgesic effectiveness and tendency for addiction parallel each other. Thus, codeine, the phenolic methyl ether of morphine (MeO— at 3 instead of HO—), used as a local anesthetic, is both much less analgesic and less liable to cause addiction than is morphine. Heroin, the 3,6 diacetate of morphine, was originally introduced as an analgesic without a tendency to cause addiction! Pethidine (meperidine, Demerol), methadone, phenazocine, and naloxone are examples of modifications that have been in (or are under investigation for) clinical use. Pethidine is less addictive than morphine but

pethidine

methadone

phenazocine (a benzomorphan)

naloxone

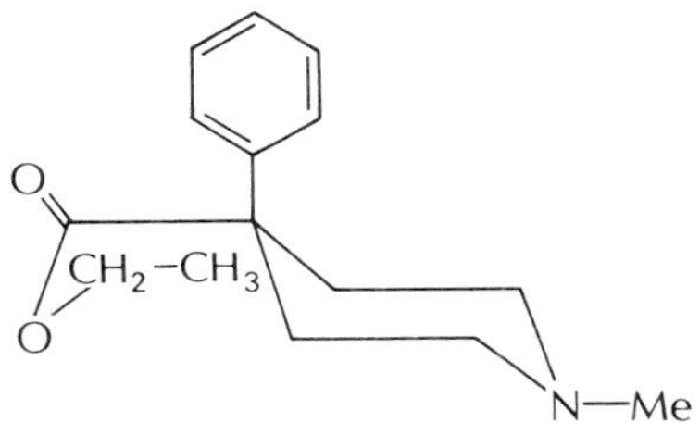

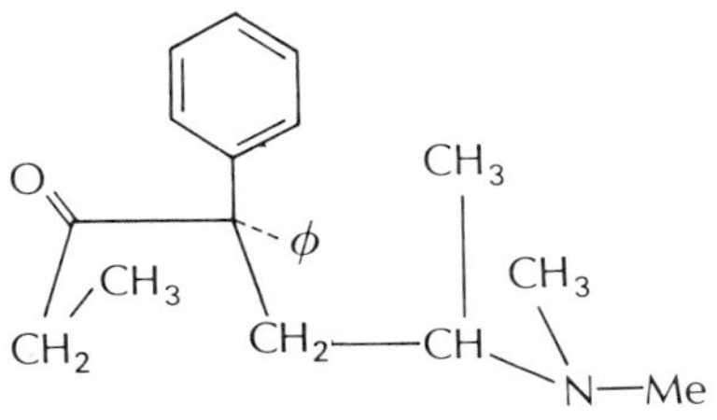

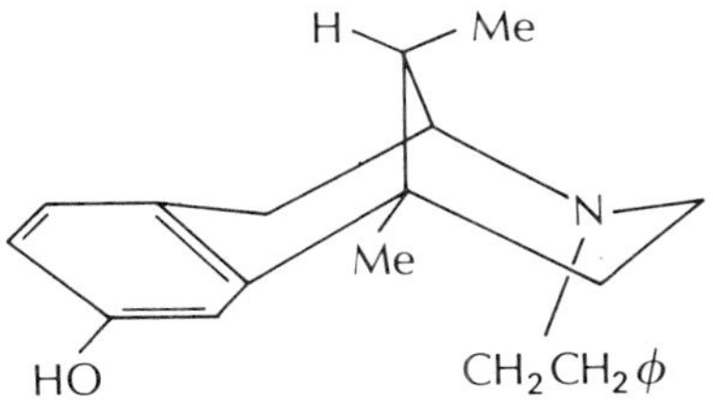

retains reasonable analgesic effectiveness. Methadone is used to replace heroin in heroin addiction. The patient may be maintained on methadone, or the methadone may be gradually withdrawn until cyclazocine (has a cyclopropylmethyl on nitrogen instead of a phenylethyl as in phenazocine) and/or naloxone treatment may be started. The latter two substances are narcotic antagonists. They act by preventing a narcotic from reaching the nervous system. Thus, even if a patient who is on a cyclazocine or naloxone regimen should try a heroin injection, he would experience no effect. The search for better analgesics and antagonists to narcotics continues. The structures, if possible, are drawn in such a way as to show similarity to morphine (e).

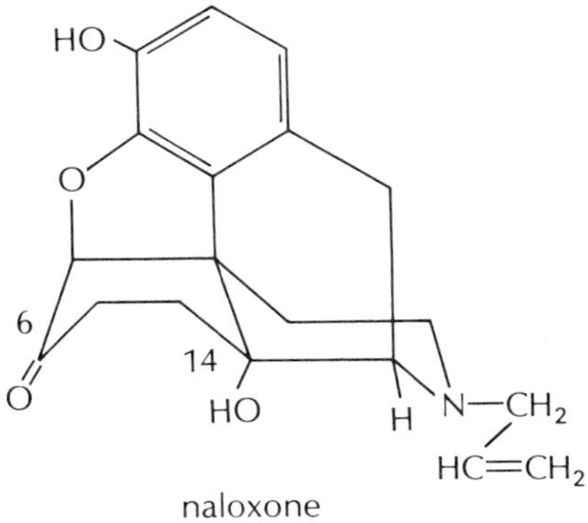

The tropane alkaloids all are derivatives of the bicyclic heterocycle tropane, or N-methyl-8-azabicyclo[3.2.1]octane. Cocaine (coca plant), atropine (belladonna), and scopolamine (scopolia) are examples.

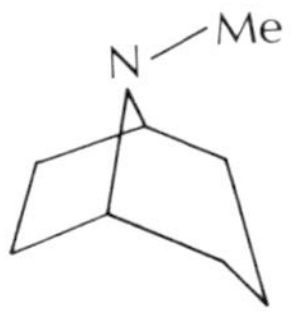

tropane 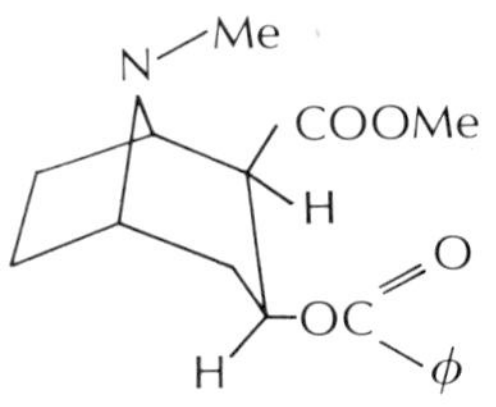   cocaine 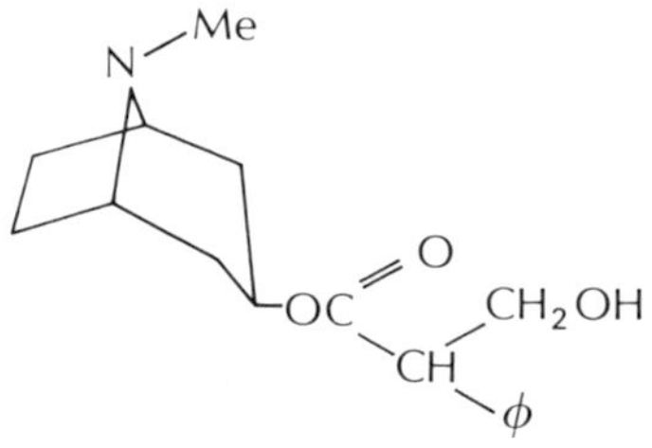   atropine

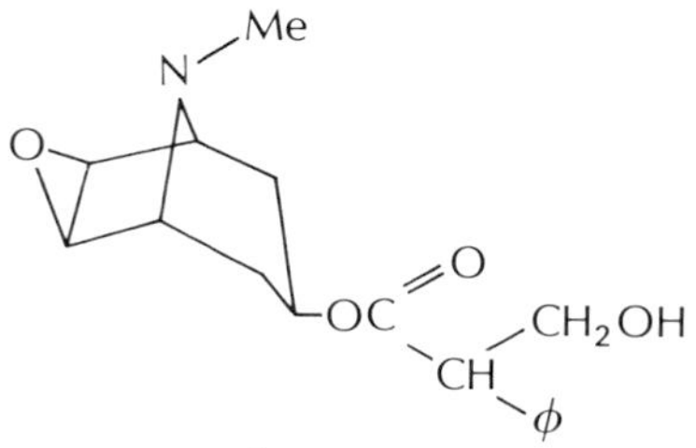

scopolamine

28.8 Nucleic acids

Discovery: DNA, RNA; structures of nucleosides and nucleotides. Nucleic acids were isolated as early as 1871 from pus. The nucleic acids occur bound to basic proteins (nucleoproteins) in the nucleus of cells. In this form, the nucleic acids are present in all living cells. When the acidity was found to be caused by phosphoric acid groups, the name nucleic acids was introduced. Although they were recognized as early as 1900, it was not until 1930 that the two classes were clearly distinguished as ribonucleic acids (RNA) and deoxyribonucleic acids (DNA). Yeast and salmon sperm are good sources of nucleic acids. DNA and RNA, although water soluble, are polymeric. RNA polymers range from as low as 25,000 to 150,000 in molecular weight (about 80 to 500 nucleotide units). DNA polymers are larger, about 6 to 16 × 10^6 in molecular weight (about 20,000 to 55,000 nucleotide units).

Acidic hydrolysis of a nucleic acid breaks the substance into the components (Eqs. 28.40). DNA and RNA differ in the sugar component and in one of the four heterocyclic

$$\text{DNA} \xrightarrow[\text{H}_2\text{O}]{\text{H}^\oplus} \text{adenine + guanine + cytosine + thymine} \\ \text{+ 2-deoxy-D-ribofuranose} + \text{H}_3\text{PO}_4$$

$$\text{RNA} \xrightarrow[\text{H}_2\text{O}]{\text{H}^\oplus} \text{adenine + guanine + cytosine + uracil} \quad (28.40)\\ \text{+ D-ribofuranose} + \text{H}_3\text{PO}_4$$

bases present: thymine in DNA, uracil in RNA. Basic hydrolysis of RNA (DNA is inert to base) breaks out only phosphoric acid, leaving a mixture of glycosides of the bases: β-ribosides, called nucleosides (Eq. 28.41). Enzymatic hydrolysis can remove a phosphate ester group,

$$\text{RNA} \xrightarrow[H_2O]{HO^{\ominus}} \text{adenosine + guanosine + cytidine + uridine} + H_3PO_4 \quad (28.41)$$

and the nucleic acid chain may be seen to consist of ribose (or deoxyribose) units linked at the 3 and 5 positions by phosphate ester groups (Eqs. 28.42). The products, specifically the

$$\text{RNA} \xrightarrow[\text{phosphodiesterase}]{H_2O} \text{5-nucleotides}$$

$$\text{RNA} \xrightarrow[\text{ribonuclease}]{H_2O} \text{3-nucleotides} \quad (28.42)$$

$$\text{DNA} \xrightarrow[\text{deoxyribonuclease}]{H_2O} \text{3-deoxynucleotides}$$

5 phosphoric acid esters, are called nucleotides as a class. The pyrimidine bases (cytosine, thymine, and uracil) are bonded β to the 1 position of ribose or deoxyribose at the 1 position of the pyrimidine. The purine bases (adenine and guanine) are bonded at the 9 position of the purine. The classes and specific names are shown, as are the structures of representative bases, nucleosides, and nucleotides.

base	ribonucleoside	ribonucleotide
adenine	adenosine	adenylic acid, A
guanine	guanosine	guanylic acid, G
cytosine	cytidine	cytidylic acid, C
uracil	uridine	uridylic acid, U

base	deoxyribonucleoside	deoxyribonucleotide
adenine	deoxyadenosine	deoxyadenylic acid, dA
guanine	deoxyguanosine	deoxyguanylic acid, dG
cytosine	deoxycytidine	deoxycytidylic acid, dC
thymine	thymidine	thymidylic acid, T

Bases

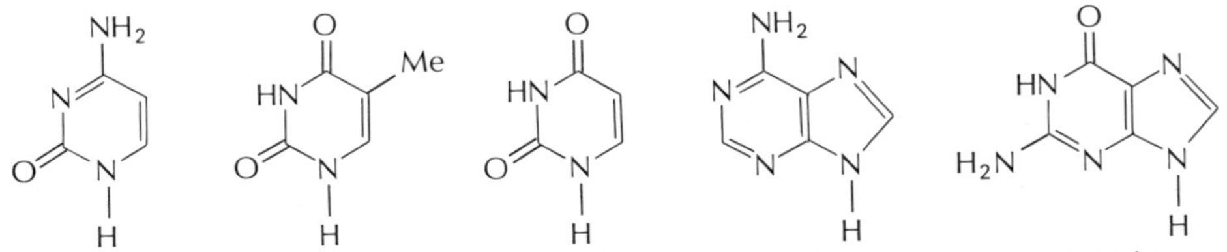

cytosine, thymine, uracil, adenine, guanine

Nucleosides

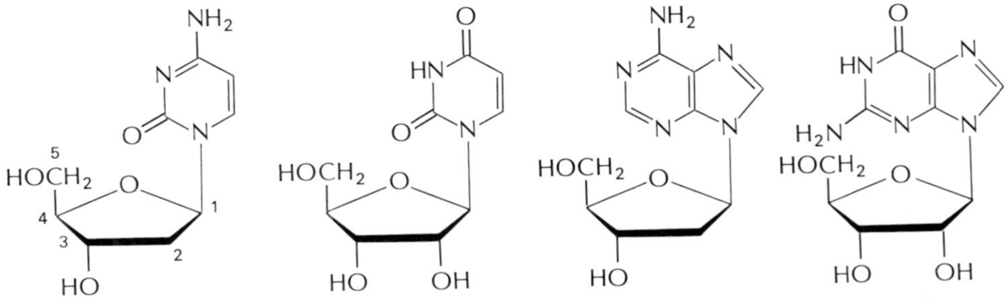

deoxycytidine, uridine, deoxyadenosine, guanosine

Nucleotides

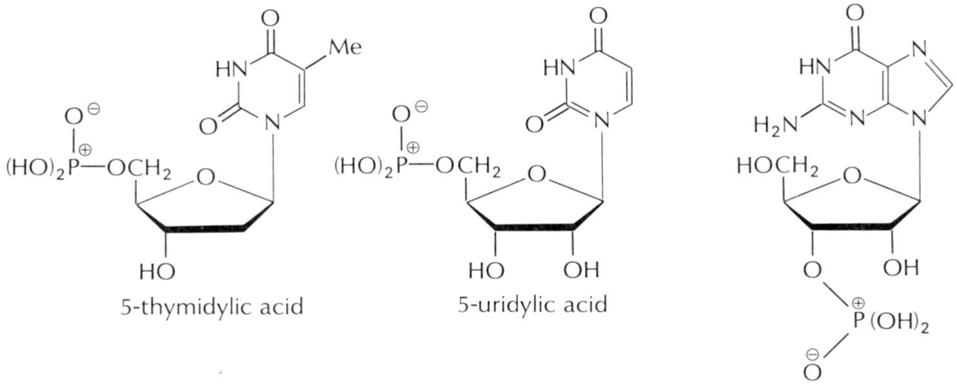

5-thymidylic acid

5-uridylic acid

3-guanylic acid

The presence of both the 3- and 5-nucleotides as hydrolytic products, along with a one-to-one ratio of phosphoric acid to ribose (or deoxyribose), leads us to the partially symbolic representation of a nucleic acid shown. We use B to represent the appropriate base. Another, more symbolic representation is to say that an RNA is a polymer consisting of A, G, C, and U units linked at the 3 position, and that DNA is a polymer consisting of dA, dG, dC, and T units linked at the 3 position (d = deoxy). Still another way is to think of an RNA as a polymer consisting of alternate ribose and phosphate ester units with varying B side chains at the 1 position of the ribose units. In a similar way, a DNA would be a polymer

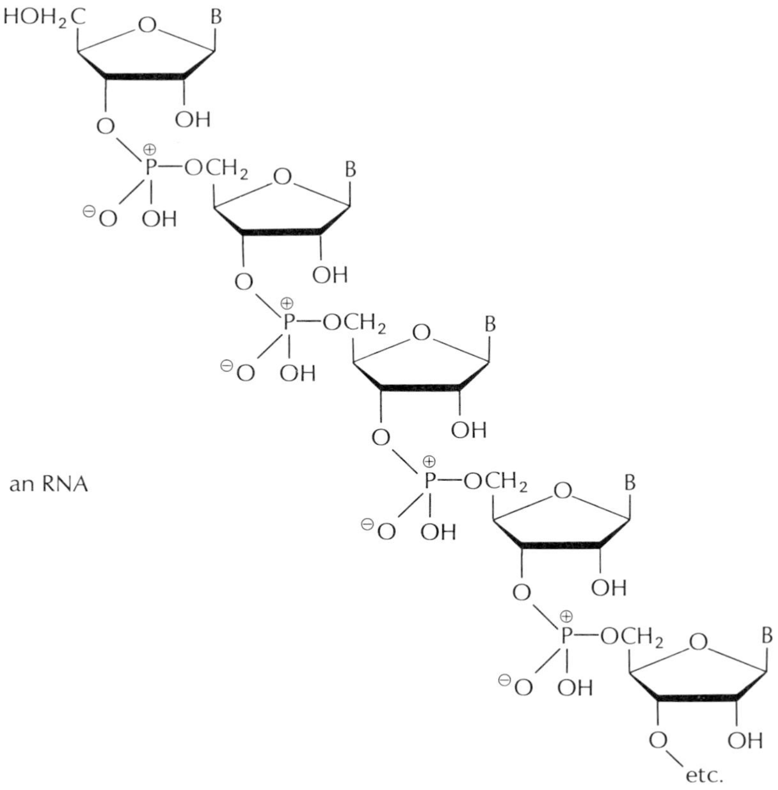

an RNA

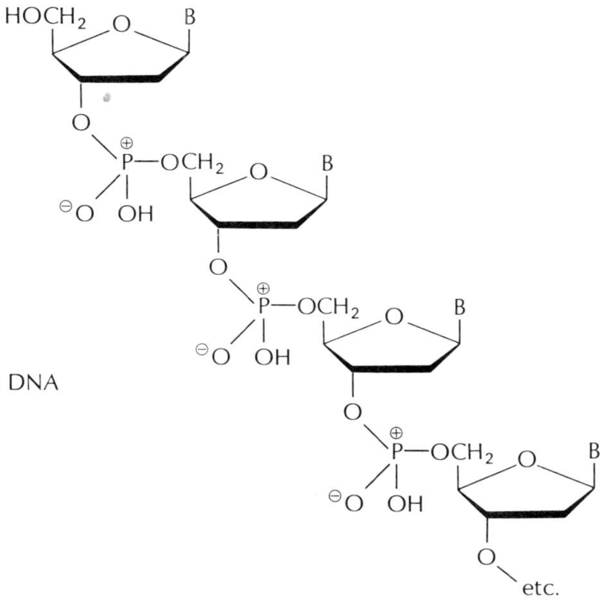

a DNA

with alternate 2-deoxyribose and phosphate ester units with varying B side chains. Knowledge of the sequence or ordering of the bases along the chains completes what is called the primary structure of the nucleic acids. We shall return to this subject later.

Secondary structure. The secondary structure of RNA and DNA is concerned with the way in which one chain may be hydrogen bonded to an adjacent chain. Referring only to DNA for the moment, it was observed that the ratios of the bases thymine to adenine and of cytosine to guanine were always one. DNA from different sources varied in other ratios of bases. For instance, the ratio of cytosine to thymine or to adenine varied with the source of the DNA. These data led to the concept of preferential hydrogen bonding of thymine to adenine and cytosine to guanine. We notice that thymine is unable to hydrogen-

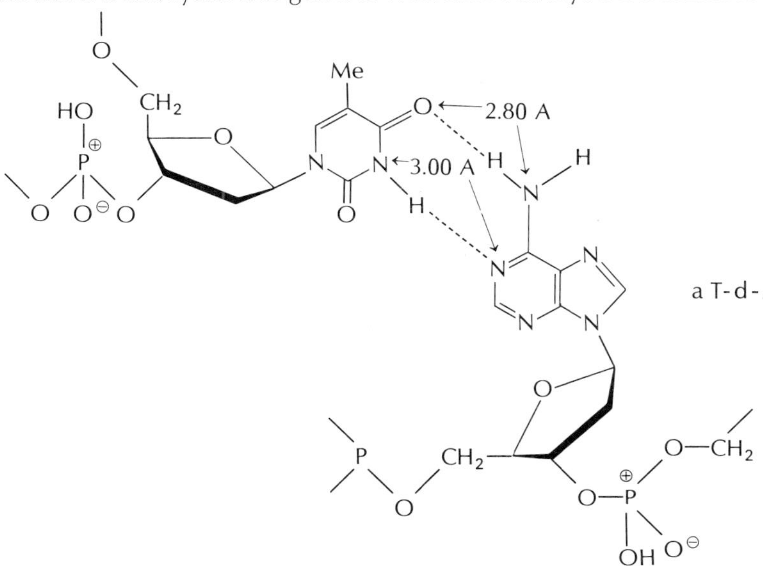

a T-d-A pair

a d-C — d-G pair

bond strongly to guanine, nor can cytidine hydrogen bond to adenine. It is difficult to illustrate the three dimensional structure that the hydrogen bond pairing of the bases imposes upon the deoxyribose-phosphate chains. However, the plane of the pair of bases is perpendicular to that of the deoxyribofuranose rings. If we imagine one chain going upward in the sequence -5,3-P-5,3-P-5,3-P-, the hydrogen bonded base pairs impose a downward sequence on the other chain.

Tertiary structure: the double helix. The tertiary structure may be thought of as follows. In DNA, base pairing causes the chains to twist into helices (coils). To a viewer at one end of the coils looking along the axis, the coils turn to the right as they move away. If we think of a long ladder that is so badly warped that the steps remain parallel to the ground as the sidepieces twist into a long double coil, we are not far off the picture. The twist is such that the steps (base pairs) are 3.4 A apart and there are ten steps per revolution of the coils, or we say that the pitch is 34 A. The diameter of the helix is 20 A. Figure 28.1 may help us to visualize the DNA double helix. At any given point along the axis, the spiral chains are 120° apart, so it should be possible to fit a third helix into the gap. A few nucleoproteins are known to have this structure, the third chain being a protein or polypeptide helix. In other nucleoproteins, the nucleic acid portion is enclosed within a sheath of protective protein.

Various RNA's; coding; the genetic code. In a given cell, the DNA is present in the nucleus of the cell as chromosomes. A gene is a segment of a chromosome and is responsible for the amino acid sequence in the synthesis of a single protein or subunit of a protein. The DNA double helix can uncoil into two separate strands. Experimentally, a solution of DNA dissociates when heated. Rapid cooling gives a solution of single strands, which slowly recombine.

There are three types of RNA. Messenger RNA (mRNA) is synthesized in the nucleus by base pairing of individual nucleotides (from cytoplasm) with single strand DNA. The mRNA (of variable size) leaves the nucleus and moves to a ribosome to complex with ribosomal RNA. Transfer RNA (tRNA), having picked up a specific amino acid in the cytoplasm of the cell (at the 3 position of adenosine at the end of the tRNA chain), arrives at the ribosomal complex and coordinates at the proper spot (codon). The amino acid is transferred to the growing polypeptide chain (protein), the tRNA moves off to find another amino acid, and the ribosomal complex rearranges to accept the next amino acid on the mRNA surface.

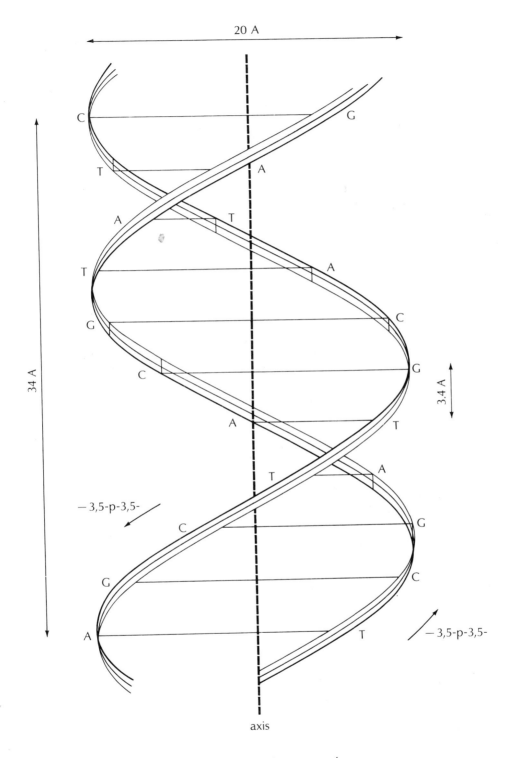

Figure 28.1 A portion of a DNA double helix, drawn to scale

926 Chapter twenty-eight

DNA is the master code from which mRNA is coded for the actual protein synthesis at the ribosomes. Thus, mRNA is the complement of a portion of a DNA strand. On the other hand, the high molecular weight ribosomal RNA is in general considered to possess a double helix structure, at least in part. The dimensions of this helix differ slightly from those of DNA. The pitch is 29 Å, with 10 to 11 base pairs per cycle. Transfer RNA is thought to possess a clover-leaf pattern. Transfer RNA is more soluble than the other RNA's (it is sometimes called sRNA) and has a relatively constant molecular weight of about 80 nucleotide units. There are also a number of unusual nucleotides in a given tRNA. The unusual nucleotides are certain N-methyl adenines and guanines and dihydrouracil as bases; and some unusual glycosides are inosine, thymine riboside, and ψ-uridine (in which uracil is bonded at the 5 position to the 1 position of ribose). In RNA, adenine pairs with uracil instead of with

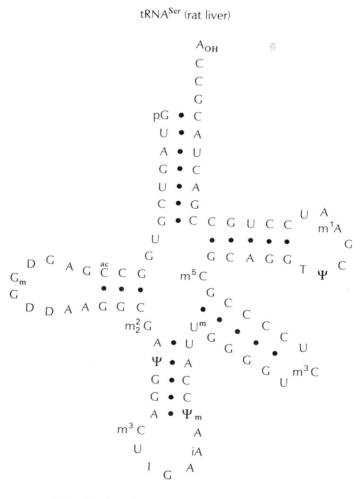

Figure 28.2 Nucleotide sequence of a rat liver serine tRNA. Modified nucleosides: acC, N^4-acetylcytidine; D, dihydrouridine; Gm, 2'-O-methylguanosine; m^2_2G, N^2-dimethylguanosine; ψ, pseudouridine; m^3C, N^3-methylcytidine; I, inosine; iA, isopentenyladenosine; ψm, 2'-O-methylpseudouridine; Um, 2'-O-methyluridine; m^5C, 5-methylcytidine; T, ribothymidine; m^1A, N^1-methyladenosine

thymine as in DNA. Thus, tRNA from rat liver for the amino acid serine is drawn in symbolic form as shown in Figure 28.2. The A-C-C- unpaired end of the chain is the free 3-hydroxy end and is present in all tRNA. The 5-hydroxy end of the chain usually has a phosphate group ending. The AGI triplet (reading from the 3-hydroxy or A-C-C- end of the chain) at the bottom of the figure is the anticodon for the serine codon of U-C-(U,C, or A). Thus this tRNA will carry only serine to the ribosomal complex for inclusion in the polypeptide chain.

DNA carries the genetic information in terms of anticodon triplets, which are replicated in reverse form as codons in mRNA. The sequence of codons in mRNA is then transferred to the growing polypeptide chain on the ribosomal-mRNA complex by selection of the proper tRNA-amino complex for incorporation in the chain. By extensive and ingenious experimentation, the genetic code for twenty amino acids was worked out (Table 28.3). Further details will not be considered here, but references are given in the bibliography.

Table 28.3 Genetic code triplets (codons) in messenger-RNA (direction of reading: free 5-terminal to free 3-terminal)[a]

First letter		Second letter				Third letter
		U	C	A	G	
U		UUU } Phe UUC } UUA } Leu UUG }	UCU } UCC } Ser UCA } UCG }	UAU } Tyr UAC } UAA STOP UAG STOP	UGU } Cys UGC } UGA STOP UGG Tryp	U C A G
C		CUU } CUC } Leu CUA } CUG }	CCU } CCC } Pro CCA } CCG }	CAU } His CAC } CAA } GluN CAG }	CGU } CGC } Arg CGA } CGG }	U C A G
A		AUU } AUC } Ileu AUA } AUG Met	ACU } ACC } Thr ACA } ACG }	AAU } AspN AAC } AAA } Lys AAG }	AGU } Ser AGC } AGA } Arg AGG }	U C A G
G		GUU } GUC } Val GUA } GUG }	GCU } GCC } Ala GCA } GCG }	GAU } Asp GAC } GAA } Glu GAG }	GGU } GGC } Gly GGA } GGG }	U G A C

[a] From I. T. Millar and H. D. Springall, *A Shorter Sidgwick's Organic Chemistry of Nitrogen*, The Clarendon Press, Oxford, 1969. © Oxford. M + S, Table 1, p. 560.

28.9 Heterocyclic metal coordination compounds

It has long been known that trace amounts of metals are essential to life processes. We shall look at a few of the heterocyclic metal coordination compounds that occur in living organisms.

Porphin systems: aromaticity. The most widely used heterocyclic ring in nature for embedding a metallic ion is porphine ($C_{20}H_{14}N_4$). With various substituents upon the

porphine

numbered positions and nearly always conjugated with a protein, a porphine ring appears in heme with Fe, in cytochrome c with Fe, in chlorophyll with Mg, and in protoporphyrin IX with Mn, to name just a few examples. Porphine, a planar pentacyclic system, appears to be made up of four pyrrole rings joined together by —CH= groups at the α and α' positions of the pyrrole rings. The various positions and identification of the rings are shown. If we do not count the 3,4 and 7,8 double bonds and count around the porphine ring, we find 18 atoms and 9 conjugated double bonds. Thus we have a $4n + 2$ system with $n = 4$. Porphine is aromatic, and an equivalent resonance structure may be written. The resonance energy has been calculated as about 400 kcal. Porphine is only slightly basic, but it forms very stable salts by incorporation of a metallic ion (Fe $^{2+}$, Mg $^{2+}$, or Cu $^{2+}$) in the center of the ring by displacement of the two protons. The presence in outer space of a tetrabenzoporphine magnesium, coordinated with pyridine rings above and below the porphine ring plane, has been suggested by optical spectra.

Hemoglobin. Hemoglobin varies somewhat from source to source (human, horse, sheep, etc.) and is the oxygen-carrying conjugated protein of blood. It is about 94% globin (the protein portion) and 6% heme (the colored substituted porphine, the prosthetic group). A typical formula is $(C_{738}H_{1166}N_{203}O_{208}S_2Fe)_4$, with a mw of about 68,000. Each red blood cell contains about 250×10^6 hemoglobin molecules. Hemoglobin consists of four distinct protein chains with four identical prosthetic groups, called heme. Two of the chains are identical, with 141 amino acids, and are called α chains; the remaining two are identical, with 146 amino acids, and are called β chains. Each of the four groups is called myoglobin. Thus, we may think of hemoglobin as consisting of two α-myoglobins and two β-myoglobins held together in a more or less spherical shape by hydrogen bonding (Fig. 28.3). This association is called quaternary structure. The tertiary structure is the conformation of each myoglobin. The secondary and primary structures were described in Chapter 21.

If the hemoglobin from 1 liter of beef blood is heated in 4 liters of glacial acetic acid with 1 g of NaCl and allowed to cool, it gives a precipitate of hemin, the oxidized prosthetic group (about 4 g). Hemin is 1,3,5,8-tetramethyl-2,4-divinylporphine-6,7-dipropionic acid ferrichloride. Replacement of chlorine with hydroxy (KOH) gives hematin. Reduction of hematin under alkaline conditions gives heme, in which Fe is in the $2\oplus$ state, as it is in hemoglobin. Both heme and hematin are unstable, especially heme, which tends to take up oxygen by coordination with Fe. Removal of the ferrous ion in heme and replacement by two protons gives the octasubstituted porphin, known as protoporphyrin IX. The dimethyl ester of protoporphyrin IX with Mn $^{3+}$ in the center of the ring and the Mn in 6 coordination, bonded vertically to a protein on one side and to water on the other side of the planar ring, is a catalyst for the release of oxygen from the leaves of vegetation.

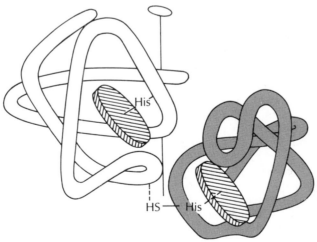

Figure 28.3 Hemoglobin molecule, as deduced from X-ray diffraction studies, showing tertiary and quarternary structure. The hemoglobin molecule is made up of two identical α chains (light blocks) and two identical β chains (dark blocks). Each chain enfolds a heme prosthetic group, and the oxygen-binding site on one group is marked by O_2. The drawing shows how closely these chains fit together in approximately tetrahedral arrangement. The model is built up from irregular blocks that represent electron density patterns at various levels in the molecule.

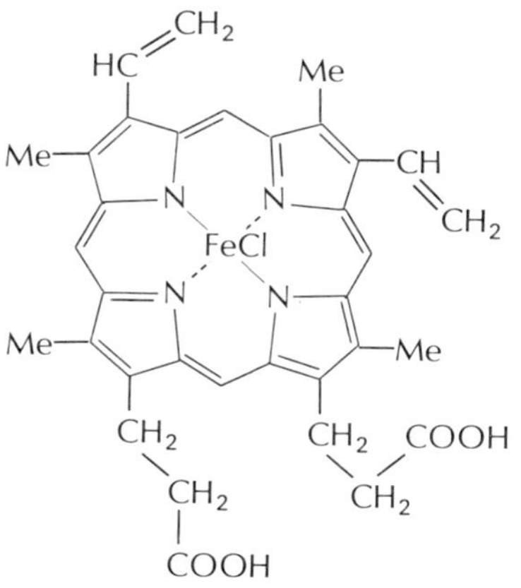

hemin

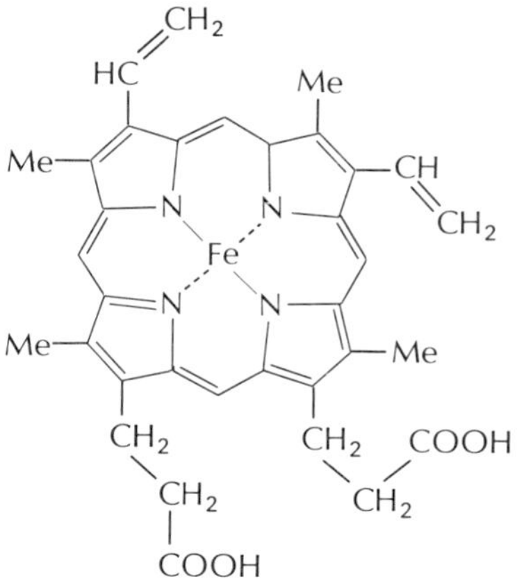

heme

Let us return to hemoglobin for a moment. Oxidation of the enclosed iron II to iron III gives <u>methemoglobin,</u> which is physiologically inactive. Hemoglobin, with four hemes present, is able to coordinate one molecule of oxygen with each heme unit. The oxygen enters the sixth coordinate position about each iron II ion. (We probably should not refer to the atom of iron as iron II, because it is so firmly bound by four of the six d^2sp^3 hybrid bonds in the center of the porphine ring; however, the name helps us to remember that the replacement of two protons has given an uncharged substance.) The fifth position is occupied by a nitrogen atom in the imidazole ring of histidine, an amino acid in the α or β chain involved. When oxygen is absent, the sixth position about the iron atom is thought to be occupied by a water molecule that is hydrogen bonded to another histidine in the α or β chain.

Myoglobin also will take up and release oxygen, depending upon the partial pressure of oxygen at the moment at a given locality. Hemoglobin, however, shows a remarkable effect. Once one of the four sites is occupied by O_2, the adjacent peptide chain is pushed aside to some extent. The change in conformation causes the chain to shift enough to allow oxygen easy access to a second site, and so on. In effect, the hemoglobin molecule expands in going over to <u>oxyhemoglobin,</u> which may be symbolized as $Hb(O_2)_4$. The arterial blood, fresh from oxygenation in the lungs, flows off to dispense oxygen to regions of low O_2 partial pressure, and the blood cells return to the lungs bearing the less firmly bound CO_2 for release in the lungs. Certain other substances are more firmly bound to hemoglobin than is oxygen. For example, inhalation of carbon monoxide gives <u>carboxyhemoglobin,</u> which is effectively removed from the oxygen-carbon dioxide transport system. If exposure is severe enough (or long enough), the individual becomes oxygen starved. Inhalation of HCN (or ingestion of NaCN or KCN) gives the even more stable <u>cyanohemoglobin,</u> with similar symptoms and results.

Hemocyanin, the blue-pigmented protein of high molecular weight, has copper instead of iron in a porphine center. It is the oxygen carrier in the blood of crustaceans.

28.9 Heterocyclic metal coordination compounds   931

Cytochromes.   The cytochromes, located in the mitochondria of cells, mediate the reduction of O_2 to water in the cell, a process that involves at least eight steps. At least four cytochromes are used in the chain of reactions, which is known as oxidative phosphorylation. Three moles of ATP (adenosine triphosphate) are formed from three moles of ADP for each mole of water formed. The process also may be thought of as electron transport to oxygen.

An exact structure is known for only one cytochrome, a cytochrome c from horse. The molecule contains a modified heme known as heme c, which is linked by four bonds to the protein chain. R_1 is cysteine at 17, R_2 is cysteine at 14, Hist is histidine at 18, and Meth is methionine at 80 in the protein chain. Other cytochromes are known to have still different heme groups, and some have copper instead of iron at the center of the porphine system.

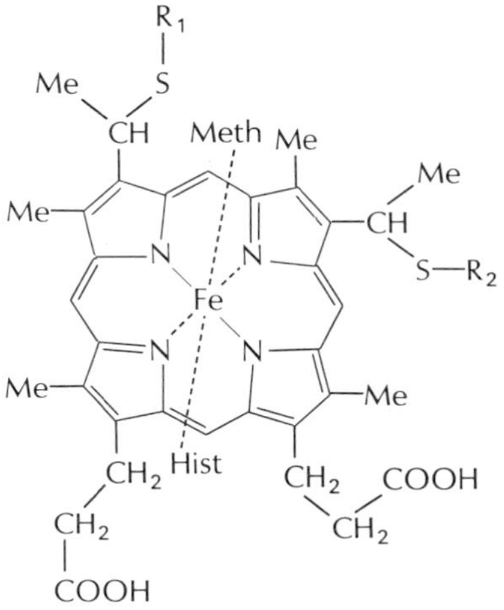

a cytochrome c

Certain enzymes, notably peroxidases, also possess an iron-containing substituted porphine as the prosthetic group attached to a protein.

Chlorophylls: phorbin.   Chlorophyll, the green pigment of plants, exists in four forms, all with a modified porphine ring and magnesium in the center of the ring system. The porphine ring system with the 7,8 double bond hydrogenated and with the formation of ring V by —CH_2CH_2— between positions 6 and γ is called phorbin. Chlorophyll a is named as the magnesium complex salt of 1,3,5,8-tetramethyl-4-ethyl-2-vinyl-9-oxo-10-carbomethoxyphorbin-7-propionic acid phytyl ester. Chlorophylls b and d differ from a as shown. Chlorophyll c is like a, except that c has a 7,8 double bond and the —CH_2CH_2COO phytyl is replaced by —CH=CHCOOH. (For uv of chlorophyll, see Sec. 10.7.)

Chlorophyll a looks on paper as if we could make it from protoporphyrin IX by insertion of Mg^{2+}, hydrogenation of the 7,8 double bond and the vinyl group at 4, closure of ring V by use of the propionic acid group at 6, oxidation of position 9 to a keto group, and esterification of the acid groups with methanol and with phytol.

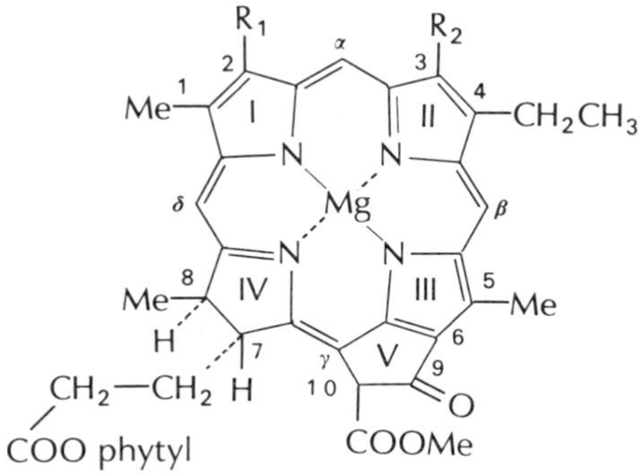

chlorophyll a, $R_1 = -CH=CH_2$; $R_2 = -CH_3$
b, $R_1 = -CH=CH_2$; $R_2 = -CHO$
c, see text
d, $R_1 = -CHO$; $R_2 = -CH_3$

Chlorophylls a and b both are present in all terrestrial plants. Chlorophylls a and c are present in marine algae, sea weeds, and so on. Chlorophyll d is relatively rare. Certain bacteria have another form called bacteriochlorophyll, which is like chlorophyll a except that the 3,4 double bond is hydrogenated. Occasionally, we find certain other names: pheophytin a (chlorophyll a from which Mg has been removed with oxalic acid), pheophorbide (the diacid obtained by hydrolysis of the ester groups of a chlorophyll or a pheophytin), and pheoporphyrin (a pheophorbide with a 7,8 double bond).

Corrin systems: vitamin B_{12}. Another ring system, corrin, was devised in nature for vitamin B_{12}. Corrin ($C_{19}H_{22}N_4$) differs from porphine in having one less carbon atom

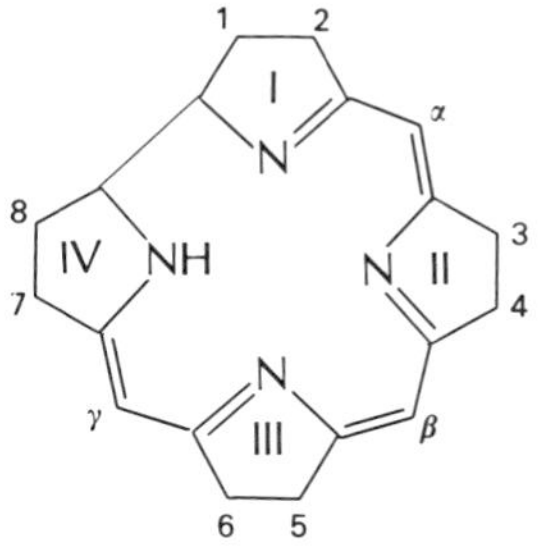

corrin
a   b

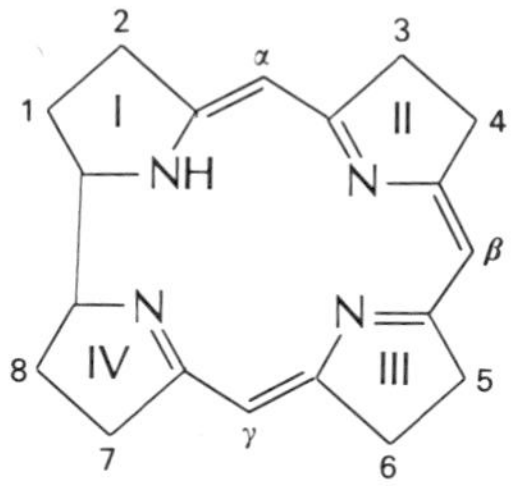

(the δ) and in being partially hydrogenated (six double bonds instead of eleven). Corrin is written as a or b, a with the N—H bond in ring IV, b with the N—H bond in ring I, with a rearrangement of the conjugated system. At first glance, it appears that a and b are tautomers. Actually, they are identical: only the numbering differs, as may be seen by exchange of the labels I and IV and of II and III, followed by turning the molecule over.

Vitamin B_{12} was isolated in 1948 from liver. The hygroscopic, dark-red crystalline substance now is produced by culture of suitable microorganisms. The structure of this complicated molecule ($C_{63}H_{88}O_{14}N_{14}PCo$) was not worked out until 1955, and then largely by X-ray work. Later on, it was found that the isolated substance actually existed in the body as coenzymes with various groups bonded to cobalt. The B$_{12}$ coenzyme, most prevalent of the coenzymes, is shown with a 5-deoxyadenosyl grouping at the sixth position on cobalt. The coenzymes decompose rapidly in the presence of light, and the presence of cyanide ion was found helpful in preserving the remainder of the system—hence the presence of $CN^{\ominus}$ at the sixth position (in place of the 5-deoxyadenosyl group) in the vitamin. The coenzyme is unusual in having a cobalt atom at the center of the ring and in having a C—Co bond.

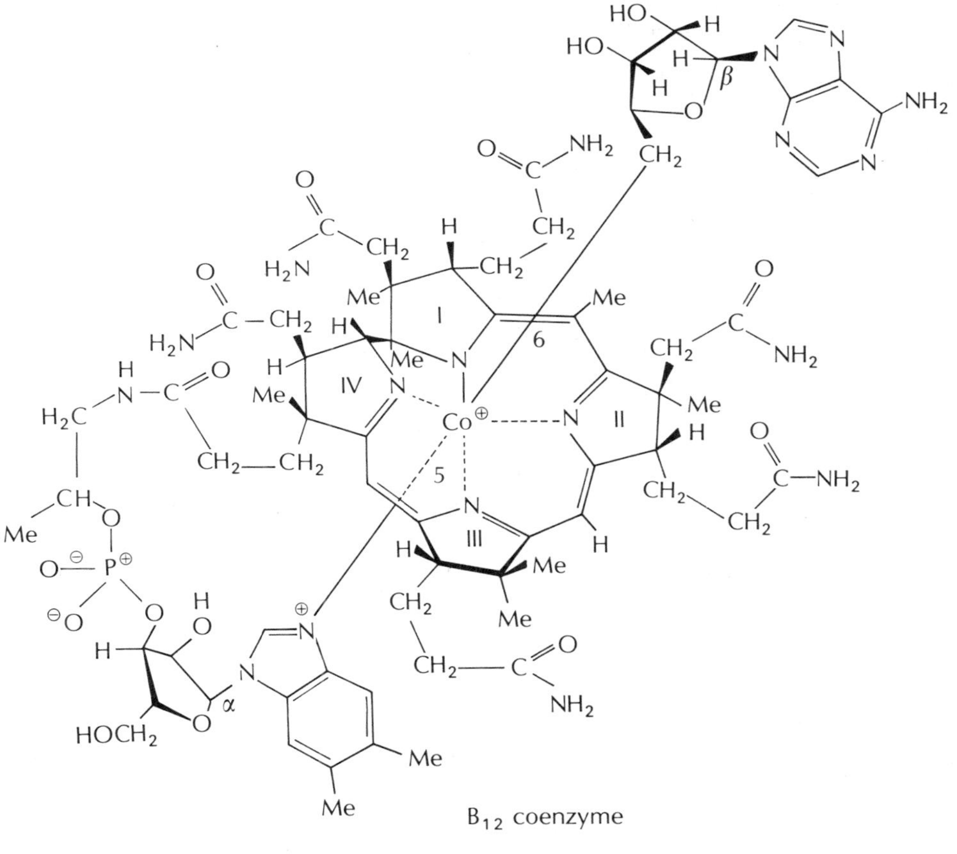

B$_{12}$ coenzyme

The molecule without a group at the sixth position is called cobalamin. Thus, the vitamin is also named as cyanocobalamin. Cobalamin from which the fifth bond to Co has been broken and the 5,6-dimethylbenzimidazole group removed is called cobamide. Hydrolysis of the —CONH$_2$ groups (side chains at 1,2,3,4,6, and 8) of cobamide gives cobamic acid. Removal of ribose and phosphoric acid (hydrolysis) from cobamic acid gives corphinic acid. Removal of CH$_3$CHOHCH$_2$NH$_2$ from corphinic acid by hydrolysis gives the heptacarboxylic acid (Co still present) known as corphyrinic acid. □ □

28.10 Summary

Through the first six sections of this chapter we have maintained a fairly logical presentation along with some of the chemistry involved. The seventh section, on nucleic acids, is centered on structure. The eighth section is little more than a listing of structures of arbitrarily selected substances.

Enough material has been given so that you can pursue a given topic into the more specialized literature available. Along the way we have touched on biochemistry and molecular biology. In any case, we have shown the variety of structures, reactions, and biological properties encompassed by the heterocyclic compounds.

Again, you should prepare a summary suited to your particular objectives.

28.11 Problems

28.1 Substance A ($C_9H_{15}NO$, mp 54°) was isolated from the bark of a pomegranate tree. A is basic, gives an oxime, and does not react with Fehling's solution. A reacts with two moles of benzaldehyde in the presence of base to give B ($C_{23}H_{23}NO$). Drastic oxidation of A gives a single product, C ($C_9H_{15}NO_4$). Heating C in MeOH with H_2SO_4 present gives D ($C_{11}H_{19}NO_4$). Exhaustive methylation (which required only one MeI) of D, followed by heating with Ag_2O, gives E ($C_{12}H_{21}NO_4$). Methylation of E, followed by heating with Ag_2O, gives F ($C_{10}H_{14}O_4$). Ozonization of F gives two moles of OHCCOOMe and one mole of $OHCCH_2CH_2CHO$. A, heated with $ZnHg_x$ and concentrated HCl, gives G. Methylation of G, followed by heating with Ag_2O, gives a single product, H ($C_{10}H_{19}N$). Methylation of H, followed by heating with Ag_2O, gives a mixture of isomers, I and J (both C_8H_{12}). Drastic oxidation of I gives a single product, but drastic oxidation of J gives a mixture of products.

Give structural formulas of the lettered substances.

28.2 Name the following compounds:

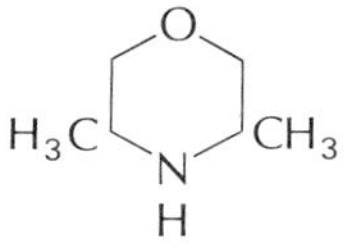

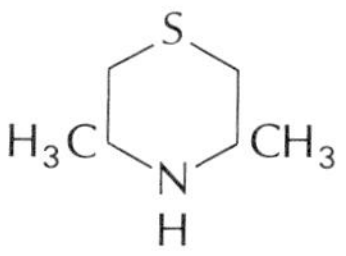

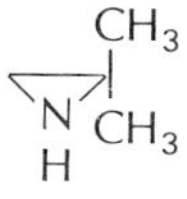

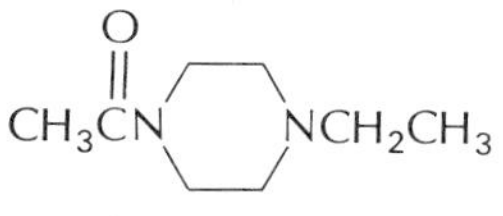

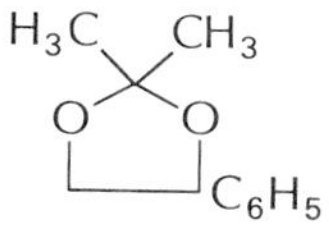

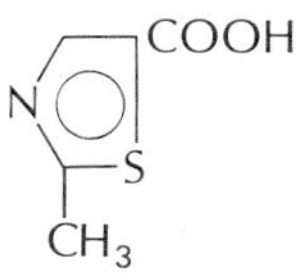

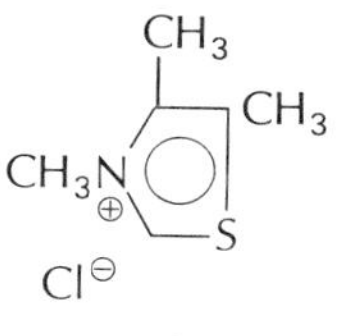

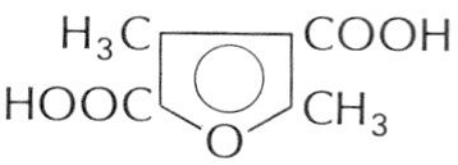

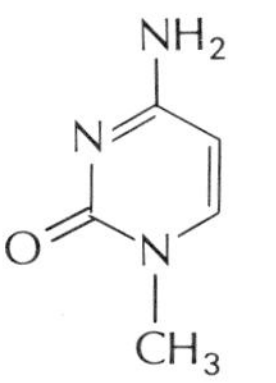

m. 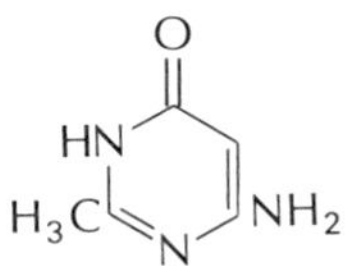 [structure: 4-amino-2-methyl-6-oxo-pyrimidine, HN–C(=O) with H₃C–C=N–C(NH₂)]

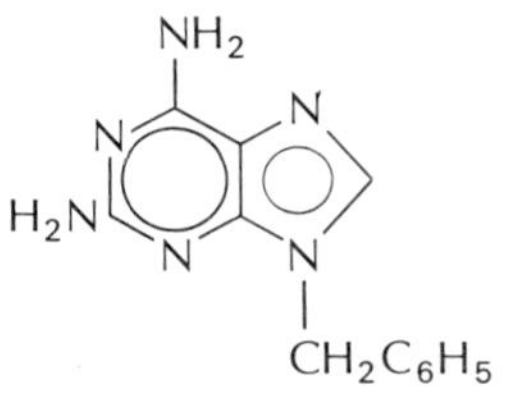

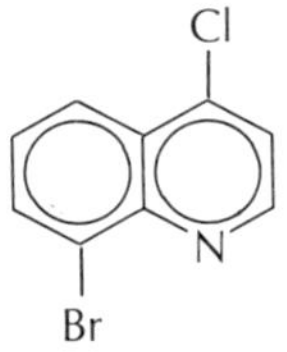

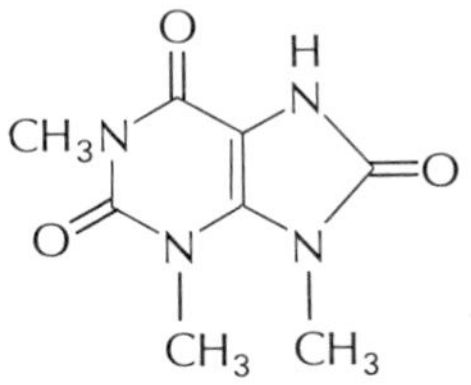

n. [structure: 4-chloro-8-bromoquinoline]

o. [structure: 5-hydroxytryptamine — HO-indole with CH₂CH₂NH₂]

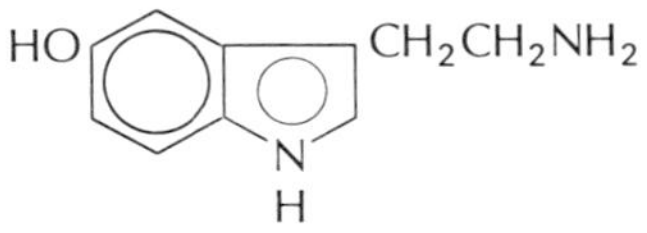

p. 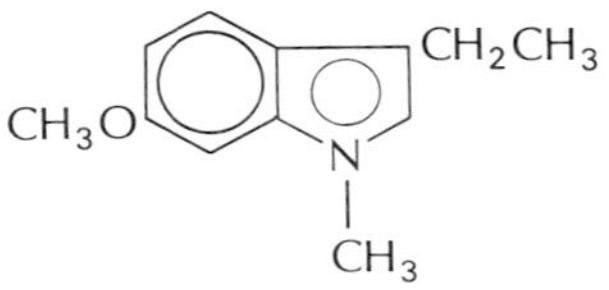 [structure: 5-methoxy-1-methyl-3-ethylindole, CH₃O on benzene, CH₂CH₃ at 3-position, N-CH₃]

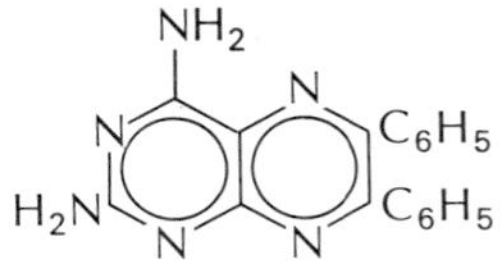

q. 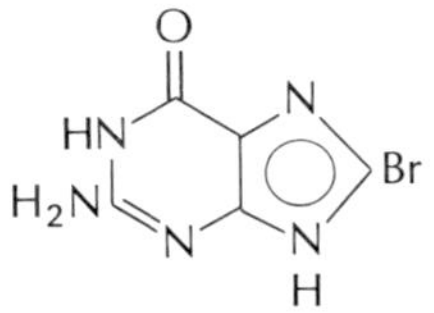 [structure: purine-like with HN–C(=O), H₂N–C=N, and Br on imidazole ring]

r. [structure: adenine with NH₂ at 6-position, H₂N at 2-position, CH₂C₆H₅ on N-9]

s. [structure: caffeine-like, CH₃N and two other N-CH₃ groups on xanthine skeleton]

t. [structure: pteridine with NH₂ at 4, H₂N at 2, C₆H₅ groups at 6 and 7]

u. 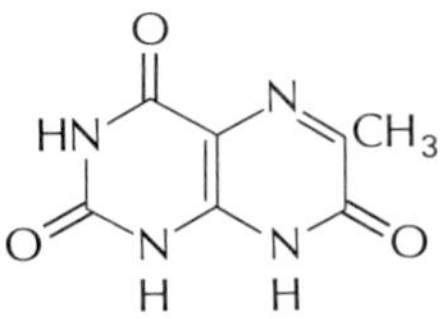 [structure: pteridine/xanthine hybrid with HN–C(=O), O=C–NH, and N–CH₃, with C=O]

28.3 Write the product(s) in each of the following reactions.

a. $CH_3COCH_2COCH_3 + NH_2NHC_6H_5 \longrightarrow$

b. $CH_3CO(CH_2)_2COCH_3 + CH_3NH_2 \longrightarrow$

c. $C_6H_5\overset{O}{\underset{\|}{C}}CHClCH_3 + NH_2\overset{S}{\underset{\|}{C}}NH_2 \longrightarrow$

d.  [o-nitrophenyl-CH₂COOCH₃] $\xrightarrow[25°]{H_2, Ni}$

e. 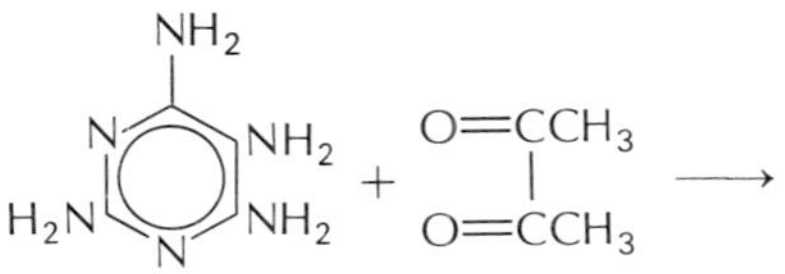 [2,4,6-triaminopyrimidine] + $O=CCH_3$ / $O=CCH_3$ (biacetyl) $\longrightarrow$

f.  [2,3-diaminopyrazine] + HCOOH $\xrightarrow[180°-200°]{HCONH_2}$

g. 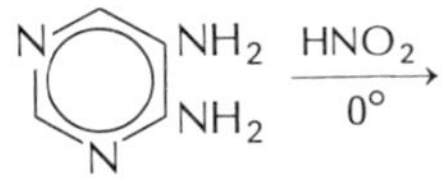 [2,3-diaminopyrazine] $\xrightarrow[0°]{HNO_2}$

h. 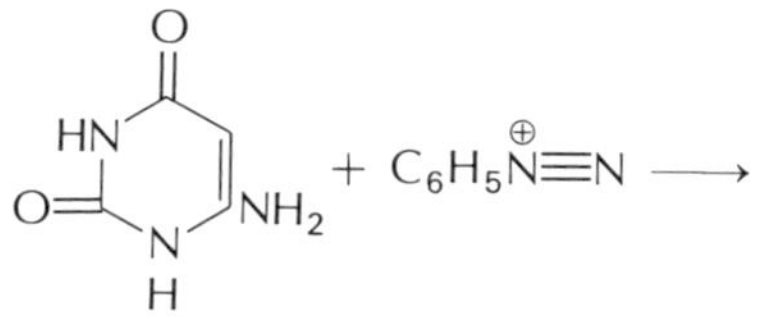 6-aminouracil + $C_6H_5\overset{\oplus}{N}\equiv N$ ⟶

i. $(CH_3OOCCH_2CH_2)_2NCH_2C_6H_5 \xrightarrow{NaOCH_3} A \xrightarrow[\Delta]{H_3O^{\oplus}} B$

j. 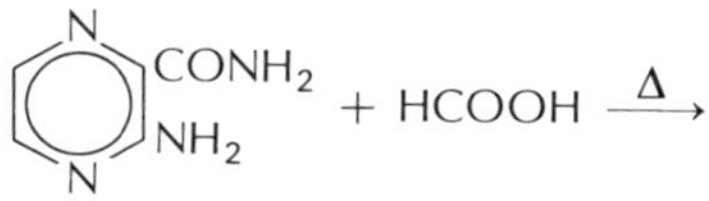 2-amino-pyrazine-3-carboxamide + HCOOH $\xrightarrow{\Delta}$

k. 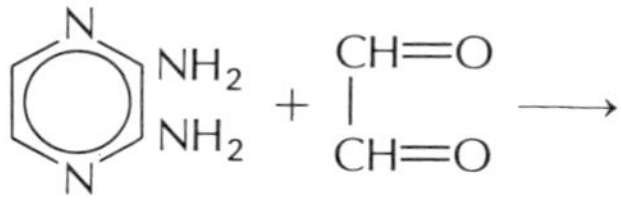 2,3-diaminopyrazine + OHC–CHO ⟶

l. 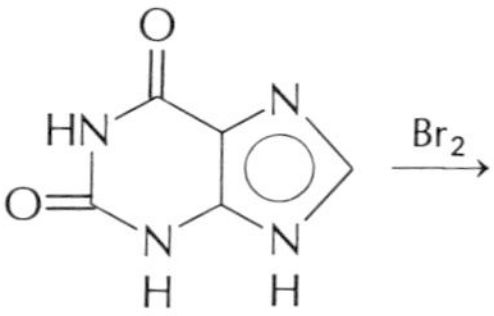 xanthine $\xrightarrow{Br_2}$

m. 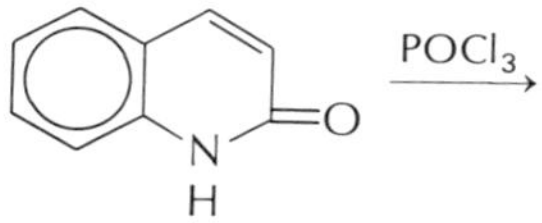 2-quinolone $\xrightarrow{POCl_3}$

n.  2,5-dimethylfuran + $\underset{CCOOCH_3}{\overset{CCOOCH_3}{|||}} \xrightarrow{100°}$

o. 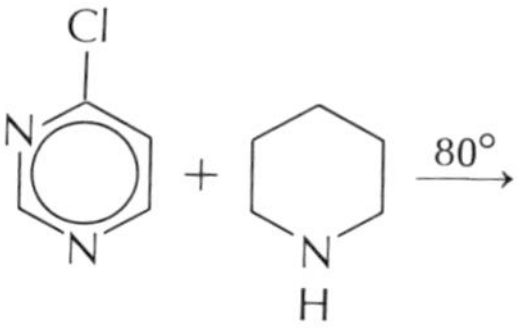 4-chloropyrimidine + piperidine $\xrightarrow{80°}$

p. 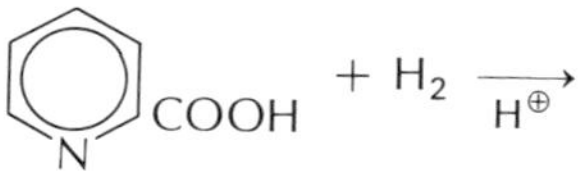 pyridine-2-COOH + $H_2 \xrightarrow{H^{\oplus}}$

q. $\underset{NH_2\cdot HCl}{\underset{|}{CH_3CHCOCH_3}}$ + KCNS ⟶ A $\xrightarrow{Ni}$ B

r. 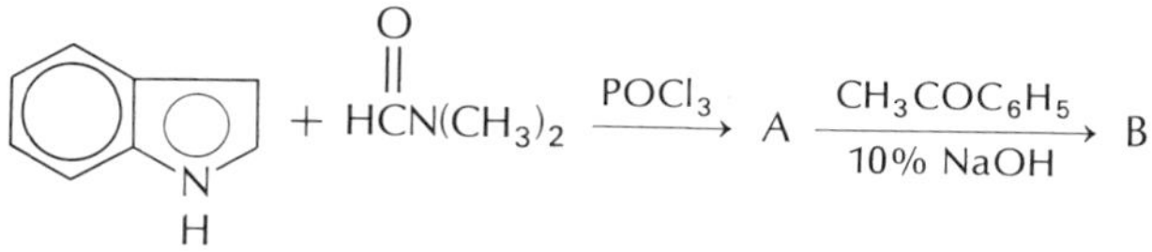 indole + $HCN(CH_3)_2$ (with C=O) $\xrightarrow{POCl_3}$ A $\xrightarrow[10\% \text{ NaOH}]{CH_3COC_6H_5}$ B

s. $\underset{\underset{NH_2}{\overset{NH}{||}}}{CH_3CNH_2}$ + $\underset{\underset{CN}{|}}{CH_2CN}$ $\xrightarrow{NaOCH_3}$

t. $NH_2\underset{\underset{S}{\|}}{C}NH_2 + CH_2(COOC_2H_5)_2 \xrightarrow{NaOC_2H_5}$

u. $NH_2\underset{\underset{NH}{\|}}{C}NH_2 + CH_3COCH_2COCH_3 \xrightarrow{NaOCH_3}$

v. 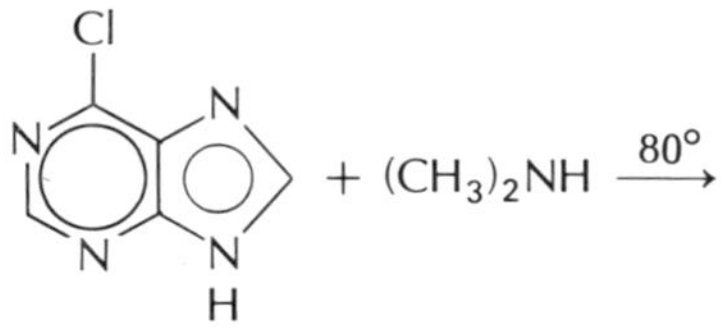 [6-chloropurine] + $(CH_3)_2NH \xrightarrow{80°}$

w. 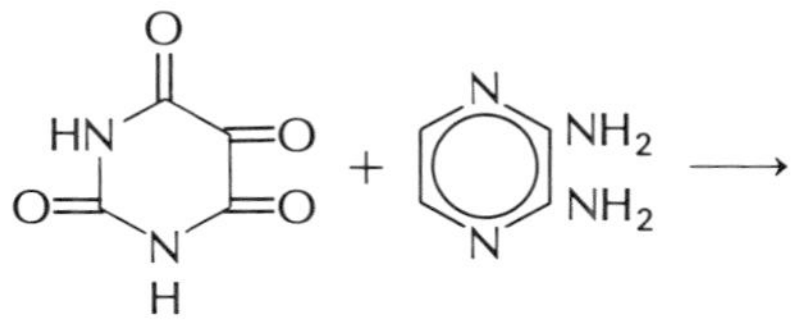 [alloxan] + [2,3-diaminopyrazine] $\longrightarrow$

x. 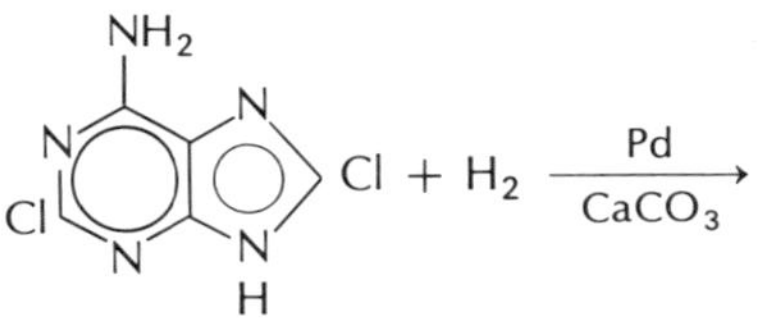 [2,6-dichloro-9H-purin-6-amine, with NH$_2$ at 6 and Cl at 2] + $H_2 \xrightarrow{Pd}{CaCO_3}$

y. 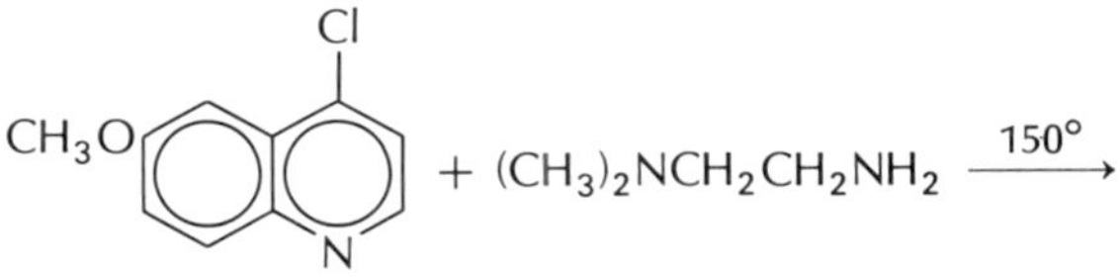 [4-chloro-8-methoxyquinoline] + $(CH_3)_2NCH_2CH_2NH_2 \xrightarrow{150°}$

z. 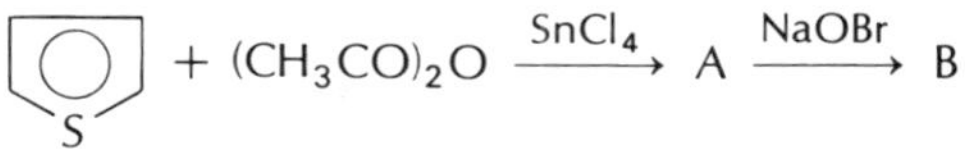 [thiophene] + $(CH_3CO)_2O \xrightarrow{SnCl_4} A \xrightarrow{NaOBr} B$

aa. 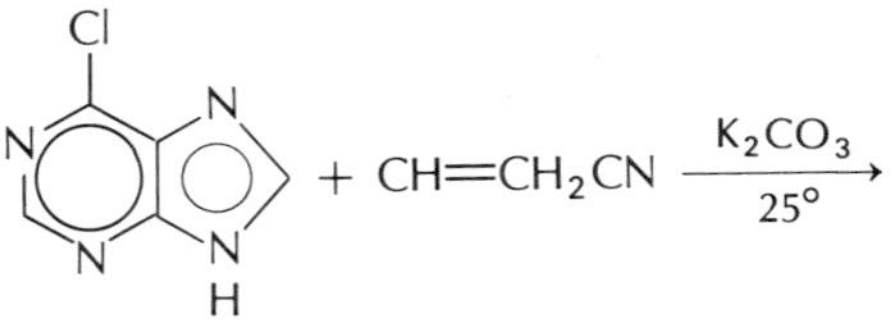 [6-chloropurine] + $CH\!=\!CH_2CN \xrightarrow[25°]{K_2CO_3}$

bb. 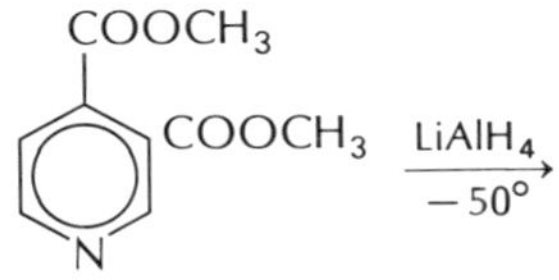 [pyridine-3,4-dicarboxylic acid dimethyl ester] $\xrightarrow[-50°]{LiAlH_4}$

cc. $C_6H_5\underset{\underset{O}{\|}}{C}CH_2CH_2CH_2CH_2NH_2 \cdot HCl \xrightarrow{OH^\ominus}$

28.12 Bibliography

Of the textbooks previously mentioned, Allinger (Chap. 7, ref. 1), Baker (Chap. 21, ref. 1), and Hendrickson *et al.* (Chap. 7, ref. 9) not only are recent and up to date, but also are readable and relatively complete. Fieser and Fieser (Chap. 7, ref. 7), Geissman (Chap. 7, ref. 8), and Noller (Chap. 7, ref. 11) are older but more comprehensive. Cason (Chap. 7, ref. 4) is excellent but restricted essentially to pyridine, quinoline, and isoquinoline chemistry. Gilman (Chap. 23, ref. 3, vols. 2, 3, and 4), though old, is excellent in coverage of older work. Norman (Chap. 22, ref. 4) is good on synthesis. Millar and Springall (Chap. 20, ref. 1) probably is the best single reference, although Bennett and Frieden (Chap. 21, ref. 3) and Organic Chemistry of Life (Chap. 21, ref. 4) also are very good.

Some specific articles of special interest are listed here.

1. MacConnell, J. G., M. S. Blum, and H. M. Fales, Alkaloid from Fire Ant Venom: Identification and Synthesis, *Science* 168, 840 (1970).
2. Synthetic Drugs Used and Abused, *C & EN* Nov. 2, 1970, p. 26.
3. Kirby, G. W., Biosynthesis of the Morphine Alkaloids, *Science* 155, 170 (1967).
4. Chemists Hone Tools to Combat LSD Abuse, *C & EN* Mar. 11, 1968, p. 32.
5. Farnsworth, Norman R., Hallucinogenic Plants, *Science* 162, 1086 (1968).
6. Chemical Mutagens, *C & EN* June 2, 1969, p. 54.
7. Cheek, F. E., S. E. Newell, and M. Joffe, Deceptions in the Illicit Drug Market, *Science* 167, 1276 (1970).
8. Hallucinogens and Narcotics Alarm Public, *C & EN* Nov. 9, 1970, p. 44.
9. Dishotsky, N. I., W. D. Loughman, R. E. Mogar, and W. R. Lipscomb, LSD and Genetic Damage, *Science* 172, 431 (1971).
10. Lennard, Henry L., Leon J. Epstein, and Mitchell S. Rosenthal, The Methadone Illusion, *Science* 176, 881 (1972).
11. Hammond, A. L., Narcotic Antagonists: New Methods to Treat Heroin Addiction, *Science* 173, 503 (1971).
12. Hayaishi, O., and M. Nozaki, Nature and Mechanisms of Oxygenases, *Science* 164, 389 (1969).
13. Soll, D., Enzymatic Modification of Transfer RNA, *Science* 173, 293 (1971).
14. Gustafson, T., and M. I. Toneby, How Genes Control Morphogenesis, *American Scientist* 59, 452 (1971).
15. Stadtman, T. C., Vitamin B_{12}, *Science* 171, 859 (1971).
16. Olson, J. M. The Evolution of Photosynthesis, *Science* 168, 438 (1970).
17. Bailar, J. C., Jr., Some Coordination Compounds in Biochemistry, *American Scientist* 59, 586 (1971).
18. Shulman, R. G., S. Ogawa, K. Wüthrich, T. Yamane, J. Peisach, and W. E. Blumberg, The Absence of "Heme-Heme" Interactions in Hemoglobin, *Science* 165, 251 (1969).
19. Perlman, D., Antibiotics—A Status Report with Prognostications, *Chemical Technology* 1, 540 (1971).

20. Jones, R. G., Antibiotics of the Penicillin and Cephalosporin Family, *American Scientist* 58, 404 (1970).
21. Horwitz, S. B., S. C. Chang, A. P. Grollman, and A. B. Borkovec, Chemosterilant Action of Anthramycin A: A Proposed Mechanism, *Science* 174, 159 (1971).
22. Albuquerque, E. X., J. W. Daly, and B. Witkop, Batrachotoxin: Chemistry and Pharmacology, *Science* 172, 995 (1971).
23. Helgeson, J. P., The Cytokinins, *Science* 161, 974 (1968).

For the naming of heterocycles, see the following.

24. Definitive Rules for Nomenclature of Organic Chemistry, *J. Amer. Chem. Soc.* 82, 5545 (1960).

These rules are also available in:

25. Weast, Robert C., ed., *Handbook of Chemistry and Physics*. Cleveland: The Chemical Rubber Co., annual editions.
26. The Chemical Abstracts usage is most conveniently found in the almost 100-page introduction to the subject index of Volume 56, for January to June, 1962.

The synthesis of vitamin B_{12} is described briefly in two separate articles. The massive joint effort involved ninety-nine people led by Professor Albert Eschenmoser of the Eidgenösische Technische Hochschule in Zurich and Professor R. B. Woodward at Harvard University.

27. Maugh, Thomas H., II, Vitamin B_{12}: After 25 Years, the First Synthesis, *Science* 179, 266 (1973).
28. Krieger, James H., Vitamin B_{12}: The Struggle Toward Synthesis, *C & EN*, Mar. 12, 1973, p. 16.

28.13 Charles C. Price: From antimalarials to involvement

Professor Charles C. Price, Ph.D. at Harvard in 1936, moved to Illinois where he remained until he went to Notre Dame in 1946 as head of the department of chemistry. In 1954 he moved to the University of Pennsylvania, again as head of the department, a task he relinquished in 1965. At present he is the Benjamin Franklin Professor of Chemistry at Penn. Professor Price has been active in research and teaching and has received many honors and awards, not the least of which was the presidency of the American Chemical Society for 1965. Here he describes some of his activity outside the laboratory.

While there have been many fascinating experiences in chemistry that have shaped my career, perhaps the most important was one that directed a modest fraction of my time and energy to the problems flowing from the increasingly important interactions of science and society. During World War II, much of my attention had been devoted to chemical problems involved in new drugs for treating malaria; the basic chemistry of polymerization to form synthetic rubber; and the behavior, detection, analysis, and removal of chemical warfare agents in water.

By 1947 the Cold War was getting under way, and scientists were again being called to Washington for consultation on problems of war and preparation for war. After a gloomy week of such discussions with a dismal feeling of the inevitability of more war, this time involving vast nuclear destruction, I spent the weekend at our family farm near Sellersville, Pennsylvania. While there, I heard a lecture by Congressman Judd (R., Minn.) on "World Law." My brother gave me a copy of a book by Emery Reves entitled "The Anatomy of Peace," which I read on the train returning to Notre Dame. The message of the latter was that the organization of the League of Nations was totally inadequate to the task of maintaining peace. The same was true for the United Nations. The need was not for a weak debating society but for the tried and true civilized principles of law and order under government. The message from both Judd and Reves was that this must now be world federal government.

This weekend encounter with the ideas of Judd and Reves turned me from disconsolate gloom about the inevitability of nuclear war and total disinterest in the political process to dedicated devotion to peace through political action for replacing the war system of international organization by a civilized system of law and order under a strengthened United Nations. Within three years, I had organized the World Federalists in Indiana, had lobbied for the cause, and had run for the U.S. Senate nomination in Indiana. My devotion to the cause has continued through service as chairman of the Federation of American Scientists, president of the United World Federalists, and board chairman of the Council for a Livable World.

The most powerful forces opposed to this goal are here in the U.S.A. I remain convinced that the only alternative to nuclear extinction of civilized human society by the products of the accelerating nuclear arms race so strongly stimulated by U.S. militarists is a civilized system of law and order in a demilitarized world. There is little hope that science can indeed become the benefactor of mankind rather than his exterminator unless serious and successful efforts are devoted to this cause. The lessons of thermodynamics indicate that an increase in order and organization can only occur with an investment of energy. The more people who are willing to invest such energy, the better chance there may be to reverse the all too natural path toward disintegration and destruction.

The whole scientific study of the nature and origin of life underlines the persistent theme of organization, order, and cooperative phenomena—from the union of elementary particles to form atomic nuclei, to covalent bonding, to association of molecules into the functioning units of living cells, to the organization of living cells into living organisms, to the use of human intelligence to create human society. In death there is disintegration and disorder. In life there is organization and cooperation. We ignore this clear message at our mortal peril.

29 Second-Row-Element Organic Compounds

From time to time we have seen various compounds of sulfur and phosphorus. Now we shall bring together the chemistry of sulfur, phosphorus, and silicon in an organized fashion. We already have covered some of the organic chemistry of the other second-row elements, Na, Mg, Al, and Cl. We shall spend more time with the sulfur compounds than with those of phosphorus and silicon.

29.1 Second-row-element organic compounds   941
   1. Introduction.   2. Pairing of S and O, P and N, Si and C.   3. Third-shell atomic orbitals.   4. Empty 3d orbitals: expansion of a shell.   5. Dimethyl sulfide, sulfoxide, and sulfone: back bonding.   6. Lack of 3p-3p π bonds: exceptions.   7. The sp^3d, sp^3d^2, and dsp^2 hybrids.   8. Summary. Exercises.

29.2 Thiols, sulfides, sulfonium ions, and disulfides   945
   1. Definitions; analogs.   2. Thiols: properties.   3. Thiols: reactions.   4. Thiols: preparation.   5. Sulfides: properties and reactions.   6. Sulfides: preparation.   7. Disulfides.   8. Occurrence of sulfur compounds in proteins and metabolic products. Exercises.

29.3 Sulfoxides and sulfones   951
   1. Naming.   2. Reactions.   3. Quaternary salts.   4. Preparations. Exercises.

29.4 Sulfenic, sulfinic, and sulfonic acids and their derivatives   954
   1. Sulfenic acids.   2. Sulfinic acids.   3. Sulfonic acids; esters; chlorides.   4. Sulfonate ion as a leaving group; chloramine-T; saccharin.   5. Synthesis. Exercises.

29.5 Organophosphorus compounds   957
   1. Naming; tautomerism.   2. Phosphines.   3. Rearrangement; ylides.   4. Tetracoordinate phosphorus: Arbusov reaction and rearrangement.   5. Radical addition of phosphines to alkenes.   6. Phosphate esters. Exercises.

29.6 Organosilicon compounds   962
   1. Bonding to silicon; naming.   2. Bond energies; nucleophilic substitution.

3. *Substitution: stereochemistry.* 4. *Radical and silene reactions.*
5. *Manufacture of chlorosilanes; silyl Grignard reagents.* 6. *Silanols, silanediols, siloxanes, and silicones. Exercises.*

29.7 Summary 967

29.8 Bibliography 967

29.9 H. Harry Szmant: Surprising sulfur radicals 968a

29.1 Second-row-element organic compounds

Introduction. A second-row element is one of the eight elements in which the process of filling the third shell is occurring: Na, Mg, Al, Si, P, S, Cl, Ar. The first three are typical metallic elements, and we have seen the organometallic compounds in several places. (An element linked to carbon through oxygen or sulfur, such as RONa, $(RS)_2Mg$, or $(RO)_3Al$, is not thought of as an organometallic compound.) In this chapter, we assemble the organic chemistry of sulfur, phosphorus, and silicon compounds.

Pairing of S and O, P and N, Si and C. The pairs sulfur and oxygen, phosphorus and nitrogen, silicon and carbon, all have the same valence shell arrangement of electrons. However, the second-row elements are larger because the third shell is being formed over a core of filled first and second shells. This fact immediately suggests to us that σ bonds are longer and of smaller energy than are the corresponding σ bonds of the first-row elements. However, as we might have suspected, things are not so simple. The second-row partner is less electronegative than is the first-row partner: F, 4.0, Cl, 3.0; O, 3.5, S, 2.5; N, 3.0, P, 2.1; and C, 2.5, Si, 1.8 (Pauling scale of electronegativities). The effect is to increase the contribution of ionic structure to the resonance hybrid for bonding to a more electronegative element or to decrease contribution of ionic structure for bonding to a less electronegative element. For example, the bond energy of C—O is 86 kcal and that of Si—O is 108 kcal, an increase of 22 kcal. For N—O (53 kcal) and P—O (95 kcal), the increase is 42 kcal. In the other direction, N—H (93 kcal) and P—H (76 kcal) show a decrease of 17 kcal; and O—H (111 kcal) and S—H (83 kcal) show a decrease of 28 kcal. Although the difference in electronegativity helps us to make a correlation, it is not a complete explanation of the bond energies.

Third-shell atomic orbitals. The third-shell orbitals, 3s and 3p, have two nodes instead of the single node characteristic of the 2s and 2p orbitals, as shown (not drawn to scale). The extra nodes make the hybrid orbitals $3sp$, $3sp^2$, and $3sp^3$ difficult to illustrate, but the general result, in terms of bond angles, is the same as for hybridization of the second shell. Thus, SiF_4 and SiH_4 are tetrahedral, and PH_3 resembles NH_3 except that inversion is much slower.

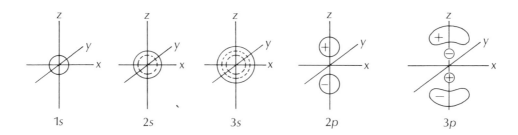

Empty 3d orbitals: expansion of a shell. The third shell possesses five d orbitals, which are not filled in by the atoms of the second row. The atomic d orbitals do not begin to be occupied until after the 4s orbital is filled (K, Ca). Filling in by 10 electrons gives the first of the long sequences in the periodic table (Sc, Ti, V, Cr, Mn, Fe, Co, Ni, Cu, Zn). However, the empty d orbitals of the second-row elements impart to them an ability to accommodate more than eight electrons. Ordinarily, the filling in of a d orbital of a second-row atom requires energy. Thus, expansion of the usual octet does not occur unless it is part of a bonding situation that releases energy.

Dimethyl sulfide, sulfoxide, and sulfone: back bonding. Let us illustrate what we mean by looking at dimethyl sulfide, sulfoxide, and sulfone, CH_3SCH_3, CH_3SOCH_3, and $CH_3SO_2CH_3$. The sulfide, like dimethyl ether, may be assumed to have two σ bonds to sulfur, the two remaining sp^3 orbitals each containing electron pairs. Now sulfur, unlike the more electronegative oxygen, may be oxidized to form a coordinate covalent bond to oxygen, as in DMSO. In the process, sulfur acquires a formal positive charge, and oxygen, a negative charge, just as in the formation of a tertiary amine oxide. Sulfur, however, has

$$Me_3\overset{\oplus}{N}—\overset{\ominus}{\underline{\overline{O}}}| \qquad Me_2\overset{\oplus}{S}—\overset{\ominus}{\underline{\overline{O}}}|$$

a remaining sp^3 orbital occupied by an electron pair, and oxygen (assuming sp hybridization) has pairs of electrons occupying the remaining sp orbital and the two p orbitals. If we assume that the S—O σ bond is the z axis, then the empty d_{xz} orbital of sulfur is in position to overlap the filled p_x orbital of oxygen. Likewise, the empty d_{yz} orbital is in position to overlap the filled p_y orbital of oxygen. Overlap leads to what is termed a d-p π bond, and there are two such d-p π bonds, one in the xz plane, the other in the yz plane.

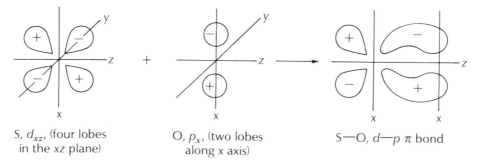

S, d_{xz}, (four lobes in the xz plane)

O, p_x, (two lobes along x axis)

S—O, d—p π bond

The overlapping leads to partial donation of each electron pair from each oxygen p orbital into a d orbital of sulfur. If the donation were complete (as in the formation of the S—O σ bond), we would find sulfur sharing a total of ten electrons and oxygen sharing six electrons (the remaining two in the *sp* orbital not being used for bonding). This situation would require a negative charge on sulfur and a positive charge on oxygen. If we postulate only partial overlap in the two *d-p* π bonds, just sufficient to average out at eight electrons shared by sulfur and four electrons shared by oxygen, then each atom would be neutral. Actually, dimethylsulfoxide has a large dipole moment (3.9 D), larger than that of acetone (2.9 D) but less than that of trimethylamine oxide (5.0 D). Thus, we conclude that the back bonding is not of great magnitude, because the dipole moment indicates only a modest diminution of positive charge on sulfur and of negative charge on oxygen. Nevertheless, any back bonding will contribute to an increase in bond energy. Although we have no means of comparing the S—O bond, we can now compare the bond energies of P—O and N—O (95 and 53 kcal) and of Si—O and C—O (108 and 86 kcal). What we are trying to say is that Si and P, like S, are capable of back bonding with a first-row element that has a filled *p* orbital and that this back bonding leads to higher bond energies.

Most authors use $Me_2S=O$ to represent DMSO, with adequate cautions that $\searrow S=O$ is not a *p-p* π bond representation, even though that is what it looks like. We shall use $Me_2\overset{\oplus}{S}-\overset{\ominus}{O}$. We also must remember that we are using it only as a symbol to represent a σ bond with back bonding present. In condensed form, we shall simply write Me_2SO or MeSOMe.

Now, what about dimethylsulfone? We may represent partial back bonding as shown in resonance structure symbolism, and we shall use *a* as the single symbolic structure. In condensed form we simply write Me_2SO_2 or $MeSO_2Me$.

Back bonding of the *d-p* π type is possible not only with S, but also with P and Si; and it does not depend upon the number of σ bonds to the second-row element, because only the vacant *d* orbitals are involved. As we stated before, the participating *p* orbital of the first-row atom must be occupied.

Lack of 3p-3p π bonds: exceptions. The 3*p* orbitals seem to be reluctant to participate in π bonding of the usual *p-p* type. Certainly 3*p*-2*p* overlap is rare. If we try to make thioaldehydes and thioketones, they dimerize, trimerize, or polymerize to give cyclic thioacetals. An exception is thiobenzophenone (Eq. 29.1). However,

$$\phi CO\phi + H_2S \xrightarrow{HCl} \phi\overset{S}{\overset{\|}{C}}\phi + H_2O \qquad (29.1)$$

the compound is blue, not colorless, as benzophenone is. This fact suggests that $\phi_2\dot{C}-\dot{S}$ is a contributor to the structure as well as $\phi_2\overset{\oplus}{C}-\overset{\ominus}{S}$.

The sp^3d, sp^3d^2, and dsp^2 hybrids. The second-row elements also expand the valence shell by hybridization of the *d* orbitals with *s* and *p*. Three types of this hybridization occur frequently: trigonal bipyramidal, sp^3d; octahedral, sp^3d^2; and plane tetragonal, dsp^2. (See Figure 29.1.) Beginning with the third shell (to have d orbitals available), hybridization of five orbitals gives five equivalent sp^3d orbitals in which the central atom has ten electrons in the valence shell, as in PCl_5. The shape is that of two triangular pyramids, back to back. Atoms of the second row of the periodic table lack the d orbitals and cannot be hybridized in this manner. If six orbitals are used (sp^3d^2), twelve electrons are required and six equivalent orbitals result, each at right angles to the nearest four. The shape is that

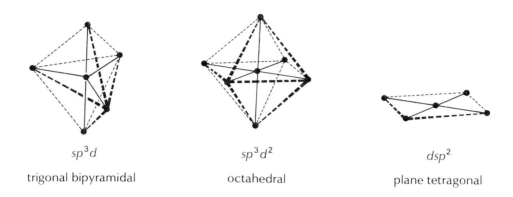

sp^3d sp^3d^2 dsp^2

trigonal bipyramidal   octahedral   plane tetragonal

Figure 29.1   Hybridization with d orbitals

of two pyramids, back to back. Fourth-row elements may utilize one of the d orbitals from the third shell to hybridize with one s and two p orbitals from the fourth shell (dsp^2), using eight electrons to give four equivalent orbitals lying in a plane, each at right angles to the nearest two. The shape is a square. Thus, sulfur forms SF_4, a sp^3d hybrid and a trigonal bipyramid, in which one of the five orbitals is occupied by an electron pair. The compound reacts with more fluorine to give SF_6, a sp^3d^2 hybrid and an octahedron, in which all six of the sulfur electrons are engaged in σ bond formation. Sulfur hexafluoride is one of the most stable and inert compounds known. Presumably, back bonding is weak in both SF_4 and SF_6 because there is no formal positive charge on sulfur to be partially neutralized by formal negative charge on any of the fluorine atoms.

The phosphorus compounds display a variety of hybridizations and back bonding. Phosphorus trichloride (PCl_3) is an sp^3 hybrid. Phosphorus oxychloride ($POCl_3$) is an sp^3 hybrid. It is like a tertiary amine oxide but has back bonding in the $\overset{\oplus}{P}$—$\overset{\ominus}{O}$ bond. Phosphorus pentachloride, ordinarily written as PCl_5, is $Cl_4P^{\oplus}PCl_6^{\ominus}$ in the crystalline state. One atom is sp^3 hybridized; the other, sp^3d^2, as in SF_6. Again, p-p π bonds are rare. An exception is phosphorobenzene, $\phi P{=}P\phi$, an analog of azobenzene.

Silicon, like carbon, prefers the sp^3 hybrid state. Free radicals and anions are known. Unlike the situation with carbon, cations and unsaturation (multiple bonds) are not known. Also, silicon is willing to expand the valence shell, as in $SiF_4 + 2F^{\ominus} \to SiF_6{}^{2\ominus}$.

Summary. Let us summarize the major points we have made about the second-row (third-shell) elements:

1. Valence shells are like those of the first-row analogs.
2. Electronegativities are less than those of the first-row analogs.

3. The 3s and 3p orbitals differ from the 2s and 2p orbitals.
4. The empty 3d orbitals are able to form d-p π bonds with filled 2p orbitals; this phenomenon is known as back bonding.
5. Hybridization of d orbitals with s and p orbitals allows penta- and hexacoordinate compounds to be formed.
6. We use $\overset{\oplus}{S}—\overset{\ominus}{O}$, $\overset{\oplus}{P}—\overset{\ominus}{O}$, and $\overset{\oplus}{Si}—\overset{\ominus}{O}$ to symbolize those instances in which back bonding is suspected.

Exercises

29.1 Work out the structures of:
   a. SO_2 and H_2SO_3
   b. SO_3 and H_2SO_4
   c. $Na_2S_2O_3$ and $Na_2S_2O_4$

29.2 Thiols, sulfides, sulfonium ions, and disulfides

Definitions; analogs. The sulfur analog of an alcohol is a thiol (RSH). The old common name, mercaptan, from "mercury capture," is still in use also. A sulfide (RSR') is the analog of an ether. The systematic name utilizes thio- in place of oxy-. Thus, methoxybenzene ($CH_3O\phi$) becomes methylthiobenzene (ϕSCH_3). A disulfide (RSSR') is the analog of a peroxide (ROOR'). The sulfonium ions ($R_3S^\oplus$) are much more stable than are the analogous oxonium ions ($R_3O^\oplus$).

Thiols: properties. Thiols have lower boiling points than those of the corresponding alcohols, even though the molecular weights are higher. The extreme case is H_2S (bp −61°) compared with water (bp 100°). The cause lies in the weakened ability of the RS—H bond to engage in hydrogen bond formation.

The smaller thiols have intense, noxious odors, and many occur in nature. Propanethiol is present in the odor of freshly sliced onions, 2-propenethiol (allyl mercaptan) is present in garlic, and butanethiol is a component of the skunk's defensive armament. Thiols and other sulfur-containing compounds occur in most petroleum, the amount varying from one oil field to another. These substances must be removed, not only to improve the odor, but also to avoid sulfur poisoning of catalysts used in the various refining processes and to avoid the production of sulfur dioxide during the combustion of fuels.

Thiols are more acidic than are alcohols, probably because the RS—H bond is longer than the RO—H bond. Thus it is easier to ionize, even though the RO—H bond has more ionic character (electronegativity difference) to begin with. The $RS^\ominus$ anion, then, is a weaker base than $RO^\ominus$. However, $RS^\ominus$ is a much better nucleophile than $RO^\ominus$ even though the C—S bond energy, 65 kcal, is considerably less than that of a C—O bond, 86 kcal. However, $RS^\ominus$ is more polarizable.

Thiols: reactions. Thiols undergo reactions completely analogous to those of the alcohols in many cases (Eqs. 29.2): anion formation, esterification with acid chlorides and

$$RSH + NaOH \longrightarrow RS^{\ominus} + Na^{\oplus} + H_2O$$

$$RSH + Hg^{2+} \longrightarrow (RS)_2Hg + 2H^{\oplus}$$

$$RSH + ArCOCl \longrightarrow ArC\overset{O}{\underset{SR}{\diagdown}} + HCl \quad (29.2)$$

$$2RSH + CH_3COCH_3 \underset{}{\overset{H^{\oplus}}{\rightleftharpoons}} \overset{H_3C}{\underset{H_3C}{\diagdown}}C\overset{SR}{\underset{SR}{\diagdown}} + H_2O$$

$$RS^{\ominus} + MeI \longrightarrow RSMe + I^{\ominus}$$

anhydrides, acetal formation (even better than that with alcohols), and ether (thioether or sulfide) formation.

Other reactions that we might expect to go do not occur or take a different course (Eqs. 29.3). The C—S bond is difficult to break by nucleophilic displacement, either S_N1 or

$$RSH + HBr \not\longrightarrow RBr + H_2S$$
$$RSH \xrightarrow{\text{oxidation}} RSO_3H \quad (29.3)$$

S_N2. Upon oxidation, a thiol reacts at the sulfur atom, passing through several oxidation states before arriving at the sulfonic acid stage. Oxidation may proceed not only by initial rupture of the S—H bond (BE 83 kcal, much weaker than O—H, 111 kcal), but also by S—O bond formation (Eqs. 29.4). Actually, because the intermediates also are easy to oxidize, it is

$$RSH \begin{cases} \xrightarrow{-H\cdot} R\underline{S}\cdot \longrightarrow R\underline{S}-\underline{S}R \xrightarrow{[O]} RS\overset{|\overline{O}|^{\ominus}}{\underset{|\underline{O}|^{\ominus}}{\overset{(2+)}{-}}}\underline{S}R \xrightarrow{[O], H_2O} RS\overset{|\overline{O}|^{\ominus}}{\underset{|\underline{O}|^{\ominus}}{\overset{(2+)}{-}}}OH \\ \xrightarrow{[O]} \underset{R\underline{S}H}{\overset{|\overline{O}|^{\ominus}}{|\overset{\oplus}{}}} \xrightarrow{[O], H_2O} \end{cases} \quad (29.4)$$

possible to use oxidation of thiols only for the preparation of disulfides and sulfonic acids (Eqs. 29.5). The formation of a disulfide plays an essential role in several biological reactions, as we shall see when we take up the disulfides.

$$R\underline{S}H \xrightarrow[\substack{\text{EtOH or} \\ CH_3COOH}]{I_2} R\underline{S}^{\oplus} + HI + I^{\ominus} \xrightarrow{R\underline{S}H} R\underline{S}-\overset{\oplus}{\underset{H}{S}}R \xrightarrow{-H^{\oplus}} R\underline{S}-\underline{S}R$$

$$RSH \xrightarrow[\text{or KMnO}_4]{H_2O_2 \text{ or HNO}_3} RSO_3H \quad (29.5)$$

Under free-radical conditions, thiols add to alkenes (Eq. 29.6).

$$R\underline{S}H \xrightarrow[h\nu]{\text{peroxide}} R\underline{S}\cdot \xrightarrow{H_2C=CHCH_2CN} R\underline{S}CH_2-\dot{C}HCH_2CN$$
$$\downarrow R\underline{S}H$$
$$R\underline{S}\cdot + R\underline{S}CH_2-\overset{H}{\underset{|}{C}}HCH_2CN \quad (29.6)$$

29.2 Thiols, sulfides, sulfonium ions, and disulfides

Thiols: preparation. Thiols may be prepared in several ways (Eqs. 29.7).

$$RX + {}^{\ominus}SH \xrightarrow{NaSH} RSH + X^{\ominus} \qquad a$$

$$RCH=CH_2 + HS\overset{O}{\underset{\|}{C}}CH_3 \xrightarrow[h\nu]{peroxide} RCH_2CH_2S\overset{O}{\underset{\|}{C}}CH_3 \xrightarrow{H^{\oplus}, H_2O} RCH_2CH_2SH \qquad b \qquad (29.7)$$

$$RX + S=C\underset{NH_2}{\overset{NH_2}{\diagup}} \xrightarrow{-X^{\ominus}} RS-C\underset{NH_2}{\overset{\overset{\oplus}{NH_2}}{\diagup\!\!\!\!=}} \xrightarrow{NaOH} RSNa + H_2NCN \text{ (polymer)} \qquad c$$

$$Me_2C=CH_2 + H_2S \xrightarrow{H_2SO_4} Me_2\underset{SH}{\overset{|}{C}}-CH_3 \qquad d$$

$$\phi MgBr + S_8 \longrightarrow \phi SMgBr \xrightarrow{work\text{-}up} \phi SH \qquad e$$

$$RX + KSCN \longrightarrow RSCN \xrightarrow{H_2O} RSH + HCNO \qquad f$$

$$RSSR \xrightarrow[H_2SO_4]{Zn} 2RSH \qquad g$$

$$\phi SO_2Cl \xrightarrow[H_2SO_4]{Zn} \phi SH \qquad h$$

Reaction a, like the alkylation of amines, suffers from the defect that the thiol also forms an anion and enters into S_N2 reaction to produce a sulfide (RSR) as a by-product. Reaction b is Equation 29.6 with CH_3COSH instead of RSH. Hydrolysis of the thioester gives the thiol. Reaction c uses thiourea and has the advantage of not giving any by-product sulfide. Reaction d is electrophilic addition of a proton to an alkene, followed by the strong nucleophile H_2S. Reaction e is similar to that of a Grignard reagent with O_2 (S_8 is crystalline sulfur, an eight-membered ring). Reaction of RX with KSCN (f) is another S_N2 reaction that avoids the production of sulfides. Both g and h are simple reduction reactions. Catalytic reductions do not go at all well because of the well-known tendency of sulfur compounds to poison hydrogenation catalysts.

Sulfides: properties and reactions. Sulfides show the usual increase in bp with increasing mw and thus have higher bp than those of the ether analogs, because there is no hydrogen bonding to complicate things. Like ethers, sulfides tend to be inert except for susceptibility to oxidation and for reactions that form sulfonium salts. Sulfides also are odoriferous substances.

A sulfide is a good nucleophile (Eq. 29.8) and displaces halogen to give sulfonium

$$\underset{Et}{\overset{Me}{\diagdown}}\underline{S}\diagup + BrCH_2COOH \longrightarrow \underset{Et}{\overset{Me}{\diagdown}}\overset{\oplus}{S}-CH_2COOH + Br^{\ominus} \qquad (29.8)$$

salts. These salts are much more stable than are oxonium salts, but they are unstable to heat. It is interesting that a sulfonium ion is nonplanar and stable enough to allow resolution into enantiomers if the three groups differ. Thus, unlike amines, which invert so rapidly that resolution is impossible, the second-row elements invert very slowly and may be said to be configurationally stable as sp^3 hybrids.

In an inert solvent, a sulfide forms a halogen-substituted sulfonium ion, but in acetic acid with some water present, the sulfonium ion undergoes a displacement and subsequent oxidation to give a sulfonyl chloride (Eqs. 29.9).

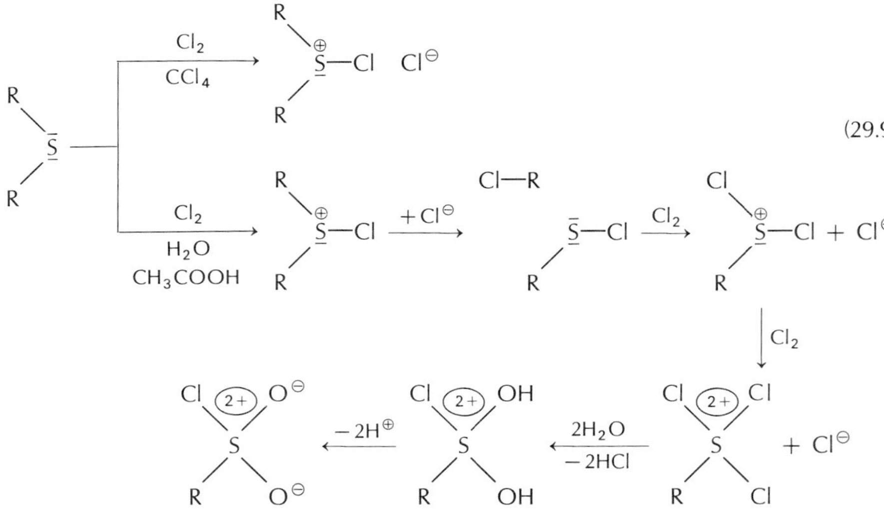

$$(29.9)$$

Episulfides are more reactive than are open-chain sulfides, just as epoxides are more reactive than open-chain ethers (Eqs. 29.10).

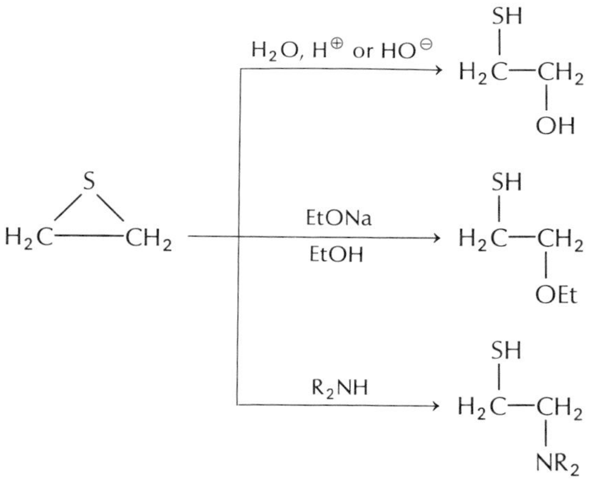

$$(29.10)$$

Sulfides, with the calculated amount of oxidizing agent (H_2O_2 or $NaIO_4$), give sulfoxides, whereas excess H_2O_2, peracid, or $KMnO_4$ give sulfones (Eqs. 29.11).

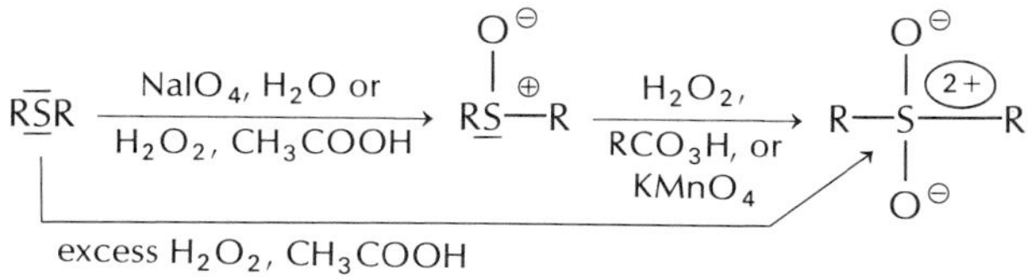

$$(29.11)$$

29.2 Thiols, sulfides, sulfonium ions, and disulfides

Sulfides: preparation. Sulfides may be prepared by several routes (Eqs. 29.12).

$$RSH \xrightarrow[R'X]{NaOH} RSR' + NaX$$

$$RSH \xrightarrow[NaOH]{(MeO)_2SO_2} RSMe + Na_2SO_4$$

$$\phi SH \xrightarrow[\substack{h\nu \\ \text{peroxide}}]{R_2C=CH_2} R_2CH-CH_2-S\phi \qquad (29.12)$$

$$RSOR \xrightarrow{HI} RSR + I_2 + H_2O$$

$$RSOR \xrightarrow{R'_3P} RSR + R'_3PO$$

$$RSO_2R \xrightarrow{LiAlH_4} RSR$$

An unusual reaction is what appears to be an aromatic nucleophilic substitution of an unactivated aryl halide by the cuprous salt of a thiol (Eq. 29.13). However, the

$$RSCu + \phi Br \xrightarrow{\Delta} RS\phi + CuBr \qquad (29.13)$$

presence of cuprous ion (or ion pair) causes us to suspect the intervention of a free-radical reaction involving single electron transfers.

The popular antiseptic for external use, thimerosal (or merthiolate sodium) may be regarded as a relative of salicylic acid in which the *ortho*-hydroxy group has been replaced by a thiol group. The thiol is converted into a mercurial sulfide with ethylmercuric chloride (Eq. 29.14). □ □

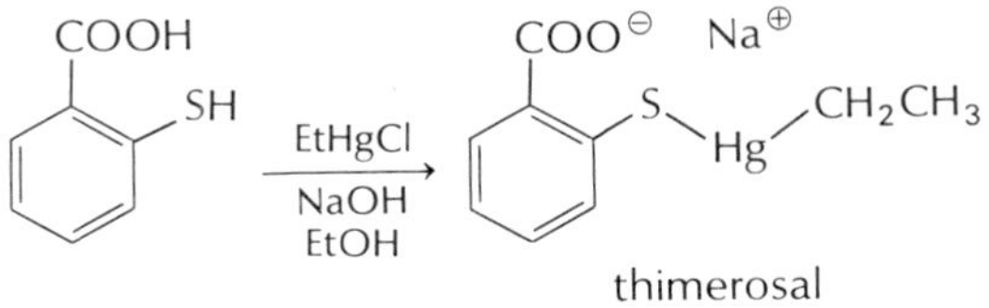

$$\underset{\text{}}{\text{(o-HOOC-C}_6\text{H}_4\text{-SH)}} \xrightarrow[\substack{NaOH \\ EtOH}]{EtHgCl} \underset{\text{thimerosal}}{\text{(o-}^{\ominus}\text{OOC-C}_6\text{H}_4\text{-S-Hg-CH}_2\text{CH}_3)\ Na^{\oplus}} \qquad (29.14)$$

Disulfides. Disulfides undergo reactions that are either oxidations or reductions (Eqs. 29.15). The S—S bond, at 54 kcal, is much stronger than the O—O bond (35 kcal) of peroxides. Thus disulfides are much less reactive than peroxides when rupture of the S—S bond (as compared to the O—O bond) is required.

The best method for the preparation of disulfides is the iodine oxidation of thiols (Eq. 29.5).

$$R-S-S-R \begin{cases} \xrightarrow{Zn, H_2SO_4} 2RSH \\ \xrightarrow{Na, ether} 2RSNa \\ \xrightarrow[-20°]{Cl_2} 2RSCl \xrightarrow[\text{excess}]{Cl_2} R\overset{\oplus}{S}Cl_2\ Cl^{\ominus} \\ \xrightarrow{KMnO_4 \text{ or } HNO_3} 2RSO_3H \end{cases} \qquad (29.15)$$

Occurrence of sulfur compounds in proteins and metabolic products. Disulfides, thiols, sulfides, and sulfonium salts all are essential components of various metabolic pathways and of the structures of proteins. Two of the amino acids are cysteine, with a β-thiol group, and cystine, the related disulfide. Insulin, the hormone that regulates the

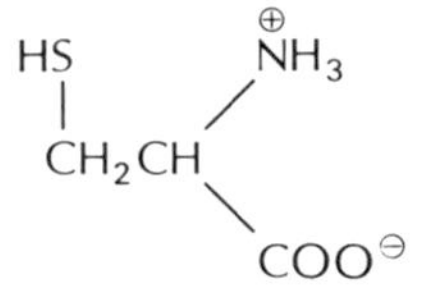

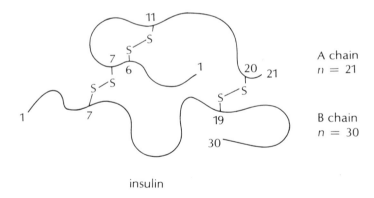

cysteine,
2-amino-3-mercaptopropanoic acid

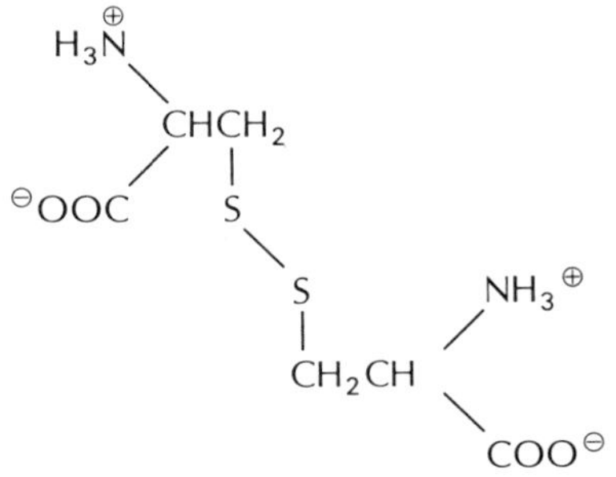

cystine

metabolism of sugar, is one of the smallest proteins, with only fifty-one amino acids in the chain. The molecule contains six cysteine groups, cross-linked as three cystine groups. Insulin consists of two chains. The A chain of twenty-one amino acids has cysteines at positions 6, 7, 11, and 20; the B chain of thirty amino acids has cysteines at positions 7 and 19. The cysteine at A 6 is paired with the one at A 11 to form a cystine bridge (or loop) in the A chain. The two chains are linked by cystine at positions A 7 to B 7 and A 20 to B 19.

insulin

Wool, hair, skin, and feathers consist mainly of keratin, a polymeric protein with a high percentage of sulfur, which is present as cystine units. The cystine units provide loops and cross-linking (as shown in insulin), which give structural rigidity, but not without flexibility. Treatment of hair with reducing agents breaks the cystine disulfide bonds and causes limpness in the hair. If the hair is shaped, curled, or straightened and treated with an oxidizing agent, many of the cystine units are regenerated in the new position. The hair tends to retain the "set." This process is the basis of the permanent wave.

Some other naturally occurring compounds are shown. Thioctic acid is required for the growth of microorganisms. (Mammals are able to synthesize the substance as needed.) Taurine is a metabolic product derived from cysteine. Methionine is a methyl transfer substance, as are the sulfonium ion and dimethylpropiothetin. Ergothionene is present both in blood and in ergot.

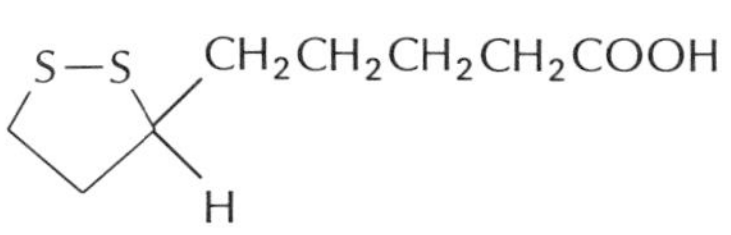

thioctic acid
(lipoic acid)

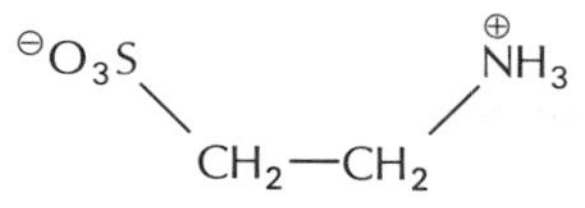

taurine

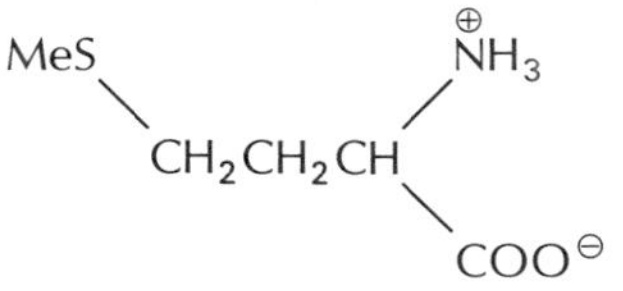

methionine

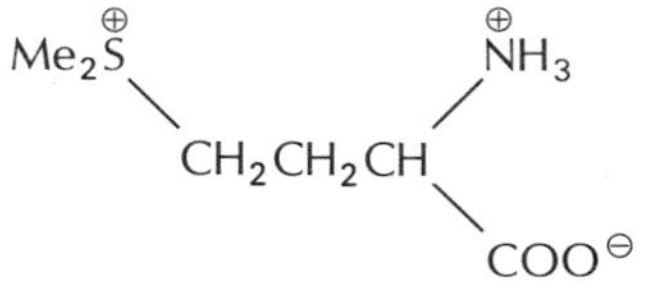

methyl sulfonium ion
of methionine

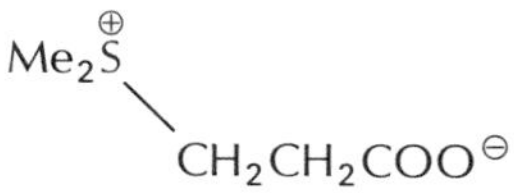

dimethylpropiothetin

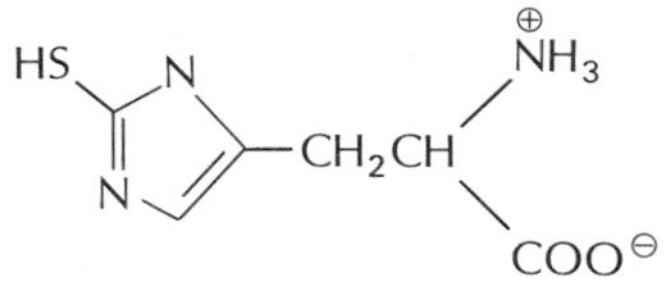

ergothionene

Exercises

29.2 Starting with iPrBr and other reagents as may be needed, show how to prepare:
- a. iPrSH
- b. iPrS(iPr)
- c. $(iPr)_3S^{\oplus}Br^{\ominus}$
- d. iPrSS(iPr)
- e. $iPrSO_2Cl$
- f. $iPrSO_3H$
- g. $iPrSCH_2CH_2CH_3$

□ □

29.3 Sulfoxides and sulfones

Naming. Sulfoxides and sulfones may be considered oxidation products of sulfides with one or two $\overset{\oplus}{S}{-}\overset{\ominus}{O}$ d-p π bonds (Sec. 29.1). They are named in the same manner as are ketones. Thus MeCOEt is ethyl methyl ketone, MeSOEt is ethyl methyl sulfoxide, and $MeSO_2Et$ is ethyl methyl sulfone. Under the systematic rules, the naming differs from that of ketones. The smaller RSO— or RSO_2— groups are seen as alkylsulfinyl or alkylsulfonyl substituents on the larger alkyl chain. Thus, MeCOiPr is 3-methyl-2-butanone, but MeSOiPr is 2-methylsulfinylpropane, and $MeSO_2iPr$ is 2-methylsulfonylpropane.

Reactions. A sulfoxide is easily oxidized to a sulfone (Eq. 29.11). It is so easily reduced to a sulfide that it may be used as a mild oxidizing agent on primary and secondary alcohol sulfonates to give aldehydes and ketones (Eq. 29.16). The postulated mechanism indicates the formation of a sulfur ylide, (a) from the alkoxysulfonium salt.

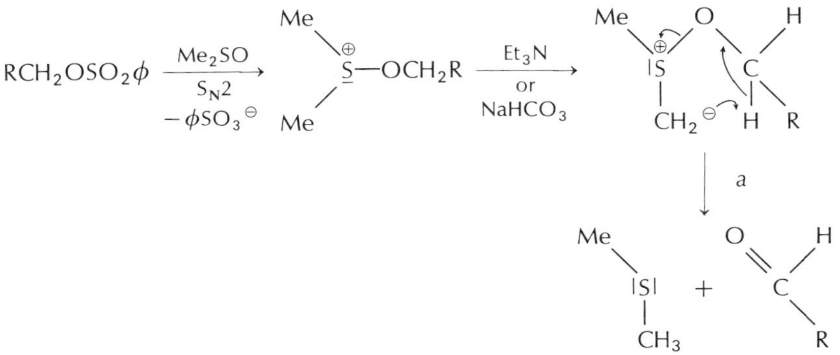

$$RCH_2OSO_2\phi \xrightarrow[\substack{S_N2 \\ -\phi SO_3^{\ominus}}]{Me_2SO} \underset{Me}{\overset{Me}{\underset{|}{S^{\oplus}}}}-OCH_2R \xrightarrow[\text{or NaHCO}_3]{Et_3N} \begin{array}{c} Me \\ \diagdown \\ |S^{\oplus}| \\ | \\ CH_2^{\ominus} \end{array} \begin{array}{c} O \quad H \\ \diagdown / \\ C \\ / \diagdown \\ H \quad R \end{array}$$

$$\downarrow a \qquad (29.16)$$

$$\begin{array}{c} Me \\ \diagdown \\ |S| \\ | \\ CH_3 \end{array} + \begin{array}{c} O \quad H \\ \diagdown\diagup \\ C \\ \diagdown \\ R \end{array}$$

The α hydrogens in sulfoxides, sulfones, and sulfonium salts all are reactive enough to be removed by a base to give analogs of enolate ions (from sulfoxides and sulfones) or ylides (from sulfonium salts). For example, dimethylsulfoxide is acidic enough (pK_a = 31; compare: acetone, 20; acetonitrile, 25; dimethylsulfone, 23) to react with strong bases. The anion, sometimes called <u>dimsyl</u>, in DMSO is very effective for the production of the anions of other weak acids (Eqs. 29.17) or to give a crossed

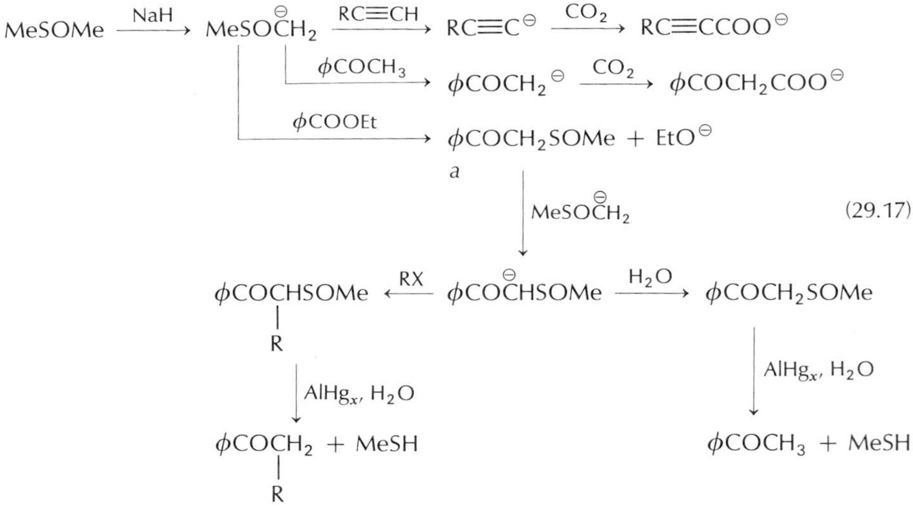

$$MeSOMe \xrightarrow{NaH} MeSO\overset{\ominus}{C}H_2 \xrightarrow{RC\equiv CH} RC\equiv C^{\ominus} \xrightarrow{CO_2} RC\equiv CCOO^{\ominus}$$

$$\phantom{MeSOMe \xrightarrow{NaH} MeSO\overset{\ominus}{C}H_2} \xrightarrow{\phi COCH_3} \phi COCH_2^{\ominus} \xrightarrow{CO_2} \phi COCH_2COO^{\ominus}$$

$$\phantom{MeSOMe \xrightarrow{NaH} MeSO\overset{\ominus}{C}H_2} \xrightarrow{\phi COOEt} \phi COCH_2SOMe + EtO^{\ominus}$$
$$ a$$
$$\bigg\downarrow MeSO\overset{\ominus}{C}H_2 \qquad (29.17)$$

$$\phi CO\overset{|}{\underset{R}{C}}HSOMe \xleftarrow{RX} \phi CO\overset{\ominus}{C}HSOMe \xrightarrow{H_2O} \phi COCH_2SOMe$$

$$\bigg\downarrow AlHg_x, H_2O \qquad\qquad\qquad\qquad \bigg\downarrow AlHg_x, H_2O$$

$$\phi CO\overset{|}{\underset{R}{C}}H_2 + MeSH \qquad\qquad\qquad \phi COCH_3 + MeSH$$

Claisen condensation. The β-ketosulfoxide a may be alkylated and reduced to give ketones. The overall effect is the conversion of an ester (no α hydrogens, ϕCOOEt) into ketones (ϕCOCH$_3$ or ϕCOCH$_2$R).

Sulfones undergo the mixed Claisen condensation as well (Eq. 29.18). The

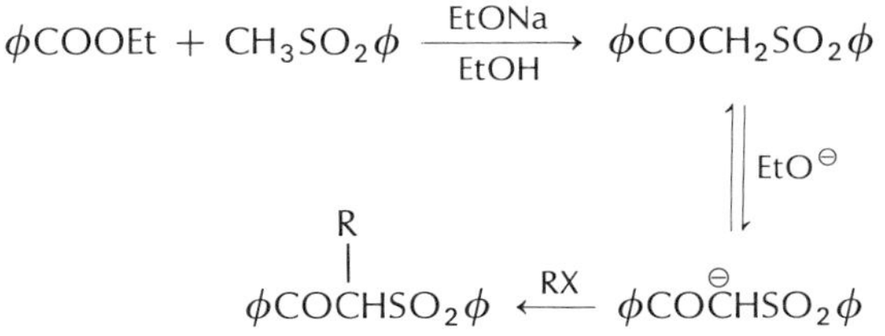

$$\phi COOEt + CH_3SO_2\phi \xrightarrow[EtOH]{EtONa} \phi COCH_2SO_2\phi$$

$$\Updownarrow EtO^{\ominus} \qquad (29.18)$$

$$\phi CO\overset{|}{\underset{R}{C}}HSO_2\phi \xleftarrow{RX} \phi CO\overset{\ominus}{C}HSO_2\phi$$

β-ketosulfone also is acidic and may be alkylated. Sulfones are not easy to reduce to sulfides, although they undergo the Raney nickel reduction.

The α,β-unsaturated sulfoxides and sulfones give Michael addition reactions just as α,β-unsaturated ketones do.

Quaternary salts. A sulfoxide forms a quaternary sulfonium salt with alkyl halides by a slow reaction at sulfur (instead of at oxygen as in Eq. 29.16). An example is shown in Equation 29.19. The α hydrogens in this ion are even more acidic than those in the

$$Me_2SO + MeI \xrightarrow[\text{2-4 days}]{\text{reflux}} Me_3\overset{2+}{S}-O^{\ominus} \; I^{\ominus} \quad (29.19)$$

$Me_2\overset{\oplus}{S}-OCH_2R$ ion. Solution in D_2O leads to a stepwise replacement of all the α hydrogens. Decomposition by heating then gives a neat method for the preparation of deuterated DMSO and CD_3I (Eq. 29.20). The sulfonium ion also reacts to give an ylide, which is useful

$$(CH_3)_3\overset{2+}{S}-O^{\ominus}I^{\ominus} \xrightarrow{D_2O} (CD_3)_3\overset{2+}{S}-O^{\ominus} \; I^{\ominus} \xrightarrow{\Delta} (CD_3)_2SO + CD_3I \quad (29.20)$$

for adding carbene to double bonds (Eq. 29.21).

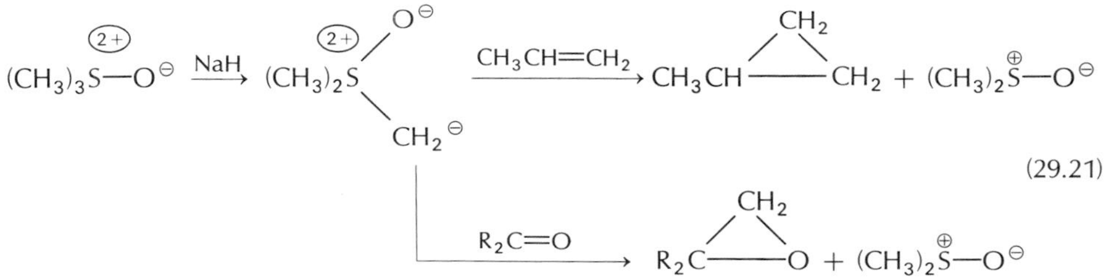

$$(CH_3)_3\overset{2+}{S}-O^{\ominus} \xrightarrow{NaH} (CH_3)_2\overset{2+}{S}\begin{matrix}O^{\ominus}\\ \diagdown \\ CH_2^{\ominus}\end{matrix} \xrightarrow{CH_3CH=CH_2} CH_3CH\overset{CH_2}{\underset{}{\diagup\!\!\!\diagdown}}CH_2 + (CH_3)_2\overset{\oplus}{S}-O^{\ominus}$$

$$\xrightarrow{R_2C=O} R_2C\overset{CH_2}{\underset{}{\diagup\!\!\!\diagdown}}O + (CH_3)_2\overset{\oplus}{S}-O^{\ominus} \quad (29.21)$$

Preparations. Sulfoxides may be prepared in several ways (Eqs. 29.22), and sulfones usually are made by the oxidation of sulfides or sulfoxides (Eq. 29.11).

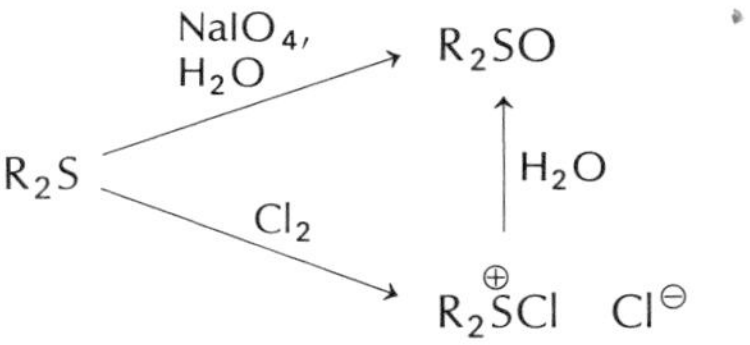

$$R_2S \begin{matrix} \xrightarrow{NaIO_4, H_2O} & R_2SO \\ & \uparrow H_2O \\ \xrightarrow{Cl_2} & R_2\overset{\oplus}{S}Cl \; Cl^{\ominus} \end{matrix} \quad (29.22)$$

$$RMgX + SOCl_2 \longrightarrow R_2SO + MgXCl$$

Exercises

29.3 Complete the equations:
 a. $\phi CH_2OSO_2\phi + Me_2SO \longrightarrow A \xrightarrow{Et_3N} B$
 b. $\phi COCH_2CH_3 + Me_2SO \xrightarrow{NaH} A \xrightarrow{CO_2} B$
 c. $t\text{BuCOOEt} + Me_2SO \xrightarrow{NaH} A \xrightarrow{\phi CH_2Cl} B \xrightarrow[H_2O]{AlHg_x} C$
 d. $Me_2SO + MeI \longrightarrow A \xrightarrow{NaH} B \xrightarrow{\phi CH=CH\phi} C$
 e. $CH_3CH_2CH_2CH_2Br \xrightarrow[\text{ether}]{Mg} A \xrightarrow{SOCl_2} B \xrightarrow{KMnO_4} C$

29.4 Sulfenic, sulfinic, and sulfonic acids and their derivatives

We have encountered sulfonic acids frequently in preceding chapters, especially the aromatic sulfonic acids and derivatives. The names <u>sulfenic</u>, <u>sulfinic</u>, and <u>sulfonic acids</u> designate sulfur acids of increasing oxidation state.

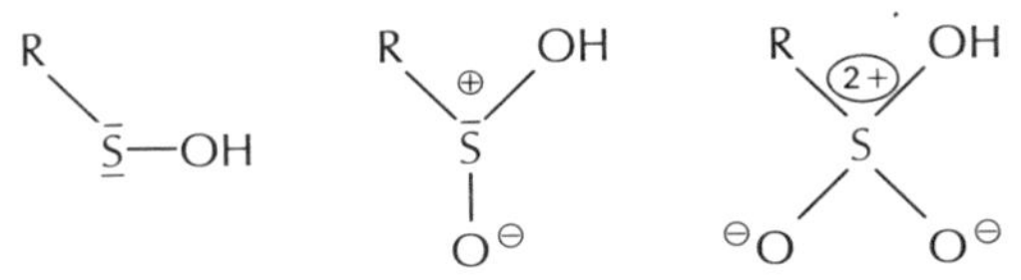

$$\underset{\text{sulfenic acid}}{R-\underline{S}-OH} \qquad \underset{\text{sulfinic acid}}{\overset{R}{\underset{O^{\ominus}}{\overset{\oplus}{S}}}\overset{OH}{}} \qquad \underset{\text{sulfonic acid}}{\overset{R}{\underset{\ominus O}{\overset{(2+)}{S}}}\overset{OH}{\underset{O^{\ominus}}{}}}$$

Sulfenic acids. Sulfenic acids are so reactive that only a few are known (one example is anthraquinone-α-sulfenic acid).

They have been postulated as reaction intermediates and are thought to disproportionate very quickly as shown in Equation 29.23. However, derivatives are more stable (the acid chlorides, RSCl; esters, RSOR′; amides, RSNR′$_2$; etc.). Some

$$3\text{RSOH} \longrightarrow \text{RSSR} + \text{RSO}_2\text{H} + \text{H}_2\text{O} \qquad (29.23)$$

frequently used derivatives are trichloromethylsulfenyl chloride and 2,4-dinitrobenzenesulfenyl chloride. They react with numerous classes of compounds to give solid derivatives. Addition to an alkene is shown in Equation 29.24, in which Ar = 2,4-dinitrophenyl.

$$\text{ArSSAr} \xrightarrow{\text{Cl}_2} \text{ArSCl} \xrightarrow{\text{H}_2\text{C}=\text{CHR}} \text{ArS}\diagdown_{\text{CH}_2-\text{CH}}\diagup^{R}_{\text{Cl}} \qquad (29.24)$$

Sulfinic acids. Sulfinic acids (RSO$_2$H) are more stable than the sulfenic acids, but they too undergo a slow disproportionation to give sulfonic acids and sulfonic esters of thiols (Eq. 29.25). The derivatives and salts of sulfinic acids are less reactive than are the free

$$2\text{RSO}_2\text{H} \xrightarrow{\text{slow}} \text{RSOH} + \text{RSO}_3\text{H}$$
$$\downarrow \text{RSO}_2\text{H, fast} \qquad (29.25)$$
$$\text{RSO}_2-\text{SR} + \text{H}_2\text{O}$$

acids. A sulfinic acid may be oxidized to a sulfonic acid (O$_2$ in air, H$_2$O$_2$, HNO$_3$, etc.) or reduced (Zn + H$_2$SO$_4$) to a thiol. Sulfinic acids may be synthesized by the reaction of a Grignard reagent with SO$_2$ or by the reduction of a sulfonyl chloride with Zn or with Na$_2$SO$_3$ (Eqs. 29.26).

29.4 Sulfenic, sulfinic, and sulfonic acids and their derivatives

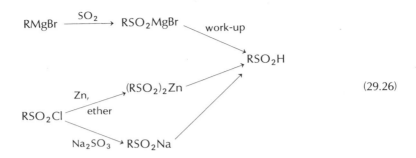
(29.26)

Sulfonic acids; esters; chlorides. Sulfonic acids, like sulfuric acid, are strong acids: trifluoromethanesulfonic acid (F_3CSO_3H) is one of the strongest acids known. They are insoluble in nonpolar solvents and very soluble in water, and they tend to crystallize as hydrates. The presence of one or more sulfonic acid groups in a dye molecule is useful in enhancing water solubility of the dye. Sulfonic acids are nonvolatile. The acid chlorides often are volatile enough to be purified by distillation.

Sulfonic acid esters cannot be prepared by direct esterification and must be synthesized from the sulfonyl chloride (Eq. 29.27; also the amides, Hinsberg test, Sec. 20.2).

$$RSO_3H \xrightarrow{SOCl_2} RSO_2Cl \begin{array}{c} \xrightarrow[\text{NaOH}]{R'OH} RSO_2OR' \\ \xrightarrow[\text{NaOH}]{R'NH_2} RSO_2NHR' \end{array}$$
(29.27)

Sulfonyl chlorides are less reactive than acid chlorides, probably because it is easier for the incoming nucleophile to approach the less hindered sp^2 hybridized carbon of an acid chloride than it is to approach the sp^3 hybrid sulfur of a sulfonyl chloride. Also, the carbon atom of an acid chloride may have less trouble in going over to the sp^3 hybrid intermediate than the sulfur in a sulfonyl chloride has in going from sp^3 to sp^3d to become pentacoordinate. The problem is even worse with hydroxide ion, because sulfonyl chlorides are insoluble in water.

Sulfonate ion as a leaving group; chloramine-T; saccharin. The acid strength of sulfonic acids is reflected in the ease with which sulfonate ions depart from sulfonate esters (Eqs. 29.28). The reactions of sulfonate esters with Grignard reagents and with $LiAlH_4$

$$CH_3CH_2OSO_2R \begin{array}{c} \xrightarrow{HO^\ominus, H_2O} CH_3CH_2OH + {}^\ominus OSO_2R \\ \xrightarrow[R'OH]{R'O^\ominus} CH_3CH_2OR' + {}^\ominus OSO_2R \\ \xrightarrow{NH_3} CH_3CH_2\overset{\oplus}{N}H_3 + {}^\ominus OSO_2R \\ \xrightarrow{R'MgBr} CH_3CH_2R' + {}^\ominus OSO_2R + {}^\oplus MgX \\ \xrightarrow{LiAlH_4} CH_3CH_3 + {}^\ominus OSO_2R \end{array}$$
(29.28)

are particularly useful for converting alcohols into hydrocarbons without the danger of rearrangements or other isomerizations.

Some of the sulfonyl chlorides and esters have been used so often that trivial names have been coined, such as mesyl chloride for CH_3SO_2Cl and ethyl mesylate for ethyl methanesulfonate (CH_3SO_2OEt), tosyl chloride for p-toluenesulfonyl chloride, and brosyl chloride for p-bromobenzenesulfonyl chloride.

Sulfonamides have been discussed in Section 22.2. The sulfonamides of ammonia and primary amines react with NaOCl to give chlorosulfonamides. That from p-toluenesulfonamide loses the remaining proton to form a sodium salt known as chloramine-T (Eq. 29.29). The dry salt is stable, but solution in water causes a

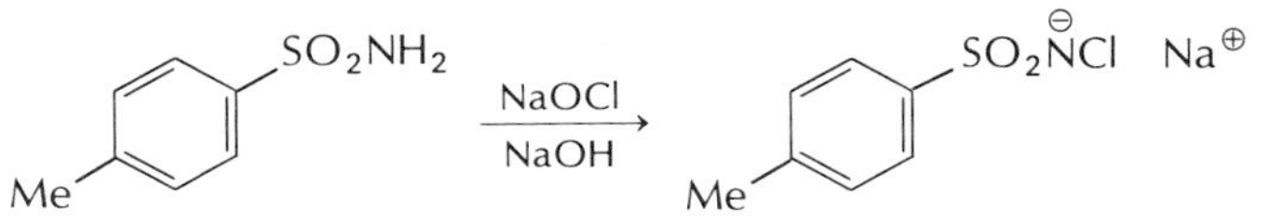

$$\text{Me–C}_6\text{H}_4\text{–SO}_2\text{NH}_2 \xrightarrow[\text{NaOH}]{\text{NaOCl}} \text{Me–C}_6\text{H}_4\text{–SO}_2\overset{\ominus}{\text{N}}\text{Cl} \ \ \text{Na}^{\oplus} \quad (29.29)$$

reversal of the reaction to give back the sulfonamide and a solution of NaOCl with its antiseptic properties.

Saccharin, the sweetening agent about 300 to 500 times stronger than sucrose, is the imide of o-sulfobenzoic acid. It usually is used as the sodium salt, which is more soluble than the imide itself. Two syntheses are shown in Equations 29.30.

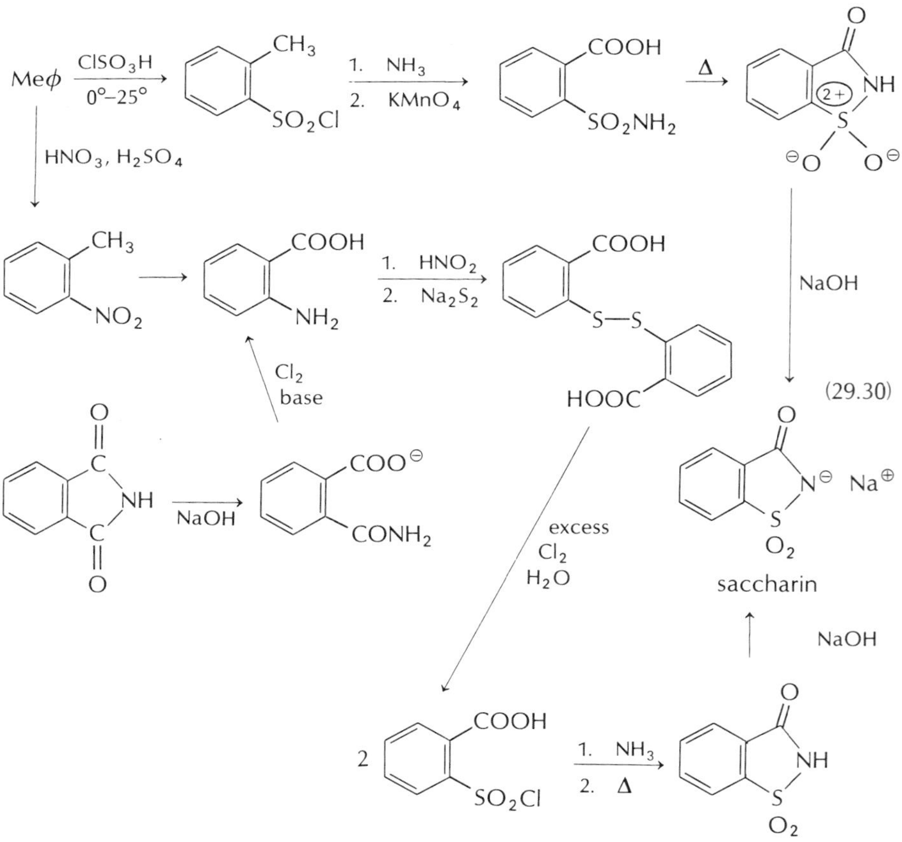

(29.30)

Synthesis. The preparation of aromatic sulfonic acids and derivatives has been covered in previous chapters. Methods applicable for the synthesis of alkanesulfonic acids are shown in Equations 29.31.

$$\begin{array}{c} \text{RSH} \\ \text{RSSR} \end{array} \xrightarrow[\text{H}_2\text{O}]{\text{KMnO}_4} \text{RSO}_3\text{H}$$

$$\text{RSSR} \xrightarrow[\text{H}_2\text{O}]{\text{Cl}_2,\text{ excess}} 2\text{R}\overset{\oplus}{\text{S}}\text{—Cl} \;\; \text{Cl}^{\ominus} \xrightarrow{\text{Cl}_2} \text{RS}\overset{(2+)}{\diagup}\overset{\text{Cl}}{\diagdown}\text{Cl} \;\; 2\text{Cl}^{\ominus} \xrightarrow[-\text{HCl}]{\text{H}_2\text{O}} \text{RSO}_3\text{H}$$

$$\text{RX} + \text{Na}_2\text{SO}_3 \xrightarrow{200°} \text{RSO}_3^{\ominus}\text{Na}^{\oplus} \xrightarrow[\text{EtOH}]{\text{HCl}} \text{RSO}_3\text{H} + \text{NaCl (insoluble in EtOH)}$$

$$\text{RCH}{=}\text{CH}_2 + \text{NaHSO}_3 \xrightarrow[\Delta]{\text{peroxide}} \text{RCH}_2\text{CH}_2\text{SO}_3^{\ominus}\text{Na}^{\oplus}$$

(29.31)

Exercises

29.4 Complete the conversions:
  a. $\phi\text{SO}_3\text{H} \longrightarrow \phi\text{SO}_2\text{H}$
  b. $\phi\text{SO}_3\text{H} \longrightarrow \phi\text{SO}_2\text{OCH}_2\text{CH}_2\phi$
  c. $\phi\text{CH}_2\text{CH}_2\text{COOH} \longrightarrow \phi\text{CH}_2\text{CH}_2\text{CH}_3$
  d. $\text{PrOH} \longrightarrow \text{Pr—Pr}$

29.5 Organophosphorus compounds

Naming; tautomerism. Any brief description of the organophosphorus compounds is complicated from the start by the naming system used. Phosphine (PH$_3$), the analog of ammonia (amine), is simple enough; and as long as we stay with alkyl and aryl substitution only, the naming is analogous to that of the amines. Thus, EtPH$_2$ is ethylphosphine, EtPHϕ is ethylphenylphosphine, Et$_2$Pϕ is diethylphenylphosphine, and Me$_4$P$^{\oplus}$ Cl$^{\ominus}$ is tetramethylphosphonium chloride.

As soon as we begin introducing HO— groups, we find differences from the corresponding nitrogen compounds. Thus H$_2$POH, the analog of hydroxylamine, does not exist as such, but tautomerizes to H$_3\overset{\oplus}{\text{P}}$—$\overset{\ominus}{\text{O}}$, phosphine oxide. In short, an organophosphorus compound (or a phosphorus acid) spontaneously tautomerizes to the more stable tetracoordinated sp^3 structure from the sp^3 but tricoordinate structure if a hydrogen is available for the shift. The ending -ous is used to designate those acids that are tricoordinate (-ite for esters and salts), and the ending -ic is used for the tetracoordinated acids (-ate for esters and salts). Those acids with two hydrogens bound to phosphorus are designated by the penultimate syllable -in-; those with one phosphorus-hydrogen bond, -on-; and none, -or-.

	tricoordinate, -ous	tetracoordinate, -ic
no P—H, -or-	phosphorous acid, $(HO)_3P$ or H_3PO_3	phosphoric acid, $(HO)_3\overset{\oplus}{P}-\overset{\ominus}{O}$ or H_3PO_4
one P—H, -on-	phosphonous acid, $(HO)_2PH$ or H_3PO_2	phosphonic acid, $(HO)_2\overset{\oplus}{P}H-O^\ominus$ or H_3PO_3
two P—H, -in-	phosphinous acid, $HOPH_2$ or H_3PO	phosphinic acid, $HO\overset{\oplus}{P}H_2-O^\ominus$ or H_3PO_2

Now we can show how the system works. We note that phosphonic acid is the more stable tautomer of phosphorous acid and that phosphinic acid is the more stable tautomer of phosphonous acid. The compound H_3PO_2 is also known as hypophosphorous acid, and it is the substance used to reduce diazonium salts. The complete set of names is required for those compounds that cannot tautomerize: esters, trimethyl phosphorite ($P(OMe)_3$) and dimethyl phosphonate ($(MeO)_2PHO$); diethyl phosphonite ($(EtO)_2PH$) and ethyl phosphinate ($EtOPH_2O$); and alkyl- or aryl-substituted acids, phenylphosphonic acid ($\phi P(OH)_2O$) and dimethylphosphinic acid ($Me_2P(OH)O$). Replacement of P—H by P—R does not change the classification for naming. The halogen compounds are named as though they were acid chlorides of the various acids or as substituted phosphines: Me_2PCl is dimethylphosphinous chloride or chlorodimethylphosphine. Finally, the hypothetical PH_5 is called phosphorane so that such compounds as pentaphenylphosphorane ($P\phi_5$) may be named. (However, PCl_5 is phosphorus pentachloride.) Most of the so-called pentavalent phosphorus compounds are tetracoordinate and are named as quaternary phosphonium salts ($R_4P^\oplus X^\ominus$).

Phosphines. The phosphines are relatively stable compounds that range from basicity less than that of the analogous amines (PH_3 has pK_a of -13.5, NH_3 of 9.2) to more basic compounds as substitution is increased. The lower nucleophilicity makes it impossible to alkylate phosphine in the manner in which ammonia and amines may be alkylated by reaction with an alkyl halide. However, if we convert a phosphine into the conjugate base, alkylation may be carried out (Eqs. 29.32). Trialkylation cannot be effected by this scheme,

$$H_3P \xrightarrow{H_2N^\ominus} H_2P^\ominus \xrightarrow{RX} H_2PR + X^\ominus$$
$$RPH_2 \xrightarrow{H_2N^\ominus} R\overset{\ominus}{P}H \xrightarrow{R'X} RR'PH + X^\ominus \qquad (29.32)$$

because the anion $RR'P^\ominus$ is basic enough to remove a proton from NH_3. A trialkylphosphine is a better nucleophile than is a tertiary amine and reacts to give a phosphonium salt (Eq. 29.33). The reaction of Grignard reagents with PCl_3 is of use in synthesis of the various

$$Me_3P + ClCH_2\phi \longrightarrow Me_3\overset{\oplus}{P}CH_2\phi \quad Cl^\ominus \qquad (29.33)$$

substituted chlorophosphines, which may be reduced to phosphines (Eqs. 29.34).

$$RMgX \xrightarrow{PCl_3} RPCl_2 \xrightarrow{NaBH_4} RPH_2$$
$$2ArMgX \xrightarrow{PCl_3} Ar_2PCl \xrightarrow{NaBH_4} Ar_2PH \qquad (29.34)$$
$$ArMgX \xrightarrow{PCl_3} ArPCl_2 \xrightarrow{2RMgX} ArPR_2$$

Rearrangement; ylides. Whenever possible, a tricoordinate phosphorus compound tries to go over to a tetracoordinate compound during reaction. Although the P—H bond energy (77 kcal) is less than that of N—H (93 kcal), and although P—C (63 kcal) is weaker than N—C (73 kcal), the P—O bond (as in $P(OR)_3$), at 95 kcal, is stronger than either N—O (53 kcal) or C—O (86 kcal). Also, $\overset{\oplus}{P}$—$O^{\ominus}$ (about 140 kcal) is almost as strong as N=O (145 kcal). Both are weaker than C=O in aldehydes and ketones (176 to 179 kcal). Thus, for $2PH_3 + O_2 \rightarrow 2H_3\overset{\oplus}{P}$—$O^{\ominus}$, we calculate $\Delta H = -161$ kcal. The large negative enthalpy, along with a relatively small energy of activation, means reactivity. Phosphine and the simpler phosphines are spontaneously flammable in air. In attempting to become tetracoordinate, phosphorus would prefer to bond to oxygen. For example, let us look at the hydrolysis of a dialkylphosphinous chloride and of an alkylphosphinous dichloride (Eqs. 29.35) to give a dialkylphosphine oxide (not the dialkylphosphinous acid, R_2POH) and an

$$R_2PCl \xrightarrow{H_2O} R_2\overset{\oplus}{P}\overset{H}{\underset{Cl}{\diagdown}} \xrightarrow[-H^{\oplus}]{H_2O} R_2P\overset{H}{\underset{Cl}{\diagdown}}{-}OH \xrightarrow{-Cl^{\ominus}} R_2\overset{\oplus}{P}\overset{H}{\underset{OH}{\diagdown}} \xrightarrow{-H^{\oplus}} R_2\overset{\oplus}{P}\overset{H}{\underset{O^{\ominus}}{\diagdown}}$$
$$a$$

(29.35)

$$RPCl_2 \xrightarrow[-H^{\oplus}]{H_2O} RP\overset{H}{\underset{Cl\ Cl}{\diagdown}}{-}OH \xrightarrow{-HCl} R\overset{\oplus}{P}\overset{H}{\underset{Cl}{\diagdown}}{-}O^{\ominus} \xrightarrow{\text{repeat}} R\overset{\oplus}{P}\overset{H}{\underset{OH}{\diagdown}}{-}O^{\ominus}$$
$$b$$

alkylphosphinic acid (not an alkylphosphonous acid). Note that we have not hesitated to write pentacoordinate intermediates such as *a* and *b*, in which phosphorus expands to accommodate ten electrons by sp^3d hybridization.

Possibly the most used of the phosphines is triphenylphosphine ($\phi_3 P$), which, though neutral, is still a good nucleophile and reacts with alkyl halides to form phosphonium salts. These salts have only one source of α hydrogens for reaction with bases to give the useful phosphorus ylides (Eq. 29.36). Reaction of the ylide with an aldehyde or ketone (the Wittig

$$\phi_3P + MeX \longrightarrow \underset{X^{\ominus}}{\phi_3\overset{\oplus}{P}CH_3} \xrightarrow{\phi Li} \phi_3\overset{\oplus}{P}{-}\overset{\ominus}{C}H_2 \xrightarrow[-H^{\oplus}]{H_2O} \underset{OH}{\phi_3\overset{}{P}{-}\overset{\ominus}{C}H_2} \xrightarrow{H_2O} \underset{\phi\ \ \ OH}{\phi_2\overset{}{P}\diagup^{CH_3}_{\diagdown}}$$

$$\Big\downarrow O=CR_2 \qquad\qquad\qquad\qquad \begin{array}{l}1.\ -\phi^{\ominus}\\ 2.\ -H^{\oplus}\end{array}\Bigg\downarrow$$

(29.36)

$$\underset{O^{\ominus}}{\phi_3\overset{\oplus}{P}}\ +\ \overset{CH_2}{\underset{CR_2}{\|}} \quad \underset{a}{\overset{\phi_3\overset{\oplus}{P}{-}CH_2}{\longleftarrow}\ \ {}^{\ominus}O{-}CR_2} \qquad\qquad \underset{b}{\phi_2\overset{\oplus}{P}\diagup^{CH_3}_{\diagdown O^{\ominus}}}$$

reaction) gives an alkene and triphenylphosphine oxide (*a*). Reaction with water gives benzene and diphenylmethylphosphine oxide (*b*). The ylide is named as a phosphorane, or methylenetriphenylphosphorane.

Trisubstituted phosphines possess stable pyramidal configurations, and if the three groups are nonidentical, the phosphine may be resolved into the enantiomers. (Contrast tertiary amines, which invert so rapidly that resolution is impossible unless the molecule is so constructed (Troger's base, Sec. 20.2) that inversion cannot take place.) Like quaternary ammonium salts and tertiary amine oxides, the phosphonium salts and tertiary phosphonium oxides also may be resolved if no two groups are identical. Pentacoordinate phosphoranes may exist as a trigonal bipyramide (PCl_5) or as a pyramid ($\phi_5 P$), depending upon which of the five d orbitals is used in forming the sp^3d hybrid.

Phosphines react with chlorine to become dichlorophosphoranes (R_3PCl_2) or chlorophosphonium chlorides ($R_3\overset{\oplus}{P}Cl\ Cl^{\ominus}$), depending upon the electron-withdrawing or electron-donating character of the R groups.

Tetracoordinate phosphorus: Arbusov reaction and rearrangement. The tetracoordinate phosphorus compounds also are oxygen hungry and react, when possible, to increase the number of P—O and $\overset{\oplus}{P}$—$O^{\ominus}$ bonds. An example is the ylide in the Wittig reaction or upon hydrolysis.

Another example is the <u>Arbusov reaction,</u> in which a trialkyl phosphite reacts with an alkyl halide to give a phosphonium salt as the first step (Eq. 29.37). The second

$$(EtO)_3P \xrightarrow[-Br^{\ominus}]{RBr} (EtO)_2\overset{\oplus}{P}\diagup^{R}_{\diagdown OCH_2\text{—}CH_3} \xrightarrow{Br^{\ominus}} (EtO)_2\overset{\oplus}{P}\diagup^{R}_{\diagdown O^{\ominus}} + CH_2\text{—}Br \quad\quad (29.37)$$
$$\mid$$
$$CH_3$$

step (S_N2) is the displacement of diethyl alkylphosphonate by $Br^{\ominus}$. If the product alkyl halide (EtBr here) is more reactive than RBr (for example, iBuBr), then it enters into the reaction and mixtures result. On the other hand, if only a little of a less reactive RBr is used, then the major reaction becomes that of EtBr, and the product is diethyl ethylphosphonate. In such a case, the reaction is called the <u>Arbusov rearrangement</u> (Eq. 29.38).

$$(EtO)_3P \xrightarrow[\substack{\text{small}\\\text{amount}}]{iPrBr} (EtO)_2\overset{\oplus}{P}\diagup^{Et}_{\diagdown O^{\ominus}} \quad\quad (29.38)$$

If we return to Equations 29.35 for a moment and consider what would happen if we carry out an alcoholysis instead of a hydrolysis, we find a parallel to the Arbusov reaction, as seen in Equation 29.39, in which phosphorus expands the valence shell to a sp^3d hybrid (a) as an intermediate or transition state. Also by analogy, the reaction of an alcohol with PCl_3, $POCl_3$, or PCl_5 to give an alkyl halide proceeds by a similar mechanism.

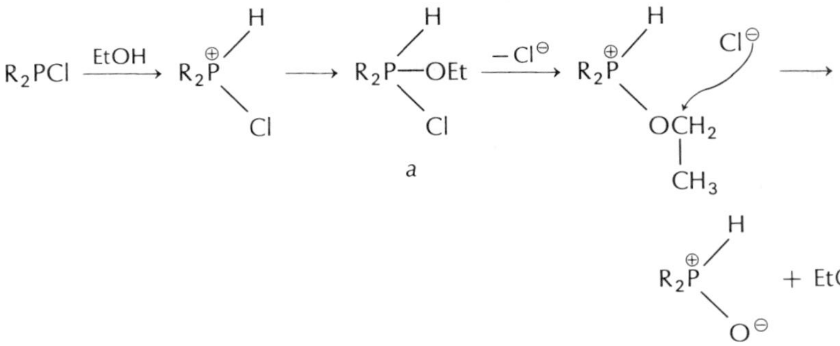

$$R_2PCl \xrightarrow{EtOH} R_2\overset{\oplus}{P}\!\!\begin{array}{c}H\\\diagup\\\diagdown\\Cl\end{array} \longrightarrow R_2P\!\!\begin{array}{c}H\\|\\-OEt\\|\\Cl\end{array} \xrightarrow{-Cl^{\ominus}} R_2\overset{\oplus}{P}\!\!\begin{array}{c}H\\\diagup\\\diagdown\\OCH_2\\|\\CH_3\end{array} \overset{Cl^{\ominus}}{\curvearrowleft} \longrightarrow$$

(29.39)

$$R_2\overset{\oplus}{P}\!\!\begin{array}{c}H\\\diagup\\\diagdown\\O^{\ominus}\end{array} + EtCl$$

Radical addition of phosphines to alkenes. The weakness of the P—H bond (77 kcal) is displayed by the susceptibility of these bonds to homolytic cleavage and participation of the phosphorus radicals in free-radical additions to alkenes (Eq. 29.40).

$$EtCH=CH_2 \xrightarrow[\substack{\text{peroxide}\\\Delta}]{R_2PH} EtCH_2-CH_2-PR_2 \qquad (29.40)$$

As is usual with radical additions, the direction is anti-Markovnikov. The reaction is not restricted to phosphines: phosphonous and phosphinous esters and phosphonic and phosphinic acids and esters also react.

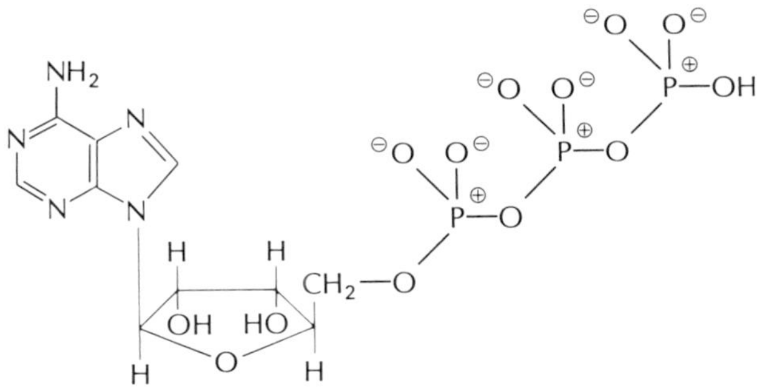

adenosine triphosphate
(ATP)
as the trinegative ion, pH 7

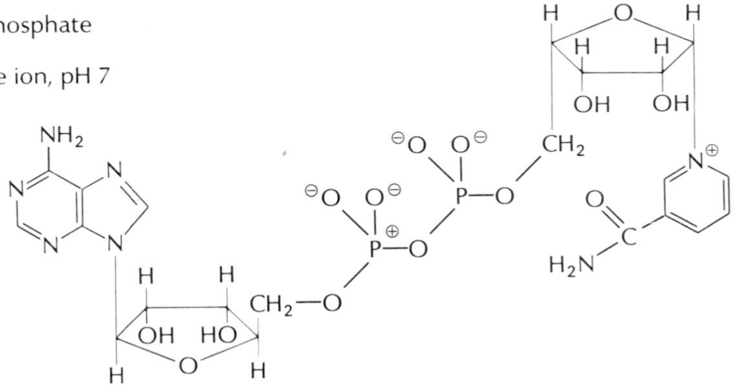

nicotinamide adenine dinucleotide
(NAD)
as the dinegative ion, pH 7

Phosphate esters. Phosphate esters really are not organophosphorus compounds, which are defined as those compounds with a P—C bond. Nevertheless, phosphate esters are universally present in cellular metabolism, so we shall bring them in. Phosphoric acid tends to form polyphosphoric acids and ions, generally called pyrophosphoric acids. At pH 7, close to that of blood, hydrolysis of a phosphate link (anhydride bond) has ΔG of about -7.4 kcal. This kind of reaction, when coupled with others in the cell that require energy, forms part of the energy-transfer system. A triphosphate and a diphosphate are shown. The di- and monophosphates of adenosine are abbreviated as ADP and AMP. Here we shall only point out that the phosphates help to carry out certain enzymatic reactions by furnishing energy (Eq. 29.41).

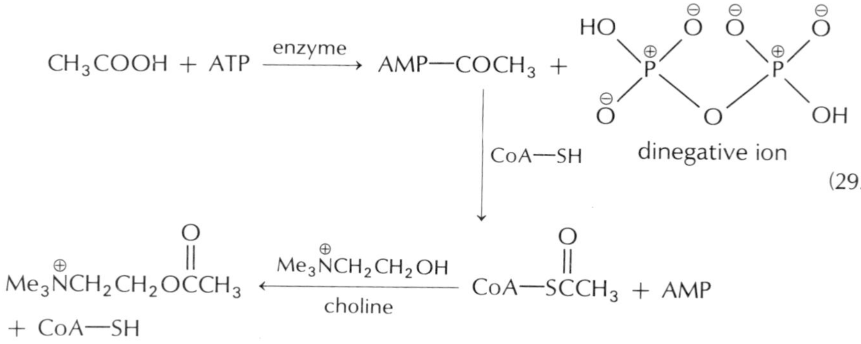

$$CH_3COOH + ATP \xrightarrow{enzyme} AMP\text{—}COCH_3 + \text{dinegative ion} \quad (29.41)$$

$$\downarrow CoA\text{—}SH$$

$$Me_3\overset{\oplus}{N}CH_2CH_2OCCH_3 + CoA\text{—}SH \xleftarrow{Me_3\overset{\oplus}{N}CH_2CH_2OH \text{ choline}} CoA\text{—}SCCH_3 + AMP$$

Exercises

29.5 Name:
  a. EtOPϕ_2
  b. $(HO)_2\overset{\oplus}{P}\text{—}O^{\ominus}$ with ϕ
  c. Et$_3$PO$_4$
  d. (EtO)$_3$P
  e. ϕ_3PCl$_2$
  f. Et$_4$P$^{\oplus}$ Br$^{\ominus}$

29.6 Show how to prepare:
  a. EtPH$_2$
  b. $\phi_3\overset{\oplus}{P}CH_2\phi$ Cl$^{\ominus}$
  c. Et$_2\overset{\oplus}{P}H\text{—}O^{\ominus}$
  d. $\phi_2C\text{=}CH_2$
  e. $(EtO)_2\overset{\oplus}{P}(O^{\ominus})Et$
  f. $\phi CH_2CH_2PEt_2$

29.6 Organosilicon compounds

Bonding to silicon; naming. In contrast with the organophosphorus compounds, which display numerous examples of $\overset{\oplus}{P}\text{—}O^{\ominus}$ d-p π bonding, the silicon compounds seem to utilize such back bonding to a much smaller extent.

This lack leads to considerable simplification in naming. The parent hydride, silane (SiH$_4$) is the analog of methane. Disilane (H$_3$Si—SiH$_3$) is the analog of ethane. The weak Si—Si bond (only 53 kcal as compared to C—C, 83 kcal) does not allow extensive chains to be built up. Silane and disilane both are spontaneously flammable in air by oxidation. The Si—O bond is very strong (108 kcal, as compared to C—O, 86 kcal). Silica (SiO$_2$) is not O=Si=O, like O=C=O, because silicon does not form any p-p π bonds. Rather, silica is the polymeric anhydride of Si(OH)$_4$, each atom of silicon (sp^3 hybrid) being bonded to four oxygen atoms, and each oxygen atom being bonded to two atoms of silicon. The polymer may take any of several crystalline forms, such as quartz (amethyst, rock crystal, agate, flint, sand, etc.), or less crystalline, hydrated forms (opal, silica gel, silicic acids, etc.).

The lack of π bonding ability by silicon means that there are no silicon analogs of the alkenes, alkynes, arenes, aldehydes, ketones, acids, acid chlorides, amides, and so on. Alkyl-, aryl-, and halogen-substituted silanes are named as substituted silanes. A silanol is the analog of an alcohol, and silanediol is the silicon system analog of a hydrate of an aldehyde or ketone. Silanols and silanediols may be bonded to other groups, as in the analogous esters, ethers, siloxanes, and ketals. A few examples are trimethylsilyl benzoate (ϕCOOSiMe$_3$), methoxytrimethylsilane (MeOSiMe$_3$), hexamethyldisiloxane (Me$_3$SiOSiMe$_3$), and dimethoxydimethylsilane ((MeO)$_2$SiMe$_2$). The naming, we see, is like the methane nomenclature used for some organic compounds, such as triphenylmethane for ϕ_3CH.

The organosilicon compounds have mp and bp not too greatly different from those of the analogous carbon compounds. Notable exceptions are the trimethylsilyl ethers (methoxytrimethylsilane, MeOSiMe$_3$, bp 56°; and methyl *tert*-butyl ether, MeOCMe$_3$, bp 106°), which are much more volatile.

Bond energies; nucleophilic substitution. Silicon, being less electronegative than carbon or hydrogen, forms σ bonds with carbon and with hydrogen that are polarized so that silicon is partially positive and carbon and hydrogen are partially negative. Tetramethylsilane (TMS, Me$_4$Si) has twelve identical hydrogen atoms. In nmr, they absorb so far upfield (shielding) that the substance is used as the standard from which nmr spectra are measured downfield. The halogens, oxygen, and nitrogen, all more electronegative than silicon (as compared to carbon), all form strong σ bonds with silicon. However, part of the extra strength must be caused by some contribution of d-p π bonding. For example, the dipole moment of MeCl (1.87 D) is greater than that of H$_3$SiCl (1.28 D) even though the Si—Cl bond is longer than a C—Cl bond. Thus $\text{Si}{=}\text{Cl}^{\oplus}$ must contribute to the bonding, the pair of electrons in a chlorine 3p orbital giving partial overlap with an empty silicon 3d orbital. The Si—F bond is particularly strong (135 kcal), so much so that SiF$_4$ reacts with 2F$^\ominus$ to give the SiF$_6^{2-}$ anion, in which silicon has expanded to a sp^3d^2 hybrid to accommodate twelve electrons.

Substitution reactions by nucleophiles take place with great rapidity on the partially positive silicon. The departing group may be a halogen, an alkoxy, an alkyl group, or a hydride ion. Some examples are shown in Equations 29.42. In a, the anion from a Grignard reagent (or RLi) simply displaces a halogen. In contrast, if no better leaving group is available, hydride is replaced (*f*). In *b*, the lack of S$_N$1 reactivity and the reduced steric hindrance of silanes is shown. No instance of ϕ_3Si$^\oplus$ or any other silicon cation has been shown: silicon is unable to delocalize the positive charge into the aromatic rings (poor Si-p–C-p overlap,

$$CH_2=CHCH_2\overset{\frown}{\overset{\ominus}{M}g\overset{\oplus}{C}l} + H_3Si-Cl \longrightarrow CH_2=CHCH_2SiH_3 + MgCl_2 \qquad a$$

$$\phi\overset{\frown}{\overset{\ominus}{C}H_2\overset{\oplus}{M}gCl} + \phi_3Si-Cl \longrightarrow \phi CH_2Si\phi_3 + MgCl_2 \qquad b$$

$$LiAlH_4 + 4Me_3Si-Cl \longrightarrow 4HSiMe_3 + LiCl + AlCl_3 \qquad c$$

(29.42)

$$4\overset{\frown}{Bu\ Li^\oplus} + Si(OMe)_4 \longrightarrow Bu_4Si + 4Me\overset{\ominus}{O}\ Li^\oplus \qquad d$$

$$HO^\ominus + Me_3\overset{\frown}{Si-CH_2}\phi \longrightarrow HOSiMe_3 + {}^\ominus CH_2\phi \xrightarrow{H_2O} CH_3\phi \qquad e$$

$$Li\overset{\oplus}{M}e^\ominus + \phi_3\overset{\frown}{Si-H} \longrightarrow MeSi\phi_3 + LiH \qquad f$$

Sec. 29.1) nor into alkyl groups by hyperconjugation. In c, hydride ion is the nucleophile. In d, an alkoxy is displaced by $R^\ominus$ of a metal organic compound. In e, a hydroxide ion is shown displacing an alkyl (benzyl) in which the negative charge is delocalized.

Substitution: stereochemistry. The enhanced reactivity of silicon may be ascribed to the ease with which the valence shell may be expanded to sp^3d hybridization. (The energy requirement is smaller than that of sp^3 carbon going to sp^2 to accommodate entering and leaving groups with a single p orbital.) Although most displacements on silicon proceed with inversion, examples of retention are common. Silicon, like carbon, is capable of chirality if the four attached groups are nonidentical. Optical activity is the means of diagnosing retention or inversion.

Particularly striking is the S_N2 reactivity of bridgehead silanes (Eqs. 29.43), with enforced retention.

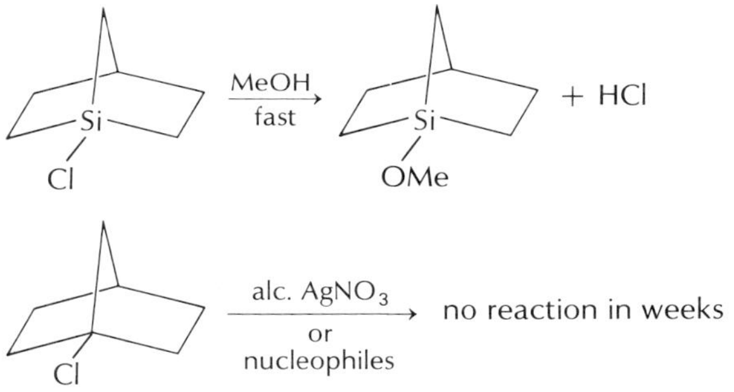

(29.43)

For retention to occur on silicon, attack by the nucleophile may be at the rear or at the side. The intermediate (not a transition state as in S_N2 attack on carbon, it has been hypothesized) may shift from one trigonal bipyramidal form to another by either of two routes (Eqs. 29.44). The vibratory transformation of one trigonal bipyramid into another is called <u>pseudorotation</u>. The bridgehead silanes require side attack, but back-side attack upon open-chain silanes also may proceed with retention of configuration.

Electrophilic reagents may attack a neighboring carbon atom adjacent to silicon, after which the usual very fast nucleophilic reaction on silicon takes place (Eqs. 29.45).

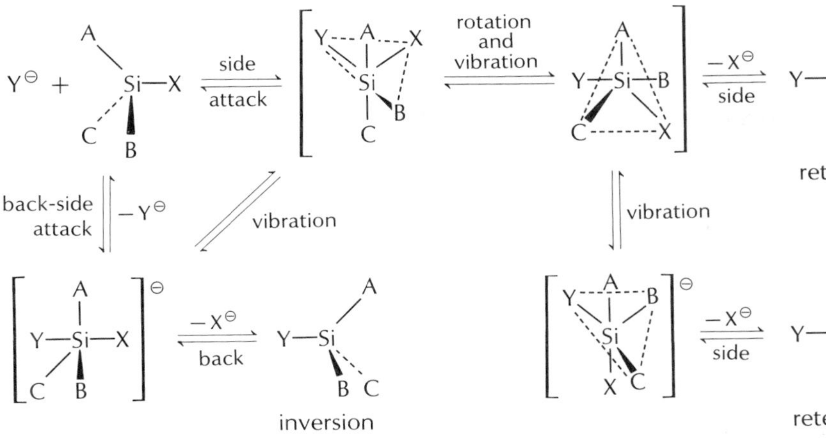

$$Y^\ominus + \underset{\underset{B}{C}}{\overset{A}{Si}}-X \underset{\text{side attack}}{\rightleftharpoons} \left[\underset{\underset{B}{C}}{\overset{Y\cdots\overset{A}{|}\cdots X}{Si}}\right] \underset{\text{vibration}}{\overset{\text{rotation and vibration}}{\rightleftharpoons}} \left[\underset{C\cdots X}{\overset{A}{\underset{|}{Si}}-B}\right] \underset{\text{side}}{\overset{-X^\ominus}{\rightleftharpoons}} \underset{\underset{B}{C}}{Y-\overset{A}{\underset{|}{Si}}}$$

retention (29.44)

back-side attack $\|-Y^\ominus$ vibration   vibration

$$\left[\underset{\underset{B}{C}}{Y-\overset{A}{\underset{|}{Si}}-X}\right]^\ominus \underset{\text{back}}{\overset{-X^\ominus}{\rightleftharpoons}} Y-\underset{B\ C}{\overset{A}{Si}} \qquad \left[\underset{X\ C}{\overset{Y\cdots\overset{A}{|}\cdots B}{Si}}\right]^\ominus \underset{\text{side}}{\overset{-X^\ominus}{\rightleftharpoons}} \underset{C\ B}{Y-\overset{A}{\underset{|}{Si}}}$$

inversion   retention

$$\phi SiMe_3 \quad \underset{HClO_4}{\overset{Br_2}{\nearrow\searrow}} \quad \begin{array}{c}\text{[Br, SiMe}_3\text{ arenium]} \longrightarrow \text{PhBr} + BrSiMe_3 \\ \text{[H, SiMe}_3\text{ arenium]} \overset{H_2O}{\underset{-H^\oplus}{\longrightarrow}} \text{PhH} + HOSiMe_3\end{array} \qquad (29.45)$$

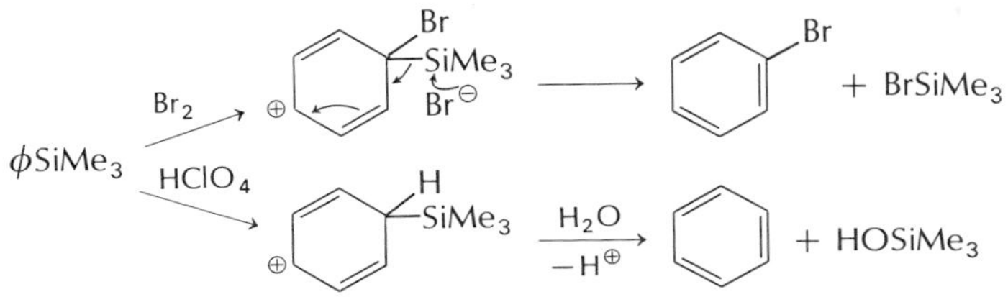

Radical and silene reactions. The Si—H bond is weak, only 76 kcal, and compounds containing these bonds undergo free-radical reactions with ease (Eq. 29.46). The abstraction of a chlorine atom from ϕCl looks surprising until we remember the

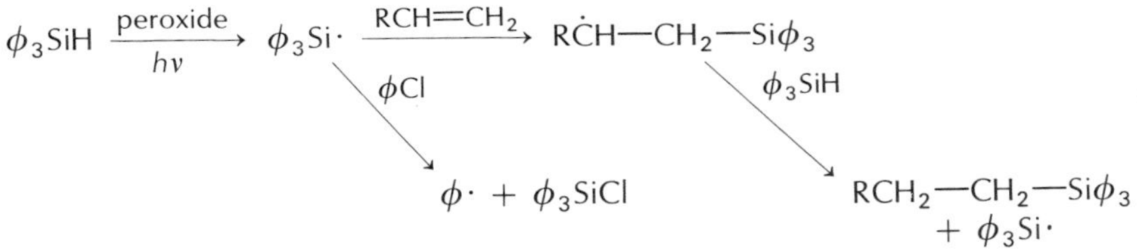

$$\phi_3 SiH \xrightarrow[h\nu]{\text{peroxide}} \phi_3 Si\cdot \xrightarrow{RCH=CH_2} R\dot{C}H-CH_2-Si\phi_3 \xrightarrow{\phi_3 SiH} RCH_2-CH_2-Si\phi_3 + \phi_3 Si\cdot$$

$$\phi_3 Si\cdot \xrightarrow{\phi Cl} \phi\cdot + \phi_3 SiCl \qquad (29.46)$$

enhanced strength of Si—X bonds (91 kcal for Si—Cl, 81 kcal for C—Cl) and that partial Si—Cl bonding in the transition state contributes to a lowering of the activation energy.

Silenes (carbene analogs) also are known. Like carbenes, they give addition reactions (to alkenes, etc.) and insertion reactions (Eqs. 29.47).

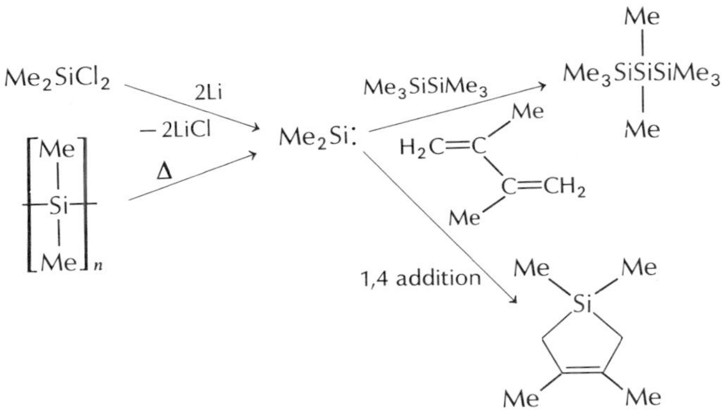

$$Me_2SiCl_2 \xrightarrow[-2LiCl]{2Li} Me_2Si: \xrightarrow{Me_3SiSiMe_3} Me_3SiSi(Me)_2SiMe_3$$

with also $[-Si(Me)_2-]_n \xrightarrow{\Delta}$ giving $Me_2Si:$, which reacts with $H_2C=C(Me)-C(Me)=CH_2$ by 1,4 addition to give the cyclic silacyclopentene with Me groups. (29.47)

Manufacture of chlorosilanes; silyl Grignard reagents. Commercially, the organosilicon compounds originate from silica (Eqs. 29.48). Silicon tetrachloride and other poly-

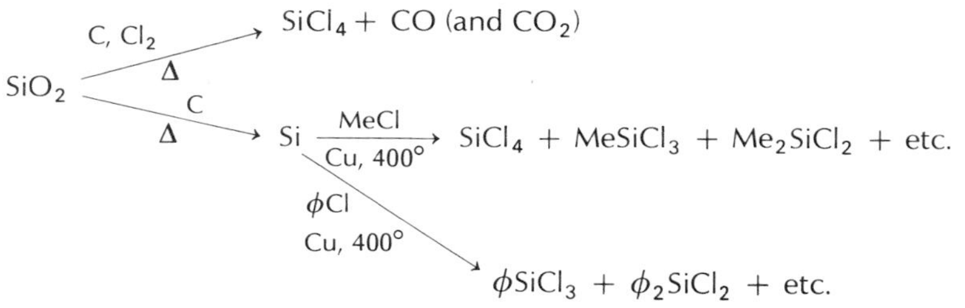

$$SiO_2 \begin{array}{l} \xrightarrow{C, Cl_2, \Delta} SiCl_4 + CO \text{ (and } CO_2) \\ \xrightarrow{C, \Delta} Si \begin{array}{l} \xrightarrow{MeCl, Cu, 400°} SiCl_4 + MeSiCl_3 + Me_2SiCl_2 + \text{etc.} \\ \xrightarrow{\phi Cl, Cu, 400°} \phi SiCl_3 + \phi_2 SiCl_2 + \text{etc.} \end{array} \end{array} \quad (29.48)$$

chlorosilanes are so reactive that they fume in the air ($nSiCl_4 + 2nH_2O \rightarrow (SiO_2)_n + 4nHCl$). The other silanes then are made by substitution reactions (Eqs. 29.42, 29.45, 29.46, 29.47). Trimethylchlorosilane (Me_3SiCl) is useful for converting alcohols of high mw or di- or triols, all of which have high bp, into the much more volatile silyl ethers (Me_3SiOR) for use in gas-liquid chromatography and in mass spectroscopy.

Silyl halides react to form Grignard and lithium reagents, which give the usual metal-organic reagent reactions (Eq. 29.49).

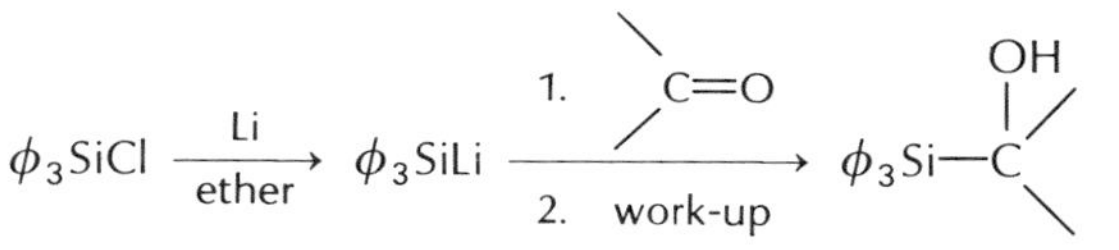

$$\phi_3SiCl \xrightarrow{Li, \text{ether}} \phi_3SiLi \xrightarrow[\text{2. work-up}]{1. \diagdown C=O\diagup} \phi_3Si-C\diagdown^{OH}_{\diagup} \quad (29.49)$$

Silanols, silanediols, siloxanes, and silicones. In general, silanols are more acidic than alcohols, but they undergo the usual reactions of alcohols. However, in the presence of acids or bases, silanols easily condense to give a siloxane (ether; Eq. 29.50). The silanediols

$$2\phi_3SiOH \xrightarrow[HO^\ominus]{H^\oplus \text{ or }} \phi_3SiOSi\phi_3 + H_2O \quad (29.50)$$

and triols react even more readily to give chains and cross-linked siloxane polymers (Eqs. 29.51). Originally, these compounds were thought to be ketonic (with $\diagdown Si=O \diagup$) and were called <u>silicones</u>. The name has stuck. Silicone oils are made from silanediols with a

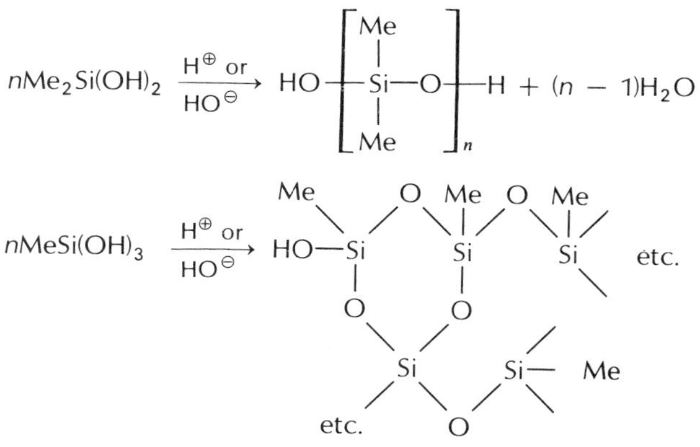

$$nMe_2Si(OH)_2 \xrightarrow[HO^\ominus]{H^\oplus \text{ or }} HO\left[\begin{array}{c} Me \\ | \\ Si-O \\ | \\ Me \end{array}\right]_n H + (n-1)H_2O$$

$$nMeSi(OH)_3 \xrightarrow[HO^\ominus]{H^\oplus \text{ or }} \text{HO—Si—O—Si—O—Si etc.} \quad (29.51)$$

(cross-linked siloxane structure with Me and O bridges, etc.)

tiny amount of trimethylsilanol present, which acts to cap the end —OH groups and to regulate the mw. Silicone rubbers and solid polymers are made with less or more silanetriol

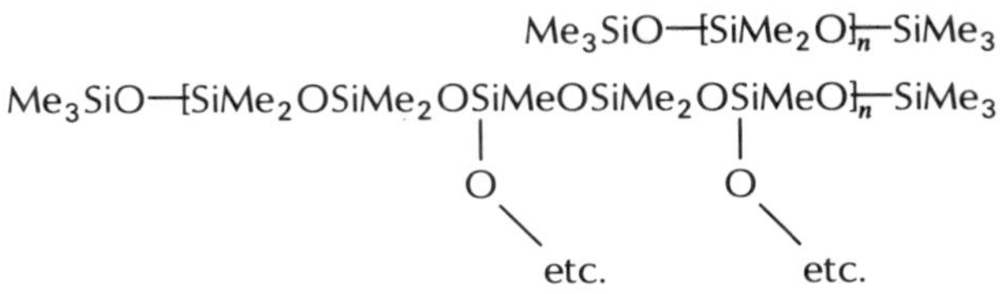

$$Me_3SiO\text{−}[SiMe_2O]_n\text{−}SiMe_3 \quad \text{silicone oil}$$

$$Me_3SiO\text{−}[SiMe_2OSiMe_2OSiMeOSiMe_2OSiMeO]_n\text{−}SiMe_3 \quad \text{silicone rubber}$$

with O\etc. branches

mixed with silanediol to provide a few or many cross-links. Silicones are used as water-repellant coatings (fabrics, glass), foam inhibitors, special-purpose rubbers, polishes, paints, electrical insulation, and so on.

Exercises

29.7 Starting with any of the products of Equations 29.48, show how to prepare:
  a. Me_2SiH_2
  b. $\phi CH_2 SiMe_3$
  c. $MeSi(OMe)_3$
  d. $\phi CH_2 CH_2 Si\phi_3$
  e. $CH_3 CH\text{=}CHCH_2 SiMe_3$
  f. 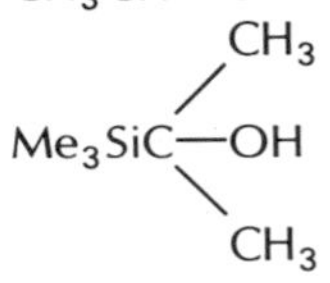 $Me_3 SiC(CH_3)_2\text{—}OH$

29.7 Summary

Aside from expanding our knowledge of the chemistry of sulfur, phosphorus, and silicon, we have included something of the essential parts played by sulfur and phosphorus in living organisms.

Our excursion into the chemistry of the second-row elements may be briefly summarized.
1. The presence of the *d* orbitals leads to differences between the corresponding first- and second-row elements in bond energies, in reactivity, in lack of π bonding, and in valence-shell expansion.
2. Steric effects are much less of a factor in determining reactivity and stability.
3. The organosulfur and organophosphorus compounds are much more versatile in type and reactions than are the organosilicon compounds.

29.8 Bibliography

Only the more recent textbooks have material on sulfur, phosphorus, and silicon. The books by Allinger *et al.* (Chap. 7, ref. 1), Hendrickson *et al.* (Chap. 7, ref. 9), Noller (Chap. 7, ref. 11), and Roberts and Caserio (Chap. 7, ref. 13) are good starting points.

We mentioned saccharin as a sweetening agent. For those who may be interested in the subject of flavor and odor, an excellent introduction is the following article.

1. Hornstein, I., and R. Teranishi, The Chemistry of Flavor, *Chemical and Engineering News* Apr. 3, 1967, pp. 93–108.

29.9 H. Harry Szmant: Surprising sulfur radicals

Professor H. Harry Szmant, Ph.D. at Purdue in 1944, spent two years in industry (Monsanto) before going to Duquesne in 1946, where he remained until 1956. From 1956 until 1960 he was at Oriente University in Cuba as head of the department of chemistry. Political upheaval there caused another move, this time to the University of Puerto Rico, and the Puerto Rico Nuclear Center where he stayed until he went to the University of Detroit in 1968 as chairman of the department, the position he holds today. In spite of the numerous upsets and moves, he has been a productive research worker; and he has an interesting story to tell us.

Twenty years ago, while teaching at Duquesne University in Pittsburgh, I was investigating the cryoscopic behavior of organic sulfur compounds in sulfuric acid. In the course of this work one of my graduate students, now Dr. Ronald L. Lapinski, set out to prepare the apparently simple compound phenethyl disulfide by the oxidation of the readily available phenethyl mercaptan. To our great surprise the oxidation of the thiol with peroxides or iodine refused to yield clean samples of the disulfide. Attempted purification by distillation, admittedly in the presence of air, gave a mixture of liquid products and a white solid. The elementary analysis of the latter revealed the presence of only one sulfur but two oxygen atoms per molecule, and this and other evidence suggested a structure of a beta-hydroxy sulfoxide $C_6H_5-CH(OH)CH_2-SO-CH_2-CH_2-C_6H_5$.

The puzzle was clarified by a literature search when we found that Professor Morris Kharasch at the University of Chicago had described the reaction of thiols, oxygen, and styrene-like olefins to give beta-hydroxy sulfoxides and had suggested the radical chain mechanism:

$$RSH + O_2 \longrightarrow RS^\cdot \ (+HO_2^\cdot)$$

$$RS^\cdot + R'-CH=CH_2 \longrightarrow R'-\overset{\cdot}{C}H-CH_2-SR$$

$$R'-\overset{\cdot}{C}H-CH_2-SR + O_2 \longrightarrow R'-\underset{\underset{O_2}{|}}{CH}-CH_2-SR$$

$$R'-\underset{\underset{O_2}{|}}{CH}-CH_2-SR + RSH \longrightarrow R'-\underset{\underset{O_2H}{|}}{CH}-CH_2-SR + RS^\cdot$$

$$R'-\underset{\underset{O_2H}{|}}{CH}-CH_2-SR \longrightarrow R'-\underset{\underset{OH}{|}}{CH}-CH_2-SO-R$$

Dr. Lapinski and I synthesized our new beta-hydroxy sulfoxide from phenethyl mercaptan, styrene, and oxygen; ascribed its accidental formation to the thermal decomposition of phenethyl disulfide; published our results; and, for the moment, forgot about this unplanned detour in our research.

About six years and several unrelated publications later, I found myself at the Nuclear Center in Puerto Rico and interested in starting some relatively simple synthetic organic projects in tune with the character of the institution and its sponsor, the U.S. Atomic

Energy Commission. In view of the then active interest in radiation protective agents and the reports that certain amino sulfur compounds, as well as dimethyl sulfoxide, exhibited radiation protective action, my research assistant, now Dr. Juan Rigau, and I set out to assemble a variety of beta-hydroxy sulfoxides by means of the "cooxidation reaction" stored away in my memory bank. Among others, we looked at the reaction of indene, molecular oxygen, and aromatic thiols.

Reports in the literature claimed the formation of the trans addition products (two isomeric trans-2-phenylsulfinyl-1-indanols that differed only in the configuration at the sulfur atom), but the careful separation technique employed by Juan Rigau demonstrated the presence of considerable amounts of one of the cis isomers. This observation contradicted the generally accepted idea that the homolytic addition of thiols to olefins was trans-stereospecific, and it had a bearing on the controversial problem of the existence of "bridged radicals."

Two M.S. and Ph.D. theses and a couple of publications later, we realized that the steroselectivity of the "cooxidation reaction" depended on the nature of the substituents in the aromatic thiols and the temperature at which the reaction was carried out. Especially curious was the now well-substantiated observation that the cis addition tended to increase at both relatively low and high reaction temperatures ($-25°$ and $+65°C$, respectively). Thus, nearly twenty years after the accidental encounter with the cooxidation reaction of thiols and olefins, we are about to publish our conclusions with regard to its reaction mechanism. We now believe that there exist two different mechanistic paths for this "simple reaction"—a radical chain reaction and a homolytic nonchain reaction—and that both mechanisms can give rise to the originally unexpected cis addition products.

The moral of this story is that one never can tell where an accidental observation in the research laboratory will lead a chemist if circumstances allow one to follow a trail staked out by curiosity.

30 Synthetic Polymers

Man-made polymers are numerous and have varying properties. Their large molecular weights make purification of products difficult, and progress in the field was slow until it was realized that very pure starting materials must be used. Efforts to replace the naturally occurring polymers (rubber, cotton, wool, and silk) with synthetic polymers of superior properties have been partially successful. On the other hand, many synthetic substances, such as polyethylene, the epoxies, and Teflon, have found use because they have properties that none of the natural substances possess. The synthetic polymer industry today is the major sector of industrial organic chemistry in volume and value of products.

30.1 Introduction   970
  1. Natural and synthetic polymers; early attempts at synthesis.   2. Definitions.
  3. Effects of cross-linking: diamond.

30.2 Structure and properties   972
  1. Stereochemistry: four classes of thermoplastics.   2. Temperature effects; molding.   3. Plasticizers or copolymerization?

30.3 Vinyl and diene polymers   974
  1. Introduction.   2. Radical polymerization: initiation and initiators; propagation.   3. Chain length; control; chain transfer reagents.
  4. Effects of substituents.   5. Cationic polymerization: butyl rubber.
  6. Anionic polymerization.   7. Ziegler-Natta catalysts: coordination polymerization.   8. Examples.

30.4 Condensation polymers   979
  1. Definitions.   2. Polyamides: the nylons.   3. Polyesters.   4. Polyethers; epoxy polymers.   5. Phenol-formaldehyde resins.   6. Polyurethanes.
  7. Formaldehyde-urea and formaldehyde-melamine resins.

30.5 Variations   990
  1. Ion exchange resins.   2. Block and graft polymers.   3. Ladder polymers.
  4. Polysulfides; polysiloxanes.

30.6 Summary   992

970   Chapter thirty

30.7   Problems   992

30.8   Bibliography   993

30.9   John E. Leffler: Why peroxides? It worked!   933a

30.1   Introduction

Natural and synthetic polymers; early attempts at synthesis. Many different polymeric substances have been devised in nature for specific uses. We already have seen the naturally occurring polymers cellulose, rubber, starch, lignin, proteins, and nucleic acids, to mention only a few. Man has developed methods for modifying a few of these polymers to improve certain properties. The vulcanization of rubber and the manufacture of rayon are examples. These modified polymers are not our major interest: most of our attention will be given to the synthetic polymers. Polymers may be described as macromolecules.

Interest in the preparation of synthetic polymers probably arose in the first place as a synthetic challenge as soon as it became apparent that rubber was a polyisoprene. Numerous attempts to polymerize dienes were made, but with limited success. Germany was able to produce a low-quality synthetic rubber from dimethylbutadiene during World War I. Progress during the 1920's and 1930's was such that the synthetic GRS rubber of World War II is still a major item of the tire industry.

The production of rigid polymers began with Bakelite in 1909. Bakelite was the trade name for a polymer produced by the reaction of formaldehyde with phenol. The first successful synthetic polymer for the production of fibers was nylon, introduced in 1938. Today, synthetic polymers are replacing metals and other inorganic materials for many structural and mechanical applications. Other uses, for which no other substance will do, have been found for other synthetic polymers: for example, the nonsticking surfaces of polytetrafluoroethylene (Teflon).

Although we are still unable to duplicate the marvelous ability of nature to synthesize polymers, methods have been developed that have allowed the preparation of synthetic polymers that rival the natural ones, as well as some that have properties unknown to nature. However, a polymer with these unnatural characteristics is likely to be nonbiodegradable. The result is a serious disposal problem to which a solution is urgently needed.

Definitions. We may think of mer (Greek) as meaning unit. Thus, a single unit is a monomer, a multiunit structure is a polymer, and dimer, trimer, tetramer, and so on have obvious meanings. An oligomer is a substance smaller than a polymer: the term does not denote a specific number of units.

A homopolymer is made up of many identical monomers. A copolymer is made up of two or more different kinds of monomeric units. Polystyrene is an example of a homo-

$$-CHCH_2CHCH_2CHCH_2CHCH_2CHCH_2- \qquad \text{polystyrene}$$
$$\phi \quad \phi \quad \phi \quad \phi \quad \phi$$

$$\diagdown \diagup (CH_2CH=CHCH_2)_6 \diagdown \diagup (CH_2CH=CHCH_2)_6 \diagdown \qquad \text{GRS rubber}$$
$$CHCH_2 \qquad\qquad CHCH_2$$
$$\phi \qquad\qquad\qquad \phi$$

polymer, and GRS rubber (government rubber, styrene, also known as SBR rubber) is an example of a copolymer (one unit of styrene to an average of six units of butadiene).

We often hear the word plastic used as a synonym for a synthetic polymer. Strictly speaking, the word is a contraction of thermoplastic, which refers to the class of synthetic polymers that soften, become pliable, and ultimately liquefy (or decompose) as the temperature is raised. The other class of synthetic polymers is called thermosetting. These polymers do not soften, but become more polymeric with increase in temperature, do not melt, and eventually decompose. A plasticizer is a substance that is added to a polymer to modify the properties of the polymer: to increase softness, cause a pleasant feel to the touch, or increase flexibility (or decrease brittleness). Resin refers to amorphous substances, either thermoplastic or thermosetting, that are used as components of varnishes and adhesives and of mixtures used for additional polymerization in molding, vulcanization, and the like.

The uses of synthetic polymers may be classified broadly as fibers, sheets, elastics, and rigid formed articles.

Effects of cross-linking: diamond.   As we saw in Section 26.8, raw rubber is a thermoplastic consisting of long polymeric chains of isoprene. Because of the *cis* or Z double bond structure, it tends to be noncrystalline. The chains prefer to entangle than to be located in a regular repeating array. We saw that cross-linking of the chains by sulfur gave greater dimensional stability.

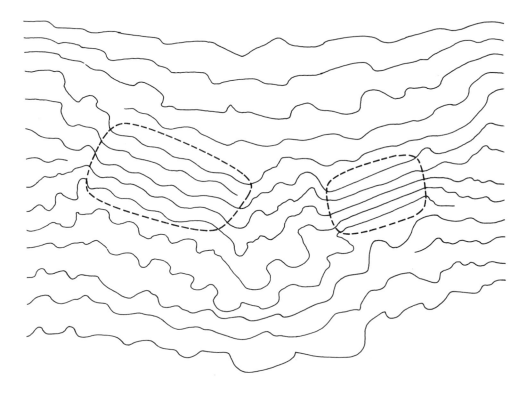

Figure 30.1   Schematic two-dimensional representation of two crystalline regions in an otherwise amorphous polymer with random arrangement of chains

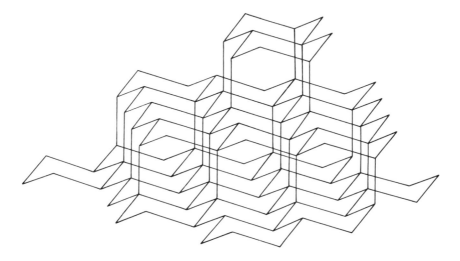

Figure 30.2   Diamond: partial structure shows only a few completed rings

As we saw in Chapter 21 when we examined protein structures, chains may assume a variety of conformations, especially if hydrogen bonding is possible. If cross-linking is not present, as in polyethylene, nonbonded forces are present and operate to form some regions in the polymer that may be considered crystalline (Figure 30.1).

If cross-linking is present, the crystalline regions increase in size, the polymer becomes more rigid and more resistant to temperature effects, and solubility decreases. The ultimate in cross-linking is reached in diamond, in which each atom of carbon is bonded to four other carbon atoms to form a completely crystalline array of tetrahedrally arranged atoms. Diamond is the hardest of all substances. It will scratch every other substance. It is relatively brittle, despite its rigidity, and is subject to cleavage by impact. Synthetic diamonds may be considered the ultimate in polymerization: each diamond is a single polymeric molecule (Fig. 30.2).

30.2   Structure and properties

Stereochemistry: four classes of thermoplastics.   The properties of a linear polymer depend upon the relative proportions of amorphous and crystalline regions. The interchain forces vary both in type (pole-dipole, dipole-dipole, hydrogen bonding), and with conformations as dictated by the monomers from which the polymer is formed.

A special terminology has been developed for use with polymers derived from 1-alkenes, as shown for polypropylene. In the extended chain, if the methyl groups occur on only one side of the chain, the polymer is described as isotactic. Isotactic polypropylene is so crystalline that the substance is used to make fibers (filaments) useful in the manufacture of carpets for outdoor use. A random arrangement of the methyl groups in the polymer chain is called atactic. Atactic polypropylene is relatively soft and rubbery. If the methyl groups alternate in position from one side of the chain to the other, the polymer chain is described as syndiotactic. A syndiotactic chain is intermediate in crystallinity and in properties.

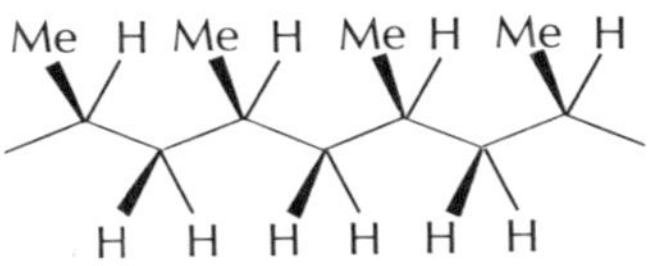

isotactic

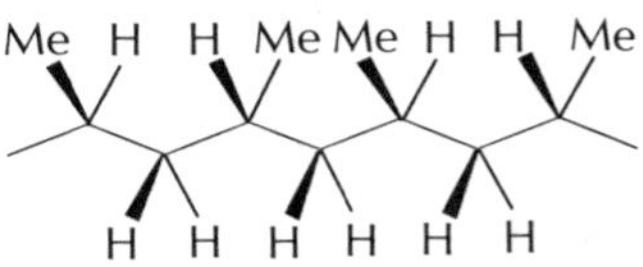

syndiotactic

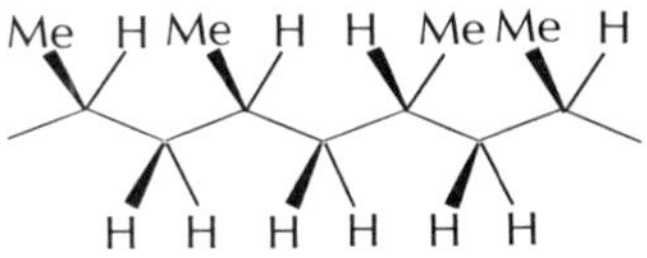

atactic

Elastic polymers usually have substituents located at random along a chain. The substituents increase the difficulty of formation of crystalline regions, and the substance is amorphous. As we saw in Section 26.8, stretching such a substance causes the chains to align and the substance to become more crystalline. The crystalline regions are not stable (because of the bumps on the chains), and the polymer reverts to the original amorphous state when the stretching force is released.

Thermoplastics, then, may be divided into four classes:

1. Amorphous
   Tensile strength is low; plastic flow is present.
2. Crystalline, unoriented
   Crystallites (crystalline regions) are randomly oriented; melting temperature is relatively sharp.
3. Crystalline, oriented
   A crystalline polymer which, as a fiber or sheet, has been subjected to cold drawing (stretching), which causes the crystallites to be aligned; strong interchain forces preserve the alignment; tensile strength is high.
4. Elastomers
   Crystalline tendencies obstructed so that the normal, unstretched state is amorphous; stretching induces crystalline regions which, however, are not stable enough to remain crystalline when the stretching force is released; plastic flow is low.

Temperature effects; molding. The effect of increasing temperature varies with the class of thermoplastic involved. An amorphous polymer may be thought of as resembling glass. Increasing the temperature has little effect until the onset of the glass temperature, above which the viscosity decreases rapidly (a glass may be worked above the glass temperature even though it is not completely molten) until at the melting temperature the polymer is fluid. The range between the two temperatures decreases as the crystallinity increases. Thermoplastics are molded at temperatures above the melting temperature. The shape of the mold is retained by the cooled molded plastic article. Injection molding is a method that allows rapid production of small plastic articles.

Chain branching of a polymer chain tends to increase the amorphous character of the polymer.

Cross-linking, as we have seen, increases crystallinity, decreases elasticity, increases the melting temperature, and decreases glass-like behavior. If cross-linking is extensive, the melting temperature may exceed 400°, at which point decomposition usually sets in (thermal rupture of bonds).

Plasticizers or copolymerization? Some polymers, such as polyvinyl chloride ($-(CH_2CHCl)_n-$) tend to be excessively brittle. This tendency has been combatted by the introduction of plasticizers into the molten polymer before molding or extrusion as sheets. Various liquid esters of phthalic acid (dibutyl and dioctyl phthalate) and of phosphoric acid (tricresyl phosphate) are used to decrease the "glassiness." These esters are relatively odorless, are high boiling, and have low vapor pressures at room temperature. Nevertheless, these plasticizers slowly evaporate and articles or sheets containing them inevitably lose flexibility and become brittle with age.

A better solution of the problem is use of a copolymer. Copolymerization of vinyl chloride with vinyl acetate gives a softer, more flexible product than polyvinyl chloride, and the need for a plasticizer is reduced. However, it is not always possible to find a copolymer that will produce the needed effect without deleterious influences upon other properties of the polymer.

30.3 Vinyl and diene polymers

Introduction. Vinyl compounds and dienes may be caused to polymerize by four different mechanisms: radical, cation, anion, and coordination. Radical polymerization was the first to be employed on a large scale and is still the most popular. The other three types of polymerization all are rather specialized in usefulness but are valuable nevertheless.

Radical polymerization: initiation and initiators; propagation. Radical polymerization of vinyl and diene monomers, as well as copolymerizations, may be carried out in bulk, in solution, or in the form of aqueous suspensions or emulsions. An aqueous suspension polymerization may be carried out so as to form the polymer into small beads. Emulsion polymerization is used to prepare a latex of the polymer. Free radical polymers are atactic.

The first step is the formation of free radicals, which initiate the polymerization itself (Eqs. 30.1) as shown for a vinyl monomer.

$$Y-Y \xrightarrow[\text{or } h\nu]{\text{heat}} 2Y\cdot \qquad \text{initiation}$$

$$R\dot{C}H=CH_2 + Y\cdot \longrightarrow R\dot{C}H-CH_2Y$$

$$R\dot{C}H=CH_2 + R\dot{C}H-CH_2Y \longrightarrow R\dot{C}H-CH_2-\overset{R}{\underset{|}{C}H}-CH_2Y \qquad (30.1)$$

$$RCH=CH_2 + R\overset{R}{\underset{|}{\dot{C}}}HCH_2CHCH_2Y \longrightarrow R\dot{C}H-CH_2-\overset{R}{\underset{|}{C}}HCH_2\overset{R}{\underset{|}{C}}HCH_2Y$$

propagation, etc.

Initiators that have been used are peroxides, azo compounds, peracids, and ketones. Heat alone will cause a reactive monomer such as styrene to polymerize. We may vary the rate of decomposition of the initiators by changing the temperature (or the intensity of the light, if photodecomposition is used). If the rate of decomposition of the initiator is slow, the number of growing chains is relatively small, and the chains are free to grow to maximum length. If the rate of decomposition of the initiator is fast, the number of growing chains is large, competition between the chains for the dwindling number of monomer molecules is intense, and the chains are shorter in length. Consequently, we should use a small concentration of initiator.

30.3 Vinyl and diene polymers

Although maximum chain length is usually a sought-after goal, there are instances in which shorter chains give rise to desirable properties. Flexibility, for example, is increased by a more amorphous polymer, which is favored by shorter rather than longer chains. A few examples of initiators are shown in Equations 30.2. Thus, Y in Equations 30.1 may be

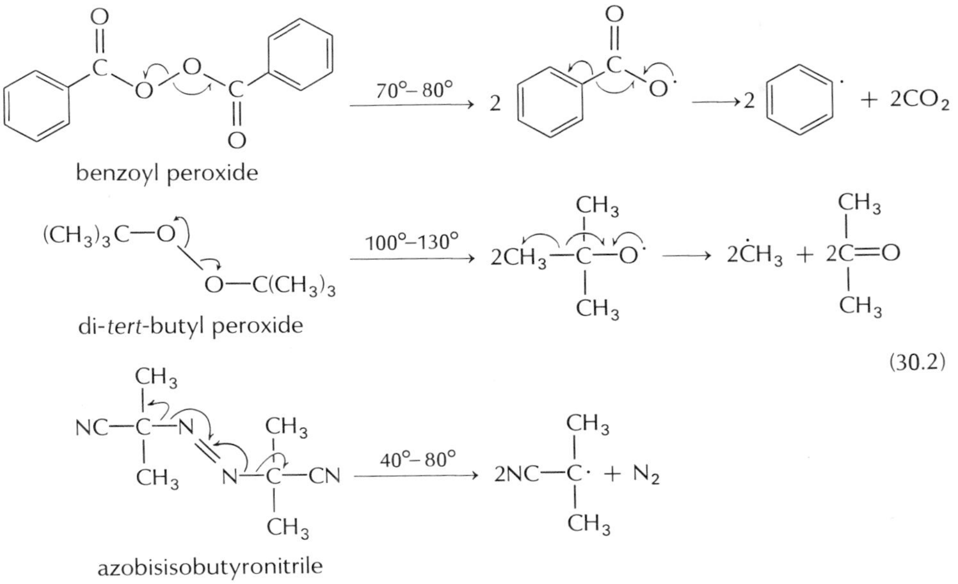

benzoyl peroxide $\xrightarrow{70°-80°}$ 2 PhC(O)O· $\longrightarrow$ 2 Ph· + 2CO$_2$

di-*tert*-butyl peroxide $(CH_3)_3C-O-O-C(CH_3)_3$ $\xrightarrow{100°-130°}$ 2 $(CH_3)_3C-O·$ $\longrightarrow$ 2 ĊH$_3$ + 2 (CH$_3$)$_2$C=O

azobisisobutyronitrile $(CH_3)_2C(CN)-N=N-C(CN)(CH_3)_2$ $\xrightarrow{40°-80°}$ 2 NC-Ċ(CH$_3$)$_2$ + N$_2$

(30.2)

phenyl (from benzoyl peroxide), methyl (from di-*tert*-butyl peroxide), or 2-cyanopropyl (from azobisisobutyronitrile).

Propagation of a chain is an exothermic reaction. The loss of a π bond requires 63 kcal; the formation of a σ bond releases 83 kcal. Thus, the addition of each vinyl monomer to a growing chain results in the release of 20 kcal of enthalpy. Very fast polymerization is avoided because the cooling requirement is excessive.

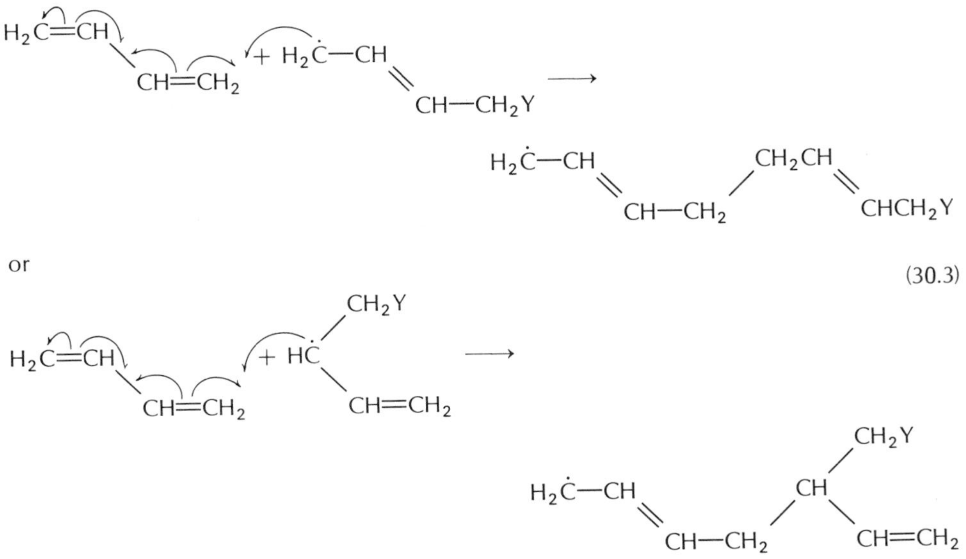

$$H_2C=CH-CH=CH_2 \;+\; H_2\dot{C}-CH=CH-CH_2Y \longrightarrow H_2\dot{C}-CH=CH-CH_2-CH_2CH=CHCH_2Y$$

or

$$H_2C=CH-CH=CH_2 \;+\; H\dot{C}(CH_2Y)-CH=CH_2 \longrightarrow H_2\dot{C}-CH=CH-CH_2-CH(CH_2Y)-CH=CH_2$$

(30.3)

Chain length; control; chain transfer reagents. Vinyl polymerization (Eq. 20.1) gives rise to very long chains. For example, the vinyl chloride polymer polyvinyl chloride has mw of about 0.6 to 1.5 × 10^6. Division by the mw of vinyl chloride, 62.5, gives 9600 to 24,000, the number of monomeric units per chain. For comparison, cellulose contains 3000 to 40,000 glucose units. Diene polymers do not usually have such long chains because of the possibility of chain branching (Eq. 30.3) and increased probability of chain-termination reactions.

As we saw in our discussion of free-radical halogenation of alkanes (Sec. 12.5), termination reactions influence the length of chains in chain reactions. In halogenation, the reaction chain can be measured in terms of moles of reactant consumed or moles of product formed per atom of the initiating halogen. In polymerization, we measure the chain length in terms of moles of monomer consumed per mole of initiating radical. The principal termination reactions in polymerization differ somewhat from those of free-radical halogenation. Termination occurs by coupling, disproportionation, and chain transfer (Eqs. 30.4).

$$
\begin{array}{c}
\text{Y}(\text{CH}_2\text{CH})_m\text{CH}_2\dot{\text{C}}\text{H} + \text{H}\dot{\text{C}}\text{CH}_2(\text{CHCH}_2)_n\text{Y} \longrightarrow \quad\text{coupling} \\
\quad\;\; | \qquad\qquad\;\; | \qquad\; | \qquad\; | \\
\quad\;\; \text{R} \qquad\qquad\;\; \text{R} \qquad\; \text{R} \qquad\; \text{R}
\end{array}
$$

$$\text{Y}(\text{CH}_2\text{CH})_m\text{CH}_2\text{CH}-\text{CHCH}_2(\text{CHCH}_2)_n\text{Y}$$

$$
\begin{array}{c}
\quad\;\;\; \text{H} \\
\quad\;\;\; | \\
\text{Y}(\text{CH}_2\text{CH})_m\dot{\text{C}}\text{H}-\text{CH} + \text{H}\dot{\text{C}}\text{CH}_2(\text{CHCH}_2)_n\text{Y} \longrightarrow \quad\text{disproportionation}
\end{array}
$$

$$\text{Y}(\text{CH}_2\text{CH})_m\text{CH}=\text{CH} + \text{H}_2\text{CCH}_2(\text{CHCH}_2)_n\text{Y}$$

(30.4)

$$\text{Y}(\text{CH}_2\text{CH})_m\text{CH}_2\dot{\text{C}}\text{H} + \text{H}-\text{SR}' \longrightarrow \text{Y}(\text{CH}_2\text{CH})_n\text{CH}_2\text{CH}_2 + \dot{\text{S}}\text{R}'$$

chain transfer $\qquad\qquad\qquad$ $\downarrow \text{H}_2\text{C}=\text{CHR}$

etc. $\longleftarrow$ R'SCH$_2$CHCH$_2\dot{\text{C}}$H $\xleftarrow{\text{H}_2\text{C}=\text{CHR}}$ R'SCH$_2\dot{\text{C}}$H

Coupling and disproportionation each lead to termination of two chains. A coupling product contains a chain of $m + n + 2$ units and has two Y groups from the initiator. The disproportionation products contain chains of $m + 1$ and $n + 1$ units, each of which has one Y group. One is an alkene; the other, an alkane. Disproportionation is more likely to occur during a diene polymerization because of the allylic type of activation in loss of hydrogen (Eq. 30.5).

Chain transfer was illustrated by the use of a mercaptan. (Dodecyl mercaptan is a popular chain-transfer reagent.) An effective chain-transfer reagent competes well against remaining monomer for reaction with a growing chain. The loss of hydrogen to the growing chain terminates that chain. However, the new radical RS· initiates a new chain: hence the term chain transfer.

30.3 Vinyl and diene polymers

$$\text{Y(CH}_2\text{CH=CHCH}_2)_m\text{CH}-\text{CH=CH}_2 \quad + \quad \text{H}_2\text{C=CH}-\text{CH=CH}_2 \longrightarrow \text{CHCH}_2(\text{CH}_2\text{CH=CHCH}_2)_n\text{Y} \quad + \quad \text{H}_2\dot{\text{C}}\text{CH=CH}_2$$

$$\downarrow$$

$$\text{Y(CH}_2\text{CH=CHCH}_2)_m\text{CH=CH-CH=CH}_2 \quad + \quad \text{CHCH}_2(\text{CH}_2\text{CH=CHCH}_2)_n\text{Y} \quad \text{with} \quad \text{H}_2\text{CCH} \quad \text{H}$$

(30.5)

Chain transfer occurs at times internally, during polymerization. Propagation, as we have seen, proceeds by addition to give the more stable radical, in the order tertiary > secondary > primary. Thus, only the polymerization of ethylene proceeds by way of primary radicals. These primary radicals can remove a secondary hydrogen from an already formed chain or from a growing chain to give a secondary radical. The chain is terminated, but another is initiated by the secondary radical and grows as a branch. Polyethylene, made by free-radical polymerization (200°, 1500 atm, O_2 initiator), has considerable branching of chains and is about 50% crystalline, 50% amorphous.

If a substance such as hydroquinone is present, the chain is terminated, but the new radical is a poor initiator. The effect is to prevent or inhibit the polymerization until all of the hydroquinone has been consumed. Such substances are known as <u>inhibitors.</u> Small amounts of an inhibitor are added deliberately to reactive monomers like styrene to prevent inadvertent polymerization (that is, to prolong shelf life or storage time). Conversely, most monomers must be pure to produce satisfactory polymers. Impurities generally act as inhibitors (no polymer is formed) or as chain-transfer reagents (polymer has low mw).

Effects of substituents. The success of vinyl (and diene) polymerization depends upon our ability to modify monomers so as to synthesize polymers with useful properties. As we have seen, ethylene is the most difficult of all alkenes to polymerize. Any substituent on ethylene gives rise to a secondary radical and enhances the polymerization. Thus, $H_2C=CHCH_3$ (or $H_2C=CHR$) and $H_2C=CH\phi$ (or $H_2C=CHAr$) polymerize with ease. If the substituent is one that causes additional stabilization of the secondary radical to be achieved (the benzylic radical, $H_3C\dot{C}H\phi$, or $-CH_2\dot{C}H\phi$), polymerization is still easier to bring about. Other examples are $H_2C=CHOR$, $H_2C=CHOCOCH_3$, $H_2C=CHCN$, $H_2C=CHCOOH$ or ester, and $H_2C=CHCl$. In general, however, disubstitution tends to disfavor polymerization, especially if the double bond is nonterminal, as in $CH_3CH=CHCH_3$, $CH_3CH=CHCOOH$, $\phi CH=CH\phi$, and so on. (There are many exceptions.) If the substituents are at one end of the double bond and are not too large (steric hindrance to addition), polymerization is still possible. Examples are $H_2C=CCl_2$ and $H_2C=C(CH_3)COOCH_3$. Trisubstitution of the double bond prevents polymerization unless the substituent is fluorine ($F_2C=CF_2$ to give <u>Teflon</u> and $F_2C=CFCF_3$ with $H_2C=CF_2$ to give <u>Viton</u>).

Cationic polymerization: butyl rubber. Cationic polymerization was mentioned in Section 13.2 (Eq. 13.25). If the alkene monomer is substituted so as to stabilize the intermediate cation, chain lengths of about 1000 units may be achieved (Eqs. 30.6). Isobutylene

$$H_2O + AlCl_3 \rightleftharpoons [HOAlCl_3]^{\ominus}H^{\oplus}$$

$$Me_2C=CH_2 + [HOAlCl_3]^{\ominus}H^{\oplus} \longrightarrow [Me_2\overset{\oplus}{C}-CH_3][HOAlCl_3]^{\ominus} \quad \text{initiation}$$

$$Me_2C=CH_2 + [Me_2\overset{\oplus}{C}-CH_3][HOAlCl_3]^{\ominus} \longrightarrow$$
$$[Me_2\overset{\oplus}{C}-CH_2-CMe_2-CH_3][HOAlCl_3]^{\ominus} \quad \text{propagation}$$

$$[Me_2\overset{\oplus}{C}CH_2(CMe_2CH_2)_nCMe_2CH_3][HOAlCl_3]^{\ominus} \xrightarrow{-H^{\oplus}}$$

$$\begin{array}{l} Me_2C=CH(CMe_2CH_2)_nCMe_2CH_3 + [HOAlCl_3]^{\ominus}H^{\oplus} \quad \text{termination} \\ \text{or} \quad +H_2O \quad (30.6) \\ \quad\quad\quad\quad\quad\quad\quad OH \\ \quad\quad\quad\quad\quad\quad\quad | \\ \longrightarrow Me_2CCH_2(CMe_2CH_2)_nCMe_2CH_3 + [HOAlCl_3]^{\ominus}H^{\oplus} \end{array}$$

$$[Me_2\overset{\oplus}{C}CH_2(CMe_2CH_2)_nCMe_2CH_3][HOAlCl_3]^{\ominus} + Me_2C=CH_2$$
$$\downarrow \text{chain transfer}$$
$$Me_2C=CH(CMe_2CH_2)_nCMe_2CH_3 + [Me_2\overset{\oplus}{C}-CH_3][HOAlCl_3]^{\ominus}$$

may be polymerized with $AlCl_3$ or BF_3 at $-100°$ in relatively dilute solution in MeCl. The low temperature is necessary to favor the formation of few and longer chains and to favor propagation over termination. The addition of about 2% of isoprene to the isobutylene gives a partially cross-linked copolymer called <u>butyl rubber.</u> Butyl rubber is relatively non-permeable to gases and is used in the manufacture of inner tubes, balloons, and so on.

Other substituents that give polymerizable monomers by cation formation are ϕ—, —OR, and —OCOR. On the other hand, substituents such as —CN and —COOR resist polymerization by this method.

Anionic polymerization. Anionic polymerization is possible for those monomers that form relatively stable anions. Substituents such as ϕ—, —CN, or —COOR stabilize anions, whereas —OR, —X, or —·R destabilize anions. Practically, anionic polymerization is used for diene polymerization. <u>Coral rubber</u> is a *cis*-polyisoprene with properties very similar to those of natural rubber (Eq. 30.7). The initial ion pair, formed at the surface of the finely dispersed lithium, remains on the surface to give the dianion (presumably). Subsequent addition (at either end of the dianion) also occurs on the surface of the metal. The growing stereospecific (all-*cis*) chain loops away off the surface, allowing more isoprene to have access to the anionic reactive sites, which cling to the surface. The reaction is carried out at 200° in a well-stirred hydrocarbon solvent with 1 part of lithium to about 1000 of isoprene.

Ziegler-Natta catalysts: coordination polymerization. Ziegler-Natta catalysis of polymerization is a heterogeneous process that, like the lithium metal catalyzed reaction, may proceed by a variation of an anion mechanism. The catalyst is prepared by combining one to three moles of Et_3Al or iBu_3Al with $TiCl_3$, $TiCl_4$, or vanadium salts. Presumably the stereospecific polymers (isotactic polypropylene and *cis*-polyisoprene for example) are formed on the surface of the catalyst by proper coordination of the monomer with the growing chain. The process might be better known as <u>coordination polymerization.</u>

$$\text{H}_2\text{C}=\text{C}\overset{\text{CH}_3}{\underset{\text{CH}=\text{CH}_2}{\Big\langle}} + \text{Li} \longrightarrow \text{H}_2\text{C}=\text{C}\overset{\text{CH}_3}{\underset{\dot{\text{C}}\text{H}-\overset{\ominus}{\text{C}}\text{H}_2}{\Big\langle}} \text{Li}^{\oplus}$$

$$\updownarrow$$

$$\underset{\overset{\ominus}{\text{H}_2\text{C}}\quad 2\text{Li}^{\oplus}\quad \overset{\ominus}{\text{C}}\text{H}_2}{\overset{\text{H}_3\text{C}}{\Big\langle}\text{C}=\text{CH}} \quad \overset{\text{Li}}{\longleftarrow} \quad \text{H}_2\dot{\text{C}}-\text{C}\overset{\text{CH}_3}{\underset{\text{CH}-\overset{\ominus}{\text{C}}\text{H}_2}{\Big\langle}} \text{Li}^{\oplus}$$

(30.7)

$$\Big\downarrow \underset{\text{H}_2\text{C}}{\overset{\text{H}_3\text{C}}{\Big\langle}}\text{C}-\text{CH}\overset{}{\underset{\text{CH}_2}{\Big\langle}}$$

$$\underset{\overset{\ominus}{\text{H}_2\text{C}}\quad \text{Li}^{\oplus}}{\overset{\text{H}_3\text{C}}{\Big\langle}\text{C}=\text{CH}}\overset{}{\underset{\text{CH}_2-\text{H}_2\text{C}}{\Big\langle}} \underset{\text{Li}^{\oplus}}{\overset{\text{H}_3\text{C}}{\Big\langle}\text{C}=\text{CH}}\overset{}{\underset{\overset{\ominus}{\text{C}}\text{H}_2}{\Big\langle}} \xrightarrow{\text{etc.}} \Big[\underset{\text{H}_2\text{C}}{\overset{\text{H}_3\text{C}}{\Big\langle}}\text{C}=\text{CH}\overset{}{\underset{\text{CH}_2}{\Big\langle}}\Big]_n$$

Examples. In Table 30.1 are listed some typical vinyl and diene polymers with the monomers used to make them, method of polymerization, and some physical properties and uses.

A few other copolymers should be mentioned. SBR rubber is made in an aqueous soapy emulsion of butadiene and styrene (3 to 1 by weight, 5.8 to 1 by moles) to which enough methanol has been added to allow operation at $-10°$ (cold rubber). EPT rubber is a copolymer of $\text{H}_2\text{C}=\text{CH}_2$ and $\text{CH}_3\text{CH}=\text{CH}_2$ with a little diene present to provide cross-linking: hence, it is a terpolymer. A Ziegler-Natta catalyst is used. EPT rubbers are also known as EPDM (ethylene, propylene, diene monomers) rubbers.

30.4 Condensation polymers

Definitions. Under this heading we include all polymers made from nonvinyl or nondiene monomers. We shall find several instances of addition polymers as well as condensation polymers. Let us define a condensation polymer as being formed from monomers by loss of simple substances, usually water. For example, a polyether can be made from ethylene glycol (Eqs. 30.8). Practically, a longer chain can be made by addition (cationic or

$$n\text{HOCH}_2\text{CH}_2\text{OH} \xrightarrow{\text{H}^{\oplus}} \text{HO}(\text{CH}_2\text{CH}_2\text{O})_n\text{H} + (n-1)\text{H}_2\text{O}$$
condensation

(30.8)

$$n\text{H}_2\text{C}\overset{\text{O}}{\overset{}{\diagup\diagdown}}\text{CH}_2 \xrightarrow[\text{trace of } \text{H}_2\text{O}]{\text{BF}_3} \text{HO}(\text{CH}_2\text{CH}_2\text{O})_n\text{H}$$
addition

Table 30.1 Representative synthetic thermoplastic vinyl and diene polymers[a]

Monomer(s)	Formula	Type of polymerization	Physical type	T_g °C	T_m °C	Trade names	Uses
ethylene	$CH_2\!=\!CH_2$	radical (high pressure Ziegler-Natta	semicrystalline	≪ 0	110	polyethylene	film, containers, piping, etc.
vinyl chloride	$CH_2\!=\!CHCl$	radical	crystalline atactic, semi-crystalline	−120 80	130 180	polyvinyl chloride, Geon	film, insulation, piping, etc.
vinyl fluoride	$CH_2\!=\!CHF$	radical	atactic, semi-crystalline	45		Tedlar	coatings
vinyl chloride and vinylidene chloride	$CH_2\!=\!CHCl$ $CH_2\!=\!CCl_2$	radical	crystalline	variable		saran	tubing, fibers, film
chlorotrifluoroethylene	$CF_2\!=\!CFCl$	radical	atactic, semi-crystalline	≪ 0	210	Kel-F	gaskets, insulation
tetrafluoroethylene	$CF_2\!=\!CF_2$	radical	crystalline	≪ −100	330	Teflon	gaskets, valves, insulation, filter felts, coatings
propylene	$CH_2\!=\!CHCH_3$	Ziegler-Natta	isotactic, crystalline	−20	175		fibers, molded articles
hexafluoropropylene and vinylidene fluoride	$CF_2\!=\!CFCF_3$ $CH_2\!=\!CF_2$	radical	amorphous	−23		Viton	rubber articles[b]
isobutylene	$CH_2\!=\!C(CH_3)_2$	cationic	amorphous	−70		Vistanex, Oppanol butyl rubber	pressure-sensitive adhesives inner tubes
isobutylene and isoprene	$CH_2\!=\!C(CH_3)_2$ $CH_2\!=\!C(CH_3)CH\!=\!CH_2$	cationic	amorphous				
chloroprene	$CH_2\!=\!C(Cl)CH\!=\!CH_2$	radical	amorphous	−40		neoprene	rubber articles[b]
isoprene	$CH_2\!=\!C(CH_3)CH\!=\!CH_2$	Ziegler-Natta or Li	amorphous (cis-1,4)	−70	28	natural rubber, Ameripol, Coral rubber	rubber articles

monomer	structure	method	properties	trade name	temp	uses
styrene	$CH_2=CHC_6H_5$	radical	atactic, semi-crystalline	Styron, Lustron	85 <200	molded articles, foam
vinyl acetate	$CH_2=CHO_2CCH_3$	Ziegler-Natta radical	isotactic amorphous		100 40	230 adhesives
vinyl alcohol	$(CH_2=CHOH)^c$	by hydrolysis of polyvinyl acetate	crystalline	polyvinyl acetate polyvinyl alcohol	dec.	water-soluble adhesives, paper sizing
vinyl butyral	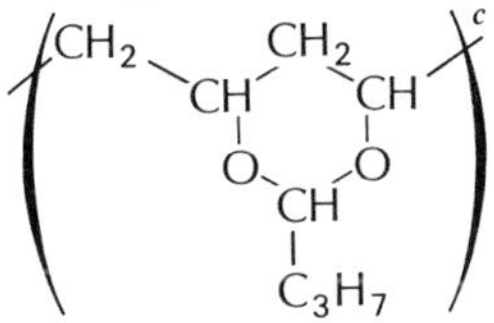 (cyclic acetal structure with C_3H_7)	by reaction of polyvinyl alcohol and butyraldehyde	amorphous	polyvinyl butyral		safety-glass laminate
acrylonitrile	$CH_2=CHCN$	radical	crystalline	Orlon, Acrilan	100 >200	fiber
methyl methacrylate	$CH_2=C(CH_3)CO_2CH_3$	radical	atactic amorphous	Lucite, Plexiglas	105	coatings, molded articles
		anionic	isotactic crystalline		115 200	
		anionic	syndiotactic, crystalline		45 160	
N-vinylpyrrolidone	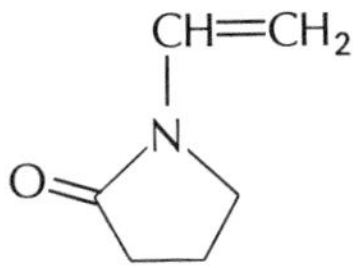 $CH=CH_2$ attached to N of pyrrolidone ring	radical	amorphous	P.V.P.		colloidal solution as blood plasma substitute, hair sprays

[a] From J. D. Roberts and M. C. Caserio, *Basic Principles of Organic Chemistry*, copyright © 1964, W. A. Benjamin, Inc., Menlo Park, California.
[b] Used particularly where ozone resistance and resistance to solvents is needed.
[c] These monomers are not the starting materials used to make the polymers, which are actually synthesized from polyvinyl acetate.

anionic) polymerization of ethylene oxide. Nevertheless, we shall classify such polymers as condensation polymers whether they are produced by condensation or by addition.

Polyamides: the nylons. The first synthetic polymer suitable for fiber formation was nylon 66, or polyhexamethylene adipamide, invented in 1935 and made by heating an equimolar mixture of hexamethylenediamine with adipic acid (Eq. 30.9). The mixture is

$$nH_2N(CH_2)_6NH_2 + nHOOC(CH_2)_4COOH \xrightarrow{280°} H{-}[HN(CH_2)_6NHC(=O)(CH_2)_4C(=O){-}]_n{-}OH + (n-1)H_2O \quad (30.9)$$

prepared by crystallization of the 1:1 salt of the diamine with adipic acid. The reaction is initially carried out under pressure to prevent loss of the monomers and is completed under reduced pressure to remove the last traces of water. The molten polymer is extruded as a ribbon, which, after cooling, is sliced up into pellets. The substance has a relatively high mp of 250°, and it may be molded into products of precise dimensions (pulleys, gears, etc.). Fibers are made by extrusion of molten material through tiny holes into a downward flowing stream of air to form filaments (melt-spinning). When they are cool, drawing the filaments out to three or four times the original length (cold drawing) increases the tensile strength by alignment of the chains so that interchain hydrogen bond formation and increased crystallinity may occur.

The deserved popularity of nylon 66 led to many attempts to find a better product or a more economical process. No better product has been found, but nylon 6, from ε-aminocaprolactam is cheaper (Eqs. 30.10) and is used as tire cord, carpet filament, and molded articles.

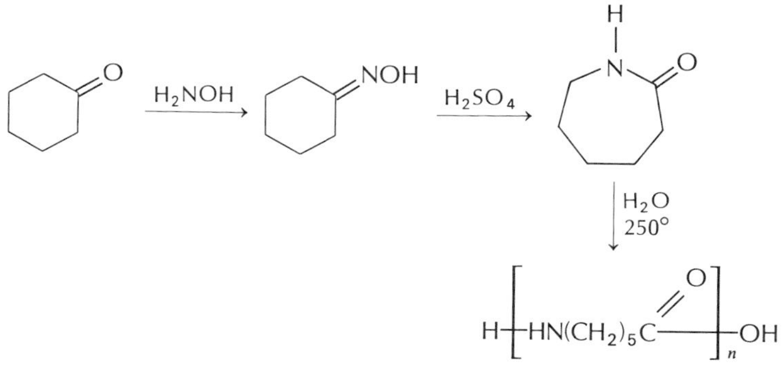

$$\text{cyclohexanone} \xrightarrow{H_2NOH} \text{cyclohexanone oxime} \xrightarrow{H_2SO_4} \text{caprolactam} \xrightarrow[250°]{H_2O} H{-}[HN(CH_2)_5C(=O){-}]_n{-}OH \quad (30.10)$$

Difficulties in the production of nylon 4 seem to have been overcome by the use of CO_2 as the initiator (Eq. 30.11). Nylon 4, with more polar groups per unit of length along the chain, is more hydrophilic than nylons 66 or 6 and approaches cotton in hydrophilicity. It has been reported to be more comfortable when fabricated into clothing than are other synthetics on hot, humid days and to be less susceptible to static electricity build-up on dry days.

Related to the nylons are polyamides, with a basis of aromatic rings instead of aliphatic open chains. Nomex, a thermally resistant substance, is the polyamide made from

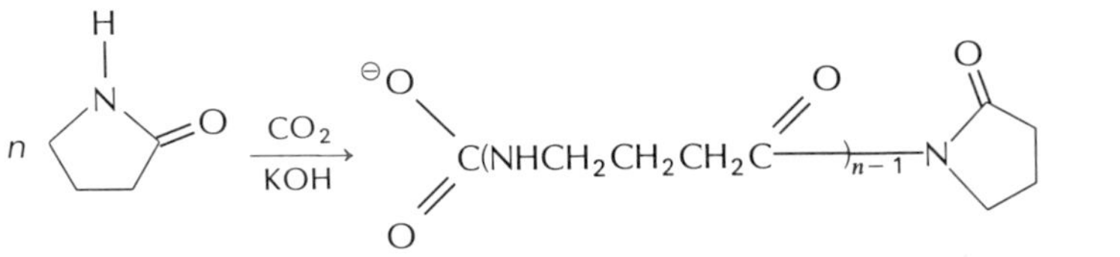

$$\text{n} \underset{\text{H}}{\overset{\text{H}}{\underset{|}{N}}}\!\!\!\!\!\!\!\!\!\bigcirc\!\!=\!\!O \xrightarrow[\text{KOH}]{\text{CO}_2, {}^{\ominus}\text{O}} \overset{O}{\underset{\parallel}{C}}(\text{NHCH}_2\text{CH}_2\text{CH}_2\overset{O}{\underset{\parallel}{C}}\!)_{n-1}\!\text{N}\!\!\bigcirc\!\!=\!\!O \qquad (30.11)$$

m-phenylene diamine and the double acid chloride of isophthalic acid (Eq. 30.12). Suits made of Nomex are flame resistant and are used by race drivers and others who may be subjected to fires. If the rings are linked by *para* instead of *meta* substitution, the mp is increased, and solubility is decreased to such an extent that spinning from a melt or from solution becomes extremely difficult (thermal decomposition sets in). Many variations of the aromatic polyamide types have been made and tested (rings other than benzene, such as heterocycles, fused systems, etc.).

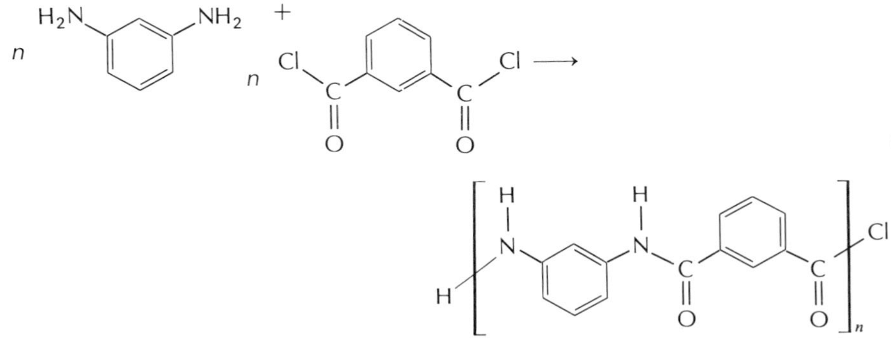

(30.12)

Polyesters. The other large class of compounds used primarily for fiber production is the polyester group of polymers. Polyethylene terephthalate (Dacron, Terylene, Fortrel, Mylar film), or PET, was patented in 1946 and has found many uses in competition with the nylons. Two methods of preparation are used (Eqs. 30.13). The first method may be thought

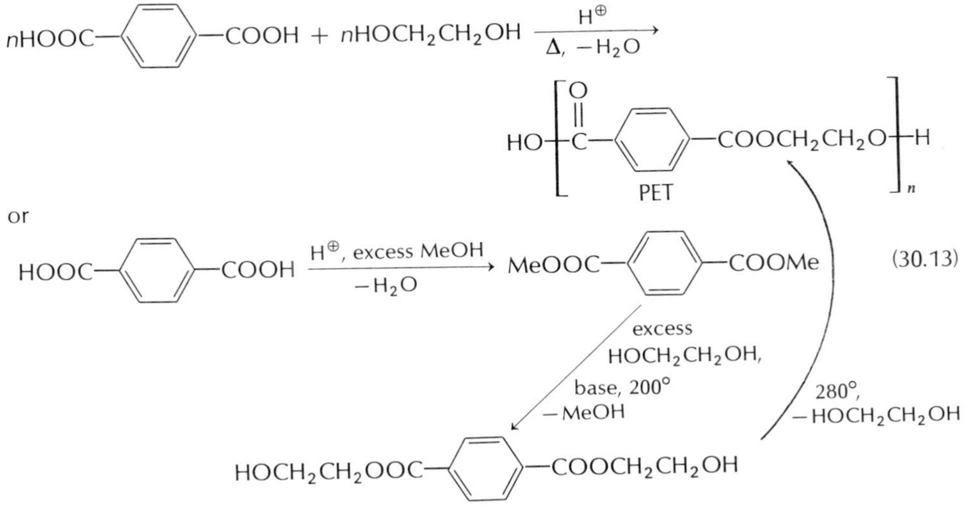

$$n\text{HOOC}\!-\!\!\bigcirc\!\!-\!\text{COOH} + n\text{HOCH}_2\text{CH}_2\text{OH} \xrightarrow[\Delta, -\text{H}_2\text{O}]{\text{H}^{\oplus}}$$

$$\text{HO}\!\!-\!\!\!\left[\overset{O}{\underset{\parallel}{C}}\!\!-\!\!\bigcirc\!\!-\!\text{COOCH}_2\text{CH}_2\text{O}\right]_n\!\!\!-\!\text{H}$$
PET

or

$$\text{HOOC}\!-\!\!\bigcirc\!\!-\!\text{COOH} \xrightarrow[-\text{H}_2\text{O}]{\text{H}^{\oplus}, \text{ excess MeOH}} \text{MeOOC}\!-\!\!\bigcirc\!\!-\!\text{COOMe} \qquad (30.13)$$

$$\xrightarrow[-\text{MeOH}]{\substack{\text{excess} \\ \text{HOCH}_2\text{CH}_2\text{OH}, \\ \text{base, 200}°}} \text{HOCH}_2\text{CH}_2\text{OOC}\!-\!\!\bigcirc\!\!-\!\text{COOCH}_2\text{CH}_2\text{OH} \xrightarrow[-\text{HOCH}_2\text{CH}_2\text{OH}]{280°}$$

of as a simple esterification driven to completion by removal of water by heat. The second method makes use of base-catalyzed ester interchange, in which methanol is removed from the mixture followed by removal of ethylene glycol by heat. The product has mp 270° and is subjected to melt spinning to produce fiber or to slit extrusion to produce film. Although many other combinations of a dialcohol with a diacid to give a linear polymer have been examined, the only other commercially successful polyester fiber is Kodel (Kodel is thought to be the polymer produced from 1,4-bis(hydroxymethyl)cyclohexane and terephthalic acid); however, the terephthalate ester of 1,4-butanediol shows promise of success.

The polycarbonate polymers (Lexan) obtained from bisphenol A (see Eq. 24.28) are useless for fiber manufacture. (The geminal methyl groups do not allow enough crystallinity to be developed.) These polymers may be molded to give glass-clear articles that have high impact strength (resistance to shattering) and high electrical resistance. Two processes for polymerization are in use (Eqs. 30.14), ester interchange and acid chloride.

$$\text{(30.14)}$$

Cross-linked polyesters were known earlier and arise by reaction of a diacid (or higher acid) with a triol (or higher) to have a mismatch of the number of functional groups. The reaction of phthalic acid with glycerol to give a glyptal resin illustrates the principle (Eq. 30.15). The esterification of the primary hydroxy groups of glycerol by the phthalic anhydride proceeds relatively rapidly to give the mono- and diesters. Thereafter, normal esterification proceeds at a much slower rate, especially so at the secondary hydroxy groups, to form a highly branched network. In practice, the initial heating is carried only to the point of formation of a solid, but still soluble, resin. After the resin is applied to a surface as a solution, the esterification is completed by baking to give a glossy, infusible, hard coating.

The general term for these substances is alkyd resins. All are thermosetting: that is, heat serves only to complete the polymeric three-dimensional network. Acids that have been used are maleic and succinic anhydrides, isophthalic acid, and the double anhydride of 1,2,4,5-benzenetetracarboxylic acid, among others. Polyalcohols other than glycerol that have been used are 2-hydroxymethyl-2-methyl-1,3-propanediol and pentaerythritol (Eq. 16.49).

Polyethers; epoxy polymers. Polyethers may be formed from several types of monomer. The simplest of the polyethers is polyoxymethylene, made by passage of formaldehyde into a heptane solution of $\phi_3 P$ (Eq. 30.16). Unless the end groups are modified, the polymer is unstable to heat and depolymerization (unzipping) will occur. The use of

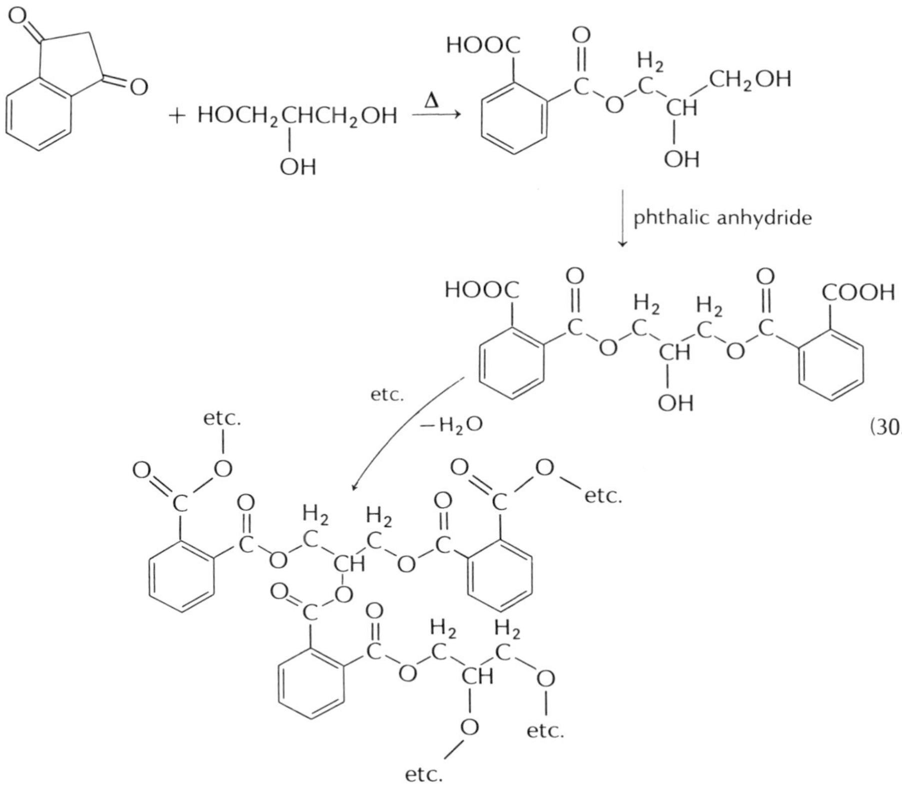

(30.15)

acetic anhydride and sodium acetate caps the end groups as the less easily removed acetate ester groups. The product is known as Delrin, a crystalline, tough substance of low solubility except in phenol (surprising in view of the large number of ether links). Polyoxymethylene also may be classified as a polyacetal.

$$\phi_3P + HCHO \longrightarrow \phi_3\overset{\oplus}{P}-CH_2O^{\ominus} \xrightarrow{HCHO} \phi_3\overset{\oplus}{P}-CH_2O-CH_2O^{\ominus}$$

$$CH_3COO\!-\!\!\!\!+\!CH_2O\!\!\!+\!\!\!_n\!CCH_3 \xleftarrow[-\phi_3PO]{(CH_3CO)_2O \atop CH_3COONa} \phi_3\overset{\oplus}{P}\!\!+\!CH_2O\!\!+\!\!\!_n\!H$$

etc. (30.16)

Ethylene oxide may be polymerized by either a cationic or an anionic mechanism. If a Lewis acid catalyst is used, a trace of water also is needed (Eq. 30.17). An anionic catalyst may be any base, such as (iPrO)$_3$Al (Eq. 30.18). Commercially, a very pure strontium carbonate is used at about 100°. The polymer, known as Carbowax, is water soluble and is used as a thickening agent (increases viscosity) and as a nonionic surfactant (lowers surface tension).

The epoxy resins are complex polyethers and polyamines made as shown in Equations 30.19 from excess epichlorohydrin and bisphenol A to give a, a relatively low molecular weight polyether chain that ends in unreacted epoxide rings. When the epoxide-ended chains (a) are mixed with trifunctional amine such as diethylenetriamine, a relatively slow cross-linking occurs as the basic amino groups react to open the epoxide rings. The final cross-linked polymer (b) has excellent adhesive properties. If inert fillers such as glass fiber are added to the mixture at the cross-linking stage, very stable and rigid structures may be molded.

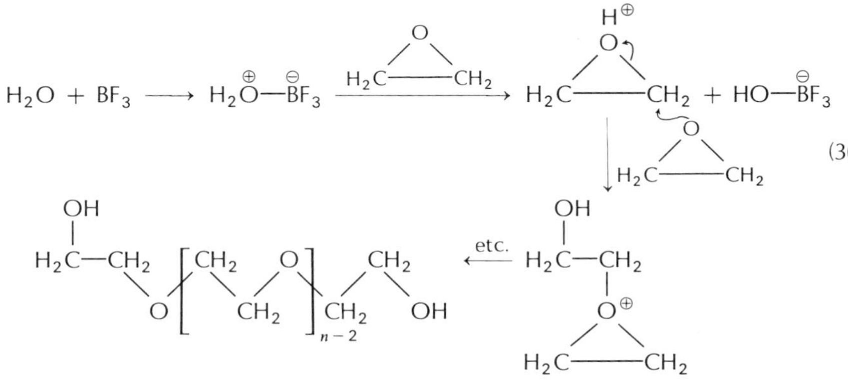

$$H_2O + BF_3 \longrightarrow H_2\overset{\oplus}{O}-\overset{\ominus}{B}F_3 \xrightarrow{H_2C-CH_2 \text{ (epoxide)}} H_2C\underset{\overset{\oplus}{O}H}{\overline{}}CH_2 + HO-\overset{\ominus}{B}F_3$$

$$\downarrow H_2C-CH_2 \text{ (epoxide)}$$

$$\underset{H_2C-CH_2}{\overset{OH}{|}}\left[\underset{CH_2}{\overset{CH_2O}{\diagdown\diagup}}\right]_{n-2} \underset{CH_2}{\overset{CH_2}{\diagdown}}OH \xleftarrow{\text{etc.}} \underset{H_2C-CH_2}{\overset{OH}{|}}\underset{H_2C-CH_2}{\overset{O^{\oplus}}{|}} \quad (30.17)$$

$$(iPrO)_2Al-OCH(Me_2) \xrightarrow{H_2C-CH_2 \text{ (epoxide)}} Me_2CHOCH_2CH_2O^{\ominus}$$

$$\downarrow H_2C-CH_2 \text{ (epoxide)}$$

$$iPrO+CH_2CH_2O+_nAl(OiPr)_2 \xleftarrow{\text{etc.}} Me_2CHOCH_2CH_2OCH_2CH_2O^{\ominus} \quad (30.18)$$

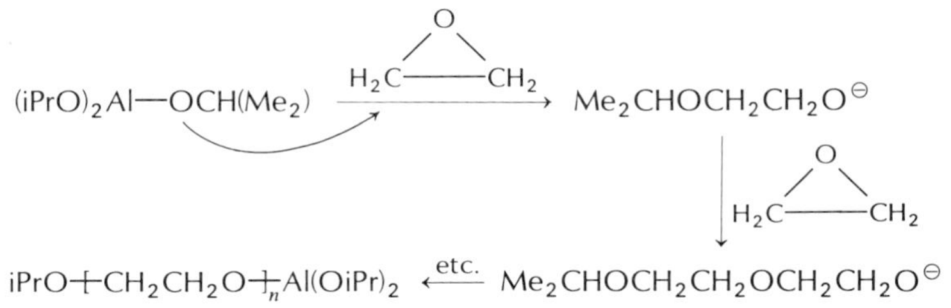

Phenol-formaldehyde resins. The first successful synthetic polymer was that made by the condensation of phenol with formaldehyde under either cationic or anionic catalysis (Eqs. 30.20).

The condensation under acidic conditions contains no ether links and is called a novolac (a). Under basic conditions, some ether formation occurs and the product is called a resole (b). Usually the condensation is carried out to partial completion. The novolac or resole is mixed with fillers and pigments, and the mixture is heated until partially melted and is molded. Heating the mold completes the cross-linking to give the final infusible thermoset product, originally known as Bakelite (now a trade name for a variety of polymers).

Polyurethanes. Other nitrogen-containing polymers are the polyurethanes, the urea-formaldehyde resins, and the melamines.

The polyurethanes are made by the reaction of an aromatic diisocyanate (such as 2,4-diisocyanotoluene or 4,4'-diisocyanobiphenyl) with a low molecular weight (1000 to 2000) polyester or polyether that has free hydroxy groups at the ends of the chains (Eqs. 30.21). Many variations of the scheme are possible and are used, including some cross-linking if a little of a polyfunctional acid, alcohol, or isocyanate is added.

Polyurethane foams are produced by the deliberate addition of water to the last step, along with some excess diisocyanate. The carbon dioxide liberated inside the mold is retained as bubbles in the hardened mass. The amount of carbon dioxide formed determines whether a relatively dense or relatively light foamed article is the result (Eq. 30.22).

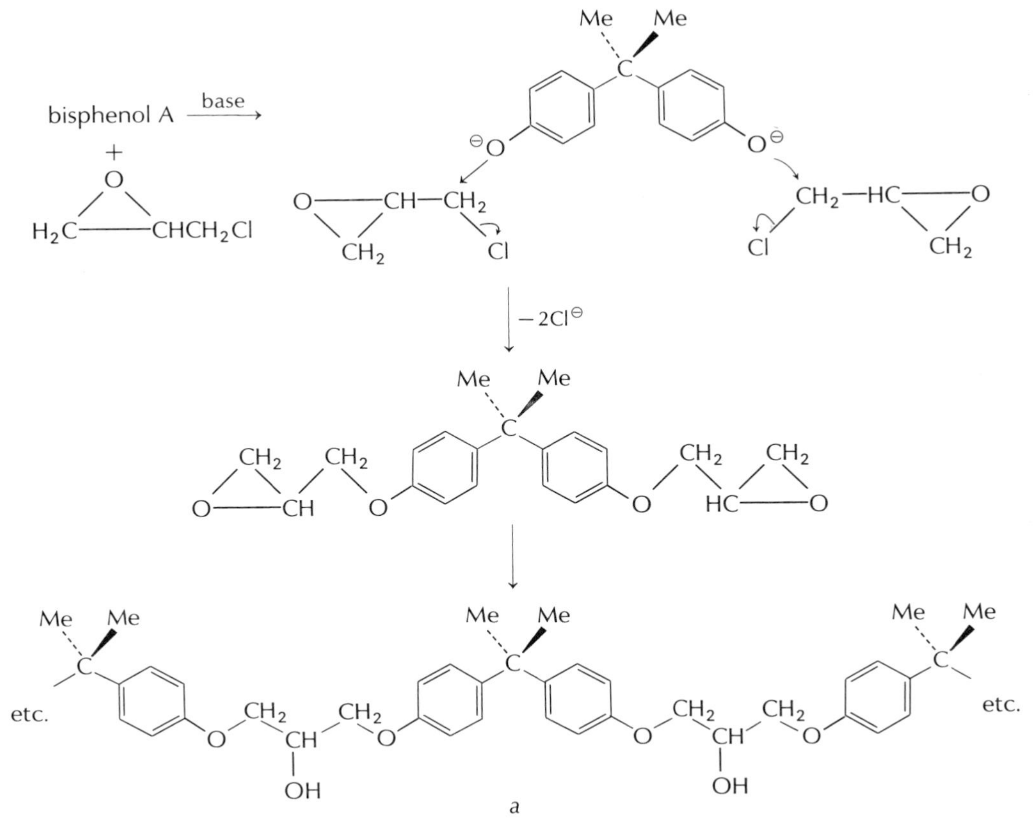

30.4 Condensation polymers

(30.19)

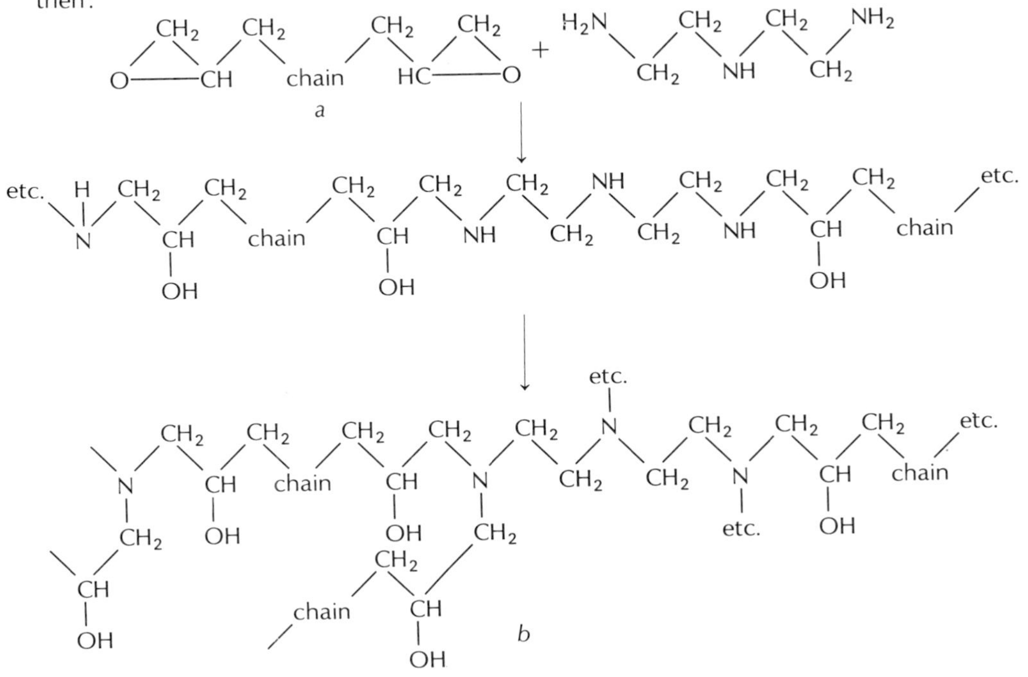

(30.20)

(30.21)

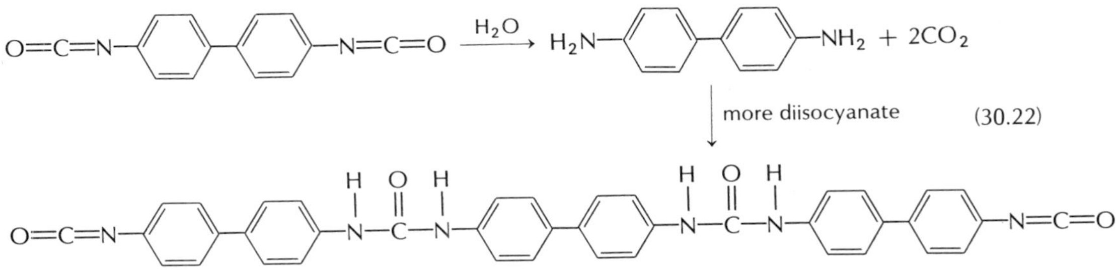

30.4 Condensation polymers

$$O=C=N-\text{C}_6\text{H}_4-\text{C}_6\text{H}_4-N=C=O \xrightarrow{H_2O} H_2N-\text{C}_6\text{H}_4-\text{C}_6\text{H}_4-NH_2 + 2CO_2$$

↓ more diisocyanate (30.22)

$$O=C=N-\text{Ar}-\text{Ar}-\underset{H}{N}-\underset{O}{\overset{\parallel}{C}}-\underset{H}{N}-\text{Ar}-\text{Ar}-\underset{H}{N}-\underset{O}{\overset{\parallel}{C}}-\underset{H}{N}-\text{Ar}-\text{Ar}-N=C=O$$

Formaldehyde-urea and formaldehyde-melamine resins. The reaction of formaldehyde with a primary amine may lead to polymeric chains (Eq. 30.23). If R is —CONH$_2$ (urea), extensive cross-linking is possible to give urea-formaldehyde resins. If the primary amine is trifunctional, an even more complex network may be made (Eq. 30.24), such as the melamine formaldehyde polymer known as Melmac.

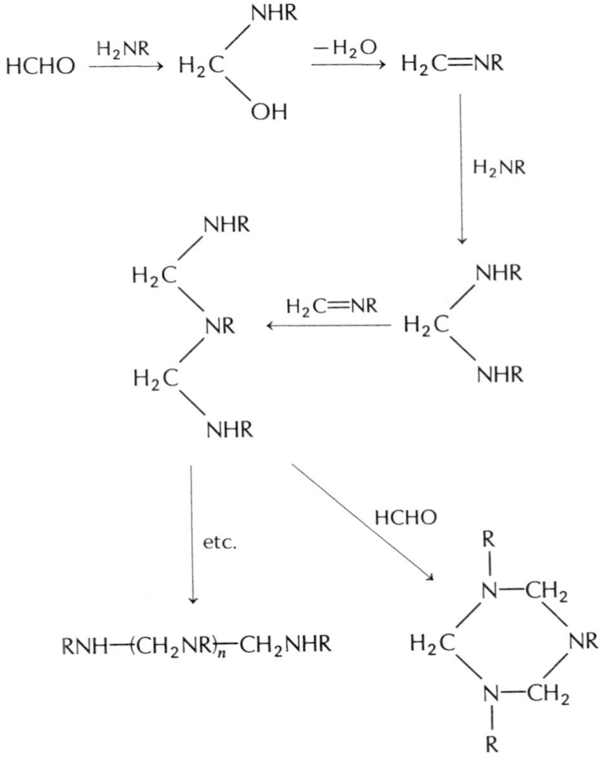

$$\text{HCHO} \xrightarrow{H_2NR} H_2C\underset{OH}{\overset{NHR}{\diagup}} \xrightarrow{-H_2O} H_2C=NR$$

↓ H_2NR

$$H_2C\underset{NHR}{\overset{NHR}{\diagup}} \xleftarrow{H_2C=NR} H_2C\underset{NHR}{\overset{NHR}{\diagup}}$$

(30.23)

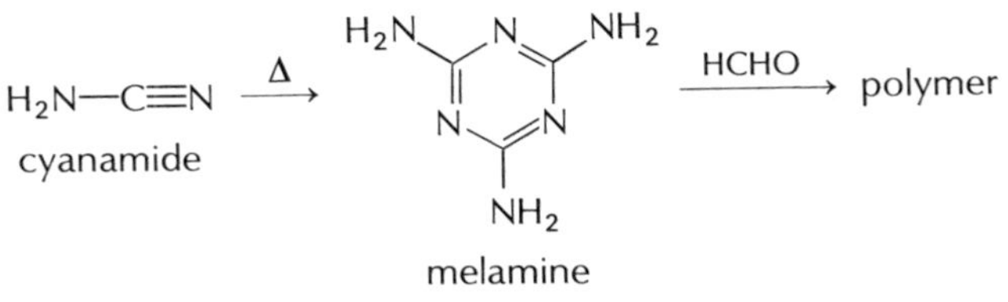

↓ etc.   ↘ HCHO

$$RNH-(CH_2NR)_n-CH_2NHR \qquad \underset{\substack{N-CH_2 \\ | \\ R}}{\overset{\substack{R \\ | \\ N-CH_2}}{H_2C\diagup\diagdown NR}}$$

$$H_2N-C\equiv N \xrightarrow{\Delta} \text{melamine} \xrightarrow{HCHO} \text{polymer} \qquad (30.24)$$

cyanamide

melamine (triazine with three NH$_2$ groups)

30.5 Variations

Ion exchange resins. Ion exchange resins may be tailor-made from certain polymers. The original cation exchangers were the naturally occurring inorganic substances known originally as green sands and later as zeolites, which were complex silicates with aluminum present: $\overline{CaAl_2Si_6O_{16}}(H_2O)_8$ and $\overline{Na_8Al_6Si_6O_{24}}Cl$, for example.

One method of preparation of an organic polymeric cation exchanger is sulfonation (on the surface of the tiny beads) of the exposed *para* positions of polystyrene. The polystyrene is partially cross-linked (to obtain a more stable, higher melting product) by the inclusion of some *p*-divinylbenzene during the polymerization. These beads may be used in two ways.

The most popular use is in water softening, whereby the cations (Ca^{2+}, Mg^{2+}, etc.) present in hard water are replaced by sodium ions as the water passes through a bed of the resin (Eq. 30.25). Sooner or later, the resin loses capacity to retain the cations (the

$$\text{Resin-}SO_3^{-}Na^{+} \;\xrightarrow[Na^{+}]{Ca^{2+}}\; \text{Resin-}SO_3^{-} \quad Ca^{2+} + 2Na^{+} \quad (30.25)$$

resin is spent), but it is easily regenerated if a saturated NaCl solution is passed into the bed. The high concentration of sodium ion shifts the equilibrium back to the left. After standing for a few minutes, the NaCl solution (which now contains the Ca^{2+}, Mg^{2+}, and so on formerly held by the sulfonate ions on the surface of the resin) is flushed out and the bed is available for fresh service as a water softener.

More expensive than simple water softening but cheaper than distillation is deionization. Deionization requires two modified polystyrenes, and the process is easily understood if we consider the two to be present in two separate tanks (or beds). For some purposes, the resins are mixed and a single bed or tank is used.

The cation exchanger now is used as shown in Equation 30.26. The hard water

$$\text{Resin-}SO_3^{-}H^{+} \;\xrightarrow[\text{HCl, then }H_2SO_4]{Ca^{2+} \text{ and other cations}}\; \text{Resin-}SO_3^{-} \quad Ca^{2+}\, 2H^{+} \quad (30.26)$$

emerges from the bed with all the metallic ions removed and replaced by hydrogen ions. The acidic effluent then is transferred to the anion exchanger. When the acidic cation exchanger bed is spent, it is regenerated by a hydrochloric acid flush (to remove Ca^{2+} and other ions as soluble chloride salts: $CaSO_4$ is relatively insoluble) followed by moderately concentrated sulfuric acid to complete the process.

The anion exchanger contains a bed of polystyrene beads that have been modified as shown in Equation 30.27. Chloromethylation (see Eq. 16.24) of the beads produces the benzylic chloride surface, which readily gives the quarternary amine salt surface. The bed is activated by a flush with dilute sodium hydroxide. The acidic effluent from the cation exchanger, in passing through the anion exchanger bed, experiences a replacement of the

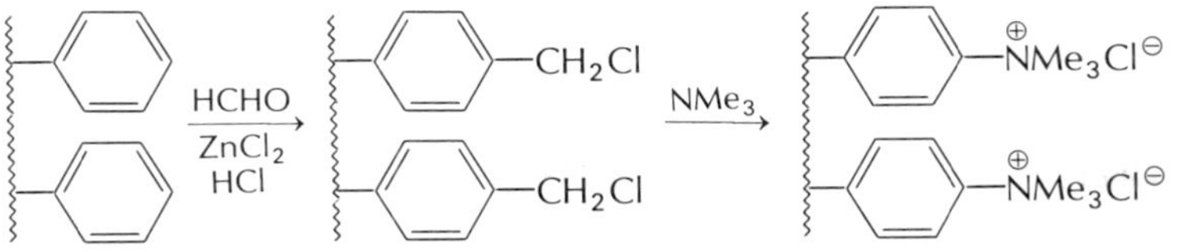

$$\text{(polystyrene)} \xrightarrow[\text{HCl}]{\text{HCHO, ZnCl}_2} \text{(chloromethylated)} \xrightarrow{\text{NMe}_3} \text{(quaternary ammonium)} \quad (30.27)$$

anions present ($HCO_3^{\ominus}$, $HSO_4^{\ominus}$, etc.) by hydroxide ion (Eq. 30.28). The hydroxide ions neutralize the hydrogen ions to form water, and the effluent deionized water compares favorably with distilled water in purity. Some of the cation exchange resins have carboxylic acid groups present instead of sulfonic acid groups.

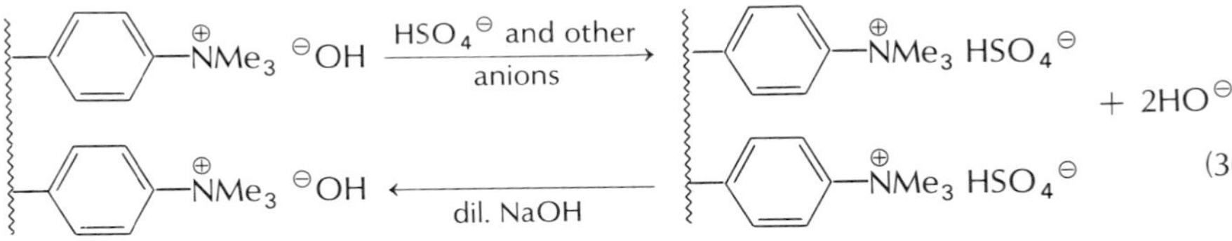

$$\text{resin-NMe}_3^{\oplus} \, {}^{\ominus}OH \xrightarrow[\text{anions}]{HSO_4^{\ominus} \text{ and other}} \text{resin-NMe}_3^{\oplus} \, HSO_4^{\ominus} + 2HO^{\ominus}$$

$$\text{resin-NMe}_3^{\oplus} \, {}^{\ominus}OH \xleftarrow{\text{dil. NaOH}} \text{resin-NMe}_3^{\oplus} \, HSO_4^{\ominus} \quad (30.28)$$

Block and graft polymers. Block and graft polymers are products made by operations other than direct polymerization. A block polymer is made by combination of two or more polymer chains so that the resulting polymer has properties somewhat modified from those of either constituent polymer. For example, if we consider a polyester with carboxylic acid end groups and a polyethylene oxide, the two may be combined by continued esterification (Eq. 30.29). Block polymers also can be made from vinyl and diene chains.

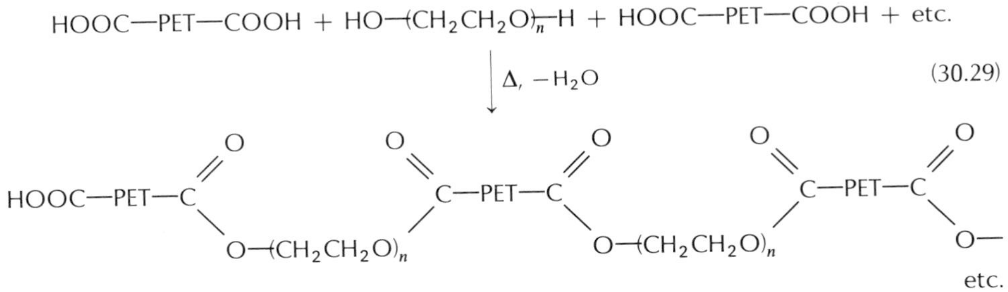

$$HOOC-PET-COOH + HO-(CH_2CH_2O)_n-H + HOOC-PET-COOH + \text{etc.}$$

$$\downarrow \Delta, -H_2O \quad (30.29)$$

$$HOOC-PET-C(=O)-O-(CH_2CH_2O)_n-C(=O)-PET-C(=O)-O-(CH_2CH_2O)_n-C(=O)-PET-C(=O)-O- \text{ etc.}$$

A graft polymer may be made if a polymer is subjected to X rays while it is in the presence of a vinyl or diene monomer. The exposure results in occasional fission of a bond in the polymer chain, and the radical that results serves as an initiator for growth of a new vinyl or diene chain.

Ladder polymers. A ladder polymer refers to those products in which the polymer chain may be thought of as a stiff ribbon instead of a chain of σ bonds. One (not a commercial product) is shown in Equation 30.30. A polymer like this one, with an almost completely aromatic system, has been sought because of outstanding thermal stability and resistance to acids and bases. However, these polymers have poor resistance to light.

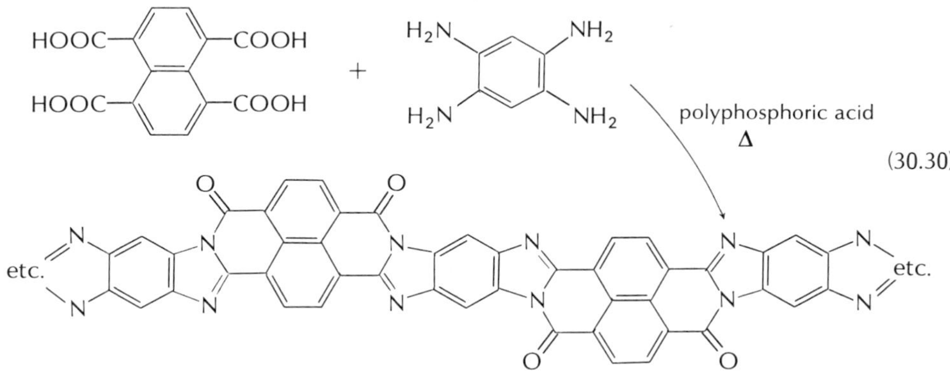

(30.30)

Polysulfides; polysiloxanes. Treatment of a dichloroalkane with sodium polysulfide (Na$_2$S solution with added sulfur) gives a latex that is rubberlike but of low tensile strength (Eq. 30.31). This polymer is but one of several Thiokols.

$$ClCH_2CH_2Cl + Na_2S_x \longrightarrow -CH_2CH_2-S_x-CH_2CH_2-S_x-CH_2CH_2- \quad (30.31)$$

The siloxanes (described in Sec. 29.6) may be oils or solids, depending upon the extent of cross-linking. The properties are remarkably resistant to change with temperature. Thus, the viscosity of a silicone oil varies little with extremes of temperature.

30.6 Summary

With this chapter we complete our journey from the simple organic compounds to the complex synthetic macromolecules of organic chemistry. Over and over, we have reluctantly left topics to go on to the next with the complaint that we have barely scratched the surface. Now, with more than the usual reason for doing so, we repeat our complaint, this time about polymer chemistry. It has been said that the last half of the twentieth century will become known as the time of the plastics revolution, just as former times were known as the iron age or the industrial revolution.

Because of the tremendous production of synthetic polymers that now exists, we must give belated consideration to the disposal problem of plastic waste, be it discarded nylon or polyester garments, polyethylene bottles, or any of the number of plastic articles to be found in every home and store.

Organic chemistry has been the source of much that is useful. With thought toward use and disposal, organic chemistry can continue to be not only a major scientific discipline for the study of nature, but also a servant of mankind.

30.7 Problems

30.1 Nylon 66 can be synthesized from butadiene starting with 1,4 addition of Cl$_2$. Show the steps.

30.2 Suggest why Nylon 45 does not make good fibers; consider the conformations of the interacting polymer chains.

30.3 1-Butene can be polymerized to an isotactic polymer but isobutene cannot. Why?

30.4 Write a generalized structure for each of the polymers obtained in the following reactions:

a. Styrene $\xrightarrow{X\cdot}$

b. Styrene $\xrightarrow[\text{TiCl}_4]{(C_2H_5)_3Al}$

c. Styrene + p-divinylbenzene $\xrightarrow{X\cdot}$

d. Maleic anhydride + styrene $\xrightarrow{X\cdot}$

e. Vinyl acetate $\xrightarrow{X\cdot}$ A $\xrightarrow{NaOH}$ B

f. $O=C=N-(CH_2)_4N=C=O + HO(CH_2)_2OH \longrightarrow$

30.5 Some monomers are made from the commercial chemicals indicated. How would you perform these required transformations? (More than one step may be required.)

a. Benzene → styrene
b. Acetylene → vinyl acetate
c. Acetylene → vinyl chloride
d. Acetone → methyl methacrylate
e. Vinylacetylene → chloroprene (2-chlorobutadiene)
f. p-Xylene → dimethyl terephthalate
g. o-Xylene → phthalic anhydride
h. Cyclohexanol → adipic acid

30.8 Bibliography

The general textbooks are spotty in coverage of polymers. Perhaps the best of the older texts is Fieser and Fieser (Chap. 7, ref. 7), Chapter 35. Of the newer books, Butler and Berlin (Chap. 7, ref. 3), Chapter 34, is more detailed than many. Of the paperbacks, Stille (Chap. 7, ref. 21) is excellent, though brief. Another is given here.

1. Mandekern, L., *An Introduction to Macromolecules.* New York: Springer-Verlag, New York, 1972. A recent paperback, not too advanced.

Frequent articles appear concerning one facet or another of polymer chemistry. A few suggestions are given.

2. Brown, A. E., and K. A. Reinhart, Polyester Fiber: From Its Invention to Its Present Position, *Science* 173, 287 (1971).

3. Peters, E. M., and J. A. Gervasi, Nylon 4—The Bridge Between Synthetic and Natural Fibers, *Chemical Technology* 2, 16 (1972).

4. Preston, J., Fibers That Resist High Temperature, *Chemical Technology* 1, 664 (1971).

30.9 John E. Leffler: Why peroxides? It worked!

Professor John E. Leffler earned his Ph.D. at Harvard in 1948. After a one-year stay at Cornell, and then at Brown, he went to Florida State University in 1950 and has remained there since, becoming professor in 1959. He has been active in several facets of mechanistic organic chemistry. Here he describes a piece of research work that went as expected!

Although organic chemistry from one point of view is merely the study of covalently bonded molecules, in fact it exists on an inherently higher level of organization than the physics and quantum mechanics that make up its theoretical foundations. Such a metasystem is more than the sum of its parts, and its development is essentially unpredictable. Just as radar sets and digital computers were not obvious to persons contemplating Maxwell's equations, the existence of, say, chlorophyll would be quite a surprise to someone acquainted only with electrons, nuclei, and the hydrogen molecule.

What first attracted me to organic chemistry, beyond my boyhood fascination with laboratory apparatus, was the opportunity that organic chemistry offers for the design, or rather, invention of ingeniously different molecules. This of course led to an interest in organic reactions as tools for synthesis.

Why does a scientist engage in a particular kind of research at a particular time? In my own case an interest in reactions generated an interest in what makes the reactions tick—the reactive intermediates involved and the mechanisms of the reactions. However, a reaction mechanism is not just something to be elucidated. Reaction mechanisms are subject to the question "Why?" as well as the question "What?" and, like molecules of ingenious structure, they can be invented as well as uncovered. The question "Why?" about a particular reaction path always has the same answer: the mechanism of a reaction is the process that is distinctly faster than all the other processes that might compete with it. To invent a new reaction or to change the mechanism of a reaction would seem to require only a knowledge of the factors affecting reaction rates. One simply selects a set of parameters that should favor the desired process. Of course, given the state of the art, a considerable amount of luck is needed as well; but a motive is provided for still another research interest: the theory of reaction rates and substituent or medium effects.

The carboxy-inversion reaction of diacyl peroxides is an example of the application of simple reaction rate theory that actually worked. A more usual result is the discovery of something new and welcome to the researcher, but quite unlike what was expected.

At that time diacyl peroxides were known to decompose by homolysis to acyloxy radicals. This mechanism was changed to a polar one by supplying an electron-withdrawing substituent at one para position of a benzoyl peroxide molecule and an electron-releasing substituent at the other para position, and running the reaction in a polar solvent. It is called the carboxy-inversion reaction, partly because the name is descriptive, and partly because of my own aversion to "name" reactions. Thionyl chloride was chosen as the solvent because it is polar, dry, and happened to be on the shelf. Since the initial example, the carboxy-inversion reaction has been observed in aliphatic peroxides and has been used as a method of synthesizing alcohols. It has also been the subject of several investigations of its mechanism. A detail that is still not yet fully understood is the exact stage in the process at which the paths leading to ion pairs and to radical pairs diverge from each other.

Selected Answers to Exercises and Problems

Chapter 2

Exercises

2.2b. CH_3OOCH_3 (CH_3OOH is methyl hydroperoxide); f. $CH_3OCH_2CH_3$. 2.3b. aminoethane or ethylamine or ethanamine; g. chloroethanoic acid (the common name is chloroacetic acid). 2.8c. $CH_3O^{\ominus} + CH_3-Br \rightarrow CH_3OCH_3 + Br^{\ominus}$.

Problems

2.3b. $HCOOH + CH_3O^{\ominus} \rightarrow HCOO^{\ominus} + CH_3OH$. 2.4c. methyl nitrite; f. methyl cyanide. 2.5a. $CH_3OH \xrightarrow[H_2SO_4]{K_2Cr_2O_7} CO_2 + H_2O \ (+ \ Cr_2(SO_4)_3 + K_2SO_4)$ (an oxidation).

Chapter 3

Exercises

3.1d. 2-aminoethanoic acid (the acid takes precedence over $-NH_2$ for the lower number; common name is glycine); e. methyl 2,2-dichloroethanoate. 3.5a. $CH_3CH_2OH + HBr \xrightarrow{25°} CH_3CH_2Br + H_2O$; c. $CH_3CH_2Br + HO^{\ominus} \xrightarrow[H_2O, \ 25°]{acetone,} CH_3CH_2OH + Br^{\ominus}$. 3.6a. $I-I + CH_3-MgBr \rightarrow I^{\ominus} + I-CH_3 + \overset{\oplus}{M}gBr$; b. $CH_3O-H + CH_3CH_2-MgBr \rightarrow CH_3O^{\ominus} + CH_3CH_3 + \overset{\oplus}{M}gBr$. 3.9a. $H_2C=CH_2 + Cl_2 \xrightarrow[CH_3COONa]{CH_3COOH} CH_3COOCH_2CH_2Cl$; c. $H_2C=CH_2 + Br_2 \xrightarrow[NaOH]{H_2O} HOCH_2CH_2Br$. 3.10d. $H_2C-CHCH_3 \xrightarrow{\Delta} CH_2=CHCH_3 + Br^{\ominus}$ (with Br and Zn on the reactant, $BrZn^{\oplus}$ byproduct). 3.11a. $CH_3CH_2OH + Na_2Cr_2O_7 + H_2SO_4 \xrightarrow{40°} CH_3COOH + Cr_2(SO_4)_3 + Na_2SO_4 + H_2O$.

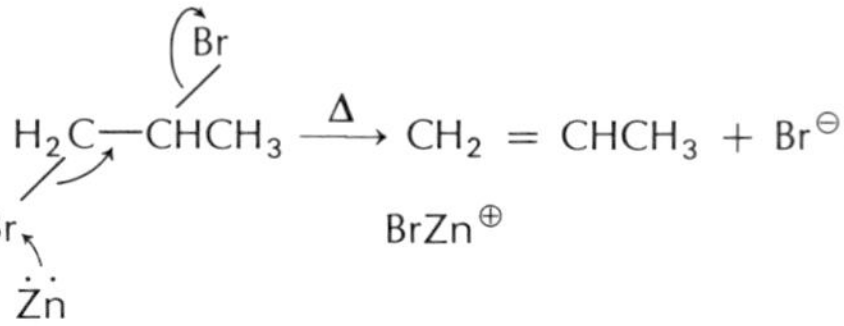

Problems

3.1c. $CH_3OH \xrightarrow{HBr} CH_3Br \xrightarrow[\text{ether}]{Li} CH_3Li \xrightarrow{I_2} CH_3I + LiI$.   3.2c. $\Delta H° = [(-66.4)] - [(12.5) + (-68.3)] = -66.4 - 12.5 + 68.3 = -10.6$ kcal.   3.3b. $CH_3CHO \xrightarrow[25°]{H_2, Pt} CH_3CH_2OH$ (or use $NaBH_4$, or $LiAlH_4$);   f. $H_2C{=}CH_2 \xrightarrow[\substack{CH_3COOH \\ CH_3COONa \\ 25°}]{Cl_2} CH_3COOCH_2CH_2Cl$.

3.4a. $CH_3COOH + LiAlH_4 \longrightarrow (CH_3CH_2O)_4LiAl \xrightarrow[HCl]{dil.} CH_3CH_2OH$; $CH_3COOH + H_2 \xrightarrow{Pt} NR$;   g. $H_2C{=}CHCHO + LiAlH_4 \longrightarrow \xrightarrow[HCl]{dil.} H_2C{=}CHCH_2OH$ (most) $+ CH_3CH_2CH_2OH$ (some); $H_2C{=}CHCHO + 2H_2 \xrightarrow{Pt} CH_3CH_2CH_2OH$.

Chapter 4

Exercises

4.1c. $C_nH_{2n}O$;   d. C_nH_{2n-2}.   4.3b. 3-bromo-2,2-dimethylbutane;   e. 2,5-dichloro-3,4-diethyl-2,3-dimethylhexane.   4.5a. $H_2C{=}C\begin{smallmatrix}CH_3\\ \\C{=}CH_2\\H_3C\end{smallmatrix}$ d. $Cl_3CC{\equiv}C{-}\underset{\underset{CH(CH_3)_2}{|}}{C}{=}CH_2$

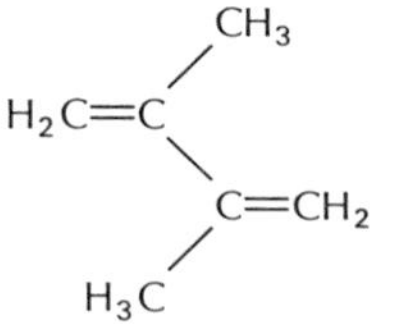

4.6a. 6-methyl-4-neopentyl-2,4-heptadiene;   c. 4-hexen-3-one.

4.7b. [structure: bromocyclopentene with vinyl and butyl substituents]   d. 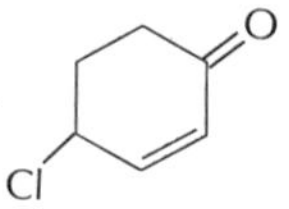 [structure: 4-chlorocyclohex-2-enone]   4.8b. 1-(1-buten-1-yl)-1,3-cyclopentadiene or 1-(1,3-cyclopentadien-1-yl)-1-butene;   d. 4-allyl-3-methyl-1-cyclohexanol.

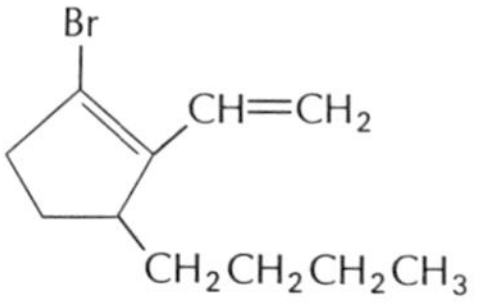

Problems

4.1b. C_3H_3Br becomes C_3H_4 or C_nH_{2n-2}. The molecule contains four fewer hydrogen atoms than are needed for saturation, or two double bonds, one double bond and one ring, two rings, or a triple bond. Possibilities are allene, cyclopropene, and propyne. Substituting a bromine into each gives:

1. $H_2C{=}C{=}CHBr$, 1-bromoallene;   2. $HC\overset{CHBr}{\underset{}{\triangle}}CH$, 3-bromocyclopropene;   3. $HC\overset{CH_2}{\underset{}{\triangle}}CBr$, 1-bromocyclopropene;   4. $BrCH_2C{\equiv}CH$, 3-bromopropyne;   5. $CH_3C{\equiv}CBr$, 1-bromopropyne.
4.2a. 2-methyl-3-hexanone;   f. 6-hydroxy-4-hexynal.   4.3c. $BrCH{=}CHCH_2CH_2CHO$;
f. 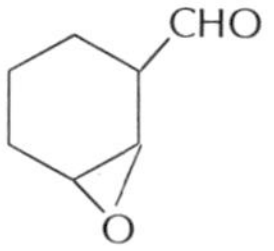 [epoxycyclohexane-CHO structure]   i. $CH_3CH_2CBrCH_3$ [cyclobutyl]   4.4a. 3-methyl-2-butenal;   e. ethyl 3-hydroxy-4,5-dimethyl-4-cyclopentene-1-carboxylate.   4.5c. [cyclopentyl]$-CH_2CH(CH_3)_2$;   f. $HOOCCH{=}CHCOOH$.

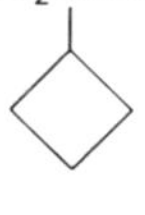

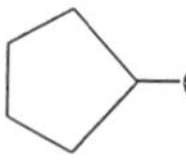

4.6d. 2-methylpentane.   4.7d. 1-*tert*-butyl-3-methylcyclobutane;   g. 2-isobutyl-1-hexene.
4.9c. $CH_3\underset{\underset{CH_3}{|}}{C}HCH{=}\underset{\underset{CH_2CH_3}{|}}{C}CH_2C{\equiv}CH$

Chapter 5

Exercises

5.2b.

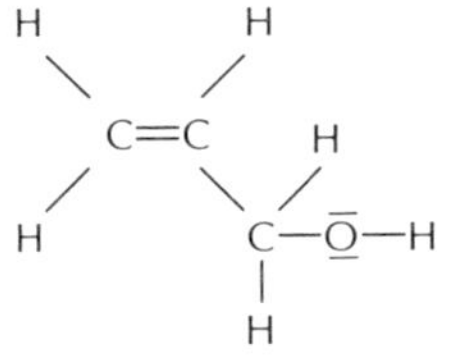

$$H_2C=CH-CH_2-\underline{\bar{O}}-H$$

5.3c. Three $(\sigma 1s2sp^3)^2$ filled MO's; one $(\sigma 1s2sp^2)^2$ filled MO; one $(\sigma 2sp^3 2sp^2)^2$ filled MO; one $(\sigma 2sp^2 2sp)^2$ filled MO, or $(\sigma 2sp^2 2p)$; one $(\pi 2p 2p)^2$ filled MO for CH_3CHO.

5.4a. $H_2C=CH-C(H)=\underline{\bar{O}}|$ ↔ $H_2C=CH-C(H)-\underline{\bar{O}}|^{\ominus}_{\oplus}$ ↔ $H_2\overset{\oplus}{C}-CH-C(H)=\underline{\bar{O}}|^{\ominus}$

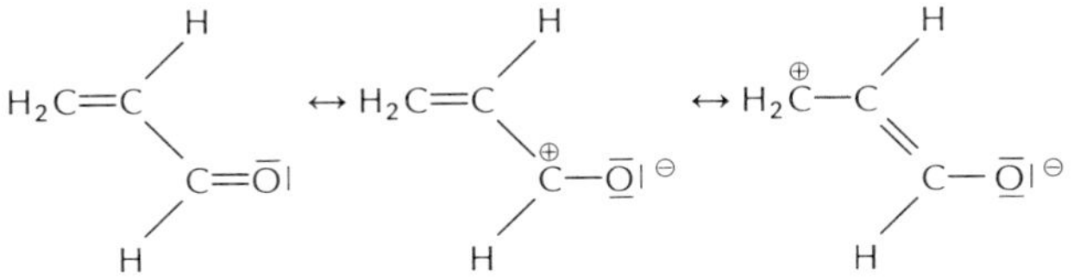

Other structures contribute very little.

5.6a. $CH_3-C(H)=O$; $C-H = 1.09$ Å; $C-O = 1.21$ Å has been reported; $\angle(H-C-CH_3) = 120°$; CH_3- is tetrahedral.

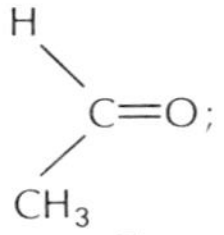

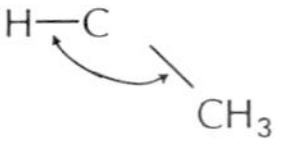

5.7a. $H-C\overset{\underline{\bar{O}}-H\cdots I\underline{\bar{O}}}{\underset{\underline{\bar{O}}|\cdots H-\underline{\bar{O}}}{}}C-H$ d. $\underset{Cl}{\overset{Cl}{>}}\!Al\cdots I\underline{\bar{O}}-CH_3$ with H

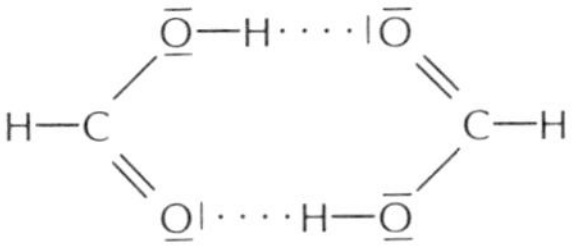

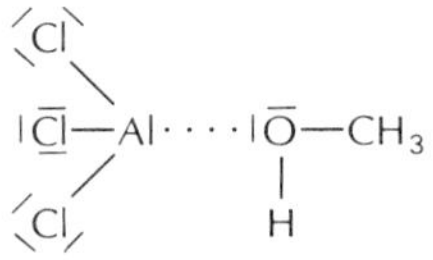

5.9. The dipole moments result mostly from the contribution of the ionic structure (b). The methyl group, with a +I effect, favors b rather than a. Replacement of the hydrogen atoms of formaldehyde (HCHO) increases the contribution of b to the resonance hybrid and μ will increase.

Thus: $C=\underline{\bar{O}}|$ ↔ $\overset{\oplus}{C}-\underline{\bar{O}}|^{\ominus}$
       a   b

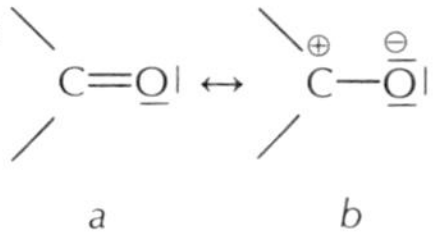

5.11a. $\dot{C}H_3 + H\cdot + \ddot{O}$ ↑ ↓
       $CH_4 + \tfrac{1}{2}O_2 \rightarrow CH_3OH$

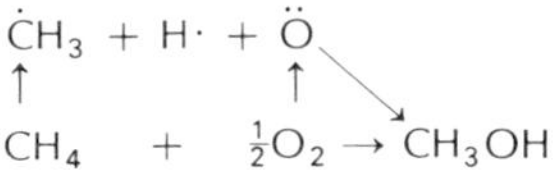

$\Delta H° = BE(C-H) + \tfrac{1}{2}BE(O=O) - BE(C-O) - BE(O-H) = (99) + \tfrac{1}{2}(119) - (86) - (111) = 99 + 59.5 - 86 - 111 = -38.5$ kcal from BE's; $\Delta H° = \Delta H°_f(CH_3OH) - \Delta H°_f(CH_4) - \Delta H°_f(\tfrac{1}{2}O_2) = (-48.1) - (-17.9) - (0) = -48.1 + 17.9 = -30.2$ kcal from $\Delta H°_f$'s. The BE method gives an answer that is too negative by 8.3 kcal.

Problems

5.1b. $NH_3 > CH_3NH_2 > (CH_3)_2NH > (CH_3)_3N$ or $1.5 > 1.3 > 1.0 > 0.6$ D. Bond angle spreading is the major cause.

5.4a.
$$\begin{array}{c} I\underline{\bar{O}} \\ \| \\ H_2C\diagdown\overset{C}{\diagup}|\underline{\bar{O}}|^{\ominus} \\ | \quad\quad H \\ H_2C\diagup\underset{C}{\diagdown}|\underline{\bar{O}}| \\ \| \\ I\underline{O} \end{array}$$

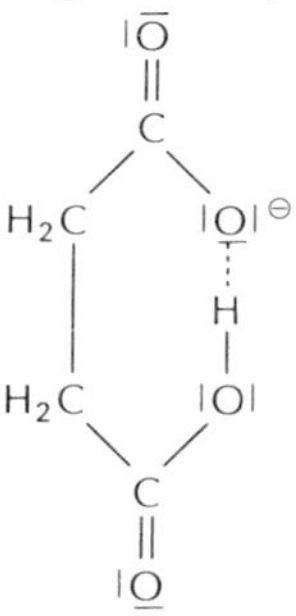

5.7a. $\dot{C}H_3 + HO\cdot + H\cdot + Cl\cdot$ ↑ ↑
$CH_3OH + HCl \longrightarrow CH_3Cl + H_2O$;
$\Delta H° = BE(C-O) + BE(H-Cl) - BE(C-Cl) - BE(O-H) = (86) + (103) - (81) - (111) = 86 + 103 - 81 - 111 = -3$ kcal; or, $\Delta H° = BDE(CH_3-OH) + BE(H-Cl) - BDE(CH_3-Cl) - BE(O-H) = (91.5) + (103) - (83.5) - (111) = 91.5 + 103 - 83.5 - 111 = 0$. The actual value (from $\Delta H°_f$'s) is -7.2 kcal. Thus neither the BE nor the BDE value is very good for this reaction.

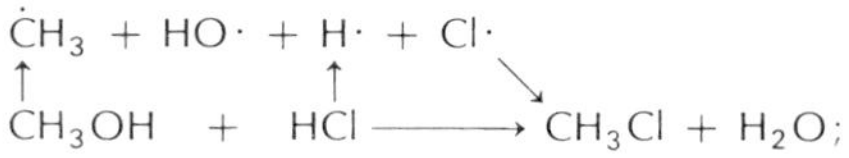

5.8b.

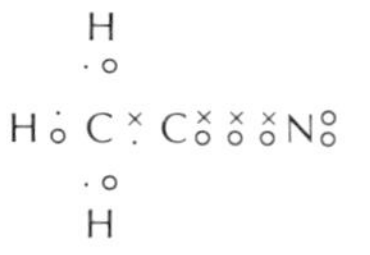

H–Ö–C::C::::N:ö:
     |
     Ö
     |
     H

g.

Hö·
  C::·C::öÖ:
Hö·

5.9c. Br—Br, none; e. 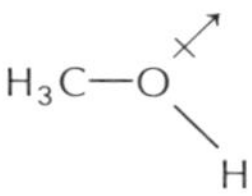 H₃C—O(↗, ↘H)

Chapter 6

Exercises

6.2a. [cyclohexane with H₃C, H on one carbon and OH, H on another]

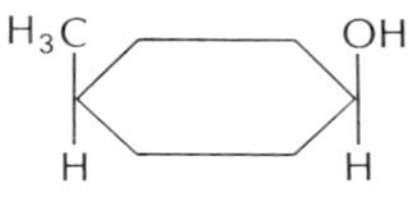

b. 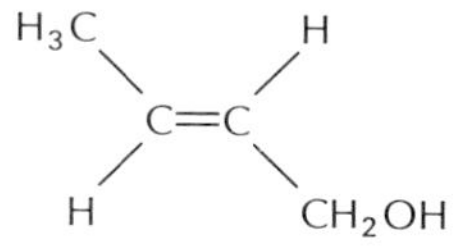 [alkene: H₃C and H on one carbon; H and CH₂OH on the other, C=C]

6.3a. [structure: HOOC–C(OH)(CH₃)–H with (R) label]

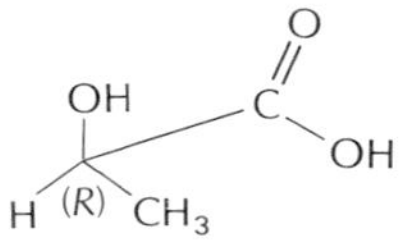

d. 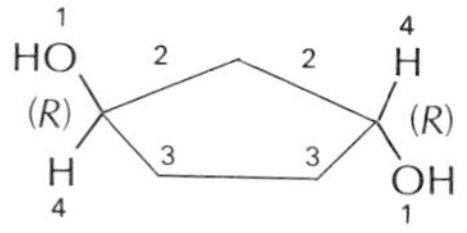 [cyclopentane with HO(1), H(2), (R) at C2; H(3); H(4), OH(1), (R) at C4 — numbered 1,2,3,4]

Problems

6.3. $CH_3\overset{*}{C}HCl\overset{*}{C}HCl\overset{*}{C}HOH\overset{*}{C}HOHCH_3$; $n = 4$, so there are $2^4 = 16$ stereoisomers, of which 8 are (+) and 8 are (−); 8 racemic pairs.   6.6g. Two isomers, one cis, the other trans; no chiral centers.

Chapter 7

Exercises

7.3b. The conditions favor E2 > S_N2, so $(CH_3)_3CCHClCH_3 \xrightarrow[\text{EtOH}]{\text{EtO}^\ominus} (CH_3)_3CCH=CH_2$.

7.8c. $BrCH_2CH_2CH_2CH_2Br \xrightarrow[80°]{\text{alc. KOH},\,\Delta} H_2C=CHCH_2CH_2Br + BrCH_2CH_2CH=CH_2$. Both product structures represent the same substances, and if excess base is used, $H_2C=CHCH_2CH_2Br \longrightarrow H_2C=CHCH=CH_2$.

Problems

7.3b. To prepare a 1-alkene, we must use an E2 reaction (to prevent rearrangement) on a primary alkyl halide. Thus:

$(CH_3)_2CHCH=CH_2 \xleftarrow[80°]{\text{alc. KOH}} (CH_3)_2CHCH_2CH_2Cl \xleftarrow[80°]{SOCl_2} (CH_3)_2CHCH_2CH_2OH$
↑
$(CH_3)_2CHCH_2OH \xrightarrow[80°]{SOCl_2} (CH_3)_2CHCH_2Cl \xrightarrow[\text{ether}]{Mg} (CH_3)_2CHCH_2MgCl + HCHO$.

Thionyl chloride is used to prepare the alkyl chlorides to avoid rearrangements.   7.4a. $CH_3CH_2CH_2Br$ (primary) is faster than $CH_3CHBrCH_3$ (secondary) in an S_N2 reaction.   7.5a. $(CH_3)_2CBrCH_2CH_3$ (tertiary) is much faster than $(CH_3)_2CHCHBrCH_3$ (secondary) in an S_N1 reaction. 7.6a. $(CH_3)_2CHCHClCH_3$ is faster than $(CH_3)_3CCHClCH_3$ in an E2 reaction because the product, $(CH_3)_2C=CHCH_3$, is more stable than $(CH_3)_3CCH=CH_2$ and the activation energy is lower. 7.7a. $(CH_3)_2COHCH_2CH_3$ (tertiary) is faster than $(CH_3)_2CHCHOHCH_3$ (secondary).
7.8b. $CH_3OC(CH_3)_2CH_2CH_3$; an S_N2 reaction.   d. $CH_3CH=CHCH_3$; an E2 reaction.

Chapter 9

Exercises

9.1. Tertiary alcohols have no C—H bond at c. Therefore, the answer list should contain the five tertiary alcohols in Table 9.1. For nonreaction at d, one has to search the table for alcohols with no C—H bond on a carbon adjacent to a C—OH bond. There is only one: 2,2-dimethyl-1-propanol.

9.3. There are two ways of having the ester labeled:

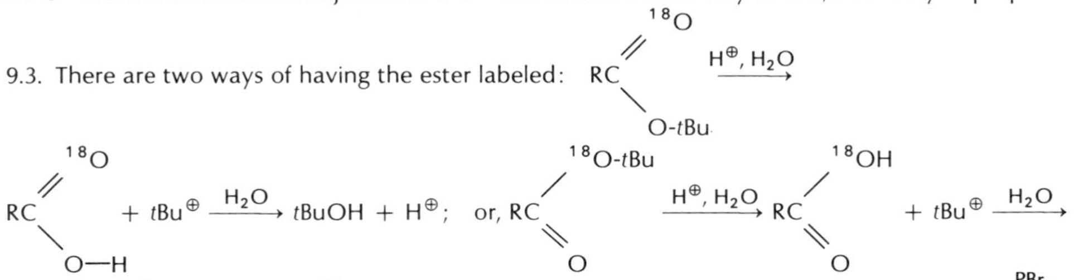

$$RC(=^{18}O)(O\text{-}tBu) \xrightarrow{H^{\oplus}, H_2O}$$

$$RC(=^{18}O)(O\text{-}H) + tBu^{\oplus} \xrightarrow{H_2O} tBuOH + H^{\oplus}; \quad \text{or,} \quad RC(=^{18}O\text{-}tBu)(=O) \xrightarrow{H^{\oplus}, H_2O} RC(^{18}OH)(=O) + tBu^{\oplus} \xrightarrow{H_2O}$$

$tBuOH + H^{\oplus}$. Either way, ^{18}O remains in the RCOOH.

9.4c. $(CH_3)_2CHCH_2OH \xrightarrow{PBr_3} (CH_3)_2CHCH_2Br$.

9.6c. cyclopentyl—OH + CrO$_3$ $\xrightarrow{\text{dil. } H_2SO_4}$ cyclopentanone + Cr$_2$(SO$_4$)$_3$ + H$_2$O.

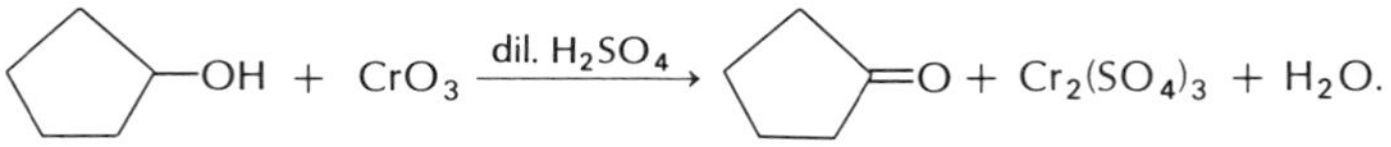

9.8d. $(CH_3)_3CMgX + H_2C\overset{O}{-\!\!-}CH_2 \longrightarrow (CH_3)_3CCH_2CH_2OH$ (after hydrolysis).

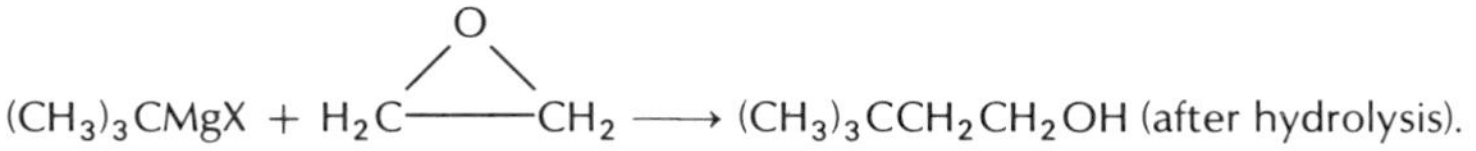

9.9c. $(CH_3)_2CHCHOHCH(CH_3)_2$; d. $(CH_3)_2CHCHOHCH(CH_3)_2$; j. 1-isopropylcyclopentanol (cyclopentane with OH and CH(CH$_3$)$_2$).

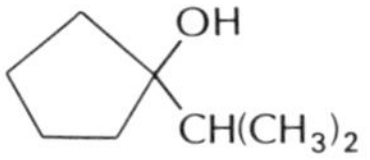

9.10b. $CH_3CH_2CHO \xrightarrow[\text{dil. } CH_3COOH]{NaBH_4, 25°, \text{ cat.}} CH_3CH_2CH_2OH$; d. $(CH_3)_3CCHO \xrightarrow[\text{dil. NaOH}]{HCHO, 40°}$
$(CH_3)_3CCH_2OH + HCOONa$; f. $(CH_3)_2CHCOOH \xrightarrow[\text{2. dil. HCl}]{\text{1. LiAlH}_4, 25°, \text{ ether}} (CH_3)_2CHCH_2OH$.

Problems

9.1a. $CH_3CH_2COCH(CH_3)_2 \xleftarrow[CH_3COOH]{CrO_3} CH_3CH_2CHOHCH(CH_3)_2 \xleftarrow{(CH_3)_2CHMgBr} CH_3CH_2CHO$

$\uparrow$ Mg | ether $\qquad$ Cu, Δ or CrO$_3$, etc.

$(CH_3)_2CHOH \xrightarrow{PBr_3} (CH_3)_2CHBr \qquad CH_3CH_2CH_2OH$
start $\qquad\qquad\qquad\qquad\qquad\qquad$ start

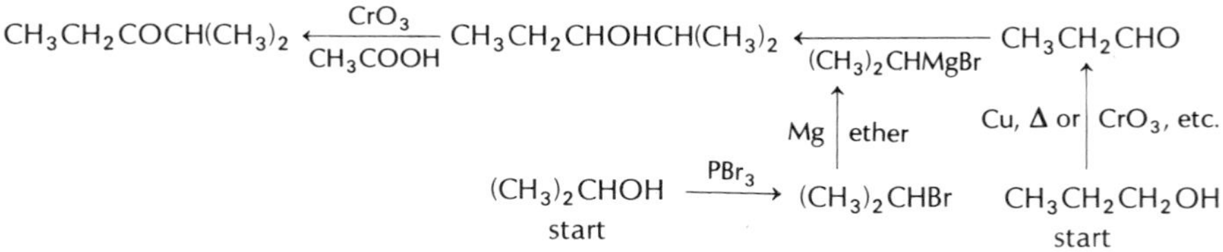

c. $CH_3OCH_2CH_2OH \xleftarrow[CH_3I]{25°} NaOCH_2CH_2OH \xleftarrow{Na} HOCH_2CH_2OH \xleftarrow[0°]{KMnO_4} H_2C=CH_2$

$\uparrow$ HI $\qquad\qquad\qquad\qquad\qquad\qquad\qquad\qquad$ H$_3$PO$_4$ | 180°

CH$_3$OH $\qquad\qquad\qquad\qquad\qquad\qquad\qquad\qquad$ CH$_3$CH$_2$OH
start $\qquad\qquad\qquad\qquad\qquad\qquad\qquad\qquad$ start

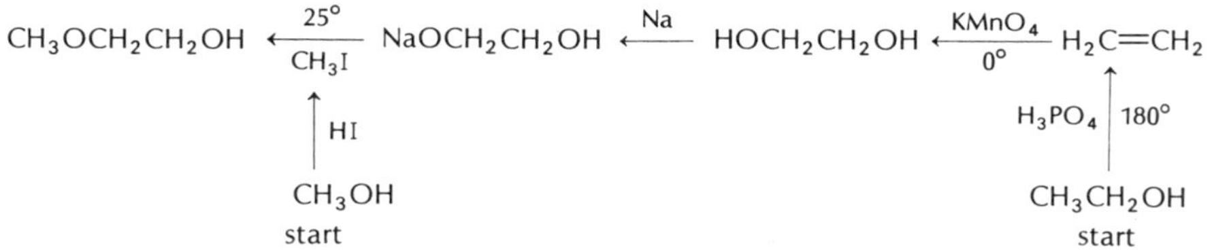

d. 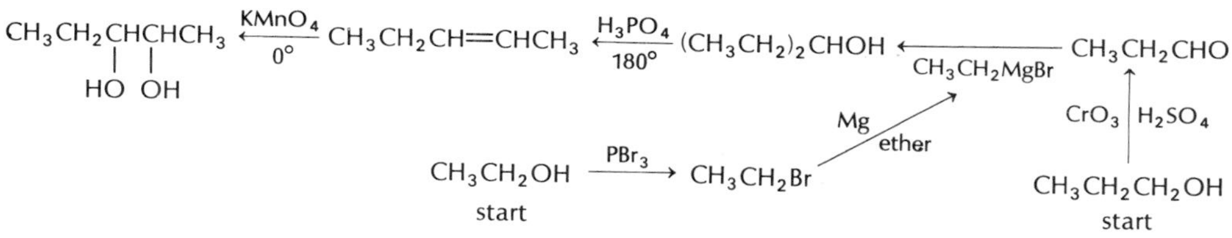 $CH_3CH_2CHCHCH_3$ $\xrightarrow{KMnO_4}{0°}$ $CH_3CH_2CH=CHCH_3$ $\xrightarrow{H_3PO_4}{180°}$ $(CH_3CH_2)_2CHOH$ $\xleftarrow{CH_3CH_2MgBr}$ CH_3CH_2CHO
     | |
    HO OH

CH_3CH_2OH $\xrightarrow{PBr_3}$ CH_3CH_2Br $\xrightarrow[\text{ether}]{Mg}$

$CH_3CH_2CH_2OH$ $\xrightarrow{CrO_3 | H_2SO_4}$
start   start

9.2a. $(CH_3)_2CHCH_2Cl$ $\xrightarrow[80°]{\text{alc. KOH}}$ $(CH_3)_2C=CH_2$ an E2 reaction; h. $(CH_3)_2CHCH_2Cl$ $\xrightarrow[\text{2. HCHO}]{\text{1. Mg, ether}}$ $(CH_3)_2CHCH_2CH_2OH$. 9.6b. 4-chloro-2-ethyl-5-fluoro-1-pentanol; e. 1-bromo-2-chloro-4-tert-butylcyclohexane. 9.7c. (cyclohexane with OCH₃) d. (cyclohexane with H, D)

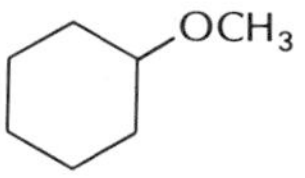

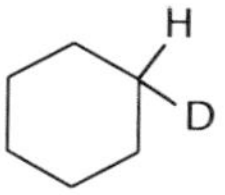

Chapter 11

Exercises

11.2 The formula, $C_3H_5Cl_3$, indicates a saturated system. The nmr indicates three protons alike at 133 Hz (δ 2.22 ppm) and two alike at 240 Hz (δ 4.00 ppm). Therefore, CH_3- and $-CH_2-$ are isolated from each other. Using $-CCl_2-$ to separate the groups gives $CH_3-CCl_2-CH_2-$. Using the lone chlorine atom left over to complete the molecule gives $CH_3-CCl_2-CH_2Cl$, 1,2,2-trichloropropane. In confirmation, the hydrogens in $-CH_2-$ are far downfield because of the location between two electron-withdrawing groups, $-CCl_2-$ and $-Cl$. The methyl group also is downfield more than usual because of the adjacent $-CCl_2-$ group.   11.5. CH_3CHBr_2 (1,1-dibromoethane).

Problems

11.1b. About 227 nm.   11.2a. (cyclohexadiene) $\xrightarrow[\text{cat.}]{2H_2}$ (cyclohexane)   11.3c. H—O—CH₂—C≡N with frequencies 3250, 2900, 1400, 2250.

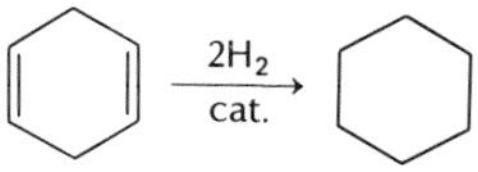

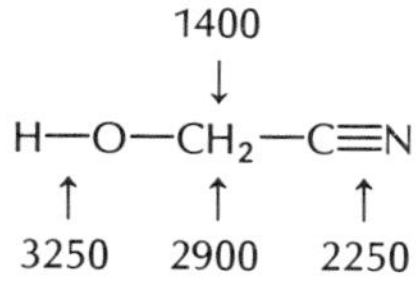

11.4a. The aldehyde bands at 2740 and 2840 cm⁻¹.   11.7a. $ClCH_2CH_2COOH$ has two triplets, and $CH_3CHClCOOH$ has a doublet and a quartet.

Chapter 12

Exercises

12.1 Cis-4-tert-butyl-1-cyclohexanol and trans-4-tert-butyl-1-cyclohexanol both have internal planes of symmetry that pass through tBu, C_4, C_1, and HO. Therefore, both are meso. The big, bulky tert-butyl group is so large that it is practically frozen into an equatorial position.

Problems

12.1. We need to form a double bond at the end of the chain. If we could persuade a halogen atom to substitute for a primary hydrogen, alcoholic KOH would remove HBr to give the product; but it is not feasible to get from $(CH_3)_2CHCH=CH_2$ to $BrCH_2CH(CH_3)CH=CH_2$ in good yield. We must use the alternative route. $(CH_3)_2CHCH=CH_2 \xrightarrow{NBS} (CH_3)_2CBrCH=CH_2$

$$(CH_3)_2CBrCH=CH_2 \xrightarrow[\text{KOH}]{\text{alc.} \;\; \Delta} CH_2=C(CH_3)-CH=CH_2$$

12.6e. $CH_3CH_2C{\equiv}CH + CH_3CH_2MgBr \xrightarrow{ether} CH_3CH_2C{\equiv}CMgBr + CH_3CH_3$.

Chapter 13

Exercises

13.1b. $(CH_3)_2C(OH)CH_2CH_2CH_3$.   13.2c. $CH_3COCH_2CH_2CH_3$ and $CH_3CH_2COCH_2CH_3$, about a 50:50 mixture.   13.5c. $tBuC{\equiv}CH \xrightarrow[\text{peroxide}]{HBr} tBuCH=CHBr$.   13.6b. $(CH_3)_2CHCH=CHCH_3$, *cis* or *trans*.

13.7d. Aldehydes require an ozonization.

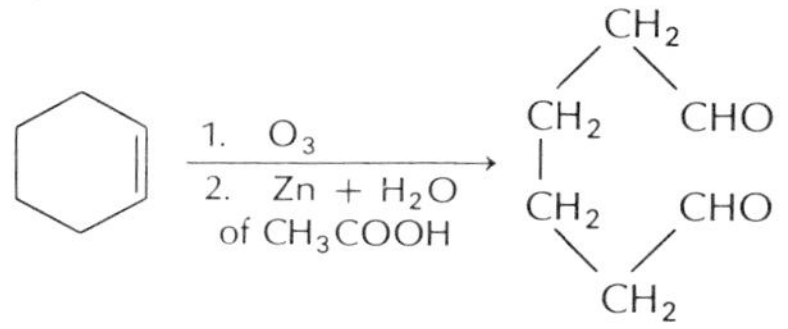

cyclohexene $\xrightarrow[\text{2. Zn + H}_2\text{O of CH}_3\text{COOH}]{\text{1. O}_3}$ OHC-CH$_2$CH$_2$CH$_2$CH$_2$-CHO

13.14a. As has been mentioned several times, $RMgX + R'X \longrightarrow RR' + MgX_2$ does not give good yields unless $R'X$ is allylic. Therefore, the more basic $HC{\equiv}CNa$ must be used. $HC{\equiv}CH \xrightarrow{NaNH_2} HC{\equiv}CNa \xrightarrow{PrCl} PrC{\equiv}CH$

Problems

13.3c. In this instance, a good answer, because few steps are needed, is:

$(CH_3)_2C=CHCH_2CH_3 \xleftarrow[\text{warm}]{H_3PO_4} (CH_3)_2C(OH)CH_2CH_2CH_3 \longleftarrow$

$\quad CH_3COCH_3 \xrightarrow[H_2SO_4]{K_2Cr_2O_7} CH_3CHOHCH_3$

$\quad CH_3CH_2CH_2MgBr \xleftarrow[\text{ether}]{Mg} CH_3CH_2CH_2Br \xleftarrow{PBr_3} CH_3CH_2CH_2OH$

The tertiary alcohol is easily dehydrated and unlikely to rearrange.

h. $CH_3CH_2CH_2CH=O + O=CH_2$.

13.4f. $CH_3CH_2CHCH_3$ with Br substituent.

13.7.

(A) cyclopentene

(B) *trans*-1,2-dibromocyclopentane

(C) pentanedial-like structure: $H_2C(CH_2CH=O)CH_2-CH=O$

(D) $H_2C(CH_2C(=O)OH)CH_2-C(=O)-OH$

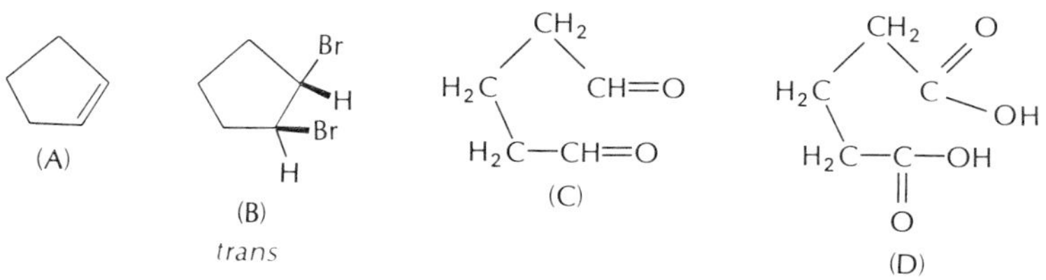

Chapter 14

Exercises

14.1b. [structure: anthracene with Br at position 1 and Br at position 9/10]

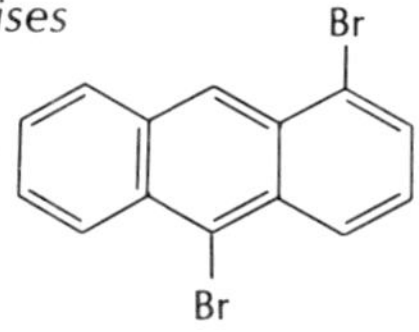

14.2e. 2-bromo-1,4-dichlorobenzene.

14.3. $H_2\overset{\oplus}{C}-CH=CH-CH=CH_2 \longleftrightarrow H_2C=CH-CH=CH-\overset{\oplus}{C}H_2$

$\qquad\qquad\searrow \qquad\qquad\nearrow$
$\qquad H_2C=CH-\overset{\oplus}{C}H-CH=CH_2$

14.8. Cl—[benzene ring]—COCH$_3$.

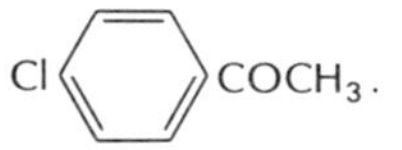

14.11. As shown in Section 14.4, a magnetic field induces a ring current in an aromatic system of such magnitude as to cause a marked downfield shift for protons attached to the ring. Thus, azulene shows a complex nmr with δ 7.11 to 8.32 ppm, definitely in the aromatic region.

14.13c. $\phi_2C=CH_2 \xrightarrow{Br_2}{MeOH}$ $\phi_2\overset{\oplus}{C}CH_2Br \longrightarrow \phi_2CCH_2Br.$
$\qquad\qquad\qquad\qquad\qquad\qquad\qquad\qquad\qquad\qquad\quad |$
$\qquad\qquad\qquad\qquad\qquad\qquad\qquad\qquad\qquad\quad MeO$

14.15a. $\phi_2CHCH=CH_2 \xrightarrow[H_2SO_4]{K_2Cr_2O_7} \phi_2C=O + 2CO_2$.

Problems

14.2f. A pentavalent carbon! This structure was given once by a student in answer to a question about what factors helped to stabilize this anion:

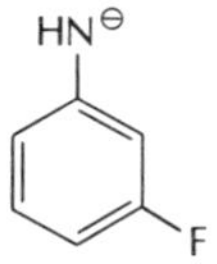

[structure: meta-fluoro phenyl with HN⊖]

14.3c. $\longrightarrow$
$\qquad\quad CH_3$
$\qquad\quad |$
$\quad CH_3\underset{\oplus}{C}-CH=CH_2$ is more stable than
$\qquad\quad CH_3$
$\qquad\quad |$
$\quad CH_2=C-\underset{\oplus}{C}H-CH_3$

$\qquad\qquad \downarrow Br^{\ominus}, -80°$

$\qquad\quad CH_3$
$\qquad\quad |$
$\quad CH_3C-CH=CH_2$
$\qquad\quad |$
$\qquad\quad Br$

14.4b. [diene + dienophile structures with CH$_3$O, CH$_2$, HC, CHO groups] $\longrightarrow$ [cyclohexene product with CH$_3$O and CHO], is the final step.

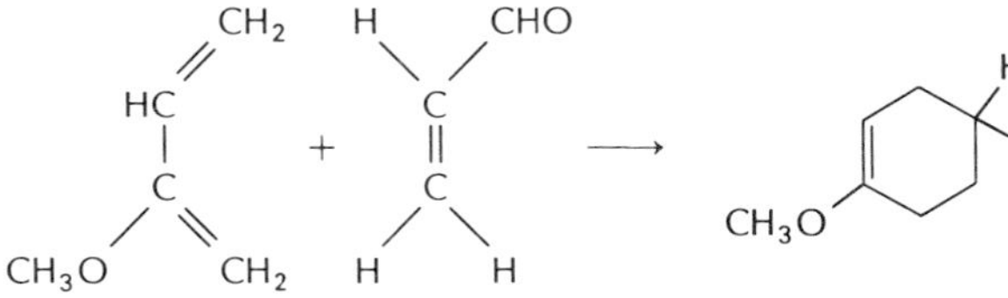

Chapter 15

Exercises

15.1e. [benzene with NH$_2$ (top), Cl, Br, NO$_2$]

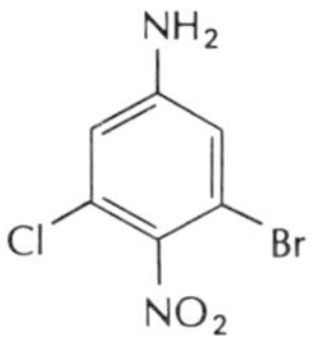

15.2e. 4-ethoxyisobutyrophenone (or para) or 1-(4-ethoxyphenyl)-2-methyl-1-propanone.

1002 Selected answers

15.3d. [biphenyl with ring A (positions 6,5,1) and ring B (positions 2,3,4) with CH₃ at 2 and Br at 4] Ring B is more activated than ring A, and the 4 position has the least steric hindrance.

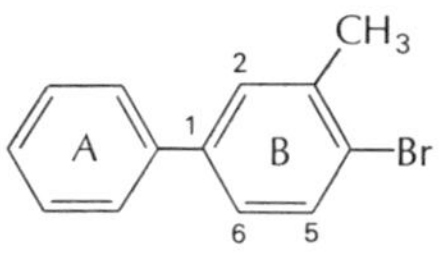

i. 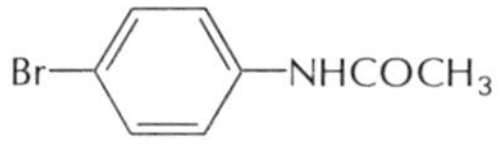 Br—⌬—NHCOCH₃

15.4a. [1-methyl-4-nitronaphthalene]

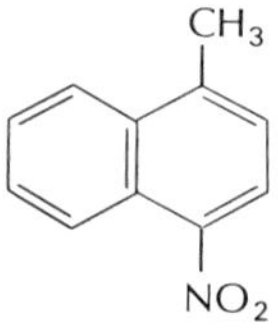

c. 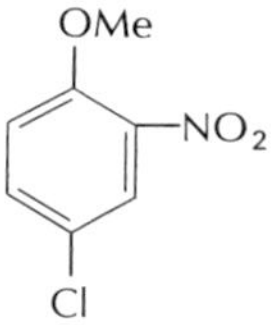 [benzene with OMe, NO₂ ortho, Cl para]

15.5b. 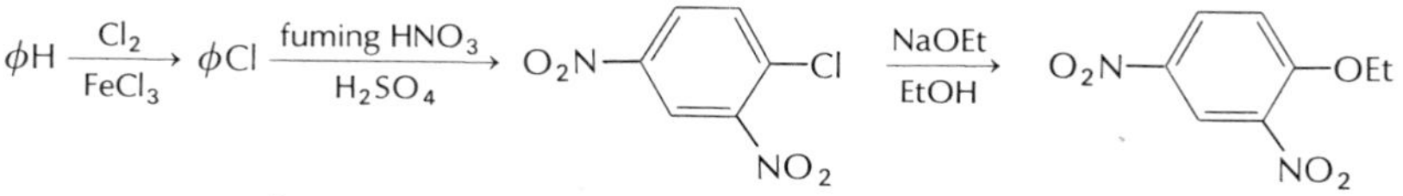 φH $\xrightarrow{Cl_2, FeCl_3}$ φCl $\xrightarrow{\text{fuming } HNO_3, H_2SO_4}$ O₂N—⌬—Cl (with NO₂) $\xrightarrow{\text{NaOEt, EtOH}}$ O₂N—⌬—OEt (with NO₂)

15.6c. φCH₂CH₂CH₃ $\xrightarrow{Br_2, h\nu}$ φCHBrCH₂CH₃ $\xrightarrow{\text{alc. KOH}}$ φCH=CHCH₃ $\xrightarrow{Br_2, CH_3COOH}$ φCHBrCHBrCH₃.

Problems

15.3a. A, Cl less deactivating than COOCH₃.   e. B, —OCOCH₃ activates, —COOCH₃ deactivates.

15.4h. [3-chloronitrobenzene]   j. 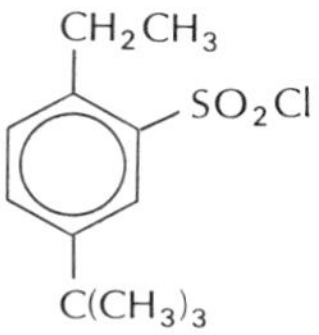 [benzene with CH₂CH₃, SO₂Cl, C(CH₃)₃ substituents]

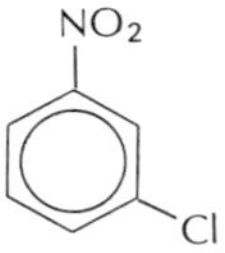

15.5d. [mechanism: anisole + Cl—Cl → arenium ion intermediate → 4-chloroanisole]

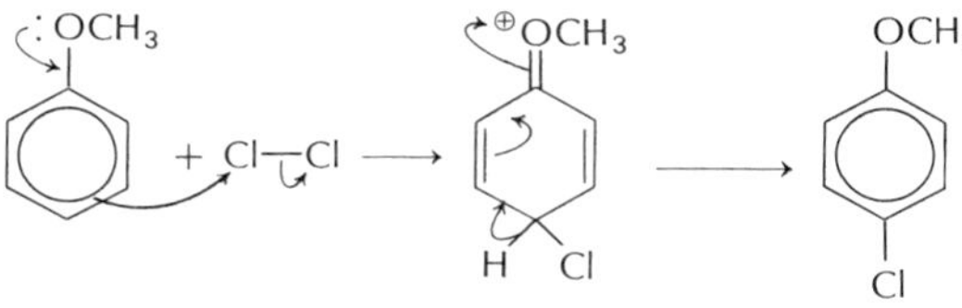

Chapter 16

Exercises

16.3. φCOCH₂CH₂CH₃. Note the strong ir band at about 1690 cm⁻¹, indicative of C=O conjugated with C=C or φ—. Likewise, note the nmr separation of the aromatic protons because of the effect of C=O on the ring.

16.7c. O=C(CH₃)(CH₂—CH₂)CHO $\xrightarrow{\text{base}}$ [cyclopentenone] (see Equation 16.50b);

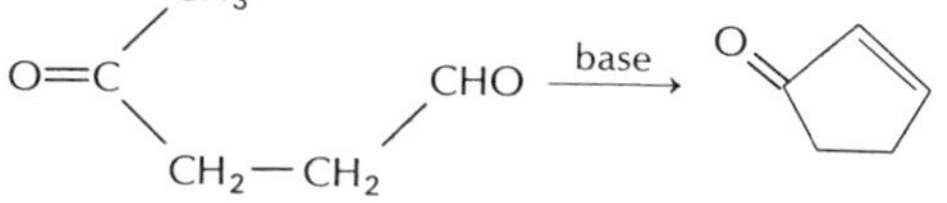

f. Claisen-Schmidt condensation (Eq. 16.47). φCHO + CH₃COiPr $\xrightarrow{\text{base}}$ φCH=CHCOiPr.

16.8a. CH₃CH₂CHO $\xrightarrow[\text{hot}]{\text{base}}$ CH₃CH₂CH=C(CH₃)—CHO $\xrightarrow{\text{cat. } H_2}$ CH₃CH₂CH₂CH(CH₃)CH₂OH.

16.11b. $\phi CH_2CO\phi \xrightarrow[\text{Wolff-Kishner}]{\text{Clemmensen or}} \phi CH_2CH_2\phi$.   16.12a. $\phi CH_2CH_2CHO \xrightarrow[25°]{\text{NaOH}}$

$\phi CH_2CH_2CHOHCHCHO$ (Eq. 16.40, aldol condensation).
$\hspace{2.5cm}|$
$\hspace{2.2cm}\phi CH_2$

Problems

16.3a. 4-phenoxy-2-butenal; c. 2,2-dimethyl-4-ethylcyclohexanone.

16.5a. $CH_3CH=CHCO\overset{\ominus}{O}NH_4^{\oplus}$ b. $C_6H_5\underset{\underset{COOH}{|}}{C}=CHCH_2OH$

c. 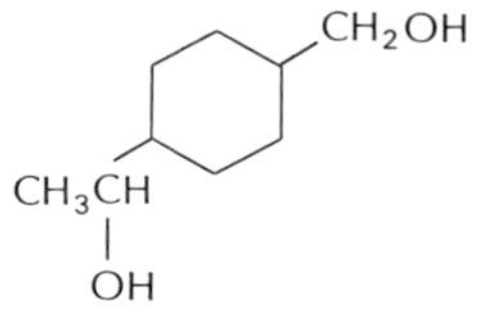 (cyclohexane with CH_2OH and $CH_3\underset{\underset{OH}{|}}{CH}$ substituents) + $HOCH_2CH_2C_6H_5$

16.8c. A; less hindrance; product is $CH_3CHOHCH_2CH=O$.   16.9b. $(CH_3)_2\underset{\underset{CH_2OH}{|}}{C}CHO$

f. 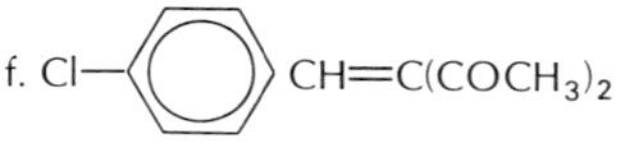 $Cl-\langle\bigcirc\rangle-CH=C(COCH_3)_2$ 16.10a. $^{\ominus}O\overset{O}{\overset{\|}{P}}OCH_2\underset{\underset{OH}{|}}{CH}\underset{\underset{HO}{|}}{CH}-\underset{\underset{OH}{|}}{CH_2}COCOOH$

Chapter 17

Exercises

17.1a. $\phi CH_2COOH \xrightarrow[Br_2]{P} \phi CHBrCOOH \xrightarrow[EtOH]{H^{\oplus}} \phi CHBrCOOEt$

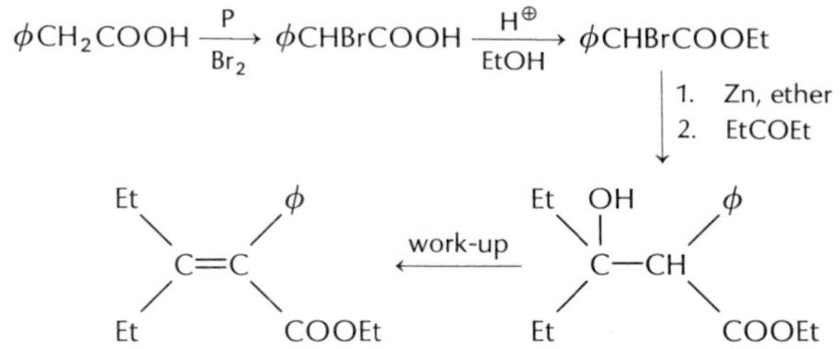

$\hspace{9cm}\downarrow \begin{array}{l}1.\ \text{Zn, ether}\\2.\ EtCOEt\end{array}$

$\underset{Et}{\overset{Et}{\diagdown}}C=C\underset{COOEt}{\overset{\phi}{\diagup}} \xleftarrow{\text{work-up}} \underset{Et}{\overset{Et\ OH}{\diagdown}}\underset{}{\overset{|}{C}}-\underset{COOEt}{\overset{\phi}{CH}}$

17.4h. $CH_3-\langle\bigcirc\rangle-CHO + (CH_3CH_2CO)_2O \xrightarrow[\Delta]{CH_3CH_2COONa} CH_3-\langle\bigcirc\rangle-\underset{\underset{}{}}{CH}=\overset{\overset{CH_3}{|}}{C}COOH$

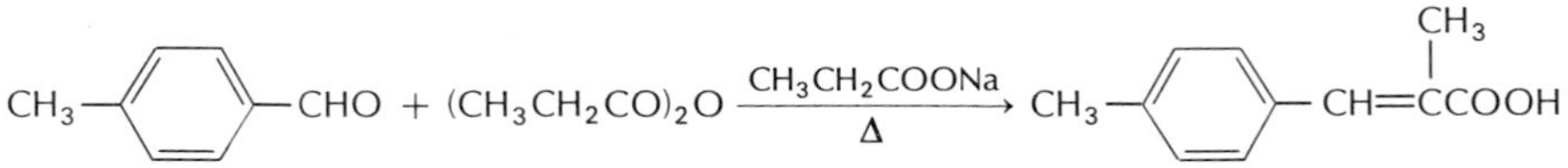

17.7a. $\phi CH_2CH_2COOEt + CH_3MgI \longrightarrow \phi CH_2CH_2\underset{\underset{CH_3}{|}}{\overset{\overset{CH_3}{|}}{C}}OH$ 17.9c. $HCOOEt + \underset{\underset{CH(CH_3)_2}{|}}{CH_2}COOEt \xrightarrow{EtO^{\ominus}}$

$\underset{\underset{}{}}{\overset{\overset{CH(CH_3)_2}{|}}{OHCCHCOOEt}}$ 17.10b. $tBuCHO + \underset{\underset{}{}}{\overset{\overset{\phi}{|}}{CH_2}}COOEt \xrightarrow{EtO^{\ominus}} tBuCH=\overset{\overset{\phi}{|}}{C}COOEt$

17.12c. $\phi COCl \xrightarrow{MeOH} \phi COOMe$.

Problems

17.3a. $\overset{\delta-}{Cl}-\overset{\delta+}{CH_2}CH_2\overset{O}{\overset{\|}{C}}\diagdown_{O}^{\ominus}$ Electron attraction by δ^+ less because of distance of δ^+ from negative charge. $CH_3CHClCOOH$ is the stronger acid.

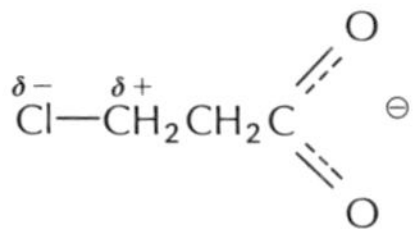

17.4e.

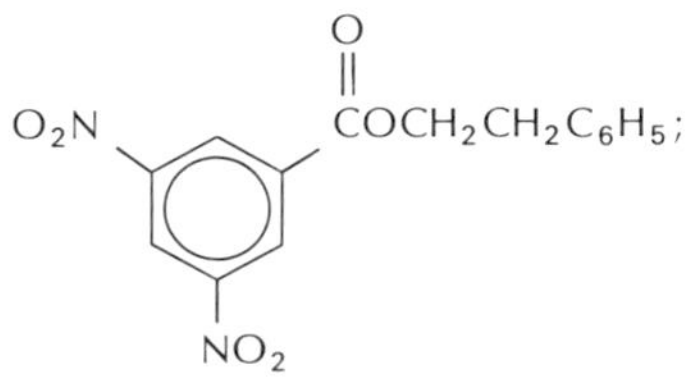

O_2N—(3,5-dinitrophenyl)—$\overset{O}{\underset{\|}{C}}OCH_2CH_2C_6H_5$ (with NO_2 at other position); m. $CH_3(CH_2)_2\overset{O}{\underset{\|}{C}}-\overset{\ominus}{C}H-\overset{O}{\underset{\|}{C}}C_6H_5$ 17.5c. $\xrightarrow{NH_2CH_2C_6H_5}$

17.8a. A; $Cl^{\ominus}$ is a better leaving group than $^{\ominus}OCH_3$, which is a strong base;   h. A; less hindrance in transition state.

Chapter 18

Exercises

18.1c. $CH_3CHOHCH_2COOH \xrightarrow{warm} CH_3CH=CHCOOH$;

e. $CH_3CHOHCH_2CH_2CH_2COOH \xrightarrow{\Delta}$ [δ-valerolactone, H₃C on ring with O and C=O]

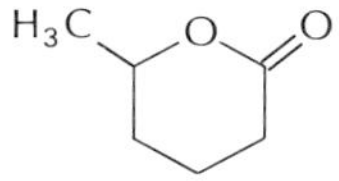

18.2c. Transesterification;

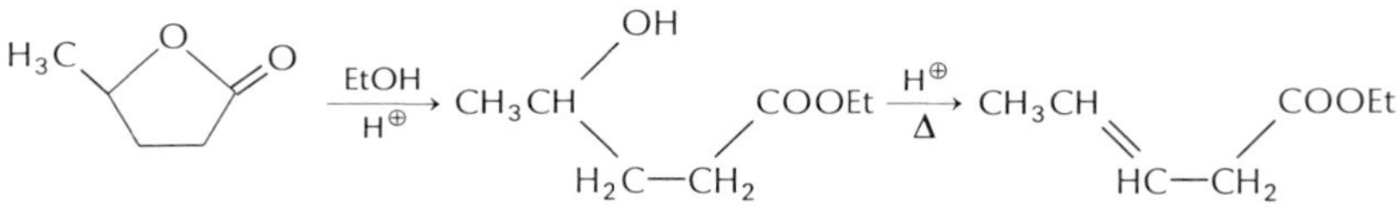

[H₃C-γ-butyrolactone] $\xrightarrow[H^{\oplus}]{EtOH}$ $CH_3\overset{OH}{\underset{H_2C-CH_2}{CH}}\diagdown COOEt$ $\xrightarrow[\Delta]{H^{\oplus}}$ $CH_3CH=\overset{}{\underset{HC-CH_2}{}}\diagdown COOEt$

18.4b. $CH_3CH_2CH=CHCOOEt + H_2C(COOEt)_2 \xrightarrow{EtO^{\ominus}} CH_3CH_2\diagdown CH-CH_2COOEt$, with $HC(COOEt)_2$

c. $H_2C=CHCOOEt + H_2C=CHCH=CH_2 \longrightarrow$ [cyclohexene with H and COOEt]

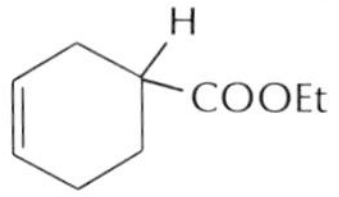

18.5a. $\phi H +$ [succinic anhydride] $\xrightarrow{AlCl_3} \phi COCH_2CH_2COOH \xrightarrow{NaBH_4} \phi CHOHCH_2CH_2COOH$ $\xrightarrow{\Delta} \phi CH=CHCH_2COOH$

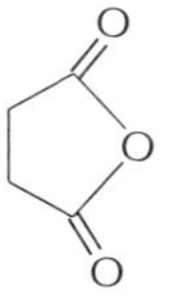

18.8a. AAE $\xrightarrow[2.\ EtBr]{1.\ EtO^{\ominus}}$ $CH_3COCHCOOEt$ with CH_2CH_3 $\xrightarrow[2.\ H^{\oplus}, \Delta]{1.\ dil.\ NaOH}$ CH_3COCH_2 with CH_2CH_3

18.9a. $\phi COCl + KCN \longrightarrow \phi COCN \xrightarrow[H_2O]{H^{\oplus}} \phi COCOOH$.   18.2d. $CH_3CH=CHCN$.

Problems

18.2d. $CH_3CH=CHCN$.   18.3a. Right; ring closure to a six membered ring is favored; internal aldol;   b. Right; CCl_3 group destabilizes $CH=O$;   c. Right; product has more stable sp³, little hindrance.

Selected answers 1005

Chapter 19

Problems

19.2a. HNO$_3$ gives *meso* dicarboxylic acid from A;   c. A gives positive reducing test with Fehlings' solution, also mutarotates;   h. A gives reducing test with Tollen's reagent.   19.3c. Br$_2$—H$_2$O, then Ruff degradation;   g. HNO$_3$.   19.4b. L-lyxose;   d. D-lyxose.

Chapter 20

Exercises

20.1c. [cyclohexyl-C(NHMe)(H)] + φCOCl + NaOH ⟶ [cyclohexyl-C(N(Me)(COφ))(H)] + NaCl + H$_2$O

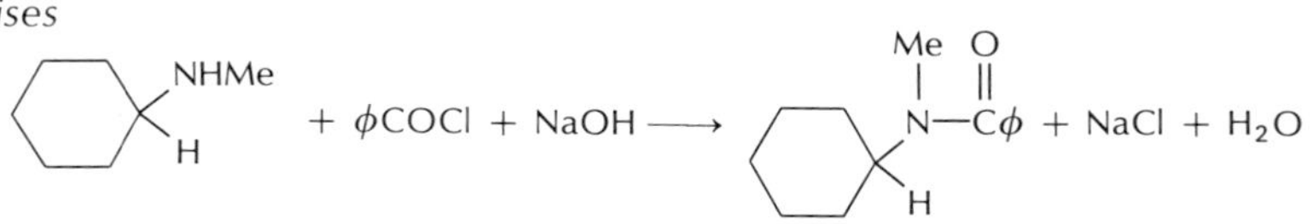

k. [cyclohexyl-C(NH$_2$)(H)] $\xrightarrow{\text{PrI}}$ [cyclohexyl-C(NHPr)(H)] $\xrightarrow{\text{EtI}}$ [cyclohexyl-C(N(Et)(Pr))(H)] $\xrightarrow{\text{MeI}}$ [cyclohexyl-C(N$^+$(Et)(Pr)(Me))(H)] I$^-$

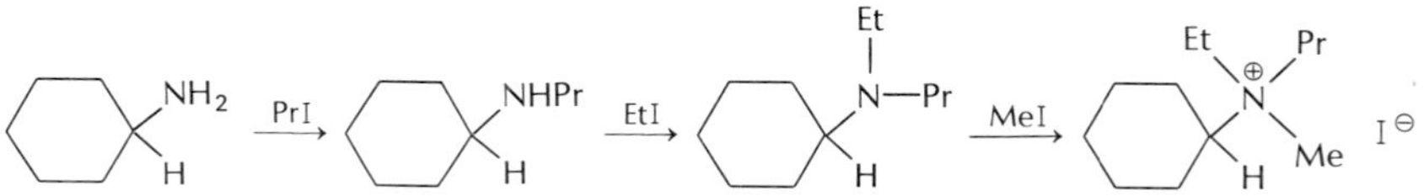

(Four different groups about the nitrogen are needed.)   20.2b. tBuOH $\xrightarrow[\text{H}_2\text{SO}_4]{\text{NaCN}}$ tBuNHCHO $\xrightarrow[\text{H}_2\text{O}]{\text{NaOH}}$ HCOONa + tBuNH$_2$;   c. Me$_2$CHC≡N $\xrightarrow[\text{ether}]{\text{LiAlH}_4}$ Me$_2$CHCH$_2$NH$_2$.   20.3d. CH$_3$COCl + Et$_2$NH ⟶ CH$_3$CONEt$_2$ + HCl;   g. H$_3$$^+NCH_2CH_2CH_2COO^-$ $\xrightarrow{\Delta}$ [pyrrolidinone] + H$_2$O.

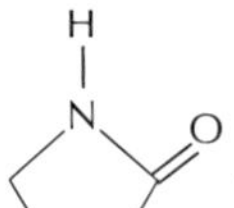

20.4a. NCCH$_2$CH$_2$CN + EtOH $\xrightarrow{\text{H}_2\text{SO}_4}$ EtOOCCH$_2$CH$_2$COOEt + (NH$_4$)$_2$SO$_4$;

b. NCCH$_2$COOEt + H$_2$C=CHCN $\xrightarrow{\text{base}}$ (NC)(EtOOC)CHCH$_2$CH$_2$CN;

c. H$_2$C=CHCH$_2$Cl + NaCN $\xrightarrow{\text{DMSO}}$ H$_2$C=CHCH$_2$CN + NaCl.
20.10b. EtOOCCHN$_2$ + CH$_3$CH$_2$COOH ⟶ CH$_3$CH$_2$COOCH$_2$COOEt + N$_2$.

Chapter 21

Exercises

21.6. His reasoning was good, but he did not go far enough. Although units 17–20 were necessary for the 1–20 peptide activity, it does not follow that only a 17–20 tetrapeptide would show activity. By continuing to build to the left from Arg—Arg—Pro—Val, it would be of great interest to find how many units are required for activity to appear.

Problems

21.6a. B + HNO$_2$ ⟶ N$_2$; A + HNO$_2$ ⟶ N-nitroso Pro, no N$_2$;   e. B gives positive ninhydrin test.   21.8. Phe. Asp. Pro. Phe. Tyr. Glu.

Chapter 22

Exercises

22.3b. The hydroxylamino group must be produced last because an attempt to reduce the nitro of m-nitrophenylhydroxylamine would reduce both groups and m-phenylenediamine would be formed.

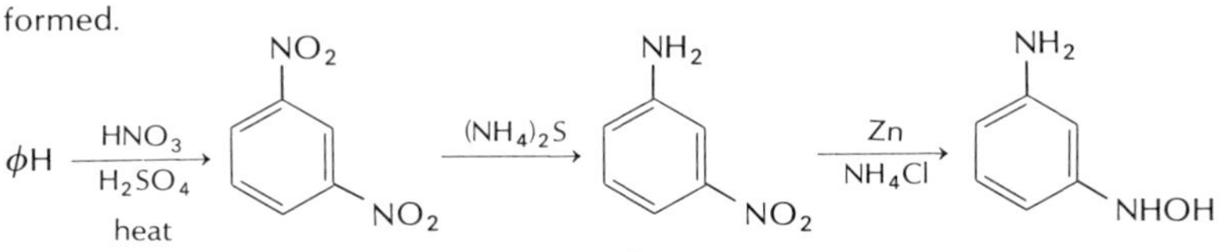

$\phi H \xrightarrow[\text{heat}]{HNO_3, H_2SO_4} m\text{-}C_6H_4(NO_2)_2 \xrightarrow{(NH_4)_2S} m\text{-}NO_2\text{-}C_6H_4\text{-}NH_2 \xrightarrow[NH_4Cl]{Zn} m\text{-}HOHN\text{-}C_6H_4\text{-}NH_2$

22.10b. $\phi H \xrightarrow[FeBr_3]{Br_2} \phi Br \xrightarrow{HNO_3, H_2SO_4} p\text{-}Br\text{-}C_6H_4\text{-}NO_2 \xrightarrow[NH_4Cl]{Zn} p\text{-}Br\text{-}C_6H_4\text{-}NHOH$

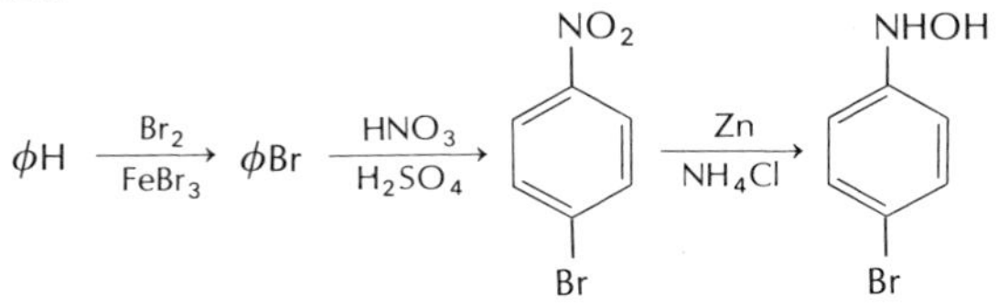

d. 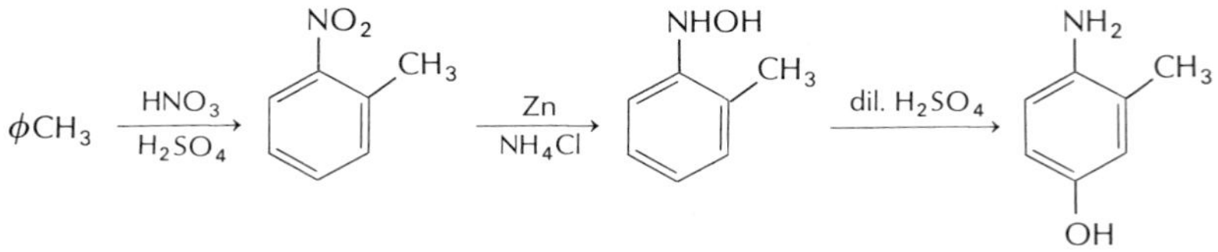 $\phi CH_3 \xrightarrow{HNO_3, H_2SO_4} o\text{-}NO_2\text{-}C_6H_4\text{-}CH_3 \xrightarrow[NH_4Cl]{Zn} o\text{-}HOHN\text{-}C_6H_4\text{-}CH_3 \xrightarrow{dil. H_2SO_4} \text{4-amino-3-methylphenol}$

22.12a. $\phi H \xrightarrow[FeBr_3]{Br_2} \phi Br \xrightarrow{HNO_3, H_2SO_4} p\text{-}Br\text{-}C_6H_4\text{-}NO_2 \xrightarrow[DMSO]{NaCN} p\text{-}NC\text{-}C_6H_4\text{-}NO_2 \xrightarrow{HCl, H_2O} p\text{-}HOOC\text{-}C_6H_4\text{-}NO_2$

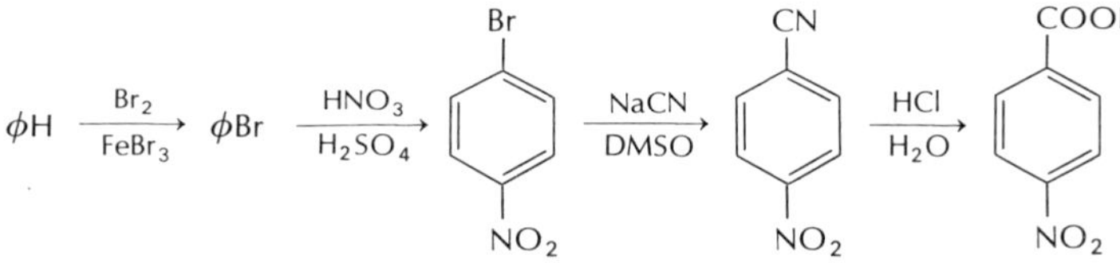

Problems

22.2a. (1) HBr, NaNO$_2$, (2) Cu$_2$Br$_2$; c. 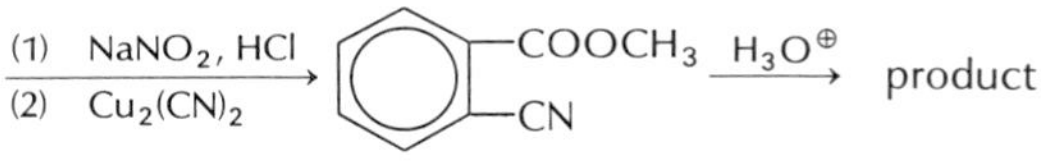 $\begin{array}{l}(1)\ NaNO_2, HCl \\ (2)\ Cu_2(CN)_2\end{array} \rightarrow o\text{-}NC\text{-}C_6H_4\text{-}COOCH_3 \xrightarrow{H_3O^\oplus} \text{product}$

Chapter 23

Exercises

23.5c. $\phi CH_3 \xrightarrow[FeBr_3]{Br_2} p\text{-}Br\text{-}C_6H_4\text{-}CH_3 \xrightarrow[\Delta]{Cu} CH_3\text{-}C_6H_4\text{-}C_6H_4\text{-}CH_3$.

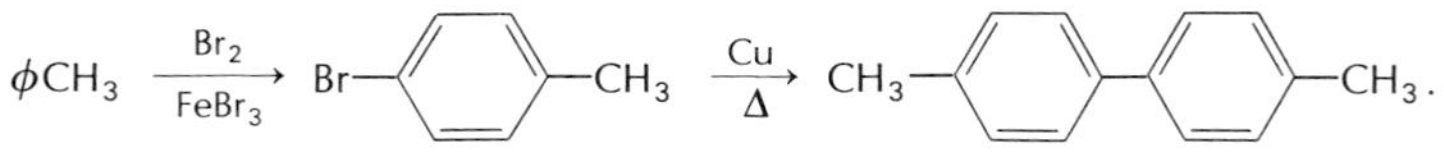

Chapter 24

Exercises

24.2. A methoxyanthracene has mw = 208, so the sample of 0.0208 g is 0.0001 mole. The reactions are: $C_{14}H_9OCH_3 \xrightarrow[\text{boil}]{HI, H_2O} C_{14}H_9OH + CH_3I \xrightarrow[\text{EtOH}]{AgNO_3} AgI + CH_3OEt + HNO_3$. The precipitate is AgI, mw = 235. Therefore 0.0001 mole of AgI is 0.0235 g. The advantage of the Zeisel methoxy determination lies in the high equivalent weight of AgI as compared with only 31 for CH$_3$O—.

24.5c. [structure: 4-methyl-2-nitrosophenol ⇌ quinone monooxime tautomer]

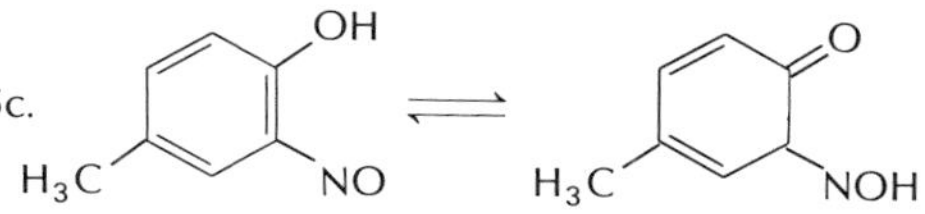

f. 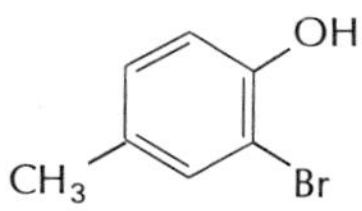 [structure: 2-bromo-4-methylphenol]

j. 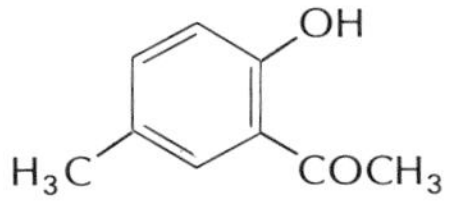 [structure: 2-acetyl-4-methylphenol]

n. 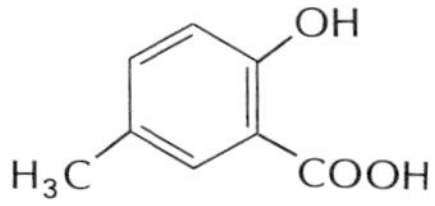 [structure: 5-methylsalicylic acid]

24.10a. [1,4-benzoquinone] + HCN ⟶ [2-cyano-4-hydroxyphenol-type intermediate] $\xrightarrow{H^{\oplus}/H_2O}$ [salicylic acid derivative]

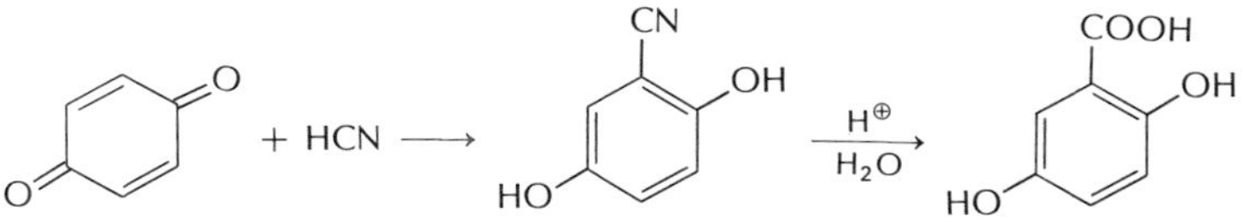

24.11a. $\phi Me \xrightarrow[h\nu]{Cl_2} \phi CH_2Cl \xrightarrow{NaCN} \phi CH_2CN \xrightarrow{LiAlH_4} \phi CH_2CH_2NH_2$

c. 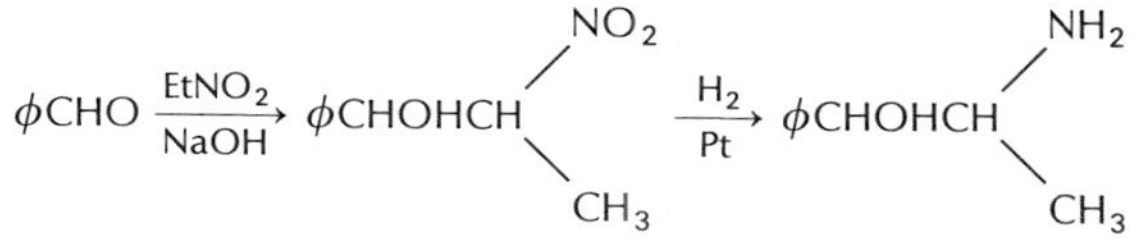 $\phi CHO \xrightarrow[NaOH]{EtNO_2} \phi CHOHCH(NO_2)CH_3 \xrightarrow[Pt]{H_2} \phi CHOHCH(NH_2)CH_3$

Chapter 25

Exercises

25.1c. [naphthalene] $\xrightarrow[H_2SO_4]{HNO_3}$ [1-nitronaphthalene] $\xrightarrow[HCl]{Sn}$ [1-naphthylamine]

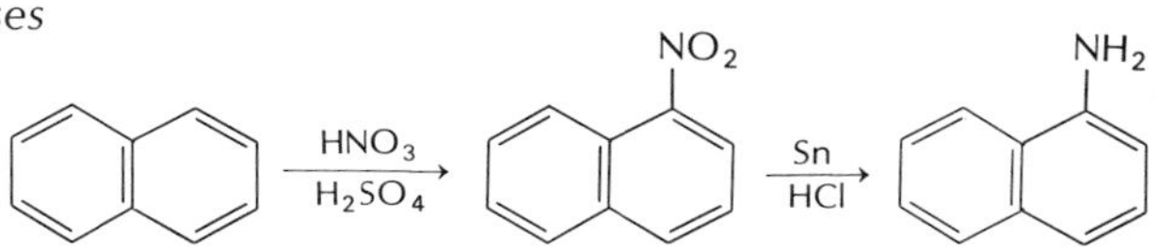

25.4b. [m-xylene] $\xrightarrow[\phi H]{AlCl_3, \text{phthalic anhydride}}$ [keto-acid intermediate] $\xrightarrow[100°]{H_2SO_4}$ [dimethylanthraquinone]

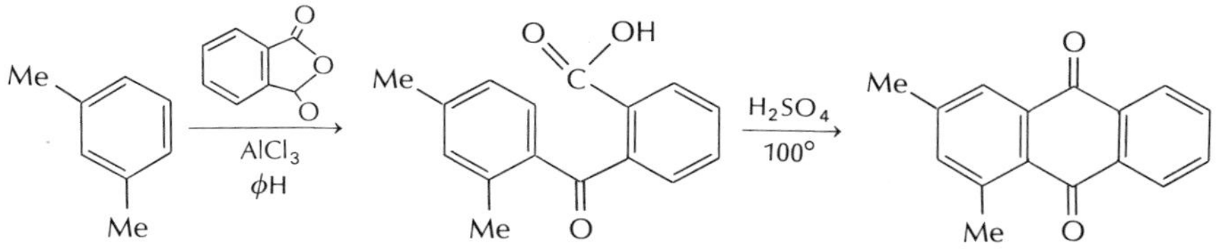

25.8b. The dimethyl compound racemizes faster.

Chapter 27

Exercises

27.4. [estrone structure] $\xrightarrow[\text{excess}]{NaC\equiv CH}$ [17α-ethynyl estradiol structure]

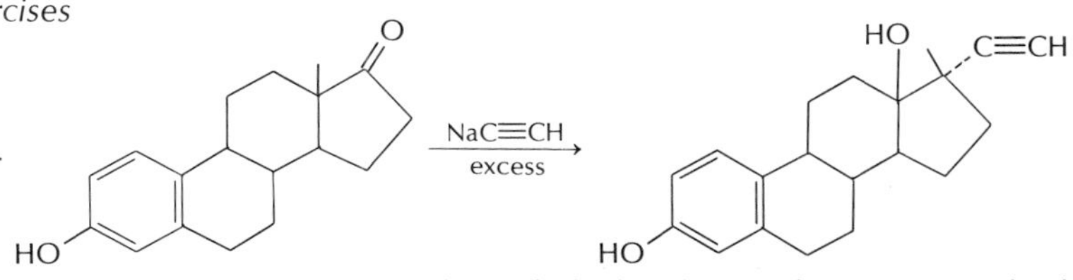

Excess NaC≡CH is needed because the acidic hydroxyl group destroys one mole of the reagent and a second mole is needed to add to the keto (attack from the α side). 27.6a. Butadiene.

Chapter 28

Exercises

28.1a. Me–[epoxide]–Me (or *trans*)

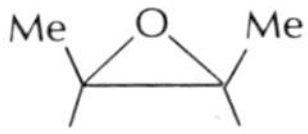

c. 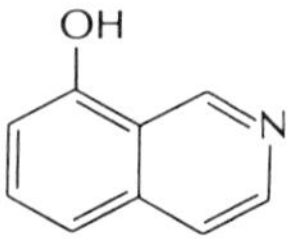 8-hydroxyisoquinoline structure

28.2c. AAE $\xrightarrow{\phi NHNH_2}$ CH$_3$C(=N-NH-ϕ)-CH$_2$-COOEt $\xrightarrow[-EtOH]{warm}$ pyrazolone (CH$_3$C=N-N(ϕ)-C(=O)-CH$_2$ ring)

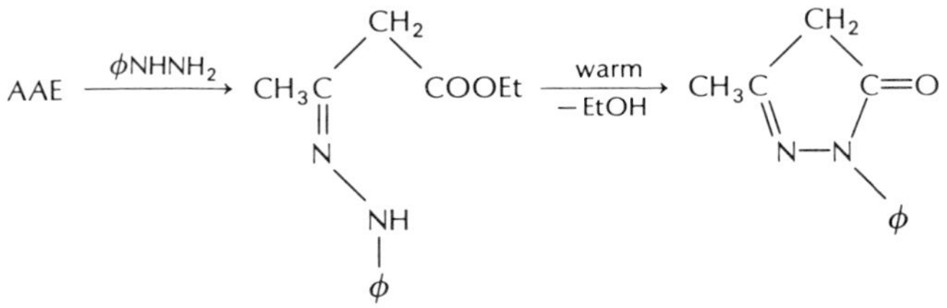

Problems

28.2a. 3,5-dimethylmorpholine; g. 3,4,5-trimethylthiazolium chloride; m. 4-amino-6-hydroxy-2-methylpyrimidine. 28.3a. 1-phenyl-3,5-dimethylpyrazole (H$_3$C at 3, CH$_3$ at 5, C$_6$H$_5$ on N1).

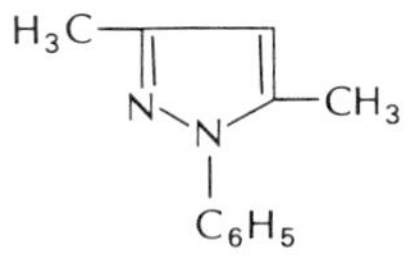

e. 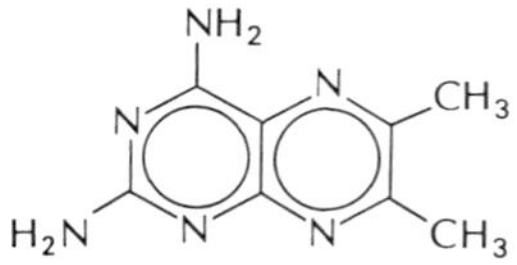 2,4-diamino-6,7-dimethylpteridine structure

Chapter 29

Exercises

29.2a. iPrBr $\xrightarrow[ether]{Mg}$ iPrMgBr $\xrightarrow{S_8}$ iPrSH; or iPrBr $\xrightarrow{KSCN}$ iPrSCN $\xrightarrow{H_2O}$ iPrSH;

e. iPrBr $\xrightarrow{H_2S/NaOH}$ iPrS(iPr) $\xrightarrow[CH_3COOH]{Cl_2, H_2O}$ iPrSO$_2$Cl. 29.3b. ϕCOCH$_2$CH$_3$ $\xrightarrow[NaH]{Me_2SO}$ ϕCO$\overset{\ominus}{C}$HCH$_3$ $\xrightarrow{CO_2}$ ϕCOCH(COOH)CH$_3$

d. Me$_2$SO $\xrightarrow{MeI}$ Me$_3$S–O$^\ominus$ (2+) $\xrightarrow{NaH}$ Me$_2$S$^{(2+)}$(–O$^\ominus$)(–$\overset{\ominus}{C}$H$_2$) $\xrightarrow{\phi CH=CH\phi}$ ϕHC–CHϕ (cyclopropane with CH$_2$)

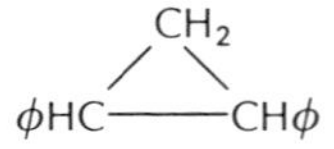

29.4b. ϕSO$_3$H $\xrightarrow{SOCl_2}$ ϕSO$_2$Cl $\xrightarrow[NaOH]{\phi CH_2CH_2OH}$ ϕSO$_2$OCH$_2$CH$_2\phi$; 29.5a. ethyl diphenylphosphinite; b. phenylphosphonic acid; d. triethyl phosphite (or phosphorite).

Index

ABC reaction, **23**, **143**
ABC system, **284**, **410**
AB₂C system, **410**
A₂BC₂ system, **410**
A₂B₂C₃ system, **280**, **281**
Abietic acid, **860**
A₂B₂M₂ system, **282**, **284**
Abnormal addition, **781**
A₂B₂ pattern, **410**
Absolute entropy, **58**
Absolute specificity, **725**, **726**
Absorbance, **244**
Absorptivity (extinction coefficient), **244**; biological uses, **263**
AB system, **271**
Acenaphthene: physical properties, **444**, **841**; structural formula, **442**
Acenaphthenequinone, reactions with organometallic compounds, **822a**
Acenaphthoquinone, **841**
Acenaphthylene, **841**
Acetaldehyde, **68**; metaldehyde, preparation from, **492**; nmr spectrum, **286**, **287**; paraldehyde, preparation from, **492**; physical properties, **36**, **480**, **481**; preparation from acetylene by hydration, **48**, **354**; preparation from ethanol, **53**; production figures, **378**; reactions, addition of water, **48**, **49**; aldol condensation, **504–505**; haloform reaction, **503**; hydration, **490**; oxidation, **53**; trimerization, **492–493**; reactions with, acetophenone, **508**; base, **504–505**; formaldehyde, **508**; sodium hydroxide, **507**; steric effects, **494**
Acetals, preparation, **491**
Acetamide, **649**; absorption peaks, **261**; physical properties, **666**
Acetamide, N-substituted, reactions, **662**
p-Acetamidobenzenesulfonyl chloride, preparation, **751**
Acetamido group, **752**
Acetamidomalonic ester, **701**

Acetamidomalonic ester synthesis, **701** ff.
Acetanilide: basicity, **748**; reactions, **748**, **749**
Acetate ion: metabolic reactions and, **867**; resonance structure, **99–100**
Acetic acid: absorption peaks, **261**; acidity, **547**; autoprotolysis constant, **22**; biosynthesis of terpenes, steroids, and fatty acids from, **867–872**; dimerization, degree of, **546**; IUPAC naming, **18**; physical properties, **36**, **538**, **543**; pK$_a$, **108**, **695**; preparation, **549**; production figures, **378**; reactions, decarboxylation, **553**; esterification, **207–208**; Kolbe electrolysis, **382**; reactions with, ethanol-methanol mixture, **194–195**; sodium hydroxide, **382**; toluene, **455**; resonance structure, **99**; solvent effect of on nucleophilic substitution, **153**; use as solvent, **53**, **375**
Acetic anhydride: production figures, **378**; solvent effect of on nucleophilic substitution, **153**; use, **548**
Acetoacetic acid: ketosis, **487**; physical properties, **71**
Acetoacetic ester, **570**; infrared spectrum, **602**; isoxazolone, preparation from, **894**; nmr spectrum, **602**, **603**; physical properties, **71**; γ-pyrone, preparation from, **896**; reactions, acylation, **600**; alkylation, **598**; cleavage, **597** ff.; condensation and cyclization, **896**; enolization, **600–604**; haloform reaction, **604**; Hantzsch synthesis, **899–901**; reactions with, dihalides, **600**; hydrazine, **893**; sodium ethoxide, **598**
Acetoacetic ester synthesis, **597–600**, **605**
2-Acetocyclohexanone,

preparation, **607**
Acetohydrazide, **649**
Acetohydroxamic acid, **649**
Acetone, **68**; as catalyst for oxidation of alcohols, **218–219**; farnesol, preparation from, **855**; infrared spectrum, **488**; ketosis, **487**; physical properties, **69**, **480**, **481**; preparation from pyroligneous liquid, **521**; production figures, **378**; reactions, addition of enolate ion to, **506**; aldol condensation, **507**; hydration, **491**; reactions with ethyl magnesium bromide, **494**; solvent effect of on nucleophilic substitution, **153**, **155**; ultraviolet spectrum, **259**; use as solvent for acetylene, **385**
Acetone azine, **649**
Acetone hydrazone, **649**
Acetone oxime, **649**
Acetone semicarbazone, **649**
Acetonitrile, **671**; solvent effect of on nucleophilic substitution, **153**
Acetophenone, **441**; physical properties, **481**; reactions with acetaldehyde, **508**
Acetoxime, preparation, **677**
Acetylacetone: nmr spectrum, **602**, **603**; physical properties, **509**; stabilization, **509**
N-Acetylamino acid, enzymatic hydrolysis, **704**
Acetylation, **560**; toluidines, **750**; use in peptide synthesis to protect amino groups, **709–710**
Acetyl azide, **649**
Acetyl chloride: absorption peaks, **261**; physical properties, **36**; reduction, **53–54**
Acetylcholine esterase, **726**
Acetylene: covalent bonding in, **5**; double bond, **45**; farnesol, preparation from, **855**; haloalkanes and haloalkenes, preparation from, **777–778**; molecular orbital theory, **94–95**; nmr spectrum, **287**; physical

Acetylene (cont.)
  properties, **36, 331**; preparation, **384–385**; preparation from vinyl chloride, **42**; reactions, addition of acid to, **47, 353**; catalytic hydrogenation, **55**; hydration, **48, 354**; oxidation by permanganate, **55**; polar addition of halogen, **46**; reactions with nucleophiles, **331–332**; reactions with peracids, **48**; reactivity, **48**; uses, **385, 386**
Acetylenedicarboxylic acid, physical properties, **594**
Acetylenic Grignard reagents, **332**
Acetylenic protons, **284 ff.**
N-Acetylglycine, condensation with aldehyde, **702, 704**
O-Acetylhydroxylamine, **649**
Acetylsalicylic acid, **104**
Acid anhydrides: acids, preparation from, **558**; carbonyl stretching frequency, **545**; infrared spectra, **253**; physical properties, **543, 544**; preparation, **565–566**; preparation from acid chlorides, **561**; preparation from acids, **548–549**; reactions, acylation, **511**; addition, **562**; alcoholysis, **580**; condensation, **563–564**; exchange reactions, **562**; Friedel-Crafts, **454, 562**; hydrolysis, **558**; olysis, **561 ff.**; Perkin reaction, **563**; reduction, **562–563**; reactions with alcohols, **212–213**
Acid anhydridolysis, **551**
Acid azides, use in peptide synthesis, **711**
Acid-base equilibria, **20 ff.**; alcohols, **206**; between ammonia and alkylammonium ion, **41**; carbonyl group, **484–485**; S_N2 reactions and, **24**
Acid-base reactions, **19, 20–23**
Acid bromides: Hell-Volhard-Zelinsky reaction, **555–556**; preparation, **562**
Acid chlorides: acids, preparation from, **558**; of amino acids, **700**; carbonyl stretching frequency, **545**; infrared spectra, **253**; physical properties, **543, 544**; preparation, **565**; preparation from acids, **547–548**; of pyridine carboxylic acids, **899**; pyruvic acid, preparation from, **605**; reactions, acylation, **511**; addition, **562**; alcoholysis, **580**; exchange reactions, **562**; Friedel-Crafts, **454, 562**; hydrogenolysis to yield aldehydes, **229–230**; hydrolysis, **558**; olysis, **560 ff.**; reduction, **562–563**; reduction with sodium borohydride, **230**; Rosenmund reduction, **529**; reactions with, alcohols, **212–213**; cadmium reagents, **523**; lithium aluminum hydride, **562**; sodium azide, **683, 684**; sodium borohydride, **562**; thiols, **945, 946**; ylides, **523–524**
Acid chloridolysis, **551**
Acid cleavage, of acetoacetic ester, **597–598**
Acid derivatives, **551**; aldehydes, preparation from, **529–530**; α-hydrogen activity, **555**; preparation, **547–551**
Acid fluorides, preparation, **562**
Acid halides: Hell-Volhard-Zelinsky reaction, **555–556**; IUPAC naming, **78**
Acid hydrazides, **770**
Acidic protons, **287**
Acid iodide, preparation, **562**
Acidity: dimerization and, **546**; substituents, effect of on, **547**
Acidolysis, of esters, **567**
Acids, **20**; aliphatic (see Aliphatic acids); aromatic (see Aromatic acids); carbonyl group as, **485**; carbonyl stretching frequencies, **544–545**; carboxylic (see Carboxylic acids); conversion to, acid anhydrides, **548–549, 565**; acid chlorides, **547–548, 565**; amides, **551**; dicarboxylic (see Dicarboxylic acids); formation during drastic oxidation of alkenes, **373**; formation during ozonization of alkenes, **367**; hydroxy acids (see Hydroxy acids); infrared spectrum, **253**; IUPAC naming, **78, 81**; keto acids (see Keto acids); mass spectra, **304**; physical properties, **53, 538, 544**; pK_a, **108–109**; preparation, **557–560**; preparation by haloform reaction of carbonyl compounds, **504**; racemates, resolution, **137–138**; reactions, addition to alkenes and alkynes, **47–48**; addition to ketene, **565**; decarboxylation, **552–555**; deuterium exchange, **555**; dimerization, **546**; esterification, **207–213, 549–550**; halogenation, **555–556**; Hell-Volhard-Zelinsky reaction, **555–556**; racemization, **555**; reduction, **53, 558**; reactions with, diborane, **230–231**; enol ester, **565**; hydrazine, **676**; hydrogen peroxide to form peracids, **368**; hydroxylamine, **676**; lithium aluminum hydride, **230, 231**; oximes, **678**; resonance theory, **999–100**; unsaturated (see Unsaturated acids)
Acids, aliphatic. See Aliphatic acids, decarboxylation
Acids, aromatic. See Aromatic acids
Acids, carboxylic. See Carboxylic acids
Acids, dicarboxylic. See Dicarboxylic acids
Acids, inorganic, esterification, **209–211**
Acids, unsaturated. See Unsaturated acids
Acid strength: of 1-alkynes, **332**; pK_a, expressed by, **22**
Acinitro form, **680**
Acridine, structural formula, **889**
Acrilan, **981**
Acrolein, Skraup synthesis using, **909**
Acrylic acid, physical properties, **69, 594**
Acrylonitrile: preparation, **674**; industrial, **346**; reactions, Michael reaction, **512, 671**; reactions, polymerization, **981**
ACTH. See Adrenocorticotropic hormone (ACTH)
Activated charcoal, **444**
Activated complex, **183**
Activation, of aromatic rings, **452 ff.**
Activation energy, **186 ff.**; halogenation of methane, **336–337**
Acyclic diterpenes, **857**
Acyclic sesquiterpenes, **855**
Acyclic terpenes, **847–850**
Acylase, **704**
Acylation, acetoacetic ester, **600**; aldehydes and ketones, **510–511**; benzene, **469**; hydrazones, **678**; naphthalene, **827, 828**; phenanthrene, **837**; phenols, **798**
Acyl azides, preparation, **683–684**
Acylium ions, **295, 297**
Acyloxy radicals, **993a**
1,2 Addition, **421 ff.**; ketones, unsaturated, **522**; 1-phenyl-butadiene, **426**; quinones, **822a**
1,4 Addition, **421 ff.**; ketones, unsaturated, **522**; Michael reaction, **511**; nitriles, **671**; quinones, **812, 822a**
Addition-elimination reactions, **4, 19, 26–28, 62**; Clemmensen reduction, **381–382**; Wolff-Kishner reduction, **381**
Addition reactions, **32, 45–50**; acid anhydrides, **562**; acid chlorides, **562**; acids, to alkenes and alkynes, **47**; aldehydes and ketones, **490–501**; alkenes, **954**; amines, to carbonyl group, **653**; anthracene, **833**; aromatic compounds, **416**; aromatic nitroso compounds, **768–769**; aromatic ring, **458–459**; benzoquinone, **812**; 1,3-butadiene, **421**; carbon-carbon double bond, **45–46**; carbon-oxygen double bond, **48–49**; competition with allylic substitution, **363–364**; electrophilic (see Electrophilic

addition reactions); esters, **570**; free-radical, to alkenes, **49–50, 961**; free-radical, to alkynes, **49–50**; homolytic (see Homolytic addition reactions); of HCl to 1,3-pentadiene, **419–420**; naphthalene, **830–831**; nucleophilic (see Nucleophilic addition reactions, hydrocarbons); peracid, **48**; phenanthrene, **837**; polar, to ethylene by hydrogen, **46**; preparations using acids, **559**; aldehydes and ketones, **522–524**; alkanes, **381–382**; alkenes, **383**; amides, **669**; esters, **582**; nitriles, **674**; quinones, **812**; silenes, **965**; styrene, **424–426**; thiols, **946**; unsaturated acids, **595**; of water to formaldehyde, **27**
Adenine, **911**; in DNA and RNA, **920, 921**; preparation from uric acid, **912**; structural formula, **921**
Adenosine, in RNA, **921**
Adenosine monophosphate, in enzymatic reactions, **962**
Adenosine diphosphonucleotide (ADP): in enzymatic reactions, **962**; in photosynthesis, **643–644**
Adenosine triphosphonucleotide (ATP): hydrolysis, **147**; in photosynthesis, **643–644**; trinegative ion, structural formula, **961**; use in enzymatic reactions, **962**
Adiabatic process, **56–57**
Adipates, Dieckmann condensation, **572**
Adipic acid: cyclic ketones, preparation from, **554**; nylon, preparation from, **982**; physical properties, **542**; preparation, **560**; preparation by malonic ester synthesis, **577**
ADP. See Adenosine diphosphonucleotide (ADP)
Adrenal cortex, **877**
Adrenal hormones, **877–879**
Adrenaline, structural formula, **806, 816**
Adrenocorticotropic hormone (ACTH), **706**; amino acid sequence in, **723**
A factor, **186**
Agar, **641**
Agaropectin, **640**
Agarose, **641**
Aglycone, **881, 882**
Air pollution, **365–367, 391b**
Alanine, **691, 692**; physical properties, **70**; pK_a, **695**; preparation, **701**; structural formula, **692**
Alanylglutamylleucine, **705**
Alcohol dehydrogenase, **219**
Alcohols: acid-base properties, **206**; alkyl halides, preparation from, **143, 169–170**; azeotropes with water, **206–207**; for beverages, **640**; ethers, preparation from, **214–216**; hydrocarbons, preparation from, **956**; hydrogen bonding, **204–205**; infrared spectra, **250–252**; IUPAC naming, **17, 78**; mass spectra, **304, 305**; physical properties, **53, 201–203, 261**; preparation, by addition reactions, **227–229**; from aldehydes and ketones, **172–174**; from alkenes by hydration, **227**; from alkyl halides, **171**; Cannizzaro reaction, **231–232**; commercial, **233, 353**; Meerwein-Ponndorf-Verley reduction, **231**; by oxidation-reduction reactions, **229–232**; by substitution reactions, **225–226**; reactions, addition to carbonyl group, **491**; addition to alkynes, **346–347**; Chugaev reaction, **223–224**; dehydration, **192**; dehydration to yield alkenes, **171–172, 223**; deuterium exchange, **205**; elimination, **158**; esterification, **207–213**; Fischer esterification, **208** ff.; fragmentation, **304**; Friedel-Crafts reaction, **454**; hydrogenolysis, **380**; Meerwein-Ponndorf-Verley reduction, **218–219**; methylation, **318b**; nucleophilic substitution, **154, 156**; Oppenauer oxidation, **218–219**; oxidation, **53, 372**; oxidation to yield carbonyl compounds, **528–529**; at oxygen-hydrogen bond, **204–205**; $S_N 1cA$ etherification, **214–215**; Williamson ether synthesis, **214**; reactions with, acid chlorides and acid anhydrides, **212**; active metals, **205–206**; hemiacetals, **618, 619**; sodium borohydride, **238a**; thionyl chloride, **211**; steroidal, **318c**; terpenes, **848, 850**; uses, **203**
Alcohols, primary: alkyl halides, preparation from, **169–170**; infrared spectra, **251**; preparation, from esters, **570**; from ethylene oxide and a Grignard reagent, **226**; from formaldehyde, **173**; by trialkylboron decomposition, **358**; reactions, dehydration, **171, 172, 223**; oxidation, **216–219**; oxidation to yield aldehydes, **528**; reactions with, acid chlorides and acid anhydrides, **213**; aldehyde, **491**; phosphorus compounds, **170**
Alcohols, secondary: alkyl halides, preparation from, **169–170**; infrared spectra, **251**; preparation, from an aldehyde, **173, 496**; from alkenes by hydration, **227**; from esters, **570**; from formic ester, **227**; from a ketone, **174**; reactions, dehydration, **171**; dehydration to yield alkenes, **223**; esterification, **208**; oxidation, **216–219**; reactions with, acid chlorides and acid anhydrides, **213**; active metals, **205**; aldehydes, **491**; phosphorus compounds, **170**
Alcohols, tertiary: alkyl chlorides, preparation from, **143, 169**; infrared spectra, **251**; preparation, **471**; from alkenes by hydration, **227**; from esters, **570**; from ketone, **173, 227, 496**; reactions, dehydration, **170, 171**; dehydration to yield alkenes, **223**; esterification, **208, 209**; oxidation, **219**; transesterification, **510–511**; reactions with, acid chlorides and acid anhydrides, **213**; active metals, **205**; aldehyde, **491**; phosphorus compounds, **170**; thionyl chloride, **170**
Alcohol sulfonates, aldehydes and ketones, preparation from, **951–952**
Alcoholysis: acid anhydrides, **561**; esters, **567, 568**; esters, preparation by, **580–581**
Aldehyde group, **484**; introduction by Gattermann-Koch reaction, **520**; reactions, **490** ff.
Aldehydes, **30**; alcohols, preparation from, **172–174**; aromatic (see Aromatic aldehydes); formation during ozonization of alkenes, **367**; glucose, **614** ff.; β-hydroxy acids, preparation from, **593**; infrared spectra, **253, 261, 487**; IUPAC naming, **18, 78**; ketones, comparison with, **68**; mass spectra, **304**; nitriles, preparation from, **674**; nmr spectra, **287**; occurrence, **480**; physical properties, **53, 382, 480–481**; preparation, from acid chloride by hydrogenolysis, **229–230**; by addition reactions, **522–524**; from alcohols by oxidation, **216** ff.; from esters, **570**; Gattermann synthesis, **798**; from α-hydroxy acids by decarbonylation, **590**; from imine by hydrolysis, **739**; by oxidation, **216** ff., **527–529**; by rearrangement reactions, **524–527**; by reduction, **529–530**; by substitution, **520–522**; reactions, addition, **490–501**; addition of Grignard reagents, **227**; aldol condensation, **504–509**; Baeyer-Villiger reaction, **515–516**; benzoin condensation, **494, 513–514**;

Aldehydes (cont.)
Cannizzaro reaction, **231–232**;
Claisen-Schmidt condensation,
507–509; Clemmensen
reduction, **381–382**;
condensation, **702, 704**;
haloform reaction, **503–504**;
halogenation, **501–504**;
Hantzsch synthesis, **899, 901**;
Knoevenagel reaction, **512–513**;
Meerwein-Ponndorf-Verley
reduction, **231**; mixed aldol
condensation, **574**; mixed
Claisen condensation, **574**;
oxidation, **53, 517–518**; Perkin
reaction, **563**; reduction, **53,
518–519**; reduction with sodium
borohydride, **230**; reductive
alkylation, **662**; Reformatsky
reaction, **495–496**; substitution,
501–504; tetramerization,
492–493; trimerization,
492–493; Wittig reaction,
497–498, 959; Wolff-Kishner
reduction, **381**; reactions with,
acetoacetic ester, **899–901**;
acetylenic Grignard reagent,
332; alcohols, **491–492**;
aluminum ethoxide, **582**;
amines, **498–499**; *tert*-butyl
perester, **566**; hydrogen cyanide,
493–494; phenols, **800–802**;
phosphorus ylide, **959**;
sodium bisulfite, **439**; sodium
borohydride, **238a**; terpenes,
848, 850; unsaturated (*see*
Unsaturated aldehydes, reactions
with Grignard reagents)
Aldehydes, aromatic. *See*
Aromatic aldehydes
Aldehydes, unsaturated. *See*
Unsaturated aldehydes, reactions
with Grignard reagents
Aldehydic proton, **286, 287**
β-Aldehydo ester, preparation, **607**
Aldimene. *See* Ethylidenimine
Aldol condensations, **504–509**;
Claisen-Schmidt condensation,
507–509; mixed, **507–509,
574**; of α-picoline, **898**;
preparation, **504**
Aldopentoses, **618**
Aldoses, **622–626**; preparation,
617; Ruff degradation, **618**;
stability, **632**
D-Aldoses, **625–626**
Aldosterone, structure, **877**
Aldoxime acetate, **617**
Aldoximes, dehydration, **674**
Aldrin, **784**
Algin, **640–641**
Alginic acid, **641**
Aliphatic acids, decarboxylation,
553
Alizarin, **408**
Alkaloids, **916–920**; nmr spectra,
318b; racemates, resolution
with, **137–138**
Alkanediol, **368–369**

Alkane ion radical, **302**
Alkanes, **37, 68**; fluorinated, **340**;
homologous series, **66**; naming,
IUPAC, **16, 18, 72, 74**; naming,
old scheme, **75–76**; physical
properties, **320 ff.**; preparation,
379–382, Clemmensen
reduction, **381–382**; preparation,
Wolff-Kishner reduction, **381**;
reactions, autoxidation,
catalyzed, **366**; carbene
insertion, **334–335**; deuterium
exchange, **375**; halogenation,
335–340; homolytic attack, **335**;
nucleophilic substitution, **331,
342**; oxidation, **365**; reactions
with electrophiles, **334**;
reactions with free radicals, **459**
Alkanesulfonic acids,
preparation, **957**
Alkanols. *See* Alcohols
Alkene dihalides, dehalogenation,
351
Alkene ion radical, **302**
Alkenes, **37, 68**; alkynes,
preparation from, **52**;
contaminated with alkanes,
purification, **351**; coupling
constants, **289**; formation in
mass spectrometry, **298**;
homologous series, **67**;
hyperconjugation, **166**; infrared
spectra, **248–250**; mass spectra,
301; negatively substituted, **347**;
nmr spectra, **287**; physical
properties, **165, 320 ff.**;
preparation, **383–384**; from
alcohols, **171–172, 223–224**;
from alkyl halides, **172**; as
by-product of Clemmensen
reduction, **382**; Cope reaction,
383–384; from esters by
pyrolysis, **578**; from halides by
elimination, **161, 162, 164**;
Wittig reaction, **498, 959**;
reactions, addition of acids
to, **47–48**; addition of *tert*-butyl
lithium,**346**; addition of
dichlorocarbene, **804**; addition
of hydrogen cyanide, **346**;
addition of proton, **947**;
addition of thiols, **946**; alkylation,
354–355; autoxidation,
catalyzed, **366**; catalytic
hydrogenation, **55**; cationic
polymerization, **977, 978**;
Diels-Alder reaction, **429–430**;
dimerization, **354**; displacement,
358–359; epoxidation, **367–370**;
fragmentation, **297**; free-radical
addition, **49–50, 961**; Friedel-
Crafts, **454**; halogenation in
polar solvents, **349–351**;
homolytic addition, **360–364**;
homolytic attack, **335**;
hydration, **227, 353–354, 358**;
hydroboration, **238b**;
hydroboration to yield alcohols,
227–229; isomerization by

bases, **333, 342**; isomerization
by phosphoric acid, **352–353**;
Markovnikov addition, **347**;
oxidation, **367–374**; oxidation
by permanganate, **54**;
ozonization, **367, 373**; peroxide
effect, **361–362**; reduction,
358, 374–377; reactions with,
diborane, **238a**; hydrazine, **377**;
hypochlorous acid, **370**; osmium
tetroxide, **370–371**; peracids,
48; potassium permanganate,
371, 372; strong acid, **352**;
sulfenic acid derivative, **954**;
Saytzeff rule, **164–165, 166,
169**; stability, **165–166**;
stereochemistry, **123 ff.**
1-Alkenes, polymers, derived
from, **972**
Alkenium ion, **302**
Alkenyl protons, **288, 289**
Alkoxide ion, in Williamson
ether synthesis, **214**
Alkoxides, preparation from
alcohols, **205**
Alkyd resins, **984**
tert-Alkyl amine, preparation, **661**
N-Alkylaniline, **754**;
alkylquinolines, preparation
from, **908**
Alkylation, **354–355**; acetoacetic
ester, **598**; amines, **653**;
benzene, **468–469**; diketones,
509–510; Friedel-Crafts, **468**;
hydrazones, **678**; nitriles,
671; phenols, **797–798**;
phosphines, **958**
Alkylbenzene, Wurtz-Fittig
reaction, **780–781**
Alkylborane, **227**
Alkyl bromides: mass spectrometry,
298; preparation from alcohols,
170; reactions with hydroxide
ion, **154**; reactions with sodium
iodide, **170**; substitution, other
halides from, **170–171**
Alkyl carbonium ion, **302**
Alkyl cation, **149**; planarity, **151**
Alkyl chlorides: mass spectrometry,
298; preparation from alcohols,
169–170; reactions with sodium
iodide, **170**; substitution, other
halides from, **170–171**
Alkyl chlorosulfite ester, **211**
1-Alkyl-3,4-dihydroisoquinoline,
910
Alkyl group: alkenes, stability of
related to, **165–166**; esters,
reactions, **578–579**; inertness,
331; infrared spectrum, **250**;
IUPAC naming, **72, 74, 75**
Alkyl halides: alcohols, preparation
from, **225**; mass spectra, **298,
303–304**; physical properties,
261; preparation, **335 ff.,
349 ff.**; reactions, Arbusov
reaction, **960**; elimination, **158,
161**; esterification, **550**; E_2
reactions with ethoxide ion,

169; Friedel-Crafts reaction, 454; Gabriel synthesis, 660; nucleophilic substitution (S_N1), 149–151, 158; nucleophilic substitution (S_N2), 39–41, 143–144, 147–149; reduction, 380; substitution, 170–171; Wurtz reaction, 380; reactions, with, acetoacetic ester, 598; enolate ion, 509; phenols, 797–798; trialkyl phosphite, 960; zinc, 495–496; transition state, 151; ylides, preparation from, 497

Alkyl halides, primary, Wurtz reaction, 380

Alkyl halides, secondary, reactions, 380

Alkyl iodides: preparation from alcohols, 170; reactions with silver nitrate, 681

Alkyl lithium compounds, 44; reactions with unsaturated aldehydes, 522

Alkylmagnesium halide, preparation, 44

Alkyl mercury compounds, 455

O-Alkyl S-methyl xanthate, 224

Alkylnaphthalenes, preparation, 832

Alkyl nitrates, 681

Alkyl nitrites, 681; reactions, acetamidomalonic ester synthesis, 701 ff.; reactions with amides, 667–668

Alkylphosphinic acid, preparation, 959

Alkylphosphinous chloride, hydrolysis, 959

Alkylquinolines, preparation, 908

Alkyl radicals, stability, 339

Alkylsulfinyl group, 951

Alkylsulfonyl group, 951

Alkyne ion radical, 302

Alkynes, 37; analytical test for 1-alkynes in, 332; infrared spectra, 250; physical properties, 320; preparation, 384–386, from alkenes, 52; preparation, from keto ylide by decomposition, 524; reactions, addition of acids, 47–48; addition of alcohols, 346–347; addition of alkynes, 355–356; addition of tert-butyl lithium, 346; addition of hydrogen cyanide, 346; bromination, in polar reactions, 351; catalytic hydrogenation, 55; epoxidation, 369–370; free-radical additions to, 49–50; homolytic addition, 360–361; homolytic attack, 335; hydration, 354; isomerization by bases, 333; Markovnikov addition, 347, 351; oxidation by permanganate, 54, 55; ozonolysis, 367; reduction, 374–377; reactions with hydrazine, 377; reactions with strong acid, 352; stereochemistry, 125

1-Alkynes, acidity, 331–333; reactions, 332–333; with tert-butyl lithium, 346

Alkynol, preparation, 332

Allene formula, 77; physical properties, 69

Allenic system, 393

Alliin, structural formula, 698

Allocinnamic acid, physical properties, 594

Alloxan, 915; preparation, 914

Alloxazine, 914

Allyl alcohol, physical properties, 69

Allylamine, physical properties, 650, 651

Allylation, phenols, 800

Allylbenzene, physical properties, 444

Allylbenzylmethylphenyl ammonium iodide, 658

Allyl group, 77

Allyl halides, reactions, 424

tert-Allylic alcohol, dehydration, 434

Allylic alcohols: hydrogenolysis, 380; oxidation to carbonyl compounds, 528

Allylic carbanion, resonance, 333

Allylic cations, 420, 423

Allylic ion, 421

Allylic rearrangement, 781

Allylic resonance, 333, 342, 426

Allylic substitution: Bromination by N-bromosuccinimide, 340–341; competition with addition, 363–364

Allyl mercaptan. See 2-Propenethiol

Allyl phenyl ethers, Claisen rearrangement, 799–800

Allyl radical, 422, 423

Alpha elimination, 31, 32

Alumina, use in dehydration reactions, 52

Aluminum: electronic distribution, 88; reactions with alcohols, 205; use to reduce carbonyl compounds, 519

Aluminum alkoxides, 205, 206, 211; as catalyst in Oppenauer oxidation, 218

Aluminum bromide-ammonia complex, 106

Aluminum chloride, reactions with amino group, 753; use as catalyst in Friedel-Crafts reactions, 454; use in Gattermann-Koch reaction, 520

Aluminum chloride-ammonia complex, 106

Aluminum ethoxide, Tishchenko reaction with aldehydes, 582

Aluminum iodide-ammonia complex, 106

Aluminum isopropoxide, 206

Ameripol, 980

Amides, 664 ff.; acidity, 666; carbonyl stretching frequency, 545; conversion to acids, 551, 558; cyclic (see Lactams); infrared spectra, 667; IUPAC naming, 78; physical properties, 543, 544, 664, 666; preparation, 568, 668–669; preparation from ketoxime and phosphorus pentachloride, 664; reactions, amidolysis, 568, 569; bromination, 663; dehydration, 668, 674; elimination, 668; Hofmann rearrangement, 662–663, 668; hydrolysis, 558, 667; olysis, 667; reduction, 668; reactions with nitrous acid, 667–668

Amides, N-substituted, preparation, 661

Amides, substituted, preparation, 663

α-Amido acid chlorides, reactions, 711, 712

Amidolysis, 551; of amides, 568, 569

Amine group, halogenation, 752

Amine oxides, 384; Cope elimination, 659; preparation, 383

Amine oxides, tertiary, 659; preparation, 897

Amine salts, 652; oxidation, 659; spectra, 651

Amines: alkaloids, resolution of racemates with, 137; aromatic (see Aromatic amines); basicity, 651–652; infrared spectra, 651; IUPAC naming, 18, 78; mass spectra, 651; nmr spectra, 651; occurrence in nature, 649–650; physical properties, 650–651; preparation, from amides by reduction, 668; Beckmann rearrangement, 663–664; Gabriel synthesis, 660–661; Hofmann rearrangement, 662–663; nucleophilic substitution, 41–42; olysis reactions, 661; rearrangement, 662–664; reduction, 661–662; reductive alkylation, 662; substitution, 660–661; reactions, addition to carbonyl group, 653; alkylation, 652–653; Cope reaction, 383–384; exhaustive methylation, 167; halogenation, 653; Hinsberg reaction, 653–654; oxidation, 659; reactions with, carbonyl group, 498–499; esters, 568, 569; nitrous acid, 654–656; phosgene, 772; stereochemistry, 658; structure, 650

Amines, aromatic. See Aromatic amines

Amines, primary, 650; aromatic, 754, 759; identification by Hinsberg test, 656; preparation, 662, 678, from diazoalkenes, 683; preparation from hydrazone, 679; reactions,

Amines, primary (cont.)
  addition of dichlorocarbene, **804**; alkylation, **653**; aminolysis, **653**; dehydrogenation, **674**; reactions with carbonyl group, **498–499**; reactions with nitrous acid, **656**
Amines, secondary, **650**; aromatic, **754, 759**; identification by Hinsberg test, **656**; preparation, **662**; reactions, **653**; reactions with nitrous acid, **655, 656**
Amines, tertiary, **650**; aromatic, **754–755, 759**; identification by Hinsberg test, **656**; as leaving group, **168**; reactions, **653**; reactions with alkyl halide, **656**; reactions with methyl iodide, **166–167**; use as catalyst in reaction between alcohols and acid chlorides, **213**
Amino acid ester, **700**
Amino acids, **669, 691** ff.; abbreviations for, **692–694, 705**; carbon-14 dating, **138–139**; detecting using nonhydrin, **840**; in DNA, **446**; infrared spectra, **697**; physical properties, **695–697**; preparation from imides, **663**; preparation from lactams, **670**; reactions, **700–701**, decarboxylation, **555**; reactions on heating, **669–670**; resolution of racemic mixtures, **704**; sequence of in peptides, determination, **708–709**; synthesis, **701–704**; transfer RNA and, **924**
α-Amino acids, preparation by Strecker synthesis, **675**
β-Amino acids, reactions on heating, **669, 670**
γ-Amino acids, reactions on heating, **669, 670**
δ-Amino acids, reactions on heating, **669, 670**
α-Aminoadipic acid, structural formula, **698**
Amino alcohol, rearrangement, **655, 656**
α-Amino alcohols, diaryl substituted, deamination, **535b**
Amino alkyl malonic ester, **701**
p-Aminobenzenesulfonic acid. See Sulfanilic acid
p-Aminobenzoic acid, structural formula, **745**
ε-Aminocaprolactam, nylon 6, preparation from, **982**
Aminodesoxybenzoin, reaction with phenylmagnesium bromide, **535b**
2-Amino-6,8-dihydroxypurine, preparation, **911**
2-Aminoglutaric acid. See Glutamic acid
Amino groups: in biochemical systems, **499**; determination in proteins, **463**; preparation from nitro group by reduction, **741**; reaction with aluminum chloride, **753**; substituent effect, **747**
Amino ketone, preparation of imidazoles from, **893**
Aminolysis reactions, **653**; amides, preparation by, **668**; esters, **568, 569**
Aminomalonic ester, **701**
2-Amino-3-mercaptopropanoic acid. See Cysteine
α-Amino nitrile, **675**
Aminophenols: oxidation to yield quinones, **813**; preparation, **769, 796, 797**
2-Aminopropionic acid. See Alanine
α-Aminopropiophenone-phenyl-^{14}C, reaction with phenylmagnesium bromide, **535c**
Aminopurines, **911**
Aminopyridines, **899**
2-Aminoquinoline, **907**
2-Aminosuccinic acid. See Aspartic acid
2-Amino-4,6,7-trihydroxypteridine. See Leucopterin
Ammonia, autoprotolysis constant, **22**; bonding in, **5, 647**; donor-acceptor complexes with, **106**; molecular structure, **650**; production figures, **378**; reactions, alkylation, **653**; reactions, halogenation, **653**; reactions with, benzaldehyde, **499**; esters, **568, 569**; ethyl halide, **41**; formaldehyde, **499**
Ammonium cyanate, urea, preparation from, **2**
Ammonium sulfide, reactions with m-dinitro aromatic compounds, **742**
Ammonolysis, of esters, **568, 569**
Amobarbital, **903**
Amorphous polymers, **973**
AMP. See Adenosine monophosphate, in enzyme reactions
Amphetamine, **816**; structural formula, **815**
amphi-, **824**
AM system, **271**
AM$_2$X$_3$ pattern, **279–280**
Amylopectin, **639**
Amylose, **639**
Amytal, **903**
Analgesics, **104**
Analytic testing: 1-alkynes, **332**; unsaturation, **351**
Androsterone, **875**; structural formula, **876**
Anethole, structural formula, **441**
Angelic acid, physical properties, **594**
Angstrom (unit), **241**
Angular momentum quantum number, **87**
Anhydrides. See Acid anhydrides
Anhydro sugar, **641**
Aniline: conjugation, **745**; dipole moment, **747**; phenol and cresols, preparation from, **810**; physical properties, **745**; preparation, **664**; preparation from nitrobenzene by reduction, **740, 741**; reactions, nitration, **747–748**; oxidation to yield quinones, **813**; sulfonation, **751**; reactions with crotonaldehyde and sulfuric acid, **908**; reactions with sodium nitrite, **754**; structural formula, **442**; ultraviolet spectrum, **745**
Anilines, substituted: pK$_a$, **746**; quinolines, substituted, preparation from, **908**
Anilinium ion, **741**
Anion exchanger, **990–991**
Anionic polymerization, **978**
Anionic surfactants, **541**
Anisaldehyde, physical properties, **481**
Anisic acid, structural formula, **442, 539**
p-Anisidine, structural formula, **442**
Anisole, bromination, **449**; structural formula, **441**
Anomers, **615**
Anthracene: occurrence in nature, **833**; physical properties, **407, 444**; reactions, addition, **833**; bromination, **833**; chlorination, **458**; reduction, **835**; substitution, **833**; resonance energy, **825**; structural formula, **396**
Anthrahydroquinone, **835**
Anthranilic acid, reactions, **764**; structural formula, **539**
Anthranol, **834, 835**
Anthraquinone, **834–836**; preparation, **836**; reactions, **812–813**; reactions, nitration, **835**; reduction, **834–835**; sulfonation, **835–836**
Anthrone, **834, 835**
anti-, **395**
Antibacterial agents, hexachlorophene, **472**
Antibiotics, **641, 699**; chloramphenicol, **817**; fumagillin, **612a**
Antibodies, **641, 724**
Antibonding orbital, **91**
Anticodon triplets, **927**
anti conformer, **123**
Antifreeze, **203**
Antigen, **641, 724**
Anti-Markovnikov addition, **228, 347**
Antioxidants, vitamin E, **819**
Antiseptic agent, **790**
Apoenzyme, **725**
Apple, ester responsible for fragrance of, **542**
Aprotic solvents, **22, 41, 463**; nucleophilic substitution, effect on, **153, 155, 156**
Aqueous suspension

polymerization, **974**
Araban, **639**
Arabinaric acid, preparation, **623**
Arabinogalactans, **639**
Arabinose, preparation, **618**; reactions, **622, 623**; structure, proof of, **622–624**
Arachidic acid, physical properties, **538**
Arbusov reaction, of trialkyl phosphite, **969**
Arbusov rearrangement, **960**
Arenes, **440** ff.; hyperconjugation, **457**; polycyclic, **458, 840**; preparation, **468–472**; reactions, benzyne pathway, **463–465**; electrophilic substitution, **452–458**; Friedel-Crafts reaction, **454**; Gattermann-Koch reaction, **520**; Gattermann reaction, **521**; halogenation, **452**; homolytic substitution, **458–460**; Houben-Hoesch reaction, **521**; mercuration, **454–455**; nitration, **452–453**; nucleophilic substitution, **460–465**; oxidation, **465–467**; substitution, **445–451**; sulfonation, **453–454**; sources of, **467–468**; substituent effects, **448–451, 455** ff., **460**
Arenes, substituted. See Benzenes, substituted
Arginine, **691, 692, 697**; structural formula, **692**
Argol, **593**
Argon, electronic distribution, **88**
Aromatic acids: carbonyl stretching frequencies, **545**; decarboxylation, **554**
Aromatic aldehydes: aldol condensation, **507**; Perkin reaction, **563**
Aromatic amines, **744** ff.; basicity, **745**; conjugation, **745**; mass spectra, **745**; preparation, **755**; primary, **754, 759**; reactions, **745, 747** ff., coupling, **758, 759**; electrophilic substitution, **747–749**; Friedel-Crafts reaction, **753**; halogenation, **752**; oxidation, **753–754**; sulfonation, **751**; reactions with nitrous acid, **655, 754**; secondary, **754, 759**; substituents in, **760**; tertiary, **754–755, 759**; ultraviolet spectra, **745**
Aromatic compounds: in cigarette smoke, **444**; coal tar as a source of, **467–468**; deshielded protons in, **408**; disubstituted, **441–442**; heterocyclic (see Heterocyclic compounds); infrared spectra, **404–407**; mass spectra, **410–411**; nmr spectra, **408–410**; nomenclature, **396–397**; reactions, catalytic hydrogenation, **418**; oxidation, **416–417**; ozonization, **416**; reduction, **418–419**; stability, **416**; ultraviolet spectra, **407**; uses, **472**
Aromaticity, **402**; furan, **890–891**; heterocycles, **890**
Aromatic protons, chemical shift, **408**
Aromatic rings, addition of free radicals to, **458–459**
Arrhenius A factor, **186**
Arrhenius equation, **186, 188**
Arrhenius plots, **186–188**
Arsenic acid, esterification, **211**
Arsenous acid, esterification, **211**
Artificial flavors, **542**
Arylacetone, **442**
Aryl alkyl ether, hydrolysis, **792, 793**
Aryl alkyl ketones, oxidation, **518**
Aryl N-arylcarbamate, preparation, **771**
Aryl bromide, preparation, **763**
Aryl ethers, hydrolysis, **792**
Aryl halides, reactions, **460**; aromatic nucleophilic substitution, **949**; Ullmann reaction, **780**
Arylhydrazines, **770**; preparation, **758**
Arylhydrazones, **770**
Aryl isocyanates, **770–772**
Aryl lithium compounds, reactions with unsaturated aldehydes, **522**
Arylnaphthalenes, preparation, **832**
Aryl nitriles, **772**
Arylolysis, **562**
Ascorbic acid (vitamin C): preparation, **629–630**; preparation, sorbitol as intermediate, **616**; structural formula, **592**
Asparagine, structural formula, **692**
Aspartate transcarbamylase, **148**
Aspartic acid, **691, 692, 696**; hydrogen bonding with, **720**; ionization, **696–697**; as photosynthesis product, **556–557**; physical properties, **71**; reactions with carbamyl phosphate, **148**; structural formula, **692**
Asphalt, **379**
Aspirin, **104, 793–794**
Astrochemistry, **28**
Astronomy, microwave spectroscopy, use in, **245**
Asymmetric carbon atom, **128**
Asymmetric catalysts, **8**
Asymmetric molecules, **8**
Asymmetry, **127** ff.
Atactic polymers, **972, 973**
Atomic orbitals: hybridization, **93**; hydrogen atom, **87, 88**; shapes, **89–90**; third shell, **941, 942**
Atomic structure, **87–90**
ATP. See Adenosine triphosphonucleotide
Atromid, **792**
Atropine, **920**
Autoprotolysis constant, **22**
Autoxidation: alkenes, **369**; arenes, **466–467**; drying of paint, **540**; rancidity, **540**
Auxins, **391a, 904**
Axial conformation, **324–326**
Axis of symmetry, **126**
AX_2 pattern, **271**
AX_6 pattern, **271, 280**
A_2X_3 pattern, **271, 276**
AX system, **271, 272**
A_4X_2 system, **282**
Aza-, **887**
Azacyclohexane. See Piperidine
Azelaic acid, **541**; physical properties, **542**
Azeotropes, alcohol-water, **206–207**
Azetidine, structural formula, **888**
Azide ion, **683**; S_N2 reaction with butyl bromide, **155–156**
Azides, **683–684**
Azimuthal quantum number, **87**
Azine. See Pyridine
Azines: preparation, **500, 677**; reactions, heat, instability to, **678**; reduction, **682**; Wolff-Kishner reduction, **678**
Azlactone, **700**
Azlactone synthesis, **702, 703**
Azoalkanes, **682**
Azobenzene, preparation from nitrobenzene, **740, 741**
Azobisisobutyronitrile, **682, 974**
Azocine, structural formula, **888**
Azo compounds, reduction to form amines, **661**
Azo compounds, substituted, preparation, **758–759**
Azoethane, **649**
Azole. See Pyrrole
Azolidine. See Pyrrolidine
Azomethane, **682**
Azoxybenzene, preparation, **740, 741**
Azulene, **856–857**; structural formula, **396**

Back bonding, **943**
Back-side attack, **159**
Back-side collision, **144**
Backus, J. G., **10a**
Bacteriochlorophyll, **932**
Baeyer-Villiger reaction, **515–516**; with α-diketone, **565**
Bakelite, **800, 970, 986**
Balata, **863**
Banana oil, **543**
Barbital, **903**
Barbiturates, **903**
Barbituric acid, preparation, **569, 903**
Bartlett, P. D., **64a, 343a, 646a**
Barton, D. H. R., **64b**
Base pairing, **446**
Base peak, **298**
Bases, **20**; aromatic, **446**; carbonyl group as, **485**; in DNA and RNA, **920** ff.; nucleophiles, comparison with, **24**; racemic, resolution, **138**; reactions,

Bases (cont.)
catalysis of addition to carbonyl group by, **493;** isomerization of alkenes by, **333, 342;** Wolff-Kishner, reduction, **381;** reactions with, acetaldehyde, **504–507;** halobenzenes, **463–465;** hydrazine, **676;** hydroxylamine, **676;** oximes, **678**
Bases, aromatic, **446**
Base strength (basicity): amines, **652;** aromatic amines, **745;** aromatic nitro compounds, **737;** pK_a, expressed by, **22;** pyridine, **894**
Bathochromic shift, **407**
Batrachotoxin, **882–883;** structural formula, **883**
Bayer's strain theory, **322**
Beckmann rearrangement, **663–664**
Bed form, **324**
Beer-Lambert law, **244**
Bees, queen bee substance, **607**
Behenic acid, physical properties, **538**
Belladonna, **920**
Benzal chloride, structural formula, **440**
Benzaldehyde, nmr spectrum, **409, 410;** physical properties, **481;** preparation from toluene, **466;** reactions with ammonia, **499;** reactions, benzoin condensation, **494;** steric effects, **494;** structural formula, **440, 480**
Benzaldoxime, isomeric forms, **500**
Benzal radical, **440**
Benzanthracene, structural formula, **841**
Benzedrex, **816**
Benzene, **396;** charge-transfer complexes, **446;** deshielded protons in, **408;** physical properties, **400–401, 407, 443;** preparation by disproportionation of cyclohexane, **433;** reactions, acylation, **469;** alkylation, **468–469;** benzyne pathway, **463–465;** bromination, **446–447, 459;** catalytic hydrogenation, **418;** chlorination, **459;** dinitration, **750;** drastic oxidation, **416, 417;** free-radical addition of chlorine, **784;** Friedel-Crafts reaction, **454, 469;** halogenation, **452, 468;** hydrogenation, **400;** inertness to oxidation, **416;** nitration, **452–453, 468;** ozonization, **416;** sulfonation, **453–454, 468, 469, 808;** reactions with, electrophilic reagents, **452–458;** iodine, **446;** sodium amide, **463;** succinic anhydride, **831;** reactivity, **458;** resonance energy, **825;** resonance structure, **403;** resonance theory, **445–446;** solvent effect of on nucleophilic substitution, **153;** stability, **394, 400, 416;** substituent effects, **448–451;** syntheses using, **469 ff.;** ultraviolet spectrum, **407**
Benzene ring, **424;** migrating, **526;** notation for, **404**
Benzenes, hydrogenated, **397**
Benzenes, substituted, **440 ff., 894–904;** infrared spectra, **404, 405;** reactions, hydrogen abstraction, **459;** hydrogenation, **416, 418;** oxidation, **416;** reduction, **418–419;** ultraviolet spectra, **407**
Benzenesulfonyl chloride, use in Hinsberg reaction, **653**
Benzenonium ion, **446**
Benzhydrol, structural formula, **440**
Benzhydryl cation, **424**
Benzhydryl chloride, structural formula, **440**
Benzhydral halides, reactivity, **781**
Benzhydryl radical, **440**
p-Benzhydryltetraphenylmethane, preparation, **782, 783**
Benzidine rearrangement, **772**
Benzidrine, **815**
Benzil: preparation, **514;** reactions with hydroxide ion, **515;** reactions with phenylhydrazine, **618, 619**
Benzilic acid, physical properties, **589**
Benzilic acid rearrangement, **515;** phenanthrenequinone, **837, 838**
Benzofuran. See Coumarone, structural formula
Benzoic acid, degree of dimerization, **546**
Benzoic acid: physical properties, **538, 544;** pK_a, **797** preparation from toluene, **466;** structural formula, **440**
Benzoin, preparation, **513–514**
Benzoin condensation, **494, 513–514**
Benzoylation, **560**
Benzophenone, **441;** physical properties, **481**
Benzopinacols, substituted, **526**
Benzo[a]pyrene, structural formula, **444**
Benzo[b]pyridine. See Quinoline
Benzo[c]pyridine. See Isoquinoline
Benzo[a]pyrrole, structural formula, **904**
Benzo[b]pyrrole. See Indole
Benzo[c]pyrrole, structural formula, **904**
Benzoquinone, **433, 811;** Diels-Alder reaction, **836;** preparation, **807**
Benzo radical, **440**
Benzothiophene. See Thianaphthene
Benzo trichloride, structural formula, **440**
Benzoyl chloride, olysis reactions, **561;** reactions with carboxylic acids, **547, 548, 551;** reactions with naphthalene, **827;** Schotten-Baumann reaction, **561**
N-Benzoylglycine, condensation with aldehyde, **702**
Benzoyl peroxide, as free-radical source, **974**
Benzpyrene: carcinogenic activity, **841;** structural formula, **842**
Benzyl–, **404**
Benzyl acetate, use in perfume, **208**
Benzyl alcohol: preparation by chloromethylation, **500;** preparation from toluene, **466;** reactions, oxidation to carbonyl compound, **528;** reactions with phosgene, **710;** structural formula, **440;** use in perfume, **208**
Benzylamine, **744;** physical properties, **650, 651**
Benzyl benzoate, use in perfume, **208**
Benzyl cation, resonance structure, **423**
Benzyl chloride, preparation by chloromethylation, **500**
Benzyl esters, hydrogenolysis, **710**
Benzyl halides: Grignard reagents from, **781;** reactions, **424**
Benzylic cation, **410**
Benzylic hydrogens, **466, 467**
Benzylic resonance, **426**
Benzyloxycarbonyl chloride, preparation, **710**
Benzyloxycarbonyl group. See Carbobenzoxy group, use in peptide synthesis as protecting group
Benzyl radical, **440, 459**
Benzyne, **463–465;** preparation from anthranilic acid, **764**
Beri-beri, **731**
Beryllium, electronic distribution, **88**
Beta bond, **410**
Beta elimination, **31, 32;** dehalogenation, **51;** dehydrohalogenation, **51, 52;** nucleophilic substitution, competition between, **42, 51, 156–158**
Betaine, **497**
Bicarbonate-carbonic acid buffer system, **21–22**
Bicyclic compounds, with five-membered heterocycle, **904–906;** with six-membered heterocycle, **906–910**
Bicyclic terpenes, **852**
Bicyclo[2.2.1]heptane. See Norbornane
Bile acids, **880–881**
Bimolecular reactions, **193**
Bimolecular reduction, **518–519**
Biochemistry, intermolecular bonding, **107**

Bio-organic chemistry, 7
Biotin, **540;** structural formula, **729;** use, **729**
Biphenyl: physical properties, **444;** reaction with succinic anhydride, **832**
2,2'-Biphenyldisulfonic acid, **844**
Biphenyls, substituted, preparation by Ullmann reaction, **780**
Bird cage hydrocarbon, **328, 329**
Bis-, **74**
Bisabolene, preparation, **855, 856**
Bischler-Napieralski synthesis, **910**
2,4-Bis(hydroxymethyl)phenol, **800**
Biphenol A, reaction with eipchlorohydrin, **985, 987**
Bisulfate ion, **47**
Biuret, **649**
Blackstrap molasses, **233**
Block polymers, **991**
Blood, artificial, **340;** hemoglobin as transport system in, **930;** oxygen transport in, **196;** pH control in, **21;** sickle-cell anemia, **724**
Blood group substances, **641–642**
Boat form, **324**
Boiling, thermodynamics, **56, 57**
Boiling point: acids, **542–544;** alcohols, **201** ff.; hydrocarbons, nonaromatic, **320** ff.; thiols, **945**
Bond, **4**
Bond angle, **101, 102;** coupling constant, effect on, **288–289**
Bond dissociation energy, **111** ff.
Bond distance, **90, 101**
Bond energy, **110–111**
Bond fixation, **825;** anthracene, **832;** phenanthrene, **837**
Bond force constant, **114**
Bonding orbital, **91**
Bond moment, **105**
Bonds, molecular orbitals, **90–96;** polarizable, **105**
Borane. *See* Boron hydride (borane)
Boric acid, esterification, **210, 211**
Bornane, structural formula, **852**
Borneol, preparation, **853**
Bornyl chloride, **853**
Boron: covalent bond radii, **102;** electronegativity, **103;** electronic distribution, **88**
Boron carbide, **385**
Boron hydride (borane), **227;** electrophilic nature, **357;** reactions, **227–229, 356–358;** reactions with, acid chlorides, **562;** amides, **668;** ethyl bromide, **331;** hydrazine, **661**
Boron trifluoride, reaction with sodium borohydride to yield diborane, **227**
Boron trifluoride-ammonia complex, **106**
Bouveault-Blanc reduction, **510**
Bromide ion, reaction with methanol, **24**
Bromination: acenaphthene, **841;** aldehydes and ketones, **502** ff.; alkanes, **339;** alkenes, free-radical, **360–361;** alkenes, in polar solvents, **349–350;** allylic, of N-bromosuccinimide, **340–341;** amide, **663;** anisole, **449;** anthracene, **833;** arenes, **452, 466;** arenes, prf effect, **456;** benzene, **446–447, 459;** carboxylic acids, **555–556;** cholesterol, **873;** esters, **577;** free-radical, arenes, **466;** furan, **892;** Hell-Volhard-Zelinsky reaction, **555–556;** hyperconjugation and, **457;** methane, thermodynamics, **337;** naphthalene, **825, 826;** β-naphthol, **829;** m-nitroacetophenone, **470;** nitrobenzene, **738;** m-nitroethylbenzene, **470;** phenanthrene, **838;** phenols, **797;** polycyclic compounds, **458;** pyridine, **896;** pyrrole, **892;** quinoline, **907;** thiophene, **892;** toluene, **470**
Bromine: covalent bond radii, **102;** electronegativity, **103;** isotopes, **299;** mass spectrum, **303;** reactions, addition to acetylene, **46;** reactions, addition to ethylene, **46;** reactions with, amide, Hofmann rearrangement, **662–663;** benzene, **446–447;** ethylmagnesium bromide, **43;** methane, **26**
Bromine water, reaction with glucose, **615**
p-Bromoacetophenone, nitration, **470**
α-Bromo acid, amino acids, preparation from, **701**
Bromoalkanes, **777**
Bromoamide, preparation, **663**
9-Bromoanthracene, preparation, **833**
Bromoarenes, **777**
m-Bromobenzal chloride, preparation, **470**
Bromobenzene, preparation, **446–447**
1-Bromobutane, **67**
2-Bromobutane, **67;** reaction with ethoxide ion, **169**
2-Bromo-2-butene, formation by E2 elimination, **162**
1-Bromo-2-chloroethane, preparation, **46**
1-Bromo-3-chloropropane, nmr spectrum, **282–284**
2-Bromo-2-chloro-1,1,1-trifluoroethane. *See* Halothane
Bromocyclobutane, structure, **79**
Bromo-enol, **555–556**
α-Bromoesters, succinic acids, substituted, preparation from, **576**
2-Bromoethyl acetate, preparation, **46**
p-Bromoethylbenzene, nitration, **470**
Bromoform, preparation, **503**
Bromomalonic esters, reactions, **578;** succinic acid, preparation from, **577**
Bromomethane, physical properties, **13;** reactions with, hydroxide ion as nucleophile, **23;** reactions with water, **24**
1-Bromo-2-methylpropane, **67**
2-Bromo-2-methylpropane, **67**
2-Bromonaphthalene, preparation by Bucherer reaction, **767**
1-Bromo-2-naphthol, preparation, **829**
m-Bromonitrobenzene, preparation, **738**
4-Bromo-3-nitro-1-ethylbenzene, preparation, **470**
Bromonium ion, **350**
2-Bromooctane, S_N1 reactions, **159**
9-Bromophenanthrene, preparation, **838**
p-Bromophenol, **797**
1-Bromopropane, isomers, **67**
2-Bromopropane, isomers, **67**
m-Bromopropiophenone, preparation, **469**
3-Bromoquinoline, **907**
N-Bromosuccinimide, **668;** reactions, bromination, allylic, **340–341;** reactions with peptide bonds, **708**
m-Bromotoluene, preparation from toluene, **765–766**
o-Bromotoluene, preparation, **766–767**
Bromotrifluoromethane, uses, **37**
8-Bromoxanthine, trimethylation, **912, 913**
Brønsted-Lowry acid, **20**
Brønsted-Lowry base, **20**
Brønsted-Lowry theory, **20**
Brown, H. C., **238a**
B_{12} coenzyme, **933**
Bucherer modification, **702**
Bucherer reaction, **755–756;** of sulfonic acid group, **836**
Bufotalin, **882**
Bufotenine, **905;** structural formula, **906**
Butadiene: conformation, **395;** infrared spectrum, **404;** preparation, commercial, **433–434;** properties, **398–399;** reactions, catalytic hydrogenation, **423;** reactions, cycloaddition with maleic anhydride, **429;** resonance, **398;** ultraviolet spectrum, **407**
1,3-Butadiene: absorption peaks, **261;** addition, **421**
Butane, **66;** conformation, **120, 121, 123;** halogenation, **339;** physical properties, **70;**

Butane (cont.)
preparation, **423**
1,4-Butanediol, physical properties, **71**
2,3-Butanediol, asymmetry, **130** ff.; physical properties, **71**
Butanedione, **416**
Butanetniol, **253, 945**
1-Butanol (butyl alcohol): physical properties, **70, 202, 480**; reactions, with hydrobromic and sulfuric acids, **143**; reactions with oxidation, **221–222**
2-Butanol (sec-butyl alcohol): asymmetry, **128, 129**; haloform reaction, **503**; mass spectrum, **304, 305**; physical properties, **70, 202**
2-Butanone, physical properties, **70, 480**
1-Butene, **67**; isomerization with phosphoric acid, **352**; physical properties, **70, 398**; preparation, **166, 167**; stereochemistry, **123–124**
2-Butene, **67**; stereochemistry, **123** ff., **162**
cis-2-Butene: physical properties, **70, 322, 400**; preparation, **352**; reactions, bromination in polar solvents, **349**; reactions, carbene addition, **355**; stability, **165**
trans-2-Butene: bromination in polar solvents, **349–350**; E$_2$ reactions, **162**; physical properties, **70, 322**; preparation, **352, 423**; stability, **165**
3-Butenoic acid, **595**
3-Buten-2-ol, physical properties, **70**
3-Butene-2-one. See Vinyl methyl ketone
1-Butoxybutane, etherification, **215**
tert-Butoxycarbonyl azide, preparation, **711**
tert-Butoxycarbonyl group, effect on peptides, **711**
Butter yellow, **773**
Butyl alcohol. See 1-Butanol (butyl alcohol)
sec-Butyl alcohol. See 2-Butanol
tert-Butyl alcohol. See 2-Methyl-2-propanol
Butylamine: physical properties, **70, 650**; reactions with nitrous acid, **655**
sec-Butylamine, physical properties, **70**
tert-Butylamine, physical properties, **70, 650**
N-tert-Butylaniline, reaction with calcium hypochlorite, **752**
tert-Butylbenzene, bromination, **457**; physical properties, **444**; structural formula, **440**
Butyl bromide: physical properties, **70**; preparation, **143**; reactions, E2 elimination, **166, 167**; nucleophilic substitution, **143, 144, 148, 155–156**; S$_N$2, with azide ion, **155–156**
sec-Butyl bromide: nucleophilic substitution, **159**; physical properties, **70**; preparation, **339**
tert-Butyl bromide, physical properties, **70**; preparation, **339**; reactions, elimination (E1), **161**; reactions with sodium ethoxide, **214**
tert-Butyl ethyl ether, reaction, **661**
sec-Butyl chloride, preparation, **339**
tert-Butyl chloride: preparation, **143, 339**; reaction with sodium ethoxide, **214**; reactions, hydrolysis, mechanism, **193, 194**; nucleophilic substitution, **157**; solvolysis, **158**
tert-Butyldimethylsulfonium ion, nucleophilic substitution, **157**
Butylene, physical properties, **70**
tert-Butyl esters, alcoholysis, **568**
tert-Butyl ethyl ester, preparation, **214**
Butyl group, formula, **75**
tert-Butyl group: bromination, **456**; infrared spectrum, **250**
tert-Butyl lithium, addition to alkenes and alkynes, **346**
tert-Butyl mesitoate, preparation, **580**
(Z)-(4R)-3-[(S)-sec-Butyl]-4-methyl-2-hexenoic acid, **394**
tert-Butyl perester, reaction with aldehyde, **566**
Butyl rubber, **978, 980**
1-Butyne: free-radical chlorination, **360**; physical properties, **70**
2-Butyne: physical properties, **70**; preparation, **385**
Butyraldehyde: physical properties, **70, 481**; preparation from 1-butanol by oxidation, **222**
Butyric acid: acidity, **547**; dimerization, degree of, **546**; physical properties, **70, 538**; preparation from 1-butanol by oxidation, **221–222**
n-Butyrophenone, physical properties, **481**

c–, **395**
Cadaverine, **555**
Cadmium reagents, reactions with acid chlorides, **523**
Caffeine, preparation, **912, 913**
Calciferol. See Vitamin D$_2$
Calcium acetate, pyrolysis, **521**
Calcium carbide, reaction with water to yield acetylene, **384–385**
Calcium hypochlorite, reaction with N-tert-butylaniline, **752**
Calvin, M., **646a**
Camphane, **854**
Camphene, **854**; preparation, **853**
Camphene hydrochloride, preparation, **853**
Camphor, **854**
Camphorquinone, preparation, **854**
Cancer, cigarette smoking and, **444**
Cannabidiol, **817**; structural formula, **818**
Cannizzaro reaction, **231–232, 506**; mixed, **508**
Capric acid, physical properties, **538**
Caproaldehyde, physical properties, **481**
Caproic acid, physical properties, **538**
Caprylic acid, physical properties, **538**
Carbamide. See Urea
Carbamidine. See Guanidine
N-Carbamylaspartic acid, **148**
Carbamyl phosphate, reaction with aspartic acid, **148**
Carbanions, **46**; nucleophilicity, **331**; preparation, **345–346**
Carbazole, structural formula, **889**
Carbene, **334**; addition to double bonds, **953**
Carbene addition, **355–356**
Carbene-carbene rearrangements, **438b**
Carbene insertion, into alkanes, **334–335**
Carbides, **385**
Carbobenzoxy group, use in peptide synthesis as protecting group, **710**
Carbocyclic compounds, **38**; hydrocarbons, **68**
Carbohydrates, **613** ff.; metabolism, **146–147, 593, 644–645**; structure, **42**
β-Carboline, **917**
Carbon: atomic structure, **88, 322**; covalent bond formation with, **4, 5, 13**; covalent bond radii, **101, 102**; divalent, **343b**; as energy source, **365**; isotopes, **299**; mass spectrometry, **298–299**; molecular orbital theory, **93**
Carbon-14 dating, **138–139**
Carbon black, effect on properties of rubber, **863**
Carbon-carbon double bond: in allylic position, **380**; in cyclic compounds, **68**; donor-acceptor complexes containing, **106**; IUPAC rules for compounds containing, **77** ff.; nonrotation around, **123**; reactions, addition, **45–46, 62**; reactions, catalytic hydrogenation, **55**; reactions with quinones, **812**; reducing agents effect of on, **54**;
Carbon-carbon single bonds, infrared spectrum, **250**
Carbon dioxide: atmospheric, **365**; atomic structure, **15**; bond energies, **111**; as combustion

product, **29**; physical properties, **13**; preparation from carbonic acid, **31**; vibrational modes, **247**
Carbon disulfide, **777**
Carbon-halogen bond: electrophilic attack on by borane, **331**; infrared spectrum, **250**
Carbon-hydrogen bond: in 1-alkynes, polarization, **332**; oxidation, **216–219**
Carbonic acid, **17**; loss of carbon dioxide, **739**; physical properties, **13**
Carbonium ions, **46, 149–152**; aromatic, **424**; hyperconjugation, **166**; IUPAC naming, **149**; planarity, **151**; preparation, by acetolysis, **64b**; from alkane reacting with an electrophile, **334**; from diazonium ion, **655**; reactions, **151–152, 160**, addition to nitrile group, **661**; reactions, elimination (E1), **156–158**; in S_N1 and E1 reactions, **156–157**; stability, **339**
Carbon monoxide: atmospheric, **365**; physical properties, **13**; preparation from formic acid, **30–31**
Carbon-nitrogen double bond, **522**
Carbon-nitrogen triple bond, **522, 558, 671**
Carbon-oxygen bond: of alcohols, reactions, **214**; in phenols, reactions, **795**
Carbon-oxygen double bond, **522**; bond energies, **111**; donor acceptor complexes containing, **106**; ground state, **482**; infrared spectrum, **253**; ketones, **68**; reactions, addition, **45–46, 62**; reactions, hydrogenation, **229–230**; resonance structures, **482–483**
Carbon-oxygen single bonds, infrared spectrum, **250–251**
Carbon tetrachloride: commercial preparation, **777**; physiological effect of, **819**
Carbonyl bond, **261**
Carbonyl chloride. See Phosgene
Carbonyl compounds, preparation by addition reactions, **522–524**; preparation, by oxidation, **527–529**; by rearrangement, **524–527**; by reduction, **529–530**; by substitution, **520–522**; reactions; acylation, **510–511**; addition, **490–501**; addition of amines, **653**; addition of enolate ion, **506, 508**; addition of hydrazine, **677**; addition of hydrogen cyanide, **675**; addition of phenylhydroxylamine, **769**; aldol condensation, **504–509**; benzoin condensation, **513–514**; Knoevenagel reaction, **512–513**; Michael reaction, **511–512**; mixed condensations, **574**; oxidation, **517–518**; reduction, **518–519**; reductive alkylation, **662**; Reformatsky reaction, **495–496**; substitution, **501–504**; Wittig reaction, **497–498**; reactions with, allylic Grignard reagent, **781**; amine group, **498–499**; ammonia, **499**; Grignard reagent, **495, 781**; hydrogen cyanide, **493–494**; sodium bisulfite, **493**
β-Carbonyl compounds, **672**; preparation by C-acylation, **511**
Carbonyldiimidazole, preparation of esters using, **581**
Carbonyl group, **483**; in esters, **566** ff.; in glucose, **614, 616**; infrared spectrum, **487–488**; ultraviolet spectrum, **488**
Carbonylic ions, **295, 297**
Carbonyl stretching frequency, **544–546**
Carbostyril. See 2-Quinolone, preparation
Carbowax, **985**
Carboxyhemoglobin, **930**
Carboxy-inversion reaction, **993a**
Carboxyl group, carbonyl stretching frequency, **545**
Carboxylic acids: conversions, to acid anhydrides, **548–549, 565**; to acid chlorides, **547–548, 565**; to amides, **551**; IUPAC naming, **18**; occurrence, **537**; physical properties, **538, 542–544**; racemic, resolution, **137–138**; reactions, addition to ketene, **565**; decarboxylation, **552–555**; deuterium exchange, **555**; dimerization, **546**; esterification, **549–550**; halogenation, **555–556**; Hell-Volhard-Zelinsky reaction, **555–556**; racemization, **555**; reduction, **555**; reduction of alkenes, **358**; reactions with enol ester, **565**
Carboxypeptidase, **708**
Carcinogens, **446**; aromatic nitro compounds, **773**; diethylstilbestrol, **472**; polycyclic aromatic compounds, **841**
Cardiac glycosides, **881–882**
Δ^3-Carene, structural formula, **852**
Caro's induline blue, **407**
β-Carotene, **262, 861**
Carotenes, **860–862**
β-Caryophyllene, structural formula, **856**
CAT, **270**
Catalysts, enzymes, **148–149, 190–191**; poisoning of, **375**; role in catalytic hydrogenation of alkenes, **374**
Catalytic dehydrogenation, alcohols, **217**
Catalytic hydrogenation, **53, 55, 646a**; acetylene, **55**; alkenes and alkynes, **374–376**; aromatic compounds, **418**; cholesterol, **873**; dienes, **423**; formaldehyde, **29**; nitrogen compounds, **662**; preparation of alcohols by, **229–230**; tetralin, **830**
Catechol, **805–806**; preparation, **810**; structural formula, **441, 790**
Catecholamines, **816**
Cation exchangers, **990**
Cationic polymerization, **977–978**
Cationic surfactants, **541**
Cations, **45, 46**. See also Carbonium ions
Cell, evolution of, **8**
Cell membrane, **107**
Cellobiose, **632**
Cellophane, **638**
Cellulose, **638, 976**; hydrolysis, **532**
Cellulose acetate, **638**
Cellulose xanthate, **638**
Cerotic acid, physical properties, **538**
Cesium, electronegativity, **103**
Chain branching, **973, 976**
Chain breaking, **336, 337**
Chain length, **976**
Chain reactions: free-radical, **45**; free-radical addition, **49–50**; homolytic displacement on hydrogen, **44–45**
Chain transfer, **976, 977**
Chair form, **324, 325**
Charge transfer complex, **106, 446**; of aromatic nitro compounds, **739–740**
Chaulmoogric acid, structural formula, **539**
Chelates, **486**
Chelation, **509**
Chemical resolution, **704**
Chemical shift reagents, **270–271**
Chemical shifts, **268, 287–288**; benzenes, substituted, **408, 410**; protons, effect on, **272**
Chenodesoxycholic acid, structural formula, **881**
Chichibabin reaction, **897**; isoquinoline, **909**; quinoline, **907**
Chicle, **863**
Chirality, **126**; biphenyl, **842–843**
Chitin, **638**
Chloral, **49**
Chloral hydrate, **786**; preparation, **49**
Chloramine-T, **956**
Chloramphenicol, **817**
Chlordane, **784**
Chlorination: aldehydes and ketones, **502** ff.; alkanes, **338–339**; alkanes, commercial, **339, 340**; alkenes, free-radical, **360**; alkenes, in polar solvents, **350**; alkynes, free-radical, **360**; arenes, **452**, free-radical, **466**; arenes, steric effects, **456**; benzene, **459**; carboxylic acids, **555–556**; ethane, **44–45**;

Chlorination (cont.)
　methane, **26, 340;** methane, thermodynamics, **337;** β-naphthol, **828;** neopentane, **339;** phenols, **797;** phosphines, **960;** propane, **338;** toluene, **470;** xylene, **466**
Chlorine: covalent bond radii, **102;** electronegativity, **103;** electronic distribution, **88;** free-radical addition of to benzene, **784;** isotopes, **299;** mass spectrum, **303**
Chloroacetic acid: degree of dimerization, **546;** pK_a, **109, 695**
Chloroacetone, infrared spectrum, **488**
Chloroacetylene: physical properties, **36;** resonance effect, **109**
Chloroalkanes, **776**
N-Chloramine, **653**
p-Chloroaniline, preparation, **769**
Chlorobenzene: preparation of phenols using, **809;** reactions, Dow process, **809;** dinitration, **743;** Raschig process, **809;** structural formula, **440**
2-Chloro-1,3-butadiene, preparation, **355**
1-Chloro-2-butene, preparation, **421**
3-Chloro-1-butene, preparation, **421**
2-Chlorocyclopentanone, structure, **291**
3-Chloro-2,2-dimethylbutane, reactions, **151–152**
Chloroethanes, commercial preparation, **777–778**
Chloroethylenes, commercial preparation, **777–778**
Chloroform: commercial preparation, **777;** solvent effect of on nucleophilic substitution, **153**
Chloroformic acid, **547**
Chloromethane: bonding in, **14;** physical properties, **13;** preparation, **26**
Chloromethylation, **500–501**
Chloromycetin, **817**
3-Chloro-2-nitrotoluene, preparation, **457**
3-Chloro-4-nitrotoluene, preparation, **457**
5-Chloro-2-nitrotoluene, preparation, **457**
Chloronium ion, **350**
4-Chloro-2-pentene, preparation, **421**
o-Chlorophenol, **797**
p-Chlorophenol, **797**
2-(p-Chlorophenoxy)-2-methylpropionic acid, **792**
Chlorophosphines, substituted, **958**
Chlorophosphonium chlorides, preparation, **960**
Chlorophyll, **857, 928, 931–932;** physical properties, **262**

Chloroprene, copolymerization, **980**
1-Chloropropane, preparation, **338**
2-Chloropropane.
　See Isopropyl chloride
3-Chloropropanoic acid, pK_a, **109**
Chlorosilane, manufacture, **966**
N-Chlorosuccinimide, **341**
Chlorosulfonamides, preparation, **956**
m-Chlorotoluene, nitration, **457**
o-Chlorotoluene, alkali fusion, **810**
p-Chlorotoluene, alkali fusion, **810**
Chlorotrifluoroethylene, polymerization, **980**
1-Chloro-2,4,6-trinitrobenzene, reactions, **461**
Cholestane, **867**
Cholesterol: biosynthesis from squalene, **871;** conversion to cholic acid in the body, **880;** dehydroepiandrosterone, preparation from, **879;** human health and, **540;** occurrence in body tissues, **872;** reactions, **873;** structure, **872**
Cholic acid, **880;** structural formula, **539, 881**
Choline, **732**
Chondroitin, **641**
Chromates, use as oxidizing agents, **53**
Chromic acid, esterification, **210**
Chromium compounds, use to oxidize aldehydes, **517**
Chromium trioxide, **217;** for oxidation of alcohols, **217–218, 222**
Chromophore, **261**
Chromosomes, **446, 924**
Chromyl chloride, **217**
Chrysanthemic acid, cyclopentenyl ester, **606;** structural formula, **539**
Chrysene, structural formula, **841**
Chugaev reaction, **223–224, 383, 579**
Chymotrypsin, **708**
Cigarette smoke, aromatic compounds in, **444**
Cinchomeronic acid, **909**
Cinchonine, **918**
Cinchoninic acid, preparation, **908**
Cinnamaldehyde, structural formula, **480**
Cinnamic acid: decarboxylation, **554;** physical properties, **594**
Cinnoline, structural formula, **889**
Circular dichroism, **134**
cis-, **69, 124–125;** EZ naming system, use with, **394;** RS- rules, **129**
cis Attack, **370–371**
cis Effect, **535b**
cis Elimination, **162, 164**
Cisoid, **395**
cis-trans Isomerism: E2 reactions, **162;** human vision, **125**

Citraconic acid: physical properties, **594;** stereochemistry, **596**
Citral, **850;** ionones, preparation from, **860, 852;** phytol, preparation from, **858;** synthesis, **849**
Citric acid, physical properties, **589**
Citric acid cycle, **645**
Citronellal, **848, 850;** reactions with acid, **850**
Citronellol, **848, 850**
Citrovorum factor, **914**
Citrulline, structural formula, **698**
Claisen condensation: acetyl-CoA with carbon dioxide, **868;** crossed, **573;** esters, **570–574;** mixed, **574**
Claisen rearrangement, **799–800**
Claisen-Schmidt condensation, **509–510**
Clemmensen reduction, **381;** ketones, **470**
Clofibrate, **792**
Coal, **467;** as energy source, **377, 378**
Coal tar, **467**
Coal tar dyes, **407–408**
Cobalamine, **731, 933**
Cobaltic fluoride, **339–340**
Cobaltous fluoride, **339–340**
Cobamide, **933**
Cocaine, **920**
Codeine, **919**
Codons, **924;** in messenger RNA, **927**
Coenzyme A, **212;** in biosynthesis of terpenes, **867 ff.;** structural formula, **730**
Coenzyme Q, **819**
Coenzymes, **725, 728;** vision, need in, **858–859**
Cofactor, **725**
Cohen S. G., **64a**
Coke, **467**
Cold drawing, **973, 982**
Cold water soap, **541**
Collagen, **690;** amino acid composition, **699, 700**
Collins, C. J., **535a**
Color, definition, **243**
Color blindness, **243**
Combustion, **28–29, 53;** hydrocarbon isomers, comparison, **76**
Competitive inhibition, **726, 727**
Competitive reactions: substitution and elimination, **42, 51, 156–158, 189, 191;** thermodynamic control, **194–195**
Complexing agents, cyclic polyethers, **806**
Concerted addition, **357**
Concerted process, **51**
Concerted reactions, **428–429**
Condensation polymers, **979, 982–989**
Condensation reactions: acetyl-CoA with carbon dioxide,

868; acid anhydrides, 563–564; aldol condensation, 504–509; aromatic nitro compounds, 738–739; Claisen condensation, 570–574; Claisen-Schmidt condensation, 507–509; crossed Claisen condensation, 573; crossed Dieckmann condensation, 573; Dieckmann condensations, 572; mixed aldol condensations, 507–509, 574; mixed Claisen condensation, 574; nitroarenes, 739; silanols, 966
Configuration, 124; optical activity and, 135
Conformation, 119, 124; configuration and, 130–131; conjugated systems, 395; cycloalkanes, 324–326; stability, 132
Conformers, 119
Congruence, 126
Coniferyl alcohol, 789; structural formula, 814
(+)-Coniine, 916
Conjugate acid, 20
Conjugate base, 20
Conjugated protein, 690, 725
Conjugated systems, 1,2 and 1,4 addition, 426; conformation, 395; dienes, 393–394; Hückel $4n + 2$ rule, 399–400; infrared spectra, 261–262, 404–407; mass spectrometry, 410–411; nmr spectra, 408, 410; reactions, 427; ultraviolet spectra, 407
Conjugation: aromatic amines, 245; biphenyl, 842; with carbonyl group, 483; carbonyl stretching frequency, effect on, 545–546; chromophores, effect on, 261; dienes, 393–394; extended, 261–262; infrared spectrum, 253
Conservation of orbital symmetry, 429
Constitution, 2
Conversion, 217
Coordinate covalent bond, 384
Coordinate polymerization, 978
Coordination chemistry, 646a
Coordination compounds, heterocyclic, 927–933
Cooxidation reaction, 968b
Cope reaction, 383–384, 659
Coplanarity, 162, 395
Copolymerization, 974
Copolymers, 970
Copper chromite, 376
Coprostane, 867
Coprostanol, 873
Coral rubber, 978 ff.
Coronene, structural formula, 842
Corphinic acid, 933
Corphyrinic acid, 933
Corrin, 932
Corticosterone, structure, 877
β-Corticotropin, 723

Cortin, 877
Cortisone: structure, 877; synthesis, 878–879
Cosmic rays, 242
Cotton, 638
Coumarone, structural formula, 889
Coupling constant, 271–272, 288–290; s character, 331
Coupling reactions: aromatic diazonium compounds, 758–759; furan, 892; phenols, 796–797; pyrrole, 892; as termination reaction, 976; thiophene, 892
Covalent bond radius, 101
Covalent bonds, 4, 13–14; electronegativity, 103; potential-energy diagram, 113–115; resonance structures, 98
Cracking, 384
Cracking (hydrocarbons), 355
Crawford, H. M., 822a
Cream of tartar, 593
Cresol: pK_a, 791; preparation, 810; structural formula, 441
Crossed aldol condensation, nitroalkanes, 680
Crossed Cannizzaro reactions, 232
Crossed Claisen condensation, 573, 599; ethyl formate, 607
Crossed Dieckmann condensation, 573
Crosslinking: in polymers, 971 ff.; rubber, effect on properties, 863
Crotonaldehyde: physical properties, 71; preparation, 506; reactions with aniline and sulfuric acid, 908
Crotonic acid, 595; physical properties, 71, 594
Crotyl alcohol, physical properties, 70
Crystalline polymers, 973
C-terminal amino acid, determination in peptides, 707–708
Cumene: physical properties, 444; preparation of phenols from, 809; structural formula, 440, 441
Cumulative bonds, 393
Cupric ion, use to oxidize aldehydes, 518
Cuprous bromide, reactions with diazonium sulfate, 763
Cuprous chloride: reactions with diazonium chloride, 762; reactions with 1-alkyne, 332
Cuprous salts: Sandmeyer reaction, 762; use in Ullmann reaction, 780
Curtin, D. Y., 535b
Cyanamide, 649; polymerization, 989
Cyanates, 684
Cyanides, 671 ff.
Cyanoacetamide, addition to β-diketone, 901–902
Cyanoacetic ester, 672; amino

acids, preparation using, 702
Cyanocobalamine, 933
Cyanoethylation, 512, 671
Cyanogen chloride, reaction with Grignard reagent, 674
Cyano group, reactions, 674; addition of Grignard reagent, 522
Cyanohemoglobin, 930
Cyanohydrin, 513; preparation, 493, 675; preparation with glucose, 616–617; reactions, 674–675; reactions, hydrolysis to yield α-hydroxy acid, 593
2-Cyano-2-propyl radicals, 682
Cyclazocine, 919
Cyclic anhydrides: preparation, 549, 554; reactions, alcoholysis, 561; reactions, reduction, 562, 582
Cyclic (ring) compounds, 38. See also under Cycloanhydrides. See Cyclic anhydrides. cis- and trans- configurations, 125; 125; conformation, 430; cycloaddition, 428 ff.; diesters, 589, 590; fused benzenoid ring systems, 396–397; highly strained, 328–330; Hückel $4n + 2$ rule, 399–400; IUPAC naming, 79–81; ketones, 493, 521; β-lactone, 591; multiple substituents, nomenclature, 395–396; nitrogen-containing, preparation, 662; polycyclic, 458; preparation, 508, 662; substituted, 326
Cyclic diamides, preparation, 669
Cyclic diesters, preparation, 589, 590
Cyclic diterpenes, 857, 859
Cyclic elimination mechanism, 383; Cope elimination, 659
Cyclic esters, β-lactone, 591
Cyclic ethers, 68
Cyclic hemiacetals, preparation, 786
Cyclic hydrazide, 701
Cyclic ketals, preparation, 607
Cyclic keto esters, preparation, 572
Cyclic ketones, Baeyer-Villiger reaction, 593; preparation, 554, 600
Cyclic photophosphorylation, 643
Cyclic terpenes, 850–852
Cyclic thioacetals, 943
Cyclization reactions: acyclic terpenes, 850; aldehydes, 492–493; indane ring system from, 573; ketene with formaldehyde, 591; keto acids, 591, 592; lycopene, 861; thiourea, 707; Thorpe-Ziegler reaction, 672, 673; unsaturated acids, 591, 592
Cycloaddition reactions, 428–431
Cycloalkanes: IUPAC naming, 79–80; physical properties, 320 ff.; preparation, 379 ff.; reactions, 362; reactions, halogenation, 339–340

Cycloalkenes: IUPAC naming, 79–80; peroxide effect, 362; physical properties, 320 ff.; 327, 328
Cycloalkynes: IUPAC naming, 79–80; physical properties, 320 ff.
Cyclobutadiene, 396; stability, 394
Cyclobutadiene dication, 400
Cyclobutane, 68; chemical shift, 288; conformation, 324; physical properties, 71, 321, 323, 327; reactions, 362; substituted, 326
Cyclobutanol, physical properties, 71, 202
Cyclobutanone, physical properties, 71
Cyclobutene: chemical shift, 288; physical properties, 71, 321, 328
Cyclodecane, physical properties, 321, 323
cis-Cyclodecene, physical properties, 321
Cyclodecyne, physical properties, 321
Cyclododecane, heat of combustion, 323
trans-Cyclododecene, 328
Cycloheptane, 330; heat of combustion, 323; physical properties, 321
Cycloheptanol, physical properties, 203
Cycloheptatrienylidene, 438a
Cycloheptene, physical properties, 321
Cyclohexadiene, heat of hydrogenation, 400; preparation, 426
Cyclohexadiene, substituted, 782
2,5-Cyclohexadienones, preparation by Reimer-Tiemann reaction, 804
Cyclohexane, 68; conformation, 324–326; heat of combustion, 323, 324; heat of hydrogenation, 400; nmr spectrum, 326; oxidation to adipic acid, 560; physical properties, 321; preparation from cyclohexene by disproportionation, 402; production figures, 378; reactions with free radicals, 459; structure, 64b; uses, 375, 378
Cyclohexane, substituted, 326
Cyclohexanecarboxylic acid, physical properties, 538
Cyclohexanol: oxidation to adipic acid, 560; physical properties, 203; preparation from phenol, 805
Cyclohexanone: infrared spectrum, 254, 488; reactions, mixed aldol condensation with ethyl acetate, 607; reactions, oxidation to adipic acid, 560; steric effects, 494

Cyclohexatriene. See Benzene
Cyclohexene, heat of hydrogenation, 400; dipole moment, 322; physical properties, 321, 327; reactions, decomposition, 296, 297; disproportionation to yield benzene, 401–402, 433; epoxidation, 369
Cyclohexenedione, tautomerism with resorcinol, 807
Cyclohexenone: preparation, 509; Robinson annelation reaction, 512
Cyclohexylamine, physical properties, 650
Cyclohexyl esters, pyrolysis, 579
Cyclohexylpropylamine, 816
Cyclononane, 330; physical properties, 321, 327
cis-Cyclononene, physical properties, 321
trans-Cyclononene, physical properties, 321
Cyclononyne, 328; physical properties, 321
Cyclooctane, 330; heat of combustion, 323; physical properties, 321
Cyclooctanol, physical properties, 203
Cyclooctatetraene, 396; hydrogenation, 400; reactions with lithium, 400
Cyclooctene, heat of hydrogenation, 400; physical properties, 321
Cyclooctyne, 328
Cyclopentadiene, reactions, 430, 840–841; reactions with Grignard reagent, 399; ultraviolet spectrum, 407
Cyclopentadiene anion, 399
Cyclopentadienes, substituted, 889–894; preparation, 890–891; reactions, 891–893
Cyclopentadienone, substituted, preparation, 508
Cyclopentadienyl anion, resonance, 890
Cyclopentadienyl cation, 399, 410
Cyclopentane, 68; conformation, 324; deuterium exchange, 375; heat of combustion, 323; physical properties, 321, 327; substituted, 326
Cyclopentanol, physical properties, 202
Cyclopentanone, infrared spectrum, 488; steric effects, 494
1,2-Cyclopentanophenanthrene, structural formula, 866
Cyclopentene: dipole moment, 322; physical properties, 321, 327–328; trans addition, 348
Cyclopentenone, preparation, 508
Cyclopentenyl esters, 606
Cyclopropane, 68, 322; chemical shift, 288; heat of combustion,

323, 324; physical properties, 69, 321, 331; reactions, 362; strain energy, 322
trans-1,2-Cyclopropane, 355
Cyclopropanol, physical properties, 69, 202
Cyclopropene, chemical shift, 288; physical properties, 69
Cyclopropylmethanol, physical properties, 71
Cyclotetradecane, physical properties, 327; heat of combustion, 323
p-Cymene: physical properties, 444; preparation, from camphor and phosphorus pentachloride, 854; from limonene, 850, 851; structural formula, 441
Cysteine, 691, 692; structural formula, 692, 950
Cystine, 694; structural formula, 950
Cytidine, in RNA, 921
Cytochrome c, 928; structural formula, 931
Cytochromes, 931
Cytosine, 904; in DNA and RNA, 920, 921; structural formula, 921

2,4-D, 791, 792
Dacron, 983
Dakin reaction, 810
Dark reactions (photosynthesis), 644
DCC. See Dicyclohexylcarbodiimide (DCC)
DDT, 783
Deactivation, of aromatic rings, 452 ff., 456
Deamination, of 1,1-diphenyl-2-amino-1-propanol with nitrous acid, 535a
Debye (unit), 104
Decalin, preparation, 830
Decarbonylation: α-hydroxy acids, 590; α-keto acids, 597
Decarboxylation, 522, 552–555; malonic acid, 597
Degradation, glucose, 617–618; Ruff degradation, 618; Wohl degradation, 617–618
Degree of dimerization, 546
Degree of ionization, 546
Dehalogenation, 51; alkene dihalides, 351
Dehydration reactions, 30, 52; alcohols, 171–172, 192, 223; tert-allylic alcohol, 434; dicarboxylic acids, 554; β-hydroxy acids, 590–591; S_N1cA ether formation, 215
Dehydroacetic acid, preparation, 894
Dehydrochlorination, 30
Dehydroepiandrosterone, preparation from cholesterol, 879
Dehydrogenation, 30; primary amines, 674; primary and secondary alcohols, 216 ff.

Dehydrohalogenation, **51–52**; alkynes, preparation by, **385**; ethers, preparation by, **40**
Deionization, as water softening process, **990**
Delocalization, **403**
Delocalization energy, **97, 400**; aromatic systems, **419**
Delrin, **493, 985**
Demerol, **104, 919**
Denaturation, **715**
Density, nonaromatic hydrocarbons, **320**
Deoxyadenosine, structural formula, **921**
Deoxycytidine, structural formula, **921**
Deoxynucleic acids. See DNA
3-Deoxynucleotides, **921**
2-Deoxy-D-ribofuranose, **920**
Deoxyribonucleic acid. See DNA
Deoxyribonucleosides, **921**
Deoxyribonucleotides, **921**
Deoxyribose, in DNA, **921**
DES. See Diethylstilbestrol (DES), carcinogenic effect
Deshielding, **267**
Desoxycholic acid, structural formula, **881**
Detergents, **540–541**
Deuterium, in solvents for nmr spectroscopy, **270**
Deuterium exchange: alcohols, **205**; alkanes, **375**
Dexamethasone, structural formula, **878**
Dexedrine, **816**
Dextran, **641**
Dextrins, **641**
Dextrose. See Glucose
Diabetes, ketosis in, **487**
Diacetone alcohol: preparation by aldol condensation, **506**; reactions, **507**
Diacetylmorphine, **104**
Diacids. See Dicarboxylic acids
Diacyl peroxides, carboxy inversion reaction, **993a**
N,N-Dialkylaniline, nitrosation, **754**
Dialkylborane, **227**
Dialkylphosphine oxide, preparation, **959**
Dialkylphosphinous chloride, hydrolysis, **959**
Dialkyl sulfite, preparation, **211**
Diallylamine, physical properties, **650**
Diamagnetic effect, **287**
Diamer, **131**
Diamines, oxidation to yield quinones, **813**
p,p'-Diaminobiphenyl. See Benzidine
Diamond, crosslinking in, **972**
Diaryl ketones, reduction, **519**
Diastase, **640**
Diastereoisomers, **131**
Diastereomers, **131, 136**; optical activity, **133** ff.; racemates, resolution, **137–138**
Diatomic molecule, vibrational modes, **247**
Diaxial diols, **369**
1,2-Diazabenzene. See Pyridazine
1,3-Diazabenzene. See Pyrimidine
1,4-Diazabenzene. See Pyrazine
1,2-Diazine. See Pyridazine
1,3-Diazine. See Pyrimidine
1,4-Diazine. See Pyrazine
Diazine, **902–904**
Diazo acid, **757**
Diazoalkanes, **683**
Diazoamino compound, preparation, **759**
Diazoate ion, **757**
Diazo compounds, preparation from hydrazones, **678, 679**
Diazocyanide, preparation, **757**
Diazoethane, **649**
Diazo group, replacement reactions, **762–763**
Diazohydroxide, **655**
1,2-Diazole. See Pyrazole
1,3-Diazole. See Imidazole
Diazoles, **893**
Diazomethane, **683**; reactions, esterification, **550**; methylation of alcohols, **318b**; with phenol, **791**
Diazonium chlorides: physical properties, **758**; reactions with cuprous salts, **762, 763**
Diazonium compounds, aromatic, **757** ff.; properties, **757–758**; reactions, coupling, **758–759**; free radical, **762–763**; reduction, **758, 763, 764**; Sandmeyer reaction, **762**; Schiemann reaction, **762**
Diazonium fluoroborate, **762**; preparation, **758**; Schiemann reaction, **762**
Diazonium ions, **655, 757, 758**; coupling, **758–759**; stability, **754**; substituents in, **760**
Diazonium sulfate: physical properties, **758**; reactions with cuprous salts, **763**
Diazotate ion, **759**
Diazotization, **759, 761**
Dibenzanthracene: carcinogenic activity, **841**; structural formula, **842**
Dibenzo-18-crown-6, **806**
Dibenzofuran. See Diphenylene oxide, structural formula
Dibenzyl ketone, condensation with α-diketone, **508**
Diborane, **227–228**; reactions with, acids, **230–231**; alkenes, **238a**, oximes, **678**
Dibromobenzene, nmr spectrum, **410**
meso-2,3-Dibromobutane, **349**; E2 elimination, **162**
5α,6β-Dibromocholestanol, **873**
5β,6α-Dibromocoprostanol, **873**
9,10-Dibromo-9,10-dihydro-anthracene, preparation, **833**
Dibromofluorescein, mercuration, **803**
2,4-Dibromo-5-methylphenol, structural formula, **442**
1,2-Dibromo-2-methylpropane, nmr spectrum, **268**
1,5-Dibromonaphthalene, preparation, **825, 826**
1,8-Dibromonaphthalene, structural formula, **443**
9,10-Dibromophenanthrene, structural formula, **443**
Di-tert-butyl ketone. See Hexamethylacetone
Di-tert-butylnaphthalenes, **822a**
Di-tert-butyl peroxide, as free-radical source, **974**
Dibutyl phthalate, use as plasticizer, **974**
β-Dicarbonyl compounds: base cleavage, **510**; enolate ions, **509**
Dicarboxylic acids: conversion to cyclic ketones, **511**; physical properties, **542**; preparation, **559–560**; reactions, heat, effect of, **554**; reactions, with lithium aluminum hydride, **555**
Dichloroacetic acid, pK_a, **109, 695**
Dichloroalkane, reaction with sodium polysulfide, **992**
3-5-Dichlorobenzaldehyde, structural formula, **442**
Dichlorocarbene, **804**
Dichlorocyclohexadiene, preparation, **459**
1,t-2-Dichloro-r-1-cyclopentane-carboxylic acid, **395**
Dichlorodifluoroethane, uses, **778**
1,2-Dichloroethane, production figures, **379**
(2Z,2E)-3,5-Dichloro-2,4-hexadienoic acid, **394**
2,4-Dichlorophenoxyacetic acid (2,4-D), **791, 792**
Dichlorophosphoranes, **960**
1,3-Dichloropropane, nmr spectrum, **282, 283**
Dicobalt octacarbonyl, **346**
Dicoumarol, **820**
2,2'-Dicyano-2,2'-azopropane. See Azobisisobutyronitrile
o-Dicyanobenzene. See Phthalonitrile
Dicyclohexylcarbodiimide (DCC): preparation of esters using, **581**; use in peptide synthesis, **711**
Dicyclohexyl-18-crown-6, **806, 807**
Dicyclohexylurea, **581**
Dicyclopentadiene, **430, 431**
Dieckmann condensation, **572**; crossed, **573**
Dieldrin, **784**
Diels-Alder reaction, **429–430**; anthracene, **834**; benzynes, **464, 465**; furan, **891, 892**; hydroquinone, **807**; isoprene, **850**; preparation of anthraquinone by, **836**;

Diels-Alder (cont.)
  quinones, **812–813**
Diene polymerization, **976, 977;**
  anionic, **978**
Diene polymers, **976, 980–981**
Dienes, conjugated, **393–394;**
  preparation, **432–434;** reactions,
  1,2 and 1,4 additions, **419–426;**
  Diels-Alder reaction, **429–430;**
  Diels-Alder reaction with
  benzyne, **464–465;** oxidation,
  427; ozonization, **427;**
  reduction by active metals, **427**
Dienolate ion, preparation, **506**
Dienophile, **429**
Diequatorial conformers, **369**
Diesters: preparation from cyclic
  anhydrides, **562–563,** reactions,
  Claisen condensation, **572;**
  reactions, Dieckmann
  condensation, **572**
N,N-Diethylacetamide, **649**
Diethylamine, **649;** physical
  properties, **36, 650**
Diethylbenzene, preparation, **468**
Diethylcarbodiimide, **649**
Diethyl carbonate, mixed
  condensations with, **599–600**
Diethylene glycol, use as
  solvent, **381**
Diethyleneglycolmethyl ether.
  See Diglyme
Diethyl ether: preparation, **40;**
  by sulfuric acid method,
  215–216
Diethyl ethylphosphonate,
  preparation, **960**
Diethyl glutarate, reactions, **573**
1,1-Diethylhydrazine, **649**
Diethyl maleate, cycloaddition
  with cyclopentadiene, **430, 431**
Diethyl malonate. See Malonic
  ester (diethyl malonate)
Diethyl oxalate, mixed
  condensations with, **600**
Diethyl oximinomalonate, **701**
Diethyl phthalate, condensation
  with dimethyl sulfoxide, **840**
Diethylstilbestrol (DES),
  carcinogenic effect, **472**
Diethyl succinate, Dieckmann
  condensation, **572**
Diethyl tartrate, reactions with
  lead tetraacetate, **605**
Digestive enzymes, **149**
Digitalis, **881**
Digitoxigenin, structural
  formula, **882**
Diglyme, **357, 358**
Dihalides: reactions with
  acetoacetic ester, **600;** reactions,
  hydrolysis, **370;** reactions,
  malonic ester synthesis
  using, **577**
1,1-Diahloethane, preparation, **47;**
  reactions, **51, 52**
Dihydrazone. See Azine
9,10-Dihydroanthracene,
  preparation, **834**

1,2-Dihydroazine. See
  1,2-Dihydropyridine
1,4-Dihydronaphthalene,
  preparation, **419, 830**
1,2-Dihydro-2-oxa-
  4,5-diazanaphthalene,
  structural formula, **887**
9,10-Dihydrophenanthrene,
  structural formula, **397**
1,2-Dihydropyridine, structural
  formula, **888**
Dihydroxanthopterin, **913**
Dihydroxyacetone, **628;**
  preparation, **631, 632**
Dihydroxybenzenes, **805** ff.;
  oxidation to yield quinones,
  813; preparation, **810**
4,4'-Dihydroxydiphenylmethane,
  800
3,4-Dihydroxymandelic acid, in
  DOPA metabolism, **817**
3,5-Dihydroxypentanoic acid. See
  Mevalonic acid
3,4-Dihydroxyphenylamines. See
  Catecholamines
Diimide, use for reducing alkenes
  and alkynes, **377**
Diisocyanate, aromatic, preparation
  of polyurethanes from, **986, 988**
4,4'-Diisocyanobiphenyl,
  preparation of polyurethanes
  from, **986**
2,4-Diisocyanotoluene, preparation
  of polyurethanes from, **986**
Diisopropyl fluorophosphate, **726**
Diketene, preparation of farnesol
  from, **855**
Diketones, internal aldol
  condensation, **600**
α-Diketones: preparation, **514;**
  reactions, **677;** Baeyer-Villiger
  reaction, **565;** benzilic acid
  rearrangement, **515;**
  condensation with dibenzyl
  ketone, **508**
β-Diketones, **509–510;**
  preparation, **574;** reactions,
  677; addition of
  cyanoacetamide, **901–902;**
  Knoevenagel reaction, **512–513**
δ-Diketones, preparation from
  acetoacetic ester by
  halogenation, **600**
Diketopiperazine, preparation,
  669, 700
Dimeric ion, **354**
Dimerization: acids, **543, 546;**
  alkenes, **354**
10,11-Dimethoxystrychnine.
  See Brucine
Dimethoxysuccinic acid, **620**
N,N-Dimethylacetamide: physical
  properties, **666;** use as
  solvent, **463**
Dimethylacetylene, physical
  properties, **70**
Dimethylamine: physical
  properties, **650;** preparation
  by nucleophilic substitution, **41**

Dimethylammonium halide,
  preparation, **42**
N,N-Dimethylaniline: physical
  properties, **768;** structural
  formula, **442**
1,5-Dimethylanthracene,
  structural formula, **443**
2,3-Dimethylbutane,
  conformational stability, **132**
3,3-Dimethyl-2-butanol:
  Chugaev reaction, **224;**
  physical properties, **202**
2,3-Dimethyl-1-butene, mass
  spectrum, **308**
3,3-Dimethyl-1-butene,
  preparation by Chugaev
  reaction, **224**
3,4-Dimethyl-1-cyclobutene,
  ring opening, **429**
1,2-Dimethylcyclopropane, **355**
3,3-Dimethylcyclopropene,
  physical properties, **331**
N,N-Dimethylformamide (DMF),
  666; solvent effect of on
  nucleophilic substitution,
  153, 155; structure, **156;**
  use as solvent, **463**
Dimethyl β-ketosulfone,
  reactions, **953**
Dimethyl β-ketosulfoxide,
  reaction, **952**
2,6-Dimethylnaphthalene,
  preparation, **832**
2,3-Dimethyl-1,4-naphthoquinone,
  reactions with organometallic
  compounds, **822a**
2,6-Dimethyl-1,5,7-octatriene.
  See Ocimene
3,7-Dimethyl-1,3,7-octatriene.
  See Ocimene
Dimethyl phthalate, ester
  interchange, **549**
2,2-Dimethyl-1-propanol,
  physical properties, **202**
Dimethylpropiothetin, **950;**
  structural formula, **951**
2,4-Dimethylpyridine,
  preparation, **901**
6,7-Dimethyl-9-(D-1-ribityl)
  isoalloxazine. See Riboflavin
Dimethyl sulfide, atomic orbitals
  in, **942**
Dimethyl sulfone, atomic orbitals
  in, **942, 943**
Dimethyl sulfoxide: atomic
  orbitals in, **942, 943;** reactions,
  952; reactions, condensation
  with diethyl phthalate, **840;**
  solvent effect of on
  nucleophilic substitution,
  153, 155; structure, **156;**
  use as solvent, **463**
Di-O-methyltartaric acid. See
  Dimethoxysuccinic acid
1,3-Dimethylxanthine. See
  Theophylline
3,7-Dimethylxanthine. See
  Theobromine
Dimsyl, **952**

Dinitration: benzene, **750**; chlorobenzene, **743**
Dinitriles, Thorpe reaction, **672**
2,4-Dinitroaniline, reactions with ammonium sulfide, **742**
m-Dinitro aromatic compounds, reactions with ammonium sulfide, **742**
2,4-Dinitrobenzaldehyde, preparation, **739**
m-Dinitrobenzene: nitration, **743**; preparation, **738**
2,4-Dinitrobenzenesulfenyl chloride, **954**
6,6'-Dinitro-2,2'-biphenyldicarboxylic acid, chirality, **843**
2,4-Dinitrochlorobenzene, **500**; preparation by dinitration of chlorobenzene, **743**
6,6'-Dinitrodiphenic acid, chirality, **842**
2,4-Dinitrofluorobenzene, reactions with proteins, **463**
Dinitrogen tetroxide, preparation, **654**
Dinitrogen trioxide: preparation, **654**; reactions with amines, **654–656**
1,8-Dinitronaphthalene, preparation, **825, 826**
2,4-Dinitrophenol: pK_a, **791**; preparation from 2,4-dinitrochlorobenzene, **743**
2,4-Dinitrophenylhydrazine, preparation, **500, 758**
2,4-Dinitrophenylhydrazones, reactions, **678**
Dioctyl phthalate, use as plasticizer, **974**
Diols, **370** ff.; oxidation, **528–529**; preparation, **562–563**
1,2-Diols, periodic acid, specific for, **632, 634**
Diosgenin, synthesis of cortisone from, **878**
Dioxane, structural formula, **888**
Dioximes, preparation, **677**
1,4-Dioxin, structural formula, **888**
Dioxyphenylalanine. See L-DOPA
Dipentene. See Limonene
Diphenic acid, preparation, **837**
trans-1,2-Diphenyl-1,2-acenaphthenediol, **822a**
Diphenylamine, basicity, **745–746**
1,1-Diphenyl-2-amino-1-propanol, deamination with nitrous acid, **535a**
Diphenyl disulfide, **343b**
Diphenylene oxide, structural formula, **889**
Diphenylmethane, physical properties, **444**
Diphenylmethyl halides, reactivity, **781**

Diphenylmethylphosphine oxide, preparation, **959**
2,3-Diphenyl-1,4-naphthoquinone, reactions with organometallic compounds, **822a**
Diphosphorylation, of thiamine, **904**
Dipole moments: aldehydes and ketones, **483**; amines, **650**; aromatic amines, **747**; aromatic nitroso compounds, **768**; definition, **104, 105**; heterocycles, **890**; hydrocarbons, **320, 322**
Disaccharides, **632–637**
Disilane, **963**
Disorder, second law of thermodynamics, **56–57**
Displacement reactions: alkenes, **358–359**; silicon, **964**
Disproportionation: limonene, **850, 851**; preparation of acids using, **559**; sulfenic acids, **954**; as termination reaction, **976**; tri(p-alkylphenyl)methyl radicals, **782**
Dissymmetry, **126** ff.
Distillation, of coal tar, **467**
Disubstitution, of naphthalene, **828**
Disulfide ion, nucleophilic aromatic substitution, **751**
Disulfides, **949**; preparation, **946**
Disulfonation, of phenol, **743**
Diterpenes, **847**
1,2-Dithioglycerol (BAL), **73**
Divalent carbon, **343b**
DNA (deoxyribonucleic acid), **446; 920**; double helix, **924, 925**; preparation from carbamyl phosphate and aspartic acid, **148**; reactions, **920–921**; secondary structure, **923–924**; structural formula, **922, 923**; tertiary structure, **924**
DNA bases, **446**
Dodecyl mercaptan, **976**
Donor-acceptor complexes, **106–107**
L-DOPA, **815** ff.
Dopamine, **815** ff.
d orbitals: hybridization, **943**; in second-row elements, **942**
Double bonds, **4, 15**; See also Carbon-carbon double bond; Carbon-oxygen double bond in allylic position, **380**; bond distance, **327**; bond force constant, **114**; in 1,3-butadiene, **398**; conjugated, **393, 395**; in cyclic compounds, **68**; IUPAC names for compounds containing, **77** ff.; nmr spectrum, effect on, **228, 289**; nonrotation around, **123**; ozone to locate, **367**; protons around, **284**; reactions, addition, **45–46**; during drastic oxidation, **373**;

nucleophilic addition, **345–346**; rings containing, IUPAC naming, **80**
Double helix, **446**; of DNA, **924, 925**
Dow process, **809**
d-p π bond, **942**
Drastic oxidation, alkenes, **372** ff.; benzene, **416, 417**; dienes, **427**; quinoline, **907**
dsp^2 hybrids, **944**
Durene, physical properties, **443**
Duroquinone, reactions with organometallic compounds, **822a**
Dyes, **8, 761**; coal tar dyes, **407–408, 468**
Dynamic stereochemistry, **119**

E–, **394**
Ebonite, **863**
Ecdysone, **883**
Eclipsed conformation, **119, 375**; hydrocarbons, **123**
Edman degradation, **707**
Ei mechanism, **383**
Einstein (unit), definition, **241**
Elastomers, **973**
Electrocyclic reactions, **429**
Electromagnetic radiation, **241–242**
Electronegativity, **103**
Electronic energy, **245, 258**
Electron pair, **4, 13**
Electron pair acceptor. See Lewis acid
Electron pair donor. See Lewis base
Electron paramagnetic resonance (epr), **294**
Electrons, atomic orbital theory, **87** ff.; delocalized, **97**; mass spectrometry, **295**; molecular orbital theory, **90–96**; in photosynthesis, **643**; in resonance structures, **98–99**; ring current, **408**; valence bond theory, **96–100**
Electron spin resonance (esr), **7, 294**
Electron volt, **241**
Electrophiles, **25, 43, 159**; reactions, alkanes, hydride removal from, **334**; reactions, with double bonds, **45**
Electrophilic addition reactions: alkenes, **947**; alkenes, with hypochlorous acid, **370**; hydrocarbons, nonaromatic, **347–356**; Markovnikov's rule, **347**
Electrophilic attack, arenes, **445, 446**
Electrophilic substitution reactions, **25, 43**; anthracene, **833**; anthraquinone, **835–836**; arenes, **452–458**; aromatic amines, **747–749**; aromatic nitro compounds, **738**;

Electrophilic (cont.)
benzene, reaction of bromine with, **446–447;** control of, **468–469;** cyclopentadienes, substituted, **891–892;** Friedel-Crafts reaction, **454;** heterocycles, preparation by, **890–891;** hydrocarbons, nonaromatic, **334–335, 342;** indene, **840;** isoquinoline, **909;** naphthalene, **826;** phenanthrene, **837;** phenols, **795 ff.;** pyridine, **896–897;** quinoline, **907**
Elimination reactions, **27, 51–52;** beta elimination, **51, 52;** configuration and conformation, effect on, **162;** dehalogenation, **51;** dehydration, **52;** dehydrohalogenation, **51–52;** methanol, **29;** nucleophilic substitution, competition between, **42, 51, 156–158;** preparations using, **171–174;** of alkenes, **383;** of diazomethane, **683;** of nitriles, **674;** Saytzeff rule, **164–165, 166, 169;** stereochemistry, **161–164;** structure, effects, **162;** of water from weak acids, **31**
Emden-Meyerhof glycolysis, **645**
Emulsifying agents, **541**
Emulsin. See β-Glucosidase (emulsin)
Emulsion polymerization, **974**
Enamine-nitrile, **672**
Enanthic acid, physical properties, **538**
Enantiomers, **126;** helices, **132;** naming, **128–130;** optical activity, **133 ff.;** racemates, resolution, **137–138**
Endergonic reaction, **57**
End-group analysis, of peptides, **706–708**
endo-, **431**
Endopeptidases, **708**
Endothermic reactions, **56**
Endrin, **784**
Energy: bond energy, **110–115, 145;** first law of thermodynamics, **56;** of a light beam, **240**
Energy-reaction-coordinate diagram, **189–190;** Fischer esterification, **209**
Enol acetate, **510**
Enolate ions, **485 ff.;** addition to carbonyl group, **506, 508;** preparation by Michael reaction, **512;** stabilization, **509**
Enol ester, reaction with an acid, **565**
Enolization, of acetoacetic ester, **600–604;** in Hell-Volhard-Zelinsky reaction, **555–556**

Enol-nitrile, **672**
Enols, **485 ff.;** phenols, **790;** reactions, **602, 604**
Enovid ketone, **876;** structural formula, **877**
Entgegen, **394**
Enthalpy, **56 ff.;** halogenation of methane, **336–337**
Enthalpy of activation, **184;** enzyme catalysis, effect on, **191**
Enthalpy of formation, **110**
Entropy, **56 ff.;** evolution of life and, **61;** in rubber band, **863**
Entropy of activation, **184;** enzyme catalysis, effect on, **191**
Entropy units, **56–57**
Enzymes, **148–149, 708, 725–732;** energetics of enzyme catalysis, **190–191;** metabolic, for polysaccharides, **640**
E1cA reactions: dehydration, **172;** rearrangements in, **166**
E1 eliminations: Saytzeff rule, **166, 169;** stereochemistry, **164**
E1 reactions, **156–158**
Eosin, **802**
EPDM rubber, **979**
Ephedrine, **816**
Epichlorohydrin, reaction with bisphenol A, **985, 987**
Epimers, **615;** interconversion, **624;** steroids, **867**
Epinephrine, **816;** function, **723;** structural formula, **806**
Episulfides, reactions, **948**
Epoxidation reactions, alkenes, **367–370**
cis-Epoxide, **368**
trans-Epoxide, **368**
Epoxides, **38, 368;** infrared spectrum, **251;** malonic ester synthesis of unsaturated acids using, **526;** preparation, **48, 367 ff.;** preparation from halohydrin, **785;** reactions with Grignard reagent to yield alcohols, **225–226;** reactions with lithium aluminum hydride, **230, 231**
Epoxides, unsaturated. See Oxirene
1,2-Epoxypropane. See Propylene oxide
Epoxy resins, **985 ff.**
EPT rubber, **979**
Equatorial conformation, **325–326**
Equilenin, **875;** structural formula, **876**
Equilibrium, **181 ff.;** in ether formation, **40**
Equilibrium constant, **181 ff.;** acid-base reaction, **21**
Ergosterol, **873, 879;** vitamin D, preparation from, **874–875**
Ergot alkaloids, **917**
Ergothionene, **950;** structural formula, **951**

Erythrose, preparation, **631, 632**
Esr. See Electron spin resonance (esr)
Essential oils, **391a, 847**
Ester group, infrared spectrum, **253**
Esterification, **207–213, 549–550;** acid-catalyzed, **207–208;** amino acids, **700;** bimolecular, of α-hydroxy acid, **589–590;** block polymers, preparation by, **991;** polyethylene terephthalate, preparation by, **984;** thiols, **945, 946**
Ester interchange: preparation of polycarbonates by, **984;** preparation of polyethylene terephthalate by, **984**
Esterolysis, **551**
Esters, in biology, **212;** carbonyl stretching frequency, **545;** of inorganic acids, **209–211;** IUPAC naming, **38;** mass spectra, **304;** physical properties, **542 ff.;** of physiological interest, **606–607;** preparation, from acid chlorides and acid anhydrides, **212–213;** by addition, **582;** from alcohols by acid-catalyzed esterification, **207–208;** by Baeyer-Villiger reaction, **515–516;** by olysis, **580–582;** by oxidation, **582;** by reduction, **582;** by substitution, **582;** preparation of acids from, **557;** reactions, **566–579;** reactions, acidolysis, **567;** addition of Grignard reagents, **227;** alcoholysis, **567, 568;** of alkyl group of, **578–579;** ammonolysis, **568, 569;** Bouveault-Blanc reduction, **570;** bromination, **577;** Claisen condensation, **570–574;** crossed Claisen condensation, **573;** crossed Dieckmann condensation, **573;** hydrogenolysis, **570;** hydrolysis, **64b, 567;** mixed aldol condensation, **574;** mixed Claisen condensation, **574;** pyrolysis, **383, 578–579;** reduction, **529, 570;** reduction with sodium borohydride, **230, 238a;** reactions with, Grignard reagents, **570;** phosphorus pentachloride, **570;** thionyl chloride, **570;** ylides, **570**
Esters, phenolic, preparation, **792**
Esters, 2,β-unsaturated, preparation, **574**
Estradiol, **875;** structural formula, **876**
Estriol, structural formula, **876**
Estrone, **875;** structural formula, **876**
Ethanal. See Acetaldehyde
Ethane, **66;** congruence,

126–127; covalent bonding in, 5; formation during halogenation of methane, 335, 336; molecular orbital theory, 93 ff.; physical properties, 36, 331; pK_a, 332; preparation by catalytic hydrogenation of acetylene, 55; preparation by Kolbe electrolysis, 382; reactions, homolytic substitution on hydrogen, 44–45; reactions, oxidation, 53, 365; reactions with chlorine, 45; reactions with free radicals, 459; stereochemistry, 119, 120, 123, 126–127; structure, 35; vibrational modes, 247
Ethanoic acid. See Acetic acid
Ethanol, 233; autoprotolysis constant, 22; azeotrope with water, 206–207; biosynthesis by brewer's yeast, 219; formation during Claisen condensation, 571; infrared spectrum, 251, 252; nmr spectrum, 276–280; physical properties, 36, 202; preparation, by addition of acid to ethylene, 47; preparation, industrial, 233; production figures, 378; reactions, dehydration, 52; estrification with acetic acid, 208; ethoxide ions from, 51–52; haloform reaction, 503; oxidation, 53; reduction of aromatic compounds in, 419; reactions with, hydroxide ion, 40; sodium, 205; sulfuric acid to yield diethyl ether, 215–216; solvent effect of on nucleophilic substitution, 153–154; use as solvent, 375
Ether, physical properties, 70. See Ethyl ether
Ethers, 38; cyclic, 68; infrared spectrum, 250–251; IUPAC naming, 17; mass spectra, 304; physical properties, 261; preparation, from alcohols, 214–216; by nucleophilic substitution, 39–40; from phenols, 791; reactions, 798
Ethoxide ion, E2 reaction with alkyl halides, 169; reactions with, ethyl bromide, 40, 192, 193; ethyl bromide, thermodynamics, 184–185; methyl bromide, thermodynamics, 184–185; methyl iodide, 40, 192; use in elimination reactions, 51
Ethoxyamine, 649
1-Ethoxybutane, mass spectrum, 304, 305
Ethoxyethane, physical properties, 70
Ethoxy group, introduction of, 471
2-Ethoxy-2-methylpropane. See tert-Butyl ethyl ether
N-Ethylacetamide, 649
Ethyl acetate: absorption peaks, 261; infrared spectrum, 307; physical properties, 69; preparation, 194–195; reactions with methyl ketones, 574; reactions with sodium borohydride, 238a; solvent effect of on nucleophilic substitution, 153; use as solvent, 375
Ethyl acetoacetate. See Acetoacetic ester
Ethylamine, 649; physical properties, 36, 650; reduction of substituted benzenes with, 419; use as solvent, 376
Ethylammonium halide, preparation, 42
Ethylaniline, structural formula, 442
Ethyl azide, 649
Ethylbenzene: physical properties, 443; reactions with free radicals, 459; structural formula, 440
Ethyl benzoate, nmr spectrum, 408, 410
Ethyl bromide, 67, 333; nmr spectrum, 276; physical properties, 36; preparation by electrophilic substitution, 43; reactions, nucleophilic substitution, 143, 148, 158–159; reactions, Williamson ether synthesis with, 214; reactions with, ethoxide ion, 40, 192, 193; ethoxide ion, thermodynamics, 184, 185; metallic lithium, 43–44
Ethyl 3-bromopropanoate, mass spectrum, 303, 304, 306
Ethyl carbamate. See Urethane
Ethylcarbylamine, 649
Ethyl cation, 151
Ethyl chloride: free-radical reactions with chlorine, 45; nmr spectrum, 276, 277; physical properties, 36; resonance effect, 109
Ethyl cinnamate, preparation by mixed Claisen condensation, 574
Ethyl cyanate, 649
Ethyl cyanoacetate, reaction with urea, 911
Ethyldiimide, 649
Ethylene, 67; covalent bonding in, 5; double bond in, 45; ethylene oxide, preparation from by catalytic conversion, 369; haloalkanes and haloalkenes, preparation from, 777–778; molecular orbital theory, 94, 95; occurrence in nature, 330; physical properties, 36, 331; pK_a, 332; preparation from ethanol by dehydration, 52; production figures, 378; reactions, addition of acid to, 47; catalytic hydrogenation, 55; halogenation, 49–50; hydration to form primary alcohols, 227; oxidation by permanganate, 54; polar addition to by halogen, 46; polymerization, 977, 980; reaction with tert-butyl lithium, 346; resonance theory, 445; stereochemistry, 123–124
Ethylene diamine tetraacetic acid (EDTA), 76
Ethylene glycol: preparation of polyethers from, 979; production figures, 378; uses, 203
Ethylene oxide, 68; physical properties, 36; preparation, 48; preparation, commercial, 369; production figures, 378; reactions, with Grignard reagent to yield alcohols, 225–226; reactions, polymerization, 985; uses, 37
Ethylenic protons, 284, 286, 287
Ethyl esters, physical properties, 542, 543
Ethyl ether: infrared spectrum, 250–251; IUPAC naming, 17; preparation, 568
Ethyl ether-ammonia complex, 106
Ethyl p-ethoxybenzoate, preparation, 471–472
Ethyl fluoride, physical properties, 36
Ethyl formate, crossed Claisen condensation, 607
Ethyl group: formula, 75; nmr spectrum, 280
Ethyl halide: preparation by addition of acid to ethylene, 47; reactions with ammonia, 41; reactions, nucleophilic substitution, 40 ff.
Ethylhydrazine, 649
N-Ethylhydroxylamine, 649
Ethylidenimine, 649
Ethyl iodide: nmr spectrum, 276; physical properties, 36; reactions with methoxide ion, 40; Williamson ether synthesis with, 214
Ethyl isocyanate, 649
Ethyl isonitrile. See Ethylcarbylamine
Ethyl 3-ketobutanoate. See Acetoacetic ester
Ethyl lithium, 44
Ethyl magnesium bromide: reactions with acetone, 495; reactions with bromide, 43
Ethyl malonate, physical properties, 69
Ethyl methyl ether: IUPAC naming, 18; physical properties, 36, 69; preparation, 40, 791; solvent effect of on nucleophilic substitution, 153

Ethyl methyl ketone: physical properties, **70**; steric effects, **494**
3-Ethyl-2-methyl-1-pentene, mass spectrum, **301, 302**
N-Ethyl-N-methylpropylamine, **649**
Ethyl nitrate, **649**
Ethyl nitrite, **649**
Ethyl oleate, work of Subba Rao, **238a**
5-Ethyl-5-phenylbarbituric acid. *See* Phenobarbital
Ethyl phenyl ketone, **440**
Ethyl propionate, physical properties, **69**
Ethyl radical, **45**
Ethyl stearate, reactions with sodium borohydride, **238a**
Ethyl sulfate, preparation, **210**
Ethylsulfuric acid, preparation from ethylene, **47**
Ethyl 2,4,6-trimethylcinnamate, preparation, **574**
17α-Ethynylestradiol, **876**; structural formula, **877**
E2cA reactions, of $EtOH_2^+$ with weak bases, **215**
E2 reactions, **158**; configuration and conformation, effect on, **162**; Saytzeff rule, **169**; structure, effect on, **162**
Eugenol, structural formula, **806**
Evolution, **7–8**; thermodynamics and, **61**
Exchange reactions, **548**; acid anhydrides, **562**; acid chlorides, **562**; amides, **551**; transesterification, **549**
Excited state, **260**
Exclusion principle, **87**
Exergonic reaction, **57**
Exhaustive methylation, **167** ff.; quaternary ammonium hydroxides, **657**
exo-, **431**
Exopeptidases, **708**
Exothermic reaction, **56**
External-mirror symmetry, **126**
Extinction coefficient. *See* Absorptivity (extinction coefficient)
EZ naming system, **394–395**

$4n + 2$ rule, **399–400**
FAD (flavin adenine dinucleotide), **914**; structural formula, **915**
Far infrared radiation, **242, 243**
Farnesol, preparation, **855**; preparation of phytol from, **858**; preparation of squalene from, **860, 861**; reactions, **855–856**
Farnesyl pyrophosphate, **869, 870**
Far ultraviolet radiation, **241** ff.
Fats, **537**; in foods, **540**; reactions, alkaline hydrolysis, **540–541**; reactions, digestion, **146–147**; solubility, **107**; structure, **42**
Fatty acids, biosynthesis, **872**
Fatty acid synthetase, **872**
FDNB. *See* 1-Fluoro-2,4-dinitrobenzene
Feedback inhibition, **148–149**
Fehling's solution, reaction with glucose, **615**
Female hormones, **875, 876**
Fenton's reagent, **618**
Ferric chloride, complexes with phenols, **794**
Fibrin, **820**
Fibrinogen, **820**
Fibroin, amino acid composition, **700**
Fibrous proteins, chain geometry, **715**
Fillers, in rubber, **863**
Fingerprint region (spectroscopy), **250**
First law of thermodynamics, **56**
First-order reactions, **192, 193**
Fischer, E., **622** ff.
Fischer esterification, **208** ff.; acid anhydrides, **561–562**
Fischer-Hepp rearrangement, **754**
Fischer indole synthesis, **906**
Fischer planar projection, **621, 622**
Fixative (perfume), **208**
Flammability, **29**
Flavine adenine dinucleotide. *See* FAD
Flavin mononucleotide. *See* FMN
Flavin oxidation-reduction enzymes, **263**
Fluorene, **840**; physical properties, **444**; preparation by benzilic acid rearrangement, **515**; structural formula, **396, 442, 840**
Fluorenol, preparation, **838**
Fluorenone: preparation, **838**; structural formula, **840**
Fluorescein, **802**
Fluorination: of alkanes, commercial, **339–340**; of alkenes, in polar solvents, **350**; of arenes, **452**; free-radical, **361**; of methane, **335, 337**
Fluorine, covalent bond radii, **102**; electronegativity, **103**; electronic distribution, **88**; hydrogen bonding to, **105–106**; isotopes, **298**; reaction with methane, **26**
Fluoroalkanes, **776–777**
Fluorobenzenes, preparation, **780**
Fluoroboric acid, methylation of alcohols using, **318b**
Fluorocarbons, **340, 779**
1-Fluoro-2,4-dinitrobenzene (FDNB), reactions with peptides, **706**
Fluoromethane, physical properties, **13**
Fluorothene, **779**
FMN (flavin mononucleotide), **731, 914**; structural formula, **915**
Folic acid, **913, 914**; noncompetitive inhibition with, **726**; preparation, **914**; structural formula, **729**; uses, **729**
Forbidden transitions, **245**
Formaldehyde, bond energies, **111**; covalent bonding in, **5, 14**; in interstellar clouds, **28**; IUPAC naming, **17**; molecular orbital theory, **95–96, 260, 261**; multiple bond in, **26**; paraformaldehyde from, **492–493**; physical properties, **13, 29, 481**; primary alcohols, preparation from, **173**; production figures, **378**; reactions, addition of water to, **27**; chloromethylation, **500–501**; condensation with phenol, **986, 988**; hydrogenation, catalytic, **29**; oxidation, **53**; polymerization, **492–493**; reduction, **19**; reactions with, acetaldehyde, **508**; ammonia, **499**; phenols, **800**; as reducing agent, **30**; sp^2 hybridization, **15**
Formaldehyde-melamine polymers, **989**
Formaldehyde-urea polymers, **989**
Formalin, **13, 492**
Formamide, physical properties, **666**
Formic acid: acidity, **547**; autoprotolysis constant, **22**; bond energies, **111**; dimerization, degree of, **546**; IUPAC naming, **17**; pH, **20–21**; pK_a, **108**; physical properties, **13, 29, 538, 543**; preparation by crossed Cannizzaro reaction, **232**; reactions, elimination of water from, **31**; reactions, Gattermann-Koch reaction, **520**; reactions with cyclohexene and hydrogen peroxide, **369**; reactions with hydrogen peroxide, **368**; redox behavior, **30**; solvent effect of on nucleophilic substitution, **153**; as solvent in nucleophilic substitution, **144, 147**
Formyl chloride, **213**
2-Formyl-1-methylpyridinium chloride oxime, **727**
Fortrel, **983**
Four-carbon plants, **556–557**
Four-center transition state, **357**
Four-membered ring, **591**
Fourth-row elements, **944**

Fragmentation, **295, 297**; of alcohols, **304**; fragment weights, commonly occurring, **301, 302**
Franck-Condon principle, **260**
Free energy, **57, 58, 181 ff.**
Free energy of activation, **184, 186**
Free-radical polymerization: azoalkanes used in, **682**; vinyl polymers, preparation by, **974**
Free-radical reactions, **20, 26**; addition to alkenes, **49–50, 961**; addition to alkynes, **49–50**; aldehyde with *tert*-butyl perester, **566**; aromatic ring, **458–459**; carbene insertion, **334**; chain reaction, **45, 336**; diazonium compounds, **762–763**; halogenation, alkenes, **360–361**; arenes, **466**; carboxylic acids, **555**; methane, **336**; organosilicon compounds, **965**; phosphines, addition to alkenes, **961**; thiols, addition to alkenes, **946**
Free radicals, **46**; triphenylnethyl, **782**
Free-radical trap, **337**
Free rotation, **121**
Freezing, thermodynamics, **56**
Freons, **37**
Frequency, **240**
Friedel-Crafts reaction, **454, 468**; aromatic amines, **753**; aryl compounds, **562**; benzene, **469**; chloromethylation, **500–501**; esters, **570**; furan, **892**; naphthalene, **827–828**; nitrobenzene, **738**; phenanthrene, **837**; phenols, **797 ff.**; polycyclic compounds, **458**; pyrrole, **892**; quinoline, **907**; thiophene, **892**
Fries rearrangement, **798–799**
Front-side collision, **144**
Fructose, **627, 628**; preparation, **631, 632**
Fuchsin dye, **8, 407**
Fulvalene, **438a**
Fumagillin, structure, **612a**
Fumaric acid: ozonization, **604**; physical properties, **542, 594**; stereochemistry, **596**
Fuming sulfuric acid, **453, 738**
Fungicides, phenyl mercury, **455**
Furan-, **618**
Furan, dipole moment, **890**; preparation, **607**; preparation, commercial, **892**; reactions, acidic polymerization, **891**; Diels-Alder reaction, **464, 891, 892**; electrophilic substitution, **891, 892**; reduction, catalytic, **892**; reactions with benzyne, **464**;

resonance, **890**; resonance structures, **402, 403**; structural formula, **888**
Furanose ring system, **618**
Fused ring systems, **841–842, 888–889**; nomenclature, **386–397, 888**; oxidation, **410**
Fusel oil, **233, 640**

Gabriel synthesis, **660–661, 701**
Galactans, **639**
Gallic acid, **807**; structural formula, **539**
Gamma globulin, **724**
Gamma rays, **242**
Gammexane, **459, 783–784**
Gas constant, **186**
Gas oil, **379**
Gasoline, **76, 379**; alkylation, use in preparation of, **355**; pollutants from combustion of, **366**
Gattermann-Koch reaction, **520**
Gattermann reaction, **521**; aldehyde synthesis using, **897**; furan, **892**; pyrrole, **892**
Gauche conformer, **123**
Geminal, **288**
Gene, **924**
Genetic code, **8, 927**
Genetic diseases, **149, 724**
Gentianose, **637**
Gentobiose, **634**
Geometrical isomerism, **69**
Geon, **980**
Geranial, **848, 850**; synthesis, **850**
Geraniol, **848**
Gibbs free energy, **57**
Gigahertz (unit), **294**
Glass temperature, **973**
Globular proteins, **690**
D-Glucaric acid, preparation, **616**
D-Glucitol. See Sorbitol
Gluco-, **615**
Gluconic acid, preparation from glucose, **615**
α-D-Glucopyranose, **614**
β-D-Glucopyranose, **614**
4-O-(β-D-Glucopyranosyl)-D-glucopyranose. See Cellobiose
β-D-Glucosamine, polymeric, **638**
Glucose, **614–622**; metabolism, **644–645**; physical properties, **614**; preparation, **631, 632**; reactions, **622**; reactions,; decomposition by zymase, **233**; fermentation, **640**; oxidation, **61, 615–616, 623**; reduction, **616, 630–632**; Wohl degradation, **617–618**; reactions with, Fehling's solution, **615**; nitric acid, concentrated, **616**; Tollens' reagent, **615**; stereochemistry, **614–615**; structural formulas, **620–622**
α-D-Glucose, **614**; structural formula, **620–622**
β-D-Glucose, **614**; reactions

with periodic acid, **632–634**; structural formula, **620–622**
Glucose-1β-6-glucose-1α-2β-fructose, **637**
Glucose-6-phosphate, **644, 645**
α-Glucosidase (maltase), **618, 620**
β-Glucosidase (emulsin), **620**
Glutaconic acid, physical properties, **594**
Glutamic acid, **691, 692, 696**; biochemical function, **499**; hydrogen bonding, **720**; structural formula, **692**
D-Glutamic acid, structural formula, **698**
Glutamic amide, preparation, **702, 703**
Glutamine, **691, 692**; structural formula, **692**
Glutaric acid, dehydration, **554**; preparation, **559**; preparation by malonic ester synthesis, **577**; physical properties, **542**
meso-Glycaric acid, preparation, **623**
(±)-Glyceric acid, physical properties, **589**
Glycerol: physical properties, **70**; preparation, **540**; reactions with phthalic acid, **984, 985**; reactions, Skraup synthesis using, **909**; uses, **203**
Glycerol-1-phosphate, **212**
Glycerol trinitrate, **681**
Glycerose, preparation, **631, 632**
Glycine, **692**; pK$_a$, **695**; preparation, **701**; structural formula, **669, 692**
Glycitols, **642**
Glycocholic acid, alkaline hydrolysis, **880**
Glycogen, **640**
Glycol: formation during oxidation of ethylene, **54**; physical properties, **36**
Glycolic acid, physical properties, **589**
Glyconic acid: preparation by Ruff degradation, **618**; reaction with pyridine, **624**
Glycosides: of nucleic acid bases, **920–921**; saponin, **881**
Glycuronic acids, **640**
Glyoxal: physical properties, **36**; preparation, **416**
Glyoxylic acid, reactions, **597**
Glyptal resin, **984**
Goiter, **459–460**
Graft polymers, **991**
Gramine, **904, 905**
Green sands, **990**
Grignard reagents, **44**; allylic, **781**; benzylic, **781**; preparation of alcohols using, **172–173, 225–226**; preparation of halides using, **171**; silicon-containing, **966**
Grignard reactions: with acid

Grignard (cont.)
chlorides and acid anhydrides, 562; addition to aldehydes, ketones, and esters, 227; addition to cyano group, 522; addition to α,β-unsaturated ketones, 522; with 1-alkyne, 332; with cadmium chloride, 523; with carbonyl group, 495; with cyanogen chloride, 674; with cyclopentadiene, 399; with enols, 602, 604; with esters, 570; with nitriles, 671; with phosphorus trichloride, 958; with quinones, 822a; with sulfonyl esters, 955–956; with sulfur dioxide, 954, 955
Ground state, 260
Group specificity, 725, 726
GRS rubber, 970, 971
Guaiazulene, structural formula, 856
Guaiol, structural formula, 856
Guanidine, 649; physical properties, 650; reactions with ethyl cyanoacetate, 911; resonance structures, 652
Guanine, 911; in DNA and RNA, 920, 921; preparation from uric acid, 912; structural formula, 921
Guanosine: in RNA, 921; structural formula, 921
3-Guanylic acid, structural formula, 922
Guiacol: reactions, 810; structural formula, 806
Gum arabic, 641
Gums, 641
Gutta-percha, 863

Haagen-Smit, A. J., 391a
Half-chair form, 324
Hallucinogenic drugs, 905
β-Haloaldehydes, reactions, 786
Haloalkanes, 776–780; commercial preparation, 777–778
Haloalkenes, commercial preparation, 777–778
Haloarenes, 780–785; preparation, 780; reactions, 781 ff.
Haloàrylamines, 797
Halobenzenes, preparation, 468; reactions with strong base, 463
β-Haloethers, preparation, 785
Haloform, reactivity, 343b
Haloform reactions, 503–504; acetoacetic ester, 604
Halogenation, 976; aldehydes and ketones, 501–504; alkanes, 335–340; alkenes, 49; alkenes, homolytic, 360–361; alkynes, in polar solvents, 351; amines, 653; ammonia, 653; anthracene, 458; arenes, 465–466; aromatic amines, 752–753; benzene, 452, 468; carboxylic acids, 555–556; enols, 602; ethylene, 46; free-radical, arenes, 466;

free radical, carboxylic acids, 555; methane, 26; methane, mechanism, 335–337; naphthalene, 825, 826; phenanthrene, 458; phenols, 797; polycyclic compounds, 458
Halogen compounds, organic, 776 ff.
Halogens, substituent effect on arenes, 462; substituent effect on benzene, 449, 456; substituent effect on reactivity of S_N1 and S_N2 mechanisms, 343a
Halohydrin, 370; preparation, 785; reactions, 785
Halothane, use, 37
Hantzsch synthesis, 899–901
Hapten, 641
Hard water, ion exchange resins to soften, 990
Harmala alkaloids, 917
Harman, 917
Hassel, O., 64b
Haworth formulation, 620, 621
Heat, first law of thermodynamics, 56; standard state reactions, 58
Heat of combustion, cycloalkanes, 322, 323
Heat of formation, of atomic states, 110–111
Heat of hydrogenation, alkenes, 165; benzene, 400–401; butadiene, 398–399
Heavy metal poisoning, 76
Heavy metals, organic compounds containing, 76
Helium: electronic distribution in, 88; molecular orbital theory, 91
Helix, 133
α-Helix, 718–720
Hell-Volhard-Zelinsky reactions, 555–556; α-hydroxy acids, preparation by, 593
Hematin, 928
Heme, 715, 716, 928; structural formula, 930
Hemiacetals, 491, 614; cyclic, 786; reactions with alcohol, 618, 619
Hemicelluloses, 638–639
Hemiketal, 491
Hemin, 928; structural formula, 930
Hemlock, 916
Hemocyanin, 930
Hemoglobin, 928–930; abnormal, 724; oxygen transport, 196; structural organization levels, 715, 716, 721
Hemophelia, 820
Heparin, 641, 820
(2Z,5E)-2,5-Heptadienoic acid, 395
2,6-Heptadione, condensation, 509
Heptaldehyde, physical properties, 481
Heptane, octane rating, 76
1-Heptanol, physical properties, 203
2-Heptanone, 253
Heptapeptide, amino acid

sequence in, 708
Heptonic acids, 617
D-Heptose, preparation, 617
Heptyl group, formula, 75
Heroin, 104, 916
Hertz (unit), 240
Heterocycle, 886
Heterocyclic compounds, 38, 68, 403, 868 ff., 920–927; alkaloids, 916–920; benzenes, substituted, 894–904; cyclopentadienes, substituted, 889–894; fused ring compounds, 841–842; metal coordination compounds, 927–933; naming, 886–888; pteridines, 910, 913; purines, 910–912
Heterogeneous equilibrium, second law of thermodynamics, 56–57
Heterolytic reactions, 19
Hexabromobenzene, preparation, 459
Hexachlorocyclohexane, 783; preparation, 459
Hexachlorocyclopentadiene derivatives, 784
Hexachloroethane. See Perchloroethane
Hexachlorophene, use as antibacterial agent, 472
1,5-Hexadiene: heat of hydrogenation, 398; reactions, 393
2,4-Hexadiene, 429
2,4-Hexadienoic acid. See Sorbic acid
2,5-Hexadione, condensation, 508
Hexafluoroacetone, 488
Hexafluorocyclopropane, 778
Hexafluoroisopropyl alcohol, pK_a, 785
Hexafluoropropylene, copolymerization, 980
Hexahydrocyclohexanes, 643
Hexamethylacetone, infrared spectrum, 488
Hexamethylbenzene, physical properties, 443
Hexamethylenediamine, polymerization with adipic acid, 982
Hexamethylene tetramine, 499
Hexane: preparation by carbene insertion into pentane, 334; solvent effect of on nucleophilic substitution, 153
Hexanoic acid, preparation, 622
1-Hexanol, physical properties, 202
2-Hexanone, physical properties, 481
3-Hexanone, physical properties, 481
1,3,5-Hexatriene: absorption peaks, 261; ultraviolet spectrum, 407
2-Hexene, ozonization, 367
3-Hexene: hydroboration isomerization during, 357;

ozonization, 367
Hexyl group, formula, 75
Hexylresorcinol, preparation, 798
Hine, J., 343a
Hinsberg reaction, 653–654
Hippuric acid, condensation with aldehyde, 702
Histamine, structural formula, 893
Histidine, 692, 694; hydrogen bonding with, 720; structural formula, 692, 893
Hoesch ketone synthesis, 798
Hoffman's violet, 407
Hofmann elimination, 383
Hofmann rearrangement, 662–663; amides, 668
Hofmann rule, 166 ff.
Holoenzyme, 725
Homogeneous hydrogenation, 376
Homolog, 18
Homologous series, 18, 68; alkanes, 66
Homolytic addition reactions, 49; dienes, 422–423; hydrocarbons, 360–364
Homolytic displacement reactions, 43 ff., 49
Homolytic reactions, 19–20; addition, 49, 360–364, 422–423; displacement, 43 ff., 49; substitution, 26, 43–45, 335–342, 458–460
Homolytic substitution reactions, 26, 43–45; arenes, 458–460; hydrocarbons, 335–341, 342
Homopolymer, 970
Hooke's law, 114
Hormones, 6, 722–723
Hot carbonium ion, 655
Houben-Hoesch reaction, 521
α-Humulene, structural formula, 856
Hyaluronic acid, 641
Hybridization, 14, 93; d orbitals, 943–944
Hydration: acetaldehyde, 490; acetone, 491; alkenes, 227, 343–354; alkynes, 354
Hydrazides: of amino acids, 708; reactions with nitrous acid, 684
Hydrazines, 676; preparation from hydrazones, 678–679; reactions, addition of to carbonyl group, 677; oxidation, 682; oxidation, copper-catalyzed, 377; reduction to form amines, 661; Wolff-Kishner reduction, 381; reactions with, acetoacetic ester, 893; aldehydes and ketones, 500; peptides, 708
Hydrazinolysis, 660–661
Hydrazobenzene: preparation from nitrobenzene, 740, 741; reactions, 769; reactions, rearrangement to benzidine, 772
Hydrazones: preparation, 381, 500, 677; reactions, 678–679; reactions, Wolff-Kishner reduction, 678
Hydride ion, 25, 53; nucleophilicity, 331; in reaction of alkane and an electrophile, 334
Hydride transfer, 354–355
Hydrindene. See Indane
Hydriodic acid, 170; reactions with monosaccharides, 622
Hydroboration, 227–229, 356–358; alkenes, 238b; ethyl oleate, 238b
Hydrobromic acid, reaction with oxonium ion, 24
Hydrocarbons, isomers, comparison of properties, 76; IUPAC naming, 16; in nature, 330; physical properties, 76, 261, 320–330; preparation, 956; preparation, petroleum refining, 379; reactions, electrophilic addition, 347–356; electrophilic substitution, 334–335; fragmentation, 297; homolytic addition, 360–364; homolytic substitution, 335–341; hydroboration, 356–358; nucleophilic addition, 345–347; nucleophilic substitution, 331–333; oxidation, 365–374; reduction, 374–376; spectroscopy, 331
Hydrocarbons, aromatic. See Arenes
Hydrocarbons, carbocyclic, 68
Hydrocarbons, cyclic. See Cycloalkanes; Cycloalkenes
Hydrocarbons: halogenated, 36, 37; preparation, 335–340; use as insecticides, 783–784
Hydrocarbons: substituted, IUPAC naming, 16–17; reactions with borane, 331
Hydroelectric power, 379
Hydrogen: covalent bond formation with, 4, 5, 13; covalent bond radii, 101; electronegativity, 103; heat of formation of atom, 110; interatomic distances, 122, 123; isotopes, 298, 299; molecule, molecular orbital theory, 90, 92; molecule, valence bond theory, 96–97; orbitals, 87, 88
α-Hydrogen, 485; activation by C≡N, 671; aldol condensation, 505; in esters, 570 ff.; reactivity in acids, 555
Hydrogen abstraction, in side chains of aromatic compounds, 459
Hydrogenation, alkenes, 165; benzene, 400; benzenes, substituted, 416; butadiene, 398–399; catalytic (see Catalytic hydrogenation); cyclooctatetraene, 400; formaldehyde, catalytic reduction to methanol, 29; industrial, 375; phenanthrene, 837; preparation of alcohols by, 229–230
Hydrogen bonding: acids, 543; alcohols, 204–205; α-helix stabilization by, 720; infrared spectrum, 251; proteins, 715, 718
Hydrogen bonds, 105–106
Hydrogen bromide, peroxide effect, 361–362
Hydrogen chloride: covalent bonding in, 13; reactions, addition to 1,3-butadiene, 421; addition to carbonyl bond, 490; addition to 1,3-pentadiene, 419–420
Hydrogen cyanide: reactions with alkenes and alkynes, 340; reactions with carbonyl group, 493–494
Hydrogen fluoride: covalent bonding in, 5; valence bond theory, 97
Hydrogen ion, pH, 20–21
Hydrogenolysis, 380; esters, 570
Hydrogen peroxide, 48; decomposition, liver catalase, 190; reactions with acid to form peracid, 368; reactions with amines, 659
Hydrogen sulfide, bond angle, 102
Hydrolysis, 42; acid anhydrides, 558; acid chlorides, 558; amides, 558; biochemistry, 146–147; tert-butyl chloride, mechanism, 193, 194; esters, 567; nitriles, 558; preparation of acids using, 557
Hydronium ion, 27, 28
Hydroperoxides: IUPAC naming, 78; reactions with alkenes, 369
Hydrophilic compounds, 107
Hydrophobic compounds, 107
Hydroquinone, 433, 807; chain transfer, effect on, 977; structural formula, 441, 790
Hydroxide ion, as nucleophile in reaction with bromomethane, 23
Hydroxyacetaldehyde, 55
Hydroxyacetic acid, preparation by oxidation of acetylene, 55
Hydroxy acids: physical properties, 589; preparation, 593; reactions, 589–592
α-Hydroxy acids, preparation, 593 reactions, 589–590
β-Hydroxy acids, dehydration, 590–591 preparation, 593
γ-Hydroxy acids, 591 preparation, 593
δ-Hydroxy acids, 591
Hydroxyaldehyde, reaction with phenylhydrazine, 618, 619
o-Hydroxyaldehyde, preparation, 803
β-Hydroxyaldehyde, preparation, 506
Hydroxyarylaldehydes, preparation

Hydroxyarylaldehydes (cont.)
    by Reimer-Tiemann reaction, 803
Hydroxyarylketones, preparation
    by Fries rearrangement, 798–799
p-Hydroxyazo compounds, 797
o-Hydroxybenzaldehyde.
    See Salicylaldehyde
p-Hydroxybenzaldehyde, physical
    properties, 481
m-Hydroxybenzoic acid, pK_a, 797
o-Hydroxybenzoic acid, pK_a, 797
p-Hydroxybenzoic acid: loss of
    carbon dioxide, 739; pK_a, 797
Hydroxybenzyl alcohols, 800
3-Hydroxybutanal. See Aldol
3-Hydroxy-2-butanone,
    structure, 6
β-Hydroxybutyric acid: ketosis,
    487; physical properties, 589
γ-Hydroxybutyric acid, physical
    properties, 589
3-Hydroxydecanoic acid, 591
3-Hydroxyhexanoic acid, 591
Hydroxyhydroquinone, structural
    formula, 790
Hydroxyketone, reaction with
    phenylhydrazine, 618, 619
α-Hydroxyketones, benzoin, 514
Hydroxylamine, 676–677;
    reaction with aldehydes and
    ketones, 500; reaction,
    reduction, 661
Hydroxylamine, disubstituted,
    preparation, 768
Hydroxylamino compounds,
    aromatic, physical
    properties, 768
Hydroxyl group, of phenol, 743
Hydroxylic protons, 279, 286
Hydroxylysine, structural
    formula, 698
2-Hydroxy-4-methylbenzoic acid,
    structural formula, 442
1-Hydroxy-6-methyl-2-naphthoic
    acid, structural formula, 443
Hydroxymethyl radical, 25
3-Hydroxy-2-naphthoic acid,
    preparation by Kolbe
    reaction, 803
α-Hydroxy nitriles. See
    Cyanohydrins
3-Hydroxyoctanoic acid, 591
11α-Hydroxyprogesterone,
    878, 879
Hydroxyproline, structural
    formula, 698, 893
Hydroxypurines, 911
2-Hydroxypyridine. See
    2-Pyridol
2-Hydroxyquinoline, 907
Hydroxyquinones, preparation, 810
β-Hydroxy sulfoxide, 968a
5-Hydroxytryptophan, 905
Hyperconjugation, 150, 166, 457
Hyperconjugation structure, 150
Hyperventilation, 22
Hypochlorous acid: electrophilic
    addition to alkene, 370;
    reactions with isobutylene, 350

Hypophosphorous acid, 958;
    reactions with aromatic
    diazonium compounds, 763–764
Hypoxanthine, preparation from
    uric acid, 912
Hypsochromic effect, 261

Identification, 6
Imidazoles, 893; preparation, 894
Imides, 664 ff.; IUPAC naming,
    78; physical properties, 666;
    preparation, 668–669; reactions,
    668, Hofmann rearrangement,
    663; reactions, reduction, 662
Imides, N-substituted,
    preparation, 661
Imine-nitrile, 672
Imines, 500; formation in Stephen
    reduction, 529; IUPAC naming,
    78; preparation, 498–499, 662;
    reactions, aminolysis, 662;
    reactions, hydrolysis, 739;
    stereochemistry, 658–659
Imino ester salt, 583
Iminonitrile, 672
Immune system, 641, 723–724
Indane: preparation, 573;
    structural formula, 840
Indene: physical properties, 444;
    reactions, 840; structural
    formula, 396, 442, 840
Index of refraction, 133
Indicators, 760–761
Indigo, 408
Indole, 904; preparation by
    Fischer synthesis, 906;
    structural formula, 889
Indoleacetic acid, 391a, 791,
    904–905
Induced polarization, 105
Inductive effect, 108–109
Induline blue, 407
Infrared radiation, 241 ff.
Infrared spectra, acetoacetic ester,
    602; acid derivatives, 253;
    acids, 253; alcohols, 250–252;
    aldehydes, 253; alkenes,
    248–250; alkyl groups, 250;
    amides, 667; amines, 651;
    amino acids, 697; benzenes,
    substituted, 404, 405;
    carbon-oxygen bonds, 250–253;
    carbonyl stretching frequency,
    487–488, 544–546; conjugated
    systems, 261–262, 404–407;
    ethers, 250–251; ketones, 253;
    nitriles, 671; nitroalkanes, 680;
    nitro group, 737
Infrared spectroscopy, 7, 244;
    theory, 246–248
Inhibition, 725, 726
Inhibitors, 725, 977
Initiation, 45
Initiators, in free-radical
    polymerization, 974, 975
Injection molding, 973
Inositols, 643
Insecticides, 459, 783–785
Insertion reactions: carbene

insertion into alkanes, 334–335;
    of organosilicon compounds, 965
Insulin, 723; structure, 950
Intensity (light), 241
Intermolecular forces, 105–107
Internal plane of symmetry, 126 ff.
Internal salt, 669
Internuclear distances, 121–123
Inulin, 640
Inversion, 636; in 1,2-shifts, 535b
Invert sugar, 636
Iodide ion, reaction with methyl
    chloride, 41
Iodination: aldehydes and ketones,
    520 ff.; alkenes, in polar
    solvents, 350–351; arenes, 452;
    benzene, 446; free-radical, 361;
    methane, thermodynamics,
    337; protein side chain, 459
Iodine: covalent bond radii, 102;
    electronegativity, 103; goiter
    from lack of, 459; isotopes, 298;
    physical properties, 446;
    reactions, 452
Iodized salt, 460
Iodoform, 504
Iodomalonic ester, preparation of
    succinic acid from, 577
Iodomethane, physical
    properties, 13
Ion collector, 295, 299
Ion exchange resins, 990–991
Ionic bond, 13, 96–97, 104
α-Ionone, 850, 852
β-Ionone, 850, 852; vitamin A,
    preparation from, 857
Ionones, 850
Ion pairs, 159–160
Ion radical, mass spectrometry,
    294 ff.
γ-Irone, 850
Irones, 850
Iso–, 74
Isoalloxazine, 914
Isoborneol, preparation, 853
Isobornyl chloride, 853
Isobutane, 66; halogenation, 339;
    IUPAC name, 74; physical
    properties, 70
Isobutyl alcohol (2-methyl-1-
    propanol), physical properties,
    70, 202
Isobutylamine, physical
    properties, 70
Isobutyl bromide, physical
    properties, 70
Isobutyl cation, 151, 152
Isobutyl chloride, preparation, 339
Isobutylene: dipole moment, 322;
    physical properties, 70;
    reactions, copolymerization,
    980; hydride transfer, 354–355;
    polymerization, 977–978, 980;
    reactions with chlorine under
    oxygen at cryogenic
    temperatures, 360; reactions
    with hypochlorous acid, 350
Isobutylene oxide, physical
    properties, 71

Isobutyl group, formula, **75**
Isobutyraldehyde: infrared spectrum, **254**; physical properties, **70**
Isobutyramide, physical properties, **666**
Isobutyric acid: infrared spectrum, **307**; physical properties, **70, 538**
Isobutyrophenone, **441**
Isocrotonic acid, physical properties, **594**
Isocyanates, **684**; preparation, **663, 772**
Isoelectric point, **696**
Isohexane, IUPAC name, **74**
Isohexyl group, formula, **75**
Isolated double bonds, **393**
Isoleucine, **693**
Isomerism: geometrical isomerism, **124**; structural isomerism, **38–39**
Isomerization: alkenes, **333, 342, 352–353**; during hydroboration, **357**
Isomers, **39, 67** ff.
Isoniazid, **899**
Isonicotinic acid, **899**
Isonitriles, preparation, **804**
Iso-octane, **76, 355**
Isopentane, IUPAC name, **74**
3-Isopentenyl pyrophosphate, **870**
Isopentyl group, formula, **75**
Isopentyl nitrite, **681**
Isophthalic acid: physical properties, **542**; structural formula, **442**
Isoprene: copolymerization, **980**; preparation, **434**
Isoprene rule, **847, 848**; exceptions, **850**
Isopropenyl group, formula, **77**
Isopropyl alcohol (2-propanol): azeotrope with water, **207**; as catalyst in reduction of alcohols, **218–219**; haloform reaction, **503**; infrared spectrum, **251**; nmr spectrum, **280–282**; physical properties, **69, 202**; production figures, **378**; reactions with sodium, **205**
Isopropylamine, physical properties, **69**
Isopropyl bromide: nucleophilic substitition, **143, 144, 148**; reaction with hydroxide ion, **154**
Isopropyl cation, **151**
Isopropyl chloride (2-chloropropanol), **338**; physical properties, **69**
Isopropylcyclopentane, structure, **80**
Isopropyl group: formula, **75**; infrared spectrum, **250**; nmr spectrum, **280**
Isopropyl methyl ether. See 2-Methoxypropane
7-Isopropyl-1-methylphenanthrene. See Retene
Isopulegol: preparation, **850,**

851; reduction, **850, 851**
Isoquinolines, **906, 908–909**; alkaloids, **918**; preparation by Bischler-Napieralski synthesis, **910**; reactions, **908–909**; structural formula, **889**
Isotactic polymers, **972, 973**
Isotopes, in mass spectrometry, **299–300**
Isoxazole: preparation, **677**; structural formula, **888**
Isoxazolones, preparation, **894**
Itaconic acid, physical properties, **594**
IUPAC rules, **16–18, 37–38**; abbreviation symbols, **74** ff.; acid halides, **78**; acids, **78**; alcohols, **78**; aldehydes, **78**; alkanes, **72, 74**; alkyl cations, **149**; alkyl group, **72, 74, 75**; amides, **78**; amines, **78**; arenes, **441** ff., carboxylic acids, **81**; complex substituents, **72**; cyclic compounds, **79–81, 395** ff.; double and triple bonds, compounds containing, **77–79**; EZ systems, **394**; heterocycles, **886** ff.; hydroperoxides, **78**; imides, **78**; imines, **78**; ketones, **78**; ketoses, **627–628**; location number, **77**; longest chain, **72**; multiplying prefix, **72, 74**; nitriles, **78**; parent chain, **77**; peroxides, **78**; position numbers, **37–38**; prefixes, **72–73, 78**; reactive substituents, **78–79**; rings with double bonds, **80, 396**; structural formula, derivation from IUPAC name, **73–74**; substituents, **72–73, 75, 78–79**; suffixes, groups designated by, **77–79**; thiols, **78**

Jasmin, **208**
Johnson, W. S., **318a**
Jones, W. M., **438a**
Juglone, **789**
Juvenile hormone, **607**

Kekule structure, **404**
Kel-F, **779, 980**
Keratin, **690, 950**; amino acid composition, **699**
Ketals, **630**
Ketenes, addition of HX, **565**
Ketimine, **649**; preparation, **522**
Keto acids, **597** ff.; preparation, **604–605**; reactions, **597–604**
α-Keto acids, **597**; preparation, **604–605**
β-Keto acids, preparation, **605, 672**; reduction, **593**
γ-Keto aldehyde, **601**
Ketoaldehydes, reactions, **677**
β-Ketobutyronitrile, **672**
trans-9-Keto-2-decenoic acid, **607**
α-Keto esters, preparation **605**
β-Keto esters: preparation, **599**;

reactions, **677**; reaction, hydrolysis, **605**
α-Ketoglutaric acid, **499**
Keto group, substituent effect on benzene, **450–451**
2-Keto-L-gulonic acid, **630**
Ketols, steroidal, **318c**
Ketone cleavage, acetoacetic ester, **597**
Ketone radical ions, McLafferty rearrangement, **301, 303**
Ketones, **68**; aromatic. See Aromatic ketones; formation during ozonization of alkenes, **367**; infrared spectra, **253**; **487**; ion radicals from, **295**; IUPAC naming, **78**; mass spectra, **304**; occurrence, **480**; physical properties, **382**, **480–481**; preparation, by addition reactions, **522–524**; from alcohols by oxidation, **216** ff.; from alkenes by drastic oxidation, **373**; from alkynes by tautomerization, **354**; from dimethyl β-ketosulfoxide, **952**; by Friedel-Crafts reaction, **454, 753**; by Houben-Hoesch reaction, **521, 798**; by Michael reaction of a Grignard reagent, **523**; by oxidation, **216** ff., **527–529**; by pinacol rearrangement, **519, 524–527**; by rearrangement reactions, **524–527**; by reduction, **529**; by semipinacol rearrangement, **655, 656**; by substitution, **520–522**; reactions, acylation, **510–511**; addition, **490–501**; addition of Grignard reagents, **227**; aldol condensation, **504–509**; Baeyer-Villiger reaction, **515–516**; Claisen-Schmidt condensation, **507–509**; Clemmensen reduction, **381–382, 470**; elimination, tertiary alcohols from, **172, 173**; between enolate ion and carbonyl group, **508**; halogenation, **501–504**; Meerwein-Ponndorf-Verley reduction, **231**; Michael reaction, **511–512**; mixed aldol condensation, **574**; mixed Claisen condensation, **574**; oxidation, **517–518**; rearrangements, **515–516**; reduction, **518–519**; reduction with sodium borohydride, **230**; reductive alkylation, **662**; Reformatsky reaction, **495–496**; substitution, **501–504**; Wittig reaction, **497–498, 959**; Wolff-Kishner reduction, **381**; reactions with, alcohols, **491–492**; amines, **498–499**; Grignard reagents, **495**; hydrogen cyanide, **493–494**; phenols, **800–802**; phosphorus

Ketones (cont.)
 ylide, **959;** sodium bisulfite,
 493; sodium borohydride, **238a;**
 syn- and *anti-* isomers of oxines
 of, **500;** terpenes, **848, 850;**
 ultraviolet spectra, **261**
Ketones: aromatic, nomenclature,
 441; preparation, by
 Gattermann-Koch reaction, **520;**
 preparation, from diacids, **521**
Ketones, α,β-unsaturated,
 addition of Grignard reagents,
 522; preparation, **506**
Keto-nitrile, **672**
7-Ketooctanoic acid,
 preparation, **607**
2-Ketopropanal, **416**
α-Ketopropionic acid. *See*
 Pyruvic acid
Ketoses, stability, **632**
D-Ketoses, **627–628**
Ketosis, **487**
α-Ketosuccinic acid. *See*
 Oxaloacetic acid
Ketoxime, reaction with
 phosphorus pentachloride, **664**
Keto ylide, **524**
Kharasch, M., **968a**
Kidneys, pH of human blood,
 control by, **21**
Kiliani-Fischer synthesis, **617**
Kinetic energy, **245**
Kinetic isotope effect, **447**
Kinetic resolution, **704**
Knock-out drops, **49**
Knoevenagel reaction, **512–513**
Kodel, **984**
Kohler, E. P., **64a**
Kolbe electrolysis, **382**
Kolbe reaction, **803**
Kraft method, **814**
Krebs citric acid cycle, **645**

γ-Lactam, preparation, **665,
 669, 670**
δ-Lactam, preparation, **669,
 670**
Lactams, **664–670**
Lactase, **635**
Lactic acid, physical properties,
 69, 589
Lactide, preparation, **589, 590**
Lactones, hydrolysis, **593;**
 preparation, **562, 786;**
 preparation from glyconic
 acid, **624**
β-Lactones, **591**
γ-Lactones: phenolphthalein,
 800–802; preparation, **591–592,
 665;** preparation, from glucose,
 615, 617; reduction, **617**
δ-Lactones: preparation, **591;**
 from glucose, **615**
Lactones, unsaturated, **604**
Lactose, **634**
Ladder polymers, **991–992**
Lamb, A. B., **64a**
Landsteiner, Karl, **641**
Lanosterol, biosynthesis from
 squalene, **871**
Lanthionine, structural
 formula, **698**
Lapinski, R. L., **968a**
Latex, **862**
Laudanosine, **918**
Lauric acid, physical
 properties, **538**
L-DOPA, **815** ff.
Lead tetraacetate: reactions with
 diethyl tartrate, **605;** use as
 oxidizing agent for diols,
 528–529
Leaving group, **24**
Lecithin, **107, 732**
Leffler, J. E., **993a**
Lemon grass oil. *See* Geraniol
Leucine, **693**
Leucine aminopeptidase, **708**
Leucopterin, **913**
Leveling effect, **22**
Levulinic aldehyde,
 preparation, **862**
Levulose. *See* Fructose
Lewis acid, **20, 106**
Lewis base, **20, 106**
Lexan, **984**
Libration, **121**
Light, **133, 241** ff.
Lignin, **814–815;**
 composition, **789**
Lignoceric acid, physical
 properties, **538**
Ligroin, **379**
Limonene, **850, 851**
Lindane, **459, 783–784**
Linstead, R. P., **646a**
Lipids, **537;** phosphatidic acid,
 formation of, **212;** stability, **107**
Lipoic acid (thioctic acid), **950;**
 structural formula, **729;** use, **729**
Lithium: electronic distribution,
 88; reactions, with
 cyclooctatetraene, **400;**
 with ethyl bromide, **43–44;**
 reduction of substituted
 benzenes with, **419;** use in
 reduction of alkenes and
 alkynes, **376**
Lithium aluminum hydride,
 reactions, **230;** reactions,
 with acid anhydrides, **562;**
 with acid chlorides, **562;**
 with amides, **668;** with
 aromatic nitro compounds, **741;**
 with carboxylic acids, **555;**
 with esters, **570;** with
 hydrazones, **679;** with hydroxy
 acids, **592;** with hydroxylamine,
 661; with nitrogen compounds,
 661, 662; with ozonides, **367;**
 with sulfonyl esters, **955–956;**
 use as reducing agent, **53, 54;**
 use in substitution reaction of
 sulfonate esters, **381**
Lithium chloride, as added ion
 in addition reactions, **46**
Lithium diethylamide,
 Thorpe-Ziegler reaction, **673**
Lithium methylphenylamide,
 Thorpe-Ziegler reactions, **673**
Lithium reagents: reactions with
 acid chlorides and acid
 anhydrides, **562;** reactions with
 esters, **570;** reactions with
 quinones, **822a**
Lithocholic acid, structural
 formula, **881**
Liver catalase, **190**
Lobry de Bruyn-van Eckenstein
 rearrangement, **630–632**
L-Sorbose, **629–630**
L-Thyronine, **459**
Lucite, **981**
Luminal, **903**
Lumisterol, preparation, **874**
Lung cancer, cigarette smoking
 and, **444**
Lungs, lipids in, **107;** pH of
 human blood, control by,
 21–22
Lustron, **981**
Lyate ion, **22**
Lycopene, **860;** structural
 formula, **861**
Lyonium ion, **22, 206**
Lysergic acid, **905;** structural
 formula, **906, 917**
Lysergic acid diethylamide,
 structural formula, **665**
Lysine, **691, 693, 697;**
 reactions, decarboxylation, **555;**
 reactions, ionization, **697;**
 structural formula, **693**

McLafferty rearrangement,
 301, 303
"Magic acid," **151**
Magnesium: electronic distribution,
 88; reactions with alcohols, **205;**
 use to reduce carbonyl
 compounds, **519**
Magnesium alkoxides, **205, 206**
Magnesium carbide, **385**
Magnetic field: mass spectrometry,
 295 ff.; nuclear magnetic
 resonance, **265** ff.
Magnetic quantum number, **87**
Malaprade's reagent, **529**
Malathion, **727–728**
Male hormone, **875, 876**
Maleic acid: ozonization, **604;**
 physical properties, **542, 594;**
 stereochemistry, **596**
Maleic anhydride, alcoholysis, **561**
Malic acid: biological conversion
 to oxaloacetic acid, **219;** as
 photosynthesis product,
 556–557; physical properties,
 71, 589
Malonic acid: acidity, **547;**
 physical properties, **69, 542;**
 preparation, **559;** reactions,
 decarbonylation, **597;** reactions,
 decarboxylation, **554, 575, 597**
Malonic ester (diethyl malonate):
 infrared spectrum, **254;**
 phthalimidomalonic ester

synthesis using, **701**;
preparation, **559**, **574–578**, **674**;
reactions with ethoxide ion,
574–575; reactions with
urea, **903**
Malonic ester synthesis, **605**
Malonitrile, **672**
Malonyl-CoA, in biosynthesis of
terpenes, **868**
Maltase. See α-Glucosidase
Maltose, **635**
Mandelic acid, physical
properties, **589**
Manganese dioxide, as oxidizing
agent, **528**
Mannan, **639**
Mannich reaction, **904–905**
Mannitol, **643**
D-Mannoheptulose, **628**
Mannose: detection, **643**;
preparation, **631**, **632**;
reactions, **622**, **623**
Margaric acid, physical
properties, **538**
Margarine, **540**
Marijuana, **817**
Markovnikov's rule, **347**
Martius yellow, **407**
Maslow, A., **64b**
Mass spectra, acids, **304**; alcohols,
304; aldehydes, **304**; alkenes,
301; alkyl halides, **303–304**;
amines, **651**; aromatic amines,
745; conjugated systems,
410–411; esters, **304**; ethers,
304; ketones, **304**; nitro group,
737; phenols, **790**
Mass spectrometer, **299–300**
Mass spectrometry, **7**; theory,
294–299
Mauve dye, **8**, **407**
Mauveine, **468**
Meat tenderizer, **42**
Mechanism, definition, **7**, **27**
Meerwein-Ponndorf-Verley
reduction, **206**, **218–219**, **231**
Melamines, **989**
Melibiose, **635**
Melmac, **989**
Melting point: alcohols, **201** ff.;
hydrocarbons, **320**
Melt spinning, **982**
Menadione. See 2-Methyl-
1,4-naphthoquinone
p-Menthane, preparation from
limonene, **850**, **851**
Menthol, preparation, **850**, **851**
Meperidine. See Pethidine
Mer, **970**
Mercaptans. See Thiols
2-Mercaptobenzothiazole, use in
rubber processing, **863**
Mercerization, **638**
Mercuration, arenes, **454–455**;
furan, **892**; nitrobenzene, **738**;
phenols, **803**; pyrrole, **892**;
thiols, **946**; thiophene, **892**
Mercuric acetate, reactions with
phenols, **803**

Mercuric acid, reactions with
arenes, **454–455**
Mercurochrome, **802**, **803**
Mercury compounds, organic,
76, **455**
Merthiolate sodium, **949**
Mesaconic acid: physical
properties, **594**;
stereochemistry, **596**
Mescaline, **807**, **815**; structural
formula, **808**
Mesitylene: physical properties,
443; structural formula, **441**
Mesityl oxide, **507**
meso-, **69**, **131**, **132**
Messenger RNA, **924**, **926**;
codons, **927**
meta-, **441**
Meta attack, **450**
Metabolism: acetate ions in, **867**;
phosphate esters in, **962**
Metaformaldehyde. See Trioxane
Metal complexes, **646a**
Metal coordination compounds,
927–933
Metaldehyde, **492**
Metal hydrides, preparation of
aldehydes with, **529–530**
Metal organic compounds:
displacement of metals from, **43**;
containing heavy metals, **76**
Metanilic acid, preparation, **751**
Meta protons, chemical shift, **408**
Methacrylic acid, physical
properties, **71**, **594**
Methadone, **104**, **919**
Methanal. See Formaldehyde,
bond energies
Methanamine. See Methylamine
Methane: bond dissociation
energies, **111**; bonding in, **647**;
bond polarity, **108**; conversion
to haloalkanes and haloalkenes,
777; covalent bonding in, **5**,
14; physical properties, **13**,
331; preparation from
carbides, **385**; preparation,
thermodynamics of, **59**;
reactions, chlorination, **26**, **340**;
combustion, thermodynamics,
56; halogenation, **26**, **335–337**;
with nucleophiles, **25**; oxidation,
29, **365**; stereochemistry, **119**;
vibrational modes, **247**
Methanides, **385**
Methanoic acid. See Formic acid
Methanol, **233**; acid-base equilibria
in water, **24**; autoprotolysis
constant, **22**; bond energies,
110–111; covalent bonding in,
5; nmr spectrum, **279**; physical
properties, **13**, **202**; pK$_a$, **332**;
preparation, industrial, **233**;
preparation by oxidation of
methane, **29**; production
figures, **378**; reactions,
elimination of hydrogen, **29**;
reactions, oxidation, **19**, **53**;
reactions with, bromide ion, **24**;

tert-butyl alcohol, S$_N$1cA
method to form ethers, **214–215**;
sodium, **205**; solvent effect
of one nucleophilic
substitution, **153**
Methemoglobin, **930**
Methine proton, **288**
Methionine, **691**, **693**, **950**;
structural formula, **693**, **951**
Methoxide ion, **24**; reaction with
ethyl iodide, **40**
Methoxybenzoic acid, pK$_a$, **797**
Methoxyethane. See Ethyl methyl
ether
Methoxyethanol, physical
properties, **69**
p-Methoxy group: effect on
hydrolysis, **710**; substituent
effect of on benzene, **449**, **450**
Methoxymethane. See Dimethyl
ether
1-Methoxypropane, physical
properties, **70**
2-Methoxypropane, physical
properties, **70**
Methoxy radical, **26**
N-Methylacetamide, physical
properties, **666**
Methyl acetate, preparation,
194–195
4-Methyl-1-acetonaphthone,
structural formula, **443**
Methylacetylene. See Propyne
Methylamine: covalent bonding
in, **5**; IUPAC naming, **17**;
pH, **21**; physical properties,
13, **650**; reactions with methyl
halide, **41**
Methylation, of alcohols, **318b**
Methyl benzoate, reactions with
sodium methoxide, **568**
Methyl bromide: physical
properties, **13**; preparation,
thermodynamics, **337**; reactions,
with ethoxide ion,
thermodynamics, **184–185**;
reactions, nucleophilic
substitution, **143**, **148**
2-Methyl-1-butanol, physical
properties, **202**
2-Methyl-2-butanol, physical
properties, **202**
3-Methyl-1-butanol, physical
properties, **202**
3-Methyl-2-butanol, physical
properties, **202**
3-Methyl-2-butanone. See
Methyl isopropyl ketone
2-Methyl-3-buten-2-ol, nmr
spectrum, **285**, **286**
Methyl tert-butyl ether,
preparation, **214–215**
Methyl tert-butyl ketone. See
Pinacolone
Methyl cation, **149** ff.
Methyl cellobioside, reactions
with periodic acid, **634**;
methylation, **632**
Methyl chloride, bond polarity,

Methyl chloride (cont.) **108**; physical properties, **13**; preparation, commercial, **777**; preparation, thermodynamics, **337**; reactions with chlorine, free-radical, **45**; reactions with iodide ion, **41**
Methylcholanthrene, carcinogenic activity, **841**; structural formula, **842**
5-Methylchrysene, structural formula, **444**
1-Methylcyclohexene, **358**
3-Methylcyclohexene, **358**
4-Methylcyclohexene, **358**
1-Methylcyclopentene, *trans*-addition, **348**
Methylcyclopropane: physical properties, **71**; structure, **79**
Methyldiazonium cation, **318b**
Methylene, vibrational modes, **247–248**
Methylene chloride, commercial preparation, **777**
Methylene group, **66**
Methylene iodide, malonic ester synthesis using, **577**
Methylene protons, **276, 288**
Methylenetriphenylphosphorane, **959**
Methyl esters, alcoholysis, **568**; physical properties, **542, 543**; preparation, **550**
Methyl ethyl ester. *See* Ethyl methyl ether
Methyl ethyl ketone: Claisen-Schmidt condensation, **507**; physical properties, **481**
Methyl fluoride: physical properties, **13**; preparation, thermodynamics, **337**
Methylformamide, physical properties, **666**
Methyl formate, **38**; physical properties, **36, 37**
Methyl gentiobioside, reaction with periodic acid, **634**
Methyl glucosides, **618**; reactions with periodic acid, **632–634**
Methyl group: inductive effect, **108**; infrared spectrum, **250**; removal of by haloform reaction, **504**
Methyl halides, nucleophilic substitutions: effect of halogen substituents on, **343a**; preparation, mechanism, **335–337**; reactions, with methylamine, **41**; reactions, nucleophilic substitutions, **40–41**
trans-3-Methyl-2-hexenoic acid, **607**
Methyl iodide: physical properties, **13**; preparation, thermodynamics, **337**; reactions with ethoxide ion, **40**; reactions with sodium ethoxide, **192**
Methyl isobutyl ketone, physical properties, **481**

Methyl isopropyl ketone, halogenation, **504**; mass spectrum, **295, 297**
Methyl ketones: haloform reaction, **503**; reactions, mixed aldol condensation, **574**; mixed condensation, **579–580**; with sodium bisulfite, **493**
Methylmagnesium iodide, electrophilic substitution, **25**
Methyl mercury chloride, **76**
Methyl methacrylate, polymerization, **981**
7-Methyl-3-methylenyl-1,6-octadiene. *See* Myrcene
α-Methylnaphthalene: oxidation, **831**; structural formula, **443**
β-Methylnaphthalene: oxidation, **831**; preparation, **832**
2-Methyl-1,4-naphthoquinone (menadione), **820**; preparation, **831**
5-Methyl-1,4-naphthoquinone, preparation, **831**
N-Methyl-N-nitrosamide, preparation, **668**
N-Methyl-N-nitroso-p-toluenesulfonamide, **683**
Methyl oleate, reactions, **385–386**
Methyl orange, **760**; preparation, **761**
Methyloxonium ion, **24**
2-Methylpentane, preparation by by carbene insertion into pentane, **334**
3-Methylpentane: preparation by carbene insertion into pentane, **334–335**; synthesis, **174–175**
2-Methylpentanoic acid, **576**
2-Methyl-1-pentanol, physical properties, **202**
2-Methyl-2-pentanol, physical properties, **202**
3-Methyl-3-pentanol, physical properties, **202**
4-Methyl-1-pentanol, physical properties, **202**
4-Methyl-2-pentanone, mass spectrum, **304, 306**
4-Methyl-1-pentene, chemical shifts, **288**
4-Methyl-2-pentenoic acid, **595**
4-Methyl-3-pentenoic acid, **595**
3-Methyl-1-pentyn-3-ol, nmr spectrum, **285–287**
2-Methyl-1-propanol. *See* Isobutyl alcohol
2-Methyl-2-propanol (*tert*-butyl alcohol): conversion into *tert*-butyl chloride, mechanism, **193**; physical properties, **70, 202**; reactions, esterification, **209**; S_N1cA method to form ethers, **214–215**; with sodium, **205**
2-Methylpropene, **67**
Methyl propyl ether. *See* 1-Methoxypropane
6-(2-Methylpropyl)undecane,

IUPAC name, derivation, **73**
Methyl protons, **276, 279, 286, 288**; chemical shift, **408**
3-Methyl-5-pyrazoline, preparation, **893**
2-Methylpyridine. *See* α-Picoline
N-Methyl-2-pyrrolidone, **666**
2-Methylquinoline, preparation, **908**
Methyl radical, **26, 339**
Methyl red, preparation, **761**
Methyl sulfate, preparation, **210**
2-Methylsulfinylpropane, **951**
2-Methylsulfonylpropane, **951**
N-Methyl-1,2,3,4-tetrahydro-papaverine. *See* Laudanosine
Methyl *p*-tolyl ketone, infrared spectrum, **407**
Methyl valerate, preparation, **550**
Methylvinylcarbinol, physical properties, **70**
Mevaldic acid, **868**
Mevalonic acid, biosynthesis, **868**; physical properties, **589**; squalene, biosynthesis from, **869, 870**
Micelles, **107**
Michael addition, nitriles, **671**
Michael reaction, **511–512**; acrylonitrile, **512**; malonic acid derivatives, **578**; nitroalkanes, **680**
Micrometer (unit), **241**
Microwave radiation, **242**
Microwave spectroscopy, **244, 245**
Migrating benzene ring, **526**
Mild oxidation, **371**
Milk intolerance, **635**
Mirror, **30**
Mirror image, **8**
Mixed acetals, **618**
Mixed acid anhydrides, **548**; preparation, **567–568**
Mixed aldol condensation, **574**; cyclohexanone with ethyl acetate, **607**
Mixed Claisen condensation, **574**; sulfones, **952–953**
Molar absorptivity, **244**
Molding, of thermoplastics, **973**
Molecular orbitals, in benzene, **403, 404**
Molecular orbital theory, **90–96**; of conjugation, **398–399**; valence bond theory, comparison with, **97–98**
Molecular rotation, **134, 135**
Molecular weight, by mass spectrometry, **298**
Monomer, **970**
Monosaccharides, **615, 622**; synthesis, **630–632**
Monoximes, preparation, **677**
Morphine, **104, 918, 919**
Morpholine, structural formula, **888**
Morse curve, **113**
Multiple bond, **26**
Multiplicity, **260**

Murphy's law, **10a**
Muscalure, **366**
Mutagens, **446**
Mutarotation, **614**
Mutations, **8**
Mycomycin, **538**
Mylar, **983**
Myoglobin, **928, 930;** structure, **720, 721**
Myosin, **690;** amino acid composition, **699**
Myrcene, structural formula, **848**
Myristic acid, physical properties, **538**

n-, **75**
$n + 1$ rule, **272–274**
NAD, spectrophotometry, **263**
NADH, use by biological systems, **868**
Naloxone, **919**
Nanometer (unit), **241**
Naphthacene, structural formula, **841**
Naphthalene: occurrence in nature, **824;** physical properties, **407, 444;** preparation from benzene and succinic anhydride, **831–832;** reactions, addition, **830–831;** bromination, **825, 826;** disubstitution, **828;** Friedel-Crafts reaction, **827–828;** nitration, **825, 826;** oxidation, **831;** reduction, **830;** substitution, **825–830;** sulfonation, **826–827;** 2 position, **767;** reactions with sodium in dimethoxyethane, **418–419;** reactions with succinic anhydride, **839;** resonance energy, **825;** resonance structures, **403;** structural formula, **396**
Naphthalenes, common names, **443**
Naphthalene-α-sulfonic acid, **827**
Naphthalene-β-sulfonic acid, **827;** fusion, **828;** preparation, **755**
Naphthalic anhydride: reactions, **841;** structural formula, **841**
Naphthenic acid salts, **540**
α-Naphthoic acid, structural formula, **443**
β-Naphthol: Kolbe reaction, **803;** preparation, **755, 756, 828;** reactivity, **828–830;** structural formula, **443**
α-Naphthoquinone: Diels-Alder reaction, **836;** preparation, **831**
β-Naphthylamine, **767, 828;** preparation by Bucherer reaction, **755–756**
α-Naphthyl isocyanate, **771**
Narcotics, **919**
Natural gas, **365, 377**
Near ultraviolet radiation, chromophores, **261**
Negative-ion mass spectrometry, **300**
Neighboring proton effect, **272**
Nembutal, **903**

Neon, electronic distribution, **88**
Neopentane, IUPAC name, **74;** reactions, chlorination, **339;** reactions, deuterium exchange, **375**
Neopentyl cation, **149**
Neopentyl group, formula, **75**
Neopentyl halides, nucleophilic substitution, **148**
Neoprene, **355, 980**
Neoretinal-b, **858, 859**
Neosynephrine, **816**
Neral, **848, 850**
Nerol, **848**
Nerve poison, **726**
Neryl pyrophosphate, **869, 870**
Niacin (nicotinic acid), commercial preparation, **899;** structural formula, **539, 729**
Niacinamide (vitamin B$_3$), **899;** structural formula, **729**
Nicotinamide adenine dinucleotide (NAD), **730, 961**
Nicotinamide adenine dinucleotide phosphate (NADP), in photosynthesis, **643–644**
Nicotine, structure, **444, 916**
Nicotinic acid. See Niacin
Ninhydrin, preparation, **840**
Nitrate ester, preparation, **681**
Nitrate ion, resonance structure, **99**
Nitration, acetanilide, **749;** aniline, **747–748;** anthraquinone, **835;** arenes and substituted aromatic compounds, **742–744;** benzene, **452–453, 468;** p-bromoacetophenone, **470;** p-bromoethylbenzene, **470;** m-chlorotoluene, **457;** furan, **892;** naphthalene, **825, 826;** nitrobenzene, **738;** phenanthrene, **458;** phenols, **796;** polycyclic compounds, **458;** propane, gas-phase, **681;** pyridine, **896;** α-pyridone, **901;** pyrrole, **892;** quaternary phenyltrimethylammonium ion, **748;** quinoline, **907;** thiophene, **891, 892;** toluene, **469;** toluidine, **750**
Nitrene, preparation, **663**
Nitric acid, mixed acid, **743;** production figures, **378;** reactions, esterification, **210, 211;** reactions, oxidation of tetramethylglucose with, **620;** reactions with, arenes, **456;** glucose, **616;** sulfuric acid, **452;** resonance structure, **99**
Nitrile group, addition of carbonium ion to, **661**
Nitriles, **671–674;** acids, preparation from, **558;** amides, preparation from, **668;** esters, preparation from, **583;** infrared spectra, **671;** IUPAC naming, **78;** physical properties, **671;** preparation, **668, 673–674;** preparation by Wohl degradation

of glucose, **617;** reactions, **671–673;** reactions, alcoholysis, **580–581;** Hoesch ketone synthesis using, **798;** hydrolysis, **558;** reduction, **662;** Stephen reduction, **529;** reactions with sodium borohydride, **238a;** ultraviolet spectra, **671**
Nitriles, aromatic, **772;** preparation, **465**
Nitriles, conjugated, reactions, **671**
Nitrilium group, preparation, **661**
p-Nitroacetanilide, preparation, **748**
m-Nitroacetophenone, bromination, **470**
Nitro-acinitro tautomerism, **680**
Nitroalkanes, **679–681;** preparation, **678**
Nitroaniline, dipole moment, **747;** nucleophilic substitution, **749;** preparation, **742, 748 ff.;** preparation from benzene, **750;** ultraviolet spectra, **745**
α-Nitroanthraquinone, preparation, **835**
Nitroarenes, condensation reactions, **739**
Nitrobenzene, **737;** boiling point, **737;** dipole moment, **747;** reactions, bromination, **738;** mercuration, **738;** nitration, **738;** reduction, **740, 768–769;** sulfonation, **738;** ultraviolet spectrum, **737;** use as solvent, **737–738**
m-Nitrobenzenesulfonic acid, preparation, **738**
p-Nitrobenzoic acid, preparation, **469**
p-Nitrochlorobenzene, **460–461**
Nitro compounds, reduction of to form amines, **661**
Nitro compounds, aromatic, **737–744;** physical properties, **737;** preparation by nitration, **742–744;** reactions, **738–742;** reactions, condensation, **738–739;** electrophilic substitution, **738;** oxidation-reduction, intramolecular, **742;** reduction, **740–742;** reactions with ammonium sulfide, **742;** spectra, **737**
Nitroethane, **649**
m-Nitroethylbenzene, bromination, **470**
Nitrogen, covalent bond formation with, **4, 5;** covalent bond radii, **102;** electronegativity, **103;** electronic distribution, **88;** hydrogen bonding to, **105–106;** isotopes, **299;** molecular orbital theory, **91**
Nitrogen compounds, **648 ff.;** aromatic, **736 ff.;** reactions, **661;** stereochemistry, **658–659**

Nitrogen dioxide, role in smog, **50**
Nitrogen oxides, **647**; in smog, **391b**
Nitrogen-oxygen bond, **384**
Nitrogen rule (spectroscopy), **298**
Nitroglycerine, **681**
Nitro group, **679**; effect on physical properties of aromatic compounds, **737**; formula, **75**; infrared spectrum, **737**; mass spectra, **737**; reactions, **738** ff.; substituent effect on arenes, **460**
p-Nitro group: effect on carbobenzoxy protecting group, **710**; substituent effect on arenes, **462, 463**
Nitromethane, solvent effect of on nucleophilic substitution, **153**
α-Nitronaphthalene, reactions, **824**
β-Nitronaphthalene, preparation, **755**
Nitronic acid, **680**
Nitronium ion, **211, 452**
m-Nitrophenol, pK_a, **791**
o-Nitrophenol: physical properties, **796**; pK_a, **791**; preparation by nitration of phenol, **743**
p-Nitrophenol: pK_a, **791**; preparation by nitration of phenol, **743**; preparation from p-nitroaniline, **749**
3-Nitrophthalic acid, **824**
4-Nitropyridine, preparation, **897, 898**
5-Nitroquinoline, **907**
8-Nitroquinoline, **907**
Nitrosamides, use, **668**
Nitrosation: aromatic amines, **754**; phenols, **796**; secondary aromatic amines, **754**; tertiary aromatic amines, **754–755**
Nitrosoalkanes, **679**
N-Nitroso-N-alkylaniline, **754**
Nitrosoamine: preparation, **655**; secondary, **656**
N-Nitrosoamines, preparation, **754**
Nitrosobenzene: physical properties, **768**; preparation from nitrobenzene, **740**; reactions, **768**; reactions, reduction, **740**; reactions with phenylhydroxylamine, **740, 741**
Nitroso compounds: aromatic, addition reactions, **768–769**; physical properties, **768**
p-Nitrosodimethylaniline, physical properties, **768**
Nitrosoethane, **649**
Nitroso group, **75, 754**; addition reactions, **768–769**
Nitrosonium ion, **211, 759**; preparation, **654, 757**
Nitrosophenol, **796**
Nitrotoluene: intramolecular oxidation-reduction, **742**; physical properties, **737**
Nitrotoluidines, preparation, **750**
Nitrous acid, acetamidomalonic ester synthesis, **701** ff.; identification of amines with, **656**; preparation, **654**; reactions, in aqueous solution, **654**; deamination of 1,1-diphenyl-2-amino-1-propanol, **535a**; esterification, **210, 211**; reactions with, amides, **667–668**; amines, **654–656**; aromatic amines, **754**; hydrazide, **684**
Nmr. See Nuclear magnetic resonance
No-bond resonance, **150**
Nomenclature, aromatic compounds, **396–397, 441** ff.; EZ system, **394**; IUPAC system, **16–18, 37–38, 72–80**; oxa-aza system, **886–887**
Nomex, **982–983**
Nonadecanoic acid, physical properties, **538**
Nonanoic acid, infrared spectrum, **254**
Nonbonded repulsion, **119–121**
Noncompetitive inhibition, **726, 727**
Noncyclic photophosphorylation, **643–644**
Nonplanar ring, **400**
Nonpolar covalent bonds, **96**
Nonproductive reaction, **336**
Nonstereoselective reaction, **360**
Noradrenalin. See Norepinephrine
Norbornane, **430–431**
Norbornene, **430–431**
Norephedrine, **816**
Norepinephrine, **789, 816**; mental health and, **816**; metabolism, **817**; structural formula, **816**
Normal–, **75**
Novolac, **986**
n State, **261**
N-Terminal amino acid, determination in peptides, **706–707**
Nuclear magnetic field, **265**
Nuclear magnetic resonance (nmr) spectra, acetoacetic ester, **602, 603**; acetylacetone, **602, 603**; alkaloids, **318b**; amides, **667**; amines, **651**; conjugated systems, **408–410**; hexaphenylethane, **782**; phenolic protons, **790**
Nuclear magnetic resonance spectrometer, **269–270**
Nuclear magnetic resonance spectroscopy, **7, 265–291, 318a**; of biomolecules, **720–722**
Nuclear magnetic resonance theory, **265–269**
Nuclear power, **378**
Nuclear spin quantum number, **265**
Nucleic acids, **920–927**; absorbance, **263**
Nucleophile, **23**

Nucleophilic addition reactions, hydrocarbons, **345–347**
Nucleophilic aromatic substitution, diazonium ion, **762**; nitro compounds, **738**
Nucleophilic displacement reactions, arenes, effect of nitro group on, **460**; halogen, **51**; thiols, **946**
Nucleophilic reactions, organosilicon compounds, **964, 965**
Nucleophilic substitution reactions, **23–25, 39–42, 143–149**; arenes, **460–465**; elimination, competition with, **42, 51, 156–158**; halogen substituents, effect on, **343a**; hydrocarbons, **331–333, 342**; hydrolysis, **146–147**; organosilicon compounds, **963–964**; pyridine, **897**; solvent polarity, **153–156**; stereochemistry, **158–160**; steric effects and, **40–41, 147**; uric acid, **911**
Nucleoproteins, double helix, **924**
Nucleosides, **921**; structural formulas, **921**
Nucleotides, **921**; structure, **922**
3-Nucleotides, **921**
5-Nucleotides, **921**
Nucleus, spin orientation, **266**
Numbering, fused ring systems, **396**
Nylon, **970, 982**; structure, **6**
Nylon, **4, 982**
Nylon, **6, 982**
Nylon, **66, 982**

Ocimene, structural formula, **848**
Octafluorocyclobutane, use, **778**
Octahedral hybridization, **944**
Octamethyl cellobiose, **632**
Octane: mass spectrum, **300–301**; preparation by Wurtz-Fittig reaction, **781**
Octane rating, **76**
1-Octanol, physical properties, **203**
Oil of mirbane, **737**
Oils, **537**; alkaline hydrolysis, **540–541**; in foods, **540**
Oil spills, **320**
Old Yellow Enzyme, **263**
Olefins, **49**; Saytzeff rule, **164–165, 166, 169**
Oleomargarine, **540**
Oligomer, **970**
Oligopeptide, **690**
Oligosaccharide, **632**
Olysis reactions, **551–552**; acid anhydrides, **561**; acid chlorides, **560** ff.; esters, **567**; nitriles, **671**; preparation of amides by, **668**; preparation of amines by, **661**; preparation of esters by, **580–582**; Schotten-Baumann reactions, **561**

Opium, **104, 918–919**
Oppenauer oxidation, **206, 218–219**
Opsin, **858**
Optical activity, **133–136**
Optical inversion, **159**
Optical isomerism, **136**
Optical isomers, **133**
Optical purity, **136**
Oral contraceptives, **876**
Orange blossom oil. *See* Nerol
Oranges, ester responsible for fragrance of, **543**
Order of reaction, **192, 193**
Organomercury compounds. *See* Mercury compounds, organic
Organophosphorus compounds, **957–962**
Organosilicon compounds, **962–967**; manufacture, **966**
Organosulfur compounds, cryoscopic behavior in sulfuric acid, **968a**; disulfides, **949**; episulfides, **948**; occurrence in proteins and metabolic products, **950–951**; sulfenic acid, **954**; sulfides, **947–949**; sulfinic acids, **954–955**; sulfones, **951–953**; sulfonic acids, **955**; sulfoxides, **951–953**; thiols, **945–947**
Organ transplant, **724**
Orientation, substituted arenes, **451**
Orlon, **981**
Ornithine, structural formula, **698**
Orthanilic acid, preparation, **751**
ortho-, **441**
Ortho attack, **450**
Ortho effect, in substituted anilines, **747**
Ortho esters, **582–583**
Ortho protons, chemical shift, Organosulfur compounds,
Osazone, from fructose, **628**
Osazones, preparation, **618, 619**
-ose, **615**
Osmium tetroxide, use in diol formation, **370–371**
Overtone (spectroscopy), **248**
Oxa-, **887**
Oxa-aza naming system, **886–887**
Oxalic acid: acidity, **547**; physical properties, **36, 37, 542**; preparation, **559**; preparation by glucose oxidation, **616**; reactions, decarboxylation, **554**; reactions, oxidation by permanganate, **55**
Oxaloacetic acid, **597**; as photosynthesis product, **556–557**; preparation from malic acid, **219**
Oxalyl chloride, reactions with carboxylic acids, **547, 548**
Oxamide, physical properties, **666**
1,3-Oxaza-4-cyclohexene, structural formula, **887**
1,2-Oxazole. *See* Isoxazole
1,3-Oxazole, structural formula, **888**
Oxazoles, **893**; preparation, **894**
Oxepin, structural formula, **888**
Oxetane. *See* Trimethylene oxide
Oxidation reactions, **32, 53–55**; acenaphthene, **841**; alcohols, primary and secondary, **216–217**; aldehydes and ketones, **517–518**; alkenes, **358**; alkenes, by permanganate, **54**; alkynes, by permanganate, **54–55**; amines, **659**; anthracene, **834**; arenes, **465–467**; aromatic amines, **753–754**; benzene, **416**; camphor, **854**; dienes, **427**; disulfides, **949**; glucose, **61, 615–616**; hydrazine, **682**; hydrocarbons, **365–374**; indene, **840**; α-keto acids, **597**; lactose, **634**; methane, **29**; monosaccharides, **622, 623**; naphthalene, **831**; α-nitronaphthalene, **824**; oximes, **678**; phenanthrene, **837, 838**; phenethyl mercaptan, **968a**; phenols, **805**; picolines, **899**; preparations using, acids, **557**; aldehydes and ketones, **527–529**; esters, **582**; quinones, **813**; sulfones, **953**; pyridine, **897, 898**; quinoline, **907–909**; side-chain oxidation, **416**; sulfinic acid, **954**; sulfoxide, **951**; tetramethylglucose, **620**; thiols, **946, 949**
Oxidation-reduction reactions, **19, 28–31, 32, 63**; balancing, **220–222**; intramolecular, in o-nitrotoluene, **742**; preparation of alcohols by, **229–232**; spectrophotometry, **263**
Oxidative dehydrogenation, **433**
Oxidizing agents, **30, 53**
Oxime acetate, **617**
Oximes: preparation, **500, 663–664**; reactions, **678**; tautomerization with nitroso group, **679**
Oximinoquinone: preparation, **796**; reduction, **796**
2H-Oxin. *See* 2H-Pyran
4H-Oxin. *See* 4H-Pyran
Oxirene, **369**
Oxolane. *See* Tetrahydrofuran, dipole moment
Oxole. *See* Furan
Oxonium ions, **206, 945**; reactions, **24**
Oxyacetylene torch, **385**
Oxygen, chain reaction termination by, **50**; covalent bond formation with, **4, 5**; covalent bond radii, **102**; electronegativity, **103**; electronic distribution, **88**; hydrogen bonding to, **105–106**; isotopes, **299**; molecular orbital theory, **93**; polarization of pi bond in formaldehyde, **26–27**; substituent effect on arenes, **462**; substituent effect on halogenation of methane, **335, 336**; transport in blood, **196**
Oxygen effect, **50**
Oxygen-hydrogen bond, in alcohols, reaction with active metals, **205–206**; in phenols, reactions at, **790–794**
Oxyhemoglobin, **930**
Oxytocin, **705**
Ozone, in smog, **366**
Ozonide, **367**
Ozonization, **367**; alkenes, **373**; benzene, **416**; conjugated systems, **427**; dienes, **427**; maleic or fumaric acids, **604**; rubber, **862**; o-xylene, **416**

Paints, **540**
Palladium, use as catalyst, **55**
Palmitic acid, physical properties, **538**
Pantothenic acid: human need for, **540**; structural formula, **730**; use, **730**
Papain, **708**
para-, **441**
Para attack, **450**
Paraformaldehyde, preparation, **492**
Paraldehyde, preparation, **492–493**; uses, **492**
Paramagnetic effect, **287**
Para protons, chemical shift, **408**
Parathion, **727–728**
Parchment paper, **638**
Parent peak, **298**
Partial double-bond fixation, **825**
Partial rate factor, **455–456**
Pauli's exclusion principle, **87**
Pectic acids, **640**
Pectins, **640**
Pelargonic acid, physical properties, **538**
Pellagra, **730**
Penicillin, **149**
Penicillin G, structural formula, **665**
Pentacene, structural formula, **842**
Pentacoordinate intermediate, esterification, **210**
Pentadecanoic acid, physical properties, **538**
1,3-Pentadiene, addition of hydrogen chloride, **419–420**; infrared spectrum, **404**
1,4-Pentadiene, heat of hydrogenation, **398**
2,3-Pentadiene, stereochemistry, **128**
Pentadienyl radical, **459**
Pentadienyl resonance, **445**
Pentadienyl system, **450, 460**
Pentaerythritol, preparation, **508**
Pentamethylbenzene, physical properties, **443**
Pentane, carbene insertion

Pentane (cont.)
reaction, 334–335; physical properties, 480
Pentanoic acid, preparation, 622
1-Pentanol, physical properties, 202
2-Pentanol, physical properties, 202
3-Pentanol, physical properties, 202
2-Pentanone, physical properties, 481
3-Pentanone, physical properties, 481
cis-2-Pentene, allylic bromination, 341
4-Pentenoic acid, 575
Pentobarbital, 903
Pentose, synthesis, 645
Pentyl alcohol, reduction of aromatic compounds in, 419
Pentyl group, formula, 75
Pepsin, 725
Peptide bond, 690; absorbance, 263; formation, 711
Peptides, 690, 705–713; end-group analysis, 706–708; hydrogen bonding in, 718; reactions, 707 ff.; structural formulas, deduction from analytical results, 798; synthesis, 709–713
Peracetic acid, reactions with alkenes, 48
Peracids, 368; preparation, 522; reactions, addition to alkenes, 48; with amines, 659; Baeyer-Villiger reaction, 515–516
Perchloric acid, 375
Perchloroethane, 778
Perchloroethylene, 778
Perfluorinated liquids, 340
Perfluoroisobutylene, 778
Perfluoromethyl cyclohexane, 340
Performic acid, 368; preparation, 552
Perfume, benzyl alcohol in, 208
Perhydro-, 397
Perhydroazine. See Piperidine
Perhydro-2,3a-thiazaindane, structural formula, 887
peri-, 824
Pericyclic reactions, 428–429
Periodic acid, determination of sugar structure using, 532, 634 ff.; as oxidizing agent for diols, 528–529
Periodic table, atomic structure and, 88
Perkin reaction, 563, 574
Peroxidase, 931
Peroxide effect, 361–362
Peroxides, carboxy inversion reaction, 993b; IUPAC naming, 78
Peroxyacyl nitrates, 366
Peroxy radical, 336; in chain reactions, 50

Perylene, structural formula, 842
Pethidine, 104, 919
Petrochemicals, 379
Petroleum, 377–379
Petroleum coke, 379
Petroleum ether, 379
PGA. See Folic acid
pH, 20 ff.
Phenanthrene, 837–839; physical properties, 444; preparation, 838–839; reactions, 458, 837–838; resonance energy, 825; structural formula, 396
Phenanthrenequinone, 837, 838
Phenazine, structural formula, 889
α-Phenethyl alcohol, haloform reaction, 503
Phenethyl disulfide, preparation, 968a
Phenethyl mercaptan, oxidation, 968a
p-Phenetidine, preparation, 769; structural formula, 442
Phenetole, structural formula, 441
Phenobarbital, 903; preparation, 569
Phenol, from chlorobenzene, 809; keto-enol tautomerism, 487; pK_a, 791; preparation by autoxidation, 466, 467; from cumene, 809–810; reactions, condensation with formaldehyde, 986, 988; reactions, ketonization, 790, 791; structural formula, 790
Phenol coefficient, 790
Phenol-formaldehyde resins, 986, 988
Phenolic esters, reactions with aluminum chloride, 799–800
Phenolphthalein, preparation, 800–802
Phenols, 440; complexes with ferric chloride, 794; as enols, 790; mass spectra, 790; occurrence in nature, 789; physical properties, 790; preparation, by diazotization, 761; by sulfonation, 808–809; reactions, acylation, 798; alkylation, 797–798; allylation, 800; bromination, 797; Bucherer reaction, 755–756; at carbon-oxygen bond, 795; chlorination, 797; coupling, 758, 759, 796–797; disulfonation, 743; Friedel-Crafts reactions, 797 ff.; Gattermann reaction, 521, 798; halogenation, 797; Houben-Hoesch reaction, 521; Kolbe reaction, 803; nitration, 743, 796; nitrosation, 796; oxidation, 805; oxidation to yield quinones, 813; at oxygen-hydrogen bond, 790–794; reduction, 805; Reimer-Tiemann reaction, 803;

replacement of hydroxy, 795; on the ring, 795–805; sulfonation, 795; Williamson ether synthesis, 791; reactions with, aldehydes and ketones, 800–802; diazomethane, 791; formaldehyde, 800; phthalic anhydride, 800; substituents in, 760
Phenoxide ion, 791; attack on carbonyl carbon, 800; coupling, 759; reactivity, 796
Phenoxyacetic acids, preparation by Williamson ether synthesis, 791
Phenylacetaldehyde, 208; physical properties, 481
Phenylacetic acid, physical properties, 538
Phenylacetonitrile, 607
Phenylalanine, 691, 693; structural formula, 698
Phenyl azide, preparation, 762
p-Phenylazo group, effect on peptides, 711
1-Phenylbutadiene, 1,2 addition, 426
Phenylcarbene, rearrangement, 438a
Phenyl cations, 460
Phenyldiazonium chloride, preparation, 754
m-Phenylene diamine, preparation of Nomex from, 983
Phenylephrine, 816
Phenyl esters, 797; Fries rearrangement, 799
Phenyl ethers, 797; Claisen rearrangement, 779–780; Gattermann reaction, 521, 798; Houben-Hoesch reaction, 521
2-Phenylethylamine, 815; Bischler-Napieralski synthesis using, 910; preparation, 910
Phenylhydrazine, 499–500; reactions with benzil, 618, 619; reactions with hydroxy aldehyde or hydroxy ketone, 618, 619
Phenylhydrazone, Fischer indole synthesis using, 906; preparation, 500; reactions, 678
Phenylhydroxylamine, instability, 769–770; physical properties, 768; preparation from nitrobenzene, 740; reactions, addition to carbonyl group, 769; with nitrosobenzene, 740, 741; rearrangement, 769
Phenyl isocyanate, 771
Phenyl isothiocyanate, reaction with peptides, 707
Phenylmagnesium bromide, reactions with aminodesoxybenzoin, 535b
Phenyl mercury, as fungicide, 455
N-Phenyl-β-naphthylamine, use

in rubber processing, 863
1-Phenylpropanol, 440
3-Phenylpropanol, 440
1-Phenyl-2-propanol, 441
Phenylpropiolic acid, 594
Phenyl radical, 440
Phenyls, substituted, 442
trans-2-Phenylsulfinyl-1-indanols, 968b
1-Phenyl-1-p-tolylethylene, preparation, 471
Pheophorbide, 932
Pheophytin a, 932
Pheoporphyrin, 932
Pheromones, 253, 366, 789; queen bee substance, 607
Phloroglucinol, 808; preparation, 810; structural formula, 790
Phorbin, 931
Phorone, preparation, 507
Phosgene (carbonyl chloride), 213; reactions with amines, 772; reactions with benzyl alcohol, 710
Phosphate esters, 962
Phosphatidic acid, preparation, 212
Phosphatidyl choline. See Lecithin
Phosphatidyl serine, 732
Phosphines, 958; free-radical addition to alkenes, 961
Phosphines, substituted, 960
Phosphinic acid, 958, 961
Phosphinic esters, 961
Phosphinous acid, 958
Phosphinous esters, 961
Phosphogluconate oxidative pathway, 645
3-Phosphoglyceraldehyde. See Triose phosphate
Phosphoglyceric acid (PGA), 556–557; in photosynthesis, 644
Phosphonic acid, 958, 961
Phosphonic esters, 961
Phosphonium oxides, tertiary, 960
Phosphonium salts, 958 ff.
Phosphonous acid, 958
Phosphonous esters, 961
Phosphoranes, 959, 960
Phosphoric acid, addition to alkenes, 353–354; esterification, 211; isomerization of alkenes with, 352; use in dehydration reactions, 52
Phosphorobenzene, 944
Phosphorous acid, esterification, 211
Phosphorus, electronegativity, 103; electronic distribution, 88; isotopes, 298
Phosphorus compounds, inorganic, 944
Phosphorus compounds, organic. See Organophosphorus compounds, cryoscopic behavior in sulfuric acid
Phosphorus oxychloride: atomic structure, 944; reactions with alcohols, 170, 211
Phosphorus pentachloride, 944; reactions with, alcohols, 170, 211; carboxylic acids, 547, 548; esters, 570; ketoxime, 664
Phosphorus trichloride, atomic structure, 944; reactions with, alcohols, 170, 211; carboxylic acid, 547, 548; Grignard reagents, 958
Phosphorus ylides, 497; preparation, 959; Wittig reaction, 959
Photochemical smog, 50
Photon, 241
Photophosphorylation, 643–644
Photosynthesis, 240, 643–644; acid products, 556–557
Phthaleins, 802
Phthalic acid, physical properties, 542; reactions with glycerol, 984, 985; structural formula, 442
Phthalic anhydride, preparation from furan, 891; preparation of anthraquinone from, 836; reactions, alcoholysis, 561; reactions, Perkin reaction, 564; reactions with naphthalene, 827; reactions with phenol, 800
Phthalimide, Gabriel synthesis using, 660–661
Phthalimidomalonic ester, synthesis, 701
Phthalocyanine, 646a
Phthalonitrile, 646a
Phthiocol, preparation, 812
Physical organic chemistry, 7
Physostigmine, structural formula, 745
Phytic acid, 643
Phytol, preparation from chlorophyll, 857
Phytyl bromide, preparation of α-tocopherol from, 818
3-Phytyl-2-methylnaphthoquinone. See Vitamin K_1
Pi bond, 15, 91, 93, 287, 393; nonrotation around, 123; polarization, 27; reactions, cis addition of hydrogen to, 358; oxidation, 372, 373; reduction 374–377; resonance theory and, 100; silicon, 963
α-Picoline, aldol condensation, 898
β-Picoline, 898, 899
Picolinic acid, 899
Pi complex, 106–107, 446; aromatic nitro compounds, 740; hyperconjugation, 457
Picrates, identification of tertiary amines using, 652
Picric acid (2,4,6-trinitrophenol): charge-transfer complexes with, 739–740; pK_a, 791; preparation, 461; preparation, from phenol by nitration, 743
Pi electron cloud, nucleophilic, 45
Pimelates, Dieckmann condensation, 572
Pimelic acid, physical properties, 542; preparation of cyclic ketones from, 554
Pinacol, dehydration, 524; preparation, 519
Pinacolone, preparation, 519, 524
Pinacol rearrangement, 519; aldehydes and ketones, preparation by, 524
Pinacols, unsymmetrical, 526
Pineapple, ester responsible for fragrance of, 543
α-Pinene, 852; structural formula, 852; synthetic camphor, preparation from, 854
Piperidine: preparation from pyridine by reduction, 897; structural formula, 887, 888
2-Piperidone, physical properties, 666
Piperine, structure, 665, 916
Piperonal, physical properties, 481
Pi state, 261
Pi* state, 261
Pituitary gland, 723
pK_a, 22, 538, 541, 542, 546, 547, 589, 594, 695–697
pK_w, 21
Planck's constant, 240
Plane-polarized light, 133
Plane tetragonal hybridization, 944
Plastic, 971
Plasticizers, 971, 974
Platinic chlorides, quantitative determination of amines with, 652
Platinum, use as catalyst, 55
Pleated structure, polypeptides, 715, 718
Plexiglas, 981
Point of symmetry, 126 ff.
Polar bond, 108–109
Polarimeter, 134
Polarity, 104; inductive effects, 108–109; of solvents, nucleophilic substitution and, 153–156
Polarizability, 26–27, 105
Polarization, 13; carbon-hydrogen bond in 1-alkynes, 332
Polarized light, optical activity, 133–136
Polio virus, 721
Pollution, organic heavy metal compounds, 76
Polyacetal, 985
Polyacrylonitrile, 981
Polyamides, 664, 982–983
Polyamino acids, peptides, 705
Poly-L-arabinoside, 639
Polyatomic molecule, vibrational modes, 247
Polycarbonates, 984
Polychloroalkanes, 777

Polychlorosilanes, 966
Polychlorotrifluoroethylene, 980
Polycyclic compounds, aromatic, 823 ff.; biphenyl, 842–844; from coal tar, 839–842; phenanthrene, 837–839; reactivity, 458
Polyesters, 983; block polymers from, 991
Polyethers, 806–807, 984–985; preparation, 979
Polyethylene, 972, 977, 980
Polyethylene oxide, 985; block polymers from, 991
Polyethylene terephthalate, 983–984
Polyfluoro compounds, 778–779
Poly-D-galactoside, 639
Polygalacturonic acids, 640
Polyglucoside, starch, 639, 640
Polyglycolide, preparation, 590
Polyhalogenation, 797
Polyhalophenols, 791
Polyhexamethylene adipamide, 982
Polyisoprene, 862–863, 977, 978
Poly-D-mannopyranoside, 639
Polymeric esters, 590, 591
Polymerization, aldehydes, 492–493; anionic, 978; cationic, 977–978; coordinate, 978; free-radical, 974; furan, 891; propagation, 975; pyrrole, 891; termination, 976; vinyl, 977
Polymers, 862, 969 ff.; alkyd resins, 984; block and graft polymers, 991; condensation polymers, 979, 982–989; epoxy resins, 985–986; ion exchange resins, 990–991; ladder polymers, 991–992; melamine-formaldehyde polymers, 989; phenol-formaldehyde resins, 986; polyamides, 982–983; polycarbonates, 984; polyesters, 983–984; polyethers, 984–985; polysiloxanes, 992; polysulfides, 992; polyurethanes, 986; urea-formaldehyde polymers, 989; vinyl polymers, 974
Polymethylene oxides, 68
Polymethyl methacrylate, 981
Polynitro compounds, use, 739–740
Polyoxymethylene, preparation, 984, 985
Polypeptides, 690; α-helix, 718–720; preparation, 712; proteins, 715–722
Polypropylene, 980; atactic, 972, 973; isotactic, 972, 973; stereochemistry, 972–973; syndiotactic, 972, 973
Polysaccharides, 107, 637–642
Polysiloxanes, 992; preparation, 966
Polystyrene, 970–971, 981; ion exchange resins from, 990,

991; structural formula, 970; use in peptide synthesis, 713
Polysulfides, 992
Polyterpenes, 862–863
Polytetrafluoroethylene, 778, 980
Polyunsaturated acids, 540
Polyurethane foam, 986, 989
Polyurethanes, 986, 988, 989
Polyvinyl acetate, 981
Polyvinyl alcohol, 981
Polyvinyl butyral, 981
Polyvinyl chloride, 973, 975, 980
Polyvinyl fluoride, 980
Polyvinylpyrrolidone, 981
Poly-D-xyloside, 639
p orbitals, 89, 90; in benzene, 400
$3p$ orbitals, 943
$3p$-$2p$ overlap, 943
Porphine systems, 928
Potassium alkyl xanthate, preparation, 223, 224
Potassium tert-butoxide, 205, 206
Potassium dichromate, reaction with alkenes, 372
Potassium hydrogen tartrate, 593
Potassium iodide, nucleophilic aromatic substitution, 762
Potassium nitrite, 654
Potassium permanganate, oxidation of alcohols, 218, 222; aldehydes, 517–518; reactions with alkenes, 371, 372; use as oxidizing agent, 53 ff.
Potassium phthalimide, 660
Potassium xanthate, preparation, 224
Potential energy, conformation of cyclohexane, 325; conformation of ethane and butane, 119–121; of a covalent bond, 113–115; energy-reaction-coordinate diagram, 189, 190; transition state, 145, 146
Potential-energy well, 113
Prednisone, structural formula, 878
Pre-exponential factor, 186
Prefixes, IUPAC rules for naming, 72–73, 78
Price, C. C., 939a
Primary carbon atom, 66
Primary carbonium ion, 352
Primary compounds. See under specific classes of compounds
Primary hydrogen atom, allylic, 341
Primary structure, nucleic acids, 923; polypeptides, 715
Principal quantum number, 87
Priority numbers, 394
Progesterone, 878, 879; structural formula, 876
Proline, 691, 693; proteins, effect on tertiary structure of, 720; structural formula, 693, 893
Propadrine, 816

Propagation, 45, 977; in vinyl polymerization, 974
Propane, 66; conformation, 123; physical properties, 69; reactions, chlorination, 338; reactions, nitration, gas phase, 681
Propanedial, 509
Propanenitrile. See Propionitrile
Propanethiol, 945
1,2,3-Propanetriol. See Glycerol
Propanoic acid, pK_a, 109
1-Propanol (propyl alcohol): azeotrope with water, 206–207; physical properties, 69, 202; reactions, etherification of, 215; with sodium, 205
2-Propanol. See Isopropyl alcohol
Propene, isomers, 67; microwave spectrometry, 331
2-Propenethiol, 945
Propionaldehyde, 68; physical properties, 69, 481
Propionamide, physical properties, 666
Propionic acid: degree of dimerization, 546; physical properties, 69, 480, 538; reduction of alkenes using, 358
Propionitrile, 649
Propionitrile oxide, 649
Propiophenone, 441; physical properties, 481
Propyl alcohol. See 1-Propanol
Propylamine, physical properties, 69, 650
n-Propylbenzene, physical properties, 443
Propyl cation, 151
Propyl chloride, physical properties, 69
Propylene, dipole moment, 320, 322; infrared spectra, 248, 250; physical properties, 69; pK_a, 333; polymerization, 980; production figures, 379; stereochemistry, 123–124
Propylene glycol, physical properties, 69
Propylene oxide, nmr spectrum, 283, 284; physical properties, 69
Propyl group, formula, 75; nmr spectrum, 280
2-Propylidenimine. See Ketimine
Propyl iodide, nmr spectrum, 280–282
2-Propylpiperidine, preparation, 916
Propyne (methyl acetylene), physical properties, 69, 322, 331
Prostaglandin E$_3$, structural formula, 608
Prostaglandins, 608
Prosthetic group, 690, 725
Proteins, 664, 690, 715–722; absorbance, 263; amino acids in, 699–700; antibody-antigen defense mechanism,

723–724; biosynthesis, role of DNA in, **924, 926;** conformation, **972;** enzymes, **725–732;** hormones, **722–723;** hydrogen bonding in, **715, 718;** reactions, digestion, **148–149;** with 1-fluoro-2,4-dinitrobenzene, **463;** hydrolysis, **691;** resonance, **715;** solubility, **107;** structure, **42;** sulfur-containing compounds in, **950**
Proteolytic enzymes, **708**
Prothrombin, **820**
Protic solvent, **22;** nucleophilic substitution, effect on, **156**
Proton acceptor, **20**
Proton donor, **20**
Proton magnetic resonance spectroscopy (pmr), **265**
Protons, chemical shift, **272;** shielding, **267–268**
Proton transfer, pK_a and, **22**
Protoporphyrin IX, **928**
Pseudoionone, **850, 852**
Pseudorotation, **964**
Psilocin, **905;** structural formula, **906**
Pteridines, **910, 913;** reactions, **913;** structural formula, **889**
Pteridines, substituted, **913–915**
Pteroic acid, structural formula, **745**
Pteroylglutamic acid. *See* Folic acid
Purine, **910–912;** preparation from uric acid, **912;** structural formula, **889, 910**
Purine bases, **921**
Purines, **910–912**
Purines, substituted, **911;** preparation from uric acid, **911, 912**
PVP, **981**
Pyran, **894, 895**
2H-Pyran, structural formula, **888**
4H-Pyran, structural formula, **888**
Pyrazine, **902;** structural formula, **888**
Pyrazole, **677;** structural formula, **888**
Pyrazoles, **893**
Pyrene, physical properties, **444;** structural formula, **442, 842**
Pyrethrin, **606–607**
Pyridazine, **902;** structural formula, **888**
Pyridine, **402, 896 ff.;** reactions, **897 ff.;** reactions with acid chlorides, **561;** reactions with glyconic acid, **624;** solvent effect of on nucleophilic substitution, **153;** structural formula, **888**
Pyridine carboxylic acid, preparation, **899**
2,3-Pyridinedicarboxylic acid. *See* Quinolinic acid
3,4-Pyridinedicarboxylic acid. *See* Cinchomeronic acid

Pyridines, substituted: preparation, **899–902;** reactions, **898–899**
2-Pyridinol. *See* 2-Pyridol
2-Pyridol, preparation, **897**
Pyridols, preparation, **899**
α-Pyridone, **897;** preparation, **901**
Pyridoxal. *See* Vitamin B_6
Pyridoxine, preparation, **901–902**
Pyridoxol, preparation, **901–902**
Pyrimidine, **902;** purines, preparation from, **911;** structural formula, **888**
Pyrimidine bases, **921**
Pyrimidines, substituted, **903;** preparation, **672, 911**
Pyrocatechol, **805;** structural formula, **441**
Pyrocatechol violet, **802**
Pyrogallol, **807;** preparation, **810;** structural formula, **790**
Pyroligneous liquid, preparation of acetone from, **521**
Pyrolysis, tartaric acid, **605**
Pyrolytic elimination, **383**
α-Pyrone, **895**
α-Pyrone, **895**
Pyrophosphate group, **869**
Pyrophosphoric acids, **962**
Pyrrole: dipole moment, **890;** occurrence in nature, **892;** reactions, **891, 892;** resonance, **890;** resonance structures, **402, 403;** structural formula, **888**
Pyrrolidine: dipole moment, **890;** structural formula, **888**
Pyrrolidines, occurrence in nature, **892**
2-Pyrrolidone, physical properties, **666**
Pyruvic acid, **597;** in glucose metabolism, **645;** preparation, **605;** reactions with Tollens' reagent, **597**
Pyrylium salt, **894**
Qualitative analysis, **6;** 1-alkynes, test for, **332**
Quantitative analysis, **6**
Quantum, **241**
Quantum numbers, **87, 245**
Quartet-triplet, **280**
Quaternary ammonium hydroxides: preparation, **657;** reactions, decomposition by S_N2 reaction, **167;** exhaustive methylation, **657;** Hofmann rule, **166 ff.**
Quaternary ammonium salts: preparation, **653, 656–657;** reactions, **657;** stereochemistry, **658**
Quaternary ions, reactivity and stability, **167**
Quaternary nitrosoamino ion, **754, 755**
Quaternary phenyltrimethylammonium ion, nitration, **748**
Quaternary phosphonium salts, **958**

Quaternary structure, hemoglobin, **928, 929;** polypeptides, **715, 720, 721**
Quaternary sulfonium salts, **953**
Queen bee substance, **607**
Quinaldinic acid, preparation, **907–908**
Quinazoline, structural formula, **889**
Quinic acid, **811**
Quinine, **918;** laboratory synthesis, **468**
Quinoid resonance structures, **800, 802**
Quinoline: alkaloids, **918;** preparation by Skraup synthesis, **908;** properties, **907;** reactions, **907–908;** structural formula, **889**
Quinolines, **906 ff.**
5-Quinolinesulfonic acid, **907**
8-Quinoline sulfonic acid, **907**
Quinolinic acid, preparation, **907**
2-Quinolone, preparation, **907**
Quinones, **433, 811–813;** coenzyme Q, **819;** preparation, **813;** reactions, **812;** vitamin K_1, **819–920**
Quinoxaline, structural formula, **889**

r-, **395**
R-, **128–130, 132;** use with EZ naming system, **394**
Racemates, **135 ff.**
Racemization, biphenyls, **843–844;** carbon-14 dating, **138–139;** esterification, **209**
Radiant energy, **240, 241;** analytical techniques using, **7**
Radicals, **26, 295;** stability, **339**
Radioactive labeling, in Fischer esterification with ^{18}O, **209;** with ^{14}C in rearrangement, **535b;** in steroids, **867–870**
Raffinose, **637**
Raman spectroscopy, **7, 257**
Rancidity, **540**
Raney nickel, reactions with thiophene, **892**
Raschig process, **809**
Rate constant, **184;** prf, **455;** transition state theory, **184 ff.**
Rate-determining step, **193**
Rate of reaction, **184;** Arrhenius equation, **186;** stoichiometry, **192**
R configuration, **128–130, 132, 394**
Reaction specificity, **725, 726**
Reactivity, **399**
Rearrangement reactions, aldehydes and ketones, **515–516;** amino alcohol, **655, 656;** Baeyer-Villiger reaction, **515–516;** Beckmann rearrangement, **663–664;** benzidine rearrangement, **772;** benzilic acid rearrangement, **515;** Claisen rearrangement,

Rearrangement (cont.)
799–800; cumene
hydroperoxide, 809–810;
in E1cA reactions, 166;
Fischer-Hepp rearrangement,
754; Fries rearrangement,
798–799; Hofmann
rearrangement, 662–663;
Lobry de Bruyn-van Eckenstein
rearrangement, 630–632; in
mass spectrometer, 297–298;
organophosphorus compounds,
959; phenylcarbene, 438a;
phenylhydroxylamine, 769;
pinacol rearrangement, 524;
preparations using, of acids,
559; of aldehydes and ketones,
524–527; of amides, 669;
of amines, 662–664
Receptor protein, 726
Reciprocal centimeter, 240
Redox reactions. *See*
Oxidation-reduction reactions
Red phosphorus,
Hell-Volhard-Zelinsky
reaction, 555–556
Reducing sugar, 615, 632
Reduction reactions, 32, 53–54;
acid anhydrides, 562–563;
acid chlorides, 562; acids, 555;
aldehydes and ketones,
518–519; alkenes, 358,
374–377; alkynes, 374–377;
amides, 668; anthracene,
834; anthraquinone, 834–835;
aromatic compounds, 418–419;
aromatic diazonium
compounds, 758, 763–764;
aromatic nitro compounds,
740–742; azine, 682;
azoalkenes, 682;
Bouveault-Blanc reduction,
570; camphor, 854;
Clemmensen reduction,
381–382; conjugated systems,
427–428; disulfides, 949;
esters, 570; glucose, 616;
hydrazones, 678–679;
hydrogenation, catalytic,
29–30, 55; hydroxy acids, 592;
p-hydroxyazo compounds,
797; indene, 840; isopulegol,
850, 851; isoquinoline, 909;
monosaccharides, 622;
naphthalene, 830; nitriles, 671;
nitroalkanes, 681;
α-nitronaphthalene, 824–825;
oximes, 678; oximinoquinone,
796; phenanthrene, 837;
phenols, 805; preparations
using, of aldehydes, 529–530;
of amines, 661–662; of
esters, 582; pyridine, 897;
quinoline, 907; Rosenmund
reduction, 529; Stephen
reduction, 529; sulfinic acid,
954; sulfonyl chloride, 954,
955; sulfoxide, 951;
Wolff-Kishner reduction, 381;

work of Subba Rao, 238a
Reductive alkylation, 662
Refining, petroleum, 379
Reflection symmetry, 126
Reformatsky reaction, 495–496,
523, 786; β-hydroxy acids,
preparation by, 593
Refractive index, 240;
hydrocarbons, 320
Reimer-Tiemann reaction, 803
Relative rate constant, 455
Relaxation, 267
Replacement reaction,
transition state, 144–146
Repulsion, nonbonded, 119–121
Reserpine, 917
Resin, 971
Resole, 986
Resolution, of racemates,
137–138
Resolving agent, 137
Resonance, 27, 97–100; allylic,
333; benzene, 445–446;
butadiene, 398; proteins, 715;
steric inhibition, 747
Resonance effect, 109; I and R
effects, 451
Resonance energy, 97; aromatic
system, 419; heterocycles, 890;
phenanthrene, 837; polycyclic
aromatic compounds, 825
Resonance hybrids, 100
Resonance stabilization energy,
protein, 715
Resonance structures, 98–99;
anthracene, 832; benzyl cation,
423; 1,3-butadiene, 398;
carbon-oxygen double bond,
482–483; cyclopentadienyl
anion, 890; dimethylsulfone,
943; furan, 890; guanidine,
652; naphthalene, 824;
phenanthrene, 837; phenol,
791; pyridine, 896; pyrimidine,
902; pyrrole, 890; quinoid,
800, 802; sigma complexes,
449; thiophene, 890
Resonance theory, aromatic
compounds, 402–403
Resorcinol, 807; Friedel-Crafts
reaction, 798; preparation, 810;
structural formula, 441, 790
Respiration, pH of human blood,
control by, 21
Retene, preparation, 860
Retention, 964
Retinal, 857–859
Retinene, preparation, 528
Retinol, 125
Reverse aldol condensation,
2-acetocyclohexanone, 607;
monosaccharides, 632
Reverse Claisen
condensation, 604
Rhodopsin, 125, 858, 859
Ribloflavin (vitamin B_2), 913,
914; preparation, 915;
structural formula, 665, 731;
use, 731

Ribonucleic acids. *See* RNA
Ribonucleosides, 921
Ribonucleotides, 921
Ribose, in RNA, 921
β-Ribosides, 920–921
Ribosomal RNA, 924
D-Ribulose, naming, 627
Ribulose diphosphate, reactions
in photosynthesis, 644
Rigau, J., 968a
Right-handed helix, 133
Right-hand rule, 246
Ring compounds. *See* Cyclic
(ring) compounds
Ring current, 408
Ring puckering, 322
Ritter reaction, 661; amides,
preparation by, 669
RNA (ribonucleic acid), 920;
stuctural formula, 922
Roberts, J. D., 318a, 343a
Rochelle salt, 593
Rosenmund reduction, 529
Rose oil. *See* Geraniol
Rotamers, 119; conjugated
systems, 395
Rotation: free rotation, 121;
nonrotation about double
bonds, 123–124
Rotational energy, 245, 258
(*RS*) rules, 128–130
Rubber, 862–863, 980; synthetic,
970; morphology, 971;
vulcanization, 863
Ruff degradation, 618
Rum, ester responsible for
fragrance of, 543
Ruzicka, L., 391a

s-, 395
S-, 128–130, 132, 394
Saccharine, preparation, 956
Safrole, structural formula, 806
Salicylaldehyde, physical
properties, 481
Salicylates, 793
Salicylic acid, 793; complex with
ferric chloride, 794;
preparation by Kolbe reaction,
803; structural formula,
442, 539
Salol, 793
Sandmeyer reaction, 762, 780
Sapogenins, 881
Saponification, 537, 567
Saponin, 881
Saran, 980
Sarcosine, structural
formula, 698
Saturation, 267
Saytzeff rule, 164 ff.
SBR rubber, 971; preparation, 979
Schaeffer, H. J., 535b
Schiemann reaction, 762, 780
Schiff base, 498 ff.
Schizophrenia, sweat odor
specific to, 607
Schotten-Baumann reaction, 561
s-cis form, 395

S configuration, **128–130, 132**;
  EZ naming system, use
  with, **394**
Scopolamine, **920**
Scurvy, **630**
Sebacic acid, physical
  properties, **542**
sec-, **74**
Secondary carbon atom, **67**
Secondary carbonium ion, **352**
Secondary hydrogen atom: allylic,
  341; in halogenation of
  alkanes, **338**
Secondary structure, of DNA,
  923–924; polypeptides,
  715, 718
Second law of thermodynamics,
  56–57
Second-order reactions, **192, 193**
Second-order resonance, **150**
Second-row elements, **941**
D-Sedoheptulose, **628**
Selection rules, **245**; ultraviolet
  spectra, **260**
Selectivity, in halogenation of
  alkanes, **339**
Self-condensation, of nitriles, **672**
Semicarbazide, **649**; reactions
  with aldehydes and
  ketones, **500**
Semicarbazone: preparation, **500**;
  reactions, **678**
Semidines, **772**
Semipinacol rearrangement, **655**
Septet-doublet pattern in nmr, **280**
Sericine, **690**
Serine, **691, 693**; decarboxylation,
  732; physical properties, **70**;
  pK_a, **695**; preparation, **702,
  703**; structural formula, **693**
Serine tRNA, **926, 927**
Serotonin, **905**; dimethylation,
  905; mental health and, **816**;
  structural formula, **816**
Serum albumin, amino acid
  composition, **699**
Sesquiterpenes, **391a, 847,
  855–857**
S_E2 reactions, **25, 32, 43, 62**
Sex hormones, **875–877**
Shielding, **267**
Shielding constant, **267**
S_H2 reactions, **26, 32, 62**;
  displacment on halogen, **43–44**
1,2-Shifts, **535b**
Sickle-cell anemia, **724**
Side-chain oxidation, **416**
Sigma bonds, **13** ff., **91, 287**;
  resonance theory and, **100**;
  of second-row elements, **941**
Sigma complex, **446, 450**; and
  hyperconjugation, **457**
Sigma *meta* complex, **450**
Sigma state, **261**
Sigma* state, **261**
Sigmatropic rearrangements, **429**
Silanediols, **963, 966, 967**
Silanes, **963, 964**
Silanetriols, **966, 967**

Silanols, **963, 966**
Silenes, **965**
Silica, atomic structure, **963**
Silicic acid, esterification, **211**
Silicon, atomic structure, **963,
  964**; electronegativity, **103**;
  electronic distribution, **88**;
  isotopes, **299**
Silicon carbide, **385**
Silicon compounds, **944**
Silicon compounds, organic.
  See Organosilicon compounds
Silicone oils, **966–967**;
  viscosity, **992**
Silicone rubbers, **967**
Silicones, **966**
Silicon tetrachloride, **966**;
  reactions with alcohols, **211**
Siloxane, preparation, **966**
Siloxane polymers. *See*
  Polysiloxanes
Silver ion, use to oxidize
  aldehydes, **518**
Silver nitrate: reactions with
  1-alkyne, **332**; reactions with
  formaldehyde, **30**
Silver nitrite, reactions with
  alkyl iodide, **681**
Silver oxide, use in exhaustive
  methylation, **167**
Silyl ethers, **966**
Silyl Grignard reagents, **966**
Silyl halides, reactions, **966**
Sinapyl alcohol, structural
  formula, **814**
Single bond, **13**; in
  1,3-butadiene, **398**
Singlet carbene addition,
  355–356
Singlet oxygen, in smog, **366**
Singlet state, carbene, **355**;
  ultraviolet spectra, **260**
Skatole, **905**
Skewed conformation, **119**
Skraup synthesis, **908**
Smog, **50, 366, 367, 391b**
S_N1cA substitution reaction,
  154–155, 158; with tertiary
  alcohols, **214–215**
S_N1 reactions, **147, 151**; halogen
  substituents, effect on, **343a**;
  mechanism and steric effects,
  149–151; racemization,
  159–160; solvent polarity
  and, **153** ff.; transition
  state, **144–146**
S_N1 replacement, of nitrogen in
  aromatic compounds, **762**
S_N1 solvolysis, haloarenes and
  allylic halides, **781**
S_N2cA reactions, esterification,
  210; substitution, **154–155, 158**
S_N2 displacement, of bisulfate ion
  by ethanol, **215**; of iodide
  ion, **166–167**; on oxygen, **384**
S_N2 inversion, **158–159**
S_N2 reactions, **23, 32, 62, 147** ff.;
  acid-base equilibria, **24**; alkyl
  halides, **39–41**; 1-alkyne

halides, **332**; aryl halides, **460**;
between epoxide and Grignard
reagent, **225–226**; halogen
substituents, effect on, **343a**;
preparations using, alkanes,
380; ethers, **39–41**; nitriles,
673–674; solvent polarity and,
154, 156; stereochemistry,
158 ff.; transition state,
144–146; Williamson ether
synthesis, **214**
Soaps, **540–541**; cleaning
  action, **107**
Soda pulp method, **814**
Sodium: electronic distribution,
  88; reactions with, alcohols,
  205; dienes, **423**; naphthalene,
  418–419, 830; thiophenes,
  892; uses, for reduction of
  alkenes and alkynes, **376**;
  uses, to reduce carbonyl
  compounds, **519**
Sodium acetate: as added ions
  in addition reactions, **46**;
  reactions with sodium
  hydroxide, **382**
Sodium alkoxides, **211**
Sodium amide, **52**; isomerization
  of alkenes by, **333**; reactions
  with, 1-alkyne, **332**; benzene,
  463; pyridine, **897**
Sodium azide: nucleophilic
  aromatic substitution, **762**;
  reaction with acid chloride,
  683–684
Sodium
  bis(2-methoxyethoxy)aluminum
  hydride, **529**; reactions with
  esters, **570**
Sodium bisulfite: reactions with
  carbonyl group, **493**; use to
  prevent benzoin condensation
  of benzaldehyde, **494**
Sodium borohydride: preparation
  of alcohols using, **230**;
  reactions with, acid anhydrides,
  562; acid chlorides, **562**;
  amides, **668**; aromatic nitro
  compounds, **741**; boron
  trifluoride to yield diborane,
  227; esters, **570**; uses, as
  catalyst during ozonization of
  alkenes, **367**; in catalytic
  hydrogenation, **376**; as
  reducing agent, **53, 54**; work
  of Subba Rao, **238a**
Sodium bromoacetate, reactions
  with sodium cyanide, **559**
Sodium *tert*-butoxide, reaction
  with ethyl bromide, **214**
Sodium chloride, ionic bonding
  in, **13**
Sodium chloroacetate, reaction
  with cyanide, **559**
Sodium cyanide, reaction with
  sodium bromoacetate, **559**;
  use in benzoin condensation,
  513–514
Sodium dichromate, use as

Sodium dichromate (cont.)
oxidizing agent, 53
Sodium ethoxide, reaction with
tert-butyl chloride, 214;
reactions with methyl
iodide, 192
Sodium fluoroacetate, 776
Sodium formate: decomposition,
pyrolytic, 382; fusion, 559
Sodium iodide, 351; reactions
with alkyl halides, 170
Sodium methoxide, reactions
with methyl benzoate, 568
Sodium nitrite, 654; reaction
with aniline, 754
Sodium oxalate, preparation, 382
Sodium phenoxide, Kolbe
reaction, 803
Sodium polysulfide, reaction with
dichloroalkane, 992
Sodium thiophenoxide, 343b
Solvating cations, 463
Solvation, 159
Solvents, for nmr spectroscopy,
270–271; nucleophilic
substitution, polarity of,
153–156
Sorbic acid (2,4-hexadienoic
acid), 394, 595; physical
properties, 594
s orbital, 89
Sorbitol, 616
β-Sorbofuranose, 630
L-Sorbose, 629–630
sp^3d hybrids, 944
sp^3d^2 hybrids, 944
Specific extinction. See
Absorptivity (extinction
coefficient)
Specificity (enzymes), 725
Specific rotation, 134
Spectrograph, 243
Spectrometer, 243
Spectrometry, 243
Spectrophotometer, 243–244
Spectroscopy, 7, 239 ff.
Spermine, 650
sp hybrid, 93
sp^2 hybrid, 93; in
formaldehyde, 15
sp^3 hybrid, 93
Spin decoupling, 276, 279
Sphingolipid, 732
Spin lattice relaxation, 267
Spin orientation, 265, 266
Spin quantum number, 87
Spin-spin coupling, 271–276
Spin-spin relaxation, 267
Splitting patterns, 410
Spontaneous combustion, 540
Spontaneous reaction, 182;
thermodynamics, 56, 57
sp orbital, covalent bond
radius, 101
sp^2 orbital, 14, 93; covalent
bond radius, 101
sp^3 orbital, 14, 93; covalent
bond radius, 101
Squalene, 860; biosynthesis,
869, 870; biosynthesis of
lanosterol and cholesterol
from, 871; preparation,
860, 861
Stability, 399; benzene, 400, 416;
styrene, 424; o-xylene, 416
Stabilization, of enolate ion, 509
Stabilization energy, aromatic
systems, 419
Staggered conformation, 119;
hydrocarbons, 123
Standard absolute entropy of
formation, 58, 59
Standard enthalpy of formation,
58, 59, 181
Standard enthalpy of reaction, 58
Standard entropy of formation, 181
Standard entropy of reaction, 58
Standard Gibbs free energy, 58
Standard Gibbs free energy of
formation, 181
Standard state, 57, 181
Standard-state reactions, 58
Stannic chloride, reactions with
imines, 529
Stannous chloride, reduction of
nitrile with, 529
Starch, 639, 640
Static stereochemistry, 119
Steady state, 194
Stearic acid, physical
properties, 538
Stearolic acid, preparation, 386
Stereochemistry, 4, 64b, 119;
amines, 658; elimination
reactions, 161–164; glucose,
614–615; imines, 658–659;
nucleophilic substitution,
158–160; polymers, 972;
quaternary ammonium salts,
658; substitution reactions of
organosilicon compounds,
964; trans addition to
unsymmetrical alkenes, 348;
unsaturated acids, 596
Stereoisomerism, 69, 124, 162;
optical isomerism, 136
Stereospecificity, singlet carbene
addition, 356
Steric effects, arenes, substitution
reactions, 456; cyanohydrins,
494; infrared spectra, effect on,
488; nucleophilic substitution,
147; Saytzeff elimination, 169;
S_N2 reactions, 40–41
Steric hindrance, 41; tert-butyl
group, 822a
Steroidal alcohols, methylation,
318c
Steroid ketols, 318c
Steroids, 866 ff.; adrenal hormones,
877–879; batrachotoxin,
882–883; bile acids, 880–881;
biosynthesis, 871; bufotalin,
882; cardiac glycosides,
881–882; ecdysone, 883; insect
hormones, 883; sapogenins,
881; sex hormones, 875–877;
sterols, 872–873; toad poisons,
882–883; vitamin D, 874–875
Sterols, 872–873
Sterulic acid, structural
formula, 539
Stigmasterol, 873, 879
trans-Stilbene: physical properties,
444; preparation, 471; structural
formula, 441; ultraviolet
spectrum, 407
Stilbestrol, 875; structural
formula, 876
Stimulants, 912–913
Stipitatic acid, structural
formula, 539
Stoichiometry, 192
Strain energy, cycloalkanes,
322, 324
s-trans form, 395
Strecker synthesis, 675
Streptomycin, 642
Strophanthidin, structural
formula, 882
Strophanthin, 882
Structural formula, derivation from
IUPAC name, 73–74
Structural isomerism, 38–39, 67 ff.
Structural stereochemistry, 119
Structure, 2 ff.
Strychnine, structural formula, 917
Strychnos alkaloids, 917
Styrene: physical properties, 444;
reactions, addition, 424–426;
reactions, polymerization, 981;
ultraviolet spectrum, 407
Styron, 981
Subba Rao, B. C., 238a
Suberic acid, physical
properties, 542
Substituents, acidity, effect on,
547; in arenes, 457; in arenes,
electrophilic substitution reaction
rates, 448–451; nucleophilic
substitution, 460–461; partial
rate factor, 455–456; steric
effects, 456; in aromatic amines,
746, 760; cooxidation reactions,
effect on, 968b; in diazonium
ion, 760; hydrolysis of esters,
effect on, 567; IUPAC rules for
naming, 72–73, 75, 78–79;
naphthalene, effect on, 828;
in phenols, 760; polymerization,
effect on, 977; on a ring, 395–396
Substitution reactions, 4, 19;
aldehydes and ketones,
501–504; allylic, competition
with addition, 363–364;
anthracene, 833; anthraquinone,
835–836; arenes, 445–451;
aromatic amines, 747–749;
aromatic nitro compounds, 738;
as competing reaction during
elimination, 51, 156–158, 189,
191; cyclopentadienes,
substituted, 891–892;
electrophilic (see Electrophilic
substitutions reactions);
Friedel-Crafts reaction, 454;
homolytic (see Homolytic

substitution reactions); naphthalene, **825–830**; nucleophilic (*see* Nucleophilic substitution reactions); organosilicon compounds, **963–964**; oxidation by, **365–367**; phenanthrene, **837**; phenols, **795 ff.**; preparations using, **169–171**; of alcohols, **225–226**; of aldehydes and ketones, **520–522**; of alkanes, **379–380**; of alkynes, **386**; of amines, **660–661**; of esters, **582**; of imides, **669**; of nitriles, **673–674**; of silanes, **966**; pteridine, **913**; pyridine, **896–897**; quinoline, **908**; uric acid, **911**; *o*-xylene, **826**
Substrate, **725**
Succinate dehydrogenase, noncompetitive inhibition with, **726**
Succinic acid: acidity, **547**; dehydration, **554**; physical properties, **71**, **542**; preparation, **559**; preparation by malonic ester synthesis, **576**, **577**
Succinic acids, substituted, preparation from α-bromo esters, **576**
Succinic anhydride, physical properties, **71**; reactions, alcoholysis, **561**; reactions, hydrolysis, **558**; reactions with, benzene, **831**; biphenyl, **832**; naphthalene, **827**, **839**; toluene, **832**
Succinimide, **649**; physical properties, **666**; pK_a, **666**
Succinonitrile, preparation, **674**
Sucrose, **636**
Suffixes, IUPAC rules for, **77–79**
Sugar alcohols, **642–643**
Sugars, **107**, **625**, **627**; disaccharides, **632–637**; monosaccharides, **614–632**; synthesis, **630–632**; trisaccharides, **637**
Sulfa drugs, **751**
Sulfanilamide, **751**
Sulfanilic acid: physical properties, **751**; preparation from aniline, **751**
Sulfates, preparation, **210**
Sulfenic acids, **954**
Sulfides, **947–949**; preparation, **949**
Sulfide-sulfate pulp method, **814**
Sulfinic acids, **954–955**
Sulfite pulp method, **814**
Sulfonamides, **956**
Sulfonamido group, **752**
Sulfonaphthalenes, **802**
Sulfonate esters, reactions, **955–956**; substitution reaction using lithium aluminum hydride, **380**
Sulfonation: aniline, **751**; anthraquinone, **835–836**; aromatic amines, **751**; benzene, **453–454**, **468**, **469**; camphor, **854**; furan, **892**; naphthalene, **826–827**; nitrobenzene, **738**; phenols, **795**; polycyclic compounds, **458**; polystyrene, **990**; preparations using, ion-exchange resins, **990**; preparations using, phenols, **808–809**; pyridine, **896**; pyrrole, **892**; quinoline, **907**; thiophene, **891**, **892**
Sulfones, **951–953**; preparation, **948**
Sulfonic acid chlorides, **955**
Sulfonic acid esters, **955**
Sulfonic acid group, Bucherer reaction, **836**
Sulfonic acids, **453**, **954**, **955**; preparation, **946**
Sulfonium ions, **945**, **953**; reactions, **948**
Sulfonium salts, preparation, **947**
Sulfonyl chlorides, **956**; atomic structure, **955**; preparation, **948**; reactivity, **752**; reduction, **954**, **955**; sulfonic acid esters, preparation from, **955**
Sulfonyl halide, preparation, **454**
Sulfoxides, **951–953**; preparation, **948**, **953**; quaternary sulfonium salts, preparation from, **953**
Sulfur, atomic orbitals in, **942**; covalent bond radii, **102**; electronegativity, **103**; electronic distribution, **88**; isotopes, **299**; use in vulcanization of rubber, **863**
Sulfur compounds, inorganic, **944**
Sulfur compounds, organic. *See* Organosulfur compounds, cryoscopic behavior in sulfuric acid
Sulfur dioxide, in smog, **367**
Sulfur hexafluoride, **944**
Sulfuric acid, autoprotolysis constant, **22**; mixed acid, **743**; reactions, addition to alkenes, **353**; reactions, esterification, **207**, **210**; reactions with ethanol, **215–216**; reactions with nitric acid, **452**; uses, in catalytic hydrogenation, **375**; in dehydration, **52**; in diazotization, **761**; in etherification, **215–216**; in oxidation, **53**
Sulfurous acid, esterification, **211**
Sulfur-oxygen bond, **942–943**
Sulfur ylide, **951**, **952**; reactions, **953**
Sunlight, **240**
Superposability, **126 ff.**
Surfactants, **541**
Suspension polymerization, **974**
Sweep coils, **269**
Symmetry, **126–133**
syn-, **395**
Syndiotactic polymers, **972**, **973**
Synephrine, **816**
Szmant, H. H., **968a**

2,4,5-T, **791**, **792**
t-, **395**
cis-Tachysterol, preparation, **874**
trans-Tachysterol, preparation, **874**
Tannic acid, **807**
Tannin, **807**
Tar, **379**
Tarbell, D. S., **612a**
Tariric acid, **538**
Tartaric acid, absolute configuration, **625**; physical properties, **71**, **589**; pyrolysis, **605**; racemates, resolution, **138**
Taurine, **950**; structural formula, **951**
Taurocholic acid, alkaline hydrolysis, **880**
Tautomer, **485**
Tautomerism, acetoacetic ester, **600–604**; organophosphorus compounds, **957–958**
Tedlar, **980**
Teflon, **37**, **340**, **778–779**, **977**, **980**
Teflon 100, **779**
Temperature: entropy, effect of on, **58**; polymers, effect of on, **973**
Terephthalic acid: physical properties, **542**; structural formula, **442**
Terminal alkynes, reactions, **376**
Termination reactions, **45**; in vinyl polymerization, **976**
Terpenes, **846 ff.**; acyclic, **847–850**; bicyclic, **852–854**; biosynthesis from acetic acid, **867 ff.**; cyclic, **850–852**; diterpenes, **857–860**; polyterpenes, **862–863**; sesquiterpenes, **855–857**; tetraterpenes, **860**; triterpenes, **860**
α-Terpineol, preparation, **853**
tert-, **74**
Tertiary alcohols, reactions during drastic oxidation, **373**
Tertiary alkyl halides, reactions, **380**
Tertiary amine oxide, reactions, **384**
Tertiary amines, Cope reaction, **383–384**
Tertiary carbon atom, **67**
Tertiary carbonium ion, **352**; dimerization, **354**
Tertiary hydrogen atom, allylic, **341**
Tertiary structure, DNA, **924**; hemoglobin, **928**, **929**; polypeptides, **715**, **720**
Terylene, **983**
Testosterone, **875**; methylation, **318c**; structural formula, **876**
Tetraalkoxyborate ion, **230**
Tetrabenzoporphine magnesium, **928**
2,3,7,8-Tetrachlorodibenzo-*p*-dioxin, **792**
Tetrachloroethylene, commercial preparation, **777**

Tetrachloromethane, **14**; physical properties, **29**
Tetracyanoethylene, **430, 671**
Tetraethylammonium chloride, **649**
Tetraethyl lead, **76**
Tetrafluoroethylene, polymerization, **980**
Tetrahaloethane, dehalogenation, **51**
Tetrahedral arrangement, **14**
Tetrahedral intermediate, **213**; esterification, **207**
trans-10a,6a-(-)-Δ⁹-Tetrahydrocannabinol. See Δ⁹-THC
Tetrahydrofuran: dipole moment, **890**; physical properties, **71**; solvent effect of on nucleophilic substitution, **153**; structural formula, **888**
1,2,3,4-Tetrahydroisoquinoline, preparation, **909**
1,2,3,4-Tetrahydronaphthalene. See Tetralin
Tetrahydro-1,4-oxazine. See Morpholine
1,2,3,4-Tetrahydrophenanthrene, oxidative dehydrogenation, **433**
1,2,3,4-Tetrahydroquinoline, preparation, **907**
Tetrahydrothiophenes, **892**; dipole moment, **890**
Tetrakis-, **74**
Tetralin (1,2,3,4-tetrahydronaphthalene): preparation, **419, 830**; oxidative dehydrogenation, **433**; reaction with naphthalene, **827–828**; structural formula, **397**
α-Tetralone, reactions, **832**
Tetramerization, of aldehydes, **492–493**
Tetramethylammonium chloride, physical properties, **70**
2,3,5,6-Tetramethyl-1,4-benzoquinone. See Duroquinone
Tetramethylene oxide, **68**
Tetramethylethylene, stereochemistry, **123–124**
Tetramethylglucose, preparation, **620**
2,2,5,5-Tetramethyl-3-hexene, stability, **154**
2,3,5,6-Tetramethyl-4-nitroaniline, dipole moment, **747**
Tetramethylsilane: atomic structure, **963**; as nmr reference standard, **268**
Tetraterpenes, **860** ff.
Δ⁹-THC, **817**; structural formula, **818**
Theobromine, **913**
Theophylline, **913**
Thermodynamics, **56–61**; 1,2 and 1,4 addition reactions, **421–422**; bond dissociation energy, **111** ff.; bond energies, **110** ff.; competing reactions, **194–195**; energy-reaction-coordinate diagrams, **189–190**; enzyme catalysis, **190–191**; equilibria and free energy, **181–183**; heat of formation of atomic states, **110–111**; life, evolution of and, **61**; steady state, **194**; stoichiometry, **192**; transition state theory, **183–186**
Thermodynamic functions of state, **57**
Thermoplastic polymers, **862, 971, 973, 980–981**
Thermosetting polymers, **971**
Thia-, **887**
Thiamine hydrochloride (vitamin B₁): preparation, **903**; structural formula, **73**; use, **731**
Thiamine pyrophosphate, **731**
Thianaphthene, structural formula, **889**
Thiazole, preparation, **894**
1,3-Thiazole, structural formula, **888**
Thiazoles, **893**
Thiazolone, substituted, **707**
Thimann, K. V., **391a**
Thimerosal, **949**
Thio-, **945**
Thioaldehydes, **943**
Thiobenzophenone, preparation, **943**
Thioctic acid. See Lipoic acid
Thioether, preparation, **946**
Thiohydantoin, substituted, **707**
Thioketones, **943**
Thiokols, **992**
Thiole. See Thiophene, dipole moment
Thiols, **945–947**; IUPAC naming, **78**; preparation, **947**; reactions, **945–946**; reactions, iodine oxidation, **949**
Thionyl chloride, reactions with acid, **212**; reactions with, alcohols, **170, 172, 211**; amides, **668**; carboxylic acids, **547, 548, 565**; esters, **570**
Thiophene, dipole moment, **890**; occurrence in nature, **892**; reactions, **891, 892**; resonance, **890**; structural formula, **888**
Thiourea, **947**
Thiourea, substituted, **707**
Third law of thermodynamics, **58**
Thorpe reaction, **672**
Thorpe-Ziegler reaction, **672–673**
Three-center reaction, **334**
Threonine, **691, 693**; physical properties, **71**; structural formula, **693**
Thrombin, **820**
Thromboplastin, **820**
5-Thymidylic acid, structural formula, **922**
Thymine, **904**; in DNA, **920, 921**; structural formula, **921**
Thyroid gland, **459**
L-Thyronine, **459**
Thyroxine, **459–460**; structural formula, **699**
Tiglic acid, physical properties, **594**
Tigogenin, **881**
Tilorone hydrochloride, **792**
Tin, reaction with aromatic nitro group, **741, 742**
Tischenko reaction, **232, 582**
Toad poisons, **882–883**
Tobacco mosaic virus, **721**
α-Tocopherol, synthesis, **818, 819**
Tollens' reagent, reaction with glucose, **615**; reaction with α-keto acid, **597**
m-Tolualdehyde, physical properties, **481**
o-Tolualdehyde, physical properties, **481**
p-Tolualdehyde, physical properties, **481**
Toluene: chemical shift, **408**; infrared spectrum, **406**; physical properties, **443**; preparation of m-bromotoluene from, **765–766**; reactions, bromination, **457, 470**; chlorination, **466, 470**; free-radical chlorination, **466**; Gattermann aldehyde synthesis using, **798**; nitration, **469**; partial rate factor, **455**; sulfonation, **469**; trinitration, **743**; reactions with acetic acid, **455**; reactions with succinic anhydride, **832**; structural formula, **440**; symbols for, **404**; syntheses using, **469** ff., **765**; ultraviolet spectrum, **407**
o-Toluenesulfonic acid, preparation, **469**
p-Toluenesulfonic acid: alkali fusion, **810**; preparation, **469**
p-Toluenesulfonyl chloride, use in Hinsberg reaction, **653**
m-Toluidine, structural formula, **442**
p-Toluidine, preparation of m-cresol from, **810**
Toluidines: preparation of phenol and cresols from, **810**; reactions, **750**
Tosylhydrazone, **438a**
trans-, **69, 124–125**; EZ naming system, use with, **394**; RS rules, **129**
trans addition, **348** ff.
trans elimination, **162, 164, 176**
Transesterification, **549**
Transfer RNA, **924, 926**
Transition moment, **245**
Transition state, **41, 144–146, 183–186**; alkyl halide, unimolecular ionization, **151**; coplanarity, **162**; S_N2 reaction, **158**; thermodynamics, **185–186**
Translational energy, **245**
Transmission effect, **739**
Transmittance, **244**
Transoid, **395**
Trehalose, **637**
Triacylborane, **230**

Trialkylborane, preparation of alcohols from, **228**
Trialkyl boron: hydrolysis, **358**; reactions with alkaline hydrogen peroxide, **358**–**359**
Tri(*p*-alkylphenyl)methyl radicals, disproportionation, **782**
Trialkylphosphine, **958**; S_N2 reaction with alkyl halide, **497**
Trialkyl phosphite, reactions with alkyl halides, **960**
Triallylamine, physical properties, **650**
Triarylmethyl system, **783**
Triarylphosphine, S_N2 reaction with alkyl halide, **497**
Triatomic molecule, vibrational modes, **247**
1,3,5-Triazines, **684**
1,1,2-Tribromoethane, nmr spectrum, **271**, **273**
Trichloroacetaldehyde, **49**
Trichloroacetic acid, pK_a, **109**
1,2,4-Trichlorobenzene, nmr spectrum, **409**, **410**
1,2,4-Trichlorocyclopentane, **395**
Trichloroethylene, **778**
Trichloromethane: bonding in, **14**; physical properties, **29**
Trichloromethylsulfenyl chloride, **954**
2,4,5-Trichlorophenoxyacetic acid (2,4,5-T), **791**, **792**
Trichloropurine, preparation from uric acid, **912**
cis-9-Tricosene. See Muscalure
Tricresyl phosphate, use as plasticizer, **974**
Tricyclic diterpenes, **860**
Tricyclo[5.2.1.0^{2,6}]3,8-decadiene. See Dicyclopentadiene
Tridecanoic acid, physical properties, **538**
Triethylamine, **649**; physical properties, **650**, **651**; uses in esterification, **211**
Triethylamine oxide, **649**
Triethylbenzene, preparation, **468**
Triethyl orthoformate, preparation, **583**
Trifluoroacetic acid, reactions with hydrogen peroxide, **368**
2,2,2-Trifluoroethanol, **785**
Trifluoromethanesulfonic acid, **955**
Trifluoroperacetic acid, **368**; oxidation of oximes using, **678**
Trigonal bipyramidal hybridization, **944**
Trihalomethyl group, **503**
Trihydroxybenzenes, **805** ff.; preparation, **810**
2,6,8-Trihydroxypurine. See Uric acid
2,4,6-Trihydroxypyrimidine. See Barbituric acid
Triiodothyronine, **459**; structural formula, **699**
1,2,3-Triketoindane. See Ninhydrin
Trimerization, of aldehydes, **492**–**493**; of formaldehyde, **493**
Trimesic acid, physical properties, **542**
Trimethoxyglutaric acid. See Trimethylxylaric acid
Trimethylacetic acid, acidity, **547**; physical properties, **538**
Trimethylamine, physical properties, **69**, **650**
Trimethylammonium methylide, **649**
Trimethylcarbonium ion, **149**
Trimethylchlorosilane, **966**
Trimethylene glycol, physical properties, **69**
Trimethylene oxide (oxetane), **68**; physical properties, **69**; structural formula, **888**
Trimethylethylene, stereochemistry, **123**–**124**
2,3,5-Trimethylhexane, IUPAC name, derivation, **73**
2,4,6-Trimethylpyridine, preparation, **901**
2,4,6-Trimethyl-1,3,5-trioxane. See Paraldehyde
1,3,7-Trimethylxanthine. See Caffeine
Trimethylxylaric acid, **620**
Trinitration, toluene, **743**
Trinitroanisole, reactions, **461**–**462**
1,3,5-Trinitrobenzene, **461**
Trinitrophenetole, reactions, **462**
2,4,6-Trinitrophenol. See Picric acid
2,4,6-Trinitrotoluene (TNT), preparation from toluene, **743**
Triose phosphate, formation in photosynthesis, **644**
Trioxane: chloromethylation using, **500**; preparation, **492**
Triozonide, **416**
Triphenylamine, basicity, **746**
Triphenylene, structural formula, **842**
Triphenylmethane, physical properties, **444**
Triphenylmethane dyes, **802**
Triphenylmethanol, structural formula, **440**
Triphenylmethyl anion, **572**
Triphenylmethyl carbonium ion, **424**
Triphenylmethyl chloride: structural formula, **440**; Wurtz reaction, **782**
Triphenylmethyl halides, **781**
Triphenylmethyl radical, **440**
Triphenyl orthothioformate, **343b**
Triphenylphosphine, **959**
Triple bonds, **4**; bond distance, **327**
Triple bonds, bond force constant, **114**; conjugated, **393**; in cyclic compounds, **68**; IUPAC names for compounds containing, **77** ff.; reactions, addition-elimination, **62**; catalytic hydrogenation, **55**; nucleophilic addition, **345**–**346**; reducing agents, effect of on, **54**; stereochemistry, **125**
Triplet carbene addition, **355**–**356**
Triplet oxygen, in smog, **366**
Triplet state, carbene, **355**; ultraviolet spectra, **260**
Tris-, **74**
Trisaccharides, **637**
Triterpenes, **847**, **860** ff.
Triton B, **512**
Tröger's base, **658**
Tropane alkaloids, **920**
Tropylium cation, **410**
Trypsin, **708**
Tryptamine, structural formula, **916**
Tryptophan, **693**, **694**, **905**; structural formula, **693**, **916**
Tschugaeff reaction. See Chugaev reaction
"Tub" conformation, **400**
Tuberculostearic acid, **538**
Turpentine, **852**
Twinned bonds, **393**
Twist boat form, **324**
Tyramine, **815**
Tyrosine, **691**, **693**, **815**; hydrogen bonding with, **720**; metabolism, **817**; structural formula, **693**, **815**

Ullmann reaction, **780**
Ultraviolet radiation, **241** ff.
Ultraviolet spectra, aromatic amines, **745**; carbonyl group, effect on, **488**; conjugated systems, **407**; nitriles, **671**; nitrobenzene, **737**; purine, **911**
Ultraviolet spectroscopy, **244**; conjugation, effect of, **261**–**262**; theory, **258**–**261**; uses, **262**–**263**
Undecanoic acid, physical properties, **538**
10-Undecenoic acid, **591**
Unimolecular fragmentation, **295**
Unimolecular reactions, **193**
Unsaturated acids: physical properties, **595**–**596**; of physiological interest, **607**–**608**; preparation, **596**; preparation, by malonic ester synthesis, **576**; reactions, **595**–**596**; stereochemistry, **596**
α,β-Unsaturated acids: preparation, **596**; reactions, addition, **595**; reactions, heating of, **553**–**554**
β,γ-Unsaturated acids: addition reactions, **595**; heating of, **553**
Unsaturated aldehydes, reactions with Grignard reagents, **522**
α,β-Unsaturated aldehydes: Michael reaction, **511**–**512**; preparation, **506**
α,β-Unsaturated ketones, diazomethane ring enlargement, **318c**
Unsaturation, **55**; nmr spectrum to determine, **290**–**291**
Uracil, **904**; in RNA, **920**, **921**; structural formula, **921**

Urea, **649**; barbituric acid, preparation from, **903**; IUPAC naming, **17**; physical properties, **13**; reactions, with ethyl cyanoacetate, **911**; reactions, exchange reaction with acid, **551**; Wohler synthesis, **2**
Urea-formaldehyde polymers, **989**
Urethane, **649**; preparation, **663**
Uric acid, **911**; alloxan, preparation from, **914**; purines, substituted, preparation from, **911, 912**
Uridine, in RNA, **921**; structural formula, **921**
5-Uridylic acid, structural formula, **922**

Vaccination, **724**
Valence bonds, **96–100**
Valence shell, **4, 14, 90**
Valence state, **93**
n-Valeraldehyde, physical properties, **481**
Valeric acid, esterification, **550**; physical properties, **538**
Valine, **694**; preparation, **675**; sickle-cell anemia, relation to, **724**; structural formula, **694**
Van der Waals forces, **105**
Van der Waals radii, **121**
Vanillin, physical properties, **481**; preparation from softwood sulfite spent liquor, **814–815**; structural formula, **480, 806**
Vanillylmandelic acid, in DOPA metabolism, **817**
Van Romburgh, P., **391a**
Van't Hoff's law, **110, 183**
Vapor state, ultraviolet absorption in, **407**
Varnishes, **540**
Veratric acid, structural formula, **806**
Veratrole, reactions, **810**
Veronal, **903**
Vetivazulene, structural formula, **856**
β-Vetivone, structural formula, **856**
Vibrational energy, **245, 258**; in infrared spectroscopy, **246–247**
Vibrational quantum number, **113**
Vibratory motion, **101, 113**
Vicinal, **288**
Vinyl acetate, polymerization, **981**
Vinylacetylene, addition of HCl to, **355**; preparation, **355**
Vinyl alcohol: preparation from acetylene, **48**; reactions, polymerization, **981**; reactions, rearrangement, **48, 55**
Vinyl bromide, **779**
Vinyl butyral, polymerization, **981**
Vinyl chloride, **778, 779, 780**; physical properties, **36**; polymerization, **980**; preparation by addition of HCl to acetylene, **353**; production figures, **378**; resonance effect, **109**
Vinyl esters, transesterification, **510–511**
Vinyl ethers, preparation, **346–347**

Vinyl fluoride, polymerization, **980**
Vinyl group, **37, 77**
Vinyl halides, **779**; dehydrohalogenation, **385**; inertness, **42, 52**; preparation, **47**
Vinylic absorption, **404**
Vinylidene chloride, **778**; copolymerization, **980**
Vinylidene fluoride, copolymerization, **980**
Vinyl ketone, preparation, **600**
Vinyl methyl ketone (3-butene-2-one): absorption peaks, **261**; Robinson annelation reaction, **512**
Vinyl monomers, polymerization, **974**
Vinylog, **796**
Vinylogy, **739**
Vinyl polymerization, **977**
Vinyl polymers, **974 ff., 980–981**
N-Vinylpyrrolidone, polymerization, **981**
Viscose, **638**
Viscose rayon process, **638**
Visible light. See Light
Visible spectra, conjugated systems, **407**
Visible spectroscopy, **244**; conjugation, effect of, **261–262**; theory, **258–261**; uses, **262–263**
Vision, chemistry of, **858–859**
Vistanex, **980**
Visual purple, **858**
Vital force, **2**
Vitamin A, vision, role of in, **858**
Vitamin A$_1$, **125**; oxidation by manganese dioxide, **528**; preparation, **857, 859**; preparation from lycopene, **862**; structural formula, **859**
Vitamin A$_2$, **857**; structural formula, **859**
Vitamin B$_1$. See Thiamine hydrochloride
Vitamin B$_2$. See Riboflavin
Vitamin B$_3$. See Niacinamide
Vitamin B$_5$. See Pantothenic acid
Vitamin B$_6$ (pyridoxal), **499**; structural formula, **731**
Vitamin B$_{12}$, **932–933**
Vitamin B$_c$. See Folic acid
Vitamin B complex, **643, 728, 729, 732, 913**
Vitamin C. See Ascorbic acid
Vitamin D$_1$, **874**
Vitamin D$_2$, **874, 875**
Vitamin D$_3$, **874, 875**
Vitamin E, **818–819**; preparation, **857, 858**
Vitamin H. See Biotin
Vitamin K, **819–820**
Vitamin K$_1$, **820**; preparation, **858**
Vitamin K$_2$, **820**
Vitamin M. See Folic acid
Vitamins, **728–732**
Viton, **779, 977, 980**
Vulcanization, **863**

Wagner-Meerwein rearrangement, **152, 466**; of halohydrin, **785**

Waste recycling, enzyme catalysis, use in, **190**
Water, acid-base equilibrium, **22**; autoprotolysis constant, **22**; azeotropes with alcohols, **206–207**; bond angle, **102**; bonding, **647**; covalent bonding, in, **5**; preparation from H_2 and O_2, thermodynamics, **58**; reactions, addition (see Hydration); reactions, nucleophilic substitution, solvent effect of on, **153–154**; reactions with lithium aluminum hydride, **230**; reactions with quaternary ions, **167**; vibrational modes, **247**
Water softening, ion exchange resins, **990**
Wavenumber, **240**
Waxes, **537**
Went, F., **391a**
Wetting agents, **541**
Williamson ether synthesis, **214**; ortho esters, preparation by, **582–583**; of phenols, **791**
Winstein, S., **64b**
Wintergreen, ester responsible for fragrance of, **542**
Wittig reaction, **497–498**; with phosphorus ylides, **959**
Wohl degradation, **617–618**
Wolff-Kishner reduction, **381, 678**
Wood, **814**
Woodward and Hoffman rules, **429**
Wurtz-Fittig reaction, **780–781**
Wurtz reaction, **380, 782**

Xanthates, **223**
Xanthenes, **802**
Xanthines, **912, 913**
Xanthins, **861**
Xanthophylls, **861**
Xanthopterin, **913**
X-ray radiation, **241 ff.**
Xylan, **639**
Xylene, chlorination, **466**
m-Xylene: infrared spectrum, **406**; physical properties, **443**; structural formula, **441**
o-Xylene: infrared spectrum, **406**; physical properties, **443**; reactions, ozonization, **416**; reactions, substitution, **826**; stability, **416**; structural formula, **441**
p-Xylene: infrared spectrum, **406**; physical properties, **443**; structural formula, **441**

Yield, **217**
Ylides, **497**; phosphorus ylides, **497, 959**; reactions with acid chlorides, **523–524**; reactions with esters, **570**
Yohimbine alkaloids, **917**

Z-, **394**
Zeolites, **990**
Zero-point energy, **113**
Ziegler-Natta catalysts, **978**
Zinc: in dehalogenation reactions,

51; reaction with alkyl halides, **495–496**
Zinc amalgam, use in Clemmensen reduction, **381**
Zinc dust, as catalyst after ozonization of alkene, **367**
Zinc oxide, effect on properties of rubber, **863**
Zusammen, **394**
Zwitterion, **669**
Zymase, **233, 640**

Some selected enthalpies and entropies of formation[a]

Elements	ΔH°_f kcal	ΔS°_f eu	Simple oxides and hydrides	ΔH°_f kcal	ΔS°_f eu
Br_g	26.8	41.8	CH_{4g}	−17.9	44.5
Br_{2g}	7.5	58.6	NH_{3g}	−11.0	45.9
Br_{2l}	0.0	36.4	HO_g	10.1	43.9
C_g	170.9	37.8	H_2O_g	−57.8	45.1
$C_{graphite}$	0.0	1.4	H_2O_l	−68.3	16.7
Cl_g	28.9	39.5	H_2O_{2g}	−32.5	55.6
Cl_{2g}	0.0	53.3	HF_g	−64.2	41.5
F_g	18.9	37.9	HCl_g	−22.1	44.6
F_{2g}	0.0	48.4	HBr_g	−8.7	47.5
H_g	52.1	27.4	HI_g	6.2	49.3
H_{2g}	0.0	31.2	HS_g	32.0	46.8
I_g	25.5	43.2	H_2S_g	−4.8	49.2
I_{2g}	14.9	62.3	PH_{3g}	2.2	50.2
I_{2s}	0.0	27.8	CO_g	−26.4	47.2
N_g	113.0	36.6	CO_{2g}	−94.1	51.1
N_{2g}	0.0	45.8	CS_{2g}	27.6	56.8
O_g	59.6	38.5	N_2O_g	19.5	52.6
O_{2g}	0.0	49.0	NO_g	21.6	50.3
S_g	66.4	40.1	NO_{2g}	8.1	57.5
S_{2g}	30.8	54.5	N_2O_{4g}	2.3	72.7
$S_{s,\ rhombic}$	0.0	7.6	SO_{2g}	−71.0	59.3
			SO_{3g}	−94.4	61.2
Carbon compounds (gases)					
CH_3Cl	−19.6	55.8			
CH_2Cl_2	−21	64.6	Carbon compounds (gases)		
$CHCl_3$	−24	70.7	$HC{\equiv}CH$	54.2	48.0
CCl_4	−25.5	73.9	$H_2C{=}CH_2$	12.5	52.5
CF_4	−218	62.5	CH_3CH_3	−20.2	54.9
$COCl_2$	−52.5	67.8	CH_3CH_2OH	−56.6	67.6
CH_3OH	−48.1	57.3	CH_3CH_2SH	−11.0	70.8
CH_3SH	−3.7	60.9	CH_3SCH_3	−9.0	68.3
$HCHO$	−27.7	52.3	CH_3CHO	−39.7	63.2
$HCOOH$	−90.4	59.4	CH_3COOH	−103.8	67.6
CH_3NC	35.9	58.8	CH_3CN	21.0	58.2

[a] H°_f is the standard enthalpy of formation at 25°C, 1 atm; and ΔS°_f is the absolute entropy at 25°C, 1 atm. Values are for one mole of the indicated substance.

Relationship of ΔG°, log K_{eq}, and K_{eq} at 25°C

ΔG°, kcal	log K_{eq}, or $-pK_{eq}$	K_{eq}
30	−21.989	1.03×10^{-22}
20	−14.659	2.19×10^{-15}
10	−7.330	4.68×10^{-8}
1	−0.733	1.85×10^{-1}
0	0	1.00
−1	0.733	5.41
−10	7.330	2.14×10^7
−20	14.659	4.56×10^{14}
−30	21.989	9.74×10^{21}

Some characteristic infrared frequencies

Frequency, cm^{-1}	Group
3600	O—H, monomeric alcohol or acid
3300	≡C—H or hydrogen bonded alcohol
3050	=C—H
2950	C—H, CH$_2$, and CH$_3$
2840 and 2740	C—H in aldehyde, —CHO
2600	O—H, hydrogen bonded acid
2230	—C≡N or R—C≡C—R
2120	RC≡CH
1820 and 1760	C=O in acid anhydride
1800	C=O in acid chloride
1740	C=O in ester^a or dilute acid
1720	C=O in aldehyde, ketone, or dimeric acid
1660	C=C
1600 and 1400	RCO$_2^-$ in salts
1475	CH$_2$
1475 and 1380	CH$_3$
1410, 990 and 910	RCH=CH$_2$
1410 and 890	R$_2$C=CH$_2$
1395 and 1370	(CH$_3$)$_3$C, unequal
1385, 1370, and 1160	(CH$_3$)$_2$CH, equal
1360	tertiary O—H
1340	tertiary C—H
1300	primary and secondary O—H
1300 and 960	trans—RCH=CHR
1250	multiple CH$_2$ or CH$_3$
1250	epoxide
1200	tertiary C—O, alcohol, ether, or ester
1200	C—F, variable
1160	(CH$_3$)$_2$CH
1100	secondary C—O
1050	primary C—O
810	R$_2$C=CHR
700	C—Cl
690	cis-RCH=CHR
550	C—Br
500	C—I

^{a} An ester also has two strong bands, C—O, in the 1250–1050 cm^{-1} region.

Some chemical shift values (δ, ppm)

R	CH$_3$—	CH$_3$—CH$_2$—		CH$_3$—CH$_2$—CH$_2$—			(CH$_3$)$_2$CH—	
RF	4.3	1.6	4.5	0.9	1.8	4.3		4.8
RCl	3.0	1.5	3.5	0.9	1.8	3.3		4.0
RBr	2.7	1.7	3.4	1.0	1.9	3.3	1.7	4.2
RI	2.2	1.7	3.2	1.0	1.9	3.2		4.2
ROH^a	3.3	1.2	3.6	0.9	1.6	3.6	1.2	3.9
ROR'	3.2	1.1	3.4	1.0	1.6	3.6	1.0	3.6
R'COOR	3.6	1.3	4.1	0.9	1.6	4.2	1.3	5.0
RCH=CH$_2$^{b}	1.6	1.0	1.9	0.9	1.3	1.9	0.9	2.4
RC≡CH^c	1.8	1.1	2.1	0.9	1.5	2.2		2.8
RCHO^d	2.2	1.1	2.3	1.0	1.7	2.4		2.4
RCOR'	2.2	1.0	2.5	0.9	1.5	2.3	1.1	2.5
RCOOH^e	2.1		3.2	1.0	1.5	2.4	1.2	2.7
RCOOR'	2.0	1.2	2.2	1.0	1.5	2.3		2.5
RNH$_2$^{f}	2.3	1.0	2.5	0.9	1.4	2.6		2.9

^{a} Hydroxylic protons exhibit strong temperature and concentration effects, δ 0.5–δ 6.0.
^{b} See Figure 11.15 for alkenyl proton shifts.
^{c} Acetylenic protons appear at δ 2.4–2.8.
^{d} Aldehydic protons appear at δ 9.7–10.0.
^{e} Acidic protons appear at δ 10–δ 13.
^{f} Amine protons appear at δ 0.5–δ 3.0.

Bond energies (BE)[a]

Diatomic molecules

H—H	104	F—F	37	H—F	135
O=O	119	Cl—Cl	58	H—Cl	103
N≡N	226	Br—Br	46	H—Br	87
C=O[b]	257	I—I	36	H—I	71

Polyatomic molecules

C—H	99	C—C	83	C—F	116
N—H	93	C=C	146	C—Cl	81
O—H	111	C≡C	200	C—Br	68
S—H	83	C—N	73	C—I	51
P—H	76	C=N	147	C—S	65
N—N	39	C≡N	213	C=S[c]	128
N=N	100	C—O	86	N—F	65
O—O	35	C=O[d]	192	N—Cl	46
S—S	54	C=O[e]	166	O—F	45
N—O	53	C=O[f]	176	O—Cl	52
N=O	145	C=O[g]	179	O—Br	48

[a] These bond energies in kcal/mole at 25° are from J. D. Roberts and M. C. Caserio, *Basic Principles of Organic Chemistry*, New York: Benjamin, 1964, by permission.
[b] Carbon monoxide.
[c] For carbon disulfide.
[d] For carbon dioxide, acids, esters.
[e] For formaldehyde.
[f] For other aldehydes.
[g] For ketones.

Some bond dissociation energies (BDE)[a]

	—H	—F	—Cl	—Br	—I	—OH	—NH$_2$	—CH$_3$
CH$_3$—	104	108	83.5	70	56	91.5	79	88
CH$_3$CH$_2$—	98	106	81.5	69	53.5	91.5	78	85
CH$_3$CH$_2$CH$_2$—	98	106	81.5	69	53.5	91.5	78	85
(CH$_3$)$_2$CH—	94.5	105	81	68	53	92	77	84
(CH$_3$)$_3$C—	91	—	78.5	63	49.5	90.5	77	80
CH$_3$C(=O)—	87.5	119	83.5	—	52.5	109	96	82
CH$_2$=CH—	103	—	84	—	—	—	—	92

[a] S. W. Benson, *J. Chem. Educ.* 42, 502 (1965). BDE = $\Delta H°$, 25°C, 1 atm, gas phase, the energy to break the bond between the fragment listed at the left and the atom or fragment listed at the top. Example:

$$\text{BDE} = 69 \text{ kcal for } CH_3CH_2\frown Br \longrightarrow CH_3CH_2\cdot + Br\cdot.$$

Some proton spin coupling constants (J, Hz)

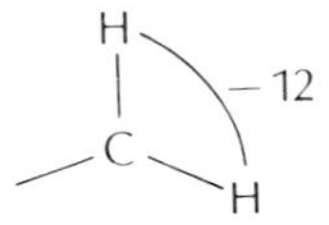

J varies with the bond angle: 105°	−20
110°	−12
115°	−7
120°	−3
125°	0
130°	+2

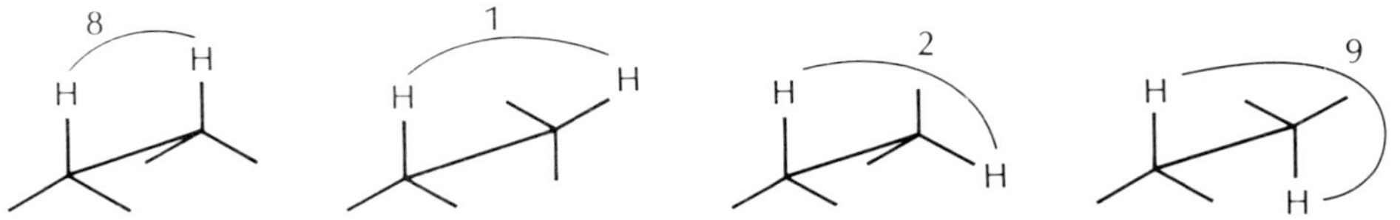

if rotation is free, $J = 7$ Hz

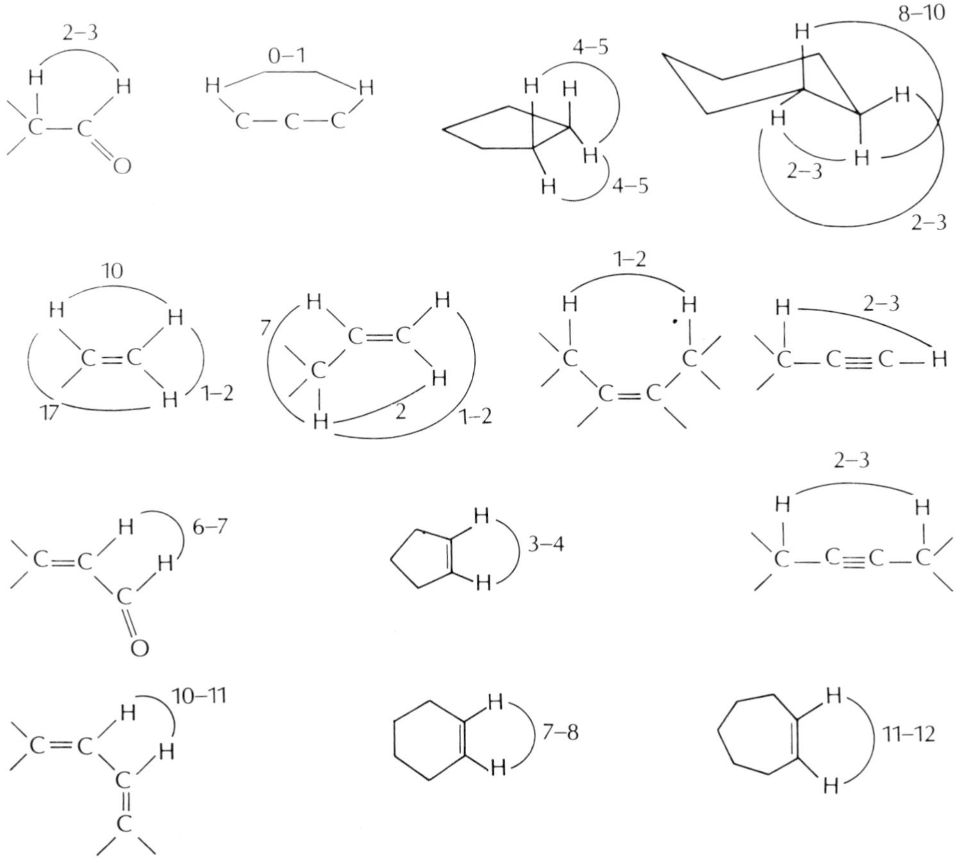